The Wetland Book

C. Max Finlayson • G. Randy Milton
R. Crawford Prentice • Nick C. Davidson
Editors

The Wetland Book

II: Distribution, Description, and Conservation

Volume 1

With 603 Figures and 125 Tables

Editors
C. Max Finlayson
Institute for Land, Water and Society
Charles Sturt University
Albury, New South Wales, Australia

UNESCO-IHE, Institute for Water Education
Delft, The Netherlands

R. Crawford Prentice
Nature Management Services
Cambridge, UK

G. Randy Milton
Department of Natural Resources
Kentville, Nova Scotia, Canada

Nick C. Davidson
Institute for Land, Water and Society
Charles Sturt University
Albury, New South Wales, Australia

Nick Davidson Environmental
Wigmore, UK

ISBN 978-94-007-4000-6   ISBN 978-94-007-4001-3 (eBook)
ISBN 978-94-007-4002-0 (print and electronic bundle)
https://doi.org/10.1007/978-94-007-4001-3

Library of Congress Control Number: 2017937719

© Springer Science+Business Media B.V., part of Springer Nature 2018

This work is subject to copyright. All rights are reserved by the Publisher, whether the whole or part of the material is concerned, specifically the rights of translation, reprinting, reuse of illustrations, recitation, broadcasting, reproduction on microfilms or in any other physical way, and transmission or information storage and retrieval, electronic adaptation, computer software, or by similar or dissimilar methodology now known or hereafter developed.

The use of general descriptive names, registered names, trademarks, service marks, etc. in this publication does not imply, even in the absence of a specific statement, that such names are exempt from the relevant protective laws and regulations and therefore free for general use.

The publisher, the authors and the editors are safe to assume that the advice and information in this book are believed to be true and accurate at the date of publication. Neither the publisher nor the authors or the editors give a warranty, express or implied, with respect to the material contained herein or for any errors or omissions that may have been made. The publisher remains neutral with regard to jurisdictional claims in published maps and institutional affiliations.

Printed on acid-free paper

This Springer imprint is published by the registered company Springer Science+Business Media B.V. part of Springer Nature.
The registered company address is: Van Godewijckstraat 30, 3311 GX Dordrecht, The Netherlands

Foreword: The Wetland Book

The venerable lineage of encyclopedic publishing can be traced back to Pliny the Elder's *Naturalis Historia*, which contained chapters on water and aquatic life. Although our terminology regarding and understanding of the aquatic environment has evolved over the past two millennia, one constant has been the need for a multidisciplinary approach to examining these areas. Using an encyclopedic model, this multidisciplinary book builds on an ancient format and adapts it for a modern audience. In this way, *The Wetland Book* builds on a long tradition of scholarly publishing and presents invaluable information for its modern audience.

Wetlands have been around longer than the traditions associated with academic publishing. Wetland management and wise use have been practiced by indigenous cultures in many forms for millennia, and that ancient knowledge about wetlands was often curated and passed down orally or in traditional systems and forms. In modern times, the pressures and threats to wetlands are vastly different in their scope and magnitude. The forms of governance and administration that respond to these pressures and threats have also changed, particularly in their scale as it has been recognized that management takes place at the level of countries and river basins, rather than simply at the local level.

Internationally, wetland conservation, management, and wise use are promoted through the Ramsar Convention on Wetlands. The countries that have signed onto the Ramsar Convention have recognized the imperatives to work with stakeholders and decisionmakers beyond the traditional wetland community and to incorporate wetlands into policy-making in other sectors such as water, energy, agriculture, and health. Indeed, in 2008 at the 10th Conference of the Contracting Parties of the Ramsar Convention, the Changwon Declaration was adopted, which contains key messages for wetland conservation, management, and wise use addressed to planners; policymakers; elected officials; managers in the environmental, land, and resource-use sectors; educators and communicators; economists; and health workers. *The Wetland Book* offers a base of knowledge that is intended to reach a similarly broad audience.

The editors and contributing authors to *The Wetland Book* have long experience and deep understanding of wetland science and management. Many have worked with the Ramsar Scientific and Technical Review Panel (STRP), the Convention's scientific advisory body, over the years. This collection of people provides a

repository of knowledge that can help meet the challenge of learning about and understanding the value of protecting and managing wetlands.

Making this knowledge more easily accessible, however, has always been difficult. There are physical limitations to how much we can pick a person's brain, and there are limitations to how much a wetland manager out in the field, perhaps with little technical support, can search for, read, and review scientific and traditional knowledge to find answers to pressing questions. Thus, the encyclopedic style of publication remains a viable format for accessing high levels of expertise, including expertise from distant locations, with similar landscape and ecological characteristics. *The Wetland Book* provides an in-depth level of knowledge in the form of a handbook to assist those seeking information on the many facets of wetland management.

Of course, reading *The Wetland Book* will not make an individual an expert in all aspects of wetland science, wise use and governance, a feat which no one publication can deliver. Instead, a truly useful publication should offer an individual the vocabulary to support further inquiry and to find knowledge that is locally, regionally, nationally, or even internationally applicable. It should also allow a reader to know who to ask and what questions to pursue when she or he needs more knowledge to solve a research question or particular management problem. *The Wetland Book* delivers this foundation through two volumes – Vol. 1: Structure and Function, Management, and Methods and Vol. 2: Distribution, Description, and Conservation.

We highly recommend *The Wetland Book*; it provides an unparalleled source of knowledge about wetlands by building on the ancient form of the encyclopedia, revitalized by new technologies for distribution and access. We are also proud to see that many of those who have contributed to the Ramsar Convention over many years or even decades have also contributed their knowledge and wisdom to *The Wetland Book*. Given our personal association with the Convention, we also recognize the incredible contribution that the Convention has made to wetland knowledge and look forward to further contributions.

Chair, Scientific & Technical Review Panel Ramsar Convention on Wetlands 2005–2012	Heather MacKay
Chair, Scientific & Technical Review Panel Ramsar Convention on Wetland 2012–2018	Royal C. Gardner

Preface

The Wetland Book is a hard copy and online production that provides an unparalleled collation of information on wetlands. It is global in scope and contains 462 chapters prepared by leading wetland researchers and managers. The wide disciplinary and geographic scope is a particular feature and differentiates *The Wetland Book* from the existing wetland literature. The editors have compiled *The Wetland Book* from contributions supplied by authors from many countries and disciplines. Combined, these chapters represent a global source of knowledge about wetlands. Given the number of chapters and the scope of the content, it has been published as two separate books.

The bibliographic detail of the two books is given below. Book II with 170 chapters covers the distribution, description, and conservation of wetlands.

The Wetland Book II: Distribution, Description, and Conservation: edited by Finlayson CM, Milton GR, Prentice RC and Davidson NC.

Its companion book, published separately, with 292 chapters is:

The Wetland Book I: Structure and Function, Management, and Methods: edited by Finlayson CM, Everard M, Irvine K, McInnes RJ, Middleton BA, van Dam AA and Davidson NC

The Wetland Book was developed following discussions with wetland experts from the Scientific and Technical Review Panel of the Ramsar Convention on Wetlands and from the Society for Wetland Scientists. These experts pointed to the rapidly expanding literature on wetlands and enthusiastically proposed the development of a comprehensive information resource aimed at supporting the trans- and multidisciplinary research and practice, which is essential to wetland science and management. They were also seeking an information resource that would both complement and extend the existing literature and in particular provide a compendium of knowledge with contributions from authors around the world.

Aware that wetland research was on the rise and that wetland researchers and practitioners often needed to work across disciplines, *The Wetland Book II* has been prepared to serve as a first port of call for those interested in the key information

about wetlands and their conservation. This was done to allow individuals and multi- and transdisciplinary teams to search for particular terms and subjects, access further details, and read overviews of topics selected by the editors and expert authors. The content provides a global coverage of wetland knowledge with chapters provided by leading wetland experts with knowledge that spans local and regional issues to the wider body of science that is needed to assist practitioners and enable students to come to grips with one of the world's most diverse and important set of ecosystems. This is especially important as these ecosystems are under increasing pressure in many parts of the world as degradation from human development continues at an alarming rate and are in need of more effective management and restoration. It draws heavily on the knowledge compiled through the formal processes of the Ramsar Convention and associated programs and extends that contained in the seminal global assessment of wetlands undertaken through the Millennium Ecosystem Assessment.

Book II is structured in sections covering the diversity of wetland types, natural and anthropogenic drivers of wetland change, and regional compilations of individual wetlands and wetland complexes. Detailed overview chapters typically describe the diversity within each wetland type, its distribution, extent (current and historical), ecosystem services, biodiversity, and threats and future challenges. Regional contributions follow a general format describing the basic ecology of the system; uniqueness; distribution; biodiversity and species adaptations; ecosystem services with an emphasis on importance to dependent peoples where appropriate; conservation status and management; and threats and future challenges. The coverage is based on a mix of wetland types and geographic extent and distribution.

Given that *The Wetland Book* constitutes a remarkable information resource, we warmly convey our special thanks to the many authors who gave up their time and shared their knowledge of wetlands to support this effort – and also for their patience while the large number of chapters they have generously provided were collated and edited. We are proud to have worked with them to produce this book. With the benefit of their unstinting efforts and incredibly rich knowledge, *The Wetland Book II* provides a comprehensive source of information for wetland researchers, students, and practitioners. It specifically provides a much needed information resource to support the many efforts to ensure the wise use of wetlands globally. It has also not only drawn on but also extended the expert guidance and advice that the Ramsar Convention's Scientific and Technical Review Panel has for almost 25 years provided for governments and wetland experts alike. In this respect, the foreword provided by the past and present chairs of the Panel is particularly appreciated. In providing the foreword, they have reflected on the wealth of knowledge collated by wetland experts from around the world who have worked tirelessly to provide government officials with the knowledge base needed to ensure the conservation and wise use of wetlands.

As editors for *The Wetland Book II*, we personally compliment the many authors for their incredible contributions to the most comprehensive compendium of knowledge about wetlands ever assembled. In particular, we acknowledge their unstinting efforts to compile the many chapters and work with the authors to produce *The*

Wetland Book. Their knowledge and efforts are matched by their willingness to share the collated knowledge that is now contained in *The Wetland Book*.

The publishers are thanked for their foresight in developing the concepts that led to *The Wetland Book* and for providing both a hard copy and online version, with the latter being available for future updating. We recommend *The Wetland Book* to all those interested in the growing international scientific knowledge about the functioning and management of these incredibly valuable but threatened ecosystems.

Institute for Land, Water and Society C. Max Finlayson
Charles Sturt University
Albury, NSW, Australia

UNESCO-IHE, Institute for Water Education
Delft, The Netherlands

Department of Natural Resources G. Randy Milton
Halifax, NS, Canada

Nature Management Services R. Crawford Prenctice
Cambridge, UK

Institute for Land, Water and Society Nick C. Davidson
Charles Sturt University
Albury, NSW, Australia

Nick Davidson Environmental
Wigmore, UK

Contents

Volume 1

Section I Introduction **1**

1 **Wetlands of the World** 3
   G. Randy Milton, R. Crawford Prentice, and C. Max Finlayson

Section II Diversity of Wetlands **17**

2 **Wetland Types and Distribution** 19
   C. Max Finlayson, G. Randy Milton, and R. Crawford Prentice

3 **Estuaries** 37
   Graham R. Daborn and Anna M. Redden

4 **Estuarine Marsh: An Overview** 55
   Ralph W. Tiner and G. Randy Milton

5 **Seagrasses** 73
   Frederick T. Short, Cathy A. Short, and Alyssa B. Novak

6 **Mangroves** 93
   C. Max Finlayson

7 **Major River Basins of the World** 109
   Carmen Revenga and Tristan Tyrrell

8 **Freshwater Lakes and Reservoirs** 125
   Etienne Fluet-Chouinard, Mathis Loïc Messager, Bernhard Lehner, and C. Max Finlayson

9 **Salt Lakes** 143
   C. Max Finlayson

10 **Tidal Freshwater Wetlands: The Fresh Dimension of the Estuary** 155
   Aat Barendregt

11	**Freshwater Marshes and Swamps** .... C. Max Finlayson	169
12	**Papyrus Wetlands** .... Julius Kipkemboi and Anne A. van Dam	183
13	**Tropical Freshwater Swamps (Mineral Soils)** .... Wim Giesen	199
14	**Peatlands** .... C. Max Finlayson and G. Randy Milton	227
15	**Peat** .... Richard Lindsay and Roxane Andersen	245
16	**Peatland (Mire Types): Based on Origin and Behavior of Water, Peat Genesis, Landscape Position, and Climate** .... Richard Lindsay	251
17	**Arctic Peatlands** .... Tatiana Minayeva, Andrey Sirin, Peter Kershaw, and Olivia Bragg	275
18	**Mires** .... Richard Lindsay	289
19	**Blanket Mire** .... Richard Lindsay	295
20	**Blanket Bogs** .... Richard Lindsay	303
21	**Lagg Fen** .... Richard Lindsay	309
22	**Karst Wetlands** .... Gordana Beltram	313
23	**Subterranean (Hypogean) Habitats in Karst and Their Fauna** ... Boris Sket	331
24	**Groundwater Dependent Wetlands** .... Ray H. Froend, Pierre Horwitz, and Bea Sommer	345

Section III Natural and Anthropogenic Drivers of Wetland Change **357**

25	**Natural and Anthropogenic Drivers of Wetland Change** .... Susan M. Galatowitsch	359
26	**Wetland Losses and the Status of Wetland-Dependent Species** ... Nick C. Davidson	369

27	**Alien Plants and Wetland Biotic Dysfunction** .... C. Max Finlayson	383
28	**Ecological Conditions and Health of Arctic Wetlands Modified by Nutrient and Contaminant Inputs from Colonial Birds** .... Mark Mallory	391
29	**Lake Chilika (India): Ecological Restoration and Adaptive Management for Conservation and Wise Use** .... Ajit Kumar Pattnaik and Ritesh Kumar	397
30	**Saemangeum Estuarine System (Republic of Korea): Before and After Reclamation** .... Nial Moores	405
31	**Peatlands and Windfarms: Conflicting Carbon Targets and Environmental Impacts** .... Richard Lindsay	413
32	**Kakagon (Bad River Sloughs), Wisconsin (USA)** .... Jim Meeker and Naomi Tillison	427
33	**Seagrass Dependent Artisanal Fisheries of Southeast Asia** ...... Richard K. F. Unsworth and Leanne C. Cullen-Unsworth	437
34	**Great Barrier Reef (Australia): A Multi-ecosystem Wetland with a Multiple Use Management Regime** .... Jon Brodie and Jane Waterhouse	447
35	**Qa'a Azraq Oasis: Strengthening Stakeholder Representation in Restoration (Jordan)** .... Fidaa F. Haddad	461
36	**Fishponds of the Czech Republic** .... Jan Pokorný and Jan Květ	469
37	**Makgadikgadi Wetlands (Botswana): Planning for Sustainable Use and Conservation** .... Jaap Arntzen	487
38	**Seagrass Recovery in Tampa Bay, Florida (USA)** .... Holly Greening, Anthony Janicki, and Ed T. Sherwood	495
Section IV	**North America, Greenland, and the Caribbean**	**507**
39	**Wetlands of Greenland** .... Christian Bay	509
40	**Peatlands of Continental North America** .... Dale H. Vitt	515

41	**Boreal Wetlands of Canada and the United States of America** . . .	521
	Beverly Gingras, Stuart Slattery, Kevin Smith, and Marcel Darveau	
42	**Yukon-Kuskokwim Delta: Yukon River Basin, Alaska (USA)**	543
	Frederic A. Reid and Daniel Fehringer	
43	**The Peace-Athabasca Delta: MacKenzie River Basin (Canada)**	549
	Jeffrey Shatford	
44	**Copper River Delta, Alaska (USA)**	557
	Frederic A. Reid, Daniel Fehringer, and Richard G. Kempka	
45	**Fraser River Delta: Southern British Columbia (Canada)**	565
	Anne Murray	
46	**The Mississippi Alluvial Valley (USA)**	577
	J. Brian Davis	
47	**Coastal Wetlands of Manitoba's Great Lakes (Canada)**	591
	Dale Wrubleski, Pascal Badiou, and Gordon Goldsborough	
48	**Coastal Wetlands of Lake Superior's South Shore (USA)**	605
	John Brazner and Anett Trebitz	
49	**The Bay of Fundy and Its Wetlands (Canada)**	621
	Graham R. Daborn and Anna M. Redden	
50	**San Francisco Bay Estuary (USA)**	637
	Beth Huning and Mike Perlmutter	
51	**Vernal Pools of Northeastern North America**	651
	Elizabeth A. Colburn and Aram J. K. Calhoun	
52	**Pocosins (USA)**	667
	Curtis J. Richardson	
53	**Prairie Pothole Region of North America**	679
	Kevin E. Doherty, David W. Howerter, James H. Devries, and Johann Walker	
54	**Playa Wetlands of the Great Plains (USA)**	689
	Anne Bartuszevige	
55	**Wetlands of California's Central Valley (USA)**	697
	Frederic A. Reid, Daniel Fehringer, Ruth Spell, Kevin Petrik, and Mark Petrie	
56	**The Everglades (USA)**	705
	Curtis J. Richardson	

Volume 2

Section V Central and South America 725

57 Amazon River Basin 727
Florian Wittmann and Wolfgang J. Junk

58 Mangroves of Colombia 747
Jenny Alexandra Rodríguez-Rodríguez, Paula Cristina Sierra-Correa, Martha Catalina Gómez-Cubillos, and Lucia Victoria Licero Villanueva

59 Ciénaga Grande de Santa Marta: The Largest Lagoon-Delta Ecosystem in the Colombian Caribbean 757
Jenny Alexandra Rodríguez-Rodríguez, José Ernesto Mancera Pineda, Laura Victoria Perdomo Trujillo, Mario Enrique Rueda, and Karen Patricia Ibarra

60 Lake Fuquene (Colombia) 773
Mauricio Valderrama, María Pinilla-Vargas, Germán I. Andrade, Eugenio Valderrama-Escallón, and Sandra Hernández

61 The Paraná-Paraguay Fluvial Corridor (Argentina) 785
Priscilla G. Minotti

62 The Pantanal: A Brief Review of Its Ecology, Biodiversity, and Protection Status 797
Wolfgang J. Junk and Catia Nunes da Cunha

63 The Paraná River Delta 813
Patricia Kandus and Rubén Darío Quintana

64 Wetlands of Chile: Biodiversity, Endemism, and Conservation Challenges 823
Alejandra Figueroa, Manuel Contreras, and Bárbara Saavedra

65 Seagrasses of Southeast Brazil 839
Joel C. Creed, Mariana V. P. Aguiar, Agatha Cristinne Soares, and Leonardo V. Marques

66 Rio de la Plata (La Plata River) and Estuary (Argentina and Uruguay) 847
Claudio R. M. Baigún, Darío C. Colautti, and Tomás Maiztegui

67 Conchalí Lagoon: Coastal Wetland Restoration Project (Chile) 857
Manuel Contreras, F. Fernando Novoa, and Juan Pablo Rubilar

68 Bahía Lomas: Ramsar Site (Chile) 865
Carmen Espoz, Ricardo Matus, and Diego Luna-Quevedo

| 69 | Patagonian Peatlands (Argentina and Chile) | 873 |

Rodolfo Iturraspe

Section VI Europe **883**

| 70 | **Danube River Basin** | 885 |

Paul Csagoly, Gernant Magnin, and Orieta Hulea

| 71 | **Lower Danube Green Corridor** | 897 |

Paul Csagoly, Gernant Magnin, and Orieta Hulea

| 72 | **Danube, Drava, and Mura Rivers: The "Amazon of Europe"** | 903 |

Paul Csagoly, Gernant Magnin, and Arno Mohl

| 73 | **Danube Delta: The Transboundary Wetlands (Romania and Ukraine)** | 911 |

Grigore Baboianu

| 74 | **Rhine River Basin** | 923 |

Daphne Willems and Esther Blom

| 75 | **Volga River Basin (Russia)** | 933 |

Harald J. L. Leummens

| 76 | **Volga River Delta (Russia)** | 945 |

Harald J. L. Leummens

| 77 | **European Tidal Saltmarshes** | 959 |

Nick C. Davidson

| 78 | **Tipperne Peninsula and Ringkøbing Fjord (Denmark)** | 973 |

Hans Meltofte, Preben Clausen, and Ole Thorup

| 79 | **Wadden Sea (Denmark)** | 983 |

Karsten Laursen and John Frikke

| 80 | **Estuaries of Great Britain** | 997 |

Nick C. Davidson

| 81 | **The Wash Estuary and North Norfolk Coast (UK)** | 1011 |

Nick C. Davidson

| 82 | **Wetlands of the Norfolk and Suffolk Broads (UK)** | 1023 |

Andrea Kelly

| 83 | **Blanket Mires of Caithness and Sutherland: Scotland's Great Flow Country (UK)** | 1039 |

Richard Lindsay and Roxane Andersen

| 84 | **Karst Wetlands in the Dinaric Karst** | 1057 |

Rosana Cerkvenik, Andrej Kranjc, and Andrej Mihevc

Contents

85 **Turloughs (Ireland)** 1067
Kenneth Irvine, Catherine Coxon, Laurence Gill,
Sarah Kimberley, and Steve Waldren

86 **The Macrotidal Bay of Mont-Saint-Michel (France): The
Function of Salt Marshes** 1079
Loïc Valéry and Jean-Claude Lefeuvre

87 **The Inner Danish Waters and Their Importance to
Waterbirds** 1089
Ib Krag Petersen and Rasmus Due Nielsen

Section VII Mediterranean Basin, Middle East, and West Asia ... 1099

88 **The Camargue: Rhone River Delta (France)** 1101
Patrick Grillas

89 **Ebro Delta (Spain)** 1113
Carles Ibáñez and Nuno Caiola

90 **Doñana Wetlands (Spain)** 1123
Andy J. Green, Javier Bustamante, Guyonne F. E. Janss,
Rocio Fernández-Zamudio, and Carmen Díaz-Paniagua

91 **Axios, Aliakmon, and Gallikos Delta Complex
(Northern Greece)** 1137
Despoina Vokou, Urania Giannakou, Christina Kontaxi, and
Stella Vareltzidou

92 **The Philippi Peatland (Greece)** 1149
Kimon Christanis

93 **Peatlands of the Mediterranean Region** 1155
Richard Payne

94 **The Hula Wetland (Israel)** 1167
Richard Payne

95 **Coastal Sabkha (Salt Flats) of the Southern and Western
Arabian Gulf** 1173
Ronald A. Loughland, Ali M. Qasem, Bruce Burwell, and
Perdana K. Prihartato

96 **Lake Seyfe (Turkey)** 1185
Serhan Cagirankaya and Burhan Teoman Meric

Section VIII Africa 1197

97 **Congo River Basin** 1199
Ian J. Harrison, Randall Brummett, and Melanie L. J. Stiassny

98	**Zambezi River Basin**	1217
	Matthew McCartney, Richard D. Beilfuss, and Lisa-Maria Rebelo	
99	**Zambezi River Delta (Mozambique)**	1233
	Richard D. Beilfuss	
100	**Nile River Basin**	1243
	Matthew McCartney and Lisa-Maria Rebelo	
101	**Nile Delta (Egypt)**	1251
	Mohamed Reda Fishar	
102	**Baro-Akobo River Basin Wetlands: Livelihoods and Sustainable Regional Land Management (Ethiopia)**	1261
	Adrian Wood, J. Peter Sutcliffe, and Alan Dixon	
103	**Bahr el Ghazal: Nile River Basin (Sudan and South Sudan)**	1269
	Asim I. El Moghraby	
104	**Machar Marshes: Nile Basin (South Sudan)**	1279
	Yasir A. Mohamed	
105	**The Mayas Wetlands of the Dinder and Rahad: Tributaries of the Blue Nile Basin (Sudan)**	1287
	Khalid Hassaballah, Yasir A. Mohamed, and Stefan Uhlenbrook	
106	**The Sudd (South Sudan)**	1299
	Lisa-Maria Rebelo and Asim I. El Moghraby	
107	**Rugezi Marsh: A High Altitude Tropical Peatland in Rwanda**	1307
	Piet-Louis Grundling, Ab P. Grootjans, and Anton Linström	
108	**Banc d'Arguin (Mauritania)**	1319
	Antonio Araujo and Pierre Campredon	
109	**Bijagos Archipelago (Guinea-Bissau)**	1333
	Pierre Campredon and Paulo Catry	
110	**Kilombero Valley Floodplain (Tanzania)**	1341
	Lars Dinesen	
111	**Lakes Baringo and Naivasha: Endorheic Freshwater Lakes of the Rift Valley (Kenya)**	1349
	Reuben Omondi, William Ojwang, Casianes Olilo, James Mugo, Simon Agembe, and Jacob E. Ojuok	
112	**Lake Turkana: World's Largest Permanent Desert Lake (Kenya)**	1361
	William Ojwang, Kevin O. Obiero, Oscar O. Donde, Natasha J. Gownaris, Ellen K. Pikitch, Reuben Omondi, Simon Agembe, John Malala, and Sean T. Avery	

113	Soda Lakes of the Rift Valley (Kenya) 1381
	Simon Agembe, William Ojwang, Casianes Olilo, Reuben Omondi, and Collins Ongore
114	Okavango Delta, Botswana (Southern Africa) 1393
	Lars Ramberg
115	Peatlands of Africa 1413
	Piet-Louis Grundling and Ab P. Grootjans
116	Peatland Types and Tropical Swamp Forests on the Maputaland Coastal Plain (South Africa) 1423
	Althea T. Grundling, Ab P. Grootjans, Piet-Louis Grundling, and Jonathan S. Price

Volume 3

Section IX Northern and East Asia 1437

117	Lena River Basin (Russia) 1439
	Victor Degtyarev
118	Lena River Delta (Russia) 1451
	Victor Degtyarev
119	Nidjili Lake: Lena River Basin (Russia) 1457
	Victor Degtyarev
120	Taiga-Alas Landscape in the South of the Central Yakutian Lowland: Lena River Basin (Russia) 1463
	Victor Degtyarev
121	The Middle Aldan River Basin: A Key Migration Corridor for the Eastern Population of the Siberian Crane Within the Lena River Basin (Russia) 1471
	Victor Degtyarev
122	Yenisei River Basin and Lake Baikal (Russia) 1477
	Nick C. Davidson
123	Amur-Heilong River Basin: Overview of Wetland Resources ... 1485
	Evgeny Egidarev, Eugene Simonov, and Yury Darman
124	Daurian Steppe Wetlands of the Amur-Heilong River Basin (Russia, China, and Mongolia) 1499
	Eugene Simonov, Oleg Goroshko, and Tatiana Tkachuk
125	Sanjiang Plain and Wetlands Along the Ussuri and Amur Rivers: Amur River Basin (Russia and China) 1509
	Thomas D. Dahmer

126	**Zhalong Wetlands (China)**	1521
	Liying Su	
127	**Highland Peatlands of Mongolia**	1531
	Tatiana Minayeva, Andrey Sirin, and Chultemin Dugarjav	
128	**Yangtze River Basin (China)**	1551
	Cui Lijuan, Zhang Manyin, and Xu Weigang	
129	**Poyang Lake, Yangtze River Basin, China**	1565
	James Harris	
130	**Huang He (Yellow River) River Basin (China)**	1575
	Cui Lijuan, Zhang Manyin, and Xu Weigang	
131	**Current Status of Seagrass Habitat in Korea**	1589
	Kun-Seop Lee, Seung Hyeon Kim, and Young Kyun Kim	
132	**Hokkaido Marshes (Japan)**	1597
	Satoshi Kobayashi	

Section X Central and South Asia 1603

133	**High Altitude Wetlands of Nepal**	1605
	Lalit Kumar and Pramod Lamsal	
134	**Anzali Mordab Complex (Islamic Republic of Iran)**	1615
	Masoud Bagherzadeh Karimi	
135	**Bujagh National Park (Islamic Republic of Iran)**	1625
	Sadegh Sadeghi Zadegan	
136	**Fereydoon Kenar, Ezbaran, and Sorkh Ruds Ab-Bandans**	1635
	Sadegh Sadeghi Zadegan	
137	**Lake Parishan (Islamic Republic of Iran)**	1647
	Ahmad Lotfi	
138	**Lake Uromiyeh (Islamic Republic of Iran)**	1659
	Ahmad Lotfi	
139	**Shadegan Wetland (Islamic Republic of Iran)**	1675
	Ahmad Lotfi	
140	**Mesopotamian Marshes of Iraq**	1685
	Curtis J. Richardson	
141	**Indus River Basin Wetlands**	1697
	Rab Nawaz, Ali Dehlavi, and Nadia Bajwa	
142	**Wular Lake, Kashmir**	1705
	Ritesh Kumar	

143	Wetlands of the Ganga-Brahmaputra Basin Ritesh Kumar and Kalpana Ambastha	1711
144	Saline Wetlands of the Arid Zone of Western India Malavika Chauhan and Brij Gopal	1725
145	The Transboundary Sundarbans Mangroves (India and Bangladesh) Brij Gopal and Malavika Chauhan	1733
146	Wetlands of Mahanadi Delta (India) Ritesh Kumar and Pranati Patnaik	1743

Section XI Southeast Asia 1751

147	Tropical Peat Swamp Forests of Southeast Asia Susan Page and Jack Rieley	1753
148	Wetlands of the Mekong River Basin: An Overview Peter-John Meynell	1763
149	Tonle Sap Lake: Mekong River Basin (Cambodia) Colin Poole	1785
150	Tram Chim: Mekong River Basin (Vietnam) Triet Tran and Jeb Barzen	1793
151	Transboundary Mekong River Delta (Cambodia and Vietnam) Triet Tran	1801
152	U Minh Peat Swamp Forest: Mekong River Basin (Vietnam) Triet Tran	1813
153	Sembilang National Park: Mangrove Reserves of Indonesia Marcel J. Silvius, Yus Rusila Noor, I. Reza Lubis, Wim Giesen, and Dipa Rais	1819
154	Wetlands of Berbak National Park (Indonesia) Wim Giesen, Marcel J. Silvius, and Yoyok Wibisono	1831
155	Danau Sentarum National Park (Indonesia) Wim Giesen and Gusti Z. Anshari	1841
156	Wetlands of Tasek Bera (Peninsular Malaysia) R. Crawford Prentice	1851
157	Intertidal Flats of East and Southeast Asia John MacKinnon and Yvonne I. Verkuil	1865
158	Seagrass in Malaysia: Issues and Challenges Ahead Japar S. Bujang, Muta H. Zakaria, and Frederick T. Short	1875

Section XII Australia, New Zealand, and Pacific Islands 1885

159 **Murray-Darling River Basin (Australia)** 1887
     Jamie Pittock

160 **Macquarie Marshes: Murray-Darling River Basin (Australia)** ... 1897
     Rachael F. Thomas and Joanne F. Ocock

161 **The Coorong: Murray-Darling River Basin (Australia)** 1909
     Peter Gell

162 **Kati Thanda: Lake Eyre (Australia)** 1921
     Richard T. Kingsford

163 **Myall Lakes (Australia)** 1929
     Brian G. Sanderson and Anna M. Redden

164 **Australia's Wet Tropics Streams, Rivers, and Floodplain Wetlands** 1941
     Richard G. Pearson

165 **Wetlands of Kakadu National Park (Australia)** 1951
     C. Max Finlayson

166 **Groundwater Dependent Wetlands of the Gnangara Groundwater System (Western Australia)** 1959
     Ray H. Froend, Pierre Horwitz, and Bea Sommer

167 **Seagrass Meadows of Northeastern Australia** 1967
     Robert G. Coles, Michael A. Rasheed, Alana Grech, and Len J. McKenzie

168 **Wetlands of New Zealand** 1977
     Karen Denyer and Hugh Robertson

169 **New Zealand Restiad Bogs** 1991
     Beverley R. Clarkson

170 **Atolls of the Tropical Pacific Ocean: Wetlands Under Threat** ... 2001
     Randolph R. Thaman

Index of Keywords 2027

About the Editors

C. Max Finlayson
Institute for Land, Water and Society
Charles Sturt University
Albury, NSW, Australia
UNESCO-IHE
Institute for Water Education
Delft, The Netherlands

Max Finlayson is an internationally renowned wetland ecologist with extensive experience internationally in water pollution, agricultural impacts, invasive species, climate change, and human well-being and wetlands. He has participated in global assessments such as those conducted by the Intergovernmental Panel for Climate Change, the Millennium Ecosystem Assessment, and the Global Environment Outlook 4 and 5 (UNEP). Since the early 1990s, he has been a technical adviser to the Ramsar Convention on Wetlands and has written extensively on wetland ecology and management. He has also been actively involved in environmental NGOs and from 2002 to 2007 was president of the governing council of global NGO Wetlands International.

He has worked extensively on the inventory, assessment, and monitoring of wetlands, in particular in wet tropical, wet-dry tropical, and subtropical climatic regimes covering pollution, invasive species, and climate change. His current research interests/projects include the following:

- Interactions between human well-being and wetland health in the face of anthropogenic change, including global change and the onset of the Anthropocenic era
- Vulnerability and adaptation of wetlands/rivers to climate change, including changing values and trade-offs between uses and users, considering uncertainty and complexity
- Integration of ecologic, economic, and social requirements and trade-offs between users of wetlands with an emphasis on developing policy guidance and institutional changes

- Environment and agriculture interactions and policy responses/outcomes, and collaboration between stakeholders and policymakers
- Wetland restoration and construction, including the use of artificial wetlands for waste water treatment and the generation of multiple values
- Landscape change involving wetlands/rivers and land use (agriculture and mining) and implications for wetland ecosystem services and benefits for local people

He holds the following associated positions:

- Scientific Expert on the Scientific and Technical Review Panel, Ramsar Convention on Wetlands, Triennium 2016–2018
- Ramsar Chair for the Wise Use of Wetlands, UNESCO-IHE, Delft, The Netherlands (2014–2018)
- Visiting Professor, Institute for Wetland Research, China Academy of Forestry, Beijing, China
- Editor-in-Chief, Marine and Freshwater Research, CSIRO Publishing
- Chair, Environmental Strategy Advisory Panel, Winton Wetlands Restoration (Australia)

He has contributed to over 300 journal articles, reports, guidelines, proceedings, and book chapters on wetland ecology and management. He has contributed to the development of concepts and methods for wetland inventory, assessment and monitoring, and undertaken many site-based assessments in many countries.

G. Randy Milton
Department of Natural Resources Kentville
NS, Canada

Randy Milton is the manager for the Ecosystems and Habitats Program with Nova Scotia's Department of Natural Resources in Canada. Randy is an ecologist and Certified Wildlife Biologist® with 35 years' experience in public and industry conservation and environmental management, especially with freshwater and coastal wetlands and forest ecosystems. He has maintained an involvement in regional and national wetland conservation efforts since the early 1990s, as well as internationally first as a volunteer with WWF (Indonesia) in the mid-1980s and subsequently as a technical advisor to the Ramsar Convention on Wetlands (2000–2015), a contributing author to the Millennium Ecosystem Assessment, and a member (2005–2014) of the International Plan Committee for the North American Waterfowl Management Plan. He has a M.Sc. from Acadia University (Canada), where he is currently an adjunct professor, and he is also an adjunct research associate at the Institute for Land, Water and Society, Charles Sturt University, Australia.

About the Editors

R. Crawford Prentice
Nature Management Services
Cambridge, UK

Crawford Prentice is an independent consulting ecologist based in Cambridge, England. He has some 30 years of biodiversity conservation experience and has led global, regional, and national programs and projects mainly in Asia, the CIS countries, and Europe. He studied Zoology at the University of Aberdeen and subsequently completed his M.Sc. in Aquatic Resource Management at Kings College, University of London.

Much of his professional life has concerned the conservation and management of wetlands and migratory waterbirds, with early beginnings at the Wildfowl and Wetlands Trust and the International Waterfowl and Wetlands Research Bureau (IWRB) analyzing International Waterfowl Census data, evolving to conservation program management at the Asian Wetland Bureau and IWRB, and leading a bilateral aid project for the integrated management of Malaysia's first Ramsar site in the 1990s. During the next decade, he worked with the International Crane Foundation on the design and implementation of the UNEP/GEF Siberian Crane Wetland Project, helping to strengthen management of seven million hectares across 16 wetland sites in four countries for this flagship species and other biodiversity. He remains a project associate with ICF, contributing to climate change adaptation planning for key wetland nature reserves in north-eastern China and wider efforts for crane conservation. Currently, Crawford conducts consultancy assignments for the preparation, implementation, and evaluation of GEF projects on integrated ecosystem management of wetlands, mountains, and tropical forests.

Nick C. Davidson
Institute for Land, Water and Society
Charles Sturt University
Albury, NSW, Australia
Nick Davidson Environmental
Wigmore, UK

Nick Davidson was the deputy secretary general of the Ramsar Convention on Wetlands from 2000 to 2014, with overall responsibility for the convention's global development and delivery of scientific, technical, and policy guidance and advice and communications as the Convention Secretariat's senior advisor on these matters. He has long-standing experience in, and a strong commitment to, environmental sustainability supported through the transfer of environmental science into policy-relevance and decision-making

at national and international scales. Nick currently works as an independent expert consultant on wetland conservation and wise use.

Nick has over 40 years' experience of research on the ecology, assessment, and conservation of coastal and inland wetlands and the ecophysiology and flyway conservation of migratory waterbirds, with a 1981 Ph.D. from the University of Durham (UK) on this topic, and continues to publish on these issues. Prior to his Ramsar Convention post, he worked for the UK's national government conservation agencies on coastal wetland inventory, assessment, information systems, and communications and as international science coordinator for the global NGO Wetlands International.

He is an adjunct professor at the Institute of Land, Water and Society, Charles Sturt University, Australia; was presented with the Society of Wetland Scientist's (SWS) International Fellow Award 2010 for his long-term contributions to global wetland science and policy; chairs the SWS's Ramsar Section; is an associate editor of the peer-reviewed journal *Marine & Freshwater Research*; is a member of several IUCN Commissions and their task forces (World Commission on Protected Areas (WCPA), Species Survival Commission (SSC), and Commission on Ecosystem Management (CEM)); and is an honorary fellow of the Chartered Institution of Water and Environmental Management (CIWEM).

Contributors

Simon Agembe Department of Fisheries and Aquatic Sciences, University of Eldoret, Eldoret, Kenya

Mariana V. P. Aguiar Departamento de Ecologia, Universidade do Estado do Rio de Janeiro, Rio de Janeiro, Brazil

Kalpana Ambastha Wetlands International South Asia, New Delhi, India

Roxane Andersen Environmental Research Institute, University of the Highlands and Islands, Thurso, UK

Germán I. Andrade Universidad de Los Andes, Bogotá, Colombia

Gusti Z. Anshari Center for Wetlands People and Biodiversity, University Tanjungpura, Pontianak, W. Kalimantan, Indonesia

Antonio Araujo MAVA – Fondation pour la Nature, Dakar, Senegal

Jaap Arntzen Centre for Applied Research, Gaborone, Botswana

Sean T. Avery Kenya Wetlands Biodiversity Research Team, Nairobi, Kenya

Grigore Baboianu Danube Delta Biosphere Reserve Authority, Tulcea, Romania

Pascal Badiou Institute for Wetland and Waterfowl Research, Ducks Unlimited Canada, Stonewall, MB, Canada

Claudio R. M. Baigún Instituto de Investigación e Ingeniería Ambiental (3iA), Universidad Nacional de San Martín, San Martín, Buenos Aires, Argentina

Nadia Bajwa Indus Ecoregion, WWF, Karachi, Pakistan

Aat Barendregt Environmental Sciences, Copernicus Institute, Utrecht University, Utrecht, The Netherlands

Anne Bartuszevige Playa Lakes Joint Venture, Lafayette, CO, USA

Jeb Barzen Private Lands Conservation, Spring Green, WI, USA

Christian Bay Department of Bioscience, Faculty of Science and Technology, Institut for Bioscience, Aarhus University, Roskilde, Denmark

Richard D. Beilfuss International Crane Foundation, Baraboo, WI, USA

College of Engineering, University of Wisconsin, Madison, WI, USA

Gordana Beltram Ministry of the Environment and Spatial Planning, Environment Directorate, Nature Conservation Unit, Ljubljana, Slovenia

Esther Blom WWF-Netherlands, Zeist, The Netherlands

Olivia Bragg School of the Geography, University of Dundee, Dundee, UK

John Brazner Wildlife Division, Nova Scotia Department of Natural Resources, Kentville, NS, Canada

Jon Brodie Catchment to Reef Research Group, TropWATER, Centre for Tropical Water and Aquatic Ecosystem Research, James Cook University, Townsville, QLD, Australia

Randall Brummett Environment and Natural Resources Department, World Bank, Washington, DC, USA

Japar S. Bujang Department of Biology, Faculty of Science, Universiti Putra Malaysia, Serdang, Selangor Darul Ehsan, Malaysia

Bruce Burwell E-Map Department, Saudi Aramco, Dhahran, Ash Sharqiyah, Saudi Arabia

Javier Bustamante Department of Wetland Ecology, Estación Biológica de Doñana (EBD-CSIC), Seville, Spain

Serhan Cagirankaya Wetlands Division, Ministry of Forests and Water Affairs, General Directory of Nature Conservation and National Parks, Ankara, Turkey

Nuno Caiola IRTA, Aquatic Ecosystems Program, Sant Carles de la Ràpita, Catalonia, Spain

Aram J. K. Calhoun University of Maine, Orono, ME, USA

Pierre Campredon Conseiller Technique, IUCN Guinea Bissau, Bissau, Guinea-Bissau

Paulo Catry MARE – Marine and Environmental Sciences Centre, ISPA – Instituto Universitário, Lisbon, Portugal

Rosana Cerkvenik Park Škocjanske jame, Divača, SI, Slovenia

Malavika Chauhan Himmotthan Society, Dehradun, Uttarakhand, India

Kimon Christanis Department of Geology, Sector of Earth Materials, University of Patras, Rio-Patras, Greece

Beverley R. Clarkson Landcare Research, Hamilton, New Zealand

Preben Clausen Department of Bioscience, Aarhus University, Kalø, Denmark

Darío C. Colautti Instituto de Limnología de La Plata "Raúl Ringuelet", La Plata, Argentina

Elizabeth A. Colburn Harvard Forest, Harvard University, Petersham, MA, USA

Robert G. Coles Centre for Tropical Water and Aquatic Ecosystem Research, James Cook University, Cairns and Townsville, QLD, Australia

Manuel Contreras Centro de Ecología Aplicada (CEA), Santiago, Región Metropolitana, Chile

Catherine Coxon University of Dublin, Trinity College, Dublin, Ireland

Joel C. Creed Departamento de Ecologia, Universidade do Estado do Rio de Janeiro, Rio de Janeiro, Brazil

Paul Csagoly Earthly Communications, Ottawa, ON, Canada

Leanne C. Cullen-Unsworth Sustainable Places Research Institute, Cardiff University, Cardiff, UK

Graham R. Daborn Acadia Centre for Estuarine Research, Acadia University, Wolfville, NS, Canada

Thomas D. Dahmer Ecosystems Ltd., Yau Tong, Kowloon, Hong Kong, China

Yury Darman WWF Russia Amur Branch, Vladivostok, Russia

Marcel Darveau Ducks Unlimited Canada, Quebec, QC, Canada

Nick C. Davidson Institute for Land, Water and Society, Charles Sturt University, Albury, NSW, Australia

Nick Davidson Environmental, Wigmore, UK

J. Brian Davis Department of Wildlife, Fisheries and Aquaculture, Mississippi State University, Mississippi State, MS, USA

Victor Degtyarev Siberian Division of Russian Academy of Sciences, Institute for Biological Problems of Cryolithozone, Siberian Branch, Russian Academy of Sciences, Yakutsk, Russia

Ali Dehlavi Indus Ecoregion, WWF, Karachi, Pakistan

Karen Denyer National Wetland Trust, Cambridge, Waikato, New Zealand

James H. Devries Ducks Unlimited Canada, Stonewall, MB, Canada

Carmen Díaz-Paniagua Department of Wetland Ecology, Estación Biológica de Doñana (EBD-CSIC), Seville, Spain

Lars Dinesen Biologist, European Representative, Scientific and Technical Review Panel of the Ramsar Convention, Jyderup, Denmark

Alan Dixon Institute of Science and the Environment, University of Worcester, Henwick Grove, UK

Kevin E. Doherty United States Fish and Wildlife Service, Bismarck, ND, USA

Oscar O. Donde KMFRI, Lake Turkana Research Station, Lodwar, Kenya

Chultemin Dugarjav Institute of General and Experimental Biology Mongolian Academy of Sciences, Ulaanbaatar, Mongolia

Evgeny Egidarev Pacific Geographical Institute FEB RAS \ WWF Russia Amur Branch, Vladivostok, Russia

Asim I. El Moghraby Sudanese National Academy of Sciences, Khartoum, Sudan

Carmen Espoz Centro Bahía Lomas, Facultad de Ciencias, Universidad Santo Tomás, Santiago, Chile

Daniel Fehringer Ducks Unlimited, Inc., Rancho Cordova, CA, USA

Rocio Fernández-Zamudio Department of Wetland Ecology, Estación Biológica de Doñana (EBD-CSIC), Seville, Spain

Alejandra Figueroa Head Natural Resources and Biodiversity, Ministry of Environment, Santiago, Chile

C. Max Finlayson Institute for Land, Water and Society, Charles Sturt University, Albury, NSW, Australia

UNESCO-IHE, The Institute for Water Education, Delft, The Netherlands

Mohamed Reda Fishar Inland Water and Aquaculture Branch, National Institute of Oceanography and Fisheries (NIOF), Cairo, Egypt

Etienne Fluet-Chouinard Center for Limnology, University of Wisconsin-Madison, Madison, WI, USA

John Frikke The Danish Wadden Sea National Park, Rømø, Denmark

Ray H. Froend Centre for Ecosystem Management, School of Science, Edith Cowan University, Joondalup, WA, Australia

Susan M. Galatowitsch Department of Fisheries, Wildlife and Conservation Biology, University of Minnesota, Saint Paul, MN, USA

Peter Gell Water Research Network, Federation University Australia, Ballarat, VIC, Australia

Urania Giannakou Department of Medical Laboratory Studies, School of Health and Medical Care, Alexander Technological Educational Institute of Thessaloniki, Thessaloniki, Greece

Wim Giesen Euroconsult Mott MacDonald, Arnhem, AK, The Netherlands

Laurence Gill University of Dublin, Trinity College, Dublin, Ireland

Beverly Gingras Ducks Unlimited Canada, Edmonton, AB, Canada

Gordon Goldsborough Department of Biological Sciences, University of Manitoba, Winnipeg, MB, Canada

Martha Catalina Gómez-Cubillos Universidad Nacional de Colombia, sede Caribe, Santa Marta, Colombia

Brij Gopal Centre for Inland Waters in South Asia, Jaipur, Rajasthan, India

Oleg Goroshko Daursky Biosphere Reserve, Chita, Russia

Natasha J. Gownaris School of Marine and Atmospheric Sciences, Stony Brook University, Stony Brook, NY, USA

Alana Grech Department of Environmental Sciences, Macquarie University, Sydney, NSW, Australia

Andy J. Green Department of Wetland Ecology, Estación Biológica de Doñana (EBD-CSIC), Seville, Spain

Holly Greening Tampa Bay Estuary Program, St. Petersburg, FL, USA

Patrick Grillas Tour du Valat, Research Institute for Mediterranean Wetlands, Le Sambuc, Arles, France

Ab P. Grootjans Centre for Energy and Environmental Studies, University of Groningen, Groningen, The Netherlands

Institute of Water and Wetland Research, Radboud University Nijmegen, Nijmegen, The Netherlands

Althea T. Grundling Water Science Programme, Agricultural Research Council – Institute for Soil, Climate and Water, Pretoria, Gauteng, South Africa

Department of Geography and Environmental Management, University of Waterloo, Waterloo, ON, Canada

Applied Behavioural Ecology and Ecosystem Research Unit, University of South Africa, Pretoria, South Africa

Piet-Louis Grundling Centre for Environmental Management, University of the Free State, Bloemfontein, South Africa

Fidaa F. Haddad Regional Dryland, Livelihoods and Gender Program, IUCN Regional Office for West Asia (ROWA), Amman, Jordan

James Harris International Crane Foundation, Baraboo, WI, USA

Ian J. Harrison Center for Environment and Peace, Conservation International, Arlington, VA, USA

Khalid Hassaballah UNESCO-IHE Institute for Water Education, Delft, The Netherlands

Hydraulics Research Center, Wad Medani, Sudan

Sandra Hernández Fundación Humedales, Bogotá, Colombia

Pierre Horwitz Centre for Ecosystem Management, School of Science, Edith Cowan University, Joondalup, WA, Australia

David W. Howerter Ducks Unlimited Canada, Stonewall, MB, Canada

Orieta Hulea WWF Danube-Carpathian Programme, Bucharest, Romania

WWF Danube-Carpathian Programme, Vienna, Austria

Beth Huning San Francisco Bay Joint Venture, Fairfax, CA, USA

Carles Ibáñez IRTA, Aquatic Ecosystems Program, Sant Carles de la Ràpita, Catalonia, Spain

Karen Patricia Ibarra Instituto de Investigaciones Marinas y Costeras "José Benito Vives de Andreis" (INVEMAR), Santa Marta, Colombia

Kenneth Irvine UNESCO-IHE Institute of Water Education, Delft, The Netherlands

Rodolfo Iturraspe Universidad Nacional de Tierra del Fuego, Ushuaia, Tierra del Fuego, Argentina

Anthony Janicki Janicki Environmental, Inc., St. Petersburg, FL, USA

Guyonne F. E. Janss Department of Wetland Ecology, Estación Biológica de Doñana (EBD-CSIC), Seville, Spain

Wolfgang J. Junk Instituto Nacional de Ciência e Tecnologia em Áreas Úmidas (INCT-INAU), Universidade Federal de Mato Grosso (UFMT), Cuiabá, MT, Brazil

Patricia Kandus Instituto de Investigación e Ingeniería Ambiental, Universidad Nacional de San Martín, San Martín, Provincia de Buenos Aires, Argentina

Masoud Bagherzadeh Karimi Department of Environment, Study Center for Environment, Wetlands and National Parks, Tehran, Iran

Andrea Kelly Broads Authority, Norwich, Norfolk, UK

Richard G. Kempka Ducks Unlimited, Inc., Rancho Cordova, CA, USA

The Climate Trust, Portland, OR, USA

Peter Kershaw University of Alberta, Edmonton, AB, Canada

Seung Hyeon Kim Department of Biological Sciences, Pusan National University, Busan, South Korea

Contributors

Young Kyun Kim Department of Biological Sciences, Pusan National University, Busan, South Korea

Sarah Kimberley University of Dublin, Trinity College, Dublin, Ireland

Richard T. Kingsford Centre for Ecosystem Science, School of Biological, Earth and Environmental Sciences, UNSW Australia, Sydney, NSW, Australia

Julius Kipkemboi Department of Biological Sciences, Egerton University, Egerton, Njoro, Kenya

Satoshi Kobayashi Faculty of Economics, Kushiro Public University, Kushiro, Hokkaido, Japan

Christina Kontaxi Administration of Environment and Spatial Planning, Region of Central Macedonia, Thessaloniki, Greece

Andrej Kranjc Slovenian Academy of Sciences and Arts, Ljubljana, SI, Slovenia

Lalit Kumar Ecosystem Management, School of Environmental and Rural Science, University of New England, Armidale, NSW, Australia

Ritesh Kumar Wetlands International South Asia, New Delhi, India

Jan Květ Faculty of Science, University of South Bohemia České Budějovice, České Budějovice, Czech Republic

Czech Academy of Sciences, CzechGlobe, Institute for Global Change Research, Brno, Czech Republic

Pramod Lamsal Ecosystem Management, School of Environmental and Rural Science, University of New England, Armidale, NSW, Australia

Karsten Laursen Department of Bioscience, Aarhus University, Aarhus, Denmark

Kun-Seop Lee Department of Biological Sciences, Pusan National University, Busan, South Korea

Jean-Claude Lefeuvre Department of Ecology and Biodiversity Management, National Museum of Natural History, Paris, France

EA 7316 Biodiversity and Land Management, University of Rennes 1, Rennes, France

Bernhard Lehner Department of Geography, McGill University, Montreal, QC, Canada

Harald J. L. Leummens Water and Nature, Heerlen, The Netherlands

Cui Lijuan Institute of Wetland Research, Chinese Academy of Forestry, Beijing, China

Richard Lindsay Sustainability Research Institute, University of East London, London, UK

Anton Linström Wet Earth Eco-Specs, Lydenburg, South Africa

Ahmad Lotfi Conservation of Iranian Wetland Project, Department of Environment, Tehran, Iran

Senior Irrigation Engineers, Member of Board of Directors, Pandam Consulting Engineers, Tehran, Iran

Ronald A. Loughland Environmental Protection Department, Saudi Aramco, Dhahran, Ash Sharqiyah, Saudi Arabia

I. Reza Lubis Wetlands International Indonesia, Bogor, Indonesia

Diego Luna-Quevedo WHSRN Executive Office- Manomet, Plymouth, MA, USA

John MacKinnon University of Kent, Canterbury, UK

Gernant Magnin WWF – Netherlands, Freshwater Programme, Zeist, The Netherlands

Tomás Maiztegui Instituto de Limnología de La Plata "Raúl Ringuelet", La Plata, Argentina

John Malala KMFRI, Lake Turkana Research Station, Lodwar, Kenya

Mark Mallory Coastal Wetland Ecosystems, Biology Department, Acadia University, Wolfville, NS, Canada

José Ernesto Mancera Pineda Universidad Nacional de Colombia, Bogotá, Colombia

Ciudad Universitaria, Mexico City, Mexico

Zhang Manyin Institute of Wetland Research, Chinese Academy of Forestry, Beijing, China

Leonardo V. Marques Departamento de Ecologia, Universidade do Estado do Rio de Janeiro, Rio de Janeiro, Brazil

Ricardo Matus Centro de Rehabilitación de Aves Leñadura, Punta Arenas, Chile

Matthew McCartney Ecosystem Services, International Water Management Institute, Vientiane, Lao People's Democratic Republic

Regional Office for Southeast Asia and the Mekong, International Water Management Institute, Vientiane, Lao People's Democratic Republic

Len J. McKenzie Centre for Tropical Water and Aquatic Ecosystem Research, James Cook University, Cairns and Townsville, QLD, Australia

Jim Meeker Formerly of: Northland College, Ashland, WI, USA

Jim Meeker: deceased

Hans Meltofte Department of Bioscience, Aarhus University, Roskilde, Denmark

Burhan Teoman Meric Wetlands Division, Ministry of Forests and Water Affairs, General Directory of Nature Conservation and National Parks, Ankara, Turkey

Mathis Loïc Messager School of Aquatic and Fishery Sciences, University of Washington, Seattle, WA, USA

Peter-John Meynell ICEM – International Centre for Environmental Management, Hanoi, Vietnam

Andrej Mihevc Karst Research Institute, Research Centre of the Slovenian Academy for Science and Arts, Postojna, Slovenia

G. Randy Milton Nova Scotia Department of Natural Resources, Kentville, NS, Canada

Tatiana Minayeva Wetlands International, Ede, The Netherlands

Priscilla G. Minotti Instituto de Investigación e Ingeniería Ambiental (3iA), Universidad Nacional de San Martín, San Martín, Provincia de Buenos Aires, Argentina

Yasir A. Mohamed Hydraulic Research Center, MoWRIE, Wad Medani, Sudan

Department of Integrated Water Systems and Governance, UNESCO-IHE, DA, Delft, The Netherlands

Faculty of Civil Engineering and Applied Geosciences, Water Resources Section, Delft University of Technology, GA, Delft, The Netherlands

Arno Mohl WWF Austria, Vienna, Austria

Nial Moores Birds Korea, Busan, Korea (Republic of)

James Mugo Kenya Marine and Fisheries Research Institute, Naivasha, Kenya

Anne Murray Nature Guides BC, Delta, BC, Canada

Rab Nawaz Indus Ecoregion, WWF, Karachi, Pakistan

Rasmus Due Nielsen Department of Bioscience, Aarhus University, Kalø, Denmark

Yus Rusila Noor Wetlands International Indonesia, Bogor, Indonesia

Alyssa B. Novak Boston University, Earth and Environment, Boston, MA, USA

F. Fernando Novoa Centro de Ecología Aplicada (CEA), Santiago, Región Metropolitana, Chile

Catia Nunes da Cunha Instituto Nacional de Ciência e Tecnologia em Áreas Úmidas (INCT-INAU), Universidade Federal de Mato Grosso (UFMT), Cuiabá, MT, Brazil

Depto Botânica e Ecologia/Núcleo de Estudos Ecológicos do Pantanal (NEPA), Instituto de Biociências, UFMT, Cuiabá, MT, Brazil

Kevin O. Obiero KMFRI, Lake Turkana Research Station, Lodwar, Kenya

Joanne F. Ocock Water and Wetlands Team, Science Division, NSW Office of Environment and Heritage, Sydney, NSW, Australia

Centre for Ecosystem Science, School of Biological, Earth and Environmental Sciences, UNSW, Sydney, NSW, Australia

Jacob. E. Ojuok Kenya Marine and Fisheries Research Institute, Kisumu, Kenya

William Ojwang Kenya Marine and Fisheries Research Institute (KMFRI), Mombasa, Kisumu, Kenya

WWF-Kenya, Nairobi, Kenya

Casianes Olilo Kenya Marine and Fisheries Research Institute (KMFRI), Mombasa, Kisumu, Kenya

Reuben Omondi Department of Applied Aquatic Sciences, Kisii University, Kisii, Kenya

KMFRI, Kisumu Research Centre, Kisumu, Kenya

Collins Ongore Kenya Marine and Fisheries Research Institute (KMFRI), Mombasa, Kisumu, Kenya

Susan Page Department of Geography, University of Leicester, Leicester, UK

Pranati Patnaik Wetlands International South Asia, New Delhi, India

Ajit Kumar Pattnaik Chilika Development Authority, Bhubaneswar, Odisha, India

Richard Payne Environment, University of York, Heslington, York, UK

Richard G. Pearson College of Science and Engineering, James Cook University, Townsville, QLD, Australia

Laura Victoria Perdomo Trujillo Universidad Nacional de Colombia, Bogotá, Colombia

Mike Perlmutter Environmental Services Division, City of Oakland Public Works Department, Oakland, CA, USA

Ib Krag Petersen Department of Bioscience, Aarhus University, Kalø, Denmark

Mark Petrie Ducks Unlimited, Inc., Rancho Cordova, CA, USA

Kevin Petrik Ducks Unlimited, Inc., Rancho Cordova, CA, USA

Ellen K. Pikitch School of Marine and Atmospheric Sciences, Stony Brook University, Stony Brook, NY, USA

Contributors

María Pinilla-Vargas Fundación Humedales, Bogotá, Colombia

Jamie Pittock Fenner School of Environment and Society, The Australian National University, Canberra, Australia

Jan Pokorný ENKI, o.p.s., Třeboň, Czech Republic

Colin Poole Wildlife Conservation Society, Phnom Penh, Cambodia

R. Crawford Prentice Nature Management Services, Histon, Cambridge, UK

Jonathan S. Price Department of Geography and Environmental Management, University of Waterloo, Waterloo, ON, Canada

Perdana K. Prihartato Environmental Protection Department, Saudi Aramco, Dhahran, Ash Sharqiyah, Saudi Arabia

Ali M. Qasem Environmental Engineering Div./ Environmental Protection Dept, Saudi Aramco, Dhahran, Saudi Arabia

Rubén Darío Quintana Instituto de Investigación e Ingeniería Ambiental, Universidad Nacional de San Martín, San Martín, Provincia de Buenos Aires, Argentina
Consejo Nacional de Investigaciones Científicas y Técnicas (CONICET), Buenos Aires, Argentina

Dipa Rais Wetlands International Indonesia, Bogor, Indonesia

Lars Ramberg Uppsala, Sweden

Michael A. Rasheed Centre for Tropical Water and Aquatic Ecosystem Research, James Cook University, Cairns and Townsville, QLD, Australia

Lisa-Maria Rebelo Water Futures, International Water Management Institute, Vientiane, Lao People's Democratic Republic
Regional Office for Southeast Asia and the Mekong, International Water Management Institute, Vientiane, Lao People's Democratic Republic

Anna M. Redden Acadia Centre for Estuarine Research, Acadia University, Wolfville, NS, Canada

Frederic A. Reid Ducks Unlimited, Inc., Rancho Cordova, CA, USA

Carmen Revenga The Nature Conservancy, Arlington, VA, USA

Curtis J. Richardson Nicholas School of the Environment, Duke University Wetland Center, Durham, NC, USA

Jack Rieley School of Geography, University of Nottingham, Nottingham, UK

Hugh Robertson Freshwater Section, Department of Conservation, Nelson, New Zealand

Jenny Alexandra Rodríguez-Rodríguez Instituto de Investigaciones Marinas y Costeras "José Benito Vives de Andreis" (INVEMAR), Santa Marta, Colombia

Juan Pablo Rubilar Minera Los Pelambres, Santiago, Chile

Mario Enrique Rueda Instituto de Investigaciones Marinas y Costeras "José Benito Vives de Andreis" (INVEMAR), Santa Marta, Colombia

Bárbara Saavedra Wildlife Conservation Society, Bronx, NY, USA

Sadegh Sadeghi Zadegan Ornithological Unit, Wildlife Bureau, Department of Environment, Tehran, Iran

Brian G. Sanderson Acadia Centre for Estuarine Research, Acadia University, Wolfville, NS, Canada

Jeffrey Shatford Ministry of Forests, Lands and Natural Resource Operations, Resource Management Objectives Branch, Victoria, BC, Canada

Ed T. Sherwood Tampa Bay Estuary Program, St. Petersburg, FL, USA

Cathy A. Short Lee, NH, USA

Frederick T. Short Department of Natural Resources and the Environment, Jackson Estuarine Laboratory, University of New Hampshire, Durham, NH, USA

Paula Cristina Sierra-Correa Instituto de Investigaciones Marinas y Costeras "Jose Benito Vives de Andreis" (INVEMAR), Santa Marta, Colombia

Marcel J. Silvius Wageningen, The Netherlands

Eugene Simonov Rivers without Boundaries Coalition, Dalian, China

Daursky NNR, Dalian, China

Andrey Sirin Center for Protection and Restoration of Peatland Ecosystems, Institute of Forest Science, Russian Academy of Sciences, Moscow, Russia

Laboratory of Peatland Forestry and Amelioration, Institute of Forest Science, Russian Academy of Sciences, Moscow, Russia

Boris Sket Oddelek za biologijo, Biotehniška fakulteta, Univerza v Ljubljani, Ljubljana, Slovenia

Stuart Slattery Ducks Unlimited Canada, Stonewall, MB, Canada

Kevin Smith Ducks Unlimited Canada, Edmonton, AB, Canada

Agatha Cristinne Soares Departamento de Ecologia, Universidade do Estado do Rio de Janeiro, Rio de Janeiro, Brazil

Bea Sommer Centre for Ecosystem Management, School of Science, Edith Cowan University, Joondalup, WA, Australia

Ruth Spell Ducks Unlimited, Inc., Rancho Cordova, CA, USA

Melanie L. J. Stiassny Division of Vertebrate Zoology, Ichthyology Department, American Museum of Natural History, New York City, USA

Liying Su International Crane Foundation, Baraboo, WI, USA

J. Peter Sutcliffe Independent Consultant, University of Huddersfield, Huddersfield, UK

Randolph R. Thaman The University of the South Pacific, Suva, Fiji

Rachael F. Thomas Water and Wetlands Team, Science Division, NSW Office of Environment and Heritage, Sydney, NSW, Australia

Centre for Ecosystem Science, School of Biological, Earth and Environmental Sciences, UNSW, Sydney, NSW, Australia

Ole Thorup Amphi Consult, Ribe, Denmark

Naomi Tillison Natural Resources Director, Bad River Band of Lake Superior Chippewa, Odanah, WI, USA

Ralph W. Tiner Institute for Wetlands and Environmental Education and Research, Inc., Leverett, MA, USA

Tatiana Tkachuk Daursky Biosphere Reserve, Zabaikalsky State University, Chita, Russia

Triet Tran International Crane Foundation, Baraboo, WI, USA

University of Natural Science, Vietnam National University, Ho Chi Minh City, Vietnam

Anett Trebitz Mid-Continent Ecology Division, United States Environmental Protection Agency, Duluth, MN, USA

Tristan Tyrrell Tentera, Montreal, QC, Canada

Stefan Uhlenbrook UNESCO-IHE Institute for Water Education, Delft, The Netherlands

Delft University of Technology, Delft, The Netherlands

Richard K. F. Unsworth College of Science, Wallace Building, Swansea University, Swansea, UK

Mauricio Valderrama Fundación Humedales, Bogotá, Colombia

Eugenio Valderrama-Escallón Fundación Humedales, Universidad El Bosque, Bogotá, Colombia

Loïc Valéry Department of Ecology and Biodiversity Management, National Museum of Natural History, Paris, France

EA 7316 Biodiversity and Land Management, University of Rennes 1, Rennes, France

Anne A. van Dam Aquatic Ecosystems Group, Department of Water Science and Engineering, UNESCO-IHE Institute for Water Education, Delft, The Netherlands

Stella Vareltzidou Axios-Loudias-Aliakmonas Management Authority, Chalastra, Thessaloniki, Greece

Yvonne I. Verkuil Conservation Ecology Group; Groningen Institute for Evolutionary Life Sciences (GELIFES), University of Groningen, Groningen, The Netherlands

Lucia Victoria Licero Villanueva Instituto de Investigaciones Marinas y Costeras "Jose Benito Vives de Andreis" (INVEMAR), Santa Marta, Colombia

Dale H. Vitt Department of Plant Biology and Center for Ecology, Southern Illinois University, Carbondale, IL, USA

Despoina Vokou Department of Ecology, School of Biology, Aristotle University of Thessaloniki, Thessaloniki, Greece

Steve Waldren University of Dublin, Trinity College, Dublin, Ireland

Johann Walker Ducks Unlimited, Great Plains Region, Bismarck, ND, USA

Jane Waterhouse Catchment to Reef Research Group, TropWATER, Centre for Tropical Water and Aquatic Ecosystem Research, James Cook University, Townsville, QLD, Australia

Xu Weigang Institute of Wetland Research, Chinese Academy of Forestry, Beijing, China

Yoyok Wibisono Wetlands International – Indonesia Programme, Bogor, Indonesia

Daphne Willems WWF-Netherlands, Zeist, The Netherlands

Florian Wittmann Institute of Floodplain Ecology, Karlsruhe Institute of Technology - KIT, Rastatt, Germany

Adrian Wood Business School, University of Huddersfield, Huddersfield, UK

Dale Wrubleski Institute for Wetland and Waterfowl Research, Ducks Unlimited Canada, Stonewall, MB, Canada

Muta H. Zakaria Department of Aquaculture, Faculty of Agriculture, Universiti Putra Malaysia, Serdang, Selangor Darul Ehsan, Malaysia

Section I

Introduction

Wetlands of the World

G. Randy Milton, R. Crawford Prentice, and C. Max Finlayson

Contents

Introduction	4
Book Structure	4
Overview of Wetland Types	5
Natural and Anthropogenic Drivers of Wetland Change	8
Regional Compilations of Wetland Articles	9
Future Challenges	14
References	14

Abstract

The Wetland Book is a first "port-of-call" reference work on the description, distribution and conservation issues for wetlands globally. The contributions in the work summarize key concepts, orient the reader to major issues, and aid in further research on issues by multidisciplinary and transdisciplinary teams. Here we introduce the book which includes articles broadly covering Ramsar defined inland and coastal wetlands structured in sections on diversity of wetland types, natural and anthropogenic drivers of wetland change, and regional compilations of individual or wetland complexes. Overview articles typically describe the

---

G. R. Milton (✉)
Nova Scotia Department of Natural Resources, Kentville, NS, Canada
e-mail: gordon.milton@novascotia.ca

R. C. Prentice
Nature Management Services, Histon, Cambridge, UK
e-mail: crawford.prentice@gmail.com

C. M. Finlayson
Institute for Land, Water and Society, Charles Sturt University, Albury, NSW, Australia

UNESCO-IHE, The Institute for Water Education, Delft, The Netherlands
e-mail: mfinlayson@csu.edu.au

© Crown 2018
C. M. Finlayson et al. (eds.), *The Wetland Book*,
https://doi.org/10.1007/978-94-007-4001-3_182

diversity within the wetland type, its distribution, extent, ecosystem services, biodiversity, and threats and future challenges. Regional contributions adhere to a general format describing the basic ecology of the system; uniqueness; distribution; biodiversity and species adaptations; ecosystem services with an emphasis on importance to dependent peoples where appropriate; conservation status and management; and threats and future challenges. A general conclusion that can be drawn from all articles is that wetlands are still under significant threat to loss and degradation from human impacts including climate change and concerted action is required among all levels of society in the management and conservation of these dynamic ecosystems.

Keywords
Wetland diversity · Anthropogenic effects · Natural effects · Drivers of change · Ramsar · Management · Multidisciplinary · Transdisciplinary

Introduction

This work developed from conversations with the Secretariat of the Ramsar Convention of Wetlands on the status of wetlands globally, and the desire to provide a relatively comprehensive compilation that complemented existing resources. Produced as an online and hardcopy publication in two books, the *Wetland Book*: Book 1 – Structure and Function, Management, Methodology; and Book 2 – Distribution, Description, and Conservation will support the work of students, transdisciplinary researchers, natural resource managers and agency staff, engineers, planners, policy advisors, NGOs, and environmental consultants.

Book 2 is a first "port-of-call" reference work on the description, distribution, and conservation issues for wetlands globally. The intent of these selected contributions by experts is to summarize the key concepts, orient the reader to the major issues, and aid in further research on issues by multidisciplinary and transdisciplinary teams. Articles do not always have an exhaustive listing of primary references but draw upon multiple sources of information including primary and tertiary literature, and government and agency publications where appropriate.

Book Structure

The book is structured in sections with articles covering the diversity of wetland types (23), natural and anthropogenic drivers of wetland change (14), and regional compilations of individual or wetland complexes (131). Detailed overview articles typically describe the diversity within the wetland type, its distribution, extent (current and historical), ecosystem services, biodiversity, and threats and future challenges. Regional contributions adhere to a general format describing the basic ecology of the system; uniqueness; distribution; biodiversity and species adaptations; ecosystem services with an emphasis on importance to dependent peoples where appropriate; conservation status and management; and threats and future challenges. Authors do, however, place emphasis on certain topics, reflecting their specific interests. Cross-references are provided to help navigate between related articles.

Overview of Wetland Types

This book could never be fully comprehensive in its coverage of wetlands throughout the world, but the editors have strived to include a broad coverage of the inland and coastal wetlands captured by the Ramsar definition that includes "...areas of marsh, fen, peatland or water, whether natural of artificial, permanent or temporary, with water that is static or flowing, fresh, brackish or salt, including areas of marine water the depth of which at low tide does not exceed six metres." (Ramsar Convention Secretariat 2010). Finlayson et al. (2018) introduce the section on wetland diversity and distribution. The section includes overviews on coastal and inland wetland types (e.g., estuarine marsh, peatlands) and wetland systems (e.g., river basins, estuaries), with numerous cross-references to related articles in the regional sections. They note the terminology in use globally to describe wetlands is diverse and that the definitions and inventory methods that have been applied have been limiting factors in estimating their global extent, as outlined by a number of studies. The authors suggest that processes adopted by Contracting Parties of the Ramsar Convention on Wetlands for wetland inventory would provide a more consistent data set for ensuring the wise use of wetlands globally.

The Ramsar wetland definition includes rivers, and the overview by Revenga and Tyrrell (2018) notes the hydrological, geological, and topographical uniqueness of river basins and their freshwater wetlands and discusses their delineation and biodiversity. Direct services to society are immense including withdrawals for domestic water and irrigation, supporting inland fisheries, and underpinning numerous energy production options. More challenging to quantify are the indirect benefits in water purification, storing of pollutants, and ground water recharge. River systems are, however, impacted by land-based activities and alterations to reaches and flow regimes to provide societal needs. Water withdrawals are expected to increase with projected human demographic growth and urbanization, and maintaining sufficient flows for ecological processes with a changing climate will be a struggle, particularly in water-stressed river basins. Drawing upon global species assessments, freshwater species are at higher risk than terrestrial species with nearly one in three threatened with extinction. Habitat loss and degradation in particular are key factors driving this threat. Despite their importance, the authors comment that the conservation, planning, and policy-making for freshwater ecosystems have not been a primary concern and conclude with a discussion on their inadequate consideration in the design of protected areas (see also Juffe-Bignoli et al. 2016). The sections on regional contributions typically include overview articles on key river basins that cross-reference articles within the basin.

Fluet-Chouinard et al. (2018) introduce lakes, ponds, and lacustrine environments, their formation, geographic distribution, and differing estimates of their global extent. The authors comment on the importance of morphology and hydrology as well as climate and the surrounding landscape in influencing many of the ecological features of lakes. Lakes can be categorized in a variety of ways including being permanent or ephemeral or on differences in their thermal stratification and mixing regime, productivity, and nutrient availability which also affect their biotic diversity and use patterns. A subset of this broader classification are salt lakes that are generally endorheic and mostly confined to semi-arid and arid regions as permanent (e.g., Lake Turkana) or ephemeral (e.g., Kati Thanda – Lake Eyre) water bodies. Biota tends to decrease with increasing salinity and there is considerable faunal endemism; cyanobacteria often play a key role in energy and biogeochemical cycles (Finlayson 2018a). Whether fresh or salt, human activities, land use changes, and a changing climate in many regions are modifying the physical, chemical, and biological processes in lake ecosystems and affecting the multitude of ecosystem benefits they provide.

As a wetland type, peatlands are widespread and could represent more than a third of global wetlands (Finlayson and Milton 2018) with a very diverse typology based upon landscape position, climate, water sourcing, and peat source (see: Lindsay 2018; Minayeva et al. 2018a, b). Peatlands are particularly prevalent in the mid-high latitudes of North America and Eurasia (see regional accounts) and some tropical regions where extremely high rainfall counterbalances high evapotranspiration (see Page and Rieley 2018). The overview provides an introduction to the importance of hydrology in describing peatlands and factors influencing their biota which is generally less rich than nearby ecosystems but supporting many regionally or globally endemic species highly adapted to particular conditions. Drawing from the review by Silvius et al. (2008), Finlayson and Milton (2018) summarize the main benefits people receive from peatlands and the important role of peatlands in the global climate cycle. Direct (e.g., draining, land conversion) and indirect (e.g., air pollution, infrastructure development) impacts are estimated to have degraded 16% of the global peatland resource. Retention and restoration of peatlands are developing as a climate change mitigation measure and are eligible in the voluntary carbon market.

The overviews on mineral soil-based freshwater swamps and marshes consider nontidal forested and nonforested wetlands generally (Finlayson 2018b) and more specifically papyrus *Cyperus papyrus* marshes (regionally called swamps, Kipkemboi and van Dam 2018) and tropical freshwater swamps (Giesen 2018). While widespread and the subject of numerous articles and compilations on wetlands regionally and globally, and inherently difficult to separate when discussing large wetland complexes, the combined extent of these systems has been estimated at $3,694 \times 10^3$ km^2. Marshes and swamps exhibit a great diversity and variability of vegetation and habitat types due to underlying geomorphic and landform characteristics (also see Beltram 2018), hydrological regime (e.g., groundwater dependent, Froend et al. 2018), and water quality. The biota supported by these ecosystems are a significant but as yet undetermined percentage of the 126,000 but incomplete listing of animal species recognized globally. Trees and herbaceous vegetation exhibit structural adaptations (e.g., aerenchyma and suberin and lignin deposition in roots) as well as physiological, phenological, and reproductive adaptations to help them survive in standing water. There is an array of ecosystem services provided by swamps and marshes of which provisioning of fresh water and supporting food security of local communities are identified as especially important. Marshes and swamps are threatened by a similar suite of natural and human pressures as other wetland types with the expectation of further declines in their ecological condition and services without an increased emphasis and implementation of a holistic approach to wetland conservation.

Tidal freshwater wetlands occur in locations along rivers experiencing a tidal pulse but where fresh water dominates the conditions and processes. The overview by Barendregt (2018) draws attention to this wetland type that is underrepresented in the scientific literature due in-part to reduced opportunities for study from historic loss given their location in sites associated with infrastructure development for coastal ports and habitation. Barendregt (*op.cit.*) discusses physical and chemical processes and the importance of salinity and the hydrological regime (especially constancy of river discharge) in structuring vegetation communities into distinct zones. Faunal diversity is not high, but the gradient in salinity is important for adjusting the physiological equilibrium of diadromous species during migration to and from the sea. Global distribution and extent is unknown but is believed to be greatly reduced from historical amounts due to human impacts. Physical processes of sedimentation and natural vegetation growth are re-established with return of dynamic flooding in restoration of tidal fresh water wetlands. Unless tidal freshwater wetland can migrate upstream in response to sea-level rise unimpeded by infrastructure or topography, they are at risk of further loss.

Coastal wetland ecosystems are described in a series of overview articles that discuss estuaries generally (Daborn and Redden 2018), and more specifically estuarine marshes (Tiner and Milton 2018), seagrasses (Short et al. 2018), and mangroves (Finlayson 2018c). These are dynamic and relatively physically stressed systems subject to wide variations in salinity and tidal inundation with typically low biodiversity but high abundance and productivity. Biological communities reflect the physical processes that dominate each estuary determined by the interaction *inter alia* between estuary morphology, substrate, tidal range, duration to air exposure, degree of fresh and salt water mixing, and nutrient and sediment inputs. Seagrasses are widely distributed along temperate and tropical coasts and estuaries, while estuarine marshes and mangroves are the dominant plant communities of the intertidal zone at middle and higher latitudes (temperate and polar regions) and tropics, respectively. Global estimates of seagrass, estuarine marsh, and mangroves are 177,000 km^2, 68,000 km^2 (minimal), and 138,000 km^2, respectively. The articles describe the diversity and zonal variation in species distribution and the principal ecosystem services provided of which carbon sequestration increasingly is being recognized in climate change mitigation strategies. Human-based threats to the extent and quality of these ecosystems are similar and include land conversion for infrastructure and food supply, pollution, and overexploitation of resources. A warming global climate will increase relative sea-level rise. The long-term stability of these ecosystems depends upon maintaining their position in the tidal frame by either horizontal migration landward and/or vertically through sedimentation. This is becoming increasingly difficult due to infrastructure designed to serve the more than one third of the global human population residing coastally. The pressures on coastal ecosystems and their importance to society require increased emphasis on water basin management, restoration, and implementation of laws and policies that support effective management.

Natural and Anthropogenic Drivers of Wetland Change

Wetlands are dynamic systems whose ecological character is influenced by both natural processes and anthropogenic activities that can variously interact at multiple spatial and temporal scales. Under anomalous conditions, these can act as agents or drivers of wetland change triggering a cascade of ecosystem effects. The 14 articles in this section discuss the underlying demographic and economic drivers of historical and ongoing wetland loss and degradation, approaches to disentangle cause and effect relationships and prioritize proximate causes of wetland change for remedial action, and the need to formulate effective strategies to ensure the sustainable use of wetlands. Examples here and regionally describe how a lack of coordination and communication among community sectors can exacerbate a deteriorating situation. Although restoration of degraded wetlands can be as simple as plugging a drainage ditch, in many instances it is a long-term, complex, multistakeholder affair that engages wetland dependent local communities, resource use and infrastructure sectors, and multiple levels of government.

In her review, Galatowitsch (2018) lists the six proximate drivers identified in the Millennium Ecosystem Assessment that have caused the most extensive wetland degradation and loss: infrastructure development, land conversion, water withdrawal, eutrophication and pollution, over harvesting and overexploitation, and invasive species. The first three drivers typically result in wetland loss or significant alterations to the hydrology and degradation of wetland quality that are difficult to reverse because of trade-offs with other anthropogenic uses in the surrounding landscape. Economic and human population growth are implicated by Davidson (2018) as the underlying drivers responsible for 62–75% loss of global wetland area since 1900 AD. Average rates of loss have declined since 1990 for inland wetlands but have remained rapid for coastal wetlands. Losses have been especially pronounced for intertidal and other coastal wetlands in Asia during the last 50 years (MacKinnon and Verkuil 2018; Moores 2018), and seagrasses are being degraded and in decline regionally and globally *inter alia* from infrastructure development, pollution, and sedimentation (Brodie and Waterhouse 2018; Short et al. 2018; Unsworth and Unsworth 2018).

Increased economic development and human population growth can bring about changes in a wetland's species structure and nutrient and energy dynamics in response to overexploitation of resources, excessive nutrient (e.g., phosphorus and nitrogen) inputs, and introduction of alien plants that become invasive (Galatowitsch 2018; Finlayson 2018d, e), and these can operate separately from or synergistically with the drivers above. Restoring degraded wetlands requires a structured approach to identify the interlinkages among the drivers, stressors (physio-chemical changes), ecological effects, and wetland attributes of greatest concern (Galatowitsch 2018). Haddad (2018) describes natural resource management as being "...composed of two interacting decision-making subsystems: a 'horizontal' land/water use system and a 'vertical' human activity system." The former addresses the land/water/user interface with spatial and temporal bounds, while the latter deals with the interplay between the different sectors and levels of community structure. Management approaches to restore degraded wetlands or plan for their sustainable management are given in accounts for Chilika Lake (Pattnaik and Kumar 2018), Qa'a Azraq Oasis (Haddad 2018), Great Barrier Reef (Brodie and Waterhouse 2018), and Makgadikgadi wetlands (Arntzen 2018).

Regional Compilations of Wetland Articles

The globe was divided into nine regions in which experts were requested to contribute articles on river basins and their wetlands, landscapes in which wetlands are a dominant feature, globally and regionally significant wetlands, or strong examples of the sustainable use of wetlands by local communities. As described above, authors were given broad guidance on the topics to be covered in the articles, yet there was also a high degree of flexibility and depth in the coverage of the topics. The 140 articles that can be specifically linked to a region are relatively evenly balanced among regions (Table 1), and variation is in-part a function of area and expert availability to contribute. The locations of these wetland accounts in the regions are shown in Plates 1, 2, and 3.

Table 1 The distribution of articles describing specific wetlands, wetland complexes, or wetland landscapes by regional compilation or in the section on natural and anthropogenic drivers of wetland change

Regions	Descriptive wetland accounts	Wetland change accounts	Total
North America, Greenland, and the Caribbean	18	2	20
Central and South America	13	0	13
Europe	17	1	18
Mediterranean Basin, Middle East, and West Asia	9	1	10
Africa	20	1	21
Northern and East Asia	16	1	17
Central and South Asia	14	1	15
South East Asia	12	1	13
Australia, New Zealand, and Pacific Islands	12	1	13
Total	131	9	140

Plate 1 The locations of wetlands, wetland complexes, or wetland landscapes covered in this book within North America, Greenland, Caribbean, Central and South America, and Pacific Islands. Numbers refer to the accompanying Legend. Note: some numbers appear more than once to represent their wide distribution. Map prepared by G. R. Milton. (*1* Arctic Peatlands, *2* Greenland Wetlands, *3* Boreal Forest Wetlands, *4* NA Continental Peatlands, *5* Yukon-Kuskokwim Delta, *6* Copper River Delta, *7* Peace-Athabasca Delta, *8* Fraser River Delta, *9* Manitoba's Great Lakes, *10* Prairie Pothole Wetlands, *11* Kakagon (Bad River Sloughs), *12* Lake Superior's South Shore, *13* Bay of Fundy, *14* Vernal Pools - northeastern NA, *15* Pocosins, *16* Playa Wetlands, *17* California's Central Valley, *18* San Francisco Bay Estuary, *19* Mississippi River Alluvial Valley, *20* Seagrass - Tampa Bay, *21* The Everglades, *22* Karst Wetlands - Yucatan Peninsula, *23* Ciénaga Grande de Santa Marta, *24* Mangroves of Colombia, *25* Lake Fuquene, *26* Amazon River Basin, *27* Pantanal, *28* Paraná Paraguay Fluvial Corridor, *29* Rio de la Plata and Estuary, *30* Paraná River Delta, *31* Seagrasses of Southeast Brazil, *32* Wetlands of Chile, *33* Conchalí Lagoon, *34* Patagonian Peatlands, *35* Bah¡a Lomas Ramsar site, *135* Atolls of the Tropical Pacific Ocean)

1 Wetlands of the World

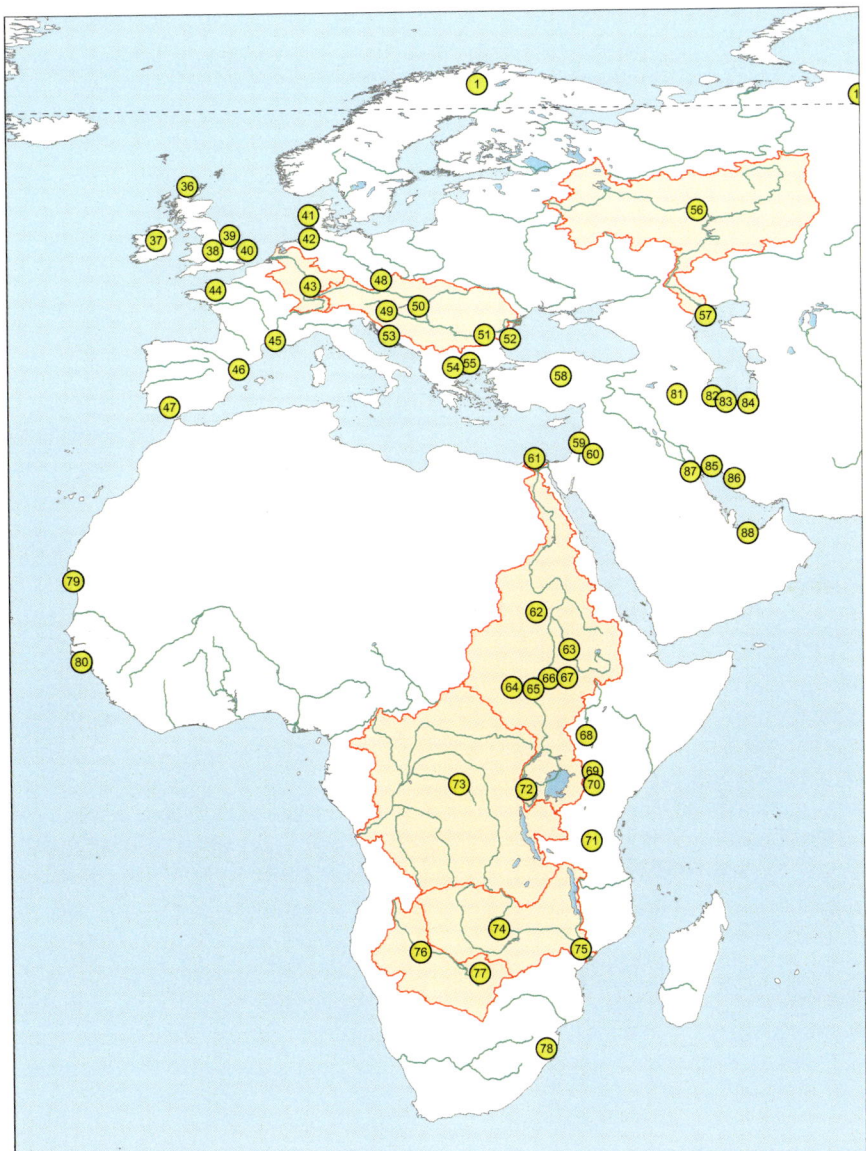

Plate 2 The locations of wetlands, wetland complexes, or wetland landscapes covered in this book within Europe, the Mediterranean Basin, Middle East, and Africa. Numbers refer to the accompanying Legend. Note: some numbers appear more than once to represent their wide distribution. Map prepared by G. R. Milton. (*1* Arctic Peatlands, *36* Caithness and Sutherland, *37* Turloughs, *38* Esturaries of Great Britain, *39* Wash Estuary, *40* Norfolk and Suffolk Broads, *41* Tipperne Peninsula and Ringkøbing Fjord, *42* Danish Wadden Sea, *43* Rhine River Basin, *44* Macrotidal Bay of Mont-Saint-Michel, *45* The Camargue, *46* Ebro Delta, *47* Doñana Wetlands, *48* Fish Ponds Czech Republic, *49* Danube River Basin, *50* Danube, Drava, and Mura Rivers, *51* Lower Danube Green Corridor, *52* Danube Delta, *53* Dinaric Karst Wetlands, *54* Axios, Aliakmon and Gallicos

River basin overviews provide the framework for most of the regions with descriptions of the interrelationship of wetlands and human communities occurring at the site level. Although there are gaps in overview coverage of a number of river basins (e.g., Mississippi, MacKenzie, Niger, Ob, and Tigris-Euphrates), articles do cover the diversity of wetland types and the major issues within each region. The article by Gingras et al. (2018), for example, on boreal wetlands has a geographic coverage that includes major portions of drainage basins for the northern half of North America: MacKenzie, Yukon, and Nelson rivers as well as the extensive lowland wetlands of Hudson and James Bay. Similarly the southern half of the potholes (Doherty et al. 2018), playas (Bartuszevige 2018), and alluvial valley (Davis 2018) are key wetland areas in the Mississippi River drainage.

Wetlands described in each of this book's articles are typically grouped into one or more types, considered together they describe the complexity and variability within and among wetlands even when "similarly" classified. Wetland heterogeneity is an outcome *inter alia* of the ecological components, topographical setting, hydrological regime, and processes interacting temporally and spatially within landscapes directly or indirectly impacted by natural and anthropogenic disturbances. Regardless of the protection level accorded to wetlands at the site or landscape level (e.g., Ramsar site, National Park or Reserve), wetlands are open systems subject to stochastic variation and uncertainty in response to external events. Many of the wetlands discussed in this book are situated within landscapes that are heavily modified by humans for agriculture and industrial use and urban living with development of infrastructure that alters hydrological regimes (e.g., dams, canals, dykes), extracts water for agriculture or human use, and inputs of point and nonpoint sources of nutrients and contaminants. While wetland loss has been significant due to drainage and land conversion, wetland degradation has also occurred when the adaptive capacity is exceeded leading to a regime shift in the system's components, processes, and ecosystem services. An underlying threat to all wetlands is the uncertain impacts of climate change over and above existing anthropogenic stresses. Particular concerns are the resilience of marine wetland communities to cope with ocean acidification and warming episodes, coastal systems' capacity to respond to accelerated sea-level rise, that of inland wetlands in drier climates to cope with warming and drying trends, and the bleak long-term outlook facing the rivers, wetlands, and large human populations dependent upon glacial melt water supply.

Plate 2 (continued) Delta Complex, *55* Philippi Peatland, *56* Volga River Basin, *57* Volga River Delta, *58* Lake Seyfe, *59* Hula Wetland, *60* Qa'a Azraq Oasis, *61* Nile Delta, *62* Nile River Basin, *63* Mayas of the Dinder and Rahad, *64* Bahr el Ghazal, *65* The Sudd, *66* Machar Marshes, *67* Baro-Akobo River Basin, *68* Lake Turkana, *69* Lakes Baringo and Naivasha, *70* Soda Lakes of the Rift Valley, *71* Kilombero Valley Floodplain, *72* Rugezi Marsh, *73* Congo River Basin, *74* Zambezi River Basin, *75* Zambezi River Delta, *76* Okavango Delta, *77* Makgadikgadi Wetlands, *78* Maputaland Coastal Plain, *79* Banc d'Arguin, *80* Bijagós Archipelago, *81* Lake Uromiyeh, *82* Anzali Mordab Complex, *83* Bujagh National Park, *84* Fereydoon Kenar Ramsar Site, *85* Shadegan Wetland, *86* Lake Parishan, *87* Mesopotamian Marshes, *88* Coastal Sabkha)

1 Wetlands of the World

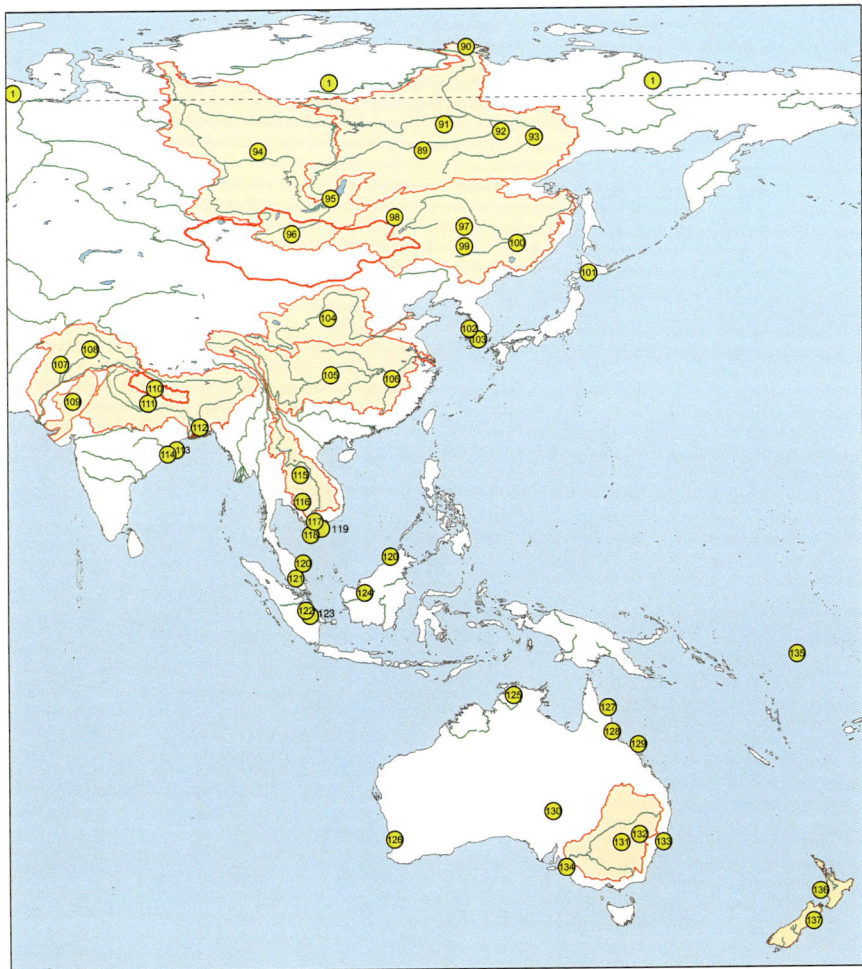

Plate 3 The locations of wetlands, wetland complexes, or wetland landscapes covered in this book within Northern, East, West, Central, South and South-east Asia, Australia, New Zealand, and Pacific Islands. Numbers refer to the accompanying Legend. Note: some numbers appear more than once to represent their wide distribution. Map prepared by G. R. Milton. (*1* Arctic Peatlands, *89* Lena River Basin, *90* Lena River Delta, *91* Nidjili Lake, *92* Taiga-Alas Landscape, *93* Middle Aldan River Basin, *94* Yenisei River Basin, *95* Lake Baikal, *96* Highland Peatlands of Mongolia, *97* Amur-Heilong River Basin, *98* Dauria Steppe Wetlands, *99* Sanjiang Plain, *100* Zhalong Wetlands, *101* Hokkaido Marshes, *102* Saemangeum Estuarine System, *103* Korea Seagrass, *104* Huang He River Basin, *105* Yangtze River Basin, *106* Poyang Lake, *107* Indus River Basin, *108* Wular Lake, Kashmir, *109* India Arid Zone Saline Wetlands, *100* Nepal High Altitude Wetlands, *111* Ganga-Brahmaputra River Basin, *112* Transboundary Sundarbans Mangroves, *113* Mahanadi Delta, *114* Chilika Lake, *115* Mekong river Basin, *116* Tonle Sap Lake, *117* Tram Chim, *118* U Minh, *119* Transboundary Mekong River Delta, *120* Malaysian Seagrass, *121* Tasek Bera, *122* Berbak Nature Reserve, *123* Sembilang National Park, *124* Danau Sentarum National Park, *125* Kakadu National Park, *126* Gnangara Groundwater System, *127* NE Australia Seagrass Meadows, *128* Australia's wet tropics streams, *129* Great Barrier Reef, *130* Kati Thanda - Lake Eyre,

Future Challenges

All of the wetlands discussed in this book are located in countries that are Contracting Parties to the Ramsar Convention on Wetlands "...that provides the framework for national action and international cooperation for the conservation and wise use of wetlands and their resources" (www.ramsar.org). A number of articles have described approaches to minimize and mitigate impacts, and restore wetland ecosystems that necessarily engage sectors of society in their implementation. However, the general conclusion that is evident from all articles is that wetlands are still under significant threat to loss and degradation from human impacts. In responding to their Ramsar commitments, Contracting Parties should plan for wetland resilience, adaptability, and transformability (as per Walker et al. 2004; Hodgson et al 2015). The use of a transdisciplinary approach (e.g., Complex Systems Science, see Filotas et al 2014; Parrott and Quinn 2016) in the analyses of the many interacting facets of these dynamic systems will improve the understanding of wetland responses to natural environmental, anthropogenic, biological, and social changes in order to develop strategies for wetland wise use.

References

Arntzen J. Makgadikgadi wetlands (Botswana): planning for sustainable use and conservation. In: Finlayson CM, Milton GR, Prentice RC, Davidson NC, editors. The wetland book II: distribution, description, and conservation. Dordrecht: Springer; 2018.

Barendregt A. Tidal freshwater wetlands, the fresh dimension of the estuary. In: Finlayson CM, Milton GR, Prentice RC, Davidson NC, editors. The wetland book II: distribution, description, and conservation. Dordrecht: Springer; 2018.

Bartuszevige A. Playa wetlands of the Great Plains (USA). In: Finlayson CM, Milton GR, Prentice RC, Davidson NC, editors. The wetland book II: distribution, description, and conservation. Dordrecht: Springer; 2018.

Beltram G. Karst wetlands. In: Finlayson CM, Milton GR, Prentice RC, Davidson NC, editors. The wetland book II: distribution, description, and conservation. Dordrecht: Springer; 2018.

Brodie J, Waterhouse J. Great Barrier Reef (Australia): a multi-ecosystem wetland with a multiple use management regime. In: Finlayson CM, Milton GR, Prentice RC, Davidson NC, editors. The wetland book II: distribution, description, and conservation. Dordrecht: Springer; 2018.

Daborn GR, Redden AM. Estuaries. In: Finlayson CM, Milton GR, Prentice RC, Davidson NC, editors. The wetland book II: distribution, description, and conservation. Dordrecht: Springer; 2018.

Davidson NC. Wetland losses and the status of wetland-dependent species. In: Finlayson CM, Milton GR, Prentice RC, Davidson NC, editors. The wetland book II: distribution, description, and conservation. Dordrecht: Springer; 2018.

Plate 3 (continued) *131* Murray-Darling River Basin, *132* Macquarie Marshes, *133* Myall Lakes, *134* The Coorong, *135* Atolls - Tropical Pacific Ocean, *136* New Zealand Wetlands, *137* NZ Restiad Bogs)

Davis JB. The Mississippi alluvial valley (USA). In: Finlayson CM, Milton GR, Prentice RC, Davidson NC, editors. The wetland book II: distribution, description, and conservation. Dordrecht: Springer; 2018.

Doherty KE, Howerter DW, Devries JH, Walker J. Prairie pothole region of North America. In: Finlayson CM, Milton GR, Prentice RC, Davidson NC, editors. The wetland book II: distribution, description, and conservation. Dordrecht: Springer; 2018.

Filotas E, Parrott L, Burton PJ, Chazdon RL, Coates KD, Coll L, et al. Viewing forests through the lens of complex systems science. Ecosphere. 2014;5(1):1–23.

Finlayson CM. Salt lakes. In: Finlayson CM, Milton GR, Prentice RC, Davidson NC, editors. The wetland book II: distribution, description, and conservation. Dordrecht: Springer; 2018a.

Finlayson CM. Freshwater marshes and swamps. In: Finlayson CM, Milton GR, Prentice RC, Davidson NC, editors. The wetland book II: distribution, description, and conservation. Dordrecht: Springer; 2018b.

Finlayson CM. Mangroves. In: Finlayson CM, Milton GR, Prentice RC, Davidson NC, editors. The wetland book II: distribution, description, and conservation. Dordrecht: Springer; 2018c.

Finlayson CM. Alien plants and wetland biotic dysfunction. In: Finlayson CM, Milton GR, Prentice RC, Davidson NC, editors. The wetland book II: distribution, description, and conservation. Dordrecht: Springer; 2018d.

Finlayson CM. Wetlands of Kakadu National Park (Australia). In: Finlayson CM, Milton GR, Prentice RC, Davidson NC, editors. The wetland book II: distribution, description, and conservation. Dordrecht: Springer; 2018e.

Finlayson CM, Milton GR. Peatlands. In: Finlayson CM, Milton GR, Prentice RC, Davidson NC, editors. The wetland book II: distribution, description, and conservation. Dordrecht: Springer; 2018.

Finlayson CM, Milton GR, Prentice C. Wetland types and distribution. In: Finlayson CM, Milton GR, Prentice RC, Davidson NC, editors. The wetland book II: distribution, description, and conservation. Dordrecht: Springer; 2018.

Fluet-Chouinard E, Message ML, Lehner B, Finlayson CM. Freshwater lakes and reservoirs. In: Finlayson CM, Milton GR, Prentice RC, Davidson NC, editors. The wetland book II: distribution, description, and conservation. Dordrecht: Springer; 2018.

Froend RH, Horwitz P, Sommer B. Groundwater dependent wetlands. In: Finlayson CM, Milton GR, Prentice RC, Davidson NC, editors. The wetland book II: distribution, description, and conservation. Dordrecht: Springer; 2018.

Galatowitsch SM. Natural and anthropogenic drivers of wetland change. In: Finlayson CM, Milton GR, Prentice RC, Davidson NC, editors. The wetland book II: distribution, description, and conservation. Dordrecht: Springer; 2018.

Giessen W. Tropical freshwater swamps (mineral soils). In: Finlayson CM, Milton GR, Prentice RC, Davidson NC, editors. The wetland book II: distribution, description, and conservation. Dordrecht: Springer; 2018.

Gingras B, Slattery S, Smith K, Darveau M. Boreal wetlands of Canada and the United States of America. In: Finlayson CM, Milton GR, Prentice RC, Davidson NC, editors. The wetland book II: distribution, description, and conservation. Dordrecht: Springer; 2018.

Haddad FF. Qa'a Azraq Oasis, Strengthening stakeholder representation in restoration (Jordan). In: Finlayson CM, Milton GR, Prentice RC, Davidson NC, editors. The wetland book II: distribution, description, and conservation. Dordrecht: Springer; 2018.

Hodgson D, McDonald JL, Hosken DJ. What do you mean 'resilient'? Trends Ecol Evol. 2015;30(9):503–6.

Juffe-Bignoli D, Harrison I, Butchart S, Flitcroft R, Hermoso V, Jonas H, et al. Achieving Aichi biodiversity target 11 to improve protected areas performance and conserve freshwater biodiversity. Aquat Conserv Marine Freshw Ecosyst. 2016;26 Suppl 1:133–51.

Kipkemboi J, van Dam AA. Papyrus wetlands. In: Finlayson CM, Milton GR, Prentice RC, Davidson NC, editors. The wetland book II: distribution, description, and conservation. Dordrecht: Springer; 2018.

Lindsay R. Peatland (mire types): based on origin and behaviour of water, peat genesis, landscape position and climate. In: Finlayson CM, Milton GR, Prentice RC, Davidson NC, editors. The wetland book II: distribution, description, and conservation. Dordrecht: Springer; 2018.

MacKinnon J, Verkuil YI. Intertidal flats of East and Southeast Asia. In: Finlayson CM, Milton GR, Prentice RC, Davidson NC, editors. The wetland book II: distribution, description, and conservation. Dordrecht: Springer; 2018.

Minayeva T, Sirin A, Chultemine D. Highland peatlands of Mongolia. In: Finlayson CM, Milton GR, Prentice RC, Davidson NC, editors. The wetland book II: distribution, description, and conservation. Dordrecht: Springer; 2018a.

Minayeva T, Sirin A, Kershaw P, Bragg O. Arctic peatlands. In: Finlayson CM, Milton GR, Prentice RC, Davidson NC, editors. The wetland book II: distribution, description, and conservation. Dordrecht: Springer; 2018b.

Moores N. Saemangeum estuarine system (Republic of Korea): before and after reclamation. In: Finlayson CM, Milton GR, Prentice RC, Davidson NC, editors. The wetland book II: distribution, description, and conservation. Dordrecht: Springer; 2018.

Page S, Rieley J. Tropical peatswamp forests of South East Asia. In: Finlayson CM, Milton GR, Prentice RC, Davidson NC, editors. The wetland book II: distribution, description, and conservation. Dordrecht: Springer; 2018.

Parrott L, Quinn N. A complex systems approach for multiobjective quality regulation on managed wetland landscapes. Ecosphere. 2016;7(6):1–17.

Pattnaik AK, Kumar R. Lake Chilika (India): Ecological restoration and adaptive management for conservation and wise use. In: Finlayson CM, Milton GR, Prentice RC, Davidson NC, editors. The wetland book II: distribution, description, and conservation. Dordrecht: Springer; 2018.

Ramsar Convention Secretariat. Designating Ramsar sites: strategic framework and guidelines for the future development of the List of Wetlands of International Importance, Ramsar handbooks for the wise use of wetlands, 4th ed., vol. 17. Gland: Ramsar Convention Secretariat; 2010.

Revenga C, Tyrrell T. Major river basins of the world. In: Finlayson CM, Milton GR, Prentice RC, Davidson NC, editors. The wetland book II: distribution, description, and conservation. Dordrecht: Springer; 2018.

Short FT, Short CA, Novak A. Seagrasses. In: Finlayson CM, Milton GR, Prentice RC, Davidson NC, editors. The wetland book II: distribution, description, and conservation. Dordrecht: Springer; 2018.

Silvius M, Joosten H, Opdam S. Peatlands and people, Chapter 3. In: Parish F, Sirin A, Charman D, Joosten H, Minayeva T, Silvius M, Stringer L, editors. Assessment on peatlands, biodiversity and climate change: main report. Kuala Lumpur; Global Environment Centre; 2008. p. 20–38. Joint publication with Wetlands International, Wageningen.

Tiner RW, Milton GR. Estuarine marsh. In: Finlayson CM, Milton GR, Prentice RC, Davidson NC, editors. The wetland book II: distribution, description, and conservation. Dordrecht: Springer; 2018.

Unsworth RKF, Cullen-Unsworth LC. Seagrass dependent artisanal fisheries of Southeast Asia. In: Finlayson CM, Milton GR, Prentice RC, Davidson NC, editors. The wetland book II: distribution, description, and conservation. Dordrecht: Springer; 2018.

Walker B, Holling CS, Carpenter SR, Kinzig AP. Resilience, adaptability and transformability in social-ecological systems. Ecol Soc. 2004;9(2):5. Accessed from http://www.ecologyandsociety.org/vol9/iss2/art5.

Section II

Diversity of Wetlands

Wetland Types and Distribution

2

C. Max Finlayson, G. Randy Milton, and R. Crawford Prentice

Contents

Introduction 20
Extent and Distribution of Wetlands 26
Wetland Inventory 31
References 32

Abstract

Wetlands have been defined by referring to particular features of the vegetation, the water quality, water regimes, soils, and landform features, and the presence of particular fauna. This is reflected in the large number of terms in use to describe wetlands. While many maps and estimates have been produced we still lack adequate information about the extent and distribution of many wetlands. Further wetland inventory is required to describe and map wetlands in many parts of the world. The Ramsar Convention has supported such initiatives and encouraged the adoption of comparable methods.

Keywords

Classification · Definition · Distribution · Extent · Inventory · Wetland types

---

C. M. Finlayson (✉)
Institute for Land, Water and Society, Charles Sturt University, Albury, NSW, Australia

UNESCO-IHE, The Institute for Water Education, Delft, The Netherlands
e-mail: mfinlayson@csu.edu.au

G. R. Milton
Nova Scotia Department of Natural Resources, Kentville, NS, Canada
e-mail: gordon.milton@novascotia.ca

R. C. Prentice
Nature Management Services, Histon, Cambridge, UK
e-mail: crawford.prentice@gmail.com

© Crown 2018
C. M. Finlayson et al. (eds.), *The Wetland Book*,
https://doi.org/10.1007/978-94-007-4001-3_186

Introduction

Wetlands have been defined in many different ways, generally by referring to particular features of the vegetation (macrophytic and microphytic), the water quality (including salinity, nutrients, and color), water regimes, soils, and landform features, and also in relation to the presence of particular fauna, such as waterbirds. This is reflected in the large number of terms in use globally to describe wetlands (some of which are listed in Table 1), often having an historical and regional or continental origin (Mitsch and Gosselink 2015). A selection of these wetland types is shown in Plate 1.

Many countries have different definitions for wetlands with one of the broadest definitions having been adopted by the Ramsar Convention on Wetlands, namely: "... wetlands are areas of marsh, fen, peatland or water, whether natural of artificial, permanent or temporary, with water that is static or flowing, fresh, brackish or salt, including areas of marine water the depth of which at low tide does not exceed six metres." (Ramsar Convention Secretariat 2010). Wetland definitions generally include the basic attributes of geologic/geomorphic, hydrologic, and biotic features and processes (Semeniuk and Semeniuk 2011) and are used to construct wetland typologies (e.g., Cowardin et al. 1979; Brinson 1993; Warner and Rubec 1997). Semeniuk and Semeniuk's (2011) typology however uses only landform types and water regimes as the basis to describe the global diversity of inland wetlands into 22 primary categories of wetlands. The typology that accompanies the Ramsar definition incorporates the three attributes to describe 42 habitat types across coastal/marine and inland environments and includes rivers, caves and reefs, as well as the more traditionally considered marshes and swamps (Ramsar Convention Secretariat 2010).

Differences in definitions have caused some difficulty in determining the extent of wetlands although considerable effort has been directed towards wetland inventory and mapping in parts of the world (Mitsch 1994; Finlayson and van der Valk 1995a; Finlayson et al. 1999; Mitsch and Gosselinck 2015). A number of the many notable collations of information on wetlands are described, in brief, in Table 2. These provide an indication of the extent of information that has been collated and also the variety of approaches to collating and combining such information. They further complement the many national and subnational inventories that have been undertaken, although there are still significant gaps in the coverage of wetland distribution and description globally (Mitsch 1994; Finlayson and van der valk 1995a, b; Finlayson et al. 1999) with many countries not having completed national wetland inventories (Finlayson 2012).

Many authors have identified gaps in the inventory and knowledge of wetlands, as well as the limited specific guidance that is available for conducting inventories that can then be used as a basis for the effective management of wetlands globally. The IV International Wetlands Conference (Columbus, Ohio, USA) in 1992 encouraged the development of guidelines for national wetland inventories given the gaps that existed in the global information base for wetlands (Finlayson and van der Valk 1995a). The Ramsar Convention on Wetlands has also supported the development of

2 Wetland Types and Distribution

Table 1 Some common names for wetlands in English (taken largely from information contained in papers in Finlayson and van der Valk 1995b and from Mitsch and Gosselink 2015). While usage of the term "wetland" has increased in recent decades, there is a diversity of terms, including in other languages that have been used to describe wetland ecosystems, and also including those in the Ramsar Convention's definition and classification of wetland types (Ramsar Convention Secretariat 2010)

Billabong	Australian term that is loosely used to describe lagoons and remnant pools in stream channels.
Bog	A peat-accumulating wetland that has no significant inflows or outflows and supports acidophilic mosses, particularly sphagnum. Includes domed and blanket bogs.
Bottomland	Lowlands, usually forested, along streams and rivers, usually on alluvial floodplains that are periodically flooded.
Carr	Wetland dominated by woody vegetation
Delta	Fan-shaped accumulation of alluvial sediments usually at mouth of a river. Inland deltas also occur
Estuary	Semi-enclosed water bodies connected with the open sea and influenced by tides and within which sea water is variously diluted with fresh water derived from land drainage
Fen	A peat-accumulating wetland that receives some drainage from surrounding mineral soil and usually supports marsh vegetation
Flark	Permanently or temporarily flooded depression containing sparse peat forming vegetation in peatlands
Intertidal flat	Flat or gently inclined nonvegetated coastal wetlands in estuaries or low energy marine environments with regular exposure and flooding by tides
Lagg	Marginal stream or swamp fringing a domed bog
Lake	Large body of water surrounded by land
Lagoon	Term frequently used to denote deep-water enclosed or partially opened aquatic system, especially in coastal delta regions
Loch	Irish and Scottish term for a lake or sea inlet. Also spelt lough. A lochan is a small loch
Mangrove	Shrubs and trees generally found in low wave energy intertidal environments in the tropics and subtropics
Marsh	A frequently or continually inundated wetland with emergent herbaceous vegetation in a mineral soil substrate
Mire	Synonymous with any actively peat-accumulating wetland
Moor	Synonymous with any peatland. A "highmoor" is a raised bog, while a "lowmoor" is a peatland in a basin that is not elevated above its perimeter
Muskeg	North American term for large expanses of peatlands or bogs
Peatland	A generic term for a wetland with a naturally accumulated layer of partly decayed plant matter (peat) greater than 30 cm deep under permanent water saturation
Playa	Flat bottomed depression in arid- to semi-arid regions with a distinct wet and dry seasons.
Polje	Flat-floored enclosed depressions with karstic drainage which may be dry or occur as seasonally flooded wetland or permanent lakes
Pond	Small pool of still water
Pothole	Shallow marsh-like ponds, particularly as found in the USA and Canadian prairie provinces.

(*continued*)

Table 1 (continued)

Reed swamp	Marsh dominated by *Phragmites* (common reed); term used particularly in Europe.
Rhyne	UK term for a drainage ditch or canal used to convert coastal wetlands into wet pasture
Riparian System	Ecosystems with a high water table because of proximity to an aquatic system, usually a stream or river.
Sabkha	A term for evaporative coastal salt flats where infrequent supratidal inundation and evaporation of saline groundwater drawn upward by capillary action forms a well-defined salt crust
Salt marsh	Wetlands dominated by salt-tolerant plants in the mid to upper intertidal zone, generally above mean sea level and above tidal flats, subject to frequent tidal flooding
Slough	A swamp or shallow lake system in the northern and midwestern United States. A slowly flowing shallow swamp or marsh in southeastern United States
Swale	Elongated depression between dunes or coastal ridges roughly parallel to the coast
Swamp	Wetlands dominated by trees of shrubs. In Europe forested fens and areas dominated by reeds (*Phragmites*) are also called swamps
Tarn	Small lake in the mountains, often with no significant tributaries
Tidal freshwater marsh	Marshes along rivers and estuaries close enough to the coastline to experience significant tides by nonsaline water
Turlough	Shallow depressions in a Carboniferous karst landscape subject to periodic flooding mainly from groundwater
Vernal Pool	Shallow temporary pools with seasonal flooding
Vlei	Southern African term for a shallow, seasonal, or intermittently flooded lake
Wet Meadow	Waterlogged grassland but without standing water for most of the year
Wet Prairie	Similar to a marsh but with water levels usually intermediate between a marsh and a wet meadow

national wetland inventories as a basis for the wise use of wetlands globally and in 1999 reiterated the value of completing comprehensive national inventories as a basis for supporting the management and restoration of wetlands (Ramsar Convention 1999). The Convention further supported the development of standardized protocols for wetland inventory, such as the Mediterranean wetland methodology (Tomas Vives 1996) and those outlined in Finlayson and van der Valk (1995b) and the later protocol for wetland inventory in Asia (Finlayson et al. 2002). The development of such protocols and in particular advances in the application of remote sensing to wetland mapping and inventory (Rosenqvist et al. 2007; Rebelo et al. 2009) have provided a strong technical basis for determining the distribution and ecological state of wetlands globally.

The following text contains a generalized description of the distribution of wetlands globally and a comment on the directions of current wetland inventory. This complements the detailed text and information on wetland types in other articles in this book.

Plate 1 (continued)

Plate 1 (continued)

2 Wetland Types and Distribution

Plate 1 (continued)

Plate 1 Examples of wetlands: (**a**) Intertidal sand and mud flats, Minas Basin (Canada); (**b**) *Spartina alterniflora* low salt marsh Southern Bight Minas Basin Ramsar Site (Canada); (**c**) sabkha flats Dubai (UAE.); (**d**) saline flats Danube River delta (Romania); (**e**) coastal Everglades, Florida (USA); (**f**) mature mangrove forest Santubong River estuary, Sarawak (Malaysia); (**g**) riverine floodplain, Wisconsin River (USA); (**h**) Pantanal riverine marsh and forest, Rio Claro (Brazil); (**i**) vernal pool in temperate forest Nova Scotia (Canada); (**j**) Cypress forest showing pneumatophores, Florida (USA); (**k**) wetland complexes in boreal forest (Canada); (**l**) lowland tropical dipterocarp peat swamp forest with encroaching *Pandanus helicopus* clumps, Tasek Bera (Malaysia); (**m**) periodically flooded savanna in the Pantanal, upper Paraguay River (Brazil); (**n**) *Phragmites* marsh at Zhalong Marsh, Heilongjiang (China); (**o**) *Phragmites* marsh in lower reaches of the Danube River delta (Romania); (**p**) dense cattail *Typha* sp. at Delta marsh, Manitoba (Canada); (**q**) prairie pothole marshes in the Missouri Coteau, southern Saskatchewan (Canada); (**r**) swamp restoration along Tua Marina River, South Island (New Zealand); (**s**) cotton grass in transitional mire in Onon-Balj National Park (Mongolia); (**t**) Restiad peatland plants: *Empodisma minus* (wire rush), *Gleichenia dicarpa* (tangle fern), and *Sphagnum* moss in peatland at Awarua-Waituna Ramsar Site, Southland (New Zealand); (**u**) domed *Sphagnum* and *Carex* bog in highlands Cape Breton Island (Canada); (**v**) Picea mariana (black spruce) peatland, Nova Scotia (Canada); (**w**) palsa mires in Bolshezemelskaya tundra, Nenets okrug (Russia); (**x**) tundra lakes at Kytalyk Nature Reserve, Sakha Republic (Russia); (**y**) Waikoropupu Springs, a karst wetland, Takaka, South Island (New Zealand); (**z**) Yellow Water lagoon, Kakadu National Park (Australia). Photo credits: (**a**) M.F. Elderkin ©; (**b–e, g, h, j, m, o, p, r, u, y**) G.R. Milton ©; (**f, l, n, x**) C. Prentice ©; (**i, v**) J. Brazner ©; (**k, q**) Ducks Unlimited Canada ©; (**t**) K. Denyer, National Wetland Trust of New Zealand ©; (**s, w**) T. Minayeva ©; (**z**) C.M. Finlayson © Rights remain with the authors

Extent and Distribution of Wetlands

The global extent of wetlands has been estimated by a number of authors with the outcomes dependent on the definition of wetlands that was used and the accuracy of the methods. Much of this information now provides a historical perspective on wetlands although the emphases and the level of detail vary greatly. There is an increasing amount of information on individual wetlands or types of wetlands including comparative studies of large sites such as that provided by Junk et al. (2006) for a number of important wetlands, or by Lukacs and Finlayson (2010) for the wetland expanses that occur across northern Australia, or for peatland globally by Joosten (2010).

Many estimates of the extent of wetland globally have not adopted the definition used by the Ramsar Convention and hence cover a smaller range of habitats. When

Table 2 Regional and global collations of information on wetlands (note–compilations on specific wetland types are listed in Table 4)

Authors	Year	Title	Coverage
Scott and Carbonell	1986	A directory of Neotropical wetlands	Distribution, extent, and features of wetlands in 45 countries or territories. Used the Ramsar definition of wetlands although coral reefs and exclusively marine systems were excluded.
Burgis and Symoens	1987	African wetlands and shallow water bodies	Detailed information on many of the largest and well known wetlands and shallow water bodies in Africa
Scott	1989	A directory of Asian wetlands.	Distribution, extent, and features of wetlands in 24 countries. Used the Ramsar definition of wetlands although coral reefs and exclusively marine systems were excluded
Hughes and Hughes	1992	A directory of African wetlands	Distribution, extent, and features of wetlands in 48 countries. Covers lakes, ponds, swamps, marshes, bogs, riverine and lacustrine floodplains, pans and wadis, coastal saltmarshes, and mangrove swamps.
Scott	1993	A directory of wetlands in Oceania	Distribution, extent and features of wetlands in 25 countries or territories. Used the Ramsar definition of wetlands although coral reefs and exclusively marine systems were excluded
Whigham et al.	1993	Wetlands of the world: inventory, ecology and management	Distribution, extent, ecology and management of wetlands in nine countries or regions. Focused on inland and coastal wetlands with country or regional definitions and classifications used.
Scott	1995	A directory of wetlands in the Middle East	Distribution, extent and features of wetlands in 13 countries. Used the Ramsar definition of wetlands although coral reefs and exclusively marine systems were excluded
Finlayson and van der Valk	1995b	Classification and inventory of the World's wetlands	Global overview of the extent of wetland inventory and classification, in particular the extent of information on wetland types in different parts of the world.

(*continued*)

Table 2 (continued)

Authors	Year	Title	Coverage
Gopal et al.	2000	Biodiversity in wetlands: assessment, function and conservation. Volume 2	Biodiversity, functions, and conservation of wetlands in different regions and important sites.
Polunin	2008	Aquatic ecosystems: trends and global prospects	Structure and function of 21 ecosystem types, from those on land (freshwater and saline) to those in the deep oceans.
Joosten	2010	The Global Peatland CO_2 Picture: Peatland status and drainage related emissions in all countries of the world	Information on the area of peatlands and carbon dioxide emissions in all countries

comparing such estimates, it is important to ascertain exactly what is being included and the accuracy of the estimates or measurements. Finlayson et al. (1999) pointed out in a summary of 233 information sources for wetland inventories that many did not include such statements. It is also necessary to check the currency of the estimates and whether or not they account for the rapid rates of wetland loss globally. For example, only about 10% of the wetlands data in the USA National Wetland Inventory are reported to be current based upon source materials less than 10 years old (Stout et al. 2007) and updated at a rate of 2% per year (https://www.fws.gov/wetlands/nwi/overview.html Accessed: 13 July 2016). This is particularly important given that Davidson (2014) has reported that on the basis of information taken from 189 published sources that some 64–71% of wetlands have been lost since 1900 AD. Davidson (2014) further reported that the evidence for such changes was patchy, in particular for Africa, the Neotropics, and Oceania, with the USA, China and Europe having more comprehensive records. The USA has possibly the most comprehensive wetland inventory and analysis of status and trends program (Dahl 2011; Dahl and Stedman 2013).

Global estimates of the extent of wetlands have varied as have the corresponding estimates of the area of the world's landmass that comprises wetlands (see discussion in Mitsch and Gosselink 2015). Despite the large effort to describe and map wetlands globally, a comprehensive and accurate map of all wetlands types does not exist. The collation undertaken on behalf of the Ramsar Convention in 1999 was acknowledged to be an underestimate while providing data for a large number of wetland types (Table 3). This analysis has not been updated, despite calls by the Convention for this to be done and a central repository for such information secured (Ramsar Convention 1999). Nor has the veracity of the underlying data sources been investigated and compared to other data sources.

The Global Review of Wetland Resources and Priorities for Wetland Inventory undertaken in support of the Ramsar Convention (Finlayson et al. 1999) used a bottom-up approach, based on an assessment of available wetland inventories, to estimate the global wetland area to be in excess of 1,280 million hectares, a figure

Table 3 Estimates of global wetland area (in million hectares) by Ramsar regions

Region	Global Review of Wetland Resources (Finlayson et al. 1999)	Global Database of Lakes, Reservoirs and Wetlands (Lehner and Doll 2004)
Africa	121–125	131
Asia	204	296
Europe	258	26
Neotropics	415	159
North America	242	287
Oceania	36	28
Total	**c. 1,280**	**917**

much larger than other estimates, and from this aspect in need of further investigation, but still considered to be an underestimate, given identified gaps in coverage. This estimate included inland and coastal wetlands, near-shore marine areas, and artificial wetlands such as rice paddies and reservoirs and rice paddies. While it considered some 233 sources of information, it also recommended that further information was required, in particular for intermittently flooded inland wetlands, peatlands, artificial wetlands, seagrasses, and coastal flats. Estimates of the extent of peatlands have been provided more recently based on an extensive collation of information on individual countries (Joosten 2010) with a global estimate of around four million square kilometers, comprising almost a third of the above estimate of all wetlands globally.

Lehner and Doll (2004) compiled information from a number of digital maps and databases to produce a Global Lakes and Wetlands Database (Fig. 1) and provide estimates of the extent of wetlands globally. The regional estimates extracted from Lehner and Doll (2004) and Finlayson et al. (1999) are presented in Table 3. This shows large discrepancies between the values for Europe and the Neotropics in particular with those collated by Finlayson et al. (1999) being much higher than those obtained from the digital databases. The reasons for these discrepancies have not been investigated in detail. Subsequent analyses have also shown further information on the extent of wetlands in South America (Junk et al. 2006). As wetland inventories are also being produced in Chile (Fernández 2018), Colombia (Ricaurte et al. 2015), and Argentina (Benzaquén et al. 2013), it is expected that more information about the extent of wetlands in the Neotropics will soon be available.

A number of estimates of the extent of land area inundated by freshwater have been prepared from Earth Observation sources, including from (Prigent et al. 2001) and Chen et al. (2015). While these analyses do not equate directly with the extent of wetlands without further analysis of landforms and consideration of issues with temporal variation and spatial resolution, they do provide an overview of the importance of water in the terrestrial environment and the increasing capability of such technology. There also may be differences in areal estimates given differences in the ecosystems that are defined and included in the analyses.

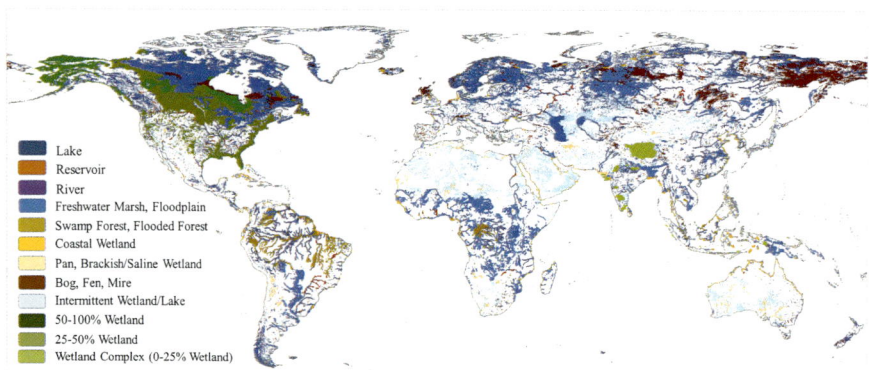

Fig. 1 Global distribution of wetlands (Reprinted from Journal of Hydrology, Vol 296, B. Lehner and P. Döll, Development and validation of a global database of lakes, reservoirs and wetlands, page 16, 2004, with permission from Elsevier)

Table 4 Estimates of area of wetland types globally

Wetland type	Area km^2	Reference
Tropical freshwater swamps	1 460,000	Multiple sources in Giessen 2018
Freshwater lakes	2.7–5,000,000	Multiple sources in Fluet-Chouinard et al. 2018
Reservoirs larger than 10 ha	284,700	Lehner et al. 2011
Salt/brackish marsh	67,580	Schuyt and Brander 2004 as reported by Tiner 2013
Unvegetated sediment	457,880	
Mangroves	121,120	
Mangroves	137,760	Giri et al. 2011
Seagrasses	177,000	Green and Short 2003; Spalding et al. 2003
Peatlands	4,000,000	Joosten 2010

Similar data are being used to provide estimates of the extent of other wetland types, as shown in Table 4. These data are from different sources and based on different approaches for determining the areal estimates. In some instances, they are not mutually exclusive which makes it difficult to derive an estimate of the total area of wetlands globally from such sources.

The Ramsar Convention has a database containing information on internationally important wetlands, known as Ramsar Sites, including some 2,241 sites that contain wetlands and cover 215,240,661 ha (www.ramsar.org accessed 20 June 2016). This represents 17% or 23% of the area of wetlands globally (based on data supplied in Table 3), although the total area of the Ramsar Sites includes an undetermined amount of nonwetland area. The information on Ramsar Sites has been supplied by the 169 countries who have signed the Convention, using a standard recording process. This database is also supported by the Ramsar Site Information Service which enables interrogation of the data from individual sites or countries. The wetland types within each Ramsar Site have been described using the Ramsar

Wetland Classification which is a listing of wetland types, not a systematic classification scheme (Semeniuk and Semeniuk 1997), that were considered by the Convention (Ramsar Convention Secretariat 2010).

Despite efforts to encourage the development of national wetland inventories, the Convention reported in 2012 that only 37% of its then 160 Contracting Parties (countries) had a national wetland inventory (Finlayson 2012). The IV International Wetlands Conference in 1992 also noted the absence of a suitable repository for global wetland information (Finlayson and van der Valk 1995a) – a situation that does not seem to have changed. Hence, while the Ramsar Convention records what countries have or are undertaking a national wetland inventory, these are not collated nor incorporated into a single global overview of wetlands. Access to the national wetland inventories recorded by the Convention is through the reporting countries.

A series of wetland continental-scale directories were produced in the 1980s and 1990s (see Table 2) and included individual volumes for the Neotropics (Scott and Carbonell 1986), Asia (Scott 1989), Oceania (Scott 1993), and the Middle East (Scott 1995) using a standardized format. A similar format was used for an Australian directory (Usbank and James 1993) with a later version being produced online and with fewer data fields. A different format was used for a directory of African wetlands (Burgis and Symoens 1987; Hughes and Hughes 1992). The data from the Neotropics, Asian, and Oceanian directories were scanned and made available on the internet by the International Water Management Institute (Rebelo et al. 2009) with the sub-Saharan African data being used in a subsequent analysis of the contribution of wetlands to agriculture (Rebelo et al. 2010).

Wetland Inventory

The Global Review of Wetland Resources comprised an assessment of the extent of wetlands in seven regions with a key conclusion that there was insufficient knowledge about the extent and condition of the global wetland resource (Finlayson et al. 1999). It was further concluded that it was not possible to make reliable overall estimates of the size of the wetland resource globally or regionally. Recommendations from the Global Review focused on the need for further national inventory programs and the inclusion of information on the location and extent of wetlands and their ecological features. This included recommendations on the collection of core data, including that for the area and boundary of the wetland, the location and geomorphic setting, and a general description of the soil, water regime and quality, and the characteristic biota. The Ramsar Convention adopted the basic recommendations from this review, although, as mentioned above, there are still many gaps in the global information base. A shortened and updated version of the specific recommendations from the Review are presented below as adapted in the light of further inventory, including through the use of remote sensing approaches. Examples of such effort are presented in the above text.

1. Countries lacking a national wetland inventory should undertake one, using, where feasible, approaches that are comparable with other inventories, noting that many Earth Observation technologies are now available.
2. Quantitative studies of wetland loss and degradation are required for many parts of the world particularly, given known recent rates of wetland loss and degradation.
3. Further inventory should focus on a core data set that can be used to describe the location and size of each wetland and its major features, including variations in area and the water regime.
4. Further information should be collected about the use of wetlands, threats to wetlands, management approaches, and values and benefits of wetlands.
5. Each inventory should contain a clear description of its purpose and the type of information that is collected and where this is stored.
6. The Ramsar Convention should further encourage the development and dissemination of models for improved globally applicable wetland inventory.
7. The Ramsar Convention should further support efforts to develop a central repository for inventory information, and the meta-data that are used to describe the inventories should be published for greater accessibility.
8. Further support is required for a global review of wetland resources and to develop procedures for regular updating and publishing of inventory information on wetlands as a contribution to understanding the state of the World's wetlands.

While these recommendations include specific reference to the Ramsar Convention, it is noted that the countries that have signed the Convention are responsible for its implementation. The Convention has been able to support international efforts to undertake wetland inventory using Earth Observation approaches (e.g., Rosenqvist et al. 2007; Fernandez-Prieto and Finlayson 2009; Mackay et al. 2009) and make these sources available to others. Given the gaps and a lack of consistency between many approaches for wetland inventory, the processes adopted by the Convention could greatly assist in closing such gaps and providing a more consistent data set for ensuring the wise use of wetlands globally.

References

Benzaquén L, Blanco DE, Bó RF, Kandus P, Lingua GF, Minotti P, Quintana RD, Sverlij S, Vidal L. Inventario de los Humedales de Argentina. Sistemas de Paisajes de humedales del Corredor Fluvial Paraná- Paraguay. Secretaria de Ambiente y Desarrollo Sustentable de la Nación GEF 4206 PNUD ARG 10/003. 2013. 376 p.

Brinson MM. A hydrogeomorphic classification for wetlands. vicksburg (MS): U.S. Army corps of Engineers Waterways Experiment Station; 1993. Wetlands Research Program Technical report WRP-DE-4.

Burgis MJ, Symoens JJ, editors. African wetlands and shallow water bodies. Paris: ORSTOM; 1987.

Chen J, Chen J, Liao A, Cao X, Chen L, Chen X, He C, Han G, Peng S, Lu M, Zhang W, Tong X, Mills J. Global land cover mapping at 30 m resolution: a POK-based operational approach. J Photogramm Remote Sens. 2015;103:7–27.

Cowardin LM, Carter V, Golet C, LaRoe ET. Classification of wetlands and deepwater habitats of the United States. Washington, DC: U.S. fish and Wildlife Service; 1979. FWS/OBS-79/31.

Dahl TE. Status and trends of wetlands in the conterminous United States 2004 to 2009. Washington, DC: U.S. Department of the Interior; Fish and Wildlife Service. 2011. 108 p.

Dahl TE, Stedman SM. Status and trends of wetlands in the coastal watersheds of the conterminous United States 2004 to 2009. US Department of the Interior, Fish and Wildlife Service & National Oceanic and Atmospheric Administration, National Marine Fisheries Service. 2013. Available from: http://www.habitat.noaa.gov/pdf/Coastal_Watershed.pdf

Davidson NC. How much wetland has the world lost? Long-term and recent trends in global wetland area. Mar Freshw Res. 2014;65:934–41.

Fernández AF. Wetlands of Chile. Biodiversity, endemism and conservation challenges. In: Finlayson CM, Milton GR, Prentice C, Davidson NC, editors. The wetland book II: distribution, description, and conservation. Dordrecht: Springer; 2018.

Fernandez-Prieto D, Finlayson CM. Special issue: The GlobWetland symposium–looking at wetlands from space. J Environ Manage. 2009;90(7):2119–286.

Finlayson CM. Forty years of wetland conservation and wise use. Aquatic Conserv Marine Freshw Ecosyst. 2012;22:139–43.

Finlayson CM, van der Valk AG. Wetlands classification and inventory: a summary. Vegetatio. 1995a;118:185–92.

Finlayson CM, van der Valk AG, editors. Classification and inventory of the world's wetlands. Advances in Vegetation Science 16. Dordrecht: Kluwer; 1995b. 192 p.

Finlayson CM, Davidson NC, Spiers AG, Stevenson NJ. Global wetland inventory–status and priorities. Mar Freshw Res. 1999;50:717–27.

Finlayson CM, Begg GW, Howes J, Davies J, Tagi K, Lowry J. A manual for an inventory of Asian wetlands (version 1.0). Wetlands International Global Series 10. Kuala Lumpur: Wetlands International; 2002. 72 p.

Fluet-Chouinard E, Message ML, Lehner B, Finlayson CM. Freshwater lakes and reservoirs. In: Finlayson CM, Milton GR, Prentice C, Davidson NC, editors. The wetland book II: distribution, description, and conservation. Dordrecht: Springer; 2018.

Giessen W. Tropical freshwater swamps (mineral soils). In: Finlayson CM, Milton GR, Prentice C, Davidson NC, editors. The wetland book II: distribution, description, and conservation. Dordrecht: Springer; 2018.

Giri C, Ochieng E, Tieszen LL, Zhu Z, Singh A, Loveland T, Masek J, Duke N. Status and distribution of mangrove forests of the world using earth observation satellite data. Glob Ecol Biogeogr. 2011;20:154–9.

Gopal B, Junk WJ, Davis JA. Biodiversity in wetlands: assessment, function and conservation, vol. 2. Leiden: Backhuys Publishers; 2000, 311 p.

Green EP, Short FT, editors. World atlas of seagrasses. Berkeley: University of California Press; 2003, 324 p.

Hughes RH, Hughes S. A directory of African wetlands. Nairobi: UNEP; 1992. Joint publications with IUCN, Gland/WCMC, Cambridge, UK.

Joosten H. The global peatland CO_2 picture: peatland status and drainage related emissions in all countries of the world. Ede: Wetlands International; 2010, 36 p.

Junk WJ, Brown M, Campbell IC, Finlayson CM, Gopal B, Ramberg L, Warner BG. Comparative biodiversity of large wetlands: a synthesis. Aquat Sci. 2006;68:400–14.

Lehner B, Döll P. Development and validation of a global database of lakes, reservoirs and wetlands. J Hydrol. 2004;296:1–22.

Lehner B, Liermann CR, Revenga C, Vörösmarty C, Fekete B, Crouzet P, Döll P, et al. High-resolution mapping of the world's reservoirs and dams for sustainable river-flow management. Front Ecol Environ. 2011;9:494–502.

Lukacs GP, Finlayson CM. An evaluation of ecological information on Australia's northern tropical rivers and wetlands. Wetl Ecol Manag. 2010;18:597–625.

MacKay H, Finlayson CM, Fernandez-Prieto D, Davidson N, Pritchard D, Rebelo L-M. The role of Earth Observation (EO) technologies in supporting implementation of the Ramsar Convention on Wetlands. J Environ Monit. 2009;90:2234–42.

Mitsch WJ, editor. Global wetlands: old world and new. Amsterdam: Elsevier; 1994.

Mitsch WJ, Gosselink JG. Wetlands. 5th ed. Hoboken: Wiley; 2015.

Mitsch WJ, Gosselink JG, Anderson CJ, Zhang L. Wetland ecosystems. Hoboken: Wiley; 2009.

Polunin NVC, editor. Aquatic ecosystems: trends and global prospects. Cambridge, UK: Cambridge University Press; 2008.

Prigent C, Matthews E, Aires F, Rossow WB. Remote sensing of global wetland dynamics with multiple satellite data sets. Geophys Res Lett. 2001;28:4631–4.

Ramsar Convention Secretariat. Designating Ramsar sites: strategic framework and guidelines for the future development of the List of Wetlands of International Importance, Ramsar handbooks for the wise use of wetlands, 4th ed., vol. 17. Gland: Ramsar Convention Secretariat; 2010.

Ramsar. Priorities for wetland inventory. Resolution VII.20. 7th Meeting of the conference of the contracting parties to the convention on wetlands (Ramsar, Iran, 1971); 1999 May 10–18, San José. Gland: Ramsar Convention Secretariat; 1999.

Rebelo L-M, Finlayson CM, Nagabhatla N. Remote sensing and GIS for wetland inventory, mapping and change analysis. J Environ Manage. 2009;90:2144–53. doi:10.1016/j.jenvman.2007.06.027.

Rebelo L-M, McCartney MP, Finlayson CM. Wetlands of Sub-Saharan Africa: distribution and contribution of agriculture to livelihoods. Wetl Ecol Manag. 2010;18:557–72.

Ricaurte LF, Patiño JE, Arias-G JE, Ó Acevedo O, Restrepo D, Jaramillo Villa U, Flórez-Ayala C, Estupiñán-Suárez L, Aponte C, Rojas S, Ignacio Vélez J, Duque S, Núñez-Avellaneda M, Lasso C, Darío Correa I, Rodríguez Rodríguez A, Duque Nivia AA, Restrepo S, Cleef A, Manrique O, Moreno EP, Vilardy Quiroga SP, Finlayson M, Junk W. Tipos de humedales de Colombia – Sistema de clasificación de humedales. In: Un pais humedales, Jaramillo Villa Ú, Cortés-Duque J, Florez-Ayala C, editors. Colombia Anfibia, vol. 1. Bogota: Instituto de Investigacion de Recursos Biologicos Alexander von Humboldt; 2015. p. 118–21.

Rosenqvist A, Finlayson CM, Lowry J, Taylor DM. The potential of spaceborne (L-band) radar to support wetland applications and the Ramsar Convention. Aquat Conserv Marine Freshw Ecosyst. 2007;17:229–44.

Schuyt K, Brander LM. "The economic values of the world's wetlands". Future flooding and coastal erosion risks. 2004. WWF, Gland, Switzerland.

Scott DA. A directory of Asian wetlands. Gland/Cambridge, UK: IUCN; 1989.

Scott DA. A directory of wetlands in Oceania. Slimbridge: IWRB; 1993. 444 p. Joint publications with AWB, Kuala Lumpur.

Scott DA. A directory of wetlands in the Middle East. Gland: IUCN; 1995. 560 p. Joint publication with IWRB, Slimbridge.

Scott DA, Carbonell M. A directory of Neotropical wetlands. Gland: IUCN; 1986.

Semeniuk V, Semeniuk CA. A geomorphic approach to global classification for natural inland wetlands and rationalization of the system used by the Ramsar Convention–a discussion. Wetl Ecol Manag. 1997;5(2):145–58.

Semeniuk CA, Semeniuk V. A comprehensive classification of inland wetlands of Western Australia using the geomorphic-hydrologic approach. J R Soc West Aust. 2011;94:449–64.

Spalding M, Taylor M, Ravilious C, Short F, Green E. Global overview: the distribution and status of seagrasses. In: Green EP, Short FT, editors. World atlas of seagrasses. Berkeley: University of California Press; 2003. p. 5–26.

Stout DJ, Kodis M, Wilen BO, Dahl TE. Wetlands layer – national spatial data infrastructure: a phased approach to completion and modernization. US Fish and Wildlife Service. 2007. Available from: https://www.fws.gov/wetlands/Documents/Wetlands-Layer-National-Spatial-Data-Infrastructure-A-Phased-Approach-to-Completion-and-Modernization.pdf. Accessed 13 July 2016.

Tiner RW. Tidal wetlands primer: an introduction to their ecology, natural history, status, and conservation. Amherst: University of Massachusetts Press; 2013.

Tomas Vives P, editor. Monitoring Mediterranean wetlands: a methodological guide. MedWet publication. Slimbridge: Wetlands International; 1996. Joint publication with IUCN, Lisbon.

Usbank S, James R. A directory of important wetlands in Australia. Canberra: Australian Nature Conservation Agency; 1993, 687 p.

Warner BG, Rubec CDA, editors. The Canadian wetland classification system. Waterloo: Wetlands Research Center, University of Waterloo; 1997.

Whigham DF, Dykyjova D, Hejny S, editors. Wetlands of the world: inventory, ecology and management, vol. 1. Dordrecht: Kluwer; 1993.

Estuaries

3

Graham R. Daborn and Anna M. Redden

Contents

Introduction	37
Classification of Estuaries	39
Dynamics of Estuaries	41
Productivity and Biodiversity	44
Ecosystem Services	48
Threats and Future Challenges	50
References	53

Abstract

Estuaries are among the most productive and valuable ecosystems on the planet, offering a wide array of ecosystem services to human beings and wildlife. They are also the among the most impacted by human activities.

Keywords

Estuarine classification · Salt wedge · Partially and well mixed estuaries · Fjords and lagoons · Tidal rhythms · Productivity · Biodiversity · Ecosystem services · Human impacts

Introduction

Estuarine environments are coastal features in which freshwater from the land interacts with saline water from the ocean. They are distinct from most other water systems on Earth, such as the freshwater lakes and rivers and the saltwater oceans,

---

G. R. Daborn (✉) · A. M. Redden (✉)
Acadia Centre for Estuarine Research, Acadia University, Wolfville, NS, Canada
e-mail: graham.daborn@acadiau.ca; anna.redden@acadiau.ca

because they are relatively physically stressed environments, experiencing wide variations in salinity, temperature, clarity, and turbulence. Physical stresses often lead to low biodiversity but high abundance and productivity for those species able to tolerate the conditions. Most of the world's largest cities and more than half of the human population are found near coastal estuaries, supported by environmental services such as food, shelter, and transportation. As a consequence of human activities, however, estuarine environments are also among the most heavily impacted ecosystems.

What is an estuary? Cameron and Pritchard (1963) defined an estuary as "a semi-enclosed body of water having a free connection with the open sea and within which sea water is measurably diluted with fresh water derived from land drainage." This definition would include a number of coastal environments, such as river mouths, deltas, lagoons, fjords, bays, and straits, all of which may be places where fresh- and salt water meet. Kjerfve (1994), on the other hand, suggested that the term should be restricted to "an inland river valley or section of coastal plain drowned as a result of [. . .] sea level rise and containing seawater measurably diluted by land drainage. . . ..". For the purposes of this article, we adopt the broader definition and thus include a variety of coastal environments in which seawater interacts with freshwater derived from land drainage, because in all of these, wetlands may be an important feature.

Oceanic water movements such as tides and longshore currents are important influences on many estuaries. Where the regional tide is small, as in the Baltic and Mediterranean Seas, Gulf of Mexico, and parts of the Indian and Pacific Oceans, the area of interface between fresh- and salt water may be very limited. Large rivers, such as the Mississippi, Nile, and Ganges, that carry a great deal of sediment from the land and enter a coastal environment with a small tide are likely to form a delta at the mouth with extensive wetlands, within the channels of which salt water and freshwater may become mixed. Large estuaries are found in regions of the world where major rivers empty into a nearby ocean that is subject to strong tidal movements, such as the Atlantic Ocean. Where the oceanic tide is large or where the river mouth is relatively wide, salt water may be pushed a great distance upstream by the rising tide, only to move seaward again as the sea level falls, creating an extensive zone in which seawater and freshwater mix, and therefore a longer estuary.

The essence of an estuary derives from the interactions between the land-based freshwater and the coastal water into which it flows. These interactions are determined by a number of factors: the geomorphology of the estuary; the seasonal variations in river flow, precipitation, and temperature; and temporal variations in wind and tidal movements. All of these factors may be greatly modified by human activities such as land "reclamation," dredging, seawall or dam construction, fishing, aquaculture, or transportation. The result is that each estuary has a number of unique features that tend to be reflected in the animal and plant life that inhabits it, the migratory species that visit it, its productivity, and its biodiversity. In spite of this variability, there are a number of useful ways to classify estuaries. These are outlined in the following section.

Classification of Estuaries

Estuarine environments may be classified in a number of ways, according to their origin and resulting morphology, the balance between precipitation and evaporation, the tidal range, and the degree to which salt water and freshwater are mixed (Fig. 1). All of these classifications reflect the interaction between geological history, geographic location, and sea-level variations.

One simple classification is based upon the relative amount of freshwater entering the estuary from the land and the amount of water that is lost from the estuary by evaporation (Perkins 1974). In a *positive estuary*, more freshwater enters than evaporates in a given time, so that the salinity throughout the estuary is generally lower than the adjacent ocean. In a *negative estuary*, evaporation exceeds freshwater inflow, leading to higher salinities. This criterion tends to separate estuaries in temperate zones, where river inflow is relatively high and consistent and evaporation is low, from tropical systems, where for much of the year, river input is low and evaporation is higher. In the latter case, salinities in the estuary may be much greater than that of the ocean.

A more extensive classification is based upon the origin of the estuary. Pritchard (1952) suggested four basic types of estuaries: *drowned river estuaries*, *fjords*, *bar-built estuaries*, and *tectonic estuaries*. Each has been formed in a different manner. *Drowned river estuaries* result from rising sea level which has invaded a coastal river mouth: examples include the Chesapeake Bay (USA), the Saint Lawrence estuary (Canada), and the Severn Estuary (UK). *Fjords* are the result of glacial scouring of the landscape, leaving a deep, u-shaped valley that commonly has a shallow entrance at the mouth formed by the original glacier's terminal moraine: examples are found in Norway (e.g., Hardangerfjord), Alaska (Misty Fjord), and New Zealand (Milford Sound). *Bar-built estuaries* result when sandbars are built up across the mouth of a stream by waves or longshore ocean currents, the sandbar restricting the outflow of freshwater and the inflow of seawater caused by tidal movements. Examples of bar-built estuaries are found in the Gulf of Mexico, the Netherlands, and South Africa. Sometimes, the closure of the estuary may become permanent, creating a *coastal lagoon*. In other cases the closure may be intermittent or seasonal: the estuary is blocked from the sea when the river flow is low and becomes open when the river is in flood. Lagoons are common coastal features in many parts of the world, occupying about 13% of coastal areas (Kjerfve 1994). *Tectonic estuaries* result when coastal land sinks as a result of geological faults or folding: examples are the Bay of Fundy (Canada) and San Francisco Bay (USA). The formation of these different types of estuaries affects the manner in which fresh- and salt water interact and has profound influence on the substrate whether rocky, sandy, or muddy and on their biological features.

Pritchard (1955, 1967) suggested that a more useful classification should be based upon the way in which river flow and tidal influences interact in an estuary. Tidal movements in an estuary tend to cause the fresh- and salt water to become mixed

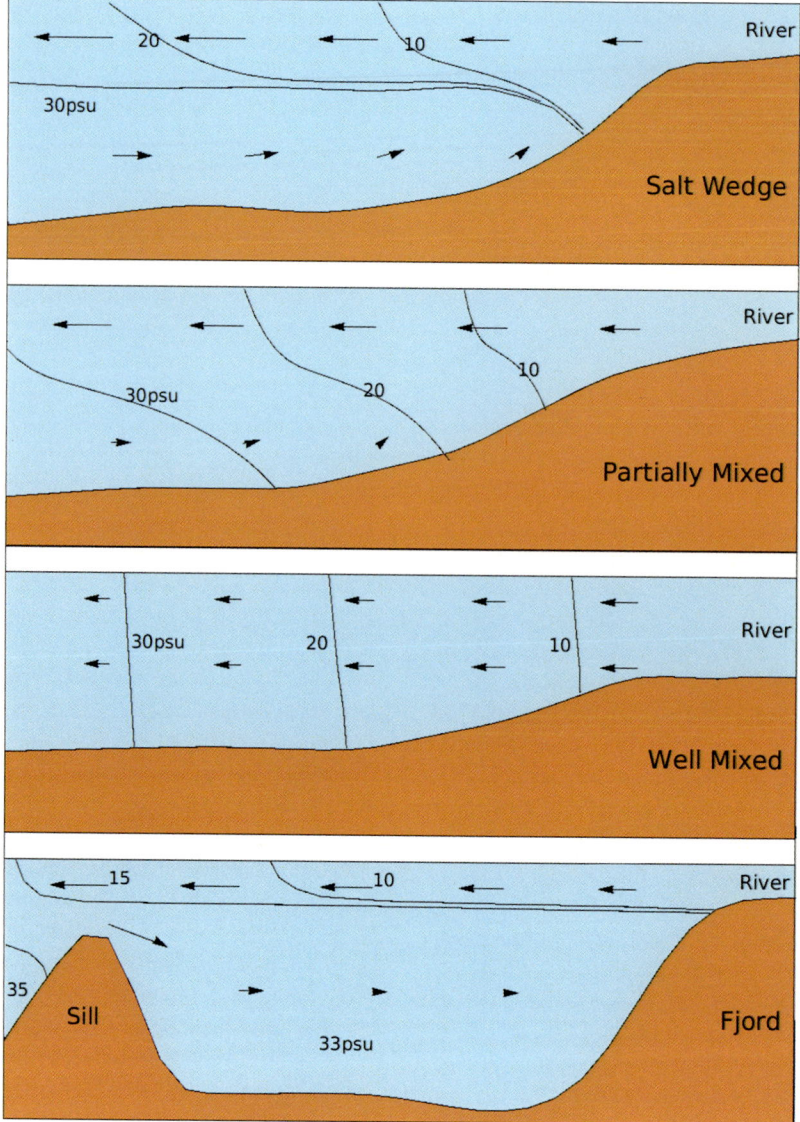

Fig. 1 Estuarine classification according to circulation patterns and salinity structure. The numbers indicate typical salinity levels (psu); the *lines* are isohalines, and *arrows* show general circulation patterns. Vertical salinity distribution varies with estuary type (Adapted from https://realfmrt.wordpress.com/2008/06/24/estuary-and-types/)

together, despite their different densities. In a *salt wedge estuary*, the less dense river water flows out as a layer over the more dense seawater, with a sharp horizontal boundary (termed the *pycnocline*) in between. This situation occurs in a fjord, where the shallow sill at the mouth restricts the influence of the tide. It also occurs in other kinds of estuaries where river flow is very high and tidal movements are small or where tidal influence is restricted by the building of causeways or barrages. At the other extreme, where tides are more dominant, freshwater is continuously mixed with seawater, producing a *vertically homogeneous estuary*, in which the salinity of the water at any one location is the same from surface to bottom but varies progressively along the length of the estuary. An intermediate condition, a *partially mixed estuary*, occurs where the river and tidal forces are more equal, so that salinity at the surface is lower than that near the bottom, but no sharp boundary exists between them. Many drowned river estuaries may oscillate between a mixed and a salt wedge condition with seasonal changes in river flow.

Another useful classification is based upon the *tidal range* (i.e., the vertical distance between the high water level and the low water level) experienced in an estuary. Although the gravitational forces causing the tides are acting directly upon the oceans, the actual tide experienced in an estuary is affected by the estuary's shape and size. The global oceanic tide averages about 1 m in range. Estuarine environments in which the mean tidal range is less than 2 m are referred to as *microtidal*; those between 2 and 4 m are *mesotidal*; and those more than 4 m are *macrotidal* estuaries. Extremely high tides occur in some cases as a result of resonance between the natural period of the estuary (a function of its depth and length) and the natural period of the oceanic tide. Examples of such *hypertidal* estuaries (mean tidal range more than 6 m) are the Bay of Fundy and Ungava Bay (Canada, $\sim$ 12 m), Severn Estuary (UK, $\sim$ 11 m), and Cook Inlet (USA, $\sim$ 9 m).

Dynamics of Estuaries

Estuaries are environments that are subject to continuous changes on a variety of time scales: short-term changes on the scale of hours are associated with the flood and ebb of tidal movements; the lunar cycle, from new moon to full moon, produces weekly variations in tidal height; and variations in precipitation and temperature over the adjacent land result in seasonal changes over the year. At the longest scale, land subsidence and rising sea levels affect the penetration of salt water into the landscape – a process of continuing change over millennia.

One of the most important influences on an estuarine environment is that of the tides. Tides are caused by the gravitational effects of the Moon and Sun acting upon the waters of the Earth and the Earth itself. Although the Moon is much smaller than the Sun, it is so much closer to the Earth that it exerts roughly twice the gravitational effect of that of the Sun. At the new moon and full moon phases, the gravitational effects of the Moon and Sun act along the same axis, so that the net effect on the ocean tide is relatively large, whereas at the first and third quarter phases of the lunar cycle, the direction of pull by the Moon is at a wide angle from the Sun, and the tidal ranges are consequently smaller. This results in an oscillation between large tides, called *spring tides*, that occur twice each month at times of full and new moons and smaller, *neap tides*, that occur in the intervening weeks (see Fig. 2). The direction of the Moon's travel is the same as that of the Earth's rotation, so that a complete orbit

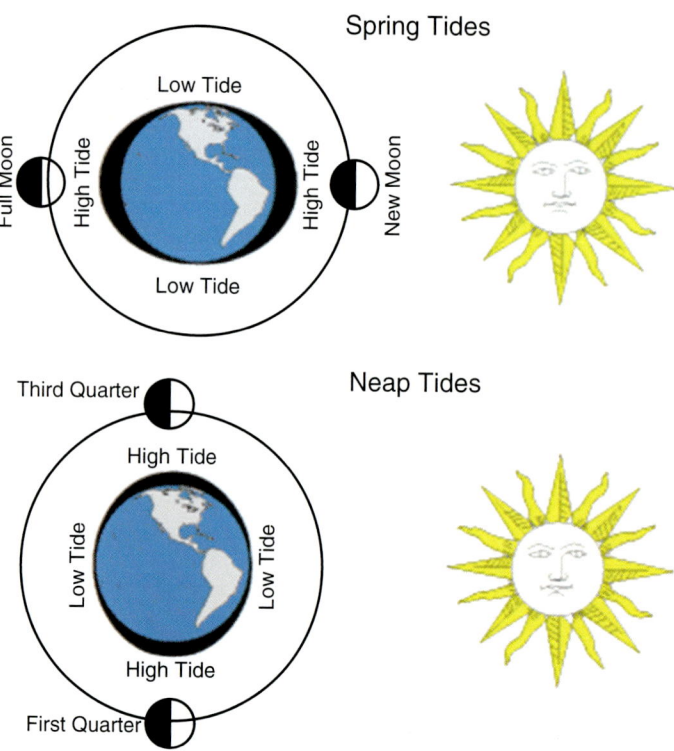

Fig. 2 The spring-neap cycle of the tides (Source, http://thebritishgeographer.weebly.com/coastal-processes.html)

takes about 24 h and 52 min; for this reason, the time of high tide on 1 day is approximately 50 min later than the corresponding tide of the day before.

If there were no large continents to block the flow of water in the oceans, there would be two tides per day (actually per 24.87 h) everywhere. In this *semidiurnal* condition, the two tides would be approximately of the same range. However, the presence of the continents produces complex patterns in water movements, so that in some places (e.g., much of the Pacific coast), one tide tends to be notably larger than the other of the same day, (referred to as a *mixed semidiurnal tide*), and in other places, there may be only a single tide per day (a *diurnal tide*, e.g., the Gulf of Mexico).

The Moon's progression around the Earth follows an ellipse rather than a circle, so that during each month, the distance between the Moon and Earth changes, varying the amount of gravitational attraction and therefore the range of the tide. When the Moon is closest to the Earth (known as *perigee*), the tidal range is increased, and when the Moon is further away (*apogee*), the tides are smaller. Similarly, the Earth's orbit around the Sun is elliptical, resulting in variations in the solar gravitational effect over the course of the year.

A number of longer-term variations in tidal movements result from the interactions between Sun, Moon, and Earth associated with oscillations in the planes of the lunar and terrestrial orbits. Important among these are two cycles, the Saros and Nodal cycles, which have roughly 18-year periods. Although they produce only small changes in the tidal range at any point, these have been shown to have significant effects on the productivity of some macrotidal estuaries (e.g., Cabilio et al. 1987). Many other cycles occur, including a postulated 1,800-year cycle (e.g., Keeling and Whorf 2000).

Water movements in an estuary caused by river flow, tides, or wind effects are important and highly variable factors affecting the biology and productivity. The speed, orientation, and persistence of water currents are determined by the shape and depth of the estuary, tidal range, and river inflow and in large estuaries are influenced by wind. Water flowing from the river generally slows down as it enters the estuary because of the increase in channel width; as a consequence, particulate material such as sediment or organic debris tends to settle out, particularly at the head of the estuary. In addition, flood and ebb currents are often dissimilar: the flood is often shorter than the ebb, because the rising tide must overcome the initial outflow of river water before it can move upstream, and as a result, flood tide currents are often stronger than those of the ebb tide. This asymmetry between flood and ebb currents may have profound effects on the movement of materials, especially sediments and larval stages of organisms (Dyer 1972). For example, stronger flood currents may resuspend sediments from the bottom and move them upstream where they may resettle during high tide. These two features result in many estuaries acting as effective traps for suspended material derived both from the land and the sea, and the upper part of the estuary may therefore accumulate nutrients and contaminants that are associated with the particles. In temperate estuaries, the lower river flows during summer months may lead to extensive accumulations of sediments in the upper reaches of the estuary, which are then washed seaward again when the river has high flows.

Wind-driven waves also may scour out deposited sediments, even redistributing them to such an extent that the channels through which water flows change on a periodic basis. The importance of waves in an estuary depends upon the size of the estuary and its orientation relative to the wind direction. When the wind blows along the long axis of the estuary as the tide is rising, it may add considerably to the amount of water entering the estuary and therefore to the height that the tide will reach. Erosion of shorelines is especially severe when wind-driven waves occur at high tide.

An important determinant of the dynamic changes occurring in an estuary is related to the nature of the substrate, whether it is rocky, sandy, or muddy, for example. This in turn is related to the origin and past history of the estuary. Estuaries which are bordered and underlain by resistant bedrock, such as granites or basalts, are not susceptible to change by the variations in tidal currents or waves. In contrast, estuaries fed by rivers that have a large sediment load, or those that are surrounded by more erodible rock (e.g., sandstones or siltstones), are likely to have a substrate that is composed of finer, mobile sediments such as sands and muds. These more mobile sediments are deposited in the estuary, particularly near the head, but may be periodically resuspended and redistributed as the strength of tidal currents or waves varies. In such an estuary, seasonal changes in channels and bars are common. The stability of the substrate is a major determinant of the kind of biological life to be found there.

Productivity and Biodiversity

Estuaries are among the most biologically productive ecosystems that we know. This may at first seem surprising, because, as described above, they are physically stressed systems and because our experience of estuaries is often limited to those which have been substantially modified by humans. Nonetheless, the tendency of estuary processes to "trap" sediments derived from the land leads to an accumulation of important nutrients, such as nitrogen and phosphorous, which are essential for plant growth. These nutrients are in organic matter brought down the river or in from the sea, or adsorbed onto particles of sediment, and become deposited wherever the sediments settle. In addition, estuaries are often shallow, so that light may penetrate to the bottom, and experience higher temperatures than the adjacent sea both from direct insolation and from the heat contributed by river waters.

The biological communities that inhabit estuaries are often notably different from those found in the neighboring sea or freshwaters in ways that reflect the physical processes that dominate each estuary. Organisms that occupy the water constitute the *plankton*, those that have limited swimming ability or float, and the *nekton* – those capable of swimming more or less independently. Organisms associated with the bottom substrate include the *benthos*, animals which live in or are fixed to the bottom; the *epibenthos*, those that move around on or just above the bottom; and plants such as seaweeds attached to rocky substrates or salt marsh plants associated with soft sediments. Finally, other important animals associated with estuaries are the marine birds and shorebirds (Fig. 3).

The *plankton* of estuaries consists mostly of small, even microscopic, plants (the *phytoplankton*) and animals (the *zooplankton*) that drift with the water currents. In a vertically mixed or partially mixed estuary, the plankton is mostly derived from forms that are similar to or the same as the plankton of the ocean nearby: a mix of diatoms and dinoflagellates, protozoans, copepods, ctenophores, coelenterates, and the larvae of numerous species that, as adults, are part of the benthos (e.g., clams, polychaetes) or the nekton (e.g., fish). The composition of the plankton varies along the course of a mixed estuary because being marine in origin, the species may have limited ability to survive in the lower salinities associated with the inner estuary (Beadle 1972; Perkins 1974). Those that can tolerate lower salinities, however, may become extremely numerous near the head of the estuary, where food is abundant and competition less. Because nutrients are often readily available in an estuary, phytoplankton production is potentially higher than the nearby ocean, but because many mixed estuaries also have high turbidity, light may be too low below the very surface waters, and thus phytoplankton productivity may be light limited. In a salt

Fig. 3 Semipalmated sandpipers *Calidris pusilla* feeding in the intertidal zone of the upper Bay of Fundy in late summer. This is an important feeding area for shorebirds migrating between their summer breeding grounds in the Arctic and wintering grounds in South America. The attraction is vast numbers of benthic crustaceans and polychaetes inhabiting intertidal zones (Photo credit: Mark Elderkin © Rights remain with the author)

wedge estuary, the surface waters are essentially fresh, and because rivers have no true plankton community, the upper layer will be primarily occupied by organisms derived from the river, most of which will die as they encounter salty water in the estuary (if they haven't already been eaten by the nekton). Below the pycnocline of a salt wedge estuary, there may be a planktonic community derived from the marine environment, but if the fresh upper layer is deep or turbid, there may be insufficient light for primary production to occur. In such a case, the plankton is comprised primarily of animals.

An important component of the plankton is comprised of the larval stages of benthic and nektonic animals. Benthic organisms release eggs, sperm, or larvae into the water, often at particular stages of the tide (e.g., ebb or flood or on particular days of spring or neap periods) with the result that their passive dispersal may lead to settlement in other appropriate habitats within the estuary (Wildish and Kristmanson 1997). Nekton animals such as fish may also release larval stages that are planktonic at first, and spawning is often timed and located to ensure that the larvae are retained within the estuary, drifting on tidal currents. When they have matured sufficiently to become active swimmers, many of these species may move out to sea to grow.

Members of the plankton play a varied but important role in the trophic pathways of an estuary. In less turbid waters, where phytoplankton can flourish, many planktonic animals feed directly on the phytoplankton and become food in turn for larger predatory plankton or nekton. In more turbid waters, however, dominant species of zooplankton include some that consume dead organic matter and even fine sediments being brought down from the land or moved around by the tide, stripping off the microbial flora commonly associated with particulate matter. Thus, both living and nonliving materials become incorporated into the estuarine food chain. Fluctuations in plankton numbers in temperate estuaries occur on a seasonal basis in response to temperature, salinity, and nutrient changes. In addition, the sporadic influx of plankton predators such as jellyfish (Cnidaria), comb jellies (Ctenophora), and planktivorous fish (e.g., menhaden *Brevoortia* spp. and herring *Clupea* spp.) from the nearby ocean can cause rapid declines in abundance of small plankton that may significantly impact larval fish being reared in the estuary.

For *nektonic* animals, which have a capacity for movement that is independent of water currents, many estuaries constitute rich feeding and spawning habitats. The *nekton* consists of larger animals such as squid, fish, turtles, and marine mammals. Some of these species are resident within the estuary for their whole lives, but many estuaries also accommodate seasonal visitations by fish and mammals that migrate between estuaries and the open ocean on a regular basis. These migrants include *anadromous* fish (e.g., salmon, Salmonidae; shad, Alosinae; and smelts, Osmeridae), which spawn in freshwater but feed and grow in the marine environment, and *catadromous* species (e.g., eels, galaxiid fish and some crabs) that spawn in salt water and grow in freshwater environments. For these migratory species, the estuary may be merely a pathway between these environments, or it may play a significant role as a feeding ground as well.

The greatest diversity of animals in estuaries is often found among those species that live on or close to the bottom – the *benthos*. This diversity reflects the great variations in physical and chemical conditions that are typical of estuarine environments. Those that live attached or burrowed into the substrate (the *infauna*) frequently exhibit a strong association both with the salinity of the overlying water and the characteristics of the substrate itself. In mixed estuaries, where salinities are higher in the outer part of the estuary, many marine species are able to live, and benthic diversity is generally high. Further upstream, however, as the salinity of bottom waters decreases, the diversity of species also goes down: very few species are well adapted to tolerate the relatively low and frequently changing salinity of the water. As a result, there tends to be a horizontal zonation of benthic organisms along the length of a mixed estuary. In a salt wedge estuary, however, this pattern is not evident because salinity within the salt wedge is essentially the same throughout the estuary. For such estuaries, therefore, the same benthic species tend to be found along the length of the estuary, and the primary determinant of their pattern of distribution is the physical nature of the bottom. Benthic communities in a fjord, which is a special case of salt wedge estuary, are often impoverished. Although the fjord may be deep, the sill at its mouth restricts tidal influence to times of higher tides, and introduced seawater tends to settle behind the sill and flows upstream without dilution. Oxygen levels are often lower than saturation in this lower level because the water is not replenished frequently, and degradation of organic matter settling from the surface freshwater uses up available oxygen. For benthic organisms this may be a principal limiting factor.

A major determinant of benthic community composition in all estuaries is the nature of the substrate. Distinctly different species are encountered attached to rocks from those inhabiting sandy bottoms or muddy areas. For areas of the bottom that are

Fig. 4 The intertidal sand and mud flats of the upper Bay of Fundy. As in most estuaries, productivity of intertidal flats is very high. In this system, polychaetes dominate the sediment macrofauna and are fed on by migrating shorebirds that stop over to feed in the late summer. Harvesters of bait worms (bloodworm *Glycera dibranchiata*) can be seen in the mid-intertidal zone (Photo credit: Glenys Gibson © Rights remain with the author)

continuously covered with water (i.e., the *subtidal zone*), the distribution of species reflects the substrate to a large extent, and benthic communities have often been classified according to their characteristic substrate (e.g., Eltringham 1971; Perkins 1974). These patterns are overlain by the salinity and mixing pattern factors described above.

In the *intertidal zone*, where the bottom substrate is alternately covered and exposed as the tide rises and falls, the characteristics of the substrate are combined with the challenges of exposure, producing a marked vertical zonation. This is most obvious on rocky shores, where distinct bands of lichens, barnacles, mollusks, and seaweeds occur, reflecting the varying tolerance of different species to exposure to the air. Although less visible, a similar zonation may exist in the intertidal zone of muddy estuaries (Fig. 4), where the pattern is confounded by variations in the characteristics of the soft sediments (sand, mud) resulting from tidal movements and wave action (e.g., Eltringham 1971; Dyer 1972).

In addition to the resident and migratory fish fauna, estuaries are often critical habitat for many birds and mammals that migrate into them from great distances in order to feed or breed. These movements may be primarily seasonal, e.g., marine waterfowl and mammals that breed in the Arctic or tropics and winter in temperate estuaries (e.g., Hicklin 1987), or more weather determined, such as the irregular movements of waterfowl and waders between European estuaries in response to local conditions (e.g., Davidson et al. 1991; Berthold et al. 2001). Such movements occur for several reasons: to take advantage of the high productivity and relatively low diversity of estuarine communities, the relative absence of competitors and/or predators, or because the habitat is only seasonally available. Nonetheless, such movements represent biological connections between estuaries that may be widely dispersed and otherwise separated from each other (Daborn 2007).

The primary producers in estuaries vary considerably, depending upon the shape, the mixing characteristics, water clarity, and nutrient status. Most estuaries tend to have a relatively high nutrient status compared with the nearby ocean because of the estuarine "trap" effect of circulation patterns; consequently, the primary production potential is high, although it may be limited by light availability. A few situations (e.g., fjords) may be extremely nutrient limited, because of low supply in incoming water, lack of circulation, and small tidal influence. Estuaries with clear waters, such as those with rocky shorelines and relatively small tides, may have both a diverse phytoplankton and an important marine macrophyte community consisting of seaweeds (e.g., *Fucus, Ascophyllum*) and/or sea grasses (e.g., *Zostera*). In addition to their role in primary production, seaweeds and sea grasses provide habitat for numerous invertebrates and refuge and spawning habitat for many fish. In turbid estuaries, on the other hand, phytoplankton tends to be light limited and impoverished. Where there are rocky shorelines, the macrophyte community may be similar to clearer estuaries, although growth rates and productivity will likely also be reduced by light limitations. If, however, the substrate is soft as a result of erodible shorelines or the high input of sediment from the river, then primary production may be entirely different, based upon salt marshes in the upper part of the intertidal zone and microscopic diatoms (etc.) on muddy or sandy shores. Some large estuaries, such as the Bay of Fundy (Canada) and the Severn Estuary (UK), may exhibit both production systems: a seaweed-phytoplankton system in the outer regions where the shoreline is rocky and the water clear and the marsh-mudflat system in inner regions where the substrates are softer (Fig. 5).

Ecosystem Services

Because of their complex interactions with both the land and the ocean, estuaries provide a wide variety of "environmental services" that are of great significance to humans. "Environmental" or "ecosystem services" have been defined as "the benefits people obtain from ecosystems" (Millennium Ecosystem Assessment 2005). First and foremost of these is *food*. The high potential productivity of an estuary is expressed in fish and shellfish, marine birds, mammals, and seaweeds, and humans have utilized these for millennia. Although with the advent of seagoing vessels attention shifted to offshore banks where larger stocks of marine fish were present, estuarine sources were always extremely important to coastal populations, and with the collapse of many offshore stocks, estuarine habitats have once again become of major importance: an increasing proportion of human food is now produced through

Fig. 5 Wetlands, tidal creeks, and intertidal flats of the hypertidal Minas Basin, Bay of Fundy. The vegetation is largely salt marsh, dominated by *Spartina alterniflora* (Photo credit: Mark Elderkin © Rights remain with the author)

aquaculture of finfish, shellfish, and marine plants, mostly in estuarine locations. Aquaculture utilizes several of the assets of estuaries, including its intrinsic productivity, the natural circulation of the water, and protection from oceanic storms (etc.) but may be limited by contamination and other forms of human degradation of estuary conditions.

In a healthy estuary, marine plants of marshes and shores play significant roles in both *water treatment* (through uptake of nutrients and contaminants) and in *coastal protection*. Estuarine marshes and sea grass beds induce settling of sediments from the water, increasing water clarity and diminishing *shoreline erosion* by waves. Since they adapt to rising sea levels, they provide a natural, low maintenance protection in spite of climate changes. They are also now being increasingly recognized for their *carbon sequestration* role in mitigation of greenhouse gases and climate change.

For many decades, an estuarine service has been in the form of *wastewater treatment*: wastes have been discharged without prior treatment into estuaries because the natural dynamic processes (circulation, oxygenation, filtration) and the disinfectant effects of salt water have been considered sufficient to mitigate human health threats.

Raw materials obtained from estuaries include sand and aggregate for construction purposes, salt, and hydrocarbons (oil, gas, methane hydrates). In modern times, estuarine plants and animals are being investigated for biological compounds that have nutritional or pharmacological benefits. In addition to hydrocarbons, the potential for *energy production* from tidal and wave movements and offshore wind is becoming of increasing interest because these sources are renewable and often close to human settlements needing energy.

Additional services from estuaries include *tourism* and *recreation*. The total economic value associated with estuarine ecosystem services has never been adequately calculated, but is thought to be very large (Barbier et al. 2011).

Threats and Future Challenges

The close association between humans and estuaries has resulted in many estuaries being significantly altered or degraded by human activities. Major threats to estuaries include:

- *Nutrient enrichment* resulting from land use practices and the natural "trapping" processes of the estuary, which may lead to development of algal blooms (some of which may have human health implications) or declines in oxygen availability, high turbidity, etc.
- *Waste disposal*, including the dredging of channels and dumping of land or estuary-based dredge spoils and human wastes
- *Shoreline modifications* associated with shoreline protection (e.g., seawalls, harbors, etc.) or *land "reclamation"* – infilling of marshes and other parts of the shoreline for residential, industrial, or agricultural purposes (e.g., the Netherlands "polders," Republic of Korea Saemangeum seawall)
- *Dredging* of channels and harbors to accommodate vessels of ever-increasing size in estuarine ports

In addition to these specific activities, the pressure on estuaries is increasing around the world as a result of the steady movement of peoples from the center of continents to the coasts. Coastal development of residential and industrial areas (example shown in Fig. 6) and construction of highways are increasingly acting to limit the potential of estuaries to adjust to long-term changes such as rising sea level. This *"estuarine squeeze"* will continue to be an important modifying factor affecting the evolution of each estuary.

Tidal barriers or seawalls have been constructed in numerous estuaries for the purposes of land reclamation (conversion to dry land and freshwater reservoirs) and as transportation connectors across rivers (e.g., Avon River estuary, Bay of Fundy) or to offshore islands (e.g., Saemangeum Estuarine System). Such structures are generally permanent and allow only minimal tidal exchange via sluice gates. The consequences are grave for natural intertidal wetlands and associated biota, leading to loss of biodiversity, die-off of shellfish and other biota, and decline in shorebird populations. Such barriers can also result in massive sediment deposits and the formation of new wetlands, as shown in Fig. 7.

Estuarine intertidal areas, including tidal mudflats, tidal marshes, and mangroves, are nutrient rich and highly productive ecosystems. They provide spawning areas and food for numerous species, including crustaceans, forage fishes, and their

Fig. 6 Aerial photo of Agua Hedionda Lagoon, Carlsbad, California. This system features extensive industrial development, protective barriers, and residential land use (Photo credit: Poseidon Water © Rights remain with the author)

predators. In addition to serving as nursery grounds, they also serve as significant shorebird staging areas where large bird populations rest and refuel on mudflat infauna (e.g., crustaceans, polychaetes). However, a steady rise in coastal development and destructive processes (e.g., tidal barriers, agricultural pollutants, mining activities, urbanization) has been impacting greatly on both the extent and productivity of intertidal habitats, with intertidal area losses greater than 50% in many parts of the world (e.g., East and Southeast Asia). Such losses reduce regional productivity and biodiversity and negatively impact fisheries and the populations of birds and marine mammals. In recent decades, national and international conservation efforts are helping to combat further losses, largely through the Convention on Wetlands (establishment of Ramsar Sites) and the Convention on the Conservation of Migratory Species of Wild Animals.

Despite habitat and species conservation efforts, nations are increasingly looking to utilize the natural resources of coastal environments, including estuaries. This includes power generation with the use of hydroelectric dams and the harvesting of renewable tidal energy from high-flow macrotidal systems. Tidal energy developments that involve the construction of tidal barrages or seawalls now exist in some countries as a means to generate power that is domestic, reliable, and renewable and to reduce carbon emissions. Such developments, however, have led to losses of intertidal habitat and to changes in the flow regime and sediment dynamics (i.e., erosion, transportation, deposition). Tidal barriers also restrict the available

Fig. 7 Avon River estuary, Bay of Fundy, as modified by the 1970 construction of a tidal barrier to allow roadway transportation and the protection of low-lying areas. This photo shows accumulated marine sediment (over a 45-year period) and the formation of an expansive and highly productive salt marsh (*Spartina alterniflora*) on the marine side. When the sluice gates are closed (as they are here), the barrier creates lake-like conditions upriver of the causeway (Photo credit: Tim Reed © Rights remain with the author)

passageways for diadromous fishes, with severe consequences for fish survival when forced to pass through turbines (Dadswell and Rulifson 1994). Recent new approaches involve stand-alone, in-stream turbine designs that do not require a tidal barrier and thus have the potential to extract tidal energy with a much reduced environmental impact. Testing of several in-stream tidal turbine designs is currently underway in Canada, France, and the UK.

Population growth and associated socioeconomic drivers in coastal regions over many centuries have resulted in alterations to estuarine systems, including wetland infilling, drainage, and conversion, with negative consequences to wetland-dependent species (migratory and endemic). Davidson (2014) estimates that, globally, >83% of wetlands have been lost since 1800 AD. Associated with significant losses in wetlands are observed declines in migratory birds and other biota. The overall health of the world's remaining wetlands, many of which continue to be degraded, remains to be assessed.

Recognition of the critical roles played by estuaries (and their wetlands) over human history, of the natural dynamic processes that lead to the ecosystem services they provide, and of the increasing pressure of human development is leading to greater focus on the issues of estuarine management. Although treatment of wastewater before release to the estuarine environment is beginning to moderate the effects of nutrient enrichment and address some of the human health issues, the rates of new development associated with increased global transportation and residential development continue to cause degradation of estuarine environments. The recent rise of aquaculture in Western countries may alleviate some of the fisheries stress on offshore stocks but leads to challenging conflicts with existing resource uses (e.g., fisheries) and facilities in the nearshore environment, as well as raising animal health issues. Conservation of rare or threatened species and habitats is seen as having increasing importance, leading to the creation of protected areas and limitations on human resource uses within them.

But the most enduring and challenging management issues are derived from the intrinsic connections that estuaries have between them and with both the land and the ocean. Understanding the manner, in which migratory fish, birds, and mammals use estuarine environments that are distant from one another, is leading to recognition that coastal estuaries represent an intrinsic network of habitats upon which many migratory species depend and that need to be protected. To address this, networks of reserves are being created that often involve international cooperation (e.g., the Western Hemisphere Shorebird Reserve system).

On the seaward side, the rate of sea-level rise appears likely to increase with global warming. This will increasingly affect the dynamic processes that drive estuarine productivity and, as the "estuarine squeeze" becomes more intense, may threaten many of the goods and services that estuaries can provide. But on the landward side also, the connection between the estuary and its watershed remains a poorly recognized fact. Estuarine circulation and productivity reflect the interplay between tidal movements and the freshwater input from the rivers. Over the last century, many of the major river systems such as the Mississippi, Nile, and St. Lawrence have been extensively modified by impoundments of water for hydroelectric production or irrigation. These dams affect the seasonal outflow of the river, decreasing the spring or rainy season floods by storing water, and then releasing it progressively so that the inflow to the estuary is more evenly distributed over the year. The effects of this can be profound, modifying all of the dynamic interactions of the estuary. This has created the need for management to consider all land-based activities in watersheds that enter estuaries in terms of their cumulative effect on estuarine environments. To manage these coastal ecosystems, it is necessary to remember that "the coastal zone begins at the head of the watershed," no matter how far from the coast that may be.

References

Barbier ED, Hacker SD, Kennedy C, Koch EW, Stier AC, Silliman BR. The value of estuarine and coastal ecosystem services. Ecol Monogr. 2011;81:169–93.

Beadle LC. Physiological problems for animal life in estuaries. In: Barnes RSK, Green J, editors. The Estuarine environment. London: Applied Science Publishers; 1972. p. 51–60.

Berthold P, Bauer H-G, Westhead V. Bird migration: a general survey. Oxford: Oxford University Press; 2001.

Cabilio P, DeWolfe DL, Daborn GR. Fish catches and long-term tidal cycles in Northwest Atlantic Fisheries: a nonlinear regression approach. Can J Fish Aquat Sci. 1987;44:1890–7.

Cameron WH, Pritchard DW. Estuaries. In: Hill MN, editor. The sea, vol. 2. New York: Wiley; 1963. p. 306–24.

Daborn GR. Homage to Penelope: unraveling the ecology of the Bay of Fundy system. In: Poehle GW, Wells PG, Rolston SJ. Challenges in environmental management in the Bay of Fundy – Gulf of Maine. Proceedings of the 7th Bay of Fundy Workshop, St. Andrews, 24–27 Oct 2006. Wolfville: Bay of Fundy Ecosystem Partnership; 2007. Technical report no. 3.

Dadswell MJ, Rulifson RA. Macrotidal estuaries: a region of collision between migratory animals and tidal power development. Biol J Linn Soc. 1994;51:93–113.

Davidson NC. How much wetland has the world lost? Long-term and recent trends in global wetland area. Mar Freshwat Res. 2014;65:934–41.

Davidson NC, Laffoley Dd'A, Doody JP, Way LS, Gordon J, Key R, Drake CM, Pienkowski MW, Mitchell R, Duff KL. Nature conservation and estuaries in Great Britain. Peterborough: Nature Conservancy Council; 1991.

Dyer KR. Sedimentation in estuaries. In: Barnes RSK, Green J, editors. The Estuarine environment. London: Applied Science Publishers; 1972. p. 10–32.

Eltringham SK. Life in mud and sand. London: English Universities Press; 1971.

Hicklin PW. The migration of shorebirds in the Bay of Fundy. Wilson Bull. 1987;99:540–70.

Keeling CD, Whorf TP. The 1,800-year oceanic tidal cycle: a possible cause of rapid climate change. Proc Natl Acad Sci. 2000;97:3814–9.

Kjerfve B, editor. Coastal Lagoon processes, Elsevier Oceanography Series, vol. 60. Amsterdam: Elsevier Science Publishers B.V.; 1994.

Millennium Ecosystem Assessment. Ecosystems and human well-being: current state and trends. Coastal systems. Washington, D.C.: Island Press; 2005.

Perkins EJ. The biology of Estuaries and coastal waters. London: Academic; 1974.

Pritchard DW. Estuarine hydrography. In: Landsberg HE, editor. Advances in geophysics, vol. 1. New York: Academic; 1952. p. 243–80.

Pritchard DW. Estuarine circulation patterns. Proc Am Soc Civ Eng. 1955;81:717.

Pritchard DW. What is an estuary? Physical standpoint. In: Lauff GH, editor. Estuaries. Washington, D.C.: American Association for the Advancement of Science; 1967. p. 3–5. Publication 83.

Wildish D, Kristmanson D. Benthic suspension feeders and flow. Cambridge: Cambridge University Press; 1997.

Estuarine Marsh: An Overview

4

Ralph W. Tiner and G. Randy Milton

Contents

Introduction 56
Estuarine Marsh Formation 58
Geographic Distribution and Extent 58
Plant Species Diversity 60
Ecosystem Services 61
Threats and Future Challenges 64
References 69

Abstract

Estuarine marshes commonly called "salt and brackish marshes" are tidal wetlands associated with the world's estuaries where salinities range from well above sea strength to nearly freshwater. Subject to frequent tidal flooding, plant communities are dominated by halophytic (salt-tolerant) herbs, subshrubs, and/or succulent-leaved shrubs. Not uniformly distributed along the world's sea coasts, tidal marshes tend to be the dominant plant community of the intertidal zone at middle and higher latitudes. The global extent of estuarine marshes is not well documented and this contributes to conservative estimates of their soil carbon stores. Most regions report significant historical and on-going loss of estuarine

R. W. Tiner (✉)
Institute for Wetlands and Environmental Education and Research, Inc., Leverett, MA, USA
e-mail: ralphtiner83@gmail.com

G. R. Milton
Nova Scotia Department of Natural Resources, Kentville, NS, Canada,
e-mail: gordon.milton@novascotia.ca

© Springer Science+Business Media B.V., part of Springer Nature 2018
C. M. Finlayson et al. (eds.), *The Wetland Book*,
https://doi.org/10.1007/978-94-007-4001-3_183

marshes by 1) human developments that in-part reflect a shift from an agrarian to industrial society and 2) natural events. Economic and cultural values set by society determine how estuarine marshes functions that yield many benefits to people and the estuarine aquatic ecosystem are valued. Estuarine marshes are increasingly being recognized among the world's most valuable ecosystems and, given their location between land and the sea, are especially vulnerable to human development and the effects of climate change.

Keywords

Brackish marsh · Climate change effect on estuarine marsh · Coastal wetland · Ecosystem services - estuarine marsh · Estuarine marsh · Plant diversity - estuarine marsh · Salt marsh · Tidal wetland · Wetland distribution - estuarine marsh · Wetland functions - estuarine marsh · Wetland threats - estuarine marsh · Wetland trends - estuarine marsh

Introduction

Estuarine marshes commonly called "salt and brackish marshes" are tidal wetlands associated with the world's estuaries where salinities range from well above sea strength to nearly freshwater (oligohaline brackish marshes). Dominated by halophytic (salt-tolerant) herbs, subshrubs, and/or succulent-leaved shrubs, these coastal wetlands are subject to frequent flooding by tides. These plant communities grow in the mid- to upper intertidal zone, generally above mean sea level and above tidal flats. They typically form in sheltered estuaries where currents and wave action are not strong enough to uproot seedlings and where ample sediments are deposited to support their growth and expansion. Such places include the mainland side of barrier islands, deltas, and shorelines of semi-enclosed bays and coastal rivers. Estuarine wetlands can also form in open shallow seas where tides are minimal, e.g., along the northeastern shore of the Gulf of Mexico (USA).

Within estuaries, salt marsh (dominated by the most salt-tolerant plants) occurs closer to the ocean than brackish marsh occupying the intertidal zone along coastal rivers whose salinity decreases with distance from the sea. While they are most characteristic of temperate and polar regions, salt marshes also occur in subtropical arid regions and in tropical climates often on the landward side of mangroves. They may include patches of and sometimes large nonvegetated or sparsely vegetated areas (salinas or salt flats), especially in arid and tropical climates. They also may include numerous shallow pools of open water. Figure 1 shows some examples around the world. This article provides an introduction to the world's estuarine marshes. More detailed information can be gathered from three publications (i.e., Chapman 1977; Adam 1990; Tiner 2013). See the senior author's contribution on "Hydrology of Coastal Wetlands" for a discussion of the hydrology of salt and brackish marshes.

Fig. 1 Examples of estuarine marshes: (**a**) Salt marsh at the head of St. Ann's Bay, Nova Scotia (Canada); (**b**) salt panne in the high marsh along St. Lawrence estuary, Quebec (Canada); (**c**) *Spartina alterniflora* low salt marsh, South Carolina (USA); (**d**) strongly brackish marsh dominated by *S. patens* with *S. alterniflora* along creekbank, New Hampshire (USA);

Estuarine Marsh Formation

Many of today's estuarine marshes are established in river valleys that have been submerged by rising sea levels since the last glacial epoch. The majority likely began forming around 6,000 years ago when the rate of sea-level rise became more gradual (Tiner 2013). As sea level continues to rise, estuarine marsh vegetation eventually colonizes coastal plain lowlands, replacing maritime forests that have succumbed to increased salt stress – a process called marine transgression or more commonly salt marsh migration. Humans have also played an important role in the formation of tidal marshes. Salt marshes have formed in front of diked marshes or on the more sheltered side of riprap piers where tides and currents combine to deposit sufficient sediments for marsh colonization. In fact, many present-day marshes in parts of Europe (e.g., Netherlands, Denmark, and Germany) are the by-product of land reclamation of former tidal marshes and shallow embayments, forming in front of embankments constructed to create land for grazing and agriculture (Davy et al. 2009). Deforestation of coastal watersheds and poor agricultural practices have also aided tidal marsh formation by introducing more sediments into coastal waters that eventually settle out, thus raising estuarine substrates to levels suitable for colonization by hydrophytes. Any process, whether natural or human induced, that increases sedimentation in coastal waters may eventually promote estuarine marsh formation.

Geographic Distribution and Extent

Tidal marshes are not uniformly distributed along the world's sea coasts. They tend to be the dominant plant community of the intertidal zone at middle and higher latitudes (in temperate and polar regions), with the tropics dominated by various species of mangroves (Fig. 2). In tropical regions, however, tidal marshes may be found either in front of the mangals (e.g., Brazil; West 1977) or more typically on less frequently flooded more saline soils behind them (e.g., salinas or salt flats, Saenger et al. 1977).

Although salt marsh can be relatively easily delineated using remote-sensing imagery, the global extent of estuarine marshes is not well documented for many regions of the world (Adam 2002). Estimates ranging from 0.14 million km^2 (Duarte et al. 2008), to 0.22 million km^2 (Laffoley and Grimsditch 2009), and 0.4 million km^2 (Woodwell et al. 1973 cited in Duarte et al. 2005) are developed from a limited database. While acknowledging their estimates are very conservative, Schuyt and

Fig. 1 (continued) (**e**) salina with *Batis maritima* behind black mangrove swamp, Florida (USA); (**f**) *Juncus roemerianus* brackish marsh, Georgia (USA); (**g**) *Salicornia* flats Camargue (France); (**h**) *Phragmites australis* with invasive *Spartina alterniflora*, Chongming Dongtan Ramsar Site (China) (Photo credits: (**a–f**) R.W. Tiner ©; (**g**) C.M. Finlayson ©; (**h**) L. Young ©. Rights remain with the author)

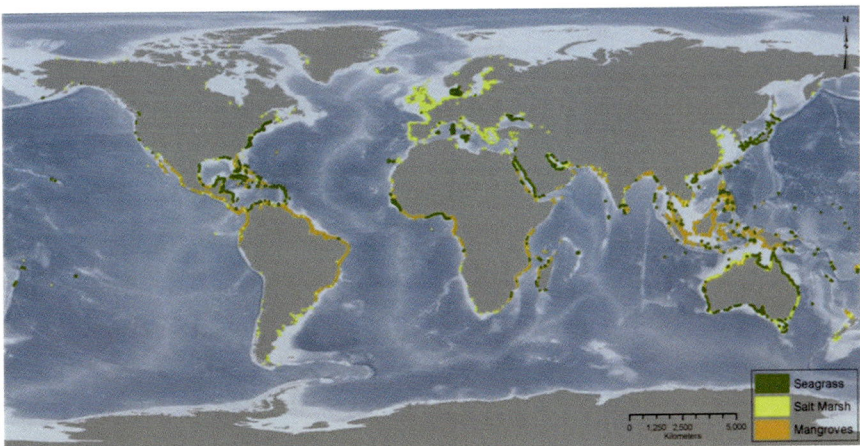

Fig. 2 Global distribution of tidal marshes, mangroves, and sea grasses (From: Pendleton et al. 2012; Fig. 1. https://doi.org/10.1371/journal.pone.0043542.goo1 licensed under Creative Commons CCO public domain dedication)

Table 1 Estimated tidal wetland area by continent. Unvegetated sediment wetlands are mostly from the coastal zone

Continent	Salt/brackish marsh (hectares)	Unvegetated sediment (hectares)	Mangrove (hectares)
North America	2,575,000	16,906,000	510,000
Latin America	1,707,000	9,223,000	4,224,000
Europe	500,000	2,374,000	Not present
Asia	1,027,000	8,011,000	1,439,000
Africa	487,000	4,632,000	3,686,000
Australasia	461,000	4,641,000	2,253,000
Total	6,758,000	45,788,000	12,112,000

Sources: Schuyt and Brander 2004: Table 9, © 2004 Living Water: the economic values of the world's wetlands, with permission WWF - World Wide Fund For Nature

Brander (2004) estimated tidal wetland area by continent using a database of nearly 3,800 wetland sites from around the world. They predicted that salt and brackish marshes minimally cover 67,580 km^2 (Table 1).

Information on the historic extent and losses of tidal marshes is extremely limited even in countries where wetlands have received much attention. Estimates of global declines in coastal wetland area do not differentiate between wetland types and are reported to be 46–50% since the beginning of the eighteenth century and 62–63% through the twentieth century (Davidson 2014). The relative rate of loss has accelerated in the twentieth and twenty-first centuries (Davidson 2014), and the Global Wetland Extent Index estimates an almost 50% decline between 1970 and 2008 (Leadley et al. 2014).

Although there are considerable regional differences, most regions report significant historical losses that are continuing due to both human and natural events. In their examination of North American wetlands, Bridgham et al. (2006) reported that the best estimate of "original" tidal wetlands in the USA was one for the 1950s and Valiela et al. (2009) provide a brief review of salt marsh loss in the contiguous United States. Various regions of North America have experienced different degrees of marsh loss in part related to their extent and the intensity and nature of development: Hudson Bay 63% (high uncertainty), Canadian Maritimes 64%, US North Atlantic 38%, US South Atlantic 12% (high uncertainty), Gulf of Mexico 18%, and Pacific Coast 93% (Gedan and Silliman 2009). In southern Sinaloa (Mexico), Camacho-Valdes et al. (2014) reported a 10–14% loss between 2000 and 2010. A study of changes in 12 of the world's largest estuaries reported a 67% loss since the onset of human settlement (Lotze et al. 2006), while Bridgham et al. (2006) arbitrarily used a 25% loss estimate for the world's tidal marshes outside of North America. Bailey and Pearson (2007) reported rapid declines of over 50% for some areas of salt marsh along the central southern British coast between 1971 and 2001, while Burd (1992) estimated losses of between 10–44% between 1973 and 1988 in 11 southeast England estuaries. An excess of 708,000 ha of land reclaimed from Chinese salt marshes (Yang and Chen 1995) and major rapid losses are reported for East and Southeast Asia (Mackinnon et al. 2012 and references within).

Plant Species Diversity

While climate is a major factor affecting plant distribution globally in tidal marshes and other habitats, many site-specific factors affect vegetation patterns within individual salt marshes (Fig. 3; see Tiner 2013 for details). Physical factors such as

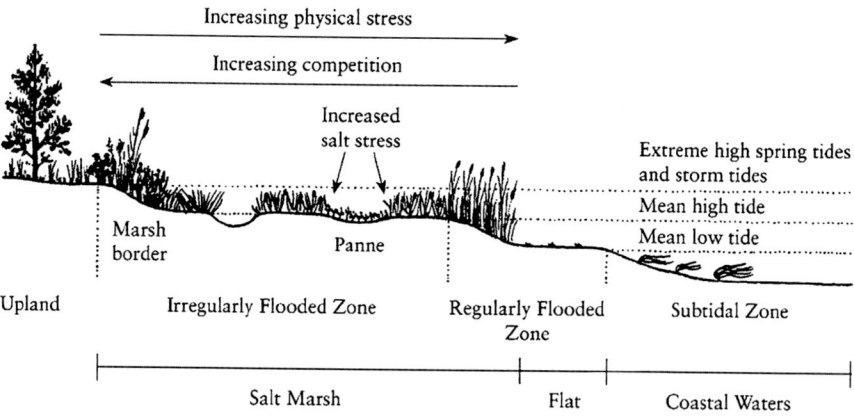

Fig. 3 In the salt marsh, increased physical stress occurs from the upland border to the water's edge, while biological competition increases in the opposite direction (From Tiner 2013; Fig. 4.15 copyrighted, permission from author)

salinity, inundation, soil saturation, and anaerobic soil conditions are more limiting in the marsh proper, while biological competition is greatest at the marsh-upland border where stress from the physical factors is low. The greatest salt stress in these marshes is in salt pannes – slight depressions where salinities can exceed 100 ppt. Here the most salt-tolerant species are found and are reduced in stature while in places even they are eliminated from growing due to the extreme salinity (i.e., salt barrens). Within the marshes, two general zones have been identified based on the frequency and duration of tidal inundation: low marsh (a zone that is flooded at least once daily for most of the year) and the high marsh (a zone that is flooded less often by the tides). A third zone – the upper marsh border – represents the highest elevations of the high marsh that, of course, are flooded less often and for shorter durations than the rest of the high marsh. Plant diversity is typically highest in this zone where salt water stress is much less than in the rest of the marsh. Where they occur in the tropics, the more extensive salt marshes (salinas) appear to be inland of the mangroves where more saline soils limit species diversity.

Species diversity is low in the Arctic, where *Puccinellia* and *Carex* are the predominant genera. Some common genera found in salt marshes elsewhere around the world include *Carex*, *Triglochin*, *Plantago*, *Glaux*, and *Hordeum* (boreal); *Spartina*, *Distichlis*, *Juncus*, and *Limonium* (temperate); and *Batis*, *Sesuvium*, *Cressa*, *Paspalum*, and *Sporobolus* (tropical) (see Chapman 1960, 1977; Beeftink 1977; Vierick et al. 1992; Thannheiser and Holland 1994; Adam 2002; Bortolus et al. 2009; Saintilan 2009; Tiner 1999, 2009, 2013 for more details). Members of the goosefoot family (Chenopodiaceae; especially *Sarcocornia*, *Salicornia*, *Suaeda*, and *Atriplex*) are also among other prominent halophytes characterizing the world's salt marshes.

Salinity in the intertidal zone along coastal rivers decreases with distance from the sea. Decreased salt stress allows more opportunities for colonization by plants with intermediate and low salt tolerances as opposed to the salt marsh proper where plants are exposed to the highest salinities. Interestingly, many plants restricted to the upland border in temperate salt marshes often become dominant species in brackish marshes further upriver. Consequently, some folks might consider the upper salt marsh border to be a brackish meadow, especially when it encompasses more than a narrow band.

Within salt marshes, especially in arid or tropical regions, there may be places where extreme salinities occur, in some cases more than three times that of sea water. The more extensive of these areas may be called salinas, salt barrens, or salt flats as they are often sparsely vegetated. It is here where the most salt-tolerant plants live: *Salicornia* spp., *Suaeda* spp., *Sagina maritima*, *Parapholis strigosa*, *Plantago coronopus*, and *Pottia heimii*, among others (Beeftink 1977; Tiner 1999, 2009, 2013).

Ecosystem Services

Historically tidal marshes have been filled for ports, and commercial, industrial, and residential development, and diked and drained for agriculture the world over (e.g., Gedan et al. 2009; Valiela et al. 2009; Silliman et al. 2009; Tiner 2013). Salt marshes

Table 2 Ecosystem services provided by estuarine marsh

Type of service	Examples
Supporting	Nutrient cycling, soil formation, primary production (photosynthesis), water cycling through wetlands
Provisioning	Food (e.g., fish, shellfish, plants), livestock grazing, aquaculture/fishing, fiber, genetic material for plant breeding, plants for natural medicines, ornamental products, animal products for fashion, etc.
Regulating	Flood regulation, coastal storm surge mitigation, climate regulation (carbon sequestration), erosion control, water quality
Cultural	Aesthetics, spiritual, educational, recreational/ecotourism, inspiration, sense of place

were highly valued by the agrarian societies since their livelihood largely depended on wetlands, and the lack of woody vegetation in salt marshes facilitated their conversion for agricultural uses. At the time of the European settlement of North America, salt marshes were arguably the most valuable natural resource in the colonies. Land prices in coastal agrarian communities were affected by the availability of salt marshes which was recognized as prized land for grazing and harvesting salt hay (an acre of salt marsh could produce enough hay to get one cow through the winter), as well as flat non-forested land that could be cultivated after diking and drainage (Hatvany 2001). However, with a change in culture to an industrial society that needed to support an ever-increasing population, values changed, technology facilitated conversion of forests to agriculture land, and it became easier to fill these areas for ports and other developments. Salt marshes became viewed by the societal leaders and the public at large as worthless lands, even a public nuisance in some regions as they produced hordes of mosquitoes that carried malaria and other life-threatening diseases. Consequently, tidal wetlands were ditched, drained, filled, or manipulated for a wide range of purposes.

While wetland functions continue to operate whether or not they are valued by society, economic and cultural values are set by society, and they may change as society changes. Tidal salt marshes provide many functions that yield many benefits to people and the estuarine aquatic ecosystem. The functions and values that they and other wetlands offer have been referred to as "ecosystem services" with the emphasis, often, on the ones of value to people (Reid et al. 2005). Table 2 lists some of the ecosystem services attributed to estuarine marsh. According to commonly used measures, the human well-being of coastal inhabitants is on average much higher than that of inland communities (Reid et al. 2005). The connection between people and the coastal waters and their wetlands is and has been an essential ingredient of this well-being.

In recognizing the ecosystem services provided by coastal wetlands in general, Laffoley and Grimsditch (2009) highlight the significant role of coastal wetlands including salt marsh in national and international climate change mitigation. Uncertainties in the global extent of salt marsh resulted in a conservative estimate by Chmura et al. (2003) of at least 430 Tg of carbon stores in the upper 50 cm of tidal salt marsh soils. Recent studies on carbon sequestration in tidal salt marsh have reported mean carbon burial rates that are probably an order of magnitude higher than terrestrial forests (Chmura 2013) but which can range from less than 20 to more than 1,700 g C m^{-2}years^{-1} (McLeod et al. 2011). In October 2013, the Intergovernmental Panel on Climate Change (IPCC) adopted the *2013 Supplement to the 2006 Guidelines for National Greenhouse Gas Inventories*: *Wetlands* (*Wetlands Supplement*) of standards to assess and report on emissions from organic soils and wetlands including tidal marshes (IPCC 2014a). Moreover, salt marsh restoration activities are eligible for certification as greenhouse gas emission reduction projects in the voluntary carbon markets using existing standards, e.g., the Verified Carbon Standard VM0033 (http://database.v-c-s.org/methodologies/methodology-tidal-wet land-and-seagrass-restoration-v10).

The attribution of these ecosystem services to salt marsh is changing how they are valued by society. Formerly the value of a good or service was based on what someone might pay for a physical marketable product harvested from the wetland or, in terms of land, what one could sell it for or earn from it by growing crops, raising livestock, harvesting timber, or extracting minerals. Conventional valuation approaches undervalue natural resources that are not harvested and sold for profit, ignoring functions that benefit society at large (e.g., shoreline and flood protection, water quality renovation, recreation, and aesthetics), and help maintain a healthy estuarine ecosystem.

The conventional approach began to change in the 1970s with development of an economic valuation system that included an analysis of the life support role that tidal wetlands play (Gosselink et al. 1973, 1974). This was perhaps the beginning of what eventually emerged as the discipline of ecological economics that attempts to value the nonmarket values of natural resources – "natural capital" and "environmental services." An American Northeast example of the use of these modern assessment techniques valued New Jersey's estuarine wetlands at over $6,000 per acre annually, for a total resource value between $1.1 and $1.2 billion (Costanza et al. 2006; Mates 2007), and storm protection by US tidal wetlands alone produce benefits worth $23.2 billion per year (Costanza et al. 2008). Using more sophisticated methods than previously available to estimate the global economic value of tidal wetlands (Costanza et al. 1997), resource economists have determined that the world's tidal wetlands (marsh and mangroves) may produce environmental services worth about $24.8 trillion annually which recognizes the immense storm protection, erosion control, and waste treatment values of these systems (Costanza et al. 2014).

Costanza et al. (2014) comment on how the valuation of ecosystem services can be used for a diverse group of purposes over multiple time and spatial scales including raising awareness and policy analyses. They caution against perceiving the valuations as exchange values rather than use or nonuse values associated with nonmarketable public goods or common pool resources. Valuation of ecosystem services is rather an approach to "…build a more comprehensive and balanced picture of the assets that support human well-being and human's interdependence with the well-being of all life on the planet." It is another "tool" to inform more balanced decision-making in achieving a healthy, productive ecosystem that supports human well-being and self-sustaining coastal fisheries and provides the vital links for migratory waterfowl, wading birds, and shorebirds moving between breeding grounds and winter habitat, without putting any species at risk of extinction.

Threats and Future Challenges

Their location at the interface of land and sea and the fact that the most populated areas of the world tend to be concentrated along the world's oceans have rendered tidal marshes especially vulnerable to the development and to the effect of climate change on sea level. Salt marshes are subjected to both natural and often intense human disturbances that affect their extent and quality (see Table 3). Their topographical position and features in close proximity to areas favored by human settlement and industrial patterns are a key determinant in the threats and challenges faced by salt marsh (Gedan et al. 2009; Adam 2016). Population pressure on coastal resources continues and in many developing regions is increasing. Nearly half of the world's major cities (having more than 500,000 people) are located within 50 km of a coastline, and more than a third of the global population resides coastally on approximately 4% of the land surface at densities 2.6 times larger than the density of inland areas (Reid et al. 2005). Despite environmental regulations and recognition/valuation of the ecosystem services, processes and functions provided by salt marsh, these systems remain under threat of loss or environmental degradation due to infrastructure development to provide for a growing and more affluent global population and accidental or purposeful discharges of industrial and domestic wastes (Adam 2016).

The introduction of new species to an ecosystem, whether by accident or deliberate, that become invasive is seen as a major threat to biodiversity and especially those species that are transformative in changing the ecological character of the ecosystem over substantive areas (Richardson et al. 2000). Estuaries and thus salt marshes are particularly exposed to invasive introductions by marine shipping from ballast release or as fouling organisms on ships' hulls (Adam 2016), e.g., the burrows of the Australian burrowing isopod *Sphaeroma quoyanum* introduced by ships to California in the mid-nineteenth century have altered sediment shear strength and increased erosion rates of US west coast salt marsh (Adam et al. 2008). Large numbers of introduced plants have been recorded in salt marsh (Adam 2002), but a smaller number of perennial species have major impacts. *Spartina* species and hybrids have been deliberately introduced to stabilize mudflats and create new marsh on North America's west coast, England, Europe, Australasia, and China with unintended results including loss of habitat for wading birds, outcompeting native plant species and displacing infauna through alteration of the physical structure of the marsh environment (Adam et al. 2008; Gedan et al. 2009; Adam 2016). *Juncus acutus*, native to the Mediterranean, is aggressively colonizing and displacing the native *J. kraussii* in eastern Australia (Adam 2002), and a number of species of *Tamarix* (native to Asia and Africa) introduced to North America are invading and converting herbaceous salt marsh to woodland (Adam 2002, 2016;

4 Estuarine Marsh: An Overview

Table 3 Some impacts to tidal wetlands from (A) natural processes or activities and (B) human-induced disturbances (Source: Tiner 2013)

(A) Natural processes or activities	Possible impacts
Sea level rise and coastal plain subsidence	Submergence (loss) of wetland, change from vegetated wetland to tidal flat/open water, change from tidal flat to open water, change in shorelines, landward migration of marshes if suitable space is available, increase in salinity with changes in plant communities
Coastal processes (wave action, currents, and tides)	Erosion, sedimentation, wetland loss or gain, smothering of vegetation (e.g., from deposition of tidal wrack or overwash sediments), and changing shorelines
Deltaic soil compaction	Same as for sea-level rise and coastal plain subsidence, except migration and salinity changes
Hurricanes and other storms	Sedimentation (including overwash deposits), erosion, vegetation impacts, wetland loss or gain, saltwater intrusion, and changing shorelines
Ice scour	Erosion, vegetation removal, creation of open water in marsh
Grazing by animals (e.g., waterfowl, fur-bearers, and invertebrates)	Loss of vegetation (denuded areas) and vegetation changes
Insect infestations	Loss of vegetation and vegetation changes
Disease outbreaks	Loss of vegetation and vegetation changes
Beaver dams	Reduce tidal flowage to tidal freshwater wetlands, and vegetation changes
Fire from lightning	Vegetation impacts
Droughts	Brown marsh syndrome, vegetation changes
(B) Human-induced disturbances	**Possible impacts**
Filling for development (port, industrial, commercial, etc.)	Loss of wetland
Disposal of dredged material or garbage	Loss of wetland or change in plant community depending on amount of spoil
Construction of jetties and groins	Loss of wetland, changes in wetland type, altered hydrology, shoreline changes, and changes in wildlife use
Armoring shorelines	Loss of wetland and stops landward migration of tidal wetland
Construction of docks	Shading of vegetation and disturbance to wildlife
Dredging for marinas and residential development (canal development)	Loss of wetland and degraded water quality
Drainage for mosquito control	Altered hydrology, increased salinity, vegetation changes, and local subsidence
Diking (for many purposes)	Altered hydrology, vegetation changes, localized subsidence, loss or diminished estuarine exchange, and loss of wetland
Road and railroad crossings	Loss of wetland, altered hydrology, and vegetation changes
Installation of tide gates	Altered hydrology, salinity changes, and changes in vegetation and aquatic life

(continued)

Table 3 (continued)

Nonpoint source pollution from farms, lawns, etc.	Eutrophication vegetation changes, and changes in aquatic life
Discharge of industrial or municipal wastewater	Water pollution, fish kills, degraded water quality, increased algal blooms, hypoxia, and changes in vegetation and aquatic life
Oil spills	Vegetation die-off, changes in aquatic life, substrate contamination, death for oiled wildlife, fish kills, and water pollution
Deepening channels and dredging canals for navigation	Increased salt water intrusion, altered hydrology, shoreline erosion, vegetation changes, and change in aquatic life
Damming of tidal rivers for various purposes	Reduction in sediment load, loss of wetland, altered hydrology, altered salinity regimes, vegetation changes, and changes in aquatic life (e.g., fish migration)
Water withdrawals	Altered hydrology, possible subsidence, altered salinity regimes, and corresponding changes in vegetation and aquatic life
Diversion of river flows	Altered hydrology, increased salinity and corresponding changes in aquatic life
Groundwater withdrawals	Subsidence and changes in vegetation
Oil and gas withdrawals	Possible subsidence, corresponding changes in vegetation and aquatic life, and pollution from spills and leaks (see oil spills)
Prescribed burning for wildlife management	Loss of soil organic matter and vegetation changes
Timber harvest	Vegetation changes, may facilitate conversion to other uses
Log storage	Topographic changes, vegetation changes, habitat degradation
Marine aquaculture	Change in current patterns, disturbance to water birds, possible loss of wetland (via conversion to open water)
Harvest of baitworms	Disturbance to water birds and changes in invertebrate density
Plant introductions	Invasive species replacement of native flora and altered hydrology
Animal introductions	Replacement of native fauna
Grazing by livestock	Soil compaction and changes in vegetation
All-terrain vehicle traffic	Vegetation impacts, disturbance and destruction of nests and young birds
Beach renourishment	Change in sand composition and changes in invertebrate usage
Noise and light pollution (nighttime)	Unknown effect on wildlife, probable displacement of sensitive species; latter disturbance disorients nesting sea turtles
Plant collection	Loss or reduction of local populations (overharvest, e.g., sea lavender)
Nitrogen inputs from runoff	Change in vegetation, plant productivity, and aquatic life
Spraying of pesticides	Aquatic organism kills

Notes: With human impacts, changes are mostly one-directional (i.e., losses). Note that where a significant change in vegetation occurs, a change in wildlife use will likely occur

Gedan et al. 2009). *Phragmites australis* is indigenous but rarely dominant in North America's upper salt marsh of North America, but an introduced European strain that is more salt tolerant now forms extensive near monocultures, displacing other native species (Adam et al. 2008; Tiner 2013; Adam 2016).

The IPCC's 5th Assessment Report (2014b) is unequivocal that warming of the global climate system is occurring and primarily human induced and that climate changes are already underway and projected to grow (e.g., sea-level rise, changing precipitation patterns, longer growing season, rise in temperature, longer ice-free season for oceans and fresh water bodies, earlier snow melt, and changes in river flows). Changes in temperature and precipitation patterns will significantly affect the life cycles of many plants and animals worldwide. Examples of these changes have already been observed such as earlier spring migrations and northward range expansions of many North American birds (Hitch and Leberg 2007; McDonald et al. 2012; Auer and King 2014), and mangrove species have expanded their range poleward on at least five continents by replacing salt marsh (Saintilan et al. 2014). Although temperature is the key delimiting factor that predicates comprehensive mangrove replacement of salt marsh-dominated ecosystems, mangrove will be favored by environmental factors associated with global climate change including elevated sea level, elevated CO_2, and higher temperatures (Saintilan et al. 2014).

Temperature changes will also affect plant community composition. As an example, while rising sea levels may create more panne habitat (water-retaining depressions) in New England (USA) high marshes (Warren and Niering 1993), panne forbs may be replaced by salt hay grass *Spartina patens* prior to the formation of new pannes, thereby precluding forb colonization of the new pannes (Gedan and Bertness 2009). Warming might tend to favor C4 plant species, but C3 plants because of differences at the cellular level have been shown to respond more positively to elevated CO_2 (Ainsworth and Long 2005). Salt marshes contain a mixture of C3 and C4, and this differential response to elevated CO_2 may favor compositional shifts to C3 plants (Gedan et al. 2009) although other ecological factors can influence the ratio of C3–C4 plants (Adam 2002). Lengthening of the growing season in middle and northern latitudes will likely increase productivity of plants and greater incorporation of organic material into salt marsh soils (Kirwan et al. 2009; Langley et al. 2009).

The effect of global warming on rising sea level has enormous consequences and poses the greatest risks to the physical integrity of tidal wetlands as well as to coastal communities and Pacific island nations and territories. Thermal expansion of the ocean in combination with accelerated melting polar and glacial ice causes a rise in sea level referred to as "eustatic sea-level rise." Local and regional conditions (e.g., subsidence and tectonic activity) may affect the position of the land relative to sea level. The combined effects of eustatic sea-level rise and local land elevation changes produce what is called "relative sea-level rise". Relative sea-level rise is greater than eustatic sea-level rise where land subsides and is less where land rises (uplifts) (e.g., Hudson Bay Lowlands, Riley 2011). The latter situation often results in "negative relative sea-level rise." Today, sea-level rise poses the greatest threat to the tidal marshes worldwide (e.g., Valiela et al. 2009; Tiner 2013).

The long-term stability of coastal wetlands in response to sea-level rise depends upon maintaining their position relative to the tidal frame by either horizontal migration landward and/or vertical migration through sedimentation (McFadden et al. 2007). A potential consequence of sea-level rise will be "drowning" of salt marsh and "coastal squeeze" where landward margins affected by steep topographical gradients or coastal defense structures prevent their horizontal migration (Adam 2016). Hoozemans et al. (1993) and Nicholls et al. (1999) using a similar scenario of a 1-m global sea-level rise in combination with human activities estimated coastal wetland (salt marsh, mangrove, intertidal areas) losses of 55% and 46%, respectively, based upon a limited 1990 dataset (~300,000 km^2) of coastal Ramsar Wetlands of International Importance. Nicholls et al.'s (1999) additional analyses of this dataset employing scenarios with a 38-cm global sea-level rise by the 2080s estimated global cumulative coastal wetland loss of 36–70% partitioned between human (25–57%) and relative sea-level rise (6–22%) impacts. Spencer et al. (2016) reevaluated these earlier broad-scale assessments of coastal wetland vulnerability to sea-level rise by incorporating improved data on global coastal wetland stocks, greater understanding of the main drivers of change, and further development of the Dynamic Interactive Vulnerability Assessment Wetland Change Model. The model's algorithm follows MacFadden et al.'s (2007) consideration of three key drivers that control wetland-type response over different time horizons to (1) local sea-level rise relative to tidal range, (2) opportunity to horizontal migration, and (3) sediment supply. Their results are however consistent with the earlier studies. Evaluating the effects of restrictions to horizontal migration capacity (dike construction) and sediment supply, comparable sea-level rise scenarios of 29 and 110 cm give coastal wetland loss estimates of 32–40% and 53–60%, respectively, by the 2080s. Since coastal wetlands are dynamic resources that move both landward and seaward in response to long-term changes in sea level, the development of low-lying areas adjacent to tidal wetlands and construction of hardened shoreline (bulkheads, seawalls, and riprap) to protect those properties prevents the natural landward migration of salt marshes accompanying rising sea level. Its impact will undoubtedly be exacerbated by the likely societal response of shoreline armoring to protect private property. This has serious implications for the future of today's tidal marsh.

The passage of environmental laws and policies has greatly improved the status of coastal wetlands in many countries (Tiner 2013). Moreover there are increasing numbers of examples at local and estuary scales of activities to restore salt marsh functions where there has been historical loss (Balletto et al. 2005; Adam et al. 2008; Tiner 2013). Nonetheless, even with existing regulations, policies and international agreements (e.g., Ramsar Convention on Wetlands) in effect, there remain serious threats to coastal wetlands particularly in developing tropical countries. In all countries, laws can be changed by legislative bodies, and the effectiveness of regulations in protecting wetlands can be weakened by the courts, lack of enforcement, or "flexibly" applied to permit further loss of salt marsh to human developments (Adam 2002; Tiner 2013).

References

Adam P. Saltmarsh ecology. Great Britain: Cambridge University Press; 1990.
Adam P. Saltmarsh in a time of change. Environ Conserv. 2002;29(1):39–61.
Adam P. Saltmarsh. In: Kennish MJ, editor. Encyclopedia of estuaries. Dordrecht: Springer Science +Business Media; 2016. p. 515–35. doi:10.1007/978-94-017-8801-4.
Adam P, Bertness MD, Davy AJ, Zedler JR. Saltmarsh. In: Polunin NVC, editor. Aquatic ecosystems. Cambridge, UK: Cambridge University Press; 2008. p. 157–71. Chapter 11.
Ainsworth EA, Long SP. What have we learned from 15 years of free-air CO_2 enrichment (FACE)? A meta-analytic review of the responses of photosynthesis, canopy properties and plant production to rising CO_2. New Phytol. 2005;165:351–72.
Auer SK, King DI. Ecological and life-history traits explain recent boundary sifts in elevation and latitude of western North American songbirds. Glob Ecol Biogeogr. 2014;23:867–75.
Baily B, Pearson AW. Change detection mapping and analysis of salt marsh areas of central southern England from Hurst castle Spit to Pagham Harbour. J Coast Res. 2007;23(6): 1549–64.
Balletto JH, Heimbuch MV, Mahoney HJ. Delaware Bay salt marsh restoration: mitigation for a power plant cooling water system in New Jersey, USA. Ecol Eng. 2005;25:204–13.
Beeftink WG. Salt marshes. In: Barnes RSK, editor. The coastline. London: Wiley; 1977. p. 93–121.
Bortolus A, Schwindt E, Bouza PJ, Idaszkin YL. A characterization of Patagonian salt marshes. Wetlands. 2009;29:772–80.
Bridgham SD, Megonigal JP, Kellet JK, Bliss NB, Trettin C. The carbon balance of North American wetlands. Wetlands. 2006;26:889–916.
Burd F. Erosion and vegetation change on the saltmarshes of Essex and orth Kent between 1973 and 1988. Research and survey in nature conservation No. 42. Peterborough, UK: Nature Conservancy Council; 1992. Available from: http://jncc.defra.gov.uk/pdf/Pubs92_Saltmarshes_of_Essex_&_North_Kent_1973-1988_PRINT.pdf
Camacho-Valdez V, Ruiz-luna A, Ghermandi A, Berlanga-Robles CA, Nunes PALD. Effects of land use changes on the ecosystem service values of coastal wetlands. Environ Manag. 2014;54:852–64.
Chapman VJ. Salt marshes and salt deserts of the world. New York: Interscience; 1960.
Chapman VJ, editor. Wet coastal ecosystems. Amsterdam: Elsevier Scientific; 1977.
Chmura GL. What do we need to assess the sustainability of the tidal salt marsh carbon sink. Ocean Coast Manag. 2013;83:25–31.
Chmura GL, Anisfeld S, Cahoon D, Lynch J. Global carbon sequestration in tidal, saline wetland soils. Global Biogeochem Cycles. 2003;17:1–12.
Costanza R, dArge R, de Groot R, Farber S, Grasso M, Hannon B, Limburg K, Naeem S, Oneill RV, Paruelo J, Raskin RG, Sutton P, van den Belt M. The value of the world's ecosystem services and natural capital. Nature. 1997;387:253–60.
Costanza R, Wilson M, Troy A, Voinov A, Liu S, D'Agostino J. The value of New Jersey's ecosystem services and natural capital. Burlington: Gund Institute for Ecological Economics, University of Vermont; 2006. Available from: http://www.state.nj.us/dep/dsr/naturalcap/nat-cap-2.pdf. Accessed 23 June 2016.
Costanza R, Pérez-Maqueo O, Martinez ML, Sutton P, Anderson SJ, Mulder K. The value of coastal wetlands for hurricane protection. Ambio. 2008;37:241–8.
Costanza R, de Groot R, Sutton P, van der Ploeg S, Anderson SJ, Kubiszewski I, Farber S, Turner RK. Changes in the global value of ecosystem services. Glob Environ Chang. 2014;26:152–8.
Davidson NC. How much wetland has the world lost? Long-term and recent trends in global wetland area. Mar Freshw Res. 2014;65:934–41.
Davy AJ, Bakker JP, Figueroa ME. Human modification of European salt marshes. Chapter 16. In: Silliman BR, Grosholz ED, Bertness MD, editors. Human impacts on salt marshes: a global perspective. Berkeley: University of California Press; 2009. p. 311–35.

Duarte CM, Middelburg J, Caraco N. Major role of marine vegetation on the oceanic carbon cycle. Biogeosciences. 2005;2:1–8.

Duarte CM, Dennison WC, Orth RJW, Carruthers TJB. The charisma of coastal ecosystems: addressing the imbalance. Estuar Coasts. 2008;31:233–8.

Gedan KB, Bertness MD. Experimental warming causes rapid loss of plant diversity in New England marshes. Ecol Lett. 2009;12:842–8.

Gedan KB, Silliman BR. Patterns of salt marsh loss within coastal regions of North America. In: Silliman BR, Grosholz ED, Bertness MD, editors. Human impacts on salt marshes: a global perspective. Berkeley: University of California Press; 2009. p. 253–83.

Gedan KB, Silliman BR, Bertness MD. Centuries of human-driven change in salt marsh ecosystems. Ann Rev Mar Sci. 2009;1:117–41.

Gosselink JG, Odum EP, Pope RM. The value of the tidal marsh. Gainesville: Urban and Regional Development Center, University of Florida; 1973. Work paper no. 3.

Gosselink JG, Odum EP, Pope RM. The value of the tidal marsh. Baton Rouge: Center for Wetland Resources, Louisiana State University; 1974. Publication no. LSU-SG-74-03.

Hatvany MG. 'Wedded to the Marshes': salt marshes and socio-economic differentiation in Early Prince Edward Island'. Acadiensis. 2001;30(2):40–55.

Hitch AT, Leberg PL. Breeding distributions of North American bird species moving north as a result of climate change. Conserv Biol. 2007;21:534–9.

Hoozemans FMJ, Marchand M, Pennekamp HA. Delft hydraulics. Sea level rise: a global vulnerability assessment: vulnerability assessment for population, coastal wetlands and rice production on a global scale. 2nd ed. Delft: Delft Hydraulics; 1993.

IPCC. In: Hiraishi T, Krug T, Tanabe K, Srivastava N, Baasansuren J, Fukuda M, Troxler TG, editors. 2013 supplement to the 2006 IPCC guidelines for national greenhouse gas inventories: wetlands. Geneva: IPCC; 2014a. Available from: http://www.ipcc-nggip.iges.or.jp/public/wetlands/index.html. Accessed 29 June 2016.

IPCC. In: Core Writing Team, Pachauri RK, Meyer LA, editors. Climate change 2014: synthesis report. Contribution of working groups I, II and III to the fifth assessment report of the Intergovernmental Panel on Climate Change. Geneva: IPCC; 2014b. 151 p.

Kirwan ML, Guntenspergen GR, Morris JT. Latitudinal trends in *Spartina alterniflora* productivity and the response of coastal marshes to global change. Glob Chang Biol. 2009;15:1982–9.

Laffoley D, Grimsditch GD. The management of natural coastal carbon sinks. Gland: IUCN; 2009.

Langley JA, McKee KL, Cahoon DR, Cherry JA, Megonigal JP. Elevated CO_2 stimulates marsh elevation gain, counterbalancing sea-level rise. Proc Natl Acad Sci. 2009;106(15):6182–6. doi:10.1073/pnas0807695106.

Leadley PW, Krug CB, Alkemade R, Pereira HM, Sumaila UR, Walpole M, Marques A, Newbold T, Teh LSL, van Kolck J, Bellard C, Januchowski-Hartley SR, Mumby PJ. Progress towards the Aichi biodiversity targets: an assessment of biodiversity trends, policy scenarios and key actions. Montreal: Secretariat of the Convention on Biological Diversity; 2014. CBD Technical Series No. 78. Available from: https://www.cbd.int/doc/publications/cbd-ts-78-en.pdf. Accessed 24 June 2016.

Lotze HK, Lenizan HS, Bourque BJ, Bradbury RH, Cooke RG, Kay MC, Kidwell SM, Kirby MX, Peterson CH, Jackson JBC. Depletion, degradation, and recovery potential of estuaries and coastal seas. Science. 2006;312:1806–9.

MacKinnon J, Verkuil YI, Murray N. IUCN situation analysis on East and Southeast Asian intertidal habitats, with particular reference to the Yellow Sea (including the Bohai Sea). Occasional paper of the IUCN species survival commission no. 47. Gland/Cambridge, UK: IUCN; 2012. ii + 70 pp.

Mates W. Valuing New Jersey's natural capital: an assessment of the economic value of the State's natural resources. Part 1: overview. Trenton: New Jersey Department of Environmental Protection; 2007. Available from: http://www.state.nj.us/dep/dsr/naturalcap/nat-cap-1.pdf. Accessed 23 June 2016.

McDonald KW, McClure CJW, Rolek B, Hill GE. Diversity of birds in eastern North America shifts north with global warming. Ecol Evol. 2012;2:3052–60.

McFadden L, Spencer T, Nicholls RJ. Broad-scale modelling of coastal wetlands: what is required? Hydrobiologia. 2007;577:5–15.

Mcleod E, Chmura GL, Björk M, Bouillon S, Duarte CM, Lovelock C, Salm R, Schlesinger W, Silliman B. A blueprint for Blue carbon: towards an improved understanding of the role of vegetated coastal habitats in sequestering CO_2. Front Ecol Environ. 2011;9:552–60.

Nicholls RJ, Hoozemans FMJ, Marchand M. Increasing flood risk and wetland losses due to global sea-level rise: regional and global analyses. Glob Environ Chang. 1999;9:S69–87.

Pendleton L, Donato DC, Murray BC, Crooks S, Jenkins WA, Sifleet S, Craft C, Fourqurean JW, Kaufman JB, Marbà N, Megonigal P, Pidgeon E, Herr D, Gordon D, Baldera A. Estimating global "Blue Carbon" emissions from conversion and degradation of vegetated coastal ecosystems. PLoS One. 2012;7(9):e43542. doi:10.1371/journal.pone.0043542.

Reid WV, Mooney HA, Cropper A, Capistrano D, Carpenter SR, Chopra K, Dasgupta P, Dietz T, Duraiappah AK, Hassan R, Kasperson R, Leemans R, May RM, McMichael AJ, Pingali P, Samper C, Scholes R, Watson RT, Zakri AH, Shidong Z, Ash NJ, Bennett E, Kumar P, Lee MJ, Raudsepp-Hearne C, Simons H, Thonell J, Zurek MB. Ecosystems and human well-being: a framework for assessment. Washington, DC: Island Press; 2005.

Richardson DM, Pyšek P, Rejmánek M, Barbour MG, Panetta FD, West C. Naturalization and invasion of alien plants: concepts and definitions. Divers Distrib. 2000;6:93–107.

Riley JL. Wetlands of the Ontario Hudson Bay Lowland: a regional overview. Toronto: Nature Conservancy of Canada; 2011. 156 p. app.

Saenger P, Specht MM, Specht RL, Chapman VJ. Mangal and coastal salt-marsh communities in Australasia. Chapter 15. In: Chapman VJ, editor. Wet coastal ecosystems. Amsterdam: Elsevier Scientific; 1977. p. 293–345.

Saintilan N, editor. Australian saltmarsh ecology. Collingwood: CSIRO Publishing; 2009.

Saintilan N, Wilson NC, Rogers KL, Rajkaran A, Krauss KW. Mangrove expansion and salt marsh decline at mangrove poleward limits. Glob Chang Biol. 2014;20:147–57.

Schuyt K, Brander L. The economic values of the World's Wetlands. Gland: World Wildlife Fund; 2004. Joint publication with Institute for Environmental Studies, Vrije Universiteit, Amsterdam I the Netherlands.

Silliman BR, Grosholz ED, Bertness MD, editors. Human impacts on salt marshes: a global perspective. Berkeley: University of California Press; 2009.

Spencer T, Schürch M, Nicholls RJ, Hinkel J, Vafeidis AT, Reef R, McFadden L, et al. Global coastal wetland change under sea-level rise and related stresses: The DIVA Wetland Change Model. Glob and Plan Chan. 2016;139:15–30.

Thannheiser D, Holland P. The plant communities of New Zealand salt meadows. Glob Ecol Biogeogr Lett. 1994;4:107–15.

Tiner RW. Field guide to coastal wetland plants of the Southeastern United States. Amherst: University of Massachusetts Press; 1999.

Tiner RW. Field guide to tidal wetland plants of the Northeastern United States and neighboring Canada. Vegetation of beaches, tidal flats, rocky shores, marshes, swamps, and coastal ponds. Amherst: University of Massachusetts Press; 2009.

Tiner RW. Tidal wetlands primer: an introduction to their ecology, natural history, status, and conservation. Amherst: University of Massachusetts Press; 2013.

Valiela I, Kinney E, Culbertson J, Peacock E, Smith S. Global losses of mangroves and salt marshes. In: Duarte CM, editor. Global loss of coastal habitats. Rates, causes, and consequences. Fundación BBVA; 2009. p. 109–42. http://www.fbbva.es/TLFU/dat/DE_2009_Global_Loss.pdf

Viereck LA, Dyrness CT, Batten AR, Wenzlik KJ. The Alaska vegetation classification system. Portland: USDA Forest Service, Pacific Northwest Research Station; 1992. General Technical Report PNW-GTR-286.

Warren RS, Niering WA. Vegetation change on a Northeast tidal marsh: interaction of sea-level rise and marsh accretion. Ecology. 1993;74:96–103.

West RC. Tidal salt-marsh and mangal formations of Middle and South America. Chapter 9. In: Chapman VJ, editor. Wet coastal ecosystems. Amsterdam: Elsevier Scientific; 1977. p. 193–213.

Woodwell GM, Rich PH, Mall CSA. Carbon in estuaries. In: Woodwell GM, Pecari EV, editors. Carbon in the biosphere. US AEC. 1973. 22(1):240.

Yang SL, Chen JY. Coastal salt marshes and mangrove swamps in China. Chinese J Oceanol Limnol. 1995;13:318–24.

Seagrasses

5

Frederick T. Short, Cathy A. Short, and Alyssa B. Novak

Contents

Introduction	74
Global Distribution	79
Elements of Ecology	82
Biota	85
Ecosystem Services	85
Threats and Future Challenges	88
Conclusions	90
References	90

Abstract

Seagrass meadows are a critical component of the coastal marine environment worldwide, providing some of the most economically and environmentally valuable ecosystem services of any marine habitat. These marine angiosperms form extensive meadows that store carbon, improve water quality, provide food and habitat, and act as biological indicators. These unique marine flowering plants are found mainly in clear, shallow estuaries and coastal waters where they propagate both sexually and vegetatively, with 72 species worldwide. They provide habitat for juvenile fish and shellfish and are eaten by sea turtles, dugong and manatee as

F. T. Short (✉)
Department of Natural Resources and the Environment, Jackson Estuarine Laboratory, University of New Hampshire, Durham, NH, USA
e-mail: fred.short@unh.edu

C. A. Short
Lee, NH, USA
e-mail: cathyashort@gmail.com

A. B. Novak
Boston University, Earth and Environment, Boston, MA, USA
e-mail: abnovak@bu.edu

© Springer Science+Business Media B.V., part of Springer Nature 2018
C. M. Finlayson et al. (eds.), *The Wetland Book*,
https://doi.org/10.1007/978-94-007-4001-3_262

well as waterfowl. Seagrasses grow both intertidally and subtidally in all the tropical and temperate ocean. Despite their importance, seagrass meadows are experiencing high rates of loss globally due to direct threats such as sedimentation, eutrophication, dredging and aquaculture, as well as diffuse threats such as water quality losses and climate change. All seagrass species have been evaluated for the IUCN Red List of Threatened and Endangered Species, with 14% at elevated risk of extinction. Strong science-based management and regulatory strategies are needed to maintain and increase seagrass habitats, as well as build their resilience to stressors in a globally changing environment.

Keywords

Seagrass · Angiosperm · Carbon storage · Coastal habitat · Submerged marine flowering plants · Vascular plants · Estuarine vegetation · Zostera

Introduction

Seagrasses are submerged flowering plants growing on all continents except Antarctica in coastal and estuarine environments; they are found from the intertidal to 90 m deep. Extensive seagrass areas are often referred to as seagrass beds or meadows, ranging from a few square meters to hundreds of square kilometers. Although there are relatively few species of seagrass (72 species in 6 families and 14 genera), the complex physical structure and high productivity of these ecosystems enable them to support a considerable biomass and diversity of associated species. Seagrass meadows provide ecosystem services that rank among the highest on Earth. Their direct monetary value is substantial since highly valued commercial and artisanal/subsistence catches such as prawns and fish are dependent on seagrass ecosystems. Seagrasses themselves are a critically important food source for dugong, manatee, sea turtles, and waterfowl. Many other species of fish and invertebrates, including sea horses, shrimps, and scallops, utilize seagrass for part of their life cycles, often for breeding or as juveniles. Seagrasses are considered to be one of the most important shallow marine ecosystems to humans, binding sediments, filtering coastal waters, and providing some protection from coastal erosion (Hemminga and Duarte 2000). There is a growing awareness of "seagrass blue carbon," referring to the fact that seagrasses sequester and store carbon in their roots and sediments. Although seagrasses represent only a small area (0.2% of the oceans' surface), it is estimated that they store 20% of oceanic blue carbon. Seagrasses are among the world's most threatened ecosystems and yet are little known because they are usually submerged and not easily seen.

Seagrasses are not true grasses. Although they are all monocotyledons, they do not have a single evolutionary origin, but are a polyphyletic group (Table 1). They live where sufficient light reaches the seafloor for these rooted plants to support photosynthesis (Fig. 1). Most have flattened leaf blades (with the exception of the genus *Syringodium* and some *Phyllospadix*, both with cylindrical leaves), elongated or strap-like leaves (with the exception of the genus *Halophila*), and an extensive

Table 1 Seagrass species, families, and the bioregions (see Fig. 4) where they are found

Species	Family	Bioregion
Amphibolis antarctica	Cymodoceaceae	6
Amphibolis griffithii	,,	6
Cymodocea angustata	,,	5
Cymodocea nodosa	,,	1, 3
Cymodocea rotundata	,,	5
Cymodocea serrulata	,,	5
Halodule beaudettei[a]	,,	2
Halodule bermudensis[a]	,,	2
Halodule ciliata	,,	2
Halodule emarginata	,,	2
Halodule pinifolia	,,	5
Halodule uninervis	,,	5
Halodule wrightii	,,	1, 2, 3, 4, 5
Syringodium filiforme	,,	2, 3
Syringodium isoetifolium	,,	5, 6
Thalassodendron ciliatum	,,	5, 6
Thalassodendron pachyrhizum	,,	6
Enhalus acoroides	Hydrocharitaceae	5
Halophila australis	,,	6
Halophila baillonii	,,	2
Halophila beccarii	,,	5
Halophila capricorni	,,	5
Halophila decipiens	,,	2, 3, 4, 5, 6
Halophila engelmannii	,,	2
Halophila euphlebia	,,	4
Halophila hawaiiana	,,	5
Halophila johnsonii[a]	,,	2
Halophila minor[a]	,,	5
Halophila nipponica	,,	4
Halophila ovalis	,,	4, 5, 6
Halophila ovata[a]	,,	5
Halophila spinulosa	,,	5
Halophila stipulacea	,,	2, 3, 5
Halophila sulawesii	,,	5
Halophila tricostata	,,	5
Thalassia hemprichii	,,	5
Thalassia testudinum	,,	2
Species	**Family**	**Bioregion**
Posidonia angustifolia	Posidoniaceae	6
Posidonia australis	,,	6
Posidonia coriacea	,,	6
Posidonia denhartogii	,,	6

(*continued*)

Table 1 (continued)

Species	Family	Bioregion
Posidonia kirkmanii	"	6
Posidonia oceanica	"	3
Posidonia ostenfeldii	"	6
Posidonia sinuosa	"	6
Ruppia cirrhosa[a]	Ruppiaceae	3
Ruppia filifolia	"	6
Ruppia maritima[a]	"	1, 2, 3, 4, 5, 6
Ruppia megacarpa	"	6
Ruppia polycarpa	"	6
Ruppia tuberosa	"	6
Lepilaena australis	Zannichelliaceae	6
Lepilaena marina	"	6
Phyllospadix iwatensis	Zosteraceae	4
Phyllospadix japonicus	"	4
Phyllospadix scouleri	"	4
Phyllospadix serrulatus	"	4
Phyllospadix torreyi	"	4
Zostera asiatica	"	4
Zostera caespitosa	"	4
Zostera capensis	"	5, 6
Zostera caulescens	"	4
Zostera chilensis[a]	"	6
Zostera geojeensis[a]	"	4
Zostera japonica	"	4, 5
Zostera marina	"	1, 3, 4
Zostera muelleri	"	5, 6
Zostera nigricaulis[a]	"	6
Zostera noltii	"	1, 3
Zostera pacifica	"	4
Zostera polychlamys[a]	"	6
Zostera tasmanica[a]	"	6

[a]Species status under review

system of roots and rhizomes. The genus *Halophila* is a unique lineage of seagrass species with small paddle-shaped blades most often in pairs (Fig. 2).

Seagrasses descended from terrestrial plants that reentered the ocean between 100 and 65 million years ago. The development of different seagrass lineages occurred at least three different times during evolution as determined by chloroplast DNA profiles (Waycott et al. 2006). The seagrasses were first scientifically noted in 1753 when Carolus Linnaeus described the species *Zostera marina*. The ecological value of seagrass was detailed in 1898 by C. G. J. Petersen through the construction

Fig. 1 Seagrass *Zostera marina* (eelgrass) intertidal meadow in Portsmouth Harbor, New Hampshire, USA (Photo credit: D. Porter © Rights remain with the author)

Fig. 2 Seagrass *Halophila australis* in the Southern Ocean off Perth, Australia (Photo credit: FT Short © Rights remain with the author)

of a trophic pyramid with *Z. marina* as the base and higher trophic levels supporting codfish and brant at the apex (Milne and Milne 1951). Cornelis den Hartog (1970) of the Netherlands wrote a seminal book on the subject, *Seagrasses of the World*, providing detailed taxonomy. The first International Seagrass Biology Workshop was held in 1973, bringing seagrass scientists together from around the world. Present-day seagrass activities include studies of genetics, formation of the World Seagrass Association in 2000 (Coles et al. 2014), publication of the *World Atlas of Seagrasses* (Green and Short 2003) and other seagrass volumes, and evaluation of all seagrass species via the International Union for the Conservation of Nature (IUCN) Red List.

Typically, seagrasses grow in areas dominated by soft substrates such as sand or mud, but some species can be found on more rocky substrates (e.g., *Phyllospadix* spp.). Seagrasses require high levels of light, more than other marine plants, because of their complex belowground structures which include considerable amounts of non-photosynthetic tissues and which require transport of oxygen to maintain a rhizosphere around the roots in anoxic sediments. Thus, although they have been recorded to 90 m depth in clear waters (Short et al. 2007), they are more generally restricted to shallow waters due to the rapid attenuation of light with depth. Seagrasses form extensive monospecific stands or areas of mixed species, creating a unique marine ecosystem or biotope. Seagrasses can also grow in isolated patches or as part of a habitat mosaic with corals, mangroves, bivalve reefs, rocky benthos, salt marshes, or bare sediments. Generally it is the larger seagrass beds in developed parts of the world which have been the subject of intensive study and mapping worldwide (Green and Short 2003). Although typically permanent over periods of decades to centuries, seagrass systems can be highly dynamic, moving into new areas and disappearing from others over relatively short timeframes.

Seagrasses themselves are an important standing stock of organic biomass, relatively stable near the equator and with broad intra-annual variation in subtropical and temperate regions. Primary productivity of these ecosystems is usually enhanced by other primary producers, including macroalgae and epiphytic algae. The abundant detrital plant material of seagrass beds is the basis of many food webs. Additionally, the complex three-dimensional structure of the seagrass bed provides shelter and cover while binding sediments and altering the patterns and strength of water currents.

Seagrasses are indicators of estuarine and coastal health worldwide (Orth et al. 2006; Waycott et al. 2009). Growing rooted in place, seagrass integrates the influences of the environmental conditions that it experiences within an estuarine system, and therefore seagrass health status acts as a barometer of impacts and changes. Seagrass beds alter their distribution and biomass in response to changing water quality, nutrient inputs, and light levels. Seagrass change can be measured at the plant population level or by examining differences in plant physiology and chemistry, functioning as an indicator of ecosystem health.

In the past century, along the developed east coast of the USA, it is estimated that up to 50% of all seagrass habitat has been lost, and the prospects for recovery in most of this area are low due to loss of water clarity, severe coastal alterations, and heavy use (Green and Short 2003). Seagrasses worldwide are in decline, and huge losses of seagrass have occurred due to coastal development and anthropogenic pressures of pollution, land cover change, and direct physical impacts. Nevertheless, there are still vast areas of seagrass in the world's nearshore environments. A global assessment of available literature and reports (Waycott et al. 2009) found that 2,640 km^2 of seagrass was documented as lost between 1980 and 2004 or 1.5% of worldwide seagrass area. Seagrass loss rates increased from a median of 0.9% year^{-1} before 1940 to 7% year^{-1} since 1990. Seagrasses are one of the most threatened ecosystems on Earth with loss rates comparable to those of mangroves, coral reefs, and tropical rainforests.

Global Distribution

Seagrasses are widely distributed along temperate and tropical coastlines of the world (Fig. 3). Virtually no seagrass species is fully investigated as to global occurrence or genetic diversity across its range. While the global species diversity of seagrasses is low, seagrass species can have ranges that extend for thousands of kilometers of coastline. The northern and southern hemispheres share ten seagrass genera, having only one unique genus each. There are roughly the same number of temperate and tropical seagrass genera as well as species. *Ruppia maritima*, only recently classified as a seagrass, is one of the most widely distributed of all flowering plants on Earth, occurring in both tropical and temperate zones and a wide variety of habitats.

For all species of seagrass, distribution is a product of combined plant sexual reproduction and clonal growth, influenced by dispersal and environmental

Fig. 3 Seagrass distribution worldwide source (Updated from Green and Short (2003), with permission)

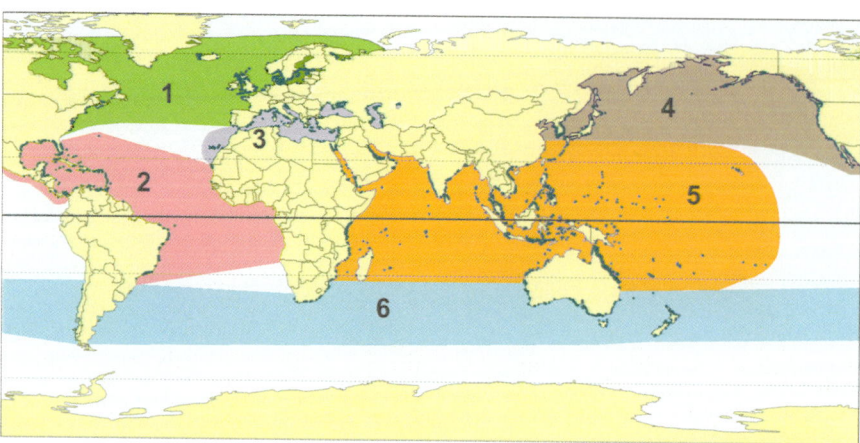

Fig. 4 Seagrass bioregions: *1* Temperate North Atlantic, *2* Tropical Atlantic, *3* Mediterranean, *4* Temperate North Pacific, *5* Tropical Indo-Pacific, and *6* Temperate Southern Oceans (Short et al. 2007) (Reprinted from Journal of Experimental Marine Biology and Ecology, 350 (1–2), Short F, Carruthers T, Dennison W, Waycott M, Global seagrass distribution and diversity: A bioregional model, 3-20, 2007, © with permission from Elsevier)

limitations (Spalding et al. 2003). Many seagrass populations are highly clonal, largely relying on asexual reproduction for population maintenance. Other seagrasses produce large numbers of sexual propagules or vary their reproductive strategies depending on environmental conditions. All seagrass species are capable of asexual reproduction, producing modular units (ramets) through horizontal rhizome growth that may be physiologically independent but are genetically identical to the parent plant (genet). Seagrasses are also capable of sexual reproduction by producing fruits and seeds or, in some cases, viviparous seedlings.

Determining the area of seagrass habitat around the globe is difficult because less than a quarter of the world's seagrasses have been mapped. The majority of seagrass information consists of observations of seagrass for specific locations with no determination of meadow size (Green and Short 2003). Globally, total seagrass area is estimated to be 177,000 km^2 based on actual mapped areas and inference of unmapped areas where seagrass occurrence has been documented (Spalding et al. 2003).

There are six global seagrass bioregions: four temperate and two tropical. The temperate bioregions include the Temperate North Atlantic, the Temperate North Pacific, the Mediterranean, and the Temperate Southern Oceans (Fig. 4, Short et al. 2007). All four temperate bioregions contain species of the genera *Zostera*. The two tropical bioregions are the Tropical Atlantic and the Tropical Indo-Pacific, both supporting mega-herbivore grazers, including sea turtles, dugong, and manatee. The Tropical Atlantic bioregion has clear water with a high diversity of seagrasses on reefs and shallow banks, dominated by *Thalassia testudinum*. The vast Tropical Indo-Pacific has the highest seagrass diversity in the world, with as many as 14 species growing together on reef flats although seagrasses also occur in very deep waters.

The Temperate North Atlantic has low seagrass diversity, the major species being *Zostera marina*, or eelgrass, typically occurring in estuaries and lagoons. Eelgrass is the most studied seagrass species worldwide and was severely impacted by the wasting disease in the 1930s in this region. In Europe, *Zostera noltii* is found in the shallow intertidal, generally inshore of *Z. marina*. On the temperate coasts of North America and Europe, *R. maritima* occurs in both brackish estuarine areas and hypersaline salt marsh pools.

The Temperate North Pacific has high seagrass diversity (15 species) with *Zostera* species in estuaries and lagoons as well as *Phyllospadix* species in the surf zone and four species which occur only at the region's boundary. Eelgrass, *Z. marina*, occurs around the Pacific Rim from Japan, Korea, and China to the northern Bering Sea and down to the Gulf of California, with other *Zostera* species centered in East Asia. *Phyllospadix* is found on both sides of the Pacific, with two species in Asia and three species in North America, all of which have a modified rhizome, attaching to rocks in the high-energy surf zone and allowing them to inhabit exposed coasts.

The Mediterranean bioregion has clear water with vast meadows of moderate diversity of both temperate and tropical seagrasses, dominated by the endemic and deep-growing species, *Posidonia oceanica*, which forms a "matte" of root and rhizome material which may be several meters deep and thousands of years old. There are other species in the Mediterranean (Table 1) including *Halophila stipulacea* which invaded from the Indian Ocean through the Suez Canal.

Extensive meadows of low-to-high diversity temperate seagrasses are found in the Temperate Southern Oceans bioregion, dominated by various species of *Posidonia* and *Zostera*. Species of *Ruppia* and *Zostera* occur on all the continents within this circumpolar region: South America, Africa, and Australia. A total of 18 seagrass species occur, of which 4 are found only at the region's boundaries; seagrasses are found in a wide range of habitats including surf zones. The southernmost seagrass in the world is *Ruppia maritima*, collected from the Straits of Magellan by Lucia Mazzella, a renowned seagrass ecologist (V. Zupo pers. com.).

The Tropical Atlantic, with ten seagrasses, is dominated by three species, *Thalassia testudinum*, *Syringodium filiforme*, and *Halodule wrightii*, which can occur in single species stands but often occur intermixed or sequentially in ecological succession. The region is predominantly a carbonate environment, occupying areas from shallow back reefs to broad deep sandbanks. *Halophila johnsonii*, a species on the US Federal Threatened Species List, is limited in distribution to the east coast of Florida (Virnstein and Hall 2009), but may actually be a form of *Halophila ovalis* (Waycott et al. 2006). The invasive seagrass *Halophila stipulacea* appeared in the Caribbean in 2002, first discovered off the island of Grenada, and, by 2014, had spread throughout the Greater and Lesser Antilles (Willette et al. 2014).

The vast Tropical Indo-Pacific bioregion extends from the east coast of Africa to the eastern Pacific Ocean with 24 species, more than any other region. In the Western Pacific, there is a trend of decreasing species numbers from west to east with (north of the equator) 14 species in the Philippines, 10 in Micronesia, and 2 in the Marshall Islands and (south of the equator) 12 species in Papua New Guinea and Vanuatu, 6 in

Fiji, and 2 in Western Samoa and French Polynesia (Green and Short 2003). In many parts of the region, seagrasses are found inshore of the coral reef. Typically, the three species *Thalassia hemprichii*, *Syringodium isoetifolium*, and *Halodule uninervis* dominate reef platforms and are found throughout the western two thirds of the region. *Enhalus acoroides*, a unique Indo-Pacific species, has flowers that are fertilized on the water surface by floating pollen, limiting its distribution to intertidal and shallow waters. *Thalassodendron ciliatum* has a flexible, woody stem and can withstand wave action. High seagrass diversity typifies the western side of the Tropical Indo-Pacific, with 12–14 species currently identified for the tropical areas of the Indian Ocean.

Elements of Ecology

The 1930s eelgrass "wasting disease" incurred an extreme, though natural, loss of *Zostera marina* (Table 2), which almost disappeared from both sides of the North Atlantic (Green and Short 2003). The pathogen causing this dieback was a marine slime mold (*Labyrinthula zosterae*), but the reasons for its eruption in the 1930s have yet to be determined. Plant leaves developed black necrotic streaks which spread rapidly in higher-salinity waters; plants in low-salinity areas survived the disease. Eelgrass began recovering in the 1950s although it has never fully regained its former distribution. A recurrence of this disease in the 1980s destroyed much of the eelgrass in the Great Bay Estuary (New Hampshire, USA). A somewhat similar mass mortality of turtle grass, *Thalassia testudinum*, in Florida Bay may also have been linked to the pathogen *Labyrinthula* sp., as may recent seagrass declines on the US west coast.

Threats to seagrasses from global climate change include increases in sea surface temperature, sea level, CO_2 concentrations, storm events, and levels of ultraviolet-B (UV-B) radiation (Short et al. 2016). Increased maximum annual seawater temperature in the Mediterranean has led to increased seagrass mortality, and the Mediterranean Institute for Marine Studies expects *Posidonia oceanica*, the dominant seagrass in the Mediterranean, to decline by 90% and become a functionally extinct ecosystem by 2050. The impacts of climate change on other seagrass species are uncertain. Scientists anticipate a shift in the geographic and depth distribution of species depending on their tolerance to different climate stressors, with the possibility of species extinction. Increasing CO_2 from anthropogenic activities is causing increased ocean acidification, adversely affecting shellfish but promoting photosynthetic carbon uptake in seagrasses, which may locally buffer ocean acidity. Increased storms throughout the world exacerbate many seagrass stressors because storm runoff brings pollutants into coastal waters.

The phenomenon of leaf reddening in seagrasses was first seen in Australia and has since been shown to be prevalent in seagrasses growing in clear, shallow waters with high light intensities (Novak and Short 2010). The response of seagrasses to UV-B radiation varies by species and can result in red-to-purple coloration of seagrass leaves (Fig. 5). The red coloration is caused by the accumulation of

Table 2 Documented major losses of seagrasses, from both natural and human impacts, as well as recoveries

Location	Date	Seagrass loss or gain	Cause
Natural impacts			
North Atlantic	1930s	90% eelgrass loss in 5 years	Wasting disease
Great Bay, NH, USA	1980s	80% eelgrass loss in 2 years	Wasting disease
Prince William Sound, AK, USA	1960s	Local loss in uplift areas	Earthquake
Puerto Morelos, Mexico	1990s	Decreased growth	Hurricane
Hervey Bay, Australia	1990s	1000 km^2 loss	Cyclone
Banda Aceh, Indonesia	2006	Local loss in uplift areas	Seismic uplift
Anthropogenic impacts			
Baltic Sea, Denmark	1901–1994	75% loss	↓Water quality
Tampa Bay, FL, USA	1950–1963	50% loss bay-wide	Nutrient load/↓water clarity
Tauranga, New Zealand	1959–1996	90% decline	Sedimentation
Weihai, China	1950s	95% (>100 km^2) loss	Kelp aquaculture
Chesapeake Bay, USA	1970s	Massive eelgrass decline	↓Water quality
Cockburn Sound, Australia	1970s	>66% loss	Nutrient loading/pollution
French Mediterranean	1970s	40% decline	Pollution/↓water quality
Lake Grevelingen, Netherlands	1970–1980s	34 km^2 loss	Nutrient loading
Wadden Sea, Netherlands	1970–1980s	100% loss	Pollution/turbidity
Westernport, Australia	1973–1984	85% decline	Sedimentation
Waquoit Bay, MA, USA	1980s	85% eelgrass loss in 6 years	Nitrogen loading
Florida Bay, USA	1980s	40 km^2 seagrass die-off	Unknown
Indian River Lagoon, FL, USA	1980s	30% decline	Nutrient load/↓water clarity
Maquoit Bay, ME, USA	1990s	10% (53 ha) loss in 8 years	Trawling
Sabah, Malaysia	2003–2005	67% loss per year	Sediment loading
Middle Placencia Lagoon, Belize	2003–2007	91% loss	Shrimp pond effluent

(*continued*)

Table 2 (continued)

Location	Date	Seagrass loss or gain	Cause
Nova Scotia, Canada	1986–2010	98% (35 ha) loss in 24 years	Green crab uprooting
Casco Bay, ME, USA	2001–2013	56% (19 km^2) loss in 12 years	Green crab uprooting
Indian River Lagoon, FL, USA	2011–2012	130 km^2 loss	↓Water clarity
Morro Bay, CA, USA	2012	1.4 km^2 loss in 6 years	Under investigation (wasting disease/↓water clarity)
Seagrass recovery			
Great Bay, New Hampshire, USA	2016	80% loss of bay-wide biomass	Eutrophication
Tampa Bay, FL, USA	1982–2012	48 km^2 revegetation	↑ Water clarity
Mondego Bay, Portugal	1997–2002	1.4 km^2 revegetation	↑ Water clarity
Delmarva Coastal Bays, USA	1999–2012	19 km^2 gain	Seed planting

Source: based on data from Short and Wyllie-Echeverria 1996; Green and Short 2003; Waycott et al. 2009, and findings from K. Merkel, C. McCarthy, H. Neckles, B. Virnstein, R. Orth

Fig. 5 Shallow intertidal *Cymodocea rotundata* with reddened leaves, Hat Yai National Park, Thailand (Photo credit: FT Short © Rights remain with the author)

anthocyanins which act as a sunscreen during periods of high light. Climate models predict global warming will diminish atmospheric ozone leading to increased levels of UV radiation in the tropics and southern latitudes over the next century. Seagrass leaf reddening is expected to increase in these regions as it augments plant resilience to increasing UV-B levels by protecting photosynthetic mechanisms from damage.

For the first time, the probability of extinction has been determined for all of the world's seagrass species under the categories and criteria of the International Union for the Conservation of Nature (IUCN) Red List of Threatened Species (Short et al. 2011). The 4-year effort involved international seagrass experts; compilation of data on species' status, populations, and distribution; and review of the biology and ecology of each of the world's seagrass species. Ten seagrass species are at elevated risk of extinction (14% of all seagrass species), with three species qualifying as endangered due to their very small ranges and/or severe human impacts (Table 1). The other seven species are ranked as vulnerable, including *Halophila beccarii*, which lives in mangrove habitat that is itself declining rapidly due to shrimp aquaculture. Seagrass species loss and degradation of seagrass biodiversity will have serious repercussions for marine biodiversity and the human populations that depend upon the resources and ecosystem services that seagrasses provide.

Biota

Seagrass habitats have higher levels of diversity than adjacent non-vegetated areas (Table 3). Many of the species that have been recorded in seagrasses are also found in other ecosystems, although some are dependent on seagrass ecosystems for at least a part of their life cycles (Gillanders 2006). Such seagrass-dependent species range from epiphytic algae and invertebrates attached to seagrass leaves to the large seagrass-grazing sea turtles, manatee (Fig. 6), and dugong (Valentine and Duffy 2006). The seagrass leaf surfaces provide substrate for fish and snails to deposit their eggs, while the dense shoots create reproductive habitat for sea horses and pipefish. Seagrasses are nursery habitats; juvenile mussels, scallops, and shrimps all find protection and food in the seagrass canopy above the seafloor. Young fish and crabs, including juvenile flounder and cod, shelter in seagrass habitat.

Ecosystem Services

Seagrasses support marine food webs and provide essential habitat for many coastal species, playing a critical role in the equilibrium of coastal ecosystems and human livelihoods. Seagrasses are important in fishery production, in sediment accumulation and stabilization, and in water filtration of nutrients and suspended sediments. Seagrass ecosystem services are estimated to have an annual global value of US$500 billion (Costanza et al. 2014). The ecosystem service value of seagrass, measured on an areal basis, has been estimated to be three times more than coral reefs and ten times more than tropical forests.

Table 3 Major taxonomic groups found in seagrass ecosystems with some specifics. **Bold text** are organisms consumed or used by humans

Taxonomic group	Notes
Bacteria	Includes purple sulfur
Fungi	Includes *Plasmodiophora*
Diatoms (Bacillariophyta)	Includes *Cocconeis*, *Nitzschia*
Blue-green algae (Cyanophyta)	
Red algae (Rhodophyta)	Includes ***Gracilaria***, ***Chondrus***, calcareous species (also epiphytic)
Brown algae (Phaeophyta)	Includes ***Padina***
Green algae (Chlorophyta)	Notably ***Ulva***, *Halimeda*, and *Caulerpa*
Protozoa	Includes *Foraminifera* and the marine slime mold *Labyrinthula*
Sponges	Includes epiphytic and free-standing species
Cnidarians	Includes epiphytic hydrozoans, sea anemones, solitary corals, and *Scleractinia* such as *Pavona*, *Psammocora*, *Porites*, *Pocillopora*, *Siderastrea*
Polychaetes	Including rag worms (nereids)
Ribbon worms	
Sipunculid worms	***Sipunculus*** (peanut worm)
Flatworms	
Crustaceans	Includes amphipods, isopods, and many **decapod crustaceans** (crabs, stomatopods, and commercially important shrimp and lobster)
Bivalve mollusks	Some **scallops, oysters, and clams**, also many boring species
Gastropod mollusks	A broad range including *Conus*, *Cypraea*, and commercially important species of ***Strombus***
Cephalopod mollusks	**Squid and cuttlefish** often found over seagrass areas, **octopus**
Bryozoans	Epiphytic on seagrass and rocks
Echinoderms	A range of commercially important **holothurian** species and ophiuroids are widespread, but also **asteroids** and **echinoids** (spoon worm)
Tunicates	Ascidians
Fish	All **fish** groups, but including the commercially important Haemulidae (**grunts**), Siganidae (**rabbitfish**), Lethrinidae (**emperors**), Lutjanidae (**snappers**), Bothidae (**left-eye flounders**), Syngnathidae (**pipefish and sea horses**); many of the latter, which are used in the aquarium trade and Chinese medicine trade, are considered threatened
Reptiles	Notably the **green turtle** *Chelonia mydas*
Birds	Notably **brant** (geese) and other migrating **waterfowl** and wading birds
Mammals	Notably the sirenian species dugong ***Dugong dugon*** and manatee (*Trichechus manatus*, *Trichechus senegalensis*)

Source: updated from World Atlas of Seagrasses (Green and Short 2003, with permission)

Fig. 6 Antillean manatee *Trichechus manatus* in Belize (Photo credit: H Jiwa © Rights remain with the author)

Dugongs, manatees, sea turtles, geese such as brant, and some herbivorous fish eat seagrass directly. "Dugong trails" are seen in seagrass beds in Thailand, where these endangered animals have grazed. Some rare and endangered sea horses use seagrass as their preferred habitat. Seagrass beds include algae and invertebrates, which serve as food for transient and resident fish. Seagrass meadows are rich in benthic epifauna and infauna. Seagrass habitats shelter and attract numerous species of breeding animals. Fish use the three-dimensional seagrass canopy as a nursery as well as a feeding ground, often moving from coral reefs or mangroves to feed in seagrass. Shrimp settle in seagrass meadows in their post-larval stage and remain there until they become adults. The seagrass detrital food web is based on decomposing seagrass leaves and roots which connect trophically to feed top-level marine consumers. Human subsistence fishers depend on foraging seagrass areas, which often provide most of the protein in their diet (Cullen-Unsworth et al. 2014).

Seagrasses are a natural filter. They both trap and bind sediment. Their leaves catch suspended materials that are brought to the seagrass meadows with the currents, clearing the water. The complex seagrass root-rhizome structure then holds these sediments in place. Seagrasses prevent coastal erosion by attenuating waves and decreasing sediment resuspension. The relatively rapid uptake of nutrients both by seagrasses and their epiphytes removes nutrients from the water column. Once removed, these nutrients can be released only slowly through a process of decomposition and consumption.

Seagrasses have relatively high productivity and biomass compared to other oceanic habitats and most terrestrial ecosystems. The estimated average net primary production of seagrass is about 2.7 g dry weight/m^2/day, far more than macroalgal communities (1 g dry weight/m^2/day) or phytoplankton (0.35 g dry weight/m^2/day)

(Duarte and Chiscano 1999). Seagrass photosynthesis oxygenates otherwise hypoxic sediments, providing O_2 that sustains microbial activity in the sediments, facilitating nutrient recycling from dead organic matter. Seagrasses also constitute an important carbon sink, a repository of "blue carbon," due to their slow decomposition rates and enrichment of stored sediment carbon.

Threats and Future Challenges

Historical losses in seagrass abundance and distribution have had substantial impacts on marine organisms and ecosystems as well as humans. Seagrasses all over the world were used in numerous ways in the past when this resource was in greater abundance. For example, leaves collected from the wrack were used as thatching for roofs: *Zostera marina* in the Netherlands through the 1800s and *Phyllospadix japonicus* in northern China for generations (Fig. 7). In China, whole villages are still mostly thatched with seagrass, but it is old seagrass collected in the 1960s; since the development of kelp aquaculture, floating over the large meadows of the seagrass *Phyllospadix japonicus*, the seagrass wrack has disappeared (Table 2). With the loss of seagrass, Chinese villagers are forced to use "inferior" tile roofs that are "colder in

Fig. 7 A roof made of the dried seagrass *Phyllospadix japonicus* in northern China, a housing practice for generations (Photo credit: FT Short © Rights remain with the author)

the winter and too hot in the summer" (Short pers. obs.). Similarly, the vast meadows of eelgrass, *Zostera marina,* that once existed in the Canadian Maritimes supported an industry involving the collection of eelgrass wrack from the beach which was dried and used in a commercial home insulation product (the "Cabot's Quilt," Cabot 1986). The 1930s eelgrass wasting disease, which was responsible for a loss of 90% of eelgrass on both sides of the Atlantic, knocked back the habitat, and it has never fully recovered (Milne and Milne 1951), in part due to degraded nearshore conditions. More recently, there are places like Waquoit Bay, Massachusetts, where all the eelgrass has been lost since the 1980s due to eutrophication; despite decades of resource management, conditions are still not adequate for seagrass recovery. There is a shifting baseline: as memory of the presence of seagrass is lost, and in the face of a general lack of awareness of its importance and distribution, the impetus to protect and restore the habitat is lost as well.

Present-day seagrass threats can be localized, such as dredge and fill, aquaculture, or shoreline hardening, or they may be more generalized and widespread, such as turbidity and phytoplankton blooms driven by nutrient loading. Turbidity (suspended sediments) in the water column from sediment loading is one of the primary stressors of seagrasses globally (Short et al. 2011). Turbidity reduces water clarity, thereby limiting the light reaching seagrass plants and reducing seagrass photosynthesis and growth. Main contributors to sediment loading of coastal waters are agriculture, which promotes soil erosion, and urbanization, creating impervious surfaces and rapid runoff.

Dense phytoplankton blooms in ocean waters derive primarily from high nitrogen concentrations that fuel phytoplankton primary production and eutrophication (McGlathery et al. 2007). In nearshore areas, excess nutrients are derived from many human sources and promote the growth of competitive algae, both seaweeds and phytoplankton. Nitrogen loading from wastewater treatment facilities, watershed runoff, agriculture, and loss of natural buffers result in excess nutrients which cause phytoplankton and seaweed blooms, shading seagrasses and causing seagrass declines worldwide (Table 2; Short and Wyllie-Echeverria 1996).

Ocean temperature in coastal waters is critical to seagrass species distribution, more than latitude per se. The distribution of temperate and tropical seagrass species worldwide is bounded by ambient temperature conditions. Temperate seagrass species are limited when coastal ocean temperatures exceed 25 °C, while tropical species are primarily found in areas above 25 °C. The primary effects of increased temperature on seagrasses are to enhance growth at the low end of a species' temperature range and to metabolically stress the plants at the high end of their temperature tolerance range (Björk et al. 2008). Salinity level also creates limits to seagrass distribution; most species survive in a salinity range from 5 to 45 psu.

In a few places, gains in seagrass area and biomass have been achieved when threats are ameliorated although these do not change the overall trend of global seagrass loss (Table 2). In many cases, resource management to reduce eutrophication, sedimentation, or direct human impacts is the first step, sometimes followed by seagrass restoration in the form of transplanting or seeding. In Tampa Bay, USA, investment in reduction of nutrient inputs created clearer water allowing natural expansion and recovery of 48 km^2 of seagrass area in 20 years (Table 2). In the lagoons off the Delmarva Peninsula on the east coast of the USA, a large area of unvegetated bottom was planted via seed scattering for a gain of 52 km^2 of seagrass meadow.

Conclusions

Seagrasses are beautiful and important coastal systems that provide habitat, food, shelter, and nursery areas to biota in the world's oceans. They also contribute filtering, sediment stabilizing, and other ecosystem services worldwide. Many coastal people depend on seagrass habitat for subsistence fishing and gleaning to provide their major source of protein. Seagrass in populated areas is declining as rapidly as any oceanic habitat with rapid coastal development and waste discharge the major impacts. A fifth of all species have a heightened risk of extinction. Seagrasses must be restored and preserved. To stop and then reverse the decline of seagrasses, a powerful combination of reduced exploitation, increased conservation and monitoring, and improved nearshore water clarity is needed globally.

References

Björk M, Short F, Mcleod E, Beer S. Managing seagrasses for resilience to climate change. Gland: IUCN; 2008. 56pp.
Cabot S. Memories of the Cabot's Quilt. Yankee, November 1986.
Coles R, Short F, Fortes M, Kuo J. Twenty years of seagrass networking and advancing seagrass science: the international seagrass biology workshop series. Pac Conserv Biol. 2014;20:8–16.
Costanza R, de Groot R, Sutton P, van der Ploeg S, Anderson SJ, Kubiszewski I, Farber S, Turner RK. Changes in the global value of ecosystem services. Glob Environ Chang. 2014;26:152–8.
Cullen-Unsworth LC, Nordlund LM, Paddock J, Baker S, McKenzie L, Unsworth RKF. Seagrass meadows globally as a coupled social-ecological system: implications for human wellbeing. Mar Pollut Bull. 2014;83:387–97.
den Hartog C. The seagrasses of the world. Verh K Ned Akad Wetenschappen Afdeling Natuurkunde. 1970;59:1–275.
Duarte CM, Chiscano CL. Seagrass biomass and production: a reassessment. Aquat Bot. 1999;65:159–74.
Gillanders BM. Seagrasses, fish and fisheries. In: Larkum AWD, Orth R, Duarte C, editors. Seagrasses: biology, ecology and conservation. Dordrecht: Springer; 2006. p. 503–36.
Green EP, Short FT, editors. World atlas of seagrasses. Berkeley: University of California Press; 2003. 324 pp.
Hemminga MA, Duarte CM. Seagrass ecology. Cambridge, UK: Cambridge University Press; 2000. 298 pp.
McGlathery KJ, Sundback K, Anderson IC. Eutrophication in shallow coastal bays and lagoons: the role of plants in the coastal filter. Mar Ecol Prog Ser. 2007;348:1–18.
Milne LJ, Milne MJ. The eelgrass catastrophe. Sci Am. 1951;184:52–5.
Novak AB, Short FT. Leaf reddening in seagrasses. Bot Mar. 2010;53:93–7.
Orth RJ, Carruthers TJB, Dennison WC, Duarte CM, Fourqurean JW, Heck Jr KL, Hughes AR, Kendrick GA, Kenworthy WJ, Olyarnik S, Short FT, Waycott M, Williams SL. A global crisis for seagrass ecosystems. Bioscience. 2006;56:987–96.

Short FT, Wyllie-Echeverria S. Natural and human-induced disturbance of seagrasses. Environ Conserv. 1996;23:17–27.

Short FT, Dennison WC, Carruthers JTB, Waycott M. Global seagrass distribution and diversity: a bioregional model. J Exp Mar Biol Ecol. 2007;350:3–20.

Short FT, Polidoro B, Livingstone SR, Carpenter KE, Bandeira S, Bujang JS, Calumpong HP, Carruthers TJB, Coles RG, Dennison WC, Erftemeijer PLA, Fortes MD, Freeman AS, Jagtap TG, Kamal AHM, Kendrick GA, Kenworthy WJ, La Nafie YA, Nasution IM, Orth RJ, Prathep A, Sanciangco JC, van Tussenbroek B, Vergara SG, Waycott M, Zieman JC. Extinction risk assessment of the world's seagrass species. Biol Conserv. 2011;144:1961–71.

Short FT, Kosten S, Morgan P, Malone S, Moore G. Present and future impacts of climate change on submerged and emergent wetland plants: a review. In: Aquat. Bot. special issue: "40 years of Aquatic Botany"; 2016. http://dx.doi.org/10.1016/j.aquabot.2016.06.006.

Spalding M, Taylor M, Ravilious C, Short F, Green E. Global overview: the distribution and status of seagrasses. In: Green EP, Short FT, editors. World atlas of seagrasses. Berkeley: University of California Press; 2003. p. 5–26.

Valentine JF, Duffy JE. The central role of grazing in seagrass ecology. In: Larkum AWD, Orth R, Duarte C, editors. Seagrasses: biology, ecology and conservation. Dordrecht: Springer; 2006. p. 463–501.

Virnstein RW, Hall LM. Northern range extension of the seagrasses *Halophila johnsonii* and *Halophila decipiens* along the east coast of Florida, USA. Aquat Bot. 2009;90:89–92.

Waycott M, Procaccini G, Les DH, Reusch TBH. Seagrass evolution, ecology and conservation: a genetic perspective. In: Larkum AWD, Orth R, Duarte C, editors. Seagrasses: biology, ecology and conservation. Dordrecht: Springer; 2006. p. 25–50.

Waycott M, Duarte CM, Carruthers TJB, Orth RJ, Dennison WC, Olyarnik S, Calladine A, Fourqurean JW, Heck Jr KL, Hughes AR, Kendrick GA, Kenworthy WJ, Short FT, Williams SL. Accelerating loss of seagrasses across the globe threatens coastal ecosystems. Proc Natl Acad Sci. 2009;106:12377–81.

Willette DA, Chalifour J, Debrot AOD, Engel MS, Miller J, Oxenford HA, Short FT, Steiner SCC, Védie F. Continued expansion of the trans-Atlantic invasive marine angiosperm *Halophila stipulacea* in the Eastern Caribbean. Aquat Bot. 2014;112:98–102.

Mangroves

C. Max Finlayson

Contents

Introduction	94
Global Distribution	95
Ecological Features	98
Biota	100
Ecosystem Services	101
Threats and Future Challenges	102
References	107

Abstract

Mangroves are trees or shrubs that are generally found in intertidal environments in the tropics and subtropics. Most but not all mangroves are found in intertidal environments along deltaic coasts, lagoons, and estuarine shorelines. Under optimal conditions they form extensive and productive forests, reaching 30 m in height, with scattered and dwarf shrubs occurring under less optimal conditions. The origin of the word mangrove is uncertain, but it is commonly used interchangeably to refer to an individual plant or to an assemblage of plants. Mangroves are an ecological rather than a taxonomic assemblage with different numbers of species reported by different authorities. Largely based on whether or not the individual species that are exclusively or nonexclusively found in mangrove communities. The mangrove communities support many other organisms and provide many benefits to people.

C. M. Finlayson (✉)
Institute for Land, Water and Society, Charles Sturt University, Albury, NSW, Australia

UNESCO-IHE, The Institute for Water Education, Delft, The Netherlands
e-mail: mfinlayson@csu.edu.au

Keywords
Mangroves · Mangal · Biodiversity · Coastal wetlands · Coastal change · Aquaculture · Pollution · Climate change

Introduction

Mangroves are trees or shrubs that are generally found in intertidal environments in the tropics and subtropics where low wave energy and shelter enable the deposition of fine particles that support their establishment and growth. Exceptions have been reported, including landlocked occurrences in Australia, in Madagascar, and in the Caribbean and some Pacific islands, although most mangroves are found in intertidal environments along deltaic coasts, lagoons, and estuarine shorelines (Saenger 2002). There is a rich and diverse literature on mangroves with overviews covering their biology being provided by a number of sources including Alongi (2002) and Saenger (2002). The information that follows is largely derived from these sources, as well as others as indicated. Under optimal conditions mangroves form extensive and productive forests, reaching 30 m in height, with scattered and dwarf shrubs occurring under less optimal conditions. The origin of the word mangrove is uncertain, but it is commonly used interchangeably to refer to an individual plant or to an assemblage of plants. The word mangal is also used to describe the assemblage of plants. Mangroves are an ecological rather than a taxonomic assemblage with some 70–84 species, many of which are not closely related phylogenetically. Different authorities report different numbers of plant species, largely based on whether or not the individual species that are exclusively or nonexclusively found in mangrove communities. The mangrove communities support many other organisms and provide many benefits to people.

All mangroves have structural and functional characteristics that enable them to thrive in the environments that occur at the interface of the land and sea, including morphological and ecophysiological adaptations such as aerial roots (Fig. 1), viviparous embryos, tidal dispersal of propagules, rapid rates of canopy production, and highly efficient nutrient retention mechanisms. They are generally seen as salt-tolerant species, although their ability to withstand regular inundation by both tidal and freshwater varies greatly. Although able to establish on a variety of substrates, the most extensive are associated with mud and muddy soils. Similarly, many, but not all, are associated with low-energy macro-tidal conditions. The range of conditions that suit mangroves are outlined in many publications, including that by Saenger (2002) who considered mangroves as plants that predominantly grow in intertidal areas along tropical and subtropical shorelines and are able to tolerate high-salt conditions and anoxia, with propagules that can be dispersed by tides and currents.

The global distribution, status and trends, biodiversity and ecosystem services, and threats and management issues for mangroves are described based largely on information contained in reviews by Alongi (2002), Saenger (2002), Spalding et al. (1997, 2010), Van Lavieren et al. (2012), and van Bochove et al. (2014) and references therein.

Fig. 1 Mangrove adaptations (**a**) pneumatophores (Photo credit: N Davidson © Rights remain with author) and (**b**) prop roots (Photo credit: C Prentice © Rights remain with author)

Global Distribution

The global distribution and areal extent of mangroves has been estimated by various means on many occasions, with uncertainty or inaccuracies associated with many efforts and the resultant data (Spalding et al. 1997; Alongi 2002). Estimates of the areal extent have been based on coarse spatial resolution satellite data, compilations of geospatial and statistical data from multiple sources, or calculated using published literature, with inconsistencies across space and time. Efforts to compile global figures from local or regional mapping exercises have also suffered from differences in scale and classification approaches. As a consequence, global estimates have ranged from 110,000 to 240,000 km^2. In an effort to overcome the limitations of past efforts, Giri et al. (2011) produced a global mangrove map for the year 2000 based primarily on archived Landsat data. This resulted in a global estimate of 137,760 km^2 with 42% of this area in Asia, 20% in Africa, 15% in North and Central America, 12% in Oceania, and 11% in South America (Fig. 2). While mangroves were recorded in 118 countries, some 75% of the total area was in 15 countries (Table 1) with the largest areas found in Indonesia, Australia, and Brazil. These data do not provide an indication of the species diversity or the quality or health of the mangroves, nor do they include all small patches; higher resolution analyses are needed to further elucidate such detail.

These analyses confirmed that mangroves are largely distributed along tropical and subtropical coastlines with a northern limit of 31–32° N and a southern limit of 33–39° S. Their global distribution is predominantly tropical and influenced by major ocean currents and the 20 °C seawater winter isotherm. In a general sense, the area of mangrove decreases with increasing latitude, with the Sundarbans in Bangladesh and India, the largest area of mangroves globally, located between 20 and 25° N providing an exception. Similarly, species diversity, height, and biomass decrease with increasing latitude. In addition to the Sundarbans, well-developed mangroves occur in the Mekong Delta, the Amazon, and coastal Madagascar, Papua New Guinea, and Southeast Asia. Floristically they have been divided into two biogeographic regions, (i) a diverse Indo-West Pacific flora extending from East Africa to Polynesia and (ii) a far less diverse Atlantic East Pacific flora in the Americas and West and Central Africa. There is very little overlap in plant species between the biogeographic regions, with the exception of the fern *Acrostichum aureum*. The highest species diversity is in the Indo-Malesian region.

Many authors have reported losses of mangroves although these data can also be questioned with Alongi (2002) noting "While it is clear that large tracts of mangroves have been either severely degraded or destroyed worldwide, most data is apocryphal, reflecting inaccurate surveys, unsubstantiated claims or old estimates not based on empirical measurements. . . .." Information collated and summarized by Alongi (2002), including that contained in the global mangrove atlas (Spalding et al. 1997) and produced by the International Society for Mangrove Ecosystems, showed that most countries with mangroves have at different times and for different reasons, such as the expansion of coastal development and aquaculture, lost significant areas. Historical records suggest that in 1980, there were 188,000 km^2 of mangroves and possibly more than 200,000 km^2 before that (Spalding et al. 1997; FAO 2007). Van Lavieren et al. (2012) report that globally about 20% of mangroves had been lost between 1980 and 2005 with >20% losses in the Asian and Pacific regions, followed by Central America, and with limited losses in East Africa, with only an 8% decline, although these data varied with individual countries. A small number of countries, such as Papua New Guinea, Australia, and Belize, have very little substantial change, while Cuba has regained mangrove forests due to restoration efforts. Other publications

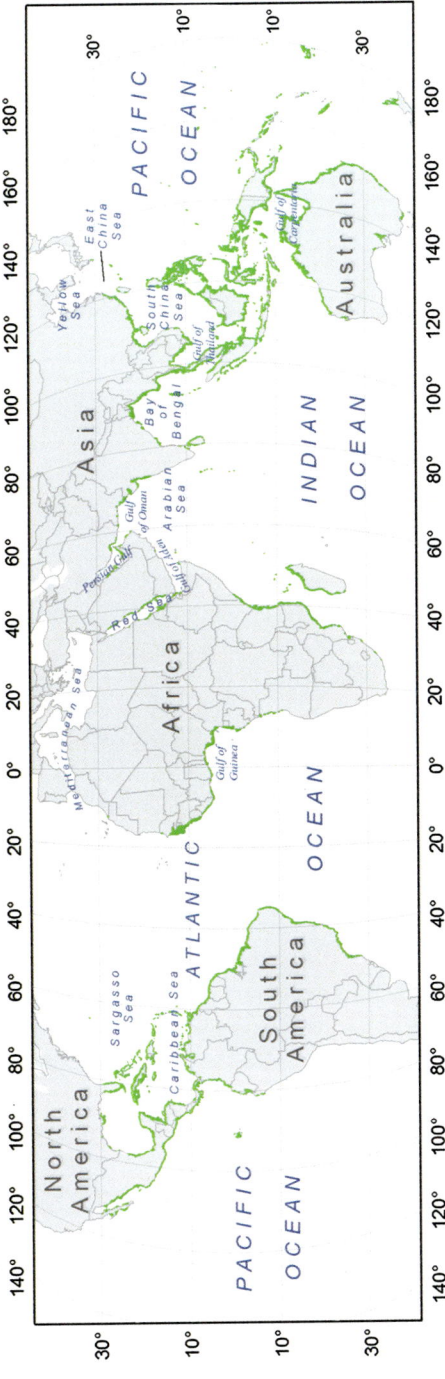

Fig. 2 Global distribution of mangroves (Adapted from Giri et al. 2011 using data supplied by UNEP-WCMC)

Table 1 Area of mangrove in individual countries (Information from Giri et al. 2011)

Country	Area (hectares)	%
Indonesia	3,112,989	22.6
Australia	977,975	7.1
Brazil	962,683	7.0
Mexico	741,917	5.4
Nigeria	653,669	4.7
Malaysia	505,386	3.7
Myanmar	494,584	3.6
Papua New Guinea	480,121	3.5
Bangladesh	436,570	3.2
Cuba	421,538	3.1
India	368,276	2.7
Guinea Bissau	338,652	2.5
Mozambique	318,851	2.3
Madagascar	278,078	2.0
Philippines	263,137	1.9

have estimated a global loss of 50% although the basis for these data was often not fully explained or justified (Dodd and Ong 2008). Overall the rate of loss has been declining, although with considerable variation between countries.

Differences in data sources emphasize the need to carefully assess such data when making general statements about national or regional rates of loss. Gains in mangrove area were also reported, although these predominantly involved the replanting of monocultures of fast-growing species with slower-growing species generally not being replaced. This is also important as mangroves are not stable – they respond to changing conditions and exhibit both seaward and landward expansion or colonization, such as that detected in various locations (Lucas et al. 2002, 2014; Krauss et al. 2014). In response to issues raised about the uncertainty associated with mangrove inventory and change analyses, Lucas et al. (2014) have developed a Global Mangrove Watch to routinely monitor and report on local to global changes in the extent of mangroves. This initiative has a number of aims including to (i) map the progression of change within or from existing global baselines of the extent of mangroves, (ii) determine likely losses and gains of tree species diversity, and (iii) validate maps of changes in the extent of mangroves and describe the causes and consequences of change.

Ecological Features

Information on the ecology of mangroves is largely taken from Alongi (2002) who provides a summary of the ecological features of mangrove systems, in particular those concerning the conditions that influence the distribution and zonation of mangrove species. While mangroves are generally found across the intertidal range from mean sea level to the highest spring tide, they also in many instances

Fig. 3 Mangrove zonation in northern Australia (Photo credit: CM Finlayson)

exhibit a conspicuous zonation pattern (Fig. 3). However, it can be difficult to make generalizations about zonation as this can result from a number of factors, including salinity, soil type and chemistry, nutrient levels, physiological tolerances, predation, and competition. The interactions between these factors are complex, and major changes in zonation have been recorded in some mangroves. Lucas et al. (2002), for example, found changes in zones in mangroves in northern Australia from 1950 to 1991, possibly in response to hydrological changes that resulted in a landward extension of saline conditions, with changes in landward and seaward zones.

The zonation within a particular mangrove can be affected by cyclones and other storms, lightning, tsunami, and floods, and once changed it can take decades to recover. Mangroves also become more susceptible to diseases and pests when stressed by changes in salinity, tidal inundation, sedimentation, and the introduction of pollutants, and damage from storms. Pests can have a severe impact on mangrove forests, destroying wood and leaves, and in some cases, greatly reduce their viability. The dynamics of change in mangrove forests is poorly understood but comprises a cycle of natural mortality and regeneration. Temporal and spatial variations within mangrove forests are commonly regulated by competition between individuals and species for light, space, and soil nutrients, which are often patchily distributed within individual stands, and can result in a progressive decline in the density of growing trees.

Food webs in mangroves comprise both marine and terrestrial components. The abundance and biomass of organisms living in mangroves – in the canopy, on or beneath the forest floor, and in associated waterways – often vary seasonally in relation to rainfall and spatially in response to a variety of factors that are often the same as those regulating the distribution and density of trees. The structure and function of mangrove food webs are driven by the production of carbon fixed mostly by the trees and by the flow of dissolved and particulate organic matter within the mangrove and with adjacent areas.

Trees and bacteria dominate the biomass and productivity of mangrove forests. The structure and function of mangrove food webs are driven by the production of carbon fixed mostly by the trees and by the flow of dissolved and particulate organic matter within the forest and adjacent tidal waters. Some estimates of the carbon fixed in mangroves show that 9% is consumed by herbivores, 40% is decomposed and recycled, 30% is exported, and 10% is stored in the sediments, although other measurements imply that more may be stored in the wood and decomposed or stored in the sediments and exported. Within the forests, a suite of decomposers directly or indirectly consumes a proportion of the litter, including crabs that retain the litter and recycle nutrients within forest, disturb the soil to stimulate microbial decomposition, or, in some cases, prey on propagules to influence the distribution, abundance, and succession of tree species. Only a few studies have constructed nutrient mass balances for entire mangrove ecosystems. The specific age and structure of a mangrove forest can influence the rates and pathways of nutrients and energy flows, as can interactions with the surrounding environments, and whether or not these have been disturbed. In many cases, mangroves can tolerate high levels of nitrogen and phosphorus inputs that can stimulate further production or even shift the forest from depending mostly on mangrove-derived organic matter to the use of nutrients derived from phytoplankton or macroalgal blooms. The continued input of nutrients could cause large changes in the dynamics of mangroves, including the distribution and dominance of species.

Biota

Mangrove communities are inhabited by a variety of other species in addition to the main trees or shrubs that are generally seen as the dominant biotic component. Mangals provide a range of habitats for animals, above and below water, including the canopy, trunks and aerial roots of the trees or shrubs, and the water and substrate.

Above the water, the trees and canopy provide important habitat for birds, insects, mammals, and reptiles. Below the water, the roots are overgrown by epibionts, including tunicates, sponges, algae, and bivalves. The soft substratum forms habitat for infaunal and epifaunal species, while the space between roots provides shelter and food for motile fauna such as prawns, crabs, and fishes. Nagelkerken et al. (2008) provide a review of the main groups of animals in mangrove habitats, covering sponges, meiofauna and macrofauna (epifauna and infauna), prawns, insects, fishes (bony fishes and elasmobranchs), amphibians, reptiles, and birds, but not zooplankton, tunicates, or mammals. The review points out the uneven nature of much of the information about different fauna groups, as well as regionally. A summary of this information is provided in Table 2.

Further information on the biota that inhabits mangroves is available in recent reviews and databases. These include information on threatened and vulnerable species in the IUCN Red List of Threatened Species (iucnredlist.org), as well as the Mangrove Reference Database and Herbarium (www.vliz.be/vmdcdata/mangroves) which contains information on mangrove coverage, taxonomy, and their

Table 2 Summary of information about mangrove fauna (Derived from Nagelkerken et al. 2008)

Fauna group	Description
Sponges	Generally restricted to prop roots hanging over tidal channels; can form distinctive high-biomass communities; lower species diversity than in adjacent subtidal habitats; 3–147 species in Caribbean mangroves
Meiofauna	Metazoans numerically dominant in sediments; nematodes and harpacticoid copepods dominate hard body forms; Turbellaria dominate soft body forms; the number of species of copepods and nematodes varies between locations
Macrofauna	Variety of benthic invertebrates with marked zonation; separated into epifauna including gastropods, crabs, and bivalves and infauna including crabs, pistol crabs, polychaetes, and sipunculids
Prawns	Inhabit mangroves when they are inundated by tides; few species exclusively associated with mangroves; penaeids can dominate the epibenthos
Insects	More diverse in Indo-West Pacific compared to Atlantic East Pacific; comprise herbivorous types that feed on leaves, etc., saproxylic and saprophagous types that feed on decaying organic matter, and parasitic and predatory types that prey on other animals
Elasmobranchs	Estuaries contain euryhaline and obligate freshwater species that feed in mangroves and shelter from larger predators
Fishes	Indo-West Pacific region has high diversity (>600 species), with diversity declining with latitude from core area in southeast Asia; mudskippers have an amphibious existence; freshwater species comprise more of fish fauna in tropical Atlantic than in Indo-West Pacific or East Pacific
Amphibians and reptiles	Little is known about amphibians and freshwater turtle use of mangroves; crocodilians often found in mangroves; many lizards from geckos to iguanas
Birds	Many species known to use mangroves; few are mangrove specialists

characteristics and the Global Mangrove Database and Information System (glomis.com) which comprises a database of mangrove literature.

Ecosystem Services

The Millennium Ecosystem Assessment (MEA 2005) and the abovementioned key references have highlighted the importance of mangroves for people both traditionally for a wide variety of products and more recently for commercial purposes. In addition, there are many benefits from regulatory services associated with coastal protection, although many of these have not received sufficient acknowledgment in coastal and marine planning and development or have only been recognized once they have been degraded or lost. van Bochove et al. (2014) have noted that the flow and provision of ecosystem services does not simply depend on the presence of mangroves but also on the species composition and other ecological factors, as well as policies and the sociocultural context, and the level of dependence of local communities on mangroves for their livelihoods and well-being.

The benefits include the provision of wood for cooking and heating and for carpentry and building purposes, including the production of furniture and boats and the construction of bridges and fish traps, as well as the production of dyes and resins, medicines, and alcohol. They are also very important sites for the gathering and local small-scale cultivation of shellfish, finfish, and crustaceans. Despite the importance of mangroves as a food source for local people, the adoption of commercial production practices, often driven by forces originating outside the local communities, has increasingly placed this at risk. Examples include the non-sustainable exploitation of wood products and aquaculture, whereas sustainable production, such as that reported for wood harvest from the Matang Mangrove Forest Reserve in Perak, Malaysia, can be achieved and provide many benefits for local communities. They are also recognized as valuable nursery areas for fish and other organisms that inhabit water adjacent to the mangroves. While mangroves are important as nursery areas for fish, this depends on the species being fished, the location, and time scale, with estimates of the amount of commercial catch explained by the presence of mangroves or estuaries ranging from 20% to 90% (Nagelkerken et al. 2008).

The regulatory services provided by fringing mangroves include stabilization of the shoreline, although recent evidence has shown major changes in mangrove zonation that may reflect or result in changes in the seaward and landward boundaries of mangroves and adjacent habitats, and protection from storms, flood surges, and tsunamis. The latter has received more attention in recent years following the high levels of damage and loss of life associated with a number of large events, although the extent of protection varies enormously, with the events themselves also causing a lot of disruption to the mangroves (Alongi 2008). The capacity of mangroves to absorb nutrients, heavy metals, and other toxic substances has been widely investigated, although the assimilative capacity for most is unknown and likely to vary depending on the type of effluent input and biophysical features of the individual location. Further, an excess of these substances can lead to adverse change and loss of this service.

The importance and quality of the ecosystem services provided by mangroves varies greatly in relation to the biophysical characteristics of the mangrove, taking into account many differences in tidal ranges and inundation patterns, the substrates, and the species assemblages and structure. In recent years the importance of mangroves as a carbon store has been increasingly recognized, with large amounts being stored, often for long periods of time, belowground in the soil and roots of mangrove species (Alongi 2012).

Threats and Future Challenges

The continued protection of mangroves has assumed more attention in recent years with substantial areas now within protected areas (Table 3), although the effectiveness of these areas is uneven. Further, the proportion of mangroves within protected areas in some countries is still low, and losses continue (van Bochove et al. 2014).

Table 3 Proportion (%) of mangroves contained within protected areas (From van Bochove et al. 2014)

Region	Percentage	Region	Percentage
Australasia	38	Central and South America	60
East Africa	38	South Asia	66
Middle East	57	Southeast Asia	28
North America	67	West and Central Africa	32
Pacific Ocean	13		

Although threats to mangroves include natural events such as large storms and tsunamis, the influence of human activities is expected to exceed these, including overexploitation and conversion to other uses, and in particular, the direct and indirect impacts of further coastal development (Alongi 2002; Dodd and Ong 2008; van Bochove et al. 2014). A regional overview of threats to mangroves is provided in Table 4. As large areas of mangroves are in developing countries that are undergoing rapid industrial development, the latter is of considerable concern. With a large and increasing proportion of people in such countries inhabiting the coastal zone, the threat of conversion of mangroves to other land uses is increasing, including land claim for tourism facilities, expansion in port facilities, and urban infrastructure. The expansion of coastal tourism has seen an expansion in hotels and other facilities, including marinas, with infilling of mangroves as well as the construction of coastal walls and tidal barriers. The discussion that follows draws on the information presented in more detail in the Alongi (2002), Dodd and Ong (2008) and van Bochove et al. (2014).

The loss of mangroves for the construction of intensive aquaculture ponds is one of the greatest threats to mangrove forests worldwide. Previous analyses have suggested that almost half of mangrove loss is due to shrimp or other aquaculture. The list of direct and indirect problems caused by aquaculture to mangroves includes, as described by Alongi (2002), the immediate loss of mangroves, blockage of tidal creeks, alteration of natural tidal flows, alteration of the groundwater table, increase in sedimentation rates and turbidity in natural waters, release of toxic wastes, overexploitation of wild seed stocks, development of acid sulfate soils, reduced water quality and introduction of excess nutrients, and alteration of natural food chains. The culturing of shrimps in ponds constructed by clearing mangroves has resulted in the loss of large areas of mangroves and been accompanied by the introduction of nutrients and antibiotics. Fishponds have been constructed in parts of Indonesia and the Philippines for many years initially on a shifting cultivation pattern with the loss of large areas of mangroves and the development of acid sulfate soils (Alongi 2002). The latter could be treated by the addition of lime that encouraged the expansion of shrimp ponds in many countries given the profits that could be made. Expansion of shrimp aquaculture has been constrained by the advent of disease as a consequence of poor hygiene and overcrowded conditions in the ponds.

Changes in the hydrology of mangroves have occurred as a result of upstream activities reducing freshwater inflows and as a result of changes to the tidal flows.

Table 4 Regional overview of the main threats to mangroves (Adapted from Van Lavieren et al. 2012)

Region	Development			Land conversion			Oil/gas		Pollution	Over harvesting	Over fishing	Sedimentation	Storms	Poor planning	Disease
	Coastal	Urban	Tourism	Agriculture	Aquaculture	Land reclamation	Extraction	Spills							
North and Central America	X	X	X	X	X				X				X		
South America				X	X										
West and Central Africa		X		X			X		X						
East Africa				X	X	X			X	X		X			
Middle East	X	X	X			X		X		X		X		X	
South Asia				X	X			X							
Southeast Asia	X	X		X	X	X	X		X	X	X	X	X	X	X
East Asia	X	X	X	X	X	X			X	X	X			X	X
Australasia	X	X				X		X	X				X	X	
Pacific Ocean	X	X	X						X	X	X	X			

The Indus Delta in Pakistan is a case where dams have reduced the inflow of freshwater and resulted in decreased sediment transport and increased salinity with a loss of mangroves. The regression of the shoreline as a consequence of reduced sediment supply is expected to exacerbate the problem mangroves will face as a consequence of projected sea level rises. Reduced flows into coastal waters may result in longer residence times and less flushing of nutrients and other pollutants that would otherwise accumulate in mangroves and adjacent areas. The construction of barriers that prevent tidal interchanges, including roads without sufficient provision for inward and outward movement of tidal water, can result in hypersaline conditions and the decline of mangroves.

The impact of pollutants on coastal wetlands, including mangroves, from both point and diffuse sources, is well known and widely documented. While an influx of nutrients may promote the growth and productivity of mangroves, other pollutants such as heavy metals can cause major adverse impacts on other organisms in the mangrove, while others such as oil may adversely affect the entire ecosystem. Nutrients could also lead to increased algal production and changes in the trophic balances in the mangroves. Eutrophication is expected to be an ongoing problem.

It is also expected that adverse impacts from aquaculture, mining, housing, and industrial encroachment and overexploitation will continue and some will probably increase as coastal settlements and port facilities are further developed. Alongi (2008) considers many past and current abuses to be irreversible. Aquaculture is expected to increase, and despite the adoption of more sustainable methods, it is expected to still result in the loss of mangrove resources. Deforestation through multiple means remains a major threat to the survival of mangroves with reforestation efforts not expected to reverse the loss.

It is expected that human-induced climate change will exacerbate existing pressures on mangroves through a rise in relative sea level and seawater temperature, a possible increase in the frequency and magnitude of extreme weather events and associated elevated storm surges and wave height (Van Lavieren et al. 2012). A summary of expected climate change related impacts on mangroves is provided in Table 5. While mangroves can respond to changes in sea levels, there are many factors that will determine whether they can adapt to the projected rises over the next century, including rates of sedimentation, below-ground biomass production, and groundwater levels, as well as available space on the landward side.

Van Lavieren et al. (2012) recommend the establishment of overarching legal and policy frameworks that can support effective mangrove management, including those that identify and accept the rights of ownership, access, and use of mangrove forests; establish national level policy and legislation for mangroves and ensure that laws and regulations are enacted and enforced; enhance human, technical, legal, and financial capacity for mangrove management at different levels; and ensure that measures, including subsidies and other incentives that lead to mangrove degradation or loss, are removed. Further, they recommended a range of management measures and tools to maximize the benefits and help

Table 5 Expected impacts of global climate change on mangrove ecosystems (Adapted from Van Lavieren et al. 2012)

Parameter	Predicted change	Expected impact
Sea level rise	Increase 18–79 cm between 1999 and 2099	Predictions of loss range from 30% to extinction. Mangroves may migrate landward dependent on the rate of sea level rise and vertical accretion, slope and available space. Adaptation to changes in tidal range and sediment supply will likely be species dependent. Zonation and species composition of plants and animals will change as erosion, and flooding at the seaward front occurs and be magnified along low-lying coasts (especially on small islands) and where landward movement is limited and in the arid tropics where rates of sediment supply and mangrove growth are low. Mangroves have the capacity to adapt to sea level rises through vertical accretion, with outcomes dependent on the rates of change
Air temperature	Increased 0.74 °C between 1906 and 2005. Predicted further 2–4 °C increase within the next 100 years	Latitudinal limits for some species could expand. Community composition could change in response to increases in photosynthesis, respiration, litter, microbial decomposition, floral and faunal diversity, growth and reproduction, but declining rates of sediment accretion. At local and regional scales, changes in the community composition could also occur if there are changes in the salinity regime, and a change in primary production if the ratio of precipitation to evaporation is altered
Atmospheric carbon dioxide	CO_2 levels have increased from 280 to 370 ppm by volume in 1880 to nearly 370 in 2000, with pH of the oceans increasing in acidity by 25%. Despite large variation it is expected that further CO_2 levels will increase by 2100, possibly a doubling or tripling	Responses will be difficult to predict but rates of photosynthesis and salinity, nutrient availability, and water-use efficiency will likely change. There will likely be no or little change in canopy production, but species patterns within estuaries are likely to change based on species-specific responses to the interactive effects of rising CO_2, sea level, temperature, and changes in local weather patterns. Elevated CO_2 could alter competitive abilities, thus altering community composition along salinity-humidity gradients

(*continued*)

Table 5 (continued)

Parameter	Predicted change	Expected impact
Storms	Intensity and frequency of storms will increase	Expected defoliation, uprooting, tree mortality, and increased stress from altering mangrove sediment elevation due to soil erosion, deposition, and compression. Recovery from storm damage can be slow. Severity of impact expected to vary regionally and based on individual events
Precipitation	Rainfall is predicted to increase nearly 25% by 2050 and the intensity of rainfall events will also increase. Changes are expected to be uneven with significant increases at higher latitudes and decreases in most subtropical regions	Both regional and local patterns of mangrove growth and distribution may be affected. Increased intensity of rainstorms is likely to influence erosion and other physical processes in catchments and tidal wetlands. Increased rainfall may increase diversity and enhance growth and coverage of previously non-vegetated areas. Reduced rainfall may lead to reduced diversity and productivity of mangroves and increases in adjacent salt marsh and salt flat areas

secure the future of mangroves and the people who rely on them. These included restoration efforts to recover lost mangrove forests and their ecosystem services, involvement of local communities in management, sustainable mangrove forestry practices, sustainable aquaculture practices, protected areas for mangrove biodiversity, cohesive management plans for entire countries and ecological units, mangrove ecotourism to generate income and employment for local communities and use multilateral environmental agreements along with national legal measures, to support mangrove management.

References

Alongi DM. Present state and future of the world's mangrove forests. Environ Conserv. 2002;29:331–49.

Alongi DM. Mangrove forests: resilience, protection from tsunamis, and responses to global climate change. Estuar Coast Shelf Sci. 2008;76:1–13.

Alongi DM. Carbon sequestration in mangrove forests. Carbon Manage. 2012;3:313–22.

Dodd RS, Ong JE. Future of mangrove ecosystems. In: Polunin NVC, editor. Aquatic ecosystems: trends and global prospects. Cambridge: Cambridge University Press; 2008. p. 172–87.

FAO. The world's mangroves 1980–2005. A thematic study prepared in the framework of the Global Forest Resources Assessment. FAO Forestry Paper 153, Rome; 2007.

Giri C, Ochieng E, Tieszen LL, Zhu Z, Singh A, Loveland T, Masek J, Duke N. Status and distribution of mangrove forests of the world using earth observation satellite data. Glob Ecol Biogeogr. 2011;20:154–9.

Krauss KW, McKee KL, Lovelock CE, Cahoon DR, Saintilan N, Reef R, Chen L. How mangrove forests adjust to rising sea level. New Phytol. 2014;202:19–34.

Lucas RM, Ellison JC, Mitchell A, Donnelly B, Finlayson M, Milne AK. Use of stereo aerial photography for quantifying changes in the extent and height of mangroves in tropical Australia. Wetl Ecol Manage. 2002;10:161–75.

Lucas R, Rebelo L-M, Fatoyinbo L, Rosenqvist A, Itoh T, Shimada M, Simard M, Souza-Filho PW, Thomas N, Trettin C, Accad A, Carreiras J, Hilarides L. Contribution of L-band SAR to systematic global mangrove monitoring. Mar Freshw Res. 2014;65:589–603.

MEA (Millennium Ecosystem Assessment). Ecosystems and human well-being: wetlands and water synthesis. Washington, DC: World Resources Institute; 2005.

Nagelkerken I, Blaber SJM, Bouillon S, Green P, Haywood M, Kirton LG, Meynecke J-O, Pawlik J, Penrose HM, Sasekumar A, Somerfield PJ. The habitat function of mangroves for terrestrial and marine fauna: a review. Aquat Bot. 2008;89:155–85.

Saenger P. Mangrove ecology, silviculture and conservation. Dordrecht: Kluwer; 2002.

Spalding M, Blasco F, Field C. World mangrove atlas. Okinawa/Japan: International Society for Mangrove Ecosystems; 1997.

Spalding M, Kainuma M, Collins L. World atlas of mangroves. Oxford: Earthscan; 2010.

van Bochove J, Sullivan E, Nakamura T, editors. The importance of mangroves to people: a call to action. Cambridge: United Nations Environment Programme World Conservation Monitoring Centre; 2014.

Van Lavieren H, Spalding M, Alongi D, Kainuma M, Clüsener-Godt M, Adeel Z. Securing the future of mangroves. A policy brief. UNU-INWEH, UNESCO-MAB with ISME, ITTO, FAO, UNEP-WCMC and TNC. London: Earthscan; 2012. p. 319.

Major River Basins of the World

Carmen Revenga and Tristan Tyrrell

Contents

Introduction	110
Defining and Mapping River Basins	111
The Biodiversity of Major River Basins	112
Ecosystem Services and the Role that River Basins Play in Human Well-being	115
Threats to Major Rivers	117
Protection of Rivers and Other Aquatic Habitats	120
References	121

Abstract

River basins are the hydrological units of the planet and as such play a critical role in the natural functioning of the Earth. While there have been advances in mapping river basins to facilitate analyses and inform basin management, information on river flow volumes, condition of freshwater habitats and species, as well as the degree of impact on rivers and freshwater biodiversity are still lacking for many parts of the world. This lack of information and, mostly a lack of appreciation for the intrinsic value and role that rivers and wetlands play in the well-being of millions of people, makes their conservation and sustainable management challenging in the face of development pressures and climate change. This chapter

C. Revenga (✉)
The Nature Conservancy, Arlington, VA, USA
e-mail: crevenga@tnc.org

T. Tyrrell
Tentera, Montreal, QC, Canada
e-mail: tristan@tentera.org

© Springer Science+Business Media B.V., part of Springer Nature 2018
C. M. Finlayson et al. (eds.), *The Wetland Book*,
https://doi.org/10.1007/978-94-007-4001-3_211

summarizes current status and trends in freshwater biodiversity, identifies threats to aquatic ecosystems, provides information of the ecosystem value of wetlands, and makes the case for an improved, and more integrated management and conservation of wetlands.

Keywords

River basins · Freshwater · Spatial data · Biodiversity · Ecosystem services · Values · Threats · Protection · Conservation

Introduction

River basins are the hydrological units of the planet and as such play a critical role in the natural functioning of the Earth; they are highly dynamic systems linked by topography, geology, and climate. The constant flow of water from rain and snow and ice melt over the continents has formed and reshaped the landscape for millennia – steep mountains are cut by fast cascading streams, major floodplain rivers wear away the continents carrying nourishing sediments that shape and replenish the coastal zone, and surface water percolates through permeable portions of the earth's surface to create systems of caves and tunnels where water collects and forms underground aquatic habitats.

Each river basin encompasses the entire area drained by a major river system or by one of its tributaries, and both the landscape and the river network that runs through it give each basin its unique characteristics. No two river basins are exactly the same, nor is a single river the same throughout the year, each is characterized by a unique combination of geographical location, climate, basin connectivity and the timing, volume, and duration of flow. Many riverine species are adapted to, and synchronized with, specific river flow patterns, such as spring peak floods or summer low flows. These patterns cue species to reproduce, disperse, migrate, feed, and hide from predators, and alterations to them can disrupt animal and plant life cycles and ecological processes with consequences for both people and nature.

While the majority of the world's rivers drain into the ocean or into another river, there are exceptions. These exceptions are known as endorheic or closed basins; some of the largest inland seas and wetlands in the world like the Caspian Sea or the Okavango Delta are closed basins. Rain that falls onto the endorheic basin flows downstream to form seasonal wetlands where water evaporates or seeps into the ground like in the Okavango Delta or forms a lake or inland water body like in the case of the Caspian Sea.

Proximity to water bodies has been a preference for the establishment of human settlements for millennia. Society has used rivers for transport and navigation, water supply, waste disposal, and as a source of food, and as a consequence we have heavily altered waterways putting the ecological functioning and the biodiversity of freshwater ecosystems at risk. Given the importance of freshwater ecosystems in sustaining human well-being, it is surprising how little we know about their changing condition, their dependent species, or the roles that these species play in sustaining ecological functions within river basins.

Defining and Mapping River Basins

River basins can be categorized in different ways, from basin size and flow volume to climatic conditions and political or administrative boundaries. This last category is often used in the field of water resources management, as managing river basins that cross international boundaries requires a high degree of international cooperation. There are 263 river basins that are shared between two or more countries; 40% of the world's population lives in one of these transboundary river basins, which collectively drain over 45% of the world's land surface (UN Water 2009; TFDD 2012). The river basin with the highest number of riparian nations is the Danube, which runs through eighteen countries. Other major transboundary river basins with eight or more riparian nations include: the Amazon, Congo, Niger, Nile, Lake Chad, Rhine, and Zambezi (TFDD 2012).

Other ways of categorizing river basins, such as basin size, flow volume, mean annual discharge, or other physical metrics, require delineations of catchments and river networks on a consistent basis for the globe. Tabular inventories of major river systems usually including data on drainage area, river length, and discharge are available, but they do not allow for spatial analyses or overlays with existing geospatial data layers such as land cover, population, or climatic zones. It also makes global and even regional analyses and comparisons extremely biased given that only the largest rivers tend to be represented in these tabular inventories. Many of these tabular inventories also suffer from out-of-date information, differences in definitions of the extent of a river system, as well as differences in the time period or location for the measurement of discharge (Revenga and Kura 2003).

Spatial databases with point, vector, and polygon data depicting surface water bodies and related information also exist, usually in the form of gazetteers, such as the Digital Chart of the World (DCW) (ESRI 1993). The great advantage of these databases is they allow for geospatial analysis and the integration of other data layers such as population, road infrastructure or land use. The disadvantage is that many of these databases tend to be of coarse resolution. The Digital Chart of the World, one of the first comprehensive database of this type developed in the 1990s, for example, includes a layer of rivers and lakes at 1:1 million resolution derived from the United States Defense Mapping Agencies' (now the National Geospatial-Intelligence Agency) Operational Navigation Charts whose information dates from the 1970s to the 1990s (Birkett and Mason 1995). While the DCW is considered consistent across the globe, its positional accuracy varies considerably between regions, and the database does not distinguish between natural rivers and artificial canals; the DCW has not been updated since 1992.

HYDRO 1K is another frequently used spatial hydrographic data set at one-kilometre resolution. Developed by the United States Geological Survey's

Eros Data Center (USGS 2000), it used to be the most comprehensive global database delineating basin boundaries. HYDRO 1 K, however, also has considerably inaccuracies with the location of actual river networks and most of the basins are not named, which makes some analyses difficult. Since 2013, HydroBASINS has become the most consistent and comprehensive global database in use for river basin analysis. Derived from HydroSHEDS a high-resolution elevation data obtained during a Space Shuttle flight for NASA's Shuttle Radar Topography Mission (SRTM), HydroSHEDS, which stands for Hydrological data and maps based on SHuttle Elevation Derivatives at multiple Scales, contains geo-referenced data on stream networks, watershed boundaries, drainage directions, and ancillary data layers such as flow accumulation, distances, and river topology information (http://hydrosheds.org) HydroBASINS (Lehner and Grill 2013) for the first time allows us to perform basic analyses, like calculating how many basins of each size there are in the world or deriving the upstream catchment of any point in a hydrologic network. When coupled with flow data, these global spatial drainage maps have additional applications, like the potential to generate stream hydrographs or classification of rivers by flow volume. Both modeled and observed discharge data for the world's rivers can be used in combination with these global drainage maps, and while observed data is preferable it is not always available on a consistent basis at global or even regional scales.

The World Meteorological Organization's Global Runoff Data Centre (GRDC) compiles and maintains a database of observed river discharge data from gauging stations worldwide. Coverage and reliability of hydrological information obtained through measurements in the GRDC varies from country to country, and although this is the best global database currently available, the number of operating stations has significantly declined in some regions since the 1980s – meaning the discharge data for some rivers have not been updated in decades (Revenga and Kura 2003). Modeled flow and discharge data are available from different rainfall-runoff models of varying complexity; many are derived for a single river basin and when possible calibrated with observed flow data; global models rely on flow calculations generated with input data from Global Climate Models.

Better and more reliable information on actual stream and river discharge, and the amount of water withdrawn and consumed at river basin scale, would increase our ability to manage freshwater resources more efficiently and set conservation measures for ecosystems and species. However, much effort and financial commitment would have to be made to restore hydrological stations and collect this type of information on a consistent basis.

The Biodiversity of Major River Basins

Freshwater ecosystems have a higher concentration of species relative to their area than terrestrial or marine biomes. Wetland ecosystems such as lakes, streams, rivers, and marshes occupy just 0.8% of the Earth's surface but support 6% of all described species, including 35% of all vertebrates (Dudgeon et al. 2006). More than 15,700 of the world's fish species, representing a quarter of all living vertebrates, live in fresh waters (Carrizo et al. 2013), with new species described each year. Between 1992 and 2011, an average of 333 new fish species, both marine and freshwater, were described each year (Eschmeyer and Fong 2012). Information on all fish species, including fresh and marine species, is available in The Catalogue of Fishes, a regularly updated searchable database (Eschmeyer et al. 2016). This is the most comprehensive database maintained since the 1980s where all freshwater fish species described are listed.

While information on freshwater fish is more accessible and available, we still lack species information for many parts of the world, particularly for the tropics and especially for invertebrates, microbes, and aquatic plants. However, available data do provide a picture of species richness patterns across the world's river basins. Large basins with high discharge volumes tend to have more fish species, and as in the terrestrial and marine realms, tropical systems tend to have higher levels of species richness. River basins with the highest levels of fish species richness include the Amazon, Orinoco, Tocantins, and the Paraná in South America; the Congo, the Niger Delta, and the Ogooue in Africa; and the Yangtze, Pearl, Brahmaputra, Ganges, Mekong, Chao Phraya, Sittang, and Irrawaddy in Asia. Lake basins with the highest levels of fish richness and endemism include the great lakes of Africa's Rift Valley: Malawi, Tanganyika, and Victoria. Smaller river basins that are also relatively high in fish species include systems in the Southeast of the United States like the Tennessee, Mobile, and Alabama Rivers as well as Sanaga, Nyong, Ntem, and Benito rivers draining into the Gulf of Guinea in West Africa, the Kapuas River basin in Borneo, and small coastal basins of the Eastern Malay Peninsula, Eastern Central Sumatra, Western India, and Japan (FEOW 2008).

Species richness patterns for other taxonomic groups follow a similar pattern than that for freshwater fish; amphibian richness is consistently high in tropical Asian and South American river basins and in most of the tropical basins in Africa, including coastal basins in Madagascar (FEOW 2008). Australia's eastern coastal drainages and the Murray-Darling basin also have high numbers of amphibians. Freshwater turtles, which comprise more than 240 species, are most concentrated in warm and humid climates, but a few basins have particularly high species numbers including the Apalachicola, Mobile, and lower Mississippi Rivers in Southeastern United States and the Sittang and Irawaddy Rivers in Southeast Asia (FEOW 2008). Freshwater dependent bird species are also predominantly present in tropical basins; river catchments with more than 125 freshwater-dependent bird species include those with important large permanent wetlands: the lower Paraná River in South America, the Lake Victoria and Upper Nile sub-basin in Africa and the Ganges, Sittang, Salween, Chao Phraya, and the Song Hong (Red) River in Asia (Hoekstra et al. 2010). Finally freshwater dependent mammal richness is concentrated in the Amazon, Magdalena, and Paraguay-Paraná basins, much of the western and central Africa river basins, Lake Victoria, and the upper Nile (Hoekstra et al. 2010).

No other major global habitat type or group of species has demonstrated such large and rapid declines. More than 50% of wetlands, from coastal marshes to peatlands, in North America, Europe, Australia, and New Zealand were drained or

Table 1 Number and percentage of assessed freshwater fish species under each Red List Category

Red List category	No. of species	% of assessed species
Extinct (EX)	66	1.1
Extinct in the Wild (EW)	8	0.1
Critically Endangered (CR)	422	6.9
Endangered (EN)	474	7.7
Vulnerable (VU)	914	14.9
Near Threatened (NT)	280	4.6
Least Concern (LC)	2,881	47.0
Data Deficient (DD)	1,083	17.7
Total	**6,128**	**100.0**

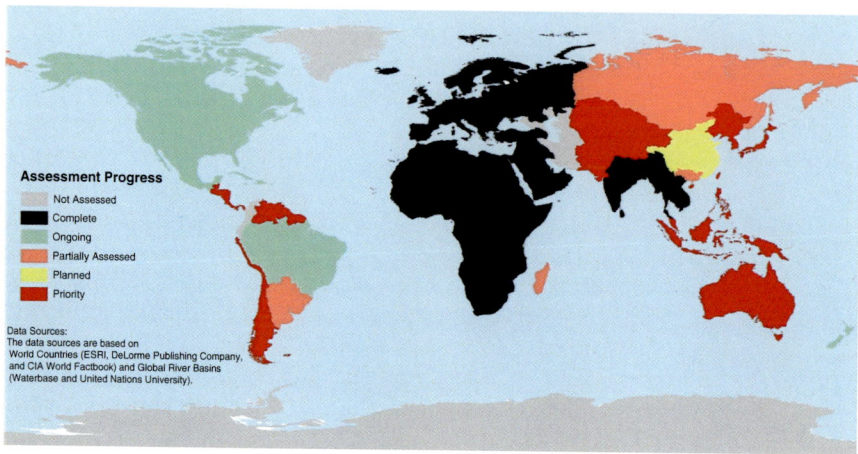

Fig. 1 IUCN Red List Assessment Progress for Freshwater Fishes. Redrawn from Carrizo et al. (2013) with updates from Smith et al. (2014), Garcia et al. (2015) and future assessment priority areas (see: http://www.iucn.org/theme/species/our-work-ssc/our-work/freshwater-biodiversity/river-bank [Accessed: 26 June 2016])

converted to other uses such as agriculture and urban development during the twentieth century (MA 2005), and while we do not have comprehensive data for other parts of the world many tropical and subtropical wetlands, particularly mangrove, swamp forests, and large-river floodplains, have been degraded or completely lost.

In terms of species, approximately one in three of all freshwater species are threatened with extinction, a higher degree of risk than for terrestrial species, and this risk is generalized across all regions of the world (Collen et al. 2014; IUCN 2015). While not all freshwater fish species have been evaluated for threat under the IUCN Red List, since 2002, there has been considerable progress with 6,128 species – or 39% – of species (Table 1) assessed by region (Fig. 1) (Carrizo et al. 2013; Smith et al. 2014; Garcia et al. 2015). These regional assessments suggest that freshwater fish are not faring any better than other taxonomic groups. In Europe 200 out of 522 species of freshwater fish (Kottelat and Freyhof 2007) and one in four in East Africa had joined the Red List (Darwall et al. 2005), and 54% of Madagascar's native freshwater fish species are threatened (Darwall et al. 2005). In total of the 27 completely assessed families of fishes, 11 families have more than 50% of their species categorized as threatened, including eight families with threat levels affecting 84–100% of their species; among this imperiled group are paddlefishes, river loaches, Valencia toothcarps, sawfishes, rainbow fishes, and sturgeons (Carrizo et al. 2013). Freshwater turtles, in particular, seem to be at an even higher risk than other taxonomic groups, with nearly half of the species classified as threatened with extinction, primarily due to hunting for food and traditional Chinese medicine and for the pet trade (Turtle Conservation Coalition et al. 2011; Collen et al. 2014). Among the most threatened are freshwater cetaceans, such as the Yangtze River dolphin *Lipotes vexillifer* which is now considered to be functionally extinct (Allen et al. 2012). Overall, the top three threats driving these numbers are habitat loss and degradation, water pollution, and exploitation (Collen et al. 2014); habitat loss and degradation, in particular, effects more than 80% of species (Collen et al. 2014). Despite the fundamental ecological and economic importance to humans, freshwater ecosystems have not been of primary concern in conservation, planning, or policy-making.

Ecosystem Services and the Role that River Basins Play in Human Well-being

In addition to providing habitat for freshwater species, wetlands deliver a wide range of ecosystem services that contribute to human well-being, such as fish and fiber, water supply, sediment retention, water purification, climate and flood regulation, energy generation, coastal protection, tourism, and recreational and cultural opportunities. While we all depend on freshwater systems in one form or another, an estimated 1.5–3 billion people depend on groundwater, often recharged through wetlands, for their drinking water (MA 2005).

The consumption of safe water for personal use, such as drinking or hygiene, is a primary development goal, and despite having met it at the global level with over two billion more people having gained access to improved drinking water sources since 1990, challenges remain for an estimated 663 million people who still lack access to safe drinking water (WHO and UNICEF 2015). In addition, 2.4 billion people still lack access to a toilet and nearly 80% of the world's population is at risk of suffering from water stress (Vorosmarty et al. 2010; WHO and UNICEF 2015). While developed nations can offset the risk of water stress through investment in water technology, poor nations are not able to do so putting more people at risk of water shortages.

River basins provide water and nutrients for agricultural activities, and therefore are key to maintaining our food security. To produce enough food to satisfy a person's daily dietary needs takes about 3,000 l of water – about one litre per calorie – while only about two to five liters are required for drinking. Predictions indicate that, in the future, more people will require more water for food, livestock, and fish. Globally, 70% of water withdrawals are used for irrigation, even though in some regions like Asia and the Middle East and North Africa this percentage is closer to 80% (WWAP 2009). In comparison, industrial use (including energy production) and domestic water withdrawal are 20% and 10%, respectively (WWAP 2009). Globally, about 82% of cropland relies on rainwater, and only about 18% of cropland is irrigated; however, this small percentage produces 40% of the agriculture production (Darghouth et al. 2008). Asia has the largest irrigated area (70% of the global), mostly in China and India (Comprehensive Assessment of Water Management in Agriculture 2007).

The world's rivers and lakes support globally important inland fisheries, providing 33% of the world's fish catch and employing over 60 million people, of whom over half are women (UNEP 2010). The consumption of fish from inland waters is especially important for human nutrition in Asia and parts of Africa (UNEP 2010). Over one fifth of Africa's population regularly consumes fish, and nearly half of this comes from inland waters. China, Bangladesh, India, and Myanmar are the top inland fish producing nations (UNEP 2010). Specific river basins, like the Mekong, are particularly important; the Tonle Sap in the lower Mekong supports 40 million rural farmers who are also engaged in fishing at least seasonally (Sverdrup-Jensen 2002) and produces 2.1 million metric tons of wild fish with a retail value of between US$4.2 and US$7.6 billion in retail markets (Hortle 2009). Lake Victoria Basin in Africa is another example; the basin produces more than one million metric tons of fish annually (Balirwa 2007) with an annual export value of exceeded US$300 million in 2007 for Nile Perch alone (LVFO 2008). As the human population continues to grow, the demand for fish both from capture fisheries and aquaculture will increase putting further pressure on freshwater ecosystems.

Water provision is a key ecosystem service and underpins numerous energy production options. Since traditional biomass energy is relatively inexpensive and more accessible than other energy sources, it plays a vital role in supporting economically poor populations, and if threatened, can greatly exacerbate poverty. Therefore, the proper functioning of freshwater ecosystems is vital. In addition to supporting biomass production and its use in hydropower installations, water is also used in oil and gas production and for cooling in electricity generation. In this manner, even industrial regions and wealthy populations are not immune to the effects of ecosystem deterioration on energy services, with the degradation of ecosystem services leading to a loss of capital assets (Williamson and McCormick 2008).

The indirect benefits derived from wetlands are equally valuable to humans, despite challenges in quantifying such contributions in monetary value. For instance, flood attenuation through wetlands and floodplains has been shown to be more cost-effective than building levies or paying to repair any damage caused. Since Hurricanes Katrina and Rita in 2005, the State of Louisiana in the United States has assigned 37% of revenues from new oil and gas projects to coastal protection and restoration, following research which suggests that an investment of US$10–15 billion in restoring the Mississippi Delta could generate the equivalent of US$62 billion by avoiding losses from storm damage and reduced ecosystem functions while gaining additional ecological benefits (Batker et al. 2010).

Wetlands play an important role in purifying water by filtering out and storing pollutants in their sediments, soils, and vegetation, and the natural ability of wetland systems to filter water has been used to treat wastewater from industry and mining as well as sewage. For instance, some floating plants, such as water hyacinth *Eichhornia crassipes*, can absorb heavy metals such as iron and copper. At least one third of the sewage from the city of Kolkata, India – which has a population of ten million people – is effectively treated by the East Kolkata marshes. The economic value of the purification function of wetlands can be huge (Ramsar Convention Secretariat 2010). In 1997, New York City found that it could avoid spending US$3–8 billion on new wastewater treatment plants, plus an additional US$700 million in annual operating costs, by investing US$1.5 billion in land purchase and conservation management measures to protect wetlands in the watershed to achieve the same outcome (Ramsar Convention Secretariat 2010).

Physical and economic water scarcity and limited or reduced access to water are major challenges facing society and are key factors limiting economic development in many countries. Many water resource developments undertaken to increase access to water have not given adequate consideration to harmful trade-offs with other services provided by wetlands. The projected continued loss and degradation of wetlands will reduce the capacity of wetlands to mitigate impacts and result in further reduction in human well-being especially for poorer people in lower-income countries, where technological solutions are not as readily available. At the same time, demand for many of these services will increase.

Threats to Major Rivers

River basins are subject to impacts from land-based activities occurring in the catchments as well as direct modifications to river reaches and flow regimes in the river network. Because of our dependence on freshwater, proximity to water bodies has been a preference for the establishment of human settlements for millennia. As a consequence we have heavily altered waterways to fit our needs by building dams, levies, canals, and water transfers. Loss and degradation of habitats, physical alteration, water withdrawals, overharvesting, pollution, and the introduction of nonnative species have all increased in scale and impact in the last century.

Of the many ways in which humans alter rivers, dams are probably the most widespread and significant in their impact. Today, the world's rivers are dotted with more than 50,000 large dams (Berga et al. 2006). Dams provide unquestionable benefits – from water supply to power generation – but they disrupt the hydrological cycle profoundly, suppressing natural flood cycles, disconnecting rivers from their floodplains, disrupting fish migrations, changing the temperature and velocity of in-stream flow, altering the deposition of sediments and essential nutrients to estuaries downstream, and flooding critical feeding and breeding habitats for many aquatic and terrestrial species. A 2005 analysis (Nilsson et al. 2005) that looked at

292 of the largest river systems in the world found that more than 60% of them were either strongly or moderately affected by dams. In addition, the livelihoods of populations living downstream of dams may be affected by dam-induced alterations of river flows through disruption of natural riverine production systems – especially fisheries, flood-recession agriculture, and dry-season grazing (Richter et al. 2010).

Another major impact that humans have on rivers is the amount of water we withdraw from them for human uses. By 2025, at least 3.5 billion people are projected to live in water-stressed river basins (Revenga et al. 2000). This per capita water supply calculation, however, does not take into account the coping capabilities of different countries to deal with water shortages. For example, high-income countries may be able to cope to some degree with water shortages by investing in desalination or reclaimed wastewater. Nearly three billion additional urban dwellers are forecast by 2050. While cities struggle to provide water to these new residents, they will also face equally unprecedented hydrologic changes due to global climate change. Currently, 150 million people live in cities with perennial water shortage, defined as having less than 100 l per person per day of sustainable surface and groundwater flow within their urban extent (McDonald et al. 2011). By 2050, demographic growth will increase this figure to almost one billion people, and climate change could increase water shortages for an additional 100 million urbanites (McDonald et al. 2011). Freshwater ecosystems in river basins with large urban populations and with insufficient water will likely struggle to maintain sufficient river flows to ecological processes (McDonald et al. 2011).

Overexploitation of resources is also a key driver in species extinction, although for some fisheries in particular the overexploitation took place decades ago, and new introduced nonnative fisheries and stocking programs have replaced the once plentiful wild and native fish populations and masked their decline. There is ample evidence of overfishing in once valuable and iconic fisheries – from salmon stocks in the Pacific Norwest of the United States to the Murray cod *Maccullochella peelii* in Australia's Murray-Darling basin and the sturgeons of the Caspian Sea. Overfishing has been a significant factor in the decline of the European eel *Anguilla anguilla* fishery, which used to employ 25,000 fishermen; eel stocks have steadily declined over the last 30 years with figures showing the number of new eel juveniles entering European rivers down to 1–5% of former levels and showing no sign of recovery even after eel stocking programs have been in place for years (ICES 2011). Even fisheries from relatively less altered rivers such as the Mekong in Southeast Asia have rapidly declined in the last decade; the Mekong giant catfish *Pangasianodon gigas* is considered critically endangered (Hogan 2013). Overexploitation is also the leading cause in the imperilment of other species, such as turtles. Recent assessments indicate that of the 263 species of freshwater and terrestrial turtles 45% are threatened, and one has already gone extinct (Turtle Conservation Coalition et al. 2011) mostly driving by overexploitation for the food and for use in traditional medicine and to a lesser degree by the pet trade; Asian and Malagasy species are at the highest risk (Turtle Conservation Coalition et al. 2011).

The intentional or accidental introduction of nonnative species is also a leading cause of species extinction in freshwater systems. Exotic species affect native fauna through predation, competition, disruption of food webs, and the introduction of diseases. The spread of exotic species is a global phenomenon, especially with increasing aquaculture, shipping, and global commerce. Worldwide, two thirds of the freshwater species introduced into the tropics and more than 50% of those introduced to temperate regions have become established (Welcomme 1988). Many introductions are usually done to enhance food production and recreational fisheries. But these introductions in most cases have hastened the decline of native fish. In Africa's Lake Victoria, for example, the introduction of Nile perch into the lake and its increase in abundance in the 1980s led to the disappearance of more than 100 of the native haplochromine cichlid fish species (Witte et al. 1992).

Finally, the IPCC 5th Assessment Report (Jiménez Cisneros et al. 2014) provides abundant evidence that freshwater systems will likely be strongly impacted by climate change and that the expected changes to water quantity and quality resulting from increased temperatures and altered rainfall patterns will significantly affect food production and availability, increasing the vulnerability of the rural poor particularly in arid and semiarid areas of the developing world. Key direct implications of warming temperatures and melting snowpack and glaciers are warmer waters, greater runoff during a shortened melt season, and longer low-flow seasons. The threats to biodiversity emerge as phenological disturbances, habitat loss, harmful temperatures, water chemistry changes, and more (Poff et al. 2002). Many freshwater animal and plant species rely on specific melting events to cue behaviors such as migration, spawning, and seed release. For instance, a fish species may require long winter seasons to allow its young to mature before the rush of spring floods. If those floods come too early, the young will be washed away by the intense flood currents and be more susceptible to predation. While cold-water dependent species such as native salmonids are projected to disappear from large portions of their current geographic range (Poff et al. 2002), as warming causes water temperature to exceed their thermal tolerance limits; behavioral or life-history adaptations may allow other species, particularly warm-water species, to persist in altered meltwater flow regimes and temperatures, and some (e.g., carp, bass) may even expand their range (Poff et al. 2002). Unlike terrestrial species and even aquatic insects with aerial or wind-dispersed life stages, freshwater species are limited in their ability to move or shift their ranges to other suitable habitats as hydrological regimes change, as most are confined to individual basins and generally unable to move from one unconnected catchment to another. This migratory adaptive behavior is even more challenging today, as we have excelled at building dams and other infrastructure, altering potential migratory river corridors that may have otherwise allowed species to relocate and adapt. Climate change will also affect how we manage and operate existing water infrastructure. And a conclusion by the IPCC 5th Assessment Report Working Group II suggests that current practices are not robust enough and that because of the cross-cutting nature of water management, many other policy arenas will be impacted, from energy production to food and water security (IPCC 2014).

Protection of Rivers and Other Aquatic Habitats

Despite the values and degree of imperilment of freshwater ecosystems, protection and conservation of inland waters has come at a slow pace, especially when compared to the protection of terrestrial and more recently marine ecosystems. For decades protected areas have been a key conservation strategy for terrestrial and marine habitats, but for inland aquatic habitats, such as rivers, lakes, marshes, and groundwater aquifers, this has not been the case. The traditional approach of fencing off an area for its protection and management is not only unlikely but impractical when it comes to protecting linear aquatic habitats such as rivers. Many terrestrial protected areas award only partial protection to aquatic habitats and important issues such as longitudinal and lateral connectivity and maintenance of essential hydrological processes (i.e., environmental flows) are not considered. Moreover, it is not uncommon to find that rivers have been used for the sole purpose of demarcating a terrestrial protected area boundary, instead of integrating them into the protected area's conservation and management objectives. A review by Abell et al. (2007) points out that many freshwater habitats and species have not benefited as much from terrestrial protected areas largely because they have not been designed or managed for aquatic ecosystems and are especially vulnerable to pressures occurring outside the protected area.

The only clear exception is the protection of aquatic habitats with a defined surface area, such as lakes, lagoons, marshes, and other wetlands. Under the Convention on Wetlands, also known as the Ramsar Convention – so named for the city in Iran where it was signed, signatory countries have protected 2,231 coastal and inland wetlands with a total surface area of nearly 215 million hectares (www.ramsar.org; Accessed: 16 March 2016). For surface water bodies, thus, establishing protected areas continues to be a customary conservation approach; but to be effective, their design and management has to also include protection, or at a minimum, sound management of important features and processes that are essential to sustaining the wetlands being protected. These key features include rivers or groundwater aquifers that feed into the particular lake or wetland but are often located outside the protected area boundary.

Protecting linear features like rivers and streams by means of a protected area is not always feasible. Only a hand full of countries including the United States, Canada, Australia, and New Zealand currently afford certain protections to entire rivers or portions of rivers. The United States and Canadian systems, where the concept of "wild river" originated, have been the model followed by other countries in establishing river protection frameworks. Undoubtedly, rivers, particularly the middle reaches and floodplains, are underrepresented in the global protected areas network, but figures showing exactly how underrepresented they are have been hard to come by. A key limitation has been the lack of inventory and assessment of freshwater ecosystems and their condition, even at national level; a limitation that has hindered the tracking of progress of individual countries towards their commitments under the Convention on Biological Diversity (CBD). The Parties to the CBD in 2004 agreed to "establish and maintain comprehensive, adequate and representative systems of protected inland water ecosystems within the framework of integrated catchment/watershed/river-basin management" (decision VII/4, CBD Conference of the Parties 2004). One of the key targets of the CBD's 2011–2020 Strategic Plan for Biodiversity developed in 2010 is to effectively conserve at least 17% of each of the world's ecological regions by 2020 for terrestrial and inland water ecosystems and 10% for marine and coastal areas.

The assumption that terrestrial protected areas protect the freshwater ecosystems that lie within them remains ingrained in the thinking and approach of most nations. Until recently there was no global spatially explicit database of river networks at enough resolution to allow for a detailed estimation of how much protection – at least in terms of coverage – terrestrial protected areas convey to waterways. In 2010, however, WWF in collaboration with McGill University and UNEP-WCMC released an assessment of the extent of coverage of rivers by protected areas by calculating both the length of rivers that lie within protected areas and the length of rivers that have their entire upstream catchments under protection (WWF 2010). The WWF analysis shows that 69% of the world's rivers are without protected areas in their upstream catchment and that smaller headwater streams are more protected at catchment level than are larger rivers. The analysis therefore points to opportunities to increase the representation of freshwater ecosystems in the global protected area network. The analysis also shows important regional differences. Central and South America lead the way in river protection, both at river length within protected areas and upstream catchment levels. Asia, North America, and Europe/Middle East have fewer river miles and upstream catchment areas protected (WWF 2010). While this first estimate of river coverage does not assess if protected areas are indeed protecting freshwater processes and species, it does provide a sense of how many lotic ecosystems could be protected if management of the existing protected area network were tailored to address freshwaters.

Protected areas can and do work for freshwaters, but the way we design and manage them is essential. Protecting freshwater ecosystems requires new approaches that incorporate the connectivity of freshwater systems and their key processes, such as spring peak flows that cue aquatic species to migrate, reproduce, feed, etc. Many of these approaches already exits but are usually not considered within a protected area framework. Zoning rules, such as establishing riparian buffers to control runoff, bank erosion, and provide shade, are in effect in many basins but are not per se protected areas. Water recharge areas or fishing reserves are other examples of protection that contribute to the conservation of freshwater habitats. Protected area design needs to start including these approaches and protected area management needs to start looking beyond the terrestrial park boundary if we are to reduce the rate of loss of freshwater species and the habitats upon which they depend.

References

Abell R, Allan JD, Lehner B. Unlocking the potential of protected areas for freshwaters. Biol Cons. 2007;134:48–63.

Allen DJ, Smith KG, Darwall WRT, compilers. The status and distribution of freshwater biodiversity in Indo-Burma. Cambridge/Gland: IUCN; 2012.

Balirwa J. Ecological, environmental and socioeconomic aspects of the Lake Victoria's introduced Nile perch fishery in relation to the native fisheries and the species culture potential: lessons to learn. Afr J Ecol. 2007;45(2):120–9.

Batker D, de la Torre I, Costanza R, Swedeen P, Day J, Boumans R, Bagstad K. Gaining ground. Wetlands, hurricanes and the economy: the value of restoring the Mississippi River Delta. Tacoma: Earth Economics; 2010.

Berga L, Buil JM, Bofill E, De Cea JC, Garcia Perez JA, Manueco G, Polimon J, Soriano A, Yague J. Dams and reservoirs, societies and environment in the 21st century. London: Taylor and Francis Group; 2006 (1).

Birkett CM, Mason IM. A new global lakes database for a remote sensing program studying climatically sensitive large lakes. J Great Lakes Res. 1995;21(3):307–18.

Carrizo SF, Smith KG, Darwall WRT. Progress towards a global assessment of the status of freshwater fishes (Pisces) for the IUCN Red List: application to conservation programmes in zoos and aquariums. Int Zoo Yb. 2013;47:46–64.

Comprehensive Assessment of Water Management in Agriculture. Water for food, water for life: a comprehensive assessment of water management in agriculture. London/Colombo: Earthscan/International Water Management Institute; 2007.

Collen B, Whitton F, Dyer EE, Baillie JEM, Cumberlidge N, Darwall WRT, Pollock C, Richman NI, Soulsby AM, Böhm M. Global patterns of freshwater species diversity, threat and endemism. Glob Ecol Biogeogr. 2014;23:40–51.

Darghouth S, Ward C, Gambarelli G, Styger E, Roux J. Watershed management approaches, policies, and operations: lessons for scaling up. Water sector board discussion paper series. Washington, DC: World Bank; 2008 (11).

Darwall W, Smith K, Lowe T, Vié JC. The status and distribution of freshwater biodiversity in Eastern Africa. IUCN–Species Survival Commission Freshwater Biodiversity Assessment Programme. Gland: IUCN; 2005.

Dudgeon D, Arthington AH, Gessner MO, Kawabata Z-I, Knowler DJ, Lévêque C, Naiman RJ, Prieur-Richard A-H, Soto D, Stiassny MLJ, Sullivan CA. Freshwater biodiversity: importance, threats, status and conservation challenges. Biol Rev. 2006;81:163–82.

Eschmeyer WN, Fricke, R, van der Laan R, editors. Catalog of fishes online database. San Francisco: California Academy of Sciences; 2016. Available at: http://www.calacademy.org/scientists/projects/catalog-of-fishes. Accessed 23 Apr 2016.

Eschmeyer WN, Fong JD. Species of fishes by family/subfamily. 2012. San Francisco: California Academy of Sciences. Available at: http://researcharchive.calacademy.org/research/ichthyology/catalog/speciesbyfamily.asp. Accessed 23 Apr 2016.

ESRI (Environmental Systems Research Institute). Digital chart of the world 1:1 Mio. Redlands: ESRI; 1993.

FEOW. Freshwater ecoregions of the world; www.feow.org. Washington, DC/Arlington: WWF and TNC; 2008.

García N, Harrison I, Cox N, Tognelli MF, compilers. The status and distribution of freshwater biodiversity in the Arabian Peninsula. Gland/Cambridge, UK/Arlington: IUCN; 2015.

Hoekstra JM, Molnar JL, Jennings M, Revenga C, Spalding MD, Boucher TM, Robertson JC, Heibel TJ, Ellison K. The atlas of global conservation: changes, challenges, and opportunities to make a difference. Berkeley: University of California Press; 2010.

Hogan Z. *Pangasianodon gigas*. The IUCN Red List of threatened species 2013: e.T15944A5324699. 2013. doi:10.2305/IUCN.UK.2011-1.RLTS.T15944A5324699.en. Accessed 16 Mar 2016.

Hortle KG. Fisheries of the Mekong River Basin. In: Campbell IC, editor. The Mekong: biophysical environment of a transboundary river. New York: Elsevier; 2009. p. 199–253.

ICES. ICES Eel Stock Advice for 2012. Online at http://www.ices.dk/sites/pub/Publication%20Reports/Advice/2011/2011/eel-eur.pdf

IPCC. Summary for policymakers. In: Climate change 2014: impacts, adaptation, and vulnerability. Part A: global and sectoral aspects. Contribution of Working Group II to the Fifth Assessment Report of the Intergovernmental Panel on Climate Change [Field CB, Barros VR, Dokken DJ,

Mach KJ, Mastrandrea MD, Bilir TE, Chatterjee M, Ebi KL, Estrada YO, Genova RC, Girma B, Kissel ES, Levy AN, MacCracken S, Mastrandrea PR, White LL, editors]. Cambridge/New York: Cambridge University Press; 2014.

IUCN. The IUCN Red List of threatened species. Version 2015-4. 2015. http://www.iucnredlist.org. Accessed 16 Mar 2015.

Jiménez Cisneros BE, Oki T, Arnell NW, Benito G, Cogley JG, Döll P, Jiang T, Mwakalila SS. 2014. Freshwater resources. In: Climate change 2014: impacts, adaptation, and vulnerability. Part A: global and sectoral aspects. Contribution of Working Group II to the Fifth Assessment Report of the Intergovernmental Panel on Climate Change [Field CB, Barros VR, Dokken DJ, Mach KJ, Mastrandrea MD, Bilir TE, Chatterjee M, Ebi KL, Estrada YO, Genova RC, Girma B, Kissel ES, Levy AN, MacCracken S, Mastrandrea PR, White LL, editors]. Cambridge/New York: Cambridge University Press; 2014.

Kottelat M, Freyhof J. Handbook of European freshwater fishes. Cornol/Berlin: Kottelat/Freyhof; 2007.

Lehner B, Grill G. Global river hydrography and network routing: baseline data and new approaches to study the world's large river systems. Hydrol Process. 2013;27(15):2171–86. Data is available at www.hydrosheds.org.

LVFO. State of fish stocks. Jinja, Uganda: Lake Victoria Fisheries Organization; 2008.

McDonald RI, Green P, Balk D, Fekete BM, Revenga C, Todd M, Montgomery M. Urban growth, climate change, and freshwater availability. Proc Natl Acad Sci U S A. 2011;108(15):6312–7.

Millennium Ecosystem Assessment (MA). Ecosystems and human well-being: wetlands and water synthesis. Washington, DC: World Resources Institute; 2005.

Nilsson C, Reidy CA, Dynesius M, Revenga C. Fragmentation and flow regulation of the world's large river systems. Science. 2005;308(5720):405–8.

Poff NL, Brinson NM, Day Jr JW. Aquatic ecosystems and global climate change: potential impacts on inland freshwater and coastal wetland ecosystems in the United States. Arlington: Pew Center on Global Climate Change; 2002.

Ramsar Convention Secretariat. Water purification wetland ecosystem services: factsheet 5. Gland: Ramsar Convention Secretariat; 2010.

Revenga C, Brunner J, Henninger N, Kassem K, Payne R. Pilot analysis of global ecosystems: freshwater systems. Washington, DC: World Resources Institute; 2000.

Revenga C, Kura Y. Status and trends of inland water biodiversity. CBD Technical Papers Series No. 11. Montreal: Secretariat of the Convention on Biological Diversity; 2003.

Richter BD, Postel S, Revenga C, Scudder T, Lehner B, Churchill A, Chow M. Lost in development's shadow: the downstream human consequences of dams. Water Altern. 2010; 3(2):14–42.

Smith KG, Barrios V, Darwall WRT, Numa C, editors. The status and distribution of freshwater biodiversity in the Eastern mediterranean. Cambridge, UK, Malaga/Gland: IUCN; 2014.

Sverdrup-Jensen S. Fisheries in the Lower Mekong Basin: status and perspective. MRC Technical Paper No.6. Phnom Penh: Mekong River Commission; 2002.

TFDD (Transboundary Freshwater Dispute Database). Department of Geosciences at Oregon State University. 2012. http://www.transboundarywaters.orst.edu/database/DatabaseIntro.html

Turtle Conservation Coalition. Rhodin AGJ, Walde AD, Horne BD, van Dijk PP, Blanck T, Hudson R, editors. Turtles in trouble: the world's 25+ most endangered tortoises and freshwater turtles – 2011. Lunenburg: IUCN/SSC Tortoise and Freshwater Turtle Specialist Group, Turtle Conservation Fund, Turtle Survival Alliance, Turtle Conservancy, Chelonian Research Foundation, Conservation International, Wildlife Conservation Society, and San Diego Zoo Global; 2011.

WHO, UNICEF. World Health Organization and UNICEF Joint Monitoring Programme (JMP). Progress on drinking water and sanitation, 2015 update and MDG assessment. 2015. http://water.org/water-crisis/water-sanitation-facts/

UNEP. Blue harvest: inland fisheries as an ecosystem service. Penang: WorldFish Center; 2010.

UN Water. World Water Day 2009 FAQs. 2009. Available from: http://www.unwater.org/wwd09/faqs.html. Accessed 16 Mar 2016.

USGS (U.S. Geological Survey). HYDRO1k elevation derivative database. Sioux Falls: USGS EROS Data Center; 2000.

Vorosmarty CJ, McIntyre PB, Gessner MO, Dudgeon D, Prusevich A, Green P, Glidden S, Bunn SE, Sullivan CA, Reidy Liermann C, Davies PM. Global threats to human water security and river biodiversity. Nature. 2010;467:555–61.

Welcomme RL. International introductions of inland aquatic species. Food and Agriculture Organization of the United Nations (FAO) Technical Series Paper 294. Rome: FAO; 1988.

Williamson L, McCormick N. Energy, ecosystems and livelihoods: understanding linkages in the face of climate change impacts. Gland: IUCN; 2008.

Witte F, Goldschmidt T, Wanink J, van Oijen M, Goudswaard K, Witte-Mass E, Bouton N. The destruction of an endemic species flock: quantitative data on the decline of the haplochromine cichlids of Lake Victoria. Enviorn Biol Fish. 1992;34:1–28.

WWAP (World Water Assessment Program). The United Nations world water development report 3: water in a changing world. Paris/London: UNESCO/Earthscan; 2009.

WWF. Increasing protection of river catchments. Factsheet. Washington, DC: World Wildlife Fund; 2010.

Freshwater Lakes and Reservoirs

8

Etienne Fluet-Chouinard, Mathis Loïc Messager, Bernhard Lehner, and C. Max Finlayson

Contents

Introduction 126
Lake Area 127
Lake Volume and Residence Time 128
Geographic Distribution 130
Ecological Features 131
Biota 133
Ecosystem Services 135
Threats and Future Challenges 137
References 139

Abstract

Lakes, ponds, and lacustrine environments are characterized by standing water in depressions of the landscape left by glacier scouring or impoundments along river networks. Depending on their origin and climatic conditions, lakes can be ancient

E. Fluet-Chouinard (✉)
Center for Limnology, University of Wisconsin-Madison, Madison, WI, USA
e-mail: fluetchouina@wisc.edu

M. L. Messager (✉)
School of Aquatic and Fishery Sciences, University of Washington, Seattle, WA, USA
e-mail: messager.mathis@gmail.com

B. Lehner (✉)
Department of Geography, McGill University, Montreal, QC, Canada
e-mail: bernhard.lehner@mcgill.ca

C. M. Finlayson (✉)
Institute for Land, Water and Society, Charles Sturt University, Albury, NSW, Australia

UNESCO-IHE, The Institute for Water Education, Delft, The Netherlands
e-mail: mfinlayson@csu.edu.au

© Springer Science+Business Media B.V., part of Springer Nature 2018
C. M. Finlayson et al. (eds.), *The Wetland Book*,
https://doi.org/10.1007/978-94-007-4001-3_201

(e.g., African rift lakes) or very young (e.g., recent landslides). The age of most of today's lakes are in the range of tens of thousands of years old as they were created during the glacier recession of the last ice age (e.g., lakes on or near the Canadian Shield). A handful of largest lakes contain a major proportion of the world's lake water volume whereas small lakes are more numerous and account for most of lakes surface area. Over time, the natural evolution of lakes is to gradually fill from autochthonous and allochthonous material and sediments. Freshwater lakes, like other surface water ecosystems, offer a wide array of benefits for people, such as water provision, fisheries, flood attenuation, and recreational purposes, and play an important role in global biogeochemical cycles, including water, carbon, and nutrient balances. Overexploitation of freshwater resources threatens the capacity of these ecosystems to provide these benefits in the future.

Keywords

Lakes · Reservoirs · Volume · Residence time · Lentic ecosystem · Biota · Ecosystem services

Introduction

Lakes, ponds, and lacustrine environments are characterized by standing water forming in depressions of the landscape, such as those left by glacier scouring or impoundments along river networks. The geomorphology of each lake and its contributing catchment determine the input of nutrients and material as well as the residence time of water in the lake (Wetzel 2001). Globally, the most important factor in the creation of lakes is the erosional and depositional effect of glacier ice movement. Lakes can also be formed by tectonic, volcanic, or riverine activity, as well as landslides, wind erosion, dissolution of limestone, or biological activity such as beaver dams (Hutchinson 1957). Depending on their origin and climatic conditions, lakes can be ancient (e.g., African rift lakes) or very young (e.g., recent landslides), but today most lakes are in the range of tens of thousands of years old as they were created during the glacier recession of the last ice age (e.g., lakes on or near the Canadian Shield). Over time, the natural evolution of lakes is to gradually fill from autochthonous and allochthonous material and sediments. Freshwater lakes, like other surface water ecosystems, provide a wide array of benefits for people, such as water provision, fisheries, flood attenuation, and recreational purposes, and play an important role in global biogeochemical cycles, including water, carbon, and nutrient balances. Yet overexploitation of freshwater resources threatens the capacity of these ecosystems to provide these benefits in the future.

For millennia, humans have created reservoirs and artificial lakes by impounding streams and digging shallow ponds for purposes including water supply, irrigation, flood protection, and hydropower production. Large impoundments (dam height >15 m) and reservoirs have been constructed at a rapid rate during the twentieth century, growing from 5000 in 1950 to 50,000 by the end of the century (Berga et al. 2006). By the end of the twentieth century, perceptions of dams had changed, as evidence of their social and environmental cost accumulated, resulting in a brief hiatus in dam building (Rosenberg et al. 1997). Renewed interest in the construction of dams and reservoirs is growing in regions where hydropower potential is still untapped such as along the Mekong River (Zarfl et al. 2015).

Lake Area

Globally, the cumulative surface area of natural lakes of at least $0.01\ km^2$ (1 ha) has been reported to range between 2.7 and $5 \times 10^6\ km^2$ or 1.8–3.3% of the world's land area (some higher estimates include reservoirs), with most estimates tending toward the lower end of the range (Table 1). The perimeter, i.e., shoreline length of lakes greater than $0.1\ km^2$, has been estimated as $7.2 \times 10^6\ km$, though this type of measure is highly sensitive to the mapping resolution (Messager et al. 2016).

Estimates of the abundance and surface area of larger natural lakes are reasonably well constrained by remote sensing inventories and cartographic surveys. Data on smaller lakes are less consistent, but their number is higher than for larger ones. Small lakes are substantially more biologically active than larger lakes due to the importance of shore and littoral processes, as well as generally possessing proportionally shorter life spans (Downing 2010; Lewis 2011). Thus, knowledge of the distribution and abundance of lakes across size classes is critical to upscale biogeochemical processes that are related to lake size, for instance, carbon sequestration (Winslow et al. 2015).

The abundance and cumulative importance of lakes smaller than the detection limit of the standard global observation methods have been estimated by extending the size-abundance distribution of observable lakes to unobserved size classes

Table 1 Recent estimates of global lake and reservoir abundance, cumulative area, and volume

Source	Smallest lake size	Number (million)	Area ($10^6\ km^2$)	Volume ($10^3\ km^3$)
Downing et al. 2006	$\geq 0.01\ km^2$	27[a,b]	3.55[a,b]	
	$\geq 0.001\ km^2$	304 [a b]	4.2[a,b]	
Ryanzhin et al. 2010	$\geq 0.01\ km^2$		2.69[a]	179.6[a]
Lewis 2011	$\geq 0.01\ km^2$		3.10[a,b]	
McDonald et al. 2012	$\geq 0.01\ km^2$		3.53[b]	
Raymond et al. 2013	$\geq 0.001\ km^2$		2.74[a,b]	
Verpoorter et al. 2014	$\geq 0.01\ km^2$	27	4.76	
	$\geq 0.002\ km^2$	117	5	
Messager et al. 2016	$\geq 0.1\ km^2$	1.43[a]	2.67[a]	181.9[a]
	$\geq 0.01\ km^2$	21.2[a,b]	3.22[a,b]	182.9[a,b,c]

[a]Excludes man-made reservoirs
[b]Based on extrapolation method to small lakes, all other estimates are from data
[c]Including all lakes $\geq 0.00001\ km^3$

Fig. 1 Dams have been constructed across many wetlands and floodplains such as the Pongolapoort Dam in South Africa (*left*) and the Burdekin Falls Dam in Australia (*right*) (Photo credit: CM Finlayson © Rights remain with the author)

(e.g., Lehner and Döll 2004; Downing et al. 2006). The number of lakes in each size class is assumed to follow a predictable scaling law, and extrapolation with a Pareto distribution suggests that small lakes constitute an important cumulative fraction of global lake area (Messager et al. 2016). Scaling laws have been questioned for their robustness and reliability for very small lakes, leaving some uncertainty around their contribution (Seekel and Pace 2011). In addition to variations in statistical distribution, divergences among estimates can also be due to differences in the discrimination between natural lakes and reservoirs and confusion among waterbody types in remote sensing products (Messager et al. 2016).

Through the construction of reservoirs and impoundments, humans have increased the Earth's water surface and storage (Fig. 1). Like natural lakes, only the area of large reservoirs is reliably estimated, and the abundance of numerous small reservoirs must be inferred from observable ones (Downing et al. 2006; Lehner et al. 2011). A database of the largest reservoirs (excluding regulated lakes), extrapolated with a Pareto distribution to smaller reservoirs, estimates that 76,000 reservoirs larger than 10 ha add a surface area of 284.7×10^3 km^2, thus increasing the world's natural terrestrial surface water area by 6.8% (Lehner et al. 2011).

Lake Volume and Residence Time

Our knowledge of the water volume stored in natural lakes is more uncertain than that of their area due to the technical and operational challenges of measuring bathymetry over broad scales (Messager et al. 2016). For lakes of a given area, their bathymetry can be modeled from surrounding shoreline slope and extended to all size classes with a statistical model (Sobek et al. 2011). Estimates of global volume tend to converge around 180×10 km^3 of water for all natural lakes of at least 10 ha in size, equivalent to 0.8% of total global nonfrozen terrestrial water stocks (Messager et al. 2016). Natural lake volume is dominated by a few large lakes,

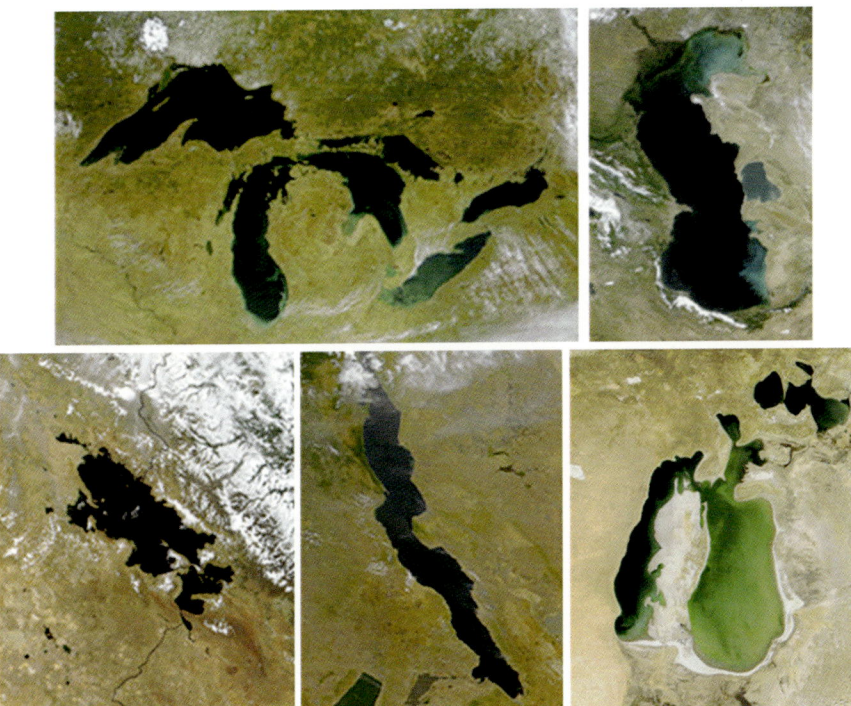

Fig. 2 Satellite imagery of some of the largest lakes: Laurentian Great Lakes (*top left*) including Lake Superior, Caspian Sea (*top right*), Lake Titicaca (*bottom left bottom center*), and Aral Sea (*bottom right*) in 2001 (Source: NASA-Visible Earth)

with 40.2% of the global lake volume contained in the Caspian Sea (Fig. 2) and 33.2% in the next four largest lakes (in decreasing order): Baikal, Tanganyika, Superior (Fig. 2), and Malawi.

The combined storage of reservoirs larger than 10 ha, extrapolated from the largest human-made reservoirs contained in the Global Reservoir and Dam database (GRanD; Lehner et al. 2011), amounts to 7,260 km^3 which represents 3.2% of the total volume of natural lakes and reservoirs of the world. Reservoir volume can be estimated from surface area and the height of the dam. Similar to lakes, reservoir volume is dominated by the largest reservoirs (dam height >15 m) accounting for 75% of the total global storage capacity (Lehner et al. 2011).

The volume of lakes and their rates of in- and outflow define the hydraulic residence time (i.e., water age) in lakes. The median residence time for lakes ≥10 ha has been estimated at 451 days, with a mean of 1,834 days or ~5.02 years (Messager et al. 2016). Short residence times are generally associated with smaller lakes and can vary with respect to the lake's position in the river network. The volume-weighted mean residence time is much larger at 275 years due to the dominant influence of the few largest lakes on Earth (Messager et al. 2016).

Geographic Distribution

Areas of high lake density are predominantly found within the glaciated extent of the Last Glacial Maximum, with the highest densities in Northern Canada, Alaska, Scandinavia, and Russia in addition to pockets in alpine regions of the Rockies, Andes, and Himalayas (Fig. 3). Other areas of high lake density not associated with glaciation are concentrated in the African Rift Valley, as well as in floodplains of the Amazon basin and regions of coastal China (Messager et al. 2016). The latitudinal distribution of intermediate and large lakes' density peaks between 40° and 50° North and can partly be attributed to the Great Lakes and Caspian and Aral Seas (Lehner and Döll 2004; Verpoorter et al. 2014). The geographic distribution of the density of small unmapped lakes is related to surface runoff processes and the regional wetness of an area, with lake densities increasing up to runoff amounts of 1,000 mm year^{-1} and then decreasing due to highly erosional landscapes (Downing et al. 2006).

While the extent and distribution of large intermittent and ephemeral water bodies, such as Lake Eyre in Australia or Lake Eyasi in Tanzania, is well known, the global prevalence of smaller intermittent and ephemeral lakes is a challenge to study and may in part be in or excluded in the numbers presented here. Indeed, most databases from satellite imagery only provide a snapshot in time of the Earth's water cover, limiting the development of a comprehensive survey of these highly seasonal geographic features. Moreover, the definition of ephemeral lakes is inconsistent across regions, complicating their assessment based on regional maps. Overall,

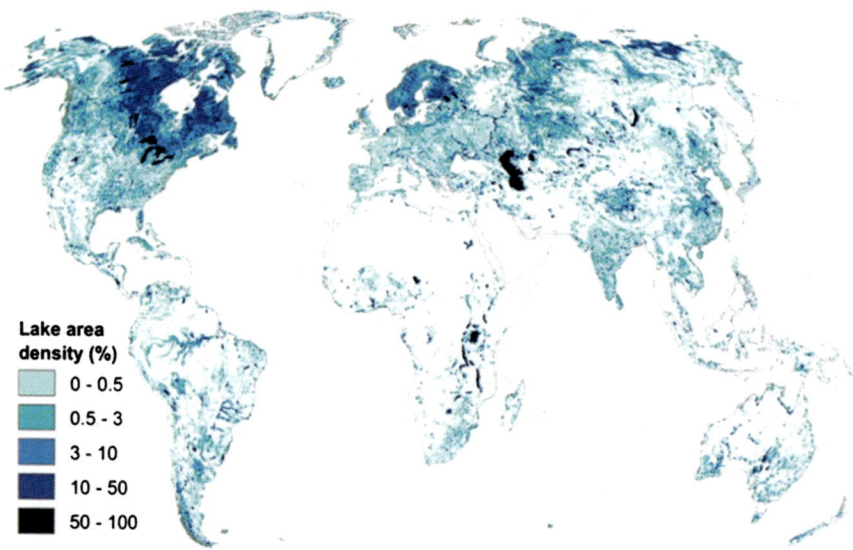

Fig. 3 Distribution of global lake area density (including reservoirs) calculated as percent area covered by lakes within a 25 km radius (From Messager et al. 2016 – Licensed under a Creative Commons Attribution 4.0 International License: https://creativecommons.org/licenses/by/4.0/

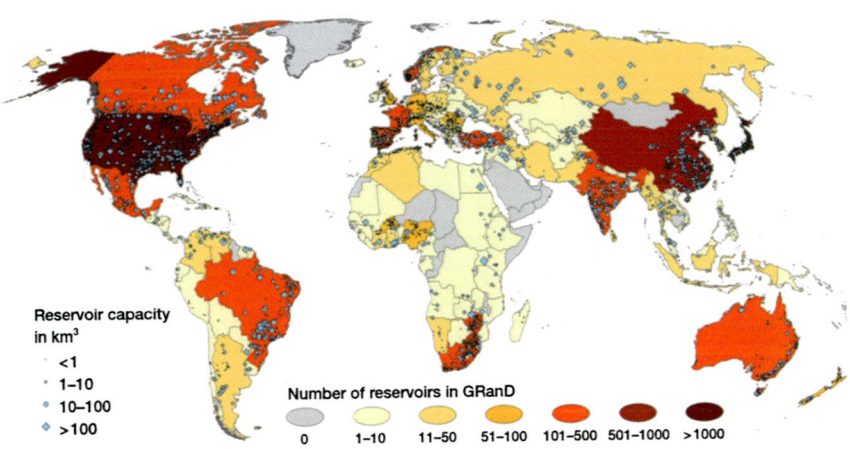

Fig. 4 Global distribution of large reservoirs across nations (Reprinted with permission by John Wiley and Sons, Ltd. From Lehner et al. 2011. High-resolution mapping of the world's reservoirs and dams for sustainable river-flow management. Frontiers in Ecology and the Environment. © the Ecological Society of America.)

these lakes tend to be found at high elevations or in arid regions and are often endorheic (i.e., without an outlet). The salinity of ephemeral and intermittent lakes is variable, as these water bodies tend to hold freshwater when recently flooded and then gradually become saltier as water evaporates.

The largest reservoirs are concentrated in the USA, Russia, Canada, Brazil, China, and other nations that have developed their water resource infrastructure during the twentieth century (Fig. 4). More than 3,700 major hydropower dams are either planned or under construction and are concentrated in developing countries and emerging economies of South East Asia, South America, and Africa (Zarfl et al. 2015).

Ecological Features

Many of the ecological features of lakes are related to their morphology and hydrology. Morphological details influence, for example, the nutrient loading rates, chemical mass, thermal stability and heat content, and biological productivity of a lake. The hydrology is also entwined with the morphology as well as the climate and form of the surrounding landscape. These features change with time and have been heavily influenced by human activities. A general description of these features has been derived from Hutchinson (1975), Wetzel and Likens (1991), and Moss (2010).

The water level of a lake is dependent on the difference between the input and output of water as well as the overall volume. Lakes may be filled from direct precipitation, runoff from streams or rivers, or from groundwater sources. Ephemeral lakes are filled at irregular time intervals and may remain dry for long periods, extending to decades in parts of the arid zones. Water can be lost from a lake due to evaporation, as well as surface or groundwater flows, or increasingly extracted for human usage. In addition, water storage has increased with dams and barrages of different shapes and sizes being used to impound water both on rivers and across floodplains. Water inflows to lakes can also be reduced by construction of dams and impoundments built upstream of the lake, for instance, Lake Poyang in Jiangxi province whose five tributaries were heavily impounded (Liu et al. 2009; Harris 2018) or Lake Urumiyeh, Iran, where reduced water inflows from regulation combined with long-term drought have caused catastrophic ecological change from hypersalinity and drying (Lotfi 2018a).

Lakes are categorized according to their thermal stratification and mixing regime or on their productivity and nutrient availability. The termal mixing regime in lakes follows general patterns differing between lakes in temperate, warm-temperate, and tropical zones. Some of the simplest mixing regimes occur in the temperate zone having snowmelt, runoff and little evaporation, with more variation within the tropics between the wet and dry tropics. Classification based on changes in the thermal stratification pattern is determined by the frequency with which a thermocline, a steep temperature gradient between the surface and deeper layers of water, establishes. The thermocline forms with more warming of the surface layers relative to the deeper, cooler layers and can be disrupted when the upper layers cool or when winds or inflowing water mixes the water in the lake. Differences in the frequency of mixing have resulted in lakes being broadly classified as monomictic, dimictic, or polymictic, although mixing patterns can be more complex. Lakes can also be classified based on their nutrient status, as being oligotrophic with low concentrations of nutrients and generally clear water and low plant or phytoplankton production, mesotrophic with increasing nutrient concentrations and production, or eutrophic with higher nutrient concentrations, turbid water, and high plant production and possibly blooms of phytoplankton. Various scales have been proposed to describe these classes, though important variation exists between lakes depending on their geomorphology, catchment features, and the impact of human activities. There is also evidence that some lakes can undergo critical transitions between stable alternate ecological states in response to increasing nutrient loading, for instance, from clear water with submerged macrophytes to turbid water with phytoplankton blooms (Scheffer and Carpenter 2003) in shallow temperate lakes (Capon et al. 2015).

The concentration of major ions in lakes varies in accordance with the geology but also in response to the historical and current climates, age of the lake, and volcanic activity. Much of the land mass of the northern hemisphere was glaciated until recent geological times, with higher ionic concentrations in lakes that formed after the glaciers melted, but then declined as the catchments were leached. The amount of organic matter in lakes also increased since the last glaciation, as the catchments became vegetated and accumulated organic matter. Where glaciation or volcanic activity have occured, the soils are leached and soluble ions largely transported to the oceans. The ionic content is also influenced by the balance between precipitation and evaporation and whether the basin is closed or open with ions exported to the oceans. The concentrations of ions affect the plant and animal communities, for example, with the amounts of bicarbonate and carbon dioxide influencing which plants and phytoplankton will predominate and calcium affecting the predominance of snails and large crustacean species. Evaporation and the accumulation of sodium chloride leading to increased salinity also greatly influence the plants and animals.

The nutrients phosphorus and nitrogen are particularly important for plant growth and the production in a lake, with some plants being able to fix nitrogen and increase nitrogen availability in the sediments and water in a lake. The cycling of nitrogen in particular is greatly influenced by the oxidation and reduction conditions in the lake and with exchanges between the sediment and the water. The primary production in a lake is also influenced by a number of other variables, including the irradiance and light attenuation, temperature, and the mixing of the water. Most of the primary production in a lake is restricted to the upper photic zone where sufficient light is available for photosynthesis whether by phytoplankton or higher plants. The balance between photosynthesis and respiration in lakes can change seasonally and also daily with a decline in the hydrogen ion (pH) and dissolved oxygen concentrations overnight. The seasonal changes are strongly interrelated with the weather and the hydrological conditions.

Biota

The biota found in freshwater lakes varies greatly across the variety of age, size, and depth of lakes as well as their geographic distribution ranging from high latitude or altitudes to those in the tropics. The abovementioned ecological features also affect the occurrence of particular species and communities with seasonal and decadal variation and both resident and migratory species. The communities in ephemeral lakes are highly dynamic and often include transitory visitors or species that can complete their life cycles in short time periods. The fluctuating water levels in many lakes are a major determinant of their use by both resident and migratory species. Seasonal fluctuations are common whether in response to temperature changes, including freezing and thawing, or in response to rainfall and flooding.

The migration of waterbirds is perhaps one of the more visible seasonal patterns, with movement of large numbers of migratory species, including ducks, swans and geese, colonial waterbirds, and shorebirds/waders, within and between continents in response to climatic changes, in search of breeding and feeding habitat. Many large and charismatic animals are found in lake habitats (Fig. 5). There is a dynamic relationship with the salinity of ephemeral lakes affecting the occurrence and life cycles of many species. Major movements of fish species between rivers and lakes also occur, as well as movement within lakes. The aquatic vegetation is generally most dominant in the shallower water with the more herbaceous species, in particular, expanding and contracting seasonally in response to water level fluctuations (Fig. 6). Many woody species can thrive along the shorelines or tolerate periods of

Fig. 5 Lakes support many large and charismatic species including crocodile species, waterbirds, and mammals (Photo credit: CM Finlayson © Rights remain with the author)

Fig. 6 Lakes contain many species of plants, including many types of water lilies such as *Nymphaea* species (*left*) and *Nelumbo nucifera* (right), with many subjects to seasonal fluctuations in water levels (Photo credit: CM Finlayson © Rights remain with the author)

inundation. Some free-floating species, such as the water hyacinth *Eichhornia crassipes*, can spread across the open water in response to variations in winds and currents. The phytoplankton are also highly diverse and can also be highly mobile and moved by winds and currents and vertical mixing.

Deep and ancient lakes that have undergone long periods of isolation can have high rates of endemism, as exemplified by the speciation of cichlids in the African Rift Valley lakes or amphipods in Lake Baikal in Russia. Moss (2010) reports that Lakes Victoria and Malawi may each contain about 700 fish species, mostly cichlids, whereas the bigger Lake Tanganyika only about 250 species. Some of the species in Lake Malawi have highly specialized feeding behavior. The endorheic Lake Chilwa near to Lake Malawi, which dries out frequently, only contains 12 species and all generalist feeders. The variability between lakes and the specialism that has occurred make these lakes important biodiversity foci.

Ecosystem Services

The importance of freshwater lakes for ecosystem services has received increased recognition in recent years, although the range of services may not have been treated equally (Finlayson and D'Cruz 2005; Vörösmarty et al. 2005). The provisioning services, especially the supply of freshwater for domestic supplies, irrigated agriculture, and hydropower, have been recognized with lakes and reservoirs being incredibly important. This includes reservoirs that have been specifically created for human purposes but often with adverse effects on other ecosystems, including riverine floodplains which themselves may contain freshwater lakes. Although not all of the water stored in lakes and reservoirs is directly available for consumption or use by people, their global storage illustrate the importance of these supplies among surface water stores. Other important ecosystem services include the supply of fish, including those from aquaculture, as well as the regulating services associated with the mediation of flooding and retention of sediments and nutrients and local climates.

Vörösmarty et al. (2005) provide a summary of freshwater usage globally but do not specifically separate information for lakes. However, the overall trends identified are seen as applicable, most notably increased extraction of freshwater for human consumption, including excessive levels from many ecosystems and across all inhabited continents. Further, a large proportion of all withdrawals comes from groundwater sources, mainly in response to the demand of water for irrigation. The importance of water for irrigation is considerable, including for agricultural land around the edges of many lakes with potential to impact their ecology if overabstraction occurs, such as has occurred in Lake Parishan, Iran (Lotfi 2018b), and Lake Turkana (Ojwang et al. 2018) as two examples among many. Irrigation schemes can be large and take large quantities of water or be much smaller and support small village communities (Fig. 7). Changes in the availability of freshwater also lead to changes in the flow regime and water balance in lakes, as well as changes in the transportation and deposition of sediments and chemical pollutants.

In some cases, severe modification or degradation of lake habitats can lead to loss of ecosystem characteristics supporting lake biodiversity, such as the disruption or fragmentation of migratory routes for various species accessing the lakes. Lakes support a large biodiversity with both resident and migratory species, such as the waterbirds that make use of the many habitats across permanent lakes as well as

Fig. 7 Water abstraction for irrigation purposes from a small reservoir in Zimbabwe (Photo credit: CM Finlayson © Rights remain with the author)

ephemeral systems. The importance of the latter is probably less known than that of permanent lakes given their variability and rapid changes in response to flooding and drying. The trade-offs between ecosystem services from lakes and reservoirs can be immense, with changes in fish or waterbird populations in individual lakes following changes to the water regime, such as from the construction of dams or diversions of water, being widely documented. Finlayson and D'Cruz (2005) provide an overview of such trade-offs in freshwater ecosystems. Ephemeral lakes can be particularly valuable for local populations, both through the supply of freshwater and fish and also through grazing or even cropping in the seasonally wet-dry margins or lake beds. The extent of inland fisheries is not well documented, but they are seen as immense and even critical for some communities for subsistence purposes, in spite of increasing commercial activity and widespread aquaculture.

Lake ecosystems also offer services: aesthetic, artistic, educational, cultural, and spiritual values and increasingly for recreation (Finlayson and D'Cruz 2005). The scenic attraction of many lakes is well known and has existed for centuries, often associated with spiritual and cultural uses. Recreational uses, both active and passive, are expected to increase in response to changing lifestyles in many countries.

Costanza et al. (2014) estimated that the value of ecosystem services from rivers and lakes in 2011 was $12,512 (2007$/ha/y), an increase of $785 from 1997. In comparison, the value for swamps/floodplains was $25,681 with a decrease of $1,340 from 1997. Despite regular discussion about valuation methods and the quality of the underlying data quality, lake ecosystems are extremely important for people in many parts of the world.

Threats and Future Challenges

Lake ecosystems in many regions are undergoing physical, chemical, and biological changes as a result of human activity and climate change. Lake water withdrawal or diversion combined with changing rainfall patterns has led to falling water levels in many regions. Examples such as the Aral Sea (Micklin 2007) and Lake Chad (Gao et al. 2011) have substantially decreased in area over the past decades. Increased loading of nutrients, particularly phosphorous, enhances productivity and can lead to eutrophic conditions and impaired water quality. Nutrient influx can shift the phytoplankton community to bloom-forming cyanobacteria whose decomposition can result in oxygen depletion and potential fish die-offs (Bennett et al. 2001).

Widespread introduction of invasive species has greatly contributed to the global homogenization of the fauna in lentic habitats (Rahel 2007). Individual invasive species can dramatically alter communities; for instance, the introduction of Nile perch to Lake Victoria is suspected to have contributed to the extinction of numerous native cichlids (Witte et al. 1992). Similarly, the zebra mussel invasion in the Laurentian Great Lakes region has fundamentally altered food webs and biogeochemistry (Strayer 2008). Further complicating the management of exotic species, food webs composed of nonnative species often support important recreational and commercial fisheries. There are also a number of plants that have invaded lakes, including the shrub *Mimosa pigra* that can establish and quickly spread thickets around lake edges and in marshy areas or the floating water hyacinth *Eichhornia crassipes* (Fig. 8).

Fig. 8 *Mimosa pigra* and *Eichhornia crassipes* can quickly spread in lake habitats (Photo credit: CM Finlayson © Rights remain with the author)

Climate change manifested through warming temperatures and more variable precipitations will have uneven and diverse consequences on the physical, chemical, and biological conditions of the world's lake ecosystems. Observation networks of surface temperatures reveal that lakes with the most rapidly warming trends are globally widely distributed, and warming trends exhibit spatial heterogeneity mediated by local characteristics of the lakes (O'Reilly et al. 2015). The duration of periods of ice cover has been reported to shorten over many lakes (Magnusson et al. 2000) and has a particular significance for high-latitude lakes losing perennial ice cover (Mueller et al. 2009). Warming surface conditions can lead to stronger thermal stratification of lakes and altered mixing regimes that reduce upwelling of nutrient-rich waters and affect planktonic activity (Carpenter et al. 2011). Less frequent mixing of oxygen-rich surface waters can lead to depletion of oxygen in the deeper hypolimnion portion of lakes which in turn can severely impact the cold-water fish depending on those thermal refugia (Ficke et al. 2007). Warmer waters are expected to favor blooming cyanobacteria (some toxic) and accelerate anthropogenic eutrophication despite increased nutrient sequestration from stronger stratification (Magnuson 2002; Pearl and Paul 2012). Changes in rainfall and inflows will also alter the hydrological balances that in many instances influence the distribution and growth of many plant and animal species, including migratory species that depend on established flow patterns to complete their life cycles. Permanently lowered lake levels could expose more shoreline, possibly deteriorating littoral (near-shore) zones, and isolate fringing wetlands (Poff et al. 2002). The exact impact of climate change will be determined by the ecological community in each lake and how the effects of climate interact with other lake stressors (Woodward et al. 2010). Due to the sensitivity and rapid response of some of their conditions, lakes are apt sentinels for climate change by offering a range of metrics for monitoring (Williamson et al. 2009; Adrian et al. 2009).

Land-use changes have had long-term impacts on large lakes and may in some cases be irreversible. Land-use change involving forestry and agriculture has resulted in soil erosion and mobilization of sediments and nutrients, resulting in eutrophication. These changes have been associated with increased human population pressure and include the clearing of riparian vegetation, the disruption of stream flow, and the decline of migratory fish species, including the disappearance of the Atlantic salmon *Salmo salar* from the Lake Ontario basin by 1900. Eutrophication has increased in many lakes, being more apparent in the shallower waters but extending to the deeper parts. It can be reduced where both point and diffuse sources are treated or reduced, but for many lakes, it is still occurring. The African Great Lakes have been variously affected by eutrophication, in particular Lake Victoria, where water clarity and dissolved oxygen have declined while cyanobacteria blooms have increased. Lake Baikal does not seem to have suffered the same fate, although there are concerns over nutrient and organic matter inputs. Pollution has become a major concern in many large lakes and was considered a factor in the decline of fisheries in the St Lawrence Great Lakes. More recently these inputs have been decreased, but problems are still occurring in the African Great Lakes and Lake Baikal. Overfishing is also widespread and is still a threat to the African Great Lakes which contain many endemic fish species. Biological invasion has had adverse impacts, including the introduction of the Nile perch *Lates niloticus* and tilapia species in Lake Victoria. The sea lamprey *Petromyzon marinus* and dreissenid mussels *Dreissena polymorpha* and *D. bugensis* have wrought major changes in the St Lawrence Great Lakes, as did the later introduction of euryhaline Ponto-Caspian species.

Large lakes may face a wide array of stressors that can have important cumulative impact. For instance, an assessment of the St Lawrence Great Lakes stressors shows that high-stress areas are driven by spatial concordance of different combinations of threats, highlighting the importance of targeted restoration planning (Allan et al. 2013). The ecological state of small lakes is particularly threatened by eutrophication, acidification, ultraviolet radiation, introduced species, toxic pollutants, water abstraction and transfer, and recreational activities (Moss et al. 2009). Most attention has focused on eutrophication and pollution, with point sources increasingly being addressed in many developed countries but with less success in the case of diffuse sources, such as those from agricultural activities.

The future of many lakes increasingly attracting attention, especially those that are used for supplying freshwater for domestic consumption and the production of food and increasingly those affected by the expansion of hydropower. As more water is directed toward these purposes, the ecological character and ecosystem services obtained from freshwater lakes are expected to change adversely (Finlayson and D'Cruz 2005; Vörösmarty et al. 2005).

References

Adrian R, O'Reilly CM, Zagarese H, Baines SB, Hessen DO, Keller W, Livingstone DM, Sommaruga R, Straile D, Van Donk E, Weyhenmeyer GA. Lakes as sentinels of climate change. Limnol Oceanogr. 2009;54(6):2283.

Allan JD, McIntyre PB, Smith SD, Halpern BS, Boyer GL, Buchsbaum A, Burton GA, Campbell LM, Chadderton WL, Ciborowski JJ, Doran PJ. Joint analysis of stressors and ecosystem services to enhance restoration effectiveness. Proc Natl Acad Sci. 2013;110(1):372–7.

Beeton AM. Large freshwater lakes: present state, trends, and future. Environ Conserv. 2002;29 (1):21–38.

Bennett EM, Carpenter SR, Caraco NF. Human impact on erodable phosphorus and eutrophication: a global perspective increasing accumulation of phosphorus in soil threatens rivers, lakes, and coastal oceans with eutrophication. BioScience. 2001;51(3):227–34.

Berga L, Buil JM, Bofill E, De Cea C, Manueco G, Polimon J, Soriano A, Yague J, editors. Dams and reservoirs, societies and environment in the 21st century. Proceedings of the International Symposium on Dams in Societies of the 21st Century; 18 June 2006; Barcelona. London: Taylor and Francis Group; 2006.

Capon SJ, Lynch JJ, Bond N, Bruce C, Chessman BC, Jenny D, Davison N, Finlayson CM, Gell PA, Hohnberg D, Humphrey C, Kingsford RT, Nielsen D, Thomson JR, Ward K, MacNally

R. Regime shifts, thresholds and multiple stable states in freshwater ecosystems; a critical appraisal of the evidence. Sci Total Environ. 2015;534:122–30.

Carpenter SR, Stanley EH, Vander Zanden MJ. State of the world's freshwater ecosystems: physical, chemical, and biological changes. Annu Rev Environ Resour. 2011;36:75–99.

Costanza R, de Groot R, Sutton P, van der Ploe S, Anderson SJ, Kubiszewski I, Farber S, Turner RK. Changes in the global value of ecosystem services. Glob Environ Chang. 2014;26:152–8.

Downing JA. Emerging global role of small lakes and ponds: little things mean a lot. Limnetica. 2010;29:9–24.

Downing JA, Prairie YT, Cole JJ, Duarte CM, Tranvik LJ, Striegl RG, McDowell WH, et al. The global abundance and size distribution of lakes, ponds, and impoundments. Limnol Oceanogr. 2006;51:2388–97.

Ficke AD, Myrick CA, Hansen LJ. Potential impacts of global climate change on freshwater fisheries. Rev Fish Biol Fish. 2007;17(4):581–613.

Finlayson CM, D'Cruz R. Inland water systems. In: Hassan R, Scholes R, Ash N, editors. Ecosystems and human well-being: current state and trends: findings of the condition and trends working group. Washington, DC: Island Press; 2005. p. 551–83.

Gao H, Bohn TJ, Podest E, McDonald KC, Lettenmaier DP. On the causes of the shrinking of Lake Chad. Environ Res Lett. 2011;6(3):034021.

Harris J. Poyang Lake, Yangtze River Basin, China. In: Finlayson CM, Milton GR, Prentice RC, Davidson NC, editors. The wetland book II: distribution, description, and conservation. Dordrecht: Springer; 2018.

Hutchinson G. A treatise on limnology: vol. I. Geography, physics and chemistry. New York: Wiley; 1957.

Hutchinson GE. A treatise on limnology: vol III limnological botany. New York: Wiley; 1975.

Lehner B, Döll P. Development and validation of a global database of lakes, reservoirs and wetlands. J Hydrol. 2004;296:1–22.

Lehner B, Liermann CR, Revenga C, Vörösmarty C, Fekete B, Crouzet P, Döll P, et al. High-resolution mapping of the world's reservoirs and dams for sustainable river-flow management. Front Ecol Environ. 2011;9:494–502.

Lewis Jr WM. Global primary production of lakes: 19th Baldi Memorial Lecture. Inland Waters. 2011;1(1):1–28.

Liu J, Zhang Q, Xu C, Zhang Z. Characteristics of runoff variation of Poyang Lake watershed in the past 50 years. Trop Geogr. 2009;29(3):213–8.

Lotfi A. Lake Uromiyeh and its satellite wetlands, Iran. In: Finlayson CM, Milton GR, Prentice RC, Davidson NC, editors. The wetland book II: distribution, description, and conservation. Dordrecht: Springer; 2018a.

Lotfi A. Lake Parishan, Iran. In: Finlayson CM, Milton GR, Prentice RC, Davidson NC, editors. The wetland book II: distribution, description, and conservation. Dordrecht: Springer; 2018b.

Magnuson JJ. Future of adapting to climate change and variability. In: McGinn NA, editor. Fisheries in a changing climate. Bethesda: American Fisheries Society; 2002. p. 283–7.

Magnuson JJ, Robertson DM, Benson BJ, Wynne RH, Livingstone DM, Arai T, Assel RA, Barry RG, Card V, Kuusisto E, Granin NG. Historical trends in lake and river ice cover in the northern hemisphere. Science. 2000;289(5485):1743–6.

McDonald CP, Rover JA, Stets EG, Striegl RG. The regional abundance and size distribution of lakes and reservoirs in the United States and implications for estimates of global lake extent. Limnol Oceanogr. 2012;57(2):597–606.

Messager ML, Lehner B, Grill G, Nedeva I, Schmitt O. Estimating the volume and age of water stored in global lakes using a geo-statistical approach. Nature Communications, 2016;7. doi: 10.1038/ncomms13603.

Micklin P. The Aral Sea disaster. Annu Rev Earth Planet Sci. 2007;35:47–72.

Moss B. Ecology of freshwaters: a view for the twenty-first century. Oxford: Wiley-Blackwell; 2010.

Moss B, Hering D, Green AJ, Aidoud A, Becares E, Beklioglu M, Bennion H, Boix D, Brucet S, Carvalho L, Clement B, et al. Climate change and the future of freshwater biodiversity in Europe: a primer for policy-makers. Freshwat Rev. 2009;2(2):103–30.

Mueller DR, Van Hove P, Antoniades D, Jeffries MO, Vincent WF. High Arctic lakes as sentinel ecosystems: cascading regime shifts in climate, ice cover, and mixing. Limnol Oceanogr. 2009;54(6 Part 2):2371–85.

O'Reilly CM, Sharma S, Gray DK, Hampton SE, Read JS, Rowley RJ, Schneider P, Lenters JD, McIntyre PB, Kraemer BM, Weyhenmeyer GA, et al. Rapid and highly variable warming of lake surface waters around the globe. Geophys Res Lett. 2015;42(24):10,773–81.

Ojwang WO, Obiero KO, Donde OO, Gownaris N, Pikitch EK, Omondi R, Agembe S, Malala J, Avery ST. Lake Turkana, the world's largest permanent desert lake. In: Finlayson CM, Milton GR, Prentice RC, Davidson NC, editors. The wetland book II: distribution, description, and conservation. Dordrecht: Springer; 2018.

Paerl HW, Paul VJ. Climate change: links to global expansion of harmful cyanobacteria. Water Res. 2012;46(5):1349–63.

Poff NL, Brinson MM, Day JW. Aquatic ecosystems and global climate change. Arlington: Pew Center on Global Climate Change; 2002. p. 44.

Rahel FJ. Biogeographic barriers, connectivity and homogenization of freshwater faunas: it's a small world after all. Freshw Biol. 2007;52(4):696–710.

Raymond PA, Hartmann J, Lauerwald R, Sobek S, McDonald C, Hoover M, Butman D, Striegl R, Mayorga E, Humborg C, Kortelainen P. Global carbon dioxide emissions from inland waters. Nature. 2013;503(7476):355–9.

Rosenberg DM, Berkes F, Bodaly RA, Hecky RE, Kelly CA, Rudd JW. Large-scale impacts of hydroelectric development. Environ Rev. 1997;5(1):27–54.

Ryanzhin SV, Subetto DA, Kochkov NV, Akhmetova NS, Veinmeister NV. Polar lakes of the world: current data and status of investigations. Water Resour. 2010;37(4):427–46.

Scheffer M, Carpenter SR. Catastrophic regime shifts in ecosystems: linking theory to observation. Trends Ecol Evol. 2003;18(12):648–56.

Seekell DA, Pace ML. Does the Pareto distribution adequately describe the size-distribution of lakes? Limnol Oceanogr. 2011;56(1):350–6.

Sobek S, Nisell J, Fölster J. Predicting the volume and depth of lakes from map-derived parameters. Inland Waters. 2011;1(3):177–84.

Strayer DL. Twenty years of zebra mussels: lessons from the mollusk that made headlines. Front Ecol Environ. 2008;7(3):135–41.

Verpoorter C, Kutser T, Seekell DA, Tranvik LJ. A global inventory of lakes based on high-resolution satellite imagery. Geophys Res Lett. 2014;41(18):6396–402.

Vörösmarty CJ, Leveque C, Revenga C. Fresh water. In: Hassan R, Scholes R, Ash N, editors. Ecosystems and human well-being: current state and trends: findings of the condition and trends working group. Washington, DC: Island Press; 2005. p. 167–207.

Wetzel RG. Limnology: lake and river ecosystems. Gulf Professional Publishing; 2001.

Wetzel RG, Likens GE. Limnological analyses. New York: Springer; 1991.

Williamson CE, Saros JE, Vincent WF, Smold JP. Lakes and reservoirs as sentinels, integrators, and regulators of climate change. Limnol Oceanogr. 2009;54(6 Part 2):2273–82.

Winslow LA, Read JS, Hanson PC, Stanley EH. Does lake size matter? Combining morphology and process modeling to examine the contribution of lake classes to population-scale processes. Inland Waters. 2015;5(1):7–14.

Witte F, Goldschmidt T, Wanink J, van Oijen M, Goudswaard K, Witte-Maas E, Bouton N. The destruction of an endemic species flock: quantitative data on the decline of the haplochromine cichlids of Lake Victoria. Environ Biol Fishes. 1992;34(1):1–28.

Woodward G, Perkins DM, Brown LE. Climate change and freshwater ecosystems: impacts across multiple levels of organization. Philos Trans R Soc Lond B Biol Sci. 2010;365(1549):2093–106.

Zarfl C, Lumsdon AE, Berlekamp J, Tydecks L, Tockner K. A global boom in hydropower dam construction. Aquat Sci. 2015;77(1):161–70.

Salt Lakes

9

C. Max Finlayson

Contents

Introduction	144
Global Distribution	145
Ecological Features	148
Biota	149
Threats and Future Challenges	151
Ecosystem Services	153
References	153

Abstract

Salt lakes are widespread and are found under a range of conditions, including cold and hot temperatures. They are, however, mostly confined to semi-arid to arid regions where evaporation exceeds precipitation. They are generally permanent or temporary bodies of water with salinities greater than 3 g L^{-1} and lacking any recent connection to the marine environment. The salinity level used to demarcate them from freshwater lakes is somewhat arbitrary, given the large variation in salinity that can occur in many lakes. Many salt lakes can be dry seasonally or for longer periods and exhibit dry saline lake beds, and even hypersaline conditions with salt crusts across the surface. Salt lakes support biota that have physiological and biochemical mechanisms that enable them to tolerate high salt levels, and are highly sensitive to even small changes in the climate. The most conspicuous invertebrate animals are crustaceans, although a variety of other non-crustacean groups also occur. The vertebrates of saline lakes comprise mainly fish and birds. Many salt lakes have

---

C. M. Finlayson (✉)
Institute for Land, Water and Society, Charles Sturt University, Albury, NSW, Australia

UNESCO-IHE, The Institute for Water Education, Delft, The Netherlands
e-mail: mfinlayson@csu.edu.au

© Springer Science+Business Media B.V., part of Springer Nature 2018
C. M. Finlayson et al. (eds.), *The Wetland Book*,
https://doi.org/10.1007/978-94-007-4001-3_255

been degraded as a consequence of human activities especially from the construction of dams and diversion of surface inflows, increased salinization, and other catchment activities, such as mining, pollution, the introduction of 205 exotic species, and human-induced climate change.

Keywords

Saline lakes · Ephemeral lakes · Crustaceans · Degradation

Introduction

Salt lakes are widespread and occur on all continents and are found under a range of conditions, including cold and hot temperatures. They are, however, mostly confined to semi-arid to arid regions where evaporation exceeds precipitation (Williams 1998). They are generally defined as "permanent or temporary bodies of water with salinities greater than 3 g L^{-1} and lacking any recent connection to the marine environment." The salinity level used to demarcate them from freshwater lakes is somewhat arbitrary, especially given the large variation in salinity that can occur in many lakes in response to the occurrence or absence of precipitation and runoff into the lakes (Williams 2002). Many salt lakes can be dry seasonally or for longer periods and exhibit dry saline lake beds (Fig. 1), and even hypersaline conditions (Fig. 2) with salt crusts across the surface, such as many of those found in southwestern Australia.

Fig. 1 Lake Bahi in Tanzania is shallow with large areas of exposed saline flats (Photograph L-M Rebelo © Rights remain with author)

Fig. 2 Hypersaline lakes with thick crusts of salt are common in south-western Australia (Photograph CM Finlayson © Rights remain with author)

The global distribution, status and trends, biodiversity and ecosystem services, and threats and management issues for salt lakes are described based largely on information contained in review articles by Williams (1998, 2002) and Jellison et al. (2008) and references therein. There are many terms used to describe particular types of salt lakes, such as saline playa lake, saline pan, salina, sabka, liman, and soda, alkali and bitter lakes, based on morphometry, chemical composition, location, or region. To avoid confusion with the specific features of individual types of lakes the generic term of "salt lake" is used while recognizing that this covers a large range of aquatic ecosystems that occur in many biomes in all continents and exhibit a large range of conditions.

Global Distribution

Salt lakes are geographically widely distributed (Fig. 3) and include many large lakes, such as the largest lake in the world, the Caspian Sea; lakes at high altitudes, such as those found on the Tibetan Plateau and the Altiplano (high plain) of South America; and the lowest lake in the world, the Dead Sea. They vary from large lakes that never dry, such as the Caspian Sea, Lake Turkana in eastern Africa and Mono Lake (USA), although their water levels may fluctuate over long time periods, to those that only fill episodically, such as Lake Eyre (Australia), and are otherwise comprised of extensive dry salt pans, to many small lakes that fill at irregular intervals, or that fill annually after rainfall. In semiarid regions they may lack surface water in the dry season, but fill annually after rainfall. The water levels of many have

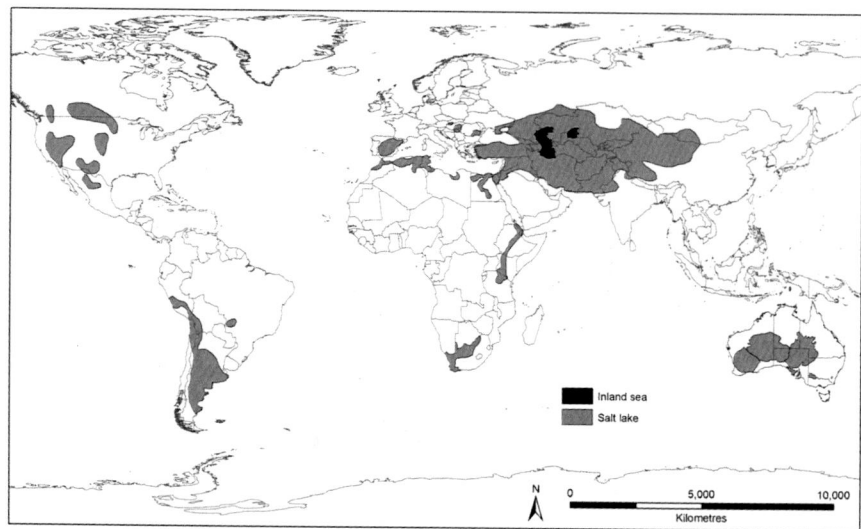

Fig. 3 Generalized distribution of salt lakes (Adapted from Williams 2002)

been drawn down by water diversions, largely for irrigated agriculture, with the Aral Sea being a widely known and disastrous example (MEA 2005). The Caspian Sea has also undergone fluctuations in water levels over the last century, attributed to short-term variability in rainfall in the catchment of the Volga River, as well as a consequence of the construction of large dams along the River (Finlayson et al. 1993).

Salt lakes generally form the terminus of a closed (endorheic) drainage system, with the balance between hydrological inputs (surface and groundwater flows, and precipitation over the lake) and outputs (evaporation and seepage) allowing the permanent or temporary persistence of a body of water, conditions that most often occur in arid and semiarid regions. As the inflows contain large concentrations of salts, these accumulate in the lake as the water evaporates. Many in eastern Africa in particular are very alkaline and known as soda lakes.

They occur under a range of temperatures, from very hot to cold, and are most abundant in semi-arid (200–500 mm precipitation per year) and arid (25–200 mm) regions that comprise approximately one-third of the Earth's land surface. Evaporation is high and generally exceeds rainfall. They are more episodic or ephemeral with increasing annual evaporation (0 cm to >200 cm); the greater the net evaporation, the more concentrated the salt. Long-term changes in climate have had major impacts on many salt lakes with palaeo-limnological studies showing fluctuations in lake volume and changes in salinity and ecology, over long time periods. Changes in the climate over annual or within-year time periods also have a considerable impact on salt lakes, notably the globally important circulation pattern involving the ocean and the atmosphere, the El Niño-Southern Oscillation (ENSO) that affects the climate of many arid and semi-arid regions. Any change to the length, intensity, and frequency of ENSO can have a profound effect on the climates of arid and semi-arid regions and the salt lakes within them.

Locally, salt lakes can dominate the landscape, whether on the basis of their size, or by the occurrence of large numbers of small lakes, many of which may be dry for long periods and exhibit hypersaline conditions. The global volume of salt lakes has been estimated as 85,000 km^3 compared to 105,000 km^3 for freshwater lakes, although 75% of this volume is contributed by the Caspian Sea. As the latter distorts such comparisons, a better view of the importance of salt lakes comes from considering them within individual catchments or landscapes and how they interact with elements in that landscape.

The distribution of salt lakes on a continental basis is described below, based on information provided by Williams (1998):

North America – salt lakes are common in six distinct regions: the Great Plains, North-western plateau, Great Basin, mid-continental, South-western, and Chihuahuan region of northern Mexico.

South America – salt lakes occur in two main regions, the Altiplano and its northern Peruvian extension, and the pampas. Salt ponds are numerous along the coast of Venezuela and on offshore islands.

Europe – salt lakes and solar ponds occur in southern and south-eastern Europe, with its warmer and drier climate. Saline water bodies in the west and north-west result from saline springs, salt mining, or have marine associations.

Africa – salt lakes are widespread through the countries bordering the Mediterranean Sea, are more scattered across the Sahara and in eastern Africa occur mostly in the Great Rift and the Western Rift Valleys.

Asia – salt lakes occupy an extensive region stretching from Turkey to China. In China, five regions may be identified where salt lakes are common: Inner Mongolia, Qaidam Basin, north Qinghai, Qinghai-Tibetan plateau, and Sinkiang.

Australia – salt lakes include permanent and intermittent lakes in the south-east and south-west, with episodic salt lakes in the arid interior.

Antarctica – mostly located along the east coast with a few being intermittent and dry in summer. The permanent lakes range from being continuously frozen to never freezing because of their high salinities.

General information on the distribution of salt lakes has been summarized in Williams (1998, 2002). Many specific studies of individual lakes have occurred, including those for some of the larger lakes, such as the Great Salt Lake and Mono Lake (USA), although a comprehensive global inventory with detailed documentation of the ecological features is not available. The destruction of the Aral Sea is one case where the demise of a large and productive system has been widely publicized and presented in many global assessments and overviews (MEA 2005).

Ecological Features

The salinity of salt lakes varies greatly among lakes and temporally (Williams 2002). For many large permanent salt lakes, the salinity may not vary greatly over long periods of time, as shown by the Caspian Sea with a salinity around 12 g L^{-1}, and prior to being degraded by human activities in recent decades, the Aral Sea had a salinity over many decades of 10 g L^{-1} and the Dead Sea 200 g L^{-1}. Episodically filled large lakes generally have salinities in the range of less than 50 to more than 300 g L^{-1} as they dry after having filled. Small lakes that are filled and dry seasonally may have equally variable salinities within an annual cycle. The major ions in salt lakes are generally the monovalent sodium (Na^+) and chloride (Cl^-), although the divalent ions, magnesium (Mg^{2+}), calcium (Ca^{2+}), and sulfate (SO_4^{2-}), may be dominant in some. In fresh water lakes in contrast, the monovalent bicarbonate (HCO_3^-) and divalent carbonate (CO_3^{2-}) are important.

Salt lakes support biota that is extremely ancient in an evolutionary sense (stromatolites and thrombolites, for example, Fig. 4), have physiological and biochemical mechanisms that enable them to tolerate high salt levels, and are highly sensitive to even small changes in the climate. The biota in moderately saline lakes (less than 10 g L^{-1}) comprises salt tolerant taxa of freshwaters, but as the salinity increases these are replaced by those found only in salt lakes. In all salt lakes the biota need to have adaptations to osmotic stress, whereas those in temporary lakes also need to have adaptations to survive desiccation. The biota tends to decrease with increasing salinity, although there are many exceptions. The long-held view that salt lake fauna were cosmopolitan is not now supported; there is considerable fauna endemism. In addition

Fig. 4 Thrombolites in a salt lake in south-western Australia (Photograph M Beemster © Rights remain with author)

to salinity determining the distribution and abundance of the biota, the decreased solubility of oxygen in water with increasing salinity may also be a factor.

Biota

Many genera of Cyanobacteria are found in salt lakes, and often play a key role in energy and biogeochemical cycles. They occur in microbial mats and in the water column. Few are as tolerant to salinity as the halobacteria that also occur in salt lakes, but some can tolerate salinities of more than 100 g L^{-1}. Stromatolites and thrombolites, which are confined to coastal marine waters and salt lakes, comprise an assemblage of bacterial microbial mats and sediment, and are among the oldest known fossils, some being about 3,000 Ma old. All major groups of true algae found in fresh waters also occur in salt lakes with microscopic forms in the water column and in algal mats on the bottom of lakes. Most occur in lakes with salinities ranging from 3 to 20 g L^{-1}, with relatively few in hypersaline lakes (>50 g L^{-1}). The most tolerant is *Dunaliella salina* which is often the only photosynthetic organism in highly saline lakes.

The diversity of macrophytes found either partially or totally immersed in salt lakes is not high. The widespread common reed, *Phragmites australis*, is found around the margins of many lakes and the submerged species, *Potamogeton pectinatus,* in shallow water, but not generally in hypersaline conditions. More species are found around the edges of the less saline lakes. In Australia, species of *Lepilaena* and *Ruppia* can occur in totally submerged conditions with salinities well above 100 g L^{-1}. The North American emergent species, *Scirpus maritimus*, has also been found in hypersaline lakes. Many salt tolerant angiosperms, particularly from the Chenopodiaceae, are found growing beyond the edge of salt lakes. Also on saline soils bordering salt lakes are some trees whose roots are clearly able to survive in saline conditions.

The most conspicuous invertebrate animals are crustaceans, although a variety of other non-crustacean groups also occur. There are many Protozoa, particularly ciliates, but including also other sorts of protozoan, and the Rotifera, are frequently encountered. One of the most common and cosmopolitan species of salt lakes is the rotifer *Brachionus plicatilis*. Of crustaceans, by far the best known is *Artemia*, the brine shrimp. Other crustaceans, though less well known, may be equally important in some salt lakes. They include representatives of all the free-living copepod groups, cladocerans, ostracods, and amphipods. Several groups of insects occur. *Ephydra*, the brine-fly, is frequently present in certain regions, while other dipterans may be common in particular lakes. Non-dipteran insect groups are well represented in lakes of moderate salinity by species of corixid and beetles, although the mayflies and stoneflies are less common.

The vertebrates of saline lakes comprise mainly fish and birds. Amphibians, reptiles, and mammals are not common except for a few species in moderately saline lakes. The Caspian seal is an example. A large number of fish species has been recorded from permanent salt lakes with salinities less than 50 g L^{-1}. Most of these

Fig. 5 Flamingos in Lake Nakuru, Kenya (Photograph CM Finlayson © Rights remain with author)

also occur in freshwater lakes. A few of these fish can tolerate salinities as high as 100 g L^{-1}, but the number of species actually confined to inland salt waters seems to be limited. *Oreochromis alcalicus grahami*, the soda lake fish of East Africa, is one. Many fish are also found in temporary salt lakes, but as they cannot withstand desiccation, they originate from streams and rivers that have filled the lake.

The best known birds associated with, and indeed confined to, salt lakes are flamingos (*Phoenicopterus*) with five species found in central and South America, Asia, Africa, and southern Europe (Fig. 5). These species feed on either Cyanobacteria, such as *Spirulina* and other phytoplankton, or small invertebrates. A number of other birds, including many long-distance migratory species, are also closely associated with salt lakes. A few of these, like the flamingos, are only rarely found elsewhere (e.g., the banded stilt *Cladorhynchus leucocephalus*), but most are also to be found at freshwater localities and use salt lakes in a facultative way for feeding, nesting, and refuge. Nonetheless, salt lakes are often critically important for supporting viable populations of these species. Mono Lake in California, for example, plays a critical role in the survival of several bird species with the eared grebe *Podiceps nigricollis*, Wilson's phalarope *Phalaropus tricolor*, and California gulls *Larus californicus* more or less dependent on the seasonally abundant invertebrates that occur in this lake. Many other species feed, nest, or find refuge at Mono Lake, a situation repeated across the many flyways used by these species. Many other lakes, such as Mar Chiquita, Argentina, and Lake Turkana, Kenya, seasonally support large numbers of migratory shorebirds, and lakes in inland Australia periodically support large numbers of waterbirds after the irregular flooding that occurs after episodic rain events, at times at immense distances upstream of the lakes.

Threats and Future Challenges

Many salt lakes have been degraded as a consequence of human activities especially from the construction of dams and diversion of surface inflows, increased salinization, and other catchment activities, such as mining, pollution, the introduction of exotic species, and human-induced climate change (Williams 2002). These activities have resulted in adverse changes to the ecological character of lakes, in particular the water regime and water quality, and the decline or even loss of biota.

The most important activity causing adverse change in large permanent salt lakes is the construction of dams and the diversion of freshwater inflows away from the lakes to support irrigated agricultural and other human needs. Diversions invariably cause a rapid decrease in lake level and volume, and an increase in salinity (Williams 2002). The most dramatic impacts have been seen on the Aral Sea in Central Asia – before 1960 the annual volume of inflows from the main rivers was 56 km^3 which was reduced to 4.2 km^3 in 1990 with a >15 m fall in water level. Other examples reported by Williams (2002) include Mono Lake, USA, where water levels have fallen 15 m since 1920; Pyramid Lake, Nevada, USA, about 21 m since 1910; the Dead Sea, Israel/Jordan, 8 m since 1980; Qinghai Hu, China, about 10 m since 1908; and Lake Corangamite, Australia, 3 m since the 1960s. Many other salt lakes have been adversely affected by dams and diversions, and many more are threatened, including the large Lake Turkana in eastern Africa which gets most of its surface water from a single river, as the pressure on water supplies and development increases. Similarly, the sea level in the Caspian Sea has been affected by changes in flows along the Volga River (Finlayson et al. 1993).

The salinity of many of these lakes has also increased from 10 to more than 80 g L^{-1} in the Aral Sea, from 48 to about 90 g L^{-1} in Mono Lake, from 3.75 to more than 5.5 g L^{-1} in Pyramid Lake, from 200 to 340 g L^{-1} in the Dead Sea, from 5.6 to 12 g L^{-1} in Qinghai Hu, and from 35 to around 50 g L^{-1} in Lake Corangamite. The effect of such changes depends in part on the original salinity, being greatest when this was low, and least when it was hypersaline. As an example, the 20 g L^{-1} increase in Lake Corangamite led to the almost complete disappearance of fish, amphipods, snails, and *Ruppia*, with consequent effects on birds that used the lake. In contrast the more than 100 g L^{-1} increase in the Dead Sea had little effect on the biota. The impacts in the Aral Sea have been near catastrophic with free-living macro-invertebrate species declining from almost 200 to less than 30, and fish from 34 to 5, as well as the loss of emergent vegetation in coastal communities. There has also been a loss of formerly important breeding and stopover habitat for migratory waterbirds in the deltas of the main rivers. Efforts to restore the water regime of the Aral Sea have been proposed, although these have been insufficient and are unlikely to succeed while irrigation continues and even expands.

The effects of falling water levels in salt lakes may include changes to the local climate, additional dust blown from exposed lake beds, falling groundwater levels, and the loss of islands, and consequently other effects. The changes in the

Aral Sea include the shorelines retreating some 100–150 km and exposing some 45,000 km^2 of former seabed and creating a salty desert that generates an estimated 100 million tonnes of salt-laden dust, and reducing the relative humidity from 40% to 30% (MEA 2005). Ultimately, the diversion of inflows can lead to the complete loss of the lake, as in Winnemucca Lake in Nevada and Owens Lake in California, or be so severe as to make restoration prohibitively expensive or near impossible.

The salinity of salt lakes can also increase as a consequence of secondary salinization after the widespread clearance of the natural vegetation and other land-use changes within catchments. This results in salt moving toward the surface as the water table rises following a decrease in the amount of underground water transpired by deep-rooted plants. Evaporation can then concentrate the salt on the surface where it may be leached and added to the inflows to lakes. The effects are not confined to salt lakes – freshwater lakes are also susceptible. As secondary salinization can affect large areas of land in semiarid and arid regions, its effects can be more extensive than those from flow diversions, resulting in changes in the natural character of many lakes and the replacement of less salt tolerant species with those that are more tolerant. The extent of secondary salinization has been widely investigated in Australia, but also occurs in other countries where similar landscape changes have occurred, largely in responses to demands for land for grazing and agriculture.

Despite salt lakes being the termini of closed hydrological systems, this has not prevented the discharge of a wide variety of pollutants to rivers flowing into them or to the lakes directly. This occurs with the realization that such substances will accumulate in the terminal systems with well-known adverse outcomes. Williams (2002) reports that almost the whole range of pollutants discharged to fresh waters is also discharged to salt lakes or their inflows. This includes sewerage discharges and agricultural runoff, and domestic and industrial garbage. Mining near salt lakes has resulted in pollution by metals, with examples occurring in Bolivia, and Lake Maryut, Egypt, which has high concentrations of tin in its sediments. High concentrations of organochloride residues are found in some Rift Valley salt lakes in Kenya. The effect of these pollutants on salt lakes is assumed to be much the same as on freshwater lakes, although the interactions with salinity and desiccation have not often been specifically determined.

In many cases fish have been introduced into moderately saline lake, such as those in parts of Canada, Bolivia, and Australia, although not all attempts have been successful. In some cases the fish populations became self-sustaining and of commercial value, such as in the Aral Sea where at least 21 species were introduced, mostly from the Caspian, Baltic, and Azov seas, and Chinese lakes, although they have now been decimated or eliminated by rising salinity. The Caspian also had many species introduced. In Australia most introduced fish cannot breed in the salt lakes but are maintained by stocking. Similarly, many invertebrates have been introduced to moderately saline lakes, such as the Aral and Caspian Seas. The latter occurred mainly after the opening of the Volga-Don canal in 1954 with some populations acclimatizing with the ctenophore *Mnemiopsis leidyi* and the coelenterate *Aurelia aurita* considered likely to have significant impacts on fish populations.

The impact of climate change on salt lakes could be significant given that their ecological character is related to the balance between evaporation, rainfall, and temperature. Relatively small changes in these parameters could cause large changes to the ecological character of salt lakes. This is likely to result in pronounced changes in inland salt lakes in semiarid basins with high evaporation and usage by people, such as Lake Urumiyeh (Urmia) and others in Central Asia. Increased aridity is expected to lead to increased salinity and ultimately to desiccation. Increased rainfall and freshwater runoff from glaciers would lead to decreased average salinities and in extreme cases to the conversion of closed to open drainage systems. Rising sea levels is likely to flood many coastal salt lakes, such as those occurring on small oceanic islands.

Ecosystem Services

The benefits provided to people by salt lakes through ecosystem services include the provision of salt (sodium chloride) from shallow solar ponds after evaporation of sea water, salt harvesting from inland lakes, or from evaporation of saline groundwater pumped to the surface. In addition, Cyanobacteria from saline lakes, in particular *Spirulina* species, have attracted attention as a source of proteins and amino acids. The green alga *Duniella* is also of interest as a source of glycerol, beta-carotene, and protein. *Artemia* larvae are also used for aquaculture to raise fish, crustaceans, and mollusks. Moderately saline lakes are important for fisheries, especially for local people, such as those from many African lakes, and from a conservation viewpoint their importance for migratory waterbirds is extremely high. Many salt lakes in many parts of the world have significant aesthetical and cultural values, as well as recreational values, including those associated with tourism to the Rift Valley soda lakes in Africa, the high altitude lakes in South America, and the large usually dry inland lakes in Australia. In some cases these values are associated with the wildlife habitats the salt lakes provide, such as those for large numbers of migratory waterbirds. Williams (2002) also points to the intrinsic scientific values that these lakes provide given the many adaptations that their biota have to high salinity. The benefits that people derive from these services have largely not been quantified, but are likely to be under threat given the generally declining status of many salt lakes.

References

Finlayson CM, Chuikow YS, Prentice, RC, Fischer W, editors. Biogeography of the Lower Volga, Russia: an overview. IWRB Special Publication No 28, Slimbridge; 1993.

Jellison R, Williams WD, Timms B, Alcocer J. Salt Lakes: values, threats and future. In: Polunin NVC, editor. Aquatic ecosystems: trends and global prospects. Cambridge: Cambridge University Press; 2008. p. 94–110.

MEA (Millennium Ecosystem Assessment). Ecosystems and human well-being: wetlands and water synthesis. Washington, DC: World Resources Institute; 2005.

Williams WD. Management of inland saline waters. Guidelines of Lake Management No 6. Shiga: International Lake Environment Committee and United Nations Environment Programme; 1998.

Williams WD. Environmental threats to salt lakes and the likely status of inland saline ecosystems in 2025. Environ Conserv. 2002;29:154–67.

Tidal Freshwater Wetlands: The Fresh Dimension of the Estuary

10

Aat Barendregt

Contents

Introduction	156
The Fresh Part of the Estuary Within the Gradient from River to Sea	157
The Gradient Water–Land Within the Tidal Freshwater Wetlands	158
Distribution and Human Impact	160
Distribution	160
Human Impacts	161
Physical and Chemical Processes	162
Flora	163
Fauna	165
Threats and Future Challenges	166
Global Warming and Sea Level Rise	167
References	167

Abstract

In the upper reaches of estuaries where rivers end, the constant input from the river creates permanent fresh water conditions even though tidal energy is still present. The physical, chemical and biological conditions differ from the brackish part of the tidal area due to the interaction of the tidal wave with the river ecosystem. Just like in the brackish estuary, tidal freshwater wetlands exhibit gradient variation from tidal flats to higher elevated forests. Chemical processes and accretion are prominent, creating wetlands with high turnover. Many tidal freshwater wetlands have been historically reclaimed for human use as cities, ports, and for agriculture. A diverse flora and fauna is represented and biomass in tidal freshwater wetlands is

A. Barendregt (✉)
Environmental Sciences, Copernicus Institute, Utrecht University, Utrecht, The Netherlands
e-mail: a.barendregt1@uu.nl

mostly very high. Because this system is at the interface of salt – fresh conditions, tidal freshwater wetlands are particularly susceptible to sea level rise and global warming.

Keywords

Salinity · Hydrology · River · Vegetation · Zonation · Marsh · Fauna · Fish · Insect · Bird · Mammal · Accretion · Nitrification · Methanogenesis · Sea level rise · Restoration

Introduction

Many papers have described the physical, chemical, and biological conditions in brackish estuaries including the salt marshes. However, most estuaries have another dimension rarely examined where a river with a continuous input of freshwater dominates the conditions and processes in the wetlands. The tidal pulse is equally prominent in this fresh section but most processes and all species are representative of this nonsaline dimension. The brackish section in the estuary is merely the transition between the fresh and salt parts in the same tidal system; many estuarine scientists have neglected the freshwater dimension, and it remains unfamiliar and underrepresented in the scientific literature.

Tidal freshwater wetlands (TFWs) are available for scientific research close to many principal cities. The fresh dimension of the estuary was historically where ships from the sea or ocean made port and where inland river navigation ended and thus the perfect location to start harbors and concomitant urbanization. Principal cities and rivers are linked worldwide to these wetlands. Portland (Columbia), San Francisco (Sacramento), New York (Hudson), Philadelphia (Delaware), Washington DC (Potomac), New Orleans (Mississippi), Hamburg (Elbe), Rotterdam (Rhine), Antwerp (Scheldt), Bordeaux (Gironde), and London (Thames) are examples of cities in the middle of tidal freshwater wetlands. Many wetlands were drained, infilled, and reshaped for economic reasons without paying attention to the ecological values in this ecosystem.

Inattention to the importance of an ecosystem by both the scientific and economic disciplines can result in its loss worldwide. Many tidal freshwater wetlands have been lost to human infrastructure and use, but fine examples are still available to researchers to describe the characteristics of this system. The following text provides a summary of our understanding (mostly from the USA and Europe) of aspects of freshwater tidal wetland ecosystem function and characteristics, known distribution, basic physical and chemical processes, represented flora and fauna, and the future outlook especially with sea level rise. The first descriptions of TFWs with global distribution, biodiversity, and ecosystem functions are by Simpson et al. (1983) and Odum et al. (1984). Recent full reviews are by Barendregt et al. (2009), Conner et al. (2007), and Barendregt and Swarth (2013) and references therein.

The Fresh Part of the Estuary Within the Gradient from River to Sea

The estuary has been classically subdivided to include three parts: the salt section with saline water (polyhaline), the brackish section (mesohaline), and the fresh section with permanently fresh water. Between the fresh and real brackish sections there is, however, an intermediate section with reduced salinity (oligohaline). The energy from the tidal wave that flows upstream and not the water itself is responsible for the mixing of the fresh river and saline sea water to become brackish. This mixing is only partly due to horizontal current flow; there is a vertical mixing processes in the zone where the bidirectional flow of heavier saline tidal sea water meets the overlaying discharging freshwater.

Freshwater hydrology dominates the components in tidal freshwater wetlands, even more so than in other wetlands. Current and tidal water are always present in the aquatic parts, including the sediments and nutrients that are transported with the water. In the semiterrestrial marshes and swamps, this water floods very regularly, causing permanent wet soil conditions irrespective of season. Although sea level induces a fixed water level in the estuary, tidal amplitude creates important daily fluctuations within the limits of high and low tide level, influenced in a 4-week period from the constellation of the moon. Occasionally storm tides elevate sea level and can create irregular differences up to some meters. The effect of fluctuating river water input to water levels in the estuary is mostly buffered by the fixed sea level. The river ecosystem itself is also characterized by water level fluctuations, but the time dimension is different. Periods with high river levels will cause river forelands to be permanently flooded for weeks or months but during a dry period the same areas will be drained for weeks or months. In the latter situation, the groundwater table in terrestrial parts will become subsurface, with soil aeration occurring as a next step.

Tidal freshwater wetlands survive due to the continuous input of fresh river water. When this input diminishes, brackish water intrudes into the fresh system and will with time replace the characteristic freshwater plants and animals with salt-tolerant species. The variation in salinity appears to be the major environmental factor for this change rather than the salinity concentration itself. In the gradient fresh saline, the species diversity is highest in freshwater, diminishes in the brackish section, and increases again with higher salinity (McLusky and Elliot 2004). Input of salinity changes the chemical processes within the wetland, e.g., the reduction of organic matter under fresh conditions takes place with methanogenesis, whereas in brackish conditions sulfate reduction is the basic process (Weston et al. 2011).

Tidal freshwater wetlands thus occur over a limited extent of an estuary. The diurnal tidal wave with permanently high water levels differentiates them from the rivers; the permanent presence of freshwater differentiates them from the other parts of the estuary. Two extra elements have to be added to indicate the special position of these wetlands: the sediments and the nutrients. The rivers transport all residuals from overland flow and human activities to the sea, including clay particles, organic components, the natural and human waste, and the free nutrients. The tidal freshwater wetlands are the section where many suspended particles flocculate and produce mud. It can be the area where huge sedimentation occurs; at least it is muddy. Moreover, in this system the nutrients are never limiting. It creates conditions in the fresh section where water, nutrients, and minerals are always present, perfect for biomass production. Within the range of wetlands, this series of vegetation produces high values up to 2 or 3 kg/m^2 dry weight a year.

The Gradient Water–Land Within the Tidal Freshwater Wetlands

Perpendicular to the gradient from fresh to salt, the tidal freshwater wetlands offer a variety of subsystems (Fig. 1), from the deep tidal channel to the upper terrestrial. The basic ecological factor is the elevation of the subsystem relative to the mean high water table; it indicates how many hours the subsystem will be flooded and be deprived of atmospheric oxygen. Hydrology is the main discerning variable. The different subsystems will be discussed to illustrate the variation within the ecosystem.

The very dynamic nature of the aquatic (subtidal) subsystem is a function of the current which is directly related to the dimension of the tidal wetlands and the tidal fluctuation. The potential exists for dynamic currents – for example, a tidal creek that fully floods a wetland of 1 km^2 with 1 m water has to discharge a million m^3 within a few hours. However, most wetlands are a mosaic of flats and higher marshes which reduces the volume and current of water entering and leaving the system. The deep water in the permanent channel might provide perfect conditions for benthic and

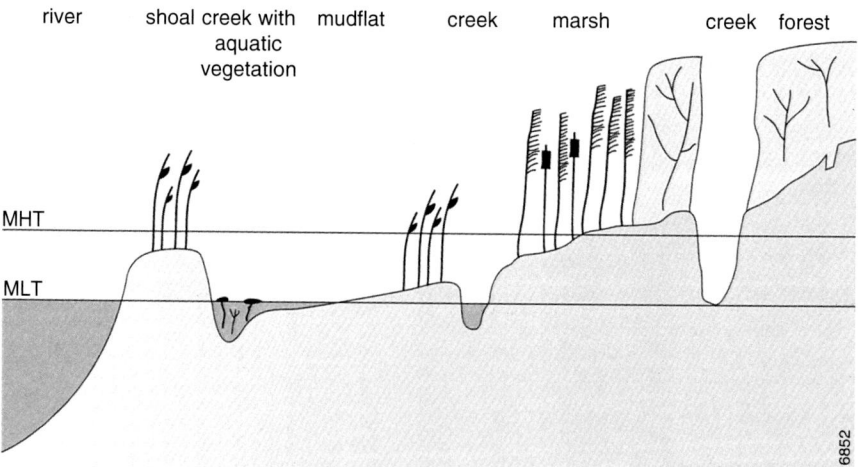

Fig. 1 Cross section from the channel to the high marsh/forest. *MHT* mean high tide level, *MLT* mean low tide level (From: Barendregt et al. 2009, © with permission of publisher)

Fig. 2 Tidal freshwater wetland in USA, Maryland (Jug Bay) – tidal creek with *Nuphar lutea* and *Zizania aquatica* (Photo credit: A. Barendregt © copyright remains with author)

pelagic species. The benthic species may, however, be confronted with water with high concentration of suspended matter creating unconsolidated sediment conditions. Next to the channel there will be shallow permanent water where dynamics will be reduced and conditions for aquatic plants and benthic fauna are better able to survive.

Higher in elevation is the intertidal subsystem, where tidal water floods for many hours a day. When the current is restricted, sediments can precipitate to form mud flats low in organic matter. Locations low in elevation will be exposed for only a few hours a day and become open mud flats. At higher elevations in the intertidal subsystem, the numbers of hours underwater is reduced and the first semiterrestrial plants can survive. In the USA yellow spatterdock *Nuphar lutea* and arrow arum *Peltandra virginica* grows in patches; in Europe bulrush *Schoenoplectus* spec. can be present.

One step higher in the elevation will be the real marshes. Where the tidal water still floods for some hours, helophyte species will dominate (Fig. 2), e.g., wild rice *Zizania aquatica* and cattail *Typha* spec. in the USA and reed *Phragmites australis* in Europe. Just higher in the gradient are the high freshwater marshes that will be flooded briefly at high tide under average conditions and for a longer period on spring tides. The soil is rich in organic matter. In the USA these high marshes are extremely rich in species including *Bidens laevis*, *Polygonum arifolium*, *P. punctatum*, and *Hibiscus moscheutos*. In Europe vegetation diversity is also high, e.g., with *Senecio paludosa* and *Calystegia sepium*.

Many locations of the highest elevated (supratidal swamps) subsystem in the tidal freshwater wetlands have a soil that is robust enough to facilitate a vegetation dominated with shrubs and trees. In between the shrubs many plant species are represented. In Europe many of the marshes were managed to grow willows (*Salix* spec.) for wood production, so a change in nature to an organized willow plantation (osier bed). The zone above these shrubs and trees is too high to be flooded and thus not a wetland for this definition.

Distribution and Human Impact

Two basic conditions need to be present to facilitate tidal freshwater wetlands: (1) a tidal wave and (2) permanent riverine freshwater. Global distribution of TFWs appears related to the constancy of river discharges that do not fall below 10–15% of maximum. When the input of freshwater decreases below this minimum value during a dry or freeze-up period, excessive saline water temporarily enters the system and TFWs cannot develop or persist. Most rivers in the temperate zones discharge at least this minimum in the dry season and potentially support a tidal freshwater wetland (Barendregt and Swarth 2013). Closer to the equator differences between seasonal flows become more extreme. Although the ecological conditions in the landscape are the same, mangroves replace TFWs in the tropical and subtropical regions. Some major tropical rivers, e.g., Amazon and Congo River, discharge freshwater constantly and in volumes that can freshen the ocean for many km from the coast. Although a tidal wave enters the Amazon for 800 km, indicating conditions for tidal freshwater wetlands may occur, none have been up to now.

Distribution

Basic knowledge on the global distribution and abundance of tidal freshwater wetlands is lacking but beginning to accumulate. North American TFWs occur where rivers meet the ocean along the west and east coasts. The most southerly reported TFWs on the west coast are in San Francisco Bay; and many parts of the extensive 7.5 million ha Yukon-Kuskokwim delta wetlands (Alaska) are tidal and should contain TFWs. Along the east coast, TFWs are well distributed from New Jersey (possibly Canada) to Florida, with a concentration around Chesapeake Bay and in North Carolina. In the Gulf of Mexico, the tidal wave is limited; the extensive tidal wetlands in the delta of the Mississippi are partly impacted by fluctuations in water level from the wind. Excluding Alaska, 200,000 to 1.5 million of ha with tidal freshwater marshes and swamps are estimated to occur in the USA.

Greatly reduced from the amount that would have occurred hundreds of years ago, ca. 15,000 ha with marshes, willow plantations, and swamps are reported in Europe. Most major rivers in Western Europe, e.g., Elbe, Weser, Rhine/Meuse, Scheldt, and Gironde, and those in the UK offer conditions for tidal freshwater wetlands. In the temperate region of South America, at least extensive TFWs are reported from the De la Plata Estuary (Parana River) in Argentina (Kandus and Malvarez 2004). This ecosystem has been reported but not described in Japan and China; and it is likely that the TFWs are well represented in the north of Russia and Siberia and the estuaries of the major rivers in China in which tidal ecosystem can be discerned from Google Earth.

Human Impacts

Tidal freshwater wetlands create for humans a location where many activities have historically taken place. The ecosystem is very rich in fish, waterfowl, and fur-bearing animals to harvest; the marshes are very productive with the input of nutrients from flooding; and freshwater is in abundance. Harvesting of firewood and plant materials for construction was common, e.g., in Europe the bulrush *Schoenoplectus* spec. was harvested to make mats and chairs, and willow *Salix spp.* used for basketwork grew abundantly in osier beds. Native Americans were present in this system 12,000 years ago (Dent 1995) and in Europe a human presence is well documented 5,000 years ago (Van de Noort 2004) and possibly longer (Early Pleistocene). In ancient times settlements were not permanent but this changed with organized societies. In Europe the first permanent settlements arose primarily during the medieval period and locally even earlier in the Roman period. Because flooding was a serious problem that affected infrastructure and the economy of these early settlements established adjacent to tidal freshwater wetlands, the inhabitants started to shape their own environment. First they created small dikes to protect the houses and fields from flooding by closing tidal creeks; and beginning about 1,000 years ago, humans started to reclaim the freshwater marshes for highly valued agriculture fields as an outcome of the dynamic sedimentation within the system.

During last 200 years this process has intensified within and adjacent to tidal freshwater wetlands. Cities have expanded in Europe as well as North America, especially those with harbors close to the ocean or sea. Most of the world's population lives close to the sea for economic reasons, starting at and with the sea harbors. The flat areas of the coastal region, especially in or next to the estuary, were well suited to support the housing, trading infrastructure, and harbors required by the increased economic activity. As a result, many thousands of ha with tidal freshwater wetlands have been occupied, and our present society is generally not aware of the extent of this past activity, e.g., the extensive harbors of Hamburg and Antwerp, and the Ronald Reagan airport in Washington DC, are reclaimed from this ecosystem.

During the twentieth century, pollution in the estuaries has increased primarily from industrial and non-treated human wastewater from the big cities upstream of the estuary. Negative effects on the ecosystems in the Elbe and Rhine Rivers were being reported beginning in the 1930s. Heavy metals and pesticides further deteriorated the ecosystem that was already suffering from extreme levels of nutrient inputs and intensification in human management. Fish stocks have been depleted from overfishing, many estuaries are closed from the sea, and shipping lanes in the main harbors have been deepened many meters by dredging. During the last 25 years, the situation has, however, begun to improve with a reduction in pollution, and the tenets of environmental awareness have become engrained within society. With improvements in water quality, the first recoveries of aquatic plants and benthic fauna have been reported, and many city inhabitants on estuaries are able now to recreate in restored tidal freshwater marshes (Baldwin 2004).

Physical and Chemical Processes

Water is the medium of transport and determinant of most chemical and physical conditions in tidal freshwater wetlands. Hydrology in estuaries is complex. The diurnal tides cause enormous flows in water with sometimes extreme current. Normal tidal amplitude in wetlands bordering oceans ranges from 1 to 2.5 m, with exceptions up to 6 m in funnel-shaped estuaries. This amplitude is the same in the freshwater sections, but the amplitude will decrease upstream in the river. On average the tidal wave comes 40–200 km inland from the sea, depending on the dimensions of the estuary and the river. The tidal amplitude differs by tide due to the neap and spring tidal cycle and is also affected by the quantity of discharge from the river. The ebbing period is longer than the flood period in the freshwater section of the estuary, resulting in an asymmetrical tidal curve. Moreover, hydrology in the larger estuary is influenced by the rotation of the earth, creating the difference in the flood stream along one shore and the ebb stream along the other and forming banks and shoals in the middle of the channel. Due to high water velocity, these sedimentary banks and shoals are mostly composed of the heavier sand component; plant growth further reduces the current. This sand originates from the river basin; the sea is merely a sink.

The production of muddy sediments is prominent in the freshwater section of the estuary, and their development is the result of sedimentation of suspended material, exported from the river catchment, stimulated through a chemical process of flocculation. Tidal movements create an interface without current between the heavier saline water in the deeper channels and the fresh river water flowing to the sea above it. At this interface, ions from the saline water start to attract organic ions from the river water, resulting in large flocs that will settle and accumulate at locations without current. A second process is that diatoms in the water are the starting points to flocculate suspended matter. This sedimentation creates tidal mud flats and later, after accumulation, in accretion in tidal marshes. Production of marsh vegetation adds many roots and litter to this mud and produces a soil rich in organic matter.

One of the major ecosystem services of these tidal freshwater marshes is wastewater treatment – input of polluted river water from upstream and the export of purified surface water from the marshes into the brackish estuary. Many nutrients and metals are bound to and accumulate with the suspended matter, e.g., phosphates binding with iron and calcium and fixation of nitrogen to organic matter. Heavy metals and pesticides can be stored during the same process. Silica is also recycled in these wetlands (Struyf et al. 2005).

Organic matter in wastewater discharged into the river is greatly reduced and disappears from the system as atmospheric CO_2. As a result of diurnal fluctuations in the water table, the intertidal system is high in oxygen at low tide but poor in oxygen at high tide with flooding. These fluctuations stimulate nitrification and denitrification of the nitrogen, finally ending in the discharge of atmospheric N_2. The reduction of organic waste and the change in nitrogen status are chemical processes performed by bacteria. Permanent wet soil with the groundwater level always at the soil surface results in non-oxide conditions with only the first millimeters with oxygen. This condition does not fluctuate seasonally as with other wetlands and is characteristic of tidal freshwater wetlands as result of a constant sea level, regular flooding of the marshes and swamps, and resistance of clay soils to water flow. The reduction of organic matter under reduced but fresh conditions takes place with the process of methanogenesis (Weston et al. 2011).

In summary, compared to other wetland types, the processes within tidal freshwater wetlands are very fast in terms of hydrological disturbance, sedimentation, and chemical or biological processes. However, at the same time, permanent high water tables in the marshes characterize the system, indicating an ecosystem with permanent oxygen-poor conditions. Another characteristic is that the system is eutrophic, and high nutrient concentrations can be reduced through both chemical and biological processes. The plant and animal species that are present can survive the flooding stress and the high nutrient levels.

Flora

There is a distinct zonation in the vegetation, linked to the flooding frequency explained by the elevation of the location and the flooding levels. This variation in subsystems perpendicular to the shoreline in the gradient from the tidal channel to the upper terrestrial parts is reported from North and South America and Europe. Although the species might be different for each area, the morphology remains comparable (Fig. 3). Aquatic plants are not common in tidal creeks as water current and turbidity prevents their establishment and development. Only in parts isolated from these pressures can aquatic plants develop. The mud flats flooded for longer than half of the tide are too dynamic and unconsolidated to allow plant growth.

Vegetation in the low marsh gradient can survive and grow although experiencing flooding for many hours because of aerenchymous tissues in stems; rhizomes transport oxygen below the water surface. In Europe and South America these might be species from the genus *Schoenoplectus*, and in North America some prominent species are *Nuphar lutea*, *Peltandra virginica*, and *Pontederia cordata*. Many helophytes are found in a somewhat higher elevated but still frequently flooded zone. In Europe some dominant species are *Phragmites australis*, *Typha angustifolia*, and *T. latifolia*; many herbs also occur. In North America the same *Typha* species are present as well as *Zizania aquatica* but these can be replaced to the south by *Zizaniopsis miliacea* and *Cladium mariscus*. Many other herbs and grasses can be added to this zone. In South America the species *Zizaniopsis bonariensis* and *Panicum grumosum* are represented.

The high marsh is represented where normal flooding is present for only a shorter period each day, but the groundwater table is still almost at the surface. In North America this vegetation has an extremely high diversity with many species of *Polygonum* and *Bidens*; in Europe it is mostly *Phragmites* with many herbs, e.g.,

Fig. 3 Tidal freshwater wetlands in the Netherlands (Oude Maas) with sequence showing zonation with increasing elevation – *Schoenoplectus* sp., *Phragmites* sp., and *Salix* sp. at high tide level (Photo credit: A. Barendregt © copyright remains with the author)

Symphytum officinale, and frequent vines. In South America *Schoenoplectus giganteus* and *Eupatorium tremulum* dominate in these marshes. In these same high-marsh locations, a swamp can also develop with many species such as *Cephalanthus occidentalis* in the northern USA and *Taxodium distichum* more to the south. In Europe many *Salix* species are present and in South America *Erythrina crista-galli* can dominate. Higher in the gradient, the flooding will be restricted to storm tides and in most cases there is a riparian forest.

Many wetland species are typical and widespread but locally some rare or threatened species are linked to tidal freshwater wetlands. They all have in common the ability to withstand flooding and the shortage in soil oxygen. Many wetland species are typically perennial. In Europe almost all species are perennial; in North America some annual species (e.g., *Zizania* and *Polygonum*) are also well represented. Research indicated that the seeds of many plants are transported by water (buoyancy), creating optimal conditions for distribution in flooding areas, although seeds of some other species (e.g., *Zizania*) fall directly from the parent plant on the soil. Just by the dynamics in the tidal wetlands, the seed banks appeared to be very rich in species, but not all species can germinate in the tidal conditions. Some species are invasive in the wetlands, with local impact on native biodiversity. In North America the problem is mainly by *Phragmites australis*; in Europe this might be with *Impatiens glandulifera*.

The tidal freshwater subsystems in all continents have in common that the yearly production in aboveground biomass is very high compared to other systems. The belowground biomass is also high (comparable with aboveground) as can be expected from vegetation with many perennial species. An aboveground production of 10–20 ton/ha/year appears to be normal and extremes up to 40 ton are reported. This level of production is not unexpected in a system with abundant and ever-present water, rich in nutrients. However, a 5-year study on effects of nutrient additions indicated that nitrogen might be limiting production (Ket et al. 2011). Other research indicated plants, especially annual species, responded positively to the addition of nitrogen. On marshes where flooding with nutrient-rich river water is present on a diurnal basis, the limitation in nutrients might, however, be of minor value compared to the stress from flooding with absence of oxygen in the soil. However, most species have a specialized root system to survive this stress.

Fauna

The faunal diversity of tidal freshwater wetlands is not extremely high, but the diversity in subsystems creates niches for many species. Compared to plants, the aquatic component is important for the fauna, the water itself (pelagic) and the soil underwater (benthic). For pelagic fish species, tidal freshwater wetlands are really an interface between the saline sea and freshwater rivers. Most inland (river) species are present in the fresh part of the estuary in addition to species from the brackish estuary entering the freshwater parts, thereby increasing the species diversity. Diadromous fish species use the gradient in salinity associated with freshwater tidal wetland habitat during their migration to and from the sea to adjust their physiological (salinity) equilibrium. A well-known example is the salmon, but many other species can be mentioned, e.g., the American shad *Alosa sapidissima* in North America and the twaite shad *A. fallax* in Europe.

Many fish species feed on invertebrates, e.g., zooplankton that in turn feed on phytoplankton. The tidal freshwater habitat is rich in both; diatoms are especially important to the food web. Crustaceans (amphipods), copepods, and insects (chironomids) together with the frequent oligochaetes (tubificids, naidids) are the most important faunal groups in the subtidal habitats with densities typically 10,000 individuals/m^2. These groups of species are also present in the intertidal habitat (mud flats, low marsh) of the tidal freshwater wetlands. Mollusks are present with many snail and bivalve species (sphaeriids and unionids).

Many of these species (especially tubificids and chironomids) are also represented in the elevated marshes and swamps; and many insects can survive the flooding conditions. Terrestrial gastropods, beetles, spiders, and springtails (Collembola) are important groups in biomass and quantities. Due to the irregular flooding frequency, aquatic species can be present just above mean high tide level and at the interface between water and land; only a select (characteristic) series of invertebrate species prefer this zone. Most terrestrial invertebrates behave optimally far above mean high tide level although depending on a species' tolerance they can be present closer to the tidal flooding zone.

European tidal wetlands are generally poor in amphibians and reptiles, whereas the freshwater section in North American estuaries is comparatively rich. Frogs and toads are frequent and turtles and snakes are not uncommon. Small mammals (mice, shrews) are frequent in both continents in the higher elevated marshes and swamps. Diversity in larger mammals is higher in North America, e.g., with many beaver *Castor canadensis*, muskrat *Ondatra zibethicus*, and otter *Lontra canadensis*.

Bird abundance and diversity is high in tidal freshwater wetlands, in both Europe and North America, although species differ. Food that includes seeds, insects, and small fishes is abundant in the wetlands and contributes to this bird abundance; the wetlands provide habitat for many specialized bird species. European reed marshes, swamps, or tidal forests are especially important, whereas in North America the high marshes and swamps contain the highest diversity of species. Waterfowl (from ducks to rails) find optimal conditions in the flooding zones. The diversity and close juxtaposition of the subsystems provide especially favorable conditions for nesting. The high abundance of fish species creates fine conditions for bitterns and herons to forage and nest. Raptors such as the osprey *Pandion haliaetus* in North America and the European hen harrier *Circus cyaneus* can be common locally. Many non-nesting birds use the wetlands during the migration period or in winter to find food. Benthivorous waders and many duck species that feed on small invertebrates (e.g., northern shoveler *Anas clypeata*) are frequent in this ecosystem. Due to the current in channels, the water is seldom frozen in winter which offers great opportunities for migrating waterfowl.

Threats and Future Challenges

Ecosystem services have been receiving more societal attention in North America and Europe in recent years. The prominent role that the tidal wetlands have in recycling of water with the reduction of nutrients and organic waste is no longer neglected. Another consideration is tidal wetlands can prevent or reduce economic damage from flooding by storing a great quantity of storm water. There are examples (e.g., close to Antwerp; Meire and Van Damme 2005) where tidal freshwater wetlands are installed for these ecosystem service benefits. Because tidal wetlands are typically close to cities, many people wish to avail themselves of the recreational opportunities these wetlands can provide, e.g., walking trails that incorporate education.

Experience from North America and Europe with the restoration of tidal freshwater wetlands indicates that there are few problems to restoring the system. Excavation, creation of new creeks, and storage of dredged sediment are elements of the activities. In the USA restoration often involves the planting of the desired species, while in Europe the general policy is species will migrate from surrounding tidal areas and species presence will be a response to the harsh flooding conditions. Monitoring of restoration projects has informed that early colonizing species dominate in the initial years but typical tidal freshwater species become established and dominate after a decade. The physical processes with sedimentation and natural vegetation growth are quickly reestablished due to the dynamic flooding system rich with sediments and nutrients.

Global Warming and Sea Level Rise

During the last millenniums, estuaries have continuously changed as a result of sea level rise, but present predictions indicate a more rapid change that might impact the otherwise robust system of tidal freshwater wetlands. A number of interlinked factors might result in changes to water levels, river discharge, and salinity.

The rising water table in the sea will affect shorelines and flood tidal freshwater wetlands more frequently. However, the freshwater section is very rich in sediments and impacts may be limited if accretion is able to keep pace with sea level rise. This process was probably present during former rises in sea level. The vertical accretion currently varies in time and space; and within hundreds of meters, sedimentation can range from 0 to 40 mm/year (Neubauer 2008). Partial loss of the accretion occurs during the winter period when vegetation is absent. A potential response to sea level rise is geographical migration of tidal freshwater wetlands upstream into the river valley. This migration will be affected by a couple of factors. Estuaries are typically funnel shaped and less space is available for wetland migration upstream into the river; and the area upstream is in most cases already occupied with housing and industry, and for economic reasons unavailable for wetland establishment.

Climate change might result in drier summer periods and less river discharge into the estuary. The consequence will be that freshwater input will be reduced and the level in the river decreases. Because the sea level is maintained, brackish water can flow further upstream. This will affect the tidal freshwater wetlands in its most vulnerable aspect, the freshwater component. Increased salinity will impact plant and animal species; many of which cannot tolerate the pulses in salinity and thus changing the competition equilibriums in species composition from a fresh to a brackish marsh. Far more significant are the impacts of salinity on the biogeochemical processes (Weston et al. 2011; Jun et al. 2013). The mineralization of organic matter in the fresh wetlands through methanogenesis will be replaced by a more efficient sulfate reduction with sulfur from brackish water. Saltwater intrusion stimulates the microbial mineralization of organic matter. As the soil in the marshes consists of clay and organic matter (in high marshes more than 50% organic matter), the marshes will shrink and will be flooded more frequently. At the same time, the storage of carbon will be disturbed and emission of CO_2 into the atmosphere will be stimulated.

References

Baldwin AH. Restoring complex vegetation in urban settings: the case of tidal freshwater marshes. Urban Ecosyst. 2004;8:125–37.
Barendregt A, Swarth CW. Tidal freshwater wetlands: variation and changes. Estuar Coasts. 2013;36:445–56.
Barendregt A, Whigham DF, Baldwin AH, editors. Tidal freshwater wetlands. Leiden: Backhuys Publishers; 2009. Joint publication. Weikersheim: Margraf Publishers.
Conner WH, Doyle TW, Krauss KW, editors. Ecology of tidal freshwater forested wetlands of the Southeastern United States. Dordrecht: Springer; 2007.
Dent Jr RJ. Chesapeake prehistory: old traditions, new directions. New York: Plenum Press; 1995.

Jun M, Altor AE, Craft CB. Effects of increased salinity and inundation on inorganic nitrogen exchange and phosphorus sorption by tidal freshwater floodplain forest soils, Georgia (USA). Estuar Coasts. 2013;36:508–18.

Kandus P, Malvárez AI. Vegetation patterns and change analysis in the lower delta islands of the Paraná River (Argentina). Wetlands. 2004;24:620–32.

Ket WA, Schubauer-Berigan P, Craft CB. Effects of five years of nitrogen and phosphorus additions on a Zizaniopsis miliacea tidal freshwater marsh. Aquat Bot. 2011;95:17–23.

McLusky DS, Elliott M. The estuarine ecosystem: ecology, threats and management. Oxford: Oxford University Press; 2004.

Meire P, Van Damme S. Special issue: ecological structures and functions in the Scheldt estuary: from past to future. Hydrobiologia. 2005;540:1–278.

Neubauer SC. Contributions of mineral and organic components to tidal freshwater marsh accretion. Estuar Coast Shelf Sci. 2008;78:78–88.

Odum WE, Smith III TJ, Hoover JK, McIvor CC. The ecology of tidal freshwater marshes of the United States east coast: a community profile. Washington, DC: U.S. Fish and Wildlife Service; 1984. FWS/OBS-83/17.

Simpson RL, Good RE, Leck MA, Whigham DF. The ecology of freshwater tidal wetlands. Bioscience. 1983;34:255–9.

Struyf E, Van Damme S, Gribsholt B, Meire P. Freshwater marshes as dissolved silica recyclers in an estuarine environment (Schelde Estuary, Belgium). Hydrobiologia. 2005;540:69–77.

Van de Noort R. The Humber wetlands, the archeology of a dynamic landscape. Macclesfield: Windgather Press; 2004.

Weston NB, Vile MA, Neubauer SC, Velinsky DJ. Accelerated microbial organic matter mineralization following salt-water intrusion into tidal freshwater marsh soils. Biogeochemistry. 2011;102:135–51.

Freshwater Marshes and Swamps

C. Max Finlayson

Contents

Introduction	169
Global Distribution	171
Ecological Features	173
Biota	173
Ecosystem Services	177
Threats and Future Challenges	178
References	180

Abstract

Freshwater swamps and marshes comprise a large variety of nontidal forested and non-forested wetlands. They have hydric soils and do not accumulate large amounts of peat (noting that the definition of peatlands comprises wetlands with at least 30% dry mass of dead organic material and greater than 30 cm deep). A swamp is dominated by trees and a marsh by emergent herbaceous plants, with both containing a wide variety of submerged and floating-leaved plants. However, the terms have not been used consistently in different parts of the world and many wetlands may contain both treed and non-treed components.

Keywords

Swamps · Marshes · Papyrus · Phragmites · Sustainable

C. M. Finlayson (✉)
Institute for Land, Water and Society, Charles Sturt University, Albury, NSW, Australia

UNESCO-IHE, The Institute for Water Education, Delft, The Netherlands

e-mail: mfinlayson@csu.edu.au

Introduction

Freshwater swamps and marshes comprise a large variety of nontidal forested and non-forested wetlands. Tidal freshwater marshes and salt marshes have been considered in separate articles. Freshwater swamps and marshes have hydric soils and do not accumulate large amounts of peat (noting that the definition of peatlands comprises wetlands with at least 30% dry mass of dead organic material and greater than 30 cm deep). A swamp is dominated by trees and a marsh by emergent herbaceous plants, with both containing a wide variety of submerged and floating-leaved plants. However, the terms have not been used consistently in different parts of the world (Thompson and Finlayson 2001; Mitsch et al. 2009), and many wetlands may contain both treed and non-treed components.

There may also be local preferences for one or other term when describing or naming wetlands, such as papyrus swamps that are dominated by the sedge *Cyperus papyrus* (Fig. 1) or reed swamps that are dominated by *Phragmites australis* and not tree species. There are also differences in opinion over what comprises a tree versus a shrub or what density of trees is needed to describe a wooded or forested environment, a situation well-known in forestry and forest ecology. Thus, it can at times be difficult to class particular wetlands as either a swamp or a marsh or even to agree how they should be called. Given this situation, freshwater swamps and marshes are considered together in this article, with a general distinction being that swamps are dominated by trees and marshes by herbaceous plants. Peatlands are not included, although many of these may have other structural or ecological similarities to swamps or marshes; these issues illustrate some of the general difficulties that surround the classification of wetlands.

The variety of freshwater swamps and marshes has been outlined in many articles and compilations about wetlands both globally and regionally. The continental wetland directories compiled largely in the 1990s by a number of nongovernmental organizations provide an extensive listing of the major wetlands of the world including many swamps and marshes. A summary of these sources is provided by

Fig. 1 Papyrus swamps (Photo credit: CM Finlayson © copyright remains with the author)

Finlayson and van der Valk (1995) – they comprise an invaluable historical information source for wetlands in Africa, Asia, Oceania, and the Neotropics, including information on their biodiversity and management. Further information sources for swamps and marshes have been collated through compendia such as those produced by Whigham et al. (1993), Mitsch (1994), McComb and Davis (1998), Polunin (2008), and Mitsch et al. (2009). Further information on freshwater systems and their species is provided in the extensive data resource contained in the impressive information source that underpins the Freshwater Ecoregions of the World (Abell et al. 2008). This provides a global biogeographic regionalization of the Earth's freshwater biodiversity and synthesis of data on the threats these face (http://www.feow.org/accessed 26 March 2016). Species compilations have also been provided by Balian et al. (2008) and through the freshwater biodiversity assessments undertaken by IUCN (see http://www.iucn.org/about/work/programmes/species/our_work/about_freshwater/what_we_do_freshwater/).

Given the extent of freshwater swamps and marshes and the inherent difficulties that are faced when attempting to separate these in complex wetland ecosystems, the following text largely addresses a number of large non-peatland freshwater wetlands that contain both marshes and swamps as examples. These include the Pantanal (South America), Okavango (Africa), Kakadu (Australia), Everglades (USA), and Tonle Sap (Cambodia) which were included in a comparative analysis that drew on separately published articles for each wetland (see Junk et al. 2006 and references therein). Information from other sources is used to complement the information provided by these examples, especially in relation to management issues (Fig. 2).

Global Distribution

Swamps and marshes are diverse and occur from the temperate zone to the tropics although their structure and ecological interactions vary enormously (Junk et al. 2006). The continental wetland directories provide information on individual wetlands and their distribution, but did not provide detailed maps showing their location and areal extent. Lehner and Döll (2004) provided estimates of the areal extent of freshwater marshes as $2,529 \times 10^3$ km^2 and swamp forest as $1,165 \times 10^3$ km^2 which comprised 1.9% and 0.9%, respectively, of the global land area. The generalized distribution of these wetlands is shown in Fig. 3. Giessen (2018) has summarized the distribution of tropical freshwater swamps in the following manner "......usually occur in vast flat floodplains and vary from the *Papyrus* and grassland swamps of the Sudd along the Nile River in South Sudan, to the grasslands of the Kafue flats in Zambia, savanna-like wetlands of southern New Guinea, backwater *billabongs* and feather- and fan palm swamps of northern Australia, wet palm savannas of South America, tall freshwater swamp forests in coastal plains and along major rivers in Southeast Asia, Central Africa and Amazonia, and complexes of woodland, forest and herbaceous vegetation of the Pantanal, in South America."

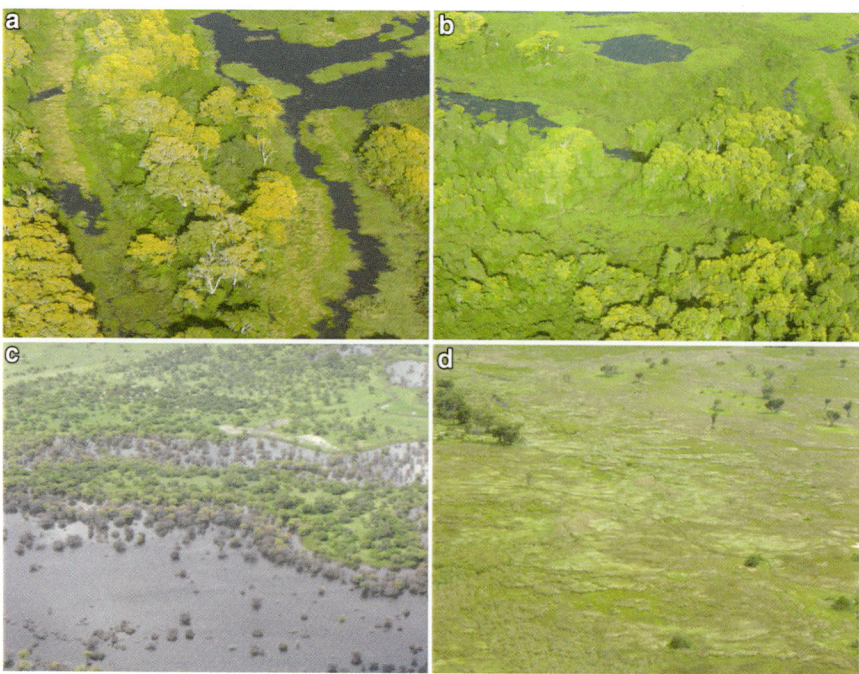

Fig. 2 Aerial view of floodplain swamp and marsh vegetation in (**a**) and (**b**) Pantanal, Brazil, (**c**) and (**d**) Okavango, Botswana (Photo credit: CM Finlayson © copyright remains with the author)

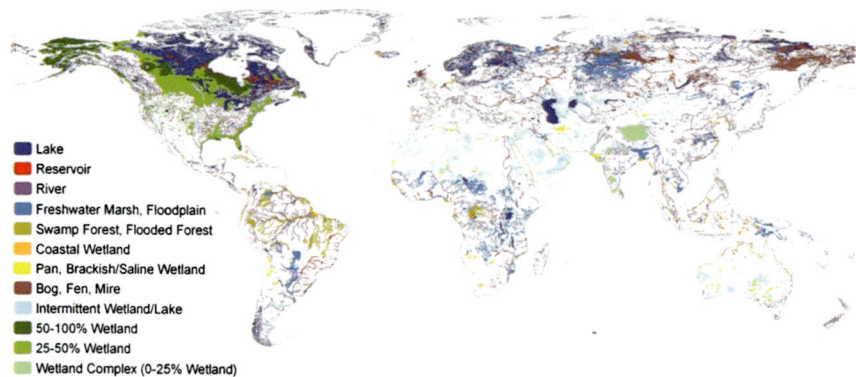

Fig. 3 Distribution of inland water systems including freshwater swamps and marshes (Reprinted from Journal of Hydrology, Vol 296, B. Lehner and P. Döll, Development and validation of a global database of lakes, reservoirs and wetlands, page 16, 2004, with permission from Elsevier)

Ecological Features

The vegetation and habitats found within swamp and marsh complexes vary considerably and can be diverse. Typically, marshes occur in areas that are frequently or continuously inundated with water and associated with mineral soils. Dominant plant species include reeds, rushes, grasses, and sedges. Swamps that may be inundated permanently or intermittently with water are dominated by trees or shrubs, and those associated with rivers typically have inorganic substrates, although others with little or no connection to flowing streams may accumulate small amounts of peat. Junk et al. (2006) provide a list of major habitats and vegetation types in five large wetland complexes which illustrates the range of wetland types or habitats that occur including lakes, non-flooded areas, and tidally affected wetlands, related to the underlying geomorphic and landform characteristics and water regimes (Table 1). It is also likely affected by differences in classifications that have been used in different sites or regions, as well as the gradations that occur between different wetlands across large landscapes.

The Freshwater Ecoregions of the World (Abell et al. 2008) provides maps of ecoregions with similar habitat types based on their biological, chemical, and physical characteristics. This includes swamps and marshes, but does not specifically display them.

The water quality of freshwater swamps and marshes also varies considerably, including annual, seasonal, and diurnal variations, as has been shown for the Kakadu freshwater wetlands (Finlayson et al. 2006). The pH and electrical conductivity of the five large wetlands assessed by Junk et al. (2006) and summarized in Table 1 provides a guide to the range of water quality in swamps and marshes, although these values cannot be seen as representative of all such wetlands given the wide range of conditions that exist. In some cases, the general descriptor of "freshwater wetland" is taken to include nonmarine brackish or even saline inland waters, especially where these are affected by flooding from freshwaters as a result of direct rainfall and surface flows.

The water regimes of swamps and marshes also vary considerably, from permanently inundated to ephemeral, including some that may be subject to long periods of drought. The importance of water regimes in wetlands has been widely discussed in the general wetland literature in reference to ecological processes such as nutrient cycling and supporting the life cycles of resident and migratory biota. The influence of the water regime for floodplain wetlands, including marsh or swamp ecosystems, is increasingly recognized with the frequency of flooding or pulsing as well as the timing, duration, and depth all having a major impact on the ecological processes and biological responses (Thompson and Finlayson 2001).

Biota

The diversity of biota in swamps and marshes is immense (Fig. 4), as illustrated in recent ecoregional descriptions (Thieme et al. 2005; Abell et al. 2008), catalogs of freshwater species (e.g., Groombridge and Jenkins 2000; Balian et al. 2008), and species assessments, in particular the regional assessments undertaken by the IUCN

Table 1 Major habitats and vegetation types, hydrological regime, and water quality (pH and electrical conductivity) from five large freshwater wetland complexes (Derived from Junk et al. 2006)

Wetland complex	Habitat and vegetation types	General wetland types	Hydrological regime	Water quality		Trophic status (whole system)
				pH	Electrical conductivity ($\mu S\ cm^{-1}$)	
1. Everglades	Sloughs, wet prairies, cypress swamps, mangrove swamps, periodically flooded and non-flooded tree islands, non-flooded pinelands, canals	Marshes, swamps, canals	Monomodal pulse, 0.5–1.0 m mean amplitude	6.8–8.5 south 6.5–7.7 north	550–1000 south 70–900 north	Oligotrophic P limited
2. Pantanal	Shallow lakes, rivers, channels, flooded savanna, flooded and non-flooded forests	Lakes, rivers/channels, marshes, swamps	Monomodal pulse, 3.5 m mean amplitude	6.5–7.4 in main tributary 5.5–7.5 up to 9.8 in flood plain	43–52 in main tributary 10–240 in flood plain >5000 in Salinas	Mesotrophic
3. Okavango	Shallow lakes, rivers, channels, permanent swamps, seasonally flooded and non-flooded savannas	Lakes, rivers/channels, swamps	Monomodal pulse, 1.85 m mean amplitude	6.3–7.4 in main tributary 6.4–8.4 in flood plain	40–50 in main tributary >5–10 times	Mesotrophic
4. Tonle Sap	Large lake, small lakes, rivers, periodically flooded shrubland, floodplain forest, rice field	Lakes, rivers, swamps, rice fields	Monomodal pulse, 8.2 m mean amplitude	7.5–8.5 in main tributary 7.5–9.3 in flood plain	3.5–18	Meso- to eutrophic
5. Kakadu	Shallow lakes, rivers including estuaries, channels, seasonally flooded savanna, mangroves, salt flats	Lakes, rivers/channels, marshes, swamps, salt flats	Monomodal pulse, 2–5 m mean amplitude	3.5–6.5	20–100 wet season ca. 1000 dry season	Oligo- to mesotrophic

Fig. 4 Variety of species found in freshwater swamps and marshes (Photo credit: CM Finlayson © copyright remains with the author)

in Europe and the Mediterranean, Africa, and parts of Asia (http://www.iucn.org/about/work/programmes/species/our_work/about_freshwater/resources_freshwater/ accessed 27 March 2016). These assessments covered freshwater ecosystems and were not confined to swamps and marshes, including rivers and also freshwater lakes that may comprise a mix of open water and areas dominated by trees or herbaceous vegetation. Other articles in this volume also illustrate the diversity of the animals and plants with particular mention being made of the overview on tropical freshwater swamps (Giessen 2018).

Balian et al. (2008), noting that their assessment of freshwater animal diversity was incomplete, considered 126,000 animal species or 9.5% of animal species

Table 2 Estimates of species numbers in five large freshwater wetland complexes (Derived from Junk et al. 2006)

Taxa	Everglades	Pantanal	Okavango	Tonle Sap	Kakadu
Algae		337	>50[a]	123	700[a]
Herbaceous wetland plants		248	Ca 350	34	75
Herbaceous terrestrial plants	1,033 (total plant species)	900	Ca 800	114	126
Woody plants		756	Ca 180	70	21
Total aquatic invertebrates	590 (terrestrial and aquatic species)			167[a]	900
Bivalves		23	6		2
Snails		5	16		12
Fishes	432 (includes marine species)	263	71	149	62
Amphibians	38	96	33	2[a]	26
Reptiles	60	40	64	24	30
Birds	349	390	444	230	107
Mammals	76	130	122	11	7

[a]Strongly underestimated

recognized globally. They determined that 60% of these were insects, 14% vertebrates, 10% crustaceans, 5% arachnids, and 4% mollusks. They also assessed 2,614 aquatic macrophytes, about 1% of the total number of vascular plants. The data from the African assessments (Darwall et al. 2011), covering 4,989 species of fishes, mollusks, dragonflies and damselflies, and crabs from 7,079 river or lake sub-catchments, were incorporated into the Freshwater Biodiversity Browser which comprises an interactive map as a basis for exploring the biodiversity of Africa's freshwaters (http://www.iucn.org/about/work/programmes/species/our_work/about_freshwater/what_we_do_freshwater/pan_africa_freshwater_ba/freshwater_biodiversity_browser/accessed 27 March 2016). These data indicate the high diversity of these ecosystems given that freshwater ecosystems cover only 0.01% of the total surface of the Earth.

Junk et al. (2006) point out that while comparisons were made between the vertebrate and higher plant species diversity of the five large freshwater wetlands they assessed, the data for lower plants and animals were insufficient for a comparative analysis. A comparison of species numbers from different plant and animal taxa is provided in Table 2. The habitat diversity in these wetlands supported a high diversity of vertebrates and higher plants; and although the habitats themselves were not categorized as swamp or marsh, both occurred in the wetlands. The species composition was to some extent related to the biogeography of the respective regions with the high diversity of large ungulate species found in Africa being apparent in the Okavango Delta and the high fish species diversity found in South America evident in the Pantanal. Except for the Everglades, the number of endemic species was low, although there were large populations of endangered or rare species. Even in the cases of these well-investigated wetlands, there are data gaps that limit the effectiveness of comparative analyses.

Giessen (2018) points out that there are many different vegetation types in tropical freshwater swamps with four main types recognized: (i) herbaceous swamps, (ii) shrub swamps, (iii) savanna/woodland wetlands, and (iv) swamp forests. Further, there are gradients across many wetlands from herbaceous swamps in deep or almost permanently flooded areas, grading into shrub swamps in areas less permanently flooded, and then savanna wetlands or swamp forest in much less permanently flooded areas.

Mitsch et al. (2009) provide further information on the biota of swamps and marshes, in particular those from North America. This includes a variety of swamp types, such as deepwater bald cypress–tupelo swamps (*Taxodium distichum* and *Nyssa aquatica*), pond cypress–black gum (*T. distichum* var. *nutans* and *N. sylvatica* var. *biflora*), and Atlantic white cedar *Chamaecyparis thyoides* swamps along the eastern seaboard and to red maple *Acer rubrum* and bottomland hardwood forests, including gum and oak species (*Nyssa* and *Quercus* species) and bald cypress *Taxodium distichum* along rivers in wetter areas. The bottomland forests are among the most dominant riparian ecosystems in the USA and are found across floodplains in the eastern and central USA and in particular in the southeast.

Tree species in these swamps have adaptations, including knees, adventitious roots, wide buttresses and stilt roots, fluted trunks, the development of aerenchyma, and deposition of suberin and lignin in roots that help them to survive in standing water. The productivity of these species is linked to the water regime with both flooding and drying resulting in lower productivity. The productivity of oligotrophic marshes is not high, but many tropical marshes and swamps have high levels of net primary production and rapid turnover of organic material with *Echinochloa* and *Cyperus papyrus* swamps being among the most productive (see comparative figures in Giessen 2018).

Ecosystem Services

The importance of ecosystem services derived from freshwater wetlands has been espoused in many recent publications since the Millennium Ecosystem Assessment (2005) drew attention to both their value and the extent of loss. Freshwater swamps and marshes in many parts of the world provide an array of production, regulating, and cultural services for many people. Their importance for freshwater for both domestic and agricultural purposes is well-known. Finlayson et al. (2013) consider the importance of floodplain wetlands, including swamps and marshes, for the supply of freshwater for people, to be one of their most important benefits, alongside those services that specifically support the food security of local communities (Fig. 5) and reduce rural poverty, such as through capture fisheries and sustainable aquaculture. However, in many cases the relative importance of different wetlands for types of food production and food security has not been determined or is highly variable, as found for swamps and marshes in sub-Saharan Africa (Rebelo

Fig. 5 Livelihood activities in African marshes – (**a**) cattle grazing, (**b**) reed cutting, (**c**) vegetable gardening, and (**d**) fishing (Photo credit: CM Finlayson © copyright remains with the author)

et al. 2010). The situation is similar for the important regulatory services provided by the hydrological cycle of such wetlands, which can make substantial contributions to groundwater recharge and discharge and flood prevention or amelioration, water purification, and retention of nutrients and sediments.

McCartney et al. (2014) when discussing the then proposed UN Sustainable Development Goals noted that "The key to sustainable development is achieving a balance between the exploitation of natural resources for economic development and conserving ecosystem services that are critical to everyone's wellbeing and livelihoods." This is particularly the case for swamps and marshes that have been lost and degraded for centuries and yet are critical for human development and the reduction of poverty. Costanza et al. (2014) in a review of the value of ecosystem services from different biomes show a reduction since their earlier analyses in 1997 in the benefits derived from swamps/floodplains along with a substantial loss of the area of these ecosystems.

Threats and Future Challenges

The threats facing freshwater swamps and marshes have been described in general assessments of wetland biodiversity and ecosystem services, such as that provided by the Millennium Ecosystem Assessment (2005). These include increasing

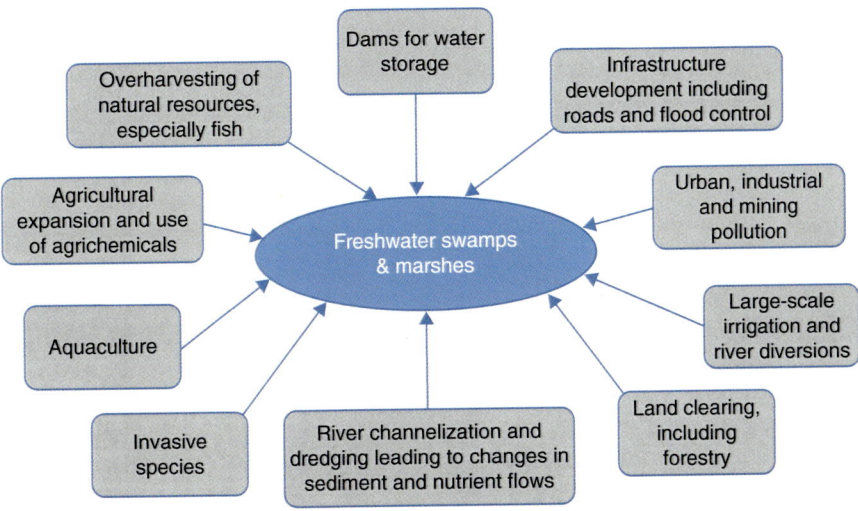

Fig. 6 Pictorial representation of the threats facing wetlands

pressures from infrastructure development for managing water supplies, land conversion for agriculture and urban or industrial development, water withdrawal for agriculture, and increasingly for domestic purposes, pollution from a variety of substances, overharvesting of timber and fish, and the introduction of invasive alien species. In addition, the threat from global climate change is expected to increase, which in many instances will exacerbate the abovementioned pressures rather than add a new dimension (Junk et al. 2013). In particular, it will change the precipitation patterns and hence the supply of water to swamps and marshes, including increased flooding in some areas, and less in others, as well as changes in the frequency and intensity of storms with fundamental changes to the life cycles of many species. Disruptions to the water regime, including the extent and timing of inflows, are expected to change the vegetation structure and wider biodiversity of swamps and wetlands. In many urban areas, seasonally drying wetlands have become permanently inundated or even converted to lakes with little fringing vegetation, whereas in rural areas the storage and movement of water have converted swamps or marshes to lakes with more regular water or deprived them of water.

The pictorial representation of threats facing wetlands was based on information supplied in the Millennium Ecosystem Assessment (MEA 2005) to depict the pressures facing wetlands generically, including for freshwater swamps and marshes (Fig. 6).

The literature on such pressures is immense, and the general situation for many wetlands is now well-known, as shown by other articles in this volume addressing swamps and marshes. Many of the swamps and marshes described in this volume also face significant threats under these categories. The five large wetlands assessed by Junk et al. (2006) add to the literature reporting the same suite of threats rather than adding anything very different.

Unless there is a major change in the way our landscapes and natural resources are managed, we can expect further decline in the ecological condition of freshwater swamps and marshes. Given increasing demands for freshwater, there is likely to be increased pressure on many swamps and marshes, although this is likely to be uneven given the extent of resource development in any particular locality and the capacity of local communities and governments to implement effective restoration efforts. With the extent of ongoing wetland loss and degradation (Davidson 2014; Gardner et al. 2015) and the seeming reticence of governments to effectively implement countermeasures, such as those agreed through the Ramsar Convention (Finlayson 2012), it is likely that further loss and degradation will continue. In this respect many governments that are signatories to the convention have yet to develop measures to address the impacts of climate change on wetlands (Finlayson 2013).

Unless a holistic and greatly increased approach to wetland conservation, including restoration, is developed, we can expect more freshwater swamps and marshes and their biota and ecosystem services to be lost with subsequent adverse consequences for human well-being and livelihoods. The potential benefits that can accrue when such wetlands are restored have been shown through the restoration and wetland management efforts that have occurred in many parts of Europe and North America, with the restoration of the prairie potholes being well-known. Similarly, the examples presented by Junk et al. (2006) demonstrate both what actions are needed and some of the successes; however, Junk et al. (2013) then caution against complacency when climate change is added.

References

Abell R, Thieme M, Revenga C, Bryer M, Kottelat M, Bogutskaya N, Coad B, et al. Freshwater ecoregions of the world: a new map of biogeographic units for freshwater biodiversity conservation. Bioscience. 2008;58:403–14.

Balian EV, Leveque C, Segers H, Martens K, editors. Freshwater animal diversity assessment. Dordrecht: Springer; 2008.

Costanza R, de Groot R, Sutton P, van der Ploe S, Anderson SJ, Kubiszewski I, Farber S, Turner RK. Changes in the global value of ecosystem services. Glob Environ Chang. 2014;26:152–8.

Darwall WRT, Smith KG, Allen DJ, Holland RA, Harrison I, Brooks EGE, editors. The diversity of life in African freshwaters: under water, under threat. An analysis of the status and distribution of freshwater species throughout mainland Africa. Cambridge/Gland: IUCN; 2011.

Davidson NC. How much wetland has the world lost? Long-term and recent trends in global wetland area. Mar Freshw Res. 2014;65:934–41.

Finlayson CM. Forty years of wetland conservation and wise use. Aquat Conserv Mar Freshwat Ecosyst. 2012;22:139–43.

Finlayson CM. Climate change and the wise use of wetlands – information from Australian wetlands. Hydrobiologia. 2013;708:145–52.

Finlayson CM, van der Valk AG. Wetlands classification and inventory: a summary. Vegetatio. 1995;118:185–92.

Finlayson CM, Lowry J, Bellio MG, Walden D, Nou S, Fox G, Humphrey CL, Pidgeon R. Comparative biology of large wetlands: Kakadu National Park. Aust Aquat Sci. 2006;68:374–99.

Finlayson CM, Davis JA, Gell PA, Kingsford RT, Parton KA. The status of wetlands and the predicted effects of global climate change: the situation in Australia. Aquat Sci. 2013;75:73–93.

Gardner R, Barchiesi S, Beltrame C, Finlayson CM, Galewski T, Harrison I, Paganini M, Perennou C, Pritchard DE, Rosenqvist A, Walpole M. State of the world's wetlands and their services to people: a compilation of recent analyses. Gland: Ramsar Convention Secretariat; 2015.

Giessen W. Tropical freshwater swamps (Mineral soils). In: Finlayson CM, Milton GR, Prentice RC, Davidson NC, editors. The wetland book II: distribution, description, and conservation. Dordrecht: Springer; 2018.

Groombridge B, Jenkins MD. Global biodiversity: earth's living resources in the 21st century. Cambridge: World Conservation Press; 2000. 246 p.

Junk WJ, Brown M, Campbell IC, Finlayson CM, Gopal B, Ramberg L, Warner BG. Comparative biodiversity of large wetlands: a synthesis. Aquat Sci. 2006;68:400–14.

Junk WJ, An S, Finlayson CM, Gopal B, Květ J, Mitchell SA, Mitsch WJ, Robarts RD. Current state of knowledge regarding the world's wetlands and their future under global climate change: a synthesis. Aquat Sci. 2013;75:151–67.

Lehner B, Döll P. Development and validation of a global database of lakes, reservoirs and wetlands. J Hydrol. 2004;296:1–22.

McCartney M, Finlayson CM, de Silva S. Sustainable development and ecosystem services. In: van der Bliek J, McCornick P, Clarke J, editors. On target for people and planet: setting and achieving water–related sustainable development goals. Colombo: International Water Management Institute; 2014. p. 29–32.

McComb AJ, Davis JA, editors. Wetlands for the future. Adelaide: Gleneagles Publishing; 1998.

MEA (Millennium Ecosystem Assessment). Ecosystems and human well-being: wetlands and water synthesis. Washington, DC: World Resources Institute; 2005.

Mitsch WJ, editor. Global wetlands: old world and new. Amsterdam: Elsevier; 1994.

Mitsch WJ, Gosselink JG, Anderson CJ, Zhang L. Wetland ecosystems. Hoboken: Wiley; 2009.

Polunin NVC, editor. Aquatic ecosystems: trends and global prospects. Cambridge: Cambridge University Press; 2008.

Rebelo L-M, McCartney MP, Finlayson CM. Wetlands of Sub-Saharan Africa: distribution and contribution of agriculture to livelihoods. Wetl Ecol Manag. 2010;18:557–72.

Thieme ML, Abell R, Stiassny ML, Skelton P, Lehner B, Teugels GG, Dinerstein E, Kamden Tohan A, Burgess N, Olson D. Freshwater ecoregions of Africa and Madagascar. Washington, DC: Island Press; 2005.

Thompson J, Finlayson CM. Wetlands. In: Warren A, French JR, editors. Conservation and the physical environment. London: Wiley; 2001. p. 147–78.

Whigham DF, Dykyjova D, Hejny S, editors. Wetlands of the world: Inventory, ecology and management, vol. 1. Dordrecht: Kluwer; 1993.

Papyrus Wetlands

12

Julius Kipkemboi and Anne A. van Dam

Contents

Introduction	184
Occurrence of Papyrus Marshes	184
Main Features of Papyrus Marshes	186
Propagation and Expansion	187
Root Architecture and Associated Features	187
Growth, Productivity, Biomass, and Water Transport	188
Zonation	189
Conservation and Management Status	190
Ecosystem Services	190
Biodiversity in Papyrus Marshes	190
Human Dependence on Papyrus Wetlands	192
Threats and Future Challenges	193
References	195

Abstract

Papyrus *Cyperus papyrus* is tropical wetland sedge that can grow up to a height of 5–6 m under optimal conditions. It is dominant vegetation in many wetlands in central, southern and eastern Africa, the Nile valley and in some parts of the Mediterranean in the Middle East and southern Europe. Papyrus is one of the most productive wetland sedge and is structurally and physiologically adapted to

---

J. Kipkemboi (✉)
Department of Biological Sciences, Egerton University, Egerton, Njoro, Kenya
e-mail: jkipkemboi@egerton.ac.ke

A. A. van Dam (✉)
Aquatic Ecosystems Group, Department of Water Science and Engineering, UNESCO-IHE Institute for Water Education, Delft, The Netherlands
e-mail: a.vandam@unesco-ihe.org

© Springer Science+Business Media B.V., part of Springer Nature 2018
C. M. Finlayson et al. (eds.), *The Wetland Book*,
https://doi.org/10.1007/978-94-007-4001-3_218

permanently and seasonally flooded wetlands. It propagates through both sexual and asexual reproduction. Wetlands dominated by papyrus are characterized by variable biotic diversity with 187 documented species in various African wetlands. These ecosystems also provide diverse ecosystem services ranging from provisioning, regulating and cultural that support livelihoods especially in sub-Saharan Africa. Like in many global wetlands, papyrus marshes are threatened by anthropogenic activities such as excessive biomass harvesting and conversion to agriculture and human settlement and climate change.

Keywords
Tropical wetlands · Macrophytes · Biodiversity · Ecosystem services · Threats · Conservation

Introduction

Papyrus *Cyperus papyrus* is a tropical wetland sedge that can grow up to 5–6 m under optimal conditions. Very often the term "swamps" is used for papyrus-dominated wetlands, yet these areas are dominated by herbaceous vegetation rather than trees. For consistency we will use the term "marshes" to refer to wetlands dominated by herbaceous vegetation and "papyrus marshes" for wetlands dominated by *C. papyrus*.

Papyrus marshes are predominantly found in riverine and lacustrine areas of tropical Africa and often exhibit high biomass productivity compared to other aquatic macrophytes and terrestrial herbaceous vegetation (Jones and Muthuri 1997; Boar et al. 1999; Mnaya et al. 2007). Papyrus utilizes the C_4 photosynthetic pathway and is characterized by high assimilation of CO_2 and solar energy into biomass (Jones 1987). Jones (1988) assigns a high quantum yield at low levels of incident light and efficient use of nitrogen by C_4 plants compared to those with C_3 pathways. This high biomass translates into numerous uses of the vegetation by the communities living adjacent to these wetlands. Papyrus wetlands are also important habitats for fish and other forms of aquatic and semiterrestrial wildlife.

Occurrence of Papyrus Marshes

Papyrus is considered native to central and eastern Africa and the Nile valley and is found from southern (Okavango), central (Zaire swamps in the present Democratic Republic of Congo), and eastern Africa to Egypt and Ethiopia and some parts of the Middle East (Israel and Syria) and southern Europe (Sicily). Major extensive papyrus wetlands occur in the lacustrine and floodplain wetlands of the White Nile River basin along the shoreline of Lake Victoria (East Africa), the largest tropical lake and second largest freshwater lake in the world; Uganda's Lake Kyoga wetland

Fig. 1 A riverine papyrus marsh in Eastern Uganda (Photo credit: J. Kipkemboi © Rights remain with the author)

complex; the shores of Lake Albert to the Sudd; Bahr el Ghazal; and the Sudd proper and associated upstream wetlands including the Machar marshes of the Sobat River and inflowing subbasins (Baro-Akobo-Pibor). The riparian ecosystem along the White Nile is also dominated by papyrus. Also significant are a number of western Ethiopia river floodplains, which are characterized by papyrus and other emergent macrophytes such as *Phragmites* and *Typha* along the water courses (Rebelo and McCartney 2012). Papyrus is also widely distributed along the River Nile in Egypt. The Nile Delta was dominated by papyrus until the early eighteenth century, but due to regulated flooding and human activities, it has been reduced to remnant stands along some of the delta lakes and lagoons.

The Sudd in the floodplains of the Nile River in South Sudan is the single most extensive papyrus marsh. Reports of the size of the papyrus-dominated zone are variable. The Jonglei Investigation Team (Jonglei Investigation Team. The Equatorial Nile project and its effects on the Anglo-Egyptian Sudan. Khartoum: Report to the Sudan Government (unpublished); 1954.) estimated the permanent swamp at 2,800 km^2, while Mefit-Babtie (1980) gave an estimate on 16,600 km^2 (Sutcliffe and Parks 1987 and citations therein) as did Chan and Eagleson (1980). An estimate of 7,000 km^2 provided by Dumont 2009 (cited in Zaroug et al. 2012) differs significantly from the approximately 40,000 km^2 permanent swamp estimate by Gaudet and Eagleson (1984 – cited in Zaroug et al. 2012).

In eastern Africa, papyrus occurs along the shores of Lake Victoria and in various valley swamps in the five East African countries. Papyrus is the dominant or codominant vegetation in many of the wetlands that cover 11–13% of Uganda's total land surface. Extensive monoculture stands are also found in most riverine marshes (Fig. 1) along rivers in eastern Uganda that drain into the Lake Kyoga system. Extensive systems include the Mpologoma, Nabajuzzi, and Namatala wetlands. Papyrus also dominates many wetlands in western Uganda's Busheyi, Kabala,

Kisoro, Ntungamo, and Rukungiri districts. Occurrence ranges from littoral vegetation of satellite lakes such as Lake Nabugabo and adjacent lakes, to the Rwambeita swamps and valley swamps along the rivers common in western Uganda and extending to Rwanda.

In Kenya, papyrus wetlands are found along Lake Victoria where it forms large stands in the floodplain wetlands of the Nyando, Yala, Sondu Miriu, Nzoia, and Sio rivers, most of which also have isolated upstream riverine marsh stands. Papyrus wetlands are also found in Kahawa swamp in Nairobi, along the riverine marshes in Nyahururu near Thompson falls, Molo river mouth at the shores of Lake Baringo, Tana River, and Loboi swamp near Lake Bogoria. Extensive lacustrine papyrus stands are found along Lake Naivasha's shoreline.

In Rwanda and Burundi papyrus is widespread in many valley swamps. Extensive areas are found in the Kanyaru-Akagera system, the Lake Kivu shoreline, and the Nyabarongo (or Nyawarungu) river at the Rwanda/Burundi border. In Tanzania, extensive areas exist along Lake Victoria and the associated floodplains such as the Mara river wetland and the Kagera river systems, with dense stands fringing either side of the river. Other extensive areas are found in the major river basins such as the Moyowosi/Malagarasi system in which floating mats occasionally detach, causing blockages in the river channels. The permanent swamps of the Pangani river system, including the associated lacustrine environments such as the Nyumba ya Mungu reservoir, are dominated by papyrus (Denny P, Bailey RG. A biological survey of Nyumba ya Mungu Reservoir, Tanzania, July–September 1974. Dar Es Salaam: Report for the Tanzanian Government (unpublished). 1976.). Papyrus also dominates the permanent swamps of the Ruaha/Rufiji system, Rukwa, Burugi/Ikimba, Eyasi/Yaida, and Natron. Along Lake Tanganyika, papyrus occurs in the delta swamps.

Within the Blue Nile, the Lake Tana system and its associated river systems (the Fogera and Dembia floodplains) have papyrus. In the Ethiopian Rift valley lakes, Lake Awassa and Lake Zwai have extensive papyrus stands on their shoreline. Other significant papyrus marshes occur in southern Africa along the Zambezi river system, particularly in the Okavango, Bangweulu, and Kafue flats. In central Africa, major papyrus-dominated areas are found in Lake Upemba and its basin in the Democratic Republic of Congo and in the Lake Chad system. Papyrus distribution is limited by altitude and disappears around 2,300 m above sea level (McClanahan and Young 1996). There is inadequate information on the actual extent of papyrus marshes in Africa. Thompson (1985) provided an estimate of 20,000 km^2, while Chapman et al. (2001) give an estimate of 85,000 km^2.

Main Features of Papyrus Marshes

The success and dominance of papyrus over other wetlands plants in permanently saturated and seasonally flooded wetlands can be attributed to its phytosociological and structural adaptations.

Propagation and Expansion

A papyrus marsh starts as a small clump from which rhizomes grow and branch in different directions. New juvenile culms develop behind the tip of the rhizome, while older culms at the aging part of the rhizome senesce gradually. The unique features of papyrus marshes are its propagation and expansion patterns and the ability to form floating mats. Mature culms (stage IV according to the growth stages proposed by Jones and Muthuri 1997) produce seeds from the racemous inflorescence. Propagation can occur sexually and asexually. Propagation by sexual means entails seed dispersal, germination, and seedling establishment. Due to seed distribution potential by different dispersal mechanisms, propagation by seed may lead to faster expansion over a larger area. Sexual propagation also allows exchange of genetic material among individuals and hence promotes genetic diversity. However, seed germination and establishment are highly sensitive stages in plant development, and their success is highly dependent on substrate characteristics, particularly soil moisture and organic content. Vegetative propagation and associated expansion occur through the growth of rhizomes, commonly in one direction but occasionally with branching and the ability to spread in all directions. Since little is known about interspecific competition, successional stages, or the dynamics of the dominant propagation method, these processes merit further long-term investigation.

Gaudet (1977) studied in detail the successional events in a drawdown marsh along the shores of Lake Naivasha. Hydric organic soils are ideal for the development of papyrus seeds transported by animals or water currents. Once the seeds germinate, the young plants thrive and grow quickly within saturated nutrient-rich organic soils and silts (Fig. 2), becoming distinctly different from other *Cyperaceae* as the culm height increases and inflorescence develops. Papyrus has an exceptional ability to respond to nutrients through root recruitment and growth in length as well as biomass accumulation compared to most other emergent vegetation (Kipkemboi et al. 2002). The ability to respond to nutrients in the interstitial water gives papyrus a competitive advantage over other emergent macrophytes. Although it is common to observe monoculture stands of papyrus with few herbs and shrubs, the climax stage may also occur as a mixed stand. Stable climax communities are uncommon in easily accessible areas due to human activities such as burning, livestock grazing, and harvesting of aboveground biomass.

Root Architecture and Associated Features

Papyrus occurs under a wide hydrological gradient from firmly rooted in the substratum to loosely anchored and easily detached floating mats and even islands (Azza et al. 2000). Its tolerance to flooding enables papyrus to dominate in fringing zones and permanently saturated areas while retaining its presence in seasonally saturated areas. Papyrus possess aerenchymatous tissue extending from the stem to the rhizomes and roots which contributes to buoyancy and enables the plant to form

Fig. 2 A young papyrus seedling (Photo credit: J. Kipkemboi © Rights remain with the author)

floating mats in low energy sheltered areas (Azza 2006). Roots occur at the rhizome nodes and play a role in anchoring the plant as well as in nutrient uptake. In floating mats, the roots dangle in the water column.

Papyrus mats are a result of intertwining rhizomes that can be secure in relatively firm soils towards the edge of the swamp substratum or may in muddy zones give rise to floating mats whenever the water level rises. As floating islands, these mats occasionally attain considerable size, covering open water bodies such as lakes. The roots on the rhizome mats play a significant role in determining the plant response to nutrient availability, especially in wetlands receiving wastewater (Kipkemboi et al. 2002). Papyrus roots have been found to be associated with nitrogen fixing bacteria (Mwaura and Widdowson 1992).

Growth, Productivity, Biomass, and Water Transport

Papyrus exhibits high growth rates compared to many other wetland plants, attaining heights up to 3.5–4 m within 4 months (Thenya 2006), and is capable of high biomass productivity and a higher standing biomass than many emergent macrophytes. Values of 14–21 g/m^2/day have been reported for Lake Naivasha (Muthuri et al. 1989). Although highly variable, values of 1.4–8.5 kg/m^2 are commonly reported for aboveground biomass, while 9–12 kg/m^2 have been reported when belowground (rhizomes and roots) are included (Muthuri et al. 1989; Boar et al. 1999; Kipkemboi et al. 2002; Mnaya et al. 2007; Terer et al. 2012 and references therein). Annual net productivity is about 5 kg/m^2/year (Jones and Muthuri 1997). The nitrogen and phosphorus in the plant biomass is variable in the different plant parts. Mean (dry matter) values for nitrogen and phosphorus range from 0.7 to 1.16% and from 0.13 to 0.16%, respectively (Gaudet and Muthuri 1981; Muthuri and Jones 1997; Boar 2006). A high growth rate and biomass imply greater carbon sequestration and storage.

Papyrus is also known for its high water fluxes. Based on a study in a papyrus marsh at the Lake Victoria shoreline, Saunders et al. (2007) reported a daily vapor flux from the papyrus canopy through evapotranspiration of as high as 4.75 kg/m^2 of water per day. This estimate is about 25% greater than loss through evaporation from open water. High water loss in papyrus marshes has also been reported for The Sudd in South Sudan. For instance, Sutcliffe and Parks (1989) indicate that an annual evaporation of 2,100 mm is reasonable, while Mohamed (2005) estimated evaporation of 1,460 and 1,935 mm per year in the dry and wet years, respectively. From these studies, it can be concluded that papyrus marshes play an important role in moisture circulation and influence local and regional hydrological cycle and climate.

Zonation

Plant zonation is a distinct feature in papyrus marshes. Emergent macrophytes can occur as monospecific bands in littoral wetlands. Species forming these bands reflect their ability to tolerate waterlogged conditions and water that sometimes rises above the substrate. There is a clear zonation of papyrus, *Typha* sp., *Phragmites* sp. and *Miscanthus* sp. from the saturated to the less saturated zones. In complex associations in large swamps, a mosaic of monospecific pockets of individual macrophyte stands is formed. Mixed stands occur where monospecific stands meet. In Lake Victoria, both in the littoral zone and at river mouths, it is common to find a fringe of papyrus, *Vossia cuspidata* and *Pennisetum purpureum* at the swamp/open water edge. In the mouths of large rivers draining into Lake Victoria, such as the Mara and Nyando rivers, *Vossia* usually occurs as floating mats and gradually gives way to fringing stands of papyrus which may be partly rooted, while *Typha, Phragmites* and other *Cyperus* spp. usually occur on the outer and often less wet parts of the swamp. At the interface zone in Lake Victoria, the papyrus fringes are connected to the floating mats of water hyacinth *Eichhornia crassipes* or, in some areas, rooted floating-leaved water lily *Nymphaea* spp.

Conservation and Management Status

Most papyrus wetlands do not have any conservation or protection status. In many East African wetlands, the management is determined by the livelihood activities of communities living around them. In many African countries, land tenure in wetlands is often unclear although the majority of the wetlands are treated as belonging to the state. Because of a lack of wetland policies or weak implementation strategies where policies do exist, most livelihood and development activities in wetlands are unregulated. As a result, many papyrus marshes are threatened by overexploitation.

Even within conservation areas, e.g., Ramsar sites, the policies for ensuring sustainable utilization are inadequate. Many of these wetlands have been converted to agricultural crop production areas. Owino and Ryan (2007) reported a loss of up to 50% of papyrus coverage in selected wetlands in the eastern shores of Lake Victoria adjacent to Winam Gulf from 1969 to 2000; and this was projected to be a loss of nearly 80% by 2020 if the trend continued unchecked. About a 70% loss of papyrus wetland has also been reported for Lake Naivasha over a period of 35 years (Boar et al. 1999). Lack of quantitative inventories is a major hindrance to monitoring losses of wetlands in many African countries.

Ecosystem Services

The high productivity and multiple uses of papyrus biomass correlate with the numerous products, functions, and uses provided by these ecosystems. Virtually every part of the plant, from rhizomes and culms to the umbel, can be used although this varies from community to community. In many instances, the rhizomes are used for fuel wood, while the culms are mainly used for craft making. At the tender stage, the umbel forms good fodder for livestock, while mature umbels are used by some communities for making brooms for sweeping. Besides these provisioning ecosystem services, papyrus marshes deliver important regulating services. They have been used for wastewater treatment (Chale 1985; Kansiime and Nalubega 1999; Mburu et al. 2015), and their carbon storage was estimated at 88 t C/ha in biomass and 640 t C/ha in the peat under permanently flooded papyrus (Saunders et al. 2014). Table 1 summarizes the common ecosystem services of papyrus wetlands.

Biodiversity in Papyrus Marshes

Floral and faunal diversity in papyrus marshes is variable (Chapman et al. 2001). Although these wetlands are characterized by the dominance of papyrus, they host a wide range of true aquatic macrophytes and semiterrestrial plants. One common feature in papyrus marshes is the presence of monoculture stands and discrete zones or pockets of plants of similar life forms. Apart from *Cyperus papyrus*, the other dominant emergent macrophytes in eastern Africa wetlands are *Typha domingensis*, *Phragmites mauritanius*, and *Miscanthus violaceus* (Denny 1985; Kipkemboi et al. 2002).

Table 1 Common ecosystem services of papyrus marshes in sub-Saharan Africa

Category	Examples
Provisioning services	
Biomass production	House construction, thatching, mat making, paper making, crafts and furniture, fuel wood, livestock fodder
Food production	Fish, wild game, seasonal crops
Clay mining	House walls smoothening
Regulating services	
Water quality improvement	Wastewater purification in natural and constructed wetlands
Hydrological regulation	Water storage and loss
Climate regulation	Evaporative cooling, carbon sequestration
Biodiversity and habitat	Wildlife habitat, tourism
Cultural services	
Education	Scientific insights on wetland biogeochemistry, hydrology and ecological studies
Ecotourism	Bird watching and other wild game observation areas.

Very few other plants are found growing within mature papyrus stands, perhaps because of the competitive advantage the sedge has over other emergent vegetation. There is a distinct difference in floral diversity between permanent and seasonally flooded areas. The seasonally flooded areas tend to be richer in species due to human disturbance and because most species occurring in this zone are predominantly opportunistic terrestrial herbs and not necessarily aquatic macrophytes (Rongoei et al. 2014).

Papyrus develop long culms supporting a 50 cm diameter umbel characterized by hundreds of cylindrical rays radiating from the tip of the culm. This canopy effectively shades the vegetation underneath. Occasionally, *Cynodon* sp. and *Leersia hexandra* occur entangled on the rhizomes of the fringe vegetation and may be moved to the open water with detached floating mats. Within the papyrus-dominated zones, many other species occur. Common species include climbers such as *Mikania* sp., *Ipomoea cairica*, and several other *Ipomoea* species, herbs such as *Commelina* sp., ferns (e.g., *Thelypteris striata*) and mosses underneath (Denny 1985). The pools between rhizomatous rafts may have floating plants such as *Lemna gibba* and *Azolla* spp. (Gaudet 1977). Other species that have been recorded in papyrus marshes are *Pyncbostacbys defixxifolia*, *Polygonum salicifolium*, *Cyperus dives*, and *Egna* sp. (Boar et al. 1999; Rongoei et al. 2014).

One of the key ecological values of papyrus marshes is habitat for a number of wetland endemic and some terrestrial fauna. Among the common residents in papyrus wetlands are representatives of a number of vertebrate groups (amphibians, reptiles, fish, birds, and mammals), insects, and a wide diversity of aquatic invertebrates. The lower water-saturated region is suitable for fish, worms, and larval stages of insects, while the canopy is ideal for perching birds. Papyrus marshes provide habitats for several avian species. Some common avian species are given in Table 2.

Table 2 Common bird species in papyrus and papyrus edge swamps (List compiled from species reported in Maclean et al. 2006)

Common name	Scientific name
Papyrus gonolek	*Liniarius mufumbiri*
Greater swamp warbler	*Acrocephalus rufescens*
Carruther's cisticola	*Cisticola carunthesi*
Papyrus yellow warbler	*Chloropeta gracilostris*
Papyrus canary	*Serinus koliensis*
Swamp flycatcher	*Muscicapa aquatica*
Crowned crane	*Balearica regulorum*
Grey heron	*Ardea cineria*
Shoebill	*Balaeniceps rex*
Hamerkop	*Scopus umbreta*
Black headed weaver	*Ploceus cucullatus*
African pied kingfisher	*Ceryle rudis*

In eastern Africa, several fish families can be found in papyrus marshes and lake fringes, such as Protopteridae, Clariidae, Anabantidae, Mormyridae, Bagridae, Shilbeidae, Cichlidae, Mochokidae, and Mastacembelidae. Fish species diversity in papyrus marshes includes both non-air- and air-breathing fish. High organic production in papyrus marshes, especially in areas where the water is above the substrate, creates a requirement for adaptation to hypoxia (Chapman et al. 1999). The air-breathers can tolerate the low oxygen conditions and include lungfish *Protopterus aethiopicus*, catfish *Clarias* spp., *Ctenopoma muriei*, and some water-breathing fish such as *Barbus neumayeri*, *Oreochromis* spp., and *Labeo* spp. Fish diversity is not necessarily related to the high biomass productivity of the sedge but rather to the different microenvironments in the papyrus wetland suitable for spawning, foraging, and refuge.

Although fish and birds are some of the widely studied animals in papyrus marshes, there is a plethora of other animal species that add to the faunal diversity (Table 3). Among these are mammals such as the swamp antelopes: sitatunga *Tragelaphus spekei* and the kobs *Kobus leche*. In some areas, hippos *Hippopotamus amphibius* associate with papyrus marshes. Otters (*Aonyx* spp.) occur especially where the water level is usually above the soil surface and there are fish. In the majority of tropical swamps and marshes, reptiles and amphibians are common among the vegetation. Among the commonest reptiles are the African python *Python sebae* and the monitor lizard *Varanus* sp., while amphibians are represented by frogs and toads. There is also a wide diversity of insects in their larval and adult stages.

Human Dependence on Papyrus Wetlands

Due to the diversity of functions and use values, papyrus wetlands support millions of people both directly and indirectly. The nature of dependence on wetlands for livelihoods and ecosystem benefits varies from place to place. The goods and services derived from papyrus marshes are intricately intertwined with culture,

12 Papyrus Wetlands

Table 3 Species richness in papyrus wetlands

Category	Study area	No. of species	Source
Herbaceous vegetation	Lake Victoria	34	Chapman et al. 2001
	Kibale forest	36	Chapman et al. 2001
Trees	General east Africa	8	Chapman et al. 2001
Amphibians	Lake Nabugabo	6	Behangana and Arusi 2004
Fish	Sondu Miriu	28	Gichuki et al. 2001
	Lake Victoria	30	Balirwa 1998
	Lake Nabugabo	18	Chapman et al. 2001
Avian	Select Ugandan wetlands	6	MacLean et al. 2006
Mammals		No data	
Reptiles		No data	
Invertebrates	Nyando	14	Mwagona 2013
Microbial diversity (bacteria)	Floating papyrus mat along the Nile in Egypt	7	Rifaat et al. 2002

prevailing economic conditions and livelihood options, and cultural interest. Papyrus wetlands attract various interest groups such as crop farmers, livestock herders, fishermen, craft makers and cottage industries, knowledge institutions, and conservationists. Compared to other emergent macrophytes, papyrus has by far the highest number of uses (described above). In some papyrus wetlands, cottage industries are the most reliable source of income. Farmers and fishermen are by far the largest direct beneficiaries of papyrus wetlands in many sub-Saharan African countries.

In the Lake Victoria and the Lake Kyoga ecosystems, papyrus wetlands are important fish nursery areas. These wetlands support the livelihoods of millions of rural populations; however, many of them are threatened by conversion to rice production. In Nyando wetland in Kenya, some local communities have initiated restoration of the papyrus vegetation for producing biomass for craft making (Fig. 3).

Threats and Future Challenges

Wetland loss globally is alarming (Davidson 2014), and papyrus wetlands are no exception. Like many other wetlands, they are primarily threatened by human population growth and associated anthropogenic activities such as unchecked urban development, human settlement, conversion to agriculture, water abstraction and diversion, pollution, introduction of alien species, and overexploitation. These threats are now aggravated by climate change. In many parts of sub-Saharan Africa, land use change and pressure from agriculture is the major threat. The increasing

Fig. 3 A regenerating papyrus wetland replanted by a local community in Nyando floodplain (Photo credit: J. Kipkemboi © Rights remain with the author)

demand for food by the fast growing population coupled with uncertainly of terrestrial rainfed agriculture due to climate change vagaries has led to more papyrus marshes being drained for agriculture. Papyrus wetlands are also affected by excessive biomass harvesting (Terer et al. 2012). Harvesting compromises aboveground plant attributes such as biomass production, culm density and height, as well as shoot regeneration. There are differences in opinion on recommended harvesting frequency (ranging between 6 and 12 months; Osumba et al. 2010; Terer et al. 2012). Burning of wetlands is also common in many areas to reduce the amount of senesced biomass and to allow fresh biomass to regenerate as pasture for livestock grazing. In other places, burning may be associated with clearing wetland margins for seasonal agriculture or for game hunting. Whatever the reason, burning disrupts the climax community and re-sets community succession. Frequent burning may also reduce the resilience of the wetland vegetation. In some wetlands, invasive species have been reported. In the case of papyrus marshes, some of these species include climbers, such as *Ipomoea* spp., and trees, such as the ambatch tree *Aeschynomene* sp. Hydrological modification is by far the most serious threat to any wetland as this affects the key ecosystem properties. In many parts of sub-Saharan Africa, hydrologic modification at the landscape and within the wetlands is seen as a threat to wetland habitat integrity (Maclean et al. 2014). Damming of the upstream rivers may compromise floodplain wetlands, whereas localized drainage in wetlands can completely alter the key properties of the ecosystem.

The challenges facing papyrus wetlands in Africa cannot be separated from other general environmental resources in the region. The trade-off between livelihood demands from the human population living adjacent to papyrus marshes and sustainable use of wetlands is poorly understood in many wetlands and will remain a challenge for many African countries for decades to come (van Dam et al. 2013; Zsuffa et al. 2014). Lack of appropriate legislative framework to guide sustainable use of wetlands is an impediment in countries in Africa. Whereas some countries have developed or are in the processes of developing appropriate policies, implementation remains a challenge in an environment where poverty and direct dependence on wetland use values by the local communities is high (Nasongo et al. 2015). Policies are often not accompanied with matching allocation of financial and human resources. More often than not, there is also inconsistency in policies and differences in institutional priorities in many countries. More research in ecology, governance, institutional, and socio-economic aspects of papyrus wetlands is required for sustainable utilization and successful conservation of these ecosystems (van Dam et al. 2014).

References

Azza NGT. The dynamics of shoreline wetlands and sediments of Northern Lake Victoria. [dissertation]. Wageningen: Wageningen University and UNESCO-IHE Institute for Water Education; 2006. 170 p.

Azza NGT, Kansiime F, Nalubega M. Differential permeability of papyrus and *Miscanthidium* root mats in Nakivubo swamp. Uganda Aquat Bot. 2000;67:169–78.

Balirwa JS. Lake Victoria wetlands and the ecology of the Nile Tilapia *Oreochromis niloticus* Linné. [dissertation]. Rotterdam: A.A. Balkema Publishers; 1998. 247 p.

Behangana M, Arusi J. The distribution and diversity of amphibian fauna of Lake Nabugabo. Afr J Ecol. 2004;42 Suppl 1:6–13.

Boar RR. Responses of a fringing *Cyperus papyrus* L. swamp to changes in water level. Aquat Bot. 2006;84(2):85–92.

Boar RR, Harper DM, Adams CS. Biomass allocation in *Cyperus papyrus* in a tropical wetland, Lake Naivasha, Kenya. Biotropica. 1999;31(3):411–21.

Chale FMM. Effects of *Cyperus papyrus* L. swamp on domestic wastewater. Aquat Bot. 1985;23:185–9.

Chan SO, Eagleson PS. Water balance studies of the Bahr El Ghazal Swamp. Cambridge, MA: Department of Civil Engineering, Massachusetts Institute of Technology; 1980.

Chapman LJ, Chapman CA, Branzeu DA, Mclaughlin B, Jordan M. Papyrus swamps, hypoxia, diversification: variation among populations of *Barbus neumayeri*. J Fish Biol. 1999;54:310–27.

Chapman LJ, Balirwa J, Bugenyi FWB, Chapman C, Chrisman TL. Wetlands of East Africa: biodiversity, exploitation and policy perspectives. In: Gopal B, Junk WJ, Davis JA, editors. Biodiversity in wetlands: assessment, function and conservation, vol. 2. Leiden: Backhuys Publishers; 2001. p. 101–31.

Davidson NC. How much wetland has the world lost? Long-term and recent trends in global wetland area. Mar Freshw Res. 2014;65(10):934–41.

Denny P. Wetland vegetation and associated plant lifeforms. In: Denny P, editor. The ecology and management of African wetland vegetation. Dordrecht: Dr. W. Junk Publishers; 1985. p. 1–18.

Dumont HJ. The Nile: origin, environments, limnology and human use. Dordrecht: Springer; 2009.

Gaudet JJ. Natural drawdown on Lake Naivasha, Kenya, and the formation of papyrus swamps. Aquat Bot. 1977;3:1–47.

Gaudet SC, Eagleson PS. Surface area variability of the Bahr el Ghazal swamp in the presence of perimeter canals. Cambridge, MA: Massachusetts Institute of Technology; 1984.

Gaudet JJ, Muthuri FM. Nutrient relationships in shallow water in an African lake, Lake Naivasha, Kenya. Oecologia. 1981;49:109–18.

Gichuki J, Guebas FD, Mugo J, Rabuor CO, Triest L, Dehairs F. Species inventory and the local uses of the plants and fishes of the Lower Sondu Miriu wetland of Lake Victoria, Kenya. Hydrobiologia. 2001;458:99–106.

Jones MB. The photosynthetic characteristics of papyrus in a tropical swamp. Oecologia. 1987;71:355–9.

Jones MB. Photosynthetic responses of C3 and C4 wetland species in a tropical swamp. J Ecol. 1988;76:253–62.

Jones MB, Muthuri FM. Standing biomass and carbon distribution in a papyrus (*Cyperus papyrus*) swamp on Lake Naivasha, Kenya. J Trop Ecol. 1997;13(3):347–58.

Kansiime F, Nalubega M. Wastewater treatment by a natural wetland: the Nakivubo Swamp Uganda: processes and implications. [dissertation]. Delft: IHE-Delft and Delft University of Technology; 1999. 300 p.

Kipkemboi J, Kansiime F, Denny P. The response of *Cyperus papyrus* (L.) and *Miscanthidium violaceum* (K. Schum.) Robyns to eutrophication in natural wetlands of Lake Victoria, Uganda. Afri J Aquat Sci. 2002;27:11–20.

Maclean IMD, Hassal M, Boar RR, Lake IR. Effects of disturbance and habitat loss on papyrus-dwelling passerines. Biol Conserv. 2006;131:349–58.

Maclean IMD, Bird JP, Hassal M. Papyrus swamp drainage and the conservation status of their avifauna. Wetl Ecol Manag. 2014. doi:10.1007/s11273-013-9335-1.

Mburu N, Rousseau DPL, van Bruggen JJA. Use of macrophyte *Cyperus papyrus* in wastewater treatment. In: Vymazal J, editor. The role of natural and constructed wetlands in nutrient cycling and retention on the landscape. Dordrecht: Springer; 2015. p. 293–314.

McClanahan TR, Young TP, editors. East African ecosystems and their conservation. New York: Oxford University Press; 1996.

Mefit-Babtie Sr. Range ecology study, livestock investigation and water supply. First interim report. Khartoum: National Council for the Development of the Jonglei Canal Area (Sudan); 1980.

Mnaya B, Asaeda T, Kiwango Y, Ayubu E. Primary production in papyrus (*Cyperus papyrus* L.) of Rubondo Island, Lake Victoria, Tanzania. Wetl Ecol Manag. 2007;15:269–75. doi:10.1007/s11273-006-9027-1.

Mohamed YA. The Nile hydroclimatology: impact of the Sudd wetland. [PhD dissertation]. Delft: UNESCO-IHE Institute for water Education/Delft University of Technology; 2005. 129 p.

Muthuri FM, Jones MB. Nutrient distribution in a papyrus swamp. Lake Naivasha, Kenya. Aquat Bot. 1997;56:35–50.

Muthuri FM, Jones MB, Imbamba SK. Primary productivity of papyrus (*Cyperus papyrus*) in a tropical swamp; Lake Naivasha, Kenya. Biomass. 1989;18:1–14.

Mwagona PC. Determination of macroinvertebrate community structure along different habitat types in Nyando wetland. [dissertation]. Njoro: Egerton University; 2013. 59 p.

Mwaura FB, Widdowson D. Nitrogenase activity in the papyrus swamps of Lake Naivasha, Kenya. Hydrobiologia. 1992;232:23–30.

Nasongo SA, Zaal F, Dietz T, Okeyo-Owuor JB. Institutional pluralism, access and use of wetland resources in the Nyando Papyrus Wetland, Kenya. J Ecol Nat Environ. 2015;7(3):56–71.

Osumba JJL, Okeyo-Owuor JB, Raburu PO. Effect of harvesting on temporal papyrus (*Cyperus papyrus*) biomass regeneration potential among swamps in Winam Gulf wetlands of Lake Victoria Basin, Kenya. Wetl Ecol Manag. 2010;18(3):333–41.

Owino AO, Ryan PG. Recent papyrus swamp habitat loss and conservation implications in western Kenya. Wetl Ecol Manag. 2007;15:1–12.

Rebelo L-M, McCartney M. Wetlands of the Nile Basin: distribution, functions and contribution to livelihoods. In: Awulachew SB, Smakhtin V, Molden D, Peden D, editors. The Nile River Basin: water, agriculture, governance and livelihoods. Abingdon: Routledge – Earthscan; 2012. p. 212–28.

Rifaat HM, Márialigeti K, Kovács G. Investigations on rhizoplane Actinobacteria communities of papyrus (*Cyperus papyrus*) from an Egyptian wetland. Acta Microbiol Immunol Hung. 2002;49(4):423–32.

Rongoei PJ, Kipkemboi J, Kariuki ST, van Dam AA. Effects of water depth and livelihood activities on plant species composition and diversity in Nyando floodplain wetland, Kenya. Wetl Ecol Manag. 2014;22(2):177–89.

Saunders MJ, Jones MB, Kansiime F. Carbon and water cycles in tropical papyrus wetlands. Wetl Ecol Manag. 2007;15:489–98. doi:10.1007/s11273-007-9051-9.

Saunders MJ, Kansiime F, Jones MB. Reviewing the carbon cycle dynamics and carbon sequestration potential of Cyperus papyrus L. wetlands in tropical Africa. Wetl Ecol Manag. 2014;22(2):143–55.

Sutcliffe JV, Parks YP. Hydrological modelling of the Sudd and Jonglei Canal. Hydrol Sci. 1987;32(2):143–59.

Sutcliffe JV, Parks YP. Comparative water balances of selected African Wetlands. J Hydrol Sci. 1989;34(1,2):49–62.

Terer T, Triest L, Muthama MA. Effects of harvesting *Cyperus papyrus* in undisturbed wetland, Lake Naivasha, Kenya. Hydrobiol. 2012;680(1):135–48.

Thenya T. Analysis of macrophyte biomass productivity, utilization and its impacts on various ecotypes of Yala swamp, Lake Victoria basin, Kenya. In: Vlek PLG, Denich M, Martius C, Rodgers C, editors. Ecology and development series no. 48. Gottingen: Cuvillier Verlag; 2006. 207 p.

Thompson K. Emergent plant of permanent and seasonally flooded wetlands. In: Denny P, editor. The ecology and management of African wetland vegetation. Dordrecht: Dr. W. Junk Publishers; 1985. p. 43–107.

van Dam AA, Kipkemboi J, Rahman MM, Gettel GM. Linking hydrology, ecosystem function, and livelihood outcomes in African papyrus wetlands using a Bayesian Network model. Wetlands. 2013;33(3):381–97.

van Dam AA, Kipkemboi J, Mazvimavi D, Irvine K. A synthesis of past, current and future research for protection and management of papyrus (*Cyperus papyrus* L.) wetlands in Africa. Wetl Ecol Manag. 2014. doi:10.1007/s11273-013-9335-1.

Zaroug MAH, Sylla MB, Giorgi F, Eltahir EAB, Aggarwal PK. A sensitivity study on the role of the swamps of southern Sudan in the summer climate of North Africa using a regional climate model. Theor Appl Climatol. 2012. doi:10.1007/s00704-012-0751-6.

Zsuffa I, van Dam AA, Kaggwa RC, Namaalwa S, Mahieu M, Cools J, Johnston R. Towards decision support-based integrated management planning of papyrus wetlands: a case study from Uganda. Wetl Ecol Manag. 2014;22(2):199–213.

Tropical Freshwater Swamps (Mineral Soils) 13

Wim Giesen

Contents

Diversity and General Description	200
Distribution and Area	202
Dominant Vegetation Types	203
Introduction	203
Herbaceous Swamps	204
Shrubland Swamps	205
Savanna Swamps	206
Swamp Forests	207
Ecosystem Services	208
Buffering Water Levels	208
Buffering of Water Quality	209
Productive Systems	210
Biodiversity	210
Carbon Storage	213
Dependent Peoples	213
Threats and Trends	215
Reclamation and Conversion	216
Damming (On-Site) and Flooding	216
Damming (Off-Site) and LU Changes Upstream	217
Pollution and Erosion	218
Fires	219
Overutilization of Resources	219
Invasive Species	220
Climate Change	220
References	221

W. Giesen (✉)
Euroconsult Mott MacDonald, Arnhem, AK, The Netherlands
e-mail: wim.giesen@mottmac.com

© Springer Science+Business Media B.V., part of Springer Nature 2018
C. M. Finlayson et al. (eds.), *The Wetland Book*,
https://doi.org/10.1007/978-94-007-4001-3_4

Abstract

Freshwater swamps on mineral soils are frequently or (almost) continuously inundated wetlands characterized by emergent vegetation. In the tropics, these areas usually occur in vast flat floodplains maintained by incoming floodwaters, with some fed by groundwater, and characterized by a cycle of seasonal flooding and desiccation. Data on exact areas of freshwater swamps is lacking and unreliable for many tropical countries, partly because of their vastness, seasonal variation in extent, changes in area over time, and extensive conversion and reclamation. Four main physiognomic types of freshwater swamps can be distinguished: (i) herbaceous swamps, (ii) shrub swamps, (iii) savanna/woodland wetlands, and (iv) swamp forests. Tropical freshwater swamps support a wide biological diversity and provide ecosystem services that benefit local dependent communities. Anthropogenic threats to tropical freshwater swamps are highly variable and depend very much on local circumstances and the type of wetland. Climate change is expected to impact tropical freshwater swamps via the increased incidence of drought and tropical storms and to lesser extent sea level rise.

Keywords

Tropical freshwater swamps on mineral soils · Ecosystem services · Biodiversity · Threats and trends

Diversity and General Description

Freshwater swamps on mineral soils are frequently or (almost) continuously inundated wetlands characterized by emergent vegetation. In the tropics these areas usually occur in vast flat floodplains and vary from the *Papyrus* and grassland swamps of the Sudd along the Nile River in South Sudan; to the grasslands of the Kafue flats in Zambia, savanna-like wetlands of southern New Guinea; backwater *billabongs* and feather and fan palm swamps of northern Australia, wet palm savannas of South America; tall freshwater swamp forests in coastal plains and along major rivers in Southeast Asia, Central Africa, and Amazonia; and complexes of woodland, forest, and herbaceous vegetation of the Pantanal, in South America. Most are floodplain swamps maintained by incoming floodwaters, but some are fed by groundwater.

Most of these freshwater swamps are characterized by a cycle of seasonal flooding and desiccation that promotes mineralization of organic matter, especially in warm tropical lowlands, and prevents the accumulation of peat. Because of the rise and fall in water levels, tropical floodplain wetlands display large differences between wet and dry season area (Table 1). Floodplain wetlands often consist of complexes that include rivers and streams, (semi-) permanent bodies of open water, permanent freshwater swamps, and a wide array of seasonal freshwater swamps. The latter may vary from almost permanently flooded swamps dominated by herbaceous (e.g., *Papyrus,* grasslands) or shrubby species (e.g., *Barringtonia, Sesbania*) to savannas and tall swamp forest that may be flooded for only a few weeks or months each wet season.

Table 1 Wet and dry season inundation of tropical freshwater wetlands

No	Wetland	Country/region	Maximum (ha)	Minimum (ha)
1	Danau Sentarum NP	Indonesia	150,000	<1,000
2	Kakadu NP	Northern Territory, Australia	195,000	<10,000
3	Haor region, Sylhet	Bangladesh	800,000	<50,000
4	Okavango	Botswana	1,390,000	250,000
5	Tonlé Sap	Cambodia	1,500,000	250,000
6	Kafue flats	Zambia	2,800,000	1,300,000
7	Trans-Fly plains	Indonesian Papua and PNG	5 Mha	<0.5 Mha
8	Sudd	South Sudan	>10 Mha	3 Mha
9	Pantanal	Brazil (62%), Bolivia (20%), and Paraguay (18%)	18–25 Mha	<1 Mha
10	Amazon basin	Bolivia, Brazil, Colombia, Ecuador, Guyana, Peru, Venezuela	80 Mha	40 Mha

Danau Sentarum (Giesen and Aglionby 2000), Kakadu (Finlayson et al. 2006), Haor Region, Sylhet (Scott 1989), Okavango (Ramberg et al. 2006), Tonlé Sap (Campbell et al. 2006), Kafue Flats (Tockner and Stanford 2002), Trans-Fly plains (WWF website), Sudd (see ▶ Chap. 106, "The Sudd (South Sudan)" by Rebelo and El-Moghraby, this volume), Pantanal (Scott and Carbonell 1986), and Amazon basin (Melack and Hess 2010)

Tropical floodplain wetlands are usually formed in lower courses of rivers, where the slope decreases and rivers become sluggish and tend to overflow their banks and spill into the floodplain. They can also form far inland in so-called pseudo-lower courses that may be formed due to tectonic uplift changing the courses of rivers (e.g., upper Kapuas River, west Borneo; Kafue River, Zambia; Okavango delta, Botswana; Tonlé Sap, Cambodia).

In contrast to peat swamps, which are predominantly fed by rainwater, freshwater swamps on mineral soils generally have higher amounts of dissolved materials, including nutrients, brought in by streams, rivers, and groundwater. The pH is usually close to neutral (i.e., usually 6–9), and productivity is high, given nutrient availability. As a result, bacteria and fungi are active, and turnover rates of organic matter are high (Mitsch and Gosselink 2000). Nutrients and sediments are deposited by flood events, but this can be highly variable over time and even within one wetland during a single flood event (Dezzeo et al. 2000). Vegetation also impacts (ground-)water quality, and deep-rooted trees such as *Acacia xanthophloea* in wetlands of semiarid Africa act as evapotranspirational pumps, selectively removing water and causing the subsurface concentration of solutes and electroconductivities 15–20 times higher than elsewhere on the floodplain (Humphries et al. 2011).

Distribution and Area

Data on exact areas of freshwater swamps is lacking and unreliable for many tropical countries, partly because of their vastness (especially in Africa and South America), but also because of their seasonal variation in extent, changes in area over time (e.g., due to medium- to long-term climate changes), and extensive conversion and reclamation (esp. in Asia; see "Dependent Peoples," below). Table 2 gives areas of tropical freshwater wetlands, and where these are unknown (which is in most countries with large expanses), areas of regularly inundated floodplains as a proxy. The total area worldwide is about 146 million hectares, with 38% occurring in Africa, 33% in South

Table 2 Area of tropical freshwater wetlands (nonlake, on mineral soils)

Asia		Africa	
Indonesia	11,500,000	DR Congo	15,000,000
Bangladesh	6,000,000	Congo	7,500,000
Pakistan	4,000,000	Chad	4,200,000
Vietnam	3,700,000	South Sudan	4,000,000
Cambodia	3,500,000	Central African Rep.	3,200,000
Burma	3,100,000	Zambia	2,900,000
Thailand	2,100,000	Nigeria	2,600,000
India	1,200,000	Tanzania	2,400,000
Malaysia	700,000	Cameroon	2,000,000
Philippines	300,000	Mali	1,800,000
Sri Lanka	100,000	Ethiopia	1,350,000
subtotal	36,200,000	Botswana	1,200,000
		Angola	1,000,000
South America		Mozambique	1,000,000
Brazil	27,200,000	Kenya	1,000,000
Colombia	1,500,000	Uganda	1,000,000
Ecuador	700,000	Gabon	900,000
French Guiana	100,000	Senegal	350,000
Paraguay	5,700,000	Somalia	350,000
Peru	600,000	Malawi	300,000
Suriname	800,000	Ghana	250,000
Venezuela	11,400,000	Rwanda	200,000
subtotal	48,000,000	Benin	150,000
		Burkina Faso	150,000
Australasia		Guinea	100,000
Papua New Guinea	6,300,000	Niger	100,000
Australia	700,000	subtotal	55,000,000
subtotal	7,000,000		
		Grand total	146,200,000

Source of data: Asia: Scott (1989, 1993); South America: Scott and Carbonell (1986), Africa: Hughes and Hughes (1992), PNG: Osborne (1993), Australia: Finlayson and Von Oertzen (1993)

America, 24% in Asia, and 5% in Australasia. These are maximum historic areas, as actual figures are expected to be lower and dropping. These contrast with figures on freshwater swamps worldwide (i.e., temperate and tropical combined) provided by Mitsch and Gosselink (2000) that vary from 89.4 Mha (Rodin et al. 1975) to 100.8 Mha (Matthews and Fung 1987).

Dominant Vegetation Types

Introduction

In the humid tropics (i.e., with minimum monthly rainfall of 100 mm; e.g., Amazonia, Congo basin, Sumatra, and Borneo), freshwater swamps that are not permanently flooded are usually dominated by tall, woody vegetation (i.e., freshwater swamp forest). Similarly, areas that are subjected to more prolonged flooding have stunted woody or shrubby vegetation (Fig. 1), while permanently flooded areas are dominated by herbaceous vegetation. This pattern seems true in most areas: in nutrient rich "varzea" whitewater floodplain swamps in Amazonia, bushes and grasses occur in low-lying areas that are subject to prolonged flooding, while forest occurs in the higher areas (Scott and Carbonell 1986).

In regions with a distinct dry season and/or lower rainfall, freshwater swamps may be dominated by woody vegetation of lower stature (e.g., shrub wetlands) or that is more sparse (e.g., savanna wetlands). Some freshwater swamps are subject to regular fires, as fuel that accumulates during the wet season becomes inflammable during dry periods, e.g., the *Melaleuca-Banksia* savannas of the Trans-Fly region in New Guinea (Paijmans 1976), *Shorea balangeran* swamp forests of Borneo (Van Steenis 1957; Dennis et al. 2000), and palm savannas of Amazonia (Junk 1993). Species of seasonal

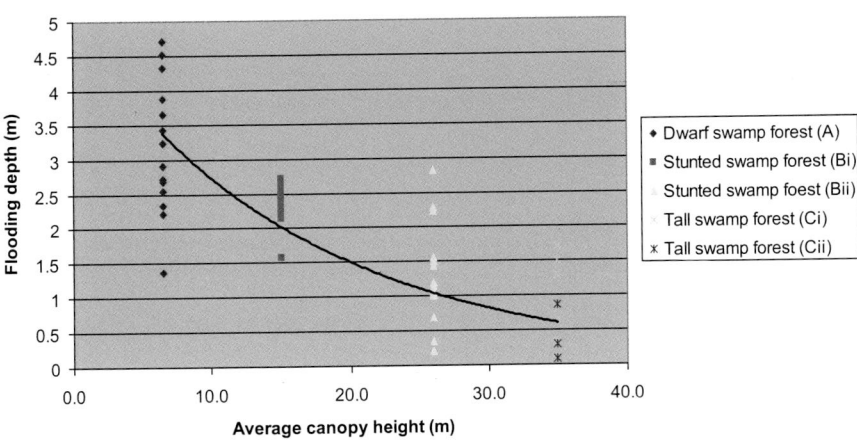

Fig. 1 Flooding and swamp forest canopy height, Danau Sentarum, Borneo (Data from Giesen (1996); © Wetlands International – Indonesia Programme, with permission)

herbaceous swamps are often annuals that cope with periods of desiccation by surviving as seeds. Woody species of freshwater swamp forests display many adaptations to regular and often prolonged flooding, and these adaptations may be structural, physiological, phenotypical, and reproductive. As summarized by Parolin et al. (2004):

- *Structural adaptations* include enlargement of lenticels, adventitious roots, plank-buttressing and stilt rooting, development of aerenchyma, and deposition of suberin and lignin in roots.
- *Physiological adaptations* include a progressive and pronounced reduction in leaf physiological activities such as photosynthetic assimilation, transpiration, and stomatal conductance.
- *Phenological adaptations.* Water logging causes reduced water conductance and consequently a water deficit in the crown, often resulting in leaf senescence and/or shedding, although some may even keep leaves below water for months without apparent damage. This periodicity in growth is often reflected in growth rings of adult trees.
- *Reproductive adaptations.* In certain tree species seed production coincides with the flood pulse to be dispersed by water flow and fish. Seeds/fruit often show morphological adaptations that enhance flotation, like spongy tissues or large air-filled spaces.

Although many hundreds of highly different vegetation types occur, four main physiognomic types of freshwater swamp can be distinguished, namely (i) herbaceous swamps, (ii) shrub swamps, (iii) savanna/woodland wetlands, and (iv) swamp forests. In many areas, gradients can be seen across wetlands, with herbaceous swamps in deeply or almost permanently flooded areas, grading into shrub swamps in areas less deeply or permanently flooded, and merging into savanna wetlands or swamp forest in much less permanently flooded parts. In the Amazonian basin, about 13% of the 800,000 km^2 of wetlands (=22% of total basin) consist of herbaceous swamps, while 78.9% has woody vegetation (Melack and Hess 2010). In East Africa, by contrast, most freshwater swamps consist of herbaceous swamps, and the huge Sudd wetland, for example, consists >90% of grass- and sedgelands (Denny 1993a).

Herbaceous Swamps

Herbaceous swamps can vary from *Nelumbo nucifera* swamps of Southeast Asia, to dense reedbeds dominated by *Phragmites karka*, *Saccharum spontaneum*, and *Typha* species, rooted or floating beds of 4–5 m tall *Cyperus papyrus* swamps in Africa, and mixed grass- or sedge-dominated swamps with a wide variety of species. In Southeast Asia and New Guinea the latter swamps include *Cyperus* and *Digitaria* species, *Echinochloa colonum*, *Leersia hexandra*, *Leptochloa chinensis*, *Panicum conjugatum*, *P. repens*, *Phragmites karka*, *Rhynchospora corymbosa*, *Saccharum spontaneum*, *Scirpus* species, and *Thoracostachyum sumatranum*, along with herbs such as *Aniseia martinicensis, Hanguana malayana, Impatiens* species, *Ludwigia hyssopifolia, Merremia hederacea, Polygonum barbatum,* and *P. celebicum* and ferns such as *Stenochlaena palustris* and *Blechnum indicum* (Paijmans 1976; Giesen 1992). In Amazonia, herbaceous swamps consist of a wide range of sedges, including *Cyperus articulatus, C. prolixus, Eleocharis minima, E. plicarhachis, Rhynchospora* species, *Scirpus californicus, S. giganteus,* grasses such as *Paspalidium paludivagum, Paspalum repens, P. virginatum, Hermathia, Panicum,* and *Axonopus* species, and herbs including *Pontederia lanceolata, Sagittaria* species, and *Ludwigia* species. In the Pantanal, the most numerous families include Fabaceae (240 species), Poaceae (grasses: 212 species), and Cyperaceae (sedges: 91 species; Desbiez et al. 2011). In tropical Africa, *Cyperus papyrus* sedge swamps and *Miscanthidium violaceum* grass swamps include sedges such as *Cyperus haspan, C. involucratus, C. nitidus, C. polystachyos, Fimbristylis* species, *Fuirena umbellata, Rhynchospora subsubquadrata,* and *Scirpus inclinatus,* and grasses such as *Andropogon canaliculatus, Echinochloa stagnina, Hyparrhenia diplandra, Leersia hexandra, Loudetia phragmitoides, Panicum parvifolium, P. subalbidum, Paspalum commersonii, P. strigosum,* and *Vossia cuspidata,* along with species such as *Typha australis* (Lind and Morrison 1974; Denny 1993a, b; John et al. 1993). Herbaceous swamps of Northern [http://wetlandinfo.derm.qld.gov.au/wetlands/ScienceAndResearch/ConceptualModels/Palustrine] Australia include species such as *Eleocharis* and *Fimbristylis dichotoma* sedges, grasses such as *Cynodon dactylon, Eragrostis* species, *Leersia hexandra, Leptochloa fusca, Oryza meridionalis,* and *Pseudoraphis spinescens,* and herbs such as *Calocephalus platycephalus* (Finlayson and Von Oertzen 1993). At Kakadu, there is very marked year-to-year variation in herbaceous composition on the floodplains, driven by variation in the timing and duration of wet season rains (McGregor et al. 2010).

Shrubland Swamps

Shrubland swamps are also extremely variable and can consist of floating mats of shrubs (e.g., dominated by *Aeschynomene indica, Sesbania javanica* in Southeast Asia), to dense woody shrubland with a wide range of species. Swamp shrubland around Tonlé Sap Lake in Cambodia includes species such as *Barringtonia acutangula, Bridelia cambodiana, Brownlowia paludosa, Capparis micrantha, Cissus hexangularis, Quisqualis indica, Croton mekongensis, C. krabas, Dalbergia pinnata, Gardenia kambodiana, Gmelina asiatica, Phyllanthus taxodiifolius, Popowia diospyrifolia, Sesbania javanica, Stenocaulon kleinii, Terminalia cambodiana,* and *Vitex holoadenon.* At Danau Sentarum in West Borneo common shrubland species include *Barringtonia acutangula, Carallia bracteata, Croton* cf. *ensifolius, Garcinia borneensis, Gardenia tentaculata, Ixora mentanggis, Pternandra teysmanniana, Memecylon edule, Syzygium claviflora,* and *Timonius salicifolius* (Giesen 2000). Shrub wetlands are not that extensive in Amazonia, where the cover about 2,490 km^2, or about 0.3% of all wetlands in the basin (Melack and Hess 2010). Shrub wetlands of the South American Pantanal are often dominated by *Vochysia divergens*, along with *Gallesia* species, *Astronium graveolens* and *Ocotea s*pecies (Ramsar data sheet on Pantanal). In Northern Australia, shrubland swamps are generally not recognized (e.g., Finlayson and Von Oertzen 1993), although *Barringtonia acutangula* and *Cassia* species are common shrubby elements of these wetlands, often lining watercourse and billabongs (Braithwaite et al. 1989), and vast areas have been invaded by the South American invasive alien shrub *Mimosa pigra* (Braithwaite et al. 1989; Cowie and Werner 1993). Shrub wetlands are also not recognized in descriptions of African wetlands (Denny 1993a), but numerous shrub species can form important elements, including *Abutilon longicuspe*, *A. mauritanum*, *Azima tetracantha*, *Cadaba farinosa*, *Capparis erythrocarpos*, *C. sepiaria*, *Celtis wrightii*, *Erythrococca bongensis*, *Grewia densa*, *G. tenax*, *Hibiscus calophyllus*, *Maerua kirkii*, *M. subcordata*, *Maytenus undata*, *Pavonia zeylanica*, *Phyllanthus ovalifolius*, *Salacia erecta*, *Salvadora persica*, *Tamarix nilotica*, and *Uvaria welwitschii* (Hughes and Hughes 1992).

Savanna Swamps

Savanna swamps are in most cases associated with seasonal fires. This does not mean that all regularly burnt wetlands end up as savannas, i.e., grass- or sedgelands with dispersed woody vegetation, as a fairly closed canopy may still occur if fires are infrequent or if species are particularly fire tolerant (e.g., in Northern Australia, McGregor et al. 2010; Indonesian Borneo, Van Steenis 1957; Giesen 2000).

In South America, savanna wetlands covered by nearly monospecific stands of palms, including *Mauritia flexuosa* and *Copernicia alba*, occur in the upper Negro River in Colombia, the upper Amazon in Peru, and the Llanos de Moxos (Melack and Hess 2010), while palm savannas with *Copernica australis*, *Mauritia flexuosa*, *Manicaria saccifera*, *Raphia taedigera*, and *Euterpe oleracea* occur in Pantanal do Mato Grosso. Savanna vegetation in Roraima (Amazonia) is dominated by grasses such as *Aristida setifolia*, *Axonopus aureus*, *Andropogon angustatus*, *Mesosetum liliiforme*, *Paspalum setllatum*, *P. Melanospermum*, along with palms such as *Mauritia flexuosa* and trees such as *Virola* and *Sumphonia* species (Junk 1993). Other savannas in Humaita (Amazonia) are dominated by the palms *Orbignya martiana* or *Mauritia armata*, along with grasses such as *Schizachyrium brevifolium* or sedges such as *Cyperus haspan* and herbs such as *Rhynchanthera grandiflora* and *Montrichardia arborescens* (Junk 1993).

In the Sudd swamps of East Africa, wooded wetlands (which form a minor element) are dominated by *Acacia seyal* and *A. siberiana*, along with *Balanites aegyptiaca* (see ▶ Chap. 106, "The Sudd (South Sudan)" by Rebelo and El-Moghraby, this volume). Savanna wetlands are otherwise generally only a minor element in African rangelands, and not described by Lind and Morrison (1974), Hughes and Hughes (1992) or Denny (1993a, b). Savanna (and forested) wetlands of Northern Australia are dominated by paperbark trees, often the same species as on New Guinea, including *Melaleuca cajuputi*, *M. leucadendron*, *M. nervosa*, and *M. viridiflora*, along with a wide variety of grasses (Finlayson and Von Oertzen 1993). In Southeast Asia, savanna swamps may consist of palm swamps, with species such as *Livistona saribus* and *Pholidocarpus sumatranus*, or secondary swamps dominated by *Melaleuca cajuputi* that emerge after mixed swamp forest is degraded by fires (Giesen 1989). In New Guinea, savanna and woodland wetlands occur with *Campnosperma, Carallia brachiata, Mitragyna ciliata, Nauclea coadunata, Syzygium*, and *Timonius* species and, in the southern part of the island, species of *Melaleuca, Acacia*, and *Tristania*; most common are *Melaleuca cajuputi, Melaleuca leucadendron*, and *Melaleuca viridiflora* (Osborne 1993).

Swamp Forests

The largest areas of freshwater swamp forest are found in South America, particularly in Amazonia where two main types of swamp forest are recognized: seasonally flooded forest along nutrient rich whitewater rivers (varzea) and swamp forests along nutrient poor black water rivers (igapo). Varzea forests may be flooded for 50–210 days per year, with depths up to 10–15 m (Parolin et al. 2004). Common tree species of the igapo are *Calophyllum brasiliense* and *Piranhea* species, while common varzea species include *Crudia amazonica, Senna reticulata, Gustavia augusta, Pouteria glomerata, Rheedia brasiliensis, Symmeria paniculata*, and *Tabernaemontana juruana*. Tree species abundance may be as high as 172 species per hectare in late successional high varzea forest (Wittmann et al. 2004). Floristic composition and tree species richness in varzea forests is strongly related to the height and the duration of annual floods and local geomorphology, which leads zonation along the flood gradient. Many high-varzea (flooded 1–2 m) trees are dense-wood species, such as *Virola calophylla, Calophyllum brasiliense*, and *Ocotea cymbarum*, while low varzea species (flooded 3–7 m) include *Laetia corymbulosa, Pseudobombax munguba, Calycophyllum spruceanum*, and *Tabebuia barbata* (Parolin et al. 2004).

Freshwater swamp forests of Central Africa (e.g., Congo basin) are characterized by species such as *Newtonia devredii* and *Uapaca guineensis*, along with *Acioa deweverei, Berlinia grandiflora, Dacryodes edulis, Dalbergia louisii, Dialium* species, *Epinetrum villosum, Eriocoelum microspermum, Homalium molle, Mitragyna stipulosa, Parinari glabra*, and *Symphonia globulifera* (Hughes and Hughes 1992). In East African freshwater swamps, characteristic tree genera are *Caraipa, Mitragyna, Nauclea, Pandanus, Spondianthus, Symphonia, Uapaca*, and *Voacanga*, which occur along with *Phoenix* and *Raphia* palms (Denny 1993a).

South Asian swamp forests include *Myristica* swamp forests (Kerala, India), creeper swamp forests with *Drimycarpus racemosa, Eugenia formosum, Machilus gamblei, Magnolia griffithii, Vatica lancaefolia*, and climbers such as *Calamus leptospadix, C. tenuis*, and *Cissus* and *Uncaria* species (along the Brahmaputra River, India), seasonal swamp forests such as those dominated by *Barringtonia acutangula, Dillenia indica*, or *Syzygium cumini* (Gopal and Krishnamurthy 1993). Swamp forests around Tonlé Sap Lake in Cambodia are 7–15 m tall and dominated by *Barringtonia acutangula* and *Diospyros cambodiana*, accompanied by climbers *Combretum trifoliatum*, *Breynia rhamnoides*, and *Acacia thailandica* and trees including *Crataeva nurvala*, *C. roxburghii*, *Terminalia cambodiana*, and *Coccoceras anisopodum*. Stunted swamp forest in West Borneo is characterized by trees 8–15(26) m such as *Diospyros coriacea*, *Vatica* cf. *umbronata*, and *Mesua hexapetalum*, along with *Cleistanthus sumatranus*, *Crudia teysmannii*, *Fordia splendissima*, *Garcinia bancana*, *Homalium caryophyllaceum*, *Ilex cymosa*, *Microcos* cf. *stylocarpa*, and *Xanthophyllum affine*. A subtype formed by repeated burning is dominated by *Shorea balangeran*. Tall swamp forest in West Borneo has trees 25–30 (–35) m and key species include *Dryobalanops abnormis*, *Hopea mengerawan*, and *Tristaniopsis obovata*, along with *Calophyllum* species, *Dichilanthe borneensis*, *Gluta pubescens*, *Gluta wallichii*, *Ilex cymosa*, *Shorea balangeran*, *Teysmanniodendron sarawakanum*, and *Vatica ressak* (Giesen 1987, 1996, 2000).

Mixed swamp forests have been described from various locations on New Guinea and common species include *Alstonia scholaris*, *Campnosperma brevipetiolata*, *Carallia brachiata*, *Dillenia alata*, *Garcinia dulcis*, *Gmelina* species, *Hopea novoguineensis*, *Myristica hollrungii*, *Pandanus tectorius*, *Rhus taitensis*, *Stemonurus* species, *Terminalia copelandii*, and *Vatica papuana*. *Campnosperma brevipetiolata* may be locally dominant, especially on peaty soils, and can form stands up to 35(–50) metres tall. Other typical swamp forest types are sago swamp dominated by *Metroxylon sagu*; *Pandanus* swamps with a range of pandan species including *Pandanus hollrungii*, *P. hysterix*, *P. kaernbachii*, *P. lauterbachii*, *P. leiophyllus*, *P. scabribracteatus*, and *P. tectorius*; and swamps dominated by *Erythrina*, *Barringtonia*, or *Leptospermum* (Eden 1974; Paijmans 1976).

Swamp forests of Northern Australia include palm swamps dominated by *Archontophoenix alexandrae* and *Licuala ramsayi*; swamp forests dominated by paperbark trees, including *Melaleuca cajuputi*, *M. leucadendron*, *M. nervosa*, and *M. viridiflora* (Finlayson and Von Oertzen 1993); and floodplain swamps dominated by *Eucalyptus camaldulensis*, often with species such as *Barringtonia acutangula*, *Bauhinia cunninghamii*, *Casuarina cunninghamiana*, *Ficus coronata*, *F. racemosa*, *Melaleuca argentea*, *M. bracteata*, *Nauclea orientalis*, *Planchonia caeya*, and *Tristania suaveolens* (Beadle 1981). On the whole, *Melaleuca* forests occur where disturbance by fire and/or floodwater is too great for rain forest to persist, rendering them the wetland analogue to the eucalypts that usually dominate well-drained portions of the north Australian environment (Franklin et al. 2007).

Ecosystem Services

Buffering Water Levels

The rise and fall of water levels in freshwater swamps may be significant. At Tonlé Sap the average difference between wet and dry season levels is 8.2 m (Campbell et al. 2006), while at Danau Sentarum it may be as much as 12 m

(Giesen 2000) and in parts of the Amazon even 15 m (Parolin et al. 2004). Not all floodplains are this deeply flooded, and much of the Kafue flats is inundated by less than 1 m of water, and much of the Sudd is flooded by 1–2 m. The temporary storage of water in floodplain wetlands modulates the flow in connected rivers. Freshwater swamps along the upper course of the Kapuas River (Indonesia's longest river) in West Borneo act as a buffer, as one quarter of peak floods are siphoned off, and in the dry season up to 50% of upper Kapuas River waters may consist of water draining from the lakes and swamp forests (Giesen and Aglionby 2000). Similarly, water stored in Lake Tonlé Sap and surrounding flooded forests contributes an estimated 16% of the dry season (December–May) flow in the Mekong River and its distributaries downstream of Phnom Penh (Campbell et al. 2006). Emergent vegetation slows water flow because of its hydraulic resistance, and clearing of vegetation can speed up flows and contribute to both increased flooding and desiccation. At the same time, the rise and fall of water levels in a floodplain wetland has many variables, and the period during which a wetland is flooded or waterlogged is termed flooding regime or hydroperiod.

Buffering of Water Quality

Water is lost from freshwater swamps due to evaporation and transpiration (via emergent plants), and in parts of the Sudd this is estimated to be as much as 50% (Sutcliffe 1974). Freshwater swamps may also buffer water quality, and studies along the Tanzanian shores of Lake Victoria indicate that peripheral *Cyperus papyrus* swamps removed almost half of the suspended solids and more than 90% of inorganic-P from waters passing through these swamps (Arcadis Euroconsult 2001). Floating water hyacinth *Eichhornia crassipes* has been found to remove 49% of nitrogen and 11% of total phosphorous (Mitsch 1977, in Mitsch and Gosselink 2000). Nutrient removal aspects depend much on vegetation type, residence time, volumes, and concentrations, but nevertheless these positive impacts have lead to the development of treatment wetlands that are specifically constructed to mimic natural floodplain swamps, in this case by improving water quality of wastewaters (e.g., Kadlec 1994; Mitsch 1994; Cole 1998; Kadlec et al. 2000; Tanner 2001). Not all swamps respond in the same way, and many are net exporters of organic matter (e.g., papyrus marshes, Gichuki et al. 2001; tropical floodplain forests, Mata et al. 2012). Nutrients stored in swamp vegetation are partitioned between above- and belowground parts, and this changes with the season. In the growing season (e.g., shortly after the wetland is freshly inundated), nutrients are mobilized from roots and rhizomes towards shoots. As senescence sets in at the end of the growing season (e.g., as marshes dry out), some of the nutrients are transferred back to belowground parts, but most are lost in litter and are subsequently decomposed (Mitsch and Gosselink 2000).

Productive Systems

The rise and fall of waters in floodplain wetlands often results in highly productive systems, as predictable seasonal rains provide replenishing floodwaters that may deposit nutrient-rich sediment and stimulate nutrient turnover which, in turn, substantially boosts both primary and secondary productivity. At the same time there are shifts in the types of primary producers and associated food webs (Pettit et al. 2011). Species that occur in both wetland and terrestrial habitats are often more productive in the former setting; production of litter by *Melaleuca quinquenervia* trees, for example, was found to be significantly higher in seasonally flooded swamps (9.3 t/ha.year) than in nonflooded areas (7.5 t/ha.year) in southern Florida (Van et al. 2002).

Many freshwater swamps in the floodplains support a diverse and abundant fish fauna (Lowe-McConnell 1987; Agostinho and Zalewski 1995; Winemiller and Jepsen 1998). Floodplain wetland fisheries peak as floodwaters recede, and fish migrating back to main bodies of water are often intercepted by a barrage of nets, traps, and eager fisherfolk. Welcomme (1979) established that average fisheries production (in metric tons/year) in floodplain wetlands worldwide was 3.83 times the area, expressed in km^2; in other words, floodplain fisheries produce almost 40 kg of fish for every hectare flooded. This seems low, but as waters recede fish concentrations will be much higher, resulting in high yields per unit of fisherfolk effort.

Temperate freshwater swamps have a net primary productivity (NPP) in the range of 500–6,000 g/m^2.year, averaging at about 2,100 g/m^2.year (Mitsch and Gosselink 2000), while tropical freshwater swamps range from 50 g/m^2.year for oligotrophic *Eleocharis* marshes in Belize to 10,220 g/m^2.year for *Echinochloa polystachya* swamps in the Brazilian Amazon (Table 3). The tropical figures average at 2,500+ g/m^2.year, which is higher than that for temperate swamps, but the figures provided in Table 3 scarcely cover the wide range of habitats encountered. In any case, the *Echinochloa* swamps (Morison et al. 2000) and *Cyperus papyrus* swamps (with a NPP of 6280 g/m^2.year; Jones and Muthuri 1997) are among the most productive swamps in the world and form very important carbon sinks.

Biodiversity

Tropical freshwater swamps support a wide biological diversity (Table 4), and due to their productivity, this diversity is often (seasonally) also in great numbers and concentrations. Fisheries of Lake Tonlé Sap (Cambodia) and Danau Sentarum (West Borneo) are highly productive and key to this industry in their region. The Kafue Flats (Zambia), Kakadu NP (Northern Australia), Okavango (Botswana), Sudd (South Sudan), Trans-Fly (Papua), and Pantanal (Brazil, Bolivia, Paraguay) support both a high avifauna diversity and seasonally (very) large numbers of water birds.

Many freshwater swamps also support unique species. Danau Sentarum, for example, supports 12–26 endemic fish species, the largest inland population of the Borneo endemic proboscis monkey *Nasalis larvatus,* and a number of endemic trees and palms including *Dichilanthe borneensis, Rhodoleia* spec. nov.*, Dicoelia*

Table 3 (Aboveground) productivity of tropical swamp vegetation

Wetland type	Name and location	Productivity g/m².year	Reference
Submerged macrophytes (*Hydrilla, Potamogeton*)	Ganges plain, India	106–266	Shardendu and Ambasht (1991)
Oligotrophic *Eleocharis* marshes	Belize, Central America	50–565	Rejmánková (2001)
Melaleuca quinquenervia swamps	Southern Florida, USA[a]	930[b]	Van et al. (2002)
Floodplain *Melaleuca*	Northern Australia	700–1,500[b]	Finlayson (2005)
Floodplain grasslands	Northern Australia	500–2,100	Finlayson (2005)
Scirpus giganteus	De la Plata River, Argentina[a]	1,514	Pratolongo et al. (2005)
Floodplain swamp	Southern Florida, USA[a]	1,607	Mitsch and Gosselink (2000)
Tropical freshwater swamps	Worldwide	1,500–3,000	Aselmann and Crutzen (1989)
Cyperus papyrus swamps	Lake Victoria, Kenya	6,280	Jones and Muthuri (1997)
Echinochloa polystachya	Brazilian Amazon	9,490–10,220	Morison et al. (2000)

[a]Subtropical
[b]Using litterfall as a proxy

beccariana, Eugeissona ambigua, and *Plectocomiopsis triquetra* (Giesen and Aglionby 2000; Giesen 2000). The Kafue lechwe *Kobus leche kafuensis* is endemic to the Kafue Flats and is one of three different races of lechwe, or marsh antelope, specifically adapted to living in wetlands (Schelle and Pittock 2005). The Sudd has an endemic grass *Suddia sagittifolia*; the endemic Nile lechwe, *Kobus megaceros*; and the rare swamp deer or sitatunga, *Tragelaphus spekei* (Riak 2006). Kakadu's fauna includes an endemic family of shrimp, the *Kakaducarididae*, comprising two endemic genera, *Leptopalaemon* and *Kakaducaris*; an endemic isopod genus, *Eophreatoicus*; and three endemic fish species, *Craterocephalus marianae, Syncomistes butleri*, and *Pingalla midgleyi* (Finlayson et al. 2006). The Trans-Fly region supports key species such as the New Guinea pig-nosed turtle *Carettochelys insculpta*, the critically endangered Fly River leptomys *Leptomys signatus*, dusky pademelon *Thylogale bruinii*, and Fly River trumpet-eared bat *Kerivoula muscina*. The Trans-Fly is also the only habitat of the New Guinea marsupial cat or quoll *Dasyurus albopunctatus* and the bronze quoll *Dasyurus spartacus* (Heads 2002; wwf website 2012).

Not all major freshwater swamps have unique species but are unique because of sheer numbers. The Okavango Delta, for example, has no endemic species but has a wide variety of large mammals occurring locally in high numbers, including African elephant, hippo, and white rhinoceros, which are the main attraction for the tourism industry (Ramberg et al. 2006). During the dry season, Kakadu floodplains are used intensively by up to 2 million water birds, including large concentrations of geese and ducks (Finlayson et al. 2006). Swamp forests of Tonlé Sap have some of the

Table 4 Species diversity in tropical freshwater swamps

Wetland	Size km²	Number of species per group					
		Phanerogams	Fish	Amphibians	Reptiles	Birds	Mammals
Danau Sentarum	1,500	262	250	<10	30–50	237	55
Kakadu	1,950	319	62	25	127	280	60
Kafue Flats	6,500	?	67	?	?	450	83
Tonle Sap	12,500	200	149(–500)	18	30–50	210	18
Trans-Fly	13,000	?	105	25	102	360	50
Okavango	28,000	500+	71	33	64	444	122
Sudd	100,000	350	100	?	?	470	100
Pantanal	250,000	1,863	263	41	113	463	132

References: Danau Sentarum (Giesen and Aglionby 2000), Kakadu (Friend and Cellier 1990; Finlayson et al. 2006), Kafue flats (Hughes and Hughes 1992; Robinson et al. 2002; Schelle and Pittock 2005), Tonle Sap (Campbell et al. 2006; Davidson 2006), Trans-Fly (WWF website; IUCN 1995; Heads 2002; Allison 2006), Okavango (Ramberg et al. 2006), Sudd (Riak 2006; see ▶ Chap. 106, "The Sudd (South Sudan)" by Rebelo and El-Moghraby, this volume), Pantanal (Alho 2008)

Table 5 Carbon sequestration in freshwater swamps

Wetland type	Carbon sequestration gC/m^2.year	Reference
Northern bogs	45	Cao et al. 1996
Net wetlands worldwide	118	Mitsch et al. 2012
Flow-through freshwater wetland, Ohio	124–160	Mitsch et al. 2010
Flow-through freshwater wetland, Costa Rica	250–260	Mitsch et al. 2010
Tropical swamps	820	Cao et al. 1996
Papyrus swamps, Kenya	1,600	Jones and Humphries 2002
Papyrus swamps, Kenya	2,510	Muthuri et al. (1989) in Saunders et al. 2007
Papyrus swamps, Uganda	3,090	Saunders et al. 2007
Echinochloa polystachya swamp, Brazil, Central Amazon	3,970	Piedade et al. (1991), in Saunders et al. 2007

largest breeding colonies of large water birds in Southeast Asia, including endangered species such as *Anhinga melanogaster, Threskiornis melanocephalus, Pelecanus philippensis, Mycteria cinerea, M. leucocephala, Leptoptilos dubius,* and *L. javanicus*; the colony of *Pelecanus philippensis* is even the largest in the world, with 25% of the world population at this one site (Campbell et al. 2006). The Sudd annually supports among others about 300,000 open bill stork *Anastomus lamelligerus*, 100,000 cattle egret *Bubulcus ibis*, 100,000 spur-wing goose *Plectropterus gambensis*, 20,000 black crowned crane *Balearica pavonina*, and 1.7 million glossy ibis *Plegadis falcinellus* (Riak 2006).

Carbon Storage

Various studies indicate that carbon sequestration in tropical freshwater wetlands is significantly higher than in similar temperate habitats (80% higher, according to Mitsch et al. 2010), but figures range widely (Table 5). However, all swamps also emit methane, which is a powerful greenhouse gas, about 25 times as much as CO_2. Although tropical swamps on the whole do not appear to emit more methane than temperate swamps (Table 6), they do not have 25 times more CO_2 sequestration than methane emissions and can be considered by some to be sources of climate warming (Mitsch et al. 2012).

Dependent Peoples

Freshwater swamps on mineral soils are often very productive, and the combination with seasonal abundance of (potable) water and waterways suited for transport makes these systems attractive for people, who in many cases have inhabited these

Table 6 Methane emission by freshwater swamps

Wetland type	Methane emission gC/m².year	Reference
Northern wetlands	11	Cao et al. 1996
Australian billabongs	12–22	Sorrell and Boon (1992), in Mitsch et al. 2012
Created marshlands, Ohio	30	Mitsch et al. 2012
Amazon basin wetlands	30	Melack et al. (2004) in Mitsch et al. 2012
Flow-through freshwater wetland, Costa Rica	33	Mitsch et al. 2012
Temperate wetlands	41	Cao et al. 1996
Tropical wetlands	54	Cao et al. 1996
Natural freshwater wetland, Ohio	57	Mitsch et al. 2012
Freshwater wetlands, Lousiana	3–225	Delaune and Pezeshki (2003), in Mitsch et al. 2012
Isolated floodplain wetlands, Costa Rica	220–263	Mitsch et al. 2012

areas for millennia. Livelihoods are often based on (combinations of) floodplain fisheries (see "Ecosystem Services"), livestock grazing, hunting (birds, reptiles, amphibians, mammals), foraging in floodplain vegetation (fruits, seeds, vegetables), and planting of crops as waters recede (e.g., vegetables, recession rice). Receding floodwaters lead to very high concentrations of fish (e.g., Tonlé Sap, Lamberts 2001; Danau Sentarum, Dudley 2000), and productive herbaceous swamps can attract large numbers of birds (e.g., water birds in the Sudd; magpie geese at Kakadu NP, Australia) and mammals (e.g., Kafue flats, Okavango), in turn attracting fisherfolk and hunters. In Amazonia, freshwater swamp forests produce more edible fruit than nearby dryland "terra firma" forest types, and are more productive than hunting or cattle ranching (Phillips 1993). Various wild grass species with edible seeds such as *Oryza rufipogon, Leersia hexandra*, and various *Panicum* species are typical for these habitats, and it is highly likely that rice *Oryza sativa* is domesticated from a wild stock – possibly *O. rufipgon* – originally found in tropical freshwater swamps (Nesbitt 2005).

As is evident from various programs (e.g., Wetlands International 2009), human health and well-being is closely related to the health of the wetlands, and this is particularly evident in tropical freshwater swamps. During the past decades, many traditional freshwater swamp systems have succumbed to overutilization and degradation of natural resources, often due to a combination of various causes. A number of cases below illustrate the trends.

Danau Sentarum NP, West Borneo, consists of a series of floodplain lakes surrounded by large areas of swamp forest. The area supports a large traditional fishing industry, which in the 1990s was utilized by over 6,500 fisherfolk inhabiting 39 villages in and adjacent the Park. Although illegal, swamp forests are heavily utilized as well, both for construction timber and for a wide variety of nontimber forest products such as rattan and honey. Exploitation levels appear to have been sustainable until about two to three decades ago; since then, however, the resource base has been steadily eroding, with fish catches declining and forest area dwindling. The main reasons are complex, involving the influx of immigrants, increased nonadherence to local customary law, population increase, increased access to external markets, and a steady development of adjacent areas (including logging and development of oil palm plantations; Giesen and Aglionby 2000).

Kafue Flats, Zambia. These wetlands are flooded for 1–7 months per year, and when dry, most of the grasslands are grazed by cattle owned by local herdsman. The annual flooding largely prevents any other type of traditional land use, but in the dryer parts some corn is grown. Fishing villages are found on the higher levees along the main river and its tributaries. Traditional inhabitants of the flats live in mostly permanent villages, while migrants from other areas usually occupy semipermanent villages that have to be abandoned during high floods. When commercial fishing became important in the 1950s people from the other fishing tribes moved onto the flats and now seem to be replacing the original inhabitants. People in the area originally practiced shifting agriculture, but increasing population and habitat deterioration have caused more people to be crowded on less land resulting in constant use of the same land and subsequent deterioration of the soil (Smardon 2009). The wet season is a lean period, and hunting of animals, which might have filled the gap, is no longer available to most people. Tight control over access to the floodplain is held by the Litunga, the traditional king of Barotseland (which includes the flats), and the annual migration with the flood is celebrated in a special ceremony.

In the *Kakadu region* of Northern Australia, freshwater wetlands have always been key habitats for food and other resources for Aboriginal people and continue to be critically important today. Kakadu freshwater wetlands, which consist of grasslands, sedgelands, and *Melaleuca* swamps, have developed from mangroves over the past 10,000 years due to falling sea levels, and hence the entire history of these ecosystems has been in association with people (Finlayson and Woodroffe 1996). Aboriginal use of fire to manage these wetlands, including the thinning of dense stands of the native grass *Hymenachne acutigluma* in order to promote a variety of other food resources, has been widespread and has formed and transformed these habitats over time (McGregor et al. 2010).

Threats and Trends

Threats to freshwater swamps on mineral soils are highly variable, and depends vary much on local circumstances and the type of wetland, and may be external (i.e., from outside the wetland), internal, or both. Where population pressures are high and availability of arable land is low (e.g., South and Southeast Asia), reclamation and conversion form main threats. Pressures may be high elsewhere (e.g., high water usage or industries in upstream areas), leading to problems such as drying out or pollution of the wetland. Various types of threats are dealt with briefly below.

Reclamation and Conversion

Reclamation and conversion of tropical freshwater swamps is one of the main causes of loss of these habitats, especially in South and Southeast Asia. In Indonesia, only 45% of the original 11.5 Mha of freshwater swamp forest remained in 1985 (Silvius et al. 1987), varying from 6.3% on Java and 97% in Indonesian Papua, and it is expected that only about a quarter remains (2012). Scales of reclamation are highly variable and may vary from a few (hundred) hectares encircled by a low bund (e.g., in the *haor* region in northeastern Bangladesh), to 10–20,000 ha development schemes (e.g., for transmigrants) in Sumatra, Indonesia, to 100,000 ha in South Sudan where excavation of the Jonglei Canal was started in 1980 (but postponed in 1983 following the civil war) to drain a large part of the Sudd swamps (see AI. Jonglei Canal – impact on the Sudd by Rebelo and El-Moghraby 2012b, this volume). Loss of fertility due to lack of floods and loss of crops and human lives following breaching of dikes have led to discussions on the wisdom of conversion (e.g., Hughes et al. 1994) but have not slowed the overall trend of conversion.

Damming (On-Site) and Flooding

Freshwater swamps may be dammed to contain water for irrigation or drinking purposes, or for promoting lake/pond fisheries. At Beung Boraphet, in Nakhon Sawan Province, Thailand, a freshwater swamp of 64,000 ha was dammed in 1930 in order to develop a lake fishery. The swamp has now largely disappeared apart from some minor fringes, and instead there now exists a 13,000 ha lake (Jintanugool and Round 1989; in Scott 1989). The construction of a small dam at the outlet of Tasek (Lake) Chini in Pahang, West Malaysia, destroyed parts of swamp forest (Shuhaimi-Othman et al. 2007), and similar plans have proposed on various occasions for Danau Sentarum, West Borneo (Giesen 1987, 1996). Permanent flooding can kill a swamp forest, as most tree species need a period during which their roots and the soil is well aerated – the exact mechanism is not well understood but seems to involve both lack of oxygen and sulpfide toxicity in anoxic permanently flooded soils. Excess water expelled from the Duri Streamflood oil fields near Duri, Riau province, Indonesia, into a small stream lead to permanent inundation of a stand of several hundred hectares of swamp forest, and dying off of trees within 2 years. An independent investigation discovered that the cause was not oil or thermal pollution (as was expected) but was simply due to permanent flooding stressing the vegetation (Northwest Hydraulic Consultants et al. 1994). In swamp forest at Songkram, Nakhon Phanom province Thailand, permanent flooding caused by bund construction lead to stress of the vegetation, leading to shedding of leaves, production of many aerial roots, and eventually death of trees (pers. obs. June 2006).

Damming (Off-Site) and LU Changes Upstream

Upstream damming of waters is a threat to many freshwater swamps, as regulation and modification of river flow reduces the area of flooding, and hence input to, and total area of freshwater swamps. Few large rivers in the tropics remain unregulated, especially in drier regions where water regulation is perceived as more acutely needed, especially for irrigation and supply of potable water. The major impact of large-scale dam development on river is often an increase in dry season flows and a decrease in wet season flows downstream.

Water diversion in the upper parts of the Indus River in Pakistan has resulted in serious threats to the lower riparian ecosystems and local communities, and impacts include lack of replenishing of wetlands, overexploitation of ground water resources, and degradation of quality of water (Nasir and Akbar 2012). Due to the Sélingué reservoir and irrigation works, water levels in the Inner Niger Delta have been lowered by 20–25 cm, and as a result the inundated area of the Inner Niger Delta in Mali has decreased by 900 km^2. A further three dams are planned and are likely to have further impacts on inundated area, livestock grazing, fisheries, and livelihoods (Wetlands International 2012).

Hydropower has also contributed to widespread regulation of rivers, with resulting impacts on freshwater swamps. In the Mekong River Basin, 30 hydropower dams of 10 MW or more existed by 2011, and at least another 13 are planned up to 2020. This will reduce the area of land subject to seasonal inundation around Tonlé Sap Lake. Planned development of two large dams in China alone could result in an increase in dry season minimum water levels in Tonlé Sap Lake of about 0.3 m and a decrease in wet season maximum water levels of about 0.4 m, leading to an 11% decrease in inundated area from 10,620 to 9,164 km^2 (Campbell et al. 2006). The construction of the Lower Kafue Gorge dam in 1971 and the Itezhi-tezhi dam in 1978 on the Kafue River in Zambia significantly affected the flooding regime of the Kafue Flats. As a result, large parts of these freshwater (grassland) swamps were no longer flooded and were invaded by shrubs, lowering the grazing value of the area to livestock. At the same time, populations of Kafue lechwe antelope *Kobus leche kafuensis* declined from 90,000 in the 1970s to 37,000 in the late 1990s. Over 70 dams are planned for Brazil's Amazonian region alone, and this is expected to have major impacts on freshwater wetlands and their biota (Pringle et al. 2000).

Widespread land use changes in upstream areas are also sufficient to alter stream water quality and hydrology and impact downstream freshwater wetlands. In the arid Loitokitok region of southern Kenya, Githaiga et al. (2003) found land use changes to be "potentially devastating ecological, cultural and socio-economic consequences," while land use changes modeled for the Niger and Chad rivers in West Africa showed that total deforestation would increase annual stream flow by

35–65%, depending on location in the basin, although forests occupy only a small portion (<5%) of the total basin area (Li et al. 2007). A 3,400 km canal ("Hidrovia") linking waterways and wetlands in Brazil-Paraguay has been on the drawing table for decades (Gottgens et al. 1998, 2001), but if implemented is expected to have major impacts on key wetlands such as the Pantanal.

Pollution and Erosion

Pollution occurs in many forms and is invariably associated with development upstream and along fringes of freshwater swamps. The most common form of pollution is eutrophication, or an overabundance of nutrients (mainly N and P). These nutrients often stem from point sources of organic waste (e.g., markets, slaughterhouses, sewerage outlets), or from diffuse sources such as from fertilized cropland and paddocks. In freshwater swamps, mild eutrophication can lead to increased productivity (e.g., fisheries) and not be directly negative. However, mild conditions are usually short-lived and in most cases eutrophication leads to increased plant growth, an overabundance of floating species such as *Salvinia molesta, Eichhornia crassipes,* and *Pistia stratiotes,* algal growth, and anoxic conditions (leading to mortality of aquatic species such as fish). High levels of eutrophication can cause harmful algal blooms (e.g., cyanobacteria, *Phaeocystis*), whereby toxins are released into waters so that drinking or bathing can be hazardous. Mining in the catchment of a freshwater swamp can also result in chemical pollution. Leachate from mining tailings is often very acid (e.g., from copper mines it is commonly pH 3–4; Pond et al. 2005) and, depending on the type of mine, can also include a variety of toxic heavy metals (e.g., Cd, Cu, Pb, Hg, Zn, Sn). Gold mining is a special case, as it may attract illegal and unregulated mine operators that make use of hazardous techniques, and associated forms of pollution include mercury and cyanide (Irwin et al. 1997; UNEP 2011).

Land use changes (e.g., deforestation, unsustainable agricultural practices) in the catchment of a tropical freshwater swamp can lead to enhanced erosion, and significant increases in sediments entering a wetland. This can lead to (much) lowered light penetration in water, eutrophication, and smothering of submerged (components) of the vegetation. Over time siltation can contribute to reduction in size of the swamp, depending on the volume ratio between wetland and incoming sediments. Chalan Beel in Bangladesh, for example, was originally 107,000 ha at the turn of the twentieth century, but this rapidly silted up following developments upstream, and by the late 1980s only 26,000 ha remained (Scott 1989). Freshwater swamps of Doon valley in Uttarakhand (Northern India) have been much affected by siltation caused by deforestation in the Himalayas (Sharma et al. 2010). Siltation can also form new swamps in deltas that occur where incoming streams deposit silt as they enter lakes, for example, at the mouth of the Sondu River, on the Kenyan shore of Lake Victoria in the 1990s (Kairu 2001).

Fires

Some freshwater swamps are subject to regular fires, as fuel that accumulates during the wet season becomes inflammable during dry periods, e.g., the *Melaleuca-Banksia* savannas of the Trans-Fly region in New Guinea (Paijmans 1976) and South and Central Kalimantan (Indonesian Borneo; Giesen 1989, 2009), *Shorea balangeran* swamp forests of Borneo (Van Steenis 1957), and palm savannas of Amazonia (Junk 1993). *Melaleuca* forests occur where disturbance by fire and/or floodwater is too great for rain forest to persist, rendering them the wetland analogue to the eucalypts that dominate well-drained portions of the north Australian environment (Franklin et al. 2007). Fire is described as an all-important factor in controlling vegetation patterns at Danau Sentarum in West Borneo (Giesen 1996, 2000), and apparently has a long history (Anshari et al. 2004). In Northern Australia, aboriginals have used fire to manage wetlands, including for the thinning of dense stands of the native grass *Hymenachne acutigluma* in order to promote a variety of other food resources; the use of fire has been widespread and has formed and transformed these habitats over time (McGregor et al. 2010). Fires play an important role in modifying the swamp forest habitat around Tonlé Sap in Cambodia, and during the dry months plumes of smoke are a common sight throughout the floodplain (Campbell et al. 2006). Fire also plays a key role on African freshwater swamps such as the Okavango, where about a quarter of the total area can burn by the end of the dry season (Heinl et al. 2007). Pollen analysis from cores in *Mauritia* palm swamp covering a period of 32,400–3,500 years B.P. shows that fires began to play a role in these wetlands from 10,500 years B.P. onwards (Ferraz-Vicentini and Salgado-Labouriau 1996), linked to human activity in the area.

Overutilization of Resources

Most tropical freshwater swamps are open-access and many are heavily utilized, and this combination has led to overutilization and degradation. Overfishing is particularly common, especially in areas where traditional systems are replaced by new techniques (e.g., gill-nets) or where there is an influx of new-comers (e.g., at Tonlé Sap). Many resources are often taken from one area, for example, at Danau Sentarum NP, West Borneo, exploitation includes fish (Dudley 2000), rattan (Peters and Giesen 2000), honey (Mulder et al. 2000), timber (Giesen 1987), turtles (Walters 2000), and mammals and birds (Jeanes and Meijaard 2000). Exploitation of freshwater swamps can be high, and the exploitation of watersnakes (esp. *Enhydris enhydris*) at Tonlé Sap, for example, is the world's largest snake hunting operation, with an estimated 3.8 million snakes being collected per annum (Brooks et al. 2007). Overexploitation of timber resources and subsequent deforestation, often in combination with land reclamation and conversion, plays a major role in Southeast Asia and Amazonia. In Indonesia, most of the original 11.5 Mha of freshwater swamp forest has disappeared, or degraded from mixed, tall primary forest to secondary forests dominated by *Melaleuca* (Giesen 1993).

Invasive Species

Invasive alien species have had significant impacts in various tropical freshwater swamps, and there are numerous examples throughout. The introduction of the rusa deer *Cervus timorensis* to Indonesian Papua at Merauke in 1928 lead to an extensive spread of this species to most of the Trans-Fly part of the island. This led to major changes to the local ecosystem, including the reduction of tall swamp grasses and consequent ceasing of breeding of the Australian pelican and magpie goose, reduction of the *Phragmites* reed species, and the extensive spread of *Melaleuca* into grasslands (Marshall and Beehler 2007). Similarly, Asian water buffalo introduced in Northern Australia have a negative effect on woody cover, particularly where they change hydrological conditions and cause salinization (Bowman et al. 2009). Threats to Kakadu wetlands are also posed by other invasive animals such as cane toad *Bufo marinus,* crazy ants *Anoplolepis gracilipes* and nonindigenous fish (Bradshaw et al. 2007). Alien mammals introduced to South American palm swamps include axis deer *Axis axis*, fallow deer *Dama dama*, European hare *Lepus europaeus*, and wild boar *Sus scrofa,* and subsequent habitat modification may have contributed to the local extinction of species such as collared peccary *Tayassu tajacu*, the giant anteater *Myrmecophaga tridactyla*, and tamandua *Tamandua tetradactyla* (WWF website 2012). Removing the alien species does not always result in a quick recovery, as forest patches at Kakadu NP do not show short-term responses to buffalo removal, although in long-term absence of buffalo, the forest tends to develop in complexity by increasing in species richness (Michels et al. 2012).

Invasive alien plants can also pose serious threats to freshwater swamps. Floating invasive species such as water hyacinth *Eichhornia crassipes, Salvinia molesta*, and *Pistia stratiotes* can choke waterways, increase evapotranspiration, cause anoxia and fish mortality, and form ideal breeding areas for vectors of diseases such as malaria and bilharzia. Giant mimosa *Mimosa pigra* can form impenetrable thickets and outcompete native species and has become a serious problem in many swamps outside its native South America, including Northern Australia (Cowie and Werner 1993), Indonesia (Giesen 1989), Cambodia (Campbell et al. 2006), and South Sudan (http://ramsar.wetlands.org/Portals/15/SUDAN.pdf).

Climate Change

Climate change is expected to impact tropical freshwater swamps via sea level rise, and the increased incidence of drought and tropical storms. Sea level rise will affect near-coastal freshwater swamps and may already be impacting freshwater ecosystems of Kakadu NP, Northern Australia, where *Melaleuca* forests that were previously impacted by high densities of feral buffalo, may have continued to decline because of salinization driven by sea level rise (Bowman et al. 2009). Drought plays a major direct role in shaping species distributions and abundance in tropical forests, including swamp forest, and is also associated with increased fire risk in tropical systems (Hannah et al. 2002) circulation. Water that evaporates from extensive tropical wetlands falls as rain over a much wider area (e.g., all of South America plus the southwestern Atlantic Ocean for Amazon basin evaporation), and the drying out of these areas is likely to influence climate over large regions (Carpenter et al. 1992).

References

Agostinho AA, Zalewski M. The dependence of fish community structure and dynamics on floodplain and riparian ecotone zone in Parana River, Brazil. Hydrobiologia. 1995;303:141–8.

Alho CJ. Biodiversity of the Pantanal: response to seasonal flooding regime and to environmental degradation. Braz J Biol. 2008;68(4 Suppl):957–66.

Allison A. 2006. Reptiles and amphibians of the Trans-Fly Region, New Guinea. WWF South Pacific Programme & WWF PNG Madang Office, Madang, Papua New Guinea. Contribution No. 2006–039 to the Pacific Biological Survey, 52 pp.

Anshari GAP, Kershaw A, Van Der Kaars S, Jacobsen G. Environmental change and peatland forest dynamics in the Lake Sentarum area, West Kalimantan. Indones J Quat Sci. 2004;19(7):637–55.

Arcadis Euroconsult. Buffering capacity of wetlands study (BCWS) Final Report, vol. 2 main report. Arnhem: Lake Victoria Environmenal Management Project (LVEMP). United Republic of Tanzania and World Bank; 2001, 183 pp.

Aselmann I, Crutzen P. Global distribution of natural freshwater wetlands and rice paddies, their net primary productivity, seasonality and possible methane emissions. J Atmos Chem. 1989;8:307–58.

Beadle NCW. The vegetation of Australia. Cambridge/London/New York/New Rochelle/Melbourne/Sydney: Cambridge University Press; 1981, 690 pp.

Bowman DMJS, Prior LD, Williamson G. The roles of statistical inference and historical sources in understanding landscape change: the case of feral buffalo in the freshwater floodplains of Kakadu National Park. J Biogeogr. 2009;37:193–9.

Bradshaw CJA, Field IC, Bowman DMJS, Haynes C, Brook BW. Current and future threats from non-indigenous animal species in Northern Australia: a spotlight on World Heritage Area Kakadu National Park. Wildl Res. 2007;34(6):419–36.

Braithwaite RW, Lonsdale WM, Estbergs JA. Alien vegetation and native biota in tropical Australia: the impact of *Mimosa pigra*. Biol Conserv. 1989;48:189–210.

Brooks SE, Allison EH, Reynolds JD. Vulnerability of Cambodian water snakes: initial assessment of the impact of hunting at Tonle Sap Lake. Biol Conserv. 2007;139:401–14.

Campbell IC, Poole C, Giesen W, Valbo-Jorgensen J. Species diversity and ecology of Tonle Sap Great Lake, Cambodia. Aquat Sci. 2006;68:355–73.

Cao M, Marshall S, Gregson K. Global carbon exchange and methane emissions from natural wetlands: application of a process-based model. J Geophys Res. 1996;101(D9):14399–414.

Carpenter SR, Fisher SG, Grimm NB, Kitchell JF. Global change and freshwater ecosystems. Annu Rev Ecol Syst. 1992;23:119–39.

Cole S. The emergence of treatment wetlands. Environ Sci Technol. 1998;32:218–23.

Cowie ID, Werner PA. Alien plant species invasive in Kakadu National Park, tropical Northern Australia. Biol Conserv. 1993;63:127–35.

Davidson PJA. The biodiversity of the Tonle Sap Biosphere Reserve. 2005 status review. Technical report of the UNDP/GEF-funded Tonle Sap Conservation Project. Phnom Penh: UNDP; 2006. 76 pp.

Dennis R, Erman A, Meijaard E. Fire in the Danau Sentarum landscape: historical, present perspectives. Borneo Res Bull. 2000;31:123–37.

Denny P. Wetlands of Africa: introduction. In: Whigham DF, editor. Wetlands of the world: inventory, ecology & management. Volume I, Handbook of vegetation science, vol. 15/2. Dordrecht/Boston/London: Kluwer Academic Publishers; 1993a. p. 1–31.

Denny P. Eastern Africa. In: Whigham DF, editor. Wetlands of the world: inventory, ecology & management. Volume I, Handbook of vegetation science, vol. 15/2. Dordrecht/Boston/London: Kluwer Academic Publishers; 1993b. p. 32–46.

Desbiez ALJ, Santos SA, Alvareza JM, Tomas WM. Forage use in domestic cattle (*Bos indicus*), capybara (*Hydrochoerus hydrochaeris*) and pampas deer (*Ozotoceros bezoarticus*) in a seasonal Neotropical wetland. Mamm Biol. 2011;76:351–7.

Dezzeo N, Herrera R, Escalante G, Chacón N. Deposition of sediments during a flood event on seasonally flooded forests of the lower Orinoco River and two of its black-water tributaries, Venezuela. Biogeochemistry. 2000;49:241–57.

Dudley RG. The fishery of Danau Sentarum. Borneo Res Bull. 2000;31:261–306.

Eden MJ. The origin and status of savanna and grassland in Southern Papua. Trans Inst Br Geogr. 1974;63:47–110.

Ferraz-Vicentini KR, Salgado-Labouriau ML. Palynological analysis of a palm swamp in Central Brazil. J South Am Earth Sci. 1996;9(3/4):207–19.

Finlayson CM. Plant ecology of Australia's tropical floodplain wetlands: a review. Ann Bot. 2005;96:541–55.

Finlayson CM, Von Oertzen I. Northern (tropical) Australia. In: Whigham DF, editor. Wetlands of the world: inventory, ecology & management. Volume I, Handbook of vegetation science, vol. 15/2. Dordrecht/Boston/London: Kluwer Academic Publishers; 1993. p. 195–243.

Finlayson CM, Woodroffe CD. Wetland vegetation. In: Finlayson CM, Von Oertzen I, editors. Landscape and vegetation ecology of the Kakadu Region, Northern Australia. Dordrecht: Kluwer Academic Publishers; 1996. p. 81–112.

Finlayson CM, Lowry J, Bellio MG, Nou S, Pidgeon R, Walden D, Humphrey C, Fox G. Biodiversity of the wetlands of the Kakadu Region, Northern Australia. Aquat Sci. 2006;68:374–99.

Franklin DC, Brocklehurst PS, Lynch D, Bowman DMJS. Niche differentiation and regeneration in the seasonally flooded *Melaleuca* forests of Northern Australia. J Trop Ecol. 2007;23:457–67.

Friend GR, Cellier KM. Wetland herpetofauna of Kakadu National Park, Australia: seasonal richness trends, habitat preferences and the effects of feral ungulates. J Trop Ecol. 1990;6(2):131–52.

Gichuki J, Triest L, Dehairs F. The use of stable carbon isotopes as tracers of ecosystem functioning in contrasting wetland ecosystems of Lake Victoria, Kenya. Hydrobiologia. 2001;458:91–7.

Giesen W. Danau Sentarum wildlife reserve: inventory, ecology and management guidelines. Bogor: WWF/PHPA; 1987, 284 pp.

Giesen W. Vegetation of the Sungai Negara wetlands. In: Zieren M, Permana T, editors. Proceedings of the workshop on "Integrating wetland conservation with land-use development, Sungai Negara, Barito Basin, Indonesia"; 1989 Mar. Banjarbaru, South Kalimantan; 1989.

Giesen W. Checklist of Indonesian freshwater aquatic herbs (including an introduction to freshwater aquatic vegetation). PHPA/AWB Sumatra Wetland Project Report No. 27, Bogor; 1992, 38 pp.

Giesen W. The state of natural wetlands in Sumatra. Implications for conservation, and the general trend in Indonesia. Case study presented at the Workshop on Tropical Environmental Management: Biodiversity for Sustainable Development in SE Asia. Wallace Research University, Dumoga Bone, North Sulawesi, 8–18 Feb 1993; 39 pp.

Giesen W. Habitat types of the Danau Sentarum Wildlife Reserve, West Kalimantan, Indonesia. UK-Indonesia Tropical Forest Management Programme. Project 5 Conservation. Work Plan Activity B.1.1. For Wetlands International Indonesia Programme/Ministry of Forestry – PHPA, Bogor; 1996, 97 pp.

Giesen W. Flora and vegetation of Danau Sentarum: unique lake and swamp forest ecosystem of West Kalimantan. Borneo Res Bull. 2000;31:89–122.

Giesen W, Aglionby J. Introduction to Danau Sentarum National Park, West Kalimantan, Indonesia. Borneo Res Bull. 2000;31:5–28.

Githaiga JM, Red R, Muchiru AN, Van Dijk S. Survey of water quality changes with land use type in the Loitokitok area, Kajiado District, Kenya, LUCID working paper series, vol. 35. Nairobi: International Livestock Research Institute; 2003, 28 pp.

Gopal B, Krishnamurthy K. Wetlands of South Asia. In: Whigham DF, editor. Wetlands of the world: inventory, ecology & management. Volume I, Handbook of vegetation science, vol. 15/2. Dordrecht/Boston/London: Kluwer Academic Publishers; 1993;15:345–414.

Gottgens JF, Fortney RH, Meyer J, Perry JE, Rood BE. The case of the Paraguay-Paraná waterway ("Hidrovia") and its impact on the Pantanal of Brazil: a summary report to the society of wetlands scientists. Wetl Bull. 1998;15:12–8.

Gottgens JF, Perry JE, Fortney RH, Meyer JE, Benedict M, Rood BE. The Paraguay–Paraná Hidrovía: protecting the Pantanal with lessons from the past. BioScience. 2001;51(4):301–8.

Hannah L, Midgley GF, Millar D. Climate change-integrated conservation strategies. Glob Ecol Biogeogr. 2002;11:485–95.

Heads M. Regional patterns of biodiversity in New Guinea animals. J Biogeogr. 2002;29:285–94.

Heinl M, Frost P, Vanderpost C, Sliva J. Fire activity on drylands and floodplains in the southern Okavango Delta, Botswana. J Arid Environ. 2007;68:77–87.

Hughes RH, Hughes JS. A directory of African wetlands. With a chapter on Madagascar by G. Bernacsek. Gland/Nairobi/Cambridge, UK: IUCN – The World Conservation Union/ UNEP – The United Nations Environment Programme/WCMC – The World Conservation Monitoring Centre; 1992, xxxiv + 820 pp.

Hughes R, Adnan S, Dalal-Clayton DB. Floodplains or floodplans: a review of approaches to water management in Bangladesh. London/Dhaka: International Institute for Environment and Development (IIED)/Research and Advisory Services (RAS); 1994.

Humphries MS, Kindness A, Ellery WN, Hughes JC, Bond JK, Barnes KB. Vegetation influences on groundwater salinity and chemical heterogeneity in a freshwater, recharge floodplain wetland, South Africa. J Hydrol. 2011;411:130–9.

Irwin RJ, Van Mouwerik M, Stevens L, Basham W. Environmental contaminants Encyclopedia: mercury. National Park Service. Fort Collins, Colorado, USA: Water Resources Division, Water Operations Branch; 1997. 108 pp

IUCN. The Fly River catchment, Papua New Guinea – a regional development assessment. Published in collaboration with the Department of Environment and Conservation, Boroko, Papua New Guinea. Gland/Cambridge, UK: IUCN; 1995, x + 86 pp.

Jeanes K, Meijaard E. Danau Sentarum's wildlife, part 1: biodiversity value and global importance. Borneo Res Bull. 2000;31:150–229.

John DM, Lévêque C, Newton LE. Western Africa. In: Whigham DF, editor. Wetlands of the world: inventory, ecology & management. Volume I, Handbook of vegetation science, vol. 15/2. Dordrecht/Boston/London: Kluwer Academic Publishers; 1993. p. 47–78.

Jones MB, Humphries SW. Impacts of the C4 sedge *Cyperus papyrus* L. on carbon and water fluxes in an African wetland. Hydrobiologia. 2002;488:107–13.

Jones MB, Muthuri FM. Standing biomass and carbon distribution in a papyrus (*Cyperus papyrus* L.) swamp on Lake Naivasha, Kenya. J Trop Ecol. 1997;13(3):347–56.

Junk WJ. Wetlands of tropical South America. In: Whigham DF, editor. Wetlands of the world: inventory, ecology & management. Volume I, Handbook of vegetation science, vol. 15/2. Dordrecht/Boston/London: Kluwer Academic Publishers; 1993. p. 679–739.

Kadlec RH. Wetlands for water polishing: free water surface wetlands. In: Mitsch WJ, editor. Global wetlands old world and new. Ohio State University conference, 1992 Sep 13–18. Amsterdam/ Lausanne/New York/Oxford/Shannon/Tokyo: Elsevier publishing; 1994, p. 335–349.

Kadlec R, Knight R, Vymazal J, Brix H, Cooper P, Haberl R. Constructed wetlands for pollution control processes, performance, design and operation. ISBN: 9781900222051; 2000, 156 pp.

Kairu JK. Wetland use and impact on Lake Victoria, Kenya region. Lakes & Reservoirs: Research and Management. 2001;6:117–25.

Lamberts D. Tonlé Sap fisheries: a case study on floodplain gillnet fisheries, RAP Publication 2001/ 11. Bangkok: FAO; 2001, 101 pp.

Li KY, Coe MT, Ramankutty N, De Jong R. Modeling the hydrological impact of land-use change in West Africa. J Hydrol. 2007;337:258–68.

Lind EM, Morrison MES. East African vegetation. London: Longman; 1974, 257 pp.

Lowe-McConnell RH. Ecological studies in tropical fish communities, Cambridge tropical biology series. Cambridge: Cambridge University Press; 1987, 387 pp.

Marshall AJ, Beehler BM. The ecology of Papua. Parts 1 & 2. Singapore: Periplus Editions; 2007, 784 + 768 pp.

Mata DI, Moreno-Casasola P, Madero-Vega C. Litterfall of tropical forested wetlands of Veracruz in the coastal floodplains of the Gulf of Mexico. Aquat Bot. 2012;98:1–11.

Matthews E, Fung I. Methane emission from natural wetlands: global distribution, area, and environmental characteristics of sources. Global Biogeochemical Cycles. 1987;1:61–86.

McGregor S, Lawson V, Christophersen P, Kennett R, Boyden J, Bayliss P, Liedloff A, McKaige B, Andersen AN. Indigenous wetland burning: conserving natural and cultural resources in Australia's World Heritage-listed Kakadu National Park. Hum Ecol. 2010;38:721–9.

Melack JM, Hess LL. Remote sensing of the distribution and extent of wetlands in the Amazon Basin. In: Junk WJ et al., editors. Amazonian floodplain forests: ecophysiology, biodiversity and sustainable management, Ecological studies, vol. 210. Dordrecht, Heidelberg, London & New York: Springer Science + Business Media B.V; 2010. p. 43–59.

Michels GH, Vieira EM, Nogueira de Sá F. Short- and long-term impacts of an introduced large herbivore (Buffalo, *Bubalus bubalis* L.) on a neotropical seasonal forest. Eur J Forest Res. 2012;131:965–76.

Mitsch WJ. The nonpoint source pollution control function of natural and constructed riparian wetlands. In: Mitsch WJ, editors Global wetlands old world and new. Ohio State University conference, 1992 Sep 13–18. Amsterdam/Lausanne/New York/Oxford/Shannon/Tokyo: Elsevier publishing; 1994, p. 351–361.

Mitsch WJ, Gosselink JG. Wetlands. 3rd ed. New York/Chichester/Weinheim/Brisbane/Singapore/Toronto: Wiley; 2000, 920 pp.

Mitsch WJ, Nahlik A, Wolski P, Bernal B, Zhang L, Ramberg L. Tropical wetlands: seasonal hydrologic pulsing, carbon sequestration, and methane emissions. Wetl Ecol Manag. 2010;18:573–86.

Mitsch WJ, Bernal B, Nahlik M, Mander U, Zhang L, Anderson CJ, Jørgensen SE, Brix H. Wetlands, carbon, and climate change. Landsc Ecol. 2012. Online submission, doi: 10.1007/s10980-012-9758-8, 15 pp.

Morison JIL, Piedade MTF, Müller E, Long SP, Junk WJ, Jones MB. Very high productivity of the C4 aquatic grass *Echinochloa polystachya* in the Amazon floodplain confirmed by net ecosystem CO_2 flux measurements. Oecologia. 2000;125:400–11.

Mulder V, Heri V, Wickham T. Traditional honey and wax collection with *Apis dorsata* in the Upper Kapuas Lake Region, West Kalimantan. Borneo Res Bull. 2000;31:246–60.

Nasir SM, Akbar G. Effect of River Indus flow on low riparian ecosystems of Sindh: a review paper. Rec Zool Surv Pakistan. 2012;21:86–9.

Nesbitt M. Grains. In: Prance G, Nesbitt M, editors. The cultural history of plants. New York/London: Routledge; 2005. p. 45–60, 452 pp.

Northwest Hydraulic Consultants, EVS Environmental Consultants and Asian Wetland Bureau. Preliminary hydro-ecological investigation of Duri Canal and swamp forest near Rantaubais. Prepared for Duri Steamflood Project and PT. Caltex Pacific Indonesia. Edmonton/Montreal/Bogor: NHC/EVS/AWB; 1994, 95 pp.

Osborne PL. Wetlands of Papua New Guinea. In: Whigham DF, editor. Wetlands of the world: inventory, ecology & management. Volume I, Handbook of vegetation science, vol. 15/2. Dordrecht/Boston/London: Kluwer Academic Publishers; 1993. p. 305–44.

Paijmans K, editor. New Guinea vegetation. Canberra: CSIRO and Australian University Press; 1976, 213 pp.

Parolin P, Ferreira LV, Albernaz ALKM, Almeida SS. Tree species distribution in Varzea forests of Brazilian Amazonia. Folia Geobot. 2004;39:371–83.

Peters CM, Giesen W. Balancing supply and demand: a case study of rattan in the Danau Sentarum National Park, West Kalimantan, Indonesia. Borneo Res Bull. 2000;31:138–49.

Pettit NE, Bayliss P, Davies PM, Hamilton SK, Warfe DM, Bunn SE, Douglas MM. Seasonal contrasts in carbon resources and ecological processes on a tropical floodplain. Freshw Biol. 2011;56:1047–64.

Phillips O. The potential for harvesting fruits in tropical rainforests: new data from Amazonian Peru. Biodivers Conserv. 1993;2:18–38.

Pond AP, White SA, Milczarek M, Thompson TL. Accelerated weathering of biosolid-amended copper mine tailings. J Environ Qual. 2005;34(4):1293–301.

Pratolongo P, Vicari R, Kandus P, Malvárez I. A new method for evaluating net aboveground primary production (NAPP) of *Scirpus Giganteus* (Kunth). Wetlands. 2005;25(1):228–32.

Pringle CM, Freeman MC, Freeman BJ. Regional effects of hydrologic alterations on riverine macrobiota in the New World: tropical–temperate comparisons. BioScience. 2000;50(9):807–23.

Ramberg L, Hancock P, Lindholm M, Meyer T, Ringrose S, Sliva J, Van As J, VanderPost C. Species diversity of the Okavango Delta, Botswana. Aquat Sci. 2006;68:310–37.

Rejmánková E. Effect of experimental phosphorus enrichment on oligotrophic tropical marshes in Belize, Central America. Plant and Soil. 2001;236:33–53.

Riak KM. Sudd area as a Ramsar site: biophysical features. Key documents of the Ramsar Convention information sheet on Ramsar wetlands. Environmental workshop event co-sponsored by the United Nations Environment Programme (UNEP); 2006 Oct 31. Juba; 2006.

Robinson CT, Tockner K, Ward JV. The fauna of dynamic riverine landscapes. Freshw Biol. 2002;47:661–77.

Rodin LE, Bazilevich NI, Rozov NN. Productivity of the world's ecosystems. In: Reichle DE, Franklin J, Goodal DW, editors. Productivity of the world's ecosystems, Washington, DC; 1975. p. 13–26.

Saunders MJ, Jones MB, Kansiime F. Carbon and water cycles in tropical papyrus wetlands. Wetl Ecol Manag. 2007;15:489–98.

Schelle P, Pittock J. Restoring the Kafue Flats. A partnership approach to environmental flows in Zambia. In 8th International River Symposium; 2005 Sep. Brisbane; 2005, 10 pp.

Scott DA. A directory of Asian wetlands. For WWF – World Wide Fund for Nature, IUCN – The World Conservation Union, ICBP – International Council for Bird Preservation and IWRB – International Waterfowl and Wetlands Research Bureau. Gland: IUCN; 1989, 1181 pp.

Scott DA. A directory of wetlands in Oceania. For IWRB – International Waterfowl and Wetlands Research Bureau, AWB – Asian Wetland Bureau, SPREP – South Pacific Regional Environment Programme and Ramsar Convention Bureau. Slimbridge (UK) IWRB and J+Kuala Lumpur, Malaysia (AWB): AWB & IWRB; 1993, 444 pp.

Scott DA, Carbonell M. A directory of neotropical wetlands. Cambridge, UK: IUCN Conservation Monitoring Centre; 1986, 684 pp.

Shardendu, Ambasht RS. Relationship of nutrients in water with biomass and nutrient accumulation of submerged macrophytes of a tropical wetland. New Phytol. 1991;117(3):493–500.

Shuhaimi-Othman M, Lim C, Mushrifah I. Water quality changes in Chini Lake, Pahang, West Malaysia. Environ Monit Assess. 2007;131:279–92.

Sharma N, Joshi SP, Pant HM. Restoration of Mothronwala Fresh water Swamp of Doon valley, Uttarakhand. 2010. http://www.sciencepub.net/researcher/research0207/06_3464research0207_53_55.pdf

Silvius MJ, Steeman APJM, Berczy ET, Djuharsa E, Taufik A. The Indonesian wetland inventory. A preliminary compilation of existing information on wetlands of Indonesia. Bogor: PHPA, AWB/INTERWADER, EDWIN; 1987, 2 vols, 121 & 268 p's, & maps.

Smardon RC. Chapter 4, The Kafue Flats in Zambia, Africa: a lost floodplain? In: Smardon RC, editor. Sustaining the world's wetlands. New York: Springer; 2009. p. 93–123.

Sutcliffe JV. A hydrological study of the southern Sudd region of the upper Nile. Hydrol Sci Bull. 1974;19:237–55.

Tanner CC. Plants as ecosystem engineers in subsurface-flow treatment wetlands. Water Sci Technol. 2001;44(11):9–17.

Tockner K, Stanford JA. Riverine flood plains: present state and future trends. Environ Conserv. 2002;29(3):308–30.

UNEP. Global forum on artisanal and small-scale gold mining. Meeting in 7–9 Dec 2010, Manila. Final report: 2011, 18 pp.

Van Steenis CGGJ. Outline of vegetation types in Indonesia and some adjacent regions. In Eighth Pacific Science Conference, 16–28 Nov. 1953, Quezon City, The Philippines, vol. IV Botany; 1957, p. 61–97.

Van TK, Rayachhetry MB, Center TD, Pratt PD. Litter dynamics and phenology of *Melaleuca quinquenervia* in South Florida. J Aquat Plant Manag. 2002;40:22–7.

Walter O. A study of hunting and trade of freshwater turtles and tortoises (order Chelonia) at Danau Sentarum. Borneo Res Bull. 2000;31:323–35.

Welcomme RL. Fisheries ecology of floodplain rivers. London/New York: Longman Publishers; 1979, 317 pp.

Wetlands International. Planting trees to eat fish. Field experiences in wetlands and poverty reduction. Wageningen: Wetlands International; 2009, 144 pp.

Wetlands International. Impact of dams on the people of Mali. Wageningen: Wetlands International, The Netherlands: 2012. 12 pp.

Winemiller KO, Jepsen DB. Effects of seasonality and fish movement on tropical river food webs. J Fish Biol. 1998;53(Supplement A):267–96.

Wittmann F, Junk WJ, Piedade MTF. The várzea forests in Amazonia: flooding and the highly dynamic geomorphology interact with natural forest succession. For Ecol Manage. 2004;196:199–212.

Peatlands

14

C. Max Finlayson and G. Randy Milton

Contents

Introduction	228
Classification of Peatlands	230
Global Distribution	230
Ecological Features	233
Biota	235
Ecosystem Services	237
Regulation Services	237
Production Services	238
Threats and Future Challenges	240
References	242

Abstract

Peatlands are ecosystems that are characterized by the accumulation of organic matter that is derived from decaying plant material under permanent water saturation. They have been defined to include areas of land with a naturally accumulated layer of peat, formed from carbon-rich dead and decaying plant material under waterlogged and low oxygen conditions, generally seen as comprising at least 30% dry mass of dead organic material and greater than 30 cm deep. They can develop under a wide range of vegetation types in fresh and saline water, including sphagnum, sedges, reed beds, and shrubs and trees in wet

---

C. M. Finlayson (✉)
Institute for Land, Water and Society, Charles Sturt University, Albury, NSW, Australia

UNESCO-IHE, The Institute for Water Education, Delft, The Netherlands
e-mail: mfinlayson@csu.edu.au

G. R. Milton
Nova Scotia Department of Natural Resources, Kentville, NS, Canada
e-mail: gordon.milton@novascotia.ca

© Springer Science+Business Media B.V., part of Springer Nature 2018
C. M. Finlayson et al. (eds.), *The Wetland Book*,
https://doi.org/10.1007/978-94-007-4001-3_202

woodland and mangroves. At the ecosystem level, the shape, size, and type of peatlands are determined by the climate and geomorphology as well as the quantity and quality of the water. Peatlands are widespread globally, although there are major gaps in data. Many are under threat from drainage and land conversion with loss of biodiversity and valuable ecosystem services. They are important for carbon storage and provide opportunities for the mitigation of climate change.

Keywords
Peatland · Peat · Biodiversity · Carbon · Agriculture · Drainage

Introduction

Peatlands are ecosystems that are characterized by the accumulation of organic matter that is derived from decaying plant material under permanent water saturation. Joosten and Clarke (2002) defined peatlands to include areas of land with a naturally accumulated layer of peat, formed from carbon-rich dead and decaying plant material under waterlogged and low oxygen conditions, generally seen as comprising at least 30% dry mass of dead organic material and greater than 30 cm deep. They can develop under a wide range of vegetation types in fresh and saline water, including sphagnum, sedges, reed beds, and shrubs and trees in wet woodland and mangroves. Joosten (2004) provides a description of peatland types and their distribution and a discussion on the differences in the terms used to describe them. At the ecosystem level, the shape, size, and type of peatlands are determined by the climate and geomorphology as well as the quantity and quality of the water (Joosten 2004; Lindsay 2018a).

In Eurasia, but not as widely in other places, the term mire is used to refer to ecosystems that are actively accumulating peat to differentiate them from the more general term peatland that also includes ecosystems where peat is not actively being accumulated (Joosten and Clarke 2002). This definition differs from some earlier definitions that instead emphasized that mires comprised wetlands that supported vegetation that normally formed peat. As such it does not require proof that peat was actively being formed, but drew on the knowledge that certain types of vegetation were particularly associated with peat formation, such as peat mosses *Sphagnum* spp., cotton grasses *Eriophorum* spp., sedges, and, in the tropics, swamp forest trees (such as *Calophyllum* spp. and *Cratoxylum* spp.) (Löfroth 1994). Peatlands where the accumulation of peat has stopped, for example, as a consequence of drainage, are no longer mires.

The literature about peatlands is diverse, given the many ecosystems that contain peat, including forested and non-forested forms, in extremely cold as well as tropical latitudes and high altitudes (Fig. 1). This diversity has led to a rich vocabulary with many local and technical terms and classifications being used to describe peatlands (Joosten 2004). Despite many differences over time,

Fig. 1 Treed domed bog and basin fen in temperate North America (*upper*), papyrus peatland in Africa (*middle*), and peatlands in the Rougerai, China (*lower*) (Photo credits: GR Milton (*upper*), NC Davidson (*middle*), CM Finlayson (*lower*) © Rights remain with the authors)

the widely accepted definition of a peatland is based on the presence of accumulated layers of peat, formed from carbon-rich dead and decaying plant material under waterlogged conditions, comprising at least 30% dry mass of dead organic material greater than 30 cm deep, and not covered by a mineral soil layer. Ecosystems with less than 30 cm of accumulated peat are not considered in this overview.

Classification of Peatlands

The term peatland has different meanings to people in different disciplines, resulting in some confusion over the nature of peat and peatlands (Charman 2002; Joosten 2004). As noted by Lindsay (2018b) "There is no *single* classification which represents the complete or definitive peatland classification system." Moreover, the plethora of terms provides for a rich vocabulary but possibly also some uncertainty when comparing between different cultural and linguistic categorizations.

Charman (2002) provided a detailed description of the classification of peatlands based initially on their hydro-morphological features and then further subdivided on the basis of vegetation, water chemistry, and peat stratigraphy. The Tope System is a further hierarchical classification that combines the mire and vegetation categories of traditional approaches with additional features such as interconnectedness of mires, surface patterns, and micro-topography (Ivanov 1981; Wells and Zoltai 1985; Moen 1985; Lindsay 2018a).

Although there are a number of definitions of peatlands (Joosten 2004) based on peat genesis, descriptions of the soil, biodiversity and carbon storage, and also on their exploitation potential, many classifications or typologies have as their basis the origin and behavior of the water that leads to waterlogging. The term ombrotrophic is used to describe peatlands that receive water directly from rainwater and minerotrophic for those supplied by mineral groundwater. The terms bog and fen have increasingly been used respectively to describe peats that are higher than their surroundings and waterlogged by rainwater and those that occur in landscape depressions and are waterlogged largely by excess ground or surface water. In tropical regions, bogs and fens are typically covered by rainforest and are often called peat swamp forest (Page et al. 1999; Lähteenoja and Page 2011).

Global Distribution

Peatlands occur in many countries and could represent more than a third of global wetlands, although such comparisons are reliant on the definition of a peatland and the adequacy of the underlying inventory data. The database on peatland inventory and distribution that is maintained by the International Mire Conservation Group (Joosten 2004) provides the basis for the following description of the global distribution of peatlands. Peatland ecosystems are most abundant in areas with excess moisture (Fig. 2). As a consequence they are widespread over large parts of the mid-high latitudes of North America and Eurasia and include forested and non-forested types, as well as the widespread cryosols, including polygon and palsa peatlands that have developed under the influence of permafrost. In more temperate climates, there are blanket bogs that are extensively developed in oceanic or highland areas, condensation peatlands on screes and boulders, and raised bogs. In some tropical regions, peatlands have developed where there the high evapotranspiration has been counterbalanced by extremely high rainfall. At very high latitudes and altitudes, the accumulation of peat may be limited by the shorter growing season

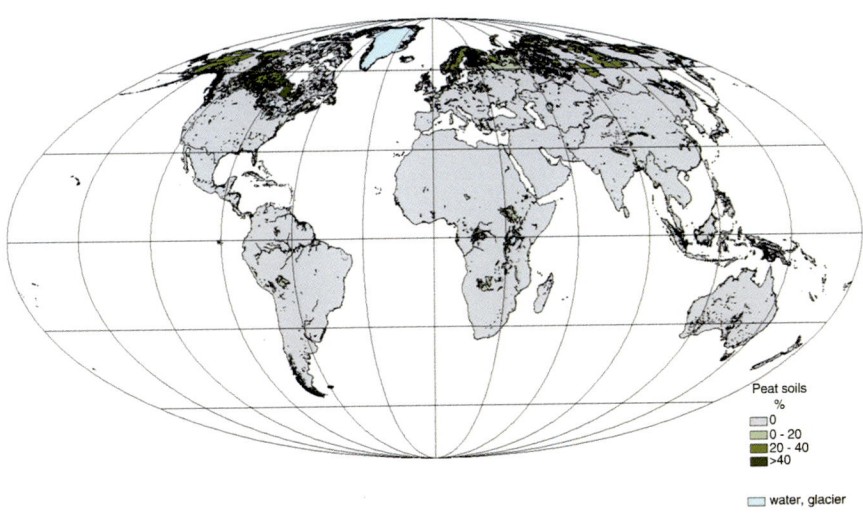

Fig. 2 Distribution of peatlands (Unpublished report by Schurmann and Joosten 2008, with permission of author)

and low plant productivity. High-altitude peatlands include the paramos that are concentrated in northwestern South America, while cloud forests and elfin woodlands generally occur in the humid tropics. Mangroves are generally confined to coastal systems, at times extending inland along macro-tidal estuaries.

Joosten (2010) used multiple sources of data in providing an areal estimate of around four million square kilometres while also pointing to gaps in data coverage and a degree of uncertainty about the actual extent of peatlands. Part of the uncertainty about the extent of peatlands comes from incomplete data, but also because global and national compilations have been hampered by the use of different definitions and classifications (Montanarella 2014). Charman (2002) also provided an areal estimate of more than four million square kilometres with an expectation that estimates of the area covered by tropical peatlands were expected to increase as further and more robust investigations were conducted.

Despite limitations in the data, it is apparent that certain countries and regions are particularly important for the area of peatlands they contain (Table 1). An overview of the distribution of freshwater peatlands is shown in Fig. 1 using data compiled from multiple sources by Joosten (2004, 2010). These data only cover peatlands with the peat layers being more than 30 cm deep, hence excluding wetlands with shallower peats. The largest areas are found in the northern hemisphere, especially in the boreal zone with 1,375,690 km^2 in Russia and 1,133,926 km^2 in Canada (Table 1). Estimates of peatlands in tropical regions range from 275,424 to 570,609 km^2, although most of the data are from pre-1990 sources, and there has been extensive destruction of some of these in recent years (Hooijer et al. 2010). The data uncertainties behind many estimates for peatland areas and distribution have been discussed in the International Mire Conservation Group database (see Joosten 2004).

Table 1 Countries with the largest area of peatlands with carbon stocks and emissions from degraded peatlands in 2008

Country	Area of peatland km^2	Peat carbon stock in 2008 Mton C	Emissions from degrading peat in 2008 Mton CO$_2$ year^{-1}
Russia	1,375,690	137,555	161
Canada	1,133,926	154,972	5
Indonesia	265,500	54,016	500
USA[a]	223,809	29,167	67[b]
Finland	79,429	5,294	50
Sweden	65,623	5,000	15
Papua New Guinea	59,922	5,983	20
Brazil	54,730	5,440	12
Peru	49,991	998	NA
China	33,499	3,224	77
Sudan	29,910	1,980	NA
Norway	29,685	2,230	NA
Malaysia	26,685	5,431	48
Mongolia	26,291	NA	NA
Belarus	22,352	1,306	41

Data from Joosten (2010)
NA information not available
[a]Conterminous states plus Alaska
[b]Not including Alaska

The data contained in Table 1 are taken from a summary of the information collated in the International Mire Conservation Group Global Peatland Database (IMCG-GPD www.imcg.net/gpd/gpd.htm) and also reported by Joosten (2010). The database is regularly being updated as further data becomes available; there are many gaps in the global inventory of peatlands, for example, for many countries in Africa and South America. The data is largely for freshwater peatlands with some ecosystems, such as mangroves, salt marshes, cloud and elfin forests, and paramos, not always being widely considered as peatlands and/or because of an absence of information. Data in submarine peatlands are not included. While the largest areas of peatland and carbon stock are found in Canada, Russia, Indonesia, and the USA, in 2008 the largest emissions from degrading peatland were in Indonesia, Russia, China, and Belarus.

In compiling these data, it was noted that peatland inventory in many countries and globally was unsatisfactory with many countries not having sufficient information about their peatland resource (Joosten 2004, 2010). The deficit in information in some places was due in part to the increasing use of remote sensing for wetland inventory with inherent difficulties in determining the "presence of peat" based on land cover attributes generally assessed in such investigations. While there is a reasonable correlation between vegetation cover and structure and the presence of peat, the variety of peatlands, varying from forest and shrubland to reeds, open grassland, and moss stretches, makes it difficult to categorically determine the presence of peatland on this basis. It is furthermore extremely difficult to ascertain the depth of the peat layer without conducting intensive field sampling with consequent severe limitations on the estimation of peat stocks (Joosten 2009; Montanarella 2014).

Data from newer or emerging sources may be able to overcome some of the deficiencies in estimating the extent and changes in the extent of some peatlands, for example, the recently developed Global Mangrove Watch (see Lucas et al. 2014). In addition to remotely sensed areal estimates, detailed ground-truthing field surveys are required to assess peat depth and verify remote sensing data (Montanarella 2014). Obtaining a consistent international overview of peatlands was seen by Joosten (2010) as being complicated by a number of issues, including differences in typologies, the scale of analyses, changes in the extent of peat over time, and differences and errors in the use of units of measure.

Ecological Features

Peatlands can be categorized in many ways based on the origin and behavior of the water they contain, the genesis of the peat, and geomorphic features, including the position in the landscape, the vegetation cover, and the climate. Many such classifications have been used as a basis for summarizing the main ecological features of peatlands as well as showing the relationships between individual peat systems and their wider ecological functions in particular landscapes (Runkle and Kutzbach 2014). Charman (2002) points out that different classifications have been used to categorize the importance of the water cycle for the formation and maintenance of peatlands and to describe the role of peatlands in the global carbon cycle. Alongside these efforts there has been considerable archeological interest in peatlands as the waterlogged and low oxygen conditions can preserve an archive of long-term human and environmental history.

As explained by Charman (2002) the water balance in a peatland consists of the inflow, outflow, and storage of water (Fig. 3) with an excess of water needed to prevent the complete decomposition of the plant biomass. The main inputs are from precipitation, surface runoff, or groundwater, with outflows coming from runoff, seepage to the groundwater, and/or interception and evapotranspiration. The balance between these flows differs greatly between peatlands with precipitation being the only flow into ombrotrophic mires, whereas for many fens, direct precipitation is likely to be less important than surface runoff or groundwater. Estimating the water budget of a peatland can prove complex, especially where groundwater plays an important role given inherent difficulties with measuring or calculating groundwater flows.

As outlined by Charman (2002) and Rydin and Jeglum (2006), the rate at which water moves within a peatland is determined using Darcy's law that relates the hydraulic conductivity of the peat to the hydraulic gradient. In general, dense peats

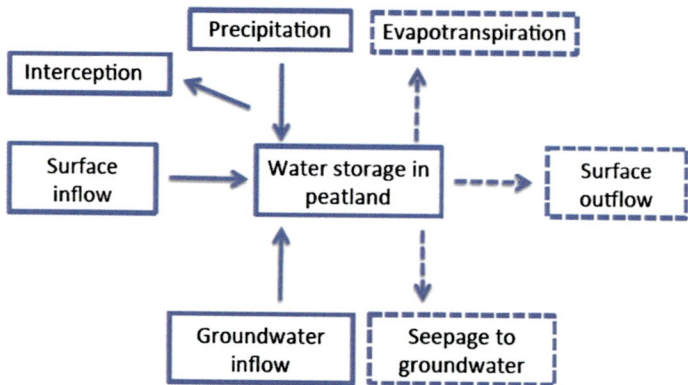

Fig. 3 Water balance in a peatland showing the inflows (*full lines*) and outflows (*dotted lines*) (Adapted from Charman 2002)

with a fine peat matrix and small pore spaces generally have low hydraulic conductivity and impeded water movement. Peats with larger and less compact fragments generally have larger pore spaces and higher hydraulic conductivity. *Sphagnum* peats are generally less permeable than peats composed of higher plants, including those derived from *Carex* species or *Phragmites*. The movement of water through a peatland is also affected by the structure and bulk density of the peat with many peatlands having an upper aerated layer, known as the acrotelm with loose, living vegetation, and an anoxic lower layer, or catotelm with brown or black and denser peat. The acrotelm is also known as the "active" layer in a peatland where most growth and decay occur, with lower rates of growth and decay in the catotelm that comprises most of the volume of the peatland. The hydraulic conductivity can vary with depth and the upper layers or acrotelm often but not always having higher values than the catotelm. Further, the distinction between these layers in peat swamp forests is generally less clear than it is in temperate and boreal peatlands.

Charman (2002) further reports that the movement of water in both fens and bogs can be complex with regional groundwater influencing not only the water movement but also the chemistry and vegetation. The depth of the water table is one of the most important factors in most peatlands, but is not necessarily an accurate measure of the amount of water held in a peatland as the peatland itself may rise and fall in relation to the surrounding land. For many peatlands surface runoff is the predominant outflow and is strongly linked to changes in the water table.

The chemistry of the water and peat within a particular peatland is determined primarily by the quality of the inflows and the chemical changes that occur within the system, including those that influence the decay processes (Charman 2002). Peatlands that depend on precipitation are typically nutrient poor and acidic, although the composition of the precipitation may vary and the amount of precipitation can affect the flux of solutes in the peatland. Surface water runoff in many cases reflects the chemistry of rainwater but is also influenced by the surrounding land and vegetation. The chemistry of the groundwater is affected by the geology and surrounding soils and the rate of flow. The chemistry is also greatly affected by the degree of aeration with saturated peat being anaerobic which leads to an increase in the reduction potential (or redox) of the peatland and an increase in bacterial activity that can result in the reduction of nitrate and the release of nitrous oxide to the atmosphere, or under highly anaerobic conditions the reduction of carbon dioxide and the production of methane. The reduction of manganese and iron can result in the production of toxic substances. The acidity of the peatland affects the reduction potential in an inverse manner. In general, bogs that receive much of their water from rainfall are more acidic than fens which are watered by surface or groundwater, although there are substantial seasonal changes and variation between peatlands.

The major nutrients nitrogen, phosphorus, and potassium are often in low supply in many peatlands and can limit plant growth and production. Many peats also absorb cations. *Sphagnum* species are particularly efficient at removing cations from solution in oligotrophic and ombrotrophic mires and creating more acidic conditions. These conditions affect the growth of plants in peatlands, with many needing specific adaptations to cope with low oxygen availability, mobilization of toxic elements, low nutrient availability, and high acidity. The traditional view was that the mineratrophic-ombrotrophic and acidic conditions were the most important control on plant growth in peatlands, although there is also support for the importance of acidity and nutrient gradients for describing peatlands (Charman 2002).

Biota

The ecological features of peatlands, in particular low oxygen availability, mobilization of toxic elements, low nutrient availability, and acidity, greatly influence the biota that they support, with many species needing special adaptations to enable them to survive. At the same time, the diversity of peatlands means that they support important biological diversity and provide refuge for some of the rarest and most unusual species of wetland-dependent flora and fauna (Joosten and Clarke 2002; Charman 2002; Minayev and Sirin 2012). In general and as reported in a number of publications (see summary in Minayev et al. 2008), the species richness in peatlands is lower than that in nearby ecosystems, but they do support a number of highly characteristic and obligate species, many of which may be regionally or globally endemic and/or highly adapted to particular conditions and having disjunct, azonal distribution patterns. In addition to these obligate peatland species, there are others that are only found in peatlands on a seasonal basis or at particular stages of their life cycles. The complexity and harshness of many peat environments also leads to a high level of intraspecific diversity that reflects biogeographic differences and the island character of many peatlands. In addition, trans-specific diversity, the expression of similar morphological or functional traits by different species, is a typical adaptive mechanism in peatlands.

The biotic diversity of peatlands has been unevenly investigated, as shown through a review undertaken by Minayeva et al. (2008). The diversity of microorganisms in peatlands is expected to be high, but has not been widely studied despite the role of such organisms in the biogeochemical cycles within peatlands, including nutrient cycling, peat accumulation and decomposition, and the release of carbon. Bacteria are found throughout the peat layers with the largest numbers being in the surface layers where nitrifying and aerobic bacteria dominate, with anaerobic species more dominant in deeper and anaerobic layers. Peatlands also contain a wide variety of fungi and fungi-like Protista with their diversity being closely related to the diversity of substrates. They include saprophytes of trees, herbs, and mosses and their litter, coprotrophs, mycorrhiza-forming species, and parasites and can vary seasonally. Similarly, the diversity of algae and algae-like Protista in peatlands is expected to be high. The diversity of lichens in peatlands is usually low compared to other ecosystems.

In contrast, peatlands provide favorable conditions for mosses and liverworts and contribute considerably to the diversity of bryophyte species. *Sphagnum* mosses are the most important peat-forming plants in ombrotrophic bogs, whereas a small number of these species create the special hydraulic conditions that lead to the formation of raised bogs. The diversity of vascular plants depends on biogeographical factors and climate, with their diversity being lower than that of bryophytes, algae, and fungi. There is though a lot of phenotypic variation in vascular plants, as shown through changes in, for example, the size of roots and shoots and the weight, shape, and number of seeds, as well as phenological changes (multiple flowering, changes in growth season) and ecological growth forms, such as that shown in *Pinus sylvestris* which ranges from tree to dwarf tree to shrub, depending on site conditions. *Sphagnum* species are widespread in many temperate and high northern latitude peatlands, but species belonging to the family Restionaceae characterize temperate oceanic South Pacific islands, and in tropical areas the peatlands are often vegetated by rainforest, the peat material is woody, and the peat surface is covered by a thick leaf litter layer.

Invertebrate diversity in peatlands can be high. Besides insects there are many other invertebrate species in peatlands, including Cladocera and Rotifera. A large number of species use peatlands at different stages in their life cycles. Water bodies in temperate peatlands are often poor in fish species, except where they are connected to other wetlands or streams. On the other hand, tropical peatlands can contain many fish species. Polymorphism is not uncommon, as shown in parts of Russia where peatland lakes generally only contain perch *Perca fluviatilis* with a high level of morphological differentiation. Although amphibians and reptiles are not widespread in peatlands, they do support different stages of the life cycle of many species. They also provide feeding, breeding, and shelter for numerous bird species, of which many are entirely dependent on them. A few bird species depend on peatlands throughout their life cycles, including black-throated diver *Gavia arctica*, black grouse *Lyrurus tetrix*, greylag goose *Anser anser*, and many others for only part of their lives, for breeding or feeding. Very few mammals are known to be entirely dependent on peatlands, although many make use of them, such as reindeer, and in the tropics they provide a haven for species such as the Sumatran tiger *Panthera tigris sumatrae*, Malayan tapir *Tapirus indicus*, and orangutan *Pongo pygmaeus* and *P. abelii*.

Ecosystem Services

The importance of peatlands for humans is well known with a long history of cultural and livelihood linkages for substantial numbers of people in many countries. Under waterlogged conditions peatlands preserve a unique paleoecological record, including valuable archeological remains, and records of environmental contamination. They support human needs for food, fresh water, shelter, warmth, and employment from the tropics to the arctic. In fact, the use of peat for fuel and horticultural purpose has resulted in large-scale drainage and extraction and the degradation of peatlands in many countries over many years (Silvius et al. 2008; Rieley 2014). In recent years the drainage and burning of tropical peat forests in parts of Southeast Asia have received a lot of attention given the smoke haze that has created health problems for many people and led to the loss of livelihood opportunities for local people impacted national and regional economies, in addition to the destruction of habitat and increased threats to the biodiversity (Harrison et al. 2009; Johnston et al. 2012; Marlier et al. 2013). These outcomes illustrate the need for balance in the use of peatlands in order to sustain the benefits that can accrue to many people. Silvius et al. (2008) provides an overview summarized here of the importance of peatlands for people with the main benefits grouped under the headings of regulation and production services.

Regulation Services

In recent years their importance as global carbon stores and sinks has come to the fore as the global community addresses measures to mitigate climate change (Hooijer et al. 2010; Joosten 2010; Joosten et al. 2012). The important role played by peatlands in the global climate cycle through the storage of atmospheric carbon and being a source of greenhouse gases including carbon dioxide, methane, and nitrous oxide has been widely described (Sirin and Laine 2008). The amount of carbon stored in peatlands, despite uncertainties in the data, exceeds that of global vegetation and may be of similar magnitude to the atmospheric carbon pool (Turetsky et al. 2015). At the same time, peatlands can emit methane. The drainage and use of peatlands for agriculture and forestry have resulted in the oxidation of peat and the release of carbon dioxide to the atmosphere (Joosten and Couwenberg 2008). Fires in drained and degraded peat forests in Indonesia are estimated to have released in 1997 the equivalent of 13–40% of the 6.4 GtCyear^{-1} global annual (2002) emissions from fossil fuels (Page et al. 2002); and CO_2 emissions from peatland drainage in Southeast Asia is contributing the equivalent of 1.3–3.1% of current global CO_2 emissions from the combustion of fossil fuels (Hooijer et al. 2010). Peatlands used for agriculture can release nitrous oxide. Thus, the way in which peatlands are managed can determine if they provide significant benefits or disadvantages for mitigating the impacts of climate change (see Biancalani and Avagyan 2014).

Peatlands also contribute to the regulation of local climates through evapotranspiration and the associated alteration of heat and moisture conditions. The influence is generally greater in warmer or drier climates than in colder or more humid. As a consequence the climate is generally cooler and more humid in areas with extensive peatlands. In the boreal the drainage of peatlands can lead to a reduction in minimum temperatures and a shortening of the yearly frost-free period and can be reversed by subsequent afforestation. Similarly, the conversion and drainage of large areas of peatlands in mountains along the Uganda/Rwanda border may have resulted in increased local temperatures.

Peatlands have many important roles in the surface and groundwater balance in catchments in terms of water storage, water quality, and flood and drought regulation, particularly at a local and regional level. They contain 10% of the global freshwater volume and are significant in maintaining freshwater quality and hydrological integrity of many river valleys. They play an important role in maintaining permafrost and preventing desertification. They have the ability to purify water by removing nutrients and other substances. Large peatlands can have a major impact on the surface- and groundwater regime by storing water under flood conditions and releasing it during dry periods. Riparian peatlands can store floodwaters and mediate the velocity and volume of downstream discharges. Coastal peat swamps can provide a buffer between salt and fresh water, mitigating saline intrusion into coastal lands. The water storage and retention function of peatlands are locally important for drinking water and for irrigation. In regions where catchment areas are largely covered by peatlands, as well as in drier regions, they can play an important role in maintaining water supplies for drinking and irrigation.

Production Services

The agricultural production of peatlands without intensive management is generally low. The high groundwater table, low bulk density, high acidity, low availability of nutrients, and their subsidence after drainage generally limit agriculture production in peatlands. Conventional agriculture interventions such as drainage, fertilizing, tilling, compaction, and subsidence eventually result in a decline in agricultural productivity in peatlands. Much of the small-scale agricultural encroachment in tropical peatlands is linked to severe poverty, whereas large-scale encroachment is mainly due to the development of palm oil plantations. The development of agriculture at this scale is relatively recent in tropical areas, whereas in many temperate areas it has occurred over centuries and eventually led to the degradation of many peatlands. Poverty-induced agricultural encroachment is still occurring in many peatlands, and while providing essential resources for local people, such activities are not generally sustainable.

Peatlands have been used for forestry in many part of the world, especially with coniferous species, such as black spruce *Picea mariana* and Scotch pine *Pinus sylvestris*. Extensive commercial forestry has been established in many countries with the growth of commercially useful species being limited by waterlogging. Forestry from naturally forested peatlands also occurs throughout Scandinavia and Canada. The afforestation of open peatlands which is especially common in the blanket and raised mires in oceanic western Europe results in fundamental changes to the physical and hydrological conditions. Tropical peat forests include substantial quantities of commercial tree species and yield some of the most valuable tropical timbers, such as ramin *Gonystylus bancanus* and agathis *Agathis dammara*. Also in Malaysia, logging of peat swamp forest plays a very important role in the economy, especially in Sarawak which has major peat swamp forest reserves.

Peat as an energy source is nowadays only important for regional or domestic socioeconomic reasons and only contributes marginally to global energy use (Fig. 4a). It is still used widely as a growing medium in horticulture and as a soil conditioner, particularly in glasshouse horticulture for the cultivation of young plants and pot plants, and for the growing of vegetable crops (Fig. 4b). Peatlands also contain many plant species that have been utilized for food, fodder, construction, and medicine. One of the oldest and most widespread utilization of wild peatland plants is their use as straw and fodder for domestic animals (Fig. 5). A further important use, especially in the temperate and boreal zones of Eurasia, is the collection of wild edible berries and mushrooms.

Fig. 4 (**a**) (*left*) Harvesting of peat (turf) for home fuel use, known as "turf cutting" in Ireland. The blocks of turf are placed upright or "footed" to aid in drying (Photo credit: D. Stroud © Rights remain with photographer). (**b**) (*right*) Mechanical harvesting of peat in New Zealand for the horticultural industry (Photo credit: CM Finlayson © Rights remain with the author)

Fig. 5 Peatlands in the Rougerai, China. Land uses include grazing and tourism (Photo credit: CM Finlayson © Rights remain with the author)

Tropical peat swamp forests provide a wide range of products, such as edible fruits, vegetables, medicinal and ritual plants, construction material (wood, rattans, bamboo), fibers and dyeing plants, firewood, and traded products like rattans, timber, and animals. Important timber species are ramin and meranti *Shorea* sp. Both timber and non-timber forest products are central to the well-being and livelihood of local indigenous communities. Socioeconomic studies indicate that in Indonesia local community livelihoods may depend for over 80% on the peat swamp forest rather than on agriculture. They provide cash income to supplement daily expenses or are a "safety net" in time of need and represent an essential part of subsistence culture and heritage.

Peatlands may also be significant for hunting and fishing, including for coyote, raccoon, mink, and lynx, and game species such as grouse, ducks, geese, and moose. Wild reindeer (caribou in North America) are hunted for meat for local markets as well as for subsistence with an estimated 250,000 people in the Eurasian Arctic dependent on reindeer as a major food source. Caribou meat and hides are marketed in Canada on a small scale, and in both Alaska and Canada, caribou are hunted recreationally, generating income for guides, outfitters, and the service industry. Peatland waters harbor many fish species, which in some regions, in addition to providing an important protein source to local communities, can be an object of sports fishing, generating income in sales of fishing equipment and licenses. Tropical black water fish diversity is extremely high. Black water species are attractive to sport fishing and the often very colorful species are attractive for the aquarium industry.

Threats and Future Challenges

Covering only 3% of the world's land area, Joosten and Couwenberg (2008) report that 65 M ha (16%) of the global peatland resource is degraded. Human pressures on peatlands are both direct, through drainage; land conversion, for example, for oil palms and oil sands; excavation; and inundation, and indirect, as a result of air pollution, water contamination, water removal, and infrastructure development. Management of peatlands has included modest levels of intervention for habitat maintenance as well as more substantial efforts for restoring peatlands or even recreating those that have been fully degraded or lost.

Many peatlands have been used for grazing, cutting of vegetation for fodder, and the removal of peat for fuel. Management of relatively undisturbed peatlands for conservation purposes generally involves restricting access or minimal intervention, or on seminatural sites low-level disturbance may be retained to ensure particular successional conditions. This is the case for many of the fens in western Europe that have been created or modified by human activity and require management to maintain their current ecological values. Charman (2002) provides the following classification of peatland based on their level of disturbance and recognizing that disturbance hundreds of years ago could alter the development of particular peatlands (Table 2). Major disturbance includes the afforestation of open peatlands

Table 2 Classification of peatlands based on their level of disturbance (Adapted from Charman 2002)

Level of disturbance	Features of the disturbance
Natural	No influence by people; the ecology and hydrology unaffected by humans. Development has proceeded naturally
Minor disturbance	Some influenced by people in the past or very minor recent disturbance. Peatlands retain their structure and function
Moderate disturbance	Past or present disturbance altered the structure and function of the peatland
Major disturbance	The structure and function of the peatland have been altered, resulting in changes in the species and hydrology in the peatland
Artificial	Almost completely destroyed the original peatland; remaining peatland is the result of restoration efforts

with both drainage and tree planting leading to changes in the form and function of the peatland. This is possibly nowadays most prevalent in Southeast Asia where peat forests are being drained, burned, and converted to oil palm plantation (Page et al. 2002). The extraction of peat has also led to the removal of most of the biotically active part of the peatland, often leaving only a thin layer of peat.

The possibilities for managing disturbed peatlands can be grouped as follows: (i) prevent further degradation and restore the peatland to its natural state, (ii) maintain the disturbance to ensure the status quo to retain existing values, and (iii) allow the disturbance to continue and restore once the economic benefit of such activities has been realized. The latter in particular has become a contentious issue in some areas where conservation of peatlands has assumed a high priority especially as once peatlands are destroyed they release large amounts of carbon and are not easily restored.

In response to the degradation of peatlands, the Ramsar Convention has adopted detailed *Guidelines for Global Action on Peatlands* including the following activities (Ramsar 2002; Joosten 2004):

– Establishing a global database of peatlands
– Detecting changes and trends in the quantity and quality of peatlands
– Developing and promoting education, training, and public awareness programs
– Reviewing national networks of peatland PAs
– Developing and implementing peatland management guidelines and actions plans
– Establishing regional centers of expertise and research networks
– Stimulating international cooperation on research and technology transfer

More recently guidance has been provided to limit the loss of carbon from peatlands and to encourage their retention and restoration as part of climate change mitigation measures (Joosten et al. 2012; Biancalani and Avagyan 2014). Approaches to the restoration of northern peatlands are well established (Quinty and Rochefort 2003) and rapidly evolving for tropical systems (Wösten et al. 2008;

Page et al. 2009; Graham 2013). The Intergovernmental Panel on Climate Change (IPCC) has prepared standards to assess and report on emissions from organic soils and wetlands, including the rewetting of peatlands (IPCC 2014). Moreover, peatland restoration or protection activities are eligible for certification as greenhouse gas emission reduction projects in the voluntary carbon markets using existing standards, e.g., the Verified Carbon Standard (http://www.v-c-s.org). These climate change mitigation measures are particularly important given the past loss of peatlands globally and the more recent degradation of tropical peatlands (Hooijer et al. 2010; Joosten et al. 2012).

References

Biancalani R, Avagyan A, editors. Towards climate-responsible peatlands management. Rome: Food and Agriculture Organization of the United Nations (FAO); 2014. p. 117.

Charman DJ. Peatlands and environmental change. West Sussex: Wiley; 2002. p. 301.

Graham LLB. Restoration from within: an interdisciplinary methodology for tropical peat swamp forest restoration in Indonesia. [dissertation]. Leicester: Department of Geography, University of Leicester; 2013.

Harrison ME, Page SE, Limin SH. The global impact of Indonesian forest fires. Biologist. 2009;56:156–63.

Hooijer A, Page S, Canadell JG, Silvius M, Kwadijk J, Wösten H, Jauhiainen J. Current and future CO_2 emissions from drained peatlands in Southeast Asia. Biogeosciences. 2010;7:1505–14.

IPCC. 2013 Supplement to the 2006 IPCC Guidelines for National Greenhouse Gas Inventories: Wetlands, IPCC, Switzerland; 2014.

Ivanov KE. Water movement in mirelands. London: Academic; 1981.

Johnston FH, Henderson SB, Chen Y, Randerson JT, Marlier M, DeFries RS, Kinney P, Bowman DMJS, Brauer M. Estimated global mortality attributable to smoke from landscape fires. Environ Health Perspect. 2012;120(5):695–701.

Joosten H, The global peatland CO2 picture: peatland status and drainage related emissions in all countries of the world. Greifswald University & Wetlands International, Ede, The Netherlands; 2009.

Joosten H. The global peatland CO_2 picture: peatland status and drainage related emissions in all countries of the world. Ede: Wetlands International; 2010. p. 36.

Joosten H. The IMCG global peatland database; 2004. www.imcg.net/gpd/gpd.htm. Accessed 7 June 2016.

Joosten H, Clarke D. Wise use of mires and peatlands. Totness: NHBS/International Mire Conservation Group and International Peat Society; 2002.

Joosten H, Couwenberg J. Peatlands and carbon, Chapter 6. In: Parish F, Sirin A, Charman D, Joosten H, Minayeva T, Silvius M, Stringer L, editors. Assessment on peatlands, biodiversity and climate change: main report. Kuala Lumpur: Global Environment Centre; 2008. p. 99–117. Joint publication with Wetlands International, Wageningen.

Joosten H, Tapio-Biström ML, Tol S, editors. Peatlands-guidance for climate change mitigation through conservation, rehabilitation and sustainable use. Rome: Food and Agriculture Organization of the United Nations; 2012. p. 114. Joint publication with Wetlands International, Wageningen.

Lähteenoja O, Page S. High diversity of tropical peatland ecosystem types in the Pastaza-Marañón basin, Peruvian Amazonia. J Geophys Res. 2011;116:G02025. https://doi.org/10.1029/2010JG001508.

Lindsay R. Peatland/mire category descriptions based on origin and behaviour of water, peat genesis, landscape position and climate. In: Finlayson CM, Milton GR, Prentice RC, Davidson NC, editors. The wetland book II: distribution, description, and conservation. Dordrecht: Springer; 2018a.

Lindsay R. Peatland classification. In: Finlayson CM, Davidson NC, Everard M, Irvine K, McInnes RJ, Middleton BC, van Dam AA, editors. The wetland book I: structure and function, management and methods. Dordrecht: Springer; 2018b.

Löfroth M. European mires – an IMCG project studying distribution and conservation. In: Grünig A, editor. Mires and man: mire conservation in a densely populated country – the Swiss experience. Birmensdorf: Swiss Federal Institute for Forest, Snow and Landscape Research; 1994. p. 281–3. Available from: www.wsl.ch/dienstleistungen/publikationen/pdf/420.pdf. Accessed 14 Apr 2015.

Lucas R, Rebelo LM, Fatoyinbo L, Rosenqvist A, Itoh T, Shimada M, Simard M, Souza-Filho PW, Thomas N, Trettin C, Accad A. Contribution of L-band SAR to systematic global mangrove monitoring. Mar Freshw Res. 2014;65(7):589–603.

Marlier ME, DeFries RS, Voulgarakis A, Kinney PL, Randerson JT, Shindell DT, Chen Y, Faluvegi G. El Nino and health risks from landscape fire emissions in Southeast Asia. Nat Clim Chang. 2013;3:131–6.

Minayeva TY, Sirin AA. Peatland biodiversity and climate change. Biol Bull Rev. 2012;2:164–75.

Minayeva T, Couwenberg J, Cherednichenko O, Grootjans A, van Duinen G-J, Olivia Bragg O, Giesen W, Nikolaev V, van der Schaaf S, Grundling P-L. Peatlands and biodiversity, Chapter 3. In: Parish F, Sirin A, Charman D, Joosten H, Minayeva T, Silvius M, Stringer L, editors. Assessment on peatlands, biodiversity and climate change: main report. Kuala Lumpur: Global Environment Centre; 2008. p. 60–98. Joint publication with Wetlands International, Wageningen.

Moen A. Classification of mires for conservation purposes in Norway. Aquil Ser Bot. 1985;21:95–100.

Montanarella L. Mapping of peatlands, Chapter 4. In: Biancalani R, Avagyan A, editors. Towards climate-responsible peatlands management. Rome: Food and Agriculture Organization of the United Nations (FAO); 2014. p. 19–21.

Page SE, Rieley JO, Shotyk ØW, Weiss D. Interdependence of peat and vegetation in a tropical peat swamp forest. Philos Trans R Soc Lond B Biol Sci. 1999;354:1885–97.

Page SE, Siegert F, Rieley JO, Boehm H-DV, Jaya A, Limin S. The amount of carbon released from peat and forest fires in Indonesia in 1997. Nature. 2002;420:61–5.

Page S, Hoscilo A, Wösten H, Jauhiainen J, Silvius M, Rieley J, Ritzema H, Tansey K, Graham L, Vasander H, Limin S. Restoration ecology of lowland tropical peatlands in Southeast Asia: current knowledge and future research directions. Ecosystems. 2009;12:888–905.

Quinty F, Rochefort L. Peatland restoration guide. 2nd ed. Québec: Canadian Sphagnum Peat Moss Association; 2003. Joint publication with New Brunswick Department of Natural Resources and Energy.

Ramsar. Resolution VIII.17 Guidelines for Global Action on Peatlands. 8th Meeting of the Conference of the Contracting Parties to the Convention on Wetlands (Ramsar, Iran, 1971) Valencia, Spain, 18–26 Nov 2002. Available from: http://www.ramsar.org/sites/default/files/documents/pdf/res/key_res_viii_17_e.pdf. Accessed 23 Jan 2016.

Rieley J. Utilization of peatlands and peat, Chapter 5. In: Biancalani R, Avagyan A, editors. Towards climate-responsible peatlands management. Rome: Food and Agriculture Organization of the United Nations (FAO); 2014. p. 6–11.

Runkle RK, Kutzbach. Peatland characterization, Chapter 1. In: Biancalani R, Avagyan A, editors. Towards climate-responsible peatlands management. Rome: Food and Agriculture Organization of the United Nations (FAO); 2014. p. 22–6.

Rydin H, Jeglum J. The biology of peatlands. Oxford: Oxford University Press; 2006. 343 p.

Schumann M, Joosten H. Global peatland restoration manual. Greifswald: Institute of Botany and Landscape Ecology, Greifswald University; 2008. Available from: http://www.imcg.net/media/download_gallery/books/gprm_01.pdf. Accessed 10 June 2016.

Silvius M, Joosten H, Opdam S. Peatlands and people, Chapter 3. In: Parish F, Sirin A, Charman D, Joosten H, Minayeva T, Silvius M, Stringer L, editors. Assessment on peatlands, biodiversity and climate change: main report. Kuala Lumpur: Global Environment Centre; 2008. p. 20–38. Joint publication with Wetlands International, Wageningen.

Sirin A, Laine J. Peatlands and greenhouse gases, Chapter 7. In: Parish F, Sirin A, Charman D, Joosten H, Minayeva T, Silvius M, Stringer L, editors. Assessment on peatlands, biodiversity and climate change: main report. Kuala Lumpur: Global Environment Centre; 2008. p. 118–38. Joint publication with Wetlands International, Wageningen.

Turetsky MR, Benscoter B, Page S, Rein G, van der Werf GR, Watts A. Global vulnerability of peatlands to fire and carbon loss. Nat Geosci. 2015;8:11–4.

Wells DE, Zoltai S. Canadian system of wetland classification and its application to circumboreal wetlands. Aquil Ser Bot. 1985;21:45–52.

Wösten JMH, Rieley JO, Page SE. Restoration of tropical peatlands. Wageningen: Altera – Wageningen University and Research Center; 2008. 252 p. Joint publication with the EU INCO – RESTORPEAT Partnership.

Peat   15

Richard Lindsay and Roxane Andersen

Contents

Introduction	246
Peat Soils - a Historical Record	247
Carbon Storage	249
The Great Challenge	250
References	250

Abstract

Peat is a soil consisting of semi-decomposed plant material which accumulates in situ as a result of waterlogging. The percentage of mineral matter contained within such soils can vary from as little as 2% by weight to as much as 30%, though even this upper limit is more an agreed convention than any strict biological threshold. Peat can be generated from a wide range of plant materials under various forms of waterlogging. The orderly nature of peat layers, which may attain depths of 40 m or more and which may have accumulated over periods as long as 100,000 years, offer much of interest to palaeobotanists and archaeologists, particularly as the processes which result in the preservation of plant material also preserve other objects, such as human remains. The carbon stored in the world's peatlands exceeds that which is stored in all the world's vegetation.

Keywords

Bog · Bog butter · Carbon · Fen · Histosol · Mire · Mosses · Peat · Peatland · Plants · Pollen · Preservation · Soil · Tollund Man · Von Post · Waterlogging

R. Lindsay (✉)
Sustainability Research Institute, University of East London, London, UK
e-mail: r.lindsay@uel.ac.uk

R. Andersen
Environmental Research Institute, University of the Highlands and Islands, Thurso, UK
e-mail: roxane.andersen@uhi.ac.uk

© Springer Science+Business Media B.V., part of Springer Nature 2018
C. M. Finlayson et al. (eds.), *The Wetland Book*,
https://doi.org/10.1007/978-94-007-4001-3_274

Introduction

Peat is a distinctive soil type because it contains an unusually high proportion of semi-decomposed plant-derived organic matter and large amounts of water. The plant matter is produced in situ by the prevailing vegetation cover. The dead plant material produced by the vegetation cover only undergoes partial decomposition because it is waterlogged. Organic matter requires large quantities of oxygen for aerobic decomposition, but oxygen diffuses through water $10,000\times$ slower than it does through air, and thus within the waterlogged mass of decomposing plant material, the oxygen supply is rapidly depleted (Clymo 1983). Anaerobic decomposition is very much slower than aerobic decomposition but a steady supply of material is produced by fresh growth, which may thus exceed the capacity of the decomposition processes to break down all of this supply, leading to accumulation of the remaining material in the form of a waterlogged deposit known as "peat." The soil classification term "histosol" refers to the fact that a peat soil consists of plant tissue – "*histos*" meaning "tissue" in Greek.

Where the waterlogging which induces peat formation is also associated with the transport of mineral soil sediments (e.g., a river flood plain or a mangrove swamp), the peat may contain a significant proportion of mineral matter. Although there is a continuum ranging from "pure peat" containing almost no mineral matter to organo-mineral soils which are largely mineral with a relatively small proportion of organic matter, peat soils are generally distinguishable from non-peat soils by their evident organic content. While to some extent arbitrary, a widely accepted definition of peat states that peat is a soil with at least 30% organic plant matter which has accumulated in situ (Joosten and Clarke 2002) and has a thickness of 30 cm or more.

Under waterlogged conditions almost any vegetation can form peat, but the constraints imposed by waterlogging restrict the types of vegetation which can survive for a sufficient length of time to generate significant thicknesses of peat. Even so, across the globe the range of vegetation which does form peat is remarkably varied and consists of almost all plant life-forms, from mosses, rushes, and herbaceous flowers to tropical forest trees (Fig. 1). Troels-Smith (1955) provides a classification of materials which may form peat, thereby giving an idea of the range of plant components which may be encountered in a peat deposit.

As plant remains are slowly degraded, the resulting peat undergoes a number of physical transformations which have been captured in the von Post humification index or scale. The scale, devised by the Swedish botanist L. von Post, divided peat into ten classes from H1 (undecomposed plant matter) to H10 (completely decomposed plant matter). A simplified classification of peat separates it into fibric (least decomposed plant material with intact fibers), hemic (partially decomposed plant material), or sapric (highly decomposed plant material). The von Post scale was never formally published, and thus a variety of "von Post scales" have been used over the years by various authors, with Clymo (1983) perhaps giving a scale which approximates most closely to the original ideas of von Post. Indeed anyone interested in exploring the detailed behavior of peat soils either as a biophysical material or a challenge to engineering will find an enormous amount of valuable information in Clymo (1983) and Hobbs (1986).

Fig. 1 Root mass of the southern hemisphere wire rush *Empodisma minus* on a patterned mixed mire, Fraser Island, Australia. The root mass closely resembles the fibrous nature of peat formed by *Sphagnum* bog moss and possesses many of the same properties of water retention and ability to preserve material (Picture credit: Richard Lindsay © Rights remain with the author)

Peat Soils - a Historical Record

Peat soils have been of considerable interest to paleobotanists and archeologists for more than a century because the accumulated peat provides a layered, preserved record of what has happened both to the peatland and to the surrounding landscape (Fig. 2). The plant remains which make up the peat provide a picture of the conditions which prevailed on the peatland over millennial timescales, but the peat also preserves any material which falls onto or into the peatland. Consequently pollen grains from the surrounding landscape and even from further afield permit a picture to be built up of regional landscape histories and past climates, even for before written records, based on changes in the local or regional vegetation which left its pollen signal in the peat (e.g., Pennington 1970; Santos et al. 2000). Other objects have been found preserved in peat, from "bog butter" (see Bog Butter website), ancient trackways, even precious objects apparently placed in the bog as votive offerings to human bodies placed in the bog after ritual sacrifice (Glob 1969; Fischer 2012) – see Fig. 3.

Fig. 2 Exposed peat face on Shetland Mainland, northern Scotland, displaying very clear layering of the peat accumulated over several millennia (Picture credit: Richard Lindsay © Rights remain with the author)

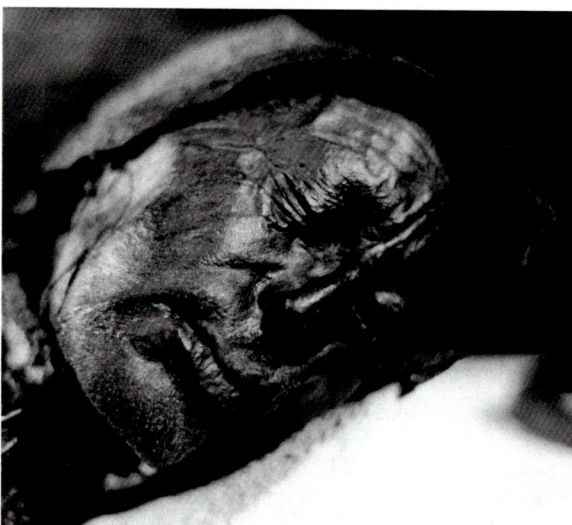

Fig. 3 The face of Tollund Man, an Iron Age individual who was apparently sacrificed by being ritually killed in three ways then buried in a peat bog in Denmark, before being discovered by peat diggers in the 1950s (Picture credit: Sven Rosborn via Wikimedia Commons)

Carbon Storage

In recent years peatlands have attracted great interest because of the plant carbon stored in peat deposits. This carbon is typically stored in undisturbed peatlands for several millennia and is thus locked up for timescales much longer than that normally associated with even the most ancient of forests. In addition, peat soils across the globe are estimated to have an average peat thickness of about 2 m, although some peat deposits are more than 40 m deep and have been accumulating for 100,000 years or more (Ramsar website for Nakaikemi-shicchi) – see Fig. 4. The carbon density of peat soils per square meter thus exceeds many forests because, unlike a forest where there are gaps between trees, there are few if any gaps in a peat deposit. Indeed the most recent figures for soil carbon storage suggest that the peatlands of the world may hold as much as half the

Fig. 4 Nakaikemi-shicchi peatland, formed within a deep basin just outside the city of Tsuruga, Fukui Prefecture, western Japan. The site has been used as a rice paddy for many years, but beneath the paddy field is some 40 m of peat, and now the site is being managed for nature conservation, with more than 70 species of dragonfly already recorded for the site (Picture credit: Richard Lindsay © Rights remain with the author)

world's soil carbon of around 3,000 Pg C (a pentagram = 10^{15} grams) or 1500 Gt C, which is equivalent to 3× the quantity of carbon stored in *all* the world's vegetation and 1.5× the amount of carbon held in the atmosphere (Scharlemann et al. 2014).

The Great Challenge

Despite this rising interest in peat and peatlands, it is undoubtedly the case that many areas of peatland, whether "archaic peat soils" which have lost their peat-forming vegetation or living "mires" which still support a vegetation capable of forming peat, are not being recognized as such and are being described as some other form of ecosystem. When a peatland is identified as something other than a peatland, it is generally managed in a way which is damaging to the peat deposit and thus ultimately to many of the fundamental features and ecosystem service benefits which characterize a functioning peatland. Recognizing the existence and extent of the world's peatlands remains one of the great challenges of ecosystem and soil science, given that peatlands are already known to be the most extensive form of the world's terrestrial wetlands and wetlands as a whole are vanishing at an ever-increasing rate (Davidson 2014). The danger is that peatlands are vanishing before we even knew we had them, and so their loss goes completely unrecorded.

References

Bog Butter website. http://nordicfoodlab.org/blog/2013/10/bog-butter-a-gastronomic-perspective. Accessed 9 Mar 2015.
Clymo RS. Peat. In: Gore AJP, editor. Ecosystems of the world 4A. Mires: swamp, bog, fen and moor. Amsterdam: Elsevier; 1983. p. 159–224.
Davidson NC. How much wetland has the world lost? Long-term and recent trends in global wetland area. Mar Freshw Res. 2014;65:934–41.
Fischer C. Tollund man – gift to the gods [English edition]. Stroud: The History Press; 2012.
Glob PV. The bog people: iron age man preserved [English edition]. London: Faber and Faber; 1969.
Hobbs NB. Mire morphology and the properties and behaviour of some British and foreign peats. Q J Eng Geol. 1986;19:7–80.
Joosten H, Clarke D. Wise use of mires and peatlands. Totness/Devon: NHBS/International Mire Conservation Group and International Peat Society; 2002.
Pennington W. Vegetation history in the North West of England: a regional synthesis. In: Walker D, West RG, editors. Studies in the vegetational history of the British Isles. Cambridge: Cambridge University Press; 1970. p. 41–80.
Ramsar website for Nakaikemi-shicchi. https://rsis.ramsar.org/ris/2057. Accessed 9 Mar 2015.
Santos L, Romani JRV, Jalut G. History of vegetation during the Holocene in the Courel and Queixa Sierras, Galicia, northwest Iberian Peninsula. J Quat Sci. 2000;15(6):621–32.
Scharlemann JPW, Tanner EVJ, Heiderer R, Kapos V. Global soil carbon: understanding and managing the terrestrial carbon pool. Carbon Manag. 2014;5(1):81–91. doi:10.4155/cmt.13.77.
Troels-Smith J. Characterisation of unconsolidated sediments. Geological Survey of Denmark. IV. Series Vol. 3, No. 10. København: Forlag and Sandal; 1955.

Peatland (Mire Types): Based on Origin and Behavior of Water, Peat Genesis, Landscape Position, and Climate

16

Richard Lindsay

Contents

Minerotrophic Mires (Fens)	252
Fens Waterlogged by a Level, Relatively Static Groundwater Table ("Topogenous")	252
Fens Waterlogged by a Level but Markedly Fluctuating Groundwater Table	253
Fens Waterlogged by Groundwater Moving Downslope ("Soligenous")	254
Fens Waterlogged by Channeled Surface-Water Flow	258
Mixed Mires Waterlogged by Both Minerogenous and Ombrogenous Water	261
Ombrotrophic Mires (Bogs)	264
Raised Bogs Domed Solely as a Result of Peat Accumulation	264
Intermediate Raised Bogs: Shaped Partially by Underlying Landform	267
Blanket Bogs: Shaped Predominantly by Underlying Landform	269
"Occult Precipitation" Mires	271
References	272

Abstract

Mires, or peat-forming systems, have traditionally been recognised as falling into two broad peat-forming types – minerotrophic fens fed by groundwater or collected surface water, and ombrotrophic bogs fed exclusively by direct precipitation. Different types of fen can then be distinguished based on sources of water and rates of water supply. In contrast, ombrotrophic bogs can be classified according to their morphology, position within the landscape and consequent developmental history. A few 'mixed' or 'intermediate' mire types can also be identified.

Keywords

Aapa · Basin · Bog · Blanket bog · Blanket mire · Climate · Cloud forest · Condensation · Estuarine · Floodplain fen · Flush · Groundwater · Immersion

---

R. Lindsay (✉)
Sustainability Research Institute, University of East London, London, UK
e-mail: r.lindsay@uel.ac.uk

© Springer Science+Business Media B.V., part of Springer Nature 2018
C. M. Finlayson et al. (eds.), *The Wetland Book*,
https://doi.org/10.1007/978-94-007-4001-3_279

mire · Ladder fen · Minerotrophic · Mire · Mixed mire · Ombrotrophic · Occult · Palsa · Patterned fen · Peatland fen · Percolation · Polygon mire · Raised bog · Raised mire · Topogenous · Saddle · Schwingmoor · Soligenous · Spring · Spur · Transition · Valley mire · Valley bog · Valley fen · Valley side · Watershed

Mires are defined as wetlands that support vegetation that is normally peat forming (Löfroth 1994; European Commission 2007) and is also used as a general term for all natural and seminatural peatland ecosystems (Wheeler and Proctor 2000). Although landform and climate are important influences, the source of water is the key factor in distinguishing two broad classes of mire ecosystem. Minerotrophic fens are peat-forming systems dependent upon groundwater (topogenous) or surface water (soligenous) flows enriched by solutes from the bedrock, subsoil, or adjacent surfaces. They display a great variety of forms due to the various ways in which landscape influences water and solute supply. Ombrotrophic bogs are systems in which the peat-forming vegetation is above the influence of the mineral groundwater table and is entirely dependent upon precipitation to supply its water and solutes.

Hydromorphology is one of the most widely used approaches to mire classification beyond the basic distinction of minerotrophy and ombrotrophy (e.g., Moore and Bellamy 1974; Botch and Masing 1983; Nature Conservancy Council 1989; Succow and Jeschke 1990; Lindsay 1995). Using a combination of established typologies together with concepts from the hydro-genetic matrix proposed by Joosten and Clarke (2002), mire systems can be categorized in a great variety of ways based on origin and behavior of water, peat genesis, landscape position, and climate.

Minerotrophic Mires (Fens)

Fens Waterlogged by a Level, Relatively Static Groundwater Table ("Topogenous")

Open-Water Transition Fen (Figs. 1 and 2)
In enclosed basins where the groundwater table forms a level and relatively static water body, peat-forming vegetation may establish at the fringes of the water body and gradually infill the basin with peat.

"Schwingmoor" Fen (Fig. 3)
In deep, steep-sided basins such as kettle holes where the groundwater table forms a level and relatively static water body, peat-forming vegetation may establish at the fringes of the water body and proceed to form a floating raft which may eventually completely cover the water body beneath. On some occasions part of a raft may become detached and drift around the lake as a floating island (e.g., Mizorogaike shicchi, Kyoto, Japan – Shimizu 1986).

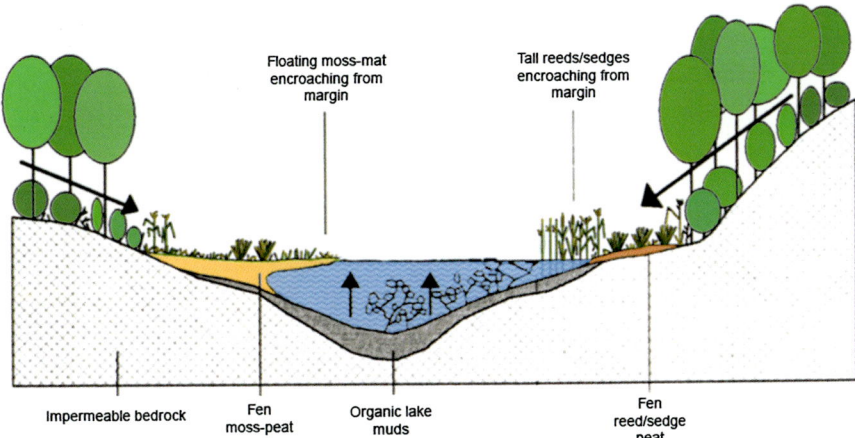

Fig. 1 Open-water transition mire (Adapted from Steiner (1992) with kind permission)

Fig. 2 Open-water transition mire beginning to infill a shallow loch (lake) in Sutherland, NW Scotland (Photo credit: Richard Lindsay © Rights remain with the author)

Fens Waterlogged by a Level but Markedly Fluctuating Groundwater Table

Karst Immersion Mire

In limestone regions, certain karst sinkholes fill with water as the associated underground cave system floods after rain (e.g., turloughs in Ireland). Fen systems are sometimes associated with such features.

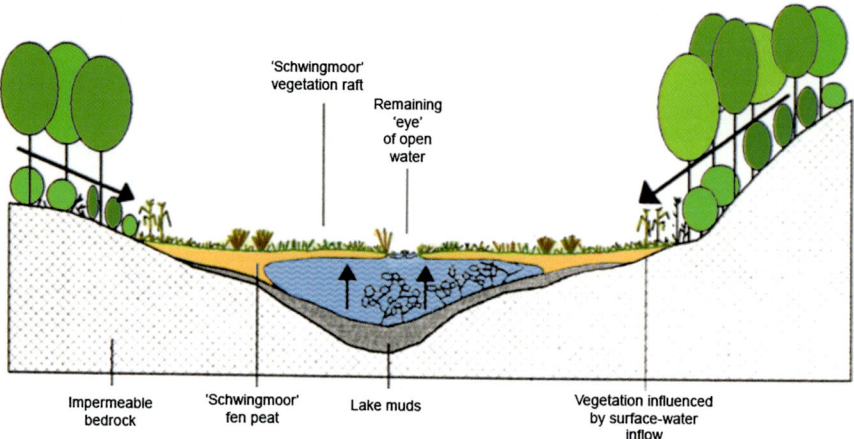

Fig. 3 "*Schwingmoor*" fen (Adapted from Steiner (1992) with kind permission)

Tidal Immersion Mire

Peat may sometimes accumulate under conditions of waterlogging provided at least in part by marine water (e.g., mangroves). Such "thalassogenic" mires are therefore subject to the rise and fall of the tides.

Fens Waterlogged by Groundwater Moving Downslope ("Soligenous")

Floodplain Fen (Fig. 4)

Where rivers have developed a floodplain, the riverbank typically develops a slight "lip" or levee which is overtopped during flood events. The levee then limits the flow of floodwaters back into the river channel. The water thereby trapped on the floodplain continues to move downslope in the direction of the river channel gradient but does so relatively slowly, thereby causing extended waterlogging and creating peat-forming conditions. The slow movement of the floodwaters downslope sometimes leads these systems to be classed as topogenous rather than soligenous fens. Oxbow lakes formed by abandoned meanders typically add open-water transition mires, while the complex nature of the underlying alluvial sediments often means that artesian spring mires also occur within the floodplain fen.

Valley Mire or Valley Fen (Often Erroneously Called "Valley Bog")

Valley mires form across flat-bottomed valleys which lack a major river channel but still nevertheless receive significant quantities of water, provided there is a relatively impermeable base. The flat nature of the valley bottom may be a natural feature of the landform or may be a result of peat accumulation along the valley floor (Fig. 5). Characteristically, valley fens possess a central, but often somewhat diffuse, zone of

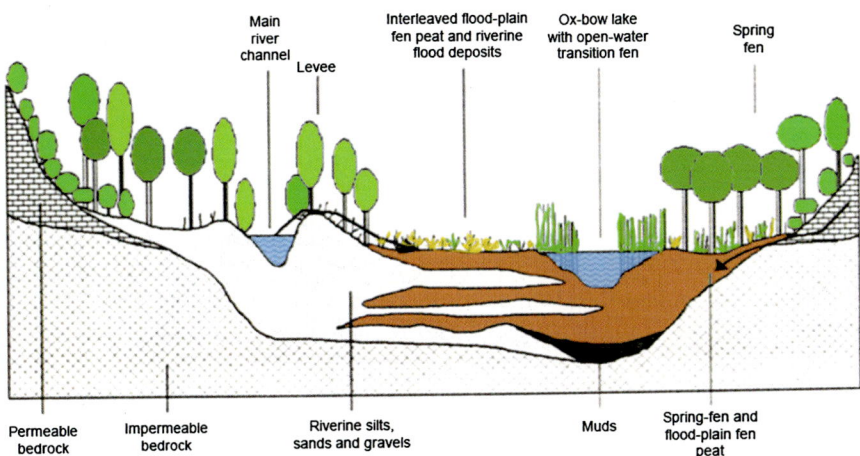

Fig. 4 Floodplain fen (Adapted from Steiner (1992) with kind permission)

Fig. 5 Small valley mire just north of Inverness, northern Scotland. The central water track running through the relatively flat mass of accumulated peat can just be seen, while the deep nature of the peat can be judged from the steepness of the valley sides, composed of glacial till and dominated by *Calluna vulgaris* (Photo credit: Richard Lindsay © Rights remain with the author)

water movement (Fig. 6), while much of the valley floor is waterlogged by water seeping from the valley side.

The rather slow and diffuse supply of water and thus of solutes to much of the valley floor often means that while the central zone of water movement and the margins of the valley mire may be relatively solute-rich, the majority of the valley floor is supplied slowly with water from which much of the solute supply has been removed by the more enriched zones of vegetation. There is often thus a zone which

Fig. 6 The zone of water movement through Thursley National Nature Reserve, a large valley mire formed across the greensands of Surrey, southern England. *Schoenus nigricans* tussocks dominate the central zone of water movement, but beyond this the vegetation is dominated by *Sphagnum papillosum, S. tenellum, Molinia caerulea, Eriophorum angustifolium*, and *E. vaginatum* (Photo credit: Richard Lindsay © Rights remain with the author)

lies between the central zone of seepage and the mire margins which is relatively poor in solutes and may, in some cases, closely resemble ombrotrophic bog. This explains the erroneous description of "valley bog" sometimes given to such systems – erroneous because the water supply of a valley fen is at least as dependent on surrounding catchment (watershed) conditions as it is on atmospheric inputs and is usually heavily influenced by activities which occur within the catchment (watershed).

Artesian Spring Mire/Fen

Where a catchment contains semipermeable sediments, particularly across valley floors, groundwater from upper parts of the catchment may be forced to the surface along the valley floor to form waterlogged areas which become dominated by peat-forming vegetation. Often these artesian spring fens occur within a broader matrix of valley fen or floodplain fen and are not therefore immediately obvious, but they may even emerge through the dome of an otherwise ombrotrophic bog. Abrupt changes in water chemistry or water temperature within a localized area may indicate the presence of an artesian fen system within a larger peatland complex.

Surface-Flow Spring Mire/Fen (Figs. 7 and 8)

When a porous sediment sits on an impermeable sediment or rock and the junction of these two is exposed at the ground surface, typically on a valley side, groundwater emerges at the surface and flows downslope across the ground surface. A surface-flow spring mire forms where such waterlogging gives rise to peat formation, but the major part of the water flows in a diffuse way over the surface of the peat.

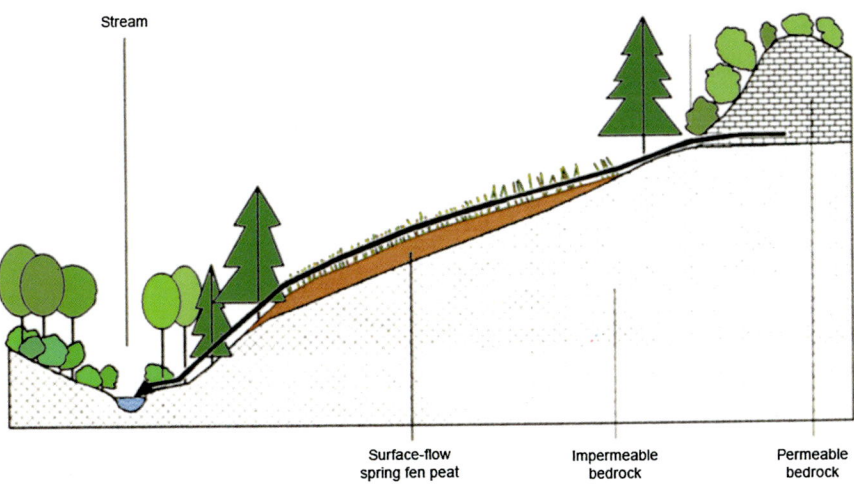

Fig. 7 Surface-flow spring mire (Adapted from Steiner (1992) with kind permission)

Fig. 8 A small percolation mire in the center of the image and a smaller spring mire with a surface-flow mire flowing out of the bottom of the image, Glen Roy, Highlands of Scotland (Photo credit: Richard Lindsay © Rights remain with the author)

Percolation Mire/Fen (Figs. 8 and 9)

Formed under the same conditions as a surface-flow spring mire, a percolation mire develops where spring-derived waterlogging gives rise to peat formation, and the major part of the water seeps through the body of the peat. In such cases the rate of

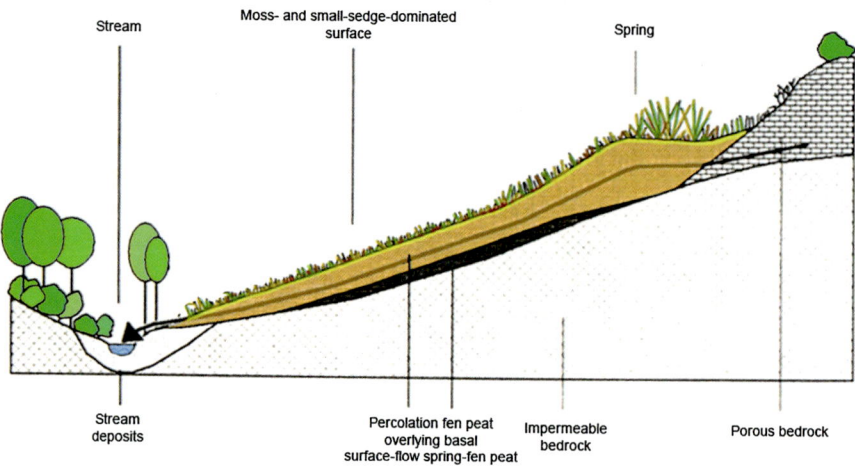

Fig. 9 Percolation mire (Adapted from Steiner (1992) with kind permission)

springwater supply tends to be rather slow compared with surface-flow spring mires. The peat is instead relatively undecomposed (un-humified) which permits more rapid water seepage through the peat (i.e., it has a higher hydraulic conductivity) than is the case for surface-flow spring mires which tend to have rather dense peat. Consequently such percolation mires are often dominated by bryophytes (particularly *Sphagnum*) rather than sedges or grasses, and the water percolates between the poorly decomposed stems and branches of the bryophyte carpet. One of the characteristic features of percolation mires is that flowing water is rarely visible at the surface. Water flowing out from a percolation mire may then become a surface-flow mire.

Fens Waterlogged by Channeled Surface-Water Flow

Surface-Flow "Flush" with a Central Zone of Water Movement (Fig. 10)

Landform often dictates that surface waters come together to form a distinct, though generally narrow, zone of surface seepage which may have no direct link with groundwater. Such zones may occur within a mineral landscape or within a peat-dominated landform (e.g., as a "flush" in blanket mire or as the headwaters of an endotelmic stream in a raised bog). The characteristic feature is that there is a central zone of water flow or seepage, while the geomorphological feature in which the water gathers is too small to merit the term "valley." Larger examples of surface-flow flushes occupying the floors of distinct valleys would instead be classed as "valley mires."

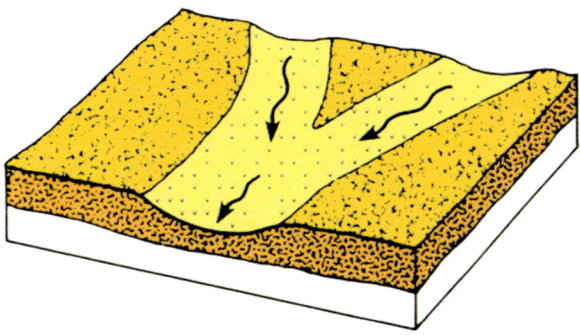

Fig. 10 Surface-flow "flush" with a central zone of water movement (Adapted from NCC (1989) with kind permission)

Patterned Fens with No Central Watercourse

Where groundwater or surface-water flow is not sufficient throughout the year to maintain a central watercourse through the mire, a series of ladderlike ridges often develops across the line of water flow in a somewhat counterintuitive pattern, apparently blocking water flow. Water seeps instead through the ridges from one pool to the next, and at times of high flow (e.g., snowmelt, rainy season), there may also be some water seepage across the surface of the intervening ridges but never enough to cut a distinct water channel. These mires, formed in what are sometimes extremely large zones of water collection, are thus characterized by an *absence* of a central water channel and the *presence* of often wide pools or hollows separated by narrow ridges, this whole "ribbed" pattern being *aligned at right angles* to the direction of water seepage.

Such patterned mires are common in boreal regions where they can form spectacular features within the landscape, consisting of large, shallow, elongated pools or hollows which may span the whole width of the mire, separated by sinuous ridges or "strings" which may be no more than 1 m wide but which may run for 0.5 km between pools, oriented at right angles to the direction of water flow (Fig. 11).

Similar patterns can, however, be found in subtropical and tropical regions and in both hemispheres. Thus while the strings of boreal fens are generally formed by species of the bog moss *Sphagnum*, in subtropical Fraser Island off the coast of Queensland, the strings are formed by the wire rush *Empodisma minus*, although the patterns formed are virtually identical to *Sphagnum*-dominated boreal patterned fens of the northern hemisphere (Fig. 12).

In more oceanic temperate regions where there is neither a period of significant snowmelt (a major feature of boreal regions) nor a distinct "wet season" (a characteristic feature of subtropical and tropical regions), the patterning of fen systems tends to be more subdued. These more oceanic examples have been described as "ladder fens" (Lindsay et al. 1988) after ladderlike fen structures were described for the Atlantic Boreal Wetland Region of Canada (Zoltai and Pollett 1983) – see Fig. 13 – although Charman (1993), who has studied several Scottish examples, favors the more generic term "patterned fens" for these systems.

Fig. 11 Patterned fen, Poplar/Nanowin Rivers Park Reserve, Manitoba, Canada. The green ovals are "pine islands" on raised mineral ground, while the ribbed pattern of the fen consists of brown bands and expanses, which are shallow pools, and narrow pale "ribbons" which are the raised ridges of peat-forming *Sphagnum*-dominated vegetation on which it is possible to walk (Image © Google Earth and Digital Globe, 2015)

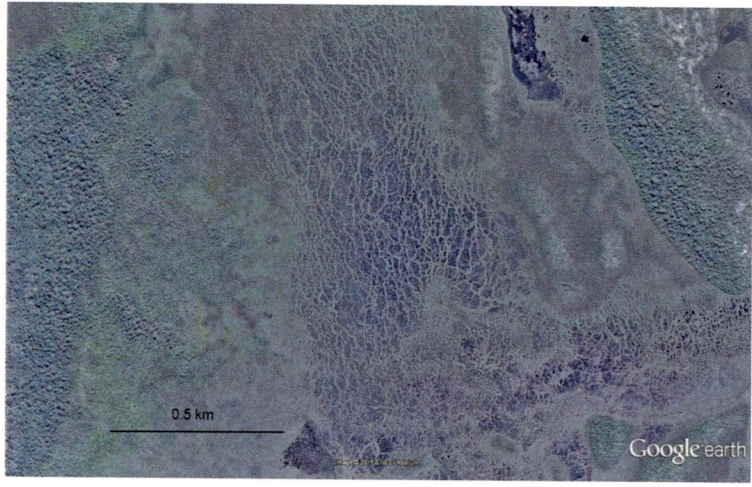

Fig. 12 Patterned fen near Moon Point, Fraser Island, Queensland, Australia. On the right-hand edge of the image are ancient high dunes covered with Eucalypt forest. In the center is the patterned fen, with shallow pools displayed as dark gray/blue bands and expanses, and narrow pale "ribbons" which are the raised ridges of peat-forming *Empodisma*-dominated vegetation on which it is possible to walk (Image © Google Earth and CNES/Astrium, 2015)

Fig. 13 Patterned "ladder" fen within the blanket mire landscape of the Flow Country, Sutherland, Northern Scotland. The patterned fen occupies the central foreground of the image. Water flows from the pale *Molinia caerulea*-dominated emergence zone located to the left of the patterned fen then seeps through the fen moving toward the right of the image across the line of ladderlike ridges before entering the stream course which exits the image on the extreme right (Photo credit: Steve Moore © Rights remain held *in-trust* by Richard Lindsay)

Mixed Mires Waterlogged by Both Minerogenous and Ombrogenous Water

Polygonal Fens of Permafrost Regions

In arctic and subarctic regions, an impermeable layer formed by permafrost sits beneath a thin layer of peat which thaws in the summer months but which is subject to frost heave and formation of ice wedges which break the peat surface up into a polygonal mosaic. Waterlogging is caused by the trapping of snowmelt water within the raised edges of each polygon. Each polygon may then give rise to its own small, enclosed fen system. The source of water is melted snow because the groundwater is frozen within the permafrost layer, but frost heave may also bring fragments of mineral subsoil to the surface if the peat is relatively thin, thereby providing the possibility of some mineral enrichment.

Snowmelt-Patterned Mixed Mires ("Aapa Mires" – Ruuhijärvi 1983)

Where the narrow ridges of patterned fen reach heights of 50 cm or more, preventing snowmelt water flow over the ridges and forcing all the water instead to percolate through the peat of the ridges, these ridges may in effect become narrow ribbons of ombrotrophic bog sitting within a matrix of wide minerotrophic fen hollows and pools. The distinction between the ombrotrophic environment of the narrow ridges and the minerotrophic environment of the pools ("flarks") is most pronounced in summer and autumn once the main flow from snowmelt has ceased. Such high ridges are often dominated by *Sphagnum fuscum* in northern boreal regions.

Fig. 14 Oblique aerial view of possible mixed mire near Moon Point, Fraser Island, Queensland, Australia. This mire lies adjacent to the patterned fen illustrated in Fig. 12, part of which can be seen in the top left of the image. At the bottom of the image, a vehicle track can be seen separating the mire from the ancient high-wooded sand dune just off the bottom of the image, though an arm of this dune can be seen extending in from the right of the image. The marked difference in pattern between this mire and the patterned mire shown in Fig. 12 suggests that a different set of hydrological processes may be in play here (Photo credit: Richard Lindsay © Rights remain with the author)

Wet-Season Subtropical/Tropical Patterned Mixed Mires (Figs. 14 and 15)

In tropical and subtropical regions where there is a wet season and a dry season, the ridges of patterned mires may display a complex relationship between minerotrophic groundwater supply and ombrotrophic conditions. Pools may spend part of the year dominated by groundwater supply, but then in the wet season the majority of their water supply may be derived from direct precipitation, depressing the groundwater to depths below the rooting zone. The intervening ridges, meanwhile, may be sufficiently high to prevent significant minerotrophic influence of the rooting zone and consequently rely largely or wholly on precipitation inputs during both dry and wet seasons.

Palsa Mires

In arctic regions just outside the continuous permafrost zone, the insulating properties of peat can give rise to formation of a permafrost ice core within a hummock. Expansion as the water freezes causes the hummock to become even more elevated which results in a thinner snow cover and more exposure to freezing conditions

Fig. 15 Ground view of the possible mixed mire shown in Fig. 14. The pools may receive water from the groundwater table, but the ridges, composed almost entirely of *Empodisma minus* peat, may be largely (or wholly?) ombrotrophic, relying on precipitation inputs during the wet season (Photo credit: Richard Lindsay © Rights remain with the author)

during the winter, causing further development of permafrost within the mound. Meanwhile the cover of peat-forming vegetation, which is typically *Sphagnum*, insulates the permafrost core during the summer months. A positive feedback loop thus develops, leading to steady growth in height and girth of the mound as the ice core steadily accumulates more water into the permafrost core to form a *palsa*, a mound which may rise as high as 7 m above the surrounding fen vegetation. The peat-forming vegetation on the sides and summit of the palsa is largely dependent upon direct precipitation inputs for water and is thus ombrotrophic. Palsas typically form a mazelike network within extensive fen systems, and the humidity provided by the fen may help to support the low-elevation ombrotrophic conditions of the palsa network.

Infilled Basin Mires Transitioning to Ombrotrophic Conditions

When an open-water transition mire has infilled its basin with peat or a "Schwingmoor" mire has completely covered the basin with a floating mat, hummock-forming bryophytes (generally *Sphagnum*) may begin to form small features which rise above the groundwater table. As these begin to coalesce and create areas which are increasingly divorced from groundwater influence, the mire surface displays a mosaic of minerotrophic, semi-ombrotrophic, and ombrotrophic conditions. This is the *Übergangsmoor* of Weber (1907) and characterized by Du Rietz's (1954) indicators of the groundwater-rainwater transition boundary (*Mineralbodenwasser-zeigergrenze*). It is commonly called "transition mire" and is now formally termed as such in the European Union Habitats Directive (European Commission 2007).

In addition, man-made "transition mires" may also be found in areas of raised bog which have been extensively and deeply cut for domestic fuel to the extent that the original raised dome has vanished except perhaps for a few scattered upstanding balks. Where the cutaway areas have recolonized with mire vegetation, this often displays a character normally associated with "Schwingmoor"-type mire with extensive rafts of, for example, *Sphagnum fallax*, while scattered remnant balks may support a highly modified version of bog vegetation. Particularly where such cutover bog is a long way from obvious habitation, it is tempting to ascribe such places to a natural "transition mire" type, but careful examination of aerial photographs will often show evidence of a rectilinear shape to these remnant balks, revealing the true nature of such cutover sites.

Ombrotrophic Mires (Bogs)

Ombrotrophic mires rely wholly on direct precipitation for its water and solute supply and are restricted to those climatic localities where evapotranspiration losses are low, and precipitation inputs of various kinds are relatively frequent such that the ecosystem is capable of retaining this moisture to the extent that the highest part of the bog surface remains constantly waterlogged.

Raised Bogs Domed Solely as a Result of Peat Accumulation

These might be regarded as "true" raised bogs because their domed morphology is determined wholly by peat accumulation. In general this gives them a somewhat symmetrical profile, and therefore any surface patterning tends toward a concentric arrangement of features. They are found throughout the range of climatic zones capable of supporting bog formation but tend to be the dominant bog type where the landform is flat and the climate is not sufficiently cool and humid to give rise to blanket mire. Raised bogs thus occur throughout the temperate and boreal zones in both hemispheres but also extend as far as the tropical zone in the form of tropical peat swamp forests (Dommain et al. 2014).

Floodplain Raised Bog (Fig. 16)

Within expanses of floodplain fen, areas of more stagnant water tend to provide conditions which favor dominance of peat-forming species which are tolerant of the more acidic and anaerobic environment created by waterlogged plant decomposition. Such species (e.g., *Sphagnum* spp.) are typically somewhat resistant to decomposition because they have a low nutrient content and thus tend to accumulate peat. Where climatic conditions provide sufficiently regular precipitation, aided by the high humidity maintained by the presence of the surrounding floodplain fen, these areas may accumulate peat which rises above the floodwaters to form domed raised bogs. A distinctive feature of these floodplain raised bogs is that their basal layers typically

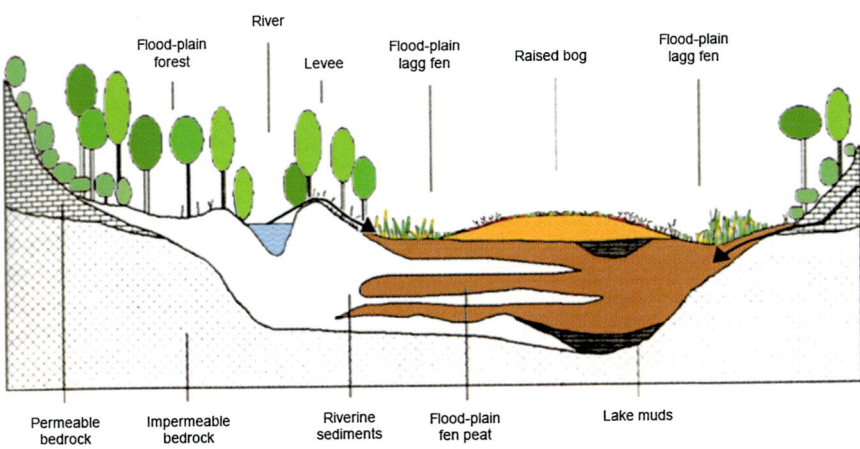

Fig. 16 Raised bog formed within a floodplain, with extensive lagg fen surrounding the raised bog dome (Adapted from Steiner (1992) with kind permission)

contain layers of riverine alluvium deposited during times of flood. This can result in quite complex hydrological properties within lower parts of the raised dome.

These bogs can occur in regions which do not, on the face of it, provide sufficient precipitation for ombrotrophic peat development, but the hydrological balance is tipped in favor of bog formation by the consistently high humidity of the valley floor. In addition, these and the next form of raised bog – estuarine raised bog – occur throughout the whole range of raised bog distribution, from the arctic to the tropics. In the arctic they are almost invariably *Sphagnum* dominated, whereas in the tropics, the peat is formed from the roots, branches, trunks, and leaves of peat swamp forest trees. In the southern hemisphere, the options are even more varied, with rushlike Restionaceae and "cushion plants" such as *Donatia fascicularis* also forming raised bogs.

Estuarine Raised Bog

Raised bogs formed on estuarine floodplains typically display complex basal deposits similar to that seen in floodplain raised bogs, but in this case the deposits may consist of interleaving layers of peat, alluvial sediment from riverine flood events and marine sediments deposited during marine transgressions. Again, the presence of extensive floodplain fen may assist the development of ombrotrophic conditions in otherwise unsuitable climate regions through maintenance of high humidity across the valley floor.

Basin Raised Bog (Fig. 17)

This is the "classic" raised bog, formed initially by the infilling of a shallow basin with peat through a process of terrestrialization, but then, because there is sufficiently regular precipitation throughout the year (including fog and dew – Weber 1902, pp. 68–69), accumulation of waterlogged peat continues even though this causes it to rise above the local groundwater table. Eventually this may form a dome of peat which rises 10 m or more above the local landscape. In drier climatic regions, ombrotrophic conditions may not be able to develop without the initial terrestrializing phase of a basin fen and may not extend as far toward the climatic limits of bog formation as floodplain raised bogs because they are not supported to the same extent by the high local humidity resulting from extensive surrounding tracts of floodplain fen.

Basin raised bogs are found from the arctic to the tropics as well as in both hemispheres and display much the same degree of variation in character as floodplain or estuarine raised bogs. However, as a consequence of not being limited to the lowland settings associated with all estuarine raised mires and many floodplain raised mires, basin raised bogs can benefit from the higher precipitation levels, lower temperatures, and higher cloud cover typical of ground lying at higher elevations. Their distribution can thus often reflect orographic influences and be associated with hill ground or montane regions.

Schwingmoor Raised Bog

Where a floating mat of peat-forming vegetation develops over a steep-sided basin such as a kettle hole lake and there is sufficiently regular precipitation, this mat can continue to accumulate peat even though the peat rises above the influence of the lake water, eventually rising to form a domed raised bog. A lens of lake water is typically trapped beneath the dome of peat. Testing the basin profile and identifying the presence of a water lens (if present) are the only sure way of identifying such raised bogs, although the immediate profile of surrounding mineral ground can sometimes give a good indication that the bog has formed in a deep basin.

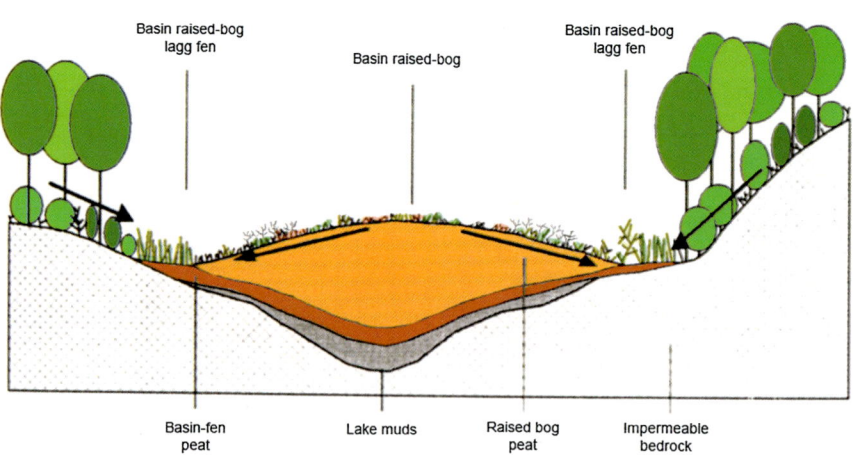

Fig. 17 Basin raised bog with a more limited lagg fen zone than that normally associated with floodplain raised bog or estuarine raised bog (Adapted from Steiner (1992) with kind permission)

Some sites described as "Schwingmoor raised bog" are actually transition mires in which small elements within the floating raft may be semi-ombrotrophic, but much of the raft and its vegetation are still influenced by groundwater. Perhaps even more commonly, cutover bogs may be erroneously assigned to this type where they have a significant extent of remaining upstanding balks surrounded by extensive areas of "Schwingmoor"-type vegetation in the cutover areas, particularly where the balks are much slumped and overgrown with modified bog vegetation and thus not immediately recognizable as peat-cutting faces.

Intermediate Raised Bogs: Shaped Partially by Underlying Landform

Where the landform is not flat and the climate is sufficiently humid to compensate for runoff losses from sloping surfaces but not so regularly humid that the entire landscape becomes cloaked in peat, bogs may form which owe as much to the underlying topography as to the depth of accumulated peat. Although their overall morphology may be somewhat complex, the arrangement of their surface features tends toward an eccentric arrangement.

It is often asserted that these more complex bogs are somewhat oceanic in character. In Britain and Ireland, they are associated with oceanic regions, but this is possibly more to do with the arrangement of lowland and upland landscapes in these two islands than with the climate because significant upland massifs occur along the oceanic west coast in both cases. "Intermediate" raised bogs which are shaped significantly by underlying landform are, however, also found in the relatively continental parts of Sweden and Finland (Sjörs 1983; Ruuhijärvi 1983), as well as from the White Sea coastal region to the West Siberian Basin (Botch and Masing 1983).

These intermediate raised bogs occupy the same positions in the landscape as blanket bog systems (see ▶ Chap. 20, "Blanket Bogs") and can thus mostly be classified based on landform and hydromorphology as various forms of watershed (water-divide), spur, saddle, and valley side mire.

Watershed or Spur Raised Bog : Ridge Raised Bog (Moore and Bellamy 1974), Semi-Confined Bog (Hulme 1980), or Atlantic Raised Bog (Goodwillie 1980)

While expansion by wetting (or "paludification" of) the adjacent ground is a feature of all raised bogs, where the climate encourages more intense paludification, this may enable the peat to climb over mineral ridges and perhaps join with neighboring lenses of peat, thereby forming a shape determined in part by peat accumulation and partly by the underlying landform. On some occasions adjacent raised bogs may fuse across quite-marked mineral ridges, ultimately swallowing them to form a shape with a complex morphology which owes something to peat accumulation and something to the underlying landform. This is particularly the case in watershed (water-divide) and spur raised mires.

Fig. 18 Valley side eccentric raised bog complex in the Andorra Valley, Tierra del Fuego, Argentina (Photo credit: Richard Lindsay © Rights remain with the author)

Valley Side Eccentric Raised Bog

Raised bogs with a distinctly eccentric profile, often displaying a striking and eccentric surface pattern, appear to fall into two distinct groups. The first group is largely associated with oceanic conditions. Bogs of this first type are often located on the lower slopes of a hill toward the valley floor. They are found in, for example, Maine on the Atlantic fringes of the USA (Davis and Anderson 1991), in the White Sea coastal region of Russia (Botch and Masing 1983), the Atlantic Boreal region of Newfoundland and New Brunswick in Canada, Tierra del Fuego of Chile and Argentina (Fig. 18), the west coast of Ireland, and arguably the northwest coast of Scotland (although here it might be argued that they form part of a blanket mire complex). The second type of eccentric raised bog appears to be more associated with large-scale snowmelt and occurs in, for example, the central Swedish uplands (Sjörs 1983), while in Finland it ranges from South Central parts almost to the northern border of the country (Ruuhijärvi 1983).

Saddle Raised Bog ("Sattel Hochmoor")

Saddle raised bogs are almost the converse of ridge raised bogs, in that the initiation point of peat formation for a saddle raised bog is on a broad watershed (water-divide) ridge lying between two summits rather than beginning in adjacent basins and extending to cross the intervening ridge at a subsequent stage of development. Saddle raised bogs are typically associated with cooler highland environments, but their distribution ranges from boreal regions to the tropics, and are thus found, for example, in the Austrian Alps (Steiner 2005) and in the highlands of Costa Rica (Fig. 19). The peat is deepest on the ridge crest and then extends some way downslope either side of the ridge before giving way to non-peat habitat further down the slope.

Intermediate Raised/Blanket Mire

On the fringes of blanket mire regions, it is possible to find bog complexes which are discrete, distinct entities and which are too small to describe as a blanket mire

Fig. 19 Saddle raised bog in the highlands of Costa Rica (Photo credit: Richard Lindsay © Rights remain with the author)

landscape but which consist of several mire units which have fused together, often though not exclusively across the varied topography of a small outlying plateau. These complexes usually have characteristics of both raised bog and blanket mire and may thus be classed as intermediate raised/blanket mire.

Blanket Bogs: Shaped Predominantly by Underlying Landform

Blanket *mire* is more of a landscape concept than a specific mire type. It applies to those circumstances where the landscape as a whole is draped with peat, from the summits of gently rolling hills to the valley floors. Blanket *bog* on the other hand refers to those parts of the blanket mire landscape which are ombrotrophic. It is important to highlight at this juncture, however, the fact that the extensive and varied nature of such ground means that its underlying peatland nature is often overlooked, and the ground may therefore be assigned to a range of non-peat habitat types. This is particularly the case when the term "moorland" is used because the word embraces a wide range of upland or highland environments and has no explicit link to the presence of peat. Moorland is not the only example, however. Areas of subalpine cloud forest and even some examples of tropical peat swamp forest may, for example, be more appropriately regarded as blanket bog or at least as intermediate raised/blanket mire because they have a significant peat thickness and their morphology is largely determined by the underlying landform. Appropriate and sustainable management of such sites is only possible when the significance of the peat-forming nature of the ecosystem is acknowledged.

While complex landforms cloaked with varying thicknesses of peat may have some areas which are somewhat undifferentiated in terms of hydromorphology, in general it is possible to partition a blanket mire landscape into a closely fitting mosaic of hydromorphological types consisting of minerotrophic fen or ombrotrophic bog. The bogs can be categorized into a small number of basic hydromorphological types, although combinations and transitions between these types can add complexity to the classification process. The various types are illustrated under "*blanket bog.*"

Watershed (Water-Divide) Blanket Bog

Broad hill summits in climate regions which favor blanket mire development tend to become cloaked with a thick layer of peat which has no possible source of waterlogging other than by direct precipitation input. As these areas are often the highest features in the local landscape, they attract more cloud cover and are therefore also recipients of more occult precipitation (in the form of cloud fog) than other parts of the landscape. The characteristic feature of such bogs is that their margins fall away downslope on all sides to merge with the general hillslope. If these systems have a marked surface pattern, this usually involves a concentric arrangement of features.

Saddle Blanket Bog

A watershed (water-divide) ridge may possess more than one summit, and in climate regions which favor blanket mire formation, the saddle which lies on the ridge between two such peat-dominated watershed summits may form its own ombrotrophic bog which is waterlogged only by direct precipitation. Such saddle mires have two types of marginal zone. In the first type, the bog margin is fed by water from the summits, and this generally occurs on the two margins closest to the two watershed summits. The other type of margin falls away downhill on either side of the water divide forming the "flaps" of the saddle. If these systems have a marked surface pattern, this tends to be displayed as a concentric arrangement of features on the central crest of the saddle mire with a tendency toward eccentricity on the "flaps."

Spur Blanket Bog

Where slopes descending from watershed (water-divide) summits in places level out to form a plateau or even a basin in mid-slope within an overall blanket mire landscape, distinct ombrotrophic bog units can form on such ground. These "spur" mires are characterized by a margin which on one side is fed by water from upper slopes, while the remainder of the margin falls away downslope like the margins of a watershed bog. If these systems have a marked surface pattern, this tends to be concentric or mildly eccentric.

Valley Side Blanket Bog

Toward the valley floor, the gradient of the hillslope generally diminishes and presents further opportunities for the formation of deep peat within a blanket mire landscape. Where this deep peat forms an ombrotrophic bog rather than a valley mire or floodplain fen, the lower margin is often delimited by a river or significant stream course. The upper margin is fed by water from the upper slopes of the valley. Rather counterintuitively, because the system is a bog rather than a fen, the deepest peat (forming the crown of the bog) is found close to the upslope margin of the bog, with the peat thinning downslope toward the valley floor. Not surprisingly, if these systems have a clearly defined surface pattern, this tends to be markedly eccentric in its arrangement of features.

"Occult Precipitation" Mires

One hydrological condition which is not catered for within the conceptual matrix assembled by Joosten and Clarke (2002) is the possibility that mist, fog, and dew ("occult precipitation" or hidden precipitation) alone can create sufficiently waterlogged conditions for peat to form. Under certain circumstances these sources of moisture can indeed give rise to substantial peat formation.

Condensation Mires (Fig. 20)

One example of peat formation as a result of occult precipitation occurs where large boulder fields cover the steep sides of valleys, and cold air descending from glaciers further up the mountainside is funneled through gaps between the rocks to emerge lower down the valley. Here the cold air comes into contact with warmer air at lower altitude, and the resulting drop in temperature causes moisture in the warmer air to condense, creating an almost continuous dewfall which is sufficient to support vigorous *Sphagnum* growth and accumulation of a substantial peat thickness over the boulder field (Steiner 2005).

Fig. 20 Condensation mire near Klammhöhe/Hochschwab, Austria (Photo credit: Richard Lindsay © Rights remain with the author)

Cloud Forest/Cloud-Affected Forest Mires

Cloud forests or "cloud-affected forests" (Mulligan and Burke 2005) typically rely more on occult precipitation inputs and low evapotranspiration as a result of regular cloud cover than they do on rainfall per se and generally possess a distinctive organic soil layer (Bruijnzeel and Hamilton 2000). Within the lower and upper montane cloud forest zones, this organic layer may be no more than 10–20 cm thick, but the depth often increases markedly toward the subalpine cloud forest zone to the point where these low-growth cloud forests become, in effect, peatland ecosystems with a low tree and scrub cover. Such peatland systems are potentially important by virtue of their carbon stores but also because they may play an important part in regulating water supplies to lower parts of the drainage basin. These subalpine cloud forest mires, and the "micromires" which occur on the branches of trees within the other cloud forest zones, also provide a rich and particularly distinctive range of epiphytic biodiversity (and potentially overlooked carbon store).

References

Botch MS, Masing VV. Mire ecosystems in the U.S.S.R. In: Gore AJP, editor. Mires: swamp, bog, fen and moor (Ecosystems of the world 4B). Regional studies. Amsterdam: Elsevier Scientific; 1983. p. 95–152.

Bruijnzeel LA, Hamilton LS. Decision time for cloud forests. IHP Humid Tropics Programme series no. 13. Paris: UNESCO; 2000. Available to download from http://www.unep-wcmc.org/resources-and-data/decision-time-for-cloud-forests. Accessed 3 Apr 2015.

Charman DJ. Patterned fens in Scotland: evidence from vegetation and water chemistry. J Veg Sci. 1993;4:543–52.

Davis RB, Anderson DS. The eccentric bogs of Maine: a rare wetland type in the United States. Technical bulletin 146. Orono: Maine Agricultural Experiment Station; 1991. Available from http://library.umaine.edu/MaineAES/TechnicalBulletin/tb146.pdf. Accessed 11 Mar 2015.

Dommain R, Couwenberg J, Glaser PH, Joosten H, Suryadiptura INN. Carbon storage and release in Indonesian peatlands since the last deglaciation. Quat Sci Rev. 2014;97:1–32.

Du Rietz GE. Die Mineralbodenwasserzeigergrenze als Grundlage einer natürichen Zweigliederung der nord- und mitteleuropäischen Moore [The mineral soil water boundary indicator as the foundation for a natural two-part division of northern and central European peatlands]. Vegetatio. 1954;5-6:571–85.

European Commission. Interpretation manual of European Union habitats. Brussels: European Commission DG Environment, Nature and Biodiversity; 2007. Available from http://ec.europa.eu/environment/nature/legislation/habitatsdirective/docs/Int_Manual_EU28.pdf. Accessed 9 Mar 2015.

Goodwillie R. European peatlands, Nature and environment series, no. 19. Strasbourg: Council of Europe; 1980.

Hulme PD. The classification of Scottish peatlands. Scott Geogr Mag. 1980;96:46–50.

Joosten H, Clarke D. Wise use of mires and peatlands. Totness: NHBS/International Mire Conservation Group and International Peat Society; 2002. Available from: http://www.imcg.net/media/download_gallery/books/wump_wise_use_of_mires_and_peatlands_book.pdf. Accessed 9 Mar 2015.

Lindsay RA. Bogs: the ecology, classification and conservation of ombrotrophic mires. Battleby, Perth: Scottish Natural Heritage; 1995. Available from: http://roar.uel.ac.uk/3594/. Accessed 11 Mar 2015.

Lindsay RA, Charman DJ, Everingham F, O'Reilly RM, Palmer MA, Rowell TA, Stroud DA. The flow country: the peatlands of Caithness and Sutherland. Peterborough: Nature Conservancy Council; 1988.

Löfroth M. European mires – an IMCG project studying distribution and conservation. In: Grünig A, editor. Mires and man: mire conservation in a Densely Populated Country – the Swiss experience. Birmensdorf: Swiss Federal Institute for Forest, Snow and Landscape Research; 1994. p. 281–3. Available from www.wsl.ch/dienstleistungen/publikationen/pdf/420.pdf. Accessed 14 Apr 2015.

Moore PD, Bellamy DJ. Peatlands. London: Elek Science; 1974.

Mulligan M, Burke S. DFID FRP Project ZF0216 global cloud forests and environmental change in a hydrological context. Final Report. London: UK Department for International Development (DFID); 2005. Available to download from www.ambiotek.com/cloudforests/cloudforest_finalrep.pdf. Accessed 2 Mar 2015.

Nature Conservancy Council (NCC). Guidelines for selection of biological SSSIs. Peterborough: Nature Conservancy Council; 1989. Available from: http://jncc.defra.gov.uk/page-2303. Accessed 11 Mar 2015.

Ruuhijärvi R. The Finnish mire types and their regional distribution. In: Gore AJP, editor. Ecosystems of the world 4b. Mires: swamp, bog, fen and moor. Amsterdam: Elsevier; 1983. p. 47–67.

Shimizu Y. Species numbers, area, and habitat diversity on the habitat islands of Mizorogaike pond. Japan Ecol Res. 1986;1:185–94.

Sjörs H. Mires of Sweden. In: Gore AJP, editor. Mires: swamp, bog, fen and moor (Ecosystems of the world 4B). Regional studies. Amsterdam: Elsevier Scientific; 1983. p. 69–94.

Steiner GM. Österreichischer Moorschutzkatalog. [Austrian Mire Conservation Catalogue]. Grüne Reihe des, Bundesministeriums für Umwelt, Jungend und Familie, Band 1. Graz: Verlag Ulrich Moser; 1992.

Steiner GM. Die Moorverbreitung in Österreich/Distribution of mires in Austria. In: Steiner GM, editor. Moore – von Siberien bis Feuerland/Mires – from Siberia to Tierra del Fuego, Stapfia 85. Linz: Oberösterreichische Landesmuseen; 2005. p. 55–96.

Succow M, Jeschke L. Moore in der Landschaft – Enststehung, Haushalt, Lebewelt, Verbreitung, Nutzung und Erhaltung der Moore [Peatlands in the landscape – formation, ecology, biodiversity, distribution, use and conservation of peatlands]. Leipzig: Urania-Verlag; 1990.

Weber CA. Über die Vegetation und Entstehung des Hochmoors von Augstumal im Memeldelta [Vegetation and development of the raised bog of Augstumal in the Memel delta]. Berlin: Verlagsbuchhandlung Paul Parey. In: Couwenberg J, Joosten H, editors. C.A. Weber and the Raised Bog of Augstumal. Tula: International Mire Conservation Group/PPE "Grif & K"; 1902.

Weber CA. Die grundlegenden Begriffe der Moorkunde *[Basic peatland concepts]*, Zeitschrift für Moorkultur und Torfverwertung, vol. 5. Wien: Jahrgang; 1907.

Wheeler BD, Proctor MCF. Ecological gradients, subdivisions and terminology of north-west European mires. J Ecol. 2000;88:187–203.

Zoltai SC, Pollett FC. Wetlands in Canada: their classification, distribution and use. In: Gore AJP, editor. Ecosystems of the world 4B. Mires: swamp, bog, fen and moor. Regional Studies. Amsterdam: Elsevier Scientific; 1983. p. 245–68.

Arctic Peatlands

17

Tatiana Minayeva, Andrey Sirin, Peter Kershaw, and Olivia Bragg

Contents

Introduction	276
Arctic Peatland Distribution and Diversity	277
Threats and Future Challenges	286
References	287

Abstract

Arctic peatlands are mire ecosystems distributed across the vast northern edges of the Eurasian and North American continents and the islands and coastal areas of the Arctic and far northern Atlantic and Pacific Oceans. Arctic peatlands are mostly represented by "frozen" mires with much of their organic deposits remaining frozen throughout the year, whose water regime and other characteristics are strongly dependent on permafrost.

Primary production in arctic peatlands is low, but the tendency for peat to accumulate is enhanced by the low decomposition rates. The peat layer in arctic

---

T. Minayeva (✉)
Wetlands International, Ede, The Netherlands
e-mail: tatiana.minayeva@wetlands.org; tania.minajewa@gmail.com

A. Sirin (✉)
Center for Protection and Restoration of Peatland Ecosystems, Institute of Forest Science, Russian Academy of Sciences, Moscow, Russia
e-mail: sirin@ilan.ras.ru; sirin@proc.ru

P. Kershaw (✉)
University of Alberta, Edmonton, AB, Canada
e-mail: gkershaw@ualberta.ca

O. Bragg (✉)
School of the Geography, University of Dundee, Dundee, UK
e-mail: o.m.bragg@dundee.ac.uk

© Springer Science+Business Media B.V., part of Springer Nature 2018
C. M. Finlayson et al. (eds.), *The Wetland Book*,
https://doi.org/10.1007/978-94-007-4001-3_109

peatlands is not thick, and only in rare cases does the peat exceed 4–6 m in depth. Three main permafrost processes are responsible for the variability of arctic frozen peatlands: thermokarst, frost heaving, and cracking. Arctic peatland diversity and distribution across the landscape depends on the climate conditions, permafrost presence, and hydrology-connected landscape dynamics, and adheres to a characteristic pattern which is beneficial to consider in inventory, mapping, or planning of wetland management.

The ecosystem diversity of arctic peatlands is presented by paludified shallow peatlands; "frozen" peatland types such as polygon mires, peat plateaus, and palsa mires; and "nonfrozen" peatland types like patterned string fens, raised bogs, riparian mires, coastal tundra, and some types of coastal marsh.

Arctic ecosystems are characterized by low species diversity, and typical species are highly specialized and intimately linked to their habitats. The short growing season limits annual production and the ecological niche capacity of these species.

Arctic peatlands are highly integrated ecosystems which are extremely fragile to both natural and human-induced perturbations.

With climate change, the arctic region will continue to warm more rapidly than the global mean. Peat, as a thermo-isolating material, plays a crucial role in minimizing permafrost thaw. However, climate change-induced peatland degradation introduces positive feedback with permafrost thaw. Accompanying rapid in situ peat decomposition is an increase in greenhouse gas (GHG) emissions, specifically methane from saturated peat water and that which was formerly bound in frozen permafrost.

Development in such areas often ignores the special hydrological and ecological characteristics that are central to the productivity of these areas. Increased development of the oil and gas industry and its supporting transport infrastructure significantly fragments the landscape and disrupts its hydrology. Even traditional land uses such as reindeer herding are being industrialized. The resulting changes in peatland status will in turn restrict use of the land by the indigenous people who have traditionally depended on peatlands for food including herded reindeer, game, and fish.

Thus, there is an urgent need to promote sustainable practices.

Keywords

Climate change · Palsa mires · Peat plateau · Permafrost · Polygon mires · Positive feedback · Shallow peatlands · Thermokarst · Thermo-isolation · resistance

Introduction

Arctic peatlands are mire ecosystems (i.e., peatlands where peat is currently being formed and accumulating – Joosten and Clarke 2002) distributed across the vast northern edges of the Eurasian and North American continents and the islands and coastal areas of the Arctic and far northern Atlantic and Pacific Oceans. The Arctic is a vast and diverse region, usually delimited at its southern boundary by the Arctic

Circle, but differing definitions include not only tundra but also the northernmost part of the boreal zone and the forest-tundra (CAFF 2002, 2013).

Arctic peatlands are mostly represented by "frozen" mires with much of their organic deposits remaining frozen throughout the year and in which the water regime and other characteristics are strongly dependent on permafrost (permanently frozen subsoil). However, in areas in which permafrost is sporadic or absent –valleys, deltas, and coastal areas (Zhang et al, 2008) – nonfrozen peatlands more typical of the boreal zone may be common. Frozen peatlands are also found far to the south of the Arctic, especially in continental Asia, associated with isolated and sporadic permafrost. Peatlands similar to arctic peatlands are also reported for the Southern Hemisphere, but their distribution is limited by available land area.

Compared with more temperate locations, primary production (by photosynthesis) in arctic peatlands is low (although the shortness of the summer is partially compensated by long daylight hours), but the tendency for peat to accumulate is enhanced by the low decomposition rates. The effectiveness of decomposer organisms is severely impeded under arctic conditions, being active only during the part of the year when the peat is not frozen. This is just a portion of the already short summer, because seasonal thawing is delayed by the heat input required to change ice into water. Once they become locked into the permafrost, plant remains are often still recognizable to species level for millennia.

The peat layer in arctic peatlands is not thick, and only in rare cases does the peat exceed 4–6 m in depth (Vasilchuk et al. 1983; Peteet et al. 1998; Andreev et al. 2001; Pitkanen et al. 2002). Peatlands by definition require a minimum peat depth and it is difficult to distinguish arctic peatlands from the much more common shallow peatlands. Although together (hereafter referred to as peatlands) they represent the most widespread wetland types in the Arctic, it is only relatively recently that they have they been the focus of heightened scrutiny. Attention is now being directed toward arctic peatlands and their specific origin and practical requirements due to the pronounced effects of climate change in high latitudes, related thawing of permafrost, shifts in traditional land use, and industrial development.

Arctic Peatland Distribution and Diversity

Wetlands, as defined by the Ramsar Convention on Wetlands, are the dominating land cover type in the Arctic, covering more than 60% of the area (Minayeva and Sirin 2009). There are two main factors responsible for the vast distribution of wetlands in the Arctic: presence of large north flowing rivers within ancient wide valleys still experiencing natural flood regimes and the presence of permafrost in all watersheds and terraces providing a permanent source of water due to the freeze-thaw processes.

Arctic peatland diversity and distribution depends on the climate conditions and permafrost and hydrology-connected landscape dynamic. The distribution and diversity of wetlands across the arctic landscape adheres to a characteristic pattern (Fig. 1) that it is beneficial to consider in inventory, mapping, or planning of wetland

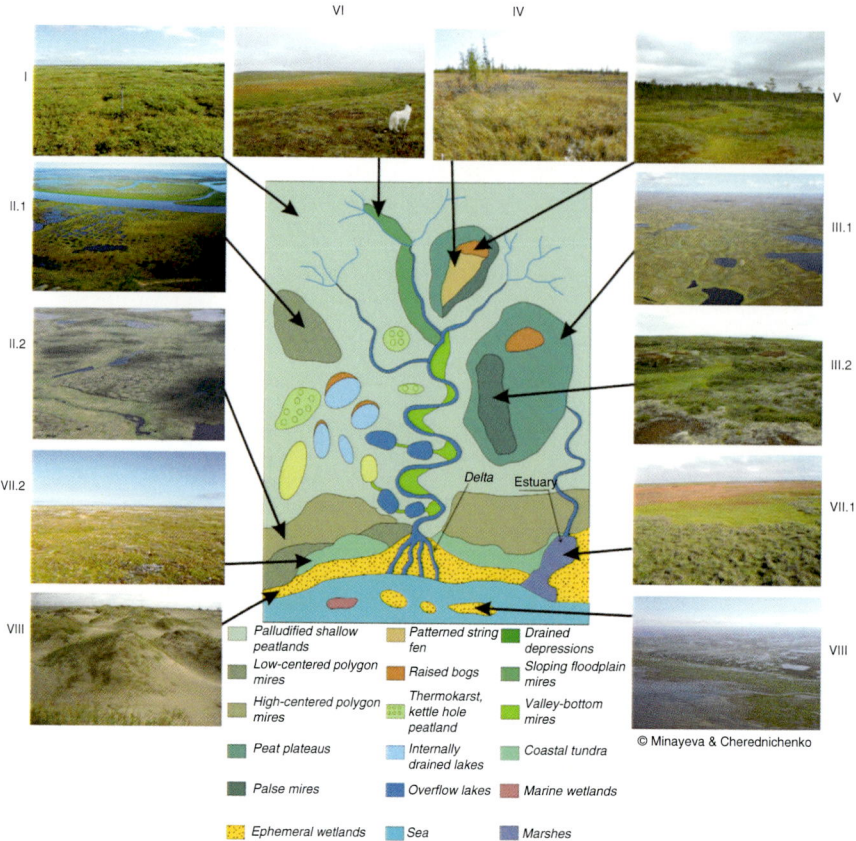

Fig. 1 Arctic wetland diversity and their relative position in the landscape: *I* Paludified shallow peatlands; *II* Polygon mires (with and without thermokarst ponds); *II.1* Low-centered polygon mires; *II.2* High-centered polygon mires; *III* Peat plateaus and Palsa mires (with and without thermokarst ponds); *III.1* Peat plateau mires; *III.2* Palsa mires; *IV* Patterned string fen; *V* Raised bogs; *VI* Riparian mires; *VII* Coastal wetlands; *VII.1* Saline marshes; *VII.2* Coastal tundra; *VIII* Ephemeral wetlands (Image credit: T. Minayeva and O. Cherednichenko © Rights remain with the authors)

management. The terrestrial wetlands of the Arctic are mostly peatlands (Tarnocai and Zoltai 1988) in which the long-term rate of organic matter production exceeds the rate at which it decomposes and include some freshwater (e.g., riparian mires and drained depressions) as well as brackish marshes. This coarse description of arctic wetland diversity is coincident with both the Ramsar classification of wetland types (Ramsar Convention Secretariat 2010) and the CAFF (2013) approach to ecosystem consideration (terrestrial, freshwater, marine).

Three main permafrost processes are responsible for the variability of arctic frozen peatlands: thermokarst, frost heaving, and cracking. Thermokarst includes all changes in topography originating from the processes of permafrost thawing, particularly negative forms of relief such as *inter alia* depressions, sinks, funnels, and gullies. Frost heaving results in different forms of positive relief such as palsas, mounds, and hummocks. Cracking is the key process forming the polygonal morphology of land surface widely present in the Arctic. All three processes can combine at different spatial levels to form specific types of arctic peatlands, e.g., mounds of palsa mires are formed by frost heaving in which peat heads are often cut by cracks and fen depressions are created by thermokarst.

Paludified Shallow Peatland (Tundra)

This is the most widespread terrestrial wetland type on the interfluves of the Eurasian Arctic, forming vast areas of shallow peatland overlying sandy or loamy soils. It occurs more sporadically in North America, where it is often replaced by peat plateaus (see Fig. 1). The peat layer can be partly degraded or decomposed; and the profile can contain alternating layers of peat and mineral material, which indicate that the peatland has periodically been re-covered by mineral soil. In the Russian Arctic, shallow peatlands make up more than half of the peat covered area (Vompersky et al. 2005).

The lighter sandy soils are associated with alluvial (riverine, lacustrine or oceanic) processes and are colonized by dwarf shrub-lichen vegetation. Patches with more snow host mainly fruticose lichens belonging to the genus *Cladonia*. Areas with less snow have species of *Cetraria* and *Alectoria*.

The heavier loamy soils are associated with diluvium (sediment deposited by flood water). They help to retain water in shallow peat layers, leading to increased dominance of both green mosses (in the southern tundra zone) and *Sphagnum* (in northern areas and uplands), cottongrass, sedges, and dwarf shrubs.

The shallow peat tundra on coastal or highland plateaus can have features resembling polygon structures, and here the cracks between polygons can host green mosses, dwarf birch, and willow. Degraded areas that retain some peat or an organic soil layer are covered by crustose lichens (*Gymnomitrion concinnatum* as well as members of the genera *Ochrolechia* and *Pertusaria*). The main colonizers of degraded tundra with sandy soil are psammophyte species, for example, in Nenetsky Okrug, the lichen *Sphaerophorus globosus*, the mosses *Racomitrium lanuginosum* and *Polytrichum piliferum*, and the vascular plants *Armeria maritima*, *Juncus trifidus*, *Festuca ovina*, *Diapensia lapponica*, and *Loiseleuria procumbens*.

Polygon Mires (With or Without Thermokarst Ponds)

Polygon landscapes arise as a consequence of thermal contraction and expansion processes in the frozen active layer and near-surface permafrost (Mackay 2000). Tension cracks propagate from the surface through the frozen active layer into the underlying permafrost in winter. Before they close in summer, snowmelt water migrates into the open cracks and freezes. As the cycle repeats annually, the original vertical ice veins are augmented and ice wedges develop. Their spacing, thickness, and depth will vary with the nature of the host materials. Initially, the growing ice wedges force host sediments upward to form polygon shoulders and create low-center polygons. When they occur in peat – a soft, unstable, and very wet substance – the effects are accentuated, forming deep ice-filled cracks which soon become secure water sources for maintenance of the mire, and in some cases for peat formation. There are two main types of polygon mire, namely, low-centered (concave) and high-centered (mounded) (Tarnocai and Zoltai 1988). Successful functioning of the mire system depends upon a complex construction of different types of peat, permafrost, and mineral soil (Fig. 2). This mire type is an integrated dynamic system driven by structural (vegetation, soil) and permafrost processes, so that even partial physical disturbance or destruction can unpredictably shift the equilibrium of the whole system. Peat thickness depends on the landscape position. The peat layer is usually 1–2 m and sometimes 3–5 m thick on watersheds, on old terraces, and in deep erosion gulleys. In valleys and on small terraces it may be only 0.2–0.5 m thick.

Polygon mires are not rich in biodiversity and host no more than 30 species of vascular plants, 30 species of mosses, and 45 species of lichens, but they represent a unique ecosystem type. Low-centered polygons have sedge-cotton grass-moss vegetation typical of peatland hollows, their shoulders offer better-drained conditions and often support shrubs like dwarf birch (*Betula nana*) and Labrador tea (*Ledum palustre* ssp. *decumbens*), and the ice in the trenches is covered by *Sphagnum* peat and *Sphagnum* carpets. The tops of high-centered polygons are usually covered by cloudberry, dwarf shrubs, and some mosses which remain sparse because they do not compete successfully with the dwarf shrubs, especially *Empetrum nigrum*. In the High Arctic and on coastal terraces, the caps of high-centered polygons are covered by lichens with sparse dwarf shrubs.

Polygon mires occur in the High Arctic. In North America they are found on the arctic coastal plains of Alaska and in the Canadian Mackenzie Delta. In Eurasia, there are relic mounded polygons along the shore of the White Sea, both types occur regularly on the shore terraces along the Barents Sea coastline, and farther east they are found on the terraces of large inland rivers. They are abundant in eastern Yamal and on the Gydansky Peninsula; and they cover immense parts of the Yenisey lowlands in Taymyr, the Yana-Indigirka and Kolyma lowlands, and the Lena delta. Relic polygon mires have also been described in Northern Sakhalin and even the Amur River delta.

Peat Plateau and Palsa Mires

The palsa mires make up the second group of peatlands whose genesis and maintenance is related to permafrost (Pyavchenko 1955; Novikov et al. 2009). The mounds are not created by peat formation but by heaving of the underlying mineral ground; the hummocks consist of fen peat (formed from mesotrophic plant communities), but their present vegetation is oligotrophic; the underlying mound of mineral material contains ice lenses; and the upper limit of permafrost is close to the ground surface in palsa mounds and deeper or absent in the intervening depressions (Fig. 3). Numerous authors have suggested mechanisms for palsa formation on the basis of their own analyses of this general structure (Tyuremnov 1976; Seppälä 1986; Tarnocai and Zoltai 1988; Kujala et al. 2008; Vasilchuk et al. 2008).

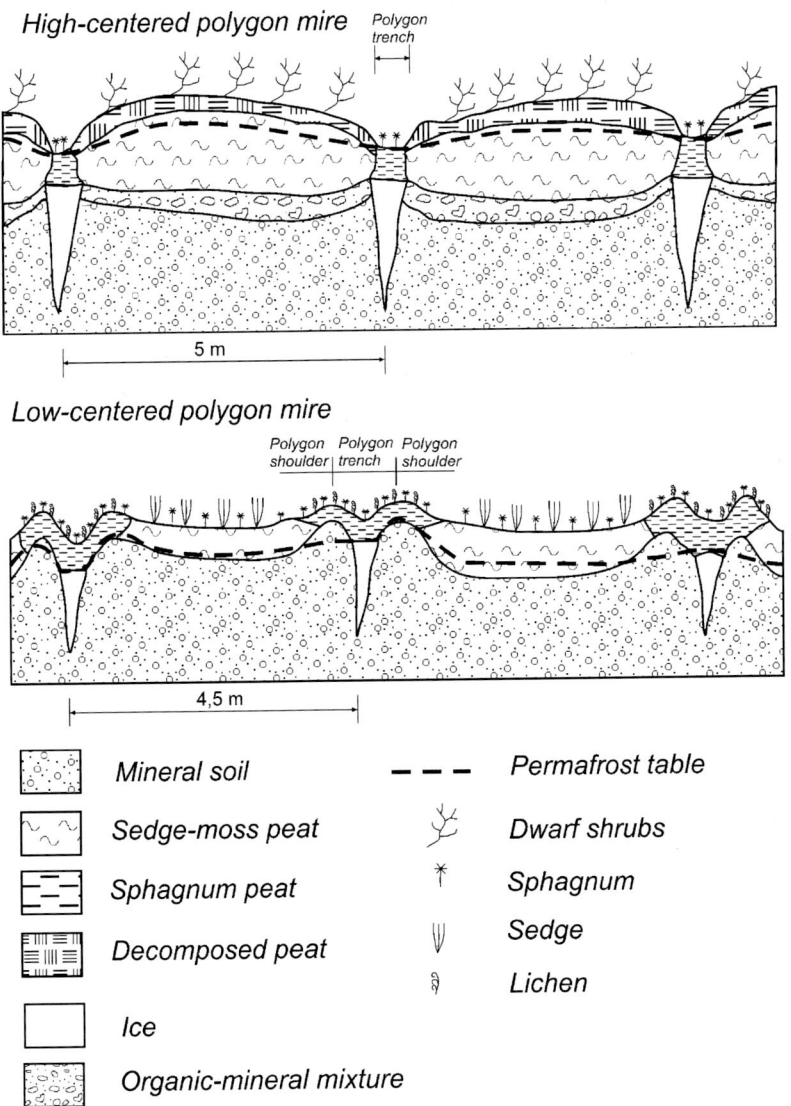

Fig. 2 Cross-sections of high-centered polygon and low-centered polygon mire (Adapted by T. Minayeva and O. Cherednichenko after Tarnocai and Zoltai 1988; © 1988, Minister of Supply and Services Canada, with permission of Environment and Climate Change Canada)

Palsa mires differ from polygon mires in that the permafrost-related processes in palsas operate below the surface rather than from the surface, and peat plays a more active role in creating the palsa structure. In polygon mires, the role of peat becomes pertinent when the basic structure is already in place – peat begins to form on the edges of preexisting ice wedge troughs and gradually modifies the hydrological and temperature regimes thereafter. In palsa mires, the peat is originally formed in

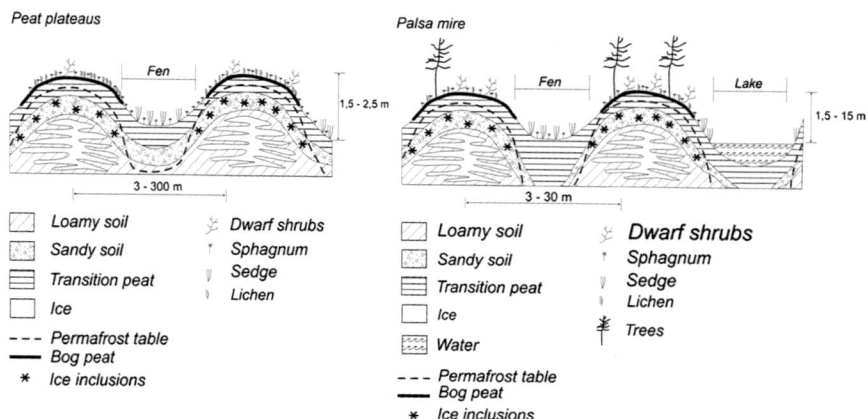

Fig. 3 Structural characteristics of peat plateau and palsa mires. Although both have the same basic structure, the indicated scale of the peat plateau/palsa is more typical for Russia. In North America, plateaus can cover tens of km^2 and the fens are only a minor part of the landscape. In this case the horizontal scale would be >300 m while the vertical would be 2.5 m (Adapted by T. Minayeva and O. Cherednichenko after Konstantinova 1963 cited in Tyuremnov 1976)

flooded sedge-moss plains (i.e., on flat areas). When a mire is present, the drier layer of peat at the surface protects ice from thawing and enhances freezing processes in the wetter mineral soil beneath. During the warm part of the year, the mineral soil becomes saturated with water, which subsequently freezes. The peat provides insulation which reduces heat gain in summer. In locations where the surface relief was originally uneven (e.g., in a hummock), ice lenses accumulate over many years, stacked on top of one another in the mineral core to progressively create the palsa mound. Thus, the mound-hollow relief of palsa mires is created by a frost-heaving process. Fine-textured mineral sediments are more susceptible to frost heaving than coarse ones, and the structure of the mineral soil beneath the peat is the main factor in determining how pronounced this relief becomes. True palsa mires with large convex mounds, separated by vast hollows and occasional ponds, are characteristic for regions with fine-textured mineral soils. "Peat plateaus" form where access to water is limited, either due to low permeability of the mineral sediments or because continuous permafrost permits water to migrate into the ice lenses only around the perimeters of the features. In either case the plateaus cannot attain heights of more than ~2.5 m because water can be drawn into ice lenses to grow superposed palsa-like features. When "palsas" on peat plateaus are sufficiently extended (up to 300 m in diameter), polygon structures can appear in their central parts, forming polygonal peat plateaus.

Whatever their sizes and shapes, the mounds of northern peat plateau and palsa mires are usually dominated by dwarf shrubs and lichens, whereas in the southern tundra their vegetation consists of *Sphagnum* and green mosses (mainly *Dicranum elongatum*, *Polytrichum juniperinum*, and *Sphagnum fuscum*). Treed palsas have also been described. Where the peat has become degraded it might be covered by scale lichens (*Icmadophila ericetorum*, *Ochrolechia frigida*, *Omphalina hudsoniana*) and form an effective springboard for invasions of new species from the south. The hollows have sedge-*Sphagnum* vegetation and often contain thermokarst ponds.

The degradation of palsas is described in numerous publications and is highly significant in the present context. As a palsa mound grows, the vegetation cover of its top gradually degenerates; this is a natural physical process. Following breakdown of the vegetation cover, the peat layer starts to degrade so that its thermal insulation function is gradually lost and the permafrost begins to thaw. In some cases, all of the supporting ice disappears leaving a roundish patch of bare peat in the tundra or a pond. The full cycle of palsa formation and degradation can be completed within a few decades.

Patterned String Fen (Aapa Mire) and Raised Bogs

These two wetland types occur in the arctic, subarctic, and boreal zones and are not directly connected with permafrost, although permafrost can play a role in certain stages of their formation. At first glance the patterns of hummock-hollow complexes in raised bogs look similar to those in patterned string fens. However, there are important differences between them in terms of structure and thickness of the peat layer, which lead to differences in hydrology and vegetation patterns. In patterned string fen, the dominant peat type is fen peat, with bog peat being found only on the hummocks (Botch and Masing 1979) (Fig. 4).

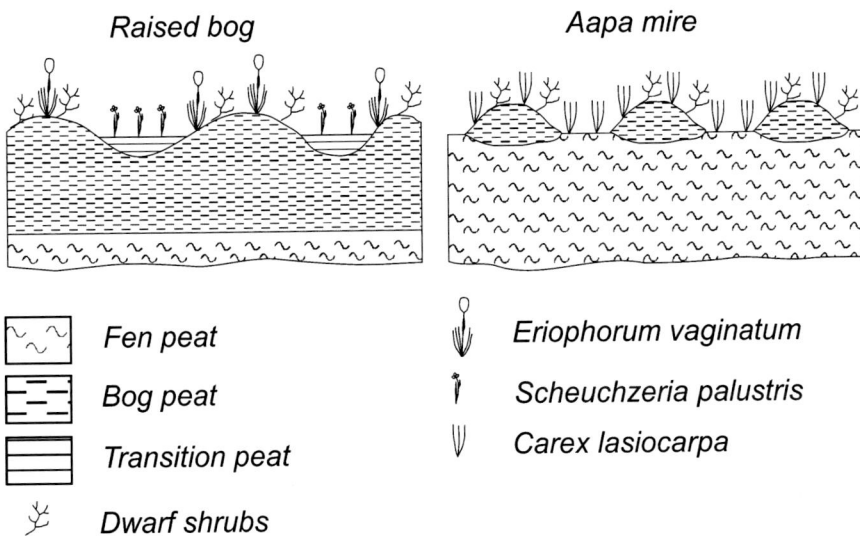

Fig. 4 Comparison of the structures of raised bog and aapa mire microtopes (Image credit: T. Minayeva and O. Cherednichenko © Rights remain with the authors; after Kats 1971 based on the description provided by Botch and Masing (1979))

Patterned string fen, or aapa mire, is widespread in the Eurasian Arctic and Subarctic, and it occurs in North America where it is known as ribbed or stringed fen. The main distinguishing feature of an aapa mire massif is its concave or inclined form, which means that it receives both direct rainfall and water with a higher mineral content as inflow from the surrounding land (Laitinen et al. 2007). The peat layer of aapa mire is shallow (0.8–1.5 m), so that deep-rooted plants can also access mineral-enriched water in the soil beneath. The typical landscape position of aapa mire is on gently sloping ground with permanent surface flow, which is a key factor in the formation of hummock-hollow complexes. The hummocks and hollows are usually well defined, and both host mesotrophic species. The patterns they form at landscape level depend on morphology. Aapa mires are nonfrozen peatlands, but seasonal ice in hummocks can persist throughout the summer under certain conditions. The strings or hummocks are largely composed of *Sphagnum* peat. The difference in peat characteristics between hummocks and hollows is more pronounced in northern ribbed fens than in southerly examples (Zoltai et al. 1988).

Raised bogs occur mostly in the boreal belt, where they can be recognized as separate mire landscapes. In the arctic and subarctic zones, raised bogs occur only as elements within peatlands and wetlands of other types which occupy lake depressions, relic water channels, etc., mostly in areas with thawed permafrost. Raised bogs have typical vegetation patterns made up of hummocks with ericaceous and non-ericaceous dwarf shrubs, sedges, cottongrass and *Sphagnum* mosses, and mesotrophic hollows with *Sphagnum* and small sedge carpets.

Thermokarst Kettle Hole Peatlands

Thermokarst kettle hole peatlands or alases are typical arctic wetland ecosystems. A kettle hole is a distinctive steep-sided depression formed by the thawing of permafrost, which may contain a thermokarst lake or an ecosystem in lake succession such as a floating mat or true peatland. The vegetation depends on the stage of development and can include sedges, hypnaceous mosses, or *Juncus*.

Drained Depressions (Syn. Khasyry)

Thermokarst processes very often lead to the drainage and drying-out of lakes and ponds. The resulting "former lakes" have the local name "khasyry" (which means "dry lake") in Nenents. These ecosystems are unique for the Eurasian Arctic insofar as peatland development and peat formation is relatively recent. Permafrost is not a factor in their genesis and function and for this reason they lack patterning. Their vegetation is species diverse, consisting of tall sedges (*Carex aquatilis*, *C. rariflora*, *C. rotundata*) and *Sphagnum* communities (*Sphagnum lindbergii*, *S. girgensohnii*, *S. squarrosum*, *S. fimbriatum*, *S. angustifolium*, *S. warnstorfii*) with forbs which can include some extremely rare species; for example, the Red Data Book species *Ranunculus pallasii* in the case of Nenetsky. In wetter places, *Sphagnum* mosses are replaced by green mosses like *Warnstorfia exannulata*, *Limprichtia revolvens*, *Calliergon stramineum*, *Sanionia uncinata*, *Mnium* spp., *Meesia triquetra*, and *Paludella squarrosa* and, more typically at mesotrophic sites, vascular plants like *Calamagrostis neglecta*, *Comarum palustre*, and *Epilobium palustre*. Occasional hummocks are present, and these host oligotrophic species (ericoid shrubs and *Sphagnum* mosses). In valleys of the southern tundra and the forest-tundra, khasyry are often colonized by pine or larch trees which grow very rapidly.

Riparian Mires
Deltas play a significant role in the ecology of the Arctic, as buffers for the impact of changing river flow. Deltas can contain all wetland types; from ephemeral dunes and sandy spits near beaches, to valley-bottom mires with sedges and *Hypnum* moss in oxbows and along low riverbanks with or without gravel embankments. Sloping floodplain fens which are dependent on the fluvial regime are also regarded as riparian mires.

Sloping floodplain fens, and especially valley-bottom fens, have homogeneous vegetation structure and highly organic soils including peat deposits. Their vegetation consists of willows, tall sedges, and mosses. These are the most productive of all arctic ecosystems and so have very high restoration potential.

Coastal Wetlands
Some coastal marshes, especially brackish and freshwater marshes, may contain a peat layer in their latest stages of development. Sedimentation of silt and accumulation of organic matter (vegetation remains) lead to a gradual rise of the marsh surface above sea level and, consequently, to changes in the periodicity and duration of seawater influence. A typical chronosequence in Eurasia would be, as the surface rises, pioneering associations of *Puccinellia phryganodes* and *Carex subspathacea* on unstable sand and silt giving way to medium-level marsh communities of *Carex subspathacea* with dicotyledonous grasses (*Potentilla egedii*, *Plantago schrenkii*, *Arctanthemum hultenii*) and associations dominated by *Calamagrostis deschampsioides* and *Carex glareosa*. In the final stage of succession, more elevated features – where seawater effects are limited to wave splash and occasional inundation by surges – are colonized by stable upper marsh communities dominated by *Festuca richardsonii* and *Salix reptans*, with *Rhodiola rosea*, *Parnassia palustris*, and other halophyte and salt-tolerant tundra species including mosses. When the moss cover develops, the marsh surface begins to rise again due to peat accumulation. Similar sequences occur in North America, often when different species of the same genera occupy the same niches as the Eurasian species.

Coastal Eurasian tundra (high-level marsh) resembles regular tundra, but is influenced by salt water. It is initially dominated by halophytes but eventually common tundra species like *Carex subspathacea*, *Calamagrostis deschampsioides*, *Carex glareosa*, *Festuca richardsonii*, *Salix reptans*, *Empetrum hermaphroditum*, *Rhodiola rosea*, *Parnassia palustris*, and *Comarum palustre*, and mosses belonging to the genera *Bryum* and *Drepanocladus* predominate. These Eurasian species have North American equivalents that fill the same niches. In some places, marshes are seriously affected by waterfowl populations.

Threats and Future Challenges

Arctic peatlands are highly integrated ecosystems which are extremely fragile to both natural and human-induced perturbations. Although their status has not yet been described comprehensively in the scientific literature, certain trends are clearly evident. These are dominated by direct and indirect effects of climate change arising from global warming, which has multiple and sometimes subtle implications for arctic peatlands (Minayeva and Sirin 2010).

In recent years the southern limit of permafrost moved significantly northward (AMAP 2011). Long-term temperature monitoring in the European part of the Russian Arctic (Nenets) shows a permanent temperature increase in permafrost upper layers. Loss of permafrost in Quebec has been attributed to the insulating effect of increased snowfall since the late 1950s rather than to temperature, which did not rise until the late 1990s, and has been accompanied by new peat accumulation on thawed areas (paludification) and in thermokarst ponds (terrestrialization) (Payette et al. 2004). Thus, small changes in weather conditions can cause abrupt changes in the direction of peatland system development. The distinctive polygonal patterns and palsa mounds of permafrost peatlands can exist only where the ground is permanently frozen. The thicker snow cover of the progressively milder arctic winters (with increased precipitation) already threatens the persistence of these remarkable peatland systems (Zuidhoff 2002; Hofgaard 2003; Kershaw 2003; Fronzek et al. 2006). Moreover, it is anticipated that trees and other boreal species will colonize arctic peatlands as the northern treeline migrates to higher latitudes in response to rising summer temperatures (CAFF 2013). This will not only affect biodiversity but also reduce albedo (surface reflectivity), further enhancing warming of the atmosphere.

In locations such as the High Arctic, where low temperatures currently limit primary production and thus peat growth, nonfrozen peatlands are likely to expand in topographically suitable locations with rising temperature. Peatlands in floodplains and lake basins are particularly susceptible to the increasingly dynamic river flow regimes that are expected as the intensity of rainfall and droughts continues to increase. The biota of surface water bodies are in turn vulnerable to changes in the load of dissolved and/or particulate organic matter (DOC, POC) in drainage water from any peatlands within their catchments that are degrading due to any cause.

The arctic region will continue to warm more rapidly than the global mean (IPCC 2014) with climate change, and this alone is expected to transform arctic peatlands through loss of permafrost. Peat as a thermo-isolating material plays a crucial role in minimizing permafrost thaw. However, climate change-induced peatland degradation introduces positive feedback with permafrost thaw. Accompanying rapid *in situ* peat decomposition is an increase in greenhouse gases (GHG) emissions, specifically methane from saturated peat water and that which was formerly bound in frozen permafrost.

The vast undisturbed peatlands of the arctic and subarctic zones are among the last remaining wilderness and natural resource areas of the world. Development in such areas often ignores the special hydrological and ecological characteristics that are central to the productivity of these areas. Although the Ramsar Convention, Convention on Biological Diversity, and UNFCCC have acknowledged that special action to conserve peatlands is urgently required, peatlands are still underrepresented in conservation strategies and seldom recognized as specific targets for management. For example, there is an urgent need for arctic peatland restoration technologies which, in order to be effective, must be designed specifically for permafrost systems.

Traditional and sustainable uses of arctic peatlands, such as grazing, hunting, and berry-picking have been carried forward into recent times. New advances in technologies have provided the means to overcome challenges presented by the harsh arctic environment, leading to increased development of the oil and gas industry and its supporting transport infrastructure which significantly fragments the landscape and disrupts its hydrology. Even traditional land uses such as reindeer herding are being industrialized, and the increased human presence means that wild mammals and birds are increasingly threatened by recreational hunting. There is thus an urgent need to promote sustainable practices.

Arctic ecosystems are characterized by low species diversity, and typical species are highly specialized and intimately linked to their habitats. The short growing season limits annual production and the ecological niche capacity of these species, so that communities have low resistance to disturbance and extremely limited potential for natural recovery. Thus impacts of increased industrial development will in turn reduce arctic peatland ecosystem diversity and thus their biodiversity value. The resulting changes in peatland status will in turn restrict use of the land by the indigenous people who have traditionally depended on peatlands for food including herded reindeer, game, and fish.

References

AMAP. Snow, water, ice and permafrost in the Arctic (SWIPA): climate change and the cryosphere. Oslo: Arctic Monitoring and Assessment Programme (AMAP); 2011.

Andreev AA, Klimanov VA, Sulerzhitsky LD. Vegetation and climate history of the Yana River lowland, Russia, during the last 6400 yr. Quat Sci Rev. 2001;20:259–66.

Botch MS, Masing VV. Mire ecosystems of the USSR. Leningrad: Nauka; 1979.

CAFF. Arctic flora and fauna: status and conservation. Helsinki: Edita; 2002.

CAFF. Arctic Biodiversity Assessment. Statistics and trends in Arctic biodiversity: synthesis. Conservation of Arctic Flora and Fauna, Arctic Council. Iceland: Akureyri; 2013. 128 pp.

Fronzek S, Luoto M, Carter TR. Potential effect of climate change on the distribution of palsa mires in subarctic Fennoscandia. Climate Res. 2006;32:1–12.

Hofgaard A. NINA Project Report 21: effects of climate change on the distribution and development of palsa peatlands: background and suggestions for a national monitoring project. Trondheim: Norwegian Institute for Nature Research; 2003.

IPCC. Climate change 2014 synthesis report. 2014. https://www.ipcc.ch/pdf/assessment-report/ar5/syr/SYR_AR5_LONGERREPORT.pdf. Last visited 1 Mar 2015.

Joosten H, Clarke D. Wise use of mires and peatlands – background and principles including a framework for decision-making. Devon: International Mire Conservation Group and International Peat Society, NHBS Ltd.; 2002.

Kats NY. Bolota Zyemnogo Shara [The Mires of the Globe] Moscow: Nauka; 1971. 295 p. [In Russian]

Kershaw GP. Permafrost landform degradation over more than half a century, Macmillan Pass/ Caribou Pass region, NWT/Yukon, Canada. In: Phillips M, Springman SM, Arenson LU, editors. Proceedings of the eighth international conference on permafrost. Zurich: Swets & Zeitlinger Publishers; 2003. p. 543–8.

Konstantinova GS. On the cryogenic formations in the region of Bolshoy Khantay rift. In: Efimov AI, editor. The permafrost deposits in the different regions of USSR. Moscow: Academy of Sciences USSR; 1963. p. 112–20 (In Russian).

Kujala K, Seppälä M, Holappa T. Physical properties of peat and palsa formation. Cold Reg Sci Technol. 2008;52:408–14.

Laitinen J, Sakari R, Antti H, Teemu T, Raimo H, Tapio L. Mire systems in Finland – special view to aapa mires and their water-flow pattern. Suo. 2007;58(1):1–26.

Mackay JR. Thermally induced movements in ice wedge polygons, western Arctic coast: a long term study. Géogr Phys Quat. 2000;54:41–68.

Minayeva T, Sirin A. Wetlands – threatened Arctic ecosystems: vulnerability to climate change and adaptation options. In: UNESCO, editor. Climate change and Arctic sustainable development: scientific, social, cultural and educational challenges. Paris: UNESCO; 2009. p. 80–7.

Minayeva T, Sirin A. Arctic peatlands. In: Arctic biodiversity trends 2010 – selected indicators of change. Akureyri: CAFF International Secretariat; 2010. p. 71–4. http://www.arcticbiodiversity.is/images/stories/report/pdf/Arctic_Biodiversity_Trends_Report_2010.pdf

Novikov SM, Usova LI, Moskvin YuP, editors. Hydrology of wetlands of the permafrost zone in West Siberia. St. Petersburg: VVM; 2009 (In Russian).

Payette S, Delwaide A, Caccianiga M, Beauchemin M. Accelerated thawing of subarctic peatland permafrost over the last 50 years. Geophys Res Lett. 2004;31:L18208. doi:10.1029/2004GL020358.

Peteet D, Andreev A, Bardeen W, Mistretta F. Long-term Arctic peatland dynamics, vegetation and climate history of the Pur-Taz region, Western Siberia. Boreas. 1998;27:575–86.

Pitkanen A, Turunen J, Tahvanainen T, Tolonen K. Holocene vegetation history from the Salym-Yugan Mire Area, West Siberia. Holocene. 2002;12:353–62.

Pyavchenko NI. Bugristyje torfyaniki (Palsa mires). Moscow: Academy of Sciences of the USSR Publishing House; 1955 (In Russian).

Ramsar Convention Secretariat. Designating Ramsar sites: strategic framework and guidelines for the future development of the list of Wetlands of International Importance, Ramsar handbooks for the wise use of wetlands, vol. 17. 4th ed. Gland: Ramsar Convention Secretariat; 2010.

Seppälä M. The origin of palsas. Geogr Ann. 1986;A68:141–7.

Tarnocai C, Zoltai SC. Wetlands of Arctic Canada. In: National Wetlands Working Group, editor. Wetlands of Canada, Ecological classification series, no. 24. Ottawa: Enironment Canada and Polyscience Publications; 1988. p. 29–53.

Tyuremnov SN. Peat deposits. Moscow: Nedra; 1976 (In Russian).

Vasilchuk YuK, Petrova EA, Serova AK. Some features of the Yamal paleogeography in Holocene. Bull Comm Quat Stud. 1983;52:73–89 (In Russian).

Vasilchuk YuK, Vasilchuk AK, Budantseva NA, Chizhova JN. Palsa of frozen peat mires. Moscow: Moscow University Press; 2008. p. 571.

Vompersky SE, Sirin A, Tsyganova OP, Valyaeva NA, Maikov DA. Peatland and paludified lands of Russia: attempt at analysis of spatial distribution and diversity. Izv RAS Geogr. 2005;5:39–50.

Zhang T, Barry RG, Knowles K, Heginbottom JA, Brown J. Statistics and characteristics of permafrost and ground-ice distribution in the Northern Hemisphere. Polar Geogr. 2008;31:47–68.

Zoltai SC, Tarnocai C, Mills GF, Veldhuis H. Wetlands of Subarctic Canada. In: National Wetlands Working Group, editor. Wetlands of Canada, Ecological classification series, no. 24. Ottawa: Enironment Canada and Polyscience Publications; 1988. p. 57–96.

Zuidhoff FS. Palsa decay in relation to weather conditions. Geogr Ann. 2002;84:103–11.

Mires

18

Richard Lindsay

Contents

Introduction to Mires	289
Defining a Mire	290
Conclusion	291
References	293

Abstract

The term 'mire' is now widely accepted as the appropriate term for peatlands which still display the features associated with active peat formation. It is, however, difficult to determine whether any given site is actually accumulating peat at any specific moment in time. The presence of vegetation which is normally peat forming has thus been proposed and been widely adopted as a pragmatic means of identifying the presence of mires – i.e., peatland systems still capable of accumulating peat.

Keywords

Mire · Moor · Bog · Fen · Moss · Carr · Swamp · Marsh · Myr · Peatland · Godwin · Tansley · Peat formation · Active raised bog · Active blanket bog · European Commission · Wetland · Sedges · Sphagnum · Cotton grass · Swamp-forest · Calcium · Suo · Spring mire · International Mire Conservation Group · IMCG · Mangroves · Carbon

---

R. Lindsay (✉)
Sustainability Research Institute, University of East London, London, UK
e-mail: r.lindsay@uel.ac.uk

Introduction to Mires

In his comprehensive review of British and Irish vegetation, Tansley (1939) discussed the meaning and use of the German word "*Moor*" and highlighted the fact that it does not translate directly into the English word "moor" because this word is applied in English usage to all unenclosed land, whether peat, rock, or mineral soils. He therefore elected to use the terms "bog," "moss," "fen," "carr," "swamp," and "marsh" to describe various peat-forming communities. It was Godwin (1941) who first proposed that the term "mire," a word closely related to the Swedish and Norwegian word "myr," should be used as a general term for all natural and semi-natural peatland ecosystems (Wheeler and Proctor 2000). With the passage of time this has led to its use as an all-embracing term for ecosystems associated with peat formation (e.g., Gore 1983). In recent years, however, interest in the carbon-sequestering role of peatlands and the desire to distinguish relatively natural "active" bogs from damaged moribund bogs has resulted in a particular focus on whether a peatland is *currently* accumulating peat. Joosten and Clarke (2002) therefore define a mire as "*a peatland where peat is currently being formed,*" but in a footnote they acknowledge the difficulty of determining whether peat *is* at the present moment being formed.

Defining a Mire

The question of defining the concept of a mire was explored extensively at an International Mire Conservation Group (IMCG) field symposium in Switzerland in 1990, with the result being the following definition: "*A mire is a wetland supporting at least some vegetation which is normally peat-forming*" (Löfroth 1994). This definition has two advantages. Firstly, it does not require proof of current peat formation but instead draws on the long-recognized fact that certain types of vegetation are particularly associated with peat formation. Thus *Sphagnum*-dominated communities (Fig. 1), stands of cotton grass *Eriophorum* spp., sedges and, in the tropics, even roots and branches of swamp-forest trees such as *Calophyllum* spp. and *Cratoxylum* spp. (Page et al. 1999) are well documented as forming peat (Troels-Smith 1955). Secondly, the definition acknowledges that there are certain circumstances where typical peat-forming vegetation does not form peat, mainly because of high calcium levels in the water supply, as in, for example, certain *Schoenus ferrugineous* spring mires (Fig. 2). In all respects, these non-peat systems are what one would normally recognize as a mire, they merely lack peat. Joosten and Clarke (2002) use the Finnish term "*suo*" to combine these peat-less and peat-forming mires into an all-embracing category, but this still does not resolve the challenge of proving that a mire is actually forming peat at the present time.

The use of vegetation which is "normally peat-forming" as an indicator of "living" peatland systems has since been adopted by the European Commission in its definition of "active raised bogs" and "active blanket bogs" for the purposes of the Habitats Directive (European Commission 2007). This decision was undoubtedly guided by a

Fig. 1 Carpet of *Sphagnum* bog moss (*S. fallax*, *S. capillifolium*, *S. magellanicum*), one of the main contributors to global peat formation, together with a few small stems of crowberry (*Empetrum nigrum*) (Picture credit: Richard Lindsay © Rights remain with the author)

recognition that disputes over protection of peatland sites under the Habitats Directive might eventually find their way to the European Court of Justice, with the prospect of proving in law that a mire was currently forming peat. The requirement instead that a site be shown to possess "*a significant area of vegetation that is normally peat forming*" offers a more reasonable and achievable legal test. Taking the IMCG's suggested definition of a mire, therefore, it is possible to identify where "mires" fit within the overall global picture of wetland and dryland habitats (Fig. 3).

Conclusion

What constitutes a "vegetation that is normally peat-forming" will depend on one's location on the globe, but it is important to recognize that *any* vegetation which gives rise to a significant thickness (30 cm or more) of moist or wet semi-decomposed vegetation is a "peat-forming vegetation" and the system can therefore be classed as

Fig. 2 Calcareous surface-water spring mire characterized by *Schoenus ferrugineous* and the absence of peat, Latvia (Picture credit: Richard Lindsay © Rights remain with the author)

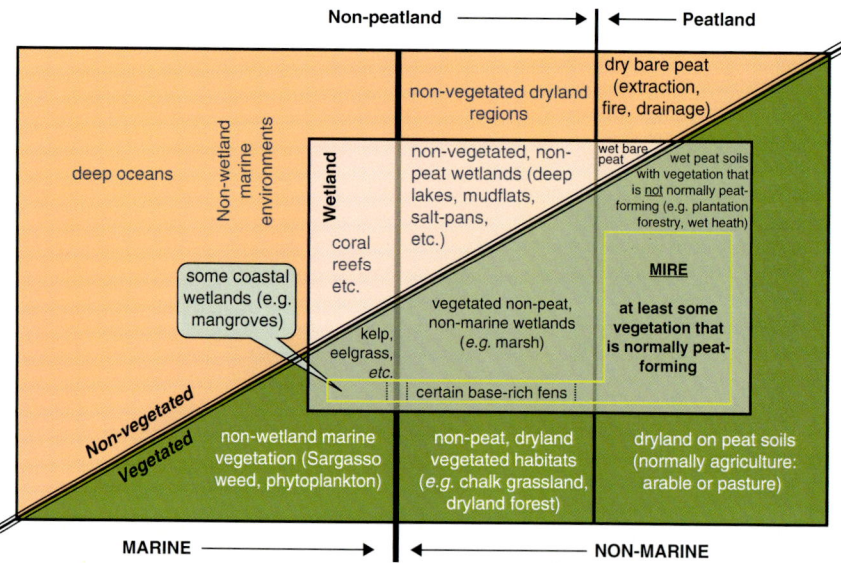

Fig. 3 Diagram to show the relationship between mires (according to the suggested IMCG definition), peatlands, wetlands, and dryland habitats (Figure credit: Richard Lindsay © Rights remain with the author)

a mire. This may mean that various habitats which have perhaps more conventionally been classified as some other type of system should be recognized first and foremost as "mires," capable of or actually accumulating significant quantities of carbon which is then stored often on millennial timescales and with hydrological and physical characteristics which are peculiar to peat.

References

European Commission. Interpretation manual of European Union habitats, Nature and biodiversity. Brussels: European Commission DG Environment; 2007.

Godwin H. The factors which differentiate marsh, fen, bog and heath. Chron Bot. 1941;6:11.

Gore AJP. Introduction. In: Gore AJP, editor. Ecosystems of the world 4A. Mires: swamp, bog, fen and moor. General studies. Amsterdam: Elsevier; 1983. p. 1–34.

Joosten H, Clarke D. Wise use of mires and peatlands. Totnes/Devon: NHBS/International Mire Conservation Group and International Peat Society; 2002.

Löfroth M. European mires – an IMCG project studying distribution and conservation. In: Grünig A, editor. Mires and man: mire conservation in a densely populated country – the Swiss experience. Birmensdorf: Swiss Federal Institute for Forest, Snow and Landscape Research; 1994. p. 281–3. Available from www.wsl.ch/dienstleistungen/publikationen/pdf/420.pdf. Accessed 14 Apr 2015.

Page S, Rieley JO, Shotyk ØW, Weiss D. Interdependence of peat and vegetation in a tropical peat swamp forest. Philos Trans R Soc Lond B. 1999;354:1885–97.

Tansley AG. The British islands and their vegetation. Cambridge: The University Press; 1939.

Troels-Smith J. Characterisation of unconsolidated sediments. Geological survey of Denmark. IV. Series Vol. 3, No. 10. København: Forlag and Sandal; 1955.

Wheeler BD, Proctor MCF. Ecological gradients, subdivisions and terminology of north-west European mires. J Ecol. 2000;88:187–203.

Blanket Mire

19

Richard Lindsay

Contents

Introduction to Blanket Mires 296
Blanket Mire Landscapes 297
References 301

Abstract

Blanket mire forms where the climate remains relatively cool and humid throughout the year and where losses, either through run-off or evapotranspiration, are not high. The resulting continuous saturation of the ground results in peat accumulation across entire landscapes, giving rise to a complex of ombrotrophic bogs and minerotrophic fens. The distinction between a raised mire landscape and a blanket mire landscape is that, in the former, the shape of the terrain is largely determined by peat accumulation whereas, in the latter, the shape of the peat-cloaked terrain is determined more by the topography of the underlying mineral-ground than by the thickness of accumulated peat. Blanket mires are usually considered to be associated with temperate oceanic regions but examples of orographic blanket mire can also be found in other parts of the world, although these examples are often described as other habitats such as 'wet alpine heath'.

Keywords

Blanket mire · Blanket bog · Mire · peatland · Bog · Fen · Climate · Ombrotrophic · Minerotrophic · Orographic · Wet alpine heath · Flow Country · Scotland · Ruwenzori · Falkland Islands · Malvinas · Rain days · Occult precipitation · Mist · Fog · Dew · Tierra del Fuego

R. Lindsay (✉)
Sustainability Research Institute, University of East London, London, UK
e-mail: r.lindsay@uel.ac.uk

© Springer Science+Business Media B.V., part of Springer Nature 2018
C. M. Finlayson et al. (eds.), *The Wetland Book*,
https://doi.org/10.1007/978-94-007-4001-3_272

Introduction to Blanket Mires

When relatively cool, humid climates prevail throughout the year as a result of oceanic or orographic influences, peat can form directly by the paludification of a mineral substrate, particularly if that substrate is base-poor and occurs in association with slopes of less than 30–35° (Tansley 1939; Lindsay 1995). Broad watershed summits, gentle hill-slopes, and wide valley bottoms may become entirely cloaked by the resulting mantle of peat. The topography of the peat mantle is largely determined by the topography of the underlying mineral base, but the varying thickness of the peat, thinner on steeper slopes, deeper on shallow gradients and in valley bottoms, results in an overall smoothing of the smaller-scale undulations within the landscape.

Where the mantle develops across broad summits and other level areas, the peat surface soon becomes entirely dependent upon direct precipitation inputs for its water and mineral supply and thus it becomes ombrotrophic bog. In parts of the landscape with moderate gradients, here too the peat surface may become ombrotrophic as the peat becomes thicker and thereby alters gradients. Other parts of the peat mantle function as water-collection zones and thus represent minerotrophic fen systems which gather surface and subsurface run-off from the various units of ombrotrophic bog and direct this water into streams, rivers, or lakes. The peat mantle as a whole therefore represents a mosaic of differing peatland systems which are interconnected in a variety of ways (Figs. 1 and 2). Although

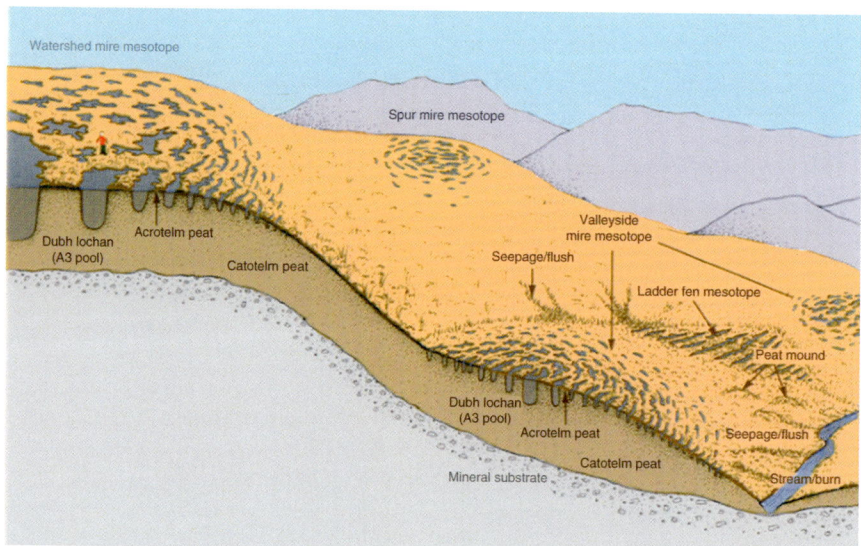

Fig. 1 Illustration of inter-relationships between different mire units within a typical example of blanket mire in the north of Scotland. Descriptions of the differing features are provided in Lindsay et al. (1988) and Lindsay (1995, 2010) (Figure credit: Adapted from Lindsay (1995) with kind permission from the Scottish Natural Heritage)

Fig. 2 Example of a blanket mire landscape from the western part of the Flow Country, Sutherland, north-west Scotland. In the foreground is a ladder fen, beyond which are several watershed bog units with pool patterning. Beyond those units are lochs (lakes), with Loch Loyal and Ben Loyal forming the backdrop (Photo credit: Steve Moore © Rights remain held *in-trust* by Richard Lindsay)

such landscapes are often referred to as blanket bog, the term is misleading because it fails to recognize the important part played by the various fen components within this interconnected landscape (Fig. 3). The more appropriate term is therefore blanket mire (Moen 1985), because a mire is any system which is normally peat forming, whether bog or fen.

Blanket Mire Landscapes

Blanket mire landscapes are generally thought of as being restricted to the oceanic fringes of temperate and boreal regions in both northern and southern hemispheres. The "type" site which best displays the nature and scale of the blanket mire habitat is probably the Flow Country of Caithness and Sutherland in northern Scotland (Lindsay et al. 1988). Blanket mire can, however, also occur in alpine and subalpine zones where orographic clouds and fogs (caused when masses of air are forced up the sides of elevated land formations and condense through increased elevation cooling) create cool, humid conditions. Blanket mire can thus be found, for example, on the upper slopes of the Ruwenzori Mountains in equatorial central Africa (Lindsay et al. 1988) or as scattered examples within the Central Alps (Steiner 2005).

Fig. 3 A blanket mire landscape from the eastern part of the Flow Country with many interlinking mire units, some with pool patterning. Fen systems provide the linkages between systems, sometimes in association with stream courses (Photo credit: Richard Lindsay © Rights remain with the author)

In general, blanket mire develops in regions which have a greater annual water surplus than is typical for raised mire systems (Tansley 1939), partly because the hill-slope gradients associated with blanket mire are often quite steep (Fig. 4). It is widely believed that a minimum mean annual rainfall of around 1,000 mm or more is necessary for blanket mire development (Tansley 1939; Pearsall 1956; Lindsay et al. 1988; Doyle 1997). However, blanket mire is extensive in the Falkland Islands (Malvinas) despite a mean annual rainfall of only 608 mm (climate-data.org website) at Port Stanley, and extensive development of blanket mire also occurs in the Caithness District of northern Scotland (Fig. 3) despite a mean annual rainfall of around 810 mm (Met Office website). The important climatic requirement for blanket mire formation is the overall pattern of water balance throughout the year, which integrates frequency of precipitation inputs with temperature, humidity, cloud cover, windspeed, and vegetation cover. It is believed that around 160 "wet days" (at least 1 mm of rain within a 24 h period) or 200 "rain days" (at least 0.25 mm within a 24 h period) provide the necessary level of precipitation frequency for blanket mire development (Lindsay et al. 1988; Doyle 1997). Significantly, however, this balance must also take into account occult precipitation (i.e., that which arrives by processes normally not recorded by a standard rain gauge) such as mist, fog, and dew (Weber 1902 p.68–69), which can contribute 10–20% of the annual precipitation in blanket mire regions and as much as 50% on individual

Fig. 4 Blanket mire landscape on the island of Mainland, Shetland Isles, northern Scotland. The blanket mire has been much dissected by domestic peat cutting and erosion caused by heavy trampling and grazing as a result of very high sheep numbers in the past. Nonetheless, the whole landscape still retains an almost-continuous mantle of peat (Photo credit: Richard Lindsay © Rights remain with the author)

days (Price 1992). A robust and reliable climate model for blanket mire formation and long-term growth has not yet been constructed.

One striking feature of British and Irish blanket mires is their almost wholly treeless nature even along stream lines (Fig. 5), whereas blanket mire landscapes in areas such as Tierra del Fuego, British Colombia, and Newfoundland have greater or lesser degrees of woodland cover on mineral outcrops, steep slopes, and along water courses (Fig. 6). The almost complete absence of trees (other than artificial conifer plantations) in British and Irish blanket mire landscapes is probably a result of grazing pressure from sheep and associated burning practices rather than because of any fundamental character of blanket mire. Indeed Tallis (1995) and Lindsay (2010) have implicated Neolithic woodland clearance from steeper slopes as one possible cause for the erosion which is a widespread feature of British and Irish blanket mires, although burning, domestic peat cutting, trampling as a result of large sheep numbers, aerial pollution since the start of the Industrial Revolution and widespread drainage are all thought to have been more important factors in recent centuries (Lindsay et al. 2014). It is perhaps no coincidence that the Falkland Islands (Islas Malvinas) in the South Atlantic also have treeless, eroding blanket mires which are subject to sheep grazing.

Fig. 5 Aerial view of blanket mire in the central part of the Flow Country, northern Scotland, with almost total absence of trees even along the streamlines. Conifer plantations can just be seen in the distance to the top left of the view, but otherwise the scenery is treeless apart from some very small dwarf shrub *Salix* and *Sorbus* in sheltered pockets along the streamlines (Photo credit: Steve Moore © Rights remain held *in-trust* by Richard Lindsay)

In contrast with such treeless examples, there is also a potentially compelling argument for including at least some areas of tropical and semitropical cloud forest within the overall concept of blanket mire. Precipitation input patterns are similar to those associated with blanket mire, although perhaps with a greater proportion of precipitation occurring as fog rather than rain (Mulligan and Burke 2005), and in some cases, particularly in "Elfin forests" there is a significant acidic peaty layer covering the forest floor to the point where aluminium toxicity resulting from the low pH becomes a limiting factor – just as it does in typical bog systems (Bruijnzeel and Hamilton 2000). Cloud forests as a whole are not well documented or described but are being lost from some parts of the tropics at an increasingly rapid rate. As with blanket mires, their importance in providing base-flows to downstream communities during dry periods has been recognized, but their potential contribution to the global soil-carbon pool in the form of long-term peatland carbon stores remains largely unstudied, particularly as they also have "peat" formed as an epiphytic deposit on branches and even tree trunks. If they are identified and managed purely as forests rather than as peatland forest systems, there is a danger that a fundamentally important part of these cloud forest "blanket mire" ecosystems will be harmed or lost, along with their many associated ecosystem services.

Fig. 6 View from a hill summit looking across part of the Mitre Peninsula, Tierra del Fuego, Argentina. Steep-sided mineral mounds and steep hillsides are covered with trees, as are streamlines, but apart from these the peat cover is largely continuous and treeless (Photo credit: Richard Lindsay © Rights remain with the author)

Although there are also some features which suggest affinities with blanket mire, tropical peat swamp forests are best considered as essentially raised mire systems. Successive changes in sea levels have left some raised mires somewhat perched and connected with raised mires that have formed during times of lower sea level, but the systems are fundamentally raised mires (Dommain et al. 2011, 2014). However, peatlands in the interior of places such as Indonesia are not well studied and it may be that there are examples which have even closer affinities with blanket mire. The overlap between tropical peat swamp forest and tropical montane cloud forest might be a fruitful area for further research.

References

Bruijnzeel LA, Hamilton LS. Decision time for cloud forests. IHP Humid Tropics Programme Series No. 13. Paris: UNESCO; 2000. Available to download from: http://www.unep-wcmc.org/resources-and-data/decision-time-for-cloud-forests. Accessed 3 Apr 2015.

Climate-data.org website (for Stanley, Falkland Islands) http://en.climate-data.org/location/714892/. Accessed 14 Feb 2015.

Dommain R, Couwenberg J, Joosten H. Development and carbon sequestration of tropical peat domes in south-east Asia: links to post-glacial sea-level changes and Holocene climate variability. Quat Sci Rev. 2011;30:999–1010.

Dommain R, Couwenberg J, Glaser PH, Joosten H, Suryadiputra INN. Carbon storage and release in Indonesian peatlands since the last deglaciation. Quat Sci Rev. 2014;97:1–32.

Doyle G. Blanket bogs: an interpretation based on Irish blanket bogs. In: Parkyn L, Stoneman RE, Ingram HAP, editors. Conserving peatlands. Wallingford: CAB International; 1997. p. 25–34.

Lindsay RA. Bogs: the ecology, classification and conservation of ombrotrophic mires. Perth: Scottish Natural Heritage; 1995.

Lindsay RA. Peatbogs and carbon: a critical synthesis to inform policy development in oceanic peat bog conservation and restoration in the context of climate change. Commissioned report to the Royal Society for the Protection of Birds. 2010. http://www.rspb.org.uk/Images/Peatbogs_and_carbon_tcm9-255200.pdf (low resolution) http://www.uel.ac.uk/erg/PeatandCarbonReport.htm (high resolution: downloadable in sections). Accessed 2 Feb 2015.

Lindsay RA, Charman DJ, Everingham F, O'Reilly RM, Palmer MA, Rowell TA, Stroud DA. In: Ratcliffe DA, Oswald P, editors. The flow country: the peatlands of Caithness and Sutherland. Peterborough: Nature Conservancy Council; 1988.

Lindsay R, Birnie R, Clough J. Peat bog ecosystems: weathering, erosion and mass movement of blanket bog. Technical report. Edinburgh: International Union for the Conservation of Nature; 2014. Available to download from: http://hdl.handle.net/10552/3986. Accessed 2 Mar 2015.

Met Office website (for Wick Airport, Caithness) http://www.metoffice.gov.uk/public/weather/climate/gfmu99nxj. Accessed 14 Feb 2015.

Moen A. Classification of mires for conservation purposes in Norway. Aquil Ser Bot. 1985;21: 95–100.

Mulligan M, Burke S. DFID FRP Project ZF0216 global cloud forests and environmental change in a hydrological context. Final Report. London: UK Department for International Development (DFID); 2005. Available to download from www.ambiotek.com/cloudforests/cloudforest_finalrep.pdf. Accessed 2 Mar 2015.

Pearsall WH. Two blanket-bogs in Sutherland. J Ecol. 1956;44(2):493–516.

Price JS. Blanket bog in Newfoundland. Part 2. Hydrological processes. J Hydrol. 1992;135: 103–19.

Steiner GM. Die Moorverbreitung in Österreich/Distribution of mires in Austria. In: Steiner GM, editor. Moore – von Siberien bis Feuerland/Mires – from Siberia to Tierra del Fuego, Stapfia 85. Linz: Oberösterreichische Landesmuseen; 2005. p. 55–96.

Tallis JA. Blanket mires in the upland landscape. In: Wheeler BD, Shaw SC, Fojt WJ, Robertson RA, editors. Restoration of temperate wetlands. Chichester: Wiley; 1995. p. 495–508.

Tansley AG. The British Islands and their vegetation. Cambridge: The University Press; 1939.

Weber CA. On the vegetation and development of the raised bog of augstumal in the Memel Delta. (1902) In: Couwenberg J, Joosten H, editors. C.A. Weber and the raised bog of Augstumal. Tula: "Grif & K"; 2002. p. 52–270.

Blanket Bogs

20

Richard Lindsay

Contents

Introduction to Blanket Bogs	304
Blanket Bog Mesotopes	305
References	308

Abstract

Blanket bog – a term first coined in 1935 – refers to areas of regular precipitation and generally cool climates where *blanket mire* develops as a mantle of peat smothering entire landscapes, *blanket bog* referring specifically to those parts of this peat-draped landscape which are entirely rain-fed (ombrotrophic) bogs. The blanket bog components in such a landscape predominate but are linked by areas of minerotrophic fen or other types of wetland system. A number of differing approaches can be used to describe individual blanket bog units but perhaps the most universally applicable is their geographic position within the landscape. The global extent of blanket bog may be larger than originally thought because certain habitat formations normally thought of as other habitat types may be more appropriately classified as blanket bog.

Keywords

Blanket bog · Blanket mire · Bog · Cloud forest · Cover moss · Fen · Fog · Godwin · Hollows · Mesotope · Minerotrophic · Mire · Mist · Ombrotrophic · Osvald · Pools · Precipitation · Rain · Ridges · Saddle · Scotland · Spur · Surface patterning · Tierra del Fuego · Valleyside · Water-divide · Waterlogging · Watershed

R. Lindsay (✉)
Sustainability Research Institute, University of East London, London, UK
e-mail: r.lindsay@uel.ac.uk

© Springer Science+Business Media B.V., part of Springer Nature 2018
C. M. Finlayson et al. (eds.), *The Wetland Book*,
https://doi.org/10.1007/978-94-007-4001-3_19

Introduction to Blanket Bogs

The term "blanket bog" was first used in 1935 by Dr H. Godwin to describe the mantle of peat which cloaks many oceanic temperate landscapes such as the north and west of Britain and the west of Ireland, in preference to the name "cover moss" suggested by Prof. H. Osvald (Osvald 1949; Pearsall 1950). It is Dr Godwin's term which has since gained widespread acceptance, but the term is not internally consistent and has therefore given rise to ambiguity and confusion from its earliest usage to present-day scientific literature. "Bogs" are ombrotrophic systems relying entirely on direct precipitation for their water and solute supply. Unfortunately since its inception the term "blanket bog" has generally been applied to the entire peat-dominated landscape (Tansley 1939), despite the fact that such landscapes contain a whole mosaic of peat-forming systems. Some landscape elements, such as broad watershed (water-divide) summits, may indeed be ombrotrophic bog systems (Fig. 1), but often many of the key interconnecting peatland elements are minerotrophic fens, each with a distinctive hydrology, chemistry, biodiversity, and carbon balance. It is therefore more accurate and less ambiguous to refer to the whole peat-draped landscape as "blanket *mire*" because the term "mire" means *any* peat forming system, whether bog or fen (Moen 1985). The term "blanket bog" can then be applied solely to those parts of the blanket mire landscape which are truly ombrotrophic.

The most distinctive features which separate blanket bogs from raised bogs are, firstly, that the shape of most blanket bog systems is determined largely by the topography of the underlying mineral base (Fig. 2), whereas in a raised bog the surface morphology is almost entirely determined by the depth of accumulated peat. Secondly, lagg fens are a relatively uncommon feature in blanket bogs because the bog peat characteristically fuses with the wider mantle of peat which becomes progressively thinner downslope or upslope, giving a diffuse and ill-defined peat margin rather than forming the distinct peat-mineral interface zone which is so characteristic of many

Fig. 1 Extensive watershed (water-divide) plateau dominated by blanket bog, Isle of Lewis, Outer Hebrides, Scotland. To the *left* (west) is the Atlantic Ocean which brings precipitation mostly in the form of rain, mist, or fog on a regular and frequent basis throughout the year. The plateau is completely treeless, as are the surrounding slopes (Photo credit: Richard Lindsay © Rights remain with the author)

Fig. 2 Ombrotrophic blanket bog within a broader blanket mire landscape, western Sutherland, NW Scotland. The peat has accumulated directly over the ice-molded bedrock of ancient Lewisian gneiss which provides no nourishment to the peat-forming vegetation cover. Frequent precipitation maintains waterlogged conditions and supplies a limited quantity of solutes (Photo credit: Richard Lindsay © Rights remain with the author)

raised bogs. Thirdly, blanket bogs are generally associated with higher water surpluses than raised bogs because the slopes typically found on blanket bogs cause more rapid runoff than is normally encountered on raised bogs. Consequently a larger water surplus is required in order to maintain sufficient waterlogging for peat formation to be possible across such sloping terrain (Lindsay 1995).

Blanket Bog Mesotopes

Individual blanket bog units, or mesotopes (Fig. 3; also refer to "Peatland classification *In* The Wetland Book I"), are defined according to their position within the landscape, giving rise to a small number of core bog types, namely, watershed (water-divide), saddle, spur (a level area formed partway down a slope), and valleyside (formed at the foot of a hillslope). Each has a distinctive morphology and pattern of surface-water flow lines (Lindsay 2010) – see Fig. 4. Intermediate forms between these main types provide further variety. Blanket bogs formed in areas of large water surplus typically display an extensive microtope surface patterning consisting of pools, hollows, and ridges, particularly where gradients are gentle (Tansley 1939; Pearsall 1956; Smart 1982; Lindsay et al. 1988; Belyea 2007) – see Figs. 5 and 6. Where gradients become steeper or in regions of smaller

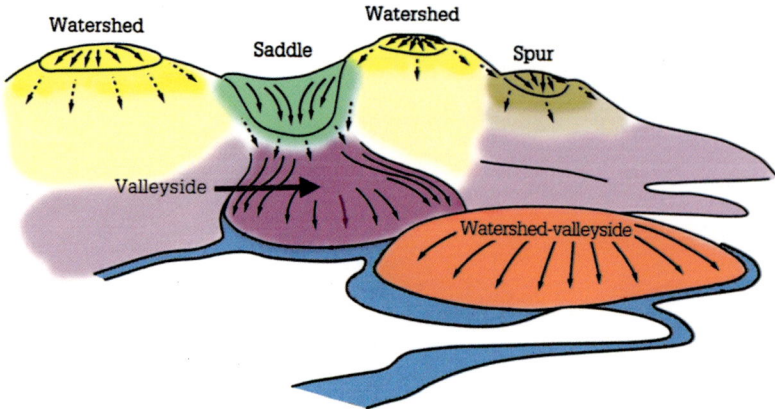

Fig. 3 Blanket bogs tend to occupy a few characteristic landform positions within the landscape. Being the locations where peat accumulates most readily, these tend to be the deepest areas of peat and form distinct bog hydromorphological units or "mesotopes" (see "Peatland Classification"). Such landform locations are also where bogs which are intermediate between raised bog and blanket bog (see "Peatland Classification") tend to form in climates which are not sufficiently humid for blanket mire development (Picture credit: Adapted from Lindsay et al. (1988) with kind permission of the Joint Nature Conservation Committee (JNCC))

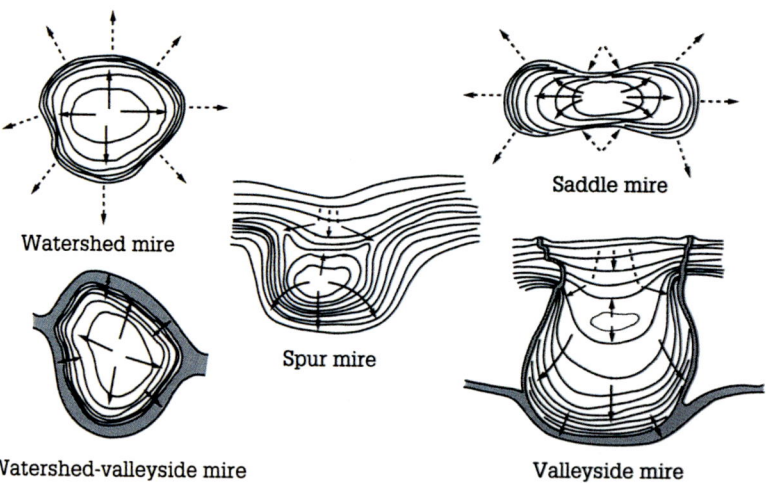

Fig. 4 The differing characteristic patterns of surface-water seepage ("flow lines" – Ivanov 1981) as seen from above for the main blanket bog hydromorphological mesotopes. Flow lines are oriented and drawn at right angles to contours and to the long axis of any surface patterning (Picture credit: Amended from Lindsay et al. (1988) with kind permission of the Joint Nature Conservation Committee (JNCC))

Fig. 5 Open-water pool complex dominating watershed (water-divide) blanket bog at Blar nam Faioleag National Nature Reserve in the Flow Country of northern Scotland (Photo credit: Richard Lindsay © Rights remain with the author)

Fig. 6 Deep blanket bog formed on a high watershed plateau (water-divide) in Tierra del Fuego, Argentina. The deep peat of the plateau expanse is characterized by a well-developed pattern of pools and ridges and an absence of trees. Towards the edge of the plateau where the gradient becomes more marked, stunted individuals of southern beech (*Nothofagus* spp.) form part of the bog community. Some way down the slope where the gradient increases significantly, the peat thins and *Nothofagus* woodland becomes the dominant vegetation cover of the steep hillslope (Photo credit: Richard Lindsay © Rights remain with the author)

water surplus, surface patterning becomes more subtle and may consist largely of hummocks and ridges with few or no aquatic elements (Lindsay et al. 2014). Certain areas of tropical cloud forest may also represent a hitherto overlooked form of blanket bog (see cloud forest description in Bruijnzeel and Hamilton (2000) and ▶ Chap. 19, "Blanket Mire").

References

Belyea LR. Climatic and topographic constraints on the abundance of bog pools. Hydrol Process. 2007;21(5):675–87.

Bruijnzeel LA, Hamilton LS. Decision time for cloud forests. IHP Humid Tropics Programme series no. 13. Paris: UNESCO; 2000. http://www.unep-wcmc.org/resources-and-data/decision-time-for-cloud-forests. Accessed 3 Apr 2015.

Ivanov KE. Water movement in mirelands. London: Academic; 1981.

Lindsay RA. Bogs: the ecology, classification and conservation of ombrotrophic mires. Perth: Scottish Natural Heritage; 1995.

Lindsay RA. Peatbogs and carbon: a critical synthesis to inform policy development in oceanic peat bog conservation and restoration in the context of climate change. Commissioned report to the Royal Society for the Protection of Birds. 2010. http://www.rspb.org.uk/Images/Peatbogs_and_carbon_tcm9-255200.pdf. (low resolution). http://www.uel.ac.uk/erg/PeatandCarbonReport.htm. (high resolution: downloadable in sections). Accessed 2 Feb 2015.

Lindsay RA, Charman DJ, Everingham F, O'Reilly RM, Palmer MA, Rowell TA, Stroud DA. The flow country: the peatlands of Caithness and Sutherland. Ratcliffe DA, Oswald P, editors. Peterborough: Nature Conservancy Council; 1988.

Lindsay R, Birnie R, Clough J. IUCN UK Committee Peatland Programme, Briefing note no. 2 – Peat Bog ecosystems: structure, form, state and condition. Edinburgh: IUCN UK National Committee; 2014. http://www.iucn-uk-peatlandprogramme.org/sites/www.iucn-uk-peatlandprogramme.org/files/2%20Biodiversity%20final%20-%205th%20November%202014.pdf. Accessed 4 Apr 2015.

Moen A. Classification of mires for conservation purposes in Norway. Aquilo Seria Botanica. 1985;21:95–100.

Osvald H. Notes on the vegetation of British and Irish mosses. Acta Phytogeographica Suecica. 1949;26:1–62.

Pearsall WH. Mountains and Moorlands. London: Collins; 1950.

Pearsall WH. Two blanket-bogs in Sutherland. J Ecol. 1956;44:493–516.

Smart JP. Stratigraphy of a site in the Munsary Dubh Lochs, Caithness, northern Scotland: development of the present pattern. J Ecol. 1982;70:549–58.

Tansley AG. The British islands and their vegetation. Cambridge, UK: The University Press; 1939.

Lagg Fen

21

Richard Lindsay

Contents

Introduction to Lagg Fens 309
Water Level and Lagg Fens 311
References 311

Abstract
The fen which typically surrounds a raised bog is often considered to be of marginal interest, but this lagg fen plays a major part in the overall hydrology of the bog. Its loss can have substantial long term impacts on the bog in terms of hydrology, morphology and biodiversity.

Keywords
Peat · Bog · Mire · Raised · Lagg · Fen · Subsidence · Settlement · Drying · Drainage · Morphology · Biodiversity · Peatland · Margin · Groundwater

Introduction to Lagg Fens

Ombrotrophic raised mires form domes of rain-fed peat which sit on the local groundwater table like a water droplet on a horizontal sheet of glass, rising as much as 10 m above the surrounding landscape. Precipitation falling onto the dome is lost through evapotranspiration, enters long-term water storage within the dome, or moves by gravity-driven seepage through the surface acrotelm (living surface layer of a peat bog) to the margin of the dome where it merges

R. Lindsay (✉)
Sustainability Research Institute, University of East London, London, UK
e-mail: r.lindsay@uel.ac.uk

© Springer Science+Business Media B.V., part of Springer Nature 2018
C. M. Finlayson et al. (eds.), *The Wetland Book*,
https://doi.org/10.1007/978-94-007-4001-3_257

with the local groundwater table. The zone of contact, where runoff from the ombrotrophic bog mixes with mineral-enriched groundwater, is termed the lagg fen (Fig. 1).

For a circular raised bog, which approximates to an oblate hemispheroid in shape, an increase in radius and/or height will result in a larger increase in surface area compared to the associated increase in its circumference. Consequently the lagg fen is likely to receive substantially increased volumes of runoff from the expanding bog surface. This large quantity of runoff is then capable of waterlogging adjacent mineral ground, giving rise to peat formation through paludification, thus resulting in further lateral expansion of the raised mire system. If the raised mire has formed in a relatively steep-sided basin with rising slopes surrounding the bog, the lagg fen zone will be narrow and lateral expansion will be limited, but if the bog has formed in a shallow basin beyond which there is relatively level or gently sloping terrain, the lagg fen zone may become 1–2 km wide, and subsequent expansion of the bog may be both rapid and considerable (Ivanov 1981, pp. 17–20). Where raised mires have formed within extensive floodplain fens, the lagg fen merges with the surrounding floodplain fen system and becomes less recognizable as a distinct feature.

The lagg fen is a minerotrophic mire habitat in its own right, though it is obviously dependent on the presence of the adjacent ombrotrophic bog system. Its character is determined by the nature of soils and bedrock in the surrounding landscape and generally provides a marked ecological contrast to the extreme

Fig. 1 Lagg fen zone of Teiči Nature Reserve, Latvia, viewed from fire-watch tower. The raised mire is to the right, with low scrub on the sloping rand of the bog and virtually treeless open bog to the far right across the main mire expanse. The lagg zone is visible in the center of the photograph as water more than 1 m deep with floating rafts of *Sphagnum*. To the left is a forest on mineral ground but which is partially flooded and becoming increasingly paludified as the bog expands to the left of the picture (Photo credit: Richard Lindsay © Rights remain with the author)

nutrient and solute poverty of the adjacent bog environment. The fact that it represents a major aquatic transition zone which may support a range of vegetation types from tall emergent macrophytes to low moss carpets means that the invertebrate assemblage of a lagg fen can often show remarkable diversity and unusual mixtures of species (Howie and Tromp van-Meerveld 2011; Thorne and Hatfield Moors Conservation News website).

Although the lagg fen is often the most extensive truly aquatic zone of a natural raised bog, across much of Western Europe and North America, the bog margin is often the driest part of a raised bog because in many cases the lagg fen has either been converted into a deep drain in order to prevent paludification of adjacent agricultural land or has been entirely replaced by agricultural fields which are subject to underdrainage and the new bog margin is represented by a dry peat face cut into the main dome of the bog. The beneficial effect of draining the lagg fen for those contemplating the exploitation of raised bogs was known as far back as the seventeenth century (King 1685), but this drained condition of the margin is now so widespread that it is almost considered "normal," and many restoration efforts focus on rewetting efforts on the remnant dome without addressing the fundamental problem of the drained lagg fen.

Water Level and Lagg Fens

The lagg fen is a fundamental part of raised bog hydrology because it establishes the basal level on which the "groundwater mound" of a raised bog sits (Ingram 1982). Lowering the water level in the lagg fen zone in effect lowers the basal level for the entire raised bog. This does not mean that the whole groundwater mound is immediately lowered by the same amount, but the lowered water table at the margin establishes a hydrological gradient which causes primary consolidation, secondary compression, and oxidative wastage of the peat in the zone immediately adjacent to the drained lagg fen (Lindsay et al. 2014). This in turn leads to progressive subsidence across the dome until a new dome profile is established – a process which may take several centuries (Bragg 1995). Raised mire restoration techniques which take this process into account seek to reestablish a lagg fen at a position which gives the remnant dome the greatest potential for minimizing further subsidence (Steiner 2005).

References

Bragg OM. Towards and ecohydrological basis for raised mire restoration. In: Wheeler BD, Shaw SC, Fojt WJ, Robertson RA, editors. Restoration of temperate wetlands. Chichester: Wiley; 1995. p. 305–14.

Howie SA, Tromp van-Meerveld I. The essential role of the Lagg in raised bog function and restoration: a review. Wetlands. 2011;31:613–22.

Ingram HAP. Size and shape in raised mire ecosystems: a geophysical model. Nature. 1982;297:300–3.

Ivanov KE. Water movement in Mirelands (trans: Thompson A, Ingram HAP). London: Academic; 1981.

King W. On the bogs and loughs of Ireland. Philos Trans. 1685;15:948.

Lindsay R, Birnie R, Clough J. Impacts of artificial drainage on peatlands. IUCN UK Committee Peatland Programme Briefing Note No.3; 2014. http://www.iucn-uk-peatlandprogramme.org/sites/www.iucn-uk-peatlandprogramme.org/files/3%20Drainage%20final%20-%205th%20November%202014.pdf. Accessed 2 Feb 2015.

Steiner GM. Zum Verständis de Ökohydrologie von Hochmooren. In: Steiner GM, editor. Mires from Siberia to Tierra del Fuego [Moore von Siberien bis Feuerland]. Linz: Land Oberöstereich, Biologiezentrum/Oberösterreichische Landesmuseen; 2005. p. 27–39.

Thorne and Hatfield Moors Conservation News website. 2013 Oct 17. https://thmcf.wordpress.com/tag/lagg-fen/. Accessed 2 Feb 2015.

Karst Wetlands

22

Gordana Beltram

Contents

Introduction 314
Hydrology 314
Ecosystem Services 316
Conservation and Wise Use 317
Diversity of Karst Wetlands and Their Ecosystem Services 319
    Karst Wetlands in the Temperate Zone 320
    Karst Wetlands in Tropics/Subtropics 322
Threats and Challenges 327
References 328

Abstract

Karst is a term used for characteristic landscapes, landscape features, and phenomena developed in water soluble and porous carbonate rocks on surface and underground. The name originates from the stony limestone region Kras/Carso on the border of Slovenia and Italy where it was first studied. Carbonate rocks outcrop across some 13.2% of world's land area, yet subterranean carbonate rocks with karst groundwater circulation extend to an even larger area, estimated to about 14% of world's land area. Surficial karst wetlands function interdependently with their subterranean "counterparts" to maintain their ecological character. Karst wetlands occur in all regions of the world but their complexity, values, and ecosystem services are still not fully recognized and many are threatened by economic developments (tourism developments, agriculture, fisheries) that depend on the same wetland resources.

G. Beltram (✉)
Ministry of the Environment and Spatial Planning, Environment Directorate, Nature Conservation Unit, Ljubljana, Slovenia
e-mail: gordana.beltram@gov.si

© Springer Science+Business Media B.V., part of Springer Nature 2018
C. M. Finlayson et al. (eds.), *The Wetland Book*,
https://doi.org/10.1007/978-94-007-4001-3_203

> **Keywords**
>
> Karst · Carbonate · Poljes · Ponors · Cenotes · Sinkholes · Caves · Hydrology · Subterranean

Introduction

Karst is a term used for characteristic landscapes, landscape features, and phenomena developed in water soluble and porous carbonate rocks on surface and underground. The name originates from the stony limestone region Kras/Carso on the border of Slovenia and Italy where it was first studied. It is often referred to as "classical" Kras/Karst. In global terms, carbonate rocks outcrop across some 13.2% of world's land area (Fig. 1), yet subterranean carbonate rocks with karst groundwater circulation extend to an even larger area, estimated to about 14% of world's land area (Williams 2008).

In the context of karst systems, wetlands cannot be delineated only on the basis of surface features since they function interdependently with their subterranean "counterparts" to maintain their ecological character. While the diversity of wetlands can be recognized on the surface in karst areas, this chapter will stress the interlinkages with subterranean hydrological systems so obvious on karst poljes in karst springs, ponors, and other karst phenomena (Cerkvenik et al. 2018; See also http://www.speleogenesis.info/directory/glossary/).

Hydrology

Wetlands and karst systems have some key factors in common. They both depend on water and the hydrological system of the catchment area. Both are closely associated with the hydrological cycle as water passes into or is stored within, flows through and emerges from karst or wetland areas. The quality and quantity of water, in combination with other factors, are crucial for maintaining the form and structure of karst and also for maintaining the ecological character of wetlands. Both are extremely fragile systems. Changes in one component can affect the ecosystem quality and considerably change its character (Beltram 2004).

The complexity of karst hydrology (Fig. 2) is best described in some recent studies with examples from different parts of the world (Ford and Williams 2007; Stokes et al. 2010; Bauer-Gottwein et al. 2011; Knez et al. 2011b; Prelovšek 2012). Very briefly, the rain water percolates through the fractures of permeable carbonate bedrock and reappears in the subterranean cavities, voids, and caves. Water flow paths (inflow, through flow, and outflow of water) are often difficult to follow and can vary from tiny conduits to underground rivers. Concentrated inflows of water from allogeneic sources sink underground at swallow holes (swallets, stream-sink, sinkholes, or ponors) that can be either vertical or lateral. The outflow of water in the karst springs are among the most remarkable. Reversing springs in karst areas that can be sinks at low water and springs at high water level are known as estavelles (Ford and Williams 2007). While the karst surface is usually dry, surface karst

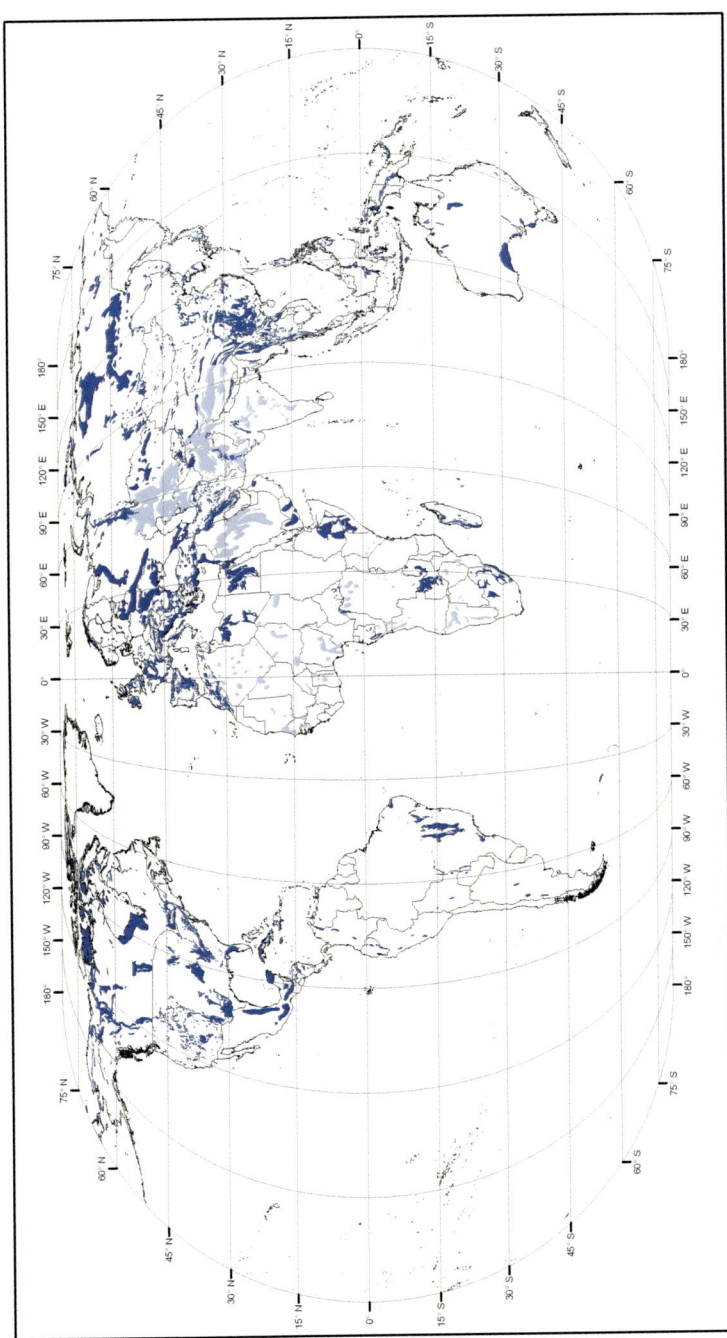

Fig. 1 Global distribution of carbonate outcrops. *Solid color* indicates that carbonates are relatively continuous; *pale color* depicts areas in which carbonates are abundant but not continuous (World Map of Carbonate Rock Outcrops v3.0 – prepared by Paul Williams and Yin Ting Fong, http://web.env.auckland.ac.nz/our_research/karst/#karst6 – Used with permission)

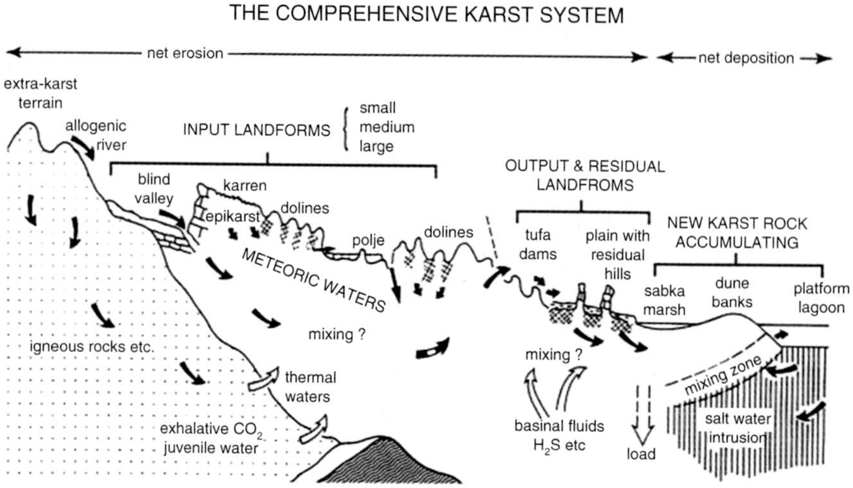

Fig. 2 The comprehensive karst system schematically presents a full range of karst phenomena from input to output margins (© Reproduced with Permission: Ford D. and Williams P. Karst Hydrogeology and Geomorphology, 2nd ed. Chichester (UK): John Wiley and Sons, Ltd; 2007. Fig. 1.2, p. 3)

wetlands are strongly dependent on geomorphology and subterranean karst hydrology including the subterranean wetlands in the caves. In active cave systems water and its characteristics are responsible for the intensity of karstification processes and formation of speleothems (General term for all cave mineral deposits, embracing all stalactites, flowstone, flowers, etc.).

Ecosystem Services

Karst systems and associated wetlands have important socioeconomic values, including maintenance of general hydrological balance and flood storage, drinking water supply, and water for grazing animals or agriculture. Karst aquifers and wetland systems remain a significant source of drinking water all over the world and may play an especially vital role in ensuring adequate water supplies for human communities in generally dry surface landscapes. It has been estimated that 20–25% of the world's population depends on karst waters and in some areas up to 50% of drinking water derives from karst areas either in springs or pumped from karst aquifers (Ford and Williams 2007). In addition, subterranean wetlands have many natural values including as habitats of endemic species of fauna to supporting life in cave systems. Increasingly, the importance of these areas is recognized for tourism development; archaeology and natural history; and places of research, education, and raising awareness of the importance and fragility of subterranean hydrological systems and their surface counterparts/components. In the past, caves were closely linked with the livelihoods of local people for storage, hiding or shelter, burial grounds or sacred places of worship, for health treatment, providing mineral resources, and as sources of drinking water (Mihevc 2005).

All the wetland values and ecosystem services are further considered in the section discussing different karst wetlands.

Conservation and Wise Use

To conserve and maintain karst wetlands, sound knowledge of their main characteristics and understanding of the functioning and processes that form and sustain these systems is essential. Recognizing the interlinkage and interaction of karst surface and subterranean hydrological systems, as well as their ecosystem services, it is crucial to conserve and sustain surface and subterranean wetlands within their catchments.

A key consideration is that strict protection has not always been the best solution for conserving dynamic and unstable ecosystems, and consequently, appropriate management of the whole *catchment area* (including both surface water and ground water) is essential (Watson et al. 1997). The Ramsar Convention introduced and elaborated on the "wise use" of wetlands and strongly advocates the "catchment approach." The holistic ecosystem approach also stresses the interdependence of all wetlands in the catchment. For karst wetlands, being surface and subterranean, this is of critical importance bearing in mind that in some karst areas water and wetlands might be hidden from sight in the subterranean world.

The complexity of karst wetlands, their values, and ecosystem services are still not fully recognized. However, in 1999, the Ramsar Convention on Wetlands with Resolution VII.13 (http://www.ramsar.org/sites/default/files/documents/library/key_res_vii.13e.pdf) recognized the principal wetland conservation values of "karst and other subterranean hydrological systems" to include (Ramsar Convention Secretariat, 2010):

- Uniqueness of karst phenomena/functions and functioning
- Interdependency and fragility of karst systems and their hydrological and hydrogeological characteristics
- Uniqueness of these ecosystems and endemism of their species
- High biodiversity and the importance for conserving particular taxa of fauna and flora

Although more than 2,231 Ramsar sites have been designated as of March 2016 from around the globe and at different altitudes, just over 100 have been identified as including surface and subterranean karst wetlands and other subterranean hydrological systems (Table 1). Furthermore, in less than 30 of these sites the *karst* "subterranean hydrological system" has been recognized as one of the main wetland types according to Ramsar Classification System of Wetland Type. Additionally stressing their conservation values or uniqueness, a few sites are also listed as UNESCO World Heritage Sites and/or Biosphere Reserves under the UNESCO MAB Programme. Two of these which are in Europe include Škocjan Caves, a karst underground water

Table 1 Percent distribution of carbonate outcrop globally and regionally (Adapted from http://web.env.auckland.ac.nz/our_research/karst/#karst6. Accessed 20 Mar 2016); number of designated Ramsar sites by karst wetland type, global and regional area, and percent of maximum carbonate outcrop within Ramsar sites (Ramsar Information Data Base: http://www.ramsar.org/sites-countries/the-ramsar-sites. Accessed 20 Mar 2016)

Region[a]	Land area km^2	Percent maximum carbonate outcrop	Ramsar site area with karst km^2	Ramsar percent of maximum carbonate outcrop	Ramsar type Zk(a)[f]	Ramsar type Zk (b)[g]	Ramsar type Zk (c)[h]
Global[b]	133,448,089	13.2	46,371	0.26	16	93	3
North America[c]	22,229,293	18.3	34,113	0.8	11	33	
South America	17,792,882	2.1	329	0.09	0	2	
Europe[d]	6,125,842	21.8	8,564	0.64	5	36	1
Russian Federation[e]	20,649,781	19.3	0	0.0	0	0	
Africa	30,001,574	10.1	2,570	0.08	0	7	1
Middle East and Central Asia	11,129,676	23.0	15	0.0006	0	3	
East and Southeast Asia	15,638,629	10.8	705	0.04	0	8	1
Australasia	9,611,377	6.2	75	0.01	0	4	

[a]Regions, land area, and percent maximum carbonate outcrop from http://web.env.auckland.ac.nz/our_research/karst/#karst6. Accessed 20 Mar 2016
[b]Excluding Antarctica, Greenland, and Iceland
[c]Including Caribbean, excluding Greenland
[d]Including Iceland and Russian Federation
[e]Including Armenia, Azerbaijan, Georgia, Kazakhstan, Kyrgyzstan, Russia, Turkmenistan, Uzbekistan
[f]Zk(a) – marine and coastal karst wetland type (Karst or other subterranean hydrological systems)
[g]Zk(b) – inland karst wetland type (Karst or other subterranean hydrological systems)
[h]Zk(c) – human-made: karst or other subterranean hydrological systems

cave system developed in the area of the "classical" Karst/Kras, and the Baradla-Domica cave system shared between Hungary and Slovakia, and there are more sites with multiple designations considered with individual sites. Moreover, Mexico is an outstanding example of a country recognizing the importance of karst subterranean wetlands and taking advantage of the Ramsar approach. In 2003 and 2004, Mexico designated a number of Ramsar sites in the Yucatan Peninsula to show the importance of the karst wetlands and the need for their protection. Subsequently, more karst wetlands were designated worldwide but still almost one fourth of all karst Ramsar sites in the world are designated in the Yucatan Peninsula.

Although karst wetlands are well distributed across the planet, there are more karst wetlands worldwide that should be considered for inclusion on the Ramsar List, in order to, *inter alia*, better integrate them into catchment management, use them wisely, and safeguard their resources. The outstanding karst phenomena in South East Asia, including the three provinces in the south of China, have been included on the World Heritage List. Karst wetland areas in the region are, however, less well represented on the Ramsar List, although the Asia-Pacific Karst forum on Karst Ecosystems and World Heritage in 2001 stressed that many areas meet Ramsar criteria and provide opportunity for complementing the World Heritage nominations (Wong et al. 2001).

Some of the most typical karst wetlands are outlined herewith to provide some best examples of inland and coastal surface-subterranean interlinkages. Data and information was extracted from the Ramsar Site Database of the List of Wetlands of International Importance (Ramsar Site Information Service) maintained by Wetlands International: (http://ramsar.wetlands.org/Database/AbouttheRamsarSitesDatabase/tabid/812/Default.aspx), as well as other relevant sources as indicated. Across Europe and North America karst areas are well documented, and increasingly more data are available from other parts of the world (see http://www.speleogenesis.info/index.php).

Diversity of Karst Wetlands and Their Ecosystem Services

Karst landscapes vary around the world mainly as a result of lithology, geological structure, and climate, taking into account that precipitation is a key factor in karst evolution. Due to different climates, broad differences in karst landscapes occur between humid tropics/subtropics (such as karst of monsoonal SE Asia), the hot deserts (arid and semiarid Australia), the humid temperate zone (the Dinaric Karst), and cold high altitude regions (karst of Canadian Rockies and Siberia). Yet karstification processes are most intensive in the humid tropics and temperate zone (Williams 2008). In dry areas, underground stream systems can also be human-made for irrigation or water supply purposes (e.g., Algeria). Additionally, water caves can be found in other rocks as well as ice (glaciers), but such examples have not been considered in this context.

Although karst wetlands occur in all regions of the world, only some most characteristic examples are considered herein particularly from the hydrological and biodiversity aspects and their ecosystem services, with a particular focus on the subterranean-surface interface of wetlands. Some typical characteristics of surface karst wetlands are additionally discussed in the Dinaric Karst (Cerkvenik et al. 2018), while an overview of subterranean biodiversity, karst habitats, and their fauna is considered separately by Sket (2018).

There are still many areas and topics to be researched to fully understand the functioning of different karst systems. Research in karst areas continues to provide new findings (Filippi and Bosák 2013). International scientific unions (such as the International Geological Union, the International Association of Geomorphologists, the International Association of Hydrogeologists, the International Union of Speleology (http://www.uis-speleo.org/), and the International Union of Biological Sciences) are some key sources of objective authoritative scientific information on different areas and sites. Their research is often conveyed at conferences and through publications (http://speleogenesis-info.isth.info/index.php).

Karst Wetlands in the Temperate Zone

Almost 22% of Europe's territory has exposed carbonate rock outcrops (see: http://web.env.auckland.ac.nz/our_research/karst/#karst6. Accessed 20 Mar 2016) and karst areas are well researched and documented. The term karst originated, and karst studies were first conducted, in Europe in the "classical" Karst and along the Adriatic Sea, in the Dinaric Karst. Additionally, Moravian karst in the Czech Republic and the Baradla-Domica cave systems are representative of some of the best karst wetlands in the temperate zone.

In Slovenia where almost half of the territory has carbonate rocks, *Škocjanske jame* (Škocjan Caves – multiple designations: Ramsar Site 1999, World Heritage Site 1986, a protected regional park 1996, UNESCO MAB site 2004) is a karst underground water cave system that developed in the area of the "classical" Karst/Kras (Kranjc 1997). The main hydrological characteristics are the extremely high fluctuations in the ground water level, moving water currents fed by rainwater, and pools of stagnant water. The underground river has a discharge oscillating between 0.050 m^3/s and more than 400 m^3/s. Water oscillations result in ground water level fluctuations of more than 130 m in the researched parts of the cave system. The area holds typical karst phenomena and karst features developed at the contact between impermeable and permeable limestone rocks. Leaving the surface, the Reka River flows through the cave system for almost 5 km and through a siphon sinks even deeper and continues underground for some 40 km to reach the surface again at the Timavo karst springs in the Adriatic coastal area in Italy. In addition to contributing to the recharge of the karst aquifer, the cave system provides habitat to numerous endemic and endangered animal species. Škocjan Caves are the second most visited cave in Slovenia with guided tours, a natural trail and cycling paths on the surface, and tourism is the main economic activity in the area (https://rsis.ramsar.org/ris/991).

The karst catchment of the Ljubljanica River is a typical complex karst wetland area in the Slovenian part of the Dinaric Karst (Fig. 3). It includes a series of surface intermittent lakes on karst poljes (Planinsko polje, Cerkniško polje) and caves with underground rivers (Križna jama, Postojnska jama, Planinska jama) that well represent

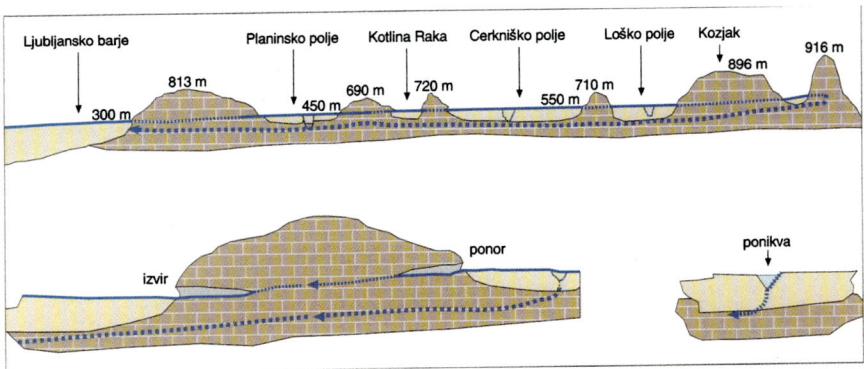

Fig. 3 A cross section through the inner karst poljes, ponors, and springs – the sinking system of the Ljubljanica River (From Kunaver et al. 1995. Obča geografija za 1. Letnik srednjih šol. Fig. 2.61, p. 42. © with permission of Državna založba Slovenije)

the interaction and interdependency between the surface and subterranean wetlands belonging to a common hydrological system (Beltram and Sket 2010). Over a relatively short distance there are numerous surface and subterranean karst phenomena holding remarkable geomorphological, hydrological, ecological, and socioeconomic values for the entire area. On the poljes, especially Cerkniško polje and Planinsko polje, more than 300 bird species have been recorded. Only 11 fish species are known to live in the waters of the area, but some of them are uniquely adapted to the intermittent character of karst lakes. The most important value of the area is the extremely rich and endemic aquatic and terrestrial subterranean fauna in the Postojna-Planina cave system (Sket 2005). Being a short distance from the capital of Slovenia, the area provides an important supply of drinking water, water purification, flood control, agricultural land, forestry, tourism, and recreation. *Lake Cerknica and its environs* became a wetland of international importance in 2006 (https://rsis.ramsar.org/ris/1600).

The Baradla-Domica cave system (multiple designations: Transboundary Ramsar Site 2001, UNESCO MAB site, World Heritage Site), shared between Hungary and Slovakia, is a typical and, with its 25 kms, the largest subterranean hydrological system of the karst plateau in the territory of the two countries. The site is characterized by a permanent subterranean stream, lakes, and wetlands rich in subterranean fauna and dripstone formations. Its natural functioning plays a substantial hydrological, biological, and ecological role. The extended underground world of the Aggtelek and Slovak Karst provides a habitat for troglobite, troglophile, and trogloxene animals including endemic species (e.g., *Niphargus aggtelekas*), as well as species first described from this region. Its cultural importance is recognized in some very significant archaeological remains of Paleolithic and Neolithic occupancy. The most important archaeological sites are the settlements of the Neolithic Bükk culture both inside and at the cave entrance, with charcoal drawings unique in Central Europe. In addition to many artifacts in the cave system, a burial ground with human skeletons has been discovered in Baradla Cave (Sabol 2004). The importance of the karstic springs was recognized by local people already in the Middle Ages.

Consequently, the cave system is nowadays important for tourism, with guided tours by boat and on foot, a visitors' center, and a transborder nature trail (https://rsis.ramsar.org/ris/1051, https://rsis.ramsar.org/ris/1092).

There are more systems that have been researched across the temperate zone in Canada (Stokes et al. 2010) and USA (longest known cave system – the Mammoth Cave), other parts of Europe and Asia where the deepest cave was found in the West Caucasus Mountains (Krubera or Voronja Cave, 2197 m deep, http://www.caverbob.com/wdeep.htm).

Karst Wetlands in Tropics/Subtropics

In addition to an example from the Island of Bermuda and from Australia, two major areas of the tropical karst are considered here, Yucatan Peninsula and South East Asia. However, it has to be noted that there are more areas with these important systems of inland and coastal wetlands.

The Mid-Atlantic island of Bermuda is in the subtropics, but with a mild climate, and supports rich biodiversity, extensive coral reefs, small mangrove swamps, and karst caves (Anderson et al. 2001). The volcanic basement rocks are covered with limestone. There are more than 150 known caves, many of which contain anchialine pools (tidal, saltwater pools with subterranean connection to the ocean) in their interior. There are extensive labyrinths of passageways, underwater caves with endemic species, and large speleothems that must have formed during Ice Age periods. Most of the caves have inland entrances but often extend to the sea. There is a system of drowned caves with a number of endemic species, while cave mouths and sinks are also important for threatened and endemic plants and ferns (http://environment.bm/caves/?rq=Bermuda%20biodiversity%20country%20study. Accessed 20 Mar 2016). Additionally, fossils of existing and extinct species of birds, tortoise, and invertebrates have been found in these caves. Stygobitic (aquatic cave-adapted) species identified are largely threatened by constructions and quarrying, water pollution, solid waste dumping, and littering as well as vandalism (http://www.tamug.edu/cavebiology/BeCKIS/hotspot.html. Accessed 20 Mar 2016).

Mexico's Yucatan Peninsula has one of the most extensive karst aquifer systems worldwide (Bauer-Gottwein et al. 2011) with numerous caves, underground rivers, and karst features that extend to Guatemala and Belize. The most notable karst feature is the cenote, a term used by the Mayans for any subterranean chamber that contains permanent water, mentioned in several designated Ramsar sites (Cuba and Mexico). While some cenotes are vertical, water-filled shafts, others are caves that contain pools and underwater passageways in their interior. The world's largest known underwater caves are underground rivers carrying water to the Caribbean Sea along Yucatan's east coast. Remarkable features include unique, spring-fed groundwater-dependent ecosystems. In Riviera Maya there are a number of underwater cave systems that are important for their biodiversity and hydrology and are becoming increasingly better researched and understood. In the whole area of the Quintana Roo several hundreds of kilometers of cave passages have been explored and the number is constantly increasing. Horizontal passage development in most of the caves occurs along the halocline separating fresh from marine waters. The halocline is found at a depth of approximately 10 m in caves along the coast but increases to 60–70 m depths in caves as much as 60 km or more inland. Biological investigations of these caves have revealed the presence of rich stygobitic fauna that are primarily marine relict species. Tourism development depends on the natural values, but water extraction and water pollution are the main threats to these fragile karst wetlands. The complex hydrology of the Yucatan Peninsula has been thoroughly studied by Bauer-Gottwein et al. (2011). Additionally, the diversity of ecosystems and ecosystem services is summarized for three Ramsar sites from the Peninsula (see Ramsar Information Database: http://www.ramsar.org/sites-countries/the-ramsar-sites): Reserva de la Biosfera Los Petenes (Ramsar site 2004) is an area in the northwest coast of the Yucatan, located north of the city of Campeche. In 2008, Los Petenes-Ría Celestún site was included in the World Heritage tentative list due to the particular wetland characteristics of its ecosystem, consisting of islands of low seasonally flooded and/or mangrove forests associated with underwater springs from cenotes. It is a coastal corridor of wetlands with rich floristic and faunistic diversity, particularly avifauna represented by 304 species. The marine part of the site is significant for the seagrass beds, while inland the landscape is dominated by particular plant communities called *petenes* or islands of vegetation in shallow, muddy, brackish marshes fed by surrounding fresh water springs. Consequently, they appear as islands of ring-shaped rainforest areas. Giving rise to plant associations such as mangroves, including samples of low-flooded rainforest and tall rainforest, *petenes* are dependent on the quantity of fresh water nourishing them, the type of soil and microtopography. These plant communities found only in the Yucatan Peninsula, Florida and Cuba are of special interest and importance (http://whc.unesco.org/en/tentativelists/5396 and https://rsis.ramsar.org/ris/1354).

Anillo de Cenotes (Ramsar site 2009) is a complex of 99 cenotes (or sinkholes) that occur in an approximately 5 km wide belt with a radius of 90 km in the most northwestern part of the Yucatan peninsula, a zone of high permeability with regional-scale fracture zone (meaning that flow paths of the karst aquifer can extend from some 10 to 100s of km). This is the outline ring of an impact crater caused by a large meteor 65 million years ago, the Chicxulub Impact Crater. It is some 200 km in diameter with its impact center just off the Yucatán coastline. As a result, the surface layers of the Earth's crust were fractured and the ring configuration of the aquifer outcrops was formed, a worldwide unique water system known as Anillo de Cenotes ("ring of cenotes"). It is a complex system of steep-walled cenotes with large chambers and underground passages extending below the water table. The area is important for its rich biodiversity, particularly as resting grounds for waterbirds during their migration to the South and habitats of a number of endangered and threatened species such as the fish species *Ogilbia pearsei*, *Ophisternon infernale*, and *Poecilia velifera*. Additionally, it is important for endemic species of reptiles, such as the Yucatan box turtle *Terrapene carolina yucatana*, amphibians like the Yucatan mushroomtongue salamander *Bolitoglossa yucatana*, and birds, such as the Ridgway's rough-winged swallow *Stelgidopteryx ridgwayi*, the Yucatan jay

Cyanocorax yucatanicus, and the black catbird *Melanoptila glabirostris*. The area is also contributing to the Yucatan aquifer and providing freshwater for species and human use. However, due to human interference the main threats to the area include introduction of invasive alien species and environmental changes caused by tourism, including extensive extraction of water for tourist resorts, and lack of water management (Bauer-Gottwein et al. 2011; https://rsis.ramsar.org/ris/2043).

Sian Ka'an (Origin of the Sky – multiple designations: Ramsar Site 2003, World Heritage Site 1987, UNESCO MAB site 1986) is a wetland area along the 120 km barrier reef east coast of the Yucatán peninsula. It comprises tropical forests, mangroves, and marshes as well as the marine area. There are two large shallow bays surrounded by mangroves and numerous cenotes in a landscape of tropical deciduous forests. Typical forested island petenes are some tens of meters to up to a kilometer in diameter. The complex hydrological system provides good living conditions for a remarkably rich flora and fauna, including 320 bird species and five neotropical felines. Endangered species in the area include green *Chelonia mydas*, loggerhead *Caretta caretta*, hawksbill *Eretmochelys imbricata*, and leatherback *Dermochelys coriacea* turtles, American and Belize crocodiles *Crocodylus acutus* and *C. moreletii*, Baird's tapir *Tapirus bairdii*, jaguar *Panthera onca*, puma *Puma concolor*, American manatee *Trichechus manatus*, and sperm whales *Physeter catodon*. It is an archaeological site and part of the Riviera Maya being used as a tourism destination. Therefore, the main threats are due to overfishing, increasing tourism developments, forest fires, and introduction of invasive alien species. A management plan has been developed to guide activities and zoning as well as to involve local communities, governmental representatives, academia, and nongovernmental organizations in management. Several research activities have been undertaken, and a community program has been initiated to train locals as tourist guides in order to have tourism benefit local communities. However, in spite of multiple international protection and national Wildlife Protection Area designations, surface and subterranean wetlands are continuously being threatened mainly due to mass tourism and accompanying activities (https://rsis.ramsar.org/ris/1329).

Southeast Asia is a significant karst area with globally recognized karst phenomena including limestone towers on the surface and big stream caves. Karst wetlands are developed from China through Southeast Asia, particularly Vietnam, Laos, Thailand, Malaysia, and Philippines. Subterranean stream systems are mostly developed in horizontal caves. In China carbonate rocks cover some 910,000 km^2 with thousands of caves, mostly within an almost continuous carbonate outcrop of 500,000 km^2 in the south of the country. Of the 3,319 caves discovered, 60% are in southern China. The three provinces in the south of China are particularly recognized for their karst landscapes (Yunnan, Guizhou, and Guanxi Zhuang). The different karst phenomena and its hydrological characteristics have been described for the South China Karst by Knez et al. (2011b). In the Shilin area, there are estimated to be more than 100 caves in a region of high cave development in Yunnan Province, and over 70 caves were explored between 1998 and 2001. The majority of these are underground river caves contributing to 80% of the documented caves (Liu and Zhou 2011).

The northwestern part of the province includes the Three Parallel Rivers National Park in which the upper reaches of the three largest East Asian rivers flow in different directions to reach the sea: the (Chang Jiang) Yangtze River, the Lancang Jiang (Mekong) River, and the Nu Jiang (Salween) River close to the Tibetan high plateau. Shuilian Cave (the Water Curtain) in the area of the Chang Jiang River is laying within the large Three Parallel Rivers National Park and since 2003, a UNESCO World Heritage Site. It is one of the several karst springs occurring above the bottom of the river valley. The Shuilian cave has two entrances 70 m above the river, and the water flowing from the lower entrance forms a splashing, curtain-like waterfall falling several tens of meters down into the river (Knez et al. 2011a).

Further north and upstream in the Yangtze River catchment the Napahai Wetland Ramsar site (designated in 2004) is located at about 3,260 masl in the Chinese province of Yunnan. It is a seasonal karst wetland composed of different habitat types. The lake outflow is through the underground hydrological system flowing into the Jinsha River in the upper reaches of the Yangtze. The wetland is of particular significance as wintering site and staging post for numerous wintering birds, supporting over 70,000 birds annually and over 1% of the population of the black-necked crane *Grus nigricollis*. Birdwatching and ecotourism are importantly contributing to the local economy of the region. Overgrazing and logging in the neighboring areas can threaten the ecosystem services, but on the other hand, engaging local communities in the management of the reserve and activities can raise their awareness and improve their livelihoods (https://rsis.ramsar.org/ris/1440).

In Vietnam there are to date three UNESCO World Heritage Sites characterized also by typical karst landscapes and landforms including water caves (http://whc.unesco.org/en/list/). *The Ba Be National Park* and Ramsar site (designated in 2011) is not among them, although it is important for its hydrological characteristics and biodiversity. It is a natural mountain lake consisting of three lakes surrounded by Limestone Mountains between 500 and 600 masl. It is an important site for globally threatened species including Francois' langur *Trachypithecus francoisi*, big-headed turtle *Platysternon megacephalum*, and endemic species such as the Vietnamese salamander *Paramesotriton deloustali*. It is also an Important Bird Area particularly due to supporting more than the 1% threshold population of the endangered white-eared night heron *Gorsachius magnificus*, a species which also has a very restricted habitat range. Hydrologically it gathers waters from three main rivers and it substantially feeds the Nang River and its catchment. Ba Be Lake has an important provisioning service providing water for local use in addition to providing numerous ecosystem services such as flood control and tourism destination. Archaeological evidence found in the area's caves indicates human habitation since the late Pleistocene some 20,000 years ago. Human pressures are the main threats to the karst system and include infrastructure development, hunting of birds, pollution, forest loss for agriculture, and tourism development (https://rsis.ramsar.org/ris/1938). Another remarkable Vietnamese karst area is the Bai Tu Long National Park located next to the Holong Bay World Heritage Site. It is a marine protected area in the northeast coastal zone, covering an area of 15,783 ha and one of the protected areas with the highest biodiversity in Vietnam including both, terrestrial and aquatic zones.

The ecosystems included in the protected area (rain broad leaves forest, limestone forest, littoral forest, coral area, and shallow waters) provide habitat for over 1,900 flora and fauna species. Its rich biological and cultural diversity plays an important role in the livelihoods of local communities – providing opportunities for ecotourism, aquaculture development, as well as offshore fishing. The park also serves as a buffer for storm surges that affect coastal settlements (ASEAN Centre for Biodiversity: http://chm.aseanbiodiversity.org/and http://vietnamnationalpark.org/2743/bai-tu-long-national-park/).

Tower karst is also well developed in southern Thailand. The Ao Phang-Nga Marine National Park and ASEAN Heritage Park is one of the most frequently visited marine national parks in Thailand. It is also a coastal area recognized for its rich folklore, prehistoric rock arts, and natural beauty. The park extends over an area of about 400 km^2 consisting of a coastal forest and a series of karst limestone towers flooded by the sea to form some islands with high cliffs, rock overhangs, caves, coral gardens, and scrubs (ASEAN Centre for Biodiversity: http://chm.aseanbiodiversity.org/).

Borneo has more caves important for their biodiversity value, particularly huge colonies of bats and swifts. The *Gunung Mulu* National Park (Borneo, Malaysia) contains large underground rivers and almost 300 km of explored caves (Clearwater Cave in Sarawak).

Puerto Princesa Subterranean River National Park is another nice example of a complex karst wetland system combining different wetland types. The protected area is located on the island of Palawan (Philippines). It is an example of surface subterranean as well as inland coastal interfaces and a distinctive site bringing together different ecosystems, including surface karst landscape in limestone with a complex cave system, mangrove forests, lowland tropical rainforests, and freshwater swamps. It is particularly rich in biodiversity with some 800 plant and 233 animal species, including some 15 endemic species of birds (https://rsis.ramsar.org/ris/2084). Additionally, it is a habitat of numerous colonies of bats and swallows and their guano production together with the biological material brought into the cave by the river develops a complex trophic network. One of the special characteristics is the Cabayugan River draining the area of Mt. Bloomfield before disappearing under Mount Saint Paul and flowing some 8.2 km underground to reach the sea. Thus, the surface water is diverted to the sea through huge flooded galleries. Due to opening directly to the sea, the tides affect a large part of the cave and reach as far as 6 km inside the mountain. The lower reaches of the river system may be considered a classic example of an underground estuary (De Vivo et al. 2013). Together with the Babuyan River, the longest river in Palawan, it is one of the main water sources for local communities for domestic and agricultural uses. In addition to its natural values the site has remarkable cultural (archaeological) and socioeconomic significance. Increasingly, it is a major ecotourism destination, and community-based sustainable ecotourism has been started to involve the local communities in park management as well as to generate income. Human activities, namely, tourism, forest clearing, and agriculture are considered the main threats in the site. The area has been a Ramsar site since 2012, in addition to a number of other significant designations including UNESCO World Heritage Site (1999, http://whc.

unesco.org/en/list/652) and being part of a large Palawan Biosphere Reserve (1990, http://www.unesco.org/new/en/natural-sciences/environment/ecological-sciences/biosphere-reserves/asia-and-the-pacific/philippines/palawan/).

Australia has several significant karst areas and Piccaninnie Ponds (South Australia) is a Ramsar site (2012) with karst wetlands located in one such area. The site is a complex of different wetland types fed by the surface runoff from groundwater discharge from the series of karst springs. The very clear water rises from deep below, and the largest and the deepest spring is more than 110 m deep. In addition to a series of these rising springs, there are also several substantial groundwater springs in the beach area. The geomorphic and hydrological features of the site produce a complex and biologically diverse ecosystem and wetland types (karst wetland, fen wetland, and beach springs) that support rich biodiversity including some internationally important species of fauna and flora such as the globally threatened Australasian bittern *Botaurus poiciloptilus* and orange-bellied parrot *Neohpema chrysogaster*. The karst springs also support unique macrophyte and algal associations, with macrophyte growth to 15 m depth mainly due to the spring water clarity. Human activities include recreation, tourism, snorkeling, camping, and research, while the surrounding areas are used for livestock grazing. Some main threats to the wetlands may be caused by ground water extraction in the surrounding area, introduction of invasive alien species, and climate change (https://rsis.ramsar.org/ris/2136).

There are more remarkable karst wetlands of high biodiversity value and providing numerous ecosystem services that have been designated all over the world and which can be explored on the Ramsar Site Information Service (https://rsis.ramsar.org/). For example, Madagascar's Lake Tsimanampetsotsa (Ramsar site 1998) with its underground caves and rivers bound to the lake on its eastern site is an important source of clean water supply and maintains general hydrological balance in addition to its rich biodiversity. In India, Renuka Wetland (Ramsar site 2005) with freshwater springs fed by a small stream from the lower Himalayan is home to many animal species, including ichthyofauna and birds. But it has also a highly religious significance, visited by thousands of pilgrims and tourists.

Threats and Challenges

Karst ecosystems are very fragile and particularly subterranean ones as they are dependent on energy flows transmitted by water. Although the subterranean systems are, in many, cases still well preserved, they are very sensitive to increasing development pressures on the surface directly above or in the river catchment and thus are becoming endangered. Tourism development is considered one of the main threats to karst wetlands, particularly in the tropics. The pressures to these fragile systems are both direct (visitors to caves and wetlands, researchers) and indirect including pollution (e.g., dumping of solid waste, sewage), development of infrastructure, invasive alien species, water extraction, and retention in reservoirs.

In general terms, many "living" karst areas are wetlands, surface, or subterranean and provide valued ecosystem services. Yet, many are threatened by economic developments (tourism developments, agriculture, fisheries) that depend on the same wetland resources. Therefore, the main challenge for the future is to safeguard their ecosystem services and reduce the development pressures. Furthermore, much of the subterranean world remains to be explored or researched, making it even more important for conservation, protection, and wise use of wetland resources. Following the Ramsar Convention objectives, subterranean and complex karst wetland systems can be maintained and protected and further included in the List of Wetlands of International Importance. Moreover, the convention is being implemented in 169 countries and is an important mechanism to promote maintenance and sustainable use of karst water resources by raising awareness to reduce development pressures and to conserve their values for nature and human development.

References

Anderson C, De Silva H, Furbert J, Glasspool A, Rodrigues L, Sterrer W, Ward J. Bermuda biodiversity country study. Bermuda Zool Soc. 2001:103. Available from: http://static1.squarespace.com/static/501134e9c4aa430673203999/t/55393543e4b0b0c454552cb2/1429812547370/Bermuda+Biodiversity+Country+Study.pdf. Accessed 30 Apr 2016.

Bauer-Gottwein P, Gondwe BRN, Charvet G, Marin LE, Rebolledo-Vieyra M, Merediz-Alonso G. Review: The Yucatán Peninsula karst aquifer. Mex Hydrol J. 2011;19:507–24.

Beltram G. Ramsar sites – wetlands of international importance. In: Gunn J, editor. Encyclopedia of caves and karst science. London: Fitzroy Dearborn; 2004. p. 1322–6.

Beltram G, Sket B. Subterranean habitats as wetlands of international importance. In: Abstract Book, 20th International Conference on Subterranean Biology, 2010. 29 Aug–3 Sept; Postojna; 2010. p. 79.

Cerkvenik R, Kranjc A, Mihevc A. Karst wetlands in the Dinaric karst. In: Finlayson CM, Milton GR, Prentice RC, Davidson NC, editors. The wetland book II: distribution, description, and conservation. Dordrecht: Springer; 2018.

De Vivo A, Piccini L, Forti P, Badino G. Some scientific features of the Puerto Princesa underground River: one of the 7 wonders of nature (Palawan, Philippines). In: Fillipi M, Bosák P, editors. Proceedings of the 16th international congress of speleology, July 21–28, 2013, Brno, vol. 3. Praha: Czech Speleological Society; 2013. p. 35–41.

Filippi M, Bosák P, editors. Proceedings of the 16th international congress of speleology, July 21–28, 2013, Brno. Praha: Czech Speleological Society; 2013.

Ford D, Williams P. Karst hydrogeology and geomorphology. 2nd ed. Chichester: Wiley; 2007, 576 p.

Knez M, Kogovšek J, Kranjc A, Liu H, Slabe T, Petrič M. Shuilian cave in the upper region of the Chang Jiang River. In: Knez M, Liu H, Slabe T, editors. South China karst II, Carsologica, vol. 12. Ljubljana: Postojna Karst Research Institute ZRC-SAZU, ZRC Publishing; 2011a. p. 125–39.

Knez M, Liu H, Slabe T, editors. South China karst II, Carsologica, vol. 12. Ljubljana: Postojna Karst Research Institute ZRC-SAZU, ZRC Publishing; 2011b. 237 p.

Kranjc A, editor. Kras: Slovene classical karst. Ljubljana: Postojna Karst Research Institute ZRC-SAZU; 1997. 254 p.

Kunaver J, Drobnjak B, Klemenčič MM, Lovrenčak F, Luževič M, Pak M, Senegačnik J, Buser S. Obča geografija za 1. letnik srednjih šol. Ljubljana: DZS; 1995.

Liu H, Zhou Y. Characteristics of the cave development in the Shilin area. In: Knez M, Liu H, Slabe T, editors. South China karst II, Carsologica, vol. 12. Ljubljana: Postojna Karst Research Institute ZRC-SAZU, ZRC Publishung; 2011. p. 99–111.

Mihevc A, editor. Kras: Voda in življenje v kamniti pokrajini – Water and Life in a Rocky Landscape, Project Aquadapt. Ljubljana: ZRC SAZU; 2005. 564 p.

Prelovšek M. The dynamics of the present-day speleogenetic processes in the stream caves of Slovenia, Carsologica, vol. 15. Postojna – Ljubljana: Karst Research Institute ZRC-SAZU; 2012. 145 p.

Ramsar Convention Secretariat. Designating Ramsar sites: strategic framework and guidelines for the future development of the list of wetlands of international importance. In: Ramsar handbooks for the wise use of wetlands, vol. 17. 4th ed. Gland: Ramsar Convention Secretariat; 2010.

Sabol M. Aggtelek caves, Hungary – Slovakia: archeology. In: Gunn J, editor. Encyclopedia of caves and karst science. London: Fitzroy Dearborn; 2004. p. 28–9.

Sket B. Kaj so podzemeljska mokrišča in zakaj so vredna varstva – What are subterranean wetlands and why are they worth conserving. In: Beltram G, editor. Novi izzivi za ohranjanje mokrišč v 21. stoletju: Ramsarska konvencija in slovenska mokrišča. Ljubljana: Ministrstvo za okolje in prostor; 2005. p. 19–25.

Sket B. Subterranean (hypogean) habitats in karst and their fauna. In: Finlayson CM, Milton GR, Prentice RC, Davidson NC, editors. The wetland book II: distribution, description, and conservation. Dordrecht: Springer; 2018.

Stokes T, Griffiths P, Ramsey C. Karst geomorphology, hydrology and management. In: Pike RG, Redding TE, Moore RD, Winkler RD, Bladon KD, editors. Compendium of forest hydrology and geomorphology in British Columbia, Land management handbook, vol. 66. Victoria: Government Publications Services/BC Government and FORREX; 2010. p. 373–400. www.for.gov.bc.ca/hfd/pubs/Docs/Lmh/Lmh66.htm

Watson J, Hamilton-Smith E, Gillieson D, Kiernan K, editors. Guidelines for cave and karst protection. Gland/Cambridge, UK: IUCN; 1997, 63pp.

Williams P. World heritage caves and karst. Gland: IUCN; 2008. 57 p.

Wong T, Hemilgton-Smith E, Chape S, Frederich H, editors. Proceedings of the Asia-Pacific Forum on karst ecosystems and world heritage, Guang Mulu National Park World Heritage Area, Sarawak Malaysia, 26–30 May 2001.

Subterranean (Hypogean) Habitats in Karst and Their Fauna

23

Boris Sket

Contents

Introduction	332
Classification of Subterranean (Hypogean) Habitats	332
Karst, Caves, and Interstitial	332
Environment	334
Inhabitants	335
Composition of Subterranean Fauna	336
Dinaric Karst as a World Hotspot, Postojna-Planina Cave System and Cave Vjetrenica as its Local Hotspots	338
Threats and Future Challenges	342
References	343

Abstract

Subterranean karst systems are wetlands with a generally poor biodiversity, but species typically are ecologically specialized with a high level of endemism and heightened risk of extinction. A number of diverse habitats may develop in the three zones characterized by the extent and permanence of water within the karst voids. The subterranean environment is a relatively closed space, with restricted connections to other ecosystems, and primarily dependent upon organic material dispersed in percolating water through the fissured ceiling or carried with streams entering through sinkholes. Permanent residents of subterranean habitats (troglobionts) exhibit morphological and physiological changes that may include disappearance of skin and eye pigmentation, reduction of eyes, elongation of body appendages, and elongation of the body. Beetles and crustaceans (especially

B. Sket (✉)
Oddelek za biologijo, Biotehniška fakulteta, Univerza v Ljubljani, Ljubljana, Slovenia
e-mail: boris.sket@bf.uni-lj.si

© Springer Science+Business Media B.V., part of Springer Nature 2018
C. M. Finlayson et al. (eds.), *The Wetland Book*,
https://doi.org/10.1007/978-94-007-4001-3_241

Copepoda and Amphipoda) are by far the richest groups of terrestrial and aquatic troglobionts, respectively, but diversity is still poorly described for many areas. Subterranean wetlands and their inhabitants remain poorly understood, and habitat destruction is the principal threat to hypogean fauna.

Keywords
Caves · Interstitial · Biodiversity · Endemism · Wetlands

Introduction

The surface of the karst landscape is mainly dry since its most characteristic attribute is the underground circulation of waters. For the same reason, the whole underground of karst in a moderately humid country is – if not even drowned – damp or wet. Thus, all subterranean karst habitats are wetlands. Due to severe living conditions, caused directly or indirectly by lack of light, subterranean faunas are generally diversity poor, reaching in best cases some low percent of the total local fauna; and their populations are small or sparse. Nevertheless, in some areas, cave fauna may be an important indicator of the habitat health or scientifically important for historical biogeography or evolutionary investigations. It is also sensitive throughout since subterranean species are ecologically specialized and are mainly endemics disposed to extinction. In the last six decades, this environment has been richly presented in a number of "secondary" publications (Vandel 1964; Gunn 2004; White and Culver 2012).

Classification of Subterranean (Hypogean) Habitats

A cave, let alone a cave system, may include a number of habitats, similar to any epigean batch of land. Habitats (Table 1) may be characterized by (1) the large feature physiognomy; (2) the trophic base and small scale physiognomy, which are together (in aphotic habitats) an equivalent of the supporting "plant-community composition" on the surface; and (3) biogeographical or ecological factors supporting a certain animal-community composition.

Karst, Caves, and Interstitial

Karst can be defined as an area of soluble rock where waters are mainly flowing below the surface and where characteristic epigean and hypogean karst phenomena are developed. Hypogean karst phenomena are generally and popularly called "caves," but ecologically, extensive systems of narrow fissures in the karst massifs are equally important; and numerous seemingly isolated caves may in fact be interconnected by fissures into extensive systems.

In principle, a karst void begins with precipitation falling onto a soluble rock, mainly limestone, which has been previously fractured by tectonic movement.

Table 1 Description and classification of subterranean (hypogean) habitats and associated fauna

	Habitat; present in:	Usually inhabited by
1	*Terrestrial hypogean habitats*	
1.1	**Terrestrial interstitial** (= superficial hypogean habitat, MSS)	
1.1.1	Terrestrial interstitial in carbonate (karst) territories; worldwide in karst territories	Troglobionts and edaphic species
1.1.2	Terrestrial interstitial in non-karst territories; worldwide in rocky areas	Mainly edaphic species
1.2	**Entrance (twilight) cave habitats**; worldwide in karst or volcanic territories	Trogloxene and subtroglophile invertebrate communities (Rhaphidophoridae, Opiliones, Gastropoda, etc.) Trogloxene vertebrates
1.3	**Dark cave habitats**; habitat worldwide in karst or volcanic regions	(Fauna only in temperate to tropical areas)
1.3.1	Dark cave habitats with high input of allochthonous organic matter	Rich edaphic invertebrate fauna and/or (locally) rich troglobiotic fauna
1.3.2	Energetically poor dark cave habitats	Scarce to rich troglobiotic faunas
2	*Aquatic hypogean habitats*	
2.1	**Interstitial waters** (waters in unconsolidated sediments: sand, gravel); worldwide	Characteristically small and slender troglomorphic animals
2.1.1	Marine and coastal interstitial habitats	
2.1.2	Freshwater interstitial habitats (hyporheic and phreatic); worldwide on land	
2.2	**Waters in porous rocks (cave waters)**	
2.2.1	Subterranean parts of permanent sinking rivers; worldwide in karst areas	Mixed communities of trogloxenes to troglobionts, changing along the channel
2.2.2	Cave waters with autochthonous energy resources (sulfurous, hydrocarbon); worldwide, very scarcely	Chemoautotrophic microbiota and very diverse, sometimes rich troglobiotic faunas
2.2.3	Percolation waters; worldwide in karst, less common in volcanic rock	Mostly highly specialized troglomorphic species, Crustacea predominate
2.2.4	Anchihaline habitats; in warm temperate to tropic karst or volcanic sea shores	Troglobionts of marine origin, either euryhaline or limnic
2.2.5	Thermal hypogean waters; worldwide, scarcely	Very scarce fauna of relic troglobionts (Crustacea, Gastropoda), denser only in springs

The precipitation reacts with CO_2, either from the atmosphere or from the soil above the massifs through which it passes, to form carbonic acid. Upon penetrating the fractures, the weak carbonic acid widens them by (chemical) dissolution. Through millions of years, the fractures turn into systems of wider fissures and caves. At the bottom of the massif, dammed up by surrounding insoluble rock or by the sea, the water accumulates and flows out in horizontal direction. In this basal level of the karst massif, the accumulated and horizontally moving water produces horizontal channels ending in springs. The preceding is a rudimentary description and very variable and complex systems of hypogean voids can develop under different conditions. The process is generally widening of voids combined with dissolution of the rock on its surface; the process permanently changes the shape of both the epigean and hypogean landscape and finally results in the total dissolution and disappearance of the soluble rock massif. The karst is a transient formation, but its cycle may extend through millions of years.

Thus, the cave system principally consists of three layers (usually called zones). The inferior layer, permanently filled with water, is called the saturated or phreatic zone. The uppermost layer, where the percolated water is moving vertically, is called the unsaturated or vadose zone or the aeration zone since the voids are mainly not filled with water. However the aeration zone in non-arid regions is damp or wet and for short periods can be partly filled by excessive percolated water. Between both zones there is a transitional layer of voids which are flooded only during heavy precipitations, so-called epiphreatic or the floodwater zone. A number of diverse habitats may develop in these three layers of the karst underground.

> "Vulcanokarst" are sometimes extensive volcanic landscapes, mainly lava fields, containing systems of voids. The origin of volcanic voids (e.g., "lava tubes") is very different, but their ecology and their faunas may be similar to those in karst voids. Volcanic caves in the Canary Islands are particularly rich in troglobiotic fauna (Oromi and Martin 1992).

Another system of subterranean habitats exists which is even more common outside karst than inside it. Interstitial is a three-dimensional netlike system of channels between grains of nonconsolidated sediments, such as that which occurs in sand, gravel, and rubble. The "dry" terrestrial interstitial has been known as MSS ("milieu souterraine superficielle"). The aquatic interstitial is developed as sand- or gravel banks along rivers as well as hidden below the bottom cover of large land depressions. The width of the interstitial channels is defined by the size of the sediment particles.

The extent of the subterranean space is difficult to predict. For example, more than 10,000 karst caves have been investigated in Slovenia with an area of ca 20,000 km^2. This number in fact equals the number of entrances to caves that penetrate only some meters and up to 20 km. The number of cave entrances is not necessarily an accurate indication of the extent of the subterranean space as 640 km of corridors have been measured in the Mammoth – Flint Ridge Cave System in Kentucky (USA) with less than 20 entrances.

Environment

The subterranean environment (Sket 1996a) is a relatively closed space, with restricted connections to other ecosystems. The most obvious consequence of this closure is the absence of daily light changes and permanent and total darkness.

Owing to the moderating influence of rock masses surrounding the voids, meteorological fluctuations are strongly reduced. Although commonly considered to be low, cave temperatures are in fact close to average yearly surface values, but without daily and with strongly reduced yearly fluctuations. Air circulation may be very limited resulting in very high relative air humidity, especially close to the cave floor or its walls – the layer in which the small cave animals are living. The air humidity is close to 100% and the substratum is normally permanently damp or wet. Therefore, the boundary between subterranean aquatic and terrestrial habitats is not sharply defined.

Since photoautotrophic producers ("green plants") are absent due to lack of light, and chemoautotrophs are – except in rare cases – scarce or relatively unproductive, autogenous food production is missing. Organic substances can only be imported, mainly by water, either through the fissured ceiling by percolating water or in higher quantities by sinking streams through sinkholes (swallow holes, ponors). Effective mediators of food are also subtroglophile animals, feeding outside but spending parts of their lives underground; bats (Chiroptera) are generally the most present vector for this type of organic input.

All these features of the subterranean environment are the most explicit in deeper caves. Closer to the cave entrances, in the twilight zone and in the transitional entrance zone, conditions are more similar to the epigean ones. In addition, water has a high heat capacity and in sinking rivers this heat may persist to influence the surrounding air and bottom temperature kilometers far underground.

Special conditions occur in caves at sea coasts, where the influence of the sea may be perceived. The water body in these anchihaline (or anchialine) caves is brackish with a strong vertical salinity gradient. It may be even limnic at the water surface and euhaline at depth. At the halocline there is a steep transition between salinities; this layer is usually nearly depleted of oxygen and rich in H_2S, conditions that may also occur below the halocline as the salinity and density stratification prevents mixing of the water.

Inhabitants

Karst develops under appropriate geological conditions all over the world. Wherever the karst is old enough and the climates not too cold, its subterranean habitats are inhabited by heterotrophic organisms. Owing to unfavorable living conditions and the restricted physical accessibility of the habitats, the biotic diversity is much lower than in the epigean realm. The number of species of aquatic subterranean fauna in Europe is comparatively rich, approximately 8% of the total freshwater fauna. While there is a greater number of terrestrial subterranean faunal species, their contribution to the total terrestrial fauna is comparatively much poorer.

Microbiota are present in moderate densities in the soil as well as the water. But some comparatively massive, obvious assemblages exist in caves. Important communities of autotrophic (White and Culver 2012), mainly sulfur-oxidizing microbiota, may develop in a few caves with mineral water influx. In the Movile cave (Dobrogea, Romania), microbial mats are the primary energy base for a very rich aquatic and terrestrial fauna. In the Villa Luz cave (Mexico), bacteria form mucous stalactite-like "snottites." The dense overgrowths of sprout like bacterial colonies are probably heterotrophic, e.g., a colony of *Troglogloea* in only one streamlet of Vjetrenica (Hercegovina; Kostanjšek et al. 2013). Wet walls may be covered by extensive, crumby, gaily colored (yellow, white, pink, gray, bluish) microbial colonies where Actinobacteria prevail (Pašić et al. 2009); a covering of condensed water droplets makes them shine gold or silver.

The animals occurring in caves may be ecologically classified according to the classification by Ruffo (1957) and Barr (1968), as summarized and adapted by Sket (2008). Trogloxenes or accidentals are species without any particular relation to the subterranean environment, occurring there accidentally. Subtroglophiles are regularly occurring in caves but explicitly bound for some life functions to epigean environment; most of them are feeding outside (like bats). Eutroglophiles are in principle epigean species, able to build permanent subterranean populations. Troglobionts are specialized to the subterranean life, normally not present outside. A eutroglophile species may include some troglobiotic races-subspecies. Aquatic species may be designated by the prefix stygo- instead of troglo-. Eventually, a permanent cave population changes morphologically and physiologically. The most general troglomorphies are the disappearance of skin and eye pigmentation, reduction of eyes, elongation of body appendages, and elongation (enlargement) of the body.

Composition of Subterranean Fauna

While no general review of terrestrial cave fauna exists, we can compare against numbers of aquatic troglobiotic (cave and interstitial) species globally in published accounts over the past century (Botosaneanu 1986; Sket 1996b). The numbers of aquatic troglobiotic species (or stygobionts) in the faunistically richest areas are summarized in Table 2.

The published accounts of numbers of terrestrial cave fauna have been steadily growing. Sket et al. (2004) increased the Dinaric list to ca. 480 aquatic troglobionts (stygobionts) and 790 terrestrial troglobionts. In the same period, the troglobiotic fauna of the USA and Canada included 420 aquatic and 930 terrestrial troglobionts (Peck 1998; Culver et al. 2003). However, the interstitial fauna in America has not been well studied and the usually rich group of Copepoda is believed to be unnaturally poorly represented in the lists. Thus in biotically rich karst regions, the true relation of terrestrial species number seems to be the approximately 1.5 number of the aquatic species. Recently, additional subterranean fauna were recorded in Australia, the Australian fauna of groundwaters, mainly in calcretes, being particularly plentiful (Guzik et al. 2010).

By far the richest group of terrestrial troglobionts are beetles (Coleoptera) with slightly less than 40% of species in both North America and in Dinaric karst. More than 100 species also known for of pseudoscorpions (Pseudoscorpiones), spiders (Araneae), and millipedes (Diplopoda). Other insects are disproportionately poorly represented, although in the entrance zones trogloxene insects and subtroglophile

Table 2 Numbers of aquatic troglobiotic species (or stygobionts) in the richest areas according to data in Botosaneanu (1986)

Region by Botosaneanu 1986	Surface area (km^2)	Number of aquatic troglobiotic species
Dinarides	117	388
Padano-Alpine District	49	105
W Macedonia	14	100
Rhodano-Lotharingian Province	141	167
Pyrenean-Aquitanian Province	168	200
Appalachian Highlands	1,100	152
Interior Low Plateaus	124	51
Edwards Plateau	198	41
Mexico without Yucatan	1,930	99
Cuba	114	87

crickets exhibit increased diversity. In the water, crustaceans are by far the richest group, especially by Copepoda and Amphipoda. The very low number of copepods recorded in America is believed to be certainly an artifact of insufficient searching. However, the very high number of snails (Gastropoda) in the Dinaric karst is an exception not repeated in any other area.

The faunal composition in some volcanic areas, e.g., Hawaiian Islands and in Islas Canarias, is very different than in European and the North American karsts (Oromi and Martin 1992). Peculiarities are troglobiotic planthoppers (Hemiptera: Fulgoroidea), earwigs (Dermaptera), and even moths (Lepidoptera: Noctuidae).

It is interesting to note that globally, in troglobiotically poor areas, shrimps (Crustacea: Caridea) are regularly represented as are troglobiotic fishes (but not in Europe). The only European troglobiotic vertebrate is the Dinaric salamander *Proteus anguinus* (family Proteidae). *Proteus* was the first troglobiotic animal scientifically described. A number of troglobiotic salamander species are present in North American caves, mainly of the family Plethodontidae.

It has been estimated (Guzik et al. 2010) that the troglobiotic fauna of the western half of Australia should consist of 4,140 species, of which only 10% are described and an additional 9% are not described but recognized by morphological or molecular approaches. A big part of this fauna inhabits calcretes in arid river valleys with a structurally poorly understood aquifer but probably of an aberrant karst character. The Pilbara region (502,000 km^2) has 78 described aquatic troglobionts and an estimated 500–550 undescribed species. The leading groups in the western half of the continent are Copepoda, Amphipoda and diving beetles (Coleoptera: Dytiscidae); the latter group is exceptional since in no other area in the World have more than approximately 10 species of aquatic beetles been found. As many as 98 dytiscid species are described and 510 are estimated from Australia. Of terrestrial animals, arachnids prevail among the described species, but insects and crustaceans (Isopoda: Oniscidea) are estimated to have higher species numbers.

Dinaric Karst as a World Hotspot, Postojna-Planina Cave System and Cave Vjetrenica as its Local Hotspots

Entrance parts of caves are a resting place of a number of trogloxene or subtroglophile insects that may occur in high numbers. These can include different dipterans, e.g., *Limonia* and *Culex*, and some lepidopterans occur regularly, e.g., *Triphosa* and *Scoliopteryx*. Bat colonies may be ecologically very important as food vectors, sometimes producing deep deposits of guano. Among invertebrates, cave crickets (*Troglophilus* and *Dolichopoda*, family Rhaphidophoridae) feeding outside caves have a function similar to bats as a food vector. The taxonomic composition of the troglobiotic fauna is presented in Table 3.

Beetles (Coleoptera) are the most diverse group of terrestrial troglobionts. Mainly 2–7 mm long and colored dark brown, they belong to three families: Cholevidae (Leptodirinae), Carabidae (Trechinae), and Pselaphidae. Leptodirinae, with close to 200 Dinaric species, are morphologically very diverse. The less troglomorphic

Table 3 Composition of the troglobiotic fauna in the Western Balkans (mainly Dinarides, as presented in Sket et al. (2004))

Terrestrial troglobiotic fauna	Number of species	Aquatic troglobiotic fauna	Number of species
Land planarians (Tricladida)	2	Sponges (Porifera)	1
Snails (Gastropoda)	36	Temnocephalans (Temnocephalida)	13
Spiders (Araneae)	97	Planarians (Tricladida)	12
Pseudoscorpions (Pseudoscorpiones)	100	Cnidarians (Hydrozoa)	1
Harvestmen (Opiliones)	16	Snails (Gastropoda)	148
Microwhip scorpions (Palpigradi)	3	Clams (Bivalvia)	1
Woodlice (Isopoda)	70	Nemerteans (Nemertini)	1
Centipedes (Chilopoda)	28	Polychaetes (Polychaeta)	3
Millipedes (Diplopoda)	71	Oligochaetes (Oligochaeta)	22
Diplurans (Diplura)	4	Leeches (Hirudinea)	4
Springtails (Collembola)	27	Water fleas (Cladocera)	4
Bristletails (Thysanura)	1	Copepods (Copepoda)	113
Beetles (Coleoptera)	326	Ostracods (Ostracoda)	36
Dipterans (Diptera)	2	Shrimps (Decapoda)	4
		Thermosbaenacea	2
		Bathynellacea	2
		Possum shrimps (Mysidacea)	1
		Isopods (Isopoda)	41
		Amphipods (Amphipoda)	107
		Salamanders (Amphibia)	1
Total species	783		517

Fig. 1 The beetle *Leptodirus hochenwartii* Schmidt was the first discovered and scientifically described troglobiotic invertebrate. It was described in 1832, found in the cave of Postojna, in Dinaric Slovenia (Photo credit: B. Sket © Rights remain with the author)

species are small and egg-shaped ("bathyscioid"). With increasing troglomorphy they became bigger and spindle-shaped ("pholeuonoid") progressing finally to a long, cylindrical prothorax and egg-shaped to globular abdomen ("leptodiroid"); the transformation of the trunk is accompanied by a progressive lengthening of the legs and antennae. Outside Dinaric karst, only the North American *Glacicavicola* can be characterized as leptodiroid. *Leptodirus* was the second troglobiotic animal and the first invertebrate ever discovered. Trechinae are less diverse and their troglomorphy is expressed in narrowing of the prothorax and slight elongation of appendages. The tiny Pselaphidae with their short elytra occur throughout but are represented only by a few tens of species. The big majority of beetle genera from Dinaric caves are endemic to the area and entirely limited to troglobiotic species. It was estimated that in a 20 × 20 km quadrat up to 30 or some more species of beetles may occur (Zagmajster et al. 2008; Fig. 1).

Many of the nearly 100 pseudoscorpion species have been only found in single caves; a molecular analysis would probably reduce this number remarkably. Most of the species belong to the genera *Neobisium* and *Chthonius*, and species belonging to the former are bigger than the latter; both genera are also represented in epigean habitats, but troglobiotic species are eyeless, slightly larger, more slender, and with very slender and long appendages. The number of spider species (Araneae) is similar; most belong to the rich genus *Troglohyphantes* (of the huge family Linyphiidae) or to the entirely troglobiotic and endemic genera of Dysderidae. The tiny *Troglohyphantes* inhabit and are adapted to all the chain of habitats between the forest bottoms to deep inside caves. Troglobiotic woodlice belong mainly to the family Trichoniscidae; many of them are of amphibious habits and members of the genus *Titanethes* are giants (ca. 15 mm) for this family and might be the most commonly encountered terrestrial troglobiont of the region. Of the millipedes (Diplopoda), *Brachydesmus* spp. are those most often encountered and most of these are probably soil-inhabiting animals not bound to caves. Common also are tiny springtails, represented by some troglobiotic and some trogloxene (edaphic) species. Snails are mainly minuscule *Zospeum* spp. of millimeter size category, but flat shells of some *Aegopis* spp. and relatives may reach ca 20 mm in diameter.

Two habitat types are particularly wet and can possibly be considered semi-terrestrial or even semiaquatic: ice caves and the cave hygropetric. If entrance parts of caves are properly shaped (sack-like), cold winter air penetrates comparatively deep inside to cause freezing of the percolating water. If the quantity of the ice is so big that it cannot entirely melt in the summer, this ice will grow gradually into a small subterranean glacier. The ice cave has to be sack-like, located on the shady slope of the mountain with its entrance shaded by forest. The shape of such a cave allows also comparatively abundant debris input from the surface. The vicinity of the ice mass is always wet from melting. In such conditions, we usually encounter rich assemblages of troglobiotic and eutroglophile beetles, and some species seem even to be specialized to these cold habitats (e.g., leptodiroid *Astagobius* species).

The cave hygropetric (Sket 2004) are mainly sintered walls (i.e., rocky wall coated by a layer of crystalline calcite) overflown by a thin layer of water. The most characteristic inhabitants are wading leptodirine beetles of the pholeuonoid body with some particular details in morphology: very large claws, knee-shaped mandibles, and densely setaceous mouthparts. These beetles belong to a number of genera and are probably not closely related. We know nothing about their life cycles and how they feed is not clearly understood. Although ecologically similar conditions are very common in karst caves, the cave hygropetric fauna has only been described from some places in Dinarides and in Italian SE Alps.

If we divide crustaceans into lower taxa (orders approximately), the snails are the richest group of aquatic animals in Western Balkans. These are generally animals with a ca 2 mm wide or high shell, belonging to the family Hydrobiidae. The shell may be ovoid like in *Belgrandiella*, wide conical like in some *Hauffenia*, but also totally flat like *Hadziella*, or narrowly conical like *Iglica*. All types of subterranean waters – from the narrowest fissures and interstitia to cave "lakes" – are inhabited by very numerous tiny copepods (Copepoda). The same is true for amphipods. Very important is *Niphargus*, the richest nominal freshwater genus of amphipods. Its species are very diverse, reaching from 2 to 30 mm, some 10 species inhabit epigean waters, while ca. 300 are in hypogean waters of Europe and Middle East. All species are eyeless (epigean included) and up to 9 species have been found in the same locality. In the Dinaric karst, some species inhabit brackish water of anchihaline caves. Very prominent members of Isopoda are *Monolistra* spp., rolling into a ball if disturbed, some with long spines on their backs. Their distribution is limited to karsts of Dinarides and SE Alps. Thermosbaenacea are represented by only two species, one in brackish waters of anchihaline caves and one inland. In Adriatic anchihaline caves, several *Niphargus* spp. occur in place of shrimps which are so characteristic in tropical anchihaline waters (Fig. 2).

The Dinaric cave waters are also inhabited by some "unique" species (Sket 1999). The salamander *Proteus anguinus* is the only European troglobiotic vertebrate and the first troglobiotic animal discovered. *Marifugia cavatica* is the only cave tube worm; it is related to the circum-global brackish-water inhabitant *Ficopomatus enigmaticus* and member of an otherwise marine family (Serpulidae). To a generally marine family (Bougainvilliidae) belongs also the only troglobiotic cnidarian

Fig. 2 *Typhlogammarus mrazeki* Schäferna is an aquatic amphipod occurring also in the cave hygropetric habitat (Photo credit: B. Sket © Rights remain with the author)

Velkovrhia enigmatica. Mytilopsis kusceri (synonym *Congeria k.*) and its two sister species (Bilandžija et al. 2013) are the only troglobiotic mussels (Bivalvia) in the world. *Eunapius subterraneus* seems to be the only troglobiotic freshwater sponge (Porifera: Spongillidae). The flatworm group Temnocephalida with its ca. 10 epizoic members is also unique; they occur mainly on shrimps. Interesting and very noticeable are troglobiotic atyid shrimps *Troglocaris* with eight known species and five additional subspecies, and up to three species may be present in the same cave (Sket and Zakšek 2009).

Impressive biotic formations in Dinaric cave waters are – or used to be – huge aggregations of the sessile filter-feeders *Marifugia* and *Mytilopsis*. A kind of tufa of tightly packed calcareous *Marifugia* tubes in the sink cave Crnulja was up to a meter thick with a layer of live individuals at the surface. Due to the loss of regular flooding in Herzegovinian poljes and caves, caused by hydrotechnical "ameliorations," these formations are now dead and decaying.

Two cave systems in the Dinaric karst appear to be the richest in the world with troglobiotic fauna (Culver and Sket 2000). In each cave system, approximately 100 troglobiotic species – terrestrial and aquatic ones – were found. Postojna-Planina cave system (PPCS) in Slovenia is however separated into two parts by a flooded ca 3 km corridor not yet been entirely traversed by man. A sinking river flows through this cave system which also has several entrances; therefore, in addition to troglobionts, it contains numerous non-troglobiotic inhabitants. PPCS is also crucial in the early development of speleology (scientific study of caves and other karst features) and speleobiology (biology of subterranean biota); and with the exception of *Proteus*, many "first" troglobiotic animal species were

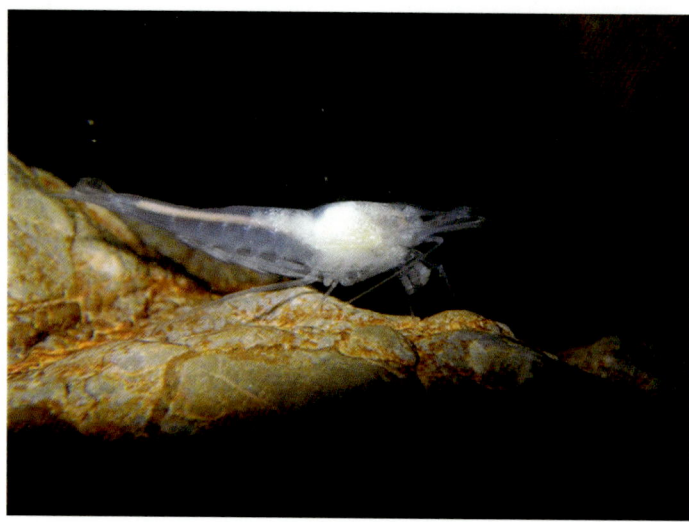

Fig. 3 The cave shrimp *Troglocaris anophthalmus* (Kollar) is one of the widely distributed and frequently seen aquatic invertebrates in the Dinaric karst (Photo credit: B. Sket © Rights remain with the author)

discovered here. PPCS continues to be the site for extensive speleobiological studies. Vjetrenica, in southern Hercegovina (Bosnia and Herzegovina) is less open to epigean habitats; its fauna is therefore nearly limited to troglobionts – a number of them is only known from this cave (Fig. 3).

Threats and Future Challenges

Globally, species are becoming extinct by human-caused environmental destruction. Hypogean fauna are scientifically interesting living treasures that are principally threatened by habitat destruction. However, also at risk are individuals and endemic populations of beetles and vertebrates of relatively low abundance and distribution that have a commercial interest to collectors. This risk can be minimized through legal protection such as occurs in Slovenia's habitat protection of all cave animals and species protection for beetles and the salamander *Proteus*. A balance must however be struck in the level of protection afforded to species as opportunities for scientific research must be preserved. Effective conservation of habitats and species is based not only upon legal statutes but also upon knowledge gained through scientific study that informs decisions.

An indirect pollution threat of special concern to subterranean fauna is particulate and dissolved organics and nitrates that enter the system through percolation or sinking streams. Slight increases in the levels of organic substances may be a desirable enrichment of food resources for troglobionts, but under certain conditions, these increases may also enable energetically more demanding surface species to invade caves and to successfully compete with troglobionts. Although sinking streams and interstitial waters can gradually be "self-purified" of organic pollutants through bacterial oxidation, this self-purification process cannot occur with nitrates. In illuminated surface waters, nitrates are being used and extracted by green plants, but nitrates in polluted waters entering the hypogean system where green plants do not exist, gradually increase in concentration along the hypogean course of the sinking river and interstitial waters. Note that a serious source of pollution may originate from soil layers lying beneath agricultural fields above interstitial or karst systems of groundwater.

Threats by hydrotechnical changes are complex. As noted above, the loss of regular flooding in Herzegovinian poljes and caves caused by hydrotechnical "ameliorations" resulted in the fatal destruction of *Marifugia* and *Mytilopsis* colonies. Adverse changes may also occur in underground water connections.

Subterranean wetlands and their inhabitants remain poorly understood and the challenge is to increase our knowledge of these systems wherever they occur globally as the basis for their conservation and protection.

References

Barr TC. Cave ecology and the evolution of troglobites. Evol Biol. 1968;2:35–102.

Bilandžija H, Morton B, Podnar M, Ćetković H. Evolutionary history of relict Congeria (Bivalvia: Dreissenidae): unearthing the subterranean biodiversity of the Dinaric Karst. Front Zool. 2013;10:5–17.

Biospéologie VA. La biologie des animaux cavernicoles. Paris: Gauthier Villars; 1964.

Botosaneanu L, editor. Stygofauna Mundi. A faunistic, distributional, and ecological synthesis of the World fauna inhabiting subterranean waters. Leiden: Brill; 1986.

Culver DC, Sket B. Hotspots of subterranean biodiversity in caves and wells. J Cave Karst Stud. 2000;62(1):11–7.

Culver DC, Christman MC, Elliott WR, Hobbs III HH, Reddell JR. The North American obligate cave fauna: regional patterns. Biodivers Conserv. 2003;12:441–68.

Gunn J, editor. Encyclopedia of caves and karst science. New York/London: Fitzroy Dearborn; 2004.

Guzik MT, Austin AD, Cooper SJB, Harvey MS, Humphreys WF, Bradford T, Eberhard SM, King RA, Leys R, Muirhead KA, Tomlinson M. Is the Australian subterranean fauna uniquely diverse? Invertebr Syst. 2010;24:407–18.

Kostanjšek R, Pašić L, Daims H, Sket B. Structure and community composition of sprout-like bacterial aggregates in a Dinaric Karst subterranean stream. Microb Ecol. 2013;66(1):5–18.

Oromi P, Martin JL. The Canary Islands subterranean fauna characterization and composition. 1992. http://www.atlantic-island.eu/documentos/Oromi Martin1992.pdf

Pašić L, Kovče B, Sket B, Herzog-Velikonja B. Diversity of microbial communities colonizing the walls of a karstic cave in Slovenia. FEMS Microbiol Ecol. 2009;71(1):50–60.

Peck SB. A summary of diversity and distribution of the obligate cave-inhabiting faunas of the United States and Canada. J Cave Karst Stud. 1998;60:18–26.

Ruffo S. Le attuali conoscenze sulla fauna cavernicola della Regione Pugliese. Mem Biogeografia Adriat. 1957;3:1–143.

Sket B. The ecology of the anchihaline caves. Trends Ecol Evol. 1996a;11(5):221–5.

Sket B. Biotic diversity of hypogean habitats in Slovenia and its cultural importance. In: Biodiversity – International Biodiversity Seminar, UNESCO, Gozd Martuljek, Proceedings; 1996b. p. 59–74.

Sket B. The cave hygropetric – a little known habitat and its inhabitants. Arch fur Hydrobiol. 2004;160(3):413–25.

Sket B. Can we agree on an ecological classification of subterranean animals? J Nat Hist. 2008; 42(21):1549–63.

Sket B. Diversity and singularity of hypogean fauna in the Dinaric region. In: Abstracts 14th International Symposium Biospeleology, Makarska; 1999. p. 21–5.

Sket B, Zakšek V. European cave shrimp species (Decapoda: Caridea: Atyidae), redefined after a phylogenetic study; redefinition of some taxa, a new genus and four new *Troglocaris* species. Zool J Linn Soc. 2009;155:786–818.

Sket B, Paragamian K, Trontelj P. A census of the obligate subterranean fauna in the Balkan Peninsula. In: Griffiths, HI & Kryštufek, B, editors. Balkan biodiversity. Pattern and process in Europe's biodiversity hotspot. Dordrecht: Kluwer Academic Publishers B.V.; 2004. p. 309–32.

White WB, Culver DC, editors. Encyclopedia of caves. 2nd ed. San Diego: Elsevier; 2012.

Zagmajster M, Culver DC, Sket B. Species richness patterns of obligate subterranean beetles in a global biodiversity hotspot – effect of scale and sampling intensity. Divers Distrib. 2008;14:95–105.

Groundwater Dependent Wetlands

24

Ray H. Froend, Pierre Horwitz, and Bea Sommer

Contents

Introduction	346
Groundwater Dependence	346
Types of Groundwater Dependent Wetlands	347
Threats and Challenges	349
References	355

Abstract

The relative importance of groundwater in the development and maintenance of hydrological and ecological character is recognized as an important feature that distinguishes so-called "groundwater-dependent" wetlands. A categorisation is presented here according to three types of groundwater dependent ecosystems (1: subterranean or cave; 2: surface expressions of groundwater discharge; and 3: ecosystems dependent on subsurface presence of groundwater). Beyond these, we recognise eight degrees of dependence based on ecology-hydrology linkages that make explicit the hydrology, biology, soil/sediment/rock relationships arranged in increasing requirement for groundwater permanence (from intermittent to seasonal to permanent, and (soil) moisture to saturation). The classification is applied to situations where groundwater wetlands have become subject to changing hydrological conditions.

Keywords

Groundwater discharge · Hypogean · Hyporheic · Stygofauna · Phreatophytes · Aquifer · Soil moisture · Groundwater abstraction · Climate change

---

R. H. Froend · P. Horwitz (✉) · B. Sommer
Centre for Ecosystem Management, School of Science, Edith Cowan University, Joondalup, WA, Australia
e-mail: r.froend@ecu.edu.au; p.horwitz@ecu.edu.au; b.sommer@ecu.edu.au

© Springer Science+Business Media B.V., part of Springer Nature 2018
C. M. Finlayson et al. (eds.), *The Wetland Book*,
https://doi.org/10.1007/978-94-007-4001-3_246

Introduction

In hydrological terms, all wetlands contain surface and/or subsurface water inflows, discharges, and storages that are spatially and temporally variable. In its broadest definition, groundwater is considered to be all water beneath the land surface; logically then it is an important hydrological component of most wetlands. So why would we specifically define some wetlands as "groundwater-dependent"?

Interest in this aspect of wetlands has largely developed from a need to understand the consequences of the direct use of groundwater or the pollution of aquifers connected to wetlands. As stated in Ramsar Convention materials (2010), many aquifers around the world are currently heavily exploited or overexploited for water (Custodio 2002), for example in Mexico, China, the Middle East, and Spain (Morris et al. 2003). Some exploitation is clearly unsustainable and has led to alteration of the hydrological regime of wetlands associated with the aquifers and significant degradation of their ecological character – examples include the Azraq wetlands, Jordan (Fariz and Hatough-Bouran 1998), and Las Tablas de Daimiel, Spain (Fornés and Llamas 2001).

The purpose of this overview is to describe the particular features of "groundwater dependent" wetlands that distinguish such wetlands from other wetlands and to describe the conditions under which some of these special hydrological-ecological linkages and processes are disrupted by human activity.

Groundwater Dependence

In the context of this article, the notion of a groundwater-dependent wetland (GDW) implies:

- Groundwater is an important contributor to the maintenance of a wetland hydrological regime.
- A change in the quantity or quality of the groundwater will impact on the state and condition of a wetland ecosystem.

We specifically refer to groundwater here as saturated unconsolidated sediment and rock strata beneath the surface that interact with a wetland (Nield 1990). The relative importance of groundwater in the development and maintenance of hydrological and ecological character is recognized as an important feature that distinguishes so-called "groundwater-dependent" wetlands. However, groundwater and wetland interaction is not limited to discharge (or inflow) into the wetland system. Wetlands interact with underlying and surrounding groundwater through both discharge and recharge pathways, either simultaneously or dynamically, depending on seasonal connection with the water table. In cases where the predominant interaction is recharge of groundwater by the wetland, the condition of the wetland is still closely related to the state of the underlying groundwater.

These relationships can be disrupted by changes to the groundwater through abstraction, pollution, and reduction in rainfall recharge, and ensuring the effective management of wetlands will require integrated management of associated surface and groundwater resources. This in turn necessitates a quantitative understanding of the surface and/or subsurface origins, pathways, and variability of flows in and out of the wetland, in order to develop groundwater abstraction strategies that minimise or prevent unacceptable levels of change in ecological character of the wetland. The quantity and quality of groundwater interacting with a wetland is as important as the spatial and temporal variability of inflow and outflow. For example, some GDWs are entirely maintained by continuous groundwater discharge while others receive relatively minor groundwater discharge critical to wetland functioning but restricted to particular times of the year.

The precise nature of interactions between groundwater and wetlands will depend on local hydrogeological conditions. The extent of interaction depends on the permeability of any rocks or sediments that lie between the wetland and the aquifer. Where impermeable strata (an aquiclude) overlie an aquifer, water cannot move vertically upwards or downwards and the aquifer is said to be "confined." In such cases, the wetland and the aquifer are hydrologically separate and exchange of water will not occur. Where rocks or sediments of low permeability (an aquitard) overlie the aquifer, interaction may occur, but the rates of movement will be slow and the amounts of water involved will be small. Where there are no overlying low permeability rocks (no aquitard or aquiclude present), the aquifer is said to be "unconfined"; here the wetland and aquifer are in direct contact and the degree of interaction can be high. (The above definitions are sourced from Ramsar Convention 2010).

The interaction between groundwater and wetlands can also vary within wetlands (e.g., along a river's course) and between individual wetlands, even ones that are close to one another (Ramsar Convention 2010). There is also groundwater within what we may recognize as the wetland ecosystem; saturated sediments beneath the water column and in the littoral zone: the hyporheic zone. Given this variability, what then distinguishes a "groundwater dependent" wetland from all other wetlands?

Types of Groundwater Dependent Wetlands

Perhaps the simplest approach to identifying different types of groundwater dependent wetlands is to use an established wetland classification scheme against which the source of water can be designated. Such is the approach taken by the Ramsar Convention (2010, reproduced here in Table 1). It is useful to demonstrate where groundwater-fed systems are and are not unique to landscape contexts. However it does not recognize habitat-related complexities associated with wetlands sourced by groundwater.

Another approach is recommended for adoption by Richardson et al. (2011) in Australia, recognizing three types of groundwater dependent *ecosystems*: Aquifer and cave ecosystems (Type 1) which provide unique habitats for living organisms

Table 1 Wetland landscape location types and hydrological subtypes (Sourced from Horwitz et al. 2008)

Landscape location	Subtype based on water transfer mechanism
Flat upland wetlands	Upland surface water-fed
Slope wetlands	Surface water-fed
	Surface and groundwater-fed
	Groundwater-fed
Valley bottom wetlands	Surface water-fed
	Surface and groundwater-fed
	Groundwater-fed
Underground wetlands	Groundwater-fed
Depression wetlands	Surface water-fed
	Surface and groundwater-fed
	Groundwater-fed
Flat lowland wetlands	Lowland surface water-fed
Coastal wetlands	Surface water-fed
	Surface and groundwater-fed
	Groundwater-fed

(e.g., stygofauna and troglofauna), including karst aquifer systems, fractured rock and saturated sediments, and including two important "subtypes": the hyporheic zones of rivers, floodplains, and coasts and the deep subsurface groundwater environment. Both provide relatively discrete and characteristic environmental conditions. For example, the latter provides more stable and dark conditions with restricted inputs of energy and low productivity. Ecosystems dependent on the surface expression of groundwater (Type 2) include wetlands, lakes, seeps, springs, river baseflow, coastal areas, and estuaries and marine ecosystems. Ecosystems dependent on subsurface presence of groundwater (Type 3) (via the capillary fringe) include terrestrial vegetation that depends on groundwater fully or on a seasonal or episodic basis in order to prevent water stress and generally avoid adverse impacts to their condition (typology text adapted from Richardson et al. 2011).

These ecosystems can be differentiated further for wetlands in particular through an assessment of the temporal and spatial relationship between groundwater and surface water under the following five conditions.

1. The timing of the requirement for the groundwater component, usually a seasonal requirement where there is a seasonal or interannual wetter or drier period.
2. The location of the water relative to the land surface: above the surface or below (hypogean or subterranean) or an intermediate category where there is a surface/subsurface exchange (this can also be related to landscape location of the wetland, as per the Ramsar Convention 2010 categorization).
3. The nature of the water movement/water retention in the wetland, where the receiving system can be standing water (like "lentic" in a lake sense) or running water (like lotic in a river sense).

4. The way that groundwater is held, which usually determines the nature of the discharge: confined systems discharge under a pressure gradient, whereas unconfined systems (which can include perched) are more diffuse (although this dichotomy is usually upheld, there are occasions where confined or unconfined conditions imperfectly categorize the nature of the discharge).
5. Nature of the groundwater discharge: from single and isolated point source discharge to diffuse discharge along a broad front. This is often a question of scale since multiple localized volumetrically limited point discharges could be interpreted as diffuse.

According to these relationships, different species and ecological communities can have different requirements for groundwater. These requirements can be explored in terms of hydrology – biology/ecology linkages shown in Table 2. Again the linkages can be arranged according to whether the requirement is temporal in nature, how and where the groundwater is expressed, and under what circumstances these linkages are *just* or even *mostly* groundwater.

Threats and Challenges

Processes that threaten groundwater dependent wetlands are by and large no different to those that threaten other wetland types (e.g., climate change, pollution, temperature changes, nutrient enrichment, water abstraction, fragmentation). Processes that disrupt the nature of the groundwater-ecology linkages discussed above (i.e., changes in the hydrological regimes that support said linkages), however, are particularly relevant to groundwater-fed wetlands, and it is these that we focus on in this review. A substantial proportion of activities that humans undertake in an effort to promote development and human well-being bring about changes in hydrological regimes, including (in no particular order):

- Climate change (increased rainfall, sea level rise, melting permafrost → increased water levels, reduced rainfall, increased temperatures → lower water levels)
- Clearing of catchments (deforestation, etc. → increased runoff, increased recharge)
- Urbanization (increased impervious surfaces resulting in increased runoff and recharge)
- Damming of rivers (increased water levels in flooded areas; decreased groundwater recharge due to runoff ending up in dam; decreased water levels downstream of dammed river)
- Discharge of water (e.g., waste water, saline land drainage, etc.) into wetlands
- Plantation forestry/agriculture/pasture for grazing animals (can result in either increased or decreased groundwater recharge)
- Groundwater abstraction (arguably the major cause of global groundwater decline)
- Drainage or infilling of wetlands (for agricultural or urban development)
- Fire

Table 2 Broad classes of Ecology-Hydrology linkages for inland and near shore wetland species, communities, and/or ecosystems. These linkages make explicit the hydrology, biology, soil/sediment/rock linkages that define a set of wetland relationships. The linkages are arranged in increasing requirement for water permanence (from intermittent to seasonal to permanent, and (soil) moisture to saturation) (Adapted from Horwitz et al. 2008, with permission Royal Society of Western Australia)

Ecology-Hydrology linkages	Relevance for groundwater dependent wetlands	Examples (species, communities and wetland ecosystems)
Requirement for seasonal soil moisture	Groundwater can maintain (through capillary action) soil moisture in seasonally dry wetlands, or the littoral zone of more permanent wetlands	Non-phreatophytic terrestrial vegetation; sedgelands, fringing wetland vegetation in seasonally drier times
Requirement for seasonally moist habitat for aestivation/drought avoidance	Groundwater can maintain (through capillary action) soil moisture in seasonally dry wetlands, or the littoral zone of more permanent wetlands	Burrowing or sheltering fauna (e.g., aestivating fish, burrowing freshwater crayfish, burrowing frogs, etc., along with any assemblage associated with their burrows)
Requirement for a seasonal or intermittent surface saturation	Where groundwater discharge is seasonal or intermittent; can occur for flow through wetlands, and some riparian and littoral habitats in lentic or lotic ecosystems, when groundwater levels only connect with wetlands at peak seasonal flows	Ephemeral wetland systems
'Terrestrial' requirement for access to groundwater table	Broad relevance, incorporating phreatophytic vegetation in otherwise terrestrial ecosystems or damplands, or in the riparian zone, to the biotic (mainly vegetation) requirements for groundwater in a wetland like a dry floodplain	Phreatophytic vegetation or floodplain vegetation, and the fauna dependent on this vegetation
Requirement for groundwater discharge to maintain a particular quality of surface water, a quality influenced by physical (i.e., surface water temperatures) and/or chemical (i.e., salinity) characteristics	Where groundwater discharge maintains a thermally and/or chemically distinct habitat. Can include groundwater fresh discharges in near shore marine or estuarine ecosystems; cool fresh discharges in warm/arid ecosystems	Stream and river fauna that are cold stenotherms and intolerant to elevated water temperatures associated with catchment flows at warmer times of the year

(continued)

Table 2 (continued)

Ecology-Hydrology linkages	Relevance for groundwater dependent wetlands	Examples (species, communities and wetland ecosystems)
Requirement for permanent surface saturation	Permanent groundwater discharge or flow-through that maintains surface waters in a wetland. Usually depends on regional aquifers	Permanent lakes, peatlands, and permanent stream and rivers, each maintained by groundwater discharges/baseflows in regions with a seasonal drought
Requirement for permanent subsurface/subterranean saturation of saturated hypogean (interstitial) spaces	Permanent groundwater discharge to maintain a water level in a subterranean wetland like a cave	A subterranean 'wetland' might be the biotic assemblage dedicated to a subterranean existence in (confined or unconfined) aquifer. Stygofauna. Rootmat communities in caves
Requirement for an exchange between surface/subsurface flows and groundwater	Hyporheic zone: where there is a spatial and temporal exchange between surface waters and subterranean waters	Hyporheic assemblages in river beds and riparian areas

The hydrological changes brought about by these activities will have varying effects on the ecology of groundwater-fed wetlands, depending on the hydrological requirements of the different types of hydrology-ecology linkages (Table 2). Because the eight types of linkages/hydrological requirements are not strict categories, rather existing on a scale of "degree of dependence" (Hatton et al. 1998), impacts of hydrological change are also to be regarded on this scale. Any significant change to the amount of groundwater discharge to wetlands can be expected to have impacts; however, changes in hydrological regimes that result in less discharge are generally more undesirable than those resulting in increased discharge. This is, on the one hand, because a wetland that becomes "more wet" is still a wetland, capable of delivering functions and services typical of wetlands, whereas one that dries up can no longer be regarded as such. Another reason is that most groundwater dependent wetlands are in semi-arid or arid zones where water is otherwise scarce (and hence the demand for groundwater high). Moreover, the reality appears to be that, on a global scale, declining groundwater reserves is one of the most pressing concerns facing the twenty-first century (Konikow and Kendy 2005; World Meteorological Organization 1997). Nevertheless, increased flows and water levels can have adverse impacts, depending on what the management objective is (Table 2).

The fact that the hydrological changes (Table 3) that threaten groundwater dependent wetlands are largely caused by human activities implies that they can (and arguably, should) be managed in order to minimize risks. The concept of adaptive management is particularly suited to dynamic systems whose responses

Table 3 Effects of hydrological change on the ecology of groundwater-dependent wetlands based on their hydrological requirements (Adapted from Horwitz et al. 2008, with permission Royal Society of Western Australia). A decrease or increase in the discharge of groundwater to wetlands can cover all temporal scales or may be expressed as seasonal deviations from the norm. In addition, hydrological changes usually have consequences for water quality which may also impact on wetland ecology. An increased fire risk applies to all of the drying scenarios and is therefore not listed in the table

Hydrological requirement	Effects of a decrease in the quantity of water supply	Effects of an increase in the quantity of water supply
Seasonal soil moisture	Reduced plant vigor as a result of reduced water availability particularly in the dry season; decreased rates of surface soil carbon and nutrient cycling. Germination failure. Potentially reduced seed set and shift in population distribution, persistence, and community composition	Depends on the amount of seasonal increase; as seasonal is by nature temporary, effects may be minimal. In certain circumstances, increased soil moisture during the warmer times of the year may benefit the proliferation of soil pathogens such as *Pythophthora cinnamomi* or disease-carrying insects such as mosquitoes
Seasonally moist habitat for aestivation/drought avoidance	Species may survive, provided moisture levels are sustained; however, if seasonal wetting fails, less frequent emergence and probably reduced reproduction may result. Weed invasion. Drop in the water table may cause land subsidence	Depends on the amount of seasonal increase; as seasonal is by nature temporary, effects may be minimal. Potential positive effects as this would increase habitat availability for these species
Seasonal or intermittent surface saturation	If inundation is less frequent or seasonality of inundation changes (e.g., inundation occurs in winter-spring instead of in summer) and/or decreased areal extent and duration of inundation: – may result in reduced frequency of plant recruitment events; gradual terrestrialization. Weed invasion. Reduced richness of wetland invertebrates. Effects on species dependent on surface water at particular times of the year to complete their life cycles (e.g., amphibians, macroinvertebrates). Drop in the water table may cause soil shrinkage and land subsidence	Depends on the amount of seasonal increase; as seasonal is by nature temporary, effects may be minimal. Potential positive effects as this would increase habitat availability for these species. However, a shift from a seasonal or intermittent system to a permanent one would bring about a significant shift in the entire ecology of the wetland and its character

(*continued*)

24 Groundwater Dependent Wetlands

Table 3 (continued)

Hydrological requirement	Effects of a decrease in the quantity of water supply	Effects of an increase in the quantity of water supply
Terrestrial to access to groundwater table	Acute drawdown and low recharge can result in mortality of overstorey and understorey plant species or local extinction of susceptible species. Less severe circumstances can result in reduced vigor of mature plants and a shift in the distribution of established juveniles. Plants and animals dependent on soil moisture (e.g., through hydraulic redistribution) and shade provided by phreatophytes could also be affected. Drop in the water table may cause land subsidence	Depends on the amount of the rise in the groundwater table. Phreatophytes can adapt to moderate changes (e.g., through 'root pruning'); however, larger, more permanent increases will affect the distribution and persistence of plant communities (i.e., a shift to more mesic communities that are adapted to tolerate wet conditions). May have positive effects on wetland associated flora and fauna
Groundwater discharge to maintain a particular quality of surface water, a quality influenced by physical (i.e., surface water temperatures) and/or chemical (i.e., salinity) characteristics	Habitat contraction and/or loss may lead to loss of gene flow, increased competition, extinctions, etc. Loss of specifically adapted stenotherms. Potential invasion by weeds and introduced pest species	Potentially positive effects due to habitat expansion
Permanent surface saturation	For example, change from permanent to temporary stream systems or from permanent to temporary lentic wetlands – in this scenario indirect effects of reduced flow/inundation on physicochemistry could be as important as the direct hydrological effects. Effects include impacts on fish community structure, survival and growth; effects on fish migration; effects on invertebrate richness and community structure; increased invertebrate drift in streams; evapoconcentration of nutrients leading to eutrophication; effects on sediments exposed to more frequent drying, potentially displacing biota.	This would result in overall higher water levels, greater areal extent of the wetland, and potentially increased flow velocity in streams. Effects include: drowning of emergent macrophytes and littoral vegetation (however, if increased inundation is more permanent, these may reestablish further upslope); formation of passive acid sulfate soils and changes in nutrient cycling; fauna and flora may be 'washed away'; increased groundwater levels could cause secondary salinization (as well as potentially acidification when these lands are drained)

(continued)

Table 3 (continued)

Hydrological requirement	Effects of a decrease in the quantity of water supply	Effects of an increase in the quantity of water supply
	Terrestrialization as more non-mesic plant species migrate towards the wetland centre. Weed invasion. Drying, heating, and cracking of sediments leading to changes in sediment structure and biogeochemistry; effects of drying on nutrient cycling; exposure of acid sulfate soils leading to acidification and disrupted nutrient cycling. Effects of low pH and associated metal toxicity on all wetland biota. Salt water intrusion into groundwater may increase salinity of the wetland. Drop in the water table may cause land subsidence. Drying and oxidation of wetland sediments may accelerate the release of greenhouse gases	
Permanent subsurface/ subterranean saturation, or saturated hypogean (interstitial) spaces	Altered patterns of carbon and nutrient cycling. Reduced ability to retreat or emerge according to life history requirement. Potential loss of habitat. Extinction of species	Potentially positive effects due to habitat expansion. Effect could be negative if increased saturation were associated with increased anoxia or pollution
Exchange between surface/ subsurface flows and groundwater	Where habitat is fixed at a certain stratigraphic level, then declines in the saturated zone will strand dependent biota resulting in local extinctions. Otherwise distributions of fauna may change according to extent of groundwater drawdown	Potentially positive effects due to habitat expansion. Effect could be negative if increased saturation were associated with increased anoxia or pollution

to environmental change are difficult to predict (Holling 1978). One consideration when deliberating over the effects of hydrological change on GWD wetlands, and whether or not these changes are acceptable, is the adaptation potential of the biological communities in these systems. For example, a groundwater fed wetland may exist in a number of alternative states corresponding to the hydrology-ecology linkage types, potentially shifting between them (except perhaps for the subterranean ones), in accordance to changes in the extant hydrological support mechanisms. Usually, however, the adaptation potential of species and natural communities is limited, and unwanted catastrophic shifts (i.e., largely irreversible ones) may occur (Gunderson 1999). In order to avoid this management actions should be adaptive and respond to previous outcomes (i.e., "learning by doing"; Walters and Holling 1990). For example, abstraction volumes can be reduced in order to prevent further shrinkage of wetland habitat. Ensuring the sustainability of groundwater and the ecosystems they support requires concerted and coordinated action by natural resource managers, politicians, and the general public. The major challenge remains the ability to balance human demands with those of the wetlands, not least because humans rely on the persistence of these water sources as much as the services provided by the wetlands themselves.

References

Custodio E. Aquifer over-exploitation; what does it mean? Hydrogeol J. 2002;10:254–77.

Fariz GH, Hatough-Bouran A. Population dynamics in arid regions: the experience of Azraq Oasis Conservation Project. In: de Sherbinin A, Dompka V, editors. Water and population dynamics. Case studies and policy implications. Washington, DC: American Association for the Advancement of Science; 1998.

Fornés JM, Llamas MR. Conflicts between groundwater abstraction for irrigation and Wetland conservation: achieving sustainable development in the La Mancha Húmeda Biosphere reserve (Spain). In: Griebler C, Danielopol D, Gibert J, Nachtnebel HP, Notenboom J, editors. Groundwater ecology. A tool for management of water resources. European Commission. Environment and Climate Programme – Austrian Academy of Sciences (Institute of Limnology). Luxembourg: Publications Office of the European Union. EUR 19887; 2001. p. 263–75.

Gunderson L. Resilience, flexibility and adaptive management – antidotes for spurious certitude? Ecol Soc. 1999;3, 7. Available from http://www.consecol.org/vol3/iss1/art7/. Accessed 29 Jan 2016.

Hatton T, Evans R, SKM. Dependence of ecosystems on groundwater and its significance to Australia. Canberra: Land and Water Research and Development Corporation; 1998.

Holling CS. Adaptive environmental assessment and management. New York: Wiley; 1978.

Horwitz P, Bradshaw D, Hopper SD, Davies PM, Froend R, Bradshaw F. Hydrological change escalates risk of ecosystem stress in Australia's threatened biodiversity hotspot. J R Soc West Aust. 2008;91:1–11.

Konikow L, Kendy E. Groundwater depletion: a global problem. Hydrogeol J. 2005;13:317–20.

Morris BL, Lawrence A, Chilton PJ, Adams B, Calow RC, Klink BA. Groundwater and its susceptibility to degradation; a global assessment of the problem and options for management. Wallingford: British Geological Survey; 2003. Report RS033 to UNEP.

Nield SP. Effects of recharge and lake linings on lake-aquifer interaction. Dissertation. Perth: University of Western Australia; 1990.

Ramsar Convention Secretariat. Managing groundwater: guidelines for the management of groundwater to maintain wetland ecological character. In: Ramsar handbooks for the wise use of wetlands, vol. 11. 4th ed. Gland: Ramsar Convention Secretariat; 2010.

Richardson S, Irvine E, Froend R, Boon P, Barber S, Bonneville B. Australian groundwater-dependent ecosystem toolbox part 1: assessment framework. Canberra: National Water Commission; 2011. Waterlines report.

Walters CJ, Holling CS. Large-scale management experiments and learning by doing. Ecology. 1990;71:2060–8.

World Meteorological Organization. Comprehensive assessment of the freshwater resources of the world. Geneva: WMO; 1997.

Section III

Natural and Anthropogenic Drivers of Wetland Change

Natural and Anthropogenic Drivers of Wetland Change

25

Susan M. Galatowitsch

Contents

Introduction	360
Assessing Wetland Change	361
Drivers of Change to Wetland Hydrology	362
Land Conversion	363
Water Withdrawals	363
Infrastructure Development	364
Water Level Management	365
Other Drivers of Wetland Change	365
Future Challenges	366
References	367

Abstract

The hydrology of wetlands is dynamic owing to daily, seasonal, and inter-annual changes in water levels caused by tides, river flooding, and/or precipitation events. The resulting water regimes are primary determinants of many wetland ecosystem attributes including soil properties, water chemistry and biotic composition. Human-caused changes to wetlands that result in anomalous water regimes usually trigger a cascade of ecological effects, including species losses and invasions and altered biogeochemical cycles. These, in turn, often cause a loss in ecosystem services. Compared to other ecosystems, rates of wetland degradation and loss have been greater, primarily due to six drivers: 1) infrastructure development, 2) land conversion, 3) water withdrawal, 4) eutrophication and pollution, 5) overharvesting and overexploitation, and 6) introduction of invasive species. Wetland degradation is often caused by multiple drivers, some of which

---

S. M. Galatowitsch (✉)
Department of Fisheries, Wildlife and Conservation Biology, University of Minnesota, Saint Paul, MN, USA
e-mail: galat001@umn.edu

© Springer Science+Business Media B.V., part of Springer Nature 2018
C. M. Finlayson et al. (eds.), *The Wetland Book*,
https://doi.org/10.1007/978-94-007-4001-3_217

are site based, while others are regional or global in scope. This makes wetland degradation difficult to reverse, even where social and institutional support is strong. However, in the past twenty years, the complexity and scale of wetland restoration has advanced, resulting in successful attempts in many different contexts around the world. In many cases, though, it is not possible to fully restore the water regime of a wetland modified by human use, so partial fixes must be accompanied by ongoing water level management to achieve desired conditions. An important future challenge is to develop and implement strategies that ensure the sustainability of wetland ecosystem services under increasing stress from climate change, increasing human population and the drivers that have historically threatened wetlands.

Keywords
Wetland loss · wetland degradation · wetland restoration · anthropogenic change · wetland hydrology · land conversion · pollution · overharvesting · invasive species · climate change

Introduction

Wetlands are typified by having some or even all of their areas alternating between being saturated or flooded and being dry. This natural ebb and flow of water in wetlands may be caused by tides, flooding of rivers, or precipitation patterns, with fluctuations in water levels occurring over time scales ranging from hours to months. Less frequent but greater magnitude changes in water levels associated with major climatic events such as major storms or droughts also contribute to the inherently dynamic nature of most wetlands. The pattern of changes in water levels, i.e., a wetland's water regime, in combination with its geomorphic setting, determines many of its ecosystem attributes, including soil properties, water chemistry, and the kinds of biota residing there (van der Valk 2012). Many wetland plants and animals, for example, possess traits that allow individuals to tolerate hydrologic fluctuations or their populations to persist through unfavorable periods, as long as changes are within the natural range of variability for that system. Changes to wetlands that result in anomalous water regimes usually trigger a cascade of ecosystem effects, including species losses and invasions and altered biogeochemical cycles.

While shifts in wetland water regimes can result from natural phenomena such as succession or geologic events, human-caused changes to wetlands are far more prevalent. In addition to altering water regimes, human actions direct pollutants into many wetlands, and in some regions, wetlands are heavily exploited for food or fiber. Wetland impacts can be the result of changes made directly to them or from indirect impacts, i.e., those resulting from modifications at landscape and global scales. Wetlands are rarely (if ever) altered by humans in only one way; most experience changes from multiple drivers. For example, wetlands within agricultural regions of the world may have altered water regimes from water withdrawals, receive pollutant-laden irrigation return water, and be dominated by introduced, invasive species.

Assessing Wetland Change

Anthropogenic alterations to wetlands are seldom recognized and addressed until degradation is extensive and adverse consequences to people serious. In order to minimize or reverse this degradation, the most important causes need to be determined. These cause-effect relationships are seldom obvious because some problems trigger multiple effects, some effects can have multiple origins, and some problems mask the effects of others. To tease apart these relationships and determine which causes should be priorities to address, adopting a formal assessment framework can be useful (Galatowitsch 2012). One typical framework is based on mapping linkages among drivers, stressors, effects, and attributes. Agents of change causing deleterious effects are called drivers. Some examples of drivers include land conversion, water diversions, overharvest of animals or plants, and introduction of toxins. Physiochemical changes to an ecosystem caused by these drivers are referred to as stressors. Water diversions, for instance, are associated with wide range of potential stressors, such as reduced duration of soil saturation or flooding, acidification, and salinization, depending on site and landscape conditions. Responses to these stressors and drivers by species or ecosystems (e.g., increased mortality of a particular species, loss of a habitat resource, spread of an invasive species) are ecological effects. Changes to ecological attributes, or components of the ecosystem that are of greatest concern, are the direct result of these ecological effects. For complex assessments, this information is often managed and presented as a diagram (Fig. 1).

The Millennium Ecosystem Assessment (2005) comprehensively considered the causes of wetland degradation worldwide, as well as the social and ecological consequence of this degradation. This assessment found that, compared to other ecosystems, the rate of wetland degradation and loss has been greater. One consequence is that the status of both freshwater and coastal wetland biodiversity is deteriorating rapidly. As importantly, wetland degradation has been implicated in the erosion of a wide variety of critical ecosystem services, such as diminished coastal reserves, impaired downstream water quality, and lost food stocks. The Millennium Ecosystem Assessment identified six drivers that have caused the most extensive wetland degradation:

- Infrastructure development
- Land conversion
- Water withdrawal
- Eutrophication and pollution
- Overharvesting and overexploitation
- Introduction of invasive species

Reversing wetland degradation must address the underlying causes of degradation (i.e., indirect drivers), which for wetlands are most often related to human population growth and increasing economic development (MEA 2005). Demographic and economic pressures often result in wetlands being used or treated in ways that are not easily avoidable resulting in tradeoffs between wetland

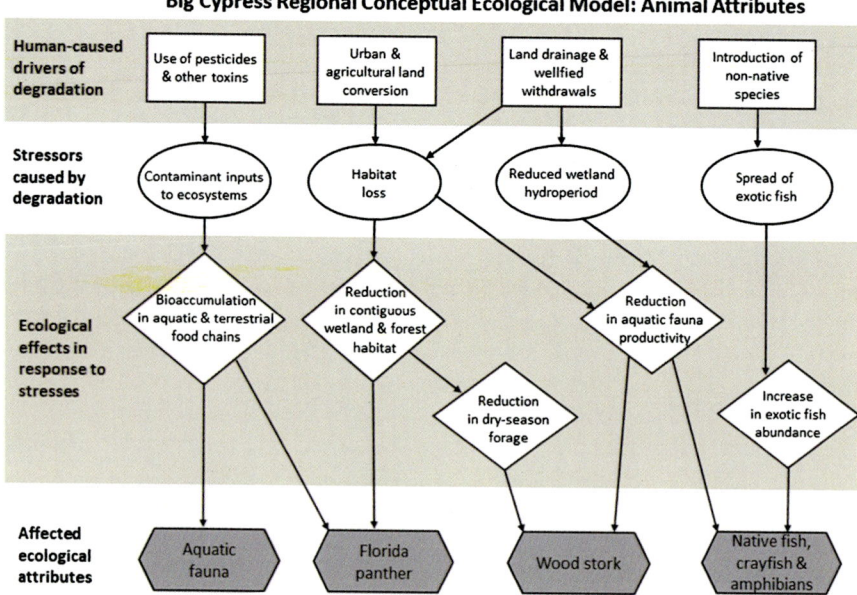

Fig. 1 Conceptual ecological models, such as this one representing key animal attributes of the big Cypress Marsh, are being used for planning ecosystem restoration in the Everglades region of south Florida. At a glance, this diagram shows that may drivers caused degradation and that each of the resulting stresses cause many ecological effects to wetland-dependent animals (after Duever 2005; Wetlands, Big Cypress regional ecosystem conceptual ecological model, 25, 2005, 843-53, Duever MJ; © 2005, Society of Wetland Scientists, with permission of Springer Science + Business Media)

protection and use. So, in order to be effective, solutions proposed for reversing wetland degradation must resolve key tradeoffs, such as those between agricultural production and water quality, land use and biodiversity, and water use and aquatic biodiversity. The entries in this section of Volume 2 provide many examples of anthropogenic changes where tradeoffs pose major challenges for reversing degradation. What these cases make clear is that wetland degradation is often caused by multiple drivers, some of which are site based, while others are regional or global in origin. Consequently, wetland impacts are often very difficult to reverse, even where there is strong social commitment. In some cases, however it has been possible to implement effective solutions and improve wetland conditions.

Drivers of Change to Wetland Hydrology

Significant changes to hydrology occur as a consequence to three of the six leading drivers of wetland degradation: land conversion, water withdrawal, and infrastructure development (MEA 2005). Land conversion of wetlands for agriculture or human settlement is mostly commonly accomplished by artificial drainage and less often by filling or pumping. Even if wetlands are not directly converted for other uses, water from wetlands may be diverted or withdrawn to serve as water supplies for agriculture or other purposes. Infrastructure engineered to limit the extent of flooding or to create overland transportation corridors also can greatly affect wetland water regimes, a problem that is particularly common in coastal and riverine wetland systems.

Land Conversion

Draining wetlands for agricultural production, construction, or peat mining is accomplished by installing subsurface conduits (called tiles) or surface ditches. Water is drawn to these ditches and tiles and carried away from the wetland, lowering the water table to beneath the level of those drainage structures. Wetlands are also drained for peat mining, a commercially significant resource for fuel and horticulture in cold climates such as Canada and Scandinavia. The peatlands are first drained by land warping – the shaping of the peat surface into convex rows – and ditching. The peat is then vacuum-harvested or stripped by cutting it into blocks. Wetland drainage is typically tied to economic incentives, either provided by governments stimulating development or in response to increases in global commodity crops. Consequently, wetland drainage losses can be rapid and extensive. For example, federal policies promoting agricultural production in the US Prairie Pothole region have been responsible for wetland losses exceeding 75% where row-crop agriculture has long been the main land use (Oslund et al. 2010). New subsidies primarily intended to stimulate biofuel production has caused a westward expansion of row-crop agriculture, triggering a recent wave of wetland losses exceeding 5,000 ha/per year (Johnston 2013).

The main challenge of reversing drainage losses is that the scale of restoration opportunities is limited compared to regional losses. For example, the most ambitious program to restore wetlands in the intensively agricultural portions of the US Prairie Pothole region (the Conservation Reserve Program) only restored about 2,700/ha over 4 years, or about 0.3% of the historic wetland extent of the area (Galatowitsch and van der Valk 1999). The technology required to reflood individual wetlands is relatively simple and affordable and so is not generally the limiting factor for restoration. Wetland drainage can often be reversed by blocking the outlets that were engineered to promote water output, for example, by plugging a ditch outlet with nonporous soil or by removing a short section of tile. Where deep ditches were required to accomplish drainage, these lines may need to be regraded in order for water to spread across the wetland rather than first filling abandoned ditch lines (see Richardson 2017. Mesopotamian Marshes of Iraq. Vol. 3).

Water Withdrawals

Actions that deliberately divert water from its normal course (i.e., stream diversions, withdrawals, groundwater pumping) alter wetland water regimes. These impacts are often most pervasive in dryland agricultural regions. The Qa'a Azraq Oasis in Jordan (see Haddad. 2017 Qa'a Azraq Oasis. This volume.), which consists of freshwater lakes, marshes, and mud flats, receives only a fraction of natural spring discharge it once did because aquifers have been seriously depleted. Agricultural development in the region was spurred by well drilling and dam building, which extracted and diverted so much water that the freshwater oasis has become highly saline, threatening water supplies for local communities, wetland biodiversity, and ecotourism. In this situation, the main driver of change is geographically dispersed, and so problems could only be addressed because a large number of stakeholders agreed to coordinate water management. The hope is that with coordination of water use, groundwater use will not exceed the bounds of what can be sustained, and freshwater aquifers will once again feed the oasis.

At an even larger scale, unsustainable dryland agricultural development in the Central Asian Desert, spanning five countries, nearly dewatered the entire Aral Sea in less than 40 years (Nilsson and Berggren 2000). Water from two main rivers flowing into the sea were diverted to irrigate cotton crops, causing water levels to dramatically recede, collapsing fisheries, and depleting supplies of potable water. Multinational agreements largely failed to restore water flows into the Aral Sea, leaving individual countries to attempt partial remedies. Kazakhstan, with support from the World Bank, built a dam to store water in the northern part of the sea and upgraded agricultural water works along the Syr Darya River, the main waterway flowing into the impoundment, to improve water use efficiency. This engineering solution has restored wetlands of the North Aral Sea, which again provide critical habitat for breeding and migratory water birds, as well as for several rare fish.

Infrastructure Development

Infrastructure such as levees associated with channelization, embankments to impound water for agriculture or aquaculture, and road beds constrain water inputs and outputs from rivers and tides. They also alter the movement of sediments and nutrients and, on ocean coasts, the balance of saltwater and freshwater. The degradation of Lake Chilika (see Pattnaik and Kumar. 2017. Lake Chilika. This volume.), a coastal lagoon in India, illustrates multiple impacts caused by infrastructure development. This lagoon was once a complex of shallow marine, brackish, and freshwater wetlands that supported a diverse fishery, served as a major wintering ground for migratory waterbirds, and provided critical habitat for several endangered species, including the Irrawaddy dolphin. Infrastructure development began along tributary rivers during colonial times but accelerated after 1950 with the installation of an extensive network of embankments and hydraulic structures used for irrigated agriculture and shrimp aquaculture within the delta. A combination of increased soil transport into the lagoon from channelized, deforested tributaries and from sediment movement along the altered coast choked the main tidal entrance from the Bay of Bengal. Reduced inputs of saltwater contributed to the collapse of the traditional fishery and the spread of invasive weeds. To restore the fishery and critical habitat for wildlife and endangered species, the government created an agency to oversee ecological restoration. Over the past 20 years, they initiated a participatory process to manage the wetland and catchment and to regulate ecotourism, established a local ferry system to reduce the road network, and reopened the mouth to the sea.

Water Level Management

It is seldom possible to fully restore the natural water regime of a wetland modified by anthropogenic change because of tradeoffs with other uses in the surrounding landscape. Often, the partial fixes must be accompanied by ongoing water level management to achieve the water regime necessary to support wetland biodiversity. A key challenge is to create hydrologic conditions, or hydropatterns, that can support the full range of species typical for that type of wetland. Hydropatterns vary in five ways that affect biotic communities and so ecosystem structure and function (Wissinger 1999):

Permanence: from permanently to temporarily inundated in most years
Predictability: from regular wet and dry phases to very sporadic wetting or drying
Phenology of inundation: the time of year when flood pulses are most likely to occur
Duration of wet and dry phases (continuous periods of inundation before drying and vice versa): from very short phases, lasting a few weeks, to very long phases, lasting most of a year (or even longer)
Harshness: the amplitude of change between wet and dry phases: from wet and dry phases marked by relatively small changes in water table elevation to dramatic changes that result in deep flooding or significant water table recession

The wetlands of Tram Chim National Park (see Tran and Barzen. 2017. Tram Chim. Vol. 3.) illustrate the challenges of water level management. These wetlands were restored by building embankments around 7,600 ha of drained agricultural lands within the Mekong Delta (Beilfuss and Barzen 1994). If the wetland does not experience strong seasonal fluctuations in water levels, wetland vegetation needed for waterbird habitat cannot be sustained. So, managers attempt to create suitable wetland hydrology by managing water inputs and outputs through sluice gates and canals. They have restored alternating wet-dry conditions in the Tram Chim wetlands by blocking water inflows at the beginning of the dry season (December). As rains returns in June, the sluice gates can be opened, allowing water levels to rise 2–3 m. The gates can then be managed so that water levels recede from October to December.

Other Drivers of Wetland Change

Increased population growth and economic development within a region invariably creates multiple drivers of wetland degradation. For example, overharvesting of fisheries, spread of introduced invasive species, and infrastructure development combined to accelerate the collapse of the Lake Chilika estuary ecosystem. Typically, the most seriously damaged attributes of an ecosystem will have been adversely affected by several stressors simultaneously. At Tram Chim National Park, waterbird populations were decimated by the loss of a food source when wetlands were drained and then impounded and by a change in habitat structure caused by the spread of an introduced shrub (*Mimosa pigra*). Nile perch have impacted the biodiversity of wetland in the Nile Basin, adding to degradation caused by agriculture and overfishing. In the Great Barrier Reef, the populations of many fish species have precipitously declined because they are overharvested and because eutrophication caused habitat impairment by stressing plant beds and living reefs. Overhunting of eggs, mammal skins, and feathers, spurred by commercial markets, decimated the wildlife of Sanjiang Plain wetlands. Siberian weasels, roe deer, and oriental storks are among the species that experienced severe losses or were extirpated. Losses in the Chinese portions of the basin have been much greater than in Russia, where human populations are much lower.

Because wetlands depend on inputs of water from the atmosphere, surface water runoff, and groundwater, eutrophication and pollution are frequently among the multiple drivers of degradation. Agricultural production (both crops and livestock), in particular, adds to nitrogen and phosphorus received by wetlands, which triggers major trophic changes, starting with overstimulation of plant growth (especially algae). In the past century (1890s–1990s), the amounts of biologically reactive nitrogen and phosphorus have increased, nitrogen by ninefold (Galloway and Cowling 2002), while changes in phosphorus have not been reliably estimated. Both phosphorus and nitrogen are readily transported in surface water runoff, within the water fraction and attached to sediment. Nitrogen also moves via groundwater (nitrates) and the atmosphere (ammonium). Pollutants, such as nitrogen, transported in the atmosphere, can be a significant driver of change to wetlands far from human settlements. In the Arctic, where pollutant levels in precipitation are relatively low, colonial seabirds serve as "biovectors," feeding on nutrient-enriched forage across vast areas and concentrating it where they nest.

Petroleum-based chemicals, such as plastics, pesticides, and a wide array of industrial products, pollute wetlands. Many synthetic compounds produced by the chemical industry have been slow to break down in the environment because few microbes can degrade them (Hinga and Batchelor 2005). For example, the half-life of polychlorinated biphenyls (PCBs) in the environment is about 13 years within a US estuary (Delaware Bay) and 34 years in long-lived mussels, which ingest toxin-infused river sediments (Hinga and Batchelor 2005).

Future Challenges

As human population increases, so too will the pressures to use wetlands to produce food and for water supplies. A key future challenge is to formulate strategies that ensure sustainable use of wetlands for these purposes, to avoid the collapse of these ecosystems and the loss of well-being to resource-dependent communities. A second challenge is linked to economic development, which globally is heavily reliant on energy sources from fossil fuel combustion. Carbon dioxide, a by-product of this combustion, is accumulating in the atmosphere, causing global warming (i.e., climate change). The vast majority of wetlands worldwide are likely to be impacted by climate change, through sea level rise, by increased incidence of extreme weather events, and by new infrastructure built to protect humans from increasingly unpredictable environmental conditions.

References

Beilfuss RD, Barzen JA. Hydrological wetland restoration in the Mekong Delta, Vietnam. In: Mitsch WJ, editor. Global wetlands: old world and new. Amsterdam: Elsevier Science B.V; 1994.

Duever MJ. Big Cypress regional ecosystem conceptual ecological model. Wetlands. 2005;25:843–53.

Galatowitsch SM. Ecological restoration. Sunderland: Sinauer Associates; 2012. p. 2012.

Galatowitsch SM, van der Valk AG. Characteristics of recently restored wetlands in the Prairie Pothole Region. Wetlands. 1999;16:75–83.

Galloway JN, Cowling EB. Reactive nitrogen and the world: 200 years of change. Ambio. 2002;31(2):64–71.

Hinga KR, Batchelor A. Waste detoxification. In: Hassan R, Scholes R, Nash A, editors. Millennium ecosystem assessment volume 1: conditions and trends working group. Washington, DC: Island Press; 2005.

Johnston CA. Wetland losses due to row crop expansion in the Dakota Prairie Pothole Region. Wetlands. 2013;33:175–82.

Millennium Ecosystem Assessment. Ecosystems and human well-being: wetlands and water synthesis. Washington, DC: World Resources Institute; 2005.

Nilsson C, Berggren K. Alterations of riparian ecosystems caused by river regulation: dam operations have caused global-scale ecological changes in riparian ecosystems. How to protect river environments and human needs of rivers remains one of the most important questions of our time. BioScience. 2000;50:783–92.

Oslund FT, Johnson RR, Hertel DR. Assessing wetland changes in the Prairie Pothole Region of Minnesota from 1980–2007. J Fish Wildl Manag. 2010;1:131–5.

van der Valk AG. The biology of freshwater wetlands. 2nd ed. Oxford: Oxford University Press; 2012.

Wissinger SA. Ecology of wetland invertebrates. Synthesis and applications for conservation and management. In: Batzer DP, Rader RB, Wissinger SA, editors. Invertebrates in freshwater wetlands of North America. Ecology and management. New York: Wiley; 1999. p. 1043–86.

Wetland Losses and the Status of Wetland-Dependent Species

26

Nick C. Davidson

Contents

Introduction 370
Long-Term Wetland Losses 372
Wetland Losses in the Twentieth and Early Twenty-First Centuries 373
Overall Natural Wetland Loss Since the Start of the Eighteenth Century 375
Status of Wetland-Dependent Species 377
References 380

Abstract

Human-kind has been draining, infilling, and converting both coastal and inland wetlands for many centuries. Recent estimates suggest that wetland losses have been as much as 87% since 1700 AD, 70% since 1900 AD, and 30% since 1970 AD. Rates of loss in the twentieth century were almost four times faster than in earlier centuries, and wetland conversion is continuing in the twenty-first century. Although rates of loss are now low or slowing in some parts of the world (e.g., Europe and North America), high losses are continuing elsewhere, especially in Asia. Not unexpectedly, the status of species dependent on wetlands is deteriorating, and at faster rates than species depending on other biomes. Although the status of migratory shorebird populations has improved in the twenty-first century in some regions (e.g., North America), it is very poor and deteriorating further elsewhere, especially in East Asia-Australasia.

Keywords

Wetland loss · Rate of loss · Wetland-dependent species · Migratory shorebirds

---

N. C. Davidson (✉)
Institute for Land, Water and Society, Charles Sturt University, Albury, NSW, Australia

Nick Davidson Environmental, Wigmore, UK
e-mail: arenaria.interpres@gmail.com

Introduction

Human-kind has been draining, infilling, and converting both coastal and inland wetlands for many centuries, for example, since at least Roman times in Europe (see also Fig. 1), for at least 2,000 years in China, and since at least the seventeenth century in North America and southern Africa. This conversion and degradation of wetlands continues, with underlying drivers being economic and human population growth, and proximate causes being conversion to first extensive and then intensive agriculture (croplands), changes in water use and availability including the downstream effects of water abstraction and major hydro-engineering schemes, increasing urbanization and infrastructure development, disease control (especially for mosquitoes), and on the coast also sea defenses, port and industrial developments, and aquaculture (Finlayson et al. 2005; van Asselen et al. 2013). The wetland losses part of this chapter is drawn largely from Davidson (2014).

Widespread (although mostly not quantified) inland and coastal wetland drainage and conversion, and particularly its impact on hunted waterfowl populations, was increasingly reported and raised as a concern since the 1920s in North America and from the early 1960s in Europe. Hoffman (1964) concluded that "in temperate regions drainage of wetlands is proceeding at an increased rate and without reference to their diverse values," and recommended the establishment of an international convention on wetlands. This led in 1971 to the establishment of the "Ramsar Convention on Wetlands" (Carp 1972), global in scope, which recognized the great value of wetlands, "the loss of which would be irreparable," to people, and which has the desire to "stem the loss and degradation of wetlands now and in the future," through the wise use of all wetlands, the designation and management of Wetlands of international importance ("Ramsar Sites") and international cooperation.

It has been widely reported that 50% (or at least 50%) of the world's wetlands have been lost (or lost since 1900), but the provenance of this figure is obscure. Its origin appears to date back to reports for a few parts of the USA in mid-1950s and then reported to a wildlife conference in Minnesota, USA, in 1981. Drawing on these reports only, Winkler and DeWitt (1985), in a paper on impacts of peat mining, then stated that "*the biggest changes in land-use since 1900 have been a 50% decrease in wetlands globally,*" introducing unsubstantiated global and temporal elements to the earlier statements. This has subsequently, in various forms, been restated in numerous other publications but without supporting data.

Davidson (2014) made a first global assessment of the published evidence for temporal and geographical trends in the extent of wetlands, and rates of wetland area change, from 189 reports in published scientific journal papers and reports. These cover a wide range of spatial scales, from a single wetland to national, regional, and global scales, and widely different time periods, from a few years to many centuries. The analysis was done to determine whether or not there is evidence to support the statements that the world has lost 50% of its wetlands, since historical times or since 1900 AD. With wetland losses known to have continued during the last quarter of the twentieth century and beyond, could such a figure also be an underestimate of overall

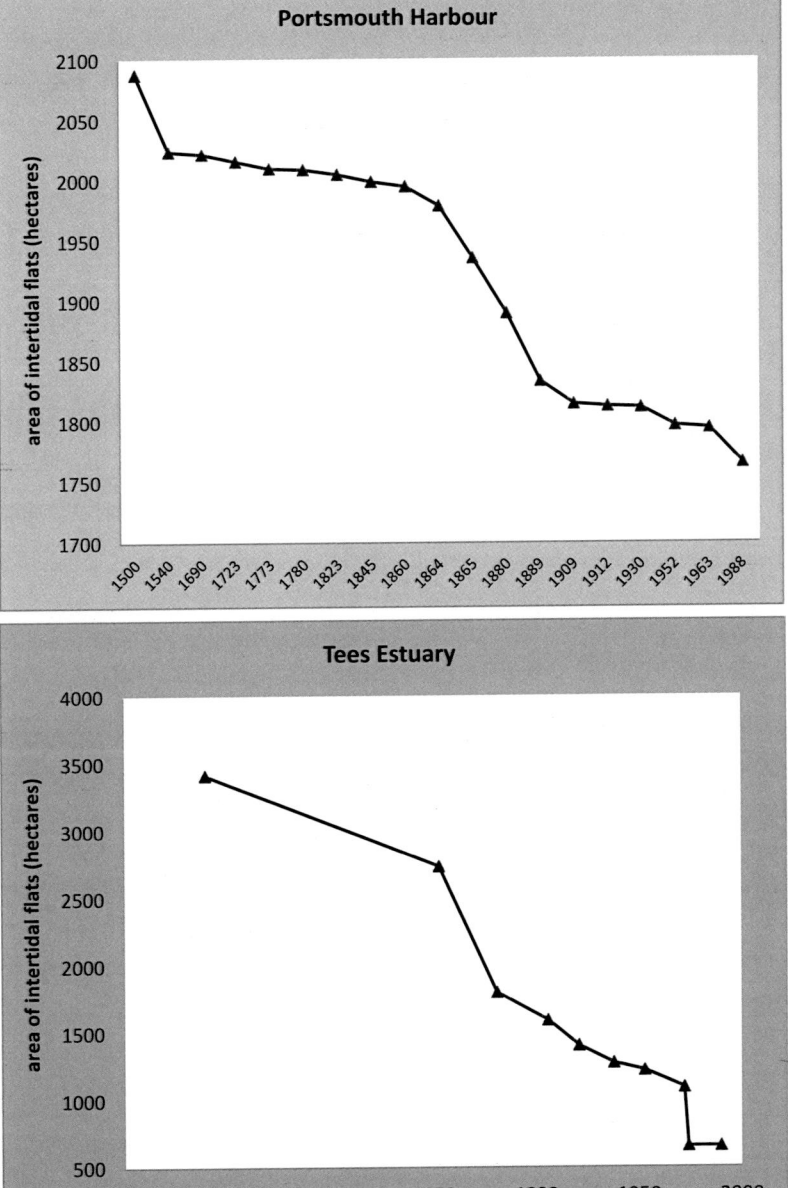

Fig. 1 Two examples of long-term progressive loss of estuarine area through the land-claim of intertidal flats and marshes in British estuaries: Portsmouth Harbour in southern England since 1500 AD and the Tees Estuary in north-east England since 1720 AD. In each case, early land claims were predominantly the embankment of saltmarshes and conversion to agriculture, and more recent land claims of saltmarsh and intertidal flats were for industrial, port, urban and infrastructure developments, and waste disposal (Drawn from data in Davidson et al. 1991. © Joint Nature Conservation Committee)

twentieth century wetland loss? Further, is there evidence that the loss of wetlands has been stemmed in the more than 40 years since the establishment of the Ramsar Convention in 1971?

The published evidence based on wetland area change is patchy and limited and there is a need to much improve this situation. There are few parts of the world yet with comprehensive assessments of wetland area change, notably exceptions being the USA, China, and Europe. Most records identified by Davidson (2014) (over 80% of the total) were for Europe, Asia, and North America, with far fewer for Africa, Neotropics, and Oceania. Some types of wetland are not well represented in the dataset: there is little information (even of baseline area from global wetland inventories and mapping) on ephemeral or intermittently flooded wetlands such as wet meadows and arid and semiarid zone shallow depressions. There is also a lack of data for areas and especially for trends in some of the world's major flooded forest areas such as the Amazon and Congo Basins (see e.g., Finlayson et al. 1999; Maltchik 2003; Keddy et al. 2009).

Reported extents and rates of loss of wetlands are partly related to the geographical scale of the assessment, with higher average values for studies at the wetland site scale than for larger scales, although the differences are not great. Why is not clear, but studies at large spatial scales are more likely to include both wetlands which have been converted and those whose area has remained unchanged. It is also possible that the research community has focussed its site-scale attention, at least in publications, on those wetlands which are known to have been partially or wholly converted and lost, and/or are under threat of further conversion, rather than those wetlands which have remained unchanged.

Long-Term Wetland Losses

All 63 records of long-term (dating from a start year prior to 1900 AD) natural wetland area change analyzed by Davidson (2014) report a loss of wetland area. The average overall long-term loss of natural wetlands was 53.5%, with more loss, on average, of inland natural wetlands (60.8%) than coastal wetlands (46.4%). Excluding site-scale records, average overall loss was 56.9%, with 59.2% for inland and 49.8% for coastal natural wetlands. This historical (long-term) loss of 54–57% of the world's wetlands exceeds the widely stated but unsubstantiated figure of 50%.

Long-term rates of loss for all types of natural wetland averaged $-0.296\% \text{ y}^{-1}$. Rates of loss were 1.75 times faster for inland wetlands ($-0.391\% \text{y}^{-1}$) than for coastal wetlands ($-0.228\% \text{y}^{-1}$). Excluding site-scale records, the average rate of loss was $-0.258\% \text{ y}^{-1}$ and was also faster for inland ($-0.342\% \text{y}^{-1}$) than coastal ($-0.298\% \text{y}^{-1}$) natural wetlands.

Long-term loss of wetlands has been reported from all regions of the world, with largest overall losses for Europe and North America. Regional average losses were: Africa 43.0%, Asia 45.1%, Europe 56.3%, North America 56.0%, and Oceania 44.3%. Average long-term rates of loss were 1.8 times faster in Europe ($-0.323\% \text{y}^{-1}$) than in

North America ($-0.181\%y^{-1}$). There were insufficient records to calculate long-term rates of loss for other regions.

Long-term rates of loss varied considerably, from -0.005 to $-1.089\%y^{-1}$, but most were in the range -0.100 to $-0.400\%y^{-1}$. Fastest long-term rates (above $-0.500\%y^{-1}$) were reported for lowland raised bogs (peatlands) in part of the United Kingdom; floodplains in parts of Germany and the USA; and freshwater marshes, coastal marshes, and saltmarshes in parts of Italy.

Wetland Losses in the Twentieth and Early Twenty-First Centuries

Rates of loss of all natural wetland types during the twentieth and early twenty-first centuries from 117 reports averaged $-1.085\%y^{-1}$, much faster than the long-term loss rate. Excluding site-scale records, the average rate of loss was $-0.901\%y^{-1}$, also much faster than the equivalent long-term rate. This pattern of faster loss rates in the twentieth and early twenty-first centuries occurred for both inland and coastal natural wetlands. Although in the twentieth/early twenty-first century the rate of loss of coastal natural wetlands was still slightly lower than that of inland natural wetlands, the relative increase in their rate of loss was greater, such that coastal wetlands were being lost 4.2 times faster and inland wetlands 3.0 times faster than in the long term.

Rates of loss of all wetlands for each of four time periods of the twentieth and early twenty-first centuries (1900–1944, 1945–1974, 1975–1989, and 1990 onwards) were considerably faster than long-term loss rates (Fig. 2): between 1.9 times (for 1900–1944) and 4.6 times (for 1945–1974). Average rates of loss were highest ($-1.363\%y^{-1}$) in the third quarter of the twentieth century, the period when there was also the highest percentage (61%) of reported losses occurring at a high ($>-1\%y^{-1}$) average annual rate. Average loss rates continued to be almost as fast ($-1.308\%y^{-1}$) into the last quarter of the century (1975–1989) but slowed after 1990 ($-0.565\%y^{-1}$).

Rates of loss were consistently higher for natural inland than natural coastal wetlands for the periods from 1900 through to the 1980s, but while the average rate of loss of inland wetlands slowed considerably after 1990, rapid losses have continued for coastal wetlands (Fig. 2). Why loss of coastal wetlands has been generally slower than inland wetlands is not clear, but it may be that the technologies needed to drain and convert inland wetlands have, until recently, been relatively easier to apply than those needed for the impoundment and infilling of often highly dynamic coastal systems (see Fig. 3).

Wetland losses occurred during the twentieth and early twenty-first centuries in all regions of the world (Fig. 4). Rates of loss of all natural wetlands were slowest in North America and fastest in the Neotropics and Asia. For inland wetlands (Africa, Asia, Europe, and North America only) rates were slowest in North America and fastest in Africa and Asia, with a similar pattern for coastal wetlands.

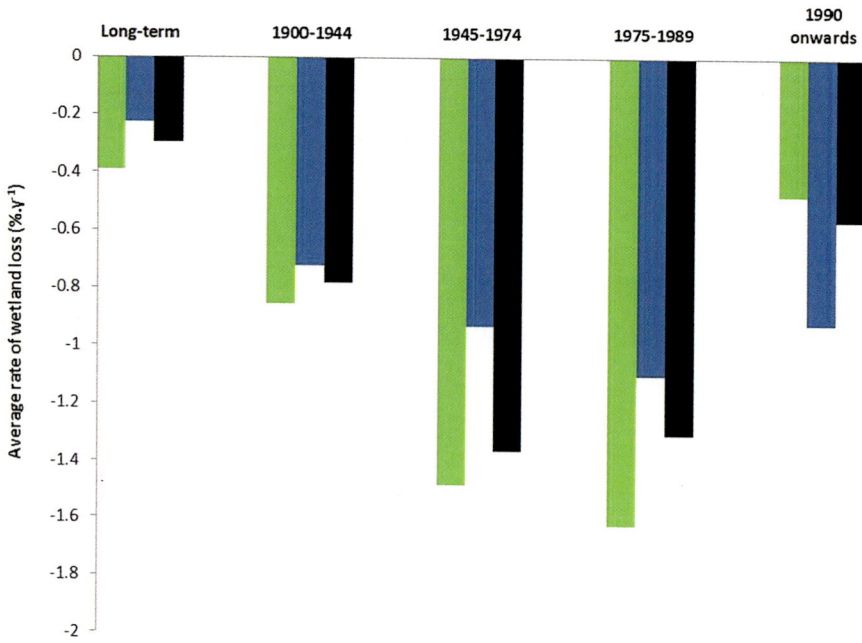

Fig. 2 Average rates of loss of inland, coastal, and all natural wetlands in different time periods (Drawn from data in Davidson 2014). *Green bars* inland wetlands; *blue bars* coastal wetlands; *black bars* all natural wetlands

Similarly, Dixon et al. (2016) report a 30% loss of wetlands in the 38 years from 1970–2008, with largest losses in Asia and Europe, and smallest in North America and Oceania.

The temporal patterns of rates of loss through the twentieth century varied regionally. In Europe, there were particularly high rates in the 1945–1974 period but much slower for 1990s onward. In North America, rates were consistently lower than in Asia and Europe and did not vary during the twentieth century. Rates of loss in Asia have been consistently high across the different periods of the twentieth century. Particularly high loss rates ($>1.5\%y^{-1}$) since 1975 have been reported for both inland and coastal wetlands in China, tropical peatswamp forest in Borneo, and inland wetlands in part of New Zealand.

Most reports of losses of natural wetlands for the twentieth/early twenty-first centuries were for less than the full time period: the average wetland loss was 38.5% for an average period of 38.6 years. Extrapolating from average rates of wetland loss for the different periods of the twentieth century, only 29.4% of wetlands present in 1900 AD may have remained by the end of the century – a loss of 70.6%, with a loss of 63.0% of coastal natural wetlands and 75.0% for inland natural wetlands. The figures are similar but slightly lower when site-scale records are excluded: an average wetland loss of 34.9% for an average period of 39.2 years, and a twentieth century 64.0% loss extrapolated from average rates of loss in each time period (62.0% for

Fig. 3 Land claim of intertidal flats can be challenging: here a truck infilling part of the Tees Estuary, north-east England in 1974 for port and industrial developments is sinking irretrievably into the mud. Much of the area infilled has subsequently remained unused for the last 40 years (Photo credit: Mike Pienkowski © Rights remain with the photographer)

coastal wetlands and 68.8% for inland wetlands). These figures of 62–75% loss of wetland area existing in 1900 AD are considerably higher than the unsubstantiated statements of 50% loss during the twentieth century.

The continuing losses reported for natural inland and coastal wetlands contrast with increases in the area human-made (artificial) wetlands (as noted earlier in Swift 1964). In Europe, open waters increased by 4.4% between 1990 and 2006, attributed mostly to creation of reservoirs and other artificial waterbodies (EEA 2010). In China, in the 30 years between 1978 and 2008 while natural inland wetlands decreased in area by 33%, artificial inland wetlands increased by 122% (Niu et al. 2012), in the USA the area of restored and created ponds increased by 12% between 1985 and 2004 (Dahl 2006) and the global area of rice paddy harvested increased by 41.5% between 1961 and 2012 (FAOSTAT http://faostat.fao.org accessed 10 July 2014).

Overall Natural Wetland Loss Since the Start of the Eighteenth Century

Although reported long-term loss of natural wetlands averaged 54–57%, overall losses may have been much greater than these values. Extrapolation from the average rates of wetland loss since the start of the eighteenth century (from Davidson

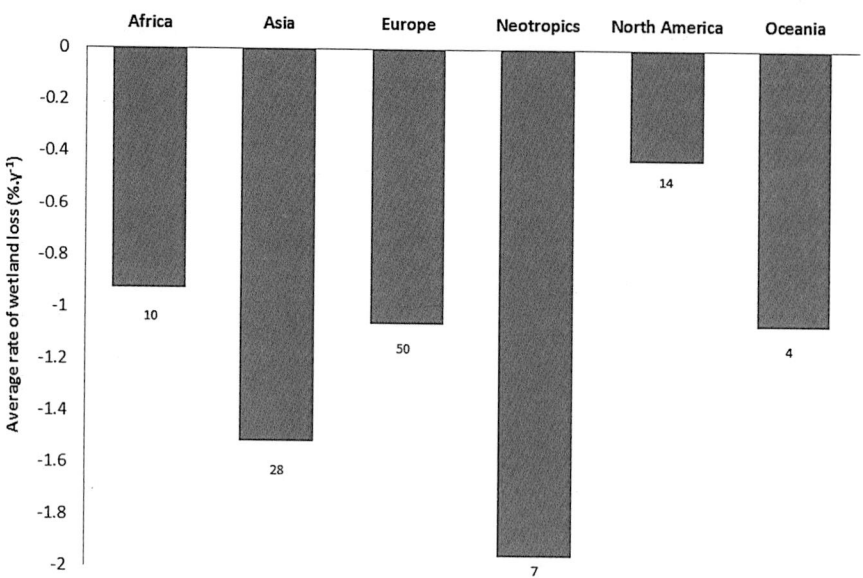

Fig. 4 Average rates of loss of all natural wetlands in different regions during the twentieth and early twenty-first centuries (Drawn from data in Davidson 2014). Numbers are the number of records for each region

2014) suggests that of the natural wetland area existing in 1700 AD, 76.3% remained in 1800 AD and 44.1% in 1900 AD, but only 13.0% at the end of the twentieth century – an overall loss of 87.0% since the start of the eighteenth century. A similar extrapolation for wetland area at the start of the nineteenth century suggests that only 16.9% remained at the end of the twentieth century – a loss of 83.1% since 1800 AD.

The patterns of wetland area change reported here raise questions about whether the governments that are Parties to the Ramsar Convention are effectively addressing the Convention's 1971 desire to "stem the loss and degradation of wetlands." The high rates of wetland loss reported for some regions, such as Asia and Europe, in the second half of the twentieth century suggest that the concerns which led to the creation of the convention were well founded. However, widespread wetland conversion and loss has continued. Although the continuing low rate of loss in North America and the considerably reduced rate of loss in Europe are encouraging signs, the continuing very high rates of loss of both inland and coastal natural wetlands in Asia are of particular concern (see MacKinnon et al. 2012; Murray et al. 2014). Finlayson (2012) presented this as a paradox whereby while there had been a substantial investment in wetland policy and information through the Ramsar Convention this had not stopped or reversed the global loss and/or degradation of wetlands: a situation that raised questions about the extent or effectiveness of national implementation of the convention.

Davidson (2014) looked only at the overall changes in the area of remaining wetlands and not at the state of those that remain, and so provides only part of the picture. As well as the rapid and continuing loss of coastal and inland wetlands, remaining wetlands continue to face severe pressures, despite many benefits of high value they provide to people (Finlayson et al. 2005; Carpenter et al. 2011; Russi et al. 2013; Costanza et al. 2014), and the many conservation/restoration successes from recent efforts at local to national to global scales. Although there has been no overall assessment of the state of health of the world's remaining wetlands, many are recognized as having deteriorated in status and to be currently degraded. In 2012, 127 national governments reporting to the Ramsar Convention indicated that the overall status of their wetlands had deteriorated in recent years in 28% of countries but had improved in only 19% (Ramsar Convention 2012).

Status of Wetland-Dependent Species

Unsurprisingly, trends in the status of wetland-dependent species broadly follow the overall patterns of continuing (and in some places accelerating) wetland losses.

The Living Planet Index (LPI), derived from trends in freshwater species and populations, declined by 37% in the 38 years from 1970 to 2008 – a larger decline than for any other biome. While it is encouraging that there was an increase of 36% in the temperate freshwater index, for tropical regions there has been a major (70%) decline (WWF 2012). The marine LPI (which includes many coastal wetland-dependent species) also declined (by 22%) over the same period. Regionally, the decline in the overall LPI has been greatest (64% since 1970) in the Indo-Pacific biogeographic realm.

Similarly, the Red List Index of globally threatened species for birds declined for freshwater species and for waterbirds between 1998 and 2004, with a faster decline for freshwater species than most other taxa (Butchart et al. 2004). There have also been major declines in the Red List status of amphibians and corals (Butchart et al. 2010).

Although the global status of waterbird biogeographic populations improved slightly overall during the period 1976–2005, more populations remain in decline (38%) than are increasing (20%) (Wetlands International 2010). As for the LPI, this global trend masks major differences in status across different regions, flyways, and taxa. While populations in Europe and North America have had relatively good, and improving, status since the mid-1970s, those depending on South America and Africa and long-distance migrants worldwide have a much poorer and declining status. The status of all types of waterbird population in the Asia-Pacific region has been, and continues to be, particularly bad.

While the status of some waterbird taxa has improved, that of others has deteriorating rapidly: the status of migratory shorebirds (Charadrii: sandpipers, plovers, and their allies) decreased by 33% over the 20 years to 2005 (Butchart et al. 2010). Their population status at the end of the twentieth century on all flyways was poor, with more declining than increasing populations (Fig. 5): it is best on the African-Eurasian East Atlantic flyway and worst on the East Asia-Australasia flyway, with over 80% of populations for which there is a trend being in decline at the end of the twentieth century. Nonmigratory (endemic) populations had a generally better status than migratory populations.

The fate of shorebird populations since the end of the twentieth century has varied between flyways (Fig. 6). For the North America/intercontinental flyway, data in Stroud et al. (2006) indicate a Population Status Index (PSI) of -0.458, with 52% of populations assessed in decline. An assessment in the mid-2000s (Morrison et al. 2006) suggests a status deterioration, with 78% of populations declining

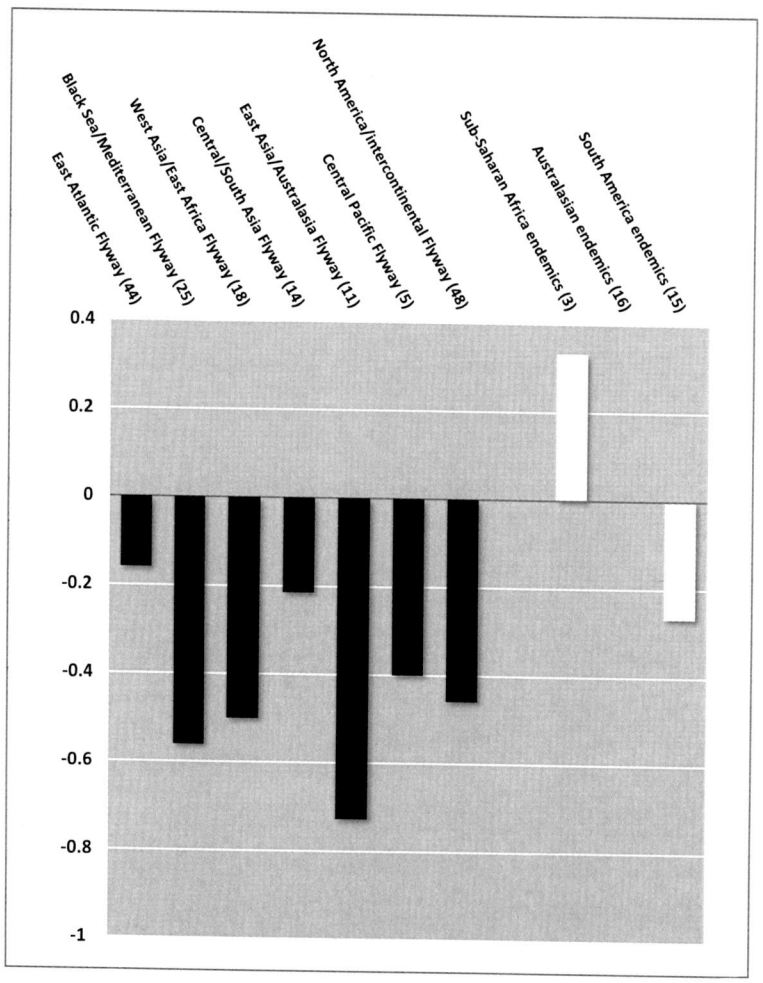

Fig. 5 The status (Population Status Index – PSI) of shorebird biogeographic populations at the end of the twentieth century (Drawn from data in Stroud et al. 2006. © Scottish Natural Heritage). *Black bars,* migratory populations; *white bars,* endemic resident populations. Numbers in parentheses are the number of populations for which a trend was available. The PSI is on a scale of +1 (all populations increasing) to -1 (all populations decreasing). For further information on methods see Butchart et al. (2010) and Wetlands International (2010). The PSI for Australasian endemics was 0

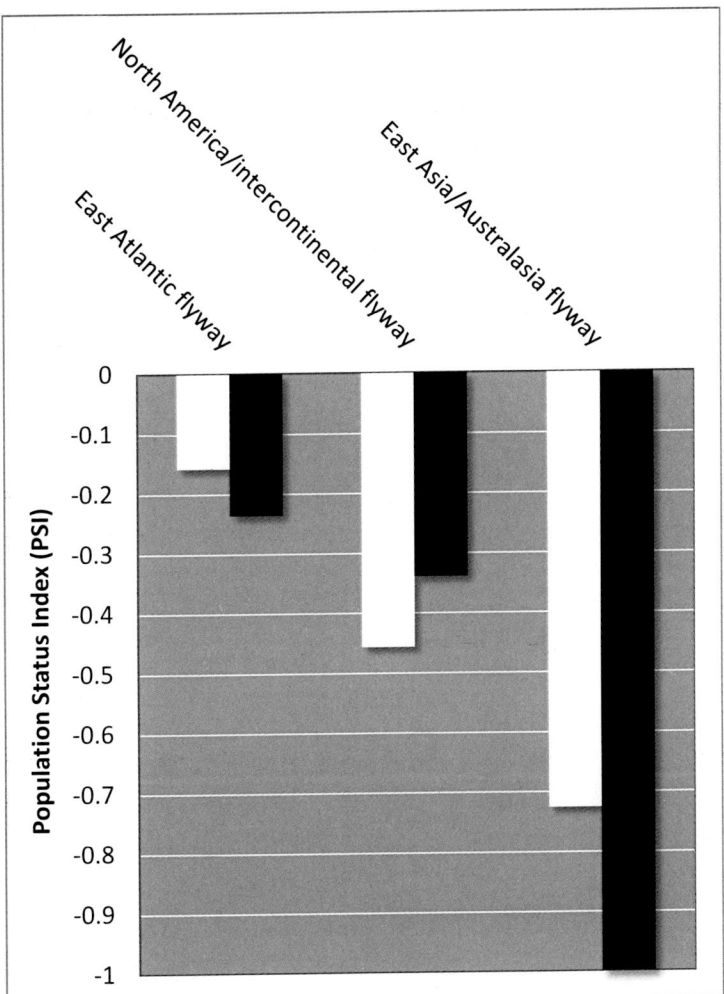

Fig. 6 Trends in migratory shorebird population status on three flyways between the end of the twentieth century (*white bars*) and the mid-2010s (*black bars*). For data sources see text. There were insufficient recent population trends available for other flyways

(PSI = −0.740). However, a reassessment for the mid-2010s (Andres et al. 2012) indicates, for the same populations, a considerable recent improvement in status, with now only 42% of populations declining (PSI = −0.340). For the East Atlantic flyway, Stroud et al (2006) reported 36% of populations decreasing (PSI = −0.159). Although there is a surprising paucity of more recent trend assessments, two (van Roomen et al. 2012; Blew et al. 2013) in the 2010s for the international Wadden Sea, the single most important wintering and migratory staging area on this flyway suggest that there has been some further deterioration of shorebird status since the end of the twentieth century: of 19 populations, 47% were in decline, with a PSI of −0.237. The status of shorebird populations depending on the East Asia/Australasia flyway was already very poor at the end of the twentieth century, with a PSI of −0.727 and 82% of populations declining. For 14 populations for which there is a 2010s trend assessment available (Waterbird Population Estimates database http://wpe.wetlands.org/; accessed 5 July 2014) all are now in decline (PSI = −1.000).

In conclusion, wetland conversion and loss in the long term was in excess of 50% and as much as 87% since the beginning of the eighteenth century; wetland loss was almost four times faster in the twentieth century than previously, with losses of up to 70% of wetlands existing in 1900 AD; conversion of coastal natural wetlands accelerated more that of inland natural wetlands in the twentieth century; and conversion and loss is continuing in all parts of the world. While low or reduced rates of wetland loss in some regions such as Europe and North America may be linked to the relatively good or improving status there of wetland-dependent species such as shorebirds, Asia is of particular concern. There the rapidly deteriorating status of shorebirds has been linked to the large-scale and rapid rate of land claim of particularly coastal wetlands in the Yellow Sea (MacKinnon et al. 2012; Murray et al. 2014), such as Saemangeum, Republic of Korea (Birds Korea – Australasian Wader Studies Group 2008) and Bohai Bay, China (Yang et al. 2011). The fate of the world's remaining wetlands and the species which depend upon them is very uncertain.

References

Andres BA, Smith PA, Morrison RIG, Gratto-Trevor CL, Brown SC, Friis CA. Population estimates of North American shorebirds. Wader Study Group Bull. 2012;119:178–94.

Birds Korea – Australasian Wader Studies Group. The 2006–2008 Saemangeum Shorebird Monitoring Program Report. Busan: Birds Korea; 2008.

Blew J, Günther K, Kleefstra R, Laursen K, Schieffarth G. Trends of migratory and wintering waterbirds in the Wadden Sea 1987/1988-2010/2011, Wadden sea ecosystem no. 31. Wilhelmshavn: Common Wadden Sea Secretariat; 2013

Butchart SHM, Stattersfield AJ, Bennun LA, Shutes SM, Akcakaya HR, et al. Measuring global trends in the status of biodiversity: red list indices for birds. PLoS Biol. 2004;2(12):e383.

Butchart SHM, Walpole M, Collen B, van Strien A, Scharlemann JPW, et al. Global biodiversity: indicators of recent declines. Science. 2010;328:1164–8. doi:10.1126/science.1187512.

Carp E, editor. Proceedings of the International Conference on the Conservation of wetlands and waterfowl. Ramsar, Iran, 30 January – 3 February 1971. IWRB: Slimbridge; 1972.

Carpenter SJ, Stanley EH, Vander Zanden MJ. State of the world's freshwater ecosystems: physical, chemical, and biological changes. Annu Rev Environ Resour. 2011;36:75–99.

Costanza R, de Groot R, Sutton P, van der Ploeg S, Anderson SJ, Kubiszewski I, Turner RK. Changes in the global value of ecosystem services. Glob Environ Chang. 2014;26:152–8.

Dahl TE. Status and trends of wetlands in the coterminous United States 1998 to 2004. Washington, DC: U.S. Department of the Interior, Fish and Wildlife Service; 2006.

Davidson NC. How much wetland has the world lost? Long-term and recent trends in global wetland area. Mar Freshw Res. 2014;65:934–41.

Davidson NC, Laffoley D'A, Doody JP, Way LS, Gordon J, Key R, Drake CM, Pienkowski MW, Mitchell RM, Duff KL. Nature conservation and estuaries in Great Britain. Peterborough: Nature Conservancy Council; 1991.

Dixon MJR, Loh J, Davidson NC, Beltrame C, Freeman R, Walpole M. Tracking global change in ecosystem area: The Wetland Extent Trends index. Biological Conservation. 2016;193:27–35.

EEA. The European environment – state and outlook 2010. Copenhagen: European Environment Agency; 2010.

Finlayson CM. Forty years of wetland conservation and wise use. Aquat Conserv Mar Freshwat Ecosyst. 2012;22:139–43.

Finlayson CM, Davidson NC, Spiers AG, Stevenson NJ. Global wetland inventory – current status and future priorities. Mar Freshw Res. 1999;50:717–27.

Finlayson CM, D'Cruz R, Davidson N. Ecosystems and human well-being: wetlands and water. Synthesis. Millennium Ecosystem Assessment. World Resources Institute, Washington, DC; 2005.

Hoffmann L. editor Proceedings of the MAR Conference organised by IUCN, ICBP and IWRB. Stes-Maries-de-la-Mer, France, 12–16 November 1962. IUCN Publications new series 3. IUCN, Morges; 1964.

Keddy PA, Fraser LH, Solomeshch AI, Junk WJ, Campbell DR, Arroyo MTK, Alho CJR. Wet and wonderful: the world's largest wetlands are conservation priorities. Bioscience. 2009;59:39–51. doi:10.1525/bio.2009.59.1.8.

MacKinnon J, Verkeuil YI, Murray N. IUCN situation analysis on East and Southeastern Asian intertidal wetlands, with particular reference to the Yellow Sea (including the Bohai Sea), Occasional paper of the IUCN Species Survival Commission no. 47. Gland/Cambridge, UK: IUCN; 2012.

Maltchik L. Three new wetlands inventories in Brazil. Interciencia. 2003;28(7):421–3.

Morrison RIG, McCaffery BJ, Gill RE, Skagen SK, Jones SL, Page GW, Gratto-Trevor CL, Andres BA. Population estimates of North American shorebirds, 2006. Wader Study Group Bull. 2006;111:67–85.

Murray NJ, Clemens RS, Phinn SR, Possingham HP, Fuller RA. Tracking the rapid loss of wetlands in the Yellow Sea. Front Ecol Environ. 2014. doi:10.1890/130260.

Niu ZG, Zhang HY, Wang XW, Yao WB, Zhou DM, et al. Mapping wetland changes in China between 1978 and 2008. Chin Sci Bull. 2012. doi:10.1007/s11434-012-5093-3.

Ramsar Convention. Report of the Secretary General on the implementation of the Convention at the global level. Ramsar COP11 DOC. 7; 2012. Available at http://www.ramsar.org/pdf/cop11/doc/cop11-doc07-e-sg.pdf. Accessed 10 Nov 2013.

Russi D, ten Brink P, Farmer A, Badura T, Coates D, et al. The economics of ecosystems and biodiversity for water and wetlands. London/Brussels/Gland: IEEP/Ramsar Secretariat; 2013.

Stroud DA, Baker A, Blanco DE, Davidson NC, Delany S, Ganter B, Gill R, Gonzales P, Haanstra L, Morrison RIG, Piersma T, Thorup O, West R, Wilson J, Zöckler C. The conservation and population status of the world's waders at the turn of the millennium. In: Boere GC, Galbraith CA, Stroud DA, editors. Waterbirds around the world. Edinburgh: The Stationary Office; 2006. p. 643–8.

Swift JJ, editor. Proceedings of the first European meeting on wildfowl conservation. St. Andrews, Scotland, 16–18 October, 1963. London/Le Sambuc: The Nature Conservancy/International Wildfowl Research Bureau; 1964.

van Asselen S, Verburg PH, Vermaat JE, Janse JH. Drivers of wetland conversion: a global meta-analysis. PLoS One. 2013;8(11):e81292. doi:10.1371/journal.pone.0081292.

van Roomen M, Laursen K, van Turnhout C, van Winden E, Blew J, Eskildsen K, Günther K, Hälterlein B, Kleefstra R, Potel P, Sxhrader S, Luerssen G, Ens BJ. Signals from the Wadden Sea: population declines dominate among waterbirds depending on intertidal mudflats. Ocean Coast Manag. 2012;68:79–88.

Wetlands International. State of the world's waterbirds. Delany S, Nagy S, Davidson N (compilers). Wetlands International, Ede; 2010.

Winkler MG, DeWitt CB. Environmental impacts of peat mining: documentation for wetland conservation. Environ Conserv. 1985;12:317–30.

WWF. Living Planet Report 2012. Gland: WWF International; 2012.

Yang H-Y, Chen B, Barter M, Piersma T, Zhou C-F, Li F-S, Zhang Z-W. Impacts of tidal land reclamation in Bohai Bay, China: ongoing losses of critical Yellow Sea waterbird staging and wintering sites. Bird Conserv Int. 2011;21:241–59.

Alien Plants and Wetland Biotic Dysfunction

27

C. Max Finlayson

Contents

Introduction 384
Population Growth and Migration 384
Methods of Invasion 386
Assessing the Impact of Alien Species 387
Methods of Control 388
Critical Lessons and Challenges 388
References 389

Abstract

The impact of alien plants on wetlands has been recognized for some time, with the displacement of native species and subsequent changes in the nutrient and energy dynamics of the wetlands. In many instances the importance of alien species was not widely recognized until it was too late and the plants had spread and invaded large areas, with *Elodea canadensis* being an example from the nineteenth century and *Spartina alterniflora* more recently. The reasons for this included: (i) a generally low level of public and institutional awareness of the problems; (ii) insufficient information about alien species and ways of controlling them; or (iii) insufficient capacity to collect information or implement control measures. The extent of wetland dysfunction caused by alien plants is now more recognized, but prevention and control can still prove difficult.

Keywords

Weeds · Alien plants · Risk assessment · Weed control

C. M. Finlayson (✉)
Institute for Land, Water and Society, Charles Sturt University, Albury, NSW, Australia

UNESCO-IHE, The Institute for Water Education, Delft, The Netherlands
e-mail: mfinlayson@csu.edu.au

© Springer Science+Business Media B.V., part of Springer Nature 2018
C. M. Finlayson et al. (eds.), *The Wetland Book*,
https://doi.org/10.1007/978-94-007-4001-3_48

Introduction

The impact of alien plants on wetlands has been recognized for some time, with the displacement of native species and subsequent changes in the nutrient and energy dynamics of the wetland being common outcomes. Unfortunately, in many instances the importance of alien species was not recognized until it was too late and the plants had spread and invaded large areas, with *Elodea canadensis* being an example from the nineteenth century and *Spartina alterniflora* more recently. The reasons for this lack of appreciation have included: (i) a generally low level of public and institutional awareness of the problems; (ii) insufficient information about alien species and ways of controlling them; or (iii) insufficient capacity to collect information or implement control measures. The extent of wetland dysfunction caused by alien plants is now more widely recognized, but prevention and control can still prove difficult.

The following definitions have been adopted as part of a global strategy on invasive alien species in order to bring some consistency to the language used to describe alien species (McNeely et al. 2001):

- An *alien species* is a species, subspecies, or lower taxon, introduced outside its normal past or present distribution; includes any part of such species that might survive and subsequently reproduce.
- An *invasive alien species* is an alien species whose establishment and spread threaten ecosystems, habitats, or species, with economic or environmental harm.
- A *weed* is a plant (not necessarily alien) that grows in sites where it is not wanted and has detectable negative economic or environmental effects.

In the text that follows, the emphasis is placed on invasive alien species that adversely affect the ecological character of wetlands, that is, their introduction results in biotic dysfunction.

Population Growth and Migration

In considering alien plants in wetlands, we need to consider the balances and checks that influence the growth and movement of specific species. When deciding whether a species is invasive we need to assess the nature of movements within its "normal" range, keeping in mind that extremes of climate can cause species to move outside of their established range. At times it can be difficult to ascertain the "native range" of some species, especially those that are highly mobile or respond to climate variability. However, there is no doubt that many species have established in nonnative habitats, with prominent examples including salvinia *Salvinia molesta* and water hyacinth *Eichhornia crassipes* (Fig. 1) that originated in South America and which are now widely spread across the tropics.

The example of Canadian pondweed *Elodea canadensis* is illustrative of the dilemma faced when considering the spread and distribution of invasive species. It is the first documented example (Sculthorpe 1967) of the explosive growth of an

Fig. 1 Water hyacinth *Eichhornia crassipes* has invaded water bodies across the tropics and subtropics (Photo credit: CM Finlayson © Rights remain with the author)

aquatic weed. It originated in North America and invaded the waterways of Europe in the late nineteenth century, growing rapidly and spreading through vegetative means and reaching a maximum density within a few months to 4 years. This density was generally maintained for up to 5 years and then declined to levels that were not considered a nuisance. The reasons for the rapid increase and subsequent decline were not determined.

Other species, such as the wetland shrub, *Mimosa pigra* (Fig. 2), can be introduced to an area, but not spread for many years before "breaking out" and rapidly invading large areas of floodplains. Experience in northern Australia and Africa illustrates that once established this plant is extremely hard to control and provides a source of invasion for nearby areas. *Egeria densa* is an example of a wetland plant that established and proliferated over many years in a newly created wetland in southern Chile but with wide population fluctuations, including a massive and contentious decline in 2004 (Delgado and Marin 2013). Other species, such as the common reed *Phragmites australis* and cattail *Typha* spp (Fig. 3), may become invasive in their home range if the environmental conditions change.

For all of the abovementioned species, the environmental determinants and drivers of change in their spatial extent and density were not widely understood with management actions being unprepared for the rate of spread, or confounded by later declines. This illustrates the need to consider the growth patterns of any alien species and to ascertain if and when it may become invasive, necessitating a combination of botanical and ecological investigations, possibly complemented by risk assessment to address the likelihood of a species becoming invasive and the consequences if it does.

Fig. 2 *Mimosa pigra* is native to Central America and has become invasive in other tropical areas (Photo credit: CM Finlayson © Rights remain with the author)

Fig. 3 Common reed *Phragmites australis* and *Typha* species are widespread and under some circumstances can become invasive (Photo credits: CM Finlayson © Rights remain with the author)

Methods of Invasion

Increasing social and economic globalization, including travel and trade, as well as agricultural and other commercial programs has accidentally or deliberately led to the introduction of alien species to wetlands. Human activities have often enabled invasive species to be transported from one locality to another or resulted in changes in habitat conditions to an extent that alien species thrive and become a nuisance.

At the same time, only a small proportion of introduced species are likely to flourish and become a serious problem. These species have traits that enable them to reproduce or establish in new habitats, or in new areas within their native range, especially if these have been altered or degraded.

Management responses need to consider both the characteristics of the particular species and those habitats that are prone to invasion. Thus, the manner in which plant propagules are transported and establish in specific habitats needs to be considered in addition to noting the occurrence of mature plants. Many wetlands are prone to invasion by propagules of species that are carried by water or by highly mobile or migratory animals. Further, low species richness in some wetland habitats may enable propagules to establish readily. The interrelationships between species diversity and habitat characteristics of wetlands, and invasion by alien species, deserve further investigation.

Assessing the Impact of Alien Species

The specific impact of an alien species on native species in wetlands includes the displacement or loss of native species, at least seasonally or in the short term, and changes to the nutrient and energetic dynamics in the wetland. Massive populations, such as those of the abovementioned floating aquatic species, salvinia and water hyacinth, have been widely recorded and documented across the tropics. Similarly, the spread and dominance of submerged species and wetland grasses have been widely recorded. In many instance though, the actual impact of such species has not been intensively investigated beyond a commentary focused on the extensive spatial distribution and biomass of the invading species.

In response to this situation, there have been many calls to more thoroughly investigate the impact of invasive species, including the use of formalized risk assessment procedures, such as those proposed by van Dam et al. (1999) and accepted by the Ramsar Wetlands Convention (http://www.ramsar.org/res/key_res_vii.10e.pdf). This involves six steps:

- Identification of the problem
- Identification of the effects
- Identification of the extent
- Identification of the risk
- Risk management and reduction
- Monitoring

An example of the application of risk assessment to a wetland invasive species is presented by Walden et al. (2004).

Risk assessments, such as those mentioned above, have mainly focused on the ecological reasons for change in wetlands, including those traits, such as long-lived or readily transported propagules, that enable plants to spread and establish.

Methods of Control

There are many control measures recommended for individual invasive species, although single solutions may not be as effective as integrated solutions. For many years, generic principles for addressing invasive and pest species have been available, and these are still valid. As an example, Mitchell (1978) proposed that aquatic weeds within Australia were addressed in a series of steps that included assessment and monitoring of infestations using systematic and standardized methods; elimination of potentially dangerous infestations at the earliest opportunity; prevention through quarantine and import regulations; consideration of potential problems during land planning exercises; increased awareness and public education; greater cooperation between government sectors and private developers; and increased knowledge of the causes of weed problems and possible control methods.

At a conceptual level, prevention, quarantine, and early intervention are attractive components of a management regime. Difficulties with implementing such approaches have occurred as a consequence of conflicts with the proponents of economic development, and also as a consequence of increased volumes of trade and transport of many goods, locally and globally. Given these factors, the management of invasive and pest species should not be considered in isolation.

In general, weed species are often tackled through a combination of physical, chemical, or biological control techniques. Early intervention and the control of satellite infestations are recommended methods, with consistent and ongoing surveillance and rehabilitation required after control. In recent years, biological control methods have been promoted as an ecologically sound approach to managing pest species, and some spectacular successes have received a lot of publicity.

In all control or management programs, it is necessary to support field assessments and actions by well-targeted monitoring and research. The former should be based on clear objectives, be constantly reviewed, and cover both the target species and the control actions. Research needs to be targeted and integrated with the management program. Further, the management program should not stop when control is effected; there may be a resurgence of the problem and possibly a need for rehabilitation of the locality or habitat in question.

Critical Lessons and Challenges

There are many lessons that can be learnt from past biotic disruption of wetlands by alien plants, including the following (Finlayson 2009):

- Prevention and early intervention are more likely to be successful or cheaper than later control and management.
- Coordinated and multisectoral approaches that incorporate effective legislative, educational, and coordination steps are more likely to result in partnerships and actions for effective management than are individualistic approaches.

- Management strategies and actions require a concerted and ongoing effort and should be based on a sound information base with the latter being compiled by ongoing and formalized risk assessment and ecological investigations.
- Control measures are generally location and species specific.
- A cost–benefit analysis should precede all control programs. It should consider the need for rehabilitation of areas where control has been successful.
- Balanced against a need for early intervention and prevention is the wisdom of deciding whether any action is needed or is likely to be successful.
- Managers should consider both the nonecological and ecological causes of biotic pressures and direct resources to address long-term strategic outcomes as well as tactical short-term outcomes.

The main challenges facing wetland managers dealing with alien plants are reflected in these lessons. The stumbling block in undertaking effective management is likely to be the vagaries of funding – funding to ensure early intervention and coordinated responses being essential.

References

Delgado M, Marin VH. Interannual changes in the habitat area of the Black-Necked Swan, *Cygnus melancoryphus*, in the Carlos Anwandter Sanctuary, Southern Chile: a remote sensing approach. Wetlands. 2013;33:91–9.

Finlayson CM. Biotic pressures and their effect on wetland functioning. In: Maltby E, Barker T, editors. The wetlands handbook. Oxford: Wiley-Blackwells; 2009. p. 667–88.

McNeely JA, Mooney HA, Neville LE, Schei P, Waage JK, editors. A global strategy on invasive alien species. Gland/Cambridge: IUCN; 2001.

Mitchell DS. Aquatic weeds in Australian inland waters. Canberra: Australian Government Printing Service; 1978.

Sculthorpe CD. The biology of aquatic vascular plants. London: Edward Arnold; 1967.

van Dam RA, Finlayson CM, Humphrey CL. Wetland risk assessment: a framework and methods for predicting and assessing change in ecological character. In: Finlayson CM, Spiers AG, editors. Techniques for enhanced wetland inventory, assessment and monitoring. Canberra: Supervising Scientist Group; 1999. p. 83–118. Supervising Scientist Report 147.

Walden D, van Dam R, Finlayson M, Storrs M, Lowry J, Kriticos D. A risk assessment of the tropical wetland weed *Mimosa pigra* in Northern Australia. Darwin: Supervising Scientist; 2004. Supervising Scientist Report 177.

Ecological Conditions and Health of Arctic Wetlands Modified by Nutrient and Contaminant Inputs from Colonial Birds

28

Mark Mallory

Contents

Introduction	391
Biotransport of Nutrients and Effects on Arctic Wetlands	392
Biotransport of Contaminants and Effects on Arctic Wetlands	393
Contaminants Move from the Marine Environment to the Terrestrial Environment	395
Future Challenges	395
References	395

Abstract

The Arctic supports relatively few but large colonies of seabirds. These birds feed in the ocean, often on contaminated food webs, then concentrate and transport nutrients and contaminants through digestion. Back at the colony, the birds release these chemicals, which can then move into terrestrial food webs, accumulating in organisms that would not normally be exposed to these concentrations. Wetlands around Arctic seabird colonies may be oases of life due to nutrient deposition from seabirds, but they also become point sources of certain contaminants.

Keywords

Seabird · Contaminant · Arctic · Biomagnification · Biotransport

Introduction

The Arctic, the circumpolar area roughly north of the treeline, is a region with a harsh climate: cold temperatures, strong winds, shallow, nutrient-poor soils underlain by permafrost, and limited light for part of the year. Consequently, the growing

---

M. Mallory (✉)
Coastal Wetland Ecosystems, Biology Department, Acadia University, Wolfville, NS, Canada
e-mail: mark.mallory@acadiau.ca

© Springer Science+Business Media B.V., part of Springer Nature 2018
C. M. Finlayson et al. (eds.), *The Wetland Book*,
https://doi.org/10.1007/978-94-007-4001-3_49

season for flora and fauna is quite short. Relatively, a few bird species are adapted to inhabit this region all year, but millions of migratory birds return to the Arctic each summer to exploit the briefly available but bountiful food supplies on the land and in the ocean before heading south in the autumn. These birds include the iconic migrations of geese, which fly principally over land, as well as seabirds, which fly principally over the ocean and out of the sight of man until they are at their nesting sites. While many species of migratory birds nest dispersed across the Arctic tundra, some species nest colonially in a small area, which may be a cliff, an island, or a defined region of tundra. These colonies are comprised of hundreds to millions of birds (Mallory and Fontaine 2004).

Although the Arctic is often thought of as clean and pristine, in part because the region is sparsely inhabited by people and there is relatively little industrial activity, in reality the area receives much contamination from pollutants emitted far to the south (AMAP 2011). These contaminants are assimilated at low trophic levels in food webs and then increase in concentration (biomagnify) at higher trophic levels (AMAP 2011). Thus, birds feeding across the tundra or ocean accumulate contaminants through their diet.

Arctic colonial birds may forage across a large region, notably for seabirds, but they return to their nesting location where they deposit nutrients and contaminants in their feces, in dropped food, in shed tissues like feathers, and through mortality (Blais et al. 2007). Migratory, colonial birds are therefore biovectors, acquiring chemicals over their vast foraging range and then concentrating them at their nesting colonies. The biotransported nutrient subsidies create oases of life in an otherwise spartan landscape (Fig. 1), but the deposition of contaminants can turn the local environment into an analogue of point-source pollution (Blais et al. 2005). By looking at sediment cores in ponds, small lakes, and wetlands (hereafter "wetlands") near the colonies, researchers can peer back in time to examine when and how conditions may have changed as a result of this phenomenon of biotransport.

Biotransport of Nutrients and Effects on Arctic Wetlands

The addition of the nutrients nitrogen and phosphorus by birds to Arctic landscapes has a pronounced effect on the biota of receiving wetlands as well as on surrounding vegetation. Algal production can be markedly higher in affected wetlands, potentially reaching eutrophic status despite their high latitude (Michelutti et al. 2009) – for example, remote, small wetlands at an eider duck colony on Southampton Island, Nunavut, Canada (64°N) resemble sewage lagoons (Mallory et al. 2006). Even in areas where bird contributions of nutrients to Arctic wetlands are less intensive, such as goose inputs on Bylot Island, Nunavut, Canada, ornithogenic effects on wetland chemistry are still detected (Côté et al. 2010). The changes to the biotic communities within the ponds are sufficient that they leave different microfossil histories in sediments (Keatley et al. 2011). By extracting and analyzing sediment cores, scientists can track the effects of changes in climate and bird contributions through time.

Fig. 1 Ponds below an Arctic seabird cliff, receiving nutrients and contaminants in runoff. Note distinctive, bright vegetation in an otherwise barren landscape. Ponds farther from the cliffs showed less influence of seabirds (Photo credit: M Mallory © Rights remain with the author)

Higher diversity and more lush growth of lichens, mosses, and herbaceous vegetation is also evident around many Arctic bird colonies (Ellis 2005), to the point where these sites can be detected as distinct by satellite data using remote sensing techniques (Blais et al. 2007). At some sites, the proliferation of vegetation at bird colonies, and the corresponding increase in terrestrial insects reliant on vegetation or emergent aquatic insects growing in eutrophic ponds, creates verdant oases of life in a polar desert, and these sites are exploited by other species such as insectivorous songbirds (Falconer et al. 2008). Effectively, these locations become local biodiversity hotspots.

Biotransport of Contaminants and Effects on Arctic Wetlands

The same process of biotransport that enhances nutrient levels near Arctic bird colonies also results in elevated levels of contaminants in the local environment (Blais et al. 2007; Evenset et al. 2007). This process is particularly evident at seabird colonies, because these species typically feed at higher trophic levels in marine food chains, and the processes of bioaccumulation and biomagnification of contaminants in these food chains mean that seabirds ingest relatively high contaminant loads compared to terrestrial, colonial birds such as geese. When the contaminants are eliminated to the local environment, many end up entering nearby wetlands where

Fig. 2 A nesting common eider *Somateria mollissima borealis* on a small island in the High Arctic. Eider faeces contribute nutrients and non-essential trace elements like lead (Pb) and manganese (Mn) to wetlands in their nesting colonies (Photo credit: M Mallory © Rights remain with the author)

Fig. 3 A nesting Arctic tern *Sterna paradisaea* on the same island as the common eider in Fig. 2. Terns defecate, drop fish and zooplankton, and many chicks die on the colony. These materials contribute nutrients and particularly persistent organic pollutants and bioaccumulating, non-essential trace elements like mercury (Hg) to wetlands in their colonies (Photo credit: M Mallory © Rights remain with the author)

they accumulate in water or sediment (Brimble et al. 2009; Michelutti et al. 2009). Dietary variation among seabirds, and the consequent types and levels of contaminants ingested (e.g. persistent organic pollutants like DDT, non-essential trace elements like mercury) can be sufficiently different that the isotopic (nitrogen, $\delta^{15}N$) and contaminant signatures left in wetlands can be used to determine what types of birds used those wetlands. For example, at a small island in High Arctic Canada, a wetland used by molluscivorous common eider ducks *Somateria mollissima* (Fig. 2), had low $\delta^{15}N$ sediment values typical of nutrient subsidies from low in marine food webs, and higher concentrations of lead and manganese, two elements often found to be relatively high in eider tissues. In contrast, a wetland on the same island used by piscivorous Arctic terns *Sterna paradisaea* (Fig. 3) had sediments with much higher $\delta^{15}N$ values, indicative of contributions from higher trophic levels in marine food webs, as well as higher concentrations of bioaccumulating mercury (Michelutti et al. 2010).

Contaminants Move from the Marine Environment to the Terrestrial Environment

A consequence of the biotransport of contaminants from the surrounding habitat to a bird colony is that they may then move into a new compartment of the environment. This is best illustrated at seabird colonies, where nearly all of the contaminants at the colony come from marine prey consumed by the birds. The release of these contaminants to local drainage means that they accumulate in receiving wetlands, but are also taken up in vegetation fed by this drainage, such as local lichens and mosses, which may have elevated contaminant levels near Arctic seabird colonies (Choy et al. 2010). Terrestrial organisms feeding on this vegetation, or feeding on biota growing in the affected wetlands then consume and assimilate these contaminants. This explains how snow buntings *Plectrophenax nivalis*, an insectivorous, migratory songbird found in abundance near some Arctic seabird colonies, could have the highest DDT concentrations of the terrestrial food web that was sampled. Snow buntings feed on emergent insects from the wetlands that receive high contaminant loadings from the nearby seabird colony; contaminants initially from marine sources end up in a terrestrial songbird (Choy et al. 2010).

Future Challenges

The issue of biotransport of nutrients and contaminants by birds has been well-studied at some Arctic locations (Blais et al. 2005; Evenset et al. 2007), but the process applies to sites around the world where migratory species concentrate for some stage of their annual cycle (e.g., Krummel et al. 2003; Roosens et al. 2007). Certainly many wetlands in temperate and tropical regions support colonial waterbirds during migration, breeding, or during the winter. The process of biotransport by colonial birds to local wetlands is a recently identified process (Blais et al. 2007), and three particular avenues of research remain to be studied. First, to what extent might receiving wetlands serve as a sink for environmental contaminants, as opposed to serving as a transfer site for these pollutants to get into terrestrial food webs? Second, by creating many artificial wetlands, notably around communities, and having these subsequently colonized by colonial waterbirds, are we affecting contaminant sources and levels for nearby flora and fauna (including municipal water supplies)? Finally, in some regions the biotransport of contaminants by biota could play a major role in local contaminant cycling, and could be incorporated into models examining compartmentalization and movement of contaminants through environments.

References

AMAP. Arctic pollution 2011. Norway: Arctic Monitoring and Assessment Programme; 2011.
Blais JM, Kimpe LE, McMahon D, Keatley BE, Mallory ML, Douglas MSV, Smol JP. Arctic seabirds transport marine-derived contaminants. Science. 2005;309:445.

Blais JM, Macdonald R, Mackay D, Webster E, Harvey C, Smol JP. Biologically mediated transport of contaminants to aquatic systems. Environ Sci Technol. 2007;41:1075–84.

Brimble SM, Foster KL, Mallory ML, MacDonald RW, Smol JP, Blais JM. High Arctic ponds receiving biotransported nutrients from a nearby seabird colony are also subject to potentially toxic loadings of arsenic, cadmium and zinc. Environ Toxicol Chem. 2009;28:2426–33.

Choy ES, Kimpe LE, Mallory ML, Smol JP, Blais JM. Biotransport of marine pollutants to a terrestrial food web: spatial patterns of persistent organic pollutants adjacent to a seabird colony in Arctic Canada. Environ Pollut. 2010;158:3431–8.

Côté G, Pienitz R, Velle G, Wang X. Impact of geese on the limnology of lakes and ponds from Bylot Island (Nunavut, Canada). Int Rev Hydrobiol. 2010;95:105–29.

Ellis JC. Marine birds on land: a review of plant biomass, species richness, and community composition in seabird colonies. Plant Ecol. 2005;181:227–41.

Evenset A, Carroll J, Christensen GN, Kallenborn R, Gregor D, Gabrielsen GW. Seabird guano is an efficient conveyer of persistent organic pollutants (POPs) to Arctic lake ecosystems. Environ Sci Technol. 2007;41:1173–9.

Falconer MC, Nol E, Mallory ML. Breeding biology and provisioning of nestling snow buntings (*Plectrophenax nivalis*) in the Canadian High Arctic. Polar Biol. 2008;31:483–9.

Keatley BK, Blais J, Douglas MSV, Gregory-Eaves I, Mallory M, Michelutti N, Smol JP. Historical seabird population dynamics and their effects on Arctic pond ecosystems: a multi-proxy paleolimnological study from Cape Vera, Devon Island, Canada. Fund Appl Limnol. 2011;179:51–66.

Krummel EM, Macdonald RW, Kimpe LE, Gregory-Eaves I, Demers MJ, Smol JP, Finney B, Blais JM. Aquatic ecology: delivery of pollutants by spawning salmon. Nature. 2003;425:255–6.

Mallory ML, Fontaine AJ. Key marine habitat sites for migratory birds in Nunavut and the Northwest Territories. Can Wildl Serv Occasional Paper. 2004;109.

Mallory ML, Fontaine AJ, Smith PA, Wiebe Robertson MO, Gilchrist HG. Water chemistry of ponds on Southampton Island, Nunavut, Canada: effects of habitat and ornithogenic inputs. Arch Hydrobiol. 2006;166:411–32.

Michelutti N, Keatley BE, Brimble S, Blais JM, Liu H, Douglas MSV, Mallory ML, MacDonald RW, Smol JP. Seabird-driven shifts in Arctic pond ecosystems. Proc R Soc Lond B. 2009;276:591–6.

Michelutti N, Brash J, Thienpont J, Blais JM, Kimpe L, Mallory ML, Douglas MSV, Smol JP. Trophic position influences the efficacy of seabirds as contaminant biovectors. Proc Natl Acad Sci U S A. 2010;107:10543–8.

Roosens L, Van Den Brink N, Riddle M, Blust R, Neels H, Covaci A. Penguin colonies as secondary sources of contamination with persistent organic pollutants. J Environ Monit. 2007;9:822–5.

Lake Chilika (India): Ecological Restoration and Adaptive Management for Conservation and Wise Use

29

Ajit Kumar Pattnaik and Ritesh Kumar

Contents

Introduction	398
Adverse Change in Ecological Character	398
Ecological Restoration	400
Restoration of Ecological Character and Removal from Montreux Record	401
Integrated Management Planning	402
Future Challenges	402
References	403

Abstract

Lake Chilika, a coastal lagoon of high biodiversity significance and base of livelihoods of 0.2 million fishers, was placed under the Montreux Record of Ramsar Convention due to adverse change in ecological character triggered primarily by changes in lagoon's connectivity with Bay of Bengal. An ecological restoration program, built on adaptive and participatory lake basin management program, implemented since 2000 has led to rapid recovery of wetland resources, particularly fisheries and a rejuvenation of aquatic biodiversity. The role of Chilika Development Authority, entrusted with design and implementation of restoration program in coordination with various sectoral agencies, is a wetland governance exemplar. The site was delisted from Montreux Record in 2002.

A. K. Pattnaik (✉)
Chilika Development Authority, Bhubaneswar, Odisha, India
e-mail: ajitpattnaik13@gmail.com

R. Kumar (✉)
Wetlands International South Asia, New Delhi, India
e-mail: ritesh.kumar@wi-sa.org

© Springer Science+Business Media B.V., part of Springer Nature 2018
C. M. Finlayson et al. (eds.), *The Wetland Book*,
https://doi.org/10.1007/978-94-007-4001-3_177

Keywords

Adaptive management · Fisheries · Livelihoods · Management planning · Montreux record

Introduction

Lake Chilika, a brackish-water coastal lagoon situated on the coast of Odisha state (India) (Fig. 1), forms the base of livelihood security for more than 0.2 million fishers and 0.4 million farmers living in and around the wetland and its catchments (Kumar and Pattnaik 2012). Spanning a peak mansoon inundation area of 1,165 km^2 and fringed by a seasonal floodplain extending to 400 km^2, Chilika is an assemblage of shallow to very shallow marine, brackish, and freshwater ecosystems exhibiting high biodiversity and harboring several endangered and endemic species. Chilika is one of only two lagoons in the world that support Irrawaddy dolphin *Orcaella brevirostris* populations, and it commonly provides winter habitat for over one million migratory birds. It has a rich flora of 726 angiosperm species, including several of economic value (Pattnaik 2003). The diverse and dynamic assemblage of fish, invertebrate, and crustacean species provide the basis of a rich fishery which generates more than US$ 17.3 million in annual revenues and contributes over 6% of the state's foreign exchange earnings. The tourism industry linked with wetland, generates more than US$ 44.2 million to various economic sectors. The wetland is also inextricably linked to the local culture and belief systems. Chilika was designated a Wetland of International Importance (Ramsar Site under the Convention on Wetlands) by the Government of India in 1981. Nalabana, a low, flat, marshy island of 15.53 km^2, was designated a wildlife sanctuary in 1987 due to its unique features as habitat for avifauna and nursery ground for fishes.

Adverse Change in Ecological Character

Chilika went through a phase of rapid degradation, particularly since the 1950s owing to increasing sediment loads from catchments and reduced connection with the sea. The Mahanadi floodplains were subject to intensive hydrological regulation during the colonial rule of the eighteenth century (D'Souza 2002). The dynamic fluvial environment of the delta was constrained by embankments and other hydraulic structures to provide a regulated water supply to agricultural fields. In 1984/85 prawn culture was introduced to Chilika as part of a supplementary income program for low-income families (Mohanty 1988). Increased demand for shrimp, especially in European markets, devaluation of the Indian rupee, and trade liberalization induced severe pressures on the use of Chilika Lake's shorelines for prawn aquaculture (Shimpei and Shaw 2009).

29 Lake Chilika (India): Ecological Restoration and Adaptive Management...

Fig. 1 Lake Chilika, Odisha inundated floodplains (Source: Kumar and Pattnaik 2012: Reprinted with permission)

Channelization of floodplains, increased agriculture, and decreasing forest cover in the direct catchments mobilized soil transport and increased the overall sedimentation of the wetland (Das and Jena 2008). Coupled with the northward littoral sediment drift, these changes led to the choking of the channel entrance from the Bay of Bengal. Average salinity within the lake declined from 13.2 ppt (parts per thousand) in 1960/61 (Jhingran and Natrajan 1966) to 1.4–6.3 ppt in 1995 (Banerjee et al. 1998). Lack of connectivity with the sea was one of the major factors leading to a decline in fisheries, with a drop in annual average landing from 8600 MT to 1702 MT between 1985/86 and 1998/99 (Mohapatra et al. 2007). The area under invasive macrophytes, primarily *Eicchornia*, increased from 20 km^2 in 1972 to 523 km^2 by October 2000 (WISA 2004). Introduction of shrimp culture, as well as an overall decline in fisheries, led to gradual breakdown of traditional resource management systems. There was an occupational displacement and loss of fishing grounds of the traditional fishing communities and resentment between traditional fishers and the immigrants (Dujovny 2009). Chilika fisheries gradually converted from a "community-managed fishery" to "contested-common," wherein those not engaged in the traditional fishery gradually exerted pressure for more fishing rights. The adverse changes in ecological character led to the inclusion of Chilika in the Ramsar Convention's Montreux Record in 1993 (a list of Wetlands of International Importance where adverse changes in ecological character have occurred, are occurring, or are likely to occur).

Ecological Restoration

Concerned over the rapid decline of Chilika, the Government of Odisha created the Chilika Development Authority (CDA) in 1991 (under the aegis of the Department of Forests and Environment) as the nodal agency to undertake measures for ecological restoration. The institutional design of the authority has as the chairperson the Chief Minister of the Government of Odisha. The governing body is constituted by the secretaries of all concerned departments, political representatives, as well as representatives of fisher communities. With financial support from the state government and Ministry of Environment and Forest (Government of India), CDA initiated several programs including treatment of degraded catchments, hydro-biological monitoring, sustainable development of fisheries, wildlife conservation, community participation, and development and capacity building at various levels. In 2000, a major hydrological intervention in the form of opening of a new mouth to the sea was undertaken based on modeling and stakeholder consultations.

CDA also initiated participatory watershed management in the western catchments to restore the vegetative cover, improve soil moisture, and enhance resources for community livelihoods. Through dedicated capacity building, conflict resolution, and trust building, CDA enabled development of watershed management plans and also provided resources for their implementation. An intensive awareness campaign on the values and functions of the wetland system, particularly amongst the fishers and school children, was undertaken in participation with civil society. A visitor center at Satapada was constructed as a hub for these activities. Specific initiatives for managing tourism by building capacity of the boatmen association were also implemented. A code of conduct for dolphin watching has also been developed. To improve connectivity in island villages, a ferry service for people and vehicles was launched benefitting more than 70,000 people. CDA has also strengthened fishing infrastructure through the construction of landing centers and jetties. Woman self help groups were organized and trained to undertake enterprises on dried fish and crab fattening.

To support systematic management, an intensive hydrological and ecological monitoring program has been put in place. These programs are coordinated through the Wetland Research and Training Center constructed on the shorelines of Chilika in 2002. Equipped with state-of-the-art facilities, the center is also a node for national and international training programs for wetland managers. Over the years, CDA has also established collaboration with over 50 organizations of international and national repute to support scientific studies related to various dimensions.

Restoration of Ecological Character and Removal from Montreux Record

The restoration initiatives created several positive impacts for the wetland ecosystem. Enhanced marine flows by opening a new mouth to the Bay of Bengal in restored the hydrological regimes and reestablished salinity regimes (Ghosh et al. 2006). Recovery of the fisheries and biodiversity was rapid. As per monitoring records of CDA, the average fish landing during 2001–2014 has remained around 13,000 MT which is more than a sevenfold increase from pre-restoration levels (Mohapatra et al. 2007). Annual censuses by CDA of Irrawaddy dolphins have indicated an increase from 89 to 158 individuals between 2003 and 2014, an increase in habitat use, improved breeding, and dispersal and decline in mortality rates. The sea grass meadows expanded from 20 km^2 in 2000 to 80 km^2 at present. There has also been a noticeable decline in the area under invasive macrophytes. Improvement of Lake Chilika's ecosystem, and in particular the increase in dolphins, has led to a resurgence of wetland tourism. The annual number of tourists visiting the wetland during 2000–2014 averaged 0.3 million – an increase of over 60% compared to arrivals during 1994–1999. In Manglajodi and adjoining villages of the northern sector, a perceptible change in community behavior is apparent. These villages, once famous for poaching of birds, are now protective of these winged visitors. Equipped with binoculars and watchtowers, these communities now provide local guides for bird-watching and CDA's staff engaged in assessing habitat and populations. Based on the positive changes noticed in the ecological character, the Ministry of Environment and Forest requested the Ramsar Convention to remove the site from Montreux

Record. Following an advisory mission in December 2001, the site was delisted in 2002 and the intervention recognized by the Ramsar Wetland Conservation Award and Evian Special Prize for "wetland conservation and management initiatives" to the Authority (Ramsar Convention Secretariat 2008).

Integrated Management Planning

Building on the knowledge base that had been developed and the outcomes of interventions, an integrated management planning process was initiated in 2008 to guide conservation and wise use of Lake Chilika. A management planning framework was developed based on assessing the status and trends in ecological character and prevailing threats. The process involved several stakeholders, particularly local communities. The plan was released in October 2012 by the chief minister, Government of Odisha. The management plan is a guide to activities of the Authority and all stakeholders.

CDA has also initiated programs for the revitalization of community based fisheries as a basis to create responsible fisheries. In July 2010, the state government established the Chilika Fishermen Central Cooperative Society (CFCCS) Ltd as the apex agency for managing Chilika fisheries. Under a pilot initiative, CDA through the Fisheries and Animal Resources Development Department is providing a revolving fund to primary fisher cooperatives to ensure fair access to credit to the member fishers. Through the support of the Marine Products Export Development Authority, CDA has also launched an initiative to provide ice boxes to the fishers so that the catch could be maintained for longer time and fishers could choose their preferred point of sale. A regulatory regime for the wetland, empowering the authority to control all detrimental fishing practices, is also in advanced stages of consideration.

Future Challenges

Lake Chilika has gradually emerged as a role model for participatory and adaptive management of wetland ecosystems. As a proactive step towards addressing emerging drivers of change, particularly those related to the coastal zone and linked to a changing climate, new research has been commissioned to assess the overall vulnerability of the wetland's ecological character and to identify suitable management response options. The wetland monitoring system is also being continually upgraded and made more sophisticated through deployment of better equipment, collaboration with expert institutions, and training of research staff. Partnerships, which are a hallmark of the CDA's institutional design, are being made more strategic. Efforts to link the management of Lake Chilika to the management of the Mahanadi River Basin and coastal zone are in advanced stages.

References

Banerjee RK, Pandit PK, Chatterjee SK, Das BB, Sengupta A. Chilika Lake: present and past. Barrackpore: CIFRI; 1998. Bulletin No. 80.

D'Souza R. Colonialism, capitalism and nature: debating the origin of Mahanadi Delta's hydraulic crisis (1803–1928). Econ Polit Weekly. 2002;37(13):1261–72.

Das BP, Jena J. Impact of Mahanadi basin development on eco-hydrology of Chilika. In: Sengupta M, Dalwani R, editors. Proceedings of Taal 2007: The 12th World Lake Conference. 28 October – 2 November 2007. Jaipur/New Delhi: Ministry of Environment and Forests; 2008. p. 697–702.

Dujovny E. The deepest cut: political ecology in the dredging of a new sea mouth in Chilika Lake, Orissa, India. Conserv Soc. 2009;7:192–204.

Ghosh AK, Pattnaik AK, Ballatore TJ. Chilika lagoon: restoring ecological balance and livelihoods through re-salinization. Lakes Reserv Res Manag. 2006;11:239–55.

Jhingran VG, Natrajan AV. Final report on the fisheries of the Chilika Lake (1957–1965). Barrackpore: Central Inland Fisheries Research Institute; 1966. Bulletin 8:1–12.

Kumar R, Pattnaik AK. Chilika – an integrated management planning framework for conservation and wise use. New Delhi/Bhubaneswar: Wetlands Internationals-South Asia/Chilika Development Authority; 2012.

Mohanty SK. Rational utilization of brackishwater resources of the Chilika lagoon for aquaculture. In: Patro SN, Sahu BN, Rama Rao KV, Misra MK, editors. Chilika: the pride of our wetland heritage. Bhubaneswar: Orissa Environmental Society; 1988. p. 36–9. 1980.

Mohapatra A, Mohanty RK, Mohanty SK, Bhatta KS, Das NR. Fisheries enhancement and biodiversity assessment of fish, prawn and mud crab in Chilika lagoon through hydrological intervention. Wetl Ecol Manag. 2007;15:229–51.

Pattnaik AK. Phytodiversity of Chilika Lake, Orissa, India [dissertation]. Bhubaneswar: Utkal Univerity; 2003.

Ramsar Convention Secretariat. *Evian project* – memorandum of understanding (translation). 2008. Available from http://www.ramsar.org/key_evian_protocol_e.htm. Accessed 7 Dec 2012.

Shimpei I, Shaw R. Linking human security to natural resources: perspective from a fishery resource allocation system in Chilika lagoon, India. Sustain Sci. 2009;4:281–92.

WISA. Socio-economic assessments of environmental flow scenarios of Chilika. New Delhi: Wetlands International – South Asia; 2004.

Saemangeum Estuarine System (Republic of Korea): Before and After Reclamation

30

Nial Moores

Contents

Introduction	406
Location	406
Reclamation Project Rationale	407
Physical Geography and Human Use Pre-seawall Closure	407
International Importance to Waterbirds Before Seawall Closure	408
History of Opposition	409
Changes to the SES following Seawall Closure	409
Saemangeum Shorebird Monitoring Program	409
Other Impacts	410
References	411

Abstract

Saemangeum is the name coined in the 1980s to promote a controversial 40,100 ha reclamation project on the west coast of the Republic of Korea (ROK). The project entails the construction of a 33.9 km long outer seawall (accredited as the world's longest man-made barrier) to impound two free-flowing estuaries in order to create 29,000 ha of land and a reclamation reservoir. In their natural state, these estuaries supported approximately 330,000–570,000 shorebirds annually and the livelihoods of 20,000–30,000 fishers and shell-fishers. Following seawall closure in 2006, there was a catastrophic decline in shorebirds supported by the site, and research found no evidence that the majority of affected birds were able to relocate to other sites in the ROK. Rather, substantial declines have been recorded at the population level in some shorebird species. Local fisheries have also been lost. Currently, construction on the inner dikes is continuing.

N. Moores (✉)
Birds Korea, Busan, Korea (Republic of)
e-mail: Nial.Moores@birdskorea.org

© Springer Science+Business Media B.V., part of Springer Nature 2018
C. M. Finlayson et al. (eds.), *The Wetland Book*,
https://doi.org/10.1007/978-94-007-4001-3_36

Keywords
Korea · Reclamation · Impacts · Shorebirds · Fisheries · Declines

Introduction

Saemangeum (alternative spelling, Saemangum) is a name originally coined in the 1980s to describe and promote a 40,100 ha reclamation project on the west coast of the Republic of Korea (ROK). The project entails the construction of 33.9 km long seawall (accredited as the world's longest man-made barrier) to impound two free-flowing estuaries (combined, the Saemangeum Estuarine System or SES). Inner dikes and walls (presently under construction) would then enable the conversion of approximately 29,000 ha of estuarine tidal flat (estimated at lowest low tide) and 11,000 ha of subtidal sea shallows into dry land and freshwater reservoirs.

The reclamation plan was developed initially in the 1980s, and construction of the outer seawall was started in 1991. Following several court cases and growing civil opposition, the last remaining gaps in the seawall were completed in April 2006. Outer seawall strengthening continued into 2008, and inner sea-dike construction started in ~2009 and was still ongoing in 2015.

The Saemangeum reclamation was the largest single coastal reclamation project in the world in the 1990s and early 2000s. It was the focus of sustained civil society opposition within the ROK especially from the late 1990s until seawall closure in 2006. Opposition focused initially on the predicted social and economic impacts on local fisheries and fishing communities and concerns over pollution. Following the ROK's accession to the Ramsar Convention in 1997 and because of the international importance of the SES in its natural or near-natural state to migratory waterbirds, the reclamation project was also the first major infrastructure project in the ROK and in the Yellow Sea Eco-region to attract sustained international concern, including reference in Ramsar Resolution IX.15 (2005). During the same period, the reclamation was actively promoted by central and provincial government. Most advertising since the mid-2000s has emphasized the reclamation's potential importance to the national economy (with designation of the area as a special economic zone) and its alleged environmental sustainability (Birds Korea 2010; Moores 2012).

Location

The Saemangeum reclamation (approximately 35°30 – 35° 50′ N and 125° 40 – 126° 45′ E) entails the impoundment of the Mangyeung (alternative spelling, Mangyeong) and Dongjin Estuaries and adjacent intertidal flats and subtidal shallows. Historically, both estuaries and the neighboring Geum Estuary formed the central part of a more or less contiguous subregion of intertidal wetland along the west coast of the ROK that extended c. 175 km from Cheonsu Bay in the north to Hampyeong Bay in the south (Moores 2012). Following on from piecemeal reclamation projects that increased in scale during the first half of the twentieth century, the whole subregion was identified as fit for reclamation by central government in 1984 in a pre-feasibility study for the 1984–2001 National Master Plan (Long et al. 1988). The area initially targeted for Saemangeum (a word incorporating part of the names of both the Mangyeung and Geum Rivers, with the additional connotation in Korean of "new big treasure") was increased between the late 1980s and the start of seawall construction in 1991 to include the outer estuarine tidal flats too. As constructed, the seawall (now with a national trunk road and a series of pocket parks running along its length) connects offshore islands and an earlier reclamation project and runs to a total length of 33.9 km. There are two sluice gates to allow tidal exchange totaling 540 m in length.

Reclamation Project Rationale

In line with the National Master Plan for Land Use and extant legislation within the Public Waters Reclamation Act (1962), the reclamation was initially proposed to create additional land and a water supply for rice agriculture. However, by 1990, the reclamation project was described as being primarily for the development of the port city of Gunsan for future trade with China (Birds Korea 2010). By 2001, agriculture was publicly given up as the primary land use and the Special Act to Promote the Saemangeum Reclamation was passed in December 2007 to legalize other development (Kim 2011). From 2010 the area was promoted as "Ariul, the Water City of Asia" with several proposed artificial islands and industrial complexes and 5,950 ha reserved as "ecological and environmental lots" (Birds Korea 2010). However, by 2012, as in 2015, less than 1,000 ha of tidal flat remained seaward of the outer seawall; almost all intertidal wetland landward of the outer seawall had become desiccated or submerged; and there was no conservation plan for the migratory shorebirds supported by the site (Moores 2012).

Physical Geography and Human Use Pre-seawall Closure

The SES was comprised of two free-flowing estuaries, divided by the Simpo headland (see Fig. 1), and was contained within seawalls constructed as part of earlier reclamation projects. In its natural and near-natural state, the SES had semidiurnal tides with a tidal range of 1.2–7.2 m. On higher spring tides, a tidal bore (+/− 1 m) moved up the Mangyeung River, and the whole system was temporarily inundated. During neap and low tide, extensive salt marshes (dominated by *Suaeda japonica*) were exposed, in addition to unvegetated tidal flat which extended 3 km north–south and up to 10 km west–east in some areas. Upstream from Simpo, most tidal flat areas were steep sided, muddy, or mud-sand mix, while most outer tidal flats were sandy with a very gentle slope.

Although high levels of pollutants were found within some of the sediments (in Long et al. 1988), the economic livelihoods of an estimated 20,000–30,000 fisherfolk depended on the system for shellfishing or commercial fishing from

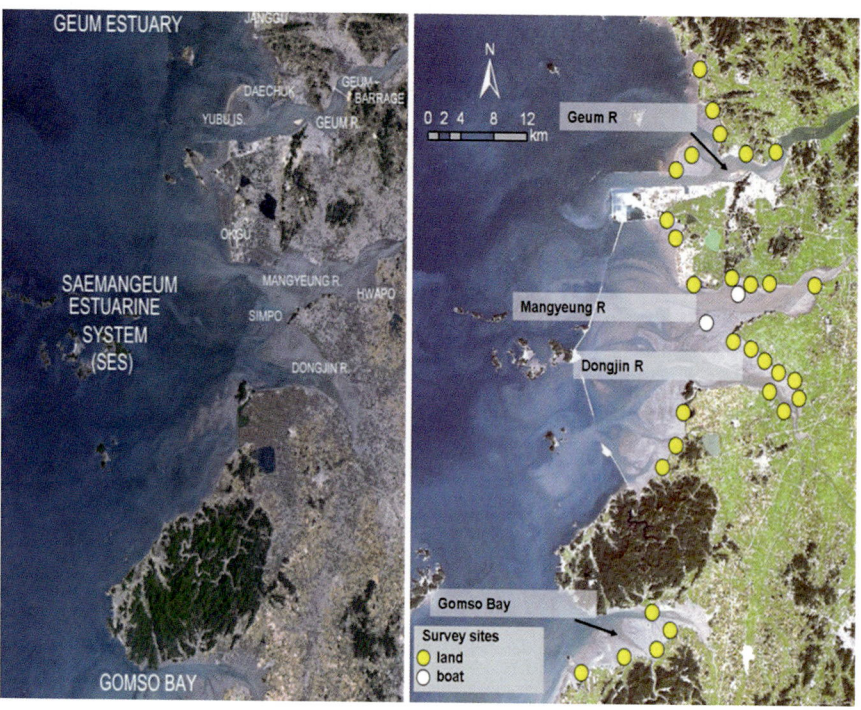

Fig. 1 The SSMP Study Region in May 1989 (*Left*) before construction started on the Saemangeum seawall and in late 2005/early 2006 (*Right*), showing SSMP count sites (Image credit: N. Moores © Rights remain with the author)

boats (Birds Korea 2010). The hinterland (through to at least 2010) contained a mix of agriculture (largely rice fields) and rural villages, with many communities economically dependent upon both shellfishing and land-based agriculture.

International Importance to Waterbirds Before Seawall Closure

At least 27 species of waterbird, including 20 species of migratory shorebird were recorded within the SES in Ramsar-defined internationally important concentrations (of either >20,000 individuals or 1% or more of a biogeographical population) between 1997 and 2003 (Moores 2012). Surveys by the national Ministry of Environment during the same period led to estimates during northward and southward migration, respectively, of 138,000 and 145,000 shorebirds supported by the Mangyeung Estuary and of 178,000 and 112,000 shorebirds supported by the Dongjin Estuary (Yi 2004). Combined, 330,000–573,000 migratory shorebirds were estimated to be supported by the SES each year. The lower estimate is based on the sum of the seasonal site maxima in Barter (2002), and the higher estimate is based on the sum of peak counts of each species at both estuaries in different years between 1997 and 2001 (Yi 2004). The SES was thus the most important known shorebird site in the whole of the Yellow Sea during northward migration and the second most important known site in the Yellow Sea during southward migration (Barter 2002).

History of Opposition

Civil society opposition to the reclamation project, led by environmental NGOs, religious leaders, leading academics, and some local community representatives, intensified during the late 1990s (Kim 2001). This was a period in which wetland conservation issues rose to prominence in the ROK due in part to the nation's accession to the Ramsar Convention. A ritual walk (the *samboilbae*) from Saemangeum to Seoul increased national awareness of the likely social and environmental impacts, as did a series of court cases which challenged the project's legal status (Kim 2011). Internationally, a growing number of environmental organizations expressed their concerns during, and subsequent to, the Ramsar Convention's 7th Conference of the Contracting Parties (COP7 – 1999). Such concerns intensified following the publication of Barter (2002) and were expressed in Ramsar Resolution IX.15, which formally requested the government of the Republic of Korea to advise the Secretary-General of the current situation concerning the seawall construction and reclamation of the Saemangeum coastal wetlands and the impact of the construction works to date on the internationally important migratory waterbird populations dependent upon these wetlands. No detailed response to this request was provided by the ROK government to the Ramsar Secretariat.

Changes to the SES following Seawall Closure

Remaining gaps in the Saemangeum seawall were closed on 26 April 2006. Tidal movement, with high tides peaking at 7 m in early 2006, was then controlled through sluice gates, with a tidal range of between 0 and 50 cm (maximum 1.3 m) during most subsequent tidal cycles. Low-lying tidal flats became submerged, and higher areas desiccated. Mass shellfish die-offs were first recorded in late April and occurred with each subsequent major influx or release of seawater (Moores 2012).

Saemangeum Shorebird Monitoring Program

As an independent scientific response to Ramsar Resolution IX.15, the ROK-based conservation organization Birds Korea and the specialist Australasian Wader Studies Group partnered to conduct the Saemangeum Shorebird Monitoring Program (SSMP). The SSMP entailed intensive shorebird counting during northward migration in 2006–2008 within both the SES and also at the adjacent Gomso Bay and Geum Estuary (combined with the SES, the "SSMP Study Region").

The SSMP also included a national shorebird count in the ROK (in May 2008) and was designed to share data with the Monitoring Yellow Sea Migrants in Australia program (in Australia).

Based only on the sum of peak counts, 181,755 shorebirds were recorded within the SES during the SSMP. Between 2006 and 2008, 16 out of the 20 most numerous shorebird species showed declines that totalled >126,000 birds. The largest declines in number were shown by great knot *Calidris tenuirostris* (>74,000), dunlin *Calidris alpina* (>37,000), and Mongolian plover *Charadrius mongolus* (5,327). The largest declines in percentage terms were shown by sanderling *Calidris alba* (96%), sharp-tailed sandpiper *Calidris acuminata* (94%), and spoon-billed sandpiper *Eurynorhynchus pygmeus* (91%). Within the SSMP Study Region as a whole, based on the sum of peak counts, there were almost 264,000 shorebirds in 2006 and 164,261 in 2008. Thus, there was a decline of almost 130,000 shorebirds within the SES and of 100,000 shorebirds within the SSMP Study Region during northward migration between 2006 and 2008. The most affected species included great knot with >92,000 lost from the SES and the two adjacent wetlands. The national survey in May 2008 failed to find evidence of displaced great knot from the SSMP Study Region at other internationally important wetlands nationwide, including all other sites known to be internationally important for the species in the ROK (Moores et al. 2008; Moores 2012). Research in Australia (yet to be published in full) also indicates a decline in numbers and adult survival of great knot and several other shorebird species after seawall closure at Saemangeum (Rogers et al. 2009; Moores 2012). The SSMP therefore became the first research program to detect trans-hemispheric impacts on shorebirds at the population level caused by reclamation on the East Asian-Australasian Flyway.

Further construction and conversion of formerly natural intertidal wetland into dry land and freshwater reservoirs at Saemangeum and other internationally important wetlands in the ROK are predicted to lead to further declines in shorebirds at the site, national, and population level (Moores 2012).

Other Impacts

The Saemangeum reclamation has already caused massive changes to the natural biological and physical conditions previously found within the SES and threatened the extinction of a recently described mollusc (Hong et al. 2007). The loss of tidal exchange and tidal flat area has led to a reduction in fisheries and shellfishing opportunity and resulted in a major loss of local livelihoods (Hahm 2004, Birds Korea 2010). Before seawall closure 50,000–90,000 t of hard clams and 1,000 t of mud octopus were collected annually in the SES (MacKinnon et al. 2012). Although robust data are presently unavailable, yields within the SES are now considered to have fallen close to zero. The reclamation has also affected patterns of erosion and sedimentation outside the outer seawall, including in the Geum Estuary to the north (Suh 2008; Lee 2010) and Gomso Bay 15 km to the south (Lee et al. 2015).

Following closure of the Saemangeum seawall, levels of suspended solids and the speed of the tidal current in Gomso Bay both increased. The direction of the tidal current changed, the area dominated by sand increased, and the sedimentation trend changed from accretion to erosion on the lower flats in the outer bay, resulting in reduced habitat for fish and shellfish, decline in water quality, and large-scale degradation of natural habitat (Lee et al. 2015). In addition, most of the sand and gravel required for reclaiming land within the Saemangeum reclamation area, estimated at 700 million cubic meters, will be supplied by dredging seabed 10–20 km off the coast (Ministry of Land, Infrastructure and Transport 2011), likely to cause further degradation to surrounding marine and coastal ecosystems.

References

Barter M. Shorebirds of the Yellow Sea: importance, threats and conservation status, Wetlands international global series 9, international wader studies 12. Canberra: Wetlands International-Oceania; 2002.

Birds Korea. The Birds Korea Blueprint 2010 for the conservation of the avian biodiversity of the South Korean part of the Yellow Sea. Busan: Birds Korea; 2010.

Hahm H-H. Ousted fishermen and Saemangeum Tideland Reclamation Project. J Korean Cult Anthropol. 2004;37(1):151–82.

Hong J-S, Yamashita H, Sato S. The Saemangeum Reclamation Project in South Korea threatens to extinguish a unique mollusc, ectosymbiotic species attached to the shell of Lingula anatina. Plankon Benthos Res. 2007;2(1):70–5.

Kim C-N. The Campaign against the Saemangeum Reclamation Project. Asia Solidarity Q. 2001;6:16–9. Published by the People's Solidarity for Participatory Democracy, Seoul.

Kim R-H. Is Ramsar home yet? A critique of South Korean laws in light of the continuing coastal wetlands reclamation. Columbia J Asian Law. 2011;24(2):437.

Lee H-J. Enhanced movements of sands off the Saemangeum Dyke by an interplay of Dyke construction and winter monsoon. In: Ishimatsu A, Lie H-J, editors. Coastal environmental and ecosystem issues of the East China Sea. Shinan County: TERRAPUB and Nagasaki University; 2010. p. 49–70.

Lee Y-K, Ryu J-H, Choi J-K, Lee S, Woo H-J. Satellite-based observations of unexpected coastal changes due to the Saemangeum dyke construction. Korea Mar Pollut Bull. 2015;97:150–9. Elsevier.

Long A, Poole C, Eldridge M, Won P-O, Lee K-S. A survey of coastal wetlands and Shorebirds in South Korea, Spring 1988. Kuala Lumpur: Asian Wetland Bureau; 1988.

MacKinnon J, Verkuil Y, Murray N. IUCN situational analysis on East and Southeast Asian intertidal habitats, with particular reference to the Yellow Sea (including the Bohai Sea). Gland: International Union for Conservation of Nature; 2012.

Ministry of Land, Infrastructure and Transport. A study on schemes to secure sand and gravel for the Saemangeum Project. Seoul: Government of the Republic of Korea (in Korean); 2011.

Moores N. The distribution, abundance and conservation of the Avian biodiversity of Yellow Sea habitats in the Republic of Korea. Doctoral thesis, University of Newcastle, Australia; 2012.

Moores N, Rogers D, Kim R-H, Hassel, C, Gosbell K, Kim S-A, Park M-N. The 2006–2008 Saemangeum Shorebird Monitoring Program Report. Birds Korea publication; 2008.

Rogers D, Hassell C, Oldland J, Clemens R, Boyle A, Rogers K. Monitoring Yellow Sea Migrants in Australia (MYSMA): North-Western Australian Shorebird Surveys and Workshops, Dec 2008; 2009.

Suh S-W. Simulation of sedimentation change around Saemangeum area. p. 203–28. In: Recent progress on the Coastal wetland restoration. Proceedings of International Symposium; 2008 Sept 8–9. Seoul National University. Hosted by Ministry of Land, Transport and Maritime Affairs; 2008.

Yi J-Y. Status and habitat characteristics of migratory shorebirds in Korea. p. 87–103. In: The Proceedings of the 2004 International Symposium on Migratory Birds, Gunsan. Published by the Ornithological Society of Korea; 2004.

Peatlands and Windfarms: Conflicting Carbon Targets and Environmental Impacts

31

Richard Lindsay

Contents

Introduction	414
Mires, Wind Farms, and Climate	415
Extent of Wind Farms on Peat	415
Wind Farm Construction on Blanket Peat	417
Turbine Towers, Turbine Bases, and Maintenance Platforms	418
Wind Farm Roads	419
Carbon Balance of Wind Farms on Peat	423
References	424
Website References	425

Abstract

A combination of landform and climate suitable for both blanket mire formation and windfarm construction means that many windfarms have been, and continue to be, constructed on peat soils. Renewable energy sources are increasingly being adopted in order to reduce carbon emissions. Meanwhile peatlands are becoming increasingly recognised globally as some of the most carbon-rich of all terrestrial habitats. When a windfarm is constructed on peat it is inevitable that some of the carbon stored in the peat will be lost through oxidation of the peat. The main source of such disturbance is the network of access roads built for construction and maintenance. The most recent research suggests that potential carbon losses resulting from windfarm construction within a natural peat bog mean that there may be no net carbon benefit from the windfarm.

R. Lindsay (✉)
Sustainability Research Institute, University of East London, London, UK
e-mail: r.lindsay@uel.ac.uk

© Springer Science+Business Media B.V., part of Springer Nature 2018
C. M. Finlayson et al. (eds.), *The Wetland Book*,
https://doi.org/10.1007/978-94-007-4001-3_50

Keywords

Blanket · Bog · Carbon · Climate · Drainage · Drying · Emissions · Energy · Mire · Oxidation · Peat · Rainfall · Roads · Settlement · Subsidence · Turbine · Wind · Windfarm

Introduction

One of the benefits of having an atmosphere cloaking our planet is that solar heating results in atmospheric circulation. This provides an energy source which is additional to direct solar irradiation and is a continually renewed source of energy which requires no mining, no exploitation of natural resources, and no complex technology for its utilization. Plants use wind energy to transport seeds at minimal cost, spiders use air currents to travel enormous distances and are thus often some of the first colonizers of new islands, while for thousands of years, ships have used the wind to cross oceans. The energy available can be considerable. The kinetic energy of a single hurricane is typically around 1.5 TW (10^{12} W) which is equivalent to half the worldwide electrical generating capacity (NOAA Hurricane Research Division website).

At a time when unsustainable use of carbon-rich fossil fuels as an energy source is altering our atmosphere in undesirable ways because of the resulting gaseous waste products, it seems entirely logical to use instead the constantly renewed winds of this same atmosphere to supply our energy needs. This assumes, however, that the methods used to harvest wind energy are not themselves responsible for significant carbon emissions. Currently, the manufacture of almost any harvesting machinery is most likely to be reliant still on the use of fossil fuels. The carbon emissions of this manufacturing process must therefore be subtracted from any carbon savings made by gathering energy from the wind. Furthermore, the potential carbon emissions associated with siting of the wind-harvesting infrastructure must also be considered, and it is this aspect which has brought the renewable industry into conflict with the principle of sustainable and the wise use of wetlands – specifically the conflict between wind farms and peatland environments.

There is an undeniable irony in the fact that some of the best locations for wind harvesting using current technology also happen to be some of the most carbon-rich environments on the planet. Globally, although peatland systems only cover around 3% of the land area, they contain more carbon than all the world's vegetation combined (Bain et al. 2011). This carbon is retained for millennial timescales in a peatland as long as it remains as a fully functioning wetland. If the peatland is damaged by, for example, drainage or burning, it will release this stored carbon back to the atmosphere. On a global scale, such damage is widespread. Consequently in order to encourage the restoration of damaged peatlands and help to prevent further damage, it has been agreed that nations committed to the Kyoto Protocol can include peatland restoration as part of their national carbon balance reporting (Smith et al. 2014). At the same time, nations are being encouraged to embrace renewable forms of energy to reduce carbon emissions, and wind energy is currently being seen as a major opportunity in terms of its potential for supplying global energy demand (Lu et al. 2009; Leung and Yang 2012).

Mires, Wind Farms, and Climate

A global map of winds (Fig. 1) makes it clear that the oceanic coastlines of temperate and boreal regions experience some of the most consistent strong wind patterns in the world. Such regions are attractive to the wind farm industry for obvious reasons, but the more rugged and steep the landscape, the greater the engineering challenges and costs of constructing and maintaining a wind farm. Gently rolling "whaleback" hills offer fewer challenges and greater rewards. These same factors of constant oceanic air masses passing over gentle hills are, however, also those which give rise to extensive peat formation because when air laden with oceanic moisture meets the land, it must shed this moisture to rise over the land. The resulting rainfall can exceed 4,000 mm annually, but, more importantly, this rainfall is distributed across much of the year. Where there is 0.25 mm or more precipitation in a day (a "rain day") on at least 200 days of the year, entire landscapes can become covered with a mantle of peat, particularly when allied to acidic bedrock or heavily leached soils (Lindsay et al. 1988; Doyle 1997). These peat-draped landscapes are termed blanket mire. The carbon store of the peat deposit, which may be more than 6 m deep in places, means that, for example, loss of only around 5% of the UK's blanket mire resource would release as much carbon to the atmosphere as all current annual UK anthropogenic greenhouse gas emissions (Bain et al. 2011).

A tension has therefore developed between, on the one hand, the need to conserve the long-term carbon stores held within peatland systems and, on the other hand, the goal of the wind industry to create the most efficient and cost-effective wind farm sites. Both find their optimal conditions in the same geographical areas, but construction and maintenance of wind farm infrastructure on a peatland inevitably involve damage to that peatland, releasing carbon from its long-term carbon store and thus to some degree negating the purpose of the wind farm.

Extent of Wind Farms on Peat

The majority of wind farms on peat occur within blanket mire landscapes. While Norway and Newfoundland have one or two examples, in 2008 Ireland had 39 wind farms on blanket mire (Renou-Wilson and Farrell 2009), Spain had 23 in the Galician blanket mires (Pontevedra Pombal et al. 2007), while 55% of installed wind farms in Scotland were on deep peat soils (Bright et al. 2008). By August 2013, Scotland had at least 578 wind farms built, in construction or proposed, associated with blanket mire habitat (Scottish Natural Heritage Wind Farm Footprint Maps website) – see Fig. 2.

Some wind farms are associated with raised mire systems or with bogs which are intermediate between raised mire and blanket mire. Indeed Renou-Wilson and

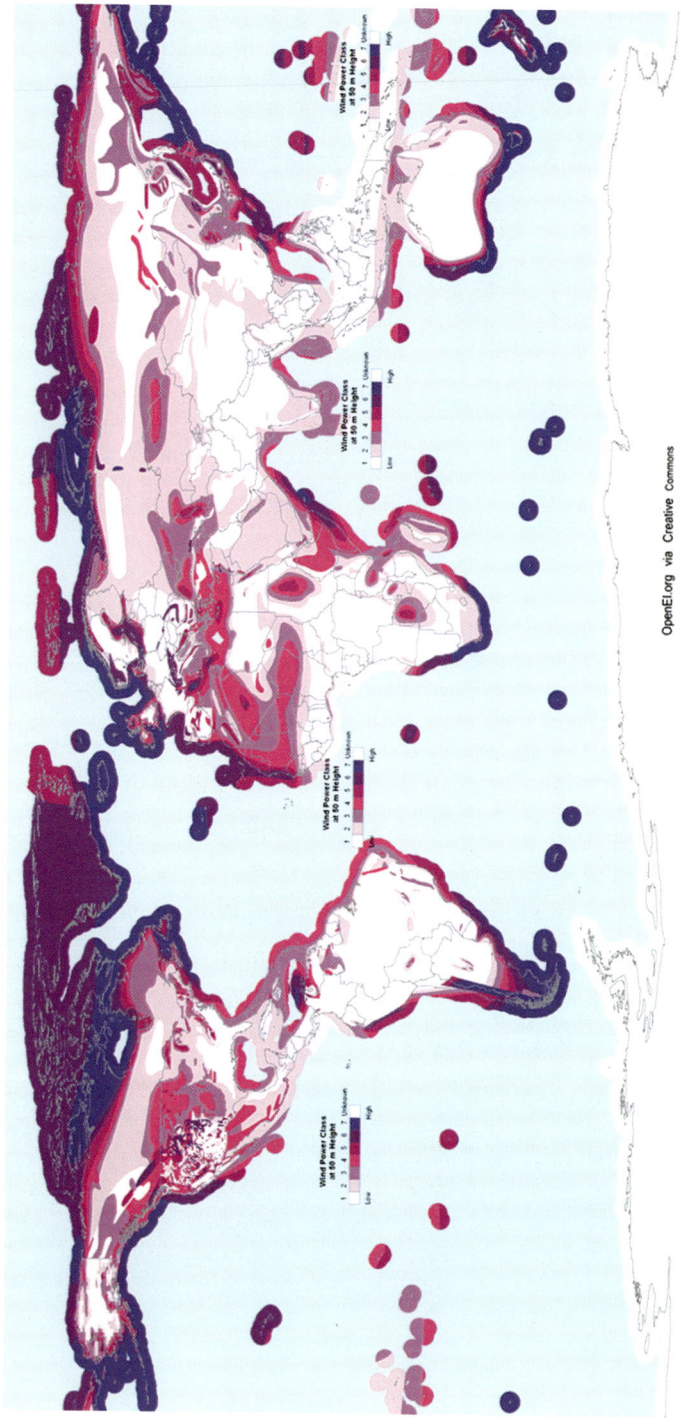

Fig. 1 Global wind resources map (Picture credit: OpenEI.org via Creative Commons (http://en.openei.org/wiki/File:Global_wind_resources_map.png#filelinks))

Fig. 2 A close view of the wind farm roads and turbines of the Whitelee wind farm (inset), the UK's largest onshore wind farm, located on Eaglesham Moor south of Glasgow, UK, constructed across an area of blanket mire which had been partly afforested some decades earlier. The forest is now being felled to make way for the wind farm, which also extends onto unplanted areas of blanket mire and consists of 215 turbines generating up to 539 MW (Photo credit: Richard Lindsay © Rights remain with the author)

Farrell (2009) recommend that pressure be taken from the relatively undamaged blanket mires of Ireland by focusing wind farm development on areas of Midland raised mire which have been worked to commercial exhaustion by peat extraction and therefore have only a thin remnant layer of peat.

The key range of impacts resulting from wind farm development on peat is thus generally those associated with blanket mire landscapes. Such landscapes offer further advantages to the wind farm developer in addition to suitable wind conditions and gentle landforms. Generally the land is considered to be of low value because the traditional economic returns from such landscapes are low. The potential to make substantial income from the land through energy production is thus extremely attractive. Furthermore, land ownership often involves only one or two large landowners, making negotiations to lease or purchase extensive tracts of land relatively simple.

Wind Farm Construction on Blanket Peat

Construction of a wind farm requires a significant range of infrastructure to be installed within the blanket mire landscape, including access and maintenance roads, a number of turbine bases, various cable channels, and a number of other ancillary elements. Impact on the peatland habitat and the carbon store cannot be avoided during installation of this infrastructure. These impacts may involve actual removal, removal and replacement of peat, compression, flooding, drainage, erosion, or even mass movement of the peat.

Possibly the most striking feature about the question of wind farm impacts is the extreme paucity of peer-reviewed evidence available from the scientific literature. While each wind farm development must produce an impact assessment as part of the planning process, little of this subsequently filters through to the scientific journals, while evidence obtained from the monitoring of impacts following construction of a wind farm is almost as sparse. Some of the construction methods now employed by the wind farm industry have only existed in their present form for a decade or so, and yet few examples of performance data for these new methods have made their way into the public domain via a process of critical scrutiny. The engineering mechanisms have been described and are available in some detail, but the impacts of these mechanisms remain largely undescribed.

Turbine Towers, Turbine Bases, and Maintenance Platforms

Turbine towers are the most visible feature of a wind farm. Their impact is therefore most strongly experienced at a cultural, psychological, and emotional level, in terms of how the turbines affect perceptions of "wilderness" and "wildland." The blanket mire landscape, for example, represents the most extensive remaining example of near-natural terrestrial habitat in the UK, offering a sense of the pre-neolithic landscape which can itself be reconstructed from the plant remains stored in the peat archive. The presence of turbine towers within such a landscape substantially impacts on that perception of a pre-neolithic scene.

To some extent it seems that birds also react to the presence of turbines, with species such as Eurasian buzzard *Buteo buteo*, hen harrier *Circus cyaneus*, and Eurasian golden plover *Pluvialis apricaria* showing marked avoidance up to 500 m away from turbines in a series of wind farms in upland Britain (Pearce-Higgins et al. 2009). For Eurasian golden plover, an Annex 1 species under the EU Birds Directive, this represented a potential 39% reduction in breeding density within 0.5 km of a turbine. In Norway, the red-throated diver *Gavia stellata* population present prior to building a wind farm on the island of Smøla ceased to use the site once the wind farm was in place and appeared to show explicit avoidance of the wind farm, being found breeding successfully in other parts of the blanket mire landscapes of the island (Bevanger et al. 2008). Sea eagles *Haliaeetus albicilla* have also shown substantially reduced breeding success within the area of the Smøla wind farm (Bevanger et al. 2008), while bird strikes between eagles and turbine blades have been a particularly contentious issue on the island, with as many as ten killed in an 18-month period.

The main direct impact on the blanket mire habitat itself is limited to the immediate vicinity of the turbine base. The cumulative impact of this therefore depends on the number of turbines to be installed. Turbine towers experience enormous forces and must therefore be placed on solid foundations embedded within the underlying mineral subsoil or bedrock. This means that the peat in the immediate vicinity of the foundation must be removed. Each turbine also generally has a maintenance platform capable of supporting a mobile crane which may be needed at some point during the life of the wind farm in order to replace blades, turbine housing, or even the whole turbine tower. This, too, requires removal of the peat and replacement with concrete to create a solid working surface. While it may be possible to use peat to backfill much of the excavation associated with construction of the turbine base, the maintenance platform cannot be covered in this way. Consequently for each turbine there is a quantity of peat which must be permanently removed, plus potentially there may be a quantity of peat which is removed but then replaced.

The peat which is removed permanently is likely to be lost to the atmosphere through oxidation fairly quickly. The peat which is replaced will have undergone significant physical disruption, and it will have shrunk to some extent because, once it is removed from the supporting bog water table, water will be lost through drying and compression. When this peat is placed back in situ, it will not be in the same condition as when it was removed. The hydrology of any area consisting of removed and replaced peat will not therefore initially resemble that of the undisturbed peat and will thus tend to undergo further degradation through settlement and oxidation. The degree to which this can be avoided depends on how quickly, if at all, it is possible to reestablish a naturally functioning hydrology and associated cover of peat-forming bog vegetation. Until this is achieved, the replaced peat will continue to oxidize and release carbon dioxide to the atmosphere.

Excavation of the peat to create a turbine base can also be a hazardous business, particularly when the peat is deep. Large quantities of peat may be deposited onto the bog surface if trucks are not continuously available to receive the excavated material. The weight of this excavated material ("arisings"), particularly if on a slope, can cause failure of the deep, nearly liquid peat and trigger mass movement in the form of a "peat slide," or avalanche of peat. One such failure occurred in Co. Galway, Ireland, in 2003, with more than 2 km of hillslope flowing continuously down the hill, across a road, and into a river system for almost a week, the material ultimately being transported for 20 km downstream (Lindsay and Bragg 2005) – see News Video of Derrybrien Bog Slide. The Irish Government was successfully prosecuted by the European Court of Justice for permitting this development to go ahead without adequate safeguards. At least one other major peat slide has occurred in Ireland associated with wind farm construction (Dykes and Jennings 2011). Such mass movements potentially involve very large losses of peat carbon to the atmosphere as the material of the slide oxidizes.

Wind Farm Roads

Construction and maintenance roads are generally the most extensive direct impact of a wind farm on a blanket mire landscape. The roads must give access to every turbine plus all other infrastructure buildings and must be sufficiently stable and usable throughout the life of the wind farm to enable, if necessary, the delivery and erection of whole replacement turbines. Given that turbines must be separated by a

Fig. 3 A wind farm road constructed across blanket mire which has developed over watershed (water divide) ridges and gentle slopes of the Xistral Mountains, Galicia, northern Spain (Photo credit: Richard Lindsay © Rights remain with the author)

distance around seven times the rotor diameter in order to minimize turbulence effects between turbines (though some research suggests that this distance may need to be doubled – Meyers and Meneveau 2012), the length of the wind farm road network is strongly influenced by the number of turbines to be installed.

The nature of the road network also depends on the location of the turbines, and as watershed ridgetops give maximum height advantage to the turbines, it is common to find lines of turbines following such ridges. In a blanket mire landscape, these broad watershed ridges are also the main location of the deepest, most extensive areas of blanket mire. Indeed in areas which are on the climatic limit of temperate-region blanket mire formation, such as northern Spain, these are almost the only locations with extensive tracts of blanket mire. In such landscapes, wind farm development tends to impact on the core of the blanket mire system (see Fig. 3). In tropical regions, some areas of cloud forest may also represent a particular type of blanket mire, and these too are affected by such road construction, but in this case the roads are often for the purposes of constructing and servicing mobile phone masts (Bruijnzeel and Hamilton 2000).

Wind farm road construction generally takes one of three forms. On shallower peat the most cost-effective solution is simply to dig the peat out to the mineral subbase and create a running surface on this subbase. This was historically the approach used by land managers prior to the advent of wind farms because there was little requirement to access the center of the deepest peats by vehicle, so the track would follow areas of thinner peat along natural contours. These excavated roads involve removal of the peat. Furthermore, because the running surface lies below the level of the surrounding peat, the road then acts, in effect, as a continuous drain cutting through the site. In addition, drains are generally dug either side of the running surface to prevent ponding of water on the running surface.

The effect of such a road is thus loss of carbon from the excavated peat, plus consolidation, and subsidence of the peat adjacent to the road because of drainage effects, thereby in effect widening the drainage effect of the road. Surface seepage can no longer pass to the expanse of peat downslope from the road. Holden (2005) has indicated that in some cases, this downslope impact may extend for several hundreds of meters. Drying of the peat adjacent to the road leads to cracking, which allows rainwater to penetrate to the base of the peat and lubricate the interface between the peat and the mineral subbase. Shallow peat is also lighter and thus more buoyant than deep peat. The root mat of the bog vegetation which binds the bog surface together has also been cut by the road line. During periods of intense rainfall, this combination of factors can lead to lifting of the peat layer from the mineral subbase during periods of intense rainfall, resulting in a peat slide, as observed in blanket mire on a number of occasions where such conditions prevail (Warburton et al. 2004).

The second approach to road construction on peat is "cut and fill," which involves excavating the peat completely to the mineral subbase then backfilling the resulting trench with crushed rock until the roadway lies a little above the level of the surrounding bog surface. This can be an expensive option because potentially large volumes of fill material must be transported to the road line and the waste peat must be disposed of somehow. Depending on the nature of the backfill material, the road may act as a hydrological barrier or as a drain. Culverts are normally run beneath the road to permit water to continue to flow downslope, but these culverts focus the water and thus increase its erosive power compared with the generally diffuse surface seepage which existed prior to the road. Some downslope areas become drier and therefore subject to cracking because they no longer receive water directed by the culverts, while other areas receive more water as a result of the culverts. Areas upslope from the road may also become ponded when the culverts are unable to cope with large volumes of water or if a culvert ceases to function. These circumstances too, over time, can give rise to conditions which are less stable than those which existed before road construction, while the peat which was excavated during road construction is also likely to be lost to the atmosphere through oxidation, or outwash downstream (Greive and Gilvear 2008).

Where peat is more than 0.5–1.0 m deep, the method of choice for the wind farm industry is a "floating road," which involves laying a geotextile mat directly onto the peat surface; then a geogrid may be laid onto which appropriately graded aggregate is poured. Another geogrid may then be added with more aggregate before finally the running surface is laid down. The aggregate fills the spaces in the geogrid tightly, thereby spreading any load across the whole width of the road. In this way it is possible to lay a road on which a 500 tonne crane can travel across the surface of a bog even when the peat is some meters deep. Perhaps surprisingly, provided the vehicle does not stop for any length of time before reaching the solidity of the concrete crane pad, the passage of such a vehicle does not in itself cause major changes to the underlying peat. This is because water cannot be expelled from the peat matrix quickly enough to have a permanent effect when such a vehicle passes, and so the peat experiences elastic compression, meaning that the peat rebounds almost completely after the vehicle has passed. The presence of the road itself, however, is a different matter.

When a peat soil is placed under a constant load, water is steadily expelled from the peat matrix causing the peat soil to compress (Hobbs 1986). In the case of a floating road, this results in long-term settlement of the road surface. For example, a road in Wales constructed across peat in 1819 by the engineer Thomas Telford is still experiencing settlement and now has a carriageway thickness of more than 2 m in places because over the years material has been regularly added in order to keep the carriageway above the peat surface (Nichol and Farmer 1998). Wind farm floating roads experience similar differential long-term settlement, and the greatest degree of settlement occurs on those sections overlying the softest, wettest peats. There is thus a tendency for the carriageway to become waterlogged in these sections, compromising road strength. Additional material may then be added to such sections in order to raise the carriageway level, but this further increases the load on the peat still further. Alternatively, drains may be added along the side of the carriageway to remove this excess water, but this results in drainage-induced subsidence of the peat beneath and adjacent to the roadway. Neither solution halts the subsidence.

Floating roads thus become increasingly significant linear features cutting through the blanket mire landscape and frequently cutting across natural flow lines. As with excavated roads, they can drain the adjacent mire if marginal drains are added, and like cut-and-fill roads, they can both pond flow upslope from the road if the road blocks

Fig. 4 Ponding of (frozen) surface water on the upslope side of a wind farm road at Scout Moor Wind Farm, southern Pennines, UK. This water would formerly have percolated downslope through the surface vegetation to provide additional water inputs to blanket mire further down the gentle hillside (Photo credit: Richard Lindsay © Rights remain with the author)

natural flow lines (see Fig. 4) and produce zones of drier and higher flow downslope as a result of culverted flow beneath the road. The long-term impact of such roads on the blanket mire landscape is unknown because there are few examples which are more than 15 years old. It should also be noted that although the turbines may be removed at the end of the wind farm life, it is generally assumed that the roads will be left as permanent features, producing a new means of easy access to the blanket mire landscape. In Spain, such increased access pressure is already being observed, with harmful consequences for the blanket mire habitat (Fraga et al. 2008).

Carbon Balance of Wind Farms on Peat

The debate about the relative carbon benefits of wind farms constructed on peat has reached such a level of intensity in the UK that the Scottish Government has commissioned the construction of a "carbon calculator." This calculator is designed to be applied to any wind farm proposal which occurs on peat, and, having taken onto account the various factors such as the carbon emissions of turbine construction, extent of road construction, degree of drainage, and numerous other factors, it then gives an indication of whether the carbon saved by harvesting wind energy are likely to exceed the carbon released during the various construction phases and likely losses from the peat during the lifetime of the wind farm (Nayak et al. 2010). The calculator also takes account of proposals by developers to restore additional areas of peatland as mitigation for any damage caused to the habitat directly affected by the wind farm development.

The calculator is intended as an aid to the planning process, allowing an informed decision to be made about whether the relative gains and losses of carbon associated with a given development proposal. In its initial form, it appeared to indicate that most wind farm developments, even if entirely on peat, would show a significant carbon benefit over the 25-year life of a wind farm. Recent refinements to the calculator, however, have since altered this picture significantly. The authors of the calculator now conclude that construction of a wind farm on an undamaged blanket mire would not result in a net long-term carbon benefit even with careful management, partly because the contribution of fossil fuels to energy production is expected to decline in the coming years (Smith et al. 2014).

The long-term effects resulting from construction and operation of a wind farm on the blanket mire environment will only come to be understood in 10–20 years time, or perhaps on even longer timescales, but even this understanding will only be possible if appropriate measurements and monitoring are undertaken now. Indeed the Scottish Government has recently initiated a research project to compare the predicted impacts set out in the Environmental Impact Statements of 11 wind farms with the impacts observed since construction. At present, the lack of published data about many aspects of potential impact suggests that the necessary research studies are not currently in place – an issue to which the wind power industry and the peatland research community should devote some serious thought.

References

Bain CG, Bonn A, Stoneman R, Chapman S, Coupar A, Evans M, Gearey B, Howat M, Joosten H, Keenleyside C, Labadz J, Lindsay R, Littlewood N, Lunt P, Miller CJ, Moxey A, Orr H, Reed M, Smith P, Swales V, Thompson DBA, Thompson PS, Van de Noort R, Wilson JD, Worrall F. IUCN UK commission of inquiry on peatlands. Edinburgh: IUCN UK Peatland Programme; 2011.

Bevanger K, Follestad A, Gjershaug JO, Halley D, Hanssen F, Johnsen L, May R, Nygård T, Pedersen HC, Reitan O, Steinheim Y. Pre- and post-construction studies of conflicts between birds and wind turbines in coastal Norway, Status Report 1st January 2008. NINA Report 355. Trondheim: Norsk institutt for naturforskning; 2008.

Bright JA, Langston RHW, Pearce-Higgins JW, Bullman R, Evans R, Gardner S. Spatial overlap of wind farms on peatland with sensitive areas for birds. Mires and Peat. 2008; 4:Article 07.

Bruijnzeel LA, Hamilton LS. Decision time for cloud forests, IHP Humid Tropics Programme Series, vol. 13. Paris: UNESCO; 2000. Available to download from: http://www.unep-wcmc.org/resources-and-data/decision-time-for-cloud-forests. Accessed 3 Apr 2015.

Doyle G. In: Parkyn L, Stoneman RE, Ingram HAP, editors. Blanket bogs: an interpretation based on Irish blanket bogs. Wallingford: CAB International; 1997. p. 25–34.

Dykes AP, Jennings P. Peat slope failures and other mass movements in western Ireland, August 2008. Q J Eng Geol Hydrogeol. 2011;44:5–16.

Fraga MI, Romero-Pedreira D, Souto M, Castro D, Sahuquillo E. Assessing the impacts of wind farms on the plant diversity of blanket bogs in the Xistral Mountains (NW Spain). Mires and Peat. 2008; 4:Article 06.

Greive I, Gilvear D. Effects of wind farm construction on concentrations and fluxes of dissolved organic carbon and suspended sediment from peat catchments at Braes of Doune, central Scotland. Mires and Peat. 2008; 4:Article 03.

Hobbs NB. Mire morphology and the properties and behavior of some British and foreign peats. Q J Eng Geol. 1986;19:7–80.

Holden J. Peatland hydrology and carbon release: why small-scale process matters. Phil Trans R Soc A. 2005;363:2891–912.

Leung DYC, Yang Y. Wind energy development and its environmental impact: a review. Renew Energy Sustain Rev. 2012;16:1031–9.

Lindsay RA, Bragg OM. Wind farms and blanket peat: the bog slide of 16th October 2003 at Derrybrien, Co. Galway, Ireland. 2nd ed. Gort: The Derrybrien Development Cooperative Limited; 2005. http://roar.uel.ac.uk/1143/. Accessed 2 Feb 2015.

Lindsay RA, Charman DJ, Everingham F, O'Reilly RM, Palmer MA, Rowell TA, Stroud DA. In: Ratcliffe DA, Oswald P, editors. The flow country: the peatlands of Caithness and Sutherland. Peterborough: Nature Conservancy Council; 1988.

Lu X, McElrow MB, Kiviluoma J. Global potential for wind-generated electricity. PNAS. 2009;106(27):10933–8.

Meyers J, Meneveau C. Optimal turbine spacing in fully developed wind turbine boundary layers. Wind Energy. 2012;15(2):305–17.

Nayak DR, Miller D, Nolan A, Smith P, Smith JU. Calculating Carbon Budgets of Wind Farms on Scottish Peatlands. Mires and Peat. 2010; 4:Article 09.

Nichol D, Farmer IW. Settlement of peat on the A5 at Pant Dedwydd near Cerrigydrudion, North Wales. Eng Geol. 1998;50:299–307.

Pearce-Higgins JW, Stephen L, Langston RHW, Bainbridge IP, Bullman R. The distribution of breeding birds around upland wind farms. J Appl Ecol. 2009;46:1323–31.

Pontevedra Pombal X, Nóvoa-Muñoz JC, Martínez Cortizas A, García Arrese AM, Nieto Olano C, Macías Vázquez, García-Rodeja E. Windfarm development on peatlands at Serras Septrionais of Galicia (NW Spain). IMCG Newsl. 2007;4:24–8. http://www.imcg.net/pages/publications/newsletter.php?lang=EN. Accessed 2 Feb 2015.

Renou-Wilson F, Farrell CA. Peatland vulnerability to energy-related developments from climate change policy in Ireland: the case of wind farms. Mires and Peat. 2009; 4:Article 08.

Smith J, Nayak DR, Smith P. Wind farms on undegraded peatlands are unlikely to reduce future carbon emissions. Energy Policy. 2014;66:585–91.

Warburton J, Holden J, Mills AJ. Hydrological controls of surficial mass movements in peat. Earth-Sci Rev. 2004;67:139–56.

Website References

News Video of Derrybrien Bog Sslide: https://www.youtube.com/watch?v=k6UMUW4IIrc. Accessed 10 Apr 2015.

NOAA Hurricane Research Division website: http://www.aoml.noaa.gov/hrd/tcfaq/D7.html. Accessed 2 Feb 2015.

Scottish Natural Heritage Wind Farm Footprint Maps (updated regularly) http://www.snh.gov.uk/planning-and-development/renewable-energy/research-data-and-trends/trendsandstats/windfarm-footprint-maps/. Accessed 14 Apr 2015.

Kakagon (Bad River Sloughs), Wisconsin (USA)

Jim Meeker and Naomi Tillison

Contents

Introduction	428
Geologic Setting	429
Wetland Setting	430
Water Level Dynamics	431
Ecology of Manomin (Wild Rice)	431
Invasion of Exotics	431
Sediment and Nutrient Loading	432
Current Status	434
Future Challenges	434
References	435

Abstract

The Kakagon/Bad River Sloughs, Honest John Lake wetland complex, is a 6,475 ha mosaic of coastal wetland and open water habitats of Lake Superior and is the homeland of the Bad River Band of the Lake Superior Tribe of Chippewa Indians. The complex is a dynamic system influenced by both natural processes and anthropogenic activities, including those occurring within the Bad River Watershed and Lake Superior. The Kakagon Sloughs support a myriad of wetland

---

Jim Meeker: deceased.

J. Meeker (✉)
Formerly of: Northland College, Ashland, WI, USA

N. Tillison (✉)
Natural Resources Director, Bad River Band of Lake Superior Chippewa, Odanah, WI, USA
e-mail: NRDirector@badriver-nsn.gov

© Springer Science+Business Media B.V., part of Springer Nature 2018
C. M. Finlayson et al. (eds.), *The Wetland Book*,
https://doi.org/10.1007/978-94-007-4001-3_229

types, and both the estuarial characteristics of the Sloughs and influence of fluctuating water levels on Lake Superior contribute to the wetland's productivity. Hydrology is a strong driver of the health of this ecosystem, and wild rice *Zizania palustris* serves as a key species indicating the health of the Kakagon Sloughs complex. The Bad River Band adaptively manages the Kakagon Sloughs complex by implementing an integrated, multifaceted approach, including controlling invasive species, improving habitat, and collaborating in watershed management efforts.

Keywords

Wild rice · Cattails · Nutrients · Sediments · Water levels · Coastal wetlands

Introduction

The Kakagon/Bad River Sloughs, Honest John Lake wetland complex (hereafter "Kakagon Sloughs" or "Kakagon Sloughs complex") is a 16,000-acre (6,475 ha) mosaic of coastal wetland and open water habitats of Lake Superior near Odanah, Wisconsin, USA (Fig. 1). The Kakagon Sloughs complex is recognized for its large size, unique features, and the many rare species it supports, such as lake sturgeon *Acipenser fulvescens*, wild rice *Zizania palustris*, English sundew *Drosera anglica*, and a regionally significant mussel bed containing rare species and species not known elsewhere in the Lake Superior basin (Epstein et al. 1997). It is the homeland of the Bad River Band of the Lake Superior Tribe of Chippewa Indians (hereafter referred to as the "Bad River Band"), sustaining generations of community members with vast Manomin (wild rice) beds and diverse fisheries and wildlife – a home known for providing plentiful foods and medicines (Fig. 2).

The Bad River Band has designated the Kakagon Sloughs as a Conservation Area, an area managed for the natural, ecological, and cultural values (Bad River Band of Lake Superior Chippewa 2001). Recognized for its biodiversity and unique features and the integrity of the ecosystem, the Kakagon Sloughs complex is also designated a Wetland of International Importance under the Ramsar Convention on Wetlands, an Important Habitat Site in the binational Lake Superior Lakewide Management Plan, and a National Natural Landmark.

The Kakagon Sloughs is a dynamic system influenced by both natural processes and anthropogenic activities, including those occurring within the Bad River Watershed and Lake Superior. To understand the current and historical state of this freshwater estuary, ecological drivers need to be examined at different spatial and temporal scales. For example, both short- and long-term lake level fluctuations are strong ecological drivers of the system. Seasonal and annual variations in the sediment and nutrient loading to the Kakagon Sloughs and the introduction and spread of nonnative species (e.g., narrow-leaf cattail *Typha angustifolia*) further drive changes to this system.

Fig. 1 Overview of the Kakagon/Bad River Sloughs, Honest John Lake wetland complex within the Bad River Reservation, Odanah, Wisconsin (Image © Rights remain with the Bad River Band of Lake Superior Tribe of Chippewa Indians)

Geologic Setting

Throughout most of the Bad River Reservation, glacial deposits, as part of the Miller Creek Formation occurring 9,500–11,500 years ago, overlay those deposited as part of the Copper Falls Formation, occurring over 11,500 years ago (Clayton 1984). Covering the majority of the lowland along Lake Superior, the

Fig. 2 Bad River tribal members harvesting wild rice in the Kakagon Sloughs in August 2013 (Photo credit: Bad River Band of Lake Superior Tribe of Chippewa Indians © Rights remain with the *Bad River Band of Lake Superior Tribe of Chippewa Indians*)

offshore clay and silt deposited by the turbidity currents flowing into glacial lakes are one component of the Miller Creek Formation (Clayton 1984); these glacial deposits are commonly referred to as the red clay plain and consist of clays interspersed with sands and silts. This heterogeneous mixture of clay and sand produces soils with little stability, which, when exposed to varying moisture conditions on steep slopes, often erodes severely (USEPA 1980). As a result of the glacial history combined with current and historic land practices, such as vast logging operations occurring in the late 1800s, nonpoint source impacts are observed in the Bad River watershed today.

Wetland Setting

The 1,000 square mile (2,590 km^2) Bad River watershed drains into the Kakagon Sloughs complex and intermixes with the waters of Lake Superior. The Kakagon Sloughs complex supports a myriad of wetland types including shore fens and bogs, sedge meadows, and marshes. Many of these wetlands are peatlands and some of the most recent peat is around 600–800 years old, and can be found in abandoned river channels and at Sand Cut, located between the older recurved spit (Brush Point) and the present day Chequamegon Point. These younger peatlands began to form, and are due to the fact that, the ancestral channel of the Bad or White rivers shifted easterly and brought more sediment to that side of the system, eventually forming a new, younger set of recurved spits (Long Island). The oldest peat was dated to be about 2,500 years ago.

Water Level Dynamics

Short-term water level changes include, in the Kakagon in particular, mini-weather-driven tide-like phenomena, called seiches, with amplitudes as high as 40 cm (more commonly 20–25 cm) that often ebb and flow in a quasi-regular fashion with six to eight cycles in one 24 h period. The long-term water levels are influenced by geological factors such as isostatic rebounding of the earth (i.e., rising of land masses depressed by the weight of ice sheets during the last glacial period) at different rates throughout the Lake Superior basin.

Water level shifts like these as well as seasonal and multiyear fluctuations are important forcing functions in these wetlands. Multiyear fluctuations, in particular, contribute to a process that ecologists term *pulse stability*. Though the wetlands have constantly changed in species abundances in the short term, over longer time scales they have remained stable, as specific wetland types, as long as the climate has remained relatively constant.

Ecology of Manomin (Wild Rice)

As research suggests (Meeker 1993, 1996), both (1) the estuarial characteristics of the Kakagon Sloughs and (2) the influence of fluctuating water levels on Lake Superior contribute to this wetland's productivity. In general, wild rice is well adapted to the riverine environs (Fig. 3). As an annual plant competing with perennials, it benefits from moving waters through both the seasonal pulse of sedimentation and the scouring action or opening of habitat. In addition, annual water level fluctuations appear to favor wild rice productivity in the long term. In contrast to the generally accepted view that stable water levels favor wild rice, flooding events, although damaging to wild rice in a given season, act to set back perennial competition and offer a more open habitat for wild rice recolonization during the subsequent drawdown years. Restoration and management that recognize these natural processes will likely be sustainable.

Invasion of Exotics

In contrast to the natural, usually cyclical, changes noted in many of Lake Superior's coastal wetlands, there also have been considerable changes in response to human activities since European settlement (Meeker and Fewless 2008). Instead of the aforementioned pulse stability, a rogue's gallery of tall, aggressive exotic plants including species such as narrow-leaf cattail, common reed grass *Phragmites*

Fig. 3 An aerial view of the Kakagon Sloughs complex during peak growing season (July 2013) shows the wild rice growing along the channels (Photo credit: Bad River Band of Lake Superior Tribe of Chippewa Indians © Rights remain with the Bad River Band of Lake Superior Tribe of Chippewa Indians)

australis, purple loosestrife *Lythrum salicaria*, and, at higher elevations, reed canary grass *Phalaris arundinacea* have invaded the sedge dominated poor fen wetlands. The ecological changes due to such exotics do not appear to be cyclic, but directional and smaller plants eventually are lost from the site. For example, transects surveyed in 1998 and again in 2005 indicated narrow-leaf/hybrid cattail *Typha x glauca* expanded by an average of 65% whereas wide leaf cattail *T. latifolia*, the species native to the Kakagon Sloughs, experienced no significant changes (Meeker 2005). The more aggressive narrow-leaf/hybrid cattails produce dense rhizome mats, encroaching on the wild rice habitat and are known to rapidly replace a diverse ecosystem with a monoculture. Declines in water levels and increase in nutrients have been shown to promote the growth of these invaders in Great Lake coastal wetlands (Meeker and Fewless 2008).

Sediment and Nutrient Loading

The Lake Superior red clay region is known for its flashy hydrologic regimes and sedimentation issues. The Bad River delivers one of the largest sediment loads to Lake Superior (Robertson 1997; Fitzpatrick et al. 2006). Erosion and sediment transport and deposition rates are accelerated in the Bad River Watershed as compared to pre-European settlement rates; the elevated rates are due to historical and current

Fig. 4 Water from the Bad River (*upper right*) flowing into the Kakagon (*lower left*) on May 24, 2013, when the Bad River was near its annual peak after receiving runoff from snowmelt and precipitation (Photo credit: Bad River Band of Lake Superior Tribe of Chippewa Indians © Rights remain with the Bad River Band of Lake Superior Tribe of Chippewa Indians)

land-use practices (e.g., timber harvests, agricultural practices) and more frequent large floods (Fitzpatrick et al. 2006). At flood stage, Bad River has been observed to flow westward to enter the Kakagon side of the system, such as in May 2013 (Fig. 4).

Sediment loading and accumulation rates vary spatially and temporally in the Kakagon Sloughs. A paleoecology study of Honest John Lake, located in the eastern portion of the Kakagon Sloughs complex, revealed that sedimentation rates increased in the early 1900s with the greatest change occurring over the last 50 years and peaking in the mid-1980s at around four times historical rates (Garrison 2011a). On the western side of the Kakagon Sloughs in Beartrap Creek Sloughs, the sedimentation rates were also higher a few decades ago and are still elevated above pre-European settlement rates (Garrison 2011b).

In addition to sediments, the Bad River transports a significant load of phosphorus during high flows to Lake Superior (Robertson 1997). In Honest John Lake, the delivery of phosphorus has increased to nearly three times the historical rate; this increase is largely driven by the increase in sediment loading to this system (Garrison 2011a). In comparison, phosphorus depositional rates in Beartrap Creek Slough have remained relatively constant in the last few decades and are lower than observed at the Honest John Lake sampling location (Garrison 2011b).

Unlike phosphorus levels, nitrogen concentrations tend to be greater at the Beartrap Creek Slough sampling location as compared to the Honest John site. Additionally, nitrogen levels have increased over the last few decades in Beartrap Creek Slough

(Garrison 2011b). In Honest John Lake, nitrogen concentrations are currently lower than at their peak in the mid-1980s, corresponding to maximum soil erosion rates, whereas phosphorus levels peaked in the early 2000s (Garrison 2011a). This difference between the recent nitrogen and phosphorus trends may be attributable to the municipal wastewater treatment discharge that previously entered Denomie Creek.

Shifts in the diatom community composition correspond with changes in water quality. In Beartrap Creek Slough, the diatom community composition indicates this area was historically mesotrophic and has remained so during the last century (Garrison 2011b). In Honest John Lake, the diatom community composition has shifted from one that is dominated by benthic diatoms to one that is dominated by planktonic diatoms, indicating an increase in phosphorus loading to the lake (Garrison 2011a).

Current Status

The condition of the Kakagon Sloughs complex is influenced by both natural processes and anthropogenic activities occurring in the Bad River watershed and in Lake Superior. The health of this coastal ecosystem is strongly driven by short-term and long-term water level fluctuations. Historic and current anthropogenic activities have resulted in the invasion of species, such as narrow-leaf/hybrid cattail, and contributed to changes in water quality. In light of its current stressors, the Kakagon Sloughs complex continues to show signs of a healthy and functioning ecosystem. For example, the abundance of juvenile lake sturgeon in the Bad River has increased over the last 15 years (Quinlan et al. 2010).

Future Challenges

Hydrology is a strong driver of the health of this wild rice ecosystem, and the Kakagon Sloughs is influenced by both the watershed draining into it and Lake Superior. As of 2013, Lake Superior water levels have been lower, with 1997 marking the most recent time water levels were greater than 10 cm above the long-term average of 183.40 m (Gronewold et al. 2013). In 2007, water levels were near or at record lows, and a rare event occurred – the Bad River Band closed the wild rice harvest season in the Kakagon Sloughs. In 2012, the harvest was closed once again due to the sparse wild rice. Multiple factors contributed to the 2012 closure, such as warmer spring water temperatures, the large runoff event occurring in June 2012, uprooting sensitive wild rice plants in the floating-leaf stage, and the low water levels in Lake Superior in the recent decade or so, shifting the overall vegetative community. As the climate changes, wild rice will continue to serve as a key species indicating the health of the Kakagon Sloughs complex.

The Bad River Band will continue to adaptively manage the Kakagon Sloughs complex by implementing an integrated, multifaceted approach, including controlling invasive species, improving habitat, and collaborating in watershed management efforts. Routine monitoring is a critical component of adaptive management, providing valuable information on the current condition and the causes and sources of change along with evaluating the success of restoration efforts. The health of the Bad River Band is intertwined with that of the Kakagon Sloughs complex. They will continue to care for this unique and diverse ecosystem because they know if they do, it will take care of them as it has for generations.

References

Bad River Band of Lake Superior Tribe of Chippewa. Integrated resource management plan. Odanah: Bad River Natural Resources Department; 2001.

Clayton L. Pleistocene geology of the Superior region, Wisconsin, Wisconsin Geological and Natural History Survey information circular, vol. 46. Madison: Wisconsin Geological and Natural History Survey; 1984.

Epstein EJ, Judziewicz EJ, Smith WA. Wisconsin's Lake Superior coastal wetlands evaluation including other selected natural features of the Lake Superior basin. Wisconsin Department of Natural Resources, Bureau of Endangered Resources; 1997. PUB ER-095 99.

Fitzpatrick F, Cahow-Scholtes K, Peppler, M. Geomorphic context for sediment sources, movement, and deposition in the Bad River, Bad River Reservation, Wisconsin. Middleton, WI: United States Geologic Survey, Wisconsin Water Science Center; 2006.

Garrison PJ. Paleoecological study of Honest John Lake, Ashland County. Wisconsin Department of Natural Resources, Bureau of Science Services; 2011a. PUB SS-1085 2011.

Garrison PJ. Paleoecological study of Bear Trap Slough, Ashland County. Wisconsin Department of Natural Resources, Bureau of Science Services; 2011b. PUB SS-1087 2011.

Gronewold AD, Clites AH, Smith JP, Hunter TS. A dynamic graphical interface for visualizing projected, measured, and reconstructed surface water elevations on the earth's largest lakes. Environ Model Software. 2013;49:34–9. Available from: http://www.glerl.noaa.gov/pubs/fulltext/2013/20130022.pdf. Accessed 29 Oct 2013.

Meeker JE. The ecology of "wild" wild-rice (*Zizania palustris* var. *palustris*) in the Kakagon Sloughs, a riverine wetland on Lake Superior [dissertation]. Madison: University of Wisconsin-Madison, Botany Department; 1993. 365 p.

Meeker JE. Wild-rice and sedimentation processes in a Lake Superior coastal wetland. Wetlands. 1996;16(2):219–31.

Meeker JE. Repeat sampling of *Typha* species in the Kakagon Sloughs and Bad River systems. Unpublished Report to Bad River Band; 2005.

Meeker JE, Fewless G. Change in Wisconsin's coastal wetlands. In: Waller DM, Rooney TP, editors. The vanishing present: Wisconsin's changing lands, waters, and wildlife. Chicago/London: The University of Chicago Press; 2008. p. 183–92.

Quinlan HR, Pratt TC, Friday MJ, Schram ST, Seider MJ, Mattes WP. Inshore fish community: lake sturgeon. In: Gorman OT, Ebener MP, Vinson MR, editors. The state of Lake Superior in 2005, Great Lakes Fishery Commission special publication 10-1. Ann Arbor: Great Lakes Fishery Commission; 2010.

Robertson DM. Regionalized loads of sediment and phosphorus to Lakes Michigan and Superior – high flow and long-term average. J Great Lakes Res. 1997;23(4):416–39.

United States Environmental Protection Agency (USEPA). Red clay project final report part II: impact on nonpoint pollution control on western Lake Superior. 1980. EPA 905/9-79-002-B.

Seagrass Dependent Artisanal Fisheries of Southeast Asia

33

Richard K. F. Unsworth and Leanne C. Cullen-Unsworth

Contents

Introduction	438
Seagrass Fisheries	438
Seagrass Fisheries and Food Security	440
Understanding Drivers of Overexploitation	441
Additional Threats to Seagrass in SE Asia	441
Over-Fishing	441
Destructive Fishing	442
Mining	442
Seagrass Extraction	443
Future Challenges	443
References	444

Abstract

Seagrass meadows are widespread and abundant across Southeast (SE) Asia providing a myriad of ecosystem services that includes artisanal fisheries which support human wellbeing though livelihood provision and food security. In addition to threats commonly associated with degradation of seagrass meadows globally, seagrass meadows across SE Asia are increasingly threatened by overexploitation of their productive fish and invertebrate assemblages. The abun-

R. K. F. Unsworth (✉)
College of Science, Wallace Building, Swansea University, Swansea, UK
e-mail: r.k.f.unsworth@swansea.ac.uk

L. C. Cullen-Unsworth
Sustainable Places Research Institute, Cardiff University, Cardiff, UK
e-mail: Cullen-UnsworthLC@cardiff.ac.uk

© Springer Science+Business Media B.V., part of Springer Nature 2018
C. M. Finlayson et al. (eds.), *The Wetland Book*,
https://doi.org/10.1007/978-94-007-4001-3_267

dance of small-scale fisheries across SE Asia means that fisheries management needs to identify socially acceptable and locally implementable controls on marine resource use that supports conservation at the same time as enhancing local livelihood interests. In this article we discuss the drivers of SE Asia seagrass meadow degradation and identify the need for increased focus on understanding socio-economic and cultural issues as on pure ecological information in conservation and fisheries management.

Keywords
Seagrass meadows · Artisanal fishery · Fisheries management · Food security · Seagrass degradation

Introduction

Seagrass meadows are widespread and abundant across Southeast (SE) Asia providing a myriad of ecosystem services. There is a growing body of evidence demonstrating the role of seagrass meadows in artisanal fisheries, this role relates to fisheries productivity which supports human wellbeing though livelihood provision and food security. As alternative food resources become scarcer or less accessible, healthy seagrass meadows within a connected seascape represent a reliable source of protein and income for coastal communities. The seagrasses across SE Asia, as in other parts of the world, remain subject to widespread threat with conservation efforts largely marginalized. Globally, degradation of seagrass meadows has been commonly associated with increased nutrient run-off, sedimentation, damage from boats, other physical impacts, and pesticide leaching (Orth et al. 2006) (See Fig. 1a, b). In SE Asia seagrass meadows are also increasingly threatened by overexploitation of their productive fish and invertebrate assemblages (Fortes 1990; Tomascik et al. 1997). This is of particular concern where seagrass meadows offer an easily accessible and abundant food source for local people.

Seagrass Fisheries

Few faunal species utilize seagrass meadows throughout their entire life, but their use of this habitat can have significant cascade effects, supporting thriving adult populations with high fisheries values (Gillanders 2006). In SE Asia, seagrass meadows have been shown to have a hidden role in fisheries productivity that was previously attributed to other habitats (Unsworth and Cullen 2010). Links between seagrass and fisheries have been demonstrated in other areas (Warren et al. 2010) and evidence of the role of seagrass in supporting food security is

Fig. 1 (**a, b**) (*top left* and *right*, respectively) impacts of excess epiphytes and macro algae suffocating seagrass; (**c**) (*middle left*) invertebrate harvesting from seagrass; (**d**) (*middle right*) abundant netting in seagrass; (**e**) (*lower left*) typical seagrass fish species caught in SE Asia; (**f**) (*lower right*) intensive static nets growing in popularity in seagrass (Photo credit: R Unsworth and C Smart © Rights remain with the author)

emerging (Unsworth et al. 2014). Important fishing grounds in their own right, seagrass meadows also play a significant role in coral reef and other fisheries productivity.

Seagrass Fisheries and Food Security

Seagrass meadows are a key component of the seascape forming important readily accessible fishing grounds (Unsworth and Cullen 2010) and critical nursery habitat (Gillanders 2006). Traditionally all tropical coastal marine fisheries are described as "coral reef fisheries." Such fisheries to some may include seagrass, but poor terminology has led to an ill appreciation for the role of seagrass habitats in supporting fisheries productivity and hence food supply (Unsworth and Cullen 2010). SE Asia's seagrass meadows support commercially important faunal species as well as species with high subsistence value. Yet, while many people are aware of the importance of coral reef fisheries in tropical habitats, seagrass fisheries are often neglected. Seagrass meadows are ideal fishing grounds because they contain abundant fish and invertebrates (see Fig. 1c, d), and their location in intertidal or shallow subtidal waters means they are easily accessible and can usually be exploited in all weather conditions by all types of fishers (de la Torre-Castro et al. 2014). For people with limited financial means this access can be a critical component of wellbeing (Cullen-Unsworth et al. 2014). In many areas the reliance of fishers on seagrass meadows, rather than coral reefs, as a fishing ground, contradicts the emphasis of fisheries management, monitoring, and conservation efforts placed primarily on coral reef habitats.

The Coral Triangle in SE Asia is considered to be the world's epicenter of marine biodiversity with 76% of all known coral species, 37% of all known coral reef fish species, 53% of the world's coral reefs, and the greatest extent of mangrove forests in the world (Coral Triangle Initiative Secretariat 2009). It also houses exceptional seagrass habitat with high conservation value (Unsworth et al. 2014). The area includes part or all of the exclusive economic zones of six countries: Indonesia, Malaysia, Papua New Guinea, the Philippines, the Solomon Islands, and Timor-Leste. Seagrass meadows at the center of the Coral Triangle have been demonstrated to support at least 50% of the fish based food supply that accounts for between 54% and 99% of daily protein intake in the area (Unsworth et al. 2014). Given that 68% of fishing activity in some areas occurs within seagrass meadows and not on coral reefs making assumptions about where fishing activity occurs and the biological production supporting fishers can lead to inappropriate management focus.

Across SE Asia, seagrass meadows are commonly harvested at low tide for subsistence foodstuffs such as small molluscs, clams, and urchins and for commercial species such as octopus and sea cucumber. Particularly as full moon approaches, the exposed inter-tidal zone at sunset becomes a hive of gleaning activity, with numerous fishers and families collecting invertebrates, trapping fish stranded in tide pools, or bringing in tidal nets. Much of this fishing is subsistence and community-based but also includes small family fishing collectives earning a basic living selling excess catch. It involves whole families, including small children (see Fig. 1c), and as a result exists as a social and recreational activity. This type of exploitation remains largely unquantified, but it can be assumed to be increasing in areas of rapid human population growth and where dependence on seafood for protein is high (Unsworth and Cullen-Unsworth 2014). With rapid population growth, continued economic expansion, and major development pressures seagrass meadows are under increasing stress, with overexploitation of herbivorous fauna a significant issue.

Understanding Drivers of Overexploitation

Due to their economic and ecological importance, and the fact that seagrass meadows are becoming increasingly degraded, management of these habitats should be a major consideration when designing local marine conservation efforts. For increased chances of success, coral reef conservation efforts must account for the inter-connectedness of ecosystems. Coral reefs rely on their connections with adjacent habitats for fish nursery grounds, supply of organic detritus and nutrients, and for fish and invertebrate feeding grounds (Harborne et al. 2006). Effective conservation and fisheries management therefore requires a thorough understanding of resource exploitation patterns and the related impacts on all marine habitats.

Conservation and fisheries management should be as much focused on understanding socioeconomic and cultural issues as on pure ecological information. Stakeholder engagement is now an accepted part of natural resource management, but to be effective strategies need to incorporate the requirements of local people and their beliefs. Throughout SE Asia, coastal marine habitats are culturally as well as economically important, with many traditional ways of life intricately associated with seagrass meadows for food, recreation, and spiritual fulfillment.

Given that vast areas of seagrass in SE Asia are present in archipelagos of small islands on reef systems, are often far from major riverine influence and urban development, and unlikely to be the focus of high trawling activity, threats to such seagrass meadows are either very limited or are dominated by other issues. Degradation in remote areas is largely a consequence of overexploitation of fish and invertebrate populations causing trophic cascades, localized pollution causing small-scale water quality issues, and localized mechanical damage. Threats from fishing activities (e.g., net-fishing) in the Tropical Indo-Pacific have been ranked higher than in other bioregions, reflecting the regional differences in the exploitation of seagrass fisheries and the nature of the fishing activities, but this was minimal in comparison to other impacts (Grech et al. 2012).

Additional Threats to Seagrass in SE Asia

Over-Fishing

In Indonesia and the Philippines, seagrass meadows are under increasing pressure from large tidal fishing nets that are laid for up to 100 m across the seagrass to catch all fish moving with the tide (see Fig. 1e, f). These tidal fishing devices fish indiscriminately, catching juveniles as well as low value species. Catch commonly comprises species from the families Emperor (*Lethrinidae*), Rabbitfish (*Siganidae*), and Parrotfish (*Scaridae*) that migrate between coral reef, mangrove, and seagrass habitats. Removing high numbers of Rabbitfish and Parrotfish may have long-term implications for coral reefs, as both families have been highlighted as playing important ecological roles, aiding coral reef resilience and recovery after bleaching (damage to corals from elevated sea temperatures) (Fox and Bellwood 2008). The juvenile catch includes fish from the Grouper, Snapper, and Wrasse families, commonly associated with coral reefs but utilizing the abundant food sources within seagrass, their success often depends on the availability of seagrass resources.

In Indonesia, the tidal fishing net catch from seagrass beds has experienced declines of up to 60% over a five-year period, which probably reflects a similar decline in the size of the fish stock (Exton 2010). Examination of satellite imagery across parts of Eastern Indonesia in particular illustrates the growing use of this intensive fishing method within intertidal seagrass areas which has knock-on effects for both seagrass and connected habitats (Unsworth and Cullen 2010).

Invertebrate seagrass fisheries are also in decline in many areas with sea cucumber fisheries in particular crashing. Even at subsistence levels, invertebrate exploitation can alter and reduce the biomass and diversity within seagrasses (Nordlund and Gullstrom 2013). Commercial exploitation of seagrass fauna is common throughout the Indo-Pacific bioregion (see Fig. 1), with seagrass meadows now largely devoid of many sea cucumber species as well as the helmet shell *Cassis cornuta* which was collected for the curio trade. Seahorses are another important component of seagrass faunal species now threatened by overexploitation (Curtis et al. 2007).

Overexploitation of fish and invertebrate populations is considered a major driver of seagrass loss (Moksnes et al. 2008). There is increasing evidence that seagrass communities are defined by top-down predator control (Eklof et al. 2009). This trend of declining predator abundance may therefore result in further cascade effects, such as loss of predators leading to increased urchin abundance with associated episodes of seagrass over-grazing, the result being a loss of cover or changes in seagrass assemblage structure (Rose 1999). The loss of key herbivorous fish populations that graze upon the epiphytes of seagrass may also lead to a loss of ecosystem resilience and potential system wide degradation (Unsworth et al. 2015).

Destructive Fishing

The use of cyanide or bleach and bombs by fishers is not restricted to coral reefs but is also used within seagrass meadows. Although quantitative evidence of this is limited, anecdotal evidence exists that this is commonly used in seagrass meadows. The impact of these activities on seagrass meadows comes largely from direct physical damage to the meadow.

Mining

Both coral and sand is mined for infrastructure construction and repair, the demand for which is increasing with an increasing population and influxes of developmental funding, and with knock-on impacts for water quality. Mining for "dead" coral or sand beneath seagrass meadows involves physical removal of large areas of seagrass to access the substrate.

Seagrass Extraction

Organically derived poison extracted from terrestrial plant material is used to remove seagrass from areas selected for static fishing gear holding pens. In some localities, seagrass is also physically removed from areas utilized for seaweed cultivation as it is believed that the seagrass causes damage and the spread of disease within the cultivated stock.

Future Challenges

To ensure that seagrass meadows remain at the forefront of management thinking (alongside coral reefs and mangroves) top-down support for bottom up action is required to protect seagrasses. Formal management plans are lacking. The value of seagrass meadows is recognized within the Convention on Biological Diversity (CBD) and Convention on the Conservation of Migratory Species (CMS) but beyond this recognition, management remains highly limited, with protection usually afforded only due to proximity with other "high profile" habitats such as coral reefs. To further support implementation of context specific and appropriate management planning, we need to understand the socio-cultural, economic, and environmental drivers that are resulting in seagrass decline.

The abundance of small-scale fisheries across SE Asia means that fisheries management needs to identify socially acceptable and locally implementable controls on marine resource use (Cohen et al. 2013). Marine protected area management needs to include partnerships between communities, civil society, and government (Gutierrez et al. 2011), so that management supports conservation at the same time as enhancing local livelihood interests. From a food security perspective, understanding food supply, its ecological origin, and the resources that support it are critical knowledge components of developing appropriate ecosystem based management actions that can foster enhancement of fisheries resources. Prohibition or controls of a specific fishing activity in one habitat type will have limited impact if those fish migrate at night into an adjacent habitat where they are readily collected by fishers. Therefore, understanding habitat links to fisheries is critical for the consideration of short-term fisheries management but is also important for understanding the vulnerability of marine systems (Folke 2006). Given the need to understand the role that different habitat types have in supporting tropical marine fisheries, the limited literature and knowledge on seagrass biodiversity across SE Asia, and the growing evidence of the role of seagrass meadows in supporting fisheries, a major future challenge is to foster widespread recognition for the role of seagrass meadows in supporting fisheries and hence local food security.

References

Cohen PJ, Cinner JE, Foale S. Fishing dynamics associated with periodically harvested marine closures. Glob Environ Chang. 2013;23(6):1702–13.

Coral Triangle Initiative Secretariat Regional plan of action; Coral Triangle Initiative on Coral Reefs, Fisheries and Food Security (CTI-CFF). Report. Interim Regional CTI Secretariat, Manado. 2009: 87 p.

Cullen-Unsworth LC, Nordlund L, Paddock J, Baker S, McKenzie LJ, Unsworth RKF. Seagrass meadows globally as a coupled social-ecological system: implications for human wellbeing. Mar Pollut Bull. 2014;83(2):387–97.

Curtis JMR, Ribeiro J, Erzini K, Vincent ACJ. A conservation trade-off? Interspecific differences in seahorse responses to experimental changes in fishing effort. Aquat Conserv Mar Freshwat Ecosyst. 2007;17:468–84.

de la Torre-Castro M, Di Carlo G, Jiddawi NS. Seagrass importance for a small-scale fishery in the tropics: the need for seascape management. Mar Pollut Bull. 2014;83(2):398–407.

Eklof JS, Frocklin S, Lindvall A, Stadlinger N, Kimathi A, et al. How effective are MPAs? Predation control and "spill-in effects" in seagrass-coral reef lagoons under contrasting fishery management. Mar Ecol Prog Ser. 2009;384:83–96.

Exton DA. Nearshore fisheries of the Wakatobi. In: Clifton J, Unsworth RKF, editors. Marine conservation and research in the Coral Triangle: the Wakatobi Marine National Park. New York: Nova Scientific; 2010.

Folke C. Resilience: the emergence of a perspective for social–ecological systems analyses. Glob Environ Chang. 2006;16(3):253–67.

Fortes MD. Seagrasses: a resource unknown in the ASEAN region. Manila: International Centre for Aquatic Living Resources; 1990.

Fox RJ, Bellwood DR. Remote video bioassays reveal the potential feeding impact of the rabbitfish *Siganus canaliculatus* (f: Siganidae) on an inner-shelf reef of the Great Barrier Reef. Coral Reefs. 2008;27(3):605–15.

Gillanders BM. Seagrasses, fish, and fisheries. In: Larkum AW, Orth RJ, Duarte CM, editors. Seagrasses: biology, ecology and conservation. Dordrecht: Springer; 2006. p. 503–36.

Grech A, Chartrand-Miller K, Erftemeijer P, Fonseca M, McKenzie L, Rasheed M, et al. A comparison of threats, vulnerabilities and management approaches in global seagrass bioregions. Environ Res Lett. 2012;7(2):024006.

Gutierrez NL, Hilborn R, Defeo O. Leadership, social capital and incentives promote successful fisheries. Nature. 2011;470(7334):386–9.

Harborne AR, Mumby PJ, Micheli F, Perry CT, Dahlgren CP, Holmes KE, et al. The functional value of Caribbean coral reef, seagrass and mangrove habitats to ecosystem processes. In: Southward AJ, Young CM, Fuiman LA, editors. Advances in Marine Biology. 2006; 50:57–189.

Moksnes PO, Gullstrom MM, Tryman K, Baden S. Trophic cascades in a temperate seagrass community. OIKOS. 2008;117:763–77.

Nordlund LM, Gullstrom M. Biodiversity loss in seagrass meadows due to local invertebrate fisheries and harbour activities. Estuar Coast Shelf Sci. 2013;135:231–40.

Orth RJ, Carruthers TJB, Dennison WC, Duarte CM, Fourqurean JW, Heck KL, et al. A global crisis for seagrass ecosystems. BioScience. 2006;56:987–96.

Rose CD. Overgrazing of a large seagrass bed by the sea urchin *Lytechinus variegatus* in Outer Florida Bay. Mar Ecol Prog Ser. 1999;190:211–22.

Tomascik T, Mah JA, Nontji A, Moosa KM. The ecology of the Indonesian Seas (part II). Hong Kong: University of Oxford Press/Periplus Editions (HK) Ltd.; 1997. p. 643–1355.

Unsworth RKF, Cullen LC. Recognising the necessity for Indo-Pacific seagrass conservation. Conserv Lett. 2010;3:63–73.

Unsworth RKF, Cullen-Unsworth LC. Biodiversity, ecosystem services, and the conservation of seagrass meadows. In: Maslo B, Lockwood JL, editors. Coastal conservation. Cambridge (UK): Cambridge University Press; 2014. p. 95–130.

Unsworth RKF, Hinder SL, Bodger OG, Cullen-Unsworth LC. Food supply depends on seagrass meadows in the coral triangle. Environ Res Lett. 2014;9(9):094005.

Unsworth RKF, Collier CJ, Cullen-Unsworth LC, McKenzie LJ, Waycott M. A framework for the resilience of seagrass ecosystems. Mar Pollut Bull. 2015;100:34–46.

Warren MA, Gregory RS, Laurel BJ, Snelgrove PVR. Increasing density of juvenile Atlantic (*Gadus morhua*) and Greenland cod (*G. ogac*) in association with spatial expansion and recovery of eelgrass (*Zostera marina*) in a coastal nursery habitat. J Exp Mar Biol Ecol. 2010;394(1–2):154–60.

Great Barrier Reef (Australia): A Multi-ecosystem Wetland with a Multiple Use Management Regime

34

Jon Brodie and Jane Waterhouse

Contents

Introduction	448
Status of Selected Species and Ecosystems of the GBR	451
Coral Reefs	451
Seagrass	451
Mangroves	452
Saltmarsh	452
Megafauna	453
Management of the GBR	454
The Management Regime	454
Management Status	456
The Future of the GBR	457
References	458

Abstract

The Great Barrier Reef (GBR) covers an area of about 350,000 km^2 on the northeastern Australian continental shelf. Biological diversity within the GBR is very high and includes coral reefs, large areas of seagrass meadows, many species of turtles, sponge gardens and high species diversity of fish, molluscs, echinoderms, sea snakes and seaweeds. The GBR also has at least 30 species of whales and dolphin and the dugong. The coral reefs of the GBR are in generally poor condition with degradation continuing. Seagrass meadows have declined recently but may recover more easily if acute stressors are removed. Dugong and turtle populations in many parts of the GBR are in very severe decline. The GBR has

---

J. Brodie (✉) · J. Waterhouse (✉)
Catchment to Reef Research Group, TropWATER, Centre for Tropical Water and Aquatic Ecosystem Research, James Cook University, Townsville, QLD, Australia
e-mail: jon.brodie@jcu.edu.au; j.waterhouse@c2o.net.au

© Springer Science+Business Media B.V., part of Springer Nature 2018
C. M. Finlayson et al. (eds.), *The Wetland Book*,
https://doi.org/10.1007/978-94-007-4001-3_46

a well-designed management system based on both regulatory measures and voluntary compliance with planning regimes but the net effect of 40 years of management has seen many habitats and species still in decline. Most of the factors leading to decline can be identified as associated with climate change, terrestrial pollutant runoff and fishing. To prevent the further decline of the GBR more stringent measures need to be implemented to reduce the impacts of terrestrial runoff and fishing as well as better global (and Australian) measures to reduce the severity of climate change.

Keywords

Corals · Seagrass · Mangroves · Stressors · Management · Future

Introduction

The Great Barrier Reef (GBR) covers an area of about 350,000 km^2 on the northeastern Australian continental shelf. It is a long, narrow system stretching 2,000 km along the coast from 25^0 S near Lady Elliot Island and as far north as Bramble Cay close to the Papua New Guinea coast at latitude 9^0 S (Fig. 1). In width it ranges from 100 km wide in the north to 200 km in the south and is generally bounded by the coast on the west and the Coral Sea on the east. The adjacent catchment area covers 400,000 km^2 (Brodie et al. 2012; Great Barrier Reef Marine Park Authority 2014).

Biological diversity within the GBR includes over 2,900 coral reefs built from over 450 species of hard coral, over one third of all the world's soft coral and sea pen species (150 species), over 43,000 km^2 of seagrass meadows including 23% of the known global species diversity, 2,000 species of sponges equaling 30% of Australia's diversity in sponges, 3,000 species of mollusks including 2,500 species of gastropods, and approximately 500 species of seaweeds (from Day 2011). It also provides habitat for six of the world's seven species of marine turtle including the largest green turtle *Chelonia mydas* breeding area in the world, one of the world's most important dugong *Dugong dugon* populations, more than 1,620 species of fish of which 1,460 are coral reef species, 630 species of echinoderms including 13% of the known global species diversity, and 14 breeding species of sea snakes including 20% of the known global species diversity. The GBR is also a breeding area for humpback whale *Megaptera novaeangliae* with at least 30 other species of whales and dolphins identified within the GBR.

The coastline is dominated by mangroves and saltmarsh interspersed with areas of low energy sandy beach and rocky shores. Immediately offshore, shallow seagrass beds are common and considerable areas of deepwater (>15 m) seagrass are found further offshore (Coles et al. 2015; Fig. 2). The GBR lagoon floor is dominated by soft-bottomed communities of algae, sponges, and bryozoans interspersed with bare sand. In the north, extensive *Halimeda* sp. algal beds occupy the deeper offshore waters, their growth stimulated by nutrient-rich water upwelling from the Coral Sea

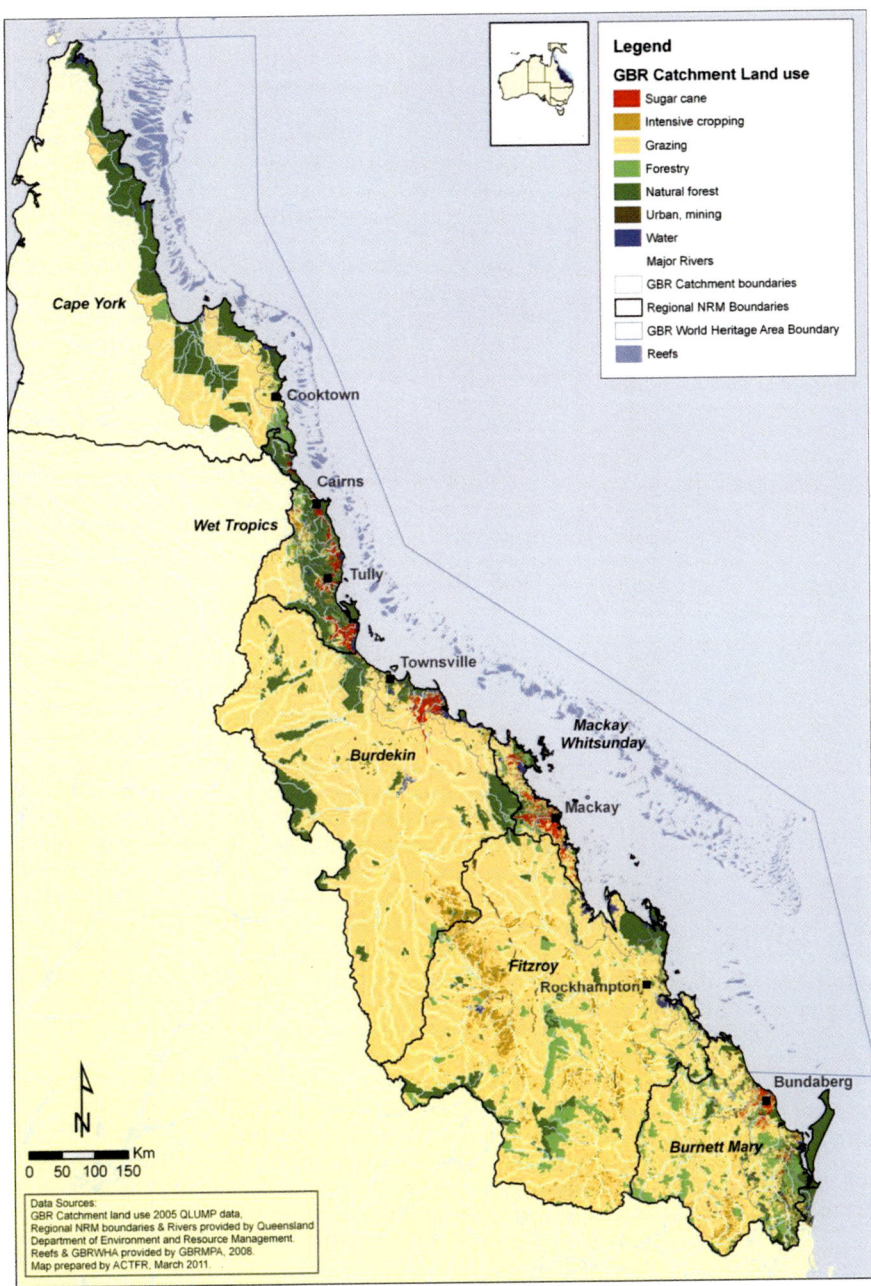

Fig. 1 The Great Barrier Reef showing the reefs, catchments and major rivers, and land uses on the catchments and major cities and towns (Map prepared by the Australian Centre for Tropical Freshwater Research (now TropWATER) for general use by staff of TropWATER, James Cook University)

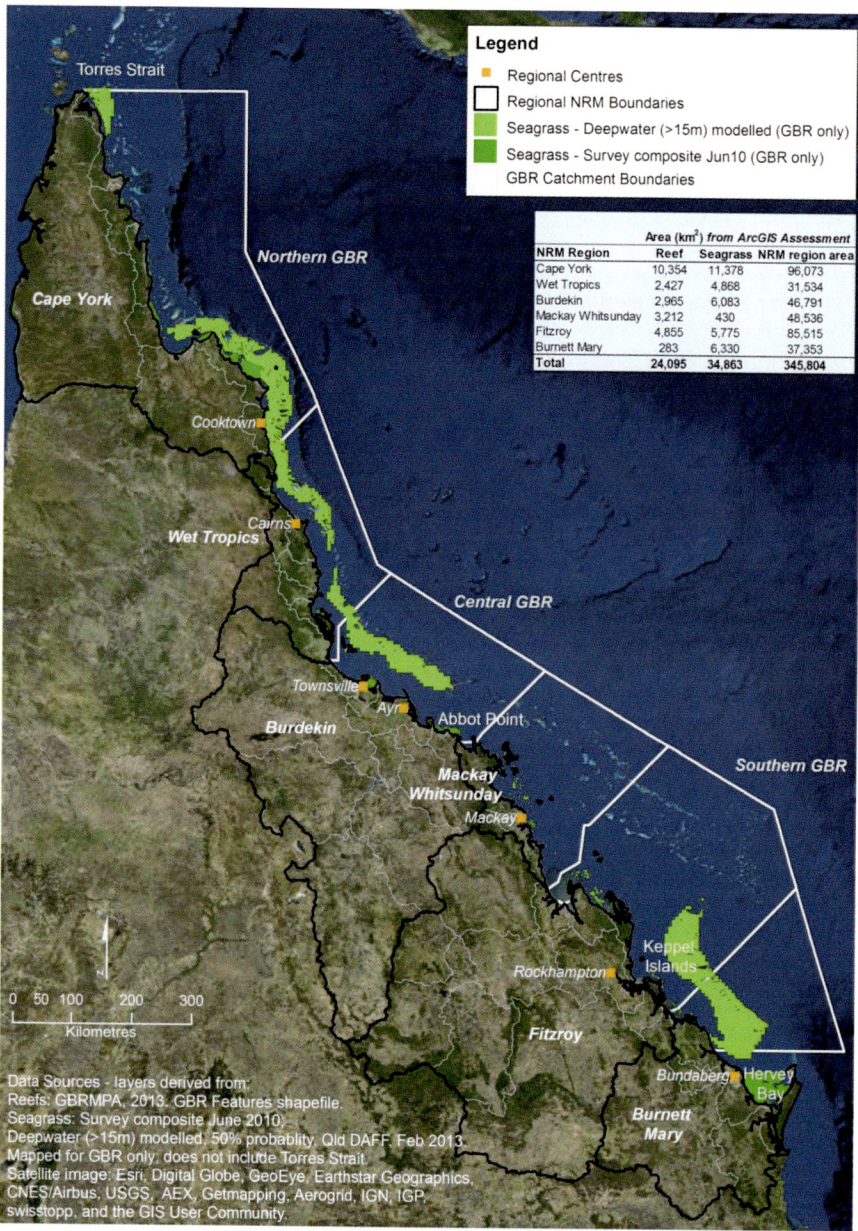

Fig. 2 Mapped extent of potential seagrass meadows and coral reefs in the Great Barrier Reef World Heritage Area. *Inset* table shows the area of coral reef, seagrass, and total area of each catchment-based natural resource management (*NRM*) region (Map prepared by J. Waterhouse, TropWATER. Data provided by the Great Barrier Reef Marine Park Authority and the Queensland Department of Agriculture, Fisheries and Forestry, 2013. Satellite image supplied by ESRI)

(Wolanski et al. 1988). The coral reefs of the GBR consist of two main types: the fringing reefs (~760 reefs) which occur on the coast and around the high islands (inner-shelf reefs), and those of the main reef (~2,200 reefs) which occupy a band on the outer part of the continental shelf (mid- and outer-shelf reefs) (Day 2011) (Fig. 2).

Status of Selected Species and Ecosystems of the GBR

Coral Reefs

Coral cover (an indicator of coral reef status) on the GBR has declined markedly in the period since the 1960s from values near 45–55% on mid- and outer-shelf reefs in the 1960s (Hughes et al. 2011) to 28% in 1986 and 14% in 2011 (De'ath et al. 2012). On the inner-shelf reefs, although our record is shorter (8 years), similar declines have been recorded (Thompson et al. 2014). Other indicators of coral "health" have also declined at different areas in the GBR, e.g., between Townsville and Cooktown coral diversity is much lower than expected and ascribed to the impact of terrestrial runoff (Thompson et al. 2014). In addition, calcification rates have declined across the central GBR with the cause mainly attributed to thermal stress on mid- and outer-shelf reefs (Great Barrier Reef Marine Park Authority 2014) with some involvement of ocean acidification, but on inner-shelf reefs terrestrial runoff has also been shown to play a part. There are many causes of the decline, often quite reef-specific, including terrestrial runoff of fine sediment (with associated nutrients) causing more turbid conditions on the inner-shelf and reducing light availability for coral growth, nutrient runoff from soil erosion and fertilizer loss causing crown of thorns starfish (COTS) outbreaks (Brodie et al. 2012), excess algal growth, enhanced sensitivity to coral bleaching, and coral diseases (Brodie et al. 2013); coral bleaching and mortality associated with climate change (Hughes et al. 2007); and increased incidence of category 4 and 5 cyclones which seems to have occurred over the last decade and is predicted to continue under climate change (Walsh et al. 2016). Rainfall extremes are also predicted to increase with an increase in the frequency of La Nina periods and hence more frequent and larger river discharge events (Great Barrier Reef Marine Park Authority 2014).

Seagrass

Seagrass health and abundance is quite variable in space and time in the GBR (Coles et al. 2015). This variability constrains our knowledge of the extent and biomass of seagrass in the GBR at any fixed time, in contrast to coral reefs which are much more structurally stable. Analysis of areas of the GBR suitable for deepwater seagrass (>15 m) using environmental factors has been carried out, and these modeled results are combined with survey results to produce an overall assessment of the likely occurrence of seagrass in the GBR (Coles et al. 2015). The combined area is shown in Fig. 2 (presented as 50% probability of occurrence).

Recently, there is strong evidence that seagrass is declining in parts of the GBR (McKenzie et al. 2015; Coles et al. 2015), particularly in the Townsville, Cairns, and Abbot Point regions and several other locations, associated with a series of severe cyclones and large river flood events (Coles et al. 2015; Devlin et al. 2012). Evidence of this decline are that 38% of sampling sites that are monitored regularly across the GBR are exhibiting shrinking meadow area, a large number of sites have reduced seagrass abundance, and many sites have limited or no sexual reproduction producing seeds that would enable rapid recovery (McKenzie et al. 2015). Degraded light regimes from increased turbidity are the driver of declining seagrass abundance in many sites. For example, the 2011 major river discharge events from many of the GBR rivers associated with the strong La Nina and the effects of category 5 tropical cyclone Yasi have had devastating effects on large areas of GBR seagrass (Devlin et al. 2012). In addition, port dredging may have severe but shorter-term effects on seagrass in locations in the GBR such as at Hay Point near Mackay, where turbidity from dredging is believed to have prevented seasonal reestablishment of deepwater seagrass, although recovery occurred in later years after the dredging ceased (York et al. 2015).

Mangroves

While mangrove forests worldwide are under increasing pressure from a range of anthropogenic threats, mangroves along the whole GBR coast are generally in excellent condition and relatively stable, with only small losses reported, mostly associated with port and urban development (Great Barrier Reef Marine Park Authority 2014). Mangrove and saltmarsh habitats cover an area of approximately 3,800 km^2 along the GBR coast. Given their importance to so many commercial fisheries, it is vital to maintain protection of this asset.

Saltmarsh

Saltmarsh in eastern Queensland has been subject to a range of development pressures due to its position at the interface between terrestrial and estuarine ecosystems, and often on private or leasehold land (Wegscheidl et al. 2015). Since European settlement, around 35 km^2 of saltmarsh has been lost in Queensland, mainly through the construction of ponded pastures, salt ponds, and urban development. The largest losses have been along the GBR coast, including in Fitzroy region (refer Fig. 2), where 17% of estuarine wetlands have been lost, primarily from the construction of tidal levees to create ponded pastures for cattle grazing (Great Barrier Reef Marine Park Authority 2014).

As is the case with mangroves, saltmarsh systems along the GBR coast are exposed to continuing threats from expanding urban and industrial coastal development (Waltham and Sheaves 2015), providing a strong imperative for better management of coastal development, especially given the poor status of many of the ecosystems in the GBR (Grech et al. 2013).

Megafauna

Dugong numbers in the GBR have declined precipitously over recent decades with numbers reducing at a rate of almost nine percent a year between 1962 and 1999. Overall, this is estimated to have reduced dugong numbers from about 72,000 in the early 1960s to 4,000 by the late 1990s. Causes of mortality include incidental netting in fish nets and shark nets, loss of seagrass habitat due to water quality impacts and coastal development, and hunting (Marsh et al. 2005). The combination of severe weather events in 2011 along the entire Queensland coast (Devlin et al. 2012) has also increased dugong mortality and the long-term effects of these events in combination with the existing stresses has yet to be assessed.

Dugong populations are in best condition in Hervey Bay at the southern end of the GBR (Coppo et al. 2014) and in the northern GBR including the Torres Strait (Sobtzick et al. 2014), while there are much lower populations left in the central and southern GBR (Grech et al. 2011). The Torres Strait region between mainland Australia and Papua New Guinea (Fig. 2) supports the largest recorded single continuous seagrass meadow in Australia (Carter et al. 2014) and is the most important dugong habitat in the world (Marsh et al. 2011). Dugongs primarily occur in a large, central area that extends south of Boigu Island to north of Badu and Moa Islands and west of Badu and Muralug Islands (Grech et al. 2011; Sobtzick et al. 2014). The northern GBR region also supports globally significant populations of dugongs (Sobtzick et al. 2014).

The population of "east Australian" humpback whales was as low as 500 animals when whaling ceased in 1963. The population in 2008 was estimated to have been more than 10,000 animals, which is about half of the estimated pre-whaling population size (Noad et al. 2008).

The population structure, distribution, range, and status of the six species of marine turtles found in the region have been well documented (Hamann et al. 2007). All six species are listed as threatened under Queensland and Federal legislation and the International Union for Conservation of Nature and Natural Resources (IUCN) Red List. Long-term census data on green turtle populations indicate that although significant declines in population size are not apparent, other biological factors such as declining annual average size of breeding females, increasing remigration interval, and declining proportion of older adult turtles in the population may indicate populations at the beginning of a decline (Hamann et al. 2007). In Queensland, the loggerhead turtle *Caretta caretta* population has been monitored annually since the late 1960s and has undergone a substantial and well-documented decline in the order of 85% in the last three decades. Long-term monitoring data collected for the eastern Australian population of flatback turtle *Natator depressus* show no signs of a declining population. No leatherback turtle *Dermochelys coriacea* nests have been reported in Queensland since 1996, despite annual nesting surveys for loggerhead turtles that use the same beaches. Mortality of green turtles increased greatly in the year after the large flood and cyclone events of 2011 (Devlin et al. 2012) with many turtles appearing to be starved. *Fibropapillomas* virus is also present in GBR green turtle populations, and studies about the possible causes are continuing.

Management of the GBR

The Management Regime

The GBR has been managed as a national Marine Park since 1975 (*Great Barrier Reef Marine Park Act, 1975*) and was listed as a World Heritage Area (WHA) in 1981 (Day 2011). The overriding objective of the legislation is the conservation of the GBR. The outer boundaries of both the GBR Marine Park (GBRMP) and the Great Barrier Reef World Heritage Area (GBRWHA) lie beyond the shelf break in the east thus enclosing a considerable area of oceanic depth water. The western boundary of the GBRWHA is the low water mark along the coast with the GBRMP boundary similar, except for a few small excluded areas along the coast. It is recognized as a global Large Marine Ecosystem, and the management system is often regarded as the best possible or best in the world and the leading contender for the best example of ecosystem-based management (Brodie et al. 2012). The GBR has been subject to an intensive management regime involving both the Australian and Queensland state governments for 40 years, focussing on managed use and ecosystem protection. The Great Barrier Reef Marine Park Authority (GBRMPA) was established as a statutory authority to manage and protect the values of the GBR in 1975. The human use impacts to be "protected against" include tourism, recreation, shipping, farm and urban pollutant runoff from the adjacent land, fishing and hunting, and climate change related environmental change. The GBRMP is not a park in the same sense as a terrestrial national park; it is a multiple use protected area and many uses, including extractive uses such as commercial fishing, can be licensed to operate with in parts of the park. The one activity which is not permitted in any part of the GBRMP is mining. Activities which occur in the GBRMP include commercial and recreational fishing, aquaculture, tourism, shipping traffic, research, and defense exercises.

Saltmarshes, mangroves, and other marine and estuarine plants in Queensland are protected through a range of legislative measures including the *Fisheries Act 1994*, Ponded Pastures Policy 2001, *Sustainable Planning Act 2009*, and *Environment Protection and Biodiversity Conservation Act 1999*. These mechanisms are vital to minimize or mitigate development impacts on saltmarsh communities; however, stronger policies, planning, and adaptation strategies are needed to ensure that buffers are set aside for anticipated landward migration of saltmarsh with sea-level rise (Sheaves et al. 2007).

The GBRMP Strategic Plan (Anon 1994) attempted to provide hierarchical steps and processes for ecologically sustainable use (ESU) and biodiversity management. It identifies the key issues or objectives in the management of the GBRMP as: (1) the maintenance of the ecology; (2) management to achieve ecologically sustainable use; and (3) maintenance of traditional, cultural, heritage, and historic values. The main 25 year objective was "to ensure the persistence of the GBRWHA as a diverse, resilient, and productive ecological system." However, the Strategic Plan failed to provide unambiguous, scientifically based targets and mechanisms for ESU and biodiversity conservation and does not attempt to define the processes by which ESU may be attained. The plan makes no attempt to identify threatened species; define limits of acceptable change to habitats, proportions of habitat which should be totally protected, or the number, size, and spatial arrangements of protected areas; or specify how representative biological communities can be identified. However, the Strategic Plan recognized that management of potentially damaging activities in the GBR, which may adversely affect conservation values, is spread among a variety of agencies, and GBRMPA attempts to maintain a level of overall coordination.

In 2003, the evidence of the relationship between land use, water quality, and declining GBR ecosystem health led to a national policy response between the Australian and Queensland governments, the Reef Water Quality Protection Plan (Reef Plan) (Department of the Premier and Cabinet 2013). The Reef Plan built on existing government policies, industry and community initiatives with a list of strategies and actions to be implemented by government, industry, and community groups. The initial goal was *"to halt and reverse the decline in water quality entering the GBR by 2013"* (i.e., within 10 years). The Plan was revised in 2009 and again in 2013, with the current long-term goal *"to ensure that by 2020 the quality of water entering the reef from broadscale land use has no detrimental impact on the health and resilience of the Great Barrier Reef."* The primary focus is to address diffuse source pollution from broadscale land use. The Plan has provided a substantial challenge for the delivery of widespread changes in land use practices and community attitudes, and while some measurable progress has been made, the rate of improvement is not likely to be enough to meet the Plan's goals and targets. To date, the governments have adopted a largely voluntary incentive-based management approach, where agricultural land holders are offered financial support and training (matched by in-kind contributions) to improve management practices. Investment to support Reef Plan implementation has increased from the governments in the last 5–8 years; however, securing resources that are commensurate with the scale of the catchment management issues has been an ongoing challenge.

In more recent times The United Nations Educational, Scientific and Cultural Organization (UNESCO) has expressed concern over the decline of the outstanding universal value of the GBRWHA, their concerns triggered by the rapid industrialization of the Queensland coastline for port, urban, and industrial development, and the recent proposals for expanded development of ports for export of unprecedented amounts of fossil fuels. In response to UNESCO's concerns, the Australian and Queensland governments drafted the Reef 2050 Long-term Sustainability Plan (Reef 2050 Plan; Commonwealth of Australia 2015). The Australian Academy of Science review of the draft Reef 2050 Plan concluded that the plan was inadequate to achieve the goal of restoring or even maintaining the diminished outstanding universal value of the GBR (Australian Academy of Science 2014). The final Reef 2050 Plan, released in March 2015, remains short-sighted given its aspiration to provide an overarching framework for the next 35 years (Hughes et al. 2015). Critically, the revised plan lacks any action on climate change, identified by scientists and the government as the key threat to the GBR owing to the impact of global warming and ocean acidification.

Management Status

Despite this impressive management system, the success of the management regime in halting the decline of many species and ecosystems in the GBR is mixed (summarized in Brodie and Waterhouse 2012; Great Barrier Reef Marine Park Authority 2014). There have been notable successes in recent times after the major rezoning of 2004 with new no-take zones showing increased fish populations (Emslie et al. 2015) but also apparent effects on COTS populations (Sweatman 2008) such that numbers of COTS are lower in the no-take zones. However, major floods have removed much of the positive effects on coral health in the no-take zones in the Keppel Island group (refer Fig. 2; Wenger et al. 2016) showing that integrated management of both fishing and terrestrial pollution (and other stresses) is needed to maintain healthy reef condition. Recent data on turtle populations indicates that numbers are declining, particularly after recent large-scale flooding. Whales (humpbacks in particular) are slowly increasing in numbers again after the cessation of most commercial whaling. There has been little loss of mangroves as a result of strong prohibitions on damaging marine plants under the Queensland fisheries legislation. However, saltmarsh has not been so well protected (Wegscheidl et al. 2015), and coastal wetlands in general are in need of further protection and restoration (Sheaves et al. 2014; Waterhouse et al. 2016).

Sewage effluent discharges from resort islands and mainland cities and towns have been improved dramatically. Strong action on shipping management for compulsory pilotage and navigation equipment may have prevented many shipping accidents but ships still manage to run onto the reef every decade or so. Water quality may have started to improve with recent programs under the Reef Plan policy addressing river pollutant discharges (Brodie et al. 2012, 2013), but the success of the Reef Plan is still uncertain, mainly due to the scale of the management effort required, the associated resource implications, and time lags in the system response from land management change. Current progress under the Reef Plan is summarized in an annual report card (e.g., Department of the Premier and Cabinet 2014). During the first years of more targeted management (2008–2013), the catchment pollutant load modeling, which removes the time lags in management practice effectiveness, indicates that the adoption of improved land management practices was estimated to have reduced loads of suspended sediment by 11%, total phosphorus by 13%, total nitrogen by 10%, and photosystem II inhibiting herbicides (e.g., atrazine and diuron) by 28% to the GBR lagoon (Department of the Premier and Cabinet 2014). However, there is still much to do and a large proportion of agricultural land in priority areas (up to 75–80% in some regions) is still managed below best management practice standards (Department of the Premier and Cabinet 2014).

Many aspects of GBR health and management are less positive; coral cover has declined greatly (De'ath et al. 2012; Thompson et al. 2014), seagrass health in the central GBR is in poor shape, dugong numbers have declined precipitously, shark populations are in serious decline (although perhaps recent management has reduced the rate of decline), many other large fish on the GBR have had large population declines (although data on many are incomplete), and the fourth wave of COTS outbreaks is in progress. Most notably, coral bleaching has become more frequent, widespread, and damaging (Great Barrier Reef Outlook Report 2014), and coral calcification has started to decline due to ocean acidification. Rapid port expansion with large-scale dredging of entry channels (Grech et al. 2013) and associated increases in shipping traffic linked to the growth of the mining industry in Australia raise significant concerns for the long-term health of the GBR (Hughes et al. 2015), with increased potential for fauna strikes and a greater risk of a major shipping incident occurring in the GBR.

The reasons for this apparent failure of effective management are complex but include the need for reasonably "certain" science before management action occurs (and the long times needed to achieve this) (Brodie and Waterhouse 2012). In addition, time lags in recovery after management action are long. For a slow breeding animal like a dugong (one calf every few years), population recovery is a very slow process. Catchment management activities such as reafforestation of riparian areas or rehabilitation of degraded land take decades to reduce erosion and river sediment loads. The likelihood of effective reduction in global greenhouse gas emissions such that temperature change associated with climate change can be limited to a maximum of two degrees also seems remote at the moment although the recent Paris agreements (http://www.cop21.gouv.fr/en/), if implemented, may limit temperature rises to at least less than a catastrophic four degrees.

The Future of the GBR

Based on the information we have outlined above, the long-term viability of the GBR in anything like its current state is in doubt (Hughes et al. 2015). Many species and ecosystems are in decline, and only a few are stable or recovering from past degradation. In addition, the heavy cyclone and flood damage of 2009–2015 and the commencement of the fourth wave of COTS outbreaks raise significant concerns for the long-term health of the GBR (Great Barrier Reef Marine Park Authority 2014). It could be argued that the system has gained some resilience through the current management interventions in water quality management (Brodie et al. 2012; Department of Premier and Cabinet 2014) and the GBRMP rezoning in 2004 which increased the level of protection in a greater proportion of the GBRMP. However, it is unlikely that this management response is adequate to prevent either a large-scale phase shift in the system or just continuing slow decline. Even successful interventions are unlikely to return the GBR to some pristine or pre-disturbance state as has been shown from experience in restoration through management in other systems.

Overall, issues such as the continuing effects of climate change including ocean warming and ocean acidification, more frequent extreme events, continued COTS outbreaks, and accelerating coastal development associated with greatly expanded port and urban development that are not managed in any strategic way raise serious concerns that recovery of many of the key species and ecosystems of the GBR is unlikely (Brodie and Waterhouse 2012). This outcome is despite the fact that the GBR remains one of the best managed coral reef systems in the world.

References

Anon. A 25 year strategic plan for the Great Barrier Reef World Heritage Area. Townsville: Great Barrier Reef Marine Park Authority; 1994. 64 p.

Australian Academy of Science. Response to the Draft Reef 2050 long term sustainability plan. Australian Academy of Science; 2014. http://go.nature.com/jz4PHT

Brodie J, Waterhouse J. A critical assessment of environmental management of the 'not so Great' Barrier Reef. Estuar Coast Shelf Sci. 2012;104–105:1–22. doi:10.1016/j.ecss.2012.03.012.

Brodie J, Bainbridge Z, Lewis S, Devlin M, Waterhouse J, Davis A, Bohnet I, Kroon F, Schaffelke B, Wolanski E. Terrestrial pollutant runoff to the Great Barrier Reef: issues, priorities and management response. Mar Pollut Bull. 2012;65:81–100. doi:10.1016/j.marpolbul.2011.12.011.

Brodie J, Waterhouse J, Schaffelke B, Johnson JE, Kroon F, Thorburn P, Rolfe J, Lewis S, Warne MSJ, Fabricius K, McKenzie L, Devlin M. Reef water quality scientific consensus statement 2013. Brisbane: Department of the Premier and Cabinet, Queensland Government; 2013.

Carter AB, Taylor HA, Rasheed MA. Torres Strait Dugong Sanctuary – deepwater seagrass monitoring 2010–2014. TropWATER Report No. 14/21. Cairns: Centre for Tropical Water & Aquatic Ecosystem Research, James Cook University; 2014. 22 p.

Coles RG, Rasheed MA, McKenzie LJ, Grech A, York PH, Sheaves M, McKenna S, Bryant C. The Great Barrier Reef World Heritage Area seagrasses: managing this iconic Australian ecosystem resource for the future. Estuar Coast Shelf Sci. 2015;153:A1–12.

Commonwealth of Australia. Reef 2050 long-term sustainability plan. Commonwealth of Australia, Canberra; 2015. http://www.environment.gov.au/system/files/resources/d98b3e53-146b-4b9c-a84a-2a22454b9a83/files/reef-2050-long-term-sustainability-plan.pdf

Coppo C, Brodie J, Butler I, Mellors J, Sobtzick S. Status of coastal and marine assets in the Burnett Mary Region. TropWATER report no. 14/36. Townsville: Centre for Tropical Water & Aquatic Ecosystem Research, James Cook University; 2014. 89 p.

Day JC. Protecting Australia's Great Barrier Reef. Solutions. 2011;2:56–66. http://www.thesolutionsjournal.com/node/846

De'ath G, Fabricius KE, Sweatman H, Puotinen M. The 27-year decline of coral cover on the Great Barrier Reef and its causes. Proc Natl Acad Sci U S A. 2012;109:17995–9. doi:10.1073/pnas.1208909109.

Department of the Premier and Cabinet. Great Barrier Reef report card 2012 and 2013. Reef Water Quality Protection Plan. Brisbane: Reef Water Quality Protection Plan Secretariat; 2014. http://www.reefplan.qld.gov.au/measuringsuccess/reportcards/assets/report-card-2012-2013.pdf

Department of the Premier and Cabinet. Reef Water Quality Protection Plan 2013. Securing the health and resilience of the Great Barrier Reef World Heritage Area and adjacent catchments. Brisbane: Reef Water Quality Protection Plan Secretariat; 2013.

Devlin M, Brodie J, Wenger A, da Silva E, Alvarez- Romero JG, Waterhouse J, McKenzie L. Extreme weather conditions in the Great Barrier Reef: drivers of change? In: Proceedings of the 12th International Coral Reef Symposium. Cairns; 2012.

Emslie MJ, Logan M, Williamson DH, Ayling AM, MacNeil MA, Ceccarelli D, Cheal AJ, Evans RD, Johns KA, Jonker MJ, Miller IR. Expectations and outcomes of reserve network performance following re-zoning of the Great Barrier Reef Marine Park. Curr Biol. 2015; 25(8):983–92.

Great Barrier Reef Marine Park Authority. Great Barrier Reef outlook report 2014. Townsville: Great Barrier Reef Marine Park Authority; 2014.

Grech A, Sheppard J, Marsh H. Informing species conservation at multiple scales using data collected for marine mammal stock assessments. PLoS One. 2011;6(3):e17993.

Grech A, Bos M, Brodie J, Coles R, Dale A, Gilbert R, Hamann M, Marsh H, Neil K, Pressey R, Rasheed MA. Guiding principles for the improved governance of port and shipping impacts in the Great Barrier Reef. Mar Pollut Bull. 2013;75(1):8–20.

Hamann M, Limpus CJ, Read MA. Vulnerability of marine reptiles in the Great Barrier Reef to climate change. In: Johnson JE, Marshall PA, editors. Climate change and the Great Barrier Reef: a vulnerability assessment. Townsville: Great Barrier Reef Marine Park Authority and Australian Greenhouse Office; 2007. p. 466–96.

Hughes TP, Rodrigues MJ, Bellwood DR, Ceccerelli D, Hoegh-Guldberg O, McCook L, Moltchaniwskyj N, Pratchett MS, Steneck RS, Willis BL. Regime-shifts, herbivory and the resilience of coral reefs to climate change. Curr Biol. 2007;17:360–5.

Hughes TP, Bellwood DR, Baird AH, Brodie J, Bruno JF, Pandolfi JM. Shifting base-lines, declining coral cover, and the erosion of reef resilience. Comment on Sweatman et al. (2011). Coral Reefs. 2011;30:653–60.

Hughes TP, Day J, Brodie J. Securing the future of the Great Barrier Reef. Nat Clim Chang. 2015;5:508–11.

Marsh H, De'ath G, Gribble N, Lane B. Historical marine population estimates: triggers or targets for conservation? The dugong case study. Ecol Appl. 2005;15:481–92. doi:10.1890/04-0673.

Marsh H, O'Shea TJ, Reynolds JR. The ecology and conservation of Sirenia; dugongs and manatees. London: Cambridge University Press; 2011.

McKenzie LJ, Collier CJ, Langlois LA, Yoshida RL, Smith N, Takahashi M, Waycott M. Marine monitoring program – Inshore Seagrass, annual report for the sampling period 1st June 2013 – 31st May 2014. Cairns: TropWATER, James Cook University; 2015. 225 p.

Noad M, Dunlop R, Cato D, Paton D. Abundance estimates of the East Australian Humpback Whale Population. Final report for the commonwealth department of the environment, heritage, water and the arts. Brisbane: University of Queensland; 2008.

Sheaves M, Brodie J, Brooke B, Dale P, Lovelock C, Waycott M, Gehrke P, Johnston R, Baker R. Vulnerability of coastal and estuarine habitats in the GBR to climate change. In: Johnson JE, Marshall PA, editors. Climate change and the Great Barrier Reef: a vulnerability assessment. Townsville: Australia Great Barrier Reef Marine Park Authority and the Australian Greenhouse Office; 2007. p. 593–620.

Sheaves M, Brookes J, Coles R, Freckelton M, Groves P, Johnston R, Winberg P. Repair and revitalisation of Australia's tropical estuaries and coastal wetlands: opportunities and constraints for the reinstatement of lost function and productivity. Mar Policy. 2014;47:23–38.

Sobtzick S, Penrose H, Hagihara R, Grech A, Cleguer C, Marsh H. An assessment of the distribution and abundance of dugongs in the Northern Great Barrier Reef and Torres Strait TropWATER Report 14/17. Townsville: Centre for Tropical Water & Aquatic Ecosystem Research, James Cook University; 2014.

Sweatman H. No-take reserves protect coral reefs from predatory starfish. Curr Biol. 2008;18(14): R598–9.

Thompson A, Schaffelke B, Logan M, Costello P, Davidson J, Doyle J, Furnas M, Gunn J, Liddy M, Skuza M, Uthicke S, Wright M, Zagorskis I. Reef rescue monitoring program. Draft final report of AIMS activities 2012 to 2013 – Inshore water quality and coral reef monitoring. Report for

Great Barrier Reef Marine Park Authority. Townsville: Australian Institute of Marine Science; 2014.

Walsh KJ, McBride JL, Klotzbach PJ, Balachandran S, Camargo SJ, Holland G, Knutson TR, Kossin JP, Lee TC, Sobel A, Sugi M. Tropical cyclones and climate change. Wiley Interdiscip Rev Clim Chang. 2016;7(1):65–89.

Waltham NJ, Sheaves M. Expanding coastal urban and industrial seascape in the Great Barrier Reef World Heritage Area: critical need for coordinated planning and policy. Mar Policy. 2015;57:78–84.

Waterhouse J, Brodie J, Audas D, Lewis S. Land-sea connectivity, ecohydrology and holistic management of the Great Barrier Reef and its catchments: time for a change. Ecohydrol Hydrobiol. 2016;16:45–57.

Wegscheidl C, Sheaves M, McLeod I, Fries J. Queensland's saltmarsh habitats: values, threats and opportunities to restore ecosystem services, TropWATER report no. 15/54. Townsville: Centre for Tropical Water & Aquatic Ecosystem Research, James Cook University; 2015. 25 p.

Wenger AS, Williamson DH, da Silva ET, Ceccarelli DM, Browne NK, Petus C, Devlin MJ. Effects of reduced water quality on coral reefs in and out of no-take marine reserves. Conserv Biol. 2016;30(1):142–53.

Wolanski EJ, Drew EA, Abel KM, O'Brien J. Tidal jets, nutrient upwelling and their influence on the productivity of the alga *Halimeda* in the Ribbon Reefs, Great Barrier Reef. Estuar Coast Shelf Sci. 1988;26:169–201.

York PH, Carter AB, Chartrand K, Sankey T, Wells L, Rasheed MA. Dynamics of a deep-water seagrass population on the Great Barrier Reef: annual occurrence and response to a major dredging program. Sci Rep. 2015;5:13167. doi:10.1038/srep13167.

Qa'a Azraq Oasis: Strengthening Stakeholder Representation in Restoration (Jordan)

35

Fidaa F. Haddad

Contents

Introduction	462
Everybody Has a Stake in Oasis Management	463
Azraq Oasis Management Initiatives: Facilitation Process to Reach Oasis Governance	465
Azraq Oasis Dialogue for Restoration: The Highland Water Forum	466
Establishing Local Committees	466
Local Water Resource Management Plan	466
Achieving Process Success: Final Words	467
References	467

Abstract

The Azraq Oasis wetland is a unique ecosystem in an arid region with immense biological, cultural, and socioeconomic value. This Ramsar Site contains three major subecosystems: a freshwater lake ecosystem; a marsh ecosystem with moderately saline waters and soils; and the *Qa'a* mudflat ecosystem, which now has highly saline waters and soils. Distribution of flora, fauna, and aquatic species varies according to each habitat. The site faces many challenges, most of which are related to the over-extraction of scarce water resources. This chapter highlights how ownership of natural resource management implies a sense of accountability for the actions undertaken for restoration management, with emphasis on stakeholder accountability for sustainable oasis use and management

---

F. F. Haddad (✉)
Regional Dryland, Livelihoods and Gender Program, IUCN Regional Office for West Asia (ROWA), Amman, Jordan
e-mail: f2hadad@hotmail.com; fida.haddad@iucn.org

toward themselves, the oasis ecosystem, and their community. It also reviews successes, challenges, and lessons learned from strengthening stakeholder platforms for oasis management in Jordan.

Keywords

Governance · Local water resource management plan

Introduction

The Azraq Oasis is an outstanding example of an oasis wetland in an arid region with a unique ecosystem and immense biological, cultural, and socioeconomic value. The Azraq area has a special historical importance; availability of water and wildlife made it a stopping point for trade caravans on the route between Hijaz and Levant. In modern times settlements in the area began at the start of the twentieth century when revolutionaries used it as a main station during the Great Arab Revolution to liberate Damascus from the Ottoman domination.

Recognized as a Ramsar Site, the Azraq Wetland Reserve (http://rscn.org.jo/content/azraq-wetland-reserve-0) contains three major sub-ecosystems: a freshwater lake ecosystem; a marsh ecosystem with moderately saline waters and soils; and the *Qa'a* mudflat ecosystem, which now has highly saline waters and soils. Distribution of flora, fauna, and aquatic species varies according to each habitat.

Azraq Oasis (Qa'a Azraq), 120 km northeast of Jordan's capital Amman, is part of the Azraq basin, which stretches from the mountains of Syria (the main water source) toward the border with Saudi Arabia. Because its water is so close to the surface, Qa'a Azraq (the main recharge area) has been inhabited since the Stone Age. Azraq Basin is one of the most important recharging groundwater basins in Jordan; it forms the largest source of high-quality ground and surface water in the country.

Azraq faces many challenges, many of which existed earlier but not to the same extent as today. In brief, they vary from increasing water salinity due to over-extraction of already scarce water resources, leading to negative impacts on human health, drought and desertification, unemployment, decreasing ecotourism, loss of natural rangelands, loss of aesthetic value in the area, drought of agricultural lands for the sake of industrial investments, and non-implementation of laws and legislations related to underground water usage. Additionally, a lack of coordination and communication among the community's sectors exacerbates the situation and increases degradation. Figure 1 shows Azraq Basin's actual water uses.

The major challenge for Azraq is agriculture. This sector consumes the largest percentage of water, demanding over 30 MCM of water per year (nearly equivalent to the entire safe yield of 24 MCM per year). This represents a critical problem in an arid area that suffers from desertification and deterioration. Agriculture not only becomes impractical in this situation but uneconomical in relation to water's value for other livelihood needs.

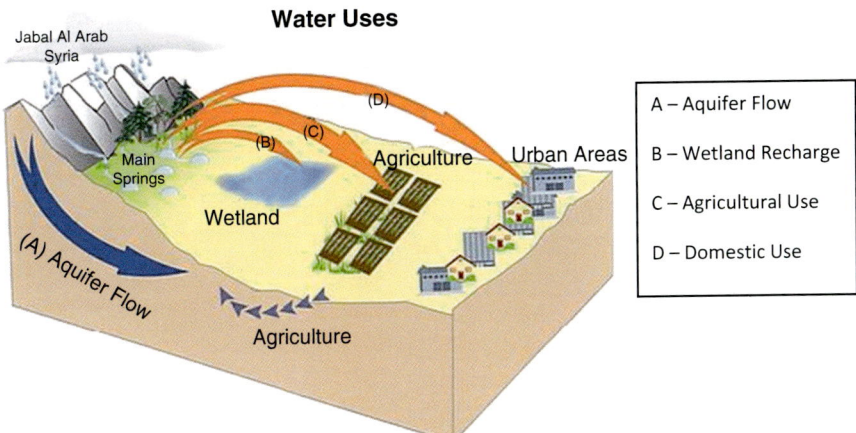

Fig. 1 Water uses in the Azraq Basin (Permission for use courtesy of IUCN – West Asia Office)

To this end, the area faced a crisis (El-Naqa 2010) in the 1990s when the natural springs and surface water disappeared, alerting those concerned that the situation had become critical and putting the drinking water supply and livelihoods of local people at stake.

However, getting the right people involved to collectively and collaboratively address this crisis has been an ongoing challenge. Initially, there was the problem of getting stakeholders from the entire aquifer, not just one sector, involved. Additionally, there was the problem of getting powerful commercial agricultural interests to abide by stakeholder decisions.

This chapter highlights how ownership of natural resource management implies a sense of accountability for the actions undertaken for restoration management with emphasis on stakeholder accountability for sustainable oasis use and management toward themselves, the oasis ecosystem, and their community. It also reviews successes, challenges, and lessons learned from strengthening stakeholder platforms for oasis management in Jordan.

Everybody Has a Stake in Oasis Management

Planning is a key element of sustainable oasis management, particularly in areas experiencing high levels of competition for limited natural resources such as water. Integrated Natural Resource management brings coordination and collaboration among individual sectors, while also fostering stakeholder participation, transparency, and cost-effective local management.

Management of natural resources is a long-term, complex, multi-stakeholder affair in which many players at many different levels have to assume responsibilities

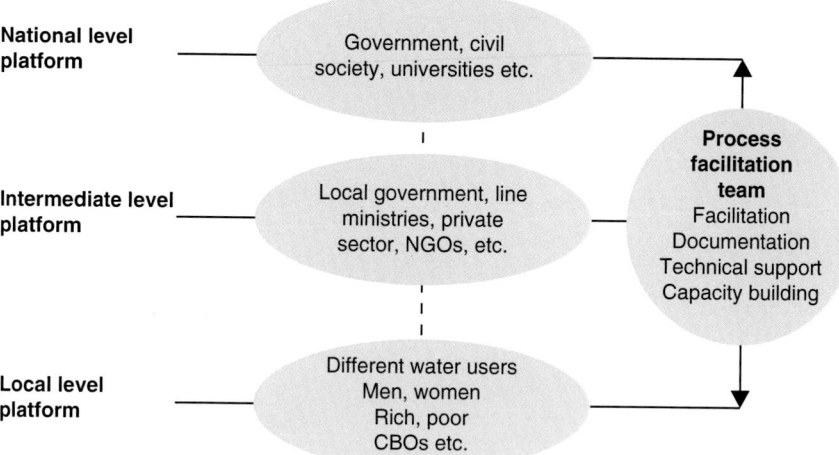

Fig. 2 The diagram illustrates the stakeholder in oasis governance at different institutional levels. The process facilitation team enhances dialogue and concentration within and between stakeholder platforms at different levels (Moriarty et al. 2007, Permission for use, courtesy of INWRDAM)

and be accountable to others (Fig. 2). Natural resource management (e.g., water management) can be seen as a system composed of two interacting decision-making subsystems: a "horizontal" land-/water-use system and a "vertical" human activity system. The horizontal system may be described as the complex interaction between land, water, vegetation, and their users within a given geographical space and time frame, taking into account down- and upstream effects. The vertical system may be described as a complex network composed of different levels of actors, from local households to governments. All these actors have their own roles, rights, responsibilities, and occasionally conflicting interests in water resource management. These two interacting systems are sometimes described as a decision-making-in-conflict system (Laban 1994). This should be considered when promoting local oasis governance (Haddad et al. 2009).

Promoting dialogue and concerted action among different stakeholders requires analysis of these stakeholders and their roles. This refers to different issues, such as forms of cooperation and coordination, information and knowledge sharing, assumed tasks and responsibilities, influence on decision-making, and interest and roles in planning and implementation but also to perceptions, political and institutional agendas, power, resistance to change, etc. In short, a detailed stakeholder analysis is an essential first step to understand how a shared groundwater resource would best be used and managed and what the implications are for the governance and social organization around such management. Completing this step allows facilitators to bring stakeholders together in a meaningful way that makes sense to all participants, where everyone feels committed to contribute to a shared vision and objectives with the oasis resource. Stakeholder Dialogue and Concerted Action

(SDCA) is an effective approach to strengthening the communication between stakeholders through different stakeholder platforms.

Azraq Oasis Management Initiatives: Facilitation Process to Reach Oasis Governance

Dealing with water scarcity, competition for water, and the ecosystem as a whole, the Jordan water manager has to find a way to fulfill needs and reduce impacts. The best way to do so is through integrated natural resource management where rewarding the ecosystem is rewarding people through sustainable use, management, and development of water, land, and related resources. This is the same as using Integrated Water Resource Management (IWRM) for planning and development activities implemented with local communities (Haddad 2011).

The Azraq situation urged many actors to make an immediate intervention to save Azraq from further deterioration. Some of these attempts had no success, while others, because they did not give enough attention to different interests and needs of stakeholders which weakened their sustainability, had short-term and limited success.

An initiative that came about as a result of failed interventions was *The Azraq Oasis Restoration project 2006–2010* (Laban 2010), a multi-donor project implemented by the IUCN West Asia regional office. The project was built on the assumption that stakeholder involvement – particularly at the intermediate and local levels – leads to improved use and management of water resources. Improved management implies taking better account of user's needs and includes collective responsibility for interventions in oasis management. The project used a participatory planning cycle for IWRM. This cycle builds on the identification of water-related problems and the development of area-specific long-term visions and strategies for water resource development. The larger process of participatory analysis, visioning, scenario building, and strategic planning was the real heart of this project.

The project was also designed to create the institutional and policy conditions under which restoration of a substantial part of the oasis/wetland could become a reality. The project approach makes evident under which policy assumptions and related scenarios such restoration is possible.

The heart of the initiative was to strengthen stakeholder representation through a joint participatory planning framework between all stakeholders at the watershed level. The initiative demonstrated how an effective planning relationship can be developed and sustained between the local community, private sector, and respective government departments for improved water resource management. Building the capacity of the different stakeholders (including end users) on local oasis management increased their participation and representation in planning and decision-making processes. Existing local community-based organizations (CBOs) were entrusted with the management and coordination of all activities in close cooperation with government institutions at the intermediate-level governorate. The Azraq district increased their accountability to contribute to the restoration and development of the Azraq Oasis and the welfare of the local communities.

As a result of the participatory methodology, all related stakeholders agreed that significant ecosystem degradation has occurred in the Azraq Oasis because of internal and external factors. Internal factors include, among others, over-pumping within the oasis area, unclear water rights, and non-sustainable agriculture, while external factors include climate change, over-pumping in the larger area of the Azraq basin (outside the oasis area), and the drilling of wells and building of dams in the upper catchment on Syrian territory. In light of these challenges, the stakeholder parties defined a shared vision for the Azraq oasis: *"The restoration of Azraq Oasis" by reducing, within 12 years, groundwater pumping down to safe yield levels in compliance with the Ministry of Water and Irrigation Policy.*

Azraq Oasis Dialogue for Restoration: The Highland Water Forum

The experiences and results in the Azraq demonstration site have been incorporated into the agenda of the Highland Water Forum (HWF – https://highlandwaterforum.wordpress.com/about/, Accessed: 3 Jan 2016), a national initiative established by the Jordanian-German Water Programme under Royal Patronage in Jordan in 2010 and hosted by the Ministry of Water and Irrigation with support from other partners. The Forum acts as a national umbrella institution under which strategies for improving ecosystem sustainability and livelihood security can be carried out. The HWF aims to enrich and strengthen the experience gained in specific pilot projects in the Azraq Basin to act as a learning platform for stakeholders.

Establishing Local Committees

In 2009, the Local Oasis Resource Management Committee was established for the Azraq basin. This committee is made up of all the existing CBOs in the Azraq area, including cooperatives, nongovernment organizations (NGOs), and women's organizations. As a means of ensuring the sustainability of this committee, a steering committee supported the group at the national level during the project implementation period.

Local Water Resource Management Plan

The results from community-led planning and pilot projects enabled the development of the Water Resource Management Plan for Azraq, which was completed in 2010. The plan promotes an integrated approach for water resources management based on water demand management and the sustainable use of nonconventional water. It emphasizes long-term access and water rights for women and underprivileged groups in target communities and participation in planning and decision-making processes to ensure that livelihoods are improved through increased effective participation in oasis management.

Achieving Process Success: Final Words

A key challenge in oasis management and better groundwater governance is breaking down sectoral barriers to planning and decision-making and providing common platforms and frameworks for the development and management of all oasis resources.

Local governance involves concepts of negotiation, fairness, and stewardship; it not only allows many voices to represent different interests but pushes them to make decisions that meet the needs of the present while accommodating the needs of the future, both for their own lifetimes and those of their descendants (Abu-Elseoud et al. 2007).

Ultimately, most groundwater resources are used at the local level, being urban or rural, for drinking water, agriculture, or other economic activities. Local user group participation in how that water is used and managed is required to ensure sustainability of the groundwater resource and efficacy of its management. Key issues here include genuine participation, transparency, empowerment, ownership over the management practices, rights, and accountability.

References

Abu-Elseoud M, Al-Zoubi R, Mizyed B, Abd-Alhadi FT, Barghout M, de la Harpe J, Schouten T. Doing things differently: stories about local water governance in Egypt, Jordan and Palestine. Amman: INWRDAM on behalf of the EMPOWERS Partnership; 2007. 99 p.

El-Naqa A. Study of salt water intrusion in the upper aquifer in Azraq Basin, Jordan. IUCN; 2010. p. 92. Available from: http://cmsdata.iucn.org/downloads/final_report_azraq_2011.pdf. Accessed 1 Feb 2016.

Haddad F. Local community participation for sustainable water resource management. The sixth IWM specialist conference on efficient use and management of water, Dead Sea- Jordan 2011.

Haddad F, Buthaina M, Laban P. Enhancing rights and local level accountability in water management in the Middle East. Chapter 4. In: Campese J, Sunderland T, Greiber T, Oviedo G, editors. Rights- based approaches: exploring issues and opportunities for conservation. Bogor: CIFOR and IUCN; 2009. p. 98–122.

Laban P. Accountability, an indispensable condition for sustainable natural resource management. In: Proceedings of the International Symposium on System-oriented Research in Agriculture and Rural Development, CIRAD-SAR, Montpellier. 1994. p. 344–9.

Laban P. "Rewarding ecosystems, rewarding people" in dryland watershed ecosystems of the West Asia and Mediterranean regions. Regional Water Resources and Dryland (REWARD) Programme, Amman: IUCN Regional Office for West Asia; 2010. Available from: https://cmsdata.iucn.org/downloads/rewarding_ecosystems_rewarding_people__reward_sharm_al_shaikh_report__final.pdf. Accessed 1 Feb 2010.

Moriarty P, Bachelor C, Abd-Alhadi F, Laban P. Fahmy H. The EMPOWERS Approach to water governance: guidelines, methods, tools. Amman: INWRDAM on behalf of the EMPOWERS Partnership; 2007. Available from: http://waterwiki.net/images/d/d2/EMPOWERS_Guidelines,_Methods_and_Tools.pdf. Accessed 1 Feb 2016.

Fishponds of the Czech Republic

36

Jan Pokorný and Jan Květ

Contents

Introduction	470
Biodiversity of Water Birds	471
Ecosystem Services	473
Management for Fish Production	476
Fishpond Vegetation Management	477
Conservation Status and Management	480
Threats and Future Challenges	483
References	483

Abstract

Fishponds are shallow water bodies the construction of which started in Central Europe in the Middle Ages. Varying in size from less than 1 hectare to several hundred hectares, fishponds have become an integrated landscape component in cultivated regions. In some of them, systems of fishponds play a principal hydrological role. These constructed wetlands provide a diversity of regulatory, provisioning, sustaining and cultural services. Fishponds are centres of high species richness. In both agricultural and forested areas, littoral zones of fishponds form ecotones providing suitable habitats for amphibians, invertebrates and namely for water birds. Presented is a case study on the role of fishponds in

J. Pokorný (✉)
ENKI, o.p.s., Třeboň, Czech Republic
e-mail: pokorny@enki.cz

J. Květ
Faculty of Science, University of South Bohemia České Budějovice, České Budějovice, Czech Republic

Czech Academy of Sciences, CzechGlobe, Institute for Global Change Research, Brno, Czech Republic
e-mail: jan.kvet@seznam.cz

© Springer Science+Business Media B.V., part of Springer Nature 2018
C. M. Finlayson et al. (eds.), *The Wetland Book*,
https://doi.org/10.1007/978-94-007-4001-3_208

water retention during the "thousand-year flood" in the Labe/Elbe river basin in 2002.

About 20,000 fishponds (180,000 hectares) existed on the territory of the present Czech Republic in the 16th century. Since then, their area has declined to 52,000 hectares of nowadays existing fishponds. Large-scale drainage, sewage discharge and intensification of fish production have resulted in water eutrophication and an augmentation of natural fish production. Its average annual amount per hectare has increased from 70 kg in the Middle Ages to the contemporary 450 kg. Adaptation of the management of fishponds to their heavy eutrophication represents a highly demanding task. The proliferation of weedy fish (some of them invasive alien species) interferes with the classical and simple scheme of top-down control of the food chain (the fish stock controls the zooplankton community which, in turn, controls the species composition and amount of phytoplankton).

In the Czech Republic, 4 out of 14 sites listed as Wetlands of International Importance under the Ramsar Convention encompass appreciable numbers and areas of fishponds. These 4 Ramsar Sites and their fishponds are briefly characterized. Discussed are also the threats and future challenges to the balance between the economy of fish production, biodiversity and hydrological and other ecosystem functions and services of these fishponds.

Keywords

Biodiversity · Ecosystem services · Water retention · Food chain · Fish production · Eutrophication · Ramsar sites · Management

Introduction

Situated in the drainage basins of three major European rivers (Danube, Labe/Elbe, and Odra/Oder), the Czech Republic has few natural water bodies. Several small natural lakes (<100 ha) at altitudes over 900 m and 118 constructed reservoirs (14,200 ha) are greatly exceeded by the 22,000 fishponds (52,000 ha) distributed across the countryside. Fishponds are mostly shallow with an average depth of about 1 m, being deepest at the water outlet through the pond dam. Varying greatly in size (<1 to 489 ha), fishponds constructed over the course of centuries have become integrated into the landscape providing ecological functions and recognized as important landscape elements that are protected to various degrees under Czech law (No. 114/92 Sb on Nature and Landscape Conservation).

Fishponds have a long history in the Czech Republic, occurring around monasteries in Silesia and Flanders in the middle ages and probably in Bohemia since the tenth century. Fishponds of several km^2 (Dvořiště, Máchovo jezero) already existed in the second half of the fourteenth century (Andreska 1987; Dykyjová 2000). By the end of the sixteenth century more than 20,000 fishponds (ca. 180,000 ha) had been constructed in Bohemia and Moravia. The Thirty Years War (1618–1648) resulted in many fishponds being drained and dams breached, and the introduction of sugar beet

Fig. 1 Fishponds constructed mostly about 500 years ago cover 10% of the Třeboň Basin Biosphere Reserve area and create an impression of a lakescape (Photo credit: J. Ševčík © Rights remain with the author)

and rising value of wheat in the early nineteenth century saw many fishponds converted into agricultural land. By the beginning of the twentieth century, fishponds in the Czech Republic had been reduced to about 50,000 ha, mostly in regions where loamy sands and/or peaty soils did not provide a solid base for arable crop farming. This is the situation in South Bohemia where the largest proportion of fishponds has remained in existence, a few have been restored, and account for about 65% of the total fishpond area (Fig. 1).

Biodiversity of Water Birds

Fishponds are centers of high species richness, supporting a diversity of animals and plants adapted to aquatic and/or wetland habitats. In agricultural regions in which the size of drained fields approaches several hundred hectares, the littoral zones of fishponds form ecotones providing refuge for various bird species and favorable conditions for the occurrence and ontogenetic development of amphibians and invertebrates. Nevertheless, egg clutches of amphibians represent an easily available food resource for fish. In general a heavy fish stock of common carp can destroy a whole amphibian egg clutch as well as reduce populations of large zooplankton or macrophytes. Amphibians need to be isolated from the fish and can find shelter in fishpond littoral ecotones, especially in their reed belts. Common reed *Phragmites australis* stands are also of crucial importance as bird nesting habitats. Dead reed

Fig. 2 Postbreeding concentration of greylag geese *Anser anser* at the Hlohovecký fishpond in South Moravia can be up to several thousand individuals (July 2013, Photo credit: P. Macháček © Rights remain with the author)

culms mostly remain standing during winter into early spring and can support bird nests. New culms reach their final size only when the young birds have abandoned their nests.

An examination into long-term trends in the abundance of waterfowl identified fishponds as their most important breeding habitat in the Czech Republic (Fig. 2). In addition to new species (e.g., mute swan *Cygnus olor*, tufted duck *Aythya fuligula*, red-crested pochard *Netta rufina*, and goldeneye *Bucephala clangula*), the populations of most water birds increased in the 100 years preceding the 1970s. Population declines of grebes, coots, ducks (several species by 30%), and black-headed gulls *Chroicocephallus ridibundus* in the 1980s have continued for other species (e.g., common teal *Anas crecca*, garganey *Anas querquedula*, northern shoveler *Anas clypeata*, tufted duck, gadwall *Anas strepera*, and common snipe *Gallinago gallinago*) as a result of changes in land use, draining of wetlands, and fishpond management (Musil et al. 2001; Musil 2006).

Fishpond management has been implicated in this decline as increasing fish stocks affect benthic and plankton communities, the extent of littoral vegetation, water transparency, and chemistry. Increased numbers of fish feeding birds (cormorants, herons) and decline of diving ducks and other species feeding on benthos and large zooplankton can be explained by increased food (fish) supply to the birds in the former case and by increased consumption of bird food resources by a dense fish stock in the latter case. Diving duck broods in fishponds of the Třeboň Biosphere Reserve (South Bohemia), for example, showed a preference for fishponds with a younger fish stock and higher water transparency (Musil 2006). The use of heavy machinery and the direct destruction of littoral vegetation in the process of pond cleaning is also a factor in the decline of many water bird populations.

Since 1990 the population of Eurasian otter *Lutra lutra* has increased with legal protection. Likewise, the reintroduction of several pairs of white-tailed eagle *Haliaeetus albicilla* to Czech fishponds in the 1980s has resulted in a population of 30–40 breeding pairs and up to 200 overwintering birds.

Ecosystem Services

Originally constructed to manage water, the first fishponds dammed slow-flowing streams in relatively flat depressions or valleys. The southern Bohemian Třeboň region and parts of the neighboring České Budějovice region, both with large swamps and bogs, were particularly advantageous for fishpond building as they offered little land suitable for agricultural use. Retained water was used to drive water wheels in mills and iron furnaces at foundries and glassworks and to pump water from open pit mines. Fishponds could also form a part of town fortifications such as the Vajgar fishpond in Jindřichův Hradec in southern Bohemia (Fig. 3). They are now considered to have a positive aesthetic impact on the landscape character and by the Czech law on nature and landscape conservation (114/92 Sb.), all fishponds are considered as "important landscape elements." These constructed wetlands indeed provide a diversity of regulating, provisioning, sustaining, and cultural services.

Fishponds provide important water regulatory services as was evident during the flood of August 2002. Daily rainfall exceeded 100 mm for several successive days and the discharge of the Vltava River in Prague exceeded its average August rate a fifty fold at 5,000 $m^3.s^{-1}$. Several districts of Prague and sections of the Prague Underground (Metro) were flooded and 50,000 inhabitants had to be quickly evacuated. Damage would have been worse if the wetland landscape of the Třeboň Basin Biosphere reserve had not retained the flood wave of the Lužnice River, an important tributary of the Vltava River. The largest fishpond, Rožmberk, with its dam holding 70 million m^3 of water, surpassed its normal water volume about thirteen times. The whole system of the Třeboň fishponds together with adjacent wetlands held about 150 million m^3 of water, retarding and moderating the flood wave by about 68 h (Lhotský 2010; Fig. 4). The area flooded in August 2002 is shown in the map (Fig. 5) and compared with the normal situation.

Water draining agricultural catchments contain relatively high amounts of mineral nutrients and suspended particles which increase the trophic status of water and thereby deteriorate its quality, especially in downstream parts of the respective catchments. Located mostly in agricultural landscapes, fishponds capture sediments and nutrients eroded from arable land. Nutrients captured in fish and plants are

Fig. 3 The Vajgar fishpond (40 ha), an impoundment on a small river, served as a component of the fortification of the castle and town of Jindřichův Hradec from the early middle ages until mid-nineteenth century. The pond provides fish, hydropower, and recreation facilities (Photo credit: Stanislav Maxa © Rights remain with the author)

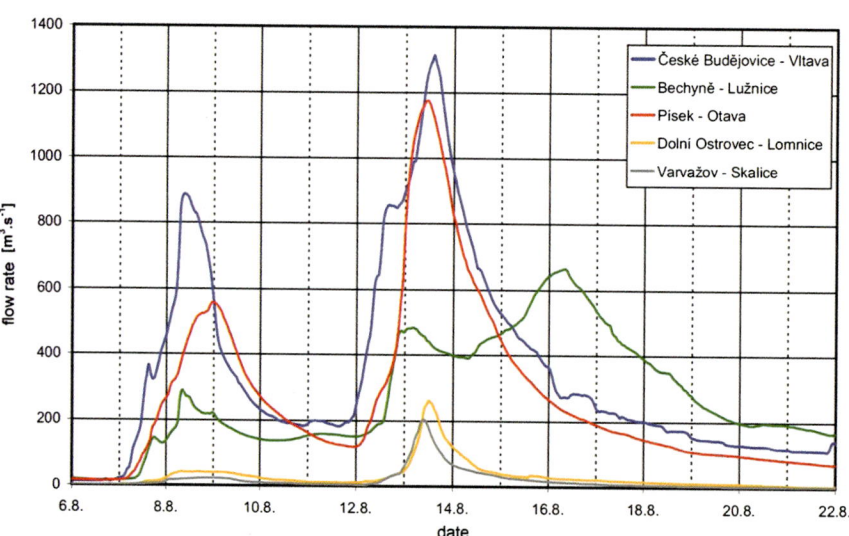

Fig. 4 Flow rate ($m^3 s^{-1}$) during the flood of August 2002 in rivers of the upper reaches of the Vltava catchment that contributed to the flood waters reaching the capital Prague and further to Dresden (Germany). The fishpond-rich landscape of the Třeboň Basin Biosphere Reserve retarded the peak flood of the Lužnice River by 3 days. (Šercl 2003 – Permission to use granted by Czech Hydrometeorological Institute)

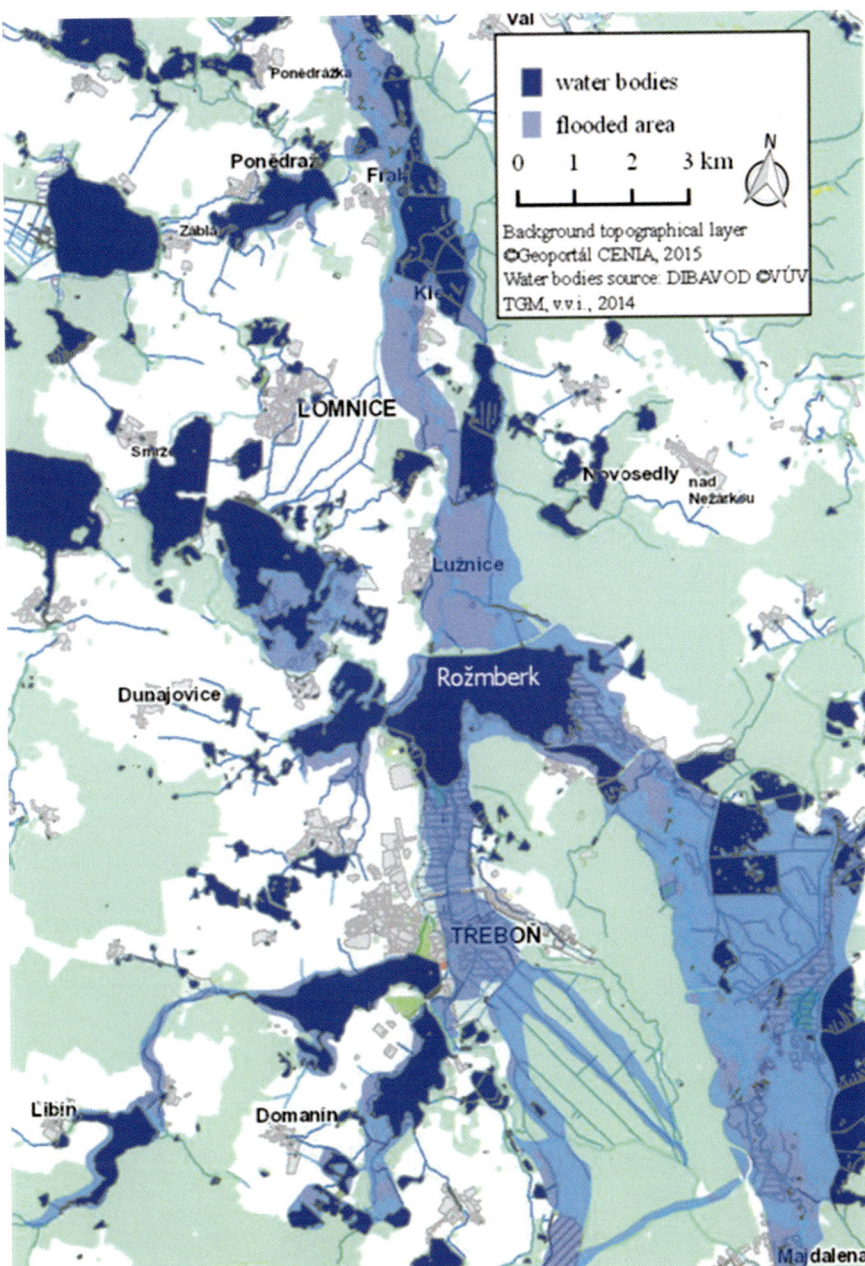

Fig. 5 The open-water area of fishponds occupies normally 10% of the area of the Třeboň Basin Biosphere Reserve whereas water covered 30% of the reserve area during the huge flood of August 2002. The dam of the large Rožmberk fishpond (constructed in 1592) held at least 70 million m^3 of water (R. Lhotský and P. Hesslerová: Permission to use granted by ENKI, o.p.s.)

removed and recycled through harvesting and consumption or composting, respectively. The bulk of the sediments is stored as mud on the fishpond bottoms and has to be occasionally dredged. The dredged material can then be composted unless it contains toxic substances (Pokorný and Hauser 2002).

Fishponds provide a source of water in case of fire, an emergency source of potable water, and groundwater replenishment for nearby wells. Many also are a source of irrigation water to support agriculture. Owing to progressively drier summers in most parts of Europe anticipated with global warming, fishponds are a centuries-old and proven way of water retention in cultivated landscapes and offer a partial adaptive response to water shortages.

Fishpond aquaculture developed through the middle ages and was an important economic activity at the end of the fifteenth century (Dykyjová 2000, Šusta 1995). The 20,000 fishponds (180,000 ha) existing in Bohemia and Moravia in the sixteenth century provided better economic returns to the nobility than did land used for agriculture until the Thirty Years War (1618–1648) severely disrupted the economy, and pond farming almost ceased to exist. The decline in pond aquaculture following the Thirty Years War lasted until the end of the nineteenth century when carp pond farming was revived (Adámek et al. 2012). Most of the fishponds are now used for extensive and semi-intensive fish farming (see below).

Selected fishponds with small nutrient loads, mainly those not fertilized with manure during the growing season, are suitable for human recreation: swimming and other water sports (sailing, rowing, etc.), sports fishing (angling), bird-watching, and hiking. The main socioeconomic service of the fishponds is the provision of employment by fishery enterprises, recreational facilities, and various other infrastructural services (e.g., accommodation or gastronomy) to tourists and other visitors to regions rich in fishponds. Fishponds also provide opportunities for scientific research on the structure and functioning of inland wetland ecosystems and various biota inhabiting them.

Agricultural areas have lost much of their earlier water-holding capacity. This loss is the result of large-scale land drainage and associated loss of soil organic carbon through mineralization of humus. Fishponds have thus almost become the only managed food-producing landscape components which conserve water in such areas. In addition, their high evaporation cools and reduces air-temperature fluctuations and increases air humidity in their wider surroundings (Hesslerová et al. 2013)

Management for Fish Production

The principal cultivated fish species has always been the common carp *Cyprinus carpio* originating from the Danube River. Other species have been added through history to the fish stock: pike *Esox lucius* (sixteenth century), tench *Tinca tinca* (late sixteenth century), pikeperch *Stizostedion lucioperca* (eighteenth century), coregonids (*Coregonus lavaretus*, 1882; *C. peled*, 1950), and herbivorous fish, mainly the grass carp *Ctenopharyngodon idella* and silver carp *Hypophthalmichthys molitrix* about 1960 (Kubů et al. 1994). The small invasive fish *Pseudorasbora parva*, which is an efficient predator on large zooplankton and competes with common carp, was mistakenly introduced with the herbivorous fish in the 1960s. Excessive populations of *Carassius auratus* behave in a similar manner.

Bishop Jan Skála Dubravius in the sixteenth century was the first to summarize the principles of pond management for carp rearing by introducing a three-stage process: winter fallowing (drawdown) of fishponds, spawning ponds (the so-called Dubravius ponds), and nursery ponds; as well as introducing long-term rearing of breeding fish (Dubravius 1547). Special hibernation (wintering) ponds were introduced somewhat later. By the end of the sixteenth century, fishponds were yielding an average 40 kg.ha^{-1} and Bohemian and Moravian carp were highly appreciated on both domestic and foreign markets.

Until the nineteenth century, annual fish yields did not exceed 50 kg.ha^{-1}. Modern management methods were introduced and described by Šusta (1995), including the description of natural food chains leading to fish production. He initiated supplementary fish feeding and also the manuring and liming of fishponds. By the beginning of the 1930s, liming and fertilizer application to the fishponds together with fish feeding (introduced at the end of the nineteenth century) brought about a moderate increase in annual fish production to 50–100 kg.ha^{-1} (Pechar et al. 2002). From 1950 to 1980, total fish production of Czechoslovak fishponds increased from 5,000 to 16,300 metric tons, with natural food accounting for some 60% (380 kg.ha^{-1}) of annual production in a prosperous rearing pond. The desirable rate of fish production was supported with 1.9 kg of feed (mostly grain) consumed per 1 kg of fish increment (Kubů et al. 1994).

Current aquaculture in the Czech Republic is characterized by the prevalence of a relatively low-impact and semi-intensive fish farming in fishponds. Annual fish production averages around 450–500 kg.ha^{-1}, with individual farms ranging between 200 and 800 (1,000) kg.ha^{-1}, and in total fluctuating between 17 and 21 thousand tonnes for the last 20 years. Common carp is the dominant fish produced (86–88%), but polyculture stocks are an important aspect of Czech pond farming (Fig. 6). Chinese carps (grass carp, bighead *Aristichthys nobilis* and silver carp) together with traditional supplementary fish (tench) and predatory species (pike, pikeperch, European catfish, and perch) are produced in the fishponds. About 25–30% of the carp production originates from supplementary feeding (wheat and barley), but the greatest share of carp production is based on natural food – zooplankton and zoobenthos. Přikryl (1996) describes the method of optimizing the amount of additional food for fish. Moderate additional feeding of the fish stock with grain is nevertheless required to obtain desirable fish yields.

Fishpond Vegetation Management

Hrbáček (1958) described the effect of fish on zooplankton structure in oxbow lakes, and Faina (1983) developed a method to estimate the consumption of zooplankton by carp on the basis of zooplankton structure. Fishpond managers by adjusting the level of fish stocking can control zooplankton and thereby impact vegetation in the

Fig. 6 Fish harvesting from a drained fishpond (Krvavý rybník) in autumn or spring when water is cold and the fish stock does not suffer from lack of oxygen. The harvested fish are sorted into species and transported live in storage tanks to temporary holding ponds (Photo credit: M. Kotyza © Rights remain with the author)

fishpond. Competition for incoming photosynthetically active radiation (PhAR) and CO_2 is assumed to be one of the crucial processes determining the development and succession in submersed vegetation in shallow ponds with increased eutrophication (Pokorný and Björk 2010). Large *Daphnia* (phytoplankton filter feeders) are not completely consumed when the fish stock is less than ca. 350 kg.ha^{-1}. The high feeding pressure by *Daphnia* reduces the density of phytoplankton, especially small planktonic algae, at high nutrient levels. Dominance of large *Daphnia* results in an, increase of water transparency and excess nutrients. These conditions stimulate the development of submersed vegetation.

Macrophytes are, however, very often invaded by filamentous algae, mostly *Cladophora* spp., which use inorganic carbon effectively even at low concentrations, grow fast, and form dense mats throughout the water column. When the filamentous algae reach the water surface, they completely shade the water column; pH frequently increases to values above 10 near the water surface. Macrophytes serving as substrate for filamentous algal growth then die-off. The water bloom of large colonies of cyanobacteria *Aphanizomenon* can be another effect of the dominance of large cladocerans.

Fish stocks above ca. 400 kg.ha^{-1} reduce large *Daphnia* numbers and thereby indirectly stimulate the development of dense populations of chlorococcal algae of

Fig. 7 In eutrophic fishponds, a low stocking rate of zooplankton-consuming fish (especially carp) results in a high water transparency and rich development of macrophytes. Excessively dense carp stocks result in a high density of phytoplankton and cyanobacterial "water blooms." Traditional wooden outlets are typical of small ponds (Photo credit: M. Baxa © Rights remain with the author)

several hundred µg chlorophyll *a* per liter. Both carp and small "weedy" fish such as *Pseudorasbora parva* and *Carassius auratus* feed on large *Daphnia* and effectively reduce the numbers of large zooplankton which is the most effective consumer of phytoplankton. Dense populations of chlorococcal algae subsequently reduce PhAR penetration into the water (Fig. 7) thereby preventing the development of cyanobacteria such as *Aphanizomenon* and *Microcystis* at higher water temperatures, but other species of cyanobacteria may instead occur in the phytoplankton (*Limnothrix*, *Planktothrix*) later in summer (Potužák et al. 2007). When "weedy" fish are present, low biomass of carp does not result in recovery of large zooplankton species (Pokorný and Květ 2004).

Conservation Status and Management

In the Czech Republic, four of the 14 sites listed under the Ramsar Convention as Wetlands of International Importance include appreciable numbers and area of fishponds (Chytil et al. 2006). The Ramsar Site Třeboň fishponds (9,710 ha, altitude 410–430 m), situated within the Třeboň Basin Protected Landscape Area and Biosphere Reserve (www.trebonsko.ochranaprirody.cz), includes 159 fishponds ranging from approximately 1 to 490 ha adjacent to littoral and marginal wetlands (mostly reed belts, wet meadows, and swamps) and parts of the Lužnice River floodplain. Within the Biosphere Reserve, the most valuable biotopes are designated as nature reserves with compulsory conservation management plans that ban destruction of littoral vegetation and waterfowl hunting. The Ramsar Site is particularly rich in avifauna including a breeding population of white-tailed eagles successfully reintroduced in the 1970s, a century after their extinction in the Třeboň area. Only a few of the Třeboň fishponds were constructed later than the fourteenth to sixteenth centuries. Fed with water from the Lužnice River, an elaborate network of canals connects the Třeboň fishponds enabling their coordinated management (Fig. 8). The Ramsar Site is the center of occurrence of the Natura 2000 listed tiny grass *Coleanthus subtilis*. The species has a short life cycle and grows on exposed shores or bottoms of fishponds that result from partial drawdown in dry years or in the first year of the fish-rearing cycle, or from full summer drainage (Hejný 1969; Dykyjová and Květ 1978; IUCN 1996). Fishpond managers are aware of the habitat requirements of this endangered species and to some extent make allowance for its sustainable conservation. The underlying clays and sands are relatively infertile, but excessive eutrophication due to dense fish stocks, manuring, and external nutrient inputs has resulted in the Třeboň Fishponds Ramsar Site being listed on the Ramsar Convention's Montreux Record since 1994. The flora of aquatic and littoral algae, cyanobacteria, and macrophytes is still relatively diverse although some formerly common species (e.g., *Nymphaea candida* and most *Potamogeton* spp.) have become quite rare with eutrophication.

Novozámecký and Břehyňský are two ancient fishponds constructed in the fourteenth century and currently included within two National Nature Reserves that constitute a European Important Bird Area (1,348 ha, altitude 250–272 m). Situated on sandstone bedrock, the mesotrophic to eutrophic fishponds and adjacent wet meadows, alder carr, waterlogged coniferous forests, and mires are rich in water birds, littoral and aquatic macrophytes, and algae. A high diversity of rather rare algae is characteristic of relatively small mires adjacent to the Břehyňský fishpond which in some years is overgrown by excessively dense macrophyte vegetation, mostly dominated by *Myriophyllum spicatum*. This Ramsar Site is being considered for enlargement to encompass several additional small- or medium-sized fishponds and one large fishpond (Dokeský Fishpond, also called Máchovo jezero – Mácha's Lake). Although used intensely for recreation, Mácha's Lake retains relatively good water quality (with periods of cyanobacterial blooms) and high biodiversity in its remote bays and littoral vegetation belts where public access is limited.

Fig. 8 The ancient (constructed in 1508–1518) Golden Canal (Zlatá stoka), 45 km long, supplies water from the Lužnice River to about 250 fishponds (Permission to use granted by publisher – ENKI, o.p.s.)

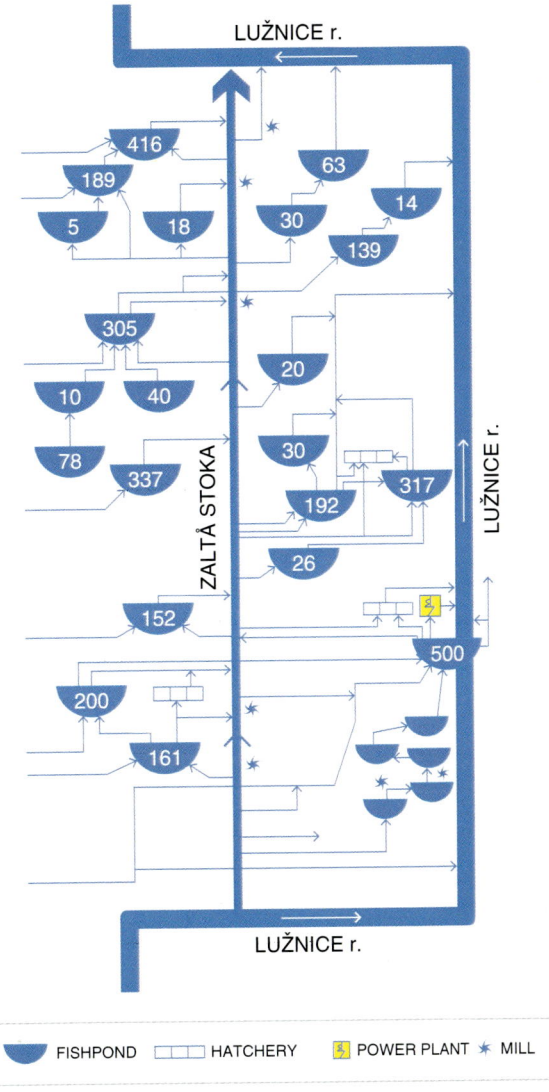

The Lednice Fishponds (www.palava.cz and www.dolnimorava.org; total area 691 ha, altitude 170–175 m) is a system of three large (>100 ha each), several medium-sized (about 50 ha each) and about 10 small fishponds dating back to the fourteenth to seventeenth centuries. They are distinguished by slightly saline water (up to 0.1% salinity mainly due to increased sulfate concentration) and solonchak soils in a few adjacent miniature continental salt marshes. The main fishpond cascade is provided with water from a small tributary (the Včelínek) in the Dyje River catchment. Only one medium-sized pond is situated directly in the floodplain of the Dyje River. The largest fishpond (322 ha) is called Nesyt (= never saturated). This name aptly describes the frequent and wide water-level fluctuations that are a distinguishing feature of the Lednice Fishponds, reflecting their location near the northern edge of the Pannonian biogeographical region with its semi-continental climatic character and unstable hydrological regime. In light of its high alkalinity, the water of these fishponds is well buffered and their biota need not adapt to wide fluctuations of pH and other hydrochemical characteristics. The vegetation includes several wetland species indicating oligohaline habitats (e.g., *Aster tripolium* ssp. *pannonicus*) and some facultatively halophytic aquatic macrophytes (e.g., *Batrachium baudotii*) and algae (e.g., *Dunaliella salina*). Sulfur bacteria occur in the reed belt of the Nesyt Fishpond. In addition to its Ramsar Site designation, the Lednice Fishponds are situated within the "Lower Morava" Biosphere Reserve and "Lednice and Valtice Area" Natural and Cultural Heritage Site of UNESCO. They are also designated an Important Bird Area under the EU's Natura 2000, and the main fishpond cascade and one adjacent salt marsh have conservation status as two separate National Nature Reserves. Controversy over management of water levels in the fishponds does, however, still occur between the authorities responsible for nature conservation and fisheries. Current water management calls for the water level to be maintained at a low watermark every fourth year in each of the large and medium-sized fishponds to protect those species of plants and invertebrates which require periodic water drawdown. Some bird populations take advantage of the drawdown as well.

The largest part of the Poodří – Odra River Floodplain Ramsar Site (www.poodri.ochranaprirody.cz; 4,427 ha, altitude 212–282 m) – is situated in the floodplain of the naturally meandering upper Odra River in northeastern Moravia close to the city of Ostrava. The Odra valley forms a section of the borderline between the Hercynian and Carpathian biogeographical regions and is in contact with the Polonian province. The system of ancient medium-sized fishponds occupies 690 ha but represents only one of several wetland types occurring in the Odra floodplain. The mostly hypertrophic fishponds, when filled during winter, suffer from oxygen deficiency and are thus unsuitable for fish hibernation. They are therefore regularly winter drained. With insufficiently dense fish stocks, the ponds provide suitable conditions for the development of excessively proliferous (with respect to fisheries management) assemblages of aquatic macrophytes (also including the protected species *Nymphoides peltata* and *Trapa natans*), which unfavorably influence populations of some other protected organisms. In such cases, the nature conservation authorities agree with the control of this vegetation by dense fish stocks reared in affected ponds at several years' intervals. In addition to providing breeding habitats for numerous bird species including white stork *Ciconia ciconia*, the Ramsar Site is situated on an important north–south migration route of water birds (e.g., Eurasian cranes *Grus grus*) and white-tailed eagles. The Eurasian otter and Eurasian beaver *Castor fiber* are also found in the site. Nationally, its conservation status is that of a Protected Landscape Area. This Ramsar Site has been on the Montreux Record since 2005 due to proposals to construct the Danube-Odra navigation canal across it.

Threats and Future Challenges

Fishponds represent a centuries-old and proven method of water retention in a cultivated landscape. These constructed wetlands are secure due to ongoing management for fish rearing and other ecosystem services. However, finding a sustainable balance between the economy of fish production, biodiversity, and hydrological and other ecological functions of the fishponds is an important pressing issue. Due to both external and internal loading most fishponds have become eutrophic to hypertrophic during the twentieth century.

With increased nutrient input by manure and agricultural runoff, stands of aquatic plants (macrophytes, chlorococcal, and blue-green algae) become denser, biomass per unit area increases, aquatic plant diversity decreases, the vertical distribution of this biomass changes, and a shortage of carbon dioxide in the fishpond water occurs as pH increases during the diurnal cycle of plant photosynthetic activity. Rapid decomposition of organic matter at the bottom results in low oxygen concentrations and release of nutrients (especially phosphorus) from the anaerobic bottom sediment. In hypertrophic fishponds, ecosystem services are impacted by excessive growth and production of algae and/or cyanobacterial water blooms, which are eventually discharged into recipient water courses.

At present, the amounts of fertilizers applied to fishponds are controlled by the Czech water-quality legislation in order to limit the negative effects of the resulting increase of fishpond eutrophication (algal blooms caused by high nutrient concentrations). Fishpond managers have to request permission from the regional office to apply manure or any other extraneous materials to a fishpond. Feeding of the fish stock is legal. Nevertheless, most Czech fishponds continue to be highly eutrophic to hypertrophic as a result of a high external nutrient load, internal load from sediments, and various management measures aimed at the highest possible marketable fish yields.

The Czech Republic is a landlocked country in which fishponds have played an important role as integral components of its hydrological network. Fishpond-rich regions can serve as positive examples of the integration of managed wetlands into agricultural landscapes (Křivánek et al. 2012). Simultaneously, protein-rich high-quality human food (fish) can be produced, biodiversity of the landscape enriched, and, last but not least, long-lasting human employment secured. Landscapes rich in wisely managed fishponds can thus serve as examples of sustainable land use. The policy public education and awarding of financial incentives should be directed towards supporting sustainable functioning of fishpond ecosystems, not solely on fish production.

References

Adámek Z, Linhart O, Kratochvíl M, Flajšhans M, Randák T, Policar T, Masojídek J, Kozák P. Aquaculture in the Czech Republic in 2012: modern European prosperous sector based on thousand-year history of pond culture. Aquac Eur. 2012;37(2):5–14.

Andreska J. Rybářství a jeho tradice (Fishery and its Tradition). Praha: Státní zemědělské nakladatelství; 1987. 208 pp.

Chytil J, Hakrová P, Vlasáková L, editors. Wetlands of the Czech Republic. 2nd ed. Prague: Ministry of Environment of the Czech Republic; 2006. 36 pp.

Dubravius I. De Piscinis ad Antonium Fuggerum. Andreas Vinglerus, Vratislaviae; 1547. 47 pp.

Dykyjová D, Květ J, editors. Pond littoral ecosystems, structure and functioning. Methods and results of quantitative research in the Czechoslovakian IBP wetland project. Berlin/Heidelberg/New York: Springer; 1978. 464 pp.

Dykyjová D. Třeboňsko, Příroda a člověk v krajině pětilisté růže. Carpio, ENKI, Třeboň, 111 pp. (Třeboňsko, Nature and Man in the Landscape of Five – Petalled Rose). In Czech with extended Engl. Summ. by S. Ridgill and M. Eiseltová, 2000. 16 pp.

Faina R. Využívání přirozené potravy kaprem v rybnících (Utilization of natural food by carp in fishponds), Metodika č.8. Vodňany: Výzkumného ústavu rybářského a hydrobiologického; 1983, In Czech 18 pp.

Hejný S. *Coleanthus subtilis* (Tratt.) Seidl. in der Tschechoslowakei. Folia Geobotanica et Phytotaxonomica, Praha. 1969;4:345–99.

Hesslerová P, Pokorný J, Brom J, Rejšková – Procházková A. Daily dynamics of radiation surface temperature of different land cover types in a temperate cultural Landscape: consequences for the local climate. Ecol Eng. 2013;54:145–54.

Hrbáček J. Typologie und Produktivität der teichartigen Gewässer. Verh int Verein theor angew Limnol. 1958;13:394–9.

IUCN, Význam rybníků pro krajinu střední Evropy. Trvale udržitelné využívání rybníků v Chráněné krajinné oblasti a biosferické rezervaci Třeboňsko (Sustainable use of Fishponds in the Třeboňsko Protected Landscape Area and Biosphere Reserve.) České koordinační středisko IUCN – Světového svazu ochrany přírody Praha a IUCN Gland, Švýcarsko Cambridge, Velká Britanie, 189 pp. In Czech with extended Engl. Summ; 1996. 30 pp.

Křivánek J, Němec J, Kopp J. Rybníky v České republice (Fishponds in the Czech Republic) MZe (Ministry of Agriculture Czech Republic). Publisher: Jan Němec – Consult.; 2012, 303pp.

Kubů F, Květ J, Hejný S. Fishpond management (Czechoslovakia). In: Patten BC et al., editors. Wetlands and shallow continental water bodies, vol. 2. The Hague: Academic Publishing; 1994. p. 391–404.

Lhotský R. The role of historical fishpond systems during recent flood events. J Water Land Dev. 2010;14:49–65.

Musil P. A review of the effects of intensive fish production on waterbird breeding populations. In: Boere GC, Galbraith CA, Stroud DA, editors. Waterbirds around the world. Edinburgh: TSO Scotland Ltd; 2006. p. 520–1.

Musil P, Cepák J, Hudec K, Zárybnický J. The long- term trends in the breeding waterfowl populations in the Czech Republic. Kostelec nad Černými lesy: OMPO & Institute of Applied Ecology; 2001. 120 pp.

Pechar L, Přikryl I, Faina R. Hydrobiological evaluation of Třeboň fishponds since the end of nineteenth century. In: Květ J, Jeník J, Soukupová L, editors. Freshwater wetlands and their sustainable future. A case study of the Třeboň Basin Biosphere Reserve, Czech Republic. Paris/Boca Raton: UNESCO/Parthenon Publishing; 2002. p. 31–62.

Pokorný J, Björk S. Development of aquatic macrophytes in shallow lakes and ponds. In: Eiseltová M, editor. Restoration of lakes, streams, floodplains, and bogs in Europe. Principles and case studies. Dordrecht/Heidelberg/London/New York: Springer Science; 2010. p. 37–43.

Pokorný J, Hauser V. The restoration of fish ponds in agricultural landscapes. Ecol Eng. 2002;18:555–74.

Pokorný J, Květ J. Aquatic plants and lake ecosystems. In: O'Sullivan PE, Reynolds CS, editors. The lakes handbook, vol. 1. Limnology and limnetic ecology. Malden/Oxford/Carlton: Blackwell Science and Blackwell Publ; 2004. p. 309–40.

Potužák J, Hůda J, Pechar L. Changes in fish production effectivity in eutrophic fishponds – impact of zooplankton structure. Aquac Int. 2007;15:201–10.

Přikryl I. Historical development of Bohemian fishpond management and its reflection in zooplankton structure (A possible criterion of biological value of ponds). In: Flajšhans M. editor. Proceedings of scientific papers to the 75th anniversary of foundation of the Research Institute of Fish Culture and Hydrobiology. Vodňany; 1996, p. 153–66.

Šercl P. Hydrologické vyhodnocení katastrofální povodně v srpnu 2002. Výzkumný ústav vodohospordářský. T.G. Masaryka; 2003.

Šusta J. Pět století rybničního hospodářství v Třeboni. Czech, translation of Šusta, J., 1889, by O. Lhotský. Carpio, Třeboň; 1995. 212 pp.

Makgadikgadi Wetlands (Botswana): Planning for Sustainable Use and Conservation

37

Jaap Arntzen

Contents

Introduction	488
Principles and Approach	488
Livelihoods	490
Resource Base	490
Resource Value	491
Choosing a Future Development Path	492
Opportunities and Challenges	493
References	494

Abstract

After the completion of the Okavango Delta Management Plan in 2008, the Botswana Government formed a partnership with private consultants led by the Centre for Applied Research to develop a similar plan for the Makgadikgadi wetlands in one year. The growing number of resource conflicts and government's desire to conserve the ecosystem, enhance sustainable development in the area and to reduce poverty formed the main reasons for the plan development. Plan development was holistic and integrated with participation of communities and leading sectors and stakeholders. The main stages were: development of an integrated framework focused on the area's main challenges; a range of specialised environmental and socio-economic studies (ecology, hydrogeology, wildlife, land use, livelihoods, economic values, policy assessment, archaeology, tourism and heritage); integration of the specialist studies into the framework and development of a range of future development scenarios; finally, selection of the preferred scenario, plan development with project activities based on the preferred scenario. Communities and other stakeholders (livestock, crop, tourism and mining sectors) actively participated

J. Arntzen (✉)
Centre for Applied Research, Gaborone, Botswana
e-mail: jarntzen@car.org.bw

© Springer Science+Business Media B.V., part of Springer Nature 2018
C. M. Finlayson et al. (eds.), *The Wetland Book*,
https://doi.org/10.1007/978-94-007-4001-3_24

throughout the plan development. This was essential in developing a shared understanding among communities and the main economic sectors and reaching consensus about the preferred scenario. The Plan is currently being implemented through the Department of Environmental Affairs, the MFMP Implementation Committee and the Makgadikgadi Wetlands Committee based in the area. The Botswana MAB Committee plans to apply to UNSECO for Biosphere Reserve status of the area.

Keywords

Makgadikgadi wetlands · Sustainable development · Integrated planning stakeholder participation

Introduction

The Makgadikgadi wetlands consist mostly of seasonal, intermittent saline/brackish lakes with four other wetland types (seasonal rivers/streams, saline pools, and salt pans used for soda ash and salt mining) (Ramsar Convention Secretariat 2007). The wetlands comprise two pan systems, i.e., the Ntetwe Pan linked with the Okavango Delta by the Boteti River and the Sua Pan fed by the Nata River (Fig. 1), which form the core of the plan area of 36,452 km^2 (agreed upon with stakeholders). Away from the pans, semi-arid rangelands dominate the landscape. The major land uses are: communal grazing, arable and residential development (primarily Tribal Land) – 19,454 km^2 (53.1%); Wildlife management areas and Protected Areas – primarily State Land – 16,366 km^2 (44.7%); mining – 763 km^2 (2.1%); and private livestock ranches and veterinary quarantine camps – 493 km^2 (1.4%). The wetland area is used for livestock production, collection of natural products, crop production, mining, eco-tourism, and adventure tourism.

Land use conflicts abound, aggravated by the fact that the area falls under three administrative districts. Increasing conflicts occur between livestock and wildlife as well as between mining and livestock, posing threats to biodiversity, damaging local livelihoods, and restricting the development potential of the wetland. Recent community-based natural resource management projects aim to combine natural resource conservation and poverty reduction.

Principles and Approach

In response to these challenges, the government initiated the development of the Makgadikgadi Framework Management Plan (MFDP). This follows the development of a management plan for the Okavango Delta (DEA 2008). The overall aim of the plan is *"to improve people's livelihoods through wise use of the wetland's natural resources"*. The formulation was guided by the following principles:

- Holistic planning will reduce intersectoral and resource conflicts.
- Future development must benefit rural livelihoods and the environment.
- Special attention is needed for vulnerable groups.

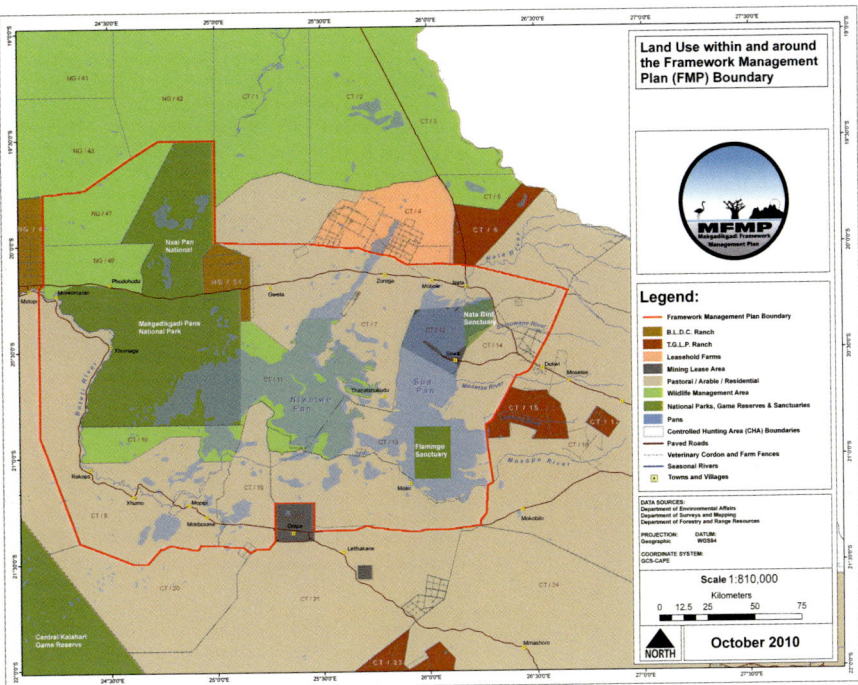

Fig. 1 Land use within and around the Framework Management Plan Boundary at Makgadikgadi wetlands (Source: © DEA and CAR 2010, Vol. 1, p. 62; with permission)

- Local stakeholder must participate in the preparation of the plan, especially regarding the future development path.
- Implementation is a shared responsibility of government, private sector, and civil society.
- Resource conservation must benefit development opportunities and livelihoods.

The MFMP approach required a clear prioritization up front (given the government stipulated MFMP preparation time of one year to address the area's challenges as soon as possible). Therefore, the following concepts were used: biodiversity hotspots, areas with tourism potential, wet spots, and heritage and archaeological sites. Firstly, all spots were identified and mapped before the ten most important spots were selected for further investigation using multi-criteria analyses.

The approach has been holistic and multidisciplinary. During the inception phase, a common environment-development framework (EDF) was developed. Then, work was done for largely disciplinary components (ecology, biology, hydrology, economics, sociology, and land use planning and policy analysis) and each component's findings were integrated back into the EDF before future development options were evaluated through scenarios and multi-criteria analysis. The plan was developed through a partnership between the government's Department of Environmental

Affairs and private consultants led by the Centre for Applied Research. This partnership combined strengths of each partner and ensured delivery on time. It also contributed to capacity building of junior government staff and a greater sense of shared ownership of the plan. Stakeholders regularly participated in workshops and communities were widely consulted. Stakeholders fully participated in the choice of future development path during a scenario workshop with the aid of a decision-support model.

Livelihoods

A village survey was carried out in eight villages to assess the main sources of livelihood in relation to wetland resources. Crop production was most commonly practiced (71.8% of the households), followed by livestock production (56.2%), (in) formal employment (42.5%), and subsidies/welfare programs (33%). This livelihood picture explains the high levels of poverty (38.5%, against a national average of 30.6%) due to low returns of crop production as well as the climatic vulnerability of crop and livestock sectors. Households attempt to increase security by engaging in multiple (3) sources of livelihood.

Natural resources proved vital to livelihoods. This is evident for agriculture but also applies to collection of veld products and tourism. Resource degradation threatens livelihoods, forcing residents to travel long distance or purchase natural resources. Households expressed the hope that the MFMP would increase employment in agriculture and tourism and diversify development as well as improve natural resources management.

Resource Base

The area is one of Botswana's biodiversity hotspots, much of it adapted to the highly variable saline and extreme conditions of the pans. The area is an important bird area with 385 different bird species, seven of which are classified as "very rare." It is one of the few African breeding sites for flamingos and pelicans. The area holds a large number of flamingos (over 170,000 lesser flamingos *Phoeniconaias minor* and 30,000 greater flamingos *Phoenicopterus roseus* (McCulloch 2003)) and hundreds of pelicans, attracted by the high productivity of the pans. Wildlife also abounds, particularly in the National Park. A few species are increasing (elephant, buffalo, gemsbok, and ostrich), while most are decreasing. Species such as zebra migrate between the Okavango Delta and the Pans. Ten biodiversity hotspots were identified based on a systematic set of factors, including threatened species, species richness, and species endemism. The ten with the highest scores were then prioritized for further analysis. Out of these ten, only five were formally protected. The rivers, Lake Xau, and a scenic escarpment at Mosu and Rysana Pan are unprotected and were prioritized for careful management.

Resource Value

The importance of economic valuation of wetlands is now widely recognized (Barbier et al. 1997; Emerton 1998). Based on the concept of total economic value, the direct use value (crop and livestock production, collection of veld products, tourism, and mining) was assessed based on focus group discussions and resource use models; the indirect use values (carbon sequestration, groundwater recharge, wildlife refuge, science, and education) were assessed based on ecological zoning and a mixture of local (e.g., wildlife numbers and tourism) and international (carbon sequestration figures and values) data.

The valuation assessment (DEA and CAR 2010) showed that both the direct and indirect use values are significant. The indirect use value is estimated to be US$28.1 million per annum (mostly carbon sequestration). The estimation of the indirect use value was based on a range of assumption given the paucity of data for the area. Indirect use values were estimated for carbon sequestration, wildlife refuge, water purification, groundwater recharge, and scientific and education values. The direct use value is US$35.7 million in terms of livelihood contributions, US$64.2 million in terms of direct economic benefits, and US$28.9 million in terms of direct and indirect economic benefits. Valuation studies need to assess both the livelihood and macroeconomic contributions as these were found to be different (Figs. 2 and 3). Agriculture and tourism clearly represent different values. While livestock and crop production together account for 18% of the livelihoods, their economic value is 6–7%. In contrast, the tourism sector contributes an estimated 7% to local livelihoods, while generating 16–26% of the economic value. Thus future tourism development should contribute more to local livelihoods. The sectors of natural resource use and mining have a more even value distribution. Use of natural resources contributes 37% to local livelihoods and 16–24% to economic values. Mining contributes virtually the same to local livelihoods (38%) but over half of the economic value.

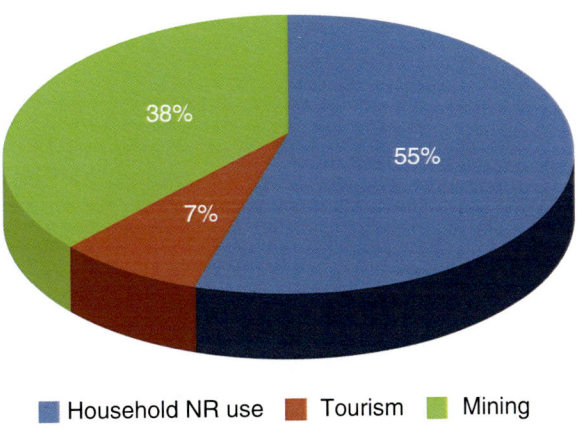

Fig. 2 Contribution of different activities to local livelihoods in the MFMP area (Source: © DEA and CAR 2010, Vol. 2, p. 38; with permission)

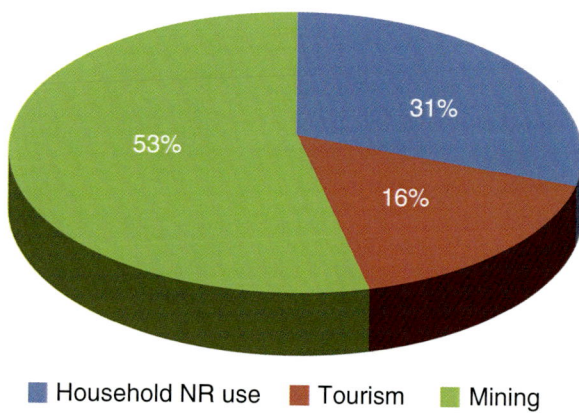

Fig. 3 Contribution of different activities to direct value added in the MFMP area (Source: © DEA and CAR 2010, Vol. 2, p. 38; with permission)

The magnitude of the indirect use value demonstrates that unsustainable natural resource management will threaten the ecosystem services and biodiversity, thus posing a significant long-term livelihood threat. The indirect use value of carbon sequestration accounts for 86% of the total indirect use value of the wetland system and is therefore the most valuable ecological service of the wetland. Wildlife (and bird) refuge and groundwater recharge value are also important, secondary indirect uses.

Choosing a Future Development Path

Initially, farmers were suspicious that the Plan would curb the livestock sector and encourage tourism. This fear disappeared during the Plan preparation and stakeholders agreed on a common preferred development path based on detailed scientific analysis, their opinions and interests, and the use of a transparent decision support system through DEFINITE multi-criteria analysis (Janssen 2001). Four aspects of sustainable development were distinguished: ecological and physical, economic, social, and institutional. Three categories of drivers of change were distinguished: 1. factors controlled by local communities; 2. factors that local communities influence but cannot control; and 3. factors beyond communities' control and influence (e.g., rainfall, climate change, etc.). The first stakeholder workshop was held to discuss the scenarios to be evaluated and the evaluation criteria. The second workshop was held to evaluate the developed scenarios and chose the preferred development path (based on the agreed weighted scores of each scenario). This was done interactively through discussions and feedback with participants. Four main scenario groups were considered: 1. Current situation; 2. Economic growth scenarios; 3. Resource preservation; and 4. Wise use based on sustainable use and Ramsar guidelines.

The outcome was that the current situation scored among the lowest and that the sustainable use scenario scored highest, followed by tourism-led development (Fig. 4). Resource preservation scored low because of the lack of development benefits. The management implications of the scenario evaluation are:

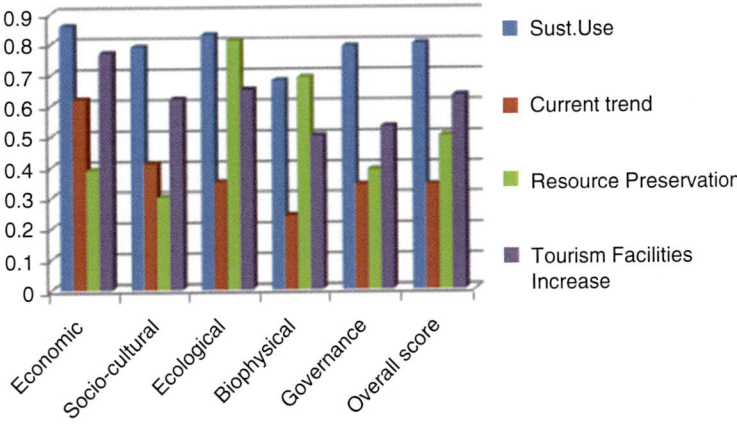

Fig. 4 Evaluation of selected development scenarios (Source: © DEA and CAR 2010, adapted, Vol. 1, p. 128; with permission)

1. There is need to move away from the current management characterized by continued conflicts, sub-optimal utilization of wetland resources, and high poverty levels.
2. National Park management needs to change; infrastructure needs to be improved, and the Park needs to generate more local benefits. This can be achieved through co-management.
3. There is potential for ecotourism and adventure tourism development. Joint venture partnership between community-based organizations and private companies would increase local benefits.
4. Conservation and utilization of heritage archaeological sites is needed to recognize and safeguard the country's cultural heritage and to create development opportunities.
5. Agricultural development should focus on better utilization of existing agricultural areas, which are underutilized. Further expansion should concentrate on areas with agricultural potential where conflicts with wildlife and tourism are minimal.

Opportunities and Challenges

The study has created opportunities to reduce sectoral conflicts by review and rationalization of the fences, comanagement of the National Park, and integrated, in-depth land use planning. In addition, poverty can be reduced by community-benign tourism development and by more productive, improved agricultural practices and management. The study showed that resource conservation can be strengthened by greater community participation and by balancing conservation and development needs. The DEA is currently leading the implementation of the MFMP and closely liaises with communities and other stakeholders in the Makgadikgadi Wetlands Committee. After a pre-feasibility study of potential Biosphere Reserve (BR) sites in Botswana (CAR 2014), The national Man and Biosphere (MAB) Committee has selected the area for Botswana's first nomination UNESCO for Biosphere Reserve status. The dossier preparation by the MAB Committee is on-going in 2016, supported by CAR (technical) and the German Federal Agency for Nature Conservation (financially). The future challenges include:

- Effective and timely implementation of the proposed activities
- Forming effective partnerships between the government, private sector, and communities
- Effective protection and management of natural resources and biodiversity, in particular hotspots (e.g., biodiversity and wet spots) and conflict spots outside protected areas.
- To complete the BR nomination dossier and to get the area accepted as a BR by UNESCO.

References

Barbier EB, Acreman M, Knowler D. Economic valuation of wetlands: a guide for policy makers and planners. Gland: Ramsar Convention Bureau; 1997.

Centre for Applied Research (2014). Prefeasibility study on the listing of biosphere reserves in botswana. Prepared for the botswana MAB committee and the german federal agency for nature conservation (BfN).

Department of Environmental Affairs (2008). Okavango delta management plan. Government of Botswana.

Department of Environmental Affairs and Centre for Applied Research. The Makgadikgadi framework management plan. http://www.car.org.bw/downloads%202010.html. Vol 1 (main report) and vol. 2 (technical reports). Report prepared for the Department of Environmental Affairs; 2010. [The Plan was developed by a core team consisting of C.Brooks, M.Ntana, P. Ruthenberg, G.McCulloch, S. Johnson and J. Arntzen (project leader)].

Emerton L. Economic tools for valuing wetlands in eastern Africa. Nairobi: IUCN; 1998.

Janssen R. On the use of multi-criteria analysis in environmental impact assessment in The Netherlands. J Multi-Criteria Decis Anal. 2001;10:101–9.

McCulloch G. The ecology of Sua pan and its flamingo populations [PhD thesis]. University of Dublin; 2003.

Ramsar Convention Secretariat. Designating RAMSAR sites: strategic framework and guidelines for future development of the list of wetlands of international importance. Gland: Ramsar Convention Secretariat; 2007.

Seagrass Recovery in Tampa Bay, Florida (USA)

38

Holly Greening, Anthony Janicki, and Ed T. Sherwood

Contents

Introduction	496
Tampa Bay Characteristics	496
Tampa Bay – Historical Loss and Stage for Recovery	498
Ecosystem Responses to Management	499
Nutrient Loadings	499
Bay Water Quality and Clarity	500
Seagrass Cover	501
Future Challenges and Opportunities	501
Conclusions	502
References	505

Abstract

In Tampa Bay, Florida, USA, reduction in wastewater nutrient loading of approximately 90% in the late 1970s resulted in rapid reduction of more than 50% of external total nitrogen loading. Continuing nutrient management actions from public and private sectors are associated with a steadily declining TN load rate since the mid-1980s – despite an increase of more than 1M people living within the Tampa Bay metropolitan area since then—and with concomitant reduction in chlorophyll-a concentrations and ambient nutrient concentrations. Seagrass extent has increased by more than 65% since the 1980s, and in 2014 exceeded the recovery goal adopted in 1996. There is evidence that Tampa Bay's successful

---

H. Greening (✉) · E. T. Sherwood (✉)
Tampa Bay Estuary Program, St. Petersburg, FL, USA
e-mail: hgreening@tbep.org; esherwood@tbep.org

A. Janicki (✉)
Janicki Environmental, Inc., St. Petersburg, FL, USA
e-mail: tjanicki@janickienvironmental.com

© Springer Science+Business Media B.V., part of Springer Nature 2018
C. M. Finlayson et al. (eds.), *The Wetland Book*,
https://doi.org/10.1007/978-94-007-4001-3_269

seagrass recovery may provide additional benefits, including buffering of global ocean acidification trends and increased carbon sequestration, both of which can be important to compensate for negative impacts of CO_2 emissions. Maintaining Tampa Bay's positive trajectory towards recovery will require continued watershed-based nutrient management and community involvement.

Keywords

Tampa Bay, Florida, USA · Seagrass recovery · Watershed-based nutrient reduction · Climate change impacts

Introduction

Prolonged and sustained human activities within an estuarine watershed can lead to increased nutrient influxes to the coast which provides a catalyst for further ecosystem degradation – a process termed cultural eutrophication (Nixon 1995; Bricker et al. 1999, 2007; NRC 2000; Cloern 2001; Duarte et al. 2013). For many coastal environments, loss of seagrass resources is a result of cultural eutrophication whereby persistent nutrient influxes cause enhanced primary production from phytoplankton and macroalgae reducing water clarity and blocking sunlight from reaching submerged aquatic vegetation. Cases of gradual estuarine eutrophication and seagrass declines have been documented include systems such as Chesapeake Bay (Boesch et al. 2001; Kemp et al. 2005; Williams et al. 2010), the Baltic Sea (Osterblom et al. 2007; Helsinki Commission 2009), and smaller systems such as Waquoit Bay, Massachusetts (Valiela et al. 1992; Valiela and Bartholomew 2014). Reversal of eutrophic conditions and recovery of lost seagrass habitats has been observed less commonly. Duarte et al. (2013) examined definitions of estuarine and coastal ecosystem recovery in published examples and found that partial (as opposed to full) ecosystem recovery often prevailed. Most successful examples involve the restoration of specific estuarine habitats rather than whole ecosystem processes: eelgrass recovery in Virginia coastal bays (Orth and McGlathery 2012) and the Danish coast (Cartensen et al. 2011; Riemann et al. 2016) and water quality and seagrass recovery in Tampa Bay, Florida (Greening et al. 2011, 2014).

Here, we provide a synopsis of seagrass recovery in Tampa Bay as more fully described in Greening et al. 2014.

Tampa Bay Characteristics

Tampa Bay is a large (water surface area of 1,036 km^2), shallow (mean depth 4 m), Y-shaped embayment located on the west-central coast of the Florida peninsula, USA (Fig. 1). The bay receives fresh water runoff from a watershed that covers an area of about 5,700 km^2. Its bathymetry has been modified by the construction and

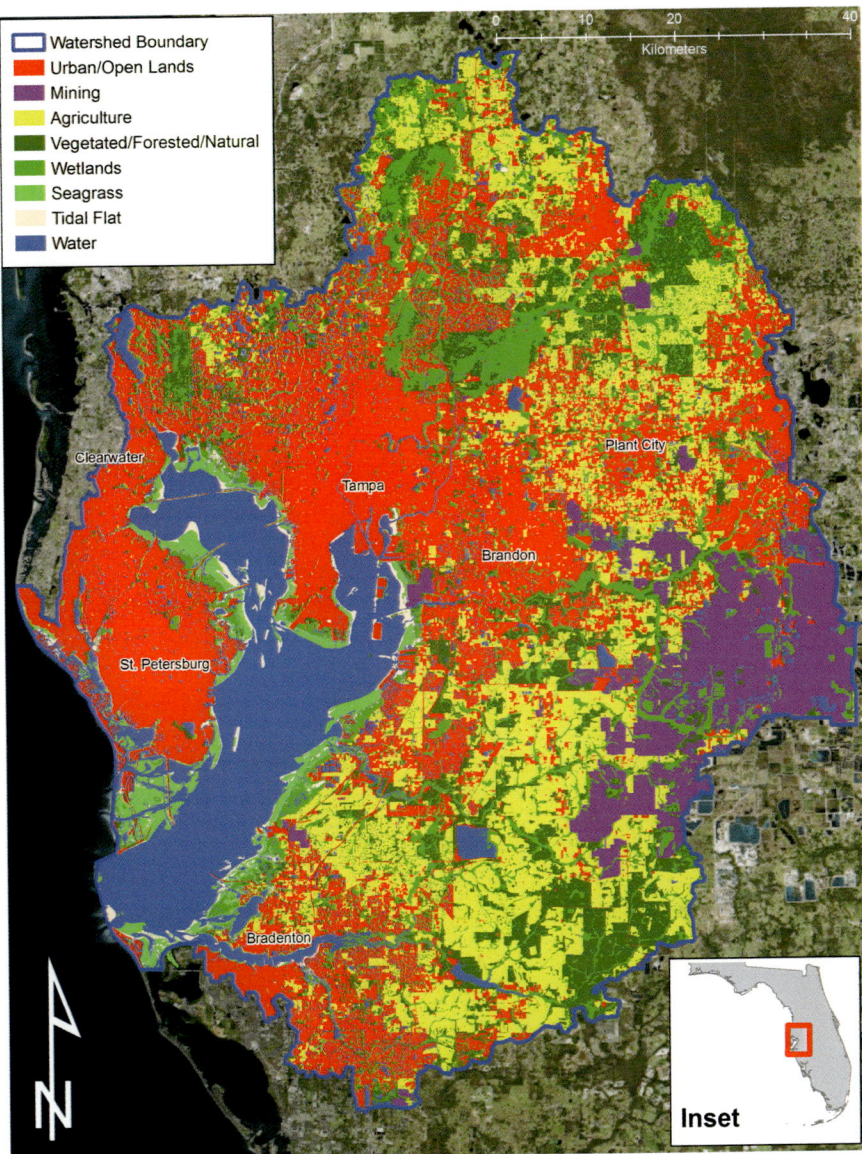

Fig. 1 Tampa Bay location and 2011 land use map of the watershed. Produced by TBEP; data source: Southwest Florida Water Management District (in public domain)

maintenance of shipping channels, dredged to depths of about 13 m. Model-based estimates of baywide residence times range from weeks to months and are primarily influenced by tides, winds, and the historic dredging alterations to bathymetry (Weisberg and Zheng 2006; Meyers et al. 2013).

Much of the land area that adjoins the bay is highly urbanized, including the cities of Tampa, St. Petersburg, Clearwater, and Bradenton, as well as numerous smaller municipalities. More than 2.3 million people currently live within the jurisdictions that directly border the bay, a number that has more than quadrupled since the early 1950s (US Census Bureau). The bay shoreline is dominated by urban land uses on its northern and western sides and by a combination of agricultural, industrial, and suburban land uses on the east, while much of the southern extent is natural shoreline. Several active port facilities with bulk commercial shipping interests and cruise terminals occur in the upper and lower bay and are an important component of the local economy.

Tampa Bay – Historical Loss and Stage for Recovery

In the late 1970s and 1980s, the degrading ecological condition of Tampa Bay became more visible to the people living along its shoreline and in the watershed. Macroalgae mats (including *Ulva* sp) were washing up on shorelines, large phytoplankton blooms were occurring, seagrass beds were disappearing, and populations of valued and visible fauna such as birds, fish, and manatees were decreasing (Yates et al. 2011; Morrison et al. 2014). These observations coincided with a rapid increase in human population, from less than 0.5 million in 1950 to 1.5 million in 1980 (Greening et al. 2014).

Several key events have occurred within the Tampa Bay region which led to the bay's recovery. Beginning in the late 1970s and continuing throughout the present recovery period, citizen engagement and pressure shaped the management efforts of both public and private entities. A major turning point for Tampa Bay occurred in the 1980s when implementation of state legislation requiring more stringent treatment standards for wastewater plants discharging to Tampa Bay was enacted. This legislation was prompted by citizen demands to local and state elected officials to improve the water quality of Tampa Bay. Upgrades to wastewater treatment plants and initiation of large-scale reclaimed wastewater programs during the late 1970s reduced the amount of total nitrogen (TN) being discharged from these sources into the bay by 90% (Johansson 2003). Implementation of advanced wastewater treatment and reuse provided the beginnings of the bay's recovery. In 1996, local and regional partners working together through the Tampa Bay Estuary Program (TBEP) adopted numerical seagrass protection and restoration goals (ca. 15,400 ha or ca. 95% of 1950s levels), and adopted numerical water transparency targets (expressed as annual mean Secchi disk depths), annual mean chlorophyll-a concentrations, and annual TN loading rates which supported attainment of the seagrass goals. The development of the goals and targets followed a multistep process (Greening and Janicki 2006; Greening et al. 2011, 2014) involving joint collaboration between public and private sectors.

To help meet the TN loading rate target, TBEP created an ad hoc public/private partnership known as the Tampa Bay Nitrogen Management Consortium (TBNMC) in 1996. The TBNMC's objective is to implement actions to meet the nutrient load targets developed for Tampa Bay. Since 1999, projects have been implemented to reduce nitrogen loads to the bay by about 270,000 kg each year (or about 7.9% of the average total Tampa Bay load from 2000 to 2011). In total, from 1992 to 2015, TBNMC participants have invested over US$450 million in projects and actions to reduce nutrient loads to Tampa Bay to meet the protective load targets for the bay. These voluntary actions and associated improvements in water quality have now been accepted by federal and state regulatory agencies as adequate to meet regulatory requirements (Greening et al. 2014).

Ecosystem Responses to Management

Nutrient Loadings

On a baywide basis, there were marked responses in the environment to the management and regulatory actions taken since the early 1980s. Most notable was the reduction in TN load from point sources (PS). Point source TN loads declined from a ca. 1976 worst case load of 5.4 million kg/year to 0.5 million kg/year by the mid-1980s. Nonpoint sources have now become the predominant TN load to Tampa Bay, contributing >57% of the total external nutrient load to the bay (Fig. 2).

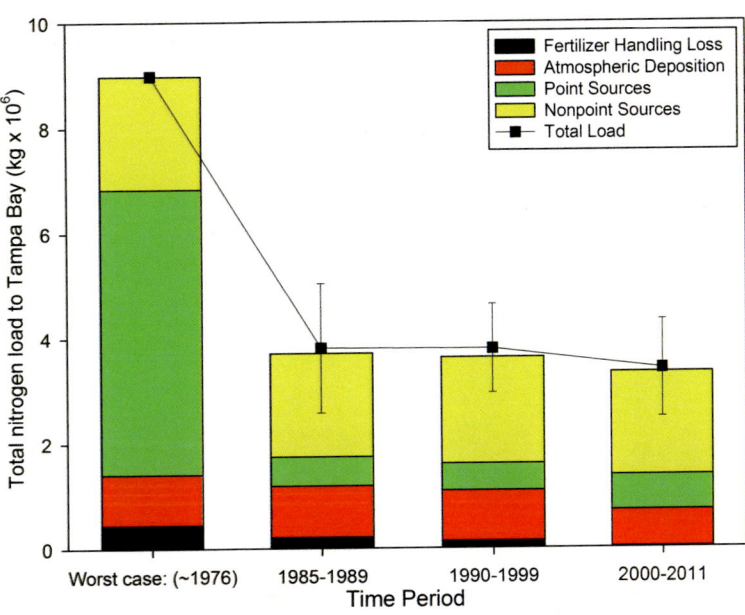

Fig. 2 Estimated annual loads of total nitrogen from various sources to Tampa Bay summarized from 1976 to 2011. Produced by TBEP; data sources from local governments and the Florida Department of Environmental Protection (in public domain)

While the human population residing in the Tampa Bay watershed from 1980 to 2011 increased by more than 1.1 million, TN load per capita continues to decrease. In the early 1980s, per capita TN load was approximately 6.6 kg/person/year. From 1985 to 1989 it was approximately 2.1 kg/person/year; from 1990 to 1999, 1.8 kg/person/year; and from 2000 to 2011, 1.3 kg/person/year. These continued per capita declines reflect an ever expanding human population, but declining TN loads entering the bay from the watershed due to implementation of multiple nutrient reduction projects and programs as outlined above. Further translated, current per capita TN load has declined by about 80% since the early 1980s (Greening et al. 2014).

Bay Water Quality and Clarity

Both TN and total phosphorus (TP) concentrations in each of the four major bay segments declined significantly following nutrient loading reductions (Greening et al. 2014). Reductions in chlorophyll-a concentration were also observed over the period of record (Fig. 3). Current chlorophyll-a concentrations have been consistently lower than historic periods with a few exceptions related to above-average rainfall conditions (e.g., 1994 and 1998 El Niño years), and to occasional summer nuisance algae blooms (*Pyrodinium bahamense*) in one segment. Concurrent with decreased nutrient loads, concentrations and chlorophyll-a concentrations, significant increasing trends in Secchi disk depth were observed in all four major bay segments (Fig. 3; Greening et al. 2014). The increasing trends in monthly mean Secchi disk depths and decreasing trends in chlorophyll-a and nutrient concentrations for each of the bay segments are all highly significant ($P < 0.001$).

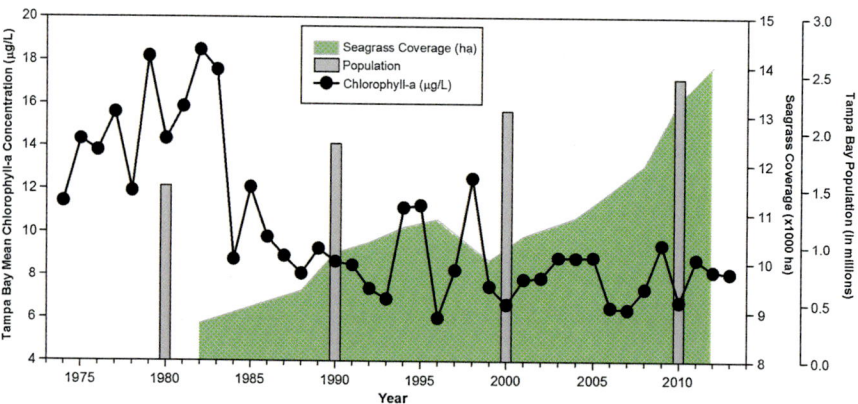

Fig. 3 Trends in mean annual chlorophyll-a concentrations and Secchi disk depth seagrass extent and watershed population estimates for Tampa Bay. Produced by TBEP; data sources: Environmental Protection Commission of Hillsborough County (in public domain); Southwest Florida Water Management District (in public domain) and US Census Bureau (in public domain)

Seagrass Cover

On a long-term basis, changes in seagrass cover in the bay have reflected changes in water quality. Analysis of aerial photographs, habitat maps, and the extent of urbanization show that seagrass cover in Tampa Bay has fluctuated markedly during the roughly six-decade period (1950–2014) for which baywide photography is available. Cover declined by an estimated 7,600 ha (or about 50%) between the early 1950s and early 1980s, during a period characterized by widespread physical impacts (e.g., dredging and filling) and increasingly poor water quality. Cover then increased by about 7,550 ha from the early 1980s through 2014, as the magnitude and frequency of physical impacts were reduced and water quality improved. Aerial photographs taken in 2014 show that the seagrass extent met and exceeded the recovery goal, and is now 16,307 ha, surpassing the original recovery goal of 15,400 ha set in 1996 (Fig. 3).

Future Challenges and Opportunities

The observed relationship between increased urbanization in coastal areas and decreased estuarine condition throughout the world has been well documented (NRC 2000; Cloern 2001; Garnier et al. 2002; Bricker et al. 2007; Duarte et al. 2013). In contrast to many coastal systems throughout the world, the water quality and seagrass improvements that started to be observed in 1985 in Tampa Bay have occurred simultaneously with an increase in the extent of urbanized land and an additional 1.1 million people (a population increase of 40%) living in the watershed during this same time period. The many projects and actions completed by public and private sectors appear to have offset nutrient loading contributions (both wastewater and nonpoint sources) typically contributed by increased population and urbanization.

Although Tampa Bay's recovery to date is on a positive trajectory, maintaining that trajectory will require continued action, assessment, and adjustment. Following evaluation of 28 long-term datasets, Carstensen et al. (2011) found that individual ecosystems exhibit unique chlorophyll-a concentration to TN relationships, such as those exhibited in Tampa Bay's different major bay segments. Carstensen et al. (2011) state that changes are likely derived from large-scale forcing associated with global change, and they imply that current chlorophyll-a and nutrient relationships cannot be used to predict future relationships. Sea level rise scenarios developed for Tampa Bay show the potential need to revise habitat restoration strategies over time, as well as nutrient management plans (Sherwood and Greening 2014).

However, the Bay's successful seagrass recovery may offer additional benefits for future resilience to impending climate change and coastal development impacts, as long as current seagrass areas are maintained and potentially expanded in the future with maintaining adequate water quality conditions. For example, Tampa Bay pH conditions have been shown to be increasing coincident with the recent expansion of seagrass areas (Fig. 4) (Sherwood et al. 2016). Increasing Tampa Bay pH trends are

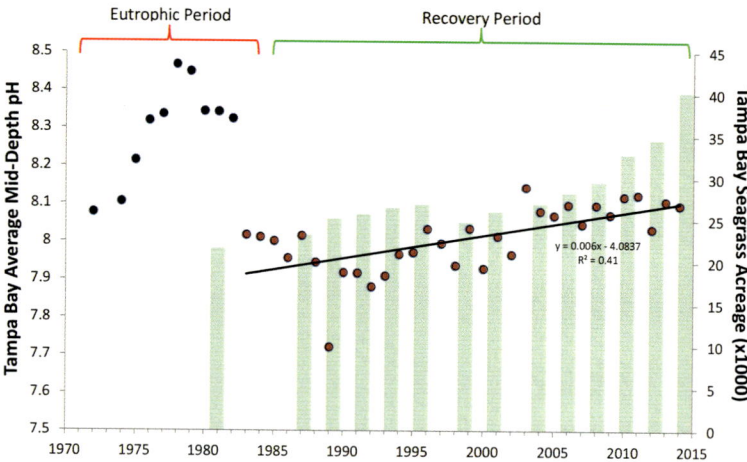

Fig. 4 Average daytime mid-depth pH (*dots*) from water quality monitoring stations in Tampa Bay relative to baywide seagrass coverage (*bars*). Produced by TBEP; data sources: Environmental Protection Commission of Hillsborough County (in public domain) and Southwest Florida Water Management District (in public domain)

contrary to global ocean acidification observations. So, an added, unforeseen ecological benefit of Tampa Bay's seagrass recovery may be the future buffering of potential ocean acidification impacts on sensitive estuarine biota, thereby making Tampa Bay a regionally significant ocean acidification refugia within the Gulf of Mexico.

Likewise, recent habitat modeling suggests that carbon sequestration benefits of Tampa Bay seagrass will be important to compensate for potential future emergent tidal wetland habitat carbon sequestration losses (Sheehan and Crooks 2016; Sherwood and Greening 2014). Sea level rise and coastal development may synergistically reduce future emergent tidal wetland habitat areas, thereby reducing their potential to sequester atmospheric CO_2. If seagrass are able to rapidly expand to newly inundated areas where emergent tidal wetlands may be lost, then Tampa Bay seagrass habitats may continue to offer "blue carbon" benefits to global efforts to mitigate the negative impacts of CO_2 emissions. Based on this scenario, it has been estimated that the suite of Tampa Bay coastal habitats, including expanding seagrass areas, may sequester up to 74 million metric tonnes of greenhouse gas emissions, an equivalent of removing 160,000 cars from Tampa Bay roads every year until 2100 (Sheehan and Crooks 2016).

Conclusions

Two broad conclusions can be drawn from the Tampa Bay example. First, under certain conditions some of the impacts of estuarine eutrophication, including loss of seagrass, can be reversed. The second conclusion is that continued watershed-based nutrient management and community involvement is critical for addressing the impacts brought about by an increasing human population, and potentially by future climate change.

Key management elements that have contributed to the observed improvements in Tampa Bay during the past several decades, as summarized in Greening et al. 2014, include:

- **Development of numeric water quality targets.** The water quality targets were developed to meet a quantitative long-term goal of restoring seagrass coverage to baseline 1950s levels. Establishing quantitative goals early in the process resulted in meaningful participation by public and private sectors in the comprehensive nutrient management strategy for Tampa Bay. The availability of long-term monitoring data, and a systematic process for using the data to evaluate the effectiveness of management actions, has allowed the Tampa Bay management community to track progress and make adaptive changes when needed. Annual reporting to the community and regulatory agencies on the attainment of water quality targets (Fig. 5) has been an essential part in continuing public and private entity engagement in the Tampa Bay nutrient management strategy.
- **Citizen involvement.** The initial reductions in TN loads, which occurred in the early 1980s, were a result of state regulations that were developed in response to citizens' demand for action. Improved water clarity and better fishing and swimming conditions were identified as primary goals by citizens again in the early 1990s, and led to development of numeric water quality targets and seagrass restoration goals. More recently, individual citizen and community actions, from pet waste management campaigns to support for reductions in residential fertilizer use, are important elements of the nitrogen management strategy in the increasingly urbanized Tampa Bay watershed.
- **Collaborative actions.** In addition to numerous other collaborative ventures that have benefitted Tampa Bay, the public/private, stakeholder-driven TBNMC, which includes more than 45 participating organizations, has implemented 500+ nutrient reduction projects since 1995. These projects have addressed stormwater treatment, fertilizer manufacturing and shipping, agricultural practices, reclaimed water use, and atmospheric emissions from local power stations, providing more than 450,000 kg of TN load reductions since 1995.
- **State and federal regulatory programs.** Regulatory requirements, such as state statutes and rules requiring compliance with advanced wastewater treatment (AWT) standards by municipal wastewater treatment facilities and stormwater treatment in residential developments, have played a key role in Tampa Bay management efforts. The technical basis and implementation plan of the Tampa Bay nitrogen management strategy was developed in cooperation with state and federal regulatory agencies, and the strategy has been recognized by them as an appropriate tool for meeting water quality standards.

Local governments, municipalities, and private businesses around the world realize the importance of a healthy watershed and bay environment to the economy and quality

Fig. 5 Average annual chlorophyll-a concentration threshold attainment for the four major bay segments. Produced by TBEP: data source Environmental Protection Commission of Hillsborough County (in public domain)

Year	Old Tampa Bay	Hillsborough Bay	Middle Tampa Bay	Lower Tampa Bay
1974	No	No	No	Yes
1975	No	No	No	Yes
1976	No	No	No	Yes
1977	No	No	No	No
1978	No	No	No	Yes
1979	No	No	No	No
1980	No	No	No	No
1981	No	No	No	No
1982	No	No	No	No
1983	No	No	No	No
1984	Yes	Yes	No	Yes
1985	No	No	No	Yes
1986	No	No	Yes	Yes
1987	No	Yes	No	Yes
1988	Yes	Yes	Yes	Yes
1989	No	Yes	Yes	Yes
1990	No	Yes	Yes	Yes
1991	Yes	Yes	Yes	Yes
1992	Yes	Yes	Yes	Yes
1993	Yes	Yes	Yes	Yes
1994	No	No	No	No
1995	No	No	No	Yes
1996	Yes	Yes	Yes	Yes
1997	Yes	Yes	Yes	Yes
1998	No	No	No	No
1999	Yes	Yes	Yes	Yes
2000	Yes	Yes	Yes	Yes
2001	Yes	Yes	Yes	Yes
2002	Yes	Yes	Yes	Yes
2003	No	Yes	Yes	Yes
2004	No	Yes	Yes	Yes
2005	Yes	Yes	Yes	No
2006	Yes	Yes	Yes	Yes
2007	Yes	Yes	Yes	Yes
2008	Yes	Yes	Yes	Yes
2009	No	Yes	Yes	Yes
2010	Yes	Yes	Yes	Yes
2011	No	Yes	Yes	Yes
2012	Yes	Yes	Yes	Yes
2013	Yes	Yes	Yes	Yes
2014	Yes	Yes	Yes	Yes

of life of their regions. Many are also facing requirements to meet federal, state, and local water quality regulations. In Tampa Bay, local communities, governments, and industries developed voluntary water quality goals and nutrient loading targets to support recovery of clear water and underwater seagrasses in the mid-1990s, and have implemented more than 500 projects since that time to help meet the nutrient loading targets developed for Tampa Bay. Their collective efforts, starting with significant wastewater point source reductions and continuing with nutrient loading reductions from atmospheric, industrial, agricultural, and community sources, have resulted in a present-day Tampa Bay which looks and functions much like it did in the relatively predisturbance 1950s period. In total > $0.6 billion has been invested by the collective management community towards Tampa Bay's overall recovery. Maintaining this progress, and potentially expanding efforts to ensure seagrass persist into the future, is paramount to make certain that the large monetary and community capital investments to date are not squandered in the future.

References

Boesch D, Brinsfield RB, Magnien RE. Chesapeake Bay eutrophication: scientific understanding, ecosystem restoration, and challenges for agriculture. J Environ Qual. 2001;30:303–20.

Bricker SB, Clement CG, Pirhalla DE, Orlando SP, Farrow DGG. National estuarine eutrophication assessment: effects of nutrient enrichment in the nation's estuaries. Special projects office and the national centers for coastal ocean science, national ocean service, national oceanic and atmospheric administration. Silver Spring; 1999.

Bricker S, Longstaff B, Dennison W, Jones A, Boicourt K, Wicks C, Woerner J. Effects of nutrient enrichment in the nation's estuaries: a decade of change. NOAA coastal ocean program decision analysis series no. 26. Silver Spring: National Centers for Coastal Ocean Science; 2007.

Carstensen J, Sánchez-Camacho M, Duarte CM, Krause-Jensen D, Marbà N. Connecting the dots: responses of coastal ecosystems to changing nutrient concentrations. Environ Sci Tech. 2011;45:9122–32.

Cloern JE. Our evolving conceptual model of the coastal eutrophication problem. Mar Ecol Prog Ser. 2001;210:223–53.

Duarte CM, Borja A, Carstensen J, Elliott M, Karawuse-Jensen D, Marbà N. Paradigms in the recovery of estuarine and coastal ecosystems. Estuar Coasts. 2013. https://doi.org/10.1007/s12237-013-9750-9.

Garnier J, Billen G, Hannen E, Fonbonne S, Videnina Y, Soulie M. Modeling transfer and retention of nutrients in the drainage network of the Danube River. Estuar Coast Shelf Sci. 2002;54:285–308.

Greening H, Janicki A. Toward reversal of eutrophic conditions in a subtropical estuary: water quality and seagrass response to nitrogen loading reductions in Tampa Bay, Florida, USA. Environ Manag. 2006;38:163–78.

Greening HS, Cross LM, Sherwood ET. A multiscale approach to seagrass recovery in Tampa Bay, Florida. Ecol Restor. 2011;29:82–93.

Greening H, Janicki A, Sherwood ET, Pribble R, Johansson JOR. Ecosystem responses to long-term nutrient management in an urban estuary: Tampa Bay, Florida, USA. Estuar Coast Shelf Sci. 2014;151:A1–16.

Helsinki Commission. Eutrophication in the Baltic Sea – an integrated thematic assessment of the effects of nutrient enrichment and eutrophication in the Baltic Sea region: Finland. Helsinki, Finland: Baltic Sea Environment Proceedings Number 115B; 2009.

Johansson JOR. Shifts in phytoplankton, macroalgae and seagrass with changing nitrogen loading to Tampa Bay, Florida. In: Treat SF, editor. Proceedings, Tampa Bay Area Scientific Information Symposium, BASIS 4; 2003 Oct 27–30; St. Petersburg. p. 31–9.

Kemp WM, Boynton WR, Adolf JE, Boesch DF, Boicourt WC, Brush G, Cornwell JC, Fisher TR, Glibert PM, Hagy JD, Harding LW, Houde ED, Kimmel DG, Miller WD, Newell RIE, Roman MR, Smith EM, Stevenson JC. Eutrophication of Chesapeake Bay: historic trends and ecological interactions. Mar Ecol Prog Ser. 2005;303:1–29.

Meyers SD, Linville A, Luther ME. Alteration of residual circulation due to large-scale infrastructure in a coastal plain estuary. Estuar Coasts. 2013. https://doi.org/10.1007/s12237-013-9691-3.

Morrison G, Greening HS, Sherwood ET, Yates KK. Management case study: Tampa Bay, Florida. Ref Module Earth Syst Environ Sci. 2014. https://doi.org/10.1016/B978-0-12-409548-9.09125-9.

National Research Council (NRC). Clean coastal waters: understanding and reducing the effects of nutrient pollution. Washington, DC: National Academies Press; 2000.

Nixon SW. Coastal marine eutrophication: a definition, social causes, and future concerns. Ophelia. 1995;41:199–219.

Orth RJ, McGlathery KJ. Eelgrass recovery in the coastal bays of the Virginia Coast Reserve, USA. Mar Ecol Prog Ser. 2012;448:173–6.

Osterblom H, Hansson S, Larsson U, Hjerne O, Wulff F, Elmgren R, Folke C. Human-induced trophic cascades and ecological regime shifts in the Baltic Sea. Ecosystems. 2007;10:877–89.

Riemann B, Carstensen J, Dahl K et al. Recovery of Danish coastal ecosystems after reductions in nutrient loading: a holistic ecosystem approach. Estuaries and Coasts. 2016;29:82–97.

Sheehan L, Crooks S. Tampa Bay blue carbon assessment: summary of findings. Technical report #07-16 of the Tampa Bay estuary program. 2016.

Sherwood ET. 2015 Tampa Bay water quality assessment. Technical report #01-16 of the Tampa Bay estuary program. 2016.

Sherwood ET, Greening HS. Potential impacts and management implications of climate change on Tampa Bay Estuary critical coastal habitats. Environ Manag. 2014;53:401–15.

Sherwood ET, Greening HS, Janicki AJ, Karlen DJ. Tampa Bay estuary: monitoring long-term recovery through regional partnerships. Reg Stud Mar Sci. 2016;4:1–11.

Valiela I, Bartholomew M. Land-sea coupling and global-driven forcing: following some of Scott Nixon's challenges. Estuar Coasts. 2014. https://doi.org/10.1007/s12237-014-9808-3.

Valiela I, Foreman K, LaMontagne M, Hersh D, Costa J, Peckol P, DeMeo-Anderson B, D'Avenzo C, Babione M, Sham C, Brawley J, Lajtha K. Couplings of watersheds and coastal waters: sources and consequences of nutrient enrichment in Waquoit Bay, Massachusetts. Estuaries. 1992;15:443–57.

Weisberg RH, Zheng L. Circulation of Tampa Bay driven by buoyancy, tides, and winds, as simulated using a finite volume coastal ocean model. J Geophys Res. 2006;111. https://doi.org/10.1029/2005JC003067.

Williams MR, Filoso S, Longstaff BJ, Dennison WC. Long-term trends of water quality and biotic metrics in Chesapeake Bay: 1986 to 2008. Estuar Coasts. 2010;33:1279–99.

Yates K, Greening H, Morrison G, editors. Integrating science and resource management in Tampa Bay, Florida. Reston: U.S. Geological Survey Circular 1348; 2011.

Section IV

North America, Greenland, and the Caribbean

Wetlands of Greenland

39

Christian Bay

Contents

Introduction 510
Wetland Type Diversity 510
Conservation Status and Management 511
Threats 512
References 513

Abstract

Wetlands are distributed throughout Greenland and occur in the lowlands. They include lakes, ponds, fens, grasslands and salt marshes. The vascular plant species diversity varies among the vegetation types and declines with increasing latitude. Fens on rich soils have the highest diversity and aquatic habitats in northernmost Greenland have only a few vascular plant species. The wetlands are dominated by graminoids from the plant families Poaceae, Cyperaceae, and Juncaceae. A continuous moss cover occurs in many habitats. Salt marshes occur on silty soil along protected coastlines and include only a few species. Selected wetlands have been designated for inclusion in the Ramsar list of Wetlands of International Importance. All wetlands in the National Park in North and Northeast Greenland are protected under the UNESCO Man and Biosphere program.

Keywords

Vascular plants · Biodiversity · Low arctic · High arctic · Habitat protection

C. Bay (✉)
Department of Bioscience, Faculty of Science and Technology, Institut for Bioscience, Aarhus University, Roskilde, Denmark
e-mail: cba@bios.au.dk

Introduction

Wetlands are known from all the bioclimatic zones of Greenland except from the polar desert zone (Bay 1997), i.e., they are distributed from the subarctic areas in the continental fjord region of South Greenland to the continental zone in high arctic Greenland. Greenland has the longest extent of all the Arctic countries, which give rise to a variation of the species composition of the plant communities. The wetlands are mostly found in the lowlands and are classified according to the Ramsar Convention into inland wetlands: lakes, ponds, fens, grasslands, and marine wetlands, i.e., salt marshes along coasts.

Lakes and ponds are found in most parts of Greenland and play a major role in the landscape of the northern part of Greenland, e.g., in Jameson Land and Hochstetter Forland in East Greenland, Lersletten, and southern Svartenhuk in West Greenland. The submerged vegetation in shallow water is dominated by *Sparganium* spp., *Potamogeton* spp., *Callitriche* spp., and *Myriophyllum* spp. Other important common species in ponds and along edges of the lakes are *Hippuris lanceolata*, *Menyanthes trifoliata*, and *Carex* spp. Characteristic for the aquatic species is that they have their main distribution outside the Arctic (Hultén 1968).

Fens occur on wet or damp soils that never dry out during the growing season. They are common along lake and pond shores, rivers and in proximity of snow banks, and in depressions in dwarf shrub heaths. The soil is not influenced by solifluction (a gradual down slope movement of surficial material related to freeze-thaw activity), which is one of the characteristic geomorphological features of the Arctic. All types are dominated by low graminoids belonging to the plant families Cyperaceae, Poaceae, and Juncaceae. A continuous moss cover occurs in many of the types.

Grasslands are dominated by species of Poaceae and *Kobresia myosuroides* and dry out during the last part of the growing season, and they have another species composition compared to fens. They are common in continental parts of West and East Greenland.

Salt marshes occur on silty soils along protected coastlines in most parts of Greenland except in the northernmost polar desert zone (Bay 1997) and cover locally only few hundred square meters. A distinct zonation is found in all parts of Greenland determined by the influence of the salty water. The outer zone is characterized by *Puccinellia phryganodes*, which is followed on less salty soil by zones dominated by *Carex subspathacea* and *Stellaria humifusa*, respectively, and the inner zone least influenced by the salty water is dominated by *Festuca rubra*.

Wetland Type Diversity

The fresh water wetlands are subdivided into a number according to their species composition and structure. The number of species varies and is dependent on the nutrients in the soil. Generally, the species diversity is reduced, when moving to the north (Fredskild 1992). Only a few aquatic species are known from lakes and ponds

Fig. 1 Low arctic fen dominated by *Carex saxatilis* with a pond covered by *Sparganium hyperboreum* in the background. Godthåbsfjorden, West Greenland (Photo credit: Christian Bay © Rights remain with the author)

in North Greenland. In addition to the widely distributed species *Ranunculus confervoides* and *Hippuris lanceolata*, *Pleuropogon sabinei*, the only high arctic aquatic plant species, occurs in fens and ponds in the northern part of Greenland.

Fens on nutrient poor soils occur mostly on acidic soils in southern parts, whereas another type occurs on basaltic and sedimentary soils in northern Greenland (Figs. 1 and 2). Characteristic species in the low arctic southern half of Greenland are *Eriophorum angustifolium*, *Carex rariflora*, *C. saxatilis*, and *Juncus biglumis* on poor soils, whereas *C. stans*, *E. triste*, and *J. castaneus* are more common to the north. In the most nutrient rich type the species diversity is higher and characterized by *Kobresia simpliciuscula*, *C. microglochin*, and *J. triglumis*. The only woody species in the fens are *Oxycoccus palustris* and *Salix arctophila*.

Conservation Status and Management

Twelve areas have been designated for inclusion in the Ramsar list of Wetlands of International Importance (Egevang and Boertmann 2001; Ramsar 2012). The National Park in North and East Greenland is protected under the UNESCO Man and Biosphere program and includes many wetland areas. The wetland areas in

Fig. 2 *Eriophorum scheuchzeri* dominated fen in the lowland, Zackenberg, Northeast Greenland (Photo credit: Christian Bay © Rights remain with the author)

North and East Greenland are in remote areas and are presently not under the influence of humans although mining activities are expected to occur in the future, whereas some of the wetland areas in West Greenland are threatened by human activities focusing on exploiting the natural resources. The local population in West Greenland does not use the wetlands directly, but as they hunt caribou *Rangifer tarandus groenlandicus* and muskoxen *Ovibus moschatus*, they are indirectly using the wetlands. These habitats constitute important foraging areas for mammals and birds. Waders and geese are dependent on the wetlands for feeding and nesting sites during their stay in the arctic summer; ducks and divers nest in close proximity to open water in the wetlands.

Threats

Exploitation of natural resource during the last decades of the 1990s and continuing into the 2000s is a potential threat to the wetlands. Environmental investigations conducted prior to mining and oil exploitation activity have focused on locating vulnerable habitats, rare, red-listed, and endemic vascular plant species and assessing the threats to the habitats. Traffic is acknowledged to impact wetland

Fig. 3 Wetlands on the west side of Jameson Land, central East Greenland (Photo credit: David Boertmann © Rights remain with the author)

wildlife, and guidelines are outlined to minimize the impact to wetlands and other vulnerable habitats by human activities connected with exploitation of natural resources.

A decline in the population size of waders in two of Greenland's Ramsar Sites due to human impacts has been recorded (Egevang and Boertmann 2001).

The large wetland in the lowlands of Jameson Land is potentially threatened (Fig. 3) by future oil exploration. It is of utmost importance that investigations in connection with Environmental Impact Assessments are carried out prior to exploitation and are focusing on the distribution and the use of the wetlands.

References

Bay C. Floristic and ecological characterization of the polar deserts zone of Greenland. J Veg Sci. 1997;8:685–96.

Egevang C, Boertmann D. The Greenland Ramsar sites, a status report. Aarhus: National Environmental Research Institute; 2001. p. 95. NERI technical report no. 346.

Fredskild B. The Greenland limnophytes – their present distribution and Holocene history. Acta Bot Fenn. 1992;144:93–113.

Hultén E. Flora of Alaska and neighboring territories: a manual of the vascular plants. Palo Alto: Stanford University Press; 1968.

Ramsar. Denmark's newest Ramsar Site in Greenland. 2012. http://www.ramsar.org/cda/en/ramsar-pubs-notes-anno-denmark/main/ramsar. Accessed 24 Jan 2016.

Peatlands of Continental North America

Dale H. Vitt

Contents

Introduction	516
Historical Development	517
Bogs and Fens	517
Peatland Landscapes	518
Threats and Future Challenges	519
References	520

Abstract

Peatlands in the USA and Canada are estimated to cover about 1.86 million km^2, most of which is located in the boreal zone with a continental climate and constitutes about 40–45% of the world's 4 million km^2 of peatland. Deep peat deposits (2–5 m) in boreal peatlands hold large stores of both carbon and nitrogen, estimated at about 33% of the global soil carbon and 10% of the world's soil nitrogen. The accumulation of carbon in peat is extremely sensitive to environmental disturbances. Changes in precipitation and temperature regimes related to climate change could act to decrease the accumulation rate, thus causing peatlands to turn from a carbon sink to a carbon source to the atmosphere. Peatlands provide a number of important ecological services, including not only significant carbon and nitrogen stores but also wildlife habitat, water filtration, and are home for a number of rare and endangered species and have been utilized and considered as culturally significant by native Americans for centuries.

D. H. Vitt (✉)
Department of Plant Biology and Center for Ecology, Southern Illinois University, Carbondale, IL, USA
e-mail: dvitt@siu.edu

Keywords

Boreal zone · Peat · Carbon · Nitrogen · Holocene · Paludification · Bogs · Fens · Climate change

Introduction

The boreal zone of North America is a mosaic of peatlands, lakes, and upland forests all adapted to exist in a climate characterized by long, cold winters and short, cool summers. Precipitation falls mostly as snow in the winter and through localized thunderstorms during the summer. Peatlands, consisting of minerogenous fens and ombrogenous bogs, are an important sink of both carbon and nitrogen, and in boreal western Canada occupy from 30% to 40% of provincial landscapes. Two of the world's largest wetlands occur in continental North America, both dominated by bogs and fens. The Hudson Bay Lowland in Quebec, Ontario, and Manitoba is the second largest peatland area in the world, while the peatlands of the Mackenzie River watershed in Alberta, Northwest Territories, and the Yukon are generally considered the third largest peatland complex (Vitt et al. 2005). It is estimated that the USA and Canada together have about 1.86 million km^2 of peatland area most of which is located in the boreal zone with a continental climate (Wieder and Vitt 2006). This is about 40–45% of the world's 4 million km^2 of peatland. Peat is the undecomposed remains of organic matter. In boreal peatlands, cold anaerobic conditions allow deep deposits (2–5 m) of peat to develop over thousands of years. These deposits hold large stores of both carbon and nitrogen, estimated at about 33% of the global soil carbon and 10% of the world's soil nitrogen (Loisel et al. 2014). Peat deposits are formed in place and hold a permanent, long-term record of the development of individual peatlands and can form significant proxies for past climate (Vitt and Wieder 2008).

Major differences exist between peatlands of northern North America and those of Siberia and Fennoscandia (Gore 1983). Although many of the ground and field layer species are similar between the New and Old World, tree and shrub species are different. For example, the dominant shrub of Eurasian bogs is *Calluna vulgaris*, whereas in North America it is *Ledum groenlandicum*. The occurrence of a tree *Picea mariana* able to tolerate the ever growing peat surface by layering of the lower branches creates peatlands that have a dense tree layer. Although *Pinus sylvestris* occurs in European and Asian bogs, it never forms dense canopies, and the bogs of Eurasia are open and have only scattered individuals of trees, similar to the scattered individuals of *Picea mariana* as it occurs in the more maritime regions of eastern Canada. Secondly, the peat depths of Siberian peatlands are much greater than those of Canada, especially those of western and northern Canada. As a result, the carbon store of Siberian peatlands tends to be greater than that of Canadian peatlands (MacDonald et al. 2006). Thirdly, the climate of western Canada is drier and more arid than eastern Canada and Eurasia. The combination of a dense, well-developed tree layer and drier climate provides for higher fire frequencies. The high incidence of fire in western Canada results in more peat lost to fire and less carbon stored in the peat deposit. This higher frequency of wildfire coupled with a more severe early Holocene climate in western Canada have prohibited peat from forming deposits in western Canada that are as deep as those elsewhere.

Historical Development

Peatlands in continental North America began to initiate soon after the retreat of glaciers, some 12,000–15,000 years ago, by primary peat formation directly on wet, mineral soils. Also, the retreating glaciers stagnated leaving isolated blocks of ice scattered on the landscape that were soon covered by eroding mineral soil. When these blocks of ice melted there remained depressions with steep sides, and these "kettle holes" soon filled with water. Surrounding these depressions, wetland vegetation developed and over time the water-filled basins were filled in by decomposing peaty materials, thus forming a peatland vegetative cover from terrestrialization. Both primary peat formation and terrestrialization were common processes in the early Holocene, and the former continues to be a common initiation process in the Hudson Bay Lowland today as isostatic rebound provides new unvegetated surfaces. However, the modern landscape of Canada and Alaska is largely the result of a third peatland initiation process that is termed paludification or the swamping of previously dry mineral soils with upland vegetation. Most of the peatland-dominated landscape of the boreal region of the continent has an ever-increasing cover of peatland landforms (Halsey et al. 1998). This paludified landscape began to develop relatively late in the Holocene in western Canada, whereas it began earlier in the east. Thus in general, peatlands are older in eastern Canada and younger in the west (Glaser and Janssens 1986). Rates of paludification, especially in the dry western portion of Canada, were cyclic, with several episodes of paludification (Campbell et al. 2000). These paludification events appear to be climate-related with higher rates of paludification associated with wetter, cooler climatic periods.

Bogs and Fens

Bogs, here defined as ombrogenous or exclusively rain-fed peatlands, are characterized by low pH's (3.0–4.5), calcium pore water concentrations of less than 5 mg/L, and a set of common species. Bogs of western and central Canada and Alaska are dominated by a tree layer of *Picea mariana*, a well-developed shrub layer usually dominated by *Ledum groenlandicum*, an herb layer of *Rubus chamaemorus*, and a ground layer of *Sphagnum fuscum, S. magellanicum,* and *S. angustifolium*. In eastern Canada, trees are smaller, more scattered, and the bogs have a more open appearance. In the east, shrubs also include species of *Kalmia* and *Andromeda*, and the ground layer consists of additional species of *Sphagnum* such as *S. capillifolium* and *S. cuspidatum*. Except for *Eriophorum vaginatum,* sedges (species of *Carex, Scirpus,* and *Eriophorum*) are infrequent in bogs. Throughout the more northern portions of the boreal zone and southern subarctic, permafrost is common in bogs, forming continuously frozen peat plateaus with intermittent unfrozen pockets (collapse scars), while farther south in the northern and mid-boreal zones bogs and occasionally fens have intermittent permafrost or frost mounds, some of which have recently melted forming wet internal lawns (Vitt et al. 1994).

Fens are minerogenous peatlands strongly influenced by the chemistry of the inflowing surface and ground waters. Poor fens are acidic, oligotrophic ecosystems having pH's between 4.5 and 5.5 along with relatively low concentrations of base cations. These fens are dominated by species of *Sphagnum* (e.g., *S. angustifolium, S. fallax, S. majus, S. jensenii, S. riparium*, and/or *S. magellanicum*). True mosses are only present as infrequent members of the *Sphagnum*-dominated ground layer. In western Canada, poor fens form on low drainage divides and at higher elevations where the inflowing waters have small amounts of base cations. In general, poor fens become more frequent eastward and become the dominant fen type on the acidic Canadian Shield region of eastern Canada. Rich fens are circum-neutral to alkaline mesotrophic ecosystems with pH's between 5.5 and 8.5 and highly variable base cation concentrations. Two types of rich fens occur – at the lower pH's (5.5–7.0) with lower concentrations of base cations moderate-rich fens have some *Sphagnum* (e.g., *S. warnstorfii* and *S. teres*) and abundant sedges (e.g., *Carix aquatilis* and *C. lasiocarpa*), while at the higher pH's (7.0–8.5) extreme-rich fens occur where sites are influenced by calcareous inflowing waters. These alkaline, extreme-rich fens are dominated by true mosses, often called brown mosses (e.g., species of *Calliergon, Hamatocaulis*, and *Tomentypnum*), abundant sedges, and may contain deposits of marl. Extreme-rich fens are most frequent along the eastern foothill of the Canadian Rockies and northward into the subarctic plain of the western NWT. Although most boreal peatlands are species poor and have rather constant flora, rich fens are home to a large variety of plant species (Vitt and Chee 1990).

Peatland Landscapes

Fens in continental areas are variable in size and form. Both poor fens and rich fens can occupy basins influenced by stagnant waters (topogenous fens) and these basin fens often have continuous peatland vegetation cover, but this plant cover can be variable, both in structure and species. In wet basins, the fens are sedge dominated, whereas in drier basins *Larix laricina* and shrubby *Betula* and/or *Salix* species characterize the sites. Other basins, especially in eastern Canada are steep sided with open water surrounded by floating and grounded mats of fen vegetation. These "kettle-hole" fens vary depending on the chemistry of the associated water body and can be *Sphagnum*-dominated poor fens or calcareous, brown moss-dominated rich fens (Vitt and Slack 1975). Basin fens are most common in the southern boreal zone and on the Canadian Shield, whereas northward where relatively flat landscapes and late melting of the winter ice provide hydrological conditions favoring the formation of large peatland complexes. The fens of these northern complexes are often patterned into a series of ridges (strings) and pools (flarks) perpendicular to the overland water flow. These patterned fens are influenced by soligenous (overland) flowing water and also by the underlying mineral topography that provides areas of water stagnation. The presence of stagnant water allows succession to more oligotrophic fens that gradually become ombrotrophic and develop into bog islands and peninsulas (Glaser et al. 1981).

Threats and Future Challenges

Permafrost thaw in bogs of the northern boreal region appears to be related to climatic changes over the past 200 or so years as climate warmed following the Little Ice Age. When frost mounds thaw in bogs, the surface collapses, the thawed area becomes very wet, and the trees and shrubs die. The collapsed area is colonized by sedges and mesotrophic species of *Sphagnum* and plant production increases. As these internal lawns succeed back to a drier hummocky bog surface, plant production is greater than decomposition and carbon is stored in the newly deposited peat, providing a short-term net carbon sink resulting from the permafrost thaw (Turetsky et al. 2002a). This reaction of boreal permafrost thaw happening in bogs of the boreal zone is quite different from the reactions of permafrost thaw in mineral soils of the arctic.

Although there is increasing disturbance in boreal peatlands from oil and gas exploration and production, reservoir creation, and peat harvesting, the overall impact of these activities is much less than the impacts from natural disturbance from wildfire (Turetsky et al. 2002b). In comparison, in the more populated areas of eastern Canada, the large-scale drainage of peatlands from agriculture and urbanization is a lesser concern in the west.

Peatlands are an important carbon store. Whereas the amount of carbon stored in peat is large, the amount of carbon currently added to the peat column on an annual basis is very small (generally estimated as 19–24 gC m^{-2} years^{-1} – Loisel et al. 2014). As a result, the accumulation of carbon in peat is extremely sensitive to environmental disturbances. Especially concerning are changes in precipitation and temperature regimes related to climate change, which could act to decrease the accumulation rate thus causing peatlands to turn from a carbon sink to a carbon source to the atmosphere. In boreal North America, peatlands occupy a significant portion of the landscape and provide a number of important ecological services, including not only significant carbon and nitrogen stores but also wildlife habitat, water filtration, and are home for a number of rare and endangered species including the woodland caribou *Rangifer tarandus caribou*. Native Americans have utilized these places for centuries and consider them as culturally significant. Community succession is a slow process, and in peatlands the fen to bog successional series is well documented and forms an important part of boreal ecology, and this is especially true for western Canada where the climate is dry and wildfire is such an important natural disturbance (Vile et al. 2011). Bogs, with their unique ombrotrophic status, also are subject to disturbances, especially fire, and their importance in providing a long-term but climate-sensitive carbon stock should not be underestimated (Wieder et al. 2008).

References

Campbell ID, Campbell C, Yu Z, Vitt DH, Apps MJ. Millennial-scale rhythms in peatlands in the western interior of Canada and in the global carbon cycle. Quatern Res. 2000;54:155–8.

Glaser PH, Janssens JA. Raised bogs in eastern North America: transitions in landforms and gross stratigraphy. Can J Bot. 1986;64:395–415.

Glaser PH, Wheeler GA, Gorham E, Wright Jr HE. The patterned peatlands of the Red Lake peatland, northern Minnesota: vegetation, water chemistry, and landforms. J Ecol. 1981;69:575–99.

Gore AJP, editor. Mires – swamp, bog, fen and moor, Ecosystems of the world 4B – regional studies. Amsterdam: Elsevier; 1983.

Halsey LA, Vitt DH, Bauer IE. Peatland initiation during the Holocene in continental western Canada. Clim Change. 1998;40:315–42.

Loisel J, Yu Z, Beilman DW, et al. A database and synthesis of northern peatland soil properties and Holocene carbon and nitrogen accumulation. Holocene. 2014;24:1028–42.

MacDonald GM, Beilman DW, Kremenetski KV, Sheng YW, Smith LC, Velichko AA. Rapid early development of circumarctic peatlands and atmospheric CH_4 and CO_2 variations. Science. 2006;314:285–8.

Turetsky MR, Wieder RK, Vitt DH. Boreal peatland C fluxes under varying permafrost regimes. Soil Biol Biochem. 2002a;34:907–12.

Turetsky MR, Wieder RK, Halsey L, Vitt DH. Current disturbance and the diminishing peatland carbon sink. Geophys Res Lett. 2002b;29(11). https://doi.org/10.1029/2001GLO14000.

Vile MA, Scott KD, Brault E, Wieder RK, Vitt DH. Living on the edge: the effects of drought on Canada's western boreal forest. In: Tuba Z, Slack NG, Stark LR, editors. Bryophyte ecology and climate change. Cambridge, UK: Cambridge University Press; 2011. p. 277–97.

Vitt DH, Chee W-L. The relationships of vegetation to surface water chemistry and peat chemistry in fens of Alberta, Canada. Vegetatio. 1990;89:87–106.

Vitt DH, Slack NG. An analysis of the vegetation of *Sphagnum*-dominated kettle-hole bogs in relation to environmental gradients. Can J Bot. 1975;53:332–59.

Vitt DH, Wieder RK. The structure and function of bryophyte-dominated peatlands. In: Goffinet B, Shaw AJ, editors. Bryophyte biology. 2nd ed. Cambridge, UK: Cambridge University Press; 2008. p. 357–92.

Vitt DH, Halsey LA, Zoltai SC. The bog landforms of continental western Canada, relative to climate and permafrost patterns. Arct Alp Res. 1994;26:1–13.

Vitt DH, Halsey LA, Nicholson BJ. The Mackenzie River basin wetland complex. In: Fraser LH, Keddy PA, editors. The world's largest wetlands: ecology and conservation. Cambridge, UK: Cambridge University Press; 2005. p. 218–54.

Wieder RK, Vitt DH, editors. Boreal peatland ecosystems. Berlin/Heidelberg/New York: Springer; 2006.

Wieder RW, Scott KD, Kamminga K, Vile MA, Vitt DH, Bone T, Xu B, Benscoter BW, Bhatti JS. Post-fire carbon balance in boreal bogs of Alberta, Canada. Glob Chang Biol. 2008. https://doi.org/10.1111/j.1365-2486.2008.01756.x.

Yu Z, Beilman DW, Jones MC. Sensitivity of northern peatland carbon dynamics to Holocene climate change. In: Baird AJ, Belyea LR, Comas X, Reeve AS, Slater LD, editors. Carbon cycling in northern peatlands. Washington, DC: American Geophysical Union; 2009. p. 55–69.

Boreal Wetlands of Canada and the United States of America

41

Beverly Gingras, Stuart Slattery, Kevin Smith, and Marcel Darveau

Contents

Introduction	522
Boreal Wetland Types	523
Factors Influencing Presence and Diversity of Boreal Wetlands	527
Boreal Wetlands Ecosystem Services	531
Potential Sources of Wetland Alteration	533
Conservation Status and Current Efforts	537
Future Challenges	539
References	540

Abstract

The Canadian and Alaskan boreal zone is one of the most water rich areas in the world, and contains an estimated combined surface water and peatland area the size of Indonesia ($\sim$1.94 million km^2). Boreal wetlands are diverse in form and function and can be classified into five major types: bogs, fens, swamps, marshes, and shallow open water wetlands. The distribution and diversity of these wetlands is primarily a function of physical drivers (e.g., climate, topography), but is also influenced by biological drivers (e.g., humans, beaver). Boreal wetlands deliver a

---

B. Gingras (✉) · K. Smith (✉)
Ducks Unlimited Canada, Edmonton, AB, Canada
e-mail: b_gingras@ducks.ca; k_smith@ducks.ca

S. Slattery (✉)
Ducks Unlimited Canada, Stonewall, MB, Canada
e-mail: s_slattery@ducks.ca

M. Darveau (✉)
Ducks Unlimited Canada, Quebec, QC, Canada
e-mail: m_darveau@ducks.ca

© Crown 2018
C. M. Finlayson et al. (eds.), *The Wetland Book*,
https://doi.org/10.1007/978-94-007-4001-3_9

variety of important ecological goods and services with regulating, provisioning, cultural, and supporting benefits that directly and indirectly benefit humans. Some of these direct benefits are the Canadian and Alaskan boreal's rich natural resources including peat, minerals, hydropower, trees, and oil and gas. However, extraction of these resources can bring a variety of landscape changes. Boreal wetland conservation efforts to address or prevent these landscape changes vary by jurisdiction but tend to fall into three groups: (a) establishing protected areas; (b) using practices that avoid or minimize net wetland loss and degradation; and (c) restoring and reclaiming to recover lost function. The Canadian and Alaskan boreal presents of the world's great conservation opportunities; however, ongoing pressure from resource development means that collaboration of all people living, studying, harvesting/ extracting, and managing the boreal is required to conserve boreal wetlands in perpetuity.

Keywords

Boreal · Wetlands · Peatland · Conservation · Ecosystem Services · Natural Resources · Landscape Change · Best Management Practices · Protected Areas · Wetland Alteration · Climate

Introduction

The extent of the boreal zone in North America has been the subject of debate, and so for the purposes of this chapter we will follow the boundaries developed by Brandt (2009) which includes most of Alaska and Canada's Yukon Territory and stretches across Canada from the province of British Columbia east to Newfoundland and Labrador (total area of 6.3 million km^2, Brandt 2009; Fig. 1). The Canadian and Alaskan boreal zone is one of the most water rich areas in the world with an estimated 850,000 km^2 of fresh surface water (Wells et al. 2011; Boggs et al. 2012). In addition, Canada and Alaska contain over one third of the world's peatlands (~1.27 million km^2) storing an estimated 160 billion tons of carbon (Bridgham et al. 2006). The majority of these peatlands are found in the Canadian and Alaskan boreal zone (e.g., 86%, Tarnocai et al. 2002). Combined, the estimated surface water and peatland area of the Canadian and Alaskan boreal zone (~1.94 million km^2) is equal in size to the country of Indonesia and exceeds the size of 234 other countries.

Boreal wetlands are diverse in form and function and provide a variety of ecosystem services. With abundant natural resources present in the Canadian and Alaskan boreal zone, a current and future challenge is to ensure resource development occurs in a manner that sustains these complex areas. Herein, we review those wetlands types, factors influencing their formation, ecosystem services they provide, and conservation issues affecting their persistence in a changing environment.

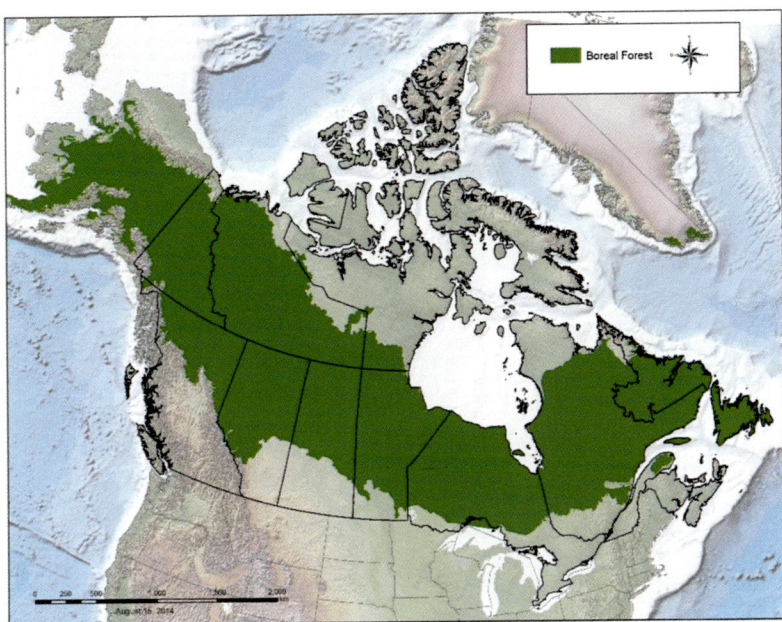

Fig. 1 Extent of the boreal forest in North America, following Brandt (2009). Brandt's boreal and boreal alpine were combined, while hemi-boreal and hemi-boreal alpine regions were omitted

Boreal Wetland Types

Boreal wetlands are functionally diverse and are influenced by and perform many ecological functions. Boreal wetlands can be classified into five major types based on soils, water table depth, water flow, nutrient availability, and vegetation communities (National Wetlands Working Group 1997):

(i) Bogs (e.g., Fig. 2) – Wetlands that have >40 cm of accumulated peat (primarily derived from decomposing *Sphagnum* moss), are hydrologically isolated from surface or groundwater interaction, and have nutrient inputs derived from precipitation (i.e., are ombrotrophic). Due to these characteristics, bogs have relatively poor nutrient availability and lower plant diversity relative to other wetland types. The water table is at or slightly below the ground surface and is raised above the surrounding terrain through capillary action in the peat. Vegetation cover is frequently dominated by *Sphagnum* spp. moss with black spruce *Picea mariana* tree and ericaceous shrub cover.

(ii) Fens (e.g., Fig. 3) – Wetlands that have >40 cm of accumulated peat (primarily derived from decomposed sedges and *Sphagnum* spp. or brown mosses) and are

Fig. 2 North American boreal bog (Photo credit: © Ducks Unlimited Canada)

Fig. 3 North American boreal fen (Photo credit: © Ducks Unlimited Canada)

hydrologically connected to surrounding areas via groundwater or surface water flow. Fens are minerotrophic (i.e., nutrients derived from ground and/or surface water flow) and as a result exhibit more diverse flora than bogs. The water table is at or above ground for most of the year and often forms characteristic surface patterns (e.g., strings, flarks, hummocks, hollows). Vegetation cover is variable, but if present, trees range from tamarack *Larix laricina* in richer sites to black spruce in poorer sites. The most common shrubs include birch *Betula* spp., willow *Salix* spp., and ericaceous shrubs *Sphagnum* spp. moss, brown mosses, and sedges *Carex* spp. are also common.

Fig. 4 North American boreal swamp (Photo credit: © Ducks Unlimited Canada)

(iii) Swamps (e.g., Fig. 4) – Wetlands found on mineral or peat soils and with tree or tall shrub (>2 m) cover influenced by minerotrophic ground or surface water. The water table is highly variable throughout the year (at or below the surface). During some of the year the water table is well below the surface creating an aeration zone in the soil that promotes root development of trees and shrubs. Swamps exhibit a range of tree, shrub, and herb vegetation cover. Stands of coniferous (e.g., black spruce; white spruce *Picea glauca*; and tamarack) or deciduous tree species (e.g., birch; balsam poplar *Populus balsamifera*; and silver maple *Acer saccharinum*) are common as well as stands of mixed coniferous and deciduous trees. Shrubs are typically dominated by willow, alder *Alnus* spp., and dogwood *Cornus* spp.

(iv) Marshes (e.g., Fig. 5) – Mineral soil wetlands that have shallow surface water tables that fluctuate dramatically during the year due to a variety of hydrologic inputs including precipitation, surface runoff, stream or river inflow, lake seches, groundwater seepage or discharge, or tidal action. Marshes are minerotrophic and are usually eutrophic due to abundant nutrients available during periods of drawdown when significant aeration of the substrate occurs. Abundant nutrients can promote significant productivity of vascular plants during the growing season and high decomposition rates of plant material afterward. Vegetation cover is primarily emergent aquatic macrophytes such as sedges, rushes, reeds, grasses, and cattails. Floating-leaves and submerged aquatic vegetation are also present. Trees and shrubs are never associated with marsh wetland types.

(v) Shallow open waters (e.g., Fig. 6) – Wetlands with abundant surface water <2 m deep and that are transitional between other wetland types and deeper (lacustrine) or moving (riverine or estuarine) water environments. Dissolved minerals and nutrients are widely variable in this wetland type and are influenced by the

Fig. 5 North American boreal marsh (Photo credit: © Ducks Unlimited Canada)

Fig. 6 North American boreal shallow open water wetland (Photo credit: © Ducks Unlimited Canada)

substrate geology, hydrology, plant communities, nutrient fluxes, and other aquatic processes common to the littoral zone of water bodies. Water levels are typically above the surface throughout the year but may experience periodic drawdowns. Vegetation, if present, is limited to floating (e.g., pond lily *Nuphar* spp.) or submerged aquatic (e.g., pondweed *Potamogeton* spp.) plants.

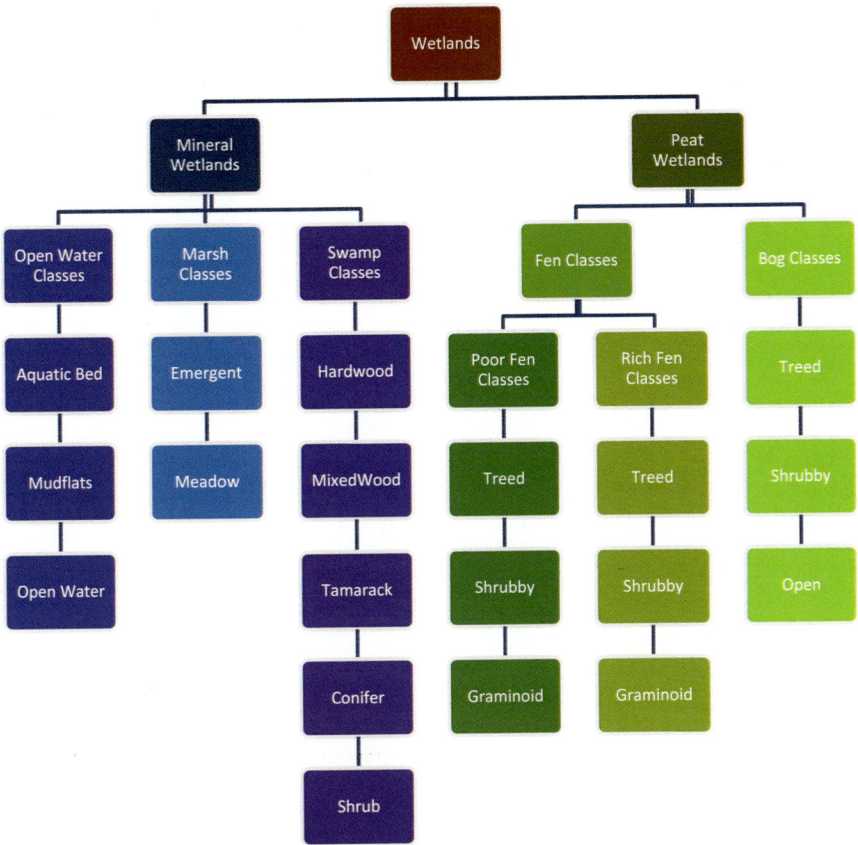

Fig. 7 The five major and nineteen minor classes of North American boreal wetlands as described in Ducks Unlimited Canada's Enhanced Wetland Classification system following Smith et al. 2007b

These wetland classes can be further broken down into subclasses that represent vegetation structure and wetland function (Smith et al. 2007b; Fig. 7). It is important to note that wetland types can grade into one another and so often create a diverse mosaic of wetlands on the landscape depending on the spatial distribution of abiotic and biotic factors discussed below (Fig. 8).

Factors Influencing Presence and Diversity of Boreal Wetlands

The distribution and diversity of wetlands in the Canadian and Alaskan boreal zone is primarily a function of physical drivers including past and present geology, topography, and climate (Carter 1997; Price et al. 2005; Brandt et al. 2013; Ménard et al. 2013), but is also influenced by biological drivers such as the presence of North American beaver *Castor canadensis* and humans on the

Fig. 8 Diversity and juxtaposition of wetlands on the landscape. Different boreal wetlands types can create a mosaic of wetland types across gradients of soils, water table depth, water flow, and nutrient availability. The *dark green* shaded area is treed rich fen, the *light green* area is graminoid rich fen, the *light blue* area is meadow marsh, and the *dark blue* area is conifer swamp (Photo credit: © Ducks Unlimited Canada)

landscape. The bedrock and surficial geology (i.e., soil type, texture, strata, depth to impermeable layer, and parent material), in conjunction with the topography of the boreal zone, influences the development of wetlands as well as the extent, permanency, and type of water flow into and out of wetlands (Devito et al. 2012). Generally, increases in elevation and slope results in a more organized and efficient drainage network and an increased surface water flow. Topography is an important driver in areas of the boreal zone with shallow soils and impermeable bedrock but has less of an influence on wetland development and diversity in areas with more water storage capacity due to deep soil deposits and permeable bedrock (Devito et al. 2012).

Much of the geology of the present day Canadian and Alaskan boreal zone is shaped by changes in climate that occurred during periods of continental glaciation and deglaciation beginning ~2.7 million BP (Brandt et al. 2013). The end of the last full glacial episode began in the Holocene period (~15,000–16,000 BP) and it was not until ~7,000 BP that the boreal zone became firmly established in northern Canada (Brandt et al. 2013). As the glaciers advanced across the boreal shield, they removed large amounts of soil, clay, rock, and gravel and created millions of depressions in the impermeable Precambrian crystalline igneous and metamorphic bedrock. These poorly drained depressions now form millions of lakes and wetlands in the shield ecozones of Canada (Taiga Shield and Boreal Shield). Glaciers also flattened the landscape of the plains ecozones of Canada (Taiga Plains, Boreal Plains, and Hudson Bay Plains) and covered much of the underlying permeable horizontal layers of sedimentary rock with deep deposits. During the retreat of the last glaciation, drainage into Hudson Bay was blocked and large lakes formed; the remnants of which left finely textured silt and clay mineral soils on much of the landscape. The flat terrain, impervious soil, and poor drainage promoted the development of wetlands throughout much of the plains, and now include some of the largest wetland complexes in the world (Warner 2004). Boreal wetlands in the Canadian and Alaskan cordillera ecozones (Taiga and Boreal Cordillera) are found in the plateaus and valleys of the mountainous sedimentary bedrock terrain. This area also experienced several cycles of glaciation; the plateaus in the region are flat to rolling with widespread deposits of glacial debris and the mountain valleys have deep glacial deposits. It is in these areas of flat terrain and poor drainage where wetlands form. In the Alaskan Boreal Interior, most of the lakes and wetlands are found on flats, lowlands, and bottomlands of low relief and were primarily formed by the hydrological processes associated with rivers and streams. Thus, the surficial geology in these areas is characterized primarily by fluvial with some aeolian and alluvial deposits overlying sedimentary bedrock. As a result, the surficial geology in the boreal zone ranges from coarse texture and porous (sands and gravels) to fine textured and impermeable (clays and silt) to a unique situation where coarse sands lay over fine clay (sandy veneer; Devito et al. 2012).

All modern boreal wetlands originate from mineral wetlands (Warner 2004). Mineral wetlands are those that form when excess water saturates mineral soils and accumulates to shallow depths. Soils of mineral wetlands (i.e., swamps, marshes, and shallow open water wetlands) have little or no organic content. Organic wetlands, including bogs and fens (collectively known as peatlands), also have a foundation of mineral soil, but over time deep layers of decomposed or compressed organics (peat) develop either by paludification (formation in the absence of open water) or by terrestrialization (natural wetland succession by infilling of ponds, lakes, or open water wetlands into marshes, swamps, and finally into peatlands;

Warner 2004). Peatland development occurs in areas where connection to ground and surface water flow is limited, movement of water is slow (Warner 2004), annual precipitation is >500 mm per year, temperatures are cool, and the water balance is positive (Halsey et al. 2000).

Climate continues to play a key role in creating the diversity of wetlands within the boreal zone as it influences the overall amount of water available in the region, primarily because of its effects on water balance through precipitation and evapotranspiration (Devito et al. 2012). Precipitation provides a direct source of water to wetlands by recharging ground and surface water (Price et al. 2005) and an indirect source by recharging soil moisture in and around wetlands (Devito et al. 2012). The relative difference between precipitation and evapotranspiration (water lost to the atmosphere) describes the amount of water available on the landscape. The climate of the Alaskan and Canadian boreal zone is generally characterized by long, cold, and dry winters and short, cool, and moist summers, but precipitation and temperature vary spatially and temporally across the region (Price et al. 2013). The climate of Canada's Boreal Shield is described as being humid, which means that precipitation (mean annual precipitation of 400–1,600 mm and temperature of −4 °C to +5.5 °C) greatly exceeds evapotranspiration. The humid climate in combination with shallower soils and impermeable bedrock geology typical of the Boreal Shield results in wetlands primarily recharged through runoff (Devito et al. 2012). In contrast, the climate of the western Boreal Plains is characterized as subhumid (mean annual precipitation of 300 –625 mm and temperature of −2 °C to +2 °C) with evapotranspiration exceeding precipitation, which results in a long-term moisture deficit. The subhumid climate in combination with deep glacial deposits and permeable bedrock typical of the western boreal plains results in wetlands heavily influenced by subsurface and ground water recharge (Devito et al. 2012). Climate in the northern boreal zone is generally cold and dry (e.g., mean annual temperature of −10 °C to −4 °C and precipitation of 200–500 mm in the Taiga Plains) and conducive to maintaining sporadic to extensive regions of peatland and permafrost.

The same conditions that can lead to the development of peatland (limited connection to ground and surface water flow, slow water movement, low annual precipitation, cool temperatures, and positive water balance) can also lead to the creation of permafrost, which is soil, rock, or sediment that remains at or below 0 °C for two or more consecutive years (Price et al. 2013). Permafrost is a prevalent feature in Canada and Alaska, covering about half the landmass of Canada and most of Alaska in either continuous regions (primarily found in the Arctic) or discontinuous patches (primarily found in the northern boreal zone). In regions with permafrost layers, water moves into the ground along a thermal gradient from warm to cold and is stored as ice within frozen soils. The presence of permafrost in the northern boreal zone has resulted in the development of peatlands as it creates a ground seal, allowing pooling of water in depressions and saturation of soils in the active layer of otherwise permeable geologic settings (Warner 2004).

Modern peatlands appeared in boreal Canada and Alaska between ~2,000 and 7,000 years after the last deglaciation (~18,000–12,000 BP; Gorham et al. 2007; Richard and Grondin 2009; Jones and Yu 2010) reaching their current extent about

2,000–3,000 BP (Halsey et al. 2000). Peatland distribution has generally been stable since that time, but current studies suggest that recent and continuing climate change could cause wetland abundance to decrease in the southwest boreal zone due to drought and decrease in the northern boreal zone due to permafrost thawing, waterlogging, and eventual drying of peatlands (Payette et al. 2004; Price et al. 2013).

Beavers are often referred to as wetland creators or engineers (Johnston 2012). They build dams that restrict water flow, increase water storage, and promote organic matter deposition, creating new beaver ponds (i.e., open water wetlands). From the time of their colonization (17,000–10, 000 BP, Gorham et al. 2007) to the arrival of European settlers in the fifteenth century, beavers greatly influenced the distribution of boreal wetlands as they built dams and created ponds wherever conditions allowed. About 60–400 million beavers were estimated to be present prior to settlement (Johnston 2012), but unregulated fur trapping gradually brought the beaver to near extinction by the end of the nineteenth century. If current patterns are any indication, over time, many of the depopulated beaver ponds in the boreal zone completed a successional transition from pond to marsh to swamp. With the recovery of beaver populations, which have doubled since the late 1980s, many of those swamps are now flooded and returning to a marsh or open water state (Naiman et al. 1988; Jarema et al. 2009).

In addition to our historic effects on beaver pond dynamics, humans play a role in the current distribution of wetlands in the Canadian and Alaskan boreal zone. We have historically and continue to impact wetlands through damming, drainage, and infilling as part of various industrial developments in the boreal zone (Warner 2004; Brandt et al. 2013; Ménard et al. 2013). Since the early 1970s, an estimated 10,000 km^2 of boreal wetlands in Canada have been drained or disturbed for industrial or agricultural development (Brandt et al. 2013). Boreal wetlands have been relatively undisturbed in Alaska (Hall et al. 1994); however, the cumulative impacts of continued resource development and human-induced climate change on the future distribution and diversity of wetlands in the boreal zone is unknown.

Boreal Wetlands Ecosystem Services

Boreal wetlands deliver a variety of important ecological, social, and economic services that directly and indirectly benefit humans. These benefits are a result of the natural biological, physical, and chemical processes that maintain and sustain wetlands. The benefits that are derived from these ecological functions include: regulating (benefits that regulate ecological functions), provisioning (benefits that supply a good or product), cultural (benefits that support artistic, social, and/or historical activities and values), and supporting (benefits that sustain ecological functions; Millennium Ecosystem Assessment 2005).

Regulating Ecosystem Services – At the local and regional scale, wetland abundance and diversity influence rainfall and temperature patterns. At the global scale, boreal zone wetlands, especially peatlands, play a key role in regulating greenhouse gases such as methane and carbon dioxide and in buffering the impacts of climate change (Smith et al. 2007a). As mentioned previously, peatlands make up a significant amount of the Canadian and Alaskan boreal zone and store immense amounts of carbon. This large storage capacity exists because the cool and anoxic conditions in much of the boreal zone causes primary production to exceed decomposition of organic material, leading to a net accumulation of carbon (Smith et al. 2007a). Although the magnitude of the fluxes is small, it is compounded by the large surface area of boreal wetlands and has resulted in large stores of carbon that are currently sequestered in boreal zone wetlands. Based on long-term average carbon accumulation rates, Canadian peatlands store and sequester between 0.025 and 0.037 Gt of atmospheric carbon per year (Smith et al. 2007a), the equivalent to the annual emissions of 19–28 million cars (United States Environmental Protection Agency 2013). Although peatlands sequester and store carbon, they also lose carbon in the form of carbon dioxide from plant respiration and aerobic peat decomposition and as methane from the anaerobic decomposition of peat. Boreal peatlands are a significant source of methane emissions, estimated to currently contribute approximately 3.2 Tg methane year^{-1} to the atmosphere in Canada (Smith et al. 2007a). Methane is 21 times more effective at trapping radiation than carbon dioxide and thus processes that could potentially increase the anaerobic decomposition of peat (e.g., large scale permafrost thaw due to climate change, see below) could dramatically increase the release of methane and have substantial impacts on global carbon and methane cycles (Price et al. 2013).

Boreal wetlands also regulate local and regional hydrology regimes. The water retention properties of boreal wetlands helps maintain water flow during droughts and floods, and the ability to regulate water flow during storm-water peaks may help to control erosion. In addition, wetlands improve water quality by trapping sediments, nutrients, anthropogenic pollutants, and accumulating peat and other biomass deposits (Smith et al. 2007a).

Provisioning Ecosystem Services – Boreal wetlands provide fresh surface water and replenish ground water supplies for industrial (e.g., petroleum extraction) use and to a lesser extent for domestic and agricultural use. In addition, some boreal wetland plants and animals offer provisioning benefits such as food (e.g., fish; wild rice *Zizania* spp., waterfowl, berries, fiddlehead ferns *Matteuccia struthiopteris* (L.), moose *Alces alces*, woodland caribou *Rangifer tarandus caribou*, and mushrooms) and are sources for timber, fuel, fodder, and fur for domestic and commercial use (Ellanna and Wheeler 1986; Smith et al. 2007a). Biochemicals within boreal biota such as the polymer sphagnan extracted from *Sphagnum* peat moss (Stalheim et al. 2009) can provide the building blocks for medicinal, therapeutic, and other organic and chemical products. Boreal wetlands may also be critical habitat for reproduction and early survival of pollinating species of insects (Millennium Ecosystem Assessment 2005).

Cultural Ecosystem Services – Boreal wetlands are an important part of Canadian and American history and culture in relation to Aboriginal peoples and early exploration and settlement by Europeans. For thousands of years hunting, trapping, and gathering have been an essential part of life for the Aboriginal peoples in the boreal zone, and these traditional activities remain an integral part of Aboriginal culture (Smith et al. 2007a). Historically, boreal wetlands and riparian areas provided relatively predictable travel corridors and stable harvest sources for food, clothing, tools, and medicine. Today, boreal ponds, lakes, and wetlands are still important economic and subsistence sources for numerous Canadian and Alaskan Aboriginal communities and have great, sometimes sacred, spiritual significance to them as well as many other citizens (Ellanna and Wheeler 1986; Smith et al. 2007a).

In addition to providing spiritual and inspirational benefits, boreal wetlands, due to their aesthetic qualities and high degree of biodiversity, provide opportunities for recreational activities that include canoeing, hunting, hiking, fishing, trapping, and bird watching. They also provide a unique "classroom" for educating people about wetland ecology and the local, regional, and global services that wetlands provide (Cimon-Morin et al. 2013).

Supporting Ecosystem Services – Boreal wetlands are rich in biodiversity and provide habitat for hundreds of species of plants and animals. For example, ~26 million waterfowl representing 35 species and ~7 million shorebirds representing 19 species are estimated to use boreal wetlands in Canada and Alaska as migratory stop over or breeding habitat (Wells et al. 2011). Many plant and animal species are considered wetland specialists (e.g., bog katydid *Sphagniana sphagnorum*) and in Canada some of these specialists are rare, threatened, or endangered (e.g., woodland caribou).

Valuation of Ecosystem Services – Placing a dollar value on all of the above mentioned ecosystem services is a considerable challenge, because only some of the provisioning services have measurable market value that is occasionally tracked. For example, wild rice, the only commercial cereal crop that grows wild in Canada, was valued at $3.5 million for the 2003 harvest of 1,013 tons (Smith et al. 2007a). *Sphagnum* peat moss is also commercially harvested and revenues equalled $170 million in Canada in 1999 (Smith et al. 2007a). Regulating, supporting, and cultural ecosystem services are difficult to value in monetary terms but preliminary estimates by Anielski and Wilson (2009) place the nonmarket value of annual: (1) carbon storage in wetlands (including peatlands) at $401.9 billion; (2) replacement cost value of peatland carbon sequestration at $383 million; (3) flood control, water filtering, and biodiversity for wetlands (excluding peatlands) at $33,724 million; and (4) flood control and water filtering for peatlands at $76,998 million.

Potential Sources of Wetland Alteration

The Canadian and Alaskan boreal forests are rich in natural resources including peat, minerals, hydropower, trees, and oil and gas. These resources are being extracted to varying degrees across the boreal forest, and extraction activities bring landscape changes ranging from complete removal or inundation of the biotic layer to clear cuts to development of complex systems of linear features (e.g., roads, pipelines, and seismic lines). Currently, it has been estimated that less than 730,000 km^2 of the boreal zone in Canada (11%) has been impacted by forestry, mining, petroleum, and hydropower (Wells et al. 2011). To present, the boreal zone in Alaska has been minimally impacted by natural resource development (Hall et al. 1994). However, given the density of wetlands, and the intensity of development in some regions of the boreal zone, it is likely that industrial activities currently intersect with wetlands and will continue to do so (Cimon-Morin et al. 2015a). These interactions may alter hydrology and fragment landscapes, influencing water regimes, wetland productivity, biodiversity, and ecosystem services (Poulin et al. 2004; Foote and Krogman 2006; Smith et al. 2007a; Slattery et al. 2011; Wells et al. 2011; Price et al. 2013; Kreutzweiser et al. 2013).

Potential local scale alterations to wetlands include:

Direct loss of wetlands – Topsoil removal during strip mining (e.g., bitumen, coal, peat, ore, diamonds) and petroleum extraction results in the complete removal and/or disruption of wetland functions. In the boreal zone of Alberta an estimated 1,500 km^2 of wetlands have or will be lost as a result of oil sand strip mining (Wells et al. 2011). Inundation has the same effect and is much less recoverable. Ninety percent of wetland loss in the boreal zone, particularly in the eastern boreal zone, is caused by flooding due to the construction of hydroelectric dams (e.g., 120 thousand ha of peatland in Quebec; Poulin et al. 2004). Conversion of wetlands for agriculture (e.g., cereal, fruit, and vegetable crop land) has been occurring since European settlement, especially in the southern boreal transition zone where as much as 73% of the zone has been converted for agricultural purposes in some areas (Hobson et al. 2002).

Damming and redistribution of water – Roads (e.g., see Fig. 9) and pipelines can create blockages to flowing systems and alter water flow patterns and rates. For example, building a road through a fen may require a ditch to redirect water towards a culvert which then creates open water where none existed. If the road is built perpendicular to the fen, water flow may become blocked resulting in the water table rising upstream and falling downstream causing changes to the vegetation and wildlife community on both sides of the road. In addition, when the water table is reduced, there is an increase in the air-filled porosity of the peat, which allows the rate of oxygen-limited decomposition to increase leading to release of large amounts of carbon dioxide. Methane may also be released as a result of flooding (Kreutzweiser et al. 2013).

Soil compaction – Physical compaction of soil, particularly in peatlands, by industrial machinery decreases its porosity and reduces its ability to conduct and retain water and potentially its capacity to store carbon. Compaction of soils due to the use of heavy machinery is a particular concern during forestry, petroleum extraction, and mining operations (Graf 2009).

Edge effects/fragmentation – The removal of vegetation during forestry (e.g., average of 165 million m^3 of harvestable timber per year from 2002 to 2011 in Canada; National Forestry Database 2013) and other industrial operations and the creation of linear features has been documented to divide once contiguous

Fig. 9 Roads are an example of landscape changes used by many industries operating in the boreal forest. While efforts are made to minimize road effects on wetlands, these are primarily directed at the most dynamic flowing systems. Often, less dynamic systems are not recognized as wetlands and so no allowance is made to account for associated hydrodynamics (Photo credit: © Ducks Unlimited Canada)

habitat into smaller, sometimes isolated fragments. The fragmentation of boreal zone wetlands and surrounding forest habitat may impact both biotic (e.g., predation) and abiotic (e.g., evaporation) processes (Boulet et al. 2000; Fahrig 2003).

Water contamination – Point source chemical or organic contamination of boreal forest wetlands can occur as a result of leaks from mine tailings ponds, pipeline ruptures, seepage and leaks from facilities, camps, or sumps, seepage and leaks during horizontal drilling, or leaks or spills during transportation. In Alberta, from 1990 to 2012, over 15,000 pipeline leaks and 880 ruptures were reported; the majority resulting in a release of less than 100 m^3 of liquid (Alberta Energy Regulator 2013a). However, some releases can be quite large (e.g., 9,500 m^3 of water containing hydrocarbons was released from a ruptured pipeline into 42 ha of peatland in northern Alberta in May 2013; Alberta Energy Regulator 2013b). Nonpoint sources of boreal forest contamination can occur as a result of erosion and runoff and as a result of atmospheric deposition of various airborne contaminants. Acid rain has resulted in large scale impacts on vegetation and wildlife in the eastern boreal zone and, with increasing industrial development, is becoming more of a concern in the western boreal zone (Kreutzweiser et al. 2013).

Water removal – Although the mining and petroleum extraction industries recycle much of the water they use, water that is removed is not returned to the original environment. Water removal can influence wetland hydrologic function on a small and large scale. For example, during strip mining, groundwater is pumped out from the deposit and surrounding areas to decrease water pressure and to prevent or slow water seepage. This process lowers the water table in the surrounding area and causes drying of nearby wetlands (Wells et al. 2011). During enhanced oil recovery, oil is either extracted by injecting water directly into remaining oil pools to increase fluid pressure or is injected as steam into oilsands deposits or heavy oil pools (Alberta Government 2013a). In Alberta, this method of oil extraction uses as much as 65 million m^3 (comparable to the volume in Canada's Lake Athabasca) of ground and surface fresh water (including wetlands) per year (Alberta Government 2013a). The impact of large scale water removal on boreal wetland ecological functions and associated ecosystem services is not clearly understood.

Water yield – Evidence suggests that the removal of vegetation through timber harvesting in watersheds with shallow soils and impermeable bedrock may increase runoff, resulting in changes to the water budgets of wetlands. Timber harvesting in peatlands with deep soil deposits and low relief may not affect runoff but may cause an increase in net radiation from the soil and in evaporation due to increased wind exposure. However, the net effects of vegetation removal on water yield are highly variable and only demonstrable when harvesting exceeds 20% of the basin area (Buttle and Murray 2011).

Permafrost alteration – Permafrost is an important source of water and carbon storage; the world's permafrost contains twice as much carbon in the form of frozen organic matter than what exists in today's atmosphere (Shaefer et al. 2012). In discontinuous permafrost regions, removal of vegetation associated with timber harvest, seismic lines, roads, and other industrial activities can warm ice-rich soils causing subsidence, altering water flows, and triggering the conversion of permafrost plateaus into bogs or channel fens. Once disturbed, permafrost is unlikely to reform, resulting in transformation of the landscape (Price et al. 2013).

Climate change – Warming has occurred in Canada's boreal zone in the last 50 years with mean annual temperatures increasing 0.5 °C in the east and 2.0 °C in the west, and precipitation increasing 10–20% in much of the boreal zone (Price et al. 2013). Evidence suggests that this trend will continue well into the next century with northern regions experiencing the most amount of warming with temperatures predicted to increase by 7.5–10.0 °C in the winter and 3.5–6.0 °C in the summer by 2100 (Price et al. 2013). In addition to the direct landscape alterations by human activities, climate change is thought to influence the amount and type of boreal wetland biodiversity, particularly in regions with permafrost. Warming of permafrost in the western boreal zone has been steadily increasing since the 1970s but warming in the eastern boreal zone has accelerated since it was first recorded in the 1990s (Price et al. 2013). In boreal Alaska, Riordan et al. (2006) observed up to ~30% decline in water surface area over a 50-year period (1950–2002), attributed to terrestrialization, encroachment of floating vegetation, and accumulation of organic matter in shallow water wetlands, resulting in the transition of these wetlands into peatlands (Roach et al. 2011). In the same region, Corcoran et al. (2009), observed substantial differences in contemporary and historic invertebrate communities of shallow open water wetlands, which they attributed to the effects of climate change. The presence and extent of similar changes to boreal wetlands in Canada are just starting to be examined (Kreutzweiser et al. 2013; Price et al. 2013).

While potential direct local scale impacts of industrial activities are starting to be better understood, only a few studies (e.g., Schindler and Smol 2006; Schindler and Lee 2010) have characterized cumulative effects at larger spatial (e.g., watershed) and temporal scales. The impact of landscape changes due to industrial activities and climate change on actual or functional loss of boreal wetlands has not been fully quantified at local or regional levels or across longer time periods (Kreutzweiser et al. 2013). Evidence exists from nonboreal wetlands that biodiversity may be negatively correlated with increased road density (Findlay and Houlahan 1997), although patterns might take years to decades to be observed (Findlay and Bourdages 2000).

In many areas, even highly altered regions, industrial development of the boreal landscape is considered in its infancy (Foote and Krogman 2006; Wells et al. 2011; Kreutzweiser et al. 2013). What this means for the future of boreal wetlands is uncertain. However, the degree of overlap between natural resources and Canadian boreal wetlands is likely large, and so there is a clear need to better understand how best to manage development in this wetland rich landscape.

Conservation Status and Current Efforts

The Canadian and Alaskan boreal zone spans nearly the width of a continent, crossing 1 state and 10 provinces and territories (Brandt 2009, Fig. 1). Wetland conservation efforts vary by jurisdiction but essentially fall into: (a) establishing protected areas; (b) using practices that avoid or minimize net wetland loss and degradation; and (c) restoring and reclaiming to recover lost wetland function. Typically, these efforts align with federal and state or provincial/territorial regulatory frameworks and corporate policies directed at environmentally sustainable land use.

Protected Areas – Across the Canadian boreal, there are or have been several forms of legislated protected area processes at the Federal (e.g., Northwest Territories Protected Areas Strategy), Territorial (e.g., North Yukon Land Use Plan), or Provincial (e.g., Quebec Plan Nord, Ontario Far North Land Use Planning Initiative) levels. Similar processes and plans are being developed or have been implemented in Alaska as well (e.g., Yukon Tanana Area Plan). These processes focus on securing large protected areas, typically in the thousands to millions of hectares. Because wetlands represent large portions of these protected areas, previous or current protected area plans by default conserve wetlands and the watersheds that support them. Indeed, conservation of wetlands and therefore conservation of the biodiversity and ecosystem services that wetlands provide (Cimon-Morin et al. 2015b) has been a core justification for protecting some areas. A challenge with these efforts is that securing protected status is often a long process (>10 years), and commitments made at the start may not be realized in the end, resulting in fewer or smaller protected areas. In addition to these protected area processes, territorial and provincial land use planning frameworks can result in the creation of special management zones, which provide varying degrees of protection for wetlands, ranging from no access for industrial development to access under defined conditions. Furthermore, green certification programs associated with the forestry industry require the conservation of high environmental value forests, which can result in the establishment of "no go zones" for certified companies. However, other land users may not recognize these areas as off-limits, and so the effectiveness of this form of protection is unknown. As of 2014, about 110,030 km^2 of Canadian boreal wetlands are in areas with permanent protection and 62,170 km^2 in areas under interim protection (Ducks Unlimited Canada, unpublished data). With an estimated 84% of Canadian boreal wetlands falling outside of permanent or interim protected areas, there is a clear need to better understand potential effects of industrial activities and to develop additional conservation approaches where existing legislation and practices are insufficient.

Best Management Practices – Best management practices may be legislatively required or voluntarily adopted practices that aim to minimize environmental and social impacts of industrial resource activities including forest harvesting, oil and gas, hydroelectric, and mining. In some cases, these practices are enforced by legislation or outlined in government or corporate policy. For example, the forestry industry has long been subject to state and provincial legislation and policies around riparian management, in block road placement, and use of pesticides, with the intent to manage impacts on open water wetlands (Lee and Barker 2005). However, these rules are primarily in place to protect fisheries habitat and often do not address peatlands or swamps. In addition, land users often do not recognize peatlands or swamps as wetlands in need of protection. This oversight can lead to considerable risk to peatlands and swamps, particularly as related to road construction. The petroleum industry is becoming increasingly subject to policies similar to that of the forest industry, with the same risks. Fortunately, partnerships are forming between conservation organizations and industrial companies to further develop practices that meet or exceed government regulations. In addition, while practices allowable under green certification programs for forestry are aimed at conservation of uplands, they still provide some measure of benefit to wetlands in that they are designed to avoid or minimize environmental impacts such as erosion and sedimentation, compaction, permafrost degradation, hydrologic alteration, and contamination. However, not all land users are subject to following these practices, making the net benefit in terms of risk to wetlands difficult to quantify.

Restoration/Reclamation – Some jurisdictions in Canada require restoration – the process of recovering disturbed lands to predisturbance ecological function, physical, chemical, and biological characteristics – or reclamation – the process of improving disturbed land to achieve land capability equivalent to the predisturbance condition or for a specified end land use – of some lands disturbed by industry. In Alberta, progressive reclamation, reclamation activities completed in stages over a specified amount of time, is a standard operating condition for all petroleum extraction operations and includes the reclamation of impacted boreal peatlands. However, in some situations and in some areas, boreal peatlands are allowed to be reclaimed as forest lands, resulting in a net wetland loss (Rooney et al. 2012, but see Jutras et al. 2006). In addition, only 0.15% of land disturbed by oil sands mining operations has been certified as reclaimed (Alberta Government 2013b). Restoration and reclamation of these wetlands is complex, and past reclamation activities in the boreal have had limited success (e.g., Graf 2009). However, restoration and reclamation of disturbed lands has become the focus of much research in Canada and is supported primarily by governments and by mining and petroleum industries.

Other Federal, Provincial, and Territorial policies – In Canada, only two of the ten jurisdictions containing portions of the boreal zone have policies specifically designed to protect wetlands (Alberta and Saskatchewan). Many other jurisdictions have policies that protect, or have the potential to protect, wetlands including land use planning frameworks and industry specific regulations as described above. Other jurisdictions are in the process of developing wetland policies. Canada has a federal wetland policy, which is applicable to all federally owned lands or projects where federal money is invested (e.g., road construction). However, the federal wetland policy is not consistently applied across Canada. In the United States, the federal government has adopted a policy of "no net loss" of wetlands, which is a shift from past federal strategies and regulations that promoted wetland conversion to current legislation and incentives that discourage or prevent wetland conversion and promote wetland protection, restoration, enhancement, and creation. Wetlands in Alaska receive protection afforded under section 404 of the federal Clean Water Act which regulates the dredging and filling of wetlands.

Future Challenges

The Canadian and Alaskan boreal zone may arguably encompass the world's largest and most important cumulative wetland and deep water areas and presents one of the world's greatest conservation opportunities. Wetlands within this zone are diverse in form and function and offer a variety of ecosystem services that are important at local, regional, and global scales. With abundant natural resources also present in this area, a future challenge will be to ensure resource development occurs in a manner that safeguards the long-term viability of these complex areas. Riparian and wetland forest ecology and management has been a favorite research domain of plant and wildlife scientists and a fertile ground for policy influence at provincial and territorial levels (Morissette and Donnelly 2010). These studies have helped develop some industrial practices focussed on sustainable management. Although great progress has been made regarding knowledge of boreal wetland distribution, ecology, management, and conservation, a lot remains to be done. Collaboration of all people living, studying, harvesting/extracting, and managing the boreal is required to conserve in perpetuity boreal wetlands. Partnerships such as the Canadian Boreal Forest Agreement (CBFA 2010) provide examples of how economic development with strong environmental stewardship objectives may be achieved.

References

Alberta Energy Regulator. Report 2012-B: pipeline performance in Alberta, 1990–2012. 2013a. http://www.aer.ca/documents/reports/R2013-B.pdf. Accessed: 4 Mar 2014.

Alberta Energy Regulator. ERCB updates volume spilled on pipeline incident near Zama City. 2013b. http://www.aer.ca/about-aer/media-centre/news-releases/news-release-2013-06-12-nr2013-13. Accessed 4 Mar 2014.

Alberta Government. Water used for oilfield injection purposes. 2013a. http://environment.alberta.ca/01729.html. Accessed 3 Mar 2014.

Alberta Government. Oil sands reclamation. 2013b. http://www.oilsands.alberta.ca/FactSheets/Reclamation_FSht_Sep_2013_Online.pdf. Accessed 7 Mar 2014.

Anielski M, Wilson S. Counting Canada's natural capital: assessing the real value of Canada's boreal ecosystems. Ottawa: Canadian Boreal Initiative; 2009.

Boggs K, Boucher TV, Kuo TT, Fehringer D, Guyer S. Vegetation map and classification: northern, western and interior Alaska. Anchorage: Alaska Natural Heritage Program, University of Alaska Anchorage; 2012. p. 88.

Boulet M, Darveau M, Bélanger L. A landscape perspective on bird nest predation in a managed boreal black spruce forest. Écoscience. 2000;7:281–9.

Brandt JP. The extent of the North American boreal zone. Environ Rev. 2009;17:101–61.

Brandt JP, Flannigan MD, Maynard DG, Thompson ID, Volney WJA. An introduction to Canada's boreal zone: ecosystem processes, health, sustainability, and environmental issues. Environ Rev. 2013;21(4):07–26.

Bridgham SD, Megonial JP, Keller JK, Bliss NB, Trettin C. The carbon balance of North American wetlands. Wetlands. 2006;26:889–916.

Buttle JM, Murray CD. Hydrological implications of forest biomass use. Final Report. Prepared for Environment Canada. 2011. https://www.trentu.ca/iws/documents/ContractK2A13-10-0033BiomassLCAHydrology_FinalReport.pdf April 4 2014. Accessed 7 Mar 2014.

Carter V. Technical aspects of wetlands wetland hydrology, water quality, and associated functions. In: National water summary–wetland resources. Washington, DC: U.S. Geological Survey; 1997. p. 35–48. U.S. Geological Survey Water-Supply Paper 2425.

CBFA. Canadian boreal forest agreement: an historic agreement signifying a new era in the boreal forest. 2010. p. 49. http://cbfa-efbc.ca/wp-content/uploads/2014/12/CBFAAgreement_Full_NewLook.pdf. Accessed 8 Mar 2014.

Cimon-Morin J, Darveau M, Poulin M. Fostering synergies between ecosystem services and biodiversity in conservation planning: a review. Biol Conserv. 2013;166:144–54.

Cimon-Morin J, Darveau M, Poulin M. Consequences of delaying conservation of ecosystem services in remote landscapes prone to natural resource exploitation. Landscape Ecol. 2015a; First Online 06 Oct 2015:1–18.

Cimon-Morin J, Darveau M, Poulin M. Site complementarity between biodiversity and ecosystem services in conservation planning of sparsely-populated regions. Environ Conserv. 2015b; Online: 1–13.

Corcoran RM, Lovvorn JR, Heglund PJ. Long-term change in limnology and invertebrates in Alaskan boreal wetlands. Hydrobiologia. 2009;620:77–89.

Devito K, Mendoza C, Qualizza C. Conceptualizing water movement in the Boreal Plains. Implications for watershed reconstruction. Synthesis report prepared for the Canadian Oil Sands Network for Research and Development. Environmental and Reclamation Research Group; 2012. p. 164.

Ellanna LJ, Wheeler PC. Subsistence use of wetlands in Alaska. In: Alaska Regional Wetland Functions – Proceedings of a Workshop. Amherst: The Environmental Institute, Univ. of Massachusetts; 1986. p. 85–103.

Fahrig L. Effects of habitat fragmentation on biodiversity. Annu Rev Ecol Evol Syst. 2003;34:487–515.

Findlay SC, Bourdages J. Response time of wetland biodiversity to road construction on adjacent lands. Conserv Biol. 2000;14:86–94.

Findlay SC, Houlahan J. Anthropogenic correlates of species richness in southeastern Ontario wetlands. Conserv Biol. 1997;11:1000–9.

Foote L, Krogman N. Wetlands in Canada's western boreal forest: agents of change. Forest Chronicle. 2006;82:825–33.

Graf M. 2009. Literature review on the restoration of Alberta's boreal wetlands affected by oil, gas, and in situ oil sands development. Prepared for Ducks Unlimited Canada. http://www.biology.ualberta.ca/faculty/stan_boutin/ilm/uploads/footprint/Graf%20Wetland_Restoration_Review%20FINAL-Small%20File.pdf. Accessed 3 Mar 2014.

Gorham E, Lehman C, Dyke A, Janssens J, Dyke L. Temporal and spatial aspects of peatland initiation following deglaciation in North America. Q Sci Rev. 2007;26:300–1.

Hall JV, Frayer WE, Wilen BO. Status of Alaska wetlands. Washington, DC: U.S. Department of the Interior, Fish and Wildlife Service; 1994. p. 36.

Halsey LA, Vitt DH, Gignac LD. Sphagnum-dominated peatlands in North America since the last glacial maximum: their occurrence and extent. Bryologist. 2000;103:334–52.

Hobson KA, Bayne EM, Van Wilgenburd SL. Large-scale conversion of forest to agriculture in the boreal plains of Saskatchewan. Conserv Biol. 2002;16:1530–48.

Jarema SJ, Samson J, McGill BJ, Humphries MM. Variation in abundance across a species' range predicts climate change responses in the range interior will exceed those at the edge: a case study with North American beaver. Global Change Biol. 2009;15:508–22.

Johnston CA. Wetland habitats of North America: ecology and conservation concerns, Chapter 12. Los Angeles: University of California Press; 2012.

Jones MC, Yu Z. Rapid deglacial and early Holocene expansion of peatlands in Alaska. Proc Natl Acad Sci U S A. 2010;107:7347–52.

Jutras S, Plamondon AP, Hökkä H, Bégin J. Water table changes following precommercial thinning on post-harvest drained wetlands. Forest Ecol Manag. 2006;235:252–9.

Kreutzweiser DP, Beall FD, Webster KL, Thompson DG, Creed IF. Impacts and prognosis of natural resource development on aquatic biodiversity in Canada's boreal zone. Environ Rev. 2013;21(4):227–59.

Lee P, Barker T. Impact of riparian buffer guidelines on old growth in western boreal forests of Canada. Forestry. 2005;78:263–78.

Ménard S, Darveau M, Imbeau L. The importance of geology, climate and anthropogenic disturbances in shaping boreal wetland and aquatic landscape types. Écoscience. 2013;20: 399–410.

Millennium Ecosystem Assessment. Ecosystems and human well-being: wetlands and water synthesis. Washington, DC: World Resources Institute; 2005.

Morissette J, Donnelly M. Riparian areas: challenges and opportunities for conservation and sustainable forest management. Edmonton: KETE Report, Sustainable Forest Management Network; 2010.

National Forestry Database. 2013. Forest products- background. http://nfdp.ccfm.org/products/background_e.php. Accessed 3 Mar 2014.

National Wetlands Working Group. The Canadian wetland classification system. In: Warner BG, Rubec CDA, editors. Waterloo: Wetlands Research Centre. 2nd ed. University of Waterloo; 1997.

Naiman RJ, Johnston CA, Kelley JC. Alteration of North American streams by beaver. BioScience. 1988;38:753–62.

Payette S, Delwaide A, Caccianiga M, Beauchemin M. Accelerated thawing of subarctic peatland permafrost over the last 50 years. Geophys Res Lett. 2004;31: Online L18208.

Price JS, Branfireun BA, Waddington JM, Devito KJ. Advances in Canadian wetland hydrology, 1999–2003. Hydrol Process. 2005;19:201–14.

Price DT, Alfaro RI, Brown KJ, Flannigan MD, Fleming RA, Hogg EH, Girardin MP, Lakusta T, Johnston M, McKenney DW, Pedlar JH, Stratton T, Sturrock RN, Thompson ID, Trofymow JA, Venier LA. Anticipating the consequences of climate change for Canada's boreal forest ecosystems. Environ Rev. 2013;21:322–65.

Poulin M, Rochefort L, Pellerin S, Thibault J. Threats and protection for peatlands in eastern Canada. Geocarrefour. 2004;79:331–44.

Richard P, Grondin P. Histoire postglaciaire de la végétation. In: Ordre des ingénieurs forestiers du Québec, editor. Manuel de foresterie. 2e éd. Québec: Éditions MultiMondes; 2009. p. 170–6.

Riordan B, Verbyla D, McGuire DA. Shrinking pongs in subarctic Alaska based on 1950–2002 remotely sensed images. J Geophys Res. 2006;111. G04002. doi:10.1029/2005JG000150.

Roach J, Griffith B, Verbyla D, Jones J. Mechanisms influencing changes in lake area in Alaskan boreal forest. Global Change Biol. 2011;17:2567–83.

Rooney RC, Bayle SE, Schindler DW. Oil sands mining and reclamation cause massive loss of peatland and stored carbon. Proc Natl Acad Sci U S A. 2012;109:4933–7.

Schindler DW, Lee PG. Comprehensive conservation planning to protect biodiversity and ecosystem services in Canadian boreal regions under a warming climate and increasing exploitation. Biol Conserv. 2010;143:1571–86.

Schindler DW, Smol JP. Cumulative effects of climate warming and other human activities on freshwaters of Arctic and Subarctic North America. J Human Environ. 2006;35(4):160–8.

Shaefer K, Lanuite H, Romanovsky V, Shuur E, Gartner-Roer I. Policy Implications of warming permafrost. United Nations Environment Programme. 2012. http://www.unep.org/pdf/permafrost.pdf. Accessed 27 Feb 2014.

Slattery SM, Morissette JL, Mack GG, Butterworth EW. Waterfowl conservation planning: science needs and approaches. In: Wells JV, editor. Boreal birds of North America: a hemispheric view of their conservation links and significance. Studies in Avian Biology (no. 41). Berkeley, CA: University of California Press; 2011. p. 23–40.

Smith C, Morissette J, Forest S, Falk D, Butterworth EW. Synthesis of technical information on forest wetlands in Canada. Research Triangle Park: National Council for Air and Stream Improvement, Inc.; 2007a. Technical Bulletin No. 938.

Smith KB, Smith CE, Forest SF, Richard AJ. A field guide to the wetlands of the boreal plains Ecozone of Canada. Edmonton: Ducks Unlimited Canada, Western Boreal Office; 2007b. p. 98.

Stalheim T, Ballance S, Christensen BE, Granum PE. Sphagnan – a pectin-like polymer isolated from *Sphagnum* moss can inhibit the growth of some typical food spoilage and food poisoning bacteria by lowering the pH. J Appl Microbiol. 2009;106:967–76.

Tarnocai C, Kettles IM, Lacelle B. Peatlands of Canada Database. Geological Survey of Canada, Open File 4002. 2002.

United States Environmental Protection Agency. 2013. Greenhouse Gas Equivalencies Calculator. 2013. http://www.epa.gov/cleanenergy/energy-resources/calculator.html#results. Accessed 4 Mar 2014.

Warner BG. Geology of Canadian wetlands. Geosci Can. 2004;31:57–68.

Wells JV, Roberts D, Lee P, Cheng R, Darveau M. A forest of blue: Canada's Boreal Forest, the world's water keeper. Washington, DC: Pew Environment Group; 2011.

Yukon-Kuskokwim Delta: Yukon River Basin, Alaska (USA)

42

Frederic A. Reid and Daniel Fehringer

Contents

Introduction	544
Hydrology	544
Wetlands	545
Biodiversity	547
Conservation Status	547
References	548

Abstract

The combined Delta of the Yukon and Kuskokwim Rivers of Western Alaska create the largest expanse of wetlands on North America's west coast. The dendritic streams and rivers mingle with a myriad of lakes and wetlands and create a lush 11 million ha delta in the poorly drained coastal plain. The intertidal mudflats of the delta cover more than 311,000 ha. The diversity and biomass of breeding waterfowl, shorebirds, and other waterbirds is among the highest in the world.

Keywords

Yukon river · Kuskokwim river · Wetlands · Coastal plain · Western alaska · Yukon-Kuskokwim Delta National Wildlife Refuge · Waterfowl · Shorebirds · Salmon

F. A. Reid (✉) · D. Fehringer
Ducks Unlimited, Inc., Rancho Cordova, CA, USA
e-mail: freid@ducks.org; dfehringer@ducks.org

Introduction

The combined Delta of the Yukon and Kuskokwim Rivers (Fig. 1) of western Alaska creates the largest expanse of wetlands on North America's west coast. The dendritic streams and rivers mingle with a myriad of lakes and wetlands and create a lush 11 million ha delta in the poorly drained coastal plain. Within the coastal plain, the two rivers converge to within 40 km of each other, but then diverge and each empty into the Bering Sea some 640 km apart from each other. At 3,185 km, the Yukon is North America's third longest river, with only the Mississippi/Missouri and Mackenzie/Peace rivers longer (Hall et al. 1994). The Yukon, which means "great river" in Gwich'in, is the largest river or watershed in Alaska and Yukon territories. At 1,150 km, the Kuskokwim River is the nation's longest free-flowing river. The intertidal mudflats of the delta cover more than 311,000 ha. The diversity and biomass of breeding waterfowl, shorebirds, and other waterbirds is among the highest in the world.

Hydrology

The headwaters of the Yukon River arise in British Columbia and Yukon territories. Major tributaries in the Yukon include the Teslin, Pelly, White, and Fortymile Rivers. Major tributaries in Alaska include the Charley, Porcupine, Black, Chandalar, Tanana, Tozitna, Nowitna, Melozitna, Koyukuk, and Innoko Rivers.

Fig. 1 Location of the Yukon Delta National Wildlife Refuge in Alaska on the coastal plain delta of the Yukon and Kuskokwim Rivers

The watershed drainage is 850,000 km² in size (Brabets et al. 2000), with 60% in Alaska and 40% in Canada, which is roughly equivalent to the size of Texas. The average flow of the river is 6,400 m³/s.

The Kuskokwim River (or Kusquqvak in Central Yup'ik) is the ninth largest river in the United States by average discharge volume at its mouth (20,420 m³/s) and 17th largest by watershed drainage area (124,000 km²) (Kammerer 1990). The river's source is at the North Fork near Medfra.

Wetlands

The majority of this ecoregion consists of tens of thousands of lakes, herbaceous wetlands (freshwater and coastal), and thermokarst features set in a matrix of peat plateaus carpeted by dwarf shrub and lichen. Low and tall shrub wetlands dominate riparian areas. Trees, including spruce, poplar, and willow, are limited to the banks of major rivers.

An earthcover mapping project was initiated in 2005 and completed in 2011, using Landsat images from summer 2002 to 2005 (Ducks Unlimited 2011). The study area consisted of 10.9 million ha, which included the Y-K Delta and the boreal forest areas of the Nulato Hills. The classification system is based on the Alaska Earth Cover scheme, derived from Viereck et al. (1992).

Forest cover types made up 6% of the area, with closed deciduous (<60% treed) and woodland needleleaf composing more than 70% of the overall class. Forest cover was found adjacent to floodplains of larger rivers and the Nulato Hills. Paper birch *Betula papyrifera* and poplar were the most common trees.

Tall/low shrub cover types made up 15% of the area, with willow *Salix* spp. and alder *Alnus* spp. dominating and sweetgale *Myrica gale* as a subdominant. Tall shrubs were found primarily in the floodplains, hills, or volcanic slopes, less so in the lowlands.

Dwarf shrub cover types made up 36% of the area with the majority found in peat lowlands. Permafrost was near the surface under most of these poorly drained lowland dwarf shrub communities. Common shrubs included dwarf birch and ericaceous shrubs such as *Ledum*, *Vaccinium*, and *Empetrum*. Herbaceous species often included *Eriophorum*, *Carex aquatilis*, *Calamagrostis*, *Rubus chamaemorus*, and *Petasites frigidus*. Dwarf shrub lichen peatland was the most extensive community in the lowlands. Vast stretches of flat to gently rolling peat plateaus covered with dwarf shrub lichen extend southward from the floodplain of the Yukon River toward the coast and from the western slopes of the Nulato Hills to the coast. Dwarf shrub peatland communities tended to occur in areas less favorable to the growth of lichen, such as in young floodplain basins, adjacent to drainages, or in river bends near the coastal plain – all areas prone to occasional flooding. Dwarf shrub peatland was also found on large hummocks or plateaus raised up out of wetland bog areas by thermokarst action. Coastal dwarf shrub was observed in the brackish zone of the coastal meadows above the tidal marshes.

Moss/graminoid class was defined as having <25% water, <25% shrub cover, and >40% herbaceous cover and made up 3% of the landcover. This community was common in drained lake basins and other depressions on saturated peat soils in the lowlands and floodplains. *Sphagnum* moss cover was often continuous.

Wet graminoid communities occurred throughout the lowlands in wet areas around the edges of lakes, in drained basins, in water tracks, and on floodplains where wetlands form in wet depressions, oxbows, and abandoned channels. This community made up 6.6% of the scene and is seasonally flooded. The dominant sedges were *Carex aquatilis* and *Eriophorum angustifolium* with other sedges present. Marsh species may be present, but not dominant.

Freshwater emergent marsh was mapped throughout the lowlands in small patches in and around lakes and in wet areas of floodplains. This class included graminoid-dominated marshes as well as forb-dominated marshes and made up 5.3% of the area. *Carex utriculata* and *Arctophila fulva* were dominant in the marsh communities of the Y-K Delta.

Coastal herbaceous communities (Fig. 2) made up 3% of the landcover. The lower coastal salt marsh is dominated by *Carex ramenskii* and/or *C. subspathacea* and is regularly inundated by tides. The upper coastal salt marsh occurs at a slightly higher elevation; this cover type is drier and is flooded less frequently, usually only at extreme high tides or storm surges. *C. rariflora* is the dominant sedge with other grasses and sedges present. Coastal graminoid occurs on upper beaches or natural levees bordering tidal sloughs. *Leymus mollis* was generally the principal grass, but other common grasses included *Poa eminens* and *Calamagrostis deschampsoides*.

Water made up 21% of the scene, while Barren (rock, gravel, sandbars) and other made up 3%.

Fig. 2 Yukon-Kuskokwim River Delta brackish marsh (Photo credit: D. Fehinger, Ducks Unlimited)

Biodiversity

The Yukon and Kuskokwim have viable salmon runs for five species of salmon. The Yukon River has one of the world's longest runs with adults reaching spawning beds in tributaries in Alaska, Yukon, and British Columbia. Villages along the Delta have historically dried, smoked, and frozen fish for both human and sled dog consumption.

The wetlands provide unique habitat for waterbirds. Historically one to two million ducks breed in the Delta. Nearly the entire population of emperor *Chen canagica* and cackling *Branta hutchinsii* geese nest in a narrow zone of coastal habitat and half of the continental population of black brant *Branta bernicla nigricans* (King and Lensink 1971). A large proportion of the Pacific flyway white-fronted goose *Anser albifrons* and tundra swan *Cygnus columbianus* populations also nest and stage on the Delta. Wrangle Island lesser snow geese *Chen caerulescens* migrate through the Delta on route to Russia.

Dramatic declines in breeding sea ducks, as well as brant and cackling geese, may be related to warmer winter conditions, less snow cover, and greater survival of predatory red fox (*Vulpes vulpes*). Population estimates of long-tailed duck have declined by nearly 50% over the past 30 years (Schamber et al. 2009). Nest success averaged 30%, while duckling survival to 30 days old averaged only 10%. Average population decline was estimated at 19% (Schamber et al. 2009). Number of spectacled eiders nesting near the Kashunuk River in the Delta has declined over 75% in the last 20 years and was linked to red fox predation (Ely et al. 1994).

The Delta provides breeding habitat for approximately 14 million shorebirds. Dominant shorebird species include bar-tailed godwits *Limosa lapponica*, dunlins *Calidris alpina*, western sandpipers *C. mauri*, northern phalaropes *Phalaropus lobatus*, red phalaropes *P. fulicarius*, black turnstones *Arenaria melanocephala*, and ruddy turnstones *A. interpres* (King and Lensink 1971). Lesser sandhill crane *Grus canadensis* and arctic *Gavia arctica* and red-throated loons *Gavia stellata* are dominant nesting species. Coastal habitats provide staging area for bristle-thighed curlews and whimbrels during late summer and in migration (King and Lensink 1971).

Riparian habitats support some of the only North American populations of primarily Eurasian breeding landbirds like yellow wagtail *Motacilla tschutschensis*, bluethroat *Luscinia svecica*, and northern wheatear *Oenanthe oenanthe*, as well as high-priority North American breeding landbirds like gray-cheeked thrush *Catharus minimus*, rusty blackbird *Euphagus carolinus*, blackpoll warbler *Setophaga striata*, and golden-crowned sparrow *Zonotrichia atricapilla* (Harwood 1999).

Conservation Status

The Yukon River or its tributaries flow through several protected lands that include Innoko, Nowitna, and Yukon Flats National Wildlife Refuges and Yukon-Charley Rivers National Preserve, before flowing into the Yukon Delta National Wildlife

Refuge. The Kuskokwim River flows into the Yukon Delta National Wildlife Refuge below Aniak.

US President Theodore Roosevelt first set aside southwestern Alaska refuge lands in 1909. President Jimmy Carter signed the Alaska National Interest Lands Conservation Act into law in 1980, officially creating the Yukon Delta National Wildlife Refuge at 77,500 km^2 and encompassing the Y-K Delta wetland complex. This refuge is the second largest in the United States, only slightly smaller in size than Arctic National Wildlife Refuge. Extraction industries are not compatible with refuge objectives and are excluded.

The Yukon River Inter-Tribal Watershed Council, a cooperative joint venture with over 70 bands of First Nations from Yukon and Alaska, oversees water quality and water rights in their traditional lands. Association of Village Council Presidents and the Tanana Chiefs Conference promote viable hunting and fishing for the health of the people and lands. The Delta itself has approximately 25,000 residents, 85% Alaskan native, either Yup'ik or Athabascan. Subsistence fishing, hunting, and gathering are a traditional way of life that continues.

Historically both rivers have minor pollution from placer gold and silver mining and waste water and debris from human settlements along the river. The Yukon River was one of the principal means of travel for gold prospectors during the 1896–1903 Klondike Gold Rush. Because of the refuge status under the Fish and Wildlife Service, extraction industries are excluded. Potential oil spills exist on coastal areas from deepwater platform accidents or shipping accidents. Transport of barge material up the rivers also is a possible means of pollution.

References

Brabets TP, Wang B, Meade RH. Environmental and hydrologic overview of the Yukon River Basin, Alaska and Canada. Anchorage: United States Department of the Interior, US Geological Survey; 2000. p. 106. Water-Resources Investigations Report 99-4204.

Ducks Unlimited Inc. Yukon delta national wildlife refuge earth cover classification user's guide. Rancho Cordova: Ducks Unlimited; 2011. p. 93. Prepared for the U.S. Fish and Wildlife Service.

Ely C, Dau C, Babcock C. Decline in a population of spectacled eiders nesting on the Yukon-Kuskokwim Delta, Alaska. Northwest Nat. 1994;75:81–7.

Hall JV, Frayer WE, Wilen WO. Status of Alaska wetlands. Anchorage: U.S. Fish and Wildlife Service; 1994. p. 32.

Harwood CM. Boreal partners in Flight Working Group, Anchorage. 1999. p. 6–7. (abstract).

Kammerer JC. Largest rivers in the United States. Reston, VA, USA: United States Geological Survey, Reston, VA, USA. 1990.

King JG, Lensink CJ. An evaluation of Alaskan habitat for migratory birds. Washington, DC: Department of the Interior, Bureau of Sport fisheries and Wildlife; 1971. p. 45.

Schamber JL, Flint PL, Grand JB, Wilson HM, Moore JA. Population dynamics of long-tailed ducks breeding on the Yukon-Kuskokwim Delta, Alaska. Arctic. 2009;62:190–200.

Viereck LA, Dryrness CT, Batten AR, Wenzlick KJ. The Alaska vegetation classification. Portland: United States Department of Agriculture, Forest Service, Pacific Northwest Research Station; 1992. p. 278. Gen. Tech. Rep. PNW-GTR-286.

43

The Peace-Athabasca Delta: MacKenzie River Basin (Canada)

Jeffrey Shatford

Contents

Introduction 550
Hydrology 550
Wetland Ecosystems 552
Biodiversity 552
Conservation Status 553
Ecosystem Services 553
Threats and Future Challenges 554
References 556

Abstract

The Peace-Athabasca Delta in northern Alberta is one of the largest freshwater deltas (>4,000 km^2) in the world. The highly complex ecosystem of lakes, wetlands and meadows supports a diverse assemblage of plants and animals. The delta is remote and continues to sustain the well-being of the First Nations communities that have depended on the area for centuries. However, large scale industrial developments upstream of the delta represent significant threats. Issues and concerns raised by stakeholders in the region have been informed by the collaborative monitoring program that has been on-going in various forms since the 1970's.

Keywords

Pulse flood · Perched basin · First Nations · Ice · Wood Buffalo · Ramsar

---

J. Shatford (✉)
Ministry of Forests, Lands and Natural Resource Operations, Resource Management Objectives Branch, Victoria, BC, Canada
e-mail: Jeffrey.Shatford@gov.bc.ca

© Crown 2018
C. M. Finlayson et al. (eds.), *The Wetland Book*,
https://doi.org/10.1007/978-94-007-4001-3_13

Introduction

The Peace-Athabasca Delta (PAD), located in the northeast corner of the Canadian province of Alberta, is one of the largest freshwater deltas in the world. The delta covers >4,000 km^2 and encompasses three large lakes and hundreds of shallow basins connected by meandering river channels. The delta stretches from the mouth of the Athabasca River on Lake Athabasca west 100 km to the far shore of Lake Claire (110° 45′ to 112° 36′ W) and 70 km north to south between 58° 17′ and 58° 55′ N (Fig. 1).

Hydrology

The Mackenzie River watershed is one of the largest watersheds in North America and the largest flowing north to the Canadian Arctic. The delta lies at the junction of the Peace and Athabasca Rivers, two large sub-basins of the MacKenzie, that

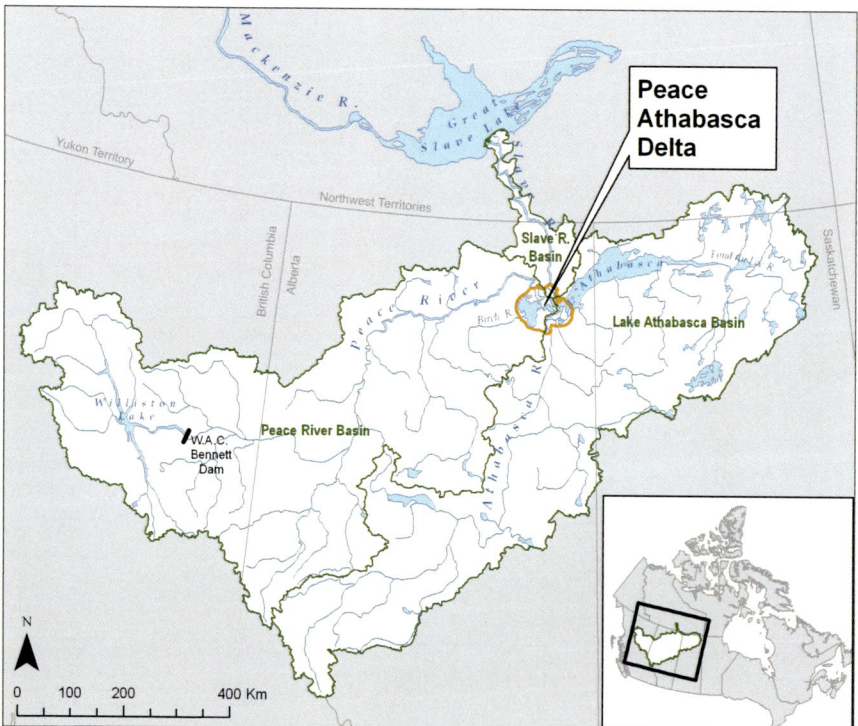

Fig. 1 The Peace-Athabasca Delta occupies the flat plain at the western edge of Lake Athabasca. Here the fluctuating water levels of the Peace and Athabasca Rivers provide the pulse of water and nutrient inputs that drive the productive lakes and wetlands of the region (© Parks Canada with permission)

intersect on the flat plain lying at the western end of Lake Athabasca. The rivers drain an enormous land area stretching from the Rocky Mountains across the foothills and plains to the Canadian Shield.

Water levels in three large central lakes are not a simple function of outflow at a constant sill elevation but are complicated by water levels on the Peace River, ice conditions during winter (November through April), and wind seiches (below) in summer. Perhaps the most complicated feature of the delta hydrology is the influence of the Peace River, which acts as a hydrologic dam on the delta outflow. In general, the delta lakes drain north to the Peace and Slave Rivers via three permanent river channels. However, during high flows in spring or midsummer, the elevation of the Peace River may exceed delta lake levels by several meters (Peters et al. 2006). As water levels on the Peace River rise, there is flow stoppage and then flow reversal on the Chenal des Quatres Fourches, Revillon Coupé, and Rivière des Rochers. These conditions may occur for a few hours or several days and lead to increased water levels in the delta lakes and adjacent wetlands. Where the hydraulic dam on the Peace is sustained for sufficient lengths of time, connectivity to smaller ephemeral channels may occur (the Claire River, Baril River, and Sweetgrass Creek) and contribute to large-scale overland flooding and the recharge of the entire delta complex.

Freezing conditions for several months of the year contribute further to the hydrological complexity of the delta. Ice formations 50–100 cm thick are common by late winter on lakes, rivers, and ponds. Ice roughness, ice grounding to lake and channel bottom, and ice dams can significantly change the outflow of water from the delta lakes and wetlands. Ice physically constrains the movement of water and during breakup in spring may form ice jams (Beltaos 2007). Such blockages result in backwater and elevated water elevations. Ice jams may occur on any of the rivers, especially where sharp bends and rocky islands constrain the movement of ice pans and ice blocks.

A seiche is the phenomena of wind and air pressure elevating water levels in the downwind portion of a water body. Water levels in Lake Athabasca, Lake Claire, and Mamawi Lake are both influenced by this condition. Wind seiche can result in the water level in the western end of Lake Athabasca rising >50 cm in a matter of days. This may push water back towards Mamawi Lake and in turn back towards Lake Claire, another type of flow reversal that can slow the flow of the Athabasca River at its mouth.

Historically, spring and summer flood events on the Peace River generated wide-scale overland flows and regular inputs of water to an extensive and complex system of wetlands, rivers, and lakes. Many of the small lakes and wetlands are situated above normal river stage and are referred to as perched basins. This portion of the flood plain is only replenished by the very highest water levels created by periodic spring or summer flooding on the Peace and Athabasca Rivers. The pulse of water and nutrients delivered to the flood plain and wetlands contribute to the productive character of the area. The recurring floods and subsequent drying periods, the so-called flood pulse, maintain the health and diversity of the ecosystem (Junk et al. 1989).

Wetland Ecosystems

The PAD contains thousands of shallow basins. The hydrological regime of each basin varies with location in the delta, elevation, and connectivity to the main flow system (delta channels and large lakes). Local basins can be permanently connected to the main flow system ("open"), connected seasonally ("restricted") or not connected at all ("isolated"). The latter two categories are also referred to as "perched" basins. The frequency, extent, depth, and duration of flood waters in each basin directly influence the structure and composition of the plant community (Timoney 2008) and the community of epiphytic diatoms (Wiklund et al. 2010). Three large shallow lakes cover approximately one third of the delta: Lake Claire, Mamawi, and Richardson. Extensive marshes dominated by open aquatic communities (*Lemnis*, *Potamogeton*) fringed by emergent vegetation (*Typha*, *Equisetum*, *Scirpus*) cover another third of the area. Wet meadows (*Carex atherodes* and *Calamagrostis* sp.) and savannah areas are also dominant features of the landscape. Portions of the delta lie within basins underlain by the Canadian Shield (granite) have considerable buildup of organic matter and intermittent permafrost. There are also remnant channels, oxbow lakes, and depressions left from scour events that remain as ponds.

Biodiversity

Despite its northern latitude, the area provides productive and valuable habitat for a wide range of species. The nutrient-rich waters and long summer days sustain high levels of primary productivity in marshes and meadows. Three hundred and sixty-four species of vascular plants are known from the area, along with 36 species of mammals, 152 species of birds, and 24 fish species (13 common species and nine rare species).

The meadows, dominated by *Carex* and *Calamagrostis*, represent some of North America's largest undisturbed grasslands and important habitat for the largest free-roaming herd of bison *Bison bison athabascae* and their main predator the timber wolf *Canis lupis occidentalis*. The delta provides breeding habitat for waterfowl and shorebirds and is a staging area for as many as one million waterfowl including tundra swans *Cygnus columbianus* and snow *Anser caerulescens*, white-fronted *Anser albifrons*, and Canada geese *Branta canadensis*. Fourteen species of ducks are known to nest in the delta and tens of thousands of shorebirds have been counted over one season (Beyersbergen 2000). Fish migrate through the delta from overwintering sites to spawning and feeding habitat in the shallow lakes (Donald and Kooyman 1977). The large shallow lakes, rich in plant growth and invertebrates, are extremely productive fish-rearing sites (Donald and Aitken 2005). Muskrats *Ondatra zibethicus* were historically found in large numbers throughout the wetlands, numbering in the tens of thousands.

Conservation Status

The delta's remote location has provided some level of protection since the human population density in the immediate area is very low. Wood Buffalo National Park was established in 1922 and in 1926 the park was extended south of the Peace River to include portions of the delta. Today, 80% of the delta lies within the national park. The remaining is on First Nations Reserve (Athabasca Chipewyan First Nations Indian Reserve 201) or Alberta Provincial Lands. In 1982, the delta was designated under the Ramsar Convention on Wetlands of International Importance. The unique character of the delta and the intact native diversity contributed to the designation of Wood Buffalo National Park as a UNESCO World Heritage Site in 1983.

The land use and management conducted by local first nations is in keeping with the traditional use (hunting, trapping, fishing, and food gathering) that has gone on for centuries. Agricultural activities have been limited to small garden plots and an earlier attempt to cultivate wild rice. Several government-led initiatives have sponsored research and monitoring of the delta as far back as the early 1970s. Initially, the impacts of the W.A.C. Bennett hydroelectric dam were at the core of the work. The list of concerns and questions has grown in step with the expansion of industrial development in the watershed (Northern River Basin Study: Gummer et al. 2006). Since 2008, the Peace-Athabasca Delta Ecological Monitoring Program (PADEMP) has worked to develop monitoring and management recommendations regarding the delta. PADEMP is comprised of representatives from 17 federal, territorial, provincial, and aboriginal governments and two nongovernment organizations.

Ecosystem Services

As the gateway to the Arctic in the Mackenzie River basin and one of the largest wetland systems in northern Canada, the Peace-Athabasca Delta provides a wide range of ecosystem services for the region. Large quantities of nutrients and sediments arrive in the delta each year, providing storage and filtering services.

The delta provides habitat of exceptional value for a range of wildlife, which has in turn attracted people to the region for thousands of years. The area continues to have great significance to the First Nations of the region who depended on this vast area rich in wild game. The continued and dependable access to local country food provides both a connection to the land, their traditions, and the ancient material and spiritual lifestyle.

The delta's large and shallow lakes provide productive nursery habitat for numerous fish species and support domestic, commercial, and sports fisheries. The walleye fishery maintains an annual quota of >40,000 kg of walleye *Sander vitreus*, while from 1950 to 2000 the commercial goldeye *Hiodon alosoides* fishery netted between 1,000 and 300,000 fish each year (Donald and Aitken 2005). The area also supports bison, moose *Alces alces*, and muskrat populations. Muskrats were once trapped by the tens of thousands by local indigenous and itinerant trappers and provided the basis of the fur trade in the region dating back to the eighteenth century. Other fur-bearing species such as lynx *Lynx canadensis*, marten *Martes americana*, and beaver *Castor canadensis* have also been important mainstays of the fur trade in the area.

The channels, rivers, and lakes are used as transportation routes, which are especially important to the community of Fort Chipewyan (1,200 people). This remote northern community has no permanent road system connecting it to the outside world. During the open water season June–October, people and goods travel via boat or barge along the Athabasca River. Historically, frozen rivers provided the major travel routes for dog sleds. While this mode of travel is now less common, the same routes are used by traditional hunters and trappers on snow machines. Transportation also takes place on a network of ice roads during the winter which includes several river crossings in the delta built up on ice bridges.

Threats and Future Challenges

The watershed upstream of the delta drains over 600,000 km^2 of northern British Columbia, Alberta, and Saskatchewan. While the human population in the vast area is relatively low (~300,000), industrial development is underway in oil and gas, hydroelectric production, forestry, and agriculture. The past century has also seen a general climate warming and projected changes in seasonality and rainfall patterns, along with projected expansion of energy development, threaten the ecological health of the watershed (Schindler and Donahue 2006).

In 1968, the W.A.C. Bennett dam was completed, comprising one of the largest hydroelectric projects in North America and significantly altering the seasonal flow regime of the Peace River. This has significantly reduced the quantity of water entering the delta lakes and basins from the Peace River watershed (Peters and Buttle 2009; Fig. 2). The speed and scale of development of the oil sands deposits in the Athabasca River region represent another large and growing threat. The oil sands in northern Alberta are one of the largest remaining oil deposits in the world. The development of this resource is dependent, in part, on the inexpensive and abundant water supply afforded by the Athabasca River. In 2011, water withdrawals approached 200 million cubic meters from the Athabasca River. By 2020, oil production is predicted to grow to more than twice the 2008 levels of 1.31 million barrels per day. The increased withdrawals of water and output of contaminants from processing and refining in the region are cumulative which together threaten to alter the dynamic and productive character of the delta wetlands. Aboriginal people with a deep connection to the area express concerns over the health of the ecosystem and question if traditional country foods are safe to eat. Uncertainties around possible contaminants in fish, waterfowl, moose, and muskrat are common. While contaminant levels in wild game remain unknown, there are documented increases in contaminant levels linked to oil extraction in the region (Kelly et al. 2009; Kurek et al. 2013).

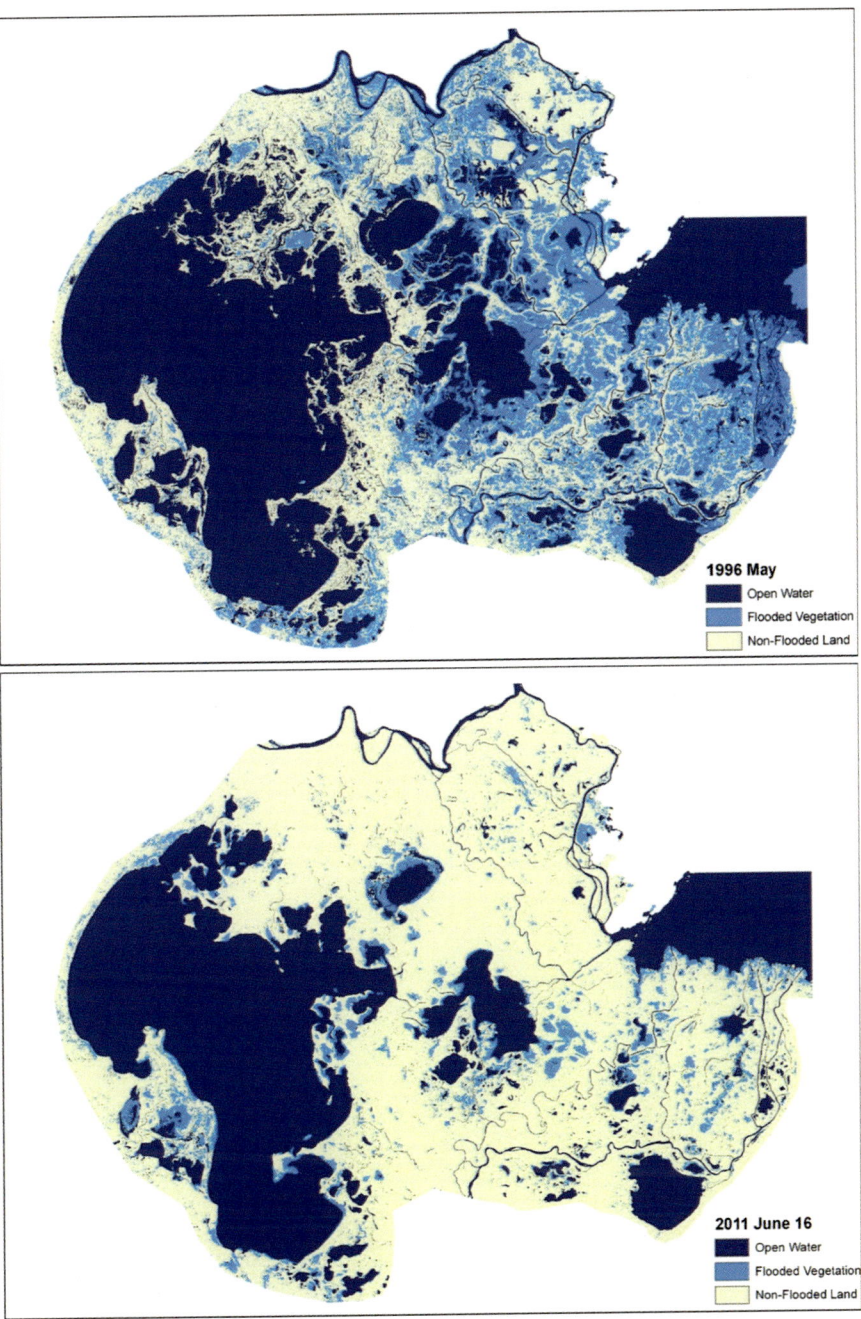

Fig. 2 Satellite imagery provides a graphic representation of the overland floods of the late 1990s (May 1996) and subsequent draw down (June 2011) to the extensive wetlands in the Peace-Athabasca Delta (© Parks Canada with permission)

References

Beltaos S. The role of waves in ice-jam flooding of the Peace-Athabasca Delta. Hydrol Proc. 2007;21(19):2548–59.

Beyersbergen GB. An investigation of migrant shorebird use of the Peace-Athabasca Delta, Alberta in 1999. Edmonton: Canadian Wildlife Service; 2000. 42 p.

Donald DB, Aitken W. Stock–yield model for a fish with variable annual recruitment. North Am J Fish Manag. 2005;25:1226–38.

Donald DB, Kooyman AH. Food, feeding habits and growth of goldeye (*Hiodon aalosoides*) in waters of the Peace-Athabasca Delta. Can J Zool. 1977;55(6):1038–47.

Gummer WD, Conly FM, Wrona FJ. Northern rivers ecosystems initiative: context and prevailing legacy. Environ Monit Assess. 2006;113:71–85.

Junk WJ, Bayley PB, Sparks RE. The flood pulse concept in river-floodplain systems. Can Spec Publ Fish Aquat Sci. 1989;106:110–27.

Kelly EN, Short JW, Schindler DW, Hodson PV, Ma M, Kwan AK, Fortin BL. Oil sands development contributes polycyclic aromatic compounds to the Athabasca River and its tributaries. Proc Natl Acad Sci. 2009;106(52):22346–51.

Kurek J, Kirk JL, Muir DCG, Wang X, Evans MS, Smol JP. Legacy of a half century of Athabasca oil sands development recorded by lake ecosystems. Proc Natl Acad Sci. 2013;110(5):1761–6.

Peters DL, Buttle JM. The effects of flow regulation and climatic variability on obstructed drainage and reverse flow contribution in a northern river-lake-delta complex, Mackenzie Basin Headwaters. River Res Appl. 2009;26(9):1065–89.

Peters DL, Prowse TD, Pietroniro A, Leconte R. Flood hydrology of the Peace-Athabasca Delta, northern Canada. Hydrol Process. 2006;20(19):4073–96.

Schindler DW, Donahue WF. An impending water crisis in Canada's western Prairie Provinces. Proc Natl Acad Sci. 2006;103(19):7210–6.

Timoney K. Factors influencing wetland plant communities during a flood-drawdown cycle in the Peace-Athabasca Delta, Northern Alberta, Canada. Wetlands. 2008;28(2):450–63.

Wiklund JA, Bozinovski N, Hall RI, Wolfe BB. Epiphytic diatoms as flood indicators. J Paleolimno. 2010;44:25–42.

Copper River Delta, Alaska (USA)

44

Frederic A. Reid, Daniel Fehringer, and Richard G. Kempka

Contents

Introduction	558
Hydrology	559
Tectonics	559
Landcover	559
Wetlands	560
Succession	561
Biodiversity	561
Conservation Status	562
Threats and Future Challenges	563
References	563

Abstract

The Copper River Delta of south-central Alaska is the largest wetland on the Pacific Coast of North America, stretching more than 75 km along the coast and up to 50 km inland. This coastal wetland complex is bordered by coastal mountains to the north (up to 2,300 m), Prince William Sound and the Gulf of Alaska to the south, and the largest glacial system of North America (Bering complex) to the east. This 252,000 ha wetland complex has minimal human

---

F. A. Reid (✉) · D. Fehringer
Ducks Unlimited, Inc., Rancho Cordova, CA, USA
e-mail: freid@ducks.org; dfehringer@ducks.org

R. G. Kempka
Ducks Unlimited, Inc., Rancho Cordova, CA, USA

The Climate Trust, Portland, OR, USA
e-mail: rkempka@climatetrust.org

© Springer Science+Business Media B.V., part of Springer Nature 2018
C. M. Finlayson et al. (eds.), *The Wetland Book*,
https://doi.org/10.1007/978-94-007-4001-3_12

disturbance, except for minor timber harvest. A 1964 earthquake of 9.2 Richter resulted in a 2–4 m uplift that modified hydrology and plant succession. This coastal delta serves as one of the most important migrational habitats for waterfowl and shorebirds on the continent.

Keywords

Wetlands · Prince William Sound · Copper river delta · Tectonics · Plant succession · Hydrology · Western sandpiper · Dunlin · Trumpeter swan

Introduction

The Copper River Delta of south-central Alaska is the largest wetland on the Pacific Coast of North America, stretching more than 75 km along the coast and up to 50 km inland (Fig. 1). This coastal wetland complex is bordered by coastal mountains to the north (up to 2,300 m), Prince William Sound and the Gulf of Alaska to the south, and the largest glacial system of North America (Bering complex) to the east. This 252,000 ha wetland complex has minimal human disturbance, except for minor timber harvest. A 1964 earthquake of 9.2 Richter resulted in a 2–4 m uplift that modified hydrology and plant succession.

Fig. 1 The Copper River Delta, Alaska

Hydrology

The headwaters of the Copper River lie in the Alaska Range to the north (above Paxson), in the Wrangell-St. Elias to the east (above Kennicott), and the Talkeetna Mountains to the west. The delta is fed by many glacial streams, including the Scott, Sheridan, Goodwin, Childs, and Miles Glaciers and the Martin River. The total length of the Copper River is approximately 460 km.

The Copper River basin is the sixth largest watershed in Alaska with an area of 62,700 km^2; however, the average discharge of the Copper River, 2.026 million m^{3}s^{-1}, ranks second behind that of the Yukon River (Brabets 1997). By volume of discharge, the Copper River is the tenth largest river in the United States. The average discharge per square kilometer of the Copper River ranks second only to the Stikine River, which also has a relatively large percentage of its area – 10% – covered by glaciers. Most of the Copper River Basin is boreal forest biome. Most of the flow of the Copper River occurs from June to September. An estimated 100 million metric tons of sediment is deposited annually on the Copper River Delta and into the Gulf of Alaska by the Copper River and associated glacial rivers (Thilenius 1990).

Annual precipitation is approximately 205 cm and average monthly temperature is 3.4 °C. Winter snowfalls can be very heavy and average 318 cm total accumulation. Young, steep mountains containing several active glaciers greatly influence this region. The lowland physiography is characterized by braided streams and glacial outwash plains. Soils are predominately high energy alluvial deposits of sand and gravel and low energy marine deposit of glacial silts and clays. Sand dunes occur in the mouth of the Copper River, deposited by wind from upriver, and as offshore barrier islands deposited by ocean currents.

Tectonics

The entire delta region is one of the most seismically active in the world. The 1964 earthquake measured 9.2 Richter and was the most recent of at least five similar coseismic uplifts of 2–3 m each that have occurred in the last 3,000+ years (Pflaker 1990). The extremely high rate of deposition of glacial sediments from the Copper River and several smaller rivers flowing from local glaciers causes an accumulation of sediments on the delta that is about equal to the rate of submergence (Reimitz 1966). Biological communities and hydrology were highly altered as the uplifted marsh began to desalinate and was transformed into an elevated freshwater wetland (Potyondy et al. 1975; Thilenius 1990; Boggs 2000).

Landcover

An analysis of multispectral satellite and airborne images (circa 2009–2011) of the Copper River Delta and adjacent areas revealed the extent of vegetation communities. A vegetation decision tree for the delta (under 150 m) was developed, field observation of over 479 sites was conducted by helicopter (Bellante et al. 2013) and a land cover image of the region was created. Analysis of land cover revealed that four upland forested communities existed. Sitka spruce *Picea sitchensis* (11,600 ha) and western hemlock *Tsuga heterophylla* (2,500 ha) types were the most common forests and encompassed just over 5% of the study area. Most were on upland sites with smaller "haystacks" or stringers in the wetland areas. Sitka spruce was also common on beach ridges and uplifted dunes. Black cottonwood *Populus trichocarpa* was widely distributed and encompassed 7,500 ha, representing 3% of the study area. The sitka spruce–black cottonwood community type is composed of mature spruce and cottonwood found on deep, well-drained alluvium of major outwash plains and floodplains. It was a minor component (3,900 ha) representing only 1.5% of the total area.

Alder and Willow communities encompassed 48,500 ha. They are widespread and, combined, were by far the most common vegetation types on the delta, encompassing nearly 20% of the project area, with alder accounting for over 14% individually. The brushlike thickets of alder *Alnus* spp., willow *Salix* spp., and mixed alder-willow occur inland along braided streams, in and around shallow wetlands, and on levees surrounding wetland basins and sloughs. Dry graminoid types, most commonly dunegrass *Leymus mollis*, represent a very minor portion of the project area (<400 ha, <0.2%) along coastal beaches.

Wetlands

Sweetgale *Myrica gale* community types encompassed 22,800 ha, over 9% of the area. Sweetgale is wide ranging and varies phenotypically in different habitats. It is found in tidally influenced areas and inland in both wet and dry sites. Understory is dominated by *Carex lyngbyei* Lyngby's sedge, while meadow horsetail *Equisetum arvense*, marsh fivefinger *Potentilla palustris*, and bluejoint *Calamagrostis canadensis* are common understory species. Sweetgale and sitka sedge *Carex sitchensis* are vegetation types found on old undisturbed portions of uplifted marshes, floodplains, and outwash plains (Boggs 2000). They typically mark the old marsh edge to tidal wetlands, created after the 1964 earthquake.

Mesic/Wet Herbaceous dominance types encompassed 5.5% of the study area (13,900 ha) and were widespread throughout the delta. This class includes wet forb types that closely correspond to the palustrine emergent class of Cowardin et al (1979). These wet forb areas are dominated by herbs and forbs such as horsetail *equisetum* spp., fivefinger, buckbean *Menyanthes trifoliata*, angelica, and sedges/grasses such as Lyngby's and sitka sedge, cottongrass *Eriophorum* spp., and pendant grass *Arctophylla* spp. More mesic types are dominated by forbs and grasses such as bluejoint, lady-fern *Athyrium filix-femina*, fireweed *Epilobium angustifolium*, iris, lupine *Lupinus spp.*, beach strawberry *Fragaria chiloensis*, and cow parsnip *Heracleum*.

Aquatic herbaceous types encompassed 3,400 ha, accounting for 1.3% of the area. This class has at least 25% floating or slightly submerged vegetation and less than 25% emergent vegetation. Pond lily *Nuphar polysepalum*, pondweed *Potamogeton* spp., starwort *Callitriche* spp., and milfoil *Myriophyllum* spp. are the most common components.

Water and bare or sparsely vegetated classes encompass 137,400 ha and account for the majority (55%) of the project area. The acreage of each of these classes is difficult to quantify because they are in constant flux as thousands of hectares of mudflats are covered and exposed daily throughout the tide cycle. Much of the water in the project area is turbid, due to heavy sediment loads transported by the flows of the Copper River. Clear water is found in the ponds, small lakes, and rivers that are isolated from glacially influenced rivers. Developed areas account for only 200 ha ($<0.1\%$) of the area.

Succession

Since the 1964 earthquake and uplift, plant succession is occurring rapidly on the old salt marsh, with much of the grass/sedge community shifting to willow and alder. Early descriptions and old photographs reveal palustrine wetlands dominated by Lyngby's sedge. A successional shift has occurred from palustrine emergent wetlands to palustrine shrub/scrub. Needleleaf trees are growing above some alder-willow stringers. Mudflats lifted above the mean high tides are slowly being colonized by vegetation (mostly *Carex lyngbyei*). Uplift of the mature tidal surface to a supra-tidal status allowed freshwater tolerant species to invade the basins and trees and shrubs displaced herbaceous vegetation on levees (Boggs and Shephard 1999). Vegetation zonation within a basin-levee complex was evident and repeated. The change in elevation and hydrography led to rapid ecological succession (Crow 1968; Thilenius 1995; DeVelice et al. 2001).

A comparison of Bellante's (2013) land cover map to a previous map produced using Landsat and SPOT imagery from 1990 and 1991 (Kempka et al. 1994) indicates significant changes in land cover composition on the delta. Needleleaf dominated forest communities increased from 3.5% of the delta in 1990 to 7.2% in 2010. Deciduous forest and shrub types have increased from 20.3% to 31.3% of the total area, wet herbaceous communities have decreased from 15% to only 5.7%, and aquatic herbaceous communities have decreased from 2.3% to 1.4% of the total area.

Biodiversity

Beaver *Castor canadensis* have followed the willows invasion of the old marsh, building dams and creating plunge pools and ponds. In addition to beavers, the Copper River Delta provides habitat for many species of fish and wildlife including four species of salmon, dolly varden *Salvelinus malma*, steelhead *Oncorhynchus mykiss*, cutthroat trout *O. clarkii*, moose *Alces alces*, wolves *Canis lupus*, geese, swans, ducks, shorebirds, and many neotropical songbirds. Understanding the succession of vegetation on the Copper River Delta has a direct effect on management of many species of fish and wildlife.

Historically, over 5–10% of all North American waterfowl annually breed in Alaska, including 70,000 swans, a million geese, and 12 million ducks (King and Lensink 1971; King 2002). As much of interior Alaska is often locked up in snow and ice in early spring and late fall, this coastal delta serves as one of the most important migrational habitats on the continent. The tidal flats have a rich habitat for ducks and shorebirds, especially at low tide.

Nearly the entire world's population of dusky Canada goose *Branta canadensis occidentalis* nests on the Copper River Delta. Prior to the 1964 earthquake (1953–1964), population estimates for the dusky Canada goose averaged 15,000 birds. With reduced tidal flooding of preferred nesting habitat after the quake (1965–1982), population estimates increased to over 21,000 birds. As vegetation succession continued on the delta (1982–1992), nesting conditions became less favorable and the average population estimate has declined to under 13,000 with a decreasing trend (Timm et al. 1979; Campbell 1990; Butler 1992). Estimates on the wintering grounds during winter 1994–1995 place population levels under 9,000 birds.

Moose were transplanted to the Copper River Delta in 1949 (MacCracken 1992), and their population had grown to an estimated 900 by 1993. Hunting of the population began in 1960 with over 100 harvested annually in recent years. Vegetation succession since the earthquake has increased the amount of forage for moose on the delta. A recent analysis indicates the delta may be able to support a much higher moose population. However, only a small portion of the delta was suitable as moose winter range due to snowfall patterns and food availability (MacCracken et al. 1997).

The Copper River Delta is important habitat to many other species of wildlife. Estimates of sixteen million shorebirds (including the entire Pacific population of western sandpipers *Calidris mauri* and dunlin *Calidris alpina*) use the mud flats and outer portions of the marsh as a critical stopover for a few weeks during spring migration. Five hundred pairs of trumpeter swans *Cygnus buccinator* (10% of the world's population) nest and rear their broods on the delta. Sandhill crane from Alaska and Wrangell Island snow geese *Chen caerulescens* from Russia forage in the wetlands as critical migration habitat.

Conservation Status

With the passage of the federal 1980 Alaska National Interest Lands Conservation Act, the Copper River Delta is to be managed by the US Forest Service under the Chugach National Forest primarily for the conservation of fish and wildlife and their habitats – ANILCA, Section 501(b). Throughout the entire US National Forest system there is only one other area with a similar congressional mandate. In 1990, the Western Hemisphere Shorebird Reserve Network recognized the delta as the first Hemispheric Site (the most important) in Alaska. Ducks Unlimited recognizes the

Copper River Delta as a critical migration and breeding area for waterfowl and part of the key coastal wetland system in Alaska. Alaska Department of Fish and Game recognizes a Copper River Delta State Game Refuge as an overlay to the Forest Service. No extraction is allowed on the delta and only minor timber harvest can be allowed either along the highway or Eyak native lands.

Threats and Future Challenges

Major threats for the Copper River Delta include oil pipeline leaks that directly impact the river or shipping accidents in Prince William Sound. The grounding of the Exxon Valdez had significant impact on marine organisms but little petroleum washed up directly on the delta. As human development expands in the boreal and arctic, proactive identification of key wetlands will be critical to making viable resource decisions on land use in the future. The Alaskan state population has tripled since the 1950s. The current human population is estimated at 740,000 and is primarily concentrated in the southeast panhandle and in a corridor between Kenai and Fairbanks. The identification and classification of wetland habitat types will allow resource agencies to understand the dynamic nature of these systems.

References

Bellante G, Goetz W, Maus P, Develice R, Riley M, Megown K. Copper River Delta existing vegetation map project. Salt Lake City: USDA Forest Service, Remote Sensing Applications Center; 2013. p. 23. RSAC-10075-RPT1.

Boggs K. Classification of community types, successional sequences, and landscapes of the Copper River Delta, Alaska. Portland: U.S. Deptartment of Agriculture, Forest Service, Pacific Northwest Research Station; 2000. p. 244. General technical report PNW-GTR-469.

Boggs K, Shephard M. Response of marine deltaic surfaces to major earthquake uplifts in Southcentral Alaska. Wetland. 1999;19:13–27.

Brabets TP. Geomorphology of the lower Copper River, Alaska. Denver: U.S. Geological Survey; 1997. 89 p. Professional Paper 1581.

Butler W. Report to the Pacific Flyway Study Committee on 1986–1992 breeding ground surveys of the dusky Canada goose on the Copper River Delta. Portland: U.S. Fish and Wildlife Serv.; 1992. 7 p. Unpubl. Rept.

Campbell BH. Factors affecting the nesting success of dusky Canada geese, *Branta canadensis occidentalis*, on the Copper River Delta, Alaska. Can Field Nat. 1990;104:567–74.

Cowardin LM, Carter V, Golet FC, Laroe ET. Classification of wetlands and deepwater habitats of the United States. Washington, DC: U.S. Department of Interior, FWS, Biological Services Program; 1979. p. 103. FWS/085-79/31.

Crow JT. Plant ecology of the Copper River Delta, Alaska [PhD dissertation]. Pullman: Washington State University; 1968. p. 120.

DeVelice RL, Delapp J, Wei X. Vegetation succession model for the Copper River Delta. Anchorage: USDA Forest Service, Chugach National Forest; 2001. p. 56.

Kempka RG, Maurizi BS, Reid FA, Logan DW, Youkey DE. Utilizing SPOT multispectral imagery to assess wetland vegetation succession in the Copper River Delta, AK. Remote Sens Mar Coastal Environ. 1994;2:529–41.

King JG. Ducks, Rampart dam and wildlife refuges in interior Alaska. Anchorage, AK: Alaska Historical Society; 2002. p. 1–22.

King JG, Lensink CJ. An evaluation of Alaska habitat for migratory birds. Washington, DC: US Dept of Interior, Bureau of Sport Fisheries and Wildlife; 1971. p. 45.

MacCracken JG. Ecology of moose on the Copper River Delta, Alaska [PhD dissertation]. Moscow: University of Idaho; 1992. p. 338.

MacCracken JG, Van Ballenberghe V, Peek JW. Habitat relationships of moose on the Copper River Delta in coastal south-central Alaska. Wildlife Monogr. 1997;136:3–52.

Pflaker G. Regional vertical tectonic displacement of shorelines in south-central Alaska during and between great earthquakes. Northwest Sci. 1990;64:250–8.

Potyondy J, Meyer M, Mace A. Ana analysis of 1964 earthquake effects upon the vegetation and hydrology of the Copper River Delta, Alaska. St. Paul: Remote Sensing Laboratory, College of Forestry, University of Minnesota; 1975. 84 p. IAFHE RSL Res. Pap. 75–6.

Reimitz E. Late Quaternary history and sedimentation of the Copper River Delta and vicinity, Alaska [PhD dissertation]. San Diego: University of California; 1966. p. 160.

Thilenius JT. Woody plant succession on earthquake uplifted coastal wetlands of the Copper River delta, Alaska. For Ecol Manag. 1990;33(34):439–62.

Thilenius JT. Phytosociology and succession on earthquake-uplifted coastal wetlands, Copper River, Alaska. Portland: U.S. Dep. Ag., Forest Service, Pacific Norwest Research Station; 1995. p. 58. Gen. technical rep. PNW-GTR-346.

Timm DE, Bromley RG, McNight DE, Rodgers RS. Management evolution of Dusky Canada Geese. In: Jarvis RL, Bartonek JC, editors. Management and biology of Pacific flyway geese. Corvallis: OSU Book Stores; 1979. p. 322–30.

Fraser River Delta: Southern British Columbia (Canada)

45

Anne Murray

Contents

Introduction	565
The Fraser River and Formation of the Delta	567
Ecological Importance	568
Wetland Diversity	569
Intertidal	569
Estuarine Marshes	569
Salt Marshes and Boundary Bay	570
Ocean	570
Salt Wedge and Plume	570
Upland Farmland	571
Bog	571
Conservation Designations	571
Threats and Future Challenges	572
References	573

Abstract

The Fraser River Delta is the largest on the west coast of Canada. Its upland, riverine, marine and bog environments are highly significant wildlife habitats, increasingly threatened by industrial and urban developments.

Keywords

Fraser river · Fraser estuary · Fraser delta · Boundary bay · Burns bog

---

A. Murray (✉)
Nature Guides BC, Delta, BC, Canada
e-mail: sanderling@uniserve.com; info@natureguidesbc.com

© Springer Science+Business Media B.V., part of Springer Nature 2018
C. M. Finlayson et al. (eds.), *The Wetland Book*,
https://doi.org/10.1007/978-94-007-4001-3_191

Introduction

The Fraser delta, in the Strait of Georgia, is a globally important ecosystem, essential habitat for wintering and migrating birds and many aquatic species, including salmonids. Birds migrating through the delta use three continents and 20 countries on the Pacific flyway (Butler and Campbell 1987). The agricultural soils and growing conditions of the delta are among the best in Canada (Klohn Leonoff et al. 1992). The Fraser River is called Stò:lo by the indigenous Halq'eméylem-speaking people that bear the same name (Carlson 2001).

The Fraser delta is located in the southwest of the province of British Columbia (BC), Canada, at 49° 10′N, 123° 05′W. The delta covers approximately 700 km^2 immediately south of the city of Vancouver and is the largest estuary on Canada's Pacific coast (Butler and Campbell 1987). It extends 30 km west from the community of New Westminster to the Strait of Georgia (part of the Salish Sea) and 22 km from the north arm of the Fraser River south to the US border at Point Roberts (Fig. 1).

Fig. 1 The Fraser River delta, British Columbia, Canada (July 20, 2000 Landsat 7 image downloaded from Geogratis.cgdi.gc.ca. Contains information licensed under the Open Government License)

The Fraser River and Formation of the Delta

The Fraser was designated a Canadian Heritage River in 1998 (http://www.chrs.ca/Rivers/Fraser/Fraser_e.php), for its ecological and cultural importance. It is the longest river in BC, flowing for 1,378 km from its headwaters in the Canadian Rockies. During its journey through mountain ranges and plateaus, it is joined by many tributaries and drains a total area of 234,000 km^2 (Fig. 2), one quarter of the province (Bocking 1997).

Following the melting of the Pleistocene ice sheets between 16 and 11,000 years BP, ocean levels adjusted, approaching their current level about 6,000 years BP. The formation of the Fraser delta began with sediment deposition near the confluence of the Pitt and Fraser rivers and aggraded southwest (Neu 1966). The river braided into channels as it flowed across the low-lying marshy landscape, and the original mouth lay to the south, discharging into the current area of Boundary Bay (Thomson 1981).

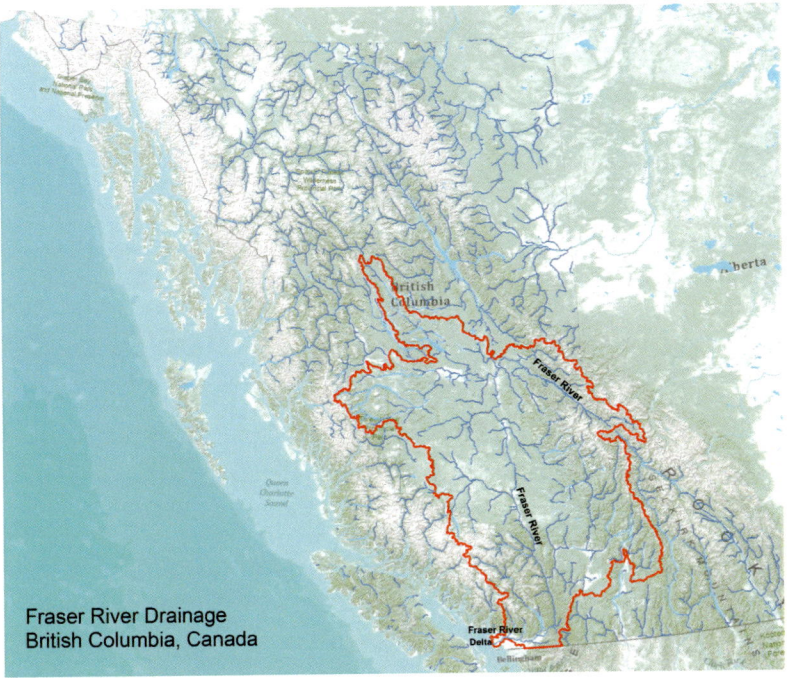

Fig. 2 Map of the Fraser River drainage, British Columbia, Canada. Basemap created using ArcGIS® software by Esri. ArcGIS® and ArcMap™ are the intellectual property of Esri and are used herein under license. Copyright © Esri. All rights reserved. For more information about Esri® software, please visit www.esri.com. Watershed from Freshwater Atlas downloaded from DataBC (www.data.gov.bc.ca). Contains information licensed under the Open Government License – British Columbia

The higher ground of the Point Roberts peninsula was an island until 2,500 years ago (Clague et al. 1998). About 6,000 years ago, the river changed direction to flow west, and Burns Bog began to develop in the center of the delta, from an accumulation of organic debris (Hebda et al. 2000). Boundary Bay was cut off from the Fraser River's flow and reformed as a shallow marine bay, bordered by salt marshes. Prior to dyking, the position of the main channel of the Fraser varied up to 3 km north or south of its current position (Church et al. 1987).

The Fraser River now has four branches as it enters the Strait of Georgia: the North Arm, Middle Arm, South (or Main) Arm, and Canoe Pass. The South Arm carries 85% of the river volume (Port Metro Vancouver 2005), with an average flow of 1.85 m/s. This increases to 2.83 m/s during the freshet in May and June, when snow melt from the interior mountains enters the river system (Ward 1980).

Larger deltaic islands formed from alluvial deposits include Lulu Island, the location of the City of Richmond; Sea Island, home to the Vancouver International Airport; and Westham Island in Delta, known for its farms and national wildlife reserve. Smaller islands line the shore of the South Arm.

The Fraser River delivers 17 million tonnes of sediment annually to the Strait of Georgia, much of which accumulates on the delta slope (City of Richmond 2016a). Two to four million tonnes of sediment are dredged from the main river channels every year, to keep shipping lanes open (Cannings and Cannings 2015). Many back sloughs and channels are silting up. On the active front, the delta progrades at a rate of 3 m per year (Kennett and McPhee 1988). However, it is thought that sand deposition may no longer be occurring through normal deltaic processes (McClaren and Ren 1995), and the expansion of marshes has slowed (Church and Hales 2007).

Areas of the delta lying below sea level are potentially at risk from storm surges and flooding. The delta is located in a seismically active part of the coast and is at risk for earthquake-induced ground liquefaction. The risk of a tsunami is considered to be low (City of Richmond 2016b).

Ecological Importance

The Fraser delta has a temperate climate and lies within the sheltered Strait of Georgia, ensuring year-round use of wildlife habitats. The delta is a critically important habitat for migratory and wintering shorebirds, waterfowl, and raptors (Butler and Campbell 1987). Several million birds stop over on migration, including tens of thousands of waterfowl – swans, geese, ducks, loons, and grebes – and up to a million shorebirds. 250,000 waterfowl spend the winter offshore and in the agricultural uplands (www.ramsar.org). The uplands around Boundary Bay and Roberts Bank host the largest density of wintering birds of prey in Canada (Butler and Campbell 1987), including falcons that follow the migrant shorebirds (Butler et al. 2003) and hawks, harriers, and owls that feed on grassland rodents, such as the abundant Townsend's vole *Microtus townsendii*. Bald eagles *Haliaeetus leucocephalus* are very common resident and wintering species. There are several colonies of the coastal race of great blue heron *Ardea herodias fannini*, a federal species of concern. The Fraser delta (Ladner) Christmas bird count regularly has the highest species count in Canada (http://netapp.audubon.org/cbcobservation/).

The estuary plays a key role in the life history of seven salmon species – chinook *Oncorhynchus tshawytscha*, coho *O. kisutch*, sockeye *O. nerka*, pink *O. gorbuscha*, chum *O. keta*, coastal cutthroat trout *O. clarkii clarkii*, and steelhead *O. mykiss* – and hundreds of genetically unique stocks. It has been estimated that the Fraser averages 800 million out-migrating juvenile salmonids a year (Environment Canada 1992), of which pink smolts average 450 million (Fraser River Aboriginal Fisheries Secretariat 2015). Federally listed species of concern include eulachon *Thaleichthys pacificus*, white sturgeon *Acipenser transmontanus*, and green sturgeon *A. medirostris*. Marine mammals, including endangered orcas or killer whales *Orcinus orca*, humpback *Megaptera novaeangliae*, gray whale *Eschrichtius robustus*, minke whale *Balaenoptera acutorostrata*, harbor seal *Phoca vitulina*, two species of porpoise, and two species of sea lion, occur in adjacent waters.

Wetland Diversity

Intertidal

Beneath the 25,700 ha sand and mudflats on Sturgeon Bank, Roberts Bank and Boundary Bay are myriads of microscopic plant eaters and filter feeders such as lugworms *Arenicola marina*, burrowing shrimp *Callianassa* sp., snails, clams, and cockles *Mollusca* sp. (Butler and Campbell 1987). Nonnative crustaceans are common. They were introduced in ballast water and by former oyster-growing operations. Biofilm occurs on muddy substrate where freshwater mingles with ocean saltwater at the mouth of the Fraser, particularly on Roberts Bank. This mucous-like mix of microbes and organic detritus is a key dietary component for small shorebirds, such as the western sandpiper *Calidris mauri* (Kuwae et al. 2008; Kuwae et al. 2012).

The tidal flats are covered by extensive growths of the native eelgrass, *Zostera marina*, and a nonnative species, *Zostera japonica*. The eelgrass meadows of Boundary Bay and Roberts Bank are the most extensive in the Strait of Georgia, covering thousands of hectares and supporting numerous species of invertebrates, algae, and vertebrate species (Short et al. 2004). Many of the local, commercially taken fish and shellfish spend at least part of their lives in eelgrass meadows.

Estuarine Marshes

Characteristic vegetation of the estuarine marshes, on the active front of the delta, includes sedge *Carex lyngbyei*, bulrush *Scirpus paludosus*, and cattail *Typha latifolia* (Butler and Campbell 1987). This marsh is important feeding habitat for wintering trumpeter swans *Cygnus buccinator* and lesser snow geese *Chen caerulescens*

caerulescens. Near the dykes, the marsh is only covered by the highest winter tides. Plants such as dunegrass *Elymus mollis*, Douglas' aster *Aster douglasii*, Canada goldenrod *Solidago canadensis*, sea-watch *Angelica lucida*, and cow parsnip *Heracleum lanatum* are typical of this backshore vegetation. Populations of Townsend's vole provide prey for resident northern harriers *Circus cyaneus*, wintering rough-legged hawks *Buteo lagopus*, and short-eared owls *Asio flammeus*.

Salt Marshes and Boundary Bay

Most of the Fraser delta salt marsh was lost when dyking occurred (Butler 2003). A remnant fringe occurs at Boundary Bay. Characteristic plants are sea asparagus *Salicornia virginica*, sea arrow-grass *Triglochin maritimum*, entire-leaved gumweed *Grindelia integrifolia*, and saltmarsh dodder *Cuscuta salina*. Hundreds of thousands of black-bellied plover *Pluvialis squatarola*, dunlin *Calidris alpina*, and other shorebirds gather to feed in Boundary Bay in season. Up to a hundred thousand dabbling ducks winter here, primarily northern pintail *Anas acuta*, mallard *A. platyrhynchos*, American wigeon *A. americana*, and green-winged teal *A. crecca* (Badzinski et al. 2008). Flocks of brant *Branta bernicla* gather offshore. The backshore habitat is similar to that described for estuarine marshes, with high populations of voles, nesting northern harriers, savannah sparrows *Passerculus sandwichensis*, and wintering hawks and owls. Other species characteristic of this marsh include short-tailed weasel *Mustela erminea* and anise swallowtail butterfly *Papilio zelicaon*.

Ocean

Within a few kilometers of the mouth of the Fraser, the Strait of Georgia drops to depths of over 300 m. The deep water is habitat for marine mammals, including endangered southern resident orca, wintering diving ducks, loons, grebes, and alcids. Gray whales often come into the shallower waters of Boundary Bay during spring and summer, and harbor seal colonies occur around the delta shores. The Strait of Georgia has commercial Dungeness crab *Metacarcinus magister* and prawn *Dendrobranchiata* sp. fisheries.

Salt Wedge and Plume

Tides in the Strait of Georgia are mixed, semidiurnal with the difference between high and low tide ranging from 3.1 to 4.8 m (Thomson 1981). At high tide, the saltwater moves up the Fraser River in a wedge that lies below the outflowing freshwater of the river. The salt wedge reaches maximum intrusion during low river discharge in winter and is at a minimum during freshet (spring snowmelt) from mid-May to July. The outgoing freshwater transports suspended sediments and is discernible as a turbid plume, which during freshet can extend across the Strait as far as Galiano Island, 30 km west of the river mouth (Thomson 1981). With salinity levels half that of the Strait of Georgia (15 psu to 30 psu), the plume is a distinct ecosystem from surrounding waters (Harrison and Yin 1998).

Upland Farmland

Agricultural land in the delta grows crops such as hay, potatoes, beans, corn, and berry fruits. Open-field farmland, together with hedgerows and ditches, provides habitat for diverse wildlife, from beneficial insects to fish, birds, and mammals. It is an essential winter habitat for birds such as trumpeter swan, lesser snow goose, dunlin, black-bellied plover, short-eared owl, and northern harrier. A few industrial-scale glasshouses have associated freshwater ponds that are used by waterfowl and shorebirds.

Bog

Burns Bog is the largest bog in the Fraser delta, and the largest, deltaic, domed bog on the west coast of North America (Environment Canada 2001). Historically covering about 4,800 ha, between the Fraser River and Boundary Bay, it has been encroached on by other land uses, and only about 29% remains relatively undisturbed (Metro Vancouver 2007). The bog formed over the course of several thousand years. Areas of the delta with a high water table and poor drainage gradually became elevated by decaying vegetation and the growth of sphagnum moss *Sphagnum flexuosum*. The domed bog is dependent on rainfall for its moisture, and the nutrient-poor, acidic conditions promote a specialized plant assemblage, uncommon on the southwestern BC coast (Hebda et al. 2000). As well as twelve species of sphagnum moss, characteristic bog plants include: Labrador tea *Ledum groenlandicum*, reindeer lichen *Cladonia rangiferina*, bog laurel *Kalmia microphylla*, cloudberry *Rubus chamaemorus*, and round-leafed sundew *Drosera rotundifolia* (Metro Vancouver 2007). Drier conditions prevail around the bog's outer, transitional region or lagg. Predominant vegetation includes shore pine *Pinus contorta* var. *contorta*, western redcedar *Thuja plicata*, red alder *Alnus rubra*, and sweet gale *Myrica gale*.

Burns Bog is also known for faunal diversity. Two locally rare dragonflies, subarctic darner *Aeshna subarctica* and zigzag darner *A. sitchensis*, have been recorded (Kenner 2000). Pacific water shrew and a rare subspecies of southern red-backed vole *Myodes gapperi occidentalis* also occur (Fraker et al. 1999). Bird species include waterfowl, songbirds, raptors, and greater sandhill crane.

Conservation Designations

The ecological value of the Fraser delta is well recognized by scientists, yet legal protection of the area is limited. There are several overlapping international conservation designations. The 20,682 ha of intertidal and marsh area in the provincial

Wildlife Management Areas outside the dykes at Boundary Bay and Sturgeon Bank, and at South Arm Marshes and Serpentine, together with the Alaksen National Wildlife Area and the multi-jurisdictional Burns Bog Ecological Conservancy Area, are collectively designated under the Ramsar Conventions as a Wetland of International Importance (Fraser River Delta – site 243: https://rsis.ramsar.org/ris/243). Roberts Bank has not yet been designated as a Ramsar Site although it meets the criteria for designation. It is however along with Boundary Bay and Sturgeon Bank designated a hemispheric site under the Western Hemispheric Shorebird Reserve Network, www.whrsn.org. In addition, the whole Fraser delta and estuary, including Boundary Bay, Roberts Bank, Sturgeon Bank, the surrounding uplands, and lower reach of the Fraser River, are designated an Important Bird Area (IBA) under the worldwide BirdLife International program (www.ibacanada.com; www.birdlife.org).

The delta also has a number of Canadian national, provincial, and regional designations which carry legislated conservation protections of varying degrees. These include the federal Alaksen National Wildlife Area that includes the George C. Reifel Migratory Bird Sanctuary; the Sea Island Conservation Area; provincial Wildlife Management Areas at Boundary Bay, Roberts Bank, Sturgeon Bank, South Arm Marshes, and Serpentine; Burns Bog Ecological Conservancy Area that is jointly owned by the federal and provincial governments, Metro Vancouver Parks and the Corporation of Delta; and three Metro Vancouver Regional Parks in the Fraser delta managed for recreation and conservation – Boundary Bay Regional Park and dyke trail, Deas Island Regional Park, and Iona Beach Regional Park.

Threats and Future Challenges

The delta has undergone considerable physical modification since the 1880s, including dyking, dredging, channel stabilization, and jetty construction. A landscape that was originally wet prairie, bog, and marsh (North et al. 1979) is now a mix of urban, industrial, and agricultural land, surrounded by dykes and crossed by numerous transportation and infrastructure networks. The original dykes were built by farmers, in order to permanently settle the land. Following major floods in 1894, the delta was comprehensively dyked and drained, eliminating 70% of the original marshes (North et al. 1979). Dykes stretch for 620 km along the shoreline and the lower reaches of the Fraser River (Bocking 1997). Water levels are regulated by ditches, pump stations, box culverts, and storm sewers, which allow for agricultural irrigation in summer.

Since the early 1960s, the wildlife-rich environment of Roberts Bank has been significantly affected by the construction and ongoing expansion of a port and a ferry terminal and their respective 4-km causeways across the tidal flats. Dominant tidal currents flow northward, parallel to the bathymetry (McClaren and Tuominem 1988). The unculverted causeways have prevented sediments from the Fraser River being deposited on southern Roberts Bank. Geophysical effects of the developments include slumping, channelization, and erosion of the delta slope (Port Metro Vancouver 2005). Biological impacts range from proliferation of nonnative species, such as *Zostera japonica*, *Spartina* sp., and purple varnish clam *Nuttallia obscurata* (Mills 1999), to bird kills from overhead wires and noise, air, and light pollution. There are additional impacts from jetties in intertidal areas at Sturgeon Bank and constraints on bird use of intertidal and marsh habitat adjacent to the Vancouver International Airport due to safety concerns.

Burns Bog was heavily impacted by twentieth-century peat mining. Today, it is ringed by major highways; the Vancouver landfill is located in the southwest corner. In 2004, 2,042 ha were protected as an Ecological Conservancy Area, under a conservation covenant, jointly owned by three levels of government. It is closed to the public. Other bogs occur in Richmond; many hectares have been converted to farmland or other land uses.

The fertile soils, level terrain, and long growing season make the Fraser delta an excellent area for agriculture. In the last 50 years, competing pressures on the land base have resulted in widespread urban, commercial, industrial, and transportation developments. The establishment of an Agricultural Land Reserve in 1973 slowed, but did not eliminate, the conversion of delta farmland to nonagricultural use. The viability of farming and maintenance of land for agriculture are now under extreme pressure, a situation with the potential to negatively impact wildlife populations.

References

Badzinski SS, Cannings RJ, Armenta TE, Komaromi J, Davidson PJA. 2008. Monitoring coastal bird populations in BC: the first five years of the Coastal Waterbird Survey (1999–2004). Bird Studies Canada. British Columbia Birds, Vol. 17. http://www.bsc-eoc.org/library/bccws5yrreport2008.pdf

Bocking RC. Mighty River. A portrait of the Fraser/Douglas & McIntyre. Vancouver, Seattle: University of Washington Press; 1997. 294 p.

Butler R. The Jade coast. The ecology of the north pacific ocean. Toronto: Key Porter Books; 2003.

Butler RW, Campbell RW. The Birds of the Fraser River delta: populations, ecology and intertidal significance. Ottawa: Canadian Wildlife Service; 1987. 73 p. Occasional Paper No. 65.

Butler RW, Ydenberg RC, Lank DB. Wader migration on the changing predator landscape. WaderStudy Group Bull. 2003;100:130–3.

Cannings R, Cannings S. British Columbia: a natural history of its origins, ecology, and diversity with a new look at climate change. Vancouver, BC: Greystone Books 2015.

Carlson KT, editor. A Stó:lo-coast Salish historical atlas. Vancouver: Douglas & McIntyre; 2001.

Church M, McLean DG, Kostachuck RA, Tassone B. Sediment transport in lower Fraser river: summary of results and field excursion guide. Environment Canada, Inland Waters Directorate, Water Resources Branch. Report IWL-HQ-WRB-SS-875. 1987. 95 pp.

Church M, Hales W. The tidal marshes of the Fraser delta. Discovery. 2007;36(1):28–33.

City of Richmond. 2016a http://www.richmond.ca/safety/prepare/city/hazards/tsunamis/tsunamistudy.htm. Accessed 18 Feb 2016.

City of Richmond. 2016b http://www.richmond.ca/safety/prepare/city/hazards/tsunamis.htm. Accessed 18 Feb 2016.

Clague JJ, Luternauer JL, Mosher DC, editors. Geology and natural hazards of the Fraser river delta, British Columbia. Natural resources Canada. Geological Survey of Canada, Bulletin 525. 1998. 270 p.

Environment Canada http://www.ec.gc.ca/media_archive/press/2001/010331_b_e.htm; 2001. Accessed 18 Feb 2016.

Environment Canada. State of the Environment for the Lower Fraser River Basin. A state of the environment report; SOE report no.92-1. Co-published by the British Columbia Ministry of Environment. EN1-11/92-1E; Canada. 1992.

Fraser River Aboriginal Fisheries Secretariat report: Forecast 2015 FRAFS Presentation FINAL (Mar 10 2015) pdf.pdf; B.C. Canada. 2015. http://frafs.ca/sites/default/files/Forecast%202015%20FRAFS%20Presentation%20FINAL%20%28Mar%2010%202015%29%20pdf.pdf. Accessed 19 Feb 2016.

Fraker M, Bianchini C, Robertson I. Burns Bog Ecosystem Review: Small Mammals. Prepared for Delta Fraser Properties Partnership and the Environmental Assessment Office in support of the Burns Bog Ecosystem Review. Report. Sidney B.C. 1999. https://a100.gov.bc.ca/appsdata/epic/documents/p60/103643488864716af42884ae640f8553d621be48be92.pdf. Accessed 18 Feb 2016.

Harrison PJ, Yin K. Ecosystem delineation in the Georgia Basin based on nutrients, chlorophyll, phytoplankton species and primary production. In: Levings CD, Pringle, Aitkens F, editors. Approaches to marine ecosystem delineation in the Strait of Georgia: proceedings of a D.F.O. Workshop, Sidney, B.C. 4–5 November 1997. West Vancouver: Fisheries and Oceans Canada; 1998. p. 3–71. Canadian Technical Report of Fisheries and Aquatic Sciences 2247.

Hebda RJ, Gustavson K, Golinski K, Calder AM. Burns bog ecosystem review synthesis report. Victoria: B.C. Environmental Assessment Office; 2000.

Kenner RD. Lower mainland dragonfly records for 1999. Discovery. 2000;29(1):22–3.

Kennett K, McPhee MW. The Fraser river estuary: an overview of changing conditions. New Westminster: Fraser River Estuary Management Program; 1988.

Klohn Leonoff Ltd, WR Hom and Associates, GG Runka Land Sense Ltd. Delta agricultural study. Canada/B.C. Agri-food regional development subsidiary program. Victoria: British Columbia Ministry of Agriculture; Fisheries and Food; 1992. Joint with Agriculture Canada.

Kuwae T, Beninger PG, Decottignies P, Mathot KJ, Lund DR, Elner RW. Biofilm grazing in a higher vertebrate: the western sandpiper, *Calidris mauri*. Ecology. 2008; 89(3):599–606. https://doi.org/10.1890/07-1442.1.

Kuwae T, Miyoshi E, Hosokawa S, Ichimi K, Hosoya J, Amano T, Moriya T, Kondoh M, Ydenberg RC, Elner RW. Variable and complex food web structures revealed by exploring missing trophic links between birds and biofilm. Ecology Letters. 2012;15:347–356. doi:10.1111/j.1461-0248.2012.01744.x/pdf. Accessed 19 Feb 2016.

McClaren P, Tuominem T. Sediment transport patterns in the lower Fraser river and Fraser delta In: Gray C, Tuominem T, editors. Health of the Fraser River Aquatic Ecosystem. Vol. 1 & 2. A synthesis of research conducted under the Fraser River action plan. Environment Canada. 1988; p. 81–92. Chapter 3.4. Available from: http://research.rem.sfu.ca/downloads/frap/S_34.pdf. Accessed 18 Feb 2016.

McClaren P, Ren P. Sediment transport and its environmental implications in the lower Fraser river and Fraser delta.; Environment Canada: Environmental Conservation; 1995. DOE FRAP-03

Metro Vancouver. Burns Bog Ecological Conservancy Area Management Plan. 2007. Report. 29 p. (& attachments).

Mills CE. 1999-present. *Nuttallia obscurata*, the purple varnish clam or the purple mahogany-clam. Electronic internet document available at http://faculty.washington.edu/cemills/Nuttallia.html. Published by the author, web page established March 1998, last updated 7 June 2004. Accessed 30 Jan 2016.

Neu HJA. Proposals for improving flood and navigation conditions in the lower Fraser River. 1966. 79 p. manuscript. Institute for Ocean Sciences; quoted in McClaren and Ren 1995.

North MEA, Dunn MW, Teversham JM. Vegetation of the southwestern Fraser Lowland 1858–1880. Vancouver: Lands Directorate, Environment Canada; 1979.

Port Metro Vancouver. Deltaport Third Berth Project Environmental Assessment, Chapter Seven: Coastal Geomorphology. 2005. Available from: http://a100.gov.bc.ca/appsdata/epic/documents/

p212/d19622/1108345982109_f26c46a9c4474ae28f98bc46cce3d3f0.pdf. Accessed 30 Jan 2016.

Short FT, Wyllie-Echeverria S, Wright N, Durance C. Eelgrass, *Zostera marina*. 2004. Available from: http://www.birdsonthebay.ca/pdf%20files/Microsoft%20Word%20-%20eelgrass_brochure_outside%20July%202012mc.pdf and http://www.birdsonthebay.ca/pdf%20files/Microsoft%20Word%20-%20Eelgrass%20Brochure%20inside%20print%20version%20revisedJuly12mc.pdf. Accessed 30 Jan 2016.

Thomson RE. Oceanography of the British Columbia coast. Ottawa: Fisheries and Oceans Canada; 1981.

Ward P. Explore the Fraser estuary. Vancouver: Lands Directorate, Environment Canada, Pacific & Yukon Region; 1980.

The Mississippi Alluvial Valley (USA)

J. Brian Davis

Contents

Introduction	578
Hydrology	578
Wetland and Aquatic Systems	580
Riverine Wetlands	580
Palustrine Emergent Wetlands	580
Bottomland Hardwood Forest	580
Palustrine Scrub-Shrub	583
Lakes	583
Reservoirs (Lacustrine)	583
Aquaculture Ponds (Lacustrine)	583
Rice Agriculture	584
Biodiversity	584
Conservation Status	585
Ecosystem Services	586
Future Challenges	587
References	588

Abstract

The Mississippi Alluvial Valley (MAV) is over 800 km long, drains about 41% of the conterminous United States, and is the largest continuous system of wetlands and aquatic habitats in North America comprising approximately 10 million ha. Elevation and hydrology primarily influence the frequency, duration, and periodicity of flooding, which in turn determine plant community composition and species distribution. Largely forested prior to the arrival of Europeans, flood control for agriculture and human settlement caused nearly 75% loss of riparian forests in the MAV by the

---

J. B. Davis (✉)
Department of Wildlife, Fisheries and Aquaculture, Mississippi State University, Mississippi State, MS, USA
e-mail: brian.davis@msstate.edu

© Springer Science+Business Media B.V., part of Springer Nature 2018
C. M. Finlayson et al. (eds.), *The Wetland Book*,
https://doi.org/10.1007/978-94-007-4001-3_236

late twentieth century, with only highly fragmented patches remaining today. However, diverse landforms and ecological communities in the MAV provide unique habitats for myriad species. Many sources of nonpoint source pollution (e.g., fertilizers, toxic chemicals, livestock waste) negatively influence water quality in the MAV. Primary crops grown in the MAV include corn, cotton, rice, and soybeans. Rice fields are especially important to diverse waterbirds during migration and winter.

Keywords

Bottomland hardwood forests · Delta · Floodplains · Hydrology · MAV · Mississippi · Mississippi River · Waterbirds · Waterfowl · Wetlands

Introduction

The Mississippi Alluvial Valley (MAV) is over 800 km long, ranges from 32 to 128 km wide, and comprises approximately 10 million ha in seven states including Arkansas, Kentucky, Illinois, Louisiana, Mississippi, Missouri, and Tennessee (Reinecke et al. 1989 [and references therein], Fig. 1). This vast floodplain begins at the convergence of the Mississippi and Ohio Rivers at Cairo, Illinois, extends to the northern Gulf of Mexico, and drains about 41% of the conterminous United States (Reinecke et al. 1989; Klimas et al. 2012). The MAV was largely forested prior to the arrival of Europeans, following which flood control for agriculture and human settlement caused nearly 75% loss of riparian forests in the MAV by the late twentieth century, with only highly fragmented patches remaining today (Gardiner and Oliver 2005; Klimas et al. 2012). Moreover, the Mississippi River was channelized and leveed for flood protection at unprecedented rates following the 1928 Flood Control Act (King et al. 2005; Oswalt 2013 [and references therein]). Primary crops grown in the MAV include corn, cotton, rice, and soybeans. Rice fields are especially important to diverse waterbirds during migration and winter (Reinecke et al. 1989; Petrie et al. 2014).

Hydrology

Climatic events of the Pleistocene (Quaternary Period) largely shaped the MAV (Reinecke et al. 1989; Klimas et al. 2012). The Mississippi River sculpted the MAV with alluvium through advancing and retreating glaciers, which also caused the Gulf Coast sea level to rise and fall. The floor of the MAV today rises gradually northward at nearly 0.1 m/km from near sea level in south Louisiana to about 100 m in southeastern Missouri (Reinecke et al. 1989).

The MAV contains six drainage subbasins including the St. Francis, Western Lowlands, Arkansas Lowlands, Yazoo, Boeuf, and Tensas. The mostly flat, broad alluvial plain with river terraces and levees contributes to the MAV being the largest continuous system of wetlands in North America. Elevation and hydrology primarily influence the frequency, duration, and periodicity of flooding, which in turn

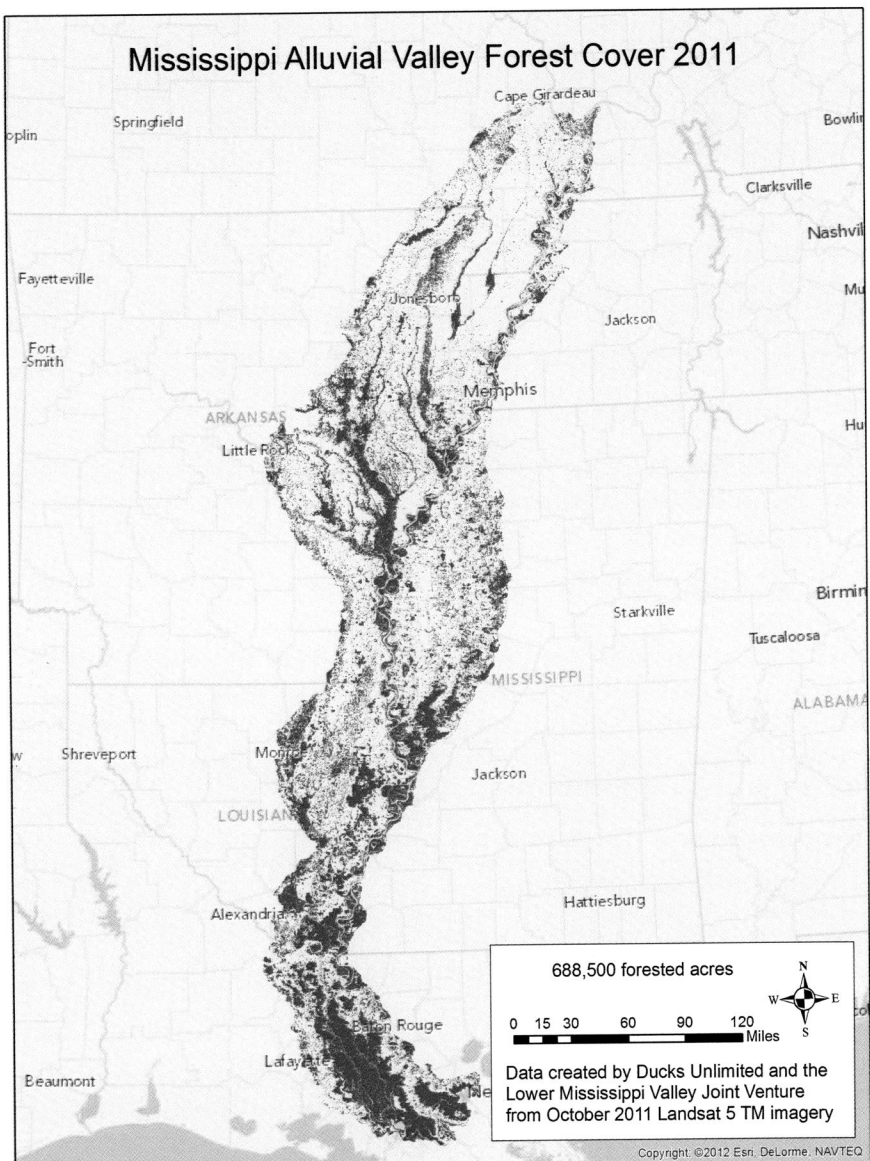

Fig. 1 Distribution and abundance (278,636 ha) of forest cover in the Mississippi Alluvial Valley (*MAV*), 2011; the MAV is contained within the forest cover perimeter (Image credit: Lower Mississippi Valley Joint Venture © with permission)

determine plant community composition and species distribution (Reinecke et al. 1989; Klimas et al. 2012).

Average annual precipitation is 137 cm in the MAV; 111 cm in Puxico, Missouri (northern MAV); and 154 cm in Baton Rouge, Louisiana (southern MAV) (Barlow and Clark 2011). Despite abundant rainfall, hydrological modifications have greatly affected flood frequency. For example, in western Mississippi, a 2-year flood, or the water level that would be reached or exceeded on average in 1 in every 2 years, originally inundated >1.8 million ha but has been reduced to approximately 415,000 ha (Reinecke et al. 1989).

Wetland and Aquatic Systems

The MAV contains a complex of wetland and aquatic habitats that include rivers, seasonal emergent wetlands, bottomland hardwood forest, scrub-shrub, natural lakes, flood control reservoirs, and aquaculture ponds (Reinecke et al. 1989). Wetlands are typically categorized as lacustrine and palustrine. Palustrine wetlands are further subdivided into emergent herbaceous, forested, and scrub-shrub wetlands. In addition to natural wetlands, post-harvested rice fields, particularly when flooded, provide important habitats to autumn migrating and wintering waterbirds.

Riverine Wetlands

Hydrofluvial processes of the Mississippi River and its tributaries formed the MAV's meandering streams, rivers, oxbow lakes, and scrub-shrub wetlands (Brown et al. 2000). River water levels are predominately driven by winter rains that cause overbank flooding of the associated floodplain, inundating forested and herbaceous plant communities and agricultural fields.

Palustrine Emergent Wetlands

Seasonally flooded wetlands generally dry out or contain exposed mudflat sometime between spring and summer (i.e., March to July) that produce annual plant communities (Fig. 2). Annual grasses and weeds such as barnyard grass (millets, *Echinochloa* spp., panic grasses *Panicum* spp., and smartweeds *Polygonum* spp. are a few species indicative of seasonal herbaceous wetland (Reinecke et al. 1989). These systems may flood anytime during the year with overbank flooding from rainfall or deliberate inundation by wetland managers from fall-spring. Wetland managers deliberately flood ($\leq$45 cm deep) these habitats to attract wetland-dependent birds, primarily in fall and winter (November to March). Combined dry mass of seeds and tubers may be $\geq$496 kg/ha in seasonal, or moist-soil, wetland impoundments (Kross et al. 2008).

Bottomland Hardwood Forest

Once covering 10 million ha, bottomland hardwood forest was the dominant ecosystem of the MAV. Extensive deforestation over the past two centuries has resulted in

Fig. 2 Seasonally flooded palustrine emergent wetland, e.g., moist-soil impoundment, flooded in fall-winter to provide high-energy natural seeds (e.g., annual wetland grasses and sedges), tubers, browse, and aquatic invertebrates for migrating and wintering waterfowl and other waterbirds (Photo credit: R.M. Kaminski © Rights remain with the photographer)

2.6 million ha of highly fragmented patches of forest as of 2005 (Conner and Sharitz 2005). Bottomland hardwood forests (Fig. 3) of the MAV are generally categorized as (1) oak-gum-cypress and (2) elm-ash-cottonwood communities (Conner and Sharitz 2005). Dominant tree species in these communities include sweetgum *Liquidambar styraciflua* L., green ash *Fraxinus pennsylvanica*, bald cypress *Taxodium distichum*, sugarberry *Celtis laevigata*, red maple *Acer rubrum*, American sycamore *Platanus occidentalis*, water tupelo *Nyssa aquatica*, eastern cottonwood *Populus deltoides*, black willow *Salix nigra*, elms *Ulmus* spp., hickories *Carya* spp., and at least nine species of oak *Quercus* spp.

Flooding of bottomland hardwood forest is dynamic and varies both temporally and spatially according to winter rains and river levels (Fig. 4). Succession of forest tree species generally occurs in three situations in the MAV: (1) on poorly drained sites at low elevations in major river bottoms (e.g., overcup oak *Quercus lyrata*, common buttonbush *Cephalanthus occidentalis*, and bald cypress), (2) on better drained higher-elevation sites in major river bottoms (e.g., sycamore, sweetgum, oak-hickory), and (3) in minor river bottoms on poorly drained sites (e.g., black willow, tupelo) and better drained sites (e.g., elm-ash, oak-hickory). Contemporary forests have increased in maples and hickories as they replace sweetgum and oaks, largely because of fire suppression and modified flooding regimes in parts of the MAV. Passage of the 2002 Security and Rural investment Acts (Farm Bill) has

Fig. 3 Cypress-tupelo brake of the Mississippi Alluvial Valley, Mississippi, USA (Photo credit: Ducks Unlimited, Inc. © Rights remain with the organization)

Fig. 4 Bottomland hardwood forest flooded during fall-winter for waterfowl hunting and conservation, Bayou Meto Wildlife Management Area, eastern Arkansas, USA (Photo credit: Ducks Unlimited, Inc. © Rights remain with the organization)

particularly spawned reforestation in the MAV through programs like the Wetland Reserve Program (WRP; see "Conservation Status" section below). More than 260,000 ha of land were enrolled in WRP in Arkansas, Louisiana, and Mississippi from 1992 to 2011 (Oswalt 2013).

Palustrine Scrub-Shrub

Densely vegetated wetland habitat dominated by common buttonbush and eastern swamp privet *Forestiera acuminate* interspersed with open water. Scrub-shrub environments provide loafing habitats for waterfowl and cover from predators for wood duck *Aix sponsa* broods and perches for other birds (Reinecke et al. 1989).

Lakes

Usually permanent deep-water (>2 m) habitat supports obligate hydrophytes and freshwater fish and provides roosting and feeding sites for some waterbirds, commonly oxbow lakes formed by abandoned meanders of the Mississippi River (e.g., Moon Lake, Mississippi).

Reservoirs (Lacustrine)

Reservoirs have earthen dams that are typically constructed by the US Army Corps of Engineers (USACE) for flood control and recreation throughout the MAV. Mississippi contains four of these reservoirs (e.g., Sardis Lake, Sardis), which are operated by the USACE. Some reservoirs provide excellent fishing opportunities for recreationists, but reservoirs are generally flooded too deeply to provide significant foraging value for most wetland-dependent birds (e.g., shorebirds and waterfowl). Some wetland birds use lake edges when water levels permit access to seeds and invertebrates or forage on mudflats when exposed in late summer-early fall (Reinecke et al. 1989).

Aquaculture Ponds (Lacustrine)

Leveed impoundments are primarily used for commercial channel catfish *Ictalurus punctatus* production and flooded ~1 m deep. Dominant bird species that use commercial ponds during fall-winter include lesser scaup *Aythya affinis*, double-crested cormorant *Phalacrocorax auritus*, and ruddy duck *Oxyura jamaicensis*, as do generalist species (e.g., northern shoveler *Anas clypeata*, American coot *Fulica americana*, great blue heron *Ardea herodias*, and great egret *Ardea alba*). As many as 150,000 birds used catfish ponds in the mid-1980s, with an average of 100,000 individuals using the ponds weekly. Some waterbirds seek catfish ponds as some impoundments may contain >50 kg/ha of macroinvertebrates and abundant small fish (Feaga 2014). Commercial aquaculture has declined in the MAV because of competition from foreign markets, infrastructure and fuel costs, and other reasons. There were 64,000 ha of commercial aquaculture ponds in Mississippi, Louisiana, and Arkansas in 2001, but only 25,000 ha were operational in those states by 2012. Drained aquaculture ponds with exposed mudflats provide important substrates for shorebirds. An estimated 1,100 ha of mudflats were available to shorebirds in MAV regions of Arkansas, Louisiana, and Mississippi in fall 2009 (Lehnen and Krementz 2013). Idle aquaculture ponds that shallowly flood (≤ 45 cm) also provide important seasonal wetland habitats for migrating and wintering birds when early succession grasses and weeds germinate and mature from exposed mudflats (Feaga 2014), similar to moist-soil impoundments.

Rice Agriculture

Commercial rice production composes >809,000 ha in MAV states of Arkansas, Louisiana, Mississippi, and Missouri, and approximately 20% of all harvested rice is flooded in the MAV. Harvested rice fields yield approximately 79 kg/ha of waste rice for wintering waterfowl, or approximately 11% of total food energy acquired by these birds during winter in the region (Petrie et al. 2014).

Biodiversity

The MAV contains diverse landforms and ecological communities that provide unique habitats for myriad species (Reinecke et al. 1989; Brown et al. 2000). Species biodiversity of the MAV is broad and includes ≥ 183 freshwater fish, 50 mammals, 45 reptiles and amphibians, and 37 mussels (Brown et al. 2000). There are three federally threatened vertebrates, piping plover *Charadrius melodus*, Louisiana black bear *Ursus americanus luteolus*, and loggerhead *Caretta caretta*; eight endangered vertebrates, Bachman's warbler *Vermivora bachmanii*, ivory-billed woodpecker *Campephilus principalis* (likely extinct), red-cockaded woodpecker *Picoides borealis*, interior least tern *Sternula antillarum athalassos*, Attwater's greater prairie chicken *Tympanuchus cupido attwateri*, gray myotis *Myotis grisescens*, West Indian manatee *Trichechus manatus*, and Kemp's ridley sea turtle *Lepidochelys kempii*; and one endangered vascular plant, pondberry *Lindera melissifolia* (Griep and Collins 2011) (US Fish and Wildlife Service, http://www.fws.gov/endangered/index.html). The manatee and sea turtle primarily associate with coastal waters of the southeastern and eastern United States. The MAV is used by approximately 60% of all species of birds in the lower 48 states (Brown et al. 2000). Wood duck and hooded merganser *Lophodytes cucullatus* nest in the MAV, and some individuals of these species remain year-round. During migration and winter, waterfowl forage primarily on acorns, wetland plant seeds and tubers, aquatic invertebrates, and waste grain (Reinecke et al. 1989; Kross et al. 2008). Wetlands containing mudflats and water depths of up to ≤ 45 cm are generally accessed by most waterbirds (Reinecke et al. 1989). Complexes of habitats are generally important to some wintering waterfowl in the MAV, for instance, habitats comprised of 47% cropland, 20% forested or scrub-shrub wetland, 20% emergent or seasonal wetland, and 13% open water contained the greatest mallard abundances at local (20 ha) and landscape (5,024 ha) scales in the Mississippi portion of the MAV in the mid-2000s (Pearse et al. 2012). Rice fields are an important agricultural resource in the winter habitat complex for waterfowl and other birds (Petrie et al. 2014).

Forests of the MAV promote avian diversity because forests form a continuous pattern of vertical structure with different seral stages, providing forest-interior species and niche specialists with essential habitat (Brown et al. 2000). Forest birds of the MAV choose habitat based on the vertical structure of hardwoods, flood regimes, and microhabitats such as vine tangles, canebrakes, Spanish moss *Tillandsia usneoides*, and scour channels (Brown et al. 2000). Shorebird use was likely limited to sandbars and riverine mudflats historically in the MAV (Brown et al. 2000). However, today's largely open, agriculturally dominated landscape of the MAV has supported ~29 species of continental shorebirds (Ranalli and Ritchison 2012). Aquatic species such as the channel catfish and red swamp crayfish *Procambarus clarkii* support commercial fisheries of local and national economic importance.

Conservation Status

A coalition of natural resource partners represented by state and federal agencies, universities, and nongovernmental organizations work to champion conservation in the MAV. The Lower Mississippi Valley Joint Venture (LMVJV) operates as a branch of the US Fish and Wildlife Service's Migratory Bird Program (LMVJV Management Board 2013). The LMVJV is one of the 22 regular (and more recently three species specific; see mbjv.org) joint ventures that were formed via the North American Waterfowl Management Plan in 1986, arguably one of the most effective wildlife management plans ever conceived. There are five working groups of the LMVJV: the Forest Resources Conservation, Landbird, Shorebird, Waterfowl, and Winter Rice Food Availability working groups. The LMVJV is the primary science entity for sustaining bird populations and related habitats in the MAV and Gulf Coastal Plain regions. To build on an already successful model, 22 Landscape Conservation Cooperatives (LCC) (http://www.fws.gov/landscape-conservation/lcc.html) were formed in the United States by the Secretary of the Interior in 2010. The LCCs are represented by myriad conservation interest groups to champion landscapes that sustain natural and cultural resources for current and future generations. Like the LMVJV, LCCs are science based and strive to understand and disseminate information on the effects of climate change and other environmental stressors at landscape scales. The Gulf Coastal Plains and Ozarks LCC, covering the ecoregion of the MAV, works across 72 million ha in 12 states, mostly in the southern United States. Restoration of wetlands, riparian forests, and other important resources in the MAV is largely guided by the LMVJV and GCPO LCC entities.

Fig. 5 Restoration of bottomland hardwood forest in the Grand Prairie region, Arkansas, USA. Note spacing between tree rows, a method to encourage sunlight in forest floor to facilitate growth of palustrine emergent herbaceous wetland plants, e.g., moist-soil vegetation, such as annual wetland grasses and sedges (Photo credit: J.Pagan © Rights remain with the photographer)

Approximately 82% (2.9 million ha) of the forested area of the MAV is privately owned. Of the remaining 18% (526,000 ha), 52% is owned by state and local governments, 31% by the US Fish and Wildlife Service, and the remainder by other federal agencies (Oswalt 2013). Preserving existing forest land is important to many constituents of the MAV, and restoration of wetlands and associated habitats from marginal agricultural lands is a priority. Some 11,000 landowners that managed >1.07 million ha have enrolled in the Wetlands Reserve Program (WRP) across the United States since the program's inception ~1992. The WRP has been significant in the MAV where 261,708 ha were reforested per 1,857 private landowner agreements in Arkansas, Louisiana, and Mississippi from 1992 to 2011 (Oswalt 2013; Fig. 5). The WRP is arguably the most significant and effective wetland and associated forest restoration program on private lands in the world. The WRP was established to restore and protect functions and values of wetlands in agricultural landscapes, emphasizing habitat for migratory birds and wetland-dependent wildlife, protection and improvement of water quality, flood attenuation, ground water recharge, protection of native flora and fauna, and educational and scientific scholarship (Faulkner et al. 2008). Examples of other conservation programs that assist private landowners in the MAV include the Conservation Reserve Program, Wildlife Habitat Incentive Program, and Environmental Quality Incentives Program.

Ecosystem Services

The WRP has generated as much as $300 M annually through the provision of ecosystem services (King et al. 2005; Jenkins et al. 2010; Oswalt 2013). When agricultural land is retired and enrolled into a WRP easement, by default the nitrogen (N) losses driven by fertilizer application, fixation, and tilling cease (Jenkins et al. 2010). Nitrate (NO_3) is the N compound that is most closely associated with the hypoxic zone in the Gulf of Mexico (see "Future Challenges" section), and WRP reforestation in the MAV provided an estimated social welfare value of $1248/ha/year in N mitigation (Jenkins et al. 2010). Relative to greenhouse gas (GHG) dynamics, BLHFs store more than eight times the level of carbon and have greater denitrification potential than agricultural fields (Faulkner et al. 2008). When retired agricultural land is reforested via WRP, the monetized net mitigation value of GHG, or the difference between WRP and agricultural sites, has ranged from $171 to $222/ha/year for the social values (Jenkins et al. 2010). By combining values of GHG mitigation, N mitigation, and waterfowl recreation, Jenkins et al. (2010) estimated the market value of these services at $1,035/ha annually for the MAV, concluding that the WRP has provided a favorable return on public investment.

Socioeconomic impacts of hunting, fishing, bird-watching, and other outdoor-related activities in the MAV are significant. In the delta of Mississippi alone, outdoor recreation generated $2.7 billion in goods and services in 2005–2006; waterfowl hunting alone provided an adjusted $86.8 million (2009 USD) in economic impacts and supported 1,139 full- and part-time jobs in Mississippi (Henderson et al. 2010).

In addition, the channel catfish aquaculture industry of Arkansas, Louisiana, and Mississippi currently generates $56 million in annual payroll that supports many rural economies, and rice agriculture contributes nearly $10 billion to the regional MAV economy. Commercial harvest of red swamp crayfish is the most valued shellfish crop in the United States. Crayfish production was valued at $127 million annually in 2008, of which >93% of production occurs on about 43,000 ha in Louisiana, some of which originate from the Atchafalaya Basin or elsewhere in the MAV (National Aquaculture Sector Overview 2011).

Future Challenges

Significant anthropogenic modifications in the nineteenth to twentieth centuries have negatively affected most flora and fauna of the MAV. Water quality is a significant indicator of the health of the Gulf of Mexico. Many sources of nonpoint source pollution (e.g., fertilizers, toxic chemicals, livestock waste) negatively influence water quality in the MAV. Excess nitrogen and phosphorous result from urban development, industry, agricultural runoff, atmospheric deposition, and other sources throughout the entire Gulf watershed (Jenkins et al. 2010). Floodplain lakes of the MAV have experienced a 50-fold increase in sedimentation rates since land clearing for agriculture (Dembkowski and Miranda 2012). Agricultural runoff, largely from America's upper Midwestern "Corn Belt," currently contributes approximately 74% of the NO_3 load transported by the Mississippi River (Rabalais et al. 2002; Howarth et al. 2002; Faulkner et al. 2008). The increase in dissolved and particulate NO_3 levels causes extensive eutrophication and hypoxia (low oxygen) or "dead zone," in the northern Gulf of Mexico (Faulkner et al. 2008). Each summer the hypoxic zone forms in the Gulf from the mouth of the Mississippi River to the Upper Texas Coast. Hypoxia is a serious environmental concern because it creates uninhabitable conditions for marine life and threatens diverse and sustainable fisheries and sensitive ecosystems. Nutrient loading is also affected by stratification (layering) of waters in the Gulf of Mexico (Jenkins et al. 2010). To reduce NO_3 and subsequent hypoxia in the Gulf, Mitsch et al. (2001) recommended that an additional 971,800 ha of lands be used for creating wetlands and restoring riparian forests throughout the entire Mississippi River Basin (Faulkner et al. 2008).

The Mississippi River Valley alluvial aquifer is the third most used aquifer in the United States. Arkansas and Mississippi are the first and second greatest users, respectively, of the aquifer, mostly for crop irrigation. In Mississippi, for example, $>3.5^{10}$ m^3 are withdrawn from the aquifer each day. In several counties in the western Mississippi Delta, an estimated cumulative loss of nearly 186 million m^3/year in water storage has occurred from 1987 to 2009. An estimated 25% reduction in water use in that geographic region of Mississippi would result in a 32% improvement across the entire Mississippi Delta, which may be an important conservation strategy to ensure adequate water supplies in the future (Barlow and Clark 2011).

References

Barlow JRB, Clark BR. Simulation of water-use conservation scenarios for the Mississippi Delta using an existing regional groundwater flow model: U. S. Geological Survey Scientific Investigations report 2011–5019, Reston; 2011.

Brown CR, Baxter C, Pashley DN. The ecological basis for the conservation of migratory birds in the Mississippi Alluvial Valley. In: Bonney R, Pashley DN, Cooper RJ, Niles L, editors. Strategies for bird conservation: the partners in flight planning process. Ithaca: Cornell Lab of Ornithology; 2000. p. 1–7. http://birds.cornell.edu/pifcapemay. Accessed 14 Oct 2013.

Conner WH, Sharitz RR. Forest communities of bottomlands. In: Fredrickson LH, King SL, Kaminski RM, editors. Ecology and management of bottomland hardwood ecosystems: the state of our understanding, Special publication No. 10. Puxico: University of Missouri-Columbia, Gaylord Memorial Laboratory; 2005. p. 93–120.

Dembkowski DJ, Miranda LE. Hierarchy in factors affecting fish biodiversity in floodplain lakes of the Mississippi Alluvial Valley. Environ Biol Fishes. 2012;93:357–68.

Faulkner SW, Barrow JRB, Keeland B, Walls S. Interim report assessment of ecological services derived from U. S. Department of Agriculture conservation programs in the Mississippi Alluvial Valley: Regional estimates and functional condition indicator models. 2008.

Feaga JS. Winter waterbird use and food resources of aquaculture lands in Mississippi [thesis]. Mississippi State: Mississippi State University; 2014.

Gardiner ES, Oliver JM. Restoration of bottomland hardwood forests in Lower Mississippi Alluvial Valley, U.S.A. In: Stanturf JA, Madsen P, editors. Restoration of boreal and temperate forests. Boca Raton: CRC Press; 2005. p. 235–51.

Griep MT, Collins B. Chapter 14: Wildlife and forest communities. The Southern Forest Futures Project Technical Report. General Technical Report 178. Asheville: USFS Southern Research Station; 2011. http://www.srs.fs.usda.gov/futures/technical-report/14.html. Accessed 20 Nov 2013.

Henderson JE, Grado SC, Munn IA, Jones WD. Economic impacts of wildlife- and- fisheries-associated recreation on the Mississippi economy: An input-output analysis. Forest and Wildlife Research Center, Research Bulletin FO398, Mississippi State University; 2010.

Howarth RW, Boyer EW, Pabich WJ, Galloway JN. Nitrogen use in the United States from 1961–2000 and potential future trends. Ambio. 2002;31:88–96.

Jenkins WA, Murray BC, Kramer RA, Faulkner SP. Valuing ecosystem services from wetlands restoration in the Mississippi Alluvial Valley. Ecol Econ. 2010;69:1051–61.

King SL, Shepard JP, Ouchley K, Neal JA. Bottomland hardwood forests: past, present, and future. In: Fredrickson LH, King SL, Kaminski RM, editors. Ecology and management of bottomland hardwood systems: the state of our understanding, Gaylord Memorial Laboratory special publication No. 10. Puxico: University of Missouri-Columbia; 2005. p. 1–17.

Klimas C, Foti T, Pagan J, Williamson M, Murray E. Potential natural vegetation maps for ecosystem restoration in the Mississippi Alluvial Valley. EMRRP technical notes collection. ERDC TN-EMRRP-ER-16. Vicksburg: U.S. Army Engineer Research and Development Center; 2012.

Kross J, Kaminski RM, Reinecke KJ, Penny EJ, Pearse AT. Moist-soil seed abundance in managed wetlands of the Mississippi Alluvial Valley. J Wildl Manag. 2008;72:707–14.

Lehnen SE, Krementz DG. Use of aquaculture ponds and other habitats by autumn migrating shorebirds along the lower Mississippi River. Environ Manag. 2013;52:417–26.

Lower Mississippi Valley Joint Venture: Management Board. 2013. www.lmvjv.org

Mitsch WJ, Day Jr JW, Gilliam JW, Groffman PM, Hey DL. Reducing nitrogen loading to the Gulf of Mexico from the Mississippi River basin: strategies to counter a persistent ecological problem. Bioscience. 2001;51:373–88.

National Aquaculture Sector Overview. United States of America. National aquaculture sector overview fact sheets. Text by Olin PG. In: FAO Fisheries and Aquaculture Department [online]. Rome. [updated 2011 Feb 1; cited 2013 Oct 14].

Oswalt SJ. Forest resources of the lower Mississippi Alluvial Valley. Gen. Tech. Rep. SRS-117. Asheville: U.S. Department of Agriculture Forest Service, Southern Research Station; 2013.

Pearse AT, Kaminski RM, Reinecke KJ, Dinsmore SJ. Local and landscape associations between wintering dabbling ducks and wetlands complexes in Mississippi. Wetlands. 2012;32:859–69.

Petrie M, Brasher M, James D. Estimating the biological and economic contributions that rice habitats make in support of North American waterfowl. Stuttgart: The Rice Foundation; 2014.

Rabalais NN, Turner RE, Scavia D. Beyond science into policy: Gulf of Mexico hypoxia and the Mississippi River. Bioscience. 2002;52:129–42.

Ranalli N, Ritchison G. Phenology of shorebird migration in western Kentucky. Southeast Nat. 2012;11:99–110.

Reinecke KJ, Kaminski RM, Moorhead DJ, Hodges JD, Nassar JR. Mississippi Alluvial Valley. In: Smith LM, Pederson RL, Kaminski RM, editors. Habitat management for migration and wintering waterfowl in North America. Lubbock: Texas Tech University Press; 1989. p. 203–47.

Coastal Wetlands of Manitoba's Great Lakes (Canada)

47

Dale Wrubleski, Pascal Badiou, and Gordon Goldsborough

Contents

Introduction	592
Coastal Wetlands	593
Coastal Wetlands of the Manitoba Great Lakes	595
Netley-Libau Marsh	595
Delta Marsh	598
Threats and Future Challenges	602
References	603

Abstract

The province of Manitoba, Canada, contains three of the world's largest freshwater lakes; Lakes Winnipeg, Winnipegosis and Manitoba. A significant feature of these lakes are their extensive coastal wetlands. A GIS-inventory found the Manitoba Great Lakes have six times more wetlands per km of shoreline than the Laurentian Great Lakes. Lake Winnipeg has a coastal wetland area of 1,404 km^2, Lake Manitoba has 564 km^2, and Lake Winnipegosis has 742 km^2. Netley-Libau Marsh (222 km^2) on Lake Winnipeg, and Delta Marsh (185 km^2) on Lake Manitoba, are believed to be the largest freshwater coastal wetlands in North America. These wetlands provide many benefits to their adjoining lakes, and provide important wildlife and fisheries habitat. However, lake-level regulation, nonpoint source nutrient pollution and invasive species are significant threats to these coastal wetlands and the ecosystem benefits they provide.

D. Wrubleski (✉) · P. Badiou (✉)
Institute for Wetland and Waterfowl Research, Ducks Unlimited Canada, Stonewall, MB, Canada
e-mail: d_wrubleski@ducks.ca; p_badiou@ducks.ca

G. Goldsborough (✉)
Department of Biological Sciences, University of Manitoba, Winnipeg, MB, Canada
e-mail: Gordon.Goldsborough@umanitoba.ca

© Springer Science+Business Media B.V., part of Springer Nature 2018
C. M. Finlayson et al. (eds.), *The Wetland Book*,
https://doi.org/10.1007/978-94-007-4001-3_190

Keywords

Coastal wetlands · Delta Marsh · Lake Manitoba · Lake Winnipeg · Lake Winnipegosis · Netley-Libau Marsh

Introduction

The province of Manitoba, Canada, located in central North America contains three of the world's largest freshwater lakes (by area). Lakes Winnipeg, Winnipegosis, and Manitoba are the remnants of glacial Lake Agassiz, which was the largest lake in North America during the last deglaciation (Mann et al. 1999). At any one time, Lake Agassiz covered at least 260,000 km^2 and had a volume of 22,700 km^3 (Leverington et al. 2000). However, over its 4,000-year history, Lake Agassiz covered an overall area of approximately 1 million km^2 (Teller and Clayton 1983).

Lake Winnipeg is the largest of the Manitoba Great Lakes at 23,750 km^2 and is the tenth largest freshwater lake in the world and sixth largest in Canada. Lake Winnipeg consists of a large, deeper north basin (mean depth 13.3 m) and a smaller, relatively shallow south basin (mean depth 9 m). Water levels on Lake Winnipeg are controlled by the Manitoba Hydro power utility, making Lake Winnipeg the third largest hydroelectric reservoir in the world (Environment Canada and Manitoba Water Stewardship 2011).

Lake Winnipegosis is the second largest of the Manitoba Great Lakes, covering an area of 5,375 km^2 with a mean depth of 3 m (Downing and Duarte 2009). Lake Winnipegosis is Canada's 11th largest lake and the 30th largest in the world. Its watershed spans 49,825 km^2. Lake Winnipegosis is the only Manitoba Great Lake whose water levels are not regulated.

Lake Manitoba is the smallest of the Manitoba Great Lakes. Like the other Manitoba Great Lakes, Lake Manitoba is very shallow with a mean depth of 5 m. At 4,625 km^2 it is Canada's 13th largest lake and the 33rd largest in the world. Lake Manitoba's watershed covers an area of approximately 79,000 km^2. However, when the Portage Diversion is in operation, which diverts flow from the Assiniboine River into Lake Manitoba, the watershed expands to cover 121,800 km^2 (Page 2011).

All three Manitoba Great Lakes are located within the Lake Winnipeg watershed. The Lake Winnipeg watershed spans nearly 1 million km^2, covering parts of Alberta, Saskatchewan, Manitoba, and Ontario in Canada and Montana, North Dakota, South Dakota, and Minnesota in the United States. The watershed is contained within the Nelson River drainage basin which flows into Hudson Bay. The northern and eastern portions of the Lake Winnipeg watershed lie mostly in the Canadian boreal forest and have vast expanses of peatlands. This is particularly the case for Lake Winnipegosis and the north basin on Lake Winnipeg. However, the dominant land cover in the Lake Winnipeg watershed is agricultural farmland which covers approximately 650,000 km^2.

Coastal Wetlands

Connection with waters of large lakes is a key feature that distinguishes coastal wetlands from other freshwater inland wetlands, and given the dynamic nature of large lakes, their coastal wetlands are also inherently dynamic (Keough et al. 1999). Most research on freshwater coastal wetlands has been conducted in the Laurentian Great Lakes where coastal wetlands have been defined as:

> ... lands transitional between terrestrial and aquatic systems where the water table is usually at or near the surface or the land is covered by shallow water. Wetlands must have one or more of the following three attributes: 1) at least periodically, the land supports predominantly hydrophytes; 2) the substrate is predominantly undrained hydric soil; and 3) the substrate is non-soil and is saturated with water or covered by shallow water at some time during the growing season of the year. Wetlands may be considered to extend lake-ward to the water depth of 2 m, using the historic low and high water levels or the greatest extent of wetland vegetation. Hydrologic connections with one of the Great Lakes may extend upstream along rivers since exchanges caused by seiches [wind tides] and longer-period lake-level fluctuations influence riverine wetlands. Wetlands under substantial hydrologic influence from Great Lakes waters may be considered coastal wetlands. (Simon and Stewart 2006)

Freshwater coastal wetlands can be further separated into three specific hydrogeomorphic systems, lacustrine, riverine, and barrier-protected (Fig. 1), based on geomorphic position, dominant hydrologic source, and current hydrologic connectivity to the lake (Albert et al. 2005). In brief, lacustrine coastal wetlands are

Fig. 1 Barrier-protected coastal wetland of Lake Winnipeg (Photo Credit: G. Goldsborough © Rights remain with the author)

controlled directly by waters of their associated lakes and are strongly affected by lake-level fluctuations, nearshore currents, seiches, and ice scour. The riverine coastal wetland class occurs along and within rivers and creeks that flow into or between Great Lakes. Water levels and fluvial processes in these wetlands are directly or indirectly influenced by coastal processes because lake waters flood back into the lower portions of the drainage system. Nearshore and onshore processes have separated barrier-protected wetlands from the Great Lakes by a barrier beach or other barrier features. These wetlands are protected from wave action but may be connected directly to the lake by a channel crossing the barrier.

The values of Great Lakes are enhanced by their associated coastal wetlands (Simon and Stewart 2006). Coastal wetlands provide numerous ecosystem services such as flood storage; sediment traps; nutrient filters; shoreline erosion buffers; habitat for plants, fish, and other wildlife; and hotspots for biodiversity (Maynard and Wilcox 1996). Additionally, due to their high productivity, wetlands are critical in the global carbon cycle (Mitsch and Gosselink 2000), and coastal wetlands are important areas for carbon sequestration (Bernal and Mitsch 2012). Coastal wetlands are situated at the interface between the watershed and lake and are particularly vulnerable to land-based stressors (Morrice et al. 2008). These stressors can impact the function of coastal wetlands and impair their capacity to provide the ecosystem services described above. In the Manitoba Great Lakes, there is an increasing development pressure that directly affects coastal wetlands and the ecosystem services they provide (Fig. 2).

Fig. 2 Residential development, visible in the foreground of this photograph, threatens the ecological integrity of coastal wetlands on Lake Winnipeg (Photo credit: G. Goldsborough © Rights remain with the author)

Coastal Wetlands of the Manitoba Great Lakes

A GIS-based inventory of coastal wetlands within 10 km of the shorelines of the Manitoba Great Lakes shows they have six times more wetlands per km of shoreline than the comparatively more well-studied Laurentian Great Lakes (Watchorn et al. 2012; Fig. 3). Lake Winnipeg has a coastal wetland area of 1,404 km^2 (0.8 km^2/km), Lake Manitoba has 564 km^2 (0.6 km^2/km), and Lake Winnipegosis has 742 km^2 (0.8 km^2/km). Riverine wetlands are the most common type on Lakes Winnipeg and Winnipegosis, whereas barrier-protected wetlands are the most common type on Lake Manitoba. When Treed Muskeg habitat in the northern regions of the lake watersheds is included in the inventory, the coastal wetland areas for lakes Winnipeg and Winnipegosis are greater by 548% and 273%, respectively, whereas the total for more southerly Lake Manitoba is greater by only 18%. Netley-Libau Marsh (222 km^2; N50.35690, W96.79201) on Lake Winnipeg and Delta Marsh (185 km^2; N50.18333, W98.31667) on Lake Manitoba are, we believe, the two largest coastal wetlands in North America. The characteristics of, and anthropogenic threats to, these important coastal wetlands are described below. Other large coastal marshes of the Manitoba Great Lakes include Lake Francis (N50.30452, W97.97242), Marshy Point (N50.55479, W98.08963) and Big Point (N50.40414, W98.55819) on Lake Manitoba, and Willow Point (N50.59843, W96.96811) and Grand Marais (N50.55806, W96.61173) on Lake Winnipeg.

Netley-Libau Marsh

Netley-Libau Marsh lies along the south shore of Lake Winnipeg, separated from the lake by a 25-km series of barrier islands (Fig. 4). The postglacial origins of the marsh have not been studied, but it likely originated when barrier beach formation along the southern shore of Lake Winnipeg isolated the southernmost part of the lake; southward migration of Lake Winnipeg due to isostatic uplift of the northern basin has eroded the barrier beach (Nielsen and Conley 1994). The marsh consists of a complex of shallow lakes, bays (lagoons), and channels, bisected by the Red River of the North that passes through the marsh on its way to Lake Winnipeg. Netley Marsh lies on the west side of the river and Libau Marsh lies on the east. Openings in the barrier islands (currently, eleven) allow water exchange between Lake Winnipeg and Netley-Libau Marsh. Strong wind setup on Lake Winnipeg can result in significant, short-term fluctuations in marsh water level, sometimes exceeding 1 m.

Netley-Libau Marsh provided resources for early aboriginal people and subsequently for European settlers who began to arrive in abundance in the early 1800s. The vicinity developed into an important recreational and agricultural area through the twentieth century. Its importance has much to do with the abundant fish and wildlife that can be found there. It has been recognized internationally as major habitat for nesting, staging, and molting waterfowl. It is also used for recreational activities such as hunting, fishing, boating, bird-watching, and ecotourism. The marsh has been recognized as an Important Bird Area (IBA). Migratory Neotropical songbirds and

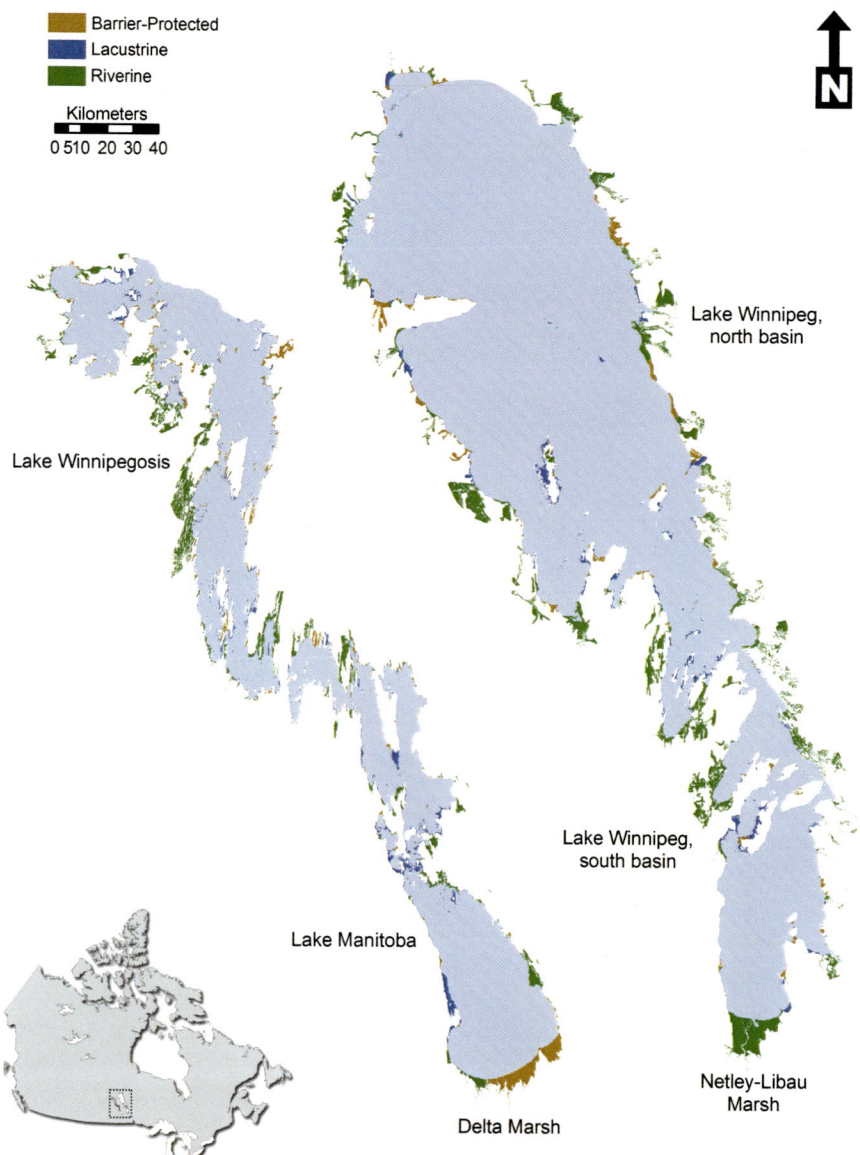

Fig. 3 Coastal wetlands of the Manitoba Great Lakes, excluding the Treed Muskeg that is abundant around the north ends of Lake Winnipeg and Lake Winnipegosis (Watchorn et al. 2012)

waterfowl use the marsh for nesting or during migration. Currently, over 90% of the marsh is publicly owned and includes a 1,073-ha Game Bird Refuge. Abundant bird species that use the marsh for nesting or during migration include Franklin's gull *Leucophaeus pipixcan*, Forster's tern *Sterna forsteri*, black-crowned night heron

Fig. 4 An aerial photograph of Netley-Libau Marsh with Lake Winnipeg visible in the upper, right and the Red River passing from left to right in the center (Photo credit: D. Wrubleski © Rights remain with the author)

Nycticorax nycticorax, western grebe *Aechmophorus occidentalis*, and several waterfowl species, including mallard *Anas platyrhynchos*, redhead *Aythya americana*, and blue-winged teal *Anas discors*. The marsh also provides spawning, nursery, and feeding habitat for fish species that are important for commercial and sport fisheries on Lake Winnipeg and the Red River (Janusz and O'Connor 1985), including northern pike *Esox lucius*, sauger *Sander canadensis*, yellow perch *Perca flavescens*, goldeye *Hiodon alosoides*, and shiners *Notropis* spp. The predominant marsh furbearer is the muskrat *Ondatra zibethicus* (Manitoba Wildlife Branch 1986).

However, the ecological integrity of Netley-Libau Marsh is under threat (Grosshans et al. 2004). Introductions of exotic species such as purple loosestrife *Lythrum salicaria* and common carp *Cyprinus carpio* have altered its structure. The Red River, which drains an enormous geographic area and supports large urban centers, is a source of nutrients and pollutants. Erosion of uplands that formerly divided the marsh into smaller units has led to the creation of large, open, windswept bays that have lost their fundamental marsh character. Dredging of a channel into Netley Lake in 1913 (Fig. 5) has permitted greater inflow of polluted water from the Red River. Habitat for fish and wildlife has been compromised through the loss of aquatic plants and uplands that provide cover, habitat diversity, and breeding sites. Changes in the marsh have been ongoing for a long period of time, and a combination of factors (e.g., lake-level regulation, Red River water quality, river dredging) are responsible.

Fig. 5 The "Netley Cut" shown here was originally excavated from the Red River into Netley-Libau Marsh in 1913 but has eroded into a channel that now redirects from one-third to one-half of the entire river flow in the marsh (Photo credit: G. Goldsborough © Rights remain with the author)

Efforts to restore marsh vegetation to Netley-Libau Marsh are predicated on the benefits that it would provide in capturing nutrients flowing through the Red River into Lake Winnipeg and reducing its eutrophication rate, providing habitat for fish and furbearing mammals, and using harvested marsh plants as a source of biofuel and fertilizer (from absorbed nutrients). At present, most of the marsh is too deep (over 2 m) for natural plant regeneration to occur. Initially, attention was placed on the creation of artificial reefs that would be planted with hybrid cattail *Typha x glauca*. It is now believed this approach would have high associated construction costs and low likelihood of successful plant regrowth. We are now investigating floating cattail bioplatforms that would grow the plants hydroponically, in theory in any depth of water so long as ambient nutrient levels are sufficient to sustain plant growth.

Delta Marsh

Delta Marsh is a large coastal wetland stretching approximately 32 km along the south shore of Lake Manitoba (Fig. 6). The marsh covers an area of between 150 and 250 km^2, depending on water levels and areas included as part of the marsh (Delta Marsh Technical Committee 1968; Shay et al. 1999; Batt 2000). It consists of a series of large and small open water bays and ponds, former river channels, flooded emergent vegetation, wet meadows, low prairie, and uplands. Average water depth in the open water areas is about 1 m, but depths of up to 3 m can be found in some of the former river channels. There are four openings in the forested barrier beach that

Fig. 6 An aerial photograph of Delta Marsh showing Lake Manitoba at the right (Photo credit: D. Wrubleski © Rights remain with the author)

allow the exchange of water between the lake and Delta: Clandeboye Channel on the far east side, Delta Channel near the middle of the marsh, and Cram and Deep Creeks on the far west side (Fig. 7). Depending on wind direction and wind setup on Lake Manitoba, flows in the connecting channels can reverse direction frequently, and water levels on the marsh fluctuate accordingly.

Much like Netley-Libau Marsh, Delta Marsh has its origins in a river depositing sand and silt into a remnant of glacial Lake Agassiz, but in this case, it was the Assiniboine River and the lake is Lake Manitoba. From approximately 4,500 to 2,000 years ago, all or part of the Assiniboine River flowed into the south end of Lake Manitoba (Last 1980). However, at approximately 2,000 years ago, the river stopped flowing northward and changed course toward the east. Wave action eroded and redistributed the river-deposited sediments and created the barrier beach that now separates Delta Marsh from Lake Manitoba.

Delta Marsh is a wetland of international importance, being designated a Ramsar wetland in 1982 (Gillespie and Boyd 1991) for its importance to waterfowl. It was designated a Manitoba Heritage Marsh in 1988 (Manitoba NAWMP Technical Committee 1988) and an Important Bird Area (IBA) in 1999. In 2006, Manitoba Conservation designated over 110 km^2 of the marsh as a Wildlife Management Area (WMA), providing some protection from industrial development. In addition, parts of the marsh are also designated as a Game Bird Refuge. Three research facilities operated on Delta Marsh and contributed significantly to our understanding of the marsh and Lake Manitoba. A major flood in 2011 resulted in the closure of the Delta

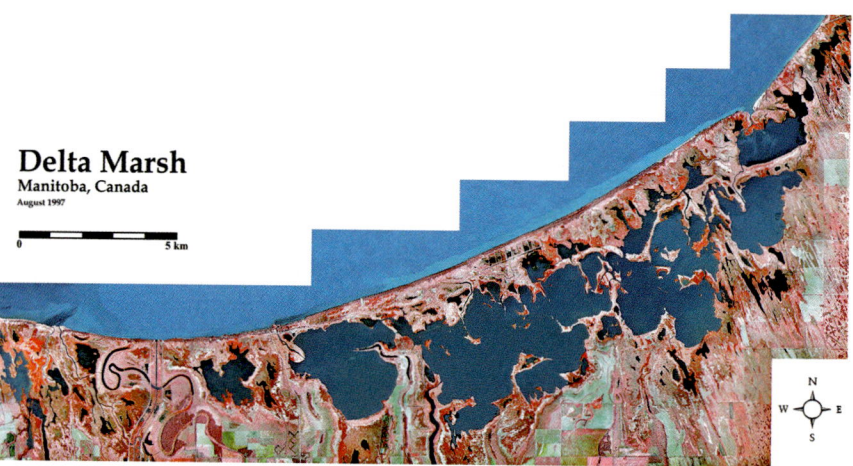

Fig. 7 Infrared color photomosaic of Delta Marsh showing the four connections between the marsh and Lake Manitoba, two on the extreme left side, one in the center, and one on the extreme right side (Photo credit: D. Wrubleski (Ducks Unlimited Canada) and G. Goldsborough (University of Manitoba) © Rights remain with the authors)

Marsh Field Station (University of Manitoba), the Delta Marsh Bird Observatory, and the Delta Waterfowl Research Station (Delta Waterfowl Foundation).

From the late 1800s, the marsh was known as an important area for waterfowl and waterfowl hunting (Fig. 8). Commercial hunting provided local employment and harvested birds were shipped to the market. Waterfowl hunting attracted a wide range of people to the marsh, including the rich and famous from around the world. Not only was the marsh important for waterfowl but also for other colonial waterbirds, including most of the same species as found in Netley-Libau Marsh. An estimated 307 species of birds have been recorded in the area, of which 136 species have been confirmed to breed (Underwood and den Haan 2000). Thirty-one species of fish use the marsh for spawning, rearing, and feeding habitat (Stewart et al. 1985; Wrubleski, unpublished data).

Starting the early 1960s, local residents and users began to report negative changes on the marsh. These changes have continued and include reduced clarity of the water column, loss of submersed aquatic vegetation, abundant populations of common carp, algal blooms, increasing dominance of the emergent plant community by the hybrid cattail and encroachment and loss of shallow ponds, loss of islands in the larger bays and ponds, and reductions in waterfowl, muskrats, and other wildlife that use the marsh. Reasons for the changes in the marsh are many and include stabilization of water levels on Lake Manitoba, abundant populations of common carp, and eutrophication of the marsh as a result of increased nutrient loading (Goldsborough and Wrubleski 2001).

Water levels on Lake Manitoba and adjacent Delta Marsh historically fluctuated within a range of 1.7 m. However, in 1961, completion of a control structure on the Fairford River, the only major outlet to Lake Manitoba, reduced fluctuations to less than 0.6 m. Stabilization of water levels has likely contributed to encroachment of hybrid cattail into shallow ponds, erosion of the islands in the larger bays, and increased

Fig. 8 The practice of waterfowl hunting has a long history at Delta Marsh and other coastal wetlands of the Manitoba Great Lakes (Photo credit: G. Goldsborough and Delta Marsh History Group (www.deltahistory.org) © Rights remain with the author)

turbidity within the larger bays due to increased wave action. In 2003, a stakeholder group, the Lake Manitoba Regulation Review Advisory Committee, recommended that the lake be allowed to fluctuate to a greater extent than allowed under the previous management plan. The province accepted this recommendation in principle. A 1-in-400-year flood in 2011 did significant damage to private property and infrastructure around the lake. On the marsh, it resulted in a significant dieback in the invasive hybrid cattail. As a result of the flood, a government-appointed stakeholder group recommended that the operating range for Lake Manitoba should be lowered by 0.15 m. Lower water levels may have benefits for recruitment of vegetation in Delta Marsh but could accelerate the encroachment by hybrid cattail.

From the early 1960s to the mid-1980s, several large-scale management plans were proposed to restore Delta Marsh. These plans primarily called for the isolation of the marsh from Lake Manitoba and to artificially manipulate water levels within the marsh. However, none of these plans were implemented. Beginning in 1997, several agencies began working on the marsh to document the extent of change and to reexamine options for improving marsh conditions (Goldsborough and Wrubleski 2001). Activities have included mapping the aquatic plants, both submersed and emergent, monitoring water quality, and documenting the large fish community of the marsh. Gillnet collections on the marsh confirmed the abundance of common carp. Subsequent experimental manipulations and controlled stocking of isolated ponds confirmed the significant effects of these large invasive benthivorous fish (Hnatiuk 2006; Parks 2006; Badiou and Goldsborough 2010; Hertam 2010).

Fig. 9 In early 2013, seven structures were constructed to exclude most large common carp from Delta Marsh as a means of restoring submersed macrophytes and other components of the coastal wetland ecosystem degraded by this invasive, benthivorous fish (Photo credit: D. Wrubleski © Rights remain with the author)

Recognizing the significant impacts of common carp on the marsh, a restoration plan was developed that called for carp exclusion. Research studies have been initiated to better understand the size structure of the large fish community and timing of the annual spring migration into the marsh. Based on this information, seven exclusion structures were constructed or modified in early 2013 (Fig. 9). These structures use temporary screens to selectively prevent common carp from accessing the marsh each spring. These structures are part of a larger restoration effort called "Restoring the Tradition at Delta Marsh." Besides excluding common carp, the project also initiated a 5-year research program to better understand the hydraulics and hydrology of the marsh, determine sources of nutrients, and examine effective control methods for hybrid cattail. This new information will inform a second phase of the restoration project to address eutrophication and hybrid cattail control.

Threats and Future Challenges

The overall threats to the Manitoba Great Lakes provide context for the specific threats mentioned above for Netley-Libau Marsh and Delta Marsh. Water quality in Lake Winnipeg has been declining steadily over the last half century, and the lake was recently declared the most threatened lake in the world. The frequency and severity of cyanobacterial blooms in Lake Winnipeg are the most noticeable symptom of the declining health of the Lake Winnipeg watershed. The ratio of watershed to lake area for all three Manitoba Great Lakes (range 9 to 40) is substantially larger and more variable relative those of the Laurentian Great Lakes (range 2 to 4). For this reason and due to the sheer size, as well as the fact that agricultural production dominates the watershed, nonpoint source nutrient pollution is suspected to be the main contributing factor in the ongoing eutrophication of Lake Winnipeg.

While water quality issues are at the forefront for Lake Winnipeg and its watershed, there are a number of other stressors that are acting on Manitoba Great Lakes and their associated coastal wetlands. Water levels on Lake Manitoba and Lake Winnipeg have been regulated since 1961 and 1976, respectively. In addition to eutrophication and water-level regulation, a number of invasive species have become established in Lake Winnipeg and its watershed. To date, 14 aquatic invasive species have been introduced to Manitoba waters: the common carp, rainbow smelt *Osmerus mordax*, white bass *Morone chrysops*, feral goldfish *Carassius auratus*, rusty crayfish *Orconectes rusticus*, spiny waterflea *Bythotrephes cederstroemi*, water flea *Eubosmina coregoni*, freshwater jellies *Craspedacusta sowerbyi*, black algae *Lyngbya wollei*, purple loosestrife, flowering rush *Butomus umbellatus*, invasive common reed *Phragmites australis*, Asian tapeworm *Bothriocephalus acheilognathi*, and koi herpesvirus (Environment Canada and Manitoba Water Stewardship 2011). Of great concern is the recent discovery of zebra mussels *Dreissena polymorpha* in the south basin of Lake Winnipeg during the fall of 2013.

References

Albert DA, Wilcox DA, Ingram JW, Thompson TA. Hydrogeomorphic classification for Great Lakes coastal wetlands. J Great Lakes Res. 2005;31 Suppl 1:129–46.
Badiou PHJ, Goldsborough LG. Ecological impacts of an exotic benthivorous fish in large experimental wetlands, Delta Marsh, Canada. Wetl. 2010;30:657–67.
Batt BDJ. The Delta Marsh. In: Murkin HR, van der Valk AG, Clark WR, editors. Prairie wetland ecology: the contribution of the Marsh Ecology Research Program. Ames: Iowa State University Press; 2000. p. 17–33.
Bernal B, Mitsch WJ. Comparing carbon sequestration in temperate freshwater wetland communities. Glob Chang Biol. 2012;18(5):1636–47.
Branch MW. Netley-Libau Marshes: resource development and management proposal, summary report. Winnipeg (MB): Manitoba Department of Natural Resources; 1986. p. 1986.
Delta Marsh Technical Committee. The Delta Marsh: it's values, problems and potentialities. Winnipeg (MB): Manitoba Department of Mines and Natural Resources; 1968.
Downing JA, Duarte CM. Abundance and size distribution of lakes, ponds and impoundments. In: Likens GE, editor. Encyclopedia of inland waters. Oxford: Elsevier; 2009. p. 469–78.
Environment Canada and Manitoba Water Stewardship. State of Lake Winnipeg: 1999 to 2007. Winnipeg (MB): Manitoba Water Stewardship; 2011.
Gillespiel DI, Boyd H. Wetlands for the world: Canada's Ramsar sites. Ottawa: Canadian Wildlife Service; 1991.

Goldsborough LG, Wrubleski DA. The decline of Delta Marsh, an internationally significant wetland in South-Central Manitoba. Sixth Prairie Conservation and Endangered Species Conference, Winnipeg. 2001. CD-ROM.

Grosshans RE, Wrubleski DA, Goldsborough LG. Changes in the emergent plant community of Netley-Libau Marsh between 1979 and 2001. Winnipeg: University of Manitoba; 2004. p. 52. Delta Marsh Field Station Occasional Publication No. 4.

Hertam SC. The effects of common carp (*Cyprinus carpio* L.) on water quality, algae and submerged vegetation in Delta Marsh, Manitoba. [master's thesis]. Winnipeg: University of Manitoba; 2010.

Hnatiuk SD. Experimental manipulation of ponds to determine the impact of common carp (*Cyprinus carpio* L.) in Delta Marsh, Manitoba: effects on water quality, algae, and submersed vegetation. [master's thesis]. Winnipeg (MB): University of Manitoba; 2006.

Janusz RA, O'Connor JF. The Netley-Libau Marsh Fish Resource. Winnipeg (MB): Manitoba Natural Resources; 1985. p. 176. Fisheries Manuscript Report 85–19.

Keough JR, Thompson TA, Guntenspergen GR, Wilcox DA. Hydrogeomorphic factors and ecosystem responses in coastal wetlands of the Great Lakes. Wetl. 1999;19:821–34.

Last WM. Sedimentology and post-glacial history of Lake Manitoba. [dissertation]. Winnipeg (MB): University of Manitoba; 1980.

Leverington DW, Mann JD, Teller JT. Changes in the bathymetry and volume of glacial Lake Agassiz between 11,000 and 9300 ^{14}C yr B.P. Quatern Res. 2000;54:174–81.

Manitoba NAWMP Technical Committee. Manitoba implementation plan of the North American waterfowl management plan. Winnipeg: Manitoba NAWMP Technical Committee; 1988.

Mann JD, Leverington DW, Rayburn J, Teller JT. The volume and paleobathymetry of glacial Lake Agassiz. J Paleolimnol. 1999;22:71–80.

Maynard L, Wilcox D. Background Report on Great Lakes Wetlands. Prepared for the state of the Great Lakes 1997 report. Chicago (IL): U.S. Environmental Protection Agency, Great Lakes National Program Office; 1996. Joint publication of Environment Canada, Toronto (ON).

Mitsch WJ, Gosselink JG. Wetlands. 3rd ed. Toronto : John Wiley and Sons; 2000.

Morrice JA, Danz NP, Regal RR, Kelly JR, Niemi GJ, Reavie ED, Hollenhorst T, Axler RP, Trebitz AS, Cotter AM, Peterson GS. Human influences on water quality in Great Lakes coastal wetlands. Environ Manag. 2008;41:347–57.

Nielsen E, Conley G. Sedimentology and geomorphic evolution of the south shore of lake winnipeg. Winnipeg (MB): Manitoba Energy and Mines, Geological Services; 1994. Geological Report GR94-1.

Page ECM. A water quality assessment of Lake Manitoba, a large shallow lake in central Canada. [master's thesis]. Winnipeg: University of Manitoba; 2011.

Parks CR. Experimental manipulation of connectivity and Common Carp; the effects on native fish, water-column invertebrates, and amphibians in Delta Marsh, Manitoba. [master's thesis]. Winnipeg: University of Manitoba; 2006.

Shay JM, de Geus PMJ, Kapinga MRM. Changes in shoreline vegetation over a 50-year period in the Delta Marsh, Manitoba in response to water levels. Wetl. 1999;19:413–25.

Simon TP, Stewart PM, editors. Coastal wetlands of the Laurentian Great Lakes: health, habitat, and indicators. Bloomington: Indiana Biological Survey; 2006.

Stewart KW, Suthers IM, Leavesley K. New fish distribution records in Manitoba and the role of a man-made interconnection between two drainages as an avenue of dispersal. Can Field Nat. 1985;99:317–26.

Teller JT, Clayton L, editors. Glacial Lake Agassiz. Geological Association of Canada Special Paper 26. 1983. p. 451.

Underwood TJ, den Haan HE. Checklist of the birds of Delta Marsh. 3rd ed. Portage la Prairie: Delta Marsh Bird Observatory; 2000.

Watchorn KE, Goldsborough LG, Wrubleski DA, Mooney BG. A hydrogeomorphic inventory of coastal wetlands of the Manitoba Great Lakes: Lakes Winnipeg, Manitoba, and Winnipegosis. J Great Lakes Res. 2012;38:115–22.

Coastal Wetlands of Lake Superior's South Shore (USA)

48

John Brazner and Anett Trebitz

Contents

Introduction	606
Inventory of Lake Superior Coastal Wetlands	607
Types of Coastal Wetlands in Lake Superior	607
Ecology of Coastal Wetlands in Lake Superior	610
Threats and Future Challenges	613
Annex	614
References	618

Abstract

There are more than two thousand coastal wetlands that encompass an area of about 215,000 ha in the Laurentian Great Lakes (LGL) of North America. Coastal wetlands in the LGL are distinguished hydrologically from nearby inland wetlands by a direct surface water connection with waters of an adjacent Great Lake. Daily, seasonal, annual and decadal lake level fluctuations exert important influences on the ecology of LGL coastal wetlands. Levels of human impacts in the LGL are generally greatest in the south, near the larger centers of population and most intense agriculture. Lake Superior is the largest (82,100 km^2) and most northern of the LGL and coastal wetlands associated with Lake Superior are among the least disturbed by human influences. The distribution of coastal wetlands along the US shoreline of Lake Superior in Wisconsin and Michigan is skewed towards the southwestern end of the lake due to differential effects of

---

J. Brazner (✉)
Wildlife Division, Nova Scotia Department of Natural Resources, Kentville, NS, Canada
e-mail: braznejc@gov.ns.ca

A. Trebitz (✉)
Mid-Continent Ecology Division, United States Environmental Protection Agency, Duluth, MN, USA
e-mail: trebitz.anett@epa.gov

© Springer Science+Business Media B.V., part of Springer Nature 2018
C. M. Finlayson et al. (eds.), *The Wetland Book*,
https://doi.org/10.1007/978-94-007-4001-3_235

isostatic adjustment following glaciation. The locations, geomorphic settings and basic characteristics of these south-shore wetlands are provided. Great Lakes coastal wetlands come in a variety of shapes and sizes. There is general agreement that there are three primary geomorphic types – lacustrine, riverine and barrier-protected with riverine and barrier-protected being most common along the US shoreline of Lake Superior. Coastal wetlands in Lake Superior are the most peat-dominated and support some of the highest levels of biodiversity among all LGL habitats. Process-oriented work indicates that Lake Superior coastal wetlands, 1) can export considerable numbers of young fish to adjacent bays and nearshore food webs, 2) have unique habitat fingerprints manifest in fish biochemical signatures that can be used to quantify wetland-nearshore interactions, and 3) structure, function, and response to anthropogenic stressors are all strongly influenced by hydrogeomorphic setting. Shoreline and watershed development, invasive species and climate change are among the most challenging factors affecting the integrity of Lake Superior and other Great Lakes coastal wetlands.

Keywords

Laurentian great lakes · Lake Superior · Coastal wetlands · Barrier-protected · Riverine · Lacustrine

Introduction

There are more than two thousand coastal wetlands that encompass an area of about 215,000 ha in the Laurentian Great Lakes (LGL) of North America (Ingram et al. 2004). The LGL include Lakes Superior, Michigan, Huron, Erie, and Ontario and spread across eight United States (US) states (Minnesota, Wisconsin, Illinois, Indiana, Michigan, Ohio, Pennsylvania, and New York) and one Canadian province (Ontario).

Coastal wetlands in the LGL are distinguished hydrologically from nearby inland wetlands by a direct surface water connection with waters of an adjacent Great Lake; although these wetlands often occur as complexes, only portions of which may have this connection (e.g., peatlands adjacent to shallow open water sites typically lack lake connectivity). Operationally LGL coastal wetlands can be defined as extending lakeward to a water depth of two meters and extending upriver to the extent influenced by seiches (sub-daily water level changes driven by wind and atmospheric pressure differentials) and lake level fluctuations on seasonal, annual, and decadal time scales (e.g., Herdendorf 1990; Keough et al. 1999). LGL coastal wetlands are, therefore, influenced by nearly continuous changes in water levels.

Levels of human impacts in the Laurentian Great Lakes are generally greatest in the south and near centers of population and agriculture. Lake Superior is the largest (82,100 km^2) and most northern of the Great Lakes, and coastal wetlands associated with Lake Superior are among the least disturbed by human influences. Development pressures and anthropogenic disturbances along the shoreline of Lake Superior are relatively low, although agricultural, forestry, commercial, and other stressors are present and significant in some locations (Danz et al. 2007).

The distribution of coastal wetlands along the US shoreline of Lake Superior in Wisconsin and Michigan is skewed toward the southwestern end of the lake. Differential effects of isostatic adjustment following glaciation (Herdendorf 1990; Wilcox 2012) resulted in the drowning of river mouths and other coastal habitats on the southwestern end of the lake (favoring development of coastal wetlands), while uplifting toward the eastern end of the lake combined with generally higher exposure to winds and waves (prevailing winds are from the west) reduced the abundance of coastal wetlands toward the east end of the lake. The presence of steeply sloped and often exposed Precambrian bedrock geology along the north shore of the lake precludes development of significant coastal wetlands along the US shore (i.e., Minnesota shoreline). This pattern changes somewhat along the Ontario shoreline in the Thunder Bay area where there is a cluster of coastal wetlands along the protected bays and islands that are common in this region (Ingram et al. 2004). Readers should refer to Ingram et al. (2004) and Cvetkovic and Chow-Fraser (2011) for more details on the location and condition of wetlands along the Canadian shoreline of Lake Superior.

Inventory of Lake Superior Coastal Wetlands

The most complete inventory of the coastal wetlands of Lake Superior was completed as part of studies coordinated by the Great Lakes Coastal Wetlands Consortium (GLCWC 2008), although earlier inventories provide some additional perspective (e.g., Herdendorf et al. 1981; Ingram et al. 2004; Moffett et al. 2006). Built upon the best coastal wetland data available and incorporating a formal classification process, the GLCWC binational inventory provides a standard reference for those interested in Great Lakes coastal wetlands. The inventory consists of wetland polygons and points, along with extensive attribute data for sites throughout the basin. All coastal wetlands are available for download (http://projects.glc.org/wetlands/inventory.html) as a single GIS coverage for the entire Great Lakes Basin.

Types of Coastal Wetlands in Lake Superior

Great Lakes coastal wetlands come in a variety of shapes and sizes and have been classified in a number of ways by different authors (e.g., Minc and Albert 1998; Keough et al. 1999; Albert et al. 2005), but there is general agreement that there are three primary geomorphic types – *lacustrine, riverine, and barrier protected*. There are a number of subforms classified under each of these main types (see Albert et al. 2005 and Wilcox 2012 for details).

Briefly, lacustrine wetlands occur mainly along shallow sloping sections of open shoreline or embayments, are dominated by robust emergent vegetation such as bulrushes (e.g., *Schoenoplectus* spp.), and have only minimal associated

Fig. 1 Duluth Minnesota Harbor in 1869 (**a**) showing extensive areas of peatland behind the barrier beach and Duluth Harbor as it looks today (**b**) with former peatlands replaced by open water or infilled and developed areas. For positional reference, a black arrow pointing to a small island (Hearding Island) is superimposed on both photos (Photo credit: (**a**): Nettleton House, Duluth, Minnesota, Paul B. Gaylord, photographer, courtesy of the University of Minnesota Duluth Kathryn A. Martin Archives and Special Collections, NEMHC Collections; Photo credit: (**b**): A. Trebitz © Rights remain with the author)

wet-meadow areas. Riverine wetlands occur along the lower reaches of drowned rivers or in river mouth deltas and can range from small bands of submerged and emergent vegetation lining a single channel to much more extensive estuarine-like complexes consisting of deep and shallow marsh within multiple braided or dendritic channels and wet meadow or swamp higher up in the riparian zone. Barrier-protected wetlands usually occur as lagoons with deep and shallow marsh habitat behind extensive barrier-beach formations with a transition to peatland complexes of fen and bog habitat immediately inland from the lagoons (Epstein et al. 1997).

Most of the coastal wetlands along the US shore of Lake Superior are of the barrier-protected lagoon or riverine geomorphic types (Epstein et al. 1997; Albert et al. 2005). There are few lacustrine coastal wetlands in Lake Superior because the high energy of the lake precludes colonization by wetland plants along most open shorelines (Albert et al. 2005). Historically, most of the coastal wetlands in Lake Superior were probably dominated by peatland-type vegetation and water chemistry, but watershed development and associated nutrient loading is thought to have resulted in a transition to more emergent marsh-type vegetation and water chemistry as well as an outright loss of coastal wetlands in some areas (Epstein et al. 1997; Figure 1a, b). Most of Lake Superior's coastal wetlands are still classified as poor fens (low-productivity, low-pH peatlands; Minc and Albert 1998), but almost all have adjacent shallow open water wetlands with well-developed, emergent, floating, and/or submerged plant communities (Figs. 2 and 3). However, there are also some richer emergent marshes (Fig. 4) that occur along riparian areas of tributary streams and in river mouth deltas (Epstein et al. 1997; Minc and Albert 1998); these are located primarily to the western end of the basin. As a result of patterns in surficial geology and human settlement, wetlands at the western end of the US shore tend to have clayey soils and higher levels of watershed urbanization and agriculture

Fig. 2 Ground-level photo of Bark Bay wetland (lagoon portion of riverine/lagoon complex). Note tree/shrub swamp as well as fen-type vegetation, tannin-stained water, exposed peat along shoreline (Photo credit: J. Brazner © Rights remain with the author)

Fig. 3 Ground-level photo of Laughing Whitefish wetland (a riverine/lagoon complex). Note mix of marsh- and fen-type vegetation. Barrier beach in background protects mouth from Lake Superior waves (Photo credit: J. Brazner © Rights remain with the author)

Fig. 4 Ground-level photo of typical productive St. Louis River Estuary wetland. Note emergent marsh-type vegetation, clay sediments (reddish color), road and railway in background, and fyke-net set in foreground to capture fish and turtles (Photo credit: J. Brazner © Rights remain with the author)

(leading to increased nutrient levels and decreased water clarity), whereas wetlands further east tend to have sandier soils and more forested and less populated watersheds (although residential development along the shoreline is increasing). Many of the coastal wetlands on Lake Superior are crossed by a road, but quite a few can still be described as remote (far from population centers or major roads), and some have no road access at all (Annex).

There are 45 prominent coastal wetlands (those greater than 10 ha in size) on Lake Superior's southern shore (Fig. 5, Annex). There are two complexes over 1500 ha, 17 sites between 100 and 600 ha and 20 sites less than 100 ha. The larger sites tend to have dendritic morphology or a combination of lagoon and riverine areas. Some of the wetlands also have smaller inundated areas that are not directly connected to Lake Superior; most prominent of these are the parallel depressions from ridge-and-swale areas (e.g., at Stockton Island Tombolo, Big Bay Lagoon, Au Train River).

Ecology of Coastal Wetlands in Lake Superior

The ecology of coastal wetlands on the US side of Lake Superior has been fairly well studied, both via efforts focused specifically on this region (primarily by researchers from the US Environmental Protection Agency's Mid-Continent Ecology Division, e.g., Sierszen et al. 2006; Morrice et al. 2009; Trebitz et al. 2011, and the Wisconsin

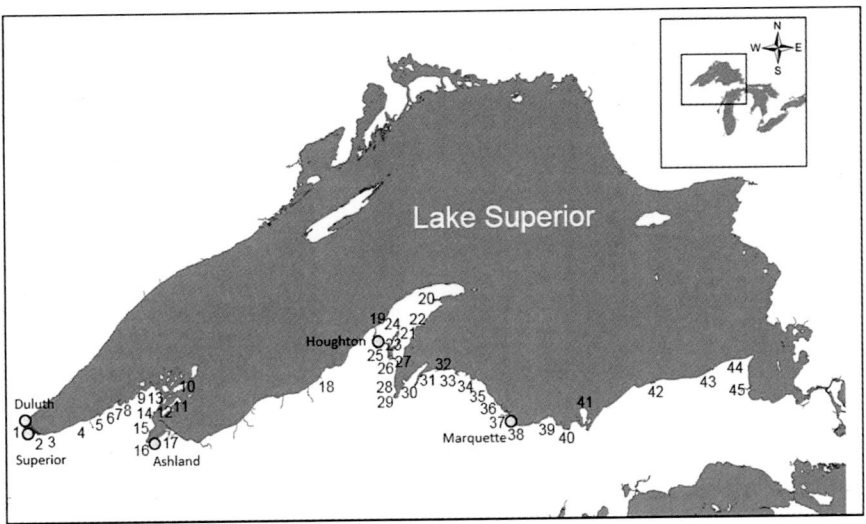

Fig. 5 Map showing approximate location of coastal wetlands along the southern shore of Lake Superior listed in Annex (numbers) and the cities mentioned in the text (labeled circles)

Department of Natural Resources, e.g., Epstein et al. 1997) and as part of broader efforts to develop indicators of ecological condition for coastal wetlands across the LGL (Great Lakes Environmental Indicator Project – Niemi et al. 2007; GLCWC 2008; Chow-Fraser laboratory at McMaster University, e.g., Seilheimer et al. 2009; Cvetkovic and Chow-Fraser 2011).

These studies have revealed some unique features associated with Lake Superior coastal wetlands. In addition to being among the least disturbed by human influences in all the Great Lakes (Danz et al. 2007; Cvetkovic and Chow-Fraser 2011) and the most peat-dominated (Epstein et al. 1997; Minc and Albert 1998), they support among the highest levels of biodiversity (Brazner et al. 2000; Trebitz et al. 2009), including many rare species (Epstein et al. 1997) and some like wild rice *Zizania palustris* that used to be common but no longer are in most of the Great Lakes Basin (Meeker 1996). While much of the biodiversity characterization effort in Lake Superior coastal wetlands has focused on fish communities (e.g., Epstein et al. 1997; Brazner et al. 2001; Trebitz et al. 2011), stable isotope studies have provided insights about entire food webs (e.g., Keough et al. 1998; Sierszen et al. 2006, 2012a), and assemblages of birds (e.g., Howe et al. 2007), amphibians (e.g., Price et al. 2005), macroinvertebrates (Kang et al. 2007), vascular plants (e.g., Johnston et al. 2007), fish (e.g., Bhagat et al. 2007), and diatoms (Reavie 2007) have all been the focus of ecological indicator studies that included coastal wetlands in Lake Superior (also see Brazner et al. 2007; GLCWC 2008). The hydromorphic determinants of aquatic habitat variability (Trebitz et al. 2005), including nutrient availability and water quality (Morrice et al. 2009), that supports biodiversity in Lake Superior coastal wetlands have also been examined.

Lake Superior coastal wetlands support over 60 species of fish, including several of importance to recreational fishing (yellow perch, *Perca flavescens*; bluegill and pumpkinseed sunfish, *Lepomis macrochirus* and *L. gibbosus*; northern pike, *Esox lucius*; black crappie, *Pomoxis nigromaculatus*; walleye, *Stizostedion vitreum*; largemouth and smallmouth bass, *Micropterus salmoides* and *M. dolomieu*) but also some nonnative fish species (e.g., common carp, *Cyprinus carpio*; Eurasian ruffe, *Gymnocephalus cernuus*; round goby, *Neogobius melanostomus*). There are at least fifteen amphibian species that use Lake Superior coastal wetlands including bullfrog *Rana catesbeiana*, pickerel frog *R. palustris*, leopard frog *R. pipiens*, mink frog *R. septentrionalis*, green frog *Lithobates clamitans*, mudpuppy *Necturus maculosus*, tiger salamander *Ambystoma tigrinum*, eastern newt *Notophthalmus viridescens*, and four-toed salamander *Hemidactylium scutatum*. Reptiles occurring in Lake Superior wetlands include painted turtle *Chrysemys picta*, snapping turtle *Chelydra serpentina*, and wood turtles *Glyptemys insculpta*, and northern water *Nerodia sipedon* and garter snakes *Thamnophis sirtalis*. A wide variety of bird species use Lake Superior coastal wetlands, including waterfowl (e.g., diving and dabbling ducks, Canada goose *Branta canadensis*) and other water-associated species (red-winged blackbird, *Agelaius phoeniceus*; rails, e.g., *Porzana carolina*; bitterns, e.g., *Botaurus lentiginosus*; herons, e.g., *Ardea herodias*; belted kingfisher, *Megaceryle alcyon*; osprey, *Pandion haliaetus*; bald eagle, *Haliaeetus leucocephalus*; common loon, *Gavia immer*) as well as many songbirds. Types of invertebrates inhabiting Lake Superior coastal wetlands include the strictly aquatic crayfishes (*Orconectes* spp.), clams (Sphaerid and Unionid spp.), snails (e.g., Gyrinus spp.), leeches (e.g., Glossiphoniidae spp.), aquatic worms (oligochaetes), water fleas (e.g., *Daphnia* spp.), and amphipods (e.g., *Gammarus* spp.) as well as immature (often aquatic) and adult butterflies (e.g., *Boloria* spp.), midges (Chironomidae spp.), dragonflies and damselflies (e.g., *Enallagma* spp.), mayflies (e.g., *Caenis* spp.), and caddisflies (e.g., *Oecetis* spp.). Among the mammals found in Lake Superior coastal wetlands are mink *Neovison vison*, otter *Lontra canadensis*, and beaver *Castor canadensis*. Characteristic vascular plants include swamp trees (tamarack, *Larix laricina*; black spruce, *Picea mariana*; white cedar, *Thuja occidentalis*; black ash, *Fraxinus nigra*; speckled alder, *Alnus incana*; willow, *Salix* spp.), wet-meadow species (sedges, *Carex* spp.; water hemlock, *Cicuta* spp.), bog associates (e.g., *Sphagnum* spp.; ericaceous shrubs; pitcher plant, *Sarracenia purpurea*; cranberry, *Vaccinium* spp.), graminoids (e.g., burred, *Sparganium* spp.; bullrush, spike rush, *Eleocharis* spp.; cattails, *Typha* spp.; giant reed, *Phragmites australis*; wild rice; horsetail, *Equisetum* spp.), and broader-leaf (e.g., pickerelweed, *Pontederia cordata*; arrowhead, *Sagittaria* spp.) emergent forms and in deeper waters, a variety of submerged and floating leaved species (e.g., water lily, *Nuphar* and *Nymphaea* spp.; water shield, *Brasenia schreberi*; pondweeds, *Potamogeton* spp.; milfoils, *Myriophyllum* spp.; coontail, *Ceratophyllum demersum*; wild celery,

Vallisneria americana; bladderworts, *Utricularia* spp.; and naiads *Najas* spp.). Invasive plants are much less prevalent in Lake Superior wetlands than in coastal wetlands of the lower Great Lakes, although purple loosestrife *Lythrum salicaria*, narrow-leaf cattail *Typha angustifolia*, Eurasian milfoil *Myriophyllum spicatum*, curly-leaf pondweed *Potomageton crispus*, and reed canary grass *Phalaris arundinacea* all occur.

Studies on ecological processes in Lake Superior coastal wetlands are more limited than those on ecological structure. Process-oriented work has shown that Lake Superior coastal wetlands can export considerable numbers of young fish to adjacent bays and nearshore food webs (Brazner et al. 2001) and have unique habitat fingerprints manifest in fish biochemical signatures that can be used to quantify wetland-nearshore interactions and wetland contributions to broader populations and food webs (Brazner et al. 2004; Sierszen et al. 2012a). Process-oriented work has also shown that Lake Superior coastal wetland structure, function, and response to anthropogenic stressors are all strongly influenced by hydrogeomorphic setting in which variable strength of and interplay among river and lake influences plays a key role (e.g., Trebitz et al. 2002; Sierszen et al. 2006, 2012a). A growing interest in evaluating wetland protection and restoration in terms of benefits to humans (e.g., fish and wildlife support, water quality improvement, recreational opportunities, shoreline protection; Sierszen et al. 2012b) is fueling a current research emphasis on ecosystem services provided by Lake Superior coastal wetlands.

Overall, there is still much research that needs to be done to better understand the ecology of these complex ecosystems, but studies suggest Lake Superior and other Great Lakes coastal wetlands in general may play a larger role in the food webs of the broader Great Lakes than their size alone suggests (Brazner et al. 2000).

Threats and Future Challenges

Shoreline and watershed development, invasive species, and climate change are among the most challenging factors affecting the integrity of Lake Superior and other Great Lakes coastal wetlands. These factors are likely to pose even greater challenges in the future. Wilcox (2012) provides an excellent summary of the threats to coastal wetlands in the Great Lakes, including those in Lake Superior, so we cover these only briefly here.

Purple loosestrife, hybrid cattail, and giant reed grass are the most aggressively invading plants affecting Lake Superior coastal wetlands. None of these are widespread in Lake Superior at this time but merit monitoring because of their potential to outcompete native plants and alter the habitat these native species provide for other biota. Several invasive animal species including Eurasian ruffe and dreissenid mussels are present in the St. Louis River Estuary, which serves as an epicenter for aquatic invasive species due to urbanization and high shipping and recreational boating traffic (Grigorovich et al. 2003). Water chemistry does not appear to be conducive for the spread of dreissenid mussels as has occurred in other Great Lakes and coastal wetlands have been shown to be a refuge from the invasive ruffe; however, continued monitoring of these species in Lake Superior coastal wetlands is warranted.

Watershed and shoreline development is by far the heaviest along the St. Louis River Estuary and will likely continue to intensify into the future. That makes it likely that water quality issues such as sedimentation and nutrient loading will continue to intensify and negatively impact coastal wetlands in the estuary. Along the rest of the US shoreline, watershed development is relatively light, although agricultural operations do result in nutrient loading that degrades Lake Superior coastal wetlands (particularly on the western end), and residential development along the entire shoreline continues to expand, with potential for septic leakage, shoreline hardening/vegetation loss, and road and boat impacts. At present, degraded coastal wetlands are found primarily in the St. Louis River Estuary (cities of Duluth MN and Superior WI), but also in the vicinity of the smaller cities of Ashland, WI, and Houghton and Marquette, MI (Fig. 5).

By far the biggest potential threat to the health of Lake Superior coastal wetlands in the immediate and long-term future is climate change and associated lower water levels that may result. Although water levels on Lake Superior are controlled to some degree at the outlet to the St. Mary's River, water levels have been largely below the long-term average since the late 1990s. Decreasing periods of ice cover and increasing water temperature and evaporation rates have also been observed during this same time period, indicating that major shifts in the physical character of Lake Superior are underway. Drier and warmer climates may reduce the flooded area of coastal wetlands and limited access by mobile species, especially larger fish, and will probably alter the typical exchange of nutrients and other materials between wetland and lake. Some coastal wetlands (e.g., lacustrine types) are able to migrate lakeward as water levels drop if bottom contours in the adjacent lake are not too steep, but barrier-protected lagoon and riverine wetlands, the most common types along Lake Superior, typically cannot. The severity of the reduction in wetland area will depend on the severity of warming that occurs but seems certain to play a major role in shaping the future ecology in these ecosystems.

Annex

Primary (>10 ha, surface connected) coastal wetlands along the US shoreline of Lake Superior arranged from west to east. Information compiled from Moffett et al. (2006), Epstein et al. (1997), and inspection of printed and online maps and imagery. Lagoon geomorphic type is short for "barrier-protected lagoon"; riverine type is subdivided into "dendritic" (deltaic or multiple arms) versus "simple riverine" (single channel). Numbers before wetland names match those mapped in Fig. 5.

Coastal wetland	Latitude	Longitude	State and county	Area (he)	Geomorphic type	Watershed setting
1. St. Louis River Estuary wetland complex	46.7029	−92.1756	MN – St. Louis, WI – Douglas	>500	Lagoon, dendritic, simple riverine	Urban to forest gradient (cities of Duluth, Superior), motorboat traffic, road/railroad nearby
2. Allouez Bay-Nemadji mouth complex	46.6924	−92.0072	WI – Douglas	310	Lagoon, dendritic	Urban (city of Superior) and forested, motorboat traffic, road/railroad nearby
3. Amnicon River mouth	46.6892	−91.8617	WI – Douglas	32	Dendritic	Mostly forested, some agric in watershed
4. Brule River mouth	46.7472	−91.6111	WI – Douglas	28	Simple riverine	Mostly forested, some agric in watershed
5. Bibon-Flag complex	46.7901	−91.3842	WI – Bayfield	376	Lagoon, dendritic	Some development (town of Port Wing, marina), road crossing, agric in watershed
6. Cranberry River mouth	46.8267	−91.2667	WI – Bayfield	14	Simple riverine	Mostly forested, some residences, road crossing, agric in watershed
7. Bark Bay complex	46.8492	−91.1956	WI – Bayfield	275	Lagoon, dendritic	Mostly forested, some residences
8. Lost Creek complex	46.8593	−91.1333	WI – Bayfield	81	Lagoon, dendritic	Mostly forested, some residences
9. Sand River mouth	46.9306	−90.9369	WI – Bayfield	95	Dendritic	Forested, remote, no road access
10. Stockton Island Tombolo	46.9258	−90.5486	WI – Bayfield	115	Lagoon	Forested, remote, no road access
11. Big Bay Lagoon (Madeline Island)	46.8069	−90.6886	WI – Bayfield	291	Lagoon	Forested, remote
12. La Pointe Wetland (Madeline Island)	46.7753	−88.7803	WI – Bayfield	34	Lagoon	Forest/urban mix (town of La Pointe, marina)
13. Raspberry Bay complex	46.9317	−90.825	WI – Bayfield	57	Dendritic	Mostly forested
14. Pike River mouth	46.7859	−90.8589	WI – Bayfield	16	Dendritic	Mostly forested, marina nearby
15. Sioux River mouth	46.7336	−90.8769	WI – Bayfield	82	Dendritic	Mostly forested, road crossing
16. Fish Creek complex	46.5864	−90.9424	WI – Bayfield	316	Dendritic, lagoon	Forested/agric mix, urban nearby (city of Ashland), road crossing

(*continued*)

Coastal wetland	Latitude	Longitude	State and county	Area (he)	Geomorphic type	Watershed setting
17. Kakagon-Bad complex	46.6361	−90.695	WI – Ashland	1949	Dendritic, lagoon	Forested, remote, no road access
18. Flintsteel-Firesteel complex	46.9305	−89.1887	MI – Ontonagon	253	Simple riverine	Forested, road crossing
19. Sevenmile Creek mouth	47.2421	−88.5786	MI – Houghton	43	Dendritic	Forested, road crossing
20. Lac La Belle complex	47.3778	−87.9789	MI – Keweenaw	516	Dendritic/lagoon	Mostly forested, some residences
21. Grand Traverse Bay complex	47.1658	−88.2469	MI – Keweenaw	277	Dendritic/lagoon	Mostly forested, some residences
22. Mud Lake-Little Traverse Bay area	47.1389	−88.2825	MI – Keweenaw	23	Dendritic	Mostly forested, some residences
23. Dollar Bay wetland	47.1524	−88.4068	MI – Houghton	82	Simple riverine, lagoon	Forested/suburban mix, road crossing
24. Torch Lake wetland	43.1524	−88.4068	MI – Houghton	478	Dendritic	Mostly forested, motorboat traffic
25. Pike River mouth	47.0183	−88.5281	MI – Houghton	36	Simple riverine	Mostly forested, some residences, road crossing
26. Sturgeon/Snake River area	47.0246	−88.4887	MI – Houghton	1538	Dendritic	Forested, motorboat traffic
27. Silver Creek/Portage River area	46.998	−88.4249	MI – Houghton	263	Dendritic	Forested, motorboat traffic
28. Sand Point area	46.7936	−88.4708	MI – Baraga	164	Lagoon	Mostly forested, some residences, road crossing
29. L'Anse Bay-Six Mile Creek	46.7537	−88.4931	MI – Baraga	202	Dendritic	Mostly forested, road crossing
30. Silver River mouth	46.8225	−88.2873	MI – Baraga	25	Dendritic, lagoon	Mostly forested
31. Lightfoot Bay	46.9	−88.1886	MI – Baraga	156	Lagoon	Forested, remote
32. Huron River mouth	46.9083	−87.0417	MI – Baraga	40	Dendritic	Forested, remote
33. Pine Lake mouth	46.8844	−87.8578	MI – Marquette	14	Dendritic	Forested, remote

34. Salmon-Trout River mouth	46.8598	−87.7658	MI – Marquette	162	Dendritic	Forested, remote, no road access
35. Iron R./Lake Independence	46.8112	−87.6559	MI – Marquette	38	Dendritic, lagoon	Forested, remote
36. Middle Isle Point area	46.5869	−87.4192	MI – Marquette	91	Lagoon	Mostly forested, road crossing
37. Dead River mouth	46.5777	−87.4003	MI – Marquette	apx. 25	Dendritic	Urban/industrial (city of Marquette), road crossing
38. Chocolay River mouth	46.5008	−87.3547	MI – Marquette	apx. 25	Dendritic	Forest/suburban mix, road crossing
39. Laughing Whitefish River mouth	46.5241	−87.0281	MI – Alger	26	Dendritic	Forested, remote
40. Au Train River mouth	46.4329	−86.8221	MI – Alger	83	Dendritic	Mostly forested, some residences, road crossing
41. Murray Bay wetland (Grand Island)	46.4722	−86.6412	MI – Alger	10	Lagoon	Forested, remote, no road access
42. East Bay/Sucker River area	46.6703	−85.9345	MI – Alger	54	Simple riverine, lagoon	Forested, remote
43. Big and Little Two-Hearted River area	46.6961	−85.4244	MI – Luce	14	Simple riverine	Forested, remote
44. Weatherhogs Creek complex	46.7578	−85.1239	MI – Chippewa	289	Dendritic, lagoon	Forested, remote
45. Tahquamenon River mouth	46.5552	−85.0297	MI – Chippewa	385	Simple riverine	Forested, road crossing

References

Albert DA, Wilcox DA, Ingram JW, Thompson TA. Hydrogeomorphic classification for Great Lakes coastal wetlands. J Great Lakes Res. 2005;31 Suppl 1:129–46.

Bhagat Y, Ciborowski JH, Johnson LB, Uzarski DJ, Burton TM, Timmermans SA, et al. Testing a fish index of biotic integrity for responses to different stressors in Great Lakes coastal wetlands. J Great Lakes Res. 2007;33(Sp. Issue 3):224–35.

Brazner JC, Sierszen ME, Keough JR, Tanner DK. Assessing the importance of coastal wetlands in a large lake context. Verh Internat Verein Limnol. 2000;26:1950–61.

Brazner JC, Tanner DK, Morrice JA. Fish-mediated nutrient and energy exchange between a Lake Superior coastal wetland and its adjacent bay. J Great Lakes Res. 2001;27:98–111.

Brazner JC, Campana SE, Tanner DK, Schram ST. Reconstructing habitat use and nursery origin of yellow perch from Lake Superior using otolith elemental analysis. J Great Lakes Res. 2004;30:492–507.

Brazner JC, Danz NP, Niemi GJ, Regal RR, Hanowski JM, Johnston CA, et al. Evaluating geographic, geomorphic and human influences on Great Lakes wetland indicators: multi-assemblage variance partitioning. Ecol Indic. 2007;7:610–35.

Cvetkovic M, Chow-Fraser P. Use of ecological indicators to assess the quality of Great Lakes coastal wetlands. Ecol Indic. 2011;11:1609–22.

Danz NP, Niemi GJ, Regal RR, Hollenhorst T, Johnson LB, Hanowski JM, et al. Integrated measures of anthropogenic stress in the U.S. Great Lakes Basin. Environ Manag. 2007;39(5):631–47.

Epstein EJ, Judziewicz EJ, Smith WA. Priority wetland sites of Wisconsin's Lake Superior Basin. Madison: Wisconsin Department of Natural Resources, Natural Heritage Inventory Program; 1997.

Great Lakes Coastal Wetlands Consortium. Great Lakes coastal wetlands monitoring plan. Great lakes commission, United States environmental protection agency, and the great lakes national program office. 2008. http://projects.glc.org/wetlands/final-report.html

Grigorovich IA, Korniushin AV, Gray DK, Duggan IC, Colautti RI, MacIsaac HJ. Lake Superior: an invasion coldspot? Hydrobiologia. 2003;499(1–3):191–210.

Herdendorf CE. Great Lakes estuaries. Estuaries. 1990;13:493–503.

Herdendorf CE, Hartley SM, Barnes MD. Fish and wildlife resources of the Great Lakes coastal wetlands within the United States, vol. 6. Washington, DC: U.S. Fish and Wildlife Service; 1981. FWS/OBS-81/02-v6.

Howe RW, Regal RR, Hanowski JM, Niemi GJ, Danz NP, Smith CR. An index of ecological condition based on bird assemblages in Great Lakes coastal wetlands. J Great Lakes Res. 2007;33(Sp. Issue 3):93–105.

Ingram J, Holmes K, Grabas G, Watton P, Potter B, Gomer T, Stow N. Development of a coastal wetlands database for the Great Lakes Canadian shoreline. Final report to: the Great Lakes Commission. WETLANDS2-EPA-03; 2004.

Johnston CA, Bedford BL, Bourdaghs M, Brown T, Frieswyk C, Tulbure M, et al. Plant species indicators of physical environment in Great Lakes coastal wetlands. J Great Lakes Res. 2007;33 (Sp. Issue 3):106–24.

Kang M, Ciborowski JHH, Johnson LB. The influence of anthropogenic disturbance and environmental suitability on the distribution of the nonindigenous amphipod, *Echinogammarus ischnus*, at Laurentian Great Lakes coastal margins. J Great Lakes Res. 2007;33(Sp. Issue 3):198–210.

Keough JR, Hagley CA, Ruzycki E, Sierszen ME. δ13C composition of primary producers and role of detritus in a freshwater coastal ecosystem. Limnol Oceanogr. 1998;43:734–40.

Keough JR, Thompson TA, Guntenspergen GR, Wilcox DA. Hydrogeomorphic factors and ecosystem responses in coastal wetlands of the Great Lakes. Wetlands. 1999;19:821–34.

Meeker JE. Wild-rice and sedimentation processes in a Lake Superior coastal wetland. Wetlands. 1996;16:219–31.

Minc LD, Albert DA. Great Lakes coastal wetlands: abiotic and floristic characterization. Lansing: Michigan Natural Features Inventory; 1998.

Moffett MF, Dufour RL, Simon TP. An inventory and classification of coastal wetlands of the Laurentian Great Lakes. In: Simon TP, Stewart PM, editors. Coastal wetlands of the Laurentian Great Lakes: health, habitat, and indicators. Bloomington: AuthorHouse; 2006. p. 17–99.

Morrice JA, Trebitz AS, Kelly JR, Cotter AM, Knuth ML. Nutrient variability in Lake Superior coastal wetlands: the role of land use and hydrology. In: Munawar M, Munawar IF, editors. State of Lake Superior: health, integrity, and management, Ecovision World Monograph Series. Burlington: Aquatic Ecosystem Health and Management Society; 2009. p. 217–38.

Niemi GJ, Kelly JR, Danz NP. Environmental indicators for the coastal region of the North American Great Lakes: introduction and prospectus. J Great Lakes Res. 2007;33(Sp. Issue 3):1–12.

Price SJ, Marks DR, Howe RW, Hanowski JM, Niemi GJ. The importance of spatial scale for conservation and assessment of anuran populations in coastal wetlands of the western Great Lakes. Landsc Ecol. 2005;20:441–54.

Reavie ED. A diatom-based water quality model for Great Lakes coastlines. J Great Lakes Res. 2007;33(Sp. Issue 3):86–92.

Seilheimer TS, Mahoney TP, Chow-Fraser P. Comparative study of ecological indices for assessing human-induced disturbance in coastal wetlands of the Laurentian Great Lakes. Ecol Indic. 2009;9:81–91.

Sierszen ME, Peterson GS, Trebitz AS, Brazner JC, West CW. Hydrology and nutrient effects on food web structure in ten coastal wetlands of Lake Superior. Wetlands. 2006;4:951–64.

Sierszen ME, Brazner JC, Cotter AM, Morrice JM, Peterson GS, Trebitz AS. Watershed and lake influences on the energetic base of coastal wetland food webs across the Great Lakes Basin. J Great Lakes Res. 2012a;38:418–28.

Sierszen ME, Morrice JA, Trebitz AS, Hoffman JC. A review of selected ecosystems services provided by coastal wetlands of the Laurentian Great Lakes. Aquat Ecosyst Health Manag. 2012b;15:92–106.

Trebitz AS, Morrice JM, Cotter AM. Relative role of lake and tributary in hydrology of Lake Superior coastal wetlands. J Great Lakes Res. 2002;28:212–27.

Trebitz AS, Morrice JA, Taylor DL, Anderson RL, West C, Kelly JR. Hydromorphic determinants of aquatic habitat variability in Lake Superior coastal wetlands. Wetlands. 2005;25:505–19.

Trebitz AS, Brazner JC, Danz NP, Pearson M, Peterson G, Tanner D, et al. Geographic, anthropogenic, and habitat influences on Great Lakes coastal wetland fish assemblages. Can J Fish Aquat Sci. 2009;66:1328–42.

Trebitz AS, Brazner JC, Tanner DK, Meyer R. Interacting watershed size and land-cover influences on habitat and biota of Lake Superior coastal wetlands. Aquat Ecosyst Health Manag. 2011;14:443–55.

Wilcox DA. Great Lakes coastal marshes. In: Batzer DP, Baldwin AH, editors. Wetland habitats of North America: ecology and conservation concerns. Berkeley: University of California Press; 2012. p. 173–88.

The Bay of Fundy and Its Wetlands (Canada)

49

Graham R. Daborn and Anna M. Redden

Contents

Introduction	622
Physical Characteristics	622
Biodiversity	625
Marshes and Tidal Flats	629
Conservation Status	632
Threats and Future Challenges	634
References	635

Abstract

The Bay of Fundy is a highly productive and biologically diverse ecosystem. Noted for the highest tides in the world, it is biologically connected to the Americas, Europe and the Arctic. Its highly productive salt marshes and tidal flats have been extensively modified by human activity, but still provide critical habitat and organic matter supporting numerous migratory species of fish, birds and marine mammals.

Keywords

Hypertidal ecosystem · Biodiversity · Salt marshes · Intertidal mudflats · Conservation issues

G. R. Daborn (✉) · A. M. Redden (✉)
Acadia Centre for Estuarine Research, Acadia University, Wolfville, NS, Canada
e-mail: graham.daborn@acadiau.ca; anna.redden@acadiau.ca

© Springer Science+Business Media B.V., part of Springer Nature 2018
C. M. Finlayson et al. (eds.), *The Wetland Book*,
https://doi.org/10.1007/978-94-007-4001-3_2

Introduction

By almost any measure, the Bay of Fundy is an extraordinary estuarine system. With the world's highest tidal range, a great diversity of habitats and organisms, and biological connections that link the Bay to the whole of the Atlantic, the Americas, and Europe, it is a complex and highly productive ecosystem that presents substantial challenges for its management and conservation. An important part of its international significance is associated with the salt marshes and tidal flats that occur primarily in the innermost portions of the bay.

The Bay of Fundy originated from a tectonic rift that formed during the break up of Pangea during the early Triassic (250–190 mya; cf. AGS 2001). An original Fundy Basin subsequently began to fill with sediments, forming the red sandstones that are now exposed in the innermost (eastern) portions of the bay. During the Early Jurassic (190–142 mya), basaltic lava flows resulting from continental breakup emerged through the crust and now form most of the floor of the bay as well as a long, resistant ridge that constitutes the south-eastern (Nova Scotia) shore. As a result of these processes, the bay has a very different character from one end to the other: the western regions are bordered by basalt and other resistant materials, forming rocky shorelines and intertidal zones, whereas the more eastern portion of the bay is bordered by more sand- and mud-stones that provide the fine sediments forming extensive intertidal flats. Because the morphology of the Bay varies substantially from one end to the other, it is convenient to recognize three major regions: the Outer Bay, the Inner Bay, and the Upper Bay (see Fig. 1). Water in the Outer Bay is exchanged with the Gulf of Maine and Atlantic Ocean on each tide and is clear, with a photic zone that extends down to 20+ m, whereas the Upper Bay is characterized by highly turbid waters and a photic zone that ranges from a few meters to a few millimeters (Brylinsky and Daborn 1987).

Physical Characteristics

The Bay of Fundy is an integral part of a complex coastal oceanographic system (referred to as the Fundy – Gulf of Maine – Georges Bank or FMG system; see Fig. 2) that also includes the Gulf of Maine, Georges and Browns Banks, and the various channels between them. The total area of the FMG system is nearly 180,000 km^2, of which the Bay of Fundy constitutes 16,000 km^2. Water entering the system on the flood tide is derived in part from northern latitudes through the cold Nova Scotia Current but may also include warmer southern waters brought north by the Gulf Stream. Periodically, large scale eddies in the Gulf Stream release "warm core rings" (see http://marine.coastal.edu/gulfstream/p5.htm) that drift north over the colder water, introducing species of tropical and subtropical origin into the FMG system. Water entering the Outer Bay during the flood tide tends to advance along the southern (Nova Scotia) shore, producing an anticlockwise circulation in the Outer and Inner Bays, which is amplified by less dense fresh water entering from the principal river (Saint John River) along the northern shore.

Fig. 1 Regions and tidal ranges of the Bay of Fundy (Adapted from Hagerman et al. 2006 and drawn by B. Sanderson)

In many ways, the FMG system responds as a unit to the semi-diurnal tidal forcing of the Atlantic Ocean. However, specific responses vary among the different regions of the FMG system as a result of past geological factors that shaped local morphology and of more recent natural and anthropogenic influences. Many dynamic oceanographic processes, such as sea level rise, channel deepening, shoreline erosion, and post-glacial land settling, result in continuing and progressive changes to the physical characteristics of the bay. Added to these natural processes are the anthropogenic effects of dam and harbor construction, dredging, marshland conversion, shoreline modifications, and changes to freshwater inputs.

As a whole, the Bay of Fundy may be considered an estuary (i.e., the sea water is measurably diluted with fresh water – see article on Estuaries this volume), but the freshwater input from rivers is very small relative to the inflow of oceanic water. Salinity ranges from 30 to 32 ‰ in the Outer and Inner Bays and 20–28 ‰ in the Upper Bay; lower salinity conditions are restricted to the mouths of inflowing rivers, but because of the great tidal range these brackish waters may extend upriver for many kilometers.

The bay has a semi-diurnal tidal system, with two full tidal cycles occurring each 24.84 h driven by the Atlantic tide. The bay becomes notably more shallow and somewhat narrower with distance from the Gulf of Maine, and because its length

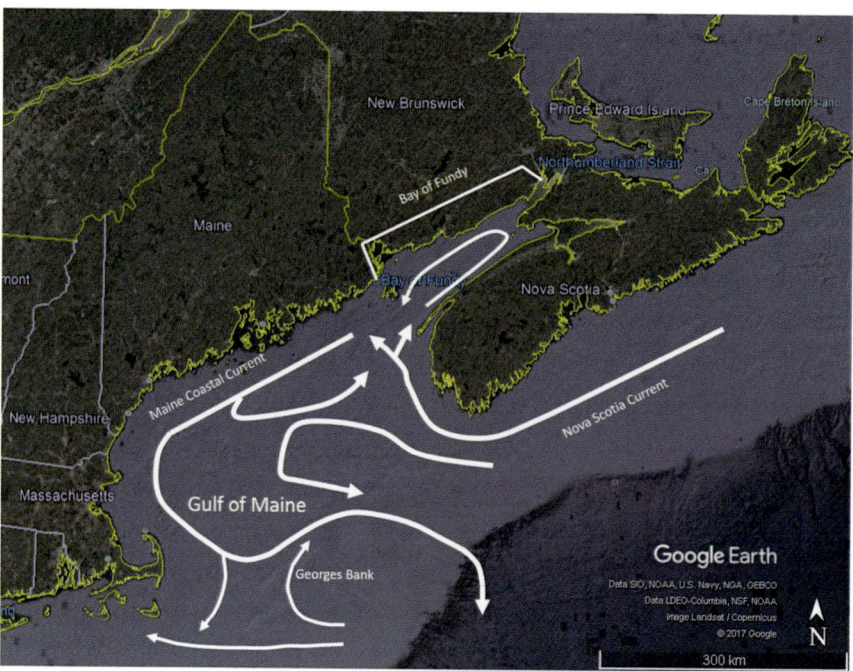

Fig. 2 Bay of Fundy-Gulf of Maine-Georges Bank system (Source: Adapted and redrawn from Gulf of Maine Area Census of Marine Life http://www.gulfofmaine-census.org/about-the-gulf/oceanography/circulation/)

(270 km) and volume are closely matched to the principal constituent of the tide (i.e., the M_2 tidal constituent, which reflects the semi-diurnal variation of the Moon's gravitational effect), it is a near-resonant system. This results in amplification of the 4–6 m tides at the mouth to the world's highest recorded tides (16+ m) in the Upper Bay. As with estuaries elsewhere, cyclical changes occur in tidal amplitude in association with lunar phases (e.g., spring-neap cycles), variations in lunar and solar distances (apogee-perigee, aphelion-perihelion), and longer tidal cycles such as the 18 year Saros and Nodal cycles (Desplanque and Mossman 2004). With the amplification of the tidal range, the effect of these cycles on water levels is magnified, especially in the Upper Bay.

Biological phenomena reflect the fact that the Bay of Fundy is fundamentally a physically driven hypertidal system. Strong tidal currents advancing over shoaling depths result in extensive vertical mixing through most of the bay, but mixing is especially pronounced at the entrance to the Outer Bay, in or near narrow channels and passages, and throughout the Upper Bay. Only in spring and summer does a central part of the Outer and Inner Bay tend to stratify (Garrett et al. 1978; Greenberg et al. 1997). Vigorous vertical mixing has important effects on biological productivity, especially through return of nutrients to surface waters where they may stimulate primary production. Upwelling zones, especially in the Outer Bay, also bring deeper-lying zooplankton to surface waters where they may become a concentrated food source for fish, birds, and baleen whales and moves surface waters down to the bottom where their contents (phytoplankton and other particles) may be accessed by benthic organisms (Daborn 1986; Emerson et al. 1986). The extent of vertical mixing varies over time with fluctuations in tidal range, producing cyclical changes in sea surface temperature and affecting rates of biological production that are detectable in fishery statistics (e.g., Cabilio et al. 1987; Campbell and Wroblewski 1986).

Current velocities on flood and ebb are >1 m s^{-1} over most of the Bay but may reach >6 m s^{-1} in narrow passages. The highest velocities scour the substrate to bedrock, leading to impoverished benthic communities. More moderate currents are associated with gravel or sand waves that occur throughout the bay or allow more productive and diverse benthic communities to occur where hard substrates are exposed (Wildish and Fader 1998).

The extreme hypertidal range (9–16m+) in the Upper Bay results in a large intertidal zone up to several kilometers in width. Because of the erodibility of nearby shorelines, the sediment contribution of rivers, and the typical flood-ebb inequality of estuaries, the intertidal habitat in the Upper Bay is dominated by sands, silts, and muds. These sediments are highly mobile, are frequently resuspended by wave and tidal action through much of the year, and may be completely reworked by ice during winter months. Ice is an important but highly variable physical force in the Upper Bay: in some winters, large sheets of pan ice may become broken and piled up into shore-grounded, sediment-laden icebergs that scour the intertidal substrate as they move. As a result, the benthic community is dominated by burrowing or epibenthic (mobile forms living at the sediment surface) animals rather than the many sessile forms that dominate rocky substrates. The regular reworking of sediments in the Upper Bay represents a physical stress that restricts the benthic fauna to pioneering species that can rapidly resettle disturbed areas (Daborn 2007). With little competition, successful species can become extremely abundant within the extensive intertidal flats, which offer a rich feeding ground for millions of fish and birds that migrate very great distances.

Biodiversity

Because of the major changes in morphology, water column stability, and substrate characteristics over the 270 km length of the Bay of Fundy, there is a great variety of habitats. Day and Roff (2000) classified regions of the Bay of Fundy according to seven physical attributes: surface and bottom water temperatures, depth, stratification, exposure, slope, and sediment type. This classification into "seascapes" (defined as: "broad, oceanographic and biophysical areas characterized by particular water-mass characteristics and sea-ice conditions" (Bredin et al. 2004)) helps to tie together the physical environmental characteristics and the biological characteristics of the bay. It distinguishes between several regional differences: shallow near-shore habitats in the Outer and Inner Bay are significantly different from deeper regions; there are differences between the northern (or New Brunswick) portions of the Outer

Bay and the southern (Nova Scotia) side; and the turbid, well-mixed region of the Inner Bay is distinct from both the clearer Outer Bay and the even more turbid Upper Bay, where substrates are dominated by finer sediments. The classification also clearly shows that habitat differences occur over short distances around the islands and banks near the mouth of the bay.

It is partly this broad diversity of habitat that gives rise to the bay's considerable biodiversity: two thirds (>2,300) of all the species known from the FMG system (c. 3,400 species) occur in the Bay of Fundy which is only 9% of the total FMG area (AECOM 2010). Some 15 different seascapes occur in the bay, ranging from deep passages with rocky substrate, to mobile sand and gravel waves, and to extensive intertidal habitats of sand or mud. Each supports a different combination of flora and fauna. The Outer Bay is deep in the center (100–200+ m), bordered by resistant rock, and floored by either bedrock or mobile sands or gravels. The food web in this outer region strongly resembles that of other coastal waters and bays: the water tends to be clear and supports a diverse plankton community dominated by diatoms (>160 spp.), dinoflagellates (at least 65 spp.), copepods, coccolithophorids, and invertebrate and fish larvae. In this Outer Bay region, phytoplankton primary production is dominant (See Fig. 3). Much of the primary production is processed in the pelagic zone (Emerson et al. 1986), supporting pelagic fish, marine birds, and baleen whales, rather than benthic communities and groundfish.

In the rocky intertidal and near-shore subtidal zones of the Outer Bay, seaweeds are abundant, providing a complex habitat for numerous species of mobile and sessile invertebrates. The seaweed community consists mainly of rockweed (e.g., *Ascophyllum nodosum*) and fucoids (*Fucus vesiculosus, F. serratus, F. edentatus,* and *F. spiralis*) within the intertidal zone and dulse *Palmaria palmata*, Irish moss *Chondrus crispus*, and kelps (*Laminaria digitata, L. longicruris, Alaria esculenta,* and *Agarum cribrosum*) in the lower littoral and sublittoral zones. Rockweed and Irish moss are harvested on a commercial scale. Numerous invertebrate species live within and attached to the fronds, and seaweed associations constitute important nursery grounds and refuge for many fish species. Although few animals, with the notable exceptions of sea urchins and some isopods, graze directly on seaweeds, the export of both dissolved and particulate organic matter from seaweed communities is thought to be an important source of nutrients to offshore waters: as much as 80% of total annual production is exported from the intertidal zone during the summer months. In addition, fragmentation of seaweeds by waves, especially during storm events, results in "rafts" of seaweeds being exported to the pelagic zone where they constitute an important floating habitat and feeding area for seabirds, various marine invertebrates, and juvenile fish (cf. Daborn and Gregory 1983). Although providing a small fraction (1–2%) of the total primary production in the Outer Bay, seaweeds provide both important organic input and critically important habitat.

On exposed subtidal bedrock with moderate currents (1–3 m s^{-1}), dense benthic populations of sessile coelenterates, sponges, mussels, tunicates, and their associated mobile fauna (including shrimps, lobsters, and crabs) occur. In higher velocity areas (>4 m s^{-1}), the bedrock is scoured and only a few species of sponges and coelenterates appear to be able to withstand the flow. The high productivity of the Outer

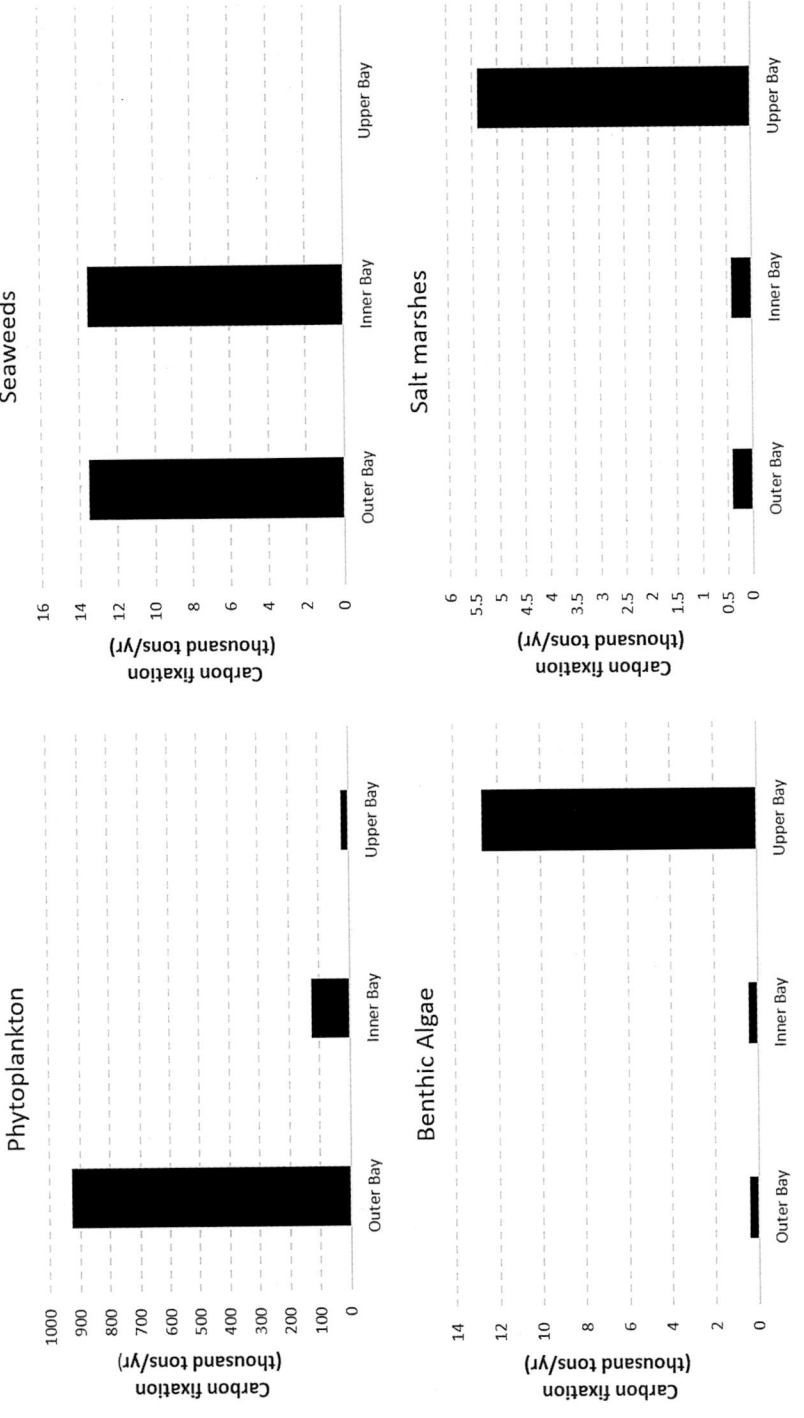

Fig. 3 Distribution of primary production in the Bay of Fundy (Drawn from data in Prouse et al. 1984)

Bay is clearly linked to extensive upwelling zones where tidal currents meet rapidly shoaling waters, which not only sustain important resident fish and fisheries but also attract large numbers of migrant fish, birds, and mammals.

In the offshore waters of the Outer and Inner Bay regions, more than half of the substrate is of highly mobile sands or gravels that form tidally controlled waves or dunes up to several meters in height and many meters in length, oriented across the direction of tidal currents. Here the benthic community is primarily composed of mobile animals such as scallops *Placopecten magellanicus* and lobsters *Homarus americanus*, with a smaller number of burrowing polychaetes, amphipods, echinoderms, and molluscs. Fisheries for lobster and scallop are the most valuable in the Bay. Over much of the Outer and Inner Bay, there are also elongated sandy ridges up to several kilometers in length, lying parallel to tidal current direction, that are capped by dense growths of horse mussels *Modiolus modiolus* which may stabilize the ridges and provide additional microhabitat for other benthic organisms (Wildish et al. 1999).

The Upper Bay, on the other hand, has very little exposed rocky substrate except for sandstone ledges that are swept by sediment-laden water on each tide, and deep channels, such as the entrance to Minas Basin where extreme currents (>6 m s^{-1}, Fig. 4) completely scour the bottom, leaving either bedrock or a boulder-scale postglacial lag. In the latter environment, benthic forms are restricted to flow-resistant sponges and occasional echinoderms, polychaetes, and molluscs that occupy refuges around large boulders (AECOM and ATEI 2013). Over the majority of the Upper Bay, sandstone ledges are overlain by various combinations of sand, silt, and mud that are highly susceptible to disturbance by waves, tidal currents, and winter ice.

Fig. 4 Minas Passage featuring strong tidal currents at Cape Split (Photo credit: C. Buhariwalla ©)

The Bay of Fundy provides habitat for a great variety of fish species (at least 120 spp. – GOMA 2014). These include some permanent residents, several anadromous species that spawn in Bay of Fundy tributaries and go to sea to grow, and others that migrate into the bay from many parts of the North Atlantic solely to feed. Several species, including herring *Clupea harengus*, cod *Gadus morhua*, haddock *Melanogrammus aeglefinus*, halibut *Hippoglossus hippoglossus*, shad *Alosa sapidissima*, and pollock *Pollachius virens*, are or have been the basis for extremely important fisheries, although populations of a number of these have declined over recent decades. Other species play significant roles as forage fish. The anadromous species utilize the marine and estuarine resources of the bay to varying degrees, some passing quickly through to offshore waters, while others spend longer foraging in different regions of the bay. The diversity of habitats existing in the Bay of Fundy is a critical feature that supports this diverse and productive group of animals.

The bay also supports more than 170 species of marine birds and up to 24 species of marine mammals. Most are migratory, traveling to the Bay from the Canadian Arctic, Europe, and the North and South Atlantic. A few including harbor porpoise *Phocoena phocoena*, harbor seal *Phoca vitulina* and grey seal *Halichoerus grypus*, Great black-backed gulls *Larus marinus*, and black duck *Anas rubripes* are year-round residents, and these are the only ones to be seen regularly in the Upper Bay. All of the other mammals including the North Atlantic right whale *Eubalaena glacialis*, humpback *Megaptera novaeangliae*, finback *Balaenoptera physalis*, sei *B. borealis*, minke *B. acuterostrata*, and long-finned pilot whale *Globicephala melaena* and more than 60 species of marine birds tend to concentrate near the major upwelling regions of the Outer Bay or follow fish (especially herring) stocks as the latter move around the Outer Bay and Gulf of Maine.

Marshes and Tidal Flats

The food web in the Upper Bay is distinctly different from that of the Outer and Inner Bay regions because high turbidity limits phytoplankton production and absence of rocky substrate provides no habitat for seaweeds and their associated fauna (see Fig. 3). As a result, a significant fraction of energy flow is heterotrophic. Primary producers include the macrophytes and blue-green algae of peripheral salt marshes and benthic diatoms occurring at the surface of intertidal flats during the summer months. Seaweeds are uncommon except for isolated patches of rockweed on exposed sandstone or boulders and some kelp in clearer waters near the mouth of the Minas Basin. In the Upper Bay, phytoplankton are light-limited because of high suspended sediments (Brylinsky and Daborn 1987); consequently, benthic diatom production occurring during low tide exposure over five summer months accounts for about 30% of the total primary production, and salt marshes now provide about 13% (Fig. 3). An unknown amount of organic matter is also derived from surrounding agricultural land through river input.

Intertidal benthic diatoms (including species of *Gyrosigma, Navicula, Pleurosigma,* and *Nitzschia*) are a high quality food source and are subject to considerable grazing pressure by deposit-feeding polychaetes and crustaceans: the latter form a major food source for fish and various waterfowl, both resident and migratory species. This intertidal community is especially important in supporting the major staging area for migratory shorebirds (which include sandpipers (*Calidris pusilla, C. minutilla*), plovers (*Charadrius semipalmatus, Pluvialis dominica, P. squatarola*), knot (*Calidris canutus*), and dunlin (*C. alpina*)) that feed mainly on mud shrimp *Corophium volutator* or polychaetes during the late summer before flying south for the winter. The arrival of about two million shorebirds from Arctic breeding grounds in July and August results in significant changes in benthic algal production and sediment erodibility of the mudflats of Minas Basin (Daborn et al. 1993).

Because of the large tidal range, about one fifth of the area of Minas Basin (1,900 km^2) is intertidal, with sandy substrates towards the low and high water marks where wave action tends to be concentrated and a dominance of finer sediments such as silts and clays in the mid-tide zone. Coarse sand deposits are inhabited by a relatively sparse group including soft-shell clams *Mya arenaria*, burrowing isopods (e.g., *Chiridotea coeca*), and various polychaetes (*Nephtys, Neanthes, Nereis* spp.). Sometimes the sand is stabilized by tube-dwelling polychaetes (e.g., *Clymenella torquata*) which construct sand-grain tubes that resist wave scour. In calmer locations, the sediment tends to have higher clay content, and the infauna is dominated by a few species of polychaete worms (*Heteromastis, Nereis, Glycera, Spiophanes* spp.) and the burrowing amphipod *Corophium volutator*. In some places, especially near to salt marshes, large numbers of the small clam *Macoma balthica* occur, and in other locations numerous mud snails such as *Ilyanassa obsoleta* occur. There appear to be dynamic competitive interactions between *Ilyanassa* and *Corophium* that may play important roles in the support of migratory shorebirds (Hamilton et al. 2003). Benthic diatom abundance is strongly seasonal, with blooms occurring in summer months when temperatures are high and daylight exposure is long. Outside of this season, the invertebrates are heterotrophic, dependent upon microbial food associated with the breakdown of organic matter originating in nearby salt marshes or derived from inflowing rivers.

Salt marshes are found around the Bay of Fundy and cover about 153 km^2 (Wrathall et al. 2013), but the vast majority that remain occur in the Upper Bay. Fundy marshes are typically two-zoned, with high marsh plants (e.g., *Spartina patens, Juncus gerardi, Plantago maritima*) occurring above mean high water, and intertidal marsh plants – mainly *Spartina alterniflora* – which dominate below mean high water (Redfield 1972). In addition to the macrophytes, Fundy marshes commonly contain blue-green algal mats in summer that at times may cover the sediment surface, and may be grazed by a number of molluscs and polychaetes. Salt marshes provide habitat for both terrestrial and marine organisms. Terrestrial species include numerous insects and spiders, and various species of waterfowl use salt marshes as breeding and over-wintering habitat. Snails and shellfish are often found within the substrate of marshes, and small fish are common to the tidal creeks and ponds associated with marshes.

Fig. 5 Bay of Fundy saltmarsh. Dominated by *Spartina alterniflora*, with height of 2 m and above-ground production < 1,800 g.m^{-2} during a 4-month growing season (Photo credit: G. R. Daborn ©)

Recent studies of salt marsh production in the Upper Bay of Fundy have indicated that, in spite of the northern latitude and short growing season, Fundy marshes (Fig. 5) rank among the most productive in North America – on a par with the Gulf of Mexico (cf. Kirwan et al. 2009). Estimates of mean above-ground annual production of *S. alterniflora* range from 500 to 900 g.m^{-2}.yr^{-1} in Texas and Louisiana (28°N), 400–600 g.m^{-2}.yr^{-1} in North Carolina (35°N), and 400–1,800 g.m^{-2}.yr^{-1} in Minas Basin (45°N) (*ibid.*; Wrathall et al. 2013). Unlike many southern marshes, the high tidal energy, open coastline, and winter ice results in most of the above-ground *S. alterniflora* production being cast up on shore where it decays or is exported into deeper water during the fall and winter months. In this way, *Spartina* becomes an important part of the detrital food chain (Gordon et al. 1985; Cranford et al. 1989). The harsh winter conditions may be the reason for the absence of insect herbivores in Fundy marshes, and so the above-ground biomass of *Spartina* is essentially ungrazed during the growing season. Consequently, the trophic cascade model of salt marsh regulation described for New England marshes (cf. Silliman and Bertness 2002; Silliman and Bortolus 2003; Bertness et al. 2008) does not appear to apply to the Upper Bay. However, a trophic cascade effect has been detected for the intertidal mudflats during the summer months when vast numbers of shorebirds reduce intertidal grazer populations, release benthic diatoms from grazing control, leading to a mid-summer bloom, and thereby changing the erodibility of the mudflats (Daborn et al. 1993).

Together, the marshes and mudflats of the Upper Bay of Fundy constitute a rich, seasonally productive system that is physically stressed by tidal exposure, strong currents, and winter ice. The result is low diversity, but very high production that supports millions of fish and marine birds, many of which migrate there to feed, as well as providing organic detritus that contributes to food webs of the Inner and Outer Bay regions. The fact that more than 80% of the marshes that existed when Europeans first began to settle the region in the early seventeenth century have since been "reclaimed" for agricultural purposes, leads to interesting speculation about the nature of the Fundy ecosystem in pre-Contact times.

Conservation Status

The productivity and diversity of habitats and species in the Bay of Fundy, the extensive migratory connections with distant ecosystems, the cumulative effects of 400 years of human influence, and the prospects of climate change underlie conservation concerns about the bay (Daborn and Dadswell 1988; Wells 1999). These concerns may be grouped under three categories: protection of species and habitats at risk; recovery of features that have been modified by human activity; and accommodation to future changes in the system as a result of both natural processes and global warming. Conservation and management efforts have incorporated international recognition of the unique features of the bay, regulatory action by two levels of government, and a substantial component of community-based management.

Twenty-three marine species that are associated with the Bay of Fundy are recognized by the Committee on the Status of Endangered Wildlife in Canada (COSEWIC). These species include five mammals, 11 fish, and seven birds (Table 1). (Criteria for assessment and designation as Endangered, Threatened, or of Special Concern may be obtained from: http://www.cosewic.gc.ca/). Programs are in place for the protection and recovery of those species classified as Endangered and are being developed for other species in the COSEWIC list. The North Atlantic right whale is one of the world's rarest species, with 2010 surveys indicating only about 450 animals remaining (see http://www.fisheries.noaa.gov/pr/species/mammals/whales/north-atlantic-right-whale.html). Prior to the expansion of the whaling industry in the eighteenth and nineteenth centuries, right whales roamed the North Atlantic but are now encountered only along the western side from Florida to the Bay of Fundy. At least half of the population appears to utilize the Outer Bay of Fundy and Roseway Basin during summer months. Because ship strikes in the Bay of Fundy region were a major cause of right whale mortality, a conservation area was established for the Outer Bay and the international ship traffic lanes were moved to minimize the overlap with right whale summer feeding areas, which are associated with major mixing zones near Grand Manan Island. This has reduced ship mortality, but significant problems remain with whale entanglement in fishing gear – especially trap lines set by lobster harvesters.

Table 1 Bay of Fundy species assessed and designated by COSEWIC as at risk (Source: http://www.cosewic.gc.ca/eng/sct0/rpt/csar_fall_2014_e.pdf)

Common name	Species	Status	Last assessed
North Atlantic right whale	*Eubalaena glacialis*	Endangered	2013
Northern bottlenose whale	*Hyperoodon ampullatus*	Endangered	2011
Harbor porpoise	*Phocoena phocoena*	Special concern	2006
Fin whale	*Balaenoptera physalus*	Special concern	2005
Sowerby's beaked whale	*Mesoplodon bidens*	Special concern	2006
Porbeagle shark	*Lamna nasus*	Endangered	2014
Atlantic salmon	*Salmo salar*	Endangered	2010
Striped bass	*Morone saxatilis*	Endangered	2012
Cusk	*Brosme brosme*	Endangered	2012
Spotted wolffish	*Anarhichas minor*	Threatened	2012
Atlantic sturgeon	*Acipenser oxyrinchus*	Threatened	2011
American eel	*Anguilla rostrata*	Threatened	2012
Atlantic cod	*Gadus morhua*	Special concern	2010
Winter skate	*Leucoraja ocellata*	Special concern	2005
Shortnose sturgeon	*Acipenser brevirostrum*	Special concern	2005
Atlantic wolffish	*Anarhichas lupus*	Special concern	2012
Piping plover	*Charadrius melodus melodus*	Endangered	2013
Roseate tern	*Sterna dougallii*	Endangered	2009
Least bittern	*Ixobrychus exilis*	Threatened	2009
Peregrine falcon	*Falco peregrinus anatum*	Special concern	2007
Harlequin duck	*Histrionicus histrionicus*	Special concern	2013
Barrow's goldeneye	*Bucephala islandica*	Special concern	2011
Yellow rail	*Coturnicops noveboracensis*	Special concern	2009

The Bay of Fundy is replete with environmentally sensitive areas, some of which have been designated for conservation or special management under international, national, and provincial programs or reserved by nongovernment organizations. There are two UNESCO Biosphere Reserves: the Fundy Biosphere Reserve in the Upper Bay of Fundy and the Southwest Nova Biosphere Reserve which includes mainland and islands of the Outer Bay. UNESCO World Heritage status has been awarded to Joggins, NS and Parrsboro, NS in the Upper Bay for their important Jurassic fossil beds, and to Grand Pré, NS, also in the Upper Bay, for the 400 year agricultural history based on marshland conversion. Three sites in the Upper Bay are recognized under the Ramsar Convention on Wetlands and form part of the Western Hemisphere Shorebird Reserve Network because of the important role played by invertebrates of the intertidal flats in supporting migratory shorebirds. In addition to these high profile international designations, there are more than 20 other sites that receive some form of protection as national or provincial ecological or historic marine sites, including wildlife reserves, marine protected areas, and bird sanctuaries (cf. Jacques Whitford 2008; AECOM 2010).

Threats and Future Challenges

In spite of the protective measures outlined above, the Bay of Fundy is very much a "lived in" ecosystem, with a variety of resource uses and human impacts that provide significant challenges for management. These include: fisheries for shellfish and finfish; aquaculture for marine plants, shellfish and fish; international shipping; tourism and recreation; channel dredging and spoil disposal; dam and causeway construction; and mining. The huge tides of the bay have naturally stimulated interests in renewable energy. For the last century proposals for generating electricity from tidal movements have been examined repeatedly, and one tidal generating station exists: the 20 MW Annapolis Tidal Generating Station that was opened in 1985 (Daborn and Redden 2009). Present renewable energy interests are based upon tidal in-stream technologies that appear to have fewer environmental risks than tidal range approaches (ATEI 2013). A major test site for large, commercial scale devices (FORCE: Fundy Ocean Research Center for Energy) has been established at the entrance to Minas Passage, and several other sites are under consideration for installation of arrays of tidal stream generators of varying size.

Managing human activities in such a diverse coastal ecosystem is challenging. Human interventions such as the modification of the bay's natural morphology by marshland conversion, dredging, damming of rivers, and causeway construction have often triggered changes in critical physical processes that may continue for years. The dynamic nature of the nonlinear processes associated with tidal movements means that even subtle changes may have significant effects over the whole system and may take years to become apparent. As more understanding of the bay ecosystem is achieved, the value of such interventions is being reassessed, and in response to pressure from local communities, efforts are under way to explore the potential for reversing the negative effects, particularly of dam and dyke construction. The Petitcodiac River Causeway, constructed in 1968, is one of three large causeways built across major estuaries entering the Bay of Fundy for highway crossing and flood control purposes. Because of the extreme turbidity of the estuary ($<30,000$ mg/L) associated with 11 m tides, rapid and massive deposition of sediment occurred on the seaward side of the dam, eliminating the tidal bore and severely impeding upstream migration of anadromous fish such as salmon, alewife, and shad. In 2010 because of public pressure the causeway was opened again, allowing tidal water to flow upstream past the city of Moncton, NB. Within months, a substantial tidal bore returned, and it is anticipated that the sediments deposited will be remobilized over time as the estuary adjusts to the change.

Appreciation of the high productivity of Fundy salt marshes is the result of research during the last decade. The progressive conversion into agricultural land of more than 80 % of the marshes over the previous four centuries provided some of the most productive farmland in Atlantic Canada, yielding large quantities of hay, grains, and vegetables. However, global competition in agriculture, shifts in demand, climate change, costs of dyke maintenance in the face of rising sea level and tidal range, and urban expansion onto dyked lands have all contributed to a re-evaluation of the benefits of exchange of marshes for dyked land. Dykes destroyed by storm surges have not always been repaired, and recent research into the process of recovery when previously dyked land has been allowed to "go out to sea" again is actively under way at the Beaubassin Research Station, a joint venture between Ducks Unlimited Canada, the Irving family and Acadia University, and by local universities.

References

AECOM. A study to identify preliminary Marine Protected Areas, Bay of Fundy Region. Report prepared for Parks Canada, Ottawa; 2010.

AECOM and ATEI. Tidal energy: strategic environmental assessment update for the Bay of Fundy. Report prepared for Offshore Energy Research Association of Nova Scotia; 2013.

AGS. The last billion years: a geological history of the Maritime Provinces of Canada, Atlantic Geoscience Society Special Publication, vol. 15. Halifax: Nimbus Publishing; 2001.

ATEI. Community and business toolkit for tidal energy development. Acadia Tidal Energy Institute Publication 2013-01. Wolfville: Acadia University; 2013.

Bertness MD, Crain C, Holdredge C, Salas N. Eutrophication and consumer control of New England salt marsh productivity. Conserv Biol. 2008;22:131–9.

Bredin KA, Gerriets SH, Van Guelpen L. Distribution of rare, endangered and keystone marine vertebrate species in Bay of Fundy seascapes. In: Wells PG, Daborn GR, Percy JA, Harvey J, Rolston SJ, editors. Health of the Bay of Fundy: assessing key issues. Proceedings of the 5th Bay of Fundy Science Workshop and Coastal Forum —Taking the Pulse of the Bay, Wolfville, Nova Scotia, May 13–16, 2002. Environment Canada – Atlantic Region, Occasional Report No. 21, Dartmouth/Sackville/New Brunswick: Environment Canada; 2004, p. 83–98.

Brylinsky M, Daborn GR. Community structure and productivity of the Cornwallis Estuary. Cont Shelf Res. 1987;7:1417–20.

Cabilio P, DeWolfe DL, Daborn GR. Fish catches and long-term tidal cycles in Northwest Atlantic Fisheries: a nonlinear regression approach. Can J Fish Aquat Sci. 1987;44:1890–7.

Campbell DE, Wroblewski JS. Fundy tidal power development and potential fish production in the Gulf of Maine. Can J Fish Aquat Sci. 1986;43:78–89.

Cranford PJ, Gordon DC, Jarvis CM. Measurement of cordgrass, *Spartina alterniflora*, production in a macrotidal estuary, Bay of Fundy. Estuaries. 1989;12:27–34.

Daborn GR. Effects of tidal mixing on the plankton and benthos of estuarine regions of the Bay of Fundy. In: Bowman MJ, Yentsch CM, Peterson WT, editors. Tidal mixing and plankton dynamics, Lecture notes in coastal and estuarine studies, vol. 17. New York: Springer; 1986. p. 390–413.

Daborn GR. Homage to Penelope: unravelling the ecology of the Bay of Fundy system. In: Poehle GW, Wells PG, Rolston SJ, editors. Challenges in environmental management in the Bay of Fundy – Gulf of Maine. Proceedings of the 7th Bay of Fundy Workshop, St. Andrews, NB, 24–27 October 2006. Bay of Fundy Ecosystem Partnership Technical Report No. 3. Wolfville: Bay of Fundy Ecosystem Partnership; 2007, p. 12–22.

Daborn GR, Dadswell MJ. Natural and anthropogenic changes in the Bay of Fundy – Gulf of Maine – Georges Bank system. In: El-Sabh MI, Murty TS, editors. Natural and man-made hazards. Dordrecht: D. Reidel Publishing Company; 1988. p. 547–60.

Daborn GR, Gregory RS. Occurrence, distribution and feeding habits of juvenile lumpfish, *Cyclopterus lumpus* L., in the Bay of Fundy. Can J Zool. 1983;64:797–801.

Daborn GR, Redden AM. A century of tidal power research in the Bay of Fundy, Canada, and the enabling role of research networks. J Ocean Technol. 2009;IV(4):1–5.

Daborn GR, Amos CD, Brylinsky M, Christian H, Drapeau G, Faas RW, Grant J, Long B, Paterson DM, Perillo GME, Piccolo MC. An ecological cascade effect: migratory shorebirds affect stability of intertidal sediments. Limnol Oceanogr. 1993;38:225–31.

Day JC, Roff JC. Planning for representative marine protected areas: a framework for Canada's oceans. Report for the World Wildlife Fund, Toronto; 2000.

Desplanque C, Mossman DJ. Tides and their seminal impact on the geology, geography, history and socio-economics of the Bay of Fundy, Eastern Canada. Atl Geol. 2004;40:1–130.

Emerson CW, Roff JC, Wildish DJ. Pelagic-benthic energy coupling at the mouth of the Bay of Fundy. Ophelia. 1986;26:165–80.

Garrett CJR, Keeley JR, Greenberg DA. Tidal mixing versus thermal stratification in the Bay of Fundy and Gulf of Maine. Atmosphere-Ocean. 1978;16:403–23.

Gordon DC, Prouse NJ, Cranford PJ. Occurrence of *Spartina* macrodetritus in Bay of Fundy waters. Estuaries. 1985;8(3):290–5.

Greenberg DA, Petrie BD, Daborn GR, FaderGB. The physical environment of the Bay of Fundy. In: Percy JA, Wells PG, Evans AJ, editors. Bay of Fundy issues: a scientific overview. Atlantic Region Occasional Report No. 8. Sackville/New Brunswick: Environment Canada; 1997, p. 11–34.

Gulf of Maine Area (GOMA). Census of marine life. 2014. http://www.gulfofmaine-census.org/about-the-gulf/biodiversity-of-the-gulf/lists/list-of-species-from-the-bay-of-fundy/#Fish

Hagerman G, Fader G, Carlin G, Bedard R. Nova Scotia tidal in-stream energy conversion (TISEC) survey and characterization of potential sites. EPRI North American Tidal Flow Power Feasibility Demonstration Project, Phase 1 – Project Definition Study, Report EPRI-TP-003 NS Rev 2; 2006. Electrical Power Research Institute, Washington, DC 20005

Hamilton DJ, Barbeau MA, Diamond AW. Shorebirds, snails and *Corophium* in the Upper Bay of Fundy: predicting bird activity on intertidal mudflats. Can J Zool. 2003;81:1358–66.

Jacques Whitford. Background report for the Fundy tidal energy strategic environmental assessment. Final Report presented to the Offshore Energy Environmental Research Association, Halifax; 2008.

Kirwan ML, Matthew L, Guntenspergen GR, Morris JT. Latitudinal trends in *Spartina alterniflora* productivity and the response of coastal marshes to global change. Glob Chang Biol. 2009;15 (8):1982–9.

Prouse NJ, Gordon DC, Hargrave BT, Bird CJ, MacLachlan J, Lakshminarayana JSS, Sita Diva L, Thomas MLH. Primary production: organic matter supply to ecosystems in the Bay of Fundy. In: Gordon DC, Dadswell MJ, editors. Update on the marine environmental consequences of tidal power development in the upper reaches of the Bay of Fundy, Canadian technical report of fisheries and aquatic sciences, Ottawa: Supply and Services Canada; vol. 1256; 1984. p. 65–95.

Redfield AC. Development of a New England salt marsh. Ecol Monogr. 1972;42:201–37.

Silliman BR, Bertness MD. A trophic cascade regulates salt marsh primary production. Proc Natl Acad Sci U S A. 2002;99(16):10500–5.

Silliman BR, Bortolus A. Underestimation of *Spartina* productivity in Western Atlantic marshes: marsh invertebrates eat more than just detritus. Oikos. 2003;101:549–54.

Wells PG. Environmental impacts of barriers on rivers entering the Bay of Fundy. Report of an *ad hoc* Environment Canada Working Group, Technical report series, vol. 334. Ottawa: Canadian Wildlife Service; 1999.

Wildish DJ, Fader GBJ. Pelagic-benthic coupling in the Bay of Fundy. Hydrobiologia. 1998;375/376:369–80.

Wildish DJ, Akagi HM, Fader GBJ. Horse mussel reef project in the Inner Bay of Fundy. In: Ollerhead J, Hicklin PW, Wells PG, Ramsey K, editors. Understanding change in the Bay of Fundy ecosystem. Proceedings of the 3rd Bay of Fundy Science Workshop, Mount Alison University, Sackville, N.B. Environment Canada, Atlantic Region Occasional Report No. 12. Sackville: Environment Canada; 1999, p. 21–2.

Wrathall C, van Proosdij D, Lundholm J. 2013. Assessment of primary productivity in the Windsor salt marsh. Report of the Environmental Science Program and Department of Geography, Saint Mary's University, Halifax; 2013.

San Francisco Bay Estuary (USA)

50

Beth Huning and Mike Perlmutter

Contents

Introduction	638
Description	638
Biodiversity	641
Ecosystem Services	643
Conservation Status	645
Threats and Future Challenges	647
References	648

Abstract

The San Francisco Bay Estuary (SFBE) is the second largest estuary in the United States, encompassing approximately 4,145 km^2 (1600 mi^2) and draining about 40% (155,400 km^2; 60,000 mi^2) of the State of California through the Sacramento and San Joaquin Rivers, which pass through the San Francisco Bay-Delta to the Pacific Ocean. The SFBE has perhaps suffered the most extensive degradation of any estuary in the United States. Many years of diking, draining, filling, pollution, and introduction of alien species have taken a great toll on the ecosystem. Although 80% of the tidal salt marshes have been lost, many of the remaining marshes are now protected, and there are large-scale restoration efforts under way to return salt evaporation ponds, agricultural areas, and some urban areas back to tidal wetlands. The SFBE was designated as a Wetland of International Importance under the Ramsar Convention on Wetlands in 2013.

B. Huning (✉)
San Francisco Bay Joint Venture, Fairfax, CA, USA
e-mail: bhuning@sfbayjv.org

M. Perlmutter (✉)
Environmental Services Division, City of Oakland Public Works Department, Oakland, CA, USA
e-mail: mperlmutter@oaklandnet.com

© Springer Science+Business Media B.V., part of Springer Nature 2018
C. M. Finlayson et al. (eds.), *The Wetland Book*,
https://doi.org/10.1007/978-94-007-4001-3_214

Keywords

Ridgeway's rail · Estuary · Shorebird · Tidal wetlands · Salt marsh · California · Ramsar · Salt marsh harvest mouse · Waterfowl

Introduction

The San Francisco Bay Estuary (SFBE) is the second largest estuary in the United States, encompassing approximately 4,145 km^2 (1600 mi^2) and draining about 40% (155,400 km^2; 60,000 mi^2) of the State of California through the Sacramento and San Joaquin Rivers, which pass through the San Francisco Bay-Delta to the Pacific Ocean (Fig. 1). The SFBE has perhaps suffered the most extensive degradation of any estuary in the United States. Many years of diking, draining, filling, pollution, and introduction of alien species have taken a great toll on the ecosystem. Although 80% of the tidal salt marshes have been lost, many of the remaining marshes are now protected, and there are large-scale restoration efforts under way to return salt evaporation ponds, agricultural areas, and some urban areas back to tidal wetlands. The SFBE was designated as a Wetland of International Importance under the Ramsar Convention on Wetlands in 2013.

SFBE was historically rimmed with tidal salt marshes, particularly in its northern and southern reaches. Despite losing one third of its size and approximately 85% of its wetlands to development, agricultural and salt flat conversion, and fill, SFBE remains critically ecologically important, accounting for 77% of California's remaining perennial estuarine wetlands (Sutula et al. 2008). SFBE is widely recognized as one of North America's most ecologically important estuaries, providing key habitat for a broad suite of flora and fauna, and a range of ecological services such as flood protection, water quality maintenance, nutrient filtration and cycling, and carbon sequestration. SFBE is home to many plant species and over 1,000 species of animals (USEPA 1999), including endemic and conservation status species, i.e., federally and state listed threatened or endangered species as well as those identified by International Union for Conservation of Nature (IUCN) or United States Fish and Wildlife Service (USFWS) as Birds of Conservation Concern.

Description

San Francisco Bay Estuary is approximately 10,000 years old and is a product of today's high sea level, which presently floods the ancient river drainage from California's Central Valley out to the Pacific Ocean through the Golden Gate (Sloan 2006). The general climate for SFBE and the surrounding region is categorized as Mediterranean featuring temperate wet winters and warm, dry summers (National Park Service 2009). Average annual rainfall in the region is 38–61 cm (15–24 in), which generally falls between November and April (Western Regional Climate Center 2009). Salinity is approximately 30 ppt near the mouth of the Golden

Fig. 1 The San Francisco Bay Estuary (Photo credit: San Francisco Estuary Institute/Aquatic Science Center ©)

Gate, whereas upstream sources in the delta are nearly fresh at 1 ppt (NOAA 2007). About 90% of the freshwater entering the Bay comes from the Sacramento/San Joaquin watershed, with the remaining 10% originating from local streams and creeks and from wastewater treatment facilities. Seasonal and year-to-year variability in freshwater inflow to the Bay have been reduced (San Francisco Estuary Partnership 2011) and is about 60% less than historic flows due to diversions of municipal water (responsible for about 9% of flow reductions) and for Central Valley agricultural uses (responsible for about 51% of flow reductions) (Sloan 2006).

SFBE water temperature variation generally follows the Pacific Ocean's, which has a cool season of upwelling from April–July, a warmer season from August–November, and a cold storm season from December–March. The Bay's temperature swings, however, are greater than the ocean's due to the Bay's shallower water and river flow inputs, making the Bay generally colder than the ocean during the winter, and warmer than the ocean the rest of the year (NOAA 2007).

The average depth of SFBE is 5.5 m (18 ft). With few exceptions, the waters are naturally deep only in parts of the Central Bay between San Francisco, Marin County, and Angel Island and in channels through Carquinez Strait (about 27.4 m; 90 ft), Raccoon Strait (18.3 m; 60 ft), and the Golden Gate (much of which is between 30.5 and 61 m; 100–200 ft) (Sloan 2006). The SFBE is at its deepest at the outer mouth of the Golden Gate, where it plunges to 107 m (350 ft) (Cohen 2000). Dredged channels provide shipping access to ports in Oakland, Richmond, and Redwood City.

Water depth changes twice daily with the tides. SFBE's tides are of unequal height. Average difference between high and low tide heights is about 1.2 m (4 ft) in the Central Bay, 1.5 m (5 ft) in northern San Pablo Bay, and about 2.1 m (7 ft) in the South Bay. These tides transport 1,603.5 Mm^3 of water, about 25% of the Bay's water volume every day (Cohen 2000).

The SFBE's floor is covered with sand, silt, or clay, along with significant quantities of oyster shell fragments. In the North Bay, channels are mostly sand, with shell fragments occurring in the southeastern and southwestern shallows of the South Bay. A few areas of bedrock rim the western part of the Central Bay and crop out at islands and a few shoreline locations. Artificial hard substrate is scattered across the Bay, including rip-rapped banks, jetties, breakwaters, seawalls, pilings, docks and piers, bridge and power line supports, and debris (NOAA 2007).

Sediment in the SFBE primarily derives from upstream watershed erosion and subsequent transport and deposition into the Bay through freshwater tributaries. The first large storm of the year carries large sediment plumes into the Bay, followed by smaller amounts of sediment during subsequent storms. Most suspended sediment is from the Sacramento and San Joaquin rivers, but sediment also comes from the Yolo Bypass, Mokelumne River, Calaveras River, Cosumnes River, and several other smaller streams. Hydraulic mining in the Sierra Nevada in the mid-late 1800s resulted in hundreds of millions of cubic meters of sediment deposition into the SFBE. Current deposition rates are much lower as hydraulic mining deposition has tapered off and additional sediment sources have been reduced (by stream flow restrictions) and trapped behind dams (NOAA 2007).

Biodiversity

Historically, wetlands accounted for 5% of California's land area, but approximately 91% have been lost, reducing their relative land cover to less than 0.5%. California retains approximately 18,000 ha (44,456 acres) of perennial estuarine wetland habitat, with 77% in the SFBE (California Resources Agency 2008). The predominant native habitats in SFBE are primarily open water of varying depths based on tidal conditions and location, tidally influenced mudflats, submerged eelgrass beds, vegetated marshes, sand and salt flats, and sandy and cobble beaches. There are also several rock islands within SFBE providing nesting habitat for colonial water birds.

Many parts of the SFBE have been altered by human activities but still provide habitat for many species. They include diked marshes, agricultural baylands (grazed and farmed), and salt ponds. Upland habitats exist adjacent to and within the SFBE in the form of transitional habitats adjacent to tidal marshes and in the form of islands. Some of the islands are manmade (fill) and some are natural and still contain native upland plant communities consisting of grasslands, shrub lands, and woodlands. Pickleweed *Sarcocornia pacifica* and cord grass *Spartina foliosa* predominate the marshes in more saline waters and bulrush *Scirpus* spp. predominates in more freshwater marshes.

Submerged aquatic plant communities of the shallow subtidal habitats and tidal flats of SFBE are important food sources for estuarine fish, invertebrates, and birds. Submerged plants such as eelgrass *Zostera marina* and certain macroalgae also provide important cover, spawning, and rearing grounds for invertebrates and estuarine fish, such as migrating salmon and Pacific herring *Clupea pallasii* (Olofson 2000). Eelgrass, surfgrass *Psyllospadix scouleri* and *P. torreyi*, widgeon grass *Ruppia maritima*, and sago pondweed *Potamegeton pectinatus* provide important nursery and foraging habitats, dampen wave energy, and aid in sediment capture (NOAA 2007).

The predominant animal communities are those associated with open water, tidal flats, and tidal marshes, as well as managed ponds. Open water provides habitat for wintering diving and sea ducks, and migratory corridors for anadromous fish to reach freshwater spawning areas. The Bay's subtidal habitats are also important for approximately 500 species (USEPA 1999) of aquatic invertebrates, while tidal flats support thousands of migratory and overwintering shorebirds and waterfowl. Tidal marshes also provide habitat for a diverse assemblage of migratory shorebirds and waterfowl, various fish species, and tidal marsh specialists such as the Ridgway's rail *Rallus obsoletus* (formerly California clapper rail) and salt marsh harvest mouse *Reithrodontomys raviventris*.

SFBE is noted for hosting more wintering and migrating shorebirds than any other estuary along the US Pacific Coast south of Alaska (Stenzel et al. 2002). For this, SFBE is recognized as a Site of Hemispheric Importance by the Western Hemispheric Shorebird Reserve Network (WHSRN). Bay-wide surveys of wintering shorebirds conducted in November from 2006 through 2008 averaged over 340,000 shorebirds, including 29 species (Wood et al. 2010). During the height of spring and fall migration, 589,000–932,000 and 340,000–396,000 shorebirds respectively were counted during surveys conducted between 1988 and 1993 (Stenzel et al. 2002). Compared to the major wetlands along the Pacific Coast, SFBE held an average of 55.7% (37.8–90.1%) of the total number of individuals of 13 key shorebird species. In particular, a significant portion of arctic-breeding dunlin *Calidris alpina* and western sandpipers *C. mauri* winter in SFBE.

SFBE is also recognized as one of 67 areas of continental significance for waterfowl by the North American Waterfowl Conservation Plan (NAWMP 2004). SFBE is the winter home for 50% of the diving ducks in the Pacific Flyway (Olofson 2000). The US Fish & Wildlife Service midwinter waterfowl counts from 1988 to 2006 document SFBE as containing 49% of the scaup *Aythya* spp. population and 43% percent of the scoters *Melanitta* spp. of the lower Pacific Flyway, from Washington State to southern California. About 99% of SFBE's scoters are surf scoters *Melanitta perspicillata*. Midwinter SFBE waterfowl surveys from 1992 to 2007 averaged 182,818 birds in mid-January (Susan Wainwright-De La Cruz, 27 May 2009, United States Geological Survey, personal communication). Additionally, from 2006 to 2009, Suisun Marsh in the eastern portion of the estuary averaged 99,649 birds. Within the larger totals of birds, there are hotspots where over 20,000 water birds regularly congregate. San Pablo Bay hosts at least 20,000 ducks in early winter until mid-January. In the East Bay, San Leandro Bay is critically important for scoters all winter but becomes increasingly more important over winter as the majority of the SFBE population moves there before migration.

The extent and diversity of SFBE fish habitats (varying salinities, substrates, water depth, etc.) make it important to over 130 species of resident and migratory marine, estuarine, and anadromous fish species (SFBCDC 2009) through many lifecycle stages. Marine species tend to use the Bay as spawning and nursery habitat, while estuarine species reside in the Bay throughout their life cycle. For anadromous chinook salmon *Oncorhynchus tshawytscha*, steelhead *Oncorhynchus mykiss*, and white sturgeon *Acipenser transmontanus*, the SFBE is a critical migratory pathway between foraging areas in the Pacific Ocean and spawning grounds upstream in the SFBE's tributary rivers (The Bay Institute 2003). SFBE is identified as Essential Fish Habitat for various fish species life stages managed under three Fisheries Management Plans of the National Marine Fisheries Service. Additionally, SFBE is designated as Habitat Areas of Particular Concern for various fish species within the Pacific Groundfish Fisheries Management Plan (NOAA 2008). SFBE supports spawning Pacific herring, which is not only a major fishery, but a source of roe forage for diving waterfowl in the Central Bay.

Eight animal species are endemic to SFBE and its associated wetlands, which also provide habitat for a number of near-endemic or range-limited species, subspecies, and races of flora and fauna. The Bay supports three endemic fish taxa: the federally endangered delta smelt *Hypomesus transpacificus*, San Francisco topsmelt *Atherinops affinis affinis*, and tule perch *Hysterocarpus traskii traskii*, in addition to four local races of chinook salmon. The federally endangered delta smelt occurs only in the San Francisco Bay-Delta Estuary. The species spends much of its lifecycle in the Sacramento and San Joaquin rivers and deltas, which feed into SFBE. Juvenile and adult smelt also may spend time adjoining northern SFBE, where they have been observed in Suisun and San Pablo bays and the Napa River (Olofson 2000). Although exact population estimates are unknown (Moyle 2002), relative population levels have been monitored for several decades by federal and state water export facilities (Bennett 2005). Counts from 2002 to 2007 showed low abundance (Armor et al. 2005).

Longfin smelt are widely but patchily distributed along North America's Pacific Coast but historically occupied only three estuaries and the lower reaches of their larger tributary rivers in California: San Francisco Bay-Delta Estuary, Humboldt Bay, and Klamath River Estuary. Presently, the largest and southernmost self-sustaining longfin smelt population is in the SFBE-Delta Estuary. The Humboldt Bay and Klamath River populations are thought to be extinct, and the small numbers of fish recently reported in the Russian River do not likely represent a self-sustaining population.

There are three recognized subspecies of tule perch, one of which occurs from the Sacramento-San Joaquin River drainage through SFBE. The range of this subspecies has contracted from its historic distribution, which formerly extended beyond the Bay to the Pajaro and Salinas rivers (Olofson 2000).

Sacramento-San Joaquin chinook salmon are grouped within four distinct races, based on the timing of adult spawning migration: winter, spring, fall, and late fall. Three of these races are presently of conservation concern and have the following status: winter run (federally and state endangered), spring run (California Class 1 qualified as threatened or endangered), and late-fall run (California listed Class 2 special concern) (Olofson 2000).

The tidal wetlands support the endemic salt marsh harvest mouse populations and the majority of the Ridgway's rail populations, both federally listed endangered species. The best mouse population estimate comes from the *Salt Marsh Harvest Mouse and California Clapper Rail Recovery Plan* published by the US Fish and Wildlife Service in 1984: "a few thousand individuals at the peak of their numbers each summer, distributed around the Bay marshes in small, disjunct populations, often in marginal vegetation and almost always in marshes without an upper edge of upland vegetation" (USFWS 1984). Ridgway's rail (Fig. 2) are now almost entirely restricted to the marshes of the SFBE, where the only known breeding populations occur. When first listed, the population was considered to be 4,200–6,000 individuals, but today's estimates appear to be about 1,200 individual birds (US Fish and Wildlife Service 2013).

Numbers of scaup, scoters, and canvasback *Aythya valisineria* vary by year and season, but SFBE hosts 44% of the Pacific Flyway diving duck population during winter months.

Ecosystem Services

SFBE wetlands provide many ecological services such as flood control, aquifer recharge, regional climate mediation, and water quality maintenance (Save the Bay 2007). Tidal marshes produce organic nutrients, sequester carbon, reduce shoreline

Fig. 2 The Ridgeway's rail, see here in pickleweed and cord grass, is the endemic rail of San Francisco Bay (Photo credit: B. Huning ©)

erosion, and provide a nursery for some fish species. The importance of the SFBE to resident and migratory species has been well documented above.

SFBE provides for a host of social and economic values through ports and industry, agriculture, fisheries, archaeological and cultural sites, recreation, and research. Open water areas of the estuary are used as shipping and ferry channels and approximately 5.35 Mm^3 (seven million cubic yards) of sediment are dredged annually to maintain shipping channels and marinas; some of the sediment is being reused to bring wetland areas back up to marsh plain elevation (the elevation within the tidal prism at which marsh vegetation can become established) during the restoration process. The industrial port of Oakland is one of the world's largest ports, ranking among the top four in the United States and 20th in the world in terms of annual container traffic (Port of Oakland 2009). Other water-related industries such as refineries, factories, and dredged material rehandling plants utilize the SFBE. In addition to water-related industries, about 1,780 ha (4,400 acres) of diked baylands in the South Bay continue to be used for salt production by Cargill Salt. Salt production in the SFBE historically occurred on nearly 16,600 ha (over 41,000 acres), but since the 1960s about 90% of these lands have been publicly acquired for conservation and restoration (SFB CDC 2008), part of the largest wetland restoration on the west coast of the United States (South Bay Salt Pont Restoration 2009). These salt ponds, both in converted and restored conditions, provide important habitat function for a wide suite of species and the restoration aims to balance the habitat needs of each species, while simultaneously providing for public access and flood protection.

Agriculture on diked baylands continues today, especially in northern San Pablo Bay where farming includes: oat hay (sometimes double-cropped with beans), dairy, row crops, vineyards, orchards, livestock, and irrigated pastureland (North Bay Water Reuse Authority 2009).

SFBE fisheries have suffered dramatic declines in the last few years. A reduced number of commercial fishing boats continue to operate in the Bay as do sport fisheries (Rogers 2009). In 2009, the Bay's last commercial fishery, herring, was shut down for the season to allow the species to recover.

Many important cultural and archaeological sites documented in the Bay include over 425 Native American shellmounds mapped by Nelson (1909). More modern historic sites include: the immigration and detention center at Angel Island, the World War II era naval shipyard at Rosie the Riveter/WW II Home Front National Historic Park, Alcatraz Island lighthouse and penitentiary, and the Presidio of San Francisco, in what is now the Golden Gate National Recreation Area. SFBE also hosts a large body of scientific research and numerous universities.

Waterfront parks and trails, such as the San Francisco Bay Trail, provide opportunity for recreation and nature appreciation, such as bird watching, along the shoreline. A variety of public and private access points and facilities are located throughout the Bay for water entry and shoreline access. These opportunities have expanded greatly in recent decades as public planning and policy has prioritized public access.

The SFBE Area is a renowned international tourist destination. Attractions such as the Golden Gate Bridge, acclaimed as one of the world's most beautiful bridges, attract an estimated nine million tourists annually. Alcatraz Island attracts more than 1.3 million visitors annually. The SFBE is a popular destination for water-oriented recreation such as boating and open water swimming. In 2013, San Francisco was the host city for the America's Cup international yacht race. Other destinations in and around the SFBE, including the city of San Francisco; Napa and Sonoma county vineyards; and national, state, and local parks; events; and festivals draw many more tourists.

Conservation Status

In addition to Ramsar and WHSRN designations, the importance of SFBE and its wetlands has been recognized by the National Audubon Society, which has designated portions of the Bay's habitats as nine distinct Important Bird Areas for the vast numbers of shorebirds, waterfowl, and endangered, threatened, and sensitive bird species populations (see: http://web4.audubon.org/bird/iba/ibaadopt.html).

The SFBE is now a major center for a vibrant habitat restoration movement. Over the past two decades, significant progress has been made to protect and enhance remaining habitats and restore as much as possible of what has been lost. A unique partnership of landowners, conservation organizations, state and federal agencies, and businesses has been formed to plan and guide the restoration. Members of this partnership, the San Francisco Bay Joint Venture (SFBJV), have established goals to protect 25,495 ha (63,000 acres), restore 14,973 ha (37,000 acres), and enhance another 14,164 ha (35,000 acres) of bay habitats that include tidal flats, marshes, and lagoons. They are also working to secure habitat values of seasonal wetlands with

Fig. 3 Breaching the dike at Cullinan Ranch in San Pablo Bay on January 6, 2015. Note the difference in marsh plain elevation between Dutchman Slough (*right*) and Cullinan Ranch (*left*), which is subsided. This former oat-hay ranch was historically tidal wetlands that had been diked from tidal action over 100 years ago and is now being restored to tidal marsh. Twice daily tidal exchanges will allow sediment to accumulate naturally (Photo credit: B. Huning ©)

protection and restoration/enhancement goals 27,114 ha (67,000 acres) (SFB Joint Venture 2001). In addition, SFBJV partners intend to protect 1619 ha (4,000 acres) of riparian corridors (Fig. 3).

Over the next 50 years, the goal is to improve the condition of the subtidal ecosystem by minimizing impacts of aquatic invasive species on subtidal habitats; protect SFBE from chronic oil spills; prevent and capture land or marine sources of trash before it enter the SFBE; identify, prioritize, and remove large sources of marine debris; increase public awareness and support for subtidal habitat protection; enhance, restore, or protect submerged aquatic vegetation, shellfish beds, sand and soft bottom habitats (SFB Joint Venture 2010). As of the beginning of 2014, more than 32,375 ha (80,000 acres) have been returned to its desired condition. (See the project tracking database at www.sfbayjv.org for updated acreage and habitat accomplishments.) The SFBJV is leading the implementation of the Baylands Goals recommendations to address climate change. An initial process is coordinating land managers and scientists in scenario planning and decision prioritization to address climate change that can lead to prioritized and integrated conservation actions to continue to restore the SFBE and accommodate species in an era of climate change (San Francisco Bay Joint Venture SF Bay Climate Decision Support Analysis, unpublished).

Threats and Future Challenges

Approximately seven million people (estimated at 7,150,739 by 2010 census) live in the nine counties surrounding SFBE. San Francisco is the original urban center of the region with an estimated population of 805,235 (SFB Area Census 2010), although San Jose is the largest city (945,942). Land use in the SFBE region in modern times has been increasing urban development and industry, agriculture (farming and grazing), and salt extraction. Specifically, nearly 50% of the estuary's watershed has been converted to agriculture; about 4% has been urbanized, and 10% is now industrial sites. Large areas of former marshlands were filled during urbanization and diked for agriculture or converted to salt evaporation ponds, thus dramatically reducing the overall acreage of tidal wetlands in the watershed today (Olofson 2000). Salt ponds provide habitat for shorebirds, and as habitats are being converted in an effort to restore the SFBE, some managed ponds are being maintained and salinity levels managed for shorebird habitat.

The habitats of SFBE face a variety of human disturbances, which can pose direct degradation impacts to habitat as well as to flora and fauna. Human access through wetlands can trample habitats and facilitate access by predators. Trail use near wetland habitats, as well as nearby boat or aircraft use, may cause animals to flush, exposing them to predators and impinging on rest and feeding behaviors.

Climate change represents a suite of challenges to SFBE such as altered species viabilities and phenologies, sea level rise, shifts in salinity content and fresh water flows, notable rises in temperature, and an increase in the severity of storms. Climate change is expected to result in sea level rise in SFBE of nearly 40 cm by mid-century and 1.4 m by the end of the century.

By mid-century, nearly 73,000 ha (180,000 acres) of SFBE shoreline will be vulnerable to flooding, and 86,200 ha (213,000 acres) will be vulnerable by the end of the century. Vulnerability within today's 100-year floodplain (the area with a one in 100 year flood probability) will increase from a one percent chance of flooding per year to a 100% chance of flooding per year by mid-century (SFB CDC 2009). Higher seas as well as more frequent and intense storm events threaten to increase shoreline damage, erosion, and inundation. Depending upon sediment availability and the rate of sea level rise, modeling has indicated that tidal marshes may be able to accrete and keep pace with higher waters, if large-scale restoration projects are completed and adjacent uplands protected. In the urbanized areas where the SFBE edges are developed, there is little opportunity for marsh transgression.

Freshwater runoff from mountain snowmelt is projected to flow earlier and more intensely in the year as warmer temperatures speed snowmelt. Resulting increased winter freshwater inputs and decreased spring and summer freshwater inflows will decrease salinities in the wet season and, by a larger degree, increase salinity levels in the dry season. Sea level rise could further drive saltwater gradients upstream in the SFBE compounding the salinity changes. These shifts in the quantity, timing, and quality of freshwater flowing into the Bay and the resulting habitat changes could cause declines of fish species and populations. The reduced freshwater inflow and resulting increase in salinity causes more salt-tolerant species to move upstream while freshwater species retreat.

Some bird species behavior, distribution, and population dynamics are susceptible to climate change (Berthold et al. 2004). Sedentary taxa of SFBE such as song sparrows *Melospiza melodia*, common yellowthroats *Geothlypis trichas*, and Ridgway's rail endemic to San Francisco tidal marshes may face additional declines if more tidal marsh habitat is impaired or lost through factors such as climate change induced sea level rise, storm damage, or associated increased invasive species pressures. Recent efforts by conservation groups and government agencies to increase acreage of tidal marsh habitat and improve tidal actions through levees might help to alleviate pressures arising from sea level rise and storm damage.

In early 2015 a science update to the Baylands Ecosystem Habitat Goals for Climate Change was released (http://baylandsgoals.org/science-update-2015/). The primary recommendations include the following: (1) Restore complete Baylands systems; (2) accelerate restoration of complete systems by 2030; (3) plan ahead for the dynamic future; (4) incerase regional coordination; (5) engage the citizenry in advocacy for baylands.

Dealing with the conservation challenges in SFBE is challenging and not simple. In spite of the prevalence of invasive species, projected climate change and sea level rise, human populations competing with the needs of wildlife, the investments in habitat protection and restoration efforts are sustaining wildlife populations and providing valuable economic services. Indicators point to the need to continue and accelerate restoration, as research and modeling have shown that the SFBE has the potential to recover endangered species and support migratory and wintering birds as well as provide for the continued economic needs of a large urban population (PRBO Conservation Science and San Francisco Bay Joint Venture 2011).

References

Armor C, Baxter R, Bennett B, Breuer R, Chotkowski M, Coulston P, Denton D, Herbold B, Larsen K, Nobriga M, Rose K, Sommer T, Stacey M. 2005. Interagency ecological program synthesis of 2005 work to evaluate the pelagic organism decline (POD) in the upper San Francisco Estuary. 2005.

Bennett WA. Critical assessment of the delta smelt population in the San Francisco Estuary, California. San Franc Estuar Watershed Sci. 2005;3:1–71.

Berthold P, Møller AP, Fiedler W. Preface. In: Møller A, Berthold P, Fiedler, editors. Birds and climate change, Advances in ecological research, vol. 35. Amsterdam: Elsevier/Academic Press; 2004. p. vii.

California State Coastal Conservancy and Ocean Protection Council, et al. San Francisco Bay Subtidal Habitat Goals Project 2010.

Cohen A. An introduction to the San Francisco Bay, save the Bay, San Francisco Estuary project, San Francisco Estuary Institute. 2000. p 4.

Moyle PB. Inland fishes of California. Revised and expanded. Berkeley: University of California Press; 2002. p. 230.

National Oceanic and Atmospheric Administration. Letter from the national oceanic and atmospheric administration to the United States Fish & Wildlife Service regarding the South Bay salt pond restoration project. 2008. http://www.southbayrestoration.org/pdf_files/Comment%20Letters/NOAA_FEIS_SBSP0001.pdf. Last accessed 21 Jan 2015.

National Oceanic and Atmospheric Administration. Report on the subtidal habitats and associated biological taxa in San Francisco Bay. 2007. p. 13.

National Park Service. A climate of contrasts. 2009. http://www.nps.gov/prsf/naturescience/climate.htm. Last accessed 21 Jan 2015.

Nelson N. Shellmounds of the San Francisco Bay region. Berkeley: University Press; 1909.

North American Waterfowl Management Plan. United States Fish & Wildlife Service, Canadian Wildlife Service, Secretaria de Medio Ambiente y Recursos Naturales. 2004. p 15.

North Bay Water Reuse Authority. North San Pablo Bay restoration and reuse project engineering and economic/financial analysis report. 2009. http://www.nbwra.org/docs/pdfs/NBWRP_Draft_Phase3_section3_part1of2.pdf. Last accessed 21 Jan 2015.

Olofson PR. Baylands ecosystem habitat goals project. Baylands ecosystem species and community profiles: life histories and environmental requirements of key plants, fish, and wildlife. Prepared by the San Francisco Bay Area Wetlands Ecosystem Goals Project. San Francisco Bay Regional Water Quality Control Board, Oakland; 2000. p. 309.

Port of Oakland. About us: revenue division. http://www.portofoakland.com/portnyou/overview.asp. Last accessed 8 Oct 2009.

PRBO Conservation Science and the San Francisco Bay Joint Venture. The state of the birds San Francisco Bay. 2011.

Rogers P. San Francisco Bay's last commercial fishery closes. Silicon Valley Mercury News. September, 5 2009. http://www.mercurynews.com/bay-area-living/ci_13278722. Last accessed 21 Jan 2015.

San Francisco Bay Area Census. 2010. http://www.bayareacensus.ca.gov/bayarea.htm. Last accessed 21 Jan 2015.

San Francisco Bay Conservation and Development Commission. 2009.

San Francisco Bay Conservation and Development Commission. San Francisco Bay Estuary. http://www.bcdc.ca.gov/bay_estuary.shtml. Last accessed 21 Jan 2015.

San Francisco Bay Conservation and Development Commission. San Francisco Bay Plan. San Francisco. 2008. p. 64.

San Francisco Bay Joint Venture. Climate adaptation decision support analysis. 2015, unpublished.

San Francisco Estuary Partnership. The State of San Francisco Bay. Oakland. 2011.

Save the Bay. Greening the Bay – financing wetland restoration in San Francisco Bay. Oakland. 2007; p. 7.

SF Bay Joint Venture. Implementation plan, restoring the estuary 2001, Baylands ecosystem habitat goals project 2000.

Sloan D. Geology of the San Francisco Bay region. Berkeley: University of California Press; 2006. p. 134.

South Bay Salt Pond Restoration Project. Project description. 2009. http://www.southbayrestoration.org/Project_Description.html. Last accessed 21 Jan 2015.

Stenzel L, Hickey C, Kjelmyr J, Page G. Abundance and distribution of shorebirds in the San Francisco Bay Area. West Birds. 2002;33:1.

Sutula M, Collins JN, Clark R, Roberts C, Stein E, Grosso C, Wiskind A, Solek S, May M, O'Connor K, Fetscher E, Grenier JL, Pearce S, Robinson A, Clark C, Rey K, Morrissette S, Eicher A, Pasquinelli R, Ritter K. California's wetland demonstration program pilot – a final report to the California resources agency. Southern California coastal water research project, technical report 572. Costa Mesa. 2008.

The Bay Institute. Ecological scorecard San Francisco Bay index. Novato: The Bay Institute; 2003.

U.S. Environmental Protection Agency. Goals project. Baylands ecosystem habitat goals. A report of habitat recommendations prepared by the San Francisco Bay area wetlands ecosystem goals project, San Francisco/S.F. Bay Regional Water Quality Control Board, Oakland. Inside front cover. 1999.

United States Fish & Wildlife Service. Recovery plan for tidal marsh ecosystems of Northern and Central California. 2013. http://www.fws.gov/sacramento/ES/Recovery-Planning/Tidal-Marsh/es_recovery_tidal-marsh-recovery.htm

United States Fish & Wildlife Service. Salt marsh harvest mouse and California clapper rail recovery plan. Portland; 1984. p. 44.

Western Regional Climate Center. San Francisco Bay area climate summaries. 2009. http://www.wrcc.dri.edu/summary/climsmsfo.html. Last accessed 21 Jan 2015.

Wood J, Page G, Reiter M, Liu L, Robinson-Nilsen C. Abundance and distribution of wintering shorebirds in San Franciso Bay, 1990–2008: population change and informing future monitoring. Grant # 2009–0179: San Francisco Bay shorebird analysis. Resources Legacy Fund. Sacramento. 2010.

Vernal Pools of Northeastern North America

51

Elizabeth A. Colburn and Aram J. K. Calhoun

Contents

Introduction	652
Landscape Distribution of Vernal Pools	652
History of Pool Studies	653
Size, Depth, and Water Quality of Vernal Pools	654
Hydrology	655
Vernal Pools as Wetlands	656
Plant Communities	657
Detritus and In-Pool Photosynthesis	657
Amphibians in Vernal Pools	658
Invertebrates in Vernal Pools	659
Conservation of Vernal Pools	662
References	664

Abstract

Vernal pools of northeastern North America are small, seasonally flooded wetlands found in forested areas. They are defined by features such as length and timing of flooding and their biological community. Most reach their maximum depth and volume each spring and draw down during the summer. Their importance as habitat for some amphibians and invertebrates, and the unique adaptations of the fauna, make vernal pools of interest for conservation, education, and research.

E. A. Colburn (✉)
Harvard Forest, Harvard University, Petersham, MA, USA
e-mail: colburn@fas.harvard.edu

A. J. K. Calhoun (✉)
University of Maine, Orono, ME, USA
e-mail: calhoun@maine.edu

© Springer Science+Business Media B.V., part of Springer Nature 2018
C. M. Finlayson et al. (eds.), *The Wetland Book*,
https://doi.org/10.1007/978-94-007-4001-3_283

Keywords

Temporary ponds, pools or waters · Seasonal pools or ponds · Forest ponds · Amphibians · Aquatic macroinvertebrates · Hydroperiod · Biodiversity · Conservation

Introduction

Vernal pools of northeastern North America are small, seasonally flooded wetlands found in forested areas. They are commonly defined by a combination of habitat characteristics, especially hydroperiod (length and timing of flooding), and biological community, especially the presence of species adapted to temporary waters. In general, vernal pools flood annually, reaching their maximum depth and volume in early spring and becoming fully or partly drawn down during the growing season. Many vernal pools dry completely by early summer, but others remain flooded until early fall, and some dry only occasionally in drought years. The annual or periodic drawdown prevents the establishment of permanent fish populations, limits the distribution of many large invertebrate predators, and contributes to decomposition of detritus by aerobic and anaerobic bacteria and fungi. Reduced predation pressure and abundant, nutritious food support a wide variety of animal life adapted to surviving seasonal inundation and drawdown. In particular, vernal pools are important breeding and nursery habitats for certain amphibians as well as a suite of pool-dependent invertebrate species. Their small size, seasonality of flooding, and unique contributions to local and regional biodiversity make vernal pools of interest for conservation, scientific research, and education.

This chapter provides a brief overview of the major physical, chemical, biological, and ecological characteristics of vernal pools, considers some of the adaptations that allow pool animals to carry out their lives in these seasonally flooded habitats, and discusses key conservation issues. Many excellent books, papers, and reports are available for those who would like more details about the ecology of vernal pools touched on here – see, e.g., Wiggins et al. 1980; Williams 1987; Semlitsch 1998; Batzer and Sion 1999; Higgins and Merritt 1999; Schneider 1999; Biebighauser 2003; Colburn 2004; Gibbons et al. 2006; Calhoun and deMaynadier 2008; and others listed in the text and at the end of the chapter.

Landscape Distribution of Vernal Pools

Temporary waters occur worldwide. Whether they occur in southern Australia, eastern Africa, the countries bordering the Mediterranean sea, central California, North American prairies, holes scoured in sandstone in desert canyons, bottomland forests of southeastern United States, or northeastern North America, temporary pools share seasonality of flooding; year-to-year variability of hydrology, chemistry, and temperature; many similarities in their biological communities and in species'

Fig. 1 Haynes Brook vernal pool in Downeast, Maine, USA, at spring thaw (Photo credit: Aram Calhoun © Rights remain with the author)

adaptations to the seasonal presence and absence of water; and conservation challenges (Colburn 2004, 2008; Calhoun et al. 2014a, b).

Vernal pools as discussed in this chapter are found in forested regions of formerly glaciated eastern North America. These pools share commonalities in their biota, most notably in the suite of amphibian species that depend on them for reproduction and also in many of the common invertebrates. (Pools in grasslands, although similar to vernal pools as defined here, support a different suite of species.)

These pools may occur in any context where depressions in the ground fill with water between fall and spring and retain water until at least late spring or early summer (Colburn 2004; Calhoun and deMaynadier 2008). Vernal pools lie in discrete depressions in forested uplands (Fig. 1), exposed bedrock hollows on mountain ridges or rocky shores, former stream channels or flood-scoured areas in floodplains, low-elevation areas within larger forested wetlands, pits formed when trees blow down, interdunal swales, and borrow pits or other areas excavated by human activities. Vernal pools often occur in clusters, especially in glacial outwash where kettlehole depressions were produced by melting ice left by the departing continental ice sheet.

History of Pool Studies

In the first half of the twentieth century, naturalists and experimental biologists reported on the biological communities and conducted research on animals of temporary forest ponds. For example, our understanding of North American invertebrate biodiversity and of the biology and ecology of temporary ponds was greatly increased by community surveys by Shelford (1913) and Kenk (1949), who documented a great diversity of phyla, classes, orders, and species of invertebrates in Michigan pools. Experiments by Libbie Hyman (University of Chicago) in the 1920s on hydras, oligochaetes, and planarians from temporary waters added to early understanding of drought survival mechanisms and developmental processes. Multi-year surveys by Ralph Dexter (Kent State University) from the 1930s through the 1960s documented year-to-year variations in distributions of fairy shrimp and clam shrimp and chronicled the loss of pool habitats in the American Mid-West over time.

Starting in the 1970s there was a great increase in scientific and conservation interest in vernal pools (Pough and Wilson 1976; Pierce 1993). This was fueled in part by new awareness of worldwide declines in amphibian populations and evidence that atmospheric deposition ("acid rain") had detrimental effects on the development of amphibian embryos in some temporary ponds in the Northeast as well as by broader concerns about the loss of local and regional biodiversity. These concerns led to applied and theoretical research focusing on conservation, especially in relation to vernal-pool-specialist amphibians (Shoop 1974). They also contributed to broader studies of vernal pool biological community composition and dynamics. During the same period, scientific curiosity about species' adaptive strategies for survival under variable disturbance regimes and about vernal pools and their inhabitants as possible aquatic models conforming to theories of island biogeography, r and K selection, and metapopulation dynamics spawned a host of research studies. Such research continued to add to theoretical understanding of vernal pools and to support conservation efforts directed toward protecting local and regional biodiversity.

Size, Depth, and Water Quality of Vernal Pools

Overall, vernal pools are highly heterogeneous (for more details see especially reviews in Colburn 2004; Calhoun and deMaynadier 2008).

Pool size and depth – Most pools are small. More than two thirds of those reported in the literature cover fewer than 0.05 ha when flooded. Fewer than one percent exceed 1.5 ha, and 6 ha is the largest reported pool area. Vernal pools also are usually shallow. The median depth of a large number of pools is 1 m, and the range is 16–300 cm. Potential implications of pool surface area for the biological community include shading by canopy trees – lower for large pools unless they are shallow enough to have trees growing within the pool basin; leaf litter inputs; depth; mixing of the water; hydroperiod (timing and duration of flooding); and extent and heterogeneity of habitat.

Water quality – Water chemistry in vernal pools covers a wide range and varies with water sources, bedrock, soils, surrounding plant community, and land use in the watershed. Reported value ranges include: pH 3.6–10.2; alkalinity 0–960; specific conductance 10–375 uS/cm; sulfate 0.5–12 mg/l; calcium 1–47 mg/l; chloride

0.1–3.0 mg/l; color 0–356 f (Vermont Wetlands Bioassessment Program 2003; Batzer et al. 2004; Colburn 2004).

Depending on a pool's size, depth, canopy cover, and water color, the water temperature may undergo dramatic diurnal fluctuations, and there may be strong thermal gradients from the surface to the bottom. These can influence the rate of embryonic development of amphibian and invertebrate eggs; affect metabolism and growth of pool animals; and, in the case of high water temperatures, reduce the amount of dissolved oxygen that the water can hold (less oxygen is available in warm water than cold).

The source of water feeding a pool can influence water temperature and chemistry. Runoff carries sediments, organic detritus, and dissolved materials transported by snowmelt and storm runoff from the watershed. Groundwater is likely to provide more constant thermal and chemical water quality than precipitation and runoff.

Hydrology

Water sources – Whether a pool is fed solely by precipitation and runoff and dries in response to evaporation from the pool surface and uptake of water by adjacent trees or if the pool is in contact with the water table and fluctuates additionally with groundwater may substantially affect the timing and duration of flooding or hydroperiod (Sobczak 1999; Sobczak et al. 2003; Leibowitz and Brooks 2008).

Hydrologic continuum for vernal pool hydroperiods – The timing and duration of flooding in vernal pools span a broad temporal continuum (Schneider 1999; Colburn 2004; Leibowitz and Brooks 2008). Highly ephemeral pools that fill following heavy rainstorms and dry within days or a couple of weeks do not remain flooded long enough to support vernal-pool-dependent species. Among pools that do support characteristic species and communities, those at the dry end of the continuum fill in spring, remain inundated for several months, and dry by early summer. Longer hydroperiod pools fill in spring but retain water through summer and sometimes into early fall. Many pools typically start to fill in fall, reach their maximum volume and depth in spring, and dry by early summer. The seasonally flooding pools that have the longest hydroperiods fill in fall and retain water through most of the following summer, with only a few weeks of drawdown in late summer before they fill again. There are also some functional vernal pools that are semi-permanently flooded, drying only during periods of drought. A particular pool may have a characteristic pattern of filling and drying over many years, but in any given year a pool may lie on the wetter or drier end of the continuum, depending on local patterns of precipitation and temperature.

Effects of hydroperiod on pool biota – Different species of pool animals are distributed across this hydrologic continuum, depending on the timing of flooding and the duration of inundation. Hydroperiod has been identified as a primary factor influencing which species can survive in a given vernal pool (Karraker and Gibbs 2009; Wiggins 1973). Hydroperiod may act both directly, with short flooding durations preventing the establishment of species that lack drought-resisting adaptations, and indirectly, with long hydroperiods allowing for greater overall species richness but excluding predation-sensitive species.

Wiggins et al. (1980) conducted a comprehensive survey of the biota from a series of pools in Ontario with differing flooding durations and evaluated the various species' adaptations for survival in the absence of water. They concluded that "vernal temporary pools," which they defined as filling in spring and drying by early summer, support species with well-developed strategies for avoiding desiccation during the long period in which the pool is without water. In contrast, they observed that species from pools with longer hydroperiods, such as "autumnal vernal pools" that fill in fall and dry by early summer, support species that can only withstand shorter periods of drying. "Permanent pools" support species that lack adaptations for withstanding drawdown. These species sometimes include predatory fish as well as invertebrate predators that can prevent successful development by vernal-pool-dependent species.

Subsequent research on a larger population of vernal pools throughout the glaciated northeast has shown that hydrologic conditions in vernal pools cover a broader continuum than discussed by Wiggins et al. (1980). Long-hydroperiod pools may provide animals with longer periods for growth and development, allowing amphibian larvae to reach larger sizes before metamorphosis, or providing invertebrates such as snails or fingernail clams with multiple opportunities to reproduce in a year (see discussions of life cycles in Colburn 2004).

During multi-year flooded periods, semi-permanent pools may support vertebrate predators including bullfrog *Lithobates catesbeiana* and green frog *L. clamitans* tadpoles (which need to overwinter before transforming the following year into frogs) (Vasconcelos and Calhoun 2006) and red-spotted newts *Notophthalmus viridescens*. Also commonly in these pools are large invertebrate predators including some dragonfly nymphs, various water bugs, and water beetle larvae. These predators can prevent most vernal-pool specialists from successfully completing their development. Such predators may be present continuously in semi-permanent pools during multi-year flooded periods, but when they are eliminated periodically by drawdown during low-water years, dormant eggs and cysts of short-hydroperiod vernal-pool species, such as fairy shrimp, may hatch and, released from predation pressure, successfully grow and reproduce. Dan Schneider (1999) has shown that hydroperiod acts as a "filter" controlling the distribution of predators whose presence limits the successful development of some species that appear to be restricted to short-hydroperiod pools but that can survive long-hydroperiod conditions in the absence of predation.

In permanent pools, predaceous fish may become established, and in their absence the presence of amphibian and invertebrate predators can limit the successful growth and development of vernal-pool-dependent amphibians and invertebrates.

Vernal Pools as Wetlands

Vernal pools fall within several categories of wetlands as defined by the US Fish and Wildlife Service (Cowardin et al. 1979) or by the Hydrogeomorphic (HGM) method (U.S. Natural Resources Conservation Service 2008). Some vernal pools support wetland vegetation and are underlain by hydric soils. Hydrophytes may be absent from pools with dense canopy cover, in deep depressions, and/or with long hydroperiods that prevent the establishment of emergent plants. Pools that occur on bedrock or with short annual hydroperiods may not be underlain by typical hydric soils.

Plant Communities

Northeastern vernal pools share many features with seasonal forest pools farther south, and with temporary waters in nonforested regions, but they are different in their general lack of unique plant communities and endemic plant species. If vegetated, they support plant species that are common in other nearby wetlands (Colburn 2004; Cutko and Rawinski 2008). As mentioned above, vernal pools in northeastern North America are of particular conservation interest because of their fauna, especially pool-dependent amphibians and crustaceans, and invertebrate species that are restricted to short-hydroperiod pools.

Detritus and In-Pool Photosynthesis

Occurring as they do in forested contexts, vernal pools receive substantial inputs of fallen leaves each autumn (Fig. 2). Leaf decomposition, nutrient cycling, and the conversion of cellulose into digestible carbohydrates and glycoproteins by aerobic and anaerobic decomposer fungi and bacteria provide abundant food for vernal pool shredders and collector-gatherer feeders and may contribute importantly to forest biogeochemistry (Bärlocher et al. 1978; Capps et al. 2014). Food quality affects the distribution and development of aquatic animals, and caddisfly larvae given choices of detritus from permanent vs. intermittently flooded habitat preferred and grew better with the vernal pool detritus (Richardson and Mackay 1984). Depending on their size and location, vernal pools may receive little or substantial amounts of

Fig. 2 (a) Southern New England vernal pool embedded in an oak forest at spring thaw and (b) drying down exposing the detritus layer of fallen leaves and woody material (Photo credit: Kevin J. Ryan © Rights remain with the author)

sunlight, and in-pool photosynthesis may be extensive or very limited. The availability of algae vs. detritus is one of many important variables affecting the relative success of different amphibian species in vernal pools (Skelly et al. 2002).

Amphibians in Vernal Pools

The importance of vernal pools for many amphibian species is central to much of the interest in these small wetlands. Amphibians are important components of local and regional biodiversity, and they are also important agents of energy exchange between forests and vernal pools and between pools and upland forests. Vernal pools are optimal breeding habitats for wood frogs *Lithobates sylvaticus* and mole salamanders in the genus *Ambystoma*, including the spotted salamander *A. maculatum*, marbled salamander *A. opacum*, small-mouthed salamander *A. tremblayi*, Eastern tiger salamander *A. tigrinum tigrinum*, and members of the Jefferson-blue-spotted salamander complex (*A. jeffersonianum* and *A. laterale*) – whose closely related species have interbred to form a variety of triploid and tetraploid unisexual hydrids (Brodman 2005). Other amphibians that often breed in vernal pools – usually at the longer-hydroperiod end of the continuum – and also breed widely in other aquatic habitats include American and Fowlers toads (*Anaxyrus* (formerly *Bufo*) *americanus* and *A. fowleri*), spring peepers *Pseudacris crucifer*, gray treefrogs (*Hyla versicolor* and *H. chrysoscelis*), and, west of the Appalachians, chorus frogs *Pseudacris* spp. Spadefoot toads *Scaphiopus holbrookii* breed most commonly in pools that fall along the ephemeral end on the continuum and often retain water for too short a time to support typical vernal pool species assemblages, but they also breed in vernal pools in some locations.

The life cycles of vernal-pool amphibians are tailored to maximize the likelihood of young successfully maturing and emerging from pools before drying (see reviews in Colburn 2004; Semlitsch and Skelly 2008). When water is first available in early spring, cold water temperatures, the possibility of a late freeze, and limited food availability pose challenges for early breeding, but the chance of a pool drying before young have completed development places limitations on how late breeding can occur, especially in shorter-hydroperiod pools. Wood frogs (Fig. 3), spring peepers, and mole salamanders typically move to vernal pools in early spring, often while snow is still on the ground and ice may be present on the pool surface. (An exception is the marbled salamander, which moves to dry pool depressions in autumn and lays its eggs on the damp or dry ground; the eggs start to develop when the pool floods.) The adult wood frogs and salamanders spend relatively few days in the pools, mating and depositing eggs, before migrating back into the upland forests where they spend most of their lives. Pools are warming rapidly in early spring, and food becomes increasingly available during the period when embryos are developing, so that tadpoles and salamander larvae find abundant food resources upon hatching (in a few weeks for wood frogs, more than a month for salamanders). Rapid development of eggs and larvae, especially in wood frogs, which often breed in shorter-hydroperiod pools than mole salamanders, maximizes the likelihood that

Fig. 3 Wood frogs spend the majority of the year in the forest away from their breeding pools. Their brown color, dark mask, and striped legs help them blend into the forest floor (Photo credit: Kristine Hoffmann © Rights remain with the author)

juveniles will be ready to leave before the pool dries. Early pool drying often results in high proportions of wood frog tadpoles failing to complete development. In the salamanders, both egg and larval development are slower than in wood frogs. After larvae reach a minimum size threshold for transformation, they may remain in pools for a period of time if conditions remain favorable for growth, as size at metamorphosis affects reproductive fitness. After juveniles leave the pools, they disperse into the upland where they will feed on invertebrates until they reach maturity. In a subsequent spring, most pool amphibians return to their larval pools when ready to breed.

Unlike the wood frogs and mole salamanders, toads and gray treefrogs start to breed later in the spring, and they tend to remain in pools for weeks at a time, the males trilling to attract mates and the females moving into pools to deposit eggs over longer periods of time and later into the spring. These animals breed in vernal pools with longer hydroperiods and also in permanent waters.

Spadefoot toads spend much of their lives burrowed into sandy soil, moving to the surface to feed on wet nights. They emerge during torrential rainstorms from spring to fall. Migrating to shallow pools, the males call with a loud, honking noise during thunderstorms, lightening reflecting off of their inflated throat pouches, and attracting females. Breeding is fast, eggs develop within days in the short-lived rainwater pools, tadpoles feed on a host of small organisms in the water, and metamorphosis is complete in just a few weeks.

Invertebrates in Vernal Pools

More than 400 species of invertebrates have been identified from northeastern vernal pools (see Wiggins et al. 1980; Williams 1987, 2006; Colburn 2004; Colburn et al. 2008). Among the most characteristic vernal pool taxa are branchiopod crustaceans including fairy shrimp *Eubranchipus* spp., clam shrimp *Lynceus brachyurus*, and cladocerans (water fleas, especially Daphniidae and Chydoridae) and water beetles (especially species of predaceous diving beetles, Dytiscidae). A host of other kinds of aquatic insects as well as aquatic earthworms, leeches, molluscs, flatworms, and even hydras also occur in vernal pools.

Invertebrates in relation to hydroperiod and seasonality – This diversity is spread across the hydrologic continuum. Long-hydroperiod pools tend to be more species-rich than short-hydroperiod pools, but the species composition is different (Schneider 1999; Brooks 2000). Collectively, the richness of species within a cluster of vernal pools with a range of hydroperiods and other habitat characteristics is greater than the number of species in the most species-rich pool, reflecting the importance of pool heterogeneity for local and regional biodiversity. A range of hydroperiods, from very short to long, is necessary to allow for the greatest richness of vernal pool invertebrate species in a particular location or region.

The composition of the aquatic invertebrate community changes as the season progresses. Early in spring one commonly finds fairy shrimp, copepods, water fleas, planarians, caddisfly larvae, mosquito and phantom midge larvae, some beetle adults that overwinter in the pool sediment, and certain dytiscid beetle larvae – known popularly as "water tigers" because of their massive, sharp jaws and their aggressive predation on fairy shrimp, tadpoles, and other pool residents. Snails and fingernail clams may be present in pools where water is not excessively acidic. As the water warms, planarians, mosquitoes, and fairy shrimp disappear; clam shrimp and different water fleas may become abundant; caddisflies pupate and fly away as brown, moth-like adults; and migratory insects including adult water beetles such as *Agabus* spp., water boatmen (Corixidae), backswimmers (Notonectidae), and other water bugs (Hemiptera, Heteroptera) fly in from permanent waters where they overwintered. In longer hydroperiod pools, especially those with some aquatic vegetation, damselfly and dragonfly nymphs may become abundant in late spring and summer, emerging progressively to feed on insects above the pool and in the adjacent forest.

Many vernal pool invertebrates are widely distributed, but other species vary with pool characteristics such as hydroperiod, forest cover, water chemistry, land use, and vegetation (Schneider 1999; Vermont Wetlands Bioassessment Program 2003).

Invertebrate life histories – As with the amphibians, the life histories of vernal pool invertebrates allow survival and reproduction in an aquatic habitat that is periodically without water, and they may also provide for dispersal to new habitats. These life cycles include strategies for withstanding pool drying within the pool basin or for escaping drying in some way. Some animals, such as planarians, form cysts that resist drying in the sediment; others including snails and fingernail clams withdraw into their shells and aestivate during drawdown; and crustaceans and many insects have eggs that lie dormant in the dry pool sediment and hatch after flooding. Post-flooding hatching may occur immediately upon water appearing in the pool, or it may be subject to other conditions such as specific water temperatures, a period of prior exposure to cold, or oxygen concentrations in the water. Some invertebrates follow the amphibian strategy of breeding and maturing in vernal pools but migrating elsewhere (often to permanent waters) as adults.

Invertebrates also have several different mechanisms for maximizing the likelihood that offspring will survive pool conditions and reach maturity. Some species produce very large numbers of eggs; even if most die, there is a chance that a few will survive to complete development. Others produce small numbers of large eggs; having access to large stores of resources helps improve an individual offspring's chances of survival. Closely related species may use alternative strategies – examples are seen in crustaceans such as copepods and in molluscs such as fingernail clams (see Wiggins et al 1980; and reviews in Colburn 2004 and Colburn et al. 2008).

Fairy shrimp are among the most classic vernal-pool-dependent animals. They (and many other pool crustaceans) have eggs that resist drying and can remain in the pool sediment for many years. When the eggs hatch, tiny shrimps emerge (seemingly by magic – hence "fairy shrimp") and can be seen swimming on their backs in vernal pools in early spring, synchronously waving a dozen feathery legs and filtering fine particles and microorganisms from the water for food. After the shrimps mature and mate, eggs fall from the females' egg pouches onto the pool bottom, and the adults die, long before salamanders and most insect predators are active in the pools. Before the eggs can hatch, they need to experience drying, chilling, and reflooding. In a semi-permanent pool, there may be no fairy shrimp for many years without drawdown and then, after a dry year when the pool bottom becomes exposed to desiccation and winter cold, fairy shrimp will appear when the pool refloods in spring. Not all eggs will hatch the first or the second – or even the third time they experience these conditions; some will remain in the sediment and hatch another time. This "bet-hedging" strategy protects the population against catastrophic losses if the pool should dry early, before the shrimp mature and mate.

Agabus is a large genus of predaceous diving beetles with several species in vernal pools. Larvae hatch from eggs in the sediment in early spring. Some weeks later, adults fly in from permanent waters where they overwintered. They mate and females leave eggs on the pool bottom. Those eggs will not hatch until the following spring. The larvae will hatch, feed, grow, pupate, and emerge as adults which disperse to permanent water when the pool dries, flying back to vernal pools the following spring to start the cycle over again. It thus takes two full years for a complete life cycle, and it includes not only the pool but also a permanent water body plus food resources in both habitats.

Water mites have even more complex life cycles. Eggs hatch into parasitic larvae that attach to discrete locations on particular species of adult insects and feed on the adult's blood as the insect flies around the pool looking for food or mates (as might be the case for a dragonfly adult) or it is carried by the adult host (such as a water beetle or water bug) to a permanent lake or pond for the winter (where it remains attached to its host) and is then returned to the original pool or dispersed to another when the adult migrates to pools the following spring. The fed larva drops into the pool and transforms into a nymph, which is a predator on the eggs of small crustaceans or of mosquito or other fly larvae. The nymph then transforms into an adult which is also predatory on the same species as the nymph. Adults mate, lay eggs, and the cycle starts over again.

Conservation of Vernal Pools

Because vernal pools are important at both pool-level (i.e., breeding habitat for specialized biotic communities) and landscape scales (e.g., genetic, hydrologic and biogeochemical reservoirs, and resting and foraging habitat for water bugs and beetles, birds, mammals, and non-pool-dependent herptiles) and because pools vary in flooding regime, geomorphic setting, and environmental quality, conservation approaches must be tailored to specific desired outcomes. Conservation strategies are further limited because these small, ephemeral wetlands largely occur on private properties. Additionally, vernal pools are hard to identify remotely (often, more than 33% are missed in pre-identification exercises), they often function as clusters of pools, and their functions depend on linkages with other pools, wetlands, and uplands (Mushet et al. 2015). We provide short guidance on conserving pool functions at the pool or species-specific scale and at the landscape scale while encouraging managers to employ all these approaches. Useful sources of additional information include Semlitsch (2002), review papers in Calhoun and deMaynadier (2008), Calhoun et al. (2014a, b), Rains et al. 2016; Cohen et al. (2016) and papers cited below.

Conservation of individual pools or species – Pool-scale conservation tools are limited to regulatory protection for a subset of exemplary pools or pools associated with rivers or lakes (this varies by state; Mahaney and Klemens 2008) and to voluntary best management practices (e.g., Calhoun and deMaynadier 2004; Calhoun and Klemens 2002). If regulated vernal pools are degraded or destroyed, mitigation may include pool creation or restoration. Restoring or creating pool habitat often fails, particularly for pool-breeding amphibians and invertebrates, due to an inability to recreate hydrology (Calhoun et al. 2014a). With the exception of regions where pool losses are very high, pool creation should be a last resort and coupled with pool preservation (Kross and Richter 2016).

Conserving habitat for individual species requires knowledge of breeding and post-breeding behaviors which varies among species and with regional context (Semlitsch et al. 2009). The biphasic life histories of vernal-pool amphibians make the adjacent terrestrial habitat an integral part of conserving pool functions (Semlitsch 2002). Similarly, the seasonal use of permanent waters by many insects of vernal pools means that a pool alone is not sufficient to meet these species' life history requirements (Colburn 2004). Although pool-breeding amphibians disperse and migrate 100s of meters between breeding pools and upland habitat (Rittenhouse and Semlitsch 2007), national and state regulations on vernal pools regulate activities within at most 10–80 m from the pool's edge (Mahaney and Klemens 2008). Stricter regulations may apply if state-listed species are documented. Within regulatory constraints, managers should work to develop mitigation or protection strategies that incorporate as much ecologically relevant habitat as possible for target species. For example, the configuration may be in the form of corridors linking and incorporating relevant habitat rather than neat circles around the pool (Baldwin et al. 2006). Given the challenge of providing meaningful post-breeding habitat for amphibians and other pool ecosystem services, we highly recommend conservation strategies that couple pool-specific conservation with broader landscape approaches.

Landscape-scale approaches to pool conservation – Because pools do not exist as isolated wetland depressions but rather as vital biological, physicochemical, and ecological integrators in a terrestrial matrix (McLaughlin et al. 2014; Cohen et al. 2016) and because many vernal pool species' life cycles extend beyond pool basins, conserving "vernal poolscapes" – complexes of vernal pools and other aquatic habitats plus associated uplands – is a preferred conservation approach. Landscape-scale conservation allows for the protection of pools spanning an array of hydrogeomorphic settings (ones that support short-to-long hydroregimes in different physical settings) that conserve a range of biogeochemical and water quality functions and support diverse biota (Mitchell et al. 2007; Marton et al. 2015). This approach also buffers changes in pool functions against changes in climate.

Landscape approaches can use stakeholder-driven energy and expertise and can be developed at scales ranging from local, low-cost, voluntary programs using citizen scientists (Jansujwicz and Calhoun 2010, 2013) to more resource-intensive regional inventory and assessments associated with research, consulting projects, or government initiatives (Lathrop et al. 2005). Development of citizen science programs is an effective strategy to develop municipal or regional pool conservation plans (Morgan and Calhoun 2014; www.vernalpools.me) and web-based reporting of vernal pools as citizens encounter them (Carpenter et al. 2011); this provides an evolving database of resources that would be too expensive for government agencies or NGOs to inventory comprehensively.

Costs of conservation at landscape scales call for collaborative approaches. For example, an incentivized approach for vernal pools in New England is being developed through collaboration among federal, state, and local governments, ecologists, the development community, land trusts, and environmental nongovernmental organizations to provide an alternative mitigation tool for vernal pools agreed upon by all parties. This tool is a local *in lieu* fee program where developers may impact wetlands in municipal growth zones in return for a fee collected to incentivize local landowners to conserve exemplary vernal pools and post-breeding habitat in municipal rural zones (Special Area Management Plan for Vernal Pools in US Army Corps of Engineers Region 1; see Levesque et al. 2016). Vernal poolscapes in this program will be targeted through partnerships with local land trusts or other conservation organizations. This innovative approach improves federal regulations by tailoring conservation of vernal pools to local needs that support both conservation and economic development. In this case discontent with the top-down "stick" of government regulations that assume one-size-fits-all served as momentum for creativity (Calhoun et al. 2014b). This type of management approach that recognizes the spatial distribution of benefits and costs and full extent of conservation costs is more likely to navigate this challenge successfully (Sunding and Terhorst 2014).

In summary, the best vernal pool conservation strategies include case-by-case regulatory approaches coupled with broader landscape-scale initiatives that capture the full suite of vernal pool ecosystem services by maximizing connectivity among wetlands and forests. Flexible conservation strategies that reduce landowner and manager costs while achieving ecological objectives will have the greatest probability of success in maintaining fully functioning landscapes. We can move toward this paradigm by tailoring conservation to local needs through stakeholder-generated solutions often coupled with government engagement at multiple levels.

References

Baldwin RF, Calhoun AJK, deMaynadier PG. Conservation planning for amphibian species with complex habitat requirements: a case study using movements and habitat selection of the wood frog *Rana sylvatica*. J Herpetol. 2006;40:442–53.

Bärlocher F, Mackay RJ, Wiggins GB. Detritus processing in a temporary vernal pool in southern Ontario. Arch Hydrobiol. 1978;81:269–95.

Batzer DP, Sion KA. In: Batzer DP, Rader RB, Wissinger SA, editors. Invertebrates in freshwater wetlands of North America. New York: Wiley; 1999.

Batzer DP, Palik BJ, Buech R. Relationships between environmental characteristics and macroinvertebrate communities in seasonal woodland ponds of Minnesota. J N Am Benthol Soc. 2004;23:50–68.

Biebighauser TR. A guide to creating vernal ponds. Morehead: U.S.D.A. Forest Service; 2003. 36 p.

Brodman R. *Ambystoma laterale*, blue-spotted Salamander. In: Lannoo M, editor. Amphibian declines: the conservation status of United States species. Berkeley: University of California Press; 2005. p. 614–16.

Brooks RT. Annual and seasonal variation and the effects of hydroperiod on benthic macroinvertebrates of seasonal forest ("vernal") ponds in central Massachusetts. Wetlands. 2000;20:707–15.

Calhoun AJK, deMaynadier PG. Forestry habitat management guidelines for vernal pool wildlife in Maine. Rye (NY): Wildlife Conservation Society Technical Paper #6; 2004.

Calhoun AJK, deMaynadier PG. Science and conservation of vernal pools in Northeastern North America. New York: CRC Press; 2008.

Calhoun AJK, Klemens MW. Pool-breeding amphibians in residential and commercial developments in the Northeastern United States. Rye (NY): Wildlife Conservation Society Technical Paper #5; 2002.

Calhoun AJK, Arrigoni J, Brooks RP, Hunter Jr ML, Richter SP. Creating successful vernal pools: a literature review and advice for practitioners. Wetlands. 2014a;34:1027–38.

Calhoun AJK, Jansujwicz JS, Bell KP, Hunter Jr ML. Improving management of small natural features on private lands by negotiating the science-policy boundary. Proc Natl Acad Sci U S A. 2014b;111:11002–6.

Capps KA, Rancatti R, Tomczyk N, Parr T, Calhoun AJK, Hunter M. Biogeochemical hotspots in forested landscapes: the role of vernal pools in denitrification and organic matter processing. Ecosystems. 2014;17:1455–68.

Carpenter L, Stone J, Griffin CR. Accuracy of aerial photography for locating seasonal (vernal) pools in Massachusetts. Wetlands. 2011;31:573–81.

Cohen MJ, Creed IF, Alexander L, Basu MB, Calhoun AJK, Craft C, D'Amico E, DeKeyser E, Fowler L, Golden HE, Jawitx JW, Kalla P, Kirkman LK, Lane CR, Lang M, Leibowitz SG, Lewis DB, Marton J, McLaughlin DL, Mushet DM, Raanan-Kliperwas H, Rains MC, Smith L, Walls SC. Do geographically isolated wetlands influence landscape functions? Proc Natl Acad Sci U S A. 2016;113(8):1978–86.

Colburn EA. Vernal pools: natural history and conservation. Blacksburg/Granville: McDonald and Woodward Publishing Co.; 2004. 426 p.

Colburn EA. Ecosystems: temporary waters. In: Jorgensen SE, Fath BD, editors. Encyclopedia of ecology. 1st ed. Oxford: Elsevier B.V; 2008. p. 3517–28.

Colburn EA, Weeks SC, Reed SK. Chapter 6. Diversity and ecology of vernal pool invertebrates. In: Calhoun AJK, deMaynadier PG, editors. Science and conservation of vernal pools in Northeastern North America. New York: CRC Press; 2008. p. 105–26.

Cowardin LM, Carter V, Golet RF, Laroe ET. Classification of wetlands and deepwater habitats of the United States. Washington (DC): U.S. Fish and Wildlife Service, Biological Services Program; 1979. Report FWS/OBS-79/31.

Cutko A, Rawinski TJ. Flora of Northeastern vernal pools. In: Calhoun AJK, deMaynadier PG, editors. Science and conservation of vernal pools in Northeastern North America. New York: CRC Press; 2008. p. 71–104.

Gibbons JW, Winne CT, Scott DE, Willson JD, Glaudas X, Andrews KM, Todd BD, Fedewa LA, Wilkinson L, Tsaliagos RN, Harper SJ, Greene JL, Tuberville TD, Metts BS, Dorcas ME, Nestor JP, Hound CAY, Akre T, Reed RN, Buhlmann KA, Norman J, Croshaw DA, Hagen D, Rothermel BB. Remarkable amphibian biomass and abundance in an isolated wetland: implications for wetland conservation. Conserv Biol. 2006;20:1457–65.

Higgins MJ, Merritt RW. Temporary woodland ponds in Michigan: invertebrate seasonal patterns and trophic relationships. In: Batzer DP, Rader RB, Wissinger SA, editors. Invertebrates in freshwater wetlands of North America. New York: Wiley; 1999. p. 279–98.

Jansujwicz J, Calhoun AJK. Protecting natural resources on private lands: the role of collaboration in land-use planning. In: Trombulak S, Baldwin RF, editors. Protecting natural resources on private lands: the role of collaboration in land-use planning. New York: Springer; 2010. p. 205–33.

Jansujwicz JS, Calhoun AJK, Lilieholm R. The Maine vernal pool mapping and assessment program: engaging municipal officials and private landowners in community-based citizen science. Environ Manag. 2013;52:1369–85.

Karraker NE, Gibbs JP. Amphibian production in forested landscapes in relation to wetland hydroperiod: a case study of vernal pools and beaver ponds. Biol Conserv. 2009;142:2293–302.

Kenk R. The animal life of temporary and permanent ponds in southern Michigan, Miscellaneous publications of the Museum of Zoology, vol. 71. Ann Arbor: University of Michigan; 1949.

Kross CS, Richter SC. Species interactions in constructed wetlands result in population sinks for Wood Frogs while benefitting Eastern Newts. Wetlands. 2016;36:385.

Lathrop RG, Montesano P, Tesauro J, Zarate B. Statewide mapping and assessment of vernal pools: a New Jersey case study. J Environ Manage. 2005;76:230–8.

Leibowitz SG, Brooks RT. Hydrology and landscape connectivity of vernal pools. In: Calhoun AJK, DeMaynadier PW, editors. Science and conservation of vernal pools in Northeastern North America. New York: CRC Press; 2008. p. 31–54.

Levesque V, Calhoun AJK, Bell KP. Turning contention into collaboration: engaging power, trust, and learning in collaborative networks. Soc Nat Resour. 2016;1:1.

Mahaney WS, Klemens MW. In: Calhoun AJK, DeMaynadier PW, editors. Science and conservation of vernal pools in Northeastern North America. New York: CRC Press; 2008. p. 192–212.

Marton JM, Creed IF, Lewis DB, Lane CR, Basu NB, Cohen MJ, Craft CB. Geographically isolated wetlands are important biogeochemical reactors on the landscape. BioScience. 2015;65:408–18.

McLaughlin DL, Kaplan DA, Cohen MJ. A significant nexus: geographically isolated wetlands influence landscape hydrology. Water Resour Res. 2014;50:7153–66.

Mitchell JC, Golet FC, Skidds DE, Paton PWC. Prioritizing non-regulatory protection of vernal pools in the Queen's River Watershed, Rhode Island. Final research report prepared for RI Department of Environmental Management. Office of Water Resources, and U. S. Environmental Protection Agency, Region 1. 2007. 91 p. http://www.dem.ri.gov/programs/benviron/water/wetlands/pdfs/queens.pdf

Morgan DE, Calhoun AJK. Maine municipal guide to mapping and conserving vernal pools. Orono: Sustainability Solutions Initiative; 2014.

Mushet DM, Calhoun AJK, Alexander LC, Cohen MJ, DeKeyser ES, Fowler L, Lane CR, Lang MW, Rains MC, Walls SC. Geographically isolated wetlands: rethinking a misnomer. Wetlands. 2015;35:423–31.

Pierce BA. The effects of acid precipitation on amphibians. Ecotoxicology. 1993;2(1):65–77.
Pough FH, Wilson RE. Acid precipitation and reproductive success of *Ambystoma* salamanders. In: Proceedings of the international symposium on acid rain and the forest ecosystem. USDA Forest Service General Technical Report 23; 1976. p. 531–43.
Rains MC, Leibowitz SG, Cohen MJ, Creed IF, Golden HE, Jawitz JW, Kalla P, Lane CR, Lang MG, McLaughlin L. Geographically isolated wetlands are part of the hydrological. Hydrol Process. 2016;30:153–60.
Richardson JS, Mackay RJ. A comparison of the life history and growth of *Limnephilus indivisus* in three temporary pools. Arch Hydrobiol. 1984;99:515–28.
Rittenhouse TAG, Semlitsch RD. Distribution of amphibians in terrestrial habitat surrounding wetlands. Wetlands. 2007;27:153–61.
Schneider DW. Snowmelt ponds in Wisconsin: influence of hydroperiod on invertebrate community structure. In: Batzer DP, Rader RB, Wissinger SA, editors. Invertebrates in freshwater wetlands of North America. New York: Wiley; 1999. p. 298–318.
Semlitsch RD. Biological delineation of terrestrial buffer zones for salamanders. Conserv Biol. 1998;12:1113–9.
Semlitsch RD. Critical elements for biologically based recovery plans of aquatic-breeding amphibians. Conserv Biol. 2002;16:619–29.
Semlitsch RD, Skelly DK. Ecology and conservation of pool-breeding amphibians. In: Calhoun AJK, DeMaynadier PW, editors. Science and conservation of vernal pools in Northeastern North America. New York: CRC Press; 2008. p. 127–48.
Semlitsch RD, Todd BD, Blomquist SM, Calhoun AJK, Gibbons JW, Gibbs JP, Graeter GJ, Harper EB, Hocking DJ, Hunter ML, Patrick DA. Effects of timber harvest on amphibian populations: understanding mechanisms from forest experiments. BioScience. 2009;59:853–62.
Shelford VE. Animal communities in temperate America: as illustrated in the Chicago region: a study in animal ecology, Bulletin (Geographic Society of Chicago), vol. 5. Chicago: Chicago University Press; 1913.
Shoop CR. Yearly variation in larval survival of *Ambystoma maculatum*. Ecology. 1974;55:440–4.
Skelly DK, Freidenburg LK, Kiesecker JM. Forest canopy and the performance of larval amphibians. Ecology. 2002;83(4):983–92.
Sobczak R. Water resources office. Barnstable: Cape Cod Commission; 1999.
Sobczak RV, Cambareri TC, Portnoy JW. Physical hydrology of selected vernal pools and kettle hole Ponds in the Cape Cod National Seashore, Massachusetts: Ground and Surface Water Interactions. Barnstable (MA) Water Resources Office, Cape Cod Commission; 2003. Report to the U. S. National Park Service, Water Rights Branch, Water Resources Division, Fort Collins, CO, and Cape Cod National Seashore, Wellfleet, MA.
Sunding D, Terhorst J. Conserving endangered species through regulation of urban development: The Case of California Vernal Pools. Land Econ. 2014;90:290–305.
U. S. Natural Resources Conservation Service. Hydrogeomorphic Wetland Classification System: An Overview and Modification to Better Meet the Needs of the Natural Resource Conservation Service. Washington (DC): US Department of Agriculture, NRCS; 2008. Technical Note No. 190-8-76.
Vasconcelos D, Calhoun AJK. Monitoring created seasonal pools for functional success: a six-year case study of amphibian responses, Sears Island, Maine. USA Wetlands. 2006;26:992–1003.
Vermont Wetlands Bioassessment Program. An evaluation of the chemical, physical, and biological characteristics of seasonal pools and northern white cedar swamps and northern white cedar swamps. Final report. Montpelier: Vermont Department of Environmental Conservation and Vermont Department of Fish and Wildlife, Nongame and Natural Heritage Program; 2003. 112 p.
Wiggins GB. A contribution to the biology of caddisflies (Trichoptera) in temporary pools, Life sciences Contributions, vol. 88. Toronto: Royal Ontario Museum; 1973. 36 p.
Wiggins GB, Mackay RJ, Smith IM. Evolutionary and ecological strategies of animals in annual temporary pools. Arch Hydrobiol Suppl. 1980;58(1,2):97–206.
Williams DD. The ecology of temporary waters. Portland: Timber Press; 1987. 205 p.
Williams DD. The biology of temporary waters. Oxford: Oxford University Press; 2006. 348 p.

Pocosins (USA)

52

Curtis J. Richardson

Contents

Introduction	668
Wetland Ecosystems	669
Geology	673
Hydrology	673
Biodiversity	674
Conservation Concerns	675
References	676

Abstract

Pocosins, an Algonquin Indian word means swamp-on-a-hill. These evergreen shrub-bog ecosystems occur on the southeastern coastal plain of the USA from Virginia to north Florida and once covered more than one million hectares in North Carolina alone. A broad definition of pocosins would include all shrub and forested bogs, as well as Atlantic white cedar *Chamaecyparis thyoides* stands and some loblolly pine *Pinus taeda* stands on flooded soils on the Coastal Plain. A stricter definition would only include the classic shrub-scrub (short pocosin) found on deep peat soils > 1-4 m and pond-pine *Pinus serotina* dominated tall pocosin on shallow peat soils < 1 m. Pocosins are rainfall driven and thus hydrologically isolated from major rivers on the landscape. However, they are often found adjacent to estuaries and have surface hydrological connections that are linked to the regional water quality and salinity gradients found in estuarine areas along the southeastern coast.

C. J. Richardson (✉)
Nicholas School of the Environment, Duke University Wetland Center, Durham, NC, USA
e-mail: curtr@duke.edu; curtr12@gmail.com

© Springer Science+Business Media B.V., part of Springer Nature 2018
C. M. Finlayson et al. (eds.), *The Wetland Book*,
https://doi.org/10.1007/978-94-007-4001-3_144

Keywords

Palustrine · Bog · Swamp · Hydrology · Peatlands

Introduction

Pocosin is an ancient Algonquin Indian word meaning "swamp-on-a-hill" (Tooker 1899). It has over 20 different spellings and is often seen on old maps and records as poquosin, poquoson, percoason, pekoson, pocoson, and pocason. Pocosins, also known as southeastern shrub bogs, are characterized by a very dense growth of mostly broadleaf evergreen shrubs with scattered pond pine *Pinus serotina*. They comprise the largest extent of true bogs in the southeastern United States (Richardson 2012). They are classified as palustrine wetland ecosystems since they are nontidal wetlands, and trees, shrubs, persistent emergents, emergent mosses, or lichens cover 30% or more of the area (Cowardin et al. 1979). Surveys have shown that pocosins occurred on >1.5 million ha along the southeastern coastal plain from Virginia (VA) to north of Florida, as far west as Alabama and once covered more than one million hectares in North Carolina (NC) alone (Richardson 1983, 2012). However, their area has been greatly diminished by agricultural development and forestry conversions to loblolly pine, *Pinus taeda*, plantations (Richardson 1983, 2012). The exact distribution of pocosins along the coastal plain has not been mapped for most states but their location as of 1960s in NC is shown in Fig. 1.

Harper (1907) characterized a pocosin as an "*...extensive, flat damp, sandy or peaty area usually remote from large streams, supporting a scattered growth of pine (mostly Pinus serotina) and a very dense growth of shrubs, mostly evergreen, giving the whole a decided heath-like aspect...*" The most holistic definitions of pocosin were proposed by Wells (1928), Woodwell (1958), and Kologiski (1977). They described pocosins as being primarily restricted to the southeastern coastal plain, occurring in broad shallow basins, drainage basin heads, and on broad flat uplands. These areas have long hydroperiods, temporary surface water, periodic burning, and soils of sandy humus, muck, or peat. These definitions suggest that pocosins are not adjacent or connected to flowing water bodies or rivers but importantly imply that pocosins are the source of fresh water on the coastal landscape. Thus understanding their role in coastal hydrology is critical to managing the region's water supply and estuarine interactions.

Only Carolina bays on the lower coastal plain have vegetation similar to pocosins (Fig. 1). Carolina bays are ovate-shaped, shallow depressions found primarily on the coastal plain of North and South Carolina and Georgia (Sharitz and Gibbons 1984; Ross 1987). However differences in size and geologic origin exist between large pocosin tracts and much smaller elliptical Carolina bays (Richardson and Gibbons 1993). Some bays exist in northern Florida and as far north as New Jersey. They were called bays by the early European explorers who found cypress, *Taxodium distichum*, hardwood trees they simply called bay trees and evergreen shrubs growing along the shore of the depression. Bays are variable in size (a few hectares up to several thousand) and are most often found on soils with layers of peat, sand, and impervious clay, and the elongated elliptical depression is often but not always found on a northwest-southeast axis, with a high sand rim to the southeast. Their origin is

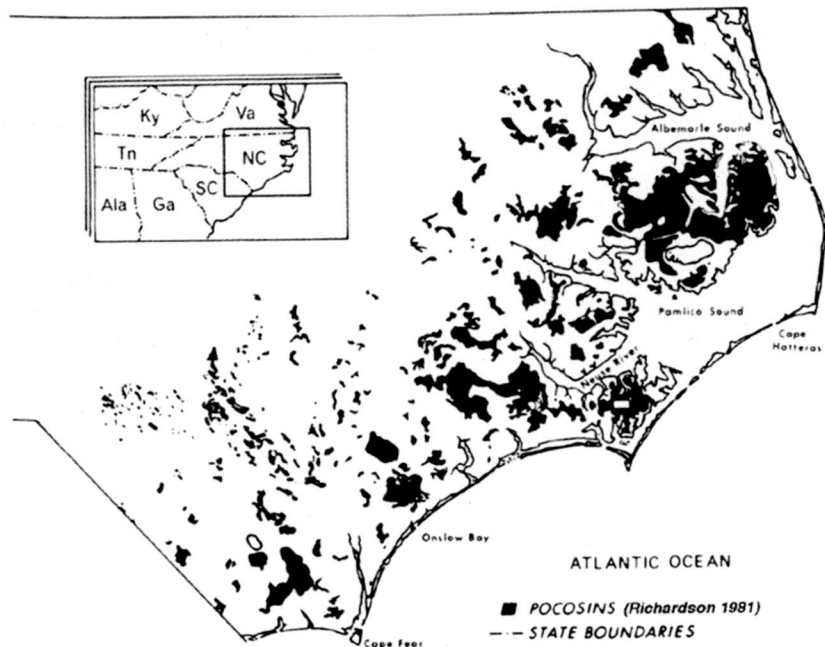

Fig. 1 A map showing the distribution of pocosin bogs and Carolina bays (mainly isolated wetlands found inland in the southern part of the state) in North Carolina around the 1960s. Total area was estimated to be 908,000 ha. The figure was compiled by Richardson (1981). The vast majority of the wetlands are found in the large expanses of pocosins. Carolina bays having pocosin-type vegetation make up the majority of the wetland type in the smaller wetlands found in the southern inland coastal counties of NC

not known, although many theories exist as to how they were formed. For a more extensive analysis of pocosins and Carolina bays, the reader is referred to Richardson (1981) and Richardson and Gibbons (1993).

Wetland Ecosystems

The pocosin communities of today are only a remnant of the vegetation of presettlement times (Christensen 1988; Christensen et al. 1988). Most of the extensive stands of Atlantic white cedar *Chamaecyparis thyoides*, bald cypress *Taxodium distichum*, pine *Pinus spp.*, and gum *Nyssa sylvatica* var. biflora were harvested during the past 200 years (Frost 1995). However, a unique mosaic of wetland ecosystems still exists across the coastal plain landscape. Distinct ecosystems, varying in vegetational composition and soil nutrient status (Table 1), occur along a topographic gradient even though local relief is generally 2 m or less. Topographic highs are occupied by nutrient-deficient, ombrotrophic shrub bogs that occur over

Table 1 Total nutrient and carbon concentrations, root-free bulk density (BD), and organic matter content (OM) at three depth intervals for short pocosin (SP), tall pocosin (TP), and bay forest (BF)

Site	Depth (cm)	BD (g/cm^3)	OM (%)	pH	C (mg/cm^3)	N (mg/cm^3)	P (mg/cm^3)	Mg (mg/cm^3)	Ca (mg/cm^3)	N: P
SP	0–5	0.027 (0.001)^{b3}	95.3 (0.3)^{a1}	3.8*	16.5 (0.6)^{b3}	0.393 (0.009)^{b3}	9.63 (0.10)^{b3}	32.5 (1.1)^{a3}	22.6 (6.6)^{a2}	46.9 (1.8)^{a**}
	5–10	0.049 (0.008)^{b2}	95.0 (0.9)^{a1}		29.2 (4.6)^{ab2}	0.793 (0.133)^{b2}	16.2 (2.7)^{b2}	52.9 (8.3)^{a3}	19.0 (2.0)^{a1}	
	10–20	0.115 (0.009)^{b1}	95.0 (0.9)^{a1}		69.7 (5.1)^{a1}	1.74 (0.14)^{b1}	36.1 (1.7)^{b1}	117 (8)^{b2}	25.2 (1.8)^{b2}	
TP	0–5	0.028 (0.002)^{b3}	93.2 (3.1)^{a1}	3.7*	15.4 (1.5)^{b3}	0.345 (0.034)^{b2}	12.5 (1.1)^{b2}	32.8 (3.9)^{a2}	50.2 (10.1)^{a1}	31.8 (3.2)^{b**}
	5–10	0.042 (0.002)^{b3}	89.7 (5.0)^{a1}		21.1 (1.3)^{b3}	0.520 (0.061)^{b2}	17.2 (1.6)^{b2}	42.1 (5.0)^{a2}	26.0 (8.0)^{a1}	
	10–20	0.132 (0.008)^{b2}	79.1 (11.0)^{ab1}		56.0 (6.8)^{b2}	1.57 (0.27)^{b1}	48.0 (4.1)^{b1}	94.2 (17.6)^{b1}	27.2 (12.2)^{ab1}	
BF	0–5	0.047 (0.007)^{a3}	81.2 (6.2)^{b1}	3.9*	23.3 (2.4)^{a3}	0.923 (0.117)^{a2}	71.2 (15.7)^{a2}	28.3 (6.1)^{a3}	43.9 (18.0)^{a1}	10.1 (1.2)^{c**}
	5–10	0.095 (0.025)^{a3}	70.3 (8.0)^{b1}		38.9 (5.5)^{a3}	1.54 (0.26)^{a2}	151 (41)^{a2}	62.8 (20.6)^{a3}	35.5 (7.4)^{a1}	
	10–20	0.265 (0.042)^{a2}	57.4 (9.1)^{b1}		89.7 (5.8)^{a2}	3.35 (0.05)^{a1}	347 (41)^{a1}	190 (35)^{a2}	69.7 (10.8)^{a1}	

Significant differences ($P < 0.05$) between sites within a depth interval are shown by different small letters, while different numbers show significant differences between depths within a site

Modified from Bridgham and Richardson 1993; (Wetlands, Hydrology and nutrient gradients in North Carolina peatlands, 13, 1993, 207–218, Bridgham SD, Richardson CJ; © 1993, the Society of Wetland Scientists, with permission of Springer Science + Business Media)

* pH values are for 0–10 cm only
** N: P at 0–30 cm only

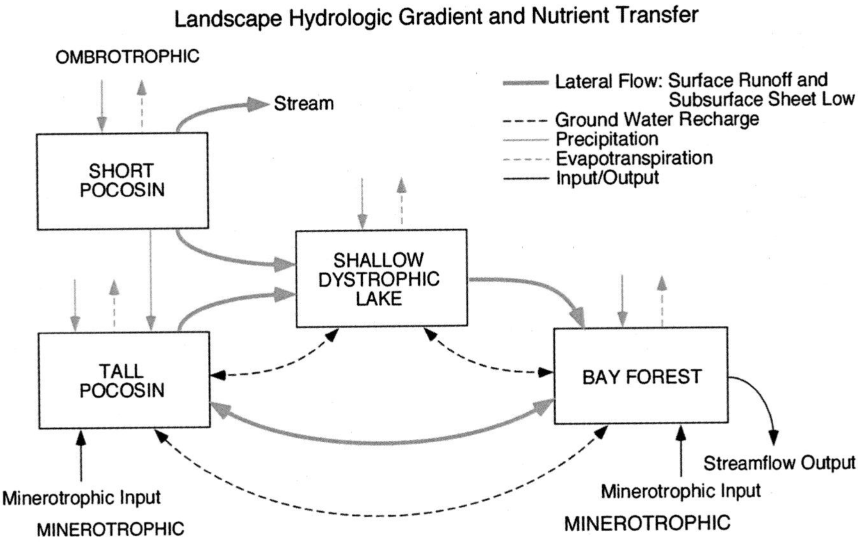

Fig. 2 A cross-sectional view of the hydrologic and nutrient gradients found on the coastal landscape for the large pocosin complexes found in the southeast USA (Fig. 14.5, Richardson, Curtis J., "Pocosins: Evergreen Shrub Bogs of the Southeast" in *Wetland Habitats of North America: Ecology and Conservation Concerns* by Darold P. Batzer (Editor), Andrew Baldwin (Editor). © 2012 by the Regents of the University of California. Published by the University of California Press – with permission)

relatively deep peat accumulations (>1 m). This lack of nutrients results in very stunted vegetation, and thus it is referred to as short pocosin.

Tall pocosins, and sometimes-shallow dystrophic lakes, border short pocosins (Fig. 2). Tall pocosins occur over shallower peat deposits (approximately 50–100 cm), have higher soil nutrient content, and exhibit greater vegetation height and aboveground biomass than short pocosins (Christensen et al. 1981, 1988). Relatively nutrient-rich gum swamps/bay forests occur along the southern margins of the lakes and along outflow stream drainages (Fig. 2). Peat depths in gum swamps/bay forest stands can range from approximately 50 cm to >150 cm of organic matter depending on the age of the stand, and soil mineral content is somewhat greater than in pocosins.

Pocosin vegetation (Fig. 3a, b) is found on mineral soils, sandy humus, and organic mucks and peats (histosols, terric, or typic Medisparist) (Bridgham and Richardson 2003; Bridgham et al. 1995). The excessive amount of annual precipitation over evapotranspiration over the past 4,000–5,000 years has resulted in massive amounts of peat accumulation, further resulting in the isolation of these ecosystems from the nutrients and minerals from the ancient marine sandy terraces found beneath (Richardson 1981, 2012). Pocosin cation chemistry is controlled by low pH and low Ca due to the siliceous parent material beneath and extremely low concentrations of phosphorus. These soils are saturated or shallowly flooded primarily during the cool seasons. On raised organic soils, precipitation is virtually the

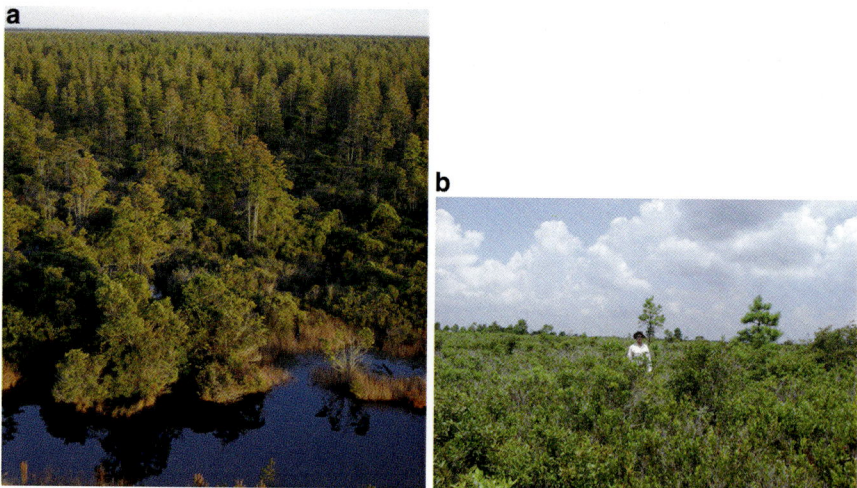

Fig. 3 (**a**) (*left*) Tall pocosin peatlands in coastal North Carolina are found on shallow peat <1 m in depth. The dominant vegetation shown is mainly pond pine *Pinus serotina*, red maple *Acer rubrum*, red bay *Persea borbonia*, and loblolly bay, *Gordonia lasianthus* (Photo credit: C. J. Richardson© Rights remain with the author). (**b**) (*right*) Short pocosin peatlands in coastal North Carolina are found on deep peat 1–4 m in depth. The dominant vegetation shown is mainly comprised of shrubs like fetterbush *Lyonia lucida*, holly *Ilex glabra, and* titi *Cyrilla racemiflora* with a few scattered pond pine *Pinus serotina*, and loblolly bay *Gordonia lasianthus* trees (Photo credit: C. J. Richardson© Rights remain with the author)

only external source of plant nutrients; the peat is thick enough to keep the underlying mineral soil out of contact with plant roots, and there is no external drainage of water into these bogs because they are the topographic high on the landscape. These peatlands are referred to as oligotrophic (i.e., having water poor in nutrients). Peat depths of greater than 3 m are found in the most extensive remaining pocosins, which are found in the Alligator River National Wildlife Refuge in northeastern NC. It has been estimated that over 529 million metric tons of peat (298 million tons of carbon) exist in NC peatlands and that pocosins comprise 82% of the peat (Richardson 2012).

The role of disturbance events as a long-term controlling factor has not been well understood and a check of peat profiles reveals that these systems are subject to intense fires, droughts, and hurricanes as evidenced by thick charcoal layers; species changes where entire mature stands of flattened Atlantic white cedar and cypress are found often buried at depths >3 m, respectively (Richardson 1981). These findings and historical studies by Wells (1928) have led to several theories how pocosins are formed and maintained in their current state. The two basic theories of pocosin succession are in sharp contrast with each other.

Wells (1928) proposed that the frequency and intensity of fire controls successional development. According to this theory, the pioneer stage is short pocosin, which succeeds to the bay forest climax within a few 100 years if disturbance is prevented. The second theory of pocosin succession (Richardson 2012) assumes that nutrient levels are the controlling factor. The limiting soil nutrient primarily is phosphorus (Richardson 1983; Walbridge 1991; Bridgham and Richardson 2003). According to this hypothesis, the successional sequence is marsh → swamp forest → bay forest → tall pocosin → short pocosin. Pollen analysis supports this latter proposed sequence in the communities. This succession has taken place over the past 5,000 years as paludification (peat expansion over the landscape) has resulted in an extensive peat-covered landscape in part of the southeastern coast and an associated gradual rise in the water table. The importance of fire versus nutrients or hydroperiod in controlling pocosin succession is, however, not clear. These factors are so interrelated that it is impossible to separate one as the primary control.

Geology

Pocosin tracts often cover hundreds of square kilometers and are found on flat, clay-based soils, in shallow basins on divides between ancient rivers and sounds on the South Atlantic coastal plain. Impeded runoff of fresh water (i.e., blocked drainage due to sediment buildup, rising sea levels, and peat formation) coupled with the milder climate since the Wisconsin Ice Age about 18,000 years ago resulted in a shift from boreal forest species to present-day southern wetland forests and evergreen shrub bog communities (Whitehead 1972, 1981).

Based on ^{14}C dating from the Dismal Swamp, a vast pocosin ecosystem located between the border of southern coastal VA and NC, the deposition of most organic clays and overlaying peat within this region began between $10,340 \pm 130$ and $8,135 \pm 160$ years before present (Daniel 1981). However, radiocarbon dates from much of the peat forest present today in the Dismal Swamp indicate ages less than 3,500 years. This information, when coupled with the changes in pollen profiles, the presence of charcoal at various depths, and the evidence of decreased rainfall (Whitehead 1972, 1981), indicates fluctuations in peat oxidation and accumulation rates, the occurrence of extensive fires in pocosin peatlands, and a dynamic peat development history.

Hydrology

Pocosins are the topographic high on the regional landscape, and as such they are the source of water for downstream areas. A graphic view of their hydrologic gradient on the landscape is shown in Fig. 2. The short pocosin is the largest area of the wetland complex, has the deepest peat, is fed primarily by rainfall (ombrotrophic), and is thus nutrient poor. It is the source of water for the region as runoff drains slowly from short pocosins to shallow dystrophic lakes or the surrounding tall pocosins. The water that flows laterally into shallow lakes or into small streams then flows to the bay forest communities comprised of species like red maple *Acer rubrum*, red bay *Persea borbonia*, loblolly bay *Gordonia lasianthus*, pond pine *Pinus serotina*, and sweetbay *Magnolia virginiana* at the downstream end of pocosin systems. These bay forest sites are more nutrient rich (minerotrophic) since they are in contact with mineral soils and have nutrient inputs from the pocosin runoff. Shallow groundwater transfers also are thought to take place among the components of the pocosin complex, but no extensive studies have been done to quantify this.

It is clear however that pocosins are the main source of freshwater on the coastal landscape where they cover large areas. The amount and timing of the runoff from these wetlands are critical to downstream flows and estuarine water quality (Richardson 1983). In natural pocosins, >90% of water output is through evapotranspiration (ET) if the rainfall is during the summer and fall but shifts to runoff if precipitation occurs in winter and spring (Richardson 2012). The water storage capacity of pocosins is thus limited in the winter and early spring months when low ET and high rainfall have caused the soil to become saturated. High ET during the summer can lower the water table 60–90 cm below the land surface, giving the wetlands extensive storm water storage capacity (Bridgham and Richardson 2003). Annual variation in rainfall and ET studied over several decades shows that runoff from pocosins can vary considerably during periods of drought or extreme rainfall. Following several years of high rainfall and reduced soil storage capacity, runoff reached nearly 75 cm/year in wet years, while runoff dropped to near 10 cm/year following low rainfall periods and higher ET (Richardson and McCarthy 1994).

Biodiversity

A comparison of short and tall pocosin community vegetation as well as bay forests in North Carolina shows that shrub species are very similar. Pocosins and bay forests also share a number of species like *Persea borbonia*, *Gordonia lasianthus*, *Cyrilla racemiflora*, and *Pinus serotina*, although the numbers and basal area vary greatly. The number of tree stems per hectare and the number of short stems are very much reduced in tall pocosins as compared to short pocosins. Herb cover is lowest in tall pocosins.

Plants that are dependent in part on pocosin-type habitat or adjacent coastal savannah are the threatened Venus flytrap *Dionaea muscipula*, dwarf fothergilla *Fothergilla gardenii*, sweet pitcher plant *Sarracenia rubra*, and the endangered white beak rush *Rhynchospora alba* (Richardson 2012). If extensive natural fires remove the peat substrate and the bay or pocosin reverts to a marsh or if the ecosystem type is developed by man, endangered plants such as white wicky *Kalmia cuneata*, arrowleaf shieldwort *Peltandra sagittaefolia*, spring-flowering goldenrod *Solidago verna*, and rough-leaf loosestrife *Lysimachia asperulaefolia* will face extinction.

Very little is known about the fauna of pocosin ecosystems. Pocosins serve as habitat for the specialized swallowtail *Papilio palamedes* and the Hessel's hairstreak butterfly *Mitoura hesseli*. They are important to the federally endangered pine barrens tree frog *Hyla andersonii* (Richardson 1981). The state (NC)-endangered eastern diamondback rattlesnake *Crotalus adamanteus* and American alligator *Alligator mississippiensis* are also found here. Pocosins are refuges for native big game species such as black bear *Ursus americanus*, white-tailed deer *Odocoileus virginianus* and smaller mammals such as the bobcat *Lynx rufus*, cottontail rabbit *Sylvilagus floridanus*, and gray squirrel *Sciurus carolinensis* (Richardson 1981). The largest population of black bear on the east coast now exists in the pocosin refuges. The federally endangered red-cockaded woodpecker *Picoides borealis* inhabits mature pond pines in pocosins. Importantly, the red wolf *Canis rufus*, one of two species of wolves in the USA, was extinct in the wild, but a successful reintroduction into the Pocosin Lakes National Wildlife Refuge has established a small population. For a more detailed analysis and listing of the animals found in pocosins, see Sharitz and Gibbons (1984) and Richardson and Gibbons (1993).

Conservation Concerns

Pocosins have come under threat many times during the past three centuries. First came the drainage of the peatlands that started in the time of George Washington in the late 1700s but really expanded during slavery as lands were converted to cotton plantations and farms in the mid-1800s (Richardson 1981). The second wave of drainage occurred during the early 1900s when logging of giant bald cypress and Atlantic white cedar trees became the main industry on the coastal plain. The next impact was the idea that peat soils would be great for farming, and by the 1980s nearly 66% of the 908,000 ha of pocosins in NC were drained in farmland or were being prepared for pine plantations (Richardson 1983). It was also at this time that pocosins were considered a great source of fuel as "peat for energy." There was a great effort by several corporations to develop peat-harvesting plans for hundreds of thousands of acres for fuel for power plants. Fortunately, it was realized that the environmental damage to the wetlands and adjacent estuaries could be massive and that the entire supply of peat for energy (reserves equal to 12 quadrillion British thermal units (BTUs) of energy) would meet less than a decade's energy requirements for NC (Richardson 2012).

A parallel wave of preservation activity started in the mid-1980s and continues to the present to preserve as much of the remaining pocosins as possible. The creation of the Alligator River National Wildlife Refuge (61,408 ha) in 1984 and in 1990 the Pocosin Lakes National Wildlife Refuge (44,482 ha) in addition to the creation of the Dismal Swamp Refuge in 1973 in coastal Virginia (44,844 ha) have resulted in the preservation of over 150,000 ha of pocosin-type wetlands under former threat of agriculture and forestry conversions. The current major threat to these coastal freshwater wetlands is the rapid rise in sea level, which according to a recent study showed that the eastern coastal region of the USA between Cape Hatteras, NC and Cape Cod, Massachusetts, is experiencing the fastest relative sea-level rise in the world, with three to four times the global average (Sallenger Jr et al. 2012). The impacts of accelerated sea-level rise, aggravated by more frequent and extreme storm surges projected by recent IPCC climate models, suggest that vast areas of coastal pocosin bogs will be destroyed by significant increases of saltwater intrusion in the coming century.

References

Bridgham SD, Richardson CJ. Hydrology and nutrient gradients in North Carolina peatlands. Wetlands. 1993;13:207–18.

Bridgham SD, Richardson CJ. Endogenous versus exogenous nutrient control over decomposition and mineralization in North Carolina peatlands. Biogeochemistry. 2003;65:151–78.

Bridgham SD, Pastor J, McClaugherty CA, Richardson CJ. Nutrient-use efficiency: a litterfall index, a model, and a test along a nutrient-availability gradient in North Carolina peatlands. Am Nat. 1995;145:1–21.

Christensen NL. Vegetation of the southeastern coastal plain. In: Barbour MG, Billings WD, editors. North American terrestrial vegetation. New York: Cambridge University Press; 1988. p. 317–63.

Christensen NL, Burchell R, Liggett A, Sims E. The structure and development of pocosin vegetation. In: Richardson CJ, editor. Pocosin wetlands: an integrated analysis of coastal plain freshwater bogs in North Carolina. Stroudsburg: Hutchinson Ross; 1981. p. 43–61.

Christensen NL, Wilbur RB, McLean JS. Soil-vegetation correlations in the pocosins of Croatan National Forest, North Carolina. Biological report 88. Washington, DC: U.S. Fish and Wildlife Service; 1988.

Cowardin LM, Carter V, Golet FC, LaRoe ET. Classification of wetlands and deepwater habitats of the United States. FWS/OBS-79/31. Washington, DC: U.S. Fish and Wildlife Service; 1979.

Daniel III CC. Hydrology, geology and soils of pocosins: a comparison of natural and altered systems. In: Richardson CJ, editor. Pocosin wetlands: an integrated analysis of coastal plain freshwater bogs in North Carolina. Stroudsburg: Hutchinson Ross; 1981. p. 69–108.

Frost C. Atlantic white cedar forests. [dissertation]. Chapel Hill: University of North Carolina; 1995.

Harper RM. A midsummer journey through the coastal plain of the Carolinas and Virginia. Bull Torrey Bot Club. 1907;34:351–77.

Kologiski RL. The phytosociology of the Green Swamp, North Carolina, Technical bulletin, vol. 250. Raleigh: North Carolina Agricultural Experiment Station; 1977.

Richardson CJ, editor. Pocosin wetlands: an integrated analysis of coastal plain freshwater bogs in North Carolina. Stroudsburg: Hutchinson Ross; 1981.

Richardson CJ. Pocosins: vanishing wastelands or valuable wetlands? Bioscience. 1983;33:626–33.

Richardson CJ. Pocosins: evergreen shrub bogs of the southeast. In: Batzer DP, Baldwin AH, editors. Wetland habitats of North America: ecology and conservation concerns. Berkeley: University of California Press; 2012. p. 189–202.

Richardson CJ, Gibbons LW. Pocosins, Carolina bays and mountain bogs. In: Martin WH, Boyce SG, Echternacht AC, editors. Biotic communities of the southeast, volume 2: lowland terrestrial communities. New York: Wiley; 1993. p. 257–310.

Richardson CJ, McCarthy EJ. Effect of land development and forest management on hydrologic response in southeastern coastal wetlands: a review. Wetlands. 1994;14:56–71.

Ross TE. A comprehensive bibliography of the Carolina bays literature. J Elisha Mitchell Soc. 1987;103:28–42.

Sallenger Jr AH, Doran KS, Howd PA. Hotspot of accelerated sea-level rise on the Atlantic coast of North America. Nat Clim Chang. 2012;2:884–8. http://doi.org/hz4.

Sharitz RR, Gibbons JW. The ecology of evergreen shrub bogs (pocosins) and Carolina bays: a community profile. FWS/OBS-82/04. Slidell: U.S. Fish and Wildlife Service Biological Services Program; 1984.

Tooker WW. The adapted Algonquin term "Poquosin". Am Anthropol. 1899;1:162–70.

Walbridge MR. Phosphorus availability in acid organic soils of the lower North Carolina coastal plain. Ecology. 1991;73:2083–100.

Wells BW. Plant communities of the coastal plain of North Carolina and their successional relations. Ecology. 1928;9:230–42.

Whitehead DR. Developmental and environmental history of the Dismal Swamp. Ecol Monogr. 1972;42:301–15.

Whitehead DR. Late-Pleistocene vegetational changes in northeastern North Carolina. Ecol Monogr. 1981;51:451–71.

Woodwell GM. Factors controlling growth of pond pine seedlings in organic soils of the Carolinas. Ecol Monogr. 1958;28:219–36.

Prairie Pothole Region of North America

Kevin E. Doherty, David W. Howerter, James H. Devries, and Johann Walker

Contents

Introduction 680
Hydrology 681
Biodiversity 682
Ecosystem Services 683
Conservation Status and Threats 684
References 686

Abstract

Located in the interior of North America straddling the US/Canada border, the Prairie Pothole Region (PPR) encompasses more than 770,000 km^2 and is one of the richest, most diverse, and unique wetland-grassland ecosystems in the world. The PPR is named for the millions of depressional wetlands called "prairie potholes" dispersed throughout the landscape. The potholes formed as subterranean masses of ice melted following the retreat of glaciers at the end of the last ice age. A majority of wetlands within the PPR are depressional and receive water by snowmelt or rain. Differences in topography, soil type, longitude, and latitude all determine the depth and ecological function of prairie potholes. Wetlands and

K. E. Doherty (✉)
United States Fish and Wildlife Service, Bismarck, ND, USA
e-mail: kevin_doherty@fws.gov

D. W. Howerter · J. H. Devries
Ducks Unlimited Canada, Stonewall, MB, Canada
e-mail: d_howerter@ducks.ca; j_devries@ducks.ca

J. Walker
Ducks Unlimited, Great Plains Region, Bismarck, ND, USA
e-mail: jwalker@ducks.org

© This is a U.S. Government work and not under copyright protection in the US; foreign copyright protection may apply 2018
C. M. Finlayson et al. (eds.), *The Wetland Book*,
https://doi.org/10.1007/978-94-007-4001-3_15

grasslands in the PPR provide vital habitat for a diverse array of plant and animal species; large populations of migratory birds including waterfowl, waterbirds, and grassland birds depend on this habitat base during the breeding season and migration. Prairie pothole wetlands confer a variety of ecological good and services to society, but wetland drainage has substantially reduced wetland abundance relative to pre-European settlement levels. Although programs have reduced the rate of loss, more work is required to slow wetland and grassland conversion.

Keywords

Adaptive management · Conservation · Energy development · Grasslands · Joint venture · Landscape conservation · Planning · Prairie Pothole Region · Wetland ecosystems

Introduction

Located in the interior of North America, the Prairie Pothole Region (PPR) is one of the richest, most diverse, and unique wetland-grassland ecosystems in the world (Baldassarre and Bolden 2006). The PPR straddles the US/Canada border and encompasses more than 770,000 km^2 including parts of five US states, the northern tier of Montana, northern and eastern North Dakota, eastern South Dakota, western Minnesota, and north-central Iowa; and three Canadian provinces, southwestern Manitoba, southern Saskatchewan, and southern Alberta (Fig. 1). The PPR is named for the millions of depressional wetlands called "prairie potholes" dispersed

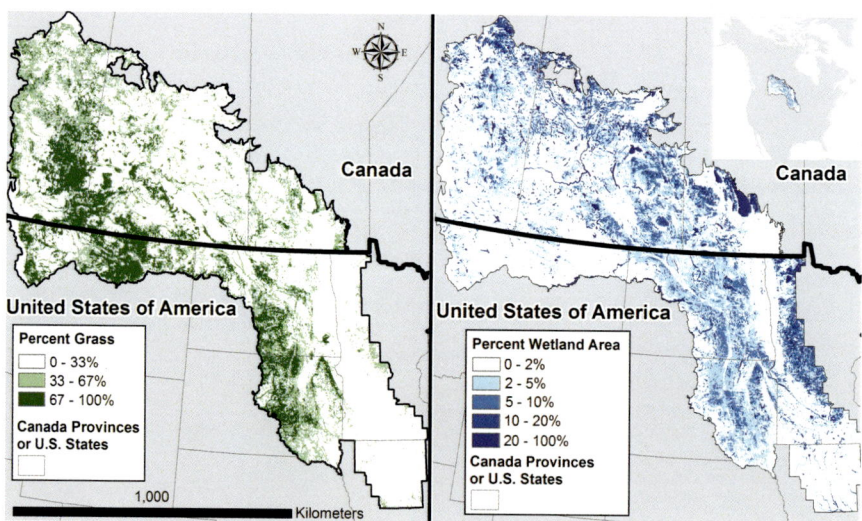

Fig. 1 Distribution of 29.2 million grassland ha and 8.3 million wetland ha (percent within 10.4 km^2 [4 mile^2] area) within the Prairie Pothole Region of North America, circa 2000 (original figure by US Fish and Wildlife Service, held in the public domain)

throughout the landscape. The vast area of the PPR ecosystem and high density of wetlands (exceeding 40/km² in some areas; Kantrud et al. 1989) make the PPR region unique.

Formed as subterranean masses of ice melted following the retreat of glaciers at the end of the last ice age, the PPR historically ranged from areas of vast grasslands in the south to grasslands interspersed with deciduous trees (parklands) in the north. The eastern portion of the PPR historically supported tall-grass prairie vegetation. Short-grass prairie dominated the western portion, with mixed-grass prairie located centrally. Parklands represent the transition from grassland to boreal forest and are interspersed with stands of aspen *Populous tremuloides* and other woody species. These gradients in vegetation reflect variation in precipitation and evapotranspiration rates across the region (Millett et al. 2009). Areas to the south and west tend to be drier than more northerly or easterly regions. The historic disturbance regime in this area was driven by wildfire, grazing by native ungulates, and drought and deluge precipitation patterns. Since settlement, wildfire has been suppressed and in many places native grasslands have been largely replaced by annual cropland. Regions of the PPR with higher annual precipitation have experienced higher rates of conversion to cropland. Accordingly, tall-grass prairies, which occur at the higher end of the moisture gradient of global grasslands, are almost extirpated (Samson and Knopf 1994).

Grasslands and wetlands within the PPR, especially the eastern PPR, are some of the most altered landscapes in the world because much of the land is privately owned, is productive as cropland, and is relatively easy to cultivate (see Fig. 4 in Hoekstra et al. 2005). Prior to European settlement, prairie pothole wetlands may have encompassed > 20% of total land area in the PPR (Euliss et al. 2006). Since settlement, up to 89% of wetlands have been lost to agricultural drainage in some parts of the PPR (Dahl 1990). Native prairie losses have exceeded rates of all other biome losses within North America (Samson and Knopf 1994). Grasslands in the PPR complement wetland resources, as many species of wetland-dependent birds nest in surrounding grasslands and grass cover is essential for successful nesting for a wide variety of ground nesting birds from passerines to waterfowl (Klett et al. 1988; Stephens et al. 2005; Winter et al. 2005). Loss of grassland, especially native grassland, has resulted in large declines in grassland bird populations, making them one of the most imperiled guilds of birds in North America (Brennan and Kuvlesky 2005).

Hydrology

A majority of wetlands within the PPR are depressional and receive water by snowmelt or rain. Differences in topography, soil type, longitude, and latitude all determine the depth and ecological function of prairie potholes (Labaugh et al. 1998). Wetlands within the PPR are classified by how long they retain water during the growing season. Temporary wetlands retain water for 1–3 weeks during the growing season, seasonal wetlands retain water for 3 weeks to 90 days, and semipermanent wetlands retain water through the growing season for many consecutive years (Stewart and Kantrund 1971). Detailed mapping of wetland types in the

US PPR indicates 3.44 million wetland basins covering 3.68 million ha in temporary (13.0% of total area), seasonal (23.7%), semipermanent (24.1%), riverine (7.7%), and permanent wetlands or lakes (31.4%) within the boundaries of the US PPR region (Fig. 1). In the Canadian PPR, detailed mapping of all individual wetlands has not been completed; however, current estimates place wetland area near 4.6 (± 1.1) million ha (Watmough and Schmoll 2007). Temporary and seasonal basins comprise an estimated 24% of this area, whereas semipermanent and permanent basins comprise 61.7 and 11.8% of wetland area, respectively (Ducks Unlimited Canada, unpublished data). Most wetlands in the PPR hold water temporarily or seasonally. In the USA, semipermanent and permanent wetlands comprise 8.6 and 1.2% of wetland basins, respectively, yet these deeper basins account for 55.5% of total wetland area. In Canada, semipermanent and permanent basins comprise 23.9 and 1.2% of wetland basins but account for 73.5% of wetland area (Ducks Unlimited Canada, unpublished data). Wetland types have different ecological values for different species. For example, northern pintail *Anas acuta*, a species that has experienced population declines since the 1970s is associated with temporary and seasonal wetlands.

The climate of the PPR is extremely variable, characterized by high interannual and regional variation in precipitation (Niemuth et al. 2010), which greatly influences the number of wetland basins in the PPR that contain water each year, water levels within those basins, and abundance of wetland-associated wildlife. Even in the absence of anthropogenic wetland drainage, wetland basin area and basin numbers are not static in the prairies. This is because large variation in precipitation causes wetland area to increase and decrease through time (Niemuth et al. 2010). Further, wetland numbers and function may change during wet/dry precipitation cycles. For example, during high precipitation years as water tables rise, many temporary wetlands are subsumed within the boundaries of seasonal or semipermanent wetlands and may function ecologically as a deeper water regime wetland (Niemuth et al. 2010). Conversely, during drought years, semipermanent wetlands may function as a temporary or seasonal wetland and temporary wetlands may remain dry for several years. Periodic drying is essential to maintain the productivity of prairie wetlands by accelerating nutrient cycling and allowing seeds of annual plants to germinate (Murkin 2000).

Biodiversity

Remaining wetlands and grasslands in the PPR provide vital habitat for a diverse array of plant and animal species, including threatened and endangered mammals (Clark 2000), fishes (Peterka 1989), amphibians (Lehtinen et al. 1999), and a variety of invertebrates (Wrubleski and Ross 2011). Most notably, large populations of migratory birds including waterfowl (Johnson and Grier 1988), waterbirds (Peterjohn and Sauer 1997), and grassland birds (Niemuth et al. 2008) depend on this habitat base for food and cover primarily during the breeding season but additionally during migration. In 2011, approximately two third of all ducks estimated in the entire Waterfowl Breeding Population and Habitat Survey area occurred within the US and Canadian PPR (Zimpher et al. 2011).

The myriad of wetlands also makes the PPR valuable to other migratory birds. For example, estimates suggest that PPR harbors 70% of the continental population of Franklin's gull *Larus pipixcan*; > 50% of the continental population of pied-billed grebe *Podilymbus podiceps*, American bittern *Botaurus lentiginosus*, sora *Porzana carolina*, American coot *Fulica americana*, and black tern *Chlidonia niger*; and 30% of the continental population of American white pelican *Pelecanus erythrorhynchos* and California gull *Larus californicus* (Beyersbergen et al. 2004) as well as ~80% of the North American population of marbled godwit *Limosa fedoa*. Remaining grasslands also support large populations of grassland birds including 91% of Baird's sparrow *Ammodramus bairdii*, 87% of Sprague's pipit *Anthus spragueii*, and 71% of chestnut-collared longspur *Calcarius ornatus* (Rich et al. 2004).

Ecosystem Services

In addition to providing habitat for a broad suite of wetland-dependent wildlife species, prairie pothole wetlands confer a variety of ecological good and services to society.

Surface water storage and flows – Prairie wetlands store water during the spring runoff and following prolonged precipitation events. Slowing runoff rates reduce annual economic losses from flood damage to infrastructure such as roads, drainage systems, and urban housing (Murkin 1998). Miller and Nudds (1996) attributed increasing magnitude of flood events in the Mississippi River, in part, to wetland losses in the upper reaches of the watershed in the PPR.

Groundwater recharge – Surface waters contained within prairie potholes interact with ground water in a variety of ways. Subsurface connectivity of water flows affects both water chemistry and a variety of biological processes. Depressional wetlands can be both areas of local recharge to groundwater at topographic highs and areas where groundwater discharges to the surface at topographic lows (Labaugh et al. 1998). Prairie wetlands can be important for recharging regional aquifers at rates ranging from 2 to 45 mm/year (van der Kamp and Hayashi 1998). In the PPR of northwest Minnesota, Cowdery et al. (2008) demonstrated that surface aquifers were recharged in significant part from seasonal and ephemeral wetlands.

Controls for contaminants, excess nutrients, and sediments – Prairie wetlands play an important role in mediating transport of a variety of contaminants and serve as sinks for excess nutrients (Murkin 1998). Agricultural applications of fertilizers and pesticides have increased substantially in the PPR since the 1960s (Crumpton and Goldsborough 1998). Accordingly, prairie potholes often receive substantial inputs of chemicals, excess nutrients, and sediments from surrounding agricultural or other industrial operations. In some cases wetlands are able to incorporate undesirable chemicals and breakdown these compounds into less toxic by-products and sediments (Goldsborough and Crumpton 1998; Murkin 1998). Excess nutrient inputs to wetlands are incorporated into wetland flora and fauna (Murkin 1998), thereby removing nitrates and phosphorous with important consequences for downstream water quality and subsequent eutrophication of receiving waterbodies (Crumpton and Goldsborough 1998). Further, wetlands reduce peak river flows allowing sediments to settle out of the water column which reduces stream turbidity. Unfortunately, increased siltation within wetlands may subject them to premature filling and impair other ecological functions (Gleason and Euliss 1998).

Greenhouse gas flux – Prairie wetlands have been documented to be important carbon stores (Gleason et al. 2009), but the same conditions that lead to the accumulation of organic carbon can also lead to the production of methane, a greenhouse gas that is more effective at trapping heat than carbon (Badiou et al. 2011). Until recently it was unknown whether prairie potholes were net sources or sinks of greenhouse gases. Badiou et al. (2011) measured the changes in soil organic carbon, and methane and nitrous oxide emissions in newly restored, long-term restored, and reference wetlands across the Canadian prairies to determine the net greenhouse gas mitigation potential associated with wetland restoration. Their research estimated that restored prairie potholes have the potential to sequester a net of approximately 3.25 Mg CO_2 equivalents ha^{-1} $year^{-1}$, after accounting for increased CH_4 emissions. Thus, restoration of wetlands within the PPR could help mitigate greenhouse gas emissions (Badiou et al. 2011).

Restoring ecological goods and services through wetland restoration – It is possible to restore many of the ecological functions associated with prairie potholes (Galatowitsch and van der Valk 1994; Begley et al. 2012). Yang et al. (2010) modeled the consequences of restoring 619 ha of wetlands lost through drainage between 1968 and 2005 within Broughton's Creek watershed in southwestern Manitoba. They estimated that peak discharge at the outlet of this 25,139 ha watershed would be reduced by 23.4%. Similarly, sediment loading would be reduced by 16.9%, while total phosphorous and nitrogen loadings would be reduced by 785 and 4,219 kg $year^{-1}$, respectively, equivalent to 23.4% of current loads. However, recent meta-analyses of 621 wetland sites through the world indicate that biological structure and biogeochemical function in restored wetlands were 26 and 23% lower, respectively, than wetlands that have not been drained (Moreno-Mateos et al. 2012). Moreno-Mateos et al. (2012) also found depressional wetlands and wetlands in colder climates, such as the PPR, are the slowest to recover full ecological functions.

Conservation Status and Threats

Wetlands – Since European settlement, wetland drainage has substantially reduced wetland abundance relative to pre-European settlement levels (Dahl 1990), yet millions still exist (Doherty et al. 2013). Historic wetland losses (state-scale estimate) across individual states ranged from 27% in Montana to 89% in Iowa (Dahl 1990). Minnesota, North Dakota, and South Dakota lost 42, 49, and 35% of their wetlands, respectively, compared to presettlement conditions (Dahl 1990). The percent of wetlands lost in the PPR portion of Minnesota is actually much higher, because state-scale estimates include many nondrained wetlands in the northern deciduous and coniferous forest biomes of Minnesota (Oslund et al. 2010). In the Canadian PPR, it is estimated that between 40% and 70% of historical wetlands have been drained for agriculture development since settlement (Environment-Canada 1986; Watmough and Schmoll 2007).

Drainage peaked across the USA during the period from the 1950s to early 1970s, when 185,346 ha of wetlands were drained annually (Dahl 2011). When compared to the 44.6 million ha of wetland area remaining across the lower 48 states in 2009, this would equate to a yearly drainage rate of 0.42% (Dahl 2011). Peaks in wetland drainage were a result, in part, of larger, more powerful farm equipment and efficiencies derived in larger crop fields (Higgins et al. 2002). Within the US PPR approximately 34.4% of wetlands are protected from conversion by legal mandate, such as federal ownership or conservation easements with private landowners. Currently, wetland protection under the US Farm Bill (conservation subtitle Swampbuster provision; Public Law 99–198) is the primary protective legislation for wetlands in agriculture landscapes (van der Valk and Pederson 2003). This leaves 65.6% of remaining wetlands having protection by farmers' voluntary participation in US farm programs (Doherty et al. 2013).

Active draining is currently occurring across the PPR, especially for shallow temporary and seasonal wetlands embedded in an agricultural matrix. In the eastern portion of the US PPR, wetland loss and drainage rates are high (Oslund et al. 2010). The documented 15% loss from 1980 to 2007 (0.57% per year) in the Prairie Coteau ecoregion in Minnesota is comparable to the nationwide peak of wetland drainage during the 1950s to early 1970s (Dahl 2011). Losses in the Prairie Coteau ecoregion in Minnesota provide further evidence for higher wetland loss rates in agriculture-dominated landscapes. Between 1985 and 2001 in the Canadian PPR, wetland area losses were 5–6% among provinces and wetland basin losses were 5–8% (Watmough and Schmoll 2007).

Grasslands – Large extents of grassland still exist, with approximately 10.7 and 18.5 million ha remaining in the PPR of the USA and Canada, respectively (Fig. 1). Conversion of grasslands for crop production continues today (Stephens et al. 2008; Rashford et al. 2011). In the Missouri Coteau region of North and South Dakota, 0.4% of grasslands (−36,540 ha) were lost per year during 1989–2003 (Stephens et al. 2008). A recent study documented 1.33% of grasslands were lost per year during 1979–1997 across the entire United States Prairie Pothole Region (Rashford et al. 2011). Major drivers of grassland conversion are soil quality and agricultural commodity prices (Rashford et al. 2011). Federal farm program subsidies also drive grassland conversion in the US PPR region (G.A.O. 2007). US Federal farm programs reduce financial risks associated with cropping marginal soils and make farming more profitable, which creates economic incentives to convert privately owned grasslands from ranching operations to agricultural cropland (G.A.O. 2007). Within the US PPR approximately 81.6% of remaining grasslands have no legal protection and are vulnerable to conversion (Doherty et al. 2013). While grassland loss has been a historical trend within the Canadian portion of the PPR, the recent trend has been toward increasing grasslands at the expense of cropland. Contributing factors include removal of grain transportation subsidies in 1995, federal and provincial programs encouraging conversion of marginal cropland, and expansion of the cattle industry increasing demand for pasture and hayland forage (Prairie Habitat Joint Venture 2008). Differences in grassland trends between the US and Canadian PPR highlight the importance of agricultural policies at the national level on individuals' land use decisions. Despite increases in overall grassland cover in Canada, native grasslands have continued to decline. Native grasslands declined by about 10% within the PPR from 1985 to 2001 (Watmough and Schmoll 2007).

Conservation Future – Many species of wildlife are adapted to the PPR's variable environment and respond to water conditions by changes in distribution and numbers, including waterfowl (Johnson and Grier 1988), waterbirds (Peterjohn and Sauer 1997), and grassland birds (Niemuth et al. 2008). In addition, reproductive effort can also be influenced by water conditions (Krapu et al. 1983). High variability complicates conservation planning and management but also provides insight into what happens when conditions are dry and what the future may hold if wetlands continue to be drained, resulting in a "permanent drought."

Protecting wetlands and grasslands from conversion is a primary step necessary for future opportunities to influence habitat quality, especially when habitat is being lost (see Figs. 3 and 4 in Doherty et al. 2013; Watmough and Schmoll 2007). Continued private landowner acceptance of conservation programs is imperative in the PPR given the amount of land privately held (Doherty et al. 2013). Focusing research on the economic and social aspects of agriculture, while specifically incorporating species-specific responses in abundance, survival, and reproduction in the PPR, may identify which agricultural practices are most favorable to wildlife yet still acceptable to private landowners (Barnes 2011). Lastly, as evidenced by the large differences in amounts of wetlands and grasslands under conservation planning scenarios in the US PPR (Doherty et al. 2013) and increases in grass cover in the Canadian PPR (Watmough and Schmoll 2007), agricultural policies that do not incentivize conversion of marginal soils, or even slow wetland and grassland conversion rates by tenths of a percent, can drastically change the future of wetlands and grasslands in the PPR.

References

Badiou P, McDougal R, Pennock D, Clark B. Greenhouse gas emissions and carbon sequestration potential in restored wetlands of the Canadian prairie pothole region. Wetland Ecol Manag. 2011;19:237–56.

Baldassarre GA, Bolden EG. Waterfowl ecology and managment. Malabar: Krieger Publishing Company; 2006.

Barnes MK. Low-input grassfed livestock production in the American West: case studies of ecological, economic, and social resilience. Rangeland. 2011;33:31–40.

Begley A, Puchniak G, Gray B, Paszkowski C. A comparison of restored and natural wetlands as habitat for birds in the Prairie Pothole region of Saskatchewan, Canada. Raffles B Zool. 2012; Suppl 25:173–187.

Beyersbergen GW, Niemuth ND, Norton MR. Northern Prairie and parkland waterbird conservation plan. Denver: Prairie Pothole Joint Venture; 2004.

Brennan LA, Kuvlesky Jr WP. North American grassland birds: an unfolding conservation crisis? J Wildl Manag. 2005;69:1–13.

Clark WR. Ecology of muskrats in prarie wetlands. In: Murkin HR, van der Vals A, Clark WR, editors. Prairie wetland ecology: the contribution of the Marsh Ecology Research Program. Ames: Iowa State University Press; 2000. p. 287–313.

Cowdery TK, Lorenz DL, Arntson AD. Hydrology prior to wetland and prairie restoration in and around the Glacial Ridge National Wildlife Refuge, Northwestern Minnesota, 2002–5. Reston: U.S. Geological Survey; 2008. Scientific investigations report 2007–5200.

Crumpton WG, Goldsborough LG. Nitrogen transformation and fate in prairie wetlands. Great Plain Res. 1998;8:57–72.

Dahl TE. Wetland losses in the United States 1780's to 1980's. Washington, DC: U.S. Department of the Interior, Fish and Wildlife Service; 1990.

Dahl TE. Status and trends of wetlands in the conterminous United States 2004 to 2009. Washington, DC: U.S. Department of the Interior, Fish and Wildlife Service; 2011. 2009.

Doherty KE, Ryba A, Stemler C, Niemuth N, Meeks W. State of the U.S. Prairie Pothole ecosystem: conservation planning in an era of change. Wildl Soc Bull. 2013;37:546–63.

Environment-Canada. Wetlands in Canada: a valuable resource. 86-4 ed. Ottawa: Lands Directorate; 1986.

Euliss NH Jr, Gleason RA, Olness A, McDougal RL, Murkin HR, Robarts RD, Bourbonniere RA, Warner BG. North American prairie wetlands are important nonforested land-based carbon storage sites. Sci Total Environ. 2006;361:179–88.

G.A.O. Agricultural conservation: farm program payments are an important factor in landowners' decisions to convert grassland to cropland. Washington, DC: U. S. Government Accountability Office; 2007. Report-07-1054.

Galatowitsch SM, van der Valk AG. Restoring prairie wetlands. Ames: Iowa State University Press; 1994. p. 246.

Gleason RA, Euliss Jr NH. Sedimentation of prairie wetlands. Great Plain Res. 1998;8:97–112.

Gleason RA, Tangen BA, Browne BA, Euliss Jr NH. Greenhouse gas flux from cropland and restored wetlands in the Prairie Pothole Region. Soil Biol Biochem. 2009;41:2501–7.

Goldsborough LG, Crumpton WG. Distribution and environmental fate of pesticides in prairie wetlands. Great Plains Res. 1998;8:73–95.

Higgins KF, Naugle DE, Forman KJ. A case study of changing land use practices in the Northern Great Plains, U.S.A.: an uncertain future for waterbird conservation. Waterbirds Int J Waterbird Biol. 2002;25:42–50.

Hoekstra JM, Boucher TM, Ricketts TH, Roberts C. Confronting a biome crisis: global disparities of habitat loss and protection. Ecol Lett. 2005;8:23–9.

Johnson DH, Grier JW. Determinants of breeding distributions of ducks, Wildlife monographs. Bethesda: Wildlife Society; 1988. p. 3–37.

Kantrud HA, Krapu GL, Swanson GA. Prairie basin wetlands of the Dakotas: a community profile. Washington, DC: U.S. Fish and Wildlife Service; 1989. Biological report 85(7.28).

Klett AT, Shaffer TL, Johnson DH. Duck nest success in the Prairie Pothole Region. J Wildl Manag. 1988;52:431–40.

Krapu GL, Klett AT, Jorde DG. The effect of variable spring water conditions on mallard reproduction. Auk. 1983;100:689–98.

Labaugh JW, Winter TC, Rosenberry DO. Hydrologic functions of prairie wetlands. Great Plains Res. 1998;8:17–37.

Lehtinen RM, Galatowitsch SM, Tester JR. Consequences of habitat loss and fragmentation for wetland amphibian assemblages. Wetlands. 1999;19:1–12.

Miller MW, Nudds TD. Prairie landscape change and flooding in the Mississippi River. Conserv Biol. 1996;10:847–53.

Millett B, Johnson WC, Guntenspergen G. Climate trends of the North American prairie pothole region 1906–2000. Climatic Change. 2009;93(1–2):243–67.

Moreno-Mateos D, Power ME, Comín FA, Yockteng R. Structural and functional loss in restored wetland ecosystems. PLoS Biol. 2012;10:e1001247.

Murkin HR. Freshwater functions and values of prairie wetlands. Great Plains Res. 1998;8:3–15.

Murkin HR. Nutrient budgets and the wet-dry cycle of prairie wetlands. In: Murkin HR, van der Vals A, Clark WR, editors. Prairie wetland ecology: the contribution of the Marsh Ecology Research Program. Ames: Iowa State University Press; 2000. p. 99–121.

Niemuth ND, Solberg JW, Shaffer TL. Influence of moisture on density and distribution of grassland birds in North Dakota. Condor. 2008;110:211–22.

Niemuth N, Wangler B, Reynolds R. Spatial and temporal variation in wet area of wetlands in the Prairie Pothole Region of North Dakota and South Dakota. Wetlands. 2010;30:1053–64.

Oslund FT, Johnson RR, Hertel DR. Assessing wetland changes in the Prairie Pothole Region of Minnesota from 1980 to 2007. J Fish Wildl Manag. 2010;1:131–5.

Peterjohn BG, Sauer JR. Population trends of black terns from the North American breeding bird survey, 1966–1996. Colonial Waterbirds. 1997;20:566–73.

Peterka JJ. Fishes in northern prairie wetlands. In: van der Valk A, editor. Northern prairie wetlands. Ames: Iowa State University Press; 1989. p. 302–15.

Prairie Habitat Joint Venture. Prairie Habitat Joint Venture Implementation Plan 2007–2008. Prairie Habitat Joint Venture. Edmonton: Environment Canada; 2008. p. 34.

Rashford BS, Walker JA, Bastian CT. Economics of grassland conversion to cropland in the prairie pothole region. Conserv Biol. 2011;25(2):276–84. http://onlinelibrary.wiley.com/doi/10.1111/j.1523-1739.2010.01618.x/full.

Rich TC, Beardmore CJ, Berlanga H, Blancher PJ, Bradstreet MSW, Butcher GS, Demarest DW, Dunn EH, Hunter WC, Inigo-Elias EE, Kennedy JA, Martell AM, Panjabi AO, Pashley DN, Rosenberg KV, Rustay CM, Wendt JS, Will TC. Partners in flight North American landbird conservation plan. Ithaca: Cornell Lab of Ornithology; 2004.

Samson F, Knopf F. Prairie conservation in North America. BioScience. 1994;44:418–21.

Stephens SE, Rotella JJ, Lindberg MS, Taper ML, Ringelman JK. Duck nest survival in the Missouri Coteau of North Dakota: landscape effects at multiple spatial scales. Ecol Appl. 2005;15:2137–49.

Stephens SE, Walker JA, Blunck DR, Jayaraman A, Naugle DE, Ringelman JK, Smith AJ. Predicting risk of habitat conversion in native temperate grasslands. Conserv Biol. 2008;22:1320–30.

Stewart RB, Kantrund HA. Classification of natural ponds and lakes in the Glaciated Prairie Region. Washington, DC: Department of the Interior, Bureau of Sport Fisheries and Wildilfe; 1971. Resource Publication 92 edition.

van der Kamp G, Hayashi M. The groundwater recharge function of small wetlands in the semi-arid northern prairies. Great Plains Res. 1998;8:39–56.

van der Valk A, Pederson R. The SWANCC decision and its implications for prairie potholes. Wetlands. 2003;23:590–6.

Watmough MD, Schmoll MJ. Environment Canada's Prairie and Northern Region habitat monitoring program phase II: recent habitat trends in the Prairie Habitat Joint Venture. Edmonton: Environment Canada, Canada Wildlife Service; 2007. Technical Report 493.

Winter M, Johnson DH, Shaffer JA. Variability in vegetation effects on density and nesting success of grassland birds. J Wildl Manag. 2005;69:185–97.

Wrubleski DA, Ross LCM. Aquatic invertebrates of prairie wetlands: community composition, ecological roles, and impacts of agriculture. In: Floate KD, editor. Arthropods of Canadian grasslands, vol. 2. Ottawa: Biological Survey of Canada; 2011. p. 91–116.

Yang W, Wang X, Liu Y, Gabor S, Boychuk L, Badiou P. Simulated environmental effects of wetland restoration scenarios in a typical Canadian prairie watershed. Wetland Ecol Manag. 2010;18:269–79.

Zimpher NL, Rhodes WE, Silverman ED, Zimmerman GS, Richkus KD. Trends in duck breeding populations, 1955–2011. Laurel: U.S. Fish & Wildlife Service, Division of Migratory Bird Management; 2011.

Playa Wetlands of the Great Plains (USA)

54

Anne Bartuszevige

Contents

Introduction	690
Formation	690
Hydrology	690
Biodiversity	692
Conservation Status	692
Threats and Future Challenges	693
References	694

Abstract

Playas are mostly small, ephemeral, recharge wetlands that are the lowest point of their own watershed. They are distributed throughout the semiarid western Great Plains, USA. Playas are biodiversity hotspots and are a primary source of recharge to the High Plains Aquifer. There are two hypotheses that explain for the formation of playas, the wind erosion hypothesis and the water dissolution hypothesis. The primary threat to playas is culturally accelerated sediment accumulation. Conversation programs in the US "Farm Bill" are the primary conservation mechanisms for playas.

Keywords

Farm bill · Great plains · High plains acquifer · Ogallala acquifer · Playas

A. Bartuszevige (✉)
Playa Lakes Joint Venture, Lafayette, CO, USA
e-mail: anne.bartuszevige@pljv.org

© Springer Science+Business Media B.V., part of Springer Nature 2018
C. M. Finlayson et al. (eds.), *The Wetland Book*,
https://doi.org/10.1007/978-94-007-4001-3_16

Introduction

Playas are mostly small, ephemeral, recharge wetlands that are the lowest point in their own watershed. They are distributed throughout the semiarid western Great Plains, USA, which includes portions of Colorado, Kansas, Nebraska, New Mexico, Oklahoma, Texas, and Wyoming (Fig. 1).

The geographic distribution of playas is almost entirely congruent with the High Plains (or Ogallala) Aquifer; they are the major source of recharge to the aquifer (Gurdak and Roe 2010). Approximately 75,000 playas have been mapped in the region (Playa Lakes Joint Venture (PLJV) unpublished data; available at www.pljv.org), but mapping and verification efforts are incomplete and on-going efforts will likely change this total. Playas range in size from <1 ha to >300 ha, with a mean size of 3.5 ha, and occupy approximately 0.4% of the surface area in the region (PLJV unpublished data).

Formation

There are two primary hypotheses for the formation of playas, although they are not mutually exclusive.

Wind erosion hypothesis (Krueger 1986; Sabin and Holiday 1995) states that playas originated from low spots on the prairie where water pooled. After the plants died, strong winds eroded exposed soils out of the basin, expanding and deepening the depression.

Water dissolution hypothesis (Osterkamp and Wood 1987) states that playas originated from low spots on the prairie where water pooled. The water flowed through the subsurface taking organic material with it, which was oxidized, releasing carbon dioxide that reacted with water to form carbonic acid. This acid dissolved the underlying calcium carbonate layer, and subsidence of the overlying soils created the playa basins.

Most current hypotheses about the formation of playas describe a combination of these two processes. Furthermore, it is likely that a different process is more prevalent in one region than another. For example, playas in Nebraska are more likely to be oblong in shape and have less soil dunes than playas in the Texas Panhandle, indicating that wind erosion may be more important than water dissolution in the formation of these playas.

Hydrology

Playas have a closed watershed and, thus, are not connected to out-flowing surface waters. Therefore, they fill through runoff from large, erratic, precipitation events and lose water through recharge to the aquifer, evaporation, and transpiration. One study in Texas estimated that playas contained water in January on average once

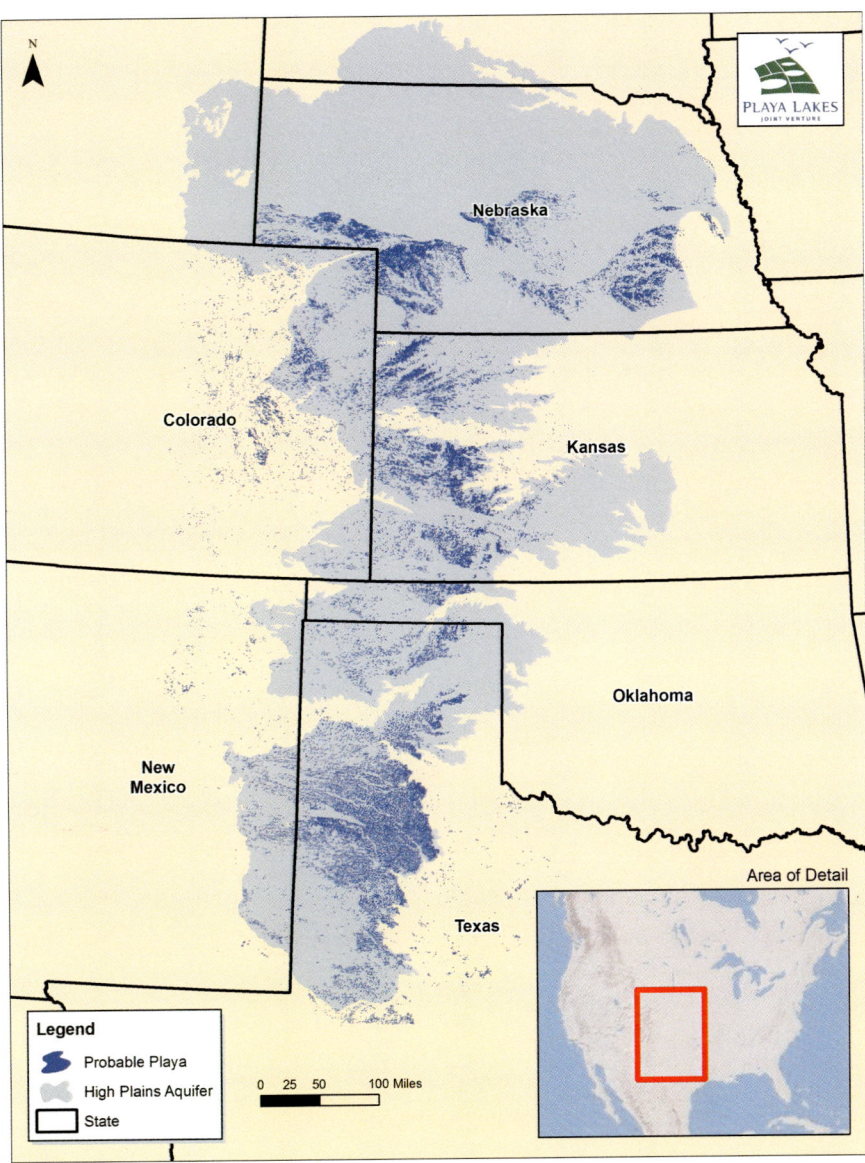

Fig. 1 Probable playas and the High Plains Aquifer in the western Great Plains, USA (Provided courtesy of Playa Lakes Joint Venture ©)

every 11 years (Johnson et al. 2011a). Playas may contain water for a few weeks to several years depending on area, volume, and amount of intervening precipitation (Smith 2003). They may also have multiple dry and wet periods in one year (Smith 2003).

Playas are a primary source of water recharge to the High Plains Aquifer. Recharge through playas is one to two orders of magnitude greater than inter-playa areas (Gurdak and Roe 2010) and happens via three different pathways. The first is through macropores that form in the hydric soil when the playa dries. Overland water flows into the playa and through macropores to the aquifer. As runoff continues and soils become saturated and swell, sealing the macropores, the playa begins to fill. When the playa fills beyond the annulus (where hydric soils meet uplands soils), water percolates through permeable soils and recharges the aquifer (second pathway). The third pathway is through micropores in the soil, including pores surrounding plant roots.

Playas have a simple physical structure. They are generally circular or oval shaped. Most do not have areas of zonation or a depth gradient, although there are exceptions among some of the larger playas. The bottom of the playa has a thick layer of hydric soils in the Randall, Lipan, Ness, and Roscoe series in the south and the Butler, Lodgepole, Fillmore, Scott, and Massie series in Nebraska playas.

Biodiversity

The extreme wet-dry fluctuation is the lifeblood of a playa ecosystem. During the dry phase, many of the wetland plants senesce or die, and macroinvertebrates aestivate in the soils. If environmental conditions are suitable, many wetland plant seeds germinate, and seedlings become established. As the playa begins to fill during the wet phase, macroinvertebrate eggs hatch. While the water is retained in the basin, wetland plants complete their lifecycle, and vertebrates use the playa for water and habitat. Tens of thousands of migrating water-dependent birds will use playas for roosting or feeding habitat.

Playas are biodiversity hotspots in this semiarid region and considered keystone ecosystems. Taxonomic diversity is 300% higher at playas than in the surrounding uplands (Smith 2003). They support approximately 185 species of birds, 350 plant species, 13 amphibian species, and 37 mammal species at some point in their lifecycle (Haukos and Smith 1994, 2004). They are especially important for migrating and wintering (in the southern region) shorebirds, waterbirds, and waterfowl in the Central Flyway. Playas in the Texas Panhandle support approximately 500,000 wintering ducks (Texas Parks and Wildlife Department 2011) and several hundred thousand sandhill cranes (*Grus canadensis*) (Smith 2003).

Conservation Status

Few playas are located on lands protected by the US state or federal governments, nor are they afforded legal protection under the US Clean Water Act. They may be considered for inclusion under the "Swampbuster" provision as part of the US Department of Agriculture policy.

Most available conservation programs are funded under the US Department of Agriculture conservation programs. However, over 90% of the playa landscape is privately owned; therefore, effective conservation involves active participation of landowners. Local conservation partnerships have emerged in the region that focus on playa conservation. In addition, states have begun to hire biologists to work with landowners and assist them in obtaining funding for playa conservation. LaGrange et al. (2011) discuss multiple specific restoration and conservation options for playas.

Threats and Future Challenges

The primary threat to playas is culturally accelerated sediment accumulation primarily from water erosion of row crop agriculture in the watersheds of playas. Due to their small size and the fact that they are often dry, farmers regularly till through playas and plant crops. The accumulated sediments fill the playa basin beyond its hydric soil-defined volume (Luo et al. 1997), thus altering the hydroperiod by reducing the ability of the playa to hold water, increasing evaporation rates, facilitating increased water loss, and supporting less avian diversity (Tsai et al. 2007; LaGrange et al. 2011). Culturally accelerated sediment accumulation also prevents germination of plants and emergence of macroinvertebrates (Jurik et al. 1994; Gleason et al. 2003), both important food resources for migrating birds. In addition, mixing of sediments with the underlying clay layer has unknown effects on playa hydrology.

Additional threats to playas include hydrological modifications such as land leveling, ditching, pits, berms, and road construction. These modifications concentrate or remove water, reducing suitable habitat for water-dependent birds. For example, shorebirds require shallow water or mudflats for foraging; concentrating water in deep pits eliminates those habitat types. In addition, road construction and pitting can disturb the playa floor by altering the hydric soil layer, which may reduce the recharge capacity of the playa. And ditching can completely eliminate the ability of a playa to hold water.

The current estimate of the number of playas in the region is approximately 75,000 (PLJV unpublished data). However, this number may be deceptive and misinforms conservation planning. Johnson et al. (2012) used remote imagery to evaluate anthropogenic impacts to playas and documented that only 0.2% of playas in the Southern Great Plains had no impacts to either the wetland or surrounding watershed. In addition, the US Department of Agriculture Natural Resources Conservation Service in Texas has undertaken a resurvey of the wetland soils that has resulted in a number of playas being reclassified as something other than a wetland (Johnson et al. 2011b). This jurisdictional reclassification reduces and may eliminate the governmental protections for these wetlands and, therefore, public money that can be spent on restoration or conservation efforts (Johnson et al. 2011b). Many playas are already filled with culturally accelerated sediment; Luo et al. (1997) predicted that at current rates of sedimentation, the remaining playas located in cropland in the Southern High Plains (Oklahoma, Texas, New Mexico) will be filled in 95 years. Johnson et al. (2012) estimated that 60% of playas in the Southern High Plains have accumulated sediments beyond their hydric soil-defined volume.

Currently, most of the funds available for conservation of playas are through programs described in and funded through a series of the US policy bills called the "Farm Bill," which is renegotiated every 5 years, but funding for these conservation programs has been declining. The current programs fall under the Conservation Reserve Program's Continuous Signup and the Wetlands Reserve Easement, both land idling programs. However, recent funding trends favor "working lands" conservation programs that alter management but retain crop or livestock production. As the link between aquifer recharge and playas becomes more widely understood, support for playa conservation and restoration is increasing.

Future research should focus on understanding the landscape patterns and processes that contribute to maintaining the ecological functioning of playas. Research that contributes to understanding of how playas function in complexes to maintain hydrology and support for migrating and wintering birds and other wetland associated wildlife is required to make effective conservation decisions. In addition, conservation programs such as buffer requirements are based on recommendations from other wetland types; effective conservation will require knowledge of applications best suited to the short-grass prairie ecoregion. Basic ecological research focused on energy flow, productivity, food webs, and resilience should also be completed. Finally, more complete understanding of the role of playas in aquifer recharge and impacts of culturally accelerated sediment on recharge rates is needed.

References

Gleason RA, Euliss Jr NH, Hubbard DE, Duffy WG. Effects of sediment load on emergence of aquatic invertebrates and plants from wetland soil egg and seed banks. Wetlands. 2003;23:26–34.

Gurdak JJ, Roe CD. Review: recharge rates and chemistry beneath playas of the High Plains aquifer, USA. Hydrogeol J. 2010;18:1747–72.

Haukos DA, Smith LM. The importance of playa wetlands to biodiversity of the Southern High Plains. Landsc Urban Plan. 1994;28:83–98.

Haukos DA, Smith LM. Plant communities of playa wetlands in the Southern Great Plains. Lubbock: Texas Tech University; 2004. Special Publication 47.

Johnson WP, Rice MB, Haukos DA, Thorpe PP. Factors influencing the occurrence of inundated playa wetlands during winter on the Texas High Plains. Wetlands. 2011a;31:1287–96.

Johnson LA, Haukos DA, Smith LM, McMurry ST. Loss of playa wetlands caused by reclassification and remapping of hydric soils on the Southern High Plains. Wetlands. 2011b;31:483–92.

Johnson LA, Haukos DA, Smith LM, McMurry ST. Physical loss and modification of southern Great Plains playas. J Environ Manage. 2012;112:275–83.

Jurik TW, Wang S-C, van der Valk AG. Effects of sediment load on seedling emergence from wetland seed banks. Wetlands. 1994;14:159–65.

Krueger JP. Development of oriented lakes in the Eastern Rainbasin Region of South Central Nebraska [Master's thesis]. Lincoln: University of Nebraska; 1986. 115 p.

LaGrange TG, Stutheit R, Gilbert M, Shurtliff D, Whited PM. Sedimentation of Nebraska's Playa wetlands: a review of current knowledge and issues. Lincoln: Nebraska Game and Parks Commission; 2011. 62 pp.

Luo HR, Smith LM, Allen BL, Haukos DA. Effects of sedimentation on playa wetland volume. Ecol Appl. 1997;7:247–52.

Osterkamp WR, Wood WW. Playa-lake basins on the Southern High Plains of Texas and New Mexico, Part 1. Hydrologic, geomorphic, and geologic evidence of their development. Geol Soc Am Bull. 1987;99:215–33.

Sabin TJ, Holliday VT. Playas and lunettes on the Southern High Plains: morphometric and spatial relationships. Ann Assoc Am Geogr. 1995;85:286–305.

Smith LM. Playas of the Great Plains. Austin: University of Texas Press; 2003. 257 p.

Texas Parks and Wildlife (TPWD). Waterfowl strategic plan: a look to the future. Austin: Texas Parks and Wildlife; 2011. 48 p. Available from: https://tpwd.texas.gov/publications/pwdpubs/media/pwd_bk_w7000_1691_07_11.pdf. Accessed 31 Jan 2016.

Tsai J-S, Venne LS, McMurry ST, Smith LM. Influences of land use and wetland characteristics on water loss rates and hydroperiods of playas in the southern High Plains, USA. Wetlands. 2007;27:683–92.

Wetlands of California's Central Valley (USA)

55

Frederic A. Reid, Daniel Fehringer, Ruth Spell, Kevin Petrik, and Mark Petrie

Contents

Introduction	698
Historical Wetland Loss	698
Wetland Types	700
Hydrologic Modifier for Seasonal Wetlands	700
Location and Ownership of Wetlands	701
Waterfowl Use	702
Threats and Future Challenges	702
References	702

Abstract

The Central Valley of California has a total area approximately 4.05 M ha. The drainages of this system is bordered by the Sierra and Coastal mountain ranges. Prior to the arrival of Europeans, the Central Valley contained more than 1.6 M ha of wetlands, chiefly seasonal in nature, based on winter and early-spring flooding. By 2006, over 83,000 ha of managed seasonal and semipermanent wetlands exist. Securing adequate surface water supplies is the most significant challenge facing public and private wetland managers in the Central Valley.

Keywords

California · Central valley · Seasonal wetland · Wintering waterfowl · Riparian and Floodplain habitats

---

F. A. Reid (✉) · D. Fehringer · R. Spell · K. Petrik · M. Petrie
Ducks Unlimited, Inc., Rancho Cordova, CA, USA
e-mail: freid@ducks.org; dfehringer@ducks.org; rspell@ducks.org; kpetrik@ducks.org; mpetrie@ducks.org

© Springer Science+Business Media B.V., part of Springer Nature 2018
C. M. Finlayson et al. (eds.), *The Wetland Book*,
https://doi.org/10.1007/978-94-007-4001-3_119

Introduction

The Central Valley (Fig. 1) stretches roughly 720 km north to south and averages 64 km across. The total area is approximately 4.05 M ha and is comprised of two smaller valleys, the Sacramento to the north and the San Joaquin to the south, and an inverted delta, where the two dominant rivers (Sacramento and San Joaquin) and the Cosumnes and Mokelumne rivers meet. The drainages of these systems are bordered by the Sierra and Coastal mountain ranges.

Prior to the arrival of Europeans, the Central Valley contained more than 1.6 M ha of wetland habitats, chiefly seasonal in nature, based on winter and early-spring flooding. Grassland and riparian habitats formed a mosaic with the wetlands and historically supported an estimated 20–40 million wintering waterfowl annually (Heitmeyer et al. 1989).

Historical Wetland Loss

The Sacramento Valley has peak runoff and discharges occurring in March and provides the dominant stream flow for the watershed. A braided system of creeks and sloughs were dominated by riparian woods and tule marshes. Peak floodwaters on the Sacramento were among the highest in the nation, with only the Mississippi River and portions of the Columbia and Ohio rivers as great (Scott and Marquiss 1984). Most wetlands have been drained for agriculture, and the land has been extensively modified with land leveling, drainage tiles, main stem levee construction, and controlled water regimes.

The San Joaquin Valley is marked by much drier conditions and less flow overall. However, when flooding does occur, water can spread extensively across the flat valley floor. Stabilization of flow and the diversion of water to agriculture have resulted in the river seasonally drying up or at least flowing below ground level for the last 40 years. Riparian and floodplain habitats have been severely degraded. The largest complex of remaining habitat in the San Joaquin Valley is the grassland complex (approximately 56,700 ha) of seasonal wetlands, grass, and riparian. In the southern portion of the Central Valley lies the Tulare Basin which is separated from the San Joaquin Basin by an uplift created by the alluvial fans of the Kings River and Los Gatos Creek. Historically large lacustrine wetlands dominated the Tulare, Kern, Goose, and Buena Vista lakes. The greater than 242,800 ha of shallow wetlands was drained and converted to barley and cotton.

Historically the Sacramento-San Joaquin Delta was comprised of nearly 100 islands with a network of sloughs, channels, and tidal freshwater marshes totaling over 283,000 ha. Large levees were constructed at the time of the 1850s gold rush, so that wetlands could be drained and crops grown. Corn, wheat, safflower, and milo dominate the crops. By the late 1980s, fewer than 7,300 ha of wetlands remained. Subsidence has occurred with the organic soils and many of the islands are now substantially below water levels of the rivers outside of the levees.

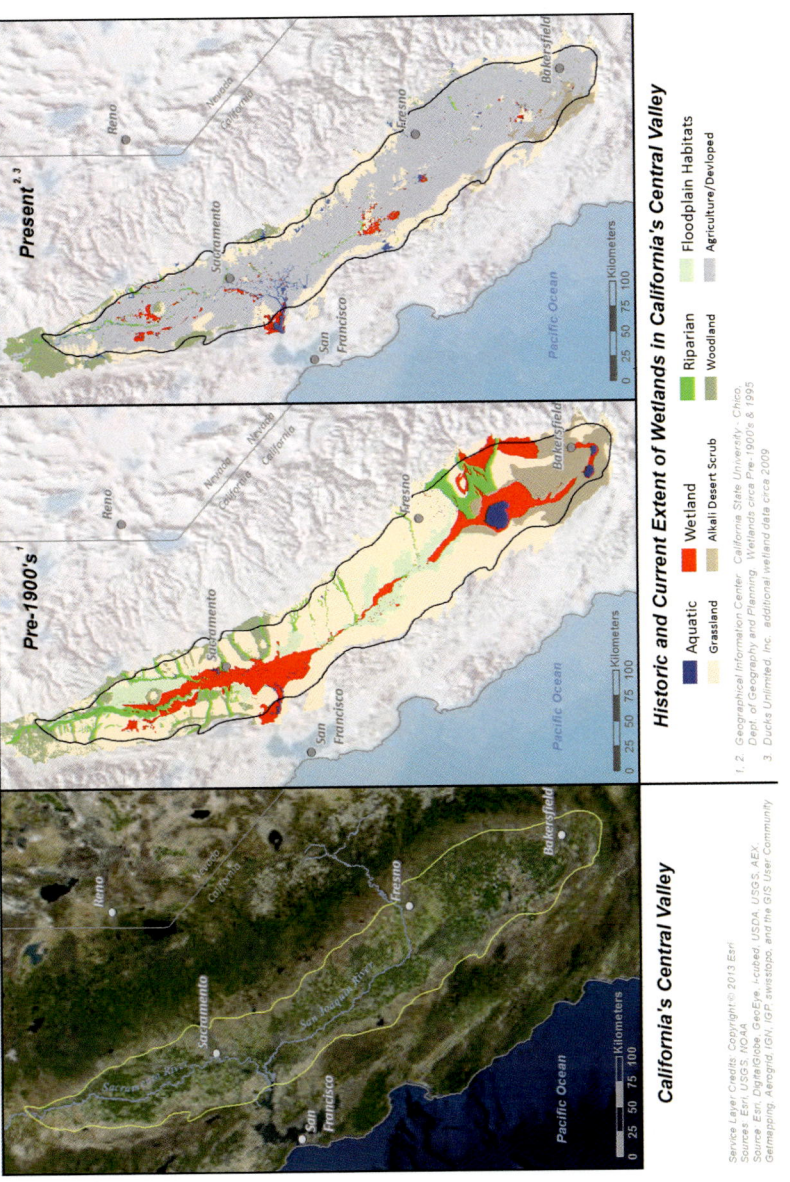

Fig. 1 California's Central Valley and historic and current extent of wetlands

Because landowners are required to repair main stem levees, agencies and conservation organizations have been reluctant to invest in wetland protection.

The Central Valley consisted of over 1.6 M ha of wetlands in the 1850s (Fig. 1), but total area had fallen to 321,740 ha by 1939. By the mid-1980s the wetland and deepwater acreage had fallen to 220,160 ha, a net loss of 32% in the 46-year period (Frayer et al. 1989). There were 227,000 ha of freshwater wetlands in 1939 and 129,100 ha in the mid-1980s, a net loss of 97,900 ha. Average annual net loss was over 2,100 ha. Almost all of the net loss came from freshwater emergent wetlands, and most of this loss consists of 93,100 ha converted to agricultural crops other than rice (Frayer et al. 1989). Satellite imagery implied that even less than 105,220 ha remained between 1986 and 1989 (Kempka et al. 1991), and many remaining wetlands were hydrologically degraded. The 2000 *Central Valley Wetland Water Supply Investigations: CVPIA 3406(d)(6)(A,B): A Report to Congress* (USFWS 2000) stated that there were only 67,113 ha of managed wetland acres in November 1996 (Kempka et al. 1997) Updating satellite data with restored wetland numbers, just 83,188 ha of managed wetlands were present in 2003 (CVJV 2006).

Wetland Types

The vast majority of Central Valley wetlands are seasonally flooded palustrine emergents (for descriptions, see Cowardin et al. 1979) which account for over 90% of all wetlands. The second most dominant type of wetland is the semipermanently flooded palustrine emergents. Estuarine emergents include seasonally flooded, semipermanently flooded, and tidal, but are restricted to the Suisun Estuary at the juncture of San Francisco Bay and western portions of the Sacramento-San Joaquin Delta. Riparian woody wetlands are restricted to the major rivers or tributaries, especially the Sacramento and Cosumnes rivers. Seasonal wetlands are typically flooded in the fall or winter, with drawdown occurring between March and May. Semipermanent wetlands are usually flooded from early fall through early July.

Hydrologic Modifier for Seasonal Wetlands

Most wetlands are managed as seasonally flooded, summer water, or permanently flooded. The former hydrologic modifier is the most common on both private and public wetlands. The later types are mostly found on public wetlands.

Seasonal wetlands are flooded from early fall through late winter or spring. Late-winter flooding or even spring flooding occurs on some wetlands within a complex to encourage vegetation diversity. Late-winter or early-spring drawdowns (January–March) encourage germination of dock, aster, and smartweeds. Drawdowns in April and May encourage germination of swamp timothy,

Table 1 Distribution of managed seasonal and semipermanent wetland areas in the Sacramento and San Joaquin Valley basins and basins connecting California's Central Valley to San Francisco Bay (Adapted from CVJV 2006)

Basin	Total wetland area	Seasonal wetland area	% private/public	Semipermanent wetland area	% private/public
Sacramento Valley basins					
American	1,517 ha	1,290 ha	100/0%	227 ha	100/0%
Butte	11,113 ha	9,446 ha	69.3/30.7%	1,667 ha	69.2/30.8%
Colusa	10,660 ha	9,061 ha	49.5/50.5%	1,559 ha	49.5/50.5%
Sutter	928 ha	789 ha	12.7/87.3%	139 ha	12.5/87.5%
Yolo	4,075 ha	3,463 ha	67.8/32.2%	612 ha	67.9/32.1%
San Joaquin Valley basins					
San Joaquin	27,435 ha	24,692 ha	76.8/23.2%	2,743 ha	76.8/23.2%
Tulare	9,088 ha	8,180 ha	33.2/66.8%	908 ha	33.2/66.8%
Basins connecting Central Valley to San Francisco Bay					
Delta	3,023 ha	2,569 ha	58.9/41.1%	454 ha	58.9/41.1%
Suisun	15,346 ha	13,044 ha	78.7/21.3%	2,302 ha	78.7/21.3%
All	83,185 ha	72,534 ha	66.5/33.5%	10,611 ha	66.6/33.3%

watergrass, and sprangletop. Drawdowns in May and June encourage cattail and tule growth and germination of cocklebur and alkali bulrush (Heitmeyer et al. 1989). Summer water management wetlands are flooded from June through March. Drawdowns in March encourage moist-soil plants, but dense stands of emergent may develop. Permanently flooded ponds hold water year-round at depths between 1 and 3 m. Periodic physical disturbance is necessary in these dense emergent stands.

Location and Ownership of Wetlands

Within the Central Valley, approximately 87.2% of managed wetland areas in 2003 were seasonal (66.5% private; 33.5% public) and 12.8% semipermanent (66.6% private; 33.3% public) (Table 1). Approximately two-thirds of all managed wetlands in the Central Valley are privately owned, most as duck clubs, and nearly 90% of all wetlands are managed on a seasonal hydrology. Sacramento Valley's five basins account for 34% of Central Valley wetland area and the San Joaquin Basin alone contains a third of the wetland areas in the Central Valley, most within the Grassland Resource Conservation District. The Suisun Basin has been called the largest single estuary on the US Pacific Coast and with the Delta Basin connects the Central Valley to the San Francisco Bay estuarine system. Seventy-seven percent of all wetland area is located in four basins: Butte, Colusa, Suisun, and San Joaquin (CVJV 2006).

Waterfowl Use

The Central Valley of California supports one of the largest concentrations of wintering waterfowl in the world despite having lost over 90% of its historic wetlands (Heitmeyer et al. 1989; Fleskes 2012). Approximately 60% of all waterfowl in the Pacific Flyway winter in the Central Valley, with a third or more of North America's northern pintail *Anas acuta*, and almost all the continental population of tule white-fronted geese *Anser albifrons elgasi* and Aleutian Canada geese *Branta canadensis leucopareia* wintering in the region (Gilmer et al. 1982). Approximately five million ducks and two million geese now winter in the Central Valley each year (Petrie et al. 2016). Birds begin arriving in the Central Valley as early as mid-August, with peak numbers occurring in mid-December to early January. Most birds have left the region by late March as they depart for their breeding grounds in Canada and Alaska (Central Valley Joint Venture Plan 2006). Most waterfowl that winter in the Central Valley rely on a mixture of natural and agricultural food resources to meet their energy needs. These foods are primarily obtained from managed seasonal wetlands and from harvested rice fields that are "winter flooded" to decompose rice straw and provide waterfowl habitat (Spell et al. 1995; Fleskes et al. 2005).

Threats and Future Challenges

Securing adequate surface water supplies for wetlands is the most significant challenge facing public and private wetland managers in the Central Valley. Most public and privately owned wetlands in the Central Valley obtain their water from the Central Valley Project, or CVP. The CVP is a federal water project that includes 20 dams and reservoirs, 11 power plants, and 500 miles of major canals. Each year, the CVP delivers about 2.3 million gallons of water for agricultural, municipal, and wildlife use (Petrik et al. 2012, 2015). Unfortunately, water supplies in the Central Valley are being increasingly constrained. Urban growth, agriculture, and the needs of endangered fish species are taxing CVP water supplies even during years of normal precipitation (Fleskes 2012). Water shortages, and the effects on wetlands, have been highlighted during the state's recent drought.

References

Central Valley Joint Venture. Central Valley Joint Venture implementation plan – conserving bird habitat. Sacramento: U.S. Fish and Wildlife Service; 2006. p. 261.

Cowardin LM, Carter V, Golet FC, Laroe ET. Classification of wetlands and deepwater habitats of the United States. Washington, DC: U.S. Dept Interior, FWS, Biological Services Program; 1979. p. 103. FWS/085-79/31.

Fleskes JP. Wetlands of the Central Valley of California and Klamath Basin. In: Batzer D, Baldwin A, editors. Wetland habitats of North America: ecology and conservation concerns. Berkeley: University of California Press; 2012. p. 357–70.

Fleskes JP, Perry WM, Yee JL, Petrik KL, Reid FA. Change in area of winter-flooded and dry rice in the northern Central Valley of California determined by satellite imagery. Calif Fish Game. 2005;91:207–15.

Frayer WE, Peters DD, Pywell HR. Wetlands of the California Central Valley: status and trends 1939 to mid-1980's. Portland: U.S. Fish and Wildlife Service, Region 1; 1989. p. 28.

Gilmer DS, Miller MR, Bauer RD, LeDonne JR. California's Central Valley wintering waterfowl: concerns and challenges. Trans North Am Wildl Nat Resour Conf. 1982;47:441–52.

Heitmeyer ME, Connelly DP, Pederson RL. The central, imperial, and Coachella Valleys of California. In: Smith LM, Pederson RL, Kaminski RM, editors. Habitat management for migrating and wintering waterfowl in North America. Lubbock: Texas Tech University Press; 1989. p. 475–505.

Kempka RG, Kollasch RP, Olson JD. Aerial techniques measures shrinking waterfowl habitat. Eugene: Geographic Information Systems; 1991. p. 48–52.

Kempka RG, Lewis K, Flint S, Shaffer B, Spell R, Lewis A, Cundiff-Gee M, Clemons S. California wetland and riparian geographic information system project. Sacramento: Ducks Unlimited; 1997. p. 37. Final Report.

Petrie M, Biddlecomb, Reid FA, Smith M. Waterfowl and the California drought. Ducks Unlimited Magazine, Sept–Oct 2015. p. 52–4.

Petrie M, Fleskes J, Wolder M, Isola C, Yarris G, Skalos D. The impact of the California drought on wintering and waterfowl habitat and food resources in the Central Valley. J Fish Wildl Manage. 2016 (In review).

Petrik KL, Carter C, Garcia F. The rice resiliency project: the Sacramento Valley ricelands spatial and economic analysis. Sacramento/Davis: Ducks Unlimited/University of California; 2012. p. 73. Final Report.

Scott LB, Marquiss SK. A historical overview of the Sacramento River. In: Warner RE, Hendrix KM, editors. California riparian systems. Berkeley: University of California; 1984. p. 51–7.

Spell R, Kempka R, Reid FA. Evaluation of winter flooding of ricelands in the Central Valley of California using satellite imagery. In: Campbell KL, editor. Versatility of wetlands in the agricultural landscape. Tampa: American Society of Agricultural Engineers; 1995. p. 357–66.

U.S. Fish and Wildlife Service. Central Valley wetlands water supply investigations: CVPIA 3406 (d)(6)(A, B): a report to congress: final report. Portland: USFWS; 2000.

The Everglades (USA)

56

Curtis J. Richardson

Contents

Introduction	706
Climate	709
Hydrology	710
Nutrients	711
Carbon Storage and Accumulation	714
Vegetation and Plant Communities	714
Sawgrass	715
Wet Prairies	715
Sloughs	716
Ponds	718
Tree Islands (Bayhead/Swamp Forests)	718
Willow Heads, Cypress Forests, Pond Apple Forests, and Hardwood Upland Hammocks	719
Periphyton	719
Future Challenges: Water Management Planning	720
References	721

Abstract

The Everglades is the largest subtropical wetland in the United States. It has been designated an International Biosphere Reserve, a World Heritage Site, and a Wetland of International Importance, in recognition of its significance to all the

---

This chapter is an updated and extracted version of some findings presented in *The Everglades Experiments: Lessons for Restoration* published by Springer in 2008 and authored by C. J. Richardson and *The Everglades: North America's subtropical wetland* authored by C. J. Richardson and published 2010 in Springer's journal on Wetlands Ecology and Management.

C. J. Richardson (✉)
Nicholas School of the Environment, Duke University Wetland Center, Durham, NC, USA
e-mail: curtr12@gmail.com; curtr@duke.edu

people of the world. However, the Everglades have undergone radical changes in both water flow and water quality over the years as the population in the state of Florida has exploded and agricultural lands have increased significantly during the past century. As a result the Florida Everglades, a peat based fen have been significantly reduced in size due to massive water drainage programs to convert these areas mainly to agriculture lands or urban areas.Together the US government and the state of Florida have spent several billion dollars to restore the water supply and ecohydrology for the remaining 50% of the Everglades, which includes native Seminole Indian Reservations. Both governments face enormous social-economic and political difficulties regarding the future allocation of water for the Everglades as the demand for water for agriculture and urban areas grows. This chapter compares and contrasts the past and current ecological conditions in the marshes, outlines the hydrologic issues facing these wetlands today as well as reviews some of the proposed solutions.

Introduction

The Everglades, originally called Pa-hay-okee ("grassy lake") by the resident Native Americans, is a 700,000 ha subtropical alkaline fen whose origin dates to around 5000 BP when the rate of sea level rise slowed and peat began to accumulate in the shallow embayment of south Florida (Gleason and Stone 1994; Porter and Porter 2002; Lodge 2005; Richardson 2008; Glaser et al 2013). The Everglades was an almost impenetrable wall of sawgrass "plains" and reptile-infested waters according to the early Spanish and American explorers (Ives 1856). The Everglades was saved from development by Marjory Stoneman Douglas's seminal 1947 book, *The Everglades: River of Grass,* which also helped establish the Everglades National Park (ENP) and prevented the Everglades from being totally drained, although vast areas of it have been destroyed over the past 100 years. Currently, 30% of the original 1,036,000-ha Everglades has been converted to agricultural and urban development, and 350,000 ha of the original area are now under state ownership as Water Conservation Areas (WCAs) 1, 2, and 3 for "flood protection, water supply, and allied purposes of navigation and fish and wildlife protection" as mandated by the 1948 US Congressional Flood Control Act (Fig. 1, Plate 1). The remaining 565,000 ha comprise the ENP. Historically, the Everglades extended from just south of Lake Okeechobee, now the Everglades Agricultural Area (EAA), to Florida Bay (26°57′N to 24°53′N) where the ENP is found. The current longitudinal width of the Everglades, although greatly narrowed by coastal development, ranges from 81°37′W to 80°13′W (Fig. 1).

The ENP is the largest federally owned peatland in the lower 48 states and is the only subtropical wetland ecosystem in the USA that is listed under the Ramsar Convention on Wetlands of International Importance. Because of its size, floral and faunal diversity, geological history, and hydrological functions on the Florida landscape, the Everglades are considered by many ecologists and conservationists to be the "sentinel wetland ecosystem" for testing the American government's resolve to

56 The Everglades (USA)

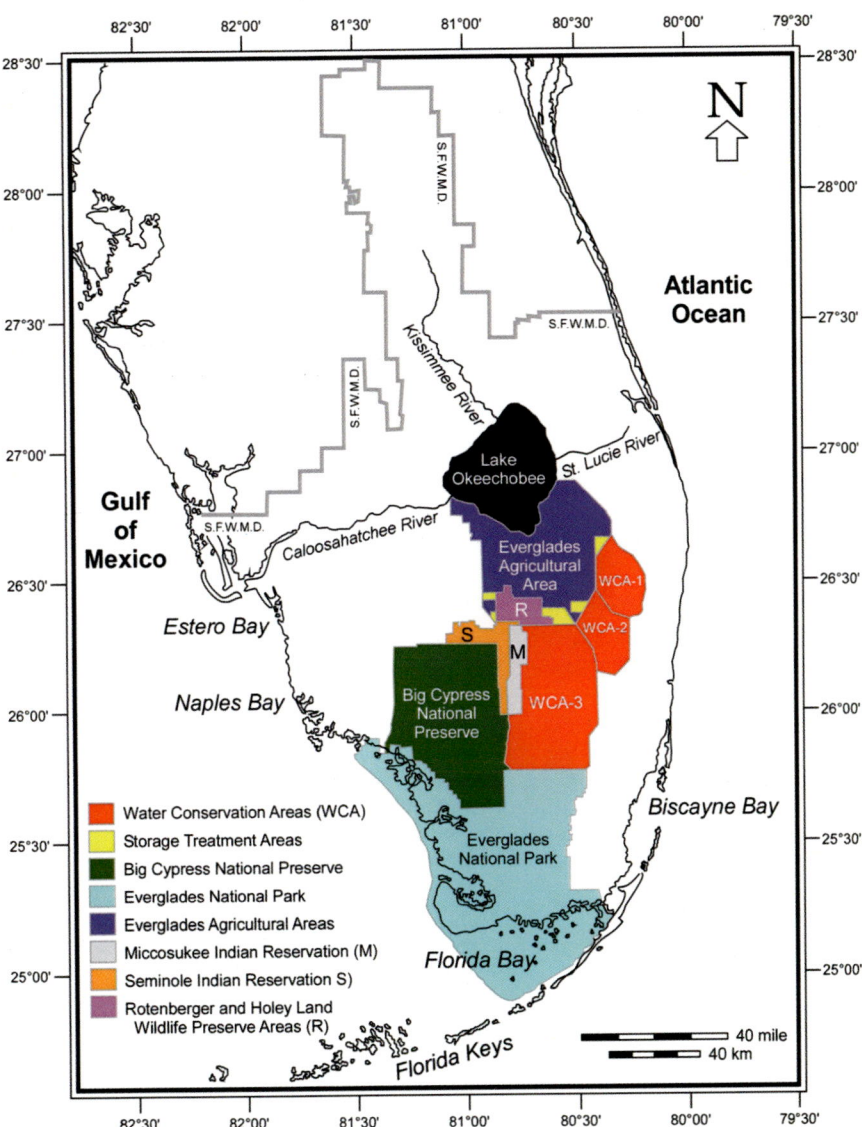

Fig. 1 A map of the current boundaries of the South Florida Water Management District (SFWMD) that shows the Kissimmee River, Lake Okeechobee, and Everglades land use complex. The Everglades is now divided into the state owned Water Conservation Areas (WCAs), Storage Treatment Areas (STAs), the Everglades National Park (ENP), and Everglades Agricultural Area (EAA). The surrounding areas are the Big Cypress National Preserve, Florida Bay, and the developed crop and urban land, a large part of which was former Everglades. The major canals and, importantly, two Indian Reservations are shown to be part of the current Everglades. (From Richardson (2010). A satellite view of these areas is shown on Plate 1.)

Plate 1 Southern Florida, Lake Okeechobee, and the Everglades visible from a Visible Earth NASA photo in 2007. The Water Conservation Areas and the Everglades Agricultural Area are clearly shown in green and brown shades just south of the lake. Note the high density of municipal areas along the east coast of Florida. A key for the areas is shown on Fig. 1

restore and maintain vast wetland areas under ever increasing urban land development pressures, agricultural runoff, and highly regulated and ever changing water management regimes (Richardson 2008; Stokstad 2008; Sandoval et al. 2016).

The Everglades, with its mosaic of wetland communities, is often referred to as a marsh or swamp; however, it is correctly identified as a fen peatland or mire by wetland ecologists (Richardson 2000; Rydin and Jeglum 2006; Richardson 2008). That is not to say that the Everglades did not start out nearly 5,000 years ago as a marsh or that some areas today have marsh habitats or tree-covered islands. However, the overall wetland complex is dominated by peat-based soils that historically formed under natural peatland hydrodynamics not present in many areas today due to extensive canal and dike systems (>2,000 km). The classification of fen or alkaline mire for the Everglades is important when one considers how different marshes and swamps are from mires in terms of their hydrologic controls, biogeochemistry, rates of peat accretion, plant and animal communities, and successional and geomorphologic development. The hydrological differences alone would greatly manifest themselves in any attempt to restore native communities and animal habitats with the wrong eco hydrologic model. Unfortunately, the terms "Everglades mire" or "peatland" by themselves do not reveal the vital and multifaceted hydrologic connections and nutrient sources that historically existed between the Everglades and surface water runoff coming from Lake Okeechobee via the Kissimmee River, the close connections of groundwater and surface waters in the region due to the karst limestone underlying the wetlands, and most importantly the seasonal influence of the key water source – rainfall on vegetatin patterns (Parker et al. 1955; Richardson 2008; Harvey and McCormick 2009; Sandoval et al. 2016).

Climate

The subtropical climate of south Florida has hot humid summers, mild winters, and a distinct wet season with 80% of the rainfall falling from mid-May through October (Richardson 2010). Harvey and McCormick (2009) report that 81% of the pre-drainage water budget for the Everglades was from rainfall, with 8% coming from Lake Okeechobee overflow, 10% from marginal runoff, and only 1% coming from groundwater. The Everglades has more in common with tropical climates in that a wet/dry season is probably more important to vegetation composition than winter/summer differences in temperature. Daily temperatures average above 27 °C from April through October in the northern part of the Everglades and from March to November in the south, but freezing temperatures do occasionally occur. The key component of climate controlling vegetation patterns and succession is the amount of precipitation. A 110-year weighted average analysis of annual rainfall over south Florida (1895–2005) shows distinct drought and heavy rainfall periods when compared to the long-term average annual rainfall of 1,320 mm per year. Evapotranspiration (ET) is also an extremely important component of the Everglades. It has been estimated that 70–100% of rainfall exits the Everglades in this way (Richardson 2008). Evapotranspiration was also recently found to be the most important driver of hydrologic flushing times in the southern Everglades (ENP). In a recent study, Sandoval et al. (2016) reported when ET was less than inflow rates even after new restoration efforts it resulted in longer flushing times for water in the southern Taylor Slough in the ENP.

These weather patterns, when combined with effects of dikes and canal drainage, have resulted in severe drying and flooding of portions of the Everglades with a resultant shift in plant communities. Annual rainfall is the main driver of hydrology, but hurricanes (sustained winds of 120 km hr^{-1}) are also an important reoccurring event ($\simeq$ every 3 years) in south Florida. Thus, extreme hydrologic events like hurricanes and droughts have also had significant effects on the water budgets for south Florida and the Everglades.

Hydrology

The role Lake Okeechobee played in supplying water to the Everglades was initially not well understood (Fig. 1). Historically, lake levels in excess of 6 m were measured in the lake in the 1850s and as late as the early 1900s, and it was reported that when lake levels exceeded 6 m water would spill over the soil bank on the southern part of the lake into the Everglades (Richardson 2010). The shallow elevation gradient of 1.57–3.16 cm km^{-1} coupled with deep overlying peat and dense native sawgrass allowed for storage of water during wet periods, slow water flow averaging 0.25 cm s^{-1}, and a gradual release of excess water during dry periods (Romanowicz and Richardson 2008). The importance of surface and ground water interactions in the Everglades was not really appreciated until the USGS report by Parker et al. (1955), who detailed studies on surface and groundwater flows and storage. Parker clearly showed for the first time the complexities of the hydrologic system that controlled the Everglades and that the extensive canal and dike system installed since the early 1900s had significantly altered water storage, surface and groundwater interactions, flow of water, and water depths throughout the Everglades.

The effects of dramatic shifts in the water flow at the landscape scale due to development of canals and ditches can be easily appreciated by comparing flows under natural conditions and current water management plans using a Minard-type diagram of the historic surface and groundwater flows (Richardson 2010) – see Fig. 2. In the diagram, the width of the lines shows the amount of water flowing along key points; the direction of flow is shown as well. Under historic conditions of no canals or dikes, a balanced and similar annual volume (~1,481 × 10^6 m^3) of water was found leaving the Kissimmee Basin flowing into the Everglades National Park (ENP). Historically, on average only 503 × 10^6 m^3 of water left Lake Okeechobee annually because of high ET rates in the lake coupled with restricted flow south due to a natural soil berm, dense sawgrass, and no direct outlets to the Caloosahatchee or St. Lucie Rivers (Fig. 1). The Central glades had approximately 1.49 × 10^6 m^3 of water, of which approximately half or 814 × 10^6 m^3 exited the Everglades to the Lower East Coast (LEC) yearly (Fig. 2). The historical total discharge for the LEC to the Atlantic was estimated to be 1,987 × 10^6 m^3 per year. By 1994 the annual Everglades water budget was highly regulated, and LEC flows dramatically doubled to 4,579 × 10^6 m^3 as freshwater water was being transported to the Atlantic Ocean via a complex series of canals and pumping stations at the expense of flows into the ENP (Fig. 2). Importantly, water inputs into the ENP were less than half of historic inputs. However,

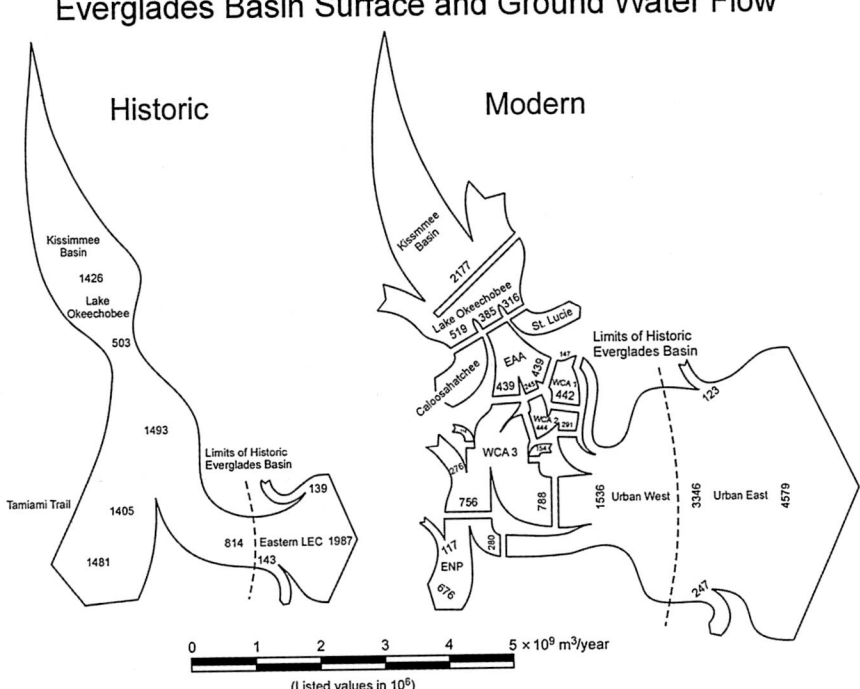

Fig. 2 Minard-type graphic of the historic (before 1880) and modern average annual water flows (based on 1993 SFWMD LEC Report data by Larsen, in Richardson 2008). The line widths are proportional to the volume of the water flows. Values are given as 10^6 m^3 per year

increased flow of freshwater from new restoration efforts during the dry season in the southern Everglades (ENP) resulted in an increase in ions, especially N and P particularly during the dry season, but at coastal sites increased inputs of fresh water decreased the inputs of brackish water from groundwater and Florida Bay resulting in a diminution of Na, Cl, and N and P concentrations flowing back into the wetlands from the Gulf (Sandoval et al. 2016).

Nutrients

Historically, the Everglades was a P limited ecosystem, which survived on nutrients primarily from rainfall, limited surface flow, and recycling within the system, especially after fire (Davis 1943; Richardson 2008). In this P-limited system, plants and algal species have evolved that can survive under TP water concentrations as low as 5–10 µg/L (Richardson and Vaithiyanathan 1995; Richardson, et al. 1999). The exception to communities evolving under low P concentrations are tree-covered islands and the vegetation around alligator holes (Davis 1943; Sklar and van der Valk 2002) as well as plant communities adjacent to Lake Okeechobee with its high historical TP concentrations >30 µg.L^{-1} (Richardson 2008). Another factor maintaining P limitations in the Everglades, unlike northern mires or marshes, is the nitrogen-fixing blue-green algae community or periphyton, found in open-water sloughs. Because of the periphyton community's high rates of nitrogen-fixation, Everglades soils are exceptionally high in nitrogen (2–4% by weight, Craft and Richardson 2008); thus, very high N: P ratios (>100) exist, further driving the system to severe P limitations (Richardson et al. 1999). While multiple studies have identified P as the primary driver of *Typha* (cattail) invasions there is no question that the cutting of canals deep into the limestone bedrock and the diking and creation of water impoundments in the northern Everglades, which began in earnest in the 1950s changed the water hardness (e.g., $CaCO_3$) and in turn community plant structure (Waters et al. 2013). The canal expansion also increased marl deposition and increased calcareous periphyton abundance as early as the 1920s according to paleoecological studies by Cooper et al. (2008) and Waters et al. (2013). This increase in calcareous periphyton before agricultural expansion and creation of impoundments in areas like WCA-2 suggest canal-derived calcium inputs and to some extent early drainage followed by deeper water retention played an important role in initiating plant, algal, and microbial community changes (Richardson 2008; Waters et al. 2013).

However, agricultural runoff from the Everglades Agricultural Area (EAA) and Lake Okeechobee (Fig. 1) significantly changed the nutrient inputs and balance in the Everglades after the 1970s because both contributed water with much higher concentrations of N and P than is typically found in rainfall and historic runoff in the Everglades (Craft and Richardson 1993; Davis and Ogden 1994; Richardson 2008). Reddy et al. (2011) provides an excellent review of phosphate loading and cycling in the Everglades and concludes that 400,000 metric tons of P is stored in the surface sediments and flocculent of lakes, rivers, and soils of the Everglades and Lake Okeechobee. Nineteen percent of this total is stored in the Everglades itself with 35% in a nonreactive form and 65% is a reactive state. A proportion (10–25%) of this phosphate leaks from the system each year (Reddy et al. 2011).

The average P concentration in water leaving the EAA farmland in the early 1990s was 150 µg.L^{-1} P, which was reduced to 115 µg.L^{-1} P in the canals and edges of WCA-1 (Richardson 2008). However, by the time surface waters reached the structures above the Everglades National Park, concentrations were 10 µg.L^{-1} P or lower. The dumping of agricultural wastewater into the WCAs and using them as a sink for excess nutrients only accomplished this reduction. The result was thousands of hectares of cattail-dominated areas of the northern Everglades with high P levels in vegetation, soils, and surface waters (Craft and Richardson 1997; Richardson 2008, Plate 2). Thus, the control of cattail expansion and community shifts now depends on best management practices (BMP) in the federally mandated regulatory P-reduction program. The major hope for reducing P loads into the Everglades is the use of Stormwater Treatment Areas (STAs) to treat EAA, upstream and Lake Okeechobee waters prior to their release. To date, six STAs covering over 16,564 ha have been built, the earliest in operation since 1994–1995 (Fig. 1). In terms of P reductions, both the BMPs and the STAs have resulted in a significant

Plate 2 A dense stand of cattail *Typha domingensis* in a heavily P enriched area in northern WCA-2A. Note: Two other invasive species water lettuce *Pistia stratiotes* and water hyacinth *Eichhornia crassipes* are shown in the foreground (From Richardson 2008)

decrease of P to the Everglades. However, EAA outflow P concentrations continue to remain high, and STA reductions have not consistently reached the low safe threshold concentrations of 10–15 µg.L^{-1} P that had been hoped for by many scientists. While P mass loadings are significantly reduced by more than 50% for the EAA, P concentrations remain too high for major improvements in the receiving waters. For example, in a 16-year study (1995–2011) Chen et al. (2012) reported that the STAs did remove 1,500 metric tons of TP from 13.6 billion m^3 of stormwater runoff from farmland and adjacent drainage basins. They found stormwater inflows to the STAs had an annual TP concentration of 143 ± 62 µg P L^{-1} and an annual phosphorus loading rate of 1.56 ± 0.91 g P m^{-2}. Unfortunately, many of the STAs exceeded their TP loading threshold of approximately 1 g P m^{-2} yr^{-1} (Richardson et al. 1997) and released annual TP concentrations of 41 ± 31 µg P L^{-1}, which is far in excess of the Everglades Forever Act national pollutant discharge elimination requirements (Germain and Pietro 2011). If the present trend continues and no additional STAs are built, the Everglades will continue to receive unacceptable concentrations and loads of P for the foreseeable future. However, the recent purchase of thousands of ha of farmland south of Lake Okeechobee provides a real opportunity to provide further nutrient reductions to help meet the established USEPA criterion of 10 µg.L^{-1} P for the Everglades by expanding the size of the STAs and developing a series of reservoirs to help store water and remove pollutants (Stokstad 2008). This plan could have significant positive consequences for the native plant communities and ecosystem structure and function if the hydrologic regime is properly restored.

Carbon Storage and Accumulation

In brief, multiple studies have shown that rates of carbon (C) storage are very low in the Everglades (Craft and Richardson 1993, 2008). Cesium-137 and Lead-210 studies have indicated that C has accumulated on average at 98 ± 13 g Cm^{-2} yr^{-1} in unenriched areas and nearly double that in P enriched areas over timespans of 26 and 100 years, respectively. These are not sustainable according to longer term Carbon-14 estimates which suggest over the past 4,600 years that C accumulation rates are closer to 12 g Cm^{-2} yr^{-1} with a sediment accretion rate of only 0.2 mm yr^{-1} in the ENP (Glaser et al. 2012). However, peat thickness decreases from north to south in the Everglades with northern areas averaging 3 m in the Loxahatchee to less than 0.5 m in the ENP (Craft and Richardson 2008). Rates of peat accretion also vary considerably depending on the timespan considered. For example, a comparison of accretion rates measured with Cesium-137 (26 years), Lead-210 (100–125 years), and Carbon-14 (5000 years) reveals mean peat accretion rates of 2, 1.6, and 0.5 mm yr^{-1}, respectively. Thus, peat accumulates at different rates in the Everglades related to nutrient and hydrologic conditions and very slowly over long time periods due to seasonal dry periods, droughts, and fire, which increases C decomposition and GHG loses via CO_2 and CH_4 efflux (Qualls and Richardson 2008).

Vegetation and Plant Communities

The successional dynamics of the Everglades are mainly controlled by the interaction of climatic patterns (droughts and rainfall) and human alterations on hydroperiod, which in turn influences fire frequency and the degree of fire intensity as well as the transfer and release of nutrients from agricultural runoff, especially P, on the landscape (Richardson 2008). In the Everglades, the dominant sawgrass communities growing typically on 1–3 m of peat are thought to be resilient to a wider range of inundation durations and depths than other species, although prolonged periods of flooding cause reduced growth (Richardson 2008). However, other communities such as pine savanna, red mangrove scrub, Bay-hardwood, and muhly grass have more pronounced physiological limitations to inundation depth and/or duration and are more restricted in their distribution (Richardson 2008; Todd et al. 2010). Within the context of more extreme cycling between wet and dry conditions, this suggests that these latter plant communities may undergo a contraction of spatial extent under future climate scenarios. With the impending sea level rise due to global climate change projected to range from 19 to 59 cm or go even higher with significant ice melt by 2090 (IPCC 2007), saltwater becomes even more of a factor as it invades further and further into the southern Everglades and alters freshwater communities (Richardson 2008). A sudden change from freshwater to saltwater conditions may accelerate oxidization of organic substrate leaving large areas of thin substrate or bare limestone bedrock with a greatly reduced potential for plant community shifts in response to climate change (Pearlstine et al. 2010; Willard and Bernhardt 2011).

A complete listing of the plant species characteristic of each vegetation community in the Everglades is given in detail in Richardson (2008). An abbreviated summary of the key ecological components of the major plant community types found in the Everglades fen is presented in the following sections. The common local names for the community types or habitats are used for comparative purposes. These habitats and their spatial distribution make the Everglades one of the more diverse peatland ecosystems in the world as shown by the historic vegetation map done by Davis in 1943 (Fig. 3).

Sawgrass

Sawgrass *Cladium jamaicense* is the dominant fen vegetation community found throughout the freshwater Everglades peatland. Sawgrass (fen-sedge) grows to 2–3 m in height on deep peat but only 0.5 m on shallow peat. It prefers sites with a fairly constant water depth of 10–20 cm (Gunderson and Loftus 1993). Its presence in the Everglades is due to its ability to survive fire, low soil nutrient content, and occasional freezing (Richardson 2008). The current diking and flooding in the Everglades has resulted in the loss of this community due to deep and fluctuating water levels. Sawgrass occurs either in almost pure stands or mixed with a wide variety of other plants, e.g., bulltongue *Sagittaria lancifolia,* maidencane *Panicum hemitomon,* pickerelweed *Pontederia cordata*, or cattail *Typha* spp (Plate 4). Davis et al. (1994) estimated that pure sawgrass-dominated areas make up only 38% of 417,000 ha of historic sawgrass-dominated areas.

Wet Prairies

Wet prairies are among the common vegetation types in the northern Everglades. Often referred to as "flats," these freshwater communities are characterized by low stature and emergent plant species, and they are found in the northern and central Everglades in conjunction with tree islands (Plate 3). Wet prairies exist on both peat and marl soil. The wet prairies in the south found on calcitic mud or marl occur on higher and drier sites but are wet 3–7 months of the year (Davis 1943; Gunderson and Loftus 1993). The water depth of these areas is generally less than sloughs but deeper than sawgrass; thus, the vegetation seldom burns. Three well-defined wet prairie associations occur in the Everglades: (1) *Rhynchospora* flats, (2) *Panicum* flats, and (3) *Eleocharis* flats. However, many other plant species may also be present on these flats, depending upon hydrological conditions, the season of the year, and soil type. Wet prairies usually dry out on an annual basis and are transition zones between sawgrass areas and sloughs.

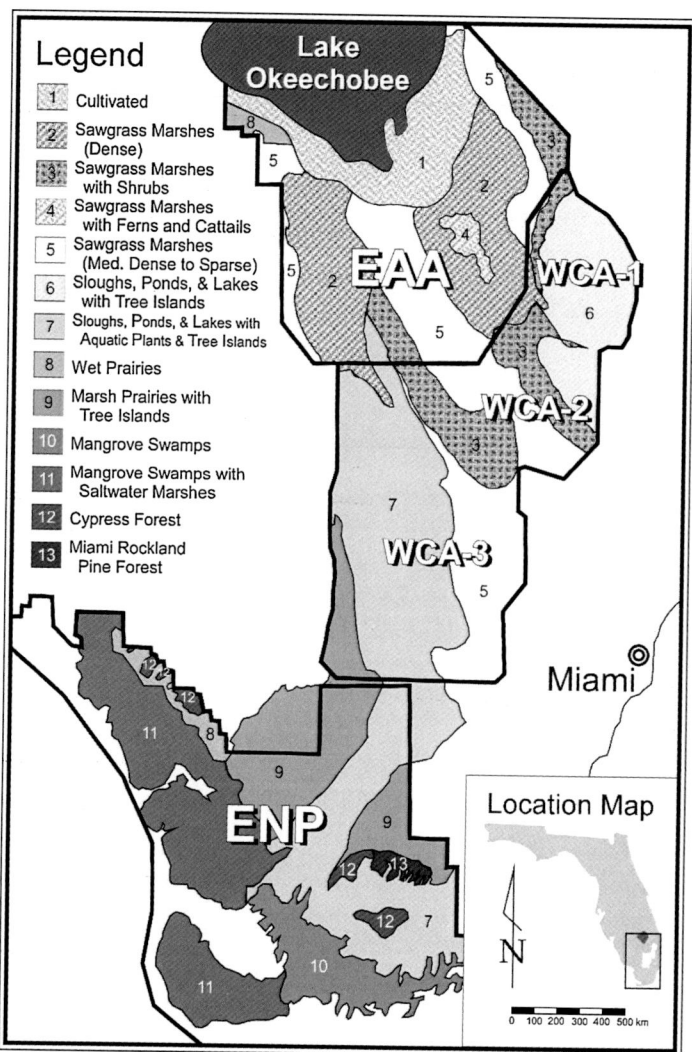

Fig. 3 Historic map showing the diversity of vegetation communities in the Everglades based on the map of J. H. Davis (1943). The map has been redrawn and simplified from the original map, and the boundaries of the current water conservation areas (WCA-1,WCA-2,WCA-3), the Everglades-National Park (ENP), and the Everglades Agricultural Area (EAA) have been added (From Richardson 2008)

Sloughs

Sloughs are among the most widespread community types in the Everglades (Plate 4). Aquatic sloughs represent the lowest elevation of the Everglades ecosystem, except for ponds. They have deep water levels averaging 30 cm annually and longer

Plate 3 Wet prairie ("often called grass flats") is a common vegetation type in the Everglades. Note the tree islands in the background (Photo credit: C.J. Richardson 2015 © Rights remain with the author)

inundation periods than other Everglades wetland communities (Gunderson and Loftus 1993). The largest pond-slough systems occur in the ENP. Sloughs are narrow drainage channels that are water-filled, or at least wet, most of the year. The "valleys" of these channels average only a few cm to 60 cm below the elevation of adjacent sawgrass areas. Not as extensive as they once were, some sloughs apparently have been replaced by either sawgrass or wax myrtle and willow stands. Cattail has also filled many of the sloughs in the natural enriched areas of the northern Everglades (Craft and Richardson 1997; Richardson 2008). Sloughs are easily recognized by their water drainage patterns and by characteristic plant species, such as white waterlily *Nymphaea odorata,* floating hearts *Nymphoides peltata,* bladderworts *Utricularia* spp., spikerushes *Eleocharis* spp., spatterdock *Nuphar lutea*, or water hyssop *Bacopa caroliniana* (Davis 1943; Gunderson and Loftus 1993).

Sloughs and wet prairies are ecologically important in the Everglades landscape. During the dry season, sloughs serve as important feeding areas and habitats for Everglades wildlife. As the higher elevation wet prairies dry out, sloughs provide refuge for aquatic invertebrates and fish. This high concentration of aquatic life, in turn, makes sloughs important feeding areas for Everglades wading bird populations. When the marsh is reflooded, the animals that have survived in the sloughs repopulate the fen as water level rises (Richardson 2008). The plant species diversity tends to be higher in sloughs and wet prairie communities

Plate 4 Sloughs or open-water areas found primarily in the northeast and south-central Everglades. Note the water lily plants (*Nymphaea odorata*) in the slough surrounded by a monoculture of sawgrass (Photo credit: C.J. Richardson 2015 © Rights remain with the author)

than in pure sawgrass and cattail communities (Craft et al. 1995). The abundance of macroinvertebrates, fish, and wading birds is also higher in sloughs than in sawgrass and cattail communities (Davis and Ogden 1994).

Ponds

Ponds are small, open-water areas that are scattered throughout most of the Everglades and represent the deepest water regime. They occur in bedrock depressions where fire has burned away the peat. Alligator activity often maintains open water in the ponds, which the locals call "alligator holes" for this reason. Ponds are wet except in the driest years; thus they are important habitats for animals, especially birds. These holes have borders of water lilies *Nymphaea* spp., spatterdock *Nuphar* spp., pickerelweed *Pontederia cordata*, and woody species such as Carolina willow *Salix caroliniana* or water primrose *Ludwigia peruviana* (Gunderson and Loftus 1993).

Tree Islands (Bayhead/Swamp Forests)

Everglades broadleaf, hardwood forests are locally called tree islands (Plate 3). The term refers to a variety of tree clusters that stand above a matrix of shorter vegetation.

Tree islands occur throughout the entire region but are most abundant in the central part of WCA-1 (Fig. 1, Richardson 2008). Tree islands may be either bayhead (swamp forests) or hammocks (upland forest) or a combination of the two (Davis 1943; Gunderson and Loftus 1993). Red bay *Persea borbonia*, swamp bay *Magnolia virginiana*, dahoon holly *Ilex cassine*, Carolina willow, and wax myrtle *Myrica cerifera* dominate the swamp forests. The large tree islands have a teardrop shape with the main axis paralleling the flow of water. The small islands ($\cong 100$ m^2) are usually round. The forests are found on the highest sites in the Everglades on a peat classified as Gandy peat (Davis 1943). The sites are wet from 2 to 6 months out of the year, but in drought conditions these systems are very susceptible to burning (Gunderson and Loftus 1993). The soil P nutrient content of tree islands is usually much higher ($>1,000$ mg kg^{-1} vs. 500 mg kg^{-1} of P) than the surrounding landscape (Richardson 2008).

Willow Heads, Cypress Forests, Pond Apple Forests, and Hardwood Upland Hammocks

These forest types comprise only a small area of the Everglades. They include interesting communities with distinct species. The pond apple forest *Annona glabra* historically existed primarily south of Lake Okeechobee in a band 5 km wide (Davis 1943). The land has been totally developed for agriculture, and now the species only exists in small, scattered stands. Willow heads exist throughout the Everglades in monotypic stands. They exist in fire-disturbed areas as well as around alligator holes. The upland hardwood hammocks are dominated by broadleaf hardwood trees of both temperate and tropical origin. Dominant trees include live oak *Quercus virginiana*, gumbo limbo *Bursera simaruba*, sabal palm *Sabal palmetto*, and strangler fig *Ficus aurea*. The cypress forests are found only in the southwestern Everglades and are dominant in the adjacent Big Cypress National Preserve. Pond cypress *Taxodium ascendens* are very short and occur as widely scattered individuals displaying very stunted growth. They are often called dwarf or hat rack cypress and seldom reach heights over 3–5 m.

Periphyton

Periphyton and algal mats are seldom thought of as "valuable" ecological resources or even listed among vegetation community species. Several authors, however, have pointed out an important role of the calcareous periphyton and algal mats characteristic of sloughs and wet prairies in the Everglades (Pan et al. 2000). Components of the periphyton/algal mat (especially diatoms) are high-quality food for some animals. Photosynthesis by the algae in sloughs can raise daytime dissolved oxygen concentrations and pH much higher (7.5 to >10) than in adjacent sawgrass marshes (Richardson 2008). Also, the calcareous periphyton deposits marl (calcitic mud), the second most common soil sediment type (190,000 ha) in the Everglades (Davis et al.

1994). Three types of calcareous periphyton are known in the southern Everglades: calcareous blue-green, calcareous diatom-rich, and calcareous green (Browder et al. 1994). A defining feature of all calcareous periphyton is their high inorganic component, no less than 49% by mass (Vymazal and Richardson 1995). Periphyton is abundant in areas of the Everglades that retain the historic oligotrophic conditions of the marsh. In these areas, periphyton biomass on an areal basis can reach values that are comparable to or higher than that of macrophytes.

Future Challenges: Water Management Planning

A Comprehensive Everglades Restoration Plan (CERP 2005; Richardson 2008) to store, pump, and flow massive amounts of water into the Everglades and create 87,668 ha of storage reservoirs was designed to restore more historic natural flow to the Everglades complex, reduce current flow to the Atlantic Ocean and the Gulf of Mexico, and increase water volume to the ENP without drowning tree islands in the northern and central WCAs (Kloor 2000). Highlights of the plan, when implemented, proposed flows and allocations that would result in a 20% reduction per year of Lower East Coast (LEC) losses to the Atlantic Ocean with new environmental water allocated to the ENP. Immediate concerns were that too little water was being allocated to the ENP, although under the plan more water is allocated than in the past (Kloor 2000). Another key concern was that moving extra water to the park would come at the expense of the central Everglades ecology. These areas would have to bear the increased flow, which in all likelihood would damage the tree island habitats (Richardson 2008) and lead to a loss of key species. Leading the objections were the Miccosukee tribe, who have over 100,000 ha of holdings in the central Everglades and view the tree islands as key to their hunting and ceremonies. The Miccosukee also worried that the extra water would be laden with excess nutrients (Kloor 2000).

Importantly, the original CERP plan (CERP 2005) is continually being altered following the concept of adaptive management, and in recent years more consideration is being given to recent scientific studies on hydrologic flow conditions and their effects. However, the correct timing and volumes of future water delivery schedules are not the only aspects of water delivery that need to be restored to maintain the original Everglades fen peatland complex. As mentioned earlier, peatland hydrodynamics needs to be taken into account in the management plans. If not, the normal successional patterns and development of the Everglades mire will forever be altered.

In the future, it appears the Everglades will be maintained mostly as a highly managed peatland, even with the planned removal of a number of dikes and canals. With only 50% of the original Everglades remaining and hundreds of control structures remaining in place, some say this is the only choice available. Peatland ecologists would argue that we have the opportunity with adaptive management to test alternative peatland restoration techniques and restore key components of the former Everglades. Harvey and McCormick (2009) suggest that water managers could, in addition to managing water depth, manipulate frequency and duration of input flows, which would better control flow velocities, water resident times, sediment settling, and biogeochemical transformations. Ultimately, it is essential that the variety and range of specific hydrologic, nutrient, and fire conditions that originally shaped the diversity of Everglades habitats be maintained. Sustaining a functioning Everglades for future generations will be a complex task given the myriad of user groups (farmers, municipalities, conservations, etc.) all vying for freshwater and land resources in a state with an ever increasing human population. A case in point is the March 2016 release of massive amounts of nutrient laden waters through the Caloosahatchee River to prevent several towns and agricultural fields from flooding, which resulted in Pine Island sound and Sanibel Island in the Gulf of Mexico region of Florida being subjected to eutrophic black waters that resulted in extensive algal blooms and anoxic water that killed fish and drove residents away. Waters in the past that might have flowed south to pollute the Everglades but can no longer be released due to Federal water quality standards.

References

Browder JA, Gleason PJ, Swift DR. Periphyton in the Everglades: spatial variation, environmental correlates and ecological implications. In: Davis SM, Ogden JC, editors. Everglades: the ecosystem and its restoration. Delray Beach: Lucie Press; 1994. p. 379–418.

CERP (Comprehensive Everglades Restoration Plan). 2005 report to congress. Available from: http://www.nrc.gov/docs/ML1219/ML12193A328.pdf. Accessed 31 May 2016.

Chen C, Meselhe E, Waldon M. Assessment of mineral concentration impacts from pumped stormwater on an Everglades wetland, Florida, USA – using a spatially-explicit model. J Hydrol. 2012;452–453:25–39.

Cooper SR, Goman M, Richardson CJ. Historical changes in water quality and vegetation in WCA-2A determined by paleoecological analyses. In: Richardson CJ, editor. The Everglades experiments: lessons for ecosystem restoration. New York: Springer; 2008. p. 321–50.

Craft CB, Richardson CJ. Peat accretion and phosphorus accumulation along an eutrophication gradient in the northern Everglades. Biogeochemistry. 1993;22:133–56.

Craft CB, Richardson CJ. Relationships between soil nutrients and plant species composition in Everglades peatlands. J Environ Qual. 1997;26:224–32.

Craft CB, Richardson CJ. Soil characteristics of the Everglades peatland. In: Richardson CJ, editor. The Everglades experiments: lessons for ecosystem restoration. New York: Springer; 2008. p. 59–74.

Craft CB, Vymazal J, Richardson CJ. Response of Everglades plant communities to nitrogen and phosphorus additions. Wetlands. 1995;15:258–71.

Davis JH. The natural features of southern Florida. Fl Geol Soc Geol Bull. 1943;25:1–311.

Davis SM, Ogden JC. Toward ecosystem restoration. In: Davis SM, Ogden JC, editors. Everglades: the ecosystem and its restoration. Delray Beach: St. Lucie Press; 1994. p. 769–96.

Davis SM, Gunderson LH, Park WA, Richardson JR, Mattson JE. Landscape dimension, composition, and function in a changing Everglades ecosystem. In: Davis SM, Ogden JC, editors. Everglades: the ecosystem and its restoration. Delray Beach: St. Lucie Press; 1994. p. 419–44.

Germain G, Pietro K. Chapter 5: STA performance and optimization. In: 2011 South Florida environmental report. West Palm Beach: South Florida Water Management District; 2011.

Glaser PH, Volin JC, Givnish TJ, Hansen BCS, Stricker CA. Carbon and sediment accumulation in the Everglades (USA) during the past 4000 years: rates, drivers, and sources of error. J Geophys Res. 2012;117:G03026.

Glaser PH, Hansen BCS, Donovan JJ, Givnish TJ, Stricker CA, Volin JC. Holocene dynamics of the Florida Everglades with respect to climate, dustfall, and tropical storms. Proc Natl Acad Sci U S A. 2013;43:17211–6.

Gleason PJ, Stone P. Age, origin and landscape evolution of the Everglades peatlands. In: Davis SM, Ogden JC, editors. Everglades: the ecosystem and its restoration. Delray Beach: St. Lucie Press; 1994. p. 149–98.

Gunderson LH, Loftus WT. The Everglades. In: Martin WH, Boyce SG, Echtemacht ACE, editors. Biodiversity of the southeastern United States: lowland terrestrial communities. New York: Wiley; 1993. p. 199–255.

Harvey JW, McCormick PV. Groundwater's significance to changing hydrology, water chemistry, and biological communities of a floodplain ecosystem, Everglades, South Florida. USA Hydrogeol J. 2009;17:185–201.

IPCC. In: Solomon S, Qin D, Manning M, Chen Z, Marquis M, Averyt KB, Tignor M, Miller HL, editors. Climate change 2007: the physical science basis. Contribution of working group I to the fourth assessment report of the intergovernmental panel on climate change. Cambridge: Cambridge University Press; 2007.

Ives JC. Memoir to accompany a military map of the peninsula of Florida south of Tampa Bay. New York: War Department; 1856.

Kloor K. Everglades restoration plan hits rough waters. Science. 2000;288:1166–7.

Lodge TE. The Everglades handbook: understanding the ecosystem. 2nd ed. Boca Raton: CRC Press; 2005.

Pan Y, Stevenson RJ, Vaithiyanathan P, Slate J, Richardson CJ. Changes in algal assemblages along observed and experimental phosphorus gradients in a subtropical wetland. USA Freshw Biol. 2000;44:339–53.

Parker GG, Ferguson GE, Love SK. Water resources of southeastern Florida, water-supply paper 1255. Washington, DC: U.S. Geological Survey; 1955.

Pearlstine LG, Pearlstine EV, Aumen NG. A review of the ecological consequences and management implications of climate change for the Everglades. J N Am Benthol Soc. 2010;29:1510–26.

Porter JW, Porter KF. The Everglades, Florida Bay, and coral reefs of the Florida Keys: an ecosystem sourcebook. Boca Raton: CRC Press; 2002.

Qualls RG, Richardson CJ. Carbon cycling and dissolved organic matter export in the northern everglades. In: Richardson CJ, editor. The Everglades experiments: lessons for ecosystem restoration. New York: Springer; 2008. p. 351–70.

Reddy KR, Newman S, Osborne TZ, White JR, Fitz HC. Phosphorus cycling in the greater Everglades ecosystem: legacy phosphorus implications for management and restoration. Crit Reg Environ Sci Technol. 2011;41(supp. 1):149–86.

Richardson CJ. Freshwater wetlands. In: Barbour MG, Billings WD, editors. North American terrestrial vegetation. New York: Cambridge University Press; 2000. p. 448–99.

Richardson CJ. The Everglades experiments: lessons for ecosystem restoration. New York: Springer; 2008.

Richardson CJ. The Everglades: North America's subtropical wetland. Wetl Ecol Manag. 2010;18:517–42.

Richardson CJ, Vaithiyanathan P. Phosphorus sorption characteristics of Everglades soils along a eutrophication gradient. Soil Sci Soc Am J. 1995;59:1782–8.

Richardson CJ, Qian S, Craft CB, Qualls RG. Predictive models for phosphorus retention in wetlands. Wetl Ecol Manag. 1997;4:159–75.

Richardson CJ, Ferrell GM, Vaithiyanathan P. Nutrient effects on stand structure, resorption efficiency, and secondary compounds in Everglades sawgrass. Ecology. 1999;80:2182–92.

Romanowicz EA, Richardson CJ. Geologic settings and hydrology gradients in the Everglades. In: Richardson CJ, editor. The Everglades experiments: lessons for ecosystem restoration. New York: Springer; 2008.

Rydin H, Jeglum JK. The biology of peatlands. New York: Oxford University Press; 2006.

Sandoval E, Price RM, Whitman D, Melesse AM. Long-term (11 years) study of water balance, flushing times and water chemistry of a coastal wetland undergoing restoration, Everglades, Florida, USA. Catena. 2016;144:74–83.

Sklar FH, van der Valk AG, editors. Tree islands of the Everglades. Dordrecht: Kluwer; 2002.

Stokstad E. Big land purchase triggers review of plans to restore Everglades. Science. 2008;321:22.

Todd MJ, Muneepeerakul R, Pumo D, Azaele S, Miralles-Wilhelm F, Rinaldo A, Rodriguez-Iturbe I. Hydrological drivers of wetland vegetation community distribution within Everglades National Park, Florida. Adv Water Resour. 2010;33:1279–89.

Vymazal J, Richardson CJ. Species composition, biomass, and nutrient content of periphyton in the Florida Everglades. J Phycol. 1995;31:343–54.

Wang H, Meselhe EA, Waldon MG. Compartment-based hydrodynamics and water quality modeling of a northern Everglades wetland, Florida USA. Ecol Model. 2012;247:273–85.

Waters MN, Smoak JM, Saunders CJ. Historic primary producer communities linked to water quality and hydrologic changes in the northern Everglades. J Paleolimnol. 2013;49:67–81.

Willard DA, Bernhardt CE. Impacts of past climate and sea level change on Everglades wetlands: placing a century of anthropogenic change into a late-Holocene context. Clim Change. 2011;107:59–80.

The Wetland Book

C. Max Finlayson • G. Randy Milton
R. Crawford Prentice • Nick C. Davidson
Editors

The Wetland Book

II: Distribution, Description, and Conservation

Volume 2

With 603 Figures and 125 Tables

Editors
C. Max Finlayson
Institute for Land, Water and Society
Charles Sturt University
Albury, New South Wales, Australia

UNESCO-IHE, Institute for Water Education
Delft, The Netherlands

R. Crawford Prentice
Nature Management Services
Cambridge, UK

G. Randy Milton
Department of Natural Resources
Kentville, Nova Scotia, Canada

Nick C. Davidson
Institute for Land, Water and Society
Charles Sturt University
Albury, New South Wales, Australia

Nick Davidson Environmental
Wigmore, UK

ISBN 978-94-007-4000-6   ISBN 978-94-007-4001-3 (eBook)
ISBN 978-94-007-4002-0 (print and electronic bundle)
https://doi.org/10.1007/978-94-007-4001-3

Library of Congress Control Number: 2017937719

© Springer Science+Business Media B.V., part of Springer Nature 2018
This work is subject to copyright. All rights are reserved by the Publisher, whether the whole or part of the material is concerned, specifically the rights of translation, reprinting, reuse of illustrations, recitation, broadcasting, reproduction on microfilms or in any other physical way, and transmission or information storage and retrieval, electronic adaptation, computer software, or by similar or dissimilar methodology now known or hereafter developed.
The use of general descriptive names, registered names, trademarks, service marks, etc. in this publication does not imply, even in the absence of a specific statement, that such names are exempt from the relevant protective laws and regulations and therefore free for general use.
The publisher, the authors and the editors are safe to assume that the advice and information in this book are believed to be true and accurate at the date of publication. Neither the publisher nor the authors or the editors give a warranty, express or implied, with respect to the material contained herein or for any errors or omissions that may have been made. The publisher remains neutral with regard to jurisdictional claims in published maps and institutional affiliations.

Printed on acid-free paper

This Springer imprint is published by the registered company Springer Science+Business Media B.V. part of Springer Nature.
The registered company address is: Van Godewijckstraat 30, 3311 GX Dordrecht, The Netherlands

Foreword: The Wetland Book

The venerable lineage of encyclopedic publishing can be traced back to Pliny the Elder's *Naturalis Historia*, which contained chapters on water and aquatic life. Although our terminology regarding and understanding of the aquatic environment has evolved over the past two millennia, one constant has been the need for a multidisciplinary approach to examining these areas. Using an encyclopedic model, this multidisciplinary book builds on an ancient format and adapts it for a modern audience. In this way, *The Wetland Book* builds on a long tradition of scholarly publishing and presents invaluable information for its modern audience.

Wetlands have been around longer than the traditions associated with academic publishing. Wetland management and wise use have been practiced by indigenous cultures in many forms for millennia, and that ancient knowledge about wetlands was often curated and passed down orally or in traditional systems and forms. In modern times, the pressures and threats to wetlands are vastly different in their scope and magnitude. The forms of governance and administration that respond to these pressures and threats have also changed, particularly in their scale as it has been recognized that management takes place at the level of countries and river basins, rather than simply at the local level.

Internationally, wetland conservation, management, and wise use are promoted through the Ramsar Convention on Wetlands. The countries that have signed onto the Ramsar Convention have recognized the imperatives to work with stakeholders and decisionmakers beyond the traditional wetland community and to incorporate wetlands into policy-making in other sectors such as water, energy, agriculture, and health. Indeed, in 2008 at the 10th Conference of the Contracting Parties of the Ramsar Convention, the Changwon Declaration was adopted, which contains key messages for wetland conservation, management, and wise use addressed to planners; policymakers; elected officials; managers in the environmental, land, and resource-use sectors; educators and communicators; economists; and health workers. *The Wetland Book* offers a base of knowledge that is intended to reach a similarly broad audience.

The editors and contributing authors to *The Wetland Book* have long experience and deep understanding of wetland science and management. Many have worked with the Ramsar Scientific and Technical Review Panel (STRP), the Convention's scientific advisory body, over the years. This collection of people provides a

repository of knowledge that can help meet the challenge of learning about and understanding the value of protecting and managing wetlands.

Making this knowledge more easily accessible, however, has always been difficult. There are physical limitations to how much we can pick a person's brain, and there are limitations to how much a wetland manager out in the field, perhaps with little technical support, can search for, read, and review scientific and traditional knowledge to find answers to pressing questions. Thus, the encyclopedic style of publication remains a viable format for accessing high levels of expertise, including expertise from distant locations, with similar landscape and ecological characteristics. *The Wetland Book* provides an in-depth level of knowledge in the form of a handbook to assist those seeking information on the many facets of wetland management.

Of course, reading *The Wetland Book* will not make an individual an expert in all aspects of wetland science, wise use and governance, a feat which no one publication can deliver. Instead, a truly useful publication should offer an individual the vocabulary to support further inquiry and to find knowledge that is locally, regionally, nationally, or even internationally applicable. It should also allow a reader to know who to ask and what questions to pursue when she or he needs more knowledge to solve a research question or particular management problem. *The Wetland Book* delivers this foundation through two volumes – Vol. 1: Structure and Function, Management, and Methods and Vol. 2: Distribution, Description, and Conservation.

We highly recommend *The Wetland Book*; it provides an unparalleled source of knowledge about wetlands by building on the ancient form of the encyclopedia, revitalized by new technologies for distribution and access. We are also proud to see that many of those who have contributed to the Ramsar Convention over many years or even decades have also contributed their knowledge and wisdom to *The Wetland Book*. Given our personal association with the Convention, we also recognize the incredible contribution that the Convention has made to wetland knowledge and look forward to further contributions.

Chair, Scientific & Technical Review Panel Ramsar Convention on Wetlands 2005–2012	Heather MacKay
Chair, Scientific & Technical Review Panel Ramsar Convention on Wetland 2012–2018	Royal C. Gardner

Preface

The Wetland Book is a hard copy and online production that provides an unparalleled collation of information on wetlands. It is global in scope and contains 462 chapters prepared by leading wetland researchers and managers. The wide disciplinary and geographic scope is a particular feature and differentiates *The Wetland Book* from the existing wetland literature. The editors have compiled *The Wetland Book* from contributions supplied by authors from many countries and disciplines. Combined, these chapters represent a global source of knowledge about wetlands. Given the number of chapters and the scope of the content, it has been published as two separate books.

The bibliographic detail of the two books is given below. Book II with 170 chapters covers the distribution, description, and conservation of wetlands.

The Wetland Book II: Distribution, Description, and Conservation: edited by Finlayson CM, Milton GR, Prentice RC and Davidson NC.

Its companion book, published separately, with 292 chapters is:

The Wetland Book I: Structure and Function, Management, and Methods: edited by Finlayson CM, Everard M, Irvine K, McInnes RJ, Middleton BA, van Dam AA and Davidson NC

The Wetland Book was developed following discussions with wetland experts from the Scientific and Technical Review Panel of the Ramsar Convention on Wetlands and from the Society for Wetland Scientists. These experts pointed to the rapidly expanding literature on wetlands and enthusiastically proposed the development of a comprehensive information resource aimed at supporting the trans- and multidisciplinary research and practice, which is essential to wetland science and management. They were also seeking an information resource that would both complement and extend the existing literature and in particular provide a compendium of knowledge with contributions from authors around the world.

Aware that wetland research was on the rise and that wetland researchers and practitioners often needed to work across disciplines, *The Wetland Book II* has been prepared to serve as a first port of call for those interested in the key information

about wetlands and their conservation. This was done to allow individuals and multi- and transdisciplinary teams to search for particular terms and subjects, access further details, and read overviews of topics selected by the editors and expert authors. The content provides a global coverage of wetland knowledge with chapters provided by leading wetland experts with knowledge that spans local and regional issues to the wider body of science that is needed to assist practitioners and enable students to come to grips with one of the world's most diverse and important set of ecosystems. This is especially important as these ecosystems are under increasing pressure in many parts of the world as degradation from human development continues at an alarming rate and are in need of more effective management and restoration. It draws heavily on the knowledge compiled through the formal processes of the Ramsar Convention and associated programs and extends that contained in the seminal global assessment of wetlands undertaken through the Millennium Ecosystem Assessment.

Book II is structured in sections covering the diversity of wetland types, natural and anthropogenic drivers of wetland change, and regional compilations of individual wetlands and wetland complexes. Detailed overview chapters typically describe the diversity within each wetland type, its distribution, extent (current and historical), ecosystem services, biodiversity, and threats and future challenges. Regional contributions follow a general format describing the basic ecology of the system; uniqueness; distribution; biodiversity and species adaptations; ecosystem services with an emphasis on importance to dependent peoples where appropriate; conservation status and management; and threats and future challenges. The coverage is based on a mix of wetland types and geographic extent and distribution.

Given that *The Wetland Book* constitutes a remarkable information resource, we warmly convey our special thanks to the many authors who gave up their time and shared their knowledge of wetlands to support this effort – and also for their patience while the large number of chapters they have generously provided were collated and edited. We are proud to have worked with them to produce this book. With the benefit of their unstinting efforts and incredibly rich knowledge, *The Wetland Book II* provides a comprehensive source of information for wetland researchers, students, and practitioners. It specifically provides a much needed information resource to support the many efforts to ensure the wise use of wetlands globally. It has also not only drawn on but also extended the expert guidance and advice that the Ramsar Convention's Scientific and Technical Review Panel has for almost 25 years provided for governments and wetland experts alike. In this respect, the foreword provided by the past and present chairs of the Panel is particularly appreciated. In providing the foreword, they have reflected on the wealth of knowledge collated by wetland experts from around the world who have worked tirelessly to provide government officials with the knowledge base needed to ensure the conservation and wise use of wetlands.

As editors for *The Wetland Book II*, we personally compliment the many authors for their incredible contributions to the most comprehensive compendium of knowledge about wetlands ever assembled. In particular, we acknowledge their unstinting efforts to compile the many chapters and work with the authors to produce *The*

Wetland Book. Their knowledge and efforts are matched by their willingness to share the collated knowledge that is now contained in *The Wetland Book*.

The publishers are thanked for their foresight in developing the concepts that led to *The Wetland Book* and for providing both a hard copy and online version, with the latter being available for future updating. We recommend *The Wetland Book* to all those interested in the growing international scientific knowledge about the functioning and management of these incredibly valuable but threatened ecosystems.

Institute for Land, Water and Society C. Max Finlayson
Charles Sturt University
Albury, NSW, Australia

UNESCO-IHE, Institute for Water Education
Delft, The Netherlands

Department of Natural Resources G. Randy Milton
Halifax, NS, Canada

Nature Management Services R. Crawford Prenctice
Cambridge, UK

Institute for Land, Water and Society Nick C. Davidson
Charles Sturt University
Albury, NSW, Australia

Nick Davidson Environmental
Wigmore, UK

Contents

Volume 1

Section I Introduction 1

1 **Wetlands of the World** 3
G. Randy Milton, R. Crawford Prentice, and C. Max Finlayson

Section II Diversity of Wetlands 17

2 **Wetland Types and Distribution** 19
C. Max Finlayson, G. Randy Milton, and R. Crawford Prentice

3 **Estuaries** 37
Graham R. Daborn and Anna M. Redden

4 **Estuarine Marsh: An Overview** 55
Ralph W. Tiner and G. Randy Milton

5 **Seagrasses** 73
Frederick T. Short, Cathy A. Short, and Alyssa B. Novak

6 **Mangroves** 93
C. Max Finlayson

7 **Major River Basins of the World** 109
Carmen Revenga and Tristan Tyrrell

8 **Freshwater Lakes and Reservoirs** 125
Etienne Fluet-Chouinard, Mathis Loïc Messager, Bernhard Lehner, and C. Max Finlayson

9 **Salt Lakes** 143
C. Max Finlayson

10 **Tidal Freshwater Wetlands: The Fresh Dimension of the Estuary** 155
Aat Barendregt

11	Freshwater Marshes and Swamps	169
	C. Max Finlayson	
12	Papyrus Wetlands	183
	Julius Kipkemboi and Anne A. van Dam	
13	Tropical Freshwater Swamps (Mineral Soils)	199
	Wim Giesen	
14	Peatlands	227
	C. Max Finlayson and G. Randy Milton	
15	Peat	245
	Richard Lindsay and Roxane Andersen	
16	Peatland (Mire Types): Based on Origin and Behavior of Water, Peat Genesis, Landscape Position, and Climate	251
	Richard Lindsay	
17	Arctic Peatlands	275
	Tatiana Minayeva, Andrey Sirin, Peter Kershaw, and Olivia Bragg	
18	Mires	289
	Richard Lindsay	
19	Blanket Mire	295
	Richard Lindsay	
20	Blanket Bogs	303
	Richard Lindsay	
21	Lagg Fen	309
	Richard Lindsay	
22	Karst Wetlands	313
	Gordana Beltram	
23	Subterranean (Hypogean) Habitats in Karst and Their Fauna	331
	Boris Sket	
24	Groundwater Dependent Wetlands	345
	Ray H. Froend, Pierre Horwitz, and Bea Sommer	

Section III Natural and Anthropogenic Drivers of Wetland Change ... **357**

25	Natural and Anthropogenic Drivers of Wetland Change	359
	Susan M. Galatowitsch	
26	Wetland Losses and the Status of Wetland-Dependent Species	369
	Nick C. Davidson	

Contents

27 Alien Plants and Wetland Biotic Dysfunction 383
    C. Max Finlayson

28 Ecological Conditions and Health of Arctic Wetlands Modified by
    Nutrient and Contaminant Inputs from Colonial Birds 391
    Mark Mallory

29 Lake Chilika (India): Ecological Restoration and Adaptive
    Management for Conservation and Wise Use 397
    Ajit Kumar Pattnaik and Ritesh Kumar

30 Saemangeum Estuarine System (Republic of Korea): Before and
    After Reclamation 405
    Nial Moores

31 Peatlands and Windfarms: Conflicting Carbon Targets and
    Environmental Impacts 413
    Richard Lindsay

32 Kakagon (Bad River Sloughs), Wisconsin (USA) 427
    Jim Meeker and Naomi Tillison

33 Seagrass Dependent Artisanal Fisheries of Southeast Asia 437
    Richard K. F. Unsworth and Leanne C. Cullen-Unsworth

34 Great Barrier Reef (Australia): A Multi-ecosystem Wetland
    with a Multiple Use Management Regime 447
    Jon Brodie and Jane Waterhouse

35 Qa'a Azraq Oasis: Strengthening Stakeholder Representation in
    Restoration (Jordan) 461
    Fidaa F. Haddad

36 Fishponds of the Czech Republic 469
    Jan Pokorný and Jan Květ

37 Makgadikgadi Wetlands (Botswana): Planning for Sustainable Use
    and Conservation 487
    Jaap Arntzen

38 Seagrass Recovery in Tampa Bay, Florida (USA) 495
    Holly Greening, Anthony Janicki, and Ed T. Sherwood

Section IV North America, Greenland, and the Caribbean 507

39 Wetlands of Greenland 509
    Christian Bay

40 Peatlands of Continental North America 515
    Dale H. Vitt

41	**Boreal Wetlands of Canada and the United States of America** . . . Beverly Gingras, Stuart Slattery, Kevin Smith, and Marcel Darveau	521
42	**Yukon-Kuskokwim Delta: Yukon River Basin, Alaska (USA)** . . . . Frederic A. Reid and Daniel Fehringer	543
43	**The Peace-Athabasca Delta: MacKenzie River Basin (Canada)** . . . . Jeffrey Shatford	549
44	**Copper River Delta, Alaska (USA)** . . . . Frederic A. Reid, Daniel Fehringer, and Richard G. Kempka	557
45	**Fraser River Delta: Southern British Columbia (Canada)** . . . . Anne Murray	565
46	**The Mississippi Alluvial Valley (USA)** . . . . J. Brian Davis	577
47	**Coastal Wetlands of Manitoba's Great Lakes (Canada)** . . . . Dale Wrubleski, Pascal Badiou, and Gordon Goldsborough	591
48	**Coastal Wetlands of Lake Superior's South Shore (USA)** . . . . John Brazner and Anett Trebitz	605
49	**The Bay of Fundy and Its Wetlands (Canada)** . . . . Graham R. Daborn and Anna M. Redden	621
50	**San Francisco Bay Estuary (USA)** . . . . Beth Huning and Mike Perlmutter	637
51	**Vernal Pools of Northeastern North America** . . . . Elizabeth A. Colburn and Aram J. K. Calhoun	651
52	**Pocosins (USA)** . . . . Curtis J. Richardson	667
53	**Prairie Pothole Region of North America** . . . . Kevin E. Doherty, David W. Howerter, James H. Devries, and Johann Walker	679
54	**Playa Wetlands of the Great Plains (USA)** . . . . Anne Bartuszevige	689
55	**Wetlands of California's Central Valley (USA)** . . . . Frederic A. Reid, Daniel Fehringer, Ruth Spell, Kevin Petrik, and Mark Petrie	697
56	**The Everglades (USA)** . . . . Curtis J. Richardson	705

Volume 2

Section V Central and South America **725**

57 **Amazon River Basin** 727
Florian Wittmann and Wolfgang J. Junk

58 **Mangroves of Colombia** 747
Jenny Alexandra Rodríguez-Rodríguez, Paula Cristina Sierra-Correa, Martha Catalina Gómez-Cubillos, and Lucia Victoria Licero Villanueva

59 **Ciénaga Grande de Santa Marta: The Largest Lagoon-Delta Ecosystem in the Colombian Caribbean** 757
Jenny Alexandra Rodríguez-Rodríguez, José Ernesto Mancera Pineda, Laura Victoria Perdomo Trujillo, Mario Enrique Rueda, and Karen Patricia Ibarra

60 **Lake Fuquene (Colombia)** 773
Mauricio Valderrama, María Pinilla-Vargas, Germán I. Andrade, Eugenio Valderrama-Escallón, and Sandra Hernández

61 **The Paraná-Paraguay Fluvial Corridor (Argentina)** 785
Priscilla G. Minotti

62 **The Pantanal: A Brief Review of Its Ecology, Biodiversity, and Protection Status** 797
Wolfgang J. Junk and Catia Nunes da Cunha

63 **The Paraná River Delta** 813
Patricia Kandus and Rubén Darío Quintana

64 **Wetlands of Chile: Biodiversity, Endemism, and Conservation Challenges** 823
Alejandra Figueroa, Manuel Contreras, and Bárbara Saavedra

65 **Seagrasses of Southeast Brazil** 839
Joel C. Creed, Mariana V. P. Aguiar, Agatha Cristinne Soares, and Leonardo V. Marques

66 **Rio de la Plata (La Plata River) and Estuary (Argentina and Uruguay)** 847
Claudio R. M. Baigún, Darío C. Colautti, and Tomás Maiztegui

67 **Conchalí Lagoon: Coastal Wetland Restoration Project (Chile)** 857
Manuel Contreras, F. Fernando Novoa, and Juan Pablo Rubilar

68 **Bahía Lomas: Ramsar Site (Chile)** 865
Carmen Espoz, Ricardo Matus, and Diego Luna-Quevedo

| 69 | Patagonian Peatlands (Argentina and Chile) | 873 |

Rodolfo Iturraspe

Section VI Europe ... 883

| 70 | Danube River Basin | 885 |

Paul Csagoly, Gernant Magnin, and Orieta Hulea

| 71 | Lower Danube Green Corridor | 897 |

Paul Csagoly, Gernant Magnin, and Orieta Hulea

| 72 | Danube, Drava, and Mura Rivers: The "Amazon of Europe" | 903 |

Paul Csagoly, Gernant Magnin, and Arno Mohl

| 73 | Danube Delta: The Transboundary Wetlands (Romania and Ukraine) | 911 |

Grigore Baboianu

| 74 | Rhine River Basin | 923 |

Daphne Willems and Esther Blom

| 75 | Volga River Basin (Russia) | 933 |

Harald J. L. Leummens

| 76 | Volga River Delta (Russia) | 945 |

Harald J. L. Leummens

| 77 | European Tidal Saltmarshes | 959 |

Nick C. Davidson

| 78 | Tipperne Peninsula and Ringkøbing Fjord (Denmark) | 973 |

Hans Meltofte, Preben Clausen, and Ole Thorup

| 79 | Wadden Sea (Denmark) | 983 |

Karsten Laursen and John Frikke

| 80 | Estuaries of Great Britain | 997 |

Nick C. Davidson

| 81 | The Wash Estuary and North Norfolk Coast (UK) | 1011 |

Nick C. Davidson

| 82 | Wetlands of the Norfolk and Suffolk Broads (UK) | 1023 |

Andrea Kelly

| 83 | Blanket Mires of Caithness and Sutherland: Scotland's Great Flow Country (UK) | 1039 |

Richard Lindsay and Roxane Andersen

| 84 | Karst Wetlands in the Dinaric Karst | 1057 |

Rosana Cerkvenik, Andrej Kranjc, and Andrej Mihevc

85 **Turloughs (Ireland)** 1067
    Kenneth Irvine, Catherine Coxon, Laurence Gill,
    Sarah Kimberley, and Steve Waldren

86 **The Macrotidal Bay of Mont-Saint-Michel (France): The
    Function of Salt Marshes** 1079
    Loïc Valéry and Jean-Claude Lefeuvre

87 **The Inner Danish Waters and Their Importance to
    Waterbirds** 1089
    Ib Krag Petersen and Rasmus Due Nielsen

Section VII Mediterranean Basin, Middle East, and West Asia ... 1099

88 **The Camargue: Rhone River Delta (France)** 1101
    Patrick Grillas

89 **Ebro Delta (Spain)** 1113
    Carles Ibáñez and Nuno Caiola

90 **Doñana Wetlands (Spain)** 1123
    Andy J. Green, Javier Bustamante, Guyonne F. E. Janss,
    Rocio Fernández-Zamudio, and Carmen Díaz-Paniagua

91 **Axios, Aliakmon, and Gallikos Delta Complex
    (Northern Greece)** 1137
    Despoina Vokou, Urania Giannakou, Christina Kontaxi, and
    Stella Vareltzidou

92 **The Philippi Peatland (Greece)** 1149
    Kimon Christanis

93 **Peatlands of the Mediterranean Region** 1155
    Richard Payne

94 **The Hula Wetland (Israel)** 1167
    Richard Payne

95 **Coastal Sabkha (Salt Flats) of the Southern and Western
    Arabian Gulf** 1173
    Ronald A. Loughland, Ali M. Qasem, Bruce Burwell, and
    Perdana K. Prihartato

96 **Lake Seyfe (Turkey)** 1185
    Serhan Cagirankaya and Burhan Teoman Meric

Section VIII Africa 1197

97 **Congo River Basin** 1199
    Ian J. Harrison, Randall Brummett, and Melanie L. J. Stiassny

98	**Zambezi River Basin**	1217
	Matthew McCartney, Richard D. Beilfuss, and Lisa-Maria Rebelo	
99	**Zambezi River Delta (Mozambique)**	1233
	Richard D. Beilfuss	
100	**Nile River Basin**	1243
	Matthew McCartney and Lisa-Maria Rebelo	
101	**Nile Delta (Egypt)**	1251
	Mohamed Reda Fishar	
102	**Baro-Akobo River Basin Wetlands: Livelihoods and Sustainable Regional Land Management (Ethiopia)**	1261
	Adrian Wood, J. Peter Sutcliffe, and Alan Dixon	
103	**Bahr el Ghazal: Nile River Basin (Sudan and South Sudan)**	1269
	Asim I. El Moghraby	
104	**Machar Marshes: Nile Basin (South Sudan)**	1279
	Yasir A. Mohamed	
105	**The Mayas Wetlands of the Dinder and Rahad: Tributaries of the Blue Nile Basin (Sudan)**	1287
	Khalid Hassaballah, Yasir A. Mohamed, and Stefan Uhlenbrook	
106	**The Sudd (South Sudan)**	1299
	Lisa-Maria Rebelo and Asim I. El Moghraby	
107	**Rugezi Marsh: A High Altitude Tropical Peatland in Rwanda**	1307
	Piet-Louis Grundling, Ab P. Grootjans, and Anton Linström	
108	**Banc d'Arguin (Mauritania)**	1319
	Antonio Araujo and Pierre Campredon	
109	**Bijagos Archipelago (Guinea-Bissau)**	1333
	Pierre Campredon and Paulo Catry	
110	**Kilombero Valley Floodplain (Tanzania)**	1341
	Lars Dinesen	
111	**Lakes Baringo and Naivasha: Endorheic Freshwater Lakes of the Rift Valley (Kenya)**	1349
	Reuben Omondi, William Ojwang, Casianes Olilo, James Mugo, Simon Agembe, and Jacob E. Ojuok	
112	**Lake Turkana: World's Largest Permanent Desert Lake (Kenya)**	1361
	William Ojwang, Kevin O. Obiero, Oscar O. Donde, Natasha J. Gownaris, Ellen K. Pikitch, Reuben Omondi, Simon Agembe, John Malala, and Sean T. Avery	

113 Soda Lakes of the Rift Valley (Kenya) 1381
    Simon Agembe, William Ojwang, Casianes Olilo,
    Reuben Omondi, and Collins Ongore

114 Okavango Delta, Botswana (Southern Africa) 1393
    Lars Ramberg

115 **Peatlands of Africa** 1413
    Piet-Louis Grundling and Ab P. Grootjans

116 **Peatland Types and Tropical Swamp Forests on the
    Maputaland Coastal Plain (South Africa)** 1423
    Althea T. Grundling, Ab P. Grootjans, Piet-Louis Grundling, and
    Jonathan S. Price

Volume 3

Section IX Northern and East Asia 1437

117 Lena River Basin (Russia) 1439
    Victor Degtyarev

118 Lena River Delta (Russia) 1451
    Victor Degtyarev

119 Nidjili Lake: Lena River Basin (Russia) 1457
    Victor Degtyarev

120 Taiga-Alas Landscape in the South of the Central Yakutian
    Lowland: Lena River Basin (Russia) 1463
    Victor Degtyarev

121 The Middle Aldan River Basin: A Key Migration Corridor for
    the Eastern Population of the Siberian Crane Within the Lena
    River Basin (Russia) 1471
    Victor Degtyarev

122 Yenisei River Basin and Lake Baikal (Russia) 1477
    Nick C. Davidson

123 Amur-Heilong River Basin: Overview of Wetland Resources ... 1485
    Evgeny Egidarev, Eugene Simonov, and Yury Darman

124 Daurian Steppe Wetlands of the Amur-Heilong River Basin
    (Russia, China, and Mongolia) 1499
    Eugene Simonov, Oleg Goroshko, and Tatiana Tkachuk

125 Sanjiang Plain and Wetlands Along the Ussuri and Amur
    Rivers: Amur River Basin (Russia and China) 1509
    Thomas D. Dahmer

| 126 | Zhalong Wetlands (China) | 1521 |

Liying Su

| 127 | Highland Peatlands of Mongolia | 1531 |

Tatiana Minayeva, Andrey Sirin, and Chultemin Dugarjav

| 128 | Yangtze River Basin (China) | 1551 |

Cui Lijuan, Zhang Manyin, and Xu Weigang

| 129 | Poyang Lake, Yangtze River Basin, China | 1565 |

James Harris

| 130 | Huang He (Yellow River) River Basin (China) | 1575 |

Cui Lijuan, Zhang Manyin, and Xu Weigang

| 131 | Current Status of Seagrass Habitat in Korea | 1589 |

Kun-Seop Lee, Seung Hyeon Kim, and Young Kyun Kim

| 132 | Hokkaido Marshes (Japan) | 1597 |

Satoshi Kobayashi

Section X Central and South Asia ... 1603

| 133 | High Altitude Wetlands of Nepal | 1605 |

Lalit Kumar and Pramod Lamsal

| 134 | Anzali Mordab Complex (Islamic Republic of Iran) | 1615 |

Masoud Bagherzadeh Karimi

| 135 | Bujagh National Park (Islamic Republic of Iran) | 1625 |

Sadegh Sadeghi Zadegan

| 136 | Fereydoon Kenar, Ezbaran, and Sorkh Ruds Ab-Bandans | 1635 |

Sadegh Sadeghi Zadegan

| 137 | Lake Parishan (Islamic Republic of Iran) | 1647 |

Ahmad Lotfi

| 138 | Lake Uromiyeh (Islamic Republic of Iran) | 1659 |

Ahmad Lotfi

| 139 | Shadegan Wetland (Islamic Republic of Iran) | 1675 |

Ahmad Lotfi

| 140 | Mesopotamian Marshes of Iraq | 1685 |

Curtis J. Richardson

| 141 | Indus River Basin Wetlands | 1697 |

Rab Nawaz, Ali Dehlavi, and Nadia Bajwa

| 142 | Wular Lake, Kashmir | 1705 |

Ritesh Kumar

143	**Wetlands of the Ganga-Brahmaputra Basin**	1711
	Ritesh Kumar and Kalpana Ambastha	
144	**Saline Wetlands of the Arid Zone of Western India**	1725
	Malavika Chauhan and Brij Gopal	
145	**The Transboundary Sundarbans Mangroves (India and Bangladesh)**	1733
	Brij Gopal and Malavika Chauhan	
146	**Wetlands of Mahanadi Delta (India)**	1743
	Ritesh Kumar and Pranati Patnaik	

Section XI Southeast Asia **1751**

147	**Tropical Peat Swamp Forests of Southeast Asia**	1753
	Susan Page and Jack Rieley	
148	**Wetlands of the Mekong River Basin: An Overview**	1763
	Peter-John Meynell	
149	**Tonle Sap Lake: Mekong River Basin (Cambodia)**	1785
	Colin Poole	
150	**Tram Chim: Mekong River Basin (Vietnam)**	1793
	Triet Tran and Jeb Barzen	
151	**Transboundary Mekong River Delta (Cambodia and Vietnam)**	1801
	Triet Tran	
152	**U Minh Peat Swamp Forest: Mekong River Basin (Vietnam)** ...	1813
	Triet Tran	
153	**Sembilang National Park: Mangrove Reserves of Indonesia**	1819
	Marcel J. Silvius, Yus Rusila Noor, I. Reza Lubis, Wim Giesen, and Dipa Rais	
154	**Wetlands of Berbak National Park (Indonesia)**	1831
	Wim Giesen, Marcel J. Silvius, and Yoyok Wibisono	
155	**Danau Sentarum National Park (Indonesia)**	1841
	Wim Giesen and Gusti Z. Anshari	
156	**Wetlands of Tasek Bera (Peninsular Malaysia)**	1851
	R. Crawford Prentice	
157	**Intertidal Flats of East and Southeast Asia**	1865
	John MacKinnon and Yvonne I. Verkuil	
158	**Seagrass in Malaysia: Issues and Challenges Ahead**	1875
	Japar S. Bujang, Muta H. Zakaria, and Frederick T. Short	

Section XII Australia, New Zealand, and Pacific Islands 1885

159 Murray-Darling River Basin (Australia) 1887
Jamie Pittock

160 Macquarie Marshes: Murray-Darling River Basin (Australia) ... 1897
Rachael F. Thomas and Joanne F. Ocock

161 The Coorong: Murray-Darling River Basin (Australia) 1909
Peter Gell

162 Kati Thanda: Lake Eyre (Australia) 1921
Richard T. Kingsford

163 Myall Lakes (Australia) 1929
Brian G. Sanderson and Anna M. Redden

164 Australia's Wet Tropics Streams, Rivers, and Floodplain
Wetlands 1941
Richard G. Pearson

165 Wetlands of Kakadu National Park (Australia) 1951
C. Max Finlayson

166 Groundwater Dependent Wetlands of the Gnangara
Groundwater System (Western Australia) 1959
Ray H. Froend, Pierre Horwitz, and Bea Sommer

167 Seagrass Meadows of Northeastern Australia 1967
Robert G. Coles, Michael A. Rasheed, Alana Grech, and
Len J. McKenzie

168 Wetlands of New Zealand 1977
Karen Denyer and Hugh Robertson

169 New Zealand Restiad Bogs 1991
Beverley R. Clarkson

170 Atolls of the Tropical Pacific Ocean: Wetlands Under Threat ... 2001
Randolph R. Thaman

Index of Keywords 2027

About the Editors

C. Max Finlayson
Institute for Land, Water and Society
Charles Sturt University
Albury, NSW, Australia
UNESCO-IHE
Institute for Water Education
Delft, The Netherlands

Max Finlayson is an internationally renowned wetland ecologist with extensive experience internationally in water pollution, agricultural impacts, invasive species, climate change, and human well-being and wetlands. He has participated in global assessments such as those conducted by the Intergovernmental Panel for Climate Change, the Millennium Ecosystem Assessment, and the Global Environment Outlook 4 and 5 (UNEP). Since the early 1990s, he has been a technical adviser to the Ramsar Convention on Wetlands and has written extensively on wetland ecology and management. He has also been actively involved in environmental NGOs and from 2002 to 2007 was president of the governing council of global NGO Wetlands International.

He has worked extensively on the inventory, assessment, and monitoring of wetlands, in particular in wet tropical, wet-dry tropical, and subtropical climatic regimes covering pollution, invasive species, and climate change. His current research interests/projects include the following:

- Interactions between human well-being and wetland health in the face of anthropogenic change, including global change and the onset of the Anthropocenic era
- Vulnerability and adaptation of wetlands/rivers to climate change, including changing values and trade-offs between uses and users, considering uncertainty and complexity
- Integration of ecologic, economic, and social requirements and trade-offs between users of wetlands with an emphasis on developing policy guidance and institutional changes

- Environment and agriculture interactions and policy responses/outcomes, and collaboration between stakeholders and policymakers
- Wetland restoration and construction, including the use of artificial wetlands for waste water treatment and the generation of multiple values
- Landscape change involving wetlands/rivers and land use (agriculture and mining) and implications for wetland ecosystem services and benefits for local people

He holds the following associated positions:

- Scientific Expert on the Scientific and Technical Review Panel, Ramsar Convention on Wetlands, Triennium 2016–2018
- Ramsar Chair for the Wise Use of Wetlands, UNESCO-IHE, Delft, The Netherlands (2014–2018)
- Visiting Professor, Institute for Wetland Research, China Academy of Forestry, Beijing, China
- Editor-in-Chief, Marine and Freshwater Research, CSIRO Publishing
- Chair, Environmental Strategy Advisory Panel, Winton Wetlands Restoration (Australia)

He has contributed to over 300 journal articles, reports, guidelines, proceedings, and book chapters on wetland ecology and management. He has contributed to the development of concepts and methods for wetland inventory, assessment and monitoring, and undertaken many site-based assessments in many countries.

G. Randy Milton
Department of Natural Resources Kentville
NS, Canada

Randy Milton is the manager for the Ecosystems and Habitats Program with Nova Scotia's Department of Natural Resources in Canada. Randy is an ecologist and Certified Wildlife Biologist® with 35 years' experience in public and industry conservation and environmental management, especially with freshwater and coastal wetlands and forest ecosystems. He has maintained an involvement in regional and national wetland conservation efforts since the early 1990s, as well as internationally first as a volunteer with WWF (Indonesia) in the mid-1980s and subsequently as a technical advisor to the Ramsar Convention on Wetlands (2000–2015), a contributing author to the Millennium Ecosystem Assessment, and a member (2005–2014) of the International Plan Committee for the North American Waterfowl Management Plan. He has a M.Sc. from Acadia University (Canada), where he is currently an adjunct professor, and he is also an adjunct research associate at the Institute for Land, Water and Society, Charles Sturt University, Australia.

R. Crawford Prentice
Nature Management Services
Cambridge, UK

Crawford Prentice is an independent consulting ecologist based in Cambridge, England. He has some 30 years of biodiversity conservation experience and has led global, regional, and national programs and projects mainly in Asia, the CIS countries, and Europe. He studied Zoology at the University of Aberdeen and subsequently completed his M.Sc. in Aquatic Resource Management at Kings College, University of London.

Much of his professional life has concerned the conservation and management of wetlands and migratory waterbirds, with early beginnings at the Wildfowl and Wetlands Trust and the International Waterfowl and Wetlands Research Bureau (IWRB) analyzing International Waterfowl Census data, evolving to conservation program management at the Asian Wetland Bureau and IWRB, and leading a bilateral aid project for the integrated management of Malaysia's first Ramsar site in the 1990s. During the next decade, he worked with the International Crane Foundation on the design and implementation of the UNEP/GEF Siberian Crane Wetland Project, helping to strengthen management of seven million hectares across 16 wetland sites in four countries for this flagship species and other biodiversity. He remains a project associate with ICF, contributing to climate change adaptation planning for key wetland nature reserves in north-eastern China and wider efforts for crane conservation. Currently, Crawford conducts consultancy assignments for the preparation, implementation, and evaluation of GEF projects on integrated ecosystem management of wetlands, mountains, and tropical forests.

Nick C. Davidson
Institute for Land, Water and Society
Charles Sturt University
Albury, NSW, Australia
Nick Davidson Environmental
Wigmore, UK

Nick Davidson was the deputy secretary general of the Ramsar Convention on Wetlands from 2000 to 2014, with overall responsibility for the convention's global development and delivery of scientific, technical, and policy guidance and advice and communications as the Convention Secretariat's senior advisor on these matters. He has long-standing experience in, and a strong commitment to, environmental sustainability supported through the transfer of environmental science into policy-relevance and decision-making

at national and international scales. Nick currently works as an independent expert consultant on wetland conservation and wise use.

Nick has over 40 years' experience of research on the ecology, assessment, and conservation of coastal and inland wetlands and the ecophysiology and flyway conservation of migratory waterbirds, with a 1981 Ph.D. from the University of Durham (UK) on this topic, and continues to publish on these issues. Prior to his Ramsar Convention post, he worked for the UK's national government conservation agencies on coastal wetland inventory, assessment, information systems, and communications and as international science coordinator for the global NGO Wetlands International.

He is an adjunct professor at the Institute of Land, Water and Society, Charles Sturt University, Australia; was presented with the Society of Wetland Scientist's (SWS) International Fellow Award 2010 for his long-term contributions to global wetland science and policy; chairs the SWS's Ramsar Section; is an associate editor of the peer-reviewed journal *Marine & Freshwater Research*; is a member of several IUCN Commissions and their task forces (World Commission on Protected Areas (WCPA), Species Survival Commission (SSC), and Commission on Ecosystem Management (CEM)); and is an honorary fellow of the Chartered Institution of Water and Environmental Management (CIWEM).

Contributors

Simon Agembe Department of Fisheries and Aquatic Sciences, University of Eldoret, Eldoret, Kenya

Mariana V. P. Aguiar Departamento de Ecologia, Universidade do Estado do Rio de Janeiro, Rio de Janeiro, Brazil

Kalpana Ambastha Wetlands International South Asia, New Delhi, India

Roxane Andersen Environmental Research Institute, University of the Highlands and Islands, Thurso, UK

Germán I. Andrade Universidad de Los Andes, Bogotá, Colombia

Gusti Z. Anshari Center for Wetlands People and Biodiversity, University Tanjungpura, Pontianak, W. Kalimantan, Indonesia

Antonio Araujo MAVA – Fondation pour la Nature, Dakar, Senegal

Jaap Arntzen Centre for Applied Research, Gaborone, Botswana

Sean T. Avery Kenya Wetlands Biodiversity Research Team, Nairobi, Kenya

Grigore Baboianu Danube Delta Biosphere Reserve Authority, Tulcea, Romania

Pascal Badiou Institute for Wetland and Waterfowl Research, Ducks Unlimited Canada, Stonewall, MB, Canada

Claudio R. M. Baigún Instituto de Investigación e Ingeniería Ambiental (3iA), Universidad Nacional de San Martín, San Martín, Buenos Aires, Argentina

Nadia Bajwa Indus Ecoregion, WWF, Karachi, Pakistan

Aat Barendregt Environmental Sciences, Copernicus Institute, Utrecht University, Utrecht, The Netherlands

Anne Bartuszevige Playa Lakes Joint Venture, Lafayette, CO, USA

Jeb Barzen Private Lands Conservation, Spring Green, WI, USA

Christian Bay Department of Bioscience, Faculty of Science and Technology, Institut for Bioscience, Aarhus University, Roskilde, Denmark

Richard D. Beilfuss International Crane Foundation, Baraboo, WI, USA

College of Engineering, University of Wisconsin, Madison, WI, USA

Gordana Beltram Ministry of the Environment and Spatial Planning, Environment Directorate, Nature Conservation Unit, Ljubljana, Slovenia

Esther Blom WWF-Netherlands, Zeist, The Netherlands

Olivia Bragg School of the Geography, University of Dundee, Dundee, UK

John Brazner Wildlife Division, Nova Scotia Department of Natural Resources, Kentville, NS, Canada

Jon Brodie Catchment to Reef Research Group, TropWATER, Centre for Tropical Water and Aquatic Ecosystem Research, James Cook University, Townsville, QLD, Australia

Randall Brummett Environment and Natural Resources Department, World Bank, Washington, DC, USA

Japar S. Bujang Department of Biology, Faculty of Science, Universiti Putra Malaysia, Serdang, Selangor Darul Ehsan, Malaysia

Bruce Burwell E-Map Department, Saudi Aramco, Dhahran, Ash Sharqiyah, Saudi Arabia

Javier Bustamante Department of Wetland Ecology, Estación Biológica de Doñana (EBD-CSIC), Seville, Spain

Serhan Cagirankaya Wetlands Division, Ministry of Forests and Water Affairs, General Directory of Nature Conservation and National Parks, Ankara, Turkey

Nuno Caiola IRTA, Aquatic Ecosystems Program, Sant Carles de la Ràpita, Catalonia, Spain

Aram J. K. Calhoun University of Maine, Orono, ME, USA

Pierre Campredon Conseiller Technique, IUCN Guinea Bissau, Bissau, Guinea-Bissau

Paulo Catry MARE – Marine and Environmental Sciences Centre, ISPA – Instituto Universitário, Lisbon, Portugal

Rosana Cerkvenik Park Škocjanske jame, Divača, SI, Slovenia

Malavika Chauhan Himmotthan Society, Dehradun, Uttarakhand, India

Kimon Christanis Department of Geology, Sector of Earth Materials, University of Patras, Rio-Patras, Greece

Beverley R. Clarkson Landcare Research, Hamilton, New Zealand

Preben Clausen Department of Bioscience, Aarhus University, Kalø, Denmark

Darío C. Colautti Instituto de Limnología de La Plata "Raúl Ringuelet", La Plata, Argentina

Elizabeth A. Colburn Harvard Forest, Harvard University, Petersham, MA, USA

Robert G. Coles Centre for Tropical Water and Aquatic Ecosystem Research, James Cook University, Cairns and Townsville, QLD, Australia

Manuel Contreras Centro de Ecología Aplicada (CEA), Santiago, Región Metropolitana, Chile

Catherine Coxon University of Dublin, Trinity College, Dublin, Ireland

Joel C. Creed Departamento de Ecologia, Universidade do Estado do Rio de Janeiro, Rio de Janeiro, Brazil

Paul Csagoly Earthly Communications, Ottawa, ON, Canada

Leanne C. Cullen-Unsworth Sustainable Places Research Institute, Cardiff University, Cardiff, UK

Graham R. Daborn Acadia Centre for Estuarine Research, Acadia University, Wolfville, NS, Canada

Thomas D. Dahmer Ecosystems Ltd., Yau Tong, Kowloon, Hong Kong, China

Yury Darman WWF Russia Amur Branch, Vladivostok, Russia

Marcel Darveau Ducks Unlimited Canada, Quebec, QC, Canada

Nick C. Davidson Institute for Land, Water and Society, Charles Sturt University, Albury, NSW, Australia

Nick Davidson Environmental, Wigmore, UK

J. Brian Davis Department of Wildlife, Fisheries and Aquaculture, Mississippi State University, Mississippi State, MS, USA

Victor Degtyarev Siberian Division of Russian Academy of Sciences, Institute for Biological Problems of Cryolithozone, Siberian Branch, Russian Academy of Sciences, Yakutsk, Russia

Ali Dehlavi Indus Ecoregion, WWF, Karachi, Pakistan

Karen Denyer National Wetland Trust, Cambridge, Waikato, New Zealand

James H. Devries Ducks Unlimited Canada, Stonewall, MB, Canada

Carmen Díaz-Paniagua Department of Wetland Ecology, Estación Biológica de Doñana (EBD-CSIC), Seville, Spain

Lars Dinesen Biologist, European Representative, Scientific and Technical Review Panel of the Ramsar Convention, Jyderup, Denmark

Alan Dixon Institute of Science and the Environment, University of Worcester, Henwick Grove, UK

Kevin E. Doherty United States Fish and Wildlife Service, Bismarck, ND, USA

Oscar O. Donde KMFRI, Lake Turkana Research Station, Lodwar, Kenya

Chultemin Dugarjav Institute of General and Experimental Biology Mongolian Academy of Sciences, Ulaanbaatar, Mongolia

Evgeny Egidarev Pacific Geographical Institute FEB RAS \ WWF Russia Amur Branch, Vladivostok, Russia

Asim I. El Moghraby Sudanese National Academy of Sciences, Khartoum, Sudan

Carmen Espoz Centro Bahía Lomas, Facultad de Ciencias, Universidad Santo Tomás, Santiago, Chile

Daniel Fehringer Ducks Unlimited, Inc., Rancho Cordova, CA, USA

Rocio Fernández-Zamudio Department of Wetland Ecology, Estación Biológica de Doñana (EBD-CSIC), Seville, Spain

Alejandra Figueroa Head Natural Resources and Biodiversity, Ministry of Environment, Santiago, Chile

C. Max Finlayson Institute for Land, Water and Society, Charles Sturt University, Albury, NSW, Australia

UNESCO-IHE, The Institute for Water Education, Delft, The Netherlands

Mohamed Reda Fishar Inland Water and Aquaculture Branch, National Institute of Oceanography and Fisheries (NIOF), Cairo, Egypt

Etienne Fluet-Chouinard Center for Limnology, University of Wisconsin-Madison, Madison, WI, USA

John Frikke The Danish Wadden Sea National Park, Rømø, Denmark

Ray H. Froend Centre for Ecosystem Management, School of Science, Edith Cowan University, Joondalup, WA, Australia

Susan M. Galatowitsch Department of Fisheries, Wildlife and Conservation Biology, University of Minnesota, Saint Paul, MN, USA

Peter Gell Water Research Network, Federation University Australia, Ballarat, VIC, Australia

Urania Giannakou Department of Medical Laboratory Studies, School of Health and Medical Care, Alexander Technological Educational Institute of Thessaloniki, Thessaloniki, Greece

Wim Giesen Euroconsult Mott MacDonald, Arnhem, AK, The Netherlands

Laurence Gill University of Dublin, Trinity College, Dublin, Ireland

Beverly Gingras Ducks Unlimited Canada, Edmonton, AB, Canada

Gordon Goldsborough Department of Biological Sciences, University of Manitoba, Winnipeg, MB, Canada

Martha Catalina Gómez-Cubillos Universidad Nacional de Colombia, sede Caribe, Santa Marta, Colombia

Brij Gopal Centre for Inland Waters in South Asia, Jaipur, Rajasthan, India

Oleg Goroshko Daursky Biosphere Reserve, Chita, Russia

Natasha J. Gownaris School of Marine and Atmospheric Sciences, Stony Brook University, Stony Brook, NY, USA

Alana Grech Department of Environmental Sciences, Macquarie University, Sydney, NSW, Australia

Andy J. Green Department of Wetland Ecology, Estación Biológica de Doñana (EBD-CSIC), Seville, Spain

Holly Greening Tampa Bay Estuary Program, St. Petersburg, FL, USA

Patrick Grillas Tour du Valat, Research Institute for Mediterranean Wetlands, Le Sambuc, Arles, France

Ab P. Grootjans Centre for Energy and Environmental Studies, University of Groningen, Groningen, The Netherlands

Institute of Water and Wetland Research, Radboud University Nijmegen, Nijmegen, The Netherlands

Althea T. Grundling Water Science Programme, Agricultural Research Council – Institute for Soil, Climate and Water, Pretoria, Gauteng, South Africa

Department of Geography and Environmental Management, University of Waterloo, Waterloo, ON, Canada

Applied Behavioural Ecology and Ecosystem Research Unit, University of South Africa, Pretoria, South Africa

Piet-Louis Grundling Centre for Environmental Management, University of the Free State, Bloemfontein, South Africa

Fidaa F. Haddad Regional Dryland, Livelihoods and Gender Program, IUCN Regional Office for West Asia (ROWA), Amman, Jordan

James Harris International Crane Foundation, Baraboo, WI, USA

Ian J. Harrison Center for Environment and Peace, Conservation International, Arlington, VA, USA

Khalid Hassaballah UNESCO-IHE Institute for Water Education, Delft, The Netherlands

Hydraulics Research Center, Wad Medani, Sudan

Sandra Hernández Fundación Humedales, Bogotá, Colombia

Pierre Horwitz Centre for Ecosystem Management, School of Science, Edith Cowan University, Joondalup, WA, Australia

David W. Howerter Ducks Unlimited Canada, Stonewall, MB, Canada

Orieta Hulea WWF Danube-Carpathian Programme, Bucharest, Romania

WWF Danube-Carpathian Programme, Vienna, Austria

Beth Huning San Francisco Bay Joint Venture, Fairfax, CA, USA

Carles Ibáñez IRTA, Aquatic Ecosystems Program, Sant Carles de la Ràpita, Catalonia, Spain

Karen Patricia Ibarra Instituto de Investigaciones Marinas y Costeras "José Benito Vives de Andreis" (INVEMAR), Santa Marta, Colombia

Kenneth Irvine UNESCO-IHE Institute of Water Education, Delft, The Netherlands

Rodolfo Iturraspe Universidad Nacional de Tierra del Fuego, Ushuaia, Tierra del Fuego, Argentina

Anthony Janicki Janicki Environmental, Inc., St. Petersburg, FL, USA

Guyonne F. E. Janss Department of Wetland Ecology, Estación Biológica de Doñana (EBD-CSIC), Seville, Spain

Wolfgang J. Junk Instituto Nacional de Ciência e Tecnologia em Áreas Úmidas (INCT-INAU), Universidade Federal de Mato Grosso (UFMT), Cuiabá, MT, Brazil

Patricia Kandus Instituto de Investigación e Ingeniería Ambiental, Universidad Nacional de San Martín, San Martín, Provincia de Buenos Aires, Argentina

Masoud Bagherzadeh Karimi Department of Environment, Study Center for Environment, Wetlands and National Parks, Tehran, Iran

Andrea Kelly Broads Authority, Norwich, Norfolk, UK

Richard G. Kempka Ducks Unlimited, Inc., Rancho Cordova, CA, USA

The Climate Trust, Portland, OR, USA

Peter Kershaw University of Alberta, Edmonton, AB, Canada

Seung Hyeon Kim Department of Biological Sciences, Pusan National University, Busan, South Korea

Young Kyun Kim Department of Biological Sciences, Pusan National University, Busan, South Korea

Sarah Kimberley University of Dublin, Trinity College, Dublin, Ireland

Richard T. Kingsford Centre for Ecosystem Science, School of Biological, Earth and Environmental Sciences, UNSW Australia, Sydney, NSW, Australia

Julius Kipkemboi Department of Biological Sciences, Egerton University, Egerton, Njoro, Kenya

Satoshi Kobayashi Faculty of Economics, Kushiro Public University, Kushiro, Hokkaido, Japan

Christina Kontaxi Administration of Environment and Spatial Planning, Region of Central Macedonia, Thessaloniki, Greece

Andrej Kranjc Slovenian Academy of Sciences and Arts, Ljubljana, SI, Slovenia

Lalit Kumar Ecosystem Management, School of Environmental and Rural Science, University of New England, Armidale, NSW, Australia

Ritesh Kumar Wetlands International South Asia, New Delhi, India

Jan Květ Faculty of Science, University of South Bohemia České Budějovice, České Budějovice, Czech Republic

Czech Academy of Sciences, CzechGlobe, Institute for Global Change Research, Brno, Czech Republic

Pramod Lamsal Ecosystem Management, School of Environmental and Rural Science, University of New England, Armidale, NSW, Australia

Karsten Laursen Department of Bioscience, Aarhus University, Aarhus, Denmark

Kun-Seop Lee Department of Biological Sciences, Pusan National University, Busan, South Korea

Jean-Claude Lefeuvre Department of Ecology and Biodiversity Management, National Museum of Natural History, Paris, France

EA 7316 Biodiversity and Land Management, University of Rennes 1, Rennes, France

Bernhard Lehner Department of Geography, McGill University, Montreal, QC, Canada

Harald J. L. Leummens Water and Nature, Heerlen, The Netherlands

Cui Lijuan Institute of Wetland Research, Chinese Academy of Forestry, Beijing, China

Richard Lindsay Sustainability Research Institute, University of East London, London, UK

Anton Linström Wet Earth Eco-Specs, Lydenburg, South Africa

Ahmad Lotfi Conservation of Iranian Wetland Project, Department of Environment, Tehran, Iran

Senior Irrigation Engineers, Member of Board of Directors, Pandam Consulting Engineers, Tehran, Iran

Ronald A. Loughland Environmental Protection Department, Saudi Aramco, Dhahran, Ash Sharqiyah, Saudi Arabia

I. Reza Lubis Wetlands International Indonesia, Bogor, Indonesia

Diego Luna-Quevedo WHSRN Executive Office- Manomet, Plymouth, MA, USA

John MacKinnon University of Kent, Canterbury, UK

Gernant Magnin WWF – Netherlands, Freshwater Programme, Zeist, The Netherlands

Tomás Maiztegui Instituto de Limnología de La Plata "Raúl Ringuelet", La Plata, Argentina

John Malala KMFRI, Lake Turkana Research Station, Lodwar, Kenya

Mark Mallory Coastal Wetland Ecosystems, Biology Department, Acadia University, Wolfville, NS, Canada

José Ernesto Mancera Pineda Universidad Nacional de Colombia, Bogotá, Colombia

Ciudad Universitaria, Mexico City, Mexico

Zhang Manyin Institute of Wetland Research, Chinese Academy of Forestry, Beijing, China

Leonardo V. Marques Departamento de Ecologia, Universidade do Estado do Rio de Janeiro, Rio de Janeiro, Brazil

Ricardo Matus Centro de Rehabilitación de Aves Leñadura, Punta Arenas, Chile

Matthew McCartney Ecosystem Services, International Water Management Institute, Vientiane, Lao People's Democratic Republic

Regional Office for Southeast Asia and the Mekong, International Water Management Institute, Vientiane, Lao People's Democratic Republic

Len J. McKenzie Centre for Tropical Water and Aquatic Ecosystem Research, James Cook University, Cairns and Townsville, QLD, Australia

Jim Meeker Formerly of: Northland College, Ashland, WI, USA

Jim Meeker: deceased

Hans Meltofte Department of Bioscience, Aarhus University, Roskilde, Denmark

Burhan Teoman Meric Wetlands Division, Ministry of Forests and Water Affairs, General Directory of Nature Conservation and National Parks, Ankara, Turkey

Mathis Loïc Messager School of Aquatic and Fishery Sciences, University of Washington, Seattle, WA, USA

Peter-John Meynell ICEM – International Centre for Environmental Management, Hanoi, Vietnam

Andrej Mihevc Karst Research Institute, Research Centre of the Slovenian Academy for Science and Arts, Postojna, Slovenia

G. Randy Milton Nova Scotia Department of Natural Resources, Kentville, NS, Canada

Tatiana Minayeva Wetlands International, Ede, The Netherlands

Priscilla G. Minotti Instituto de Investigación e Ingeniería Ambiental (3iA), Universidad Nacional de San Martín, San Martín, Provincia de Buenos Aires, Argentina

Yasir A. Mohamed Hydraulic Research Center, MoWRIE, Wad Medani, Sudan

Department of Integrated Water Systems and Governance, UNESCO-IHE, DA, Delft, The Netherlands

Faculty of Civil Engineering and Applied Geosciences, Water Resources Section, Delft University of Technology, GA, Delft, The Netherlands

Arno Mohl WWF Austria, Vienna, Austria

Nial Moores Birds Korea, Busan, Korea (Republic of)

James Mugo Kenya Marine and Fisheries Research Institute, Naivasha, Kenya

Anne Murray Nature Guides BC, Delta, BC, Canada

Rab Nawaz Indus Ecoregion, WWF, Karachi, Pakistan

Rasmus Due Nielsen Department of Bioscience, Aarhus University, Kalø, Denmark

Yus Rusila Noor Wetlands International Indonesia, Bogor, Indonesia

Alyssa B. Novak Boston University, Earth and Environment, Boston, MA, USA

F. Fernando Novoa Centro de Ecología Aplicada (CEA), Santiago, Región Metropolitana, Chile

Catia Nunes da Cunha Instituto Nacional de Ciência e Tecnologia em Áreas Úmidas (INCT-INAU), Universidade Federal de Mato Grosso (UFMT), Cuiabá, MT, Brazil

Depto Botânica e Ecologia/Núcleo de Estudos Ecológicos do Pantanal (NEPA), Instituto de Biociências, UFMT, Cuiabá, MT, Brazil

Kevin O. Obiero KMFRI, Lake Turkana Research Station, Lodwar, Kenya

Joanne F. Ocock Water and Wetlands Team, Science Division, NSW Office of Environment and Heritage, Sydney, NSW, Australia

Centre for Ecosystem Science, School of Biological, Earth and Environmental Sciences, UNSW, Sydney, NSW, Australia

Jacob. E. Ojuok Kenya Marine and Fisheries Research Institute, Kisumu, Kenya

William Ojwang Kenya Marine and Fisheries Research Institute (KMFRI), Mombasa, Kisumu, Kenya

WWF-Kenya, Nairobi, Kenya

Casianes Olilo Kenya Marine and Fisheries Research Institute (KMFRI), Mombasa, Kisumu, Kenya

Reuben Omondi Department of Applied Aquatic Sciences, Kisii University, Kisii, Kenya

KMFRI, Kisumu Research Centre, Kisumu, Kenya

Collins Ongore Kenya Marine and Fisheries Research Institute (KMFRI), Mombasa, Kisumu, Kenya

Susan Page Department of Geography, University of Leicester, Leicester, UK

Pranati Patnaik Wetlands International South Asia, New Delhi, India

Ajit Kumar Pattnaik Chilika Development Authority, Bhubaneswar, Odisha, India

Richard Payne Environment, University of York, Heslington, York, UK

Richard G. Pearson College of Science and Engineering, James Cook University, Townsville, QLD, Australia

Laura Victoria Perdomo Trujillo Universidad Nacional de Colombia, Bogotá, Colombia

Mike Perlmutter Environmental Services Division, City of Oakland Public Works Department, Oakland, CA, USA

Ib Krag Petersen Department of Bioscience, Aarhus University, Kalø, Denmark

Mark Petrie Ducks Unlimited, Inc., Rancho Cordova, CA, USA

Kevin Petrik Ducks Unlimited, Inc., Rancho Cordova, CA, USA

Ellen K. Pikitch School of Marine and Atmospheric Sciences, Stony Brook University, Stony Brook, NY, USA

María Pinilla-Vargas Fundación Humedales, Bogotá, Colombia

Jamie Pittock Fenner School of Environment and Society, The Australian National University, Canberra, Australia

Jan Pokorný ENKI, o.p.s., Třeboň, Czech Republic

Colin Poole Wildlife Conservation Society, Phnom Penh, Cambodia

R. Crawford Prentice Nature Management Services, Histon, Cambridge, UK

Jonathan S. Price Department of Geography and Environmental Management, University of Waterloo, Waterloo, ON, Canada

Perdana K. Prihartato Environmental Protection Department, Saudi Aramco, Dhahran, Ash Sharqiyah, Saudi Arabia

Ali M. Qasem Environmental Engineering Div./ Environmental Protection Dept, Saudi Aramco, Dhahran, Saudi Arabia

Rubén Darío Quintana Instituto de Investigación e Ingeniería Ambiental, Universidad Nacional de San Martín, San Martín, Provincia de Buenos Aires, Argentina

Consejo Nacional de Investigaciones Científicas y Técnicas (CONICET), Buenos Aires, Argentina

Dipa Rais Wetlands International Indonesia, Bogor, Indonesia

Lars Ramberg Uppsala, Sweden

Michael A. Rasheed Centre for Tropical Water and Aquatic Ecosystem Research, James Cook University, Cairns and Townsville, QLD, Australia

Lisa-Maria Rebelo Water Futures, International Water Management Institute, Vientiane, Lao People's Democratic Republic

Regional Office for Southeast Asia and the Mekong, International Water Management Institute, Vientiane, Lao People's Democratic Republic

Anna M. Redden Acadia Centre for Estuarine Research, Acadia University, Wolfville, NS, Canada

Frederic A. Reid Ducks Unlimited, Inc., Rancho Cordova, CA, USA

Carmen Revenga The Nature Conservancy, Arlington, VA, USA

Curtis J. Richardson Nicholas School of the Environment, Duke University Wetland Center, Durham, NC, USA

Jack Rieley School of Geography, University of Nottingham, Nottingham, UK

Hugh Robertson Freshwater Section, Department of Conservation, Nelson, New Zealand

Jenny Alexandra Rodríguez-Rodríguez Instituto de Investigaciones Marinas y Costeras "José Benito Vives de Andreis" (INVEMAR), Santa Marta, Colombia

Juan Pablo Rubilar Minera Los Pelambres, Santiago, Chile

Mario Enrique Rueda Instituto de Investigaciones Marinas y Costeras "José Benito Vives de Andreis" (INVEMAR), Santa Marta, Colombia

Bárbara Saavedra Wildlife Conservation Society, Bronx, NY, USA

Sadegh Sadeghi Zadegan Ornithological Unit, Wildlife Bureau, Department of Environment, Tehran, Iran

Brian G. Sanderson Acadia Centre for Estuarine Research, Acadia University, Wolfville, NS, Canada

Jeffrey Shatford Ministry of Forests, Lands and Natural Resource Operations, Resource Management Objectives Branch, Victoria, BC, Canada

Ed T. Sherwood Tampa Bay Estuary Program, St. Petersburg, FL, USA

Cathy A. Short Lee, NH, USA

Frederick T. Short Department of Natural Resources and the Environment, Jackson Estuarine Laboratory, University of New Hampshire, Durham, NH, USA

Paula Cristina Sierra-Correa Instituto de Investigaciones Marinas y Costeras "Jose Benito Vives de Andreis" (INVEMAR), Santa Marta, Colombia

Marcel J. Silvius Wageningen, The Netherlands

Eugene Simonov Rivers without Boundaries Coalition, Dalian, China

Daursky NNR, Dalian, China

Andrey Sirin Center for Protection and Restoration of Peatland Ecosystems, Institute of Forest Science, Russian Academy of Sciences, Moscow, Russia

Laboratory of Peatland Forestry and Amelioration, Institute of Forest Science, Russian Academy of Sciences, Moscow, Russia

Boris Sket Oddelek za biologijo, Biotehniška fakulteta, Univerza v Ljubljani, Ljubljana, Slovenia

Stuart Slattery Ducks Unlimited Canada, Stonewall, MB, Canada

Kevin Smith Ducks Unlimited Canada, Edmonton, AB, Canada

Agatha Cristinne Soares Departamento de Ecologia, Universidade do Estado do Rio de Janeiro, Rio de Janeiro, Brazil

Bea Sommer Centre for Ecosystem Management, School of Science, Edith Cowan University, Joondalup, WA, Australia

Ruth Spell Ducks Unlimited, Inc., Rancho Cordova, CA, USA

Melanie L. J. Stiassny Division of Vertebrate Zoology, Ichthyology Department, American Museum of Natural History, New York City, USA

Liying Su International Crane Foundation, Baraboo, WI, USA

J. Peter Sutcliffe Independent Consultant, University of Huddersfield, Huddersfield, UK

Randolph R. Thaman The University of the South Pacific, Suva, Fiji

Rachael F. Thomas Water and Wetlands Team, Science Division, NSW Office of Environment and Heritage, Sydney, NSW, Australia

Centre for Ecosystem Science, School of Biological, Earth and Environmental Sciences, UNSW, Sydney, NSW, Australia

Ole Thorup Amphi Consult, Ribe, Denmark

Naomi Tillison Natural Resources Director, Bad River Band of Lake Superior Chippewa, Odanah, WI, USA

Ralph W. Tiner Institute for Wetlands and Environmental Education and Research, Inc., Leverett, MA, USA

Tatiana Tkachuk Daursky Biosphere Reserve, Zabaikalsky State University, Chita, Russia

Triet Tran International Crane Foundation, Baraboo, WI, USA

University of Natural Science, Vietnam National University, Ho Chi Minh City, Vietnam

Anett Trebitz Mid-Continent Ecology Division, United States Environmental Protection Agency, Duluth, MN, USA

Tristan Tyrrell Tentera, Montreal, QC, Canada

Stefan Uhlenbrook UNESCO-IHE Institute for Water Education, Delft, The Netherlands

Delft University of Technology, Delft, The Netherlands

Richard K. F. Unsworth College of Science, Wallace Building, Swansea University, Swansea, UK

Mauricio Valderrama Fundación Humedales, Bogotá, Colombia

Eugenio Valderrama-Escallón Fundación Humedales, Universidad El Bosque, Bogotá, Colombia

Loïc Valéry Department of Ecology and Biodiversity Management, National Museum of Natural History, Paris, France

EA 7316 Biodiversity and Land Management, University of Rennes 1, Rennes, France

Anne A. van Dam Aquatic Ecosystems Group, Department of Water Science and Engineering, UNESCO-IHE Institute for Water Education, Delft, The Netherlands

Stella Vareltzidou Axios-Loudias-Aliakmonas Management Authority, Chalastra, Thessaloniki, Greece

Yvonne I. Verkuil Conservation Ecology Group; Groningen Institute for Evolutionary Life Sciences (GELIFES), University of Groningen, Groningen, The Netherlands

Lucia Victoria Licero Villanueva Instituto de Investigaciones Marinas y Costeras "Jose Benito Vives de Andreis" (INVEMAR), Santa Marta, Colombia

Dale H. Vitt Department of Plant Biology and Center for Ecology, Southern Illinois University, Carbondale, IL, USA

Despoina Vokou Department of Ecology, School of Biology, Aristotle University of Thessaloniki, Thessaloniki, Greece

Steve Waldren University of Dublin, Trinity College, Dublin, Ireland

Johann Walker Ducks Unlimited, Great Plains Region, Bismarck, ND, USA

Jane Waterhouse Catchment to Reef Research Group, TropWATER, Centre for Tropical Water and Aquatic Ecosystem Research, James Cook University, Townsville, QLD, Australia

Xu Weigang Institute of Wetland Research, Chinese Academy of Forestry, Beijing, China

Yoyok Wibisono Wetlands International – Indonesia Programme, Bogor, Indonesia

Daphne Willems WWF-Netherlands, Zeist, The Netherlands

Florian Wittmann Institute of Floodplain Ecology, Karlsruhe Institute of Technology - KIT, Rastatt, Germany

Adrian Wood Business School, University of Huddersfield, Huddersfield, UK

Dale Wrubleski Institute for Wetland and Waterfowl Research, Ducks Unlimited Canada, Stonewall, MB, Canada

Muta H. Zakaria Department of Aquaculture, Faculty of Agriculture, Universiti Putra Malaysia, Serdang, Selangor Darul Ehsan, Malaysia

Section V

Central and South America

Amazon River Basin

57

Florian Wittmann and Wolfgang J. Junk

Contents

Introduction	728
Types and Extent of Amazonian Wetland Types	729
Large-River Floodplains	729
Interfluvial Wetlands	732
Coastal Wetlands	733
Biodiversity	733
Flora	733
Fauna	735
Conservation Status	737
Ecosystem Services	738
Threats and Future Challenges	739
References	742

Abstract

Amazonian wetlands cover an area of more than two million km^2 and consist of different wetland types that vary in hydrology, water and soil fertility, and productivity. Wetlands harbor a large fraction of Amazonian biodiversity also including many endemic plant and animal species, and provide multiple ecosystem services to humans. However, few Amazonian countries have detailed wetland inventories, maps, and classification systems, and therefore also lack specific conservation and wetland management strategies. While remote and

---

F. Wittmann (✉)
Institute of Floodplain Ecology, Karlsruhe Institute of Technology - KIT, Rastatt, Germany
e-mail: f-wittmann@web.de

W. J. Junk
Instituto Nacional de Ciência e Tecnologia em Áreas Úmidas (INCT-INAU), Universidade Federal de Mato Grosso (UFMT), Cuiabá, MT, Brazil
e-mail: wjj@evolbio.mpg.de

© Springer Science+Business Media B.V., part of Springer Nature 2018
C. M. Finlayson et al. (eds.), *The Wetland Book*,
https://doi.org/10.1007/978-94-007-4001-3_83

scarcely inhabited wetland types are still in a fairly pristine stage, the conservation status of most Amazonian wetlands is at high risk because of multiple threats and in particular due to the lack of national and transnational policies regarding wetland conservation. Major threats of Amazonian wetlands include land cover change, river damming for hydropower generation, pollution, ecosystem degradation and local changes in hydrology. This trend can only be mitigated by the creation of a more holistic understanding of the benefits provided by wetlands combined with integrated, transnational conservation measures.

Keywords

Biodiversity · Ecosystem Degradation · Endemism · Hydropower Generation · Large-river floodplains

Introduction

The Amazon basin is located in the northern part of South America and comprises an area of 6.8 million km^2 (Fig. 1). Due to high annual precipitation (generally >2,000 mm year^{-1}) and flat relief in most parts of the basin, approximately 2 million km^2 or 30% of its area is covered by a complex system of different wetland types. The Amazon harbors four of the ten largest rivers on Earth in terms of water discharge and drainage area, and its river system discharges ~16–18% of the global freshwater flow (Latrubesse 2008). Amazonian wetlands vary considerably with respect to hydrology, water and soil fertility, vegetation cover, diversity of plant and animal species, and primary and secondary productivity (Junk et al. 2011). They include seasonal floodplains along large river systems, and interfluvial wetlands comprised of episodically flooded riparian areas along upland streams, permanently flooded swamps, and hydromorphic savannas. Coastal wetlands include mangroves and tidal floodplains that range from freshwater to saline conditions. Altitude wetlands occur >500 m above sea level upon northern Amazonian table mountains and the eastern slope of the Andes.

While Amazonian large-river wetlands are relatively well studied, studies of other wetland types are scarce and highly scattered in the scientific literature. Very little information exists for Amazonian altitude wetlands. Moreover, few Amazonian countries have detailed wetland inventories, maps, and classification systems and therefore also lack specific conservation and wetland management strategies. Summaries exist for the wetlands of Bolivia (Navarro and Maldonado 2002; Pouilly et al. 2004). Only recently, Junk et al. (2011) elaborated a wetland classification of the Amazon basin based on geological, hydrological, and botanical parameters. A similar approach is being developed for Colombian wetlands. However, due to high environmental variability and remote location, there are still significant knowledge gaps regarding extent, biodiversity, and ecosystem services of many Amazonian wetland types.

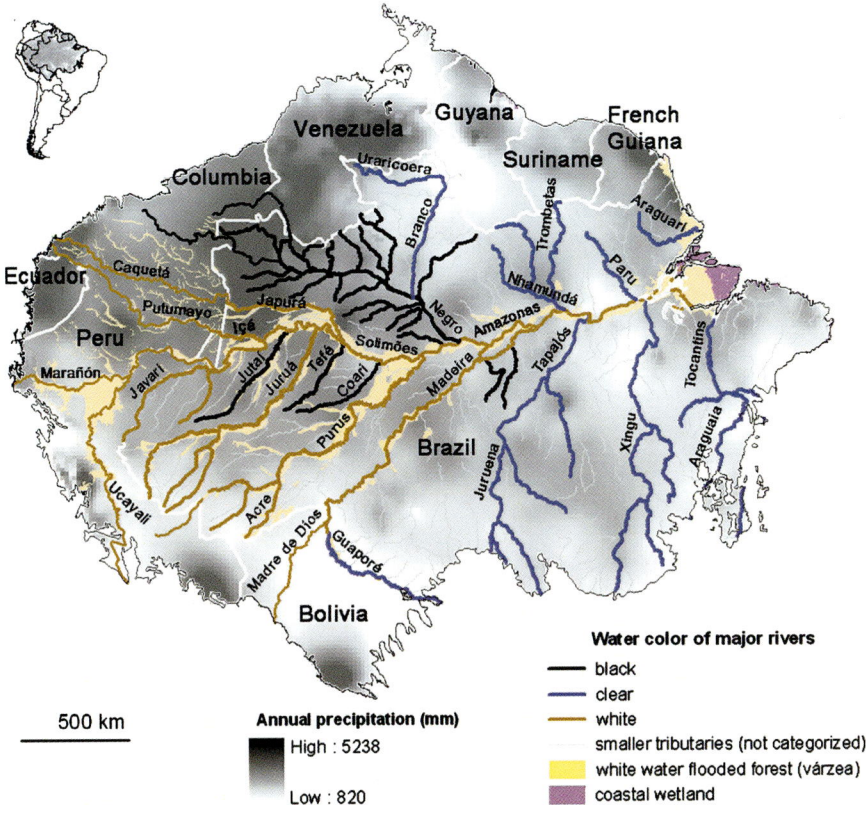

Fig. 1 The large *white-water*, *black-water*, and *clear-water* rivers of the Amazon basin (© With kind permission from Springer Science+Business Media: Junk et al. (2011), Fig. 1)

Types and Extent of Amazonian Wetland Types

Large-River Floodplains

Amazonian large-river floodplains occur along major rivers (>fifth order) and are subject to a monomodal, predictable flood pulse (Junk et al. 1989) of high amplitude (>4 m), with one low- and one high-water period of several months' duration each. These floodplains cover an area of approximately 750,000 km^2 (Table 1). Flood amplitudes are highest in the central part of the Amazon basin, where they exceed 10 m near the confluence of the Solimões and Negro Rivers and the several tributaries of the Solimões/Amazon system, such as the Madeira and Purus Rivers, and decline to 4–6 m toward the western and eastern parts of the basin, either through

Table 1 Major Amazonian wetland types and their extent

Wetland type	Area (10^3 km^2)
White-water floodplains (várzea)[a]	
Amazon main stem[b]	117.1
Madeira[c]	92.5
Marañon, including Marañon-Ucayali palm swamps	71.0
Ucayali	41.5
Purus	36.1
Caquetá-Japurá	31.9
Juruá	20.9
Putumayo-Içá	20.3
Napo	10.6
Smaller white-water rivers (Javarí, Jutai, Nanay, Jandiatuba)	14.4
Subtotal	456.3
Clear- and black-water floodplains (*igapó*)[a]	
Araguaia-Tocantins[d]	76.6
Negro[e]	119.6
Xingu	37.1
Tapajós	22.4
Trombetas[f]	7.4
Smaller black- and clear-water rivers (Abacaxis, Uatumã, Madeirinha, Paru, Coari, Curuá Una, Nhamundá, Curuá, Tefé, Maicuru, Jarauçu, Piorini, Mamuru)	39.3
Subtotal	302.4
Other wetland types	
Riparian zones along high-order rivers (baixios)[g]	1,000.0
Hydromorphic savannas not included in large river basins (Campina and Campinarana)[h]	418.7
Palm swamps not included in large river basins (buritizais–aguajales)[i]	80.3
High-elevation freshwater ecosystems[j]	70.0
Mangroves[k]	11.0
Total	2,338.7

[a]Data from Melack and Hess (2010) for the floodable area of the Amazon lowland (>500 m above sea level) at 100-m resolution
[b]Including aggregated area of basins draining directly to the mainstem floodplain
[c]Excluding the floodable area of Llanos de Moxos
[d]Data from Castello et al. (2013) excluding the floodable area of Bananal
[e]Excluding the floodable area of Negro Campinas and Campinaranas
[f]Excluding the floodable area of Trombetas savannas
[g]According to estimates on river density by Junk (1993)
[h]Data from Junk et al. (2011), including Negro River Campinas/Campinaranas, Madeira River Campinas, Trombetas savannas, Roraima/Rupununi savannas, Llanos de Moxos, Bananal, and smaller savannas interspersed in terra firme. Note that not all white-sand savannas are hydromorphic.
[i]Palm swamps on mineral and organic substrates as indicated in (Lähteenoja et al. 2009) but excluding the Marañon-Ucayali palm swamp complex
[j]Estimate for freshwater ecosystems >500 m above sea level (Andean streams, Tepuis) by Castello et al. (2013)
[k]Data from Huber et al. (1995)

increased declivity in the pre-Andean zone or through enlarged river beds near the mouth of the Amazon River into the Atlantic Ocean.

Amazonian large-river wetlands are classified by varying sediment and nutrient loads of river waters, which are traced to the geology of their catchments. The three major freshwater types within the Amazon basin, white waters, black waters, and clear waters (Fig. 1), were classified by Sioli (1954). White-water rivers, such as the Solimões-Amazon main stem and the Juruá, Japurá, Purús, and Madeira Rivers, drain the Andes or the Andean foothills. White-water rivers are rich in suspended matter and nutrients and have near-neutral pH. They carry clay minerals that are characterized by a high ion-exchange capacity, such as smectite and illite, releasing nutrients to the alluvial soils (Irion et al. 2010). The high loads of suspended solids give them a muddy (white) color, and floodplain substrates are exceptionally fertile in comparison to substrates of other Amazonian habitats (Junk et al. 2012). White-water floodplains are called *várzea* and are estimated to cover approximately 450,000 km^2 (Table 1).

Black-water rivers drain the Archaic or Precambrian formations in the North and South of the Amazon basin, such as the Guyana and Central Brazilian Shields (Fig. 1). Their catchment is covered by strongly weathered, tertiary sediments that formed kaolinitic and/or strongly podzolic soils. Their water is poor in suspended solids, is transparent, and has a dark, reddish-brown color, which derives from the accumulation of organic material from adjacent forest and leached from podzolic soils. The water is acidic, with a pH that ranges from 4 to 5. Black-water floodplains are called *igapó*, and they are generally very poor in nutrients. The largest Amazonian black-water river is the Negro River, which alone floods an area of approximately 118,000 km^2 (Melack and Hess 2010). Smaller black-water rivers in Central Amazonia are the Tefé, Coari, and Uatumã Rivers.

Clear-water rivers also drain the Archaic and Precambrian formations of the Guyana or the Central Brazilian Shields but are more frequent river types in the eastern part of the Amazon basin (Fig. 1). Although the physicochemical variability of clear waters may be relatively large (Junk et al. 2011), their water is mostly transparent to greenish colored, with low contents of electrolytes, and acidic. The most important Amazonian clear-water river is the Tapajós; other large rivers are the Trombetas, Xingu, Araguaia-Tocantins, and Branco Rivers. Due to their origin upon the cratonic shields, many clear-water rivers have rocky beds in the upper and middle reaches. The rocky beds combined with high water velocities and low loads of dissolved matter compared to other types of lowland rivers make them the preferred choice for the establishment of dams for hydropower generation. Prance (1979) classified clear-water floodplains as *igapó*, due to their floristic resemblance to black-water floodplains. However, recent studies have shown that the flora of clear-water floodplains may be distinct from black-water floodplains, as they also contain many endemic tree species as well as floral elements from white-water floodplains (Junk et al. 2011). Seasonal floodplains influenced by black- and clear-water rivers cover an area of approximately 300,000 km^2 (Table 1).

Interfluvial Wetlands

The thousands of smaller streams and creeks draining the uplands (*terra firme*) are characterized by riparian zones of varying size that episodically inundate for hours, days, or weeks according to local rainfall events. Most creeks in central and eastern Amazonia contain black waters of poor nutrient status. However, when creeks drain alluvial deposits of Andean origin, as in most parts of the western Amazon basin, and, in the central part of the basin, upon ancient *várzea* substrates, nutrient conditions might be intermediate. Irion et al. (2010) reported that nutrient contents of alluvial substrates of these *paleo-várzeas*, which were deposited during interglacial periods of high sea level, are lower than that of recent *várzea*, however still higher than that of non-flooded uplands. There is still no reliable quantification of the spatial coverage of riparian zones along Amazonian creeks, but based on the high stream density of 2 km per km^2 in most parts of the Amazon basin, Junk (1993) estimated that the total area of episodically flooded riparian zones in the Amazon basin accounts for approximately 1 million km^2. In Brazilian Amazonia, the vegetation formation of riparian zones is called *baixios*.

Amazonian permanent swamps are characterized by relatively stable water levels, with flood amplitudes of maximally 1–2 m (Junk et al. 2011). Several swamps of the Negro River and Marañon–Ucayali basins extend over several thousand kilometers squared, and their total area in the Amazon may exceed 150,000 km^2 (Lähteenoja et al. 2009). Most swamps are of intermediate to poor nutrient status, and some of them are characterized by peat formation to depths of up to 9 m (Householder et al. 2012), particularly in the western Amazon basin. In Brazilian Amazonia, permanent swamps are called *buritizais*, because of the predominance of the indicator palm species *Mauritia flexuosa* (buriti). In the Peruvian, Venezuelan, Ecuadorian, and Colombian Amazon, they are called *aguajales*, *morichales*, *moritales*, and *cananguchales*, respectively (Kahn 1991).

Amazonian white-sand savannas are characterized by rainwater-fed soils of tertiary origin, strongly leached, and of low fertility (Luizão et al. 2007). Sandy soils have an underlying hardpan (Horbe et al. 2004), and many white-sand savannas are shallowly flooded during the rainy seasons or have high groundwater tables and saturated soils (Franco and Dezzeo 1994; Junk et al. 2011). There are no reliable numbers about the extent of this Amazonian wetland type, because many white-sand savannas are small and patchily distributed in Amazonian uplands and along many black-water rivers. Other white-sand savannas, however, cover huge areas, such as the Roraima/Rupununi (~80,000 km^2), Llanos de Moxos (~117,000 km^2), and Araguaia savannas (~88,000 km^2, Dias 2014). Junk et al. (2011) estimated the total coverage of Amazonian white-sand savannas to account for at least 418,000 km^2; however, all of this area is not subject to flooding (Table 1). There are many different local names for this wetland type in the Amazon, such as *bana*, *cunurí*, *muri*, and *yevaro* in Venezuela, *varillales* in Peru, *Wallaba* in the Guyanas, and *Campina*, *Campinarana*, *Amazonian caatinga*, and *caatinga-pó* in Brazil (Takeuchi 1960; Klinge and Medina 1979; Coomes 1997).

Coastal Wetlands

A polymodal, predictable flood pulse affects most coastal tidal wetlands, where flooding and varying degrees of salinity create specific ecological conditions (Junk et al. 2011). Major habitats are mangroves, beaches, mud flats, and coastal lagoons. Because of the flat relief of the coastal area, lower reaches of the Amazonian rivers are affected by marine tides more than 100 km upriver. While only the outermost seaside part is influenced by salt waters, tidal freshwater forests along the Amazon main stem and major tributaries, such as the Xingu, Jari, and Paru Rivers, cover extensive areas. In the Amazon estuary at Marajó Island, natural levees protect inland floodplain areas from tidal pulses and create freshwater swamps. Mangroves extend along the coasts of Suriname and the Brazilian States of Amapá, Pará, and Maranhão. Their extent was estimated to be 11,000 km^2 (Huber et al. 1995).

Biodiversity

In general, wetlands harbor a large fraction of global biodiversity, and tropical wetlands in particular are considered biodiversity hotspots (Gopal et al. 2000). As demonstrated above, the complex wetland system of the Amazon basin includes several large and very different wetland types in terms of hydrology, physicochemical water characteristics, nutrient status, productivity, and hydro-ecologic connectivity. Due to the remote location of many Amazonian wetlands, databases on species numbers in most taxa are incomplete, with new species being described continuously. Moreover, wetlands attract many terrestrial species that facultatively colonize wetlands (i.e., vascular plant species) or that use wetlands temporarily or episodically as habitat and food source (i.e., migratory birds, deep-water fish, terrestrial mammals). The ephemeral nature of wetland interactions with such a wide variety of species complicates the generation of complete species lists (Wittmann et al. 2015). In the following, we summarize the state of knowledge about Amazonian freshwater biodiversity in flora and fauna.

Flora

Most Amazonian wetlands are forested. Of the approximately 5,000 tree species with valid species names occurring in the Amazon basin (Ter Steege et al. 2013), approximately 50% are able to colonize temporally or permanently waterlogged and/or inundated substrates (F. Wittmann and Amazon Tree Diversity Network (ATDN) unpublished data). Amazonian large-river floodplains harbor the most species-rich floodplain forests on Earth (Wittmann et al. 2006). In Amazonian *várzea*, more than 1,000 tree species with valid species names were described, while in Amazonian *igapó*, the number of tree species with valid names amounts to approximately 600 (Wittmann et al. 2010). In general, tree species diversity in floodplain forests increases from eastern to western Amazonia (Wittmann et al. 2006), as such reflecting the trend of increasing diversity also reported for upland forests, which is presumably coupled to the longitudinal gradient of rainfall seasonality and soil-nutrient availability (Ter Steege et al. 2003). Brackish tidal *várzea* forests near the mouth of the Amazon are relatively poor in tree species and are characterized by the predominance of the palm *Euterpe oleracea* (Queiroz et al. 2005). In contrast, western Amazonian *várzea* forests may harbor up to 160 tree species ($\geq$10-cm diameter at breast height) in a single hectare (Wittmann et al. 2006).

Of the 658 most important (abundant + frequent) Amazonian *várzea* tree species, about 11% are endemic habitat specialists, and another 20% are floodplain specialists that only facultatively occur with low-density populations in other Amazonian habitat types (Wittmann et al. 2013). For *igapó*, the low number of floristic inventories is not sufficient for comparison; however, Prance (1979), Kubitzki (1989), and Mori (2001) suggested that this floodplain type also contains elevated numbers of endemic and specialist floodplain tree species.

Numbers of tree species in other Amazonian wetland types are scattered within the scientific literature and may be highly variable at different locations. Episodically flooded forests along riparian zones of upland streams in the central and southern Amazon basin might be as tree species rich as non-flooded upland forests, with local richness often exceeding 150 species ha^{-1} ($\geq$10-cm diameter at breast height). Swamp forests are thought to be relatively tree species poor environments compared to other Amazonian habitat types (Pitman et al. 2014), but local ecotones may harbor similar high species richness than other types of wetland forest (Galeano et al. 2015), with high environmental and floristic variability on organic peat and mineral substrates (Householder et al. 2012, 2015). Amazonian white-sand savannas are relatively poor in tree species compared to the adjacent uplands. However, white-sand savannas contain unique plant communities with high proportions of endemic tree species (Fine et al. 2004). The water-saturation gradient is an important determinant of tree species richness and forest structure, with the lowest richness and lowest stature vegetation occurring in areas of longest saturation (Targhetta et al. 2015). In a review of the flora of white-sand savannas in western Amazonia, central Amazonia, and the Guyana Shield, Guevara et al. (2016) reported 1,482 tree species among 44,579 sampled individuals. However, the number of tree species that occur on strictly hydromorphic parts is yet unknown.

As the Amazon biome is predominantly covered by forest, the number of aquatic and semiterrestrial herbaceous species is comparatively low in comparison to neotropical savannas (i.e., Cerrado) and even temperate regions (i.e., Pampas) (Junk 2013). Junk and Piedade (1993) list for central Amazonian *várzeas* 388 herbaceous plant species, with 47 species considered as aquatic or palustric, of which 28 are free-floating and 19 are rooted in the sediment. While high flood pulses and turbid waters restrict the growth of submersed herbaceous plants in the lower, unshaded proportion of river floodplains, higher topographies are shaded by forest, eliminating suitable habitats. Submersed plants occur only in areas with low water-level fluctuations where light penetrates down to the bottom, such as in permanent swamps and small streams. In *várzea* floodplains, only a few aquatic species that are rooted in the sediment can establish (e.g., *Victoria amazonica*, *Nymphaea gardneriana*) where water velocities are reduced, such as in secondary channels and scroll lakes. During the low-water periods, annual semiterrestrial herbaceous species colonize the unshaded proportion of the lower topographies. In nutrient-rich and highly dynamic *várzea*, these species may achieve high population densities, and some species in particular grasses are highly productive (Piedade et al. 1991). In contrast, productivity of mostly sparsely distributed annual herbaceous species in *igapó* is generally low because of low river dynamics combined with low nutrient availability of black waters and alluvial substrates (Junk et al. 2015). Herbaceous species in *igapó* are mostly sedges (Cyperaceae). Free-floating aquatic species do not occur in blackwater floodplains and are sparsely distributed and of low productivity in most clearwater floodplains.

Fauna

Aquatic and terrestrial invertebrate species in Amazonian wetlands are numerous. There are up to 400 planktonic species, several types of species-rich benthic communities, and a very species-rich community of aquatic invertebrates that colonize floating vegetation in white-water floodplains (Junk and Robertson 1997). Most species have a short life cycle, which allows their rapid occupation of the new niches provided by the fluctuating water level (Irmler 1981). Animals also migrate or drift in the water to new habitats. Species numbers and the importance of aquatic invertebrates for food webs were reviewed by Junk and Robertson (1997). Species lists and the adaptations of terrestrial invertebrates to flood pulsing Amazonian wetlands are available, for example, from Adis (1997), Adis and Messner (1997), Adis and Junk (2002), and Franklin et al. (1997). These studies report that many terrestrial invertebrate groups have representatives that are endemic to the *várzea* and/or the *igapó*, respectively.

Aquatic vertebrates include mammals such as manatees *Trichechus inunguis*, capybara *Hydrochoerus hydrochaeris*, river dolphins (*Sotalia fluviatilis*, *Inia geoffrensis*), river otter *Lontra enudris*, and giant otter *Pteronura brasiliensis*. Reptiles include turtles (*Podocnemis* spp., *Phrynops* spp., *Peltocephalus tracaxa*, *Platemys platycephala*, *Chelus fimbriatus*, *Kinosternon scorpioides*, *Rhinoclemmys punctularia*), caimans (*Melanosuchus niger*, *Caiman crocodilus*), and aquatic snakes (*Eunectes murinus*, *Helicops* spp., and *Hydrops* spp.). Although common and rich in species, amphibians are yet poorly studied. Hödl (1977) lists 15 sympatric anuran species with synchronous breeding in the floating vegetation of Lago Janauari near Manaus (*Sphaenorhynchus* spp., *Hyla* spp., *Lysapsus* spp., *Leptodactylus* spp.).

Many terrestrial vertebrate species use floodplains, swamps, and riparian zones as temporary habitat and food source. For example, the fruits of the swamp palm species *Mauritia flexuosa* are an important part of the diet of tapirs *Tapirus terrestris* and of several species of parrots (Psittacidae) (Brightsmith and Bravo 2005; Tobler et al. 2009). Several primate species are highly dependent on flooded forests for habitat and food source, and some of them have restricted occurrence in floodplains (i.e., uakari taxa, such as *Cacajao* spp.) (Haugaasen and Peres 2005). Tapirs, peccaries *Tayassu pecari*, margays *Leopardus wiedii*, and jaguars *Panthera onca* frequently visit nutrient-rich *várzea* floodplains (Lees and Peres 2008).

In the 1,200-km-long reach of the lower Negro River, Goulding et al. (1988) collected more than 450 fish species and estimated that it contained more than 700 species. The fish fauna in the Amazon River and its floodplain is at least as large as that of the Negro River. Bayley (1983) reported more than 226 fish species in Camaleão Lake, a floodplain lake on an island in the Amazon River near Manaus with a surface area of ~2 km^2 during periods of high water. The entire Amazon basin likely hosts 2,500–3,000 fish species (Junk 2013). About half of the total species number is estimated to occur in the large rivers and their floodplains and the other half in the smaller tributaries (Junk 2007). Most of the former species have a large distribution area, whereas many species in the latter have a restricted distribution and are therefore much more vulnerable to extinction. As an example, approximately 70% of the local fish fauna is expected to be endemic to upper streams in the eastern slope of the Bolivian Andes (Abell et al. 2008).

Of the 1,042 bird species recorded in the Amazonian lowlands (without Andean foothills and the Orinoco River basin), 729 occur in Amazonian floodplains. Of the latter, 83 species are considered aquatic, and 132 occur exclusively in floodplain habitats, including sand bars to different types of floodplain forests. The remaining 514 species have been observed in the floodplains but occur mostly in the uplands (Petermann, pers. comm.). For Marchantaria Island, an island in the Amazon River floodplain near Manaus, Petermann (1997) recorded 204 bird species.

Many vertebrates have synchronized their life cycles with the hydrological cycle. For Amazonia, this was reported for mammals, reptiles, and amphibians by Junk and da Silva (1997) and in the Pantanal for birds (Petermann 2011), amphibians and reptiles (Strüssmann et al. 2011), and mammals (Tomas et al. 2011). To escape the drought, reptiles and aquatic mammals move to the river channel or to permanent water bodies, where they enter their starvation period. Amphibians, reptiles, birds, and mammals have synchronized their reproductive cycles with the flood pulse to optimize reproductive success. Fish-feeding birds reproduce during low and rising water levels, when there is ample available food. The large-river turtles (*Podocnemis* spp.) and some shore birds reproduce during the low-water period on sandy beaches, when nesting places are protected from flooding. Caimans reproduce at the end of the low-water period, when there is little danger that the habitats of the offspring will dry out. Calvation and lactation of herbivorous manatees coincide with the high-water period, when the food supply for females is greatest. The birth rate of capybaras peaks during rising water, when the animals have access to terrestrial and aquatic herbaceous plants. The breeding season of both species of river dolphins starts with the receding water level, when fish become abundant in the river channel and the food supply for females is largest. Both otter species breed year-round, with peaks during the low-water period.

Conservation Status

Compared to the situation of many other large river basins worldwide that lost extensive floodplains due to regulation for flood control, irrigation, and hydropower generation (Lehner et al. 2011; Nilsson et al. 2005), most Amazonian large-river floodplains are still in a fairly pristine stage, mainly because of their remoteness and low human population density in large parts of the basin (Junk 2013). However, the conservation status of most Amazonian wetlands is at high risk because of multiple threats and in particular due to the lack of national and transnational policies regarding wetland conservation.

Even though more than 50% of the remaining rain forest area of the Amazon basin is sheltered within some type of protected area (national parks, biosphere, ecological and fauna reserves, provincial and state parks, indigenous reserves) (Soares-Filho et al. 2010), most protection is directed to the distribution of terrestrial taxa (Peres and Terborgh 1995; Abell et al. 2007) and therefore of limited use for the integrative conservation of freshwater systems. Furthermore, many protected areas are governed by laws that allow timber extraction, mining, or the development of dams for hydropower generation (Ferreira et al. 2014). Although Brazil and Peru have created laws that prohibit deforestation in buffer zones along rivers and streams, degradation of both floodplain forests and riparian zones through unsustainable resource exploitation, deforestation, and agriculture is common.

Most Amazonian countries have established some form of environmental policy or have national programs that indirectly relate to wetlands – *inter alia* policies on water resource management, biodiversity and genetic resources, forests, sustainable development, fisheries, and water sanitation – but lack specific policies regarding wetland conservation at the national levels, a situation further exacerbated by limited infrastructure and monitoring capacity (Wittmann et al. 2015). All Amazonian countries are members of the Ramsar Convention on Wetlands. In 2015, the total area of Ramsar sites in Amazonia was 142,618 km^2, distributed over nine sites. This represents approximately 6.2% of the estimated area of Amazonian wetlands (approximately 2.3 million km^2, Table 1) and 38.2% of the total area of Ramsar sites on the South-American continent. For most wetlands and Ramsar sites, however, most contracting parties still have yet to establish a strategic monitoring system. Quantitative monitoring of environmental conditions are mostly completely lacking, hindering the identification and valuation of ecosystem services. Still, there is no transnational initiative for Amazonian wetland conservation, and only a few countries, such as Brazil and Peru, have National Ramsar or Wetland committees. Therefore, Ramsar sites often suffer from the lack of coordination at the regional and national levels, weakening cooperation among sites and complicating the formulation of specific wetland conservation strategies (Wittmann et al. 2015).

Ecosystem Services

Because Amazonian wetlands share valuable ecosystem services related to water and rain forest and in addition include coastline protection (De Groot et al. 2012), the combined value of its ecosystem services is probably one of the highest on Earth in terms of provisioning material goods, regulating biogeochemical cycles, providing habitat, and sustaining cultural practices.

Most of the rural population in Brazilian Amazonia lives in or close to fertile white-water river floodplains. They practice subsistence fishery, agriculture, and cattle ranching (Junk 2013), and use floodplain forest for hunting game animals, and the extraction of timber and non-timber forest products such as fruits, tools, construction material, and medical and ornamental resources (Wittmann and Oliveira Wittmann 2010). Together, these activities often contribute as much as two-thirds of rural household income (McGrath et al. 2008). Clear- and black-water floodplains have a much lower fertility of water and soils and a lower potential for fishery, agriculture, and timber harvest. This is also mostly valid for other Amazonian wetland types, such as palm swamps, riparian zones along upland streams, and hydromorphic savannas. Consequently, population density in these wetland types is comparatively low (Junk et al. 2011), and economic activities are mostly limited to ecotourism (*igapó*), small-scale subsistence fishery and sport fishing (*igapó*), catch of ornamental fishes (palm swamps, riparian zones), and hunting (palm swamps, riparian zones).

All Amazonian wetlands regulate biogeochemical cycles including services such as the maintenance of soil fertility, prevention against erosion, waste treatment, water purification, water flow control, discharge buffering, and mitigation of natural flooding hazards. They influence local, regional, and even global climates through fluxes of important greenhouse gases such as carbon dioxide and methane. Particularly the *várzea* is characterized by complex nitrification and denitrification processes but has an overall considerable contribution to the N budget (Kern et al. 2010). The primary productivity in *várzea* floodplains is up to five times higher than those of upland forest, mostly through highly productive forests (Schöngart et al. 2010), semiterrestrial herbaceous plants (Piedade et al. 1991), and algae (phytoplankton and periphyton) (Melack and Fisher 1990). Although *igapó* forest harbors similar aboveground wood biomass and carbon stocks compared to *várzea* forest (Schöngart et al. 2010), ecosystem productivity is much lower. The same holds true for hydromorphic white-sand savannas, although environmental variability and aboveground wood C-stocks maybe highly variable (Targhetta et al. 2015).

Habitat services include gene pool protection of an exceptionally rich flora and fauna, as well as nursery services for wetland species, terrestrial species, and deep-water species that use wetlands temporarily for habitat and food source. They further importantly contribute to the maintenance and generation of regional biodiversity, because they often support a large fraction of the regional habitat heterogeneity within the Amazon (Salo et al. 1986; Wittmann et al. 2013). For example, many endemic floodplain tree species (Wittmann et al. 2013) and endemic fish species (Junk 2007; Abell et al. 2008) diversified in Amazonian wetlands because of wetland persistence and relative environmental stability over evolutionary time scales (Wittmann et al. 2015). Similarly, the large Amazonian rivers are important dispersal barriers for many terrestrial vertebrate species and often form natural boundaries for species ranges and areas of endemism (Silva et al. 2005).

Finally, cultural services include recreation and tourism for modern cultures and numerous spiritual uses for traditional cultures. Tourist numbers in the Amazon region are constantly increasing, as exemplified in the Amazon State of Brazil, which received more than 1.1 million of tourists during 2014, a 13% increase to the year 2013 and a 44% increase compared to average tourist numbers during the years 2003–2008 (IBGE 2015).

Threats and Future Challenges

Major threats of Amazonian wetlands include (1) land cover change (mainly deforestation); (2) river damming for hydropower generation; (3) pollution through mining, agriculture, navigation, aquacultural activities, and domestic and industrial wastewaters; (4) ecosystem degradation (e.g., unsustainable timber extraction, local overfishing); and (5) local changes in hydrology (e.g., drainage of swamps for the construction of roads, transmission lines, and oil and gas pipelines, rectification of river channels in urban areas).

Deforestation, in particular due to cattle ranching and agricultural activities, has affected about 20% (approximately 1.4 million km^2) of the Amazon basin (Hansen et al. 2013). Although annual deforestation rates significantly decreased after 2005, particularly in Brazil (Nepstad et al. 2014), they show an increasing trend since 2012 (INPE 2014). Over 50% of floodplain forest in the eastern Brazilian Amazon basin was deforested by 2008 (Renó et al. 2011). Deforestation also affects riparian zones particularly in the southern Amazonian "deforestation belt," where many rivers and streams were deforested down to the stream's edge for the expansion of agricultural activities (Junk 2013). Deforestation affects wetlands in the short term because of locally altered hydrologic conditions, such as decreased evapotranspiration, and increased runoff, water temperature, and stream discharge (Macedo et al. 2013; Bleich et al. 2014), while long-term effects include reduced precipitation regimes and alterations in rainfall seasonality (Yin et al. 2014). Deforestation also increases soil erosion and alters sediment loads and water chemistry (Junk 2013). For example, deforestation along the northern Amazonian Branco River due to the expansion of the area used for cattle ranching increased sediment loads of river waters and affected the vegetation far downriver, as can be observed along the left bank of the Negro River, where formerly sparsely developed herbaceous species now show elevated stand densities and productivity (Junk et al. 2015).

River damming of Amazon rivers for hydropower generation is an increasing threat with perhaps the most disastrous consequences for the wetland biota at large scales. Dams interrupt hydrological connectivity in multiple ways; lead to the elimination or reduction of natural hydrological cycles and flood pulses; trap sediments and nutrients; change water temperatures, transparency, and chemistry in reservoirs and downriver; and act as dispersal barriers for fish and other aquatic organisms (i.e., Rosenberg et al. 1995; Nilsson and Berggren 2000; Agostinho et al. 2008; Pelicice et al. 2014). While national policies still regard hydropower as a clean energy source, Amazonian reservoirs are often located far away from where energy consumption is needed and many times operate significantly below initially designed energy outputs (Fearnside 2006). In addition, greenhouse gas emissions from tropical hydropower plants may be as high or even higher than those emitted through fossil fuel burning (Kemenes et al. 2011; Fearnside 2015), a fact still largely ignored by national governments. Although the construction of large dams is usually accompanied by studies of environmental impact, these concentrate on the floodable area of the reservoir and largely ignore environmental impacts far up- and downriver (Wittmann et al. 2015). In addition to the 48 existing river dams in Amazonia, the construction of more than 150 dams is at an advanced stage, and approximately one-third of them are estimated to have high ecological impact (Finer and Jenkins 2012). In its "Decennial Plan for Energy Expansion," Brazil plans to construct 38 new large-river dams in the Amazon region by 2023, totaling a predicted energy potential of 12,500 MW (MME 2014). Most of the new dams will be established along the South Amazonian Araguaia-Tocantins, Tapajos, and Madeira Rivers, but construction plans also exist for the Xingu, Purus and Negro Rivers (Brazil), as well as for the Ucayali, Marañon, and Napo Rivers (Peru), and Caquetá-Japurá Rivers (Colombia).

Pollution through mining, agriculture, navigation, aquacultural activities, and domestic and industrial wastewaters affects many Amazonian rivers. Hydrocarbon, iron, and bauxite mining in Brazil and Venezuela, aluminum mining in Brazil, and gold mining in all parts of the Amazon are important local threats. Besides pollution, mining activities often alter stream and river morphology by excavations and increased sediment loads. Large-scale mining also promotes deforestation; road, railway, and pipeline constructions; and the construction of river dams and transmission lines for energy production. Pollution through fertilizers and pesticides affects many Amazonian tributaries in particular where plantations of monocultures are common, such as in Southern Amazonia. Brazil is the world leader in the use of fertilizers per capita and per area, and the use of agro-toxic products is constantly increasing (IBGE 2012). Pollution through navigation occurs particularly in rivers where sea ships can navigate, such as along the Amazon main stem up to the city of Iquitos, Peru, and domestic navigation and transport through all navigable Amazonian rivers. The number of fish ponds is steadily increasing in many parts of the Amazon basin, heavily impacting and polluting headwater streams, while shrimp farms along the coastline lead to eutrophication and habitat alteration of mangroves. Finally, most Amazonian cities do not have efficient wastewater treatment, leading to the pollution of rivers and headwater streams by liquids and solids of domestic and industrial wastewaters.

The unsustainable use of resources is another threat to many Amazonian wetlands. Selective timber extraction is an economically important activity, particularly in *várzea* forests, where valuable timber species occur and where costs of logging and transport are low because timber can be rafted to the sawmills during the high-water period. Some of these timber species are commercially exploited for regional, national, and even international markets (Higuchi et al. 1994). The overexploitation of some commercially important timber species, such as *Cedrela odorata*, *Ceiba pentandra*, and *Hura crepitans*, has already led to significant reductions of their populations, in particular in eastern and central Amazonia (Wittmann and Oliveira Wittmann 2010). Overfishing, particularly near the centers of human population, combined with the preference of few commercially important fish species leads to the reduction of local fish stocks. The inland fishery potential of the Amazon basin is estimated at 900,000 t year^{-1} (Bayley and Petrere 1989), of which only one-third to one-half is actually used. Nonetheless, increasing population in most Amazonian regions and the generally low political influence of the fishery sector on long-term development strategies for Amazonian floodplains are likely to increase overfishing at larger scales (Soares and Junk 2000). Despite existing laws to protect many charismatic aquatic key animal species, the poaching of river turtles and manatees continues, even in protected areas. For example, in 2014 scientists of the Brazilian National Institute for Amazon Research (INPA, Manaus) called attention to the heavy killing of river dolphins, which are used as baits for the carnivorous catfish *Calophysus macropterus* (i.e., Rede Globo 2014). The activities are stimulated by fish merchants, who buy the catfish at very low prices and sell their filets under fantasy names at high prices to the consumer.

Although the importance of Amazonian rain forest as carbon stock, climate regulator, and its remarkable biodiversity and ecosystem services are globally recognized and have resulted in the establishment of many protective reserves, the importance of Amazonian wetlands is still poorly acknowledged by Amazonian governments and modern societies. Most Amazonian countries have not performed complete inventories, maps, and classification systems of their wetlands and its biodiversity. Many wetlands are still considered unproductive wastelands, and ecosystem services are not acknowledged by modern societies and stakeholders. Therefore, in political debates economic arguments and short-term profit through deforestation, agriculture, resource extraction, mining, pollution, and the establishment of infrastructure and hydropower generation often outweigh environmental arguments and the long-term benefits provided by pristine wetlands. Moreover, existing environmental laws regarding the conservation of wetlands are often contradictory. For example, while Brazil is a member of the Ramsar convention that stipulates "the wise use of wetland resources," it also modified its forestry laws in 2012 to decrease the permanent protection area along riparian zones, thus permitting the expansion of deforestation and agriculture into wetlands (De Sousa Jr. et al. 2011). Ongoing trends of hydropower plant establishment will stimulate large-scale wetland loss throughout the following decades, accompanied by species extinctions and the loss of multiple societal benefits. This trend can only be mitigated by the creation of a more holistic understanding of the benefits provided by wetlands combined with integrated, transnational conservation measures that will require concerted efforts from governments, scientists, and the civil society.

References

Abell R, Allan JD, Lehner B. Unlocking the potential of protected areas for freshwaters. Biol Conserv. 2007;134:48–63.

Abell R, Thieme ML, Revenga C, et al. Freshwater ecoregions of the world: a new map of biogeographic units for freshwater biodiversity conservation. BioScience. 2008;58:403–14.

Adis J. Terrestrial invertebrates: survival strategies, group spectrum, dominance and activity patterns. In: Junk WJ, editor. The Central Amazonian floodplain: ecology of a pulsing system. Berlin: Springer; 1997. p. 299–318.

Adis J, Junk WJ. Terrestrial invertebrates inhabiting lowland river floodplains of Central Amazonia and Central Europe: a review. Freshw Biol. 2002;47:711–31.

Adis J, Messner B. Adaptations to life under water: tiger beetles and millipedes. In: Junk WJ, editor. The Central Amazon floodplain: ecology of pulsing system. Berlin: Springer; 1997. p. 319–30.

Agostinho AA, Pelicice FM, Gomes LC. Dams and the fish fauna of the Neotropical region: impacts and management related to diversity and fisheries. Braz J Biol. 2008;68:1119–32.

Bayley PB. Central Amazon fish populations: biomass, production and some dynamic characteristics. [dissertation]. Halifax: Dalhousie University; 1983.

Bayley PB, Petrere Jr M. Amazon fisheries: assessment methods, current status, and management options. Can Spec Publ Fish Aquat Sci. 1989;106:385–98.

Bleich ME, Mortati AF, André T, Piedade MTF. Riparian deforestation affects the structural dynamics of headwater streams from southern Brazilian Amazonia. Trop Conserv Sci. 2014;7:657–76.

Brightsmith D, Bravo A. Ecology and management of nesting blue-and-yellow macaws (*Ara ararauna*) in *Mauritia* palm swamps. Biodivers Conserv. 2005;15:4271–87.

Castello L, McGrath DG, Hess LL, et al. The vulnerability of Amazonian freshwater ecosystems. Conserv Lett. 2013;6:217–29.

Coomes DA. Nutrient status of Amazonian caatinga forests in a seasonally dry area: nutrient fluxes in litter fall and analyses of soils. Can J For Res. 1997;27:831–9.

da Silva JMC, Rylands AB, da Fonseca GAB. The fate of the Amazonian areas of endemism. Conserv Biol. 2005;19:689–94.

De Groot R, Brander L, Van der Ploeg S, et al. Global estimates of the value of ecosystems and their services in monetary units. Ecosyst Serv. 2012;1:50–61.

De Sousa Jr PT, Piedade MTF, Candotti E. Brazil's forest code puts wetlands at risk. Lett Nat. 2011;478:458.

Dias AP. Análise especial aplicada a delimitação de áreas úmidas da planície de inundação do médio Araguaia. [master's thesis]. Cuiabá: Faculdade de Engenharia Florestal, Universidade Federal de Mato Grosso; 2014. Portuguese.

Fearnside PM. Dams in the Amazon: Belo Monte and Brazil's hydroelectric development of the Xingu River basin. Environ Manag. 2006;38:16–27.

Fearnside PM. Emissions from tropical hydropower and the IPCC. Environ Sci Pol. 2015;50:225–39.

Ferreira J, Aragão LEOC, Barlow J, et al. Brazil's environmental leadership at risk: mining and dams threaten protected areas. Science. 2014;346:706–7.

Fine PVA, Mesones I, Coley PD. Herbivores promote habitat specialization by trees in Amazonian forests. Science. 2004;305:663–5.

Finer M, Jenkins CN. Proliferation of hydroelectric dams in the Andean Amazon and implications for Andes-Amazon connectivity. PLoS One. 2012;7, e35126.

Franco W, Dezzeo N. Soils and soil-water regime in the terra firme-caatinga forest complex near San Carlos de Rio Negro, state of Amazonas, Venezuela. Interciencia. 1994;19:305–16.

Franklin E, Adis J, Woas S. The oribatid mites. In: Junk WJ, editor. The Central Amazon floodplain: ecology of a pulsing system. Berlin: Springer; 1997. p. 331–49.

Galeano A, Urrego LE, Sánchez M, Peñuela MC. Environmental drivers for regeneration of Mauritia flexuosa L.f. in Colombian Amazonian swamp forest. Aquat Bot. 2015;123:47–53.

Gopal B, Junk WJ, Davis JA, editors. Biodiversity in wetlands: assessment, function and conservation. Leiden: Backhuys; 2000.

Goulding M, Carvalho ML, Ferreira MG. Rio Negro: rich life in poor water. Hague: SPB Academic Publ. bv; 1988.

Guevara JE, Damasco G, Baraloto C. Low phylogenetic beta diversity and geographic neo-endemism in Amazonian white-sand forests. Biotropica. 2016;48:34–46.

Hansen MC, Potapov PV, Moore R, et al. High-resolution global maps of 21st century forest cover change. Science. 2013;342:850–3.

Haugaasen T, Peres CA. Primate assemblage structure in Amazonian flooded and unflooded forest. Am J Primatol. 2005;67:243–58.

Higuchi N, Hummel AC, Freitas JV, Malinowski JR, Stokes BJ. Exploração florestal nas várzeas do Estado do Amazonas: Seleção de árvores, derrubada e transporte, Proceedings of the VIII Harvesting and Transportation of Timber Products Workshop. Curitiba: IUFRO/UFPr; 1994. p. 168–93. Portuguese.

Hödl W. Call differences and calling site segregation in anuran species from Central Amazonian floating meadows. Oecologia. 1977;28:351–63.

Horbe AMC, Horbe MA, Suguio K. Tropical spodosols in northeastern Amazonas State Brazil. Geoderma. 2004;119:55–68.

Householder JE, Janovec JP, Tobler MW, Page S, Lähteenoja O. Peatlands of the Madre de Dios River of Peru: distribution, geomorphology, and habitat diversity. Wetlands. 2012;32:359–68.

Householder JE, Wittmann F, Tobler MW, Janovec JP. Montane bias in lowland Amazonian peatlands: plant assembly on heterogeneous landscapes and potential significance to palynological inference. Palaeogeogr Palaeoclimatol Palaeoecol. 2015;423:138–48.

Huber O, Gharbarran G, Funk V. Vegetation map of Guyana. Georgetown: Centre for the Study of Diversity, University of Guyana; 1995.

IBGE. Indicadores de Desenvolvimento Sustentável, Estudos e Pesquisas Informação geográfica, vol. 9. Rio de Janeiro: Instituto Brasileiro de Geografia e Estatística; 2012.

IBGE. Economia do turismo – uma perspectiva macroeconômica 2003–2009. 2015. http://www.ibge.gov.br/home/estatistica/economia/industria/economia_tur_20032009. Accessed 24 Jan 2016.

INPE. Satellite monitoring of Brazil's Amazon forest (PRODES). São José dos Campos: Brazilian National Agency for Space Research; 2014. http://www.obt.inpe.br/prodes/.

Irion G, Mello JASN, Morais J, Piedade MTF, Junk WJ, Garming L. Development of the Amazon valley during the middle to late quaternary: sedimentological and climatological observations. In: Junk WJ, Piedade MTF, Wittmann F, Schöngart J, Parolin P, editors. Amazonian floodplain forest: ecophysiology, biodiversity and sustainable management. Berlin: Springer; 2010. p. 27–42.

Irmler U. Überlebensstrategien von Tieren im saisonal überfluteten amazonischen Überschwemmungswald. Zool Anz Jena. 1981;206:26–38.

Junk WJ. Wetlands of tropical South America. In: Wigham D, Hejny S, Dykyjowa D, editors. Wetlands of the world. Dordrecht: Junk Publications; 1993. p. 679–739.

Junk WJ. Freshwater fishes of South America: their biodiversity, fisheries, and habitats – a synthesis. Aquat Ecosyst Health Manag. 2007;10:228–42.

Junk WJ. Current state of knowledge regarding South America wetlands and their future under global climate change. Aquat Sci. 2013;75:113–31.

Junk WJ, da Silva VMF. Mammals, reptiles and amphibians. In: Junk WJ, editor. The Central Amazon floodplain: ecology of a pulsing system. Ecological studies. Berlin: Springer; 1997. p. 409–17.

Junk WJ, Piedade MTF. Herbaceous plants in the floodplain near Manaus: species diversity and adaptations to the flood pulse. Amazoniana. 1993;12:467–84.

Junk WJ, Robertson B. Aquatic invertebrates. In: Junk WJ, editor. The Central Amazon floodplain: ecology of a pulsing system. Ecological studies. Berlin: Springer; 1997. p. 279–98.

Junk WJ, Bayley PB, Sparks RE. The flood pulse concept in river-floodplain systems. Proceedings of the International Large River Symposium, Ottawa. Can Spec Publ Fish Aquat Sci. 1989;106:110–27.

Junk WJ, Piedade MTF, Schöngart J, Cohn-Haft M, Adeney JM, Wittmann F. A classification of major naturally-occurring Amazonian lowland wetlands. Wetlands. 2011;31:623–40.

Junk WJ, Piedade MTF, Schöngart J, Wittmann F. A classification of major natural habitats of Amazonian white-water river floodplains (várzea). Wetlands Ecol Manag. 2012;20:461–75.

Junk WJ, Wittmann F, Schöngart J, Piedade MTF. A classification of the major habitats of Amazonian black-water river floodplains and a comparison with their white-water counterparts. Wetlands Ecol Manag. 2015;23:677–93.

Kahn F. Los nombres mas comunes de palmeras de la Amazonia. Biota. 1991;15:17–32.

Kemenes A, Forsberg BR, Melack JM. CO_2 emissions from a tropical hydroelectric reservoir (Balbina, Brazil). J Geophys Res: Biogeosci. 2011;116, G03004.

Kern J, Kreibich H, Koschorreck M, Darwich A. Nitrogen balance of a floodplain forest of the Amazon River: the role of nitrogen fixation. In: Junk WJ, Piedade MTF, Wittmann F, Schöngart J, Parolin P, editors. Central Amazonian floodplain forests: ecophysiology, biodiversity and sustainable management. Berlin: Springer; 2010. p. 281–99.

Klinge H, Medina E. Rio Negro caatingas and campinas, Amazonas States of Venezuela and Brazil. In: Specht RL, editor. Heatlands and related shrublands. Ecosystems of the world, vol. 9a. Amsterdam: Elsevier; 1979. p. 483–8.

Kubitzki K. The ecogeographical differentiation of Amazonian inundation forests. Plant Syst Evol. 1989;63:285–304.

Lähteenoja O, Ruokolainen K, Schulman L, Oinonen M. Amazonian peatlands: an ignored C sink and potential source. Glob Chang Biol. 2009;15:2311–20.

Latrubesse EM. Patterns of anabranching channels: the ultimate end-member adjustment of mega rivers. Geomorphology. 2008;101:130–45.

Lees AC, Peres CA. Conservation value of remnant riparian forest corridors of varying quality for Amazonian birds and mammals. Conserv Biol. 2008;22:439–49.

Lehner B, Liermann CR, Revenga C, et al. High-resolution mapping of the world's reservoirs and dams for sustainable river-flow management. Front Ecol Environ. 2011;9:494–502.

Luizão FJ, Luizão RCC, Proctor J. Soil acidity and nutrient deficiency in central Amazonian heath forest soils. Plant Ecol. 2007;192:209–24.

Macedo MN, Coe MT, Defries R, Uriarte M, Brando PM, Neill C, Walker WS. Land-use-driven stream warming in southeastern Amazonia. Phil Trans R Soc B: Biol Sci. 2013;368:20120153.

McGrath DG, Cardoso A, Almeida OT, Pezzuti J. Constructing a policy and institutional framework for an ecosystem-based approach to managing the lower Amazon floodplain. Environ Dev Sustain. 2008;10:677–95.

Melack JM, Fisher TR. Comparative limnology of tropical floodplain lakes with an emphasis on the Central Amazon. Acta Limnologica Brasiliensia. 1990;3:1–48.

Melack JM, Hess LL. Remote sensing of the distribution and extent of wetlands in the Amazon basin. In: Junk WJ, Piedade MTF, Wittmann F, Schöngart J, Parolin P, editors. Amazonian floodplain forest: ecophysiology, biodiversity and sustainable management. Berlin: Springer; 2010. p. 27–42.

MME. Plano decenal de expansão de energia 2023. Brasília: Ministério de Minas e Energia; MME/EPE Empresa de Pesquisa Energética; 2014.

Mori S. A Família da Castanha-do-Pará: Símbolo do Rio Negro. In: Oliveira AA, Daly DC, editors. Florestas do Rio Negro. São Paulo: UNIP, NYBG e Companhia das Letras; 2001. p. 119–42.

Navarro G, Maldonado M. Geografía ecológica de Bolivia: Vegetación y Ambientes acuáticos. Santa Cruz de la Sierra: Centro de Ecología Aplicada Simón I. Patiño, Departamento de Difusión; 2002.

Nepstad D, McGrath D, Stickler C, et al. Slowing Amazon deforestation through public policy and interventions in beef and soy supply chains. Science. 2014;344:1118–23.

Nilsson C, Berggren K. Alterations of riparian ecosystems caused by river regulation. BioScience. 2000;50:783–92.

Nilsson C, Reidy CA, Dynesius M, Revenga C. Fragmentation and flow regulation of the world's large river systems. Science. 2005;308:405–8.

Pelicice FM, Pompeu PS, Agostinho AA. Large reservoirs as ecological barriers to downstream movements of Neotropical migratory fish. Fish Fish. 2014;16:697–715.
Peres CA, Terborgh JW. Amazonian nature reserves – an analysis of the defensibility status of existing conservation units and design criteria for the future. Conserv Biol. 1995;9:34–46.
Petermann P. The birds. In: Junk WJ, editor. The Central Amazon floodplain: ecology of a pulsing system. Berlin: Springer; 1997. p. 419–52.
Petermann P. The birds of the Pantanal. In: Junk WJ, da Silva CJ, Nunes da Cunha C, Wantzen KM, editors. The Pantanal: ecology, biodiversity and sustainable management of a large neotropical seasonal wetland. Sofia: Pensoft; 2011. p. 523–64.
Piedade MTF, Junk WJ, Long SP. The productivity of the C_4 grass Echinochloa polystachia on the Amazon floodplain. Ecology. 1991;72:1456–63.
Pitman NCA, Andino JEG, Aulestia M, et al. Distribution and abundance of tree species in swamp forests of Amazonian Ecuador. Ecography. 2014;37:902–15.
Pouilly M, Beck SG, Ibenes C. Biodiversidad biológica en la Llanura de inundación del Rio Marmoré. Importancia ecológica de la dinamica fluvial. Santa Cruz de la Sierra: Centro de Ecologia Aplicada Simón I. Patiño; 2004.
Prance GT. Notes on the vegetation of Amazonia III. The terminology of Amazonian forest types subject to inundation. Brittonia. 1979;3:26–38.
Queiroz JAL, Mochiutti S, Machado SA, Galvão F. Composição florística e estrutura de floresta em várzea alta estuarina Amazônica. Floresta. 2005;35:41–56.
Rede Globo. Pescadores matam boto rosa para usar de isca na pesca de peixe. 2014. http://g1.globo.com/fantastico/noticia/2014/07/pescadores-matam-boto-rosa-para-usar-de-isca-na-pesca-de-peixe.html. Accessed 27 Jan 2016.
Renó VF, Novo EMLM, Suemitsu C, Renó CD, Silva TSF. Assessment of deforestation in the lower Amazon floodplain using historical Landsat MSS/TM imagery. Remote Sens Environ. 2011;115:3446–56.
Rosenberg DM, Bodaly RA, Usher PJ. Environmental and social impacts of large-scale hydroelectric development: who is listening? Glob Environ Chang. 1995;5:127–48.
Salo J, Kalliola R, Häkkinen L, Mäkinen Y, Niemelä P, Puhakka M, Coley PD. River dynamics and the diversity of the Amazon lowland forest. Nature. 1986;322:254–8.
Schöngart J, Wittmann F, Worbes M. Biomass and NPP of Central Amazonian floodplain forests. In: Junk WJ, Piedade MTF, Wittmann F, Schöngart J, Parolin P, editors. Amazonian floodplain forests: ecophysiology, biodiversity and sustainable management. Berlin: Springer; 2010. p. 347–88.
Sioli H. Beiträge zur regionalen Limnologie des Amazonasgebietes. Arch Hydrobiol. 1954;45:267–83.
Soares MGM, Junk WJ. Commercial fishery and fish culture of the state of Amazonas: status and perspectives. In: Junk WJ, Ohly JJ, Piedade MTF, Soares MGM, editors. The Central Amazon floodplain: actual use and options for a sustainable management. Leiden: Backhuys Publ; 2000. p. 433–61.
Soares-Filho B, Moutinho P, Nepstad D, et al. Role of Brazilian Amazon protected areas in climate change mitigation. Proc Natl Acad Sci. 2010;107:10821–6.
Strüssmann C, Prado CPA, Ferreira VL, Ribeiro RAK. Diversity, ecology, management and conservation of amphibians and reptiles of the Brazilian Pantanal: a review. In: Junk WJ, da Silva CJ, Nunes da Cunha C, Wantzen KM, editors. The Pantanal: ecology, biodiversity and sustainable management of a large neotropical seasonal wetland. Sofia: Pensoft; 2011. p. 497–521.
Takeuchi M. A estrutura da vegetação na Amazônia: III – A mata de campina na região do rio Negro. Bol Mus Paraense Emilio Goeldi. 1960;8:1–13.
Targhetta N, Kesselmeier J, Wittmann F. Effects of the hydroedaphic gradient on tree species composition and aboveground wood biomass of oligotrophic forest ecosystems in the central Amazon basin. Folia Geobotanica. 2015;50:185–205.
Ter Steege H, Pitman N, Sabatier D, et al. A spatial model of tree α-diversity and -density for the Amazon region. Biodivers Conserv. 2003;12:2255–77.

Ter Steege H, Pitman NCA, Sabatier D, et al. Hyper-dominance in the Amazonian tree flora. Science. 2013;342:325–34.

Tobler M, Janovec JP, Cornejo F. Frugivory and seed dispersal by the lowland Tapir (*Tapirus terrestris*) in the Peruvian Amazon. Biotropica. 2009;42:215–22.

Tomas WM, Cáceres NC, Nunes AP, Fischer E, Mourão G, Campos Z. Mammals in the Pantanal wetland, Brazil. In: Junk WJ, da Silva CJ, Nunes da Cunha C, Wantzen KM, editors. The Pantanal: ecology, biodiversity and sustainable management of a large neotropical seasonal wetland. Sofia: Pensoft; 2011. p. 565–97.

Wittmann F, Oliveira Wittmann A. Use of Amazonian floodplain trees. In: Junk WJ, Piedade MTF, Wittmann F, Schöngart J, Parolin P, editors. Amazonian floodplain forests: ecophysiology, biodiversity and sustainable management. Berlin: Springer; 2010. p. 389–418.

Wittmann F, Schöngart J, Montero JC, Motzer T, Junk WJ, Piedade MTF, Queiroz HL, Worbes M. Tree species composition and diversity gradients in white-water forests across the Amazon basin. J Biogeogr. 2006;33:1334–47.

Wittmann F, Schöngart J, Junk WJ. Phytogeography, species diversity, community structure and dynamics of Amazonian floodplain forests. In: Junk WJ, Piedade MTF, Wittmann F, Schöngart J, Parolin P, editors. Amazonian floodplain forests: ecophysiology, biodiversity and sustainable management. Berlin: Springer; 2010. p. 61–104.

Wittmann F, Householder E, Piedade MTF, Assis RL, Schöngart J, Parolin P, Junk WJ. Habitat specifity, endemism and the neotropical distribution of Amazonian white-water floodplain trees. Ecography. 2013;36:690–707.

Wittmann F, Householder E, Oliveira Wittmann A, Lopes A, Junk WJ, Piedade MTF. Implementation of the Ramsar convention on South American wetlands: an update. Res Rep Biodivers Stud. 2015;4:47–58.

Yin L, Fu R, Zhang Y-F, et al. What controls the interannual variation of the wet season onsets over the Amazon? J Geophys Res: Atmos. 2014;119:2314–28.

Mangroves of Colombia

Jenny Alexandra Rodríguez-Rodríguez, Paula Cristina Sierra-Correa, Martha Catalina Gómez-Cubillos, and Lucia Victoria Licero Villanueva

Contents

Introduction	748
Location Climate and Distribution	748
Wetland Diversity and Biodiversity	750
Vegetation	750
Colombian Mangrove Associated Fauna	750
Ecosystem Services	751
Conservation Status and Management	752
Threats and Future Challenges	754
References	755

Abstract

Colombia's mangrove ecosystem displays a great variety of fauna and flora over the Caribbean and Pacific coast and is a very important link between the whole coastal and marine landscape. Several studies in the country have proven its capacity to provide certain ecosystem services, such as carbon sequestration among others highly related to food security. To promote mangrove ecosystems conservation, National Environmental System of Colombia has defined an institutional framework composed by local authorities and research institutes, which had developed several laws and regulations that have worked as tools for the conservation of this ecosystem. Although changes in land use, resource

---

J. A. Rodríguez-Rodríguez (✉) · P. C. Sierra-Correa · L. V. L. Villanueva
Instituto de Investigaciones Marinas y Costeras "Jose Benito Vives de Andreis" (INVEMAR), Santa Marta, Colombia
e-mail: alexandra.rodriguez@invemar.org.co; jarodriguezrod@unal.edu.co; paula.sierra@invemar.org.co; lucia.licero@invemar.org.co

M. C. Gómez-Cubillos
Universidad Nacional de Colombia, sede Caribe, Santa Marta, Colombia
e-mail: macgomezcu@unal.edu.co

© Springer Science+Business Media B.V., part of Springer Nature 2018
C. M. Finlayson et al. (eds.), *The Wetland Book*,
https://doi.org/10.1007/978-94-007-4001-3_280

extraction, natural phenomena and accidental spills of oil are still affecting these ecosystems, it has become necessary to maintain the monitoring system recently implemented at the national level and strengthen land-use planning instruments in order to answer to vulnerability to climate and non-climate stressors.

Keywords

Neotropics · Wetlands · Mangrove management · biodiversity · ecosystem services

Introduction

Mangroves are one of the characteristic types of coastal wetlands in the tropical and subtropical intertidal zones of the world. They are formed by facultative tree-like or bushy halophytic plants (Ball and Farquhar 1984), which have in common a wide variety of morphological, physiological and reproductive adaptations that allow them to inhabit extreme environments with unstable substrate, high content of organic matter, high temperatures, wide salinity fluctuations and low oxygen concentrations (Tomlinson 1986; Hutchings and Saenger 1987). Tropical mangroves are ranked amongst the most productive natural ecosystems in the world (Lee et al. 2014) and mangrove coasts reveal important responses to environmental variability while providing structural and functional benefits (Yáñez-Arancibia et al. 2014).

Colombia is the northern-most country of South America (4° 39' N 74° 3' W) and is the only country of the region that has coastlines on both the Pacific Ocean and the Caribbean Sea. The mainly tropical climate has facilitated the development of important extensions of mangrove ecosystems, covering about 289,122.25 ha; 209,402.84 ha along the Pacific coast and 79,719.41 ha along the Caribbean coast.

Location Climate and Distribution

The Colombian Pacific coastline is highly influenced by the Intertropical Convergence Zone (ICZ). The clash of air masses with thermal differences and humidity gradients generate variable and weak winds, heavy rainfall throughout the year (3,000–8,000 mm/year) and 89% relative humidity. The tide in the Pacific has semi-diurnal features and its regimen can be classified as high mesotidal (Sanchez-Páez et al. 1997). The environment's annual average temperature is constant and ranges between 25 °C and 27 °C, rarely exceeding 30 °C, although there is a sharp difference between daytime and nighttime temperature of around 12 °C (CCCP 2002). These climatic features favor the development of lush and diverse vegetation in this area of the country, including mangrove ecosystems that reach up to 40 m and

Fig. 1 Distribution of Mangroves on the Pacific and Caribbean coasts of Colombia (Provided by: Laboratorio de Servicios de Informacion- Labsis (INVEMAR))

occupy an almost continuous fringe from the Mataje river in the south of Nariño up to an area near Cabo Corrientes (Chocó) whereupon it is discontinuous to the Panama border with small fringes in the Tribugá Gulf, Utría and Juradó Cove (Fig. 1).

Along the Caribbean coastline, the climatic regimen is markedly bimodal with rainy (August-November) and dry (December-February) periods interrupted by a transition period (March-July). The average temperature varies between 24 °C and 28 °C but there are localized microclimate effects including rainfall which can vary between 600 and 1500 mm/year. The Caribbean tide is mixed, semidiurnal, and generally less than 0.5 m amplitude (microtide). On the Caribbean shore, mangroves are limited to the narrow flooded fringes along the intertidal zone, forming patches among coastal wetlands, marshes, estuaries and river and stream mouths. The Caribbean coastal mangroves are less developed than the Pacific coast mangroves, with a maximum height of approximately 25 m. The areas of greatest coverage occur around the big river mouths, mainly on the Ciénaga Grande de Santa Marta, Dique canal, and Sinú and Atrato River delta (INVEMAR 2013; Fig. 1).

Wetland Diversity and Biodiversity

Vegetation

In Colombia there are eight true mangrove species, five of which are found on the Caribbean coast: *Avicennia germinans* and *Rhizophora mangle* are the most abundant and most heavily exploited, followed by *Laguncularia racemosa*, *Conocarpus erectus* and *Pelliciera rhizophorae*. The last species is limited to Cispata Bay in Cordoba, on the Morrosquillo Gulf and in Marirrió Bay on the Antioquia Uraba (MMA 2002). Species along the Colombian Pacific coast include those on the Caribbean coast as well as *Rhizophora harrisonii*, *R. racemosa* and *Mora oleifera* (Von Prahl et al. 1990).

Mangrove ecosystems in Colombia are associated with other types of wetlands such as brackish herbaceous plant associations along beaches and swamps comprising species such as *Typha dominguensis, Junco* sp., Araceae (e.g.,: *Montrichardia arborescenses*), ferns (e.g., *Acrostichum aureum*) and floodplain forests with dominant species such as: açai palm *Euterpe oleraceae*, "local name-guandales" (includes: kapok *Ceiba pentadra*; Panama rubber tree *Castilla elastica*; Vochysia sp.), "local name – cuangariales" (includes: *Virola* sp.; *Dialyanthera* spp.), and Sajo "local name – sajales" *Campnosperma panamensis* (Fig. 2).

Colombian Mangrove Associated Fauna

The heterogeneous landscapes surrounding Colombian mangrove wetlands include contiguous xerophytic forests, tropical dry forests and gallery forests. These provide biological corridors for a wide variety of native and migratory fauna, and strategic habitats for reproduction, feeding, refuge and transit. The mangrove forests support abundant microorganisms, invertebrates associated with roots, pneumatophores or sediments (such as polychaetes, mollucsc, arthropods, sipunculids, priapullids and bryozans), fishes, reptiles (alligators, lizards, turtles and snakes), amphibians, birds (including pelicans, flamingos, herons, falcons and eagles) as well as mammals including monkeys, anteaters and a wide variety of bats. Some species lists of Colombian mangrove-associated fauna can be viewed in Von Prahl et al. (1990), Moreno-Bejarano and Álvarez-León (2003), and Castellanos-Galindo et al. (2015). In reality, most Colombian populations of the fauna reported here are declining, although remaining in the "least concern" category of the IUCN red list (http://www.iucnredlist.org/). However, American crocodile (*Crocodylus acutus*), tarpon (*Megalops atlanticus*), agami heron (*Agamia agami*), and mammals like Caribbean manatee (*Trichechus manatus*), white-lipped peccary (*Tayassu pecari*), Guajira mouse opossum (*Marmosa xerophila*) and tiny yellow bat (*Rhogeessa minutilla*) require attention and currently are categorized as vulnerable species (VU) (Fig. 2).

Fig. 2 Colombian mangrove biodiversity. (**a**). *Pelliciera rhizophorae* in Utria National Park, Chocó, Pacific Coast; (**b**): American flamingo *Phoenicopterus ruber* at La Guajira, Caribbean Coast; (**c**): Cocoi heron *Ardea cocoi* in Cienaga Grande de Santa Marta, Magdalena, Caribbean Coast; (**d**): Cauca's Mangrove forests, Pacific Coast; (**e**): Capuchin monkey *Cebus albifrons* in Ciénaga Grande de Santa Marta, Magdalena, Caribbean coast; (**f**): Great white egret *Ardea alba* in Musichi, La Guajira, Caribbean Coast (Photo Credit **a** and **d**: JA Rodríguez-Rodríguez © rights remain with author; Photo Credit **b**, **e**, and **f**: MC Gómez-Cubillos © rights remain with author; Photo Credit **c**: LV Licero-Villanueva © rights remain with author)

Ecosystem Services

Human populations have an historical and traditional relationship with mangrove wetlands because of their strategic location on the coastline and the provisioning of a large number of ecosystem services (ES) including support, biogeochemical, geomorphological and cultural (welfare and aesthetic enjoyment) processes, among others.

Studies undertaken on San Andres island, showed that mangrove forests captured sediment to create ground up to 1.6 m depth; and *Rhizophora mangle* areas are the principal nursery grounds of important commercial sea reef fish species. The Ciénaga Grande of Santa Marta (CGSM) is considered the biggest wetland complex on the Colombian Caribbean and a wetland of high productivity that supports the maintenance of the lagoon's fish community and allows development of important socio-economic activities, e.g., fishing. In addition the CGSM provides habitat for terrestrial and aquatic species; directly improves water quality, increases dissolved oxygen in the water column and is the main source of feedstock for construction and fuel (Vargas-Morales et al. 2013). Carbon accumulation rates in the CGSM are estimated to vary between -10.8 and 65.3 t C ha^{-1} annually according to its state of conservation or degradation (Betancourt. 2013) which further highlights the large ecosystem service contribution provided by this biosphere reserve wetland. Yepes et al. (2016) estimated the potential carbon stored in the 8570.9 ha of the Integrated Management District (Distrito de Manejo Integrado – DMI) mangrove forests of Cispata Bay to be about 555,795.93 Mg. Furthermore, they suggest the equations developed in the study can be used as an alternative for the assessment of carbon stocks in above ground biomass of mangrove forests in Colombia; and can be optionally used for analysis at a more detailed scale to assess greenhouse gas emission reduction with avoided deforestation.

All along the Pacific Coast there are communities of people that live along rivers, coastlines or in swamps and mangrove areas with a livelihood dependency on these wetlands. Mangroves offer permanent refuge for fisher folks, who derive their livelihood from fishing in the mangroves and hundreds of woman are engaged in the collection of molluscs, locally known as "piangua," (*Anadara tuberculosa, A. similis, A. grandis*) coveted on the local market. Traditionally, the residents of this coastline have taken advantage of timber for house, boat and equipment construction, as well as wood for fuel (firewood or charcoal). The over-exploitation of mangrove for wood products constitutes one of the main causes of mangrove ecosystem decline in these regions of the country (Sanchez-Páez et al. 1997; Fig. 3).

Conservation Status and Management

Since 1993, a National Environmental System (Spanish acronym "SINA") is being implemented with the purpose of defining an institutional framework to protect the environment, define administrative responsibilities among different entities, and provide coordination to assure the environmental sustainability of the country. SINA includes, among others, the Ministry of Environment, Regional Environmental Authorities (Spanish acronym "CAR") and Research Institutes. Colombia was one of the first countries to implement the Jakarta Mandate under the Biodiversity Convention (https://www.cbd.int/doc/publications/jm-brochure-en.pdf; accessed 31 Dec 2015) with the formulation of a National Integrated Coastal Zone

Fig. 3 Mangroves and Society in Colombia. (**a**). Buenavista: Palaphytic village build on the Ciénaga Grande of Santa Marta, Colombian Caribbean; (**b**). Fishing task on a Valle del Cauca estuary, Pacific Coast; (**c**). Transportation on the Juribirá estuary, Chocó, Colombian Pacific (**d**). Mangrove wood pile to produce charcoal in Nariño, Colombian Pacific (Photo Credit **a**: JA Rodríguez-Rodríguez © rights remain with author; Photo Credit **b** and **d**: MC Gómez-Cubillos © rights remain with author; Photo credit **c**: O Garcés-Ordoñes © rights remain with photographer)

Management (ICZM) policy. ICZM has been implemented since year 2000 in Coastal Environmental Units planning sectors (Spanish acronym "UAC") using a specific Colombian methodology (Spanish acronym "COLMIZC"). Recently (in 2011), a law was enacted specifically to regulate and establish Coastal Environmental and Land Use Unit Plans (Spanish acronym "POMIUAC").

At the same time, mangrove forest and their associated natural resources were put under specific environmental legislation and CAR have held since 1995 a specific mandate for management, zoning and monitoring of these ecosystems. In 2002 Colombia established the "National Program for Sustainable use, Management and Conservation of Mangrove Ecosystems in Colombia," which describes guidelines for priority work. This remains the approach for the management of these wetlands and includes: zoning, planning for conservation and sustainable use, restoration and reestablishment of degraded areas, pilot projects as alternative forms of employment in the community away from felling of mangroves, update and application of rules on mangroves, an information system to aid in mangrove management and decision-makings, institutional empowerment, and monitoring of national actions. The Marine and Coastal Research Institute (INVEMAR) is a national Institution that supports SINA with scientific knowledge about these and other marine topics.

Colombian mangrove legislation includes three main zoning categories: (i) Preservation area (no-take areas, high productivity strategic locations, relevant functions and irreplaceable systems), (ii) Recovery area (areas under high pressure and degradation that require restoration, rehabilitation or special care) and (iii) Sustainable Use area (abundant natural resources and healthy mangroves that permit human uses under strict monitoring and control) (Sánchez-Páez et al. 2004). Currently 57% of mangrove areas have zoning approved by the Ministry of Environment; 23% have a management plan and more than 50% along the Pacific coast are under sustainable use. Mangroves along the Caribbean coast have a similar percentage in the preservation category.

In 31 Colombian Marine Protected Areas (MPA), mangroves are the second most highly represented ecosystem after coral reefs. Today, Colombia has 34% of its mangroves represented under different levels of protection between no-take/no-entry areas in National Natural Parks and sustainable use that involves local communities in DMIs.

"Blue carbon" and "ecosystem-based adaptation" (EBA) scientific studies are being undertaken to improve management of mangrove ecosystems for mitigation and adaptation of climate change impacts. Experiments conducted at Ciénaga Grande de Santa Marta/Cispata Bay/Malaga Bay show that healthy Colombian mangroves capture high amounts of CO_2 and methane and nitrogen emissions in mangrove sediments (Betancourt-Portela et al. 2013).

In the EBA concept, recent PhD research recognizes that the EBA approach to managing mangrove coasts in tropical areas should be presented as an addition to existing coastal planning practices, rather than as a new model itself, to improve integrated coastal zone management (ICZM). In the arena of rising sea-level, improving the resilience of mangrove coasts is a big question, and EBA within the coastal planning (design, implementation and financing) could be an answer. In developing tropical countries, such as Colombia, the mangrove services should be protected, these natural services will require huge investments that cannot be maintained over time. For Colombia, thereby, climate change planning guidelines and recommendations for strengthening capacities of SINA and National System of Disaster Prevention (in Spanish SNPAD) will be considered (Sierra-Correa and Cantera Kintz 2015).

Threats and Future Challenges

Even though Colombian mangrove ecosystems are extensive, there are nearly 30 factors contributing to the deterioration and loss of these ecosystems along the Caribbean and Pacific coasts. These factors can be consolidated into four large groups: changes in land use (mainly related to rural and urban expansion, hotel and tourism infrastructure, development of crops and livestock, and the operation of pipelines, harbors and highways); resource extraction (timber; harvesting of fish, shellfish, plant products; and artisanal and industrial mining); natural phenomena (earthquakes, meteorological events, exposure to wave action, parasites and diseases); and accidental spills of oil or other chemicals. Nationally, the departments of Nariño (29%), Chocó (15%) and Valle del Cauca (11%) located on the Pacific coastline have experienced the highest incidence of these factors on mangrove forests (INVEMAR 2015). On the Caribbean coast, a major natural hazard is erosion affecting 28% of the coast and still increasing, with consequences to not only mangroves but also urban/rural areas, as well as leading to economic losses. For these reasons, strengthening of research into ecosystem restoration and rehabilitation as well as the effective development and implementation of projects in these area are required but will be a great challenge for the country.

To achieve effective control and management of mangroves in the country, it is necessary to strengthen and maintain the monitoring system recently implemented at the national level. The development of multi-temporal land cover change studies (minimum scale of 1:25,000) is critical to quantify the changes in mangrove extent and will facilitate decision making and effective land-use planning especially for new developments on areas near wetlands. The monitoring of variables related with ecosystem function, including the periodic inventory of associated fauna at different trophic levels and their productivity rates, is certainly needed. The information derived from the implementation of this monitoring network must be stored, processed and integrated efficiently at a national level. Therefore, the strengthening of the Information System for Management of Mangroves in Colombia-SIGMA (http://sigma.invemar.org.co), a technological tool recently developed to support decision making based on quality data, is a priority.

Local coastal planning authorities in Colombia must assess site-specific vulnerability to climate (i.e., sea level rise–SLR) and non-climate (i.e., anthropogenic interventions) stressors, as well as incorporate the results into land-use planning instruments which are flexible and responsive to new needs. Multidisciplinary and multi-sectorial approaches that include local communities will be preferred over less complex linear analysis (Sierra-Correa and Cantera Kintz 2015). To reduce coastal vulnerability and enhance ecosystem resilience to projected scenarios of SLR, mainstreaming EBA (design, implement, and fund) is an emerging option in Colombia which can not only offset anticipated ecosystem losses and improve coastal planning but provide benefits beyond responding to climate change stressors (Sierra-Correa and Cantera Kintz 2015).

References

Ball MC, Farquhar C. Photosynthetic and stomatal responses of the grey mangrove, Avicennia marina, to transient salinity conditions. Plant Physiol. 1984;74:7–11.

Betancourt JM. Variación de la emisión de gases efecto invernadero en un ecosistema de manglar del caribe colombiano sometido a intervención antropogénica [master's thesis]. Cali: Faculty of Engineering of Universidad del Valle; 2013. 85 p.

Betancourt-Portela JM, Parra JP, Villamil C. Emisión de metano y óxido nitroso de lso sedimentos de manglar para la Ciénaga Grande de Santa Marta. Bol Invest Mar Cost. 2013;42(1):131–52.

Castellanos-Galindo GA, Prieto ML, Uribe C, Zapata LA. Peces de manglar del Pacífico colombiano. Cali: WWF-Colombia; 2015. p. 28.

CCCP-Centro Control Contaminación del Pacífico. Compilación Oceanográfica de la Cuenca Pacífica Colombiana. 2002. p. 124.

Hutchings P, Saenger P. Ecology of Mangroves. Brisbane: University of Queensland Press; 1987. p. 388.

INVEMAR – Instituto de Investigaciones Marinas y Costeras. Informe del estado de los ambientes y recursos marinos y costeros en Colombia: Año 2012. Serie de Publicaciones Periódicas No. 8. 2013. p. 169.

INVEMAR – Instituto de Investigaciones Marinas y Costeras. Informe del estado de los ambientes y recursos marinos y costeros en Colombia: Año 2014. Serie de Publicaciones Periódicas No. 3. Santa Marta; 2015. p. 176.

Lee SY, Primavera JH, Dahdouh-Guebas F, McKee K, Bosire JO, Cannicci S, Diele K, Fromard F, Koedam N, Marchand C, Mendelssohn I, Mukherjee N, Record S. Ecological role and services of tropical mangrove ecosystems: a reassessment. Glob Ecol Biogeogr. 2014;23:726–43.

MMA – Ministerio de Medio Ambiente. Resolución zonificación de manglares bajo la jurisdicción de CARSUCRE, CORALINA, CORPAMAG, CRA, CVS y CVC. 2002. p. 53.

Moreno-Bejarano LM, Álvarez-León R. Fauna asociada a los manglares y otros humedales en el Delta-Estuario del río Magdalena. Colombia Rev Acad Colomb Cienc. 2003;27(105):517–34.

Sanchez-Páez H, Álvarez-León R, Ariel-Guevara O, Zamora-Guzmán A, Rodríguez-Cruz H, Bravo-Pazmiño H. Diagnóstico y zonificación preliminar de los manglares del Pacifico de Colombia. Santa Fé de Bogotá DC: Ministerio del Medio Ambiente; 1997. p. 343.

Sánchez-Páez H, Ulloa-Delgado GA, Tavera-Escobar HA. Manual sobre zonificación y planificación para el manejo sostenible de los manglares. Proyecto PD 60/01 REV. 1 (F) "Manejo Sostenible y restauración de los manglares por comunidades locales del Caribe de Colombia". MAVDT, Dirección de Ecosistemas. Bogotá: CONIF. OIMT; 2004. p. 32.

Sierra-Correa PC. Ecosystem-based adaptation for improving mangrove coasts planning for climate change impact – sea level rise [Dissertation]. Cali: Faculty of Natural Science of Universidad del Valle; in press.

Sierra-Correa PC, Cantera-Kintz JR. Ecosystem-bases adaptation for improving coastal planning for sea-level rise: a systematic review for mangrove coasts. Mar Policy. 2015;51:385–93.

Tomlinson PB. The botany of mangroves. New York: Cambridge University Press; 1986.

Vargas-Morales M, Sánchez D, Amaya E, Contreras A, Sánchez-Maldonado J, Acosta A, Pérez D, Pupo L, Viloria E. Valoración integral de los principales bienes y servicios ecosistémicos provistos por los ecosistemas de manglar En: convenio interadministrativo no 57 de 2013 entre el MADS y el INVEMAR elementos técnicos y generación de capacidad para el ordenamiento, conservación y manejo de los espacios y recursos marinos, costeros e insulares de Colombia. 2013. p. 404–544.

Von Prahl H, Cantera J, Contreras R. Manglares y hombres del Pacífico colombiano. Bogotá, Colombia: Editorial Presencia; 1990. 193 p.

Yáñez-Arancibia A, Day JW, Twilley RR, Day RH. Mangrove swamps: sentinel ecosystem in front of the climatic change. Gulf Mexico Maderas y Bosques. 2014;20:39–75.

Yepes A, Zapata M, Bolívar J, Monsalve A, Espinosa SM, Sierra-Correa PC. Ecuaciones alométricas de biomasa aérea para la estimación de los contenidos de carbono en manglares del Caribe Colombiano. Biología Tropical. 2016;64(2):913–26.

Ciénaga Grande de Santa Marta: The Largest Lagoon-Delta Ecosystem in the Colombian Caribbean

59

Jenny Alexandra Rodríguez-Rodríguez,
José Ernesto Mancera Pineda, Laura Victoria Perdomo Trujillo,
Mario Enrique Rueda, and Karen Patricia Ibarra

Contents

Introduction	758
Environmental Setting	759
Water Quality Characteristics	761
Biotic Components	762
Plankton	762
Invertebrates	762
Fishes	763
Birds, Reptiles, Mammals	764
Vegetation	764
Environmental Impact, Conservation, and Management	766
Ecosystems Services	767
Threats and Future Challenges	768
References	769

J. A. Rodríguez-Rodríguez (✉) · M. E. Rueda · K. P. Ibarra
Instituto de Investigaciones Marinas y Costeras "José Benito Vives de Andreis" (INVEMAR), Santa Marta, Colombia
e-mail: alexandra.rodriguez@invemar.org.co; jarodriguezrod@unal.edu.co; mario.rueda@invemar.org.co; karen.ibarra@invemar.org.co

J. E. M. Pineda
Universidad Nacional de Colombia, Bogotá, Colombia

Ciudad Universitaria, Mexico City, Mexico
e-mail: jemancerap@unal.edu.co

L. V. P. Trujillo
Universidad Nacional de Colombia, Bogotá, Colombia
e-mail: lvperdomotr@unal.edu.co

© Springer Science+Business Media B.V., part of Springer Nature 2018
C. M. Finlayson et al. (eds.), *The Wetland Book*,
https://doi.org/10.1007/978-94-007-4001-3_126

Abstract

Ciénaga Grande de Santa Marta (CGSM), the main Colombian coastal lagoon, is situated in the south Caribbean basin and has a deltaic geomorphology. It is recognized as the most productive lagoon in the neotropics and is an important habitat for a wide variety of fish, reptiles, mammals, and birds. The fishery is one of CGSM's most important ecosystem services providing both food and income for rural communities. The mangrove forest is the most important vegetal coverage in this estuarine complex, and variations in the soil salinity levels determine the distribution, composition, and survival of four mangrove species. The construction of a coastal highway and levee road along the Magadalena River has altered the natural flow of marine and fresh waters, causing severe environmental damage to the mangrove-lagoon ecosystem. Different projects have been conducted for the rehabilitation of the CGSM. However, pressures from human population growth that overexploits ecosystem resources, a gradual expansion of the agricultural frontier, contamination by the lack of sanitary infrastructure in human settlements, waste generated by local businesses, and lack of continuity and rigor in maintaining channels continue to cause a decrease in mangrove coverage and a deterioration in the health of the forest, fauna, and quality of life or rural people.

Keywords

Neotropics · Ramsar · Lagoon · Restoration · Mangrove ecosystem · Hypersalinization

Introduction

Located on the central Caribbean coast of Colombia (Fig. 1), the lagoon complex of Ciénaga Grande de Santa Marta (CGSM) is geomorphologically part of the exterior delta of the Magdalena River, the largest river in Colombia with an annual average water discharge of 7,200 m^3 s^{-1} and a mean sediment load of 144×10^6 t year^{-1}, corresponding to a sediment yield of 560 t km^{-2} year^{-1} for the 257,438 km^2 basin (Restrepo and Kjerfve 2000). The delta and the CGSM rest on a coastal plain, which was formed by marine and fluvial sedimentary depositions during Holocene transgression and regression phases (10,000 BP) while the surrounding formations date back to the Cenozoic era. The lagoon complex (1,321 km^2) comprises the Ciénaga Grande, the Ciénaga de Pajarales, several smaller lagoons, creeks, and channels (150 km^2), and mangrove swamps. A narrow, continuous sandbar (Isla de Salamanca) borders the entire CGSM complex to the north. At the eastern end of Isla de Salamanca a 120-m-wide and 10-m deep inlet (Boca de la Barra) connects the shallow (average depth 1.2 m) Ciénaga Grande lagoon to the Caribbean Sea.

The CGSM is recognized as the most productive lagoon in the neotropics; with a primary production of 990 gC m^{-2} year^{-1} due in part to a high chlorophyll a biomass in the water column (Hernandez and Gocke 1990; Gocke et al. 2003a, b, c, 2004; Cloern and Jassby 2008, 2010) (Fig. 2). This high biological production is the foundational support for the economic livelihood of seven fishing villages existing in the interior of CGSM with a total population of approximately 20,000

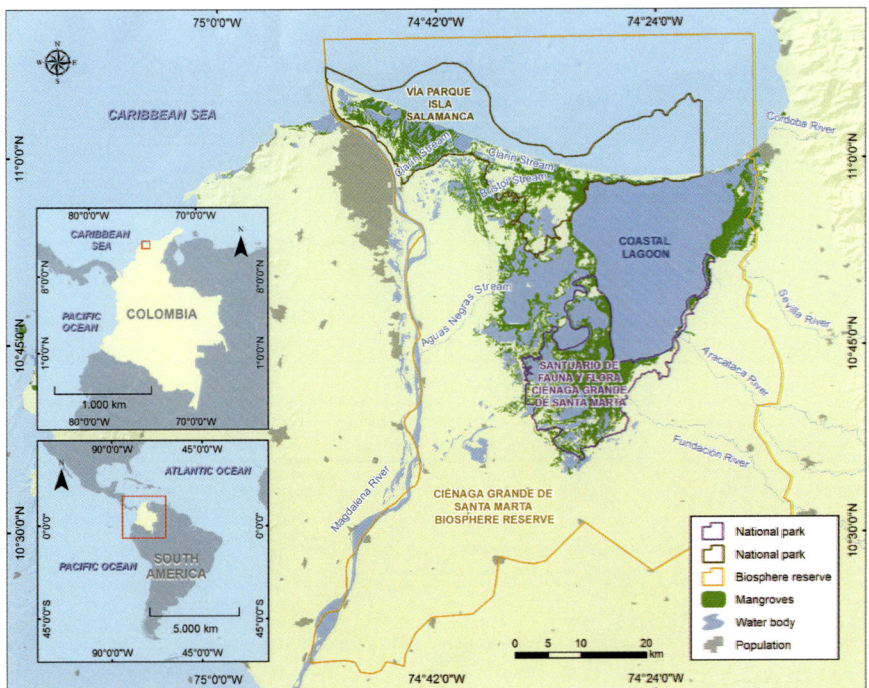

Fig. 1 Location of Ciénaga Grande de Santa Marta (Provided courtesy of: Laboratorio de Servicios de Información – Labsis (INVEMAR))

persons of which 3,200 are fishermen. The whole lagoon-delta complex experienced severe environmental stress, including freshwater diversion, large-scale death of mangroves, fish kills, water contamination, and biodiversity reduction (Botero and Salzwedel 1999; Rivera-Monroy et al. 2001). The system started to recover after an ambitious restoration plan was implemented in 1993 (PROCIENAGA project); however, conditions are again deteriorating which highlights the vulnerability of the estuarine complex.

Environmental Setting

The geomorphologic, geophysical, and biological characteristics of the CGSM identify the region as a type I setting (river-dominated, arid, micro-tidal regime Thom 1984). The climate zone is tropical arid, with a high annual water deficit of ~1,031 mm/year because evapotranspiration (1,431 mm/year) greatly exceeds precipitation (400 mm/year) (Rivera-Monroy et al. 2011). Rainfall and trade wind circulation patterns determine four seasons: (a) major dry: from December to April, (b) minor rainy: from May to June, (c) minor dry: from July to August, and (d) major rainy: from September to November (Botero and Salzwedel 1999).

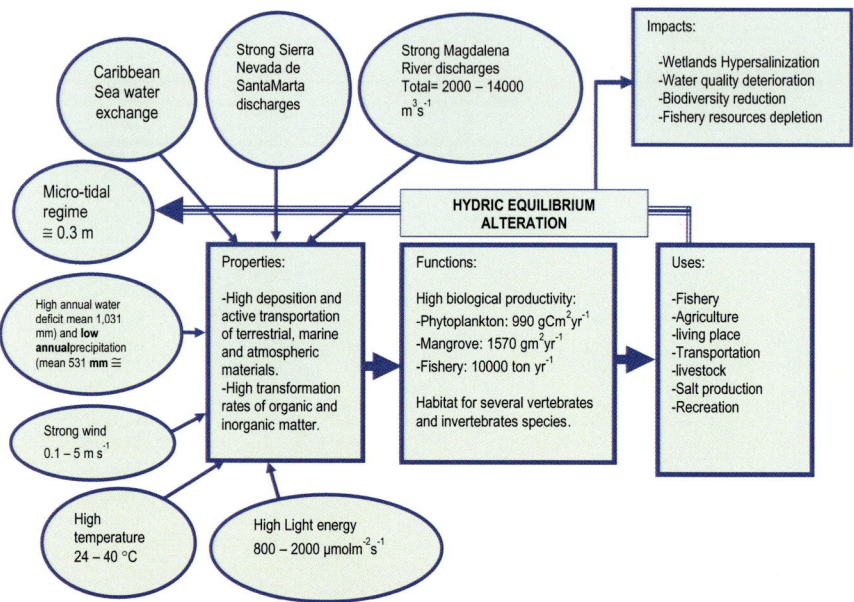

Fig. 2 Main setting of Ciénaga Grande de Santa Marta Lagoon-delta system

Northeasterly trade winds (3.2–5.4 ms^{-1}) dominate during the dry season. Average air temperature varies between 27 °C and 40 °C, and relative humidity ranges from 77% to 83%. The average annual irradiance is 2,140 cal cm^{-2}.

The CGSM is separated from the Caribbean Sea by Salamanca Island, which has an inlet approximately 100 m wide. To the west and southwest, the lagoon–delta complex is limited by the flood plain of the Magdalena River, through which five main distributaries historically carried freshwater from the river to the lagoon. To the east and southeast, the system is bounded by the Sierra Nevada de Santa Marta (SNSM), which is the highest coastal mountain range in the world (5,800 m > msl). Four major rivers with an average annual flow of ∼90 m^3 s^{-1} discharge into the CGSM from the SNSM.

Surface currents (5–3 cm s^{-1}) from the east to the west, and a microtidal amplitude (<0.5 m) characterize the central Colombian coast. According to Kjerfve (1986), the CGSM can be defined as a restricted lagoon, owing to the geomorphology of the entrance channel, which acts as a filter, and the vertical mixing of water by winds.

Hydrological patterns are a consequence of freshwater discharge by the tributaries. Prior to flood control and drainage projects in the 1970s, the Magdalena River supplied most of the freshwater to the CGSM complex. The runoff from the Magdalena River is highly seasonal, with high discharge in April, May, and October to December, and low discharge from January to March and in August. In contrast, runoff from tributary rivers (Sevilla, Aracataca, and Fundación) at the east of the CGSM is more constant. These rivers annually add a combined average freshwater flow of 19.25 m^3 s^{-1} from extensive agricultural areas at the foothills of the Sierra Nevada de Santa Marta in the east and southeast.

Circulation patterns in the lagoon are a consequence of the interaction between east-northeasterly trade winds and river discharge. Due to opposing directions of winds and freshwater runoff, the mixing of the different water masses is restrained and horizontal physicochemical gradients are thus common. During the rainy season, the increased freshwater levels in the lagoon cause an outwelling current to the open sea; during dry season, seawater enters through Boca de la Barra (Kaufman and Hevert 1973).

Sedimentary loads of the Magdalena River (see above) and those from tributaries of SNSM (180 × 10^6 t year^{-1}; Deeb-Sossa 1993) have decreased the lagoon depth at several sites from 2.3 to an average 1.5 m over the last 30 years (Bernal 1996). Muddy-clay sediments with a surficial component of detrital matter predominate in the lagoon, though in some areas gravel represents an important fraction, and sand and shell fragments of marine origin tend to increase towards the sandbar.

The CGSM can be considered a euhaline-mixohaline system, with mean annual water temperature of about 30 °C. Nevertheless, pronounced temporal and spatial salinity gradients are common, resulting from variable freshwater runoff, seawater intrusion, rainfall, and evaporation. During the dry season, salinity in the inlet is about 40‰ and may still exceed 30‰ two km up the tributaries. During the short wet season, salinity values decrease to about 15–20‰ in the center of the lagoon, but during the long wet season salinity may be close to zero even near the inlet (Botero and Mancera 1996). Furthermore, every 6 or 7 years the Magdalena River has high discharge and, for extended periods, the entire CGSM experiences salinities close to zero (Kaufman and Hevert 1973). Because of the lagoon's shallow depth and constant wind, the water column of the CGSM is normally well mixed and the resuspension of nutrients along with high light intensity and minimum variability in water temperature (25–36 °C) contribute to the lagoons's high productivity. Nevertheless, salinity stratification may occur in some areas during the rainy seasons (Rivera-Monroy et al. 2001).

Water Quality Characteristics

Hydric equilibrium alteration in the estuarine complex by the 1990s decade caused fishes and mangroves deaths. The opening of Clarin, Renegado, and Aguas Negras channels between 1996 and 1998 contributed to improvement in the water quality, promoting freshwater input to the system (PROCIENAGA project, see below) that resulted in a decrease in the salinity and chlorophyll *a* and inorganic nutrients (nitrate, nitrite, and phosphate) concentrations. There is however a significant interannual variation in seasonal mean salinity values in the CGSM. Furthermore, the distribution of salinity gradients reveal water circulation in the main lagoon has a counterclockwise pattern due to the dominant southeast to northeast flow of freshwater (Wiedemann 1973).

Since 1998, the pH value and the concentration of dissolved oxygen (DO) have fluctuated, most of the time, in the range of allowable values for preservation of flora and fauna established by Colombian law (DO > 4.0 mg/L, pH 6.5–8.5; MinSalud 1984). In the area influenced by the rivers of the Sierra Nevada de Santa Marta and particularly in the rainy season, DO has been recorded at values below the limit established in Decree 1594 of 1984 (MinSalud 1984), evidence of an increase in the content of organic matter and change in ammonia concentrations. In years with "La Niña" events, which increase the frequency and amount of precipitation, salinity decreases in most areas of the CGSM system, contrary to the years with "El Niño" events in which there are marked increases in salinity (Cadavid et al. 2011; Ibarra et al. 2014). The levels of heavy metals (Cd, Cr, Pb, Ni, Zn, Cu) have been below at-risk values for acute effects (Buchman 2008); the highest concentrations have been found with the Magdalena River, largely related with anthropogenic activities. In the case of microbiological quality, the concentrations of thermotolerant coliforms are above the allowable primary contact value (<200 NMP, MinSalud 1984) due to in-part to dwellings constructed on pilings along the rivers of the Sierra Nevada, and connection of the Clarin channel with the Magdalena river.

In spite of the improved quality in the water's physicochemical variables after the completion of waterworks, the contributions of Magdalena River, tributaries of the SNSM and the lack of sewerage systems in the communities living in pile-dwellings and those located around the estuarine complex, generate increased concentrations of organic matter that contribute to conditions of hypoxia and anoxia in the evening hours, which are, among other factors, responsible for the death of fish at certain times of the year. Similarly, the increase in low-quality micro health indicators that exceed the limits established by Colombian law has resulted in the deterioration of the water quality.

Biotic Components

Plankton

The phytoplankton community in the CGSM is comprised of 578 taxa approximately as follows: 71 (12%) Cyanobacteria; 95 (17%) centric diatoms; 189 (33%) pennate diatoms; 59 (10%) dinoflagellates; 83 (14%) Chlorophyta; 63 (11%) Euglenophyta; and 18 (3%) other groups (Vidal 2010). Some of the typical brackish water species are continuously present at high numbers and may exceed densities of 2×10^6 cells m^{-3}. The distribution of zooplankton varies between seasons and with depth. Microzooplankton and ichtyoplankton species have higher abundances during the dry season. Salinity changes and predation by ichthyoplankton influence successional changes between micro- and macrozooplankton components.

Invertebrates

Among the invertebrate fauna, mollusk are represented by approximately 98 species (66 genera and 48 families) of which 61 are of marine origin. Six species occur in mangrove sediments, *Melampus coffeus* being the most abundant among the three gastropods. The fiddler crabs *Uca rapax* and *U. vocator* are the most abundant crabs in areas of mangrove sediments. The abundance of the mangrove crab *Aratus pisonii* is controlled by salinity fluctuations, and reproductive effort and recruitment correlate with rainfall patterns. Biometric and life history traits of this crab appear to be closely related to mangrove productivity. The largest individuals of *A. penii* and the highest percentages of ovigerous females occur in arboreous mangrove habitats. Similarly, the large size and high fecundity of *Cardisoma guanhumi* can be ascribed to the high productivity of CGSM. Some other species of invertebrates are of commercial importance like the swimming crabs of the genus *Callinectes*, the oyster *Crassostrea rhizophorae*, and the conch *Melongena melongena*. However the latter two species were representative until 1996 but are now heavily depleted and are absent from current commercial catches. Since 1999, after reopening of the channels connecting the Magdalena River with the CGSM, invertebrate catches became dominated by shrimps, swimming crabs, and the clam *Polymesosa solida* (Rueda et al. 2011).

Fishes

A total of 122 bony fish species (58% with a standard length of <350 mm) belonging to 49 families, as well as 8 cartilaginous species, have been reported for the CGSM. Estuarine species represent the largest group (63% of the 122 recorded), 19 are coastal species with marine affinities and 9 freshwater species may appear seasonally; whereas the 17 remaining species are marine species that infrequently occur. Among the 32 resident species (species occurring year round with life cycles constrained within the CGSM), 24 are estuarine species widely distributed in the lagoon and near the inlet (Santos-Martínez and Acero 1991; Rueda 2001), while 8 coastal species are more or less restricted to the inlet. Species belonging to Poeciliidae, Blenniidae, Eleotrididae, and Gobiidae are true residents, whereas species like *Arius proops*, *Lutjanus cyanopterus*, *Scomberomorus brasiliensis*, *Anchoa trinitatis*, and *A. hepsetus* are frequent visitors.

Most (62%) of the resident icthyofauna, for example, Engraulididae (*Anchovia clupeoides*, *Cetengraulis edentulous*, and *Anchoa parva*), either spawn throughout the year or have two spawning peaks. Most of the species seem to spawn inside the CGSM and Boca de la Barra (estuarine resident), although some (i.e., *Mugil spp.*) spawn in adjacent marine waters (Santos-Martínez and Acero 1991). Several of the largest species (i.e., *Elops saurus*, *Tarpon atlanticus*, *Centropomus undecimalis*, and *C. latus*) appear to use the lagoon only for recruitment since the larvae and juveniles occur during the entire year (Castaño and Garzón-Ferreira 1994). Although several species (i.e., *Mugil incilis*, *Eugerres plumieri*, and *Cathorops spixii*) tolerate a broad range of salinities, interannual changes in the salinity regime not only influence their abundance but may favor the occurrence of stenohaline (*Triportheus magdalenae*, *Caquetaia krausii*, *Oreochromis niloticus*) and/or freshwater species (*Hoplias malabaricus*, *Ageneiosus caucanus*, *Sorubium lima*, *Pterygoplites undecimalis*) (Sánchez and Rueda 1999; Rueda and Defeo 2003).

Although 90 species have been recorded in the CGSM traditional fisheries to 2015, not all of them are target species for the fishery. Commercial fishing has however been an important indicator of changes in the fish community. The high environmental perturbations caused by El Niño conditions in 1992–1995, the reopening of channels in 1996–1998, and La Niña (lowering salinity) in 1999–2000 produced not just a variation in the overall community but also a change in the fish assemblage. Freshwater species (e.g., *O. niloticus*), once assumed to be a sign of ecosystem recovery, were in fact an expression of a perturbation caused by prolonged flooding (Rueda et al. 2011). Increased salinity as a result of recent El Niño conditions have favored the occurrence of marine species, including lionfish *Pterois volitans* (Viloria and Acero 2015).

Birds, Reptiles, Mammals

CGSM's mangrove forest is important as habitat for at least 26 reptile species, 19 mammal species, and 200 species of birds (Espinosa and Victoria Perdomo 2008; Campos et al. 2004). The CGSM is important to a number of IUCN listed (http://www.iucnredlist.org/) species including the manatee *Trichechus manatus* and American crocodile *Crocodylus acutus* that occur on Sevilla and Fundación rivers (Vulnerable); *Lontra longicaudis* (Near Threatened); and *Corallus cropanii* (Endangered). Snakes (e.g., *Corallus hortulanus*) and some tortoises are also common. Birds are the most diverse component in the estuarine complex and include terrestrial, aquatic, migratory, and resident species. Several of these are of global conservation concern, for example, the hummingbird *Lepidopyga lilliae* (Critically Endangered) and the northern screamer *Chauna chavaria* (Near Threatened); or nationally, for example, the flamingo *Phoenicopterus ruber*, black duck *Netta erythrophthalma*, and bronzed cowbird *Molothrus aeneus* (Renjifo et al. 2002). Additionally, the Ciénaga is an important migration stopover point for a number of families but especially waterbirds, that is, Pelecanidae, Ardeidae, Falconidae, Charadridae, Scolopacidae, Phalaropodidae, Laridae, Rynchopidae, Cuculidae, Hirundinidae, Parulidae, and Thraipidae (Campos et al. 2004; Ruíz-Guerra et al. 2012).

Vegetation

The 39,325 ha mangrove forest, sheltered from the waves around the lagoons and channels within the CGSM (Fig. 3), is the most important and most studied vegetation coverage in the estuarine complex (INVEMAR 2015). Variations in soil salinity levels determine the distribution, survival, composition, and development of the four mangrove species in CGSM: *Avicennia germinans*, *Laguncularia racemosa*, *Rhizophora mangle*, and *Conocarpus erectus* (Rivera-Monroy et al. 2001). *A. germinans* is the dominant species in the complex; it is able to withstand severe environmental conditions and uses water efficiently (Sobrado and Ewe 2006;

Fig. 3 The Ciénaga Grande de Santa Marta complex. *Rhizophora mangle* (*upper left*); Nueva Venecia's inhabitant transported herself via push pole on the water mirror of CGSM (*upper right*); *Phalacrocorax brasilianus* (*lower left*); fisherman in their daily activities on CGSM (*lower right*) (Photo credits: *upper left* and *upper right* JA Rodríguez-Rodríguez © rights remain with author; *lower left* LV Perdomo Trujillo © rights remain with author; *lower right* LV Licero Villanueva © rights remain with author)

Duque et al. 2013), and can growth in salinities up 100 psu, while *L. racemosa* and *R. mangle* withstand 85 and 70 psu, respectively.

Production, dispersion, and propagule establishment occurs at the end of the rainy season, coinciding with a reduction in salinity due to high water levels and a decrease in currents that facilitates seed retention (Lema et al. 2003). Recent (2013–2015) hydrodynamic conditions have been affected by low rain and improper channel maintenance connecting the lagoons with the Magdalena river. The physiological stress of increased salinity combined with human impacts and low natural regeneration has resulted in diminished forest health and extent.

Riparian and dry forests and freshwater marshes are also found within the delta of the Magdalena River providing habitat for native and exotic species that can tolerate high salinity levels and fresh water.

Depending upon the salinity regime, the following species can be encountered within the mangrove forest: *Pterocarpus officinalis*, *Erytroxylum cartagenense*, the fern *Acrosticum aurum*, and creeper *Rhabdadenia biflora*. Floating macrophytes are commonly found next to mangrove forests in the main water body, some strictly submerged (e.g., *Eichornia crassipes*, *Pistia stratiotes*, *Lemma minor*) and others rooted in substrate (e.g., *Nymphaeae ampla*, *Polygonum acuminatum*, *Thypa dominguensis*). On bare soils without standing water grow various types of grasses (e.g., *Urochloa mutica*, *Paspalum fasiculatum*, *Echinochloa polystachya*). Terrestrial species, intolerant to flooding and salinity, can be found south and west of the CGSM (e.g., *Anacardium excelsum*, *Sterculia apetala*, *Cavanilesia okantifolia*).

Others can grow in the dunes over the coastline on Isla Salamanca (e.g., *Melochia crenata*, *Sporobolum virginicus*, *Sesuvium portulacastrum*, *Batis maritima*).

More than 288 vegetal species have been reported within the estuarine complex (Álvarez-León et al. 2004; Espinosa and Victoria Perdomo 2008). Vegetation response to the coastal lagoon's hydrological variability over 44 years was described by Röderstein et al. (2014). They demonstrated the freshwater supply has a major influence on the composition of the vegetation in the CGSM. This is particularly pronounced in mangrove areas adjacent to freshwater tributaries. While low surface water salinity does not impair mangrove growth, freshwater plants are sensitive to increasing salinity.

Environmental Impact, Conservation, and Management

The CGSM complex has been impacted by the construction of a coastal highway and a levee road along the Magdalena River. The construction of these two structures has altered the natural flow of marine and fresh waters, respectively, resulting in severe environmental damage to the mangrove–lagoon ecosystem. Past alterations in coastal hydrology were due to diversion of freshwater associated with highway construction, agriculture expansion (banana and African palm), and illegal land appropriation. Until the early 1960s, the lagoon was surrounded by ~52,000 ha of mangrove (i.e., *Rhizophora mangle*, *Avicennia germinans*, and *Laguncularia racemosa*). Freshwater diversion from the lagoon–delta complex located in this arid coastal region resulted in hypersalinization of mangrove soils (>90 psu for 7 months a year) leading to a dieback of almost 27,000 ha of mangrove forests over a 36 year period (Botero and Mancera 1996; Botero and Salzwedel 1999; Simard et al. 2008). In addition to extensive mangrove mortality, several fish kills caused by low oxygen concentrations and eutrophication have occurred during the past 15 years (Mancera and Vida 1994; Cotes 2004).

The "Colombo-German Project for the rehabilitation of the CGSM" (PRO-CIENAGA) began in June 1992. Its main objective was improvement of the environmental and socioeconomic conditions experienced by the inhabitants of the region (Cotes 2004). To improve environmental conditions, the project sought to restore the hydrologic system by opening five naturally pre-existing tributaries (contributing a total maximum freshwater flow of 160 m^3 s^{-1}) and partially connecting the lagoon with the sea through a series of box-culverts built under the coastal highway. Freshwater from the Magdalena River started to flow through El Clarin channel into the northwestern part of the lagoon in February 1996. At the beginning of 1998, two additional channels were reopened (Aguas Negras and Renegado) each delivering a maximum freshwater flow of 60 m^3 s^{-1} from the Magdalena River to the Ciénaga de Pajarares lagoon and to the southern region of the CGSM. Although contributing with a lower water flow rate, other channels were reopened on the island of Salamanca in the northwestern sector of the CGSM.

Additionally, to ensure the sustainability of actions and results of the project, the CGSM was designated as a Ramsar wetland (1998), a UNESCO Biosphere Reserve in 2000, and an area of international importance for bird conservation (AICA) in 2001. These designations built upon preexisting national designations for the conservation and planning of the CGSM under the National System of Protected Areas of Colombia: the "Vía Parque Isla de Salamanca" (56,200 ha, 1969) and the Sanctuary of Fauna and Flora of the Ciénaga Grande de Santa Marta (26,810 ha, 1977) (Cotes 2004; Vilardy et al. 2011; http://www.parquesnacionales.gov.co/portal/es/parques-nacionales).

Since completion of the final phase of the project (1998), the Marine and Coastal Research Institute José Benito Vives De Andreis (INVEMAR), with the support of the Ministry of Environment and Sustainable Development (MADS) and the local environmental authority (CORPAMAG) have been monitoring changes in environmental conditions, as well as structural and functional changes in plant communities and fisheries resources of the estuarine complex. CORPAMAG is also part of the Ramsar Wetland Steering Committee and presides over the Board of the Biosphere Reserve. An improvement in water quality, the preservation of wildlife, and an increase in mangrove cover has been observed over the last 16 years (Ibarra et al. 2014). The success of PROCIENAGA is evident in recovery of forest system composition, structure, and functionality including natural regeneration of mangrove vegetation as a result of the restoration of the hydrological regime although significantly influenced by the climatic phenomena El Niño and La Niña (Rodríguez-Rodríguez 2015).

Ecosystems Services

The fishery is one of CGSM's most important ecosystem services providing both food and income for rural communities (Rueda and Defeo 2003; INVEMAR 2013). The CGSM supports a wide diversity of fish fauna important to small-scale fishing by approximately 3,500 artisanal fishers harvesting between 4,178 and 9,269 tonnes annually (Fig. 4a). The CGSM fishery contributes on average 35% of the Caribbean's small-scale fish catches. The species composition of the catch is mostly fish followed by crustaceans and mollusks (Fig. 4b). The level of the fish catch has varied over the period of monitoring by INVEMAR's Fishing Information System (SIPEIN) in response to the changing hydrological and salinity conditions in the CGSM. Since 2007 fish catches have declined gradually linked to reduced rainfall and increasing salinity, while mollusk harvests have been relatively constant since 2001. Although fishing has been the CGSM's main source of employment, it is important to highlight the average income for this small-scale fishery is around US$1,600 family/year (Rueda et al. 2011), which is below the poverty line. Therefore younger men prefer to develop more profitable and less demanding activities that generate a more permanent income.

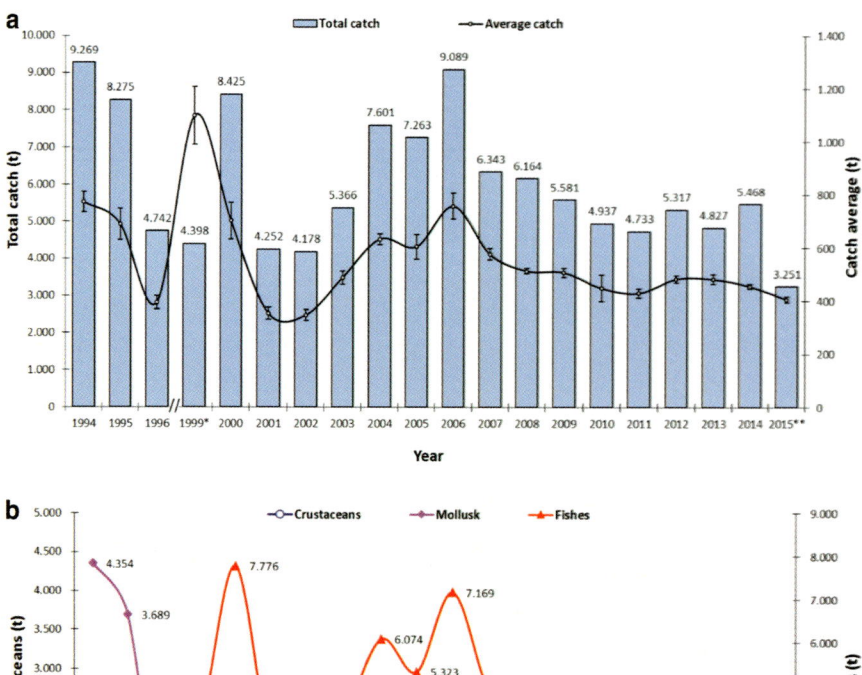

Fig. 4 (**a**) Time series of total and average catch; (**b**) Species composition (© Rights remain with the authors)

Threats and Future Challenges

Despite the actions taken to restore ecological conditions and improve the conservation status of the CGSM, pressures persist on several fronts: overexploitation of resources due to an expanding population experiencing poverty conditions; expansion of agricultural area, particularly oil palm plantations, by large landowners; lack of sanitary infrastructure in human settlements; agribusiness waste generated in areas close to channels; and lack of continuity and rigor in maintaining channels open by the PROCIÉNAGA Project (Cotes 2004; Aguilera 2011; Vilardy et al. 2011). Forest fires in 2014 and 2015 have devastated many hectares of mangrove forest, over

17 km of interconnected dikes and embankments have been illegally built to convert a large expanse of wetland into livestock pasture and agriculture production, and in 2016 a massive fish mortality was observed in the estuarine complex. The CGSM remains vulnerable and at-risk to conflicting interests that threaten the biodiversity and economy of the largest coastal wetland complex in the country. Governance gaps, little public investment, and lack of sustainable conservation management leadership from environmental authorities at the local and national levels leave the CGSM exposed to individual interests, overexploitation, and abandons local people to their fate.

Coastal erosion especially in the west of Isla de Salamanca in the last decade is another pressure. Trade winds can impart high energy wave patterns against the coast and creates instability in the coastal road that joins coastal departments with the center of the country. In addition, severe storms and windstorms that occur cyclically during the dry season months generate significant damage to the trees (PNN 2013).

A major challenge to ensure the future sustainability of this estuarine system is to improve governance using a multi- and interdisciplinary management approach that considers the ecological, social, and economic importance of the CGSM. Although climatic events significantly affect hydrological and salinity parameters within the CGSM, the proper maintenance and management of the installed hydraulic system must be guaranteed (Cotes 2004).

A strengthening of environmental monitoring and social monitoring program will serve as a basis and guide for management decision making.

References

Aguilera M. Habitantes del agua: El complejo lagunar de la Ciénaga Grande de Santa Marta. Documentos de trabajo sobre economía regional, Núm. 144. Cartagena: Cento de Estudios Regionales, CERES, Banco de la República – sucursal Cartagena; 2011. 59 p.

Álvarez-León R, De la Hoz E, Troncoso-Olivo W, Casas-Monroy O, Reyes-Forero P. La vegetación terrestre, eurihalina y dulceacuícola de la Ecorregión Ciénaga Grande de Santa Marta. In: Los Manglares de la Ecorregión Ciénaga Grande de Santa Marta: pasado, presente y futuro. Serie de publicaciones especiales. No. 11. Santa Marta: INVEMAR; 2004. p. 77–96. 236 p.

Bernal G. Caracterización geomorfológica de la llanura deltaica del río Magdalena con énfasis en el sistema lagunar de la Ciénaga Grande de Santa Marta, Colombia. Bol Invest Mar Cost. 1996;25:45–76.

Botero L, Mancera JE. Síntesis de los cambios de origen antrópico ocurridos en los últimos 40 años en la Ciénaga de Santa Marta (Colombia). Revista de la Academia Colombiana de Ciencias Exactas, Físicas, y Naturales. 1996;20:465–74.

Botero L, Salzwedel H. Rehabilitation of the Ciénaga Grande de SantaMarta, a mangrove-estuarine system in the Caribbean coast of Colombia. Ocean Coast Manag. 1999;42:243–56.

Buchman MF. Screening Quick Reference Tables (SQuiRTs). Seattle: Office of Response and Restoration Division, National Oceanic and Atmospheric Administration; 2008. 34 p. NOOA OR&R report 08-1.

Cadavid BC, Bautista PA, Espinosa LF, Hoyos AJ, Malagón AM, Mármol D, Orjuela AM, Parra JP, Perdomo LV, Rueda M, Villamil CA, Viloria EA. Monitoreo de las condiciones ambientales y los cambios estructurales y funcionales de las comunidades vegetales y de los recursos pesqueros durante la rehabilitación de la Ciénaga Grande de Santa Marta. Informe Técnico Final. Santa Marta: INVEMAR; 2011. 127 p.+ anexos.

Campos NH, Troncoso F, Blanco J. La fauna asociada a los bosques de manglar de la ecorregión Ciénaga Grande de Santa Marta. In: Garay J, Restrepo J, Casas O, Solano O, Newmark F. editores. 2004. Los manglares de la ecorregión Ciénaga Grande de Santa Marta: pasado, presente y futuro. Serie de publicaciones especiales. No. 11. Santa Marta: INVEMAR; 2004. p. 99–111. 236 p.

Castaño M, Garzón-Ferreira S. Ecología trófica del Sábalo *Megalops atlanticus* (Pisces: Megalopidae) en el área de Ciénaga Grande de Santa Marta, Caribe Colombiano. Rev Biol Trop. 1994;42(3):57–684.

Cloern JE, Jassby AD. Complex seasonal paterns of primary producers at the land-sea interface. Ecol Lett. 2008;11:1294–303.

Cloern JE, Jassby AD. Pattterns and scales of phytoplankton variability in estuarine-coastal ecosystems. Estuar Coasts. 2010;33:230–41.

Cotes G. Gestión institucional para la rehabilitación de la ecorregión Ciénaga Grande de Santa Marta y sus bosques de manglar. In: Garay J, Restrepo J, Casas O, Solano O, Newmark F. editores. 2004. Los manglares de la ecorregión Ciénaga Grande de Santa Marta: pasado, presente y futuro. Serie de publicaciones especiales. No. 11. Santa Marta: INVEMAR; 2004. p. 41–58. 236 p.

Deeb-Sossa SC. Análisis de sedimentación en algunos caños de la Ciénaga Grande de Santa Marta. Consultores, Santafé de Bogotá: Estudios y Asesorías Ing; 1993.

Duque G, Gomes M, Oliveira F, Fernandez V. Analysis of the structural variability of mangrove forests through the physiographic types approach. Aquat Bot. 2013;111:135–43.

Espinosa LF, Victoria Perdomo L. editores. Monitoreo de las condiciones ambientales y los cambios estructurales y funcionales de las comunidades vegetales y recursos pesqueros durante la rehabilitación de la Ciénaga Grande de Santa Marta: Informe Técnico, 2008. Santa Marta: INVEMAR; 2008.

Gocke K, Mancera PJE, Vallejo A. Heterotrophic microbial activity and organic matter degradation in coastal lagoons of Colombia. Rev Biol Trop. 2003a;51:85–98.

Gocke K, Mancera PJE, Vidal LS, Fonseca D. Planktonic primary production and community respiration in several coastal lagoons of the outer delta of the Rio Magdalena. Colombia Boletín de Investigaciones Marinas y Costeras. 2003b;32:125–44.

Gocke K, Meyerhofer M, Mancera PJE, Vidal LS. Phytoplankton composition in coastal lagoons of different trophic status in northern colombia determined by microscope and HPLC-pigment analysis. Boletin de Investigaciones Marinas y Costeras. 2003c;32:263–78.

Gocke K, Hernandez C, Giesenhagen H, Hoppe HG. Seasonal variations of bacterial abundance and biomass and their relation to phytoplankton in the hypertrophic tropical lagoon Ciénaga Grande de Santa Marta, Colombia. J Plankton Res. 2004;26:1429–39.

Hernandez CA, Gocke K. Productividad primaria en la Ciénaga Grande de Santa Marta, Colombia. Anales del Instituto de Investiganiones Marinas. 1990;19–20:101–19.

Ibarra KP, Gómez MC, Viloria EA, Arteaga E, Cuadrado I, Martínez MF, Nieto Y, Rodríguez JA, Licero LV, Perdomo LV, Chávez S, Romero JA, Rueda M. Monitoreo de las condiciones ambientales y los cambios estructurales y funcionales de las comunidades vegetales y de los recursos pesqueros durante la rehabilitación de la Ciénaga Grande de Santa Marta. Santa Marta: INVEMAR; 2014. 140 p.+ anexos. Informe Técnico Final.

INVEMAR. Monitoreo de las condiciones ambientales y los cambios estructurales y funcionales de las comunidades vegetales y recursos pesqueros durante la rehabilitación de la Ciénaga Grande de Santa Marta. Informe Técnico. In: Espinosa LF, Victoria Perdomo L, editors. Santa Marta; 2008.

INVEMAR. Elementos técnicos y generación de capacidad para el ordenamiento y manejo de los espacios y recursos marinos, costeros e insulares de Colombia. Código: ACT-VAR-001-013. Informe técnico final. Convenio MADS-INVEMAR No. 57. Santa Marta; 2013.

INVEMAR. Monitoreo de las condiciones ambientales y los cambios estructurales y funcionales de las comunidades vegetales y recursos pesqueros durante la rehabilitación de la Ciénaga Grande de Santa Marta. Santa Marta: Informe Técnico; 2015.

Kaufman R, Hevert F. El régimen fluviométrico del Rio Magdalena y su importancia para la Ciénaga Grande de Santa Marta. Mitt. Instituto Colombo-Alemán de Investigaciones Científicas INVEMAR. 1973;7:21–137.

Kjerfve B. Comparative oceanography of coastal lagoons. In: Wolfe DA, editor. Estuarine variability. New York: Academic; 1986. p. 63–81.

Lema LF, Polanía J, Urrego LE. Dispersión y establecimiento de las especies de mangle del río Ranchería en el período de máxima fructificación. Rev Acad Colomb Cienc. 2003;27 (102):93–103.

Mancera PJE, Vida VLA. Florecimeinto de microalgas relacionado con la mortalidad masiva de peces en el complejo lagunar Ciénaga Grande de Santa Marta, Caribe Colombiano. Anales del Instituto de Investigaciones Marinas de Punta de Betín. 1994;23:103–17.

MinSalud. Decreto No. 1594 del 26 de junio. Ministerio de Salud. Por el cual se reglamenta parcialmente el Título I de la Ley 9 de 1979, así como el Capítulo II del Título VI -Parte III- Libro II y el Título III de la Parte III -Libro I- del Decreto – Ley 2811 de 1974 en cuanto a usos del agua y residuos líquidos. 1984. 61 p.

PNN-Parques Nacionales Naturales de Colombia. Plan de manejo Santuario de Flora y Fauna de la Ciénaga Grande de Santa Marta. 2013. 222 p.

Renjifo JM, Franco-Maya AM, Amaya-Espinel JD, Kattan G, López-Lanús B. Libro rojo de aves de Colombia. Serie Libros Rojos de Especies Amenazas de Colombia. Instituto de Investigaciones de Recursos Biológicos Alexander von Humboldt y ministerio del Medio Ambiente: Bogotá; 2002. 562 p.

Restrepo J, Kjerfve B. Magdalena river: Interannual variability (1975–1995) and revised water discharge and sediment load estimates. J Hydrol. 2000;235:137–49.

Rivera-Monroy V, Mancera-Pineda JE, Twilley R, Casas O, Castañeda E, Restrepo J, Daza F, Perdomo L, Reyes P, Campos E, Villamil M, Pinto F, Cardona P, Vidal A, Troncoso W, Fonseca D, Viloria E, Sanchez G, Rojas P, Narváez J, Blanco J, Ramírez G, Henry C, Fernández J, Newmark F, Carbonó E, Hernández C, Cotes G, Sanchez H, Herrera Y, Maria A, Zuñiga R, Acosta I, Eguren A. Estructura y función de un ecosistema de manglar a lo largo de una i de restauración: El caso de la región de la Ciénaga Grande de Santa Marta. Santa Marta: University of Louisiana at Lafayette, Instituto de Investigaciones Marinas y Costeras (INVEMAR); 2001. 285 p.

Rivera-Monroy VH, Twilley RR, Mancera Pineda JE, Madden CJ. Salinity and chlorophyll a as performance measures to rehabilitate a mangrove-dominated deltaic coastal region: the Ciénaga Grande de Santa Marta–Pajarales lagoon complex, Colombia. Estuar Coasts. 2011;34:1–19.

Röderstein M, Perdomo L, Villamil C, Hauffe T, Schnetter M. Long-term vegetation changes in a tropical coastal lagoon system after interventions in the hydrological conditions. Aquat Bot. 2014;113:19–31.

Rodríguez-Rodríguez JA. Trayectorias de rehabilitación del bosque de manglar de la Ciénaga Grande de Santa Marta, luego de su reconexión con el Río Magdalena. Trabajo de Tesis como requisito parcial para obtener el título de Magister en Ciencias, Biología, Línea Biología Marina. Santa Marta: Universidad Nacional de Colombia, Sede Caribe; 2015. 94 p.

Rueda M. Spatial distribution of fish species in a tropical estuarine lagoon: a geostatistical appraisal. Mar Ecol Prog Ser. 2001;222:217–26.

Rueda M, Defeo O. Spatial structure of fish assemblages in a tropical estuarine lagoon: combining multivariate and geostatistics techniques. J Exp Mar Biol Ecol. 2003;296:93–112.

Rueda M, Blanco J, Narváez JC, Viloria EA, Beltrán C. Coastal fisheries of Colombia. In: Salas S, Chuenpagdee R, Charles A, Seijo JC, editors. Coastal fisheries of Latin America and the Caribbean. FAO Fisheries and Aquaculture Technical Paper. No. 544. Rome: FAO; 2011. p. 117–36. 430 p.

Ruíz-Guerra C, Eusse-Gonzalez D, Johnston-Gonzalez R, Castillo LF, Angulo C, Gonzalez AF. Distribución de aves acuáticas de la Ecorregión Ciénaga Grande de Santa Marta, Costa Caribe colombiana. Santiago de Cali: CALIDRIS, Asociaciones para el Estudio y la Conservación de las Aves Acuáticas en Colombia y la Dirección Territorial Caribe de Parques Nacionales de Colombia; 2012. 24 p.

Sánchez C, Rueda M. Diversity and abundance variation of dominant fish species on the Magdalena river delta, Colombia. Rev Biol Trop. 1999;47(4):1067–79.

Santos-Martínez A, Acero A. Fish community of the Ciénaga Grande de Santa Marta (Colombia): composition and zoogeography. Icthyol Explor Freshwaters. 1991;2(3):247–63.

Simard M, Rivera-Monroy VH, Mancera-Pineda JE, Castaneda-Moya E, Twilley RR. A systematic method for 3D mapping of mangrove forests based on Shuttle Radar Topography Mission elevation data, ICEsat/GLAS waveforms and field data: Application to Ciénaga Grande de Santa Marta, Colombia. Remote Sens Environ. 2008;112:2131–44.

Sobrado M, Ewe S. Ecophysiological characteristics of *Avicennia germinans* and *Laguncularia racemosa* coexisting in a scrub mangrove forest at the Indian River Lagoon, Florida. Trees. 2006;20:679–87.

Thom BG. Coastal landforms and geomorphic processes. In: Snedaker S, Snedaker J, editors. The mangrove ecosystem: research methods. Paris: UNESCO; 1984. p. 3–17.

Vidal LA. Manual del fitoplancton hallado en la Ciénaga Grande de Santa Marta y cuerpos aledaños. Bogotá: Fundación Universidad Jorge Tadeo Lozano; 2010. 384 p.

Vilardy S, González JA, Montes C. 2011. La Ciénaga Grande como un sistema socioecológico. En: Vilardy S. González JA, editores. Repensando la Ciénaga: nuevas miradas y estrategias para la sostenibilidad en la Ciénaga Grande de Santa Marta. Santa Marta: Universidad del Magdalena y Universidad Autónoma de Madrid. 228 p.

Viloria EA, Acero A. Aparición del pez león, Pterois volitans (Actinopterygii: Scorpaenidae), en la Ciénaga Grande de Santa Marta: caída del último reducto. 2015. Bol Invest Mar Cost. 2015;44(2):387–90.

Wiedemann HU. Reconnaissance of the Ciénaga Grande de Santa Marta, Colombia: physical parameters and geological history. Mitt Instituto Colombo-Alemán de Investigaciones. 1973;85–119.

Lake Fuquene (Colombia)

60

Mauricio Valderrama, María Pinilla-Vargas, Germán I. Andrade, Eugenio Valderrama-Escallón, and Sandra Hernández

Contents

Introduction	774
Location	774
Biodiversity	776
Ecosystem Services	777
Conservation Status and Management	779
Threats and Future Challenges	781
References	782

Abstract

Lake Fúquene is the second largest high altitude lake in Colombia and it is very important for the survival and prosperity of the inhabitants of its catchment area. Additionally, Fúquene harbors endemic and threatened fauna and flora, and it is a key habitat for migratory bird species. This wetland supports fish stocks, provides water for human consumption, irrigation for agriculture, flow regulation and materials for handcrafts. The lake has been subjected to years of unsustainable practices and inappropriate management policies. Climate change poses a threat to the lake's ability to provide essential environmental services and therefore

---

M. Valderrama (✉) · M. Pinilla-Vargas · S. Hernández
Fundación Humedales, Bogotá, Colombia
e-mail: mvalde@fundacionhumedales.org; mpinilla@fundacionhumedales.org; sandrahe@fundacionhumedales.org

G. I. Andrade
Universidad de Los Andes, Bogotá, Colombia
e-mail: giandradep@gmail.com

E. Valderrama-Escallón
Fundación Humedales, Universidad El Bosque, Bogotá, Colombia
e-mail: evalderr@fundacionhumedales.org

© Springer Science+Business Media B.V., part of Springer Nature 2018
C. M. Finlayson et al. (eds.), *The Wetland Book*,
https://doi.org/10.1007/978-94-007-4001-3_282

policies must assure measures towards adaptation and reduction of the vulnerability of the communities. Whole catchment policies, restoration efforts, sustainable agricultural and economic activities and sewage treatment are some of the recommendations to guarantee the ecosystem's health for future generations.

Keywords

Fúquene · Colombia · Northern Andes · Wetland management · Socio-ecological conflicts

Introduction

Lake Fúquene is one of the most important freshwater bodies of the high altitude ecosystems in the Northern Andes. Because of its uniqueness and the nature of the threats it faces, this lake is strategic for both the communities living around it and professionals involved in water management disciplines.

In this chapter, we look into the successes and failures of pursuing sustainable management of the lake and raise a discussion into actions aimed to preserve the ecosystem services it provides. Finally, we present a strategy for the adaptation of this ecosystem and its human communities to climate change.

Location

Lake Fúquene is located in South America, in the Colombian Northern Andes (5° 30′ N; 73° 50′ W), in the Ubaté River Basin (1,752 km^2 surface; Map 1). The lake has a surface area of 3,260 ha, a volume of 82.5 Mm^3, and a mean depth of 2.5 m (JICA 2000). The lake is situated in a high altitude tropical region (2,500 m above sea level) near the nation's capital Bogotá with ca. 8,000,000 citizens. The 29 km Ubaté River is the main inflowing tributary to the lake; and the only discharge is by the Suárez River which flows for 120 km before joining the Sogamoso River and eventually empties into the Caribbean Sea via the Magdalena River. The lake is part of a wetland complex that includes other small water bodies, like the lagoons of Cucunubá and Palacio.

The weather is characterized by an almost constant temperature and humidity varying between 70% and 79%. The mean rainfall in the central part of the lake is 1,030 mm per year and the average duration of sunlight is 5.3 h/day (JICA 2000). The water quality indicates eutrophic conditions (see Table 1).

Historically the lake has lost 76% of its original surface compared to the minimum level of year 1934 (Valderrama et al. 2013). The water surface has declined 29.6 ha per year between 1940 and 2009 (Franco et al. 2011b).

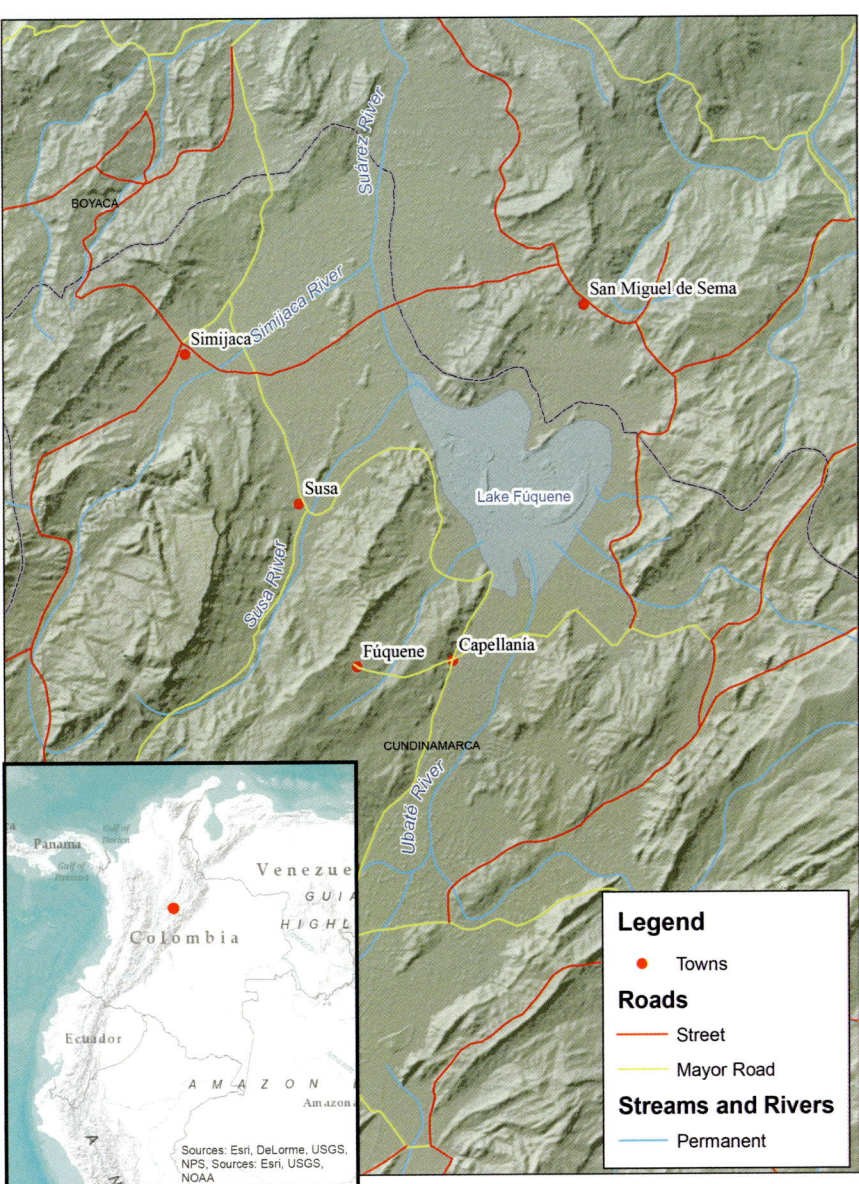

Map 1 Geographical position of Lake Fúquene and topography of the surrounding area

Table 1 Lake Fúquene water quality (Adapted from: JICA 2000; Franco et al. 2011a; Asociación de Pescadores Los Fundadores and Fundación Humedales 2011). BOD (Biological oxygen demand) and COD (Chemical oxygen demand)

Parameter	Value
Temperature	18.1 °C
Dissolved oxygen	4.8 mg/L
Total dissolved solids	47.5 ppm
Conductivity	98.5 us/cm
pH	6.78
Alkalinity	46.35 mg/L CaCO3
Hardness	77.1 mg/L CaCO4
Organic matter in sediment	16.4%
Nitrites	0.002 mg/L
Nitrates	0.21 mg/L
Phosphates	0.13 mg/L
Total phosphorus in sediments	478
BOD	2.5
COD	25.6 mg/kg mg/l mg/L

Table 2 Lake Fúquene biodiversity (Adapted from: Hilty and Brown 1986; Olivares 1969; JICA 2000; Asociación de Pescadores los Fundadores and Fundación Humedales 2011)

Group	N° spp.
Fishes	6
Macro-crustaceans	1
Birds	125
Mammals	12
Amphibians	7
Reptiles	5
Aquatic plants	17
Phytoplankton	27
Coastal plants	20
Surrounding flora	87

Biodiversity

Lake Fúquene represents a local "hotspot" of freshwater biodiversity; 307 different species have been registered at the lake and its surrounding area (Table 2).

Among the species present in the lake, it is important to highlight the catfishes *Eremophilus mutissi* and *Pygidium bogotense*, the small characid *Grundulus bogotensis* and the river crab *Neoestrengeria macropa* as they are endangered endemic species (Mojica et al. 2012). Also, there are two exotic species present: the goldfish *Carassius auratus* and the carp *Cyprinus carpio* which were introduced to the lake. Fish is still important as a food resource for local people.

Fúquene's area represents a key habitat in Colombia for globally threatened bird species and subspecies. For example, the extinct Andean grebe *Podiceps andinus* and subspecies of the ducks *Anas georgica niceforoi* and *Anas cyanoptera borreroi* were found in the lake until the 1960s. A high proportion of the bird species found in Fúquene (43 of 125 spp.) are migratory (41 boreal, 2 austral), 39 spp. are aquatic, 3 are endemic, and 5 are threatened (Renjifo et al. 2002), i.e., Apolinar's wren *Cistothorus apolinari*, Bogota rail *Rallus semiplumbeus*, spot-flanked gallinule *Porphyriops melanops*, ruddy duck *Oxyura jamaicensis*, and subtropical doradito P*seudocolopteryx acutipennis*. Other species include the blue-winged teal *Anas discors*, the cinnamon teal *Anas cyanoptera*, the American osprey *Pandion haliaetus*, the lesser scaup *Aythya affinis*, and the southern pochard *Netta erytrophtalma*. Some bird species are occasional, e.g., ruddy duck *Oxyura jamaicensis* and the endemic masked duck *Nomonyx dominicus*. Among the aquatic weeds and cattails are found the least bittern *Ixobrychus exilis* and the yellow-hooded black bird *Chrysomus icterocephalus*.

The lake provides habitat for four species of amphibians: the green-dotted tree frog *Dendropsophus labialis*, the Bogotá robber frog *Pristimantis bogotensis*, the cream-backed poison frog *Hyloxalus subpunctatus*, and Buckley's giant glass frog *Centrolene buckleyi*, and four species of reptiles: the Bogotá Anadia *Anadia bogotensis*, the striped lightbulb lizard *Riama estriata*, the water snake *Liophis epinephelus*, and the ground snake *Atractus* sp.

The vegetation of the lake is mainly composed of emerging aquatic plants such as the rush *Scirpus californicus* or the cattail *Thypha angustifolia*, floating plants including the water hyacinth *Eichiornia crassipes* and the duck weed *Lemna minor*, and the submerged Brazilian elodea *Egeria densa*. The surrounding landscape consists of a dry slope covered in subxerophytic vegetation, with important plant species such as the endemic *Agave cundinamarcensis* and small relict stands of formerly extensive neotropical oak forests *Quercus humboldtii* on wetter slopes on the east side of the lake.

Ecosystem Services

During Pre-Columbian times, Lake Fúquene was a sacred place to the Muisca (the original inhabitants of the region) who named it "Bed of the Vixen Goddess". Nowadays, the lake is still an important part of the popular traditions of the population, but its spiritual meaning has faded.

Currently, Lake Fúquene provides freshwater to ca. 105,000 rural and 76,000 urban inhabitants. The water provided is used for human consumption as well as supporting agricultural activities (e.g., dairy farming, cattle breeding, and crops) and water distribution to an irrigation district. The lake also contributes with flow regulation; plant material used for handcrafts; and supports fish stocks and habitat to endemic, migratory, and threatened bird species.

In terms of economic activities, the region is dedicated to agriculture with ca. 50 factories of different capacities producing dairy products. A continuous growth in the number of cattle has taken place over the last 10 years, from 120,232 cows (2001) to 146,218 cows (2010) (Valderrama et al. 2013). The area's per capita gross domestic product (GDP) is around US$ 2,455 (JICA 2000).

Currently only 48 of the nearly 200 fishermen reported 20 years ago remain fishing at Lake Fúquene (Fig. 1). Of these, 69% derive their income from fishing and the rest fish to supplement other activities such as cattle raising, agriculture, and sale of rush handcrafts. The average monthly income deriving from fishing activities is very low at just US$68 which represents less than one third of the country's legal monthly minimum wage.

Two native species are caught by fishermen: the catfish "*Capitán de la sabana*" *(Eremophilus mutisii)* and the crab "*Cangrejo de la sabana*" (*Neoestrengeria macropa)*, along with two introduced species: the carp and goldfish. The annual fish harvest from the lake is 19 t and valued at ca. US$ 36,000. Carp is the most frequently caught species, followed by catfish, and 36% of the total harvest is destined for the fishermen's family use, which in turn translates to a fish consumption of 53.3 kg per person per year. Fúquene clearly represents an important food source for the community, guaranteeing food security, especially when compared to the national average consumption of 6 kg of fish per person per year (Valderrama and Hernández 2007; Asociación de Pescadores Los Fundadores and Fundación Humedales 2011).

Handcrafts are another economic and social activity that employs more than 175 people and their families who sell baskets and hampers made of native plants like rush or cattail in the local markets (Hernández and Valderrama 2007).

Fig. 1 Lake Fúquene's fishermen inspecting their catch (Photo credit: Sandra Hernández © Rights remain with the author)

Conservation Status and Management

Five centuries of transformations have changed the natural conditions (i.e., climate, hydrological pulses) of the lake and associated wetlands into a problematic human regulated irrigation district (Andrade et al. 2013).

The catchment basin of Lake Fúquene has suffered an intensive process of deforestation. Only 5% of the land cover is primary and secondary forests (JICA 2000). During the last 22 years, the natural cover of the basin lost 107 ha (7%) at a rate of 5 ha per year. The resulting erosion has increased the sedimentation rate of the lake from approximately 0.4 mm per year over the last 10,000 years to 1 mm per year during the last 500 years (van der Hammen 1998). This trend has caused a decrease of 2.5 m in the depth of the lake and an estimated loss of 50% of its water storage capacity (Franco et al. 2011b).

The assessment of the lake water's physicochemical conditions and biological indicators reveals poor water quality; and in several spots of the lake these indicators reach critical status and a high level of eutrophication (Asociación de Pescadores Los Fundadores and Fundación Humedales 2011). Dissolved oxygen in the water has decreased and the levels of phosphorus and nitrogen are high. The levels of biological oxygen demand and chemical oxygen demand are high according to national standards. The availability of water is also compromised with a medium to high vulnerability index (Franco et al. 2011b). Water contamination is caused primarily by sewage, milk processing by-products, and water drainage from cattle ranches.

The situation described above is exacerbated by the expansion of invasive aquatic plants – floating water hyacinth *Eichiornia crassipes* and submerged elodea *Elodea densa* which reduce the extent of open water. During 2003–2008 the rate of increase of invasive aquatic vegetation was 64 ha per annum; and when projected to 2020–2025 suggests there will be no open water in the lake.

Environmental degradation due to low water quality and loss of habitat has affected the fishery as well. In the last 10 years fish catches declined 80% affecting food security of fishermen who in turn abandoned the fishery by 90% (Asociación de Pescadores los Fundadores and Fundación Humedales 2011). Two endangered native species have disappeared from the harvest: the small sardine *Grundulus bogotensis* and the crayfish *Neostrengeria macropa*. The principal native fish in the fishery, the catfish *E. mutisii*, is overexploited.

Another important environmental impact was the construction of a peripheral channel that was intended to address the contraction in the amount of open water, but instead caused a disruption of the hydraulic regime affecting water circulation patterns and constraining the wetland's area (Useche 2003). The channel also isolated the central body of the lake from the lakeside ecosystems. Water level fluctuations and pulses now depend on the management of the floodgates.

The Regional Autonomous Environmental Authority (CAR) has proposed a management plan for the lake (CAR and CI 2011), which has not yet been approved. Recently, the same authority defined an action plan (CAR 2015) that recommends the declaration of Lake Fúquene as a protected area under the District of Soil

Conservation category. This category prioritizes the conservation of soil and water over productive activities and thus restricts potential uses of the area, creating conflicts of interest especially with the local dairy industry. The action plan also calls for a functional assessment of the lake and suggests actions aimed to improve hydraulic conditions. These actions include the removal of invasive aquatic vegetation, the recovery of degraded soil and aquatic habitat, the management of sediments, an economic assessment of the services provided by the ecosystem, improving the connectivity to other associated wetlands, and the development of environmental education programs.

Although the Colombian Authority of Aquaculture and Fishing, AUNAP, regulates the fisheries by types of fishing gear permitted, fish catch sizes, and fishing seasons through Resolution 3,897 (2004), these regulations are no longer relevant given the lake's environmental problems. The lake has however been declared an Area of Importance for Bird Conservation (AICA) according to criteria established by Bird Life International which highlights the lake's importance and the need to apply effective management practices.

Van der Hammen (2003), Fundación Humedales (2007b), and Andrade (2010) recommended measures to improve the environmental sustainability of the lake. Among these are: increase the water level of the lake through better management of current floodgates; the establishment of protected areas in the surrounding area; restoration of native forests; and preservation of habitat for fish and birds. Additionally, Fundación Humedales has contributed to the understanding of Lake Fúquene using models of current and possible future scenarios by varying provisioning of ecosystem services and decision-making processes. Researchers at this nongovernmental organization (NGO) have also identified social and ecological barriers to achieve sustainable management of the lake, addressed the vulnerability of the basin to climate change, and proposed a strategy for climate change adaptation (Franco et al. 2011b, 2015).

The current state of Lake Fúquene is the product of a history of transformation expressed in the decrease of the lake's area and the change in structure and function of the ecosystem (depth, sediments, and introduced species, among others). Currently, the wetland system is part of an irrigation district that has an objective to improve the pastures for cattle raising and dairy production. This reality creates "threats" (in the sense of external processes on a natural area) and also changes the natural water regime of the system due to the artificial regulation of the water flow. Therefore, lake management is focused on maintaining the pastures and does not consider measures for climate change adaption. This will result over the long term in a continual reduction of the lake's open water and the transformation of the system into an extensive swamp with possible water shortages in "El Niño" (Southern Oscillation) years and floods in "La Niña" years.

The NGOs, Fundación Humedales and the Global Nature Fund (GNF), as a means to improve the lake's water quality and restore its natural attributes are building constructed wetlands for the wastewater treatment of sewage effluents from small towns in the lake's basin. Three of these treatment systems are already operating with community participation in San Miguel de Sema, Susa, and Fúquene.

Both institutions have since 2004 promoted a participatory biodiversity monitoring program.

Threats and Future Challenges

Although changes in ecosystem functions seem inevitable, a comprehensive management plan for the water bodies can still improve the provision of environmental services and preserve biodiversity. Information on the factors affecting the hydrology and provisioning of ecosystem services throughout the catchment and its contributing rivers is necessary to enhance management plans for the wetland. For example, management objectives for Lake Fúquene have shifted from drainage to provide more land for agriculture in the early twentieth century, towards the current artificially regulated irrigation system. Neither of these approaches has been successful in the protection of the natural resources. Fúquene needs a change in policy towards an ecosystem-based approach that includes the stakeholders and provides adaptation strategies to face the threats of climate change (JICA 2000; van der Hammen 2003; Andrade 2010).

The conservation and restoration of the lake and all the benefits that it provides to the communities should be the main management objective. With the inclusion of Colombia as a signatory to the Ramsar Convention of Wetlands of International Importance (Law 357 of 1997), and with Lake Fúquene as part of the Living Lakes Conservation Network (since 2000), new opportunities not yet realized for integrated resource management policies became available.

Colombia's high Andean wetlands are particularly sensitive to climate change, and it is essential that many dependent human communities adapt to that inevitability. An adaptive management plan for climate change requires acknowledgement not only of the magnitude of the climate threats but of the aspects that determine vulnerability (Franco et al. 2013).

Analyzing Lake Fúquene's vulnerability to climate change, Franco et al. (2011) proposed a desired scenario for a successful adaptive climate change strategy for this ecosystem that identifies actions to counteract the principal challenges the wetland faces: "The lake and its associated wetlands will be immersed in an active process of planning and environmental action, so that its irreversible environmental deterioration is halted and the foundations of the management plan aimed to consolidate it as a wetland of international importance are laid." It is necessary to increase the resilience of social-environmental systems undergoing environmental transformations. In order to achieve this it is essential that the ecosystem components and the interactions between them are understood, so that the outcome of management actions can be predicted with less uncertainty.

To accomplish the latter, there must be an integral planning process that directs the social welfare, economic, and human development towards the conservation, restoration, recovery, and sustainable use of the territory. Pursuing the following objectives based upon the methodology of Hallie and Luers (2006) to assess the vulnerability of social-environmental systems will aid the process:

(a) Include the entire catchment and wetland complex into the management and adaptation plans, not just Lake Fúquene
(b) Promote ecological restoration and conservation efforts in the lake
(c) Support social development and community involvement in conservation initiatives

To ensure the resilience of the territory, the quality of the water supply, and the sustainable management of soils under changing environmental conditions due to climate change, it is necessary to have a forest cover on at least 36% of the basin area in the next 20 years (Franco et al. 2011). Additionally, the activities that occur in the basin, traditional (their environmental impacts should however be reduced) or as yet unknown that might arise from the community or governmental efforts, must be directed towards sustainable use of the resources. This requires political will in the long term focused on a model of administration and management that ensures environmental security and, above all, an improvement of rural productivity, competitiveness, and the quality of life of the people.

To achieve a successful and long-term adaptation to climate change and the provision of ecosystem services in the Lake Fúquene Basin, it is important to ensure the protection of the territory as a whole. Particular attention to other wetlands in the complex is needed to guarantee the water supply, and conservation and restoration of vegetation relicts currently considered almost extinct is vital (Valderrama et al. 2013). On the other hand, a better and more efficient use of the soil and the reduction of natural hazards will improve the livelihoods of local communities that depend on the lake for their survival and income. This in turn will be reflected in the slowdown of the widespread environmental deterioration and loss of biodiversity.

References

Andrade G. Recuperación de la laguna de Fúquene, riesgos y oportunidades del CONPES- Fúquene para la conservación de la biodiversidad y los servicios ecosistémicos. Fundación Humedales. Gestión de Humedales. 2010;1(1):1–23.

Andrade G, Franco L, Delgado J. Socioecological barriers to adaptative management of Lake Fúquene Colombia. Chap. 10. In: Brebbin C, Jorgensen S, editors. Lake sustainability. Southampton: WIT Press; 2013. p. 127–35.

Asociación de Pescadores Los Fundadores and Fundación Humedales. Estado del ecosistema. Tendencias y cambios en la laguna de Fúquene a través del monitoreo participativo. In: Valderrama M, Quevedo Y, editors. Encuentro por la Laguna de Fúquene. Es posible evitar su desaparición? Diálogo por un futuro posible. Fundación Humedales. Gestión De Humedales; 2011. p. 27–48.

CAR. Plan de Manejo Ambiental PMA Complejo Lagunar de Fúquene. Plan de Acción. CAR. 2015. Available from https://www.car.gov.co/index.php. Accessed 14 Jan 2016.

CAR and Conservación International (CI). Propuesta de declaratoria como área protegida de carácter Regional del Complejo de Humedales Fúquene, Cucunubá y Palacio junto con el Plan de Manejo Ambiental. Convenio Específico de Cooperación Técnica No. 951 de 2009. Bogotá; 2011.

Franco L., Delgado J, Andrade G. Laguna de Fúquene: entender las crisis, visualizar el futuro y acordar el camino. In: Valderrama M, Quevedo Y, editors. Encuentro por la Laguna de Fúquene.

Es posible evitar su desaparición? Diálogo por un futuro posible. Fundación Humedales. Gestión de Humedales; 2011a. p. 17–26.

Franco L, Delgado J, Andrade G, Hernández S, Valderrama J. Evaluación de la vulnerabilidad y estrategia de adaptación en un complejo de humedales de la cordillera oriental colombiana: Lagunas de Fúquene, Cucunubá y Palacio. Anexo Técnico, Convenio DHS N° 131 de 2009 ECOPETROL – Fundación Humedales, Humedales Altoandinos frente al Cambio Climático Global, Bogotá; 2011b.

Franco L, Delgado J, Andrade G. Factores de la vulnerabilidad de los humedales altoandinos de Colombia al cambio climático global. Cuad Geografía Rev Colomb Geografía. 2013; 22(2):69–85.

Franco L, Ruiz C, Delgado J, Andrade G, Guzmán A. Interacciones socioecológicas que perpetúan la degradación de la laguna de Fúquene, Andes orientales de Colombia. Rev Ambiente y Desarrollo. 2015;29(37):49–66.

Fundación Humedales. Monitoreo participativo de la laguna de Fúquene. Una iniciativa para llenar el vacío de información, conocimiento y gestión ambiental. In: Franco L, Andrade G, editors. Fúquene, Cucunubá y Palacios, Conservación de la biodiversidad y manejo sostenible de un ecosistema lagunar andino. Bogotá: Instituto Alexander von Humboldt-Fundación Humedales; 2007a. p. 301–16.

Fundación Humedales. Visión de Futuro del Sistema Fúquene, Cucunubá y Palacios, Percepciones, deseos y necesidades. In: Franco L, Andrade G, editors. Fúquene, Cucunubá y Palacios, Conservación de la biodiversidad y manejo sostenible de un ecosistema lagunar andino. Bogotá: Instituto Alexander von Humboldt-Fundación Humedales; 2007b. p. 333–48.

Hallie E, Luers A. Assessing vulnerability of social-environmental systems. Annu Rev Environ Resour. 2006;31:365–94.

Hernández S, Valderrama M. El uso de la vegetación palustre en la laguna de Fúquene. In: Franco L, Andrade G, editors. Fúquene, Cucunubá y Palacios, Conservación de la biodiversidad y manejo sostenible de un ecosistema lagunar andino. Bogotá: Instituto Alexander von Humboldt-Fundación Humedales; 2007. p. 247–65.

Hilty S, Brown WL. Birds of Colombia. Princeton: Princeton University Press; 1986.

JICA. Estudio sobre el plan de mejoramiento ambiental regional para la cuenca de la Laguna de Fúquene, CAR, Resumen Ejecutivo. Bogotá: CTI Engineering Co LTD.; 2000.

Mojica I, Usma S, Alvarez R, Lasso C, editors. Libro rojo de peces dulceacuícolas de Colombia (2012). Bogotá: Instituto Alexander von Humboldt; 2012.

Olivares A. Aves de Cundinamarca. Bogotá: Universidad Nacional de Colombia; 1969.

Renjifo L, Amaya A, Amaya J, Kattan G, López B, editors. Libro Rojo de Aves de Colombia. Bogotá: Instituto Alexander von Humboldt; 2002.

Useche F. Optimización de la operación hidráulica de la laguna de Fúquene. In: Memorias del comité de expertos para la recuperación de la laguna de Fúquene. Bogotá: CAR; 2003. p. 57–65.

Valderrama MB, Hernández SB. Local participation and empowerment to application and monitoring a fish management plan in lake Fúquene Colombia. In: World Lake Vision Action Report Committee. 2007. p. 346–53

Valderrama J, Franco L, Andrade G. Del Páramo a la laguna. Conocimiento y gestión participativa de la biodiversidad asociada a humedales y el sistema hídrico de la cuenca del río y la Laguna de Fúquene, Fundación Humedales, Comunidad de la Cuenca del río Fúquene, ECOPETROL Convenio de Colaboración DHS No. 521141, Bogotá; 2013.

van der Hammen T. Bases para un plan de manejo ambiental de la cuenca hidrográfica de la laguna de Fúquene. Bogotá: CAR; 1998.

van der Hammen T. Bases para un plan de manejo de la laguna de Fúquene y su cuenca hidrográfica. In: CAR, editor. Memorias del comité de expertos para la recuperación de la laguna de Fúquene. Bogotá: CAR; 2003. p. 33–565.

The Paraná-Paraguay Fluvial Corridor (Argentina)

Priscilla G. Minotti

Contents

Introduction	786
Hydrology	788
Wetland Ecosystems	790
Biodiversity	792
Conservation Status	792
Ecosystem Services	793
Threats and Future Challenges	793
References	794

Abstract

The Paraná-Paraguay Corridor forms the main fluvial axis of De la Plata River basin, spanning free of hydroelectric dams, between 20° S and 35° S, crossing Brazil, Bolivia, Paraguay, Argentina, and Uruguay. There are big differences in precipitation and hydrological seasons which drive the pulsing regimes of its large wetlands with extended wet and dry phases. From the Pantanal wetlands in the north and ending with the Rio de la Plata Estuary in the south, the corridor connects large tropical and subtropical wetlands with temperate deltaic and coastal areas, making it a remarkable thermal, geochemical and biogeographic highway. Species richness figures for the whole area are not available although some of its major wetlands (Middle Paraná, Lower Paraná Delta) have been extensively studied. A thorough assessment of the conservation status for the

---

P. G. Minotti (✉)
Instituto de Investigación e Ingeniería Ambiental (3iA), Universidad Nacional de San Martín, San Martín, Provincia de Buenos Aires, Argentina
e-mail: priscilla.minotti@gmail.com

© Springer Science+Business Media B.V., part of Springer Nature 2018
C. M. Finlayson et al. (eds.), *The Wetland Book*,
https://doi.org/10.1007/978-94-007-4001-3_242

whole Fluvial Corridor has not been made yet. Its wetland mosaics contribute to buffer the effects of extraordinary floods and function as water reservoirs during dry seasons or spells. The Fluvial Corridor provides land settings, freshwater for domestic and industrial uses, pollution cleanup services, recreation opportunities, and local and international commerce transportation to more than 40 million urban dwellers which inhabit its shores. Major threats for the whole system are related to climate variability, river flow regulation from hydroelectrical projects and waterway development expansion, excessive runoff from agricultural and industrial mainland activities with few or no water treatments. Current economic development is heavily focused on transforming wetlands into terrestrial ecosystems for recreational urban development, year round cattle ranching and agriculture.

Keywords
Fluvial wetlands · Biodiversity corridor · Thermal corridor · De la Plata River Basin

Introduction

The Paraná-Paraguay Corridor forms the main fluvial axis of De la Plata River basin, ranking second in South America after the Amazon in length, flow, drainage area, and network of large wetlands. The fluvial axis is formed by the lower Paraguay River, which merge with the Paraná River near Corrientes city (Argentina) forming the Middle Paraná, which continues as Paraná Inferior and its delta, to finally discharge into the Atlantic Ocean through De la Plata Estuary (Fig. 1). From the Pantanal wetlands in the north and ending with the Rio de la Plata Estuary in the south, the corridor spans free of hydroelectric dams, between 20° and 35° S, crossing five countries: Brazil, Bolivia, Paraguay, Argentina, and Uruguay.

One of its unique features is this connection from tropical to temperate areas (Neiff et al. 2006). Integrating regions with different geological, ecological, and cultural history, the Paraná-Paraguay Corridor functions as a highway for thermal, geochemical, and biodiversity distribution as well as human transportation, forming the sustaining basis of different livelihoods.

Its complex geology which spans geological history since the breakup of Gondwana includes the uplift of the Andes, marine ingressions with different inland extent, accompanied with changing climates during Andean uplift and glaciations (Wilkinson et al. 2006; Iriondo 2010). The resulting geomorphology is composed of tectonic blocks at different elevations, fluvial megafans exiting from the Andean ranges and the Brazilian Shield, tributary streams having different types of valley shapes depending on their lithology, and a huge alluvial floodplain with anabranches forming the main corridor axis (Iriondo 2010; Latrubesse et al. 2005, 2011). This setting results in a series of major fluvial wetland complexes linked along its way,

61 The Paraná-Paraguay Fluvial Corridor (Argentina)

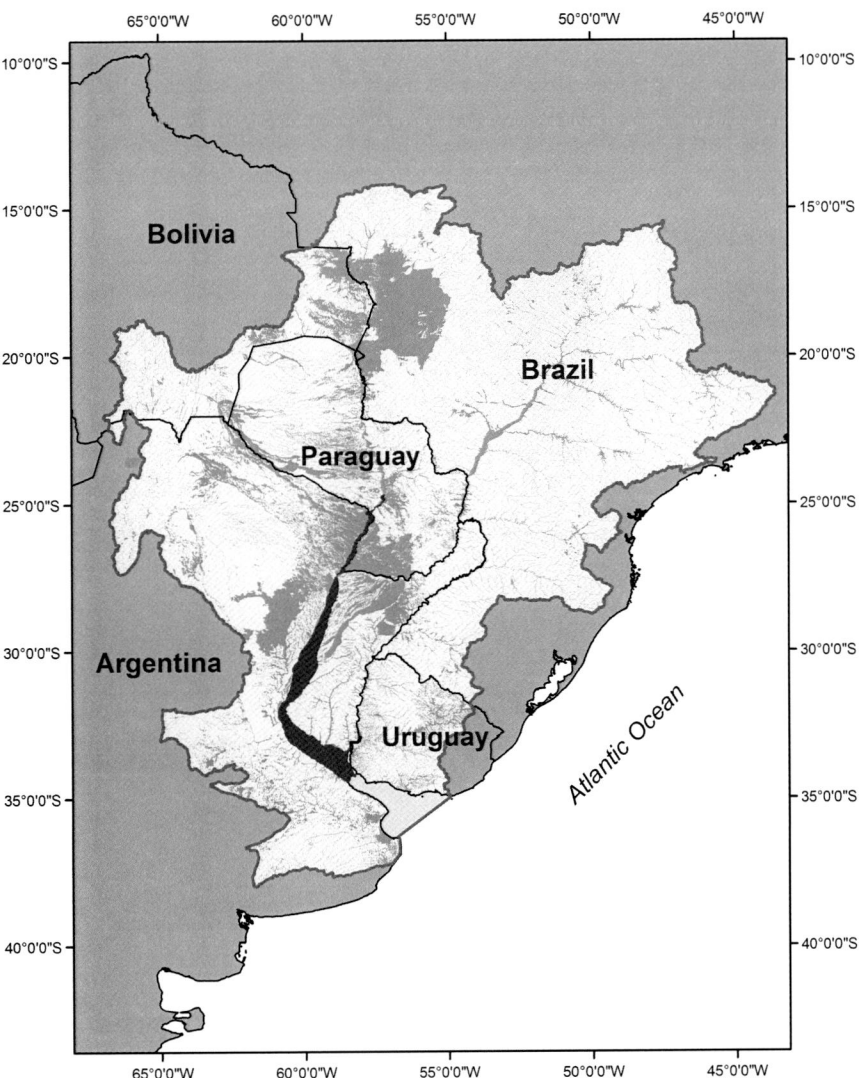

Fig. 1 Location of the Paraná-Paraguay Fluvial Corridor within De La Plata Basin (Source: Priscilla Minotti © Rights remain with the author)

which Neiff et al. (1994) have termed macroecosystems: Pantanal; Patiño-La Estrella, San Francisco- Juntas del Teuco, Bajos Submeridionales, and Bajo de los Saladillos in the Chaco-Pampean plains from the Andean margin, Ñembucú-Iberá from the Brazilian Shield side; and Middle Paraná, Paraná Delta, and De La Plata Estuary forming the southernmost network (Fig. 2).

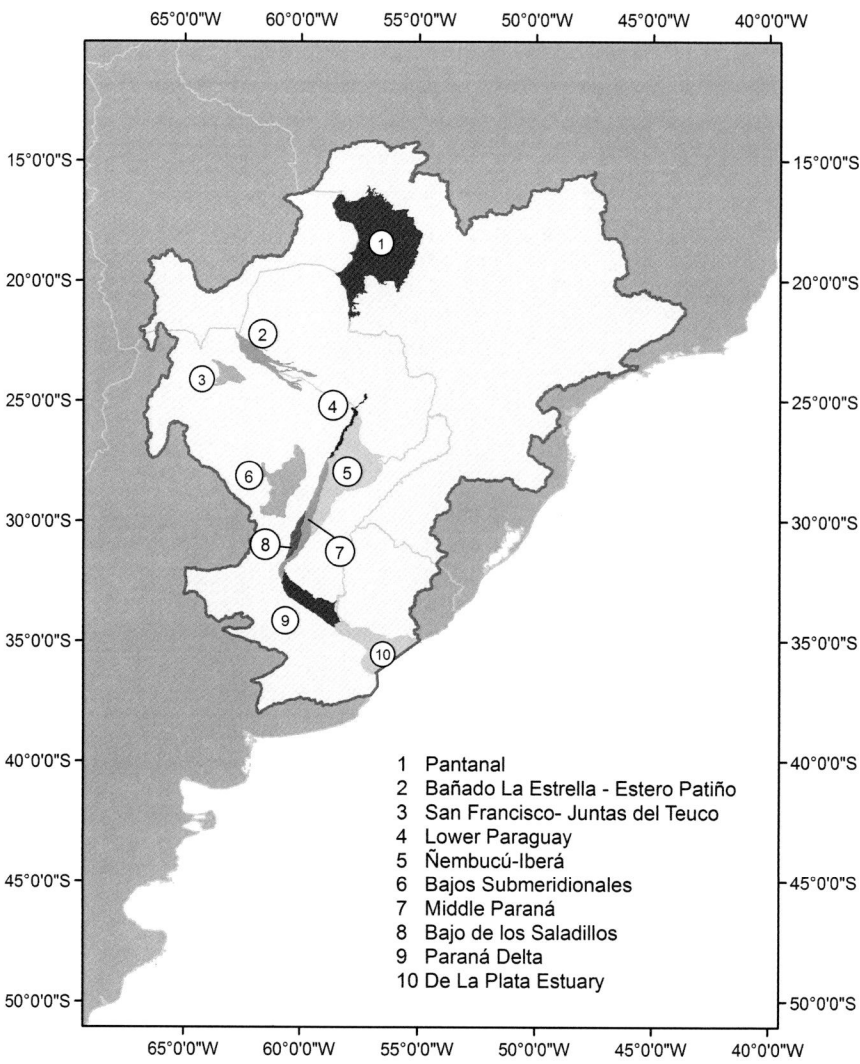

Fig. 2 Main wetland complexes related to the Paraná-Paraguay Fluvial Corridor (Source: Priscilla Minotti © Rights remain with the author)

Hydrology

There are big differences in precipitation and hydrological seasons along the Fluvial Corridor which drive the pulsing regimes of its large wetlands with extended wet and dry phases (Neiff and Malvarez 2004). In addition, the flood pulse traveling along the main fluvial axis impounds the lower courses of the tributaries, decreasing their current and even reversing their flow temporarily for considerable distances upriver (Minotti et al. 2013).

The lower Paraguay River is the outlet of the Pantanal, and its hydrograph is uncoupled from the rainfall (Hamilton 1999). The rainy season takes place mostly from late spring to summer (November–March) with a drought period of intense evapotranspiration from July to August during the mild austral subtropical winter. The regulating effect of the Pantanal wetlands is showed by a deferred unimodal flood pulse with a relatively regular and predictable flood pattern picking up between June-July (Girard 2011).

The Chaco plains tributaries are mostly autochtonous rivers formed on paleochannels from the fringe of the Pilcomayo and Bermejo megafans. Their small flows reflect the pattern of rainy season which can be monomodal in summer or bimodal at the end of spring or beginning of fall, depending on the years. During the dry or low water season from May to October, they are sustained by groundwater base flows of higher salinity (Drago et al. 2008). Bermejo River is the only one currently exiting the topographic front of the Andean range and reaching the Paraguay waters, carrying heavy loads of suspended silt and clay sediments which deposit as islands extending the Paraná deltaic front. Although its main rain input takes place during austral summer months, its hydrosedimentary pulse reaches the Paraguay River and propagates along the right margins of the Parana River between 1 and 3 months afterwards (Ramonell et al. 2000).

The high Paraná joins the Paraguay River at the Confluence, an area located between the Argentine cities of Corrientes and Paso de la Patria. The Paraná River contributes the bulk of flow, with waters of very low conductivity, sandy bed load, and currently extreme clarity as most suspended sediments are retained by more than 30 dams in its upper reaches. Although rains occur year-round, higher precipitations are concentrated during summer and reduced during winter months. Its hydrological regime is complex and unpredictable due to flow regulation by hydroelectrical dams (Neiff et al. 1994). Its current course cuts in half its megafan, with the Ñembucú swamps developed to the north and Iberá wetlands to the south. Both wetlands are currently completely disconnected from the river; its swamps and drainages are fed by precipitation with intense evapotranspiration throughout the year, with very small contributions to the Fluvial Corridor.

Rainfall distribution does not vary much along the Middle Paraná sector, with two peaks in December-January and March-April. Ordinary floods take place by the end of summer, in February-March, with a mean recurrence of 2 years. The peak sediment pulse takes place when the Bermejo waters bring its silt and clay load of suspended sediments, which compose circa 80% of the 130–135 million tons transported annually by the main Paraná channel. The main channel flows range from 12,000 to 25,000 $m^3 s^{-1}$, with a maximum of 60,000 $m^3 s^{-1}$ in the extraordinary floods of 1982–1983. The hydrology of different wetlands in this extensive and complex floodplain varies according to their relative elevation, connectivity, and distance to the main river channel (Neiff et al. 2006; Marchetti et al. 2013).

The delta region is a fluvial-coastal complex, with multimodal and unpredictable flood pulses of different source: high flows of the Parana river; local rainfall; main tributaries such as Gualeguay, Caracarañá, Tala, or Luján; flood pulses from Uruguay River; moon tides; storm surges; and even ocean wave surges coming from De La Plata Estuary (Kandus et al. 2006; Baigún et al. 2008). Extraordinary floods related to El Niño–Southern Oscillation or to the conjunction of multiple water sources may completely cover the whole region for several months, such as occurred in the 1959–1960, 1982–1983, and 1998 events.

The Fluvial Corridor ends as De la Plata Estuary, going from silty-clayish brown freshwaters in the shallow inner estuary, brackish and turbid environments in the middle section, and a completely ocean environment in the deep outer estuary. The estuary is subject to bi-daily moon tides and storm surges. Tides have amplitudes less than 1 m on the delta, which can momentarily reverse the flow of distributaries. The influence of the tide upriver depends on the Paraná water stage. If strong winds from the southeast persist several days, the storm surge can block the outlet of Paraná and Uruguay waters, increasing water levels by 3 m or more.

Wetland Ecosystems

The Paraná-Paraguay Fluvial Corridor is considered a complex mosaic of wetlands which defies classification and detailed mapping based on individual wetland types (Benzaquen et al. 2013). In the recent inventory of the Argentine portion of the Parana-Paraguay Fluvial Corridor wetland landscape systems (PNUD 2013), the corridor axis is described as composed of the following wetland landscapes or functional sets (Drago et al. 2003; Wantzen et al. 2005; Kandus et al. 2006; Minotti et al. 2013):

(i) **Main channel and anabranches**: completely aquatic, open, deep (2–40 m), and fast-flowing waters, with mobile sandy beds, which can be divided by central or lateral strips of banks and islands, presenting areas of deep scour ("pozones"). Their courses are rather unstable and can change due to lateral migration, log deposition, or avulsion. They have sandy bottoms with differences in benthic communities towards the shores but a remarkable homogeneity in the assemblages of the main channel central strip (Wantzen et al. 2014). They are considered whitewater systems as their high flow allow carrying a heavy load of suspended clay and silting sediments from the Andean ranges. Large-sized fish species as surubi *Pseudoplatystoma corruscans*, dorado *Salminus brasiliensis*, and sabalo *Prochilodus lineatus* actively migrate upstream and reproduce in the main channel, while their eggs and developing larvae float passively before entering the floodplain several hundred kilometers downstream. During flood, floating mats of camalotes (the water hyacinth *Eichornia crassipes* and *Eichornia azurea*) coming from floodplain lakes and streams enter the main channel and move downstream, colonizing new areas of the floodplain when the waters recede. Floating mats carry their load of invertebrates, plankton, fish, and even terrestrial vertebrates several hundred kilometers downstream and can even reach Rio de la Plata Estuary (Sabattini and Lallana 2007). During low waters, banks can be colonized by the floating seeds of Humbolt's willow *Salix humboldtiana* and river alder *Tessaria integrifolia* forming dense mats, which may evolve into fluvial fringe forest if the islands persist for several years.

(ii) **Floodplain channels**: are smaller channels completely aquatic, with depth and velocity changes depending on river stage, bed siltation, and floating vegetation cover. Their channel planform is stable but can dry up if completely disconnected from the main channel or anabranches. Their levees may present gallery forest or well-developed shrub patches. During low waters they are usually covered with free-floating mats of camalotes or floating fringe mats of canutillo (*Paspalum repens* or *Panicum elephantipes*) or catayzales *Polygonum* sp. Floating mats reduce flow and favor stream bed elevation due to sediment trapping and deposition and also by accumulation of decaying plants. Waters have amberine color and although relatively transparent, they are heavily shaded by the floating vegetation. If temporarily disconnected from the main channel, it can be colonized by reeds and tall sedges and the deepest parts can remain as floodplain lakes, rejuvenating with each flood.

(iii) **Floodplain lakes**: are permanent aquatic habitats of different origin, size, shape, and proportion of open water/vegetated areas, such as oxbow lakes, swamps, ponds, or depressions. Their hydrology depends on river stage, surface and hyporheic connectivity, and distance to the main channel (Neiff 2001). They are considered key habitats for biodiversity due to their function as nursery, feeding, and refuge areas for many insect and vertebrate species (Welcomme 2001). They can be functionally connected to the main river during high water, integrating them into an aquatic floodplain maze, or can be completely and permanently isolated being fed by rain or groundwater. Oxbow lakes and ponds have larger open water areas with submerged vegetation or can be covered up by floating mats. Swamps are characterized by massive stands of emergent macrophytes such as junco reed *Schenoplectus californicus*, piri *Cyperus giganteus*, guajo (*Thalia geniculata* and *T. multiflora*), and espadaña *Zizaniopsis* sp. and may dry up for some time.

(iv) **Marshlands, flooded grasslands, and savannas**: they completely dry up during the dry season and usually occupy the center of islands. Marshlands present extensive stands of tall graminoids such as cortaderia saw grass *Cyperus giganteus* or thatching grass *Colaetaenia prionitis* and junco reed in deeper and permanent waters. Palm or shrub savannas have drier physiognomies and are flooded during high waters. Flooded grasslands (locally termed "Bañados") are the first areas to be flooded due to their low topography and connection to streams. They are all subject to natural and anthropogenic fires during the dry season.

(v) **Distributaries**: aquatic and permanent lotic systems forming the secondary channels in the delta region, streaming from the main Paraná branch (Paraná Guazú) downstream to De La Plata Estuary. Their flow can be reverted by tides and storm surges, which can erode and rework their shores. Channel margins can present Humbolt's willow forests or reed beds or *S. californicus*.

(vi) **Fluvial forests**: they are located always on areas of locally higher elevations and are subject to shorter periods of flooding. Tree species have anatomical and physiological adaptations to cope with waterlogging stress and present strategies for masting or dispersing propagules during high waters, even by fish. They function as refuge and feeding areas during floods and fires. In the delta region they are mainly secondary forest or plantations, dominated by exotic tree species.

Biodiversity

A unique feature of this fluvial network is the connection from tropical to temperate areas, and fresh to brackish waters, mostly along different wet climates, making it a remarkable biogeographic highway (Neiff et al. 2006). Species richness figures for the whole area are not available as studies have focused in the different particular large wetlands such as Middle Parana (Iriondo et al. 2007) or delta (Kandus et al. 2006; Quintana et al. 2011).

The Fluvial Corridor is home to nearly 400 fish species (Sverlij et al. 2013). Most species have Brazilian lineage and are shared with the Amazon basin, having in this region their southernmost distributional range. On the delta and estuarial end, species of marine ancestry and colder waters are present and their dominance increases towards the middle and outer estuary (Baigún and Sverlij 2003; García et al. 2003; Minotti et al. 2011).

Forest, birds, mammals, and herps show a similar trend, with species whose occurrence extend more than 1200 km south of Tropic of Capricorn (Bó 2005; Oakley et al. 2006; Arzamendia and Giraudo 2009).

Conservation Status

A thorough assessment of the conservation status for the whole Fluvial Corridor has not been made yet. All wetland reptile and mammal species present are under some level of international extinction threats, while defining conservation status for other groups has been difficult as there are many knowledge gaps in terms of distribution or population levels, combined with conflicting economic and political interest in cases of important commercial species.

The Argentine portion registered circa 50 protected areas by 2012, covering more than 20,000 km^2, which include four National Parks and three Natural Strict Reserves under federal management, five Ramsar sites, and two Man and the Biosphere Programme reserves, among other provincial, municipal, and private reserves, including specific reserves for fish protection (Boscarol 2013). Although sufficient in terms of areal coverage or biodiversity representation, most conserving units are under mixed management, lacking management plans and adequate funding.

Ecosystem Services

The Fluvial Corridor provides land settings, freshwater for domestic and industrial uses, pollution cleanup services, recreation opportunities, and local and international commerce transportation to more than 40 million urban dwellers which inhabit its shores. It brings huge amounts of warm waters to a temperate region, which ameliorates winter effects. The corridor wetland mosaics contribute to buffer the effects of extraordinary floods and function as water reservoirs during dry seasons or spells.

As a biogeographical highway, it allows the maintenance of a rich biodiversity. Wetland habitats are essential to most fish and birds species because they provide spawning grounds and food sources, and main channels are used as habitats or visual clues for migratory species. This in turn sustains important sport and commercial fisheries and bird watching tourism activities of local and international importance.

The main economic island activity is extensive seasonal cattle ranching on the flooded grassland and savannas, and the drier fringe of lakes and marshlands. Many species have well-known medicinal uses and also provide flower resources for honey production, soft and hard construction materials for roof thatching, furniture, canoes, and crafts.

Threats and Future Challenges

Major threats for the whole system are related to climate variability, river flow regulation from hydroelectrical projects and waterway development expansion, and excessive runoff from agricultural and industrial mainland activities with few or no water treatments (Baigún et al. 2008).

Hydroelectrical projects are mostly located in the upper basins of the Paraná and Paraguay rivers, and their numbers and extents of impounded water continue to increase with population numbers and industry development in need of a larger energy infrastructure. As citizen awareness is easily focused on international projects, governments are switching their focus to hydropower projects completely within national boundaries.

The Paraguay-Paraná Waterway (or Hidrovía) plans to modify the Paraguay and Paraná River for large-scale water transportation. The complete project involves dredging through the Pantanal and eliminating some rock outcrops in the Paraguay River to facilitate year-round navigation of barge trains and dredging the Paraná River stretch between Buenos Aires and Santa Fé to allow the entrance and secure navigation of huge grain carrier vessels. The project has aroused international concerns in relation to the potential environmental effects, highlighting big knowledge gaps not only in biodiversity but mainly in the poor understanding of the hydrological relationships between rivers and floodplains (Hamilton 1999). Only the Ocean-Santa Fé part has been completed and operational. The only well-known impacts reported in the literature are related to the introduction and expansion of exotic Asian mollusks (Darrigan 2004).

Urban growth and agriculture intensification in the islands and mainland shores have increased, with polder-type diking, road and bridge building, excavation for navigation channels, and boat resorts, particularly in the delta region where most reclamation projects for productive activities have been made (Kandus and Minotti 2010). Current economic development is heavily focused on transforming wetlands into terrestrial ecosystems. These land-use changes result in floodplain fragmentation, wetland disconnection, exotic species introduction or invasion, increase in harmful algal blooms, and continuing loss of habitats for biodiversity.

In face of these threats, national governments together with nongovernment organizations are busy analyzing risks and developing conservation strategies for the protection of biodiversity of the whole basin wetlands, e.g., De La Plata River Treaty Program Framework (CIC 2009) and the Ramsar Regional Initiative (SRC 2010), or for areas covering in part the Fluvial Corridor such as the Gran Chaco Americano Ecoregional Evaluation (TNC 2005), the Ecological Risk Assessment for the Paraguay River (Petry et al. 2012), or FREPLATA for the Estuary (Brazeiro et al. 2003).

References

Arzamendia V, Giraudo AR. Influence of great South American rivers of the Plata basin in distributional patterns of tropical snakes: a panbiogeographic analysis. J Biogeogr. 2009;36:1739–49.

Baigún C, Sverlij, SB. Ictiofauna y recursos pesqueros del Río de la Plata interior (margen argentina). Documentos de FREPLATA. 2003. http://adt.freplata.org/documentos/archivos/Documentos_Freplata/Ictiofauna_y_recursos_pesqueros_VJCM.pdf

Baigún CRM, Puig A, Minotti PG, Kandus P, Quintana R, Vicari R, Bó R, Oldani N, Nestler J. Resource use in the Parana River Delta (Argentina): moving away from an ecohydrological approach? Ecohydrol Hydrobiol. 2008;8(1–2):245–62.

Benzaquen L, Blanco D, Bó R, Kandus P, Lingua G, Minotti P, Quintana R, Sverlij S, Vidal L. Inventario de los Humedales de Argentina. Sistemas de Paisajes de humedales del Corredor Fluvial Paraná- Paraguay. Secretaria de Ambiente y Desarrollo Sustentable de la Nación GEF 4206 PNUD ARG 10/003. 2013. 376 p.

Bó RF. Situación ambiental en la Ecorregión Delta e Islas del Paraná. In: Brown A, Martínez Ortiz U, Acerbi M, Corcuera J, editors. La situación ambiental argentina 2005. Buenos Aires: Fundación Vida Silvestre Argentina; 2005. p. 130–74.

Boscarol, N. Áreas protegidas y humedales del Corredor Fluvial Parana-Paraguay. In: Benzaquen L, Blanco D, Bó, R, Kandus P, Lingua G, Minotti P, Quintana R, Sverlij S, Vidal L, editors. Inventario de los Humedales de Argentina. Sistemas de Paisajes de humedales del Corredor Fluvial Paraná- Paraguay. Secretaria de Ambiente y Desarrollo Sustentable de la Nación GEF 4206 PNUD ARG 10/003. 2013. p. 357–72.

Brazeiro A, Acha M, Mianzan H, Gomez-Erache M, Fernandez V. Aquatic priority áreas for the conservation and management of the ecological integrity of the Rio de La Plata and its Maritime Front. Technical Report PNUD Project/GEF RLA/99/G31. 2003. 81 p.

CIC (Comité Intergubernamental Coordinador de los Países de la Cuenca del Plata). Programa Marco para la Gestión Sostenible de los Recursos Hídricos de la Cuenca del Plata, en Relación con los Efectos de la Variabilidad y el Cambio Climático 2010–2015. 2009. http://www.cicplata.org/?id=marco_docs

Darrigan G. Moluscos Invasores, en especial *Corbicula fluminea* (Almeja asiática) y *Limnoperna fortunei* (Mejillón dorado) de la región Litoral. Instituto Superior de Correlación Geológica (INSUGEO), Universidad Nacional de Tucumán. Serie Miscelánea. 2004;12:205–10.

Drago EC, de Drago IE, Oliveros OB, Paira AR. Aquatic habitats, fish and invertebrate assemblages of the Middle Paraná River. Amazoniana. 2003;17(3/4):291–341.

Drago EC, Wantzen KM, Paira AR. The Lower Paraguay river floodplain habitats in the context of the Fluvial Hydrosystem Approach. Ecol Hydrobiol. 2008;8(1):125–42.

García ML, Jaureguizar A, Protogino LC. Asociaciones de peces en el estuario del Río de la Plata. Documentos de FREPLATA. 2003. 4 p. http://adt.freplata.org/documentos/archivos/Documentos_Freplata/Asociaciones_peces_VJCM.pdf

Girard P. Hydrology of surface and ground waters in the Pantanal Floodplain. In: Junk W, da Silva C, Nunes da Cunha C, Wantzen KM, editors. The Pantanal; Ecology, biodiversity and sustainable management of a large neotropical seasonal wetland. Sofia: Pensoft Publishers; 2011. p. 103–26.

Hamilton S. Potential effects of a major navigation project (Paraguay–Parana Hidrovia) on inundation in the Pantanal floodplains. Regul Rivers Res Manag. 1999;15:289–99.

Iriondo MH. Geología del Cuaternario en la Argentina. Ed. Santa Fe, Argentina: Museo Provincial de Ciencias Naturales Florentino Ameghino; 2010. 437 p.

Iriondo MH, Paggi JC, Parma MJ, editors. The Middle Paraná River: limnology of a subtropical wetland. Berlin: Springer; 2007. 382 p.

Kandus P, Minotti P. Distribución de terraplenes y áreas endicadas en la región del Delta del Paraná. In: Blanco D, Mendez M, editors. Endicamientos y terraplenes en el Delta del Paraná: Situación, efectos ambientales y marco jurídico. Buenos Aires: Fundación para la Conservación y el Uso Sustentable de los Humedales; 2010. p. 15–27.

Kandus P, Quintana R, Bó R. Landscape patterns and biodiversity of the Lower Delta of the Parana River: landcover map. Buenos Aires: Ed. Pablo Casamajor; 2006. 48 p.

Latrubesse E, Stevaux JC, Sinha R. Tropical rivers. Geomorphology. 2005;70(3):187–206.

Latrubesse E, Cafaro E, Ramonell CG. The Chaco megafans: hydrogeomorphology of the largest coalescing megafans system in Earth. Amsterdam: American Geophysical Union-AGU-Fall meeting; 2011.

Marchetti ZY, Giraudo AR, Ramonell C, Barberis I. Humedales del Río Paraná con grandes lagunas. In: Benzaquen L, Blanco D, Bó R, Kandus P, Lingua G, Minotti P, Quintana R, Sverlij S, Vidal L, editors. Inventario de los Humedales de Argentina. Sistemas de Paisajes de humedales del Corredor Fluvial Paraná- Paraguay. Secretaria de Ambiente y Desarrollo Sustentable de la Nación GEF 4206 PNUD ARG 10/003. 2013. p. 187–98.

Minotti P, Baigun C, Brancolini F. Peces del Bajo Delta Insular: Una mirada distinta. In: Quintana RD, Villar MV, Astrada E, Saccone P, Maltzof S, editors. El patrimonio natural y cultural del Bajo Delta Insular del Río Parana: Bases para su conservación y uso sostenible. Buenos Aires: Aprendiz; 2011. p. 109–19.

Minotti P, Ramonell C, Kandus P. Regionalización del Corredor Fluvial. In: Benzaquen L, Blanco D, Bó R, Kandus P, Lingua G, Minotti P, Quintana R, Sverlij S, Vidal L, editors. Inventario de los Humedales de Argentina. Sistemas de Paisajes de humedales del Corredor Fluvial Paraná- Paraguay. Secretaria de Ambiente y Desarrollo Sustentable de la Nación GEF 4206 PNUD ARG 10/003. 2013. p. 35–90.

Neiff JJ. Diversity in some tropical wetland systems of South America. In: Gopal B, Junk W, Davis J, editors. Biodiversity in wetlands: assessment, function and conservation. Leiden: Backhuys Publishers; 2001. p. 157–86.

Neiff JJ, Malvárez AI. Grandes humedales fluviales. In: Malvárez AI, Bó RF. (comp.) Documentos del Curso Taller Bases ecológicas para la clasificación e inventario de humedales en Argentina. Buenos Aires. 2004.

Neiff JJ, Iriondo M, Carignan R. Large tropical south american wetlands: an overview. In: Link GL, Naiman RJ, editors. The Ecology and management of aquatic-terrestrial ecotones. Proceedings book, University of Washington. 1994. p. 156–65.

Neiff JJ, de Neiff ASG P, Casco SL. Importancia ecológica del Corredor Fluvial Paraguay-Paraná como contexto del manejo sostenible. In: Petean J, Cappato J, editors. Humedales fluviales de América del Sur. Santa Fe: Ediciones Proteger; 2006. p. 193–210.

Oakley LJ, Prado D, Adámoli J. Aspectos Biogeográficos del Corredor Fluvial Paraguay-Paraná. Insugeo Miscelanea. 2006; 14:245–58.

Petry P, Rodrigues S, Ramos Neto M, Matsumoto M, Kimura G, Becker M, Rebolledo P, Araujo A, Caldas de Oliveira B, da Silvia Soares, M, Gonzáles de Oliveira M, Guimarães J. Ecological risk assessment for the Paraguay River Basin: Argentina, Bolivia, Brazil and Paraguay. Brasilia: The Nature Conservancy and WWF-Brazil; 2012. 54 p.

PNUD. ARG 10/003 Ordenamiento Pesquero y Conservación de la Biodiversidad en los Humedales Fluviales de los Ríos Paraná y Paraguay, República Argentina. 2013. http://www.ambiente.gob.ar/default.asp?idseccion=299

Quintana RD, Villar MV, Astrada E, Saccone P, Maltzof S. El patrimonio natural y cultural del Bajo Delta Insular del Rio Parana: Bases para su conservación y uso sostenible. Buenos Aires: Aprendelta; 2011. 316 p.

Ramonell CG, Amsler CG, Toniolo H. Geomorfología del cauce principal. In: Paoli C, Schreider M, editors. El río Paraná en su tramo medio. Contribución al conocimiento y prácticas ingenieriles en un gran río de llanura. Santa Fe: Centro de Publicaciones de la Universidad Nacional del Litoral; 2000.

Sabattini RA, Lallana V. Aquatic Macrophytes. In: Iriondo MH, Paggi JC, Parma MJ, editors. The Middle Paraná River: limnology of a subtropical wetland. Berlin: Springer; 2007. p. 205–26.

SRC (Secretariat of Ramsar Convention). Estrategia de Conservación y Uso Sustentable de los Humedales Fluviales de la Cuenca del Plata. 2010. http://www.ramsar.org/pdf/sc/40/key_sc40_reginits_b15.pdf

Sverlij S, Liotta J, Minotti P, Brancolini F, Baigún C, Firpo Lacoste F. Los peces del Corredor Fluvial Paraná-Paraguay. In: Benzaquen L, Blanco D, Bó R, Kandus P, Lingua G, Minotti P, Quintana R, Sverlij S, Vidal L, editors. Inventario de los Humedales de Argentina. Sistemas de Paisajes de humedales del Corredor Fluvial Paraná- Paraguay. Secretaria de Ambiente y Desarrollo Sustentable de la Nación GEF 4206 PNUD ARG 10/003. 2013. p. 341–56.

The Nature Conservancy (TNC). Vida Silvestre Argentina (FVSA), Fundación para el Desarrollo Sustentable del Chaco (DeSdel Chaco) y Wildife Conservation Society Bolivia (WCS). Evaluación Ecorregional del Gran Chaco Americano/Gran Chaco Americano Ecoregional Assessment. Buenos Aires: Fundación Vida Silvestre Argentina; 2005.

Wantzen KM, Drago E, da Silva CJ. Aquatic habitats of the Upper Paraguay River-Floodplain-System and parts of the Pantanal (Brazil). Ecohydrol Hydrobiol. 2005;21:1–15.

Wantzen KM, Michael B, Marchese M, Bacchi M, Amsler M, Ezcurra de Drago I, Drago E. Sandy Rivers: a review on general ecohydrological patterns of benthic invertebrate assemblages across continents. Int J River Basin Manag. 2014;12(3):1–26.

Welcomme RL. Inland Fisheries: ecology and management. Food and Agriculture Organization of the United Nations. Oxford: Fishing News Books/Blackwell Science; 2001.

Wilkinson MJ, Marshall LG, Lundberg JG. River behavior on megafans and potential influences on diversification and distribution of aquatic organisms. J S Am Earth Sci. 2006;21:151–72.

The Pantanal: A Brief Review of Its Ecology, Biodiversity, and Protection Status

62

Wolfgang J. Junk and Catia Nunes da Cunha

Contents

Introduction	798
Geographic and Ecological Setting	799
Biodiversity	804
Conservation Status	806
Ecosystem Services	807
Threats	807
Future Challenges	810
References	810

Abstract

The Pantanal is a large wetland of ~150,000 km^2 located in the center of the South-American subcontinent. Dry and wet periods during the Upper Pliocene and Lower Pleistocene formed a geomorphically complex landscape in the Upper Paraguay Basin, covered by the Pantanal. The actual hydrological cycle of the Pantanal is characterized by a monomodal, predictable, annual flood pulse, however also with multiannual wet and dry periods. During high floods, ~110,000 km^2 of the Pantanal are flooded, but during very dry periods only ~5,500 km^2 are covered by water. The hydrological dynamics in combination

---

W. J. Junk (✉)
Instituto Nacional de Ciência e Tecnologia em Áreas Úmidas (INCT-INAU), Universidade Federal de Mato Grosso (UFMT), Cuiabá, MT, Brazil
e-mail: wjj@evolbio.mpg.de

C. Nunes da Cunha
Instituto Nacional de Ciência e Tecnologia em Áreas Úmidas (INCT-INAU), Universidade Federal de Mato Grosso (UFMT), Cuiabá, MT, Brazil

Depto Botânica e Ecologia/Núcleo de Estudos Ecológicos do Pantanal (NEPA), Instituto de Biociências, UFMT, Cuiabá, MT, Brazil
e-mail: catianc@ufmt.br

© Springer Science+Business Media B.V., part of Springer Nature 2018
C. M. Finlayson et al. (eds.), *The Wetland Book*,
https://doi.org/10.1007/978-94-007-4001-3_129

with geomorphological heterogeneity have given rise to a very large macrohabitat diversity that provides the basis for a large biodiversity. This diversity, which includes aquatic, wetland, and many terrestrial species, is evidence of the important interactions among aquatic, terrestrial, and intermittent macrohabitats. Flora and fauna are dominated by species from the surrounding savanna (cerrado biome), but there are also species from the Amazon, chaco, and dry forest biomes. However, in the Pantanal, there are very few endemic species, because paleoclimatic and hydrologic instability hindered speciation. Cattle ranchers, who actually own 90% of the land in the Pantanal have slowly modified its vegetation cover for the last 200 years. Nonetheless, habitat and species diversity have been maintained because of low-density cattle herds. In recent years, economic pressure on the ranches has increased, forcing ranchers to augment cattle production. This has resulted in accelerated deforestation, the planting of artificial pastures, the draining of swamps, and the construction of dikes. Additional, external stress factors have also affected the Pantanal. These factors include the construction of hydroelectric dams along the headwaters, changing the natural flood regime, increased sediment input from upland agroindustries into the rivers, pollution by agrochemicals, mercury input from gold mining, and liquid and solid wastes from cities along the rivers entering the Pantanal. The recently established New Forest Code (Federal Law no. 12.561/12) left large stretches of riverine wetlands along the headwaters and along the Pantanal itself unprotected. While a new law regulating the use and protection of the Pantanal is under discussion, scientists are under intense pressure by groups with economic interest in the region, who seek to undermine existing environmental laws and inhibit the implementation of further protection measures needed to confront new environmental threats.

Keywords
Pantanal · Flooded savanna · Macrohabitat · Biodiversity · Threats · Management

Introduction

The Pantanal is a large, periodically flooded savanna in the depression of the upper Paraguay River (16–20°S and 55–58°W) that extends between the old crystalline shield of Central Brazil and its transition zone to the foothills of the geologically young Andes. It is one of the best-studied South American wetlands. Extensive scientific information is provided in many papers and in several books, with the best known those of Heckman (1998), Britski et al. (1999), Pott and Pott (1994, 2000), Junk et al. (2011), Ioris (2012), and Nunes da Cunha et al. (2014).

The occupation of the Pantanal by humans dates back to ~5,000 years BP, when the climate became moister and groups of Tupi-Guarani Indians began to colonize the region. By the time the Europeans arrived, the Pantanal had long been occupied by several indigenous nations. Wars, slave raids, and the diseases introduced by Europeans quickly reduced the size of the native population. Today, in the Brazilian Pantanal, there are ~150 remaining members of the Guató nation and ~250 of the

Bororo nation. More than 90% of the Pantanal is occupied by cattle ranchers, who have slowly modified the natural vegetation cover to increase beef production. Over the course of two centuries, this process proceeded gradually such that habitat and species diversity were maintained, as was the cultural identity of the local population. During the last three decades, however, threats to the environmental integrity of the Pantanal have increased such that the future of the area has become of increasing concern to scientists, environmentalists, and segments of the local population (Junk and Nunes da Cunha 2012). Here we provide a brief review of the ecological setting, biodiversity, and ecosystem services of the Pantanal as well as its protection status and the major threats to this region.

Geographic and Ecological Setting

The Upper Paraguay depression was formed during the last compression of the Andes, during the Upper Pliocene and Lower Pleistocene (~2.5 million years ago). Sediment deposition in this depression by the Paraguay River and its large tributaries resulted in the formation of large internal deltas, with the largest one being that of the Taquari River (Fig. 1).

The Upper Paraguay Basin is situated in the Brazilian Cerrado (savanna) belt. The climate in the Pantanal is mostly hot, with a pronounced dry season lasting from May to September and a rainy season from October to April (Fig. 2). Annual rainfall decreases from 1,250 mm in the northern Pantanal, near Cáceres, to 1,089 mm in the southern part, near Corumbá. The mean monthly temperature as measured near Cuiabá varies between 27.4 °C in December and 21.4 °C in July. Short-term ingressions of subpolar air masses can lead to a drop in air temperature to 0 °C.

Dramatic climatic changes during the Quaternary led to intermittent periods of large-scale flooding and severe drought, which resulted in differences in sediment deposition and the extent of flooding. Consequently, the surface of the Pantanal is composed of sediments of different ages, some of which extend above the actual highest water level. Both these paleo-dikes, locally called *capões* and *cordilheiras*, and the paleo-deltas show different stages of erosion and internal sedimentation, after their deactivation by the tributaries. The active recent floodplain, with its characteristic features of sediment deposition and erosion, covers only a small part of the Pantanal along the main river channels. Most of the area is shallowly flooded by rainwater and by sediment-depleted water from the tributaries.

In the northern Pantanal, flooding coincides with the rainy season, whereas in the southern part, floods reach a peak about 3 months later, after the water has traversed the wetland area (Fig. 2). The downward slope from north to south is only ~2–3 cm km^{-1}, whereas from east to west it is 5–25 cm km^{-1}. The Paraguay River at Ladário, at the southern end of the Pantanal, shows a predictable monomodal flood pulse, but multi-annual extremely dry and wet periods lead to both extreme flooding and drought events (Fig. 3). Tributaries of the Paraguay River have individual flood patterns and sediment loads according to the regional rainfall. Therefore, different authors have divided the Pantanal into distinct geographic subunits.

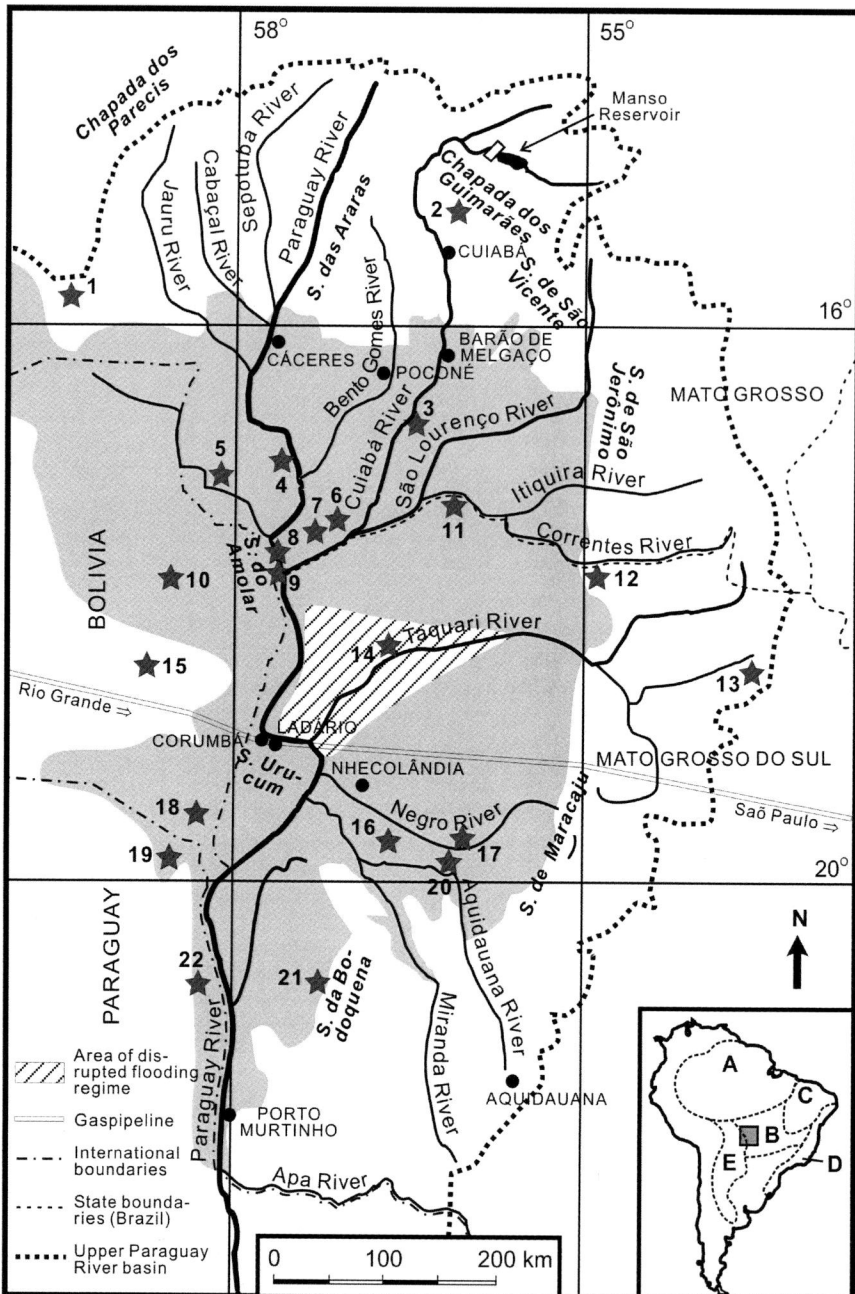

Fig. 1 Map of the Pantanal and its catchment area and the position of major protected areas. *1* Serra de Ricardo Franco State Park, *2* PN Chapada dos Guimarães, *3* RPPN-SESC Pantanal, *4* EE Taiamã, *5* Guira State Park, *6* RPPN Dorochê, *7* PN do Pantanal, *8* RPPN Acurizal, *9* RPPN Penha, *10* ANMI San Matías, *11* Fazenda Poleiro Grande Private Reserve, *12* Serra de Sonora State

Fig. 2 (**a**) Mean monthly precipitation near Cuiabá (1933–1993) and mean water level of the Cuiabá River at Cuiabá (1971–1988), northern Pantanal. (**b**) Mean monthly precipitation near Corumbá (1912–1971) and mean water level of the Paraguay River at Ladário (1979–1987), southern Pantanal (From Junk et al. 2006: Fig.2; Aquatic Sciences, Biodiversity and its conservation in the Pantanal of Mato Grosso, Brazil, 68,2006,278-309, Junk WA, Nunes da Cunha C, Wantzen KM, Petermann P, Strüssmann C, Marques MI, Adis J; © with permission of Springer Science + Business Media)

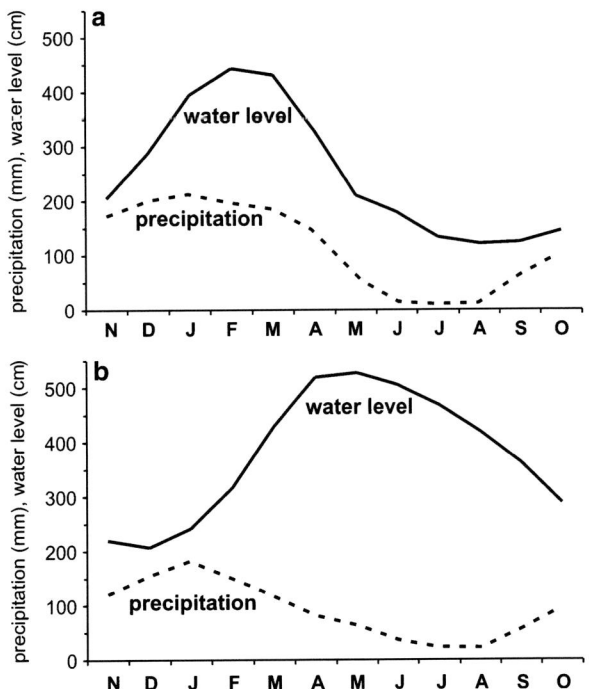

Hydrologically, the Pantanal can be categorized as a temporary wetland, subject to a predictable monomodal flood pulse (Junk et al. 1989). This wetland type is very common in tropical and subtropical regions with a strongly seasonal rainfall pattern. Large parts of these wetlands become completely dry during the low-water period and are colonized by terrestrial plant and animal species that may or may not be wetland-specific. However, these species are integral components of the wetlands because of their important contributions to bioelement cycles, food webs, primary and secondary production, community structure, and biodiversity.

Fig. 1 (continued) Park, *13* Nascentes do Rio Taquari State Park, *14* Fazenda Nhumirim Private Reserve, *15* Reserva Municipal del Valle de Tucavaca, *16* Complex of the Pantanal do Rio Negro State Park and the Private Reserves Fazendinha and Santa Sofia, *17* Fazenda Rio Negro Private Reserve, *18* PN-ANMI Otuquis, *19* PN Rio Negro, *20* Dona Aracy Private Reserve, *21* PN Serra da Bodoquena, *22* Fazenda Rancho Seguro and Tupaciara Private Reserves. *PN* national park, *RPPN* private reserve of natural patrimony, *EE* ecological station, *ANMI* national area of integrated management. For details, see Chapter 4. The small map indicates the position of the Pantanal in South America and the biomes referred to in the text. *A* Amazon forest, *B* Cerrado, *C* caatinga, *D* Atlantic Forest, *E* Chaco (From Junk et al. 2006: Fig.1; Aquatic Sciences, Biodiversity and its conservation in the Pantanal of Mato Grosso, Brazil, 68,2006,278-309, Junk WA, Nunes da Cunha C, Wantzen KM, Petermann P, Strüssmann C, Marques MI, Adis J; © with permission of Springer Science + Business Media)

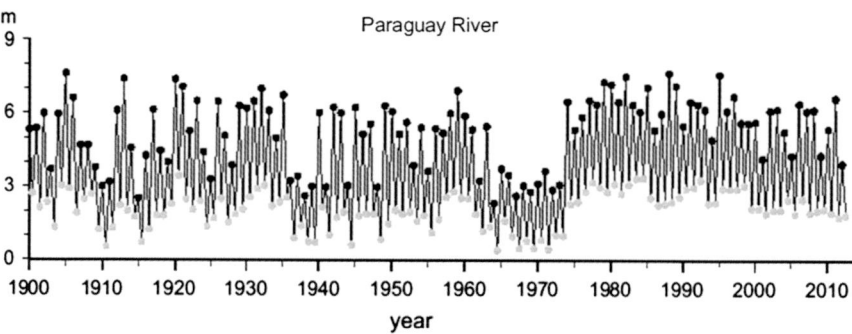

Fig. 3 Annual water-level fluctuations of the Paraquay River at Ladário from 1900 t0 2012 (Data according to DNAEE – Departmento Nacional de Águas e Energia Elétrica; adapted from Junk et al. 2006: Fig.3; Aquatic Sciences, Biodiversity and its conservation in the Pantanal of Mato Grosso, Brazil, 68,2006,278-309, Junk WA, Nunes da Cunha C, Wantzen KM, Petermann P, Strüssmann C, Marques MI, Adis J; © with permission of Springer Science + Business Media) ● = maxima; ○ = minima

The variation in flood amplitude has important consequences for the extent of the annually flooded area, because of the flat relief of the Pantanal. The reported area of the Pantanal varies between 137,000 km^2 (Hamilton et al. 1996) and 150,500 km^2 (Padovani 2010). During maximum floods, ~110,000 km^2 are flooded, during mean floods 53,000 km^2, during dry periods 11,000 km^2, and during very dry periods only about 5,500 km^2. The extensive wild fires that occur during low-water periods are an additional major stress on the plants and animals of the Pantanal.

The challenge to the delimitation of the wetland area of the Pantanal is largely due to the periodic changes in surface area in response to the hydrological pulses. Brazilian planners and the agro-industry prefer to define the wetland boundaries as following the wet areas during the low-water period. This led to modifications in the 2012 forest code (Federal Law N° 12.651). Ecologists, however, have proposed the following definition: "The extent of a wetland is determined by the border of the permanently flooded or waterlogged area, or in the case of fluctuating water levels, by the limit of the area influenced during the mean maximum flood, including, if present, internal permanently dry areas, as these habitats are of fundamental importance to the maintenance of the functional integrity and biodiversity of the respective wetland. The outer borders are indicated by the absence of hydromorphic soils and/or hydrophytes and/or specific woody species adapted to grow in periodically or permanently flooded or waterlogged soils." (Junk et al. 2014).

This definition considers that the Pantanal and other large flood-pulsing wetlands include permanently terrestrial macrohabitats, which must also be protected. However, these areas have been preferentially used for human settlements and are under intense stress from human activities such as agriculture, cattle ranching, and road construction.

Furthermore, the ecologists' definition also recognizes that flood-pulsing wetlands react strongly to extreme hydrological events, which, although they occur only every few years, readjust the wetland flora and fauna and act as important drivers of wetland dynamics. They also may lead to social and economic catastrophes for local people living in and around the wetlands. According to the IPCCs 5th Assessment Report (https://www.ipcc.ch/report/ar5/), the frequency of these events is predicted to increase. Therefore, it would be politically "wise" to consider the mean maximum flood level as the external border of flood-pulsing wetlands, to avoid heavy losses of goods and human life in the future.

The alternation of wet and dry periods in the paleoclimatic history of the Pantanal and the connected changes in sediment deposition and erosion have given rise to a very complex surface structure and to peculiarities in the physical and chemical properties of the soil. The annual changes in the Pantanal's hydrological condition are the major driver of the region's biological and biogeochemical processes. Fires that occur during the dry period pose a setback of flood-adapted plant species and especially woody species, which need decades to recolonize the affected areas. However, the permanent perturbations lead to very dynamic plant and animal communities.

For a better understanding of the complex landscape and dynamic vegetation of the Pantanal, Nunes da Cunha and Junk (2014) established a macrohabitat classification for this area that recognizes functional units, defined as "large landscape units in the floodplain, characterized by specific hydrological conditions." The six functional units are permanently aquatic, periodically aquatic, periodically terrestrial, permanently terrestrial, swampy, and artificial environments. The first five cover the entire environmental gradient and its hydrological dynamics inside the floodplains. The sixth unit comprises areas strongly modified by humans, independent of the hydrological status of these areas.

Within these large functional units, a number of macrohabitats are distinguished. Macrohabitats are defined as "landscape units in the respective wetlands, subject to similar hydrological conditions and covered by a specific and characteristic higher plants or, in its absence, subject to a similar terrestrial or aquatic environment."

The parameters used in the definition of macrohabitats are the frequency, duration, amplitude, and predictability of the floods and droughts, because they influence the occurrence, distribution, and life cycles of wetland organisms. Good indicators of the ecological conditions in macrohabitats are higher plants, because of their fixation in the soil. Herbaceous plants reflect the environmental conditions occurring over weeks, months, and a few years and woody plants those of years, decades, and centuries. Macrohabitats allow the description of plant and animal communities according to environmental conditions. They also provide the ecological basis for the sustainable management of the respective floodplains, because human activities are often concentrated within specific macrohabitats and may threaten them. Moreover, major environmental changes first become evident at the macrohabitat level, because they affect the specific vegetation units. Thus far, 57 macrohabitats have been described (Nunes da Cunha and Junk 2014), but the number can be expected to increase because large parts of the Pantanal have not yet been investigated with respect to macrohabitat classification.

Biodiversity

The Pantanal is a "hyperseasonal savanna," i.e., a savanna subject to prolonged flooding. Its southern border is the Chaco biome, its northern border the Amazon biome, its eastern border the Cerrado biome, and its western border the dry forest biome. The flora of the Pantanal is related primarily to the Cerrado biome, which is composed of different savanna types. There are very few endemic species in the Pantanal because the 6,000 years following the last intense dry period have not been enough to produce endemic species. Speciation in the Pantanal is also probably hindered by the flood pulse, which forces mobile species to move from the floodplain to the rivers and back, with water currents passively transporting propagules or less mobile species. This active and passive mobility leads to a permanent genetic exchange throughout the area and prevents speciation arising from the spatial segregation of populations.

As noted above, the macrohabitats of the Pantanal cover the entire habitat gradient, from permanently aquatic to permanently terrestrial, including a large aquatic terrestrial transition zone that is periodically wet and dry. These conditions were taken into account by Gopal and Junk (2000), who defined wetland species as "all those plants, animals and microorganisms that live in a wetland permanently or periodically (including migrants from adjacent or distant habitats), or depend directly or indirectly on the wetland habitat or on another organism living in the wetland." To be useful in practice, this comprehensive definition requires the establishment of different categories according to distinct taxonomic units: (a) residents of the wetlands proper (specific to wetlands in general, with a subgroup of endemics, and residents not specific to wetlands), (b) regular migrants from deep water habitats, (c) regular migrants from terrestrial uplands, (d) regular migrants from other wetlands (for instance, waterfowl), (e) occasional visitors, and (f) species dependent on wetland biota (for instance, epiphytes, canopy invertebrates, and parasites).

Lists of algae and aquatic and terrestrial invertebrates are thus far incomplete and restricted to a few localities only. They are not discussed here in detail. Within the Pantanal, there are approximately 144 families of Spermatophyta (division of the plant kingdom containing plants that reproduce by means of seeds) (Pott and Pott 1996): 104 are exclusively terrestrial, 21 are exclusively aquatic, and 19 include terrestrial and aquatic species. Of the 1,903 species recorded thus far, 247 are considered aquatic macrophytes or hydrophytes and 1,656 terrestrial (Pott and Pott 2000). The latter includes 900 species of grasses, herbs, vines, epiphytes, and parasites and 756 species of woody plants (shrubs, subshrubs, trees, lianas, and palms). Of 85 tree species analyzed for their environmental requirements, 26 are terrestrial and 4 occur only in habitats subjected to long-term inundation; the remaining 55 species tolerate a wide range of periodic flooding and drought conditions. The considerable morphological and physiological plasticity accounts for the co-occurrence of many terrestrial and wetland grasses as well as herbaceous plants on moist ground during the rainy season. A large seed bank in the sediments that is activated only in small portions at a time allows recolonization of the floodplain by terrestrial and aquatic grasses, sedges, and herbaceous species after flooding and droughts and acts as a safeguard against the unpredictable hydrological events that frequently occur in the Pantanal.

Vertebrates are the most well-studied animals in the Pantanal, but the available information varies between taxa and the number of recorded species continues to increase. Britski et al. (1999) listed 263 fish species belonging to 161 genera and 36 families. Characiformes, with 65 genera and 129 species, and Siluriformes, with 61 genera and 105 species, predominate, as is generally characteristic for neotropical freshwaters. However, there are still many gaps in our knowledge because many of the Pantanal's rivers and their tributaries have yet to be adequately sampled and the life history traits of the species found in these waters have not been adequately investigated.

The richness of the Pantanal herpetofauna, evidenced by the 135 species of amphibians and reptiles reported thus far, reflects its position on major faunal boundaries, such that Cerrado elements are juxtaposed or interdigitate with those from adjacent biomes. However, for the reasons noted above, in spite of the abundance and diversity of aquatic habitats, there are no strictly endemic amphibians or reptiles in the Pantanal. Rather, the recent colonization by invading faunal elements, mainly from the adjacent Cerrado, Gran Chaco, and Amazonia domains, and to a lesser extent from the Atlantic and Chiquitan forests, is still in progress. Therefore, species presently known only from peripheral, elevated habitats are found in the Pantanal wetlands. Among the reptile species of the Pantanal wetlands 49 (52%) are terrestrial species, 21 (22%) are arboreal or semiarboreal species, 12 (13%) are aquatic or semiaquatic species, and 12 (13%) are fossorial, semi-fossorial, or cryptozoic species. The percentages for the total herpetofauna of the floodplain are roughly the same: 52% terrestrial (among anurans, all bufonids, leptodactylids, and the only dendrobatid), 26% arboreal (nearly all hylid species), 12% fossorial, and 10% aquatic or semiaquatic (summarized in Junk et al. 2006).

There are 600–700 bird species in the Pantanal region but only 390 have been confirmed (summarized in Junk et al. 2006). Among the Pantanal species, 358 (97%) have also been recorded in the Cerrado, between 234 (64%) and 277 (75%) in the Chaco and Amazonia biomes, and 151 (39%) in all four biomes (include the Atlantic Forest biome) and the Pantanal. Both the Cerrado and the Pantanal lists include an important group of Southern Amazonian species that are generally found in gallery forests and in geographical proximity to Amazonia. The avifauna of the Pantanal is thus part of the Cerrado fauna.

The vast majority (286 out of 390 species) of the confirmed bird species are terrestrial and are not dependent on the wetland. Another 40 species are terrestrial but wetland dependent. Among the remaining 64 "aquatic species," the dominant groups are wading birds (Ciconiiformes egrets, herons, storks, ibises, spoonbills, 21 spp.), shorebirds (Charadriiformes sandpipers, stilts, plovers, and allies, 16 spp., including 11 Nearctic migrants), and kingfishers (Alcedinidae, 5 spp.). By contrast, waterfowl (Anseriformes, screamers, ducks, and allies, 8 spp.) are relatively poorly represented. In addition, 20 Nearctic migrants, 11 austral wintering migrants, 44 other austral migrants, and 13 nomadic species have been reported.

Similar to the birds of the Pantanal, a full record of its mammals has yet to be compiled, although there are several species lists. In particular, a thorough revision of small, species-rich taxa, such as bats and small rodents, is required. Based on an analysis of different inventories of mammals in and around the Pantanal and an extrapolation of the distribution patterns of mammals in the Brazilian Cerrado and the Argentinean Chaco, 132 mammalian species are estimated in the Pantanal (summarized in Junk et al. 2006). Of these, 91% also occur in the Cerrado, 85% in Amazonia, and 84% in the Chaco. The total number of mammal species in the Pantanal, the surrounding areas of Cerrado and Chaco and adjacent areas of Amazonia reaches 149 species, the maximum number to be expected. In all inventories, bats make up about one-third of the total species number. Most of the Pantanal's mammal species are terrestrial, which highlights the importance of its permanently terrestrial macrohabitats for the maintenance of mammalian species diversity. The rarity of contiguous dense forests most likely explains the low numbers of rodents and monkeys.

Conservation Status

Growing concern about the future of the Pantanal has led to a variety of initiatives by universities, state and government agencies, and national and international NGOs. According to a conservation assessment by the World Wide Fund for Nature and the Biodiversity Support Program, the Pantanal is "globally outstanding" (rank 1 of 4) in terms of biological distinctiveness, "vulnerable" (rank 3 of 5) in terms of conservation, and has "highest priority" (rank 1 of 4) in regional priorities for conservation action. In 1988, the Pantanal was proclaimed by the Brazilian Constitution as a National Heritage site, although this status has yet to assume practical consequences. Law no. 6040 of 2007 confers upon the *pantaneiros* the specific status of a "traditional population," which guarantees them specific rights that can be used for the sustainable management of the Pantanal. However, these rights have yet to be specified and thus are not implemented.

In 1993, the Pantanal of Mato Grosso National Park was recognized as a Ramsar Site, and in 2000 the complex of protected areas comprising the Pantanal of Mato Grosso National Park and the Private Natural Heritage Reserve (RPPN) (Doroché, Acorizal, and Penha) was officially declared by UNESCO as a World Natural Heritage Site. In the same year, the UNESCO recognized the Pantanal of the States of Mato Grosso end Mato Grosso do Sul as a World Biosphere Reserve. In 2002, the Pantanal Regional Environmental Program, associated with the United Nations University (UNU/PREP), was founded at the University of Mato Grosso, Cuiabá, and in 2008 the National Institute for Science and Technology in Wetlands (INCT-INAU) was established. Both institutions are part of a network of national and foreign institutions interested in the sustainable management and protection of the Pantanal.

Currently, in the Brazilian part of the Pantanal, there are two national parks and one ecological station under federal administration, several state parks, and an increasing number of private protected sites, including the Private Natural Heritage Reserve, administered by the NGO ECOTROPICA (http://www.ecotropica.org.br/) and by the Social Service of Commerce (http://www.sescpantanal.com.br/hotel.aspx?s=12). The total protected area is ~908,280 ha (6.4% of the Brazilian Pantanal) (http://www.imasul.ms.gov.br/setores/gerencias/unidades-de-conservacao/reserva-particular-do-patrimonio-natural-rppn/ and http://sistemas.icmbio.gov.br/simrppn/publico/rppn/MT/?nome=&proprietario=&municipio=). In the Wet Chaco and Paraguayan Pantanal, the Rio Negro National Park was expanded to 123,786 ha. In the Bolivian Pantanal, the Natural Area of Integrated Management San Matías covers 2,918,500 ha, the National Park and Area of Integrated Management Otuquis 1,005,950 ha, and the Municipality Reserve of Tucavaca 262,305 ha. These areas include flooded areas but also uplands in different proportions (http://www.fobomade.org.bo/pantanal_bolivia/conociendo.php) (Fig. 1). A research unit (Embrapa Pantanal), under the leadership of the Brazilian Agricultural Research Agency at Corumbá, provides technical assistance for agriculture and cattle ranching inside the Pantanal.

Ecosystem Services

All ecosystem services of the Pantanal are related to water. The vast Pantanal plain stores water during the rainy season and slowly delivers it to the lower sections of the Paraguay River, thereby buffering its flood amplitude. During its passage through the Pantanal, about 90% of the water evaporates, thus contributing considerably to regional water and heat balance (Ponce 1995).

The Pantanal's ecosystem services can be summarized as follows: (1) discharge buffering, (2) water purification, (3) groundwater recharging, (4) water provision, (5) maintenance of biodiversity, (6) interconnection of forest patches by riverine forests (thus promoting gene flow in forest plants and animals), (7) fish production, (8) provision of other renewable wetland products, (9) pasture for cattle ranching, (10) home to local human populations, (11) recreation for local people, and (12) ecotourism. The first six points are of high social and environmental value but little, if any, commercial value, such that each of these diverse wetland services suffers the tragedy of the commons. Thus, for the Pantanal, despite its clean water, natural beauty, and rich biodiversity, these benefits are sustained only by the traditional *pantaneiros*, who have an economic interest in doing so.

Threats

Some of the threats to the Pantanal have abated, such as the poaching of wild animals. Populations of endangered animal species have recovered because of intense control measures. But many threats, both internal and external, continue and new ones have arisen. Internal threats include: (1) changes in vegetation cover (deforestation), (2) overexploitation of pasture areas, (3) replacement of native pastures by exotic grasses, (4) a change from extensive to intensive livestock management, (5) local changes in hydrology due to the construction of roads on dikes and the drainage of swamps, (6) large-scale hydrological changes due to the dredging of river channels and the demolition of geomorphologic obstacles in the river bed of the Paraguay River to construct a large waterway (*hidrovia*, still under discussion), and (6) rectification of river channels. However, the external threats are much more dangerous and are increasing. They include:

1. Deforestation in and sediment input from the uplands
2. Input of pollutants (domestic, industrial, mining, and agrochemicals)
3. Hydrological changes due to the construction of reservoirs in the tributaries
4. Economic pressure on ranchers to intensify cattle production
5. The development of industries in the surrounding areas (e.g., the gas pipeline, mining, and smelting complex of Corumbá)
6. The introduction of exotic plants, animals, and diseases
7. The advance of agro-industry to the borders of the Pantanal, thus disrupting the buffer zone
8. The establishment of settlements of landless people along the border of the Pantanal without ensuring them sufficient area for subsistence agriculture, thus forcing the illegal use of protected wetland resources.

The many reservoirs that are under construction or that are planned in the headwaters of the Pantanal's major tributaries are a major threat to the hydrological regime of the Pantanal (Calheiros et al. 2012). Wetland-friendly approaches, such as the environmental flow assessment, have yet to be considered but should be implemented as part of a master plan that adjusts the release of water to the natural flood regime. This would allow energy production compatible with the hydrological requirements of the Pantanal and would minimize the negative environmental impacts on the region, while leaving the remaining rivers unimpeded by barriers or dams

The plan to rectify and deepen the sinuous channel of the Paraguay River to facilitate ship transport through the Pantanal (as part of the *hidrovia* project) would dramatically affect the hydrology of the entire area, with far-reaching negative consequences for the flora, fauna, and the local human population (Ponce 1995; Hamilton 1999). In 2000, the Brazilian government retreated from this plan, but private enterprises continue to construct related infrastructure despite the strong resistance of NGOs.

Cattle ranching started in the mid-eighteenth century and has expanded. Today, it is a major economic activity, with >90% of Pantanal occupied by cattle ranches. Traditional low-density cattle ranching is ecologically friendly, but increasing competition with cattle ranches on artificial pastures of the surrounding upland is forcing ranchers inside the Pantanal to expand their herds. This leads to accelerated deforestation of forested high-lying areas to provide additional pasture areas and to the substitution of native pastures by exotic grasses. The destruction of key habitats will, over the long term, severely reduce species diversity. Clearly, the effective protection of the Pantanal is only possible if the necessary measures are developed and implemented in close cooperation with the cattle ranchers.

Large-scale soybean and cotton production and intensive animal ranching in catchment areas have resulted in increased erosion and sediment deposition inside the Pantanal, with dire ecological consequences. For instance, during the last few decades, the sediment load of the Taquari River has increased considerably. Because of the low downward slope, the river has expanded beyond its former channel such that it now floods an area of $\sim 11,000$ km^2 for much longer periods than before, thus killing all shrubs and trees and transforming diversified herbaceous plant communities into monotonous swamp communities, dominated by a few aquatic species, such as *Eichhornia crassipes*, *E. azurea*, and *Oxycaryum cubense*.

Recently, several exotic animal species have become established in the Pantanal, with dangerous consequences, such as the Amazonian fish tucunaré *Cichla ocellaris*, a voracious predator that was introduced for sport fishing; the very aggressive African bee (*Apis mellifera* hybrid of *A. ligustica* and *A. scutellata*); and the Asian golden mussel (*Limnoperna fortunei*, Mytilidae). It was probably during the Paraguay War (1864–1870) that a feral population of pigs (*porco monteiro*) became established; over the years, it has been locally managed. Aerial surveys indicate about 9,800 of these animals. A few decades ago, water buffaloes *Bubalus bubalis* were introduced; they also became feral and now comprise a population of 5,100 animals.

The introduction of cattle and horses was accompanied by several animal diseases (discussed in Junk et al. 2006). Mal-de-cadeiras (*Trypanosoma evansi*) was most likely introduced by Spanish settlers in the sixteenth century and quickly affected capybara (*Hydrochoerus hydrochaeris*) populations. The parasite has also been found in coatis (*Nasua nasua*) and dogs. Since the 1930s, there have been outbreaks of foot-and mouth-disease, which have severely affected the deer population. More recently introduced diseases and parasites include the hornfly (*Haematobia irritans*), observed since 1991, equine infectious anemia (swamp fever), a retrovirus transmitted by horseflies (tabanids) and observed since 1974, and bovine trypanosomiasis (*Trypanosoma vivax*), also transmitted by tabanids and observed since 1996. Their eventual impact on the Pantanal's populations of wild animals remains to be determined.

However, the largest threat to the Pantanal (and Brazilian wetlands in general) is the lack of efficient governance. Federal and state laws are inadequate to protect and sustainably manage the area. The New Forest Code (Federal Law no. 12.561/12) and a law passed by the state of Mato Grosso for the Pantanal (Law no. 8.830, of 2008) apply wetland protection only until the borders of floods at the low-water level, which leaves large parts of the Pantanal unprotected. These laws treat the Pantanal as a mostly terrestrial ecosystem and encourage the establishment of traditional, damaging forms of agro-industry, despite the Pantanal's specific status in the Brazilian Constitution as a National Heritage site.

Future Challenges

Although the Pantanal officially has a highly protected status, the reality suggests otherwise. With the establishment of the new forest code in 2012, the Brazilian government severely undermined wetland protection, not just in the Pantanal but throughout Brazil. The Pantanal belongs to the states of Mato Grosso and Mato Grosso do Sul, and their governments are responsible for legislation ensuring the protection and management of the respective areas. However, agro-industry's large role in these states ensures its strong influence on political decisions concerning the development of the Pantanal. For example, soybean plantations begin to abut the borders of the Pantanal, without leaving a buffer zone for the floodplain. Forested areas and shrublands inside the Pantanal are being increasingly cleared for cattle pastures. The rapidly growing number of hydroelectric power plants along the tributaries of the Paraguay River have interrupted the ecological connectivity of the headwaters with the Pantanal, thus altering the hydrology of the area, with far-reaching consequences for its biota. Considering the vastness of the Pantanal, these modifications represent a slow, diffuse, and creeping process. In the absence of an acute crisis, the resulting destruction has not caught the attention of national and international audiences. Nonetheless, the negative consequences for the ecosystem are cumulative and in part irreversible. The scientific community has called for the elaboration of a master plan for the sustainable development and the protection of Brazilian wetlands in general and for the Pantanal specifically. The necessary knowledge is available (Nunes da Cunha et al. 2014), but the political will to use it is still lacking.

References

Britski HA, Silimon KZS, Lopes BS. Peixes do Pantanal – Manual de identificação. Corumbá: EMBRAPA; 1999. 184 p.

Calheiros DF, Oliveira MD, Padovani CR. Hydro-ecological processes and anthropogenic impacts on the ecosystem services of the Pantanal wetland. In: Ioris AAR, editor. Topical wetland management: the South-American Pantanal and the international experience, Ashgate studies in environmental policy and practice. Surrey: Ashgate Publishing Company; 2012. p. 29–58.

Gopal B, Junk WJ. Biodiversity in wetlands: an introduction. In: Gopal B, Junk WJ, Davis JA, editors. Biodiversity in wetlands: assessment, function and conservation, vol. 1. Leiden: Backhuys Publishers b.V.; 2000. p. 1–10.

Hamilton SK. Potential effects of a major navigation project (Paraguay-Paraná Hidrovía) on inundation in the Pantanal floodplains. Regul Rivers: Res Manag. 1999;15:289–99.

Hamilton SK, Sippel SJ, Melack JM. Inundation patterns in the Pantanal wetland of South America determined form passive microwave remote sensing. Arch Hydrobiol. 1996;137:1–23.

Heckman CW. The Pantanal of Poconé. Den Haag: Kluwer; 1998. p. 620.

Ioris AAR. Tropical wetland management: the South-American Pantanal and the international experience, Ashgate studies in environmental policy and practice. Surrey: Ashgate Publishing Company; 2012. 374 p.

Junk WJ, Nunes da Cunha C. Wetland management challenges in the South-American Pantanal and the international experience. In: Ioris AAR, editor. Tropical wetland management: the South-American Pantanal and the international experience, Ashgate studies in environmental policy and practice. Surrey: Ashgate Publishing Company; 2012. p. 315–31.

Junk WJ, Bayley PB, Sparks RE. The flood pulse concept in river-floodplain systems. Can J Fish Aquat Sci. 1989;106:110–27.

Junk WJ, Nunes da Cunha C, Wantzen KM, Petermann P, Strüssmann C, Marques MI. Biodiversity and its conservation in the Pantanal of Mato Grosso, Brazil. Aquat Sci. 2006;68:278–309.

Junk WJ, da Silva CJ, Nunes da Cunha C, Wantzen KM, editors. The Pantanal: ecology, biodiversity and sustainable management of a large neotropical seasonal wetland. Sofia-Moscow: Pensoft; 2011. 870 p.

Junk WJ, Piedade MTF, Lourival R, Wittmann F, Kandus P, Lacerda LD, Schaeffer-Novelli Y, Agostinho AA. Brazilian wetlands: definition, delineation and classification for research, sustainable management and protection. Aquat Conserv. 2014;24:5–22.

Nunes da Cunha C, Junk WJ. A Classificação dos Macrohabitats do Pantanal Mato-grossense. In: Nunes da Cunha C, Piedade MTF, Junk WJ, editors. Classificação e Delineamento das Áreas Úmidas Brasileiras e de seus Macrohabitats. Cuiabá: EDUFMAT; 2014. p. 77–122.

Nunes da Cunha C, Piedade MTF, Junk WJ, editors. Classificação e Delineamento das Áreas Úmidas Brasileiras e de seus Macrohabitats. Cuiabá: EDUFMAT; 2014. 157 p.

Padovani CR. Dinâmica espaço-temporal das inundações do Pantanal [Dissertation]. Piracicaba: Escola Superior de Agricultura "Luiz de Queiróz", Centro de Energia Nuclear na Agricultura; 2010. 174 p.

Ponce VM. Impacto Hidrologico e Ambiental da Hidrovia Paraná-Paraguai no Pantanal Matogrossense: Um estudo de referência. San Diego: San Diego State University; 1995. 132 p.

Pott A, Pott VJ. Plantas do Pantanal. Brasília: EMBRAPA; 1994. 320 p.

Pott A, Pott VJ. Flora do Pantanal – Listagem atual de Fanerógamas. In: EMBRAPA, editor. Anais II Simpósio sobre Recursos Naturais e Sócio-econômicos do Pantanal. Manejo e Conservação. Corumbá: Embrapa Pantanal; 1996. p. 297–325.

Pott VJ, Pott A. Plantas Aquáticas do Pantanal. Brasília: EMBRAPA; 2000. 404 p.

The Paraná River Delta

Patricia Kandus and Rubén Darío Quintana

Contents

Introduction	814
Geomorphology and Hydrology	815
Landscapes, Wetland Ecosystems, and Plant Communities	816
Fish and Wildlife Biodiversity	817
Ecosystem Functions and Values	817
Conservation Status	818
Threats and Future Challenges	818
References	819

Abstract

The Paraná River Delta region is a huge mosaic of wetlands. It covers over 17,500 km^2 on the final 330 km of the Lower Paraná River basin, between 60°39′W, 32°6′S south of Diamante City, and 58°30′W, 34°30′S, next to Buenos Aires City, the Capital City of Argentina. This delta is formed in a complex littoral setting that developed mainly during the last 6,000 years BP although some former processes took place during the Late Pleistocene and early Holocene periods. This region is

P. Kandus (✉)
Instituto de Investigación e Ingeniería Ambiental, Universidad Nacional de San Martín, San Martín, Provincia de Buenos Aires, Argentina
e-mail: patriciakandus@gmail.com

R. D. Quintana (✉)
Instituto de Investigación e Ingeniería Ambiental, Universidad Nacional de San Martín, San Martín, Provincia de Buenos Aires, Argentina

Consejo Nacional de Investigaciones Científicas y Técnicas (CONICET), Buenos Aires, Argentina
e-mail: mossisland2@gmail.com

© Springer Science+Business Media B.V., part of Springer Nature 2018
C. M. Finlayson et al. (eds.), *The Wetland Book*,
https://doi.org/10.1007/978-94-007-4001-3_232

characterized by a high environmental heterogeneity which provides a variety of habitats, all freshwater wetlands, for a productive and rich biodiversity and a wealth of ecological functions. As for biodiversity, while the set of species comprises more than 700 vascular plants, just a few make up the majority of plant biomass. More than 80% of the surface is covered by herbaceous plant communities. Only 4% of the delta area is occupied by different types of native forests located in the less flooded places such as levees and meander spires. As for wildlife, 50 species of mammals, 260 species of non-passerine birds, 37 reptiles and 27 amphibians have been registered in the region. In addition, 200 species of fishes have been registered too, about 60% of the known to the Argentine sector of Paraná-Paraguay River Corridor, the region with the richest freshwater fish in the country. Many wildlife and fish species found here are at the southern limit of their distribution and the Paraná River Delta is an important area for several threatened species. The Paraná River Delta wetlands provide a number of ecosystem functions (such as the reduction in water flow rate and turbulence, increased short- and long-term water retention and storage, and regulation of evapotranspiration) that ensure a good quality of life for local and neighboring areas' inhabitants. This Delta is no longer a pristine region since it has a long history of use by local people. Despite this, the region still has an important biological and functional diversity that should be preserved. There are varying degrees of threats that affect not only the conservation of wetland biodiversity but also the Delta culture. Some of these threats are regional or originate upstream, while others have local caracter (e.g., impacts of infrastructure, water pollution, urban development, and biological invasions).

Keywords

Argentina · Biodiversity · Freshwater wetlands · Landscape patterns · Paraná river delta

Introduction

The Paraná River Delta region is a huge mosaic of wetlands (Malvárez 1999). It covers over 17,500 km^2 on the final 330 km of the Lower Paraná River, between 60°39'W, 32°6'S south of Diamante City, and 58°30'W, 34°30'S, next to Buenos Aires City, the capital city of Argentina (Fig. 1). The Paraná River drains a 2,310,000 km^2 area and ranks second in South America after the Amazon, in terms of basin size, length, and water discharge (Latrubesse et al. 2005). Among the great rivers throughout the world, the Paraná is the only river that flows from tropical to temperate latitudes, where it converges with the Uruguay River into the Río de la Plata estuary. Therefore, the Paraná River Delta region is a complex floodplain having unique biogeographic and ecological characteristics in South America (Malvárez 1997; Quintana and Bó 2011).

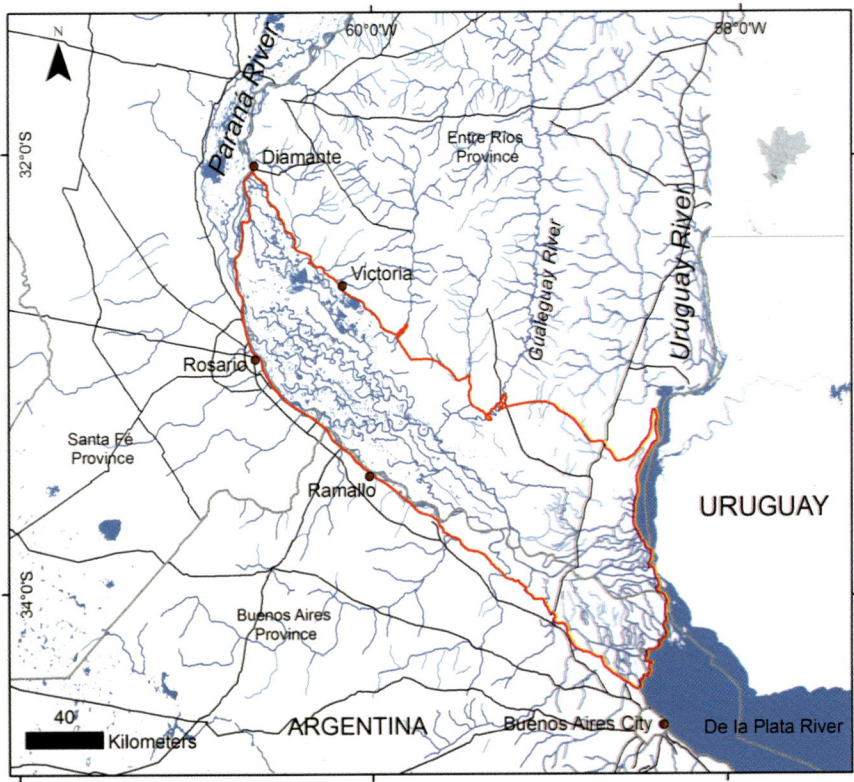

Fig. 1 Paraná River Delta region (*red line* in main frame), in the context of de la Plata basin in South America (*upper right*)

Geomorphology and Hydrology

The Paraná River Delta is formed in a complex littoral setting that developed mainly during the last 6,000 years BP although some former processes took place during the Late Pleistocene and early Holocene periods (Iriondo and Scotta 1979; Cavallotto et al. 2005). The evolution of the Delta includes a fluvial period represented by river flood deposits and terraces; a marine ingression phase with development of a sand barrier, coastal lagoons, and estuaries and well-developed sandy deposits made during a regression phase; an estuarine phase characterized by extensive tidal and delta deposits in the central area; and the present fluvial period, with fluvial bars and channels and deltaic deposits advancing into the Río de la Plata estuary (Iriondo 2004; Ramonell personal communication).

The current hydrological regime is dominated by floods from the Paraná, combined with those from Gualeguay and Uruguay rivers, tidal and storm surges from the Río de la Plata estuary, and local rainfall events, each with a distinctive hydrological signature. The Paraná River represents the main flood source in the upper and middle Delta. However, during strong "El Niño" events, the whole Delta is affected. Additionally, the alternation of relatively wet and dry decades provides an extra source of variability. The Paraná River's average water discharge can range from 16,000 m^3/s to 60,000 m^3/s. The flood pulse from the Paraná River is more irregular than other large South American rivers (Carignan and Neiff 1992). It shows several peaks in the Delta, the main ones at the end of summer (March) from High Paraná River and during winter (June–July) from the Paraguay River (its main tributary), with low flows typically by the beginning of summer.

Landscapes, Wetland Ecosystems, and Plant Communities

Minotti et al. (2013), in the frame of the Paraná-Paraguay National Wetlands Inventory, identified three main landscapes: wetlands of the Paraná River with large lagoons in the upstream area, wetlands of the littoral complex of the Lower Paraná, and wetlands of the Paraná River Delta. Based on the complex hydrogeomorphic setting, Malvárez (1999) identified eleven landscape units, later enriched to fifteen, with the contributions of Kandus et al. (2003) and Zóffoli et al. (2008). The Delta heterogeneity provides a variety of habitats, all freshwater wetlands, for a productive and rich biodiversity and a wealth of ecological functions.

While the number of vascular plants includes more than 700 species, just a few make up the majority of the plant biomass. More than 80% of the surface is covered by herbaceous plant communities (Salvia et al. 2009): emergent hydrophytic broadleaf prairies (i.e., *Polygonum* spp., *Ludwigia elegans*, *Alternanthera philoxeroides*, *Eichhornia* spp.); marshes dominated by equisetoid plants in permanently flooded areas (*Schoenoplectus californicus* and *Cyperus giganteus*); tall marshes dominated by *Scirpus giganteus* in tidal wetlands; and grasslands of tall graminoids (i.e., *Hymenachne* spp., *Panicum elephantipes*, *Coleataenia prionitis*), medium graminoids (*Echinochloa* spp., *Bromus catharticus*), and short graminoids < 0.5 m (*Cynodon dactylon*, *Luziola peruviana*, *Leersia hexandra*, *Paspalum vaginatum*) (Malvárez 1997; Kandus et al. 2003; Kandus and Malvárez 2004; Quintana et al. 2005; Morandeira and Kandus 2015; Magnano et al. 2013).

Shallow lakes are numerous and occupy more than 10% of the upper and the middle portion of the region providing habitat for a number of wildlife species, particularly migratory birds and fishes (Borro et al. 2010). Although only 4% of the delta area is occupied by native forests located in less flooded places, such as levees and meander spires, they include many different types with 26 tree species (e.g., *Acacia caven*, *Albizia inundata*, *Blepharocalyx salicifolius*, *Salix humboldtiana*, *Erythrina crista-galli*) from three biogeographic provinces: Yungas, Paranaense, and Chaco (Menalled and Adámoli 1995; Enrique 2009).

In the downstream portion of the Delta, afforestation of *Salix* spp. and *Populus* spp. covers large areas accounting for another 4% of the region. Finally, secondary succession forests mainly dominated by exotic species (*Ligustrum lucidum, L. sinense, Acer negundo*) are found in the Lower Delta next to Río de la Plata estuary.

Fish and Wildlife Biodiversity

Fifty species of mammals have been documented in the region (e.g., *Myocastor coypus, Hydrochoerus hydrochaeris, Blastocerus dichotomus, Lontra longicaudis, Leopardus geoffroyi*), as well as 260 species of nonpasserine birds (e.g., *Rhea americana, Penelope obscura, Aramides ypecaha*), 37 reptiles (e.g., *Tupinambis merianae, Phrynops hilarii, Rhinocerophis alternatus*), and 27 amphibians (e.g., *Leptodactylus latrans*) (Quintana et al. 1992; Bó and Quintana 2013; Quintana and Bó 2013). The 200 species of fishes recorded represent about 60% of those known to occur in the Argentine sector of Paraná-Paraguay River Corridor, the region with the richest diversity of freshwater fish in the country (Sverlij et al. 2013). The high biodiversity of fish functionally supports the high abundance of birds, particularly the large variety of herons (Minotti 2011b). Most fish are from Brazil, tropical and subtropical lineage, many of them common to the Amazon basin. A smaller group is formed by species of coastal or estuarine lineage (e.g., *Licengraulis grossidens, Odontesthes bonariensis*). Among the large migratory species include predators (e.g., *Pseudoplatystoma corruscans, P. reticulatus, Pseudopimelodus pati, Salminus brasiliensis*), omnivores (e.g., *Oxidoras kneri, Pterodoras granulosus*), herbivores (*Leporinus obtusidens, Schizodon borelli*), and mudsuckers like *Prochilodus lineatus* (Minotti and Kandus 2010).

Many wildlife and fish species found here are at the southern limit of their distribution; also, the Paraná River Delta is an important area for several threatened species. For instance, the IUCN Red List categorizes *Sporophila palustris, Gubernatrix cristata,* and *Argenteohyla siemersi* as endangered species, while *Blastocerus dichotomus, Anisolepis undulatus, Xolmis dominicanus,* and *Xanthopsar flavus* are listed as vulnerable ones.

Ecosystem Functions and Values

The Paraná River Delta wetlands provide a number of ecosystem functions that ensure a good quality of life for local inhabitants and from the neighbouring areas (Kandus et al. 2010; Oddi and Kandus 2011). Among the important hydrological functions provided by the system are the reduction in water flow rate and turbulence, the increased short- and long-term water retention and the storage and regulation of evapotranspiration. Different biochemical processes enhance water quality including storage, transformation, and degradation of nutrients and pollutants and salt regulation. From an ecological perspective, most herbaceous plant communities of the Delta are highly productive, sequestering carbon in soil and biomass (Pratolongo et al. 2007; Pratolongo et al. 2008; Ceballos et al. 2012) and providing forage for livestock and wildlife species (Pereira et al. 2003; Quintana et al. 1998; González et al. 2008; Magnano et al. 2013). More than 25% of wildlife species is used by local people for meat, leather, and feathers (Bó and Quintana 2013; Quintana and Bó 2013). The region offers good sport fishing throughout the year and supplies large migratory species of commercial value for export, fishmongers, and restaurants (Baigún et al. 2009).

Conservation Status

The Paraná River Delta is no longer a pristine region since it has a long history of use by local people. Despite this, the region still has an important biological and functional diversity that should be preserved. To date, the Delta has 23 conservation units, with a total area of 1,171,504 ha. However, only 1% of this area is found in units with effective conservation categories (national parks and other units with management plans). Most of the area within the conservation units under some legal protection incorporates multiple use tenure and mixed (public-private) ownership with no management plans. The "jewels" of the conservation units are "Parque Nacional Predelta La Azotea" (2458 ha; 32°7′S, 60°39′W; nowadays part of the "Delta del Paraná" Ramsar Site) and "Reserva de Biosfera Delta del Paraná" (88.624 ha; 34° 5′S, 58°30′W) (Minotti 2011a; Boscarol 2013).

Threats and Future Challenges

There are varying degrees of threats that affect not only the conservation of wetland biodiversity but also the Delta culture. Some of these threats are regional or originate upstream, while others have a local character:

- Alterations of the water regime for energy and transport infrastructure (i.e., existing and future dams and roads) and the advance of industrial agriculture at country scale.
- Construction of polders and artificial levees and changes in courses and water bodies for cattle ranch, afforestation, agriculture, and tourism.
- Urban development favored by both the lack of planning and building codes in island environments and the low cost of lands.
- A set of diffuse threats is given by the overexploitation of natural resources (e.g., commercial fishing, hunting reptiles and mammals for hides, trapping birds for sale as pets, felling trees for firewood) and the spread of invasive alien species, modifier of natural communities. They include both aquatic species (*Limnoperna fortunei, Cyprinus carpio*) as well as terrestrial ones (plants such as *Rubus ulmifolius, Ligustrum sinense*, and *Iris pseudacorus* and animals like deer axis – *Axis axis*).

References

Baigún CRM, Minotti PG, Puig A, Kandus P, Quintana R, Vicari R, Bó R, Oldani NO, Nestler J. Resource use in the Paraná River delta (Argentina): moving away from an ecohydrological approach? Ecohydrol Hydrobiol. 2009;8(2–4):77–94.

Bó R, Quintana RD. Sistema 5e – Humedales del Delta del Paraná. In: Benzaquen L, Blanco D, Bó R, Kandus P, Lingua G, Minotti P, Quintana R, Sverlij S, Vidal L, editors. Inventario de los Humedales de Argentina. Sistemas de Paisajes de humedales del Corredor Fluvial Paraná-Paraguay. Buenos Aires: Secretaría de Ambiente y Desarrollo Sustentable de la Nación; 2013. p. 297–319. ISBN 978-987-29340-0-2.

Borro MM, Morandeira NS, Salvia MM, Minotti PG, Puig A, Karszenbaum H, Kandus P. Las lagunas de la planicie aluvial del Delta del Río Paraná: clasificación multitemporal e integración con datos limnológicos. In: Varni M, Entraigas I, Vives L, editors. Hacia la gestión integral de los recursos hídricos en zonas de llanuras. Libro de Actas del 1° Congreso Internacional de Hidrología de Llanuras, Azul, Buenos Aires. 2010. p. 639–46. ISBN: 978-987-543-393-9.

Boscarol N. Areas Protegidas y Humedales del Corredor Fluvial Paraná-Paraguay. En: Benzaquen L, Blanco D, Bó R, Kandus P, Lingua G, Minotti P, Quintana R, Sverlij S, Vidal L, editors. Inventario de los Humedales de Argentina. Sistemas de Paisajes de humedales del Corredor Fluvial Paraná- Paraguay. Buenos Aires: Secretaría de Ambiente y Desarrollo Sustentable de la Nación; 2013. p. 357–72. ISBN: 978-987-29340-0-2.

Carignan R, Neiff JJ. Nutrient dynamics in the floodplain ponds of the Parana River (Argentina) dominated by the water hyacinth *Eichornia crassipes*. Biochemistry. 1992;17:85–121.

Cavallotto JL, Violante RA, Colombo F. Evolución y cambios ambientales de la llanura costera de la cabecera del río de la Plata. Rev Asoc Geol Argent. 2005;60(2):353–67.

Ceballos DS, Frangi J, Jobbágy EG. Soil volume and carbon storage shifts in drained and afforested wetlands of the Paraná River Delta. Biogeochemistry. 2012;112(1–3):359–72.

Enrique C. Relevamiento y caracterización florística y espectral de los bosques de la región del Delta del Paraná a partir de imágenes satelitales. Tesis de Licenciatura en Ciencias Biológicas. Buenos Aires: Universidad de Buenos Aires; 2009.

González G, Rossi CA, Pereyra AM, de Magistris AA, Lacarra H, Varela E. Determinación de la calidad forrajera en un pastizal de la región del Delta Bonaerense Argentino. Zootec Trop. 2008;26:223–25.

Iriondo M. The littoral complex at the Paraná mouth. Quat Int. 2004;114:143–54.

Iriondo M, Scotta E. The evolution of the Paraná River Delta. Proceedings of the 1978 International Symposium on Coastal Evolution in the Quaternary, September 11–18, 1978. Sao Paulo: Symposium Organizing Committee; 1979. p. 405–18.

Kandus P, Malvárez AI. Vegetation patterns and change analysis in the Lower Delta Islands of the Paraná River (Argentina). Wetlands. 2004;24(3):620–32. USA ISSN: 0277-5212.

Kandus P, Malvárez AI, Madanes N. Estudio de las comunidades de plantas herbáceas de las islas bonaerenses del Bajo Delta del Río Paraná (Argentina). Darwiniana. 2003;41:1–16.

Kandus PN. Morandeira N, Schivo F, editors. Bienes y Servicios Ecosistémicos de los Humedales del Delta del Paraná. Buenos Aires: Wetlands International; 2010. ISBN 978-987-24710-2-6. 28 p.

Latrubesse EM, Stevaux JC, Sinha R. Tropical rivers. Geomorphology. 2005;70(3):187–206.

Magnano A, Vicari R, Astrada E, Quintana RD. Ganadería en humedales. Respuestas de la vegetación a la exclusión del pastoreo en tres tipos de ambientes en un paisaje del Delta del Paraná. RASADEP. 2013;5:137–48.

Malvárez AI. Las comunidades vegetales del Delta del Río Paraná. Su relación con factoresambientales y patrones de paisaje [disertation]. Buenos Aires: Facultad de Ciencias Exactas y Naturales, Universidad de Buenos Aires; 1997.

Malvárez AI. El delta del río Paraná como mosaico de humedales The Parana Delta as a wetland mosaic. In: Malvárez AI, editor. Tópicos Sobre Humedales Subtropicales y Templados de Sudamérica. Montevideo: MAB-ORCYT; 1999. p. 35–53.

Menalled F, Adámoli J. A quantitative phytogeographic analysis of the species richness in forest communities of the Parana River Delta. Argent Vegetatio. 1995;120:81–90.

Minotti P. Áreas Protegidas. In: Kandus P, Minotti P, Borro M, editors. Contribuciones al conocimiento de los humedales del Delta del Río Paraná: herramientas para la evaluación de la sustentabilidad ambiental. San Martín: Universidad Nacional de General San Martín; 2011a. 32 p. ISBN 978-987-1435-35-7.

Minotti P. Biodiversidad de Peces. In: Kandus P, Minotti P, Borro M, editors. Contribuciones al conocimiento de los humedales del Delta del Río Paraná: herramientas para la evaluación de la sustentabilidad ambiental. San Martín: Universidad Nacional de General San Martín; 2011b. 32 p. ISBN 978-987-1435-35-7.

Minotti P, Kandus P. Marcos geográficos para evaluar el estado de conservación de los peces de la Cuenca del Plata en Argentina. En: Cappato J, De la Balze J V, Petean, J, Liotta J, editors. Conservación de los peces de la Cuenca del Plata en Argentina: enfoques metodológicos para su evaluación y manejo. Buenos Aires, Fundación para la Conservación y el Uso Sustentable de los Humedales. Wetlands International; 2010. p. 51–60. ISBN97898724710-5-7.

Minotti P, Ramonell C, Kandus P. Regionalización del Corredor fluvial Paraná-Paraguay. In: Benzaquen L, Blanco D, Bó R, Kandus P, Lingua G, Minotti P, Quintana R, Sverlij S, Vidal L, editors. Inventario de los Humedales de Argentina. Sistemas de Paisajes de humedales del Corredor Fluvial Paraná- Paraguay. Buenos Aires: Secretaría de Ambiente y Desarrollo Sustentable de la Nación; 2013. p. 35–90. ISBN 978-987-29340-0-2.

Morandeira NM, Kandus P. Multi-scale analysis of environmental constraints on macrophyte distribution, floristic groups and plant diversity in the Lower Paraná River floodplain. Aquat Bot. 2015;123:13–25.

Oddi J, Kandus P. Bienes y servicios de los humedales del Bajo Delta Insular. En: El Patrimonio natural y cultural del Bajo Delta Insular. Bases para su conservación y uso sustentable. In: Quintana R, Villar V, Astrada E, Saccone P, Malzof S, editors. Convención Internacional sobre los Humedales (Ramsar, Irán, 1971)/Aprendelta, Buenos Aires. 2011. p. 135–146. ISBN 978-987-27728-0-2.

Pereira J, Quintana RD, Monge S. Diets of plains vizcacha, greater rhea and cattle in Argentina. J Range Manage. 2003;56(1):13–20.

Pratolongo P, Kandus P, Brinson M. Net aboveground primary production and soil properties of floating and attached freshwater tidal marshes in the Rio de la Plata estuary. Estuar Coasts. 2007;30(4):618–26.

Pratolongo P, Kandus P, Brinson M. Net aboveground primary production and biomass dynamics of Schoenoplectus californicus (Cyperaceae) marshes growing under different hydrological conditions. Darwiniana. 2008;46(2):258–69.

Quintana RD, Bó R. Por qué el Delta del Paraná es una región única en la Argentina? In: Quintana R, Villar V, Astrada E, Saccone P, Malzof S, editors. El Patrimonio natural y cultural del Bajo Delta Insular. Bases para su conservación y uso sustentable. Buenos Aires: Convención Internacional sobre los Humedales (Ramsar, Irán, 1971)/Aprendelta; 2011. p. 42–53. ISBN 978-987-27728-0-2.

Quintana RD, Bó RF. Sistema 5d – Humedales del complejo litoral del Paraná Inferior. In: Benzaquen L, Blanco D, Bó R, Kandus P, Lingua G, Minotti P, Quintana R, Sverlij S, Vidal L, editors. Inventario de los humedales de Argentina. Sistemas de paisajes de humedales del Corredor Fluvial Paraná-Paraguay. Buenos Aires: Secretaría de Ambiente y Desarrollo Sustentable de la Nación; 2013. p. 271–96. ISBN 978-987-29340-0-2.

Quintana RD, Bó R, Merler, J, Minotti P, Malvárez AI. Situación y uso de la fauna silvestre en la región del Delta del Río Paraná (Argentina) [Wildlife use and status in the Paraná Delta region]. Iheringia Sér. Zool., Porto Alegre. 1992;73:13–33.

Quintana RD, Monge S, Malvárez AI. Feeding patterns of capybara *Hydrochaeris hydrochaeris* (Rodentia, HYDROCHAERIDAE) and cattle in the non-insular area of the Lower Delta of the Paraná River, Argentina. Mammalia. 1998;62(1):37–52.

Quintana RD, Madanes N, Malvárez AI, Kalesnik FA. Análisis de la vegetación en tres tipos de hábitat de Carpinchos en la baja cuenca del Río Paraná, Argentina. INSUGEO Miscelánea. 2005;14:183–200.

Salvia M, Karszenbaum H, Kandus P, Grings F. Datos satelitales ópticos y de radar para el mapeo de ambientes en macrosistemas de humedal. Rev Esp Teledetección. 2009;31:35–51.

Sverlij S, Liotta J, Minotti P, Brancolini F, Baigún C, Firpo Lacoste F. Los Peces del Corredor fluvial Paraná-Paraguay. En: Benzaquen L, Blanco D, Bó R, Kandus P, Lingua G, Minotti P, Quintana R, Sverlij S, Vidal L, editors. Inventario de los Humedales de Argentina. Sistemas de Paisajes de humedales del Corredor Fluvial Paraná- Paraguay. 2013. p. 341–56. ISBN 978-987-29340-0-2.

Zóffoli L, Kandus P, Madanes N, Calvo D. Seasonal and interannual analysis of wetlands in South America using NOAA AVHRR-NDVI time series: the case of the Parana Delta Region. Landsc Ecol. 2008;23(7):833–48.

Wetlands of Chile: Biodiversity, Endemism, and Conservation Challenges

64

Alejandra Figueroa, Manuel Contreras, and Bárbara Saavedra

Contents

Introduction	824
Main Types of Wetlands in Chile and Their Location	825
Andean Wetlands	827
Coastal Wetlands and Forest	829
Chilean Peatlands	830
The National Inventory and the Environmental Tracking of Wetlands in Chile	832
Approaches, Goals, and Challenges	832
Wetlands as Indicators of the Environmental Condition	832
Conservation for Sustainable Use Within and Outside Protected Areas	833
Participatory Environmental Management on Greater Island Chiloé	833
Ramsar Sites in Chile, Brief Overview	834
Disturbances and Threats to Wetland Ecosystems	836
References	836

Abstract

Chile has various types of wetlands throughout its territory, their biological diversity is low, but it concentrates high endemism (52% accounts for vascular plants, continental fish account for 55%, and amphibians account for

---

A. Figueroa (✉)
Head Natural Resources and Biodiversity, Ministry of Environment, Santiago, Chile
e-mail: afigueroa@mma.gob.cl; figueroaale2010@gmail.com

M. Contreras (✉)
Centro de Ecología Aplicada (CEA), Santiago, Región Metropolitana, Chile
e-mail: mcontreras@cea.cl

B. Saavedra (✉)
Wildlife Conservation Society, Bronx, NY, USA
e-mail: bsaavedra@wcs.org

© Springer Science+Business Media B.V., part of Springer Nature 2018
C. M. Finlayson et al. (eds.), *The Wetland Book*,
https://doi.org/10.1007/978-94-007-4001-3_247

65%). Towards the northern end of the country, endorheic (landlocked) basins are located in a hydrological network that has developed as a result of the geology of the Andes, here the Andean wetlands are unique and microbial biodiversity is unique. Towards the south of Chile, meadows, coastal wetlands, swamp, lacustrine (lakes, ponds), estuaries, forested wetlands (marshy wetlands, hualves) increase in abundance. The coastal wetlands types are principality tidal flats and marshes, lagoons, and estuarine waters. Other unique wetlands type are the peatlands, which are principally found in Chile and Argentina in South America. Chile currently does not have historical trends data for its aquatic environments and how pressures have acted on their quantity, quality, and morphological structure. The Ministry of Environment has proposed a standardized Wetland Environmental Monitoring System that is integrated and complementary to the National Inventory of Wetlands. The objective is to use the wetlands as indicators of the environmental condition of basins.

Keywords

High endemism · Andean wetlands · Coastal wetlands · Peatlands · Pressures · Environmental monitoring system · National inventory

Introduction

Chile is an elongated but relatively narrow country averaging approximately 180 km wide extending between 18° S and 56° S, a distance of approximately 4,300 km. The Pacific Ocean borders the territory on the west and, the Andean Mountain Range, with altitudes up to 6,900 m above sea level (Ojos del Salado, Atacama region) on the east. Within this area are described eight climatic zones, the most extreme zones are in the north, the hyper-arid desert zone and arid steppes occur on the highlands, and towards the southernmost end are the cold hyper humid and semi-arid steppe in Patagonia. Climate and geography result in the development of various types of wetlands. Thus, important hotspots of biodiversity are recognized in Chile (Cowling et al. 1996; Arroyo et al. 2006). Although Chile has low diversity, it concentrates high endemism; for example, 52% accounts for vascular plants, continental fish account for 55%, and amphibians account for 65% (CONAMA 2008; MMA 2016).

Watersheds in the Mediterranean zone in the northernmost end of Chile are water deficient (Banco Mundial 2011), but wetlands are an exception to the rule. Southwards, wetlands are increasingly frequent and diverse, and the human population is concentrated around headwaters, at the mouths of rivers, at the margins of lakes, ponds, estuaries, and creeks. Changes in population settlement patterns have redefined the natural dynamics of these ecosystems; changes in weather patterns pose additional challenges to preserve these ecosystems.

Main Types of Wetlands in Chile and Their Location

Towards the northern end of the country, endorheic (landlocked) basins are located in a hydrological network that has developed as a result of the geology of the Andes. Salt flat-type wetlands, Andean ponds, meadows, and highland wetlands predominate in the Puna area. The interdependence between wetlands and the aquifers that feed these systems is narrow and fragile. The biogeographic isolation of this area has favored the existence of endemic species. For example, the diversity of fish only reaches 44 species in Chile, 81% of which are endemic (Vila et al. 1999; Habit et al. 2006) and is a low number when compared to other biogeographic regions of the world.

Southward towards central Chile between 27° S and 31° S, the climatic conditions change and the environments are no longer dominated by high radiation and evaporative processes. Andean wetlands and meadows begin to predominate as salt flats and highland wetlands disappear. Transverse valley systems, ritronic and potamon habitats, and seasonal wetlands (creeks and estuaries) appear.

Towards the south of Chile, rivers are flowing, vegetation increases, and soils are oversaturated. Meadows, coastal wetlands, swamp, lacustrine (lakes, ponds), estuaries, forested wetlands (marshy wetlands, hualves), nonforested wetlands, and peatlands increase in abundance. Peatlands have their greatest expression in the southernmost region of Chile, between 39° S and 56° S (regions of Los Lagos and Magallanes).

Based on the national register of wetlands, their area is estimated to not exceed 2%. However, this is an underestimate as it excludes wetlands on oceanic islands and those (including peatlands and meadows) in Chile's southernmost area (regions of Aysen and Magallanes: see Fig. 1). An overview of wetland types in Chile and their most representative environments are presented in Table 1.

With the exception of coral reefs, tundra, and karst, all the wetlands described in the Ramsar classification system are found in Chile (see Ramsar Convention Secretariat 2010 – Annex 1). In Chile we can find various types of wetlands, some of which have been characterized by Hauenstein et al. (2004) and in some cases, their names are due to denominations given by indigenous communities, below are some descriptions;

Hualves: Rain forests vegetated with native woody Myrtaceae such as temu *Blepharocalyx cruckshanksii*, pitra *Myrceugenia exsucca*, chequén *Luma Chequén*, and tepu *Tepualia stipularis* on waterlogged poorly drained soils. These wetlands are the habitat of the huillín or river otter *Lontra provocax* and meadow shrimp *Parastacus nicoletti*. These wetlands are primarily located in grabens, creek beds, or gullies with poorly drained soils (Varela 1981; Ramírez et al. 1983; Castro 1987; San Martín et al. 1988; Solervicens and Elgueta 1994) in the central depression coastal range (Ramírez and Añazco 1982; San Martín et al. 1988) and on the island of Chiloé (41° 00′–42° 30′S).

Ñadis: Systems with thin soils, saturated or flooded only in winter. They have a waterproof layer of "fierrillo" between the organic soil and gravel substrate. Located in the central depression of south central Chile, these wetlands have poor plant and animal diversity.

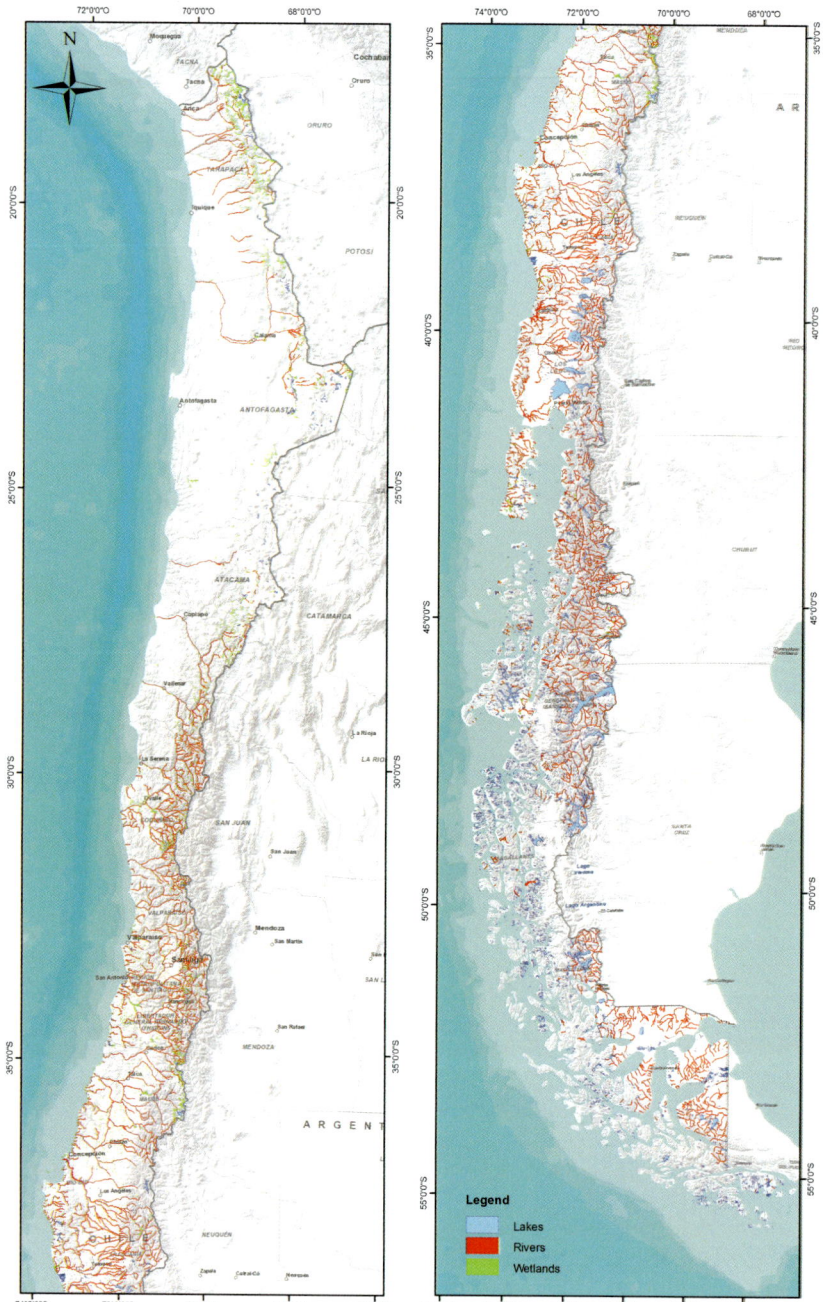

Fig. 1 Cadastre rivers, lakes and wetlands, excluding peatlands, meadows and saline pastures of Los Lagos, Aysen and Magallanes regions (Source: Ministerio de Medio Ambiente 2014 based upon MMA 2012, with permission)

Table 1 Predominant wetland ecotypes (CONAMA 2006) and some examples where they can be found

Ecotypes	Common name	Some locations in Chile
Coastal wetlands	Coastal lakes, salines lagoons, intertidal marshes, estuarine waters, intertidal salt	Lago Budi (saline coastal lake), Ramsar Site Laguna Conchalí (lagoons), Laguna Cahuil (saline lagoons and estuarine), Ramsar Site El Yali (coastal freshwater lagoons and saline lagoons), Humedal Tubul-Raqui (marshal, estuarine), Estuary del río Queule (estuarine waters), Putemún (intertidal marshes)
Inland wetlands	salar, "bofedal," "Puquios"	Salar de Atacama (salar and saline lagoons), Sitio Ramsar Surire (salar and saline lagoons), Sitio Ramsar Sistema hidrológico Soncor (saline lagoons), Salar de Huasco, Sitio Ramsar Negro Francisco and Santa Rosa (saline lagoons and salar, Andean prairie)
	Hualves, pitranto, ñadi, swamp	Wetlands of the central depression between the Maule and Araucanía (primarily) and the coastal area of Araucanía (Queule and Moncul). (forested wetlands, swamps, permanent rivers)
	Mallín, turberas (peatlands, bogs, fens), "pomponales"	Parque Karukinka, Tierra del Fuego (peatlands), Parque Nacional Chiloé (peatlands), Parque Tantauco (peatlands, bogs)
	River, lakes, streams	Río Lluta (river), Lago Chungará (andean lake), in Ramsar Site Parque Andino Juncal (rivers and streams, bogs), estero Tongoy (stream), Lago Lleu-Lleu (permanent lake)

Albúferas: Brackish lagoons, located in the coastal area, with seasonal connectivity to the sea. They are highly eutrophic, due to the salt loads coming from the sea. A representative example is the lagoon of the estuary El Yali, Valparaíso Region.

Andean Wetlands

Andean wetlands are widely distributed on the Andes Mountain Range, being prominent in the Andean highland area and Andean steppe (Fig. 2). Andean wetlands are usually located above 3,000 m asl, while high Andean wetlands typically occur around 2,300 m asl (MMA et al. 2011a). The Puna Seca is a bioregion associated with the Andean highlands characterized by intense cold, dryness, and daily temperature fluctuations, mainly endorheic basins, vegetation types defined as azonal hydric, and can include halophytes (Ahumada and Faundez 2009). These types of wetlands are flooded or partially flooded and have a strong balance with groundwater (aquifers).

Fig. 2 A peatland (bog) in Parque Nevado Tres Cruces that lies in the northern end of the southern Andean steppe, Atacama Region, and includes Laguna Santa Rosa and Laguna del Negro Francisco designated as Ramsar wetlands of international importance (Photo Credit: A. Figueroa © Rights remain with the author)

Wetlands as freshwater lakes and ponds (glacial, volcanic, and tectonic), hot springs, and geysers also occur in Puna Seca. Associated with these lakes and ponds are dense stands of flooded or semi-flooded sedges. In the subregion of the Mediterranean Andes and the Cordillera de la Araucanía (from the Biobio region to the northern region of Los Lagos), wetlands occupy the higher parts of the Cordillera de los Andes and Nahuelbuta interspersed among evergreen coniferous *Araucaria sp.* and southern beech *Nothofagus sp.* forests and steppes characterized by the genera *Festuca, Juncus, Carex,* and *Eleocharis.*

These type of wetlands (meadows, sedges, bogs) are supplied almost exclusively by groundwater and there exists a functional relationship between riparian or edge vegetation (azonal vegetation) and water courses and other bodies of water. Biological communities can be distinguished between those occurring in low salinity ecosystems (macrophytic vegetation, birds, fish, and amphibians), high salinity ecosystems (communities of flamingos, benthic microalgae, and bacteria), and ecosystems dominated by hydrophytic vegetation.

Andean wetlands are fragile ecosystems with high species endemism. Their high fragility is associated with changes in rainfall patterns and also with anthropogenic causes, such as drainage activities, overgrazing, or disturbance in the water regime. For wetlands in the dry Puna, sustained water extraction kills meadows and wetlands. Reduced flows also negatively affect the habitats of fish, amphibians, and birds.

Andean wetlands are rich in species, in response to the spatial heterogeneity, local factors that occur in different basins (e.g., water, soil, water quality), and the phenomena of geographic isolation. In general, the biological composition of wetlands is specific constituting biodiversity concentration areas in the highland region ("hot spot"). The fish fauna of Andean wetlands is endemic; it has primitive characteristics and is of great ecological significance (Arratia 1982; Vila et al. 1999, 2006). This is the case for the genus *Orestias* with six endemic species present only in the Andean wetlands (from 3,000 to 4,500 m. in altitude). Birds are another important group that depend upon Andean wetlands. Chilean flamingo *Phoenicopterus chilensis*, Andean flamingo *P. andinus*, James's flamingo *P. jamesi*, and other migratory bird species use the salt flats and lagoons for reproduction and feeding.

A remnant microbial world was recently recognized in Argentina and Chile salt mines of the Andean otherwise known to occur in certain parts of the world, e.g., Australia (Shark Bay); Mexico (Cuatro Ciénagas), and USA (The Bahamas-Yellowstone) (Farías and Conteras 2013). Among these microbial groups are distinguished cyanobacteria, microbialites, stromatolites, evaporates, and a variety the nomenclature of which varies according to its structure and composition. In general, they correspond to the microorganisms that gave life to the planet. This demands precision in terms of water resources management, particularly in endorheic basins, and agreements to limit activities in areas of ecological importance and scientific interest.

Coastal Wetlands and Forest

These wetlands are located along the continental coast and on the oceanic islands, including Chiloé Island. There are mainly three types of coastal wetlands in Chile: tidal flats and marshes, lagoons, and estuarine waters. The estuaries are numerous and occur south of latitude 35°S (subhumid zone in Maule and Biobio Regions); south of latitude 40°S, estuaries become large deltas and are characteristic southern fjords.

In the northern part of Chile are rivers with low flows, small streams, and coastal lakes. The largest estuaries located to the south of Mataquito River (Maule Region) input nutrients and sediments that benefit coastal zone productivity (Stuardo and Valdovinos 1989) and supports activities like an inshore fishery and salt works. Especially important are forested wetlands of the coastal zone between the Imperial and Toltén Rivers. In this area these forested wetlands (wooded swamps) are known locally as "hualves" (mapudungun language, Ferriere 1982) or "pitrantos," they do not receive salinity from the tides but have daily influence of these. Several authors have described the importance of these wetlands, its origin, and its vegetal composition (Ramírez et al. 1995; Hauenstein et al. 2004, 2014; Peña-Cortés et al. 2011). Principal genera in these wooded swamps include *Myrceugenia, Blepharocalyx, Luma*, and *Tepualia* and are described as having a canopy cover between 18 and

20 m in height (Ramírez et al. 1995; Correa-Araneda et al. 2011). These wetlands are the favored habitat for hullín, (*Lontra provocax*, Thomas 1908), birds as torcaza (*Araucaria Patagioenas*, Lesson) and crow swamp (Plegadis chihi, Veillot) as well as amphibians.

Coastal wetlands are under pressure by development activities in adjacent basins, particularly the central Chilean zone where most of the population lives and most of the industrial activities occur (with the exception of mining which takes place in the north of Chile). The farming activity exerts some pressure on coastal systems which extends into the river flood zones, especially near estuarine wetlands. The high input loading of nitrogen and phosphorus with the application of pesticides and herbicides negatively impacts water quality, the fishery, and marine and freshwater biodiversity.

Nutrient levels and chlorophyll "a" concentrations used to determine the trophic status indicates there is a progressive decline in the general environmental condition of coastal wetlands. In 2011 and 2013, the 68 monitored coastal systems all showed either a tendency towards eutrophication or hypereutrophication (MMA 2011b, 2012). About the Biobío Region 11 systems were evaluated and 72% were in a bad conservation state. For the coastal zone, between the Andalién and Biobío Rivers, the conservation status of 83% of 12 evaluated systems was classified bad or very bad (MMA 2011c). Chilean coasts are however also exposed to dramatic and mostly irreversible natural changes caused by earthquakes and tsunamis that produce changes in the morphology and ecological characteristics of wetlands (e.g., coastal zones and wetlands of Valdivia and Cruces Rivers, Queule River Estuary, Tubul – Raqui Estuary).

In addition, the mouths of rivers and marshes are modified by opening of the terminal bars. Summer tourism can also impact these coastal wetlands if it has not been planned. To ensure continued enjoyment and pleasant visitation of coastal wetlands by many people, coastal wetlands need protection along with appropriate access infrastructure for visitors and educational information.

Chilean Peatlands

South American peatlands are principally found in Chile and Argentina, and the largest proportion of these is in the humid zones of Patagonia where regional climate patterns favor the development of these cold temperate wetlands (Roig and Roig 2004). Although Chile does not have a detailed land registry for peatlands, the largest proportion of peatlands are distributed in the administrative regions of Los Lagos south to Magallanes and particularly south of $45°S$ (Lappalainen 1966). Joosten and Clarke (2002) estimated Chile's peatlands at no more than 10,470 km^2, while other publications consider peatland covering between 10,684 km^2 (Luebert and Pliscoff 2006) to 21,000 km^2 (1.4–2.8% of the national territory). The two southernmost provinces of Tierra del Fuego and Chilean Antarctica in the Region of Magallanes have approximately 4,900 km^2 of peatland, respectively (Ruiz and Doberty 2005). Karukinka Natural Park at nearly

Fig. 3 Peatlands in Karukinka Natural Park. Magallanes Region, Chile (Photo credit: Carlos Silva-Quintas © Rights remain with photographer)

2,800 km² is the largest privately owned protected area in southern Magallanes and contains significant areas of peat (Fig. 3), while Chiloé National Park and the private Tantauco Park on Chiloé Island (Los Lagos Region) have only minor areas of peat. Bernardo O'Higgins National Park within both the Aysén and Magallanes regions is the largest of the protected areas in Chile at 35,259 km² and peat is one of the predominant vegetation types.

Peatlands provide habitat to a diverse group of species including amphibians (e.g., marbled wood frog *Batrachyla antartandica*), birds (e.g., white-tufted grebe *Rollandia rolland*, silvery grebe *Podiceps occipitalis*, magellan goose Caiquén *Cholephaga picta*), and mammals (e.g., guanaco *Lama guanicoe*). The beaver *Castor canadensis* is however an introduced species that negatively impacts peatlands, particularly in Magallanes and Tierra del Fuego, with the construction of dams from slow growing native trees and shrubs. The dams flood wide areas resulting in changes in basin hydrology, habitats, and peatlands (Baldini et al. 2008).

Peatlands are also being negatively affected by human activities, particularly by a growing peat harvesting industry (Domínguez 2013). Although there are a number of legal standards respecting the use of inlands wetlands, an assessment by Möller and Muñoz-Pedreros (2014) concluded peatlands had among the lowest legal measures for their protection. Harvesting of this resource has a number of ecological and social consequences including the disruption of *Sphagnum* ecosystems, changes in water storage capacity affecting water supply to rural communities, exhaustion of the moss and impoverishment of the producers, carbon cycle impacts, landscape erosion, and loss of biodiversity (Zegers et al. 2006; Diaz and Delano 2012).

The National Inventory and the Environmental Tracking of Wetlands in Chile

Approaches, Goals, and Challenges

Beginning in 2011 work has been proceeding on developing a national inventory of wetlands. The first step in this undertaking was development of a national land register from which the locations and area of the different watercourses and bodies of water existing in the country could be determined (Table 2). However, full coverage has not yet been completed using the available tools and methodology for some areas and types of wetlands, such as the peatlands in Patagonia (discussed below). Although there is information for these areas included in the Forest Cadastre (CONAF – CONAMA 1999; CONAF-UACH 2011), it must be revised. The current information of the locations of Chilean aquatic ecosystems (rivers, lakes and wetlands) now is available in mapping (see Fig. 1) and an estimate of their coverage (Table 2). The detailed information for eventual consolidation in the national inventory is still underway. The national inventory is the means to implement an environmental monitoring system of wetlands and support the territorial planning at country level. The National Cadastre of wetlands is a first step.

Wetlands as Indicators of the Environmental Condition

Chile currently does not have historical trends data for its aquatic environments and how pressures have acted on their quantity, quality, and morphological structure. A monitoring and evaluation program (Early Warning System) is a prerequisite for assessing the health of aquatic systems and determining whether or not a wetland has suffered changes in its ecological character. A standardized Wetland Environmental Monitoring System integrated and complementary to the National Inventory of Wetlands and Environmental Monitoring project (MMA 2011c, 2012) is proposed for this purpose. Applying a basin approach, wetlands will be selected to collect basin and class/ecotype specific information on condition indicators, forcing factors, and threats in order to ascertain wetland environmental and ecologic condition (CONAMA 2006). Spatiotemporal considerations include where and when to measure condition indicators and anthropogenic activity specific to each wetland, i.e., defining the most sensitive area and period for each wetland (dependent on geographical location and wetland ecotype), included in the monitoring and assessment program. Pilot projects are underway in wetlands including Sitio Ramsar Laguna Negro Francisco and Sta. Rosa and The Yali National Reserve.

Conservation for Sustainable Use Within and Outside Protected Areas

Several aspects have driven the development of projects that improve the knowledge and management of the environment: the progressive request for natural spaces, the loss and alteration of wetlands, the growing use of hydric resources, and some deficiencies of regulatory instruments to protect wetland ecosystems. This improvement involves local governments, communities, and private systems through participatory environmental management and environmental monitoring of wetlands inside and outside of protected areas; two current examples of which occur in Greater Island of Chiloé and Nevado Tres Cruces National Park and Sitio Ramsar Negro Francisco, respectively.

Participatory Environmental Management on Greater Island Chiloé

The island is located between latitudes 42°–43° S and has a diversity of wetlands: marshlands, lakes, rivers, and peatlands. Towards the eastern coast of Chiloé are intertidal wetlands and salt marshes used by migratory birds and diverse fauna of invertebrates. More than the 30% of the population of the whimbrel *Numenius phaeopus* and Hudsonian godwit *Limosa haemastica* arrive every year after travelling over 14,000 km from the arctic to these coastal wetlands, making this one of the most important areas for these species in South America. Chiloé wetlands are also the habitat for the Chilean flamingo *Phoenicopterus chilensis* (Fig. 4), chorlon chilean (rufous-chested plover) *Charadrius modestus*, and the small churrete (grey-cheeked cinclodes) *Cinclodes oustaleti.*

Peatlands vegetated with Sphagnum, grasses (gramineas), and "pulvinadas" (dense, compact, and hard cushion-shaped plants, e.g., *Donatia fascicularis, Astelia pumila, Caltha dioneifolia*) are very common, the latter having recently developed over the last 100–200 years by anthropogenic processes. The vegetation in some areas corresponds to: Totora azul marsh (*Scirpo-cotuletum coronopifoliae*), Seliera marsh (*Puccinellia-Sellerietum radicantae*), and Llinto marsh (*Sarcocornio Spartinetum densiflorae*).

A project of participatory management has been developed for 11 coastal areas that involves four local governments and communities. Management plans for these 11 coastal area wetlands contain local developmental proposals that consider resource sustainability, agriculture, and wetland recovery and protection. The stakeholders define the priorities and conservation interests. The management proposal is presented before the local government in order for them to identify changes in the territorial planning and to develop agreements for land use restrictions in fragile areas and degraded zones. Sustainable tourism of coastal wetlands will be promoted using guides, interpretative tables, and signposts.

Fig. 4 Chilean flamingo *Phoenicopterus chilensis* (Photo credit: Roberto Villablanca)

Ramsar Sites in Chile, Brief Overview

Chile has 12 Ramsar sites covering an area of 359,989 ha. Eight of these are Andean wetlands (Table 3) and four are coastal (Table 4), with the exception of Lomas Bay, the second southernmost Ramsar site in the world after Argentina's Atlantic Coast Reserve of Tierra del Fuego, which includes the intertidal flats of Bahía Lomas.

The first site designated was the Santuary Carlos Anwandter (4,877 ha). It is a fluvial wetland with a marked marine influence due to the daily fluctuations of the tides and freshwater flows and affected by major natural and anthropogenic events. Located in the region of Los Rios, this wetland originated from the earthquake that hit Chile in 1960.

Since 1981, public interest in the protection of wetlands has increased. Two examples of private conservation in Chile are: Andino Juncal Park and Laguna Conchalí. Andino Juncal Park was nominated by a private owner who proposed 13,796 ha of private land be included within the boundaries of a Ramsar site for the purposes of preservation, education, and research. This is a symbolic action, considering the mining activity of our Cordillera and neighboring areas to Juncal, where glaciers, Andean prairie, streams, and watershed (header basin) are protected. Laguna Conchalí is managed by a mining company. This 52 ha site is an example of passive recovery and community education.

The remaining Ramsar sites are within the National System of Protected Areas. The demands made by society to improve the conditions of their environment have attracted increasing interest in the conservation of wetlands and in designation of Ramsar sites as a protective measure.

64 Wetlands of Chile: Biodiversity, Endemism, and Conservation Challenges

Table 2 Identified area of wetlands, rivers, and lakes by region

Región	Lakes (km²)	Rivers (km²)	Wetlands (km²)	Total (km²)	Total (ha.)
Arica y Parinacota	56	34	196	286	28,600
Tarapacá	12	14	11	137	13,700
Antofagasta	167	76	166	409	40,900
Atacama	87	12	78	177	17,700
Coquimbo	9	10	140	159	15,900
Valparaíso	15	27	24	66	6,600
Región Metropolitana	22	61	49	132	13,200
O'Higgins	6	4	101	111	11,100
Maule	115	171	138	424	42,400
Biobío	176	283	64	522	52,200
Araucanía	561	176	82	819	81,900
Los Ríos	1,050	230	13	1,293	129,300
Los Lagos*	178	232	2,036	2,446	244,600
Aysén	4,000	329	113	4,442	444,200
Magallanes Y Antártica Chilena	2,700	180	6	2,886	288,600
*Isla Grande de Chiloé	130	49	116	295	29,500
TOTAL	**9,284**	**1,887**	**3,433**	**14,604**	**1,460,400**

*Chiloé Island is part of the Los Lagos Region
Source: Ministerio de Medio Ambiente, Chile (2012)

Table 3 Andean Ramsar Sites in Chile

Ramsar Sites High Andean	Area (ha)
Laguna del Negro Francisco y Laguna Santa Rosa	62,460
Salar de Surire	15,858
Salar de Tara	96,439
Salar del Huasco	6,000
Salar de Aguas Calientes IV	15,529
Salar de Pujsa	17,397
Parque Andino Juncal	13,796
Sistema hidrologico de Soncor del Salar de Atacama	67,133

Table 4 Coastal Ramsar Sites in Chile

Ramsar Sites Coastal	Area (ha)
Laguna Conchalí	34
Reserva Nacional El Yali	520
Santuario de la Naturaleza Carlos Anwandter	4,877
Bahía Lomas	59,946

Disturbances and Threats to Wetland Ecosystems

The relationship between wetland ecosystems and people is close as much of the population is located on the coastline or the banks and mouths of rivers or estuaries from which natural resources are extracted, e.g., seafood, fish, and even salt. Wetlands are used as navigation routes and as tourist attractions. The same pattern is repeated for wetlands in the interior of the country, concentrating on the edges of rivers, estuaries, lakes, or ponds. Along with the increase in population and human activities is however alterations in the ecological integrity of wetlands with changes affecting water quality and the loss of wetlands and their biodiversity. Habitats are modified along with the landscape, thereby changing riparian and terrestrial vegetation types. Space is opened for exotic species, and habitats for fish species are lost or become isolated.

Moreover, the increased and growing demand for water extracted from aquifers and natural waterways exceeds recharge. Wetlands are also being affected by other factors including the removal of native forest, soil drainage, construction of coastal roads, or changes to the dynamics of the terminal bar connecting coastal lagoons and estuaries to the sea. Coupled with changes in rainfall patterns and retreating glaciers feeding streams and lakes in much of the Chilean territory, a complex scenario is established for the maintenance of fragile ecosystems.

The systematic fragmentation of bodies of water, rivers, and estuaries to consolidate city real estate development extends into rural areas and develops a landscape where wetlands are surrounded by people. However, in planning urban spaces the inclusion of wetlands and the goods and services they provide is not considered as a contribution to human welfare. For example, in the city of Concepcion, regional capital of the region of Biobío, the expansion and development of real estate projects has "made" more ponds. These become wetlands embedded in an urban matrix. Their viability is however not assured, because the fragmentation modifies the natural waterways and the hydrographic network (MMA 2010a). Heavy rains maintain the wetlands, but their environmental condition deteriorates.

References

Ahumada M, Faundez L, SAG. Guía descriptiva de los Sistemas Vegetacionales Azonales Hídricos terrestres de la Ecorregión Altiplánica (SVAHT). Santiago: Servicio Agrícola y Ganadero (SAG); 2009 118 p.

Arratia G. Peces del altiplano de Chile. In: Veloso A, Bustos E, editors. El Hombre y los ecosistemas de montaña. Montevideo: Ediciones Oficina Regional de Ciencia y Tecnología de la UNESCO para América Latina y el Caribe; 1982. p. 93–133.

Arroyo MTK, Marquet P, Marticorena C, Simonetti JA, Cavieres L, Squeo FA, Rozzi R, Massardo F. El Hotspot chileno, prioridad mundial para la conservación. In: Saball P, Arroyo MTK, Castilla JC, Estades E, Ladrón De Guevara JM, Larraín S, Moreno C, Rivas F, Rovira J, Sánchez A, Sierralta L, editors. Biodiversidad de Chile. Patrimonio y Desafíos. Santiago: Comisión Nacional del Medio Ambiente; 2006. p. 94–9.

Baldini A, Oltremari J, Ramírez M. Impact of american beaver (Castor canadensis, Rodentia) in lenga (Nothofagus pumilio) forests of Tierra del Fuego, Chile. Bosque (Valdivia). 2008;29 (2):162 Valdivia.

Banco Mundial. Diagnóstico de la situación de recursos hídricos de Chile. Departamento de Medio Ambiente y Desarrollo Sostenible, Región para América Latina y el Caribe. Santiago de Chile, Marzo de 2011. http://documentos.dga.cl/ADM5263.pdf

Castro C. Transformaciones geomorfológicas recientes y degradación de las dunas de Ritoque. Rev Geogr Norte Grande (Chile). 1987;23:63–77.

Comisión Nacional del Medio Ambiente, Chile (CONAMA). Protección y manejo de Humedales integrados a la cuenca hidrográfica. CONAMA y Centro de Ecología Aplicada (CEA); 2006.

Comisión Nacional de Medio Ambiente, Chile (CONAMA). Estado del Medio Ambiente en Chile, Informe país. Universidad de Chile, Instituto de Asuntos Públicos. CONAMA, PNUMA, CEPAL; 2008. p. 175–8.

CONAF-CONAMA. Catastro y Evaluación de los Recursos Vegetacionales Nativos de Chile. 1999.

CONAF – UACH. Monitoreo de Cambios, Corrección Cartográfica y Actualización del Catastro de Bosque Nativo en la XI Región de Aysén. 2011.

Correa-Araneda F, Urrutia J, Figueroa R. Knowledge status and principal threats to freshwater forested wetlands of Chile. Rev Chil Hist Nat. 2011;84:325–40.

Cowling RM, Rundel PW, Lamont BB, Arroyo MK, Arianoutsou M. Plant diversity in Mediterranean-climate regions. Trends Ecol Evol. 1996;11:362–6.

Díaz MF, Delano G. Plan de manejo sustentable y modelo de fiscalización para hume dales con predominio de musgo Pompón (Sphagnum magellanicum) en las provincias de Llanquihue y Chiloé. In: Valde's-Barrera A, Repetto F, Figueroa A, Saavedra B, editors. Actas del Taller: Conocimiento y valoración de las turberas de la Patagonia: oportunidades y desafíos. Punta Arenas: Anales Instituto Patagonia; 2012. p. 79–80.

Domínguez E. Manual de buenas prácticas para el uso sostenido del musgo *Sphagnum magellanicum* en Magallanes, Chile, vol. 256. Punta Arenas: Instituto Nacional de Investigación Agropecuaria; 2013. p. 82.

Farías ME, Contreras LM. Ecosistemas microbianos asociados a humedales altoandinos. Rev Bitácora Ecol. 2013;1:6–12.

Ferriere F. Distribución, flora y ecología de los bosques pantanosos de Mirtáceas en la Región de Los Lagos, Chile. [Licenciatura]. Valdivia: Escuela de Ingeniería Forestal, Universidad Austral de Chile; 1982.

Habit E, Dyer B, Vila I. Current state of knowledge of freshwater fishes of Chile. Gayana. 2006; 70(1):100–13.

Hauenstein E. A synoptical view of freshwater macrophytes of Chile. Gayana. 2006;70(1):16–23.

Hauenstein E, González M, Peña-Cortés F, Muñoz-Pedreros F, Muñoz-Pedreros A. Clasificación y caracterización de la flora y vegetación de los humedales de la costa de Toltén (IX Región, Chile). Gayana Bot. 2002;59(2):87–100.

Hauenstein E, González M, Peña-Cortés F, Muñoz-Pedreros A. Diversidad vegetal en humedales costeros de la Región de la Araucanía. Dirección de Investigación de la Universidad Católica de Temuco; 2004. Través de los proyectos DIUCT N°95-3-08, 97-4-02 y 99-4-04.

Hauenstein E, Peña-Cortés F, Bertrán C, Tapia J, Vargas-Chacoff L, Urrutia O. Floristic composition and evaluation of the degradation of the swampy coastal forest of temu-pitra in the Araucanía Region, Chile. Gayana Bot. 2014;71(1):43.

Joosten H, Clarke D. Wise use of mires and peatlands. Totness, Devon: NHBS/International Mire Conservation Group and International Peat Society; 2002.

Lappalainen E. Global peat resources. Finland: International Peat Society; 1966.

Luebert F, Pliscoff P. Sinopsis bioclimática y vegetacional de Chile. Santiago: Editorial Universitaria; 2006.

Ministerio de Medio Ambiente, Chile (MMA). Diagnóstico y propuesta para la conservación y uso sustentable de los humedales lacustres y urbanos principales de la región del Biobío. MMA-Centro EULA. Universidad de Concepción; 2010a.

Ministerio de Medio Ambiente, Chile (MMA). Aplicación piloto del estudio protección y manejo sustentable de humedales integrados a la cuenca hidrográfica. Humedales costeros. 99 pp. MMA-Centro de Ecología Aplicada (CEA); 2010b.

Ministerio de Medio Ambiente-Servicio Agrícola y Ganadero –Dirección General de Aguas, (MMA, SAG, DGA.). Guía para la conservación y seguimiento ambiental de humedales andinos; 2011a.

Ministerio de Medio Ambiente, Chile (MMA). Diagnóstico y propuesta para la conservación y uso sustentable de los humedales lacustres y urbanos principales de la Región del Biobío. MMA-Centro EULA. Universidad de Concepción; 2011b.

Ministerio de Medio Ambiente, Chile (MMA). Inventario Nacional de Humedales y seguimiento ambiental. Etapa 1. Santiago: MMA y Centro de Ecología Aplicada (CEA); 2011c

Ministerio de Medio Ambiente, Chile (MMA). Actualización del Catastro Nacional de Humedales y Guía Metodológica. Santiago. 2012.

Ministerio de Medio Ambiente, Chile (MMA). Especies Endemicas de Chile. 2016. Available from: http://especies.mma.gob.cl/CNMWeb/Web/WebCiudadana/especies_endemicas.aspx. Accessed 5 May 2016.

Möller P, Muñoz-Pedreros A. Legal protection assessment of different inland wetlands in Chile. Rev Chil Hist Nat. 2014;87:23. doi:10.1186/s40693-014-0023-1.

Peña-cortés F, Pincheira-ulbrich J, Bertrán C, Tapia J, Hauenstein E, Fernández E, Rozas D. A study of the geographic distribution of swamp forest in the coastal zone of the Araucanía Region. Chile Appl Geogr. 2011;31(2):545–55.

Ramírez C, Añazco N. Variaciones estacionales en el desarrollo de Scirpus californicus, Typha angustifolia y Phragmites communis en pantanos valdivianos. Chile Agro Sur. 1982;10:111–23.

Ramírez C, Ferreire F, Figueroa H. Estudio fitosociológico de los bosques pantanosos templados del sur de Chile. Rev Chil Hist Nat. 1983;56:11–26.

Ramírez C, San Martín C, San MJ. Estructura florística de los bosques pantanosos de Chile sur-central. In: Armesto JJ, Villagrán C, Arroyo MK, editors. Ecología de los bosques nativos de Chile. Santiago: Editorial Universitaria; 1995. p. 215–34.

Ramsar Convention Secretariat. Designating Ramsar sites: strategic framework and guidelines for the future development of the list of wetlands of International importance. In: Ramsar handbooks for the wise use of wetlands, vol. 17. 4th ed. Gland: Ramsar Convention Secretariat; 2010.

Roig C, Roig FA. Consideraciones generales. In: Blanco DE, de la Balze VM, editors. Los Turbales de la Patagonia Bases para su inventario y la conservación de su biodiversidad. Buenos Aires: Wetlands International - América del Sur; 2004. p. 5–21. Publicación No. 19.

Ruiz A, Doberty N, Ltda. Catastro y Caracterización de los Turbales de Magallanes. Sernageomin. BIP N°20196401-0, Chile. 2005. 123 p.

San Martín J, Troncoso A, Ramírez C. Estudio fitosociológico de los bosques pantanosos nativos de la Cordillera de la Costa en Chile central. Bosque. 1988;9(1):17–33.

Solervicens J, Elgueta M. Insectos de follaje de bosques pantanosos del Norte Chico, centro y sur de Chile. Rev Chil Entomol (Chile). 1994;21:135–64.

Stuardo J, Valdovinos C. Estuarios y lagunas costeras: ecosistemas importantes de Chile. Rev Am Y Des. 1989;5(1):107–15.

Thomas O. LXIII. On certain African and S.-American otters. J Nat Hist. 1908;1(5):387–95.

Varela J. Geología del cuaternario del área de los Vilos – Ensenada el Negro (IV Región) y su relación con la existencia del bosque "relicto" de Quebrada Quereo. Congreso Internacional de Zonas Áridas y Semiáridas. La Serena, Chile. Comunicaciones (Chile). 1981;33:17–30.

Vila I, Fuentes L, Contreras M. Peces Límnicos de Chile. Bol Museo Nac Hist Nat, Chile. 1999;48:61–75.

Vila I, Veloso A, Schlatter R, Ramirez C. Macrófitas y vertebrados de los sistemas límnicos de Chile. Coleccion: Biodiversidad. Santiago: Editorial Universitaria; Universidad de Chile; Programa Interdisciplinario de Estudios en Biodiversidad (PIEB); 2006. 186 p.

Zegers G, Larraín J, Diaz MF, Armesto J. Impacto ecológico y social de la explotación de pomponales y turberas de Sphagnum en la Isla Grande de Chiloé'. Rev Ambiente Desarrollo. 2006;22:28–34.

Seagrasses of Southeast Brazil

65

Joel C. Creed, Mariana V. P. Aguiar, Agatha Cristinne Soares, and Leonardo V. Marques

Contents

Introduction	840
Geography and Oceanographic Context	840
Seagrass Species and Distribution	840
Biology and Ecology	842
Conservation	845
Threats and Future Challenges	845
References	846

Abstract

The tropical/subtropical southeast region of Brazil comprises three coastal and one landlocked state. The region is the most highly developed in Brazil and contains the urban conglomerations of São Paulo and Rio de Janeiro; between 80-89% of native Atlantic rainforest has been lost through conversion. The 1,650 km coastline of the three coastal states represents 22.4% of the Brazilian coastline. Five species of seagrass occur in southeast Brazil. Runoff and terrestrially derived pollution decreases seawater clarity and quality. Although little known, this chapter describes the current state of knowledge on seagrass species and habitat in southeast Brazil, especially with regard to their geography and oceanographic context, species composition and distribution, biology and ecology, conservation, threats and future challenges.

Keywords

Brazil · Conservation · Distribution · Seagrass · Species composition · Review

---

J. C. Creed (✉) · M. V. P. Aguiar · A. C. Soares · L. V. Marques
Departamento de Ecologia, Universidade do Estado do Rio de Janeiro, Rio de Janeiro, Brazil
e-mail: jcreed@uerj.br; mvpaguiar@gmail.com; soaresacp@hotmail.com; vidalleo@gmail.com

© Springer Science+Business Media B.V., part of Springer Nature 2018
C. M. Finlayson et al. (eds.), *The Wetland Book*,
https://doi.org/10.1007/978-94-007-4001-3_265

Introduction

The tropical/subtropical southeast region of Brazil comprises three coastal and one landlocked state (Fig. 1). The region is the most highly developed in Brazil and contains the urban conglomerations of São Paulo and Rio de Janeiro; between 80% and 89% of native Atlantic rainforest has been lost through conversion to a range of other land uses such as urbanization, crop production, and pasture. With a population of 64 million (2014), the 1,650 km coastline of the three coastal states represents 22.4% of the Brazilian coastline. Runoff and terrestrially derived pollution decreases seawater clarity and quality. Positioned at the interface of land and sea, seagrasses thrive and are able to absorb some level of excess nutrients and other pollutants but can be stressed and decline or disappear when pollution levels and coastal development increase. Although little known, this chapter describes the current state of knowledge on seagrass species and habitat in southeast Brazil.

Geography and Oceanographic Context

Southeastern Brazil is one of five geographical regions that divide the Brazilian territory. Southeastern Brazil comprises four states; Minas Gerais is landlocked, with the three coastal states being Espírito Santo, Rio de Janeiro, and São Paulo (Fig. 1). In 2014, the Instituto Brasileiro de Geografia e Estatística reported that about 85.1 million people, equivalent to 41% of the national population, lived in southeastern Brazil.

The coastline of southeastern Brazil is variable in composition and geomorphology. Beach ridge plains dominate from the north of Espírito Santo (Doce River) to the north of Rio de Janeiro (Paraíba do Sul River). Beach rock reefs can occur nearshore. The coast alternates between regions in equilibrium, accretion, and erosion; more than 30% of the coastal area experiences erosion, with accretion occurring mainly on the coastal plains of river deltas.

The double-barrier lagoon coast has an almost east–west alignment, beaches being highly exposed to storm waves from the south; in contrast the numerous hypersaline, salt or brackish water lagoons are wave protected. Rocky shores become increasingly common in a southward direction due to the proximity of the Serra do Mar range, which extends to São Vicente in São Paulo. Rocky shores create large bays (Vitória, Guanabara, Sepetiba, and Ilha Grande), peninsulas, headlands, protected pocket beaches, and coastal islands and archipelagos with wave protected aspects. Sporadic wind-driven coastal upwelling occurs between Rio de Janeiro and São Paulo, centered on Cabo Frio ("Cape Cold").

Seagrass Species and Distribution

Four seagrass species belonging to three families (Cymodoceaceae: shoal grass *Halodule wrightii* and *H. emarginata*; Hydrocharitaceae: paddle grass *Halophila decipiens*; Ruppiaceae: widgeon grass *Ruppia maritima*) (Fig. 2)

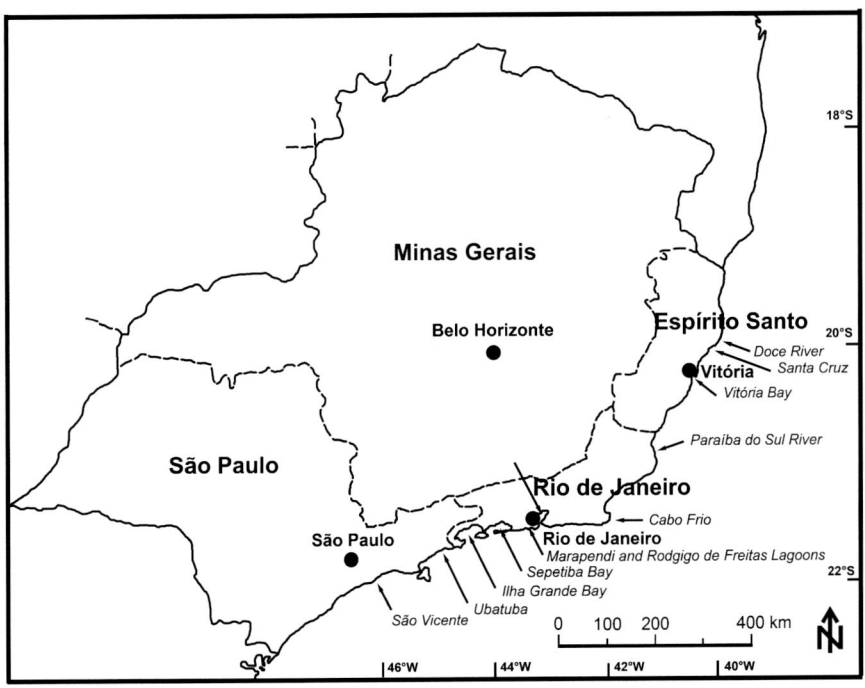

Fig. 1 Map of southeast Brazil showing states, cities and important seagrass locations

are found in the coastal waters of southeast Brazil (Oliveira Filho et al. 1983).

All of these species occur in the coastal states, though most occurrences are in Rio de Janeiro. *Halodule wrightii* is about three times more common than *H. emarginata* and ten times more common than *R. maritima* and *Halophila decipiens*. As waters are often turbid due to wave action and river discharge, seagrass meadows occur sporadically in small isolated populations to depths of 7 m in medium to fine sand and silts in four situations: (1) as sparse intertidal populations, where beach sand is sufficiently stable on the leeward side or on beachrock reefs (*Halodule wrightii*, north Espírito Santo, Fig. 3); (2) comparatively denser populations in smaller rock shore-formed embayments or estuaries in the intertidal zone (e.g., *Halodule* spp., *Halophila decipiens*, and *R. maritima* at Santa Cruz and Cabo Frio); (3) in larger bays with less turbid waters extending into the subtidal zone (*Halodule* spp., *Halophila decipiens*, e.g., Vitória Bay, Guanabara Bay, Ilha Grande Bay, Ubatuba); and (4) intertidal to shallow subtidal areas in lagoons (*R. maritima*, Southern Espírito Santo, the urban Marapendi and Rodrigo de Freitas Lagoons, Rio de Janeiro).

There is very little information about historical losses of seagrass beds but Creed (2003) reported that 16% of *Halodule wrightii* beds at Rio de Janeiro were lost in 10 years.

Fig. 2 Seagrass beds in southeast Brazil: (**a**) Shoal grass *Halodule wrightii* in Rio de Janeiro. (**b**) Paddle grass *Halophila decipiens* in Espírito Santo. (**c**) Widgeon grass *Ruppia maritima* in Rio de Janeiro (Photo credits: JC. Creed © Rights remain with the author)

Biology and Ecology

Seagrasses are flowering plants, and *Halophila decipiens* and *R. maritima* do produce flowers, fruits, and seeds in southeast Brazil. Sexual reproduction is, however, very rare in *Halodule* spp. One population of *Halodule wrightii* at Cabo Frio, for example, has been monitored for 20 years with no evidence of flower or seed production. Sexual reproduction may be latitude dependent as it is more common further north in Brazil. All four species also grow by rhizome elongation; clonal growth is the norm and allows the seagrasses to rapidly occupy large areas. Where flushing is low, seagrasses often complete with associated drift algae. Species of *Jania*, *Cladophoropsis*, *Hypnea*, *Chaetomorpha*, *Cladophora*, and *Rhizoclonium* are common competitors and when abundant can smother seagrass beds, leaving sediments anoxic.

Seasonal changes in biomass, shoot density, and shoot length are pronounced in populations of *Halodule wrightii*, with maxima in late summer and minima in late winter (Creed 1999). A comparison of ten populations distributed along the coast of the State of Rio de Janeiro revealed substantial variation in these parameters. Low

Fig. 3 *Halodule wrightii* seagrass growing on beach rock reefs in rock pools in north Espírito Santo, Brazil (Photo credit: JC. Creed © Rights remain with the author)

biomass may reflect more extreme environmental conditions. Seagrasses form the basis of the food chain for grazing animals and for the detritus food chain in situ or in adjacent areas when dead leaves are transported from the bed to rocky shores and reefs or unvegetated bottoms. Seagrass leaves also harbor a highly productive community of epiphytes.

Halodule beds in southeast Brazil support herbivores such as the sea urchins *Echinometra lucunter* and *Lytechinus variegatus* (Oliveira 1991), the green turtle *Chelonia midas*, and some crabs. One previously unknown trophic interaction between the capybara *Hydrochoerus hydrochaeris*, an aquatic rodent, and *R. maritima* was reported in Rio de Janeiro (Creed 2004). The capybaras foraged 18% of the seagrass meadow, and the impact of capybara grazing is similar to that reported for Sirenia (manatees and dugongs).

Bivalve mollusks, polychaete worms, and crustaceans such as crabs, shrimp, and amphipods are abundant detritivores within the seagrass beds. The cerith (snail) *Ceritium atratum* is a key detritivore and may form dense populations which enhance local diversity by providing microhabitat – their still-inhabited shells provide rare hard substratum for oysters and polychaete worms, and their empty shells are used by hermit crabs (Creed 2000; Fig. 4). Seagrass vegetated areas have a greater density and diversity of macrofauna than adjacent unvegetated areas (Casares and Creed 2008).

Predators such as crabs, fish, and seabirds hunt within the seagrass beds. Economically important fish species are the bluewing searobin *Prionotus punctatus*,

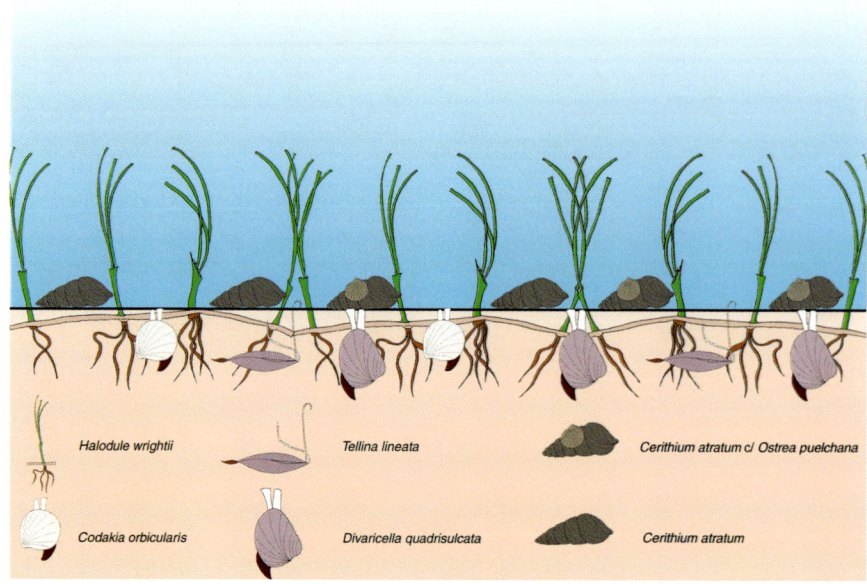

Fig. 4 Model of important species interactions in a shoal grass (*Halodule wrightii*) bed in southeast Brazil (Figure credit: © Monique Kinupp with permission)

Fig. 5 Blue crab (*Callinectes sapidus*) – a commercially important seagrass inhabitant (Photo credit: JC Creed © Rights remain with the author)

whitemouth croaker *Micropogonias furnieri*, and mullet *Mugil platanus*. Local fisheries exploit crustaceans such as blue crab *Callinectes sapidus* (Fig. 5), stone crab *Menippe nodifrons*, lobster (*Panulirus argus* and *P. laevicauda*), shrimp (*Penaeus brasiliensis* and *P. paulensis*), clams (*Anomalocardia brasiliana*, *Tagelus plebeius*, *Tivela mactroides*), volutes (*Voluta ebraea*), rock shells (*Thais haemastoma*), oyster (*Ostrea puelchana*), and cockles (*Trachycardium muricatum*) (Creed 2003).

Conservation

Seagrasses are not protected by habitat specific legislation in Brazil but are covered by resource management and conservation legislation. However, the ecosystem services provided by seagrasses justify their inclusion in specific legislation. The federal government has among other items the responsibility to "preserve and restore essential ecological processes and promote the ecological management of species and ecosystems.... preserve the diversity and integrity of genetic resources.... protect the flora and fauna." The coastal zone is recognized as a national resource by the constitution. The 1998 Environmental Crimes Law makes it a criminal offense to damage algal beds, coral reefs, and mollusk beds, but seagrasses are not mentioned (Creed 2003).

Brazil has a complex management system based upon conservation units at federal, state, and municipal government levels. Seagrasses are found in numerous federal and state Marine Protected Areas throughout the region. Unfortunately these are underfunded and policy is often not enforced.

Threats and Future Challenges

Seagrasses of the region suffer anthropogenic pressures which are concentrated near coastal cities such as Rio de Janeiro and Vitória (Creed 2003). Eutrophication caused by sewage and other organic wastes that are released untreated into the sea, landfills, dredging, vessel and anchor impacts, overfishing, and tourism are some of the recognized stressors to seagrasses in southeastern Brazil. Global warming, sea level change, and change in acidity may all impact on seagrass ecosystems.

Phillips (1992) stated that "the most basic research is needed in almost every place" regarding seagrasses. The challenge is mapping of all seagrass meadows to both provide information on their distribution and areal coverage, and a baseline for future monitoring and conservation. For example, mostly unknown seagrass meadows were found in 14 of 80 sites visited in 2012 in Espírito Santo State as part of an ongoing mapping project, and although northern Rio de Janeiro was better known, six new records were found in the 19 sites visited (JC Creed, unpublished data).

Southeast Brazil still lacks basic research of the biology and ecology of the seagrasses, especially *Halodule emarginata*. This species is listed as data deficient by the IUCN Red List (Short et al. 2011). Standardized monitoring should be expanded throughout the region in order to assess changes and mitigate impacts to seagrass area and services. At Cabo Frio, Rio de Janeiro, seagrass monitoring has been conducted seasonally since 1995 using protocols which include the

SeagrassNet Global Seagrass Monitoring Program (www.SeagrassNet.org). The seagrass bed area, biomass, and height vary seasonally as well as increasing and decreasing over an 11-year cycle related to sunspot activity (Marques et al. 2014), which demonstrate that temporal seagrass change in southeast Brazil is complex and apparently contains multiple cycles.

References

Casares FA, Creed JC. Do small seagrasses enhance density, richness, and diversity of macrofauna. J Coast Res. 2008;24:790–7.
Creed JC. Distribution, seasonal abundance and shoot size of the seagrass *Halodule wrightii* near its southern limit at Rio de Janeiro state, Brazil. Aquat Bot. 1999;65:47–58.
Creed JC. Epibiosis on cerith shells in a seagrass bed: correlation of shell occupant with epizoite distribution and abundance. Mar Biol. 2000;137:775–82.
Creed JC. The seagrasses of South America: Brazil, Argentina and Chile. In: Green EP, Short FT, editors. World atlas of seagrasses. Berkeley: University of California Press; 2003. p. 243–50.
Creed JC. Capybara (*Hydrochaeris hydrochaeris* Rodentia: Hydrochaeridae): a mammalian seagrass herbivore. Estuaries. 2004;27:197–200.
Marques LV, Short FT, Creed JC. Sunspots drive seagrasses. Biol Rhythm Res. 2014;46:63–8.
Oliveira MC. Survival of seaweeds ingested by three species of tropical sea urchins from Brazil. Hydrobiologia. 1991;222:13–7.
Oliveira Filho EC, Pirani JR, Giulietti AM. The Brazilian seagrass. Aquat Bot. 1983;16:251–67.
Phillips RC. The seagrass ecosystem and resources in Latin America. In: Seeliger U, editor. Coastal plant communities of Latin America. San Diego: Academic; 1992. p. 107–21.
Short FT, Polidoro B, Livingstone SR, Carpenter KE, Bandeira S, Bujang JS, Calumpong HP, Carruthers TJB, Coles RG, Dennison WC, Erftemeijer PLA, Fortes MD, Freeman AS, Jagtap TG, Kamal AHM, Kendrick GA, Judson Kenworthy W, La Nafie YA, Nasution IM, Orth RJ, Prathep A, Sanciangco JC, Tussenbroek BV, Vergara SG, Waycott M, Zieman JC. Extinction risk assessment of the world's seagrass species. Biol Conserv. 2011;144:1961–71.

Rio de la Plata (La Plata River) and Estuary (Argentina and Uruguay)

66

Claudio R. M. Baigún, Darío C. Colautti, and Tomás Maiztegui

Contents

Introduction	848
Main Environmental Characteristics	848
Biodiversity Patterns	849
Ecosystem Services	852
Threats and Future Challenges	853
References	854

Abstract

The La Plata River system can be defined as a funnel coastal plain tidal river with a semi-closed shelf at the mouth. La Plata River is both the world's widest freshwater system and an estuary that drains the second largest basin in South America and the fifth largest in the world. The Rio de la Plata system is shared by Argentina and Uruguay and has an area of 38,000 km^2, extends almost 300 km in length, and widens from about 40 km at the inner freshwater part to 227 km at the Atlantic Ocean boundary. The system is mainly formed by the Paraná and Uruguay rivers that provide 97% of the water discharge contributing with a mean annual flow of 16,000 and 4,000 m^3/s, respectively. The Rio de la Plata comprises three well-defined areas: the internal zone that starts at the end of the Parana Delta and is characterized by only freshwater, an intermediate or mixing zone, and an external or marine zone. Distribution and abundance of organisms are strongly associated with environmental factors, but mainly salinity gradients.

C. R. M. Baigún (✉)
Instituto de Investigación e Ingeniería Ambiental (3iA), Universidad Nacional de San Martín, San Martín, Buenos Aires, Argentina
e-mail: cbaigun@gmail.com

D. C. Colautti · T. Maiztegui
Instituto de Limnología de La Plata "Raúl Ringuelet", La Plata, Argentina
e-mail: colautti@ilpla.edu.ar; maiztegui@ilpla.edu.ar

© Springer Science+Business Media B.V., part of Springer Nature 2018
C. M. Finlayson et al. (eds.), *The Wetland Book*,
https://doi.org/10.1007/978-94-007-4001-3_243

Water and sediment contamination are the main threats affecting the ecological integrity; and conservation strategies should encompass transboundary policies and agreements oriented to balance socio-economic benefits and ecological requirements at a basin scale.

Keywords
Euryhaline species · Fine sediments · Pollution · Saline gradient · Turbidity front

Introduction

The Río de la Plata system can be defined as a funnel coastal plain tidal river with a semi-closed shelf at the mouth. La Plata River is both the world's widest freshwater system and an estuary that drains the second largest basin in South America and the fifth largest in the world. The Rio de la Plata system is shared by Argentina and Uruguay and has an area of 38,000 km^2, extends almost 300 km in length, and widens from about 40 km at the inner freshwater part to 227 km at the Atlantic Ocean boundary. The system is mainly formed by the Paraná and Uruguay rivers that provide 97% of the water discharge contributing with a mean annual flow of 16,000 and 4,000 m^3/s, respectively. The Rio de la Plata comprises three well-defined areas: the internal zone that starts at the end of the Parana Delta and is characterized by only freshwater, an intermediate or mixing zone, and an external or marine zone.

Main Environmental Characteristics

The Río de la Plata system exhibits a complex topography with a mean depth of 5 m and a maximum of 18–20 m at the mouth. The system's main environmental characteristics are defined by the presence of longitudinal thermal, saline and sediment gradients. The estuary is divided into two main areas by the submersed Barra del Indio (Del Indio bar) that crosses the river transversally at 6.5–7 m depth (Fig. 1). Sediments close to the Parana and Uruguay river mouths are mostly coarse (sand and silty sands), increasing the fraction of fine sediments (silt and clay) to the riverine-estuarine zonal boundary (FREPLATA 2005). The inner or freshwater riverine zone is defined by salinity values up to 5 practical salinity units (PSU) whereas the boundary with the estuarine zone occurs where the halocline intersects the bottom, corresponding to the bottom salinity front and boundary for the brackish estuarine water (Guerrero et al. 1997).

Turbidity varies along a gradient and is influenced by river discharges, wave effects, dredging and fishing, sedimentation, and flocculation processes. Most of the sediments are carried by the Paraná River contributing 160 million tons per year, 50% of which is silt and 28% clay (Sarubbi et al. 2006). A noticeable characteristic is the appearance of a turbidity front with high suspended sediment concentrations, typically from 50 to 300 mg l^{-1} at the boundary between the riverine and inner estuarine areas

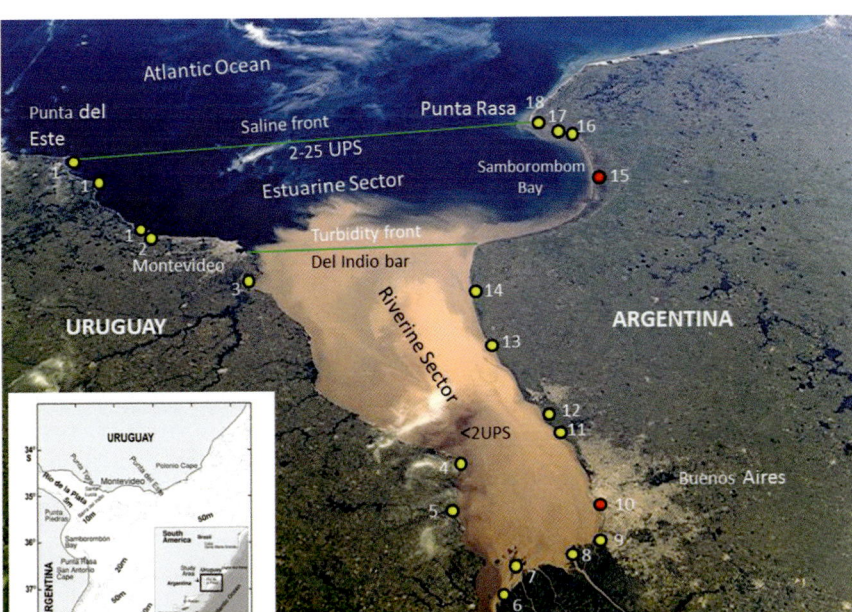

Fig. 1 The Río de la Plata system showing the main sectors and protected natural areas: *1* Parque Nacional Islas Costeras; *2* Parque Nacional F.D. Roosevelt; *3* Islas del río Santa Lucía; *4* Parque Nacional Isla San Gabriel; *5* Parque Nacional Anchorena; *6* Isla Martin García; *7* Río Barca Grande; *8* Bajos del Temor; *9* Ribera Norte; *10* Costanera Sur (Ramsar Site); *11* Selva Marginal de Hudson; *12* Punta Lara; *13* El Destino; *14* Parque Costero Sur; *15* Samborombón Bay (Ramsar site); *16* Rincón de Ajo; *17* Punta Rasa; *18* Campos del Tuyú (Image courtesy of the Earth Science and Remote Sensing Unit, NASA Johnson Space Center, NASA Photo ID ISS006-E-38952, https://eol.jsc.nasa.gov/)

(Acha and Mianzán 2003). This turbidity maximum is due to the suspended matter flocculation at the tip of the salt wedge and re-suspension of sediment due to tidal stirring (Acha et al. 2008). The estuarine waters extend to the oceanic shelf delimited by the maximum horizontal gradient of surface salinity (Fig. 2).

Biodiversity Patterns

Distribution and abundance of organisms in the Río de la Plata are strongly associated with environmental factors but mainly salinity gradients (Jaureguizar et al. 2004). Several of these species are potadromous moving seasonally to the Paraná and Uruguay rivers. A few species such as *Odontesthes* sp., *Lycengraulis grossidens*, and *Genidens barbus* show anadromous behavior migrating from the estuary or the Atlantic Ocean to spawn in the lower Parana River (Baigún et al. 2003; Baigún and Minotti 2012).

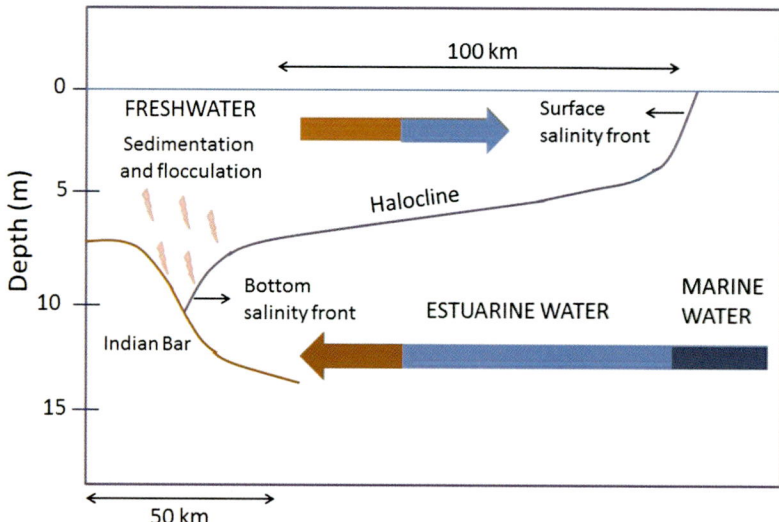

Fig. 2 Water circulation at the boundary between the riverine and estuarine zones (Adapted from Acha and Mianzán 2003, with permission)

Fish assemblage variation is related to the horizontal gradients of salinity (Jaureguizar et al. 2003). The innermost area comprises only freshwater species such as *Prochilodus lineatus*, *Saaminus brasiliensis*, *Pimelodus albicans*, *Parapimelodus valenciennsi*, *Lucipimelodus pati*, and the exotic *Cyprimus carpio*. Estuarine species are characterized by the presence of *Micropogonias furnieri*, *Brevoortia aurea*, *Macrodon ancylodon*, *Paralonchurus brasiliensis*, and *Paralichthys orbignyanus*, whereas the marine group includes *Cynoscion guatucupa*, *Conger orbignyanus*, *Discopyge tschudii*, *Paralichthys patagonicus*, *Percophis brasiliensis*, *Atlantoraja castelnaui*, *Mustelus schmitti*, *Sympterigia bonapartei*, *Stromateus brasiliensis*, *Squatina guggenheim*, *Myliobatis goodei*, and *Prionotus punctatus*. Some species, including *Micropogonias furnieri*, *B. aurea*, *Macrodon ancylodon*, and *Pogonias cromis*, spawn in the estuary taking advantage of convergent water masses that facilitate the retention of eggs and larvae within the estuary and share it as rearing habitat with other species, e.g., *Menticirrhus americanus*, *Parona signata*, *Paralichthys patagonicus*, and *Anchoa marinii* (Brazeiro et al. 2003).

In the same vein, benthonic species are distributed according to saline tolerance (Masello and Menafra 1998). For example, estuarine species are represented by *Brachiodontes darwinianus*, *Tagelus plebeious*, *Littoridina australis*, *Erodona mactroides*, *Balanus improvisus*, and *Mytella charruana*, and euryhaline species are composed of *Cyrtograpsus angulatus*, *Mactra isabelleana*, *Neanthes succinea* whereas marine species include *Mesodesma mactroides*, *Ocypode cuadrata*, *Hemipodus olivieri*, and *Emerita brasiliensis*. Occasionally cetaceans including the dolphin *Pontoporia blainvillei*, porpoises (*Phocoena dispotica*, *Tursiops gephyreus*, and *T. truncatus*), and whales (*Balaenoptera musculus*, *B. physalus*, and *B. acutorostrata*) enter the estuary.

Several important wetlands are located in the Río de la Plata such as Santa Lucía and Samborombón along the Uruguayan and Argentinian coasts, respectively (see Fig. 1). The latter is a Ramsar Site and one of the largest and most important wetlands in South America covering 140,000 ha of freshwater and brackish marshes (Lasta 1995). The area contains a diverse plant community adapted to different soil conditions. In the wet prairies, the most common species are *Eleocharis* spp., *Juncus* spp., *Paspalidium paludiphalus*, *Paspalum* spp., *Panicum* spp., *Polypogon elongatus*, and *Lolium multiflorum*. Characteristic species occupying lowland wet areas with saline soils are *Agropyrum* spp., *Spartina* spp., *Salicornia ambigua*, and *Spergularia* spp. whereas aquatic vegetation communities are composed of *Scripus californicus*, *Typha* spp., and *Zizanopsis bonariensis*. Shallow waters are inhabited by polychaetes (*Heteromastus similis; Poludora ligni*), amphipods (*Corosium insidiosum*), and bivalves (*Tagelus gibbus*). The burrowing crabs (*Chasmagnathus granulata* and *Cyrtograpsus angulatus*) are a distinctive characteristic of Samborombón Bay's intertidal zone (Fig. 3).

The south portion of the bay has been identified as a resting area for piscivorous marine birds including *Sterna hirundo*, *Thalasseus maximus*, and *T. sandvicencis* and migratory species such as *Rynchops niger*. Moreover, the bay represents the southernmost distribution for *Tachyphonus rufus*, *Icterus cayanensis*, *Synallaxis frontalis*, *S. cinerascens*, *Limnornis curvirostris*, *Cyclarhis gujanensis*,

Fig. 3 Typical crab community of Río de la Plata coastal intertidal area (Photo credit: courtesy of Alejandra Volpedo ©)

Phylloscartes ventralis, *Geothlypis aequinoctialis*, *Parula pitiayumi*, *Circus bofoni*, *Polyborus plancus*, *P. chimango*, and *Elanus leucurus*.

Common mammals found in this wetland are *Myocastor coypus*, *Hydrocaeris hydrocaeris*, *Cavia pamparum*, *Ctenomys talarum*, *Didelphys albiventris*, *Galictis cuja*, *Conepatus chinga*, and *Ozotoceros bezoarticus*. Several amphibians can be also found (e.g., *Rhinella arenarum*, *Leptodactylus ocellatus*, *Odontophrynus americanus*) along with reptiles (e.g., *Tupinambis teguxin*, *Prhynops hilarii*, *Lystrophis dorbignyi*, *Oxyrhopus rhombifer*). The coastal area of Samborombon Bay also provides rearing habitat for fish such as *Brevoortia aurea*, *Pogonias cromis*, *Micropogonias furnieri*, *Macrodon ancylodon*, and *Mugil* sp.

Ecosystem Services

The Río de la Plata River provides a number of valuable ecosystem services related to the use of the coastal and offshore areas. Commercial foreign trade is a main activity but passenger transportation between Argentina and Uruguay and recreational navigation are also important (FREPLATA 2005). An artisanal fishery based on *Prochilodus lineatus* has been an important activity formerly relevant in the inner Argentinean sector (Baigún et al. 2003) but is still practiced in Uruguay. The main fishing area for both countries is currently located in the estuary, with *Micropogonias furnieri* as the main target species (Fig. 4). An important inshore recreational and

Fig. 4 Fishing boats used in Río de la Plata River estuary (Photo credit: Claudio Baigún © rights remain with the author)

sport fishery occurs along the riverine and estuarine sectors for *Odontesthes* sp., *Genidens barbus*, and *M. furnieri* as the main target species (Colautti et al. 2003; López et al. 2012). In the coastal areas several wetlands, but particularly the Samborombón Bay, provide a wide mosaic of feeding, reproductive and rearing habitats for terrestrial and aquatic organisms.

The Río de la Plata River is a major source of water for domestic use in the most highly populated urban areas of Argentina and Uruguay. At the boundary with the Atlantic Ocean where freshwater and salt water merge, contaminants adsorbed and transported by clay particles sink and are immobilized due to flocculation thereby reducing the input of contaminated particles to the marine environment.

Threats and Future Challenges

Water and sediment contamination are the main threats affecting the ecological integrity of Río de La Plata, impacting planktonic, benthonic, and nektonic communities with population reductions, mortality, and disease that can ultimately affect human health through bioaccumulation. Coastal and urban areas, particularly in Argentina, are heavily impacted due to land use and lack of appropriate treatment of industrial and domestic sewage effluents, affecting in turn a wide spectrum of biological communities and human health. Inshore fecal bacteria levels in the water column exceed limits for safe recreational use, acceptable levels only occurring 3,000 m from the coast. Harmful algal blooms based on Ciliates, Cyanophytes, and Dinoflagellates are commonly triggered by high nutrient loads, mainly in the Argentine sector.

Waters contaminated by PCBs, PAHs, hydrocarbons, and heavy metals are found along the south coastal area of the Río de la Plata between Buenos Aires and La Plata at levels higher than those recommended for aquatic biota (PNUD 2009). Along the Uruguayan coast, however, only the Montevideo area is impacted by heavy metal (chromium and lead) inputs from sewage and tributaries (Kurucz et al. 1998).

Offshore sediments along the estuary are only slightly contaminated except at the turbidity front (see Fig. 1) where flocculation promotes the sinking of clay particles with adsorbed metal ions. Waves, currents, dredging, and bottom fishing tows can re-suspend and even alter sediment transport and result in bioaccumulation of pollutants by filter feeding and detritivorous species. Habitat structure and the diversity and abundance of benthic communities are also impacted by these activities. The fishing areas encompass many of the main spawning and rearing habitats, and without proper management the fisheries can be severely impacted as has been shown for *Micropogonias furnieri* (Ministerio de Asuntos Agrarios 2007). In turn, on the Uruguay coast *Cynocion guatucupa* and *Mustelus schmitti* have been acknowledged as overexploited (Defeo et al. 2009).

International commercial navigation can result in introductions of invasive species via transport in ballast tanks or adhering to the hull. The diversity of exotic species in de la Plata River Estuary is still low although reported species show high population abundances. In addition to *Cyprinus carpio*, and *Oreochromis niloticus*, the bivalves *Corbicula fluminea*, *C. langilleri*, and *Limnoperna fortunei* are currently widely distributed and have a high economic impact on shore structures and water pumps (Darrigran 2002). Other exotic species such as the gastropod *Rapana venosa* has also been found (Giberto et al. 2006). Brazeiro et al. (2003) stated that the Ortiz Bank, the Turbidity Front, and the Santa Lucía and Samborombón wetlands can be classified as the most highly critical areas because they have both functional relevance for the fluvio-marine ecosystem but also support important environmental risks.

Coordinated management policies are required to restore the Río de la Plata's ecological integrity, particularly in coastal and wetland areas and to maintain their ecosystem services. This will require managers to enforce appropriate water treatment for both urban and industrial effluents, improve socioeconomic conditions for people inhabiting shore areas, and develop appropriate governance institutions and processes to apply and enforce environmental laws, including sustainable fishing regulations and management policies. As the Río de la Plata system represents the end point for the second largest river basin in South America, conservation strategies should also encompass transboundary policies and agreements oriented to balance socioeconomic benefits and ecological requirements at a basin scale.

References

Acha M, Mianzán H, Gerrero R, Carreto J, Gilberto D, Montoya N, Carrignan M. An overiview of physical and ecological processes in the Rio de La Plata estuary. Continental Shelf Research; 2008;28:1579–158.

Acha M, Mianzán H. El estuario del Plata: donde el río se encuentra con el mar. Ciencia Hoy. 2003;13:10–20.

Baigún C, Sverlij SB, López H. Capítulo I. Recursos pesqueros y pesquerías del Río dela Plata interior y medio (Margen argentina)- Informe final. En: Protección Ambiental del Río de la Plata y su Frente Marítimo: Prevención y Control de la Contaminación y Restauración de Hábitats, FREPLATA, PROYECTO PNUD/GEF/RLA 99/G31, Montevideo; 2003. www.freplata.org/documentos/tecnico.asp

Baigún C, Minotti P. The current status of bagre marino (*Genidens barbus*). In: Gough P, editor. From sea to sources. The Regional Water Authority Hunze en Aa's; Netherlands; 2012. p. 220–1.

Brazeiro A, Acha E, Mianzán H, Gómez M, Férnandez V. Aquatic priority areas for the conservation and management of the ecological integrity of the Rio de la Plata and its Maritime Front. Montevideo-Buenos Aires, FrePlata, Proyecto PNUD-GEF RLA/99/631. 2003.

Colautti D, López H, Nadalin D. La pesca en el sector costero del Río de la Plata entre Punta Atalaya y Punta Piedras. In: Athor J, editor. Parque Costero del Sur. Buenos Aires: Fundación de Historia Natural Félix de Azara; 2003. p. 370–83.

Darrigran G. Potential impact of filter-feeding invaders on temperate inland freshwater environments. Biol Invasions. 2002;4:145–56.

Defeo O, Horta S, Carranza A, Lercari D, De Alava A, Gómez J, Martínez G, Lozoya JP, Celentano E. Hacia un manejo ecosistémico de pesquerías. Areas marinas protegidas en Uruguay. Montevideo: Facultad de Ciencias-DINARA; 2009.

FREPLATA. Análisis diagnóstico transfronterizo del Río de la Plata y su frente marítimo. Documento Técnico. Proyecto "Protección Ambiental del Río de la Plata y su Frente Marítimo: Prevención y Control de la Contaminación y Restauración de Hábitats" FREPLATA Montevideo, junio 2005Proyecto PNUD/GEFRLA/99/G31, 2005.

Giberto DA, Bremec CS, Schejter L, Schiariti A, Mianzan H, Acha EM. The invasive rapa whelk *Rapanavenosa* (Valenciennes 1846): status and potential ecological impacts in the Río de la Plata Estuary, Argentina-Uruguay. J Shellfish Res. 2006;25:919–24.

Guerrero RA, Acha EM, Framiñan MB, Lasta CA. Physical oceanography of the Río de la Plata Estuary, Argentina. Cont Shelf Res. 1997;17:727–42.

Jaureguizar A, Menni RC, Bremec C, Mianzan HW, Lasta CA. Fish assemblages and environmental patterns in the Río de la Plata estuary. Estuar Coast Shelf Sci. 2003;56:921–33.

Jaureguizar AJ, Menni R, Guerrero R, Lasta C. Environmental factors structuring fish communities of the Río de la Plata estuary. Fish Res. 2004;66:195–211.

Kurucz A, Masello A, Méndez S, Cranston R, Wellas PG. Calidad ambiental del Río de la Plata. In: Wells P, Daborn GR, editors. El Río de la Plata. Una revisión ambiental. Un informe de antecedentes del Proyecto Ecoplata. Nova Scotia: Dalhousie University Halifax; 1998. p. 71–86.

Lasta CA. La Bahía Samborombón: zona de desove y cría de peces. [disertación], La Plata (Argentia):Facultad de Ciencias Naturales, Universidad Nacional de La Plata; 1995.

López H, Colautti D, Baigún C. Peces y pesca en la zona metropolitana: Una perspectiva histórica. In: Athor J, editor. Buenos Aires, la historia de su paisaje natural. Buenos Aires: Fundación de Historia Natural Félix de Azara, Universidad Maimónides; 2012. p. 233–47.

Masello A, Menafra R. Comunidades macrobentónicas de la zona costera uruguaya y áreas adyacentes. In: Wells P, Daborn GR, editors. El Rio de la Plata. Una revisión ambiental. Un informe de antecedentes del Proyecto Ecoplata. Nova Scotia: Dalhousie University Halifax; 1998. p. 117–68.

Ministerio de Asuntos Agrarios de la Provincia de Buenos Aires. Distribución geográfica de los sectores de pesca utilizados por la flota comercial que operó en el Río de la Plata durante la zafra invernal de la corvina rubia *Micropogonias furnieri*; 2007. www.maa.gba.gov.ar/pesca/minagri

PNUD (Programa de las Naciones Unidas para el Desarrollo). Prevención y contaminación de la contaminación de origen terrestre en el Río de la Plata y su frente Marítimo mediante la implementación del frente estratégico de Freplata. ARG/09/G31; 2009. http://www.undp.org.ar/docs/Documentos_de_Proyectos/ARG09G46Prodoc%20I.pdf

Sarubbi A, Pittau MG, Menéndez AN. Delta del Paraná: avance del frente e incremento areal. INA. Proyecto LHA. 2006; 235.

Conchalí Lagoon: Coastal Wetland Restoration Project (Chile)

67

Manuel Contreras, F. Fernando Novoa, and Juan Pablo Rubilar

Contents

Introduction	858
Biodiversity	859
Physical Process	861
Landscape Process	862
Threats and Future Challenges	864
References	864

Abstract

Conchali Lagoon is a coastal wetland located at the northern part of Chile (31° 52.757° S; 71° 29.769° W). This ecosystem is hydrologically dynamic since its catchment receives contributions of water and salt from both rivers and the sea. During periods of high recharge, the lagoon connects to the sea across a sand barrier and waters enter into a euryhaline condition. This contrasts with a mainly freshwater lagoon, disconnected from the sea, during low recharge periods. This hydrodynamic connection/disconnection process generates inter and intra-annual patterns in the aquatic flora and fauna, such as the dominance of aquatic plants and marine algae during periods of low and high salinity, respectively. In 1998 several management actions were implemented in order to improve its ecological status, mainly for controlling local threats. Comparative studies have shown that the area has significantly improved its ecological indicators, including a progressive

M. Contreras (✉) · F. F. Novoa
Centro de Ecología Aplicada (CEA), Santiago, Región Metropolitana, Chile
e-mail: mcontreras@cea.cl; cea@cea.cl

J. P. Rubilar
Minera Los Pelambres, Santiago, Chile
e-mail: jrubilar@pelanbres.cl

© Springer Science+Business Media B.V., part of Springer Nature 2018
C. M. Finlayson et al. (eds.), *The Wetland Book*,
https://doi.org/10.1007/978-94-007-4001-3_249

increase in the number of bird species. From a management perspective, the plan implemented to control local threats was successful. However, new threats at a cathment and global scale are affecting the lagoon and they have implications to the maintenance of the lagoon's ecological character. Therefore, for the conservation of this Ramsar Site, the local management actions are by themselves insufficient and measures must be undertaken to develop a new plan for the sustainable use of water resources at a catchment scale.

Keywords
Wetland · Coastal Lagoon · Biodiversity · Watershed water balance · Eutrophication

Introduction

Conchalí Lagoon (Fig. 1, 31°52.757′ S; 71°29.769′ W) is located 4 km north of the town of Los Vilos Region of Coquimbo (Wetlands International 2004). This water body is hydrologically dynamic since its catchment receives contributions of water and salt from both rivers and the sea. During periods of high recharge, the lake connects to the sea across a sand barrier and waters enter into a euryhaline condition. This contrasts with a lagoon characterized by freshwater during low recharge conditions and disconnected from the sea. This hydrodynamic connection/disconnection process generates inter- and intra-annual patterns in the aquatic flora and fauna, such as the dominance of aquatic plants and marine algae during periods of low and high salinity, respectively. As the sand barrier opens, fish migrate from the

Fig. 1 Aerial view of Conchalí Lagoon (Photo Credit: © Centro de Ecología Aplicada, with permission)

sea toward the lagoon to feed. The lagoon is surrounded by a belt of riparian vegetation maintained by high groundwater levels generated by the presence of the lagoon, and which may be either reeds or halophyte plants depending on the salinity of the soil.

Like most coastal lagoons worldwide, Conchalí Lagoon has a high human population in its immediate surroundings using its resources and ecosystem services (see: Millennium Ecosystem Assessment 2005). During winter, local fisherman harvested the lagoon for food resources (mainly *Mugil cephalus* and *Odonthestes brevianalis*) even though Chilean law prohibits sport fishing on inland waters. Livestock were pastured year-round along the edges of the lagoon, feeding on the riparian vegetation. Hunters assisted by dogs caught birds and mammals which are fed and reproduced near the lagoon. The lagoon as a result of this human-related pressure existed in a poor ecological state, with a low species richness of terrestrial and aquatic flora and fauna.

In 1998, the mining company "Los Pelambres" bought the land where the lagoon is located and presented to the authorities the following management actions for the water body: (i) a fence around the lagoon; (ii) the creation of areas with restricted public access; (iii) implementation of an environmental education program; (iv) construction of paths; and (v) monitoring of the flora and fauna around the lagoon. In 2000 the lagoon was declared "Santuario de la Naturaleza," and in 2004 it was designated as a Ramsar Wetland of International Importance, due to the presence of eight species of conservation concern, three reproducing native fish species, 84 bird species, and a reproduction and feeding ground for migratory birds.

Studies comparing the period before the area was protected (between 1990 and 1992; Tabilo and Mondaca 2011), with those obtained after 1998 (Fig. 2) showing a progressive increase in the number of bird species. This indicates that the implementation of the management actions caused a positive change in biodiversity and ecological character, mostly due to the control of local threats and the recovery of riparian vegetation.

Biodiversity

On a regional scale, Conchalí Lagoon is recognized as the site with the highest number of bird species in the area. The 95 recorded species are divided among terrestrial (34), aquatic (20), riparian (31), and marine (9) species; 33 are described as permanent, 40 as frequent, 7 as rare, and 15 as occasional residents.

In the time series of avian abundance and richness, temporal changes can be observed with annual maxima abundance and richness occurring in autumn (March–April) and minima in summer (December–February; Fig. 3). In 2010, the lowest abundance was recorded (316 individuals), with an average richness of 37 species. However, in the last few years some aquatic species, e.g., *Fulica leucoptera*, *F. rufifrons*, *F. armillata*, *Anas platalea*, and *Egretta thula*, have

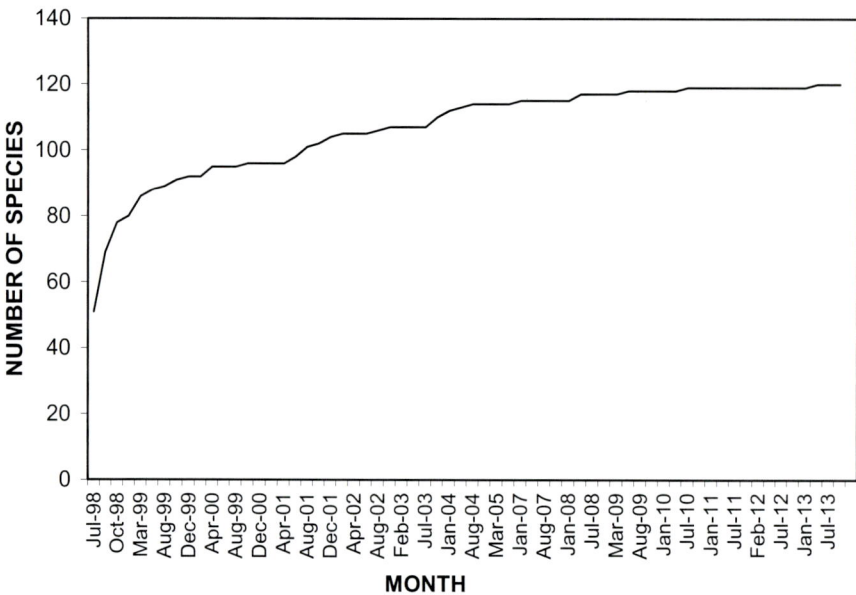

Fig. 2 Cumulative species richness curve of avifauna in Conchalí Lagoon (Data: © Centro de Ecología Aplicada, with permission)

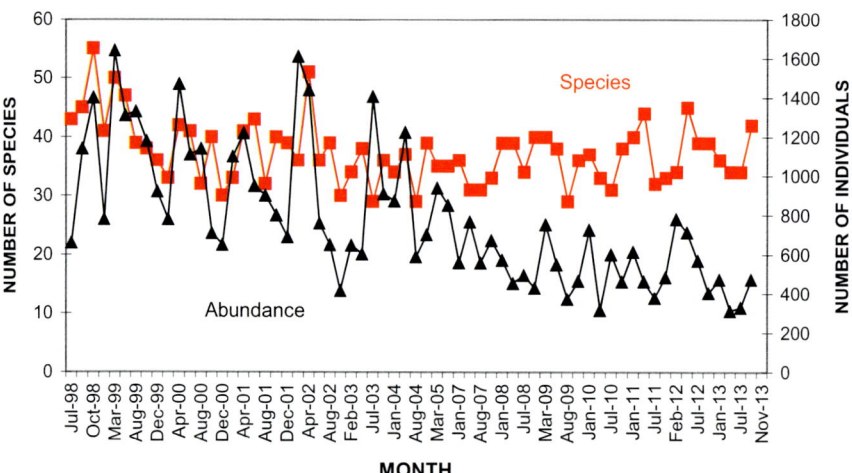

Fig. 3 Changes in the number of species and number of bird individuals observed during the period from July 1998 to November 2013 (Data: © Centro de Ecología Aplicada, with permission)

decreased in numbers. Comparing the data in two periods, before and after connection with the sea (year 2004), there was neither a significant change in the richness of birds (F = 1.52; p = 0.22) nor in the number of species (richness). However, there was a tendency of decreasing avian abundance (F = 3.65; p = 0.06).

In terms of functionality, the bird community is composed of five groups: granivores, omnivores, piscivores, insectivores, and herbivores. Herbivores, followed by insectivores, are the most abundant functional group during the period of record, indicating that the food web of Conchalí Lagoon is sustained by primary producers, and that it comprises a fundamental feeding habitat for primary consumers. Looking at the temporal changes in functional groups as an indicator of habitat condition, there were no significant changes during the observed time period. Although the response in the composition of the avifauna appears species-specific, it could still indicate changing ecological properties.

Physical Process

The discharge of the principal tributary (Pupío stream) into Conchalí Lagoon has decreased since 2004 (Fig. 4), coinciding with the closing of the connection with the sea. However, the rate of decrease does not coincide with the variation in precipitation, since discharge is near zero even during the rainy period. In 2006 and 2007 the average annual discharge was 0.054 and 0.032 m^3/s, respectively. In 2009 and 2010 precipitation was higher than 2006–2007 but average discharge was 0.011 and 0.0098 m^3/s, respectively. The decrease was sustained in the baseflow period, in which flow is dependent of contributions from aquifers located in the upper part of the catchment. Between December 2010 and May 2011, discharge was zero, except in March during which discharge averaged 0.010 m^3/s. According to seasonal variation curves of the Pupío stream (CADE –

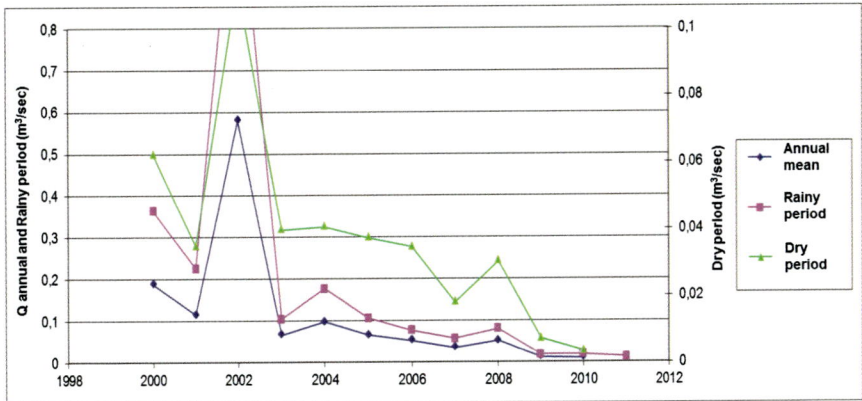

Fig. 4 Discharge of Pupío stream into Conchalí Lagoon (Data: © CADE-IDEPE 2004, with permission)

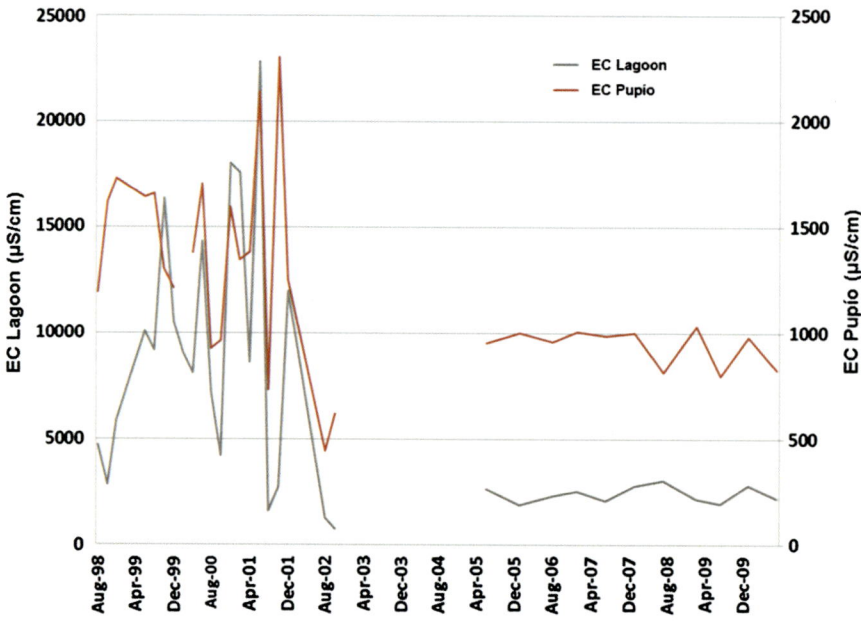

Fig. 5 Historical variation comparison of the EC (μS/cm) in Conchalí Lagoon and Pupío tributary (Data: © Centro de Ecología Aplicada, with permission)

IDEPE 2004), the average discharge measured during the 2010–2011 baseflow period has a 95% exceedance probability, while the rainy period of 2011 represents an exceedance probability in the order of 90%.

Between August 1998 and August 2002 monthly electrical conductivity (EC) values in the lagoon are highly variable but generally greater than 5,000 μS/cm with peaks reaching over 20,000 μS/cm (Fig. 5). The increase in EC and its variability is associated with events of marine water input breaking through the sand barrier separating the lagoon and the sea. These events caused changes in the internal dynamics of the lake, which can be seen in the seasonal variation and range in EC. The occurrence of maxima and minima during March and September, respectively, is related to the changes in contributions of freshwater and evaporation, which are factors that modify the lake's dilution capacity.

Landscape Process

Through air-photo interpretation, classification, and area of Conchalí Lagoon's water, dune, reed bed, and pasture habitats were determined for 2003 and 2012 (Fig. 6). Between these two dates, reed beds ("junco") increased in extent

Fig. 6 Conchalí Lagoon in (**a**) December 2003 and (**b**) December 2012. Google earth V 7.1.2.2041. (July 13, 2007). Conchali Lagoon, Chile. 31°52'45.30"S, 71°29'46.42"W, Eye alt 1.896 ft. DigitalGlobe 2003. http://www.earth.google.com (December 10, 2003). DigitalGlobe 2012. http://www.earth.google.com (December 5, 2012)

Table 1 Area (ha) of Conchalí Lagoon habitats in 2003 and 2012

Habitat	Area (ha) by year		Difference (ha)
	2003	2012	
Water	11.69	8.27	−3.42
Dunes	4.1	3.87	−0.23
Reed Bed	0.28	2.35	2.07
Pasture	0	1.31	1.31

surrounding the lagoon, dunes were displaced toward the center of the lagoon, and a pasture formed upon the sand barrier which isolates the lagoon from the sea. The area of water decreased by 3.42 ha while reed beds increased by 2.07 ha (Table 1).

Conchalí Lagoon experienced a periodic connection-disconnection process as the sand barrier was repeatedly opened and closed. In 2004 however, this dynamic stopped, and the lake has since been continuously isolated from the sea. This change resulted in a change in the trophic state of the lake from mesotrophic (in 1998) to hypertrophic (2012).

Threats and Future Challenges

Possible factors that explain the decrease in richness and abundance of the lagoon's avifauna are: (i) the decreased area of water; (ii) the advance of the dunes; (iii) the strengthening of the sand barrier separating the lagoon from the sea; (iv) the decrease in discharge from the Pupío tributary into the lagoon; and/or (v) the decrease in area covered by aquatic plants. The ecological character of the lagoon is changing, from a condition of variable estuary- and freshwater-like periods to a condition of continued freshwater, decreasing the heterogeneity of the habitat both spatially and temporally. This change in the ecological state will result in changes in the present biodiversity of Conchalí Lagoon.

From a management perspective, the plan implemented to control local threats was successful. However, threats affecting the lagoon originate at a catchment and even global scale and have implications to the maintenance of the lagoon's ecological character. Threats currently affecting Conchalí Lagoon are likely to cause, directly or indirectly, a decrease in the water recharge to the lake and thus move the ecosystem into a hypereutrophic condition.

This process, which probably affects other water bodies' level globally, is likely to be stronger in coastal water bodies due to their transitional character, changing between fresh- and marine water (estuary) conditions. This is a fundamental property which is a basis for their ecological character. As water contributions decrease due to increasing water demands within their catchments (human use and less recharge), a likely first impact to systems of this type is the loss of the dynamic pattern of sea-lagoon interchange of water and salt, initiating an accelerating process of eutrophication and biodiversity loss within the system. Therefore, for the conservation of this Ramsar Site, the local management actions are by themselves insufficient and measures must be undertaken to develop a plan for the sustainable use of water resources at a catchment scale. In addition, implications of global climate change confront us with an urgent necessity to assess the adaptability of Conchalí Lagoon to future environmental conditions and its ecological character.

References

CADE-IDEPE: Consultores en Ingeniería. Diagnóstico y Clasificación de los Cursos y Cuerpos de Aguas Según Objetivos de Calidad: Cuenca del Estero Pupío. 2004.

Millennium Ecosystem Assessment. Ecosystems and human well-being: biodiversity synthesis. Washington (DC): World Resources Institute; 2005.

Tabilo R and Mondaca V. Aves acuáticas en humedales costeros de la Región de Coquimbo, Chile. Boletín Chileno de Ornitología. 2001;8:13–7.

Wetlands International. Ficha Informativa de los Humedales de Ramsar (FIR): Santuario de la Naturaleza Laguna Conchalí. 2004. https://rsis.ramsar.org/RISapp/files/RISrep/CL1374RIS.pdf. Accessed 3 Apr 2016.

Bahía Lomas: Ramsar Site (Chile)

68

Carmen Espoz, Ricardo Matus, and Diego Luna-Quevedo

Contents

Introduction	866
A Key Site for Shorebirds	867
Studies and Monitoring Programs	868
Management Plan of Bahía Lomas	869
Threats and Future Challenges	870
References	870

Abstract

Designated as a Ramsar Wetland of International Importance in 2004 and Site of Hemispheric Importance by the Western Hemisphere Shorebird Reserve Network in 2009, Bahía Lomas is among the most important wetlands for shorebirds in the Southern Hemisphere. It is located near the end of the Southern Hemisphere, in the island of Tierra del Fuego, in Chile. The bay is known especially as the most important site in South America for the red knot *Calidris canutus rufa* and the second most important for the Hudsonian godwit *Limosa haemastica*, with significant numbers of Magellanic Oystercatcher *Haematopus leucopodus*, white-rumped sandpiper *C. fuscicollis*, two-banded plover *Charadrius falklandicus*, least seedsnipe *Thinocorus rumicivorus*, and Magellanic plover *Pluvianellus socialis*. The site has

---

C. Espoz (✉)
Centro Bahía Lomas, Facultad de Ciencias, Universidad Santo Tomás, Santiago, Chile
e-mail: cespoz@santotomas.cl

R. Matus (✉)
Centro de Rehabilitación de Aves Leñadura, Punta Arenas, Chile
e-mail: rmatusn@gmail.com

D. Luna-Quevedo (✉)
WHSRN Executive Office- Manomet, Plymouth, MA, USA
e-mail: diego.luna@manomet.org

© Springer Science+Business Media B.V., part of Springer Nature 2018
C. M. Finlayson et al. (eds.), *The Wetland Book*,
https://doi.org/10.1007/978-94-007-4001-3_248

its own Management Plan which was developed as part of a process of good governance. This Plan contains the actions, strategies, and programs necessary to achieve the effective conservation of this Ramsar Site. In this context, the Bahía Lomas Center ("Centro Bahía Lomas" in Spanish), is part of the effective conservation of Bahia Lomas, and is one of the strategic objectives of the Management Plan. The main threats for the site are related to activities that could produce oil pollution or other chemicals. Future challenges are being addressed from the "Centro Bahía Lomas" maintaining and strengthening ecological monitoring programs and the implementation of actions from the Management Plan, to ensure the conservation of this Ramsar Site and species associated with this wetland of international importance.

Keywords
Wetlands of Chile · Ramsar site · Shorebirds · Red knot

Introduction

Among the Ramsar sites to be found in Chile, Bahía Lomas (52°28′08″S, 69°22′54″W) is the most southerly, located at the eastern end of the Magellan Strait, on the northern coast of the island of Tierra del Fuego, Chile (Fig. 1).

As a tidal flat, the major characteristic is its large, 8–9 m, tidal range, which exposes an intertidal distance of about 7 km each day. The bay comprises a wide area of mud flats which are occasionally interrupted by channels (Morrison and Ross 1989) and some extensive areas of sand (Fig. 2). The bay, designated as Ramsar Wetland of International Importance in 2004 and Site of Hemispheric Importance by

Fig. 1 Bahía Lomas (Tierra del Fuego, Chile) is located in the extreme south of South America (Adapted from Image CC 21 June 2009. Penarc. https://commons.wikimedia.org/wiki/File:Bahia_Lomas_wetlands.JPG)

Fig. 2 Bahía Lomas (Tierra del Fuego, Chile). General view of the tidal flat with a mixed flock of Hudsonian godwits and red knots in summer season (Photo credit: Antonio Larrea ©)

the Western Hemisphere Shorebird Reserve Network (WHRSN) in 2009, is among the most important wetlands for shorebirds in the Southern Hemisphere.

A Key Site for Shorebirds

Bahía Lomas is a highly productive system, with an intertidal area of nearly 590 km^2. It supports a great diversity and biomass of invertebrates, such as polychaete worms, bivalves, amphipods, and isopods, and is thus a favored feeding habitat for many resident and migrant shorebirds. The bay is known especially as the most important site in South America for the red knot *Calidris canutus rufa* and the second most important location in South America for the Hudsonian godwit *Limosa haemastica*, supporting about 16% of the world population. Moreover, Bahía Lomas supports significant numbers of Magellanic oystercatchers *Haematopus leucopodus*, white-rumped sandpiper *C. fuscicollis*, two-banded plover *Charadrius falklandicus*, least seedsnipe *Thinocorus rumicivorus*, and Magellanic plover *Pluvianellus socialis* (Matus and Espoz 2012). It has been estimated that the tidal flat supports more than 60,000 shorebirds during the austral summer (Morrison and Ross 1989; Fig. 3).

Among all subspecies of the red knot *Calidris canutus rufa* – which breeds in the central Canadian arctic and migrates to wintering grounds in north temperate, tropical, and south temperate regions from Florida to Tierra del Fuego – has undergone a massive population decline over the past decade (Morrison and Ross 1989; Morrison and Harrington 1992; Harrington and Flowers 1996; Morrison

Fig. 3 Mixed flocks of Hudsonian godwit and red knot, Bahía Lomas, Tierra del Fuego, Chile. This Ramsar Site is the most important nonbreeding area in the Southern Hemisphere to *Calidris canutus rufa*, where they stay near to 5 or 6 months every year (Photo credit: Antonio Larrea©)

et al. 2004; Niles et al. 2008). During the nonbreeding season, the largest segment of the population has always occurred in southern South America. There the distribution has contracted, confining virtually the entire population to Tierra del Fuego, where numbers have decreased from 53,232 in 1986 (Morrison and Ross 1989) to 14,800 in 2008 (Niles et al. 2008). In Bahía Lomas, counts declined from 42,000 in 1986 (Morrison and Ross 1989) to near 10,000 in 2013 (R.I.G. Morrison and R. Matus, pers. comm.).

Studies and Monitoring Programs

Over time, research in Bahía Lomas has sought to promote both the scientific and technical knowledge necessary to make decisions that secure the effective management and conservation of the wetland.

Aerial counts, initiated in 1985 by Guy Morrison and Ken Ross (National Wildlife Research Center, Canada), indicate that Bahía Lomas represents the most important wintering area for red knot, *Calidris canutus rufa*, in South America (Morrison and Ross 1989). At a local level, the counts of resident and migratory birds, carried out by Ricardo Matus and Olivia Blank, have been key in defining the ecological importance of the site. At present, both aerial and terrestrial surveys are supported by the National Wildlife Research Center of Canada and "Empresa Nacional del Petróleo" (ENAP). In 2002, a ringing program of red knots began in Bahía Lomas, funded by the New Jersey Division of Endangered and Nongame Species from the United States. The program was part of a hemisphere-wide effort to understand the reasons for the population decline of *Calidris canutus rufa* over the last few decades. The researcher in charge of this program, Lawrence Niles, has been assisted by numerous other researchers both national and international. At the same time, coordinated by Olivia Blank, the first virological, bacteriological, and serological study for avian flu, Newcastle disease, and salmonella in two species of shorebirds was being carried out. In 2003, an ecological monitoring program was started, led by a team from Universidad Santo Tomás. This group of researchers has provided key information with regard to the abundance and distribution of macroinvertebrates in the mudflats, the trophic ecology of *Calidris canutus rufa*, and the physical, chemical, and biological characteristics of the sediments and water of Bahía Lomas (Espoz et al. 2008, 2011). Currently the group maintains an ecological monitoring program supported by the Manomet Center for Conservation Sciences, Universidad Santo Tomás, Ministry of the Environment, ENAP, and the Ramsar Convention. Universidad de Magallanes has also provided relevant information through research on cetaceans carried out by Jorge Gibbons (Gibbons et al. 2000, 2006) and on the ecological community carried out by Iván Cañete (Cañete et al. 2010). Alfredo Prieto, from Instituto de la Patagonia, with Mauricio Massone, from Museo de Historia Natural de Concepción, carried out studies on the presence of Selknam (i.e., native inhabitants of the island) in the region (Massone and Prieto 2005; Prieto and Gibbons 2008).

Management Plan of Bahía Lomas

In 2005, the National Commission for the Environment gave official status to the "National Strategy for the Conservation and Sustainable Use of Wetlands," and since then it has consolidated diverse coordinated actions by the National Wetlands Committee of Chile. Special attention has been given to Ramsar Sites, working on updating the information and promoting the development of the management plans of these sites, as is the case with Bahía Lomas. This Ramsar Site faces the ENAP platforms which are the main threat to the area in the case of an environmental accident. However, the designation of the area as a Ramsar Site provided an opportunity for the company to consider improving its processes and support for conservation work, in relation to natural environments, in the Magallanes region. In April of 2005, ENAP signed a collaboration agreement with CONAMA which looks at, among other actions, the development of a risk plan for the area, support and collaboration with scientists, the training of ENAP personnel, and the creation of an action plan in case of accidents. In this context, in November 2007, a participatory process began, with the access and exchange of information and the combining of forces of all interested parties, including CONAMA, ENAP, Municipality of Primavera, the universities of Santo Tomás and Magallanes, other government agencies, and those with private interests in the area. These parties form a committee for the promotion of the management of Bahía Lomas ("Comité Promotor de Manejo de Bahía Lomas") whose objective is to promote coordinated and permanent beneficial actions among all key interests to effectively conserve the Ramsar Site. In June 2010, Universidad Santo Tomás and WCS-Chile took charge of the design of the management plan for the site. Their main objectives are to identify the actions, strategies, and action programs necessary to achieve the effective conservation of Bahía Lomas. The Management Plan for Bahía Lomas was submitted in July 2011 (Espoz et al. 2011), and from this date, its implementation is ensuring the effective conservation of the diversity and biological productivity of the wetland and the sensible use of its resources on the part of all the interested parties.

Finally, within the framework of the strategic planning, the Bahía Lomas Center (Matus and Espoz 2012) fulfills an important role, serving as both a space and platform for promoting and carrying out scientific work and educational and regional development associated with conserving the Ramsar Bahía Lomas Site. This center, now administrated by Universidad Santo Tomás and the Manomet Center for Conservation Sciences (USA), is supported by the Chilean Ministry of the Environment, ENAP, and the Ramsar Convention on Wetlands. "Centro Bahía Lomas" located in Punta Arenas has been established with three main objectives: (1) to develop and carry out scientific research into migrant shorebirds, whales, ecological systems, and the culture of the indigenous Selknam people, first reported by Magellan in 1520, (2) to facilitate education and public awareness, and (3) to promote tourism and economic development associated with the conservation of Bahía Lomas.

Threats and Future Challenges

The main threats are related to activities that produce pollution by oil or other chemicals, habitat loss from mining and farming activities, and those associated with the administration and management of the area (e.g., tourism not oriented to the conservation of the site).

Future challenges are being addressed from the "Centro Bahía Lomas" maintaining and strengthening ecological monitoring programs and the implementation of actions of the management plan, to ensure the conservation of this Ramsar Site and species associated with this wetland of international importance. In this context, one of the most important action for this Ramsar Site is to obtain, by the Chilean government, the protection status of "Santuario de la Naturaleza."

References

Cañete JI, Astorga MS, Santana M, Palacios M. Abundancia y distribución espacial de *Scolecolepides uncinatus* Blake, 1983 (Polychaeta: Spionidae) y características sedimentológicas en Bahía Lomas, Tierra del Fuego, Chile. An Inst Patagonia. 2010; 38(2):81–94.

Espoz C, Ponce A, Matus R, Blank O, Rozbaczylo N, Sitters H, Rodríguez S, Dey A, Niles LJ. Trophic ecology of the Red Knot *Calidris canutus rufa* at Bahía Lomas, Tierra del Fuego, Chile. Wader Study Group Bull. 2008;115:69–76.

Espoz C, Labra F, Matus R, Ponce A, Barría I, Saavedra B, Figueroa A, Rondanelli M. Plan de manejo para el sitio RAMSAR Bahía Lomas. Santiago: Ministerio del Medio Ambiente/ Universidad Santo Tomás/Wildlife Conservation Society; 2011. 131 p.

Gibbons J, Gazitúa F, Venegas C. Cetáceos del Estrecho de Magallanes y senos Otway, Skyring y Almirantazgo. An Inst Patagonia Ser Cienc Nat. 2000;28:107–18.

Gibbons J, Capella J, Kusch A, Cárcamo J. The Southern Right Whale *Eubalaena Australis* (Desmoulins, 1822) in the Strait of Magellan, Chile. An Inst Patagonia. 2006;34:75–80.

Harrington B, Flowers C. The flight of the Red Knot. New York: WW Norton; 1996.

Massone M, Prieto A. Ballenas y delfines en el Mundo Selk'nam. Una aproximación Etnográfica. Magallanania. 2005;33(1):25–35.

Matus R, Espoz C. Centro Bahía Lomas: a new facility for shorebird research at "the end of the world". Wader Study Group Bull. 2012;119(1):69.

Morrison RIG, Harrington BA. The migration system of the red knot *Calidris canutus rufa* in the New World. Wader Study Group Bull. 1992;64(Suppl):71–84.

Morrison RIG, Ross RK. Atlas of nearctic shorebirds on the coast of South America, 2 vols. Ottawa: Canadian Wildlife Service Special Publication; 1989.

Morrison RIG, Ross RK, Niles LJ. Declines in wintering populations of Red Knots in southern South America. Condor. 2004;106:60–70.

Niles LJ, Sitters HP, Dey AD, Atkinson PW, Baker AJ, Bennett KA, Carmona RC, Clark KE, Clark NA, Espoz C, González PM, Harrington BA, Hernández DE, Kalasz KS, Lathrop RG, Matus R, Minton CDT, Morrison RIG, Peck MK, Pitts W, Robinson RA, Serrano IL. Status of the Red Knot (*Calidris canutus rufa*) in the Western Hemisphere. Stud Avian Biol. 2008;36:1–185.

Prieto A, Gibbons J. Selk'nam y cetáceos en el área norte de Tierra del Fuego. En: Análisis y diagnóstico de una ruta eco-cultural Bahía Lomas, Primera Angostura (Estrecho de Magallanes), Pali Aike y sectores asociados. Informe Final. Universidad de Magallanes. 2008.

Patagonian Peatlands (Argentina and Chile) 69

Rodolfo Iturraspe

Contents

Introduction	874
Age and Origin of Patagonian Peatlands	874
Peatland Distribution	876
Peatland Use and Conservation	879
Values and Functions	880
Threats and Future Challenges	881
References	881

Abstract

Peatlands are present in humid areas of Patagonia, reaching a latitudinal distribution along 2,000 km, from the Chiloe Island to the Tierra del Fuego archipelago. Deglaciation and postglacial processes that occurred after the Last Glacial Maximum favoured the conditions for peatlands development. In addition, the strong negative west-east humidity gradient that the Andes mountain range induces is a substantial factor for peatland distribution, determining the location of the more extensive peatland complexes at the west of the Andes and in the Tierra del Fuego archipelago. However, peatlands are also present as small minerogenic units in the oriental Andean foothill and occasionally in the driest extra-Andean Patagonia. Climatic and latitudinal gradients, in addition to the local drainage conditions gave rise to the conformation of several peatland complex types and subtypes that show differences on their dominant flora and morphological patterns. Particular regional names as "tepual", "pomponal", "mallín", "vega", etc., are utilized by the local people to identify different peatlands types.

R. Iturraspe (✉)
Universidad Nacional de Tierra del Fuego, Ushuaia, Tierra del Fuego, Argentina
e-mail: rodolfoiturraspe@yahoo.com; riturraspe@untdf.edu.ar

© Springer Science+Business Media B.V., part of Springer Nature 2018
C. M. Finlayson et al. (eds.), *The Wetland Book*,
https://doi.org/10.1007/978-94-007-4001-3_230

Main Patagonian peatland complexes, favoured by their isolated situation, remain in an almost pristine state, but peat mining is an increasing activity in both Argentina and Chile, affecting peatlands accessible by roads. Numerous minerogenic peatlands of the extra-Andean Patagonia present degradation signs produced by overgrazing and drainage to favouring cattle raising.

The Patagonian peatlands provide significant environmental functions. They constitute the main carbon sink and carbon storage in the extratropical Southern Hemisphere, so they contribute to climate change mitigation. In addition, they give support to the biodiversity, offer a tourist attraction as components of the landscape, contribute to the hydrological regulation and provide freshwater. The wise use of these wetlands is a challenge for the peatlands management in Patagonia.

Keywords

Peatland distribution · Peatland complex · Patagonian wetlands · Ecological gradients · Peatland functions · Peatland management · Peat mining · Andes · Argentina · Chile

Introduction

The largest part of South American peatlands is distributed in the humid zones of Patagonia (southern Argentina and Chile) over a longitudinal distance of close to 2,000 km (Fig. 1). The regional climate patterns favor the development of these cold temperate wetlands, especially in the west and the south of the region, where the passage of the oceanic wind determines very wet conditions.

The steep W-E climatic gradient which is induced by the Andes is a substantial factor determining differences in peatland types, scale, and distribution: small and scattered minerogenic units occur in the extra-Andean Oriental Patagonia, in contrast to the extensive mire complex at the west of the Andes and in Tierra del Fuego. Peatlands are strongly embedded within the southern temperate forest matrix, and they are intimately connected at the level of ecosystem processes (Arroyo et al. 2005).

The information about the Patagonian peatlands is very patchy and not easily accessible (Blanco and de la Balze 2004); however, new studies undertaken through since 2,000 both in Argentina and in Chile have improved the knowledge of these wetlands.

Age and Origin of Patagonian Peatlands

Most peatland locations were occupied by ice during the Last Glacial Maximum (LGM) approximately 24,500 BP. Glacial and postglacial processes molded favorable landforms for peatland development which followed the deglaciation, in the range of 9–12 ky BP (early Holocene). Older records from early deglaciated locations near the LGM termini have however been reported: 16 ky BP in southern continental Patagonia in Rubens (Markgraf and Huber 2010) and Punta Arenas

69 Patagonian Peatlands (Argentina and Chile)

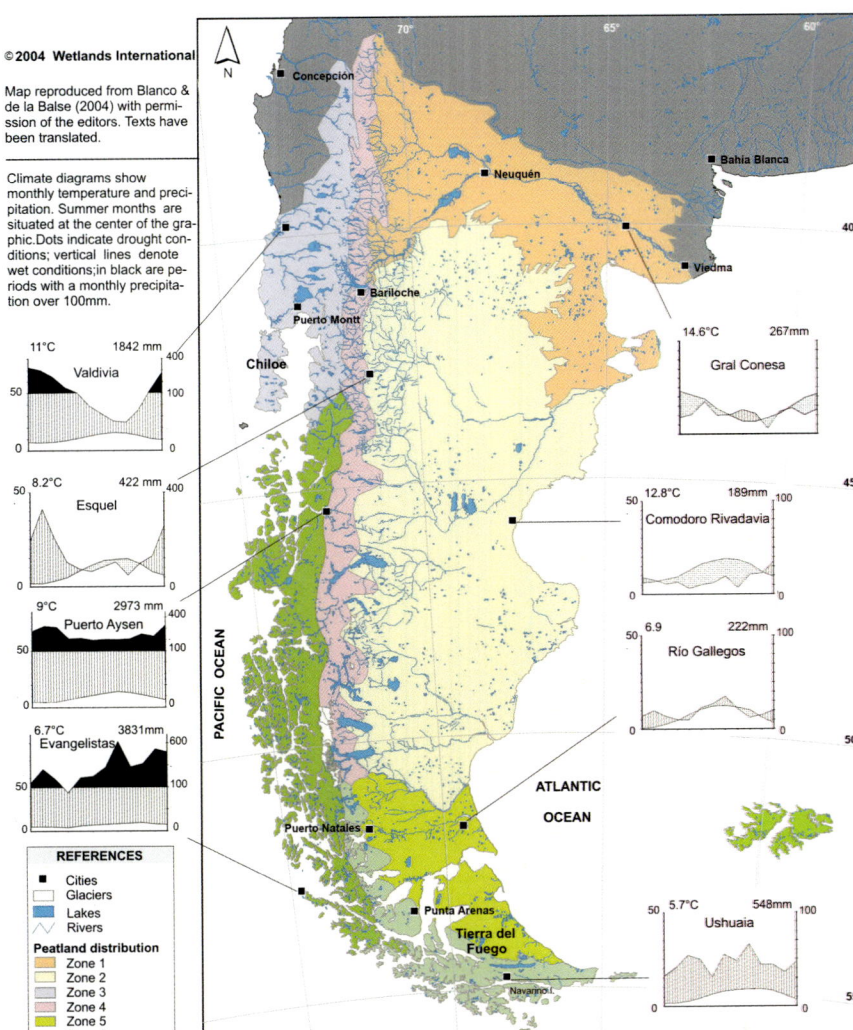

Fig. 1 Distribution of Patagonian peatlands (Malvarez et al. 2004). Reproduced and translated with permission of Blanco and de la Balse (eds). *Zone 1*, very low presence of peatlands; *Zone 2*, widely separated peatlands localized in patches in wet meadows; *Zone 3*, peatlands in land depressions or areas with more altitude in NW Patagonia; *Zone 4*, peatlands and peat meadows in the east side of the Andean region; *Zone 5*, peat meadows and occasionally peatlands in the south of the extra-Andean region; *Zone 6*, significant peatland cover. It includes the subzones *6a* (high precipitation and dominance of peatlands over the rest of ecosystems), *6b1* (high precipitation and mesic–udic soil regimes), and *6b2* (low precipitation and cryic–xeric soil regimes)

(Heusser 1995) and 17 ky BP in southern Tierra del Fuego (Markgraf and Huber 2010). Older ages have been determined in North Patagonia for successional deposits of peat and lake sediments (Villagran 2001). Palynological studies performed in peatlands provided valuable information about paleoecology and paleoclimate in Patagonia.

Peatland Distribution

In the east of the Patagonian Andes, the annual precipitation decreases from more than 2,000 to 200 mm along a less than 100 km wide transitional environment from the forest to the steppe. Peatlands are insignificant at this regional scale and are mainly localized in the Andean foothill forest environment and the environs. They occur as scattered units, mostly small fens dominated by *Cyperaceae*. *Sphagnum*-raised bogs are rare.

The extra-Andean Patagonia is covered by treeless shrub and grass steppes that give way to dwarf-shrub semideserts in the drier areas of the central plateaus (Roig 1988). The windy and dry climate prescribes a general water balance deficit; nevertheless, a widely distributed type of small wetlands, named "mallines" or "vegas" (wet meadows), provides accessible water and high primary productivity (Buono et al. 2010). These riparian ecosystems preserve biodiversity and sustain livestock production.

Table 1 Peatland types in the Magellanic moorland (Pisano 1983; Arroyo et al. 2005)

Wetland complex	Precipitation range (mm year^{-1})	Particular features	Dominant vegetation	Related forest
Ombrotrophic-raised bogs and blanket bogs	500–1,500	Wider distribution	*Sphagnum* sp.	Deciduous forest (east)
			Empetrum sp.	Deciduous forest-evergreen rain forest (west)
Juncoid	500–1,500	Poor drainage and waterlogged areas	*Marsippospermum gradiflorum* and *Sphagnum* sp.	Deciduous and evergreen forest
Cyperoid	1,000–2,000	Flat morphology	*Schoemus antarcticus, Carpha alpina, Rostkovia*, grass species	*Nothofagus antarctica, N. betuloides, N. nitida*
Cushion bogs	>2,000	Wind exposed	*Astelia pumila, Donatia fascicularis, Bolax caespitosa, Oreobulus obtusangulus*	*Nothofagus antarctica*
		Oceanic condition		*N. betuloides*

Fig. 2 *Sphagnum* bog covered by dwarf trees of *Pilgerodendron uviferum* (Guaiteca Cypress – Ciprés de las Guaitecas) near Seno Obstrucción (at south of Puerto Natales, Chile) (Photo credit: R. Iturraspe © Rights remain with the author)

Mallines occur in hollows alongside drainage networks where the water table reaches the surface. Usually they are connected to the slopes of basaltic rock plateaus, which are typical formations in the extra-Andean area (Mazzoni and Rabassa 2013). Peat accumulation can occur where the water table remains close to surface throughout the year. Peat thickness is linked to the water level during the summer period (Collantes et al. 2009).

This type of "peat wet meadow" is scarce in the eastern and more common in the western part of the extra-Andean Patagonia and in the ecotonal zone of Tierra del Fuego. The extent of mallines in the extra-Andean Patagonia reaches about 3% of the total area, but information on the proportion of those that accumulate peat is not available.

Very different bioclimatic conditions occur in the Chilean Patagonia west of the Andes. Native temperate rain forest dominates in the 10th Region de Los Lagos, but peatlands are visible landscape components, especially in Chiloe Island (41°48′S–43°28′S) where, sensu Richardson (2011), peatland extent reaches 890 km^2 (10% of the land area). These wetlands are *Sphagnum* bogs, fens, "pomponales," and "tepuales." "Pomponal" is the local name for a *Sphagnum magellanicum* ecosystem without a basal peat layer that has developed in the last 100 years following extensive forest fires.

Fig. 3 *Astelia–Donatia* cushion bog in Moat, East of Tierra del Fuego (Argentina) (Photo credit: R. Iturraspe © Rights remain with the author)

A "tepual" is a swamp forest on low and poorly drained lands with normally waterlogged soils. Tepuales are dominated by *Tepualia stipularis* and usually include Guaitecas cypress *Pilgerodendron uviferum* and winter's bark *Drimys winteri* (Veblen and Schlegel 1982). Soils are acidic, pH may reach 3.5, and peat layers are frequently present (Holdgate 1961).

The Magellanic tundra (Pisano 1983), also known as the Magellanic moorland (Moore 1979), is a wild peatland complex that extends from Golfo de Penas (48°S) to Cabo de Hornos (56°S) and covers 44,000 km^2 (Arroyo et al. 2005). Its development is the consequence of the extremely wet climate which prevails during the whole year, especially in the western islands. Guarello Island (50°21′S; 75°21′W) at 7220 mm year^{-1} is the rainiest site at the sea level (Aravena and Luckman 2009).

Pisano (1983) differentiates four wetland complex subtypes according to precipitation, drainage, and wind regimes (Table 1). In addition, Kleinebecker et al. (2007) reported that the floristic variation of Patagonian peatlands correlated with two major environmental gradients: continentality and soil water level. Peatland and forest ecosystems are strongly related by synergistic and competitive interactions. Dwarf trees of species such as Antarctic beech *Nothofagus antarctica*, Magellan's beech *N. betuloides*, or Guaitecas cypress are usually found on peatlands (Fig. 2).

The Tierra del Fuego archipelago represents the southernmost part of the Magellanic moorland. The main island, Isla Grande de Tierra del Fuego (Tierra del Fuego (TDF)), is divided between Chile (west) and Argentina (east). Oceanic conditions and the SW wind direction prevail in southern TDF, where a west-east aligned mountain range dominates the landscape. The Darwin icefield (2400 m a.s.l.), whose glacier termini reach the sea level in the fiords of both the Beagle channel and the Strait of Magallanes, is located there.

In the west of the main island and leeward of the mountains, *Sphagnum*-raised bogs occupy the upper valleys of the Grande River. *Sphagnum* peatlands occur also in the drier Argentinean side of TDF. The SW-NE precipitation gradient determines a transitional ecotonal zone from the Andes Fueguinos to the north, with mosaics of *Nothofagus* forest, grasslands, *Sphagnum* bogs, and *Cyperaceae* fens. *Sphagnum* occurrence decreases in a northerly direction and disappears at 54°S.

The main bog concentration in TDF is located in the east, in Peninsula Mitre, where the precipitation is about 700–900 mm year^{-1}. These peatlands show well-marked local features which are linked to the local hydrology, climate, relief, geomorphology, and water and soil chemical properties (Iturraspe et al. 2012). *Astelia–Donatia* cushion bogs (Fig. 3) dominate in the windy coastal lowlands under oceanic conditions, even though the precipitation is less than what has been indicated for the Pacific cushion bog environment. Similar biogeographic patterns occur in the southern part of the nearby Navarino Island.

Peatland coverage reaches 2700 km^2 in the Argentinean sector of TDF, 2400 km^2 of which it is located at the eastern extreme where peatlands are the dominant ecosystem and their cover reaches 90% in some eastern water basins (Iturraspe et al. 2012). Ruiz y Doberty Ltda (2005) estimated the peatland extent in the Chilean sector of TDF to be 946 km^2 and 3960 km^2 in the Antarctica Chilean Province.

Peatland Use and Conservation

Most of the Patagonian peatlands remain in an almost pristine state. This occurs in several extended zones in TDF, in the Chilean Pacific Archipelago, and in the southwestern uninhabited mainland areas where they are confined by the Patagonian icefield and prominent fjords. Nevertheless, in the neighborhood of inhabited territories, human activities produce negative impacts on peatlands.

Peat cutting has been growing since the decade of the 1970s. The main peat mining places are Chiloe and Magallanes in Chile and TDF in Argentina. Most of the peat production is sold to the internal market to be used as horticultural substrate. Peat mining activities are regulated by the respective mining laws of these countries. Nearly all the peat produced in Argentina comes from the TDF province. Since 2011 this state has implemented a strategy for the wise use of peatlands, establishing new policies for mire management and restricting new peat mining concessions to a defined zone which is located in the center of the province, to the east of Fagnano Lake.

There is concern about how *Sphagnum* fiber is being harvested without regulation in pomponales in the 10th Region de Los Lagos, Chile. Production of this exported product has been growing; export income from *Sphagnum* fiber increased from 0.5

to 14.8 million USD between 2001 and 2012 (Domínguez 2014). This practice has affected a large percentage of pomponales in that region. The continuity of moss extraction at the current rate and without proper management would mean the extinction of this resource in a short period of time.

Cattle raising is the main use of minerotrophic peatlands in the extra-Andean Patagonia. Since the late nineteenth century, the entire region has been affected by severe sheep grazing. Overgrazing results in the degradation and disturbance of natural biodiversity. Nevertheless, the natural capacity for restoring the original species after grazing closure is greater in wet meadows than in dry grasslands (Collantes et al. 2013).

Tourist and recreational uses of peatlands produce economic profit in TDF, Argentina. Since the 1980s, several tourist centers have become established beside peatlands in mountain valleys and provide services and support for winter sports and recreational activities on these peatlands.

Many protected areas in Patagonia, e.g., national parks (NP), natural reserves, and Ramsar sites, include peatlands. Bernardo O'Higgins NP (Chile: 35,200 km^2) is the largest of these protected areas. Alberto de Agostini NP and Laguna San Rafael NP are also large Chilean reserves that include peatlands. Several Argentinean NPs, e.g., Nahuel Huapi NP, Los Alerces NP, Los Glaciares NP, and Tierra del Fuego NP, protect extensive zones along the Patagonian Andes, but they contain considerably less peatland areas. The main bog system situated in Argentina's eastern TDF is not protected. "The Vinciguerra Glacier and related peatlands" (TDF, Argentina) is the southernmost Ramsar Site in the world, and "the Cape Horn Biosphere Reserve," located in the Chilean part of the archipelago, is the southernmost of UNESCO's Man and the Biosphere Reserve.

Values and Functions

Patagonian peatlands represent the main terrestrial carbon sink and carbon storage in the extratropical Southern Hemisphere. They play a relevant function in the global carbon cycle regulation and contribute to climate change mitigation.

Peatlands provide habitat for notable species. Although minerotrophic wetlands such as fens and wet meadows have a small extent, they hold a special role in the extra-Andean Patagonia as a valuable water source for the natural fauna providing habitat to a considerable variety of vascular plants.

Patagonian peatlands contribute to hydrological water basin regulation in a natural environment characterized by hyper contrasting climatic patterns and poor distribution of water.

Peatlands provide freshwater, enhance flood regulation capacity, improve water quality, and diminish soil erosion. Hydrological peatland functions are especially important for populated river basins. Several Patagonian cities, such as Castro, Puerto Natales, Punta Arenas, Ushuaia, and Río Grande, use surface water from rivers whose peatland systems provide an important hydrological regulation function.

Threats and Future Challenges

Problems related to the environmental condition of peatlands usually increase as the regional population increases: peat extraction, open-pit mining, drainage of peatlands, extensive cattle raising, changes in the land use, urban expansion, and new infrastructure on peatland areas are some threats whose significance change in each place.

The wise use of these peatlands is the real challenge. Joosten (2002) defines this concept as that use for which reasonable people now and in the future will not attribute blame. The concept of "wise use" incorporates complex environmental, economic, and social concerns that requires integrated decision-making and further demands that peatland values and functions should be properly recognized by the local people through increased environmental education.

References

Aravena JC, Luckman BH. Spatio-temporal rainfall patterns in southern South America. Int J Climatol. 2009;29(14):2106–20.
Arroyo MTK, Mihoc P, Pliscoff P, Arroyo M. The Magellanic moorland. In: Fraser LH, Keddy PA, editors. The World's largest wetlands. New York: Cambridge University Press; 2005. p. 425–45.
Blanco DE, de la Balze VM. Los Turbales de la Patagonia: bases para su inventario y la conservación de su biodiversidad. Buenos Aires: Wetlands International; 2004.
Buono G, Oesterheld M, Nakamatsu V, Paruelo JM. Spatial and temporal variation of primary production of Patagonian wet meadows. J Arid Environ. 2010;74:1257–61.
Collantes MB, Anchorena J, Stoffella S, Escartín C, Rauber R. Wetlands of the Magellanic Steppe (Tierra del Fuego. Argentina). Folia Geobot. 2009;44:227–45.
Collantes MB, Escartín C, Braun K, Cingolani AM, Anchorena JA. Grazing and grazing exclusion along a resource gradient in magellanic meadows of Tierra del Fuego. Rangel Ecol Manag. 2013;66:688–99.
Domínguez E. Manual de buenas prácticas para el uso sostenido del musgo *Sphagnum magellanicum* en Magallanes, Chile. Instituto de Investigaciones Agropecuarias. Centro Regional de Investigación Kampenaike. Punta Arenas. Boletín INIA; 2014. N° 276. 113 p.
Heusser CJ. Three late quaternary pollen diagrams from Southern Patagonia and their paleoecological implications. Paleogeogr Paleoclim Paleocol. 1995;118:1–24.
Holdgate MW. Vegetation and soils in the South Chilean Islands. J Ecol. 1961;49(3):559–80.
Iturraspe R, Urciuolo AB, Iturraspe RJ. Spatial analysis and description of eastern peatlands of Tierra del Fuego, Argentina. In: Heikkilä R, Lindholm T, editors. Mires from pole to pole, The Finnish Environment, vol. 38. 2012. p. 385–99.
Joosten HD, Clarke D. Wise use of mires and peatlands. Saarijärvi: IMCG-IPS; 2002. 304 p.
Kleinebecker T, Hölzer N, Vogel A. Gradients of continentality and moisture in South Patagonian peatland vegetation. Folia Geobot. 2007;42:363–82.
Malvarez AI, Kandus P, Carbajo A. Distribución regional de los turbales en Patagonia. In: Blanco DE, de la Balze VM, editors. Los turbales de la Patagonia. Bases para su inventario y la conservación de su biodiversidad. Buenos Aires: Wetlands International; 2004. p. 22–9.
Markgraf V, Huber UM. Late and postglacial vegetation and fire history in Southern Patagonia and Tierra del Fuego. Palaeogeogr Palaeoclim Palaeoecol. 2010;297:351–66.
Mazzoni E, Rabassa J. Types and internal hydro-geomorphologic variability of mallines (wet-meadows) of Patagonia: emphasis on volcanic plateaus. J South Am Earth Sci. 2013;46:170–82.

Moore DM. Southern oceanic wet-healthland (including Magellanic moorland). In: Spech RI, Goodhall DW, editors. Ecosystems of the World, vol. 9A. Amsterdam: Elsevier Science; 1979.

Pisano E. The magellanic tundra complex. In: Gore A, editor. Mires:swamp, Bog, Fen and Moor. B. Regional studies. Amsterdam: Elsevier; 1983. p. 295–329.

Richardson R. A registry of productive peat bogs in the lake region. Hemispherics Polar Stud J. 2011;2(4):249–66.

Roig FA. La Vegetación de la Patagonia. In: Correa MN, editor. *Flora Patagónica*, Colecc. Ci. INTA. 1988;8(1):48–166.

Ruiz y Doberty Ltda. Catastro y Caracterización de los Turbales de Magallanes. Sernageomin. BIP N°20196401-0, Chile. 2005. 123 p.

Veblen TT, Schlegel FM. Reseña ecológica de los bosques del sur de Chile. Bosque. 1982;4(2):73–115.

Villagran C. A model for the history of vegetation of the coastal range of central-southern Chile: Darwin's glacial hypothesis. Rev Chil Hist Nat. 2001;74:793–803.

Section VI

Europe

Danube River Basin

70

Paul Csagoly, Gernant Magnin, and Orieta Hulea

Contents

Location	886
Climate and Hydrology	887
Wetland Distribution and Diversity	888
Biodiversity	889
People	890
Threats	890
Values	892
Future Challenges	892
Box: Assessing the Basin's Restoration Potential	894
References	895

Abstract

The Danube River is 2,857 km long and it is the world's most international river basin, including territories of 19 countries. Its source lies in the Black Forest in Germany and it flows into the northwestern part of the Black Sea, splitting into three branches that form the Danube Delta (6,750 km^2). Over the last two centuries, 80% of the wetlands and floodplains in the DRB became disconnected, and many disappeared. Still, the remaining floodplains and wetlands are uniquely valuable ecosystems in European

---

P. Csagoly
Earthly Communications, Ottawa, ON, Canada
e-mail: paul.csagoly@gmail.com

G. Magnin (✉)
WWF – Netherlands, Freshwater Programme, Zeist, The Netherlands
e-mail: gernantmagnin@gmail.com; gmagnin@wwf.nl

O. Hulea (✉)
WWF Danube-Carpathian Programme, Bucharest, Romania

WWF Danube-Carpathian Programme, Vienna, Austria
e-mail: ohulea@wwfdcp.ro

© Springer Science+Business Media B.V., part of Springer Nature 2018
C. M. Finlayson et al. (eds.), *The Wetland Book*,
https://doi.org/10.1007/978-94-007-4001-3_87

and global terms, providing numerous ecosystem services to basin residents, such as flood protection, groundwater replenishment, sediment and nutrient retention, biodiversity, river-floodplain products, cultural values and climate change buffering capacity. With the implementation of the EU Water Framework Directive, major threats in the DRB, like pollution and hydromorphological alterations have been identified as well as measures to mitigate or reduce their impact, including the ambitious objective to restore lost wetlands and their valuable services and functions.

Keywords
Danube River Basin · Water Framework Directive · Water management · Wetlands restoration · Ecosystem services

Location

The Danube River is 2,857 km long and up to 1.5 km wide with depths of up to 8 m (Mandl 2009). Its source lies in Central Europe on the eastern slopes of the *Schwarzwald* (Black Forest) in Germany at 678 m above sea level. At its end, the Danube flows into the northwestern part of the Black Sea along the boundary between Ukraine and Romania (Fig. 1).

Fig. 1 The Danube River Basin, indicating key areas covered by related articles in this volume: the Danube-Drava-Mura Rivers, Lower Danube Green Corridor, and Danube Delta (Reproduced with permission from WWF)

After the Volga River Basin, the Danube River Basin (DRB) is the second largest river basin in Europe (and 21st in the world). With a total drainage area of 817,000 km², it includes 7.8% of Europe. It lies to the west of the Black Sea in Central and South-eastern Europe. To the west and northwest, the DRB borders on the Rhine river basin; the Weser, Elbe, Oder, and Vistula river basins in the north; the Dnjestr in the north-east; and in the south on the catchments of the rivers flowing into the Adriatic Sea and the Aegean Sea. Importantly, it is the world's most international river basin, including the territories of 19 countries (ICPDR 2004).

Based on its gradients, the DRB can be divided into three subregions: the Upper, Middle, and Lower Basins. The Upper Basin extends from the source of the Danube in Germany to Bratislava in Slovakia. The Middle Basin is the largest of the three subregions, extending from Bratislava to the dams of the Iron Gate gorge on the border between Serbia and Romania. The lowlands, plateaus, and mountains of Romania and Bulgaria form the Lower Basin. Before reaching the Black Sea, the river divides into three main branches, forming the Danube Delta, which covers an area of about 6,750 km² (Mandl 2009). Each subregion possesses different hydro- and geographic features.

The Danube connects with 27 large and over 300 small tributaries on its way from the Black Forest to the Black Sea. The Tisza is the longest tributary (966 km) of the Danube and the largest by catchment area (157,186 km²). Many lakes are also found in the Danube Basin, with Lake Balaton in Hungary being the largest at 605 km² (Mandl 2009).

Climate and Hydrology

The DRB shows great differences in climate due to its large extension from west to east and its diverse relief. Upper regions in the west show a strong influence from the Atlantic climate with high precipitation, while eastern regions are affected by the continental climate with lower precipitation and cold winters. The Mediterranean climate influences the area of the Drava and Sava rivers. The heterogeneity of the relief, especially the differences in the extent of exposure to the predominantly westerly winds, and the differences in altitude diversify this general climate pattern (Wise 2016). This leads to distinct landscape regions showing differences in climatic conditions (WISE) and vegetation. Precipitation ranges from under 500 mm to over 2,000 mm. This in turn has strong effects on surface runoff and discharge into streams (ICPDR 2004).

The hydrologic regime of the Danube River, in particular the discharge regime, is distinctly influenced by regional precipitation patterns. Austria shows by far the largest contribution (22.1%) to the Danube, followed by Romania (17.6%). This reflects the high precipitation in the Alps and in the Carpathian Mountains. In the upper part of the Danube, the Inn river contributes the main water volume. In the middle reach, it is the Drava, Tisza, and Sava rivers, which together contribute almost half of the total discharge that finally reaches the Black Sea (ICPDR 2004).

The Danube River has a very high water volume, with a mean annual flow of about 6,400 m³ s^{-1} into its estuary. Melting mountain snow, heavy rains and groundwater feed the river, and floods occur during warmer seasons (February until August). The Danube is particularly shallow in September and October, before a potential freezing over in January and February, although freezing does not regularly occur on an annual basis (Bogutskaya 2015).

Wetland Distribution and Diversity

The current extent of wetlands in the DRB is only a remnant of the former wetland systems. Over the last two centuries, 80% of the wetlands and floodplains in the DRB became disconnected, and many disappeared.

The *morphological* floodplain is defined as the postglacial terraces; the *active* floodplain as those within current flood protection dikes, regularly flooded on a periodic basis; and the *former* floodplain as the portion of the morphological floodplain outside the dikes, disconnected from the flood pulse. The total size of the DRB's morphological floodplain, including the Drava, Sava, and Tisza floodplains, is 79,406 km^2, equal to almost 10% of the basin. The remaining active floodplain is 15,542 km^2. For the Danube River, floodplains originally covered an area of approximately 26,524 km^2, equal to about 3.3% of the total Danube catchment area. Only 8,452 km^2 (32%) of that floodplain remains (Schwarz 2010).

At the same time, the 20% of the original wetlands that remain are highly impressive. Floodplain forests, marshlands, deltas, floodplain corridors, lake shores, and other wetlands are essential components in the basin's biodiversity and hydrology. The DRB extends into five of Europe's eight biogeographic regions: Alpine, Continental, Pannonic, Steppic, and Black Sea. Each shows characteristic wetlands, some of which are protected (ICPDR 2004). Many of the larger wetland complexes are transboundary in nature and include UNESCO and Ramsar sites, national parks, and nature reserves. A few main examples are provided below.

The transboundary *Danube Delta* (see Fig. 1) is the basin's most important wetland, covering 675,000 ha across Romania and Ukraine. It is also one of the world's largest wetlands, a UNESCO World Heritage Site, and Europe's largest remaining natural wetland (World Wide Fund for Nature -LDGC website). The complex includes three large river arms, floodplain forests, limans (estuary formed at the widening mouth of a river, where flow is blocked by a bar of sediments), inner lakes, natural and man-made channels, sand dunes, coastal biotopes, and the largest reed bed in the world at 180,000 km^2 (ICPDR 2004). Wetland ecosystems are dominated by flooded reed beds, floating reed beds, and riparian willow formations.

The *Lower Danube wetlands* are a major complex covering 600,000 ha across six countries (see Fig. 1). The Lower Danube is one of the last free-flowing stretches of river in Europe. Here, the natural dynamics of the river have formed and re-formed nearly 200 islands that are home to rich floodplain ecosystems. However, 70% of the Lower Danube's floodplains were lost over the last 200 years, while its restoration potential is some 500,000 ha (WWF 2010).

The *Drava-Mura wetlands* (see Fig. 1) cross three countries and include a 380 km long bio- and landscape corridor and a floodplain corridor of some 60,000 ha.

Austria's *Donau-Auen National Park* near Vienna consists of 11,000 ha of floodplain forests and riparian habitats and side arms and represents the last intact floodplain in the Upper Danube region. Together with the floodplains of the Morava and Dyje rivers, intersecting three countries, the overall area forms a transboundary "wetland of international importance" and trilateral Ramsar site covering 25,000 ha (ICPDR 2004).

Many other large transboundaries and national wetland complexes can also be found in the basin's Middle and Lower regions.

Biodiversity

The floodplains and wetlands of the DRB are uniquely valuable ecosystems in European and global terms, although few areas are still in their natural or even near-natural state. Along the Upper and Middle Danube, only about 10% of the basin's floodplains remain in near-natural conditions (Schwarz 2010).

The basin hosts 2,860 sites that are designated under *Natura 2000*, an EU-wide network of nature protection areas that is the centerpiece of EU nature and biodiversity policy, aimed at assuring the long-term survival of Europe's most valuable and threatened species and habitats. Two hundred thirty are found on the Danube River itself (Mandl 2009).

Over 2,000 plant species and 5,000 animal species live in or by the waters of the Danube, including approximately 100 fish species as well as important bird sanctuaries for species such as the Dalmatian pelican *Pelecanus crispus* (Mandl 2009).

The *Danube Delta* features rare fauna and flora, as well as 30 different types of ecosystems. Overall, it supports a wide variety of taxa with 2,383 plant species and 4,029 animal species, including 135 fish species and 331 bird species (Danube Delta Biosphere Reserve Authority website). WWF regards the Delta among the 200 most valuable ecological areas on earth (WWF Lower Danube Green Corridor website). In the *Lower Danube*, the islands are important elements of the Danube migration corridor – stepping stones for fish, waterbirds, and other fauna as well as flora on their journeys up and down the river. Together, the Lower Danube and Danube Delta are especially important as breeding and resting places for some 331 species of birds, including the rare Dalmatian pelican, the white-tailed eagle *Haliaeetus albicilla*, and 90% of the world population of red-breasted geese *Branta ruficollis* (Baboianu 2016).

Other examples include the *Drava-Mura wetlands* which boast migratory freshwater species and alpine pioneer species living on sand and gravel bars and islands, as well as forest species and mammals such as Eurasian otter *Lutra lutra* and Eurasian beaver *Castor fiber*. Hungary's 47,000 ha *Gemenc-Béda-Karapancsa wetlands* is an important bird site for black stork *Ciconia nigra* and white-tailed eagle. Croatia's 30,000 ha *Kopacki Rit* Ramsar Site and Nature Park boasts 100 days of flooding a year and an abundance of food and underwater vegetation, making it, after the Danube Delta, the most important fish-spawning ground along the entire Danube (ICPDR 2004).

People

The DRB is home to 83 million people with a wide range of cultures, languages, and historical backgrounds. Twenty-nine million people live in the Lower basin. Geopolitically, the basin was divided between east and west until communism ended in the former Eastern Bloc countries and Soviet Union – many of these former communist countries have since acceded to the EU, Bulgaria and Romania having been the latest in 2007.

Today, there is a wide gulf between the GDP per capita of Austria, Germany, and Slovenia and the other Danube Basin countries: the wealthiest country's GDP per capita is nearly 14 times higher than that of the poorest (Mandl 2009). Lower Danube countries continue to be far more reliant on the agricultural sector than those in the Middle and Upper Danube regions: while 9.8% of Ukrainian, 12.4% of Romanian, and 21.8% of Moldovan GDP are generated from agriculture, this share is only 1.7% for Austria, 2.4% for Germany, and 2.6% for the Czech Republic (Mandl 2009).

Most Danube Basin countries have begun to experience negative population growth rates, with only three countries – Austria, Bosnia and Herzegovina, and Slovakia – displaying marginal population growth. As populations in the Danube Basin shrink and age, this will result in changing social and consumption patterns that may, in turn, lead to a change in environmental impacts (Mandl 2009).

The basin's wetlands provide numerous ecosystem services to basin residents, such as flood protection, groundwater replenishment, sediment and nutrient retention, water purification, resilience and recovery of river ecosystems after accidents, biodiversity/habitat, river-floodplain products (e.g., wood, fish, game, reed), cultural values, recreation and tourism, and climate change buffering capacity (Schwarz 2010).

The wetlands of the Danube Delta more or less influence the living standards of some 200,000 inhabitants. They are the major source of water for industrial, agriculture (irrigation), and domestic use for local communities (Baboianu 2016).

Threats

The *Danube River Basin Management Plan* (*DRBMP*) and its *Joint Programme and Measures* (*JPM*) focus on four *Significant Water Management Issues* (*SWMIs*) that affect the overall quality of rivers and lakes as well as transboundary groundwater bodies, namely: pollution by organic substances, pollution by nutrients, pollution by hazardous substances, and hydromorphological alterations (i.e., to the physical characteristics of a water body's shape, boundaries, and content) (ICPDR Fact Sheet 2 2013a). The four issues also significantly affect wetlands.

The Plan identifies hydropower generation, navigation, and flood protection as the key water uses that cause hydromorphological alterations. These alterations result in the following key pressures of basin-wide importance: interruption of river and habitat continuity, disconnection of adjacent wetlands/floodplains, and hydrological alterations (ICPDR Fact Sheet 6 2013e).

The Plan further concluded that 80% of the former wetlands/floodplains in the DRB are disconnected largely due to the expansion of agricultural uses, and river engineering works for flood control, navigation, and power generation. For example, some 80% of Romania's floodplains were drained under agricultural intensification schemes during the 1960s and 1970s alone (ICPDR Wetlands website).

Hydrological alterations impact the status of water bodies for different reasons. These significantly reduce the flow and quantity of water and impact the water status where the minimum ecological flow of rivers is not guaranteed. In the DRB, 449 water bodies are affected by impoundments, 140 by water abstractions, and 89 by hydropeaking (rapid increase or decrease in the release of (operating) water from reservoir hydroelectric power stations that changes the flow regime in the river downstream of the plant causing artificial fluctuations of flow). The key water uses causing significant alterations through water abstractions are hydropower generation, agriculture, forestry, and public water supply, followed by cooling water, the manufacturing industry, and navigation (ICPDR Fact Sheet 6 2013e).

A number of future infrastructure projects (FIPs) for navigation, hydropower, and flood protection have been proposed for the basin, many of which may have negative impacts on water status and wetlands.

Organic, nutrient, and hazardous pollution is another key threat to DRB wetlands. Organic pollution can cause significant changes in the oxygen balance of surface waters. As a consequence, it can impact the composition of aquatic species and populations and therefore water status. Organic pollution in the DRB is mainly caused by the emission of partially treated or untreated wastewater from cities or towns, industry, and agriculture. In terms of organic pollution, water quality in the Danube ranges between moderate pollution and moderate to critical pollution (ICPDR Fact Sheet 3 2013b).

Nutrient pollution is also mainly caused by emissions from cities and towns, industry, and agriculture. Its main threat is *eutrophication* – the enrichment of water causing an accelerated growth of algae and higher forms of plant life which produce an undesirable disturbance to the balance of organisms present in the water and to the quality of the water concerned (ICPDR Fact Sheet 4 2013c).

Sources of hazardous substances are industrial effluents, storm water overflow, pesticides and other chemicals applied in agriculture as well as discharges from mining operations, and accidental pollution. For some substances, atmospheric deposition may also be a source. Manufacturing industries are responsible for large emission loads of a number of hazardous substances, including heavy metals and organic micro-pollutants. A basin-wide inventory of potential accident risk spots has been created, with some 650 risk spots recorded and 620 evaluated. As a result, a hazardous equivalent of about 6.6 million tonnes has been identified as a potential danger in the Danube catchment area (ICPDR Fact Sheet 5 2013d).

Values

During 2010, commissioned by WWF, the Institute for European Environmental Policy (IEEP) assessed how to put the key recommendations of the The Economics of Ecosystems and Biodiversity (TEEB) initiative into practice, through a case study of ecosystem services in the Danube River Basin (Tucker et al. 2010). It primarily demonstrates the potential benefits of an ecosystem service-based approach to land management while also identifying potential constraints and opportunities. On the basis of available information and expert judgment, five key types of ecosystem service were selected for assessment on the basis of their known importance in the DRB and the availability of quantitative information on their values. These are river fish production, water provisioning and purification, flood storage, climate regulation through carbon sequestration and storage, and nature-based tourism.

It was too early to accurately attach monetary values to each of these services, but, for example, nature-based tourism was preliminary estimated to be at least €711 million per year across ten DRB countries.

Future Challenges

Danube River Basin waters are situated in both EU and non-EU country states. While EU member states are obliged to comply with EU directives and regulations, non-EU member states are not. Fortunately, however, all of the DRB countries not in the EU have also agreed to comply with the EU's Water Framework Directive (WFD). Possibly the strongest water legislation in the world, the WFD requires that all EU waters reach at least "good status" by 2015 (or at the latest by 2027), including good chemical and ecological status.

The *International Commission for the Protection of the Danube River (ICPDR)* is a transnational body which serves as the platform for the implementation of all transboundary aspects of the WFD. Originally, it was established to implement the Danube River Protection Convention.

Regarding DRB wetlands and floodplains, the Danube River Basin Management Plan's (DRBMP) ambitious vision is that they are all reconnected and restored. ("The integrated function of these riverine systems ensures the development of self-sustaining aquatic populations, flood protection and the reduction of pollution in the DRB District"). In turn, significant restoration efforts are needed. In 2009, 95 wetlands/floodplains (covering 612,745 ha) with the potential to be reconnected to the Danube River and its tributaries were identified. Of this, the Joint Programme of Measures (JPM) indicated that 11 wetlands/floodplains (62,300 ha) should be reconnected by 2015 (ICPDR Fact Sheet 6 2013e). As a result of the 2009 DRBMP and its measures, by 2015 more than 50,000 ha of wetlands and floodplains were partly or totally reconnected. For the period 2015–2021, additional measures are planned to be implemented, including improvement of the morphological conditions and habitats by restoration measures in 77 water bodies, 15,130 additional ha of reconnected wetlands and floodplains, ensuring ecological flow requirements, ecological improvement of impoundments, and addressing hydropeaking in more than 60 cases (ICPDR 2015).

In some cases, efforts are proving highly successful. Europe's "most ambitious wetland protection and restoration project" is the *Lower Danube Green Corridor* (see Fig. 1). Based on an agreement signed in 2000 by four governments and facilitated by WWF, the project aimed at protecting 1 million ha of existing and new protected areas and restoring 224,000 ha of natural floodplains, including the final 1,000 km of the Danube to the Danube Delta. By 2010, the first target was surpassed with 1.4 million ha protected, while about 25% of the restoration target had been reached (WWF LDGC 2010).

To minimize further threats to the basin, the ICPDR started a cross sectoral discussion process with the navigation sector involving all relevant stakeholders and NGOs (leading to a Joint Statement) (ICPDR 2008). A similar process has also been initiated with the hydropower sector, and the "Guiding Principles on Sustainable Hydropower Development in the Danube Basin" have been developed by an interdisciplinary team and were finalized and adopted in June 2013 (ICPDR 2013f). Nonetheless, a significant number of future infrastructure projects (FIPs) (navigation, flood protection, hydropower) may have negative impacts on water status by 2021 and need to be addressed accordingly. Thirty-nine FIPs have been reported for the DRBD. Thirty-two of them are located in the Danube River itself. For eight FIPs, Strategic environmental assessments have been performed during the planning process. Further, Environmental Impact Assessments have already been performed for 20 FIPs and are intended for another 18 FIPs. Fourteen FIPs are expected to have a negative transboundary effect on other water bodies, and four FIPs are expected to provoke deterioration of water status, for which exemptions according to WFD Article 4(7) are applied (ICPDR 2015).

Regarding organic pollution reduction, the implementation of the Urban Wastewater Treatment Directive (UWWTD) in EU member states and the development of wastewater infrastructure in non-EU member states are the most important measures to reduce organic pollution in the DRBD by 2015. In general, upstream countries have almost completely achieved overall treatment efficiency; less has been accomplished in the Middle/Lower Danube countries, but extensive efforts are underway throughout the basin (ICPDR Fact Sheet 3 2013b). As a result of the 2009 DRBMP and its measures, a reduction of organic emissions from urban wastewaters by half from 2005 levels is observed (ICPDR 2015).

Regarding nutrient pollution reduction, fulfilling the UWWTD and EU Nitrates Directive for EU member states will be the main measures. For non-EU member states, the main measure is the implementation of the ICPDR Best Agricultural Practices (BAP) recommendation. Expanding the use of phosphate-free detergents would also be supportive, although regulation may be required to influence some countries and industries (ICPDR Fact Sheet 4 2013c). As a result of the 2009 DRBMP, a decline of emissions – via point and diffuse sources – of nitrogen and phosphorus by approximately 10% and 30%, respectively, leading to a significant reduction of transported nutrient loads to the Black Sea, was observed (ICPDR 2015).

For hazardous substance reduction, fulfilling the UWWTD and EU Nitrates Directive for EU member states is a key measure. For non-EU member states, the construction of 47 municipal WWTPs by 2015 will improve the situation. EU industrial and agro-industrial controls and regulations, pesticide bans, and best practices are also recommended. Regarding accident prevention, the ICPDR has taken important steps to develop an accident early warning system which sends out international warnings to countries downstream (ICPDR Fact Sheet 5 2013d). Improving wastewater treatment and industrial technologies, regulating market products and application of chemicals, and closing knowledge gaps on hazardous substances via emission inventories are the most important current activities to address hazardous substance pollution. National inventories on priority substance emissions that are currently being compiled by the countries will deliver substantial information on the emission sources (ICPDR 2015).

Overall, for all four of the Significant Water Management Issues, the proposed measures carried out by 2015 with the ICPDR and Danube countries resulted in considerable improvements as mentioned above (ICPDR 2015). The 2015–2021 DRBMP defines additional measures agreed to be implemented by 2021 that should contribute to achieving good chemical and ecological status as required by the WTD.

However, to date (ICPDR 2015) 13.742 river kilometers (rkm) (49%) is at risk or possibly at risk of achieving good ecological status, and 9.390 rkm (35%) is at risk or possibly at risk of achieving good chemical status by 2021, due to ongoing pressures by organic pollution (19%), nutrient pollution (20%), hazardous substances pollution (27%), and hydromorphological alterations (30%).

Box: Assessing the Basin's Restoration Potential

In July 2010, an *Assessment of the restoration potential along the Danube and main tributaries* was conducted for WWF (Schwarz 2010). The study analyzed 439 areas with major existing, planned, and proposed restoration projects (total area 1.38 million ha). Most of the proposed restoration areas are located in the former floodplain. The study found that the overall loss of floodplains could be reduced by 44% of their original extent through restoration efforts, with the largest potential in Romania.

For the Danube River, 196 areas (covering 810,228 ha) were identified (about 560,000 ha are already officially planned according to the ICPDR). Of this,

179,708 ha are in the active floodplain, and 630,520 ha are in the former floodplain. Some 8% lie in "near-natural" floodplains. The study found that about 24% of the former floodplain could be restored.

The study further found that some current action plans and implementation strategies provide a framework for floodplain restoration. However, these often include only a few proposals and implementation schedules in later management cycles toward 2026. It therefore recommends the clear planning of one large-scale pilot restoration project per country by the next cycle of WFD management planning in 2015–2021. As a result, if all sites proposed in this study were implemented over the long term (e.g., in the third WFD planning cycle by 2026 and beyond 2032), then it would be necessary to restore 28,000 ha each year. This number would require strong and clear political direction and funding for implementation, with an overall investment of over 6 billion €. It would further require favorable legal frameworks (e.g., clear protection of existing retention areas and no-go areas for further land development in floodplains), strong spatial planning instruments, and tight administrative and political structures that allow for transparent public participation (Schwarz 2010).

References

Baboianu G. Transboundary wetlands of the Danube Delta, Romania and Ukraine. 2016. [This volume].
Bogutskaya N, Hales J. Freshwater ecoregions of the world (418: Dniester-Lower Danube). 2015. http://www.feow.org/ecoregions/details/418. Accessed Feb 2016.
Danube Delta Biosphere Reserve Authority. Website: http://www.ddbra.ro/en/danube-delta-biosphere-reserve/danube-delta/biodiversity. Accessed Feb 2016.
ICPDR. Danube basin analysis (WFD Roof Report 2004). Vienna: ICPDR; 2004.
ICPDR. Development of inland navigation and environmental protection in the Danube river basin, joint statement on guiding principles. Vienna: ICPDR; 2008.
ICPDR. Fact sheet 2: JDS3 and the EU water framework directive. Joint Danube survey 3. Vienna: ICPDR; 2013a.
ICPDR. Fact sheet 3: organic pollution. Joint Danube survey 3. Vienna: ICPDR; 2013b.
ICPDR. Fact sheet 4: nutrient pollution. Joint Danube survey 3. Vienna: ICPDR; 2013c.
ICPDR. Fact sheet 5: hazardous substances pollution. Joint Danube survey 3. Vienna: ICPDR; 2013d.
ICPDR. Fact sheet 6: hydromorphological alterations. Joint Danube survey 3. Vienna: ICPDR; 2013e.
ICPDR. Sustainable hydropower development in the Danube Basin, Guiding Principles. Vienna: ICPDR; 2013f.
ICPDR. Danube river basin district management plan 2015–2021. Vienna: International Commission for the Protection of the Danube River; 2015.
ICPDR. Wetlands page in ICPDR website. Vienna: ICPDR. http://www.icpdr.org/main/issues/wetlands. Accessed Feb 2016.
Mandl B. The Danube river basin, facts and figures. Vienna: ICPDR; 2009.
Schwarz U. Assessment of the restoration potential along the Danube and main tributaries. Vienna: WWF; 2010.

Shepherd K. A functioning river system: incorporating wetlands into river basin management (in Danube Watch magazine). Vienna: ICPDR; 2006. http://www.icpdr.org/main/publications/functioning-river-system-incorporating-wetlands-river-basin-management. Accessed Feb 2016.

Tucker GM, Kettunen M, McConville AJ, Cottee-Jones E. Valuing and conserving ecosystem services: a scoping case study in the Danube basin. Report prepared for WWF. London: Institute for European Environmental Policy; 2010.

WISE Water Knowledge Portal. Climate and hydrology in the Danube river basin district. European Commission. http://www.wise-rtd.info/en/info/22-climate-and-hydrology-danube-river-basin-district. Accessed Feb 2016.

WWF. Lower Danube green corridor (fact sheet). WWF. September 2010. http://awsassets.panda.org/downloads/wwf_ldgc.pdf. Accessed Feb 2016.

WWF. Lower Danube green corridor (website). http://wwf.panda.org/what_we_do/where_we_work/black_sea_basin/danube_carpathian/our_solutions/freshwater/danube_river_basin/lower_danube/. Accessed Feb 2016.

Lower Danube Green Corridor

71

Paul Csagoly, Gernant Magnin, and Orieta Hulea

Contents

Location and Extent	898
Wetland Types and Biodiversity	898
Threats	900
Conservation Status and Management	901
References	902

Abstract

The Lower Danube, flowing for more than 1,000 km through Bulgaria, Romania, Ukraine and Moldova, has been identified, together with the Danube Delta as one of the most outstanding biodiversity regions in the world. It is also a region of rich cultural heritage where local livelihoods have been closely connected with the river. Besides sheltering a rich diversity of rare and endangered habitats and species, the freshwater ecosystems of the Lower Danube perform essential environmental services and provide numerous opportunities for the sustainable development of local communities.

Human interventions have destroyed more than 80% of all Danube wetlands since the turn of the last century. Dykes, drainage and irrigation systems have

---

P. Csagoly (✉)
Earthly Communications, Ottawa, ON, Canada
e-mail: paul.csagoly@gmail.com

G. Magnin (✉)
Consultant, WWF – Netherlands, Freshwater Programme, Zeist, The Netherlands
e-mail: gmagnin@wwf.nl; gernantmagnin@gmail.com

O. Hulea (✉)
WWF Danube-Carpathian Programme, Bucharest, Romania

WWF Danube-Carpathian Programme, Vienna, Austria
e-mail: ohulea@wwfdcp.ro

© Springer Science+Business Media B.V., part of Springer Nature 2018
C. M. Finlayson et al. (eds.), *The Wetland Book*,
https://doi.org/10.1007/978-94-007-4001-3_251

been built along the river and its tributaries. Floodplain habitats have been converted to arable lands or forestry monocultures. The construction of large hydro-engineering structures for navigation and hydropower generation have had a heavy impact on biodiversity. Many of the remaining wetlands and floodplain habitats in the Lower Danube still support globally significant biodiversity but are cut off from the natural river dynamics. Consequently their ecological functions and services provided are impaired and their condition is gradually deteriorating. The restoration potential of the Lower Danube Green Corridor is about 500,000 ha, which if implemented, could store a significant volume of water to reduce the flood risks and enhance biodiversity and other ecosystem services. The Lower Danube Green Corridor aims at making the Lower Danube a living River again, connected to its natural flooding areas and wetlands, reducing the risks of major flooding in areas with human settlements and offering benefits both for local economies – fisheries, tourism – and for the ecosystems along the river.

Keywords
Lower Danube Green Corridor · Biodiversity · Floodplain restoration · Ecosystem services

Location and Extent

The Lower Danube Green Corridor (LDGC) covers a total area of 18,344 km^2 and comprises three sections. The first is the Lower Danube floodplain in Romania and Bulgaria, stretching from the Iron Gate dam to the beginning of the Chilia branch of the Danube Delta, and including the morphological floodplain (between the terraces) and the active floodplain, covering 9,080 km^2. The second is the Danube Delta in Romania and Ukraine, including the Razim-Sinoie lagoon complex in Romania, covering 6,264 km^2. The final section is the Lower Prut River in Romania and the Republic of Moldova up to Costesti, covering about 3,000 km^2 (see Fig. 1).

Some 29 million people live in the Lower Danube River Basin, many of which directly benefit from the services that the river provides, from drinking water to natural resources and recreation (http://wwf.panda.org/dcpo).

Wetland Types and Biodiversity

Some 80% of the Lower Danube's floodplains are known to have been lost over the last 200 years, along with a significant corresponding amount of water retention capacity (WWF Fact Sheet 2010). Nonetheless, much of what remains is ecologically valuable. In fact, the Lower Danube, together with the Danube Delta, is one of the world's outstanding freshwater ecoregions (http://wwf.panda.org/dcpo).

The Danube floodplains, now limited to the areas between the river bank and flood protection dikes, typically include habitats such as the relics of oxbows and

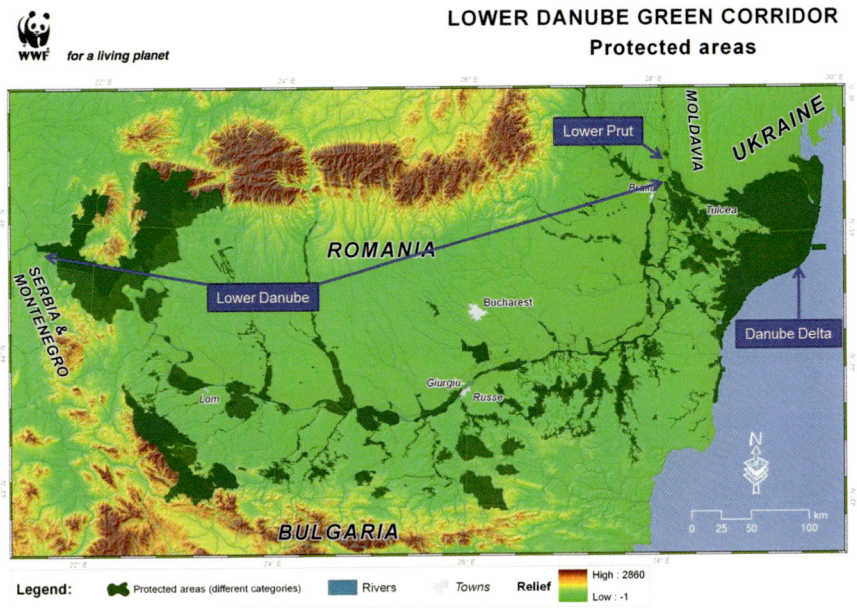

Fig. 1 The Lower Danube Green Corridor, indicating the main river sections and protected areas (Reproduced with permission from WWF)

lakes, flood channels, depressions, and small water courses. The hydrological dynamics of the river, its erosion and sedimentation processes, and periodic flooding have led to the formation of numerous islets along the border in Romania (111 islets covering 11,063 ha) and Bulgaria (75 islets covering 10,713 ha). The islets host rich floodplain ecosystems including natural floodplain forest, sand banks, marshes, and natural river channels, and are integral parts of the Danube migration corridor, essential for the distribution of many plant and animal species.

There are an impressive number of both terrestrial and aquatic habitats and species, many of them globally important, including: 55 species of aquatic macrophytes, 906 terrestrial plants, 502 insects, 10 amphibians, 8 reptiles, 56 fish, 160 birds, and 37 mammals.

The Danube's greatest jewel is the Danube Delta, Europe's largest remaining natural wetland area and, as regarded by WWF, among the 200 most valuable ecological areas on earth. A total of 5,137 species have been identified along the lower stretch of the river, including 42 different species of mammals and 85 species of fish (http://wwf.panda.org/dcpo).

The Lower Danube and Danube Delta are especially important as breeding and resting places for some 331 species of birds, including the rare Dalmatian pelican *Pelecanus crispus*, the white-tailed eagle *Haliaeetus albicilla*, and 90% of the world population of red-breasted geese *Branta ruficollis* (http://wwf.panda.org/dcpo).

Threats

The river's natural dynamics have been seriously damaged by modifications, mainly to allow the river to be navigated by large vessels (the riverbed was deepened and dredged in many places and there are numerous canals and dams) and to facilitate agriculture (floodplain conversion, irrigation). Intensified agriculture has also increased nutrient discharges to water bodies (MacKay 2006). As a result, many types of habitat have disappeared or shrunk in size, and ecological corridors critical for migration of species have been severed.

During 2010, commissioned by WWF, the Institute for European Environmental Policy (IEEP), assessed how to put the key recommendations of the The Economics of Ecosystems and Biodiversity (TEEB) initiative into practice, through a case study of ecosystem services in the Danube River Basin (Tucker et al. 2010). It primarily demonstrates the potential benefits of an ecosystem-service based approach to land management, whilst also identifying potential constraints and opportunities. On the basis of available information, and expert judgement, five key types of ecosystem service were selected for assessment on the basis of their known importance in the DRB and the availability of quantitative information on their values. These are river fish production; water provisioning and purification; flood storage; climate regulation through carbon sequestration and storage; and nature-based tourism. Three of these ecosystem services are of particular relevance in the Lower Danube and Delta region, but have been negatively affected.

The first is the fish production, especially through damming and upstream nutrient pollution affecting fish in the Delta. Fishing represents a significant livelihood for many remote communities with few alternatives sources of income. The Danube River basin preserves some of the most important sturgeon populations in the world today (beluga – *Huso huso*, Russian sturgeon – *Acipenser gueldenstaedtii*, stellate sturgeon – *Acipenser stellatus*, all critically endangered and the sterlet – *Acipenser ruthenus*, listed as vulnerable) with Romania and Bulgaria holding the only – still – viable populations of wild sturgeons in the European Union. Illegal fishing – principally for their caviar – is the main direct threat to the survival of Danube sturgeons. Habitat loss and disruption of spawning migration, largely because of physical barriers (dams), are further threats to sturgeon survival.

The second type of ecosystem service that has suffered degradation is the flood storage service. Floodplains help to reduce the impacts of flooding along the Danube. The growing frequency and severity of floods and droughts in the Danube basin – which are expected to worsen with climate change – have highlighted the value of these floodplains and called into question traditional paradigms of water management based on diking and dredging. Floodplains play an important role in recharging groundwater supplies – the source of drinking water for many people in the Danube basin. Wetlands also help purify the water of excess nutrients (WWF 2012).

Economically, the degradation of these services represents annual losses (if restoration does not occur) of US$15 million for the fishery, about US$131 million for water treatment for nitrogen and phosphorus nutrients, and US$16 million for tourism (MacKay 2006).

The biodiversity loss is in part reversible, when natural river dynamics can be restored and wetlands returned to at least a seminatural state. However, full restoration to their original state is unlikely to be feasible (MacKay 2006).

The restoration of 37 floodplain sites (areas with high restoration potential – about 225,000 ha) that make up the LDGC is estimated to cost US$ 204 million (calculated from EUR) compared to total damages costing an estimated US$ 445 million on the lower Danube from the spring 2006 floods (Schwarz et al. 2006) and likely ecosystem services earnings of US$ 125 million per year (based on an average calculation of US$ 556/ha) (Ebert et al. 2009).

In the Danube Delta, the living standards of some 200,000 inhabitants are more or less influenced by its wetlands, which provide the major source of water for industrial, agriculture (irrigation), and domestic use for local communities. They are used for navigation by both commercial and public ships and boats. The main natural resources represented by fish, reed, pasture, natural, and planted forests support local traditional economic activities. Fish are by far the most exploited resource, with an annual harvest of about 7,000 t supporting commercial fishing, family consumption, and sport fishing. Reed beds support the potential for harvesting about 40,000–50,000 t annually of reed, and pastures support the production of sheep, cattle, pigs, and horses. Agriculture is practised, both in polders for crops with cereals (i.e., wheat, barley, maize, and sunflower) and, at a smaller scale, for family needs (i.e., vegetables, fruit trees, and vineyards) (see ▶ Chap. 73, "Danube Delta: The Transboundary Wetlands (Romania and Ukraine)" by Baboianu, this volume).

Conservation Status and Management

The Lower Danube Green Corridor is Europe's "most ambitious wetland protection and restoration project" (WWF Fact Sheet 2010). It is based on an Agreement signed in 2000 by the four governments of Bulgaria, Romania, Ukraine, and Moldova, and facilitated by WWF, who pledged to work together to establish a green corridor along the entire length of the Lower Danube River (Declaration on the cooperation for the creation of a Lower Danube Green Corridor. Bucharest; 2000). The Agreement originally made a minimum commitment of 773,166 ha of existing protected areas, 160,626 ha of proposed new protected areas and 223,608 ha of areas proposed to be restored to natural floodplain. Areas would either have strict protection, some limited human activity, or be targeted for the development of sustainable economic activities. Its key goals were to reconnect natural flooded areas and wetlands, reduce the risks of floods in areas of human settlement, and offer benefits to local economies such as fisheries and tourism.

More recently, the project aimed to protect one million ha of existing and new protected areas and restoring 224,000 ha of natural floodplains, including the final 1,000 km of the Danube to the Danube Delta. By 2010, the first target was surpassed with 1.4 million ha brought under some form of protection, while about 25% of the restoration target had been achieved (http://wwf.panda.org/dcpo).

The overall restoration potential for the area has been estimated at 500,000 ha. The restoration of 100,000 ha is equivalent to an increase of 1.6 billion m^3 of additional water retention capacity (Mackay 2006).

Taken together, the economic value of the benefits and services from Danube floodplains, including flood and drought management, climate change adaptation, water purification, fish production, reed harvesting, and recreation, is estimated to be at least €500 per hectare per year (WWF 2012).

In addition, many of the wetlands in the Lower Danube are already included in the Annexes of the EU Habitats Directive which, together with the Birds Directive, forms the cornerstone of Europe's nature conservation policy.

References

Declaration on the cooperation for the creation of a Lower Danube Green Corridor. Bucharest; 2000. Official Declaration of the Ministries of Environment from Romania, Bulgaria, Moldova and Ukraine (http://www.ramsar.org/news/creation-of-a-lower-danube-green-corridor-and-gift-to-the-earth).

Ebert S, Hulea O, Wickel B. Floodplain restoration along the lower Danube (Romania, Ukraine): a climate change adaptation case study. Clim Develop. 2009;1(3):212–9.

MacKay E. Annex 2. Loss of ecosystem services provided by river basins – case study of the Danube River basin and delta (Germany and Romania). In: Kettunen M, ten Brink P, editors. Value of biodiversity – documenting EU examples where biodiversity loss has led to the loss of ecosystem services. Final report for the European Commission. Brussels: Institute for European Environmental Policy (IEEP); 2006.

Schwarz U, Bratrich C, Hulea O, Moroz S, Pumputyte N, Rast G, Bern MR, Siposs V. Floods in the Danube River basin: flood risk mitigation for people living along the Danube and the potential for floodplain protection and restoration. Working paper. Vienna: WWF Danube-Carpathian Programme; 2006.

Tucker GM, Kettunen M, McConville AJ, Cottee-Jones E. Valuing and conserving ecosystem services: a scoping case study in the Danube basin. Report prepared for WWF. London: Institute for European Environmental Policy; 2010.

WWF. Lower Danube Green Corridor (Fact Sheet). WWF; 2010. http://awsassets.panda.org/downloads/wwf_ldgc.pdf. Accessed Feb 2016.

WWF. Lower Danube Green Corridor (website). WWF; 2010. http://wwf.panda.org/what_we_do/where_we_work/black_sea_basin/danube_carpathian/our_solutions/freshwater/danube_river_basin/lower_danube/. Accessed Feb 2016.

WWF. Veritable revolution for Danube wetlands. WWF website; 2012. http://wwf.panda.org/?205554/Prava-revolucija-u-mocvarama-Dunava. Accessed Feb 2016.

Danube, Drava, and Mura Rivers: The "Amazon of Europe"

72

Paul Csagoly, Gernant Magnin, and Arno Mohl

Contents

Location	904
Wetland Types and Diversity	904
Threats	905
Conservation Status and Management	906
References	908

Abstract

Spanning Austria, Croatia, Hungary, Serbia and Slovenia, the lower courses of the Drava and Mura Rivers and related sections of the Danube are among Europe's most ecologically important riverine areas: the so-called "Amazon of Europe".

The rivers form a 700 kilometers long "green belt" connecting almost 1,000,000 hectares of highly valuable natural and cultural landscapes from all five countries and shall therefore become a symbol of unity by becoming the world's first five country Transboundary UNESCO Biosphere Reserve "Mura-Drava-Danube" (TBR MDD) as well as Europe's largest river protected area.

Despite numerous man-made changes in the past, this stunning river landscape hosts an amazing biological diversity and is a hotspot of rare natural species and

P. Csagoly (✉)
Earthly Communications, Ottawa, ON, Canada
e-mail: paul.csagoly@gmail.com

G. Magnin (✉)
WWF – Netherlands, Freshwater Programme, Zeist, The Netherlands
e-mail: gernantmagnin@gmail.com; gmagnin@wwf.nl

A. Mohl (✉)
WWF Austria, Vienna, Austria
e-mail: arno.mohl@wwf.at

© Springer Science+Business Media B.V., part of Springer Nature 2018
C. M. Finlayson et al. (eds.), *The Wetland Book*,
https://doi.org/10.1007/978-94-007-4001-3_252

habitats. A coherent network of 12 major protected areas along the rivers highlight their ecological values.

On March 2011, WWF brokered a commitment by the Ministers responsible for environment and nature conservation of Austria, Croatia, Hungary, Serbia and Slovenia to jointly establish the reserve. Whereas UNESCO officially approved the Croatian-Hungarian part of the Biosphere Reserve, Serbia, Austria and Slovenia are still in the designation process.

Despite outstanding natural features and protection status the Mura-Drava-Danube area is suffering from an ongoing degradation. A century of river channelling, the construction of dikes, the extraction of gravel and sand, as well as the construction of hydropower plants led to a loss of up to 80% of the former floodplain areas and the alteration of about 1,100 km of natural river banks and associated stretches. The situation can only improve if the characteristic natural conditions are restored. A potential study, conducted by WWF, shows that about 650 km of river banks and 120 side arms could be restored as well as 165,000 ha of new floodplains created. A series of EU Life funded restoration projects have already been implemented or are are under implementation.

Keywords

Mura-Drava-Danube · Transboundary Biosphere Reserve · TBR MDD · Amazon of Europe · River restoration

Location

The Drava River has a length of 750 km. Its most important tributary, the Mura River, is 420 km long. Both rivers originate in the Alpine Mountains with max altitude at 3,800 m a.s.l. with the Drava joining the Danube River on the edge of the Pannonian lowland at 80 m a.s.l. Together the rivers drain an area of 48,000 km^2 shared by Austria, Croatia, Hungary, Italy, and Slovenia (Schwarz and Mohl 2009).

Wetland Types and Diversity

The three rivers in their lower parts form a 700 km-long "green belt" connecting more than 1,000,000 ha of highly valuable natural and cultural landscapes from all five countries. The lower courses of the Drava and Mura rivers are extraordinary in having a 380 km-long stretch that remains free flowing without any dams, thereby retaining the unity of an original riverine landscape. Adjacent stretches of the middle Danube in excellent condition bring this figure to almost 700 km (Schwarz and Mohl 2009).

Often referred to as the "Amazon of Europe," the area hosts excellent examples of rare natural habitats such as large softwood forests, wet meadows, natural islands, gravel and sand banks, steep banks, side branches, and oxbows. The area is home to the highest density of breeding pairs of white-tailed eagles *Haliaeetus albicilla* in

Continental Europe and other endangered species such as the black stork *Ciconia nigra*, Eurasian beaver *Castor fiber*, Eurasian otter *Lutra lutra*, and two endangered sturgeon species: the ship sturgeon *Acipenser nudiventris* and sterlet *Acipenser ruthenus*. The gravel and sand banks of the Drava River provide one of the last breeding grounds for the little tern *Sterna albifrons* in inland Europe. Every year, more than 250,000 migratory waterfowl use the rivers to rest and feed (Mohl 2014). Within a 30 km-long river stretch of the Drava downstream of the Mura confluence, over 50 main types and combinations of habitats have been described (Schwarz and Mohl 2009).

For example, the well-preserved alluvial wetlands along the Danube in the trilateral area between Croatia, Hungary, and Serbia include the world famous Kopacki Rit Nature Park, covering some 23,000 ha of swampy softwood floodplains with shallow and dynamic floodplain lakes. The Danube-Drava National Park's large floodplain forests and wetlands are a major feeding and breeding ground for some 110 bird species including the grey heron *Ardea cinerea*, great cormorant *Phalacrocorax carbo* colonies, and a high density of common kingfishers *Alcedo atthis*. The area is also home to the largest population of black stork in Hungary (Mohl 2014).

Overall, the banks of the three rivers remain in a natural state over a length of about 190 km (9%), in a near-natural state over 765 km (38%) and already altered/impacted over 1,081 km (53%). However, there is a wide variation between different river sections and countries. For example, in stretches such as the Mura along the border between Austria and Slovenia, 95% of river banks are fixed by embankments, while along some stretches of the Mura and Drava in Croatia and Hungary, and the Danube between Croatia and Serbia, this figure is less than 40% (WWF 2013).

The total active floodplain area comprises 132,341 ha, representing 22% of the original extent, while 78% was lost through flood protection dikes (WWF 2013). In different countries, the loss of active floodplains varies from 66% to 90%.

Key ecosystem services provided by the extensive floodplains include reducing the risks from floods, securing favorable groundwater conditions, and purifying water. The river ecosystem is also an important area for recreation and tourism.

Threats

Until the end of the eighteenth century, the Lower Drava and Mura rivers were free flowing and wild. Over the past 200 years, the riverine landscape has seen many changes and human impacts leading to the loss of up to 80% of the former floodplain areas and the alteration of about 1,100 km of natural river banks and stretches (WWF 2013). Engineering practices, mainly for navigation and flood protection projects, continue to have a major impact on its ecological integrity, biodiversity values, and natural resources. Key threats include the channelling of natural river courses, extraction of gravel and sand from the riverbed, and hydropower dams. Furthermore, past and ongoing river regulation and sediment extraction activities have considerably multiplied and accelerated the impacts of the hydropower dams which are situated in the upstream sections (Schwarz and Mohl 2009).

River channelling leads to the deepening of riverbeds, dries out wetlands and floodplain forests, ruins natural river habitats, and threatens endangered species. This is shown in the decline of the sand martin *Riparia riparia* along the Drava from 12,000 breeding pairs in 2005 to 3,000 in 2010 (Mohl 2014). Beside the loss in biodiversity, irresponsible river management causes considerable economic damage (e.g., decreasing water levels has negative impacts on drinking water, forests, agriculture, and fish stocks). River channelling also increases the risk of floods in downstream settlement areas. Currently, the areas most affected by newly planned large-scale river channelling are the natural stretches of the Danube and Drava rivers in the border area between Croatia, Hungary, and Serbia.

A chain of 50 hydropower dams – 22 on the Drava and 28 on the Mura – has already been built along the upper and middle courses of the Drava and Mura rivers in Austria, Croatia, and Slovenia. Their impacts on the free-flowing lower stretches of both rivers are pervasive with regard to changes in hydrology. As a result of fluctuations in electricity demand, hydropower stations are set to satisfy peaks ("hydropeaking"); they, thus, often tend to work intermittently, creating sudden and rapid fluctuations in the river system, and eventually alter the river bed morphology. Planned future hydropower dams continue to threaten the Mura River in Slovenia and Drava River in Croatia.

Conservation Status and Management

The ecological importance of the Drava-Mura and Danube areas is reflected in the declaration of 12 major protected areas at the national level (see Fig. 1), such as the Danube-Drava National Park in Hungary, Kopacki Rit Nature Park in Croatia, and Gornje Podunavlje Nature Reserve in Serbia. In February 2011, the Croatian Government protected the Drava and Mura as a Regional Park, covering about 88,000 ha of valuable natural and cultural landscape. The natural values of the river system satisfy the criteria to be recognized and protected under international conventions such as the Ramsar, Bern and Bonn Conventions, as well as EU environmental legislation such as the Habitats and Birds Directives, and as part of the Europe-wide Natura 2000 network (an ecological network of protected areas in the European Union). Natura 2000 sites covering riparian habitats have already been established in Austria, Slovenia, Hungary and in Croatia.

Since 1993, NGOs have been campaigning to protect the unique landscape of the three rivers in a Transboundary UNESCO Biosphere Reserve (TBR MDD), they call "Amazon of Europe". Step-wise, public administrations and NGOs cooperate to jointly achieve the TBR. The protection of the area as a TBR is one of Europe's most ambitious nature conservation projects, covering an overall area of more than 1,000,000 ha and spanning EU member states and countries currently outside the EU (Schwarz and Mohl 2009).

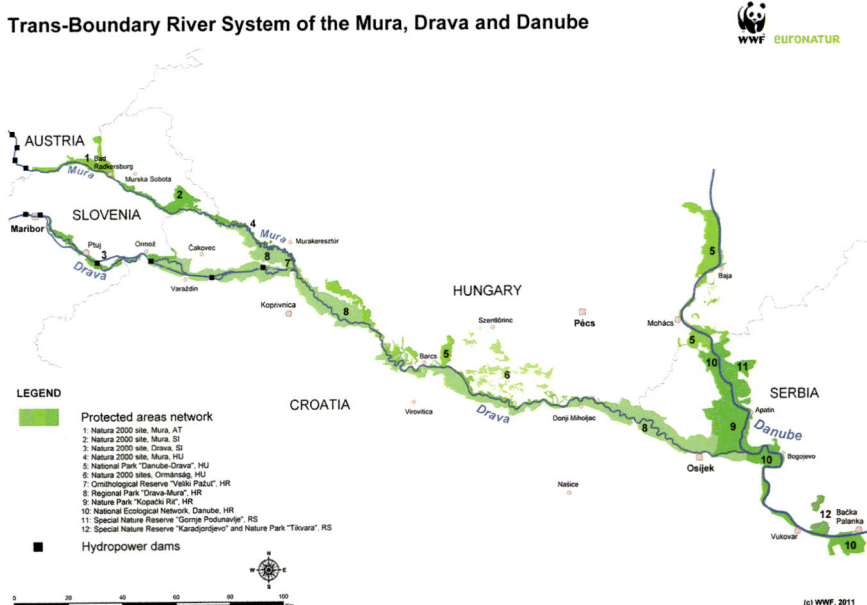

Fig. 1 Transboundary river system of the Mura, Drava, and Danube (Reproduced with permission from WWF)

In March 2011, the environment ministers of all five countries agreed to jointly protect and manage the area as a Transboundary UNESCO Biosphere Reserve. The sections within Croatia and Hungary have already been designated by UNESCO. These 630,000 ha cover more than 60% of the future 5-country area. Serbia, Austria and Slovenia are currently in the designation process. Once established, it will be Europe's largest protected river area and the world's first pentalateral (i.e., five-country) Biosphere Reserve (WWF 2013). It will also form a crucial part of the "European Green Belt" proposed by the World Conservation Union (IUCN), which aims to create an ecological network along the former Eastern Bloc countries from the Barents Sea to the Black Sea.

The central management goal of the TBR is to halt any further degradation of the riverine landscape and to begin the improvement of natural river dynamics. This will be achieved through implementing a transboundary "River Restoration Programme" for the Mura, Drava and Danube Rivers. Examples of important steps include banning further river regulation and sediment extraction activities, preventing further impacts from hydropower dams, and active restoration measures. These would further be essential steps for achieving compliance with the requirements of the EU Water Framework Directive, Flood Directive, and Habitats and Birds Directives.

The Biosphere Reserve concept defines about 300,000 ha of core and buffer zones (existing protected area network) and another roughly 700,000 ha of transition zones. The core zone is the ecological backbone of the reserve. It primarily covers the river and floodplain areas which are mostly situated within flood control dikes (Mohl 2014).

The assessment study, undertaken by FLUVIUS and conducted by WWF, of the restoration potential found that 60% of all impacted banks could be restored to highly dynamic and other near-natural banks. More than half of the destroyed banks (53% – fixed by embankments, so-called rip-rap) could be restored to natural or near-natural conditions; and 120 major side channels totalling 519 km could be reconnected to the river system. Restoration could further result in 297,244 ha of active floodplain, reducing the overall loss to 50% with reconnections of 28% of the active floodplain back to the river (WWF 2013). Restoration projects significantly reduce the further degradation of the river and floodplain areas. As already demonstrated by first mostly EU Life funded restoration efforts carried out by the MDD countries, this will safeguard the long-term survival of the characteristic habitats and species, and benefit of healthy river system for nature and local communities.

In 2010, commissioned by WWF, the Institute for European Environmental Policy (IEEP), assessed how to put the key recommendations of The Economics of Ecosystems and Biodiversity (TEEB) initiative into practice, through a case study of ecosystem services in the Danube River Basin (Tucker et al. 2010). It primarily demonstrates the potential benefits of an ecosystem-service-based approach to land management while also identifying potential constraints and opportunities. On the basis of available information, and expert judgment, five key types of ecosystem service were selected for assessment on the basis of their known importance in the DRB and the availability of quantitative information on their values. These are river fish production, water provisioning and purification, flood storage, climate regulation through carbon sequestration and storage, and nature-based tourism. It was too early to accurately attach monetary values to each of these services, but, for example, nature-based tourism was preliminarily estimated to be at least €711 million per year across ten DRB countries. Given its size and attractive nature, the Mura-Drava-Danube area would no doubt also be of significant economic value for ecotourism. As a socio-economic guiding project for the further development of the Transboundary Biosphere Reserve "Mura-Drava-Danube" (TBR MDD), WWF, in cooperation with EuroNatur, have already developed a concept of a cross-border cycle path, a unique brand for bike tourism, the so-called "Amazon of Europe" Bike Trail, which should be implemented with support of EU funds (WWF/EuroNatur 2016).

References

Mohl A. Conserving the Amazon of Europe: Mura-Drava Danube Rivers at a crossroad between protection and destruction. Vienna: WWF and Euronatur Foundation; 2014.

Schwarz U, Mohl A. Lifeline Drava-Mura 2009–2020. A plan for conserving and restoring the Drava and Mura Rivers for nature and People. Vienna: WWF and Euronatur Foundation; 2009.

Tucker GM, Kettunen M, McConville AJ, Cottee-Jones E. Valuing and conserving ecosystem services: a scoping case study in the Danube basin. Report prepared for WWF. London: Institute for European Environmental Policy; 2010.

WWF. Assessment of the River and floodplain restoration potential in the transboundary UNESCO biosphere reserve "Mura-Drava-Danube". Vienna: WWF; 2013.

WWF/EuroNatur. "Amazon of Europe" Bike Trail; From the Alps to the Pannonian plain along the largest natural river system in Central Europe. Vienna; Implementation concept: 2016. http://www.amazon-of-europe.com/en/aoe-bike-trail/

Danube Delta: The Transboundary Wetlands (Romania and Ukraine)

73

Grigore Baboianu

Contents

Introduction	912
Hydrology	912
Wetland Ecosystems	914
Biodiversity	917
Conservation Status	918
Ecosystem Services	919
Threats	920
Future Challenges	921
References	921

Abstract

The Danube Delta (4,455 km^2) is the second largest delta in Europe after the Volga Delta, shared by Romania (3,510 km^2 (79%)) and Ukraine (945 km^2 (21%)). It forms part of a large wetland region including several limans, large lakes that formed when the Danube permanently flooded the lower parts of the valleys of tributary rivers (about 6,496 km^2). The Danube River branches are the main conduits of water and sediments discharged through the Danube Delta, and a large diversity of natural, partially man-modified and anthropogenic ecosystems (30) have formed, hosting a wide variety of taxa with 7,402 species recorded to date. In addition to supporting a high level of biodiversity, the Danube Delta Region provides many ecosystem services including its important effect on water quality and nutrient retention, and provision of extensive economic and environmental benefits to the local communities (about 200,000 inhabitants) living in and around the Delta. The management of the Danube Delta should consider short and medium term needs including a wetland restoration program to increase the

---

G. Baboianu (✉)
Danube Delta Biosphere Reserve Authority, Tulcea, Romania
e-mail: gbaboianu@ddbra.ro

© Springer Science+Business Media B.V., part of Springer Nature 2018
C. M. Finlayson et al. (eds.), *The Wetland Book*,
https://doi.org/10.1007/978-94-007-4001-3_192

natural flooded area in abandoned polders, measures to reduce the impact of the more ecologically damaging economic activities including navigation and related hydrotechnical works and over-exploitation of natural resources, and transboundary cooperation.

Keywords

Danube river · Danube delta · Liman · Reed bed · Meadow · Riparian willow formation · Shrubs and herbaceous vegetation · Temperate riverine forest · Wetland restoration · Ecosystem service · Transboundary cooperation · Ramsar Site

Introduction

The Danube Delta is located in the eastern part of Europe and shared by two countries: Romania and Ukraine. At the terminus of the second largest river in Europe, the head of the Danube Delta is located near the towns of Tulcea (Romania) and Izmail (Ukraine), where the river divides into two branches: the northern branch known as the Chilia (Kilia) marks the natural border between Romania and Ukraine (120 km), and the southern branch known as the Tulcea (17 km) splits 10 km downstream and east of the town of Tulcea into the Sulina (63.7 km) and Sfântu Gheorghe (69.7 km) branches. The Danube Delta (4,455 km^2) is the second largest delta in Europe after the Volga Delta, shared by Romania (3,510 km^2 (79%)) and Ukraine (945 km^2 (21%)). It forms part of a larger wetland region including several limans (Kagul, Jalpug, Katlabuh, Kugurluj, etc.), large lakes that formed when the Danube permanently flooded the lower parts of the valleys of tributary rivers, along the left side of the Chilia (Kilia) branch (468 km^2) in Ukrainian territory (Chernichko et al. 2003), and former marine bays (1,331 km^2) and other permanent flooded areas (242 km^2), in the southern part of the delta in Romanian territory, together forming about 6,496 km^2 of transboundary wetlands (Fig. 1).

Hydrology

The Danube River branches are the main conduits of water and sediments discharged through the Danube Delta. Before splitting in the Danube Delta, the Danube River has a multiannual average annual discharge of about 6,515 m^3/s (1921–2000 period). During this period, the maximum discharge was recorded in 1970 (15,540 m^3/s) and the minimum (1,350 m^3/s) in 1921 (Bondar 1993). The Danube River transports higher volumes of water during spring and at the very beginning of winter, while the discharge is smaller in the summer-autumn period.

The Chilia (Kilia) branch carries the highest water and sediment discharge in spite of the decreasing trend recorded during recent decades: 63.8% in 1950, 63% in 1960, 60.8% in 1970, 59.1% in 1980, and 58% in 1990. The Tulcea branch carries 42% of the total discharge, with an increasing trend from 28% recorded in 1921, this

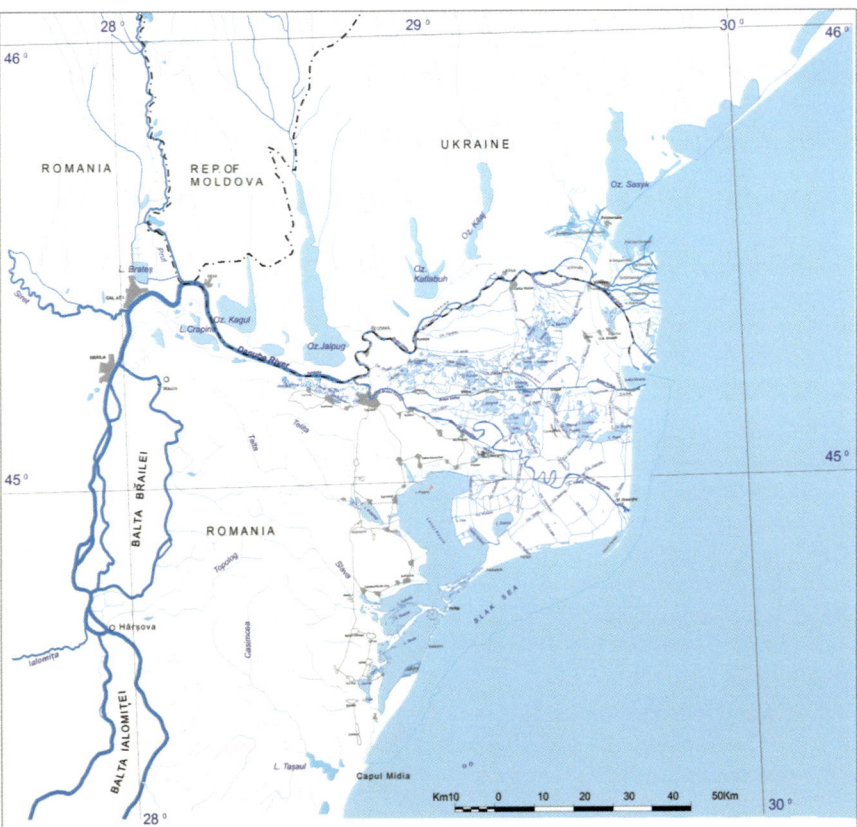

Fig. 1 Transborder wetlands in the Danube Delta, Romania/Ukraine (Map compiled in the Danube Delta National Institute (DDNI))

discharge being distributed to the Sulina branch (19.5%) and Sf. Gheorghe branch (22.5%). The recorded changes in water and sediment discharge distribution among the Danube River branches are the consequences of the main hydrotechnical works that have been implemented in the Danube Delta during the last 150 years, such as the deep navigation channel on the Sulina branch (1862–1902), and the dredging works for cutting the meanders of the Sf. Gheorghe branch (1984–1989). It is estimated that most of the total water discharge of the Danube River (an average of 6,515 m^3/s) flows into the Black Sea through the main Danube branches and only 620 m^3/s (less than 10%) flows through the Danube Delta (350 m^3/s from the Chilia branch, 170 m^3/s from the Sulina branch and 100 m^3/s from the Sf. Gheorghe branch), through the existing channels or over the levees of the Danube branches (Bondar 1993).

The creation and the evolution of the Danube Delta was and still is determined by the sediment discharge of the Danube River in front of the river mouths on contact with marine waters. The annual sediment transport has changed significantly during recent decades: the average annual suspended sediment discharge was 65.7 million tons/year (1921–1960), 41.3 million tons/year (1971–1980) and 29.2 million tons/year (1981–1990).

Wetland Ecosystems

Taking into account the morphological-hydrographical configuration of the area, its flora and fauna communities and long-term human impact, two main categories of ecosystems have been described consisting of 30 types of ecosystems, namely natural and partially man-modified ecosystems, and secondly, anthropogenic ecosystems. The first category – the natural ecosystems and partially man-modified ecosystems, comprises 23 types of ecosystems, ranging from the Danube branches to the beaches including five main groups of ecosystems: water bodies; wetlands; forests, shrubs and herbaceous vegetation; and open places with little or no vegetation. The second category – agricultural and artificial areas (anthropogenic ecosystems), includes seven types of ecosystems: agricultural lands, forest areas, poplar plantation on the river banks, fish farms and settlements – villages and towns. These are described in more detail as follows.

Water bodies, including
 (i) Running waters (lotic ecosystems), represented by the Danube branches, channels and canals with active free water circulation, and canals inside the polders with controlled or no water exchange, and whose margins are dominated by *Phragmites australis, Typha angustifolia,* and floating vegetation when the water speed is low: *Trapa natans, Nuphar lutea, Nymphaea alba, Stratiotes aloides* as well as submerged vegetation: *Potamogeton pectinatus, P. perfoliatus, P. fluitans, Elodea canadensis* and others.
 (ii) Standing freshwaters (lentic ecosystems), represented mainly by lakes with large surface area and/or active water exchange and lakes with reduced water exchange, partially covered with floating vegetation. The depth of these lakes fluctuates in terms of the time that they are influenced by the water levels of the main Danube branches: 3–4 m deep during the spring season and 1.5–2 m deep during the summer-autumn seasons. A few lakes are inside the polders with controlled water exchange and water level. The main characteristic of these lakes is the belt-like distribution of vegetation (from the center of the lakes towards their shores). The submerged vegetation forms a more less compact layer in the waterbody which is suitable for phytophilic fauna; including *Ceratophyllum demersum, Myriophyllum spicatum, M. verticillatum, Potamogeton natans, P. perfoliatus, P. pectinatus* and *P. Crispus* amongst others. The floating vegetation, occurring especially in still, sheltered marginal areas (undisturbed by boats), consists of rooted species, e.g., *Nymphoides peltata, Stratiotes aloides,* amongst free-floating species like *Salvinia natans, Azolla caroliniana, Lemna minor, L. gibba,* and *Utricularia* species.

The emergent vegetation forms a belt dominated by *Phragmites australis, Typha sp., Scirpus lacustris, Carex sp., Sagittaria sagittifolia, Polygonum amphibium, Rumex hydrolapatum, Butomus umbelatus, Thelypteris palustris, Symphytum officinalis*, etc.. The floating reed-beds (locally called "plaur") which often form free-floating islands moved by the wind are dominated by *Phragmites australis* and other species of emergent vegetation.
(iii) Standing brackish and salt waters represented by isolated lakes located in the southern part of the Razim-Sinoie lagoon system. The high level of evaporation (caused by the harsh climate) and lack of freshwater input produces high quantities of salt and sapropelic mud.
(iv) Coastal ecosystems represented by lagoons connected (naturally or controlled) to the sea, the semi-enclosed bays and coastal marine waters. These systems vary according to the respective influences of freshwater from the Danube and saltwater from the Black Sea as the connectivity between water bodies along the coast is more open or controlled. The Danube has a major influence on the coastal marine waters, resulting in a certain degree of mineralisation (brackish waters), turbidity and pollution.

Marginal vegetation (wetlands), including
(i) Flooded reed beds that cover large areas, usually in the depressions limited by fluvial or by mixed (fluvial and marine) sand ridges, with a permanent water layer varying from 0.3 to 0.5 m (during the summer and autumn seasons) to more than 1 m during the spring and early summer time. The vegetation is dominated by *Phragmites australis* accompanied by a great floristic diversity including *Typha latifolia, T. angustifolia, Scirpus radicans, S. lacustris, Carex acutiformis, C. riparia, C. elata, Equisetum palustre, Sagittaria sagittifolia, Iris pseudacorus, Glyceria aquatica, Mentha aquatica, Salix cinerea, Thelypteris palustris, Polygonum amphibium, Ranunculus lingua*, etc.
(ii) Floating reed beds, constructed of the roots (rhizomes) of *Phragmites australis* and other plants. The base of the bed (0.5–1.5 m thick), contains much undecayed organic matter and humus. The dominant plant of this ecosystem is *Phragmites australis* joined by species such as *Typha angustifolia, T. latifolia, Scirpus lacustris, S. radicans, Rumex hydrolapatum, Thelypteris palustris, Sagittaria sagittifolia, Carex* spp., *Stachys palustris, Oenanthe aquatica, Cicuta virosa, Galium palustre, Rorippa amphibia, Lythrum salicaria, Salix cinerea*, etc.
(iii) Riparian willow formations, occurring on sand ridges of fluvial origin covered by a relatively wide diversity of willow forests *(Salix alba, S. fragilis, S. pentandra, S. aurita, S. cinerea)*, where flood conditions last for a longer period. *Populus alba, Alnus glutinosa, Fraxinus angustifolia* also occur. In association with the above-mentioned species, *Amorpha fruticosa, Rubus caesius* and a layer of shorter vegetation occurs, consisting of *Equisetum palustre, Polygonum hydropiper, Stellaria aquatica, Raphanus raphanistrum, Rorippa palustris, Potentilla reptans, Symphytum officinale, Solanum dulcamara*, etc.

Forests, shrubs and herbaceous vegetation, including
  (i) Temperate riverine forests, consisting of mixed oak woods with several tree and bush species: *Quercus robur, Q. pedunculiflora, Fraxinus angustiflora, F. pallisiae, Ulmus foliacea, Populus alba, P. canescens, P. tremula, Salix alba, Tilia tomentosa, T. cordata, Malus silvestris, Pyrus pyraster, Corylus avellana, Prunus spinosa, Crataegus monogyna, Rosa canina, Rhamnus frangula, R. catharticus, Euonymus europaea, Cornus mas, C. sanguinea, Hippophae rhamnoides, Tamarix ramosissima*. Sometimes, their stems are covered by climbing plants like *Hedera helix, Vitis sylvestris, Humulus lupulus, Clematis vitalba, Calystegia sepium and Periploca graeca*.
  (ii) Shrubs and herbaceous vegetation, including very limited sites for shrubs and herbaceous vegetation on calcareous cliffs where the dominant plant species are represented by *Artemisia santonicum, Limonium gmelinii, Halimione verrucifera, Lepidium cartilagineum, Camphorosma annua, Halocnemum strobilaceum, Leuzea salina, Taraxacum bessarabicum*, etc., and steppe meadows dominated by *Thymus zygoides, Festuca callieri, Melica ciliata, Althea rosea* and *Convolvulus cantabrica*.
  (iii) Meadows on loessial plains, developed on the sand ridges of fluvial origin or marine levees. Depending on the specific environmental conditions, the vegetation is dominated in the slightly inclined, frequently flooded marginal areas, by associations with *Agrostis stolonifera* and *Trifolium fragiferum*. During summer, the halophilous species show a certain degree of salinisation *(Juncus gerardi, Suaeda maritima, Puccinella distans)*. In higher horizontal levels that generally remain unflooded, and where the phreatic water table occurs at depths of 2–3 m, apart from the above-mentioned species, *Aleuropus littoralis* (a species indicative of highly saline soils) can be found.
  (iv) Meadows on low marine levees that are up to 2 m high and largely covered by water during the spring time, supporting specific vegetation – the main factor influencing the vegetation being the distance to the Black Sea. In higher places, with half-shifting sands, the dominant species are adapted to sandy rather than saline soils. Typical species of the sandy environment include *Festuca arenicole, Ephedra distachya, Carex colchicum, Elymus giganteus, Apera spica-ventu* (ssp. maritima), *Plantago maritima, P. coronopus, Centaurea arenaria, Convolvulus persicus, Gypsophila perfoliata* and *Stachis maritima*.

Open places with little or no vegetation, including
  (i) Dunes with shifting or partially shifting sands, partially covered with vegetation, found on the beach ridge and dune complexes. Typical vegetation includes *Stipa borysthenica, S. pulcherrima, S. capillata, Convolvulus persicus, Dianthus pontederae, Silene thymifolia, Fumana procumbens, Scutellaria altissima, Trifolim dubium, Papaver maeoticum, Melampyrum arvense, Falcaria vulgaris* and *Botriochloa ischaemum*.
  (ii) Weakly consolidated coastal sand-belts that are frequently flooded by marine waters, especially during stormy weather. The vegetation is dominated by sand-loving species such as *Festuca arenicola, Ephedra distachya, Carex colchicum,*

Elymus giganteus, Apera spica-ventu (ssp. maritima), *Plantago maritima, Centaurea arenaria, Convolvulus persicus, Gypsophyla perfoliata, Plantago coronopus. Stachys maritima and Helichrysum arenarium. Hippophae rhamnoides* is found in some places.

(iii) As part of the above-mentioned coastal sand-belts, beaches are generally barren. If plants occur, they have a patchy distribution, few specimens being grouped together. In most areas the beaches are expanding. Beaches width ranges from 5 to 100 m. Generally, the narrow beach strips (washed by the sea-waves) are built of sandy matter of organic origin and pieces of shells (Gâştescu et al. 1998).

Agricultural land and artificial areas (anthropogenic ecosystems), including polders (areas that have been isolated from the surrounding hydrological system by dikes, with controlled water regime) dedicated to agriculture, forestry, fish farming, mosaic polders and abandoned ones in process of ecological restoration; as well as poplar plantations on river levees, and urban and rural settlements. The first polder in the Danube Delta was built in 1939 for agriculture, and by 1990 the total area of polders had increased to about 96,000 ha, including polders for agriculture, fish farming and forests. The man-made changes brought about through diking, drainage and use of fertilisers and pesticides, transformed the natural ecosystems into simplified systems – monocultures of cereals, poplar plantations of Euro-American species (covering 97% of the total area of forested polders) and intensive fish culture. These polders are considered to be detrimental to the natural ecological balance, causing a decrease in biodiversity (Gâştescu et al. 1998).

In conclusion, indigenous vegetation largely prevails in the Danube Delta although major reclamation works have been carried out for agriculture, fish farming and polders for planted forests. The natural marsh and aquatic vegetation are most widespread in the Danube Delta, covering 398,676 ha in the Romanian part and 35,711 ha in the Ukrainian part of the Delta (Hanganu et al. 2002).

Biodiversity

The Danube Delta supports a wide variety of taxa with 7,402 species recorded to date including, according to "Universal Taxonomic Services" and "Sistema Naturae 2000" (Brands 1989–2005): Plants (2,383 species), Animals (4,026 species), Fungi (145 species), Protozoa (429 species), Chromista (210 species) and Bacteria (209 species). This high number is due to the great variety of terrestrial, wetland and marine habitats (30), which form part of the steppic and pontic bio-regions protected within the Natura 2000 Network, and the proximity of several sub-zones of the Palearctic faunal region (e.g., Mediterranean, Pontic, Eurasian) (Management Plan of DDBR: Romanian Government 2015).

Invertebrates form by far the greatest part of the fauna in the Danube Delta, with a total of more than 3,000 species. Of these, there are some 255 species of worms, 84 species of molluscs, 223 species of crustaceans, 168 species of arachnids, 98 species of miriapods and 2,260 species of insects. To date, 38 species of invertebrates have been described for science from the Danube Delta, including a worm *Proleptonchus deltaicus,* 5 species of arachnids, and 32 species of insects such as *Isophya dobrogensis, Dialulinopsis deltaicus,* and *Homoporus deltaicus.* 26 species of insects are known only from the Danube Delta (Management Plan of DDBR: Romanian Government 2015).

The fish fauna of the Danube Delta is represented by 135 species belonging to 30 families. The majority of these are freshwater species, the others being euryhaline species that live in the Black Sea and visit the delta mainly during the breeding season (including the sturgeons: *Huso huso, Acipenser guldenstaedti, A. stellatus, A. nudiventris,* the Danube Shad *Alosa pontica,* etc.). About one third of these species are exploited by intensive commercial fisheries.

The amphibians are represented by 10 species of frogs and toads, including *Rana ridibunda, Bombina bombina, Hyla arborea, Pelobates fuscus, Bufo bufo, Bufo viridis* and two species of newts, *Triturus dobrogicus* and *T. vulgaris.* The reptiles include terrapins, lizards and snakes. All the amphibians and reptiles are protected in Europe through the Bern Convention. The Danube Delta is most famous for its birds, with a recorded total of 341 species (out of about 520 in Western Europe as a whole). The site is of global importance for breeding populations of many waterbirds such as great white pelican *Pelecanus onocrotalus,* Dalmatian pelican *P. crispus* and pygmy cormorant *Microcarbo pygmaeus.* In addition there are important colonies of Eurasian spoonbill *Platalea leucorodia,* and several breeding pairs of white-tailed eagle *Haliaeetus albicilla.* The Danube Delta region is a major migratory stop-over area in both spring and autumn for several million birds, especially ducks, white stork *Ciconia ciconia* and various birds of prey. In winter, the region also hosts large flocks of swans and geese, including at times almost the entire world population of the threatened redbreasted goose *Branta ruficollis* (Goriup et al. 2007). There are 54 species of mammals including some terrestrial species of European importance such as Eurasian otter *Lutra lutra* and Eurasian mink *Mustela lutreola,* stoat *Mustela erminea,* European wildcat *Felis silvestris,* as well as marine species: short-beaked common dolphin *Delphinus delphis* and bottle nose dolphin *Tursiops truncatus.*

Several invasive species have been recorded, especially in the Black Sea: warty comb jelly *Mnemiopsis leidyi,* soft-shell clams *Mya arenaria,* the bivalve mollusc *Scapharca inaequivalvis,* the crustacean *Balanus improvisus* and the gastropod mollusc *Rapana venosa.* The flowering shrub *Amorpha fruticosa* (Fabaceae) is invading the levees of Danube branches.

Conservation Status

In 1991, an area of 647,000 ha of wetlands was declared as a Wetland of International Importance under the Ramsar Convention (Ramsar Site) on the Romanian side of the Danube Delta region, and later on in 1995, several wetlands with a total area of 60,800 ha were declared as a Ramsar Site on the Ukrainean side of the Delta. Prior to this designation, several protected areas existed within the Danube Delta region in both sides of the border covering 626,403 ha, namely:

(i) Letea Forest Nature Reserve, 500 ha, established in 1939 (Romania);
(ii) Danube Delta Biosphere Reserve (DDBR), 580,000 ha, established in 1990, including Letea Forest Nature Reserve (Romania);
(iii) Kilia Delta Reserve (1 km wide coastal strip of the Kilia Delta), established in 1967 (Ukraine);
(iv) Dunaiski Plavni Reserve, 14,851 ha, established in 1981 (Ukraine);
(v) Danube Biosphere Reserve (DBR), 46,403 ha, established in 1998, including both the Kilia Delta Reserve and the Dunaiski Plavni Reserve (Ukraine) (Baboianu et al. (2004))

The management of the Danube Delta in Romania is mainly based on the special law, 82/1993 (updated in 2011), concerning the establishment of DDBR, the law 265/2006 for environment protection, and the law 49/2011 concerning the protection of biodiversity and protected areas. According to existing legislation, visitor access is allowed in the Danube Delta based on entrance permit, and on special permit to the core areas (strictly protected areas) for scientific purposes only. Economic activities are allowed with the authorisation of the Reserve Authority. Most of the land is state owned (82%), or owned by the local authorities (17%). Only 1% of the total land area is privately owned. Wetlands cover about 40% of the total area of the Danube Delta. The management of the Danube Delta in Ukraine is based on the Law of Nature and Reserve Fund of Ukraine 2456-XII/1992 and two Decrees of the President of Ukraine. Entry to the reserve and traditional economic activities (reed harvesting, commercial fishing, tourism, cattle grazing, etc.) in the reserve teritory (which is totally state owned) are allowed based on permits issued by the reserve administration (Baboianu et al. 2004).

Ecosystem Services

In addition to supporting a high level of biodiversity, the Danube Delta Region provides many ecosystem services. It has an important effect on water quality, and nutrient retention, especially for the Black Sea ecosystems. Moreover, it provides extensive economic and environmental benefits to the entire region: the socio-economic benefits of the wetlands to local communities living in and around the Danube Delta are very important. Practically, all aspects of the lives of the delta's 200,000 inhabitants are related to water in one way or another. The Danube River and its branches, and several canals are the major sources of water for industrial, agriculture (irrigation) and domestic use for local communities. They are also used for navigation by both commercial and public ships and vessels, boats and canoes. The main natural resources represented by fish, reed, pasture, natural and planted forests support traditional economic activities undertaken by local communities. The fishery is far the most exploited resource, with about 7,000 t per year supporting commercial, subsistence and recreational fishing, mostly consisting of fresh water species. The reed beds have the potential to produce about 40,000–50,000 t of reed per year, and the pastures support grazing sheep, cattle, pigs, and horses. The use of reed has a long history in the Danube Delta, with local people building shelters for fishermen, refuges for cattles and sheep, roofs for houses, fences for yards, etc. When used for thatched roofs elsewhere in Europe, it would imply that significant income is obtained. The paper industry used the reed as raw material for producing paper and cellulose, until the collapse of this activity in the late 1980s. While not permitted on the Romanian side, hunting is permitted on more than 60% of the Ukrainian part of the Danube Delta. Agriculture is practiced, both in polders for cereal crops (wheat, barley, maize), sunflowers, and, on a smaller scale, for family needs (vegetables, fruit trees, vineyards).

Threats

The first major human intervention in the Danube Delta started in the 19th century with measures that were taken to improve navigability and to support agricultural exploitation in the Delta. The consequences of these interventions, however, remained without major impact on the homogeneous natural integrity of the river and delta. A development program for building dams and canals necessary for establishment of agricultural polders, reed harvesting, fishing and fish farming, and silviculture started in 1960, and significantly altered the network of watercourses between the main branches, as mentioned in the section onhydrology. Many areas were cut off from the Danube's natural water level fluctuations, and their water balance and the exchange of waters in the Delta was completely disturbed. This also disturbed the Delta's interdependent ecosystems. The complex Delta transformation programme adopted in Romania in 1983 foresaw more intensive land use, which necessarily implied the erection of further dams and large-scale drainage works (Gomoiu 1996). As a consequence, the natural habitats of numerous plant and animal species were reduced and partly destroyed. The Danube Delta wetland complex with its broad reed areas (180,000 ha) which act as a filter for the ecological balance in the Black Sea, had been considerably affected. The cutting of the meander bends of the Sf. Gheorghe Branch in 2000, in Romania, was followed by the Deep Navigation Route development (Bystroe) in Ukraine, in 2004, affecting the habitats and the hydrology of the River. More than 96,000 ha of wetlands and reed beds, many of them acting as important spawning areas for cyprinids or nesting areas for birds, were drained for agriculture, fish farming, and artificial forests, as mentioned above (Agricultural land and artificial areas). The effects of the changes within the delta was amplified by the impacts of human activity outside the area: land reclamation works in the flood plains of the River Danube and its tributaries, and the construction of dams and sluices for regulating navigation and for hydropower plants, etc., eliminated the important areas of wetland habitats and reduced the sediment load reaching the delta. Wetland restoration in several polders (totalling about 15,000 ha), works for improving the water circulation through the delta (canal closures and calibration), and waste water treatment plants are some of the measures taken to mitigate human impacts.

Future Challenges

The management of the Danube Delta should take into consideration several needs for the short and medium terms. In the short term, the implementation of a wetland restoration program to increase the natural flooded area in abandoned polders for agriculture and fish farming should be continued. In addition, measures are needed to reduce the impacts of the more ecologically damaging economic activities (including navigation and related hydrotechnical works, over-exploitation of natural resources (especially fish)) and other landuses according to the carrying capacity of the ecosystems and pollution control. The living standards of local communities should be improved through the extension of drinking water supply, waste water treatment networks, waste management, green energy use, and the involvement of the local communities in the direct management of the wetlands and their resources is another urgent need.

In the short and medium terms, the development of transboundary/transnational cooperation between Ukraine and upstream riparian Danubian countries should be taken into consideration for implementing regional cooperation programs such as the *Lower Danube Green Corridor* between Romania, Bugaria, Republic of Moldova and Ukraine, the *Trilateral Agreement for joint management of transboundary protected areas* in Romania, Ukraine and Republic of Moldova, and the development of the *Network of Protected Areas of the Danube River – DANUBEPARKS* within the EU Danube River Strategy.

The joint projects implemented during the last decade in the Danube Delta Transboundary Biosphere Reserve in Romania and Ukraine, and the Lower Prut River Reserve in Romania and Republic of Moldova, were focused on developing cooperation in the Lower Danube Region to develop monitoring, biodiversity and habitat conservation, wetland restoration, public awareness and to establish the Transboundary Biosphere Reserve of the Danube Delta and Lower Prut River in Romania, Ukraine and the Republic of Moldova.

References

Baboianu G, Munteanu I, Voloshkevich O, Zhmud M, Fedorenko V, Nebunu A, Munteanu A. Transboundary cooperation in the nature protected areas in Danube Delta and Lower Prut – management objectives for biodiversity conservation and sustainable development. Dobrogea: Editura; 2004. p. 45–55, 85–94.

Brands SJ, Compiler. *Systema Naturae 2000*. Amsterdam, The Netherlands; 1989–2005. http://sn2000.taxonomy.nl/

Bondar C. Hydrology in the Danube Delta case study (Hidrologia în studiu de caz Delta Dunării). Tulcea: Analele Institutului Delta Dunării; 1993.

Chernichko J, Overmars W, Nesterenko M. A vision for the danube delta, Ukraine. Vienna/Odessa: WWF-Danube Carpathian Programme Office/Ukraine WWF-Project Office in collaboration with WWF-Netherlands; 2003.

Gâştescu P, Oltean M, Nichersu I, Constantinescu A. Ecosystems of the of the romanian danube delta biosphere reserve. Lelystad/The Netherlands: RIZA; 1998.

Gomoiu M-T. Facts and remarks on the danube delta, danube delta – black sea system under global changes impact. Bucharest-Constanta/Romania: National Institute of Marine Geology and Geo-ecology; 1996. p. 70–82.

Gomoiu M-T, Baboianu G. Some aspects concerning the ecological restoration in the Danube Delta Biosphere Reserve (DDBR). In: Analele științifice ale Institutului Delta Dunării. Tulcea: Danube Delta National Research Institute; 1992. p. 259–64. In Romanian; English summary.

Goriup P, Baboianu G, Chernichko J. The Danube Delta: Europe's remarkable wetland. Br Birds. 2007;100(4):194–213.

Hanganu J, Grigoraș I, Ștefan N, Sârbu I, Dubyna D, Zhmud E, Menke U, Drost H. Vegetation of the biosphere reserve Danube Delta with transboundary map. Lelystad/The Netherlands: RIZA; 2002.

Romanian Government (HG) Decision nr. 763/2015, Management plan and the rules of the danube delta biosphere reserve authority (Planul de management și Regulamentul Rezervației Biosferei Delta Dunării), Annex 3; 2015.

Rhine River Basin

74

Daphne Willems and Esther Blom

Contents

Introduction	924
Historical Background	924
Restoration of the River Rhine	926
Key Principles for River Restoration	927
Future Challenges	929
References	930

Abstract

With a length of 1,320 km and a river basin of 168,757 km^2, the Rhine is the largest river in Northwest Europe. On its way, the Rhine passes through eight countries: Switzerland, Liechtenstein, Austria, Germany, France, Luxembourg, Belgium, and the Netherlands. In the nineteen-eighties, the water and spatial quality of the river were very poor. A chemical disaster stimulated international cooperation at river basin level. Restoration measures started in the nineties, with success. Key principles used are Working with Nature (Room for the River), Integrated approaches, strategic partnerships and flagship species. The latter has been used to bring parties from different backgrounds together, and unite them in their efforts. The working with Nature approach involves the entire river basin: from the mountains, where water can be stored near the source in natural sponges; the middle reaches, where the Room for the River concept can be applied; and the Delta, where natural processes can keep pace with a rising sea level. Although many challenges are still there, nowadays the salmon and sturgeon are back in the Rhine.

D. Willems · E. Blom (✉)
WWF-Netherlands, Zeist, The Netherlands
e-mail: dwillems@wwf.nl; esther.blom@ark.eu

© Springer Science+Business Media B.V., part of Springer Nature 2018
C. M. Finlayson et al. (eds.), *The Wetland Book*,
https://doi.org/10.1007/978-94-007-4001-3_187

Keywords

ICBR · International cooperation · Living Rivers · River basin approach · River restoration · River Rhine · Smart Rivers · Working with nature

Introduction

With a length of 1,320 km and a river basin of 168,757 km^2, the Rhine is the largest river in Northwest Europe. The Rhine has two main sources, both in Switzerland. The resulting small streams Vorderrhein and Hinterrhein merge near Reichenau, becoming the Alpen Rhine. From there on, the Rhine has the dimensions of a river, with an average discharge of 2,490 m^3/s (Fairbridge and Herschy 1998). On its way, the Rhine passes through eight countries: Switzerland, Liechtenstein, Austria, Germany, France, Luxembourg, Belgium, and the Netherlands (Fig. 1).

Being a glacier-fed river, the discharge in the upper reaches of the Rhine is highly variable. Lake Konstanz (Bodensee) buffers this huge variation, being able to store four times the annual discharge of the inflowing rivers (Fairbridge and Herschy 1998). The outflowing river from the Bodensee is named the Hochrhein (High Rhine), subsequently known as the Oberrhein (Upper Rhine) after Basel, the Mittelrhein (Middle Rhine) after Mainz and the Niederrhein (Lower Rhine) near Bonn. Finally, the river enters the delta of the Netherlands. Just after passing the border, the Rhine splits into three branches: the Waal, Nederrijn, and IJssel; the first two heading west, to end up in the North Sea near Rotterdam, the latter heading north towards Lake IJsselmeer connected with the Wadden Sea. In the western delta, the Rhine branches form a braiding pattern with the branches of the River Meuse.

Historical Background

The River Rhine is highly influenced by human interventions (Vorogushyn and Merz 2013). Already in the Middle Ages, the first dikes had been constructed. Nowadays, the position of the river bed and the floodplains are fixed as a result of human intervention.

Besides the construction of dikes, groynes were built in the river bed to gain more agricultural land and to steer the water in the desired direction, away from the dikes. Meanders were cut off, to avoid ice dams ramming the square dikes. Straightening of the river flow, through canalization, shortened shipping distances as well, while the groynes kept this shipping route fixed. Weirs and dams were constructed to assure water availability during low water discharge and to generate hydroelectricity. At the same time, in the hinterland wetlands were drained. Drainage of the beds of valley streams has drastically reduced water storage through the natural sponge effect in the source areas of the Rhine.

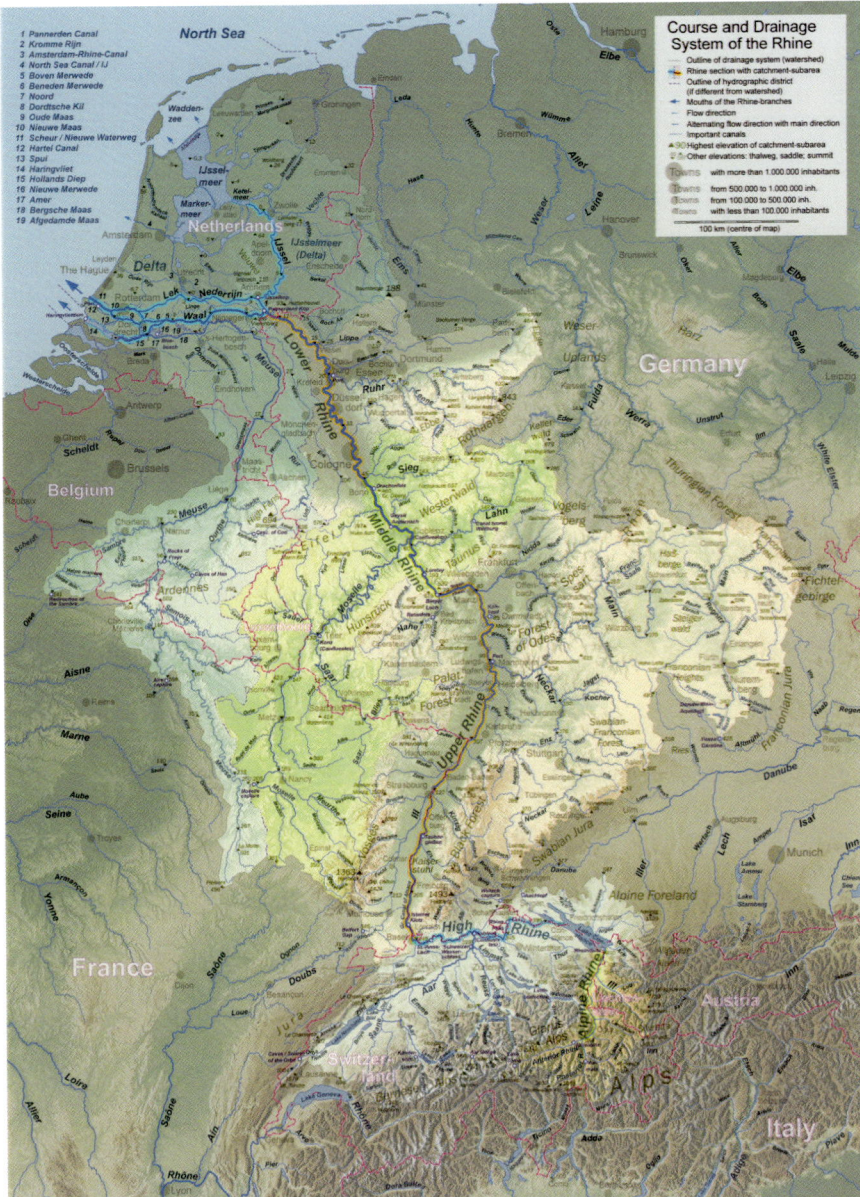

Fig. 1 Course and drainage system of the Rhine (This file is downloaded from Wikipedia at: https://commons.wikimedia.org/wiki/File:Flusssystemkarte_Rhein_04.jpg#filelinks. The file information indicates that it can be reused under Creative Commons: See https://commons.wikimedia.org/wiki/File:Flusssystemkarte_Rhein_04.jpg#filelinks)

With all these modifications, the characteristics of the river system have changed entirely (Vorogushyn and Merz 2013). The spatial variation of the river system was drastically reduced (it did not entirely disappear, nor did the natural processes), as was the influence of natural processes that formerly (re)created variety. Natural diversification provides a wide range of habitats for flora and fauna, such as deep and shallow waters, fast and slow flowing stream stretches, high and low energy sections, sand and clay soils. The modified river Rhine lacks this variation, while the weirs and dams block the river's connectivity (Brevé et al. 2014). The river discharge has become more reactive to these interventions. As a result of the increased modification and intensified use of the river, its forelands and hinterlands, problems began to arise in the second half of the twentieth century.

In 1986, a chemical disaster near Basel (Switzerland) took place. The chemicals discharged into the Rhine caused massive kills of benthic organisms and fish, particularly eels and salmonids. The disaster had an important impact on transboundary cooperation, improving the willingness for international cooperation. The International Commission for the Protection of the Rhine (ICPR), existing since 1950, began to play a more important role. The Rhine Convention (1999) was an important step in developing cooperation, and the Rhine Action Plan and currently "Rhine 2020" have provided means for achieving the targets of this convention: improve water quality, flood prevention, and ecological restoration. International cooperation has proven to be an important success factor for the recovery of the Rhine river basin (Giger 2009).

Restoration of the River Rhine

During the last decades, considerable effort has been put in the restoration of the River Rhine, resulting in the improved quality of the entire Rhine system (ICPR 2013). There have been strong improvements in water quality. An increase in the structural diversity of river banks has been achieved along more than 100 km of the mainstream of the Rhine; since 2000, bypasses have been constructed at 481 weirs. Flood levels and flood damage risks have been lowered, awareness raised, and the flood forecasting system improved.

These restoration efforts have had very positive effects. A Dutch study concerning the return of riverine species during the last 20 years in the Netherlands has shown spectacular results (Kurstjens and Peters 2012). The number of returned, formerly rare plant species in the Dutch part of the River Rhine is spectacular, including species inhabiting sandy banks and dunes (Fig. 2), pioneer species, aquatic species, and riparian forests. Mammals like Eurasian beaver *Castor fiber* and Eurasian otter *Lutra lutra*, many breeding birds, fish species of running waters, grass snake *Natrix natrix*, dragonflies, and grasshoppers all profited from the new nature areas (improved spatial variety), natural management (natural grazing), partly restored natural processes (erosion and sedimentation), and improved water quality.

Fig. 2 River dunes: restored habitat (river dunes) in the Dutch part of the Rhine (the Waal) (Photograph by Twan Teunissen/ARK Natuurontwikkeling © Rights remain with the author)

Key Principles for River Restoration

The restoration of the River Rhine has been based on several key principles. This chapter focuses on the spatial restoration principles; while measures to improve water quality and water quantity (flood prevention) are linked to this spatial restoration but not discussed in detail.

1. *Working with nature:* Instead of fighting against the forces of nature, it is wiser to work together with these powers. Restoration efforts should, therefore, be based on the features of the natural river, the "DNA" of the river, making proper use of its natural characteristics. To be able to do so, it is essential to understand the river system as a whole, from the sources to the mouth and from the main river bed to the smallest tributaries. To start with, the river should have enough space to effectuate it's natural power and perform the processes that naturally take place: only then a river can be a 'living river'. Using and restoring natural processes like erosion and sedimentation can help to build a sustainable, natural river system and provide an environment for many riverine species (www.smartrivers.nl).
2. *Integrated approaches, strategic partnerships:* The waters of the River Rhine are used for shipping, drinking water production, irrigation, energy generation, fishing, industrial cooling water, waste water disposal and recreation, amongst others. Its floodplains are also used in many ways, including mineral extraction (sand, gravel, and clay), agriculture, housing, and recreation. There are many different interests in one river, but also many possible partners to join forces with.

As an example, the concept of "Living Rivers" has been adopted by the Dutch Ministry of Infrastructure and Water Management, resulting in the national program "Room for the River" (Roomfortheriver.nl). As a result, many side channels have been restored along the Rhine branches in the Netherlands, contributing to flood safety and nature development.

3. *Flagship species:* The Rhine has several flagship species or "ambassadors" for natural restoration. The Atlantic salmon *Salmo salar* is a migratory fish that symbolises the need for maintaining the aquatic connectivity of the river, from source to sea and vice versa. The European sturgeon *Acipenser sturio* is the emblem of the connection between the River Rhine and the sea and a healthy estuarine system to grow up in (Vis and De Bruijn 2012). Providing a "face" to a target contributes towards the motivation of the different stakeholders involved.

Concepts

To restore a river system, the entire river basin has to be taken into account, from source to sea.

Storing water near the source: Marshes and peatlands are able to absorb rain water during heavy rain fall and releasing it during dryer periods (Van Winden et al. 2004). This "sponge effect" can buffer peaks and shortages in water, just like the Bodensee does. By restoring spongy nature areas, currently often heavily drained and used for agriculture, local nature development can contribute to the restoration of the entire river system. These areas can become highly valuable nature areas; the total surface of marshes and peat lands has been decimated in the last century, causing the accompanying plant and animal species to become rare. By storing water near its source in high wetlands, natural woodland and wide, boggy valleys, flood waves can be moderated and slowed down, resulting in decreased flood problems downstream.

Living Rivers: The large-scale development of side channels and peeling off the clay layer in the floodplains serves a number of ecological and economic interests (Helmer et al. 1992). In the subsoil along the Rhine, many old river beds that have been abandoned can be found. These structures can be restored. Living Rivers was applied first in the Millingerwaard, a restored floodplain along the Rhine in the Netherlands. By extracting clay and sand following the underlying geographic relief, the natural pattern and structure of the riverine landscape has been uncovered. Here, through careful planning, commercial mineral extraction companies contributed towards nature development, river safety, and a highly valued recreation area. In the Millingerwaard, nature management is achieved through the grazing of large herbivores. Cattle and horses roam the area year round (Fig. 3), and together with beavers, deer, and geese, they control the growth of vegetation and improve spatial variety, creating microhabitats for many species. The herds of animals live in natural functioning social groups.

Additionally, natural disasters are imitated from time to time, as these do not occur anymore in the highly regulated Dutch rivers. Then part of the alluvial forest is removed, the wood itself sold as biomass for green energy. These "disturbances" help to regenerate riverine pioneer conditions, leading to increased biodiversity.

Fig. 3 Koniks: grazing by large herbivores (for example Konik horses) in the restored nature areas contributes to a diversification of the landscape (Photograph by Twan Teunissen/ARK Natuurontwikkeling © Rights remain with the author)

Moreover, in this way flood safety is assured, leaving sufficient space for the river to flow during peak discharges.

The Millingerwaard has become a very popular recreational nature area, where visitors are allowed to roam around freely. The experienced freedom and nature perception is unlike any other natural park in the Netherlands.

Growing with the sea: The soil in the delta of the Netherlands is subsiding, and the sea level is rising due to climate change. People tend to look for technical solutions like higher dykes, but in the long term they will only contribute to a further distortion in land growth. An attractive alternative exists: "Growing with the sea," restoring the natural resilience of the delta (WNF and RIKZ 1996). The Netherlands is a delta by nature. This fertile area is a key driver of the economy, but it needs space for natural maintenance processes. By providing natural processes like siltation and marshland development in the space they require, it is possible to break the current vicious circle of fresh water shortages and subsiding ground.

Future Challenges

Much has been achieved in the Rhine river basin during the last decades, but much remains to be done. Spatial measures can play a very important role in the ongoing restoration of the Rhine river basin, benefiting the landscape and riverine nature values and at the same time its users, inhabitants, and visitors. These measures include the following:

First, to cope with ongoing climate changes bringing increased, heavier rainfall and longer dry periods, it is necessary to restore the hydrological resilience of the river system. Starting at the source, the restoration of natural spongy areas such as peatbogs in the middle mountains (the central hilly and mountainous part of the Rhine basin, such as the Eiffel Range) will deliver important nature areas and a more sustainable river basin.

Secondly, making the room required for riverine processes has been widely accepted in the Netherlands, and partly in Germany and Switzerland. Because of climate change effects, it is expected that both the intensity of droughts as well as periodic high discharges of the river will increase. It is highly important that the natural characteristics or "the DNA" of the river remain the starting point for the spatial interventions that are needed to cope with such expected increased variation in river discharge.

Thirdly, also in the downstream direction, a lot remains to be done. The southwestern estuary of the Rhine-Meuse system, which is the entry and exit point for many fish species of the Rhine system, requires substantial restoration. Migratory birds and fish need the brackish and nutritious estuarine conditions and an open connection to migrate between the river and the sea.

Finally, further upstream, there are barriers for migratory fishes, both in the large tributaries and in the main stream of the River Rhine. In particular, the hydropower dams of the Canal d'Alsace are not yet passable to fish. However, good examples of ecologically optimized hydropower projects are available to inform improvements (Damm et al. 2011). By installing functional fish passages and preventing the development of new barriers, it is possible that salmon will be able to access their original spawning grounds in Switzerland and Germany.

References

Brevé NWP, Vis H, Houben B, de Laak GAJ, Breukelaar AW, Acolas ML, de Bruijn QAA, Spierts I. Exploring the possibilities of seaward migrating juvenile European sturgeon *Acipenser sturio* L., in the Dutch part of the River Rhine. J Coast Conserv. 2014; 18:131–143.

Damm C, Dister E, Fahlke N, Follner K, König F, Korte E, Lehmann B, Müller K, Schuler J, Weber A, Wotke A. Auenschutz – Hochwasserschutz – Wasserkraftnutzung. Beispiele für eine ökologisch vorbildliche Praxis. Naturschutz und Biologische Vielfalt Heft 112. Bonn: Bundesamt für Naturschutz. 321 S; 2011.

Fairbridge R, Herschy R, editors. Encyclopedia of hydrology and lakes: SpringerReference. Berlin/Heidelberg: Springer; 1998.

Giger W. The Rhine red, the fish dead – the 1986 Schweitzerhalle disaster, a retrospect and long-term impact assessment. Environ Sci Pollut Res. 2009;16 Suppl 1:S98–111.

Helmer W, Litjens G, Overmars W, Barneveld H, Klink A, Sterenburg H, Janssen B. Living rivers. Zeist: WWF; 1992.

International Commission for the Protection of the Rhine (ICPR). The Rhine and its catchment, an overview – ecological improvement, chemical water quality, survey on the action plan on floods; 2013.

Kurstjens G, Peters B. Rijn in Beeld, deel 1: ecologische resultaten van 20 jaar natuurontwikkeling langs de Rijntakken (Ecological results of 20 years of nature development along the river Rhine branches). Projectgroep Rijn in Beeld. Beek-Ubbergen/Berg en Dal: Kurstjens Ecologisch adviesbureau/Bureau Drift; 2012. p. 136.

Van Winden AL, Overmars W, Braakhekke W. Storing water near the source. WWF: Nijmegen; 2004.

Vis H, De Bruijn QAA. Onderzoek naar het migratiegedrag van de Atlantische steur (Acipenser sturio) in de Rijn (Investigation on the migratory behaviour of the Atlantic Sturgeon in the Rhine). Sportvisserij Nederland: Nieuwegein, the Netherlands; 2012.

Vorogushyn S, Merz B. Flood trends along the Rhine: the role of river training. Hydrol Earth Syst Sci. 2013;17:3871–84. doi:10.5194/hess-17-3871-2013.

WNF and RIKZ. Growing with the sea; creating a resilient coastline. Zeist: Worldwide Fund for Nature (Netherlands) and National Institute for Coastal and Marine Management RIKZ; 1996.

Volga River Basin (Russia)

75

Harald J. L. Leummens

Contents

Introduction 934
Climate 935
Hydrology 936
Landscapes 937
Biodiversity 939
Human Impacts 940
Threats and Future Challenges 942
References 943

Abstract

The Volga River is the largest and longest river in Europe, with a length of 3,500 km and a basin area of about 1,360,000 km^2. The Volga River basin is located in European Russia, is inhabited by 57 million people and harbors half of the country's industrial and agricultural production. The Volga River basin is characterized by a variety of biomes: coniferous, deciduous and mixed forest, steppe, semi-desert and desert, as well as wetlands, largely conditioned by the wide variety in climate conditions. The average total annual discharge of the Volga River at Volgograd is 252 km^3 (8,076 m^3/s), showing large inter-annual variations and long-term trends of change related to climate variation. River flow and related sediment and material transport has been significantly changed due to reservoirs constructed throughout the basin, free flow conditions remain only in the river's headwaters and its most downstream section. Water resources are used in agriculture, for industrial production, transportation, energy generation and as drinking water for the population. While in the Volga basin still significant areas of

H. J. L. Leummens (✉)
Water and Nature, Heerlen, The Netherlands
e-mail: harald.leummens@gmail.com

© Springer Science+Business Media B.V., part of Springer Nature 2018
C. M. Finlayson et al. (eds.), *The Wetland Book*,
https://doi.org/10.1007/978-94-007-4001-3_88

biodiversity importance remain, with a rich and typical variety in terrestrial and aquatic flora and fauna diversity, human activities throughout the basin, mainly agriculture, industry and urbanization, have significantly impacted upon ecosystems and biodiversity. A large network of protected areas has been designated at the national and provincial level, including important wetlands, aiming to conserve the Volga basin's natural landscapes and rich biodiversity. Challenges for the future include addressing remaining and present pollution from solid and liquid, point and diffuse sources, improving water quality and water use efficiency, to be addressed through an effective and efficient integrated water resources management approach, including appropriate legal reforms, intersectoral and interregional coordination and cooperation, monitoring and enforcement, for the benefit of people and nature.

Keywords
Volga · Rivers · Europe · Regulated flow · Freshwater

Introduction

The Volga River is the largest and longest river in Europe, with a total length of 3,500 km and a watershed basin area of about 1,360,000 km^2, completely located in Western – European – Russia (Fig. 1). Although the basin covers only 8% of the Russian Federation, or 62% of European Russia, it contains 40% of the country's population, or 57 million people, an average of 42 people per km^2 (Naidenko 2003). Industrial activities in the basin account for 45% of the national output, while agriculture contributes 50% of the country's food production (Leummens 2005a). Constructed large reservoirs provide irrigation water and hydroelectric power, and the navigable waterway network, including the Moscow Canal, the Volga-Don Canal, and the Volga-Baltic Waterway, connects the White Sea, the Baltic Sea, the Caspian Sea, and the Black Sea.

The Volga River is part of the Caspian Sea closed drainage basin. The river originates at 225 m above sea level (asl) in the Valdai Hills, located halfway between Moscow and Saint Petersburg. The Volga maintains an easterly direction until the city of Kazan, dissecting the Valdai and Central Russian uplands and passing cities like Tver, Dubna, Rybinsk, Yaroslavl, and Nizhny Novgorod. Turning south, the river passes Ulyanovsk, Tolyatti, Samara, Saratov, and Volgograd, to discharge into the Caspian Sea about 200 km south of the city of Astrakhan at 28 m below sea level. Overall, 11 of Russia's 20 largest cities, including Moscow, are located in the Volga River basin. The major tributaries are the Oka, the Belaya, the Vyatka, and the Kama, each exceeding 1,000 km in length and their catchment areas each exceeding 100,000 km^2.

The Volga River basin includes a variety of biomes, from southern taiga forests and mixed coniferous-deciduous forests in the north to deserts in the southern North Caspian Lowland. Intermediate biomes include forest steppe, steppe, and semideserts. Intrazonal biomes include the Volga-Akhtuba floodplain, located between Volgograd and Astrakhan, and the Volga Delta (Litvinov et al. 2009).

Fig. 1 The Volga River basin (Image © Karl Musser, permission provided through Creative Commons Attribution-Share Alike 4.0 International Public License http://creativecommons.org/licenses/by-sa/4.0/legalcode. Accessed 15 Dec 2015)

Climate

The Volga basin is located in the Atlantic continental European climate region, conditioned by eastward drifting North Atlantic Ocean air masses. In the southern part of the basin, the climate is also influenced by continental air masses from southern Europe and Kazakhstan, causing more frequent anticyclonic weather conditions and decreasing average precipitation. From north to south, the degree of continentality increases from 30% to 80%, the average yearly precipitation decreases from 750 to 150 mm, the average snow cover depth decreases from 60 to 3 cm, and its duration from 240 days to 30 days (Fig. 2). The period with above zero air temperatures decreases from 180 days to 110 days, while the vegetation growth

Fig. 2 A typical stretch of the Lower Volga floodplain near Volgograd, showing a mosaic of wetland habitats that supports rich biodiversity (Photo credit: Harald Leummens © Rights remain with the author)

period increases from 150 days to 220 days. However, due to moisture deficits, the total productivity of (agro)-landscapes in the southern part of the basin is 2–2.5 times lower than in the northern part (Golosov and Belaev 2011).

The whole length of the Volga River can be affected by ice cover formation. Under natural conditions, a stable ice cover develops in November, persisting for 120–140 days. Following the construction of artificial reservoirs, the ice cover on average develops 3–5 days earlier, while its persistence has also increased by a few days. Also, the maximum thickness of the ice cover in reservoirs is larger than in the natural river channel. Over the last decades, the duration and thickness of the ice cover has decreased notably, related to generally warmer winter weather conditions (Golosov and Belaev 2011).

Hydrology

The hydrological network of the Volga River basin includes 151,000 streams with a total length of 574,000 km, of which about 200 directly enter the Volga main stream. Left tributaries are more numerous and carry more water than right tributaries. The largest tributaries include the Kama River from the left (4,100 m^3/s) and the Oka River from the right (1,170 m^3/s). Downstream of Kamyshin (Volgograd Province), no significant tributaries enter the main stream, and before entering the Caspian Sea, the river loses 2% of its water (Golosov and Belaev 2011). Significant riverbed slopes occur in the upper reaches of rivers in the Volga Upland and Central Russian Upland as well as on the western slopes of the Ural Mountains, where river stretches occasionally may appear mountainous. However, the majority of the rivers in the basin are characterized by slow currents and meandering channels, with floodplains rich in lakes and oxbow channels.

The average total annual discharge of the Volga River at Volgograd between 1880 and 2009 was 252 km^3 (8,076 m^3/s). Although typical large interannual variations occur, the total annual discharge shows a trend of gradual decrease until 1977, after which a trend of increase is observed. These long-term trends are assessed to be related to cyclone activity above European Russia, causing continental-scale variations in precipitation and evaporation. The cyclone activity in turn is shown to have a dependency relationship with the North Atlantic Oscillation (Leummens 2005b).

The Volga is a typical plain-type river, predominantly fed by snowmelt (60%). Additional sources of water include groundwater outflow (30%) and rainfall (10%). The specific combination of water sources contributes to the river's discharge regime under natural conditions, characterized by a distinct spring flooding period (April–July) related to snowmelt, followed by low-water periods in summer, autumn, and winter (Golosov and Belaev 2011).

Today the flow regime of the Volga main stream is dominated by a series of artificial reservoirs, constructed between 1843 and 1980, dominating the majority of the river's length. Natural hydro-morphological conditions, such as bifurcating streams dissecting terrestrial floodplains, only remain in the river's headwaters and the most downstream reaches of the Volga-Akhtuba floodplain and Volga Delta, where the downstream area is impacted by an artificially installed discharge regime. Reservoirs mainly serve to control seasonal discharge – hardly any impact on the total annual discharge is observed, although accordingly, significant changes in the intraannual hydrograph are apparent (Leummens 2005a).

Before discharge regulation, the Volga River annually transported large volumes of sediments (26 × 10^6 t) and dissolved matter (45 × 10^6 t) into the Caspian Sea, equal to 19 t/km^2/year and 33 t/km^2/year, respectively. Sediment entrapment in the reservoirs decreased the suspended sediment yield at the delta outlet to 8 × 10^6 t/year. At the same time, pollution caused a dramatic increase in dissolved material yield, to 65–70 × 10^6 t/year. Sediments mainly originate from agriculturally developed drainage basins in the forest-steppe and steppe zones, where locally, sediment yield can reach 100–200 t/km^2/year, with average values of 20–40 t/km^2/year. Sediment yield in the forest zone, less affected by agricultural developments, rarely exceeds 5–10 t/km^2/year (Golosov and Belaev 2011).

Landscapes

The Volga River basin is part of the Russian Plain, consisting of Precambrian crystalline rocks covered by a thick layer of sedimentary rocks, mainly limestone, marl, and dolomite from the Carboniferous, Permian, Triassic, Jurassic, and

Cretaceous ages. These sediments reach up to 10 km in depth in the North Caspian Lowland. Marine deposits from the Cenozoic age are only found in the southern Volga basin (Litvinov et al. 2009).

The largest part of the basin (65%) consists of lowlands (<200 m asl). Upland areas vary commonly between 200 and 250 m asl, reaching 350–400 m asl at subbasin watershed boundaries. Mountains occupy less than 5% of the basin. On average rivers in the Volga River basin are incised 50–100 m into their surroundings, in uplands occasionally reaching 150–200 m, while in lowland areas such as the Volga upper reaches, the Meshcherskaya, Oksko-Donskaya, and North Caspian Lowlands, the incision depth is significantly less (Golosov and Belaev 2011).

The Volga River basin can be subdivided into three sections according to general climatic conditions: the Upper Volga basin, from the northern basin boundaries up to Nizhny Novgorod and the confluence of the Oka tributary; the Central Volga basin, between Nizhny Novgorod and the confluence of the Kama tributary between Kazan and Ulyanovsk; and the Lower Volga basin, from the Kama confluence to the Caspian Sea. These boundaries largely coincide with the boundaries between the forest/forest-steppe and the steppe/semidesert landscapes, respectively (Golosov and Belaev 2011).

The Upper Volga basin is characterized by a high cover of forests, varying according to their location from middle taiga fir (*Abies* spp.)-spruce (*Picea* spp.) and pine (*Pinus* spp.) forests in the North, via southern taiga fir-spruce intermixed with broadleaf and grassy forest, to broadleaf lime (*Tilia* spp.)-oak (*Quercus* spp.) and maple (*Acer* spp.)-lime-oak forests in the South. A significant part of the drainage in the Upper Volga is occupied by swampy lowlands, such as Mologo-Sheksninskaya, Kostromskaya, Balakhninskaya, and Meshcherskaya, characterized by forested and open upland peat bogs (FEOW 2012). In the past, many peatlands were drained for agriculture, forestry, and peat mining and subsequently left abandoned, giving rise to wind and water erosion, fires, and the emission of large amounts of carbon dioxide.

The Central Volga basin includes the forest, forest-steppe, and steppe biomes, varying from spruce and northern broadleaf forest in the North, via mixed meadow-steppe broadleaf and pine forest in the central part, to herb-feather grass (*Stipa* spp.) steppe in the South. After the construction of reservoirs, hardly any floodplain ecosystems were conserved (Litvinov et al. 2009). Peatlands in this zone typically include valley fens and swamps still occurring on the floodplains of larger tributary rivers, while in valleys of smaller rivers, peatlands have been almost totally destroyed as a consequence of the long and intensive use by man. Remaining peatlands are increasingly important temporary and permanent habitats for fauna species, providing food, shelter, and breeding grounds in stressed agro-landscapes (Minaeva et al. 2009).

The Lower Volga basin is characterized by herb-feather grass, fescue (*Festuca* spp.)-feather grass, and desert-like wormwood (*Artemisia* spp.)-fescue-feather grass steppes. Floodplain vegetation remains on river islands as well as downstream of the Volgograd reservoir – the Volga-Akhtuba floodplain and the Volga Delta (Litvinov et al. 2009).

Lakes in the Volga basin include both natural lakes and artificial reservoirs. Natural lakes in the upper reaches include lakes like Seliger, Sterzh, Penno, and Volgo; lakes in the middle reaches include Nero, Pleshcheyevo, Galichskoye, and

Chukhlomskoye, while downstream from Volgograd floodplain lakes occur in-between numerous branches and former riverbeds (FEOW 2012).

Biodiversity

Faunal diversity in the Volga basin is characterized by 12 amphibian species, 22 reptile species, 100 mammal species, and 356 bird species. Of amphibians, the most common and widely distributed include frogs of the genus *Ranidae* and two species of toads (*Bufo* spp.). Widespread but less numerous are two species of *Triturus*. Of the reptiles, especially numerous are lizards (10 species, mainly *Lacerta* and *Eremias*) and snakes (12 species), while turtles are represented only by one species (bog turtle – *Emys orbicularis*). Mammals mostly include rodents (44 species) and predators (especially small ones), 22 species; even-toed ungulates (*Artiodactyla*, 5 species) and double-toothed rodents (*Lagomorphs*, 3 species) are few in number. The largest bird order is the passerines, within which the most common families include the finches (*Fringillidae*), Old World warblers (*Sylviidae*), and thrushes (*Turdidae*) (Rozenberg 2011).

No unified overview of the flora of the Volga basin exists, although numerous publications are dedicated to the floral diversity in one or several of the federal districts, administrative provinces, or landscape zones in the basin. A comprehensive list of rare flora species of the Volga basin is currently being argued for (Senator and Saksonov 2014).

Under natural conditions, the Volga River was home to 69 fish species belonging to 23 families, the most diverse being *Cyprinids* (36 species), *Percids* (9 species), and *Salmonids* (8 species). Based on occurrence, four groups are distinguished: residential omnipresent, e.g., pike *Esox lucius*, roach *Rutilus rutilus*, bream *Abramis brama*, and carp *Cyprinus carpio*, occurring along the whole Volga river; residential-site specific, e.g., river lamprey *Lampetra fluviatilis*, trout *Salmo trutta*, taimen *Hucho taimen*, grayling *Thymallus thymallus*, and minnow *Phoxinus phoxinus*, found locally at suitable stretches and tributaries; semi-migratory (semi-anadromous), e.g., Caspian roach *Rutilus rutilus caspius*, Caspian Sea sprat *Clupeonella cultriventris*, and Caspian shemaya *Alburnus chalcoides*, migrating between the Caspian Sea and the Lower Volga region; and migratory (anadromous), e.g., beluga *Huso huso*, Russian sturgeon *Acipenser gueldenstaedtii*, stellate sturgeon *Acipenser stellatus*, ship sturgeon *Acipenser nudiventris*, black-backed shad *Alosa kessleri*, and Caspian salmon *Salmo trutta caspius*, annually swimming from the Caspian Sea to spawning areas in the Upper Volga basin.

Discharge regulation related to the construction of the cascade of dams and reservoirs impacted on fish communities all along the Volga River, but specifically the Middle and Lower Volga sections of the basin. While the dams were equipped with fish passes, these were poorly designed and functionally ineffective, and as a result especially Caspian sturgeon migrants were limited to the Lower Volga for natural spawning. Today, the area of remaining spawning grounds is assessed as 325 ha, down from 3390 ha before discharge regulation (KaspNIIRKh 2008), while once they swam as far as Rzhev (Russian sturgeon), Tver (beluga), as well as the Oka and Kama tributaries (black-backed shad). Although numerous hatcheries were constructed to support the sturgeon population, some of which are still in operation, sturgeon populations have declined significantly, especially since the early 1990s. While quantitative data are disputed, assessments state that surgeon populations in the Lower Volga and northern Caspian have undergone at least a 15-fold reduction in the last 30 years (Bologov 2013). While the Caspian littoral states strive to enforce limitations on catch volumes as well as trade in caviar – since 1998 international trade has been regulated under CITES (the Convention on International Trade in Endangered Species of Wild Flora and Fauna) – the illegal catch of sturgeons has thrived (Berkeliev 2002). As a result, all sturgeon species are considered to be threatened, largely assessed as critically endangered (beluga, Russian sturgeon, stellate sturgeon, ship sturgeon, Persian sturgeon *Acipenser persicus*), while sterlet *A. ruthenus* is assessed as vulnerable (IUCN 2013).

At present, the fish diversity in the Middle and Upper Volga regions predominantly includes lake-type (*Limnophilous*) species, although small residual populations of previously migrating species remain. Efforts were made to increase diversity by the direct introduction of nonnative species, many of which did not develop self-reproducing populations. Fish communities resembling those before flow regulation remain only in the Volga headwaters and in the Lower Volga – the Volga-Akhtuba floodplain and Volga Delta. Along the whole length of the river, fish populations have also been significantly impacted by overfishing and poaching (Litvinov et al. 2009).

The Volga River basin still maintains significant areas of biodiversity importance, despite being dominated by anthropogenic activities. About 2,000,000 ha of land are formally protected under federal legislation as Strictly Protected Nature Reserves and National Parks (OOPT of Russia 2013), while at the provincial level, many more areas have been designated. Protected areas include wetlands of importance to related species of flora and fauna, specifically peatlands in the northern forest zone, freshwater lakes and rivers in the Middle Volga zone, and the freshwater Volga Delta in the Lower Volga zone. The Volga Delta in particular supports millions of migratory water birds during spring and autumn migration, largely ducks, geese, and swans, in addition to their local breeding populations. No comprehensive overview of the number and populations of endangered species exists specifically for the Volga basin.

Human Impacts

The Volga basin and river are actively used for agriculture and fisheries, navigation, industrial production, and generation of clean energy. The Volga basin is home to about 57 million inhabitants of Russia, carries 70% of the country's inland water traffic, accounts for half of the inland fish stocks, and supports half of the country's agricultural production (Kashchenko and French 2005). In total, 39 administrative provinces of Russia are wholly or partly part of the basin, as well as two administrative provinces of Kazakhstan (Naidenko 2003). Yearly about 20 km^3 of wastewater enters the Volga River, including oil, nutrients and heavy metals, from

Fig. 3 The Volgograd hydro-electric station barrage, the construction and operations of which continue to have a major impact upon the hydrological regima and ecological functions of the Lower Volga (Photo credit: Harald Leummens © Rights remain with the author)

residential, industrial, and agricultural sources. Volga surface water sources provide 85% of the basin population's water needs. Meanwhile, only 15% of treated wastewater meets established national standards (Kashchenko and French 2005).

Discharge regulation of the Volga started in 1943 with the construction of the Verkhnevolshskoe reservoir. By 1980, 12 major dams were constructed along the course of the Volga River, guided by the large-scale expansion of economic activities in the Volga basin (Fig. 3). The largest storage lakes extend over hundreds of kilometers, their width in general not exceeding several tens of kilometers. The reservoirs cover an area of about 26,000 km^2 and have a maximum storage capacity of about 190 km^3, of which about 90 km^3 is useable (Leummens 2005a). Currently the total installed power generating capacity is 11,369 MW, the average total annual energy production is 40,658 million kWh, or almost 4% of the total production in Russia and 22% of renewable energy sources. Additionally, numerous small dams with or without small hydropower stations operate on Volga tributaries (Naidenko 2003).

The main strategic goals for discharge regulation as defined in the 1930s included establishing a unified waterway for shipping in European Russia, obtaining cheap energy, solving water problems in the agricultural sector, and decreasing the negative effects of extreme flooding by means of intra-annual redistribution. The plans also targeted community drinking water supply problems as well as the need for technical water in industry. The envisioned reduced income from fisheries was expected to be compensated through the development of fisheries in the reservoirs. Today's revised exploitation rules stress the principle of integrated use of water resources and the need for a yearly approach to management which involves the filling of reservoirs in spring and acceptable winter water level decreases. For the three downstream reservoirs, the spring discharge regime should comply with fisheries' and agricultural demands in the Lower Volga; from summer to autumn, the regime should provide for navigation; and only in winter should the regime focus on energy production (Leummens 2005a).

Threats and Future Challenges

The main challenges for the Volga River basin in the near future focus on handling the legacy of the former Soviet Union. With the Volga's surface water resources being of critical importance as drinking water resources for the population, improving its quality remains of key importance. While industrial production as well as agriculture decreased after the breakup of the Soviet Union, pollution sources were largely not cleaned up, and pollutants continue to enter surface waters, from point as well as diffuse sources. At the same time, the aging of the water supply and sewerage system infrastructure provides increased stress due especially to nutrient outflows from leakages and poorly functioning wastewater treatment facilities. Meanwhile, recent years have been characterized by a recurrent growth in industrial as well as agricultural production, giving rise to potential new sources of pollution as typical control and enforcement remain suboptimal. In parallel, proper landfilling, including recycling, remains a significant challenge in the Volga basin. Accordingly, pollution maintains a high pressure on the health of the population, worsened by the fact that about 40% of the population lives at or below the subsistence level.

Moreover, the inefficient use of natural resources and especially water creates challenges for the future. While water levels in the reservoirs of the Volga basin are suboptimal for power generation, the discharge regime has a significant negative impact on aquatic biodiversity, especially in the river's unregulated sections. Irrigation water losses are significant due to poor infrastructure maintenance.

The pressure on natural ecosystems and especially wetlands from natural as well as man-made causes will continue into the future. In addition to stress caused by past and present industrial water pollution, the reinvigorated agricultural sector will likely be accompanied by an increase in the use of fertilizers, herbicides, and pesticides, potentially accumulating in wetland sinks throughout the basin. An expanding middle class will likely further affect biodiversity, through hunting and fishing tourism, while the high level of poverty will continue to exert a significant subsistence level pressure on the use of natural resources.

In order to improve the water resources in the Volga basin, an effective and efficient integrated approach is needed. Although the institutional structure has been modernized, and basin management approaches are being applied, intersectoral and interregional coordination and cooperation need to be further strengthened, supported by appropriate legislative-institutional reforms as well as enforcement mechanisms. The provision of sufficient and appropriate information from monitoring programs for effective decision-making on land and water resources remains an additional challenge.

References

Berkeliev T. Main ecological problems of the Caspian Sea. Moscow: Center for Nature Protection; 2002. Available from: www.biodiversity.ru. Accessed 30 Jan 2016. Russian.

Bologov P. Will the Caspian Sea change into the "Dead Sea". 16 April 2013. Available from: Lenta.ru. Accessed 30 Jan 2016. Russian.

FEOW Freshwater Ecoregions of the World. Volga-Ural. 2012. Available from: http://www.feow.org/ecoregions/details/410. Accessed 30 Jan 2016.

Golosov V, Belayev V. The Volga River basin report. UNESCO International Sediment Initiative. Moscow (Russia): Lomonosov Moscow State University, Faculty of Geography. Case study to: Sediment issues and sediment management in large river basins. UNESCO Office in Beijing & IRTCES. International Sediment Initiative Technical Documents in Hydrology. 2011. 145 p. Available from: http://www.irtces.org/isi/isi_document/2010/ISI_Case_Study_Volga.pdf. Accessed 30 Jan 2016.

IUCN. IUCN red list of threatened species. Version 2013.1. Available from: www.iucnredlist.org. Accessed 28 July 2013.

Kashchenko O, French M. The Volga River: Russia's strained lifeline. Cabri Volga Brief, December 2005, issue 1, p. 1.

KaspNIIRKh. Inventory of Sturgeon spawning areas in the Lower Volga river. Astrakhan: Caspian Scientific Research Institute for Fisheries; 2008. 139 p. + maps Russian.

Leummens H, editor. Integrated impact analysis of Volga river discharge regulation on floodplain and delta ecosystems. Astrakhan: UNESCO/ROSTE; 2005a. 33 p.

Leummens H, editor. Conservation of wetland biodiversity in the Lower Volga Region – integrated assessment report. Moscow: UNDP/GEF; 2005b. 43 p.

Litvinov AS, Mineeva NM, Papchenko VG, Korneva LG, Lazareva VI, Shcherbina GK, Gerasimov YV, Dvinskikh SA, Noskov VM, Kitaev AB, Alexevnina MS, Presnova EV, Seletkova EB, Zinov'ev EA, Baklanov MA, Okhapkin AG, Shurganova GV. Volga River Basin. In: Tockner K, Robinson CT, Uehlinger U, editors. Rivers of Europe. Amsterdam: Academic; 2009. p. 23–57. 728 p.

Minayeva T, Sirin A, Bragg O, editors. A quick scan of peatlands in Central and Eastern Europe. Wageningen: Wetlands International; 2009. 132 p.

Naidenko VV. The Great Volga in the new millennium – from environmental crisis to sustainable development. Nizhniy Novgorod: Promgrafika Publishing House; 2003. 423 p.

OOPT of Russia. Protected areas of Russia – information system. Moscow: Center for Nature Protection; 2013. Available from: http://oopt.info/index.php?page=1. Accessed 25 July 2013.

Rozenberg GS. Prognosis of the state and management of biodiversity in the Volga basin towards reaching sustainable development. 2011. Russian. Available from: http://rudocs.exdat.com/docs/index-173299.html. Accessed 30 Jan 2016.

Senator SA, Saksonov SV. The Red Book of the Volga basin in implementing the principles of sustainable development. Povolzhkiy Environ J. 2014;1:38–49. Russian.

Volga River Delta (Russia)

76

Harald J. L. Leummens

Contents

Introduction	946
Hydrology	949
Wetland Ecosystems and Biodiversity	949
Land Use	953
Conservation Status	954
Ecosystem Services	955
Threats	955
Natural Environmental Variability	955
Discharge Regulation	956
Unsustainable Exploitation of Natural Resources	956
Future Challenges	956
References	957

Abstract

The Volga delta, globally the tenth largest delta by size, is located in the southeastern part of European Russia, in the Astrakhan Province. The delta is part of the North Caspian Lowland, a flat sedimentary basin largely located below oceanic sea level. The dynamic location, structure and sedimentary composition of the actual Volga Delta are influenced by two systems: the Volga River and the Caspian Sea. The present-day delta developed on young Holocene sediments, exhibiting a classic triangular form. The delta comprises two distinct landscape sub-regions: the terrestrial delta and the shallow waters of the fore-delta, separated by the transitory *kultuk* zone (freshwater bays with slow flowing water and many small islands). Typical ecosystems include wet and dry meadows, back-swamps, reed and reedmace dominated communities, freshwater bays with

---

H. J. L. Leummens (✉)
Water and Nature, Heerlen, The Netherlands
e-mail: harald.leummens@gmail.com

© Springer Science+Business Media B.V., part of Springer Nature 2018
C. M. Finlayson et al. (eds.), *The Wetland Book*,
https://doi.org/10.1007/978-94-007-4001-3_29

islands, and open water. The global importance of the Volga Delta for wetland biodiversity is widely recognized. The delta occupies a strategic position on 3 important flyways, and supports a large diversity of common as well as globally threatened flora and fauna species, specifically birds, mammals as well as highly valuable sturgeon species. The wetlands' natural resources have long supported the local population, providing products like waterfowl, fish, caviar, plant materials and freshwater. During the 20th century, the Volga Delta wetlands and their biodiversity became subjected to intense and increasing pressures from human activities. Since the 1960s the Volga River flow is regulated, its hydrological regime changed towards reduced interannual and seasonal variation. In the terrestrial part of the delta, large areas of land were converted for irrigated agriculture, many of which today remain abandoned and degraded. Annually flooded meadows are intensively used for grazing and haymaking. More recent, urbanization and infrastructure caused increased disturbance, fragmentation and degradation of wetlands, augmented by unsustainable use of terrestrial and aquatic natural resources, including legal and illegal hunting and fishing for commercial, livelihood and touristic purposes. Flow regulation and diking has affected the delta ecosystems' natural capability to adapt to dynamic environmental conditions, especially variable river flow and sea levels. Nature conservation is pursued in the State Nature Biosphere Reserve "Astrakhansky", several State Nature Reserves and Nature Monuments. Since 1975, the southern part of the Volga Delta is designated as wetland of international importance under the Ramsar Convention, expanded in 2009 to include the "Western Ilmen Area". Within and beyond protected areas, institutional and individual capacities for successful management of biodiversity conservation remain weak, inhibiting effective enforcement of land use restrictions. The challenge towards successful conservation of the Volga Delta's wetland biodiversity is to install an adaptive holistic management approach to integrated sustainable use and conservation of its natural resources, linked to wetlands in the wider landscape to account for natural environmental variability.

Keywords

Volga river · Delta wetlands · Europe · Regulated flow · Freshwater ecosystems

Introduction

The Volga Delta is located in the southeastern part of European Russia (Fig. 1), in the North Caspian Lowland, a flat large sedimentary basin largely located below oceanic sea level, its surface monotony only interrupted by local salt domes and Paleozoic and Mesozoic outcrops (Butorin et al. 2008). The Caspian Lowland developed as a result of a complex pattern of Caspian Sea transgressions and regressions since the Late Pliocene. With a terrestrial surface of 23,000 km², the Volga Delta is globally the tenth largest delta by delta plain size. The current delta has no paleo-analogues, as with varying sea level the location of the delta wandered throughout the North

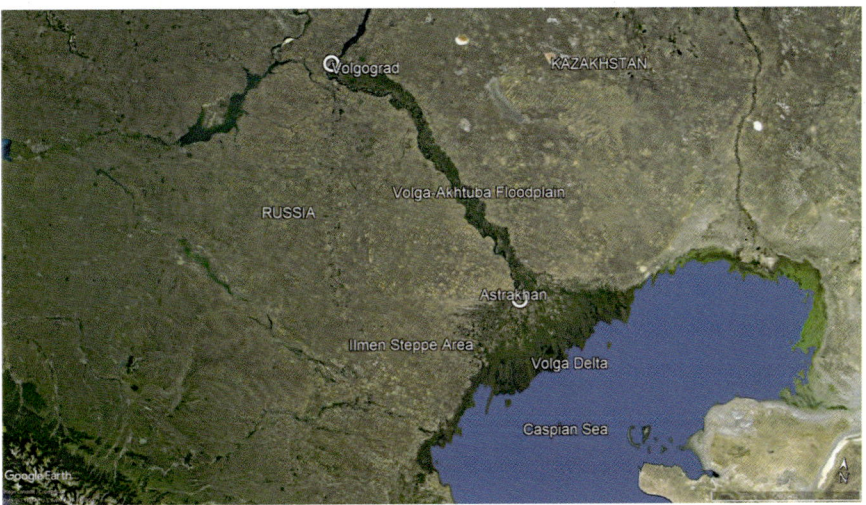

Fig. 1 Location of the Volga Delta in south-east European Russia (Source: Google Earth, visited 13 December 2017)

Caspian Lowland (Kroonenberg et al. 1997). The relief and deposits of the present-day delta are very young formations, dating from the Late Holocene. The location, structure, and sedimentary composition of the actual Volga Delta are subjected to continuous changes under the influence of two systems: the Volga River and the Caspian Sea (Arsenyev 2011).

The climate is characterized by a long hot summer with frequent strong easterly winds and low relative humidity and a cold winter with little snow, conditions typical for the strong continental Eastern European climatic province. The average monthly temperatures vary between −7 °C in January and 25 °C in July; absolute air temperatures vary between −40 °C and +45 °C. The frost-free period lasts for 235–260 days; the total solar radiation reaches 100–115 kcal/cm^2; the average yearly hours of sunshine amount to 2,265 h, and the total temperature sum above 10 °C to 3,400 °C. The yearly precipitation averages 200 mm, with large inter-annual variations observed. Precipitation between April and October exceeds that of November-March by 1.5 times. Summer precipitation is often of a stormy character, although biologically effective rainfall pulses – showers that exceed 5 mm per day in volume, enough to stimulate biological processes, particularly growth and reproduction – are rare. A stable snow cover develops only in 40% of all winters, the prevailing snow cover being 1–5 cm. The maximum number of days with stable snow cover is 42–50 days. Yearly atmosphere evaporative demand varies from 1,100 mm in the south-west to 920 mm in the eastern part of the delta, being strongly seasonal, rising to 175–200 mm during the summer months (Leummens 2005a).

The Volga Delta wetlands comprise two distinct subregions: the terrestrial delta and the shallow waters of the fore-delta, separated by the transitory *kultuk* zone

Fig. 2 MSR image of the Volga Delta, 27 August 2001 (Source: © NASA's Earth Observatory, in public domain, accessed 20 October 2012)

(freshwater bays with slow-flowing water and many small islands). The delta exhibits a classic triangular form; the distance from the apex to the Caspian Sea is about 120 km, its width at the coast about 200 km. The apex is located 50 km north of the city of Astrakhan, where the Volga main stream branches into the Buzan and Bakhtemir. The delta becomes increasingly dissected in a south-easterly direction, with finally more than 800 small and large streams entering the Caspian Sea. Except for the most eastern part, located in Kazakhstan, the Volga Delta is located within the boundaries of Astrakhan Province of Russia (UNDP/GEF 2000). The delta is very flat, the height from the apex to the coast decreasing from -22 m to -27 m asl. or 4 cm per km (Kroonenberg et al. 1997). In addition to the terrestrial section, the Volga Delta also shows a well-developed subaqueous component, the surface ratio between them being 1.97 (Coleman et al. 2008; Fig. 2).

The global importance of the Volga Delta for wetland biodiversity is widely recognized. Occupying a strategic position on three important flyways, the delta supports at least 15 globally threatened migratory bird species during various stages of the migration cycle. In addition, four threatened and highly valuable sturgeon species depend on the region for spawning and feeding as well as at least 20 endemic subspecies of fish. Additional ecological importance is given to the area by its geographical location and structure, being one of the few riverine north–south land corridors crossing the extended dry semidesert and steppe area of southern Russia and Kazakhstan. Occasionally, the wetlands serve as a feeding area for saiga antelopes (*Saiga tatarica*) from the Kalmykian steppes (Leummens and Menke 2008).

Hydrology

The Volga discharge averaged 252 km^3/year (8,076 m^3/s) over the period 1880–2009, showing significant interannual variation. The period 1880–1935 was the period of unregulated flow (average 265 km^3/year), 1935–1960 (average 235 km^3/year) of semiregulated flow, and since 1960 completely regulated flow (Zemlyanov 2010). The regulated flow period subdivides into the period of lower than average Volga river discharge, also the period of Caspian Sea level fall (1961–1977; average 224 km^3/year), and a period of higher than average Volga river discharge, during which Caspian Sea level rise was observed (1978–2009; average 265 km^3/year).

In total 12 large artificial reservoirs regulate the water flow along the Volga River (Zemlyanov 2010). The most downstream reservoir is located at Volgograd, 300 km upstream of the Volga Delta apex. As a result, the interannual variation in total yearly discharge has been significantly reduced (Zemlyanov 2010), causing lower lands to be flooded more frequently, while higher areas became excluded from flooding (Golub and Kuzmina 1997). Also the seasonal distribution of river flow has been altered: the traditional spring flooding period (April-July) became less pronounced, with reduced peak levels, shortened duration, earlier peak date, faster rise and fall of the water level, and reduced overall volume, while the winter discharge (December-March) notably increased, with a doubling of total volume and occurrence of flooding events. The ratio between the discharge volume during the flooding period and the winter period reduced to 1.6 from 4.5, the average maximum spring discharge declining from 33,250 m^3/s to 28,000 m^3/s (Leummens and Menke 2008). Meanwhile the discharge in the summer-autumn (August-November) low water period increased only slightly (Leummens 2005b) (Figs. 3 and 4).

Wetland Ecosystems and Biodiversity

The terrestrial part of the Volga Delta shows a typical structure of back-swamps and meadows surrounded by low levees along channels. Small channels may be overgrown with bur-reed (*Sparganium erectum*), pondweeds (*Potamogeton perfoliatus, P. lucens*), *Limnanthemum peltata*, white water-lily (*Nymphaea alba*), and other emergent and submerged plants.

Large areas in the upper and middle part of the delta mainly consist of meadows. Conditioned by the duration of flooding and soil salinity, three types are distinguished.

- At the lowest topographical level, yearly flooded for several months, aquatic and marsh plants dominate, including species like reed canary-grass (*Phalaroides arundinacea*), fine-leafed water-dropwort (*Oenanthe aquatica*), spike-rush (*Eleocharis palustris*), common water-plantain (*Alisma plantago-aquatica*), common reed (*Phragmites australis*), blue water-speedwell (*Veronica anagallis-*

Fig. 3 Typical "kultuk" freshwater bays in the Volga Delta, showing riparian vegetation belts interchanging with open water, featuring populations of colonial waterbirds such as egret species (Photo credit: Harald Leummens © Rights remain with the author)

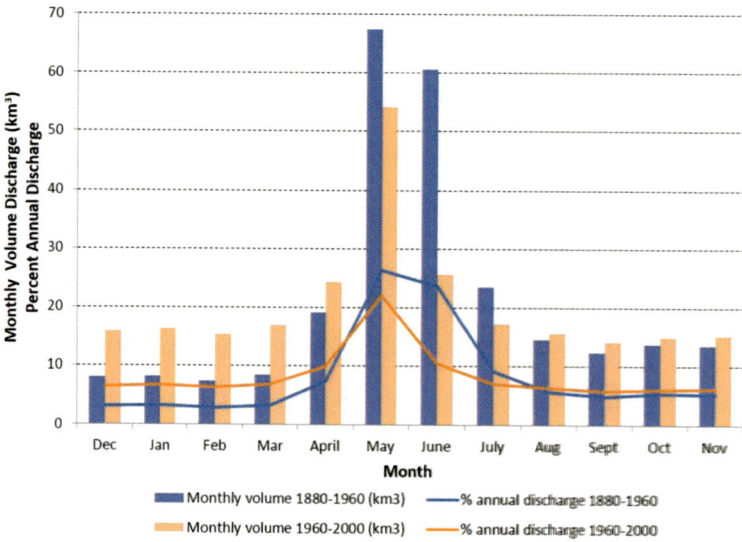

Fig. 4 Total monthly and seasonal Volga River discharge volumes at Volgograd before and after regulation (Source: Leummens 2005b)

Fig. 5 Despite hydrological flow regulation, a dynamic seasonal flood regime is maintained in the Volga floodplain downstream of Volgograd, in support of ecological functionality (Photo credit: Harald Leummens (c) Rights remain with the author)

aquatica), slender-tufted sedge (*Carex acuta*), and amphibious bistort (*Persicaria amphibia*). Salt-tolerant species are absent, as salts are leached by flooding water.
- At the intermediate level, flooded yearly for 1–2 months, mesophytic species dominate, including couch grass (*Elytrigia repens*), holy grass (*Hierochloe odorata*), wood small-reed (*Calamagrostis epigeios*), and bermuda-grass (*Cynodon dactylon*).
- At the highest level, not yearly flooded or flooded up to 1 month, the marsh and mesophytic species are replaced by the dominance of typical salt-tolerant species, like bog-rush crypsis (*Crypsis schoenoides*), *Crypsis aculeata*, *Suaeda confusa*, glasswort (*Salicornia europaea*), sea aster (*Tripolium pannonicum*), sagittate-leafed orache (*Atriplex calotheca*), and tamarisk (*Tamarix ramosissima*).

Meadows in the Volga Delta differ from more northern types in the increased occurrence of salt-tolerant species, like *Bolboschoenus maritimus*, marsh mallow (*Althaea officinalis*), broad-leaved pepperwort (*Lepidium latifolium*), common liquorice (*Glycyrrhiza glabra*), and hedge hog liquorice (*Glycyrrhiza echinata*). Both the lowest and intermediate levels also show flowering plants like British fleabane (*Inula britannica*), marsh spurge (*Euphorbia palustris*), angular garlic (*Allium angulosum*), gratiola (*Gratiola officinalis*), common meadow-rue (*Thalictrum flavum*), and strawberry clover (*Amoria fragifera*).

In the southern part of the terrestrial delta, dense monospecific communities of lesser reedmace (*Typha angustifolia*) and common reed occur. Other important species-poor communities are dominated by slender reed mace (*Typha laxmannii*) and bur-reed. Towards the central part of the delta the monospecific communities, characterized by surface water throughout the year, are replaced by more species-rich reed communities in spring-flooded areas, including species like hedge bindweed (*Calystegia sepium*), swamp sedge (*Carex acutiformis*), *Dipsacus gmelinii*, marsh woundwort (*Stachys palustris*), and marsh spurge. Relative small areas (50–300 ha) of softwood gallery forests with white willow (*Salix alba*) and almond willow (*Salix triandra*) occur on low levees (UNDP/GEF 2000), although typically affected – crown-thinning, reduced growth rates, dying – by increased water levels (Rusanov 2003).

The *kultuk* area is a zone of freshwater bays with slow-flowing water and a large number of small islands, extending along the southern edge of the terrestrial delta, varying in width from one to several km. The bays are between 0.4 and 1.3 m deep at low water. Large permanently submerged parts of the bays are covered with emergent and submerged plants, representing different stages of succession. Typical are communities combining submerged aquatic species like pondweeds, *Chara* spp., rigid horn-wort (*Ceratophyllum demersum*), Eurasian water-milfoil (*Myriophyllum spicatum*), holly-leafed naiad (*Najas marina*), and wild celery (*Vallisneria spiralis*) with species like duckweed (*Lemna* spp.), *Salvinia natans*, frogbit (*Hydrocharis morsus-ranae*), and greater bladderwort (*Utricularia vulgaris*) as well as floating species like white water-lily, candid water-lily (*Nymphaea candida*), yellow water-lily (*Nuphar lutea*), fringed water-lily (*Nymphoides peltata*), caltrop (*Tribulus terrestris*), broad-leaved pond-weed (*Potamogeton natans*), and amphibious bistort.

Along the border of the open water zones species like arrow-head (*Sagittaria sagittifolia*) and flowering rush (*Butomus umbellatus*) occur. Typical also are monospecific common reed formations, with a coverage varying between thin (20–40%) and massive stands (90–100%), as well as beds of bur-reed with a cover of 50–70%. The *kultuk* zone leads into the fore-delta, comprising an extensive shallow area with a large number of low islands. Water depths vary between 1.0 and 1.7 m at low water. Habitats mainly resemble those wetter habitats of the *kultuk* area (UNDP/GEF 2000). The *kultuk* area and the northern part of the fore-delta are also the main regions where the endangered lotus lily (*Nelumbo caspica*) occurs.

The variety of dynamic landscapes within a small area makes the *kultuk* zone and the fore-delta the ecologically most interesting and important parts of the Volga Delta. The southern part of the terrestrial delta, the *kultuk* zone, and the fore-delta developed in their present form in response to Caspian Sea level decrease since the early twentieth century. Sea level rise since 1978 caused large changes, mainly in the fore-delta and *kultuk* zone; extended areas covered by common reed and bur-reed suffered from increased water levels, as did the most southern part of the gallery forests. Meanwhile, the surface area covered with lotus lily strongly increased, from 2 to 20,000 ha, to stabilize at the present-day area of 10,000 ha (UNDP/GEF 2000).

The Volga Delta serves as habitat for numerous rare and endangered species of animals – mammals as well as nesting and migratory birds – included in the IUCN

Red List of Threatened Species, the Red Books of the Russian Federation and the Astrakhan Province. Globally threatened species occurring in the Volga Delta include birds – Dalmatian pelican (*Pelecanus crispus*), red-breasted goose (*Branta ruficollis*), lesser white-fronted goose (*Anser erythropus*), marbled teal (*Marmaronetta angustirostris*), white-headed duck (*Oxyura leucocephala*), greater spotted eagle (*Aquila clanga*), eastern imperial eagle (*Aquila heliaca*), Pallas' fish eagle (*Haliaeetus leucoryphus*), saker falcon (*Falco cherrug*), lesser kestrel (*Falco naumanni*), Siberian crane (*Grus leucogeranus*), great bustard (*Otis tarda*), sociable lapwing (*Vanellus gregarious*), slender-billed curlew (*Numenius tenuirostris*) (Rusanov 2010); mammals – Russian desman (*Desmana moschato*), giant blind mole rat (*Spalax giganteus*), European mink (*Mustela lutreola*), European marbled polecat (*Vormela peregusna*) (Arsenyev 2011). The southern part of the Volga Delta site is known as a mass molting area for ducks – the total number of birds migrating through these wetlands during autumn 2009 was estimated as 4,617,720 birds (Arsenyev 2011).

The Volga River plays a significant role in the reproduction of anadromous, semianadromous, and freshwater fish species living in the Volga-Caspian basin. The Lower Volga and North Caspian region supports highly productive spawning, nursery, and feeding areas for 76 species and 47 subspecies of fish, more than half of which are endemic and 40 are of commercial value (UNDP/GEF 2000). Key common species include carp (*Cyprinus carpio*), bream (*Abramis brama*), pikeperch (*Stizostedion lucioperca*), catfish (*Silurus glanis*), caspian roach (*Rutilus rutilus spp. caspicus*), pike (*Esox lucius*), chub (*Leuciscus cephalus*), rudd (*Scardinius erythrophthalmus*), tench (*Tinca tinca*), and Caspian herring (*Alosa kessleri*). In addition, a number of rare species occur: fringebarbel sturgeon (*Acipenser nudiventris*), Volga shad (*Alosa kessleri volgensis*), inconnu (*Stenodus leucichthys leucichthys*), Caspian trout (*Salmo trutta caspius*), Caspian vimba (*Vimba vimba persa*), and Caspian barbel (*Barbus brachycephalus caspicus*) (Arsenyev 2011). All sturgeon species are included as threatened species in the IUCN Red List of Threatened Species – Russian sturgeon (*Acipenser gueldenstaedti*), beluga sturgeon (*Huso huso*), stellate sturgeon (*Acipenser stellatus*), and sterlet (*Acipenser ruthenus*), and their trade is regulated by the Convention on International Trade in Endangered Species of Wild Fauna and Flora (CITES) (Arsenyev 2011).

Land Use

The population in Astrakhan Oblast is 1,016,500 of which 530,900 (Rosstat 2014) live in the city of Astrakhan, in the northern part of the Volga Delta. Densely populated rural areas (>50 persons per km^2) occur in the wide vicinity of Astrakhan city and along the Bakhtemir in the western part of the delta.

In the northern and central part of the Volga Delta, large areas were converted into irrigated agricultural fields during the Soviet era, protected against flooding by dikes. Although following the economic turmoil after the breakup of the Soviet Union many of them were abandoned, in recent years agricultural production is again on the rise. Still however agriculture suffers from the lack of capital investment to replace obsolete, costly, and ineffective irrigation installations. Also, part of the previously irrigated lands is degraded by salinity, resulting from poor irrigation practices in the past. Typical crops under irrigation include potato, tomato, and other vegetables as well as grains. The remaining yearly flooded meadows are intensively used for grazing and haymaking, providing fodder for local cattle as well as for herds in the surrounding semiarid steppe zone. Fish resources of the Volga River and northern Caspian Sea, including the globally endangered sturgeon species, are of significant importance to the local economy as well as for the livelihood of the local population (Losev et al. 2006).

In contrast, the southern part of the terrestrial delta, the *kultuk* zone, and the fore-delta are hardly used for agriculture (UNDP/GEF 2000). Traditionally these areas are used for organized and unorganized tourism. In recent years, activities related to large-scale hunting and fishing tourism rapidly expanded, related to the increase financial capacities mainly of people in central Russia (Arsenyev 2011).

Conservation Status

The State Nature Biosphere Reserve "Astrakhansky" (ASBR) was established in 1919 (IUCN Category 1a) and covers a total area of 66,800 ha, divided over three sections. Its buffer zone – IUCN category VI – covers an additional 31,000 ha. The ASBR obtained the status of "Biosphere Reserve" in 1984. All three clusters of the Astrakhansky Reserve lie within the lower zone of the Volga Delta and its shallow coastal waters (Butorin et al. 2008).

Several State Nature Reserves (*Zakaznik*) – IUCN category IV – were established in the Volga Delta, covering an area of 18,200 ha. Additionally four State Nature Monuments – IUCN category III – were defined, mainly to protect breeding colonies of *Pelecaniformes* and *Ciconiiformes* as well as fish spawning areas (Arsenyev 2011).

The southern part of the Volga Delta was designated a wetland of international importance by the Government of the USSR on 25 December 1975, confirmed by the Government of the Russian Federation on 13 September 1994. On 14 October 2009, by joint Decree of the Government of Astrakhan Province and the Ministry for Natural Resources and Environment of the Russian Federation, the area of the Ramsar Site was increased from 800,000 ha to 1,122,500 ha. The Ramsar Site includes part of the Important Bird and Biodiversity Areas "Volga Delta" (1,150,000 ha) and "Western Ilmen Area" (598,145 ha). The Government of the Astrakhan Province recently designated 20 sites of extensive waterbird nesting areas within the Ramsar Site, covering 25,653 ha, towards strengthening conservation measures (Arsenyev 2011).

As part of the UNDP/GEF Project "Conservation of wetland biodiversity in the Lower Volga" science-based recommendations on the permissible recreational load on the natural wetland systems in the Lower Volga have been elaborated (Arsenyev 2011).

Ecosystem Services

The wetlands of the Volga-Akhtuba floodplain and Volga Delta provide a range of ecological functions and economically valuable goods and services of importance at the local, regional, national, and international level. The occurrence of such an extended area of wetland habitats in a semiarid region, as well as their relatively intact ecological and hydrological functioning, are important actual considerations for biodiversity and landscape conservation and their integration with the sustainable use of wetland resources (Losev et al. 2006)

The wetlands' natural resources have long supported the local population, providing products like waterfowl, fish, caviar, and plant materials. The abundance of water is of critical importance not only for irrigation but also for the extensive meadows in the delta, providing fodder resources for cattle in the wider surroundings outside the Volga Delta.

Due to the combination of pressures arising from the increasing human population's use of natural resources, the direct loss and transformation of wetlands following diking and sea level changes, pollution, the regulation of natural river water regimes, as well as tourism developments, the wetlands' biodiversity values are now under intense and increasing pressures (Leummens and Menke 2008).

Threats

Natural Environmental Variability

The region's rich wetland biodiversity is an expression of its varied and dynamic aquatic resources. Under natural conditions, the wetlands of the Lower Volga – location, area, and environmental conditions – are subjected to large inter-seasonal, annual, as well as long-term fluctuations in Volga river discharge. The long-term variation in discharge is considered to be related to cyclone activity above European Russia and the Caspian, causing continental-scale variations in precipitation and evaporation. The Volga River discharge also has a profound effect on the Caspian Sea's water level, since it is a closed sea and the Volga provides 80% of its inflow. Although no human-induced threat, the continuous hydrology-induced changes of the Volga Delta's location and habitat quality are reflected in increasing/decreasing numbers, and the presence/absence, of aquatic and terrestrial plant and animal species, including fish recruitment (Leummens and Menke 2008).

Sea level rise between 1978 and 1995 resulted in a loss of shallow aquatic habitats, in the drowning of land and swamp vegetation, and changes in the feeding and breeding conditions for many mammals and water birds. Habitat loss could not be compensated for further inland, because dikes protecting agricultural fields were not relocated and no agricultural fields have been restored to wetlands. As a result, some habitats important for wetland biodiversity have decreased significantly (Leummens and Menke 2008).

Discharge Regulation

As a result of reservoir construction, the traditional spawning grounds of migratory sturgeon species became inaccessible, and their area reduced from 3,400 ha along the whole Volga course and its tributaries to 430 ha within the Volga-Akhtuba floodplain south of Volgograd. The changes in volume, amplitude, and duration as well as temperature regime of the spring flooding caused negative effects on the natural reproduction of fish and survival rate of fish fry, especially of meadow-spawning species (Leummens 2005a).

Unsustainable Exploitation of Natural Resources

In recent years, the expansion of urban areas and the transportation network (roads, pipelines, and canals) caused an increased disturbance, fragmentation, and degradation of wetland habitats. Around urban areas, particularly Astrakhan, both urban and recreational housing development has increased its pace, related to increased living standards. Equally, commercial tourism and recreation increased, with popular activities including angling, hunting, boating, and swimming, as well as berry and mushroom picking, and picnicking. The consequences include increased disturbance, littering, trampling, soil erosion, and habitat degradation (Leummens 2005a; Losev et al. 2006). As a response, impact stress norms for recreation in the region's wetlands have been established (Arsenyev 2011).

Meanwhile, especially in rural areas, harsh economic conditions continue, characterized by unemployment and poor payments, causing poverty and subsistence use of natural resources, including poaching (Leummens and Menke 2008). Artificial fires continue to affect seminatural grasslands, being a cost-effective way of meadow vegetation clearance and stimulation of fresh growth. Especially meadows at the highest topographical level often show a significant degradation of the natural vegetation and soil erosion, caused by overgrazing and trampling of the topsoil, especially during the flooding period, when these areas serve as cattle refuge areas. In all meadow types, excessive grazing pressure stimulates the appearance of resistant species like rough cocklebur (*Xanthium strumarium*), tartarian orache (*Atriplex tatarica*), and red goosefoot (*Chenopodium rubrum*) (UNDP/GEF 2000).

Future Challenges

Key factors adversely impacting on the natural wetland conditions in the Volga Delta include illegal use of terrestrial and aquatic natural resources (poaching), environmental pollution, Caspian Sea level rise, reed fires, and weakly regulated recreation activities (Arsenyev 2011). Despite attention in recent years given to expanding the area with recognized protection status, largely multiple use protected areas, enforcement of related land use restrictions remains difficult. In protected areas, directorates have limited technical and staff capacity to monitor infringements (Leummens 2005a).

Taking into account that the Volga Delta wetlands show a natural dynamic equilibrium, their location and features related to global climate conditions, Volga river discharge, and Caspian Sea level, the challenge towards successful conservation of the delta's wetland biodiversity is to install an adaptive holistic management approach to integrated sustainable use and conservation of its natural resources. Specifically, this management approach should be linked to areas beyond the Volga Delta to include the Volga-Akhtuba floodplain as well as the steppe wetlands. Both the floodplain and steppe wetlands at times serve as "escape" refuge areas, especially for migratory birds and animals, when habitat conditions in the delta wetlands worsen, either due to man-made or natural causes. For a successful and sustainable conservation of wetland biodiversity, it is critical that the community as well as decision makers come to recognize that ecosystem services provided by the Volga Delta wetlands, both locally and regionally, and the benefits that are currently gained can be lost when the wetlands collapse due to inappropriate (over) use. For this, information and awareness need to be increased and cause-effect relations made palatable for nonexperts, in order to enable decisions being made favoring the long-term provision of nature's benefits to the community. The important role that protected areas can play in reaching this understanding should be better recognized.

References

Arsenyev A. Ramsar Information sheet on Ramsar Wetlands, revised. Astrakhan: Service for Natural Resources Management & Environmental Protection, Government of the Astrakhan Province; 2011. 29 p.

Butorin A, Glasov P, Gorbunov A, Lychagin M, Maxakovsky N, Rusanov G. The Volga Delta – potential world heritage site. Moscow: Natural Heritage Protection Fund/Institute of Geography, Russian Academy of Sciences; 2008. 40 p.

Coleman JM, Huh OK, DeWitt Braud JR. Wetland loss in world deltas. J Coast Res. 2008;24:1–14.

Golub VG, Kuzmina EG. The communities of cl. Querco-Fagetea Br.-Bl. & Vlieger in Vlieger 1937 of the lower Volga Valley/Feddes Repertorium. 1997;108(3–4): 205–18.

Kroonenberg SB, Rusakov GV, Svitoch AA. The wandering of the Volga delta: a response to rapid Caspian sea-level change. Sediment Geol. 1997;107:189–209.

Leummens H, editor. Conservation of wetland biodiversity in the Lower Volga Region – integrated assessment report. Moscow/Astrakhan/Volgograd/Elista: UNDP/GEF project "Conservation of wetland biodiversity in the Lower Volga Region"; 2005a. 43 p.

Leummens H, editor. Integrated impact analysis of Volga river discharge regulation on floodplain and delta ecosystems. Astrakhan/Volgograd: UNESCO/ROSTE publication; 2005b. 33 p.

Leummens H, Menke U. ECRR – Addressing practitioners. Mestre (Venezia): European Centre for River Restoration, c/o Centro Italiano per la Riqualificazione Fluviale (CIRF); 2008. 60 p.

Losev GA, Laktionov AP, Afanasyev VE, Leummens H. Flora of the Lower Volga valley (the Volga-Akhtuba floodplain and the Volga Delta) – an annotated list of wild vascular plant species. Lelystad: RIZA Netherlands Institute for Inland Water Management & Waste Water Treatment; 2006. 170 p. Report No. 2006.039X.

Rosstat. Regions of Russia – Main features of administrative regions of Russia. Federal State Statistical Service of the Russian Federation. 2014. http://www.gks.ru. Accessed 20 Nov 2015. (in Russian).

Rusanov GM, editor. Structural ecosystem changes in the Astrakhan Biosphere Reserve caused by a rise of the Caspian Sea level. Astrakhan: Volga Publishers; 2003 (in Russian).

Rusanov GM. Birds of the Lower Volga. UNDP/GEF Project "Conservation of wetland biodiversity in the Lower Volga Region". Astrakhan; 2010. 389 p.

UNDP/GEF. Analytical paper – Biodiversity conservation on the Lower Volga region. Annex 1 to the UNDP/GEF project proposal "Conservation of wetland biodiversity in the Lower Volga region". Wetlands International – Africa, Europe, Middle East, Wageningen/Moscow; 2000. 44 p.

Zemlyanov IV, editor. Analysis of ecological impacts due to the exploitation of the Volgograd reservoir on wetland biodiversity in the Lower Volga Region. Moscow: Federal Service for Hydro-meteorology and Environmental Monitoring, Zubov State Oceanographic Institute; 2010. 675 p. (in Russian).

European Tidal Saltmarshes

Nick C. Davidson

Contents

Introduction	960
Geographic Distribution and Extent	960
Biodiversity	961
Conservation Status	966
Threats and Future Challenges	968
References	971

Abstract

Tidal vegetated marshes (saltmarshes) are widespread around all the coastlines of Europe, from the Mediterranean and Black Seas in the south to the Arctic Ocean in the north. Largest areas of European saltmarsh are on the Atlantic and North Sea coasts, particularly in the many estuaries around the coast of Great Britain and the international Wadden Sea (Netherlands, Germany and Denmark). Saltmarsh vegetation diversity is highest amongst these Atlantic saltmarshes and also in the Mediterranean. European saltmarshes are important as spawning and nursery areas for fish, and for their breeding and wintering waterbird populations. Although much of the saltmarsh resource is covered by multiple international and national nature conservation designations, its conservation status is generally rated unfavourable. Much saltmarsh has been embanked or

---

N. C. Davidson (✉)
Institute for Land, Water and Society, Charles Sturt University, Albury, NSW, Australia

Nick Davidson Environmental, Wigmore, UK
e-mail: arenaria.interpres@gmail.com

© Springer Science+Business Media B.V., part of Springer Nature 2018
C. M. Finlayson et al. (eds.), *The Wetland Book*,
https://doi.org/10.1007/978-94-007-4001-3_275

infilled in past centuries for agriculture, and urban and industrial developments, leading to a "coastal squeeze" which, with rising sea levels, has increased marsh erosion, threatening sea defences and increasing flood risk.

Keywords

Tidal · Vegetated marsh · Saltmarsh, vegetation diversity · Nature conservation · Land-claim · Managed realignment

Introduction

This chapter describes the resource of tidal vegetated marshes growing around the coasts of Europe. In addition to these intertidal saltmarshes, some inland areas of Europe support similar halophytic vegetation communities growing on saline soils. These include areas of naturally saline soils (e.g., the Hungarian *puszta*) and small areas (e.g., in parts of the United Kingdom) where commercial extraction of underground salt deposits has led to saline surface water. Some of the data in this chapter, including on saltmarsh areas, are derived from previously unpublished information compiled by the author.

Geographic Distribution and Extent

Tidal saltmarshes are widely distributed around almost all the coast of Europe, from the Mediterranean and Black Seas in the south to the Arctic Ocean in the north (Fig. 1). Many of the largest saltmarshes occur in the numerous sheltered estuaries and bays with large tidal ranges around the coasts of the United Kingdom and the southern North Sea. Large saltmarshes also occur in parts of the Mediterranean Sea, notably in southern France and on the Adriatic and Ligurian coasts of northern Italy.

There are no comprehensive estimates of the total area of European saltmarshes. The most extensive information dates from the 1980s (e.g., Dijkema 1984; Burd 1989; Hecker and Tomas-Vives 1995) but lacks area figures for countries including Albania, Ireland, Italy, and Spain. Although Schuyt and Brander (2004) give "a conservative estimate" of 500,000 ha (cited as being derived from information in CCRU (2003)), this may be a considerable overestimate. Area data from the 1980s compiled for 13 countries, including most of those considered to have the largest European saltmarsh areas, provide a total of only 128,435 ha. Given that saltmarsh area in Europe is also considered to be generally decreasing (see below), European saltmarsh extent seems unlikely to exceed 150,000 ha in total area.

By country, the largest areas (1980s data) of saltmarsh in Europe (Fig. 2) are in Great Britain (44,400 ha, 34.7% of the total area), Germany (18,900 ha, 14.7%), Denmark (15,150 ha, 11.8%), France (14,900 ha, 11.6%), and The Netherlands (13,000 ha, 10.1%). The largest contiguous saltmarsh area (35,450 ha, 27.6% of the total) is in the international Wadden Sea (The Netherlands, Germany, and Denmark).

Fig. 1 The geographical distribution of coastal saltmarshes in Europe (Derived from multiple sources including Dijkema (1984), Heliotis (1988), Burd (1989), Hecker and Tomas-Vives (1995), J.P. Doody pers. comm.)

Biodiversity

Europe's coastal saltmarshes display a large variety of habitat types, with types of vegetation varying geographically, latitudinally, and vertically across the intertidal zone. The EUNIS habitat classification (EEA 2012), a hierarchical comprehensive pan-European system designed to facilitate the harmonized description and collection of data across Europe through the use of criteria for habitat identification, under category A2.5 *Coastal saltmarshes and saline reedbeds* divides saltmarsh habitats into five broad types depending on their location in the intertidal zone. From upper to lower tidal levels these are: saltmarsh driftlines (A2.51); upper saltmarshes (A2.52); mid-upper saltmarshes and saline and brackish reed, rush and sedge beds (A2.53); low-mid saltmarshes (A2.54); and pioneer saltmarshes (A2.55).

Under these broad saltmarsh types, the classification identifies 50 different habitat types. Some of these types are further divided into a number of habitat subtypes. Taking subtypes into account there are 98 different categories of coastal saltmarsh

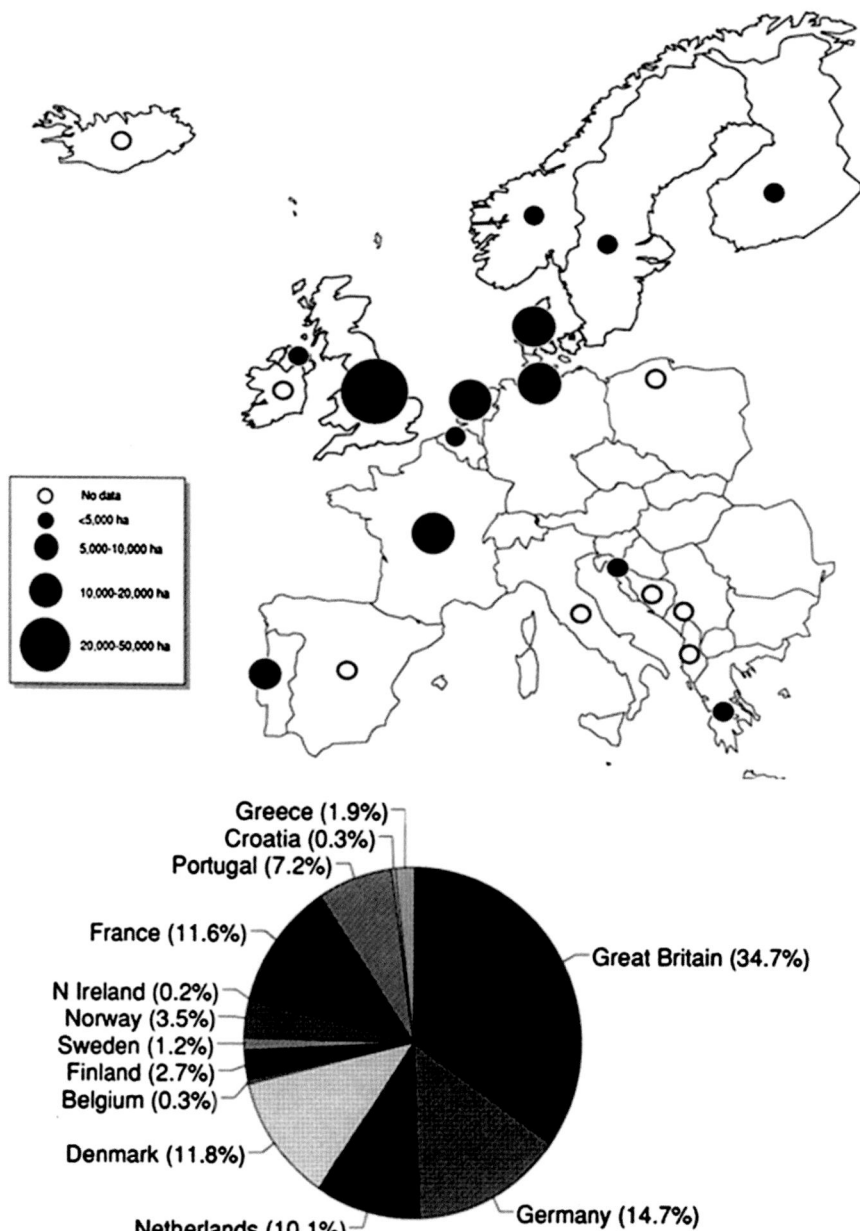

Fig. 2 The national distribution of saltmarsh area in Europe in the 1980s (Derived from multiple sources including Dijkema (1984), Heliotis (1988), Burd (1989) and Hecker and Tomas-Vives (1995))

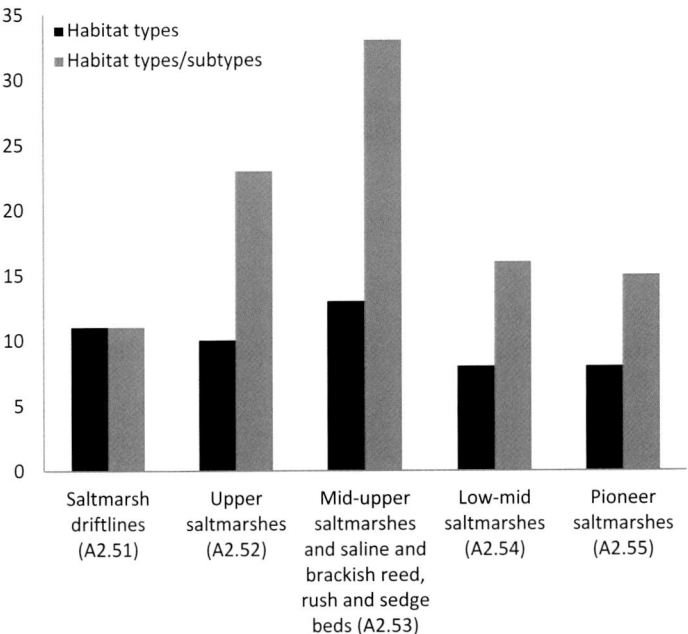

Fig. 3 The number of EUNIS habitat classification saltmarsh types and subtypes identified for different intertidal saltmarsh zones in Europe (Source: EUNIS habitat classification 2012. http://eunis.eea.europa.eu/habitats-code-browser.jsp?expand=B,A,A2,A2.5#level_A2.5)

vegetation recognized across Europe. The most diverse saltmarsh zone is mid-upper saltmarshes, followed by upper saltmarshes (Fig. 3).

Geographically, European saltmarshes are by far the most vegetationally diverse in Atlantic coastal regions, with over 60 habitat types/subtypes recognized (Fig. 4). Many parts of these coastlines have large areas of intertidal flats and marshes (see Fig. 2) in regions of high tidal range and salinities varying from marine to brackish. There is also a high diversity of types of saltmarsh habitat on the northern shores of the Mediterranean Sea. Lowest diversity of saltmarsh types occur in the Black Sea and Arctic regions of Europe (Fig. 4).

Other, more simplified, classifications of European saltmarsh vegetation are also used. For example, four main general types of saltmarsh are often recognized in western and northern Europe, a zonation across increasing intertidal elevation: pioneer marsh, low marsh, upper or high marsh, and driftline or transitional marsh (Environment Agency 2007). Typical plants of these saltmarsh zones are:

- *Pioneer marsh*. The commonest plants of these pioneer marshes are annual marsh samphire *Salicornia* spp., cord-grass *Spartina*, and sea aster *Aster tripolium*. Others include annual seablite *Sueda maritima* and, in the more brackish upper reaches of estuaries, common saltmarsh grass *Puccinelia maritima*, sea club-rush *Scirpus maritimus*, and orache *Atriplex* spp.

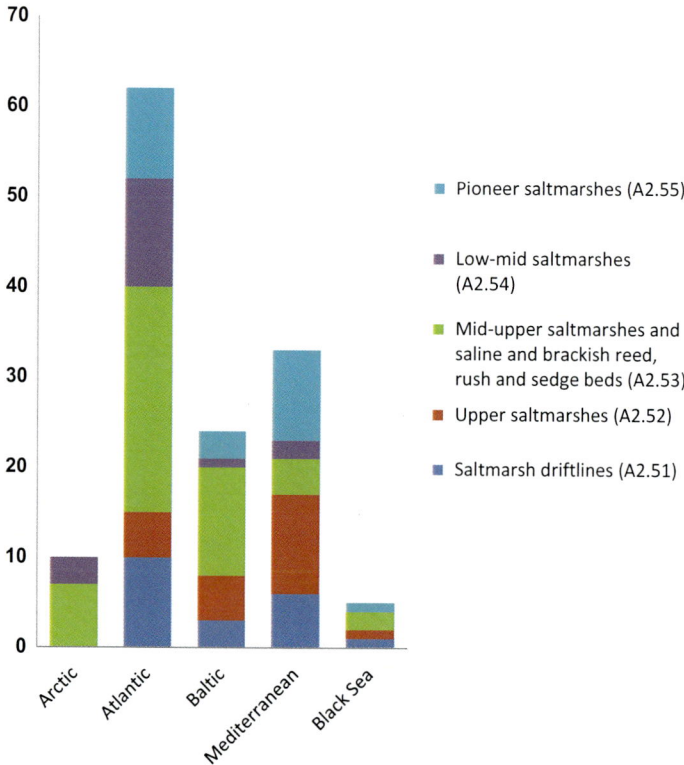

Fig. 4 The number of EUNIS habitat classification saltmarsh types identified for different coastal regions of Europe (Source: EUNIS habitat classification 2012. http://eunis.eea.europa.eu/habitats-code-browser.jsp?expand=B,A,A2,A2.5#level_A2.5)

- *Low marsh.* This zone is characterized by the dominance of two species: sea purslane *Halimione portulacoides* and common saltmarsh grass. Ungrazed marshes are generally dominated by sea purslane, with common saltmarsh grass more abundant on livestock-grazed marshes. A more diverse, herb-rich plant community occurs in the higher level parts of low marsh, with species such as common sea-lavender *Limonium vulgare* (Fig. 5), thrift *Armeria maritima*, sea arrowgrass *Triglochin maritima*, sea plantain *Plantago maritima*, greater sea-spurrey *Spergularia media*, common scurvy-grass *Cochlearia officinalis*, and sea milkwort *Glaux maritima*.
- *Upper (high) marsh.* The zone is usually dominated by grasses and rushes. The most characteristic plants of upper marsh are saltmarsh rush *Juncus gerardii* and red fescue *Festuca rubra*. Others can include (depending on the latitude of the marsh) common couch grass *Elytrigia repens*, sea couch grass *Agropyron pungens*, hard-grass *Parapholis strigose*, and creeping bent *Agrostis stolonifera*.
- *Transitional (driftline) marsh.* Widespread communities around the highest tide level include those dominated by rushes and reeds, such as sea club-rush *Scirpus*

Fig. 5 Common sea-lavender *Limonium vulgare* characteristically forms dense meadows in low-mid level saltmarshes on the Atlantic coasts of Europe (Photo credit: Nick Davidson © Copyright remains with the author)

maritimus, sea rush *Juncus maritimus*, and, in the less saline areas, common reed *Phragmites australis*. A number of scarce saltmarsh plants also occur in this zone, including matted sea-lavender *Limonium bellidifolium*, rock sea-lavender *Limonium binervosum*, sea heath *Frankenia laevis*, and the bushy perennial shrubby seablite *Suaeda vera*.

Although western and southern Europe has only a few truly estuarine fish species (sea bass *Dicentrarchus labrax*, the goby *Pomatoschistus microps*, and three mullet species (*Chelon labrosus*, *Liza ramada*, and *L. aurata*)), at least 100 fish species occur in European estuaries, and the tidal creeks of European tidal saltmarshes contain fish assemblages with high proportions of juvenile individuals and so can provide significant refugia as spawning and nursery areas for life-cycle stages vulnerable to predation (Kneib 1997; Mathieson et al. 2000).

European saltmarshes are of major importance for their assemblages of breeding shorebirds (Charadrii), notably common redshank *Tringa totanus*, lapwing *Vanellus vanellus*, Eurasian oystercatcher *Haematopus ostralegus*, and common ringed plover *Charadrius hiaticula*, and in some areas dunlin *Calidris alpina schinzii*. The saltmarsh-breeding bird assemblage also typically includes other species such as mallard *Anas platyrhynchos*, skylark *Alauda alauda*, reed bunting *Emberiza schoeniclus*, and meadow pipit *Anthus pratensis*, and sometimes also yellow wagtail *Motacilla flava* and colonial-breeding gulls and terns (Laridae). In Great Britain, densities of breeding common redshank are much higher on some saltmarshes than on inland wetlands, and an estimated 45% of the population of common redshank breeding in Great Britain and 7% of the northwest and central European population breeds on British saltmarshes (Davidson et al. 1991; Brindley et al. 1998). Of concern is that many of the important breeding shorebird populations of saltmarshes in Britain and the international Wadden Sea are in long-term decline (Brindley et al. 1998; Koffijberg et al. 2015).

Short grass-dominated saltmarshes in western Europe, and particularly in the United Kingdom and the international Wadden Sea, are vitally important spring feeding areas for several Arctic-breeding geese species, notably barnacle goose *Branta leucopsis*, dark-bellied brent goose *B. bernicla bernicla*, and white-fronted goose *Anser albifrons* (Bos et al. 2005). Rich feeding areas at this time of year permit the geese rapidly to accumulate fat and protein reserves for migration and breeding, and feeding conditions in spring are closely linked with breeding success.

Conservation Status

Much of the coastal saltmarsh in Europe is covered by multiple national and international conservation designations. There are 219 Wetlands of International Importance (Ramsar Sites) (covering a total area of 6.63 million hectares) designated in Europe under the Ramsar Convention on Wetlands which include coastal saltmarshes (Fig. 6). These include all major UK estuarine saltmarsh systems, and

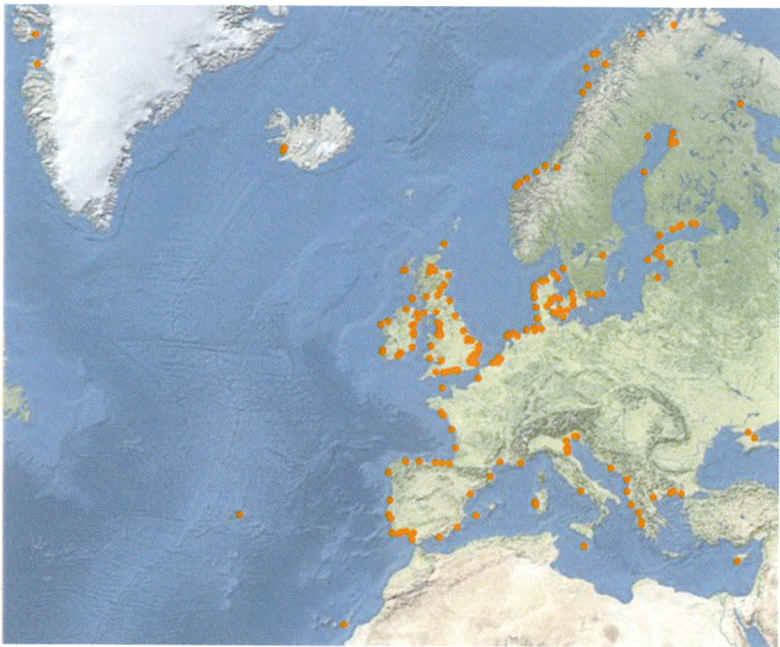

Fig. 6 The location of the 219 European Ramsar Sites which include areas of saltmarsh (Source: Ramsar Sites Information Service (RSIS) (Accessed 10 March 2016))

much of the extensive saltmarsh areas of the international Wadden Sea, as well as saltmarshes on the Norwegian coast, and around the Baltic and Mediterranean Seas. The Wadden Sea was also inscribed in 2009 (extended in 2014) as a UNESCO World Heritage Site for the global importance of its natural coastal ecosystems as the largest unbroken system of intertidal sand and mud flats in the world.

The European Union's 1992 Directive on the conservation of natural habitats and of wild fauna and flora (Council Directive 92/43/EEC), in its Annex I (Natural Habitat Types of Community Interest whose Conservation requires the Designation of Special Areas of Conservation), lists two such saltmarsh habitats: "Atlantic and continental salt marshes and salt meadows" and "Mediterranean and thermo-Atlantic salt marshes and salt meadows." Nineteen EU countries with coastlines have designated 1,212 Natura 2000 network sites (under the EU's Birds and Habitats Directives) which contain saltmarshes (EEA 2015). The largest numbers (79% of the total) of these Natura 2000 sites are in Spain, France, Italy, Denmark, Sweden, Ireland, and the United Kingdom.

Several European saltmarsh areas are also included in Biosphere Reserves, declared under the UNESCO Man and Biosphere Programme. These include the Po Delta (Italy), the Camargue (France), Dublin Bay (Ireland), and the Wadden Sea in Germany and the Netherlands.

However, despite the many nature conservation designations covering European saltmarshes, the conservation status of many parts of the resource is not good. Under the EU Habitats Directive Article 17, EU member states are required to report periodically on the conservation status of different habitats, including saltmarshes. Very few report that their saltmarsh habitats are in "favorable conservation status."

For the most recent period (2007–2012) (EIONET 2015):

- Salicornia *and other annuals colonizing mud and sand* (Natura 2000 habitat code 1310), a widespread lower saltmarsh habitat around all European coasts, have "unfavorable bad" status on Atlantic region coasts (except for "favorable" status reported for Denmark), and "unfavorable inadequate" on Mediterranean and Baltic coasts.
- Spartina *swards* (Habitat code 1320), widespread along Atlantic coasts, but also occurring locally in the northern Adriatic, have "unfavorable bad" status for most of the Atlantic region except for Germany and Denmark ("favorable" status).
- *Atlantic salt meadows* (Habitat code 1330) distributed along the coasts of the Atlantic and the western Baltic in the upper part of the intertidal zone have "unfavorable" conservation status except for Spain (status "unknown").
- *Mediterranean salt meadows* (Habitat code 1410), saltmarshes in the Mediterranean basin dominated by *Juncus* rushes especially sea rush *Juncus maritimus* have "unfavorable bad" status throughout the European coasts of the Mediterranean.
- *Mediterranean and thermo-Atlantic halophilous scrubs* (Habitat code 1420), salt tolerant scrub growing on saline muds in the Mediterranean region (Fig. 7) and also in particularly warm sites (e.g., south facing) in other European regions have "unfavorable bad" status on Mediterranean coasts, and "unfavorable inadequate" status elsewhere.

Fig. 7 Mediterranean salt-tolerant scrub often develops around abandoned salt pans (salines), here in the Bay of Cadiz, Spain (Photo credit: Nick Davidson © Rights remain with the author)

Many pressures and threats to these saltmarsh habitats were reported, with the most frequent including altered coastal dynamics, urbanization and other land-claim and livestock grazing, as well as rising sea levels.

Threats and Future Challenges

Many parts of Europe's coastal saltmarshes have been progressively embanked and converted to agricultural land for centuries, for example, around the shores of the Wadden Sea and the estuaries of the United Kingdom (see e.g., Davidson et al. 1991 for examples from Great Britain). Mostly in the southern parts of Europe, natural saltmarshes have been converted to salines (salt pans) for the artisanal and commercial production of sea salt. More recently, saltmarshes have been extensively infilled for urban, industrial, port, tourism, and infrastructure developments. As a response to the devastating 1953 storm surge and flooding in the southern North Sea, the Netherlands undertook major coastal protection engineering works in the Dutch Delta region, including the complete closure of some tidal inlets, with consequent loss of saltmarshes in these areas. Although no overall figures for area losses exist, most land claims have particularly affected upper saltmarsh parts of European and coastal systems.

Although such land claim of saltmarshes and other parts of coastal areas has diminished in recent years, its long-term legacy has been to "fix" the location of the upper tidal limit of many parts of the European soft coast. With progressively rising sea levels linked to climate change, this has led to a "coastal squeeze," whereby the width of the intertidal shoreline, including its saltmarsh, is narrowing and steepening, increasing saltmarsh erosion. For example, Burd (1992) estimated that in 11 estuaries in southeast England 20.3% of saltmarsh was lost through erosion between 1973 and 1988, an average rate of 1.35% year^{-1}. Much of the loss was of lower pioneer marsh vegetation, and over the same period there was very little saltmarsh accretion. Since saltmarshes are important in dissipating wave energy as a natural coastal defense, such continuing erosion increases the risk of breach of sea defenses and flooding during storm events.

Cord-grass *Spartina* in European saltmarshes has an interesting history (Doody 1984; Davidson et al. 1991). Originally there was a single *Spartina* species native to the UK, small cord-grass *S. maritima*. Sometime before 1870, the native American smooth cord-grass *S. alterniflora* was accidentally introduced into Southampton Water in southern England. Hybridization with the native species then produced the fertile hybrid common cord-grass *S. anglica*. This species, now the most widespread in European saltmarshes (Fig. 8), showed the ability to colonize open mudflats rapidly and to lower tidal levels than other species. As a consequence, in the twentieth century it was extensively planted in western Europe (and elsewhere in the world) to increase shoreline stabilization and coast protection (Hubbard and Stebbings 1967). However, the spread of *S. anglica* over open mudflats has also raised nature conservation concerns over the loss of feeding grounds of internationally important wintering populations of waterbirds, which led to several attempts to control *Spartina* by spraying with herbicides. More recently there has been *Spartina* "die-back" occurring in parts of its range, notably in southern England. Although not fully understood, this is thought to be a natural phenomenon and is now causing concerns over reduced coastal defense capacity of such marshes.

Many areas of saltmarsh in western Europe are grazed by domestic livestock. The intensity of such grazing has major impact on the type of saltmarsh vegetation that results. With high grazing densities, saltmarshes have lower plant species diversity, reduced structural diversity (with lower invertebrate diversity and lower densities of breeding birds), but provide important winter feeding areas for grazing ducks and geese (Doody 1987).

Since the early 1990s concerns over the continuing losses of saltmarsh in northwest Europe, including in relation to the requirements of the EU Habitats Directive to maintain such habitats and compensate for losses, have led to an increasing number of "managed realignment" (sometimes called "managed retreat") schemes. These involve the deliberate breaching of artificial seawalls to allow the reinstatement of tidal inundation over areas of low-lying land claimed in past centuries for agriculture. With the maintenance and heightening of seawalls becoming increasingly cost-ineffective, this approach has the co-benefit of reinstating the role of saltmarshes in providing natural sea defenses. Since sea levels have risen since these areas were originally embanked, such schemes often also involve major works to raise surface levels in the areas to be inundated to intertidal levels, so that saltmarsh and intertidal mud and sandflats can develop.

Wolters et al. (2005) record 29 such deliberate managed realignment schemes in northwest Europe being undertaken up to 2003 (and a further six planned), in the

Fig. 8 Common cord-grass *Spartina anglica often* forms extensive swards over low-level saltmarshes around the coasts of Europe (Photo credit: Nick Davidson © Rights remain with the author)

UK, Netherlands, Belgium, and Germany, with the majority being around the shores of the southern North Sea and a few in the southern Baltic. More recently there have been further major such realignment schemes initiated, notably the 600 ha scheme at Wallasey Island in the Crouch-Roach Estuary in southeast England.

However, there are a number of concerns over the effectiveness of such realignment schemes in achieving their objectives. Atkinson et al. (2004) found that although 5 years after initial breaches in the seawalls wintering shorebird communities within managed realignment sites were broadly similar to those of surrounding mudflats there were some notable exceptions, largely a consequence of the extent of colonization of different benthic invertebrates upon which the birds feed. Mossman et al. (2012) found that although halophytic plant species colonized realignment sites rapidly, attaining species richness similar to nearby reference marshes after 1 year, the plant community of managed realignment sites was significantly different from reference sites, with early successional species remaining dominant, even on the high marsh. They also found that, in other sites where seawalls had accidentally breached up to a century ago, there still remained differences in plant communities, with the characteristic perennials sea lavender, sea arrowgrass, sea plantain, and thrift remaining relatively rare but the shrub sea

Fig. 9 Reestablished saltmarsh vegetation, dominated by sea purslane *Halimione portulacoides* after accidental breaching of seawalls on the Blyth Estuary, eastern England (Photo credit: Nick Davidson © Rights remain with the author)

purslane more abundant (Fig. 9). Mossman et al. (2012) concluded that currently marshes created by managed realignment do not satisfy the requirements of the EU Habitats Directive.

References

Atkinson PW, Crooks S, Drewitt A, Grant A, Rehfisch MM, Sharpe J, Tyas CJ. Managed realignment in the UK – the first 5 years of colonization by birds. Ibis. 2004;146:101–10.

Bos D, Loonen MJJE, Stock M, Hofeditz F, van der Graaf J, Bakker JP. Utilisation of Wadden Sea salt marshes by geese in relation to livestock grazing. J Nat Conserv. 2005;13:1–15.

Brindley E, Norris K, Cook T, Babbs S, Forster Brown C, Massey P, Thompson R, Yaxley R. The abundance and conservation status of redshank *Tringa totanus* nesting on saltmarshes in Great Britain. Biol Conserv. 1998;86:289–97.

Burd F. The saltmarsh survey of Great Britain, Research & survey in nature conservation, vol. 17. Peterborough: Nature Conservancy Council; 1989.

Burd F. Erosion and vegetation change on the saltmarshes of Essex and north Kent between 1973 and 1988. Research & survey in nature conservation, no. 42. Peterborough: Nature Conservancy Council; 1992.

Cambridge Coastal Research Unit (CCRU). Global wetland database developed for the Dynamic and Interactive Assessment of National Regional and Global Vulnerability of Coastal Zones to Climate Change and Sea-Level Rise (DINAS-COAST) project. Potsdam: Potsdam Institute for Climate Impact Research; 2003.

Davidson NC, Laffoley D'A, Doody JP, Way LS, Gordon J, Key R, Drake CM, Pienkowski MW, Mitchell RM, Duff KL. Nature conservation and estuaries in Great Britain. Peterborough: Nature Conservancy Council; 1991. 422 p.

Dijkema KS. Salt marshes in Europe, Nature and environment series, vol. 30. Strasbourg: European Committee for the Conservation of Nature and Natural Resources; 1984.

Doody P, editor. Spartina anglica in Great Britain. A report of a meeting held at Liverpool University on 10th November 1982, Focus on nature conservation, vol. 5. Huntingdon: Nature Conservancy Council; 1984.

Doody JP. Botanical and entomological implications of saltmarsh management in intertidal areas. RSPB Symposium. Sandy: Royal Society for the Protection of Birds; 1987.

Environment Agency. Salt marsh management manual, R&D technical report SC030220. Bristol: Environment Agency; 2007.

European Environment Agency. EUNIS Habitat Classification. 2012. Available from http://www.eea.europa.eu/themes/biodiversity/eunis/eunis-habitat-classification. Accessed 28 Mar 2016.

European Environment Agency. 2015. Available from Natura 2000 data – the European network of protected sites. http://www.eea.europa.eu/data-and-maps/data/natura-6. Accessed 16 Mar 2016.

European Topic Centre on Biological Diversity (EIONET). Habitat assessments at EU biogeographical level. 2015. Available from http://art17.eionet.europa.eu/article17/reports2012/habitat/summary/. Accessed 16 Mar 2016.

Hecker N, Tomas-Vives P, editors. The status of wetland inventories in the Mediterranean Region. IWRB publication 38. Slimbridge: International Waterfowl & Wetlands Research Bureau; 1995 146 p.

Heliotis FD. An inventory and review of the wetland resources of Greece. Wetlands. 1988;8:15–31.

Hubbard JCE, Stebbings RE. Distribution, date of origin and acreage of *Spartina townsendii* (s.l.) marshes in Great Britain. Proc Bot Soc Br Isl. 1967;7:1–7.

Kneib RT. The role of tidal marshes in the ecology of estuarine nekton. Oceanogr Mar Biol Annu Rev. 1997;35:163–220.

Koffijberg K, Laursen K, Hälterlein B, Reichert G, Frikke J, Soldaat L. Trends of breeding birds in the Wadden sea 1991–2013, Wadden sea ecosystem, vol. 35. Wilhelmshaven: Common Wadden Sea Secretariat, Joint Monitoring Group of Breeding Birds in the Wadden Sea; 2015.

Mathieson S, Cattrijsse A, Costa MJ, Drake P, Elliot M, Gardner J, Marchand J. Fish assemblages of European tidal marshes: a comparison based on species, families and functional guilds. Mar Ecol Prog Ser. 2000;204:225–42.

Mossman HL, Davy AJ, Grant A. Does managed coastal realignment create saltmarshes with 'equivalent biological characteristics' to natural reference sites? J Appl Ecol. 2012;49:1446–56.

Schuyt K, Brander L. Living waters: the economic values of the world's wetlands. Gland: World Wildlife Fund; 2004. Available from http://vinc.s.free.fr/IMG/wetlandsbrochurefinal.pdf. Accessed 28 Mar 2016.

Wolters M, Garbutt A, Bakker JP. Salt-marsh restoration: evaluating the success of de-embankments in north-west Europe. Biol Conserv. 2005;123:249–68.

Tipperne Peninsula and Ringkøbing Fjord (Denmark)

78

Hans Meltofte, Preben Clausen, and Ole Thorup

Contents

Introduction	974
Physiochemical Conditions in the Lagoon	974
Biodiversity	977
Submerged Vegetation	977
Benthos and Fish	977
Birds	978
Ecosystem Services	980
Conservation Status	981
Threats and Future Challenges	981
References	982

Abstract

The establishment of the Tipperne Reserve (55° 52′ N, 8° 14′ E) in the brackish lagoon Ringkøbing Fjord in Denmark dates back to 1898 when the breeding birds were protected on 6 km^2 of brackish meadows in the southern end of the lagoon. In 1928 the Reserve was significantly expanded to also include about 18 km^2 of shallow flats with prohibition of hunting. From the same year, monitoring of waterbirds and other biodiversity was initiated and has continued ever since. Today, the Reserve presents almost 90 years of continuous bird data together with more randomly collated data on benthos, soil invertebrates and year round water levels in the fjord as well as on the ever expanding meadows. Changes during this

H. Meltofte (✉)
Department of Bioscience, Aarhus University, Roskilde, Denmark
e-mail: mel@bios.au.dk

P. Clausen
Department of Bioscience, Aarhus University, Kalø, Denmark

O. Thorup
Amphi Consult, Ribe, Denmark

© Springer Science+Business Media B.V., part of Springer Nature 2018
C. M. Finlayson et al. (eds.), *The Wetland Book*,
https://doi.org/10.1007/978-94-007-4001-3_221

long time period have been dramatic, both for the breeding birds and for the staging waterbirds. First of all, eutrophication led to an almost total die off of submerged vegetation around 1979 and mass reductions in associated waterbird numbers that was followed by several attempts to recover good ecological conditions in the fjord. Not until recently have these efforts seem to have been successful.

Keywords

Tipperne Reserve · Ringkøbing Fjord · Ramsar site · Special Protection Area · Denmark · Wetlands · Waterbirds · Biodiversity · Eutrophication · Long-term monitoring

Introduction

The core area of the Ramsar Site of Ringkøbing Fjord is made up of the state-owned Tipperne Reserve (55° 52′ N, 8° 14′ E). It constitutes the northern quarter of a low peninsula with surrounding waters in the 285 km^2 coastal lagoon of Ringkøbing Fjord in West Jutland, Denmark (Fig. 1). It covers 7 km^2 of brackish meadows and marshland surrounded by 18 km^2 of primarily shallow water (Fig. 2). The reserve was established as a hunting-free bird sanctuary in 1928, and from the very beginning, numbers of breeding as well as staging birds were monitored each year (Madsen 1985; Meltofte 1987; Thorup 1998; Meltofte and Clausen 2011, 2016). The present account is primarily based on these major analyses.

The remaining part of the peninsula is a similar area of privately owned brackish meadows, marshlands, and heathland primarily used for grazing, hunting, reed cutting, and birding. Ringkøbing Fjord is separated from the North Sea by a 32 km long and 1–2 km wide natural sand isthmus. The fjord is state property, while most of the surrounding coastal fringes are privately owned.

Physiochemical Conditions in the Lagoon

The water level and salinity of the lagoon is regulated by a sluice to the North Sea, but the regulation has varied considerably through the years, and so has the outflow of sediments and nutrients from the main river, Skjern Å, into the lagoon.

Originally, the lagoon was an open inlet with the same salinity as the North Sea, but during the eighteenth and the nineteenth centuries, expanding sandbars from the south and north closed off the inlet from the sea, thus forming a brackish lagoon with an outlet to the southwest of Tipperne.

After a number of different man-made outlets had existed from 1845 onwards, the present channel and sluice at Hvide Sande was established in 1931. But even since then, the administration of the sluice has varied so much – for political reasons – that living conditions in the lagoon have changed considerably concerning both annual patterns of water level and salinity.

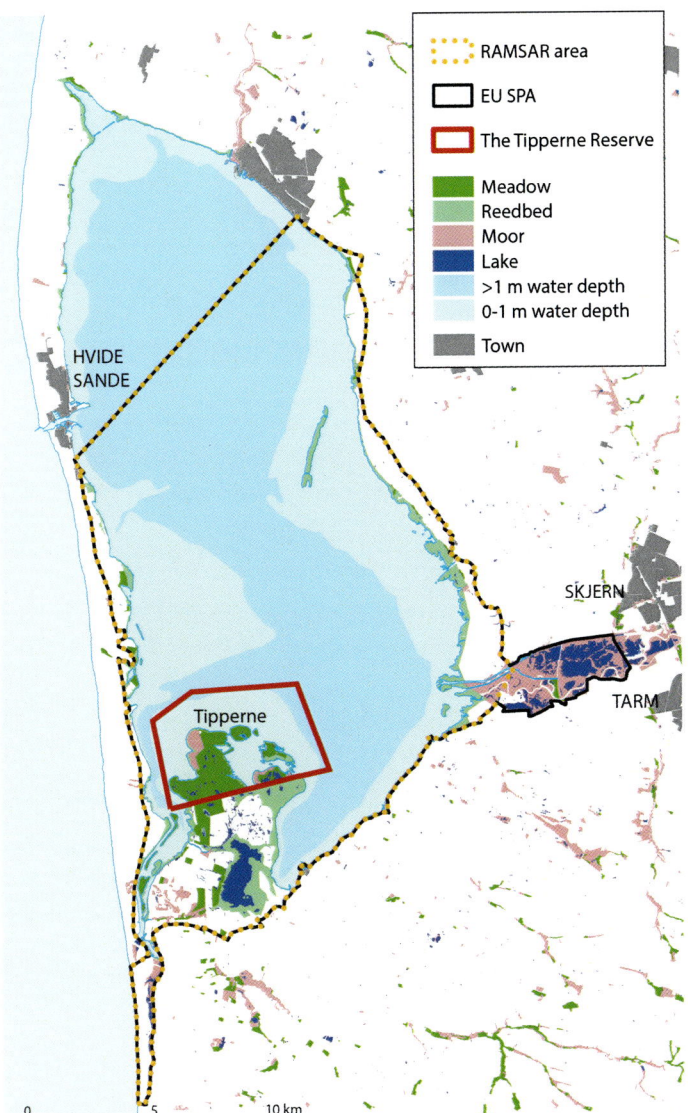

Fig. 1 Map of Ringkøbing Fjord with international conservation regulations. These include the Danish Ramsar Site no. 2, which is almost identical to Special Protection Area under the EU Birds Directive no. 43. The Tipperne Reserve is framed in *red* (Translated from Meltofte and Clausen 2011; Reproduced with permission by Dansk Orn. Foren. Tidsskr.)

Fig. 2 The Tipperne Reserve is the core area of the Ringkøbing Fjord Ramsar Site and EU Special Protection Area. It covers 7 km² of brackish meadows and marshland surrounded by 18 km² of primarily shallow water. Ever since 1928 birdlife and other biodiversity in the reserve has been monitored from a field station manned year round or most of the year at least (Photo credit: courtesy of Erik Thomsen ©)

Another important change to the environment of the lagoon was the drainage in the mid-1960s of the large delta of Skjern Å together with increased use of agricultural fertilizers in the catchment area; this resulted in heavily increased sedimentation and eutrophication of the lagoon (see below). A large part of the delta was reestablished as wetlands during 1999–2002, but the effect on the inflow of nutrients has been limited.

Large parts of the lagoon are shallow (Fig. 1). The area with less than 0.5 m depth – and hence appropriate for most nondiving waterbirds – makes up 72 km² or 25% of the lagoon. Furthermore, 63 km² or 22% has depths of between 0.5 m and 1.5 m and thereby accessible for foraging swans and Eurasian coot *Fulica atra*, provided vegetation is present.

The inflow of salt water from the North Sea is a very important factor for the ecological conditions in the lagoon. From 1967 onwards, the intention was that the salinity of the lagoon should vary between 5 ‰ and 10 ‰, a management practice that led to significantly less salt and more stagnant water conditions in the lagoon, which together with the sedimentation and eutrophication from the 1960s onwards severely reduced the water quality. To compensate for this, the desired salinity was raised to 8–15 ‰ in 1987 and further to a summer salinity of 12–15 ‰ and a winter salinity of at least 8 ‰ from late 1995 onwards.

Biodiversity

Submerged Vegetation

Before the sluice in Hvide Sande was established in 1931, the submerged vegetation around Tipperne was dominated by salt-tolerant species of tasselweeds *Ruppia* spp. In 1956, i.e., 25 years after the displacement of the outlet to the North Sea from southwest of Tipperne to Hvide Sande, the dominant species around Tipperne were fennel pondweed *Potamogeton pectinatus* and spiral tasselweed *Ruppia cirrhosa* together with other brackish water species. In both periods, extensive vegetation cover was found to a depth of about 2 m.

In 1972, the increased freshwater conditions had resulted in brackish water-crowfoot *Batrachium baudotii* being almost as abundant as spiral tasselweed, and the first indications of eutrophication were evident, including a decline in the Charophyte community. The amount of submerged vegetation peaked in 1978, when the dry biomass on the reserve was 762 metric tons. During this and the following year most of the plants became overgrown with epiphytic algae, and 2 years later most plants had disappeared, leaving a bottom covered with mats of algae. In 1984, the dry biomass of the submerged vegetation was only 161 t. At that time, however, a recovery had begun, so that the dry biomass was 474 t in 1985 and 581 t in 1986, but plants only grew at depths of 1 m or less.

This situation – with varying amounts of submerged vegetation at relatively low depths – continued until 1995, when more salt water was led into the lagoon. This induced a regime shift resulting in disappearance of almost all vegetation (Petersen et al. 2008); a situation that the vegetation only to some degree had recovered from by 2009–2010. Since then there has been a return of submerged vegetation on the shallows in the fjord as evidenced from transect surveys as well as aerial images (see an orthophoto from 2014 in Meltofte and Clausen 2016).

Benthos and Fish

Due to the brackish water, the benthic fauna is relatively poor. Furthermore, as a result of the variation in the inflow of salt water to the lagoon and the increasing eutrophication during the last half century, the relative abundance among species has varied considerably.

Throughout the study period, the benthic fauna of the shallow flats has been dominated by the rag worm *Hediste diversicolor*, mud shrimp *Corophium volutator*, and varying species of mud snails *Hydrobia* spp. Furthermore, hypoxia-tolerant sludge worms *Tubifex costatus/Tubificoides benedii* have been common since the 1980s.

Rag worms, mud shrimps, and mud snails certainly have increased in density and biomass from the 1930s to the 1990s, in parallel with the increased sedimentation and eutrophication. The biomass of mud shrimps increased tenfold from the 1930s to the 1970s, mud snails tripled from the 1930s to the early 1990s, and rag worms increased fourfold up to the late 1990s. However, all three species subsequently declined, so that the total biomass of invertebrates on the mudflats peaked in the mid-1990s.

Below the vegetated zone, the benthic fauna is dominated by two rag worms, *Hediste diversicolor* and *Marenzelleria viridis*, the latter being an invasive species found in the lagoon at least since 1990. However, since more salt water was let into the lagoon from 1995 onwards, soft-shell clams *Mya arenaria* have been dominating with up to 85–95% of the biomass after having been almost absent for decades and have been the driving force behind the observed regime shift. Rag worms and crustaceans have just about doubled in biomass in deep water during the same period.

As for the benthos, varying sampling methods make it difficult to quantify the development in the fish fauna of the lagoon. However, based on available data together with specialist evaluations, it appears that the commercially important fish populations were stable from the 1930s up to 1970, declined in the 1980s with the eutrophication collapse and decreased further with the increased salt water intake from 1995 onwards – perhaps with a slight increase in the most recent years.

Birds

Following the deterioration of the food resources, especially the almost total die-off of the submerged vegetation around 1979–1980, numbers of most of the waterfowl have now fallen below the Ramsar Convention's criteria for wetlands of international importance, i.e., numbers of dabbling ducks – e.g., Eurasian widgeon *Anas penelope* (Fig. 3) and common teal *Anas crecca* – decreasing from annual maximum numbers generally between 25,000 and 50,000 ducks up until the late 1970s to a few thousand during 2001–2009. However, since then numbers of ducks have almost recovered to the level from the 1970s following widespread recovery of the submerged vegetation (Fig. 4). Other species, such as mute swan *Cygnus olor* and Eurasian coot occurred with peak numbers of 7,550 and 22,700, respectively, during the mass growth of submerged vegetation in the 1970s, but have largely disappeared since then, albeit with some recovery of swan numbers in recent years. Fish- and benthos-eating species have reacted according to similar large changes in their food resources.

However, for many of the waterfowl, the large numbers staging in the reserve in autumn were not only influenced by food resources within the area but also by intensive hunting in adjacent areas. During the first decades, the Tipperne Reserve was the only large shooting-free area in this part of the country, and the majority of the waterfowl in West Jutland could be concentrated within a few shooting-free areas. Especially during the 1980s and 1990s, an expanded network of shooting-free reserves was established all over Denmark, resulting in greatly increased numbers of staging waterfowl in the country (Clausen et al. 2004, 2013). In relation to Tipperne, this meant that the birds had many alternative safe staging areas, when the submerged vegetation disappeared almost completely in 1995.

Fig. 3 Eurasian widgeon is one of the most numerous dabbling ducks staging in the Tipperne Reserve, and the species has shown marked numerical fluctuation in relation to feeding possibilities in the form of submerged vegetation (Photo credit: courtesy of Jens Kristian Kjærgaard ©)

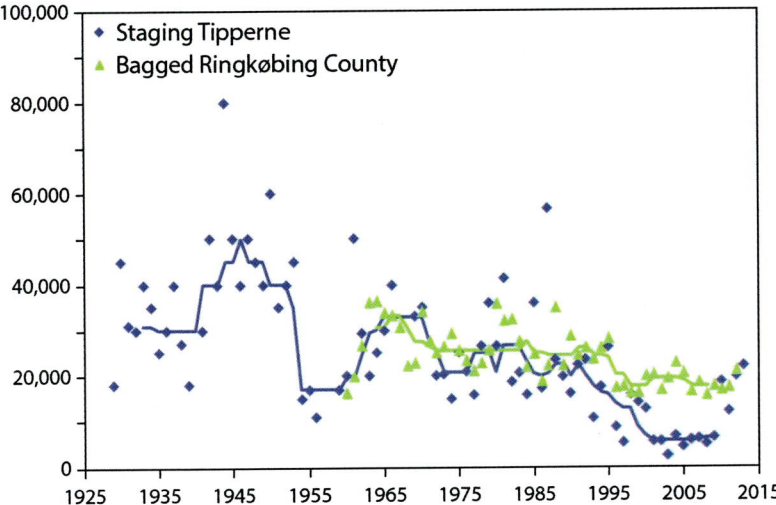

Fig. 4 Annual maximum numbers of autumn staging dabbling ducks (all species) on the Tipperne Reserve during 1929–2015. The bag of dabbling ducks shot in Ringkøbing County 1959–2012 is also shown. The lines give sliding 9 year medians (Data are from Meltofte and Clausen (2011), national reserve monitoring programs (Clausen et al. 2013) and the national game bag statistics)

In contrast to most ducks, several wader species have benefitted from the increased sedimentation of the fjord and the increasing densities of benthic invertebrates. This particularly concerns common snipe *Gallinago gallinago*, spotted redshank *Tringa erythropus* and common greenshank *Tringa nebularia*, whose numbers have multiplied on the Tipperne Reserve. Dunlin *Calidris alpina*, bar-tailed godwit *Limosa lapponica*, Eurasian golden plover *Pluvialis apricaria* are the most numerous staging wader species with numbers in the thousands.

During spring and autumn, hundreds or even thousands of pink-footed geese *Anser brachyrhynchus* and dark-bellied brent geese *Branta b. bernicla* utilized the meadows on Tipperne up until around the mid 1990s, but their numbers have declined since then. Instead, thousands of barnacle geese *Branta leucopsis* now occur in the reserve. The pink-footed geese are still around the lagoon, but nowadays primarily feed on agricultural fields east of the lagoon and roost on the Klægbanken reserve, whereas the brent geese most likely have declined in local numbers in response to a general population decline (Clausen et al. 2013).

The northern part of the Tipperne Peninsula constitutes one of the most important breeding areas for meadow breeding shorebirds in Europe, and is one of very few sites in temperate North and Northwest Europe with remaining and stable numbers of ruff *Philomachus* [now *Calidris*] *pugnax* and the Baltic biogeographic population of dunlin *C. a. schinzii*. Favorable breeding conditions for ruff, dunlin, and other shorebirds like black-tailed godwit *Limosa limosa*, common redshank, Eurasian curlew *Numenius arquata*, northern lapwing *Vanellus vanellus*, and common snipe are safeguarded in the Tipperne Nature Reserve by keeping the freshwater table as high as possible and by fine-tuned land use with cattle grazing and mowing, adjusted annually according to an evaluation of bird distribution, breeding success, and vegetation development.

In the past, there have been periods with large numbers of breeding colonial species like pied avocet *Recurvirostra avosetta* and Arctic tern *Sterna paradisaea*. At present, they are only breeding in small numbers because of the presence of predators like red fox *Vulpes vulpes* and peregrine falcon *Falco peregrinus*, which prevent successful breeding of colonial nesting birds (Thorup and Bregnballe 2015).

No mammals occur in any significant numbers in the fjord, but Eurasian otter *Lutra lutra* is now reoccurring after having been locally extinct for decades (Elmeros et al. 2006).

Ecosystem Services

Ringkøbing Fjord is used quite intensively for fishing, hunting, and recreation. The total catches of the eight most important fish species for human consumption (European flounder *Platichthys flesus*, herring *Clupea harengus*, eel *Anguilla anguilla*, whitefish *Coregonus lavaretus*, European smelt *Osmerus eperlanus*, roach *Rutilus rutilus*, perch *Perca fluviatilis*, and northern pike *Esox lucius*) reached about 1,500 t per year until a decrease began in the 1980s. The exact hunting bag taken in and around the lagoon is unknown, but it is reasonable to assume that the majority of dabbling ducks taken in the former Ringkøbing County were shot in or near Ringkøbing Fjord. The bag numbered around 30,000 ducks in the 1960s and 1970s, but was much reduced between 1995 and 2009, followed by an upsurge in numbers harvested that reflects duck numbers in the lagoon area (Fig. 4). Water sports like wind and kite surfing are very popular and well regulated.

Conservation Status

The Tipperne Reserve and Ringkøbing Fjord are among the best protected areas in Denmark. The breeding birds were protected already in 1898, and the present reserve was established in 1928, with a ban on hunting and regulation of public access.

Except for the northernmost corner (Fig. 1), Ringkøbing Fjord and the Tipperne Peninsula has been a Ramsar Site since 1978. In 1983, the area was also declared a Special Protection Area for birds under the EU Birds Directive and is designated for no less than 11 breeding and 24 staging bird species due to a unique richness and diversity of habitats and birdlife in the area – a total number of designations only exceeded by the much larger Danish part of the Wadden Sea. The majority of these 30 species are waterbirds, which until a few decades ago prospered from rich food resources in the form of submerged vegetation, invertebrates, and fish. Fortunately, there are signs of recovery in recent years, most likely as a response to improved water quality.

Furthermore, all of Ringkøbing Fjord, including the reserve and all islands, was protected as a nature reserve in 1985, when also the area protected from hunting and different kinds of disturbing boating activities was expanded considerably. Also, in 1987, hunting from motor boats was banned in the entire lagoon. Finally, it was declared an EU Special Area for Conservation in 1998.

Threats and Future Challenges

It is clear that the lagoon still suffers from eutrophication, primarily originating from admission of nutrients from arable land in the catchment area. Indeed, the reduction of this nutrient admission resulting from three national Action Plans for the Aquatic Environment (see Riemann et al. 2016) seems to be a main precondition for the reestablishment of a healthy lagoon with a submerged vegetation and a fauna of staging waterbirds resembling those found before the ecosystem collapsed around 1979–1980. On top of this comes the regulation of the salinity in the lagoon, which at present aims at a relatively high salinity of 12–15 ‰ in May–September in order to enable soft-shell clams to reproduce. Because of the eutrophication, soft-shell clams are considered necessary for securing a water transparency that allows submerged vegetation to grow to depths of more than 1.5 m. Since this high salinity induced a regime shift in the lagoon and killed the original submerged vegetation in 1995, and new more salt-tolerant plants have been slow to move in, but as reported above there are recent signs of significant numbers of dabbling ducks returning to the lagoon. Thus, in the future the lagoon may once again attract high and internationally important numbers of waterfowl under the present sluice management regime, provided that the inflow of nutrients to the fjord is under control, so that the fjord can fully recover.

References

Clausen P, Bøgebjerg E, Hounisen JP, Jørgensen HE, Petersen IK. Reservatnetværk for vandfugle. En gennemgang af udvalgte arters antal og fordeling i Danmark 1994–2001. Faglig Rapport fra DMU. 2004;490:1–142 [in Danish].

Clausen P, Holm TE, Laursen K, Nielsen RD, Christensen TK. Rastende fugle i det danske reservatnetværk 1994–2010. Del 1: Nationale resultater. Videnskabelig rapport fra DCE. 2013;72:1–118 [in Danish].

Elmeros M, Hammershøj M, Madsen AB, Søgaard B. Recovery of the Otter *Lutra lutra* in Denmark monitored by field surveys and collection of carcasses. Hystrix It J Mam. 2006;17:17–28.

Madsen J. The goose populations at the Tipperne peninsula, western Jutland, Denmark. I: occurrence and population trends, 1929–1983. Dansk Orn Foren Tidsskr. 1985;79:19–28 [in Danish with English summary].

Meltofte H. The occurrence of staging waders Charadrii at the Tipperne reserve, western Denmark, 1928–1982. Dansk Orn Foren Tidsskr. 1987;81:1–108 [in Danish with extensive English summary].

Meltofte H, Clausen P. The occurrence of swans, ducks, Coot and Great Crested Grebe in the Tipperne Reserve 1929–2007 in relation to environmental conditions in the brackish lagoon, Ringkøbing Fjord, Denmark. Dansk Orn Foren Tidsskr. 2011;105:1–120 [in Danish with extensive English summary].

Meltofte H, Clausen P. Trends in staging waders on the Tipperne Reserve, western Denmark, 1929–2014 with a critical review of trends in flyway populations. Dansk Orn Foren Tidsskr. 2016;110:1–72.

Petersen JK, Hansen JW, Laursen MB, Clausen P, Carstensen J, Conley DJ. Regime shift in a coastal marine ecosystem. Ecol Appl. 2008;18:497–510.

Riemann B, Carstensen J, Dahl K, Fossing H, Hansen JW, Jakobsen HH, Josefson AB, Krause-Jensen D, Markager S, Stæhr PA, Timmermann K, Windolf J, Andersen JH. Recovery of Danish coastal ecosystems after reductions in nutrient loading: a holistic ecosystem approach. Estuar Coasts. 2016;39:82–97.

Thorup O. The breeding birds on Tipperne 1928–1992. Dansk Orn Foren Tidsskr. 1998;81:1–192 [in Danish with extensive English summary].

Thorup O, Bregnballe T. Pied Avocet conservation in Denmark – breeding conditions and proposed conservation measures. Dansk Orn Foren Tidsskr. 2015;109:134–44.

Wadden Sea (Denmark)

79

Karsten Laursen and John Frikke

Contents

Introduction	984
A Tidal Ecosystem	984
Physical Conditions	984
Sediments	986
Nutrients	987
Biodiversity	988
Plants and Benthos	988
Fish	988
Breeding Birds	989
Migratory Birds	989
Sea Mammals	993
Ecosystem Services	993
Conservation Initiatives and Status	993
Threats and Future Challenges	994
References	995

Abstract

The Danish part of the Wadden Sea comprises 1,500 km^2, and makes up about one tenth of the whole Wadden Sea area. In contrast to the other parts of the Wadden Sea the water quality in the Danish section is not influenced by large rivers, but mostly inputs from the North Sea. During the period 1990–2010 nitrogen in the water body has been nearly halved. The increase in the number of houting, an endemic fish

K. Laursen (✉)
Department of Bioscience, Aarhus University, Aarhus, Denmark
e-mail: kl@dmu.dk; kl@bios.au.dk

J. Frikke (✉)
The Danish Wadden Sea National Park, Rømø, Denmark
e-mail: jofri@danmarksnationalparker.dk

© Springer Science+Business Media B.V., part of Springer Nature 2018
C. M. Finlayson et al. (eds.), *The Wetland Book*,
https://doi.org/10.1007/978-94-007-4001-3_135

species, indicates it may now be secure due to intensive river restoration. Breeding birds are numerous, but the populations of many species have decreased recently. Millions of water birds stay during spring and autumn, and the site is of international importance to 20 species. The harbour seal is the most numerous of the sea mammals, and the small population of grey seal is increasing in number. The Danish part of the Wadden Sea is well protected, but still under pressure from changes in land use and recreational activities.

Keywords
Breeding birds · Houting · Intertidal flats · Marine mammals · Migratory birds · Nutrients · Sediments

Introduction

The Danish part of the Wadden Sea (Fig. 1) is considered the largest and most natural wetland in the country; and the trilateral Danish-German-Dutch Wadden Sea makes up one of the largest coherent tidal ecosystems in the world. The Wadden Sea region includes the whole coastal area from Den Helder in the Netherlands to Blåvands Huk in Denmark. It is a strip of coastal wetland some 10–30 km wide and more than 500 km long, bordering the North Sea. With an area of about 1,500 km^2, including 300 km^2 of land, the Danish part makes up about one-tenth of the trilateral area. The tidal amplitude in this section is 1.5–1.8 m.

The major part of the Danish Wadden Sea is dominated by natural, dynamic processes, with minor anthropogenic influences. It is situated in the transition zone between sea and land, and the forces of wind and water lead to the formation and erosion of the typical landscape elements of the area – tidal flats, sandbanks, salt marshes, and islands.

As one of its most obvious biological characteristics, the area is known as the most important locality for waders in Denmark and is also important to large numbers of other waterbird species, such as geese, dabbling ducks, and terns. In total, it is estimated that 10–12 million waterbirds are passing through the Wadden Sea on an annual basis, and for many of the species migrating along the East Atlantic Flyway, the area is of vital importance for food uptake and improving body condition.

A Tidal Ecosystem

Physical Conditions

Compared with other wetlands in Denmark, the Wadden Sea is special in view of its position as a wetland on the "edge" of the North Sea, sheltered by the peninsula of Skallingen, some high sandbanks, and the barrier islands of Fanø and Rømø. In addition to the dynamic ecosystem, the landscape is also remarkable in terms of

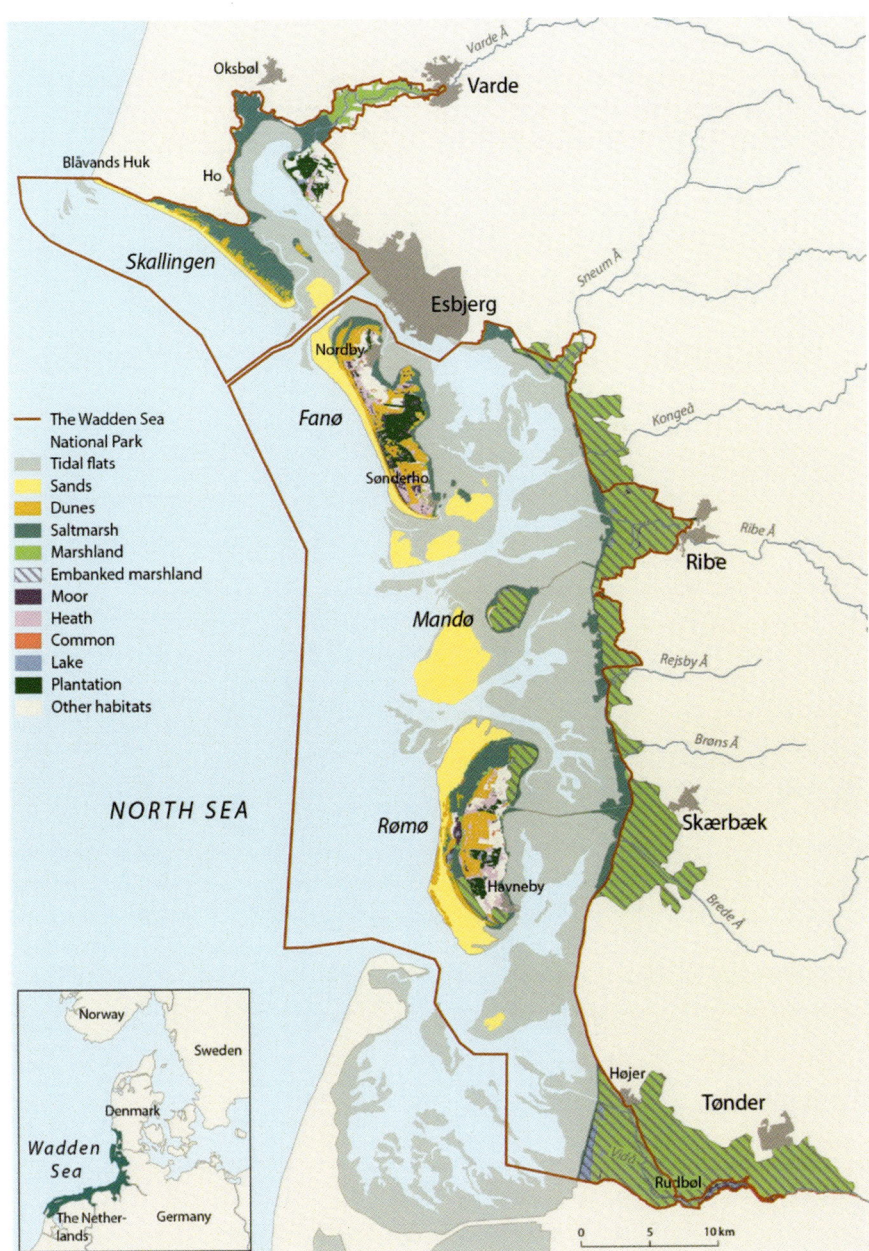

Fig. 1 The Danish Wadden Sea with indication of habitat types. The border of the Wadden Sea National Park, established in 2010, is shown (With permission from the Wadden Sea National Park)

Fig. 2 Due to the tide, extensive areas of highly productive sand and mud flats are exposed twice a day – here a lugworm flat in Lister Dyb tidal area in the southern part of the Danish Wadden Sea (Photo Credit: John Frikke © Rights remain with the author)

cultural history, housing, and dike construction that are influenced by the close proximity of the sea.

The most pristine and dynamic landscapes and natural conditions are found to the west, while to the east, the most human-influenced landscapes and ecosystems occur. Toward the mainland coast, the natural marine environment is replaced by human-modified freshwater meadows and regulated watercourses. The most crucial element of this difference is the sea dikes, which in the Danish Wadden Sea were built mainly in the early 1900s. Only one river, the River Varde in the very northern part, is not regulated by dikes or locks and forms an original tidal estuary.

Sediments

Due to the significant tidal range and the relatively low water level, the Danish Wadden Sea is the only site in Denmark where extensive tidal flats of mud, sand-mixed mud, and sand occur. Almost 80% of the Wadden Sea behind the peninsula of Skallingen and the barrier islands of Fanø and Rømø are exposed at low tide (Fig. 2).

The environment is mainly dominated by inputs from the North Sea, since no large rivers flow into the site. This is in contrast to the situation in the German and the Dutch parts of the Wadden Sea, where the enormous freshwater outlets from large parts of Central Europe via large rivers, such as the Rhine, the Weser, and the Elbe, have a major impact on the sediment and nutrient conditions in the Wadden Sea.

Large amounts of sediments (sand and clay) enter the Danish Wadden Sea from the North Sea. Since the last Ice Age, sea level rise and sedimentation have interacted with isostatic lowering of the landscape and local processes in determining the actual water level. In two large tidal basins in the northern (Grådyb) and southern (Lister Deep) parts, the rate of sedimentation exceeds erosion and isostatic lowering, which seems to be a general trend in the intertidal area in the Danish Wadden Sea. This means that the seabeds are continuously rising at a rate faster than the rise in sea level (Ingvardsen 2006a, b). For waterbird species depending on exposed tidal flats for feeding, this balance between the rising sea level and the landscape lowering processes is crucial. It implies that the tidal flats will remain and that the Danish Wadden Sea also will exist in the future.

However, the area and distribution of the major sediment types are constantly changing. The tidal flats are highly dynamic habitats, and creeks formed by tidal movements are changing over the years; and a year of sedimentation can be eroded and resuspended during a single storm (Pejrup et al. 1997). Increasing amounts of soft sediment fractions (clay) are aggregated in the inner parts of the tidal area, and the size of the area covered by this sediment type will continue to expand (Pejrup et al. 1997). This process will favor ecological communities that are associated with soft sediment types.

Nutrients

In contrast to the Wadden Sea in Germany and the Netherlands, the water quality in the Danish section is not influenced by large rivers, and the four tidal basins, Grådyb, Knudedyb, Juvre Dyb, and Lister Dyb, receive only freshwater inputs from areas with relatively low population densities in southwestern Denmark. The main sources are small rivers like Varde Å, Kongeå, Ribe Å, Brede Å, and Vidå, and in general the Danish Wadden Sea is considered to be dominated by coastal North Sea water (de Jong et al. 1999).

Between 1990 and 2010, the concentration of nitrogen in the water body decreased from 550 to 300 µg/l, but most of the decrease has occurred since 1999. The concentration of nitrite-nitrate decreased too, but in contrast to nitrogen, the reduction was most pronounced from 1990 to 1995. Phosphate concentration decreased only slightly, while chlorophyll a concentration has been stable (van Beusekom et al. 2009).

Despite the decreasing trends in nutrient concentration, the Danish Wadden Sea is assessed, according to the trilateral Wadden Sea targets, as a "problem area," which means that the concentrations are higher than the threshold values given in the "Wadden Sea Plan" (van Beusekom et al. 2009).

Biodiversity

Plants and Benthos

Plants and animals in the Danish Wadden Sea are adapted to live in an environment which changes constantly between wet and dry conditions and in some cases between saline and freshwater conditions. Extreme combinations of these conditions may last for long periods being affected by changing weather conditions.

Both the sublittoral and the littoral zone are, in some places, inhabited by two species of eelgrass, the common eelgrass *Zostera marina* and the dwarf eelgrass *Zostera noltii*. Closer to land, in the transition zone, pioneer plant species with high salt tolerance occur, such as salicornia *Salicornia europaea* and spartina *Spartina alterniflora x maritime*. The plant community changes above the mean high water level and continuous plant cover takes over and forms the typical habitat of salt marsh. Here, the species composition changes according to ground elevation, with those species that are most adapted to freshwater conditions at the highest levels. On the landward side, the salt marsh plant communities are in many places restricted by sea dikes, which separate the salty environment from the freshwater environment. However, in some salt marsh and dune areas on the islands, on the peninsula of Skallingen and in the estuary of the River Varde, a natural transition zone is found that is uninfluenced by dikes and other human infrastructure.

The benthic diatoms and phytoplankton are the main primary producers on the tidal flats. They provide a net primary production of about $100 \text{ g C m}^{-2} \text{ y}^{-1}$, which is about the average production level of most estuaries and terrestrial habitat types (Wolff 1983). The species composition of phytoplankton is extremely rich, whereas the number of macrophyte species is poor, as a consequence of the instability of the substrate and of the high water turbidity. The Danish Wadden Sea is also an extremely unpredictable environment for invertebrate animals. However, at least 1,250 benthic species are found, and of these there are about 400 macrobenthic species. These include well-known species, such as blue mussel *Mytilus edulis*, cockle *Cerastoderma edule*, lugworm *Arenicola marina*, and rag worm *Nereis diversicolor*, all of which are important food items for fish and bird species (Wolff 1983). The highest production of benthos is found in the middle part of the tidal flats between the mean high and mean low water level. The distribution of benthic species is strongly related to sediment type and water coverage. Corophium *Corophium volutator* and rag worm are found in soft sediments with medium to low degree of water coverage, whereas cockles live in sandy sediments with high water coverage.

Fish

The Danish Wadden Sea is an important nursery ground for commercial fish species, especially flatfish that later move to the North Sea. Large numbers of migratory fish species pass through the Danish Wadden Sea to the river systems, of these the eel

Anguilla anguilla and Atlantic salmon *Salmo salar* are the most important species commercially and recreationally. The houting *Coregonus oxyrhynchus*, an endemic species of whitefish closely related to the trouts, was historically widespread in rivers along the entire Wadden Sea coast. However, it has been extirpated from the German and Dutch areas. Due only to recent intensive river restoration, the remnant number of houtings has been increasing in most of the Danish rivers flowing into the Wadden Sea. This rare species may now be more secure with its recent spread within the Danish section.

Breeding Birds

Dunlin *Calidris alpina* and ruff *Philomachus pugnax* breed in small numbers, and these species are both rare and among the most threatened breeding birds in the entire Wadden Sea. More numerous and widespread are northern lapwing *Vanellus vanellus*, common redshank *Tringa totanus*, and Eurasian oystercatcher *Haematopus ostralegus*, which are territorial, breeding species that prefer salt marshes, wet meadows, and in some cases fields (Thorup and Laursen 2008).

Breeding birds on the mainland suffer from high predation pressure, mainly caused by red foxes *Vulpes vulpes* and other mammalian predators. Thus the highest densities of breeding birds are found on the islands, especially on those without foxes and/or humans. For example, the highest breeding density of black-tailed godwits *Limosa limosa* is found on the island of Mandø. The small uninhabited island of Langli (in the northern section, east of the peninsula of Skallingen) holds more than 60% of all colonial breeding birds in the Danish Wadden Sea, especially gulls and terns (Thorup and Laursen 2013).

The populations of many species of breeding birds are decreasing, and in general the most endangered are those species breeding in beaches and in freshwater meadows, due to high recreational activity on the beaches during summer and intensive agricultural management of the meadows in the landside of the seawalls (Koffijberg et al. 2009).

Migratory Birds

The dunlin is the most numerous migratory waterbird species in the Danish Wadden Sea. In the first part of October, when juvenile birds arrive from Scandinavian and Russian breeding grounds and join the adults that have already arrived, about 160,000 dunlins are present. At this time of the year, about three-quarter million waterbirds can be counted in the Danish Wadden Sea, feeding and building up body reserves to continue their southward migration (Fig. 3). Also, large numbers of dabbling ducks and geese are staging, especially along the mainland coast and in the extensive marshlands behind the sea dikes. Northern pintail *Anas acuta*, one of the dabbling duck species breeding in Finland and Russia, has increased during the last two decades. In recent years the number of pintail in the Danish section of the

Fig. 3 At rising water several wader species are often seen in flocks numbering tens of thousands of individuals. Here it is red knots gathering at the coast of the island of Mandø (Photo Credit: John Frikke © Rights remain with the author)

Wadden Sea makes up about 12% of the flyway population during autumn (Laursen and Frikke 2013). The Danish Wadden Sea is also very important for the light-bellied brent goose *Branta bernicla hrota*, a subspecies numbering only 7,600 individuals, which breeds on Svalbard. Most of these birds stay in Denmark during autumn and spring, and a large proportion arrive in the Danish Wadden Sea in late August and early September (Laursen and Frikke 2013).

A number of several Arctic breeding species, e.g., red knot *Calidris canutus*, bar-tailed godwit *Limosa lapponica*, and dark-bellied brent goose *Branta b. bernicla*, peak in late spring (in May or June). The departure date of most of these species from the Wadden Sea is quite stable from year to year, although late compared to others. These species must time their arrival on the Arctic breeding grounds as snow begins to melt. This is in contrast to another arctic breeder, the barnacle goose *Branta leucopsis*, which has recently prolonged its stay in the Danish Wadden Sea by up to a month during spring (Laursen and Frikke 2013).

A regular monitoring program for the entire Wadden Sea has shown that of 40 waterbird species in the Danish part, the numbers of 11 species have increased, 8 species have been stable, 15 species have declined, and 6 species have fluctuated in numbers (Fig. 4). The site is of international importance for 20 species of geese, ducks, and waders, meaning that at least 1% of the entire biogeographical population of each species are using the area on a regular basis (Fig. 5). For nine of the 20 species, 2–10% of the population is staging in the Danish Wadden Sea, and for three, more than 10% use the site.

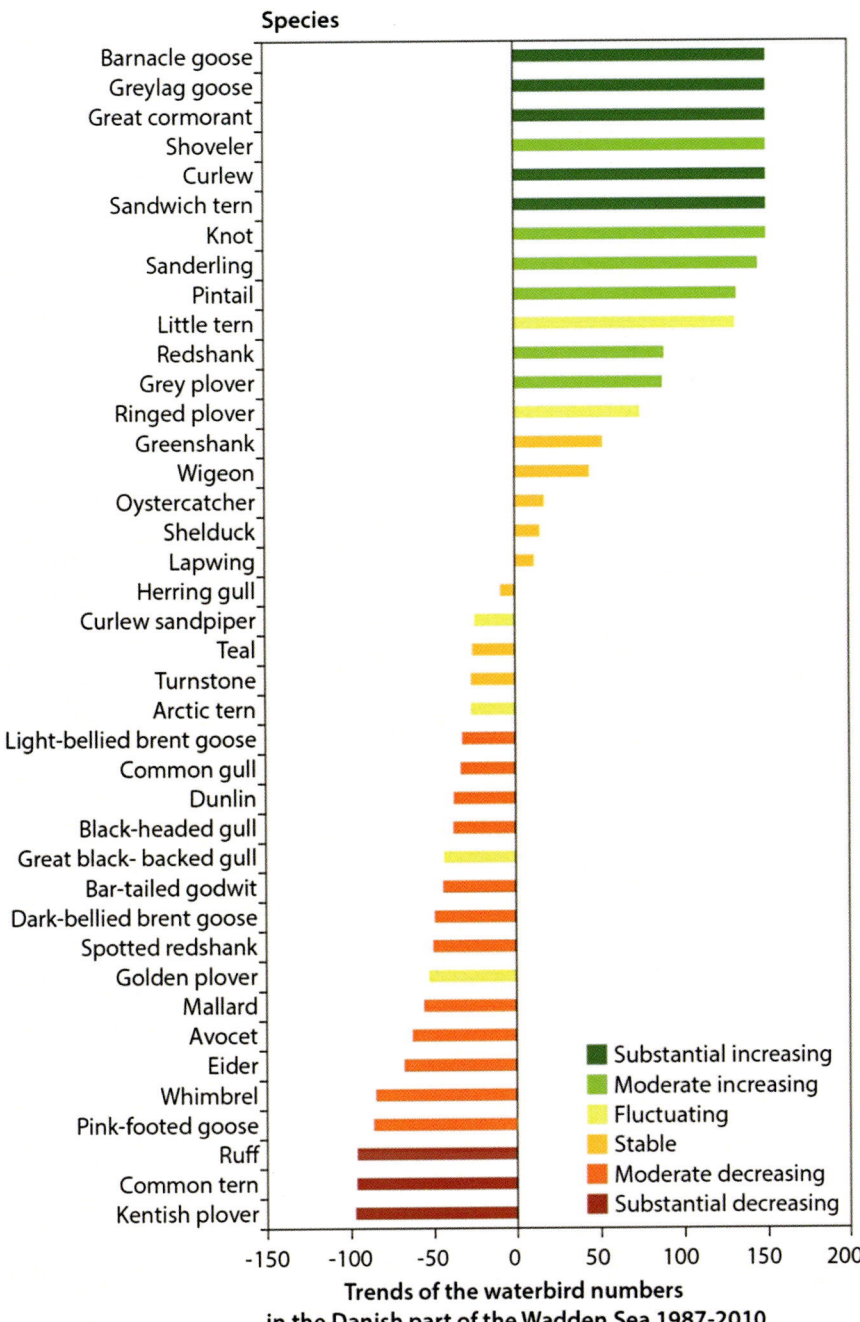

Fig. 4 Changes in population numbers (%) in the Danish Wadden Sea 1987/88-2009/10 (eider from 1992/93) arranged by decreasing order (From Laursen and Frikke 2013, with permission from the Danish Ornithological Society)

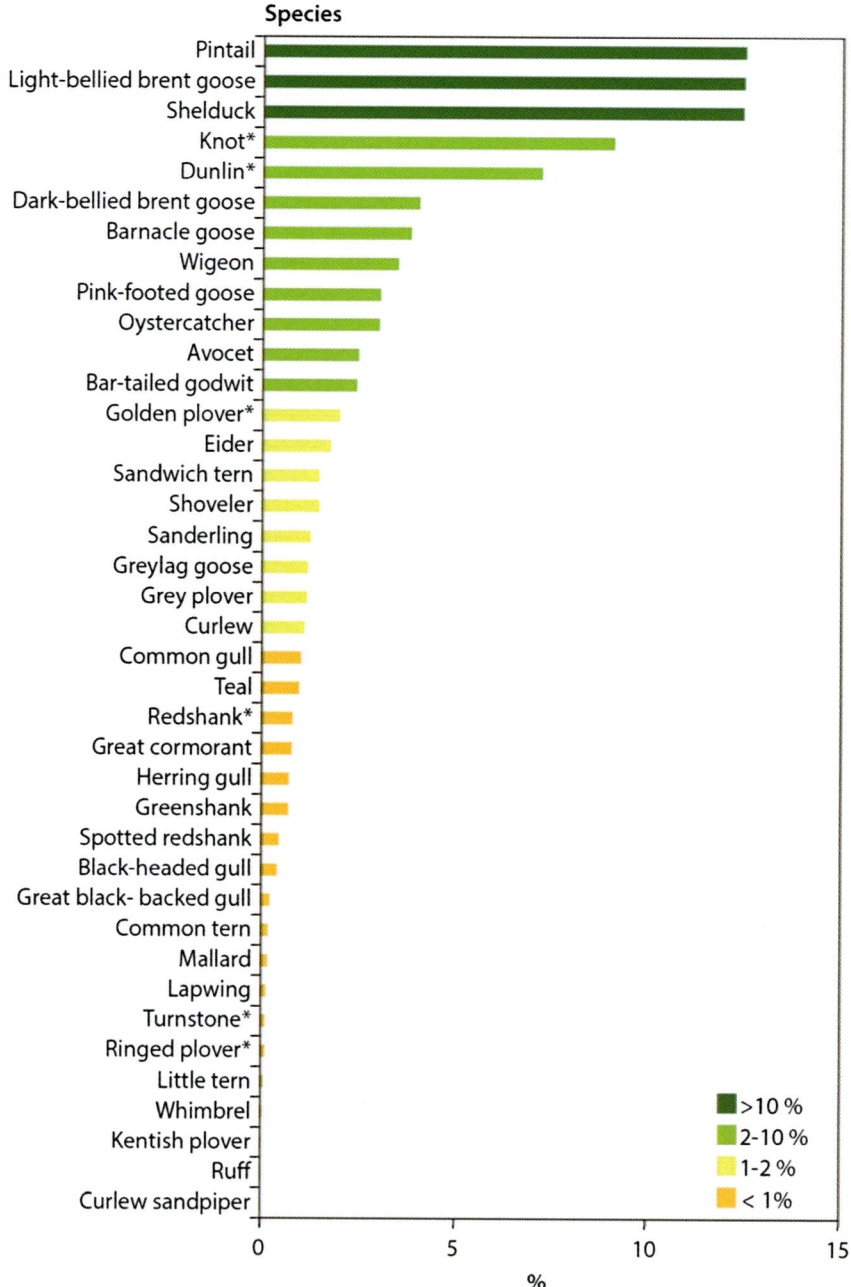

Fig. 5 International importance of the Danish Wadden Sea for 39 waterbird species (arranged by decreasing values) calculated as counted numbers in percent of the flyway population size. * Species with more subpopulations involved (From Laursen and Frikke 2013, with permission from the Danish Ornithological Society)

Sea Mammals

Three species of sea mammals occur regularly in the Danish Wadden Sea, the harbor seal *Phoca vitulina*, the gray seal *Halichoerus grypus*, and the harbor porpoise *Phocoena phocoena*. With a population of more than 4,000 individuals, the harbor seal is by far the most numerous, and they rest in groups on sandbanks in the western part alongside the deeps between the barrier islands. They feed in the Wadden Sea, but satellite transmitters have revealed that they also undertake regular sorties of several hundred kilometers in the North Sea. The number has increased since 1975, but suffered setbacks in 1988 and in 2002 due to epizootic outbreaks that killed a major part of the population (Reijnders et al. 2009). The species has since recovered (Hansen 2015).

There are up to 80 gray seals, and their numbers have increased recently. They rest in mixed groups with the harbor seals, and so far breeding has only been observed in very few cases.

The number of harbor porpoises is difficult to determine. They occur in the North Sea off the Wadden Sea, but are regularly observed in the deeps between the Wadden Sea islands. Hence, the Danish Wadden Sea is not considered to be of great importance to this species.

Ecosystem Services

About 200,000 people are permanent residents in the Danish Wadden Sea region, and the annual number of tourism overnight stays is about seven million (Brandt and Wollesen 2009). Potentially, this could create a large recreational pressure on the Wadden Sea ecosystem, but in fact most of the guests spend their days on the sandy beaches on the west coast of the islands of Fanø and Rømø. Here, between 2,200 and 3,900 persons are regularly observed in July and August.

Several other outdoor activities take place in the area, e.g., walking, fishing, sailing, bird watching, hunting, and guided tours. Compared to the popular beaches, only a few hundred persons take part in these activities per autumn (Laursen et al. 1997). Most of these non-beach-related activities occur around and north of the city of Esbjerg and on the northern part of Fanø and near Højer in the southern part.

Sailing activity is most popular from May to August, when up to 120 vessels are seen regularly. The numbers of motorboats and sailing boats have been rather stable for several years, whereas the number of windsurfers, and recently kitesurfers, has increased considerably. These activities are restricted to special sites of Hjerting north of Esbjerg and on the west side of the islands of Fanø and Rømø.

Conservation Initiatives and Status

The Danish Wadden Sea has always been a popular hunting ground, and one of the first conservation steps was to regulate this activity. This started in 1939 in the southern part, when the little island of Jordsand and its surroundings to the south to the Danish-German border and an area east of Fanø (Albue Bugt) were closed for hunting. In 1948 and in 1978, areas around the causeways to Rømø and to Mandø were closed for public access and consequently also closed to hunting. These initiatives were followed by increasing numbers of waterbirds, especially south of the Rømø Dam.

A big step in terms of nature protection occurred in 1979, when the entire Danish Wadden Sea became a wildlife reserve, with the extension of restrictions on hunting and public access over a larger area and establishment of seal reserves. In 1982, conservation was extended to include the habitat types, and the area became a combined wildlife and nature reserve. As a result, more activities were forbidden, e.g., changes to the tidal flats and mechanical excavation of cockles and other benthic organisms. In the following year, the entire Danish Wadden Sea was designated as an Important Bird Area according to the EEC Birds Directive. In 1992, all hunting activity was forbidden between the Wadden Sea islands and the mainland coast and in some salt marsh areas on the islands and along the mainland coast. It is expected that the governmental notice for the Wildlife and Nature Reserve Wadden Sea will be revised in 2016 by the Danish Nature Agency.

In 1988, 2,570 ha of embanked marshes in Tøndermarsken were protected by law – the largest coherent grassland in Denmark and very important to breeding and migratory waterbirds. Recent initiatives for protection of the Danish Wadden Sea culminated in 2010, when the Wadden Sea National Park was established, which includes the sea territory, the islands, and the several larger marshland areas behind the mainland seawalls (Fig. 1). In 2014 this was followed by a designation of the Danish Wadden Sea as a World Heritage Area, and with the Dutch and German designations in 2009, the entire trilateral Wadden Sea area is now a coherent IUCN World Heritage Site.

Threats and Future Challenges

As a result of national conservation initiatives during the last three decades, in particular the EU Birds and Habitat Directives and initiatives by the Trilateral Wadden Sea Cooperation, the Danish Wadden Sea is now well protected, and it has gained large national awareness. However, there are still things that can be improved.

Monitoring of breeding birds since the beginning of the 1980s has revealed decreasing trends, and the breeding areas are shrinking due to intensified land use, which includes plowing, drainage, and intensive management of former wet permanent grasslands (Laursen and Thorup 2009). Increased predation by red foxes and other predators is another problem for the birdlife, and mitigation actions are necessary to increase breeding success in a number of species. Reduced grazing by sheep and cattle also constitutes a problem in several habitats like cultural grasslands and salt marshes, as a lower grazing pressure is followed by overgrowth by higher plants, bushes, and trees, which reduce the quality of the habitats for lower plant communities and for a number of breeding birds.

Invasive species is also a serious point of concern. In the terrestrial environment, invasive mammals like the Asian raccoon dog *Nyctereutes procyonoides* and the North American mink *Neovison vison* are well known as predators of ground-nesting birds such as gulls, terns, and waders, but it is especially in relation to the unique marine ecosystem that invasive species in the past two to three decades has been a growing problem. To mention the most serious examples, the Pacific oyster *Crassostrea gigas* has invaded and spread in the Wadden Sea since the early 1980s, the Asian shore crab *Hemigrapsus sanguineus* reached France in 1999 and invaded the Wadden Sea shortly after 2000, and since 2007 a comb jellyfish (Ctenophora) native to western Atlantic coastal waters, *Mnemiopsis leidyi*, has invaded the water column of all parts of the Wadden Sea. They are all predators that compete for food with native Wadden Sea species and are a serious problem for species like the blue mussel.

New forms of recreational activities are also introduced from time to time, and activities like kayaking, kiteflying, hang gliding, and other forms of recreational flying may become threats to birds and seals in the Danish Wadden Sea. Further, the general pressure from recreational activities is predicted to increase due to the immense public relation, which is related to the designations of both the national park and the world heritage.

References

Brandt AC, Wollesen A. Tourism and recreation. Quality Status Report. Wilhelmshaven: Common Wadden Sea Secretariat; 2009. Thematic Report No. 34.

De Jong F, Bakker JF, van Berkel CJM, Dankers NMJA, Dahl K, Gätje C, Marencic H, Potel P. Wadden Sea Quality Status Report. Wilhelmshaven: Common Wadden Sea Secretariat; 1999. Wadden Sea Ecosystem No. 9.

Hansen JW, editor. Marine areas, NOVANA. Scientific report from DCE, Aarhus University [In Danish]. 2015.

Ingvardsen SM. Morfologisk udvikling i Vadehavet, Grådybs Tidevandsområde og Skallingen. Kystdirektoratet, Transport- og Energiministeriet. 2006a. 86 p. [In Danish].

Ingvardsen SM. Morfologisk udvikling i Vadehavet, Juvre Dybs Tidevandsområde. Kystdirektoratet, Transport- og Energiministeriet. 2006b. 88 p. [In Danish].

Koffijberg K, Dijksen L, Hälterlein B, Laursen K, Potel P, Schrader S. Breeding birds. Thematic Report No. 18. In: Marencic H, de Vlas J, editors. Quality Status Report 2009. Wilhelmshaven: Common Wadden Sea Secretariat; 2009. Wadden Sea Ecosystem No. 25.

Laursen K, Frikke J. Staging waterbirds in the Danish Wadden Sea 1980–2010. Dan Ornitol Foren Tidsskr. 2013;107(1):1–184. (In Danish with an English summary).

Laursen K, Thorup O. Breeding birds in the Danish Wadden Sea Region 1983–2006, assessment of SPAs. Dan Ornitol Foren Tidsskr. 2009;103(2):77–92.

Laursen K, Salvig J, Frikke J. Vandfugle i relation til menneskelig aktivitet i Vadehavet 1980–1995. Faglig rapport fra DMU, nr. 187. Miljø- og Energiministeriet, Danmarks Miljøundersøgelser 1997. [In Danish]

Pejrup M, Larsen M, Edelvang K. A fine-grained sediment budget for the Sylt-Rømø tidal basin. Helgoländer Meeresun. 1997;51:253–68.

Reijnders PJH, Brasseur SMJM, Borchardt T, Camphuysen K, Czeck R, Gilles A, Jensen LF, Leopold M, Lucke K, Ramdohr S, Scheidat M, Siebert U, Teilmann J. Marine mammals.

Thematic Report No. 20. In: Marencic H, de Vlas J, editors. Quality Status Report 2009. Wilhelmshaven: Common Wadden Sea Secretariat; 2009. Wadden Sea Ecosystem No. 25.

Thorup O, Laursen K. Status of breeding Oystercatcher Haematopus ostralegus, Lapwing Vanellus vanellus, Black-tailed Godwit Limosa limosa and Redshank Tringa totanus in the Danish Wadden Sea in 2006. Dan Ornitol Foren Tidsskr. 2008;102(2):255–67.

Thorup O, Laursen K, Ynglefugle i Vadehavet. Notat fra DCE – Nationalt Center for Miljø og Energi. Aarhus: Aarhus Universitet; 2013.

Van Beusekom JEE, Bot PVM, Carstensen J, Goebel JHM, Lenhart H, Pätsch J, Petenati T, Raabe T, Reise K, Wetsteijn B. Eutrophication. Thematic Report No. 6. In: Marencic H, de Vlas J, editors. Quality Status Report 2009. Wilhelmshaven: Common Wadden Sea Secretariat; 2009. Wadden Sea Ecosystem No. 25.

Wolff WJ, editor. Ecology of the Wadden Sea, vol. 1. Rotterdam: A.A. Balkema; 1983.

Estuaries of Great Britain

80

Nick C. Davidson

Contents

Introduction	998
The Size and Diversity of British Estuaries	998
Biodiversity Importance of British Estuaries	1001
Estuarine Habitats and Communities	1001
Fish	1002
Waterbirds	1002
Conservation Status of British Estuaries	1004
Land-Claim of British Estuaries	1004
Threats and Future Challenges for British Estuaries	1005
Annex 1	1007
References	1008

Abstract

For its relatively small geographical size, Great Britain (comprising England, Scotland, and Wales and their associated islands) is uniquely well endowed with estuaries, and these vary greatly in their geomorphologic origins, size, shape, extent of freshwater influence, tidal range, and their variety of coastal and marine habitats. They form a major component of the British natural environment and are of major significance for wetland biodiversity conservation and for the many ecosystem services they provide to people.

Keywords

Estuary · Tidal flats · Saltmarshes · Great Britain · Waterbirds · Land claim · Coastal squeeze · Managed realignment

N. C. Davidson (✉)
Institute for Land, Water and Society, Charles Sturt University, Albury, NSW, Australia

Nick Davidson Environmental, Wigmore, UK
e-mail: arenaria.interpres@gmail.com

© Springer Science+Business Media B.V., part of Springer Nature 2018
C. M. Finlayson et al. (eds.), *The Wetland Book*,
https://doi.org/10.1007/978-94-007-4001-3_3

Introduction

For its relatively small geographical size, Great Britain (comprising England, Scotland, and Wales and their associated islands) is uniquely well endowed with estuaries, and these vary greatly in their geomorphologic origins, size, shape, extent of freshwater influence, tidal range, and their variety of coastal and marine habitats. They form a major component of the British natural environment and are of major significance for wetland biodiversity conservation and for the many ecosystem services they provide to people (see also UK National Ecosystem Assessment 2011).

The information summarized here is derived largely from a major late 1980s review of the distribution, features, importance of, and pressures on British estuaries (Davidson et al. 1991; summarized in Davidson 1991) and its underlying data, which remains the most contemporary national assessment. This review included all parts of the coast covered by an inclusive definition of "estuary" as "a partially enclosed area at least partly composed of soft tidal shores, open to saline water from the sea and receiving freshwater from rivers, land run-off or seepage." Included were all parts of the British coastal zone with an intertidal channel or shoreline length of greater than 5 km.

Estuaries included were classified into nine estuary types: fjord, fjard, ria, coastal plain estuary, bar-built estuary, complex estuary (with characteristics of more than one other type), barrier beach, linear shore, and embayment. The only major estuary type which does not occur on the coast of Britain is the delta.

Further information on each estuary (derived from the 1991 review) is provided in the seven-volume *An inventory of UK estuaries* (Buck 1996–1997), and regional overviews are provided in the 16-volume *Coasts and seas of the United Kingdom* (Barne et al. 1995–1997).

The Size and Diversity of British Estuaries

There are 155 estuaries around the coast of Britain (Fig. 1). In addition there are a further eight estuaries wholly or partly in the Northern Ireland part of the United Kingdom of Great Britain and Northern Ireland. Their total area in Britain is almost 530,000 ha, of which over 303,400 ha are intertidal flats and marshes, with vegetated saltmarshes being 42,350 ha (14%) of that intertidal area – so much of the intertidal area of British estuaries is formed of unvegetated intertidal mud and sand flats. British estuaries have a total of 2,450 km of main tidal channels and a shoreline of just over 9,000 km: almost half of the total shoreline length of the British coast.

The most common types of estuaries in Britain are bar-built (47) and coastal plain (35) systems (Table 1). Because they are often large, coastal plain estuaries (35%) and embayments (25%) form the largest proportions of the British estuarine area.

Estuaries are distributed around all parts of the British coast, but most and the largest are on the southeastern and western shores of England, with fewer and

Fig. 1 The location and area (hectares) of British estuaries (From Davidson et al. 1991). The location and names of all 155 British estuaries are provided in Annex 1. © Joint Nature Conservation Committee

generally smaller estuaries on the rockier coastlines of Scotland and Wales. The largest estuaries are the Wash (66,600 ha) in eastern England and the macrotidal Severn Estuary (55,700 ha) in southwest England and Wales, which has the second largest tidal range in the world (after the Bay of Fundy in Canada).

Largest intertidal areas are found in Morecambe Bay (33,750 ha) in northwest England (Fig. 2), the Wash (29,770 ha), and the Solway Firth (27,550 ha) in

Table 1 The numbers of British estuaries of each estuary type and their percentage contribution to the total area of British estuaries

Estuary type	No. of estuaries	Percentage of total estuarine area
Fjord	6	2
Fjard	20	5
Ria	15	3
Coastal plain	35	35
Bar-built	47	6
Complex	10	18
Barrier beach	2	2
Linear shore	7	4
Embayment	13	25

Fig. 2 Morecambe Bay in northwest England has the largest intertidal area of any estuary in Britain (Photo credit: Nick Davidson© Rights remain with the author)

northwest England/southwest Scotland. Although many British estuaries are individually small (80 – 61.5% of the total – each have an intertidal area of less than 500 ha), their overall contribution to the diversity of estuarine resource and its wildlife importance is high.

The number and variety of British estuaries is unrivaled in Europe, and together they form about 28% of the total estuarine area of c. 1,895,000 ha on the Atlantic seaboard of western Europe. This is the largest national estuarine area in Europe, although the single largest contiguous estuarine area in Europe is the c. 764,000 ha of the Wadden Sea behind the North Sea barrier islands of the Netherlands, Germany, and Denmark, forming about 40% of the total estuarine area in Western Europe.

Biodiversity Importance of British Estuaries

Estuarine Habitats and Communities

Even small British estuaries are typically composed of a mosaic of four to nine major habitat types (subtidal, intertidal mudflats, intertidal sandflats, saltmarshes, shingles, rocky shores, coastal lagoons, sand dunes, and coastal wet grasslands, the last often having been converted from formerly intertidal habitats). Tidal flats occur in all, and saltmarshes and subtidal areas in almost all, British estuaries.

Saltmarshes larger than 0.5 ha occur on 135 estuaries, with saltmarsh plant communities (Fig. 3) being most diverse in southern and eastern England, where they include plants such as sea purslane *Halimione portulacoides* in low-mid marsh and sea lavender *Limonium* spp. and shrubby sea blight *Sueda fruticosa* in mid-upper marsh. Cord-grass *Spartina townsendii* swards now occur in 82 British estuaries and dominate the lower saltmarsh zone especially in southern and western England. First appearing in Southampton Water in the late nineteenth century, it has spread, both naturally and through planting for shoreline stabilization, but is now dying back in much of southern England.

Sand dunes are associated with 55 British estuaries, often being a major force in shaping the estuary through the formation of estuary-mouth spits. Of the seven nationally important shingle structures in Britain, five are associated with estuaries. As in the case of sand dunes, some shingle structures are a major influence on the geomorphological development of the estuary, for example, Orford Estuary. Often

Fig. 3 A diverse British natural upper saltmarsh community with sea purslane *Halimione portulacoides* and shrubby sea blight *Sueada fruticosa* on the North Norfolk Coast estuary in eastern England (Photo credit: Nick Davidson© Rights remain with the author)

associated with such shingle structures is the scarce and highly vulnerable habitat of coastal saline (or hypersaline) lagoons: about 83% of the area of British saline lagoons is associated with 37 estuaries in England and Wales, and they support a highly specialized flora and fauna often of very local distribution.

Aquatic estuarine communities in British estuaries are diverse, with 17 hard-shore and 16 soft-shore communities recognized. While each hard-shore community typically occurs either intertidally or subtidally, most soft-shore communities occur in both situations. Although diverse, hard-shore estuarine communities are generally small in area and restricted to the outer parts of a few estuaries. They are most diverse in the estuaries of southwest England and south Wales, and parts of Scotland.

Soft-shore communities are more widespread, with five occurring in over 20% of British estuaries and two occurring in over 80% of estuaries. One is a muddy sand community in areas of variable or normal salinity, dominated by lugworms *Arenicola marina*, but intertidally also with abundant cockles *Cerastoderma edule*, Baltic tellins *Macoma balthica*, and polychaete worms. The other is a mud community typical of more sheltered areas of variable or reduced salinity, with a benthic fauna dominated by bivalve mollusks and worms.

Other estuarine communities are important because of their rarity. These include the maerl beds of the Fal Estuary, Helford Estuary, and Milford Haven in western Britain; a sand or muddy sand community dominated by razor shells *Ensis* spp. in a few southwest English, Welsh, and western Scottish estuaries; and the rich fauna of a muddy gravel community in outer estuaries in south and southwest England.

Fish

Eighteen British fish species are considered estuarine, with five dependent on estuaries throughout their life cycles and seven others moving between estuaries and fresh or marine waters, including sea *Petromyzon marinus* and river *Lampetra fluviatilis* lampreys, salmon *Salmo salar*, sea trout *S. trutta*, and eel *Anguilla anguilla*. The sheltered waters of major estuaries such as Plymouth Sound, the Humber Estuary, and the Wash are important spawning and nursery areas for flatfish, and at least 32 estuaries in southern and western England and Wales support sea bass *Dicentrarchus labrax* nursery areas.

Waterbirds

The network of British estuaries is of major national and international importance as migratory staging and wintering areas for migratory waterbirds, chiefly wildfowl (ducks, geese, and swans) and waders (shorebirds), from a vast range of breeding areas from northern Canada to Siberia. In mid-winter in the 1980s over 1,740,000 waterbirds depended on British estuaries – 62% of the British wintering waterbird population and over 10% of the relevant international populations. Of these, there were 581,000 wildfowl (38% of the British and over 10% of the northwest European populations) and almost 1,159,000 waders (90% of the British and over 15% of the

East Atlantic Flyway populations). While waterbirds are widely distributed, the biggest concentrations are on the largest estuaries, notably the Wash and North Norfolk Coast, Morecambe Bay, and the estuaries of Essex and north Kent.

The wader assemblage is dominated by three species (together forming almost three quarters of all wintering waders): dunlin *Calidris alpina*, red knot *Calidris canutus*, and oystercatcher *Haematopus ostralegus*; over half the wildfowl are wigeon *Anas penelope*, dark-bellied brent geese *Branta bernicla bernicla*, and shelduck *Tadorna tadorna*.

British estuaries are of particular international importance for the large proportions of some waterbird populations they support in winter (Fig. 4), notably among waders red knot (67%), common redshank *Tringa totanus* (55%), bar-tailed godwit *Limosa lapponica* (50%), *C.a.alpina* subspecies of dunlin (27%), and oystercatcher (26%). Among wildfowl, British estuaries support over 75% of the small Svalbard-breeding population of light-bellied brent geese *Branta bernicla hrota*, over 50% of dark-bellied brent geese and 100% of the Svalbard population, and 70% of the greenland population of barnacle geese *Branta leucopsis*.

Many wintering waterbird populations reached peak numbers in Britain in the late 1990s. Since then some population sizes have leveled off, or in the case of waders, declined – by 11% since a peak in 2000–2001 (Eaton et al. 2012). Bigger declines have been noted for west coast than east coast estuaries, and there is also evidence of distribution shifts within northwest Europe, perhaps as a response to recent milder winters meaning that fewer birds are now moving on to British estuaries (Maclean et al. 2008). However, for some species such as dunlin and redshank, the declines may also reflect genuine declines in breeding populations (Eaton et al. 2012).

Fig. 4 A flock of wintering red knots *Calidris canutus* on the Wash estuary, eastern England. Britain's estuaries are of major importance for red knots, supporting two thirds of the *C. c. islandica* subspecies which breeds in Arctic Canada and Greenland (Photo credit: Nick Davidson © Rights remain with the author)

Conservation Status of British Estuaries

Much of the British estuarine resource is recognized as of major nature conservation importance. Under national legislation, about one quarter of the total area of Sites of Special Scientific Interest (SSSIs) is estuarine, associated with 136 British estuaries. Significant parts of the estuarine resource are also recognized as internationally important, under two mechanisms: Natura 2000 sites and Wetlands of International Importance (Ramsar Sites). Natura 2000 sites are designated under two European Union Directives: Special Protection Areas (SPAs) under the 1979 "Birds Directive" and Special Areas of Conservation (SACs) under the 1992 "Habitats Directive." Ramsar sites are designated for their international importance as wetlands under the Ramsar Convention on Wetlands, the intergovernmental treaty addressing the conservation and wise use of wetlands worldwide. Sixty-six of the 128 Ramsar sites designated by the United Kingdom in England, Scotland, and Wales are estuarine, with a further seven estuaries designated in Northern Ireland (as at July 2013). Overall, 68 (44%) British estuaries or parts of them have been recognized as internationally important as Ramsar sites and/or Natura 2000 sites.

Land-Claim of British Estuaries

People have been converting and modifying natural estuaries in Britain for at least a millennium, since Roman times. Initially conversions were predominately for agriculture (with such land-claims continuing up to the 1970s) but increasingly in recent centuries also for ports and industry, and urban and infrastructure developments. Land-claim has affected at least 85% of British estuaries and has removed over 25% of the intertidal area from many estuaries and over 80% from some such as the Blyth, Tees (Fig. 5), and Tyne estuaries in eastern England.

The construction of linear sea defenses has had a major impact on British estuaries, with 85% of estuaries having some artificial embankments restricting natural tidal flows. Such sea defenses are particularly extensive along the low-lying coasts of southeastern England and with rising sea levels contribute to an increasing "coastal squeeze" since the natural inland migration response of intertidal systems is curtailed. One consequence has been extensive and continuing erosion and loss of saltmarshes especially in southern and eastern England, with, for example, 25% of Essex saltmarsh estimated to have been lost in the last quarter of the twentieth century (Covey and Laffoley 2002) and losses continuing at an estimated 100 ha per year.

Much of the past agricultural land-claim created freshwater coastal wet grasslands used for stock grazing, which now have considerable biodiversity importance. But much of this grazing marsh has subsequently been converted for intensive crop production and urban and industrial developments: between 30% and 70% of such marshes associated with different southeast England estuaries have been lost since 1930.

Fig. 5 The Tees estuary in northeast England has lost over 80% of its intertidal area since the eighteenth century, through land-claims for agriculture and more recently for port and industrial developments, but still provides many benefits to people (Photo credit: Nick Davidson © Rights remain with the author)

The largest area (47,000 ha) has been claimed from the Wash since Roman times. On just 18 of Britain's estuaries a total of at least 89,000 ha have been claimed: 37% of their former area and an almost 25% loss of the overall estuarine resource. Although mostly small scale, in the late 1980s further land-claims were underway – 123 land-claims in progress affecting 45 (29%) estuaries. Two thirds of these land-claims were for rubbish and spoil disposal, transport schemes, housing and car parks, and marinas, with at least 1,100 ha under claim, 62% of which was for rubbish and spoil disposal. Further development proposals in the late 1980s, mostly for urban development, marinas, and barrages, if implemented would have led to further estuarine losses from 55 British estuaries.

Rates of estuarine land-claim were low before the seventeenth century but accelerated in the eighteenth century, again in the late nineteenth century, and then again in the second half of the twentieth century, when losses averaged 0.3% per year, but have slowed greatly since then – in large part as a consequence of their protected status under international treaties (see above).

Threats and Future Challenges for British Estuaries

Although rates of British estuarine habitat loss appear to have slowed in the early part of the twenty-first century, and the major environmental and ecosystem services importance of British estuaries has been increasingly recognized in decision-making, some land-claims have continued, and changes in land-use planning, energy, and transport policies may lead to further estuarine habitat loss. For example, in the late

1990s an amenity barrage constructed to create a freshwater lake as part of the urban regeneration of Cardiff Docks impounded 200 ha of tidal flats and marshes of the Taff and Ely estuaries – a side arm of the internationally important Severn Estuary. But some estuarine restoration projects have also been initiated.

With a potentially major impact on the whole of the Severn Estuary, in the 1980s long-standing proposals for a barrage across the mouth of the estuary, primarily for tidal power generation, were resurrected and the subject of a major study and impact assessment, but the plans were rejected on economic and environmental grounds. Subsequently such plans have been revisited, following a 2007 report from the UK's Sustainable Development Commission (SDC) which supported such a barrage but noted that full compliance with the EU Birds and Habitats Directives would be vital, as would be a long-term commitment to create compensatory habitats on an unprecedented scale (Sustainable Development Commission 2007). Although a 2-year feasibility study revisiting much of the assessment work of the late 1980s was started, further work on this project was abandoned in 2010. But in 2011 further barrage proposals were again under discussion, which might also lead to proposals for trials of technology on smaller estuaries such as the Mersey and Duddon Estuaries in northwest England, both designated as internationally important Ramsar sites.

In a similar pattern to the Severn barrage proposals, a new Thames Estuary Airport has been proposed at various times since the 1970s. Several locations for a new airport have been proposed including Maplin Sands, Foulness on the north side of the estuary; Cliffe, Kent, and the Isle of Grain on the south side (all designated as internationally important wetlands); and on artificial islands to be created in the middle of the estuary. None of these proposals has been implemented, but some continue to be under review.

As a response to both "coastal squeeze" and the increasingly uneconomic costs of maintaining and raising sea defenses defending agricultural land, an increasing number of "managed realignment" projects have been done or are underway, mainly on eastern English estuaries. This is in the context that the UK's biodiversity action plan (JNCC 1997) aims to prevent net losses to the area of saltmarsh present in 1992. It is therefore a requirement that all losses in marsh area must be compensated by replacement habitat with equivalent biological characteristics. This equates to the need to restore approximately 140 ha of saltmarsh habitat per year in the UK.

Through breaching of seawalls, these realignments seek to reinstate tidal flows and restore saltmarshes and tidal flats in former intertidal areas. The restorations can be costly and challenging because rising sea levels since sea defense construction mean than land levels inside the realignments are often lower than those needed to support saltmarsh vegetation following tidal inundation. To date approximately 400 ha of salt marsh have been restored by managed realignment, but there is increasing evidence that while some biodiversity returns rapidly, saltmarsh plant and benthic communities can take at least hundreds of years to return towards their natural state (Atkinson et al. 2001; Mossman et al. 2012).

Annex 1

The locations and names of all 155 British estuaries covered in this chapter, from Davidson et al. (1991). © Nature Conservancy Council.

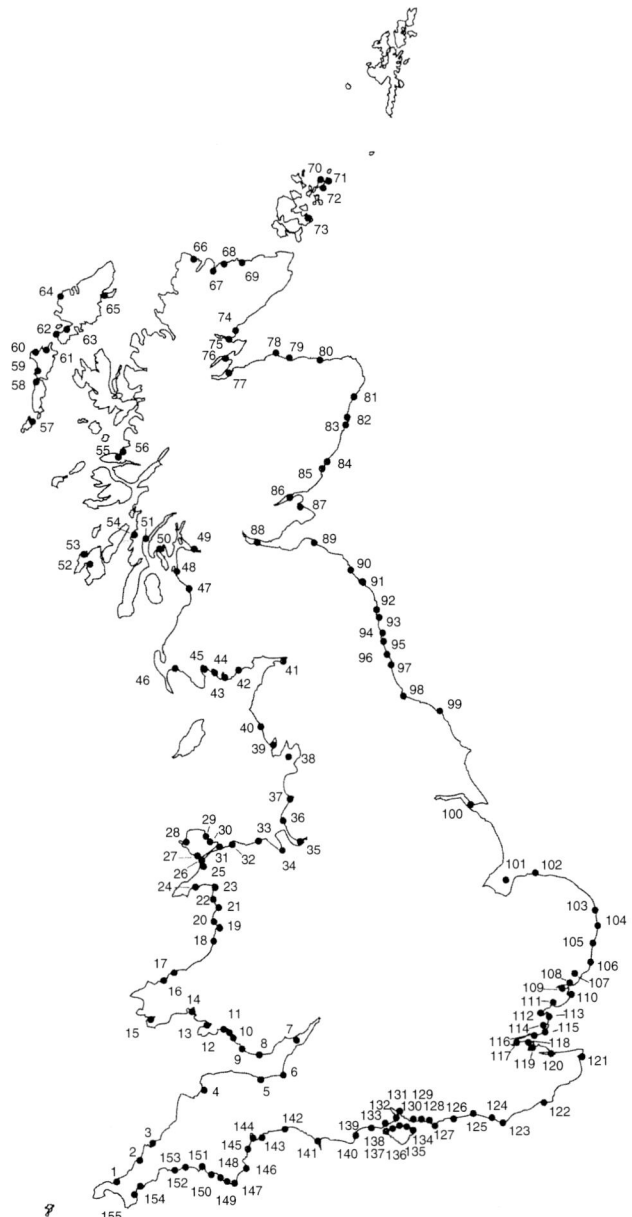

#	Name	#	Name	#	Name
1	Hayle Estuary	57	Tràigh Mhór	112	Blackwater Estuary
2	Gannel Estuary	58	Bagh Nam Faoilean	113	Dengie Flat
3	Camel Estuary	59	Oitir Mhór	114	Crouch-Roach Estuary
4	Taw-Torridge Estuary	60	Tràigh Vallay	115	Maplin Sands
5	Blue Anchor Bay	61	Oronsay	116	Southend-on-Sea
6	Bridgwater Bay	62	Scarista	117	Thames Estuary
7	Severn Estuary	63	Tràigh Luskentyre	118	South Thames Marshes
8	Thaw Estuary	64	CamusUig	119	Medway Estuary
9	Ogmore Estuary	65	Laxdale Estuary	120	Swale Estuary
10	Afan Estuary	66	Kyle of Dumess	121	Pegwell Bay
11	Neath Estuary	67	Kyle of Tongue	122	Rother Estuary
12	Tawe Estuary & Swansea Bay	68	Torrisdale Bay	123	Cuckmere Estuary
13	Loughor Estuary	69	Melvich Bay	124	Ouse Estuary
14	Carmarthen Bay	70	Otters Wick	125	Adur Estuary
15	MilfordHaven	71	Cata Sand	126	Arun Estuary
16	Nyfer Estuary	72	Kettletoft Bay	127	Pagham Harbour
17	Teifi Estuary	73	Deer Sound & Peter's Pool	128	Chichester Harbour
18	Aberystwyth	74	Loch fleet	129	Langstone Harbour
19	Dyfi Estuary	75	Dornoch Firth	130	Portsmouth Harbour
20	Dysynni Estuary	76	Cromarty Firth	131	Southampton Water
21	Mawddach Estuary	77	Moray Firth	132	Beaulieu River
22	Artro Estuary	78	Lossie Estuary	133	Lymington Estuary
23	Traeth Bach	79	Spey Bay	134	Bembridge Harbour
24	Pwllheli Harbour	80	Banff Bay	135	Wootton Creek & Ryde Sands
25	Foryd Bay	81	Ythan Estuary	136	Medina Estuary
26	Traeth Melynog	82	Don Estuary	137	Newtown Estuary
27	Cefni Estuary	83	Dee Estuary (Grampian)	138	Yar Estuary
28	Alaw Estuary	84	St Cyrus	139	Christchurch Harbour
29	Traeth Dulas	85	Montrose Basin	140	Poole Harbour
30	Traeth Coch	86	Firth of Tay	141	The Fleet & Portland Harbour
31	Traeth Lavan	87	Eden Estuary	142	Axe Estuary
32	Conwy Estuary	88	Firth of Forth	143	Otter Estuary
33	Clwyd Estuary	89	Tyninghame Bay	144	Exe Estuary
34	Dee Estuary & North Wirral	90	Tweed Estuary	145	Teign Estuary
35	Mersey Estuary	91	Lindisfarne & Budle Bay	146	Dart Estuary
36	Alt Estuary	92	Alnmouth	147	Salcombe & Kingsbridge Estuary
37	Ribble Estuary	93	Warkworth Harbour	148	Avon Estuary (Devon)
38	Morecambe Bay	94	Wansbeck Estuary	149	Erme Estuary
39	Duddon Estuary	95	Blyth Estuary (Northumberland)	150	Yealm Estuary
40	Esk Estuary (Cumbria)	96	Tyne Estuary	151	Plymouth Sound
41	Solway Firth	97	Wear Estuary	152	Looe Estuary
42	Rough Firth & Auchencairn	98	Tees Estuary	153	Fowey Estuary
43	Dee Estuary (Dumfries)	99	Esk Estuary (Yorkshire)	154	Falmouth
44	Water of Fleet	100	Humber Estuary	155	Helford Estuary
45	Cree Estuary	101	The Wash		
46	Luce Bay	102	North Norfolk Coast		
47	Garnock Estuary	103	Breydon Water		
48	Hunterston Sands	104	Oulton Broad		
49	Clyde Estuary	105	Blyth Estuary (Suffolk)		
50	Ruel Estuary	106	Ore-Alde-Butley		
51	Loch Gilp	107	Deben Estuary		
52	Tràigh Cill-a-Rubha	108	Orwell Estuary		
53	Loch Gruinart	109	Stour Estuary		
54	Loch Crinan	110	Hamford Water		
55	Kentra Bay	111	Colne Estuary		
56	Loch Moidart				

References

Atkinson PW, Crooks S, Grant A, Rehfisch MM. The success of creation and restoration schemes in producing intertidal habitat suitable for waterbirds. English Nature Research Report 425. Peterborough: English Nature; 2001.

Barne JH, Robson CF, Kaznowska SS, Doody JP, Davidson NC, editors. Coasts and seas of the United Kingdom. Peterborough: Joint Nature Conservation Committee; 1995–1997. (Sixteen regional volumes: 3 published in 1995, 5 in 1996, 8 in 1997. Each 200–300 pp.)

Buck AL, editor. An inventory of UK estuaries. 7 volumes. Peterborough: Joint Nature Conservation Committee; 1996–1997.

Covey R Laffoley D d'a. Maritime state of nature. Report for England; Getting onto an Even Keel. Peterborough: English Nature; 2002.

Davidson NC. Estuaries, wildlife and man. A summary of nature conservation and estuaries in Great Britain. Peterborough: Nature Conservancy Council; 1991. 20 pp.

Davidson NC, Laffoley Dd'A, Doody JP, Way LS, Gordon J, Key R, Drake CM, Pienkowski MW, Mitchell RM, Duff KL. Nature conservation and estuaries in Great Britain. Peterborough: Nature Conservancy Council; 1991. 422 pp.

Eaton MA, Cuthbert R, Grice PV, Hal, C, Hearn RD, Holt CA, Knipe A, Marchant J, Mavor R, Moran N, Mukhida F, Musgrove AJ, Noble DG, Oppel S, Risely K, Small C, Stroud DA, Toms M, Wotton S. The state of the UK's birds 2012. Sandy: RSPB, BTO, WWT, CCW, JNCC, NE, NIEA and SNH; 2012. 40 pp.

Joint Nature Conservation Committee. UK Biodiversity Action Plan. 1997. See: http://jncc.defra.gov.uk/default.aspx?page=5155

Maclean IMD, Austin GE, Rehfisch MM, Blew J, Crowe O, Delany S, Devos K, Deceuninck B, Günther K, Laursen K, van Roomen M, Wahl J. Climate change causes rapid changes in the distribution and site abundance of birds in winter. Glob Chang Biol. 2008;14:2489–500.

Mossman HL, Davy AJ, Grant A. Does managed coastal realignment create saltmarshes with 'equivalent biological characteristics' to natural reference sites? J Appl Ecol. 2012. doi:10.1111/j.1365-2664.2012.02198.x.

Sustainable Development Commission. Turning the tide, Tidal Power in the UK. London: SDC; 2007. 154 pp. www.sd-commission.org.uk/data/files/publications/Tidal_Power_in_the_UK_Oct07.pdf

UK National Ecosystem Assessment. The UK National Ecosystem Assessment Technical Report. Cambridge, UK: UNEP-World Conservation Monitoring Centre; 2011. 1466 pp.

The Wash Estuary and North Norfolk Coast (UK)

81

Nick C. Davidson

Contents

General Description	1012
Biodiversity Importance	1015
Conservation Status	1017
Land-Claim and Conversion	1018
Threats and Future Challenges	1019
References	1021

Abstract

The tidal embayment of The Wash and its associated barrier island system of the North Norfolk Coast extending eastwards along the north coast of East Anglia is the largest estuarine area in the United Kingdom. Their total area is 729.5 km^2 (12.5% of the UK estuarine area and about 3.5% of estuarine area of northwest Europe), with intertidal flats and marshes covering 35.6 km^2 (10.2% of the UK intertidal area). This coast is of major importance for breeding seals, for migratory waterbirds, supporting the largest nonbreeding waterbird population in the UK, and as a nursery area for fish. Almost the entire area is covered by multiple international, national and local nature conservation designations. The intertidal areas of The Wash have been progressively embanked and converted to farmland since at least Roman times – at about 470 km^2 since Saxon times the largest area claimed from any UK estuary. A major shellfishery was overexploited in the 1980s and 1990s, leading to declines in internationally important waterbirds. Some parts of The Wash are now the subject of managed realignment projects aiming to reinstate tidal flats and marshes.

N. C. Davidson (✉)
Institute for Land, Water and Society, Charles Sturt University, Albury, NSW, Australia

Nick Davidson Environmental, Wigmore, UK
e-mail: arenaria.interpres@gmail.com

© Springer Science+Business Media B.V., part of Springer Nature 2018
C. M. Finlayson et al. (eds.), *The Wetland Book*,
https://doi.org/10.1007/978-94-007-4001-3_20

Keywords

Estuary · Tidal flats · Saltmarsh · Barrier Island · Seals · Migratory waterbirds · Land-claim · Ramsar Site · Natura 2000

General Description

The tidal embayment of The Wash and its associated barrier island system of the North Norfolk Coast extending eastwards along the north coast of East Anglia (Fig. 1) is the largest estuarine area in the United Kingdom (Buck and Donaghy 1997). For further information on these, and all other, estuarine areas in the UK, see

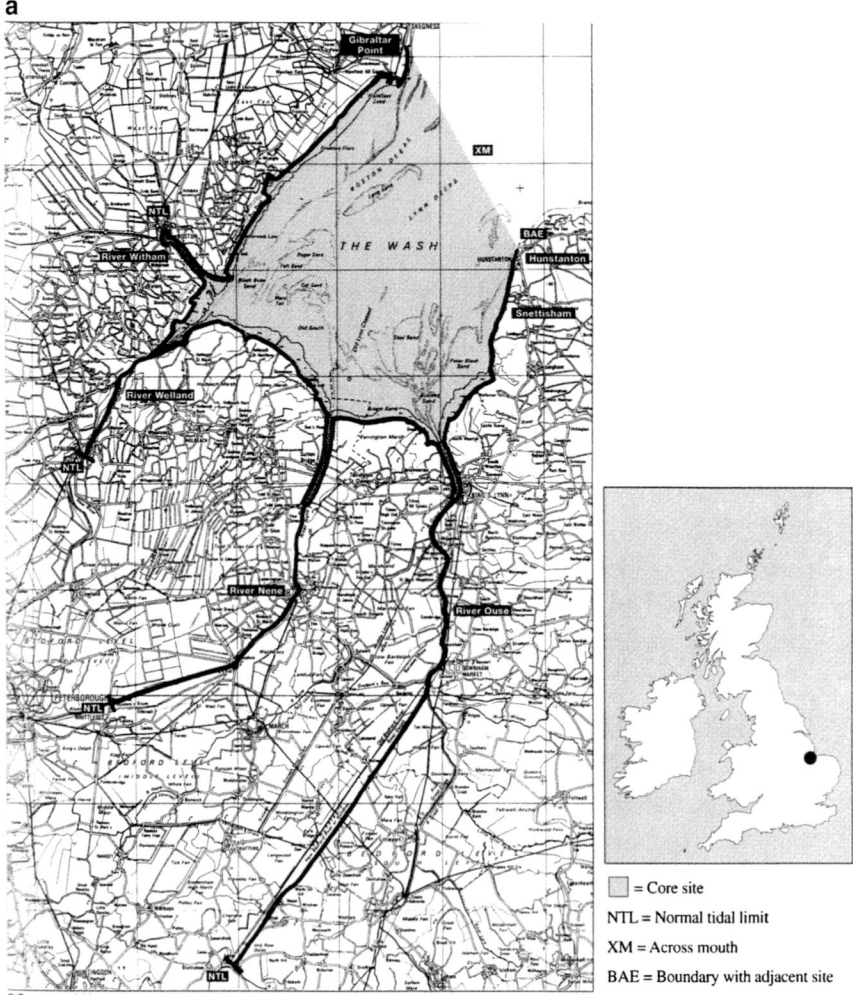

Fig. 1 (continued)

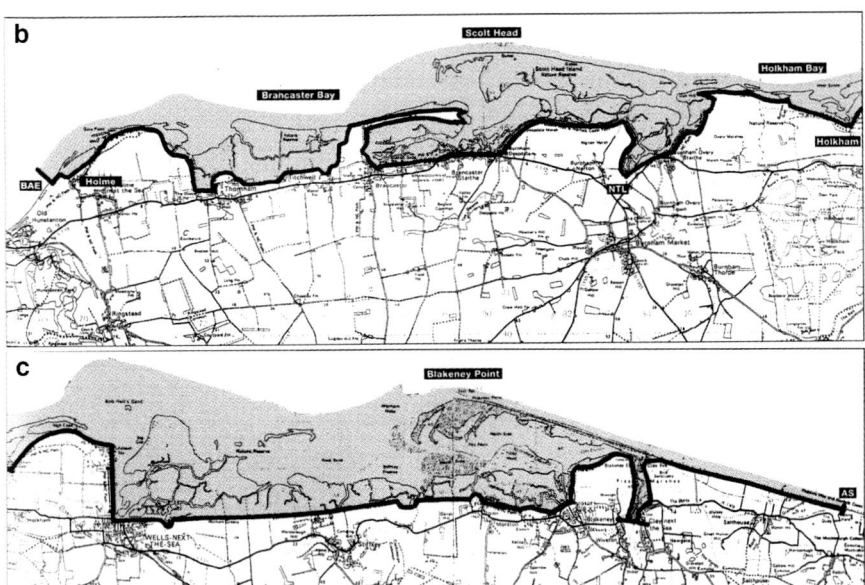

Fig. 1 Maps of The Wash and North Norfolk Coast estuarine area: (**a**) The Wash. (**b**) North Norfolk Coast – west. (**c**) North Norfolk Coast – east. These areas are contiguous (From Buck (1996). © Nature Conservancy Council. © Joint Nature Conservation Committee)

Davidson et al. (1991) and Buck (1997). The total area of The Wash and North Norfolk Coast is 729.5 km^2 (12.5% of the UK estuarine area and about 3.5% of estuarine area of northwest Europe), with intertidal flats and marshes covering 35.6 km^2 (10.2% of the UK intertidal area).

Much of the intertidal area, especially of The Wash (85.8% of its intertidal area), is unvegetated sand flats and mud flats (Fig. 2). Vegetated saltmarshes cover 64.5 km^2 (18% of the intertidal zone), are a particularly important component (37.7% of intertidal area) of the North Norfolk Coast system, and form 14.2% of The Wash intertidal area. There are important and diverse lime-rich sand dune systems at Gibraltar Point at the northwest edge of The Wash, and at Holme and Holkham, and topping the shingle barrier island of Scolt Head and the cuspate shingle spit of Blakeney Point, on the North Norfolk Coast. Scolt Head and Blakeney Point are the major geomorphological features of the North Norfolk Coast and are both actively accreting westwards. There is a small stretch of chalk and sandstone cliffs and rocky shore at Hunstanton at the northeast corner of The Wash.

On the North Norfolk Coast, behind shingle ridges and artificial earth seawalls, there are wet grazing marshes, reedbeds, and a number of small coastal saline lagoons, which are scarce in the UK.

Sediment in The Wash is of largely marine origin, from the east from the North Norfolk Coast, and from offshore glacial banks to the north. There is little input of sediment to The Wash from the four largely channelized rivers (Ouse, Nene,

Fig. 2 The intertidal mud and sand flats of The Wash are among the largest in the UK (Photo credit: Nick Davidson © Rights remain with the author)

Welland, and Witham) which flow into it across the large former wetlands of The Fens from the south and west. These rivers are tidal for a long distance inland – about 60 km for the River Ouse.

There are diverse human uses of the estuarine area. Those on the North Norfolk Coast are predominantly low intensity tourism and recreational activities, notably walking, bird-watching, and water-based pursuits including sailing, windsurfing, water-skiing, and canoeing (Fig. 3). There is also cultivation and use of natural resources including cultivation of oysters and mussels and traditional harvesting of cockles and marsh samphire *Salicornia* spp., reeds (for roofing), and wildfowling. There is little industrial activity, aside from a small port at Wells-next-the-Sea.

On The Wash, there are similar recreational activities around parts of the shoreline, and wildfowling and bait digging is extensive. There are dock and port facilities at Boston (River Welland) on its west coast and King's Lynn (River Ouse) in the southeast of the embayment, and small ports near the mouths of the other two rivers. Commercial fisheries are of major importance, with, in the 1980s, landings forming 60% of mussels, 12–40% of cockles, 22–25% of brown shrimps, and 100% of pink shrimps of England and Wales totals, but the extent of this exploitation of mussels and cockles was considered unsustainable, shellfish harvests have declined, and there has been major impact on intertidal habitats and their wildlife (see Threats and Future Challenges below).

Fig. 3 Tourism and recreation are important human uses of the North Norfolk Coast: seal-watching at Blakeney Point (Photo credit: Nick Davidson © Rights remain with the author)

Biodiversity Importance

The Wash and North Norfolk Coast are of major national, European, and global wildlife importance.

The sandbanks of The Wash support the largest breeding population of harbor seals *Phoca vitulina* in the UK. Following two outbreaks of phocine distemper virus (PDV) in the late 1990s and early 2000s, the population has now recovered to about 3,500 adults by the early 2010s (Thompson, undated). There is also a large (of about 500 adults) mixed colony of harbor and gray seals *Halichoerus grypus* on Blakeney Point (North Norfolk Coast).

Both The Wash and the North Norfolk Coast are of major international importance for their nonbreeding waterbird populations. The Wash supports the largest nonbreeding waterbird population in the UK, averaging 363,450 birds in recent years (5-year mean from 2007/2008 to 2011/2012), and the North Norfolk Coast the fourth largest, averaging 183,809 birds (Austin et al. 2014). These estuaries hold 16 waterbird biogeographic populations of international importance (>1% of their biogeographic population) (Table 1). During the nonbreeding season, The Wash and North Norfolk Coast support the largest populations in the UK of seven waterbird species: pink-footed goose *Anser brachyrhynchus*, dark-bellied brent goose *Branta b. bernicla*, gray plover *Pluvialis squatarola*, red knot *Calidris canutus*, sanderling *Calidris alba*, black-tailed godwit *Limosa limosa*, and bar-tailed godwit *Limosa lapponica*.

Table 1 Nonbreeding waterbird populations of international importance (>1% of their biogeographic population) in The Wash and the North Norfolk Coast. Numbers are the average population size for the most recent 5-year period available: 2007/2008–2011/2012 (Sources: Austin et al. (2014) and WeBS online: www.bto.org/webs)

Species	The Wash	North Norfolk Coast	Total	Percent of biogeographic population
Pink-footed goose *Anser brachyrhynchus*	39,893	53,283	93,176	26.6
Dark-bellied brent goose *Branta bernicla bernicla*	16,480	6,472	22,952	9.6
Shelduck *Tadorna tadorna*	5,705	1,074	6,779	2.3
Pintail *Anas acuta*	547	562	1,109	1.8
Oystercatcher *Haematopus ostralegus*	20,177	4,102	24,279	3.0
Ringed plover *Charadrius hiaticula*	1,409	1,310	2,719	3.7
Golden plover *Pluvialis apricaria*	24,544	4,849	29,393	3.2
Gray plover *Pluvialis squatarola*	10,482	1,785	12,267	1.4
Lapwing *Vanellus vanellus*	17,296	10,398	27,694	1.4
Red knot *Calidris canutus islandica*	134,468	57,063	191,531	42.6
Sanderling *Calidris alba*	2,973	1,096	4,069	3.4
Dunlin *Calidris alpina*	25,421	3,585	29,006	2.2
Black-tailed godwit *Limosa limosa*	8,922	801	9,723	15.9
Bar-tailed godwit *Limosa lapponica*	14,934	4,324	19,258	16.0
Common curlew *Numenius arquata*	9,259	2,359	11,618	1.4
Common redshank *Tringa totanus*	7,072	1,944	9,016	3.8

These estuaries are of particularly major importance to nonbreeding pink-footed geese, supporting over one quarter of their nonbreeding population: these birds roost on the estuaries and disperse during daytime to feed on agricultural land inland of the coast; and Nearctic-breeding *islandica* red knots: in autumn and early winter The Wash and North Norfolk Coast currently hold over 40% of their nonbreeding population, with birds feeding on the tidal flats and roosting at high tide on shingle banks and upper saltmarshes around the shore. These estuaries also support over 15% of the mostly Icelandic-breeding population of black-tailed godwits and the northern European and Siberian-breeding population of *lapponica* bar-tailed godwits, and almost 10% of Siberian-breeding dark-bellied brent geese (Fig. 4).

The Wash is a major nursery area for fish including plaice *Pleuronectes platessa*, cod *Gadus morhua*, and common sole *Solea solea*. The wet dune slacks of the sand

Fig. 4 A flock of wintering red knots on The Wash, which supports over 40% of the nonbreeding population of the *C. c. islandica* subspecies which breeds in arctic Canada and Greenland (Photo credit: Nick Davidson © Rights remain with the author)

dune systems of Gibraltar Point (north-west Wash) and North Norfolk Coast hold around 10% of the UK population of nationally rare natterjack toads *Bufo calamita*.

Botanically, the saltmarshes (Fig. 5) of the North Norfolk Coast are one of the finest and most diverse saltmarsh systems in the UK, with a number of nationally scarce saltmarsh communities and uncommon species, including the nationally rare matted sea lavender *Limonium bellidifolium* growing at the transition between sand dunes and saltmarsh on the North Norfolk Coast, and Rock Sea Lavender *L. binervosum* at Gibraltar Point.

Conservation Status

Almost the entire Wash and North Norfolk Coast systems are covered by multiple national and international wildlife and landscape conservation designations, and significant parts of the area are managed as nature reserves.

Their international importance is recognized by both being designated Wetlands of International Importance (Ramsar Sites) under the Ramsar Convention on Wetlands. Under European Union legislation, both are Special Protection Areas (SPAs) under the EU Birds Directive, as are the Ouse Washes and the Nene Washes: freshwater wetland areas adjacent to the tidal reaches of rivers flowing into The Wash. All these areas, and also Gibraltar Point, are Special Areas for Conservation

Fig. 5 Common sea lavender *Limonium vulgare* carpeting the important and diverse saltmarshes of the North Norfolk Coast in July (Photo credit: Nick Davidson © Rights remain with the author)

(SACs) under the EU Habitats and Species Directive. Part of the North Norfolk Coast is also a UNESCO Biosphere Reserve.

Under UK national legislation, all the estuarine area is covered by Sites of Special Scientific Interest (SSSIs), within which there are six National Nature Reserves. Four (Holme Dunes, Scolt Head Island, Holkham and Blakeney) are on the North Norfolk Coast, and there are two in The Wash (about 15% of The Wash embayment, and Gibraltar Point). Landscape designations include the Wash and North Norfolk Coast Heritage Coast and the Norfolk Coast area of outstanding natural beauty (AONB).

Nongovernmental organizations (NGOs) manage a number of nature reserves: on The Wash, County Wildlife Reserves at Frampton Marsh, Snettisham and the Ouse Washes, Royal Society for the Protection of Birds (RSPB) reserves at Frampton Marsh, Snettisham, Nene Washes and Ouse Washes, and a Wildfowl and Wetlands Trust reserve on the Ouse Washes. On the North Norfolk Coast, there are County Wildlife Reserves at Cley and Salthouse Marshes, Holme Dunes, and Scolt Head Island; an RSPB at Titchwell; and there is a Bird Observatory at Holme. The National Trust owns considerable parts of the North Norfolk Coast.

Land-Claim and Conversion

Some formerly intertidal areas of the North Norfolk Coast have been converted to freshwater grazing marshes (grasslands). In part this was through the construction of earth banks from the seventeenth century onwards, but parts – notably around the eastern part of the coast, also because of westwards movement of shingle banks isolating these areas from tidal influence. Up to the seventeenth century, Cley-next-the-Sea on the shores of the River Glaven was one of Britain's most important ports, until the river silted up.

Land-claim and wetland conversion has been progressive on The Wash for many centuries, since Roman times (first to fifth century AD) (Fig. 6a) when The Wash was the northern tidal part of the much larger Fenland Basin with its very extensive freshwater and brackish marshes to the south. Almost all of these alluvial and peat soil marshes were drained and poldered for high-quality agricultural land in the seventeenth century. Along the tidal shores of the Wash there have been successive land-claims for agriculture since Saxon times (fifth to tenth century AD) until the late 1970s (Fig. 6b). Each land-claim involved construction of seawalls (earth banks) along the outer part of the tidal saltmarsh. Subsequently, new saltmarsh accreted outside each of these banks, leading to subsequent saltmarsh land-claim episodes. With sea level rising relative to land levels in eastern England, each successive seawall is at higher altitude than the earlier ones, so that the land behind early seawalls is now several meters below current sea level.

These estuarine land-claims on The Wash are the largest areas recorded for any UK estuary, at about 470 km^2 since Saxon times. Of this, much (about 320 km^2) has occurred since the seventeenth century, 30 km^2 of which were claimed in the twentieth century (Davidson et al. 1991). Since the low tide mark of the Wash has been largely static for at least the twentieth century, the effect of these land-claims has been a "coastal squeeze" of a progressively narrowing width and reduction in remaining area of tidal flats and marshes and, with further saltmarsh accretion outside the most recently constructed walls, a reduction in the area of intertidal sand and mud flats of about 3% between 1971 and 1985 (Doody 1987).

Threats and Future Challenges

Although shellfish harvesting (Fig. 7) has been an important aspect of natural resource exploitation on The Wash for centuries, from the 1980s onward there was increasing concern over the impact of its scale and increased mechanization on both the shellfish stocks themselves, which had led *inter alia* to the removal of all long-established mussel beds, reduced cockle densities (Dare et al. 2004), and the closure of the mussel fishery in 1994, and the impact on the internationally important waterbird populations depending on these internationally important tidal flats (Atkinson et al. 2003). The overexploitation of shellfish (mussels and cockles) in the 1980s and 1990s in The Wash was associated with nonbreeding population declines in mollusk-specialist feeders oystercatcher, red knot, and shelduck and, in severe winters, with major mortality of oystercatchers and reduced juvenile recruitment into the population of oystercatchers and red knots (Atkinson et al. 2003, 2010).

As a response to both "coastal squeeze" and the increasingly uneconomic costs of maintaining and raising sea defenses defending agricultural land, an increasing number of "managed realignment" projects have been done or are underway in the UK, in the context that the UK's biodiversity action plan (JNCC 1997) aims to

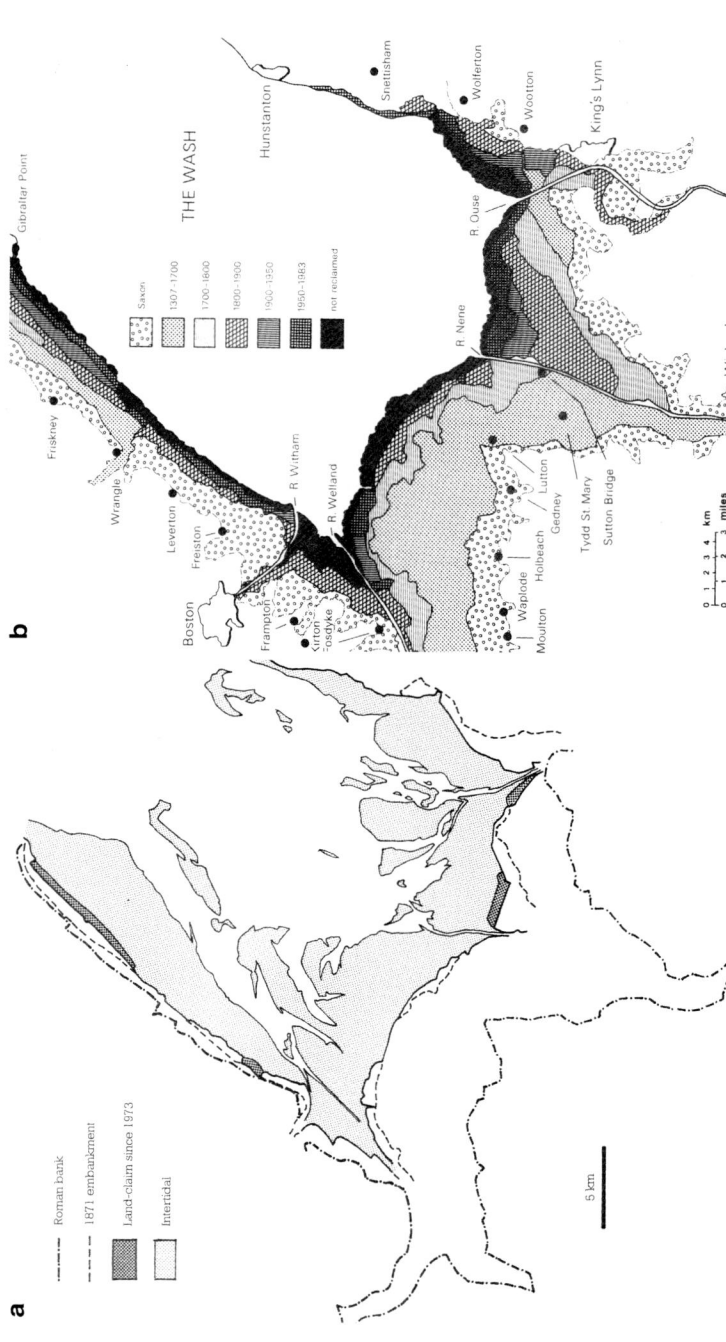

Fig. 6 Progressive land-claim on The Wash. (**a**) The overall pattern since Roman times. (**b**) The pattern of agricultural land-claim since Saxon times (From Davidson et al. (1991), adapted from Doody (1987). © Nature Conservancy Council. © Joint Nature Conservation Committee)

Fig. 7 Hand raking for cockles has been practiced on the North Norfolk Coast for centuries (Photo credit: Nick Davidson © Rights remain with the author)

prevent net losses to the area of saltmarsh present in 1992. These are particularly on eastern English estuaries, including The Wash (Mossman et al. 2012). There, one of the larger "managed realignment" schemes is at Freiston Shore on the western side of The Wash, where the seawall was breached in 2000 and is leading to the reestablishment of 0.7 km^2 of tidal saltmarsh.

As sea level continues to rise, the remaining earth bank seawalls become increasingly vulnerable to high tides and storms. A recent major North Sea storm surge (in December 2013) caused extensive damage to tidal systems along the North Norfolk Coast, including breaching such walls at Blakeney with saline water inundation of freshwater marshes. Whether it will prove viable to rebuild these walls to protect such freshwater marsh nature reserves (which were originally natural tidal saltmarshes) remains a matter of considerable debate.

References

Atkinson PW, Clark NA, Bell MC, Dare PJ, Clark JA, et al. Changes in commercially fished shellfish stocks and shorebird populations in the Wash, England. Biol Conserv. 2003;114:127–41.

Atkinson PW, Maclean IMD, Clark NA. Impacts of shellfisheries and nutrient inputs on waterbird communities in the Wash, England. J Appl Ecol. 2010;47:191–9.

Austin GE, Read WJ, Calbrade NA, Mellan HJ, Musgrove AJ, Skellorn W, Hearn RD, Stroud DA, Wotton SR, Holt CA. Waterbirds in the UK 2011/12: the Wetland bird survey. Thetford: British Trust for Ornithology; 2014.

Buck AL, Donaghy A. An inventory of UK estuaries. Volume 5. Eastern England. Peterborough: Joint Nature Conservation Committee; 1997.

Buck AL, editor. An inventory of UK estuaries. Peterborough: Joint Nature Conservation Committee; 1996–1997. 7 volumes.

Dare PJ, Bell MC, Walker P, Bannister RCA. Historical and current status of cockle and mussel stocks in The Wash. Lowestoft: Centre for environment, fisheries and aquaculture science (CEFAS); 2004.

Davidson NC, Laffoley D'A, Doody JP, Way LS, Gordon J, Key R, Drake CM, Pienkowski MW, Mitchell RM, Duff KL. Nature conservation and estuaries in Great Britain. Peterborough: Nature Conservancy Council; 1991. 422pp.

Doody JP. The impact of 'reclamation' on the natural environment of the Wash. In: Doody JP, Barnett B, editors. The Wash and its environment, Research and survey in nature conservation, vol. 7. Peterborough: Nature Conservancy Council; 1987. p. 165–72.

Joint Nature Conservation Committee. 1997. UK Biodiversity Action Plan. http://jncc.defra.gov.uk/default.aspx?page=5155

Mossman HL, Davy AJ, Grant A. Does managed coastal realignment create saltmarshes with 'equivalent biological characteristics' to natural reference sites? J Appl Ecol. 2012. doi:10.1111/j.1365-2664.2012.02198.x.

Thompson D. Distribution and abundance of harbor seals (*Phoca vitulina*) during the breeding season in the Wash and along the Essex and Kent coasts. Unpublished report to Natural England covering surveys carried out in 2004 to 2011; undated. http://www.dassh.ac.uk/dataDelivery/filestore/1/0/0_e74b5c443587fc8/100_7a6b8bf8c592c5d.pdf

Wetlands of the Norfolk and Suffolk Broads (UK)

82

Andrea Kelly

Contents

Introduction	1024
Extent (Current and Historical)	1024
Biodiversity	1027
Habitats	1027
Species	1030
Conservation Status	1032
Management	1032
Ecosystem Services	1034
Dependent Peoples	1035
Future Challenges	1035
References	1036

Abstract

The Broads is renowned as one of the UK's premier wetlands and its third largest system for inland navigation. A unique and globally important landscape, it has been shaped and nurtured by its inhabitants since at least Roman times. The Broads Authority executive area encompasses an area of 303 km^2 in Norfolk and North Suffolk, nestled between the urban areas of Norwich to the west and Great Yarmouth and Lowestoft to the east, with a short coastal strip and an estuary at Breydon Water. The Broads lies at the bottom end of the much larger Broadland Rivers Catchment, with water flowing through or under it and out to sea.

This low-lying, mainly open and undeveloped wetland landscape is an interconnected mosaic of rivers, shallow lakes, fens, drained marshland, wet woodland, saltmarshes, intertidal mudflats and various coastal formations.

A. Kelly (✉)
Broads Authority, Norwich, Norfolk, UK
e-mail: andrea.kelly@broads-authority.gov.uk

© Springer Science+Business Media B.V., part of Springer Nature 2018
C. M. Finlayson et al. (eds.), *The Wetland Book*,
https://doi.org/10.1007/978-94-007-4001-3_18

Water, not surprisingly, is the vital element that links everything in this landscape, and its careful and integrated management is central to everything in this plan.

Each habitat has its own distinctive characteristics and hosts a wealth of species, many rare and some unique to the Broads within the UK. The importance of the area is borne out by a range of national and international designations in recognition of its landscape, nature conservation and cultural features.

The Broads is also a dynamic, living landscape. Over the centuries its natural, cultural and built features have been shaped by the way peat diggers, traders and merchants, reed cutters and thatchers, farmers and fishermen have lived and worked. The shallow lakes referred to as 'broads' originated as great pits dug for peat to provide fuel during medieval times. Around the 14th century, these peat diggings flooded and became part of an extensive communication network for transporting fuel, building materials including reed for thatch, and livestock and their products, especially wool.

Looking ahead, the Broads will continue to be influenced and shaped by environmental, social, economic, technological and political change. Some of the biggest challenges facing this easterly, low-lying freshwater wetland are likely to come from the projected more rapid changes to the climate, together with sea level rise. Other significant changes in global, national and regional economies, patterns in leisure and tourism, demands on food and energy resources, and population growth and demands for housing and infrastructure in the East of England will also have an impact on the landscape and communities of the Broads.

While we cannot predict what the Broads will look like in 50 or 100 years' time, understanding and responding now to the challenges ahead will help us to plan a longer term future that maintains the area as a unique, special and valued landscape for generations to come, even if it does not stay the same as it is now. As part of the UK National Parks family and global network of protected landscapes, the Broads has a vital role to play in demonstrating how wetland resources can be managed sustainably for the benefit of both nature and people.

Keywords

Wetland · Water · Lake restoration · Fen management · Peat digging · Eutrophication control · Ecosystem services

Introduction

Extent (Current and Historical)

The area of the Norfolk and Suffolk Broads is renowned as one of the UK's premier wetlands, a 303 km^2 mix of coastal and inland wetland habitats comprising six rivers and an estuary flowing into the North Sea (Fig. 1). The inhabitants of the Broads have navigated and managed this ever-changing landscape for thousands of years.

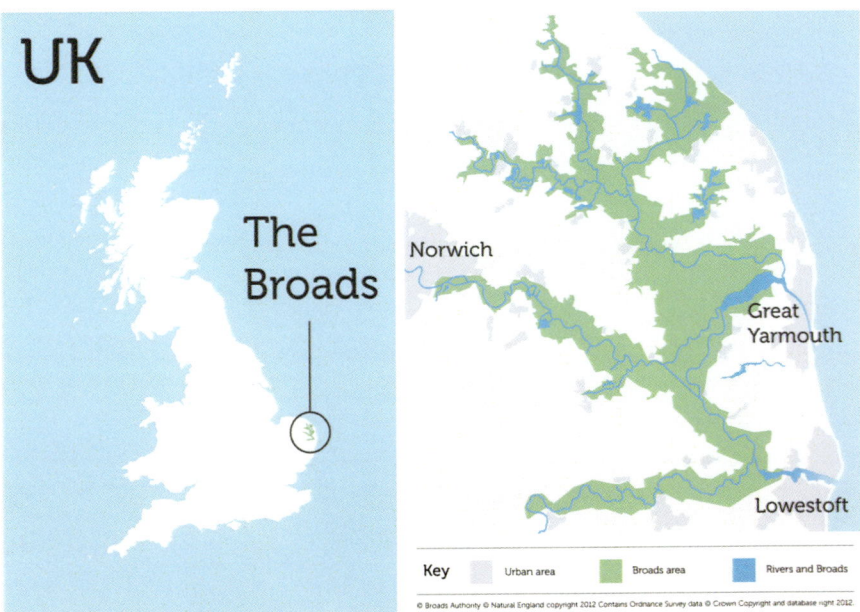

Fig. 1 Map showing the Norfolk and Suffolk Broads and their location in the UK (*inset*) (Crown Copyright 2012, Contains public sector information licensed under Open Government License v3.0)

The area of the Broads sits between the urban areas of Norwich, Great Yarmouth, and Lowestoft, including a short coastal strip and the UK's most easterly tidal estuary (Breydon Water). The low-lying, mainly open and undeveloped, landscape of the Broads comprises an interconnected wetland mosaic of rivers, broads, fens, marshes, and woodland rich in rare habitats and species, 66 of which depend on the Broads for their survival in the UK (Panter et al. 2011). Designated as part of the UK national park family in 1989, the Broads make up 10% of the protected landscapes in the UK.

Back in the first century CE, the Broads wetland covered a vast area, as a result of sea level being about 0.5 m higher than today. The large eastern estuary was defended by the Romans against raids from the continent. Around 1,500 years ago, sea levels dropped and marsh areas became mainly freshwater. Between the ninth and thirteenth centuries deep peat soils, formed in the marsh environment from decomposing vegetation, were excavated on an industrial scale for fuel. By the fourteenth century, peat cutting had to be abandoned when the pits left by the cutting process filled with water when water levels started rising again. These flooded holes became what we now know as the broads. The Broads name came from the term describing a "broadening" of the river.

The rivers and broads of Norfolk and Suffolk formed the principal routes for transporting goods such as fuel, building materials including reed for thatch, livestock, and wool until the advent of the railways in the mid-nineteenth century and

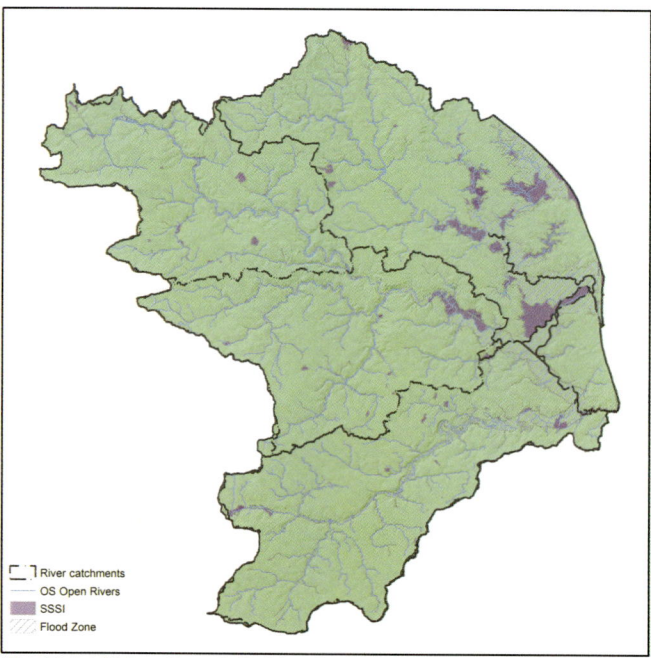

Fig. 2 Map of the Broadland catchment area (Crown Copyright and database right 2017, Contains public sector information licensed under Open Government License v3.0)

motor vehicles in the twentieth century which brought most river-borne commerce to an end. Nonetheless, the waterways continued to be used as a popular tourist destination for boating holidays, with 200 km of navigable rivers and open waterbodies to be explored and enjoyed.

Almost half of the Broads habitats are large expanses of drained grazing marsh, created over the centuries by drainage attempts that were enhanced in the eighteenth century by drainage mills with sails, followed by steam power and diesel, then electric pumps. This increase in drainage efficiency also increased productivity in the marshes, providing cattle grazing free from flood water. Farmers were also now able to produce crops. This change in land usage created new habitats from the network of grass grazing marshes, newly created water ditches, and riverbanks. The peat-based fens were left undrained but still provided crops of marsh hay, and reed and sedge for thatching buildings.

The Broads Authority (equivalent to a national park authority) and national conservation organizations own only around 11% of the Broads, all other land is in private ownership.

The Broadland rivers catchment area is more than ten times bigger than the Broads (Fig. 2). Water that falls in this area runs, drains, percolates, or is pumped into rivers and ultimately flows through the Broads and out to sea at Great Yarmouth. A Broadland Catchment Partnership has been formed to work together for healthier water and wetlands in the wider area and Broadland Catchment Plan provides the framework for partnership working (Punchard 2014).

The Broads, being low-lying and located at the bottom of the catchment, are significantly affected both by tidal influences and by what happens further upstream. The upstream areas of land are 80% arable and generally steeper sloping, and present a greater risk of runoff and pollution from land-based sources. Therefore upstream planning helps focus partners' efforts and resources towards the headwaters rather than just in the areas immediately around the Broads.

Biodiversity

Habitats

The Broads area contains a wide range of internationally and nationally important habitats: open water, calcareous fen, carr woodland, and grazing marsh (Fig. 3).

The waterbodies include 122 km of navigable river and 63 shallow lakes, some of which are also navigable. The majority of the large shallow lakes are supplied by rivers that drain a large agricultural and highly populated catchment area, which consequently have been degraded by eutrophication. However, the water quality of the majority of the 63 isolated shallow lakes has been restored by the Broads Authority and partners over the past 30 years (Kelly 2008).

Considerable further restoration is required to deliver status required by the European Water Framework and the European Natura 2000 sites for the rivers and lakes in the Broadland area. Over 90% of rivers still fail to meet European Water Framework Directive targets due to factors including physical modification, water quantity, phosphate, dissolved oxygen, and fish populations. With three quarters of the lake area connected to the river network, the management of the upstream rural and urban land is important in lowering the nutrient status and in providing adequate flows of fresh water (Kelly 2008; Punchard 2014).

In some areas of the Broads, water levels are too high for agriculture or too low for wildlife and amenity, while recent droughts have resulted in a lack of water availability for agriculture, wildlife, and public garden use. Another issue is that heavy rainfall running off rural and urban areas causes surface water and river flooding in specific locations. Tidal surges continue to threaten lives, property, farmland, coarse fish populations, and important freshwater wildlife habitats.

The Broads area contains three quarters of the UK's calcareous fen, which is a European Union (EU) priority habitat. Fens are the first stage in the natural succession from open water to wet woodland. The fens are the richest area for biodiversity in the Broads, supporting over 250 plant species, many of which are so rare that they cannot be found elsewhere in lowland Britain, such as the fen orchid *Liparis loeselii*, intermediate stonewort *Chara intermedia*, crested buckler fern *Dryopteris cristata*, and greater water parsnip *Sium latifolium*. Areas of mature wet woodland and some younger scrub occur in a mosaic around the fen (Fig. 4).

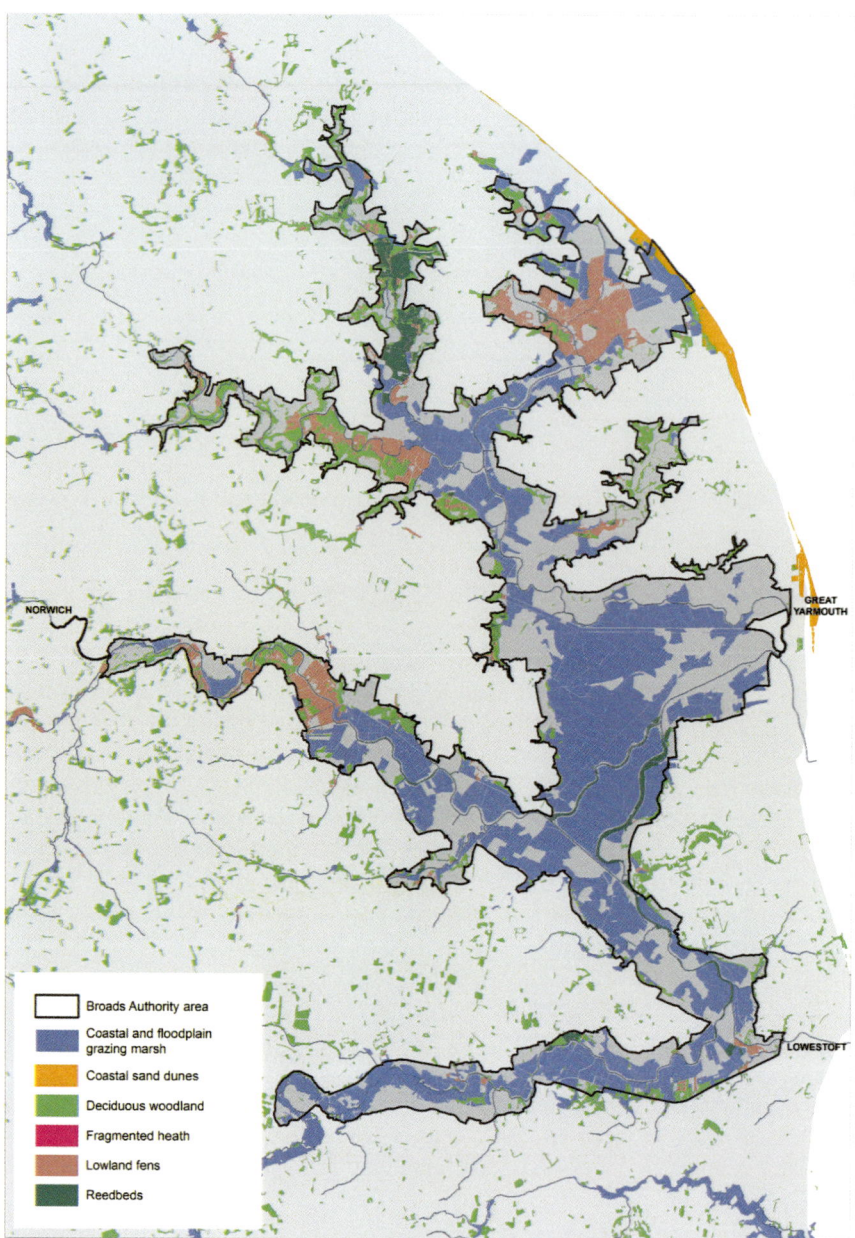

Fig. 3 Map showing the distribution of habitats in the Broads (Crown Copyright and database right 2017, Contains public sector information licensed under Open Government License v3.0)

Fig. 4 Fen habitat in the Broads (Photo credit: Ian Aitkin © Rights remain with the author)

Significant areas of mature wet carr woodland which are left unmanaged support a diversity of invertebrates, and the refuge and food provided by the alder trees attract redpolls *Carduelis flammea* and other small birds. Alder and willow *Salix cinerea* are the most common canopy species, along with downy birch present in patches with ash and pedunculate oak in drier stands. Holly *Ilex* sp., rowan *Sorbus* sp., buckthorn Rhamnus sp., and guelder rose *Viburnum opulus* also occur occasionally. The field layer is related to the preceding swamp and fen communities, with a small woodland influence. Greater tussock-sedge *Carex paniculata* or lesser pond sedge *C. acutiformis* usually form a major component of the field layer. Wet woodlands are rich in invertebrate species and provide an important habitat for many associated animal and plant groups: mammals such as otters, water voles, and bats.

The largest area of seminatural habitat in the Broads is the coastal grazing marsh which supports thousands of wintering water birds, breeding waders, and a host of rare wildlife in the water-filled ditches.

The Broads include the UK's most easterly estuary, Breydon Water. Breydon supports internationally important numbers of wintering and passage birds, an average of over 105,000 birds being present in recent years. This is the eighth most important site for waterfowl in Great Britain and has the densest concentrations of wintering birds in the UK. The area has a long history of protection and was one of the first areas to be protected in this country, with a Breydon Water Wild Bird Protection Society being formed in 1889.

Breydon Water reserve and adjacent areas is internationally important for pink-footed goose *Anser brachyrhynchus*, Eurasian wigeon *Anas penelope*, Eurasian teal *Anas crecca*, northern shoveler *Anas clypeata*, pied avocet *Recurvirostra avosetta*, Eurasian golden plover *Pluvialis apricaria*, northern lapwing *Vanellus vanellus*,

Fig. 5 Eurasian cranes (Photo credit: Nick Upton © Rights remain with the author)

black-tailed godwit *Limosa limosa*, and common tern *Sterna hirundo*. It is also holds nationally important numbers of Bewick's swan *Cygnus columbianus*, greater white-fronted goose *Anser albifrons*, dunlin *Calidris alpina*, ruff *Philomachus pugnax*, and common redshank *Tringa totanus*. Additionally, it is locally important for breeding numbers of common shelduck *Tadorna tadorna* and a host of other species which require large areas of undisturbed mud flats on which to feed and rest.

Species

The diversity of plants in the Broads is far exceeded by the animal life. The vast majority of rarities are invertebrate species such as the shining ram's horn snail, *Pseudamnicola confuse*, Norfolk hawker *Aeshna isosceles*, Bure long-legged fly *Dolichopus nigripes*, and swallowtail butterfly *Papilio machaon*.

Four resident birds which are commonly seen in the fens are Eurasian bittern *Botaurus stellaris*, Eurasian crane *Grus grus*, bearded tit *Panurus biarmicus*, and marsh harrier *Circus aeruginosus*. In the 1800s, Eurasian bitterns were hunted for food and by collectors and their steady decline, compounded by loss of habitat, resulted in only 15 males in eastern England by 1995. Today there are around 20 pairs in the Broads as a result of protection and habitat restoration (Fig. 5).

Numerous migratory bird species make use of the Broads as summer and winter visitors and are seasonally resident. The Broads also provide an essential wetland stopover as a feeding area for passage migrants. Winter visitors include waders, geese, and swans, while summer visitors include Eurasian cuckoo *Cuculus canorus* and many other passerines that breed in the wetlands.

Fig. 6 Fen raft spider (Photo credit: Ross Hoddinott © Rights remain with the author)

Almost half of the Broads habitat consists of large expanses of drained grazing marsh. The increase in drainage efficiency and productivity in the marshes to provide cattle grazing free from flood water has resulted in poorer quality habitat for birds. However, much of the summer grazed pasture still provides important habitat for vast flocks of wintering birds – Eurasian wigeon, Eurasian teal, gadwall *Anas strepera*, greater white-fronted geese, and Bewick's swans – as well as breeding birds – common redshank, northern lapwing, and short-eared owls *Asio flammeus*.

A biodiversity audit (Panter et al. 2011) showed that 11,067 taxa (species aggregates, species and designated subspecies or varieties) have been recorded in the Broads, of which 1,519 are priorities for conservation, including 403 beetle species, 251 true flies, and 179 moth species. These include 26% of all UK Biodiversity Action Plan species (Natural England 2014), 13% of all UK Red Data Book species, and 17% of all nationally notable or scarce species. The Broads also provide habitat for 85% of the UK bird species – and 94% of the birds – Amber list species (Eaton et al. 2009; IUCN 2014). Sixty-six species were identified as Broads specialities, of which 31 are entirely or largely restricted to the Broads within the UK ($\geq 80\%$ of UK range, assessed as 10 km squares, or population size) and a further 35 have a primary stronghold in the region ($\geq 50\%$ of UK range or population size). These include the swallowtail butterfly, holly-leaved naiad *Najas marina*, and the fen raft spider *Dolomedes plantarius* (Fig. 6).

The audit showed that since 1950 an average of 10 species have been lost per decade in the Broads. In addition, the status of 356 species is unknown, and it is likely that a similar situation would be found in most English wetlands. However, it is worth noting that species with no recent record were more often associated with dry or damp ecotonal habitats than wetland habitats. This indicates that much of the species loss has occurred around the margins of the wetland, as agriculture has intensified.

Sixty-three percent of conservation priority species require fully freshwater conditions and are considered to be unlikely to tolerate brackish influence. Many of these are invertebrates that may find it difficult to move away from saline conditions resulting from increased coastal flooding predicted to result from climate change.

Conservation Status

The importance of the Broads as one of Europe's finest wetlands is reflected by the fact that many sites in the Broads are internationally, nationally, or locally designated for conservation value. Around 25% of the Broads are covered by international designations, forming part of the Natura 2000 site network under the EU Habitats and Birds Directive, including sites listed under the Ramsar Convention of Wetlands of International Importance. Twenty-eight sites, lying wholly or partly within the Broads Authority executive area, are nationally important. In addition, a growing number of local nature reserves (currently 84) provide refuge for wildlife and are greatly valued by local people and visitors.

Management

Most of the wetland habitats in the Broads require management to retain and enhance biodiversity. The shallow lakes and rivers have been subject to over 30 years of restoration. This has included phosphate removal from waste water that has decreased nutrient loading to some waterways by 90%. Techniques such as sediment removal and fish population manipulation to gain clear water conditions that can stimulate water plant growth have been carried out (Kelly 2008; Moss et al.1996; Madgwick 1999; Kelly and Harris 2010). Considerable improvements have been made in water quality and habitats as a result; however, many of the larger lakes that are connected to rivers remain in unfavorable ecological condition (Figs. 7 and 8).

To protect the grazing marshes from saline tides, a major investment program for flood alleviation was implemented from 2000 to 2010. Despite this, many marshes have shown a decline in breeding bird numbers since the 1990s but have the potential for increases in both bird use and ditch quality for plants and invertebrates if the incentives provided to farmers under the European Union Common Agricultural Policy were sufficient to stimulate their participation.

The fens have also been the subject of considerable study and restoration (Moss 2001; Phillips 1992; Phillips et al. 2005; Kelly 2008; George 1992; Broads Authority 1997). Declines in species and habitat area have been arrested with extensive scrub removal. New fen management techniques have been developed, as well as expanding traditional reed cutting practices. Commercial reed and sedge cutting retains a traditional and economically viable open fen landscape. Within a mosaic of larger areas of longer rotation cutting (every four to seven years) there is no significant impact on biodiversity (Harding et al. 2010). Most of the fens in the Broads are supplied by the base-rich river water; however, some groundwater-

Fig. 7 Barton Broad showing restored clear water areas (Photo credit: Mike Page © Rights remain with the author)

Fig. 8 Trinity Broads with the North Sea behind (Photo credit: Mike Page © Rights remain with the author)

Fig. 9 A restored wetland in the Broads (Photo credit: Mike Page © Rights remain with the author)

dependent fens require further study to determine the possible impact of water abstraction (Fig. 9).

Ecosystem Services

Management in the Broads makes a significant national and regional contribution to mitigating and adapting to climate change through flood control, water conservation, carbon conservation, biodiversity conservation, and promoting sustainable farming. The Broads draft climate change adaptation plan (Broads Authority 2011b) is in the process of being revised. Work is ongoing through a "Broads Community" initiative to engage decision makers, stakeholders, and local communities and give them an opportunity to share information and ideas. It looks at predicted changes in climate and sea level rise and considers the impacts on people and the natural environment.

Wetland management in the Broads involves taking a landscape-scale approach to the water catchment boundary. This involves not just looking at biodiversity opportunities but also other environmental, social, and economic factors such as climate change and flood management, food production, and health and well-being.

Economic values for the Broads ecosystem services have been calculated, for example, £18 million per year for biodiversity alone, £17 million for water supply, and over £300 million for tourism expenditure (Kelly 2008). These costed benefits are high compared to the cost of managing these natural resources to retain good condition. The principles of securing critical conservation objectives while taking account of the economic, recreational, social, and cultural needs of the community are central to the strategic management plan for the Broads (Broads Authority 2011a).

Dependent Peoples

With 850,000 permanent residents living on the doorstep and within the water catchment of the Broads, the needs of both people and wildlife are integral to the area. The sectors that depend most on the natural resources of the Broads are tourism, including accommodation and food and drink industry, boating and fishing, farming on the grass marsh and arable land, and water companies that extract water from lakes, rivers, and groundwater. In 2011 there were 7.4 million visitors to the Broads alone, resulting in an estimated visitor spend of £469 million and supporting over 6,000 jobs. About 8,500 jobs in the catchment rely on farming. Energy and life sciences are also important components of the local economy.

Local partnerships, for example, with tourism businesses, landowners, boaters, anglers, reed cutters, graziers, and community groups, as well as the involvement of local volunteers, help people understand and value the cultural and natural heritage. One of the challenges going forward is to widen the support for sustainable management of the Broads from funders and sectors that directly benefit from a quality natural environment.

The community of people with a shared and passionate interest in the Broads is widespread. There is a complex interrelationship between conserving and enhancing the area's natural beauty, wildlife, and cultural heritage, promoting opportunities for the understanding and enjoyment of these special qualities and protecting the interests of navigation, and the Broads Plan (Broads Authority 2011a) considers many of these complexities and links. It is a plan not just for the Broads Authority but for the Broads in its entirety, involving many organizations and individuals in the management and care of this precious and fragile landscape.

Future Challenges

The Broads low-lying coastal wetland habitats include some of the most threatened places in the UK in the light of water pollution, water abstraction, invasive species, climate change, sea level rise, and pressure from recreational water use (Fig. 10).

Sea level rise, exacerbated by global warming, is the biggest risk to the Broads in the long term. This would result in saline intrusion further up the rivers during spring tides and storm surges, drought, and flooding events. However, short-term evidence shows that lack of management intervention needed to retain the diversity of wetland habitats and pollution still remain significant concerns. For example, scrub removal and new solutions found for creating sustainable harvesting (such as bioenergy from the biomass) need incentivizing if we are not to lose botanically rich fens.

Tackling issues around water quality, water shortage, flooding, and wildlife habitat is important to many individuals and organizations locally and internationally, as is

Fig. 10 Hickling Broad and the North Sea behind (Photo credit: Mike Page © Rights remain with the author)

supporting recreation, tourism, agriculture, and dependent industries. Improvements within the catchment, particularly over the past 30 years, are due to the dedicated effort of many individuals, groups, and organizations. Despite this excellent work, there is much more to be done.

References

Broads Authority. Fen management strategy. Norwich: Broads Authority; 1997. www.broads-authority.gov.uk.

Broads Authority. Broads plan 2011: a strategic plan to manage the Norfolk and Suffolk Broads. Norwich: Broads Authority; 2011a.

Broads Authority. Draft Broads climate change adaptation plan. Norwich: Broads Authority; 2011b.

Eaton MA, Brown AF, Noble DG, Musgrove AJ, Hearn R, Aebischer NJ, Gibbons DW, Evans A, Gregory RD. Birds of conservation concern 3: the population status of birds in the United Kingdom, Channel Islands and the Isle of Man. Br Bird. 2009;102:296–341. http://www.rspb.org.uk/Images/BoCC_tcm9-217852.pdf.

George M. The land use, ecology and conservation of Broadland. Chichester: Packard Publishing; 1992.

Harding M, et al. Fen plant communities of broadland results of a comprehensive survey 2005–2009; Norwich: Broads Authority; 2010. http://www.broads-authority.gov.uk/__data/assets/pdf_file/0009/416376/Broads_Fen_plant-communitites.pdf

IUCN. The IUCN red list of threatened species. Version 2014.2. http://www.iucnredlist.org. Downloaded in March 2014.

Kelly A. Lake restoration strategy. Norwich: Broads Authority; 2008. www.broads-authority.gov.uk.

Kelly A, Harris J. Aquatic plant monitoring in the broads. In: Hurford C et al., editors. Conservation monitoring in freshwater habitats: practical guide and case studies. Dordrecht/London: Springer Science+Business Media B.V; 2010. doi:10.10007/978-1-4020-9278_1.

Madgwick FJ. Restoring nutrient-enriched shallow lakes: integration of theory and practice in the Norfolk Broads, UK. Hydrobiologia. 1999;409:1–12.

Moss B. The Broads. London: Collins; 2001.

Moss B, Madgwick J, Phillips G. A guide to the restoration of nutrient-enriched shallow lakes. Norwich: Broads Authority; 1996.

Natural England. The UK biodiversity action plan priority species. 2014. http://jncc.defra.gov.uk/page-5717

Panter C, Mossman H, Dolman PM. Biodiversity audit and tolerance sensitivity mapping for the broads. Broads Authority Report. Norwich: University of East Anglia; 2011. http://www.broads-authority.gov.uk/__data/assets/pdf_file/0020/412922/Broads-Biodiversity_audit_report.pdf

Phillips G. A case study in restoration: shallow eutrophic lakes in the Norfolk Broads. In: Harper ED, editor. Eutrophication of freshwaters. London: Chapman and Hall; 1992. p. 251–78.

Phillips GL, Kelly A, Pitt JA, Sanderson R, Taylor E. The recovery of a very shallow eutrophic lake, 20 years after the control of effluent derived phosphorus. Freshwater Biol. 2005;50:1628–38.

Punchard N. Broadland catchment plan. Norwich: Broads Authority; 2014. www.broads-authority.gov.uk.

Blanket Mires of Caithness and Sutherland: Scotland's Great Flow Country (UK)

83

Richard Lindsay and Roxane Andersen

Contents

Introduction	1040
Physical Setting	1040
Climate	1041
Mire Development	1043
Vegetation, Surface Pattern, and Birdlife	1043
Other Animal Life of the Flow Country	1047
Land-Use Pattern	1047
Conservation Status: Forestry, Conservation, and the "Battle for the Flow Country"	1049
Threats and Future Challenges	1051
References	1054

Abstract

The Flow Country of northern Scotland dominates the northernmost part of mainland Britain. It is thought to be one of, if not the, largest more-or-less continuous expanses of blanket mire in the world. It extends from the coast of Caithness in the east to the foot of the mountain chain which dominates western parts of Sutherland, covering some 400,000 ha in all. The mire systems of the Flow Country occur in a wide variety of forms as well as displaying distinct east-west gradients in their vegetation and surface microtopography. These mires also support internationally important breeding bird populations, including many species more usually associated with the northern tundra. The area has a long

---

R. Lindsay (✉)
Sustainability Research Institute, University of East London, London, UK
e-mail: r.lindsay@uel.ac.uk

R. Andersen
Environmental Research Institute, University of the Highlands and Islands, Thurso, UK
e-mail: roxane.andersen@uhi.ac.uk

© Springer Science+Business Media B.V., part of Springer Nature 2018
C. M. Finlayson et al. (eds.), *The Wetland Book*,
https://doi.org/10.1007/978-94-007-4001-3_193

history of human occupation and land use but the treeless peat-dominated landscape, which developed around 6200 BP, remained largely unchanged until after World War 2 when some experimental conifer plantations were established on the deep peat. These were then followed in the 1980s by a huge expansion of conifer plantation across the peatlands, leading to a major battle between conservation bodies and the forest industry. The afforestation was eventually halted but only after some 70,000 ha had been planted or was scheduled for planting. Substantial efforts are now being devoted to restoring these plantations to open bog. Meanwhile windfarm developments pose a new threat. Despite these challenges, the area is under consideration for possible World Heritage status.

Keywords

Blanket · Bog · Britain · Caithness · Conifer · Flow · Forestry · Forsinard · Mire · NCC · Peat · Plantations · Pools · RSPB · Scotland · Sutherland · Tundra · Windfarm

Introduction

The Flow Country of northern Scotland is the most extensive and varied expanse of treeless blanket mire in the UK and is thought to be one of, if not the, largest continuous expanses of blanket mire in the world. It contains a wide range of mire types in which the mire patterns and vegetation assemblages display a marked east–west gradient in their form and composition. The area also supports internationally important breeding bird populations which are more usually associated with the northern tundra. The area became the focus of a major battle between conservation bodies and the forest industry over large-scale planting of the area with conifers. Afforestation was eventually halted but now windfarm developments pose a new threat. The area is under consideration for possible World Heritage status.

Physical Setting

The Flow Country dominates a 400,000 ha ice-scoured plain forming the northernmost part of the Scottish mainland. This plain extends across the whole County of Caithness in the east and approximately two thirds of the County of Sutherland towards the west (Fig. 1). Much of the area is underlain by Old Red Sandstone or Moine schists with some felsic igneous intrusions, but the whole is overlain by glacial till which is generally highly leached and base-poor, though it becomes markedly calcareous in eastern Caithness where seabed material was deposited by the east–west movement of the last glaciation. The plain is gently inclined from NE to SW rising from sea level to 350 m and is interrupted in a few places by isolated high hills, none of which rises above 1,000 m. Apart from these few interruptions, the plain is almost entirely blanketed by a mantle of peat which in places reaches

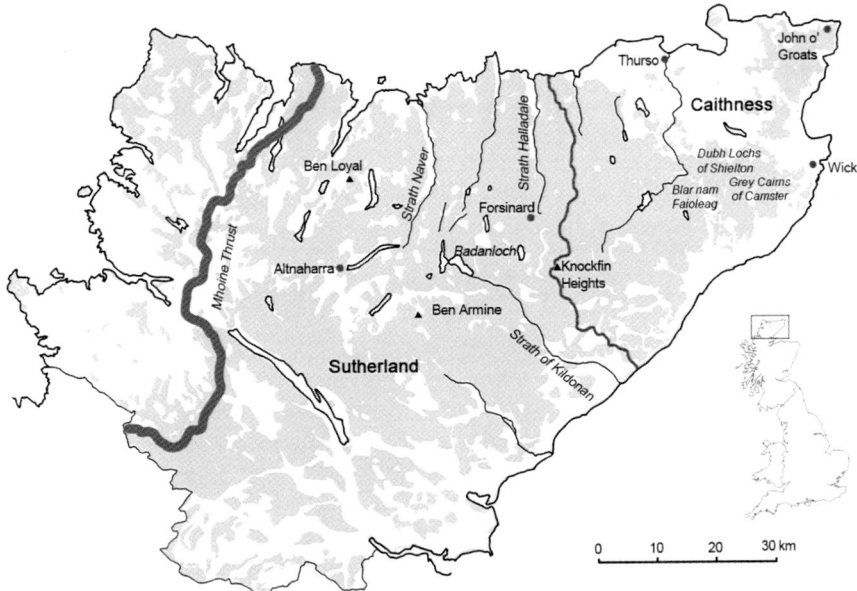

Fig. 1 Caithness and Sutherland, northern Scotland, showing named locations and features. The extent of peat is shown in *pale gray*, the county boundary running through Knockfin Heights is shown as a *solid gray line*, while the Moine Thrust in western Sutherland is shown as a *thicker solid gray line*. Named hill peaks are shown as *black triangles* (Figure credit: Adapted from Lindsay et al. (1988) with kind permission of the Joint Nature Conservation Committee (JNCC))

depths of more than 6.5 m. It is traversed by only two main roads in Caithness. In Sutherland there are five roads, mostly single track. The immensity of this open landscape and the sense of solitude it engenders have been likened to that of the Arctic tundra, and indeed the wildlife it supports is not so different from that of the tundra (Ratcliffe 2000). The key difference between the two areas is that the tundra is waterlogged because it has an impermeable permafrost layer, whereas the Flow Country has no permafrost but is instead waterlogged because precipitation consistently exceeds evapotranspiration, thereby resulting in widespread paludification of the ground and development of a landscape almost entirely cloaked by blanket mire.

Climate

The climate of the Flow Country varies significantly across the inclined plain from NE to SW and also from the plain to the scattered broad high plateaux. In the far north-eastern corner of Caithness the climate is hyperoceanic humid hemiboreal (Lindsay et al. 1988), with an annual rainfall of around 815 mm but with never less than 49 mm in any month. This is distributed across the year in the form of some 166 "wet days" per annum with at least 10 wet days per month or between 225 and

Fig. 2 Aerial view across the summit of Knockfin Heights in midwinter, looking north-west across the Flow Country. The pool system on the summit of Knockfin Heights can be seen to be at least partially frozen, while the pool systems at lower altitude were unfrozen (Photo credit: Steve Moore © Rights remain held *in-trust* by Richard Lindsay)

250 "rain days" per annum (Lindsay et al. 1988). A "wet day" is one where at least 1 mm precipitation is recorded in a 24 h period, whereas a "rain day" is one where at least 0.25 mm is recorded in a 24 h period. The precipitation pattern is accompanied by a maximum mean monthly temperature of 16 °C and a minimum mean monthly temperature of 1 °C (Metoffice data for– Wick Airport, http://www.metoffice.gov.uk/public/weather/climate/gfmeffeh7). Higher elevation but decreased oceanic influence give western parts of the plain (which lie in the rain shadow of the Moine Thrust line of high hills dominating the west coast) a euoceanic extremely humid southern boreal and lower oroboreal climate (Birse 1971) with a total annual rainfall of around 1196 mm and at least 60 mm of precipitation in the driest month, spread across 196 wet days per annum, with the driest month having at least 12 wet days per month. Maximum mean monthly temperature is 18 °C and minimum mean monthly temperature is −1.3 °C (Metoffice data for– Altnaharra, http://www.metoffice.gov.uk/public/weather/climate/gfkgdgj2j), but the minimum absolute temperature can fall as low as −27 °C whereas minimum recorded temperatures in the east fall no lower than −13 °C because the land is at lower altitude and benefits more from the moderating maritime influence. Peat-covered high-level plateaux such as Knockfin Heights (Fig. 2) and Ben Armine, lying at altitudes of more than 400 m, experience a much harsher euoceanic extremely humid orohemiarctic climate (Birse 1971)

highlighted by the presence of northern boreal/subarctic spider assemblages (Lindsay et al. 1988).

Mire Development

Detailed studies by Charman (1992) of the vegetational history of the Flow Country have shown that a post-glacial community of *Betula* and *Juniperus* scrub transformed into *Betula-Corylus* (or possibly *Myrica*) open woodland around 9000 BP. This open woodland appears to have declined from around 6200 BP onwards, being steadily replaced by an ericaceous peat-forming vegetation. There was then a brief expansion of *Pinus*, possibly representing a single generation of trees, starting around 4250 BP and vanishing again by 3400 BP. The blanket mire landscape then became almost entirely treeless by 1470 BP.

Loss of virtually all woodland cover, even as scattered scrub, may be attributable to grazing pressure from deer and from possible human actions such as the introduction of sheep, as there is ample evidence of a substantial community presence between neolithic and iron age times which was sufficiently well resourced to be capable of constructing large chambered tombs and fortified towers or "brochs." Examples of blanket mire from other parts of the world where there is much lower or indeed no pressure from sheep grazing tend at least to have scrubby woodland growth along streamlines (e.g., Nova Scotia, Tierra del Fuego), which makes the almost complete absence of such scrub in the Flow Country all the more striking. *Betula*, *Salix*, *Alnus*, and *Sorbus* are all present but only as low, heavily-browsed stems no taller than the accompanying mire vegetation. In the event of reduced pressure from sheep grazing, streamline scrub woodland might develop and thereby re-establish a landscape closer to that which existed 3,000 years ago.

Vegetation, Surface Pattern, and Birdlife

The range of climatic conditions and landforms found across the Flow Country together give rise to distinctive forms of blanket mire habitat characterized by particular surface patterns and vegetation assemblages. The following descriptions are largely based on the detailed accounts of the Flow Country provided by Stroud et al. (1987) and Lindsay et al. (1988). In their analysis of the Flow Country mire systems, Lindsay et al. (1988) identified seven natural blanket bog types, six types of damaged bog generally involving some form of erosion, two types transitional between bog and fen, and 12 types of fen community. The more calcareous fen systems were associated with the marine-derived glacial tills of eastern Caithness. The most extensive features across the Flow Country as a whole proved to be the various forms of eroded bog, which reflects the poor condition of blanket bog generally in the UK as a result of burning, drainage, atmospheric pollution, and trampling resulting from overstocking of sheep. The remarkable feature of the Flow Country, however, is the extent of ground which is not degraded. The natural

Fig. 3 Large pool patterns, with some pools several meters deep, formed across the blanket mire landscapes of eastern Caithness at the Dubh Lochs of Shielton ("dubh loch" meaning "black loch" – i.e., a water body formed within the peat) (Photo credit: Steve Moore © Rights remain held *in-trust* by Richard Lindsay)

variation in microtope pattern and vegetation across the expanse of the Flow Country can therefore be readily seen within a common background vegetation assemblage of *Sphagnum papillosum*, *S. capillifolium*, *Erica tetralix*, *Calluna vulgaris*, *Eriophorum angustifolium*, and *Narthecium ossifragum*.

In the east, broad watersheds are dominated by large pool systems which show affinities to the "bog lake" raised bog complexes which occupy much of the west Siberian basin (Fig. 3). The largest pools at the center of the pattern complex may be three or four meters deep, vertical sided, and largely devoid of vegetation, but these pools are important roosting sites for species such as the greenland white-fronted goose (*Anser albifrons flavirostris*) and breeding sites for red-throated diver (*Gavia stellata*). Occupying the ground between pools are broad *Sphagnum*-rich ridges with scattered hummocks. Beneath a short dwarf shrub layer, species such as *Listera cordata* and *Vaccinium microcarpum* may be found. These broad ridges are favored breeding areas for dunlin (*Calidris alpina*) and greenshank (*Tringa nebularia*), which feed on the pool margins and in the smaller, shallower hollows of pure *Sphagnum cuspidatum* which surround the main pool complex. Where the bog lies

Fig. 4 Pool, ridge, and hummock system at Badanloch Bogs, eastern Sutherland, in the heart of the Flow Country. The large green hummock is the moss *Racomitrium lanuginosum*, which becomes a more frequent component of the natural bog vegetation in western parts of the Flow Country, but which is also a common feature of eroding bog in all parts of the Flow Country (Photo credit: Richard Lindsay © Rights remain with the author)

on a distinct slope, the pools and hollows are narrower and shallower, and the narrow intervening ridges are characterized by a *Sphagnum-Eriophorum vaginatum* community.

In central parts of the Flow Country, the ridge vegetation of some bog systems takes on a rather more "continental" character, with *Betula nana, Arctostaphylos uva-ursi*, and *Sphagnum fuscum* featuring significantly within the vegetation of the ridges (Fig. 4). In other cases, the vegetation shows signs of a distinct oceanic influence, with *Potentilla erecta, Polygala polygonifolius*, and the oceanic liverwort *Pleurozia purpurea* becoming increasingly frequent. The small *Carex pauciflora* forms distinctive stands in low *Sphagnum* hummocks. In hollows and shallow pools, stands of *Eleocharis multicaulis* grow within a mixed carpet of *Sphagnum cuspidatum* and *S. auriculatum*. Some of the softest examples of blanket bog are to be found in these central mires, displaying a simple repeated pattern of *Sphagnum*-rich high ridge, low ridge, and hollow in which *S. papillosum* is often the dominant species. For Eurasian golden plover (*Pluvialis apricaria*), this central part of the Flow Country is almost as important as the eastern bog-pool systems, but dunlin tend to favor only the soft low-relief sites (Stroud et al. 1987) while common greenshank are even more selective. Merlin (*Falco columbarius*), hen harrier (*Circus cyaneus*), and short-eared owl (*Asio flammeus*) can all be seen hunting over these central and eastern mire systems.

Towards the west, the Flow Country becomes markedly oceanic in its vegetation composition, as indicated by the presence in the far west of hyperoceanic species such as *Sphagnum skyense* and *Schoenus nigricans* growing on blanket bog. *Trichophorum cespitosum* and *Molinia caerulea* form prominent components of

Fig. 5 A drought-sensitive pool in the far west of the Flow Country, with mud sedge (*Carex limosa*) and bog bean (*Menyanthes trifoliata*). The pool complex as a whole has many shallow mud-bottom hollows in which the leaves shed by purple moor grass (*Molinia caerulea*) collect to form a relatively firm base to the hollow (Photo credit: Richard Lindsay © Rights remain with the author)

the vegetation sward, along with low cushions of *Campylopus atrovirens* and soft low-ridge mats and hummocks of *Racomitrium lanuginosum*. By way of contrast, in the east, *R. lanuginosum* only occurs as large hummocks or erosion haggs (isolated upstanding hummocks of original bog surface surrounded by erosion gullies) within severely eroded bog systems, while *M. caerulea* is entirely restricted to the margins of stream courses and distinct zones of seepage.

Some deep permanent pools still occur as part of the bog pattern in these western mires, but they are comparatively rare. Pool systems are generally dominated by shallower drought-sensitive pools, often characterized by *Carex limosa* growing as emergent open stands (Fig. 5). Very shallow drought-sensitive pools and mud-bottom hollows become frequent aquatic features because much *Molinia* litter is continually blown into these water-filled elements of the pattern. The semi-decomposed litter creates a relatively firm base for *Drosera intermedia*, which forms distinctive red mats within mud-bottom hollows, together with occasional stands of *Rhynchospora fusca*, while the looser litter in drought-sensitive pools mixes with *Sphagnum auriculatum* to create a framework within which *Utricularia vulgaris* forms an open network. There are few bog pools here which are large enough for red-throated divers, but in these central and western parts of the Flow

Country the greater number of larger lochs (lakes) with a mineral base offer more nesting opportunities for black-throated divers (*Gavia arctica*), while the greater frequency of rock outcrops and high hills provide nesting opportunities for peregrine falcons (*Falco peregrinus*) and even golden eagles (*Aquila chrysaetos*) – see "Blanket mire" Fig. 2. In former times it seems that substantial gull colonies may have been a feature of some large remote bog systems, because the Gaelic name *Blar nam Faoileag* given to some sites means "moor of the seagulls," and a few bog systems continue to support breeding colonies of common gull (*Larus canus*).

Other Animal Life of the Flow Country

The largest British land mammal, the red deer (*Cervus elaphus*) occurs in large numbers throughout the Flow Country, but large herds are more commonly encountered in central and western parts. Occasionally, they will wallow in the loose peat of drought-sensitive pools, either to cool off or to combat parasites, but their main impact is trampling of the soft bog surface when they gather in large numbers. Roe deer (*Capreolus capreolus*) are also often seen as solitary individuals or pairs on the fringes of the main peatland expanses but rarely if ever as more than a family group. Otters (*Lutra lutra*) occur in the larger rivers and around the coasts but may occasionally visit pool systems presumably in search of wader nests for their eggs, because there are no fish in any of the bog pools. Water voles (*Arvicola amphibius*) have colonies in some pool systems and along stream courses. The common toad (*Bufo bufo*) and smooth newt (*Lissotriton vulgaris*) are often encountered, as is the common lizard (*Zootoca vivipera*), but the common frog (*Rana temporaria*) is rarely seen within the bog systems.

The Flow Country is important for invertebrates (Lindsay et al. 1988), particularly relict arctic-alpine species, both for those which occur within the blanket mire habitat and for species in streams and rivers which are supplied by run-off from the blanket mire systems. The bog pool systems support the core national population of the whirligig beetle *Gyrinus opacus*, which is scarce throughout Britain and Europe as a whole. They also support one of only two known populations for the British Red Data Book caddis fly *Nemotaulius punctatolineatus*. Freshwaters associated with the blanket mire systems are the only known location in Britain for the British Red Data Book beetle *Oreodytes alpinus*, which is otherwise restricted to northern Scandinavia and Siberia. These waters are also home to one of the few remaining strong British populations of the freshwater pearl mussel (*Margaritifera margaritifera*), a species listed as "vulnerable" in the IUCN Red Data Book.

Land-Use Pattern

The presence of neolithic chambered tombs has already been noted, the most spectacular examples of which are the reconstructed Grey Cairns of Camster, pointing to a significant human population which was capable of more than simple hunter-gathering at least 5,000 years BP. Their effect on the blanket mire landscape cannot be determined with any certainty, but they may have used fire to drive red deer, they may have cut riparian trees for fuel or for tools, and they almost certainly cut peat as a source of fuel for heating and cooking (given the limited availability of woodland – Charman 1992).

Documentary evidence of land-use patterns within the blanket mire landscape is scarce until the 1600s, by which time a number of clan families were established along the main river valleys of Strath Naver, Strath Halladale, and Strath of Kildonan as well as in the more fertile "shelly till" ground of eastern Caithness. Breakdown of the traditional clan system, particularly after the failure of the 1745 rebellion and defeat at Culloden, just south of Inverness, meant that by the late 1700s most people who worked the land were, in effect, tied farm laborers. New animal husbandry and breeding methods, in part pioneered by Sir John Sinclair, a family member of the Earls of Caithness, stimulated a huge interest in sheep farming amongst the landlords of the Highlands. In many cases, this led to wholesale eviction of tenant families in what has become known as the Highland Clearances. Some of the worst examples of these evictions occurred in Strath Naver, owned by Sinclair's neighbor the Duke of Sutherland, in the heart of the Flow Country (Prebble 1963).

The effect of this enormous and often tragic social upheaval on the blanket mire landscape is difficult to determine, but it seems certain that whatever low-intensity management had been undertaken until that point must have largely ceased. In their place, more widespread grazing by sheep and potentially more intensive management (essentially burning to produce fresh plant growth) for the newly fashionable sports of deer stalking (shooting) and grouse shooting became the norm. A railway was constructed through the Flow Country in the early 1870s in part to bring guests for this new sport. Not only did the sporting management involve burning the blanket mire landscape to encourage fresh growth of heather (*Calluna vulgaris*) for the red grouse (*Lagopus lagopus scoticus*) and of cotton grass (*Eriophorum vaginatum*) for red deer but sparks from the steam train caused regular fires along the line which then burnt across the surrounding bog.

This picture remained largely unchanged until 1980, with sheep numbers across the blanket mire landscapes underpinned by government support mechanisms and what were now called "sporting estates" managing the same land for grouse and red deer, but estate staff numbers were dwindling due to high post-war staff costs. Burning management was therefore becoming increasingly difficult to carry out in a controlled way. Fires begun by steam trains may have ceased once the railway line through the Flow Country was electrified but in their place were many more "managed" fires which simply ran away from the small estate management team. This accumulation of impacts over a 200-year period explains the widespread occurrence of damaged, eroding bog observed by Lindsay et al. (1988) during their surveys in the early 1980s. The early 1980s were, however, also to see one of the most dramatic changes to the Flow Country blanket mire landscape since its initial development some 6,000 years ago.

Conservation Status: Forestry, Conservation, and the "Battle for the Flow Country"

The scale and significance of the Flow Country has only come to be appreciated relatively recently. Indeed the area was not referred to as "the Flow Country" until the 1980s. Prior to this, the area was known only through descriptions of a few individual sites (e.g., Tansley 1939; Pearsall 1956), although there had been some recognition in the 1960s and 1970s that the area was remarkable for the sheer extent of blanket mire habitat (Ratcliffe 1964). Given this extent and the variety of the blanket mire habitat which is now known to occur there, it might seem curious that so few studies were carried out in the area during the early and middle decades of the last century – a period which was, in many ways, a golden age for mire ecology. However, the road network north of Inverness (the nearest city) was not good, accommodation was limited, and the peat-dominated interfluves were so large that the centers of most mire units involved walking for considerable distances across difficult ground (see Lindsay 2015). Mire systems further south were easier to investigate, and the very isolation of the area appeared to afford it a degree of protection from any major development threat.

Consequently, by the start of the 1980s a limited number of blanket mire sites in the "far north" had been identified and notified by the Great Britain government wildlife agency (the Nature Conservancy Council – NCC) as valuable wildlife sites (Site of Special Scientific Interest – SSSI) on the basis of published research papers and limited survey, thereby providing a degree of statutory protection to a very small proportion of the Flow Country. In 1980, however, tax accountants had identified an attractive investment opportunity in forestry support funding which meant that individuals earning enough to be in the "super-tax" bracket could offset most of the costs of establishing new conifer plantations against tax. The trees did not even have to grow, because most of the financial advantages were in the setup costs. If land was cheap – and Flow Country land was very cheap – the tax advantages were enormous. As a result, between 1980 and 1988 thousands of hectares of blanket mire in the heart of the Flow Country were plowed and planted. By 1988 almost 70,000 ha had been planted or scheduled for planting (Fig. 6).

Concurrently with this new development, the NCC had coincidentally been surveying the area for its peatland and bird interest because it was recognized as the last great unsurveyed tract of Britain. As these surveys continued, sometimes quite literally in front of the forestry plows, it became evident that this was an extraordinary place both for its peatland habitats and its breeding bird populations (Lindsay 2015). Alarm was growing across a wide spectrum of interests about the effect of these forestry incentives across much of upland Britain, but attention was beginning to focus on the particularly dramatic changes occurring in the far north. At the time, the area did not even have a name and so was referred to simply as "the peatlands of Caithness and Sutherland," though Ratcliffe's (1964) description of the area as a "great flow country" ("flow" being a northern word for a flat expanse of wet bog) was increasingly being adopted as a formal name (e.g., RSPB 1985).

Fig. 6 An aerial view of forestry plowing in eastern Sutherland during the mid-1980s. The wettest areas of pool system were left unplowed because the machinery could not traverse the softest, wettest peatland areas, but drains were dug as far as possible into such areas. Note the proximity of the plowing to water courses (Photo credit: Steve Moore © Rights remain held *in-trust* by Richard Lindsay)

Forest interests were powerful and the financial opportunities were considerable, but at the same time it was becoming clear that the "Flow Country" was an area of global significance. The resulting battle for the Flow Country has been described as "one of nature conservation's more desperate battles" (Ratcliffe 2000) and the outcome had enormous consequences (Stroud et al. 2015). The Royal Society for the Protection of Birds (RSPB) coordinated a major campaign targeting the financial system causing the damage and arguing for protection of the area (Bainbridge et al. 1987). The International Mire Conservation Group (IMCG) held its 2nd field symposium in the Flow Country to assess the global status of the area and, having had the opportunity to see the area for itself and witness the scale of damage being caused by afforestation, drew up and sent to the Prime Minister, the Right Honorable Margaret Thatcher, a formal Resolution confirming the global importance of the area and calling for a halt to the afforestation. Meanwhile the NCC surveys had been completed and the reports were published (Stroud et al. 1987; Lindsay et al. 1988). These reports then formed the basis of a government decision whereby half the area identified as internationally important would be protected under domestic (and ultimately EU) wildlife legislation. At the same time, the tax incentive driving forest expansion was removed.

Stroud et al. (2015) describe the aftermath of that particularly difficult battle, but the most positive result was the raising of general awareness about the importance of the area and the high profile which it has since come to enjoy. Forestry expansion was halted, and in recent years large programs of bog restoration involving the felling of the 1980s plantations have been established through EU, UK Heritage

Fig. 7 A view of Forsinard Estate, eastern Sutherland, now owned by the Royal Society for the Protection of Birds (RSPB), who are felling the conifers planted in the 1980s and restoring the area back to actively growing blanket mire habitat. The *grey furrows* are areas of former plantation which have now been felled, while some areas of plantation still await felling (Photo credit: Richard Lindsay © Rights remain with the author)

Lottery, NGO, and government funding, with the RSPB playing a lead role in this work (Fig. 7). The futility of the whole saga is perhaps best highlighted by the fact that some of the same individuals originally paid to drain the bogs and plant the trees are now being paid to fell the trees and fill the drains.

The whole area approved for statutory conservation designation as a Site of Special Scientific Interest (SSSI) has subsequently been designated as a Special Area for Conservation (SAC) under the European Union's Habitats Directive which now provides it with strong protection against potential damaging activities, backed up, where necessary, by the European Court of Justice.

Threats and Future Challenges

Resolution of the forestry issue does not, however, mean that the Flow Country as a whole is now protected. Only half of the internationally important area was afforded legal conservation protection, which left substantial areas without any form of environmental protection (Fig. 8). While the proportion of the Flow Country designated as a composite SAC site has strong protection, the unprotected areas are now being exploited by a new and in some ways equally dramatic threat – namely

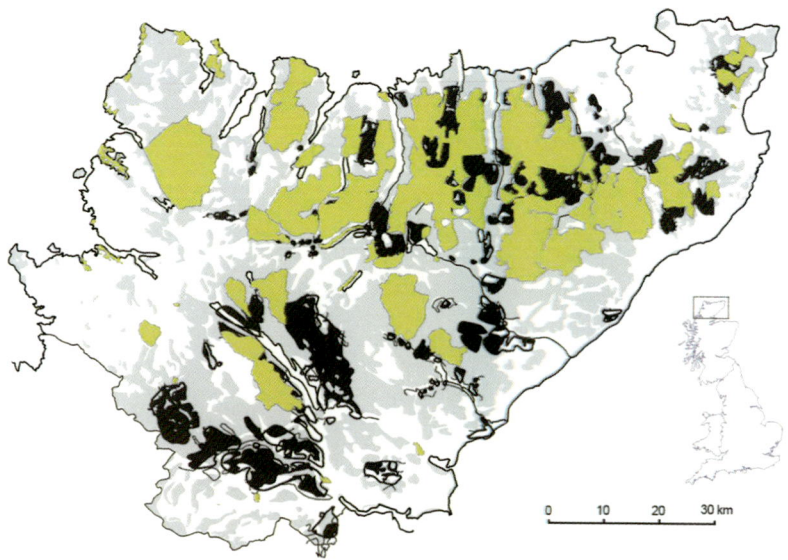

Fig. 8 Map of the Flow Country showing the peat soil (*pale gray shading*), areas planted with conifers or designated for planting (*shaded black*) and areas now designated as Special Areas of Conservation under the EU Habitats Directive (*khaki shading*). Note that in some places SAC designation overlays areas which were planted or designated for planting because the mire unit (or mesotope – see "Peatland Classification") to be protected as an SAC already had some planting within the unit. These trees have since been felled and restoration management is being undertaken in an effort to minimize the long-term effects of such damage to the mire systems (Picture credit: Adapted from Lindsay et al. (1988) with kind permission of the Joint Nature Conservation Committee (JNCC))

windfarm development. Some of what used to be the wildest, most tundra-like views are already now interrupted by the tall turbine towers and blades of windfarms, and the industry is keen to expand further into the heart of the Flow Country. There are currently (2015) two proposals which will lie close to one of the original bogs first surveyed by Pearsall in the 1950s right in the heart of the Flow Country (Pearsall 1956).

The Flow Country is being used increasingly as the focus for research which seeks to shed light on changes which are occurring throughout the blanket mires and wider upland environments of the UK. The Flow Country supports populations of various bird species which are of national and international conservation importance but which have recently shown declines across the UK, ground-nesting waders such as the dunlin, the common greenshank, and the Eurasian golden plover being of particular concern (Douglas et al. 2011). The causes of the declines remain unclear but land-use change, including forestry (Wilson et al. 2014) and wind farms (Pearce-Higgins et al. 2012), has been implicated. Indeed the profound changes brought about by forestry, drainage and other land-uses have been felt across a wide spectrum of Flow Country biodiversity, from higher organisms to micro-organisms like desmids (Goodyear 2014). The same threats could also impact species populations of major economic interest such as the Atlantic salmon (*Salmo salar*).

Studies are now being undertaken within the Flow Country to assess whether restoration can bring back assemblages typical of the remaining open areas. A recent study by Gilbert (2013) has demonstrated that more ticks were found in forestry plantation than in open peatlands, but that restoration reduced tick numbers and thus also reduced the potential for transmission of tick-borne diseases. Other studies have shown that bog restoration by tree removal and drain blocking alters micro-nutrient levels (Muller and Tankéré-Muller 2012) and DOC concentrations (Vinjili 2012) in streams and rivers. This, in turn, could have potential implications for species such as the salmon which use these same rivers to spawn. In such cases, assessing impacts and informing best practice for land management is critical given the economic significance of, for example, wild fishing in the Highlands.

Beyond their importance for biodiversity and water quality, peatlands have also been explicitly recognized through international climate change agreements under the Kyoto Protocol for the significant carbon stores they hold. The Flow Country is the UK's single largest terrestrial carbon store. However, when degraded, peatlands can switch from carbon sink to carbon source and thus fuel climate change. In Scotland, it is estimated that restoration could provide up to 2.7 Mt CO_2-eq savings per year (Chapman et al. 2012). Restoring the Flow Country peatlands could thus contribute meaningfully to meeting Scotland's climate mitigation targets, and a team of researchers is currently investigating the potential scale of this contribution.

Many forestry plantations in the Flow Country are now coming to the end of their first rotation (Fig. 9). This is at a time when national forestry targets, GHG

Fig. 9 Forestry which is approaching first-rotation felling age in the Forsinard area of the Flow Country (Photo credit: Richard Lindsay © Rights remain with the author)

reductions targets, and biodiversity targets all need to be met. Unfortunately, the actions currently being taken to meet these targets are often conflicting. For example, restocking of a plantation on blanket bog in order to meet forestry targets prevents restoration of that blanket bog habitat to meet biodiversity targets. It is therefore critical to assess which sites – if any – are suitable for restocking, which ones should be restored and how, and which ones should be conserved and protected.

Current research in the Flow Country therefore seeks to characterize the wider implications of land-use changes on ecosystem service provision, including potential co-benefits in ecological and economic terms. The conflicts between conservation and development in the Flow Country, as in many other places, highlight the need for an integrated and holistic understanding of peatland functions and a better appreciation of how their responses to disturbances could impact society. Launched in 2012, the Flow Country Research Hub works to achieve this by facilitating a coordinated scientific approach and by providing a platform where science, policy, and practitioners can work together (Andersen et al. 2014). Such a holistic approach will be more vital than ever in the future because the international value of this extraordinary blanket mire landscape has now been sufficiently well recognized for the Flow Country to be placed on the tentative list for designation as a World Heritage Site.

References

Andersen R, Payne R, Ratcliffe J, Gaffney P. Restoration of afforested peatlands in Scotland. Proceedings from In the Bog: the ecology, landscape, archaeology and heritage of Peatlands, 3–4th Sept, Sheffield; 2014.

Bainbridge IP, Minns DW, Housden SD, Lance AN. Forestry in the flows of Caithness and Sutherland, Royal Society for the Protection of Birds Conservation topic paper 18. Sandy: RSPB; 1987.

Birse EL. Assessment of climatic conditions in Scotland. 3. The bioclimatic sub-regions, Soil survey of Scotland bulletin no. 4. Aberdeen: Macaulay Institute for Soil Research; 1971.

Chapman S, Artz R, Donnelly D. Carbon savings from peat restoration. ClimateXChange enquiry, 1205-02, 1–17. 2012

Charman DJ. Blanket mire formation at the Cross Lochs, Sutherland, northern Scotland. Boreas. 1992;21:53–72.

Douglas DJ, Bellamy PE, Pearce-Higgins JW. Changes in the abundance and distribution of upland breeding birds at an operational wind farm. Bird Study. 2011;58:37–43.

Gilbert L. Can restoration of afforested peatland regulate pests and disease? J Appl Ecol. 2013;50:1226–33.

Goodyear, E. Quantifying the Desmid diversity of Scottish Blanket Mires. Doctoral dissertation, University of the Highlands and Islands, Scotland. 2014.

Lindsay RA, Charman DJ, Everingham F, O'Reilly RM, Palmer MA, Rowell TA, Stroud DA. The flow country: the peatlands of Caithness and Sutherland. Peterborough: Nature Conservancy Council; 1988.

Lindsay R. A letter to Derek Ratcliffe. In: Thompson D, Birks J, Birks H, editors. Nature's conscience: the life and legacy of Derek Ratcliffe (wildlife and people). Peterborough: Langford Press; 2015.

Muller FL, Tankéré-Muller SP. Seasonal variations in surface water chemistry at disturbed and pristine peatland sites in the Flow Country of northern Scotland. Sci Total Environ. 2012;435:351–62.

Pearce-Higgins JW, Stephen L, Douse A, Langston RH. Greater impacts of wind farms on bird populations during construction than subsequent operation: results of a multi-site and multi-species analysis. J Appl Ecol. 2012;49:386–94.

Pearsall WH. Two blanket-bogs in Sutherland. J Ecol. 1956;44(2):493–516.

Prebble J. The Highland clearances. London: Secker and Warburg; 1963.

Ratcliffe DA. Mires and bogs. In: Burnett JH, editor. Vegetation of Scotland. Edinburgh: Oliver and Boyd; 1964. p. 426–78.

Ratcliffe DA. In search of nature. Leeds: Peregrine Books; 2000.

Royal Society for the Protection of Birds (RSPB). Forestry in the flow country – the threat to birds. Sandy: RSPB; 1985.

Stroud DA, Reed TM, Pienkowski MW, Lindsay RA. Birds, bogs and forestry: the peatlands of Caithness and Sutherland. Peterborough: Nature Conservancy Council; 1987.

Stroud D, Reed T, Pienkowski M, Lindsay R. E. The flow country: battles fought, war won, organisation lost. In: Thompson D, Birks J, Birks H, editors. Nature's conscience: the life and legacy of Derek Ratcliffe (wildlife and people). Peterborough: Langford Press; 2015.

Tansley AG. The British islands and their vegetation. Cambridge: The University Press; 1939.

Vinjili S. Landuse change and organic carbon exports from a peat catchment of the Halladale River in the Flow Country of Sutherland and Caithness, Scotland. Doctoral dissertation, University of St Andrews, Scotland. 2012.

Wilson JD, Anderson R, Bailey S, Chetcuti J, Cowie NR, Hancock MH, Quine CP, Russell N, Leigh S, Thompson D. Modelling edge effects of mature forest plantations on peatland waders informs landscape-scale conservation. J Appl Ecol. 2014;51:204–13.

Karst Wetlands in the Dinaric Karst

84

Rosana Cerkvenik, Andrej Kranjc, and Andrej Mihevc

Contents

Introduction	1058
Conservation Status of Karst Wetlands in the Dinaric Karst	1058
Examples of Karst Wetlands in the Dinaric Karst	1060
Poljes	1060
Uvalas and Dolines	1060
Blind and Pocket Valleys	1061
Lakes	1062
Water Caves – Karst Springs and Ponors	1062
Future Challenges	1063
References	1064

Abstract

Dinaric karst is the largest continuous karst area in Europe and among the largest in the world extending over 60,000 km^2 from the Soča (Isonzo) River in the northwest to Skadar (Shkodra) Lake in the southeast, and between the Pannonian Basin and the islands or coast of the Adriatic Sea. Parts of the Dinaric karst have been internationally listed as Ramsar Sites and World Heritage properties as

---

R. Cerkvenik (✉)
Park Škocjanske jame, Divača, SI, Slovenia
e-mail: rosana.cerkvenik@psj.gov.si

A. Kranjc (✉)
Slovenian Academy of Sciences and Arts, Ljubljana, SI, Slovenia
e-mail: andrej.kranjc@sazu.si

A. Mihevc (✉)
Karst Research Institute, Research Centre of the Slovenian Academy for Science and Arts, Postojna, Slovenia
e-mail: mihevc@zrc-sazu.si

© Springer Science+Business Media B.V., part of Springer Nature 2018
C. M. Finlayson et al. (eds.), *The Wetland Book*,
https://doi.org/10.1007/978-94-007-4001-3_240

well as receiving different national level designations. Classified into two major groups: karst depressions (including poljes, blind and pocket valleys, uvalas, lakes, and dolines) and water caves (including ponors and karst springs), water that reaches the surface from underground can fluctuate as much as 200 m. In this chapter, the different types of Dinaric Karst wetlands and some of the challenges affecting their hydrology are briefly described.

Keywords
Dinaric Karst · Karst phenomena · Karst hydrology

Introduction

Dinaric karst is the largest continuous karst area in Europe and belongs among the largest in the world. It extends over 60,000 km^2 from the Soča (Isonzo) River in the northwest to Skadar (Shkodra) Lake in the southeast, and between the Pannonian Basin and the islands or coast of the Adriatic Sea. It extends over a distance of around 650 km and is up to 150 km wide. It is named after Mount Dinara and the Kras (Carso) Plateau (Mihevc et al. 2010); (Fig. 1).

Researches into the phenomena, which are now described as karst, were first conducted in the Kras region, known by the names Kras (Slovenian), Karst (German), and Carso (Italian). The first scientific explorations that produced theories on the development of karst, karst morphology, and hydrology were carried out in the Dinaric karst (Gams 1974). Many karst landforms were first scientifically described in this area; their locally used names became standard in international terminology (some of them are: dolina, uvala, polje, ponor, kamenitza, hum).

Karst wetlands in the Dinaric karst can be classified into two major groups: karst depressions (including poljes, blind and pocket valleys, uvalas, lakes, and dolines) and water caves (including ponors and karst springs). The levels of water that reaches the surface from underground can fluctuate as much as 200 m. Underground water flows through narrow channels which are not accessible to humans.

Conservation Status of Karst Wetlands in the Dinaric Karst

Individual parts of the Dinaric karst have been listed under one or several international designations, as well as protected at different national levels (e.g., national, regional and nature parks, nature monuments). There are also some examples of overlapping national and international designations.

At present, there are three individual World Heritage properties in the Dinaric karst, namely, Škocjanske Jame (Slovenia), Plitvićka Jezera (Croatia), and Durmitor (Montenegro). All those areas or parts of them also cover some karst wetlands.

At present, the following karst wetlands have been designated as Ramsar Sites in the Dinaric karst: River Buna (Albania), poljes Hutovo Blato and Livanjsko Polje (Bosnia and Herzegovina), Neretva River Delta and Vransko Jezero

84 Karst Wetlands in the Dinaric Karst

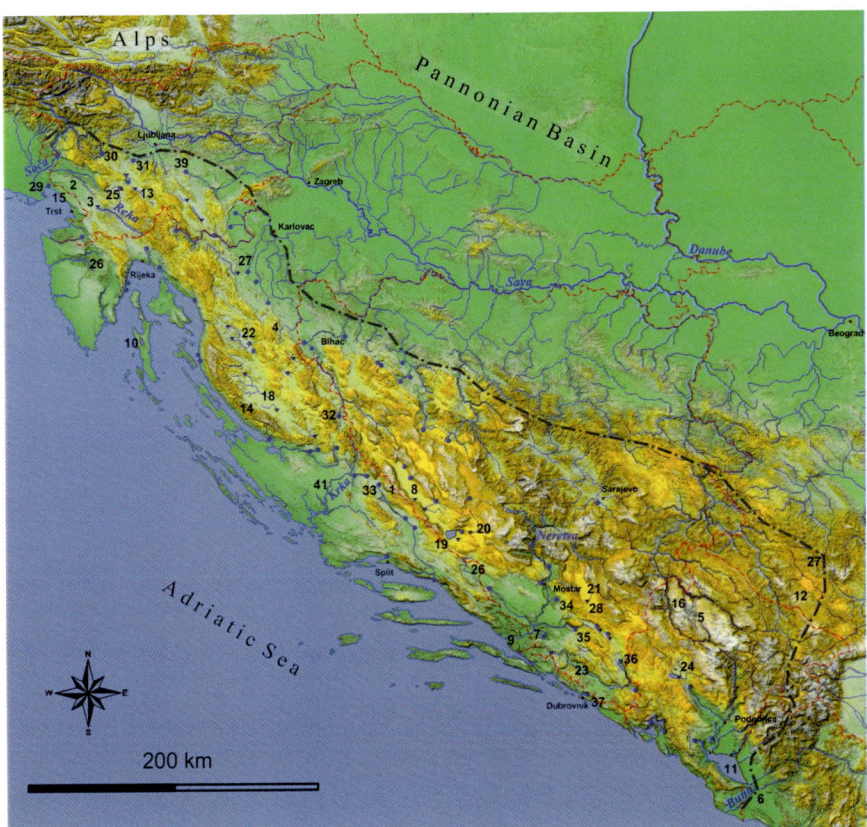

Fig. 1 Dinaric Mountain Range and important karst sites mentioned in the text. The gray dashed line marks the extent of the Dinaric Karst and red lines the state borders. *1* Mount Dinara, *2* Kras Plateau. UNESCO sites: *3* Škocjanske Jame, *4* Plitvićka Jezera, and *5* Durmitor. Ramsar Sites: *6* River Buna, *7* Hutovo Blato, *8* Livanjsko Polje, *9* Delta of Neretva River, *10* Vransko Jezero, *11* Skadarsko Jezero (Shkodra Lake), *12* Pestersko Polje, *13* Cerknisko Polje, and *3* Škocjanske Jame. Biosphere reserves: *14* Velebit, *15* Miramare, *16* Tara River Basin, *2* Kras Plateau, and *17* Golija-Studenica. Important karst features, first poljes: *18* Lićko Polje, *8* Livanjsko Polje, *19* Buško Blato, *20* Duvanjsko Polje, *21* Nevesinjsko Polje, *22* Gacko Polje, *23* Popovo Polje, and *24* Nikšićko Polje. Sinking points of rivers: *3* Škocjanske Jame, *25* Pivka, *26* Foiba Cave, *27* Dobra River, and *28* Zalomka River. Karst springs of rivers: *29* Timavo (Timava), *30* Divje Jezero, *31* Ljubljanica, *32* Una, *33* Cetina, *34* Buna, *35* Bregava, *36* Trebišnjica, and *37* Ombla. Karst lakes in collapsed doline: *38* Crveno Jezero; in uvala: *39* Hrastov Dol; lakes *40* Vransko Jezero, *4* Plitvićka Jezera, and lakes of Krka River *41*. Blue dots mark large springs and triangles are large sinking points of rivers (© A. Mihevc, Rights remain with author. Cartography is based on SRTM 90m Digital Elevation Data)

(Croatia), Skadarsko Jezero (Shkodra) (transboundary area in Albania and Montenegro), Pestersko Polje (Serbia), and Cerkniško Jezero and Škocjanske Jame (Slovenia).

Some of the karst wetlands have also been designated as biosphere reserves, namely, Velebit in Croatia, Miramare in Italy, Tara River Basin in Montenegro, Kras in Slovenia, and Golija-Studenica in Serbia. Many karst wetlands are also Natura 2000 sites under EU Directives.

Examples of Karst Wetlands in the Dinaric Karst

Poljes

Poljes are the largest flat-floored enclosed depressions in karst. Polje has a flat floor in the rock or covered by unconsolidated sediments such as alluvium, a closed basin with steeply rising marginal slope on at least one side, and karstic drainage. The surface of the polje's bottom can range from a few km^2 to over 500 km^2. There are about 130 poljes in the Dinaric karst; about 44 of them are recognized as outstanding examples. Dinaric poljes are polygenetic features. The most important factors for their development are geological and hydrological characteristics (Bonacci 1987). They are generally elongated along the strike in a northwest–southeast direction. There are several variants of poljes – ranging from dry to seasonally flooded poljes (seasonal karst lakes) and from permanently flooded poljes (karst lakes) to cryptodepressions – "hidden depressions," lakes, whose surface is above sea level and their bottoms below it.

The international term polje originated in the karst of Bosnia and Herzegovina where they are the main karst features (Kranjc 2013). The dimensions of polje vary considerably. They are developed at all elevations, between 20 and 1,200 m above sea level. The largest poljes are Lićko Polje (460 km^2) and Livanjsko Polje (402 km^2), but the majority of poljes are smaller, covering only few 10 km^2 (Milanović 2004b).

The main characteristic of poljes that influences flora, fauna, ecosystems, and land use is the water fluctuations at the bottom of poljes. In natural conditions, floods are normal phenomena for many poljes. In some poljes like Cerkniško Polje and Buško Jezero they last about half a year.

Poljes are very important for human settlement. In their surroundings the best, and often the only, arable land in the Dinaric karst is available. There is always some water available and floods are regular and limited to the lower parts of poljes, so all of the largest settlements are located on the edges of the poljes (Fig. 2).

Uvalas and Dolines

Both terms – doline and uvala – derive from the Dinaric karst. A doline is a karst depression in circular to subcircular plan form, varying in diameter from a few meters to around 1 km. Their sides range from a few to several hundred meters deep. They are formed by various processes, mostly by dissolution, collapse, and subsidence (Ford and Williams 2007). The bottoms of the dolines can be dry, but some have stagnant or running water. The most outstanding examples of these wetlands are near Imotski and in Škocjanske Jame. Crveno Jezero near Imotski is a 528 m deep collapse doline, and its bottom extends 10 m below sea level. The lake is, on average, 285 m deep and fluctuates ± 35 m. The diameter of the lake is about 200 m (Bonacci and Roje-Bonacci 2000; Bonacci et al. 2013; Bonacci and Andrić 2014). In the vicinity there is also Modro Jezero collapse doline which is 190 m deep and can be completely dry in summer (Milanović 1979). In front of the terminal ponor in

Fig. 2 Cerkniško Polje (Photograph by: Valentin Schein, archive of the Park Škocjan Caves. © Rights remain with the author)

Škocjanske jame there are two collapsed dolines, with depths of up to 165 m and water level fluctuations of up to 72 m.

Uvalas are large (in kilometer scale) karst closed depressions of elongated or irregular plan form. Their bottoms are undulating or pitted with dolines, seldom flattened by colluvial sediments. Uvalas are always situated above the karst water table (Ćalić 2009). Although uvalas are usually dry, there are some examples where the water table occasionally rises to the bottom of uvalas. Such examples are Vagan Mazin near Gračac which is permanently hydrologically active and Hrastov Dol near Ivančna Gorica (Gams 2004).

Blind and Pocket Valleys

Blind valleys appear on the contact karst, a karst phenomenon formed by the influence of the surface rivers. The term is familiar in Slovenia and Croatia, where the karst contacts noncarbonate rocks, and specific relief forms and phenomena have developed (Roglič 1957). Contact karst features are limited only to the area influenced by sinking rivers which can form large water caves behind the ponor (a hole or opening in the bottom or side of a depression where a surface stream or lake flows either partially or completely underground into the karst groundwater system). The best examples are blind valleys in Matarsko Podolje on the Istria Peninsula (Mihevc 1994). Pocket valleys are the reverse phenomena of blind valleys. They extend headwards into the calcareous massive with a karst spring at the contact. Blind and pocket valleys also represent karst wetlands where water can fluctuate considerably.

Lakes

Typical wetland areas in the Dinaric karst are permanent lakes, formed either in karst depressions or by natural dams. Water in the lakes can be fresh, saline, or brackish. In some cases, there are karst springs at the bottoms of lakes. The largest lake in the Balkan Peninsula is Skadar (Shkodra) Lake, a cryptodepression, with an area ranging from 370 to 530 km^2. Another cryptodepression, Vransko Jezero on the Cres Island, is an important source of fresh water with the average volume of 220 million m^3. Typical karst wetlands are also lakes formed by travertine or calcareous tufa dams. Such examples are Plitvička Jezera (Bonacci and Roje-Bonacci 2004) with a total of 16 lakes and representing an integral part of the Korana River and lakes on the Krka River in Dalmatia.

Water Caves – Karst Springs and Ponors

Hydrological conditions in the karst landscape are characterized by underground runoff, which enabled the development of special hydrological phenomena, vertical percolation, and feeding of waters into major underground conduits. Rivers that flow from the surrounding nonkarst terrains sink when they reach karst ground. Water resurfaces again in big karst springs at the edge of karst terrain or in the bottoms of poljes. Surface water forms are different types of ponors of large rivers (which vary in size considerably and extend from alluvial sinks to large ponor caves, e.g., more than 20 km long Postojnska Jama); large springs including submarine springs (vrulja), coastal springs (of fresh and brackish water), karst alluvial springs, and large and deep vauclusian springs; estavelles; and occasionally flooded bottoms of poljes. A vauclusian spring is an ascending karst spring where important quantities of water originate from a deep phreatic zone. An estavelle or inversac is a ground orifice in karst, which, depending on weather conditions and season, can serve either as a swallow hole or as a spring.

One of the largest sinking rivers is the Reka River, with a mean discharge of 8 m^3/s (and maximum discharge of over 300 m^3/s), that sinks on the edge of the Kras Plateau in Škocjanske Jame. Large sinking rivers also include the Pivka River that sinks in the Postojnska Jama, Pazinska Reka in Istria that sinks in the Foiba Cave, Dobra River which sinks near Ogulin, and Zalomka that sinks in the Biogradski Ponor on the Nevesinjsko Polje. Large ponors are essential features of poljes. Underground connections between sinking streams and nearby poljes often form a complex network, including surface and underground bifurcations.

Karst springs in the Dinaric karst (Fig. 3) are among the largest in the world, both as regards their specific discharge as well as their dimensions. The largest karst springs are Trebišnjica, Ombla, Ljubljanica, Buna, Bunica, Timavo (Timava), and Bregava which can reach the flow rate over 100 m^3/s (Milanović 1979; Ford and Williams 2007). The depth of the deepest springs surpasses 200 m; such are Una and Divje Jezero.

Fig. 3 Cetina River Spring (Photograph by: Andrej Mihevc. © Rights remain with the author)

Distances between some ponors and springs are considerable in the Dinaric karst. Some examples of this are the Reka (Timavo) and Zalomka. Reka River flows from Škocjanske jame towards the Timavo (Timava) Spring in the distance of over 35 km, and permanent Zalomka River flows underground over a distance of more than 60 km from Gatačko Polje and Nevesinjsko Polje through Cerničko Polje, Fatničko Polje, Dabarsko Polje, Ljubijsko Polje, and Ljubomirsko Polje to Popovo Polje and finally to the Ombla Spring at the Adriatic Coast and springs along the Neretva River (Milanović 2006).

Important karst wetlands have also been formed in the estuaries of the rivers that cross the Dinaric karst, for example, the Zrmanja, Krka, Cetina, and Neretva along the Adriatic Coast.

Future Challenges

Precipitation in the Dinaric karst is in general high, but most of the water runs off vertically. Surface water forms are rare features and thus very precious. Several karst springs serve as water supplies, and additional and important water resources are lakes of fresh water as, for example, Vransko Jezero on the Cres Island. Due to these particular karst characteristics, water sources are very vulnerable. The highest risk to karst wetlands in the Dinaric karst is pollution of surface and underground water, ponors and karst springs by waste water from households, agriculture, and industry, by spills of hazardous materials, and illegal waste disposal. Regulation of poljes has been conducted for centuries to achieve either permanent flooding or drainage.

Fig. 4 Regulation of the Trebišnjica River (Photograph by: Andrej Mihevc. © Rights remain with the author)

Cerkniško Polje is an example where the first plans for drainage were mentioned by Hacquet in 1778, while plans were later produced for building a dam and creating a permanent lake. Several smaller land reclamation efforts have been conducted as well as other interventions, but the lake, nevertheless, remains in its relatively natural state. On the other hand, there are some examples of intense use and regulation of large poljes (e.g., Ličko Polje, Gacko Polje, Duvanjsko Polje, Nikšićko Polje, and Popovo Polje) where in the 1950s and 1960s, hydroelectric systems were built. The hydrological regime has been completely changed in these poljes. The flooding of 60 km long Popovo Polje under natural conditions reached a height of 40 m in the lowest section of the polje, and the average yearly flood duration was 253 days, while today the Trebišnjica River flows for around 60 km in a concrete channel (Fig. 4) (Milanović 1979). The largest water storage facilities in the Dinaric karst are Bileća reservoir on the Trebišnjica River (1.28 billion m^3), Nikšićko Polje (220 million m^3), and Buško Blato reservoir (800 million m^3) (Milanović 2004a).

References

Bonacci O. Karst hydrology. With special reference to the Dinaric karst. Berlin/Heilderberg: Springer; 1987.

Bonacci O, Roje-Bonacci T. Interpretation of groundwater level monitoring results in karst aquifers: examples form the Dinaric karst. Hydrol Process. 2000;14:2432–8.
Bonacci O, Roje-Bonacci T. Plitvice Lakes, Croatia. In: Gunn J, editor. Encyclopedia of caves and karst science. New York: Fitzroy Dearborn; 2004.
Bonacci O, Andrić I, Yamashiki Y. Hydrology of Blue Lake in the Dinaric karst. Hydrol Processes. 2014;28:1890.
Bonacci O, Željković I, Galić A. Karst rivers' particularity: an example from Dinaric karst (Croatia/Bosnia and Herzegovina). Environ Earth Sci. 2013;70:963.
Ćalić J. Uvala- contribution to the study of karst depressions (with selected examples from Dinarides and Carphato-Balkanides) [dissertation]. Nova Gorica: University of Nova Gorica; 2009.
Ford DC, Williams PW. Karst geomorphology and hydrology. Chichester: Wiley; 2007.
Gams I. Kras. Ljubljana: Slovenska matica; 1974.
Gams I. Kras v Sloveniji v prostoru in času. Ljubljana: Založba ZRC SAZU; 2004.
Kranjc A. Classification of closed depressions in carbonate karst. In: Shroder JF, editor-in-chief, Frumkin A, volume editor. Treatise on geomorphology, vol. 6, Karst geomorphology. San Diego: Academic; 2013; p. 104–11.
Mihevc A. Contact harst of Brkini hills. Acta Carsologica. 1994;23:99–108.
Mihevc A, Prelovšek M, Hajna NZ, editors. Introduction to the Dinaric karst. Postojna: Inštitut za raziskovanje krasa ZRC SAZU (Karst Research Institute at ZRC SAZU); 2010.
Milanović PT. Hidrogeologija karsta i metode istraživanja. Trebinje: Hidroelektrarne na Trebišnjici, Institut za korištenje i zaštitu vode na kršu; 1979.
Milanović PT. Dams and reservoirs on karst. In: Gunn J, editor. Encyclopedia of caves and karst science. New York: Fitzroy Dearborn; 2004a.
Milanović PT. Dinaride Poljes. In: Gunn J, editor. Encyclopedia of caves and karst science. New York: Fitzroy Dearborn; 2004b.
Milanović PT. Kars istočne Hercegovine i Dubrovačkog priobalja (Karst of Eastern Herzegovina and Dubrovnik Littoral). Beograd: Asocijacija speleoloških organizacija Srbije; 2006. p. 362.
Roglič J. Zaravni u vapnencima. Geografski Glas. 1957;19:103–34.

Turloughs (Ireland)

85

Kenneth Irvine, Catherine Coxon, Laurence Gill, Sarah Kimberley, and Steve Waldren

Contents

Introduction	1069
Wetland Description	1069
Hydrology	1070
Biodiversity in the Flooded Phase	1071
Biodiversity in the Dry Phase	1072
Threats and Management Challenges	1073
Conservation Status	1075
References	1075

Abstract

Turloughs are shallow depressions in a Carboniferous karst landscape, subject to periodic flooding, mainly from groundwater. They are most prevalent in the west of Ireland, although similar sites are found in Wales, Spain, Slovenia and eastern Canada. Hydrological pathways can connect several turloughs over large areas, which can fill and empty over short periods. The hydrology of turloughs can act as either a flow-through model, where inflow and outflow occur simultaneously, or a surcharged tank model, where the turlough acts as overflow storage for the underlying karst flow network, leading to greater residence time than in the through-flow model.

K. Irvine (✉)
UNESCO-IHE Institute of Water Education, Delft, The Netherlands
e-mail: K.Irvine@unesco-ihe.org

C. Coxon · L. Gill · S. Kimberley · S. Waldren
University of Dublin, Trinity College, Dublin, Ireland
e-mail: cecoxon@tcd.ie; Laurence.Gill@tcd.ie; kimbersj@tcd.ie; swaldren@tcd.ie

© Springer Science+Business Media B.V., part of Springer Nature 2018
C. M. Finlayson et al. (eds.), *The Wetland Book*,
https://doi.org/10.1007/978-94-007-4001-3_256

During the inundation phase, turloughs resemble and function like shallow lakes, but typically have no or few fish, that promotes persistence of large-bodied plankton and rare invertebrates. Amphibians, however, are common. During the winter, turloughs provide important habitat and feeding areas for wildfowl. Overall, phytoplankton tend to be dominated by cryptophytes, together with small diatoms. Shallow depth and wind-induced mixing results in suspension of algae that are predominantly associated with the vegetation on the turlough floor. Many turloughs are phosphorus-limited, although impact from greater intensification of agriculture is leading to increased phytoplankton populations in many sites. Being part of a karstic limestone catchment results in generally high calcium and carbonate concentrations of between 100 and 250 mg l^{-1} $CaCO_3$. Some turlough catchments arising on acidic, non-limestone rocks overlain by upland blanket peat can have high concentrations of dissolved organic matter, and colour reaching over 80 mg l^{-1} PtCo.

The duration of flooding results in a characteristic zonation of turlough plant communities, but it is difficult to develop generalised models of vegetation zonation. Areas with long-duration flooding or permanent pools show characteristics of wetland communities, while the less-frequently inundated upper parts of a turlough show a transitions to grassland, scrub and woodland. During the dry phase, turloughs are often used as pastures for low intensity grazing, and are renowned for their diverse semi-wetland plant communities and associated invertebrates. Some turloughs dry out completely during dry summer periods, while others can retain some surface water throughout the year. Oligotrophic turloughs tend to have vegetation dominated by sedges with associated herbs, while those that are subjected to higher nutrient concentrations, particularly phosphorus, tend to be dominated by grasses and herbs.

Both the terrestrial and flooded phase contain ecological communities of national and international importance. Seasonal transitions between surface water and terrestrial systems results in a variable and species-rich ecotone and wetland mosaic habitat. This contributes to high conservation value, and many are protected as Special Areas of Conservation under European legislation. Yet, the hydrology of approximately one third of turloughs over 10 hectares had been irreversibly altered by drainage by the mid-1980s, and continuing pressures include removal of scrub and woodland, and reseeding of grassland with *Lolium perenne*, and direct application of fertiliser, or bulldozing the turlough basin to remove stones, with obvious degradation of vegetation communities. In recent years, there has been greater attention to their conservation, although many remain vulnerable to both localised and wider catchment pressures.

Keywords

Turlough · Ephemeral · Biodiversity · Ireland · Karst · Groundwater · Grazing

Introduction

Turloughs (derived from the Irish *tur loch* meaning *dry lake*) are shallow depressions in a Carboniferous karst landscape that are subject to periodic flooding, mainly from groundwater. Although internationally rare ephemeral wetlands, they are common in the limestone geology of the west of Ireland. While this article focuses on Irish turloughs, similar habitats are found in Wales, Spain, Slovenia, and eastern Canada (Sheehy Skeffington et al. 2006; Sheehy Skeffington and Scott 2008). Owing to strong natural environmental gradients, turloughs are difficult to classify. However, under the EU Water Framework Directive (CEC 2000), they are considered as groundwater-dependent terrestrial systems.

Reviews of turlough geomorphology, hydrology, ecology, and conservation importance can be found in Coxon (1986) and Sheehy Skeffington et al. (2006). Summaries of turlough wetland plant and freshwater habitats and communities are found in Goodwillie and Reynolds (2003), and comparisons of Irish vegetation in lakes, fens, and turloughs in the Burren limestone plateau in Proctor (2010). Duigan and Kovach (1991) provide extensive descriptions of chydorid cladoceran communities across a number of turloughs. A comparison with small temporary karst lakes in the Slovenian Pivka mountains is found in Sheehy Skeffington and Scott (2008).

Wetland Description

During the inundated phase, typically in winter, turloughs resemble and function like shallow lakes, but contain only low density, or no, fish populations and harbor characteristic plant and animal communities. During the dry phase, they are often used as pastures for low-intensity grazing and are renowned for their diverse semi-wetland plant communities and associated invertebrates. Some turloughs dry out completely during dry summer periods, while others can retain some amount of surface water throughout the year. Oligotrophic turloughs tend to have vegetation dominated by various sedges with associated herbs, while those that are subjected to higher nutrient concentrations, particularly phosphorus, tend to be dominated by grasses and herbs (Sharkey 2012).

The duration of flooding results in a characteristic zonation of turlough plant communities (Praeger 1932; Procter 2010; Sharkey 2012). However, the variability of flooding regimes across turloughs makes it difficult to develop generalized models of vegetation zonation. Areas with long-duration flooding or permanent pools show characteristics of wetland communities, while the less-frequently inundated upper parts of a turlough show transitions to grassland, scrub, and woodland communities. However, there can be considerable variation in community structure across different sites, often mediated by other factors, including nutrient status, grazing regime, and land use.

The karst limestone region covering the midwestern parts of Ireland provides the physical setting for a large number of turloughs, fed largely by rising groundwater through *swallow holes* or *estavelles*. These are conduits to flow linked to subterranean karst geomorphology that can be sinks or springs depending on water level (Coxon and Drew 1986, 2000; Naughton et al. 2012). While tending to be shallow water bodies during the flood period, depths of over 10 m are also recorded. The upper limits of flooding can also be highly variable, sometimes causing inundation of roads and houses. The main flooded period is usually between 5 and 9 months, with high interannual variation and periods of early and late emptying and refilling depending on local rainfall.

Being part of a karstic limestone catchment results in generally high calcium and carbonate concentrations of between 100 and 250 mg l^{-1} $CaCO_3$ across flooded turloughs. Carbonate deposition during the flooding season may result in coatings of calcite crystals on vegetation and tufaceous crusts (Coxon 1994). Some turlough catchments contain sinking streams arising on acidic, non-limestone rocks overlain by upland blanket peat, and turloughs in this situation can have high concentrations of dissolved organic matter and color reaching over 80 mg l^{-1} PtCo (Cunha Pereira et al. 2010).

Phosphorus concentration in turlough water is a major determinant of vegetation community and species distribution. Even though this process is mediated by soil character, vegetation patterns are more closely linked to total phosphorus in water than in soils. The likelihood of phosphorus transmission from the zones of contribution to the turlough will reflect the bedrock and quaternary geology. The most important factor is the presence of point recharge to the limestone aquifer by sinking streams, where phosphorus in the stream water will reach the turlough by rapid conduit flow with little attenuation (Kilroy and Coxon 2005). Closed depressions (dolines) in the karst landscape are also potential routes for phosphorus entry, depending on the thickness of soil and sediment and geomorphology of the doline (Mellander et al. 2013).

Hydrology

The hydrological functioning of turloughs can be described using two general conceptual models: the flow-through system and the surcharged tank. The principal difference between the two models is the timing and interaction between inflow and outflow. In the flow-through model, both inflow and outflow occur simultaneously and largely independently within the turlough basin, resulting in constant flow of groundwater through the turlough. In the surcharged tank model, the turlough acts as overflow storage for the underlying karst flow network, accumulating excess groundwater that cannot be accommodated owing to insufficient capacity, and with greater residence time than in the through-flow model. It is feasible that both types of hydrological conceptualization apply to turloughs in the Irish landscape (Naughton et al. 2012).

Hydrological pathways to turloughs can connect over large areas, linking several turloughs, and estimating the hydrological *zone of contribution* has inherent uncertainty. The groundwater catchment can be markedly different from that apparent from surface catchment topology. Strong connectivity with karst groundwater also makes turloughs highly dynamic, filling and emptying often over short time periods. The duration of flooding and the speed of filling and emptying are also associated with sediment character, with turloughs dominated by sand and silt/clay tending to have shorter flood durations than those with predominance of peat or marl sediments.

Biodiversity in the Flooded Phase

The hydroperiod has major effects on both terrestrial and aquatic communities, strongly influencing the distribution of vascular plant species and the zonation of vegetation communities within turlough basins (Praeger 1932; Proctor 2010; Sharkey 2012), seasonal succession of phytoplankton (Cunha Pereira et al. 2011), and invertebrate compositions (Porst et al. 2012).

Overall, phytoplankton tend to be dominated by cryptophytes, together with small diatoms. Shallow depth and high amount of wind-induced mixing result in a large component of algae that, although predominantly associated with the vegetation on the turlough floor, are frequently suspended in the water column (Cunha Pereira et al. 2011). The highly variable and dynamic nature of turloughs leads to dominance of fast growing "r-selected" algal species, and, although restricted mainly to the winter period, turloughs can be as productive as permanent lakes (Cunha Pereira et al. 2011). Many turloughs are phosphorus limited and show a similar relationship between total phosphorus and chlorophyll *a* to permanent lakes, although this is less apparent in the more highly colored turloughs where light limitation tends to reduce overall phytoplankton biomass (Cunha Pereira et al. 2010).

Zooplankton taxa of turloughs include large-bodied cladocerans, reflecting low predation pressure from fish. This likely also explains the occurrence of a number of rare species such as *Tanymastrix stagnalis*, *Diaptomus castor* (Grainger 1966, 1991), and *Eurycercus glacialis* (Duigan and Frey 1987). The benthic macroinvertebrates include a number of rare and threatened species or those of conservation importance (Porst and Irvine 2009a; Williams and Gormally 2009). Examples include the Dytiscidae diving beetles *Graphoderus bilineatus*, *Hygrotus quinquelineatus*, and *Agabus labiatus*, the Hydrophilidae beetle *Berosus signaticollis*, and the temporary pond adapted Lymnaeidae mud snail, *Omphiscola glabra*. The mossy edges of turlough ecotones house a number of rare beetle species (Foster et al. 1992). The turlough environment is a rich habitat for dragonflies, including notable rarities such as *Lestes dryas* (Nelson and Thompson 2004).

While fish may be rare in turloughs, amphibians are found frequently, likely also benefitting from the low fish presence. During the winter, turloughs provide important habitat and feeding areas for wildfowl (Cabot 1999).

Biodiversity in the Dry Phase

As vegetation communities show distinct zonation in relation to the duration of flooding, the matrix of vegetation in the dry phase is largely dependent on species tolerance to flooding and inundation. The upper communities with short duration flooding may contain characteristic limestone grassland communities or woodland dominated by ash (*Fraxinus excelsior*) and buckthorn (*Rhamnus cathartica*). Woody species generally do not occur in the more frequently flooded zones of turloughs and those that do tend to be small, such as creeping willow (*Salix repens*), although the development in turloughs of carr woodland dominated by various willow trees (*Salix* sp.) is uncommon. Instead, communities dominated by sedges (*Carex* spp., especially *C. flacca*, *C. hostiana*, and *C. panicea*) tend to prevail in the more oligotrophic turloughs, while those with higher nutrient status communities are dominated by herbs such as silverweed (*Potentilla anserina*) and creeping buttercup (*Ranunculus repens*). Fully aquatic species often occur in semipermanent shallow pools in the base of turloughs (Fig. 1), such as various pondweeds (*Potamogeton* spp.) and white water lily (*Nymphaea alba*).

A characteristic feature of many of the plant species, particularly those that occur in zones with longer duration flooding, is their ability to grow amphibiously. Species

Fig. 1 A small permanent pool in Carran turlough, County Clare, Ireland, at low water level in midsummer. The pool, which may have formed from cutting and removal of fen peat, contains *Nymphaea alba* (foreground) and *Menyanthes trifoliata* (middle distance); the surrounding vegetation is mostly dominated by sedges (*Carex* spp.). This location would typically be submerged beneath 3–4 m of water in winter, with flooding extending into the *Rhamnus/Fraxinus* woodland in the background (Photo: S. Waldren)

such as curled dock (*Rumex crispus*) and lesser spearwort (*Ranunculus flammula*) may extend their leaf stalks in floodwater such that leaves float on the water surface, which enable an oxygen supply to shoots and roots through aerenchymatous tissue. When floodwaters recede, normal leaf growth is resumed. Other species survive the flooded period as dormant rhizomes (e.g., *Potentilla anserina*) or seed and grow rapidly when flooding recedes. In some normally terrestrial species such as *Ranunculus repens*, genetically distinct populations occur in turloughs which differ morphologically and physiologically from typical populations and can tolerate periodic complete inundation to several meters (Lynn and Waldren 2001, 2002, 2003). Other species, such as autumn hawkbit (*Leontodon autumnalis*) and amphibious bistort (*Polygonum amphibium*), show markedly different leaf morphologies when growing submerged or above floodwater.

Most studies of turlough invertebrate diversity have focused on the aquatic phase. However, the adults of many of the insects that have aquatic larval phases become obvious features of turlough biodiversity during the dry phase. This is particularly true of Odonata, for example, with characteristic species such as the scarce emerald (*Lestes dryas*) and ruddy darter (*Sympetrum sanguineum*). Terrestrial ground beetle (Carabidae) diversity studied in a few turloughs (see Ní Bhriain et al. 2002; Moran et al. 2012) has revealed the presence of several rare or threatened species, and their distribution in relation to vegetation community, depth and duration of flooding, and grazing intensity.

Some wetland birds breed in turloughs depending on vegetation type and substrate moisture (see Goodwillie 1992), including water rail (*Rallus aquaticus*), Eurasian curlew (*Numenius arquata*), common snipe (*Gallinago gallinago*), common redshank (*Tringa totanus*), northern lapwing (*Vanellus vanellus*), coot (*Fulica atra*), mallard (*Anas platyrhynchos*), black-headed gull (*Chroicocephalus ridibundus*), and occasionally dunlin (*Calidris alpina*) and little grebe (*Tachybaptus ruficollis*). Many waders (Charadriidae, Scolopacidae) and some waterfowl (Anatidae) utilize the emergent and shallowly flooded zones of turloughs in spring and autumn.

Threats and Management Challenges

Altering the hydrology of turloughs affects all features of a turlough. Many turloughs have been drained, with water diversions to alleviate local flooding. The impact of land drainage on groundwater resources is particularly acute in karst areas owing to the unique characteristics of karst aquifers (Sheehy Skeffington et al. 2006), and the hydrological regime of approximately one third of turloughs over 10 ha had been irreversibly altered by drainage by the mid-1980s (Coxon 1987; Drew and Coxon 1988).

As grazing systems, turloughs have been subject to agricultural improvement, particularly in the upper, less-frequently flooded zones. This has often resulted in removal of scrub and woodland and reseeding of grassland with *Lolium perenne*. In some cases, agricultural improvement has involved direct application of fertilizer to

Fig. 2 Skaghard turlough, part of the Cooloorta/Skaghard/Travaun complex, County Clare, Ireland. This is an oligotrophic turlough surrounded by limestone pavement, with the vegetation dominated by sedge (Cyperaceae) communities with very little grazing by livestock. Shown at early summer flood level with the emergent *Cladium mariscus* in the foreground and middle distance, water level will recede further to leave small pools with extensive deposits of marl, whereas in winter the flood level extends into the surrounding limestone pavement (Photo: S. Waldren)

the turlough vegetation, or bulldozing the turlough basin to remove stones, with obvious degradation of vegetation communities, removing seminatural limestone grasslands, scrub, and marginal woodland.

Overall, grazing regime within a turlough is important for its biodiversity (Sheehy Skeffington et al. 2006). Grazing also modifies the vegetation, but its effects seem to be linked with nutrient status. Most oligotrophic turloughs have little grazing pressure, with plant communities dominated by less palatable sedges (Fig. 2). In more nutrient-enriched turloughs, grazing intensity is often higher and becomes more important in determining vegetation structure, with both undergrazing and overgrazing leading to reductions in species diversity. Trampling by grazing animals disturbs the moist and typically shallow soils, allowing recruitment of some specialist annual plants more or less restricted to turloughs, but high grazing pressure may compact and damage soil structure, eliminating some sensitive species from the turlough sward. Increased cover of scrub and woodland may occur in the upper zones of turloughs when grazing is reduced, and in more mesotrophic turloughs reduction in grazing pressure often leads to the dominance of a small number of taller herbs and grasses in the zones of longer duration flooding.

Localized input of nutrients can occur by overland flow, through-flow, and through the surface layers of karst. Turloughs close to intensively farmed grassland can be vulnerable to nutrient enrichment. Many turloughs are surrounded by sloping, fertilized grassland providing a route for localized nutrient impacts.

While long recognized as intriguing ecosystems, only recently has there been integrated work that has attempted to link their hydrochemistry with the ecology of their terrestrial and flooded phases and the response of plant, algal, and macroinvertebrate communities to hydrology, hydrochemistry, substrate, and land use. Hydroperiod exerts a strong influence on the community structure of macroinvertebrates (Porst and Irvine 2009a, b; Porst et al. 2012), the distribution of plants (Sharkey 2012), and the terrestrial phase grazing regime (Kimberley et al. 2012). The recorded continuum of turlough hydrological regimes (Naughton et al. 2012) concurs with the biotic distinctiveness of turloughs.

Conservation Status

Both the terrestrial and flooded phase contain ecological communities of national and international importance. Seasonal transitions between surface water and terrestrial systems result in a variable and species-rich ecotone and wetland mosaic habitat. This unique transitional status has resulted in high conservation value and protected status as priority habitats under the European Habitats Directive (CEC 1992). Many are protected as Special Areas of Conservation under European legislation, but effective management is difficult because they can be subject to pressures that arise some distance from their location. Further work is needed to better understand their hydrological and nutrient pathways and response of the biotic communities to these and more localized effects.

References

Cabot D. Ireland: a natural history, Collins New Naturalist Library. London: Collins; 1999.

CEC (Council of the European Communities). Directive 2000/60/EC of the European Parliament and of the Council – Establishing a framework for Community action in the field of water policy. Official Journal of the European Communities L327. 2000. p. 1–73.

CEC (Council of the European Communities). Council Directive of 21 May 1992 on the conservation of natural habitats and of wild fauna and flora (92/43/EEC). Official Journal of the European Communities, No. L 206/35. 1992.

Coxon CE. An examination of the characteristics of turloughs using multivariate statistical techniques. Ir Geogr. 1987;20:24–42.

Coxon CE. Carbonate deposition in turloughs (seasonal lakes) on the western limestone lowlands of Ireland, I: present day processes. Ir Geogr. 1994;27:14–27.

Coxon CE. A Study of the hydrology and geomorphology of turloughs [Unpublished PhD thesis]. Trinity College Dublin; 1986.

Coxon CE, Drew DP. Groundwater flow in the lowland limestone aquifer of eastern Co. Galway and eastern Co. Mayo, western Ireland. In: Paterson K, Sweeting MM, editors. New directions in Karst. Norwich: Geo Books; 1986. p. 259–79.

Coxon C, Drew D. Interdependence of groundwater and surface water in lowland karst areas of western Ireland: management issues arising from water and contaminant transfers. In: Robins NS,

Misstear BDR, editors. Groundwater in the Celtic regions: studies in hard rock and quaternary hydrogeology, Special Publication, vol. 182. London: Geological Society; 2000. p. 81–8.

Cunha Pereira H, Allott N, Coxon C. Are seasonal lakes as productive as permanent lakes? A case study from Ireland. Can J Fish Aquat Sci. 2010;67:1291–302.

Cunha Pereira H, Allott N, Coxon C, Naughton O, Johnston P, Gill L. Phytoplankton of turloughs (seasonal karstic Irish lakes). J Plankton Res. 2011;33:385–403.

Drew DP, Coxon CE. The effects of land drainage on groundwater resources in karstic areas of Ireland. In: Yuan D, editor. Proceedings of the 21st I.A.H. Congress. Beijing: Geological Publishing House; 1988. p. 204–9.

Duigan CA, Frey DG. *Eurycercus glacialis*, a chydorid cladoceran new to Ireland. Irish Nat J. 1987;22:180–3.

Duigan CA, Kovach WL. A study of the distribution and ecology of littoral freshwater chydorid (Crustacea, Cladocera) communities in Ireland using multivariate analysis. J Biogeogr. 1991;18:267–80.

Foster GN, Nelson BH, Bilton DT, Lott DA, Merritt R, Weyl RS, Eyre MD. A classification and evaluation of Irish water beetle assemblages. Aquat Conserv Mar Freshwat Ecosyst. 1992;2:185–208.

Goodwillie R. Turloughs over 10 hectares: vegetation survey and evaluation. Unpublished report for the National Parks and Wildlife Services. Dublin: Office of Public Works; 1992.

Goodwillie R, Reynolds JD. Turloughs. In: Otte ML, editor. Wetlands of Ireland: distribution, ecology, uses and economic value. Dublin: University College Dublin Press; 2003. p. 130–4.

Grainger JNR. *Diaptomus castor* in Co. Meath. Irish Nat J. 1966;15:211.

Grainger JNR. The biology of *Tanymastix stagnalis* (L.) and its survival in large and small temporary water bodies in Ireland. Hydrobiologia. 1991;212:77–82.

Kilroy G, Coxon C. Temporal variability of phosphorus fractions in Irish karst springs. Environ Geol. 2005;47:421–30.

Kimberley S, Naughton O, Gill L, Johnston P, Waldren S. The influence of flood duration on the surface on the surface soil properties and grazing management of karst wetlands (turloughs) in Ireland. Hydrobiologia. 2012;692:29–40.

Lynn DE, Waldren S. Morphological variation in populations of *Ranunculus repens* from the temporary limestone lakes (Turloughs) in the west of Ireland. Ann Bot. 2001;87:9–17.

Lynn DE, Waldren S. Physiological variation in populations of *Ranunculus repens* L. (Creeping Buttercup) from the temporary limestone lakes (turloughs) in the west of Ireland. Ann Bot. 2002;89:707–14.

Lynn DE, Waldren S. Survival of *Ranunculus repens* L. (Creeping Buttercup) in an amphibious habitat. Ann Bot. 2003;91:75–84.

Mellander P-E, Jordan P, Melland A, Murphy P, Wall D, Mechan S, Meehan R, Kelly C, Shine O, Shortle G. Quantification of phosphorus transport from a karstic agricultural watershed to emerging spring water. Environ Sci Technol. 2013;47:6111–9.

Moran J, Gormally M, Sheehy SM. Turlough ground beetle communities: the influence of hydrology and grazing in a complex ecological matrix. J Insect Conserv. 2012;16:51–69.

Naughton O, Johnston PM, Gill LW. Quantifying groundwater flooding in Ireland: the hydrological characterisation of ephemeral karst lakes (turloughs). J Hydrol. 2012;470–471:82–97.

Nelson B, Thompson R. The natural history of Ireland's dragonflies. Belfast: The National Museums and Galleries of Northern Ireland; 2004.

Ní Bhriain B, Sheehy Skeffington M, Gormally M. Conservation implications of land use practices on the plant and carabid beetle communities of two turloughs in Co. Galway, Ireland. Biol Conserv. 2002;105:81–92.

Porst G, Irvine K. Distinctiveness of macroinvertebrate communities in turloughs (temporary ponds) and their response to environmental variables. Aquat Conserv-Mar Freshwat Ecosyst. 2009a;19:456–65.

Porst G, Irvine K. Implications of the spatial variability of macroinvertebrate communities for monitoring ephemeral lakes. An example from turloughs. Hydrobiologia. 2009b;636:421–38.

Porst G, Naughton O, Gill G, Johnston P, Irvine K. Adaptation, phenology and disturbance of macroinvertebrates in temporary water bodies in temporary water bodies. Hydrobiologia. 2012;696:47–62.

Praegar RL. The flora of the turloughs: a preliminary note. Proc R Ir Acad. 1932;41:37–45.

Proctor MCF. Environmental and vegetational relationships of lakes, fens and turloughs in the Burren. Biol Environ. 2010;110B:17–34.

Sharkey N. Turlough vegetation communities – links with hydrology, hydrochemistry, soils and management [Unpublished PhD thesis]. Trinity College Dublin; 2012.

Sheehy Skeffington M, Scott NE. Do turloughs occur in Slovenia? Acta Carsologica. 2008;37:291–306.

Sheehy Skeffington M, Moran J, O'Connor A, Regan E, Coxon CE, Scott NE, Gormally M. Turloughs – Ireland's unique wetland habitat. Biol Conserv. 2006;133:265–90.

Williams CD, Gormally MJ. Spatio-temporal and environmental gradient effects on the mollusc communities in a unique wetland ecotone (turloughs). Wetlands. 2009;29:854–65.

The Macrotidal Bay of Mont-Saint-Michel (France): The Function of Salt Marshes

86

Loïc Valéry and Jean-Claude Lefeuvre

Contents

Introduction 1080
Hydrology and Geomorphology of Salt Marshes 1080
Plant Diversity, Salt Marsh Structure, and Primary Production 1081
Animal Diversity and Salt Marsh Functions 1082
Recent Changes in Salt Marsh Functioning and Potential Loss of Ecosystem Services 1084
Conservation Status 1086
Future Challenges 1086
References 1087

Abstract

A site recognized for its exceptional biological richness by many international designations, the Mont-Saint-Michel Bay (northwestern France) has the largest undivided tidal salt marshes in Europe. Owing to their extent and the heterogeneity of environmental conditions that characterize them, they are home to a relatively rich and diverse flora and fauna – given the particularly stressful abiotic factors of such type of habitat – and play a critical role in the ecological functioning of the Bay at the origin of many ecosystem services. However, the functions and services provided by these salt marshes are being profoundly modified by the invasion of sea couch grass, for which nitrogen enrichment of coastal waters from watersheds seems to be the main driver.

L. Valéry (✉) · J.-C. Lefeuvre (✉)
Department of Ecology and Biodiversity Management, National Museum of Natural History, Paris, France

EA 7316 Biodiversity and Land Management, University of Rennes 1, Rennes, France
e-mail: loic.valery@gmail.com; jeanclaudelefeuvre@wanadoo.fr

© Springer Science+Business Media B.V., part of Springer Nature 2018
C. M. Finlayson et al. (eds.), *The Wetland Book*,
https://doi.org/10.1007/978-94-007-4001-3_137

Keywords

Salt marsh functioning · Plant and animal diversity · Ecosystem services · Eutrophication · *Elymus* invasion

Introduction

Mont-Saint-Michel Bay is located within the angle of the Normandy-Breton Gulf formed by the Cotentin (Normandy) and Brittany and is open to the English Channel over a distance of 20 km (48° 36′–48° 44′ N and 1° 52′–1° 22′ E). Delimited by the cliffs of Carolles to the east and by the rocky spur of Cancale to the west, it consists of a coastal sedimentary basin covering an area of about 600 km². It is also the estuary of three main small coastal rivers (the Couesnon, the Sée, and the Sélune) (Fig. 1).

Hydrology and Geomorphology of Salt Marshes

The input of freshwater by the coastal rivers is negligible (less than 1%) relative to the volume of seawater in the bay. Therefore, its hydrology is largely dominated by the tides. The exceptional tidal range of the bay (14.50 m for the spring tide per equinox) is the 5th largest in the world. At low water, the intertidal zone represents a surface area of about 250 km².

The tidal flow is slowed down very little due to the shallow slope of the intertidal zone, which is rarely above 3 ‰ and can even be less than 1 ‰. This is the main cause of water agitation and turbidity, as the oceanic swell only enters the bay with difficulty. The asymmetry of the ebb and flow of the tides arriving at the interior of

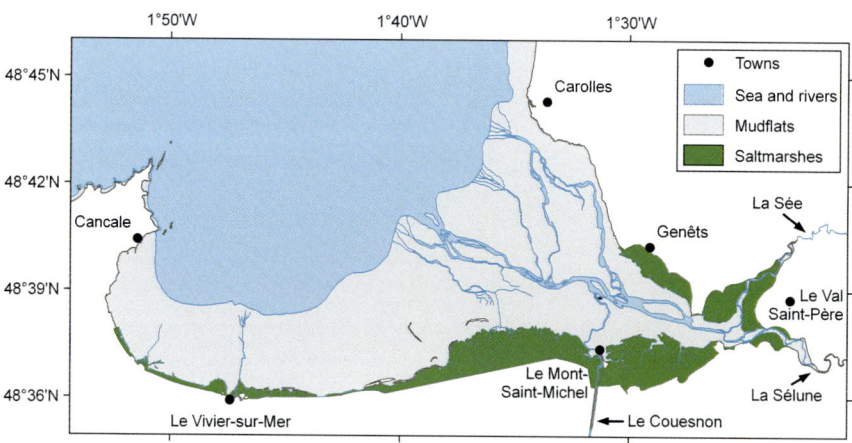

Fig. 1 Mont-Saint-Michel Bay, northwestern France

the bay results in the inevitable silting up of the intertidal zone. Sediment inputs have been estimated to be between 1 and 1.5 million m³ per year (Verger 2005).

On the higher part of the intertidal zone, these inputs have thus resulted in the formation of the largest undivided salt marsh area in Europe. These now occupy an area of about 4,200 hectares and have increased at a mean rate of 16 ha year^{-1} during the last 30 years.

These salt marshes are only flooded twice a day during the spring tides. At the end of the 1960s, only 10% of the tides covered them. The active sedimentation processes in the bay mean that it is possible to hypothesize that the frequency of tidal flooding has decreased even more.

Plant Diversity, Salt Marsh Structure, and Primary Production

After a survey of the flora of all of the Atlantic coastal salt marshes, Mont-Saint-Michel Bay was considered the most interesting area in France from the point of view of the rarity and biological value of its species and plant communities (Géhu 1979).

This remarkably rich flora (about 70 species and 30 plant associations) is mainly explained by the large area of the salt marshes and the diversity of environmental conditions (substrate, micro-topography, freshwater input, salinity, etc.).

The vegetation associations have a spatial distribution related to the physico-chemical nature of the sediment and the frequency and length of time that they are flooded (Fig. 2). This relatively clear zonation occurs in bands parallel to the mudflats (Guillon 1984). Classically, one can distinguish (i) a pioneer zone

Fig. 2 The salt marshes surrounding the Mont-Saint-Michel: in the foreground, the so-called "Great Eastern Salt Marsh"; beyond the dike road, the "Great Western Salt Marsh" (Photograph courtesy of André Mauxion, © rights remain with the photographer)

colonized by *Salicornia dolichostachya* and, less frequently, *Salicornia fragilis*, *Suaeda maritima*, and *Spartina anglica*; (ii) a low marsh zone with *Puccinellia maritima*, *Aster tripolium*, and *Suaeda maritima*; (iii) a middle marsh zone dominated by the shrub species *Atriplex portulacoides*; and (iv) a high marsh zone, generally characterized by the presence of grasses *Festuca rubra* and *Elymus athericus* (Guillon 1984; Bouchard 1996). A large proportion of these marshes, used for sheep grazing, has lost its zonation and has been transformed into short grasslands (the salt meadows).

This zonation can also be disturbed by the proximity of channels or variations in the micro-topography within the salt marshes. Furthermore, species and plant community diversity is higher in the upper part of the salt marshes due to weaker abiotic constraints (e.g., lower salinity, shorter periods of immersion, and periodic freshwater inputs) which could induce a patchy organization of the plant groups (Guillon 1984).

Functionally, the primary production of the salt marshes is very high: the most productive plants (*Atriplex portulacoides*, *Elymus athericus*, and *Festuca rubra*), situated in the middle and high parts of the marsh, each produce between 2 and 3 kg dry matter $m^{-2} years^{-1}$ and supply a total of more than 90% of salt marsh production (Bouchard 1996; Valéry et al. 2004).

Animal Diversity and Salt Marsh Functions

The Mont-Saint-Michel Bay salt marshes contain an abundant and diverse invertebrate fauna. The spatial distribution of species varies mainly as a function of the salinity, immersion frequency, as well as the structure and composition of the plant communities (Fouillet 1986). This explains the spatial distribution of nearly 70 species of carabid beetles (Carabidae). The Trechinae are best represented by the *Pogonus*, *Emphanes*, *Philochthus*, *Metallina*, and *Tachys* genera. The Harpalinae, *Dichirotrichus gustavii*, is also very abundant, whereas the Cicindelinae of the *Cicindela* genus are concentrated in smaller zones (Fouillet 1986; Pétillon et al. 2008).

The spider population consists of about 60 species, most of which are not specific to saline areas. Three main families that have been identified are (i) the Lycosidae (especially the *Pardosa* and *Arctosa* genera), (ii) the Tetragnathidae (mainly the *Pachygnatha* genus), and (iii) the Linyphiidae (essentially the *Erigone* and *Oedothorax* genera) (Fouillet 1986; Pétillon et al. 2008).

Large populations of the crustacean, *Orchestia gammarella* (an amphipod), live under the well-developed vegetation cover, especially that of *Atriplex portulacoides*, which provides a particularly suitable habitat (Fouillet 1986). Associated with other decomposers and detritivores (e.g., Nematoda) as well as numerous fungi and bacteria (Bouchard 1996; Lucas 1997), this crustacean participates actively in the decomposition of organic matter. Most of the latter is exported into the coastal marine water (outwelling) in a dissolved and particulate form (Lefeuvre et al. 1994; Créach 1995; Troccaz 1996), and the decomposers thus play a key role in the intersystem exchanges (Bouchard 1996). The exported organic matter

Fig. 3 Dark-bellied brent geese grazing on a *Puccinellia maritima* sward: a high heritage value species wintering every year in the bay (Photograph courtesy of André Mauxion, © rights remain with the photographer)

favors the development of the microphytobenthos (mainly diatoms) on the mudflats and partly explains the abundance of filter feeders (e.g., polychaetes and bivalves) (Meziane 1997). Two of the latter are grown commercially in marine farms: oysters (6,000 tonnes per year) and mussels (12,000 tonnes per year) (Lefeuvre 2010).

The intertidal and subtidal zones of the bay are exploited by nearly a hundred species of fish belonging to about 40 families (Laffaille 2000). The bay represents the nursery zone for numerous species, like rays (the *Raja* genus), some Pleuronectiformes (e.g., common sole *Solea solea*, European plaice *Pleuronectes platessa*, European flounder *Platichthys flesus*, brill *Scophthalmus rhombus*), some Gadidae (e.g., pouting *Trisopterus luscus*), as well as some Clupeidae (e.g., *Sprattus*, *Clupea*, and *Sardina* genera).

Nearly a dozen species take advantage of the high tide to feed on the diatoms (e.g., mullets (*Liza spp.* and *Mugil spp.*), gobies (*Pomatoschistus spp.*)), or amphipods (e.g., sea bass juveniles (*Dicentrarchus labrax*)) in the drainage channels and sometimes even in the salt marshes. Therefore, these species participate in the outwelling process (Laffaille 2000).

As it is located on the East Atlantic Flyway, Mont-Saint-Michel Bay is a remarkably rich zone from an ornithological point of view. About 130 species, belonging to about 40 families, can be seen regularly. Depending on the species, this site provides successively during the year wintering grounds, a pre- and postnuptial staging area, summering grounds, and a molting site. During very cold periods, this bay represents a climatic refuge for all of the Nordic birds, in particular the Anatidae (e.g., common goldeneye *Bucephala clangula* and long-tailed duck *Clangula hyemalis*) and the waders (e.g., Eurasian oystercatcher *Haematopus ostralegus* and Eurasian curlew *Numenius arquata*) (Schricke 1983; Eybert et al. 1999).

The salt marshes play an important role for many species. In the winter, a few Anatidae (i.e., dark-bellied brent goose *Branta bernicla bernicla*, Eurasian wigeon *Anas penelope*, and Eurasian teal *Anas crecca*) feed exclusively on the grass *Puccinellia maritima* and seeds of *Suaeda maritima* and *Salicornia spp.* (Schricke 1983) (Fig. 3). Among the passerines, the common starling *Sturnus vulgaris*, common linnet *Carduelis cannabina*, Eurasian skylark *Alauda arvensis*, several pipit species (*Anthus spp.*), Lapland longspur *Calcarius lapponicus*, and snow bunting *Plectrophenax nivalis* also find food there (seeds and/or invertebrates). Finally, a few birds of prey (diurnal and nocturnal) are observed regularly while hunting: the western marsh harrier *Circus aeruginosus* and hen harrier *Circus cyaneus*, merlin *Falco columbarius*, common kestrel *Falco tinnunculus*, peregrine falcon *Falco peregrinus*, and short-eared owl *Asio flammeus* (Eybert et al. 1999).

During the breeding period, many passerines nest in the salt marshes (e.g., Eurasian skylark *Alauda arvensis*, certain pipit species, and bluethroat *Luscinia svecica*). The common quail *Coturnix coturnix* use the tall dense meadows of *Festuca rubra* and *Elymus athericus* for nesting (Eybert et al. 1999), and the Kentish plover *Charadrius alexandrinus* use the shellfish beds.

For the mammals, the abundance of inland species like the wild boar *Sus scrofa* and European hare *Lepus europaeus* must be highlighted, for whom the salt marshes form a very favorable habitat.

Recent Changes in Salt Marsh Functioning and Potential Loss of Ecosystem Services

Since the beginning of the 1990s, the salt marshes of Mont-Saint-Michel Bay have been undergoing major changes in their vegetation cover due to the invasion of the sea couch grass *Elymus athericus* (Fig. 4). This indigenous grass characteristic of the high marshes has continued expanding its area by spreading toward the middle and low marshes: overall, between 1984 and 2013, the proportion of the salt marsh area covered by this species has increased from 3% to 45%, corresponding to a gain of 1,800 ha. By forming dense monospecific stands, this species eliminates the

Fig. 4 Invasion of salt marshes by the sea couch grass (Photograph courtesy of André Mauxion, © rights remain with the photographer)

Fig. 5 Spread of sea couch grass (*background*) in the middle marsh at the expense of sea purslane (*foreground*) (Photograph courtesy of Jean-Claude Lefeuvre, © rights remain with the photographer)

characteristic zonation described above, to the detriment of the common salt marsh grass *Puccinellia maritima* and the sea purslane *Atriplex portulacoides* in particular (Valéry et al. 2004; Valéry and Radureau 2014) (Fig. 5).

The most plausible hypothesis to explain this phenomenon is coastal eutrophication by nitrogen. On the one hand, intensification of agriculture since the beginning of the 1970s in all of the coastal river catchment areas on the eastern side of the bay is probably the origin of an increase in nitrogenous nutrients in the coastal waters: between 1970 and 2000, the nitrate content of these rivers has increased from less than 10 mg L^{-1} to more than 30 mg L^{-1}, a level that has since stabilized (Valéry et al. 2016). On the other hand, the increase in nitrogen availability has allowed *Elymus athericus* to synthesize osmoticum (molecules biosynthesized by halophilous plants allowing them to resist salt stress) more abundantly (Leport et al. 2006).

The invasion of sea couch grass has several effects on the structure and functioning of the ecosystem, resulting especially in a large reduction of essential ecological functions:

(i) Overall reduction of land-sea exchanges due to increased sedimentation in the salt marshes, whose elevation increases more quickly (Valéry 2006).
(ii) Qualitative and quantitative modifications of the outwelling function (Valéry et al. 2004; Valéry 2006).
(iii) Reduction of the densities of certain invertebrate species specific to salt marshes (especially spiders and carabid beetles) (Pétillon et al. 2008).
(iv) Reduction of the nursery role for fish, due to the decrease in the density of their favorite prey, *Orchestia gammarella* (Laffaille et al. 2005).
(v) Reduction of the carrying capacity for the dark-bellied brent goose (Valéry 2006; Valéry et al. 2008), Eurasian wigeon, and Eurasian teal which do not eat sea couch grass, except at the young shoot stage (Schricke 1983). In certain highly invaded sectors, plant cover closure is such that, in general, the birds have no access, which very strongly limits the interest of the salt marshes for these migratory birds.

Fig. 6 The production of salt-meadow lambs: a flagship activity of Mont-Saint-Michel Bay (Photograph courtesy of Jean-Claude Lefeuvre, © rights remain with the photographer)

From an agronomic and socioeconomic point of view, the progression of this plant species, which is grazed very little by sheep, also results in a reduction of grazing area and constitutes a threat to the pastoral sheep activities (Fig. 6), which were awarded the trademark of controlled designation of origin (Appellation d'Origine Contrôlée) for lamb production in 2009 (a status given to local products whose quality is certified by a specific governmental body).

Conservation Status

The bay forms an exceptional heritage, recognized by numerous classification and protection measures. The most significant are:

(i) The classification of Mont-Saint-Michel and its bay as a World Heritage Site by UNESCO in 1979 based on cultural and natural criteria.
(ii) The inscription of Mont-Saint-Michel Bay as a Ramsar Site in 1994, underlining the biological value of this coastal wetland.
(iii) The inclusion of the whole site into the "Natura 2000 network" under the European "Birds" (1979) and "Habitat, Fauna and Flora" (1992) directives.
(iv) The bay is recognized nationally as a "listed and registered site" (according to the 1930 Act): 18,000 ha of terrestrial and maritime zones have thus been protected since 1987. There is also a hunting and wildlife reserve (3,000 ha, 1974).

Future Challenges

The creation of a large Natural National Park, efforts to improve river hydrology (including the removal of two hydroelectric dams), and the projected improvement of the physicochemical quality of water should result, in the midterm, in an overall improvement of the ecological functioning of this bay which is subjected to strong anthropogenic pressures.

References

Bouchard V. Production et devenir de la matière organique dans un marais salé européen en système macrotidal (Baie du Mont-Saint-Michel) [dissertation]. Université Rennes 1, Rennes; 1996. 209 p.

Créach V. Origines et transferts de la matière organique dans un marais littoral (Baie du Mont-Saint-Michel): utilisation des compositions isotopiques naturelles du carbone et de l'azote [dissertation]. Université Rennes 1, Rennes; 1995. 156 p.

Eybert M-C, Geslin T, Le Dréan-Quenec'hdu S, Schricke V. Etudes de l'avifaune. Rétablissement du caractère maritime du Mont-Saint-Michel. Syndicat mixte pour le rétablissement du caractère maritime du Mont-Saint-Michel. Université Rennes 1; 1999. 79 p.

Fouillet P. Evolution des peuplements d'Arthropodes des schorres de la baie du Mont-Saint-Michel. Influence du pâturage ovin et conséquences de son abandon [dissertation]. Université Rennes 1, Rennes; 1986. 330 p.

Géhu J-M. Etude phytocoenotique analytique et globale de l'ensemble des vases salées et prés salés et saumâtres de la façade atlantique française. Rapport de synthèse, Ministère de l'Environnement, Mission des Etudes; 1979. 514 p.

Guillon L-M. Les schorres de la baie du Mont-Saint-Michel – Unités de végétation et facteurs du milieu, Rapport CEE Environnement, Université de Rennes 1, Muséum National d'Histoire Naturelle, Ecole Pratique des Hautes Etudes, Paris; 1984. 78 p.

Laffaille P. Relations entre l'ichtyofaune et les marais salés macrotidaux : l'exemple de la baie du Mont-Saint-Michel [dissertation]. Université Rennes 1, Rennes; 2000. 202 p.

Laffaille P, Pétillon J, Parlier E, Valéry L, Ysnel F, Radureau A, Feunteun E, Lefeuvre J-C. Does the invasive plant *Elymus athericus* modify fish diet in tidal salt marshes? Estuar Coast Shelf Sci. 2005;65:739–46.

Lefeuvre J-C. Histoire et écologie de la baie du Mont-Saint-Michel. Editions Ouest-France; 2010. 271 p.

Lefeuvre J-C, Bertru G, Burel F, Brient L, Créach V, Gueuné Y, Levasseur J, Mariotti A, Radureau A, Retière C, Savouré B, Troccaz O. Comparative studies on salt marsh processes: Mont-Saint-Michel Bay, a multi-disciplinary study. In: Mitsch WJ, editor. Global wetlands: old world and new. Amsterdam: Elsevier Science; 1994. p. 215–34.

Leport L, Baudry J, Radureau A, Bouchereau A. Sodium, potassium and nitrogenous osmolyte accumulation in relation to the adaptation to salinity of *Elytrigia pycnantha*, an invasive plant of the Mont-Saint-Michel bay. Cah Biol Mar. 2006;47:31–7.

Lucas F. Activité et structure de la communauté bactérienne des sédiments colonisés par *Nereis diversicolor* (O.F. Müller) – Baie du Mont-Saint-Michel [dissertation]. Université Rennes 1, Rennes; 1997. 196 p.

Meziane T. Le réseau trophique benthique en Baie du Mont-Saint-Michel : intégration de la matière organique d'origine halophile à la communauté à *Macoma balthica* [dissertation]. Université Rennes 1 – Muséum National d'Histoire Naturelle, Rennes; 1997. 182 p.

Pétillon J, Georges A, Canard A, Lefeuvre J-C, Bakker J-P, Ysnel F. Influence of abiotic factors on spider and ground beetle communities in different salt-marsh systems. Basic Appl Ecol. 2008;9:743–51.

Schricke V. Distribution spatio-temporelle des populations d'Anatidés en transit et en hivernage en baie du Mont-Saint-Michel en relation avec les activités humaines [dissertation]. Université Rennes 1, Rennes; 1983. 299 p.

Troccaz O. Evolution de la dynamique d'un marais salé : processus fonctionnels internes et relations avec le milieu côtier – La Baie du Mont Saint-Michel [dissertation]. Université Rennes 1, Muséum National d'Histoire Naturelle, Rennes; 1996. 106 p.

Valéry, L. Approche systémique de l'impact d'une espèce invasive – Le cas d'une espèce indigène dans un milieu en voie d'eutrophisation [dissertation]. Muséum National d'Histoire Naturelle, Paris; 2006. 276 p.

Valéry L, Radureau A. Evolution des marais salés de la baie du Mont-Saint-Michel – Analyse cartographique 1984–2013. Rapport Université de Rennes 1 et Centre Régional d'Etudes Biologiques et Sociales, Rennes; 2014. 15 p.

Valéry L, Bouchard V, Lefeuvre J-C. Impact of the invasive native species *Elymus athericus* on carbon pools in a salt marsh. Wetlands. 2004;24:268–76.

Valéry L, Radureau A, Lefeuvre J-C. Spread of the native grass *Elymus athericus* in salt marshes of Mont-Saint-Michel bay as an unusual case of coastal eutrophication. J Coast Conserv. 2016; in press. Available from: doi:10.1007/s11852-016-0450-z [Accessed 14th August 2016].

Valéry L, Schricke V, Fritz H, Lefeuvre J-C. A synthetic method to assess the quality of wintering sites for the Dark-bellied Brent Goose *Branta bernicla bernicla* – The case study of the salt marsh of Vains in the Mont-Saint-Michel Bay, France. Vogelwelt. 2008;129:221–5.

Verger F. Marais et estuaires du littoral français. Paris: Belin Editions; 2005. 335 p.

The Inner Danish Waters and Their Importance to Waterbirds

87

Ib Krag Petersen and Rasmus Due Nielsen

Contents

Introduction	1090
Physiochemical Conditions	1091
Benthos	1092
Birds and Mammals	1092
Conservation Status	1094
Threats, Uses, and Future Challenges	1095
References	1095

Abstract

The inner Danish waters comprise an area of 50,000 km^2. The area is very important to a range of species of waterbirds breeding further north in Scandinavia or further northeast in Russia and Siberia as the inner Danish waters are ice-free in most winters. The shallow nearshore areas are important to mute swan, Eurasian coot while the near-shore, slightly deeper areas are important to a range of seaducks most importantly common eider and common scoter. Open sea areas are important to red-throated and black-throated diver and several species of alcids. The largest number of waterbirds are present during migration and winter but some species also molt in the area during late summer. During migration dabbling ducks and geese are present in their thousands in coastal areas. Harbor seal and harbor porpoise are the most numerous marine mammals occurring in the area. Climate change might be affecting the distribution of waterbirds with several species showing an increase in numbers in the northeastern part of their wintering range with e.g. increasing numbers of wintering smew in the inner Danish waters.

I. K. Petersen (✉) · R. D. Nielsen
Department of Bioscience, Aarhus University, Kalø, Denmark
e-mail: ikp@bios.au.dk; rdn@bios.au.dk

Surveys of waterbirds have been conducted in Denmark since the 1960s, when the aerial survey method was introduced. For many years the total count method was used, but since 2000 the line transect sampling method, allowing for fine scale estimation of densities, has been used intensively.

Keywords

Waterbirds · Wetlands · Monitoring · Line transect · Sampling method · Molt

Introduction

The inner Danish marine area comprises the western part of the Baltic Sea and its "estuary" with fjords and sounds into the North Sea. It covers an area of around 50,000 km^2 of which which more than 23,000 km^2 have water depths of less than 20 m (Fig. 1). The area is characterized by a mild winter climate with mean midwinter temperatures around 0 °C, leaving the area free of ice in most winters.

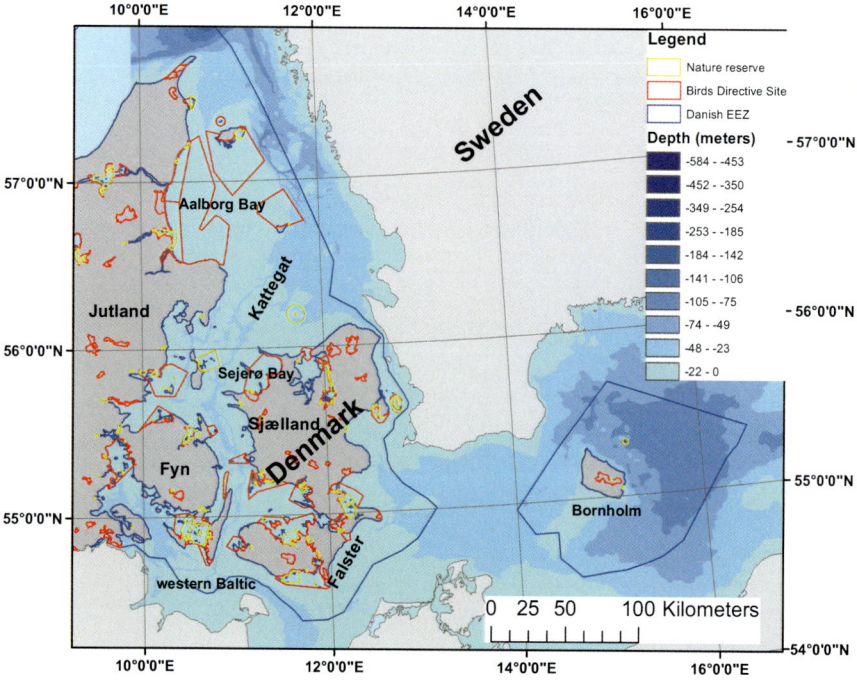

Fig. 1 The inner Danish waters comprise the area east of Jutland all the way to the island of Bornholm. Birds Directive sites are the most important protected sites for waterbirds in the inner Danish waters. National nature reserves cover smaller areas and are primarily found within the boundaries of the Birds Directive sites. The Danish EEZ boundary is indicated (produced from data in public domain)

The combination of mild winter climate and huge area of shallow sea creates very attractive wintering and staging habitats for numerous species of waterbirds. Millions of waterbirds (Laursen et al. 1997) migrate into these milder areas from Scandinavian and Russian/Siberian breeding grounds where winter conditions are too harsh. As such, the inner Danish waters provide an internationally very important wetland for wintering waterbirds. At the same time, the inner Danish waters offer good breeding options for waterbirds such as great cormorant, *Phalacrocorax carbo*; mute swan, *Cygnus olor*; and common eider, *Somateria mollissima*. Over the late summer, greylag geese, *Anser anser*, mute swan, and common scoter, *Melanitta nigra* undergo moult in the area, during which time these birds are particularly vulnerable to human disturbances.

The marine landscape of the inner Danish waters provides a large variety of habitats. Numerous fjords and sounds with complex topography create sheltered habitats, favored by dabbling ducks, geese, swans, and mergansers (*Mergus* spp.), while exposed coastlines are favored by several species of diving ducks. In the open water areas, divers (Gaviidae), diving ducks, and auks (Alcidae) winter in significant numbers.

The inner Danish waters are monitored as part of the national monitoring program every third winter and every sixth summer (Pihl et al. 2013). Since the late 1960s, these surveys have been conducted as land-based total counts or from an aircraft. From 2000, aerial line transect survey sampling over large open water areas was introduced to provide abundance estimates and distribution patterns in wider parts of the Danish waters (Petersen et al. 2006; Petersen and Nielsen 2011).

Physiochemical Conditions

The salinity in the inner Danish waters decreases from northern Kattegat toward the south, becoming brackish in the southern parts of the Danish Baltic.

One of the major environmental problems in the inner Danish waters is eutrophication, but nutrient inputs from land have reduced by ca. 50% for nitrogen (N) and 56% for phosphorous (P) since 1990 as a consequence of legislation regulating agricultural practices among other things (Riemann et al. 2016). The maximal depth of at which eelgrass, *Zostera marina*, occurs decreased during the period 1989–2008, but from 2010 onward, eelgrass has expanded toward deeper water depth (Riemann et al. 2016). The reductions in N and P have resulted in significant and parallel declines in nutrient concentrations and initiated a shift in the dominance of primary producers toward reduced phytoplankton biomass and increased cover of macroalgae in deeper waters (Riemann et al. 2016). Events of anoxic situations is an annually reoccurring problem in the inner Danish waters, mainly in September and October, with varying spatial distribution between years (Jensen et al. 2012). The anoxic events mainly occur at water depths of more than 15 m. Thus, the overlap between high concentrations of diving ducks and anoxic areas is limited.

During the last 40 years the mean water temperature in Danish waters has increased by more than 1 °C.

Benthos

The higher benthic fauna comprises more than 400 species, with the largest diversity in areas of high salinity. In shallow water, species such as blue mussel, *Mytilus edulis*, and soft-shell clam, *Mya arenaria*, dominate. Both the number of benthic species and their abundance have decreased since the mid-1990s with no apparent reason (Hjort and Josefson 2010; Riemann et al. 2016). Meanwhile, since the 1980s the invasive American razor clam, *Ensis americanus*, has spread from the south, northward along the west coast of Jutland, and into inner Danish waters. This clam species provides an important source of food to diving ducks, in particular common eider and common scoter.

Birds and Mammals

The sheltered, near shore areas of the inner Danish waters are of great international importance for a range of different waterbirds. Fjords and sheltered bays harbor the majority of wintering mute swan, greylag geese; common goldeneye, *Bucephala clangula*; red-breasted merganser, *Mergus serrator*; and Eurasian coot, *Fulica atra*, in Denmark (Petersen et al. 2006, 2010). Increasing numbers of white-tailed eagle, *Haliaeetus albicilla*, forage on the large numbers of wintering and molting waterbirds in these habitats (Ehmsen et al. 2011). During the migration periods, many species of dabbling ducks and geese pass through wetlands, particularly in the southern parts of Denmark.

The number of mute swan wintering in Denmark fluctuates with winter severity. In cold winters large numbers are observed, with reduced numbers counted in subsequent winters due to a higher mortality rate in the cold years (Pihl 2000). Total numbers (41,777–72,130 birds) vary considerably but with fairly stable numbers from 1970 to 2008 (Pihl et al. 2013). The majority of birds winter in sheltered bays and fjords in the central part of the inner Danish waters. Molting mute swans are generally found in the same habitats as wintering birds, and total numbers appear to have been stable or slightly increasing during the period 1968–2006 (Pihl et al. 2013). Eurasian coot is found in the same habitats as mute swan, often in dense groups ranging from a few hundred to several thousands. Total wintering numbers are dependent on winter severity and fluctuate accordingly. The overall trend seems stable (Pihl et al. 2013). In mild winters, mute swan and coots also winter in numbers in larger inland lakes. Large numbers of Eurasian wigeon, *Anas penelope*, use the inner Danish waters as a staging area during migration in spring (March–April) and autumn (September–November). They mainly feed on eelgrass and other ditch grasses, *Ruppia* sp. Numbers of greylag geese have increased throughout the years with birds also wintering in significant numbers over the last 20 years (Pihl et al. 2013). The majority of greylag geese occur on meadows and fields or in sheltered bays and fjords. Most birds are found in the fjords of southern Denmark. In the more exposed parts of shallow fjords and bays, large numbers of red-breasted merganser and common goldeneye can be found, with the majority of wintering birds found in

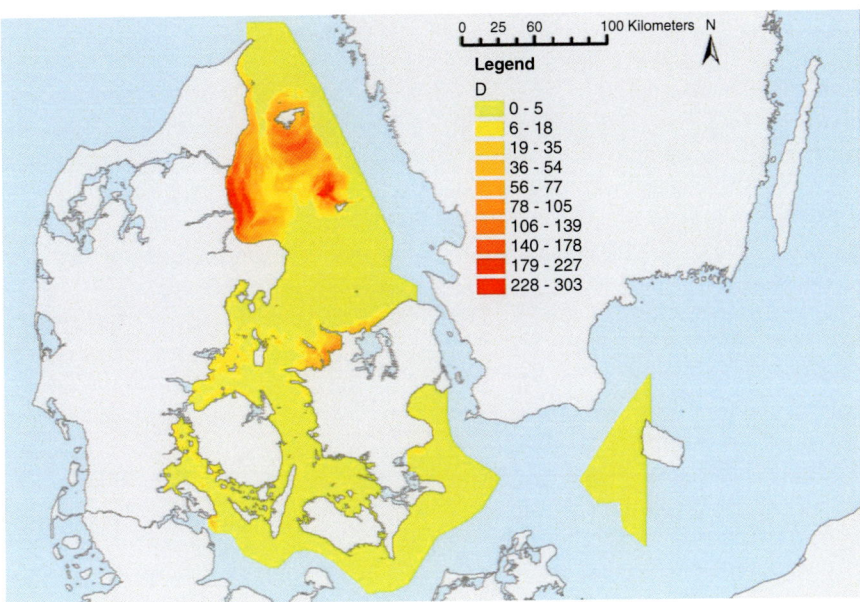

Fig. 2 The modelled distribution of 401,339 common scoters in the inner Danish waters in the winter of 2008 (Petersen and Nielsen 2011) (© IK Petersen and RD Nielsen, rights remain with authors)

the inner Danish waters. Numbers of red-breasted merganser have declined since the 1970s with around 54,000 birds estimated to winter in 2008. Common goldeneye shows stable or slightly increasing numbers over the same period, with an estimated total of 65,000 birds wintering in 2008 (Petersen and Nielsen 2011).

Large numbers of diving ducks winter and molt in the open parts of the inner Danish waters. Several species occur in numbers of international importance. Up to 83% of the entire biogeographical population of common eider winter in the inner Danish waters. The number of wintering common eider increased from 430,000 around 1970 (Joensen 1973) to 730,000 in 1992 (Pihl et al. 1992) and have since declined to 630,000 in 2004 (Petersen et al. 2006) and 502,000 in 2008 (Petersen et al. 2010) of which the vast majority winter in the inner Danish waters. The only other important Danish wintering area for common eider is the Wadden Sea (18,000 birds in 2008). Most common eiders winter in the western part of the Baltic with only few birds east of Falster (indicated on Fig. 1). The majority (67–69%) of common eider occurs at water depths of less than 10 m (Petersen et al. 2010) where they primarily forage on blue mussels. The number of molting common eider has declined from an estimated 250,000 around 1970 (Joensen 1973) to 110,000 birds in 2006 (Peterdsen and Nielsen 2011) with a geographical distribution similar to that of the wintering distribution. Common scoter and velvet scoter, *Melanitta fusca*, winter in high numbers, with the former being by far the most abundant. The total number of wintering common scoter in the inner Danish waters was estimated to be 401,000 in 2008 (Fig. 2, Petersen and Nielsen 2011). The survey method and coverage have changed through the period, and estimates of 200,000 wintering birds around 1970 (Joensen 1974) may have been underestimates. The majority of common scoters wintering in the inner Danish waters are found in the northern parts of inner Danish waters (Fig. 2). The preferred water depth was between 2 and 10 m, where they mainly forage on *Spisula subtruncata*. An estimated 35,000 common scoter molted in the inner Danish waters in 2008 (Petersen and Nielsen 2011), which was a decline as compared to 150,000 molting birds recorded in the 1960s (Joensen 1973). Molting common scoters are mainly found in Aalborg Bay and Sejerø Bay.

Other estimates of species wintering in significant numbers in the open parts of the inner Danish waters from the 2008 national midwinter survey (Petersen and Nielsen 2011) include 6,000 red- and black-throated divers, *Gavia stellata/G. arctica*; 28,000 long-tailed duck, *Clangula hyemalis;* and 76,000 razorbills/common guillemots, *Alca torda/Uria aalge*.

Changes in winter distribution and phenology of greylag goose; tufted duck, *Aythya fuligula*; common goldeneye; smew, *Mergellus albellus*; goosander, *Mergus merganser*; and several species of dabbling duck in northern Europe could be a result of increasing winter temperatures and a shift toward more northerly and easterly wintering areas (Clausen et al. 2013; Lehikoinen et al. 2013; Jordán-Pavón et al. 2015).

Three species of marine mammals regularly occur in the inner Danish waters. Two species of pinnipeds occur with harbor seal, *Phoca vitulina*, being common throughout and grey seal, *Halichoerus grypus*, more locally distributed but steadily increasing in numbers in recent years (Søgaard et al. 2015). Harbor porpoise, *Phocoena*, is the only frequent cetacean and is commonly found in the central and western part of the area with fever animals present in the eastern part of the area. The population of harbor porpoises in the western Baltic (including the German part), Belt Seas, and Kattegat (including the Swedish part) was estimated to be 40,475 animals in 2012 (Viquerat et al. 2014).

Conservation Status

In 1983, 111 Special Protection Areas (SPAs) were declared under the EU Birds Directive, the majority of which are situated in marine parts of the inner Danish waters. These SPAs were revised and finalized in 1994, covering a total area of 9,400 km^2. In 2004, a second revision of the Danish SPAs resulted in enlargement of three sites and the designation of a new site in Danish waters with the specific aim to cover important common scoter molting and wintering areas. Denmark also designated a total of 27 sites under the Ramsar Convention. All Danish Ramsar Sites are also designated as SPAs under the EU Birds Directive and cover a total of 7,400 km^2.

Hunting regulations in designated areas within the SPAs have in the period from 1994 to 2010 resulted in increased numbers of staging and wintering waterbirds in the inner Danish waters (Clausen et al. 2013).

Threats, Uses, and Future Challenges

Human utilization of the marine areas is intense. Water sports, fishing activities, and hunting are prominent recreational activities (Laursen et al. 2016). A commercial fishery also plays an important role in the inner Danish waters.

In the winter season 2011/2012, a total of 750,000 waterbirds was harvested in Denmark. Of these, 63% were mallards, *Anas platyrhynchos*, which to a large extent derive from a captive bred stock. A total of 43,200 common eiders were also harvested (Asferg 2012).

Bycatch in fishing gear has an impact on some waterbird species in the Baltic (Zydelis et al. 2009). In the western parts of the Baltic, common eider was the most abundant species in the bycatch statistics with ca. 600 birds caught annually within that region in the period 2001–2003 (Degel et al. 2010).

The construction of large offshore wind farms in the inner Danish waters began with the arrival of the twenty-first century. There are currently five large wind farm areas (>25 turbines) and several smaller installations. The location of offshore wind farms is currently limited to fairly shallow areas, often coinciding with important areas for staging/wintering and/or migratory waterbirds. Comparisons of pre- and post-construction distribution and abundances of common scoters and long-tailed ducks around Danish offshore wind farms have demonstrated a distribution impact on these species, with significant impacts out to a distance of up to 5 km from the wind farm footprint (Petersen et al. 2011, 2014). The plans for further development of offshore wind farms in Denmark are prominent, and spatial planning is thus essential in order to reduce potential impacts on the important concentrations of waterbirds in Danish marine areas.

References

Asferg T. Vildtudbyttestatistik for jagtsæsonen 2011/2012. – Report from DCE, Aarhus University. 2012. p. 7. http://dce.au.dk/fileadmin/dce.au.dk/Udgivelser/Vildtudbyttestatistik_2011_12.pdf.

Clausen P, Holm TE, Laursen K, Nielsen RD, Christensen TK. Rastende fugle i det danske reservatnetværk 1994–2010. Del 1: Nationale resultater. Aarhus Universitet, DCE – Nationalt Center for Miljø og Energi, 118 s. Videnskabelig rapport fra DCE – Nationalt Center for Miljø og Energi nr. 2013; 72. http://dce2.au.dk/pub/SR72.pdf.

Degel H, Petersen IK, Holm TE, Kahlert T. Fugle som bifangst i garnfiskeriet. Estimat af utilsigtet bifangst af havfugle i garnfiskeriet i området omkring Ærø. DTU Aqua-rapport nr. 2010;227–2010:56.

Ehmsen E, Pedersen L, Meltofte H, Clausen T, Nyegaard T. The occurrence and reestablishment of White-tailed Eagle and Golden Eagle as breeding birds in Denmark. Dansk Orn Foren Tidsskr. 2011;105:139–50.

Hjort M, Josefson AB. Marine områder 2008. NOVANA. Tilstand og udvikling i miljø- og naturkvaliteten. Danmarks Miljøundersøgelser, Aarhus Universitet, 136 s. –Faglig rapport fra DMU nr. 2010;760. http://www.dmu.dk/Pub/FR760.pdf.

Jensen PN, Boutrup S, Fredshavn JR, Svendsen LM, Blicher-Mathiesen G, Wiberg-Larsen P, Jerring R, Hansen JW, Nielsen KE, Ellermann T, Thorling L, Holm AG. Vandmiljø og Natur 2011. NOVANA. Tilstand og udvikling – faglig sammenfatning. Aarhus Universitet, DCE –

Nationalt Center for Miljø og Energi, 102 s. Videnskabelig rapport fra DCE – Nationalt Center for Miljø og Energi nr. 2012;36. http://www.dmu.dk/Pub/SR36.pdf.

Joensen AH. Moult Migration and Wing-feather Moult of Seaducks in Denmark. Danish Rev Game Biol. 1973;8(4):1–42.

Joensen AH. Waterfowl populations in Denmark 1965-73. Danish Rev Game Biol. 1974;9(1):206.

Jordán-Pavón D, Fox AD, Clausen P, Dagys M, Deceuninck B, Devos K, Hearn R, Holt CA, Hornman M, Keller V, Langendoen T, Lawicki L, Lorentsen SH, Luigujoe L, Meissner W, Musil P, Nilsson L, Paquet J-Y, Stipniece A, Stroud D, Wahl J, Zenatello M, Lehikoinen A. Climate-driven changes in winter abundance of a migratory wanterbird in relation to EU protected areas. Diversity Distrib. 2015;21:571–82.

Laursen K, Kaae BC, Bladt J, Skov-Petersen H, Rømer JK, Clausen P, Olafsson A, Draux H, Petersen IK, Bregnballe T, Nielsen RD. Fordeling af vandorienterede friluftsaktiviteter og vandfugle i Danmark. Aarhus Universitet, DCE – Nationalt Center for Miljø og Energi, 66 s. – Teknisk rapport fra DCE – Nationalt Center for Miljø og Energi nr. 2016;81. http://dce2.au.dk/pub/TR81.pdf.

Laursen K, Pihl S, Durinck J, Hansen M, Skov H, Frikke J, Danielsen F. Numbers and distribution of waterbirds in Denmark 1987-1989. Danish Rev Game Biol. 1997;15(1):181.

Lehikonen A, Jaatinen K, Vähätalo AV, Clausen P, Crowe O, Deceuninck B, Hearn R, Holt CA, Hornman M, Keller V, Nilsson L, Langendoen T, Tománková I, Wahl J, Fox AD. Rapid climate driven shifts in wintering distributions of three common waterbird species. Global Change Biology. 2013;19:2071–81.

Pihl S. Vinterklimaets indflydelse på bestandsudviklingen for overvintrende kystnære vandfugle i Danmark 1987–1996. DOFT. 2000;94(2):73–89.

Pihl S, Laursen K, Hounisen JP, Frikke J. Landsdækkende optælling af vandfugle fra flyvemaskine, januar/februar 1991 og januar/marts 1992. Danmarks Miljøundersøgelser. Faglig rapport fra DMU. 1992;44:42.

Pihl S, Clausen P, Petersen IK, Nielsen RD, Laursen K, Bregnballe T, Holm TE, Søgaard S. Fugle 2004–2011. NOVANA. Aarhus Universitet, DCE – Nationalt Center for Miljø og Energi, 188 s. Videnskabelig rapport fra DCE – Nationalt Center for Miljø og Energi nr. 2013; 49. http://www.dmu.dk/Pub/SR49.pdf.

Petersen IK, Pihl S, Hounisen JP, Holm TE, Clausen P, Therkildsen O, Christensen TK. Landsdækkende optællinger af vandfugle, januar og februar 2004. Danmarks Miljøundersøgelser. 76 s. Faglig rapport fra DMU nr. 2006; 606. http://www.dmu.dk/Pub/FR606.pdf.

Petersen IK, Nielsen RD, Pihl S, Clausen P, Therkildsen O, Christensen TK, Kahlert J, Hounissen JP. Landsdækkende optælling af vandfugle i Danmark, vinteren 2007/2008. Danmarks Miljøundersøgelser, Aarhus Universitet. 70 s. Faglig rapport fra DMU nr. 2010; 785. http://www.dmu.dk/Pub/FR785.pdf.

Petersen IK, Nielsen RD. Abundance and distribution of selected waterbird species in Danish marine areas. Report commissioned by Vattenfall A/S. National Environmental Research Institute, Aarhus University, Denmark. 2011. p. 62

Petersen IK, MacKenzie ML, Rexstad E, Wisz MS, Fox AD. Comparing pre- and post-construction distributions of long-tailed ducks Clangula hyemalis in and around the Nysted offshore wind farm, Denmark: a quasi-designed experiment accounting for imperfect detection, local surface features and autocorrelation. CREEM Technical Report, no. 2011–1, University of St Andrews. 2011.

Petersen IK, Nielsen RD, Mackenzie ML. Post-construction evaluation of bird abundances and distributions in the Horns Rev 2 offshore wind farm area, 2011 and 2012. Report commissioned by DONG Energy. Aarhus University, DCE – Danish Centre for Environment and Energy. 2014. p. 51

Riemann B, Carstensen J, Dahl K, Fossing H, Hansen JW, Jakobsen HH, Josefson AB, Krause-Jensen D, Markager S, Stæhr PA, Timmermann K, Windolf J, Andersen JH. Recovery of Danish

Coastal Ecosystems after reductions in nutrient loading: a holistic ecosystem approach. Estuaries Coasts. 2016;39:82–97.

Søgaard B, Wind P, Bladt JS, Mikkelsen P, Wiberg-Larsen P, Galatius A. Teilmann J. Arter 2014. NOVANA. Aarhus Universitet, DCE – Nationalt Center for Miljø og Energi, 74 s. Videnskabelig rapport fra DCE – Nationalt Center for Miljø og Energi nr. 2015; 168. http://dce2.au.dk/pub/SR168.pdf.

Viquerat S, Herr H, Gilles A, Peschko V, Siebert U, Sveegaard S, Teilmann J. Abundance of harbour porpoises (*Phocoena phocoena*) in the western Baltic. Belt Seas Kattegat Mar Biol. 2014;161:745–54.

Zydelis R, Bellebaum J, Österblom H, Vetemaa M, Schirmeister B, Stipniece A, Dagys M, van Eerden M, Garthe S. Bycatch in gillnet fisheries – An overlooked threat to waterbird populations. Biological Conservation. 2009;142:1269–81.

Section VII

Mediterranean Basin, Middle East, and West Asia

The Camargue: Rhone River Delta (France)

Patrick Grillas

Contents

Introduction	1102
Hydrology	1102
Wetland Ecosystems	1104
Biodiversity	1106
Conservation Status	1108
Ecosystem Services	1108
Threats and Future Challenges	1109
References	1110

Abstract

The Camargue, the delta of the Rhône River, is the second largest delta in the Mediterranean. Its hydrology is strongly modified by artificial flows of water during spring and summer. Water management enhances the natural gradients of salinity and hydromorphy, factors that determine the existence of diverse ecosystems (sand dunes, *Salicornia* steppes, fresh and saline wetlands). These ecosystems perform numerous ecological functions and provide important ecosystem services. Overall, the Rhône delta is one of the most important wetlands in the Mediterranean basin in terms of its area, hydrology, and biodiversity. Although the Camargue is known under numerous nature conservation designations, it is increasingly difficult to counteract the mounting pressures from global change. A revised perspective on nature conservation is needed for the long-term protection of wetlands, its functions, and values.

P. Grillas (✉)
Tour du Valat, Research Institute for Mediterranean Wetlands, Le Sambuc, Arles, France
e-mail: grillas@tourduvalat.org

© Springer Science+Business Media B.V., part of Springer Nature 2018
C. M. Finlayson et al. (eds.), *The Wetland Book*,
https://doi.org/10.1007/978-94-007-4001-3_21

> **Keywords**
>
> Hydrology · Biodiversity · Pressures · Ecosystem services · Conservation · Management

Introduction

The Camargue is the delta of the Rhône River by the Mediterranean Sea (Fig. 1). It extends over 1,500 km² and is the second largest delta in the Mediterranean (with the Pô delta in Italy), far behind the Nile delta (22,000 km²) in Egypt. The Rhône river is 805 km long, rising in Switzerland (the Rhône Glacier), and its catchment area extends over 96,000 km².

The delta has been created mostly over the last 10,000 years (Flandrian transgression) by the sediment deposited by the river, with major extensions during the "Little Ice Age" (fourteenth to nineteenth centuries) due to the heavy loads of sediments resulting from both the cold wet climatic conditions and the high level of deforestation of the catchment areas. Since the middle of the nineteenth century, the Rhône River has been dramatically transformed. Today, the load of coarse sediment in the river has been strongly reduced through complete embankment and channelization, gravel extraction (300,000 m³ in the last 30 years), the construction of more than 120 reservoirs in the Northern Alps catchment areas, and 18 run-of-the-river hydropower plants along the Rhone itself. The present regression of the coastline in the delta (about 4 m/year regression) results from the interaction between the reduction of the sediment load and marine processes (Maillet et al. 2006) enhanced by a rise in the sea level (2 mm/year). It is, however, heterogeneous in space with a few areas in accretion and some with fast regression. The mouth of the Rhône River is now divided into two branches, which divides the delta into three parts (from west to east: Petite Camargue, Grande Camargue, and Plan du Bourg) with disconnected hydrology.

The delta has been occupied by human populations at least since the Copper and Bronze Ages. The population is today about 40,000 inhabitants, with lower density in the central part of the delta (Grande Camargue: 10 inhabitants/km²). The delta is used for agriculture, salt production, tourism, hunting, fishing, industry, etc. and wetland conservation. The development of these activities since the middle of the twentieth century has led to major changes in land use and the loss of natural habitats converted into industrial and agricultural land (Tamisier 1990).

Hydrology

With a Mediterranean climate, the rainfall (600 mm/year with wide interannual variations) is mainly concentrated from autumn to early spring, with a very dry summer. For about 150 years, most of the Rhône delta has been polderized and disconnected by embankments from natural flows of water from the river and from the sea. Therefore, the hydrology of the delta relies to a great extent on artificial

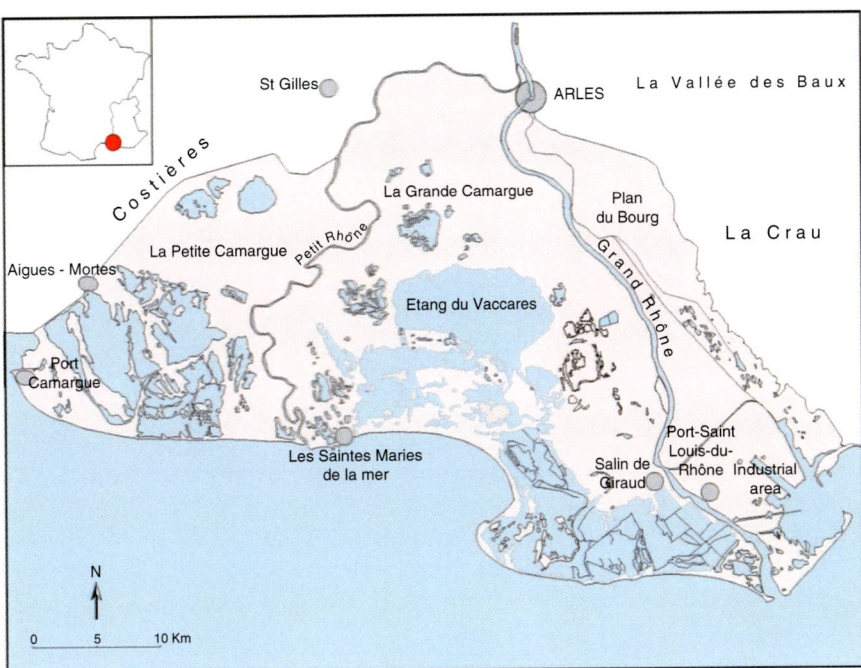

Fig. 1 The Camargue, Rhône delta, Southern France (© Tour du Valat, with permission) (Modified from Sandoz 1996)

flows through pumping. About 250–400 Mm³ per year of freshwater are introduced into the central part of the delta (Grande Camargue) by more than 150 pumping stations (mainly for rice production) and 360 Mm³ by rainfall (Chauvelon 2013). Similarly, 125 Mm³ sea water are pumped for salt production and about 9 Mm³ enter the lagoon system naturally. Most of the pumping is done during the dry season (April to September) which has a strong impact on natural ecosystems, enhancing water levels in lagoons and preventing salinity from rising in summer. Evapotranspiration exports to the atmosphere more than 500 Mm³ of the freshwater that enters the delta, most of the remaining part (180 Mm³) is pumped back into the Rhône River and only 22 Mm³ are exported to the sea through the lagoons. There are three channels between the lagoon complex and the sea but only one, near Saintes-Maries-de-la-Mer, is really efficient for water exchanges.

Many pollutants are found in the water in the delta (Roche et al. 2000, 2003), those produced by industry and intensive agriculture in the Rhône valley (e.g., PCBs and PAHs) accounting for only 8% of the total occurrence in water analyses. The current trend with regard to water quality in the river is a decrease in the concentration of heavy metals, nitrogen, and phosphorus and conversely an increase in the concentrations of pesticides and more particularly herbicides. Most of the pesticides found in the delta are however of local origin, rice cultivation alone accounting for 55% of the total occurrence (Comoretto et al. 2007, 2008).

Wetland Ecosystems

The ecosystems in the Camargue are distributed along a double gradient of hydrology and salinity. Low salinity and short flooding periods are mostly found in the northern part of the delta and, conversely, high salinity and long flooding periods in its southern tip. However, geomorphological features (former beds of the river, levees) account for local discrepancies in this general pattern and water management tends both to exacerbate the hydrological and salinity gradients and disrupt hydrological cycles with large inputs of water in summer for rice production in the north and salt production in the south. Brackish ecosystems are often restricted to protected areas and especially the central lagoons. In this double gradient, a wide diversity of ecosystems can be observed:

Coastal dunes cover extensive areas in the south of the delta. The dunes are low in elevation, exposed to strong winds and are under pressure from sea level rises and shortage of sediment. They are very rich in species with different communities from the upper beach, white dunes (*Ammophila arenaria, Pancratium maritimum, Euphorbia peplis*, etc.), dune slacks (*Scirpus* spp., *Linum maritimum, Juncus* spp., *Limonium* spp., etc.), and grey dunes (*Pinus* spp., *Juniperus turbinata, Vulpia* spp., etc.).

Mediterranean coastal lagoons: They occur along the whole of the coastline at the southern end of the delta covering more than 20 km^2. These lagoons are brackish with salinity fluctuating through the seasons and in between years in phase with the fluctuation of rainfall and the extent of rice cultivation, with a strong impact on their plant and fish communities (Poizat et al. 2004; Charpentier et al. 2005). Although mostly flooded with permanent water, the shallowest lagoons can dry up in late summer. The vegetation of permanent brackish lagoons is dominated by *Zostera noltei, Ruppia cirrhosa, Chaetomorpha linum*. In temporarily flooded brackish lagoons, the vegetation is dominated by *Ruppia maritima* and *Althenia filiformis* with some rare species such as *Lamprothamnium papulosum* and *Tolypella salina*. The lagoons host specialized fish species (e.g., *Atherina boyeri*) with amphihaline (e.g., *Anguilla anguilla*), marine (*Sparus auratus, Dicentrarchus labrax*, etc.), and/or freshwater species (e.g., *Cyprinus carpio, Carassius carassius*) (Poizat et al. 2004). These lagoons are important habitat for waterbirds in every season, e.g., wintering for thousands of ducks, feeding habitat for crowds of migrating waders and breeding habitats for terns, gulls, greater flamingo *Phoenicopterus roseus*, etc. (Tamisier and Isenmann 2004).

Many of the lagoons (22,000 ha) have been transformed into salt pans at the SE and SW extremities of the delta. The salinity gradient ranges from 37 PSU to crystallization of salts (360 PSU). These transformed lagoons are dried up in winter and flooded from spring to autumn when evapotranspiration is highest. At intermediate salinity, the salt pans contain high concentrations of invertebrates, especially *Artemia spp*. Because of the summer flooding, the presence of islands and high food biomass, these lagoons are intensively used by waterbirds for reproduction and feeding (Britton and Johnson 1987; Sadoul et al. 1998; Béchet and Johnson 2008).

Temporarily flooded marshes occupy a wide range of depressions from former shallow lagoons to former river beds. Their salinity varies naturally from brackish to almost fresh according to the geomorphological origin of the depression where they stand (river bed/lagoons) and to the distance from the sea. They usually flood in autumn with heavy rainfall and remain flooded until next spring or early summer. When ungrazed, these marshes are dominated by helophytes, mainly *Phragmites australis*, *Bolboschoenus maritimus*, and fringed by *Juncus maritimus* belts. During the flooding period (autumn to late spring) they are occupied by dense and diverse beds of submerged macrophytes, e.g., *Ranunculus*, *Zannichellia*, *Chara*, *Tolypella*, *Callitriche*, and *Potamogeton* (Grillas 1990). These marshes contain rich communities of zooplankton and macro-crustaceans (Waterkeyn et al. 2008). They are often grazed by local breeds of cattle and horses and intensively used by wintering waterfowl for feeding and roosting.

Many of the temporarily flooded marshes have been transformed for hunting purposes through large inputs of freshwater. This management results in highly productive permanently flooded freshwater marshes (or with a short dry period often in spring) dominated by entangled patches of reedbeds (*P. australis*) and dense beds of submerged macrophytes, e.g., *Potamogeton pectinatus*, *P. pusillus*, *Myriophyllum spicatum*, and *Najas marina* (Tamisier and Grillas 1994; Aznar et al. 2003). They are often colonized by *Ludwigia* spp., two invasive exotic species. They are very rich in fish (including many exotic species), amphibians, and invertebrates and are intensively used by waterfowl as breeding and feeding habitats, especially ardeids and ducks (Tamisier and Isenmann 2004).

Salt marshes occupy large areas mostly in the south of the delta on saline soils often rich in clay and silt. Located at the interface between aquatic and terrestrial habitats, salt marshes are often flooded in winter by shallow water. They are dominated by different species according to the duration of flooding and the salinity, from the most flooded towards the driest habitats: *Salicornia gr. europaea*, *Salicornia perennis*, *Arthrocnemum macrostachyum*, *Salicornia fruticosa*. Ephemeral non-halophyte species grow in early spring (e.g., *Bellis annua*, *Hutchinsia procumbens*, *Myosurus minimus*, and *Cardamine hirsuta*) at the highest elevations in salt marshes benefiting from the reduced salinity of the topsoil after winter rains. These salt marshes are used by wintering waterfowl for feeding and breeding habitats for waders (e.g., common redshank *Tringa totanus*) and a few species of passerine (spectacled warbler *Sylvia conspicillata*, yellow wagtail *Motacilla flava*).

Grasslands occupy the highest topographic location and are rarely flooded by intense rainfall events and their structure and species composition varies with elevation and two related factors: frequency of flooding and salinity. At lower topographical level, they are occupied by very diverse plant communities dominated by annual species (*Trifolium* spp., *Bellis annua*, *Crepis* spp.,

Medicago spp., *Bromus* spp., *Allium chamaemoly*, *Iris spuria*, *Limonium* spp., etc.). At higher locations, these grasslands are less diverse and are dominated by perennial species (e.g., *Brachypodium sylvaticum*) and are locally colonized by shrubs (*Phillyrea angustifolia*). These grasslands are very rich in invertebrates (e.g., grasshoppers), lizards, and snakes and are intensively used by domestic grazers which contribute to maintaining the diversity of the plant communities.

Riparian forests are restricted to small patches along the Rhône River. Most of their potential habitats have been transformed for agriculture and the remaining patches have been intensively used for wood until the mid-twentieth century. Degraded communities of the riparian forest also exist in small patches along the hundreds of kilometers of irrigation and drainage canals in the delta. These forests are dominated by *Populus alba*, *Fraxinus angustifolia*, and *Ulmus minor* with a rich understorey with *Quercus* spp., *Ficus carica*, *Viburnum tinus*, *Crataegus monogyna*, etc. The riparian forests are important habitats for mammals (European badger *Meles meles*, red fox *Vulpes vulpes*, European pine marten *Martes martes*, red squirrel *Sciurus vulgaris*), etc.) and birds (black kite *Milvus migrans*, woodpeckers, etc.).

Freshwater lagoons are found in the north-west of the delta with three main water bodies covering about 12 km^2. Receiving drainage water form agriculture (rice), these lagoons are eutrophic and are fringed by large reedbeds. The reedbeds are very important breeding habitats for the eight European species of heron and ibis which make their nest in the reeds or on the shrubs (*Tamarix gallica*) within the reedbed. The reed is harvested in winter. The center of the lagoons is occupied by sparse beds of *Potamogeton* spp. and *Myriophyllum spicatum* in Scamandre and Charnier lagoons and by dense beds of charophytes in Crey lagoon. These lagoons are used by large numbers of wintering ducks and breeding herons (Tamisier and Isenmann 2004) and are used for hunting wildfowl.

Biodiversity

The Camargue is very rich in species, notably waterbirds, which represent the most emblematic group of the delta with one of the most important breeding colonies of greater flamingo in the Mediterranean basin, large numbers of wintering ducks, migrating waders, etc. The Camargue is one of the largest wetlands in the Mediterranean basin, located directly on one of the two main migration routes used by millions of birds. The biodiversity of the Camargue includes more than birds, however, and is highly diversified because of the diversity and extent of the habitats. Within the Camargue are found about a thousand of species of plants ranging from marine and lagoon species (e.g., *Zostera noltei*, *Lamprothamnium papulosum*) to terrestrial species on dunes, grasslands, etc. and from halophytes (e.g., *Salicornia* spp.) to non-salt tolerant species (e.g., *Nuphar luteum*). A large number of exotic species are to be found, especially in agricultural areas, including a few invasive wetland species (*Ludwigia* spp., *Baccharis halimifolia*, *Amorpha fruticosa*, *Heteranthera* spp. notably), 75 species of fish (including 13 exotic species) (Rosecchi et al. 1997), 43 species of mammals including 5 exotic species (Poitevin et al. 2010), 15 species of reptiles (1 exotic: *Trachemys scripta subsp elegans*), and 10 amphibians. Among the invertebrates, Odonata and crustaceans are abundant, with respectively 43 and more than 150 species.

Biodiversity is more than a matter of the number of species, however. For centuries, the biodiversity has been constantly changing as a result of the profound changes that the delta has undergone. Three species of mammals have disappeared in the previous centuries because of hunting pressure (red deer *Cervus elaphus* in the fifteenth Century, wolf *Canis lupus* in the nineteenth century, and Eurasian otter *Lutra lutra,* which is currently recolonizing habitats in Southern France and has been found in the Camargue on a few occasions). A few species of mammals have recently appeared (e.g., common genet *Genetta genetta*) while others are declining, e.g., rabbit *Oryctolagus cuniculus*, garden dormouse *Eliomys quercinus*. The rabbit, which until recently was proliferating in the dry parts of the delta, is now on the verge of extinction in the Camargue because of the successive introduction of two diseases, myxomatosis in the 1970s and viral hemorrhagic disease caused by a calicivirus introduced into France in 1986. Similarly, the populations of European eel *Anguilla anguilla* have collapsed in the last decades (Bevacqua et al. 2007), which could partly be explained by the introduction of an exotic parasite, *Anguillicola crassus* (Lefebvre et al. 2003). Bird populations are also experiencing changes with new breeding species appearing in the delta in the last decades such as the cattle egret *Bubulcus ibis* in the 1970s, and the purple swamphen *Porphyrio porphyrio*. In the last 40 years (1970–2010), the populations of vertebrates have undergone contrasting changes (Galewski et al. 2011). Populations of waterbirds have undergone strong increases in size while the populations of other groups of vertebrates have decreased by 40%. The increase in the populations of waterbirds may be explained by a combination of several factors, with impacts at different scales: the protection of species in Europe, the increase of protected areas in the Camargue, the management of water bodies with consequent increased duration of flooding, decreased salinity and eutrophication, and the addition to the local fauna of exotic species which are intensively used by herons (fish species, and Louisiana crayfish *Procambarus clarkii* notably) (Poulin et al. 2007).

Although there are no consolidated data, it is likely that while the biomass of their vegetation has increased, the biodiversity of the wetlands has been declining as a result of eutrophication and their increasing homogeneity. Intensification of management for waterbirds (mostly for hunting) through the construction of embankments and large inputs of freshwater into the wetlands has led to changes in the abundance and species composition of the aquatic vegetation (Tamisier and Grillas 1994; Aznar et al. 2003). It is thought that the populations of invertebrates have also been strongly altered by the changes in hydrology and salinity and by the increased connectivity with permanent water bodies leading to the introduction of fishes (Pont et al. 1991; Waterkeyn et al. 2009).

Conservation Status

There are numerous protected areas, and 17 conservation designations are recognized in the Camargue. The strongest conservation status is held by the four nature reserves, the first being founded in 1927 (the first in France):

- Camargue National Nature reserve (13,117 ha), established in 1927
- Tour du Valat Volunteer Nature reserve (1,070 ha), established in 1984; it was enlarged to 1,844 ha and upgraded to the status of Tour du Valat Regional Nature reserve in 2008
- Scamandre Regional Nature Reserve (146 ha), established in 1994
- Vigueirat National Nature reserve (919 ha), established in 2011

A regional park was created in 1970, now covering 1,350 km^2 in the center of the delta. The Camargue is covered by two Ramsar Sites, the first created in 1986 within the boundaries of the Natural Regional Park, the second (37,000 ha) was created in 1996; it is contiguous to the former Ramsar Site on the west side and extends beyond the delta. A Man & Biosphere Reserve was created in 1977 and extends today over 1,930 km^2, covering the whole geomorphological delta. The Conservatoire du Littoral, a public land holding body aiming at preserving coastal ecosystems by purchasing them, owns numerous estates for a total surface area of more than 20,000 ha, including 14,000 ha in nature reserves. There are, in addition, several Natura 2000 sites, a European Union status aiming at the preservation of selected habitats and species, covering most of the delta.

This diversity of labels and the complexity of legal statuses are difficult to understand for the layman and do not guarantee a strong conservation status outside the nature reserves. Therefore, despite the numerous existing conservation statuses and labels, although the strong decline in the areas of wetlands in the 1960s–1980s is now past, the loss of natural habitat in the Camargue is still continuing at an estimated rate of 0.1%/year (Perennou 2009).

Ecosystem Services

The Camargue performs a number of important ecological functions including habitat provision, the maintenance of biodiversity, water purification, and nutrient cycling (Isenmann 2004). On the basis of these functions, the Camargue provides important ecosystem services (Mathevet 2000; EEA 2010).

The high primary productivity of the area supports provisioning services including agricultural production (especially rice), natural habitats support extensive grazing of domestic herbivores (mostly cattle), and the freshwater marshes support hunting, fishing, and reed harvesting. Commercial hunting has become a key activity contributing towards the maintenance of other land uses (Mathevet and Mesléard 2002). In addition, saline lagoons support intensive salt production in the southern part of the delta. Camargue wetlands provide regulating services including water purification, flood extension, and by the evapotranspiration of water make a very significant contribution to the water balance of the delta, decreasing the amount of water pumped from the delta for drainage.

Cultural services are also important, supported by the well-known image of the Camargue in terms of the natural environment and picturesque traditional activities. Widespread husbandry of Camargue horses and cattle and associated bull-games, traditional fishing and reed-cutting, and religious events in Saintes-Maries-de la Mer attract numerous visitors.

Threats and Future Challenges

Although the Camargue benefits from many labels and types of protection status (see above), it is facing numerous threats and the pressures are increasing. The delta is particularly exposed to sea-level rise (2 mm/year) because it is disconnected from its tributaries and the load of coarse sediment transported by the river has drastically declined. Furthermore, subsidence resulting from the compaction of the recently deposited sediment (last few thousand years) increases the need for soil accretion to compensate for the sea level rise and climate change, with more frequent storms contributing to coast erosion. These combined processes are already currently transforming coastal ecosystems, increasing their hydromorphy and salinity and threatening human activities. The natural ecosystems and the biodiversity are also threatened by intensive agriculture introducing large amounts of pesticides (including illegal types) added to those transported by the Rhône River and which ultimately find their way into the protected areas (Roche et al. 2000, 2003; Comoretto et al. 2007, 2008). In addition, the intensification of agriculture affects the landscape by decreasing interstitial habitats between parcels with consequent homogenization. Conversely, the abandonment (or conversely the intensification) of some extensive uses of the wetland such as grazing could lead to biodiversity losses.

The industrial area fringing the delta in the southeast, in addition to threatening rare habitats and species by its development, emits large amounts of pollutants through the industrial activity and truck traffic (heavy metals, HAPs) which are conveyed by the frequent SE winds directly onto the wetlands.

The large amounts of freshwater introduced into the wetlands have contributed to their eutrophication, to a decrease in their salinity, and to favor the development of exotic invasive species of plants (e.g., *Ludwigia* spp., *Baccharis halimifolia*) and animals including many fish (e.g., *Siluris glanis*, *Gambusia affinis*, etc.), crustaceans (*Procambarus clarkii*), birds (African sacred ibis *Threskiornis aethiopicus*), and mammals (coypu *Myocastor coypu*).

The nature conservation objective for the delta in the large scale planning in the 1960s is being progressively weakened by the increasing pressure from the neighboring industrial (east) and touristic (west) poles of development.

Overall, the Rhône delta is one of the most important wetlands in the Mediterranean basin in terms of its area, hydrology, and biodiversity. It has been much transformed over the centuries and more intensively in the last two centuries with almost complete embankment and intensification of human activities. In this process, the biodiversity has remained outstanding, constantly changing in phase with land use and climate changes. Sound management is needed for the conservation of the wetland, its functions, and values, now threatened by global change.

References

Aznar JC, Dervieux A, Grillas P. Association between aquatic vegetation and landscape indicators of human pressure. Wetlands. 2003;23(1):149–60.

Béchet A, Johnson AR. Anthropogenic and environmental determinants of Greater Flamingo *Phoenicopterus roseus* breeding numbers and productivity in the Camargue (Rhone delta, southern France). Ibis. 2008;150:69–79.

Bevacqua D, Melia P, Crivelli AJ, Gatto M, De Leo D. Multi-objective assessment of conservation measures for the European eel (*Anguilla anguilla*): an application to the Camargue lagoons. ICES J Mar Sci. 2007;64:1483–90.

Britton RH, Johnson AR. An ecological account of a Mediterranean salina: the Salin de Giraud, Camargue (S. France). Biol Conserv. 1987;42:185–230.

Charpentier A, Grillas P, Lescuyer F, Coulet E, Auby I. Spatio-temporal dynamics of a *Zostera noltii* community over a period a fluctuating salinity in a shallow coastal lagoon, southern France. Estuar Coast Shelf Sci. 2005;64:307–15.

Chauvelon P. La gestion des eaux dans l'ile de Camargue. In: Blondel J, Vianet R, Barruol G, editors. L'encyclopédie de la Camargue: nature et culture du delta du Rhône. Paris: Buchet Chastel Ecologie Editions; 2013. p. 39–41.

Comoretto L, Arfib B, Chiron S, Höhener P. Pesticides in the Rhône river delta (France): basic data for a field-based exposure assessment. Sci Total Environ. 2007;380:124–32.

Comoretto L, Arfib B, Talva R, Chauvelon P, Pichaud M, Chiron S, Höhener P. Runoff of pesticides from rice fields in the Ile de Camargue Rhône river delta, France: field study and modeling. Environ Pollut. 2008;151:486–93.

European Environment Agency. Ecosystem accounting and the cost of biodiversity losses – The case of coastal Mediterranean wetlands. ISSN Technical report series; 2010. p. 1725–2237. doi:10.2800/39860.

Galewski T, Collen T, McRae B, Loh J, Grillas P, Gauthier-Clerc M, Devictor V. Long-term trends in the abundance of Mediterranean wetland vertebrates: from global recovery to localized declines. Biol Conserv. 2011;144:1392–9.

Grillas P. Distribution of submerged macrophytes in the Camargue in relation to environmental factors. J Veg Sci. 1990;1:393–402.

Isenmann P. Les oiseaux de Camargue et leurs habitats. Une histoire de cinquante ans 1954–2004. Paris: Buchet– Chastel Ecologie Editions; 2004.

Lefebvre F, Acou A, Poizat G, Crivelli AJ. Anguillicolosis among silver eels: a 2-year survey in 4 habitats from Camargue (Rhône delta, South of France). Bull Fr Pêche Piscic. 2003;368:97–108.

Maillet GM, Vella C, Provansal M, Sabatier F. Connexions entre le Rhône et son delta (partie 2): évolution du trait de côte du delta du Rhône depuis le début du XVIIIe siècle. Géomorphologie: relief, processus, environnement [on line], 2/2006|2006, online 01 juillet 2008, consulted on 01 octobre 2012. doi:10.4000/geomorphologie.559. http://geomorphologie.revues.org/559

Mathevet R. Usages des zones humides camarguaises: Enjeux et dynamique des interactions Environnement/Usagers/Territoires [Unpublished PhD dissertation]. Lyon: University Jean Moulin; 2000.

Mathevet R, Mesléard F. The origins and functioning of the private wildfowling lease system in a major Mediterranean wetland: the Camargue (Rhone delta, southern France). Land Use Policy. 2002;19(4):277-86.

Perennou C. La Camargue au fil du temps – Evolution récentes et perspectives. Arles: Tour du Valat Edition; 2009.

Poitevin F, Olivier A, Bayle P, Scher O. Mammiféres de camargue. In: Regard du Vivant & Parc Naturel Régional de Camargue Editions. France: Castelnau-le-Lez & Arles; 2010. p. 232.

Poizat G, Rosecchi E, Chauvelon P, Contournet P, Crivelli AJ. Long-term fish and macro-crustacean community variation in a Mediterranean lagoon. Estuar Coast Shelf Sci. 2004;59(4):615-24.

Pont D, Crivelli AJ, Guillot F. The impact of three-spined sticklebacks on the zooplankton of a previously fish-free pool. Freshw Biol. 1991;26:149-63.

Poulin B, Lefebvre G, Crivelli AJ. The invasive red swamp crayfish as a predictor of Eurasian bittern density in the Camargue. France J Zool. 2007;273(1):98-105. doi:10.1111/j.1469-7998.2007.00304.

Roche H, Buet A, Jonot O, Ramade F. Organochlorine residues in European eel (Anguilla anguilla), crusian carp (Carassius carassius) and catfish (Ictalurus nebolosus) from Vaccarès lagoon (French National Reserve of Camargue) – effects on some physiological parameters. Aquat Toxicol. 2000;48:443-59.

Roche H, Buet A, Tidou A, Ramade F. Contamination du peuplement de poissons d'un étang de la Réserve Naturelle Nationale de Camargue, le Vaccarès, par des polluants organiques persistants: Report of the National Reserve of Camargue. [Contamination by persistent organic pollutants of the fish community of the Vaccarès lake, French National Reserve of Camargue]. Revue d'Ecologie. 2003; 58(1):77-102.

Rosecchi E, Poizat G, Crivelli AJ. Introduction de poissons d'eau douce et d'écrevisses en Camargue. Historique, origine et modifications des peuplements. Bull Fr Pêche Piscic. 1997;344/345:221-32.

Sadoul N, Walmsley JG, Charpentier B. Salinas and nature conservation. In: Crivelli AJ, Jalbert J, editors. Conservation of Mediterranean wetlands No. 9. Arles: Station Biologique de la Tour du Valat; 1998.

Sandoz A. Proposition d'une méthodologie adaptée au suivi de l'occupation du sol d'une zone humide aménagée: application au bassin du Fumemorte (Grande Camargue, France). Thèse doctorat: sciences géographiques et de l'aménagement, Université d'Aix-Marseille I; 1996. 167 p.

Tamisier A. Camargue. Milieux et paysages, évolution de 1942 à 1984. Arles: Editions de l'Association ARCANE; 1990.

Tamisier A, Grillas P. A review of habitat changes in the Camargue. An assessment of the effects of the loss of biological diversity on the wintering waterfowl community. Biol Conserv. 1994;70:39-47.

Tamisier A, Isenmann P. Milieux et paysages de Camargue. Description, statuts fonciers, et statuts de protection, changements quantitatifs et qualitatifs, pollution. In: Isenmann P, editor. Les oiseaux de Camargue et leurs habitats. Paris: Collection Ecologie, Buchet/Chastel Editions; 2004.

Waterkeyn A, Grillas P, Vanschoenwinkel B, Brendonck L. Invertebrate community patterns in Mediterranean temporary wetlands along hydroperiod and salinity gradients. Freshw Biol. 2008;53:1808-22.

Waterkeyn A, Grillas P, De Roeck EM, Boven L, Brendonck L. Assemblage structure and dynamics of large branchiopods in Mediterranean temporary wetlands: patterns and processes. Freshw Biol. 2009;54(6):1256-70.

Ebro Delta (Spain)

89

Carles Ibáñez and Nuno Caiola

Contents

Introduction	1114
Wetland Types and Biodiversity	1115
Wetland Use and Ecosystem Services	1117
Conservation Status and Management	1118
Threats and Future Challenges	1119
References	1120

Abstract

The Ebro Delta is located in the Western Mediterranean (Catalonia, NE Spain), and the delta plain has an area of 320 km^2. Up to 80% of the delta area has been reclaimed (250 km^2), mostly for rice agriculture (210 km^2), and there is only 56 km^2 of wetlands left. At the level of species the delta stands out for its ornithological and ichthyological fauna, as well as for its halophilous vegetation. A significant amount of these habitats and species are very scarce in the European and Mediterranean context. During the last century agriculture became the main human activity of the Delta and nowadays rice fields play a crucial role in its economy and its ecology. The ecological functioning of the Ebro Delta at present is largely dependent on and affected by human activities because of modification of the natural hydrological regime. There have been changes in the temporal and spatial patterns of water salinity, eutrophication, pesticide pollution, and changes in the spatial and temporal patterns of sediment transport and deposition in the deltaic plain. The reduction of sediment transport of the Ebro River is about 99% of that existing prior to the construction of reservoirs in the catchment basin. Under these conditions, the delta has stopped its growth and the coast is being

---

C. Ibáñez (✉) · N. Caiola (✉)
IRTA, Aquatic Ecosystems Program, Sant Carles de la Ràpita, Catalonia, Spain
e-mail: carles.ibanez@irta.cat; nuno.caiola@irta.cat

© Springer Science+Business Media B.V., part of Springer Nature 2018
C. M. Finlayson et al. (eds.), *The Wetland Book*,
https://doi.org/10.1007/978-94-007-4001-3_145

strongly reshaped by waves, with rates of retreat as high as 10 m/year in the mouth area. Additionally, the sediment deficit and the relative sea level rise imply a loss of land elevation of the delta. The main future challenges in terms of conservation and sustainable management of wetlands in the Ebro Delta are:

- A better coordination and stronger involvement of the different administrations and stakeholders in order to plan long-term conservation goals integrating the whole delta.
- A strong program of wetland restoration in order to recover part of the coastal lagoons and marshes which were lost in the past.
- A plan for adaptation to climate change and sediment deficit, in order to reduce coastal retreat and keep pace with sea level rise by means of ecological engineering measures such as wetland restoration and enhancement of vertical accretion through the restoration of the sediment flux in the river.
- A comprehensive monitoring program measuring a set of indicators of ecological status of the main wetland habitats.

Keywords
Rice fields · Wetland loss · Sediment deficit · Sea level rise · Salinity

Introduction

The Ebro Delta is located in the Western Mediterranean (Catalonia, NE Spain), being its seaward end at 40° 43′ N of latitude and 0° 53′ E of longitude. The delta plain has an area of 320 km^2, and the two existing bays account for another 68 km^2 (Fig. 1). Up to 80% of the delta area has been reclaimed (250 km^2), mostly for rice agriculture (210 km^2), and there is only 56 km^2 of wetlands left, as well as 14 km^2 of lagoons. The delta started to undergo important human modifications by the mid-nineteenth century, with the construction of irrigation canals and the transformation of wetlands to rice fields. Other major modifications affecting the delta during the twentieth century (mostly the second half) have been the reduction and regularization of the river flow due to irrigation and damming and an increase in organic and chemical pollution. As a consequence of the nearly complete retention of sediment discharge due to dam construction, the growth at the present river mouth has stopped and the coast is retreating in many areas (Rovira and Ibáñez 2007).

The Ebro River is 928 km long and the drainage basin has a surface of 85,550 km^2, the largest in the Iberian Peninsula. The mean annual flow of the Ebro River near the mouth was 592 m^{3}s^{-1} at the beginning of the twentieth century (1914–1935); however, a continuous decreasing tendency has led to a mean value of 270 m^{3}s^{-1} (1990–2000) due to the intensification of water uses (Ibáñez and Prat 2003). Climate is continental in most of the basin, except in the Pyrenees (mountain climate) and in the coast (Mediterranean). In the delta, annual rainfall is close to 500 mm and the mean annual air temperature is around 17 °C. Tidal regime in the

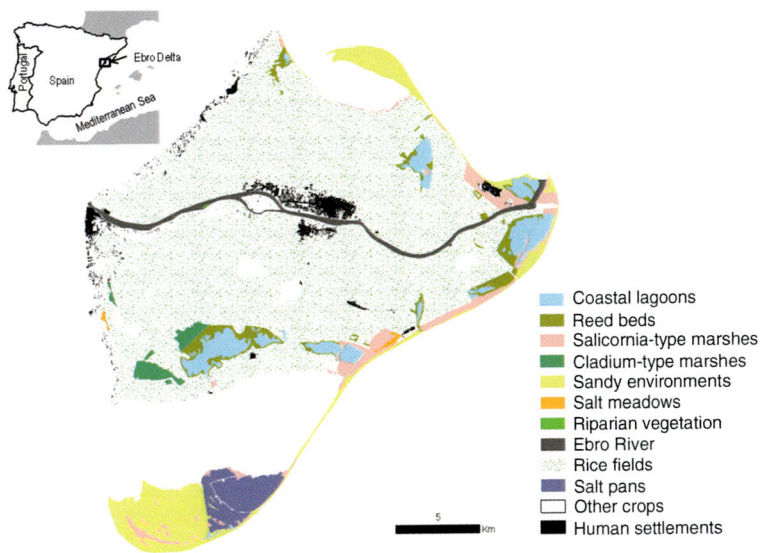

Fig. 1 Map showing the present land uses and main wetland habitats in the Ebro Delta (© IRTA, with permission)

delta is semi-diurnal and is characterized by very weak tides; mean and maximum tidal ranges are 16 and 25 cm, respectively.

Wetland Types and Biodiversity

The Ebro Delta is remarkable for its great diversity of habitats and species in a relatively small surface (33,000 ha). At the level of species the delta stands out for its ornithological and ichthyological fauna, as well as for its halophilous vegetation. A significant amount of these habitats and species are very scarce in the European and Mediterranean context. Rice fields are nowadays the sustaining bases of most of the biodiversity of this important wetland area. A total of 330 species of birds have been observed in the delta, from which 81 species breed regularly and 28 occasionally. Among breeding species 50 are aquatic, with approximately 50,000 breeding pairs and a mean population approaching 200,000 birds in January. The Ebro Delta has international importance for breeding of at least 24 species and for migration and wintering of 13 species, and occasionally for 14 more (SEO/BirdLife 1997). The delta is particularly important for aquatic birds and especially for its nesting populations of gulls and terns. Outstanding is the world's biggest colony of ca. 10,000 pairs of Audouin's Gull, a threatened species. The delta is also a very important resting, molting, and feeding area for thousands of migratory birds. In autumn, up to 20,000 herons of different species have been seen, while in the winter peak counts, 100,000 ducks, 20,000 coots, and over 30,000 waders, are present (Martínez-Vilalta 1996). The fish fauna is also very rich, with 62 fish species (45 marine/brackish and 17 freshwater

Table 1 Brief description of the main habitats of the Ebro Delta (urban areas and crops excluded) indicating the present surface and the predicted surface under natural conditions according to a habitat distribution model (From Benito et al. 2014: Table 1; Wetlands, Modelling habitat distribution of Mediterranean coastal wetands: the Ebro Delta as case study, 34, 2014, 775–85, Benito X, Trobajo R, Ibañez C; © 2014, Society of Wetland Scientists, with permission of Springer Science + Business Media)

Habitat	Description	Present area (km^2)	Predicted area (km^2)	% change
Coastal lagoons	Shallow coastal water bodies with submerged macrophytes. Salinity may vary from brackish to marine	14.08	60.97	−76.9
Reed beds	Brackish marshes dominated by *Phragmites australis*	8.62	7.59	+13.5
Salicornia-type marshes	Salt marshes with succulent shrubby species dominated by Chenopodiaceae	7.89	14.18	−44.4
Cladium-type marshes	Fresh marshes dominated by *Cladium mariscus*, affected by significant inputs of underground freshwater	3.35	32.65	−89.7
Sandy environments	Sandy beaches and dunes with or without halophilous vegetation	3.85	17.37	−77.8
Salt meadows	Meadows dominated by rushes (e.g., *Juncus maritimus* and *J. acutus*) in salty soils with occasional flooding	0.52	125.41	−99.6
Riparian vegetation	Forests along river levees composed by trees adapted to flooding events	0.02	32.29	−99.1

species) observed in the delta plain, i.e. excluding strictly marine systems. It is remarkable the presence of two endemic and endangered species of the Western Mediterranean coast or the Iberian Peninsula, the Spanish toothcarp (*Aphanius iberus*) and the Valencia toothcarp (*Valencia hispanica*). In negative terms is also remarkable the introduction of 15 exotic species and 4 translocated (native species from the Ebro basin that should not be in the delta). A total of 16 of these nonnative species were introduced during the last four decades.

Mediterranean wetland habitats were originally dominated by salt marshes (Ibañez et al. 2000), but most of them have been reclaimed for farmland, mostly to rice fields in the Ebro Delta case. However, the delta still keeps a high variety of wetland habitats whose features change according to the gradient of salinity and elevation (Table 1). Salt marshes are dominated by nonsucculent xerophilous species, which can excrete salts through salt glands, such as *Limonium* spp. or *Limoniastrum* spp. Seasonally flooded salt marshes are very diverse and they show a wide range of water and salinity regimes. In the driest areas, soil salinity can reach values so extreme that plants cannot survive and they are colonized by algal and microbial mats. When the flooding period does not include the entire summer, annual species of glasswort (*Salicornia* spp.) form dense stands. However, the most typical and widespread salt marshes are dominated by succulent shrubby species, mostly belonging to the Chenopodiaceae family (*Sarcocornia*,

Arthrocnemum, Suaeda). In the upper salt marsh, *Arthrocnemum macrostachyum* forms sparse communities, sometimes with winter annuals; this community can tolerate strong variations of water level and soil salinity. In the middle marsh, *Sarcocornia fruticosa* is dominant and forms taller shrublands with a more dense cover. Finally, the low marsh is often dominated by *Sarcocornia perennis*, a prostrate shrub much rarer than the two other chenopod species. It grows in areas at mean sea level, where water and salinity conditions are quite stable during the year.

Brackish marshes are also widespread, especially in areas influenced by freshwater inputs from the rice drainage and from the river. Temporal ponds are dominated by small submerged macrophytes (such as *Zannichellia* spp. and *Ranunculus* spp.), while permanent aquatic systems (coastal lagoons) usually are dominated by fennel pondweed (*Potamogeton pectinatus*) and, in more saline waters, by *Ruppia* spp. Presently, many coastal lagoons receive high inputs of fresh water from agricultural runoff and other human activities, so several former saline coastal lagoons have become brackish, and plant communities have completely changed (Prado et al. 2013). Reed beds are widespread around coastal lagoons. They are dominated by *Phagmites australis*, though the sea club-rush (*Scirpus maritimus*) is sometimes abundant. Brackish rush communities are found around reed beds in slightly higher elevations. Several species of halophylic rushes, like *Juncus maritimus* and *J. acutus*, and grasses (*Aeluropus littoralis* or *Paspalum vaginatum*) are dominant in these communities.

Freshwater marshes are scarce and are normally associated with underground freshwater springs from karstic zones. Permanent freshwater habitats (springs, old river channels, drainage channels) contain submerged and floating communities composed by several macrophytes, such as several species of pondweed (*Potamogeton* spp., *Myriophyllum spicatum*) and water lily (*Nymphaea alba*). In marshy environments, emerged vegetation forms a belt dominated by cattails (*Typha* spp.) and the common reed (*P. australis*). In peatland areas saw sedge (*Cladium mariscus*) is dominant. Rice fields are shallow, temporary, highly productive artificial wetlands. The vegetation can be quite complex, depending on the agricultural management, and is dominated by annual species with a pantropical distribution (*Bergia, Ammannia, Lindernia, Echinochloa*, etc.).

Wetland Use and Ecosystem Services

The economy of the Ebro Delta is essentially relying on the ecosystems services provided by the wetland habitats. The main human activities before rice cultivation were salt production, glasswort harvest for soap production, artisanal fishing and hunting, grazing, and some farming along the river levees and the inner border. The oldest economic activities recorded in the historical documents are salt production and fishing in the coastal lagoons, dated in the year 1149. All these activities depended from the exploitation of the natural resources of the Delta, and ultimately on the fluxes and pulses of water, nutrients, and sediments carried by the Ebro River. The existence of a book published in the sixteenth century in which the Ebro Delta natural resources and their use are described in detail and the delta considered a "paradise" thanks to what their wetlands provide is a nice example of the valuation of ecosystem services by preindustrial societies (SEO/BirdLife 1997).

Rice farming in the Ebro Delta was introduced in the 1860s where most of the rice paddies were carved out of salt marsh areas through the active application of fresh river water rich in sediment that was delivered to the wetlands through the system of irrigation canals. After a few years of supply of river water to the marsh area a new layer of several centimeters of fertile fluvial sediment was deposited and it allowed the start of the rice cultivation. This procedure was the basis of the culture of "rising grounds" through the introduction of fluvial sediments into the delta plain that is still deeply rooted in the local society. This system of creating farming areas through controlled sediment supply (known as the "siltation method") is a good example of ecological engineering methods based on the ecosystem services provided by the river (Ibáñez et al. 2014).

During the last century agriculture became the main human activity of the Delta and nowadays rice fields play a crucial role in its economy and its ecology. However, other activities depending on the conservation and use of wetlands, such as tourism, fishing, aquaculture, hunting, or those directly related to biodiversity conservation, have an increasing importance in the last decades. Rice production is still the main income to the local economy, with a total production approaching 120,000 Mt/year. In addition to the economic benefit rice fields are artificial wetlands providing several ecosystem services, such as the maintenance of large population of aquatic birds (attracting ecotourism), their function of filter of nutrients, and regulation of soil salinity (Ibáñez et al. 2010a). Fisheries and aquaculture is another important economic sector which benefits from the ecosystem services provided by coastal lagoons, bays, and the continental shelf. Fish captures amount ca. 7,000 Mt/year and aquaculture production approaches 5,000 Mt/year. Tourism is another economic activity of growing importance, which is mostly associated to the natural and cultural values of the Ebro Delta. The number of visitors is close to 500,000 per year and the protection status of the delta clearly represents an attractive for the touristic activity.

Conservation Status and Management

The Ebro Delta is the second most important bird area of Spain, and part of its surface (7,700 ha) is protected as a natural park, Special Protection Area (SPA) for birds and Ramsar Site. A total of 12,000 ha, including the natural park plus other wetlands and some rice fields, is included in the Natural 2000 Network of the European Union. Recently the Ebro Delta has been included in a biosphere reserve covering most of the lower Ebro Basin in Catalonia. There are a total of 18 habitats included in the 92/43/EEC Directive for the Conservation of Natural Habitats and Wild Flora and Fauna, from which two are of priority conservation and eight are locally endangered. There are about 30 species of vertebrates and 17 species of plants endangered in the Delta. The relatively most affected groups are fishes, amphibians, and aquatic reptilians, so indicating the reclamation and degradation of aquatic environments, not only due to the intensification of agricultural practices but also due to the artificialization of the hydrological system (i.e., pavement of the irrigation canals with concrete).

The ecological functioning of the Ebro Delta at present is largely dependent on and affected by human activities because of modification of the natural hydrological regime. There have been changes in the temporal and spatial patterns of water salinity, increased nutrients and organic matter concentrations, pesticide pollution, and changes in the spatial and temporal patterns of sediment transport and deposition in the deltaic plain. Rice cultivation has become the crucial element in the hydrology and ecology of the Ebro Delta, with both positive and negative implications on biodiversity conservation (Ibáñez 2000). Rice fields are artificial wetlands which are a suitable habitat for feeding and reproduction of many aquatic birds, as well as for many invertebrates, amphibians, and fishes. Rice fields also act as a biological filters and, in many cases, they link fluvial environments with lagoon and marine environments. Furthermore, rice cultivation has also important economic, cultural, and aesthetic implications for the mentioned areas.

Although rice cultivation has many positive environmental aspects, current cultivation systems have some negative impacts on wetland conservation due to the intensive use of pesticides, fertilizers, and artificial drainage systems. Consequences are, for instance, the eutrophication and chemical pollution of the lagoons and marshes of the Ebro Delta, as well as salt and water stress in some wetland ecosystems. Thus, the implementation of environmentally friendly cultivation practices in order to minimize the negative impacts of rice cultivation is a conservation priority. However, the maintenance of rice fields as a conservationist objective does not preclude the fact that the recovery of natural wetlands is also a convenient measure in many cases. Actually, in the last two decades there have been several conservation projects whose main goal was the restoration of wetland habitats. Thus, the best situation in terms of socio-ecological sustainability is that with a proper combination of rice fields and natural habitats.

Threats and Future Challenges

The environmental crisis of the Ebro Delta aggravated in the 1960s, when the mechanization and the chemical agriculture arrived and they strongly increased the impact of human activities on wetlands. Formerly, the transformation of wetlands into rice fields implied a severe reduction of natural habitats, but the fact that rice fields are wetlands in many aspects and the use of traditional techniques of farming allowed the conservation of many aquatic species, though those linked to freshwater environments were favored against the halophilous ones.

The fragmentation and small surface of the natural areas, as well as the absence of an institution for a global management and protection of the Delta, are important problems for a sustainable management of wetlands. The reduction of marshes to a narrow belt around the lagoons and along the coast diminishes the ability of these areas to contain fauna, to conserve the natural gradients of plant communities, and to accomplish the function of filtering nutrients. As a consequence, the biodiversity and productivity of the lagoons is also affected, since the ecological functioning of both environments is closely linked. The degradation of the lagoons has implied a strong reduction of fish captures leading to a situation of overfishing.

The reduction of sediment transport of the Ebro River is about 99% of that existing prior to the construction of reservoirs in the catchment basin. The construction of the Mequinensa and Riba-roja reservoirs at the end of the 60s drastically culminated this process of reduction. Under these conditions, the delta has stopped its growth and the coast is being strongly reshaped by waves, with rates of retreat as high as 10 m/year in the mouth area. Additionally, the sediment deficit and the relative sea level rise (eustatic rise + subsidence) imply a loss of land elevation of the deltaic plain. Considering an estimated mean subsidence of 2 mm/year and a eustatic sea level rise of 50 cm for the year 2100, the relative sea level rise at the end of the century is expected to be around 70 cm. This means that approximately 50% of the emerged plain will likely be under sea level, causing the formation of polders and wetland degradation due to salt stress and water logging (Ibáñez et al. 2010b).

The extension of tourism during the last decades has caused an increasing environmental impact that should be minimized through a careful planning and the development of ecotourism as an alternative to the growth of the classical tourism of sun and beach. The almost complete deforestation, the loss of traditional architecture, the excessive proliferation of power lines, and the widespread pavement of the irrigation channels are some examples of human alterations affecting both biodiversity and landscape.

The main future challenges in terms of conservation and sustainable management of wetlands in the Ebro Delta are:

- A better coordination and stronger involvement of the different administrations and stakeholders in order to plan long-term conservation goals integrating the whole delta.
- A strong program of wetland restoration in order to recover part of the coastal lagoons and marshes which were lost in the past.
- A plan for adaptation to climate change and sediment deficit, in order to reduce coastal retreat and keep pace with sea level rise by means of ecological engineering measures such as wetland restoration and enhancement of vertical accretion through the restoration of the sediment flux in the river.
- A comprehensive monitoring program measuring a set of indicators of ecological status of the main wetland habitats.

References

Benito X, Trobajo R, Ibáñez C. Modelling habitat distribution of Mediterranean coastal wetlands: the Ebro Delta as case study. Wetlands. 2014. doi:10.1007/s13157-014-0541-2.

Ibáñez C. Integrated management in the special protection area of the Ebro Delta: implications of rice cultivation for birds. In: Proceedings of the conference on the council directive on the conservation of wild birds. Brussels: European Commission; 2000. pp. 98–103.

Ibáñez C, Prat N. The environmental impact of the Spanish Hydrological Plan on the lower Ebro River and delta. Water Resour Dev. 2003;19(3):485–500.

Ibàñez C, Curcó A, Day JW, Prat N. Productivity of Mediterranean coastal marshes and estuaries: differences with macrotidal systems. In: Weinstein MP, Kreeger DA, editors. Concepts and controversies in tidal marsh ecology. Netherlands: Springer; 2000.

Ibáñez C, Curcó A, Riera X, Ripoll I, Sánchez C. Influence of rice field management practices on birds in the growing season: a review and an Ebro Delta Case Study. Waterbirds. 2010a; 33(Special Publication 1):167–80.

Ibáñez C, Sharpe PJ, Day JW, Day JN, Prat N. Vertical accretion and relative sea level rise in the Ebro Delta wetlands. Wetlands. 2010b. doi:10.1007/s13157-010-0092-0.

Ibáñez C, Day JW, Reyes E. The response of deltas to sea-level rise: natural mechanisms and management options to adapt to high-end scenarios. Ecol Eng. 2014;65:122–130.

Martínez-Vilalta A. The rice fields of the Ebro delta. In: Morillo C, González JL, editors. Management of Mediterranean wetlands. Madrid: Ministerio de Medio Ambiente; 1996. p. 173–97.

SEO/BirdLife 1997.

Prado P, Caiola C, Ibáñez C. Spatio-temporal patterns of submerged macrophytes in three hydrologically altered Mediterranean coastal lagoons. Estuar Coast. 2013;36:414–29.

Rovira A, Ibáñez C. Sediment management options for the lower Ebro River and its delta. J Soils Sediments. 2007;7(5):285–95.

Doñana Wetlands (Spain)

Andy J. Green, Javier Bustamante, Guyonne F. E. Janss, Rocio Fernández-Zamudio, and Carmen Díaz-Paniagua

Contents

Introduction	1124
Hydrology and History	1124
Biodiversity	1130
Conservation Status	1132
Ecosystem Services	1132
Threats and Challenges	1133
References	1134

Abstract

The Doñana wetlands in SW Spain constitute what remains of the original 180,000 ha of marshland in the Guadalquivir delta. The natural wetlands are protected in the National Park and World Heritage Site (54,252 ha), and include more than 3,000 temporary dune ponds fed mainly from groundwater and 30,000 ha of seasonal marshes dependent on surface flows. They are particularly famous for their avifauna, and hold more wintering waterfowl than any other European wetland. Wading birds exploit surrounding ricefields and fishfarms, and their breeding numbers are increasing. Doñana holds a high diversity of herpetofauna, mammals, invertebrates and plants, with several endemic species. The wetlands have high value for ecotourism and are the setting for the El Rocío pilgrimage. Alien fish and crayfish have had a significant impact. The quantity and quality of water entering the wetlands is under threat from groundwater extraction and other effects of agricultural and urban development.

A. J. Green (✉) · J. Bustamante (✉) · G. F. E. Janss (✉) · R. Fernández-Zamudio (✉) ·
C. Díaz-Paniagua (✉)
Department of Wetland Ecology, Estación Biológica de Doñana (EBD-CSIC), Seville, Spain
e-mail: ajgreen@ebd.csic.es; jbustamante@ebd.csic.es; guyonne@ebd.csic.es; rzamudio@ebd.csic.es; poli@ebd.csic.es

© Springer Science+Business Media B.V., part of Springer Nature 2018
C. M. Finlayson et al. (eds.), *The Wetland Book*,
https://doi.org/10.1007/978-94-007-4001-3_139

> **Keywords**
> Biological invasions · Climate change · Dune ponds · Ecotourism · Eolian sands · Temporary marshes · Waterfowl

Introduction

The Doñana Wetlands (37°N, 6°25′W) are located in southwest Spain including parts of the provinces of Huelva, Sevilla, and Cadiz (Fig. 1). Doñana contains one of the largest wetland complexes in Western Europe, lying within the delta of the Guadalquivir River. It consists of an intricate matrix of marshlands and aquifer-fed dune ponds. It is surrounded by Mediterranean scrubland, pine forests, a 30 km-long mobile dune ecosystem along the shoreline of the Atlantic Ocean, and cultivated areas (Fig. 2). Two main habitat types characterize Doñana's natural wetlands: extensive seasonal marshes and adjacent eolian sands with natural depressions which can hold over 3,000 temporary ponds in rainy years (Díaz-Paniagua et al. 2010). Examples of wetland systems and fauna characteristic of the Doñana Wetlands are illustrated in Figs. 3, 4, 5, 6, 7, 8, and 9.

The first written references to the territories of Doñana date from the fourteenth century, in a book dedicated to game hunting by King Alfonso XI. The area is named after Doña Ana de Silva y Mendoza, wife of the seventh Duke of Medina-Sidonia. In the nineteenth century, kings and nobles who used this area for hunting and leisure were joined by naturalists such as Abel Chapman and Walter J. Buck, attracted by the great diversity of fauna and the ease of obtaining new specimens for scientific collections in Northern Europe. In the middle of the twentieth century, Doñana was studied by pioneering Spanish ornithologists such as Francisco Bernis and José Antonio Valverde, who caught the interest of conservationists from abroad, such as the founders of the World Wildlife Fund. Their support led to the purchase in 1963 of what became the first protected area of Doñana, which was then managed by the largest scientific entity in Spain (the Spanish National Research Council-CSIC).

Hydrology and History

Doñana is located in the Mediterranean climatic region, with a sub-humid climate with rainy autumns and winters, hot and dry summers and mild winters. Average annual rainfall is 549 mm, and the average daily temperature ranges from 4.6 °C in January to 32.6 °C in July. Of deltaic origin, Doñana is located between the Guadalquivir River Estuary and the Atlantic Ocean. The size and depth of the wetlands varies remarkably between years, driven principally by variable rainfall. Wetland inundation starts from September onwards, although timing of the first rainfall is highly unpredictable. On average, the highest monthly rainfall occurs in November and maximum inundation levels are reached during February. In late spring, evaporation becomes the most important factor influencing water levels, and

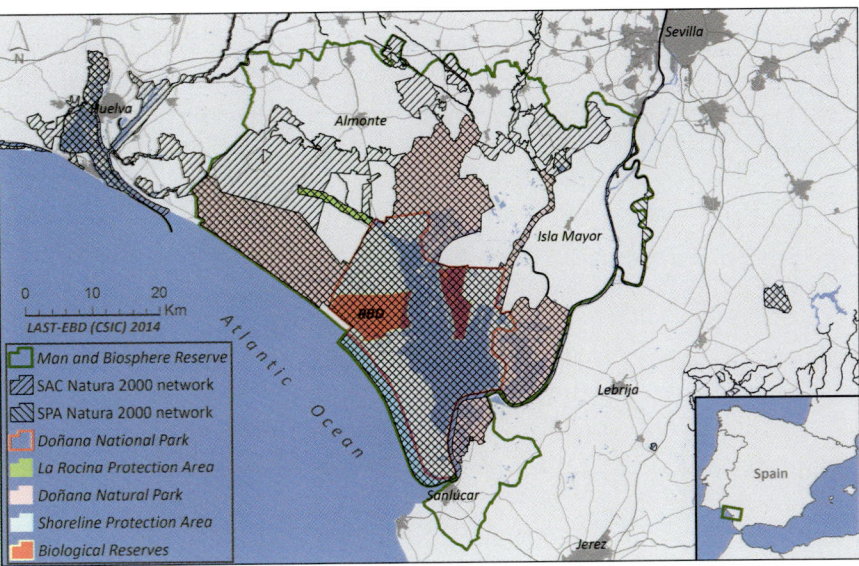

Fig. 1 Protected area limits of Doñana Natural Space at national, European, and International Scales (Figure by LAST-EBD/CSIC 2014 ©, with permission)

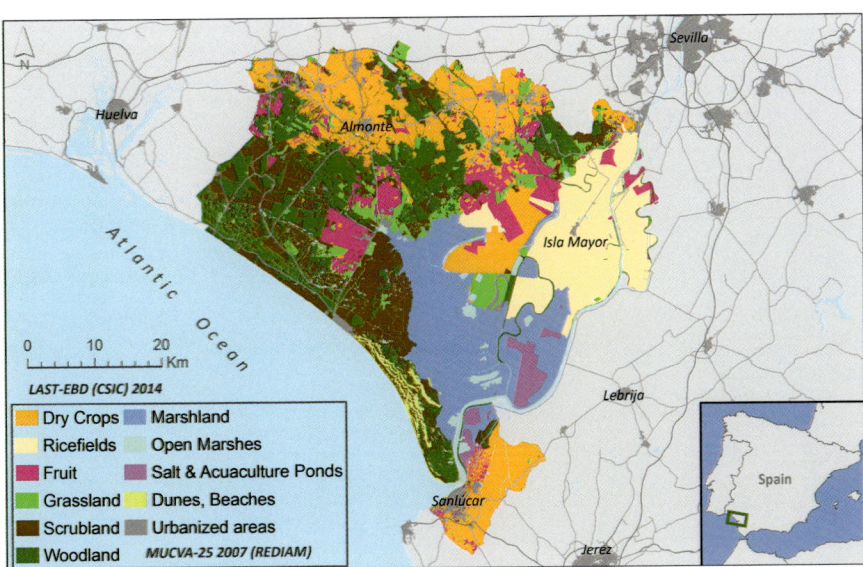

Fig. 2 Distribution of main land uses and land covers in the Doñana Natural Space and surrounding areas (Figure by LAST-EBD/CSIC 2014 ©, with permission)

Fig. 3 Flamingos, *Phoenicopterus ruber*, in the Doñana marshes (municipality of Hinojos) (Author: Héctor Garrido)

Fig. 4 Feral horses in the Doñana marshes (Sotogrande River mouth) (Author: Héctor Garrido)

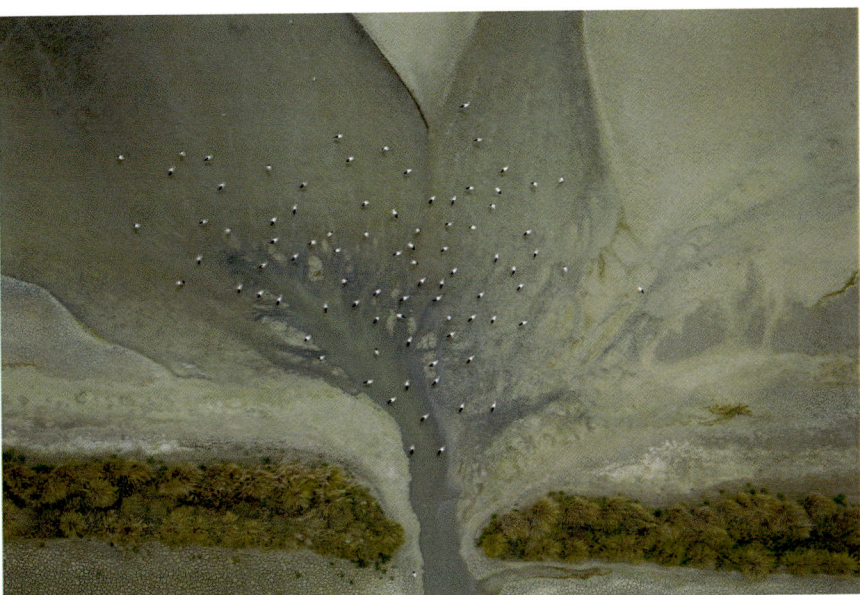

Fig. 5 White storks, *Ciconia ciconia*, in the Doñana Natural Park (Veta la Palma real estate) (Author: Héctor Garrido)

Fig. 6 Dark spreadwings, *Lestes macrostigma,* are abundant in Doñana marshes and in temporary ponds, where they lay the eggs in the stems of reeds (Author: Carmen Díaz-Paniagua)

both the marshes and ponds dry up slowly until most of the surface area is completely dry by the end of July.

Most dune ponds in Doñana are temporary, within a wide gradient of hydroperiod. Ponds are mainly filled by the rise of the water table after heavy rains, and the high interannual variation in rainfall increases the variation in hydroperiod of the ponds, which in some years may be filled in autumn, during

Fig. 7 The natterjack toad, *Bufo calamita*, is one of the most abundant amphibians of Doñana, laying long strings of eggs in shallow temporary ponds, where tadpoles may complete metamorphosis in 1–2 months (Author: Héctor Garrido)

Fig. 8 A typical temporary pond in Doñana, which usually dries out during summer (Author: Carmen Díaz-Paniagua)

Fig. 9 In Spring, Doñana ponds exhibit a dense cover of macrophytes, such as *Ranunculus peltatus* seen here in the laguna del Sopetón (Author: Carmen Díaz-Paniagua)

other years in winter, and still others in spring. This unpredictable hydrological cycle generates a high level of interannual variation in the aquatic communities present in each pond (Gómez-Rodríguez et al. 2010; Florencio et al. 2014). Only a small number of dune ponds retain water throughout the year, these being at the interface between mobile and stable dunes where the local and regional aquifers discharge. There is a high heterogeneity of size, depth, and water chemistry in the dune ponds as well as in their flora and fauna, which is mainly related to variations in geomorphology (Díaz-Paniagua et al. 2010; Florencio et al. 2014). Some natural ponds have been deepened to supply water for domestic and wild animals, converting them into permanent water bodies. These ponds play an important role in the conservation of aquatic flora and fauna, as in the case of some hydrophytes that now only persist in these permanent water bodies.

The marshes were formed by sediment deposition in the estuary of one of Spain's largest rivers, the Guadalquivir, but currently receive most of their water from a network of streams (some of which are fed by aquifer discharge) and direct rainfall. Until the eighteenth century, the Doñana marshes were largely tidal, but gradually this marine influence has reduced, and most of the wetlands currently have a continental character. Sediments from the rivers have formed an elevated platform of about 3.6 m above mean sea level. Sea water only enters a small part of the marshes at high tides. The natural marsh is largely a flat bed of clay, but slight changes in topography translate into a wide diversity of hydroperiods and vegetation types, ranging from channels "caños" and seasonal lakes "lucios" to low-lying islands or "vetas." A few flooded areas are artificially maintained around the edge of the National Park at Visitor Centres, notably at the José Antonio Valverde Centre where a major heronry has been established (Ramo et al. 2013).

From historical data, the original extent of the natural marshland has been estimated at around 180,000 ha, which has been gradually reduced to the remaining 32,000 ha and confined to the western side of the Guadalquivir river bed. Mainly since 1960, the remaining 150,000 ha have been completely drained or turned into cultivated marshlands, such as ricefields (up to 37,000 ha), fish farms (3,200 ha), and salt ponds (1,000 ha). Drier parts are cultivated with cotton, wheat, sunflower, and other crops or turned into urban areas and roads. The rivers and its meanders that formed Doñana's marshes have been greatly modified, and two of the three arms of the Guadalquivir River were closed off, while the central channel has been repeatedly dredged to allow progressively bigger ships to reach the port of Seville. Remaining rivers and streams were channelized, with dykes to prevent the flooding of fields. In the 1980s, a large dyke was built along the boundary between the marshes in the Doñana National Park and the Guadalquivir River with the aim of recovering part of the natural marshland. The presence of this dyke means that water levels become artificially high in wet winters, and the excess water is slowly drained into the river through sluice gates. The Aznalcóllar mine spill disaster in 1998 contaminated part of the area adjacent to the National Park with sludge rich in heavy metals (Taggart et al. 2006). The response by the central government was an ambitious "Doñana 2005" restoration project. The aim was to partially recover the natural dynamics of the Doñana marshland and has included the restoration of over 5,000 ha of agricultural land back to marshland by eliminating the drainage system (García-Novo and Marín-Cabrera 2006). Measures to improve water quality using water purification and treatment systems were also included, as was the development of an extensive experimental system of 96 temporary ponds (Frisch et al. 2012).

Plans to restore exchange between the Guadalquivir and the marsh inside the National Park within the framework of this restoration project were not completed before 2014.

Biodiversity

Doñana has a biodiversity that is unique in Europe. The area features a great variety of ecosystems and holds a highly diverse combination of European and Africa flora and fauna, including many globally threatened species such as the marbled duck (*Marmaronetta angustirostris*), Spanish imperial eagle (*Aquila adalberti*) and Iberian lynx (*Lynx pardinus*) as well as localized and Iberian endemics. Temporary ponds form a robust network of aquatic habitats of specialized aquatic flora and fauna, with different strategies to resist summer desiccation, whereas the Doñana marshes have many similarities with the Camargue in the Rhône river delta in France (*Parc Naturel Régional de Camargue*). At least 33 Natura 2000 habitats have been identified in Doñana, of which eight are European priority habitats (Mediterranean temporary ponds, several dune communities, wet heath *Erica* sp., and Mediterranean salt steppes and pseudo-steppe habitats).

Including all habitats, over 1,300 vascular plants have been identified in the Doñana region, 170 of which are endemic species and 60 are threatened (applying IUCN criteria). Nearly 400 species are typical of Doñana wetlands (marsh and dune ponds), with almost 23 species exclusive to the Iberian Peninsula (e.g., *Juncus emmanuelis; Rorippa valdes-bermejoi*). At least 24 species are threatened at national and six at international level (e.g., *Avellara fistulosa, Hydrocharis morsus-ranae, Micropyropsis tuberosa, Caropsis verticillato-inundata*). Three main groups of hydrophytes can be differentiated for the vegetation of temporary ponds: (1) nonstrict aquatic plants (*Agrostis stolonifera, Paspalum paspalodes, Cynodon dactylon, Mentha pulegium, Baldellia ranunculoides*); (2) wetland species occupying the borders of ponds (*Juncus maritimus, Eleocharis palustris, Eleocharis multicaulis*); and (3) submerged and floating macrophytes in the deep zones (*Juncus heterophyllus, Ranunculus peltatus, Isolepis fluitans, Myriophyllum alterniflorum, Callitriche obtusangula*) (Díaz-Paniagua et al. 2010). In the marshes, two main vegetation groups are found: the lower marsh is dominated by the emergent bulrushes *Bolboschoenus* (*Scirpus*) *maritimus* and *Schoenoplectus* (*Scirpus*) *litoralis*, whereas the higher marsh is a salt-marsh dominated by the glasswort *Arthrocnemum macrostachyum* and common woodrush *Juncus subulatus (*Espinar et al. 2002). The microflora is not well studied but includes diatom species not recorded elsewhere (Blanco et al. 2013).

Regarding vertebrates, exceptional numbers of birds are recorded. Due to its strategic location between the continents of Europe and Africa and its proximity to the Strait of Gibraltar, Doñana's large expanse of seasonal freshwater marshes is a breeding ground as well as a transit point for hundreds of thousands of European and African birds (aquatic and terrestrial) and hosts many species of migratory waterbirds during the winter. Over 300 different species of birds may be sighted there annually, of which about half are local breeding species and over 60 are localized or in decline at a European level (such as the crested coot *Fulica cristata*). Many are waterbirds (e.g., *Ardeola ralloides, Plegadis falcinellus, Platalea leucorodia, Phoenicopterus ruber*), Doñana being one of the most important sites in Europe and the Mediterranean region for breeding waterbirds (Martí and del Moral 2003), with major increases in the number of colonial waterbirds in recent decades (Ramo et al. 2013). Eight of the regularly occurring bird species are nationally threatened and five are globally threatened (IUCN criteria), such as marbled duck and white-headed duck (*Oxyura leucocephala*). Doñana meets the Ramsar Convention's 1% criterion for international importance for at least 25 species of wintering waterbirds. Many of these species are highly dependent on the ricefields and fish farms, and this is likely to explain why these species have tended to increase their population size (Rendón et al. 2008). Although originally protected because of its spectacular wetlands and waterbird concentrations, Doñana is also famous for its birds of prey, which are also highly dependent upon the aquatic systems (Sergio et al. 2011). The Doñana coast is also of considerable importance for wintering and roosting seabirds (e.g., Audouin's gull *Ichthyaetus audouinii*). In the last century, species like *Turnix sylvaticus* and *Otis tarda* became extinct.

In Doñana, 27 continental fish species have been identified (e.g., *Mugil cephalus, Atherina boyeri*). Seven species are exotic, these being highly dominant (e.g., *Gambusia holbrooki, Cyprinus carpio*). Nearly all native species are classified as Vulnerable or Endangered using IUCN criteria (e.g., *Anguilla anguilla*), one species is endemic (*Aphanius baeticus*), and *Acipenser sturio* is extinct. Amphibians reach high abundance and diversity in this area: six out of a total of twelve species are Iberian endemics. A characteristic dwarfism has been reported for the Doñana populations of some amphibians (*Triturus pygmaeus, Lissotriton boscai* and *Pelobates cultripes*). Among the aquatic reptiles, there is a high abundance of the aquatic turtles *Mauremys leprosa* and *Emys orbicularis* that mainly inhabit the more permanent ponds, both considered threatened at a national level (IUCN criteria). *Natrix maura* and *Natrix natrix* are also present in the area. Mammals include both wild and domestic species. The Iberian lynx (*Lynx pardinus*) is the most emblematic predator in the area, and the Eurasian otter (*Lutra lutra*), genet cat (*Genetta genetta*), and other carnivores are also present. The Eurasian water vole *Arvicola sapidus* is characteristic of the temporary ponds. Ancient breeds of cattle (Mostrenca cow) and horses (Retuertas' horse) feed in the marshlands.

There is a high diversity of aquatic macroinvertebrates, with more than 110 species of aquatic Coleoptera (including ibero-african endemics such as *Hygrotus lagari, Hydroporus lucasi*, and *Cybister tripunctatus africanus* as well as rare species such as *Rhantus hispanicus* and *Haliplus andalusicus*), 19 heteropterans, and seven large branchiopods (Millán et al. 2003; Florencio et al. 2014). Doñana is considered a hotspot for Odonata, with 42 species recorded, including eleven threatened species, among which the vulnerable *Lestes macrostigma* find optimal habitats in temporary ponds and marshes (Florencio and Díaz-Paniagua 2012). The zooplankton are relatively well studied, and include 48 cladocerans, 20 cyclopoids, 13 diaptomids, 8 harpacticoids (Fahd et al. 2009), 20 ostracods (Alcorlo et al. 2014), and 74 rotifers (García-Novo and Marín-Cabrera 2006). There are species of Rhabdocoela flatworms that have not been recorded elsewhere (Van Steenkiste et al. 2011).

Conservation Status

In 1969, the Doñana Biological Reserve and several surrounding estates were declared a National Park, enclosing nearly 38,000 ha. In 1974, park management was handed over from the Doñana Biological Station (EBD-CSIC) to the authorities with responsibility for environment and nature conservation (Institute of Nature Conservation, Ministry of Agriculture). EBD-CSIC has remained in charge of research coordination since then. In 1978, the National Park was extended to over 50,000 ha and buffer zones of over 25,000 ha were delimited. In 1989, these buffer zones were declared a "Natural Park" by the regional government (Andalucía). An important extension of the Natural Park followed in 1997, and the latest extension of the National Park occurred in 2004. Doñana is the third biggest National Park in Spain and the largest wetland protected as a National Park. The National and Natural Park together now cover over 110,000 ha (Fig. 2). The Doñana Natural Space ("Espacio Natural de Doñana" END) is the official name used to refer collectively to these two protected areas since 2006.

Doñana is covered by four international protection designations. The Doñana Biosphere Reserve was declared in 1980 and then occupied both the National Park and buffer zones (about 77,000 ha). It was extended in 2013 to nearly 270,000 ha, including nearly all municipalities of the region (Fig. 2). Doñana was included in the Natura 2000 Network as a Special Protection Area for Birds (Birds Directive) in 1987 and as Site of Community Importance (Habitats Directive) in 1997, being extended to include adjacent areas in 2006. Today nearly 145,000 ha of the area are included in the Natura 2000 Network (Fig. 2), covering the whole Doñana protected area and nearby water bodies as well as pine forests. The area was designated as a Ramsar wetland in 1982, first covering the National Park and being extended subsequently to include the Natural Park in 2005. In 1994, Doñana National Park was designated as a UNESCO World Heritage Site.

Ecosystem Services

There have been several studies on the ecosystem services provided by the Doñana wetlands, their monetary estimation and criteria for their valuation (Martín-López et al. 2007a, b, 2009), and their spatial distribution (Palomo et al. 2013, 2014). Also, there has been a preliminary review of the ecosystem services provided by waterbirds (Green and Elmberg 2014). In terms of the cultural values provided by birds, Doñana is clearly critical for the maintenance of many populations across the Western Palearctic. The seed dispersal service provided by birds dependent on Doñana is also of great importance.

The greater Doñana ecosystem is used to produce strawberries that are exported all over Europe, as well as large quantities of rice. The deliberate introduction of the Louisiana red swamp crayfish (*Procambarus clarkii*) starting in 1974 aimed to promote commercial exploitation of this species (Habsburgo-Lorena 1978), although this introduction has had serious negative consequences for the rest of the fauna. Crayfish are extracted and mainly exported to other parts of the Iberian Peninsula, Europe, and the United States. Hunting is still permitted in limited areas outside the National Park, including small game (rabbits, partridge, thrushes, etc.) and waterfowl. Limited grazing of livestock is permitted throughout the National Park. Other natural resources exploited in the area are pine kernels, honey, and charcoal. The marine areas off the shores of Doñana are hugely productive, and traditional shellfish exploitation on the beaches of the National Park is permitted under licence.

Doñana attracts birdwatchers and other ecotourists from all over Europe and beyond. The beaches are used mainly by national tourists and day trippers from Seville during weekends and summer holidays. The El Rocío pilgrimage is of huge cultural importance and is strongly linked with the natural values of Doñana. The major pilgrimage, with participation of nearly one million people, takes place in the week before Pentecost, while during the rest of the year thousands of pilgrims visit the shrine of El Rocío each weekend.

Threats and Challenges

The most important threats to Doñana are related to human activities in the surrounding areas (Fernández-Delgado 1997). The ecosystem has been under constant threat from the drainage of the marshes, the gradual intensification of agricultural production (including groundwater extraction from thousands of wells and use of fertilizers and other chemicals), and the expansion of tourist facilities along the coast. These activities affect both the amount and quality of the water available for the marsh and dune pond ecosystems. Groundwater extraction has led to the shortening of hydroperiods, loss of some of the most important seasonal ponds, and has reduced the flow into the Rocina, a major stream feeding the marshes (Guardiola-Albert and Jackson 2011; Manzano et al. 2013; WWF 2009, 2013). Agricultural and urban pollution has led to a major increase in phosphorus loading of the marsh since 1990, causing eutrophication (WWF 2012). In the absence of better management of the catchment area and aquifer, climate change will exacerbate these problems (Guardiola-Albert and Jackson 2011; Green et al. in press). Owing to the low altitude of the area, sea level rise poses a long-term threat to the freshwater marsh and temporary ponds. Finally, the planned dredging in the Guadalquivir River to allow even bigger ships to reach Sevilla port is likely to cause a major change in the hydrology of the area, including salinization of the lower part of the river and an increase in the risk of invasive species entering the Doñana wetlands.

Invasive species represent a major problem, including the crayfish *Procambarus clarkii*, the aquatic fern *Azolla filiculoides*, exotic fish species (*Ameiurus melas*,

Gambusia holbrooki), and pathogens like *Phytophthora cinnamomi* (affecting the roots of cork oaks). The eutrophication of the marsh (Espinar et al. 2015) and the rainfall pattern along the hydrological cycle (Fernández-Zamudio 2011) have played a major role in the invasion by *Azolla*. Some exotic species like the raccoon dog *Nyctereutes procyonoides*, American sliders *Trachemys scripta*, and the North American ruddy duck *Oxyura jamaicensis* have been eradicated locally but are still present in other areas of Spain or Europe. The presence of emergent infectious diseases whose vectors are associated with wetlands (migrating birds and mosquitos: West Nile, plasmodium) is being monitored but is not yet of concern (Vázquez et al. 2011).

Herbivory may influence the population dynamics of endangered plant species, but little research has been dedicated to this phenomenon. Ungulates (e.g., *Sus scrofa* and cattle) affect ground breeding bird species, reducing breeding success significantly. Human-induced wildlife mortalities (power lines, road kills, illegal hunting, poisoning) and impacts related with tourists and pilgrimages are managed, reducing their impacts. The isolated position of Doñana, surrounded by intensive agricultural areas (greenhouses, orange groves, etc.), a highway connecting the main cities of the area (Huelva and Seville), and the Guadalquivir River, creates dispersal problems for many terrestrial vertebrates. A corridor connecting Doñana with more northern areas has been partially created along the Guadiamar River but needs to be extended in order to become effective.

References

Alcorlo P, Jiménez S, Baltanás A, Rico E. Assessing the patterns of the invertebrate community in the marshes of Doñana National Park (SW Spain) in relation to environmental factors. Limnetica. 2014;33(1):189–204.

Blanco S, Alvarez-Blanco I, Cejudo-Figueiras C, Espejo JMR, Barrera CB, Becares E, del Olmo FD. The diatom flora in temporary ponds of Doñana National Park (southwest Spain): five new taxa. Nord J Bot. 2013;31:489–99.

Díaz-Paniagua C, Fernández-Zamudio R, Florencio M, García-Murillo P, Gómez-Rodríguez C, Siljeström P, Serrano L. Temporary ponds from the Doñana National Park: a system of natural habitats for the preservation of aquatic flora and fauna. Limnetica. 2010;29:1–18.

Espinar JL, García LV, García-Murillo P, Toja J. Submerged macrophyte zonation in a Mediterranean salt marsh: a facilitation effect from established helophytes? J Veg Sci. 2002;13: 831–40.

Espinar JL, Díaz-Delgado R, Bravo-Utrera MA, Vilà M. Linking *Azolla filiculoides* invasion to increased winter temperatures in the Doñana marshland (SW Spain). Aquat. Invasions. 2015;10 (1): 17–24.

Fahd K, Arechederra A, Florencio M, Leon D, Serrano L. Copepods and branchiopods of temporary ponds in the Doñana Natural Area (SW Spain): a four-decade record (1964–2007). Hydrobiologia. 2009;634:219–30.

Fernández-Delgado C. Conservation management of a European natural area: Doñana National Park, Spain. In: Meffe GK, Carroll CR, editors. Principles of conservation biology. 2nd ed. Sunderland, MA: Sinauer Associates; 1997. p. 458–67.

Fernández-Zamudio R. Plantas acuáticas del Parque Nacional de Doñana: aspectos ecológicos y biología de una especie exótica. Ph.D. dissertation, University of Seville (Spain). 2011.

Florencio M, Díaz-Paniagua C. Presencia de *Lestes macrostigma* (Eversmann, 1836) (Odonata: *Lestidae*) en las lagunas temporales del Parque Nacional de Doñana (SO España). Boletín de la Sociedad Aragonesa de Entomología. 2012;50:579–81.

Florencio M, Diaz-Paniagua C, Gómez-Rodríguez C, Serrano L. Biodiversity patterns in a macroinvertebrate community of a temporary pond network. Insect Conserv Diver. 2014;7:4–21.

Frisch D, Cottenie K, Badosa A, Green AJ. Strong spatial influence on colonization rates in a pioneer zooplankton metacommunity. PLoS One. 2012;7(7):e40205.

García-Novo F, Marín-Cabrera C. Doñana: water and biophere. Doñana 2005 project, Guadalquivir Hydrologic Basin Authority, Spanish Ministry of Environment, Madrid; 2006.

Gómez-Rodríguez C, Díaz-Paniagua C, Bustamante J, Portheault A, Florencio M. Inter-annual variability in amphibian assemblages: implications for diversity assessment and conservation in temporary ponds. Aquat Conserv Mar Freshwat Ecosyst. 2010;20:668–77.

Green AJ, Elmberg J. Ecosystem services provided by waterbirds. Biol Rev. 2014;89:105–22.

Green AJ, Alcorlo P, Peeters ETHM, Morris EP, Espinar JL, Bravo MA, Bustamante J, Díaz-Delgado R, Koelmans AA, Mateo R, Mooij WM, Rodríguez-Rodríguez M, van Nes EH, Scheffer M. In press. Creating a safe operating space for wetlands in a changing climate. Frontiers in Ecology and the Environment.

Guardiola-Albert C, Jackson CR. Potential impacts of climate change on groundwater supplies to the Doñana Wetland, Spain. Wetlands. 2011;31:907–20.

Habsburgo-Lorena AS. Present situation of exotic species of crayfish introduced into Spanish continental waters. Freshw Crayfish. 1978;4:175–84.

Manzano M, Custodio E, Lozano E, Higueras H. Relationships between wetlands and the Doñana coastal aquifer (SW Spain). In: Ribeiro L, Stigter TY, Chambel A, de Melo MTC, Monteiro JP, Medeiros A, editors. Groundwater and ecosystems. CRS Press/Balkema, Leiden; 2013. p. 169–82.

Martí R, del Moral, JC. Atlas de las aves reproductoras de España. Madrid, Dirección General de Conservación de la Naturaleza-Sociedad Española de Ornitología; 2003.

Martín-López B, Montes C, Benayas J. Influence of user characteristics on valuation of ecosystem services in Doñana Natural Protected Area (south-west Spain). Environ Conserv. 2007a;34:215–24.

Martín-López B, Montes C, Benayas J. The non-economic motives behind the willingness to pay for biodiversity conservation. Biol Conserv. 2007b;139:67–82.

Martín-López B, Gómez-Baggethun E, Lomas PL, Montes C. Effects of spatial and temporal scales on cultural services valuation. J Environ Manage. 2009;90:1050–9.

Millán A, Hernando C, Aguilera P, Castro A, Ribera I. Los coleópteros acuáticos y semiacuáticos de Doñana: reconocimiento de su biodiversidad y prioridades de conservación. Bol Soc Entomol Aragon. 2005;36:157–64.

Palomo I, Martín-López B, Potschin M, Haines-Young R, Montes, C. National Parks, buffer zones and surrounding lands: mapping ecosystem service flows. Ecosyst Serv. 2013;4:104–116.

Palomo I, Martín-López B, Zorrilla-Miras P, Del Amo DG, Montes C. Deliberative mapping of ecosystem services within and around Doñana National Park (SW Spain) in relation to land use change. Reg Environ Change. 2014;14(1):237–51.

Ramo C, Aguilera E, Figuerola J, Máñez M, Green AJ. Long-term population trends of colonial wading birds breeding in Doñana (SW Spain) in relation to environmental and anthropogenic factors. Ardeola. 2013;60:305–26.

Rendón MA, Green AJ, Aquilera E, Almaraz P. Status, distribution and long-term changes in the waterbird community wintering in Doñana, south-west Spain. Biol Conserv. 2008;141:1371–88.

Sergio F, Blas J, Lopez L, Tanferna A, Diaz-Delgado R, Donázar JA, Hiraldo F. Coping with uncertainty: breeding adjustments to an unpredictable environment in an opportunistic raptor. Oecologia. 2011;166:79–90.

Taggart MA, Figuerola J, Green AJ, Mateo R, Deacon C, Osborn D, Meharg AA. After the Aznalcollar mine spill: Arsenic, zinc, selenium, lead and copper levels in the livers and bones of five waterfowl species. Environ Res. 2006;100:349–61.

Van Steenkiste N, Tessens B, Krznaric K, Artois T. Dalytyphloplanida (Platyhelminthes: Rhabdocoela) from Andalusia, Spain, with the description of four new species. Zootaxa. 2011;2791:1–29.

Vázquez A, Ruiz S, Herrero L, Moreno J, Molero F, Magallanes A, Sánchez-Seco MP, Figuerola J, Tenorio A. West Nile and Usutu viruses in mosquitoes in Spain, 2008–2009. Am J Trop Med Hyg. 2011;85:178–81.

WWF. Environmental flows in the marsh of the National Park of Doñana and its area of influence. Synthesis report. 2009. Available from http://awsassets.wwf.es/downloads/synthesis_report_final_ecological_flows_1.pdf

WWF. Contaminación del agua en Doñana. Evaluación de los vertidos sin depurar de los municipios de Almonte, Rociana del Condado y Bollullos Par del Contado (Comarca de Doñana, Huelva). 2012. Available from http://awsassets.wwf.es/downloads/informe_vertidos.pdf

WWF. Evaluación del estado del aquifero 2011–2012. 2013. Available from http://awsassets.wwf.es/downloads/informe_wwf_estado_acuifero_2011_2012__2_.pdf

Axios, Aliakmon, and Gallikos Delta Complex (Northern Greece)

91

Despoina Vokou, Urania Giannakou, Christina Kontaxi, and Stella Vareltzidou

Contents

Introduction	1138
Hydrology and History	1138
Biodiversity	1140
Wetland Ecosystems, Vegetation Zones, and Flora	1140
Fauna	1140
Conservation Status	1143
Natura 2000 Habitats	1144
Ecosystem Services	1144
Threats and Challenges	1146
References	1146

Abstract

This is a complex wetland area of northern Greece encompassing the mouths of rivers, riverbeds and lagoons, in close proximity to Thessaloniki, the second largest city and the second most important harbour in the country. It offers major ecosystem services and despite this proximity and the major hydraulic works in the

---

D. Vokou (✉)
Department of Ecology, School of Biology, Aristotle University of Thessaloniki, Thessaloniki, Greece
e-mail: vokou@bio.auth.gr

U. Giannakou
Department of Medical Laboratory Studies, School of Health and Medical Care, Alexander Technological Educational Institute of Thessaloniki, Thessaloniki, Greece

C. Kontaxi
Administration of Environment and Spatial Planning, Region of Central Macedonia, Thessaloniki, Greece

S. Vareltzidou
Axios-Loudias-Aliakmonas Management Authority, Chalastra, Thessaloniki, Greece

© Springer Science+Business Media B.V., part of Springer Nature 2018
C. M. Finlayson et al. (eds.), *The Wetland Book*,
https://doi.org/10.1007/978-94-007-4001-3_253

20th century, it is still an area of beauty, of high biodiversity value, and home to many important species, primarily birds and fish. More than 290 bird species, of which some are globally threatened, and more than 80 species of fish including a Greek endemic are reported from the area. There are several designations for this important delta complex: a Ramsar Site, a national park, and a number of Natura 2000 sites and wildlife refuges. Current threats make the management of the Axios, Aliakmon and Gallikos Delta complex a challenging goal.

Keywords

Axios · Conservation · Estuaries · Ecosystem services · Lagoon · Mussel culture · Riparian forest · Thermaikos gulf · Thessaloniki · Threatened species · Natura 2000 sites

Introduction

This is a major wetland area of Greece, in close proximity to Thessaloniki, the second largest city and the second most important harbor in the country. It includes the mouths of the rivers Axios, Aliakmon, Gallikos, and Loudias that all discharge into Thermaikos Gulf, the two old riverbeds of the Axios, and the Kalochori Lagoon (Fig. 1).

At 380 km long, the Axios is the most important of the rivers as it is the main freshwater contributor and sediment supplier of Thermaikos Gulf (Kourafalou et al. 2004). For most of its length, it flows within the Former Yugoslav Republic of Macedonia (FYROM); only the last 80 km and 10% of its watershed are within Greece. The Aliakmon, 350 km long, flows entirely within the Greek territory and so does the far shorter Gallikos (65 km). The Loudias, 40 km long, is nowadays only a drainage channel.

Major changes took place in the twentieth century resulting in heavy losses of the original size and wilderness of this complex wetland area. Despite these and its proximity to a city of more than one million inhabitants, it is still an area of beauty, of high biodiversity value, and home to many important species, primarily birds and fish.

Administratively, it belongs to three prefectures and four municipalities: Delta and Halkidon municipalities belonging to Thessaloniki Prefecture, Alexandreia belonging to Imathia Prefecture, and Pydna-Kolindros to Pieria Prefecture.

Hydrology and History

Natural processes had been shaping the area and its wetlands until the beginning of the twentieth century. Between 1929 and 1936, major hydraulic works took place: drainage of the floodplains, sand quarrying, canalization, land reclamation, and construction of levees and weirs. The Axios River was displaced from its natural course in order to reduce the siltation of Thessaloniki bay and port, the main river channels were artificially realigned, the nearby Giannitson Lake and the connected Loudias swamps were drained, and the Loudias River itself, stretching to the sea, became a drainage channel. Due to repeated subsidence and in order to protect the

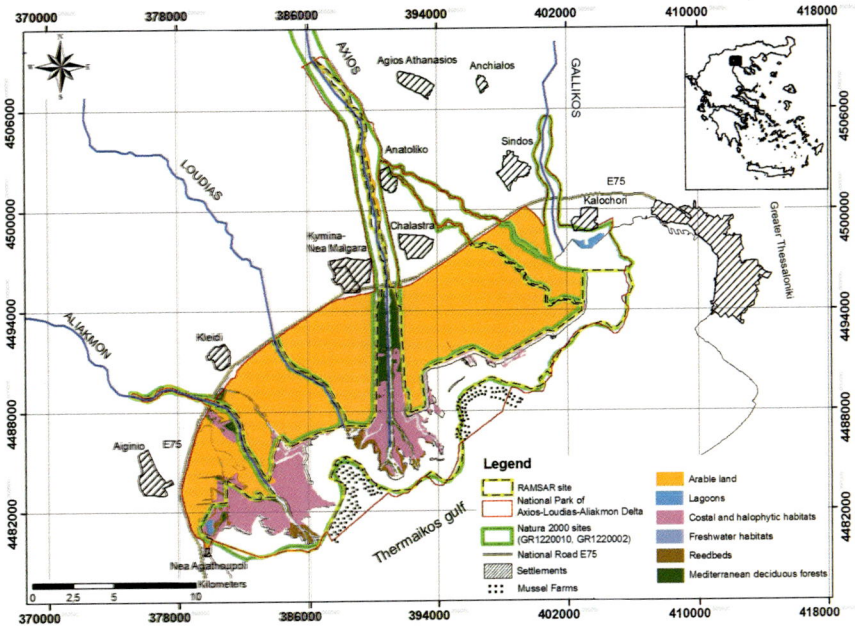

Fig. 1 The Axios, Aliakmon, and Gallikos Delta complex. Its location, in the north of Greece, is shown in the *inset*. Given are the borders of the protected area under different protection schemes, at the national, European, and international level, as well as the major habitat types both natural and man-made (© Axios Loudias Aliakmonas Management Authority, with permission)

mainland, embankments were also constructed in the mid-1960s and onward. All these produced an artificial coast in the western part of Thermaikos Gulf and led to the creation of Kalochori Lagoon. With these modifications, the wetland landscape took its current form.

Thermaikos Gulf is rather shallow, not exceeding 50 m in depth. Continuous deposition of sediment transported by the discharging rivers formed a large deltaic complex, subjected continuously to major changes. Well-known coastal places like Pella, the famous ancient Macedonian city, are now inland. Deposited material off the port of Pella encircled part of the sea creating a lagoon, which was completely cut off from the sea by the fifth century AD, thus becoming an inland lake, the Giannitson Lake. By the twentieth century, the lake covered a large part of the plain of Thessaloniki that was created by the increasing river sediment discharge and the continental uplift that caused the gradual regression of the sea. At that time, the Axios River, to the east of its current location, formed its delta south of Kalochori (Fig. 1) threatening with its sediments the port of Thessaloniki. The Axios was diverted and consequently the city did not share the same fate as Pella. Yet in present times, most of the suspended particulate matter is deposited on the seabed of Thermaikos Gulf, near the delta and pro-delta regions (Kourafalou et al. 2004).

Biodiversity

This complex deltaic area hosts coastal, inland, riparian, and agricultural ecosystems and a rich and important associated flora and fauna (Dafis et al. 1996; RSIS 1998; YPEHODE 1999–2001; Smardon 2009).

Wetland Ecosystems, Vegetation Zones, and Flora

More than 300 plant species and subspecies have been reported from the area. These are associated with the following six major vegetation types, the distribution of which is mainly influenced by soil humidity and salinity:

(i) Halophytic vegetation. Widespread in sites directly influenced by the sea, thus producing saline environments, this vegetation type is dominated by *Halocnemum strobilaceum* with *Salicornia europaea, Halimione portulacoides*, and *Arthrocnemum* and *Limonium* species having a considerable contribution.
(ii) Tamarisk (*Tamarix*) scrubland. Found along the rivers, it is of high importance for birds like herons that use it as breeding and nesting refuge.
(iii) Rush (*Juncus*) meadows. Occurring in areas of low salinity, it is adversely affected by grazing.
(iv) Reed beds. Associated with the slow flowing or stagnant waters of river mouths and by the riverbeds and drainage canals, they are of great importance for birds. *Phragmites australis, Bolboschoenus maritimus*, and *Typha latifolia* are dominant species.
(v) Hydrophytic vegetation. Found in wet areas, in the shallow freshwater ponds, irrigation canals, drainage ditches, and rice fields, it represents an important part of the wetland vegetation. Pondweed (*Potamogeton* species) and duckweed (*Lemna* species) are the dominant plants, and *Ruppia maritima* is an ecological indicator of shallow brackish waters; the alien water fern (*Azolla filiculoides*) is quite common.
(vi) Riparian forest. Found along the riverbanks and on the many islets in the riverbeds and dominated by poplars, willows, and alder, it has well-formed understory vegetation. It regulates the rivers' flow and is of high importance for the fauna providing shelter and nesting sites.

Fauna

Lying on one of the main south–north migratory routes of birds in Europe, the Axios, Aliakmon, and Gallikos Delta complex is famous for its avifauna. Thousands of birds stop over to rest and feed before continuing their journey or overwinter and breed. The long coastline, the shallow waters, the salt marshes and mudflats, the

Table 1 Bird species, for which one site of the area was selected to be included in the European Natura 2000 network, and their extinction risk status at national and global scales (Legakis and Maragou 2009; IUCN 2013); *CR* critically endangered, *EN* endangered, *VU* vulnerable, *NT* near threatened

Scientific name	Common name	Classification	
		Global scale	National scale
Tadorna tadorna	Common shelduck		VU
Platalea leucorodia	Eurasian spoonbill		VU
Ixobrychus minutus	Little bittern		
Nycticorax nycticorax	Night heron		NT
Ardeola ralloides	Squacco heron		VU
Egretta garzetta	Little egret		
Pelecanus crispus	Dalmatian pelican	VU	VU
Phalacrocorax pygmeus	Pygmy cormorant		
Haematopus ostralegus	Eurasian oystercatcher		
Himantopus himantopus	Black-winged stilt		
Recurvirostra avosetta	Pied avocet		VU
Charadrius alexandrinus	Kentish plover		
Limosa limosa	Black-tailed godwit	NT	
Numenius tenuirostris	Slender-billed curlew	CR	CR
Glareola pratincola	Collared pratincole		
Larus genei	Slender-billed gull		VU
Larus melanocephalus	Mediterranean gull		EN
Sterna nilotica	Gull-billed tern		VU
Sterna albifrons	Little tern		NT
Calandrella brachydactyla	Short-toed lark		

lagoons, the riparian forests, the reed beds, and also the rice fields offer shelter and breeding, feeding, and resting grounds to birds. More than 290 species have been reported from the area. Recently, a total of 124 migratory birds have been recorded with 103 nesting there.

Twenty-one of the bird species are included in the IUCN Red List of Threatened Species, 74 are listed in the Greek Red Data Book (Legakis and Maragou 2009), and 101 are listed in Annex I of the EU Birds Directive (2009). The bird species for which one site of the area was selected to be included in the European Natura 2000 Network are given in Table 1 along with their extinction risk status at the national and global scale (for more information regarding this network and the related European directives, see the following section "Conservation Status").

The area is of global importance for the globally threatened Dalmatian pelican (*Pelecanus crispus*), the ferruginous duck (*Aythya nyroca*), the greater spotted eagle (*Aquila clanga*), the black-tailed godwit (*Limosa limosa),* the slender-billed curlew (*Numenius tenuirostris*), and the Eurasian curlew (*Numenius arquata*); also, as a breeding ground for three herons (*Nycticorax nycticorax, Egretta garzetta,* and *Ardeola ralloides*), for two waders, the pied avocet (*Recurvirostra avosetta*) and the collared pratincole (*Glareola pratincola*), for the Mediterranean gull (*Larus melanocephalus*) and the gull-billed tern (*Sterna nilotica*); and as a wintering area for the pygmy cormorant (*Phalacrocorax pygmeus*). At the European scale, it is important for two waders (*Charadrius alexandrinus, Glareola pratincola*) and two terns (*Sterna nilotica, S. albifrons*); also, for the wintering of common shelduck (*Tadorna tadorna*) and the breeding of the Eurasian spoonbill (*Platalea leucorodia*) and the little bittern (*Ixobrychus minutus*).

During spring and summer, nine heron species can be seen feeding in the rice fields, salt marshes, reed beds, riverbanks, and canals along with spoonbills, cormorants, waders, and terns. During the cold months, thousands of ducks and geese are found in the delta along with many birds of prey including eagles (*Aquila clanga, Haliaeetus albicilla*), falcons (*Falco columbarius, F. peregrinus*), buzzards, and harriers. A variety of shore birds also occur, most of which are seen during spring and autumn migration periods.

Notably, the area hosts the only breeding colony in Greece of the slender-billed gull (*Larus genei*) and the largest breeding colonies of the terns *Sterna sandvicensis, S. nilotica*, and *S. hirundo*. The riparian forests of the Axios, in particular, host one of the largest heron colonies in Europe (Kazantzidis and Goutner 2008); overall, seven heron species nest in the area. The endangered glossy ibis (*Plegadis falcinellus*), the white-tailed eagle (*Haliaeetus albicilla*), and the spur-winged lapwing (*Vanellus spinosus*) have all made remarkable comebacks after many decades of absence from the area.

Among animals other than birds that occur there, the following are of European importance: two mammals, the European ground squirrel (*Spermophilus citellus*) and the European otter (*Lutra lutra*), one amphibian (*Triturus carnifex*), three reptiles (*Testudo graeca, Emys orbicularis*, and *Mauremys caspica*), two fish (*Aphanius fasciatus* and *Rhodeus sericeus amarus*), and four invertebrate species (*Lycaena dispar, Lindenia tetraphylla, Cerambyx cerdo*, and *Morimus funereus*). All of these species are listed in Annex II of the EU Habitats Directive (1992) (see the following "Conservation Status" section).

Apart from the three reptiles mentioned above, the area is also home to the dice snake (*Natrix tessellata*) and the nose-horned viper (*Vipera ammodytes*). Regarding mammals, in addition to the two species above, there are also wolves (*Canis lupus*), red foxes (*Vulpes vulpes*), European badgers (*Meles meles*), beech martens (*Martes foina, M. martes*), weasels (*Mustela nivalis vulgaris*), hedgehogs (*Erinaceus concolor*), shrews (*Crocidura leucodon, C. suaveolens*), Nathusius' pipistrelle (*Pipistrellus nathusii*), hares, mice, and rats. The invasive coypu (*Myocastor coypus*) having escaped from fur farms is a potential threat to the area's ecology.

An impressive stock of water buffalos belonging to the Mediterranean subtype of the river buffalo can be seen in the Gallikos estuary. They can use roughage of variable nutritional value, have high disease resistance, and can contribute to the conservation of wetland ecosystems. Also, wild horses live in the Axios delta and in some of the river's islets; descendants of animals once used by farmers have become members of the local fauna and a visitors' attraction.

Parts of the rivers and the deltas and also the shallow sea waters serve as spawning grounds for fish populations of the Thermaikos Gulf and the North Aegean Sea. The variety of aquatic ecosystems of the delta complex supports a fish fauna of more than 80 species, including the Greek endemic *Alburnoides bipunctatus thessalicus*.

Conservation Status

The Axios, Aliakmon, and Gallikos delta complex is designated a protected area at the local, regional, and global scales.

Being of international importance, this wetland was designated as a Ramsar Site in 1975 (the Ramsar Site boundaries are shown in Fig. 1).

At the European scale, it forms part of the Natura 2000 network with two sites, designated after the Birds Directive (Special Protection Area, SPA) and the Habitats Directive (Special Area of Conservation, SAC); these are GR1220010 Delta Axiou–Loudia–Aliakmona–Alyki Kitrous (SPA) and GR1220002 Delta Axiou–Loudia–Aliakmona–Evryteri Periochi–Axioupoli (SAC) (EEA 2013a, b). The Birds and Habitats Directives form the cornerstone of Europe's nature conservation policy, which is built around the Natura 2000 network of protected sites and the strict system of species protection. The purpose of this network that includes protected sites in every member state of the European Union is to maintain or restore the natural habitats and species of Europe to a favorable conservation status in their natural range. Annex I of the Birds Directive lists the bird species of European importance; Annex II of the Habitats Directive lists the other species and Annex I the habitat types of European importance. All these habitats and species are targets of the Europe-wide conservation efforts; hence, these Annexes are referred to when describing the biodiversity of the delta complex.

At the national scale, the wetland complex forms part of the National Park of Axios–Loudias–Aliakmon Delta; this has an area of around 338 km^2 including the Alyki Kitrous Lagoon, located to the south, beyond the borders of the map in Fig. 1. The Park is organized in zones of decreasing order of protection: (i) strict nature reserves, (ii) nature reserves, and (iii) a peripheral zone. Several activities are allowed in the latter, while the first is under strict protection. According to the Greek legislation (Law 3937/2011 for the protection of biodiversity), the Natura 2000 sites along with a preexisting group of protected areas, the wildlife refuges, fall within a special category, the habitat/species management areas; these may or may not have additional designations, such as strict or simple nature reserves. Included in this category are three wildlife refuges: the Aliakmon Delta/Kleidi (12 km^2) in Imathia Prefecture, designated in 1981; the Axios Delta/Chalastra (21.7 km^2) in Thessaloniki Prefecture, designated in 1988; and the Aliakmon Delta/Stergiou (23.8 km^2) in Pieria Prefecture, designated in 1997.

The Ramsar Site corresponds to around 35% of the Park, whereas the Natura 2000 sites almost coincide with the Park, covering 99% of its area.

The management authority of the Axios–Loudias–Aliakmon Delta, first appointed in 2003, is based in Chalastra (Fig. 1). It implements a management plan that includes guarding of the area, monitoring biodiversity and threats, hosting and contributing to other research activities, setting conservation objectives, and

Table 2 Habitat types of Annex I of the EU Habitats Directive that are found in the Axios, Aliakmon, and Gallikos Delta complex

Code	Description
1130	Estuaries
1150	Coastal lagoons
1160	Large shallow inlets and bays
1210	Annual vegetation of drift lines
1310	*Salicornia* and other annuals colonizing mud and sand
1410	Mediterranean salt meadows (*Juncetalia maritimi*)
1420	Mediterranean and thermo-Atlantic halophilous scrubs (*Sarcocornetea fruticosi*)
3150	Natural eutrophic lakes with *Magnopotamion* or *Hydrocharition*-type vegetation
3280	Constantly flowing Mediterranean rivers with *Paspalo–Agrostidion* species and hanging curtains of *Salix* and *Populus alba*
6420	Mediterranean tall humid grasslands of the *Molinio-Holoschoenion*
92A0	*Salix alba* and *Populus alba* galleries
92D0	Southern riparian galleries and thickets (*Nerio-Tamaricetea* and *Securinegion tinctoriae*)

applying conservation measures in cooperation with the other public and private players, providing consultation, informing the authorities empowered by the state to enforce the law on illegal activities, and implementing communication events and activities; the latter aim primarily at changing the way that people understand, value, and use the natural resources of the protected area. It also runs the existing infrastructure. This includes an information center at Chalastra, where environmental education programs are offered; a fully equipped observation post and a thematic pavilion at Nea Agathoupoli, where programs specifically designed for children are offered; and an environmental park with observation posts, trails, and rest places at the Gallikos River. To improve its effectiveness, the management authority is guided by the Open Standards of the Conservation Measures Partnership (CMP 2010).

Natura 2000 Habitats

The natural habitat types, as coded and described in Annex I of the Habitats Directive, in the Axios, Aliakmon, and Gallikos Delta complex, are presented in Table 2. The habitat type "1150 Coastal lagoons," which is found in Gallikos and Aliakmon estuaries, is a priority habitat type for the European Union.

Ecosystem Services

The complex wetland system formed by the Axios, Aliakmon, Gallikos, and Loudias rivers, and their streams and irrigation channels, provides many services; prominent among these are water supply, rice, and mussel production.

Extensive drillings for groundwater extraction in the past had a serious impact on its quantity. Today, all municipalities and communities in the area, including the city of Thessaloniki, satisfy their needs for drinking water from the Aliakmon river. But the most important use of water is to cover the irrigation needs of the plain of Thessaloniki. To do so, major works had to take place to store and transfer water. Operation of the Axios dam, providing annually around 430×10^6 m^3 of water began in 1962, whereas construction works to transport approximately 360×10^6 m^3 of water from the Aliakmon to the Axios via a canal started after 1975. During dry years, the quantity used for irrigation is around $695–700 \times 10^6$ m^3 annually. During non-dry years, this may rise up to 900×10^6 m^3. Arable land covering some 4,300 km^2, of which 2,000 km^2 are located on the plain of Thessaloniki, the largest deltaic plain in Greece, is served from April to September by collective irrigation networks. The General Organization for Irrigation Improvements distributes the water, and many smaller local organizations control its supply to cultivated areas. Water from the Aliakmon and the Axios flows into 320 km long canals to reach the fields. After its use, with the aid of drainage channels, pumps, water tanks, and underground pipes, it eventually drains into the Thermaikos Gulf.

Rice cultivation was introduced in the 1950s, in a small area of 20 km^2, after the remediation of the saline land along the lower riverbeds and the river mouths, which decreased the pH from 9–10 to 7.5–8. Forty years later, the area of rice-cultivated land had expanded to 120 km^2, a sixfold increase. Today, 120,000 t – 67% of the annual rice production of Greece is being cultivated in the river mouths and the land surrounding the riverbeds of the Axios, Loudias, and Aliakmon, in an area covering 142.5 km^2 (Hellenic Statistical Authority 2011).

Agricultural and drinking water demands reduce freshwater availability. This has impacts on the wetland complex and its biodiversity, on the Thermaikos Gulf, and also on productive activities. The unique convergence of rivers with currents continuously moving large volumes of freshwater and providing large amounts of nutrients has resulted in the country's largest mussel (*Mytilus galloprovincialis*) production. Located off the Axios, Gallikos, and Aliakmon river mouths, 90% of the mussel farms of Greece produce about 80–90% of the annual national harvest (Theodorou et al. 2011). Mussel farming is strongly dependent on continuous river flows. Low freshwater supply and/or quality can severely affect mussel farming. There is, therefore, a conflict of interest between mussel and rice farmers due to the direct contribution of agricultural activities to freshwater pollution with agrochemicals and to eutrophication phenomena in the deltaic and coastal area (Karageorgis et al. 2006).

Several other wetland products and services of direct use are offered by the Axios, Aliakmon, and Gallikos Delta, such as fish, meat, provision of fodder, and other agricultural commodities. Apart from these, it is of great value as a reservoir of biodiversity and also as a recreation area offering aesthetic and spiritual benefits, particularly to the inhabitants of the nearby big city of Thessaloniki. Microclimatic regulation, flood control, groundwater replenishment, water purification, soil formation, primary production, and nutrient cycling are other major ecosystem services offered by this important wetland.

Threats and Challenges

There are several threats encroaching upon the integrity and environmental values of this complex delta area (Vareltzidou and Strixner 2009). The most serious ones are caused by

(i) Its proximity to a major national road and a city of more than one million inhabitants, resulting in high pressures for land use changes, i.e., from agricultural to urban and commercial uses;
(ii) The excessive removal of freshwater, primarily for irrigation purposes
(iii) The discharge of inadequately treated urban and industrial sewage to the freshwater systems
(iv) Unsustainably conducted activities including hunting, environmentally hazardous sand extraction, intensification of mariculture and excessive use of agrochemicals that are both associated with harmful algal blooms
(v) The inadequately managed solid waste and the uncontrolled dumpsites

In addition to the above, sea level rise and an increase in the frequency and duration of droughts due to climate change are expected, in the medium to long term, to have serious impacts on the hydrology of the area, its species and habitat types, and the ecosystem services that it offers.

The major challenge for those involved in the management of the important Axios, Aliakmon, and Gallikos Delta complex is how to withstand the pressures from aggressive urbanization so that it retains its critical role as a biodiversity reservoir and maintains its valuable life-sustaining natural processes while remaining an important supplier of freshwater and food products from both the land and the sea. Raising public awareness, making ecological knowledge common knowledge, and pressing for integration of this knowledge into all development plans of the nearby area is a goal to be pursued not only locally, by the management authority of the national park, but by the entire conservation community of the country.

References

CMP (Conservation Measures Partnership). Open standards for the practice of conservation; 2010. Available from: http://www.conservationmeasures.org/wp-content/uploads/2013/05/CMP-OS-V3-0-Final.pdf

Dafis S, Papastergiadou E, Georghiou K, Babalonas D, Gorgiadis T, Papageorgiou M, Lazaridou Th, Tsiaoussi V. Directive 92/43/EEC The Greek habitat project NATURA 2000: an overview. Life Contract B4-3200/94/756, Commission of the European Communities DG XI. The Goulandris Natural History Museum – Greek Biotope/Wetland Centre; 1996.

EEA (European Environmental Agency). Natura 2000 network viewer – standard data form for GR1220002 Delta Axiou-Loudia-Aliakmona-Evryteri Periochi-Axioupoli; 2013a [updated 2014 Feb 07]. Available from: http://natura2000.eea.europa.eu/natura2000/SDF.aspx?site=GR1220002

EEA (European Environmental Agency). Natura 2000 network viewer – standard data form for GR1220010 Delta Axiou-Loudia-Aliakmona-Alyki Kitrous; 2013b [updated 2014 Feb 07]. Available from: http://natura2000.eea.europa.eu/Natura2000/SDF.aspx?site=GR1220010

EU Birds Directive. Directive 2009/147/EC of the European Parliament and the Council of 30 November 2009 on the conservation of wild birds (codified version of the Directive 79/409/EEC of 2 April 1979 as amended); 2009. Available from: http://eur-lex.europa.eu/legal-content/EN/TXT/?uri=CELEX:32009L0147

EU Habitats Directive. Council Directive 92/43/EEC of 21 May 1992 on the Conservation of natural habitats and of wild fauna and flora; 1992. Available from: http://eur-lex.europa.eu/legal-content/EN/TXT/?uri=CELEX:31992L0043

Hellenic Statistical Authority. Agricultural statistics of Greece, year 2006; 2011. Available from: http://dlib.statistics.gr/Book/GRESYE_02_0903_00076.pdf

IUCN (International Union for Conservation of Nature and Natural Resources). The IUCN red list of threatened species; 2013. Available from: http://www.iucnredlist.org/

Karageorgis AP, Kapsimalis V, Kontogianni A, Skourtos M, Turner KR, Salomons W. Impact of 100-year human interventions on the deltaic coastal zone of the inner Thermaikos Gulf (Greece): a DPSIR framework analysis. Environ Manage. 2006;38:304–15.

Kazantzidis S, Goutner V. Abundance and habitat use by herons (Ardeidae) in the Axios delta, northern Greece. J Biol Res-Thessalon. 2008;10:129–38.

Kourafalou VH, Savvidis YG, Koutitas CG, Krestenitis YN. Modelling studies on the processes that influence matter transfer on the Gulf of Thermaikos (NW Aegean Sea). Cont Shelf Res. 2004;24:203–22.

Legakis A, Maragou P, editors. The red data book of threatened animals of Greece, Athens. Greece: Hellenic Zoological Society; 2009.

RSIS (Ramsar Sites Information Service). Ramsar information sheet for Axios – Loudias – Aliakmon Delta; 1998 [updated 1998 June 01]. Available from: http://ramsarsites.wetlands.org/reports/infosheet.cfm?siteref=3GR007

Smardon RC. The Axios River Delta – Mediterranean wetland under Siege. In: Sustaining the world's wetlands: setting policy and resolving conflicts. New York: Springer; 2009. p. 57–92.

Theodorou JA, Viaene J, Sorgeloos P, Tzovenis I. Production and marketing trends of the cultured Mediterranean mussel *Mytilus galloprovincialis* Lamarck 1819, in Greece. J Shellfish Res. 2011;30:859–74.

Vareltzidou S, Strixner L. Recommended strategic plan for maintaining favourable conservation status of Natura 2000 areas in the Axios Delta (2009–2013). Axios-Loudias-Aliakmonas Delta Management Authority, Greece; 2009.

YPEHODE (Hellenic Ministry for the Environment Physical Planning and Public Works). Identification and description of habitat types in areas of interest for the conservation of nature. Sub-programme 3 – protection of the natural environment of the operational programme for the environment (measure 3:3); 1999–2001.

The Philippi Peatland (Greece)

92

Kimon Christanis

Contents

Introduction 1150
Climate and Hydrology 1150
The Peatland 1150
Uses 1153
Threats and Future Challenges 1153
References 1153

Abstract

The Philippi peatland is located in the intermontane Drama basin in northern Greece. The mire occupied an area of 274 km^2 before drainage in the 1940s. A drilling project carried out in the 1960s proved that the thickness of the peat alternating with other limnic and limnotelmatic sediments reaches up to 190 m. Plans to exploit the upper part of the peatland for power generation, were canceled due to the lack of social acquiescence. After drainage the area started being cultivated with maize, tobacco and sugar beet but the intensive cultivation caused severe soil subsidence and flooding.

Keywords

Cultivation · Greece · Peat · Peatland · Philippi · Power generation · Subsidence

K. Christanis (✉)
Department of Geology, Sector of Earth Materials, University of Patras, Rio-Patras, Greece
e-mail: christan@upatras.gr

© Springer Science+Business Media B.V., part of Springer Nature 2018
C. M. Finlayson et al. (eds.), *The Wetland Book*,
https://doi.org/10.1007/978-94-007-4001-3_147

Introduction

The Philippi peatland is located in the southwestern part of Drama basin, an intermontane basin in Eastern Macedonia, northern Greece, covering c.700 km^2 (Fig. 1). The basin surface lies at altitudes ranging from 42 (in the southern part) to 200 m (in the northern part) above sea level, whereas the surrounding mountains rise to over 2,000 m a.s.l. Before drainage in the 1940s, the Philippi mire covered 274 km^2, of which 98 km^2 were permanently flooded, 89 km^2 water-saturated, and more than 87 km^2 periodically flooded (Melidonis 1969).

The area is well known due to three main reasons: (a) in the north of the mire, the battle between the armies of Octavian and Mark Antony on one side and Brutus and Cassius, Julius Caesar's assassins on the other side took place in 43 B.C.; (b) at the town of Philippi, the first Christian church on European ground was founded by the Apostle Paul in 49 or 50 A.D.; and (c) the area hosts the thickest (c. 190 m) peat deposit known worldwide (Melidonis 1981).

Climate and Hydrology

Despite the proximity to the Mediterranean Sea, the climate shows central-European features with cold winters and warm summers. The annual average precipitation ranges from 500 to 700 mm. The mean temperature is 4 °C in January and 24.5 °C in July (Panilas 1998).

According to Panilas (1998), the central and southern part of the Drama basin has a shallow unconfined aquifer overlaying impermeable limnic and telmatic Pleistocene sediments (deposited in lakes and mires). The water level lies between 1 and 1.5 m beneath the current surface of the peatland. Furthermore, there is a karstic aquifer in the basin margins, which generally shows no connection with the basin aquifer except at the southwestern margins.

Several streams drain the northern and central part of the Drama basin into the Angitis River, a tributary of the Strymon River, which in turn flows into the Aegean Sea (Fig. 1). Several ditches and channels were dug between 1931 and 1944 in order to drain the southern part of the basin, namely, the former Philippi mire, to protect the local people from malaria and gain agricultural land; these also discharge into the Angitis River.

The Peatland

After World War II, the Philippi mire was successfully drained; the water table dropped 1–2 m beneath the former mire surface, and the area was cultivated mainly with maize, tobacco, and sugar beet. In the late 1950s, i.e., after only 15 years of

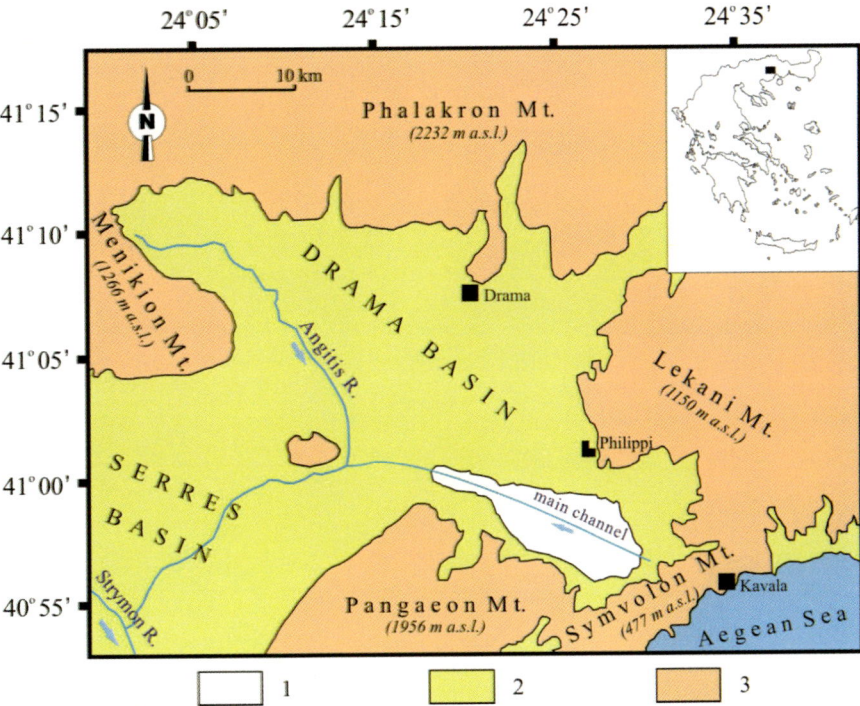

Fig. 1 Sketch map of the Drama basin showing the Philippi peatland (*1* Philippi peatland, *2* Neogene and Tertiary sediments, *3* pre-Neogene basement) (© K Christanis, rights remain with the author)

intense cultivation, the topsoil suffered subsidence of more than 3.5 m in the central part of the mire (Fig. 2); this made the road infrastructure over the former mire useless.

The first step taken to address this problem was to drill through all the peat layers and reach the inorganic ground. This, however, revealed a peat thickness exceeding 20 m (van der Molen and Smits 1962). Following this result, the Greek Geological Survey (IGME) carried out a drilling exploration campaign in the Philippi plain in 1964–1965. The thickness of the peat alternating with other limnic and limnotelmatic sediments such as clay, calcareous and detrital mud, as well as with millimeter-to-meter-thick volcanic tephra layers, proved to be around 190 m at the central part of the drained area (Melidonis 1981). Thus, the entire peat sequence forms the thickest known deposit in the world, i.e., Philippi represents the longest record of almost continuous peat formation.

The peat has accumulated over the last one million years (Mommersteeg et al. 1995) or 1.35 million years (Tzedakis et al. 2006). It is derived from reeds and sedges. With increasing depth, the peat appears to be increasingly coalified, and beneath a depth of around 120 m, it can be considered lignite (Teichmüller 1968).

Fig. 2 Due to its better foundation, the level of the bridge (*left*) over a drainage ditch remained stable or subsided less in comparison to the road; this made the transport system over the drained mire useless (photo taken in 1979) (Photo credit: Kimon Christanis © Rights remain with the author)

These deeper telmatic facies (sediments or sedimentary rocks accumulated in mires) may extend also to the central part of the Drama basin (in the north of the Philippi peatland; see Fig. 1), where peat accumulated until the Eemian interglacial period, resulting in the formation of a lignite deposit with 570 Mt of proven reserves (on dry basis) (Broussoulis et al. 1991).

The proven peat reserves amount to approximately 4.3 Gm^3 with a gross calorific value ranging between 13.2 and 20.4 MJ/kg on dry basis (Melidonis 1981).

The uppermost part of the peat sequence up to a depth of 15 m consists mainly of well-humified Upper Weichselian-Holocene peat, alternating with some thin layers of limnic/limnotelmatic sediments. The peat has been derived mainly from Cyperaceae, namely, *Cladium mariscus* and various *Carex* species (Christanis 1983). Two volcanic tephra layers of up to 4 cm in thickness, and one of 25 cm thickness (even thicker in places), have been proven to derive from the volcanic fields of Thera in the Hellenic arc and the Campanian Province of Italy, respectively (Christanis 1983; Seymour et al. 2004). This upper sequence includes 300 Mm^3 bulk peat constituting about 7.5% of the entire Philippi deposit, capable of supporting power plants of 375 MW installed capacity (Christanis 1987).

Uses

Plans to exploit the upper 15 m thick peat layers for power generation were canceled in 1972 due to the lack of social acquiescence and the subsequent protests. Due to the same reason, one pilot plant producing milled peat for agricultural and horticultural uses on a 10 ha field since 1990 ceased operation in 1993. Today, the peatland is exclusively used for agricultural use except for a small spa located at the northeastern margin that uses peat and mud for balneotherapeutic purposes (the treatment of disease by bathing).

Threats and Future Challenges

The drainage and subsequent intensive agricultural cultivation of the peatland, which through plowing disturbs the top layer and exposes fresh peat to the atmosphere every year, has caused severe surface subsidence that nowadays exceeds 7 m in the central part of the peatland. This, in turn, has resulted in the flooding of large areas and the destruction of crop yields, which necessitate the deepening of the ditches and channels from time to time. The exposure of the organic matter under the dry and hot climatic conditions of Greece results in its oxidation and mineralization, and occasionally also self-combustion. Furthermore, the application of agrochemicals such as fertilizers and pesticides represents a severe threat to the local fauna and results in eutrophication of the drainage channels and ditches and contamination of the Angitis River.

References

Broussoulis J, Yakkoupis P, Arapogiannis V, Anastasiadis J. Drama′s lignite deposit. IGME report [in Greek]; 1991.
Christanis K. Ein Torf erzählt die Geschichte seines Moores. Telma. 1983;13:19–32.
Christanis K. Philippi/Greece: a peat deposit awaiting development. Int Peat J. 1987;2:45–54.
Melidonis N. The peat and brown coal deposit of Philippi (Eastern Macedonia, Greece). Geol Geophys Res. 1969;33(3):1–250 [in Greek].
Melidonis N. Beitrag zur Kenntnis der Torflagerstätte von Philippi (Ostmazedonien). Telma. 1981;11:41–63.
Mommersteeg HJPM, Loutre MF, Young R, Wijmstra TA, Hooghiemstra H. Orbital forced frequencies in the 975000 year pollen record from Tenagi Philippon (Greece). Clim Dyn. 1995;11:4–24.
Panilas S. Hydrogeological questions concerning the open pit exploitation of lignite deposits. The case of the Drama lignite deposit. Ph.D. thesis: Department of Geology, University of Patras [in Greek]; 1998.
Seymour KS, Christanis K, Bouzinos A, Papazisimou S, Papatheodorou G, Moran E, Dénès G. Tephrostratigraphy and tephrochronology in the Philippi peat basin, Macedonia, Northern Hellas. Quat Int. 2004;21:53–65.

Teichmüller M. Zur Petrographie und Diagenese eines fast 200-m mächtigen Torfprofils (mit Übergängen zur Weichbraunkohle?) im Quartär von Philippi (Mazedonien). Geol Mitt. 1968;8:65–110.

Tzedakis PC, Hooghiemstra H, Pälike H. The last 1.35 million years at Tenaghi Philippon: revised chronostratigraphy and long-term vegetation trends. Quat Sci Rev. 2006;25:3416–30.

van der Molen WH, Smits H. Die Sackung in einem Moorgebiet in Nord-Griechenland (Philippi). Report, Bremen; 1962.

Peatlands of the Mediterranean Region

Richard Payne

Contents

Peatland Types, Occurrence, and Distribution 1156
Biodiversity and Ecosystem Services 1159
Human Impacts 1162
Future Challenges 1163
References 1164

Abstract

The generally arid climate of the Mediterranean region means that peatlands are typically small and scattered but these sites should not be overlooked. Peatlands are present in the vast majority of Mediterranean countries and make a disproportionately large contribution to regional biodiversity with endemic species recorded for several sites. These peatlands are often poorly documented and increasingly threatened by human activity, in particular the increasing need for agricultural land and water resources. The vast majority of sites are disturbed or destroyed but relatively natural sites do remain, particularly in mountainous regions. A greater focus on the important biodiversity and ecosystem service benefits of Mediterranean peatlands is needed, particularly in countries outside Europe.

Keywords

Agriculture · Europe · Levant · Mire · North Africa · Peatland · Wetland

R. Payne (✉)
Environment, University of York, Heslington, York, UK
e-mail: richard.payne@york.ac.uk

© Springer Science+Business Media B.V., part of Springer Nature 2018
C. M. Finlayson et al. (eds.), *The Wetland Book*,
https://doi.org/10.1007/978-94-007-4001-3_111

Peatland Types, Occurrence, and Distribution

The Mediterranean drainage basin encompasses a large area of southern Europe, western Asia, and northern Africa. Climate in the region is variable but mostly warm and dry with mild, damp winters and hot dry summers. Cooler and wetter regions occur through the mountains of the Alps, Pyrenees, and Balkans and hotter drier regions across North Africa (Fig. 1). Despite this generally arid climate peatlands of some type are present in most countries of the region.

In the western European Mediterranean countries (France, Italy, Spain) there are many peatlands, although the majority are in areas which drain to the Atlantic rather than Mediterranean. Peatlands are present, and in some areas relatively abundant in the Pyrenees, Alps, Jura, and Apennine mountains (Table 1). In the Balkans peatlands are present but scattered, most are fens located particularly at high altitudes in upland regions although ombrotrophic bogs are recorded in at least Slovenia. Peatlands are present in Turkey including lowland and upland fens and one ombrotrophic bog. In the Levant, important sites include the extensive (drained/restored) Hula peatland in Israel and the Aammiq wetland in Lebanon (Table 2). There are extensive wetlands including areas referred to as peatlands (much now drained) in Egypt's Nile Delta. Through North Africa there are scattered peatlands, particularly in the mountains of Morocco, Algeria, and possibly Tunisia, although literature is extremely sparse. There are peatlands of various types on many of the larger Mediterranean islands including Corsica, Crete, and Cyprus (Tables 1 and 2).

The region is large and diverse; the peatlands are shaped by their local environment (physical, biogeographical, and human) which makes generalization difficult, particularly as most scientists can only be personally familiar with a small proportion of the area and the published literature is sparse. However, in general terms the peatlands can be grouped as:

(a) Intramontane basin fens. These peatlands are the largest in the region, often with very deep peats. Minerotrophic peats, often intercalating with lake sediments, have accumulated over long periods of time in basins which receive water supply from the surrounding land but have restricted, or no, outflow. Vegetation typically comprises reeds and sedges, often with extensive *Phragmites australis*. Examples are found in many countries, for instance, the Hula peatland in Israel (Fig. 2) or Philippi peatland in Greece. These sites are now generally damaged or destroyed with a large proportion drained for agriculture.
(b) Coastal peatlands. These peatlands are in low-lying coastal regions where peats have accumulated in estuaries or deltas, often with marine influence. Such sites are less frequent but sometimes extensive, for instance, in the Nile Delta. Vegetation generally comprises species such as *Phragmites australis, Typha domingensis*, and *Cyperus papyrus*. Many of these sites have been drained.
(c) Upland fens. These peatlands are in mountainous regions with water supply from springs or streams, often forming in areas with impeded drainage (Fig. 3). Such sites occur in all but the most arid mountain ranges and are perhaps the most

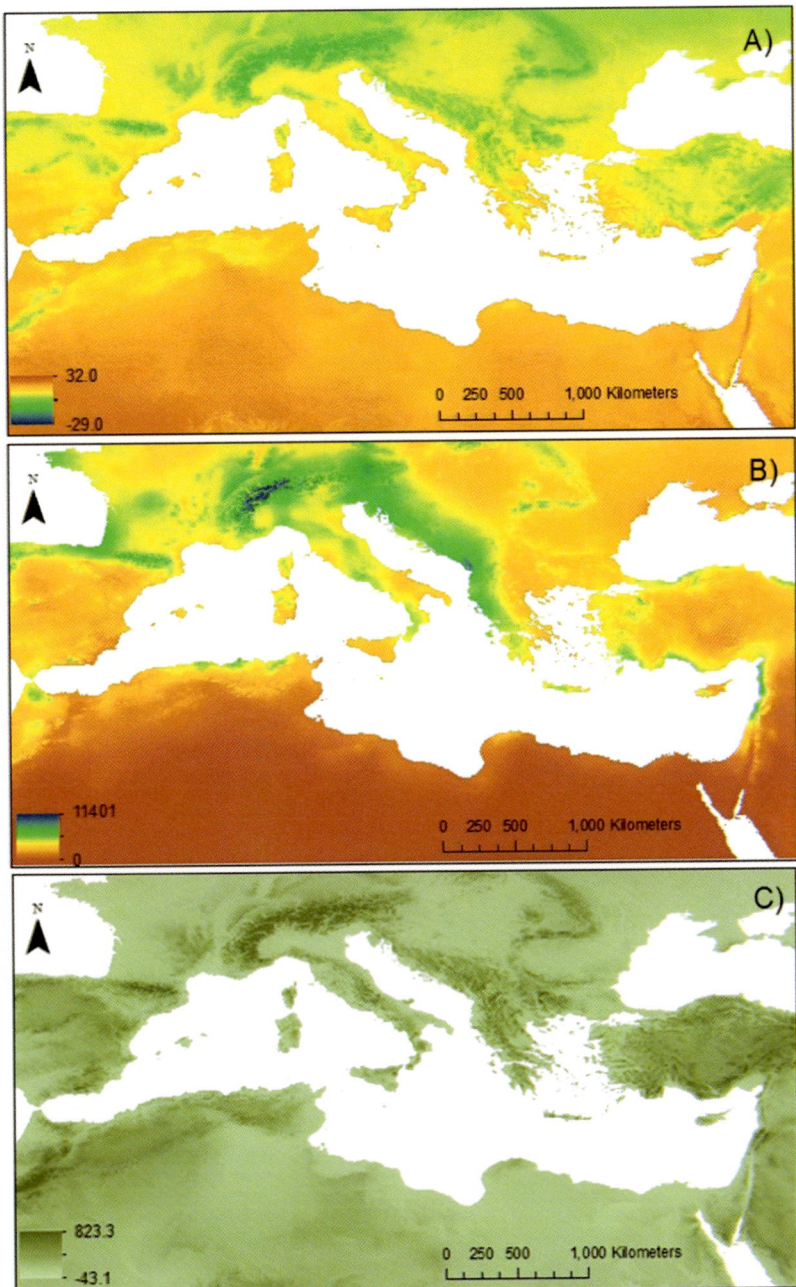

Fig. 1 Climate and physical geography of the Mediterranean basin – (**a**) temperature (°C), (**b**) Precipitation (mm), (**c**) Altitude (m). Interpolated data from the WorldClim database (Hijmans et al. 2005) produced by the author

Table 1 Peatlands in European countries directly bordering Mediterranean Sea (excluding Black Sea) (Turkey and Cyprus listed in Table 2)

Country/territory	Peatlands
Albania	European Soil Database peat area: 44 km^2, Map of organic C content, >25%: 0 km^2 (Montanarella et al. 2006) Fen plant communities recognized in national phytosociology (Dring et al. 2002) Peatland total area <100 km^2 including peatlands at Cukë, Xavë, and Malig (Lappalainen and Żurek 1996)
Bosnia and Herzegovina	European Soil Database peat area: 170 km^2, Map of organic C content, >25%: 32 km^2 (Montanarella et al. 2006) Peatland area 190 km^2 (Lappalainen and Żurek 1996) Minerotrophic peatlands in alpine and subalpine zones (Sulejman 2011)
Croatia	European Soil Database peat area: 41 km^2, Map of organic C content, >25%: 0 km^2 (Montanarella et al. 2006) Selected significant sites: Blatuša fen (1,000 ha), fen at Dubravica village, mires in the Gorski kotar region (Wetlands International 2006)
France	European Soil Database peat area: 3,157 km^2, Map of organic C content, >25%: 775 km^2 (Montanarella et al. 2006) National estimates of peatland area vary from 550 to 1,200 km^2 (Lappalainen and Żurek 1996) Peatlands are relatively widespread but much of the area is in northern France draining to the Atlantic not Mediterranean. Peatlands occur in the mountains of the Massif Central, Jura, Vosges, Pyrenees, and southern Alps (Francez et al. 1992; Muller, personal communication 2012). Fewer lowland peatlands such as Marais de Lavours in the Rhone Valley. Upland fens (Pozzine) occur in the mountains of Corsica (Gamisans and Marzocchi 1996). National mapping project currently underway
Gibraltar	None
Greece	European Soil Database peat area: 554 km^2, Map of organic C content, >25%: 0 km^2 (Montanarella et al. 2006) Peatland area 101 km^2 (Lappalainen and Żurek 1996) Peatlands mostly fens in intra-montane basins such as Phillipi, Koroni, Nissi, and Kalodiki (Botis et al. 1993), now mostly drained (Christanis 1996). Small transitional mires with *Sphagnum* occur near Bulgarian border (Papazisimou et al. 2002)
Italy	European Soil Database peat area: 292 km^2, Map of organic C content, >25%: 1 km^2 (Montanarella et al. 2006) Peatland area 300–1,200 km^2. Ombrotrophic peatlands including *Sphagnum* in upland areas, particularly in north (Lombardy, Piedmont, Veneto), valley, and coastal mires further south (Lappalainen and Żurek 1996)
Malta	None known (V. Gauci, personal communication 2012) European Soil Database peat area: 0 km^2, Map of organic C content, >25%: 0 km^2 (Montanarella et al. 2006)
Monaco	None European Soil Database peat area: 0 km^2, Map of organic C content, >25%: 0 km^2 (Montanarella et al. 2006)
Montenegro	European Soil Database peat area: 110 km^2, Map of organic C content, >25%: 0 km^2, figures for Yugoslavia as of 2006 including Montenegro, Kosovo, and Serbia (Montanarella et al. 2006)

(*continued*)

Table 1 (continued)

Country/territory	Peatlands
	Serbia and Montenegro: valley mires and lacustrine peatlands with herbs dominant, small montaine sloping mires, ombrotrophic *Sphagnum* peatlands rare (Wetlands International 2006) In all of the former Yugoslavia: 1,000 km^2, 300 Mt (Wetlands International 2006)
Slovenia	European Soil Database peat area: 78 km^2, Map of organic C content, >25%: 0 km^2 (Montanarella et al. 2006) Many peatlands in Julian Alps including ombrotrophic bogs, often forested, for instance, Pokljuka plateau (Kutnar and Martinčič 2003; Kutnar 2000) Bogs at Pohorje, fens at Bloška Planota (Beltram 2007)
Spain	European Soil Database peat area: 360 km^2, Map of organic C content, >25%: 184 km^2 (Montanarella et al. 2006) Peatland area 385 km^2 (Lappalainen and Żurek 1996) One important site is the Padul peatland (Granada) – a basin fen, which has a history of around 400 ka (Ortiz et al. 2004) Many blanket bogs and other peatlands occur in Galicia and northern Spain more generally (Martinez Cortizas and Garcia-Rodeja Gayoso 2001; Martinez Cortizas 2009a) but much of this area is in the Atlantic not the Mediterranean drainage basin. Transition mires and raised bogs occur in northern and central Spain (Martinez Cortizas 2009b, c). Small montane minerotrophic peatlands are found in the Sierra Nevada (Jiménez-Moreno and Anderson 2012). One coastal Ramsar Site includes peatland – Prat de Cabanes (Martinez Cortizas 2009d)

numerous type of peatlands in the region. Vegetation typically includes species of *Carex* and *Juncus;* in the most oligotrophic sites species of *Sphagnum* also occur. Examples include the Pozzines of Corsica and small mires in the Troodos Mountains of Cyprus (Christanis et al. 2008) and White Mountains of Crete (Atherden and Hall 1999). Such sites are often in less accessible areas and have consequently escaped human impacts to a greater extent than lowland fens.

(d) Ombrotrophic bogs. These sites are much the rarest type of peatland in the region. Sites are comparatively few and always situated at moderate to high altitude in regions with higher rainfall. Examples are found in the Alps, Pyrenees, and Soğanlı Dağ Mountains of northeast Turkey. Vegetation is *Sphagnum* dominated, and sites are in many ways similar to northern European bogs.

Biodiversity and Ecosystem Services

In contrast to other regions of the world, the peatlands of the Mediterranean are often small and frequently widely separated, essentially forming wetland islands in a generally arid landscape. Perhaps due to this disparate nature the sites include many endemic species. Some species only found in a single site include the grass *Poa asiae-minoris* in a Turkish mire (e.g., Scholz and Byfield 2000), the fish

Table 2 Details of peatlands in countries and territories of Africa and Asia bordering the Mediterranean Sea. Data is from the IMCG global peatlands database (Joosten 2004) with the exception of entries shown in italics. The literature referred to is often fragmentary, old, or secondary and for many regions should be treated with considerable caution

Country/ territory	Details of peatlands (and other wetlands and organic soils)
Algeria	290 km^2 freshwater marshes (Britton and Crivelli 1993) Peatlands in Atlas Mountains (Schokalskaja 1953) Peatlands may include Mekhada Marsh (Lappalainen and Żurek 1996) Possible peatlands in coastal regions (Schneider and Schneider 1990) No histosols. 193 km^2 gley soils (Van Engelen and Huting 2002)
Ceuta	None known
Cyprus	No histosols or gley soils (Van Engelen and Huting 2002) *Pashia Livadi and Almyrolivado Juncus-dominated small mires in Troodos Mountains* (Christanis et al. 2008) *Fassouri Phragmites-dominated wetland with saline peat*
Egypt	"Peatlands" in Nile Delta and adjacent to Red Sea (Markov et al. 1988) Manzala "peatland" 540 km^2, Burulus "peatland" 170 km^2 No histosols. 11,080 km^2 gley soils (Van Engelen and Huting 2002)
Gaza	None known No histosols or gley soils (Van Engelen and Huting 2002)
Israel	*Most significant peatland is the Hula wetland in northern Israel. Peatland drained in 1950s and now consists of oxidized drained peats, an area of mostly recently developed peats and a recent restoration project* (Payne and Gophen 2012 et seq.) *Possible peatlands in Wadi Kubani, Wadi Faliq, Zevulun Valley, and Ain Arus although some of these seem unlikely to still exist* 50 km^2 peatlands (Moore and Bellamy 1974) 48 km^2 wetlands (Lappalainen and Żurek 1996) No histosols. 93 km^2 gley soils (Van Engelen and Huting 2002)
Lebanon	*Most significant is Phragmites-dominated Aammiq wetland in the Bekaa Valley* (A Rocha 2012) *Possible montane peatlands* No histosols. 49 km^2 gley soils (Van Engelen and Huting 2002)
Libya	No peatlands known. Oasis wetlands
Melilla	None known
Morocco	Peatlands in High Atlas and Rif Mountains (Schokalskaja 1953) Peatlands may include Iriki swamp (now destroyed), Zima lake, and in the Middle Atlas (Howard-Williams and Thompson 1985) 2 km^2 freshwater marshes (Britton and Crivelli 1993) No histosols. 1140 km^2 gley soils (Van Engelen and Huting 2002) *Sedge fens. Sphagnum subnitens occurs in Krimada Fen (Larache region), Sphagnum auriculatum occurs in Rif Mountains* (Muller et al. 2011)
Syria	Formerly peatlands in the Ghab Valley (Niklewski and van Zeist 1970) No histosols. 1087 km^2 gley soils (Van Engelen and Huting 2002)
Tunisia	51 km^2 freshwater marshes (Britton and Crivelli 1993) Peatlands may include Ichkeul lakes and swamp and Sebka Kelbia Swamp (Howard-Williams and Thompson 1985) 8 km^2 wetlands (Lappalainen and Żurek 1996) No histosols. 93 km^2 gley soils (Van Engelen and Huting 2002)

(*continued*)

Table 2 (continued)

Country/territory	Details of peatlands (and other wetlands and organic soils)
Turkey	Wetland area 561 km² (Öz 1996) Peatlands (>0.3 m depth) 130 km² *Most significant site is Sphagnum-dominated Ağaçbaşi peatland very like northern-European blanket mires. Vegetation includes endemic species* (Byfield and Özhatay 1997) No histosols. 45,069 km² gley soils (Van Engelen and Huting 2002)

Fig. 2 The Hula peatland, Israel. This site is a drained and partially restored fen complex in the northern Afro-Syrian Rift. Photo shows the margins of the Hula Nature Reserve

Acanthobrama hulensis in an Israeli wetland (Crivelli 2006), and the fish *Aphanius sirhani* in a Jordanian wetland. In many more cases, a peatland may include the only examples of species occurring in a country or region. Mediterranean peatlands are important for migratory birds with many being important stopping points on routes between Europe, Asia, and Africa. Peatlands therefore contribute considerably to the diversity of this global biodiversity hotspot. Given the limited taxonomic survey work which has been conducted in these regions much unrecorded diversity, and possibly new species, remain to be discovered. This major contribution to regional biodiversity is one of the most important arguments for the conservation of these peatlands.

Ecosystem services provided by Mediterranean peatlands have been little considered and are almost entirely unquantified. In contrast to northern peatlands, the

Fig. 3 The Elatia Mires of northern Greece. These small fens occur in the Rhodope Mountains in an area with comparatively high rainfall. These mires are described by Papazisimou et al. (2002)

carbon sequestration function of Mediterranean peatlands is globally insignificant – although sometimes very deep, Mediterranean peatlands are scattered and small as a proportion of total area. Other functions are locally significant, particularly those of flood control and water purification.

Human Impacts

Very few peatlands in the region are in a pristine condition; most have been affected by human activity and many destroyed. Key threats include drainage, fire, grazing, eutrophication and, to a limited extent, peat cutting. Of these much the most serious impacts have been through drainage. Most of the large, low altitude fens have been partially or completely destroyed by drainage (e.g., Bouzinos et al. 1997); for instance, it is estimated that 87% of the peatlands in Turkey have been lost (Çayci et al. 1988; Byfield and Özhatay 1997). Motivations for drainage include control of disease, water management, and particularly the provision of new agricultural land. Drainage schemes date back to prehistoric times (Christanis 1996) but increased in number and extent from the late nineteenth century and particularly into the first half of the twentieth century as mechanization made drainage easier. Most of these drained sites are now extensively degraded with extensive loss of peatland species and oxidation and erosion of peats. Many of these drainage schemes have not lived up to initial expectations – agriculture has proved less productive than expected with large quantities of pesticides and fertilizers required, and there have been problems with subsidence due to peat oxidation and fire.

Some peatlands are affected by nutrient enrichment; for instance, the Ioannina peatland (Christanis 1996) adjacent to Lake Pamvotis in north-central Greece has been affected by sewage input into the lake from the city of Ioannina. Fire is an important agent of environmental change in Mediterranean ecosystems and a significant impact on some peatlands. Other peatlands, particularly in montane areas, are affected by grazing – in many cases the impacts of this grazing appear limited to minor trampling and nutrient enrichment but may be more serious in some sites. For instance, in the Asi Gonia peatland on Crete there is extensive trampling by sheep leading to considerable physical disturbance and some areas of bare peat. This site contains the most southerly *Sphagnum* in Europe and a unique palaeoecological archive (Atherden and Hall 1999) but does not appear to be in a favorable condition.

Peat-cutting is not widespread in the region. There is little culture of cutting and as peats are generally minerotrophic they produce poor quality fuel. Small-scale or experimental peat cutting has been carried out in several areas, for instance, the Yeniçağa fen in Turkey, the Phillipi peatland in northern Greece, and the Hula peatland in Israel. Particularly notable is peat cutting in the Sürmene Ağaçbaşı Yaylası peatland in north-eastern Turkey. This site is a remarkable ombrotrophic *Sphagnum*-dominated peatland which contains at least one endemic species and several more for which this is the only site known in Turkey (Byfield and Özhatay 1997; Payne et al. 2007). Peat has been cut for at least 30 years with very extensive disturbance across the site. Peat has been removed by digging holes a few meters across down into the peat; fortunately, this pattern of cutting means that the hydrological impact has been less severe than might be expected.

Increasingly, the main pressure on peatlands may shift to water resources with rising human populations and climate change. An example is one of the most recent episodes of catastrophic drainage of an important wetland – the Azraq wetland in Jordan. From the 1980s the water table of the site dropped drastically with water abstraction, primarily to supply the growing city of Amman. By 1993 no surface water remained, and this unique and important site was all but destroyed. Climate change projections suggest rising temperatures and reduced precipitation in the Mediterranean basin (Giorgi and Lionello 2008). As many peatlands are at the climatic margins for peat formation and human usage of water is likely to increase the continued existence of many Mediterranean peatlands is in doubt.

Future Challenges

Peatlands in the region, particularly outside Europe, are highly under-researched. Unsurprisingly, in regions where peatlands are rare there is little expertise in peatland science and consequently little focussed study of the sites. Much basic inventory work is lacking, as can be seen from the overview of published literature presented in Tables 1 and 2. Only a minority of countries have comprehensive inventories of peatlands. Basic biological data such as lists of plant taxa are available for only a very small minority of sites.

Peatland conservation across the Mediterranean is also variable. In Western Europe (France, Italy, Spain) there is often considerable protection for peatlands – many are managed by national, regional, or charitable conservation organizations and protected from disturbance. In other regions, particularly in the Middle East and North Africa there is frequently no protection at all. While some sites in developed countries are important destinations for ecotourism, particularly with respect to bird-watching, in many others there is little or no public recognition of the importance of peatlands. Much greater efforts are needed to catalog, conserve, and educate the public about the importance of Mediterranean peatlands before even more are lost.

References

A Rocha. About the Aammiq wetland. A Rocha Lebanon. 2012. http://www.arocha.org/lb-en/work/aammiq.html. Accessed 7 July 2012.

Atherden M, Hall J. Human impact on vegetation in the White Mountains of Crete since AD 500. Holocene. 1999;9:183–93.

Beltram G. Conservation and management of wetlands in Slovenia. Quark; 2007. p. 70–81. http://www.quark-magazine.com/pdf/quark07/0702CMWetlandsinSloveniaB.pdf.

Botis A, Bouzinos A, Christanis K. The geology and palaeoecology of the Kalodiki peatland, western Greece. Int Peat J. 1993;5:25–34.

Bouzinos A, Christanis K, Kotis T. The Chimaditida fen (W. Macedonia, Greece): a peat deposit lost. Int Peat J. 1997;7:3–10.

Britton RH, Crivelli AJ. Wetlands of southern Europe and North Africa: Mediterranean wetlands. In: Whigham DF, Dykyjová D, Hejný S, editors. Wetlands of the world: inventory, ecology, and management, vol. 1. Dordrecht: Kluwer; 1993. p. 129–94.

Byfield A, Özhatay N. A future for Turkey's peatlands: a conservation strategy for Turkey's peatland heritage. Istanbul: Doğal Hayati Koruma Derneği; 1997.

Çayci G, Ataman Y, Ünver I, Munsuz N. Distribution and horticultural values of the peats in Anatolia. Acta Hortic. 1988;238:189–96.

Christanis K. The peat resources in Greece. In: Lappalainen E, editor. Global peat resources. Jyskä: International Peat Society; 1996. p. 87–90.

Christanis C, Kalaitzidis S, Loizou L, Emmanouloudis D. Small mire habitats in Troodos National Forest Park, Cyprus. IMCG Newsl. 2008;1:22–3.

Crivelli AJ. Acanthobrama hulensis. In: IUCN 2010. IUCN Red List of Threatened Species, Version 2010.3; 2006.

Dring J, Hoda P, Mersinllari M, Mullaj A, Pignatti S, Rodwell J. Plant communities of Albania – a preliminary overview. Ann Bot. 2002;2:7–30.

Francez AJ, Bignon JJ, Mollet AM. The peatlands in France: localization, characteristics, use and conservation. Suo. 1992;43:11–24.

Gamisans J, Marzocchi J-F. La flore endemique de la Corse. Aix-en-Provence: Edisud; 1996.

Giorgi F, Lionello P. Climate change projections for the Mediterranean region. Glob Planet Change. 2008;63:90–104.

Hijmans RJ, Cameron SE, Parra JL, Jones PG, Jarvis A. Very high resolution interpolated climate surfaces for global land areas. Int J Climatol. 2005;25:1965–78.

Howard-Williams C, Thompson K. The conservation and management of African wetlands. In: Denny P, editor. The ecology and management of African wetland vegetation. Dordrecht: Dr. W. Junk; 1985. p. 203–30.

Jiménez-Moreno G, Anderson RS. Holocene vegetation and climate change recorded in alpine bog sediments from the Borreguiles de la Virgen, Sierra Nevada, southern Spain. Quatern Res. 2012;77:44–53.

Joosten H. IMCG Global peatlands database. 2004. www.imcg.net/gpd/gpd.htm. Accessed 7 July 2012.

Kutnar L. Spruce mire types on the Pokljuka plateau, Slovenia. Phyton. 2000;40:123–8.

Kutnar L, Martinčič A. Ecological relationships between vegetation and soil-related variables along the mire margin- mire expanse gradient in the eastern Julian Alps, Slovenia. Ann Bot Fenn. 2003;40:177–89.

Lappalainen E, Zurek S. Peat in other European Countries. In: Lappalainen E, editor. Global peat resources. Jyskä: International Peat Society; 1996. p. 153–62.

Markov V, Olenin A, Ospennikova L. World peat resources: reference book. Moscow: Nedra Publishing House; 1988.

Martinez Cortizas A. 7110. Turberas elevadas activas. Madrid: Ministerio de Medio Ambiente, y Medio Rural y Marino. Secretaría General Técnica. Centro de Publicaciones; 2009a.

Martinez Cortizas A. 7130. Turberas de Cobertor. Madrid: Ministerio de Medio Ambiente, y Medio Rural y Marino. Secretaría General Técnica. Centro de Publicaciones; 2009b.

Martinez Cortizas A. 7140. Mires de transición. Madrid: Ministerio de Medio Ambiente, y Medio Rural y Marino. Secretaría General Técnica. Centro de Publicaciones; 2009c.

Martinez Cortizas A. 71. Bases ecológicas para la gestión de turberas acidas de esfagnos. Madrid: Ministerio de Medio Ambiente, y Medio Rural y Marino. Secretaría General Técnica. Centro de Publicaciones; 2009d.

Martinez Cortizas A, Garcia-Rodeja Gayoso E. Turberas de montaña de Gailicia. Santiago de Compostela: Xunta de Gailicia; 2001.

Montanarella L, Jones RJA, Hiederer R. The distribution of peatland in Europe. Mires Peat. 2006;1:1–10.

Moore PD, Bellamy D. Peatlands. Berlin: Springer; 1974.

Muller SD, Laïla R, Saber E-R, Rifai N, Daoud-Bouattour A, Bottollier-Curtet M, Saad-Limam SB, Ghrabi-Gammar Z. Peat mosses (*Sphagnum*) and related plant communities of North Africa. II. The Tingitanean-Rifan range (northern Morocco). Nova Hedwigia. 2011;93:335–52.

Niklewski J, van Zeist W. A late Quaternary pollen diagram from northwestern Syria. Acta Bot Neerl. 1970;19:737–54.

Ortiz JE, Torres T, Delgado A, Julià R, Lucini M, Llamas FJ, Reyes E, Soler V, Valle M. The palaeoenvironmental and palaeohydrological evolution of Padul Peat Bog (Granada, Spain) over one million years, from elemental, isotopic and molecular organic geochemical proxies. Org Geochem. 2004;35:1243–60.

Őz D. Peatlands in Turkey. In: Lappalainen E, editor. Global peat resources. Jyskä: International Peat Society; 1996. p. 201.

Papazisimou S, Bouzinos A, Christanis K, Tzedakis PC, Kalaitzidis S. The upland Holocene transitional mires of Elatia forest, northern Greece. Wetlands. 2002;22:355–65.

Payne RJ, Gophen M. The Hula peatland: past, present and future. Mires Peat. 2012;9:1–2.

Payne R, Eastwood W, Charman D. The ongoing destruction of Turkey's largest upland mire by peat cutting. International Mire Conservation Group Newsletter. 2007;2007/1:5–6.

Schneider S, Schneider R. Verteilung der Moore auf der Erde. In: Gottlich Kh, editor. Moor und Torfkunde. 3rd ed. Stuttgart: Schweizerbart; 1990. p. 65–101.

Schokalskaja SJu. Die Boden Afrikas. Berlin: Akademie-Verlag; 1953.

Scholz H, Byfield A. Three grasses new to Turkey. Turk J Bot. 2000;24:263–7.

Sulejman R. Phytogeographic and syntaxonomic diversity of high mountain vegetation in Dinaric Alps (Western Balkan, SE Europe). J Mt Sci. 2011;8:767–86.

Van Engelen V, Huting J. Peatlands of the world. An interpretation of the world soil map. GPI Project 29 GPI 1. Wageningen: ISRIC; 2002.

Wetlands International. Quick scan of peatlands in Central and Eastern Europe. Amsterdam: Wetlands International and the Netherlands Ministry of Agriculture, Nature & Food Quality; 2006.

The Hula Wetland (Israel)

Richard Payne

Contents

Introduction	1168
Location, Hydrology, and Climate	1168
The Hula Wetland: Drainage and Restoration	1168
The Pre-Drainage Wetlands	1168
Destruction of the Wetland	1170
Post-Drainage and Restoration	1170
The Hula Today	1171
Biodiversity	1171
Future Challenges	1172
References	1172

Abstract

This article provides a brief introduction to the Hula wetland: a regionally significant wetland complex in the Middle East. The article describes the physical setting of the wetland and how it has been shaped by human intervention over long time-periods. The present biodiversity is briefly introduced along with a discussion of possible future threats.

Keywords

Peatland · Wetland · Levant · Middle East · Israel · Palestine

R. Payne (✉)
Environment, University of York, Heslington, York, UK
e-mail: richard.payne@york.ac.uk

© Springer Science+Business Media B.V., part of Springer Nature 2018
C. M. Finlayson et al. (eds.), *The Wetland Book*,
https://doi.org/10.1007/978-94-007-4001-3_146

Introduction

The Hula wetland (החולה in Hebrew, also Hula wetland variously transliterated as Hulah, Houla, Huli, Hooleh, and Huleh) is a drained and partially restored wetland complex in northern Israel. Although now a fraction of its former size, the Hula remains one of the most important wetlands in the Middle East. The Hula is well researched, and this short article can only provide a very brief introduction. Useful entry points into the extensive literature are the book by Dimentman et al. (1992) which describes the pre-drainage Hula and two journal special issues: *Wetlands Ecology and Management* volume 6, no.2-3 (http://link.springer.com/journal/11273/6/2/page/1) which summarizes the early years of the Hula restoration project and *Mires and Peat* volume 9 (2011–12) (http://mires-and-peat.net/pages/volumes.php) which updates this with some more recent research.

Location, Hydrology, and Climate

Situated in northern Israel, the Hula Valley lies in the northern reaches of the Afro-Syrian rift valley (33°04'N, 35°35'E) and forms part of the Jordan River catchment (Fig. 1). The Hula Valley is at around 70 m above sea level but is surrounded by higher ground, with the Golan Heights to the east, Mt Hermon to the north and the Galilee hills to the west. The Hula Valley catchment is 1,470 km^2 and lies almost entirely to the north of the wetland. Numerous springs and tributaries reach the floor of the Hula Valley; the most important water supply to the wetlands is the River Jordan, which forms from a number of tributaries slightly upstream of the Hula. Downstream, water from the Hula travels down the River Jordan to Lake Kinneret (the Sea of Galilee) and ultimately to the Dead Sea.

The climate of the Hula basin is hot, Mediterranean. Average annual rainfall is around 600 mm with variability in the period 1986–1997 of approximately 400–1,000 mm (Tsipiris and Meron 1998). Most of this precipitation falls in the winter. Average annual temperatures are around 20 °C with hot, dry summers with temperatures reaching 40 °C and cool, moist winters with temperatures rarely dropping below freezing (Hambright and Zohary 1998).

The Hula Wetland: Drainage and Restoration

The Pre-Drainage Wetlands

The Hula today bears the imprint of its history of human management, drainage, conservation, and restoration over a long period of time. Peat deposits started accumulating in the northern valley of the river Jordan around 20,000 BP, reaching depths of 8–9 m by the early twentieth century (Hambright and Zohary 1998). The wetland complex consisted of a shallow lake (Lake Hula, 15 km^2), an extensive area of *Cyperus*

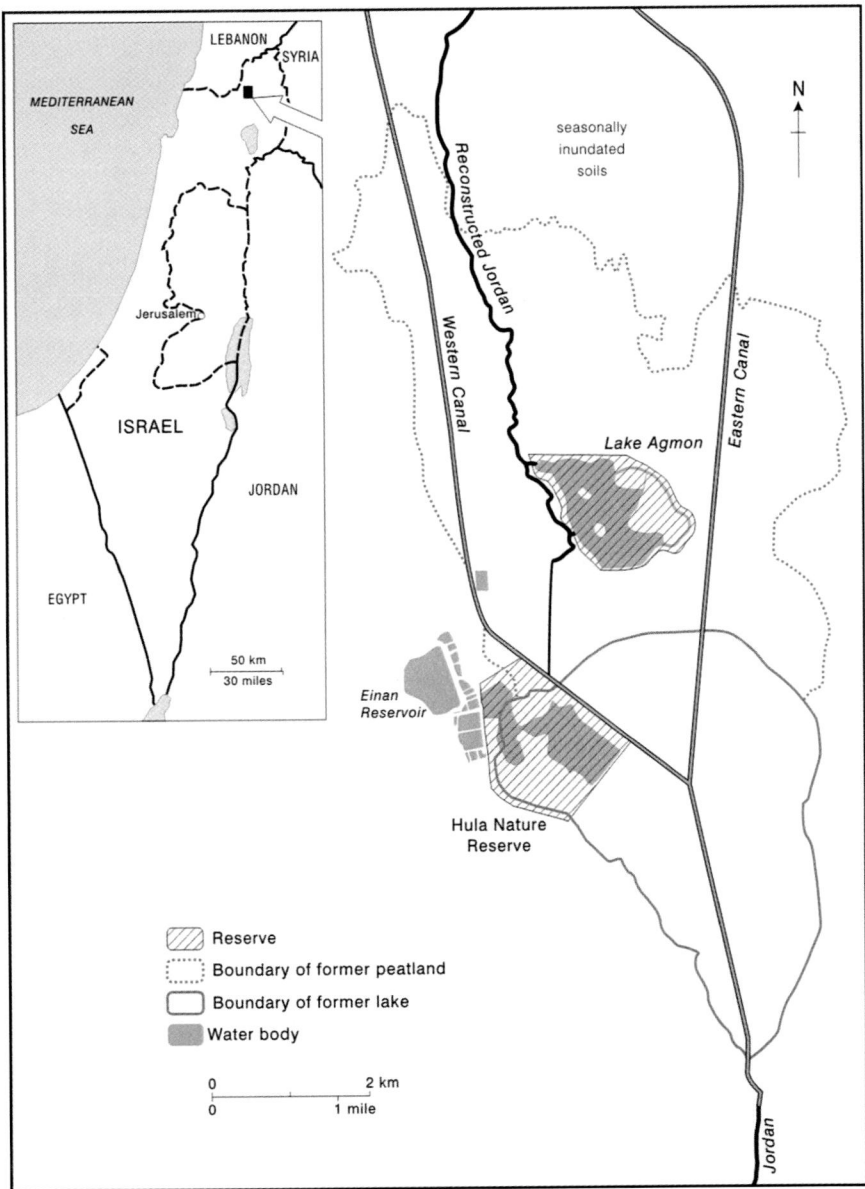

Fig. 1 Location (*inset*) and map of the Hula wetland (From Payne 2012, © with permssion from International Mire Conservation Group and International Peatland Society)

papyrus peatland, and areas of seasonally inundated inorganic soils (Fig. 1). The peatlands were a biodiversity hot spot hosting many species that are extremely rare in the region, providing an important stop for migratory birds and including several endemic species such as the fish *Acanthobrama hulensis* (Crivelli 2006) and frog

Latonia (formerly *Discoglossus*) *nigriventer* (Mendelssohn and Steinitz 1943). The American missionary William Thompson (1883) wrote that *the lake is alive with fowls, the trees with birds, and the air with bees. At all times fair, but fairest of all in early spring and at eventide...such is the Huleh.* The Hula Valley was successively part of the Ottoman Empire, the British Mandate in Palestine, and Israel, with relatively few detailed records before the mid-nineteenth century. Early accounts describe an ecosystem extensively shaped by its inhabitants, the Ghawarna. Crops including rice, maize, and millet were cultivated in areas of drained peat and other crops in the drier areas (Jones 1940). Buffalo were reared and large quantities of fish extracted (Dimentman et al. 1992). One of the most extensive industries involved the cutting of papyrus to make reed mats which were widely traded; large areas were cleared for this purpose (Crowfoot 1934; Larsson 1936). Taken together these disturbances created a mosaic of more- and less-disturbed habitats (Zohary and Orshansky 1947; Payne 2012).

Destruction of the Wetland

From at least the mid-nineteenth century, numerous people commented on the idea of draining the "wasteland" of the Hula (Payne 2012). Peripheral drainage by ditch digging has a long history, with areas of the seasonally inundated wetland being drained by the 1850s. However, the work required to drain the entire wetland was considerable and initial efforts ineffective. Drainage was finally achieved after the creation of the modern state of Israel. The aims of the Hula Drainage Project were the creation of new arable land, eradication of malaria, and release of greater water resources (Karmon 1960). From 1953 to 1958, the outflow from Lake Hula was widened and two major canals constructed through the peatland, converging toward the center of the former lake and continuing south along a newly deepened River Jordan (Fig. 1). A network of smaller canals and ditches extended from the main canals through the former peatland. A small area of the former lake and wetlands was enclosed by dykes to become the Hula Nature Reserve – Israel's first.

Post-Drainage and Restoration

The drainage project was not a great success. Decomposition and burning led to considerable loss of peat and subsidence. There were outbreaks of pests and dust storms. Agriculture required considerable application of pesticides and fertilizers. Considerable quantities of nutrients were leached downstream to Lake Kinneret, imperiling water quality in Israel's most important freshwater source. The Hula Nature Reserve proved inadequate to prevent the loss of species, with the enclosing dykes extensively leaking and much of the area drying out. Over a hundred species were lost from the area after drainage with many species lost entirely from Israel; some endemic species became extinct (Dimentman et al. 1992; Hambright and Zohary 1998). For instance, Melamed et al. (2011)

list seven wetland plant species which were abruptly lost from the site in this period: water parsnip *Berula erecta*, common frog's-bit *Hydrocharis morsus-ranae*, water clover *Marsilea minuta*, white water-lily *Nymphaea alba*, greater yellow cress *Rorippa amphibia*, hooded willow-herb *Scutellaria galericulata*, and common bladderwort *Utricularia australis*.

Due to these considerable problems, a restoration program was conceived in the 1980s. The restoration program was pragmatically based with the aims to preserve the peat soils and prevent nutrient leaching and promote ecotourism. In the early 1990s, the water table was raised with an underground partition to prevent groundwater seepage and new canals to distribute water around the site, the original path of the River Jordan was partly restored, and a new lake, Lake Agmon, was dug in the central part of the peatland. Lake Agmon is smaller (c.1 km^2) and shallower ($\sim$1 m) than the original Lake Hula.

The Hula Today

The Hula today consists of two main wetland areas: the Agmon restoration project and the Hula Nature Reserve along with the canals and rivers, and the Einan Reservoir and various peripheral ditch and spring habitats. The Agmon project has become an extremely popular tourist destination. A path network encircles the central lake with visitors touring the area by golf cart and bicycle; several bird hides are situated around the site. The lake is surrounded by reed beds with increasing cover of *Phragmites australis* and *Typha domingensis*. Into the drier areas further beyond the lake, there is considerable seasonal cover of grasses such as *Cynodon dactylon* (Henkin et al. 2011). Vegetation showed considerable variability in the early years of the restoration project with, for instance, a severe dieback event in *Typha domingensis* in 1996 (Simhayov et al. 2012). Although this is now reduced, considerable variability in vegetation composition continues (Kaplan 2012). The Hula Nature Reserve has changed considerably over time. The original area encompassed a small fraction of the original lake and adjacent papyrus peatland; however, this dried out in the following decades. Currently the vegetation consists of areas of secondary papyrus, open water, and buffalo-grazed meadow vegetation with considerable diversity in a small area.

Biodiversity

Although the biodiversity value of the Hula wetland is much reduced relative to its pre-drainage state, it remains one of the most important wetlands of the Middle East and a key nature conservation site for Israel. Although the number of species present is much lower than in previous times, it remains considerable. The site is a very important stopover on bird migration routes between Europe and Africa with millions of individuals passing through the site every year. In an encouraging sign, the Hula painted frog *Latonia nigriventer* which was believed to be extinct was rediscovered in 2011; recent research has shown it to represent a genus which was itself believed to have gone extinct in the Pleistocene (Biton et al. 2013). This perhaps gives hope that other "lost" species may eventually be recovered.

Future Challenges

Looking to the future, a key challenge for the Hula will be climate change. Water resources are already under considerable pressure in the region, and this situation is likely to be exacerbated by a rise in temperature or reduction in rainfall. Conserving wetlands in such a generally arid region is likely to be a considerable challenge. Both the Hula Nature Reserve and the Agmon project have been extremely successful locations for environmental education with very large visitor numbers; but managing the sites for both this public engagement function and for nature conservation is likely to be an increasing challenge. Finally, history shows that the Hula has been heavily impacted by the changing political landscape of the region; this situation is unlikely to change in the future.

References

Biton R, Geffen E, Vences M, Cohen O, Bailon S, Rabinovich R, Malka Y, Oron T, Boistel R, Brumfeld V, Gafny S. The rediscovered Hula painted frog is a living fossil. Nat Commun. 2013;4:1959.
Crivelli AJ. Acanthobrama hulensis. (2006) In: IUCN. IUCN red list of threatened species, Version 2010.3. IUCN; 2010.
Crowfoot GM. The mat looms of Huleh, Palestine. Palest ExplorFund QStatement. 1934;67:195–8.
Dimentman C, Bromley HJ, Por FD. Lake Hula: reconstruction of the fauna and hydrobiology of a lost lake. Jerusalem: The Israel Academy of Science and Humanities; 1992.
Hambright KD, Zohary T. Lakes Hula & Agmon: destruction and creation of wetland ecosystems in northern Israel. Wetl Ecol Manag. 1998;6:83–9.
Henkin Z, Walczak M, Kaplan D. Dynamics of vegetation development on drained peat soils of the Hula Valley, Israel. Mires Peat. 2011;9:1–11.
Jones RF. Report of the Percy Sladen expedition to Lake Huleh: a contribution to the study of the fresh waters of Palestine: the plant ecology of the district. J Ecol. 1940;28:357–76.
Kaplan D. Instability in newly-established wetlands? Trajectories of floristic change in the re-flooded Hula peatland, northern Israel. Mires Peat. 2012;9:1–10.
Karmon Y. The drainage of the Huleh swamps. Geogrl Rev. 1960;50:169–93.
Larsson T. A visit to the mat makers of the Huleh. Palest ExplorFund QStatement. 1936;68:225–9.
Melamed Y, Kislev M, Weiss E, Simchoni O. Extinction of water plants in the Hula Valley: evidence for climate change. J Hum Evol. 2011;60:320–7.
Mendelssohn H, Steinitz H. A new frog from Palestine. Copeia. 1943;4:231–3.
Payne R. A longer-term perspective on human exploitation and management of wetlands: the Hula Valley, Israel. Mires Peat. 2012;9:1–3.
Simhayov R, Litaor MI, Barnea I, Shenker M. The catastrophic dieback of Typha domingensis in a drained and restored East Mediterranean wetland: re-examining proposed models. Mires Peat. 2012;9:1–12.
Thomson WM. The land and the book: central Palestine and Phoenecia. London: T. Nelson & Sons; 1883.
Tsipris J, Meron M. Climatic and hydrological aspects of the Hula restoration project. Wetl Ecol Manag. 1998;6:91–101.
Zohary M, Orshansky G. The vegetation of the Huleh plain. Palest J Bot. 1947;4:90–104.

Coastal Sabkha (Salt Flats) of the Southern and Western Arabian Gulf

95

Ronald A. Loughland, Ali M. Qasem, Bruce Burwell, and Perdana K. Prihartato

Contents

Introduction	1174
Background	1174
Sabkha Landforms	1176
Examples of Sustainable Saline Agro-Systems Developed Within Sabkha Landforms	1180
Conclusion	1182
References	1182

Abstract

The importance of coastal saline salt flats along the southern and western Arabian Gulf coast (known as sabkha) is discussed. Sabkha landforms are being rapidly converted for industrial and urban coastal expansion. In Abu Dhabi, sabkha has also been successfully converted to productive saline agro-systems using both local mangrove (*Avicennia marina*) and saline tolerant algae (*Dunaliella salina*). The mangrove systems have developed into productive forest ecosystems

---

R. A. Loughland (✉)
Environmental Protection Department, Saudi Aramco, Dhahran, Ash Sharqiyah, Saudi Arabia
e-mail: ronloughland@gmail.com; ronald.loughland@aramco.com

A. M. Qasem
Environmental Engineering Div./ Environmental Protection Dept, Saudi Aramco, Dhahran, Saudi Arabia
e-mail: ali.qasem@aramco.com; amqnet@yahoo.com

B. Burwell
E-Map Department, Saudi Aramco, Dhahran, Ash Sharqiyah, Saudi Arabia
e-mail: bruce.burwell@gmail.com; bruce.burwell@aramco.com

P. K. Prihartato
Environmental Protection Department, Saudi Aramco, Dhahran, Ash Sharqiyah, Saudi Arabia
e-mail: perdana.karim@gmail.com; perdana.prihartato@aramco.com

© Crown 2018
C. M. Finlayson et al. (eds.), *The Wetland Book*,
https://doi.org/10.1007/978-94-007-4001-3_185

supporting local fisheries. The algae systems developed in ponds within sabkha have produced a natural pharmaceutical (beta-carotene), biomass and oil for production of biofuel. The use of sabkha landforms as a sink for heated saline brine produced by the desalination and industrial use of seawater is preferable than its current discharge to the marine environment where it is causing ecological impacts. The alteration of sabkha landforms into saline agro-systems is argued to be more productive than the permanent conversion of these same landforms for industrial or residential development.

Keywords
Arabian Gulf · Biofuels · Mangroves · Sabkha · Saline agro-systems · Saline algae

Introduction

This chapter provides information on coastal saline salt flats along the southern and western Arabian Gulf coast (known as sabkha). The extensive natural coastal sabkha ecosystems are discussed, and their incorporation as part of productive halophytic plant agro-systems is recommended. Examples from the region of the utilization of sabkha landforms for the development of sustainable mangrove (*Avicennia marina*) ecosystems and hypersaline microalgae (*Dunaliella salina*) ponds for the cultivation of pharmaceuticals and biofuels are provided. This chapter is intended to showcase the inherent natural values of the unique coastal sabkha of the Arabian Gulf and provide sustainable alternatives for its development, particularly its potential for saline agro-systems, providing nonmarine sinks for industrial saline brine currently being discharged along the Arabian Gulf coast.

Background

The Arabian Gulf is a relatively shallow sea surrounded by large arid landmasses with little rainfall and therefore minimal terrestrial runoff (Qurban et al. 2011). The main exception to this situation is the seasonal flows of the Shatt Al-Arab (Euphrates and Tigris rivers in Iraq) that drain higher elevations to the west of the Gulf. Seawater input through the Straits of Hormuz (from the Gulf of Oman) is also restricted (Sheppard et al. 2010), and sea surface evaporation is high in both summer and winter, with the resulting denser saline waters sinking and driving local current patterns. This saline-driven current phenomenon is particularly pronounced in the Abu Dhabi southern embayment between the Qatar Peninsula and the Strait of Hormuz (Sheppard and Loughland 2002).

The southern and western shoreline of the Arabian Gulf is generally flat with minor relief, resulting in vast areas of shoreline that are either intertidal or supratidal, with both being inundated by tidal and/or storm surges. One of the most unique features of the southern and western Arabian Gulf is the formation of extensive supratidal coastal saline sabkha. Sabkha is a regional term for evaporative coastal salt flats, and Abu Dhabi Emirate and the Eastern Province of Saudi Arabia have the most developed sabkha landforms (Barth and Böer 2002). The sabkhas of Abu Dhabi were some of the first to be studied in detail (Curtis et al. 1963; Evans et al. 1969) and were known as the *Unique Trucial Coast Sabkha*. The development of the *"Evan's Line,"* which was the first study transect across a coastal sabkha on the Arabian Gulf coast, and some early borehole sites southwest of Abu Dhabi City which described the structure, development, and sediment of coastal sabkha were instrumental in the early understanding of these unique coastal formations. The majority of the coastline of Abu Dhabi is an active coastal sabkha and is recognized as the largest in the world. It is around 300 km long and extends continuously for around 20 km or more inland. The hydrology of sabkha systems is very variable. At coastal sites, storms and gales may force seawater on the shallow and gently sloping southern Gulf shores, thus inundating extensive areas of coastal sabkha. Winter rainfall is often also trapped on the sabkha surface until evaporated. The sabkha may remain flooded for many months because the mineral precipitates, principally gypsum or anhydrite, seal the sabkha base forming natural sabkha ponds that increase in salinity as water evaporates.

The Abu Dhabi Gulf coast also has numerous inshore and nearshore islands, most of which are simply part of a former more extensive and continuous coastal sabkha ecosystem eroded by post-Quaternary storms and inundated by a higher sea level. Because of their proximity to marine waters and their shallow soft substrate and saline groundwater, sabkha landforms were traditionally left undeveloped. Because sabkha saline crust inhibits plant establishment and growth, they were also considered to be of little value for grazing. As a result, in most coastal areas, large sabkha landforms still persist. It has only been in recent decades that sabkha landforms adjacent to coastal communities became dumping sites for solid waste or were filled with dune sand and used for makeshift developments, despite their inadequate and corrosive foundation. Today, sabkhas are being developed at a much faster rate and are being drained (dewatered), sheet pilled, and filled with dune sand for expanding permanent developments around major towns and cities. Development of sabkha areas for commercial, residential, and industrial purposes is likely to continue at a rapid pace because the price of coastal land continues to rise.

The Arabian Gulf's entire southern and western shoreline has become increasingly developed over the past few decades and now is extensively utilized for residential, commercial, and industrial development (Loughland and Saji 2007; Loughland et al. 2011), having one of the highest proportions of coastal-based desalination, power generation, hydrocarbon, and petrochemical industries in the world. Desalination of seawater is utilized to provide potable and industrial feedwater, and the resulting marine discharge is hot saline brine. Both heat and salt are natural marine ecosystem stressors in the Arabian Gulf with most species being at their biological threshold, and only a small sustained increase above ambient conditions for water temperature and salinity can be lethal. Seawater is also used for industrial cooling, and large quantities of heated and slightly elevated saline cooling water are discharged back into the gulf annually. The volume and hence impacts of these combined discharges on the marine environment can be reduced through utilizing the same discharges as a source of saline brine for the cultivation of saline algae in nearby extensive coastal sabkha landforms.

Sabkha Landforms

Sabkhas occur along the southern and western Arabian Gulf coastline occupying the sea-land interface, where infrequent supratidal inundation and evaporation of saline groundwater that is drawn upward by capillary action saturate, forming a well-defined salt crust. The physical and chemical profile of sabkha is provided in detail by Evans and Kirkham (2002), with the surface being usually less than a meter in thickness; beneath the salt crust is a layer of precipitate soft gypsum mush, and below that is the saline groundwater. Ecologically, sabkha has limited terrestrial habitat value; however, windrows of detritus including seeds (that often germinate in adjacent sand hummocks) along their edges provide concentrated forage resources for wildlife, and naturally folding sabkha salt crust formations (Teepees, Fig. 1) can also provide refuge habitat to small animals (Loughland and Cunningham 2002). Despite its high salinity, sabkha is actually a productive surface for microorganisms. Coastal sabkha is also often coated in a stromatolitic algal mat with a reducing layer below. These mats are a mixture of cyanophytes, diatoms, and bacteria. When the sabkha surface dries out, the algal layers begin to crack and peel, and the entire sabkha surface becomes a rough mosaic of drying layers. Permanent saline pools are a notable feature of coastal sabkha; these can often also have a subterranean connection to the sea and thus fluctuate tidally.

Some of the largest coastal sabkha occur along the west coast of Abu Dhabi Emirate into the Kingdom of Saudi Arabia (KSA), with the largest being Sabkha Matti straddling both the UAE and KSA borders (Fig. 2). In Abu Dhabi, Sabkha Matti continues for 100 km inland, crossing into Saudi Arabia where it continues for a further 100 km. As a result, Sabkha Matti covers many hundreds of square kilometers. Sabkha in Abu Dhabi Emirate has been profiled and mapped (Fig. 2).

In the Eastern Province of the KSA, sabkhas were classified using remote sensing. All sabkha within 100 km of the Arabian Gulf coast was recently mapped as part of this work (Figs. 3 and 4) with sabkha area representing around 2,775 km^2 or 4.5 % of the total land area examined. The largest sabkha covered an area of around 138 km^2, and many sabkhas occurred close to existing large coastal industrial developments such as Jubail, Ras Tanura, and Dammam meaning that saline brine discharges from these industrial sites that are normally discharged into the marine environment instead could be discharged inland within sabkha formations to create sustainable saline algae agro-systems.

Where present, coastal sabkha prevents coastal flooding and erosion and is thus economically valuable. Coastal sabkha along the Arabian Gulf coast also has both aesthetic value and cultural importance as these were the same sabkhas that were extensively studied and documented as part of the early petroleum geology undertaken in the region. Sadly, most of the original Evan's Line has been lost due to the development of the sabkha southwest of Abu Dhabi City.

Fig. 1 Coastal sabkha in Abu Dhabi illustrating teepee salt crust formations. These formations are utilized by small animals (e.g., Cape hare *Lepus capensis*) for refuge (Photo by Ronald Loughland) © Rights remain with the author)

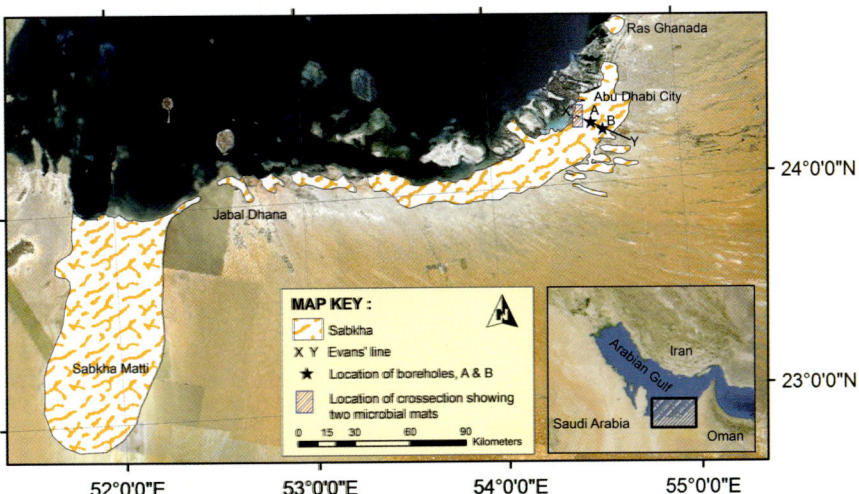

Fig. 2 The location of coastal sabkha along Abu Dhabi Emirate Coast (illustrating the Evan's Line and locations of two exploratory boreholes) (Saudi Aramco)

Fig. 3 Extent of coastal zone in the KSA (100 km) where sabkha landforms were mapped (Saudi Aramco)

Fig. 4 Extent of coastal sabkha in the Eastern Province of KSA (Saudi Aramco)

Examples of Sustainable Saline Agro-Systems Developed Within Sabkha Landforms

In Abu Dhabi Emirate, coastal sabkha has been successfully converted to productive saline agro-systems using local mangrove (*Avicennia marina*) by the Public Works Department, Private Department of the President, and other organizations such as the Emirates Heritage Club. In Abu Dhabi, early mangrove planting was undertaken around Abu Dhabi City with varied success, and in the late 1990s, the Public Works Department surveyed existing mangrove habitats using theodolite, dumpy level, and leveling staff to determine their optimal tidal range above mean sea level (Loughland and Saenger 2001). New mangrove plantation sites planned within sabkha areas were then excavated down to this optimal tidal level, and with the provision of tidal channels to allow adequate flushing and drainage, sabkha landforms were converted to lower saline productive intertidal areas which were extensively planted with local mangrove (*A. marina*). The results were exceptional, with large areas of sabkha around Abu Dhabi City, on near shore islands, and at coastal sabkha sites in central and western Abu Dhabi Emirate being transformed to sustainable tidal irrigated saline agro-systems. Invertebrates, fish, and bird species soon colonized these new plantations, and today Abu Dhabi has some of the most productive mangrove habitats in the gulf (Saenger et al. 2002; Blasco et al. 2004).

Also in Abu Dhabi Emirate, coastal sabkha has been converted to grow saline-tolerant microalgae within hypersaline agro-systems. This has been undertaken on a smaller scale than for mangroves; however, it has proven extremely productive in initial trials conducted on Futaisi Island adjacent to Abu Dhabi City. Opposite to the concept of mangrove establishment through decreasing salinity within sabkha landforms through excavations and seawater flushing, the conversion of sabkha for the development of algal systems requires increasing salinity within shallow ponds (<1 m) developed on top of sabkha by mounding soil around the edge of sabkha. The algal ponds are flooded with seawater or saline groundwater and maintained at a depth of around 15 cm. Surface evaporation results in hypersalinity of the pond water, and the resulting precipitates provide a natural seal for the base and walls of the ponds.

Once the pond water has reached high enough salinity levels (200 ppt), the pond is inoculated with saline-tolerant algae such as *Dunaliella salina*, which has naturally occurring strains in the Arabian Gulf. The algae are able to persist and thrive in the hypersaline brine; however, other organisms that normally control their growth in seawater cannot, and this results in a monoculture of the required algae species.

Dunaliella salina has been successfully grown in research scale saline agro-systems within sabkha landforms on Al Futaisi Island in Abu Dhabi. More recent trials have also been conducted at Kiran in southern Kuwait. The objective was to test the efficiencies of producing a natural pharmaceutical known as beta-carotene. Abu Dhabi trials indicated that *D. salina* algae cultivation in saline ponds is extremely productive and rivals other global sites in its annual production potential. This is due to Abu Dhabi's abundant sabkha ecosystems; warm, sunny, and dry climate; and high salinity gulf waters (Fig. 5).

Other saline-tolerant algae species (e.g., *Pleurochrysis carterae*) can also be grown in saline ponds to produce biomass and oils used for developing biofuels. Algae produce around 36 times more oil per area than that sourced from vegetable crops such as canola, corn, or sunflower. Algal oil has a similar caloric value to that of canola oil, being around 36,000 kj/kg of oil. Producing biofuel in coastal sabkha from saline-based algae in saline agro-systems makes ecological, social, and economic sense, as unlike vegetable crops, no freshwater resources are required, and no arable land is utilized. Hence, the value of terrestrial environments is not degraded, and food crops are not replaced by biofuel production. When brine effluents from coastal industry are discharged into coastal sabkha ponds, then the overall benefits are increased and the system is significantly more efficient. To increase the benefits even further, the industrial developments with respect to flue gas emissions such as CO_2 and NOX could also be utilized as a source of fertilization within the sabkha algae agro-systems to increase algae productivity and decrease overall air emissions (Negoro et al. 1991). The industrial biodiesel (diesel type 1) is possibly the most practical biofuel that can be developed from saline algae, and the demand for "green biodiesel" is expanding rapidly (Chisti 2007). Although the sabkha landforms utilized for mangroves and algae production require physical alteration, their marine interface is still maintained, and this is preferable to permanent sabkha conversion through reclamation for residential, commercial, or industrial land use.

Fig. 5 Sabkha ponds developed for the cultivation of *Dunaliella salina* in Abu Dhabi Emirate. The red color is the natural pharmaceutical beta-carotene within the algae cells that is ready for harvesting (Photo by Ronald Loughland © Rights remain with the author)

Conclusion

Saline sabkhas along the southern and western Arabian Gulf are unique coastal landforms, and when combined with the abundant saline water resources that dominate in the gulf coastal region, these landforms have huge potential for sustainable productivity utilizing indigenous halophytic plant species such as *Avicennia marina* and *Dunaliella salina*.

The long-term conservation of coastal sabkha will rely on its utilization within the overall development of the coastal zone, and lessons from the observations of natural sabkha ecosystems may provide the solutions. The utilization of coastal sabkha as a sustainable sink for industrial saline brine and incorporating various productive saline agro-systems using halophytes to absorb CO_2 and NOX emissions from these industries would mean that the productivity of the agro-system would increase and the sabkha would be viewed not only as valuable environmental sinks but also as productive systems with a substantial economic and social value.

References

Barth H-J, Böer B, editors. Sabkha ecosystems. Dordrecht: Kluwer; 2002. p. 363.

Blasco F, Saenger P, Auda Y, Aizpuru M, Loughland RA, Youssef A. Mapping main coastal habitats and mangroves. In: Loughland RA, Al Muhairi FS, Fadel SS, Al Mehdi AM, Hellyer P, editors. Marine atlas of Abu Dhabi. Abu Dhabi: Emirates Heritage Club; 2004.

Chisti Y. Biodiesel from microalgae. Biotechnol Adv. 2007;25:294–306.

Curtis R, Evans G, Kinsman DJJ, Shearman DJ. Association of dolomite and anhydrite in the recent sediments of the Persian Gulf. Nature. 1963;197:679–80.

Evans G, Kirkham A. The Abu Dhabi sabkhat. Distribution of sabkhat in the Arabian Peninsula and adjacent countries. In: Barth H-J, Böer B, editors. Sabkha ecosystems. Dordrecht: Kluwer; 2002. p. 353.

Evans G, Schmidt V, Bush RP, Nelson H. Stratigraphy and geological history of the sabkha, Abu Dhabi, Persian Gulf. Sedimentology. 1969;12:145–59.

Loughland RA, Cunningham PL. Vertebrate fauna of Sabkhat from the Arabian Peninsula: a review of Mammalia, Reptilia and Amphibia. In: Barth H-J, Böer B, editors. Sabkha ecosystems. Dordrecht: Kluwer; 2002. p. 255–66.

Loughland RA, Saji B. Remote sensing: a tool for managing marine pollution in the Gulf. In: Barth H-J, Böer B, editors. Gulf ecosystems. Dordrecht: Kluwer; 2007.

Loughland RA, Saenger P. Report to the Public Works Department of Abu Dhabi on the methods for the development of mangrove plantations in Abu Dhabi Emirate. Abu Dhabi: Department of Environmental Research, Emirates Heritage Authority; 2001

Loughland RA, Wyllie A, Al-Abdulkader K. Anthropogenic induced changes along the Gulf coast of KSA from 1967–2010. In: Piacentinj T, editor. Geomorphology. Rijeka: Intech; 2011. p. 333.

Negoro M, Shioji N, Miyamoto K, Miura Y. Growth of microalgae in high CO2 gas and effects of SOx and NOx. Appl Biochem Biotechnol. 1991;28–9:877–86.

Qurban M, Kumer K, Al-Abdulkader K, Loughland RA. Overview of the marine and coastal habitats. In: Loughland RA, Al-Abdulkader K, editors. Marine atlas of the Western Arabian Gulf. Saudi Aramco: Kingdom of Saudi Arabia; 2011. p. 455.

Saenger P, Blasco F, Loughland R, Youssef A. In: Salim Javid, de Soyza AG, editors. Research and Management Options for Mangrove and Salt Marsh Ecosystems – Proceedings of the 2nd International Symposium and Workshop on Arid Zone Environments (22–24 Dec 2001, Abu

Dhabi, UAE). Abu Dhabi: Environmental Research and Wildlife Development Agency (ERWDA); 2002. p. 196–8

Sheppard C, Loughland R. Coral mortality and recovery in response to increasing temperature in the southern Arabian Gulf. Aquat Ecosyst Health Manage. 2002;5(4):1–8.

Sheppard C, Al-Husiani M, Al-Jamali F, Al-Yamani F, Baldwin R, Bishop J, Benzoni F, Dutrieux E, Dulvy NK, Rao S, Durvasula V, Jones DA, Loughland RA, Medio D, Nithyanandan M, Pilling GM, Polikarpov I, Price A, Purkis S, Riegl B, Saburova M, Namin KS, Taylor O, Wilson S, Zaina K. The Gulf: – a young sea in decline. Mar Pollut Bull. 2010;60:13–38.

Lake Seyfe (Turkey)

96

Serhan Cagirankaya and Burhan Teoman Meric

Contents

Introduction	1186
Conservation Status	1187
Management Structure	1188
Hydrological Aspects	1188
Geological Aspects	1189
Biodiversity	1189
Habitats	1189
Flora	1189
Fish	1190
Amphibians and Reptiles	1190
Birds	1190
Mammals	1192
Cultural and Social Aspects	1192
Archaeology	1192
Past and Present Land Use	1192
Natural Resource Use	1193
Agriculture and Livestock	1193
Recreation and Tourism	1194
Wetland Management Plan	1194
Threats and Future Challenges	1194
References	1195

Abstract

Lake Seyfe is located in the tectonic depression of north-eastern Kırşehir Province and the centre of the Anatololian Region. Lake Seyfe is a Nature Conservation Site and Ramsar Site covering 10,700 ha. It is also recognized internationally

---

S. Cagirankaya (✉) · B. T. Meric (✉)
Wetlands Division, Ministry of Forests and Water Affairs, General Directory of Nature Conservation and National Parks, Ankara, Turkey
e-mail: c.serhan@gmail.com; scagirankaya@ormansu.gov.tr; teoman.meric@gmail.com

© Springer Science+Business Media B.V., part of Springer Nature 2018
C. M. Finlayson et al. (eds.), *The Wetland Book*,
https://doi.org/10.1007/978-94-007-4001-3_142

as a Key Biodiversity Area, Important Bird Area, and Important Plant Area. The site supports rare bird species such as the great bustard *Otis tarda*, Eurasian crane *Grus grus*, ruddy shelduck *Tadorna ferruginea*, and large clusters of greater flamingo *Phoenicopterus roseus*. The site was designated as a Natural Site of First Degree in 1989. It was designated as a Nature Conservation Site in 1990 and part of the area listed as a Wetland of International Importance under the Ramsar Convention in 1994. The visitor center is located at Seyfe Village and is owned by the Directorate General of the Nature Conservation and National Parks. The Lake Seyfe Ramsar Site meets three of the nine criteria of the Ramsar Convention.

There is no significant stream source at the site. Local people use the groundwater that feeds the lake for agriculture. Groundwater resources were sufficient until 2000, but have decreased substantially since 2002, and since 2007 the Ministry has forbidden any new uses of groundwater within the Seyfe basin. The water from the lake is brackish and contains sodium, so it cannot be used directly for agriculture, whereas the groundwater feeding into the lake is used for household needs as well as for irrigation.

Management Structure Lake Seyfe comes under the competency of the Directorate General of Nature Conservation and National Parks of the Ministry of Forestry and Water Affairs and the Directorate of Forestry and Water Affairs of the Provincial Kırşehir for its designation as a Ramsar Site and its status as a Nature Conservation Site. The Ministry of Environment and Urbanization is also responsible for the site as it has a conservation status as a Natural Site. The area within the boundaries of the Ramsar Site is state property, while the area outside the boundaries of the Ramsar Site consists of state, village entities, and private property.

Keywords

Wetlands · Lake · Natural resources · Ramsar · Nature conservation · Management plan

Introduction

Lake Seyfe is located in the tectonic depression of north-eastern Kırşehir Province. It is situated in Mucur District, 220 km from Ankara and 30 km from Kırşehir, which has a population of 15,000. There are six villages in the vicinity of the lake, namely, Seyfe, Gümüşkümbet, Yazıkınık, Budak, Kızıldağ, and Eskidoğanlı.

Lake Seyfe is a Nature Conservation Site and Ramsar Site covering some 10,700 ha, located at coordinates 39°12′N 034°25′E and an elevational range of 1,120–1,200 m (Fig. 1). It is also recognized internationally as a Key Biodiversity Area, Important Bird Area, and Important Plant Area (Eken et al. 2006; Çagırankaya and Meriç 2013). The site supports rare bird species such as the great bustard *Otis tarda*, Eurasian kestrel *Falco tinnunculus*, Eurasian crane *Grus grus*, ruddy shelduck *Tadorna ferruginea*, and large clusters of greater flamingo *Phoenicopterus roseus*.

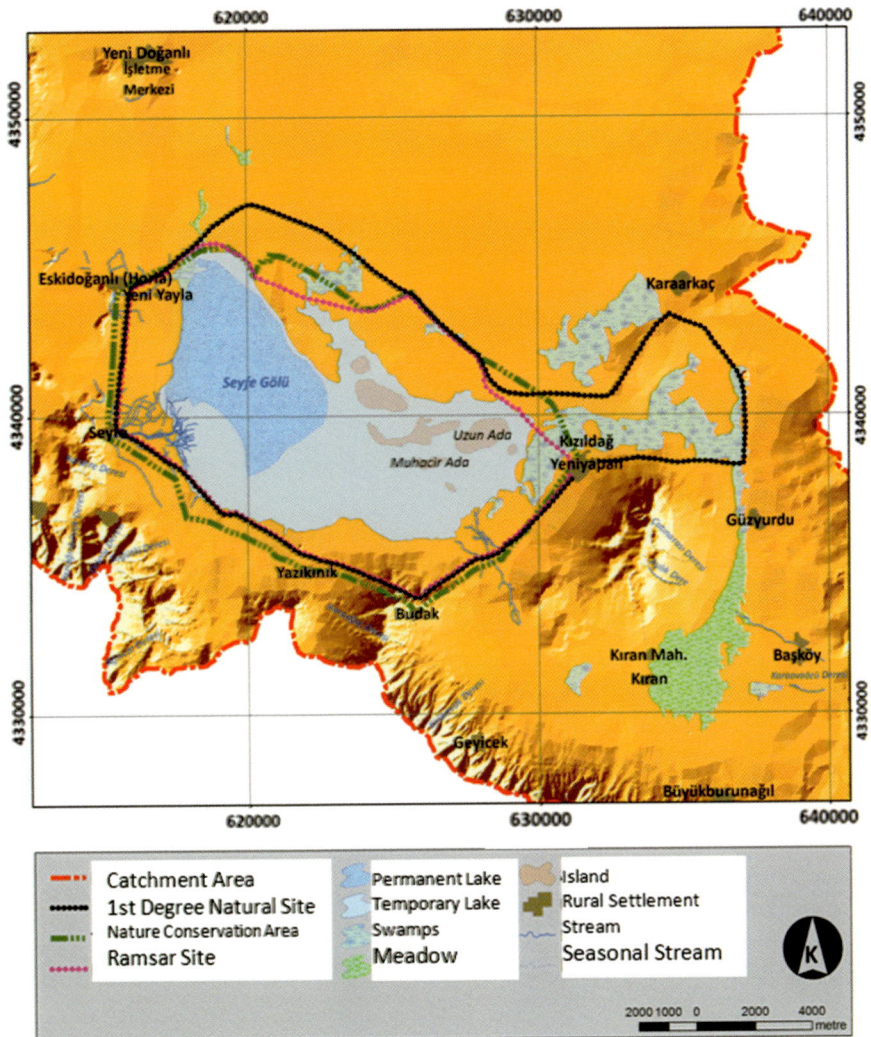

Fig. 1 Map showing the location and boundaries of conservation designations of Lake Seyfe

Conservation Status

A 23,585 ha portion of the site was designated as a Natural Site of First Degree in 1989. It was designated as a Nature Conservation Site in 1990 and part of the area listed as a Wetland of International Importance under the Ramsar Convention in 1994. The visitor center is located at Seyfe Village and is owned by the Directorate General of the Nature Conservation and National Parks. The Lake Seyfe Ramsar Site meets three of the nine criteria of the Ramsar Convention (Cagirankaya and Meric 2013). These are:

Criterion 2 (the site supports species listed in IUCN Red List categories): The site supports globally threatened species such as the great bustard, imperial eagle *Aquila heliaca*, and greater spotted eagle *Aquila clanga*.

Criterion 4 (large aggregations of waterbirds stage at the site during migration periods): greater flamingos inhabit the site in large numbers in winter (32,000 flamingos were recorded in 1987) and white storks *Ciconia ciconia* form large colonies at the site (1,300 individuals were recorded during the migration period).

Criterion 5 (20,000 waterbirds regularly inhabit the site): The highest number of waterbirds recorded between 1969 and 1970 was 152,380. A total of 32,000 waterbirds were counted during a census in 1986.

Management Structure

Lake Seyfe comes under the competency of the Directorate General of Nature Conservation and National Parks of the Ministry of Forestry and Water Affairs and the Directorate of Forestry and Water Affairs of the Provincial Kırşehir for its designation as a Ramsar Site and its status as a Nature Conservation Site. The Ministry of Environment and Urbanization is also responsible for the site as it has a conservation status as a Natural Site. The area within the boundaries of the Ramsar Site is state property, while the area outside the boundaries of the Ramsar Site consists of state, village entities, and private property.

Hydrological Aspects

The streams of Akpınar, Horla, Seyfe, and Özlühüyük feed into Lake Seyfe, Seyfe Stream being the most important source for the lake. However, the volume of water coming from Seyfe Stream into Lake Seyfe has been decreasing recently (Çelik et al. 2008).

Central Anatolia is characterized by a continental climate (Central Anatolian steppe climate). Since the site is positioned in a low-lying part of the basin, the surface area of the lake fluctuates greatly according to the amount of precipitation and the seasons. Its average depth in winter is 1 m. Since the area receives low precipitation combined with high evaporation rates, streams feeding the lake dry out. This causes the surface area of Lake Seyfe to fall to 1,560 ha and its depth to 60–70 cm, causing most parts of the lake to turn into salt swamps. Brackish and freshwater marshes, which are important for birds that are breeding or wintering in the lake, are located in the east and southeast of the temporary lake area (Reis and Yilmaz 2008).

There is no significant stream source at the site. Local people use the groundwater that feeds the lake for agriculture. Groundwater resources were sufficient until 2000, but have decreased substantially since 2002, and since 2007 the Ministry has forbidden any new uses of groundwater within the Seyfe basin. The water from the lake is brackish and contains sodium, so it cannot be used directly for agriculture, whereas the groundwater feeding into the lake is used for household needs as well as for irrigation.

Geological Aspects

Metamorphic rocks prevail in the basin. These rocks have been formed when the existing rocks, which contain remnants of organisms (i.e., sedimentary rocks), were transformed as a result of being subjected to high pressure and heat. Formations dating back to 545 B.P. and until 251 million years B.P. constitute the base of the basin. These formations which exist in the northwestern and southwestern parts of the lake consist of schist and marble. During orogenesis, granite and diorite type of intrusions occurred. With a deep-sea medium effect, having been loaded discordantly on the former base, calcareous rocks formed within the basin, which had begun to remain under water as of the Eocene period. The sea then started to become shallow, and faults were formed by the occurring intrusions. The most important fault line lies in the direction of Gümüşkent-Yenidoğanlı. The sources of water feeding into the lake surface are through the faulted zones. The basin took its current form in the Quaternary period, having been filled with silt, sand, and pebbles coming from the immediate surroundings of the plain. These loaded materials minimize the inflow of groundwater into the lake and constitute a natural seawall for the lake area. Neogene units are also recorded in the Paleozoic formations.

Biodiversity

Habitats

The Ramsar Site consists of a shallow salty lake and wet grasslands around the lake. There are areas of steppe and dry agricultural lands surrounding the wetland.

Flora

There are no trees or bushes around Lake Seyfe. The globally endangered *Centaurea pergamacea* and *Lepidium caespitosum* species occur in the steppes around the lake. Semishrub forms of *Lycium depressum* are found in the north of the lake. The name of the lake comes from Seyfe Village, which is situated to the west of the lake. In Seyfe Village, there are fruit gardens and poplar and willow groves. There are no aquatic plants because the lake is salty. In the salt marshes, there are plant species such as *Halocnemum strobilaceum*, *Salicornia prostrata*, *Salsola inermis*, *Panderia pilosa*, *Petrosimonia brachiata*, *Krascheninnikovia ceratoides*, *Camphorosma monspeliaca*, *Gypsophila perfoliata*, *Frankenia hirsuta*, *Limonium iconicum*, and *Limonium globuliferum* (Boissieri 1867–1888; Davis 1965–1988; Heywood and Tutin 1964–81).

In addition, in the channels and where the water sources empty into the lake, there are fresh water plants such as common reed *Phragmites australis*, *Sparganium erectum*, cattail *Typha angustifolia*, fennel pondweed *Potamogeton pectinatus*, and yellow iris *Iris pseudacorus*.

Fish

No fish species occur in the main lake body since the lake water is salty and contains sodium. Only two small fish species (i.e., 5–6 cm long) are found where freshwater enters and disperses into the lake, namely, *Aphanius chantra* and *Spirlinus* sp. These fish species are not commercially important but are important ecologically because they form part of the diet of pelicans and egrets (Ahıska and Karabatak 1994).

Amphibians and Reptiles

Five amphibian and 28 reptile species have been recorded at Seyfe Lake. Globally threatened species on the IUCN Red List (http://www.iucnredlist.org/) include Clarks' lizard *Lacerta clarkorum* (endangered), spur-thighed tortoise *Testudo graeca* (vulnerable), and European pond turtle *Emys orbicularis* (near threatened).

Birds

Salt marshes in the east of the lake are important feeding and breeding areas for birds. Islets in Lake Seyfe also provide breeding sites for birds. These habitats, which are rich in food resources and of different ecological character including safe islets located far from hunters, together with the large steppes, which vary gradually from salt to freshwater swamps, and the salt lake, provide an ideal breeding, feeding, and sheltering area for thousands of birds from different species. It is of special importance for rare species such as the great bustard and Eurasian crane *Grus grus* as well as for congregatory species such as the greater flamingo (Figs. 2 and 3).

Records have shown that there were 205 bird species in and around the lake in 1999. The number of birds at the lake reached peak counts during migration and during winter. The greater white-fronted goose *Anser albifrons*, common shelduck *Tadorna tadorna*, ruddy shelduck *Tadorna ferruginea*, Eurasian teal *Anas crecca*, and Eurasian coot *Fulica atra* gather in large flocks (Ayaş and Turan 2001).

Lake Seyfe is also one of the important breeding areas for waterbirds in the country. At the islets at the east of the lake, the important breeding species are the greater flamingo, great white pelican *Pelecanus onocrotalus*, Eurasian spoonbill *Platalea leucorodia*, little egret *Egretta garzetta*, red-crested pochard *Netta rufina*, black-winged stilt *Himantopus himantopus*, pied avocet *Recurvirostra avosetta*, spur-winged lapwing *Vanellus spinosus*, Mediterranean gull *Larus melanocephalus*,

Fig. 2 Greater flamingos at sunset (Ömer Çetiner© Rights remain with the author)

Fig. 3 Greater flamingos in flight (Ömer Çetiner© Rights remain with the author)

black-headed gull *Larus ridibundus*, and gull-billed tern *Sterna nilotica*. In addition to these waterbirds, white storks also gather around the lake. Hundreds of thousands of ducks also stop over at the site in autumn. Steppes around the lake also provide feeding and breeding areas for the great bustard. Local people say that the number of birds at the lake has decreased dramatically, because of drought.

Mammals

There are 31 mammal species that have been recorded around Seyfe Lake. These include the following, including their IUCN Red List categories: lesser horseshoe bat *Rhinolophus hipposideros* (least concern), long-fingered bat *Myotis capaccinii* (vulnerable), lesser blind mole rat *Spalax leucodon* (data deficient), Eurasian otter *Lutra lutra* (near threatened), greater mouse-eared bat *Myotis myotis* (least concern), common bent-wing bat *Miniopterus schreibersii* (near threatened), Anatolian squirrel *Sciurus anomalus* (least concern), gray dwarf hamster *Cricetulus migratorius* (least concern), and forest dormouse *Dryomys nitedula* (least concern).

Cultural and Social Aspects

Archaeology

Findings from archaeological excavations have shown that the first settlements in the area occurred in the Bronze Age (between 3500 and 200 BC). There are 20 hoyuks (ancient settlements) and tumulus from this age around the lake and its surroundings. Studies have shown that the people who settled at the coasts have engaged in agricultural activities throughout history. It is believed that the residents of the lake lived on hunting migratory birds and on agricultural products. Historical artifacts excavated from the hoyuks have been preserved in the Kırşehir Museum.

Past and Present Land Use

Lake Seyfe was an important area for birds in the past, especially during winter and during migration when the number of birds was high. However, the site has been adversely affected as a result of drainage channels, the unsustainable use of groundwater, and a shifting agricultural pattern toward crops requiring more water. There is a 20 cm thick salt layer on the lake which completely dries out in the summer months. Salt from this layer spreads over the agricultural lands around the lake, threatening human health and causing financial losses amounting up to millions of Turkish liras per year. The groundwater level has been decreasing as a result of use of the water in the lake, such that the depth of the wells is now 200 m.

Agricultural yields have started to decrease due to droughts in the region, and reed harvesting has also been adversely affected because it destroys the reed beds. Poplar and willow groves have also become extinct. Frost incidents have also increased as the surface of the lake has decreased and apple production has come to a standstill.

Natural Resource Use

The main occupation of the local people around the lake is agriculture and animal husbandry (Fig. 4); about 90% of the people make their living through these activities.

Agriculture and Livestock

Rain-fed agriculture is practiced in 91.7% of the basin, while the remaining area is used for irrigated agriculture. The main agricultural products are wheat, sugar beet, barley, lentils, chickpeas, beans, oats, and sunflowers. There are also orchards and groves but these are rare. There are almost a thousand caisson wells with a maximum depth of 10 m, which are used for irrigation purposes, but these are becoming harder to use because of a shortage of water. In the past 3 years, clover and trefoil cultivation has been promoted under the Agricultural Land Protection Programme (ÇATAK) to encourage a transition from irrigated agriculture to rain-fed agriculture. In four villages surrounding Lake Seyfe, a total of 1,350 ha has been included in the project. The water and humidity ratio, which was adequate to grow wheat in the past, decreased between

Fig. 4 Lake Seyfe, showing cultivated lands in the lowlands and range lands on higher slopes (Serhan Cagirankaya© Rights remain with the author)

2000 and 2007, and farmers in the region have reported that cultivating wheat has become harder compared to the past (Bozkır Çevre Derneği 2009).

Small ruminant husbandry in pasture lands has become a prominent occupation as a result of the large meadows in the basin. Livestock production activities in the region are carried out using combined facilities.

Recreation and Tourism

The historical and cultural treasures of the hoyuks and tumuli in the area, the bird population that the lake supports, and the beautiful view of the lake and its surroundings have increased the significance of the site in terms of nature tourism.

Wetland Management Plan

The wetland management plan for the site (Çagırankaya and Meriç 2010) has been implemented since 2011. The main objective of the plan is to conserve and preserve Lake Seyfe and its wetland ecosystems and to ensure the sustainable use of the area. The management plan has one main objective, five operational objectives, and 34 activities. The plan will be revised in 2016.

Threats and Future Challenges

Lake Seyfe was remarkable for its bird populations in the past, with a rich diversity and great abundance of bird species especially during winter and migration periods. Three years ago, there was a 20 cm thick salt layer covering the lake which became completely dry during the summer months and has spread over the agricultural lands around the lake, threatening human health and causing significant financial losses. The groundwater level has decreased due to the use of the springs which are feeding the lake. Nowadays, the depth of the wells is up to 200 m. Apart from the existing decrease in agricultural yield due to droughts in the region, reed harvesting has also been adversely affected due to destruction of the reed beds (Kuş Araştırmaları Derneği 1998). Moreover, poplar and willow groves have been lost due to drought conditions. While the condition of the lake is now good because of seasonal rainfall, under the increasing effects of global warming, it is anticipated that the wetland will be affected by drought (Bozkır Çevre Derneği 2009).

Recent studies on drought in Lake Seyfe (Kıymaz et al. 2011) have emphasized the importance of taking necessary drought adaptation measures. The occurrence of meteorological drought based on precipitation in the Seyfe Lake area in past years over two periods – from 1975 to 1991 (first period) and 1992–2008 (second period), have been analyzed using the Standardized Precipitation Index (SPI) method. The results of the analysis showed that the drought values of these two periods were different from each other. The values of the second period indicated mild drought during all dry periods (durations of 3, 6, 12, and 24 months), with increasing drought conditions of various intensity in comparison to the first period. Water stress is anticipated to increase in the coming years in Kırşehir Province.

In response to this threat, the Ministry of Forest and Water Affairs plan to bring irrigation water from another basin. They also plan to modernize the current irrigation system. Thus, through this project, they aim to protect water resource availability in the basin.

References

Ahıska S, Karabatak M. Lake Seyfe (Kirsehir) benthic fauna. Tr J Biology. 1994;18:61–75.
Ayaş Z, Turan L. Ornithological observations of Seyfe Lake, Kırşehir. Hacet Bull Nat Sci Eng. 2001;30:7–16.
Boissieri E. Flora Orientalis, 1–6. Geneve/Basel; 1867–1888.
Bozkır Çevre Derneği. Lake Seyfe urgent action plan. Seyfe drought area becomes a lake project; 2009.
Çagırankaya SS, Meriç BT. Lake Seyfe Wetland Management Plan 2011–2015. Ankara: Ministry of Environment and Forest, G.D. Nature Conservation and National Parks, Wetlands Division; 2010.
Çagırankaya SS, Meriç BT. Turkey's important wetlands. Ankara: Ministry of Forest and Water Affairs, G.D. Nature Conservation and National Parks, Wetlands Division; 2013.
Çelik M, Ünsal N, Tüfenkçi O, Bulat S. Assessment of water quality and pollution of the Lake Seyfe basin, Kırşehir, Turkey. Environ Geol. 2008;55:559–69.
Davis PH. Flora of Turkey and the East Aegean Islands. vols.1–10. Edinburgh; 1965–1988.
Eken G, Bozdoğan M, İsfendiyaroğlu S, Kılıç DT, Lise Y, editors. Turkey's key biodiversity areas, vol. 2. Ankara: Doğa Derneği; 2006.
Heywood VH, Tutin GT, editors. Flora Europeae. vols. I-V. Cambridge University Press; 1964–1981.
Kıymaz S, Güneş V, Asar M. Determination of drought periods for Seyfe Lake by standardized precipitation index. GOÜ Ziraat Fakültesi Dergisi. 2011;28(1):91–102.
Kuş Araştırmaları Derneği. What did they say? How have Seyfe Lake, Gavur Lake, Ereğli Marshes, Eşmekaya Marshes been dried? 1998.
Reis S, Yilmaz HM. Temporal monitoring of water level changes in Seyfe Lake using remote sensing. Hydrol Process. 2008;22:4448.

Section VIII

Africa

Congo River Basin   97

Ian J. Harrison, Randall Brummett, and Melanie L. J. Stiassny

Contents

Introduction	1200
Geographic Extent	1200
Biodiversity	1205
Ecosystem Services	1207
Threats to Biodiversity	1209
Conservation and Management Priorities	1212
References	1212

Abstract

The Congo River is the second only to the Amazon in terms of size and freshwater species diversity. The basin covers 4 million km^2. The basin has over 1,200 fish species, 400 mammal species, 1,000 bird species and over 10,000 vascular plant species. It provides about 30% of Africa's freshwater resources, and about 77 million people living in the Congo basin rely on them. The basin has remained relatively undeveloped compared to other basins in Africa, but increased political stability is allowing development, with loss of riparian habitat through deforestation, and reduction of water quality through pollution and sedimentation being some of the main threats to the freshwater ecosystems. Effective environmental

I. J. Harrison (✉)
Center for Environment and Peace, Conservation International, Arlington, VA, USA
e-mail: iharrison@conservation.org

R. Brummett
Environment and Natural Resources Department, World Bank, Washington, DC, USA
e-mail: rbrummett@worldbank.org

M. L. J. Stiassny
Division of Vertebrate Zoology, Ichthyology Department, American Museum of Natural History, New York City, USA
e-mail: mljs@amnh.org

© Springer Science+Business Media B.V., part of Springer Nature 2018
C. M. Finlayson et al. (eds.), *The Wetland Book*,
https://doi.org/10.1007/978-94-007-4001-3_92

planning is essential to ensure that resources are managed wisely and the ecosystems that provide them are adequately protected. Additional surveying and monitoring of biodiversity throughout the basin is required. It will also be important to designate additional protected areas with a focus on freshwaters.

Keywords

Congo · Cuvette centrale · Ubangi · Kasia · Blackwater · Rapids · Forest · Fisheries · Habitat loss

Introduction

The Congo River basin is the second largest river basin in the world, after the Amazon, draining an area approximately the size of western Europe. The tropical forests in the democratic republic of Congo, which accounts for the greatest part of the basin, have a surface area of more than 2.0 million km^2, and they represent about 50% of the rain forests of the African continent. The Congo basin has an extremely large diversity of freshwater species, again, globally second only to the Amazon in terms of species richness. There is a dense system of tributaries that extends throughout the basin, providing an essential navigation system through large parts of Central Africa which otherwise has a poorly developed transport network. The rivers and surrounding forests also supply many important goods and services to the people living in the basin, providing food, sources of energy (eg. firewood) and income from local trades to large industries (such as mining). However, many areas these resources are not being managed sustainably, resulting in loss of the forest habitat, and severe degradation to many other areas, including the rivers.

The basin has remained in a relatively intact state compared to many other parts of Africa, and this is partly due to the poor network of communications, and long periods of political instability. These conditions have made large parts of the basin inaccessible. Hence, while we know that the basin has a unique and rich biodiversity, our knowledge of the overall biodiversity is still relatively poor. The development of more peaceful conditions, and a better communication network, has facilitated more extensive research into the ecosystems of the basin over the last several years. However, these conditions are also promoting increased economic activity, and unsustainable management of the resources in the basin. This is resulting in loss of the forest habitat, severe degradation to many other areas, including the rivers, and increasing loss of biodiversity.

Geographic Extent

The Congo River has the largest of Africa's river basins, covering about 4 million km^2 from 09° 15′ N to 13° 28′ S and 11° 18′ E to 31° 10′ E. It occupies almost all of the Democratic Republic of the Congo (DRC), much of the Republic of the Congo, and large portions of Cameroon, the Central African Republic, Zambia, and Angola

(Fig. 1). The main channel of the river is 4,374 km long (Runge 2007) and can conveniently be divided into upper (Lualaba), middle, and lower sections.

The Congo headwaters include three main branches: the Luapula River, which drains from Lake Bangweulu to Lake Mweru in the Bangweulu-Mweru ecoregion (Fig. 2) of northeastern Zambia, and the Lufira and Lualaba rivers in the Upper Lualaba ecoregion of southeastern DRC (according to freshwater ecoregions defined by Thieme et al. (2005) and Abell et al. (2008)). The Bangweulu basin, which supplies the Luapula branch of the Congo headwaters, is characterized by several lakes, none of which are more than 10 m deep, and Lake Bangweulu is the largest (2,070 km^2). Lake Bangweulu has several characteristic sand ridges that run from the southwest to the northeast and create long sandy spits and beaches. It is bordered to the east by large grassy swamps and a floodplain (Fig. 3a, b) such that the total combined area of the wetlands is at least 13,770 km^2 (Thieme et al. 2005). Several rivers drain into the Bangweulu wetlands, of which the largest is the Chambeshi River. Lake Mweru is deeper (37 m) and larger (4,413 km^2) than Lake Bangweulu according to Thieme et al. (2005), although there are also estimates of over 5,000 km^2, and is drained by the Luvua River.

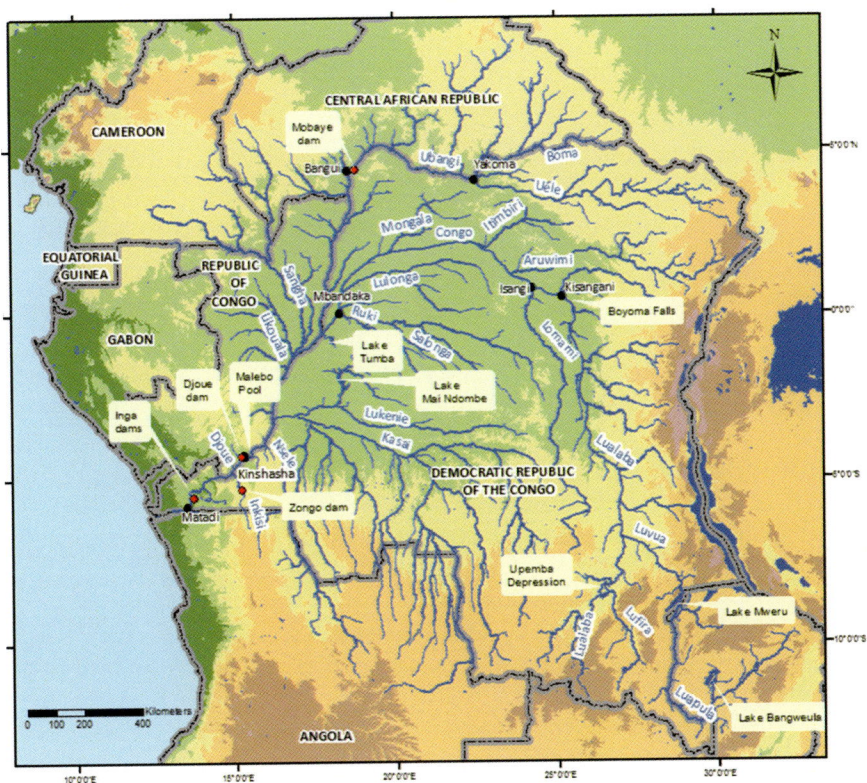

Fig. 1 Map of the Congo Basin (Topography from SRTM30 dataset. Rivers from HydroSHEDS)

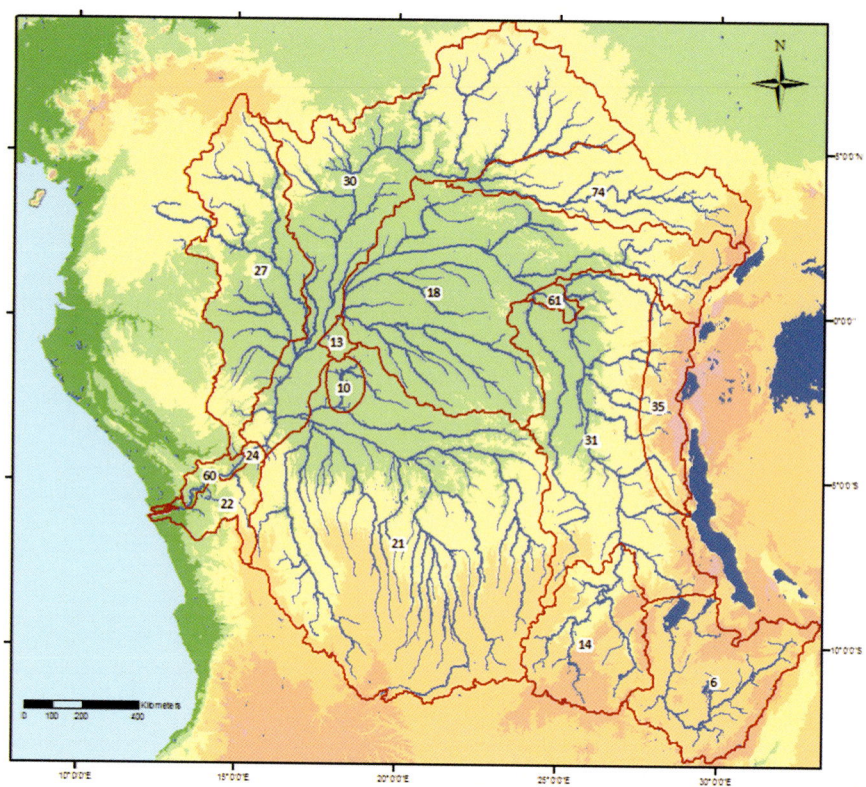

Fig. 2 Freshwater ecoregions of the Congo Basin. Ecoregions are numbered according to Thieme et al. (2005). *6* Bangweulu-Mweru, *10* Mai-Ndombe, *13* Tumba, *14* Upper Lualaba, *18* Cuvette Centrale, *21* Kasai, *22* Lower Congo, *24* Malebo Pool, *27* Sangha, *30* Sudanic Congo (Oubangi), *31* Upper Congo, *35* Albertine Highlands, *60* Lower Congo Rapids, *61* Upper Congo Rapids, *74* Uele (ecoregion boundaries from Freshwater Ecoregions of the World)

Considering the other two headwater branches, the Lufira River meets the Lualaba close to the Upemba (or Kamolondo) Depression which includes a mosaic of lakes and wetlands that seasonally extend between 8,000 and 11,840 km² (in the flood season). The Lualaba then passes over several falls and rapids and enters the Upper Congo ecoregion and is joined by the Luvua (draining Lake Mweru, see above). Within this Upper Congo ecoregion, the Lualaba runs north to Boyoma Falls (formerly Stanley Falls), just upstream of the town of Kisangani (formerly Stanleyville). Several rivers drain to the Lualaba from the Albertine Highlands ecoregion of the central Rift Valley to the west. The Lomami runs parallel and to the west of the Lualaba in the Upper Congo ecoregion and joins the Congo downstream from Kisangani, at Isangi. The Upper Congo Rapids ecoregion, upstream from Kisangani, forms a boundary between the upper and middle sections of the Congo River.

The middle section of the Congo follows a large arc, heading west and then southwest from Boyoma Falls to Malebo Pool. Most of this middle section, below

Fig. 3 (**a**) Bangweulu grassy swamps (Photo credit: Kevin Cummings © Rights remains with the MUSSEL Project http://www.mussel-project.net/); (**b**) Aquatic vegetation in Bangweulu swamp (Photo credit: Maarten Van Steenberge © Rights remains with the photographer)

the Upper Congo Rapids, is within the large Cuvette Centrale ecoregion that covers most of the DRC. This ancient continental depression is thought to have formerly been a large endorheic lake, contained by uplifting around the basin. Once the lower section of the Congo River (in the Lower Congo Rapids ecoregion) cut back through the Crystal Mountains to meet the endorheic lake, possibly in the region of Malebo Pool, the lake drained leaving the extensive, pristine forests, networks of tributaries to the main Congo channel (see below), and flooded wetlands of the Cuvette Centrale. The lakes Mai-Ndombe and Tumba (which form two discrete ecoregions at the junction of the Sudanic Congo, Cuvette Centrale, and Kasai ecoregions) are thought to be remnants of the former lake (Lévêque 1997; Thieme et al. 2008). There are many other, smaller swamps that are either permanently or seasonally flooded. These forest swamps are typically "blackwater" containing very low dissolved oxygen and high carbon dioxide concentrations, hence low pH values in the range of 4 or 5. They typically have muddy substrates overlain by large amounts of allochthonous plant materials that

Fig. 4 Main channel of Congo River between Kinshasa and Bumba, showing wide anastomosing channels (Photo credit: Jos Snoeks © Rights remains with the photographer)

fall into the swamps from the forest cover or are washed in from stream networks (especially during flooding) (Stiassny et al. 2011). The Congo main channel in the Cuvette Centrale is a deep brown color from the muddy sediment load and is characterized by wide, anastomosing channels (Fig. 4) without surface rapids. Several large rivers drain into the Congo in this middle section, significantly increasing the volume of water flowing down the main channel of the Congo. The largest of these is the Ubangi, at the north of the Congo Basin. The Ubangi originates at Yakoma where the Boma and Uele rivers meet – the drainage of the latter comprising the Uele freshwater ecoregion. The Ubangi drainage forms most of the Sudanic Congo (Ubangi) ecoregion and extends from the Central African Republic to its junction with the Congo just south of Mbandaka. The Sangha runs north-south, parallel to the lower part of the Ubangi, forming the Sangha ecoregion, draining large parts of the Republic of the Congo and extending into Cameroon. Other large rivers joining the middle Congo from the north include the Aruwimi, Itimbiri, and Mongala.

The Lomami, Lulonga, Ruki (with several large tributaries), Lukenie, and Kasai rivers drain the central and southern parts of the Congo Basin. The Kasai and its tributaries form the very large Kasai ecoregion; several of those tributaries join the Kasai from the south and drain a large part of the southern and southwestern DRC and northern Angola.

Fig. 5 View over Malebo Pool, near Kintele, Republic of the Congo (Photo credit: Ian Harrison © Rights remain with author)

The flow of the Congo slows considerably at the end of its middle section fanning out to form Malebo Pool (a discrete ecoregion), a shallow, lake-like expanse up to 28 km across with many extensive sand islands (Fig. 5). The large cities of Kinshasa (DRC) and Brazzaville (Republic of the Congo) are situated at the pool's outflow, on its southern and northern banks, respectively. The short, lower section of the Congo (498 km long) starts at this outflow and includes the Lower Congo Rapids ecoregion comprising the main channel of the Congo from below the pool to Matadi and the Lower Congo ecoregion that lies to the south of the river between Kinshasa and Banana Point. The river descends 280 m over the Lower Congo Rapids ecoregion, with at least 32 and perhaps as many as 66 cataracts (Robert 1946; Beadle 1981; Thieme et al. 2005) (Fig. 6). The largest of these are the Inga Rapids (see below). The mouth of the Congo at the Atlantic Ocean is between Banana Point in the DRC and Sharks Point in Angola. The sediment-bearing plume of the river extends over 20 km into the Atlantic, although the surface freshwater plume extends up to 800 km from the mouth (Runge 2007).

Biodiversity

The Central African rainforests of the Congo Basin and adjacent Lower Guinea (to the west of the Congo Basin; Stiassny and Hopkins 2007) have the greatest biodiversity on the African continent, harboring over 1,200 fish species,

Fig. 6 Rapids in Lower Congo (Photo credit: Uli Schliewen © Rights remain with photographer)

400 mammal species, 1,000 bird species, and over 10,000 vascular plant species (CARPE 2001; African Development Bank 2006). There is also a rich herpetofauna, with over 280 aquatic frogs and about 20 aquatic snakes, turtles, and crocodiles (Brummett et al. 2009). The middle to lower sections of the Congo and the lower part of the Ubangi have more than ten co-occurring species of turtle, which is high (only two subbasins in the world have as many as 18 or 19 co-occurring species) (Buhlmann et al. 2009). These large numbers are also likely to be underestimates because taxonomic surveys are uncommon and the flora and fauna are poorly documented for the region (Brummett et al. 2011). As new surveys are undertaken and existing museum collections reexamined, new species are frequently discovered (e.g., Lévêque et al. 2005; Frost et al. 2006). There is a clear imperative for further research and field collections, to obtain better knowledge of the freshwater species diversity and its conservation status.

Regions of greatest species richness are found along the main channel of the Congo and its two main tributaries, the Kasai and the Ubangi (Brooks 2011). This is mainly because fish species richness is very high for these rivers. The Congo River has the highest species diversity of fishes of any freshwater system in Africa and, globally, is second only to the Amazon (Lundberg et al. 2000; Teugels and Thieme 2005).

High freshwater biodiversity in the Congo Basin is likely due to the dense and extensive hydrographic network and long geomorphic stability largely unaffected by the rifting and Miocene volcanism in Eastern Africa. In the Cuvette Centrale, the large amounts of allochthonous materials (see above) provide rich nutrient resources. However, the waters have low dissolved oxygen and high carbon dioxide concentrations (see above) and restrict the habitat availability to species that are adapted to these conditions (e.g., fishes with accessory breathing organs). Although many of the rivers of the Congo Basin are low-gradient, "blackwater" systems (Thieme et al. 2005; Stiassny and Hopkins 2007), there are also some high-gradient rivers, such as those draining the Albertine Highlands. Stream width, depth, current velocity, and substrate type all affect the spatial and temporal distribution of freshwater species (Lowe-McConnell 1975; Kamdem-Toham and Teugels 1997) effectively partitioning the resource base and increasing opportunities for novel forms to evolve (Hoeinghaus et al. 2003; Dejen et al. 2006). The large amounts of intact forest cover through many parts of the Congo Basin contribute to regional rainfall (Sheil and Murkiyarso 2009), increase productivity, and support community diversity.

Hydrographic barriers between habitats prevent mixing of populations of freshwater species and promote diversification. These barriers may be waterfalls (such as those at the edges of the Congo channel) or rapids in the main channel of the Congo (Thieme et al. 2005; Brummett et al. 2009, 2011). Recent morphological and molecular analyses show that the extremely fast horizontal and vertical currents in the Lower Congo isolate fish populations along the river and laterally across it, thereby promoting diversification of populations over extremely small distances (Jackson et al. 2009; Markert et al. 2010; Lowenstein et al. 2011; Alter et al. 2015).

Malebo Pool and the Upper Congo Rapids in the vicinity of Boyoma Falls are particularly rich in species (Brooks 2011). The Lower and Upper Congo Rapids regions are especially high in species with restricted ranges; this is unsurprising because species found in these regions are generally adapted for life in the fast-flowing and turbid waters of rapids rather than the slower flowing parts of the Congo. In 2011, droughts in the Congo Basin resulted in unusually low water levels that allowed fish collections by one of us (MLJS) in parts of the Congo channel above Malebo Pool that are usually inaccessible. Interestingly, the collections included species that were previously thought to be restricted to the Lower Congo Rapids below Malebo Pool. These results suggest that the apparent restricted ranges of some species may be the result of a simple lack of knowledge, and they are more widespread than currently believed.

Ecosystem Services

The Congo provides about 30% of Africa's freshwater resources (Brummett et al. 2009), and about 77 million people living in the Congo Basin rely on them. However, the DRC, covering most of the basin, lacks adequate policies or enforcement for sustainable management and conservation of these freshwater resources.

Congo Basin fishes are an important food resource for local communities and supply markets in towns and cities throughout the basin. Upward of 20% of the population living in the rainforests of Central Africa is thought to be engaged in river fisheries (Brummett and Teugels 2004), and artisanal and traditional fisheries account for 90% of the catch from the Central African rainforests (Mino-Kahozi and Mbantshi 1997). The maximum total annual catch between 1980 and 1984 was 119,500 tons/year (not including the Ubangi) with a production value of US$47.8 million/year and an estimated potential catch of 520,000 tons/year, with a value of US$208 million/year (Neiland and Béné 2008). Fishes represent over 40% of the animal protein supply to people living in the DRC (UNEP 2010) – although this includes fisheries from the Rift Valley lakes. In the region of the Salonga National Park, in the Cuvette Centrale, fishing represents 61% of the total cash income even for households that are dependent on more than one activity (i.e., not just fishing) (Béné et al. 2009). Men and women are both involved in fishing; men tend to fish the main channels for larger fishes and they sell their catch, while women fish river margins for smaller fishes that are kept for home consumption. Fishing is a "bank in the water" for these communities (Béné et al. 2009) because it provides people with quick access to cash that is necessary to buy basic necessities and manufactured goods and for their children's education. Even so, the fishery could be better managed to reduce waste and thus take pressure off other forms of non-timber forest products, including other bushmeat (Shumway et al. 2003).

Large floodplain fisheries, such as those of the Cuvette Centrale, are highly dependent on the seasonal flood cycle because the large fish populations are sustained by seasonal access to the flooded forest. The species can feed and spawn in the flooded regions, and the floods bring extra nutrients from the forest into the aquatic ecosystem, thereby increasing their productivity.

Commercial fisheries for the aquarium trade do not appear to be a threat to native fishes in Central Africa at present (Stiassny et al. 2011). Instead, Brummett (2005) has proposed that there are opportunities for a locally managed, sustainable fishery for the aquarium trade in African freshwaters that could support rural livelihoods and integrate the community in conservation of freshwater habitats.

The river network in the Congo Basin is an important transport and trade route, supplying goods and services that are essential for local economies and livelihoods. Hydropower is another important resource supplied by the Congo River. At the end of 2005, most of the capacity for generating electricity within the Congo Basin came from hydropower, and this is a small fraction of the overall hydropower potential of the basin (Brummett et al. 2009). The largest dam project for Central Africa is in the Lower Congo Rapids ecoregion at Inga. Two dams (Inga I and II) are operational, but are working at only half their installed capacities of 358 and 1,424 MW, respectively (Brummett et al. 2009). Numerous plans have been proposed to build more dams at Inga with capacities ranging from 4,500 to 39,000 MW and then later, if realized, would represent the largest hydroelectric power generation project on the planet. An objective for these dams is the production of hydroelectric power that could be exported to other parts of Africa (SNEL 2002) and even beyond to southern

Europe. However, the environmental and social effect of the dams, associated infrastructure, and transmission lines could be very significant (Brummett et al. 2011; Winemiller et al. 2016; and see below).

Other existing dams in the Lower Congo include the 17 m Zongo Dam on the Inkisi River (built to supply power to Kinshasa) and the 36 m Djoue Dam on the Djoue River near Brazzaville (FAO 2013). The 30 m Mobaye Dam is on the Ubangi, and in the upper parts of the Congo, several large dams have been built on the headwaters of the rivers, to supply power for mining industries in the region. Three large dams of over 70 m exist on the Upper Lualaba, and smaller dams (less than 20 m) are on the Lufira.

Threats to Biodiversity

Within the basin, loss of riparian habitat through deforestation and reduction of water quality through pollution and sedimentation are two of the main threats to freshwater ecosystems (Brummett et al. 2011). Yet even by the mid-1990s, 37% of the total exploitable forest within the DRC had been designated for timber concessions (Meditz and Merrill 1994).

There are multiple causes of deforestation in addition to commercial logging: land conversion for mining, agriculture, and human habitation; and felling of trees for firewood, especially for the production of charcoal for cooking fuel that is transported, often over long distances, into towns and cities. The Congo Basin had lost an estimated 46% of its rainforest to logging and conversion to agriculture and continues to lose forested watershed at an average rate of 7% per year (Revenga et al. 1998). Deforestation is particularly severe in the Lower Congo Rapids, Upper Congo, and Kasai ecoregions (Brummett et al. 2011), and the clearing of riparian cover for charcoal production, often denuding tributaries near population centers, is an increasing threat at local and regional scales (e.g., Iyaba et al. 2013).

Conversion of forest to oil palm is especially evident in the Upper Congo ecoregion, and Chinese subsidization is anticipated to support significant oil palm expansion around Lac Tumba (Brummett et al. 2011). Highly destructive slash and burn agriculture already occurs in many parts of the basin and has denuded riparian vegetation along large sections of the entire Congo main channel. Deforestation opens up habitat for exploitation by other activities, such as mining for gold, diamonds, and other minerals. This occurs in several parts of the Congo Basin and is a particular problem in the Upper Lualaba and Kasai ecoregions (Brummett et al. 2011; Darwall and Smith 2011, Fig. 2.1; Stiassny et al. 2011). Other potential, extractive threats include oil and gas extraction. Oil in the lower part of the Kasai drainage (Shumway et al. 2003) and methane under Mai-Ndombe could attract industrial activity with potentially dire environmental impacts (Brummett et al. 2011; Darwall and Smith 2011: Fig. 2.1).

Significant additional consequences of deforestation, agriculture, and mining are erosion and increased sediment loads into freshwater systems, as described for the

Lac Tumba region by Inogwabini et al. (2006) and for the Nsele by Iyaba et al. (2013). In the Lower Congo Rapids, small-scale mining and sandstone quarrying along the river increase the turbidity and sediment deposition that are already high due to sediment runoff from the heavily deforested hillsides. The increased sediment load to the Ubangi River near Bangui has reduced river depth to the extent that shipping is impeded several months of the year (Brummett et al. 2009; Darwall and Smith 2011: Fig. 2.1). Increased surface runoff from cleared land also increases flooding risk. Loss of riparian cover and disturbances to hydrology and water quality significantly affect aquatic biodiversity (Growns and Davis 1991; Bradshaw et al. 2007, 2009; Brummett et al. 2009, 2011). Industrial development, such as mining, has been hampered by civil unrest in the DRC and neighboring countries (e.g., Rwanda, Angola), and this has reduced the potential threats posed to freshwater ecosystems. As areas restabilize (e.g., the Kasai Basin), there is high probability of resumed mining and associated impacts; therefore it is especially important that effective environmental planning is included during recovery (Thieme et al. 2005; Brummett et al. 2011).

While there are fewer dams in the Congo Basin than in many other parts of Africa, and most are in the Lower Congo Rapids ecoregion or in the headwaters (in the Upper Congo and Upper Lualaba ecoregions), additional dams are likely to be built (Winemiller et al. 2016, and see above). As currently implemented, dams have a profound impact on natural flow regimes and sediment load that determine ecosystem structure and function (Bunn and Arthington 2002). The associated power stations and other infrastructure, including clearing for power lines, roads, and housing for workers, also add significantly to deforestation, pollution, and siltation impacts on the aquatic system.

Pollution is also a serious threat from human settlement and industry in parts of the Congo Basin. Ninety-five percent of factories in the DRC discharge their waste directly into rivers and other freshwater systems (UNEP 1999). Pollution is greatest near large urban centers such as Kisangani (Thieme et al. 2005) and around Brazzaville and Kinshasa where large quantities of sewage are released into Malebo Pool and the Congo River. Lead and waste oil originating from industry, from cars, or from boat traffic add to this pollution. Significant pollution by heavy metals, particularly lead and cadmium, as far as approximately 300 km downstream from Kinshasa has been reported (Shumway et al. 2003), and the main channel of the Ubangi is affected by pollution from the city of Bangui (Brummett et al. 2011; Darwall and Smith 2011: Fig. 2.1).

Habitat disturbance is also caused by invasive species. In the Congo Basin, the greatest threats come from introduced aquatic plants that can cover vast expanses of water surface, thereby reducing available light, impacting water quality, and rendering the habitat unfavorable to many species. The most significant invasive plant species in Central Africa are *Pistia stratiotes*, *Eichhornia crassipes*, *Cyperus papyrus*, and *Lasimorpha senegalensis* (Ghogue 2011). According to Welcomme (1988), 16 fish species have been introduced to the Central African region; however, there is relatively little information on their impacts on native faunas, and, compared to many other parts of Africa, the number of exotic species is modest. Efforts to impede the introduction of additional exotic species, particularly of tilapiine cichlids, are of high priority.

Migration due to civil unrest in Central Africa has forced human populations to migrate into previously unsettled areas, and displaced populations often settle along waterways (Thieme et al. 2008) where they cut riparian forest for cooking fuel, with concomitant effects on ecosystem condition (see above). The influx of people into previously undisturbed areas, with their need to use the available natural resources for their well-being (e.g., food, housing), places pressure, particularly on national parks and reserves, which are attractive areas for relocation. Areas where this is a particular problem include the eastern parts of the Upper Congo, Uele, and Albertine Highlands ecoregions and the Sangha ecoregion near the confluences of the Sangha, Likouala, and Ubangi rivers (Brummett et al. 2011).

Overharvesting of freshwater species, although not as intense as in many other parts of Africa, has been reported for many regions. High levels of fishing occur along the Congo and its tributaries in the vicinity of Malebo Pool (supplying markets of Kinshasa and Brazzaville; Thieme et al. 2005, 2008); in the Mai-Ndombe ecoregion (Stiassny et al. 2011) and neighboring Lac Tumba ecoregion (Inogwabini et al. 2010), with declining yields and loss of large species; and in the region of Mbandaka (supplying the town's markets; Shumway et al. 2003), with a high proportion of juveniles in fisheries in the Mbandaka-Ngombe region (reported for 2003), indicating the stocks were overexploited (ERGS Research Group, cited in Thieme et al. 2005; Stiassny et al. 2011). Unregulated fishing also occurs in parts of the Cuvette Centrale (Iyaba and Stiassny 2013), Kasai, Upper Lualaba, and Bangweulu-Mweru ecoregions (Darwall and Smith 2011, Fig. 2.1; Stiassny et al. 2011). In the latter, Lake Bangweulu supports an important fishery, although it is at risk from overfishing. The most common cause of overfishing in the Congo Basin is the use of very fine mesh nets and fish poisons that indiscriminately kill individuals of all sizes for many species (Shumway et al. 2003; Inogwabini 2005; Stiassny et al. 2011) and destroy fringing vegetation and nesting areas (Mbimbi et al. 2011).

Climate change potentially creates additional layers of stress to freshwaters in Africa and is likely to compound some of the threats noted above. While it is expected that aquatic species will experience hydrological change over the next several decades, it is not clear what the ecological effects will be. This uncertainty is partly because climate change models for Africa are in their infancy, and impacts are likely to be complex and regionally specific (Schiermeier 2008; Matthews et al. 2011). Thieme et al. (2010) expected that runoff and discharge would increase in the Guinean-Congolian forests, but other regional changes have been described; for example, Inogwabini et al. (2006) describe a decrease in precipitation and increase in temperature in the latter part of the twentieth century. African ecosystems and species have a history of resilience in response to dramatic eco-hydrological changes (Matthews et al. 2011). Therefore, conservation efforts should be directed at enabling these climate-adaptive capacities in preparation for new climate regimes (see below).

Conservation and Management Priorities

The Congo Basin is rich in natural resources (see section "Ecosystem Services" above); as development proceeds across the basin, it is imperative that this is accompanied by effective environmental planning to ensure that resources are managed wisely and the ecosystems that provide them are adequately protected. Such environmental planning is dependent upon a comprehensive and reliable baseline of ecological and environmental data; without which it is impossible to identify conservation priorities or assess when natural resources are being used sustainably. However, much of the Congo Basin is understudied, and significant gaps remain in our knowledge of the distribution of species, their ecology, and the extent of threats they face. Central Africa has the highest proportion of "data-deficient" species for any part of Africa, according to IUCN's assessments of the status of the freshwater biodiversity present (Brooks 2011). Hence there is a priority to conduct more surveying and monitoring throughout the basin. This also requires increased investment in local resources to support this work, including the training of additional in-country scientists and conservation managers.

Only 3% of the Congo Basin's 5,255 subcatchments are included within existing protected areas (Linke et al. 2012); therefore it will be important to designate additional protected areas with a focus on freshwaters and their biodiversity (Harrison et al. 2016). The importance of conserving as much original habitat as possible cannot be overstated, since this is the key to ensuring the resilience of the existing biodiversity. Plans for designating protected areas must take into account the high connectivity within the dense network of rivers in the basin (Brooks 2011). There is also a need for additional resources for management of existing protected areas, especially those that are being affected by human encroachment through civil unrest (Stiassny et al. 2011).

There is a growing consensus that effective management of freshwater ecosystems requires an understanding of the environmental flow requirements that sustain the ecosystem at the same time as supplying the necessary freshwater services to people (Poff and Matthews 2013). This is especially important in preparation for adapting to the modified hydrological conditions that will result from climate change. Several management actions can help ensure species resilience to climate change (Thieme et al. 2010), most notably minimizing perturbation of water quality and maintenance of natural flow regimes. Maintaining connectivity between freshwater ecosystems and protection of spatial and thermal refugia will also be important; hence there is a need to identify optimum sites for designation as protected areas (see above).

References

Abell R, Thieme ML, Revenga C, Bryer M, Kottelat M, Bogutskaya N, Coad B, Mandrak N, Balderas SC, Bussing W, Stiassny MLJ, Skelton P, Allen GR, Unmack P, Naseka A, Ng R, Sindorf N, Robertson J, Armijo E, Higgins JV, Heibel TJ, Wikramanayake E, Olson D, Lopez

HL, Reis RE, Lundberg JG, Sahaj Pérez MH, Petry P. Freshwater ecoregions of the world: a new map of biogeographic units for freshwater biodiversity conservation. Bioscience. 2008;58:403–4.

African Development Bank Group. Eta des lieux de la gestation des resources en eau dans le basin du Congo; 2006.

Alter S, Brown B, Stiassny MLJ. Molecular phylogenetics reveals convergent evolution in lower Congo River spiny eels. BMC Evol Biol. 2015;15:224. doi:10.1186/s12862-015-0507-x.

Beadle LC. The inland waters of tropical Africa. 2nd ed. London: Longman; 1981.

Béné C, Steel E, Kambala Luadia B, Gordon A. Fish as the "bank in the water" – evidence from chronic-poor communities in Congo. Food Policy. 2009;34:108–18.

Bradshaw CJA, Sodhi NS, Peh KS-H, Brook BW. Global evidence that deforestation amplifies flood risk and severity in the developing world. Glob Chang Biol. 2007;13:2379–95.

Bradshaw CJA, Brook BW, Peh KS-H, Sodhi NS. Flooding policy makers with evidence to save forests. Ambio. 2009;38:125–6.

Brooks EGE. Regional synthesis for all taxa. In: Brooks EGE, Allen DJ, Darwall WRT, editors. The status and distribution of freshwater biodiversity in Central Africa. Gland: IUCN; 2011. p. 110–21.

Brummett RE. Freshwater ornamental fishes: a rural livelihood option for Africa? In: Thieme ML, Abell R, Stiassny MLJ, Skelton P, Lehner B, Teugels GG, Dinerstein E, Kamdem Toham A, Burgess N, Olson D, editors. Freshwater ecoregions of Africa and Madagascar: a conservation assessment. Washington, DC: Island Press; 2005. p. 132–5.

Brummett RE, Teugels GG. Rainforest rivers of Central Africa: biogeography and sustainable exploitation. In: Welcomme R, Petr T, editors. Proceedings of the second international symposium on the management of large rivers for fisheries, RAP 2004/16. Bangkok: Food and Agriculture Organization of the United Nations; 2004.

Brummett R, Tanania C, Pandi A, Ladel J, Munzimi Y, Russell A, Stiassny M, Thieme M, White S, Davies D. Water resources, forests and ecosystem goods and services. In: De Wasseige C, Devers D, de Marcken P, Eba'a Atyi R, Nasi R, Mayaux P, editors. The forests of the Congo Basin: state of the forest 2008. Luxembourg: Publications Office of the European Union; 2009. p. 1411–57.

Brummett R, Stiassny M, Harrison I. Background. In: Brooks EGE, Allen DJ, Darwall WRT, editors. The status and distribution of freshwater biodiversity in Central Africa. Gland: IUCN; 2011. p. 1–20.

Buhlmann KA, Akre TSB, Iverson JB, Karapatakis D, Mittermeier RA, Georges A, Rhodin AGJ, van Dijk PP, Gibbons JW. A global analysis of tortoise and freshwater turtle distributions with identification of priority conservation areas. Chelonian Conserv Biol. 2009;8:116–49.

Bunn SE, Arthington AH. Basic principles and ecological consequences of altered flow regimes for aquatic biodiversity. Environ Manag. 2002;30:492–507.

CARPE. Congo Basin information series; taking action to manage and conserve forest resources in the Congo Basin. Central African Regional Program for the Environment. Gland: World Wildlife Fund; 2001.

Darwall WRT, Smith KG. Assessment methodology. In: Brooks EGE, Allen DJ, Darwall WRT, editors. The status and distribution of freshwater biodiversity in Central Africa. Gland: IUCN; 2011. p. 21–6.

Dejen E, Vijverberg J, de Graaf M, Sibbing FA. Predicting and testing resource partitioning in a tropical fish assemblage of zooplanktivorous 'barbs': an ecomorphological approach. J Fish Biol. 2006;69:1356–78.

FAO. AQUASTAT – Geo-referenced database on dams in the Middle East. 2013. Available from: http://www.fao.org/nr/water/aquastat/dams/index.stm. Accessed 9 Sept 2013.

Frost DR, Grant T, Faivovich J, Bain RH, Haas A, Haddad CFB, De Sá RO, Channing A, Wilinson M, Donnellan SJ, Raxworthy CJ, Campbell JA, Blotto BL, Moler P, Drewes RC, Nussbaum RA, Lynch JD, Green DM, Wheeler WC. The amphibian tree of life. Bull Am Mus Nat Hist. 2006;297:1–370.

Ghogue J-P. The status and distribution of freshwater plants of central Africa. In: Brooks EGE, Allen DJ, Darwall WRT, editors. The status and distribution of freshwater biodiversity in Central Africa. Gland: IUCN; 2011. p. 92–109.

Growns IO, Davis JA. Comparison of the macroinvertebrate communities in steams in logged and undisturbed catchments eight years after harvesting. Aust J Mar Freshw Resour. 1991;42:689–706.

Harrison IJ, Green PA, Farrell TA, Juffe-Bignoli D, Sáenz L. Vörösmarty CJ. Protected areas and freshwater provisioning: a global assessment of freshwater provision, threats and management strategies to support human water security. Aquatic Conserv: Mar Freshw Ecosyst 2016;26 (Supplement 1):103–120.

Hoeinghaus DJ, Layman CA, Albrey Arrington D, Winemiller KO. Spatiotemporal variation in fish assemblage structure in tropical floodplain creeks. Environ Biol Fish. 2003;67:379–87.

Inogwabini B-I. Fishes of the Salonga National Park, Democratic Republic of Congo: survey and conservation issues. Oryx. 2005;39:78–81.

Inogwabini B-I, Sandokan BM, Ndunda M. A dramatic decline in rainfall regime in the Congo basin: evidence from a thirty-four year data set from the Mabali Scientific Research Centre, Democratic Republic of Congo. Int J Meteorol. 2006;31:278–85.

Inogwabini B-I, Dianda M, Lingopa Z. The use of breeding sites of *Tilapia congica* (Thys and van Audenaerde (1960) to delineate conservation sites in the Lake Tumba, Democratic Republic of Congo: toward the conservation of the lake ecosystem. Afr J Ecol. 2010;48:800–6.

Iyaba RJCM, Stiassny MLJ. Fishes of the Salonga National Park (Congo basin, central Africa): a list of species collected in the Luilaka, Salonga, and Yenge Rivers (Equateur Province, Democratic republic of Congo). Check List. 2013;9:246–56.

Iyaba RJCM, Liyandja T, Stiassny MLJ. Fishes of the N'sele River (Pool malebo, Congo basin, Central Africa): a list of species collected in the main channel and affluent tributaries, Kinshasa province, Democratic Republic of Congo. Check List. 2013;9:941–56.

Jackson PR, Oberg KA, Gardiner N, Shelton J. Velocity mapping in the lower Congo River: a first look at the unique bathymetry and hydrodynamics of the Bulu reach, west central Africa. Proc IAHR Symp River Coast Estuar Morphodynamics. 2009;6:1007–14.

Kamdem-Toham A, Teugels GG. Patterns of microhabitat use among fourteen abundant fishes of the lower Ntem River Basin (Cameroon). Aquat Living Resour. 1997;10:289–98.

Lévêque C, editor. Biodiversity dynamics and conservation: the freshwater fish of tropical Africa. Paris: ORSTOM; 1997.

Lévêque C, Balian EV, Martens K. An assessment of animal species diversity in continental waters. In: Segers H, Martens K, editors. The diversity of aquatic ecosystems. Hydrobiologia. 2005;542:39–67.

Linke S, Hermoso V, Thieme ML. Preliminary results of a freshwater biodiversity Marxan analysis for the Democratic Republic of Congo. Technical Report. Program to Reinforce the Protected Area Network; 2012. http://research.freshwaterbiodiversity.eu/downloads/DRCongo_FW_Technical_Report.pdf. Accessed 26 Apr 2016.

Lowe-McConnell RH. Fish communities in tropical freshwaters; their distribution, ecology, and evolution. London: Longman; 1975.

Lowenstein JH, Osmundson TW, Becker S, Hanner R, Stiassny MLJ. Incorporating DNA barcodes into a multi-year inventory of the fishes of the hyperdiverse Lower Congo River, with a multi-gene performance assessment of the genus *Labeo* as a case study. Mitochondrial DNA. 2011;21 (S2):1–9.

Lundberg JG, Kottelat M, Smith GR, Stiassny MLJ, Gill AC. So many fishes, so little time: an overview of recent ichthyological discoveries in freshwaters. Ann Mo Bot Gard. 2000;87:26–62.

Markert JA, Schelly RC, Stiassny MLJ. Genetic isolation and morphological divergence mediated by high-energy rapids in two cichlid genera from the lower Congo rapids. BMC Evol Biol. 2010;10:149–57.

Matthews JH, Wickel AJ, Freeman S, Thieme ML. The future of African freshwaters. Section 8.6. In: Darwall WRT, Smith KG, Allen DJ, Holland RA, Harrison IJ, Brooks EGE, editors. The diversity of life in African freshwaters: under water, under threat. An analysis of the status and distribution of freshwater species throughout mainland Africa. Gland: IUCN; 2011. p. 264–9.

Mbimbi M, Munene JJ, Stiassny MLJ. Fishes of the Kwilu River (Kasai basin, central Africa): a list of species collected in the vicinity of Kikwit, Bandundu Province, Democratic Republic of Congo. Check List. 2011;7:691–9.

Meditz SW, Merrill T. Zaire: a country study. Washington, DC: Federal Research Division, Library of Congress; 1994. Available at: https://cdn.loc.gov/master/frd/frdcstdy/za/zairecountrystud00medi_0/zairecountrystud00medi_0.pdf. Accessed 26 Apr 2016.

Mino-Kahozi B, Mbantshi M. Pollution and degradation of African aquatic environments and consequences for inland fisheries and aquaculture; the case of the Republic of Zaire. In: Remane K, editor. African Inland Fisheries, Aquaculture and the Environment. Based on proceedings of the Seminar on Inland Fisheries, Aquaculture and the Environment in conjunction with the ninth session of CIFA, Harare, Zimbabwe, December 1994. Rome: FAO; 1997. p. 99–114.

Neiland A, Béné C. Review of river fisheries valuation in West and Central Africa. In: Neiland AE, Béné C, editors. Tropical river fisheries valuation: background papers to a global synthesis. The WorldFish Center Studies and Reviews 1836. Penang: The WorldFish Center; 2008. p. 47–106.

Poff NL, Matthews JH. Environmental flows in the Anthropocene: past progress and future prospects. Curr Opin Environ Sustain. 2013;5:667–75.

Revenga C, Murray S, Abramovitz J, Hammond A. Watersheds of the world: ecological value and vulnerability. Washington, DC: World Resources Institute and Worldwatch Institute; 1998.

Robert M. Le Congo Physique. 3rd ed. Liège: Presse Universitaires de France; 1946.

Runge J. The Congo River, Central Africa. In: Gupta A, editor. Large rivers: geomorphology and management. Hoboken: Wiley; 2007. p. 293–309.

Schiermeier Q. A long dry summer. Nature. 2008;452:270–3.

Sheil D, Murkiyarso D. How forests attract rain: an examination of a new hypothesis. Bioscience. 2009;59:341–7.

Shumway C, Musibono D, Ifuta S, Sullivan J, Schelly R, Punga J, Palata J-C, Puema V. Biodiversity survey: systematics, ecology, and conservation along the Congo River. Congo River Environment and Development Project (CREDP). Boston: New England Aquarium; 2003.

Société National d'Electricité [SNEL]. Inga. Hydroelectric development on Congo river. Kinshasa: SNEL; 2002.

Stiassny MLJ, Hopkins CD. Introduction. In: Stiassny MLJ, Teugels GG, Hopkins CD, editors. Poissons d'eaux douces et saumâtres de basse Guinée, ouest de l'Afrique centrale, vol. 1. Paris: IRD Éditions; 2007. p. 31–45.

Stiassny MLJ, Brummett RE, Harrison IJ, Monsembula R, Mamonekene V. The status and distribution of freshwater fishes in central Africa. In: Brooks EGE, Allen DJ, Darwall WRT, editors. The status and distribution of freshwater biodiversity in Central Africa. Gland: IUCN; 2011. p. 27–46.

Teugels GG, Thieme ML. Freshwater fish biodiversity in the Congo Basin. In: Thieme ML, Abell R, Stiassny MLJ, Skelton P, Lehner B, Teugels GG, Dinerstein E, Kamdem Toham A, Burgess N, Olson D, editors. Freshwater ecoregions of Africa and Madagascar: a conservation assessment. Washington, DC: Island Press; 2005. p. 51–3.

Thieme ML, Abell R, Stiassny MLJ, Skelton P, Lehner B, Teugels GG, Dinerstein E, Kamdem Toham A, Burgess N, Olson D. Freshwater ecoregions of Africa and Madagascar: a conservation assessment. Washington, DC: Island Press; 2005.

Thieme M, Shapiro A, Colom A, Schliewen U, Sindorf N, Kamdem Toham A. Inventaire Rapide des Zones Humides Représentatives en République Démocratique du Congo. 2008. Available at: http://www.ramsar.org/sites/default/files/documents/library/wurc_dr-congo_inventaire2008.pdf. Accessed 26 Apr 2016.

Thieme ML, Lehner B, Abell R, Matthews J. Exposure of Africa's freshwater biodiversity to a changing climate. Conserv Lett. 2010;3:324–31.

UNEP. Regional overview of land-based sources and activities affecting the coastal and associated freshwater environment in the West and Central African region, Regional seas reports and studies, vol. 171. The Hague: UNEP/GPA Co-ordination Office and West and Central Africa Action Plan, Regional Coordinating Unit; 1999.

UNEP. Blue harvest: inland fisheries as an ecosystem service. Penang: WorldFish Center; 2010.

Welcomme RL. International introductions of inland fish species, FAO fisheries technical paper, vol. 294. Rome: FAO; 1988.

Winemiller KO, McIntyre PB, Castello L, Fluet-Chouinard E, Giarrizzo T, Nam S, Baird IG, Darwall W, Lujan NK, Harrison I, Stiassny MLJ, Silvano RAM, Fitzgerald DB, Pelicice FM, Agostinho AA, Gomes LC, Albert JS, Baran E, Petrere Jr M, Zarfl C, Mulligan M, Sullivan JP, Arantes CC, Sousa LM, Koning AA, Hoeinghaus DJ, Sabaj M, Lundberg JG, Armbruster J, Thieme ML, Petry P, Zuanon J, Torrente Vilara G, Snoeks J, Ou C, Rainboth W, Pavanelli CS, Akama A, van Soesbergen A, Sáenz L. Balancing hydropower and biodiversity in the Amazon, Congo, and Mekong. Science. 2016;351:128–9.

Zambezi River Basin

98

Matthew McCartney, Richard D. Beilfuss, and Lisa-Maria Rebelo

Contents

Introduction	1218
Wetlands in the Basin	1219
The Upper Zambezi	1221
The Middle Zambezi	1222
The Lower Zambezi	1226
The Role of Wetlands	1228
Existing Threats to Wetlands	1229
Future Challenges	1230
References	1231

M. McCartney
Ecosystem Services, International Water Management Institute, Vientiane, Lao People's Democratic Republic

Regional Office for Southeast Asia and the Mekong, International Water Management Institute, Vientiane, Lao People's Democratic Republic
e-mail: m.mccartney@cgiar.org

R. D. Beilfuss (✉)
International Crane Foundation, Baraboo, WI, USA

College of Engineering, University of Wisconsin, Madison, WI, USA
e-mail: rich@savingcranes.org

L.-M. Rebelo
Water Futures, International Water Management Institute, Vientiane, Lao People's Democratic Republic

Regional Office for Southeast Asia and the Mekong, International Water Management Institute, Vientiane, Lao People's Democratic Republic
e-mail: l.rebelo@cgiar.org

© Springer Science+Business Media B.V., part of Springer Nature 2018
C. M. Finlayson et al. (eds.), *The Wetland Book*,
https://doi.org/10.1007/978-94-007-4001-3_91

Abstract

More than 4.7% of the Zambezi River Basin is wetlands, several of which individually cover areas in excess of 1,000 km^2. The basin contains 13 Ramsar Sites and thousands of lesser known wetlands. It is estimated that 20 million people (ca. 50% of the basin population) live in the vicinity of wetlands largely because of the wide range of ecosystem services, they provide, including support to fisheries, livestock and other forms of agriculture, as well as tourism. The wetlands also support considerable biodiversity, influence the hydrology of the basin and play an important role in the economies of the riparian countries. Currently there are a number of threats to the basin wetlands, including inappropriate agricultural practices, altered hydrology due to hydropower dams and overfishing. Increased irrigation and climate change are likely to add to future stresses on wetlands. However, careful planning and management, including coordinated releases from hydropower dams, could safeguard and rejuvenate many wetlands in the basin.

Keywords

Biodiversity · Dambo · Hydrology · Livelihoods · Ecosystem services · Agriculture · fIsheries

Introduction

The Zambezi River basin is the largest river basin in Southern Africa. With a total drainage area of approximately 1.34 million km^2, it is Africa's fourth largest river after the Nile, Congo, and Niger Rivers. The main river, with a length of 3,000 km, originates in the Kalene Hills in northwest of Zambia at an altitude of 1,500 m and flows first southwest and then south before turning east to the Indian Ocean. The river has three distinct stretches: the *upper Zambezi* from its source to Victoria Falls, the *middle Zambezi* from Victoria Falls to Cahora Bassa Gorge, and the *lower Zambezi* from Cahora Bassa to the delta. Riparian countries are Zambia, Angola, Namibia, Botswana, Zimbabwe, Malawi, Tanzania, and Mozambique. Malawi and Tanzania do not have direct contact with the Zambezi River itself but are linked to it via the Shire River, which drains Lake Malawi. Other principal tributaries are the Luangwa, the Kafue, the Manyame, the Sanyati, the Chobe, and the Kabompo Rivers (World Bank 2010).

Lying between latitude 10° and 20° south and between longitude 20° and 37° east, the climate of the basin is largely controlled by the movement of air masses associated with the Intertropical Convergence Zone (ITCZ). Rainfall occurs predominantly during the summer (November to March), while the winter months (April to October) are usually dry. However, rainfall is characterized by considerable spatial and temporal variation. Droughts of several years duration have been recorded almost every decade (Tyson 1986), and floods also occur frequently. The natural flow regime of the river reflected the rainfall and was characterized by high seasonal and annual variability. The average annual discharge is approximately 130,500 million m^3 (Mm^3) (4,134 m^{3}s^{-1}).

Currently, due to the absence of large dams and water diversions, the *upper Zambezi* remains the most natural portion of the river. Further downstream, the flow is regulated by two large dams on the Zambezi main stem – Kariba and Cahora Bassa dams – as well as a number of tributary dams (most notably Kafue Gorge and Itezhi-Tezhi on the Kafue River) (Fig. 1). These were built primarily for hydropower generation (Beilfuss and dos Santos 2001). The operation of these dams has resulted in an increase in dry season flows, a delay and decrease in peak flows during the flood season, and an overall reduction in the depth and duration of floodplain inundation in the middle and lower Zambezi reaches. These changes in flow regime have had an impact on the morphology and ecology of the river and the Zambezi Delta (Beilfuss and dos Santos 2001; Nugent 1983; Ronco et al. 2010).

The basin comprises a mosaic of miombo woodland, grassland, savannah, agricultural land, and wetlands. The evolution of the basin and its major biomes and species distribution are described in Timberlake (2000).

Wetlands in the Basin

Permanent and seasonal wetlands comprising swamps, marshes, and floodplains are a major feature of the basin covering a total area of at least 63,266 km^2 (4.7% of the basin) according to Lehner and Döll (2004) (Fig. 1). However, this is certainly an underestimate because in addition to the major wetlands, (Table 1) smaller wetlands, known as dambos, are common in much of the uplands, covering up to 15% of the

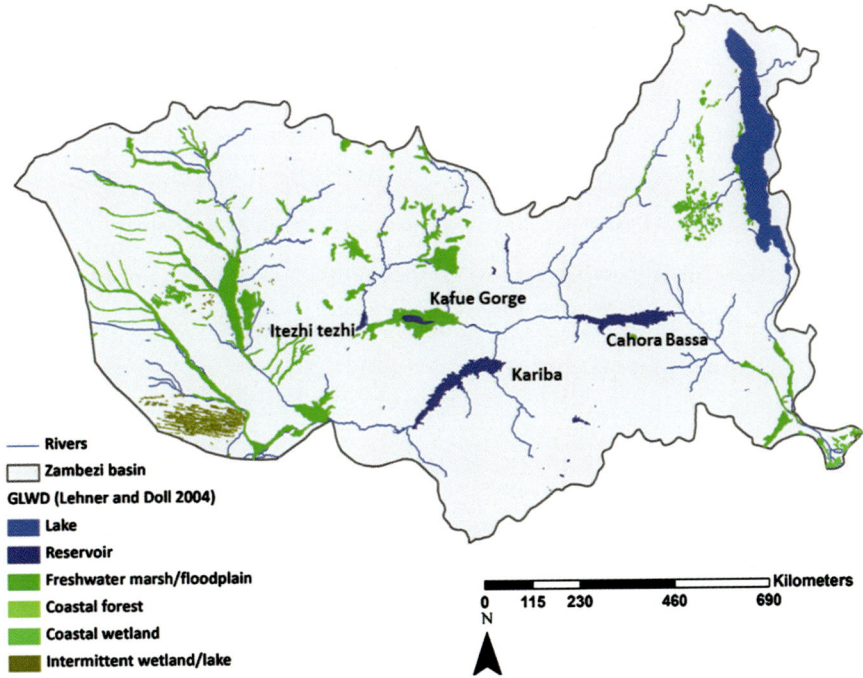

Fig. 1 Dams and major wetland areas in the Zambezi River basin

Table 1 Major wetlands in the basin (Compiled from Hughes and Hughes 1992)

Name	Location (lat and long and subbasin)	Area (km^2)	Description (e.g., wetland type)
Zambia			
Swamps of the Kabompo River	Kabompo	180	Small riparian swamps, extending in narrow strips
Swamps of the Lungue-Bungo River	The Lungue-Bungo River and two tributaries (Litapi and Lutembwe)	1,000	Large permanent swamp in the triangle of land between the two tributaries (*papyrus*, *Phragmites*, and floodplain grasslands)
Luena Flats	Luena River	897	*Papyrus* and *Phragmites* swamps with grass floodplains fed by several small streams (i.e., Nkala, Luambua, Lukuti, and Ndanda)
Nyengo Swamps	Luanginga River	700	Seasonal flood waters spread between the Luanginga, Ninda, and another tributary
Lueti and Lui Swamps	Lueti and Lui Rivers	375	Floodplain wetlands + patches of permanent swamp that merge with Barotse floodplain
Barotse Floodplain	Upper course of the Zambezi River 14°19′–16°32′S/ 23°15′–23°33′E	7,700	Floodplain wetland located on Kalahari sand
Sesheke-Maramba Floodplain	Zambezi along the northern border of the Caprivi Strip	1,500	Floodplain
Busanga Swamp	Kafue 14°05′–14°21′S/ 25°46′–25°57′E	600	Permanent shallow swamp
Lukanga Swamp	Lukanga but with spill from Kafue 14°00′–14°40′S/ 27°19′–28°00′E	2,100	Reed/*papyrus* swamp
Kafue Flats	Kafue River 15°11′–16°11′S/ 26°00′–28°16′E	7,000	Floodplain swamps and marshes located between Itezhi-Tezhi dam and Kafue Gorge dam
Zimbabwe			
Mid-Zambezi Valley and Mana Pools	Zambezi 15°36′–16°24′S/ 29°08′–30°20′E	360	Floodplain – pans and pools
Malawi			
The Shire Marshes	Shire River draining Lake Malawi 16°11′–17°05′S/ 34°59′–35°19′E	740	Two tracts of permanent swamp and lagoons in the Chikwawa and Bangula areas plus floodplain

(*continued*)

Table 1 (continued)

Name	Location (lat and long and subbasin)	Area (km^2)	Description (e.g., wetland type)
Namibia			
Cuando-Linyanti-Chobe-Zambezi (including Linyanti Swamp, eastern Caprivi wetland, Chobe Swamps)	Cuando, Linyanti (Chobe) 17°39′–18°40′S/ 23°18′–25°10′E	Total 3,930 900 (Linyanti Swamp)	Floodplain, swamps, and shallow lakes through the Caprivi Strip. Near the Chobe-Zambezi confluence in phase flooding of both rivers may inundate 1,700 km^2 of floodplain
Mozambique			
Lower Zambezi	Downstream of Tete, particularly in the vicinity of the Shire River	>325	Floodplain, swamps, and shallow lakes (e.g., Lake Mimbingue and Lake Tanie)
Zambezi Delta	Zambezi downstream of Caia	1,300	Zambezi discharges via distributaries through wide delta. Swampy floodplain and areas of mangrove forest extending up to 15 km inland along the main channels

landscape in some places (Bullock 1992). Dambos are clay-based, low-lying areas that are flooded by a combination of direct precipitation, surface runoff, and seepage from higher ground (Acres et al. 1985). They occur under a wide range of ecological conditions, and the shape and areal extent vary considerably. However, a common feature is poor drainage. The majority of dambos are characterized by vegetation communities dominated by herbaceous species, typically a large number of sedges (*Cyperacea*) and hygrophilous grasses (Hughes and Hughes 1992).

Riparian reed swamps, dominated by *Phragmites mauritianus* and *Typha domingensis*, occur along the upper courses of many tributary rivers with riverine forest occurring at lower altitudes (e.g., along parts of the Kafue and Zambezi above Victoria Falls where dense stands of *Syzygium* spp. characterize the riverbanks). Further downstream several tributaries flow into large depressions which contain permanent swamps, each of which cover tens of thousands of hectares and across which water flow is diffuse, often taking place in the absence of discrete channels. At even lower levels, the river and its tributaries have formed huge seasonally inundated floodplains.

The Upper Zambezi

Wetlands occur on many of the tributaries of the upper Zambezi (i.e., the Kabompo, the Lungue-Bungo, the Lutembwe, the Litapi, the Luena, the Luanginga, the Lueti, and the Lui), sometimes extending for many kilometers either side of the rivers. Most wetlands are essentially riparian (oxbow lakes, tall reed swamps, pans, and pools) backed by floodplain grasslands that are inundated in the wet season. Some individual wetlands are extensive (up to about 1,000 km^2), but they tend to become narrower and more discontinuous further upstream (Hughes and Hughes 1992). The most important wetland system of these upper tributaries is the vast Liuwa Plain (5,000 km^2) near the confluence of the Lungue-Bungo, Luambimba, and Zambezi Rivers, which supports the second largest migration of wildebeest (*Connochaetes gnou*) in Africa and substantial populations of endangered African painted dog (*Lycaon pictus*), gray crowned crane (*Balearica regulorum*), and wattled crane (*Bugeranus carunculatus*) (Kamweneshe et al. 2003).

The Barotse Floodplain covering an area of 7,700 km^2 is a major wetland on the upper Zambezi that is inundated to depths of 1.5–3.0 m when the flood peaks in April. As well as the main Zambezi River, the Luanginga, Luampa, Lueti, and Lui Rivers all flow into the floodplain.

The annual inundation of the floodplain significantly influences the pattern of life in Barotse, determining seasonal human and livestock migration patterns and production cycles and also making some areas inaccessible for parts of the year. The Lozi people, the native inhabitants of the Barotse, derive a range of diverse livelihoods from the floodplain, including those based on agriculture and fisheries. Every year they celebrate the flooding of the Zambezi with the Kuomboka ceremony when, toward the end of the rainy season, they make a ceremonial move to higher ground. The successful move is celebrated with traditional singing and dancing. This ceremony dates back more than 300 years (IWMI 2013). Across the floodplain, a complex network of more than 5,000 interlinked canals was built by the Barotse Royal Establishment, the traditional authority, in the late nineteenth century. These canals serve a range of different purposes including navigation, irrigation, drainage, livestock and domestic uses, and fisheries habitat. For its important biodiversity, the Barotse was designated a Ramsar Wetland of International Importance in 2007. The area has also been proposed as a World Heritage Site in recognition of both its environmental and its cultural heritage.

The Sesheke-Maramba Floodplain (also called the eastern Caprivi or southern Barotse Floodplain) occurs along the northern border of the Caprivi Strip close to the confluence of the Zambezi and the Chobe River which enters from Botswana. The floodplain is approximately 100 km in length and extends over an area of about 1,500 km^2 much of which lies in Namibia (Hughes and Hughes 1992). This floodplain is contiguous with the eastern portion of the Chobe-Linyanti floodplain system, which begins at the point where the Cuando River (the name given to the Chobe River in Angola) enters Botswana (van der Waal and Skelton 1984). The Linyanti Swamp is about 300 km^2 in area, but its size varies according to the extent of flooding in the upper Zambezi (Marshall 2000). Lake Liambezi, which periodically dries up, lies at the end of the Linyanti Swamp, has an open water surface of 100 km^2 when full, and is bordered by a swamp of 200 km^2 (van der Waal and Skelton 1984).

The Middle Zambezi

Designated in 2010, the UNESCO Middle Zambezi Biosphere Reserve comprising riverine and terrestrial ecosystems, extends from Lake Kariba (the reservoir created by the building of the Kariba dam) and the Matusadona National Park through various national parks and safari areas adjacent to the Zambezi River, including Mana Pools, Sapi, and Chewore which together are designated as a UNESCO World Heritage Site. In total it covers an area of 28,793 km^2.

Just downstream of Kariba dam, the Zambezi is joined by the Kafue River (see below) and then broadens into a braided course for some 130 km to the Mupata Gorge. The numerous streams produce many low-lying sandy islands, containing pans and pools with further pans and pools along the riverbanks. Much of the flat valley floor (ca. 12 km wide) and some river terraces used to be inundated when the river was in high flood, but now that flow is regulated by the Kariba dam, this no longer happens. However, several small tributaries (i.e., Nyamuchera, Chiruwa, Mbera, and Sapi) flow down through swampy land to the Zambezi in the vicinity of the Mana Pools. If floods in these tributaries coincide with major discharges from Kariba, substantial areas along the south bank of the Zambezi may be inundated. The area has a rich riparian flora of sedges, reeds, riverine forest, and grasslands with a clear succession based on flood tolerance (Hughes and Hughes 1992). The Mana Pools are small permanent pools that mark depressions in former river channels that have become isolated as the river moved progressively northward (Hughes and Hughes 1992). The pools have flat, grass- and reed-covered banks surrounded by forest. The main pools are permanent, deriving water from precipitation and groundwater seepage and only occasionally from flooding. The area of the pools, just a few hundred hectares, is a very small (ca 3%) portion of the total area that the Zambezi once flooded in this region.

The Kafue River is one of the major tributaries of the Zambezi, draining an area of 155,000 km^2, entirely within western Zambia. It is the most significant waterway in terms of the national economy in Zambia; most of the mining, industrial, and agricultural activities and approximately 50% of Zambia's total population are concentrated within the catchment area (Burke et al. 1994). Many permanent swamps occur on the Kafue River and its tributaries in the upper catchment. Most are narrow strips occurring on one or both sides of the river, and many are just a few hundred meters in width. However, there are several large permanent swamps: Lushwishi Swamp (100 km^2), Lufwanyama Swamp (74 km^2), Mininga Swamp (144 km^2), and an unnamed swamp on the main stream (310 km^2) (Hughes and Hughes 1992).

At a similar latitude but on different tributaries of the Kafue River, the extensive Busanga (600 km^2) and Lukanga (1,800 km^2) Swamps both lie in shallow depressions and have similar physiography. The Busanga Swamp supports a rich diversity of waterbirds but is isolated and poorly known (Beilfuss et al. 2007) (Photo 1). The Lukanga wetland, although named a swamp, actually comprises a treeless lake and marsh ecosystem – an intricate maze of reeds, pools, channels, and large bodies of open water (Kamweneshe and Beilfuss 2002; McCartney et al. 2011). The palustrine wetland covers approximately 95% of the area and includes stands dominated by reeds (*Phragmites*), mixed grass, cattail/reed mace (*Typha*), and termitaria grasslands. The lacustrine area comprises about 5% of the total wetland. It provides habitat for a wide range of terrestrial and aquatic flora and fauna, including at least 316 species of birds, including cranes, storks, ducks, geese, pelicans, herons, egrets,

Photo 1 The Busanga wetland (Photo credit: R. Beilfuss © Rights remain with the author)

Photo 2 Fishing camps in the Lukanga wetland (Photo credit: R. Beilfuss © Rights remain with the author)

and bitterns. The hydrology of the system is complex: at times of high flow, the Kafue River causes water in the Lukanga River to backup into the swamp, and during very high floods, the Kafue River itself overflows into the wetland (Seagrief 1962). It is estimated that about 60,000 people live in, or close to, the wetland (predominantly from the Lenje and Bemba tribes) and that products, derived from fishing, hunting, and agriculture, support a hinterland population of some 6.1 million people (Ramsar 2005) (Photo 2).

The broad alluvial plain of the Kafue Flats (area 7,000 km^2) lies between the Itezhi-Tezhi and Kafue Gorge dams. The river gradient through the Flats is just 0.022 m/km, and the travel time from Itezhi-Tezhi to Kafue Gorge is on average 6 weeks. Under natural conditions, the Kafue Flats flooded in the wet season (February to May) each year. Flooding usually commenced in December as a result of direct rainfall and tributary inflows, but maximum flood levels were not attained until the inflow in the main river channel reflected the heavy rainfall in the upper part of the catchment. The maximum flood arrived first in the western part of the Flats in February/March and moved slowly east, arriving at the head of the Kafue Gorge in April/May.

The Kafue Flats are one of the most biologically diverse ecosystems in Zambia. Comprising the meandering river and a complex of lagoons, oxbow lakes, abandoned river channels, marshes, levees, and floodplain grassland, they provide habitat for a wide range of birds and animals, including rare species. Over 400 bird species, including the endangered wattled crane (*Bugeranus carunculatus*), and 67 species of fish have been documented (Douthwaite 1982; Muyanga and Chipundu 1982). The Flats are home to the Kafue lechwe (*Kobus leche kafuensis*), an endemic antelope especially adapted to life in marshes (Howard and Chabwela 1986) (Photo 3). Two national parks (Lochinvar and Blue Lagoon) and associated tourist facilities were

Photo 3 Lechwe in the Kafue Flats (Photo credit: R. Beilfuss © Rights remain with the author)

established in the early 1970s. Designated as internationally important locations of high conservation value, the combined area of these parks is 830 km^2.

Traditionally, the natural resources of the Kafue Flats have been utilized in a wide variety of ways, for both commercial and subsistence purposes. It is estimated that more than 100,000 people are in some way dependent on the Flats (Scudder and Acreman 1996). Cattle grazing is a major commercial activity, and it is estimated that up to 290,000 head of cattle (10–20% of the national herd) utilize the Flats during the dry season. There is some commercial farming, primarily sugar and winter wheat. The largest producer of sugarcane is the Nkamabala Sugar Estate, owned by the Zambia Sugar Company, which presently cultivates 13,400 ha and abstracts water from the Kafue throughout the year for irrigation (McCartney and Houghton-Carr 1998). There are reeds and papyrus from which baskets and mats are woven at a subsistence level, but timber and other forest products are not common. Hunting (e.g., of lechwe), although illegal, provides an important source of protein for local people. The Flats support one of Zambia's most productive artisanal fisheries, supplying not only the floodplain communities but also urban centers such as Kafue town and Lusaka.

The Luangwa River, another major tributary, joins the Zambezi close to the border with Mozambique, just upstream of the Cahora Bassa reservoir. The upper reaches have relatively few wetlands, but there are strips of fringing reed swamps and riparian forest in places and patches of swampy forest that occur around springs in the headwaters. In its final 350 km before its confluence with the Zambezi, the floodplain broadens out and there are many oxbows and sections of abandoned channel with levees. In the rainy season, the entire floodplain, several kilometers wide in places, is completely inundated.

The Lower Zambezi

The Zambezi River enters Mozambique at Zumbo and immediately flows into the reservoir of the Cahora Bassa dam. Downstream of the dam, the river is contained within a narrow gorge until the town of Tete. Here the valley broadens and the river develops a narrow floodplain. For much of its course between Tete and the delta, the bed of the Zambezi is 1–5 km wide, and in the dry season, the river flows in several deeply incised channels. However, during the wet season, the entire bed may be one swiftly moving current (Hughes and Hughes 1992).

The most important tributary of the Zambezi in its lower course is the Shire River, which drains Lake Malawi and Malombe. Barrages near Liwonde regulate the flow for hydroelectricity generation. In its lower course before discharging into the Zambezi, the Shire flows through an extensive low-lying area, and a series of swamps extends along the river. The Shire Swamps comprise two tracts of permanent swampland (Elephant Marsh (570 km^2) and Ndinde Marsh (200 km^2)) in the Chikwawa and Bangula areas. The numerous lakes and lagoons which comprise these marshes may have connection with the anastomosing river channels only during the wet season. The marshes are virtually treeless and dominated by

Photo 4 Mangrove forest in the Zambezi Delta (Photo credit: R. Beilfuss © Rights remain with the author)

herbaceous vegetation (Hughes and Hughes 1992). The Shire Swamps support an important fishery, cattle grazing, and agriculture (irrigated cotton and sugar) in the marginal lands.

The Zambezi Delta occurs at the downstream terminus of the river, from the Zambezi-Shire confluence to the Indian Ocean. The delta is a broad, flat alluvial plain, approximately 12,000 km^2 in size. From its apex near the village of Mopeai, 120 km inland, the delta forms a large triangular area with a 200 km coastal frontage along the Indian Ocean. The Zambezi Delta is bordered to the west and north by the gently rising backslope of the African rift escarpment. The delta supports a diverse mosaic of wetland communities grading from acacia and palm savanna at the floodplain periphery, to seasonally flooded grassland, papyrus swamps, evergreen forests, and open water bodies on the low-lying plains, to mangrove forest and mud flats bordered by dunes near the coast (Beilfuss et al. 2000) (Photo 4).

The delta is an immensely productive wetland system, supporting large concentrations of African buffalo (*Syncerus caffer*), African elephant (*Loxodonta africana*), waterbuck (*Kobus ellipsiprymnus*), southern reedbuck (*Redunca arundinum*), sable antelope (*Hippotragus niger*), Lichtenstein's hartebeest (*Alcelaphus lichtensteinii*), and Livingstone's eland (*Taurotragus oryx*). Seventy-three waterbird species have been recorded, including endangered wattled crane, gray crowned crane (*Balearica regulorum*), large breeding colonies of great white pelican (*Pelecanus onocrotalus*), African openbill (*Anastomus lamelligerus*), and many other species of storks, herons, and spoonbills, and numerous Palearctic and intra-African migrants (Beilfuss et al. 2010).

Ninety-four fish species have been recorded in the lower Zambezi River, of which 55 are primarily freshwater species and mostly floodplain dependent (Bills 2000). The mangrove crab (*Scylla serrata*) and other crustaceans (portunids, etc.) are present and exploited by the local population, while prawns spawning in the delta mangroves are of great economic importance as a source of foreign revenue.

The Role of Wetlands

By affecting how water is routed and stored and evaporated, wetlands play an important role in the hydrology of the Zambezi River system. Spill from the Zambezi and its tributaries into wetlands and subsequent evaporation are major components of the basin water budget. A recent study of the effect of natural wetlands on river flow in the Zambezi basin concluded that broadly (i) floodplains decrease the magnitude of flood flows and increase low flows and (ii) headwater wetlands increase the magnitude of flood flows and decrease low flows. However, in all cases examples were found which produced contrary results and simple relationships between the areal coverage of a particular wetland type within a catchment, and the impact on the flow regime was not found. This confirms that the hydrological functions of wetlands depend to a large extent on location-specific characteristics that make it difficult to generalize (McCartney et al. 2013).

Zambezi basin wetlands support considerable biodiversity and productivity in terms of plants, large mammals, birds and fish, and other groups. Knowledge of the taxonomy of the various groups is generally good, with the exception of many invertebrate groups where even a rough indication of numbers of species present is not available. Although there are a number of species restricted to the wetlands of the Zambezi basin in a number of different groups, detailed listings are not yet available except for large mammals, birds, reptiles, amphibians, and fish (Timberlake 2000). It is estimated that there are 122 species of fish in the basin of which 25 are endemics (IUCN 2003). The lechwe antelope (with most of the global population occurring on

Table 2 Ramsar wetlands of international importance in the Zambezi basin

Country	Ramsar wetlands of international importance
Mozambique	Zambezi Delta
Zambia	Busanga Swamps
	Kafue Flats
	Luangwa Floodplains
	Lukanga Swamps
	Zambezi Floodplains
Zimbabwe	Mana Pools
	Monavale Wetland
	Cleveland Dam
	Chinhoyi Caves
	Driefontein Grasslands
	Victoria Falls
	Lakes Chivero and Manyame

Zambezi floodplains) and the wattled crane (seasonally with up to 75% of the world population in the wetlands of the Zambezi basin) are possibly the best "flagship species" for conservation of the wetlands (Timberlake 2000; Beilfuss et al. 2007).

The Zambezi wetlands play an important role in the livelihoods and well-being of many people in the basin and in the economies of the riparian countries. There are 13 Ramsar wetlands of international importance located in the basin (Table 2) which support fisheries, livestock, and other forms of agriculture, as well as tourism in addition to their considerable biodiversity value. For example, the annual gross financial value of the Barotse Floodplain is estimated to be $ 417 per household with a total annual economic value (from fish, crops, cattle, wildlife, reeds, and papyrus) of $ 12.2 million (Turpie et al. 1999). It has been estimated that the Barotse fisheries provide the bulk of the protein in the diet of about 200,000 people (Hughes and Hughes 1992). Similarly, 250,000 cattle, with a market value of $4 million, graze in the Kafue Flats wetland during the dry season each year (Seyam et al. 2001). Altogether flood recession agriculture in the major wetlands of the Zambezi is estimated to be worth US$36 million annually (Seyam et al. 2001). Many thousands of smaller, lesser-known wetlands, such as the dambos, also play a vital role in the everyday lives of poor rural communities, through the provision of clean drinking water and, because they retain extensive wet regions during the dry season, as a valuable agricultural resource in the semiarid regions of the basin (Wood et al. 2013).

Existing Threats to Wetlands

The population of the Zambezi basin is about 40 million of which 70% is rural and poor (Tumbare 2004). Currently resource overexploitation, land drainage and encroachment for agriculture, and modification of the river hydrology for large-scale hydropower and large-scale irrigation schemes are the greatest threats to the wetlands in the basin. As noted above, conversion of wetlands, including dambos, for agriculture is widespread. If conducted in an appropriate manner, it can, and does, make an important contribution to livelihoods, food security, and poverty alleviation (McCartney et al. 2010). However, with limited agricultural inputs and equipment, poor agricultural practices are prevalent resulting in wetland degradation in many places.

Currently 15 hydropower power stations are located within the basin of which by far the largest are Kariba and Cahora Bassa on the main river. The dams built in the Zambezi basin have significantly affected flow regimes, and wetlands have been greatly modified in both obvious and indirect ways through the creation of new habitats, through facilitation of distribution of species, and through reduced flooding. For example, downstream of the Kariba dam, wetlands are under extreme pressure as a result of year-round utilization by large mammals which under natural conditions would have been forced to migrate off the floodplain in the wet season. As result of this and impoverishment of the alluvium, as a consequence of silt removal from the Zambezi water before it reaches the floodplain, changes in the biotic communities of the floodplain are marked. Similarly, the ecology of the delta has been significantly altered as a consequence of flow regulation. Woody savanna and thicket species have increased in density and colonized far into the floodplain grassland mosaic. Relatively drought-tolerant grassland species have displaced flood-tolerant species in the broad alluvial floodplain, and saline grassland species have displaced freshwater species on the coastal plain. Abandoned alluvial channels are undergoing advanced stages of terrestrialization. Coastal mangrove has been replaced by saline grassland at the tidal margin. Sandbars have become stabilized and colonized by grassland and woody species (Beilfuss and dos Santos 2001).

The Itezhi-Tezhi dam was the first major dam in Africa, designed and constructed with additional storage specifically for the purpose of releasing managed floods. Flood releases were incorporated into the release regime of the dam in order to simulate the natural flooding of the Kafue Flats in March and April each year. Approximately 15% of the total live storage of the reservoir was set aside for such flood releases; a very progressive concept at the time the dam was built in the 1970s. Still the dam has undoubtedly had impacts on the Kafue Flats, many of which have been compounded by other socioeconomic changes that have occurred over the years. Substantial efforts are underway to coordinate "environmental flow" releases among Itezhi-Tezhi, Kariba, and Cahora Bassa dam operators to rejuvenate major floodplains of the middle and lower Zambezi (Beilfuss and Brown 2010; SWRSD Zambezi Basin Joint Venture 2010).

Future Challenges

Future challenges to the Zambezi basin wetlands relate to increasing population, economic development, and climate change. The population of the basin is growing rapidly, and this will inevitably increase competition for scarce resources and add to the pressure on wetlands. There are ambitious plans for additional hydropower dams and irrigation on both the main river and the tributaries, especially in Angola which has been to date a minor player in basin development. FAO (1997) estimates a potential area of 422,000 ha for formal irrigation in the basin. One likely consequence of increased water abstraction and greater flow regulation is further reduction in downstream flooding which may have significant impacts on wetland hydrology and functioning.

By changing patterns of rainfall and modifying flow regimes, climate change may also affect the basin hydrology with potentially significant impacts on flooding and hence wetlands. The Zambezi basin exhibits the worst potential effects of climate change among 11 major sub-Saharan African river basins and will experience the most substantial reduction in rainfall and runoff, according to the Intergovernmental Panel on Climate Change (IPCC). Multiple studies cited by the IPCC estimate that rainfall across the basin will decrease by 10–15% and runoff by as much as 40% or more (Beilfuss 2012). These changes will exacerbate existing problems for basin wetlands already degraded by altered and reduced flooding patterns.

Given the importance of the Zambezi wetlands not only for the livelihoods of many millions of people but also for the economies of the riparian countries, it is vital that future development planning takes into account the multiple services that they provide and incorporates measures to safeguard important ecosystem services.

References

Acres BD, Rains AB, King RHB, Lawton RM, Mitchell AJB, Rackham LJ. African dambos: their distribution, characteristics and use. Z Geomorphol. 1985;52:63–86.

Beilfuss RD. A risky climate for Southern African Hydro: assessing hydrological risks and consequences for Zambezi River Basin dams. Berkeley: International Rivers Network; 2012.

Beilfuss RD, Brown C. Assessing environmental flow requirements and tradeoffs for the Lower Zambezi River and Delta, Mozambique. Intl J River Basin Manag. 2010;8(2):127–38.

Beilfuss R, dos Santos D. Patterns of hydrological change in the Zambezi Delta, Mozambique. Working Paper #2: Program for the sustainable management of Cahora Bass Dam and the Lower Zambezi Valley. International Crane Foundation, USA and Direcção Naçional de Aguas, Mozambique; 2001.

Beilfuss RD, Dutton P, Moore D. Land cover and land use changes in the Zambezi Delta. In: Timberlake J, editor. Biodiversity of the Zambezi basin wetlands. Volume III. Land use change and human impacts. Harare: Biodiversity Foundation for Africa, Bulawayo/The Zambezi Society; 2000. p. 31–106.

Beilfuss RD, Dodman T, Urban EK. The status of cranes in Africa (2005). Proceedings of the 11th Pan African Ornithological Congress. Ostrich. 2007;78(2):175–84.

Beilfuss R, Bento C, Haldane M, Ribaue M. Status and distribution of large herbivores and birds of conservation concern in the Marromeu Complex of the Zambezi Delta, Mozambique. Mozambique: World Wide Fund for Nature; 2010, Department of Nature Conservation, Mozambique.

Bills R. Freshwater fish survey of the Lower Zambezi River, Mozambique. In: Timberlake J, editor. Biodiversity of the Zambezi Basin wetlands. Harare: Biodiversity Foundation for Africa, Bulawayo/The Zambezi Society; 2000. p. 461–85.

Bullock A. Dambo hydrology in southern Africa – review and reassessment. J Hydrol. 1992;134:373–96.

Burke JJ, Jones MJ, Kasimona V. Approaches to integrated resource development and management of the Kafue basin, Zambia. In: Kirby C, White WR, editors. Integrated river basin development. Chichester: Wiley; 1994.

Douthwaite RJ. Waterbirds: their ecology and future on the Kafue Flats. In: Howard GW, Williams GJ, editors. Proceedings of the National Seminar on environment and change: the consequences of hydroelectric power development on the utilization of the Kafue Flats. Lusaka: University of Zambia; 1982. p. 137–40.

Food and Agriculture Organization (FAO). Irrigation in Africa: a basin approach. Rome: FAO Land and Water Development Division; 1997.

Howard GW, Chabwela HN. Fauna of Zambia's major wetlands. Proc. Zambia Wetlands Seminar, University of Zambia, Lusaka; 1986.

Hughes RH, Hughes JS. A directory of African Wetlands. Gland/Cambridge, UK: IUCN/; 1992, UNEP, Nairobi, Kenya/WCMC, Cambridge, UK; 820 pp.

International Union for Conservation of Nature (IUCN). Watersheds of the world CD. Gland: IUCN; 2003.

International Water Management Institute (IWMI). Wetlands and people. Colombo: IWMI; 2013. 32pp.

Kamwenshe B, Beilfuss R. Wattled Cranes, waterbirds, and large mammals of the Lukanga Swamp, Zambia. Working paper #2 of the Zambia Crane and Wetland Conservation Project. International Crane Foundation, USA and Endangered Wildlife Trust, South Africa; 2002.

Kamwenshe B, Beilfuss R, Morrison K. Population and distribution of Wattled cranes and other large waterbirds and large mammals on the Liuwa Plains National Park, Zambia. Working paper #4 of the Zambia Crane and Wetland Conservation Project. International Crane Foundation, USA and Endangered Wildlife Trust, South Africa; 2003.

Lehner B, Döll P. Development and validation of a global database of lakes, reservoirs and wetlands. J Hydrol. 2004;296(1–4):1–22.

Marshall BE. Fishes of the Zambezi basin. In: Timberlake J, editor. Biodiversity of the Zambezi basin wetlands. Harare: Biodiversity Foundation for Africa; 2000. p. 393–460. Bulawayo/The Zambezi Society.

McCartney MP, Houghton-Carr HA. A modelling approach to assess inter-sectoral competition for water resources in the Kafue Flats, Zambia. J Ch Inst Water Environ Manag. 1998;12:101–6.

McCartney MP, Rebelo L-M, Senaratna Sellamuttu S, de Silva S. Wetlands, agriculture and poverty reduction. Colombo: International Water Management Institute (IWMI Research Report 137); 2010. doi:10.5337/2010.230. 39 pp.

McCartney MP, Rebelo L-M, Mapedza E, de Silva S, Finlayson CM. The Lukanga Swamps: use, conflicts and management. J Int Wildl Law Policy. 2011;14:293–310.

McCartney MP, Cai X, Smakhtin V. Evaluating the flow regulating functions of natural ecosystems in the Zambezi Basin. Colombo: International Water Management Institute (IWMI Research Report 148); 2013. doi:10.5337/2013.206. 51pp.

Muyanga ED, Chipundu PM. A short review of Kafue flats fishery from 1968 to 1978. In: Howard GW, Williams GJ, editors. The consequences of hydroelectric power development on the utilization of the Kafue Flats. Proceedings of the National Seminar on environment and change. Lusaka: University of Zambia; 1982. p. 105–13.

Nugent C. Channel changes in the Middle Zambezi. Zimb Sci News. 1983;17:127–9.

Ramsar. Ramsar information sheet, Ramsar sites information service, Lukanga Swamps. 2005. http://www.wetlands.org/reports/ris/1ZM0032007.pdf

Ronco P, Fasolato G, Nones M, Di Silvio G. Morphological effects of damming on the lower Zambezi River. Geomorphology. 2010;115:43–55.

Scudder T, Acreman MC. Water management for the conservation of the Kafue wetlands, Zambia and the practicalities of artificial flood releases. In: Acreman MC, Hollis GE, editors. Water management and wetlands in Sub-Saharan Africa. Gland: IUCN; 1996. p. 101–6.

Seagrief SC. The Lukanga swamps, northern Rhodesia. J S Afr Bot. 1962;28:3.

Seyam IM, Hoekstra AY, Ngabirano HHG. The value of freshwater wetlands in the Zambezi basin. Delft: IHE; 2001. 22pp.

SWRSD Zambezi Basin Joint Venture. Dam synchronisation and flood releases in the Zambezi River Basin project. Transboundary Water Management in SADC. Main Report and Four Annexes. 2010. 802 pp.

Timberlake J. Biodiversity of the Zambezi Basin. Biodiversity Foundation for Africa. Occasional Publications in Biodiversity No. 9, Bulawayo; 2000. 23 pp.

Tumbare MJ. The Zambezi River: its threats and opportunities. Paper presented at 7th river Symposium, Brisbane, 1–3 Sept 2004. http://www.archive.riversymposium.com/2004/index.php?element=Tumbare+M. Accessed 20 June 2015.

Turpie JK, Smith B, Emerton L, Barnes J. Economic value of the Zambezi Basin wetlands. Report to IUCN ROSA. Harare; 1999.

Tyson PD. Climatic change and variability in Southern Africa. Oxford: Oxford University Press; 1986.

van der Waal BCW, Skelton PH. Check list of the fishes of Caprivi. Modoqua. 1984;13:303–20.

Wood A, Dixon A, McCartney M. Wetland management and sustainable livelihoods in Africa. London: Routledge; 2013.

World Bank. The Zambezi River basin: a multi-sector investment opportunities analysis, Volume 3: State of the basin; 2010.

Zambezi River Delta (Mozambique)

99

Richard D. Beilfuss

Contents

Introduction	1234
Hydrology	1235
Wetland Ecosystems	1235
Floodplain Savanna Communities	1236
Floodplain Grassland and Swamp Communities	1236
Mangrove and Swamp Forest Communities	1237
Biodiversity	1237
Conservation Status	1238
Ecosystem Services	1239
Threats and Future Challenges	1239
References	1240

Abstract

The Zambezi River Delta on the Indian Ocean coast of Mozambique is a broad, flat alluvial plain, approximately 1.2 million ha in size. The delta includes a rich mosaic of wetland communities ranging from acacia and palm savanna on the delta periphery to seasonally flooded grassland, papyrus swamps, evergreen forests, and open water bodies on the low-lying delta plains to mangrove forest and mudflats bordered by dunes near the coast. The Zambezi Delta supports abundant wildlife, including African buffalo, African elephant, sable antelope, Lichtenstein's hartebeest, Livingstone's eland, and Endangered wild dog. Diverse waterbirds include a globally significant breeding population of Vulnerable wattled cranes, Endangered grey crowned cranes, numerous Palearctic and intra-African migrants, and immense breeding colonies of pelicans, herons,

---

R. D. Beilfuss (✉)
International Crane Foundation, Baraboo, WI, USA

College of Engineering, University of Wisconsin, Madison, WI, USA
e-mail: rich@savingcranes.org

© Springer Science+Business Media B.V., part of Springer Nature 2018
C. M. Finlayson et al. (eds.), *The Wetland Book*,
https://doi.org/10.1007/978-94-007-4001-3_195

spoonbills, and other species. The Zambezi Delta has a vital role in the local, regional, and national economy of Mozambique through the ecological goods and services it provides. Although the delta is formally designated as a protected area and *Wetland of International Importance* under the Ramsar Convention, it faces serious threats including dams, dredging, mining, agricultural conversion, and overexploitation of many resources.

Keywords

Zambezi river · Wetlands · Delta · Biodiversity · Ecosystem services · Hydropower dams · Environmental flows · African buffalo · Wattled crane · Papyrus · Mangrove

Introduction

The Zambezi Delta is situated in Mozambique at the downstream terminus of the Zambezi River, the largest river system in southern Africa, where it discharges into the Indian Ocean (Fig. 1). The delta is a broad, flat alluvial plain, approximately 1.2 million ha in size. From its apex near the village of Mopeai, 120 km inland, the delta

Fig. 1 The Zambezi Delta. Landsat ETM satellite image acquired 1.12.03 (With permission of International Crane Foundation)

forms a large triangular area with a 200 km coastal frontage along the Indian Ocean. The Zambezi Delta is bordered to the west and north by the gently rising backslope of the rift escarpment. The delta supports a diverse mosaic of grassland, palm savanna, woodland, and mangrove communities and features some of largest concentrations of wildlife found in African floodplain systems.

Hydrology

Rainfall over the delta is strongly influenced by the regional intertropical convergence zone (ITCZ), with most rainfall concentrated over the 4 month period between December and March and a prolonged dry season from May to November. Mean annual rainfall is approximately 1,150 mm but highly variable among years, with evidence of long-term cycles of relatively wet and dry years. Torrential rains and flooding may result from periodic cyclones, and severe droughts are a common phenomenon. Evaporation is high year-round and exceeds rainfall in all but the wettest months.

High flows from the Zambezi River arrive during the rainy season, with natural peak flows typically occurring between January and March. An intricate network of distributary channels conveys floodwaters from mainstem Zambezi River to the delta floodplains, and inundation of the entire delta occurs when the Zambezi River overtops its banks and spreads laterally across the floodplain. Zambezi floodwaters maintain shallow open water bodies and high water table conditions on the floodplain through most of the dry season. Groundwater contribution from the adjacent escarpment is a negligible component of the floodplain water balance but locally important for maintaining high water table conditions at the base of the escarpment.

Flooding patterns and processes in the Zambezi Delta are profoundly influenced by the upstream regulation of the Zambezi River for hydropower production, as well as local embankments constructed for roads, railroads, and flood protection. In the 28 years prior to Zambezi River regulation (1930–1958), the delta was inundated during all but one year, with an average flooding duration of 78 days above bankful discharge. Since construction of the upstream Kariba Dam (1958) and Cahora Bassa Dams (1974), flows have exceeded bankful discharge in only 54% of all years, for 26 days on average (Beilfuss 2001, 2012). As a consequence, floodplain water levels have declined by several meters, and the extent of permanent open water bodies and swamps has decreased.

Wetland Ecosystems

The Zambezi Delta is part of the Lower Zambezi Freshwater Ecoregion and includes a mosaic of wetland communities grading from acacia and palm savanna at the floodplain periphery to seasonally flooded grassland, papyrus swamps, evergreen forests, and open water bodies on the low-lying plains to mangrove forest and mudflats bordered by dunes near the coast (Beilfuss et al. 2000, 2001). Major wetland vegetation associations include:

Floodplain Savanna Communities

(i) Acacia savanna (140,000 ha): widespread at the periphery of the more elevated delta plain, rainfed, and dominated by *Acacia polyacantha* in association with *A. sieberana*, *A. xanthophloea*, *Cordyla Africana*, *Kigelia africana*, and many other tree species. Midstory vegetation is sparse, and grass growth is vigorous, with a dense cover of *Hyparrhenia dichroa*, *Ischamum afrum*, *Panicum maculatum*, and other species.

(ii) Borassus palm savanna (8,000 ha): conspicuous component of the floodplain margin on rainfed hydromorphc soils, with *Borassus aethiopum* occurring along the transitional zone from Acacia savanna to *Hyphaene* palm savanna and seasonally flooded grassland. Forms monotypic stands with increasing soil moisture, with vigorous grass growth.

(iii) Hyphaene palm savanna (86,000 ha): monotypic stands occur over vast areas on the upper delta floodplain, *Hyphaene* palm is the first woody pioneer of the seasonally flooded grassland, persisting after short-duration inundation eliminates other woody invaders. Dense understory of tussock grassland species including *Hyparrhenia rufa*, *Imperata cylindrica*, *Setaria* spp., and other species.

Floodplain Grassland and Swamp Communities

(iv) Rain flooded grasslands (115,000 ha.): bunch grasses on seasonally flooded floodplain vertisols (clay-rich soils that shrink and swell with changes in moisture content: during dry periods, the soil volume shrinks, and deep wide cracks form; during wet periods the clay absorbs water and increases in volume (swells). The alternate shrinking and swelling causes a mixing of vegetation and other surface matter into the subsoil and promotes a uniform soil profile). Grassland species are tall and rank, including various *Setaria* species, *Ischaemum afrum*, *Vetiveria nigritana*, and others depending on elevation. Together with the acacia and palm savanna communities, supports large herds of sable antelope *Hippotragus niger*, Lichtenstein hartebeest *Alcelaphus lichtensteinii*, and diverse others species of ungulates.

(v) River flooded grasslands (117,000 ha.): rhizomatous grasses and sedges on sites with prolonged inundation, on strongly expansive vertisols of the low-lying plains. Characteristic species include *Echinochloa pyramidalis*, *Cyperus digitatus*, *C. exaltatus*, *Leersia hexandra*, *Oryza longistaminus*, *Panicum maximum*, and many others. Supports vast herds of African buffalo *Syncerus caffer* and African elephant *Loxodonta africana*. Isolated patches of *Eleocharis acutangula* and *E. dulcis* are important for globally threatened (Vulnerable) wattled crane *Grus carunculatus* and other waterbirds.

(vi) Papyrus swamps (84,000 ha.): occur along deep, permanently inundated waterways of the delta; on the delta northbank a vast *Cyperus papyrus* swamp covers more than 50,000 ha of the low-lying floodplain. Floating papyrus mats support a diversity of shallow-rooted vegetation.
(vii) Phragmites reedswamps and saline grasslands (147,000 ha.): Dense stands of *Phragmites australis* occur in brackish water and saline soils near the coastal mangrove belt, occasionally with *Typha latifolia*, and often bordered by *Leersia hexandra-Cyperus digitatus* grassland on saline soils. Supports large mixed colonies of breeding waterbirds.
(viii) Open water vegetation: shallow oxbow lagoons and swamp depressions in low-lying areas throughout the floodplain grasslands, mostly permanent flooded. Colonized by a variety of floating and submerged aquatic plants, including *Nymphaea caerulea, Nymphoides indica, N. nilotica,* and *Utricularia* spp. and free-floating plants including invasives *Eichhornia crassipes, Azolla filiculoides, Pistia stratiotes,* and *Salvinia molesta*.

Mangrove and Swamp Forest Communities

(ix) Mangrove forest (100,000 ha.): part of the *East African Mangrove* ecoregion, a critically threatened and global biodiversity conservation hotspot, the delta coastline includes some of the most extensive mangrove forest on the Indian Ocean coast. Includes eight species of true mangrove, with conspicuous zonation – *Avicennia marina* and *Sonneratia alba* – are common pioneers on the exposed seaward front, with belts of *Rhizophora mucronata, Ceriops tagal,* and *Bruguiera gymnorrhiza* moving inland.
(x) Saline mudflats (45,000 ha.): large areas of hypersaline mudflats occur on the inland margin of the mangrove associations; sparsely covered with *Hibiscus tiliaceous, Salicornia* spp., and other succulents.
(xi) Evergreen swamp forest (3,000 ha): evergreen "inland mangrove" thickets of *Barringtonia racemosa* and associated species occur on muddy alluvium at the margin of tidal influence. Supports large, mixed colonies of breeding waterbirds.

Biodiversity

The Zambezi Delta is an immensely productive wetland system, featuring many species of global conservation concern. The delta supports diverse and abundant large mammals, including one of the large concentrations of African buffalo on the continent and an "aquatic" African elephant population renowned for having pink pigment on their legs associated with prolonged exposure to flooded swamps. Other large herbivore species moving seasonally between the flooded grassland and adjacent savanna and woodland include waterbuck *Kobus ellipsiprymnus*, southern reedbuck *Redunca arundinum*, sable antelope, Lichtenstein's hartebeest, and

Livingstone's eland *Taurotragus oryx*. A morphologically unique population of plains zebra *Equus quagga crawshayi*, referred to locally as "Selous zebra," occurs only in the delta region. African lion *Panthera leo*, leopard *Panthera pardus*, and globally threatened (Endangered) wild dog *Lycaon pictus* occur on the escarpment-floodplain periphery. Hippopotamus *Hippopotamus amphibious* are common in permanently flooded oxbow lagoons and other waterways (Beilfuss et al. 2010). Cape clawless otter *Aonyx capensis, in* regional decline, also occur here (Beilfuss et al. 2010).

The delta supports 73 waterbird species, including large breeding colonies of several species and numerous Palearctic and intra-African migrants (Bento and Beilfuss 1997). Species on the global Red List include wattled crane, grey crowned crane *Balearica regulorum*, and African skimmer *Rynchops flavirostris*. Thousands of pairs of great white pelican *Pelecanus onocrotalus* breed in the delta in addition to large breeding colonies of storks and herons, including African openbill *Anastomus lamelligerus*, African sacred ibis *Threskiornis aethiopicus*, grey heron *Ardea cinerea*, squacco heron *Ardeola ralloides*, African spoonbill *Platalea alba*, and *Egretta* spp.

Ninety-four fish species have been recorded in the Lower Zambezi River, of which 55 are primarily freshwater species and mostly floodplain dependent (Bills 2000).

Aquatic reptiles include the hinged terrapin *Pelusios castanoides*, Nile monitor *Varanus niloticus*, Nile crocodile *Crocodylus niloticus*, and various snakes (Branch 2000). Nineteen amphibian species are known to occur in the delta, with another five considered probable (Branch 2000).

Invertebrate studies are highly incomplete, but 18 gastropod and three bivalve mollusk species (Dudley 2000) and 25 species of Odonata (Kinvig 2000) have been recorded.

Conservation Status

Most of the south bank of the Zambezi Delta is protected through the Marromeu Buffalo Reserve, two forest reserves (*Reserva Floresta de Nhampacué* and *R.F. de Inhamitanga*), and four large hunting concessions. Other major land holdings in the delta include the large commercial agricultural lands (mainly sugar plantations) and community lands. Collectively, these lands are managed as the Marromeu Complex (1,127,200 ha) and were designated a *Wetland of International Importance* under the Ramsar Convention in 2003. In December 2015, the Government of Mozambique approved a major expansion of the Ramsar Site to include most of the north bank of the Zambezi Delta and local catchment, establishing one of the largest protected wetland complexes in the world – more than 1.2 million ha. (http://www.ramsar.org/news/mozambique%E2%80%99s-zambezi-delta-joins-the-ramsar-list).

Ecosystem Services

In addition to providing high biodiversity value, the Zambezi Delta has a vital role in the local, regional, and national economy of Mozambique through the ecological goods and services it provides. Guveya and Sukume (2008) estimate that the annual total value of ecosystem services in the Marromeu Complex ranges between US$0.93 billion and US$ 1.6 billion. Important goods and services provided by the delta include:

- Forest and woodland products (construction wood, fuel wood, wild fruits, honey, medicinal plants, and other forest and woodland resources)
- Carbon sequestration (woody species linked to voluntary carbon markets)
- Wetland products (papyrus and reeds used to make a variety of household items, palms used for palm wine, thatch grasses harvested from seasonal floodplain grasslands, and other resources)
- Grazing lands for livestock (grasslands of the floodplains, pans, and drainage lines, most notably late dry-season grazing lands supported by persistent high water table conditions)
- Nutrient-rich lands for flood-recession agriculture (floodplain agricultural lands receiving irrigation waters and nutrients from the natural ebb and flow of the mainstem Zambezi River and distributary channels)
- Riverine and floodplain freshwater fisheries (freshwater fisheries in the mainstem Zambezi, seasonal and permanent floodplain waterbodies, and pans and drainage lines on the escarpment)
- Clean and abundant freshwater (clean water supply for drinking, cleaning, bathing, and other household uses provided by surface water and groundwater recharge)
- Estuarine Penaeid shrimp fisheries (shrimp fisheries produced in mangroves and harvested off the Sofala Bank – an important source of export revenue for Mozambique)
- Storm surge and coastal erosion protection (mangroves and coastal dune vegetation that serve to stabilize coastal areas from erosion during cyclones and storm surge)
- Flood storage and mitigation (the capacity of the floodplain to store or attenuate large runoff events and reduce flood damage to settled areas)
- Diverse landscapes and wildlife for ecotourism (natural features of the landscape that contribute to the ecotourism potential of the region)
- Wildlife for sustainable trophy hunting and subsistence meat supply

Threats and Future Challenges

Although the Zambezi Delta enjoys significant institutional protection, the wetland faces a daunting range of conservation challenges. Adverse changes in the timing, volume, duration, and frequency of floodplain inundation are associated with the upstream river regulation for hydropower production and compound the impacts of locally severe droughts. Dredging and canalization of the Zambezi River is proposed to transport coal from inland mines to coastal ports. The Zambezi Delta has been long considered a prospective source for crude oil, with prospecting and drilling operations by multinational oil companies dating back to 1937, and additional prospecting is underway for natural gas and other minerals. Agro-industrial drainage and pollution from commercial sugar expansion is resulting in eutrophication of floodplain waterways (Government of Mozambique 2011). On the open floodplain and savanna, uncontrolled fires – especially in association with reduced water availability – kill many smaller antelope, destroy nests, and reduce the dry-season carrying capacity of the floodplain. Floodplain woodlands are threatened by clear cutting for smallholder shifting agriculture, unsustainable logging for export and building materials, and charcoal production.

The prolonged civil war in Mozambique, and the Zambezi Delta region in particular, had a profound impact on wildlife. The African buffalo population was reduced by 95% to fewer than 2,000 individuals and waterbuck, hippopotamus, and plains zebra declined by >98%. During the postwar period since 1994, however, most wildlife species in the delta have undergone steady population growth. The current population of African buffalo now exceeds 18,000. African elephant numbers and most large herbivores are approaching their historic population levels, and hippo have reestablished several large pods on the floodplain (Beilfuss et al. 2010). Despite these gains, illegal hunting for the bushmeat trade is increasing and threatens many smaller antelope.

The delta also faces a range of daunting socioeconomic challenges, fueled by rapid population growth and immigration. The delta population lacks adequate primary health care, housing, sanitation, and safe drinking water. Food security is a chronic challenge, driving exploitation of delta resources. The region also lacks adequate primary and secondary education opportunities and employment alternatives.

Numerous programs are underway to address these challenges, including a regional effort to restore environmental flows in the Zambezi River basin through coordinated water releases from upstream dams (Beilfuss and Davies 2000; Beilfuss and Brown 2010). The Marromeu Complex Management Plan governs the administration and management of the delta, prohibited and regulated activities, and coordinated strategies for biodiversity conservation and sustainable livelihoods among agencies, academic institutions, concession operators, community-based organizations, and nonprofit conservation and social development organizations (Government of Mozambique 2011).

References

Beilfuss R. Hydrological degradation, vegetation change, and restoration potential: the story of an African floodplain. Madison: University of Wisconsin-Madison; 2001.

Beilfuss, R. A risky climate for southern African hydro: assessing hydrological risks and consequences for Zambezi Dams. International Rivers: Berkeley, California, USA; 2012; 56 pp

Beilfuss RD, Brown C. Assessing environmental flow requirements and tradeoffs for the Lower Zambezi River and Delta, Mozambique. Intl J River Basin Manag. 2010;8:127–38.

Beilfuss R, Davies B. Prescribed flooding and wetland rehabilitation in the Zambezi delta, Mozambique. In: Streever W, editor. An international perspective on wetland rehabilitation. Kluwer, Dordrecht, The Netherlands; 2000. p. 143–58.

Beilfuss RD, Dutton P, Moore D. Land cover and land use changes in the Zambezi delta. In: Timberlake J, editor. Biodiversity of the Zambezi basin wetlands. Volume III. Land use change and human impacts. Harare/Bulawayo: Biodiversity Foundation for Africa/The Zambezi Society; 2000. p. 31–106.

Beilfuss R, Bento C, Haldane M, Ribaue M. Status and distribution of large herbivores and birds of conservation concern in the Marromeu Complex of the Zambezi delta, Mozambique. Mozambique: World Wide Fund for Nature/Department of Nature Conservation; 2010.

Bento CM, Beilfuss RD. Status and distribution of waterbirds in the Zambezi delta. Proceedings of the Workshop on the Sustainable Use of Cahora Bassa Dam and the Zambezi Valley; Songo, Mozambique; 1997.

Bills R. Freshwater fish survey of the lower Zambezi river, Mozambique. In: Timberlake J, editor. Biodiversity of the Zambezi Basin wetlands. Harare/Bulawayo: Biodiversity Foundation for Africa/The Zambezi Society; 2000. p. 461–85.

Branch WR. Survey of the reptiles and amphibians of the Zambezi Delta. In: Timberlake J, editor. Biodiversity of the Zambezi Basin wetlands. Harare/Bulawayo: Biodiversity Foundation for Africa/The Zambezi Society; 2000. p. 377–92.

Dudley C. Freshwater molluscs of the Zambezi River Basin. In: Timberlake J, editor. Biodiversity of the Zambezi Basin wetlands. Harare/Bulawayo: Biodiversity Foundation for Africa/The Zambezi Society; 2000. p. 487–526.

Government of Mozambique. General management plan for the Marromeu Complex, Mozambique: a wetland of international importance. Prepared by In: Beilfuss R, Bento C, da Silva P, editors. Mozambique: Ministry of Tourism; 2011.

Guveya E, Sukume C. The economic value of the Zambezi Delta. Maputo. Report to WWF Mozambique Country Office; 2008

Kinvig. Odonata survey of the Zambezi Delta. In: Timberlake J, editor. Biodiversity of the Zambezi Basin wetlands. Harare: Biodiversity Foundation for Africa/Bulawayo/The Zambezi Society; 2000. p. 559–64.

Nile River Basin

100

Matthew McCartney and Lisa-Maria Rebelo

Contents

Introduction	1244
The White Nile	1246
The Blue Nile	1247
The Main Nile	1248
The Role of Wetlands	1248
Existing Threats to Wetlands	1249
Future Challenges	1249
References	1250

Abstract

The Nile, the World's longest river, is well endowed with wetlands, including the Sudd, one of the World's largest. The wetlands are not only biodiversity hotspots but also vital for the livelihoods and wellbeing of people. In total there are 14 Ramsar Sites in the basin and thousands of smaller, lesser-known wetlands. Many

---

M. McCartney (✉)
Ecosystem Services, International Water Management Institute, Vientiane, Lao People's Democratic Republic

Regional Office for Southeast Asia and the Mekong, International Water Management Institute, Vientiane, Lao People's Democratic Republic
e-mail: m.mccartney@cgiar.org

L.-M. Rebelo
Water Futures, International Water Management Institute, Vientiane, Lao People's Democratic Republic

Regional Office for Southeast Asia and the Mekong, International Water Management Institute, Vientiane, Lao People's Democratic Republic

© Springer Science+Business Media B.V., part of Springer Nature 2018
C. M. Finlayson et al. (eds.), *The Wetland Book*,
https://doi.org/10.1007/978-94-007-4001-3_89

of these wetlands sustain rural communities through the provision of drinking water and by supporting fisheries, livestock and cultivation. There are a number of current threats to the basin wetlands, including inappropriate agricultural practices, overfishing, invasive species, and extraction of minerals and oil. Ambitious plans for hydropower and irrigation development, as well as rapid population increase and climate change, all pose future challenges to the sustainable management of the wetlands.

Keywords

Biodiversity · Hydrology · Livelihoods · Ecosystem services · Agriculture · Fisheries

Introduction

The Nile River in Northeast Africa is a major river basin and the Earth's longest river, flowing from south of the equator to the Mediterranean Sea (4 °S to 32 °N), a total distance of 6,718 km (Dumont 2009). However, the current river course has only been in existence for approximately 12,500 years, when the previously closed basins of both Lake Victoria and the Sudd overflowed to the north (Said 1993). The river comprises two main tributaries: the White Nile that originates as the outflow from Lake Victoria in Uganda and the Blue Nile, which originates as the outflow from Lake Tana in Ethiopia (Fig. 1). The total catchment area is 3,112,369 km^2, approximately 10% of the African continent. The longitudinal profile of the river comprises a number of flat reaches linked by steep connecting channels. It is this profile (i.e., breaks in slope and flat areas), in combination with large volumes of water, that results in a significant area of wetlands (ca. 183,000 km^2), approximately 6% of the basin.

It is estimated that there are about 115–130 fish species in the main Nile (Witte et al. 2009). However, in part because of the extreme climatic variation across the basin, the taxon richness and the degree of endemism of Nile biota are unexceptional, particularly in comparison to the adjacent Congo River Basin. In this context, the wetlands of the Nile Basin represent hotspots of biodiversity. For example, both Lake Victoria and Lake Tana support many endemic fish species (i.e., cichlids and cyprinids, respectively) (Dumont 2009), and the Sudd wetland is extremely species rich.

The Nile also provides an important pathway for African species to extend from the tropics to the Mediterranean and out into the Levant and Arabia. It is thought that this may have occurred numerous times throughout history. One example is papyrus (*Cyperus papyrus*) which has established itself in the River Jordan valley and even extends to Scilly (Dumont 2009). Reverse movements have also occurred. For example, although relatively species poor, the Ethiopian plateau is the most important, but not the only, focus of aquatic palearctic relicts, some of which are still

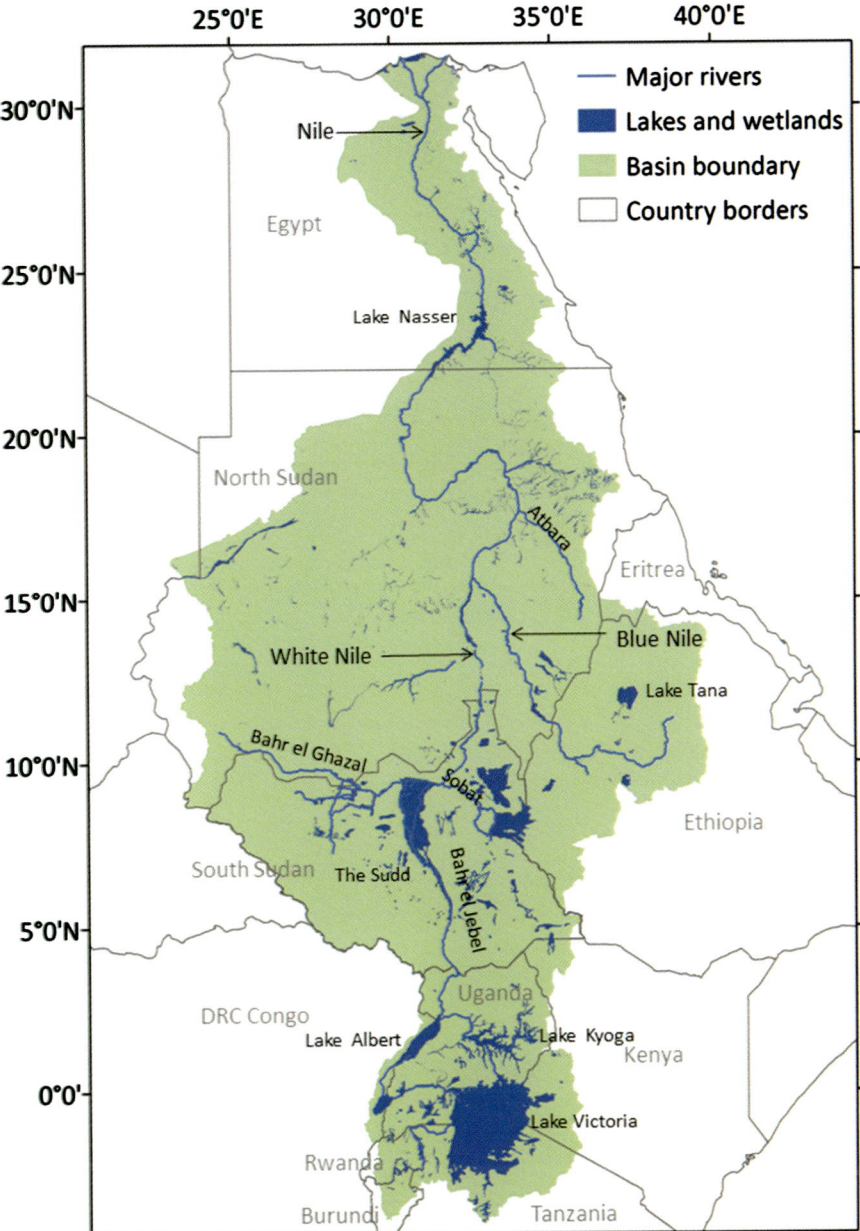

Fig. 1 Spatial distribution and areal extent of wetlands within the Nile Basin. Data are derived from the Global Lakes and Wetlands Database (Lehner and Döll 2004) and country-based Africover datasets (FAO 2002)

identical to northern populations, while others are clearly speciated, indicating more than one wave of immigration (Dumont 2009).

The White Nile

The tributary furthest from the mouth of the Nile is the Kagera River, which drains an area of 60,000 km^2 in the mountains of Rwanda and Burundi, before dropping onto the lowland plains where the river slope varies from 0.2 to 0.5 mkm^{-1}. Here the river flows through a series of lakes and swamps before discharging into Lake Victoria. A number of other tributaries drain the forested escarpment to the northeast of the lake and other tributaries drain the Serengeti plains and the swamps of Uganda. The catchment area of Lake Victoria is 184,000 km^2 and comprises 17 large river basins. However, the lake has a surface area of 68,800 km^2, and it is rainfall over the lake surface (not inflowing rivers) that provides the largest component (on average about 85%) of the input to the lake water balance (Sutcliffe 2009).

The Victoria Nile downstream of the lake comprises a single channel which discharges from the lake at Jinja and reaches Lake Kyoga about 100 km downstream. Lake Kyoga is a relatively shallow, grass-filled lake (3–7 m deep) with a surface area of 6,270 km^2 and a catchment area of 57,669 km^2 (WMO 1974). Downstream from Lake Kyoga, the Nile flows toward the western arm of the Rift Valley through a succession of level reaches and swamps, interrupted by rapids and falls before entering Lake Albert through a swamp near its northern end. Lake Albert (also known as Mobutu-Sese Seko) is located at the border of Democratic Republic of Congo and Uganda. It has a surface area of 5,300 km^2 and depth range of 0–50 m. In addition to flow from the White Nile, it receives inflow from the Semliki River which drains Lake Edward and the Ruwenzori mountains.

Downstream of Lake Albert, the White Nile (or Bahr el Jebel as it is known here) flows toward Nimule in a flat reach through swamp vegetation. At Nimule the river enters Sudan, turns to the northwest, and flows in a steeper channel toward Juba and Mongalla. Between Lake Albert and Mongalla, the river receives inflow from a number of tributaries known as the torrents, which provide seasonal and sediment-laden discharge to supplement the less variable outflow from the East African lakes. Downstream of Mongalla, the channel cannot carry the high flows occurring during the flood season, and the alluvial channels are built up above the flood plain. Excess flows leave the river via small channels through the banks, or spill over the banks at higher flows, inundating large areas on either side of the main channel, so forming the Sudd, one of the largest wetlands in the World. Due to evaporation, the outflow from the Sudd is typically less than half the inflow, and the significant seasonal variation of the inflow is damped out. Between 1907 and 1983, the mean inflow was 33 km^3 and the mean outflow was 16 km^3 (note: data records stopped in 1983 due to the onset of the civil war in Sudan).

Downstream of the Sudd, the White Nile turns east at Lake No, where the Bahr el Ghazal enters from the west. The Bahr el Ghazal drains a large area of South Sudan. The upper catchment consists of highlands with relief of up to 1,000 masl and relatively high annual rainfall (1,200–1,400 mmy^{-1}). Several seasonal tributaries converge near the confluence of Bahr el Ghazal and Bahr el Jebel. The rivers mainly leave the upper catchments in well-defined river valleys where sediments are deposited as the slopes decrease. The lower Bahr el Ghazal consists of a series of river channels, lakes, and papyrus swamps and seasonally flooded grasslands. Only about 3% of the inflow into the Bahr el Ghazal swamps is discharged into the White Nile.

The final tributary of the White Nile is the Sobat, which itself has two main tributaries, the Baro-Akobo and the Pibor. The Baro-Akobo River drains an area of 41,400 km^2 mainly in the southwestern highlands of Ethiopia where annual rainfall ranges from 1,300 to 2,370 mm. The Pibor River drains the Southern Sudan highlands and has a catchment area of 109,000 km^2 with an average annual rainfall of 950 mm. These rivers spill into the Machar marshes and other seasonally waterlogged grass swamps, which extend to its confluence with the White Nile River. A number of channels cross the marshes and discharge directly into the White Nile River downstream of the Sobat confluence. The course of the White Nile downstream of the Sobat to its confluence with the Blue Nile near Khartoum is confined to a single channel with no inflow except in exceptional years. The steady outflow from the Sudd is supplemented by the seasonal contribution of the Sobat, with each providing approximately 50% of the total flow of the White Nile at Khartoum.

The Blue Nile

Close to Khartoum the White Nile joins the Blue Nile (also known as the as the Abay in Ethiopia and the Al Bahr al-Azraq in Sudan). Two-thirds of the annual discharge of the main Nile at Khartoum is contributed by the Blue Nile. The Blue Nile drains approximately 311,548 km^2. Rainfall is largely concentrated in the months of July to October and, as a result, the river flows are highly seasonal. The upper river flows through Lake Tana, but only about 8% of the total flow, is generated upstream of the lake. The Blue Nile flows southeast from Lake Tana before looping back on itself, flowing west and then turning northwest close to the border with Sudan.

In the Ethiopian highlands, the basin is composed mainly of volcanic and Precambrian basement rocks with small areas of sedimentary rocks. The catchment is cut by deep ravines in which the major tributaries flow. The valley of the Blue Nile itself is 1,300 m deep in places. The primary tributaries in Ethiopia are the Bosheilo, Welaka, Jemma, Muger, Guder, Finchaa, Anger, Didessa, and Dabus on the left bank and the North Gojjam, South Gojjam, Wombera, and Beles on the right bank. The Blue Nile enters Sudan at an altitude of 490 masl and, just before crossing the frontier, the river enters a clay plain, through which it flows to Khartoum. The average slope of the river from the Ethiopian frontier to Khartoum is only 15 cmkm^{-1}. Within Sudan, the Blue Nile receives water from two major tributaries draining from the north, the Dinder and the Rahad, both of which also originate in

Ethiopia. At Khartoum the Blue Nile joins the White Nile to form the main stem of the Nile River.

The Main Nile

The main Nile flows north from Khartoum through the Sabaloka Gorge and is joined some 300 km north by the Atbara which drains the northern Ethiopian highlands, part of Eritrea and the lowlands in eastern Sudan. Its catchment area is 68,800 km^2 and its rainfall is concentrated in single wet season in August and September. The river is dry for much of the year. The main tributaries are the Tekeze and the Bahr el Salam.

Downstream of the confluence with the Atbara, the river flows in a series of loops through an arid region of successive cataracts, where the river flows are reduced by evaporation before entering Egypt within Lake Nasser, the reservoir of the Aswan High Dam. The mean annual flow into the reservoir is approximately 85 km^3. The area below the Aswan Dam is characterized by high aridity. The main water sources are the outflow from the Aswan High Dam and some seasonal streams (Wadi Natash, Wadi Kashao, Wadi Gurara, Wadi Elmiya, Wadi Mutula, Wadi Qena, Wadi Sheitun, Wadi el Asyut, and Wadi el Tarfa) draining the Maaza plateau between the Nile and the Red Sea. The Nile Delta catchments consists of the area, north of Cairo, where the Main Nile River bifurcates into the Rosetta and Damietta branches before draining into the Mediterranean Sea. As a result of abstractions, primarily for irrigation (estimated to cover approximately 2.5 million ha in the Nile Valley and Delta), an average of less than 10 km^3 of water flows into the sea each year.

The Role of Wetlands

As described above, wetlands play an important role in the hydrology of the Nile River system. Wetland form, function, and maintenance are governed to a large extent by the hydrological processes that occur both within them and their interaction in the catchment in which they are located. Lake and wetland storage are particularly important in the White Nile Basin, where spill from the river and its tributaries into wetlands and subsequent evaporation are major components of the catchment water budget (Sutcliffe and Parks 1999).

Throughout the Nile Basin, patterns of flow and water chemistry are significantly modified by the complex movement of water within wetlands. This in turn affects many ecosystem services upon which millions of people depend. In total there are 14 Ramsar Sites located in the basin which support fisheries, livestock, and other forms of agriculture. Many thousands of smaller, lesser-known wetlands, such as those in the headwaters of the Baro-Akobo River, also play a vital role in the everyday lives of poor rural communities, through the provision of clean drinking water, wetland sedge (*Cyperus latifolius*) which is used for thatching and handicrafts, and a range of medicinal plants (Dixon et al. 2013).

Existing Threats to Wetlands

There are a number of factors that currently threaten Nile Basin wetlands, including inappropriate agricultural practices both within and upstream, overfishing, invasive species, and extraction of minerals and oil. Conversion of wetlands for agriculture is already a widespread practice in many Nile Basin countries, including Ethiopia and Uganda, and has resulted in degradation in places. The Nile Basin most polluted wetlands are those of the Nile Delta where irrigation drainage, untreated urban wastes, and industrial effluents have reduced water quality, destroyed several forms of aquatic life, reduced the productivity of fisheries, and contaminated fish catches (UNEP 2006). The aquatic diversity of many wetlands in the basin including the Sudd, Lake Kyoga, and the Nile Delta is vulnerable to invasive species such as water hyacinth (*Eichhornia crassipes*). The introduction of Nile Perch into Lake Victoria while creating a highly productive fishing industry (over 800,000 t annually) has significantly depleted the lake biodiversity and produced an ecology that may not be sustainable in the long term (Lehman 2009).

Future Challenges

Future challenges to the Nile Basin wetlands relate to increasing population, economic development, and climate change. The population of the basin is expected to grow from 300 million (just under 20% of the total population of Africa) to approximately 550 million by 2030. This will inevitably increase competition for scarce resources and add to the pressure on wetlands (Rebelo and McCartney 2012). There are ambitious plans for dams for hydropower and irrigation on both the main rivers and the tributaries of the Blue and the White Nile. One likely consequence of water abstraction and increased flow regulation is reduced downstream flooding which may have significant impacts on wetland hydrology and functioning. In the Sudd, consideration is again being given to completing, or constructing an alternative to, the Jonglei Canal, a 360 km long channel intended to divert water around the eastern edge of the wetland in order to reduce evaporation losses and provide water for downstream irrigation (Bessière and de Savignac 2013). If built, such a canal is likely to have significant impacts on the area of the wetland and may result in the loss of biodiversity, livestock grazing areas, and fish habitat and affect the seasonal migration of cattle and wildlife (Lamberts 2009; WWF 2010). By changing patterns of rainfall and modifying flow regimes, climate change may also affect the basin hydrology with potentially significant impacts on flooding and hence wetlands.

Given the importance of the Nile wetlands not only for the livelihoods of many millions of people but also for the economies of the riparian countries, it is vital that future development planning takes into account the multiple services that the wetlands provide and incorporates measures to safeguard important ecosystem services.

References

Bessière C, de Savignac, X. Technical data for a possible Jonglei Canal. Water Storage and Hydropower Development for Africa. 2013. International Conference and Exhibition: Addis Ababa 16–18 April.

Dixon A, Hailu A, Semu T. Local institutions, social capital and sustainable wetland management: experiences from western Ethiopia. In: Wood A, Dixon A, McCartney MP, editors. Wetlands management and sustainable livelihoods in Africa. London: Routledge and Earthscan; 2013. p. 85–111.

Dumont HJ, editor. The Nile: origins, environments, limnology and human ose, Monographiae Biologicae, vol. 89. Dordrecht: Springer; 2009.

Food and Agriculture Organization of the United Nations (FAO). Africover – east Africa module: land cover mapping based on satellite remote sensing. Rome: FAO; 2002.

Lamberts E. The effects of Jonglei Canal operation scenarios on the Sudd swamps in Southern Sudan [master's thesis]. Enschede: University of Twente; 2009.

Lehman JT. Lake Victoria. In: Dumont HJ, editor. The Nile, Monographie Biolgicae, vol. 89. Dordrecht: Springer; 2009. p. 215–41.

Lehner B, Döll P. Development and validation of a global database of lakes, reservoirs and wetlands. J Hydrol. 2004;296:1–22.

Rebelo LM, McCartney MP. Wetlands of the Nile Basin: distribution, functions and contributions to livelihoods. In: Awulachew SB, Smakthin V, Molden D, Peden D, editors. The Nile River Basin: water agriculture, governance and livelihoods. Abingdon: Routledge; 2012. p. 212–28.

Said R. The River Nile: geology, hydrology and utilization. Oxford: Pergamon; 1993.

Sutcliffe J. The hydrology of the Nile Basin. In: Dumont HJ, editor. The Nile, Monographie Biolgicae, vol. 89. Dordrecht: Springer; 2009. p. 335–64.

Sutcliffe J, Parks Y. The hydrology of the Nile, IAHS special publication, vol. 5. Wallingford: IAHS Press; 1999.

United Nations Environment Programme (UNEP). Africa environment outlook 2: our environment, our wealth. Nairobi: UNEP; 2006.

Witte FM, van Oijen MJP, Sibbing FA. Fish Fauna of the Nile. In: Dumont HJ, editor. The Nile, Monographie Biolgicae, vol. 89. Dordrecht: Springer; 2009. p. 647–75.

World Meteorological Organization of the United Nations (WMO). Hydrometeorological survey of the catchments of lakes Victoria, Kyoga & Albert. Geneva: WMO; 1974.

World Wildlife Fund (WWF). Saharan flooded grasslands. Available from: www.worldwildlife.org/wildworld/profiles/terrestrial/at/at0905_full.html. Accessed June 2010.

Nile Delta (Egypt)

101

Mohamed Reda Fishar

Contents

Introduction	1252
Nile Delta Characteristics	1253
Population	1253
Climate	1254
Agriculture	1254
Fisheries	1254
Biodiversity	1254
North Coast Wetlands	1255
North Delta Lakes	1255
Threats and Future Challenges	1257
References	1258

Abstract

The Nile Delta is situated in the middle of the Egyptian Mediterranean coastline of approximately 1,000 km. The Delta shoreline consists of sandy and silty shores of greatly varying lateral configurations, depending on where the old branches of the Nile had their outlets. It is home to over 50 percent of Egypt's population of 80 million. The Nile delta is suitable for intensive agriculture and it supports 63% of the country's agricultural land. The coastline has two promontories, Rosetta and Damietta, and beaches are backed by coastal flats followed by coastal dunes and four brackish shallow lakes (from east to west: Mariut, Edku, Borullus and Manzalah). Delta and lakes are economically support both a large fishery and

M. R. Fishar (✉)
Inland water and Aquaculture Branch, National Institute of Oceanography and Fisheries (NIOF), Cairo, Egypt
e-mail: mfishar@hotmail.com

© Springer Science+Business Media B.V., part of Springer Nature 2018
C. M. Finlayson et al. (eds.), *The Wetland Book*,
https://doi.org/10.1007/978-94-007-4001-3_216

many fish farms. Lake fisheries produce 182,525 tons which represent about 12.54% of the nation's total fish production. Nile Delta is part of one of the world's most important migration routes for birds. Every year, millions of birds pass between Europe and Africa along the East Africa Flyway, and the wetland areas of Egypt are especially critical stopover sites.

Delta lakes are threatened by continuous land reclamation projects, construction of roads along the north coast, coastal erosion, soil salinization, extensive land use, pollution and degradation, and lack of appropriate institutional management systems. Nile Delta is subject also to shoreline changes resulting from erosion and accretion, subsidence, and sea level rise (SLR) resulting from climate change.

Keywords

Delta, Nile · Egypt · Biodiversity · Lakes · Threatens

Introduction

Described by Herodotus in the fifth century BC, the Nile Delta is probably the earliest known delta in the world (Said 1981). It is an arcuate delta (arc shaped), resembling a triangle or lotus flower when seen from above (Fig. 1), formed by sedimentary processes between the upper Miocene and present (Nelsen 1976; Stanley and Warne 1993) and built up by the alluvium brought by the old seven active branches of the Nile. The extinct Sebennetic branch that started during the Holocene between 7,500 and 3,000 BP (Arbouille and Stanley 1991) crossed the middle of the delta and formed its central hump that reaches its maximum extension immediately to the east of El Burullus lagoon (Orlova and Zenkovitch 1974). The Nile Delta is a typical wave dominated delta (Coleman et al. 1981) with a semidiurnal microtidal regime and maximum tidal range of 50 cm (UNESCO/UNDP 1978).

The delta was formed by the division of the Nile River into the old seven distributaries as it flowed north through the valley formed by the Nile in Upper Egypt. Those distributaries have subsequently silted up and been replaced by the present Damietta (east) and Rosetta (west) branches as the Nile River enters the delta.

The Nile Delta is situated in the middle of the Egyptian Mediterranean coastline of approximately 1,000 km (16% of the total Mediterranean coast) and supports many natural and man-made types of habitat. The delta has an onshore area of about 22,000 km^2 and about an equal area offshore down to the 200 m isobath. The delta extends some 220 km along the Mediterranean coastline and a maximum of 170 km from north to south (El-Kady et al. 2000). The southern apex of this classically shaped delta is located approximately 30 km north of Cairo. Along the Mediterranean coast, the delta is bounded on the west and east by Alexandria and Port Said, respectively.

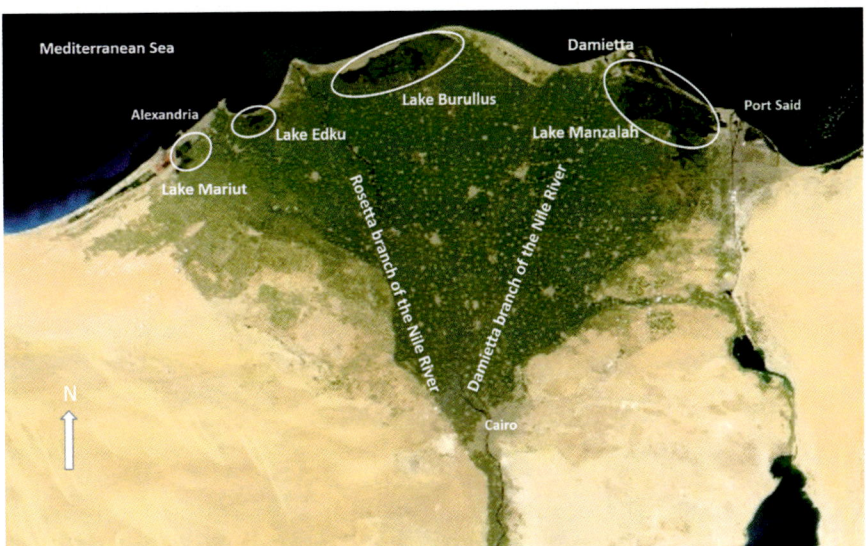

Fig. 1 The Nile Delta and north coast lakes (Adapted from: Jacques Descloitres, MODIS Rapid Response Team, NASA/GSFC. Sensor: Terra – MODIS. 19 July 2004)

The Nile Delta represents about 2.4% of the total area of Egypt; without the Nile Valley and Delta, Egypt is mainly a desert country (Fig. 1). It accounts for 63% of arable lands in Egypt, 50% of the country's industrial production, 65% of agricultural production (Hamza 2006), and 60% of fish catch in the country (El-Fishawi 1993).

Nile Delta Characteristics

Population

There are five harbors located on the coast: Idku, New Burullus, and El Gamil for fishing; while Damietta and Port Said are commercial ports. This part represents the major industrial, agricultural, and economic resource of the country. It is home to over 50% of Egypt's population of 80 million and to about 70% of the nation's industrial and commercial activities (EEAA 2009). With only 2.4% of the country's surface, the Nile Delta is very densely inhabited, with population densities up to 1,600 inhabitants per square kilometer, and strongly dependent on farming.

Alexandria and Port Said are the main economic centers of the coastal zone. These cities are vulnerable to sea level rise as a result of the low elevation of adjacent land below sea level. The Mohamed Ali Seawall, built in 1830, protects the lowland area southeast of Alexandria against inundation by water from Abu Qir Bay, and narrow strips of elevated land protect the southern area of Port Said. Many smaller towns and villages on the northern coast are also vulnerable to sea level rise.

Climate

The climate along Egypt's north coast is moderated by the influence of the Mediterranean Sea. Summers are moderately hot and humid, while the winters are moderately wet and mild. Although the northern coast receives the most rain in Egypt, the annual rainfall is low and concentrated during the winter months following a Mediterranean pattern. The coastal region includes most of the Egyptian big cities which benefit from the moderate climate. Temperatures rarely go lower than 9–10 °C in winter but in summer temperatures can exceed 40 °C. Between these two seasons, temperatures are next to 15 °C.

Agriculture

The Nile Delta is one of the oldest intensely cultivated areas on earth. The Nile Delta is suitable for intensive agriculture, and it supports 63% of the country's agricultural land (Hereher 2009). Hamza (2006) stated that approximately 85% of Egyptian water resources (mainly supplied by the Nile) are committed to irrigation of the 3.4 million hectares of cultivated lands. About 65% of these agricultural activities are within the delta area. The delta cultivates some cotton, wheat, corn, rice, citrus fruits, groundnuts, and other crops.

Fisheries

The Delta swamps are stocked with a large quantity of fishes. The lakes are also economically important as they support both a large fishery and many fish farms. Furthermore, there is a strong expansion in fish farming which is an important food source for the population. Lake fisheries produce 182,525 t which represent about 12.54% of the nation's total fish production (GAFRD 2013). Fishing is mainly done by trammel net and various primitive methods (e.g., catching by hand and collecting fishes under vegetation using a cone-shaped net). Catch is composed mainly of tilapia (mainly *Oreochromis niloticus*), catfish *Clarias gariepinus*, grass carp *Ctenopharyngodon idellus*, and mullet (mainly *Mugil cephalus*, *Liza aurata*, and *L. ramada*).

Biodiversity

The Nile Delta is part of one of the world's most important migration routes for birds. Every year, millions of birds pass between Europe and Africa along the East Africa Flyway, and the wetland areas of Egypt are especially critical stopover sites (Denny 1991). Many Palearctic bird species migrate via the delta lakes in internationally significant numbers (Goodman et al. 1989). Species that pass through the Nile Delta include white stork *Ciconia ciconia*, black stork *C. nigra*, european crane *Grus grus*, and great white pelican *Pelecanus onocrotalus*, as well as numerous birds of prey, including short-toed snake eagle *Circaetus gallicus*, booted eagle *Hieraaetus pennatus*, steppe eagle *Aquila nipalensis*, lesser spotted eagle *Clanga pomarina*, steppe buzzard *Buteo buteo*, European honey buzzard *Pernis apivorus*, and Levant sparrowhawk *Accipter brevipes*. Large numbers and a wide diversity of waterbirds, passerines, and other bird groups also pass through the country during the spring and autumn.

The lakes margins are extensively vegetated, mainly by *Phragmites* sp. and *Typha* sp. These plants are also frequent on the islands, and the invasive water hyacinth *Eichhornia crassipes* is proliferating. Agriculture production around the lakes includes date palm and sugar cane plantations, cereals and leguminous crops and has been encouraged in recent decades by the increased supply of fresh Nile water for irrigation (Dumont and El Shabrawy 2008). The water quality of all the lakes is locally affected by sewage and agrochemicals originating mainly from the southern agricultural regions.

North Coast Wetlands

The Nile Delta shoreline between Alexandria and Port Said consists of sandy and silty shores of greatly varying lateral configurations, depending on where the old branches of the Nile had their outlets. The coastline has two promontories, Rosetta and Damietta, and beaches are backed by coastal flats followed by coastal dunes and four brackish shallow lakes (from east to west: Mariut, Edku, Borullus, and Manzalah). The region is characterized by relatively low land elevation, which leaves it severely exposed to rising sea levels. In addition, it suffers from local land subsidence, compounding the effects of rising seas. Some estimates indicate that the northern delta region is subsiding at a rate that varies from about 0.60 mm/year at Alexandria to about 3.5 mm/year at Port Said.

North Delta Lakes

In their study on the Nile Delta, Stanley and Warne (1998) showed that wetlands in early nineteenth century mapping remained widespread in the northern delta. Large-scale anthropogenic alterations of wetlands began during the early nineteenth century Napoleonic-British campaigns in Egypt. Modifications included successive flooding and draining of Lake Mariut near Alexandria (De Cosson 1935), subsequent reduction of this water body by pumping, and drainage and eventual disappearance of Abu Qir lagoon by the end of the 1800s (Chen et al. 1992). Since that time, remaining lagoons have been artificially fragmented and reduced in size (Shaheen and Yosef 1978).

As the final reservoirs of Nile water before it flows into the Mediterranean, lakes Mariut, Edku, Borullus, and Manzalah are the last opportunity for Egyptians to use the Nile water (Fig. 1). The four shallow (average depth 1.10 m) lakes occupy an

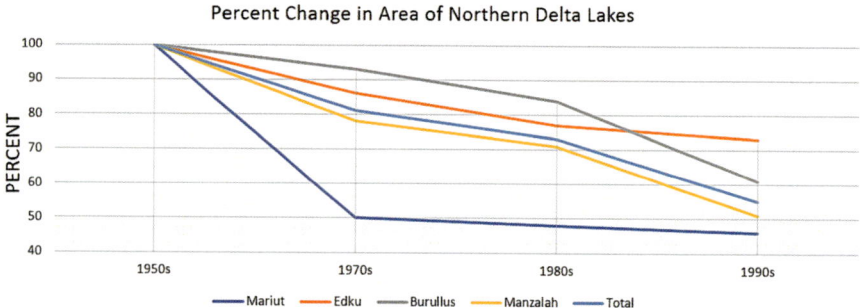

Fig. 2 Percent change in the surface area (km^2) of the northern delta lakes between the 1950s and 1990s. Surface area in the 1950s for Mariut = 136.9 km^2, Edku = 150.1 km^2, Burullus = 571.8 km^2, and Manzalah = 1274.2 km^2 (Drawn by author from data in Hamza 1999)

area of about 1,100 km^2, and their salinity changes from fresh to brackish in a seaward direction. They are connected to the Mediterranean Sea either directly or indirectly. The quantity and quality of water discharging to the lakes are determined by up-river activities, not by the requirements for activities within the lakes. The surface area of the four lakes has been reduced by approximately 50% between the 1950s and 1990s (Fig. 2).

The northern delta lakes provide a rich and vital habitat for estuarine and marine fish and their reproduction and have always been major areas of Egyptian fisheries. These lakes account for more than 75% of the fish harvested from Egyptian lakes (Mehanna 2008). The lakes are also internationally important in providing valuable habitat for several hundred thousand wintering water birds.

(a) **Lake Mariut** (31° 10′ N, 29° 55′ E) located southwest of Alexandria is the most heavily polluted lake in Egypt and with no natural outlet its water is pumped to the Mediterranean Sea (Fig. 1). The lake is artificially divided into four basins: the eastern basin or lake's proper area is 27.3 km^2 ranging in depth from 90 to 150 cm; fish farms have an area of about 4 km^2; and the southwestern and northwestern basins have a combined area of 32 km^2 (Masoud et al. 1981; Shaltout and Khlalil 2005). The lake receives a huge and continuous influx of pollutants (sewage, industrial, and agriculture wastes of Alexandria and El-Beheira Governorates). The El-Amoum drain represents the main source of water supply (60%) followed by the Noubaria canal (22%) and Qalaa drain (10%) (Kossa 2000).

(b) **Lake Edku** lies in the north of the Nile Delta, west of the Nile's Rosetta branch between 30° 8′ 30″ – 30° 23′ 00″ E and 30° 11′ 00″ – 31° 18′ 00″ N. It is a shallow brackish 126 km^2 water basin ranging in depth from 60 to 150 cm (Masoud et al. 2004). Its long axis is about 17 km, and its width varies between 5 and 11 km. The deepest part lies generally in the central and eastern regions, while the western and southern parts are more shallow (Abd Alla 1994) The lake is connected to the Mediterranean at its western extremity via Boughaz El-Maadia. Three main drains (Edku, Bousily, and Brezik) discharge agriculture drainage water to the lake.

(c) **Lake Borullus** extends along the northern fringe of the Nile Delta. It is centrally situated and second in size only to Lake Manzalah. Around the turn of the twentieth century, it had a surface area of about 600 km², but land reclamation for agriculture in its southern sector had caused it to decline to approximately 460 km² by 1974, and this decline continues to date. The long axis of the lake runs parallel to the Mediterranean shore approximately 65 km; its width varies between 6 and 16 km and averages 11 km. The lake is extremely shallow, with depths between 0.4 and 2.0 m. The deepest part is in the western sector, which is also the freshest, while the eastern sector that contains a 250 m long canal connecting Borullus to the sea (Bughaz) is shallow and saline. The lake contains several large and countless small islands, and large sectors of its water surface are invaded by emergent, floating and submerged water plants. The number of macrophytes recorded in the lake, as estimated by Shaltout and Al-Sodany (2000), is 197 species: 100 annuals and 97 perennials, including 12 hydrophytes. A canal (Brimbal) connecting Borullus to the Rosetta branch of the Nile used to supply up to half of the lake's water; after 1964 this volume rapidly declined to 5%. The lake is a Ramsar Site and it plays host to large populations of migratory and resident water birds.

(d) **Lake Manzalah** is located on the northeastern edge of the Nile Delta, separated from the Mediterranean Sea by a sandy beach ridge. It is the largest of the delta lakes and is bordered by the Mediterranean Sea on the North, the Suez Canal on the East, and the Damietta branch of the Nile on the West. The lake is generally rectangular in shape, about 60 km in length and 40 km in width, with an average depth of 1.3 m. At the beginning of the twentieth century, the lake contained approximately 1,000 islands of varying sizes and had an area of 1,698 km². By 1970 the area of the lake had diminished to 1,275 km² (El Wakeel and Wahby 1970). Land reclamation projects further reduced the lake's area to 905 km² by 1981 and 770 km² in 1988. Lake Manzalah is connected to the Mediterranean Sea through the main opening at Al-Gameel, west of Port Said, and with the Suez Canal through a small canal at Al-Gabouty near Port Said. Highly productive fresh waters of the Nile enrich the lake through three canals originating from the Damietta Branch: Al-Enania, Al-Rotma, and Al-Sufara. The Lake is also enriched by drainage water entering through five main drains (Al-Serw, Al-Gamaliah, Bahr Hadus, Ramsis, and Bahr-Bagar) connected at the south and southeastern Borders. The amount of drainage water inflow to the lake has increased from $1,399 \times 10^6 m^3$ in 1930 (Faouzi 1934) to $6,770 \times 10^6 m^3$ in 1980 (Khalil 1990) and $7,249 \times 10^6 m^3$ in 1988 (Abdel-Baky and El-Ghobashy 1990).

Threats and Future Challenges

Degradation, habitat loss, infilling, and drought have decreased the original area of all Delta lakes by more than 70%. In recent years, Delta lakes are threatened by continuous land reclamation projects, construction of roads along the north coast, coastal erosion, soil salinization, extensive land use, pollution and degradation, and lack of appropriate institutional management systems (Shaltout and Galal 2006; Zalat and Vildary 2007). Accompanying the decrease in wetland area has been a continuing decrease in the amount of open water, due not only to land reclamation but also an increase in the amount of aquatic vegetation. This is a consequence of a local customary practice of building Hoshas, basins to isolate water bodies to catch fish. At the same time there has been a deterioration of the lakes' fisheries due to over fishing and the use of illegal fishing methods.

The North Delta Lakes are under increasing threat from water withdrawal for human use, and secondarily from eutrophication, pollution, and destruction of surrounding wetlands (Birks et al. 2001). Eutrophication is a result of increased nutrient influx from agricultural drains carrying large amounts of fertilizers and pesticides into the lakes. The drains also receive industrial, agricultural, and sewage wastes that are a leading cause of the lake water's high pollution levels. Gu et al. (2013) reporting on the spatial distribution of heavy metals showed high levels in the eastern Manzala and western Edku lagoons, but an obvious low in the central Borullus lagoon. Most heavy metals showed an increasing trend with time, especially in the upper 10–15 cm of the sediment cores.

Like other deltaic regions worldwide, the Nile Delta is subject to shoreline changes resulting from erosion and accretion, subsidence, and sea level rise (SLR) resulting from climate change. Egypt is considered one of the top five countries expected to be most impacted with a 1 m SLR in the world (Susmita et al. 2007). Several general analyses of the potential impact of SLR on the Nile Delta coast have been carried out. As a result, areas of high vulnerability in the Nile Delta and possible socioeconomic impacts have been generally defined. These high-risk areas include parts of Alexandria, Beheira, Port Said, and Damietta Governorates. Several studies indicated that a large percentage of the Nile Delta is directly vulnerable to inundation and saltwater intrusion potentially impacting groundwater quality in the Nile Delta. Increasing soil salinization will lead to deterioration of crop quality and lower productivity, which will have significant socioeconomic consequences and serious implications for food security and public health. That could drive millions from their homes (El-Raey et al. 1999).

References

Abd Alla NA. Major ions and nutrient salts in the water of Lake Edku. [master's thesis]. Alexandria: Faculty of Sciences, Alexandria University; 1994. 199 p.

Abdel-Baky TE, El-Ghobashy A. Some ecological aspects of the western region of Lake Manzalah, Egypt. 1-Physical and chemical characteristics. Mansoura Sci Bull. 1990;17:166–84.

Arbouille D, Stanley DJ. Late Quaternary evolution of the Burullus lagoon region, north-central Nile delta, Egypt. Mar Geol. 1991;99:45–66.

Birks HH, Peglar SM, Boomer I, Flower RJ, Ramdani M, with contributions from Appleby PG, Bjune AE, Patrick ST, Kraiem MM, Fathi AA, Abdelzaher HMA. Palaeolimnological responses of nine North African lakes in the CASSARINA Project to recent environmental changes and human impacts detected by plant macrofossil, pollen, and faunal analyses. Aquat Ecol. 2001;35:405–30.

Chen Z, Warne AG, Stanley DJ. Late Quaternary evolution of the northwestern Nile delta between the Rosetta promontory and Alexandria, Egypt. J Coast Res. 1992;8:527–61.

Coleman JM, Robert HH, Murray SP, Salama M. Morphology and dynamic sedimentology of the eastern Nile delta shelf. Mar Geol. 1981;42:301–12.

DE Cosson A. Mareotis: being a short account of the history and ancient monuments of the Northwestern Desert of Egypt and of Lake Mareotis. London: Morrison and Gibb Ltd.; 1935. 219 p.

Denny P. Africa. In: Moser M, Finlayson M, editors. Wetlands. International waterfowl and wetlands research bureau. Oxford: Facts on File; 1991. p. 115–48.

Dumont HJ, El-Shabrawy G. Seven decades of change in the zooplankton (s.l.) of the Nile Delta lakes (Egypt), with particular reference to Lake Borullus. Int Rev Hydrobiol. 2008;93:44–61.

EEAA. National circumstances. Egypt second national communication on climate change. Cairo: EEAA; 2009. 127 p.

El-Fishawi NM. Sea level changes and their consequences for hydrology and water management. In: Proceedings of the International Workshop, Noordwijkerhout; 1993. Session 4, p. 3.

El-Kady HF, Shalotout KH, El-Shourbagy MN, Al-Sodany YM. Characterization of habitats in the northwestern part of Nile Delta. The International Conference on Biological Sciences (ICBS), Faculty of Sciences, Tanta University, vol. 1. 2000. p. 144–57.

El-Raey M, Dewidar K, El Hattab M. Adaptations to the impacts of sea level rise in Egypt. Climate Res. 1999;12:117–28.

El-Wakeel SK, Wahby SD. Hydrography and chemistry of Lake Manzalah, Egypt. Arch Hydrobiol. 1970;67:173–200.

Faouzi H. Report on the fisheries of Egypt for the year 1933. Ministry of Finance, Egypt. Cairo: Coast Guards and Fisheries Service Government Press; 1934. 120 p.

GAFRD. Fisheries statistics yearbook. Cairo: General Authority for Fish Resources Development; 2013. 93 p.

Goodman SM, Meininger PL, Baha El-Din SM, Hobbs JJ, Mullie WC. The birds of Egypt. Oxford: Oxford University Press; 1989. 548 p.

Gu J, Salem A, Chen Z. Lagoons of the Nile delta, Egypt, heavy metal sink: with a special reference to the Yangtze estuary of China. Estuar Coast Shelf Sci. 2013;117:282–92.

Hamza W. Differentiation of phytoplankton communities of Lake Mariut: a consequence of human impact. Bull Fac Sci Alex Univ. 1999;39:159–68.

Hamza W. Estuary of the Nile. In P. Wangersky (ed), Estuaries; Handbook of Environmental Chemistry. Vol. 5 H; 2006. Springer Verlage, Heidelberg, pp. 149–173.

Hereher ME. Inventory of agricultural land area of Egypt using MODIS data. Egypt J Remote Sens Space Sci. 2009;12:179–84.

Khalil MT. The physical and chemical environment of Lake Manzalah, Egypt. Hydrobiologia. 1990;196:193–9.

Kossa AA. Effect of industtrial and organic pollution on potential productivity and fish stock of Lake Mariut. [dissertation]. Cairo: Faculty of Sciences, Ain Shams University; 2000. 201 p.

Masoud AHS, Ezzat A, El-Rayis OA, Hafez HA. Occurrence and distribution of chemical pollutants in Lake Mariut, Egypt. II. Heavy metals. Water Air Soil Pollut. 1981;6:401–7.

Masoud AHS, Elewa AA, Ali AE, Mohamed EA. Metal distribution in water and sediments of Lake Edku, Egypt. Egypt Sci Mag. 2004;1:13–22.

Mehanna SF. Northern Delta Lakes, Egypt: constraints and challenges. Stuttgart: Hohenheim University; 2008.

Nelsen EV. Shore evolutions. Proceedings of Seminar on Nile Delta Coastal Process, CoRI/UNESCO/UNDP, Cairo. 1976. p. 15–59.

Orlova G, Zenkovitch V. Erosion of the shores of the Nile delta. Geoforum. 1974;18:68–72.

Said R. The geological evolution of the River Nile. New York: Springer; 1981. 151 p.

Shaheen AH, Yosef SF. The effect of the cessation of Nile flood on the hydrographic features of lake Manzala, Egypt. Hydrobiology. 1978;84:339–67.

Shaltout KH, Al-Sodany YM. Phytoecology of Lake Burullus Site. Cairo: MedWetCoast, Global Environmental Facility (GEF)/Egyptian Environmental Affairs Agency (EEAA); 2000.

Shaltout KH, Galal TM. Comparative study on the plant diversity of the Egyptian northern lakes. Egypt J Aquat Res. 2006;32:254–70.

Shaltout KH, Khlalil MT. Lake Borullus (Borullus protected area), vol. 13. Cairo: Publications of National Biodiversity Unit EEAA; 2005. 578 p.

Stanley DJ, Warne AG. Nile delta: recent geological evolution and human impact. Science. 1993;260:628–34.

Stanley DJ, Warne AG. Nile delta in its destruction phase. J Coast Res. 1998;14(3):794–825.

Susmita D, Laplante D, Meisner C, Wheeler D, Yan J. The impact of sea level rise on developing countries. A comparative analysis. World bank policy research working paper 4136, Feb 2007.

UNESCO/UNDP. Coastal protection studies. Final technical report, Paris; 1978. 1, 55 p.

Zalat A, Vildary SS. Environmental change in Northern Egyptian Delta lakes during the late Holocene, based on diatom analysis. J Paleolimnol. 2007;37:273–99.

Baro-Akobo River Basin Wetlands: Livelihoods and Sustainable Regional Land Management (Ethiopia)

102

Adrian Wood, J. Peter Sutcliffe, and Alan Dixon

Contents

Introduction	1262
Highland Wetlands	1262
Extent and Character	1262
Local Use and Knowledge	1263
Lowland Wetlands	1265
Extent and Character	1265
Local Use	1265
Coordination of Competing Land Use	1266
Threats and Future Challenges	1266
References	1267

Abstract

The Baro-Akobo system from Ethiopia, along with a major tributary the Sobat from South Sudan, contributes 48% of the flow of the White Nile where these river systems join downstream of Malakal. Within the Baro-Akobo system in Ethiopia there are wetlands at altitudes from 400 m amsl to over 2,000 m, varying in size from 1 ha to more than 1,000 ha. These wetlands provide a range of ecosystem services and play critical roles in the livelihoods of the local people.

Peter Sutcliffe: deceased.

A. Wood (✉)
Business School, University of Huddersfield, Huddersfield, UK
e-mail: a.p.wood@hud.ac.uk

J. P. Sutcliffe (✉)
Independent Consultant, University of Huddersfield, Huddersfield, UK

A. Dixon (✉)
Institute of Science and the Environment, University of Worcester, Henwick Grove, UK
e-mail: a.dixon@worcs.ac.uk

These communities have built up considerable local knowledge about these areas and have developed community management systems. These skills need to be developed and applied more rigorously to address the threats to wetlands to ensure sustainable use with catchment and wetlands managed together in a functional landscape approach.

Keywords

Livelihoods · Sustainable wetland management · Local knowledge · Community institutions · Functional landscape approach · Valley bottom wetlands · Floodplains · Ethiopia

Introduction

The Baro-Akobo system, flowing into the White Nile, covers some 76,000 km^2 in Ethiopia. It had a population of 2.4 million in 2000. The Baro and Akobo rivers originate in the highlands of southwest Ethiopia and flow west into the Sudan. Joined by Sobat in South Sudan these rivers together contribute 48% of the flow of the White Nile where these river systems join downstream of Malakal (Sutcliffe 2006a).

The Baro-Akobo basin within Ethiopia is almost equally divided between highlands (58%) and lowlands (42%). The highlands are mostly between 1,500 and 2,500 m amsl, while the lowlands are a gently sloping plain mostly between 400 and 500 m amsl. A major escarpment separates the two areas. In the highlands the predominant vegetation is Afromontane forest. In the lowlands the vegetation grades from *Acacia seyal* to *Balanites aegyptiaca* savanna woodland in east through grassland to seasonal wetlands and swamps in the west. Rainfall in the highlands is over 2,000 mm over a rainy season of 10 months, while in the lowlands it is around 600 mm. The population is concentrated in the highlands (ca. 80%), while the lowland part of the basin is sparsely populated (Sutcliffe 2006b).

The wetlands in this basin vary in character, those in the highlands being mostly quite small valley bottom features, while in the lowlands they are much more extensive floodplains supplied by river overspill. These wetlands play important roles in rural livelihoods, and the sound management of the wetlands is an important element in the development of an ecologically sensitive and sustainable use regime for this basin.

Highland Wetlands

Extent and Character

Within the highland portion of the basin there are concentrations of permanent and semipermanent wetlands. They vary in size from less than 1 ha to over 20 ha, with a few reaching 1,000 ha. *Cyperus latifolius* is the dominant vegetation in these wetlands. Most of these wetlands are in the mid-altitude plateau areas, between

Table 1 Wetland uses and beneficiaries in Illubabor Zone

Uses	Estimate of households benefiting
Social/ceremonial use of sedges	100% (including urban dwellers)
Thatching reeds	85% (most rural households)
Temporary crop guarding huts of sedges	30%
Dry season grazing	Most cattle owners (30% of population)
Water for stock	Most cattle owners (30% of population)
Cultivation	25%
Domestic water from springs	50–100%
Craft materials (palm products and sedges)	5%
Medicinal plants	100% (mostly indirectly by purchase from collectors/traditional doctors)

Source: Wood and Dixon (2002), with permission of author

1,600 and 2,400 m amsl, in the northern part of the basin in Western Wellega and Illubabor Zones. In Illubabor, wetlands cover between 2% and 4% of the area. Many of these permanent wetlands seem to be the result of blocked or impeded drainage and have considerable depths of silt deposits. Water depth is up to 2 m at the height of the flood but generally less, and often there is no standing water in the smaller ones in the dry season (Wood and Dixon 2002).

To the south of Illubabor, in the higher altitude interfluves between the Baro and Akobo drainage basins there are fewer wetlands, but in the far south, in the Akobo basin of Bench Maji zone, there are small floodplain wetlands in the lower mid-altitude zone, around 1,300 m amsl, where stream gradients are low.

Local Use and Knowledge

These highland wetlands play a critical role in the lives of many communities providing a range of ecosystem services (ESS). Springs around wetland edges are sources of drinking water, while wetlands provide sedges which are used for thatching, craft work, and floor covering on holidays (Table 1). Wetlands are also used throughout the highlands for the grazing of cattle, and cultivation is increasingly common.

In Bench Maji zone, wetlands are used for the cultivation of taro *Colocasia esculenta* in small floodplains. While traditionally this does not involve water management, such management is developing now farmers have realized that yields can be increased when flooding is extended (Sutcliffe 2006b). In Illubabor and Western Wellega, dry season cultivation of maize using drainage of permanent wetlands is common and long standing, being known as "bonee." Records of such cultivation go back to the mid-nineteenth century (McCann 1995). It seems these areas have been cultivated in the past when population density was high in the area, while the recent development of urban markets, pressure from government food security campaigns, and a serious decline in upland harvests, due to land degradation in some places, have stimulated more intensive wetland use (Wood et al. 2001). In Western Wellega because of the upland degradation wetlands account for 70% of the grain produced in some communities, while more commonly between 10% and 20% of the domestic food production comes from these areas. However, the timing of these harvests, as the hungry season approaches, makes such food particularly valuable (Dixon and Wood 2003).

Cultivation of wetlands, when drainage is involved, is not easy, and there has been extensive, and in some areas complete, loss of wetlands in the southwest highlands of Ethiopia, these areas remaining now as rough grazing (Wood and Dixon 2008). Wetland degradation is often caused by overdrainage, damage to the "plug" which has blocked the valley, intensive cattle grazing, and degradation of the catchment which reduces water storage and leads to sediment deposition in wetlands (Dixon and Wood 2003).

Not all wetlands are degraded once cultivation starts, and some have been cultivated sustainably for several decades. In fact a body of knowledge about wetland drainage, hydrological management, and soil fertility maintenance has been built up in this area (Dixon 2003). Farmers have learnt that a rotation with fallowing can help maintain the cultivation and other benefits from these areas, with the return of sedge vegetation a good indicator of recovery (Dixon and Wood 2003). In addition, during times of cultivation drain blocking in the rainy season helps recreate the natural flooding and sediment deposition which helps maintain soil fertility, while maintaining wetland patches within wetlands can also help maintain the wetland ecosystem services of these areas.

Increasingly the link between catchments and wetlands has been recognized through the work of the Ethiopian Wetlands Research Programme (EWRP) and a local NGO – Ethio Wetlands and Natural Resources Association (EWNRA). A combination of biological and physical soil conservation work in the catchments has had a combined effect of increasing upland crop yields by more than 30%, so reducing the demand to cultivate in wetlands, and apparently increasing infiltration in the catchment and the year round supply of water to the wetlands. Such measures have been so successful that some wetlands which had suffered degradation are now permanently flooded and no longer used for cultivation or grazing (PHE 2012).

Community-Based Management, Institutions, and Sustainable Wetland Use

One key to sustainable wetland management is the existence of community-based institutions which coordinate wetland management (Dixon et al. 2013). Such organizations are essential given the way in which land management within a wetland is linked through hydrological processes and other ESS. In the highlands of Southwest Ethiopia, the post of Abba Laga, "father of the water," has existed among the Oromo people for managing wetland areas, while more recently the government has supported communities to develop their own committees for managing wetland use. Such bodies coordinate drainage and plowing and also play a role in organizing guarding of the crops from pests such as monkeys and wild pigs (Dixon et al. 2009).

Lowland Wetlands

Extent and Character

Wetlands cover 2,372 km^2 of the lowland portion of the Baro-Akobo basin within Ethiopia, this being 3.5% of the basin. These wetlands are both permanent and seasonal in nature, being fed by overspill from the Baro and Akobo rivers and their tributaries. Three types of swamp are found in these lowlands: (a) *Cyperus papyrus* swamp, (b) *Typha domingensis* swamp, and (c) *Vosia cuspidata* swamp. The seasonally flooded areas are dominated by (a) *Oryza longistaminata* dominant and (b) *Echinochloa pyramidalis* (Sutcliffe 2006b).

The major concentration of wetlands is in the western part of the lowlands. Their complex ecological conditions make for rich and varied patterns of habitats, and they support very distinctive flora and fauna (Woube 1999). As a result a large part of this area is covered by the Gambella National Park (5061 km^2). This park was proposed because of the numerous large wildlife species, particularly Nile lechwe *Kobus megaceros*, white-eared kob *K. kob*, and whale-headed stork *Balaeniceps rex*. The white-eared kob migrates every year between the Sudd in South Sudan and the Gambela marshes. Some 43 species of mammals are found in the park and an IBA team recorded 230 species of birds (EWNHS/Bird Life International 1996). The park is not legally gazetted and no management plan has been prepared. There have been major incursions into the park by the government's Akobo large-scale farm in the 1980s/1990s and by foreign investment for irrigation development after 2000.

Local Use

Two ethnic groups dominate in these lowlands, each using the wetlands in different ways. The Anuak live on the high levees along the main rivers, especially the Baro, and practice flood retreat cultivation of sorghum and maize in the adjoining wetlands. They also engage in fishing. The Nuer are agropastoralists who are based on limited higher areas in the west of the lowlands away from the tsetse belt. They graze their cattle extensively across the seasonal wetlands, migrating between the drier uplands in the rains and the seasonally flooded grassland in the dry season.

In recent years there have been reports of worsening floods in the lowlands along with lower dry season flows in the main rivers (Woube 1999; Sanne van Aarst, personal communication 2014). This has forced the relocation of some villages and increased competition for the higher settlement sites. This change in river regime is thought to be due in large part to changes in land use in the highland catchments as a result of deforestation, the expansion of cultivation, and the degradation of catchments and wetlands.

Coordination of Competing Land Use

It has been recognized that there is considerable potential for large-scale irrigated agriculture in the lowlands, mostly in the area between the Baro and the Akobo along the Alwero and Gila rivers, before they enter the wetland dominated west of the sub-basin. Estimates vary but more than 500,000 ha are reported to be suitable for crops such as cotton and rice. Dams and water control for irrigation are already altering flows in the Alwero and Gila rivers, and this will impact on the wetlands to the west.

There have been a number of conflicts between the local communities, the national park, and the investors, and this is now being addressed by a projects implemented by the Horn of Africa Regional Environment Centre and Network which seek to help communities maintain their livelihoods and improve the security of these while ensuring that the national park is redefined and demarcated in a way which will meet the needs of the animals, especially their migration patterns. Land and water management to maintain the wetlands is also a part of this project to ensure that the wetlands are recharged and so support the livelihoods of the agropastoral Nuer and the grazing and migration of the wildlife.

Threats and Future Challenges

The wetlands in both the highland and lowland parts of the Baro-Akobo basin in Ethiopia are important for the range of ecosystem services they provide, especially provisioning ones for communities, support services for biodiversity, and hydrological regulation which attenuates the flow of the rivers and moderates and maintains the flood regime in the lowlands and further downstream. In this way and through the migration of wildlife these wetlands are not just of local value but also of international importance. However, these wetlands are under increasing pressure, especially from agriculture across the basin and from water extraction and diversion in the lowlands. There is also a considerable potential for hydropower development along the escarpment which would further impact on the downstream hydrological system and wetlands but also need functioning catchments and wetlands for water storage in the highlands.

These challenges suggest two key perspectives are needed in measures to manage development in this basin and to maintain the various wetlands and maximize the range of benefits from ecosystem services.

(a) The first is an integrated basin-wide approach which recognizes the functional linkages across the basin landscape and applies a functional landscape approach through policies and incentives to support critical land use in catchments and in wetlands which will help to maintain ecosystem services in a sustainable way, thereby benefiting people, the national economy, and biodiversity (e.g., www.wetlandaction.org/FLA, Accessed: 9 April 2016).

(b) Secondly, with respect to wetlands specifically, there is a need for communities and planners to recognize the full range of ecosystem services which are provided by wetlands and develop the value of all of these as far as possible. In this way, through multiple use, wetlands can be managed in ways which will make them valued and competitive parts of the landscape which communities and planners will seek to maintain (Wood and Van Halsema 2008).

These perspectives should provide critical guidance for the proposed Baro-Akobo Basin Commission which it is expected will be established as part of the water resource development planning ongoing in Ethiopia today.

References

Dixon AB. Indigenous management of wetlands: experiences in Ethiopia. Aldershot: Ashgate; 2003.
Dixon AB, Wood AP. Wetland cultivation and hydrological management in East Africa: matching community and hydrological needs through sustainable wetland use. Nat Res Forum. 2003; 27(2):117–29.
Dixon AB, Hailu A, Semu T, Taffa L. Local responses to marginalisation: human-wildlife conflicts in Ethiopia's wetlands. Geography. 2009;94(1):38–47.
Dixon AB, Hailu A, Semu T. Local institutions, social capital and sustainable wetland management: experiences from western Ethiopia. In: Wood AP, Dixon AB, McCartney M, editors. Wetland management and sustainable livelihoods in Africa. London: Earthscan; 2013. p. 85–111.
EWNHS (Ethiopian Wildlife and Natural History Society)/Bird Life International. Important bird areas of Ethiopia: a first inventory. Addis Ababa: EWNHS; 1996.
McCann J. People of the plow: an agricultural history of Ethiopia 1800–1990. Wisconsin: University of Wisconsin Press; 1995.
PHE Ethiopia. Integrated practical success stories and challenges from the field: Ethio wetlands and natural resources association. Addis Ababa: PHE Ethiopia Consortium; 2012. Available from: http://phe-ethiopia.org/pdf/Ethio_wetlands_spotlight.pdf. Accessed 9 Apr 2016.
Sutcliffe JP. Transboundary analysis of the Baro-Sobat-White Nile sub-basin. Cooperative regional assessment for watershed management. Addis Ababa: ENTRO (Eastern Nile Technical Regional Office); 2006a.
Sutcliffe JP. Transboundary analysis. Final country report – Ethiopia. Addis Ababa: ENTRO (Eastern Nile Technical Regional Office); 2006b.
Wood AP, Dixon AB. Sustainable wetland management in Illubabor Zone. Research report summaries. Huddersfield: Wetlands & Natural Resources Research Group, Huddersfield University; 2002.
Wood AP, Dixon AB. Small swamp wetlands in southwest Ethiopia. In: Wood AP, Van Halsema GE, editors. Scoping agriculture-wetland interactions: towards a sustainable multiple-response strategy. Rome: FAO; 2008. p. 65–72. FAO Water Resources Report 33.
Wood AP, Van Halsema GE, editors. Scoping agriculture-wetland interactions: towards sustainable multiple response strategy. Rome: FAO; 2008. Water Resources Report 33.
Wood AP, Abbot PG, Dixon AB. Sustainable management of wetlands: local knowledge versus government policy. In: Gawler M, editor. Strategies for the wise use of wetlands: best practices for participatory management. Wageningen: Wetlands International; 2001. p. 81–8.
Woube M. Flooding and sustainable land and water development in the lower Baro-Akobo River Basin, Ethiopia. App Geogr. 1999;19(3):235–51.

Bahr el Ghazal: Nile River Basin (Sudan and South Sudan)

103

Asim I. El Moghraby

Contents

Introduction	1270
Hydrology	1270
Wetland Ecosystems	1270
Biodiversity	1271
Insects	1272
Plankton	1273
Benthos	1273
Fish	1273
Wildlife	1273
Conservation Status	1274
Ecosystem Services	1274
Threats and Future Challenges	1275
References	1276

Abstract

Bahr el Ghazal is one five sub-basins of the River Nile. It is the only one that originates in the Sudan and is one of largest tropical wetlands in the World. It is rich in biodiversity. The area, rich in ecosystem services, has been traditionally and historically shared amicably by the Dinka Noke of Abyei and the transhumant Myesseria cattle owning tribe, for their subsistence livelihoods. After the cessation of South Sudan, in 2011, it became a contested area between the two countries. An additional threat to the integrity of the sub-basin is oil development on both sides of the borders. Understanding how the wetlands function will enable both countries to make maximum use of its values and services and would be a step in the right direction in conflict resolution.

A. I. El Moghraby (✉)
Sudanese National Academy of Sciences, Khartoum, Sudan
e-mail: asim.mog@gmail.com

© Springer Science+Business Media B.V., part of Springer Nature 2018
C. M. Finlayson et al. (eds.), *The Wetland Book*,
https://doi.org/10.1007/978-94-007-4001-3_215

Keywords

Tropical Wetlands · Biodiversity · Land use · Conflict resolution

Introduction

The source of the Nile is in Rwanda's Nyungwe forest, thence flowing northwards 6,718 km from 4° S to 31° N (Dumont 2009), making it the longest river in the world. The area of the Nile Basin (3.2×10^6 km^2, NBIS 2012) is approximately one tenth of the area of Africa. It is shared by 11 sovereign countries and is divided into eight distinct subbasins, namely, Lake Victoria, the Equatorial Lakes, the Sudd, Bahr el Ghazal, Sobat, Central Sudan, Blue Nile, and Atbara Basins (Conway and Hulme 1996). The volume of the annual discharge of the Nile has marked variations both seasonally and annually.

The Nile Basin along its middle reaches harbors three extensive wetlands in the Sudan and South Sudan. The Sudd is the largest of these, as well as being one of the largest floodplains in Africa and one of the largest tropical wetlands in the world (Rebelo and Moghraby 2018). The other two wetlands along the Nile's middle reaches are the Machar Marshes (Mohamed 2018) and Bahr el Ghazal – "the sea of antelopes."

Hydrology

Bahr el Ghazal, the largest of the Nile's eight subbasins, originates in the Nile/Congo divide and flows westward over 700 km to join with Bahr al Jabal at Lake No to become the White Nile (Fig. 1). The principal tributaries to the Bahr el Ghazal include the Lol (formed by the Boro, Sopo, Kuru, and Pongo), the Bahr al-Arab, and Jur River. The area of the watershed is around 555,000 km^2 (NBIS 2012) with a conservative estimate 9,000 km^2 and 5,000 km^2 of perennial wetlands and floodplain, respectively (Hughes and Hughes 1992). The system is composed of a myriad of streams, lakes, and wetlands with a total water budget of 12 billion meters cube (10^9 m^3) of water. Only 0.6×10^9 m^3 reach Lake No, at the confluence with Bahr el Jebel, when the White Nile starts its long journey toward the Mediterranean.

Wetland Ecosystems

Due to the flat topography and the nature of water flow, erosion, and deposition processes, a large number of back swamps (known locally as "Mayaas") were formed along the flood plains of the intricate system of lakes and water courses. Mayaas are one of the most prominent features of Bahr el Ghazal wetland. Their areas vary considerably according to the bends of the water channels (from 0.16 km^2 to 4.5 km^2). Generally Mayaas are flat with slight or no clear banks. They get filled in

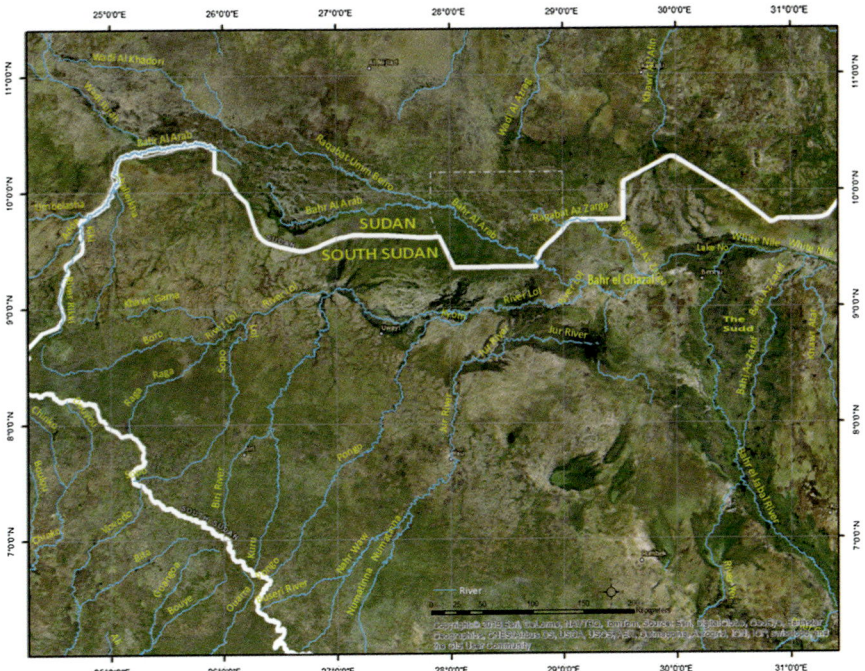

Fig. 1 Major tributaries of the Bahr el Ghazal

the rainy season by rain and/or flood waters. Flow from seasonal streams is variable in quantity, duration, and turbidity. Some Mayaas dry up soon after the end of the rainy season. Mayaas are critical to the survival of the ecosystem, their diversity, and carrying capacity.

In lean years the carrying capacity of the area is diminished and inter- and intra-tribal conflicts over resources do occur. Cyclic droughts and the repeated use of fire as a management tool have reduced the carrying capacity to a small fraction of what it was formerly. Many of the inlet streams have been blocked and Mayaas, receiving the silt-laden runoff of the parched catchment, have silted up to a varying extent.

Biodiversity

Three zones are distinguished in the wetlands: permanent swamps, intermediate swamps, and highlands. The main vegetation features of the permanent swamps are a fringing wall of reeds, a tangle of climbers, submerged and floating hydrophytes and climbers, and ferns. The vegetation is distinguished from those of "the Sudd" (Rebelo and Moghraby 2018) in that it is stunted and brownish in color.

The main species that occur in the permanent swamps include:

- On the fringe: *Cyperus papyrus*, *Vossia cuspidata*, *Phragmites karka*, and *Typha domingensis*
- Climbers: *Ipomoea aquatic*, *I. carica*, *Luffa cylindrica*, *Cissus ibuensis*, and *Vigna nilotica*
- Submerged and floating hydrophytes: *Ceratophyllum demersum*, *Najas minor*, *N. pectinata*, *Ottelia alismoides*, *O.brachyphylla*, *Polygonum* spp., *Utricularia* spp., *Vallisneria* sp., *Trapa* sp., *Potamogeton* spp., *Nymphaea lotus*, *N. micrantha*, and *Eichhornia crassipes* (water-hyacinth – an invasive species)
- Ferns: *Dryopteris gongylodes* and *Azolla nilotica*

The intermediate swamps (the *Ragaba* catena equivalent to the *Toich* "intermediate wetlands" of the Sudd) include three alternating types of habitats, connected by surface water movements. These are the *Gardud*, *Talha*, and *Fau* (Harrison and Jackson 1958).

- *Gardud* occupies the largest areas on non-cracking clay flats with immense runoff.
- *Talha* occurs on dark-cracking clay (Vertisol soils) and is typical of *Acacia seyal/ Balanitis* savannah.
- *Fau* is the smallest of the three areas, occurs on dark-cracking clay, and is subject to shallow flooding. It consists of open grassland of "tussocky perennial grasses." These are mainly *Setaria incrassate*, *Hyparrhenia rufa*, and *Vetiveria nigritana*.

Harrison and Jackson (1958) describe the Highlands as *Acacia seyal/Balanites* tall grass Savannah woodland. The dominant woody species include *Acacia seyal*, *Balanites aegyptiaca*, *Combretum glutinosum*, *Acacia nubica*, *Terminalia laxiflora*, *Boswellia papyrifera*, *Dichrostachys glomerata*, *Sclerocarya birrea*, *Boscia senegalensis*, *Leptadenia pyrotechnica*, *Piliostigma reticulate*, *Albizia amara*, and *Faidherbia albida*,

The following section describing the fauna of the Bahr el Ghazal is based on the following published sources: Girgis (1948), Talling (1957), Gronblad et al. (1958), Bishai (1962), Rzóska (1976), Monakov (1969), Moghraby (1974, 1982), Green (1977), Denny (1984), and ERM (2004).

Insects

There are at least 120 species of insects of medical, veterinary, and/or agricultural relevance. Most of these are aquatic or semiaquatic during at least one stage of their life cycle. The fauna includes some 50 species of mosquitoes, and nine tabanids. Orthoptera, Hemiptera, Lepidoptera, and Coleoptera are richly represented.

Plankton

Slowly moving and stagnant water bodies, and inside the fringing vegetation, contain more phytoplankton than the free flowing streams. The diatom *Melosira granulata* and its variety *angustissima* is the dominant species in the area. The blue-green algae *Lyngbya limnetica* and *Anabaena flos-aquae* are also prevalent.

The zooplankton is represented by a rich fauna especially under the fringe vegetation and in the adjacent water bodies. These include 17 species of Rhizopoda, 39 of Rotifera, 1 of Chonostraca, 7 of Codopoda, 27 of Cladocera, 2 of Decapoda, and 1 of Chaoborus.

It should be stressed that the subbasin is regarded by many as a hot spot of biodiversity. A good example is Lake Ambadi where Gronblad et al. (1958) identified 205 species of Desmids – 21 were new to science and 48 were new forms. Phytoplankton in Lake No is of the order of 2,300 cells per ml. The diverse invertebrate communities support the large fish populations in the wetland. Freshwater crabs of the genus *Potamonautes* occur as well as rare freshwater shrimps (e.g., *Caridina nilotica*, *Palaemon nilotica*, and *Cyclestheria hislopi*).

Benthos

The benthos has low species diversity and abundance especially in flowing streams and is largely composed of insect larvae, oligochaetes, leaches, and molluscs.

Fish

Fish is an extremely important source of protein and is a key element in the subsistence economy of the area. The swamp system and adjacent wetlands provide favorable conditions for the breeding and growth of fish. The rich macroflora is an excellent substratum for oviposition, as well as sanctuary and feeding grounds for juvenile fish.

Forty species of fish have been recorded in the area, nine of which make up 80% of the total commercial catch. These include *Distichodus spp*, *Heterotis niloticus*, *Clarias spp*, *Mormyrus caschive*, and *Lates niloticus*.

In general, the fish species found in the area have a high tolerance of low oxygen tension and an ability to use atmospheric oxygen (e.g., *Clarias*, *Polypterus*, and *Hetrotis*). In order of abundance detritus feeders are most prominent, followed by phytophagous fishes and predators. Typical planktonivores are not represented.

Wildlife

Historically the wetlands were teeming with wildlife. This is how the name Bahr el Ghazal originated. The seasonal hydrological regime has an obvious influence on the wildlife fauna. For example, sitatunga *Tragelaphus spekii* and the Nile lechwe *Kobus megaceros* can be found in the swamps even in the dry season. Elephant *Loxodonta africana*, buffalo *Syncerus caffer*, tiang *Damaliscus korrigum*, waterbuck *Kobus ellipsiprymnus defassa*, and oribi *Ourebia ourebia* extend their range to rivers' edge during the dry season. The herbivores are followed by predators including lion *Panthera leo*, caracal *Caracal caracal*, spotted and striped hyena (*Crocuta crocuta* and *Hyaena hyaena*), and jackals *Canis* spp.

The hippopotamus *Hippopotamus amphibius* are quite frequent and play a role in changing flow patterns by making deep tracks through the fringing vegetation. The spotted-necked otter *Hydrictis maculicollis* is not uncommon.

The Bahr el Ghazal wetlands harbor large populations of crocodiles *Crocodylus niloticus*. This is mainly due to the vastnesses of the area and its relative inaccessibility to hunters. Other reptiles include the Nile monitor lizard *Varanus niloticus*, soft-shelled turtle *Trionyx triunguis*, side-necked turtle *Pelomedusa subrufa*, green mamba *Dendroaspis angusticeps*, African python *Python sebae*, small-scaled asp *Atractaspis microlepidota*, puff adder *Bitis arietans*, four species of scaled vipers (*Echis* spp.), and four species of cobra (*Naja* spp.).

Birdlife is generally rich along the whole of the Nile, which constitutes one of the main routes of migration to and from Africa. The floodplains of Bahr el Ghazal provide an important breeding habitat and support the largest resident population of Shoebill stork *Balaeniceps rex* in Africa. It also acts as stopover and wintering ground for many migrant water birds including great white pelican *Pelecanus onocrotalus*, black-crowned crane *Balearica pavonina*, white stork *Ciconia ciconia*, darter *Anhinga rufa*, long-tailed cormorant *Microcarbo africanus*, herons (*Ardea* spp.), egrets (*Egretta* spp.), white-winged black tern *Chlidonias leucopterus*, as well as ibises ducks, geese, and raptors.

Conservation Status

There are neither national parks nor areas with conservation status in the Bahr el Ghazal.

Ecosystem Services

The land use pattern is seasonal. The Messerya cattle herdsmen (the pastoralist Baggara Arabs) consider the whole wetland areas as their historic summer grazing range shared willingly with the Southern Sudanese Dinka Noke. With the onset of rains, biting flies swarm and the cattle of both the Dinka and Messerya are driven out of the wetlands. The Messerya spend the rainy season on the sandy "Goz" lands, near their larger settlements. In the dry season they are sustained by the rich wetland pasture and water. The practice and precise cattle migration routes have been observed and respected by culture and tradition for generations and generations.

Threats and Future Challenges

In an administrative memo to the Ministry of Foreign Affairs, the Egyptian Government agreed in 2010 to fund a project to deepen the channels of Bahr el Arab and its tributaries (Ambassador Sharfi, Person Com). The dredging is intended to increase the volume of water flow to Egypt by draining the wetlands (see the Jonglei Canal project – Rebelo and Moghraby 2018). Although the project has not been implemented due in part to ongoing instability in the region, the degree of impacts of dredging the channels and draining the wetlands on the ecosystem services provided to the local communities is unknown. Environmental and social impact assessments should precede any such action.

The carrying capacity of the grazing range is steadily deteriorating at an accelerating rate since the droughts of the 1980s. The use of fire as a management tool has only added to the deterioration. Distances to rich pastures are progressively increasing season after season and the situation was exasperated with the outbreak of civil war in 1983 as a conflict over natural resources. The cessation of the South Sudan, in 2011, did not change the social and physical local conditions. However, it was used by both sides as a point of contention.

Most of the area lies in the heart of the newly discovered oil fields, "The Muglad Basin." With oil development came people and infrastructure. All-weather compacted raised roads constructed to connect all facets of the oil exploration, extraction, processing, and service industries have had major impacts to the ecosystem. Mirghani (2005) estimated that 579 million trees were cut from an area of more than half a million hectares in less than 5 years from the start of oil development activities. In addition to the loss of biomass and sequestered carbon, roads changed topography and impeded drainage and free flow of water in the watershed including the flow of major khores like *Wadi Shalango*, *Al Ragaba al Zarga*, and *Wadi Al Ghalla* as well as the runoff from the Nuba Mountains. Mitigation measures were not provided and culverts are installed only when and where roads are washed away by sheet flow. Traditional summer grazing areas of the Baggara were distorted while new fishing grounds were created. Although welcomed by the Nilotic tribesmen who catch fish which migrate to the edge of the wetland during the flood season to spawn, there is the concern that increased accessibility will deplete the fisheries. Crude oil sprayed on the roads as a binding agent eventually contaminates the vicinity and water table. Roads also access to loggers, charcoal makers, and poachers into formerly pristine and inaccessible habitat.

One of the major problems facing the oil industry is the disposal of "produced water" accumulating from oil operations. This "formation water" is pumped from the reservoir along with the oil. Formation produced water is the largest source of waste in the oil industry when separated from the crude oil coming out of the wells. It is not uncommon to find the ratio of oil to produced water in the order of 1:7 or greater.

Produced water contains a vast array of substances that include salts, minerals, dissolved and insoluble hydrocarbons, heavy metals (e.g., arsenic, cadmium, mercury, and lead), aromatics, phenols, cyanide, and other chemicals. Some of these occur naturally and come out from the reservoir. Others are "treatment chemicals" which are added to the crude oil for various reasons during various stages of oil production.

Oil exploration and production operations generate huge amounts of waste. Some of these wastes are considered hazardous to humans, biodiversity, and ecosystems structurally and functionally. Impacts of wastes can be readily observed in both humans and animals as frequent accidents and acute health problems. While ecosystems in some instances have the ability to remediate some of these effects in the short term, many case studies have identified environmental catastrophes 20 or 40 years after the onset of oil exploration (ECOS 2008).

Current waste management practices are far below satisfactory levels although this varies greatly between companies and from one oil development block to the next. Effective and responsible waste handling and disposal are key elements of an efficient environmental management system in order to minimize their potential harm to human health or the environment. Moreover, efficient management of wastes can reduce operating costs and potential liabilities (PIES 2005).

References

Bishai HM. The water characteristics of the Nile in the Sudan, with a note on the effect of *Echhornia crassipes* on the hydrobiology of the Nile. Hydrobiologia. 1962;20:31–9.

Conway D, Hulme M. The impacts of climate variability and future climate change in the Nile basin on water resources in Egypt. Int J Water Resour Dev. 1996;12(3):277–96.

Denny P. Permanent swamp vegetation of the upper Nile. Hydrobiologia. 1984;110:79–90.

Dumont HJ, editor. The Nile. Origin, environments, limnology and human use. Dordrecht: Springer; 2009.

ECOS. Sudan oil industry. Facts and analysis. Holland: European Coalition for Oil in the Sudan; 2008.

Environmental Resources Management (ERM). An environmental and social impact assessment study submitted to the White Nile Petroleum Operating Company. 2004 (Unpublished report).

Girgis S. A list of common fish from the upper Nile with their Shilluk, Dinka and Nuer names. Sudan Notes Rec. 1948;29:2–7.

Green J. Water birds in the Sudan in winter. Sudan Notes Rec. 1977;58:199–204.

Gronblad R, Prowse GA, Scott AM. Sudanese desmids. Acta Bot Fenn. 1958;58:1–82.

Hammerton D. The Nile River: A case history, 171–214. In: Oglesby RT, Carison CA, McCann, editors. River ecology and man. NY and London: Academic Press; 1972.

Harrison MN, Jackson JK. Ecological classification of the vegetation of the Sudan. Bull Min Agric/Sudan. 1958;2.

Hughes RH, Hughes JS. A directory of African wetlands. Gland: International Union for the Conservation of Nature; 1992. p. 820. Joint publication with United Nations Environment Programme, Nairobi (Kenya) and World Conservation Monitoring Centre, Cambridge (UK).

Hurst HE. The Nile. Constable, 2nd Edn. London; 1957.

Ibrahim AM. The Nile: Description, hydrology, control and utilization. In: Dumont HJ, Moghraby AI el, Desugi L, editors. *Limnolgy and marine biology of the Sudan*. The Hague: A Dr Junk Publishers; 1984.

Mirghani A. The balance between development of oil resources and forest cover. Khartoum: Tayba Press Seminar; 2005 (In Arabic).

Moghraby AI. The Jonglei canal-needed development or potential ecodisaster? Environ Conserv. 1982;9(2):141–8.

Moghraby AI el, Mohamed YA, Hamid A, Mustafa K. On the wetlands of the Sudan. UNEP Publications. in press.

Moghraby AI el. A note on wildlife in southern Sudan. Khartoum: Hydrobiological Research Unit/ University of Khartoum; 1974. 21st Ann. Report.

Mohamed YA. The Machar marshes, Nile Basin. In: Finlayson CM, Milton GR, Prentice RC, Davidson NC, editors. The wetland book volume 2: distribution, description and conservation. Dordrecht: Springer; 2018.

Monakov AV. The zooplankton and zoobenthos of the White Nile and adjoining waters of the Republic of the Sudan. Hydrobiologia. 1969;33:162–85.

Nile Basin Initiative Secretariat (NBIS). State of the river Nile basin. Entebbe: Nile Basin Initiative Secretariat; 2012.

PIES. Waste Management Study, *Blocks 5A and 8*. An unpublished report to the White Nile Petroleum Operating Company. 2005.

Rebelo LM, Moghraby AI el. The Sudd. In: Finlayson CM, Milton GR, Prentice RC, Davidson NC, editors. The wetland book volume 2: distribution, description and conservation. Dordrecht: Springer; 2018.

Rzóska J, editor. The Nile: biology of an ancient river. The Hague: Dr. W. Junk b.v., Publishers; 1976.

Sutcliffe JV. Parks. The Hydrology of the Nile. IAHS Special Publication no.5; 1999.

Talling JF. The longitudinal succession of water characteristics in the White Nile. Hydrobiologia. 1957;11:73–89.

Machar Marshes: Nile Basin (South Sudan)

104

Yasir A. Mohamed

Contents

Introduction 1280
The Natural Resources of the Machar Marshes 1281
    Biodiversity 1281
References 1286

Abstract

The Machar marshes are the major wetland in the Sobat basin, in South Sudan. The Sobat is a tributary of the White Nile originating from the Ethiopian highlands. The wetland is formed by the spill from the Baro River and the inflows from the eastern torrents (Yabus and Daga). The Machar marshes is an important ecosystem that supports rich flora and fauna and provides grazing land for the Nilotic tribes of South Sudan. Unlike other wetlands in Southern Sudan (e.g., Sudd), the Machar marshes is the least monitored and understood. The literature shows different estimates of the flooded area, varying between 3,350 and 20,000 km^2. This chapter provides a description of the Machar marshes, its biodiversity and the water resources system, as well the key challenges for conservation and/or development of the marshes.

Keywords

Machar marshes · Nile · Wetland · Biodiversity · South Sudan

Y. A. Mohamed (✉)
Hydraulic Research Center, MoWRIE, Wad Medani, Sudan

Department of Integrated Water Systems and Governance, UNESCO-IHE, DA, Delft, The Netherlands

Faculty of Civil Engineering and Applied Geosciences, Water Resources section, Delft University of Technology, GA, Delft, The Netherlands
e-mail: y.mohamed@hrc-sudan.sd

© Springer Science+Business Media B.V., part of Springer Nature 2018
C. M. Finlayson et al. (eds.), *The Wetland Book*,
https://doi.org/10.1007/978-94-007-4001-3_220

Introduction

The Machar wetland known as the "Machar marshes" is the largest wetland in the Baro-Akobo-Sobat subbasin (BAS), one of the major subbasins of the Nile (Fig. 1). The wetland is located in South Sudan, to the east of the famous Sudd wetland. The literature shows different wetland size varying between 3,500 and 6,500 km². Large spills of water from the Baro River, which originates from the Ethiopian highlands, find their way to the Machar marshes during the flood season from July to September. In addition, significant inflows from torrential tributaries, namely the Yabus and Dagga, end up in the marshes. Only a small percentage of the water drains

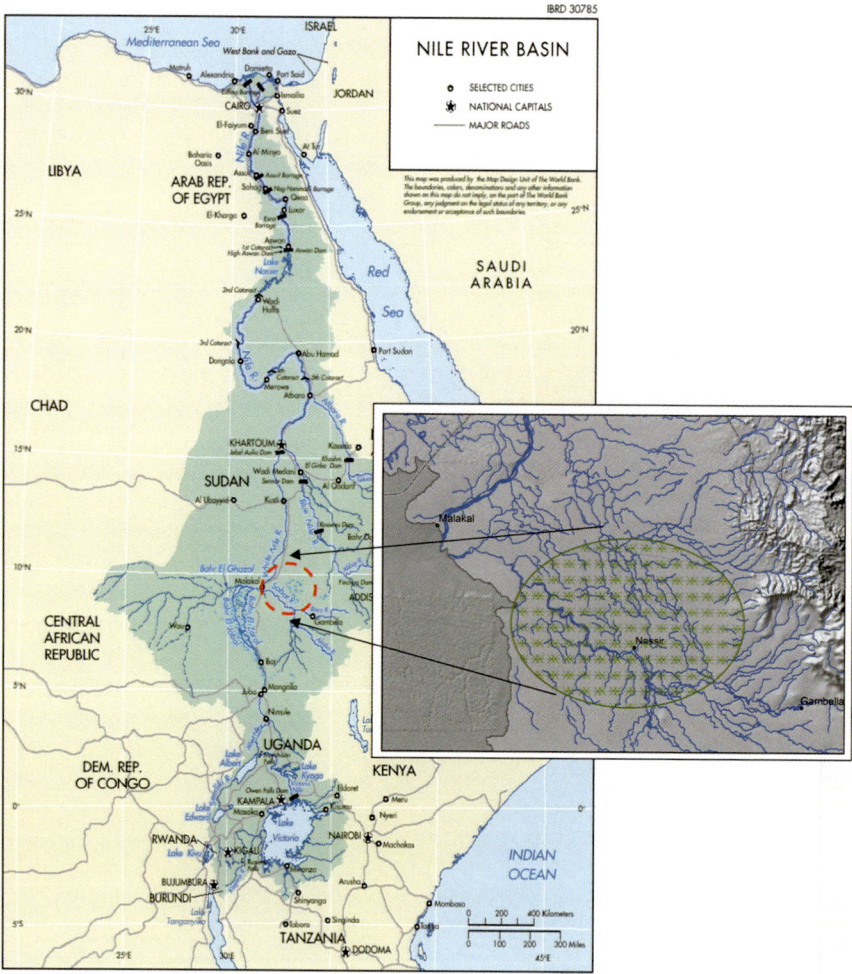

Fig. 1 Location map, showing the position of the Machar marshes (*inset*) in the Nile River Basin. (Nile River Basin Map: © Map Design Unit of the World Bank, with permission; *inset map* by the author)

downstream to the White Nile through Khor Adar. The topography of the Machar marshes is characterized by very flat slopes, of less than 1%.

The climate and vegetation cover of the Machar marshes are dominated by the seasonal flooding of the swampy plains. The temperature ranges from 20 °C to 35 °C, with extremes of 11–43 °C. The annual rainfall over the marshes is about 800 mm/year, while the actual evapotranspiration over the wider catchment around the Machar marshes is 1,300 mm/year (Mohamed et al. 2004). However, the potential evapotranspiration is much higher than actual.

The hydrology of the BAS basin is very complex and the least understood among the other subbasins of the Nile. Numerous rivers join and bifurcate along the flat slopes of the basin, in particular over the plains of South Sudan. There are very limited ground measurements of river levels and discharges. The literature shows few studies on the Machar marshes, mainly investigating the water balance of the marshes, and potential water savings (JIT 1954; El-Hemry and Eagelson 1980; Sutcliffe 1993; WaterWatch 2006). Those studies show no consensus regarding the derived estimates of wetland area, water volume, or evaporation losses.

The area of the Machar marshes is a home to hundreds of thousands of people from the Nilotic tribes – the Dinka, Nuer and Shuluk, as well as other tribes – the Morlei and Anjwak. These are nomadic tribes that migrate with their cattle to and from the "toich," the local word for grazing land. It is also a home of the unique Mangala Ghazals (ancient poetic form), named after the town of Mangala in South of Sudan. The Machar marshes have a very rich biodiversity, and the region of the marshes, as well as the larger BAS, is characterized by very limited or no infrastructural development. Accessibility to the marshes during the rainy season is only possible by air. Tall grasses and shallow water depth allow no navigation during the rainy season.

The Natural Resources of the Machar Marshes

Biodiversity

The BAS basin includes numerous wetlands, of which the Machar marshes are the largest. The Machar marshes provide habitats supporting a very rich flora and fauna. The seasonal climatic variation from the wet rainy season (June to October), to a long dry season (November to May), plus seasonal river spills into the marshes provides the marshes with their unique alternating wet/dry habitat characteristics. Three zones of land cover exist in the Machar area: permanent wetlands only in the deepest parts of the water bodies, seasonal flood plains inundated due to river spills and rainfall, and the dry areas at the fringes. While Acacia trees and scattered shrubs exist in the fringes of the marshes, papyrus and tall grass dominate the permanent water bodies. The main land use types in the Machar area are shown in Fig. 2, including grassland, savannah, open shrub land, and limited permanent wetlands.

The Machar marshes provide rich habitats supporting a very wide range of wildlife species, in particular on the seasonally flooded grasslands, including tiang

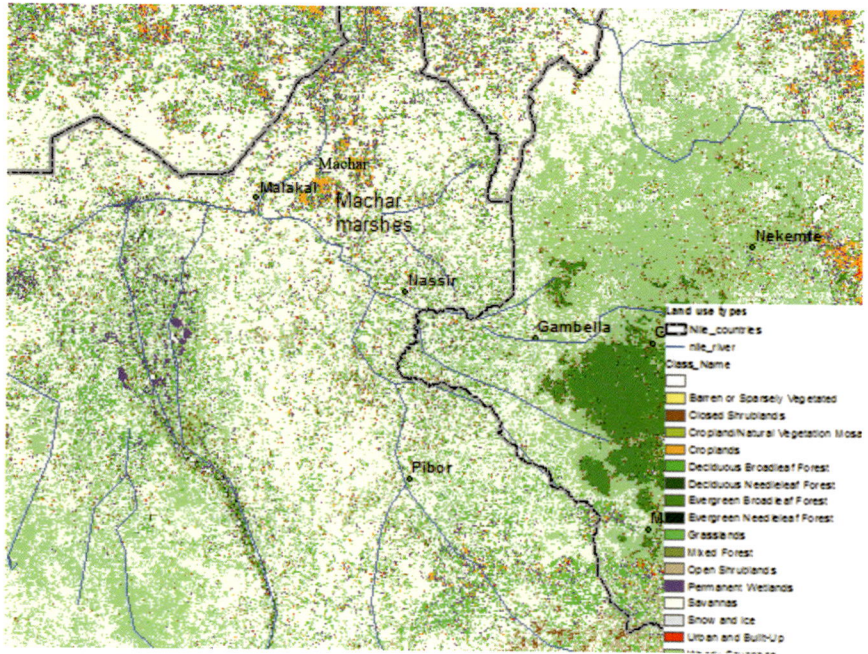

Fig. 2 Land and water use types in the Baro-Akobo-Sobat basin around Machar marshes (map by the author)

Damaliscus lunatus; white-eared *kob Kobus kob ssp. leucotis*, hartebeest *Alcelaphus buselaphus*, and oribi *Ourebia ourebi* are found in large numbers.

The ecosystem of the marshes provides essential services to the rural communities living in the area. The seasonal flood plains of the marshes are the main grazing lands for the local Nilotic Nuer, Morle, and Dinka tribes. The forest is the main source of building materials and energy for the rural communities. The livelihoods of tens of thousands of the people living alongside the marshes depend on fish resources as an important source of protein. However, available information about the natural resources, including water, land, biodiversity, and ecosystem services, remains very scarce.

Water Resources

The drainage network of the Machar marshes and BAS in general is very complex, see Fig. 3. Although clear river channels drain the high mountains of the Ethiopian Plateau flowing from east to west, the river courses become indistinct and spill at a number of places in the plains of South Sudan.

The Machar marshes are formed because of the excessive spills from the Baro River, inflows from Eastern Torrents (Yabus, Daga), and rainfall over the wetland itself. Sutcliffe and Parks (1999) indicated the occurrence of spills from the Baro River if the flow at the upstream gauging station of Gambela exceeds 1.5 km^3/month. This threshold of 1.5 km^3/month was estimated by comparing the measured

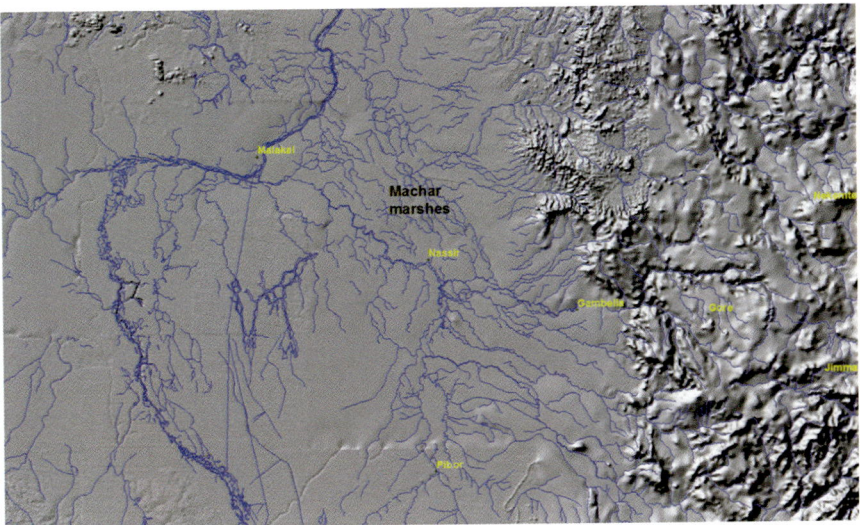

Fig. 3 The drainage system of the Baro-Akobo-Sobat basin and the Machar marshes (map by the author)

river flows at the upstream and downstream stations of Gambela and the Baro River mouth, respectively. Depending on the Baro River flows at Gambela, the spill to Machar marshes varies considerably between a minimum of 1 km^3/year up to 7 km^3/year for years with high floods. Average inflow to the Machar marshes from Yabus, Daga and other streams was estimated as 1.7 km^3/year (JIT 1954).

The computation of the water balance of the Machar marshes may give a fair idea about the water resources potential of the wetland. The main components of the water balance are: (i) spills from the Baro river, (ii) inflow from the eastern torrents, Yabus and Daga, (iii) rainfall over the marshes (Fig. 4a), (iv) evapotranspiration (ET) (Fig. 4b), and (v) outflows through Khor Adar and other streams to the White Nile, or to the lower Sobat. However, little is known about the areal extent of the Machar marshes. Consequently, quantifying the water balance has been difficult. It is not uncommon to see different definitions or delineations of the Machar wetlands and hence different results of water balance computations (see Table 1).

The large variation in the estimate of the water gain from the Machar marshes, from 1 to 8 km^3/year, indicates the high uncertainty of the available results. However, most literature agrees on 4 km^3/year as an average potential water saving from the Machar marshes.

Various studies estimated an annually flooded area (swamp and grassland) between 3,350 and 20,000 km^2. Using Landsat imagery, El-Hemry and Eagleson (1980) estimated the area of the swamps as 8,700 km^2, while FAO Africover gives only 2,913 km^2 of permanently and seasonal swamps. Hurst (1950) estimated the area of swamp as 6,500 km^2. Sutcliffe and Parks (1999) estimated the area of inundated land using a water balance model to vary between 1,500 and 6,000 km^2.

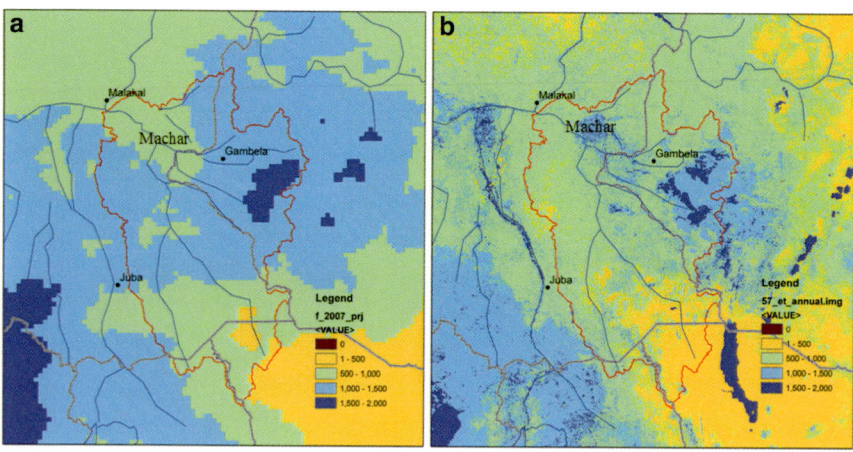

Fig. 4 Annual rainfall and actual evapotranspiration (ET_a) over the Machar area in mm/year for 2007 (graphics by the author)

Recent analysis of satellite images to assess wetland area in South Sudan showed that the Machar marshes are mostly seasonal and not permanent swamps (Mohamed et al. 2004). In fact the (actual) evaporation from the Machar marshes significantly drops during the dry season, resulting in smaller annual ET_a compared to the neighboring Sudd wetlands. The study by Water Watch (2006) indicated a wetland area of up to 12,346.7 km^2. The disagreements regarding the area of the Machar marshland reflect the current limited level of understanding.

Future Challenges

The Machar marshes are a very important seasonal wetland in South Sudan, fed by rivers originating from the Ethiopian highlands. The Machar area has not seen development in recent history, because of its harsh environment and also instability in the area. The people living in the Machar area are among the poorest in the Nile Basin. Competition for grazing land and water sources is not uncommon in the area. Regional plans for water resources development have been started by the Eastern Nile Technical Regional Office (ENTRO). This includes storage dams upstream on the Ethiopian side of the Baro River. The strong dependency of communities on the area's natural resources requires critical assessment of the impacts of future interventions in the BAS basin.

The natural resources system of the Machar marshes, and the hydrology in particular, is not clearly understood. The literature has shown different results of water resources potential in the Machar area. The major part of the river system is not gauged, and even the installed gauges at key locations have not been operational most of the time. Therefore, it is strongly recommended to create a knowledge base of the natural resources of the Machar area, to inform wetland conservation and future development planning. Because of the very difficult

Table 1 Water balance results of the Machar marshes according to different researchers

Investigator	Methodology	Spills from Baro km³/yr	Inflow from East km³/yr	Outflow to W. Nile km³/yr	Water saving potential km³/yr	Machar marshes area km²	Remarks
JIT 1954	Measurements, water balance, aerial surveys, site visits	2.8	1.7	0.1	~4	?	No evidence for area
El-Hemry and Eagelson 1980	Water balance, landsat images	3.5	5.6	0.1	~8	3500	Over-estimated
Sutcliffe 1993	Measurements, water balance	2.3	1.7	0.12	~4	3350	
WaterWatch 2006	ET from Satellite images,				Gain from Machar = 1	?	Area could not be delineated

accessibility to the Machar marshes, remote sensing techniques can be a useful source of information.

References

Africover http://www.glcn.org/activities/africover_en.jsp

El-Hemry II, Eagleson PS. Water balance estimates of the Machar Marshes. Cambridge, MA: Department of Civil Engineering, Massachusetts Institute of Technology; 1980. Report no. 260.

Jonglei Investigation Team. The Equatorial Nile project and its effects in the Anglo-Egyptian Sudan: Being the report of the Jonglei Investigation Team. Khartoum; 1954.

Mohamed YA, Bastiaanssen WGM, Savenije HHG. Spatial variability of evaporation and moisture storage in the swamps of the upper Nile studied by remote sensing techniques. J Hydrol. 2004;289:145–64.

Sutcliffe JV. Hydrological regimes and vegetation of Nile basin wetlands. In: Regional Conf. on Environmentally Sound Management of the Upper Nile Watershed (Khartoum); 1993.

Sutcliffe JV, Parks YP. The hydrology of the Nile, IAHS Special Publication No. 5. Wallingford/Oxfordshire: IAHS Press/Institute of Hydrology; 1999. http://iahs.info/bluebooks/SP005.htm

WaterWatch. Wetland evaporation in the Baro-Akobo Basin. Technical report prepared for ENSAP; 2006.

The Mayas Wetlands of the Dinder and Rahad: Tributaries of the Blue Nile Basin (Sudan)

105

Khalid Hassaballah, Yasir A. Mohamed, and Stefan Uhlenbrook

Contents

Introduction	1288
Hydrology of the Mayas	1290
Rainfall	1292
Evaporation	1293
Temperature	1293
Humidity	1293
Biodiversity	1293
Ecosystem Services	1294
Conservation Status	1294
Threats and Future Challenges	1296
References	1298

K. Hassaballah (✉)
UNESCO-IHE Institute for Water Education, Delft, The Netherlands

Hydraulics Research Center, Wad Medani, Sudan
e-mail: k.hassaballah@unesco-ihe.org; hrs_khalid@yahoo.com

Y. A. Mohamed
Hydraulic Research Center, MoWRIE, Wad Medani, Sudan

Department of Integrated Water Systems and Governance, UNESCO-IHE, DA, Delft, The Netherlands

Faculty of Civil Engineering and Applied Geosciences, Water Resources Section, Delft University of Technology, GA, Delft, The Netherlands

S. Uhlenbrook
UNESCO-IHE Institute for Water Education, Delft, The Netherlands

Delft University of Technology, Delft, The Netherlands

© Springer Science+Business Media B.V., part of Springer Nature 2018
C. M. Finlayson et al. (eds.), *The Wetland Book*,
https://doi.org/10.1007/978-94-007-4001-3_223

Abstract

Mayas wetlands forms an important ecological zone in the arid and semiarid Sodano – Saharan region in Dinder and Rahad basins. They are the most unique feature of the Dinder National Park (DNP) and one of its three major ecosystems. "Mayas" is a local name for floodplain wetlands that are found on both sides along the Dinder and Rahad Rivers. According to DNP authority, there are more than 40 mayas that are part of the rivers Dinder and Rahad ecosystems inside the DNP. They are the main source of food and water for wildlife (herbivores) especially during the dry season which extends from November to June. The mayas support large communities of wildlife and provide a refuge for a large number of migratory birds. Recently, the Dinder River has experienced significant changes of floodplain hydrology (i.e. dryness of some of the major mayas), and the reasons are not fully identified. This has significant negative implications on the mayas ecosystem functions and services. Thus, understanding the climate variability and its hydrological impacts is important for water resources management and sustainable ecosystem conservations.

Keywords

Dinder and Rahad basins · Dinder National Park · Mayas ecosystem

Introduction

"*Mayas*," a local name for riverine wetland, are found along the Dinder and Rahad tributaries of the Blue Nile basin in Sudan. They are the most important features in the Dinder National Park (DNP), as they are the main source of food and water for wildlife during the dry season from November to June (Abdel Hameed et al. 1997). The DNP (10,291 km^2) was proclaimed as a national park in 1935 following the London Convention for the Conservation of African Flora and Fauna (Dasmann 1972). It is located in the Dinder and Rahad river basin in the southeastern part of Sudan next to the Sudan-Ethiopian border, between longitude 34°30′ and 36°00′ east and latitude 11°00′ and 13°00′ North (Fig. 1).

Mayas are formed by the combination of a river with its seasonal inflow pattern flowing into a relatively flat area where the channel flow is restricted by gentle gradient, and the river channel is braided and meandering. A braided river is one of the river types that consist of a network of small channels separated by small and often temporary islands called braid bars. Braided streams occur in rivers with high slope and/or large sediment load (Schumm and Kahn 1972). A meander is a bend in a sinuous watercourse or river, forming when the water flow in a stream erodes sediments from the outer banks (concave configuration) and deposits them on the inner banks (convex). The result is a snaking pattern as the stream meanders back and forth across its down-valley axis (Julien 2002). An oxbow lake forms when a meander bend becomes cut off from the main stream.

Mayas have formed due to the meandering nature of the river on both sides of the present channel of the Dinder River. Their areas vary significantly from 0.16 to 4.5 km^2. Generally, mayas are flat with slight and/or no clear banks (Fig. 2). The

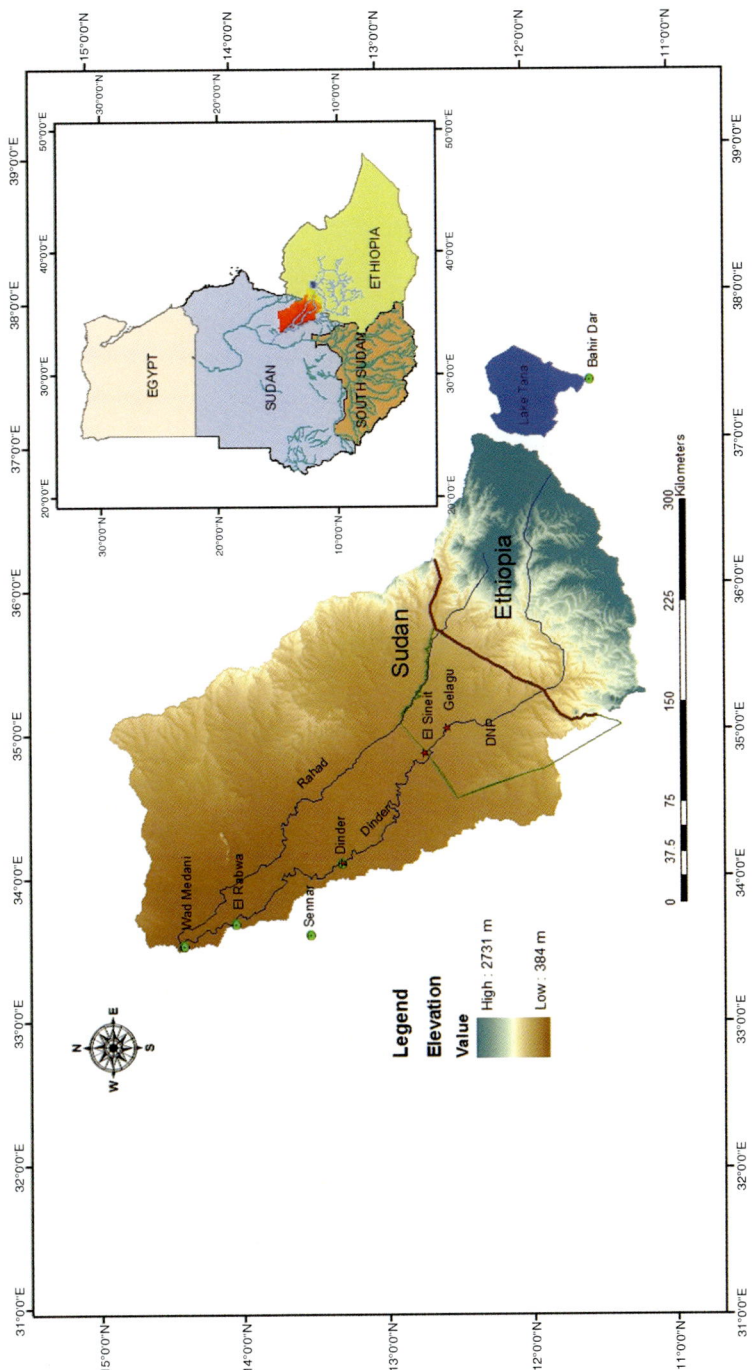

Fig. 1 Dinder National Park location and topography map (legend shows maximum and minimum elevation)

Fig. 2 Mayas – the main source of water and food for wildlife during the dry season in the DNP (Photographs by K. Hassaballah © Rights remain with the author)

feeder channel connecting the river with the mayas is also not well defined in many cases. Some mayas dry up soon after the end of the rainy season, and others remain wet throughout the dry season. Recently, the DNP authority started to manage the mayas water levels with the objective of supporting wildlife, and some of the dry mayas are kept artificially wet by pumping from groundwater.

This chapter aims to highlight the ecological importance of the mayas of the DNP and to describe the mayas ecosystem services, biodiversity and threats, with more emphasis on the hydrological threats, as well as ongoing research to improve planning for mayas ecosystem conservation.

Hydrology of the Mayas

The Dinder and Rahad River basins have a complex hydrology, with varying climate, topography, soil, vegetation, and geology as well as human interventions. The Dinder catchment area is about 37,600 km^2 and its effective catchment in Ethiopia is about 14,000 km^2. The Dinder River average flow is about 3.0×10^9 m^3/a with monthly precipitation records indicating a summer rainy season, with highest totals in the months of June–September (Block and Rajagopalan 2006). The rains during this season account for nearly 90% of total annual precipitation in the lower part near Sennar town, while in the Ethiopian highlands, approximately 75% of the annual precipitation falls during the long rainy season June-September (Shahin 1985).

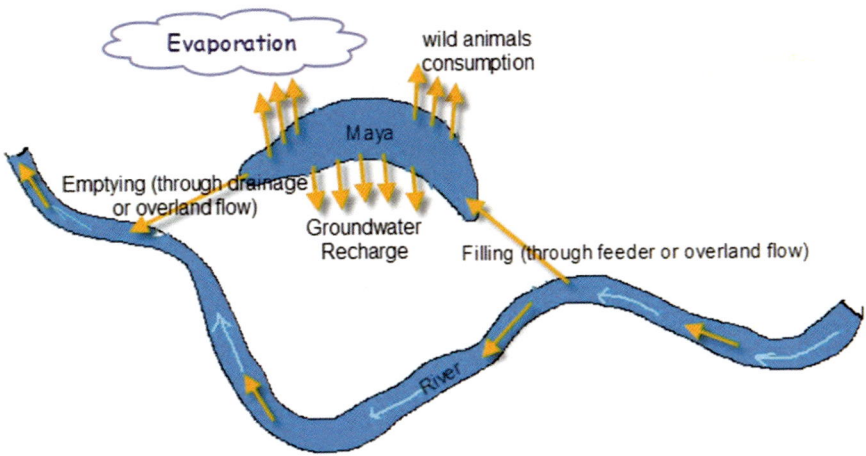

Fig. 3 Schematization of the functioning of the mayas hydrological system

Although the flow of the Dinder River is seasonal, large areas of mayas used to be inundated each year and then become dry as the spill drains or evaporates, or due to wildlife consumption, or through infiltration to recharge groundwater (Fig. 3). The ability of mayas to store significant amounts of runoff depends on the hydrology and storage capacity of individual mayas and the permeability of the subsurface. Man-made channelization of feeders to enhance the filling of mayas also complicates their hydrology.

The hydrology of mayas can be very complex. Some mayas have relatively well-defined channels while others do not. Some retain water throughout the year, while others may be dry for part of the year. Depending on the type and condition of vegetation and the amount of open water, evaporation rates will vary greatly. The mayas are also affected by surface drainage and groundwater recharge. Antecedent conditions of soil moisture and the amount of water already stored in the wetland will affect how much storage is available for runoff.

During recent years, the hydrology of the mayas has experienced significant changes. This has large implications on the ecosystem of the Dinder National Park. In particular, engineering solutions that have been undertaken in order to support wetland conservation in DNP such as channelization of the mayas feeders and excavation of mayas for removing sediment need to take account of land use and land cover changes and their impacts on runoff.

An overview of the hydrological characteristics of the Dinder and Rahad basin follows. Little baseline data has been obtained for the catchment near the border in and outside of the Sudan; therefore the climate data presented in this section has been obtained from the following sources:

http://entroportal.nilebasin.org/OSIKit/Pages/subBasin.aspx?bid=2
http://www.climate-charts.com/World-Climate-Maps.html#rain

Rainfall

The rainfall counts for 1,200 mm in the Ethiopian highlands near Lake Tana and reduces to 900 mm at the highland plateaus at the upper part of Dinder and Rahad catchment. In the middle course as in Gelagu station (inside the DNP) mean annual rainfall is reduced to less than 600 mm and further in the lower course (in Sudan) it is reduced to less than 400 mm at the village El Rabwa at the mouth of the Dinder sub-basin.

Figure 4 shows the variations in the monthly mean rainfall at the Dinder station downstream of the Dinder catchment, the Gelagu station within the mayas area inside the DNP and at Bahir Dar station further upstream of the catchment (Lake Tana). Bahir Dar is the nearest rainfall station to the upper catchment of the Dinder and Rahad with long historic records. The mean annual rainfall varied from 470 mm/a at Dinder station to 600 mm/a at Gelagu and up to 1,000 mm/a at Bahir Dar. Figure 5 shows the annual flow of Dinder River at Dinder station.

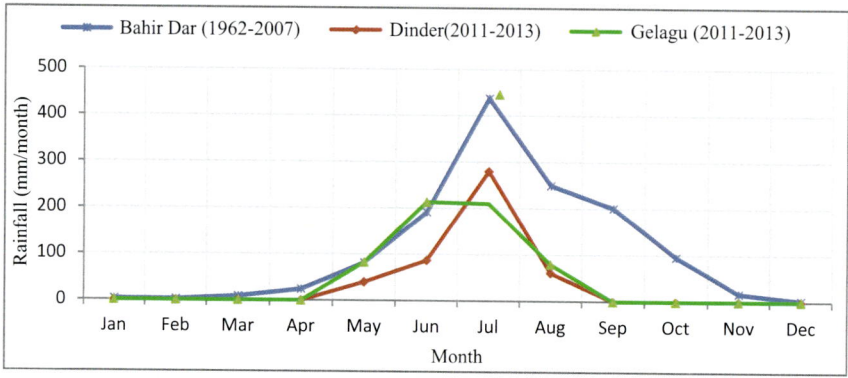

Fig. 4 The monthly mean rainfall at three different locations within the catchment

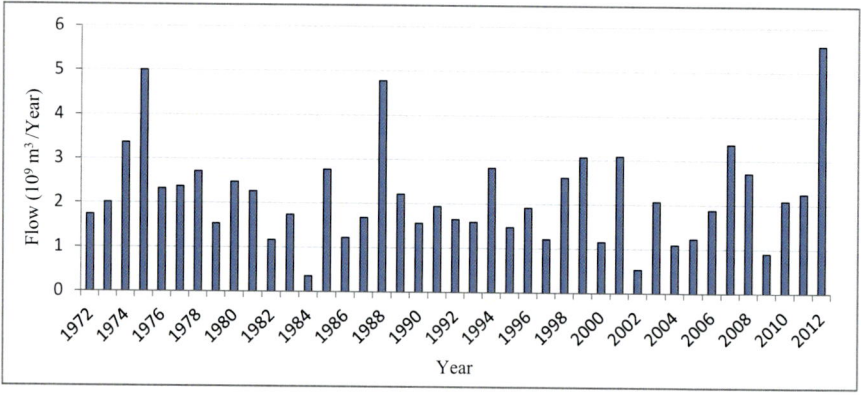

Fig. 5 Annual flow of Dinder River at Dinder station

Evaporation

Mean annual evaporation follows similar trend as that of temperature. In the highland plateau of the sub-basin, the evaporation rate is observed to be 1,150 mm/a. The low-lying area (below 1,500 masl) located at the foot of the highland plateaus, up until the border and a little beyond, experiences mean annual evaporation that ranges from 1,600 to 1,935 mm and covers some 30% of the sub-basin. Further in the Sudan lowland area at Gelagu station, evaporation is some 2,300 mm and further downstream at Dinder is observed to exceed 2,500 mm (Block 2007).

Temperature

Temperature in the highland plateau of the sub-basin is pleasant when the mean annual temperature does not exceed 20 °C. Large proportions of this highland exhibit mean annual temperatures of 18 °C. In the western low-lying area of the sub-basin, around the border, mean annual temperature is in the order of 25 °C. Further in the downstream portion of the sub-basin, around the Gelagu station, mean annual temperature is estimated to be 27 °C. In the lower course, at the mouth of the sub-basin, temperature exceeds 30 °C.

Humidity

Most likely it could be resulted from severe land and environmental degradations; nearly 80% of the sub-basin is identified to have a mean annual relative humidity of less than 55%. It is only 20% of the sub-basin with relative humidity exceeding 55%. This portion of the sub-basin is confined in the Ethiopian Plateau.

Biodiversity

The drainage system of the park includes both the Dinder and Rahad rivers and their tributaries and mayas. Dasmann (1972) classified the vegetation of DNP into four categories: wooded grassland, open grassland, woodland, and riverine forest. While, the vegetation assessments by Hakim et al. (1978) and Abdel Hameed et al. (1996) recognized three types of ecosystems, namely, the *Acacia seyal – Balanites aegyptiaca* ecosystem, the riverine ecosystem, and the mayas ecosystem. These assessments show that all three ecosystems are composed of diverse communities with relatively few species.

The mayas support large numbers of wildlife during the dry season and a smaller number during the wet season (Dasmann 1972). Yousif and Mohamed (2012) reported that waterbuck *Kobus defassa harnieri*, reedbuck *Redunca bohor cottoni*, tiang *Damaliscus korrigum tiang*, buffalo *Syncerus caffer aequinoctialis*, oribi *Ourebia ourebia montana*, roan antelope *Hippotragus equinus bakeri*, warthog *Phacochoerus aethiopicus aelinani*, and bushbuck *Tragelaphus scriptus bor* are the major herbivores that inhabit the DNP, while other animals such as baboon *Papio anubis* and red bussar monkey *Erythrocebus patas* are numerous. The major predators are lion *Panthera leo leo*, striped hyena *Hyaena hyaena dubbah*, spotted hyena *Crocuta crocuta fortis*, and black-backed jackal *Canis mesomelas*. The mayas also provide habitat for various birds such as ostrich, storks, herons, pelicans, starlings, and others.

Hakim et al. (1978) showed that the ecosystems are composed of different plant communities with relatively few species. The dominant trees in the clay plains are *Acacia seyal* and *Balanites aegyptiaca*. *Acacia fistula* is associated with *A. seyal* in areas of heavy clay which are slightly wetter than the general plain. *Combretum* sp. and *Intada sudanica* are found in drained silty soils. *Combretum hartmannianum* and *Anogeissus schimperii* are the most abundant trees along the border of Ethiopia. *Hyphaene thebaica* and *Acacia siberiana* occur along the Dinder River in the light-colored soils with varying amount of silt. *Sorghum sudanense*, *Becheropsis uniseta*, *Hyparrhenia* spp. and *Aristida plumosa* represent the dominant grasses in the park. Shrubs are limited to a few species, the most common being *Dichrostachys cinerea*.

Ecosystem Services

The DNP ecosystems provide a huge range of ecosystem services to the communities living around the park, as illustrated in Fig. 6 below. These services include purification of air and water, regulation of rainwater runoff and drought, soil formation, seed dispersal and nutrient cycling, opportunity for formal and informal education and research, climate stabilization through carbon sequestration, and moderating extremes of temperature and wind. In addition, the DNP has large potential for tourism.

The provided services can be grouped as (a) provisioning services (e.g., foods such as fish and honey, fuel wood, and medicines that used to be extracted from biota) for wildlife conservation staff and for the local communities living in the area around the DNP; (b) regulating services (e.g., flood, climate, and groundwater recharge); (c) high potential opportunity for tourism and educational services (e.g., an attractive place for local people and foreign visitors and opportunities for research and training); and (d) supporting services that maintain the conditions for life on Earth (e.g., nutrient cycling and habitat for wildlife).

Conservation Status

Wildlife management and biodiversity conservation in the DNP are challenging (Thomas et al. 2003). Yousif and Mohamed (2012) studied trends of poaching, livestock trespassing, fishing, and resource collection from 1986 to 2010 inside the DNP. This study showed that the DNP is confronted with problems such as surface water shortage during the dry season, trespassing livestock, poaching and increased

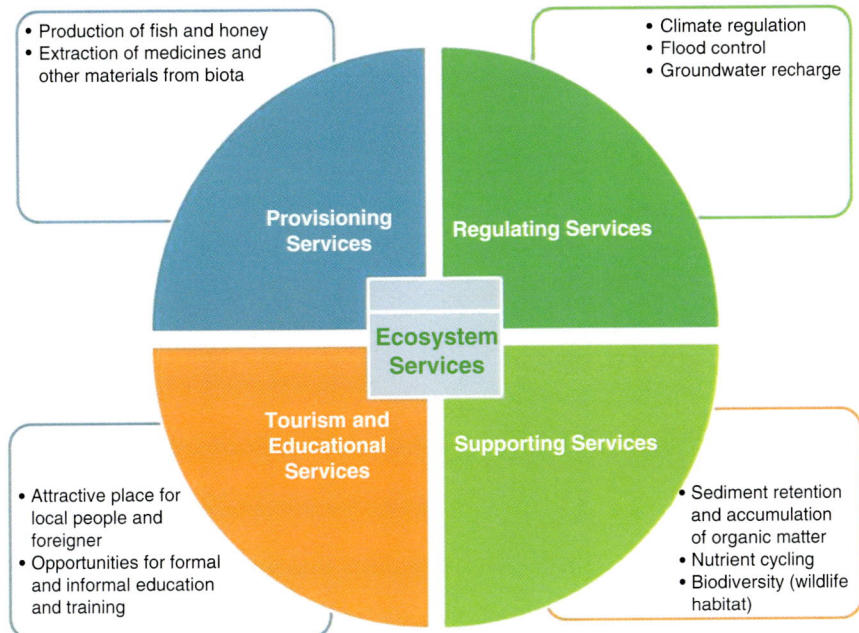

Fig. 6 Ecosystem services provided by the mayas

human settlements in surrounding areas, and consequently serious deterioration in the densities of wild animals and their related habitats. Bashir and Tong (2004) studied the protected areas in Sudan. They reported that the government of the Sudan had shown strong commitment to wildlife conservation reflected by national membership of international and regional agreements such as the African Convention, the World Heritage Convention, and the Convention on Biological Diversity. In response to the requirements of such international commitments, some efforts have been undertaken to assess the status of biodiversity and to develop an action plan by the Government of Sudan, its environmental agencies and NGOs with donor support from the United Nations Development Programme (UNDP), the World Conservation Union (IUCN), and the Global Environment Facility (GEF).

Thomas et al. (2003) studied the environmental threats and opportunities in Sudan and reported that from 2000 to 2004, the Higher Council for Environment and Natural Resources (HCENR), as part of its early efforts to accomplish its mandate, collaborated in the Sudan Country Study on Biodiversity as part of the National Biodiversity Strategy and Action Plan Project (SUD/97/G31) and with technical assistance from the IUCN and funding from the GEF implemented the Dinder National Park Project (DNPP). The main objectives of the DNPP were the conservation of biodiversity of the park and encouraging the positive interaction of the surrounding communities in the conservation and sustainable use of biodiversity in the park through the application of Biosphere Reserve concepts. The project was implemented jointly with the Wildlife Conservation General Administration (WCGA) and the Sudanese Environmental Conservation Society (SECS) and succeeded in developing a management plan for DNP involving all stakeholders adopting Biosphere Reserve concepts (Ramzy et al. 2012). In contrast, Thomas et al. (2003) reported that the outcome of this project, despite its length, was rather shallow in its treatment of wildlife, offering little more than academic or anecdotal discussions on the topic although providing considerable narrative insights into the forest, insect, agricultural, and fisheries aspects of biodiversity. Consequently, wildlife and biodiversity conservation remains an open issue until more reliable data and information can be compiled.

The Government of the Sudan, through the Ministry of Water Resources and Electricity (MoWRE), and the Eastern Nile Technical Regional Office (ENTRO) have been implementing a number of natural resource management projects financed by the World Bank and other donors. These include the Sudan Community Watershed Management Project (SCWMP) to ensure efficient water management and optimal use of resources. The SCWMP 2009–2015 sought to strengthen relevant local institutions, stakeholders, and systems in order to achieve integrated and sustainable management of watersheds in the project sites. In the Dinder area, the project aimed to reduce pressure on DNP and its biodiversity by providing alternative livelihoods to target groups encroaching on the park and by improving park management. The project's development activities were focused on promoting sustainable utilization and management of natural resources and enhancing livelihoods through the community-based project in selected target villages and communities inside and outside the park. Socioeconomic and biophysical surveys were undertaken, as well as environmental awareness campaigns which led to the election of Village Development Committees (VDCs) in 36 villages in three states (Blue Nile, Sennar, and Gedarif). The project succeeded in the rehabilitation of two mayas. To reduce the pressure on DNP and its biodiversity, the project provided alternative livelihoods to the target communities such as implementing the demarcation and plantation of animal routes within the area surrounding the park, establishment of community forest, reforestation of reserved forest, rehabilitation of range land with range seeds, and rehabilitation and construction of new hafirs (water ponds) within the animal routes to ensure enough drinking water for livestock outside the park.

Threats and Future Challenges

Although the area is relatively rich with a variety of natural resources, it is being impacted by damaging human activities such as intensive grazing, deforestation, poaching, and improper farming practices on steep slopes. These practices have posed a great threat to the sustainable management of the area as well as to its ecosystem integrity and have influenced the wildlife and plant species in the DNP and in downstream areas. The main threats facing the Dinder National Park can be summarized as the absence of proper land use practices surrounding the park, continually increasing size of human population in the Dinder area, and the trespassing of pastoralists (Ali and Nimir 2006).

During the past two decades the area of mayas inside the DNP has significantly decreased due to the temporal and spatial variations in quantities of rainfall, river discharges, and sediment deposition processes, causing many mayas to be subjected to dryness (Abdel Hameed et al. 1997). To sustain the ecosystem, some mayas are artificially watered from boreholes drilled near the mayas, and some are watered during the wet season from the Dinder River using artificial feeders. In many situations this engineering approach has led to over-engineering of the environment (through canalization and impoundment), which seriously reduces the role of the ecological processes in moderating the water cycle. Moreover, it has driven these ecological processes toward further reduction of water quality, e.g., sediment transport, or increase in secondary pollution, e.g., eutrophication (Robarts 1998). As a result, the entire ecosystem of the DNP has become threatened because it largely depends on the mayas. In addition, unlicensed mechanized rain-fed agricultural farms have been established in the areas surrounding the park within the wet season habitats of the wild animals (Abdel Hameed et al. 1999).

Areas around the park have been degraded as a result of mechanized farming and removal of the tree cover. Expansion of mechanized rain-fed agricultural farms around the park has also diminished the natural grazing area available for domestic livestock in the Dinder region (Yousif and Mohamed 2012). Large numbers of sheep and camels used to move from the Butana grasslands to the north of the park in the rainy season each year, while they returned to the banks of the Dinder and Rahad rivers and the Blue Nile during the dry season. Previously, the nomads and pastoralists used to move with their livestock over the extensive areas that are now occupied by mechanized farms or Rahad irrigation schemes bordering the northwest side of the park (Maghraby 1982). Accordingly, they are forced to forage elsewhere, and this has led to increasing invasion of the park by their livestock. Although it is prohibited, such livestock grazing within the park causes competition with wildlife for food and water.

Much of the population of wild animals migrates to habitats outside the park boundaries during the rainy season. However, many of the rainy season habitats near the borders of the park have been destroyed by agricultural activities, and the migrant wild animals are subject to increasing poaching outside the park. Animal poaching in the park during the dry season and other activities greatly affects the ecology of the area, for instance, both poachers and honey gatherers light fires throughout the park.

Recently, the DNP has experienced a serious shortage problem of surface water and green food during the dry seasons. The problem of surface water availability can be considered as a watershed management problem in which the sources of the rivers need to be studied to quantify the impacts of upstream land use on downstream areas. In addition, repeated fires within the area surrounding the mayas inside the DNP have increased the rate of soil erosion in the form of sheet flow, runoff, and silt deposition on the beds of the mayas.

Therefore, an in-depth hydrological study of the basin is crucial for planning and management of water resources as well as wider environmental management. Current research by the authors aims to contribute toward this need by assessing and understanding the interactions between the hydrology, morphology, and ecosystem of the mayas to fill a key knowledge gap for more effective ecosystem management in this important national park. It uses an ensemble of techniques tracing water and sediment fluxes from the source in the upper catchment to the sink at the mayas. This includes long-term trend analysis of hydroclimatology, field measurements, rainfall-runoff modeling, GIS and remote sensing data analysis, floodplain modeling, land-use land-cover changes, and morphological modeling of the river Dinder and its floodplain, with more focus on the mayas within the DNP.

References

Abdel Hameed SM, Hamid AA, Awad NM, Maghraby EE, Osman OA, Hamid SH. Assessment of wildlife habitats in Dinder National Park by remote sensing techniques. Albuhuth. 1996;5(1):41–55.

Abdel Hameed SM, Awad NM, ElMoghraby AI, Hamid AA, Hamid SH, Osman OA. Watershed management in the Dinder National Park, Sudan. Agric For Meteorol. 1997;84(1):89–96.

Abdel Hameed, S. M., A. A.Hamid ,N .M.Awad, E.E. Maghraby and O.A.Osman. Assessment of Watershed Problem in Dinder National Park. WRS, Unpul, report. 1999;18pp.

Ali AM, Nimir MB. Putting people first. Sustainable use of natural resource in Dinder National Park (Biosphere reserve). Khartoum. 2006. http://www.earthhore.ca/elients/WPC/English/grfx/session/PDFs/session-2Ali-Nimir.pdf{11.2006}

Bashir M, Tong F. Protected areas. 2004. http://postconflict.unep.ch/sudanreport/sudan_website/doccatcher/data/Papers%20to%20NPEM%20Workshop%201/Session%202%20papers/Paper%20(10)%20Protected%20Areas%20Management.doc

Block P. Integrated management of the Blue Nile Basin in Ethiopia: hydropower and irrigation modeling. Discussion paper; 2007.

Block P, Rajagopalan B. Interannual variability and ensemble forecast of Upper Blue Nile Basin Kiremt season precipitation. J Hydrometeorol. 2006;8:327–43.

Dasmann W. Development and management of the Dinder N. Park and its wildlife: a report to the Government of Sudan. Rome: FAO No TA 31 1 3; 1972.

Hakim SA, Fadlalla B, Awad NM, Wahab SA. Ecosystems of the vegetation of Dinder N. Park. Khartoum: WLRC; 1978.

Julien PY. River mechanics. Cambridge: Cambridge University Press; 2002.

Maghraby IE. The Dinder National Park, Environmental monitoring baseline and trend. Analysis report. ETMA/IES No. 6. Khartoum; 1982. 25 pp.

Ramzy AYE, Fawzi AMA, Reem RMS. Evaluation of the impact of Dinder National Park Project (DNPP) on the local community in Dinder Area, Sudan. J Life Sci Biomed. 2012;2(3):101–4.

Robarts RD. Factors controlling primary production in a hypertrophic lake (Hartbeespoort Dam, South Africa). J Plankton Res. 1998;6(1):91–105.

Schumm S, Kahn H. Experimental study of channel patterns. Bull Geol Soc Am. 1972;83:1755–70.

Shahin M. Hydrology of the Nile basin. Amsterdam: Elsevier Science & Technology; 1985.

Thomas C, Ejigu M, Malik Doka JTE, Ojok LI. Environmental threats and opportunities assessment. Presented to: USAID/REDSO/NPC and the USAID Sudan Task Force, Washington, DC; 2003.

Yousif RA, Mohamed FA. Trends of poaching, livestock trespassing, fishing and resource collection from 1986–2010 in Dinder National Park, Sudan. J Life Sci Biomed. 2012;2(3):105–10.

The Sudd (South Sudan)

Lisa-Maria Rebelo and Asim I. El Moghraby

Contents

Introduction 1300
Hydrology 1300
Wetland Ecosystems 1300
Biodiversity 1302
Conservation Status 1303
Ecosystem Services 1304
Threats and Future Challenges 1304
References 1305

Abstract

The Sudd is one of the largest floodplains in Africa and one of the largest tropical wetlands in the world. It is located in South Sudan, and forms part of the White Nile, or Bahr el Jebel river system, which originates in the African Lakes Plateau. Derived from an Arabic word meaning obstacle or blockage of river channels, the Sudd is composed of a maze of wetland ecosystems. The wetland supports high levels of biodiversity. Located on the eastern flyway between Africa and Europe/Asia, the Sudd is one of the most important wintering grounds in Africa for Palaearctic migrants. While variations in rainfall contribute to annual changes in the extent of the Sudd, long-term variations in the amount of water discharged from the East

---

L.-M. Rebelo (✉)
Water Futures, International Water Management Institute, Vientiane, Lao People's Democratic Republic

Regional Office for Southeast Asia and the Mekong, International Water Management Institute, Vientiane, Lao People's Democratic Republic
e-mail: l.rebelo@cgiar.org

A. I. El Moghraby (✉)
Sudanese National Academy of Sciences, Khartoum, Sudan
e-mail: asim.mog@gmail.com

© Springer Science+Business Media B.V., part of Springer Nature 2018
C. M. Finlayson et al. (eds.), *The Wetland Book*,
https://doi.org/10.1007/978-94-007-4001-3_23

African lakes are the main source of changes in the system with impacts on habitats and plant composition in the channels, lagoons, and seasonal floodplain areas.

Keywords

Ramsar · South Sudan · Sudd · Wetland

Introduction

The Sudd is one of the largest floodplains in Africa and one of the largest tropical wetlands in the world. Derived from an Arabic word meaning obstacle or blockage of river channels, the Sudd extends 650 km and 10–40 km wide between 6° 0′–9° 8′ N and 30° 10′–31° 8′ E in South Sudan (Fig. 1). The Sudd is part of the Bahr el Jebel river system (the upper reach of the White Nile in South Sudan) which originates in the African Lakes Plateau.

Hydrology

Within the Sudd, the Bahr el Jebel bifurcates into multiple distributary channels that feed a wide swamp area. High river flows from the East African lakes occur in the rainy season (April to December) with peak flows occurring in August to September. About 50% of the 33×10^9 m^3 of water that flows into the Sudd circulates within the ecosystem and does not contribute to the outflow at Lake No (Sutcliffe and Parks 1999). The volume of discharge of the Bahr el Jebel at its confluence with the Bahr el Ghazal at Lake No is calculated as 16×10^9 m^3 as steady outflow over the year (Rebelo et al. 2012).

Permanent swamps cover approximately 30,000 km^2, but depending upon season and inflow conditions, seasonal flooding extent varies considerably to over 100,000 km^2 (Mohamed et al. 2004; Dumont 2009). In periods of high flood and rainfall (e.g., 1917–1918, 1932–1933, 1961–1964, and 1988–1989), the floodplain remains flooded well into the dry season. In contrast, during periods of low flood and rainfall (e.g., 1921, 1923, and 1984), the floodplain shrinks and the permanent swamps dry up. Long-term variations in the amount of water discharged from the East African lakes are the main source of changes in the system with impacts on habitats and plant composition in the channels, lagoons, and seasonal floodplain areas.

Wetland Ecosystems

The Sudd is composed of a maze of wetland ecosystems, grading from open water and submerged vegetation to floating fringe vegetation, seasonally inundated woodlands, rain-fed and river-fed grasslands, and floodplain scrubland (Hickley and Bailey 1987). Moving laterally from the center of the swamps, different vegetation zones are recognized, determined by elevation and frequency of inundation:

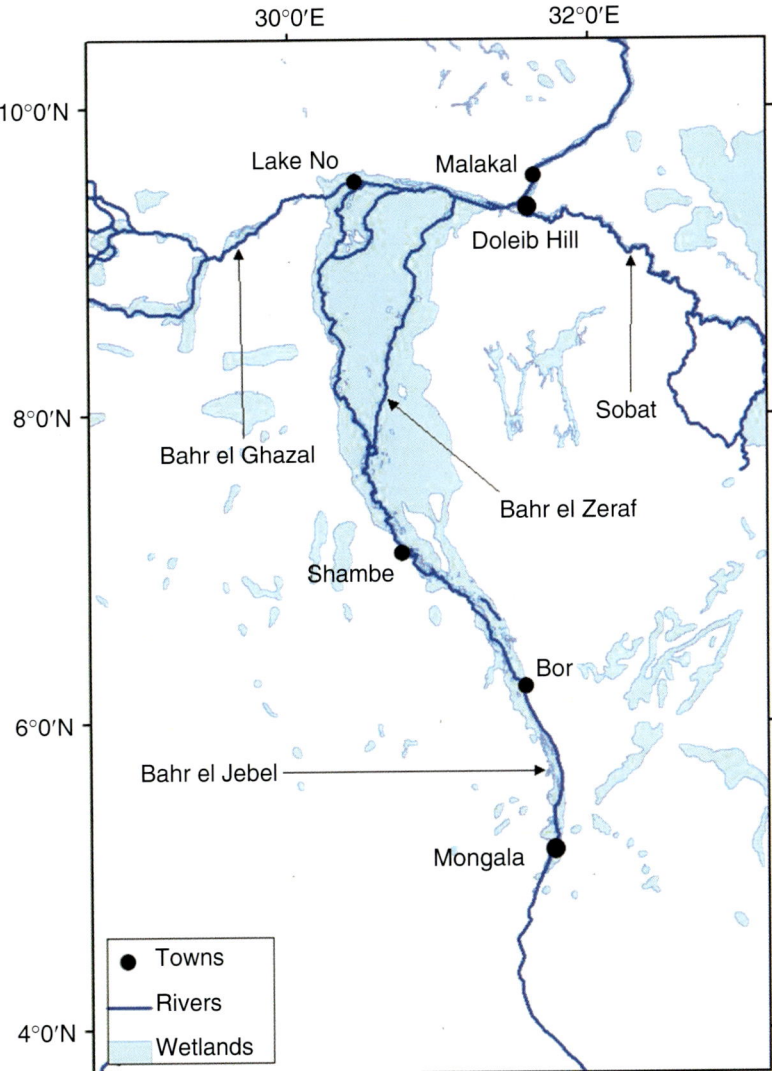

Fig. 1 The Sudd, South Sudan

(i) Open water vegetation: dominated by free-floating-leaved plants like *Eichhornia crassipes*, *Lemna gibba*, *Azolla nilotica*, and *Nymphaea lotus* as well as submerged vegetation, e.g., *Potamogeton*, *Trapa*, and *Ceratophyllum* spp., in both flowing waters and lakes. Macrophytes are prevalent in lakes but less so in the flowing turbid water of the main channel.

(ii) Permanent swamps: occurring on land that is flooded throughout the year, the most prevalent types of plants are the floating and fringe vegetation, dominated along the banks of the river by massive stands of *Cyperus papyrus*, *Vossia cuspidata*, *Phragmites karka*, and *Typha domingensis*.

(iii) Seasonally river-flooded grasslands ("toich" in the local Dinka language): during the rainy season, these are often saturated and completely covered with luxuriant growth of tall grasses. Species composition and biomass are closely linked to soil type and degree of inundation. The tall grasses are dominated by *Phragmites*, *Sorghum sudanica*, *Hyparrhenia*, and *Setaria* as well as *Oryza* and *Echinochloa*. Two grassland types are recognized: wild rice grassland dominated by *Oryza longistaminata* and *Echinochloa* grassland dominated by *Echinochloa pyramidalis*. Sedges, herbs, and other grasses such as *Sporobolus pyramidalis* are associated with the *Oryza* grassland. It is the most productive grassland type in terms of year round grazing for livestock and wildlife due to the high protein content of dead materials of wild rice grass. Within the toich there are many small seasonally flooded pools that are sources of water for domestic, livestock, and wildlife use as well as fish.

(iv) Rain-flooded grasslands: grasslands seasonally inundated either by the river or rainfall on seasonally waterlogged clay soil. The zone is made up of a comparatively well-drained portion dominated by *Echinochloa haploclada*; grassland heavily grazed by livestock dominated by *Sporobolus pyramidalis*, *Phragmites*, and *Sorghum*; and areas that are inundated by rainwater and sheet flow, a high-biomass but nutrient-poor *Hyparrhenia rufa* grassland.

(v) Floodplain woodland: occurring at a higher altitude than the grassland floodplains; this rain-fed belt consists of open mixed *Acacia* forest supporting several species of trees and shrubs and a luxuriant growth of grasses. The tree vegetation is dominated by *Acacia seyal*, *Acacia sieberiana*, and *Balanites aegyptiaca*.

Biodiversity

Biodiversity within the Sudd is high, with over 470 bird and 100 mammal species recorded. Located on the eastern flyway between Africa and Europe/Asia, the Sudd is one of the most important wintering grounds in Africa for Palaearctic migrants, providing essential habitats for millions of migrating birds (Howell and Lock 1988). During the 1980s the region was listed as supporting the highest population of shoebill storks *Balaeniceps rex* (East 1999) and the greatest numbers of antelopes in Africa (Stuart et al. 1990). Many of the antelopes undertake large-scale migrations across the Sudd, following the changing water levels and vegetation. The wetland also has a high density and diversity of aquatic plants.

The Sudd is home to over 100 species of fish, providing feeding areas on the seasonal floodplains during the flood season (Welcomme 1979). Recruitment, growth, and survival of most of the fish species take place within the various habitats of the wetland. Within the permanent swamps, invertebrate zooplankton is abundant, as are mollusks. Rare freshwater shrimps such as *Caridina nilotica*, *Palaemon nilotica*, and *Cyclestheria hislopi* are also found in the wetland. Several endangered species are found in the Sudd, including cheetah *Acinonyx jubatus*, white addax *Addax nasomaculatus*, Grévy's zebra *Equus grevyi*, various gazelles (*Gazella dama*, *G. dorcas*, *G. leptoceros*, *G. rufifrons*, *G. soemmerringii*), the Nile lechwe *Kobus megaceros*, and the African wild dog *Lycaon pictus*.

Throughout the 21 years of the Sudanese civil war (1983–2005), no surveys were undertaken within the wetland. The war not only resulted in increased levels of poaching but also halted all management activities within the protected areas in Southern Sudan. The full effect of the civil war on the biodiversity of the Sudd is currently unknown, but is expected to have resulted in a downward trend in numbers and diversity of wildlife.

Conservation Status

In 2006 an area of 5.7 million ha of the floodplains within the Sudd region was designated as a Ramsar wetland site of international importance. Prior to this designation, three protected areas existed within the Sudd region, covering 1,080,000 ha:

(i) Zeraf Game Reserve	970,000 ha, established in 1939
(ii) Shambe National Park	62,000 ha, established in 1985
(iii) Fanyikang Game Reserve	48,000 ha, established in 1939

Under the Wildlife Act of 1987 and the recently developed New Sudan Wildlife Provisions Act of 2003, all protected areas in the Sudan are not to be accessed or used without a permit. To enforce the 2003 Act, a Wildlife Force was deployed in each protected area. With the exceptions of access roads and housing for wildlife personnel, no development is allowed in any of the protected areas in Sudan. According to the recently proclaimed constitution of Southern Sudan, the government of Southern Sudan is the custodian of the land on behalf of the people of Southern Sudan. The protected areas are solely owned by the government, while ownership of the surrounding communal lands is based on customary land ownership. The wetland comprises approximately 20% protected areas and 80% communal land. Since 2006, the Sudd has been under the jurisdiction of the government of Southern Sudan under a "one country two systems" approach. However, the present status of the protected areas is not known with the recent division of Sudan into two countries.

Ecosystem Services

In addition to supporting high levels of biodiversity, the Sudd provides many ecosystem services and has a profound effect on water quality, nutrient retention, and river hydrology. It moreover provides extensive economic and environmental benefits to the entire region; the socioeconomic benefits of the wetland to the communities living in its catchment are immense. It is estimated that over one million people are almost entirely dependent on the wetland, their way of life inextricably tied to the seasonal expanding and shrinking of the swamps and the changing depth of water (Rebelo et al. 2012). The channels of the wetland are used for navigation by both commercial and public steamers as well as boats and canoes. The wetland is a major source of water for domestic use and provides building materials for the local communities. An important source of fish, it is one of the only water bodies of the Nile which is not currently overfished, and the potential yield on a sustainable basis has been estimated to be around 75,000 t per year based on a surface area of 30,000–40,000 km^2 (Witte and van Oijen 2009). Due to the lack of processing and storage facilities, commercial fisheries are yet to be exploited.

The Sudd provides a source of water and essential dry-season grazing land for livestock, the backbone of the Nilotes' economy (Sutcliffe 2009). Most of the tribes living within the Sudd catchment are nomadic and move with their large herds of cattle in response to the annual regime of the Bahr el Jebel and rainfall. Fishing is the second most important occupation of the inhabitants of the wetlands especially the Shilluk and Nuer. Subsistence hunting is also important to the Nilotes of the Sudd catchment. Crop production is not a significant occupation although some subsistence agriculture is carried out in the highland areas during the wet season. The agricultural potential of the area is limited by the vagaries of the climate, pests, weeds, and diseases.

The Sudd has very high cultural values for the Dinka, Nuer, and Shilluk communities. During periods of low flow, the wetland is a social center for the initiation of relationships and dancing leading to courtship and marriage. The wetland sustains the livestock used as dowries for marriages, religious rituals, and payment of penalties. Furthermore the wetland contains various wildlife species considered to be sacred by the Nilotes. These include the Nile lechwe (*Ontragus megaceros*), the shoebill stork (*Balaeniceps rex*), and the black crowned-crane (*Balearica pavonina*).

Threats and Future Challenges

The Sudd wetlands have come under considerable pressure during the past few years. Hunting was uncontrolled during the civil war and remains a threat due to the large number of firearms as yet uncollected. In addition, the inflow of large numbers of refugees and their cattle following the civil war has put pressure on the natural resources of the Sudd due to competition for grazing land, deforestation, and infrastructure development. The aquatic diversity of the Sudd is also vulnerable to invasive species such as the water hyacinth *Eichhornia crassipes*.

Recent discovery and exploitation of oil reserves in the Sudd threaten the diversity of the wildlife, aquatic macrophytes, and floodplains, as well as the hydrology of the intricate ecosystem (Springuel and Ali 2005). Several blocks have already been allocated to oil companies and exploration drilling is underway in the permanent swamps. Concerns surrounding the exploration and extraction of oil include disruption of water flow patterns as a result of seismic testing and diking, wetland and floodplain fragmentation due to access roads and oil exploration sites, and contamination due to oil spills and human waste.

With the signing of the Comprehensive Peace Agreement in 2005 and the end of the civil war, a major threat to the wetland was the potential completion of the Jonglei Canal (Dumont 2009). The Jonglei Canal is a project to divert inflows to the Sudd in order to reduce evaporation from the wetland, thereby gaining approximately 4.7×10^9 m^3 of water for downstream use, as well as to reclaim approximately 100,000 ha of land for agriculture. Construction of the canal started in 1980 but stopped with the onset of the Sudanese civil war in 1983, after 260 of the total 360 km had been completed. The canal was expected to reduce the water level of the swamp by 10% during flood season and by 20% during the dry season, greatly reducing the area of the *toich*. Its completion is likely to have a significant impact on climate, groundwater recharge, sedimentation, and water quality; it is also likely to result in the loss of biodiversity, fish habitats, and important grazing areas and to interfere with the seasonal migration patterns of both cattle and wildlife, all of which will have an effect on the livelihoods of the local populations. In 2008, discussions to continue the work were resumed. However following subsequent civil unrest and recent political instability in South Sudan, plans for the future of the canal are unclear.

While various other interventions are planned on the White Nile which are expected to decrease inflows to Lake Victoria, research has shown that the river flow from Lake Victoria to the Sudd is not expected to be significantly affected (Awulachew et al. 2012).

References

Awulachew SB, Demissie SS, Hagos F, Erkossa T, Peden D. Water management intervention analysis in the Nile Basin. In: Awulachew SB, Smakhtin V, Molden D, Peden D, editors. The nile river basin: water, agriculture, governance and livelihoods. Abingdon: Routledge/Earthscan; 2012. p. 292.

Dumont HJ. The nile: origins, environments, limnology and human use. Monographiae biologicae, vol. 89. Dordrecht: Springer; 2009. p. 818.

East R. African antelope database 1998. World conservation union species survival commission occasional paper 21. 1999. p. 454.

Hickley P, Bailey RG. Food and feeding relationships of fish in the Sudd swamps (River Nile, southern Soudan). J Fish Biol. 1987;30:147–60.

Howell P, Lock M. The jonglei canal: impact and opportunity. Cambridge, UK: Cambridge University Press; 1988.

Mohamed YA, Bastiaanssen WG, Savenije HH. Spatial variability of evaporation and moisture storage in the swamps of the upper Nile studies by remote sensing techniques. J Hydrol. 2004;289:145–64.

Rebelo L-M, Senay GB, McCartney MP. Flood pulsing in the Sudd wetland: analysis of seasonal variations in inundation and evapotranspiration in South Sudan. Earth Interact. 2012;16:1–19.

Springuel I, Ali OM. Nile wetlands: an ecological perspective. In: Fraser LH, Keddy PA, editors. The world's largest wetlands: their ecology and conservation. Cambridge, UK: Cambridge University Press; 2005. p. 448.

Stuart SN, Adams RJ, Jenkins MD. Biodiversity in Sub-Saharan Africa and its Islands: conservation, management and sustainable use, Occasional papers of the IUCN species survival commission, vol. 6. Gland: IUCN; 1990.

Sutcliffe J. The hydrology of the Nile Basin. In: Dumont HJ, editor. The nile: origins, environments, limnology and human use, Monographiae biologicae, vol. 89. Dordrecht: Springer; 2009. p. 335–64.

Sutcliffe J, Parks Y. The hydrology of the nile, IAHS special publication, vol. 5. Wallingford: IAHS Press; 1999. p. 179.

Welcomme RL. Fisheries ecology of floodplain rivers. London: Longman; 1979. p. 317.

Witte F, van Oijen M. Sibbing fish fauna of the Nile. In: Dumont HJ, editor. The nile: origins, environments, limnology and human use, Monographiae biologicae, vol. 89. Dordrecht: Springer; 2009. p. 335–64.

Rugezi Marsh: A High Altitude Tropical Peatland in Rwanda

107

Piet-Louis Grundling, Ab P. Grootjans, and Anton Linström

Contents

Introduction	1308
Hydrology	1310
Peat of Rugezi Marsh	1311
Biodiversity	1313
Ecosystem Services of the Rugezi Marsh	1315
Threats and Future Challenges	1316
References	1317

Abstract

The Rugezi Marsh is an extensive peatland located in a mountain valley in the north of Rwanda. It is groundwater fed and is therefore a fen with a rich biodiversity. An IUCN endangered listed orchid species, *Disa stairsii*, occurs in the system as well as three Bryophytes: *Lycopodiella cernua*, *Sphagnum cuspidatum* and *S. perichaetiale*. *Sphagnum perichaetiale* is a new species recorded for Rwanda. The peatland has been degraded by diverse activities, such as draining, cultivation, and roads crossing the wetland. Uncontrolled

---

P.-L. Grundling (✉)
Centre for Environmental Management, University of the Free State, Bloemfontein, South Africa
e-mail: peatland@mweb.co.za

A. P. Grootjans (✉)
Centre for Energy and Environmental Studies, University of Groningen, Groningen, The Netherlands

Institute of Water and Wetland Research, Radboud University Nijmegen, Nijmegen, The Netherlands
e-mail: a.p.grootjans@rug.nl

A. Linström (✉)
Wet Earth Eco-Specs, Lydenburg, South Africa
e-mail: wetearth@telkomsa.net

© Springer Science+Business Media B.V., part of Springer Nature 2018
C. M. Finlayson et al. (eds.), *The Wetland Book*,
https://doi.org/10.1007/978-94-007-4001-3_152

burning regimes due to agriculture, as well as over-exploitation of non-timber forest and wetland products further impact the system. Consequently these activities have led to the lowering of the water table, peat desiccation and establishment of terrestrial invasive vegetation species. Local communities and Rwanda-at-large are dependent on this wetland for various goods and services ranging from production of fibre, fodder, protein to water supply and hydroelectricity. Conservation and proper management of the wetland, including sustainable restoration measures, are therefore of importance.

Keywords
Africa · Degradation · High Altitude · Peatland · Tropical

Introduction

The Rugezi Marsh is located in a mountain valley in the north of Rwanda, to the east of Lake Bulera on the border with Uganda at an altitude of 2,050 m (Fig. 1). This peatland covers an area of 6,735 ha (between 1°21′30″ and 1°36′11″ south and 29°49′59″ and 29°59′50″ east). Rwanda experiences two wet seasons and two dry seasons – a major dry season from June to September, a short rainy season from October to December, a shorter dry season from January to February, and a long wet season from March to May. The mean annual precipitation is 1,142–1,309 mm (1957–1987). Annual mean temperatures range between 12 °C and 18 °C (MINALOC 2004).

Fig. 1 Location of the Rugezi Marsh, Rwanda (From Google Earth February 2016)

The original peat-forming Rugezi system (a mire) has developed into two main valleys: (i) the Kamiranzovu valley, characterized by water flowing in a north–south direction, and (ii) the Rugezi valley, which is considered as the main valley of the original mire, with water flowing in a south–north direction (MINALOC 2004). The present peatland and its catchment can be divided into three major parts (Fig. 2).

The northern part includes the Kamiranzovu valley. This region is the most degraded part of the marsh system due to drains, road crossings, agriculture, and livestock grazing. This zone used to be covered by papyrus, but this has totally changed since 1995–2000. Today isolated papyrus stands mostly occur on trampled soil with low water retention capacity. This sector has become poor in terms of biological diversity and has lost several endemic bird species due to habitat destruction.

The central part starts about 5 km south (upstream) from the confluence of the Rugezi stream with the Kamiranzovu stream (northern part) toward the split of the main valley into two tributaries. Though heavily degraded as well, natural vegetation typical of highland marshes is observed from place to place, and water levels are moderately higher. The main hydrological impact is related to a series of drains dug across the peatland. It feeds into a longitudinal drain on the eastern edge flowing toward the confluence with the Kamiranzovu stream from where it is channeled to the Ntaruka hydropower plant.

The central part of the catchment is totally transformed by cultivation activities that extend to and into the edge of the wetland and with rural housing being built against slopes steeper than 35% gradient (Hategekimana and Twarabamenye 2007, Fig. 3). This has led to increased runoff from human settlements and increased sedimentation in the marsh.

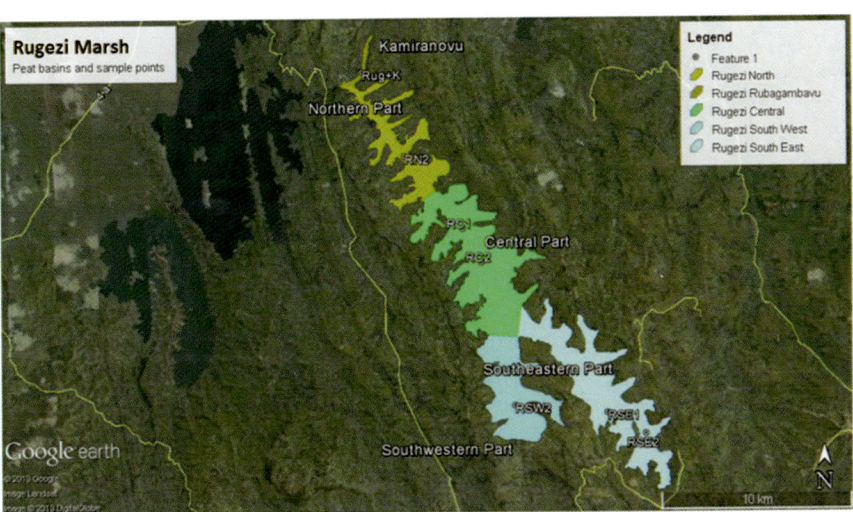

Fig. 2 The Rugezi Marsh consisting of various peat basins has been divided into three parts for discussion purposes; peat sample points are indicated. Codes of sampling points refer to codes used in Fig. 5

Fig. 3 The upper slopes in the catchment are totally transformed by cultivation, which has a pronounced effect on the low-lying wetlands (Photo credit: Anton Linström)

The southern part starts from the split in the main valley and consists of two tributaries. This part is least affected by human activities, and the water level is high, which maintains typical mire vegetation (with peat-forming capabilities). Some runoff from human settlements and agricultural land contributes to changes in the flow regime, especially the areas around the margins of the wetland. Many large water birds, such as cormorants, pelicans, bitterns, and herons, thrive in this sector, primarily due to high fish productivity and inaccessibility limiting the presence of humans inside the wetland.

The southeastern section (of the southern part) was partially drained between 1960 and 1983 due to the breaching of the eastern slope with an irrigation canal for a tea plantation project in Mulindi. Water levels fell rapidly and the whole section was reclaimed for agriculture. In 1983, the breach was blocked, and consequently the water level rose again (Hategekimana and Twarabamenye 2007). However, the restoration created water bodies and inundated the area to such an extent that large areas became flooded and wetland species typical of the original Rugezi Marsh could not recover.

The wetlands in each catchment area of the Rugezi Marsh comprise a significant portion of the catchment (Table 1). The marsh has a gentle slope (0.05%) and the combined peatland area covers 33% of the catchment.

Hydrology

The Rusumo stream originates at the confluence of the Rugezi and Kamiranzovu streams. Due to its size and accumulated peat deposits, the Rugezi Marsh plays a major role in the regulation of water flow to Bulera and Ruhondo lakes, the main sources of hydropower in Rwanda (Hategekimana and Twarabamenye 2007). The analysis of flow measurements for the Rusumo stream showed that the variation of

Table 1 The extent and gradients of the wetlands and the percentage of wetland area

Wetland (name)		Wetland size (ha)	Wetland slope	Catchment estimated size (ha)	Wetland area (as percentage of catchment)
Northern part		71	0.33	2,908	2
Central part		3,151	0.06	15,912	20
Southern part	Southeast	1,166	0.05	2,633	44
	Southwest	1,840	0.04	4,767	39
Total		**6,228**	**0.05**	**18,816**	**33**

water volume in the marsh is strongly related to rainfall (Deuse 1966). In May (rainy season) 1996, the mean discharge was 5.72 m^3/s compared to 0.43 m^3/s in August (dry season). During 1984–1986, years of less rainfall (931 mm), the mean annual discharge fell to 1 m^3/s. The groundwater contribution to the marsh is not known, and it could be a significant component of the water balance given the catchment's geology and topographical environment. Various springs, seeps, and ferrous irons on the edges of the marsh (Fig. 4) confirm groundwater discharge into the peatland.

It is, therefore, a fen and not a bog as previously indicated by Hategekimana and Twarabamenye (2007). In its natural state, the Rugezi Marsh formed a dense floating mat on a peat layer (Hategekimana 2005). However, the water levels in the marsh decreased as a result of the large amounts of water channeled to the Ntaruka hydropower plant.

Peat of Rugezi Marsh

Peat was sampled with a Russian peat corer at 0.5 m increments and was described in the field according to the von Post humification scale (Grosse-Brauckmann et al. 1977). The peat in the drier (drained) Kamiranzovu valley is much shallower than in the Rugezi valley and can be described as highly humified, while the thicker peat layers in the central and southern section were undecomposed in the surface layers, becoming progressively more humified in depth (Fig. 5). The peat in the northern part was mostly a sedge peat, while the peat in the central and southern parts was dominated by *Miscanthus violaceus* to a depth of 7 m, with some *Sphagnum* peat in the top 0.5 m in the southwestern section.

The thickness (and therefore the peat volume) in Table 2 is inferred and estimated based on field observations, as the bottom of the peat body could not be reached beyond 7–7.5 m due the limitations of the hand auger.

The average peat content per core depth for the various sections of the marsh is indicated in Table 2. The estimated volume of the peat is 5.25×10^8 m^3, and the peat carbon content for the Rugezi peatland is 2.77×10^8 m^3, about half of the 5×10^8 m^3 carbon reported by Joosten (2004). However, it is expected, given the morphology of the peat basin, that the thickness could be 50–100% greater than sampled (thicknesses of up to 20 m were reported in the IMCG Global Peatland

Fig. 4 A spring on the edge of the Rugezi Marsh with ferrous iron evident (*red* color of the water) (Photo credit: Anton Linström)

Peat Profile (Von Post Humification Scale) for the Rugezi Mire from North to South

	North			South	South East	South East	South West	
Site no:	Rug+K3	RN2	RC1	RC2	RSE1	RSE2	RSW3	
Distance (km)	2	8	12	14	24	27	21	
Depth (m)								Depth (m)
0.25	H2	H3	H4	H2	H2	H3	H1	0.25
0.5	H8	H3	H2	H2	H2	H3	H1	0.5
0.75	H8	H3	H2	H2	H2	H4	H2	0.75
1	H8	H3	H2	H2	H2	H2	H2	1
1.25	H9	H3	H2	H2	H2	H2	H2	1.25
1.5	H9	H3	H2	H7	H2	H2	H2	1.5
1.75	H9	H3	H2	H4	H3	H2	H2	1.75
2	H9	H4	H2	H6	H4	H2	H2	2
2.25	H9	H6	H2	H5	H4	H7	H2	2.25
2.5	H9	H6	H2	H5	H4	H7	H4	2.5
2.75	H9	H6	H2	H8	H4	H4	H4	2.75
3	H9	H7	H2	H8	H4	H4	H2	3
3.25	H9	H10	H2	H9	H7	H4	H2	3.25
3.5	H9	H10	H2	H9	H7	H4	H2	3.5
3.75	H10	H2	H2	H10	H7	H5	H2	3.75
4		H2	H5	H10	H9	H5	H2	4
4.25		H2	H7	H6	H9	H5	H5	4.25
4.5		H4	H4	H6	H9	H5	H5	4.5
4.75		H5	H4	H7	H9	H5	H2	4.75
5		H5	H4	H7	H9	H5	H2	5
5.25		H5	H4	H7	H9	H5	H4	5.25
5.5		H5	H4	H7	H9	H5	H4	5.5
5.75		H7	H4	H7	H6	H5	H4	5.75
6		H7	H4	H7	H6	H5	H4	6
6.25		H7	H4	H7	H6	H8	H5	6.25
6.5		H7	H4	H7	H6	H8	H5	6.5
6.75		H6	H6	H4	H6	H10	H5	6.75
7		H6	H6	H8		H10	H9	7
7.25		H9	H6	H10				7.25
7.4			H8	H10				7.4

Legend: H1, H2, H3, H4, H5, H6, H7, H8, H9, H10

Fig. 5 A diagram from north to south through the Rugezi Marsh indicating the nature of the peat (according to the von Post humification scale; Grosse-Brauckmann et al. 1977) with an increase in depth (From Linström et al. 2013, with permission of authors). The position of sampling points is indicated in Fig. 2

Table 2 Peatland area and peat volume

Locality	Area (ha)	Thickness (m)	Thickness factor	Volume (m^3)
Rugezi	6,228	–	–	–
Kamiranzovu	71	4	1	2,840,000
Rugezi main trunk	3,151	6.58	1.3	269,536,540
Rugezi southwest	1,166	6.5	1.3	98,527,000
Rugezi southeast	1,840	7	1.2	154,560,000
Subtotal	6,228			525,463,540
[a]**Kamiranzovu unmapped section**	507			
Total	6,735[a]			

[a]Wetland area reported by Hategekimana and Twarabamenye (2007) is 507 ha more than that of this study, due to the presence of clouds in our imagery in the northern part of Rugezi

Database for the Akanyaru River and Rugezi valley; Joosten 2004); thus the carbon content could be substantially higher as it is expected that the carbon content will increase with depth as the degree of humification increases.

Biodiversity

The vegetation of the marsh is dominated by species that are indicative of permanent wet conditions, such as *Miscanthidium violaceum* accompanied by *Vaccinium stanleyi*, *Erica rugegensis*, *Sphagnum* spp., *Thelypteris confluens*, and *Xyris validus* and *Nymphaea nouchali* (Fig. 6). However, stands of *Cyperus latifolius* and *Cyperus papyrus* accompanied by *Juncus oxycarpus*, *Crassocephalum* sp., *Dicrocephala*, and *Spilanthes* are still present (MINITERE 2003). An orchid species, *Disa stairsii*, which is listed by IUCN as an endangered plant, still occurs here.

The dominant fern species in the wetland is *Thelypteris confluens* with *Osmunda regalis* and *Pityrogramma aurantiaca* also occurring in places. Orchids such as *Satyrium crassicaule* and *Cynorkis anacamptoides* have been observed. Bryophytes (mosses) found in the wetland areas include three species, *Lycopodiella cernua*, *Sphagnum cuspidatum*, and *S. perichaetiale* (Fig. 7). *Sphagnum perichaetiale* is a new species recorded for Rwanda.

The catchment area has been invaded by several alien and invasive plant species that should be controlled through a management plan; further afforestation has also contributed to areas dominated by exotic trees. Exotic species include *Acacia dealbata*, *A. mearnsii*, *Eucalyptus grandis*, *Solanum* spp., and *Pennisetum clandestinum*. *Alnus acuminata* has been introduced to the wetland area to indicate the buffer zone in which no agriculture activities may take place in. This species is

Fig. 6 Open water area in the Fels outlet indicating the dominance of *Nymphaea nouchali* var. *caerulea* (Photo credit: Anton Linström)

Fig. 7 Two *Sphagnum* spp. identified in the Rugezi Marsh (Photo credit: Anton Linström)

native to the American continent and grows well in moist soil environments, such as the Rugezi Marsh.

The Rugezi Marsh was declared a Ramsar Site on 1 December 2005 (Ramsar Convention Secretariat 2012). Some 82 bird species have been recorded within the marsh and its immediate vicinity. This marsh is important for Grauer's swamp warbler (*Bradypterus graueri*), an endangered bird species endemic to the Albertine Rift region, with more than 60% of the global population found within the marsh (Byaruhanga *et al.* 2006). Rugezi hosts a high population of breeding gray-crowned

Fig. 8 An artificial channel draining the northern part of Rugezi Marsh (Photo credit: Anton Linström)

crane *Balearica rugulorum*, which is an endangered bird species (Nsabagsani 2010), and is home to other threatened bird species including white-winged scrub warbler (*Bradypterus carpalis*). The diversity and abundance of mammals have reduced significantly in recent years. The population of sitatunga *Tragelaphus spekii* (CITES and IUCN listed), the largest antelope specialized to live in wetlands, was abundant up to 1998. Currently, these antelopes are rare or have disappeared, mainly due to water reduction and hunting. Some mammals that are not hunted are still present, namely, African clawless otter *Aonyx capensis*, rodents like Inkezi *Thryonomys* sp., and many species of imbeba (Muridae family) (MINALOC 2004). The Rugezi Marsh and the lakes Bulera and Ruhondo have a very poor fish fauna possibly due to the relatively acidic water (pH between 4.6 and 6; Deuse 1966), but in some areas also due to intensive drainage activities (Hategekimana and Twarabamenye 2007). Very little information exists on amphibians and invertebrates found in the Rugezi Marsh area.

Ecosystem Services of the Rugezi Marsh

Ecosystem functions of wetlands such as the Rugezi Marsh include (i) sediment and nutrient removal of surface waters, (ii) carbon sequestration, (iii) provision of habitat for wildlife, and (iv) flow augmentation by retaining ground- and surface water in the soil and vegetation (Macfarlane et al. 2009). The northern section (including the Rubagambavu wetland) has lost many of its ecosystem services due to the large drainage channel in the middle of the wetland (Fig. 8). The wetland is now less effective in regulating stream flow and purifying water. The reduction of vegetation cover has reduced sediment retention and flood control and has led to a gradual erosion of biodiversity (REMA 2009).

The central part of the marsh (main trunk) is less intensely modified. This part of the marsh still stores large amounts of peat and maintains much biodiversity including endangered species such as Grauer's bush warbler and the gray-crowned crane. The southern sections (southwest and southeast) are largely natural, particularly the southwestern section which contributes significantly toward carbon storage and may also trap sediments and nutrients released by agricultural activities on the slopes.

The wetlands also provide local people with material they can use for housing, tools, and equipment. For instance, several *Cyperus* species (including papyrus) are exploited by the local people for fodder, baskets, mat making, roofing, ceilings, and rope making. However, extensive grass and sedge cutting can be detrimental to the integrity of the marsh. Grass and sedges are cut for feeding livestock, traditional medicine, and the making of crafts such as baskets and mats (Linström et al. 2013). Local people use the wetland for water supply and domestic use. The open water areas also created the opportunity for local people to cross the wetland via dugout canoes and contributed toward job creation. A local canoeist could earn a mean revenue of 1000 Rwandan francs per day (Hategekimana and Twarabamenye 2007).

Threats and Future Challenges

The Rugezi Marsh is extensive with clear indications of being groundwater fed and is therefore a fen. It has been degraded by diverse activities, such as draining, cultivation, roads crossing the wetland area, and uncontrolled burning regimes due to agriculture, as well as overexploitation of non-timber forest and wetland products. Many of these activities have led to the lowering of the water table in the marsh, peat desiccation, and establishment of terrestrial invasive vegetation species. Local communities and wider populations in Rwanda at large are dependent on this wetland for various goods and services ranging from production of fiber, fodder, and protein to water supply and hydroelectricity. Proper management and conservation of the marsh, including restoration measures, are therefore of importance.

For the protection of the aquatic resource in the marsh and its catchment, it is essential that buffer zones are maintained and enlarged. The primary purposes of such proposed buffer zones include (i) reducing the impacts of adjacent land uses on water resource quality, (ii) sustaining or improving the ability of the wetlands to provide goods and services to the local communities, and (iii) providing habitats and protection for aquatic and semiaquatic species (Macfarlane et al. 2009). Rewetting of desiccated areas is needed to preserve the peat, but this might have a direct impact on the livelihoods of various communities around the marsh. The creation of income-generating activities should be an integral part of the natural restoration planning. Erosion control activities in the catchment should also be increased. These could include storm water control (e.g., further terracing, water harvesting to reduce high runoff, etc.), encouragement of vegetated cover, and installation of sediment traps.

References

Byaruhanga A, Sande E, Plumptre A, Owiunji I, Kahindo C, editors. International species action plan for the Grauer's Swamp-warbler *Bradypterus graueri*. BirdLife international, Nairobi, Kenya and royal society for the protection of birds, Sandy. 2006.

Deuse P. Contribution à l'étude des tourbières du Rwanda et du Burundi. Esquisse de la végétation des tourbières du Rwanda et du Burundi. In: Treizième Rapport d' Institut pour la Recherche Scientific en Afrique. Butare; 1966. p. 53–115.

Grosse-Brauckmann G, Hacker E, Tüxen J. Moore in der Bodenkundlichen Kartierung. Telma. 1977;7:39–54.

Hategekimana S. La dégradation actuelle du Marais de Rugezi: une catastrophe écologique. Mémoire de Licence, Université Nationale du Rwanda. 2005; 115 pp.

Hategekimana S, Twarabamenye E. The impact of wetlands degradation on water resources management in Rwanda: the case of Rugezi Marsh. In: Harding B, Devisscher T, editors. Review of the economic impacts of climate change in Kenya, Rwanda and Burundi. Nairobi, Kenia: African Conservation Centre; 2007.

Joosten H. The IMCG global peatland database; chapter Rwanda. 2004. www.imcg.net/gpd/gpd.htm. Accessed 24 Jan 2016.

Linström A, Grundling PL, Pretorius L, Nsabagasani C, Rift A. A basic broad scale assessment of Rugezi Marsh and its carbon capacity: final report, 74 pp. Bloemfonteint: Centre for Wetland Research and Training (WETREST); 2013.

Macfarlane DM, Dickens J, Von Hase F. Development of a methodology to determine the appropriate buffer zone width and type for developments associated with wetlands, watercourses and estuaries, INR Report No: 400/09. Concord: Institute of Natural Resources; 2009.

MINALOC. Mission d'étude de conservation et de gestion intégrée du Marais de Rugezi. Rapport provisoire. Helpage-Rwanda, Kigali: Ministry of Local Government, Community Development and Social Affairs; 2004. 122 pp.

MINITERE. Etudes relatives à la protection intégrée et conservation des resources naturelles des zones humides critiques du Rwanda. Évaluation de la diversité biologique des zones humides, Experco des Grands Lacs. Kigali: Ministry of Environment, Lands and Mines; 2003. 78 pp.

Nsabgasani C. Status of grey crowned crane in Rugezi Marsh, Rwanda. Survey report. Kigali: Centre d'echange de Convention de la diversite biologique; 2010.

Ramsar Convention Secretariat. The list of wetlands of international importance. 2012. http://www.ramsar.org/pdf/sitelist.pdf. Accessed 24 Jan 2016.

REMA. Rwanda state of environment and outlook report. Chapter 7: water and wetlands resources. Rwanda Environment Management Authority, P.O. Box 7436 Kigali, Rwanda. 2009. http://www.rema.gov.rw/soe/

Banc d'Arguin (Mauritania)

108

Antonio Araujo and Pierre Campredon

Contents

Introduction	1320
Hydrology	1322
Coastal and Marine Ecosystems	1323
Coastal and Marine Biodiversity	1324
Benthos	1324
Fishes	1325
Sea Turtles	1325
Birds	1326
Marine Mammals	1328
Conservation Status	1328
Ecosystem Services	1328
Threats and Future Challenges	1329
References	1330

Abstract

The Banc d'Arguin refers to a vast coastal wetland located between Cap Blanc (20°46′N – 17°02′W) and Cap Timiris (19°23′N – 17°02′W) in Mauritania, West Africa. The area presently supports over one million wintering Palearctic waders and other shorebirds. It's one of the three most important hotspots along the East Atlantic Flyway between the Arctic and South Africa. Primary production in the Golfe d'Arguin depends mostly on the extensive pristine sea grass beds, but the area is also influenced by the Canary Current Upwelling. Banc d'Arguin is recognized worldwide for its marine biodiversity and especially for its globally

---

A. Araujo (✉)
MAVA – Fondation pour la Nature, Dakar, Senegal
e-mail: antonio.araujo@fondationmava.org

P. Campredon (✉)
Conseiller Technique, IUCN Guinea Bissau, Bissau, Guinea-Bissau
e-mail: Pierre.campredon@iucn.org

important wild bird populations. It holds significant cultural values and contributes to the fisheries economy in Mauritania. The extensive area of sea grass, mudflats, sand banks, channels, islands and islets are included in the Parc National de Banc d'Arguin, one of the largest marine protected areas in Africa listed as a Ramsar Site and inscribed on the World Heritage List. Threats include fishing, offshore oil and gas prospecting, rural migration to coastal areas, pollution from oil and gold mines, and climate change. Important changes are to be expected on species composition and richness if external pressures continue.

Keywords

Golfe d'Arguin · Canary Current Upwelling · Palearctic waders · Seagrass beds · Benthic macrofauna · National Park of Banc d'Arguin · Ramsar · World Heritage Site · Imraguen fishermen

Introduction

The Banc d'Arguin refers to a vast coastal wetland located between Cap Blanc (20°46′N – 17°02′W) and Cap Timiris (19°23′N – 17°02′W) in Mauritania, West Africa. The area is named after a series of underwater rocky shoals of the same nature as the coastal sandstone hillocks and low plateaus located not far inland from the coast line. The inner part of the basin known as Golfe d'Arguin (Gulf of Arguin) was a landlocked freshwater lake until ca. 8,700 years BP (Aleman et al. 2014). The lake was flooded by the sea following the last glaciation and a new marine-estuarine environment developed subsequently. The situation gradually evolved from 6,500 years BP when sea-level stabilization took effect (Aleman et al. 2014), but the area became progressively aridified, especially from 3,500 BP (Renssen et al. 2006).

The coastal shoals defend the inner part of the gulf from rough seas, thus enabling the preservation of a vast shallow coastal wetland with channels, mudflats, sand banks, islands, and islets (see Fig. 1). The coastline comprises a succession of sand beaches, bays, and rocky capes bordered to the east by the Sahara Desert. To the West, not far from the shore, some of the fifteen sandy islands and islets play an important role as the breeding grounds of several species of colonial sea and water birds. Between the coast and the islands, huge areas of intertidal mudflats interrupted by channels are the main feeding areas for over two million wintering Palearctic waders and other shorebirds. Edging the gulf to the North, the sea cliffs of Cap Blanc and Guerguerat hold the most important colony of Atlantic monk seals *Monachus monachus* in the world. The important biodiversity in the area is partly explained by its latitude. The northern edge of the gulf represents a biogeographical limit, a meeting point where several Palearctic and afro-tropical species of animals and plants coexist along their respective limits of distribution.

Primary production in the Golfe d'Arguin depends mostly on the extensive pristine seagrass beds, but the area is also influenced by the Canary current upwelling (CCU). The existence of an ancient vast hydrographic freshwater network and a

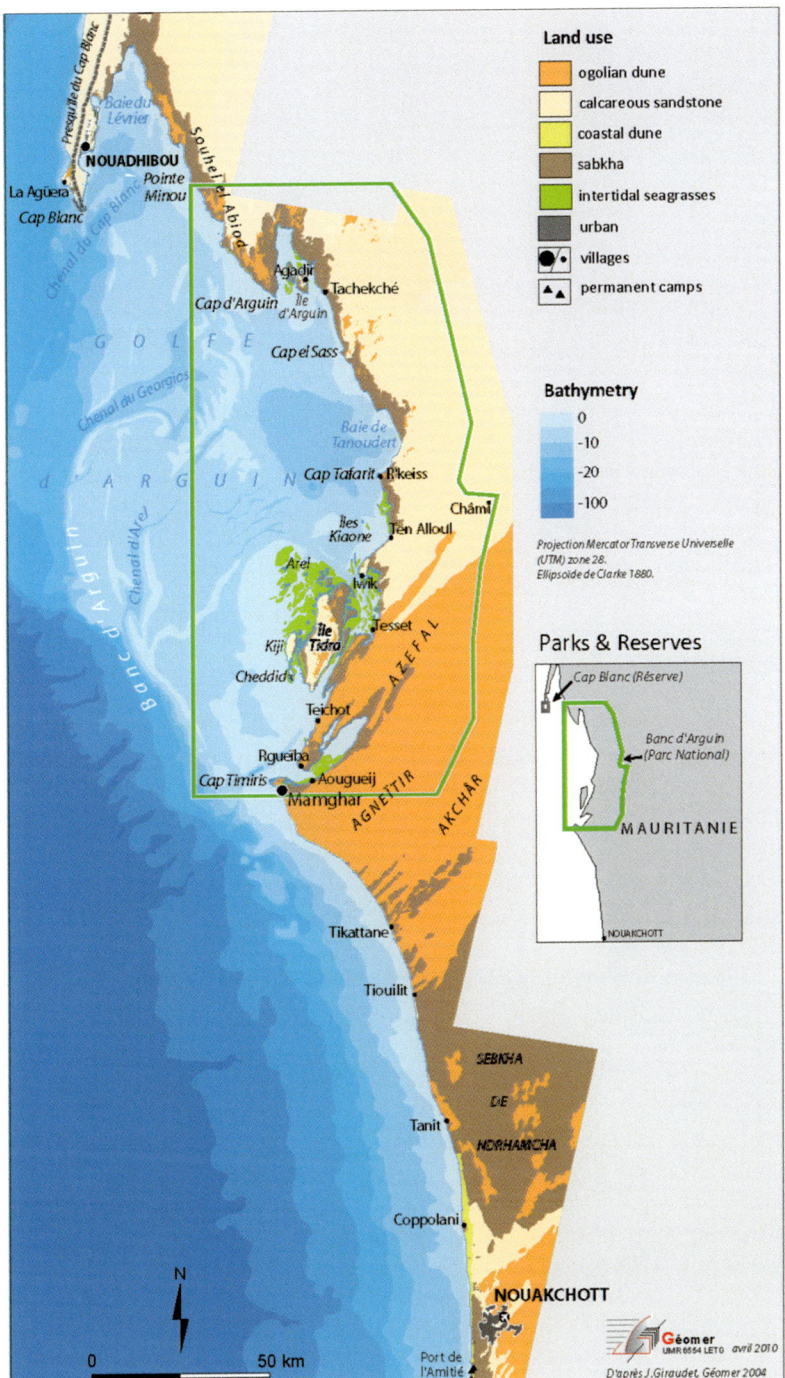

Fig. 1 The Mauritanian coastal zone from Nouadhibou to Nouakchott, including the Golfe dÁrguin and Banc d'Árguin National Park (*inset*) (Adapted from Fig. 1: Gourmelon et al. (2006), with the permission of the publisher)

tropical climate are documented by the presence of several species isolated from their present distribution areas, such as the mangrove species *Avicennia africana* and the blackchin tilapia *Sarotherodon melanotheron*. These are living witnesses of the ancient estuarine environment, which was once similar to other estuarine areas and tropical coastal lagoons still covered by mangroves further south in West Africa.

The entire area of sea grass, mudflats, channels, and islands, equivalent to half of the Golfe d'Arguin (some 6,000 km^2), is under protection. Threats around the National Park are, however, important. Nouadhibou, the second largest city in Mauritania, has an internationally important ore port only 30 km NW from the park's NW corner. Offshore oil prospecting is underway in the same areas where hundreds of international trawlers operate, not too far from the western limit of the park. An important gold mine and a new city are located close to the park's eastern border.

The area has been inhabited for thousands of years by human populations of different origins. The Imraguen are a small resident community of fishermen presently spread over eight small villages inside the marine protected area. They were already present when the first Europeans of Portuguese origin arrived in the area in the mid-fifteenth century, and they are still the only fishermen entitled to operate in the National Park using sailing boats.

Hydrology

Most of the area is characterized by shallow sea depths of less than 10 m. With the incoming tides, the offshore water masses enter Banc d'Arguin through the channels at its northwestern border and leave via its southwestern border after a residence time of 18–34 days depending on the winds. These waters, which are extremely rich in nutrients, descend into the southern canyons on their way out of Banc d'Arguin and come back to surface later, thus representing an important contribution by Banc d'Arguin to the coastal upwelling.

The tides are semi-diurnal with a mean tidal range that varies from 1 m during neap tides to 2 m during spring tides, generating currents of up to three knots in certain channels. Circulation of water masses is also influenced by north trade winds driving the CCU.

Fresh water inputs into the gulf are much reduced from historical levels. The hot air masses emanating from the Sahara cool down when they come into contact with the cold waters of the Canary current, thus stabilizing and reducing precipitation to minimum levels (on average less than 35 mm/year in Banc d'Arguin). The climate was much more humid in Mauritania 5,000 years ago when the Gulf of Arguin was still receiving important fresh water flows (deMenocal and Tierney 2012).

The Arguin Basin is interpreted as a land-locked freshwater lake during the period of post-glacial sea-level rise, corresponding to wet climatic conditions. The inner part was flooded ca. 8,700 years ago when the sea level reached the sills. The filling then corresponded to a marine–estuarine environment. The climatic aridification and the sea level stabilization from 6,500 BP onwards has resulted in the surface sediments of the Banc d'Arguin to be composed of Aeolian sand combined with a significant marine biogenic carbonate fraction (Aleman et al. 2014). The wind is the dominant key factor for continental sediment transport that feeds the gulf, building up and maintaining the mudflats and other key substrates but the global origin of these sediments, their mineralogical composition, iron content, and nutritive potentiality is not known (Valance et al. 2011).

The existence of several deltas filled with fluviatile sediments all along the coast is plausible until 4,800 BP (Aleman 2010). Thereafter the climate changed towards the present Saharan conditions, alternating arid and humid periods. Presently, the ancient wadis are filled and their basins revealed only in rainy years.

Coastal and Marine Ecosystems

The information available concerning phytoplankton shows that the influence of the CCU on primary production is relatively modest in the inner part of Banc d'Arguin (Berghuis et al. 1993; Clavier et al. 2010). Nutrient concentrations decrease from Cape Blanc in the NW to the area of seagrass beds and mudflats in the SE. The taxonomic composition of phytoplankton also changes and the food web in the intertidal flats seem to be fueled mostly by local ^{13}C enriched sources. These conclusions are supported by the distribution of fish, for which two major ecosystem realities are clearly apparent, one in the North where phytoplankton is a key issue (the Saharan coast South of Dakhla in Morocco and North Mauritania hosts one of the four most important and permanent upwelling on the planet, supporting remarkable phytoplankton blooms) and one in the south where mudflats and seagrass beds are mostly responsible for primary production. Most pelagic fish species occur in the North and most demersal sediment feeders occur in the South (Sidi MOTOS 2007; Braham CBOI 2011; 2012).

The infra-littoral zone is mostly covered by fine sand soils and on a minor scale by muddy sand substrates. The seagrass species *Cymodocea nodosa* seems to comprise the most important and extensive habitat, covering over 300 km^2 of this area. The main substrate in the intertidal zone (some 500 km^2) is muddy sand but fine sand soils are also present. The seagrass species *Zostera noltii* is dominant there, covering some 80% of the intertidal area, but *Halodule wrightii* is also present.Seagrass beds are extremely important habitats for cyanobacteria and different species of epiphytic algae.

Important fields of the algae *Caulerpa sp.* as well as some maërl (corraline algae) beds were recently detected West of Tidra and Kiji islands. The yellow-green algae *Vaucheria sp.* is also quite common in some areas and represents an important habitat for certain species of waders (Charadriiformes).

Part of the vast network of channels, sub-, and infra-tidal areas are covered by *Cymodocea nodosa* sea-grass beds where fishes find refuge from predators like dolphins, sharks, pelicans *Pelecanus onocrotalus*, and cormorants. In some of the upper areas of the inter-tidal zone settlements of small cordgrass *Spartina maritima* (southern distribution limit) and mangroves *Avicennia africana* (northern distribution limit) may be found. The mangrove trees rarely over 5 m high are sometimes used by cormorants and egrets to build their nests. The presence of diatoms along the shoreline favor the establishment of large communities of fiddler crabs *Uca tangeri* which are important prey for species like Eurasian curlew *Numenius arquata* and whimbrel *Numenius phaeopus*.

Tidra is the largest island in the Gulf (30 kms long). Other smaller islands like Kiji (8 km long), Agadir (5 km long), the Large Kiaone (800 m long), and Arel (500 m long) are important breeding places for a number of marine and shore bird species like terns, *e.g., Sterna maxim*, Eurasian spoonbills *Platalea leucorodia*, or greater flamingos *Phoenicopterus roseus* and welcome significant high tide roosts of different species of waders. Some of these sand-stone islands covered with sand have vegetation, and Tidra holds an important population of dorcas gazelle *Gazella dorcas*.

Coastal and Marine Biodiversity

Benthos

A benthic species inventory reveals the tidal flats as remarkably rich with 131 species (Wolff et al. 1993). Localized sampling throughout an annual cycle between 2011 and 2012 (Salem et al. 2014) seems to show that bivalves are much more common than gastropods. The density of the bivalve *Loripes lucinalis* may reach 5,000 individuals m^{-2}. This species dominated the assemblage (58% of total number), followed by *Dosinia isocardia* (10%), *Senilia senilis* (8%), and the gastropod *Gibbula umbilicalis* (6%). In this study, the lowest densities were reached in late spring (May) and summer (August), whereas highest densities occurred in autumn (October). The results obtained suggest that by the time migratory waders return in high numbers to Banc d'Arguin in autumn, the densities of benthic animals are at a peak. Other than the mollusks, some species of annelid worms of the taxonomic groups polychaeta and oligochaeta and isopod crustaceans are also present and important within the local food chains.

Although estimates of density provide an indication of the amount of energy available to migratory waders, a more important number for comparative purposes is ash-free dry mass (AFDM) m^{-2}. Van der Geest (2013) reported on the ornithological research undertaken in the past decades on Banc d'Arguin and several studies collected data on the benthic macrofauna biomass in the intertidal flats. These studies established that, contrary to initial expectation, food resources potentially available to shorebirds were relatively low, 7.6–28.6 g AFDM m^{-2}, compared to other tidal flats in the world where biomass values of 100 g AFDM m^{-2} are not uncommon. This value becomes even smaller, ranging between 2.9–8.9 g AFDM per m2, when the biomass represented by *Senilia senilis* is excluded.

The AFDM has been shown to vary over time. Salem et al. 2014 reported an average biomass of 32 g AFDM m^{-2} of which the West-African bloody cockle *Senilia senilis* made up three-quarters, *Loripes lucinalis* 16%, *Gibbula umbilicalis* 2%, and *Dosinia isocardia* 1%. This is twice Wolff et al.'s (1987) estimate of 14.5 g

AFDW m^{-2} of which 5.5 g was for *Senilia senilis* (formerly *Anadara senilis*). However, during the 1980s and 1990s juveniles of *Senilia* were absent from the population. Wolf and Smit (1990) commented "...The population of this species consists of mainly very old (10–30 years) individuals, which is most probably not the case for the majority of the other tidal flat species." The arrival of rains in the beginning of the 2000 decade saw an initial decrease in AFDM with the replacement of old by juvenile *Senilia*, and subsequent proportional increase in the AFDM as the juveiles grew, contributing to one third of the AFDM in 2006 (unpublished data) and further increasing to three fourths of the AFDM in 2011/2012 (Salem et al. 2014).

Fishes

Golfe d'Arguin is a critical area for an important number of fish species and with a vast marine protected area covering 11,200 km^2, the National Park of Banc d'Arguin (PNBA), is largely responsible for the sustainability of Mauritanian fisheries. Of the 145 fish species identified in the National Park, 39 occur exclusively in Mauritania within its borders.

The family Mugilidae (mullets) is well represented in the area, especially *Mugil cephalus*. This species has been sustainably fished for centuries during its southward migration by the local Imraguen population. A unique on-foot cooperative fishing technique accomplished on shallow waters along the coast has become famous due to the usual presence of dolphins hunting the mullet shoals in perfect symbiosis with the fishermen.

The area is in fact the only large estuarine ecosystem between Europe and tropical Africa and is therefore also very important for elasmobranchs (sharks, rays, and skates). Important populations of big sharks and rays like tiger shark *Galeocerdo cuvier* and saw fish *Pristis* sp. that were once abundant may well be considered extirpated in spite of all conservation measures established since 1976. Significant populations of hammerhead sharks *Sphyrna* sp., milk sharks *Rizonopriodon acutus*, eagle rays *Rhinoptera marginatai*, and guitarfish *Rhinobatos cemiculusi* are species typical of estuarine ecosystems and intertidal mudflats. The area is also very important for other species of commercially important scale fishes such as the meagre *Argyrosomus regius,* for which the only spawning area known in Mauritania is located in the Gulf. The abundance of two species of tilapia and catfish witness the brackish character of the Gulf in the past.

Sea Turtles

Several species of sea turtles occur in the Gulf, and the vast pristine seagrass beds are an irreplaceable habitat for large concentrations of green turtles *Chelonia mydas* both immature and adult. The satellite tracking of a dozen females breeding in the Archipelago of the Bijagós in Guinea-Bissau clearly showed strong links between the globally important colony of the island of Poilão, one of the most important in the

Table 1 Summary of available information on colonial breeding birds in Gulf d'Arguin (Data from Naurois 1969 (1959–1965), Campredon 1987 (1984–1985), Hafner et al. 1999 (1997–1999), Zwarts et al. (1998) and Veen and Dallmeijer no date (2004/2005/2007))

Species	Breeding pairs			
	Average	Min	Max	Tendency
White-breasted cormorant	4,688	1,400	8,130	Increasing
Long-tailed cormorant	2,012	900	2,900	Decreasing
Great white pelican	2,192	400	3,800	Increasing
Western reef egret	862	280	1,900	Decreasing
Mauritanian heron	1,738	1,070	2,400	Decreasing
Banc d'Arguin spoonbill	2,516	1,500	4,990	Stable
Greater flamingo	11,463	6,600	16,500	Stable
Slender-billed gull	1,411	870	1,780	Increasing
Grey-headed gull	35	9	70	Decreasing
Gull-billed tern	978	180	1,950	Decreasing
Caspian tern	5,691	1,800	10,900	Increasing
Royal tern	12,526	5,630	19,353	Increasing
Common tern	219	15	900	Decreasing
Bridled tern	463	48	900	Decreasing
Little tern	91	10	210	Decreasing

Atlantic for this species, and Banc d'Arguin. Turtle breeding is rare and very occasional (Godley et al. 2003; Catry et al. 2009).

Birds

The importance of the islands of the Golfe d'Arguin for birds was first highlighted by Portuguese sailors in the fifteenth century, but its ornithological significance was not revealed until the late 1950s by the French ornithologist René de Naurois (1959) in his reporting of vast colonies of seabirds and huge concentrations of waders. Since then several censuses have been organized, at the beginning of the 1960s and the 1980s (Naurois 1969; Campredon 1987), Tour du Valat at the end of the 1990s (Hafner et al. 1999), and recently in 2004, 2005, and 2007 (Veen and Dallmeijer no date). The large annual variation in the number of breeding pairs within comprehensive annual counts is still unexplained. It may be due to subregional dynamics, as most of these species also breed in neighboring countries, but a comprehensive regional count has never been performed. A minimum count of 37,340 pairs (of all colonial breeding species) was recorded in the 1960s, and the maximum number was estimated in 1997 (54,693 breeding pairs). A summary of the information collected during these counts is presented in Table 1. The Mauritanian heron *Ardea monicae* is now considered a new species by some authors (Erard et al. 1986; Isenmann 2006), and the subspecies of Eurasian spoonbill *Platalea leucorodia balsaci* breeding in Banc d'Arguin is also endemic (referred to in Table 1 as Banc d'Arguin

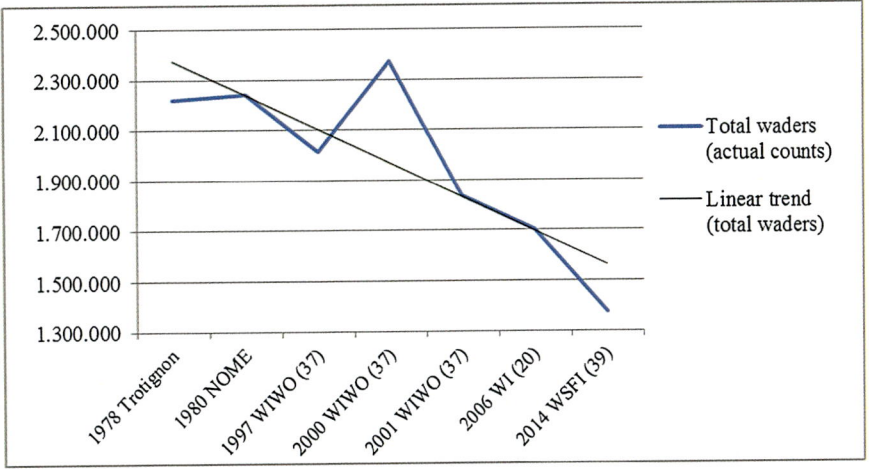

Fig. 2 Comprehensive wader counts in Banc d'Arguin between 1978 and 2014. The number of sites counted is indicated in parentheses (From Altenburg et al. (1982), Hagemeijer et al. (2000), Diawara and Overdijk (2007) and unpublished data retained by Parc National du Banc d'Arguin)

spoonbill), but some mixed pairs of *balsaci/leucorodia* have been observed. A couple of kelp gulls *Larus dominicanus* were confirmed breeding in 2004.

Few trees occur along the coast and no trees exist on the islands. Most of the birds breed on the bare soil or on low halophyte vegetation. Bridled terns *Onychoprion anaethetusi* breed inside rock holes or under loose boulders. Long-tailed cormorant *Phalacrocorax africanus* and reef egrets *Egretta gularis* may breed on the rare mangroves.

Most of the species start breeding in spring from March/April through to the end of the summer, but larger fish-eating birds (great white pelican *Pelecanus onocrotalus* and great cormorant *Phalacrocorax carbo*) breed from September/October onwards.

The Mauritanian heron breeds all around the year, but spring is the most important breeding period for this species. Royal terns are usually active breeding from March/April to November.

Since the end of the 1970s onwards, the global importance of Banc d'Arguin for migratory waders has been recognized. Several comprehensive counts performed by different research teams established total numbers slightly over two million which has decreased to less than 1.5 million birds wintering in the area by 2014 (Fig. 2).

No direct anthropogenic impacts have been noticeable in Banc d'Arguin since the 1970s, and the area is apparently unchanged. However, in spite of natural counting bias affecting different counts, an important overall decrease in the number of waders present in January is evident. The true causes of decline are not known. They may well be local due to yet unexplained natural causes or events in the arctic and changes in different key wetlands located along the flyway suffering from significant human disturbance such as the Wadden Sea.

Whether the strong evidence for food limitation in red knot *Calidris canuta* is a more general phenomenon for the other shorebirds wintering at Banc d'Arguin remains a topic for further research (Van der Geest 2013).

Marine Mammals

The two most common species of marine mammals inside Banc d'Arguin are the Atlantic humpback dolphin *Sousa teuszii* and the common bottlenose dolphin *Tursiops truncatus*. The first species is always observed in small groups but seems to be widespread in the area. The bottlenose dolphin is much more conspicuous and may form larger groups, sometimes including a few dozen individuals. Up to 100 years ago these dolphins used to fish along the shore with the Imraguen fishermen. Killer whales *Orcinus orca* are occasionally observed inside the Park close to the coast as well as immature monk seals *Monachus monachus* from the largest colony known for this species that is currently under protection, situated North of Cape Blanc some 40 km away from the northern limit of the Park. Several other species of marine mammals are commonly observed within the Gulf outside Banc d'Arguin.

Conservation Status

Half of the Golfe d'Arguin became a National Park (Parc National de Banc d'Arguin, PNBA) in 1976. With a total area of 11,200 km^2 of which half is terrestrial and the other half coastal and marine, the Park is one of the largest marine protected areas in Africa. The natural reserve of Cap Blanc is legally and administratively attached to the national park. It was created in 1986 close to the city of Nouadhibou, around Cap Blanc to protect a small population of monk seals and important roosts of seabirds.

PNBA was designated a wetland of international importance under the Ramsar Convention in 1982, and it was inscribed on the World Heritage List under Natural Criteria IX and X in 1989. The park's management policy implemented by Mauritanian authorities elicited an honorific WWF designation as a "Gift to the Earth" in 2000.

Ecosystem Services

The importance of Banc d'Arguin for marine biodiversity is recognized worldwide especially for its globally important wild bird populations. The significance of the area is otherwise recognized for its contribution to the economy of fisheries in Mauritania and the cultural values it holds, namely the Imraguen maritime culture, sailing know-how, and traditional fishing and fish preservation techniques.

The economic value of the area is difficult to determine, but some investigation has been conducted on this subject regarding PNBA. The total economic value (TEV) of the park has been estimated at slightly over 300 million euros, of which approximately one million concern direct use value (DUV) (Fernandez 2009). Nonuse value was then estimated at 80% of direct and indirect use values, and the cost-benefit analysis largely encourages keeping the legal status of the marine protected area (Binet et al. 2013).

Perfectly aware of the important ecosystem services supplied by the park, the European Union has provided annual support of one million euros to the park's budget through the Mauritania Fisheries Agreement. Although being small compared to the revenues generated to the European fleet in Mauritania, this agreement marks an important step towards the recognition of ecosystem services by marine protected areas. It signals an important turn in foreign fishing policies, opening the door to other similar initiatives elsewhere in the world. In the future, fishing agreements may well represent an important source of financing for marine biodiversity conservation and MPAs.

Threats and Future Challenges

Human stresses influencing the Golfe d'Arguin and the park are to be found both inside and outside its limits. The local community consists of some 500 families (some 1,200 people in total). These Imraguen fishermen are the only people authorized to exploit the natural resources within the park in a traditional and sustainable manner and exclusively using sailing boats whose number is limited to 114. Motorized vessels are forbidden with the exception of maritime surveillance and research which makes the marine part of the park one of the largest areas of the planet without motorboats (6,000 km^2). Nevertheless, Imraguen fisheries became commercial at the end of the 1980s, and shark populations have been targeted by the sailing fleet for more than 20 years. Big sharks that were quite abundant in the past as well as saw fish disappeared from the area during the 1990s. To mitigate the negative impacts of the Imraguen fleet on rays and sharks is one of the most important conservation objectives of the park.

External pressure has been increasing, especially since the inauguration of an important road following the terrestrial border of the park to connect the two major coastal cities in Mauritania, Nouadhibou in the north to the capital Nouakchott, some 500 km to the south.

The industrial fishing fleet that once inflicted damage to the park has been banned since 2005. However, the small-scale fleet consisting of over 6,000 motorized pirogues (mostly light wooden boats up to 10 m long, small enough to be easily taken onto land and move through shallow water) keeps putting pressure upon the park, even though the results of maritime surveillance are outstanding and the authorities have remained committed so far with respect to the legal framework. Most of the fish stocks are overexploited outside the park, and the pressure of the fishing fleet will keep increasing.

A new city has recently been founded between Nouakchott and Nouadhibou, and important gold mines are being exploited in the vicinity of the park's eastern border. Offshore oil and gas prospecting is underway not far from its western limit, and much of the pollution coming from Nouadhibou is drawn into the park by the sea currents and the wind, namely significant amounts of iron dust coming from the industrial port facilities.

As a result of important socioeconomic changes, rural populations are migrating to coastal areas where fishing is the most important economic activity. In 1960, the largest villages had 10,000 inhabitants. The population of Nouakchott was estimated to be around 140,000 in 1980, slightly over one million in 2013 and is predicted to be close to three million in 2030 (Choplin and Vincent 2014).

Last but not least, climate change impacts are underway and are particularly visible inside the park, where most of the villages are inaccessible because of the tides several times a year – this was not the case 10 years ago. Ecosystems are being impacted, and important changes are to be expected on species composition and richness. Some of the islands where colonial birds breed are already affected, losing land area, and some of the colonies are already being systematically flooded during spring tides, including the endemic subspecies of spoonbill.

References

Aleman N. Etude du remplissage sédimentaire de la partie interne du Banc d'Arguin (Mauritanie). Mémoire de Master 2ème année Environnement Océanique et Côtier. Université de Perpignan Via Domitia. UFR Sciences Exacts et Expérimentales; 2010. 44 p.

Aleman N, Certain R, Dia A, Barusseau J, Courp T. Post-glacial filling of a semi-enclosed basin: the Arguin Basin (Mauritania). Mar Geol. 2014;349:126–35.

Altenburg W, Engelmoer M, Mes R, Piersma T. Wintering waders on the Banc d'Arguin, Mauritania. Report of the Netherlands Ornithological Expedition 1980. Communication N° 6, Wadden Sea Working Group; 1982.

Berghuis EM, Duineveld GCA, Hegeman J. Primary production and distribution of phytopigments in the water column and sediments on the upwelling shelf off the Mauritanian coast (Northwest Africa). In: Wolff WJ, van der Land J, Nienhuis PH, de Wilde PAWJ, editors. Ecological studies in the coastal waters of Mauritania. Hydrobiologia. 1993;258:81–93.

Binet T, Failler P, Chavance PN, Mayif MA. First international payment for marine ecosystem services: The case of the Banc d'Arguin National Park. Mauritania Glob Environ Change. 2013;23:1434–43.

Braham CBOI. Analyse de l'activité des pêches (efforts et captures) au niveau du Parc National du Banc d'Arguin en 2010. Programme de Suivi des Pêcheries Imrague (PSPI). Rapport de l'Institut Mauritanien des Recherches Océanographiques et des Pêches. IMROP; 2011. 38 p.

Braham CBOI. Description de l'activité de pêche au niveau du Parc National du Banc d'Arguin en 2011. Programme de Suivi des Pêcheries Imrague (PSPI). Rapport de l'Institut Mauritanien des Recherches Océanographiques et des Pêches. IMROP; 2012. 34 p.

Campredon P. La reproduction des oiseaux d'eau sur le Parc National du Banc d'Arguin (Mauritanie) en 1984/85. Alauda. 1987;55:187–210.

Catry P, Barbosa C, Paris B, Indjai B, Almeida A, Limoges B, Silva C, Pereira H. Status, ecology, and conservation of sea turtles in Guinea-Bissau. Chelonian Conserv Biol. 2009;8:150–60.

Choplin A, Vincent F. Document de contexte. Nouakchott - L'avenir pour défi. Adaptation et mutation d'une ville vulnérable. Atelier international de maitrise d'oeuvre urbanine 26 Avril – 9 Mai, 2014, Nouakchott, Mauritanie. Cergy-pontoise Cedex: Les Ateliers de Cergy; 2014. 130 p.

Clavier J, Chauvaud L, Chauvaud S. Écologie benthique du banc d'Arguin. Action scientifique n° 1: Production primaire benthique des zones intertidales. Rapport de la mission du 21 janvier au 2 février 2010 à Iwik. Projet d'Approfondissement des Connaissances scientifiques des écosystèmes du golfe du Banc d'Arguin – PACOBA; 2010. 24 p.

deMenocal PB, Tierney JE. Green Sahara: African humid periods paced by Earth's orbital changes. Nat Educ Knowl. 2012;3(10):12.

Diawara Y, Overdijk O. Wader count in the Banc d'Arguin National park (Mauritania). In: Diagana CH, Dodman T, editors. Coastal waterbirds along the West African Seaboard, January 2006. Wageningen: Wetlands International; 2007.

Erard E, Guillou JJ, Mayaud N. Le Héron Blanc du Banc d'Arguin A*rdea monicae*. Ses affinités morphologiques. Son histoire. Alauda. 1986;54:161–9.

Fernandez S. Vers une estimation de la valeur économique totale du Parc National du Banc d'Arguin. [Master Développement durable dans les pays en développement et en transition]. [Clermont-Ferrand I (FR)]: Université d'Auvergne, École d'Économie; 2009.

Godley BJ, Almeida A, Barbosa C, Broderick A, Catry P, Hays G, Indjai B. Using satellite telemetry to determine post-nesting migratory corridors and foraging grounds of green turtles nesting on Poilão, Guinea-Bissau. Final Project Report, unpublished. 2003.

Gorumelon F, Robin M, Creuseveau JG, Pennober G, da Silva AS, Affian K, Hauhouot C, Pottier P. Contraintes d'utilisation des technologies de l'information géographiqu pour la gestion intégrée des zones côtières en Afrique. Vertigo. 2006;7(3):1–14.

Hafner H, Pineau O, Kayser Y, Gueye A, Sall MA, Lamarche B, Lucchesi L, Johnson A. Suivi des colonies d'oiseaux d'eau du Parc National du Banc d'Arguin (République Islamique de Mauritanie) et formation d'homologues mauritaniens. Département Ecologie des Oiseaux, Station Biologique de la Tour du Valat; 1999. Unpublished. 10 p.

Hagemeijer EJM, Smit CJ, de Boer P, van Dijk AJ, Ravenscroft N, van Roomen MWJ, Wright M. Wader and waterbird census at the Banc d'Arguin, Mauritania, January 2000. Beek-Ubergen: WIWO; 2000. 146 p. Report 81.

Isenmann P. The Birds of the Banc d'Arguin. Parc National du Banc d'Arguin – Mauritania. Arles: La fondation Internationale du Banc d'Arguin; 2006. 190 p.

Naurois R. Peuplements et cycles de reproduction des oiseaux de la côte occidentale d'Afrique, Mémoire Muséum d'Histoire Naturelle, Série A, Zoologie, vol. 56. Paris: Editions du Muséum; 1969. p. 1–312.

Naurois R. Premieres recherches sur l' avifaune des lles du Banc d'Arguin (Mauritanie). Alauda. 1959;27:241–308.

Renssen H, Brovkin V, Fichefet T, Goosse H. Simulation of the Holocene climate evolution in Northern Africa: the termination of the African Humid Period. Quat Int. 2006;150:95–102.

Salem MVA, van der Geest N, Piersma T, Saoud Y, van Gils JA. Seasonal changes in mollusc abundance in a tropical intertidal ecosystem, Banc d'Arguin (Mauritania): testing the depletion by shorebirds' hypothesis. Estuar Coast Shelf Sci. 2014;136:26–34.

Sidi MOTOS. Synthèse préliminaire des travaux scientifiques menés par l'IMROP: période 1997–2006. Rapport projet Régulation de l'Accès aux Ressources naturelles et Surveillance dans le PNBA (RARES). IMROP; 2007. 39 p.

Valance A, Dupont P, Moctar OA. Action : Transport Éolien. Projet d'Approfondissement des Connaissances Scientifiques du Banc d'Arguin (PACOBA). 2011.

Van der Geest M. Multi-trophic interactions within the seagrass beds of Banc d'Arguin, Mauritania: a chemosynthesis-based intertidal ecosystem [disertation]. Groningen: University of Groningen; 2013. 253 p.

Veen J, Dallmeijer H. Le suivi des colonies d'oiseaux piscivores sur le Banc d'Arguin, Mauritanie. Unpublished. 2004, 2005, 2007.

Wolff WJ, Duiven AG, Duiven P, Esselink P, Gueye A, Meijboom A, Moerland G, Zegers J. Biomass of macrobenthic tidal flat fauna of the Banc d'Arguin, Mauritania. In: Wolff WJ, van der Land J, Nienhuis PH, de Wilde PAWJ, editors. Ecological Studies in the Coastal Waters of Mauritania. Hydrobiologia. 1993;258:151–163.

Wolff WJ, Gueye A, Meijboom A, Piersma T, Sall MA. Distribution, biomass, recruitment and productivity of Anadara senilis (L.) (Mollusca: Bivalvia) on the Banc d'Arguin, Mauritania. Neth J Sea Res. 1987;21:243–53.

Wolff WJ, Smit CJ. The Banc d'Arguin, Mauritania, as an environment for coastal birds. ARDEA. 1990;78(1–2):17–38.

Zwarts L, Kamp J, Overdijk O, Spanje TM, Veldkamp R, West R, Wright M. Wader count of the Banc d' Aguine, Mauritania in January February 1997. Wader Study Group Bull. 1998;86:53–69.

Bijagos Archipelago (Guinea-Bissau)

Pierre Campredon and Paulo Catry

Contents

Introduction	1334
Hydrology	1335
Wetland Ecosystems	1336
Biodiversity	1337
Conservation Status	1338
Ecosystem Services	1338
Current Threats and Future Challenges	1339
References	1340

Abstract

The Bijagós archipelago (11°14′N–16°02′W) emerges from the shelf off Guinea-Bissau, not far from the mainland coast. It is the only active deltaic archipelago on the Atlantic coast of Africa. It includes 88 islands and islets, some of which are mangrove islands flooded during spring tides, and some of which have several permanently emerged portions linked by intertidal mangroves. Twenty-one of the islands are permanently inhabited by communities of the Bijagós ethnic group. The islands are separated by a network of channels, and in general are surrounded by mangroves and extensive mud and sand flats, which together represent the

---

P. Campredon (✉)
Conseiller Technique, IUCN Guinea Bissau, Bissau, Guinea-Bissau
e-mail: pierre.campredon@iucn.org

P. Catry (✉)
MARE – Marine and Environmental Sciences Centre, ISPA – Instituto Universitário, Lisbon, Portugal
e-mail: paulo.catry@gmail.com

© Springer Science+Business Media B.V., part of Springer Nature 2018
C.M. Finlayson et al. (eds.), *The Wetland Book*,
https://doi.org/10.1007/978-94-007-4001-3_158

most extensive intertidal area in Africa. Sediments originate mostly from the Corubal and Geba rivers and are deposited and moved around by a complex system of currents and wave action. The region is under the seasonal influence of the upwellings linked to the Canary current but also benefits from an important input of organic matter and nutrients via continental runoff and from the productivity of the mangroves. The archipelago and its surrounding flats and channels, covering an area of approximately one million hectares, harbors remarkable biodiversity which justifies its classification as a Biosphere Reserve (1996) and as a Ramsar Site (2014) and has prompted the creation of three marine protected areas. The traditional management of the natural resources of this area by the animistic Bijagós ethnic group is based on strong cultural and religious values and has allowed the long-term conservation of the site, for example, through the effective protection of sacred islands and islets and sacred forest patches where initiation rites take place, which remains true oasis of biodiversity.

Keywords

Deltaic archipelago · Mangrove · Intertidal flat · Upwelling · Biosphere reserve · Migratory waterbird · Fishery

Introduction

The Bijagós archipelago (11°14′N–16° 02′W) emerges from the shelf off Guinea-Bissau, not far from the mainland coast. It is the only active deltaic archipelago on the Atlantic coast of Africa (Pennober 1999; Cuq 2001; Fig. 1). According to Limoges and Robillard (1991) it includes 88 islands and islets, some of which are mangrove islands flooded during spring tides, and some of which have several permanently emerged portions linked by intertidal mangroves. Twenty-one of the islands are permanently inhabited by communities of the Bijagós ethnic group. The islands are separated by a network of channels, and in general are surrounded by mangroves and extensive mud and sand flats, which together represent the most extensive intertidal area in Africa. Sediments originate mostly from the Corubal and Geba rivers and are deposited and moved around by a complex system of currents and wave action. The region is under the seasonal influence of the upwellings linked to the Canary current but also benefits from an important input of organic matter and nutrients via continental runoff and from the productivity of the mangroves. The archipelago and its surrounding flats and channels, covering an area of approximately one million hectares, harbors remarkable biodiversity which justifies its classification as a Biosphere Reserve (1996) and as a Ramsar Site (2014) and has prompted the creation of three marine protected areas. The traditional management of the natural resources of this area by the animistic Bijagós ethnic group is based on strong cultural and religious values (Henry 1994) and has allowed the long term conservation of the site, for example through the effective protection of sacred islands and islets and sacred forest patches where initiation rites take place, and which remain true oases of biodiversity.

Fig. 1 The Bolama-Bijagós Archipelago Biosphere Reserve, Guinea-Bissau. The Biosphere Reserve encompasses the entire Bijagós archipelago described in this chapter

Hydrology

The archipelago is largely bordered by the deep (30–50 m) channels of the rivers Geba and the Rio Grande de Buba, whose orientation contribute towards the triangular shape of the delta. The islands are split by three large channel groups of shallower depth (10–30 m) which are maintained by tidal streams. The seaward extension of the continental shelf is responsible for the highest tidal ranges on the West African coast, with spring tides occasionally reaching up to 4.5 m in amplitude. The currents generated by the tides are strong (up to 78 cm/s) and the tidal flats are huge. The high turbidity of the waters highlights the fact that this is an active delta. The hydrology is influenced by the existence of currents generated by the northern and southern swells in contact with the shelf and the coastline. The influence of the Canary current is mostly felt during the dry season (December–April), while the Guinea current is more active during the rains. Trade winds reach the area and give rise to coastal upwelling, which increases overall productivity, but this influence is not strong. The mean water temperature is at its lowest around February (26.5 °C) and highest in October (30.1 °C) although sampling is incomplete (Lafrance 1994). There is considerable spatial variation, however, and measurements in March generally varied between 26 °C and 29 °C (Diouf et al. 1994). Salinity is lower during the rainy season (30.4 g/l) than during the dry season (around 35–36 g/l), when it reaches levels comparable to nearby oceanic waters (Lafrance 1994). The waters of the archipelago present a relatively low chlorophyll concentration when compared with coastal ecosystems further north, which can be explained by the relatively low concentration of nutrients. This presumably is partly compensated by a high concentration of organic matter originating from the mangroves and continental sources (Diouf et al. 1994).

Wetland Ecosystems

Channels and Shallow Shelf

These are the aquatic environments that are permanently submerged by predominantly marine waters. The seabed is mostly mobile, but here and there, particularly in the outer part of the archipelago, one can find rocky reefs, partly covered by algae (mostly *Caulerpa* sp., *Sargassum* sp.) where fishes and other sea life (including turtles) tend to congregate. Among lower trophic consumers and detritivorous species, large numbers of small pelagics, such as clupeids (*Sardinella*, *Ethmalosa*, *Ilisha*), or shallow sea shoaling fishes, such as mojarras (Gerreidae), sicklefishes (Drepaneidae), threadfins (Polynemidae), grunts (Haemulidae), and mullets (Mugilidae) provide food for abundant predators, such as *Scomberomorus tritor* (Scombridae), jacks (Carangidae, particularly *Caranx hippos*), barracudas (Sphyraenidae), catfishes (Ariidae), croakers (Sciaenidae), and various types of sharks and skates (Chondrichthyes), among others (Diouf et al. 1994; Lafrance 1994).

Intertidal Flats

With a total area often reaching c.100,000 ha (excluding mangroves), depending on the tidal state, intertidal flats represent a major marine habitat in the Bijagós. In the more sheltered areas, finer sediments are found, with mud mixed in various proportions with sand, while sand flats predominate on the seaward side of the archipelago. Flats tend not to be covered by sea grasses (although some *Halodule wrightii* can be found). The benthic fauna is rich in polychaetes, molluscs (e.g., *Senilia senilis*, *Tagelus adamsonii*, *Dosinia* sp.), and crabs (particularly *Uca tangeri*, *Callinectes* sp.), which at low tide are fed upon by birds such as ibises, palm-nut vultures, waders, and gull-billed terns (Zwarts 1988). Sandy beaches are often used as nesting sites by turtles.

Mangroves

Covering an area of 42,480 ha, mangroves represent an important habitat in the archipelago, in terms of their total area as well as their ecological roles. Mangrove tree species in the Bijagós are *Avicennia germinans*, *Rhizophora racemosa*, *Rhizophora mangle*, *Rhizophora harrisonii*, *Laguncularia racemosa*, and *Conocarpus erectus*. Mangroves harbor important biodiversity. Many waders use mangroves as roosting sites, while some, particularly whimbrel *Numenius phaeopus* and common sandpiper *Actitis hypoleucos* also forage extensively within this habitat. Some raptors are also common, particularly the palm-nut vulture *Gypohierax angolensis* (perhaps thousands of individuals), the African fish eagle *Haliaeetus vocifer* and migratory ospreys *Pandion haliaetus* from Europe that find one of their main wintering areas here. These habitats are also important for migratory passerines, particularly reed *Acrocephalus scirpaceus*, melodious *Hippolais polyglotta*, willow *Phylloscopus trochilus*, and subalpine *Sylvia cantillans* warblers (Altenburg and van Spanje 1989). Mammals are represented by manatees *Trichechus senegalensis*, African clawless otters *Aonyx capensis*, and hippopotamuses *Hippopotamus amphibius*, which similarly to the crocodiles *Crocodylus suchus* seem to have adapted to the evolution of the delta into a marine ecosystem. Mangrove forests and creeks are important nurseries and growing areas for various fishes and crustaceans.

Biodiversity

A total of 175 fish species has been recorded (see Lafrance 1994 for a first inventory) but clearly the actual number present is likely to be considerably higher, given the limited effort and low variety of sampling techniques used this far. Sharks, rays, and related species (Chondrichthyes) are particularly well represented, and the Bijagós can be seen as a sanctuary for this group, which includes two species listed by the IUCN as globally critically endangered (*Pristis pristis*, *Pristis pectinata*), six endangered, ten vulnerable, and nine near-threatened species. Most fish species are typically marine, but there are good numbers of species that are most often associated with estuaries.

Reptiles (17 species) include five species of marine turtles of which the most noteworthy is the green turtle *Chelonia mydas* with up to c.30,000 clutches in certain years, particularly in the João Vieira–Poilão Marine National Park where the largest African rookery (and the third largest in the Atlantic) is found (Catry et al. 2009). Two crocodiles (*Crocodylus suchus* and *Osteolaemus tetraspis*) are also present. Thirteen species of Anurans have also been recorded (Auliya et al. 2012).

Including terrestrial, freshwater, and marine species, a total of 282 bird species have been recorded in the Bijagós (Dodman et al. 2004) and numerous new species are still being added to the list. In an international context, the most remarkable values are related to the presence of one of the largest migratory wader (Charadriiformes) concentrations in the East-Atlantic flyway (the third most important site in this flyway, and one of the most important globally) with estimates varying between 700,000 and 900,000 birds (Zwarts 1988; Salvig et al. 1994, 1997; Dodman and Sá 2004; Dodman et al. 2004). Among the most abundant species found at this site are gray plovers *Pluvialis squatarola*, ringed plovers *Charadrius hiaticula*, curlew sandpipers *Calidris ferruginea*, red knots *Calidris canutus*, little stints *Calidris minuta*, bar-tailed godwits *Limosa lapponica*, common redshanks *Tringa totanus*, and whimbrels *Numenius phaeopus*.

The tern (Sternidae) community is also remarkable. An estimated 62,000 individuals of 10 species spend the boreal winter in Guinea-Bissau, the majority of which remain within the Bijagós archipelago (Salvig et al. 1997). More extensive surveys, including open marine waters, will likely show these estimates need an upward revision. Among the most numerous species are gull-billed *Gelochelidon nilotica*, little *Sternula albifrons*, common *S. hirundo*, sandwich *S. sandvicensis*, royal *S. maxima*, Caspian *S. caspia*, and black terns *Chlidonias niger*. Caspian and royal terns also nest regularly, sometimes in large numbers (up to 19,000 pairs of royal terns).

Among the breeding waterbirds are: common terns, gray-headed gulls *Chroicocephalus cirrocephalus*, slender-billed gulls *C. genei*, sacred ibises *Threskiornis aethiopicus*, African spoonbills *Platalea alba*, Western reef-egret *Egretta gularis*, little egrets *E. garzetta*, striated herons *Butorides striatus*, squacco herons *Ardeola ralloides*, pink-backed pelicans *Pelecanus rufescens*, long-tailed *Microcarbo africanus*, and great cormorants *P. (carbo) lucidus* (de Naurois 1966; Dodman and Sá 2004). The increasingly rare Timneh parrot *Psittacus timneh* (globally vulnerable) is still widespread, although at low densities.

Mammals are represented by 29 species. This low diversity is counterbalanced by a high abundance of some charismatic species such as the African manatee *Trichechus senegalensis*, the Atlantic humpbacked dolphin *Sousa teuszii*, the common bottlenose dolphin *Tursiops truncatus*, and the hippopotamus *Hippopotamus amphibius*. The first two of these are classified as vulnerable by the IUCN, and it is thought that the Bijagós may be one of the most important strongholds, strategic for their conservation, although quantitative data are lacking (Silva and Araújo 2001).

Conservation Status

For centuries, the archipelago has benefited from traditional conservation measures. In particular, sacred islands and islets and sacred forests were protected by restrictive access and use rules. These sites maintained important biodiversity values, as is the case with the island of Poilão, and its sea turtle rookery, or with several small islets where waterbirds nest (Campredon 2010). These traditional rules inspired the zoning of the Bolama-Bijagós Archipelago Biosphere Reserve, created in 1996. The central zones of the reserve were declared national parks: the Orango National Park (158,235 ha) and the João Vieira–Poilão Marine National Park (49,500 ha), both officially created in 2000 – although Orango National Park was managed as such since 1997; or protected with a different status: the Community Marine Protected Area of the Urok Islands (54,500 ha). The Biosphere Reserve as a whole, however, is a UNESCO label, not formally recognized in the national legislation. The archipelago is also listed, since 2014, as a Wetland of International Importance under the Ramsar Convention.

Ecosystem Services

The tidal flats and the mangroves of the archipelago play a major role in sustaining the productivity of fisheries resources. These in turn are critical for the food security of local communities. In particular, shellfish represent the main source of animal protein in many villages. The archipelago's fishery stocks are important in a national context and most fish landed in Bissau originates in the Bijagós. Total landings are estimated at 4,000–5,000 t per year (Lafrance 1994; Gonzalez 2010). The nursery areas of the archipelago may also be relevant in the wider context of the national shelf (and beyond) and contribute significantly to industrial fishing which has considerable weight in the national economy through the trading of fishing rights. The abundance of predatory fish favored the creation of several sports fishing lodges, which represent the main tourist attraction of the Bijagós. In general, the Bijagós are considered one of the main areas for the reproduction of fish in West Africa.

Current Threats and Future Challenges

The main current threats are related to fisheries and tourism. The waters of the Bijagós are exploited by migrant fishermen originating from neighboring countries (particularly Senegal) which target, among others, sharks and rays for their fins that are traded in Asian markets (Campredon and Cuq 2001). According to sports fishing club owners, there is already a clear reduction in the abundance and size of sharks in the archipelago. Sawfishes have virtually disappeared. Many fishing camps of foreign fishermen have been illegally implanted in various islands with sociocultural impacts on the local communities, and causing environmental disturbances, for example, through the cutting of (mangrove) wood used for smoking fish, or through the disturbance of nesting colonies of birds. Large fishing nets are also responsible for bycatch of endangered marine megafauna, such as sea turtles or manatees.

Tourist operators keep trying, sometimes with success, to purchase uninhabited islets, originating conflicts among traditional owners (generally local village communities) and weakening or destroying the traditional rules of access to sacred sites. Mass tourism has not developed this far, largely because of constant political instability, but the risk remains.

Longer-term threats involve the hydrocarbon sector and mining. Prospecting carried out so far and planned prospecting suggests the possibility of future oil exploitation offshore, in the immediate vicinity of the archipelago. There are also plans to build a major port in the Rio Grande de Buba for the exportation of bauxite, and the vessels using this port would cross the archipelago in close proximity of the João Vieira–Poilão Marine National Park. Environmental security compliance and impact mitigation may be problematic in a context where the balance of power between the national government and large international companies may not be equitable.

The main challenge for the future is to reinforce the management of the archipelago by involving all the stakeholders, mobilized by a common vision and having access to adequate resources for development and conservation initiatives. The Bijagós may either fall into the hands of external actors little concerned about the long-term future of the area (migratory fishermen, international tourism, mining companies, narcotic trafficking activities) or, alternatively, national stakeholders with local communities may succeed in implementing a development model that values the common heritage and natural resources, to the benefit of the sustainable development of Guinea-Bissau.

References

Altenburg W, van Spanje T. Utilization of mangroves by birds in Guinea-Bissau. Ardea. 1989;77:57–74.
Auliya M, Wagner P, Böhme W. The herpetofauna of the Bijagós archipelago, Guinea-Bissau (West Africa) and a first country-wide checklist. Bonn Zool Bull. 2012;61:255–81.
Campredon P. Mami Wata, mère des eaux. Nature et communautés du littoral ouest-africain. Ed. Actes Sud. 2010. 220 p.
Campredon P, Cuq F. Artisanal fishing and coastal conservation in West Africa. J Coast Conserv. 2001;7:91–100.
Catry P, Barbosa C, Paris B, Indjai B, Almeida A, Limoges B, Silva C, Pereira H. Status, ecology and conservation of sea turtles in Guinea-Bissau. Chelonian Conserv Biol. 2009;8:150–60.
Cuq F. Un système d'Information Géographique pour l'aide à la gestion intégrée de l'archipel des Bijagos (Guinée-Bissau). Notice de la carte, constitution et exploitation du SIG. Brest: Géosystèmes; 2001.
de Naurois R. Colonies reproductrices de spatules africaines, ibis sacrés et laridés dans l'archipel des Bijagos (Guinée portugaise). Alauda. 1966;XXXIV:257–78.
Diouf PS, Deme-Gningue I, Albaret JJ. L'Archipel des Bijagos: environment aquatique et peuplement de poissons. Unpublished Report, CRODT, ORSTOM/CECI/Ministry of Fisheries, Guinea-Bissau; 1994.
Dodman T, Sá J. Monitorização das Aves Aquáticas no Arquipélago dos Bijagós, Guiné-Bissau. Dakar: Wetlands International; 2004.
Dodman T, Barlow C, Sá J, Robertson P. Zonas Importantes para as Aves na Guiné-Bissau/Important Bird Areas in Guinea-Bissau. Dakar/Bissau: Wetlands International/Gabinete de Planificação Costeira/ODHZ; 2004.
Gonzalez JM. Enquête sur les aspects socio-économiques de la Pêche Artisanale en Guinée Bissau. Bissau: Secretaria de Estado das Pescas; 2010.
Henry C. Les Îles où Dansent les Enfants Défunts. Âge, sexe et pouvoir chez les Bijogo de Guinée-Bissau. Paris: CNRS-Éditions; 1994.
Lafrance S. Archipel des Bijagos. Ichtyofaune et éléments d'écologie marine. Bissau: CIPA; 1994.
Limoges B, Robillard M. Proposition d'un plan d'aménagement de la réserve de la biosphere de l'archipel des Bijagós. Bissau: CECI/UICN/MDRA; 1991.
Pennober G. Analyse spatiale de l'environnement côtier de l'Archipel de Bijagós (Guinée-Bissau). PhD thesis. Université de Bretagne occidentale, Institut Universitaire Européen de la Mer. 1999; 193 p.
Salvig JC, Asbirk S, Kjeldsen JP, Rasmussen PAF. Wintering waders in the Bijagós archipelago, Guinea-Bissau, 1992–1993. Ardea. 1994;82:137–41.
Salvig JC, Asbirk S, Kjeldsen JP, Rasmussen PAF, Quade A, Frikke J, Christophernsen E. Coastal waters in Guinea-Bissau – aerial survey results and seasonal occurrence on selected low water plots. Wader Study Group Bull. 1997;84:33–8.
Silva MA, Araújo A. Distribution and current status of the west African manatee (*Trichechus senegalensis*) in Guinea-Bissau. Mar Mamm Sci. 2001;17:418–24.
Zwarts L. Numbers and distribution of coastal waders in Guinea-Bissau. Ardea. 1988;76:42–55.

Kilombero Valley Floodplain (Tanzania)   110

Lars Dinesen

Contents

Introduction	1342
Hydrology	1342
Wetland Ecosystem	1343
Biodiversity	1344
Ecosystem Services	1345
Conservation Status	1346
Threats and Future Challenges	1346
References	1347

Abstract

The Kilombero floodplain at 300 m asl is one of the largest wetlands in Africa. Its headwaters originate about 40 km north of Lake Nyasa. The floodplain is about 260 km long and up to 52 km wide at its widest point, covers approximately 6,300 km^2, and comprises a myriad of rivers and seasonally flooded marshes and swamps. It has a rich flora and fauna and is of immense importance for biodiversity and contains several endemics including three species of birds and a frog. The floodplain is rich in waterbirds and traditionally in large mammals but an increasing pressure from settlements and agriculture have blocked wildlife migration routes and the fishery is in decline. The floodplain was designated a Ramsar Site in 2002 and fielding an Ramsar Advisory Mission is recommended due to a likely change in ecological status. A large plan to turn the entire southern portion of the Kilombero floodplain into a huge rice scheme is pending.

L. Dinesen (✉)
Biologist, European Representative, Scientific and Technical Review Panel of the Ramsar Convention, Jyderup, Denmark
e-mail: larsdinesen8@gmail.com

© Springer Science+Business Media B.V., part of Springer Nature 2018
C. M. Finlayson et al. (eds.), *The Wetland Book*,
https://doi.org/10.1007/978-94-007-4001-3_224

Keywords

Kilombero floodplain · Ramsar · Tanzania · Marsh · Africa

Introduction

The Kilombero floodplain at 300 m asl is one of the largest wetlands in Africa. Its headwaters originate about 40 km north of Lake Nyasa. Source streams run northeast to become the Ruhudji (Hughes and Hughes 1992) before entering the head of the great floodplain to the southwest at 7°44′–9°26′S to 35° 33′–36°36′E.

The floodplain is about 260 km long and up to 52 km wide at its widest point, covers approximately 6,300 km^2, and comprises a myriad of rivers and seasonally flooded marshes and swamps within the Kilombero and Ulanga Districts. To the west the floodplain is flanked by the ancient Udzungwa Mountains that rise to 2,600 m (Rodgers and Homewood 1982; Dinesen et al. 2001) and to the east the Selous Game Reserve and Mahenge Highlands. The flooding is caused by an extremely low plain gradient. The southern central parts of the floodplain descend only 40 m over a distance of 210 km, i.e., with a mean gradient of 1:5250 (Hughes and Hughes 1992).

The northern part of the floodplain is transversed by Msolwa and many other Udzungwa sourced rivers. At the confluence with the Ruhudji the river swings sharply southeast and leaves the floodplain as the Kilombero River. At the confluence with the Luwegu River 65 km downstream, the merged flow forms the Rufiji River which feeds the large mangroves in the Rufiji delta at the Indian Ocean (Fig. 1).

Hydrology

The Kilombero floodplain also receives water from two other large southern rivers, the Mnyera and Pitu, and together with the Ruhudji and merging smaller streams divide into a myriad of tributaries in the central part of the floodplain. Permanent or seasonal streams draining the Udzungwa Mountains in the south also flow into the floodplain. The central floodplain is divided by a number of large parallel and dynamic waterways. The seasonal change in water is huge. The plains may become totally flooded during the wet season and with the exception of river margins, oxbow lakes and the relatively few permanent marshes, swamps, and water bodies becoming almost dry during the dry months (Hughes and Hughes 1992; Baker and Baker 2002; Starkey 2002) from June and onwards.

Rainfall tends to be unimodal and very heavy, overall water levels in the Kilombero Valley tend to rise (November–April) and fall smoothly (May onwards). Flood peaks tend to occur during March–April but can happen as early as January and as late as May (RIS 2002). Moist air drifting inland from a warm Indian Ocean during the monsoon hits the old granite massive hundreds of kilometers from the coast. As the warm and moisture laden air moves upwards at the eastern Udzungwa escarpment, it cools and falls as rain covering the mountain range.

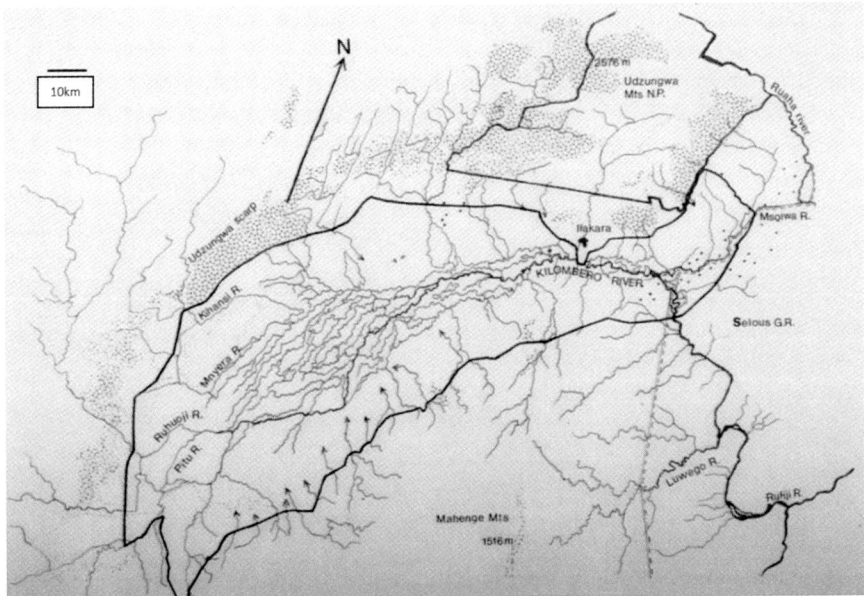

Fig. 1 Kilombero Valley Floodplain with its major rivers. The Kilombero Valley Floodplain Ramsar Site is demarcated with a solid black line encompassing the central valley. The Udzungwa Mountains National Park is shown with a similar line to the north and the Selous Game Reserve is shown with a stippled line to the east. *Dots* represent evergreen forest. Map by Jon Fjeldså

It is speculated that evaporation from the floodplain forms regional water cycles and contributes to form the mist and clouds drifting with the monsoon winds to hit the adjacent mountains. The effect of a particular wetland on local rainfall patterns, however, varies from site to site (Taylor 2010). The floodplain may be seen as a mirror to the ancient Udzungwas. Water from these old mountain streams and rivers has been collected in the floodplain presumably for millennia.

Wetland Ecosystem

The Kilombero floodplain forms part of the Zambezian flooded grasslands ecoregion in Southern and Eastern Africa (WWF 2013). This ecoregion includes other large international important African wetlands such as the Okavango Delta in Botswana, Bangweulu, Luanga and the Kafue Flats in Zambia, and the Malagarasi-Moyowosi in western Tanzania.

A number of vegetation communities are identified representing gradients away from the rivers (after Hood et al. 2002):

(i) Riverside vegetation often dominated by up to 3 m tall *Phragmites mauritianus* and flooded in the wet season to an extent that trees cannot readily establish. In the dry season it is often the last zone to dry out and is usually confined to areas near open water.

(ii) Low-lying valley grassland with the perennial grass *Panicum fluviicola* is characteristic of the interior of the floodplain and often grows in dense belts or clumps over 2 m in height.
(iii) Tall grass dominated by several tall grass species. This community is not flooded as deeply as the low-lying valley grassland due to their presence on slightly higher grounds. Trees are occasionally present.
(iv) Marginal grassland including *Echinochloa colona*, *Paspalum scrobiculatum*, and *Panicum coloratum* and occasionally with trees. The community occurs towards the edges of the floodplain and experience shallow flooding in the wet season and is often heavily grassed and burned in the dry season.
(v) Marginal woodland with the growth of shrubs and trees usually concentrated on termite mounds.
(vi) Combretaceous wooded grassland with *Combretum fragans* as a dominant species is rarely flooded at the floodplain edges.
(vii) Miombo woodland exists on higher grounds and in the foothills.

The valley can roughly be divided into two sections demarcated by the narrowing of the floodplain to 3–4 km at the town of Ifakara. The upstream section to the south is broader, more extensive, and less densely populated than the northern part. However, this may change with the establishment of rice schemes in the south (see below).

Biodiversity

The floodplain has a rich flora and fauna and is of immense importance for biodiversity and contains several endemics as well. More than 350 species of plants have been found including several rare and endangered species (Hood et al. 2002).

The floodplain is a crucial breeding and nursery ground for fish in the whole of the Rufiji Basin. *Citharinus congicus* and *Alestes stuhlmanni* are considered endemic to the Rufiji River Basin (RIS 2002). The Merera toad *Amietophrynus reesi* and Kihansi spray toad *Nectophrynoides asperginis* are only known from the Kilombero and upstream Kihansi rivers.

More than 300 birds have been recorded (Rainey et al. 2002). The Kilombero weaver *Ploceus burnieri* (Baker and Baker 1990) has a patchy distribution in less than 1,500 km^2 of seasonally flooded grassland in the valley. Two new Cisticola species have been illustrated in several books, and their distributions are restricted to the floodplain and mapped out in detail by Rannestad et al. (2015), but the formal publication and naming of these birds are still in progress (J. Fjeldså pers. com.).

The floodplain is rich in waterbirds (Baker and Baker 2002; Rainey et al. 2002). The rising and falling water level in the rivers provides a unique habitat for breeding river birds such as African skimmer *Rhynchops flavirostris* and white-headed lapwing *Vanellus albiceps* depending on the dynamics of the shifting sandbars and bank formations and occur in internationally important numbers (Baker and Baker 2002), which is also the case with the African open-bill *Anastomus lamelligerus* (RIS 2002; Dinesen unpubl.). Moreover, river birds such as the fish-eating Pel's fishing owl *Scotopelia peli*, peculiarities such as the African finfoot *Podica senegalensis*, and at least seven species of kingfishers: giant kingfisher *Megaceryle maxima*, pied kingfisher *Ceryle rudis*, malachite kingfisher *Alcedo cristata*, half-collared kingfisher *Alcedo semitorquata*, brown-hooded kingfisher *Halcyon albiventris*, striped kingfisher *Halcyon chelicuti*, and African pygmy kingfisher *Ceyx pictus* can be encountered along the Udzungwa foothill rivers or in the Kilombero floodplain.

The floodplain has traditionally been very rich in wildlife; the plain and its margins have contained a significant proportion of the world's population of the wetland-dependent puku antelope *Kobus vardonii* (Rodgers 1984; Jenkins et al. 2003). Elephants moved seasonally between protected areas on either side of the floodplain, and hippos, buffalo, lions, and a number of other large mammals are recorded (Starkey et al. 2002; Jones 2007; Bamford et al. 2009). These mammals were once abundant but are in rapid decline due to a markedly increasing human pressure from especially large- and small-scale farming, pastoralism, and fishing (Jones et al. 2012; Nindi et al. 2014). Crocodiles were abundant but are heavily hunted (Starkey et al. 2002).

Ecosystem Services

The floodplain forms the basis for one of the most important freshwater fisheries in Tanzania. Fish are exported to many major population centers and is the single most important source of meat-based protein in the local districts (RIS 2002). Fish migrate upstream to spawn, usually at the beginning of the rains in November. Peak spawning activity has been recorded in November–December, and it seems that there is a second spawning peak in March/April in the shallow water of the inundated floodplain (RIS 2002).

Agriculture is by far the most important activity for local communities. There are two major kinds of agriculture: commercial sugar-cane plantations in the north with virtually no diversity of habitat and subsistence farming in the south that includes rice, maize, and other crops (Jones et al. 2007; Bamford et al. 2009; Nindi et al. 2014). Currently rice is the most important crop, being cultivated on the majority of the farmed area in the valley. Moreover, the floodplain and its margins are important for hunting and utilization of forest products.

Kilombero contributes about two third of the water to the Rufiji (RIS 2001) and is a crucial source of nutrients and sediment to the large mangrove stands, rich mudflats, and sea grass beds at the delta. The floodplain also acts as an enormous buffer in regulating the smooth rise and fall of flooding in the Rufiji and adjacent areas.

Tourists visit the accessible parts of the floodplain, and it is a paradise for hunters and naturalists.

An assessment of the ecosystem services and the changes in these by altering the floodplain may seem to be of outmost importance (see e.g., Peh 2013 for a toolkit).

Conservation Status

Kilombero valley was designated a Game Controlled Area (GCA) since 1956 (Starkey et al. 2002) with quotas for game hunting. There are two hunting blocks following district boundaries. The conservation effect remains somewhat limited and Rodgers (1984) and Baker and Baker (2002) advocated for upgrading to a game reserve. Ulanga District Council is giving parts Wildlife Management Area status and others may be in preparation (Bamford et al. 2009; Jones et al. 2012).

In 2002, the floodplain was designated under the Ramsar Convention as a Wetland of International Importance. The Ramsar Site covers 7,967 km^2, i.e., the flooded areas and adjacent parts of the woodland margins (RIS 2002). Following article 3.2 in the convention *"each Contracting Party shall arrange to be informed at the earliest possible time if the ecological character of any wetland in its territory and included in the List has changed, is changing or is likely to change as the result of technological developments, pollution or other human interference."* Such information shall without delay be passed to the convention secretariat.

Following the threats and future challenges described below it may seem appropriate to take action regarding the changes and planned changes in the Kilombero floodplain covered by Ramsar status including fielding advisory missions to the valley.

Threats and Future Challenges

The proposed plan estimated at US$200 million to dam the major water source – the Ruhudji River – at the head of the Kilombero floodplain (Kabigumila 2002; International Rivers 2013) may have an effect on the biodiversity values. A thorough and carefully planned assessment seems logical. The damming is in the World Bank Pipeline to watch loan projects (International Rivers 2013), and some preliminary investigations have been undertaken.

Damming of Kihansi River has resulted in alteration of the largest permanent Kibasira swamp in the floodplain as well as the extinction in the wild of the endemic Kihansi spray toad (IUCN 2012). The toad has been reintroduced to Kihansi into an artificial spray zone but never recovered, and it is not clear whether it will survive in the wild (Howell et al. 2015).

Elephants and other large mammals were known to regularly cross the plain between Selous Game Reserve and the Udzungwa Mountains. Nearly all of these paths are today, however, blocked by human settlements and agriculture (Jones et al. 2007; Bamford et al. 2009) although there may be a chance of restoring some of them (Jones et al. 2012). Moreover, the fishery is in decline.

Although important for a considerable number of birds and mammals, the gallery and floodplain forests' margins are in many areas increasingly exploited (Starkey et al. 2002) and in adjacent margins within the catchment Miombo woodland is being converted to teak plantations.

The valley is widely recognized within Tanzania as one of the most fertile areas for cultivation of both cash and subsistence crops, and as a result its conversion to agriculture has been widespread and rapid. A large plan to turn the entire southern portion of the Kilombero floodplain into a huge rice scheme has been drafted but its implementation is still pending (D. Moyer pers. com). Throughout there has been rapid immigration of people over the last decades, driven by an increased demand for fertile land and jobs (RIS 2002; Bamford et al. 2009; Nindi et al. 2014) including very large numbers of seminomadic pastoralists and their cattle causing major conflict over land and water resources.

References

Baker NE, Baker EM. A new species of weaver from Tanzania. Bull Br Ornithol Club. 1990;110:51–8.

Baker NE, Baker EM. Important bird areas in Tanzania: a first inventory. Dar es Salaam: Wildlife Conservation Society of Tanzania; 2002.

Bamford A, Ferrol-Schulte D, Smith H. The status of the Ruipa corridor between the Selous Game Reserve and the Udzungwa Mountains. Frontier; 2009.

Dinesen L, Lehmberg T, Rahner MC, Fjeldså J. Biological priorities in the Udzungwa Mountains, Tanzania – based on primates, duikers and birds. Biol Conserv. 2001;99:223–36.

Hood L, Cameron A, Daffa RA, Makoti J. Botanical survey. In: Starkey M, Birnie N, Cameron A, Daffa RA, Haddelsey L, Hood L, Johnson N, Kapapa L, Makoti J, Mwangomo E, Rainey H, Robinson W, editors. The Kilombero Valley Wildlife Project: an ecological and social survey in the Kilombero Valley, Tanzania. Edinburgh: Kilombero Valley Wildlife Project; 2002.

Howell KM, Menegon M, Loader SP. Frogs without limits – amphibians daring to survive in the most peculiar places. In: Scharff N, Rovero F, Jensen FP, Brøgger-Jensen S, editors. Udzungwa tales of discovery in an East African rainforest. Trento: Natural History Museum of Denmark and MUSE – Science Museum; 2015.

Hughes RH, Hughes JS. A directory of African Wetlands. IUCN: Gland; 1992. Joint publication with IUCN, Cambridge (UK), UNEP, Nairobi (Kenya), WCMC, Cambridge (UK).

International Rivers. World Bank projects to watch. 2013. Available from http://www.internatio nalrivers.org/resources/world-bank-pipeline-projects-to-watch-3639. Accessed 28 Mar 2016.

Jenkins RK, Maliti HT, Corti GR. Conservation of the puku antelope (Kobus vardoni, Livingstone) in the Kilombero Valley, Tanzania. Biodivers Conserv. 2003;12(4):787–97.

Jones T, Rovero F, Msirikale J. Vanishing corridors: a last chance to preserve ecological connectivity between the Udzungwa and Selous-Mikumi ecosystems of Southern Tanzania. A feasibility study. Conservation International and Museo de Tridentino di Scienze Naturali; 2007. 35 p.

Jones T, Bamford, AJ, Ferrol-Schulte D, Hieronimo P, McWilliam N, Rovero F. Vanishing wildlife corridors and options for restoration: a case study from Tanzania. Tropl Conserv Sci. 2012;5 (4):463–74. Available online www.tropicalconservationscience.org

Kabigumila J. Potential impacts of the proposed Ruhudji Hydropower plant on the downstream ecology of the Kilombero Valley Floodplain. Final report submitted to SwedPower AB International; 2002.

Nindi SJ, Maliti, H, Bakari S, Kija H, Machoke M. Conflicts over land and water resources in the Kilombero Floodplain, Tanzania. Afr Study Monogr. 2014;50:173–90. (Suppl).

Peh KS-H. TESSA: a toolkit for rapid assessment of ecosystem services at sites of biodiversity conservation importance. Ecosys Serv. 2013;5:51–7.

Rainey H, Birnie N, Cameron A, Mwangomo E, Starkey M. Ornithological survey. In: Starkey M, Birnie N, Cameron A, Daffa RA, Haddelsey L, Hood L, Johnson N, Kapapa L, Makoti J, Mwangomo E, Rainey H, Robinson W, editors. The Kilombero Valley Wildlife Project: an ecological and social survey in the Kilombero Valley, Tanzania. Edinburgh: Kilombero Valley Wildlife Project; 2002. p. 41–101.

Rannestad OT, Tsegaye R, Munishi PKT, Moe SR. Bird abundance, diversity and habitat preferences in the riparian zone of a disturbed wetland ecosystem – the Kilombero Valley, Tanzania. Wetlands. 2015. doi:10.1007/s12157-015-0640-8.

RIS. Ramsar Information Sheet for Kilombero Valley Floodplain. Wildlife Division, Ministry of Natural Resources and Tourism, Tanzania; 2002. https://rsis.ramsar.org/RISapp/files/RISrep/TZ1173RIS.pdf

Rodgers WA. Status of the Puku (Kobus vardoni Livingstone) in Tanzania. Afr J Ecol. 1984;22:117–25.

Rodgers A, Homewood K. Biological values and conservation prospects for the forests and primate populations of the Udzungwa Mountains, Tanzania. Biol Conserv. 1982;24:285–304.

Starkey M, Birnie N, Cameron A, Daffa RA, Haddelsey L, Hood L, Johnson N, Kapapa L, Makoti J, Mwangomo E, Rainey H, Robinson W (eds). The Kilombero Valley Wildlife Project: an ecological and social survey in the Kilombero Valley, Tanzania. Edinburgh: Kilombero Valley Wildlife Project; 2002. 104 p.

Taylor CM. Feedbacks on convections from an African wetland. Geophys Res Lett. 2010;37:1–6.

WWF. Flooded grasslands and savannas. 2013. Available from http://worldwildlife.org/ecoregions/at0907. Accessed 5 Mar 2016.

Lakes Baringo and Naivasha: Endorheic Freshwater Lakes of the Rift Valley (Kenya)

Reuben Omondi, William Ojwang, Casianes Olilo, James Mugo, Simon Agembe, and Jacob E. Ojuok

Contents

Introduction	1350
Lake Baringo	1351
Geochemistry	1353
Wetland Ecosystems and Biodiversity	1353
Ecosystem Services	1355
Lake Naivasha	1355
Geochemistry	1356
Wetland Ecosystem and Biodiversity	1357
Ecosystem Services	1358
Threats and Future Challenge for Lakes Baringo and Naivasha	1358
References	1359

R. Omondi (✉)
Department of Applied Aquatic Sciences, Kisii University, Kisii, Kenya
e-mail: reubenomondi@yahoo.com

W. Ojwang (✉) · C. Olilo (✉)
Kenya Marine and Fisheries Research Institute (KMFRI), Mombasa, Kisumu, Kenya
e-mail: wojwang@wwfkenya.org; olilocasianes@yahoo.com

J. Mugo (✉)
Kenya Marine and Fisheries Research Institute, Naivasha, Kenya
e-mail: mugojam@yahoo.com

S. Agembe (✉)
Department of Fisheries and Aquatic Sciences, University of Eldoret, Eldoret, Kenya
e-mail: agembesimon@yahoo.com

J. E. Ojuok
Kenya Marine and Fisheries Research Institute, Kisumu, Kenya
e-mail: jojuok@yahoo.co.uk

© Springer Science+Business Media B.V., part of Springer Nature 2018
C. M. Finlayson et al. (eds.), *The Wetland Book*,
https://doi.org/10.1007/978-94-007-4001-3_133

> **Abstract**
> Lakes Baringo and Naivasha are among the Great Rift Valley lakes that were formed about 25 million years ago by violent separation of two of the earth's continental plates floating on the molten magma of its core. Lake Baringo is a freshwater lake in the eastern arm of Kenya's Great Rift Valley, with surface and catchment areas of approximately 130 and 6,820 km^2, respectively, and a mean depth of 5.9 m. The lake is shallow and is virtually a wetland in its entirety, with submerged *Ceratophyllum demersum* occurring in the deepest portions of the lake. However, larger marshes are found at the river mouths to the south and east. Lake Naivasha has a surface area of approximately 119,130 km^2, a catchment area of 3,200 km^2, and a mean depth of 4.1 m. Lake Naivasha's water level has experienced great fluctuations sometimes as much as 7 m over many years attributed to large-scale climatic influence, also causing changes in water geochemistry. The main lake has a fringing vegetation of *Cyperus papyrus* and other macrophytes and the lake body supports a diverse macrophyte community. In addition to the lakes' high biodiversity, they provide water to the local communities for domestic, irrigation, and industrial purposes. Fish in these lakes contribute to the diet of the local communities and are a source of income. Due to their unique scenery and biodiversity, the lakes are destinations for both local and international tourists. Lakes Baringo and Naivasha are faced with numerous threats and challenges including an increasing human population, deforestation and erosion, pollution, impacts associated with tourism, human–wildlife conflict, impact of invasive species, and climate change. Investments around these lakes need to be balanced against the threats to biodiversity and the human population.

> **Keywords**
> Anthropogenic activities · Biodiversity · Ecosystem services · Endorheic freshwater · Rift valley

Introduction

Lakes Baringo and Naivasha are among the Great Rift Valley lakes (Fig. 1) that were formed about 25 million years ago by violent separation of two of the earth's continental plates floating on the molten magma of its core. In addition to the lakes' high biodiversity, they provide water to the local communities for domestic, irrigation, and industrial purposes. Fish in these lakes contribute to the diet of the local communities and are a source of income. Due to their unique scenery and biodiversity, the lakes are destinations for both local and international tourists. The lakes are however threatened by anthropological activities, especially water diversion and pollution. Investments around these lakes need to be balanced against the threats to biodiversity and the human population.

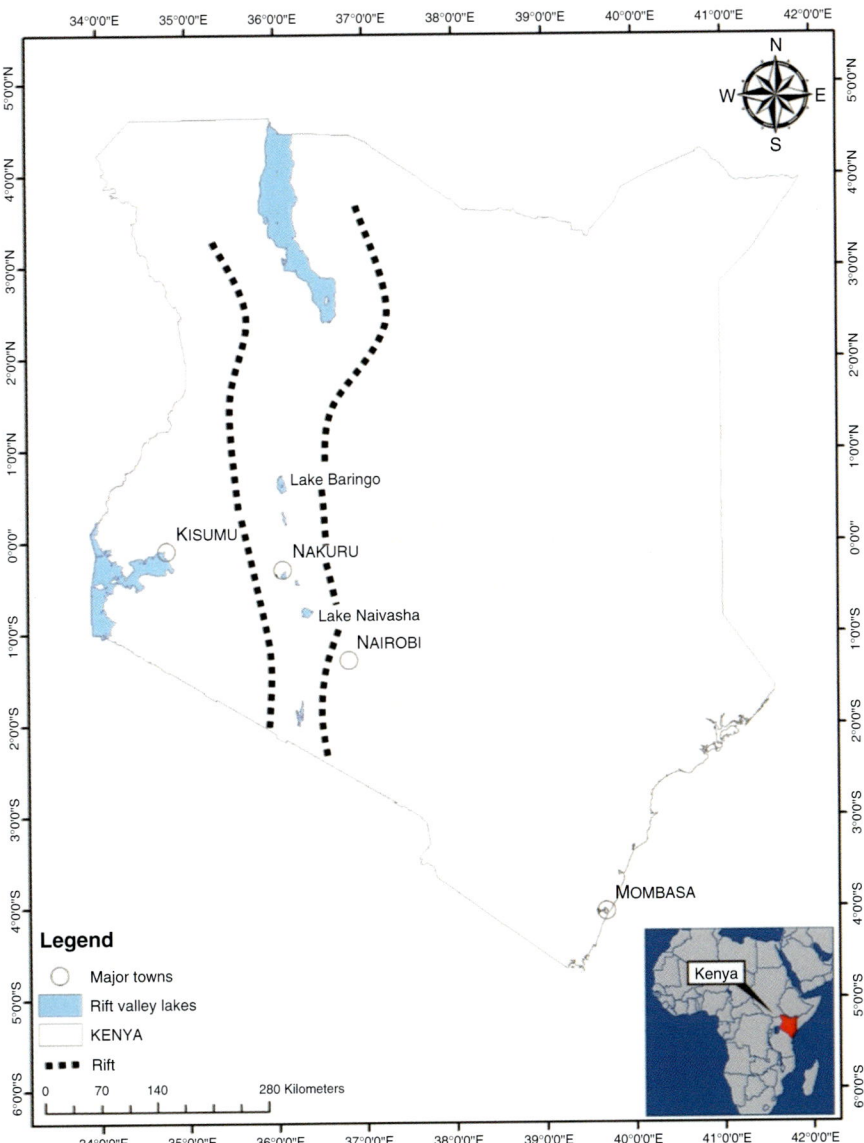

Fig. 1 Map showing the location of lakes Baringo and Naivasha, Kenya

Lake Baringo

Lake Baringo is a freshwater lake in the eastern arm of Kenya's Great Rift Valley. Located between 0°30′–0°45′N and 36°00′–36°10′ E, the lake lies approximately 60 km north of the equator at an altitude of 975 m asl (Kållqvist 1987) (Fig. 2). The

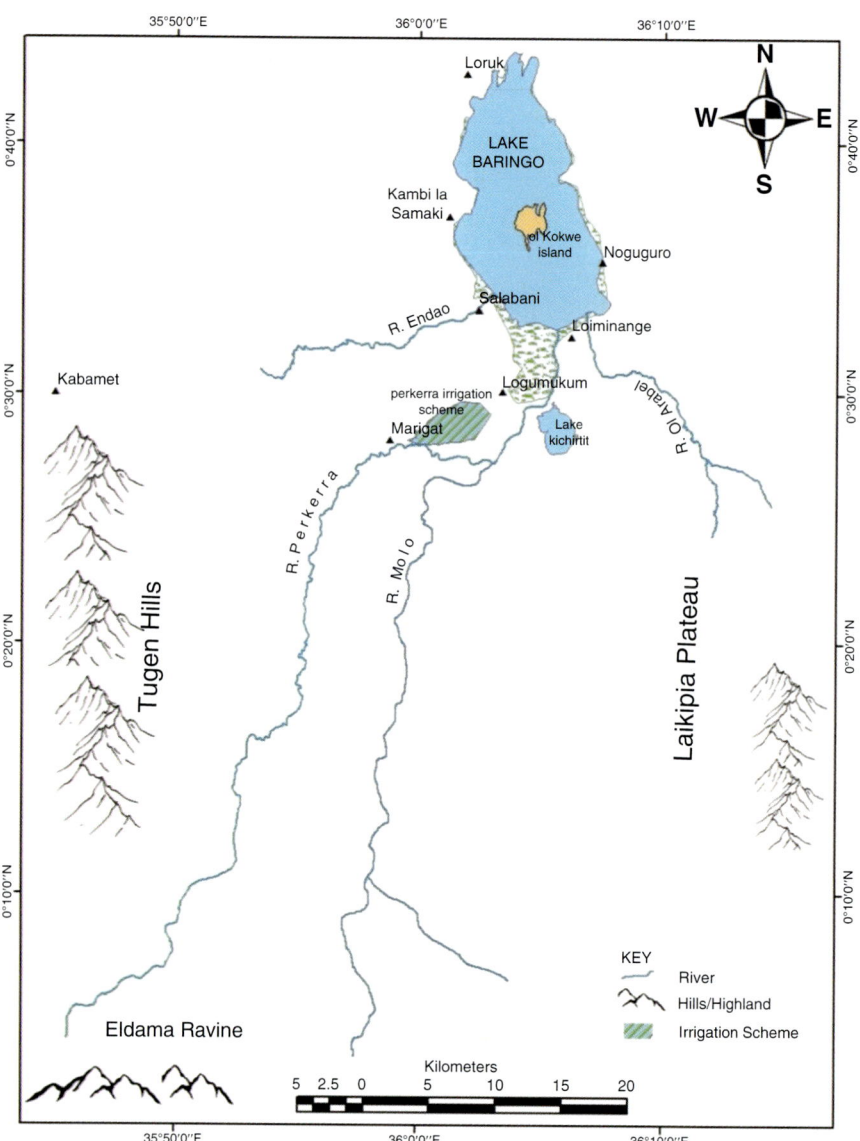

Fig. 2 Map of Lake Baringo catchment

lake has surface and catchment areas of approximately 130 km^2 and 6,820 km^2, respectively, and a mean and maximum depth of 5.9 m and approximately 10 m, respectively, at high water levels (Omondi et al. 2014). Lake Baringo waters remain fresh despite its shallow depth, a high net evaporation that characterizes the rift floor, and absence of a surface outlet. Recent hydrogeological evidence confirms that some lake water is lost by underground seepage through the fractured lake floor (Onyando et al. 2005).

The climate of the area is arid to semiarid with dry and wet seasons that are unpredictable and irregular. The dry season runs from September to February, while the rainy season is between March and August. Mean annual rainfall ranges from about 600 mm on the east and south of the lake to 1,500 mm on the western escarpment of the Rift Valley. The lake is fed by perennial rivers Molo and Perkerra and seasonal rivers Ol Arabel, Makutan, Endao, and Chemeron (Fig. 2). Damming and abstraction of water from some of these rivers have reduced the amount of water reaching the lake. Lake Baringo experiences very high annual evaporation rates of 1,650–2,300 mm (Odada et al. 2006) and its persistence depends on the inflows from rivers originating from the hilly basin.

Geochemistry

The history of water quality measurements from Lake Baringo reveals that the geochemistry is greatly influenced by the rainfall patterns and the high evaporation rate (Omondi et al. 2014). The decreasing trend of water transparency of Lake Baringo has been attributed to suspended solids brought into the lake from the catchment and resuspension of sediments by wind action (Wahlberg et al. 2003). The lowest Secchi depth of 8 cm was recorded in 1987 (Kållqvist 1987); however, recent data showed that this had risen to 25.9 cm and this was attributed to the general rise of water levels of Rift Valley lakes (Omondi et al. 2014). The lake has a mean conductivity of 577.7 $\mu S\ cm^{-1}$, while total nitrogen and total phosphorus have mean concentrations of 1,163.6 and 93.5 $\mu g\ l^{-1}$, respectively.

Wetland Ecosystems and Biodiversity

Lake Baringo was designated as a Wetland of International Importance under the Ramsar Convention on Wetlands on 10 January 2002. The lake is shallow and is virtually a wetland in its entirety, with submerged *Ceratophyllum demersum* occurring in the deepest portions of the lake. However, larger marshes are found at the river mouths to the south and east. The southern marsh, which is the most expansive, is dominated by a member of the Poaceae, *Paspalidium geminatum*, while the southeastern marsh is dominated by *Typha domingensis*. The northeastern marsh is dominated by *Aeschynomene pfundii*, while there is a low population of *P. geminatum* in the eastern bay. Other macrophytes in the lake include the free-floating *Azolla pinnata*, *Azolla nilotica*, and *Pistia stratiotes*, submerged *Ceratophyllum demersum* and *Najas horrida*, floating-leaved *Nymphaea lotus*, and emergent *Aeschynomene pfundii* (Fig. 3). During high water levels, a large swamp forms at the northeastern area within dry *Acacia tortilis* trees.

Phytoplankton in Lake Baringo is dominated by blue-green algae (Oduor 2000), zooplankton by copepods (Omondi et al. 2011), and macroinvertebrates by Mollusca

Fig. 3 Some common macrophytes in Lake Baringo. (**a**) *Paspalidium geminatum* and *Typha domingensis*, (**b**) *Aeschynomene pfundii*, (**c**) *Ceratophyllum demersum*, and (**d**) *Nymphaea lotus* (Photo credit: R. Omondi © Rights remain with the author)

(Owili et al. 2008). Lake Baringo has large populations of crocodiles (*Crocodylus niloticus*), hippopotamus (*Hippopotamus amphibious*), lizards (*Varanus* spp.), and some species of frogs (*Rana* spp.).

The lake is an Important Bird Area (IBA) with over 470 species of birds including globally important species such as lesser kestrel *Falco naumanni*, lesser flamingo *Phoeniconaias minor*, Madagascar squacco heron *Ardeola ralloides*, and pallid harrier *Circus macrourus*. A number of regionally threatened species, great crested grebe *Podiceps cristatus*, African darter *Anhinga rufa*, great egret *Ardea alba*, saddle-billed stork *Ephippiorhynchus senegalensis*, white-backed duck *Thalassornis leuconotus*, white-headed vulture *Trigonoceps occipitalis*, martial eagle *Polemaetus bellicosus*, Baillon's crake *Porzana pusilla*, and African skimmer *Rynchops flavirostris*, are also found in the lake. These are classified as least concern to near threatened species. In addition, the lake hosts over 20,000 water birds throughout the year including Palearctic migrants. Ol Kokwe Island is an important breeding site for Goliath herons *Ardea goliath*.

The fish community of Lake Baringo comprises seven species (Odada et al. 2006). These include *Aplocheliches* sp., *Barbus intermedius australis*, *B. lineomaculatus*, *Clarias gariepinus*, *Labeo cylindricus*, *Oreochromis niloticus baringoensis*, and *Protopterus aethiopicus* (Britton et al. 2006). Of these, four species, namely,

B. i. australis, *C. gariepinus*, *O. n. baringoensis*, and *P. aethiopicus*, are economically exploited. The fishery of the lake was once dominated by the endemic *O. n. baringoensis* but is presently dominated by *P. aethiopicus* introduced in 1975.

Ecosystem Services

The human communities around the lake include Pokot to the north, Ilchamus to the south, and Tugen to the east. The area is essentially a rangeland and apart from the scattered isolated pockets of dry subsistence agriculture and small irrigation farming around Marigat; the major socioeconomic activities undertaken by the communities include mostly livestock husbandry and beekeeping. Lake Baringo is an important water body nationally and internationally. The lake is of great benefit to the communities living in the basin for domestic use and for watering livestock. Fish from the lake provides food and a source income for the local communities. Lake water is used for irrigation agriculture, around Marigat, and for production of food crops like watermelon, onions, tomatoes, and maize among others. Lake Baringo is known to attract both local and international tourists for its biodiversity and boat riding. Tourism-related activities such as the expansion in hotel industry and related services provide employment and income to the local community.

Lake Naivasha

Lake Naivasha is situated between 0°40′–0°50′ and 36°15′–36°25′ E at an altitude of 1890 m above sea level (Sikes 1936). The lake has a surface area of approximately 130 km^2, a catchment area of 3,200 km^2, and a mean depth of 4.1 m (Mugo 2010) (Fig. 4). The lake's surface inflows come via rivers Gilgil and Malewa which are perennial and Karati which is seasonal. River Malewa contributes 90% of the water discharged into the lake. The waters of Lake Naivasha remain fresh despite the lack of a surface outlet, shallow depth, and high net evaporation that characterizes other lakes on the rift floor. Some waters from the lakes are lost by underground seepage through the fractured lake floor (Clark et al. 1990).

Mean air temperatures are moderate with monthly means varying from 15.9 °C to 18.5 °C. The combination of low temperatures, low relative humidity, and low rainfall makes January and February the months with the highest evaporation rates (Gaudet and Melack 1981). Rainfall is bimodal occurring in April–July and October–November for the long and short rains, respectively. Direct precipitation on the lake is minimal although occasional torrential rains are witnessed. Irregularity of the rainfall pattern is also quite common. Rainfall on the surrounding highlands quickly percolates into the ground and from there rapidly seeps through into the lake (Gaudet and Melack 1981). Rainfall in the basin is highest in the Nyandarua Mountains (1,400–1,600 mm/year^{-1}), while the lake which is located in the rain shadow receives between 500 and 700 mm/year^{-1} (Richardson and Richardson 1972).

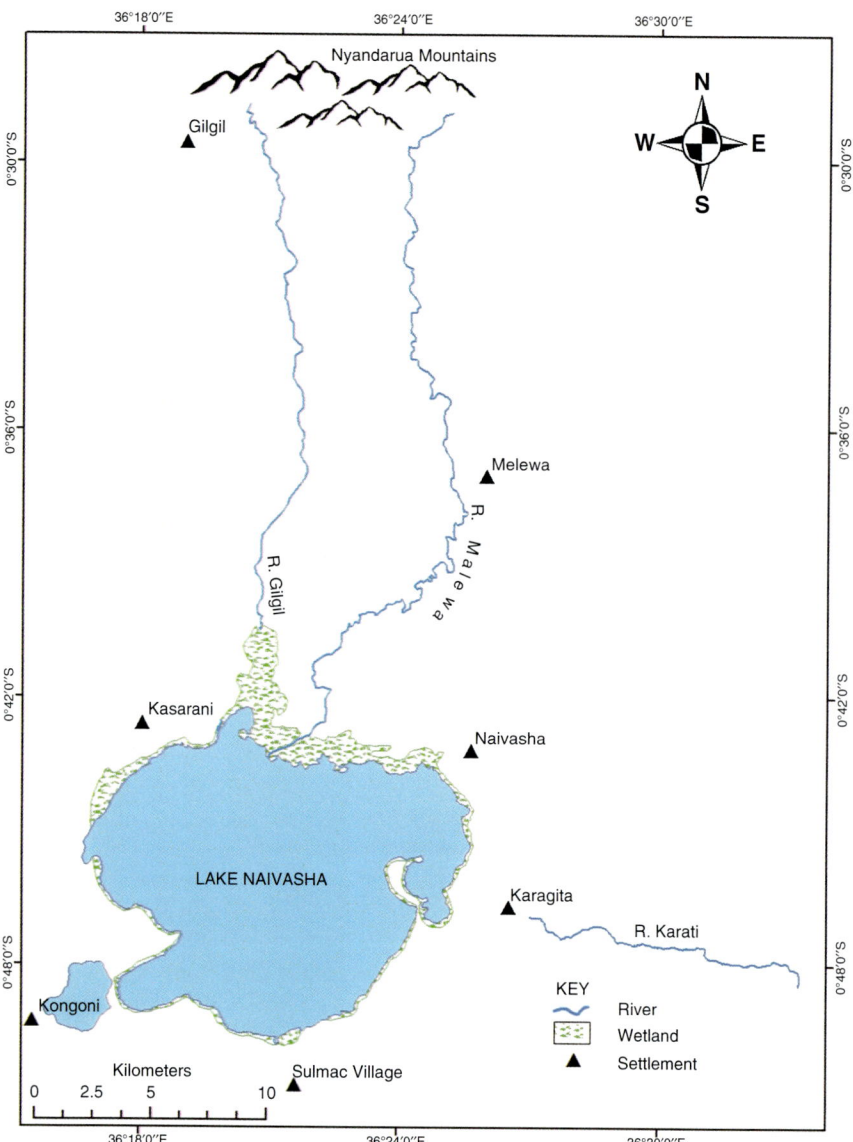

Fig. 4 Map of Lake Naivasha catchment

Geochemistry

Lake Naivasha's water level has experienced great fluctuations sometimes as much as 7 m over many years attributed to large-scale climatic influence (Vincent et al. 1979; Becht and Harper 2002). These fluctuations have been correlated with wide variation in the geochemistry of the water. Earlier studies reported on the mean ranges for various parameters: total nitrogen values (300–675 µg l^{-1}), total phosphorus (50–200 µg l^{-1}), conductivity (250–400 µS cm^{-1}), and Secchi depth (50–250 cm) (Litterick et al. 1979; Gaudet 1981; Njuguna 1982; Kitaka 1991). Recent investigations, however, report mean water and Secchi depths of 4.1 m and 48 cm, respectively, a mean conductivity of 259.2 µS cm^{-1}, and mean concentrations of total nitrogen and total phosphorous of 304 and 42.7 µg l^{-1}, respectively (Mugo 2010).

Wetland Ecosystem and Biodiversity

Lake Naivasha was designated in April 1995 as Kenya's second Wetland of International Importance under the Ramsar Convention on Wetlands (LNRA 1999). The wetland consists of the main lake covered by a fringing vegetation of *Cyperus papyrus* and other macrophytes. Macrophytes in the lake are diverse and include emergents dominated by *Cyperus papyrus*, free-floating species dominated by nonnative *Salvinia molesta* and *Eichhornia crassipes*, floating-leaved *Nymphaea caerulea*, and submerged angiosperms consisting of three species of *Potamogeton* – *P. pectinatus*, *P. schweinfurthii*, and *P. octandrus*, together with *Najas pectinata* (Harper et al. 1995). Mats of floating papyrus *Cyperus papyrus*, water hyacinth *Eichhornia crassipes*, and *Salvinia molesta* are currently the most notable features in the lake.

The phytoplankton community in Lake Naivasha is diverse with 143 described taxa but dominated by Cyanophyta (blue-green) and Bacillariophyta (diatoms) (Kalff and Watson 1986). The main zooplankton groups include Copepoda, Cladocera, and Rotifera. Copepods include the genera *Thermocyclops* and *Mesocyclops*; cladocerans reported from the lake are *Diaphanosoma*, *Simocephalus*, *Daphnia*, and *Moina*, while rotifers are *Brachionus*, *Hexarthra*, *Keratella*, *Filinia*, and *Lecane* (Uku and Mavuti 1994).

Terrestrial trees within the vicinity of Lake Naivasha consist predominantly of *Acacia xanthophloea*, which is gradually being cleared and the area converted to riparian cultivation, grasslands for intensive livestock management, and irrigated production of horticultural crops (Watson and Parker 1969).

The avifauna of Lake Naivasha is diverse, with approximately 80 resident and migratory water bird species, with large populations of the African fish eagle *Haliaeetus vocifer* and red-knobbed coot. The great crested grebe, maccoa duck, African darter, great egret, saddle-billed stork, white-backed duck, Baillon's crake, and African skimmer are all found in the Lake Naivasha region.

Lake Naivasha has historically had multiple introductions and reintroductions of various fish species (Britton et al. 2006; Ojuok et al. 2007). Introductions were purportedly desired to provide food sources and sport fishes and for mosquito control. Introduced fish species which are still being landed on the shores of Lake Naivasha include the largemouth bass *Micropterus salmoides*, redbelly tilapia *Tilapia zillii*, blue spotted tilapia (*Oreochromis leucostictus*), Nile tilapia (*O. niloticus*), and common carp *Cyprinus carpio*. Other species which have been recorded but not of commercial importance include rainbow trout *Oncorhynchus mykiss*, mosquito fish *Gambusia* sp., guppy *Poecilia reticulata*, and *Barbus amphigramma*. A crustacean, Louisiana red swamp crayfish *Procambarus clarkii*, was introduced in 1970 as food for *Micropterus salmoides*.

Lake Naivasha provides natural habitat to several resident populations of large mammals such as zebras *Equus quagga*, impalas *Aepyceros melampus*, wildebeests *Connochaetes taurinus*, and giraffes *Giraffa camelopardalis*. The lake has the largest metapopulation (approximately 600–700) of *Hippopotamus amphibious* in the Kenyan Rift Valley. The littoral zones provide crucial habitat for fish breeding and foraging by wildlife, which include hippo, waterbuck, and buffalo.

Ecosystem Services

The area surrounding the lake has a cosmopolitan population made up of various ethnic groups in Kenya. This population has continued to increase rapidly since the 1990s associated with the increase of the acreage under horticulture and floriculture products which has provided ready employment for the people. The majority of the local population is made up of Kikuyu on the lake's eastern and northeastern side who are mostly involved in farming. The pastoralist and agropastoralist Maasai community has settled toward the southern side. The majority of the workforce in the flower farms and fish and hotel industry includes the Luo, Luhya, and Kikuyu communities.

Lake Naivasha ecosystem is very rich in biodiversity and provides habitat for a wide range of terrestrial flora and fauna and aquatic organisms. Together these play an important role in sustaining ecosystem services and supporting anthropogenic activities. The beautiful sceneries of the lake and the abundant hippo and birds are popular with nature-loving tourists.

The fringing vegetation acts as a filter to organic and inorganic nutrients/material before they enter the lake. The lake which has a large surface area is an important storage facility for water. This water is utilized by various stakeholders in farming, industrial, and domestic use. (Harper et al. 1990, 1995).

Threats and Future Challenge for Lakes Baringo and Naivasha

Lakes Baringo and Naivasha are faced with numerous threats and challenges including an increasing human population, deforestation and erosion, pollution, impacts associated with tourism, human–wildlife conflict, impact of invasive species, and climate change. Although the current population of communities immediately adjacent to the lakes is 150,000, the population in the catchment areas is 900,000 people and has a $6.5\%/\text{year}^{-1}$ population growth rate which is higher than the national rate of $3.5\%/\text{year}^{-1}$ (Kenya Republic, 2010). The high human population growth rate in these riparian zones has resulted in an increasing demand for land dedicated to human settlement and agricultural. These trends related to socioeconomic demands in an area well known for its wildlife population continue to cause human–wildlife conflicts resulting in the loss of human lives and property. Poaching is an illegal activity and court battles are to be expected; alternatively human–wildlife conflicts can result in misunderstandings between land owners/users and resource managers. The increasing human populations have also led to the unregulated extraction of forest resources for medicinal purposes, fuelwood, charcoal burning, and timber and the destruction of riparian forest, coupled with foraging by livestock; these activities are severely damaging to the catchment area and ultimately lead to soil erosion and sedimentation in the lakes' riparian areas.

Other agricultural-related activities (e.g., floral industries) and urban centers are major sources of chemicals (e.g., fertilizers, pesticides, herbicides) and general wastes that pollute these freshwater resources. Periodic algal blooms and fish kills in Lake Naivasha have been attributed to these pollutants.

Recent attempts to increase aquaculture production in Kenya may result in the unintentional introduction of invasive species. Due to a lack of adequate local supply of fingerlings, there has been uncontrolled transportation of fingerlings from different regions for stocking fish in the two lake basins. Introduction of new species to these water bodies may lead to disruption of food webs. Such transfers may also result in the introduction of aquatic macrophytes, especially water hyacinth, which to date has not been reported in Lake Baringo. In addition to macrophytes, there is also the danger of increased spread of the invasive shrub *Prosopis juliflora*, from its current locations around Lake Baringo. Although there are economic gains from tourism, the industry poses some threats to the communities around lakes Baringo and Naivasha. Tourism encourages clothing styles and other practices which are inconsistent with the cultural norms of the local communities. Lakes Naivasha and Baringo are situated in semiarid areas, which put them at risk of climate change phenomenon. The increased variability on local temperature, rainfall, and wind patterns, among other climate change factors, could negatively impact lakes' morphology, hydrology, biodiversity, and socioeconomic aspects. Recent rise in the lake levels in the two lakes resulted in the destruction of infrastructures and economic losses in the riparian zones.

References

Britton JR, Ng'eno JBK, Lugonzo J, Harper D. Can an introduced, non-indigenous species save the fisheries of Lakes Baringo and Naivasha, Kenya? In: Odada EO, Olago DO, Ochola W, Ntiba M, Wandiga S, Gichuki N, Oyieke H, editors. Proceedings of the XI World Lake Conference, Nairobi, Kenya, vol. II. Tokyo: ILEC; 2006. p. 568–72.

Clark MCG, Woodhall DG, Allen D, Darling G. Geological, volcanic and hydrological controls on the occurrence of geothermal activity in the area surrounding Lake Naivasha, Kenya. Nairobi: Ministry of Energy; 1990.

Gaudet JJ, Melack JM. Major ion chemistry in a tropical African lake basin. Freshw Biol. 1981;11(40):309–33.

Harper DM, Mavuti KM, Muchiri SM. Ecology and management of Lake Naivasha, Kenya in relation to changing climate, alien species introductions and agricultural development. Environ Conserv. 1990;17:328–36.

Harper DM, Adams C, Mavuti KM. The aquatic plant communities of the Lake Naivasha wetland, Kenya: pattern, dynamics and conservation. Wetl Ecol Manag. 1995;3(20):111–23.

Kalff J, Watson S. Phytoplankton and its dynamics in two tropical lakes: a tropical and temperate zone comparison. Hydrobiologia. 1986;138(1):161–76.

Kâllqvist T. Primary production and phytoplankton in Lake Baringo and Naivasha, Kenya. Norwegian Institute for Water Research Report. Blindern, Oslo; 1987.

Kenya Republic of. Ministry of Planning and National Development: Kenya National Bureau of Statistics, Socio Economic Aspects, Nairobi. 2010.

Kitaka N. Phytoplankton productivity in Lake Naivasha [MSc thesis]. Kenya: University of Nairobi; 1991.

Lake Naivasha Riparian Association (LNRA). The Lake Naivasha management plan. Naivasha: Lake Naivasha Riparian Association; 1999.

Litterick M, Gaudet JJ, Kalff J, Melack JM. The limnology of an African lake, Lake Naivasha, Kenya. Document presented at International Conference on Tropical Limnology Nairobi; 1979.

Mugo JM. Seasonal changes in physical – chemical status and algal biomass of Lake Naivasha, Kenya [MSc thesis]. Nairobi: Kenyatta University; 2010.

Odada EO, Onyando JO, Obudho PA. Lake Baringo: addressing threatened biodiversity and livelihoods. Lakes Reserv Res Manag. 2006;11:287–99.

Oduor SO. Physico-chemical dynamics, pelagial primary production and algal composition in Lake Baringo, Kenya [MSc thesis]. Deft, Austria: IHE; 2000.

Ojuok J, Njiru M, Mugo J, Morara G, Wakwabi E. Increase dominance of common carp, *Cyprinus carpio* L. the boon or the bane of Lake Naivasha fisheries? Afr J Ecol. 2007;46(3):445–8. 2000.

Omondi R, Kembenya E, Nyamweya C, Ouma H, Machua SK, Ogari Z. Recent limnological changes and their implication on fisheries in Lake Baringo, Kenya. J Ecol Nat Environ. 2014;6(5):154–63.

Onyando JO, Kisoyan P, Chemelil MC. Estimation of potential soil erosion for River Perkerra catchment in Kenya. Water Resour Manag. 2005;19:133–43.

Owili M, Omondi R, Muli J, Ondiba R. Spatial variations in plankton, macroinvertebrates and macrophytes in Lake Baringo, Kenya. In: Muli JR, Gichuki J, Getabu A, Wakwabi E, Abila R, editors. Lake Baringo research expedition: fisheries and environmental impact. KMFRI/LABRE/Technical Report 3. 2008. p. 74–99.

Richardson JL, Richardson AE. History of an African Rift Lake and its climactic history. Ecol Monogr. 1972;42(4):499–533.

Sikes HL. Notes on the hydrology of Lake Naivasha. J East Afr Uganda Nat Hist Soc. 1936;13:73–84.

Uku JN, Mavuti KM. Comparative limnology, species diversity and biomass relationship of zooplankton and phytoplankton in five freshwater lakes in Kenya. Hydrobiologia. 1994;272(1/3):251–8.

Vincent CE, Davies TD, Beresford UC. Recent changes in the level of Lake Naivasha, Kenya, as an indicator of equatorial westerlies over East Africa. Clim Change. 1979;2:175–91.

Wahlberg HT, Harper D, Wahlberg NT. A first limnological description of Lake Kichiritit, Kenya: a possible reference site for the freshwater lakes of the Gregory Rift Valley. S Afr J Sci. 2003;99:494–6.

Watson RM, Parker ISC. The ecology of Lake Naivasha. The identification and description of some important components for a model. Unpublished report to the East African Wildlife Society; 1969. p. 49.

112

Lake Turkana: World's Largest Permanent Desert Lake (Kenya)

William Ojwang, Kevin O. Obiero, Oscar O. Donde,
Natasha J. Gownaris, Ellen K. Pikitch, Reuben Omondi,
Simon Agembe, John Malala, and Sean T. Avery

Contents

Introduction	1362
Hydrology	1363
Geochemistry (Water Quality and Nutrients)	1365
Diversity of Wetland Ecosystems	1366
Omo Delta Wetland	1367
Kerio/Turkwel Deltas	1369
Ferguson's Gulf	1370
Sibiloi/Koobi Fora Protected Area	1370
Other Protected Areas of the Lake	1370
Lake Turkana Biodiversity	1371
Ecosystem Services	1372
Riparian Communities	1373

W. Ojwang (✉)
Kenya Marine and Fisheries Research Institute (KMFRI), Mombasa, Kisumu, Kenya
e-mail: wojwang@wwfkenya.org

K. O. Obiero (✉) · O. O. Donde (✉) · J. Malala (✉)
KMFRI, Lake Turkana Research Station, Lodwar, Kenya
e-mail: kevinobiero@yahoo.com; oscinho2524@gmail.com; malalajohn@gmail.com

N. J. Gownaris (✉) · E. K. Pikitch (✉)
School of Marine and Atmospheric Sciences, Stony Brook University, Stony Brook, NY, USA
e-mail: ngownaris@gmail.com; ellen.pikitch@stonybrook.edu

R. Omondi (✉)
Department of Applied Aquatic Sciences, Kisii University, Kisii, Kenya
e-mail: reubenomondi@yahoo.com

S. Agembe (✉)
Department of Fisheries and Aquatic Sciences, University of Eldoret, Eldoret, Kenya
e-mail: agembesimon@yahoo.com

S. T. Avery (✉)
Kenya Wetlands Biodiversity Research Team, Nairobi, Kenya
e-mail: sean@watres.com

© Springer Science+Business Media B.V., part of Springer Nature 2018
C. M. Finlayson et al. (eds.), *The Wetland Book*,
https://doi.org/10.1007/978-94-007-4001-3_254

Conservation Status and Management 1375
Threats and Future Challenges 1375
References 1378

Abstract

Located in the "cradle of mankind" of the East African Rift Valley, Lake Turkana is distinguished as both the world's largest permanent desert lake and alkaline water body. With a surface area of about 7,560 km^2, Lake Turkana is a highly pulsed, variable system as a result of its closed-basin nature, arid surroundings, and its strong dependence on River Omo for the majority of its inflow, which originates as rainfall over the Ethiopian highlands. In this article we describe the lake's unique ecosystem and associated vicissitudes, diverse habitats and incredible biodiversity, and ecosystem services. Although parts of the lake and lower Omo Delta have been zoned as an international biosphere reserve, Lake Turkana and the region are facing immense threat from anthropogenic activities. A combination of external factors (hydropower dams, irrigation schemes, climate anomalies) and internal drivers (demography, economic growth) will strongly impact the Lake Turkana basin over the next decade. In turn, this will have significant negative consequences on resource productivity and the wellbeing of local communities.

Keywords

Desert Lake · Alkaline · Omo Delta Wetland · Blue-green algae · Transboundary Wetland · Hydropower Dams · Irrigation Schemes · Dependent Communities

Introduction

Lake Turkana is a unique ecosystem, distinguished as the world's largest permanent desert lake and the largest alkaline water body. Of the East African Rift Valley Lakes, Lake Turkana is the most remote (Johnson and Malala 2009) and the last of the world's great lakes to be studied (Hopson 1982). Lake Turkana occupies an arid region in East Africa, largely within northwestern Kenya, but extending into southwestern Ethiopia (Fig. 1). The lake's catchment basin covers an area of approximately 130,860 km^2. With a surface area of about 7,560 km^2, the lake is 260 km long with an average width of 30 km, a mean depth of 31 m, and a maximum depth of 114 m (ibid.). It is fed by three major rivers: the Omo, Turkwel, and Kerio. In addition, numerous small seasonal streams discharge into the lake. The Omo River, which flows continuously and is fed by precipitation from the Ethiopian Highlands, accounts for more than 90% of the lake's freshwater influx and acts as the lake's "umbilical cord" (Kolding 1992; Avery 2010). The Turkwel and Kerio Rivers provide intermittent freshwater inputs (Ricketts and Johnson 1996). Thanks to the Turkwel dam, the river's discharge is today regulated and is perennial, but sometimes all of the water is lost through the riverbed before reaching the lake (Avery 2012a). The Turkana area has been called the "cradle of mankind" due to

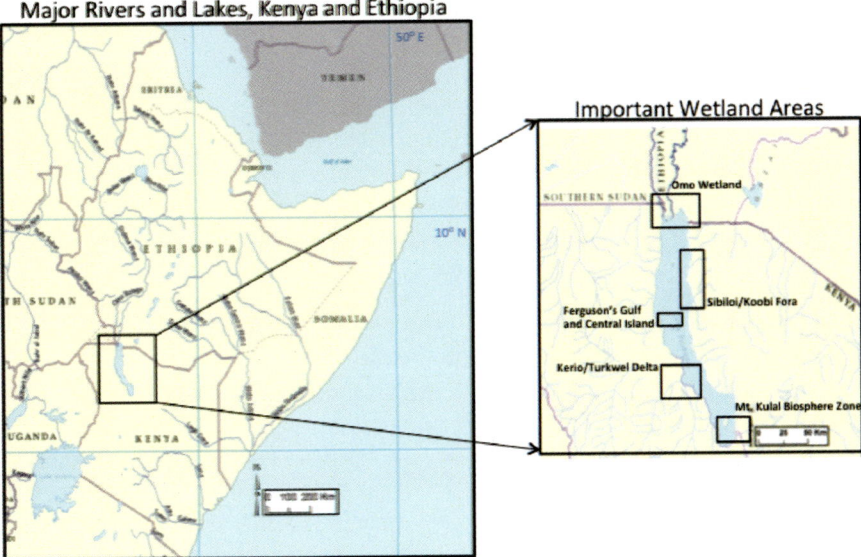

Fig. 1 Map of Kenya and Ethiopia water systems showing Lake Turkana and key areas mentioned in the text. Spatial data on lake and river centerlines were developed by Natural Earth (http://www.naturalearthdata.com/downloads/50m-physical-vectors/50m-rivers-lake-centerlines/) and base map was provided by the Environmental Systems Research Institute®

the preponderance of early hominid fossils that have been found in the region (e.g., Joordens 2011). Due to its national and global archaeological importance, Sibiloi National Park was created in 1973. In 1978, UNESCO listed Mount Kulal as a biosphere zone. In 1983 and 1985, the Central and South Island National Parks were formed, and together with Sibiloi, these were designated a UNESCO World Heritage Site in 1997. The lake is widely known as the "Jade Sea," because of its remarkable, almost incandescent color caused by the blue-green phytoplankton present on its surface.

Hydrology

Despite its large size, Lake Turkana is a highly pulsed, variable system as a result of its closed-basin nature, arid surroundings, and its strong dependence on one river for the majority of its inflow. As a result, the lake is sometimes called an "amplifier lake", i.e., "amplifies" changes in climate (Street-Perrott and Roberts 1983). The water budget of the lake is balanced between river inflows, groundwater exchanges, and evaporation losses (Avery 2010). The surface area of the lake, which receives less than 200 mm year^{-1} of rainfall, is 5.7% of its drainage area (Avery 2012a). An estimated mean evaporation rate of 2.5 m year^{-1} (Kolding 1989) requires an inflow compensation of about 600 m^3 s^{-1} or 19 km^3 year^{-1} to maintain the lake's water balance. More recent water balance modeling suggested a similar but lower inflow of

550 m^3 s^{-1} (Avery 2010). The Omo River's catchment area makes up 56.6% of the lake's drainage basin but the river contributes approximately 90% of the lake's inflow (Avery 2012b, 2013). As a result, the lake's water level fluctuations are almost entirely caused by variations in rainfall over the Ethiopian highlands. Rainfall in the lower Omo basin and over the lake is bimodal, as in the rest of Kenya, with peak rainfall in April and November. As one progresses north, the rainfall becomes unimodal, and in uppermost basin, the peak rains fall in July and August (Avery 2012a). This is the wettest part of the basin, and the lake peaks 1 month later (ibid.). At its present contemporary size, the lake has a relatively long residence time of about 12.5 years (Kolding 1992).

Data on historical and current water levels of Lake Turkana were provided by the Kenya Marine and Fisheries Research Institute (KMFRI) and obtained from TOPEX/Poseidon and other satellite records (Avery 2010, 2012a; Crétaux et al. 2011; USDA 2013). The highest lake level in recent history was recorded in the late 1800s, when levels were approximately 15 m above the zero datum of 365.4 m AMSL (Hopson 1982; Avery 2012a). Between the late 1800s and mid-1900s, the lake level dropped approximately 20 m (Avery 2012a; USDA 2013). The lake level rose 5–10 m in the 1970s and 1980s and then decreased again to current levels by 1990 (Photo 1). Over the past 25 years, the lake level has fluctuated between 360 and 365 m AMSL and has at times reached such low levels that the lake's most productive fisheries area, Ferguson's Gulf, has become dry (Avery 2012a). Ferguson's Gulf dries up when the lake level is 3.1 m lower than the Hopson zero datum, which happened most recently from 1993 to 1998, 2003 to 2008, and in 2010 and 2012 (Fig. 2). Within a given year, the lake level varies 1–1.5 m with the highest annual water levels generally occurring from September through December (USDA 2013; Fig. 3).

Photo 1 Historical lake level changes (Photo credit: W. Ojwang ©)

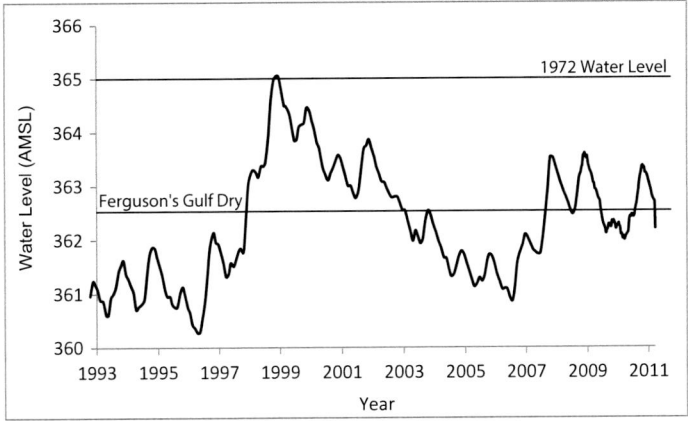

Fig. 2 Satellite altimetry data showing Lake Turkana water levels from 1993 to 2011, downloaded from http://www.LEGOS.pbs-mip.fr/soa/hydrologic/HYDROWEB (as described in Crétaux et al (2011)). The top horizontal line represents the 1972 water level and provides historical context. The bottom horizontal line represents the water levels at which Ferguson's Gulf, the lake's most productive area, dries up (Hopson 1982; Avery 2010)

Geochemistry (Water Quality and Nutrients)

Lake Turkana has average conductivity levels of 3,500 μScm^{-1}, making it a "high ion" or "Class II" lake (Talling and Talling 1965). Due to its closed-basin nature, the conductivity of the lake has been increasing by approximately 0.45 μScm^{-1} $year^{-1}$ (Hopson 1982). Current salinity levels are approximately 2.5 practical salinity units (psu) (Odada et al. 2003), with the system moving toward the limit of 3 psu defining true saline lakes. Lake Turkana is also the world's largest alkaline lake, with a pH range of 8.6–9.5 (Cohen 1986). The annual surface temperature ranges between 27.2 °C and 29.4 °C and bottom temperatures vary only 1.0 °C from 25.4 °C to 26.4 °C. Turbidity levels are high in Lake Turkana and the euphotic zone extends to only 6 m in the open lake (Källqvist et al. 1988). The lake is known for its strong southeasterly winds, which create surface water currents to the northwest and deep reverse bottom water currents (Hopson 1982). Due to these currents and the lake's relatively shallow nature, the lake is well mixed and the water is generally well oxygenated at all depths (Källqvist et al. 1988), with oxygen levels of ≥ 5 mgl^{-1} observed at all stations in a recent study (Ojwang et al. 2007).

Lake Turkana is considered to be an "allotropic riverine lake" (Jul-Larsen et al. 2003), meaning that it has a high dependence on riverine nutrient inputs (Hopson 1982; Källqvist et al. 1988). In the past, very low to trace nitrogen levels were measured (<1 mgl^{-1}) with dissolved phosphorus levels of approximately 2 mgl^{-1} (Ferguson and Harbott 1982). Recent measurements indicate much higher total nitrogen levels of 6–520 mgl^{-1} and total phosphorus levels of 0.5–140 mgl^{-1} (Ojwang et al. 2007). Although nitrogen levels are higher than they were in the past,

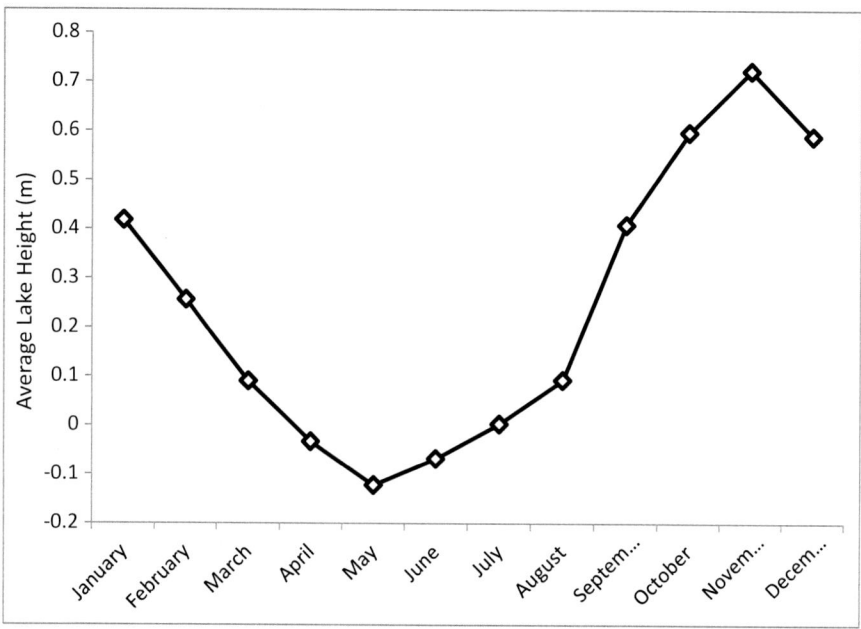

Fig. 3 Average monthly water levels (years 1993–2013) relative to a 9-year average as measured by satellite altimetry (USDA 2013, held in the public domain). Lake Turkana generally has intra-annual fluctuations of 1–1.5 m (Avery 2010)

the N:P ratio for the system is still lower than the Redfield ratio, indicating nitrogen limitation. Continued nutrient inputs from the Omo River are therefore essential to maintain the lake's productivity. The Omo River also controls the salinity and alters the turbidity of the ecosystem (Photo 2). Typical of tropical lake ecosystems, Lake Turkana exhibits relatively little seasonal variation in water temperature, and day length also varies little (Lowe-McConnell 1987). Instead, periods of high inflow from the Omo River and the resultant changes in the lake's limnological parameters act as signals for the lake's fishes to move into shallow areas or inflowing rivers to breed (Hopson 1982). The volume of the river's inflow, which controls the lake's water levels, also influences the availability and distribution of different habitat types. Sediment plumes extend up to 100 km into the lake following flood influxes (Yuretich 1979).

Diversity of Wetland Ecosystems

The most notable wetland ecosystem in Lake Turkana is the Omo Delta. Other notable wetland ecosystems within the lake include Ferguson's Gulf, Central Island National Park, and the mouths of the Kerio and Turkwel rivers (Fig. 1). In addition, the lake margins host many smaller but important wetlands, in some cases physically separate from the lake.

Photo 2 Turbid waters of River Omo flowing through delta into Lake Turkana (Photo credit: W. Ojwang ©)

Omo Delta Wetland

The Omo Delta is located at the northern tip of Lake Turkana. The delta has a complex pattern of waterways, which experience strong spatial and temporal fluctuations (Olago and Odada 2007). During the early 1970s, the Omo Delta was entirely contained within the boundary of Ethiopia. The delta increased by 500 km^2 between mid-1980s and late 1990s (Haack and Messina 1997). By the mid-2000s, the front edge of the delta had moved approximately 12 km to the south and had crossed over the Ethiopian border into Kenya, and the area of wetland vegetation had increased nearly 300%, from 117 to 334 km^2. The basis for the delta expansion is likely to be reduced lake levels and increases in sediment inflow (Avery 2010). Increases in sediment inflow are a consequence of anthropogenic influences on the river's watershed that have led to deforestation, including overgrazing (Photo 3) and clearing of land for agriculture (Haack and Messina 1997; Ayalew 2009; Avery 2010). There is a unique mode of succession taking place within the lake, with wetland vegetation largely replacing the previously water-covered region, but when the lake rises again, as it has done in 2015, these vegetated areas become inundated. While a consequent increase in faunal biodiversity in the delta area has been noted, the expansion of the delta has also attracted human population, possibly fueling the recent increase in human-wildlife conflicts (Olago and Odada 2007). The expansion of the delta southward has also increased conflicts between Ethiopian and Kenyan tribes, as Ethiopian tribes have migrated south into Kenya in order to continue fishing in the lake.

The Omo Delta and fringing riverine wetlands are characterized by dense macrophyte vegetation, dominated by *Potamogeton* spp., and the emergent grasses

Photo 3 Herds of cattle within the Omo Delta (Photo credit: W. Ojwang ©)

Paspalidium geminatum and *Sporobolus spicatus*, which occur in shallow areas. Several submerged plants, including *Ceratophyllum demersum* and *Hydrocotyle* sp., and floating plants, including *Lemna gibba*, *Nymphaea* spp., and *Ottelia ovalifolia*, have been recorded in the area (Hughes and Hughes 1992). Besides the ubiquitous acacia tree *Acacia tortilis* that dominates the landscape, there is the gingerbread palm tree *Hyphaene thebaica*, whose oval fruit is edible, and the doum palm *Hyphaene coriacea*, used for making local raft boats. The presence and imminent impact of the noxious invasive weed water hyacinth *Eichhornia crassipes* is of great concern to many conservationists, though it presumably may not survive in the semi saline lake.

Although the Omo Delta and the fringing riverine wetlands may well be considered as part of the broader Lake Turkana, they are home to some unique species that are rarely and/or hardly found in the lake proper. These include several species of mormyrids (freshwater elephant fish), *Mormyrus longirostris*, *Marcusenius victoriae*, *Marcusenius macrolepidotus*, *Mormyrus anguilloides*, *Mormyrus kannume*, *Marcusenius stanleyanus*, *Hyperopisus bebe*, and unidentified *Mormyrus* sp.; Arapaimidae (*Heterotis niloticus*), the African arowana; Gymnarchidae (*Gymnarchus niloticus*), the African knife fish (an electric fish); and Polyteridae (*Polypterus senegalus*) (Photo 4). Together with the riverine fish species, the Delta hosts representatives of more than 15 different fish families (Ojwang et al. 2011).

Other species found within the Delta include Nile crocodiles *Crocodylus niloticus* (average of eight individuals/km^{-1} along the river channel – Photo 5), several rare and endemic species of invertebrates, reptiles and amphibians, and over 128 avian species (Ojwang et al. 2011).

Photo 4 *Polypterus senegalus* inhabits the Omo Delta (Photo credit: W. Ojwang ©)

Photo 5 Nile crocodile *Crocodylus niloticus* inhabiting the mouth of R. Omo (Photo credit: W. Ojwang ©)

Kerio/Turkwel Deltas

These two smaller deltas differ greatly from each other and are associated with the seasonal Kerio and Turkwel Rivers (Fig. 1). The Kerio River runs parallel to the primary direction of the wind in the region, so its mouth is situated on a low-energy shore. It is also protected from the direct north–south wave action that arises when winds change direction and has consequently developed a dense mass of riverine-associated macrophytes. The Turkwel River, on the other hand, drains directly into the part of the bay facing the strong SE winds for which the lake is known. As a result, rooted macrophytes have failed to take hold along the banks, and there is little permanent vegetation except for the invasive thorny shrubs of *Prosopis juliflora*.

Ferguson's Gulf

Ferguson's Gulf is the most important tilapia habitat in Lake Turkana, especially for the indigenous tilapia species *Oreochromis niloticus*. The Gulf generally experiences annual water level fluctuations of 0.5–1.5 m, but has also dried up completely three times in the last 25 years (Fig. 2). The Gulf, which is approximately midway down the lake's western shoreline east of Kalokol market, is protected from the open lake's wave action and direct mixing by the Longech/Namukuse spit. The relatively calm waters of the Gulf support a different phytoplankton community from the rest of the lake, with primary production rates up to three orders of magnitude higher than in the open lake (Källqvist et al. 1988). Intensive fishing activities conducted using small-mesh beach seine, set gillnets, and purse seines are rampant in the area. The fishery is characterized by boom and bust cycles that are largely dependent on the Omo River's floods. The invasive shrub, *P. juliflora*, heavily covers the shores of the Gulf. Its thick interlocking thorny canopy blocks access to previously important fishing grounds and certain landing beaches.

Sibiloi/Koobi Fora Protected Area

Sibiloi National Park was designated a protected area under Kenyan law in 1973. The shoreline within the Park is approximately 90 km long and is characterized by several spits, muddy shorelines, some rocky shorelines, inlets, seasonal river mouths, and some lush growths of submerged and rooted macrophytes (e.g., *Potamogeton pectinatus*). Sibiloi Bay and Allia Bay are shallow regions near the Park's headquarters that boast the largest submerged beds of rooted aquatic macrophytes of Lake Turkana proper. Another important wetland habitat in Sibiloi National Park is Koobi Fora, which lies directly east of North Island, midway between the southern and northern ends of Sibiloi National Park. The wetland areas of Sibiloi National Park, which are devoid of fishing activities other than some sports fishing and cases of poaching, support the highest fish biomass in Lake Turkana. Furthermore, experimental fishing within Sibiloi National Park produced individuals larger than those caught with the same fishing gear in highly fished areas, such as those surrounding Ferguson's Gulf (Ojwang et al. 2007).

Other Protected Areas of the Lake

In 1978, UNESCO listed Mt Kulal and the southern lake area, including South Island, in its Biosphere Reserves Directory. In 1983, South Island was created as a national park in its own right, followed in 1985 by Central Island. At the south end of the lake, there are a crater lake and some small lakes that are hydraulically connected to the main lake. At Loiyangalani, on the southeastern shore, an oasis of potable springs and doum palms is the focal point for the largest human settlement on the eastern shores of the lake. Another oasis not far north is the water source for El Molo village on the lake. Similar spring-fed oases on the western lake shores at Eliye and Lobolo provide valuable sources of potable water to the local population. These springs are crucially important, as the main lake water itself is too high in fluoride for safe consumption (Avery 2010, 2012a, 2014).

Central Island is especially interesting as it includes three distinct lakes within the main lake, each with different salinities and each providing a distinct habitat for birds in particular, including lesser flamingos (Avery 2012b).

Lake Turkana Biodiversity

Phytoplankton diversity is relatively low in Lake Turkana. The phytoplankton community is dominated by the blue-green algae *Microcystis aeruginosa* and the green alga *Botryococcus braunii*. The total annual photosynthetic plankton primary production was estimated at ca. 2 kg O_2/m^2/year (Källqvist et al. 1988). The zooplankton community includes copepods, cladocerans, and protozoans, whose total production has been estimated at 216,000–540,000 tonnes dry weight per year (Hopson 1982). There are records of 50 species of benthic organisms, dominated by ostracods and insects, within the lake and Omo Delta (Cohen 1986).

Lake Turkana is home to at least 60 fish species (Froese and Pauly 2013), 12 of which are endemic. For the most part, the species found in Lake Turkana can be found elsewhere in Nilo-Sudan lake and river systems. The number of fish species in the lake is low when compared to other African lake and river ecosystems, except for Lake Albert, which has a similar fish composition and diversity. Many of the more diverse lake systems, which host hundreds of species dominated by cichlids, are older and deeper than Lake Turkana (Lowe-McConnell 1987). Endemic species of fish include small zooplanktivores (*Brycinus minutus*, *Brycinus ferox*) that form a unique mid-water scattering layer in the lake, a smaller and more pelagic species of *Lates* (*L. longispinus*) and cichlids (*Haplochromis turkanae*, *Hemichromis exsul*). Unlike in Lake Victoria and some other African Lakes, the Nile tilapia *Oreochromis niloticus* and Nile perch *Lates niloticus* are native to Lake Turkana and in fact are the highest valued species in the lake's commercial fishery.

Lake Turkana supports over 350 native and migratory bird species, making it an "Important Birdlife Area" (UNESCO 2014). The lake and Omo River, in particular, also host the world's largest remaining population of Nile crocodile (*Crocodylus niloticus*) and contain protected breeding grounds for this species as well as for hippopotamus and several venomous snake species (UNESCO 2014). Mammals sighted in the park areas and their environs include Grevy's *Equus grevyi* and Burchell's zebra *Equus quagga burchellii*, Grant's gazelle *Nanger granti*, beisa oryx *Oryx beisa*, topi *Damaliscus korrigum*, greater kudu *Tragelaphus strepsiceros*, hippopotamus *Hippopotamus amphibious*, lion (*Panthera leo*; IUCN Red List status – vulnerable), cheetah (*Acinonyx jubatus*; IUCN Red List status – vulnerable), leopard *Panthera pardus*, striped hyena *Hyaena hyaena*, wild dog (*Lycaon pictus*; IUCN Red List status – endangered), and silver-backed jackal *Canis mesomelas*. There are also four species of endemic reptiles in the region, including three species of frogs (*Bufo chappuisi*, *B. turkanae*, and *Phrynobatrachus zavattari*) and the endemic Turkana mud turtle (*Pelusios broadleyi*; IUCN Red List status – vulnerable).

Ecosystem Services

Lake Turkana's broader basin, the lake, and its fringing floodplain wetlands provide a host of hydrologic, ecological, economic, and socioeconomic services. These services include water provision for domestic and livestock use, energy (hydroelectric power) and agricultural uses, habitat for fisheries, forage for livestock, fuel, building materials (Photo 6), natural food products, climate moderating effects, as well as significant opportunities for ecotourism and preservation of cultural values. Important sites for tourism include Sibiloi National Park and the geologically active Central Island, which hosts the magnificent Crocodile, Flamingo, and Tilapia lakes, also South Island National Park. The lake, the Omo and Turkwel rivers, and associated springs are permanent water sources utilized by thousands of people, hence forming important lifelines in the region for millennia, perhaps dating back to the dawn of humankind.

Fishing has taken place on Lake Turkana for at least 10,000 years, with catches used primarily for local consumption until the emergence of the commercial fishery

Photo 6 Macrophytes used to thatch huts within the Omo Delta (Photo credit: W. Ojwang ©)

in the 1940s (Owen et al. 1982) – see Fig. 4. Although Lake Turkana may have the potential to increase food security in a region where reliance on food aid is ubiquitous, the sustainability of the fishery has not been extensively studied. Pastoralism has been the preferred livelihood of people surrounding the lake (Photo 7) for the last few thousand years, but fishing provides an important alternative and a "safety net" livelihood (Photo 8) in the region (Kaijage and Nyagah 2010). Currently, one of the largest obstacles faced by the Lake Turkana fishery is postharvest losses, which can be as high as 50% (Ojwang et al. 2007).

Riparian Communities

Lake Turkana is abutted by Turkana County on its western side and Marsabit County on its eastern side, with some of the lake's southernmost regions within Samburu County. The Turkana tribe dominates Turkana County, but minor tribes that have migrated to the lake from other regions are also present. Marsabit County has a more diverse group of tribes, including the Dassanech, the Gabra, the Rendille, the Samburu, the El Molo, and the Turkana (Kaijage and Nyagah 2010). Nearly 100,000 members of at least eight distinct indigenous ethnic groups are heavily reliant on flood-recessional farming along the Omo River (Richter et al. 2010), while about 250,000 people of various ethnicities are dependent on fishing within the lake basin.

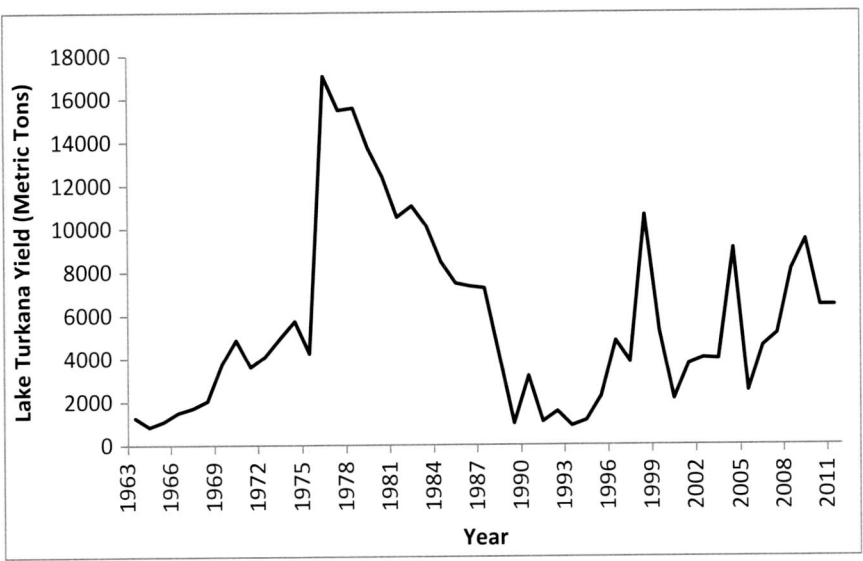

Fig. 4 Fisheries yield for Lake Turkana in metric tons 1963–2011. Fisheries data collected and provided by Kenya's Ministry of Agriculture, Livestock and Fisheries

Photo 7 Riparian settlements (typical of pastoralists) in the northern part of Lake Turkana (Photo credit: W. Ojwang ©)

Photo 8 Turkana lady amidst piles of dried fish (Photo credit: W. Ojwang ©)

Conservation Status and Management

Lake Turkana is a shared transboundary resource by Kenya and Ethiopia. The region is rich in natural resources and hosts unique endemic species. Even though parts of the lake and lower Omo Delta have been zoned as an international biosphere reserve, the protected areas are facing immense threat from anthropogenic activities. In spite of the apparent threats, there is no management plan in place to guide resource use in the region. Recent efforts toward the development of wildlife management and fisheries management plans by Kenya Wildlife Services (KWS), National Museums of Kenya (NMK), and the State Department of Fisheries, Kenya, are initiatives worth fast tracking. Otherwise the prevailing scenario of uncoordinated Lake Turkana resources management will ultimately compromise ecosystem services with drastic negative implications for development, poverty alleviation, and adaptation toward anticipated long-term environmental changes.

The natural resources and human populations within the Omo Delta, which is an "oasis" in the region, could benefit tremendously if it was considered for designation as a Ramsar Site. Other management efforts to consider include the establishment of a Lake Turkana-Omo Delta Transboundary Resource Management Committee, with members drawn from focal point ministries in Kenya and Ethiopia. Conservation and utilization of the World Heritage archaeological sites are needed to recognize and safeguard the region's cultural heritage and to create development opportunities.

Threats and Future Challenges

Human activities in the Lake Turkana basin have accelerated the rate of ecological change and increased threats to the existing natural resources. Major threats in the region include current and planned construction of dams on the Omo River; agricultural expansion and intensification through irrigation projects along the Omo River; environmental stressors including climate change, recent oil discovery, and ongoing exploration activities; associated resource use conflicts (Photo 9); and construction of Africa's largest wind power plant. The number and variety of threats facing the Turkana Basin all attest to a rapidly changing environment, which generates great concern for the future persistence and long-term viability of the important ecosystem services that have been provided over generations. Thus, a combination of external factors (hydropower dams, irrigation schemes, climate anomalies) and internal drivers (demography, economic growth) will strongly impact the Lake Turkana basin over the next decade. In turn, this will have significant negative consequences on resource productivity and the well-being of local communities.

Two dams (Gibe I and Gibe II) have been constructed along the Omo River, a third dam is under construction (Gibe III), and there are plans to build two additional

Photo 9 Insecurity and resource use conflicts (Photo credit: W. Ojwang ©)

dams in the future (Gibe IV and V) (See Fig. 5). Gibe III will be 240 m high and produce 6,400 GWh/year of energy (Asnake et al. 2009). The filling of its reservoir, which will store 11,750 m^3 of water, will lead to a reduction of 2 m in Lake Turkana's water levels (Avery 2010, 2012a; Velpuri and Senay 2012). Gibe IV's reservoir would require a similar volume of water to fill (Avery 2012a).

A minimum environmental flow and an artificial 10-day flood have been proposed for Gibe III, but it is unknown whether a flood of this duration and size will be sufficient to sustain the ecological functioning of the lake (Avery 2012a). Although a reduction in flooding is touted as a benefit by some (e.g., Asnake et al. 2009), the resultant dampening of Lake Turkana's water level fluctuations is of great ecological concern, given the importance of intra-annual fluctuations in controlling fish productivity in African lakes (e.g., Kolding and van Zweiten 2012). The amplitude of the controlled lake level fluctuations, assuming the environmental floods proposed by Asnake et al. (2009) are implemented, will be 400 mm less than the amplitude of the lake's natural fluctuations.

Associated with the Gibe dams are thousands of hectares of sugar cane and cotton plantations and their irrigation infrastructure. Currently 150,000 ha are being developed along the Omo River as part of the Kuraz Sugar Project (see Fig. 5), whose planting began in February 2013 (Ethiopian Sugar Corporation 2014). The Ethiopian government plans to increase sugar production within Ethiopia from 300,000 t in 2009/2010 to 2.25 million tons by mid-2015 (Ethiopian Sugar Corporation 2014). Additional land concessions have been awarded, mainly for cotton production. A total of 445,000 ha commercial agricultural development is planned, of which 135,285 ha has been excised from two national parks (Omo and Mago) and one wildlife reserve (Tama) (Avery 2012a, 2013, 2014).

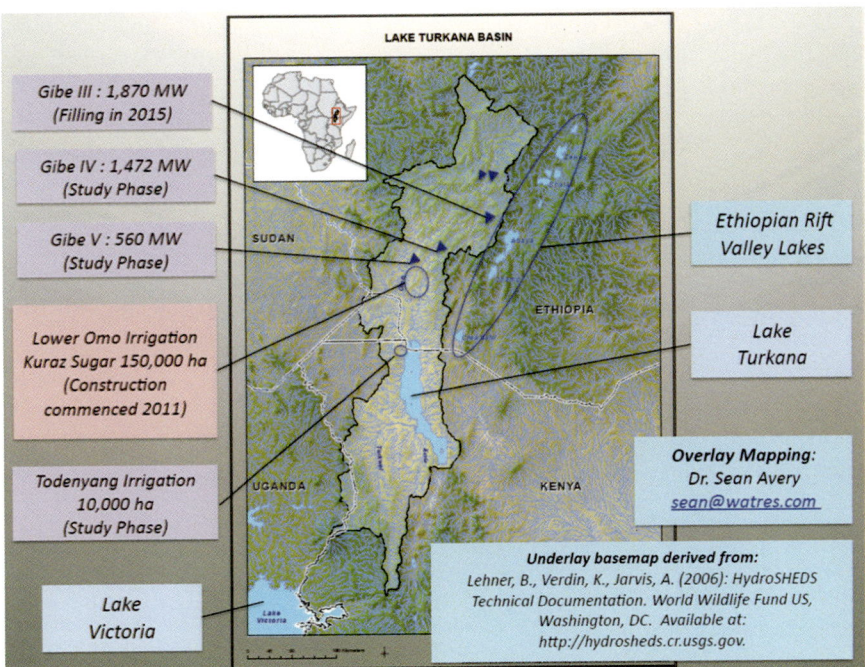

Fig. 5 Lake Turkana basin showing major development projects

The Kuraz sugar scheme will alone abstract at least 30% of the Omo River's flow (ibid.). With the development of other irrigation schemes, the abstraction can only increase, which means the inflow to the lake will diminish. If there is 33.5% reduction in the lake's Omo River inflow, the lake will drop 13 m, and its volume will reduce to 59% of its otherwise sustainable volume (ibid.). If there is 52% reduction in Omo inflow, the lake will fall 22 m, and over half the lake's volume will be lost (ibid.). A drop of as little as 10 m in lake level would reduce the lake's size from 7,560 to 5,900 km^2 and the lake's volume from 238 to 170 km^3 (Avery 2012a). Changes in the lake's water levels will also lead to changes in the shape of the lake's shorelines, which will be the most prominent in the wetland areas shown in Fig. 1 (Omo Delta, Ferguson's Gulf, Allia Bay of Sibiloi National Park, and at the Turkwel and Kerio Deltas; Avery 2010, 2012a, 2014; Velpuri and Senay 2012). Changes in the annual volume and patterns of water inflow from the Omo River will impact a variety of important parameters in Lake Turkana, including turbidity, salinity, productivity, and habitat availability. These changes will interact to influence the feeding, breeding, movement, and ultimately the population levels of fishes in the lake and therefore the lake's fishery. The Gibe Dams and irrigation schemes under construction will also have impacts beyond water inflow changes. For example, eutrophication caused by increased nutrient loads from fertilization of upstream crops or changes in turbidity due to deforestation along the Omo River will also alter the Lake Turkana ecosystem.

In addition to local development projects, the impacts of global climate change on the Lake Turkana ecosystems are projected to increase in severity. The ecosystem, which is already known for its historical environmental vicissitudes, will progressively experience even more extreme changes as a result of global climate change (e.g., Bishaw 2012). Greater efforts are therefore needed to forecast hydrodynamic and ecological responses to climate change in Lake Turkana. It is also important to understand and to strengthen local adaptation not only to improve local conditions but to avoid destabilization of the region occasioned by resource use conflicts.

In addition, the Lake Turkana region, which has historically been ignored by investors, has rapidly gained interest recently. This is attributed to the discovery of oil in the area and the ongoing exploration, which covers nearly the entire lake basin. The oil discovery is seen by many as the ultimate solution to the perennial resource use conflicts, poverty, and famine in the region. However with recent tumbling oil prices (in late 2014 and early 2015), the expected bonanza may be long delayed. Plus, if experiences from other part of Africa are anything to go by, then the usual "oil resource curse" may be unavoidable, i.e., a scenario where African countries with oil are ironically among the most economically troubled, undemocratic, and most conflict prone. Other socioeconomic activities that are usually associated with oil discovery and exploitation are also likely to have negative impacts on human well-being and biodiversity in the region.

The Lake Turkana region has also been earmarked for the development of one of the largest wind power farms in the region. The project when completed is expected to generate 310 MW. Other development activities in the area include the construction of the Lamu Port-South Sudan-Ethiopia Transport Corridor (LAPSSET), which will cut across the Lake Turkana area. These activities are not well planned and mitigated as per the requirements of Environmental Impact Assessments and will have enormous negative impacts on the resources of the region as well as potentially compromising the various ecosystem services that are currently enjoyed by the local communities and the region as a whole.

References

Asnake A, Stigliano G, Woldermarian D. Gibe III Hydroelectric Project: Environmental and Social Management Plan. Addis Ababa (Ethiopia): EEPCO (Ethiopian Electric Power Corporation); 2009.

Avery ST. Hydrological impacts of Ethiopia's Omo basin on Kenya's Lake Turkana water levels & fisheries. Nairobi (Kenya): African Development Bank; 2010.

Avery ST. Lake Turkana and the Lower Omo: hydrological impacts of Gibe III and lower Omo irrigation development, vols. I and II. African Studies Centre/University of Oxford; 2012a. http://www.africanstudies.ox.ac.uk/what-future-lake-turkana

Avery PK. Kenya's Jade Jewel in Peril from Ethiopia Plans. SWARA, April–June 2012b.

Avery ST. The impact of major hydropower and irrigation development on the world's largest desert lake. What future Lake Turkana? African Studies Centre/University of Oxford; 2013. http://www.africanstudies.ox.ac.uk/what-future-lake-turkana

Avery ST. What future for Lake Turkana and its wildlife? SWARA, Jan–Mar 2014.

Ayalew L. Analyzing the effects of historical and recent floods on channel pattern and the environment in the Lower Omo basin of Ethiopia using satellite images and GIS. Environ Geol. 2009;58:1713–26.

Bishaw Y. Evaluation of climate change impact on Omo Gibe Basin (case study of the Gigel Gibe III reservoir). Civil Engineering [Thesis]. Addis Ababa University; 2012.

Cohen AS. Distribution and faunal associations of benthic invertebrates at Lake Turkana. Kenya Hydrobiologia. 1986;141:179–97.

Crétaux JF, Jelinski W, Calmant S, Kouraev A, Vuglinski V, Bergé-Nguyen M et al. SOLS: a lake database to monitor in the near real time water level and storage variations from remote sensing data. Adv Space Res. 2011;47:1497–507.

Ethiopian Sugar Corporation. http://www.etsugar.gov.et/en/. Accessed May 2014.

Ferguson AJD, Harbott BJ. Geographical, physical and chemical aspects of Lake Turkana. In: Hopson AJ, editor. Lake Turkana: a report of the findings of the Lake Turkana Project 1972–1975. London: Overseas Development Administration; 1982. p. 1–107.

Froese R, Pauly D, editors. FishBase. World Wide Web electronic publication. www.fishbase.org; 2013.

Haack B, Messina J. Monitoring wetland changes with remote sensing: an East African example. Paper presented at: Space Technology and Applications International Formum (STAIF - 97). AIP Conference Proceedings Vol. 387: 215–20; 1997 Jan. 26–30; Albuquerque, New Mexico.

Hopson AJ, editor. Lake Turkana: a report on the findings of the Lake Turkana project 1972–1975, vols. 1–6. London: University of Stirling; 1982.

Hughes RH, Hughes JS. A directory of African wetlands. Gland/Nairobi/Cambridge: IUCN/UNEP/WCMC; 1992.

Johnson TC, Malala JO. Lake Turkana and its link to the Nile. In: Dumond HJ, editor. The Nile: origin, environments, limnology and human use. Monographiae Biologicae. 2009;89:287–304.

Joordens JCA. The power of place: climate change as driver of hominin evolution and dispersal over the past five million years. Amsterdam: Vrije Universiteit; 2011.

Jul-Larsen E, Kolding J, Overå R, Nielsen JR, van Zwieten PAM, editors. Management, co-management or no management?: major dilemmas in southern African freshwater fisheries. Case studies. Rome: Food & Agriculture Org; 2003.

Kaijage S, Nyagah N. Socio–economic analysis and public consultation of Lake Turkana Communities in Northern Kenya. Final Report. Tunis: African Development Bank; 2010.

Källqvist T, Lien L, Liti D. Lake Turkana – Limnological Study 1985–1988. Mombasa (Kenya): Norwegian Institute for Water Research (NIVA)/Kenya Marine and Fisheries Institute; 1988. 99 pp.

Kolding J. The fish resources of Lake Turkana and their environment. Cand. Scient. [Thesis]. University of Bergen and final report of project KEN-043. Oslo: NORAD; 1989.

Kolding JA. Summary of Lake Turkana: an ever-changing mixed environment. Mitt Internat Ver Limnol. 1992;23:25–35.

Kolding J, van Zwieten PA. Relative lake level fluctuations and their influence on productivity and resilience in tropical lakes and reservoirs. Fish Res. 2012;115:99–109.

Lowe-McConnell RH. Ecological studies in tropical fish communities. Cambridge, UK: Cambridge University Press; 1987.

Odada EO, Olago DO, Bugenyi F, Kulindwa K, Karimumuryango J, West K, Achola P. Environmental assessment of the East African Rift Valley lakes. Aquat Sci. 2003;65:254–71.

Ojwang WO, Abila R, Malala J, Ojuok JE, Owili M, R Omondi. Critical transboundary resource: assessment of ecological and socio-economic importance of River Omo Wetland. Project Technical report submitted to the National Council of Research and Technology, Kenya; 2011.

Ojwang WO, Asila AA, Malala JO, Ojuok JE, Othina A. The status of fishery in Lake Turkana. In: Gichuki J, Getabu A, Wakwabi E, Abila R, editors. Lake Turkana, fisheries people and the future I. Interventions for economic benefits. KMFRI/TECHNICAL REPORT/1; 2007: 122p.

Ojwang OW, Gichuki J, Getabu A, Wakwabi E, Abila R. Lake Turkana, Fisheries, People and the Future II "Intervention for Economic Benefits". Mombasa (Kenya): Kenya Marine Fisheries Research Institute; 2007.

Olago DO, Odada EO. Sediment impacts in Africa's transboundary lake/river basins: case study of the East African Great Lakes. Aquat Ecosyst Health Manage. 2007;10:23–32.

Owen RB, Barthelme JW, Renaut RW, Vincens A. Palaeolimnology and archaeology of Holocene deposits north-east of Lake Turkana. Kenya Nat. 1982;298(5874):523–9.

Richter BD, Postel S, Revenga C, Scudder T, Lehner B, Churchill A, Chow M. Lost in development's shadow: the downstream human consequences of dams. Water Altern. 2010;3:14–42.

Ricketts RD, Johnson TC. Climate change in the Turkana basin as deduced from a 4000 year long record. Earth Planet Sci Lett. 1996;142:7–17.

Street-Perrott FA, Roberts N. Fluctuations in closed-basin lakes as an indicator of past atmospheric circulation patterns. In: Street-Perrott A, Beran M, Ratcliff R, editors. Variations in the global water budge. Dordrecht: D. Reidel Pub. Co; 1983. p. 331–45.

Talling JF, Talling IB. The chemical composition of African lake waters. Int Rev gesamten Hydrobiol Hydrogr. 1965;50:421–63.

UNESCO. Lake Turkana National Parks. http://whc.unesco.org/en/list/801. Accessed May 2014.

USDA. Lake Turkana height variations. http://www.pecad.fas.usda.gov/cropexplorer/global_reservoir/gr_regional_chart.aspx?regionid=eafrica&reservoir_name=Turkana. Accessed Dec 2013.

Velpuri NM, Senay GB. Assessing the potential hydrological impact of the Gibe III Dam on Lake Turkana water level using multi-source satellite data. Hydrol Earth Syst Sci. 2012;16:3561–78.

Yuretich RF. Modern sediments and sedimentary processes in Lake Rudolf (Lake Turkana) eastern rift valley, Kenya. Sedimentology. 1979;26:313–31.

Soda Lakes of the Rift Valley (Kenya)

Simon Agembe, William Ojwang, Casianes Olilo, Reuben Omondi, and Collins Ongore

Contents

Introduction 1382
Lake Bogoria 1384
Lake Nakuru 1385
Lake Elementeita 1387
Lake Magadi 1388
Threats and Future Challenges Facing the Kenyan Soda lakes 1389
References 1391

Abstract

Soda lakes are alkaline with pH values ranging from 8 to 12 and characterized by high concentrations of principal ions such as Na^+, HCO^-, $CO3^{2-}$, and Cl^-. Kenya is endowed with many soda lakes forming part of the East African Rift Valley system and includes lakes Bogoria, Nakuru, Elementeita, and Magadi. In addition, Lake Turkana is located on the Kenya-Ethiopian border further north. These lakes are characterized by steep fault escarpments, deep gorges, canyons, and craters on the rift floor, some of which have gushing geysers and hot springs. Historically, the lakes were thought to have been one continuous system called

---

S. Agembe (✉)
Department of Fisheries and Aquatic Sciences, University of Eldoret, Eldoret, Kenya
e-mail: agembesimon@yahoo.com

W. Ojwang (✉) · C. Olilo (✉) · C. Ongore (✉)
Kenya Marine and Fisheries Research Institute (KMFRI), Mombasa, Kisumu, Kenya
e-mail: wojwang@wwfkenya.org; olilocasianes@yahoo.com; collongore@gmail.com

R. Omondi (✉)
Department of Applied Aquatic Sciences, Kisii University, Kisii, Kenya
e-mail: reubenomondi@yahoo.com

© Springer Science+Business Media B.V., part of Springer Nature 2018
C. M. Finlayson et al. (eds.), *The Wetland Book*,
https://doi.org/10.1007/978-94-007-4001-3_150

Lake Kamatian. Reconstruction of the history of the four lakes based on dated sedimentary time-series data reveal unique hydrological, ecological, and species richness trends that have fluctuated through time between alkaline and freshwater conditions. In spite of their apparent inhospitality, these soda lakes are among the most productive aquatic environments on earth and support a great diversity of species, some which are endemic, rare, and endangered. This apparently unique phenomenon has been attributed to the virtually unlimited availability of dissolved carbon dioxide. Other ecosystem services provided by the lakes include habitats for biota including the famous populations of flamingos and rare and threatened mammals, supply of water for domestic use and irrigation, thermal energy, pasture, and recreational and cultural contributions. Currently, Kenya's soda lakes face various challenges, ranging from climate change and water extraction to direct habitat modification, all of which may have significant impacts on existing biodiversity. Even though the lakes are generally referred to as soda lakes, each has its distinct qualities in terms of biogeochemical attributes.

Keywords

Alkaline lakes · Hydrochemistry · Biodiversity · Kenya · Rift valley lakes · Soda lakes

Introduction

Soda lakes are alkaline with pH values ranging from 8 to 12 and characterized by high concentrations of principal ions such as $Na+$, HCO^-, CO_3^{2-} and Cl^-. Kenya is endowed with many soda lakes forming part of the East African Rift Valley system and includes lakes Bogoria, Nakuru, Elementeita, and Magadi (Fig. 1). In addition, Lake Turkana is located on the Kenya-Ethiopian border further north and is described in a separate chapter in this volume. These lakes are characterized by steep fault escarpments, deep gorges, canyons, and craters on the rift floor, some of which have gushing geysers and hot springs (Grant et al. 1990). Historically, the lakes were thought to have been one continuous system called Lake Kamatian (Gregory 1921). Reconstruction of the history of the four lakes based on dated sedimentary time-series data reveal unique hydrological, ecological, and species richness trends that have fluctuated through time between alkaline and freshwater conditions (Grant et al. 1999).

In spite of their apparent hospitality, these soda lakes are among the most productive aquatic environments on earth and support a great diversity of species, some which are endemic, rare, and endangered (Melack and Kilham 1974; Grant et al. 1999). This apparently unique phenomenon has been attributed to the virtually unlimited availability of dissolved carbon dioxide. Other ecosystem services provided by the lakes include habitats for biota including the famous populations of flamingos and rare and threatened mammals, supply of water for domestic use and irrigation, thermal energy, pasture, and recreational and cultural contributions. Currently, Kenya's soda lakes face various challenges, ranging from climate change

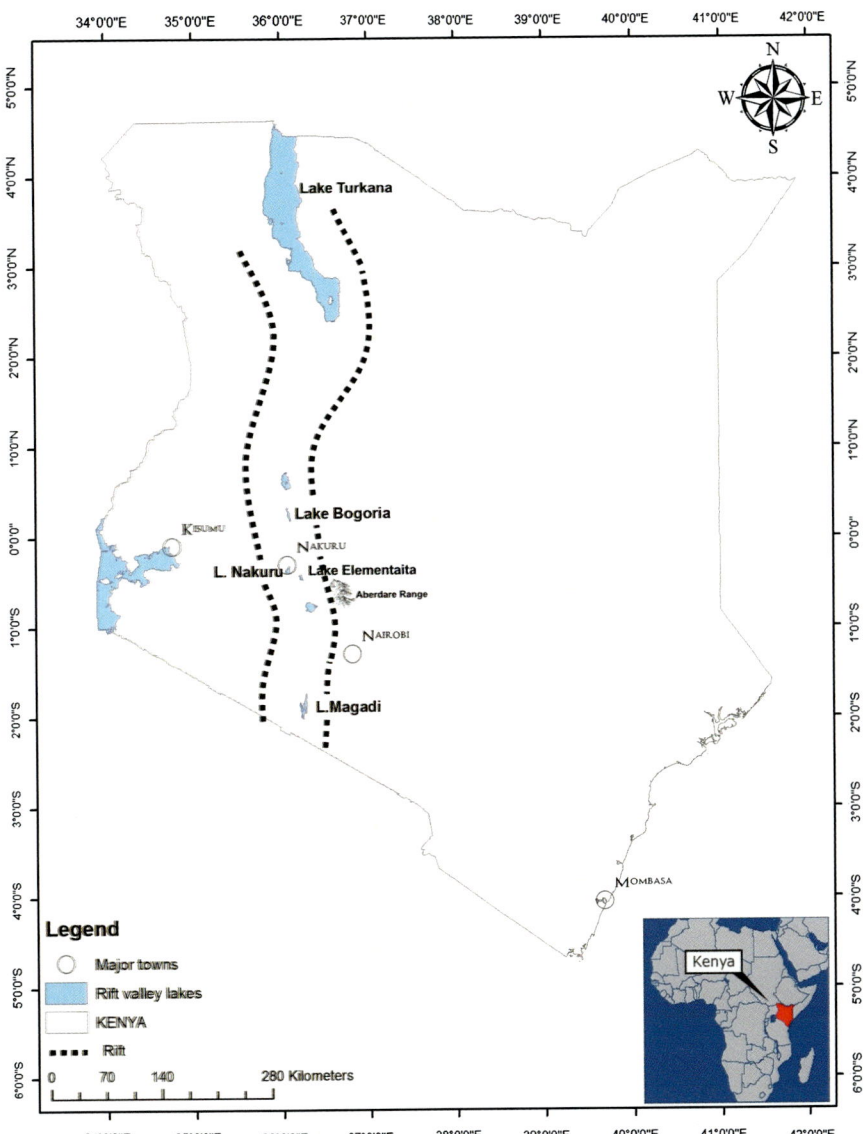

Fig. 1 Map of the Soda Lakes of the Rift Valley in Kenya (Lakes Bogoria, Nakuru, Elementeita, and Magadi). Lake Turkana (furthest north) is covered by a separate chapter in this volume

and water extraction to direct habitat modification, all of which may have significant impacts on existing biodiversity.

Even though the lakes are generally referred to as soda lakes, each has its distinct qualities in terms of biogeochemical attributes.

Lake Bogoria

Lake Bogoria (00° 15′N 36° 05′E) lies within Lake Bogoria National Park at an altitude of 990 m above sea level. The lake has a surface area of 34 km^2 and a catchment area of 930 km^2. It is shallow with an average and maximum depth of 10 m and 14 m respectively. Sediment cores indicate freshwater conditions existed at various times during the past century, and the lake level was up to about 9 m higher than its present level (Tiercelin and Vincent 1999).

Its wetland ecosystems include Lake Bogoria and shoreline, Loburu and Loboi swamps, hot spring marshes, geysers, springs, and mouths of the Emsos River and Sandai River. Some of the services provided by these ecosystems include production of cyanobacteria such as *Arthrospira fusiformis* that is an important food for lesser flamingo *Phoenicopterus minor*; bee honey, wild fruits, and vegetables which are important human food; salt lick and trona (soda ash) used for livestock fodder; Cyperus macrophytes used for making mats and baskets; and sand and stones used for buildings and murram (laterite) used for road making. It also supports wildlife such as the greater kudu *Tragelaphus strepsiceros* and birds such as the lesser flamingo. The recreational activities include archaeological and contemporary historic sites, landscape, and tourism.

In 1970, Lake Bogoria was first declared a national reserve because of its biodiversity, scenery, and hydrologic features. The status was later changed to a National Park in 1990 and subsequently designated as a Wetland of International Importance under the Ramsar Convention in 2001. The lake region has continued to have significant sociocultural, sacred, and economic (pastoral) meaning to the local ethnic groupings of Endorois and Illchamus people.

The hydrology of Lake Bogoria is primarily influenced by the inflows from the Sandai, Loboi, and Emsos Rivers during the wet season (April to August), discharges from the lake floor, and from approximately 200 alkaline hot springs nearby. The lake affects the local hydrological cycle with its evaporation and subsequent condensing as rainfall in the Mochongoi Hills and runoff from the lake's 930 km^2 drainage basin. During the April to August wet season, the increased flow from the catchment coupled with geysers along the banks of Lake Bogoria raise the water level which subsequently declines through December to March (Renault and Owen 2005). It also regulates hydrologic cycle, through evapotranspiration, condensation in Mochongoi Hills, and subsequent rainfall and runoff to Lake Bogoria drainage system of 930 km^2 and into the lake itself. During wet season in the months of April to August, the increased runoff flows from the catchments' area coupled with geysers in the banks of Lake Bogoria cause inundations and rise in the lake level, while during dry season in the months of December to March, the lake level falls (Renault and Owen 2005). Lake Bogoria has no known surface outlet and is located on a semiarid belt within a region, otherwise exhibiting a tropical climate and receiving approximately 50 mm of rainfall per month and a precipitation of 400–600 mm per year with high rates of evapotranspiration at 1500 mm per day. Temperatures range from 18 °C to 35 °C during wet and dry seasons, respectively.

Lake Bogoria National Park supports a range of wetland and terrestrial biodiversity. The hypersaline nature of Lake Bogoria provides an optimal environment for primary production of the populations of algae such as cyanobacteria. Cyanobacteria populations in the lake are dominated by *Arthrospira fusiformis* but also include other species such as *Anabaenopsis arnoldii, Anabaenopsis* spp., *Microcystis flos-aquae, Raphidiopsis* sp., and *Haloleptolyngbya alcalis* (Vareschi 1978; Kaggwa et al. 2013). Other phytoplankton groups in the lake include *Cryptomonads* and green algae, e.g., *Ankistrodesmus* sp., *Crucigenia* sp., and *Monoraphidium minutum*. Species of diatoms include *Nitzschia* sp. and *Navicula* sp., and euglenoids, *Euglena* spp. *Anabaenopsis arnoldii* and *Microcystis flos-aquae*.

Several higher plant species are found within the lake's vicinity. These include the following Poaceae species: *Dactyloctenium aegyptium, Sporobolus ioclados, Digitaria velutina,* and *Chloris virgata*; while shrubs include *Grewia tenax, G. bicolor,* and *Acacia mellifera*; and the trees *Balanites aegyptica, Combretum* spp., *Ficus* spp., *Terminalia* spp., and *Acacia tortilis*; cactus (*Opuntia sp.*); and *Prosopis juliflora*. The shoreline of the lake is vegetated with *Cyperus laevigatus, Sporobolus spicatus,* and *Cynodon dactylon*.

The zooplankton community in Lake Bogoria is dominated by the monogonont rotifer species *Brachionus plicatilis* and *Hexathra jenkinae*. Insect communities in the area include Diptera such as tsetse fly *Glossina morsitans* that causes nagana illness in cattle and sleeping sickness in humans, and odonata larvae adapted to cope with extreme alkaline conditions.

While over 120 bird species occur in Lake Bogoria, its status as a National Park is based on the presence of several key bird species. The lesser flamingo does not breed in the lake but migrates here in large numbers to feed, specifically on *Arthrospira fusiformis* (Vareschi 1984). Between the months of October and March, at least 10,000 lesser flamingos occur on the lake and decline to less than 5,000 birds during the months of April to September. Other bird species include the common ostrich *Struthio camelus*, grey heron *Ardea cinerea*, hamerkop *Scopus umbretta*, little egret *Egretta garzetta*, white-faced whistling duck *Dendrocygna viduata*, African fish-eagle *Haliaeetus vocifer*, tawny eagle *Aquila rapax*, yellow-billed stork *Mycteria ibis*, common sandpiper *Actitis hypoleucos*, African citril *Serinus citrinelloides*, and the brubru bushshrike *Nilaus afer*.

Besides birds' species, the status of Lake Bogoria as a National Park is further supported with the occurrence of more than 20 mammal species including the threatened greater kudu *Tragelaphus strepsiceros*, the East African impala *Aepyceros melampus rendilis*, and the rare vervet monkey *Cercopithecus pygerythrus* (Lake Bogoria National reserve 2007).

Lake Nakuru

The lake is situated within the Nakuru National Park at $00°20'$ S $36°05'$ E and at 1759 m above sea level. It has a surface area of 40 km^2 and a catchment area of 800 km^2 (Vareschi 1982; Tiercelin and Vincens 1987). The lake has gradually developed alkaline conditions over the last several decades, creating harsh conditions and consequent ecological limitations for various flora and fauna (Shivoga 2001).

Lake Nakuru wetland ecosystems include the lake shoreline and Baharini springs. The lake has a mean depth of 2 m, and surface water temperature ranges between 25° and 27 °C (Vareschi 1982). The wet season in the region fall between April and August, while the dry season occurs between October and March. The mean annual rainfall is 95.1 mm. Rivers draining Lake Nakuru include Njoro, Makalia, Lamuriak, and Enderit. Sources of water into the lake include rainfall, surface runoff, and treated wastewater from two sewage treatment plants in the nearby town of Nakuru. In 1997, the lake was almost completely dry, partly as a result of drought and large volumes of water being extracted upstream by farmers.

Lake Nakuru has been designated as a Bird Sanctuary since 1960, declared a National Park in 1968, as well as the first rhino sanctuary and the first Ramsar Site in Kenya in 1987 and 1990, respectively. It was identified as an Important Bird Area in 1999, and lastly in 2011, it was designated as a World Heritage Site by UNESCO.

The conservation status of Lake Nakuru signifies its importance as a habitat for key biodiversity including its signature species, the lesser flamingo, and over 56 species of mammals including the near-threatened white rhinoceros *Ceratotherium simum cottoni* and endangered Rothschild's giraffe *Giraffa camelopardalis rothschildi*, buffalo, lion, zebra, eland, and waterbuck.

The lake is world famous for its bird fauna, with over 450 species, dominated by the lesser flamingo. Home to over a million flamingos, flocks of lesser and greater flamingos *Phoenicopterus roseus* line the shores of the lake, giving it a pinkish appearance. The total flamingo count varies year to year and season to season, but frequently stands at about two million with lesser flamingo being more than greater flamingo. A cichlid fish, *Alcolapia grahami*, introduced into the lake in 1960, is the main herbivore (Vareschi 1979). Other biota found in the lake includes dense populations of Cyanobacteria *Anabaenopsis magna* and *Arthrospira fusiformis* which dominate the phytoplankton community (Finlay et al. 1987). The surrounding vegetation is mainly wooded and bushy grassland with wide ecological diversity. It has about 550 different plant species, including the unique euphorbia (*Euphorbia candelabrum*) forest and acacia (*Acacia tortilis*) woodlands.

The lake is fringed by alkaline swamps with areas of sedge and typha marsh along the river inflows and springs. The surrounding areas support a dry transitional savanna with lake margin grasslands of *Sporobolus spicatus* salt grass moving into grasslands of *Hyparrhenia hirta* and Rhodes grass *Chloris gayana* in the lower areas. More elevated areas have dry forest with *Acacia xanthophloea*, olive *Olea hochstetteri*, and *Euphorbia candelabrum* forest. The bushland is dominated by the composites, Mulelechwa *Tarchonanthus camphoratus* and *Psiadia arabica*.

Because of its incredible biodiversity, the lake's natural resources continue to contribute considerably to the country's economy through tourism activities in the area. It supports the livelihoods of cosmopolitan (i.e., Nakuru town) and diverse multiethnic communities (predominantly Kikuyus, Maasais, and Kalenjins) through provision of water for domestic and agricultural uses, recreation, employment in the tourism sector, and other values that are hardly appreciated such as local climate modulation and aesthetics.

Lake Elementeita

Lake Elementeita was formed through tectonic and volcanic activities and lies approximately 30 km south of Lake Nakuru. The lake derives its name from the Maasai word *muteita*, meaning "dust place" in reference to the dry and dusty conditions of the area, especially between January and March. It lies at 00°27′ S 36°15′ E and an altitude of 1782 m above sea level (Melack 1988; Mwaura and Moore 1991). The lake has a surface area of 20 km^2 and a catchment area of 500 km^2. Lake Elementeita is fed by three seasonal streams, the Mbaruk to the north and Kariandusi and Memeroni to the south. Some hot springs along the southern end also supply the lake with a minimal amount of water (Mwaura and Moore 1991). Annual rainfall in the area is low and highly variable, mostly concentrated between April and July for long rains and in October and November for the short rains. The water budget is strongly influenced by precipitation, evaporation, and highly seasonal small inflows. Air temperatures are highly variable, ranging from 25 °C to 32 °C during the day, while at night temperatures fall below 15 °C. The lake is extremely shallow and shows a pan-like characteristic that makes it lose water rapidly through evaporation exacerbated by rapid heating of its sediments (Oduor and Schagerl 2007).

Similar to the other soda lakes, Lake Elementeita hosts a diversity of flora and fauna. In recognition of its importance as a critical habitat, Lake Elementeita was formally recognized as a Ramsar Site in 2005. The lake hosts Kenya's only breeding colony of great white pelicans *Pelecanus onocrotalus* and great crested grebes *Podiceps cristata*. Gray-crested helmet-shrike *Prionops poliolophus* and Jackson's widowbird *Euplectes jacksoni* reside around the lake. At the southern end of the lake lies the "Kekopey" hot springs, which is a key breeding area for the alkali-tolerant tilapia, *Alcolapia grahami*. In addition, the Soysambu Wildlife Conservancy (a nonprofit organization) covering 19,424 ha, which is located within the lake and covering two-thirds of its shoreline, is also a Ramsar Site. Soysambu is home to over 12,000 large mammals including Rothschild's giraffes, giant eland *Taurotragus derbianus*, dik-diks *Madoqua* sp., cliff-dwelling klipspringers *Oreotragus oreotragus*, impalas, lions and buffalos, as well as over 450 bird species.

Lake Elementeita's littoral areas have patches of reeds. The lake is characterized by mass developments of filamentous cyanobacteria including *Arthrospira fusiformis* and *Anabaenopsis* spp. (Vareschi 1978, 1982; Melack 1988; Owino et al. 2001). Some of these algae and crustaceans are preferred prey for the visiting flamingos and other bird species, and as such the lake provides an important refuge for flamingos (*Phoenicopterus ruber*) and other birds and an important breeding ground for the pelican.

Lake Magadi

The lake gets its name from a Swahili word, meaning sodium bicarbonate, and is the southernmost in the soda lake series in Southern Kenya. It lies at 1°52′ S 36° 16′ E and 579 m above sea level and has a surface area of 104 km^2. Several thousand years ago, during the late Pleistocene to mid-Holocene, the lake had freshwater with many fish species, whose remains are preserved in the High Magadi Beds, a series of lacustrine and volcaniclastic sediments preserved in various locations around the present shoreline (Fig. 2).

Lake Magadi is mainly fed by hot springs around its edges, with relatively minor additions from the monsoonal rains. The main part of Lake Magadi is only covered by water for a short period of the year. A relatively thin layer of water accumulates during wet seasons, which then evaporates exposing a vast pan of trona (sodium sesquicarbonate) during the dry seasons. The lake is 80% covered by trona (Jones et al. 1977). Seasonal runoff, mainly from the valley floor north of the lake, is considered the major groundwater recharge (Jones et al. 1977).

Lake Magadi is a popular destination for wading birds during the dry season including flamingos, heron, pelicans, and spoonbills. The lake is one of the few places in Kenya where the chestnut-banded plover *Charadrius pallidus* can be seen regularly; and the endemic tilapiine *Alcolapia grahami* is reported by Seegers and Tichy (1999) to be the only fish species in the lake.

Other biota inhabiting the lake includes many species of the microbial alkaliphile community: *Habmonas campisalis*, *Haloalkaliphilic archaea*, *Tindallia magadii*, *Natronococcus amylolyticus*, *Methylohalomonas lacus*, *Methylonatrum kenyese*, *Amphibacillus fermentum*, *Amphibacillus tropicus*, Desulfonatronovibrio

Fig. 2 Lake Magadi – Courtesy Jacob E. Ojuok (KMFRI)

hydrogenovorans, *Halonatronum saccharophilum*, *Natronoincola histidinovorans*, *Natrionella acetigena*, *Spirochaeta alkalica*, *Spirochaeta africana*, and *Spirochaeta asiatica*.

Lake Magadi is not a protected ecosystem in Kenya (Nyamasyo and Owuor 2009). The ecosystem services in Lake Magadi include recreation (it is a tourism attraction). The lake's trona is used for glass manufacturing, fabric dyeing, and paper production. Chert is a sodium silicate sedimentary mineral rock that was discovered at Lake Magadi during the 1960s (Behr 2002). Magadi Chert with high fluoride deposits is drawn from the lake and used as a meat tenderizer by the local community, thereby ingesting high levels of fluorine, which is harmful to the bones and other body tissues. Alkaliphilic bacteria obtained from the lake are used in manufacturing of alkali-tolerant enzymes (Jones et al. 1988). The local community comprises mainly of the indigenous Maasai living within the rural hinterlands and the nearly 1,000 residents from diverse ethnic backgrounds living in the town of Magadi.

Threats and Future Challenges Facing the Kenyan Soda lakes

The soda lakes of the Rift Valley are facing myriad threats and challenges, including an increase in human population and poor land use practices, land tenure conflicts, catchment degradation, pollution, conflicts over water rights and uses, cultural erosion due to influences from tourism activities, human-wildlife conflicts, and the impacts of invasive species and climate change.

The human population density is currently 20 people per square kilometer, but the growth rate of 3.5% per annum has resulted in higher demand for settlement and agricultural land. Subdivision of land has encouraged encroachment and general reclamation of wetlands in the riparian areas surrounding the soda lakes. The latter is more apparent in semiarid areas (especially in Lake Bogoria) where agricultural activities and pastoralism are practiced in the fragile habitats.

A cultural system that glorifies high livestock numbers rather than their quality has led to an increase in the livestock, and this has led to uncontrolled grazing around lakes Bogoria and Elementeita. In addition, poor land management practices among these communities has impacted negatively on land cover. Extraction of forest resources for medicinal purpose, fuel wood, charcoal burning, and timber has led to catchment degradation and forest loss. The foregoing anthropogenic activities have led to soil erosion and consequently sedimentation in the lakes. This is more evident in the major water towers such as the Aberdare Range and Mau Escarpment, which are important in recharging aquifers and the maintenance of wetlands. It has been predicted that with continuous increase in human population and settlement within the lakes' subdrainage, the demand for water will outstrip supplies with barely minimal amounts left to support wildlife populations. Land tenure and land use systems around the soda lakes have also changed over the years, from nomadic pastoralism (surrounding lakes Bogoria, Elementeita, Magadi) to communal sedentary grazing and currently individual holdings. This has accelerated catchment degradation resulting in the drying up of rivers, encroachment of riparian areas, and subdivision of land leading to unsustainable livelihood prospects.

In addition, the general disruption of hydrological regimes, impacts on water quality and quantity due to cultivation along streams and riverbanks, and clearing of forests in the upper parts of the catchment and diversion of water to support settlement, agriculture, and industrial activities have caused serious conflicts in water rights and use.

On the other hand, the increasing use of chemicals such as fertilizers, herbicides, and pesticides in agricultural practices around the lakes has led to the accumulation of these chemicals in the lakes, thus compromising the aquatic life and the wellbeing of those whose lives depend on the waters of the soda lakes. Increased use of fertilizers in the catchment areas has also led to eutrophication of the soda lakes, hence the high primary production in these lakes. Besides the foregoing, tourism activities also contribute to the pollution of these lakes, as most of the hotels mushrooming along the shores of Lakes Nakuru and Bogoria empty their wastes into the lakes. Other sources of pollution in the lakes include effluents from municipal (e.g., Nakuru) and industrial sources (Magadi).

In spite of economic gains from tourism, the industry has continued to impact negatively on the general lifestyle of riparian communities. Tourism encourages practices which are inconsistent with traditional norms and contribute to cultural erosion. These may include influences on dress codes among other social practices.

Lakes Bogoria, Nakuru, and Elementeita are national protected reserves. However, there are numerous occasions when wild animals stray out of the park boundaries. This has resulted in the destruction of properties, and in some cases, human lives have been lost. These human-wildlife conflicts have resulted in standoffs between the wildlife managers and local communities. Conflict arises when animals from the national parks invade communities' farms, and the authorities concerned hardly compensate them adequately for their losses. The communities seek revenge by killing the animals, which results in retaliation by park managers by arresting the persons responsible for killing the animals.

There are two invasive plant species found within the proximity of the soda lakes. These include prickly pear cactus *Opuntia* sp. from the family Cactaceae and Mesquite *Prosopis juliflora* locally known as "Mathenge." The impact of these on the local environment and biodiversity is not yet apparent, but they certainly have some competitive ecological advantages over the local biota.

The soda lakes of Kenya have limited capacity to mitigate the impacts of global warming and climate change. Lake salinity levels have fluctuated over the years but clearly exhibit an increasing trend. Together with other climate change factors including the increased variability of local temperature, rainfall and wind patterns could negatively impact on the lakes' biogeochemistry with adverse effects on the biodiversity. For instance, recent reports indicate a shift in preferences and reduction of flamingo populations in some of the soda lakes as a result of environmental vicissitudes. Changes in climatic conditions may also result in trophic interruption.

References

Behr HJ. Magadiite and Magadi chert: a critical analysis of the silica sediments in the Lake Magadi Basin, Kenya. SEPM Spec Publ. 2002;73:257–73.

Finlay BJ, Curds CR, Bamforth SS, Bafort 1JM. Ciliated protozoa and other microorganisms from two african soda lakes (Lake Nakuru and Lake Simbi, Kenya). Archiv für Protistenkunde. 1987; 133 (1–2):81–91.

Grant WD, Mwatha WE. Alkaliphiles: ecology, diversity and applications. FEMS Microbiology Reviews. 1990;75:255–270.

Grant S, Grant WD, Jones BE, Kato C, Li L. Novel archaeal phylotypes from an East African alkaline saltern. Extremophiles. 1999;3:139–45.

Gregory JM. The rift valleys and geology of East Africa. London: Seeley Service; 1921.

Jones BF, Eugster HP, Rettig SL. Hydrochemistry of the Lake Magadi Basin, Kenya. Geochim. Cosmochim. Acta. 1977;41(1):53–72.

Jones BF, Eugster HP, Rettig SL. Hydrochemistry of the Lake Magadi basin, Kenya. Geochim Cosmochim Acta. 1988;41:53–72.

Kaggwa MN, Burian A, Oduor SO, Schagerl M. Ecomorphological variability of *Arthrospira fusiformis (Cyanoprokaryota)* in African soda lakes. Microbiol Open. 1–11. Published online; July 2013. 2013 doi: 10.1002/mbo3.125.

Lake Bogoria National Reserve-IMP 2007–2012. Integrated Management plan (IMP). Baringo County Council and Koibatek County Council, Baringo County Kenya. 2007, 48 p.

Melack JM. Primary producer dynamics associated with evaporative concentration in a shallow, equatorial soda lake (Lake Elementeita, Kenya). Hydrobiologia. 1988;158:1–14.

Melack JM, Kilham P. Photosynthetic rates of phytoplankton in East African alkaline saline lakes. Limnol Oceanogr. 1974;19:743–55.

Mwaura F, Moore TR. Forest and woodland depletion in the Lake Elementeita Basin, Kenya. Geoforum. 1991;22:17–26.

Nyamasyo GHN, Owour JBO. Fourth national report to the conference of parties to the convention on biological diversity. NEMA, Kenya and UNDP. 2009.

Oduor SO, Schagerl M. Phytoplankton primary productivity characteristics in response to photosynthetically active radiation in three Kenyan Rift valley saline-alkaline lakes. J Plankton Res. 2007;29:1041–50.

Owino AO, Oyugi JO, Nasirwa OO, Bennun LA. Patterns of variation in waterbird numbers on four Rift Valley lakes in Kenya, 1991–1999. Hydrobiologia. 2001;458:45–53.

Renault RW, Owen RB. The geysers of Lake Bogoria, Kenya Rift Valley, Africa. GOSA Trans. 2005;9:4–18.

Seegers L, Tichy H. The *Oreochromis alcalicus* flock (Teleostei: Cichlidae) from Lakes Natron and Magadi, Tanzania and Kenya, with descriptions of two new species. Ichthyol Expl Freshw. 1999;10:97–146.

Shivoga WA. The influence of hydrology on the structure of invertebrate communities, in two streams flowing into Lake Nakuru, Kenya. Hydrobiologia. 2001;458:121–30.

Tiercelin JJ, Vincens A (eds). Le demi-graben de Baringo-Bogoria, Rift Gregory, Kenya: 30,000 ans d'histoire hydrologique et sédimentaire. *Bulletin des Centres de Recherches Exploration-Production Elf-Aquitaine*. 1987;11: 249–540.

Vareschi E. Ecology of Lake Nakuru (Kenya). I. Abundance and feeding of Lesser Flamingo. Oecologia. 1978;32:11–35.

Vareschi E. The ecology of Lake Nakuru (Kenya). II Biomass and partial distribution of fish. Oecologia (Berlin). 1979;37:321–35.

Vareschi E. Ecology of Lake Nakuru (Kenya). III. Abiotic factors and primary production. Oecologia. 1982;55:81–101.

Vareschi E. Ecology of Lake Nakuru (Kenya). IV biomass and distribution of consumer organisms. Oecologia. 1984;61:70–82.

Okavango Delta, Botswana (Southern Africa)

114

Lars Ramberg

Contents

Introduction	1394
Social and Economic Factors	1396
The Landscape	1397
Site-Specific Features	1399
Flood Switching	1400
Desalination	1401
Biological Responses to the Flood Switching	1403
Biological Responses to the Seasonal Flood Pulse	1405
Biodiversity	1406
Ecosystem Services	1408
Threats and Future Challenges	1409
References	1410

Abstract

The Okavango Delta in Northern Botswana as well as its drainage basin in Namibia and Angola is located in the Kalahari sand basin. It is an *alluvial fan* where fine sand brought in from upstream settle in the maze of branched streams that are filled up forcing abandonment and formation of new river channels. This gives the Delta a slightly convex shape but the gradients are very small at around 1:3,000. Although almost endorheic, the Delta is a freshwater system. This is caused by removal through evaporative concentration and precipitation of Ca-Mg-bicarbonates and silica complexes in vegetated islands into which stream water is flowing. It causes the creation and growth of more than 150,000 islands that usually start as termite mounds. The remaining water with more soluble ions, mainly Na and Cl, flows into the centre of islands, concentrate further until the

L. Ramberg (✉)
Uppsala, Sweden
e-mail: lars.ramberg@gmail.com

density is becoming critically high when the water sinks down into deep groundwater due to a *density fingering process*.

An annual flood pulse arrives to the Delta in March-April caused by rains in the Angola highlands 4–5 months earlier. This causes a seasonal maximum flooding of 14,000 km^2 with large variations between years. Over historic times 28,000 km^2 has been flooded and this area is usually defined as the Okavango Delta. The Okavango fan is however 40,000 km^2 and was created over longer geologic time.

The dynamic geologic-hydrologic features create a wetland landscape under constant change with phases of flooding and desiccation from annual to intervals of decades and with all spatial scales from a few square meters to hundreds of km^2. The specific habitat units of which there are 500,000 in the Delta have however a mean 0.05 km^2. The inflowing water and the Delta itself is very nutrient poor caused by the substrate of well withered sand; and the nutrients, nitrogen and phosphorus, are mainly kept in organic matter dead and alive. Both flooding and desiccation cause therefore a release of nutrients with sudden patchy outbursts of biological productivity, recognized in the aquatic phase by the congregation of fish eating birds and in the drying phase by flood plains grazers in large numbers. A bottle neck nutritious resource is provided during the dry winter season June - August causing a grazers biomass 10 times higher than in similar savanna systems. Many of the mammal species have adopted their reproductive cycle to this unusual seasonality but there are no endemic species. The biodiversity is however fairly high caused by the dry-wet mix of habitat types.

Recently large-scale irrigation has started in the drainage basin in Angola and Namibia. This may lead to eutrophication, blue-green algal blooms, growth of denser reed-beds and expansion of the floating weed *Salvinia molesta* all with far reaching consequences for biodiversity of the fauna. It is however not unlikely that the two unique and vital processes of *flood switching* and *desalination* will be hampered as well. If so, a destruction of the Delta in its present form is likely.

Keywords

Okavango Delta · Southern Africa · Alluvial fan · Seasonal flood pulse · River avulsion · Flood switching · Desalination · Island growth · Density fingering · Carbon sequestering · Biological productivity · Biodiversity · Irrigation threats

Introduction

The maze of wetlands and streams in northern Bechuanaland – now Botswana – has been a challenge and a temptation for travelers for more than 150 years. It took a long time to realize that it all was one hydrologic system where all streams emanated from one mighty river (Mendelsohn et al. 2010) (Figs. 1 and 2). The area was very difficult to travel in – and still is – only a few hunter-gatherers lived there, plagued by endemic sleeping sickness and malaria. This enigmatic wetland with a rich aquatic

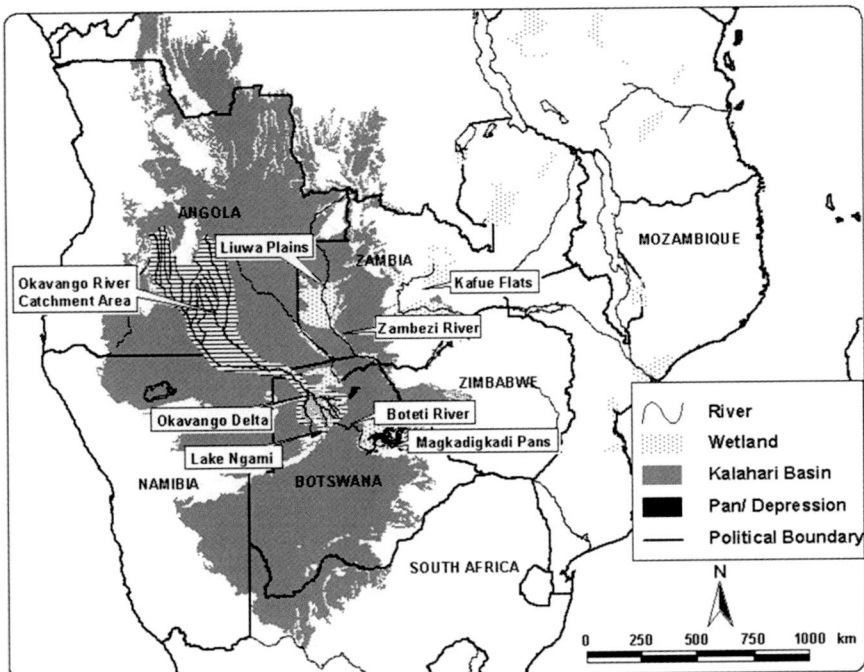

Fig. 1 The location of the Okavango Delta in the Kalahari sand basin in the interior of Southern Africa (From: Ramberg et al. 2006b. Fig. 1, p. 311. With permission of Aquatic Sciences)

and terrestrial fauna of African mammals and birds has since been a growing attraction and is now the base for a thriving tourism industry.

In the beginning of the 1900s, this bewildering river–wetland system was named the Okavango Delta by European travelers because of its network of braided and meandering streams. Fairly recently however the system was described as a *fan* (Stanistreet and McCarthy 1993) that grows vertically by sedimentation in its stream channels, while a *delta* grows horizontally in water, usually at the inflow of a river to a lake or the sea. The discovery that the Okavango Delta is actually an alluvial fan with natural processes that are completely unlike most other land forms is of profound importance for understanding its ecological functioning and management needs.

It has also been a long-standing question how the Delta can be a freshwater system when practically all inflowing water remains and probably evaporates. It ought to be a saline pan like so many found in similar settings in the Kalahari, in the East African Rift System, and in the Sahara. Only during the last decade have the remarkable processes that keep the Delta waters fresh been understood (Ramberg and Wolski 2008).

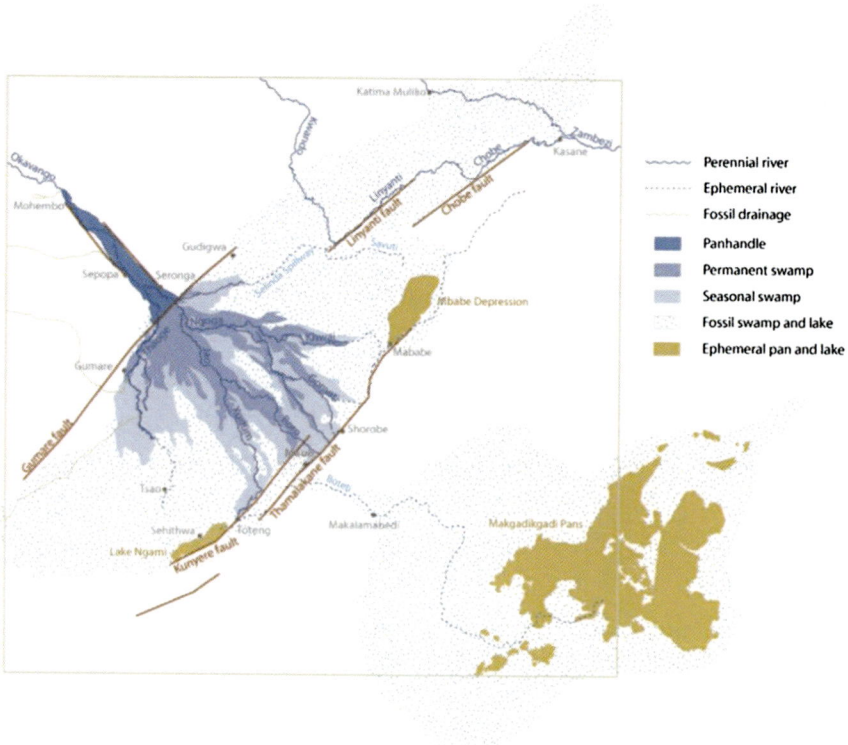

Fig. 2 The Okavango Delta with branching stream network, seasonal, and permanent swamps (From: Mendelsohn and El Obeid 2004. Fig. 4, p. 35. With permission of RAISON, Windhoek, Namibia)

Social and Economic Factors

Due to a harsh climate with low and erratic rainfalls in combination with unfertile soils, the whole Okavango drainage basin is sparsely populated. Between 2000 and 2004, the upper reaches in Angola had an estimated population of 350,000; an additional 163,000 lived in the Caprivi Strip in Namibia and 88,000 in Botswana including the Boteti River (Mendelsohn and El Obeid 2004). Most live in small villages along the river courses and along the western and southern fringes of the Delta and survive in Angola and Namibia mainly on small-scale farming, while in Botswana formal employments and government subsidies are the main livelihood sources.

Up until recently, the only significant "modern" economic activity in the river basin is that of ecotourism in the Delta (now also irrigation, see below). It is fairly well organized into the "high-cost low-density type" and is based on the abundance of African wildlife in a fascinating water landscape. It is the second largest export industry (next to mining) in Botswana and an important local employer for about 50,000 people (Mendelsohn et al. 2010).

In the high-spirited excitement following Botswana's 1964 independence, a number of water development projects supported by foreign aid were initiated in the Delta between 1970 and 1990. All of these projects failed due to lack of knowledge about the nature of the Delta system or ignorance of the local political and socioeconomic situation (Wolski et al. 2010). However, similar proposals to exploit the river water within the whole drainage basin have surfaced quite frequently during the last 40 years. Sharing of the water from the river basin between the three countries would not be without problems. The upper two states showed interest in diverting the water to more densely populated areas to use it for irrigation on more fertile soils or for hydroelectric power production. In contrast, Botswana, where the whole Delta is located with its very important tourism industry, has resisted such developments in the fear that the sensitive wetland ecosystem might be destroyed. Therefore, on the initiative of the Botswana government, the whole Delta area has been designated a Ramsar Site as well as a UNESCO World Heritage Site, and considerable efforts have been made to develop management practices and structures. There is also an administrative structure under development for management of the whole basin, the OKACOM, as an agreement between the three riparian states (op.cit.).

The diversion of river water was up to the beginning of this decade (2010) very small and did not exceed 33 Mm^3 per year which is less than 1% of the mean annual discharge of the Okavango River (Wolski et al. 2010). Large-scale withdrawals for irrigation seems to be under advanced planning in Angola and Namibia (Ramberg 2016 article 135) and some has already started. There is already by year 2010, 3,400 ha under irrigation and no more recent figures seem to be available. The off-take of river water will increase and as much as 30% of mean flow seems to be under serious consideration. In such case one year out of 20 will leave the river dry which has to be mitigated by construction of storage dams.

Before these new developments, the Okavango Delta was protected due to lack of knowledge as described above and other negative factors: the presence of the tsetse fly, which was not eradicated until 2001–2002 (Ramberg et al. 2006b), deterred early attempts in livestock farming; low agricultural potential and diseases have kept the human population low; and civil war in Angola that did not end until early 2002 prevented the implementation of a number of water development projects. The Okavango Delta was therefore until recently in an almost pristine state but has now been put on a trajectory of change.

The Landscape

The Okavango Delta is located in Botswana in Southern Africa (Figs. 1 and 2) about 400 km north of the Tropic of Capricorn. The Okavango drainage basin is part of the much larger and endorheic Makgadikgadi basin. However, most of this has no net runoff, and all inflowing water to the Delta (910–980 m a.s.l.) emanates from the

Angola highlands that rise to 1,500–1,800 m a.s.l., where the mean annual rainfall is over 1,200 mm. The highlands are the sources for the main headwaters of the Cubango and Cuito rivers with their well- developed drainage networks and topographic relief. Between 900 and 1,100 m a.s.l., the land flattens out into a plateau that includes the Delta with gradients of less than 1:3,000. The two rivers meander and follow their confluence as the Okavango River flow into the panhandle, the uppermost part of the Delta, 500 km from the headwater sources. The panhandle is a broad and flat valley with a system of seasonal and permanent swamps that widens into the Delta proper where the river loses its confinement and currently divides into five streams, each of which subdivides into secondary and tertiary distributaries (Mendelsohn et al. 2010; Wolski et al. 2010; McCarthy 2013).

At the bottom of the Delta, the flowing water is collected in the Thamalakane and Kunyere rivers located along fault lines at right angles to the distributaries (Fig. 2). Dependent upon the magnitude of the flood and which distributaries carry the most water, the outflow can take three alternative courses or any combination of them: to the eastern Mababe depression, to the western Lake Ngami, or most commonly down the Boteti River to the huge Makgadikgadi salt pan. In all cases, these are the final receiving sinks for the water.

The climate is hot and dry with an annual potential evaporation of 1,800 mm, greatly exceeding the annual rainfall of 450 mm which experiences large variations between years and falls mainly during the hot season (December–January).

The Okavango River has a unimodal flood pulse with a peak in March–April in the panhandle, 4 months after it fell as rain on the Angola highlands. The water moves as a slow wave during 6 months across the flat Delta surface, and the maximum flooded area occurs in August–September.

The runoff into the Delta amounts to an average of 9,306 Mm^3 per year and an additional (approximate) 4,800 Mm^3 fall on the Delta as rainfall. The outflow into the Boteti River only takes place during wet periods and averages 20 Mm^3 per year, which is only 2% of the inputs. Leakage from the Delta via groundwater flow to the surroundings is probably nil because the groundwater chemistry differs and the gradients are very small. Thus, some 98% of the inputs are removed from the Delta by evapotranspiration (Ramberg and Wolski 2008). A significant part of this (up to 24%) takes place through terrestrial vegetation, supplied by flood water infiltration and lateral groundwater flow of local character (Ramberg et al. 2006a).

The channel banks are usually vegetated by emergent aquatic macrophytes such as *Cyperus papyrus* and *Phragmites australis* that function as semipermeable confinements. They are often overflown, and sometimes the streams disappear as the water seeps over floodplains into other channel systems.

The vegetation can broadly be classified along a wet–dry gradient. There are permanent streams and small lakes with floating leaved plants such as *Nymphaea* spp., permanent flooded areas overgrown with emergent macrophytes like *C. papyrus* and *P. australis*, seasonal floodplains with short sedges and grasses, and occasionally flooded plains with a sparse grass cover. Fringes of riparian woodlands often with tall trees are very common close to permanent or seasonal water but are normally not flooded. Similarly, but further away from water in drier settings are large areas of wooded grasslands (Ramberg et al. 2006b; Mendelsohn et al. 2010).

There are 150,000 island-like features in the Delta that give it much of its character (Gumbricht and McCarthy 2003). They usually have a fringe of riparian woodland enclosing grassland. Sometimes, they are surrounded by water for at least part of the year. There is no subterranean geological structure that can explain their existence; seemingly they are floating on the sand. Most are small, at only a few square meters, but a few are very large at several tens of square kilometers and the largest is 700 km^2.

The sizes of the units of relevance in the Delta context are far from precise and are a matter of definition. Over a 30-year period, the area flooded at least every decade is 14,000 km^2, of which 9,000 km^2 is wetland, the rest being islands. The total area of this Okavango wetland is apportioned among the panhandle (820 km^2), permanent swamp (2,500 km^2), seasonal swamp (3,300 km^2), and occasional swamp (7,100 km^2) (flooded at least once every 10 years) (Gumbricht et al. 2004). Including the permanently dry areas that form the islands and plains that were flooded during historic times (since the 1850s), the total area of the *Okavango Delta* thus defined is 28,000 km^2 (Ramberg et al. 2006b). The *Okavango fan*, however, is much larger at about 40,000 km^2, being a result of geomorphologic processes over long time scales. In addition, the *Okavango Delta Ramsar Site* is 68,640 km^2, which includes the Delta and adjacent areas as buffer zones.

The Kalahari sand basin has been steadily filled with sediments over the past 65 million years. The major rivers here, the Zambezi, Kwando, and Okavango rivers, have changed courses in various ways and sometimes discharged into the current Delta area and Makgadikgadi pan. The present hydrologic configuration when the Okavango fan started to build is estimated to be not more than 120,000 years old. The climate has experienced large variations, and the rivers have sometimes been much larger than at present, but very dry periods have also occurred. For instance, there are fossil sand dunes to the northwest and southeast of the Delta.

Site-Specific Features

Four unusual hydrologic and geomorphologic features coincide and interact to make the Okavango Delta unique.

1. *The inflowing river has a pronounced flood pulse* that is unimodal with a peak in March–April. This is of great importance, as both erosion in the drainage basin and the transport of the eroded material in the river and primary distributaries follow an exponential function in relation to water flow (McCarthy 2013). The transport of sediments is therefore small during low and moderate river flow but very high at peak flows. A steady transport of sediments into the Delta is necessary to maintain the dynamic of river avulsions (abandonment and formation of new river channels) and flood switching (described below). It is also mainly during high flows that the streams in the Delta break their confinements

and change flow path. Finally, the area of seasonal flood plains is related to the size of the flood pulse.
2. *Its drainage basin is situated in a sand sea*, the vast Kalahari Desert, where the depth of the sand is usually 20–60 m. This is the source of the sediments that build up the fan. The sand consists mainly of grains of quartz that have been leached over millions of years. There is therefore very little clay in the river water, which is also nutrient poor. However, the Okavango River water has a high concentration of bicarbonate and silica as a result of chemical weathering of quartz and other silica minerals in the drainage basin. This process occurs with net consumption of carbon dioxide and the production of dissolved silica and bicarbonate ions (Ramberg and Wolski 2008). The process is facilitated by the high permeability of the Kalahari sand and a long exposure during underground transport pathways to the bottom of stream valleys. The Okavango River is the only known large river in Africa with bicarbonate as the dominant anion in the water, and it is very unusual worldwide with the exception of rivers in calcareous regions.
3. *It lies on a deep sand-filled rift* – an extension of the East African Rift System that cuts across from east to west. The rift is hardly visible in the Okavango Delta, which occurs on 100–270 m deep sand between the rift's north and south faults, a distance of 210 km. Apart from a few meters at the top with freshwater, the whole underlying pore volume constitutes a permanent sink for saline groundwater. The area is seismically active, small tremors are common, and a few earth quakes have been recorded (McCarthy 2013).
4. *It is subsiding* due to these seismic activities. This subsidence causes the river that flows from the panhandle into the Delta to lose its confinement and currently divide into five streams that in turn branch out in numerous rivulets (Stanistreet and McCarthy 1993).

The combination of these four specific hydrologic and geologic features enables two vital processes – flood switching and desalination – to operate in conjunction, both of which are decisive for the creation and functioning of the Okavango Delta ecosystem. Moreover, a functional flood switching phenomenon is conditional for sustainable desalination, as described below.

Flood Switching

Flood switching is instrumental for sustaining the high biological productivity of the Delta as well as its high biodiversity.

The stream channels in the Delta are banked by stands of papyrus and reeds, and water can easily seep across their banks. The water flowing into the Delta carries most sediment as a bed load of fine Kalahari sand that has been eroded upstream and which cannot be transported across vegetated banks. Water flow is reduced from 120–300 cm s^{-1} in the panhandle to 40–80 cm s^{-1} in the major channels of the fan but only 10–20 cm s^{-1} in the small distal channels. This reduction in energy causes the particulate matter to settle as sediment in the channels, ultimately leading to channel aggradation, development of vegetation blockages, and channel abandonment, as the stream finds a new path, known as the process of river avulsion (McCarthy 2013). This process forms the alluvial fan that is slightly convex with an almost perfect conical shape and its apex at the inflow from the panhandle. In aerial photos and satellite imagery, old meander patterns can be seen everywhere on the fan surface, which suggests that streams carrying sediments must have reached every part of the fan several or many times.

The gradients are very small (1:3,000) giving the impression of a flat landscape. However, a closer look reveals that on a finer scale there are irregularities on the surface such as bars, banks, and valleys, often meandering with altitude differences of a few decimeters to a meter that obviously are remnants of old hydrological systems overlaying each other. This creates an intricate flooding pattern that in turn is the reason for the mosaic-like vegetation pattern described below.

Desalination

The removal of salts within the Delta islands through desalination (Fig. 3) creates a beautiful wetland ecosystem with high productivity and biodiversity.

The seasonal flood pulse inundates large areas of seasonal and occasional swamp. It first infiltrates to fills the unsaturated volume underneath, but following that, the infiltration flow paths become almost horizontal onto the surrounding dry lands,

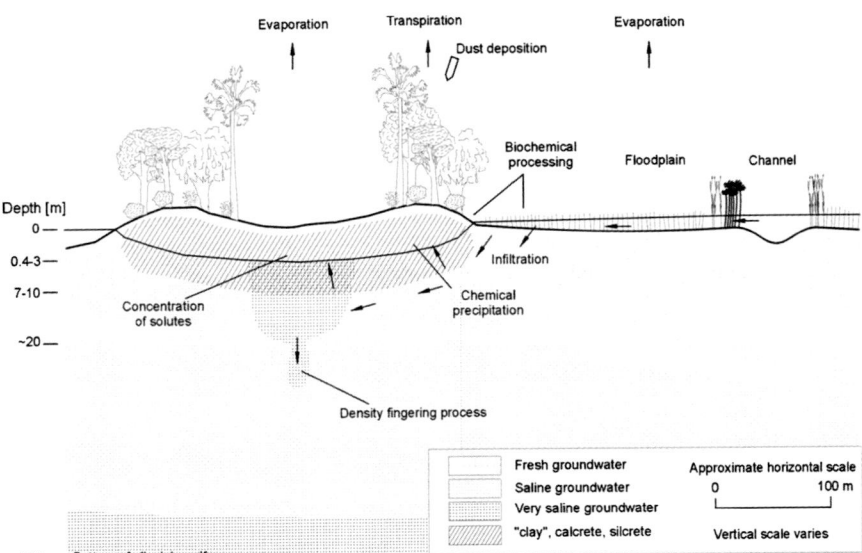

Fig. 3 The function of an island in the Okavango Delta as a desalination plant (From: Ramberg and Wolski. 2008. Fig. 1, p. 216. With permission of Plant Ecology)

Fig. 4 Aerial photo of the Okavango Delta during beginning of the flood, with sedges and grasses still dry. In the foreground to the *left* is a young island and to the *right* an older island with a well-developed fringe of riparian woodland and a central barren area that is probably salt incrusted. The central island is probably a merger of 3–4 islands. In the background lie a large number of vegetated termite mounds that might develop into new islands. Note the animal tracks in the foreground (Photo credit: Lars Ramberg © Rights remain with the author)

usually atoll-shaped islands with a fringe of riparian woodlands surrounding central lower-laying grassland (Fig. 4). The transpiration from this vegetation is high and causes an increase in concentration of solutes. Eventually calcium, magnesium, and bicarbonate ions precipitate and form calcretes, while the silica and potassium ions form clay complexes (McCarthy et al. 1993; Ramberg and Wolski 2008). This causes a volume increase and island growth horizontally and vertically. The transpiration also causes a gradient in the groundwater table, forcing a steady flow of water from the flood plains to the centers of the islands. Here, the remaining ions with high solubility such as chloride and sodium are concentrated 500–1,000 times; the salinity is high, up to 20 g dm^{-3}, as is the density. At these high concentrations and density, the solutes will become unstable in relation to the underlying water and, at times, sink to deep groundwater in a density fingering process (Bauer 2004).

The salts in the inflowing river water are removed by these two processes within the islands and are deposited in two permanent sinks, calcrete and clay minerals, and deep groundwater. It has been calculated that of the 381,000 t $year^{-1}$ total solute load at the inlet, 21,000 t leave through the outlet, 216,000 t precipitate in the Delta islands, and 144,000 t sink to the deep groundwater (Ramberg and Wolski 2008).

Thus, the islands in the Delta are the result of a growth process that initiated when termite mounds on dry areas became re-flooded. A flood switch is thus required for the island growth to start – the termites will die but their mounds will become vegetated. The islands keep on growing as long as water is coming, but there are recent observations (McCarthy 2013) that indicate that they have a limited life; when aging, the trees are dying probably due to poisoning by the accumulated salts, the transpiration will cease, and the whole desalination process will stop. For the desalination of Delta waters to be sustainable, new islands must therefore be continuously created.

Biological Responses to the Flood Switching

The stream avulsions cause major changes in the distribution of flooded areas over a wide range of space and time scales; the process where old flooded areas become dry land and dry areas become flooded is continuous. When the western distributary, the Thaoge, for instance, dried out between 1890 and 1920, several thousand square kilometers of wetlands turned into dry Kalahari savanna. At the same time, eastern distributaries developed and drowned large areas that turned into permanent swamps. Changes at smaller spatial scales often of 1–100 km^2 have since been observed in many parts of the system. The vegetation responses in such cases are very fast, with terrestrial grasses and aquatic species becoming established within a few months.

Overlaying these large switches is the seasonal flooding, with a wet phase of 4–8 months and a similarly long dry phase. It is caused by a single flood pulse inundating an area from a low of 4,500–6,000 km^2 during February to an annual high of 9,000–12,000 km^2 in August–September.

On the broadest scale, Delta habitats follow a wet–dry gradient of permanent streams, pools, and lakes, permanent swamps, seasonal and occasional flood plains, riparian woodlands, and dry savanna woodlands. Satellite imagery analyses and ground truthing identified 45 habitat types distributed among 500,000 habitat units that are on average about 5 ha in size, and thus the occurrence of each habitat type is numbered in the tens of thousands (Ramberg et al. 2006b). There are often sharp boundaries between these various habitats, and small topographic differences on the scale of centimeters makes large differences in wetness that is the foremost factor organizing plant communities in the Delta (op. cit.). Also within the various broad habitat groupings, such as flood plains and permanent swamps, the plant species can be ordinated mainly along hydrological gradients (Murray-Hudson et al. 2014).

A major effect of flood switches in the Delta is that they cause outbursts in biological productivity and migrations of animals to biological hot spots that are created by these switches – fishes and fish eaters, particularly birds, in aquatic habitats and grazing animals and their predators in terrestrial habitats (Hogberg et al. 2002; Lindholm et al. 2007; Ramberg et al. 2010). However, seasonal flooding also shows the same dynamic pattern on a smaller scale (Lindholm et al. 2007). The major reason is that these turnovers release nutrients (nitrogen and phosphorus) which have accumulated the longer an area remains either dry or flooded. Seasonal floodplains have 7–50 times higher nutrient content than permanent channels. A

Fig. 5 Pelicans and flamingoes amassed at Lake Ngami when it was re-flooded in 2004 (Photo credit: Ken Oake © Rights remain with the photographer)

flood switch will turn dry areas wet, and the released nutrients will become available to new biota. Large amounts of dead organic matter of terrestrial origin become available for heterotrophic aquatic food chains with flooding of formerly dry land (Lindholm et al. 2007). This creates biological hot spots in the Delta with very high productivity commonly identified by the aggregation of fish-eating birds. Such situations persist for only a few years as the feeding animals disperse the nutrients. For example, in 2004 birds amassed at Lake Ngami (20–50 km^2) (Fig. 5) when it was re-flooded after having been dry for 20 years and typically supporting 2,000–5,000 mainly terrestrial birds. During 2004, about 30,000 birds in total were counted, of which 6,000 were fish eaters (pelicans, cormorants, and herons) and 21,000 were geese and ducks. In 2005, the total number declined to 25,000 including fish eaters (2,000) and ducks and geese (7,000) although waders increased. Although the lake was still flooded in 2007, the total count of birds had declined to 4,000, mainly waders, while the fish eaters were almost all gone. Data was from counts by members of Birdlife Maun (www.birdlifebotswana.org.bw).

Organic matter that had accumulated as peat in wet areas will with flood switching dry and be decomposed or consumed by the Delta's common fires to become dry areas. During the dry season, floodplains in particular are often burned; on average 60% of floodplains were burned between two and ten times within a 15 year period (Heinl et al. 2007). These oxidative processes also release the nutrients. The newly created grasslands are initially very nutrient rich and attract grazers in large numbers (Table 1). The most numerous, impala and lechwe, benefit from the extreme extent of ecotones; both species graze on the floodplains, while the impala flee into the associated riparian woodlands as protection against predators the lechwe when threatened escape into the permanent waters. The overall herbivore

Table 1 Number of large mammals in the Okavango Delta in 2002, calculated for an area of 20,000 km² based on 10 aerial counts. For impala, the numbers have been corrected based on ground counts as they are difficult to see from the air. For the same reason, the numbers given for kudu and sitatunga are likely to be underestimates, but no corrections have been made (From: Ramberg et al. 2006b. Table 17, p. 329. With permission of aquatic sciences)

Species	Total number
Elephant *Loxodonta africana*	35,000
Zebra *Equus burchelli*	14,000
Warthog *Phacochoerus aethiopicus*	2,000
Hippopotamus *Hippopotamus amphibius*	2,500
Giraffe *Giraffa camelopardalis*	5,000
Wildebeest *Connochaetes taurinus*	8,000
Tsessebe *Damaliscus lunatus*	3,000
Impala *Aepyceros melampus*	140,000
Buffalo *Syncerus caffer*	60,000
Kudu *Tragelaphus strepsiceros*	300
Sitatunga *Tragelaphus spekei*	500
Red lechwe *Kobus leche*	60,000

Table 2 The number of species, density, and biomass of large herbivores in a 6966 km² study area in the Okavango Delta. The high biomass for "not grassland dependent" is caused by elephants (Modified from Bonyongo 2004)

	Grassland dependent	Not grassland dependent	Total
Number of species	8	4	12
Density no. km^{-2}	29	2	31
Biomass kg km^{-2}	6,700	5,200	11,900

biomass in the Delta is 12,000 kg km^{-2}, which is similar to that of the highly productive savannas of Eastern Africa and ten times higher than that predicted by models (Table 2). The year-round accessibility of water and the prolonged period of nutrient-rich grazing are the likely reasons (Ramberg et al. 2006b).

The overall effect of flood switches is that the Delta functions both as an effective accumulator of nutrients as dead and living organic matter and an efficient mechanism for their release and uptake in new biota. For phosphorous in particular, the Delta seems to be a closed but dynamic system with no permanent sinks and small losses to the outside.

Biological Responses to the Seasonal Flood Pulse

When the water starts to seep onto the dry flood plains, often following old channels made by hippopotamus, fishes in large numbers follow just at the water front where terrestrial insects and other animals are caught by surprise and devoured. Catfishes (Clariidae) are typically among these pioneers, but also small Cyprinidae and

Poeciliidae (topminnows) find a rich food source of zooplankton that hatch out from resting stages (Siziba et al. 2011). In addition, the drowned debris, dead grasses, and sedges provide protection and a substrate for creeping and crawling invertebrates. Fish-eating birds follow. Eighteen species of heron (Ardeidae) stalk among the emergent grasses and sedges while diving birds such as cormorants and darters hunt in the open waters of pools and channels. The floodplains are also used by spawning fish such as most Cichlidae (18 species) that build nests and guard their young in this comparatively sheltered environment (Ramberg et al. 2006b).

The oxygen levels are relatively low, reaching 3 mg l^{-1} during the day and often less than 1 mg l^{-1} during night. Many fish species have adapted to these conditions by having accessory air-breathing organs (e.g., catfishes) or have physiological adaptions to survive (e.g., Cichlidae). In addition, all Cichlidae have special strategies to provide a better oxygenated environment for eggs and fry, either by mouth brooding or by fanning and guarding eggs in nests. Other fish species build bubble foam nests that float on the surface or simply attach the eggs on vegetation to avoid the often anoxic sediment. In total, 23 species out of 71 have such adaptations in the Delta (op. cit.).

Although most fishes move back into permanent waters before the flood plains dry out, many are trapped in pools that shrink daily. This situation is used by big fish-eating birds that amass here, such as the African fish eagle, storks, and flocks of pelicans that can empty a pool of fish within a few hours in a cooperative foraging effort.

Many sedges and grasses on the flood plains pass the dry period as rhizomes and respond immediately by sprouting with the arrival of water. This attracts grazers such as zebras, wildebeests, buffaloes, and reedbucks that feed at the edge and out in the water on the fresh nutrient-rich vegetation (Fig. 6). Soon there develops a cover of new grass along the flood plain edge where impalas and warthogs feed. These green fringes in the bone-dry landscape are heavily grazed during the cold season that otherwise would be a bottleneck for the grazer guild. With the arrival of the rains in November–December, they disperse over to the much larger areas of savanna.

The flood pulse creates a different seasonality in peak food provisioning during the cold dry season compared to the African savanna landscape which usually happens during the warm rainy season. A number of mammal species – 39 out of a total of 122 – have responded to this by being winter breeders in the Delta while the same species in the Southern African region breed (drop their young) either in the summer or the year around (op. cit.). As mating must take place usually several months before the breeding, this is probably a genetically determined adaption.

Biodiversity

Wetlands do not have high biodiversity in comparison with for instance subtropical and tropical forests. However, due to the hydrological variability caused by flood switching, the Okavango Delta has a rich biodiversity that includes 1,300 identified plants, 71 fish, 31 amphibians, 64 reptiles, 444 birds, and 122 mammal species

Fig. 6 Grazing animals in the Okavango Delta are attracted in large numbers to a previously dry floodplain when the seasonal flood arrives and fresh grass is sprouting. In the foreground are three reedbucks grazing in the water and in the background a heard of wildebeests (Photo credit: Lars Ramberg © Rights remain with the author)

Table 3 Number of species in taxonomic groups of originally terrestrial origin observed in each major habitat in the Okavango Delta. The total for the taxonomic group may not equal the sum of habitats as there are some species with overlapping habitats. The total for plants is now 1,300, but not all have been classified according to habitat (From: Ramberg et al. 2006b. Table 21, p. 333. With permission of aquatic sciences)

Taxonomic group	Total number of species	Aquatic/perennial swamp	Wetland/seasonal swamp	Dry land/terrestrial
Plants	1061	205	519	704
Reptiles	64	7	5	52
Birds	444	112	57	275
Mammals	122	3	21	110

(Ramberg et al. 2006b). Compared to other large wetlands in the world, the Delta has a low number of fish species, the second highest number of plants and mammals, and the highest number of reptiles and birds (Junk et al. 2006). Flood switching is also the reason for the occurrence of most of the nonaquatic species (Table 3) in the Delta. Dry land species are generally much more common than wetland and aquatic species within each group and reflect the complex landscape of the Delta.

There are six bird species classified as "vulnerable" of which the slaty egret *Egretta vinaceigula* and the wattled crane *Bugeranus carunculatus* have 85% and

15%, respectively, of their global populations in the Delta. Healthy populations of the large African predators, notably lions, cheetahs, and wild dogs, occur in the Delta and are an important genetic resource as these species are becoming rare in many parts of their historical distribution. The black rhinoceros *Diceros bicornis* has long been locally extinct due to poaching, but there are now a few introduced white rhinoceros *Ceratotherium simum*.

Aquatic groups like fish, dragonflies, and mollusks are dominated by species that have their main distribution to the north in the Zambezi and Congo systems, and for some the Delta is the southernmost location. The dry Kalahari to the south is an effective migration barrier for aquatic species (op. cit.).

No endemic species have yet been found in the Delta. The very dry periods in the past when the Delta might not have existed could have caused endemic wetland species to become extinct. In contrast, during the very wet periods, the Okavango Delta must have been part of the huge wetland complex in the very flat region of northern Botswana and northern Zambia that includes the whole Makgadikgadi basin and the wetland systems of the Kwando, Kafue, and upper Zambezi rivers. This is in the center of the *Zambezian Phytochoria* (White 1983), a regional center of endemism with more than 50% endemic plant species. During wet periods, wetland species could probably disperse fairly easily within this whole area. It is a meaningful unit for studies of evolutionary processes and endemism, not only for wetland plants but also for animals such as the wetland antelope subfamily Reduncinae (waterbuck, reedbuck, lechwe) that here seems to have a larger number of species and subspecies than anywhere else (Cotterill 1998).

Ecosystem Services

The Delta provides important ecosystem services including freshwater to about 100,000 people, their livestock, and for small-scale agriculture withdrawal.

Sequestering of organic carbon by peat formation probably does not occur in the Delta due to flood switching in combination with frequent fires. However, there is a low content of fine particulate organic matter in cores (about 2.8% carbon), so some accumulation takes place (Huntsman-Mapila et al. 2006). The sedimentation rate in the Delta during the past 40,000 years can be estimated to be 0.1 mm year^{-1} for a sequestration of 6.4 g C m^{-2} year^{-1} or 260,000 t C year^{-1} for the whole Okavango fan. The more unusual process in the Delta of inorganic carbon sequestration can be calculated to be 80,000 t C year^{-1} from data in Ramberg and Wolski (2008) or 5.6 g C m^{-2} year^{-1} for the Okavango wetland. In total, the sequestering rate in the Okavango wetland is about 12 g C m^{-2} year^{-1}, which is an order of magnitude less than that of peat accumulating wetlands (Mitsch et al. 2010) but in the order of forest systems.

The Delta is of worldwide biological and cultural importance. The Delta has a rich biodiversity and maintains healthy populations of a number of threatened bird and animal species including the large African predators mentioned above. Tourism in the Delta is the second largest source of foreign income for Botswana and a source of local employment for about 50,000 people. Visiting tourists, now totaling around 100,000 per year, receive a unique experience. Its wildlife richness including almost all the spectacular African species in beautiful settings and the ease by which ecological and conservation principles can be illustrated makes it an outstanding destination for film makers whose productions appear frequently in television channels across the world.

Threats and Future Challenges

The most important challenge is to protect the two vital processes, flood switching and desalination, from serious threats that could affect their natural functioning and possibly cause irreparable damage to the Delta. The most serious threats are:

1. During the recent two decades, the following water development projects have been proposed: a pipeline from the Okavango River in Namibia to the capital Windhoek (Ramberg 1997) and a dam for hydroelectric power production at Popa Falls in Namibia. There are a number of earlier and similar proposed projects (Mendelsohn and El Obeid 2004; Mendelsohn et al. 2010; Wolski et al. 2010). Some of these projects had been taken to an advanced stage of planning before they encountered local and international criticism and were stopped but not permanently ruled out. Most alarming however is that irrigation has begun in Angola and Namibia and large scale schemes seems to be under advanced planning (Ramberg 2016). There is no doubt that such projects will have serious negative effects on the Delta ecosystem (see below).
2. Dredging of the primary distributaries in the Delta to improve communications and to secure water to communities in the fringe of the Delta has been proposed a number of times by representatives for the government of Botswana but will have direct negative effects on the flood switching.
3. If dams are built in the river as proposed (see above), it is likely that dissolved silica will be taken up by silica algae and eventually collect in the sediment. This will have direct negative impacts on island growth.
4. The desalination process is dependent on a high bicarbonate concentration in the Okavango River. Acidification of the drainage basin due to acid rain for instance by the escalating industrialization and building of coal-fired power plants in Southern Africa could result in a reduction of the bicarbonate concentration.
5. Increased eutrophication of the Delta, by the use of fertilizers in agriculture or by urbanization along the river, would have very complex effects. The emergent macrophyte belts along streams would become higher and denser, resulting in more vegetation blockages. The combined result would probably be fewer but bigger avulsions and flood switches. The invasive *Salvinia molesta* that is still a small problem due to the nutrient-poor waters would definitely expand. The waters would become more turbid with negative impacts on benthic production and on most fish-eating species (more than 40 species) now present that hunt by

sight. Blue-green algal blooms and with it poor water quality are other likely effects.
6. A 30% off-take of river water (the medium level scenario) in Angola and Namibia for irrigation, diversion to Windhoek and for hydro-electric power production seems to be the level under serious consideration (OKACOM 2011, see also Ramberg 2016 article 135). It must be understood that the proportionate loss of Delta environment could be much larger. The vital process of flood-switching is driven by sediment transport which takes place primarily at very high stream flows: a leveling out of flow for instance by dams will seriously hamper or stop this process all together.
7. It is predicted that the proposed water off takes will cause the river to dry out one year out of 20 (OKACOM 2011) and of course generally increase the risk for droughts in the Delta. It must be understood that the process of desalination is entirely driven by evaporating trees in the woodland fringes around the thousands of islands. If these woodlands are killed for instance by drought the desalination process will immediately cease to function and the entire Delta ecosystem will change fairly fast into a salt marsh – a probably irreversible process.

However, it is difficult to predict the magnitude of the negative effects of all these various threats. The Delta is such a unique system that management mistakes documented in systems elsewhere are of little use in informing the consequences of mismanagement here. These threats are now immediate, and they are many, they involve different economic and social actors in three countries, and they have happened elsewhere. If one or a few of them actually materialize, it would probably be enough to destroy the Okavango Delta. It is difficult to imagine a more vulnerable ecosystem.

Due to the newly started large irrigation schemes in the drainage basin three comprehensive studies are now urgent: **(1) In the drainage basin in Angola**, the causal links between geology, hydrology, sediment processes, and water chemistry. **(2) In the Delta**, the causal links between river avulsions, flooding patterns, nutrient dynamics, biological productivity, successions, and migrations. **(3) In the Delta,** the formation, growth, and decay of islands using a "population dynamic" quantitative approach. The Okavango Research Institute is the natural hub for these studies.

References

Bauer P. Flooding and salt transport in the Okavango Delta, Botswana: key issues for sustainable wetland management [dissertation]. Zurich: Swiss Federal Institute of Technology; 2004.

Bonyongo MC. The ecology of large herbivores in the Okavango Delta, Botswana [dissertation]. Bristol: School of Biological Sciences, University of Bristol; 2004.

Cotterill F. Reducine antelopes of the Zambezi basin. In: Timberlake J, editor. Biodiversity of the Zambezi wetlands. Famona, Bulawayo: Biodiversity Foundation for Africa; 1998. p. 145–99.

Gumbricht T, McCarthy TS. Spatial patterns of islands and salt crusts in the Okavango Delta, Botswana. S Afr Geogr J. 2003;85:164–9.

Gumbricht T, McCarthy J, McCarthy TS. Channels, wetlands and islands in the Okavango Delta, Botswana, and their relation to hydrological and sedimentological processes. Earth Surf Process Landf. 2004;29:15–29.

Heinl M, Frost P, VanderPost C, Sliva J. Fire activity on drylands and floodplains in the Southern Okavango Delta, Botswana. J Arid Environ. 2007;68:77–87.

Hogberg P, Lindholm M, Ramberg L, Hessen DO. Aquatic food web dynamics on a floodplain in the Okavango Delta, Botswana. Hydrobiologia. 2002;470:23–30.

Huntsman-Mapila P, Ringrose S, Mackay AW, Downey WS, Modisi M, Coetzee SH, Tiercelin J-J, Kampunzu AB, VanderPost C. Use of geochemical and biological sedimentary record in establishing palaeo-environments and climate change in the Lake Ngami basin, NW Botswana. Quat Int. 2006;148:51–64.

Junk WJ, Brown M, Campbell IC, Finlayson M, Gopal B, Ramberg L, Warner BG. The comparative biodiversity of seven important wetlands: a synthesis. Aquat Sci. 2006;68:400–14.

Lindholm M, Hessen DO, Mosepele K, Wolski P. Food webs and energy fluxes on a seasonal floodplain: the influence of flood size. Wetlands. 2007;27:775–84.

McCarthy TS. The Okavango Delta and its place in the geomorphological evolution of southern Africa. S Afr J Geol. 2013;116:1–54.

McCarthy TS, Ellery WN, Ellery K. Vegetation-induced, subsurface precipitation of carbonate as an aggradational process in the permanent swamps of the Okavango (delta) fan, Botswana. Chem Geol. 1993;107:111–31.

Mendelsohn J, El Obeid S. Okavango River. The flow of a lifeline. Cape Town: Struik Publishers; 2004.

Mendelsohn J, VanderPost C, Ramberg L, Murray-Hudson M, Wolski P, Mosepele K. Okavango Delta: floods of life. Windhoek: RAISON (Research & Information Services of Namibia); 2010.

Mitsch WJ, Nahlik A, Wolski P, Bernal B, Zhang L, Ramberg L. Tropical wetlands: seasonal hydrologic pulsing, carbon sequestration and methane emissions. Wetl Ecol Manag. 2010;18:573–86.

Murray-Hudson M, Wolski P, Brown M, Kashe K. Disaggregating hydroperiod: components of the seasonal flood pulse as drivers of plant species distribution in floodplains of a tropical wetland. Wetland. 2014;34:927–42.

Ramberg L. A pipeline from the Okavango River? Ambio. 1997;26:129.

Ramberg L, Wolski P. Growing islands and sinking solutes: processes maintaining the endorheic Okavango Delta as a freshwater system. Plant Ecol. 2008;196:215–31.

Ramberg L, Wolski P, Krah M. Water balance and infiltration in a seasonal floodplain in the Okavango Delta, Botswana. Wetlands. 2006a;26:677–90.

Ramberg L, Hancock P, Lindholm M, Meyer T, Ringrose S, Sliva J, Van As J, VanderPost C. Species diversity of the Okavango Delta, Botswana. Aquat Sci. 2006b;68:310–37.

Ramberg L, Lindholm M, Hessen DO, Murray-Hudson M, Bonyongo C, Heinl M, Masamba W, VanderPost C, Wolski P. Aquatic ecosystem responses to fire and flood size in the Okavango Delta: observations from seasonal floodplains. Wetl Ecol Manag. 2010;18:587–95.

Siziba N, Chimbari MJ, Mosepele K, Masundire H. Spatial and temporal variations in densities of small fishes across different temporary floodplain types of the lower Okavango Delta, Botswana. Afr J Aquat Sci. 2011;36:309–20.

Stanistreet IG, McCarthy TS. The Okavango Fan and the classification of subaerial fan systems. Sediment Geol. 1993;85:115–33.

White F. The vegetation of Africa. A descriptive memoir to accompany the UNESCO/AETFAT/UNSO vegetation map of Africa. Paris: Unesco; 1983. 336 p.

Wolski P, Ramberg L, Magole L, Mazvimavi D. Evolution of River Basin management in the Okavango system, Southern Africa. In: Ferrier RC, Jenkins A, editors. Handbook of catchment management. Blackwell Publishing Ltd: John Wiley & Sons Ltd...West Sussex UK; 2010. p. 457–75.

Peatlands of Africa

115

Piet-Louis Grundling and Ab P. Grootjans

Contents

Introduction	1414
Peatland Distribution	1415
Mire Types	1417
Threats and Future Challenges	1419
References	1422

Abstract

An abundance of different wetland types occur across the continent but the occurrence of peatlands is less common. Africa, the second largest landmass on earth, is a continent with diverse landscapes ranging from snow-capped mountains on the Equator (e.g., Kilimanjaro) to the world's largest desert, the Sahara in the north, and Fynbos Biome the world's richest and smallest floral kingdom located in South Africa. The variety of mires and peatlands on the world's second largest continent are an expression of its diverse landscapes and associated climatic conditions. Local communities often depend on these peatlands (as

P.-L. Grundling (✉)
Centre for Environmental Management, University of the Free State, Bloemfontein, South Africa
e-mail: peatland@mweb.co.za

A. P. Grootjans (✉)
Centre for Energy and Environmental Studies, University of Groningen, Groningen, The Netherlands

Institute of Water and Wetland Research, Radboud University Nijmegen, Nijmegen, The Netherlands
e-mail: a.p.grootjans@rug.nl

© Springer Science+Business Media B.V., part of Springer Nature 2018
C. M. Finlayson et al. (eds.), *The Wetland Book*,
https://doi.org/10.1007/978-94-007-4001-3_112

with most other wetland types) for their livelihoods. However, these peatlands are often poorly managed and mostly severely degraded and in developing areas frequently threatened by future developments. Africa's peatlands are poorly researched and inventories on where they occur and their conservation status are lacking. Many opportunities exist to do research, implement wise use practices and restoration.

Keywords
Africa · Arid · Peatland · Threats · Tropical

Introduction

Africa, the second largest landmass on earth, is a continent with diverse landscapes ranging from snowcapped mountains on the equator (e.g., Kilimanjaro) to the world's largest desert, the Sahara in the north, and the richest (and smallest) floral kingdom in the south: the fynbos biome. An abundance of different wetland types occur across the continent. These include large systems such as Sahelian floodplains (e.g., the Niger) in Western Africa and the Sudd in Sudan (Rebelo and Moghraby, this volume). The world's largest inland delta, the Okavango Delta, occurs in Botswana, and extensive tropical swamp forests occur in the central Congo Basin (Joosten et al. 2012). The lakes of the Great Rift Valley in Eastern Africa also contribute to this rich diversity of Africa's wetlands, as do the ephemeral pans in the deserts of southern and Northern Africa and the alpine mires of Lesotho, Tanzania, Rwanda, and Ethiopia (Fig. 1).

The extreme range of habitats characteristic of this continent and their rich biodiversity are threatened by cultivation, harvesting, damming, and draining by a human population that is just as diverse. Nowhere are anthropogenic impacts more evident than in the continent's peatlands with rural populations often dependent on the availability of water and organic soils for cultivation. This means that natural peat-forming systems (mires) probably are much less abundant than the estimated area of peatlands. Peatlands are wetlands that have a minimum peat depth of 30 cm, and peat has a minimum percentage of sedentarily accumulated material consisting of at least 30% (dry mass) of dead organic material (Joosten and Clarke 2002).

The African continent is primarily an uplifted plateau located in the tropical and subtropical zones (McCarthy and Rubidge 2005). The average rainfall of the continent is 652 mm/year, which is well below the global average of 860 mm/year. In many parts of Africa even the average rainfall is not very reliable and large fluctuations in groundwater level occur. Peatlands are therefore restricted to the tropics with its high rainfall, the uplands of Lesotho and Ethiopia with cool climates, and regions with sustained groundwater discharge such as the Mozambique coastal plain and the karst landscapes of southern Africa.

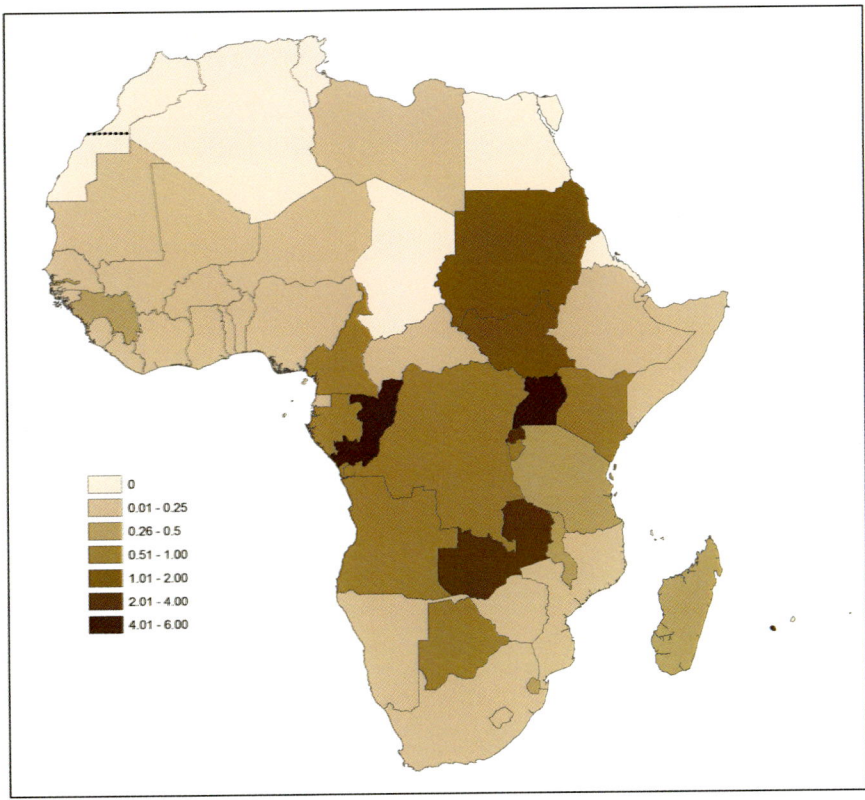

Fig. 1 Peatlands in Africa: percentage cover per country (Drawn from data in Joosten et al. 2010)

Peatland Distribution

It is estimated that as much as 50% of the world's wetlands could be peatlands, which include mires (supporting peat-forming vegetation). Peatlands occur in most African countries (Figs. 1 and 2) but often as small features in the landscape of less than a few hectares, as, for instance, the interdunal depressions in the Highveld, South Africa (Fig. 3), or the seeps and sloping mires in the mountain areas of Lesotho and Ethiopia (Fig. 4). But extensive peatlands also occur, such as the 7,000 ha Rugezi Marsh in Rwanda (Fig. 5) at the equator (Hategekimana and Twarabamenye 2007). Even more extensive mires occur in the Okavango Delta in Botswana and the Sudd catchment in Sudan (Fig. 6), but it is unknown how much peat such systems actually have. They have rather thin organic layers often associated with swamp forests, and we know little about the actual peat resources. It is estimated that the equatorial areas of Africa, such as the Gulf of Guinea and Central Africa, may contain about 14% of the global area of peat swamp forest. Estimates of

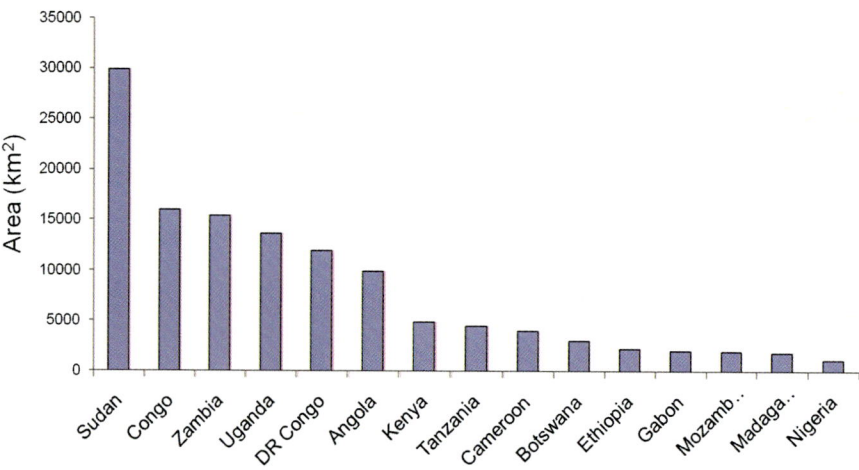

Fig. 2 Countries hosting the most extensive mires and peatlands in Africa (Adapted from Joosten 2010)

Fig. 3 Tevreden Pan during winter in South Africa: a mire formed by floating *Phragmites australis* mat in an endorheic depression in the Highveld grassland – a biome severely threatened by coal mining (Photo P.L. Grundling © Rights remain with the author)

the total peatland area in Africa vary between 4,856,500 ha (Andriesse 1988) and over 11,000,000 ha (Joosten 2010), which is less than 1% of the total land surface area of the continent.

In the wet equatorial belt of Africa, a distinction can be made between the areas flanking the Gulf of Guinea (e.g., Benin) and the Lower Congo Basin (e.g., Congo-Brazzaville), which are both large depressional areas with extensive lowland mires (Lappalainen and Zurek 1996; Joosten 2004), compared to those in eastern parts of Central Africa. In the eastern parts, peat formation occurs mainly in upland valleys following recent geological uplift, rift formation, and volcanism (e.g., in Rwanda and Burundi). In these high-altitude regions, the conditions for peat formation are more similar to temperate regions (Deuse 1966; Andriesse 1988).

Fig. 4 A sloping mire in Lesotho with flowering *Kniphofia caulescens* (Photo P.L. Grundling © Rights remain with the author)

Fig. 5 Cultivation by the raised-bed technique in the Rugezi Marsh in Rwanda leads to siltation and sediment deposition in downstream wetlands (*left*). As a consequence, such marshes lose their ability to purify water as illustrated in the *right* photo indicating high loads of suspended solids and turbidity (Photo credits: A. Linström © Rights remain with the author)

Mire Types

Original mire types in Africa reflect the drier climates of the continent; the systems are almost always groundwater fed. Fens are thus the dominant mire type, while rainwater fed bogs, which are often the dominant mire type in the temperate

Fig. 6 The Okavango Delta in Botswana – It is yet to be established to what extent parts of the Okavango Delta might contain peat (Photo: A.P. Grootjans © Rights remain with the author)

northern hemisphere, are rare. Peat-forming species include swamp forest species (e.g., *Ficus* and *Syzygium*), sedges (Cyperaceae, such as *Papyrus* and *Cladium*), grasses (e.g., *Phragmites* and *Miscanthus*), and mosses such as *Sphagnum*. Peat stratigraphy in Africa reflects a diversity of depositional environments and is very variable between sites (Hamilton and Taylor 1986; Smuts and Akiaoue 1999). A detailed description on African mire types is not available, but based on limited research in Rwanda and more extensive work in southern Africa, the following types have been noted: (i) groundwater rise mires; (ii) terrestrializing fens; (iii) sloping fens; (iv) groundwater through-flow mires, mostly treeless; and (v) flood mires. Groundwater rise mires mostly consist of relatively productive swamp forests. They are abundant in the Central Congo Basin (Joosten et al. 2012). Peat layers are usually shallow (less than 1 m). Terrestrializing fens are floating mats that may or may not produce peat. The extensive *papyrus* swamps in the Okavango Delta partly belong to this type and probably do not accumulate much peat. Sloping fens, sometimes even with *Sphagnum* species, are groundwater-fed mires. Usually the groundwater comes from springs that are sustained by groundwater that is forced to surface due to occurrence of geological faults. The systems are most frequent in mountain areas. Groundwater through-flow mires have been found in interdunal depressions along the east coast of South Africa. Flood mires are mires that are influenced by both groundwater and surface water. The surface water regularly floods these systems and may inundate them for several months.

Fig. 7 Cultivation of a papyrus fen in Kenya: note the drainage pattern (Photo: H. Joosten © Rights remain with the author)

They are usually very productive. The vegetation may consist of swamp forests, but can also be treeless.

More detailed information on a local mire typology for South Africa is presented by Grundling et al., ▶ Chap. 116, "Peatland Types and Tropical Swamp Forests on the Maputaland Coastal Plain (South Africa)."

Threats and Future Challenges

Agriculture, mining, and damming for electricity production and/or water reservoirs are some of the main threats to peatlands in Africa. Many low-lying peatlands in Africa have been converted into agricultural land (Figs. 5 and 7), while upland mires are increasingly threatened by grazing (Fig. 8). Pristine mires are under severe threat of conversion and degradation due to mining for minerals in valleys where mires occur (such as coal and diamonds), in particular in southern and Central Africa. Logging of forests and draining of swamp forests for timber and cultivation are a continuous threat in many parts of the continent. Furthermore, the damming of peatlands for hydroelectricity (e.g., by Eskom in South Africa and planned schemes in the Congo Basin) and also extraction of peat for energy (e.g., Burundi and Rwanda) pose severe threats to peatlands in Africa. Access routes in and across mires to these developments are often cause for concern (Figs. 9 and 10).

Various studies have shown that Africa is particularly vulnerable to climatic change (Hulme et al. 2001; Challinor et al. 2007). The semiarid and subhumid

Fig. 8 Livestock grazing the Berga mire in Ethiopia (Photo: M. Drummond © Rights remain with the author)

Fig. 9 The alpine fens of Lesotho are critical to water security for the southern African region, but they are often eroded and degraded due to overgrazing, partly as a result of the flooding of the lower valleys by hydroelectricity schemes (Photo P.L. Grundling © Rights remain with the author)

areas, in particular, will be most adversely affected by a shift toward a drier regime with extremely unreliable rainfall. This will not only affect the natural character of peatlands but also people will be affected, leading to increased dependence and pressure on the remaining peatlands.

Fig. 10 A dugout canoe (kano) in an artificial canal on the Rugezi mire, Rwanda. Artificial channels (dug to supply more water to the hydropower facility) drain the wetland but also provide a means of transport for local people (Photo P.L. Grundling © Rights remain with the author)

In general, wetlands are resilient, dynamic systems that change over time. However, peatlands in Africa cannot easily adapt to major changes in hydrological conditions associated with a combination of climatic change and anthropogenic impacts. It is expected that increased carbon emissions will result from these combined threats. While Africa covers 20% of the world's landmass, it only contains 3.45% of the peatland area and hosts 2.4% of its carbon, yet 10–15% of its peatlands are degraded, and it releases 4.3% of global peat CO_2 emissions (Joosten 2010). Due to drought and drainage, these peatlands also become more fire-prone in drier areas, and in Sub-Saharan Africa (South Africa excluded) peat emissions are equivalent to 25% of all fossil fuel emissions (Joosten 2010).

The variety of mires and peatlands in Africa are an expression of its diversity in landscapes and associated climatic conditions. Local communities often depend on these wetlands, where they occur in close proximity, for their livelihoods. However, these peatlands are poorly managed and mostly severely degraded and threatened by future developments. Basic inventories for Africa's peatlands are lacking and they are not well researched. Therefore, many opportunities exist to implement wise-use practices, restoration, and applied research projects.

References

Andriesse JP. Nature and management of tropical peat soils. FAO Soils Bulletin 59. Rome: FAO – Food and Agriculture Organization of the United Nations; 1988. Available from: http://www.fao.org/docrep/x5872e/x5872e04.htm. Accessed 24 Jan 2016.

Challinor A, Wheeler T, Garforth C, Craufurd P, Kassam A. Assessing the vulnerability of food crop systems in Africa to climate change. Clim Change. 2007;83(3):381–99.

Deuse P. Contribution à l'étude des tourbières du Rwanda et du Burundi. Esquisse de la végétation des tourbières du Rwanda et du Burundi. Butare: Institut pour la Recherche Scientific en Afrique; 1966. p. 53–115. Treizième Rapport d' Institut pour la Recherche Scientific en Afrique.

Hamilton A, Taylor D. Mire sediments in East Africa. London: Geological Society; 1986. Special publications 25. p. 211–7. doi:10.1144/GSL.SP.1986.025.01.

Hategekimana S, Twarabamenye E. The impact of wetlands degradation on water resources management in Rwanda: the case of Rugezi Marsh. In: Harding B, Devisscher T, editors. Review of the economic impacts of climate change in Kenya, Rwanda and Burundi. Nairobi: African Conservation Centre; 2007.

Hulme M, Doherty R, Ngara T, New M, Lister D. African climate change 1900–2100. Climate Res. 2001;17(2):145–68.

Joosten H. The IMCG global peatland database. 2004. www.imcg.net/gpd/gpd.htm. Accessed 24 Jan 2016.

Joosten H. The global peatland CO_2 picture; peatland status and drainage related emissions in all countries of the world. Ede: Wetlands International; 2010. p. 36.

Joosten H, Clarke D. Wise use of mires and peatlands: background and principles including a framework for decision-making. Totness/Devon: NHBS/International Mire Conservation Group and International Peat Society; 2002.

Joosten H, Tapio-Biström MJ, Tol S, editors. Peatlands – guidance for climate change mitigation through conservation, rehabilitation and sustainable use. 2nd ed. Rome: Food and Agriculture Organization of the United Nations; 2012. p. 114. Joint publication with Wetlands International, Wageningen.

Lappalainen E, Zurek S. Peat in other African countries. In: Lappalainen E, editor. Global peat resources. Jyskä: International Peat Society; 1996. p. 239–40.

McCarthy T, Rubidge B. The story of earth and life: a southern African perspective on a 4.6-billion-year journey. Cape Town: Struik Publishers; 2005. p. 333.

Smuts WJ, Akiaoue E. Characterization of the peats of the South Central part of the Congo. Afr Geosci Rev. 1999;6:65–70.

Peatland Types and Tropical Swamp Forests on the Maputaland Coastal Plain (South Africa)

116

Althea T. Grundling, Ab P. Grootjans, Piet-Louis Grundling, and Jonathan S. Price

Contents

Introduction	1424
Geomorphological History of the Maputaland Coastal Plain	1424
Mire Types of Maputaland Coastal Plain	1426
Peat Swamp Forests	1428
Swamp Forest on Mineral Soils	1433
Threats and Future Challenges to Mires and Swamp Forests	1433
References	1434

A. T. Grundling (✉)
Water Science Programme, Agricultural Research Council – Institute for Soil, Climate and Water, Pretoria, Gauteng, South Africa

Department of Geography and Environmental Management, University of Waterloo, Waterloo, ON, Canada

Applied Behavioural Ecology and Ecosystem Research Unit, University of South Africa, Pretoria, South Africa
e-mail: althea@arc.agric.za

A. P. Grootjans
Centre for Energy and Environmental Studies, University of Groningen, Groningen, The Netherlands

Institute for Water and Wetland Research, Radboud University Nijmegen, Nijmegen, The Netherlands
e-mail: a.p.grootjans@rug.nl

P.-L. Grundling
Centre for Environmental Management, University of the Free State, Bloemfontein, South Africa
e-mail: peatland@mweb.co.za

J. S. Price
Department of Geography and Environmental Management, University of Waterloo, Waterloo, ON, Canada
e-mail: jsprice@uwaterloo.ca

© Springer Science+Business Media B.V., part of Springer Nature 2018
C. M. Finlayson et al. (eds.), *The Wetland Book*,
https://doi.org/10.1007/978-94-007-4001-3_166

Abstract

The Maputaland Coastal Plain (MCP) is on the north-eastern seaboard of the KwaZulu-Natal Province, South Africa. It hosts a variety of wetlands and mire types that range from rare tropical swamp forest and calcareous fens to various inter-dune settings with groundwater-fed to surface water fed fens and floodplains. The MCP peatland areas are important not only for biodiversity (e.g. many endemic species) and supporting subsistence cultivation but they also host the largest and thickest peat deposits found in South Africa. The peatlands on the MCP developed in a drowned dune landscape and the peat thickness vary from 0.5 to 11 m. Threats and future challenges on the peatlands in Maputaland include the effects of land-use practices: plantations, agriculture and urbanization. Attempts to conserve swamp forests are often frustrated by the different management frameworks of provincial conservation agencies and the national government.

Keywords

Groundwater · Hydrology · Mire types · Subsistence cultivation

Introduction

The Maputaland Coastal Plain (MCP) is one of the wetter areas in South Africa with a mean annual precipitation of 900–1,000 mm. It hosts the most extensive wetlands and best developed peat deposits in South Africa and has a high biodiversity with many endemic species (Grundling et al. 1998; Smuts 1992). Maputaland, the southern tip of the Mozambique Coastal Plain (Fig. 1) in South Africa's KwaZulu-Natal Province, contains 60% of the known peat resources in South Africa, ranging in size from a few to thousands of hectares of wetland (Grundling et al. 1998). Approximately 266 peatlands occur within the MCP with peat thickness varying from 0.5 to 11 m (Grundling et al. 2000). The highest concentration of peatlands occurs close to the coast where the precipitation is highest, but larger peatlands also occur towards the western edge of Lake Sibaya (Grundling 1994) and the tributaries of the northward flowing Futi (Mkuzi-North) stream, where the rainfall varies between 600 and 800 mm pa (Grundling et al. 1998). The tropical peat swamp forests of Maputaland form about a third of the peatlands in Maputaland and are one of the most unique and rare wetland types in southern Africa (Moll 1980; Wessels 1997). Various land use practices impact negatively on the peatlands in Maputaland, including plantation, cultivation and urbanization (Grundling et al. 2013a).

Geomorphological History of the Maputaland Coastal Plain

The original rivers of this area flowed from west to east, i.e., approximately at right angles to the present coastline (Fig. 1). Most had estuaries, controlled partly by coastal sand bars, partly by a rise in sea level. Periods of transgression have resulted

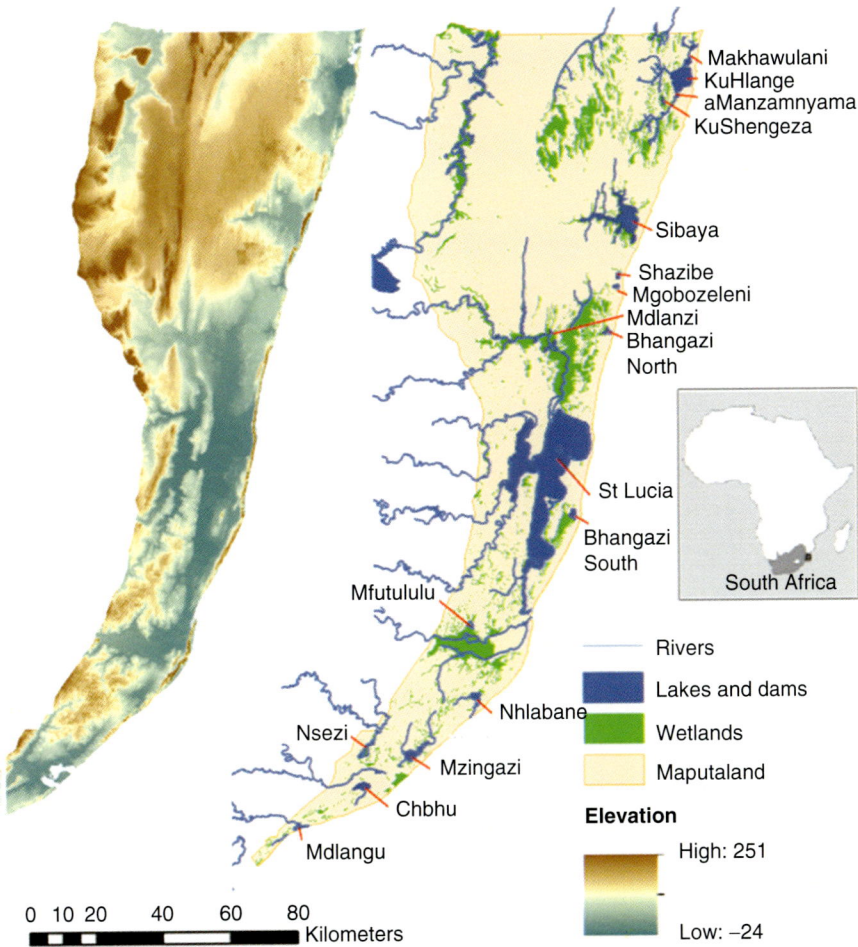

Fig. 1 Landscapes (high 250 m.a.s.l and low <50 m.a.s.l), rivers, lakes and wetlands and major peatlands of the Maputaland Coastal Plain, on the Indian seaboard of South Africa (With permission from Grundling et al. 2014)

in the drowning of the coast-parallel dune valleys allowing low energy, low-sedimentation regimes to prevail. Peat deposits have developed in this drowned dune landscape (Smuts 1992), often in association with the lakes, which are the result of damming and diversion of rivers by belts of sand dunes. These dunes originate from beach deflation (Hart 1995).

The shape of the lakes is controlled by the topography of the drowned dune landscape. Some of these lakes, such as Lake Sibaya, are up to 35 m deep (Wright et al. 2000). Some lakes are isolated and their river mouths closed by dunes up to 180 m high (northern part of the study area). Lake Sibaya and Lake Bhangazi were formed in this way (Hart 1995). Excess water from the lake seeps through the dune to reach the ocean in the form of fresh-water beach springs (Taylor et al. 2006). Lake Sibaya is the largest natural fresh water body in South Africa and 29 groundwater-fed peatlands (fens) are directly related to this lake (Grundling et al. 2000).

Lake St. Lucia represents a composite diversion of several river systems, such as the Mkuze and the Hluhluwe (Hart 1995). The diversion in the case of Lake St. Lucia has been for a distance of 80 km. The St. Lucia wetland system hosts the largest and the thickest peatlands in South Africa. They are respectively the Mkuze Delta (a floodplain with papyrus, reed and peat swamp forest) with an aerial extent of 8,800 ha and the Mfabeni Mire with a peat thickness of 10 m (Grundling et al. 2013b). The genesis of lake associated peatlands is interpreted to be related to a rise in sea level and with high groundwater levels in adjacent dune systems (Grundling et al. 2000; Grundling et al 2013b). This rise in sea level has reduced the sediment carrying capacity of the rivers. Consequently the rivers deposit the sediment within their valleys, many kilometres before reaching the sea shore. During floods, the rivers flow out of their normal river courses onto the floodplain. In many instances, these flood waters spread wide enough to submerge the lower reaches of smaller tributaries of the main rivers (Grundling et al. 2000; Ellery et al. 2012). The sediment build-up on the main floodplain, in some instances, has further dammed the lower reaches of the local tributaries, resulting in the formation of more lakes in the lower portions of these tributaries. About 15% of the peatlands in Maputaland are associated with these lakes.

Mire Types of Maputaland Coastal Plain

A variety of mire types, all fens as they are all groundwater-fed or surface water fed, can be found on the MCP, of which the following are distinguished (Fig. 2):

Sloping mires (Fig. 3) are fed by groundwater from a larger plateau that is discharging groundwater on its slopes (i.e., through seepage) to a low lying area. Usually these systems are associated with geological faults or deeply incised valleys, where erosion has exposed an aquifer. In Maputaland these are usually associated with perched aquifers formed by impermeable clay layers, or occur closer to the coast where the regional water table is intersected by incised valleys. Slope or spring mires are often associated with swamp forest components (Grundling et al. 2014).

Through flow mires (also called percolation fens) (Fig. 4) have groundwater flowing through the shallow peat layers, or more often passing over the surface of the peat. Such fens are also relatively flat and can contain both groundwater and rainwater on the surface. The peat layers can be quite thick (up to 11 m). These isolated inter-dunal mires are quite common in northern Maputaland and are usually covered by reeds and sedges. They are orientated parallel to the Pleistocene dune ridges and hydrologically linked to coastal lakes with an estuary. Examples are Siyadla, the Muzi North system in Tembe Elephant Park and the Mfabeni Mire (Grundling et al. 2013b).

Calcareous mires (Fig. 5) have deposited calcite on the peat, thus creating a very alkaline and nutrient-poor environment. The calcite originates from supersaturated

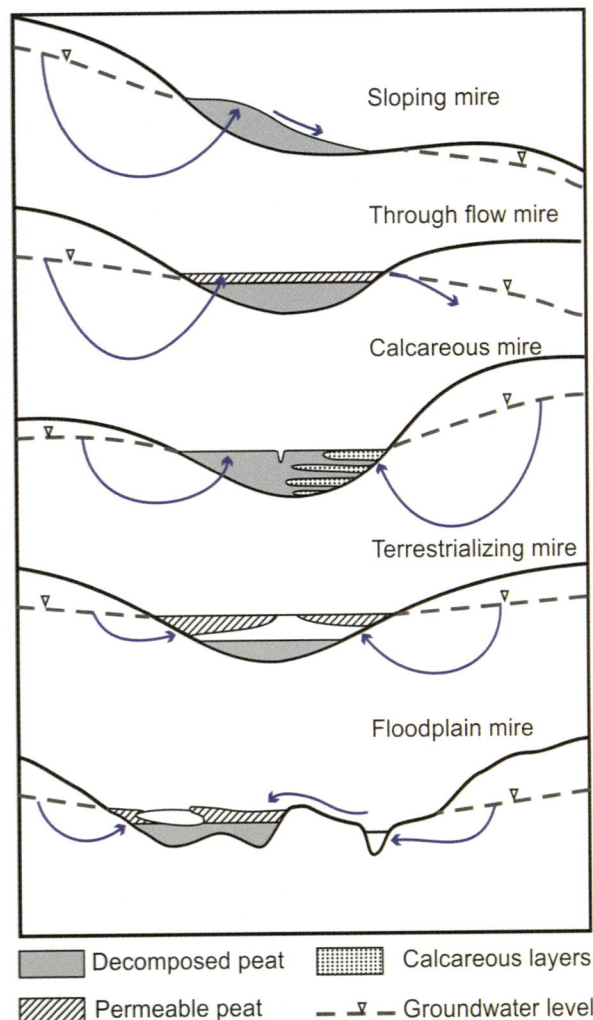

Fig. 2 Conceptual models of hydrological functioning of some Maputaland mire types (© Rights remain with the authors)

groundwater, which precipitates $CaCO_3$ when exposed to the air (Grootjans et al. 2012). Only one example of a calcareous mire has been found on the MCP (the Muzi North system in Tembe Elephant Park). Chalk (marl) deposits were observed in the peat profile where it borders a through flow mire (Grundling 2014).

Terrestrializing mires (Fig. 6) are flat and float on surface water. Peat formation can proceed because such mires do not dry out. They are mainly fed by lake water, but groundwater may also reach such fens. The floating mat mostly consists of papyrus (*Cyperus papyrus*), reeds (*Phragmites australis*) and sedges (*Fimbristylis longiculmis, Cladium mariscus*) and during the last stage of terrestrializing, swamp forest (dominated mainly by *Ficus trichopoda, Syzigium cordatum* and *Voacanga thouarsii*) colonizes the mire. Terrestrializing mires are the most common type of peatland in the Maputaland area.

Fig. 3 The Mfabeni sedge fen (*foreground*), an 11-m deep peatland with a sloping mire (the Nkazana swamp forest) comprising its western edge (*middle ground*) (Photo credit: Piet-Louis Grundling © Rights remain with the author)

Floodplain mires (Fig. 7) are fed by surface water (floods from rivers) and also by groundwater. Due to the high energy conditions experienced during flood events, much sediment is deposited on these mires. Consequently, they are more productive than the former fen types and covered by papyrus, reeds and sedges. The peat is often formed by floating mats with the peat layers practically always containing grey clay particles or thin grey clay laminae. The clay observed in this peat is deposited in the papyrus mats when the floodplain back-floods into the off-channel lakes during flood events. Some swamp forests occur in these settings as well. These peatlands are usually found at the fringes of large floodplains, where they are supplied with groundwater from the surrounding plateaus or dunes. Good examples are mires in the vicinity of the Pongolo, Mkuze, Hluhluwe, Umfolozi and Mhlatuzi rivers.

Peat Swamp Forests

Peat swamp forests (Fig. 8) are fresh-water forested wetlands that are established on peat and require a high water table with periodic saturated conditions, as a result of inundation, that creates a favourable hydrological regime for peat development (Sliva 2004). Most swamp forests occur in the proximity of lakes (Table 1) within inundated (flooded) inter-dune valleys (Grundling et al. 2000). Other important settings are seep zones, inter-dune valleys and drainage lines underlain by low-permeability sediments (Sliva 2004), whereby groundwater seepage elevates the water table sufficiently to promote peat accumulation in which the long-term annual average biomass production exceeds decay (Clymo 1983). The MCP hosts 75% of South Africa's swamp forests (Lubbe 1997) with 59% of all South African swamp forest protected within the iSimangaliso Wetland Park (Wessels 1997).

Fig. 4 A through flow mire in an inter-dune setting in the iSimangaliso Wetland Park. Below is a close-up of the gardens (Photos credit: Althea Grundling © Rights remain with the author)

Fig. 5 Aerial view of the Muzi peatland, Tembe Elephant Park – a calcareous mire. Note the elephants in the mire (Photo credit: Piet-Louis Grundling © Rights remain with the author)

Fig. 6 Terrestrializing mires (Photo credit: Retief Grobler © Rights remain with the author)

Fig. 7 Floodplain mires. *Above*, aerial view with meandering river (Photo credit: Piet-Louis Grundling © Rights remain with the author). *Below*, the Pongolo Floodplain (Photo credit: Althea Grundling © Rights remain with the author)

Fig. 8 Peat swamp forest at Twamansi (Photo credit: Retief Grobler © Rights remain with the author)

Table 1 Lakes associated with peatlands on the Maputaland Coastal Plain

Lake	Isolated, floodplain or open ended	No of peatlands	Peat thickness (m)	Peat area (ha)
Manzamnyama	Open ended	1	0.65	3.5
Bhangazi North	Isolated	16	6.2	66.5
Bhangazi South	Open ended	1	11	1460
Cubhu	Open ended	3	2	30
KuHlange	Open ended	6	6.1	46.5
KuShengeza	Isolated	2	1.1	16
Makhawulani	Open ended	3	4.7	21
Mdlangu	Open ended	1	1.65	4
Mdlanzi	Floodplain	3	4.7	248
Mfutululu	Floodplain	1	6.5	87.5
Mgobozeleni	Open ended	4	2.4	131
Mzingazi	Open ended	5	3.5	715
Nhlabane	Open ended	3	3.2	23
Nsezi	Floodplain	1	5.9	324
Shazibe	Open ended	2	2.85	150
Sibaya	Isolated	29	5.5	400
St Lucia	Open ended	7	5.9	7650
Total: 17		**88**		**10 732.5**

Swamp Forest on Mineral Soils

It has been noted extensively throughout the whole study area that swamp forests or riverine forests that are dominated by *Barringtonia racemosa* do not contain peat. Only some organic, sandy clay or dark clay material is present. *B. racemosa* only dominates parts of swamp forest peatlands where there is concentrated surface water flow. Under these conditions the heavy seed pod of *B. racemosa* can be transported. However, much of the organic material that may have accumulated is flushed out of the system at the same time. Therefore, the presence of *B. racemosa* could be an indicator of wetlands with low peat accumulation potential (Grundling et al. 2000).

Threats and Future Challenges to Mires and Swamp Forests

Pinus and *Eucaluptus* plantations both on a commercial and a woodlot scale, has severe impacts on peatlands and swamp forests on the MCP (Blackmore and Mulqueeny 1996). Since swamp forests are largely groundwater dependent the lowering of the water table by forest plantations of exotic species results in desiccation of the peat profile and eventually degradation of the swamp forest, often with accompanying peat fires (Grundling and Blackmore 1998). Soils of the coastal plain of north-eastern South Africa are very sandy, leached and low in nutrients (Watkeys et al. 1993; Sliva 2004), resulting in their low agricultural potential. The low-lying peatland areas, however, are relatively fertile, mostly swamp forests (Grundling et al. 2000). Consequently, the expansion of cultivation is the most significant land-use pressure in swamp forest of the MCP (SANBI 2006; Grundling and Grundling 2007), threatening both the ecosystem and the soil resource itself (Grundling et al. 1998). The relatively fertile peat soils with high water-holding capacity are very suitable for crops such as bananas, madumbe (taro), cassava and sweet potato as well as a variety of vegetables; these crops are instrumental in the local food supply, economy and social structure (Sliva 2004). However, the productivity of subsistence cultivation on swamp forest peat is threatened by soil degradation associated with drainage, especially carbon loss through soil oxidation and peat fires. Since the type of crop and farming practice influences water use and soil carbon store (Robinson and Alberts 1989; Jones 2006), a sustainable swamp forest use strategy is important to maintain ecological condition and prolongs productive use, essential for building food security and easing pressure on the remaining pristine systems. Sustainable management actions include the blocking of drains to maintain high water tables to reduce desiccation, oxidation and collapsing and burning of the peat. The increasing population, which is one of the most impoverished in the country (Morgenthal et al. 2004), is partly constituted by illegal cross-border migration and depends mainly on subsistence living (Grobler et al. 2004) derived primarily from swamp forest cultivation. Degradation of the swamp forest is a key symptom of complex socio-economic and ecological processes that constitute a 'difficult and complex problem' for stakeholders across the wider socio-political terrain (Van Bueren et al. 2003).

Rainfall patterns dictate the distribution of the human population in Maputaland and most communities are located on the fertile floodplains of the Pongola, Mkuze and other rivers in the western parts of Maputaland or close to the higher rainfall areas along the eastern coastal dune corridor. The higher population densities associated with the higher rainfall thus also result in more pressures on coastal swamp forests.

Evidence of abandoned raised garden beds (described as fossil gardens by Grundling 1996, Fig. 9) in seasonal wetlands in Maputaland highlights the unpredictable inundation of wetlands dependent on variable rainfall patterns. Local groundwater tables vary as much as 3 m within a season responding to large rainfall events after extensive drier periods. The stable water sources associated with swamp forests are thus another important consideration when selecting cultivation plots, which leads to greater pressure on the swamp forests. The pattern and expansion of cultivation in swamp forests is directly linked to road access and distance to informal markets (KwaZulu-Natal Top Business 2009; Grundling et al. 2013a).

Roughly 4,000 ha of peat swamp forest occurs on the coastal plain, much of which is still unclassified due to its remoteness and inaccessibility (Wessels 1997), although Sliva (2004) recognized swamp forest sites as either "pristine", "recently disturbed", "long-time recovering" or under "active gardening". Some swamp forests have been designated as Ramsar Sites, recognizing their unique character and settings. However, such attempts to conserve swamp forests are often frustrated by other management frameworks operating within different socio-spatial, political and resource jurisdictions (Sliva 2004).

The Maputaland Coastal Plain hosts a variety of mires ranging from calcareous fens to swamp forest that developed in an array of hydrogeomorphic settings. Not only do these mires contribute to the rich biodiversity of the region, they also provide the basis for subsistence farming and the only means of survival for many families. Conserving and making wise use of these systems is therefore of importance, and research to understand how they function and the need for sustainable management should be promoted.

References

Blackmore A, Mulqueeny C. Towards developing a model for the integration of afforestation and wetland conservation within the Zululand coastal plain. Natal Parks Board Annula Research Symposium: Programme and Abstracts. 1996.

Clymo RS. Peat. In: Gore AJP, editor. Ecosystems of the World 4A, Mires: swamp, bog, fen and moor. Amsterdam: Elsevier; 1983.

Ellery WN, Grenfell SE, Grenfell MC, Humphries MS, Barnes K, Dahlberg A, Kindness A. Peat formation in the context of the development of the Mkuze floodplain on the coastal plain of Maputaland, South Africa. Geomorphology. 2012;141–142:11–20.

Grobler R, Moning E, Sliva J, Bredenkamp G, Grundling, P. Subsistence farming and conservation constraints in coastal peat swamp forests of the Kosi Bay lake system, Maputaland, South Africa. Géocarrefour [En ligne]. 2004; 79/4: 316–24.

Grootjans AP, Jansen AMJ, Šefferová Stanová V, editors. Calcareous mires of Slovakia; landscape setting, management and restoration prospects. Zeist: KNNV Publishing; 2012. p. 109.

Grundling P. Peat reserves at Lake Sibaya. Pretoria (South Africa): Council for Geoscience; 1994. Report no: 1994–0122.

Grundling P. Sustainable Utilisation of Peat in Maputaland, KwaZulu-Natal, South Africa. In: Lüttig GW, editor. Abstracts of the 10th International Peat Society Congress. May 1996, Bremen. 1996; p. 6.

Grundling AT, Van den Berg EC, Pretorius ML. Influence of regional environmental factors on the distribution, characteristics and functioning of hydrogeomorphic wetland types on the Maputaland Coastal Plain, KwaZulu-Natal, South Africa. Water Research Commission Report No. 1923/1/13. ISBN 978-1-4312-0492-2. 2014.

Grundling P, Blackmore A. Peat fire in the Vasi Pan Peatland, Manzengwenya Plantation. Pretoria (South Africa): Council for Geoscience Geological Survey; 1998. Report No. 1998–0208.

Grundling P, Mazus H, Baartman L. Peat resources in northern KwaZulu-Natal wetlands Sustainable: Maputaland. Pretoria: Department of Environmental Affairs and Tourism; 1998.

Grundling P, Baartman L, Mazus H, Blackmore A. Peat resources of KwaZulu-Natal wetlands: Southern Maputaland and the North and South Coast. Pretoria (South Africa): Council for Geoscience; 2000. Report no: 2000–0132.2000.

Grundling P, Grundling AT. Natural resource baseline assessment mapping the state of the peatland swamp forests in the catchment of Kosi Bay, Greater St Lucia Wetland Park and surrounding areas. Progress report for the iSimangaliso Wetland Park Authority Wetlands International: Wetland and poverty reduction project: 14. St. Lucia, KwaZulu-Natal, South Africa. 2007.

Grundling AT, Van den Berg EC, Pretorius ML, Price JS. Assessing the distribution of wetlands over wet and dry periods and land-use change on the Maputaland Coastal Plain, north-eastern KwaZulu-Natal, South Africa. S Afr J Geomat. 2013a;2:120–39.

Grundling P, Grootjans AP, Price JS, Ellery WN. Development and persistence of an African mire; how the oldest South African fen has survived in a marginal climate. Catena. 2013b;100:176–83.

Grundling AT. Remote sensing and biophysical monitoring of vegetation, terrain attributes and hydrology to map, characterise and classify wetlands of the Maputaland Coastal Plain, KwaZulu-Natal, South Africa. [dissertation]. Waterloo (Ont.): University of Waterloo; 2014.

Hart RC. South African Lakes. In: Cowan GI, editor. Wetlands of South Africa. Pretoria: Department of Environmental Affairs and Tourism; 1995. p. 103–30.

Jones C. Carbon and Catchments: Inspiring real change in natural resource management. 'Managing the Carbon Cycle' NATIONAL Forum 22–23 November 2006.

KwaZulu-Natal Top Business Umkhanyakude District Municipality. 2009. http://www.kzntopbusiness.co.za/site/umkhanyakude-district-municipality. Accessed 27 Nov 2010.

Lubbe RA. Vegetation and flora of the Kosi Bay Coastal Forest Reserve in Maputaland, northern KwaZulu-Natal, South Africa. [master's thesis] Pretoria (South Africa): University of Pretoria; 1997.

Moll EJ. Terrestrial plant ecology. In: Burton MN, Cooper KH, editors. Studies on the ecology of Maputaland. Rhodes: Rhodes University & Wildlife; 1980. p. 52–68.

Morgenthal TL, Kellner K, Van Rensburg L. Auditing the conservation status of the natural resources in the OR Tambo and Umkanyakude ISRDS Nodes. Pretoria (South Africa): Agricultural Research Council – Institute for Soil, Climate and Water; 2004. ARC-ISCW Report No. GW/A/2003/47/1.

Robinson JC, Alberts AJ. Seasonal variation in the crop water-use coefficient of banana (cultivar 'Williams') in the subtropics. Sci Hortic. 1989;40:212–25.

SANBI (South African National Biodiversity Institute) National Wetland Inventory. South African National Biodiversity Institute – Freshwater Programme. 2006. http://bgis.sanbi.org/nwi/map.asp. Accessed 2 April 2013.

Sliva J. Maputaland – Wise Use Management in Coastal Peatland Swamp Forests in Maputaland, Mozambique/South Africa. Wageningen (NL): Wetlands International: 2004. Project No. WGP2 – 36 GPI 56.

Smuts WJ. Peatlands of the Natal Mire Complex: geomorphology and characterization. S Afr J Sci. 1992;88:474–83.

Taylor R, Kelbe B, Haldorsen S, Botha GA, Wejden B, Været L, Simonsen MB. Groundwater-dependent ecology of the shoreline of the subtropical Lake St. Lucia estuary. Environ Geol. 2006;49:586–600.

Van Bueren EM, Klijn EH, Koppenjan JF. Dealing with wicked problems in networks: analyzing an environmental debate from a network perspective. J Public Adm Res Theory. 2003;13:193–212.

Watkeys MK, Mason TR, Goodman PS. The role of geology in the development of Maputaland, South Africa. J Afr Earth Sci. 1993;16:205–21.

Wessels NG. Aspects of the ecology and conservation of Swamp Forests in South Africa. [master's thesis]. [Port Elizabeth]: Technikon; 1997.

Wright CI, Miller WR, Cooper JAG. The late Cenozoic evolution of coastal water bodies in Northern KwaZulu-Natal, South Africa. Mar Geol. 2000;167:207–29.

The Wetland Book

C. Max Finlayson • G. Randy Milton
R. Crawford Prentice • Nick C. Davidson
Editors

The Wetland Book

II: Distribution, Description, and Conservation

Volume 3

With 603 Figures and 125 Tables

Editors
C. Max Finlayson
Institute for Land, Water and Society
Charles Sturt University
Albury, New South Wales, Australia

UNESCO-IHE, Institute for Water Education
Delft, The Netherlands

R. Crawford Prentice
Nature Management Services
Cambridge, UK

G. Randy Milton
Department of Natural Resources
Kentville, Nova Scotia, Canada

Nick C. Davidson
Institute for Land, Water and Society
Charles Sturt University
Albury, New South Wales, Australia

Nick Davidson Environmental
Wigmore, UK

ISBN 978-94-007-4000-6   ISBN 978-94-007-4001-3 (eBook)
ISBN 978-94-007-4002-0 (print and electronic bundle)
https://doi.org/10.1007/978-94-007-4001-3

Library of Congress Control Number: 2017937719

© Springer Science+Business Media B.V., part of Springer Nature 2018
This work is subject to copyright. All rights are reserved by the Publisher, whether the whole or part of the material is concerned, specifically the rights of translation, reprinting, reuse of illustrations, recitation, broadcasting, reproduction on microfilms or in any other physical way, and transmission or information storage and retrieval, electronic adaptation, computer software, or by similar or dissimilar methodology now known or hereafter developed.
The use of general descriptive names, registered names, trademarks, service marks, etc. in this publication does not imply, even in the absence of a specific statement, that such names are exempt from the relevant protective laws and regulations and therefore free for general use.
The publisher, the authors and the editors are safe to assume that the advice and information in this book are believed to be true and accurate at the date of publication. Neither the publisher nor the authors or the editors give a warranty, express or implied, with respect to the material contained herein or for any errors or omissions that may have been made. The publisher remains neutral with regard to jurisdictional claims in published maps and institutional affiliations.

Printed on acid-free paper

This Springer imprint is published by the registered company Springer Science+Business Media B.V. part of Springer Nature.
The registered company address is: Van Godewijckstraat 30, 3311 GX Dordrecht, The Netherlands

Foreword: The Wetland Book

The venerable lineage of encyclopedic publishing can be traced back to Pliny the Elder's *Naturalis Historia*, which contained chapters on water and aquatic life. Although our terminology regarding and understanding of the aquatic environment has evolved over the past two millennia, one constant has been the need for a multidisciplinary approach to examining these areas. Using an encyclopedic model, this multidisciplinary book builds on an ancient format and adapts it for a modern audience. In this way, *The Wetland Book* builds on a long tradition of scholarly publishing and presents invaluable information for its modern audience.

Wetlands have been around longer than the traditions associated with academic publishing. Wetland management and wise use have been practiced by indigenous cultures in many forms for millennia, and that ancient knowledge about wetlands was often curated and passed down orally or in traditional systems and forms. In modern times, the pressures and threats to wetlands are vastly different in their scope and magnitude. The forms of governance and administration that respond to these pressures and threats have also changed, particularly in their scale as it has been recognized that management takes place at the level of countries and river basins, rather than simply at the local level.

Internationally, wetland conservation, management, and wise use are promoted through the Ramsar Convention on Wetlands. The countries that have signed onto the Ramsar Convention have recognized the imperatives to work with stakeholders and decisionmakers beyond the traditional wetland community and to incorporate wetlands into policy-making in other sectors such as water, energy, agriculture, and health. Indeed, in 2008 at the 10th Conference of the Contracting Parties of the Ramsar Convention, the Changwon Declaration was adopted, which contains key messages for wetland conservation, management, and wise use addressed to planners; policymakers; elected officials; managers in the environmental, land, and resource-use sectors; educators and communicators; economists; and health workers. *The Wetland Book* offers a base of knowledge that is intended to reach a similarly broad audience.

The editors and contributing authors to *The Wetland Book* have long experience and deep understanding of wetland science and management. Many have worked with the Ramsar Scientific and Technical Review Panel (STRP), the Convention's scientific advisory body, over the years. This collection of people provides a

repository of knowledge that can help meet the challenge of learning about and understanding the value of protecting and managing wetlands.

Making this knowledge more easily accessible, however, has always been difficult. There are physical limitations to how much we can pick a person's brain, and there are limitations to how much a wetland manager out in the field, perhaps with little technical support, can search for, read, and review scientific and traditional knowledge to find answers to pressing questions. Thus, the encyclopedic style of publication remains a viable format for accessing high levels of expertise, including expertise from distant locations, with similar landscape and ecological characteristics. *The Wetland Book* provides an in-depth level of knowledge in the form of a handbook to assist those seeking information on the many facets of wetland management.

Of course, reading *The Wetland Book* will not make an individual an expert in all aspects of wetland science, wise use and governance, a feat which no one publication can deliver. Instead, a truly useful publication should offer an individual the vocabulary to support further inquiry and to find knowledge that is locally, regionally, nationally, or even internationally applicable. It should also allow a reader to know who to ask and what questions to pursue when she or he needs more knowledge to solve a research question or particular management problem. *The Wetland Book* delivers this foundation through two volumes – Vol. 1: Structure and Function, Management, and Methods and Vol. 2: Distribution, Description, and Conservation.

We highly recommend *The Wetland Book*; it provides an unparalleled source of knowledge about wetlands by building on the ancient form of the encyclopedia, revitalized by new technologies for distribution and access. We are also proud to see that many of those who have contributed to the Ramsar Convention over many years or even decades have also contributed their knowledge and wisdom to *The Wetland Book*. Given our personal association with the Convention, we also recognize the incredible contribution that the Convention has made to wetland knowledge and look forward to further contributions.

Chair, Scientific & Technical Review Panel   Heather MacKay
Ramsar Convention on Wetlands
2005–2012
Chair, Scientific & Technical Review Panel   Royal C. Gardner
Ramsar Convention on Wetland
2012–2018

Preface

The Wetland Book is a hard copy and online production that provides an unparalleled collation of information on wetlands. It is global in scope and contains 462 chapters prepared by leading wetland researchers and managers. The wide disciplinary and geographic scope is a particular feature and differentiates *The Wetland Book* from the existing wetland literature. The editors have compiled *The Wetland Book* from contributions supplied by authors from many countries and disciplines. Combined, these chapters represent a global source of knowledge about wetlands. Given the number of chapters and the scope of the content, it has been published as two separate books.

The bibliographic detail of the two books is given below. Book II with 170 chapters covers the distribution, description, and conservation of wetlands.

The Wetland Book II: Distribution, Description, and Conservation: edited by Finlayson CM, Milton GR, Prentice RC and Davidson NC.

Its companion book, published separately, with 292 chapters is:

The Wetland Book I: Structure and Function, Management, and Methods: edited by Finlayson CM, Everard M, Irvine K, McInnes RJ, Middleton BA, van Dam AA and Davidson NC

The Wetland Book was developed following discussions with wetland experts from the Scientific and Technical Review Panel of the Ramsar Convention on Wetlands and from the Society for Wetland Scientists. These experts pointed to the rapidly expanding literature on wetlands and enthusiastically proposed the development of a comprehensive information resource aimed at supporting the trans- and multidisciplinary research and practice, which is essential to wetland science and management. They were also seeking an information resource that would both complement and extend the existing literature and in particular provide a compendium of knowledge with contributions from authors around the world.

Aware that wetland research was on the rise and that wetland researchers and practitioners often needed to work across disciplines, *The Wetland Book II* has been prepared to serve as a first port of call for those interested in the key information

about wetlands and their conservation. This was done to allow individuals and multi- and transdisciplinary teams to search for particular terms and subjects, access further details, and read overviews of topics selected by the editors and expert authors. The content provides a global coverage of wetland knowledge with chapters provided by leading wetland experts with knowledge that spans local and regional issues to the wider body of science that is needed to assist practitioners and enable students to come to grips with one of the world's most diverse and important set of ecosystems. This is especially important as these ecosystems are under increasing pressure in many parts of the world as degradation from human development continues at an alarming rate and are in need of more effective management and restoration. It draws heavily on the knowledge compiled through the formal processes of the Ramsar Convention and associated programs and extends that contained in the seminal global assessment of wetlands undertaken through the Millennium Ecosystem Assessment.

Book II is structured in sections covering the diversity of wetland types, natural and anthropogenic drivers of wetland change, and regional compilations of individual wetlands and wetland complexes. Detailed overview chapters typically describe the diversity within each wetland type, its distribution, extent (current and historical), ecosystem services, biodiversity, and threats and future challenges. Regional contributions follow a general format describing the basic ecology of the system; uniqueness; distribution; biodiversity and species adaptations; ecosystem services with an emphasis on importance to dependent peoples where appropriate; conservation status and management; and threats and future challenges. The coverage is based on a mix of wetland types and geographic extent and distribution.

Given that *The Wetland Book* constitutes a remarkable information resource, we warmly convey our special thanks to the many authors who gave up their time and shared their knowledge of wetlands to support this effort – and also for their patience while the large number of chapters they have generously provided were collated and edited. We are proud to have worked with them to produce this book. With the benefit of their unstinting efforts and incredibly rich knowledge, *The Wetland Book II* provides a comprehensive source of information for wetland researchers, students, and practitioners. It specifically provides a much needed information resource to support the many efforts to ensure the wise use of wetlands globally. It has also not only drawn on but also extended the expert guidance and advice that the Ramsar Convention's Scientific and Technical Review Panel has for almost 25 years provided for governments and wetland experts alike. In this respect, the foreword provided by the past and present chairs of the Panel is particularly appreciated. In providing the foreword, they have reflected on the wealth of knowledge collated by wetland experts from around the world who have worked tirelessly to provide government officials with the knowledge base needed to ensure the conservation and wise use of wetlands.

As editors for *The Wetland Book II*, we personally compliment the many authors for their incredible contributions to the most comprehensive compendium of knowledge about wetlands ever assembled. In particular, we acknowledge their unstinting efforts to compile the many chapters and work with the authors to produce *The*

Wetland Book. Their knowledge and efforts are matched by their willingness to share the collated knowledge that is now contained in *The Wetland Book*.

The publishers are thanked for their foresight in developing the concepts that led to *The Wetland Book* and for providing both a hard copy and online version, with the latter being available for future updating. We recommend *The Wetland Book* to all those interested in the growing international scientific knowledge about the functioning and management of these incredibly valuable but threatened ecosystems.

Institute for Land, Water and Society Charles Sturt University Albury, NSW, Australia	C. Max Finlayson
UNESCO-IHE, Institute for Water Education Delft, The Netherlands	
Department of Natural Resources Halifax, NS, Canada	G. Randy Milton
Nature Management Services Cambridge, UK	R. Crawford Prenctice
Institute for Land, Water and Society Charles Sturt University Albury, NSW, Australia	Nick C. Davidson
Nick Davidson Environmental Wigmore, UK	

Contents

Volume 1

Section I Introduction 1

1. **Wetlands of the World** 3
   G. Randy Milton, R. Crawford Prentice, and C. Max Finlayson

Section II Diversity of Wetlands 17

2. **Wetland Types and Distribution** 19
   C. Max Finlayson, G. Randy Milton, and R. Crawford Prentice

3. **Estuaries** 37
   Graham R. Daborn and Anna M. Redden

4. **Estuarine Marsh: An Overview** 55
   Ralph W. Tiner and G. Randy Milton

5. **Seagrasses** 73
   Frederick T. Short, Cathy A. Short, and Alyssa B. Novak

6. **Mangroves** 93
   C. Max Finlayson

7. **Major River Basins of the World** 109
   Carmen Revenga and Tristan Tyrrell

8. **Freshwater Lakes and Reservoirs** 125
   Etienne Fluet-Chouinard, Mathis Loïc Messager, Bernhard Lehner, and
   C. Max Finlayson

9. **Salt Lakes** 143
   C. Max Finlayson

10. **Tidal Freshwater Wetlands: The Fresh Dimension of
    the Estuary** 155
    Aat Barendregt

xi

11	**Freshwater Marshes and Swamps**	169
	C. Max Finlayson	
12	**Papyrus Wetlands**	183
	Julius Kipkemboi and Anne A. van Dam	
13	**Tropical Freshwater Swamps (Mineral Soils)**	199
	Wim Giesen	
14	**Peatlands**	227
	C. Max Finlayson and G. Randy Milton	
15	**Peat**	245
	Richard Lindsay and Roxane Andersen	
16	**Peatland (Mire Types): Based on Origin and Behavior of Water, Peat Genesis, Landscape Position, and Climate**	251
	Richard Lindsay	
17	**Arctic Peatlands**	275
	Tatiana Minayeva, Andrey Sirin, Peter Kershaw, and Olivia Bragg	
18	**Mires**	289
	Richard Lindsay	
19	**Blanket Mire**	295
	Richard Lindsay	
20	**Blanket Bogs**	303
	Richard Lindsay	
21	**Lagg Fen**	309
	Richard Lindsay	
22	**Karst Wetlands**	313
	Gordana Beltram	
23	**Subterranean (Hypogean) Habitats in Karst and Their Fauna** ...	331
	Boris Sket	
24	**Groundwater Dependent Wetlands**	345
	Ray H. Froend, Pierre Horwitz, and Bea Sommer	

Section III Natural and Anthropogenic Drivers of Wetland Change **357**

25	**Natural and Anthropogenic Drivers of Wetland Change**	359
	Susan M. Galatowitsch	
26	**Wetland Losses and the Status of Wetland-Dependent Species** ...	369
	Nick C. Davidson	

27	**Alien Plants and Wetland Biotic Dysfunction** .... C. Max Finlayson	383
28	**Ecological Conditions and Health of Arctic Wetlands Modified by Nutrient and Contaminant Inputs from Colonial Birds** .... Mark Mallory	391
29	**Lake Chilika (India): Ecological Restoration and Adaptive Management for Conservation and Wise Use** .... Ajit Kumar Pattnaik and Ritesh Kumar	397
30	**Saemangeum Estuarine System (Republic of Korea): Before and After Reclamation** .... Nial Moores	405
31	**Peatlands and Windfarms: Conflicting Carbon Targets and Environmental Impacts** .... Richard Lindsay	413
32	**Kakagon (Bad River Sloughs), Wisconsin (USA)** .... Jim Meeker and Naomi Tillison	427
33	**Seagrass Dependent Artisanal Fisheries of Southeast Asia** .... Richard K. F. Unsworth and Leanne C. Cullen-Unsworth	437
34	**Great Barrier Reef (Australia): A Multi-ecosystem Wetland with a Multiple Use Management Regime** .... Jon Brodie and Jane Waterhouse	447
35	**Qa'a Azraq Oasis: Strengthening Stakeholder Representation in Restoration (Jordan)** .... Fidaa F. Haddad	461
36	**Fishponds of the Czech Republic** .... Jan Pokorný and Jan Květ	469
37	**Makgadikgadi Wetlands (Botswana): Planning for Sustainable Use and Conservation** .... Jaap Arntzen	487
38	**Seagrass Recovery in Tampa Bay, Florida (USA)** .... Holly Greening, Anthony Janicki, and Ed T. Sherwood	495
Section IV	**North America, Greenland, and the Caribbean**	**507**
39	**Wetlands of Greenland** .... Christian Bay	509
40	**Peatlands of Continental North America** .... Dale H. Vitt	515

41	**Boreal Wetlands of Canada and the United States of America** ... Beverly Gingras, Stuart Slattery, Kevin Smith, and Marcel Darveau	521
42	**Yukon-Kuskokwim Delta: Yukon River Basin, Alaska (USA)** .... Frederic A. Reid and Daniel Fehringer	543
43	**The Peace-Athabasca Delta: MacKenzie River Basin (Canada)** .... Jeffrey Shatford	549
44	**Copper River Delta, Alaska (USA)** .... Frederic A. Reid, Daniel Fehringer, and Richard G. Kempka	557
45	**Fraser River Delta: Southern British Columbia (Canada)** .... Anne Murray	565
46	**The Mississippi Alluvial Valley (USA)** .... J. Brian Davis	577
47	**Coastal Wetlands of Manitoba's Great Lakes (Canada)** .... Dale Wrubleski, Pascal Badiou, and Gordon Goldsborough	591
48	**Coastal Wetlands of Lake Superior's South Shore (USA)** .... John Brazner and Anett Trebitz	605
49	**The Bay of Fundy and Its Wetlands (Canada)** .... Graham R. Daborn and Anna M. Redden	621
50	**San Francisco Bay Estuary (USA)** .... Beth Huning and Mike Perlmutter	637
51	**Vernal Pools of Northeastern North America** .... Elizabeth A. Colburn and Aram J. K. Calhoun	651
52	**Pocosins (USA)** .... Curtis J. Richardson	667
53	**Prairie Pothole Region of North America** .... Kevin E. Doherty, David W. Howerter, James H. Devries, and Johann Walker	679
54	**Playa Wetlands of the Great Plains (USA)** .... Anne Bartuszevige	689
55	**Wetlands of California's Central Valley (USA)** .... Frederic A. Reid, Daniel Fehringer, Ruth Spell, Kevin Petrik, and Mark Petrie	697
56	**The Everglades (USA)** .... Curtis J. Richardson	705

Volume 2

Section V Central and South America 725

57 Amazon River Basin 727
Florian Wittmann and Wolfgang J. Junk

58 Mangroves of Colombia 747
Jenny Alexandra Rodríguez-Rodríguez, Paula Cristina Sierra-Correa,
Martha Catalina Gómez-Cubillos, and Lucia Victoria Licero Villanueva

**59 Ciénaga Grande de Santa Marta: The Largest Lagoon-Delta
Ecosystem in the Colombian Caribbean** 757
Jenny Alexandra Rodríguez-Rodríguez, José Ernesto Mancera Pineda,
Laura Victoria Perdomo Trujillo, Mario Enrique Rueda, and
Karen Patricia Ibarra

60 Lake Fuquene (Colombia) 773
Mauricio Valderrama, María Pinilla-Vargas, Germán I. Andrade,
Eugenio Valderrama-Escallón, and Sandra Hernández

61 The Paraná-Paraguay Fluvial Corridor (Argentina) 785
Priscilla G. Minotti

**62 The Pantanal: A Brief Review of Its Ecology, Biodiversity, and
Protection Status** 797
Wolfgang J. Junk and Catia Nunes da Cunha

63 The Paraná River Delta 813
Patricia Kandus and Rubén Darío Quintana

**64 Wetlands of Chile: Biodiversity, Endemism, and Conservation
Challenges** 823
Alejandra Figueroa, Manuel Contreras, and Bárbara Saavedra

65 Seagrasses of Southeast Brazil 839
Joel C. Creed, Mariana V. P. Aguiar, Agatha Cristinne Soares, and
Leonardo V. Marques

**66 Rio de la Plata (La Plata River) and Estuary (Argentina and
Uruguay)** 847
Claudio R. M. Baigún, Darío C. Colautti, and Tomás Maiztegui

**67 Conchalí Lagoon: Coastal Wetland Restoration Project
(Chile)** 857
Manuel Contreras, F. Fernando Novoa, and Juan Pablo Rubilar

68 Bahía Lomas: Ramsar Site (Chile) 865
Carmen Espoz, Ricardo Matus, and Diego Luna-Quevedo

69	Patagonian Peatlands (Argentina and Chile) Rodolfo Iturraspe	873

Section VI Europe **883**

70	**Danube River Basin** Paul Csagoly, Gernant Magnin, and Orieta Hulea	885
71	**Lower Danube Green Corridor** Paul Csagoly, Gernant Magnin, and Orieta Hulea	897
72	**Danube, Drava, and Mura Rivers: The "Amazon of Europe"** Paul Csagoly, Gernant Magnin, and Arno Mohl	903
73	**Danube Delta: The Transboundary Wetlands (Romania and Ukraine)** Grigore Baboianu	911
74	**Rhine River Basin** Daphne Willems and Esther Blom	923
75	**Volga River Basin (Russia)** Harald J. L. Leummens	933
76	**Volga River Delta (Russia)** Harald J. L. Leummens	945
77	**European Tidal Saltmarshes** Nick C. Davidson	959
78	**Tipperne Peninsula and Ringkøbing Fjord (Denmark)** Hans Meltofte, Preben Clausen, and Ole Thorup	973
79	**Wadden Sea (Denmark)** Karsten Laursen and John Frikke	983
80	**Estuaries of Great Britain** Nick C. Davidson	997
81	**The Wash Estuary and North Norfolk Coast (UK)** Nick C. Davidson	1011
82	**Wetlands of the Norfolk and Suffolk Broads (UK)** Andrea Kelly	1023
83	**Blanket Mires of Caithness and Sutherland: Scotland's Great Flow Country (UK)** Richard Lindsay and Roxane Andersen	1039
84	**Karst Wetlands in the Dinaric Karst** Rosana Cerkvenik, Andrej Kranjc, and Andrej Mihevc	1057

Contents   xvii

85 **Turloughs (Ireland)** 1067
    Kenneth Irvine, Catherine Coxon, Laurence Gill,
    Sarah Kimberley, and Steve Waldren

86 **The Macrotidal Bay of Mont-Saint-Michel (France): The
    Function of Salt Marshes** 1079
    Loïc Valéry and Jean-Claude Lefeuvre

87 **The Inner Danish Waters and Their Importance to
    Waterbirds** 1089
    Ib Krag Petersen and Rasmus Due Nielsen

Section VII   Mediterranean Basin, Middle East, and West Asia ... 1099

88 **The Camargue: Rhone River Delta (France)** 1101
    Patrick Grillas

89 **Ebro Delta (Spain)** 1113
    Carles Ibáñez and Nuno Caiola

90 **Doñana Wetlands (Spain)** 1123
    Andy J. Green, Javier Bustamante, Guyonne F. E. Janss,
    Rocio Fernández-Zamudio, and Carmen Díaz-Paniagua

91 **Axios, Aliakmon, and Gallikos Delta Complex
    (Northern Greece)** 1137
    Despoina Vokou, Urania Giannakou, Christina Kontaxi, and
    Stella Vareltzidou

92 **The Philippi Peatland (Greece)** 1149
    Kimon Christanis

93 **Peatlands of the Mediterranean Region** 1155
    Richard Payne

94 **The Hula Wetland (Israel)** 1167
    Richard Payne

95 **Coastal Sabkha (Salt Flats) of the Southern and Western
    Arabian Gulf** 1173
    Ronald A. Loughland, Ali M. Qasem, Bruce Burwell, and
    Perdana K. Prihartato

96 **Lake Seyfe (Turkey)** 1185
    Serhan Cagirankaya and Burhan Teoman Meric

Section VIII   Africa 1197

97 **Congo River Basin** 1199
    Ian J. Harrison, Randall Brummett, and Melanie L. J. Stiassny

98	**Zambezi River Basin** .... 1217 Matthew McCartney, Richard D. Beilfuss, and Lisa-Maria Rebelo	
99	**Zambezi River Delta (Mozambique)** .... 1233 Richard D. Beilfuss	
100	**Nile River Basin** .... 1243 Matthew McCartney and Lisa-Maria Rebelo	
101	**Nile Delta (Egypt)** .... 1251 Mohamed Reda Fishar	
102	**Baro-Akobo River Basin Wetlands: Livelihoods and Sustainable Regional Land Management (Ethiopia)** .... 1261 Adrian Wood, J. Peter Sutcliffe, and Alan Dixon	
103	**Bahr el Ghazal: Nile River Basin (Sudan and South Sudan)** .... 1269 Asim I. El Moghraby	
104	**Machar Marshes: Nile Basin (South Sudan)** .... 1279 Yasir A. Mohamed	
105	**The Mayas Wetlands of the Dinder and Rahad: Tributaries of the Blue Nile Basin (Sudan)** .... 1287 Khalid Hassaballah, Yasir A. Mohamed, and Stefan Uhlenbrook	
106	**The Sudd (South Sudan)** .... 1299 Lisa-Maria Rebelo and Asim I. El Moghraby	
107	**Rugezi Marsh: A High Altitude Tropical Peatland in Rwanda** .... 1307 Piet-Louis Grundling, Ab P. Grootjans, and Anton Linström	
108	**Banc d'Arguin (Mauritania)** .... 1319 Antonio Araujo and Pierre Campredon	
109	**Bijagos Archipelago (Guinea-Bissau)** .... 1333 Pierre Campredon and Paulo Catry	
110	**Kilombero Valley Floodplain (Tanzania)** .... 1341 Lars Dinesen	
111	**Lakes Baringo and Naivasha: Endorheic Freshwater Lakes of the Rift Valley (Kenya)** .... 1349 Reuben Omondi, William Ojwang, Casianes Olilo, James Mugo, Simon Agembe, and Jacob E. Ojuok	
112	**Lake Turkana: World's Largest Permanent Desert Lake (Kenya)** .... 1361 William Ojwang, Kevin O. Obiero, Oscar O. Donde, Natasha J. Gownaris, Ellen K. Pikitch, Reuben Omondi, Simon Agembe, John Malala, and Sean T. Avery	

| 113 | **Soda Lakes of the Rift Valley (Kenya)** 1381
Simon Agembe, William Ojwang, Casianes Olilo,
Reuben Omondi, and Collins Ongore

| 114 | **Okavango Delta, Botswana (Southern Africa)** 1393
Lars Ramberg

| 115 | **Peatlands of Africa** 1413
Piet-Louis Grundling and Ab P. Grootjans

| 116 | **Peatland Types and Tropical Swamp Forests on the
Maputaland Coastal Plain (South Africa)** 1423
Althea T. Grundling, Ab P. Grootjans, Piet-Louis Grundling, and
Jonathan S. Price

Volume 3

Section IX Northern and East Asia 1437

| 117 | **Lena River Basin (Russia)** 1439
Victor Degtyarev

| 118 | **Lena River Delta (Russia)** 1451
Victor Degtyarev

| 119 | **Nidjili Lake: Lena River Basin (Russia)** 1457
Victor Degtyarev

| 120 | **Taiga-Alas Landscape in the South of the Central Yakutian
Lowland: Lena River Basin (Russia)** 1463
Victor Degtyarev

| 121 | **The Middle Aldan River Basin: A Key Migration Corridor for
the Eastern Population of the Siberian Crane Within the Lena
River Basin (Russia)** 1471
Victor Degtyarev

| 122 | **Yenisei River Basin and Lake Baikal (Russia)** 1477
Nick C. Davidson

| 123 | **Amur-Heilong River Basin: Overview of Wetland Resources** ... 1485
Evgeny Egidarev, Eugene Simonov, and Yury Darman

| 124 | **Daurian Steppe Wetlands of the Amur-Heilong River Basin
(Russia, China, and Mongolia)** 1499
Eugene Simonov, Oleg Goroshko, and Tatiana Tkachuk

| 125 | **Sanjiang Plain and Wetlands Along the Ussuri and Amur
Rivers: Amur River Basin (Russia and China)** 1509
Thomas D. Dahmer

126	**Zhalong Wetlands (China)**	1521
	Liying Su	
127	**Highland Peatlands of Mongolia**	1531
	Tatiana Minayeva, Andrey Sirin, and Chultemin Dugarjav	
128	**Yangtze River Basin (China)**	1551
	Cui Lijuan, Zhang Manyin, and Xu Weigang	
129	**Poyang Lake, Yangtze River Basin, China**	1565
	James Harris	
130	**Huang He (Yellow River) River Basin (China)**	1575
	Cui Lijuan, Zhang Manyin, and Xu Weigang	
131	**Current Status of Seagrass Habitat in Korea**	1589
	Kun-Seop Lee, Seung Hyeon Kim, and Young Kyun Kim	
132	**Hokkaido Marshes (Japan)**	1597
	Satoshi Kobayashi	

Section X Central and South Asia **1603**

133	**High Altitude Wetlands of Nepal**	1605
	Lalit Kumar and Pramod Lamsal	
134	**Anzali Mordab Complex (Islamic Republic of Iran)** ...	1615
	Masoud Bagherzadeh Karimi	
135	**Bujagh National Park (Islamic Republic of Iran)** ...	1625
	Sadegh Sadeghi Zadegan	
136	**Fereydoon Kenar, Ezbaran, and Sorkh Ruds Ab-Bandans** ...	1635
	Sadegh Sadeghi Zadegan	
137	**Lake Parishan (Islamic Republic of Iran)**	1647
	Ahmad Lotfi	
138	**Lake Uromiyeh (Islamic Republic of Iran)**	1659
	Ahmad Lotfi	
139	**Shadegan Wetland (Islamic Republic of Iran)**	1675
	Ahmad Lotfi	
140	**Mesopotamian Marshes of Iraq**	1685
	Curtis J. Richardson	
141	**Indus River Basin Wetlands**	1697
	Rab Nawaz, Ali Dehlavi, and Nadia Bajwa	
142	**Wular Lake, Kashmir**	1705
	Ritesh Kumar	

Contents

143	Wetlands of the Ganga-Brahmaputra Basin	1711
	Ritesh Kumar and Kalpana Ambastha	
144	Saline Wetlands of the Arid Zone of Western India	1725
	Malavika Chauhan and Brij Gopal	
145	The Transboundary Sundarbans Mangroves (India and Bangladesh)	1733
	Brij Gopal and Malavika Chauhan	
146	Wetlands of Mahanadi Delta (India)	1743
	Ritesh Kumar and Pranati Patnaik	

Section XI   Southeast Asia **1751**

147	Tropical Peat Swamp Forests of Southeast Asia	1753
	Susan Page and Jack Rieley	
148	Wetlands of the Mekong River Basin: An Overview	1763
	Peter-John Meynell	
149	Tonle Sap Lake: Mekong River Basin (Cambodia)	1785
	Colin Poole	
150	Tram Chim: Mekong River Basin (Vietnam)	1793
	Triet Tran and Jeb Barzen	
151	Transboundary Mekong River Delta (Cambodia and Vietnam)	1801
	Triet Tran	
152	U Minh Peat Swamp Forest: Mekong River Basin (Vietnam)	1813
	Triet Tran	
153	Sembilang National Park: Mangrove Reserves of Indonesia	1819
	Marcel J. Silvius, Yus Rusila Noor, I. Reza Lubis, Wim Giesen, and Dipa Rais	
154	Wetlands of Berbak National Park (Indonesia)	1831
	Wim Giesen, Marcel J. Silvius, and Yoyok Wibisono	
155	Danau Sentarum National Park (Indonesia)	1841
	Wim Giesen and Gusti Z. Anshari	
156	Wetlands of Tasek Bera (Peninsular Malaysia)	1851
	R. Crawford Prentice	
157	Intertidal Flats of East and Southeast Asia	1865
	John MacKinnon and Yvonne I. Verkuil	
158	Seagrass in Malaysia: Issues and Challenges Ahead	1875
	Japar S. Bujang, Muta H. Zakaria, and Frederick T. Short	

Section XII   Australia, New Zealand, and Pacific Islands **1885**

159 **Murray-Darling River Basin (Australia)** 1887
Jamie Pittock

160 **Macquarie Marshes: Murray-Darling River Basin (Australia)** ... 1897
Rachael F. Thomas and Joanne F. Ocock

161 **The Coorong: Murray-Darling River Basin (Australia)** 1909
Peter Gell

162 **Kati Thanda: Lake Eyre (Australia)** 1921
Richard T. Kingsford

163 **Myall Lakes (Australia)** 1929
Brian G. Sanderson and Anna M. Redden

164 **Australia's Wet Tropics Streams, Rivers, and Floodplain Wetlands** 1941
Richard G. Pearson

165 **Wetlands of Kakadu National Park (Australia)** 1951
C. Max Finlayson

166 **Groundwater Dependent Wetlands of the Gnangara Groundwater System (Western Australia)** 1959
Ray H. Froend, Pierre Horwitz, and Bea Sommer

167 **Seagrass Meadows of Northeastern Australia** 1967
Robert G. Coles, Michael A. Rasheed, Alana Grech, and Len J. McKenzie

168 **Wetlands of New Zealand** 1977
Karen Denyer and Hugh Robertson

169 **New Zealand Restiad Bogs** 1991
Beverley R. Clarkson

170 **Atolls of the Tropical Pacific Ocean: Wetlands Under Threat** ... 2001
Randolph R. Thaman

Index of Keywords 2027

About the Editors

C. Max Finlayson
Institute for Land, Water and Society
Charles Sturt University
Albury, NSW, Australia
UNESCO-IHE
Institute for Water Education
Delft, The Netherlands

Max Finlayson is an internationally renowned wetland ecologist with extensive experience internationally in water pollution, agricultural impacts, invasive species, climate change, and human well-being and wetlands. He has participated in global assessments such as those conducted by the Intergovernmental Panel for Climate Change, the Millennium Ecosystem Assessment, and the Global Environment Outlook 4 and 5 (UNEP). Since the early 1990s, he has been a technical adviser to the Ramsar Convention on Wetlands and has written extensively on wetland ecology and management. He has also been actively involved in environmental NGOs and from 2002 to 2007 was president of the governing council of global NGO Wetlands International.

He has worked extensively on the inventory, assessment, and monitoring of wetlands, in particular in wet tropical, wet-dry tropical, and subtropical climatic regimes covering pollution, invasive species, and climate change. His current research interests/projects include the following:

- Interactions between human well-being and wetland health in the face of anthropogenic change, including global change and the onset of the Anthropocenic era
- Vulnerability and adaptation of wetlands/rivers to climate change, including changing values and trade-offs between uses and users, considering uncertainty and complexity
- Integration of ecologic, economic, and social requirements and trade-offs between users of wetlands with an emphasis on developing policy guidance and institutional changes

- Environment and agriculture interactions and policy responses/outcomes, and collaboration between stakeholders and policymakers
- Wetland restoration and construction, including the use of artificial wetlands for waste water treatment and the generation of multiple values
- Landscape change involving wetlands/rivers and land use (agriculture and mining) and implications for wetland ecosystem services and benefits for local people

He holds the following associated positions:

- Scientific Expert on the Scientific and Technical Review Panel, Ramsar Convention on Wetlands, Triennium 2016–2018
- Ramsar Chair for the Wise Use of Wetlands, UNESCO-IHE, Delft, The Netherlands (2014–2018)
- Visiting Professor, Institute for Wetland Research, China Academy of Forestry, Beijing, China
- Editor-in-Chief, Marine and Freshwater Research, CSIRO Publishing
- Chair, Environmental Strategy Advisory Panel, Winton Wetlands Restoration (Australia)

He has contributed to over 300 journal articles, reports, guidelines, proceedings, and book chapters on wetland ecology and management. He has contributed to the development of concepts and methods for wetland inventory, assessment and monitoring, and undertaken many site-based assessments in many countries.

G. Randy Milton
Department of Natural Resources Kentville
NS, Canada

Randy Milton is the manager for the Ecosystems and Habitats Program with Nova Scotia's Department of Natural Resources in Canada. Randy is an ecologist and Certified Wildlife Biologist® with 35 years' experience in public and industry conservation and environmental management, especially with freshwater and coastal wetlands and forest ecosystems. He has maintained an involvement in regional and national wetland conservation efforts since the early 1990s, as well as internationally first as a volunteer with WWF (Indonesia) in the mid-1980s and subsequently as a technical advisor to the Ramsar Convention on Wetlands (2000–2015), a contributing author to the Millennium Ecosystem Assessment, and a member (2005–2014) of the International Plan Committee for the North American Waterfowl Management Plan. He has a M.Sc. from Acadia University (Canada), where he is currently an adjunct professor, and he is also an adjunct research associate at the Institute for Land, Water and Society, Charles Sturt University, Australia.

About the Editors

R. Crawford Prentice
Nature Management Services
Cambridge, UK

Crawford Prentice is an independent consulting ecologist based in Cambridge, England. He has some 30 years of biodiversity conservation experience and has led global, regional, and national programs and projects mainly in Asia, the CIS countries, and Europe. He studied Zoology at the University of Aberdeen and subsequently completed his M.Sc. in Aquatic Resource Management at Kings College, University of London.

Much of his professional life has concerned the conservation and management of wetlands and migratory waterbirds, with early beginnings at the Wildfowl and Wetlands Trust and the International Waterfowl and Wetlands Research Bureau (IWRB) analyzing International Waterfowl Census data, evolving to conservation program management at the Asian Wetland Bureau and IWRB, and leading a bilateral aid project for the integrated management of Malaysia's first Ramsar site in the 1990s. During the next decade, he worked with the International Crane Foundation on the design and implementation of the UNEP/GEF Siberian Crane Wetland Project, helping to strengthen management of seven million hectares across 16 wetland sites in four countries for this flagship species and other biodiversity. He remains a project associate with ICF, contributing to climate change adaptation planning for key wetland nature reserves in north-eastern China and wider efforts for crane conservation. Currently, Crawford conducts consultancy assignments for the preparation, implementation, and evaluation of GEF projects on integrated ecosystem management of wetlands, mountains, and tropical forests.

Nick C. Davidson
Institute for Land, Water and Society
Charles Sturt University
Albury, NSW, Australia
Nick Davidson Environmental
Wigmore, UK

Nick Davidson was the deputy secretary general of the Ramsar Convention on Wetlands from 2000 to 2014, with overall responsibility for the convention's global development and delivery of scientific, technical, and policy guidance and advice and communications as the Convention Secretariat's senior advisor on these matters. He has long-standing experience in, and a strong commitment to, environmental sustainability supported through the transfer of environmental science into policy-relevance and decision-making

at national and international scales. Nick currently works as an independent expert consultant on wetland conservation and wise use.

Nick has over 40 years' experience of research on the ecology, assessment, and conservation of coastal and inland wetlands and the ecophysiology and flyway conservation of migratory waterbirds, with a 1981 Ph.D. from the University of Durham (UK) on this topic, and continues to publish on these issues. Prior to his Ramsar Convention post, he worked for the UK's national government conservation agencies on coastal wetland inventory, assessment, information systems, and communications and as international science coordinator for the global NGO Wetlands International.

He is an adjunct professor at the Institute of Land, Water and Society, Charles Sturt University, Australia; was presented with the Society of Wetland Scientist's (SWS) International Fellow Award 2010 for his long-term contributions to global wetland science and policy; chairs the SWS's Ramsar Section; is an associate editor of the peer-reviewed journal *Marine & Freshwater Research*; is a member of several IUCN Commissions and their task forces (World Commission on Protected Areas (WCPA), Species Survival Commission (SSC), and Commission on Ecosystem Management (CEM)); and is an honorary fellow of the Chartered Institution of Water and Environmental Management (CIWEM).

Contributors

Simon Agembe Department of Fisheries and Aquatic Sciences, University of Eldoret, Eldoret, Kenya

Mariana V. P. Aguiar Departamento de Ecologia, Universidade do Estado do Rio de Janeiro, Rio de Janeiro, Brazil

Kalpana Ambastha Wetlands International South Asia, New Delhi, India

Roxane Andersen Environmental Research Institute, University of the Highlands and Islands, Thurso, UK

Germán I. Andrade Universidad de Los Andes, Bogotá, Colombia

Gusti Z. Anshari Center for Wetlands People and Biodiversity, University Tanjungpura, Pontianak, W. Kalimantan, Indonesia

Antonio Araujo MAVA – Fondation pour la Nature, Dakar, Senegal

Jaap Arntzen Centre for Applied Research, Gaborone, Botswana

Sean T. Avery Kenya Wetlands Biodiversity Research Team, Nairobi, Kenya

Grigore Baboianu Danube Delta Biosphere Reserve Authority, Tulcea, Romania

Pascal Badiou Institute for Wetland and Waterfowl Research, Ducks Unlimited Canada, Stonewall, MB, Canada

Claudio R. M. Baigún Instituto de Investigación e Ingeniería Ambiental (3iA), Universidad Nacional de San Martín, San Martín, Buenos Aires, Argentina

Nadia Bajwa Indus Ecoregion, WWF, Karachi, Pakistan

Aat Barendregt Environmental Sciences, Copernicus Institute, Utrecht University, Utrecht, The Netherlands

Anne Bartuszevige Playa Lakes Joint Venture, Lafayette, CO, USA

Jeb Barzen Private Lands Conservation, Spring Green, WI, USA

Christian Bay Department of Bioscience, Faculty of Science and Technology, Institut for Bioscience, Aarhus University, Roskilde, Denmark

Richard D. Beilfuss International Crane Foundation, Baraboo, WI, USA

College of Engineering, University of Wisconsin, Madison, WI, USA

Gordana Beltram Ministry of the Environment and Spatial Planning, Environment Directorate, Nature Conservation Unit, Ljubljana, Slovenia

Esther Blom WWF-Netherlands, Zeist, The Netherlands

Olivia Bragg School of the Geography, University of Dundee, Dundee, UK

John Brazner Wildlife Division, Nova Scotia Department of Natural Resources, Kentville, NS, Canada

Jon Brodie Catchment to Reef Research Group, TropWATER, Centre for Tropical Water and Aquatic Ecosystem Research, James Cook University, Townsville, QLD, Australia

Randall Brummett Environment and Natural Resources Department, World Bank, Washington, DC, USA

Japar S. Bujang Department of Biology, Faculty of Science, Universiti Putra Malaysia, Serdang, Selangor Darul Ehsan, Malaysia

Bruce Burwell E-Map Department, Saudi Aramco, Dhahran, Ash Sharqiyah, Saudi Arabia

Javier Bustamante Department of Wetland Ecology, Estación Biológica de Doñana (EBD-CSIC), Seville, Spain

Serhan Cagirankaya Wetlands Division, Ministry of Forests and Water Affairs, General Directory of Nature Conservation and National Parks, Ankara, Turkey

Nuno Caiola IRTA, Aquatic Ecosystems Program, Sant Carles de la Ràpita, Catalonia, Spain

Aram J. K. Calhoun University of Maine, Orono, ME, USA

Pierre Campredon Conseiller Technique, IUCN Guinea Bissau, Bissau, Guinea-Bissau

Paulo Catry MARE – Marine and Environmental Sciences Centre, ISPA – Instituto Universitário, Lisbon, Portugal

Rosana Cerkvenik Park Škocjanske jame, Divača, SI, Slovenia

Malavika Chauhan Himmotthan Society, Dehradun, Uttarakhand, India

Kimon Christanis Department of Geology, Sector of Earth Materials, University of Patras, Rio-Patras, Greece

Beverley R. Clarkson Landcare Research, Hamilton, New Zealand

Preben Clausen Department of Bioscience, Aarhus University, Kalø, Denmark

Darío C. Colautti Instituto de Limnología de La Plata "Raúl Ringuelet", La Plata, Argentina

Elizabeth A. Colburn Harvard Forest, Harvard University, Petersham, MA, USA

Robert G. Coles Centre for Tropical Water and Aquatic Ecosystem Research, James Cook University, Cairns and Townsville, QLD, Australia

Manuel Contreras Centro de Ecología Aplicada (CEA), Santiago, Región Metropolitana, Chile

Catherine Coxon University of Dublin, Trinity College, Dublin, Ireland

Joel C. Creed Departamento de Ecologia, Universidade do Estado do Rio de Janeiro, Rio de Janeiro, Brazil

Paul Csagoly Earthly Communications, Ottawa, ON, Canada

Leanne C. Cullen-Unsworth Sustainable Places Research Institute, Cardiff University, Cardiff, UK

Graham R. Daborn Acadia Centre for Estuarine Research, Acadia University, Wolfville, NS, Canada

Thomas D. Dahmer Ecosystems Ltd., Yau Tong, Kowloon, Hong Kong, China

Yury Darman WWF Russia Amur Branch, Vladivostok, Russia

Marcel Darveau Ducks Unlimited Canada, Quebec, QC, Canada

Nick C. Davidson Institute for Land, Water and Society, Charles Sturt University, Albury, NSW, Australia

Nick Davidson Environmental, Wigmore, UK

J. Brian Davis Department of Wildlife, Fisheries and Aquaculture, Mississippi State University, Mississippi State, MS, USA

Victor Degtyarev Siberian Division of Russian Academy of Sciences, Institute for Biological Problems of Cryolithozone, Siberian Branch, Russian Academy of Sciences, Yakutsk, Russia

Ali Dehlavi Indus Ecoregion, WWF, Karachi, Pakistan

Karen Denyer National Wetland Trust, Cambridge, Waikato, New Zealand

James H. Devries Ducks Unlimited Canada, Stonewall, MB, Canada

Carmen Díaz-Paniagua Department of Wetland Ecology, Estación Biológica de Doñana (EBD-CSIC), Seville, Spain

Lars Dinesen Biologist, European Representative, Scientific and Technical Review Panel of the Ramsar Convention, Jyderup, Denmark

Alan Dixon Institute of Science and the Environment, University of Worcester, Henwick Grove, UK

Kevin E. Doherty United States Fish and Wildlife Service, Bismarck, ND, USA

Oscar O. Donde KMFRI, Lake Turkana Research Station, Lodwar, Kenya

Chultemin Dugarjav Institute of General and Experimental Biology Mongolian Academy of Sciences, Ulaanbaatar, Mongolia

Evgeny Egidarev Pacific Geographical Institute FEB RAS \ WWF Russia Amur Branch, Vladivostok, Russia

Asim I. El Moghraby Sudanese National Academy of Sciences, Khartoum, Sudan

Carmen Espoz Centro Bahía Lomas, Facultad de Ciencias, Universidad Santo Tomás, Santiago, Chile

Daniel Fehringer Ducks Unlimited, Inc., Rancho Cordova, CA, USA

Rocio Fernández-Zamudio Department of Wetland Ecology, Estación Biológica de Doñana (EBD-CSIC), Seville, Spain

Alejandra Figueroa Head Natural Resources and Biodiversity, Ministry of Environment, Santiago, Chile

C. Max Finlayson Institute for Land, Water and Society, Charles Sturt University, Albury, NSW, Australia

UNESCO-IHE, The Institute for Water Education, Delft, The Netherlands

Mohamed Reda Fishar Inland Water and Aquaculture Branch, National Institute of Oceanography and Fisheries (NIOF), Cairo, Egypt

Etienne Fluet-Chouinard Center for Limnology, University of Wisconsin-Madison, Madison, WI, USA

John Frikke The Danish Wadden Sea National Park, Rømø, Denmark

Ray H. Froend Centre for Ecosystem Management, School of Science, Edith Cowan University, Joondalup, WA, Australia

Susan M. Galatowitsch Department of Fisheries, Wildlife and Conservation Biology, University of Minnesota, Saint Paul, MN, USA

Peter Gell Water Research Network, Federation University Australia, Ballarat, VIC, Australia

Urania Giannakou Department of Medical Laboratory Studies, School of Health and Medical Care, Alexander Technological Educational Institute of Thessaloniki, Thessaloniki, Greece

Wim Giesen Euroconsult Mott MacDonald, Arnhem, AK, The Netherlands

Contributors

Laurence Gill University of Dublin, Trinity College, Dublin, Ireland

Beverly Gingras Ducks Unlimited Canada, Edmonton, AB, Canada

Gordon Goldsborough Department of Biological Sciences, University of Manitoba, Winnipeg, MB, Canada

Martha Catalina Gómez-Cubillos Universidad Nacional de Colombia, sede Caribe, Santa Marta, Colombia

Brij Gopal Centre for Inland Waters in South Asia, Jaipur, Rajasthan, India

Oleg Goroshko Daursky Biosphere Reserve, Chita, Russia

Natasha J. Gownaris School of Marine and Atmospheric Sciences, Stony Brook University, Stony Brook, NY, USA

Alana Grech Department of Environmental Sciences, Macquarie University, Sydney, NSW, Australia

Andy J. Green Department of Wetland Ecology, Estación Biológica de Doñana (EBD-CSIC), Seville, Spain

Holly Greening Tampa Bay Estuary Program, St. Petersburg, FL, USA

Patrick Grillas Tour du Valat, Research Institute for Mediterranean Wetlands, Le Sambuc, Arles, France

Ab P. Grootjans Centre for Energy and Environmental Studies, University of Groningen, Groningen, The Netherlands

Institute of Water and Wetland Research, Radboud University Nijmegen, Nijmegen, The Netherlands

Althea T. Grundling Water Science Programme, Agricultural Research Council – Institute for Soil, Climate and Water, Pretoria, Gauteng, South Africa

Department of Geography and Environmental Management, University of Waterloo, Waterloo, ON, Canada

Applied Behavioural Ecology and Ecosystem Research Unit, University of South Africa, Pretoria, South Africa

Piet-Louis Grundling Centre for Environmental Management, University of the Free State, Bloemfontein, South Africa

Fidaa F. Haddad Regional Dryland, Livelihoods and Gender Program, IUCN Regional Office for West Asia (ROWA), Amman, Jordan

James Harris International Crane Foundation, Baraboo, WI, USA

Ian J. Harrison Center for Environment and Peace, Conservation International, Arlington, VA, USA

Khalid Hassaballah UNESCO-IHE Institute for Water Education, Delft, The Netherlands
Hydraulics Research Center, Wad Medani, Sudan

Sandra Hernández Fundación Humedales, Bogotá, Colombia

Pierre Horwitz Centre for Ecosystem Management, School of Science, Edith Cowan University, Joondalup, WA, Australia

David W. Howerter Ducks Unlimited Canada, Stonewall, MB, Canada

Orieta Hulea WWF Danube-Carpathian Programme, Bucharest, Romania
WWF Danube-Carpathian Programme, Vienna, Austria

Beth Huning San Francisco Bay Joint Venture, Fairfax, CA, USA

Carles Ibáñez IRTA, Aquatic Ecosystems Program, Sant Carles de la Ràpita, Catalonia, Spain

Karen Patricia Ibarra Instituto de Investigaciones Marinas y Costeras "José Benito Vives de Andreis" (INVEMAR), Santa Marta, Colombia

Kenneth Irvine UNESCO-IHE Institute of Water Education, Delft, The Netherlands

Rodolfo Iturraspe Universidad Nacional de Tierra del Fuego, Ushuaia, Tierra del Fuego, Argentina

Anthony Janicki Janicki Environmental, Inc., St. Petersburg, FL, USA

Guyonne F. E. Janss Department of Wetland Ecology, Estación Biológica de Doñana (EBD-CSIC), Seville, Spain

Wolfgang J. Junk Instituto Nacional de Ciência e Tecnologia em Áreas Úmidas (INCT-INAU), Universidade Federal de Mato Grosso (UFMT), Cuiabá, MT, Brazil

Patricia Kandus Instituto de Investigación e Ingeniería Ambiental, Universidad Nacional de San Martín, San Martín, Provincia de Buenos Aires, Argentina

Masoud Bagherzadeh Karimi Department of Environment, Study Center for Environment, Wetlands and National Parks, Tehran, Iran

Andrea Kelly Broads Authority, Norwich, Norfolk, UK

Richard G. Kempka Ducks Unlimited, Inc., Rancho Cordova, CA, USA
The Climate Trust, Portland, OR, USA

Peter Kershaw University of Alberta, Edmonton, AB, Canada

Seung Hyeon Kim Department of Biological Sciences, Pusan National University, Busan, South Korea

Young Kyun Kim Department of Biological Sciences, Pusan National University, Busan, South Korea

Sarah Kimberley University of Dublin, Trinity College, Dublin, Ireland

Richard T. Kingsford Centre for Ecosystem Science, School of Biological, Earth and Environmental Sciences, UNSW Australia, Sydney, NSW, Australia

Julius Kipkemboi Department of Biological Sciences, Egerton University, Egerton, Njoro, Kenya

Satoshi Kobayashi Faculty of Economics, Kushiro Public University, Kushiro, Hokkaido, Japan

Christina Kontaxi Administration of Environment and Spatial Planning, Region of Central Macedonia, Thessaloniki, Greece

Andrej Kranjc Slovenian Academy of Sciences and Arts, Ljubljana, SI, Slovenia

Lalit Kumar Ecosystem Management, School of Environmental and Rural Science, University of New England, Armidale, NSW, Australia

Ritesh Kumar Wetlands International South Asia, New Delhi, India

Jan Květ Faculty of Science, University of South Bohemia České Budějovice, České Budějovice, Czech Republic

Czech Academy of Sciences, CzechGlobe, Institute for Global Change Research, Brno, Czech Republic

Pramod Lamsal Ecosystem Management, School of Environmental and Rural Science, University of New England, Armidale, NSW, Australia

Karsten Laursen Department of Bioscience, Aarhus University, Aarhus, Denmark

Kun-Seop Lee Department of Biological Sciences, Pusan National University, Busan, South Korea

Jean-Claude Lefeuvre Department of Ecology and Biodiversity Management, National Museum of Natural History, Paris, France

EA 7316 Biodiversity and Land Management, University of Rennes 1, Rennes, France

Bernhard Lehner Department of Geography, McGill University, Montreal, QC, Canada

Harald J. L. Leummens Water and Nature, Heerlen, The Netherlands

Cui Lijuan Institute of Wetland Research, Chinese Academy of Forestry, Beijing, China

Richard Lindsay Sustainability Research Institute, University of East London, London, UK

Anton Linström Wet Earth Eco-Specs, Lydenburg, South Africa

Ahmad Lotfi Conservation of Iranian Wetland Project, Department of Environment, Tehran, Iran

Senior Irrigation Engineers, Member of Board of Directors, Pandam Consulting Engineers, Tehran, Iran

Ronald A. Loughland Environmental Protection Department, Saudi Aramco, Dhahran, Ash Sharqiyah, Saudi Arabia

I. Reza Lubis Wetlands International Indonesia, Bogor, Indonesia

Diego Luna-Quevedo WHSRN Executive Office- Manomet, Plymouth, MA, USA

John MacKinnon University of Kent, Canterbury, UK

Gernant Magnin WWF – Netherlands, Freshwater Programme, Zeist, The Netherlands

Tomás Maiztegui Instituto de Limnología de La Plata "Raúl Ringuelet", La Plata, Argentina

John Malala KMFRI, Lake Turkana Research Station, Lodwar, Kenya

Mark Mallory Coastal Wetland Ecosystems, Biology Department, Acadia University, Wolfville, NS, Canada

José Ernesto Mancera Pineda Universidad Nacional de Colombia, Bogotá, Colombia

Ciudad Universitaria, Mexico City, Mexico

Zhang Manyin Institute of Wetland Research, Chinese Academy of Forestry, Beijing, China

Leonardo V. Marques Departamento de Ecologia, Universidade do Estado do Rio de Janeiro, Rio de Janeiro, Brazil

Ricardo Matus Centro de Rehabilitación de Aves Leñadura, Punta Arenas, Chile

Matthew McCartney Ecosystem Services, International Water Management Institute, Vientiane, Lao People's Democratic Republic

Regional Office for Southeast Asia and the Mekong, International Water Management Institute, Vientiane, Lao People's Democratic Republic

Len J. McKenzie Centre for Tropical Water and Aquatic Ecosystem Research, James Cook University, Cairns and Townsville, QLD, Australia

Jim Meeker Formerly of: Northland College, Ashland, WI, USA

Jim Meeker: deceased

Hans Meltofte Department of Bioscience, Aarhus University, Roskilde, Denmark

Burhan Teoman Meric Wetlands Division, Ministry of Forests and Water Affairs, General Directory of Nature Conservation and National Parks, Ankara, Turkey

Mathis Loïc Messager School of Aquatic and Fishery Sciences, University of Washington, Seattle, WA, USA

Peter-John Meynell ICEM – International Centre for Environmental Management, Hanoi, Vietnam

Andrej Mihevc Karst Research Institute, Research Centre of the Slovenian Academy for Science and Arts, Postojna, Slovenia

G. Randy Milton Nova Scotia Department of Natural Resources, Kentville, NS, Canada

Tatiana Minayeva Wetlands International, Ede, The Netherlands

Priscilla G. Minotti Instituto de Investigación e Ingeniería Ambiental (3iA), Universidad Nacional de San Martín, San Martín, Provincia de Buenos Aires, Argentina

Yasir A. Mohamed Hydraulic Research Center, MoWRIE, Wad Medani, Sudan

Department of Integrated Water Systems and Governance, UNESCO-IHE, DA, Delft, The Netherlands

Faculty of Civil Engineering and Applied Geosciences, Water Resources Section, Delft University of Technology, GA, Delft, The Netherlands

Arno Mohl WWF Austria, Vienna, Austria

Nial Moores Birds Korea, Busan, Korea (Republic of)

James Mugo Kenya Marine and Fisheries Research Institute, Naivasha, Kenya

Anne Murray Nature Guides BC, Delta, BC, Canada

Rab Nawaz Indus Ecoregion, WWF, Karachi, Pakistan

Rasmus Due Nielsen Department of Bioscience, Aarhus University, Kalø, Denmark

Yus Rusila Noor Wetlands International Indonesia, Bogor, Indonesia

Alyssa B. Novak Boston University, Earth and Environment, Boston, MA, USA

F. Fernando Novoa Centro de Ecología Aplicada (CEA), Santiago, Región Metropolitana, Chile

Catia Nunes da Cunha Instituto Nacional de Ciência e Tecnologia em Áreas Úmidas (INCT-INAU), Universidade Federal de Mato Grosso (UFMT), Cuiabá, MT, Brazil

Depto Botânica e Ecologia/Núcleo de Estudos Ecológicos do Pantanal (NEPA), Instituto de Biociências, UFMT, Cuiabá, MT, Brazil

Kevin O. Obiero KMFRI, Lake Turkana Research Station, Lodwar, Kenya

Joanne F. Ocock Water and Wetlands Team, Science Division, NSW Office of Environment and Heritage, Sydney, NSW, Australia

Centre for Ecosystem Science, School of Biological, Earth and Environmental Sciences, UNSW, Sydney, NSW, Australia

Jacob. E. Ojuok Kenya Marine and Fisheries Research Institute, Kisumu, Kenya

William Ojwang Kenya Marine and Fisheries Research Institute (KMFRI), Mombasa, Kisumu, Kenya

WWF-Kenya, Nairobi, Kenya

Casianes Olilo Kenya Marine and Fisheries Research Institute (KMFRI), Mombasa, Kisumu, Kenya

Reuben Omondi Department of Applied Aquatic Sciences, Kisii University, Kisii, Kenya

KMFRI, Kisumu Research Centre, Kisumu, Kenya

Collins Ongore Kenya Marine and Fisheries Research Institute (KMFRI), Mombasa, Kisumu, Kenya

Susan Page Department of Geography, University of Leicester, Leicester, UK

Pranati Patnaik Wetlands International South Asia, New Delhi, India

Ajit Kumar Pattnaik Chilika Development Authority, Bhubaneswar, Odisha, India

Richard Payne Environment, University of York, Heslington, York, UK

Richard G. Pearson College of Science and Engineering, James Cook University, Townsville, QLD, Australia

Laura Victoria Perdomo Trujillo Universidad Nacional de Colombia, Bogotá, Colombia

Mike Perlmutter Environmental Services Division, City of Oakland Public Works Department, Oakland, CA, USA

Ib Krag Petersen Department of Bioscience, Aarhus University, Kalø, Denmark

Mark Petrie Ducks Unlimited, Inc., Rancho Cordova, CA, USA

Kevin Petrik Ducks Unlimited, Inc., Rancho Cordova, CA, USA

Ellen K. Pikitch School of Marine and Atmospheric Sciences, Stony Brook University, Stony Brook, NY, USA

María Pinilla-Vargas Fundación Humedales, Bogotá, Colombia

Jamie Pittock Fenner School of Environment and Society, The Australian National University, Canberra, Australia

Jan Pokorný ENKI, o.p.s., Třeboň, Czech Republic

Colin Poole Wildlife Conservation Society, Phnom Penh, Cambodia

R. Crawford Prentice Nature Management Services, Histon, Cambridge, UK

Jonathan S. Price Department of Geography and Environmental Management, University of Waterloo, Waterloo, ON, Canada

Perdana K. Prihartato Environmental Protection Department, Saudi Aramco, Dhahran, Ash Sharqiyah, Saudi Arabia

Ali M. Qasem Environmental Engineering Div./ Environmental Protection Dept, Saudi Aramco, Dhahran, Saudi Arabia

Rubén Darío Quintana Instituto de Investigación e Ingeniería Ambiental, Universidad Nacional de San Martín, San Martín, Provincia de Buenos Aires, Argentina

Consejo Nacional de Investigaciones Científicas y Técnicas (CONICET), Buenos Aires, Argentina

Dipa Rais Wetlands International Indonesia, Bogor, Indonesia

Lars Ramberg Uppsala, Sweden

Michael A. Rasheed Centre for Tropical Water and Aquatic Ecosystem Research, James Cook University, Cairns and Townsville, QLD, Australia

Lisa-Maria Rebelo Water Futures, International Water Management Institute, Vientiane, Lao People's Democratic Republic

Regional Office for Southeast Asia and the Mekong, International Water Management Institute, Vientiane, Lao People's Democratic Republic

Anna M. Redden Acadia Centre for Estuarine Research, Acadia University, Wolfville, NS, Canada

Frederic A. Reid Ducks Unlimited, Inc., Rancho Cordova, CA, USA

Carmen Revenga The Nature Conservancy, Arlington, VA, USA

Curtis J. Richardson Nicholas School of the Environment, Duke University Wetland Center, Durham, NC, USA

Jack Rieley School of Geography, University of Nottingham, Nottingham, UK

Hugh Robertson Freshwater Section, Department of Conservation, Nelson, New Zealand

Jenny Alexandra Rodríguez-Rodríguez Instituto de Investigaciones Marinas y Costeras "José Benito Vives de Andreis" (INVEMAR), Santa Marta, Colombia

Juan Pablo Rubilar Minera Los Pelambres, Santiago, Chile

Mario Enrique Rueda Instituto de Investigaciones Marinas y Costeras "José Benito Vives de Andreis" (INVEMAR), Santa Marta, Colombia

Bárbara Saavedra Wildlife Conservation Society, Bronx, NY, USA

Sadegh Sadeghi Zadegan Ornithological Unit, Wildlife Bureau, Department of Environment, Tehran, Iran

Brian G. Sanderson Acadia Centre for Estuarine Research, Acadia University, Wolfville, NS, Canada

Jeffrey Shatford Ministry of Forests, Lands and Natural Resource Operations, Resource Management Objectives Branch, Victoria, BC, Canada

Ed T. Sherwood Tampa Bay Estuary Program, St. Petersburg, FL, USA

Cathy A. Short Lee, NH, USA

Frederick T. Short Department of Natural Resources and the Environment, Jackson Estuarine Laboratory, University of New Hampshire, Durham, NH, USA

Paula Cristina Sierra-Correa Instituto de Investigaciones Marinas y Costeras "Jose Benito Vives de Andreis" (INVEMAR), Santa Marta, Colombia

Marcel J. Silvius Wageningen, The Netherlands

Eugene Simonov Rivers without Boundaries Coalition, Dalian, China

Daursky NNR, Dalian, China

Andrey Sirin Center for Protection and Restoration of Peatland Ecosystems, Institute of Forest Science, Russian Academy of Sciences, Moscow, Russia

Laboratory of Peatland Forestry and Amelioration, Institute of Forest Science, Russian Academy of Sciences, Moscow, Russia

Boris Sket Oddelek za biologijo, Biotehniška fakulteta, Univerza v Ljubljani, Ljubljana, Slovenia

Stuart Slattery Ducks Unlimited Canada, Stonewall, MB, Canada

Kevin Smith Ducks Unlimited Canada, Edmonton, AB, Canada

Agatha Cristinne Soares Departamento de Ecologia, Universidade do Estado do Rio de Janeiro, Rio de Janeiro, Brazil

Bea Sommer Centre for Ecosystem Management, School of Science, Edith Cowan University, Joondalup, WA, Australia

Ruth Spell Ducks Unlimited, Inc., Rancho Cordova, CA, USA

Melanie L. J. Stiassny Division of Vertebrate Zoology, Ichthyology Department, American Museum of Natural History, New York City, USA

Liying Su International Crane Foundation, Baraboo, WI, USA

J. Peter Sutcliffe Independent Consultant, University of Huddersfield, Huddersfield, UK

Randolph R. Thaman The University of the South Pacific, Suva, Fiji

Rachael F. Thomas Water and Wetlands Team, Science Division, NSW Office of Environment and Heritage, Sydney, NSW, Australia

Centre for Ecosystem Science, School of Biological, Earth and Environmental Sciences, UNSW, Sydney, NSW, Australia

Ole Thorup Amphi Consult, Ribe, Denmark

Naomi Tillison Natural Resources Director, Bad River Band of Lake Superior Chippewa, Odanah, WI, USA

Ralph W. Tiner Institute for Wetlands and Environmental Education and Research, Inc., Leverett, MA, USA

Tatiana Tkachuk Daursky Biosphere Reserve, Zabaikalsky State University, Chita, Russia

Triet Tran International Crane Foundation, Baraboo, WI, USA

University of Natural Science, Vietnam National University, Ho Chi Minh City, Vietnam

Anett Trebitz Mid-Continent Ecology Division, United States Environmental Protection Agency, Duluth, MN, USA

Tristan Tyrrell Tentera, Montreal, QC, Canada

Stefan Uhlenbrook UNESCO-IHE Institute for Water Education, Delft, The Netherlands

Delft University of Technology, Delft, The Netherlands

Richard K. F. Unsworth College of Science, Wallace Building, Swansea University, Swansea, UK

Mauricio Valderrama Fundación Humedales, Bogotá, Colombia

Eugenio Valderrama-Escallón Fundación Humedales, Universidad El Bosque, Bogotá, Colombia

Loïc Valéry Department of Ecology and Biodiversity Management, National Museum of Natural History, Paris, France

EA 7316 Biodiversity and Land Management, University of Rennes 1, Rennes, France

Anne A. van Dam Aquatic Ecosystems Group, Department of Water Science and Engineering, UNESCO-IHE Institute for Water Education, Delft, The Netherlands

Stella Vareltzidou Axios-Loudias-Aliakmonas Management Authority, Chalastra, Thessaloniki, Greece

Yvonne I. Verkuil Conservation Ecology Group; Groningen Institute for Evolutionary Life Sciences (GELIFES), University of Groningen, Groningen, The Netherlands

Lucia Victoria Licero Villanueva Instituto de Investigaciones Marinas y Costeras "Jose Benito Vives de Andreis" (INVEMAR), Santa Marta, Colombia

Dale H. Vitt Department of Plant Biology and Center for Ecology, Southern Illinois University, Carbondale, IL, USA

Despoina Vokou Department of Ecology, School of Biology, Aristotle University of Thessaloniki, Thessaloniki, Greece

Steve Waldren University of Dublin, Trinity College, Dublin, Ireland

Johann Walker Ducks Unlimited, Great Plains Region, Bismarck, ND, USA

Jane Waterhouse Catchment to Reef Research Group, TropWATER, Centre for Tropical Water and Aquatic Ecosystem Research, James Cook University, Townsville, QLD, Australia

Xu Weigang Institute of Wetland Research, Chinese Academy of Forestry, Beijing, China

Yoyok Wibisono Wetlands International – Indonesia Programme, Bogor, Indonesia

Daphne Willems WWF-Netherlands, Zeist, The Netherlands

Florian Wittmann Institute of Floodplain Ecology, Karlsruhe Institute of Technology - KIT, Rastatt, Germany

Adrian Wood Business School, University of Huddersfield, Huddersfield, UK

Dale Wrubleski Institute for Wetland and Waterfowl Research, Ducks Unlimited Canada, Stonewall, MB, Canada

Muta H. Zakaria Department of Aquaculture, Faculty of Agriculture, Universiti Putra Malaysia, Serdang, Selangor Darul Ehsan, Malaysia

Section IX

Northern and East Asia

Lena River Basin (Russia)

117

Victor Degtyarev

Contents

Introduction	1440
Factors of Paramount Importance to Ecological Conditions	1442
Biodiversity and Ecosystem Services	1443
Change in Ecological Character and Conservation	1444
Wetland Types and Distribution	1445
Conservation	1448
References	1449

Abstract

The Lena is one of the world's largest rivers, distinguished for its length (4,400 km), catchment area (2,500,000 km^2), and delta (32,000 km^2). It is located in northern Asia, belongs to the Arctic Ocean basin, and flows into the Laptev Sea. The annual water discharge fluctuates from 417 to 631 km^3, sediment runoff averages 12,000,000 tonnes annually. Orographic barriers in combination with the geographic location minimize oceanic influence on the basin and promote maximal cooling of the surface atmosphere layer of Eurasia. Permafrost and cryogenic effects are the most significant factors determining annual fluctuations in water discharge and wetland composition and distribution. This watered frozen layer on sedimentary areas produces and maintains lakes as a result of thermokarst. Especially, these effects are a significant factor in areas that experience a negative moisture budget. Taiga is the dominant biome as the basin narrows sharply toward the subarctic zone, while tundra is confined within the delta area. Owing to orographic heterogeneity and the extreme continental nature of the climate, unique landscapes occur. The Lena has an influence on Arctic Ocean ecosystem functioning and the global greenhouse gas

V. Degtyarev (✉)
Siberian Division of Russian Academy of Sciences, Institute for Biological Problems of Cryolithozone, Siberian Branch, Russian Academy of Sciences, Yakutsk, Russia
e-mail: dvgarea@yandex.ru

© Springer Science+Business Media B.V., part of Springer Nature 2018
C. M. Finlayson et al. (eds.), *The Wetland Book*,
https://doi.org/10.1007/978-94-007-4001-3_96

budget, and supports a significant proportion of fish species, including stocks of the Siberian sturgeon and salmonid species. Its basin, as a vast territory covered by light coniferous taiga, is a key area of this biome in northern Asia. Negative influences intensified since the 1950s–1960s with a wave of mineral resource development, increased availability of motorboats, and fishing with kapron nets which increased the take of fish and hunting takes. The of the basin environment is still relatively stable but this is due to the large pristine areas that still remain. The 2000s coincided with a new wave of industrial and large-scale development that incorporated special transportation, fishing, and hunting technology. This development threatens primarily pristine areas with their game animal and fish resources. In this respect, the protected areas are of great importance. In 2012, the Lena Pillars National Park was inscribed on the World Heritage List.

Keywords

Alas · Cryolithozone · Endangered species · Permafrost · Siberia · Unique climatic conditions

Introduction

The Lena (Fig. 1) is one of the world's largest rivers, distinguished for its length, catchment area, and delta. It is located in northern Asia, belongs to the Arctic Ocean basin, and flows into the Laptev Sea. The general direction of flow is northward. The river source is situated on the Baikal ridge, about 10 km from the west shore of Lake Baikal. The name "Lena" is a derivative of an aboriginal name translated as "great

Fig. 1 Lena River in its middle reaches (Photo credit: V. Degtyarev © Rights remain with the author)

river." The basin is predominantly located (65%) in Yakutia, while Irkutskaya oblast (province) and Zabaykalsky krai (a similar level of administrative unit) share about 30% in the south. The rest of its area consists of extreme western and eastern fragments belonging to three other administrative units of the Russian Federation (Mostakhov 1972).

The Lena River has a length of 4,400 km and a catchment area totaling 2,500,000 km^2 (Fig. 2). Precipitation is the river's main water source, and its spring peak discharge is 530 times greater than its winter minimum. The annual water discharge fluctuates from 417 to 631 km^3, sediment runoff averages 12,000,000 tonnes annually, and water turbidity is up to 60 g/m^3, and while annual solute runoff is 41,000,000 tonnes, the extent of mineralization is up to 100 mg/l under flood and up to 500 mg/l during the low-flow conditions. The water's maximum temperature is 19 °C in the upper and middle reaches and 14 °C in the lower reaches. The river is iced over for 6–8 months: freezing-over occurs in

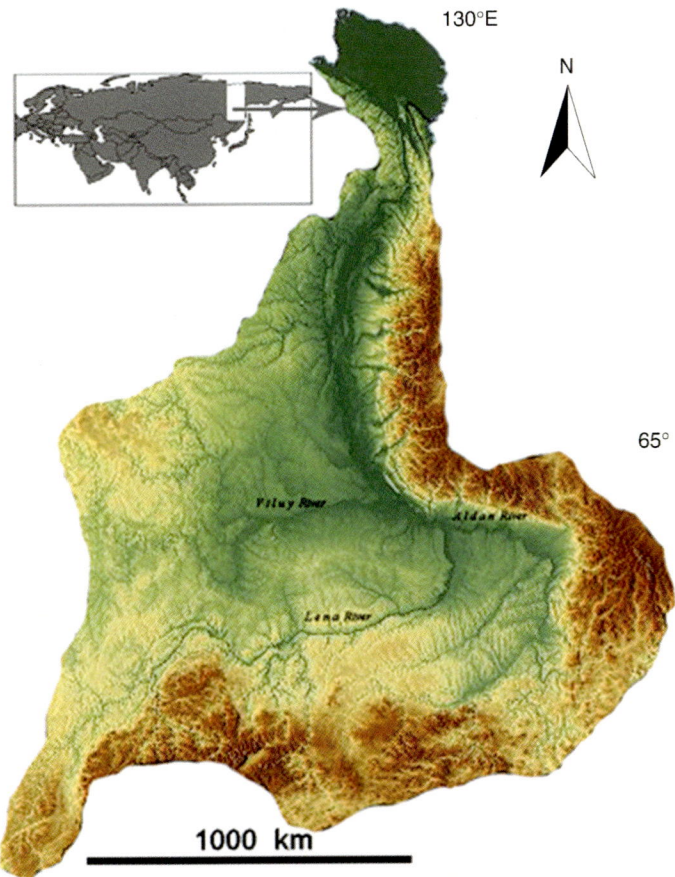

Fig. 2 Lena River Basin (V. Degtyarev © Rights remain with the author)

September–October, and spring breakup and drifting of ice occurs from late April to early June. Ice jams followed by floods (sometimes catastrophic) are frequent during the ice drifting period when the water level can rise up to 28 m. The drainage system consists of 242,000 streams, 98% of which are up to 25 km in length, while 1.7% are over 1,000 km in length. The largest tributaries (1,400–2,700 km in length, draining 8–29% of the basin area, and contributing 9–31% of the discharge) are the Vitim, Olyekma, Aldan, and Vilyuy Rivers (Mostakhov 1972, 1974).

The near heart-shaped basin extends 21° (52–73 °N) latitudinally (2,400 km) and 37° (104–141 °E) longitudinally (2,000 km) across the temperate and subarctic climatic zones. As it is situated on the Siberian platform and the Central Asian orogenic belt, the basin is orographically heterogeneous. The Central Yakutian lowland forms the central part of the basin. The Nizhnelenskaya lowland is located northward along the river. To the south, east, and north of the lowland is a system of high plateaus (the Aldanskoe, Patomskoe, etc.) and mountain ridges (e.g., the Verkhoyansky, Dzhugdzhur, Stanovoy.) rising to 2,600 m in altitude, and westward the Vilyuiskoe and Prilenskoe plateaus rise up to 1,000 m altitude. Water in the Lena's upper reaches thus flows as a torrent, whereas the middle and lower reaches are a smoothly flowing stream terminating in the large delta. The drainage density averages 0.42 km/km^2. There are 330,000 lakes totaling some 20,000 km^2 (1% of the basin) (Mostakhov 1974). A high lake density occurs in the lowlands, watershed depressions, and enlarged valleys of rivers, with 65 lakes per 100 km^2 on the Central Yakutian lowland (Matveev et al. 1989). Peatlands and mires occupy more than 5% of the basin (Andreev et al. 1987).

Factors of Paramount Importance to Ecological Conditions

The significant elevation of the watershed produces conditions for orographic isolation of the basin, especially for the lowland in its center. Orographic barriers in combination with the geographic location minimize oceanic influence on the basin and promote maximal cooling of the surface atmosphere layer of Eurasia (Balobaev 1991). Consequently, unique climatic conditions have arisen; the continental influence on the modern climate in the Central Yakutian lowland is 90–99% (Borisov 1967). The range in annual temperature is 102 °C, with mean annual temperature of −10 to −12 °C, a mean January temperature of −40 °C, and mean July temperature of 19 °C. Insolation is very high, and summer is sultry (but short). Therefore, the moisture budget is negative with an annual precipitation of 220–300 mm and evaporation of 300–420 mm. Around the Central Yakutian lowland, the continental influence diminishes slightly and the moisture budget becomes positive (Borisov 1967; Gavrilova 1972; Elovskaya and Konorovskiy 1978).

The climate generates and retains permafrost (cryolithozone) which in the Lena River Basin spreads to its southernmost extent (up to 60°N) (Matveev et al. 1989). Permafrost and cryogenic effects are the most significant factors determining annual fluctuations in water discharge and wetland composition and distribution. Forming an underlying waterproof layer, permafrost contributes to the accumulation of snowmelt runoff into the river system, lakes, mires, and other depressions. This watered (ice saturated) frozen layer on sedimentary areas produces and maintains lakes as a result of thermokarst. Especially, these effects are a significant factor in areas that experience a negative moisture budget (Soloviev 1961).

The population of the basin is less than one million people, with an average density of 0.3 persons per km^2. There are only three towns populated by 30,000–300,000 people. Regional economic development is dominated by the extractive industry with gold bearing, uranium, apatite, complex ore deposits or placers, and coal fields concentrated in the south, east, and northeast of the basin; diamond pipes and placers in the west and northwest; and oil and gas fields in the center and southwest. Energy is mainly generated by three hydroelectric and two thermal power stations. Cultivated lands total less than 200,000 ha. The rural population is employed in cattle breeding, fishery, reindeer raising, and the fur trade, considered as the traditional economy (Matveev et al. 1989). The surface transportation network is not developed. Besides outstanding biological, mineral, and water-power resources, the freshwater resources of the basin support large areas of virgin nature, picturesque surroundings, as well as geological and archeological monuments (Degtyarev 2007a).

Biodiversity and Ecosystem Services

Taiga is the dominant biome as the basin narrows sharply toward the subarctic zone, while tundra is confined within the delta area. Owing to orographic heterogeneity and the extreme continental nature of the climate, unique landscapes occur. According to its significant latitudinal range and complicated surface structure, the basin is clearly divided into 6–7 floristic or faunistic regions, which as a rule cover some areas beyond the watershed (Krivenko 2000; Troeva et al. 2010). Based on flora and fauna reviews of Yakutia (Degtyarev 2000a; Solomonov and Lukina 2009; Troeva et al. 2010) and geographic characteristics of the rest of the basin, the number of its flora species is estimated at about 6,000, including 2,500 algae, 700 lichens, 700 mosses, 900 fungi, and 1,500 higher vascular plants; the number of species of vertebrate animals is 450 (including 300 birds and 67 mammals) and arthropods 5,000 species. Besides synanthropic vertebrates, there are nonnative species: Baikal cisco *Coregonus migratorius*, common carp *Cyprinus Linnaeus*, bream *Abramis brama*, muskrat *Ondatra zibethicus*, American mink *Mustela vison*, Russian polecat *M. eversmanni*, and musk ox *Ovibos moschatus*. Species present in the basin during the late Holocene but which have since disappeared are water caltrop *Trapa natans*, Eurasian beaver *Castor fiber*, greylag goose *Anser anser*, and, presumably, great crested grebe *Podiceps cristatus*.

Vertebrates ecologically dependent on inland wetlands are 41 species of fish, 5 species of Amphibia, 4 species of mammals (water vole *Arvicola terrestris*, muskrat, American mink, and common otter *Lutra lutra*), and 145 species of birds. For the coastal area, two sea species of fish and four species of marine mammals (walrus *Odobenus rosmarus*, bearded seal *Erignathus barbatus*, ringed seal *Pusa hispida*, and polar bear *Ursus maritimus*) have been recorded (Degtyarev 2000a, b; Solomonov and Lukina 2009).

Twenty-five of the 145 bird species are casual visitors from both nearby territories (e.g., snow goose *Chen caerulescens*, Steller's sea eagle *Haliaeetus pelagicus*, marbled murrelet *Brachyramphus marmoratus*) and remote regions (e.g., eastern white pelican *Pelicanus onocrotalus*, oriental ibis *Threskiornis melanocephalus*, greater flamingo *Phoenicopterus roseus*). Thirty-two species breed only in the delta. The basin is important for the long-term global viability of a significant population segment of Middendorff's bean goose *Anser fabalis middendorffii*, Baikal teal *Anas formosa* (in recent times), falcated *Anas falcata* and harlequin duck *Histrionicus histrionicus*, white-winged scoter *Melanitta deglandi*, hooded crane *Grus monacha*, long-toed stint *Calidris subminuta*, little *Numenius minutus* and far eastern *N. madagascariensis* curlews, and black-tailed godwit *Limosa limosa*. Osprey *Pandion haliaetus*, white-tailed eagle *Haliaeetus albicilla*, loons, a majority of ducks, waders, gulls, and terns are widespread, common or numerous species. The basin composes a significant part of the East Asian-Australasian flyway supporting a diversity of migrating birds. In particular, one-third and one-fifth of the entire migration distance of Baikal teal and eastern population of the Siberian crane *Grus leucogeranus*, respectively, lie here (Degtyarev 2007a).

The Lena has an influence on Arctic ocean ecosystem functioning and the global greenhouse gas budget, and supports a significant proportion of fish species, including stocks of the Siberian sturgeon *Acipenser baerri*, taimen *Hucho taimen*, Siberian white salmon *Stenodus leucichthys*, and another ten salmonid species (Gukov 2001; Kirillov 2002). Its basin, as a vast territory covered by light coniferous taiga, is a key area of this biome in northern Asia. The light coniferous taiga with its keystone species Gmelin's larch *Larix gmelinii* plays a significant role in the global water and carbon budget (Dolman et al. 2008). The basin supports unique soil and wetland types (Degtyarev 2000b; Dobrovolsky et al. 2009), populations and significant proportion of plant and animal species, including 20 endemic plants and 8 threatened bird species listed by BirdLife International (Dolinin et al. 2000; BirdLife International 2001; Ostrovsky et al. 2002; Gaikova et al. 2010), forms an optimal environment for characteristic taiga animals like black-billed capercaillie *Tetrao parvirostris*, sable *Martes zibellina*, Siberian musk deer *Moschus moschiferus*, and elk *Alces alces* (Solomonov and Lukina 2009). The species (e.g., mammals) reveal diverse adaptations to the adverse conditions of extremely low winter temperatures.

Change in Ecological Character and Conservation

Negative environmental trends and changes in the basin, frequently affecting wetlands, have been documented during the last 40–50 years, although gold-bearing rivers in the upper reaches of the Aldan had already undergone changes before this time, and stocks of sturgeon and some white fish suffered damage from an unlimited fishery. Negative influences intensified since the 1950s–1960s with a wave of mineral resource development, increased availability of motorboats, and fishing with *kapron* nets which increased the take of fish and hunting takes. Dozens of new settlements were built in remote virgin areas. Large power stations (including hydroelectric) were constructed to provide the energy needs of the mining industry and settlements. Shipping increased dramatically. In rural areas, consolidation of agricultural enterprises and settlements were undertaken. Fishery and hunting management was not effective. Public utilities and industrial technologies were environmentally harmful. The changes had local impacts as both industry and settlements are dispersed. Water in all large streams has been assessed as contaminated, and the steady decrease in waterfowl and some other bird numbers is obvious. Irreversible changes are observed for the Vilyuy ecosystem as a result of the Viluislkaya hydropower stations construction. Efforts to improve the situation have been undertaken since the 1980s. Even with these changes, the basin environment is still relatively stable but this is due to the large pristine areas that still remain (Kirillov 2002; Degtyarev 2007a).

Wetland Types and Distribution

Wetlands of the Lena River Basin are diverse and using the Ramsar classification system have been identified as belonging to the following natural inland wetland and coastal habitats: permanent inland deltas; permanent rivers/streams/creeks; seasonal/intermittent/irregular rivers/streams/creeks; permanent freshwater lakes (over 8 ha) including large oxbow lakes; seasonal/intermittent freshwater lakes (over 8 ha) including floodplain lakes; permanent saline/brackish/alkaline lakes; permanent freshwater marshes/pools, ponds (below 8 ha), marshes, and swamps on inorganic soils, with emergent vegetation waterlogged for at least most of the growing season; seasonal/intermittent freshwater marshes/pools on inorganic soils including sloughs, potholes, seasonally flooded meadows, and sedge marshes; non-forested peatlands including shrub or open bogs, swamps, and fens; alpine wetlands including alpine meadows and temporary waters from snowmelt; tundra wetlands including tundra pools and temporary waters from snowmelt; shrub-dominated wetlands including shrub swamps, shrub-dominated freshwater marshes, shrub carr, alder thicket on inorganic soils, and freshwater; tree-dominated wetlands including freshwater swamp forests, seasonally flooded forests, and wooded swamps on inorganic soils; forested peatlands including peatswamp forests; geothermal wetlands; karst and other subterranean hydrological systems, inland, and human-made wetlands (ponds; includes farm ponds, stock ponds, small tanks); (generally below 8 ha); irrigated land, seasonally flooded agricultural land including intensively managed or grazed wet meadow or pasture; salt exploitation sites including salt pans, salines, etc.; water storage areas including reservoirs/barrages/dams/impoundments generally over 8 ha; excavations including gravel/brick/clay pits; borrow pits and mining pools; wastewater treatment areas including sewage farms, settling ponds, oxidation basins, etc.; and estuarine waters exclusively in the Lena Delta (Krivenko 2000).

By origin, lakes are categorized into six types (erosion, thermokarst, eolian, karstic, glacial, tectonic, anthropogenic, and **trapps** – massive eruptive formations specific for ancient platforms (Spiridonov 1980)). River types are graded as mountain, semi-mountain, and lowland streams. Peat-bearing depressions are represented by peatlands and mires differentiated by vegetation composition and structure. Among them is **mari** – a kind of shrubby-wooded peatland originally linked with permafrost (see Box below)

> **Mari**
> The term "mari" is generally adopted in Russian-language scientific publications to designate specific landforms (Spiridonov 1980; Kireev 1984). It has no satisfactory analogues in Germanic or Romanic languages, notably, in English (Poppe and Brown 1976). Mari is peatland (peat depth less than 30 cm) that is sparsely wooded with oppressed larch hummocky forest, occurring predominantly in conditions of permafrost that is annually thawed out to a depth of 20–40 cm (specific for the taiga zone of northeast Asia) (Kireev 1984).

Independent of origin, a lake aging in zonal conditions with a positive moisture budget develops from a highly watered basin into a blanket bog. In azonal lowland conditions characterized by a negative moisture budget, the late aging stages of an endorheic lake depression are dependent on long-term fluctuations in precipitation and thus cyclically dry and become inundated like lakes in the steppe. It is a unique Central Yakutian succession and the key point in understanding the **alas** phenomenon (Soloviev 1961; Degtyarev 2000b) – see Box below.

Lakes, rivers, and peat-bearing depressions, by regional classification of waterbird habitats, are categorized into the following inventory types: "alas lake," "bog lake," "collapsed lake," "eolian lake," "mountain lake," "large river floodplain lake," "small river floodplain lake,""large or medium lowland river," "small lowland river," "semi-mountain river," "mountain river," "brook," "wooded or shrubby peatland," and "grassy peatland." As a waterbird habitat, these wetlands as a rule form complexes divided into four groups (flat interfluve, watershed depression, large river valley, and small river valley) and nine types differentiated by wetland typological composition. Based on geomorphology and landscape structure and wetland composition, the basin (except for the delta) is regionalized into hierarchical aligned regions, provinces, and districts (Degtyarev 2000b).

> **Alas**
> According to Poppe and Brown (1976), is "a depression with gentle slopes and flat bottom; the bottom may be covered with a meadow and/or a lake." But, strictly speaking, as a landform it is an aging succession of endorheic thermokarst lakes drying up due to a negative moisture budget. It is specific to the present-day permafrost area of the Central Yakutian lowland

(continued)

Fig. 3 Eolian dune sands (Photo credit: V. Degtyarev © Rights remain with the author)

characterized by cryoaridic conditions (the combination of aridity and low temperatures, characteristic of permafrost areas). The term is used incorrectly when referring to thermokarst lakes of the tundra zone that dry up not as a result of moisture deficiency but rather due to runoff. In the cryoaridic conditions, originally different lakes can develop to alas (alas-like landform) (Degtyaryev 2007b). The term "alas" is generally adopted in Russian-language scientific publications to designate specific landforms (Spiridonov 1980). It has no satisfactory analogues in Germanic or Romanic languages and notably not in English (Poppe and Brown 1976).

The lowland region embraces the Central Yakutian and Nizhnelenskaya lowlands (altitude up to 300 m). This part of the basin is rich in wetlands and distinguished for its diversity of waterbird habitats. At the same time, it is the most populated. The lowest level is the vast floodplain area of the Lena, Aldan, and Vilyuy rivers. Flat interfluves bear the unique **alas** landscapes attributed to cryoaridic conditions, which is especially inherent in the north of the Lena and Aldan interfluve and in the Vilyuy middle reaches basin. Eolian sand dune (up to 3,500 km^2, e.g., 64°11′N, 121°59′E; Fig. 3) occurs with lakes, bogs, and **mari**. There are brine (up to 135 g/ml) lakes but meso- or eutrophic types dominate. Their hydrophyte composition is the most diverse and includes common reed *Phragmites communis*, bulrush *Scirpus lacustris*, great reedmace *Typha latifolia*, beach arrowhead *Sagittaria natans*, pondweeds, etc. This region plays a major role in support of the Lena populations of such valuable fishes as Siberian sturgeon, Siberian white salmon, Arctic *Coregonus autumnalis* and Siberian *C. sardinella* ciscos, peled *C. peled*, and muksun *C. muksun*, as well as of crucian carp *Carassius carassius* especially valued in the provincial cuisine. The region was characterized by extremely abundant breeding and migrating waterbirds until the middle of the twentieth century. The number of waterbirds has since substantially decreased. Alas-based complexes support colonies of gulls and terns and unique bird communities including gadwall *Anas strepera*, northern lapwing *Vanellus vanellus*, marsh *Tringa stagnatilis* and broad-billed *Limicola falcinellus* sandpipers, Temminck's stint *Calidris temminckii*, ruff *Philomachus pugnax*, and Richard's pipit *Anthus richardi*.

The plateaus and piedmont areas (altitude up to 1,000 m) around the lowlands are less flattened and more well drained. Considerable pristine areas still remain, notably in the westernmost parts, where vast trapps formation generated the stepped table mountains and trapps lakes (a trapps lake is a poorly investigated lake type conditionally marked out by Voskresensky (1962)). The areas are crossed by the Lena, Aldan, and Vilyuy rivers with their wide floodplains. Semi-mountainous rivers dominate, and the flat watershed areas are characterized by a depression bearing bog-lake-river-tundra complex. Large peat depressions occur rarely in the west-north. Within the basin of the Vilyuy river is a large man-made water reservoir (Viluiskoe vodohranilizshe) of 2,170 km^2 and 2,650 km of shoreline. The large rivers support stocks of Siberian sturgeon and tugun *Coregonus tugun* and in the middle and small streams of humpback white-fish *C. lavaretus*, taimen, Arctic grayling *Thymallus arcticus*, and lenok *Brachymystax lenok*. Semi-mountainous rivers and developed floodplains mostly meet ecological requirements of breeding mergansers, common goldeneye *Bucephala clangula*, Middendorff's bean goose, falcated duck, osprey, and white-tailed eagle. The region is linked by three known breeding sites of the hooded crane. The easternmost area forms a significant section of the Baikal teal and Siberian crane migration corridor. Depressions in the west-north are densely inhabited by little curlew, whimbrel *Numenius phaeopus*, and black-tailed godwit.

The mountain region embraces the eastern watershed area from Lake Baikal to the Laptev Sea. It is the most uniform and pristine part of the basin in regard to wetlands mostly consisting of mountain rivers. There is a small (11 km^2) water reservoir for the Mamakan hydroelectric power station. Lakes are oligotrophic and rare, but there are a few large (up to 80 km^2) water bodies. This region plays a determinative role in the viability of taimen, Arctic grayling, lenok, round white-fish *Prosopium cylindraceus* and Arctic char *Salvelinus alpinus* stocks, harlequin duck, and significant numbers of osprey, Middendorff's bean goose, mergansers, and Far Eastern curlew.

Conservation

The 2000s coincided with a new wave of industrial and large-scale development that incorporated special transportation, fishing, and hunting technology. This development threatens primarily pristine areas with their game animal and fish resources. In this respect, the protected areas are of great importance and include Ust-Lenskiy (73°8′N, 126°4′E), Olekminsk (59°14′N 122°14′E), Dzhugdzhursky (57°28′N,

138°19′E), Vitimskiy (57°6′N, 116°33′E), and Baikalo-Lensky (54°5′N, 107°5′E) zapovedniks (the highest federal level of protected area in Russia, usually referred to as a strict nature reserve) providing the appropriate protection regime. Up to 80 regional and local nature parks, reserves, and others sites including lakes receive nominal protection (Solomonov and Lukina 2009). Except for the Lena Delta, the Muna River Basin (67°59′N, 122°37′E), the middle Lena valley and the interfluve of the Dyanyischka and Lepiske rivers (64°48′N, 125°27′E), Nidjili lake (63°35′N, 125°9′E), Beloe lake (63°21′N, 128°58′E), the interfluve of the Aldan and Amga rivers (61°58′N, 135°14′E), and the interfluve of the Aldan and Maya rivers (59°53′11N, 134°23′E) are included in the Ramsar shadow list (Krivenko 2000). There are a number of wetlands of conservation concern (e.g., Central Yakutia areas of the taiga-alas and eolian complexes, large depressions in the Viluiy and the Chara upper reaches). In 2012, the Lena Pillars National Park was inscribed on the World Heritage List (http://whc.unesco.org/en/list/1299), based on criterion viii (outstanding examples representing major stages of the earth's history, including the record of life, significant ongoing geological processes in the development of landforms, or significant geomorphic or physiographic features).

References

Andreev VN, et al. Osnovnye osobennosti rastitelnogo pokrova Yakutskoy ASSR. Yakutsk: Izdatelstvo YaF SO AN SSSR; 1987 (In Russian). – Basic features of vegetation cover of the Yakutian ASSR. Yakutsk: YaB, SD, AS of USSR Publ; 1987.
Balobaev VT. Geotermiya mersloi zonyi litosferyi severa Evrasii. Novosibirsk: Isdatelstvo Nauka; 1991 (In Russian). – Geothermometry of permafrost of lithosphere of northern Eurasia. Novosibirsk: Nauka Publ; 1991.
BirdLife International. Threatened birds of Asia: the BirdLife International Red Data Book. Cambridge, UK: BirdLife International; 2001.
Borisov AA, Klimatyi SSSR. Moskva: Isdatelstvo Prosvyastchenie; 1967 (In Russian). – Climates of USSR. Moscow: Prosvyastchenie Publ; 1967.
Degtyarev VG. Vodno-bolotnyie ptitzyi v usloviyiah Tsentralnoyakutskoi ravninyi. Avtoreferat na soiskanie uchenoi stepeni doktora biologicheskih nauk. Moskva: Institut problem ecologii i evolutsii im. AN Severtsova RAN; 2000b (In Russian). – Waterbirds in conditions of Central Yakutian lowland. Abstract of thesis for a doctor's degree. Moscow: Institute of problems of ecology and evolution of RAS; 2000.
Degtyarev AG. Vertebral animals of Yakutia. A fourlingual taxonomic list. Yakutsk; 2000a (In Russian).
Degtyarev VG. Strategiya sohraneniya vodno-bolotnyih ugodii Yakutii. Yakutsk: Isdatelstvo YaNTs SO RAN; 2007 (In Russian). – Yakutia wetland conservation strategy. Yakutsk: YaSC, SB, RAS Publ; 2007a.
Degtyarev VG. Waterbirds in conditions of cryoaridic lowland. Novosibirsk: Nauka Publ; 2007b (In Russian).
Dobrovolsky GV, et al. Krasnaya kniga pochv Rossii: Obiektyi Krasnoy knigi i kadastra osobo tsennych pochv. – Moskva: MARS Press; 2009 (In Russian). – Red Data Book of soils of Russia: objects of Red Data Book and especially protected soils land-survey. Moscow: MARS Press Publ; 2009.
Dolinin IN, et al. Krasnaya kniga Respubliki Saha (Yakutia) Tom 1: Redkie i nakhodyaschiesya pod ugrozoy ischezoveniya vidy rasteniy i gribov. Yakutsk: NIPK Sahapoligrafizdat; 2000

(In Russian). – Red Data Book of Saha Republic (Yakutia) V.1: rare and endangered species of plants and fungi. Yakutsk: NIPK Sahapoligrafizdat Publ; 2000.

Dolman AJ, Maximov T, Ohta T. Water and energy exchange in East Siberian forest: an introduction. Agric For Meteorol. 2008;148:1913–5.

Elovskaya LG, Konorovskiy AK. Raionirovanie i melioratsiya merzlotnyih pochv Yakutii. Novosibirsk: Isdatelstvo Nauka; 1978 (In Russian). – Districting and melioration of frozen soils of Yakutia. Novosibirsk: Nauka Publ; 1978.

Gaikova OY, et al., editors. Krasnaya kniga Irkutskoi oblasti. Irkutsk: Isdatelstvo Vremya stransvii; 2010 (In Russian). – Red Data Book of Irkutskaya oblast. Irkutsk: Vremya stransvii Publ; 2010.

Gavrilova MK. Klimat Tsentralnoi Yakutii. Yakutsk: Yakutskoe knizhnoe isdatelstvo; 1972 (In Russian). – Climate of Central Yakutia. Yakutsk: Yakutian book Publ; 1972.

Gukov AY. Gidrobiologiya ustievoi oblasti reki Lenyi. Moskva: Isdatelstvo Nauchnyii mir; 2001 (In Russian). – Hydrobiology of the Lena mouth area. Moscow: Nauchnyii mir Publ; 2001.

Kireev DM. Ekologo-geografichtskie terminyi v lesovedenii. Slovar-spravochnik. Novosibirsk: Isdatelstvo Nauka; 1984 (In Russian). – Ecological and geography terms in forestry. Dictionary and handbook. Novosibirsk: Nauka Publ; 1984.

Kirillov AF. Promyislovyie ryibyi Yakutii. Moskva: Izdatelstvo Nauchnyii mir; 2002 (In Russian). – Food fishes of Yakutia. Moscow: Nauchnyii mir Publ; 2002.

Krivenko VG, editor. Wetlands in Russia. Volume 3. Wetlands on the Ramsar shadow list. Moscow: Wetlands International Global Series No. 3; 2000. In Russian.

Matveev IA, et al., editors. Atlas selskogo khozyaistva Yakutskoy ASSR. Moscow: GUGK; 1989 (In Russian). – Atlas of agriculture of the Yakutian ASSR. Moscow: GUGK; 1989.

Mostakhov SE. Reka Lena. Yakutsk; 1972 (In Russian). Yakutsk: Lena River; 1972.

Mostakhov SE. Ozernost basseina Lenyi. In Ozera kriolitozonyi Sibiri. Novosibirsk: Isdatelstvo Nauka; 1974 (In Russian). – Lake density of Lena river basin. In Lakes of Siberia permafrost. Novosibirsk: Nauka Publ; 1991.

Ostrovsky AP, et al. Krasnaya kniga Chitinskoi oblasty i Aginskogo Buryatskogo avtonomnogo okruga (rasteniya). Chita: Stil; 2002 (In Russian). – Red Data Book of Chitinskaya oblast and Buryatskii autonomous region (plants). Chita: Stil Publ; 2002.

Poppe VN, Brown RJE. Russian-English glossary of permafrost terms. Tech. Memorand. No. 117. Ottawa: Associate Committee on Geotechnical Research, National Research Council of Canada; 1976.

Solomonov NG, Lukina LM, editors. Dar planete Zemlya. Osobo ochranyaemyie prirodnyie territorii Respubliki Saha (Yakutia). Yakutsk; 2009 (In Russian). – Gift to the Earth. Yakutsk: Protected areas of Saha Republic (Yakutia); 2009.

Soloviev PA. Tsiklicheskie ismeneniya vodoobilnosti alasnyih ozer Tsentralnoi Yakutii. In Voprosyi geografii Yakutii. Yakutsk: Knizhnoe isdatelstvo; 1961. Vipusk 1 (In Russian). – Water supply cyclical changes of alas lakes of Central Yakutia linked with fluctuation of climate elements. In Yakutia geography issues. Fascicle 1. Yakutsk: Yakutian book Publ; 1961.

Spiridonov AI, editor. Chetyirehyazyichnyii entziklopedicheskii slovar terminiv po fizicheskoi geografii. Moskva: Izdatelstvo Sovetskai Enciklopedia; 1980. (In Russian). [Four-linqual encyclopaedic glossary of terms in physical geography. Moscow: Soviet Encyclopedia Publ; 1980.

Troeva EI, et al. editors. The far north: plant biodiversity and ecology of Yakutia – plant and vegetation v. 3. Springer Science + Business Media B. V. 2010.

Voskresensky SS. Geomorfologiya Sibiri. Moskva: Izdatelstvo Moskovskogo universiteta; 1962 (In Russian). – Geomorphology of Siberia. Moscow: Moscow University Publ; 1962.

Lena River Delta (Russia)

118

Victor Degtyarev

Contents

Introduction	1452
Hydrology	1452
Wetland Ecosystems	1453
Biodiversity	1454
Conservation Status	1454
Ecosystem Services	1455
Threats and Future Challenges	1455
References	1455

Abstract

The Lena delta is one of the world's largest river deltas (32,000 km^2) with average annual discharge of 513 km^3, 6,500 km of distributaries and about 30,000 lakes (predominantly of thermokarst and erosion origin). Mires consist mostly of vast areas of bogs with low-centered polygons and rarely, hillocky string bogs. Rivers, lakes, and bogs are normally interconnected by means of shallow tributaries and runoff troughs at the time of high water and spring to form a unified wetland complex. The coastal area abounds in small bays, while its lowest areas hold salty water lagoon lakes and foreland meadows. In coastal waters, materials washed away and carried over by distributaries have formed 0.5–5.0 m deep sandbars. The number of species of higher vascular plants, Bryopsida, lichens, fish, birds, and mammals is estimated at 316, 95, 51, 109, and 29 respectively. The endemic *Salvelinus jakuticus* inhabits the delta. Some species are presented by several

---

V. Degtyarev (✉)
Siberian Division of Russian Academy of Sciences, Institute for Biological Problems of Cryolithozone, Siberian Branch, Russian Academy of Sciences, Yakutsk, Russia
e-mail: dvgarea@yandex.ru

© Springer Science+Business Media B.V., part of Springer Nature 2018
C. M. Finlayson et al. (eds.), *The Wetland Book*,
https://doi.org/10.1007/978-94-007-4001-3_207

ecological forms, e.g., four forms of the muksun *Coregonus muksun* are identified to occupy various ecological niches and have marked morphological differences. In 1985, the state natural zapovednik "Ust-Lenskyi" was established with a total area of 14,330 km^2. The delta has been included in the Ramsar shadow list. Being a place of fish reproduction and fattening, it supports the long-term viability of significant populations or stocks of the Siberian sturgeon *Acipenser baeri* and 12 salmonid species. Fishing technologies applied currently in the area are accompanied by mass take of the whitebait. The salmonid harvest has decreased 2.5 times from the 1940s up to the late 1970s.

Keywords
Tundra · Alluvial terrace · Lake · Lagoon · Mire · Zapovednik · Fishery · Siberian sturgeon · Salmonid

Introduction

The Lena delta (Fig. 1) is one of the world's largest river deltas. Located in northeastern Asia at the Laptev Sea coast (63°35′N, 125°9′E, altitude 1–114 m), the delta occupies a 400 × 100 km fan-shaped area with a total area of 32,000 km^2. Scientific data on the Lena delta's natural conditions were obtained in the early nineteenth century. The first special study of the delta was ichthyological research undertaken by the State Research Institute for Lacustrine and Riverine Fishery and Chief Directorate of the Northern Sea Route in the 1930s–1940s. A comprehensive biological study was carried out by the Yakut Biology Institute (USSR Academy of Science) alongside the West Siberian Survey Agency in designing the Ust-Lensky zapovednik (the highest federal level of protected area in Russia, usually referred to as a strict nature reserve) in 1982–1983. Following the zapovednik's establishment, studies were conducted by its scientific departments and the international biological station "Lena-Nordenscheld" in cooperation with Russian and foreign academic institutions and within international projects (Labutin et al. 1985; Grigoriev et al. 2000; Gukov 2001; Kirillov 2002).

Hydrology

The annual discharge reaching the delta averages 513 km^3 with its maximum in June and minimum in winter. The river breaks down into numerous branched distributaries by Tit-Ary island, the largest and longest (more than 100 km) of which are the Arynskaya, Trofimovskaya, and Bykovskaya. The latter two channels receive 90% of the annual water runoff. Only the Bykovskaya and Olenekskaya distributaries are navigable. The ice cover forms in October and lasts 220–230 days and is up to 2.0–2.5 m thick by spring breakup which occurs in late May. Ice drift is often accompanied by jams resulting in flooding of vast wetlands and terraces above

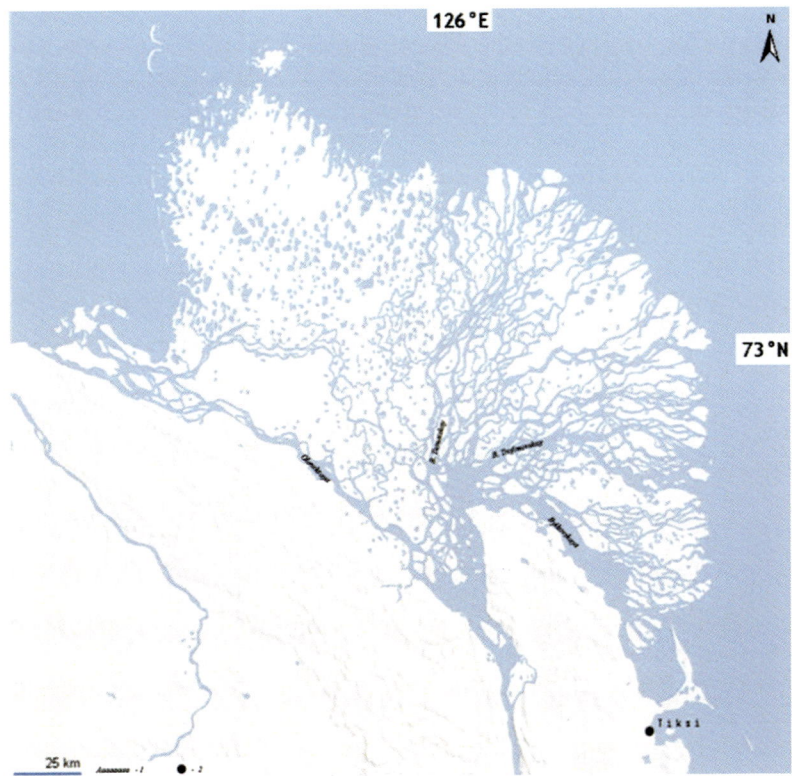

Fig. 1 Lena delta. *1* main distributaries, *2* villages (V. Degtyarev © Rights remain with the author)

flood plains. The fluctuation in water level amplitude decreases from 15 m to 6–9 m closer to the sea. Significant water level fluctuations (the maximal coastal amplitude being 2.5 m) are induced by onshore winds that often reach storm velocities. Mechanical and warming effects of the water runoff and abrasion bring about intensive erosion of delta distributary banks, formed by ice-saturated frozen rocks, and their resedimentation.

Wetland Ecosystems

The Lena delta is a lowland sloping slightly eastwards, consisting of ancient marine and late Pleistocene and Holocene alluvial terraces formed by permafrost rocks. That structure defines the distribution of the main wetness regimes: wet polygonal tundra on the first terrace, dry sparsely vegetated regime on the sandy second terrace, and the medium wetness regime of the third Yedoma terrace (Widhalm et al. 2015).

Water and mire areas include river flood plains, lake, and bog systems, as well as a coastal area. The total length of the delta distributaries reaches 6,500 km. Most of the delta is made up of shallow distributaries with slow currents and terraced valleys forming spits and sandbars in the sandy riverbed. About 30,000 lakes (predominantly of thermokarst and erosion origin), 22,000 of which are concentrated in its western part, are distributed over the delta. The greatest percentage (90%) of the lakes have an area of less than 0.25 km^2 and only 100 have a water surface exceeding 10 km^2. Mires consist mostly of vast areas of bogs with low-centered polygons and rarely, hillocky string bogs. Rivers, lakes, and bogs are normally interconnected by means of shallow distributaries and runoff troughs at the time of high water and spring snow melting to form a unified wetland complex. The coastal area abounds in small bays, while its lowest areas hold salty water lagoon lakes and foreland meadows. In coastal waters, materials washed away and carried over by distributaries have formed 0.5–5.0 m deep sandbars (Samoilov 1952; Zalogin and Rodionov 1969; Labutin et al. 1985; Grigoriev et al. 2000; Gukov 2001).

Biodiversity

The number of species of higher vascular plants, Bryopsida, and lichens occurring in the delta is estimated at 316, 95, and 51, respectively. Five types of vegetation and up to 100 plant communities have been identified. In delta waters 153 algae species have been found, and 43 fish species have been registered including marine, semi-migrating, migrating, and lacustrine species. The endemic *Salvelinus jakuticus* inhabits the delta. Some species are presented by several ecological forms, e.g., four forms of the muksun *Coregonus muksun* are identified to occupy various ecological niches and have marked morphological differences. Birds are represented by 109 species, of which 64% are water birds (mostly *Anseriformes* and *Charadriiformes*), the significant part of which find favorable conditions for breeding and molting. Mammals are represented by 29 species, of which less than 10 are resident. Historically, among marine mammals, the walrus *Odobenus rosmarus* and polar bear *Ursus maritimus* had close affinity with the delta (Labutin et al. 1985; Krivenko 2000; Gukov 2001; Kirillov 2002; Solomonov and Lukina 2009).

Conservation Status

In 1985, the state natural zapovednik "Ust-Lenskyi" was established in the Lena delta and its lower reaches with a total area of 14,330 km^2 consisting of two divided sectors ("Deltovyi" – 13,000 km^2 and "Sokol" – 1,330 km^2) and a buffer zone of 10,500 km^2. The delta has been included in the Ramsar shadow list. In the east of the delta is located the Lena delta regional reserve, established in 1996 (Krivenko 2000; Solomonov and Lukina 2009).

Ecosystem Services

The discharge of the Lena River divided among the distributaries of the delta has a significant effect on the Laptev Sea coastal zone. It results in the formation of vast freshwater and desalted zones. Being places of fish reproduction and fattening, these areas support the long-term viability of significant populations or stocks of the Siberian sturgeon *Acipenser baeri* and 12 salmonid species. Almost to the mid-twentieth century, the mode of life of indigenous peoples occupying the Lena's lower reaches was centered on utilization of the fish including their use in sled dog breeding. Currently, these stocks provide a major input into a substantial and commercial fishery in the Lena delta and its lower reaches and, correspondingly, secure employment of a significant amount of local people in the region's fish-processing industry. Up to the 1980s a significant part of the Bulunsky population of the reindeer *Rangifer tarandus* populated the delta and adjoining territories, but overhunting resulted in a 20-fold decrease and the species is currently not exploited here (Labutin et al. 1985; Krivenko 2000; Kirillov 2002).

Transportation of goods within the Lena basin and the Anabar, Yana, Indigirka, and Kolyma river basins is through the major navigable waterways of the Bykovskaya and Olenekskaya distributaries to the sea port of Tiksi, the outlet to the Northern Sea Route.

Threats and Future Challenges

Fishing practices are unsustainable as they do not adequately take into account basic scientific rationale for sustainable harvesting (current practices include overharvesting, high bycatch, high take in the desalted zone supporting the main stocks of salmonids for fattening). The salmonid harvest has decreased 2.5 times from the 1940s up to the late 1970s. Fishing technologies applied currently in the Lena delta are accompanied by mass take of the whitebait, being fished at a rate of up to 180 tonnes per 100 tonnes of conditioned fish taken. Intensive overfishing of recruitment stock makes it almost impossible to restore the delta stocks to any significant abundance values. An expected increase in navigation intensity will result in an acceleration of delta waterway erosion and levels of chemical and biological water pollution (Labutin et al. 1985; Krivenko 2000; Gukov 2001; Kirillov 2002).

References

Grigoriev MN, Rachold F, Schwamborn G. Problema proischozhdeniya i razvitiya oser ostrova Arga-Muora-Sis (severo-zapadnyii sector deltyi r. Lenyi) – rezultatyi Rossiisko-Germanskoi expeditsii 1998–99 gg. In: Zhirkov II, Pestryakova LA, Maksimov GN, editors. Mezhdunarodnaya konferentsiya "Ozera holodnyih regionov". Chast 4. Voprosyi paleoklimatologii, paleolimnologii i paleoekologi. Yakutsk: Yakut state University; 2000. p. 69–81 (In Russian). – Problem of origin and development of lakes of Arga-Muora-Sis island (north-west sector of Lena delta) – results of Russian-Germany expedition 1998–99. In: Zhirkov II, Pestryakova LA, Maksimov GN, editors. Items of paleoclimatology, paleolimnology and paleoecology. Yakutsk: Yakut state University; 2000. p. 69–81.

Gukov AYu. Gidrobiologiya ustievoi oblasti reki Lenyi. Moskva: Isdatelstvo Nauchnyii mir; 2001 (In Russian). – Hydrobiology of the Lena mouth area. Moscow: Nauchnyii mir; 2001.

Kirillov AF. Promyislovyie ryibyi Yakutii. Moskva: Izdatelstvo Nauchnyii mir; 2002 (In Russian). – Food fishes of Yakutia. Moscow: Nauchnyii mir; 2002.

Krivenko VG, editor. Wetlands in Russia. Volume 3. Wetlands on the Ramsar shadow list, Wetlands International global series no. 3. Moscow: Wetlands International; 2000 (In Russian).

Labutin YuV at al. Rastitelnyii i zhivotny mir delty reki Lena. Yakutsk: Izd-vo YaF SO AN SSSR; 1985 (In Russian). Plants and animals of the Lena river delta. Yakutsk: YaB, SD, AS of SSSR; 1985.

Samoilov IV. Ustia rek. Moskva: Isdatelstvo Geografgiz; 1952 (In Russian). – Mouths of rivers. Moscow: Geografgiz; 1952.

Solomonov NG, Lukina LM, editors. Dar planete Zemlya. Osobo ochranyaemyie prirodnyie territorii Respubliki Saha (Yakutia). Yakutsk; 2009 (In Russian). – Gift to the Earth. Protected areas of Saha Republic (Yakutia). Yakutsk; 2009.

Widhalm B, Bartsch A, Heim B. A novel approach for the characterization of tundra wetland regions with C-band SAR satellite data. Int J Remote Sens. 2015;36(22):5537–56.

Zalogin BS, Rodionov NA. Ustievyie oblasti rek SSSR. Moskva: Isdatelstvo Myisl; 1969 (In Russian). – Mouth areas of rivers of the USSR. Moscow: Myisl; 1969.

Nidjili Lake: Lena River Basin (Russia)

119

Victor Degtyarev

Contents

Introduction	1458
Hydrology	1458
Wetland Structure	1458
Biodiversity	1459
Conservation Status	1460
Ecosystem Services	1460
Threats and Future Challenges	1460
References	1461

Abstract

Nidjili Lake is the largest lake (catchment area of 1,010 km^2, a surface area of 119 km^2) in the Lena River basin. It is located in the northern Lena-Vilyuy interfluve (63°35′N, 125°90′E). The lake is believed to have formed in the late Pleistocene as a result of erosion and thermokarst processes. The maximum Lena basin values of benthos biomass and density were recorded for the dominating substratum. At least since the 1930s, the lake has experienced an intense natural eutrophication process accompanied by alterations in hydrochemical properties and corresponding simplification in fish species composition. Up to the late 1950s, the lake played a significant role in sustaining populations of common goldeneye and white-winged scoter, which nested and molted in the tens of thousands of individuals. No other area of such a small size was known in northeastern Asia to support such large accumulations of molting goldeneye. High fish productivity resulted in a high nesting density of osprey, white-tailed

---

V. Degtyarev (✉)
Siberian Division of Russian Academy of Sciences, Institute for Biological Problems of Cryolithozone, Siberian Branch, Russian Academy of Sciences, Yakutsk, Russia
e-mail: dvgarea@yandex.ru

© Springer Science+Business Media B.V., part of Springer Nature 2018
C. M. Finlayson et al. (eds.), *The Wetland Book*,
https://doi.org/10.1007/978-94-007-4001-3_184

eagle, ducks, waders, gulls, and terns. The processes of significant anthropogenic technological transformations are absent. The lake and its catchment area have not endured any man-made habitat changes and are still capable of sustaining the same significant numbers of water birds that occurred here in the mid-twentieth century if hunting (poaching), boating, and summer fishing are controlled, allowing the bird population to recover.

Keywords
Common goldeneye · High productivity · Largest lake

Introduction

Nidjili Lake (Fig. 1) in the Central Yakutian lowland is the largest lake in the Lena River basin. It is located in the northern Lena-Vilyuy interfluve (63°35′N, 125°9′E) 26 km from the Vilyuy River (altitude 134 m). The major data on it were obtained during hydrobiological and ichthyological studies in the 1930s–1960s by the State Research Institute for Lacustrine and Riverine Fishery (Averintsev 1933; Titova et al. 1966). Ornithological data are scarce and have been obtained during brief summer excursions by secondary school children under the supervision of Andreev (1987). Bird census has been conducted for the lake (excluding the catchment area) (Degtyarev and Larionov 1981), although a full-scale study of water birds in the area has not been carried out.

Hydrology

The lake has a catchment area of 1,010 km^2, a surface area of 119 km^2, a length of 33.5 km, maximum width of 5.7 km, maximum depth of 7 m, and mean depth of about 3 m. The minor waterways of Kyunkyu and Kharyia-Yurekh flow into the lake from the southwest and southeast. The lake is the source of the Sien Yurekh River which flows north, linking the lake with the Vilyuy. Lake freeze-over occurs in October to a depth of 1.7 m with more than 50% water volume decrease. Ice cover breakup continues for about a month from mid-May. The fluctuation in water level reaches 44 cm – water level rise begins in mid-May, peaks in late June, and then it drops to its minimum level in winter. Summer precipitation brings about brief water level rises of up to 2 cm (Titova et al. 1966).

Wetland Structure

The lake is believed to have formed in the late Pleistocene as a result of erosion and thermokarst processes. Its elongated shape narrows in the middle with an indented coastline, and its depression profile is asymmetric. The southern coastline is steep

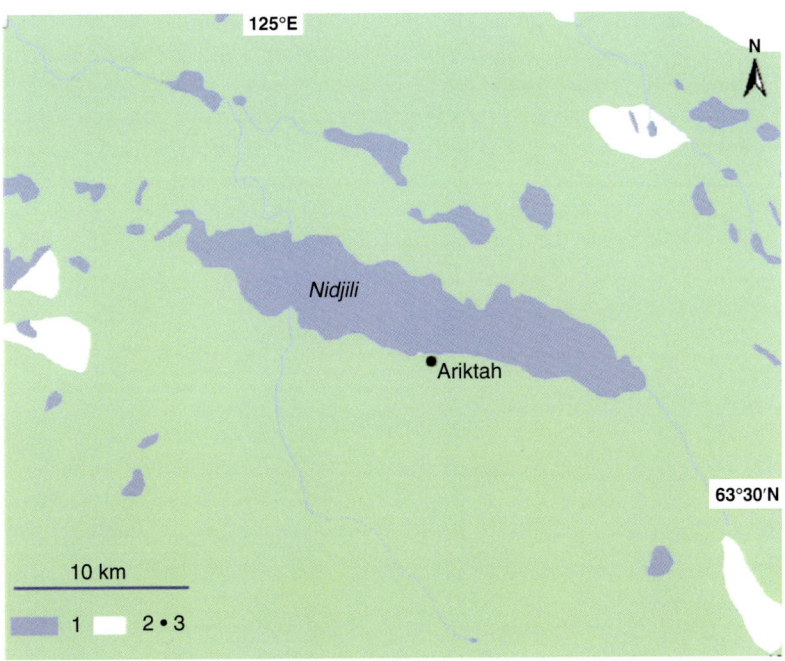

Fig. 1 Nidjili lake. *1* lake surface, *2* eolian sand dune, *3* villages (V. Degtyarev © Rights remain with the author)

(up to 1.5 m), while the northern coastline is gently sloping and swampy. The windward facing shoreline has been transformed by ice movements into a 4 m wide and 0.5 m high levee. There are two islands in the northwestern (7 ha) and northeastern tips (74 ha). The bottom is predominantly soft organic substrate. The littoral zone has a belt of emergent wetland vegetation (Titova et al. 1966). The lake basin consists of taiga where are located numerous lakes of various sizes, **mari** (see Box in ▶ Chap. 117, "Lena River Basin (Russia)") and five elongated dune sands (up to 150 km^2), 4–20 km from the lake.

Biodiversity

At least since the 1930s, the lake has experienced an intense natural eutrophication process accompanied by alterations in hydrochemical properties and corresponding simplification in fish species composition. By the 1960s, the pike *Esox lucius*, perch *Perca fluviatilis*, roach *Rutilus rutilus*, and dace *Leuciscus leuciscus* were extirpated and only the common minnow *Phoxinus percnurus* and crucian carp *Carassius carassius* currently occur in the lake. The ide *Leuciscus idus*, pike, and roach entering the lake in spring via the Sien Yurekh River die in winter if they remain (Averintsev 1933; Titova et al. 1966).

Higher hydrophyte communities include sedges, common reed *Phragmites communis*, water milfoil *Myriophyllum spicatum*, and bladderworts. In the zoobenthos, 48 species have been identified with chironomids, mollusks, gammarids, and oligochaetes prevailing. The maximum Lena basin values of benthos biomass and density were recorded for the dominating substratum (Titova et al. 1966). The dominant tree in the surrounding landscape is Cajander larch *Larix cajanderi* substituted by pine *Pinus sylvestris* in sandy soils. Four endemic plants grow in the dune sands of the Lena basin. In the lake area, 174 bird species (117 nesting) have been recorded. The muskrat *Ondatra zibethicus* is recorded as an aquatic mammal (Krivenko 2000).

Conservation Status

The lake was included in the Ramsar shadow list for the Russian Federation and the regional list of unique lakes. A special regional decree was proclaimed on the protected unique Lake Nidjili; it defines water protection zones and restrictions in its resource usage (Krivenko 2000; Solomonov and Lukina 2009).

Ecosystem Services

Up to the late 1950s, the lake played a significant role in sustaining populations of common goldeneye *Bucephala clangula* and white-winged scoter *Mellanitta deglandi*, which nested and molted in the tens of thousands of individuals (Andreev 1987). No other area of such a small size was known in northeastern Asia to support such large accumulations of molting goldeneye. High fish productivity resulted in a high nesting density of osprey *Pandion haliaetus*, white-tailed eagle *Haliaeetus albicilla*, ducks, waders, gulls, and terns. Currently, the lake is a regular staging area for water birds during seasonal migrations (Krivenko 2000). The lake supports a population of about 200 people in the village of Aryktakh primarily with a fishery for crucian carp. This species is especially valued in the provincial cuisine. However, by 2000, the annual carp harvest had dropped from 200–300 to 30–70 t (Kirillov 2002).

Threats and Future Challenges

The major factor influencing the lake ecosystem's diminished ecological condition is unsustainable utilization of its resources (Andreev 1987; Kirillov 2002). The processes of significant anthropogenic technological transformations are absent. The lake and its catchment area have not endured any man-made habitat changes – it did not undergo drastic changes following hydroengineering activities – and is still capable of sustaining the same significant numbers of water birds that occurred here in the mid-twentieth century if hunting (poaching), boating, and summer fishing are controlled, allowing the bird population to recover.

References

Andreev BN. Ptitsy Vilyuiskogo basseina. Yakutsk: Yakutskoe knizhnoe isdatelstvo; 1987 (In Russian). Birds of the Vilyui Basin. Yakutsk: Yakutian Book; 1987.

Averintsev SV. Ryibolovstvo na ozere Nidjili i na sosednih s nim ozerah. Trudyi Yakutskii nauchnoi ribochosyaistvennoi stantsii. Vypusk 2. Yakutsk; 1933 (In Russian). Fishery on Nidjili and Neighbouring Lakes. Proceedings of Yakutian research station of Fishery. Issue 2. Yakutsk; 1933.

Degtyarev AG, Larionov GP. Rasmezshenie i chislennost plastinchatokluvyih Tsentralnoi Yakutii. In: Labutin YuV, editor. Migratsii i ekologiya ptits Sibiri. Novosibirsk: Isdatelstvo Nauka; 1981 (In Russian). In: Labutin YuV, editor. Distribution and numbers of anseriformes of Central Yakutia. Migration and ecology of birds of Siberia. Novosibirsk: Nauka; 1981.

Kirillov AF. Promyislovyie ryibyi Yakutii. Moskva: Izdatelstvo Nauchnyii mir; 2002 (In Russian). Food fishes of Yakutia. Moscow: Nauchnyii mir; 2002.

Krivenko VG, editor. Wetlands in Russia. Volume 3. Wetlands on the Ramsar shadow list, Wetlands International global series, vol. 3. Moscow: Wetlands International; 2000 (In Russian).

Solomonov NG, Lukina LM, editors. Dar planete Zemlya. Osobo ochranyaemyie prirodnyie territorii Respubliki Saha (Yakutia). Yakutsk; 2009 (In Russian). Gift to the Earth. Protected areas of Sakha Republic (Yakutia). Yakutsk; 2009.

Titova KN, et al. Ozero Nidjili. Yakutsk: Yakutskoe knizhnoe isdatelstvo; 1966 (In Russian). Lake Nidjili. Yakutsk: Yakutian Book; 1966.

Taiga-Alas Landscape in the South of the Central Yakutian Lowland: Lena River Basin (Russia)

120

Victor Degtyarev

Contents

Introduction	1464
Hydrology	1464
Wetland Ecosystems	1465
Biodiversity	1468
Conservation Status	1469
Ecosystem Services	1469
Threats	1469
References	1470

Abstract

The expression of cryo-arid conditions over vast territories in modern times is best exemplified by the Central Yakutian lowland. In the study of climatology, geo-cryology, and landscapes, the Central Yakutian lowland is recognized as a unique area, characterized by specific climatic parameters and specific interrelations of permafrost and climate, predetermined by the extremely continental influence. A result has been the origin and development of unique wetland areas determined under the conditions of the arduous interaction of climatic and geo-cryological factors affecting watering of the earth's surface. The thermokarst produces depression in bulk, and the water impermeability and high ice content of the permafrost supply the depressions with water from fossil ice and surface drainage to produce lakes, while the deficit in atmospheric precipitation and high summer temperatures interact to dry up water bodies and their catchment areas. Due to the negative water budget of the surface, the Central Yakutian lowland is not included in the general scheme of modern landscape dynamics in the

V. Degtyarev (✉)
Siberian Division of Russian Academy of Sciences, Institute for Biological Problems of Cryolithozone, Siberian Branch, Russian Academy of Sciences, Yakutsk, Russia
e-mail: dvgarea@yandex.ru

© Springer Science+Business Media B.V., part of Springer Nature 2018
C. M. Finlayson et al. (eds.), *The Wetland Book*,
https://doi.org/10.1007/978-94-007-4001-3_136

permafrost area. Rather it is described as a cryo-aridic lowland due to its climatic and geo-cryological characteristics. These unique conditions are most evident in the south of the lowland in the interfluve of the Lena and Aldan rivers. To a considerable degree, modern conditions of the cryo-arid lowland simulate conditions of the periglacial zone of the Late Pleistocene glaciation in the Eurasian lowlands. Study of the modern wetlands of the Central Yakutian lowland thus allows us to hypothesize about conditions of the Eurasian Pleistocene and adaptive strategy directions of the waterbird populations and other animals that inhabited the periglacial zone.

Keywords

alas · cryo-arid conditions · cryolithozone · permafrost · thermokarst

Introduction

The expression of cryo-arid conditions over vast territories in modern times is best exemplified by the Central Yakutian lowland. In the study of climatology, geo-cryology, and landscapes, the Central Yakutian lowland is recognized as a unique area, characterized by specific climatic parameters and specific interrelations of permafrost (cryolithozone) and climate, predetermined by the extremely continental influence (up to 99%) of the climate (Abolin 1929; Borisov 1967; Gavrilova 1972). A result has been the origin and development of unique wetland areas determined under the conditions of the arduous interaction of climatic and geo-cryological factors affecting watering of the earth's surface. The thermokarst (i.e., changes in topography originating from the process of permafrost thawing) produces depression in bulk, and the water impermeability and high ice content of the permafrost supply the depressions with water from fossil ice and surface drainage to produce lakes, while the deficit in atmospheric precipitation and high summer temperatures interact to dry up water bodies and their catchment areas. Due to the negative water budget of the surface, the Central Yakutian lowland is not included in the general scheme of modern landscape dynamics in the permafrost area. Rather it is described as a cryo-aridic lowland due to its climatic and geo-cryological characteristics. These unique conditions are most evident in the south of the lowland in the interfluve of the Lena and Aldan rivers (Fig. 1). The taiga-**alas** landscape (Fig. 2) and **alas**es have long received academic attention in various fields of science, notably in works by researchers of the Academy of Sciences of the USSR and later of Russia (Degtyarev 2007; Dobrovolsky et al. 2009; Troeva et al. 2010).

Hydrology

Lakes of thermokarst and erosion origin predominate. Early-stage thermokarst lakes are supplied with permafrost mobilized by thermokarst and meteoric waters and have a positive water budget. Erosion lakes and late aging successions of

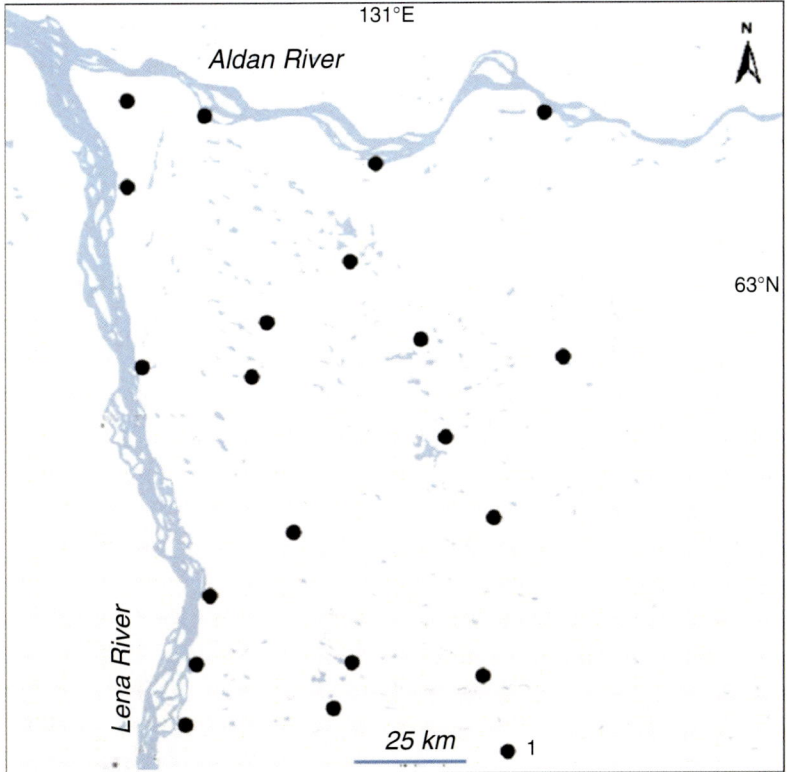

Fig. 1 The south-central Yakutian lowland (interfluve of the Lena and Aldan rivers) indicating villages (*1*) (V. Degtyarev © Rights remain with the author)

thermokarst lakes are supplied exclusively with meteoric water, and water-level fluctuations fully depend on a long-term recurrence of meteoric water. Freezing-over occurs in late September to early October. Lakes freeze through at a depth of 1.0–1.7 m and above. Freezing often reaches bottom sediments. Ice cover breakup continues up to May (Degtyarev 2007).

Wetland Ecosystems

The successional progression of thermokarst lakes commences with a collapse depression containing fallen trees or forest stands, an unstable shoreline, and underdeveloped water biocenoses. At this stage, the lake is termed a "collapse lake" (Fig. 3).

The next stage of lake evolution occurs following shore stabilization and a constant water budget. Formation of littoral sedge bogs and flattening of the

Fig. 2 Taiga-alas landscape (Photo credit: V. Degtyarev © Rights remain with the author)

Fig. 3 Collapse lake (Photo credit: V. Degtyarev © Rights remain with the author)

depression profile begin. Sedimentation, aquatic vegetation and benthos development, and introduction of fish and muskrat into the lake are of significance in the evolution of the lake ecosystem. At this stage, the lake is termed a "bog lake" (Fig. 4).

Following the termination of the thermokarst process, the lake turns into an endorheic lake. If it lacks a sufficient catchment area, the thermokarst depression begins to dry up and develops the properties of an alas (Fig. 5). From this stage on, the successional pathway of the aging thermokarst lake becomes similar to an aging erosion lake, having lost links with the river system and its catchment area.

Fig. 4 Bog lake (Photo credit: V. Degtyarev © Rights remain with the author)

Fig. 5 Alas lake (Photo credit: V. Degtyarev © Rights remain with the author)

Lake water mineralization, soil salinity, the ratio of meadow to lake area, and within the lake the ratio of emergent aquatic plant blankets and open water change with progressive drying. The succession of meadows to steppe begins with a medium-level decrease in alas depression watering. A thick growth of aquatic plants and conditions that produce algal blooms are characteristic of this state of alas. An increase in water mineralization, steppe vegetation formation in meadows, and growth of cane, reed, and sedge thickets are the evident features of a steppe closed lake. Normally, maximal levels of water mineralization do not reach critical values for water animals and plants (up to 10 g/ml). In rare cases, when the thermokarst touches upon rocks enriched with soluble salts, mineralization of water in the drying-up alas can reach extreme values (up to 135 g/ml), which

Fig. 6 Dry alas with pingo (Photo credit: V. Degtyarev © Rights remain with the author)

stresses the vegetation and reduces the density of populations of benthic organisms (Zhirkov 1982). Evolution of steppe and meadow formations in the drying-up depression results in occurrence of the xerophytic *Orthoptera*, thus enlarging the feeding base for birds.

The lacustrine stage of evolution of alases is succeeded by the dry stage (Fig. 6). The dominant vegetation cover consists of steppe meadows with soil salinization of the highest degree including patches of saline lands. Up to 10 m-high pingos are often characteristic components of dry alases. At this stage, the alas, as waterbird habitat, loses its functions as a nesting biotope while retaining its status as a major feeding habitat for *Charadriiformes*. Further, the alas moves into a regime of cyclic watering and drying-up following the pattern of closed lakes of Eurasian arid zones that correspond with long-term precipitation cycles. The alas then becomes dependent on long-term meteoric water fluctuations, cyclically drying-up and being inundated like endorheic lakes in the steppe zone. The most significant fact in understanding of the depression and cryo-aridic peculiarity of the region is the relatively continuous formation of new thermokarst depressions. As a consequence, a series of thermokarst formations, reacting differently to meteoric water alterations and differing in bird habitat quality, occur within relatively homogenous areas (Degtyarev 2007).

Biodiversity

The lowland is why the Central Yakutian floristic region differs greatly from the other floristic regions of the Lena basin in soils and flora-vegetation peculiarities. The flora of the Central Yakutian region contains about 1,000 species and subspecies of higher vascular plants. Its core is composed of azonal (30%), boreal (29%), and steppe (over 25%) zonal complexes. The vegetation of alases is composed of steppe, meadow, hygrophytic, and aquatic communities. Alases form four unique soil types (Dobrovolsky et al. 2009; Troeva et al. 2010).

Fishes are presented by the common minnow *Phoxinus percnurus* and crucian carp *Carassius carassius* that occur almost everywhere. Perch *Perca fluviatilis*, pike *Esox lucius*, and roach *Rutilus rutilus* occur only in a small number of lakes in the center of the region. Other water-related vertebrates are two amphibians, water vole *Arvicola terrestris*, muskrat *Ondatra zibethicus*, and waterbirds (38 breeding and up to 35 migrating species). The alas supports a unique bird community including species that are typical of the tundra and forest steppe: gadwall *Anas strepera*, coot *Fulica atra*, northern lapwing *Vanellus vanellus*, marsh *Tringa stagnatilis* and broad-billed *Limicola falcinellus* sandpipers, Temminck's stint *Calidris temminckii*, ruff *Philomachus pugnax*, and Richard's pipit *Anthus richardi* (Degtyarev et al. 2006; Degtyarev 2007).

Conservation Status

Four alases are included in the regional list of unique lakes (one of which also ranks among the regional protected landscape category), and three more lakes hold the status of local nature reserves (Solomonov and Lukina 2009).

Ecosystem Services

Historically, an alas landform was a determinative base for a specific form of animal husbandry of local peoples. Today, alas meadows are used intensively as livestock and horse pastures and hayfields and support a considerable part of the regional stockbreeding. Up to the 1960s, waterfowl were an important component of the local people's subsistence, and the crucian carp still fills this role. The early stages of thermokarst depressions serve as water supplies for some settlements.

To a considerable degree, modern conditions of the cryo-arid lowland simulate conditions of the periglacial zone of the Late Pleistocene glaciation in Eurasian lowlands. Study of the modern wetlands of the Central Yakutian lowland thus allows us to hypothesize about conditions of the Eurasian Pleistocene and adaptive strategy directions of the waterbird populations and other animals that inhabited the periglacial zone (Degtyarev 2007).

Threats

Major threats are related to unregulated human impacts. In most cases, settlements and cattle-breeding husbandries are located in alases. Significant degradation of the landform has resulted from intensified agricultural development. For example, overgrazing brings about exhaustion of the vegetation cover and compression of the upper soil layer. This leads to elevated salt concentrations in saline soils, transformation of dry habitat soils of light mechanical structure into dust, etc. (Dobrovolsky et al. 2009; Troeva et al. 2010). The waterfowl have long been overhunted, and their number has decreased by almost tenfold (Degtyarev 2007).

References

Abolin RI. Geobotanicheskoy i pochvennoye opisaniye Leno-Viluyskoy ravniny. In: Trudyi komissii po izucheniyu Yakutskoy ASSR, Tom 10. Leningrad: Isdatelstvo AN SSSR; 1929 (In Russian). – Geobotanical and soil description of the Lena-Viluy lowland. In: Proceedings of Commission for Study of Yakutian ASSR, vol. 10. Leningrad: AS of USSR Publications; 1929.

Borisov AA. Klimatyi SSSR. Moskva: Prosvyascthenie; 1967 (In Russian). – Climates of the USSR. Moscow: Prosvyascthenie Publ; 1967.

Degtyarev VG. Waterbirds in conditions of cryoaridic lowland. Novosibirsk: Nauka; 2007 (In Russian).

Degtyarev VG, Larionov AG, Antonov AK. Isolated boreal populations of Temminck's Stint and eastern Broad-billed Sandpiper in central East Siberia. Wader Study Group Bull. 2006;110:30–5.

Dobrovolsky GV, et al., editors. Krasnaya kniga pochv Rossii: Obiektyi Krasnoy knigi i kadastra osobo tsennyich pochv. Moskva: MARS Press; 2009 (In Russian). – Red Data Book of soils of Russia: objects of Red Data Book and especially protected soils of land-survey. Moscow: MARS Press Publications; 2009.

Gavrilova MK. Klimat Tsentralnoi Yakutii. Yakutsk: Yakutskoe knizhnoe isdatelstvo; 1972 (In Russian). – Climate of Central Yakutia. Yakutsk: Yakutian book Publ; 1972.

Solomonov NG, Lukina LM, editors. Dar planete Zemlya. Osobo ochranyaemyie prirodnyie territorii Respubliki Saha (Yakutia). Yakutsk; 2009 (In Russian). – Gift to the Earth. Protected areas of Saha Republic (Yakutia). Yakutsk; 2009.

Troeva EI, et al., editors. The far North: plant biodiversity and ecology of Yakutia, Plant and vegetation, vol. 3. Dordrecht: Springer; 2010.

Zhirkov II. Morfogeneticheskaya klassifikatsiya kak osnova razionalnogo ispolsovaniya, ochranyi i vosproizvodstva prirodnyih resursov ozer kriolitozonyi (na primere Tsentralnoy Yakutii). In: Zhirkov II, editor. Voprosyi razionalnogo ispolsovaniya i ochranyi prirodnyih resursov raznotipnyih ozer kriolitozonyi (na primere Tsentralnoy Yakutii). Yakutsk; 1982 (In Russian). – Morphogenetic classification as base of rational utilization, protection and reproduction natural resources of permafrost lakes (with an example of Central Yakutia). In: Zhirkov II editor. Issues of rational utilization and protection of natural resources of different types of permafrost lakes (with an example of Central Yakutia). Yakutsk; 1982.

The Middle Aldan River Basin: A Key Migration Corridor for the Eastern Population of the Siberian Crane Within the Lena River Basin (Russia)

121

Victor Degtyarev

Contents

Introduction	1472
Hydrology	1472
Wetland Ecosystems	1473
Biodiversity	1474
Conservation Status	1475
Ecosystem Services	1475
Threats and Future Challenges	1476
References	1476

Abstract

Spanning a submontane area of the middle Aldan River within the taiga zone lies a migration corridor for the eastern population of the Siberian crane which is ranked as a flagship species, for which its conservation requires protection of its habitats in all their diversity. The river system is formed by the middle reaches of the Aldan (one of the Lena's largest tributaries), the lower reaches of its large tributary the Maya, and numerous middle-sized and minor tributaries. The Aldan and Maya are characterized as lowland and semi-mountainous rivers, respectively. Waterways in the interfluve of the Aldan and Amga rivers are mostly of the lowland type, semi-mountain type in the interfluve of the Aldan and Maya rivers, and semi-mountain and mountain types in the right bank ributaries of the Aldan. The Middle Aldan Basin forms a significant portion of the Siberian crane and Baikal teal migration routes and supports populations of mergansers, common goldeneye, Middendorff's bean goose, falcated duck, osprey, and white- tailed eagle and a significant number

V. Degtyarev (✉)
Siberian Division of Russian Academy of Sciences, Institute for Biological Problems of Cryolithozone, Siberian Branch, Russian Academy of Sciences, Yakutsk, Russia
e-mail: dvgarea@yandex.ru

© Springer Science+Business Media B.V., part of Springer Nature 2018
C. M. Finlayson et al. (eds.), *The Wetland Book*,
https://doi.org/10.1007/978-94-007-4001-3_138

of fish species including stocks of Siberian sturgeon, tugun, humpback whitefish, and taimen. Four regional nature reserves are located within the area where one can observe migration and stopovers of the Siberian crane and Baikal teal. Currently, major threats are coincident with the rapid growth in accessibility of highly efficient hunting, fishing tackle, and transportation equipment.

Keywords
critically endangered species · flagship species · *Grus leucogeranus* · Siberia

Introduction

Spanning a submontane area of the middle Aldan River within the taiga zone (approximately between 58 and 63°N, Fig. 1) lies a migration corridor for the eastern population of the Siberian crane *Grus leucogeranus* on passage to and from their Arctic breeding grounds. Having passed over the sparsely populated Verchoyanskiy ridge (during autumn migration) or the Aldanskoe high plateau (during spring migration), the population passes over the Aldan and Amga and Aldan and Maya interfluves (i.e., an area of higher ground between two rivers in the same drainage basin). Residents in the scattered dozen or so settlements along the Aldan regularly observe the Siberian crane flocks, enjoying a rare opportunity of contemplating a species of great cultural significance as well as top conservation concern. The first data on the mass migration of the Siberian crane over the middle Aldan were obtained in the 1980s (Degtyarev and Antonov 1989), and, in the mid-1990s, satellite tracking revealed the migration route and its duration (Kanai et al. 2002). Since 2006, the Institute for Biological Problems of Cryolithozone of the Siberian Branch of the Russian Academy of Sciences has studied Siberian crane habitats and conducted regular observations of the migration within the framework of the UNEP/GEF Siberian Crane Wetland Project and the Russian Foundation for Basic Research projects (No. 11–04–00130 and 12–04–10009).

The Siberian crane is a large (2 m wingspan, 6–8 kg weight), smooth-flying (Fig. 2), musical white bird totaling no more than 4,000 individuals that is listed as a critically endangered species (IUCN 2011). In accordance with its attractive appearance, cultural significance, behavior, ecological peculiarities, population dynamics and distribution, as well as its critically endangered conservation status, the Siberian crane is ranked as a flagship species, for which its conservation requires protection of its habitats in all their diversity (Prentice et al. 2006).

Hydrology

The river system is formed by the middle reaches of the Aldan (one of the Lena's largest tributaries), the lower reaches of its large tributary the Maya, and numerous middle-sized and minor tributaries. The Aldan and Maya are characterized as lowland and semi-mountainous rivers, respectively. Waterways in the interfluve of

Fig. 1 Siberian cranes (Photo credit: V. Degtyarev © Rights remain with the author)

the Aldan and Amga rivers are mostly of the lowland type, semi-mountain type in the interfluve of the Aldan and Maya rivers, and semi-mountain and mountain types in the right bank (eastern) tributaries of the Aldan.

The main water source of the rivers is meteoric water; therefore, the most active period in the hydrological regime is spring and summer when three to four notable water peaks in the mainstream are observable. The water in the Aldan is slightly mineralized. The oxygen content is satisfactory and contamination is weak or moderate. Freezing-over occurs in late October and drifting of ice in early to mid-May. In large rivers, ice jams during spring breakup regularly occur resulting in rapid water rise, which threatens settlements and brings about bank overflows that supply wetlands positioned on the higher parts of the floodplain (State report on environment condition and protection 2008).

Wetland Ecosystems

Large rivers have mostly vast terraced valleys with huge floodplains containing numerous small- and medium-sized lakes (mostly of erosion origin), **mari** (see Box in ▶ Chap. 117, "Lena River Basin (Russia)"), bogs, marshes, and riverine sandbanks. Tributaries of large rivers often have floodplains which are in most cases largely vegetated by tussocky meadows and mari and sparse lakes. The interfluves are well drained; therefore, wetlands are absent. Lake systems of glacial origin are located in the **sandur** area of the Verchoyanskiy ridge foothills, interfluves, or sources of minor rivers, among vast maris (Degtyarev et al. 2011). A **sandur** is a submontane outwash plain formed during glacier melting (Spiridonov 1980; Ritter et al. 1995).

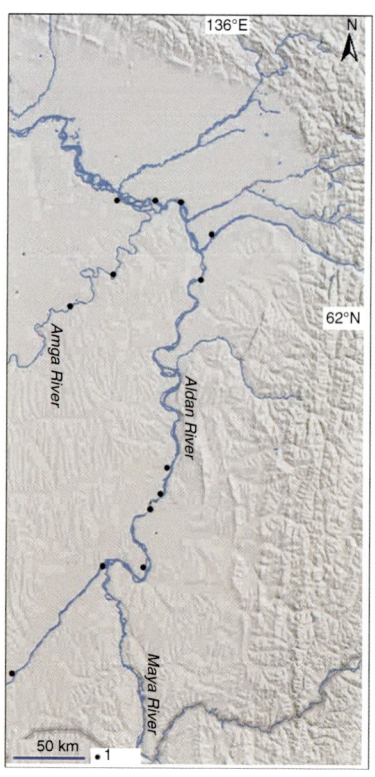

Fig. 2 Middle Aldan River and adjacent territories indicating villages

Biodiversity

The middle Aldan Basin is located at the junction of three floristic regions and is characterized by rich flora with only higher hydrophytes of about 1,500 species including 4 endemics (Troeva et al. 2010). Here, 27 fish species (including the Siberian sturgeon *Acipenser baerii* and 8 salmonids), 4 amphibian species, 1 reptile species, up to 170 bird species of birds, and 38 mammal species have been recorded.

Water-related vertebrates, besides fishes and amphibians, are the water vole *Arvicola terrestris*, muskrat *Ondatra zibethicus*, American mink *Mustela vison*, common otter *Lutra lutra*, and waterbirds (42–46 breeding and 28–35 migrating species). The migration route of the Siberian crane in this area coincides with that of the Baikal teal *Anas formosa*. The nesting sites of osprey *Pandion haliaetus*, white-tailed eagle *Haliaeetus albicilla*, Middendorff's bean goose *Anser fabalis middendorffii*, falcated duck *Anas falcata*, hooded crane *Grus monacha*, and Far Eastern curlew *Numenius madagascariensis* occur in the same areas where the Siberian crane makes its spring stopovers (Fig. 3). Since 2007, pairs of mandarin duck *Aix galericulata* have been observed occasionally (most probably casual visitors).

Fig. 3 Siberian crane flock on spring stopover (Photo credit: V. Degtyarev © Rights remain with the author)

Conservation Status

Four regional nature reserves, "Chabda" (853,736 ha) and "Kyupsky" (1,761,378 ha), "Koluma" (251,500 ha), "Kuoluma-Chappanda" (580,133 ha), and "Tomporuk" (285,600 ha) (Solomonov and Lukina 2009), are located within the area under study. The Chabda and Koluma reserves contain parts of wetlands in the interfluves of the Aldan/Amga and the Aldan/Maya rivers were included in the Ramsar Shadow List (Krivenko 2000). In all of the above mentioned nature reserves, one can observe migration and stopovers of the Siberian crane and Baikal teal; while in Chabda, there is breeding of hooded crane and osprey and breeding of Middendorff's bean goose in Tomporuk.

Ecosystem Services

The Middle Aldan Basin forms a significant portion of the Siberian crane and Baikal teal migration routes and supports populations of mergansers, common goldeneye *Bucephala clangula*, Middendorff's bean goose, falcated duck, osprey, and white-tailed eagle and a significant number of fish species including stocks of Siberian sturgeon, tugun *Coregonus tugun*, humpback whitefish *C. lavaretus*, and taimen *Hucho taimen*. Waterfowl, fish, and ungulates are an important component in the diet of local people. The Aldan floodplain is the primary area utilized by a poorly developed agriculture complex and for subsistence production.

The Aldan and Maya form a navigation route used mainly for local motorboat transportation. The rivers are navigable for limited-tonnage cargo vessels. A coal stowing cargo terminal is operating on the Aldan to ship steam coal from the Dzebariki-Haya coalfield.

Threats and Future Challenges

Subsistence fishing and hunting are having strong negative effects on the population status of game animals. Currently, major threats are coincident with the rapid growth in accessibility of highly efficient hunting, fishing tackle, and transportation equipment (multi-charge shotgun, various power multiform ammunition, satellite navigators, callers, decoys, high-performance net and hook tackle, high-speed motorboats and snowmobiles, water-jet propellers, cross-country vehicles) (Degtyarev 2007).

References

Degtyarev VG. Strategiya sohraneniya vodno-bolotnyih ugodii Yakutii. Yakutsk: Isdatelstvo YaNTs SO RAN; 2007 (In Russian). – Yakutia wetland conservation strategy. Yakutsk: YaSC, SB, RAS Publ.; 2007.

Degtyarev VG, Antonov AK. Sterh v yuzhnyih raionah Yakutii. In: Amirhanov AM editor. Redkie i nuzhdayuschiesya v ochrane zhivotnyie. Moskvwa; 1989 p. 68–69 (In Russian). –Siberian Crane in southern regions of Yakutia. In: Amirhanov AM editor. Rare animals and animals in need of protection. Moscow; 1989 p. 68–69.

Degtyarev VG et al. Preliminary assessment of the Hooded Crane status and wetlands structure along Siberian Crane eastern population flyway in the Middle Aldan River Basin. In: Ilyashenko E, Winter S editors. Cranes of Eurasia (biology, distribution, migrations, management). Issue 4. Moscow; 2011 (In Russian).

Gosudarstvennyii doklad o sosostoyanii i ohrane okruzhayuszhei sredyi Respubliki Saha (Yakutia) v 2007 godu. Yakutsk; 2008 (In Russian). – State report on environment conditions and protection in Sakha Republic (Yakutia) in 2007. Yakutsk; 2008.

IUCN 2011. IUCN red list of threatened species. Version 2011.2. www.iucnredlist.org. Downloaded 13 Mar 2012.

Kanai Y, et al. Migration routes and important resting areas of Siberian cranes (*Grus leucogeranus*) between northeastern Siberia and China as revealed by satellite tracking. Biol Conserv. 2002;106:339–46.

Krivenko VG editor. Wetlands in Russia. vol 3. Wetlands on the Ramsar shadow list. Moscow: Wetlands International Global Series No. 3; 2000 (In Russian).

Prentice C, Mirande C, Ilyashenko E, Harris J. Flyway site network development in Asia: wetlands conservation using the Siberian Crane *Grus leucogeranus* as a flagship species. In: Boere GS, Galbraith CA, Stroud DA, editors. Waterbirds around the world. Edinburgh: The Stationary Office; 2006.

Ritter DF, Kochel RC, Miller JR. Process geomorphology. Dubuque: William C. Brown; 1995.

Solomonov NG, Lukina LM editors. Dar planete Zemlya. Osobo ochranyaemyie prirodnyie territorii Respubliki Saha (Yakutia). Yakutsk; 2009 (In Russian). – Gift to the Earth. Protected areas of Sakha Republic (Yakutia). Yakutsk; 2009.

Spiridonov AI, editor. Chetyirehyazyichnyii entziklopedicheskii slovar terminiv po fizicheskoi geografii. Moskva: Izdatelstvo Sovetskai Enciklopedia; 1980. (In Russian). –Four-lingual encyclopaedic glossary of terms in physical geography. Moscow: Soviet Encyclopedia Publication; 1980.

Troeva EI et al. editor. The far north: plant biodiversity and ecology of Yakutia – plant and vegetation. vol. 3. Springer Dordrecht Heidelberg London New York: Springer Science + Business Media B. V. 2010.

Yenisei River Basin and Lake Baikal (Russia)

122

Nick C. Davidson

Contents

Introduction	1478
General Description	1479
The Yenisei River Basin	1479
Lake Baikal	1479
Biodiversity Importance	1480
The Yenisei River Basin	1480
Lake Baikal	1480
Conservation Status	1482
Human Uses	1482
Threats and Future Challenges	1484
References	1484

Abstract

Rising in northern Mongolia, the Yenisei (sometimes spelled Yenisey) River flows predominantly northwards through Russia into the Kara Sea part of the Arctic Ocean. The largest of the major Siberian rivers, its basin is the fifth largest in the world. A major and important feature of the upper part of the Yenisei River Basin is Lake Baikal, considered the deepest and oldest lake in the world, and a major centre of plant and animal endemism. Although in a remote part of the world, the basin is strongly affected by flow regulation and fragmentation from damming for particularly hydropower for the major industries along its banks exploiting the basin's rich mineral and forestry resources, and leading to the lower reaches of the river being amongst the most polluted of Arctic rivers.

N. C. Davidson (✉)
Institute for Land, Water and Society, Charles Sturt University, Albury, NSW, Australia

Nick Davidson Environmental, Wigmore, UK
e-mail: arenaria.interpres@gmail.com

© Springer Science+Business Media B.V., part of Springer Nature 2018
C. M. Finlayson et al. (eds.), *The Wetland Book*,
https://doi.org/10.1007/978-94-007-4001-3_276

Keywords
Yenisei · Lake Baikal · Arctic · Endemism · Dams · Pollution

Introduction

Rising in northern Mongolia, the Yenisei (sometimes spelled Yenisey) River then flows predominantly northwards into the Kara Sea part of the Arctic Ocean (Fig. 1). A major and important feature of the upper part of the Yenisei River Basin is Lake Baikal. Parts of the text of this chapter are derived from information provided in the World Wildlife Fund/The Nature Conservancy (WWF/TNC) *Freshwater Ecoregions of the World (FEOW)* descriptions of the Yenisei and Lake Baikal ecoregions (http://feow.org/).

Fig. 1 The Yenisei River Basin (Source: Wikimedia Commons. Use licensed under the Creative Commons Attribution-Share Alike 3.0 Unported license)

General Description

The Yenisei River Basin

Almost all (97%) of the Yenisei River Basin lies in northern Russia (Siberia). The Yenisei is the largest of the three major Siberian rivers that discharge into the Arctic Ocean (the others being the Lena to the east and the Ob to the west), with a maximum length (Selenge-Angara-Yenisei arms) of 5,940 km. At 2.58 million km^2, the Yenisei River Basin is the fifth largest in the world and drains a substantial part of central Siberia. Of this area, 1.04 million km^2 is in the (eastern) Angara sub-basin system.

The Yenisei River is often divided into three sections: the *upper Yenisei*, from the headwaters to the Tuba River confluence; the *middle Yenisei* from this confluence to the confluence with the Angara River; and the *lower Yenisei* from the Angara confluence to its discharge into the Arctic Ocean. Lake Baikal lies in the upper to middle reaches, with the Selenge River inflowing from its Mongolian headwaters and the outflow becoming the Angara River.

The Yenisei has an average depth of 14 m (maximum 24 m), and in its lower reaches (north of the inflow of the Nizhnyaya Tunguska River) it is a braided river channel between 2 and 5 km wide.

Given its northern latitude location and continental climatic influence, the Yenisei River Basin has great seasonal temperature variations. In its upper reaches summer temperatures can exceed 40° C; in the middle reaches summer temperatures are regularly above 30° C and winter temperatures below −30° C; and in its Arctic lower reaches winter temperatures can fall below −60° C. In winter its lower, northerly reaches are completely frozen, with rapid spring river icemelt, snowmelt, and rainfall leading to spring and summer flooding, and major silt deposition in the lower reaches.

Lake Baikal

Lake Baikal is the largest freshwater lake by volume (23,615.39 km^3) in Asia, the seventh largest in the world by surface area (31,722 km^2), and the deepest lake in the world (1,642 m). Lake Baikal is a long (636 km), narrow (79 km wide), crescent-shaped rift valley lake, surrounded by mountains, and whose waters are well mixed and well oxygenated throughout the water column. Among the lake's 27 islands, Olkhon (72 km long) is the third largest inland lake island in the world. The rift is active and widening, with hot springs in the surrounding area. Including some 7 km depth of sediment, the rift floor is between 8 and 11 km below the lake surface, making Lake Baikal the deepest continental rift valley in the world. At an estimated 25–30-million-years-old Lake Baikal is considered the oldest lake in the world. Very unusually for a high-latitude lake, it has not been scoured by continental ice sheets, so its undisturbed deep sediments are of major importance for their paleoecological and paleoclimatic records dating back millions of years.

Biodiversity Importance

The Freshwater Ecoregions of the World (FEOW) defines much of the Yenisei River Basin as a single ecoregion with its major habitat type being "polar freshwaters" (Bogutskaya and Hales 2015a). This ecoregion excludes Lake Baikal and lakes of the Khantayka River headwaters on the southwestern Putorana Plateau. Lake Baikal is treated as a separate freshwater ecoregion (major habitat type "large lakes") (Bogutskaya and Hales 2015b).

Terrestrially, the Yenisei River Basin is covered by 11 of WWF's terrestrial ecoregions: Taimyr-Central Siberian tundra; Yamai-Gydan tundra; East Siberian taiga; West Siberian taiga; West Siberian hemiboreal forests; South Siberian forest steppe; Sayan montane conifer forests; Sayan Alpine meadows and tundra; Trans-Baikal conifer forests; Selenge-Orkhon forest steppe; and Daurian forest steppe (see Olson et al. 2001). The upper reaches of the system are mountainous and the rivers flow through arctic-alpine and dry forested steppe ecosystems. Further downstream, the Yenisei rivers flow through boreal forests and extensive "taiga" ecosystems, the latter dominated by larch *Larix* forests. This whole region is considered as one of the centers of endemism in Siberia. Further north into the Arctic, in its lower reaches the river flows through extensive tundra ecosystems.

The Yenisei River Basin

The Taimyr reindeer *Rangifer tarandus sibiricus* herd migrates to tundra areas along the Yenisei River in winter. It is the largest reindeer herd in the world, with a population of 700,000 – one million animals during the 2000s – a major population increase since the 1950s.

The tundra systems surrounding the lower reaches of the Yenisei support major summer breeding populations of Arctic-breeding waterbirds including ducks, geese, swans (Anatidae), and shorebirds (Charadrii). Important breeding species include: red-breasted goose *Branta ruficollis*, white-fronted goose *Anser albifrons*, lesser white-fronted Goose *A. erythropus*, bean goose *A. fabalis*, and Bewick's swan *Cygnus columbianus bewickii*.

Excluding Lake Baikal and other headwater lakes, the basin supports 55 native fish species. Most are widespread Siberian species, but two are endemic to the basin: a gobionine cyprinid *Gobio sibiricus* and a grayling *Thymallus nigrescens*. Other notable species include Siberian sturgeon *Acipenser baerii*, sterlet sturgeon *A. ruthenus*, tench *Tinca tinca*, and Arctic flounder *Liopsetta glacialis*.

Lake Baikal

Lake Baikal is one of the most biodiverse lakes in the world. In part because of its great age and relative geographical isolation, Lake Baikal supports rich plant (at least 1,000 species) and animal (at least 2,500 species) diversity, many of which are

Fig. 2 Siberian sturgeon *Accipenser baerii* is native to the Yenisei River (© *Citron/CC-BY-SA-3.0*, licensed under the Creative Commons Attribution-Share Alike 3.0 Unported license)

endemic to the lake. Over 26% of plant species are endemic, as are 80% of the Lake's animal species, including the Baikal seal *Pusa sibirica*. This is one of only three seal populations which live wholly in freshwater systems.

Lake Baikal, especially the delta areas of inflowing rivers particularly the Selenge River delta, is internationally important for its breeding, molting, and migrating waterbird populations. In and around the Selenge River delta alone, the total breeding population of Anatidae (ducks, geese, swans) varies annually from 20,000 to 138,000 individuals, and the number of fledged young from 23,000 to 175,000 individuals. There is a substantial breeding population of the globally near-threatened Asian dowitcher *Limnodromus semipalmatus* as well as important breeding concentrations of northern lapwings *Vanellus vanellus* and marsh sandpipers *Tringa stagnatilis*. In autumn, the molting Anatidae population includes, in addition to locally breeding ducks, eight to ten thousand males congregating from surrounding areas. Species include Eurasian wigeon *Anas penelope*, teal *A. crecca*, falcated duck *A. falcata*, and northern pintail *A. acuta*. The area is of major importance in autumn as a migratory staging area, with an estimated five million birds passing through the area, including ducks, Siberian crane *Grus leucogeranus*, black stork *Ciconia nigra*, Bewick's swan *Cygnus bewickii*, swan goose *Anser cygnoides*, and shorebirds, notably Pacific golden plover *Pluvialis fulva*, Temminck's stint *Calidris temminckii*, and Curlew sandpiper *C. ferruginea* (Ramsar Convention 1994a; Fefelov and Tupitsyn 2004).

Although there are less than 60 native fish species occurring in Lake Baikal, over half of these are endemic. All species of deepwater sculpins, Baikal oilfish (golomyankas) – including *Comephorus baicalensis* and *C. dybowskii* which together form the largest fish biomass – and Baikal sculpins are endemic, as is the Baikal sturgeon *Acipenser baerii baicalensis*. The Baikal sturgeon occurs mostly in the northern end of Lake Baikal and spawns upstream in the Selenge River (Fig. 2).

Lake Baikal supports very diverse communities of invertebrates, especially amphipods, freshwater snails, and worms, with very high levels of endemism. All of the more than 350 amphipod taxa are endemic, as are about 78% of freshwater snails and sponges and about 80% of oligochaete worms.

Conservation Status

Lake Baikal was declared a UNESCO World Heritage Site in 1996, in recognition of the lake being an outstanding example of a freshwater ecosystem, including for its outstanding variety of endemic flora and fauna, which is of exceptional value to evolutionary science. It is also surrounded by a system of protected areas that have high scenic and other natural values. The World Heritage Site includes the lake itself, its catchment basin, the headwaters of the Angara River, and the Irkutsk water reservoir. In 2010, the area of the Putorana Plateau covered by the Putorana Nature Reserve, to the east of the Yenisei River, was inscribed on the World Heritage List as "a complete set of subarctic and arctic ecosystems in an isolated mountain range, including pristine taiga, forest tundra, tundra, and arctic desert systems, as well as untouched cold-water lake and river systems."

There are two Wetlands of International Importance (Ramsar Sites) designated in the Yenisei River Basin, both designated in 1994. One is the *Selenga Delta*, the largest river inflow delta in Lake Baikal. The Ramsar Site covers 12,100 ha of the Delta and consists of the shallow water area of Lake Baikal, streams and oxbow lakes, reedbeds, regularly flooded sedge-grass meadows, and willow shrub (Ramsar Convention 1994a). The other Ramsar Site, *Brekhovsky Islands in the Yenisei estuary* (approx. 1,400,000 ha), is in the Arctic tundra lower reaches of the Yenisei River. It is an estuarine wetland complex, made up of a network of rivers, streams, lakes, islands, floodplains, and terraces supporting various types of tundra vegetation communities (Ramsar Convention 1994b).

In addition, there are a number of nationally designated Nature Reserves in the Yenisei River Basin, including Kabansky Nature Reserve (Selenge Delta, Lake Baikal), Sayano-Shushensky Nature Reserve (upper reaches), Stolby Nature Sanctuary (middle reaches), Putorana Nature Reserve (on the Kureika River tributary of the Yenisei, middle reaches); and the Putorana Nature Reserve (Arctic plateau east of the lower reaches of the Yenisei).

Human Uses

Although in a remote part of the world, the Yenisei River Basin is rich in natural resources, including oil, coal (the richest in Russia), metal ores, and timber, which have been extensively exploited from the twentieth century onwards, with the development from the 1950s onward of major industrial complexes along the river. The Yenisei River has provided the major transport route for extraction of these resources, although the extent of shipping transport has declined considerably in recent years.

The largest city on the Yenisei River is Krasnoyarsk which, with a population of just over one million, is the third largest city in Siberia. On the Arctic lower reaches

Fig. 3 Krasnoyarskaya hydroelectric dam, one of the largest dams on the Yenisei River (© Denis Belevich releases this work into the public domain, and grants anyone the right to use this work for any purpose, without any conditions, unless such conditions are required by law)

of the river Norilsk (population 175,000) is an industrial city with mining and ore smelting its major industries. Further upstream, Abakan (population 165,000) lies at the confluence of the Abakan and Yenisei Rivers. On the Angara River, Bratsk (population around 250,000) was developed in the 1950s with the construction of the Bratsk dam and hydroelectric plant. Further upstream, Irkutsk (population about 600,000) is one of the larger Siberian cities, with aircraft manufacture and aluminum smelting its main industries. Upstream of Lake Baikal, on the Selenge River, Ulan-Ude (population 400,000) is a major railway junction on the Trans-Siberian Railway.

The Yenisei River has been extensively dammed, particularly for generation of hydropower (Stuefer et al. 2011). There are at least 64 dams and reservoirs in the Angara-Yenisei basin, with a total surface area of 482 km^2. These include the two largest hydroelectric power stations in Russia – Sayano-Shushenskaya and Krasnoyarskaya (Fig. 3). The reservoirs upstream of the confluence of the Yenisei and Angara rivers were designed for the yearly redistribution of water.

Lake Baikal supports one of the most important fisheries in Russia, the catch being primarily endemic species, the most important being a whitefish: the omul *Coregonus migratorius*. Traditional activities in the more northerly Arctic parts of the river basin include fishing, reindeer herding, and hunting of Arctic foxes.

Threats and Future Challenges

The Yenisei River is classified as "strongly affected by flow regulation/fragmentation" and is considered the most affected by anthropogenic impacts of all rivers flowing into the Arctic Ocean (Stuefer et al. 2011). As a consequence of discharges since the 1950s from industrial and military activities along its banks, the Yenisei is among the most polluted of Arctic rivers, including with heavy metals (particularly nickel and copper), and also with radioactive sediments in the middle and lower reaches of the river, discharged from a plutonium enrichment plant.

Since the 1960s until its final closure in 2014 Lake Baikal has been polluted by waste discharges from the Baykalsk Pulp and Paper Mill, built on the lake shore. There are also concerns of pollution from the discharge of untreated waste from tourist facilities around the lake. Concerns over plans to construct a major oil pipeline passing close to the Lake Baikal shore have now led to its construction on an alternative route away from the lake.

References

Bogutskaya N, Hales J. Freshwater Ecoregions of the World (FEOW). 605: Yenesei. 2015a. http://www.feow.org/ecoregions/details/yenisei. Accessed 1 Mar 2016.

Bogutskaya N, Hales J. Freshwater Ecoregions of the World (FEOW). 606: Lake Baikal. 2015b. http://www.feow.org/ecoregions/details/lake_baikal. Accessed 1 Mar 2016.

Fefelov I, Tupitsyn I. Waders of the Selenga delta, Lake Baikal, eastern Siberia. Wader Study Group Bull. 2004;104:66–78.

Olson DM, Dinerstein E, Wikramanayake E, Burgess ND, Powell GVN, Underwood EC, D'Amico JA, Itoua I, Strand HE, Morrison JC, Loucks CJ, Allnutt TF, Ricketts TH, Kura Y, Lamoreux JF, Wettengel WW, Hedao P, Kassem KR. Terrestrial ecoregions of the world: a new map of life on Earth. Bioscience. 2001;51(11):933–8.

Ramsar Convention. Selenga delta. Information sheet on Ramsar Wetlands. 1994a. https://rsis.ramsar.org/RISapp/files/RISrep/RU682RIS.pdf. Accessed 29 Feb 2016.

Ramsar Convention. Brekhovsky Islands in the Yenisei estuary. Information sheet on Ramsar Wetlands. 1994b. https://rsis.ramsar.org/RISapp/files/RISrep/RU698RIS.pdf. Accessed 29 Feb 2016.

Stuefer S, Daqing Y, Shiklomanov A. Effect of streamflow regulation on mean annual discharge variability of the Yenisei River. In: *Cold region hydrology in a changing climate*. International Association of Hydrological Sciences (IAHS) Publication. 2011;346:27–32.

Amur-Heilong River Basin: Overview of Wetland Resources

Evgeny Egidarev, Eugene Simonov, and Yury Darman

Contents

Introduction	1486
Conservation	1487
Future Challenges	1496
References	1498

Abstract

The Amur River (Heilong River in Chinese) basin is one of the largest and most diverse aquatic systems in Asia, being divided into seven distinctive freshwater ecoregions, most of them transboundary. Wetlands other than rivers and lakes are valuable components of the Amur-Heilong ecosystem. Unfortunately, no consistent comparative information is available on wetland types and their distribution in the transboundary basin due to the different approaches to land inventories in three countries. This paper describes the composition and distribution of major wetlands and current issues in Amur River basin wetland protection.

Keywords

Transboundary · Ecoregion · River · Floodplain · Lake · Reservoir · Mire · Bog · Ecological network · Hydropower · Protected area

E. Egidarev (✉)
Pacific Geographical Institute FEB RAS \ WWF Russia Amur Branch, Vladivostok, Russia
e-mail: egidarev@yandex.ru

E. Simonov (✉)
Rivers without Boundaries Coalition, Dalian, China

Daursky NNR, Dalian, China
e-mail: Simonov@riverswithoutboundaries.org

Y. Darman (✉)
WWF Russia Amur Branch, Vladivostok, Russia
e-mail: ydarman@wwf.ru

© Springer Science+Business Media B.V., part of Springer Nature 2018
C. M. Finlayson et al. (eds.), *The Wetland Book*,
https://doi.org/10.1007/978-94-007-4001-3_7

Introduction

The Amur River (Heilong River in Chinese – Fig. 1) basin is one of the largest and most diverse aquatic systems in Asia, being divided into seven distinctive freshwater ecoregions, most of them transboundary (Abell et al. 2008; Simonov 2010).

Only a limited area of the wetlands of the Amur River Basin that meet the Ramsar Convention's criteria for Wetlands of International Importance has been designated under the Convention (see Table 1). At Ramsar COP X (2008), the World Wildlife Fund (WWF) proposed the Amur-Heilong Ramsar Regional Initiative for the Amur basin as an overarching mechanism for the preservation of Amur River wetlands and to protect the network of 20 globally important wetlands already listed (see Markina et al. 2008).

Egidarev and Simonov (2008) distinguished 10 broad wetland categories in their map of the "Wetlands of the Amur River Basin" which aggregated 15 types of wetlands in the Ramsar wetland classification system (letters in brackets correspond to Ramsar classifications: see Ramsar Convention Secretariat 2010: Annex 1):

1. Rivers, deltas, and creeks (permanent flowing water) (M, L)
2. Rivers, deltas, and creeks (seasonal flowing water) (N)
3. Permanent freshwater lakes, pools, and reservoirs (O)
4. Permanent freshwater mires, swamps, fens, marshes, and bogs (Tp, W, Xf, Xp, U)
5. Permanent brackish lakes (Q)
6. Saline, brackish, and alkaline wetlands permanent marshes (SP)
7. Seasonal-intermittent lakes/pools saline, brackish, or freshwater (P) (R)
8. Seasonal/regularly inundated areas/floodplains with intermittent marshes with herb vegetation (Ts)
9. Seasonal-intermittent brackish marshes – solonchaks (pale or gray soil type found in arid to subhumid, poorly drained conditions) and alkaline flats that seasonally get some water (SS)
10. Springs (Y)

Note: Category (5) was problematic since many lakes, especially in the western part of the basin, fluctuate during the climate cycle in terms of acreage and salinity. Dalai (Hulun) Lake, for example, varies between 500 and 2,300 km^2 and saline-alkaline lake to freshwater. Furthermore, the area of wetlands in all categories can naturally fluctuate dramatically.

Due to the ambiguity of some categories, wetlands have been further aggregated into three major types (i.e., large floodplains, water bodies, and bogs and mires) to report the total area of wetland and the proportion of each wetland category within each country's portion of the Amur River Basin (see Table 2 below).

Russia with 70% has the largest proportion of wetlands within the Amur River Basin. Russia is especially rich in bogs and mires due to huge expanses of sparsely forested swamps dominated by larch (*Larix sp.*) growing on permafrost. The relatively narrow floodplains of small tributaries are not included in the totals for floodplain area shown in Table 2.

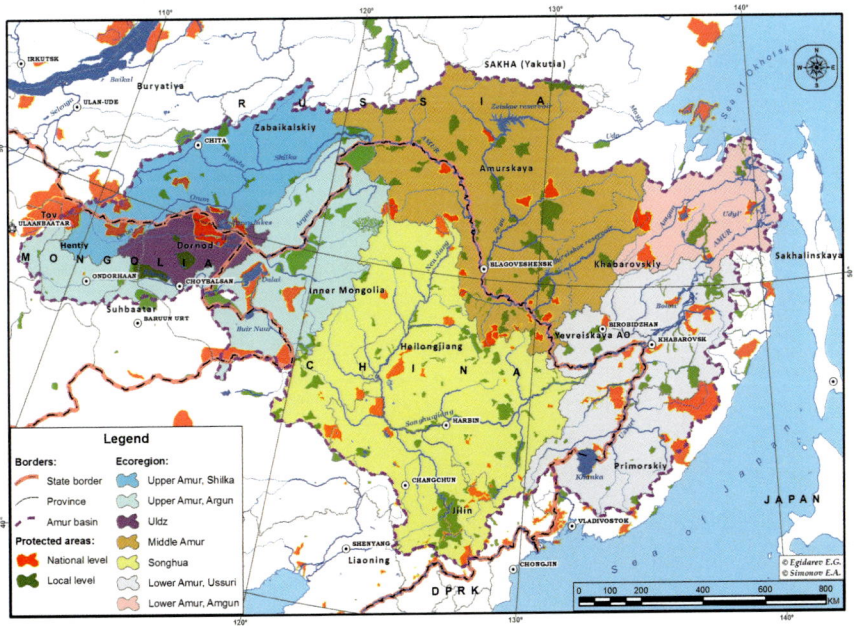

Fig. 1 Freshwater ecoregions of the Amur-Heilong River Basin divided by countries and protected areas network (E. Egidarev and E. Simonov © Rights remain with the authors)

Conservation

Although slightly more than 17% of the wetland area in the Amur River Basin is protected, the distribution of protected wetlands is very uneven and varies greatly between countries and ecoregions (Table 3). Within the borders of individual countries, China has the highest percentage of Amur River Basin wetlands in protected areas (30%), followed by Mongolia (14%) and Russia (12%). Moreover, in most ecoregions, China has a greater percentage of protected wetlands than the other countries of the basin. However, the gazetting of protected areas does not automatically ensure sufficient levels of protection. It estimated that more than half of the original wetland area has been converted to agriculture and other human uses in the Chinese part of the basin (Simonov and Dahmer 2008) – a greater percentage than all the wetlands currently found in protected areas.

In developing a conservation strategy for the Amur River Basin, it is necessary to identify important wetland regions, each possessing high natural values and unique wetland ecosystem features. This was attempted for the Russian Far East by Wetlands International (Bocharnikov 2005), and Northeast China (Li 2006). Based on a review of available literature and our own analysis, we suggest for the Amur River a basin-wide transboundary classification of major wetland-dominated regions and a brief list of conservation measures needed for each (see Table 4 and Fig. 5).

Table 1 Ramsar sites in the Amur-Heilong River Basin (see Fig. 2 for locations)

#	Name	Region	Area (ha)
Russia			
1	Torey Lakes	Zabaikalsky	172,500
2	Zeya-Bureya Plains	Amurskaya	31,600
3	Khingano-Arkharinskaya Lowland	Amurskaya	200,000
4	Bolon Lake and river estuaries	Khabarovskiy	53,800
5	Udyl Lake and river estuaries	Khabarovskiy	57,600
6	Khanka Lake	Primorskiy	310,000
China			
7	Xingkai Lake	Heilongjiang	222,488
8	Zhenbaodao	Heilongjiang	44,364
9	Dongfanghong	Heilongjiang	31,538
10	Qixing River	Heilongjiang	20,000
11	Honghe	Heilongjiang	21,836
12	San Jiang	Heilongjiang	164,400
13	Nanweng River	Heilongjiang	229,523
14	Zhalong	Heilongjiang	210,000
15	Momoge	Jilin	144,000
16	Xianghai	Jilin	105,467
17	Dalai Lake	Inner Mongolia	740,000
Mongolia			
18	Buir Lake and wetlands	Dornod	104,000
19	Mongol Daguur	Dornod	210,000
20	Khurkh-Khuiten Valley Lakes	Hentiy	42,940

Table 2 Major wetland types and forest cover of Amur-Heilong River Basin by country

Country	Total land area	Floodplain area	Water area	Bog area	Wetland area	Forests
Russia (ha)	100,888,242	3,215,520	2,258,117	22,945,485	25,455,755	71,559,739
% basin	49	63	58	73	70	64
China (ha)	88,649,511	1,781,040	1,367,755	7,723,268	9,845,716	37,870,055
% basin	42	35	35	25	27	34
Mongolia (ha)	18,630,699	58,911	204,238	636,801	880,563	1,589,816
% basin	9	1	5	2	2	1
Amur River Basin Total	208,168,452	5,055,471	3,830,110	31,305,554	36,182,034	111,019,610

The list includes the most important wetland regions in each of the freshwater ecoregions delineated in the Amur River Basin. Out of 10 major wetland regions only one, Song-Nen Plain, is confined to one country, all the rest cross national borders and require bilateral or trilateral cooperation to achieve long-term protection.

Table 3 Distribution of wetlands[a] in freshwater ecoregions and protected areas by 2011 (PAs)

Ecoregion	Country	Wetland area inside PA (ha)	Wetland area in ecoregion (ha)	Ecoregion area (ha)	% of Wetland in PA	Ecoregion % wetland
Uldz	Russia	86,694	112,921	955,097	77	12
	Mongolia	77,512	481,372	5,723,895	16	8
	China	0	12,145	374,144	0	3
Lower Amur/Ussuri	China	1,093,797	2,748,200	7,271,871	40	38
	Russia	861,763	7,183,167	26,386,745	12	27
Argun	China	545,779	1,421,076	15,143,185	38	9
	Mongolia	25,165	313,188	10,053,258	8	3
	Russia	1,881	236,554	4,901,258	1	5
Songhua	China	1,091,052	4,464,289	55,336,371	24	8
Shilka	Mongolia	21,007	86,003	2,853,546	24	3
	Russia	60,533	1,071,562	17,387,139	6	6
Middle Amur	China	253,781	1,200,007	10,523,939	21	11
	Russia	1,670,621	13,229,242	38,570,418	13	34
Lower Amur/Amgun	Russia	495,225	3,622,309	12,687,274	14	29
Amur River Basin total		**6,284,811**	**36,182,035**	**208,168,142**	**17**	**17**

[a]Wetlands are taken as the cumulative overlap of lakes and rivers, river floodplains, and bogs taken from 1:500,000 digital topographical maps and satellite images

The regions are biased towards large floodplains and lakes, paying less attention to ecosystems of headwaters. This is determined by the nature of headwater ecosystems, which are dispersed over large relatively similar areas without any concentration of wetland features. The only exception is delineation of "Mountain tributaries and larch swamps of Small Hingan Mountains" although it must be noted that many similar larch swamps are dispersed in the Russian part of the basin (see Fig. 2).

The Lower Amur-Ussuri freshwater ecoregion includes four important wetland regions, the Argun and Middle Amur freshwater ecoregions have two wetland regions each, while all other freshwater ecoregions are represented by a single wetland region. Originally, one more distinctive wetland region existed in the Zeya River headwaters in the Middle Amur freshwater ecoregion, but it was largely destroyed by the creation of a giant reservoir by Zeya Hydro and lost most of its natural character (see Figs. 3 and 4).

A large-scale ecological network for the conservation of wetlands of the Amur River Basin should consist of these 10 core areas, linked by floodplain/river corridors to ensure connectivity and supplemented by specific general rules of integrated river basin management. Rules may slightly vary in different freshwater ecoregions to reflect varying natural conditions across the basin (Fig. 5).

Table 4 Major wetland-dominated regions of the Amur-Heilong River Basin (Adapted from Simonov and Dahmer 2008 and updated)

Area (freshwater/terrestrial ecoregion)	Major PA status in Russia	Major PA status in China and Mongolia	Ramsar Sites listing	Conservation management action required
1. Lake Khanka, Songhacha River (Lower Amur-Ussuri/Suifen-Khanka ecoregion)	Khankaisky Zapovednik (NNR), MAB reserve	Xingkaihu NNR, MAB reserve	Listed in both countries, in Russia Ramsar area much larger than PA	Listing of International Ramsar Site and joint management, extension of PA along Songacha River. New agreement on fishing needed
2. Ussuri/Wusuli midflow (Lower Amur-Ussuri/Suifen-Khanka ecoregion)	Sredneussuriisky NR	Hutou NR, Dajiahe NR (south), Zhenbaodao NR, Dongfanghong wetland NR	Eligible but not listed	Protection of Russian river bank, coordinated management across agencies and borders
3. Sanjiang-Middle Amur (Lower Amur-Ussuri/Middle Amur meadows ecoregion-east section)	Amur mainstream: Zabelovsky NR (managed by Bastak Zapovednik NNR), Bolonsky Zapovednik NNR. Ussuri right bank-complementary wetlands. No sizable wetland PAs. Only Bolshekhehtsirsky (NNR) has small wetland	Dajiahe NR (north), Naolihe NR, Wusulijiang NR, Sanjiang NNR, Heixiazidao NR, Honghe NNR, Qindeli NR, Bachadao NNR	Sanjiang, Bolonsky, and Honghe listed, all the rest highly eligible, especially in Russia	Expanded protection of Russian river bank and islands, coordinated management across agencies and borders. Environmental flow norms have to be introduced into HPP reservoir management upstream. Enhanced pollution control. Ecosystem-based flood risk management needed. Developing international PA at Heixiazi/Tarabarovy Islands

(*continued*)

Table 4 (continued)

Area (freshwater/terrestrial ecoregion)	Major PA status in Russia	Major PA status in China and Mongolia	Ramsar Sites listing	Conservation management action required
4. Mountain tributaries and larch swamps of Small Hingan Mountains (Middle Amur/Manchurian mixed forest ecoregion)	Similar ecosystems also abundant in Russia, further to the north. Closest in Bastak Zapovednik in Evreiskaya Province. In Russia contains most remote salmon runs in Amur basin	Da Zhanhe NNR, Xinqing NR, Wuyilin NR	Largest best preserved wetlands of the China, part of Amur basin but no Ramsar Sites in China or Russia	Needs establishment of complex transboundary econet for forest and wetland conservation around Hinggan Gorge. Strengthening protection of mountain wetlands in China's Lesser Hingan/prohibition of gold mining in salmon rivers
5. Zeya-Bureya Plain wetlands (Middle Amur/Middle Amur meadows ecoregion-west section)	Khingansky Zapovednik (NNR), Ganukan NR, Muravievsky NR, Amurskii NR, Tom'-Tashina interfluvial protected wetland (several NR)	No sizable wetlands in China, but for several islands in Amur main channel	Two Ramsar Sites in Russia	Joint measures needed to sustain ecological flow in Amur-Heilong main channel. Environmental flow norms have to be introduced into HPP reservoir management. Ecosystem-based flood risk management needed
6. Lower Amur wetland plains and lakes (Lower Amur-Amgun/Lower Amur Valley ecoregion)	Udyl Lake NR, Mukhtel NR, Orel/Chlya Lakes, Kizi Lakes, Amgun River mouth, Evoron-Chukchagirskaya Lowland; five wildlife refuges	Confined to Russia, but most of migratory fish and waterbirds in China reserves depend on these habitats	Udyl Lake is Ramsar Site. Little percent of waterbody and floodplain protected by PAs	Establish coherent ecological network to safeguard this ecoregion, develop specific measures to protect, and

(*continued*)

Table 4 (continued)

Area (freshwater/ terrestrial ecoregion)	Major PA status in Russia	Major PA status in China and Mongolia	Ramsar Sites listing	Conservation management action required
				restore waterbirds and fish populations. Enhanced pollution control upstream
7. Argun/ Erguna River midflow (Argun/Dauria Steppe Global 200 ecoregion)	No PAs	Erguna Wetland NR, Erka NR, Huliyetu NR, no protection on the ground so far. Hui River National Nature Reserve	Highly eligible. No protected areas so far, planning underway	Establishment of continuous NR network along Argun River on both sides under DIPA (Dauria International Protected Area). Environmental flow norms have to be introduced into transboundary water management agreement. Enhanced pollution control upstream. Amendments to border agreements to prevent dyke building
8. Great lakes and wetlands of Dauria Steppe (Argun, Uldz/ Dauria Steppe Global 200 ecoregion)	Daursky Zapovednik (NNR) - Torey Lakes	Dalainur NNR: Hulun (Dalainur) and Beier (Buirnur) Lakes, Huihe NNR. Mongol Daguur Strict Reserve, no other PAs in Mongolia	Dalaihu and Daursky – Ramsar Sites, Huihe NR also highly eligible. Buir Lake listed but not included in PA in Mongolia	Develop Sino-Russia-Mongolian international Ramsar site and MAB reserve. Link Daursky and Dalai NR with Argun Midflow thus securing critical habitat

(*continued*)

Table 4 (continued)

Area (freshwater/terrestrial ecoregion)	Major PA status in Russia	Major PA status in China and Mongolia	Ramsar Sites listing	Conservation management action required
				linkages. Upgrade protection of Huihe NNR and Mongolian wetlands in Khalkh, Kherlen river basins. Prevent mining in river valleys
9. Wetlands of Song-Nen Plain (Songhua River/Nen River grasslands ecoregion)	Exclusively in China, linked to Russia by migration of cranes and other waterbirds	Zhalong NNR, Longfeng NR, Xianghai NNR, Tumuji NNR, Ke'erchin NNR, Momoge NNR	Haling and Xianghai are Ramsar Sites, while other areas are not	Restore wetland hydrology severely affected by water withdrawals and agricultural encroachment. Ecosystem-based flood risk management needed
10. Onon River and adjacent wetlands (Shilka/Dauria forest-steppe ecoregion)	Russia has only Aginskaya Steppe, Gurney Steppe, and Tassuchei NR nearby. This stretch of Onon still has sturgeon population. Area important for *Grus vipio* and *Otis Tarda*	Onon-Baldj National Park protects river valleys. Taimen is important species – base of international tourism	Khurh-Khuiten listed but not included in PA in Mongolia	Expansion of Daursky Biosphere Reserve and Sokhondo Biosphere Reserve into Onon River Valley. Upgrade protection of Mongolian wetlands Khurh-Khuiten. Prevent mining in river valleys

Acronyms used in the table: *NNR* National Nature Reserve, *NR* nature reserve, *MAB* Man and Biosphere Programme, *HPP* hydro power plant

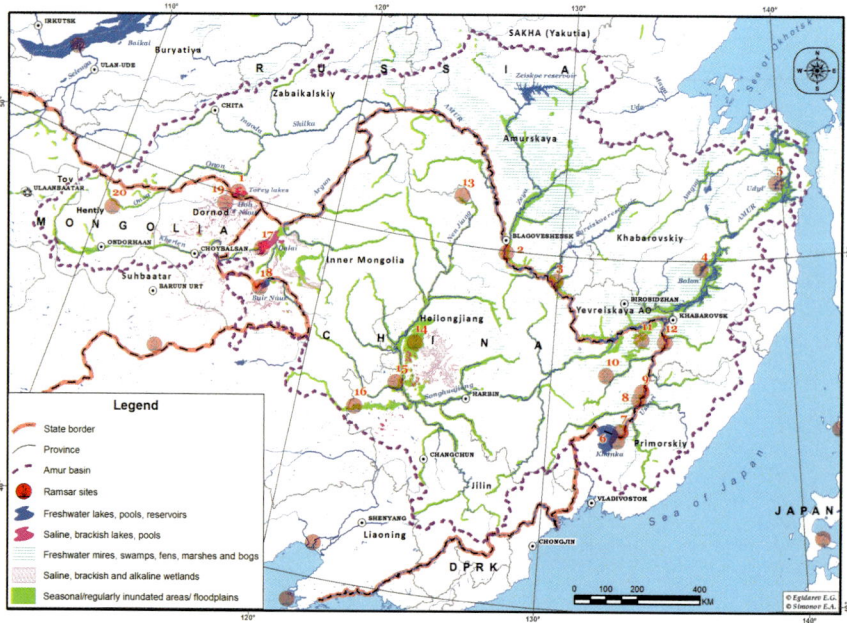

Fig. 2 Wetlands of the Amur-Heilong River Basin, refer to Table 1 for names of Ramsar Sites (E. Egidarev and E. Simonov © Rights remain with the author)

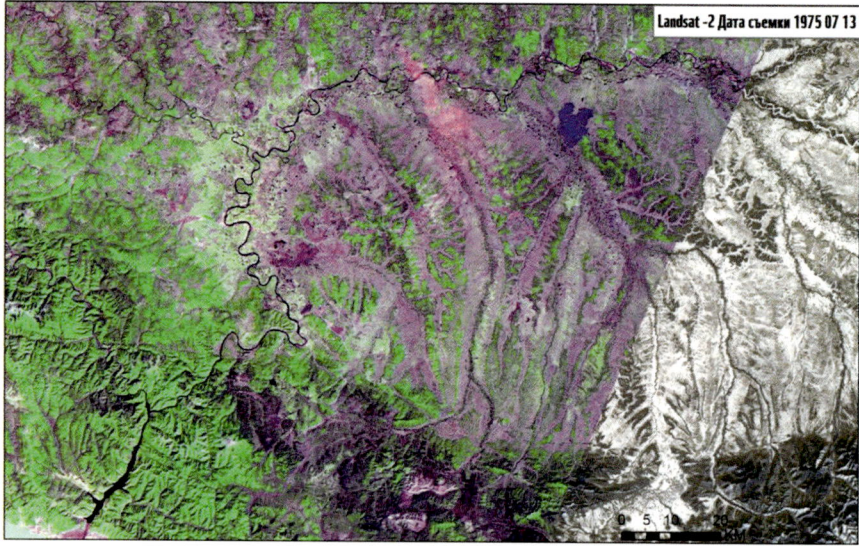

Fig. 3 Wetlands of the Zeya River headwaters before inundation (1975) (Landsat imagery courtesy of the US Geological Survey)

Fig. 4 Zeya Hydro Reservoir destroyed most natural wetlands of the Zeya headwaters (Landsat imagery courtesy of the US Geological Survey)

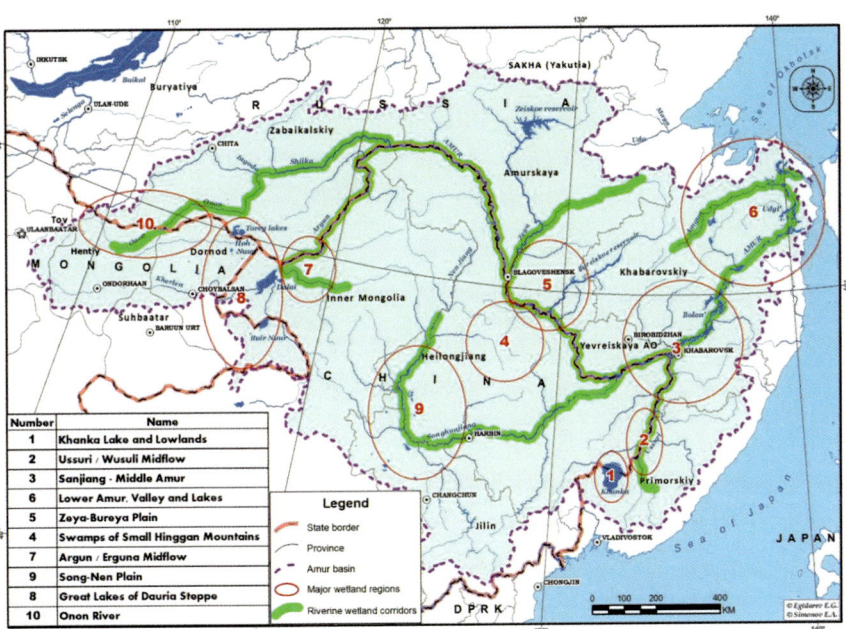

Fig. 5 Major wetland regions and wetland corridors of the Amur-Heilong River Basin (E. Egidarev and E. Simonov © Rights remain with the author)

Fig. 6 Construction of a new dyke along the transboundary Ussuri River in 2014 © Rivers without Boundaries Coalition

Future Challenges

On the transboundary Amur, wetland conservation and water management planning is burdened with numerous big differences between the three countries, their competition in protecting national borders and making use of resources, and cultural and organizational barriers hindering cooperation.

For example, for several decades China invested billions of yuan "to protect the motherland's banks with dykes" along the Amur, Argun, and Ussuri transboundary rivers without much coordination with Russia, which owns the opposing river banks. These embankments arrest the floodplain development process and reduce flooding, thus effectively degrading natural floodplain wetlands (Fig. 6).

During the 2013 large flood on the Amur River, disastrous dyke failures occurred in the counties Jiayin, Luobei, and Tongjiang, resulting in flooding of large populated areas on China's riverbank. Multiple dam failures along the Amur in 2013 prompted the emergence of new engineering projects. In June 2014, the Heilongjiang Province Water Department announced that 24.6 billion yuan ($4 billion) will be invested in construction and reinforcement of dykes (a total planned length of 2,722 km) on the major rivers of the Heilongjiang (Amur) River basin. By summer 2014 the work had already begun along the Amur and Ussuri rivers (Fig. 6), with old dykes being raised 2–5 m, with the declared objective of protecting land from floods occurring every 50–100 years. Once such embankments are built, it should be expected that up to 6,000 km^2 of natural floodplains will be isolated from the river and 25 km^3 of floodplains' natural flood retention capacity will be lost along the Chinese banks of transboundary rivers (Fig. 7). Presently this is the single most important threat to transboundary wetlands in the Amur River Basin (Simonov et al. 2015).

Additional and major threats/challenges to Amur River wetland ecosystems are listed below in decreasing order of importance (Sparling and Simonov 2014). (Note: The order reflects current Russian expert opinion and would be different if compiled by Chinese or Mongolian experts.)

- Hydropower development
- Lack of adaptive strategies for flood and drought-risk management
- Mining (especially placer gold mining)

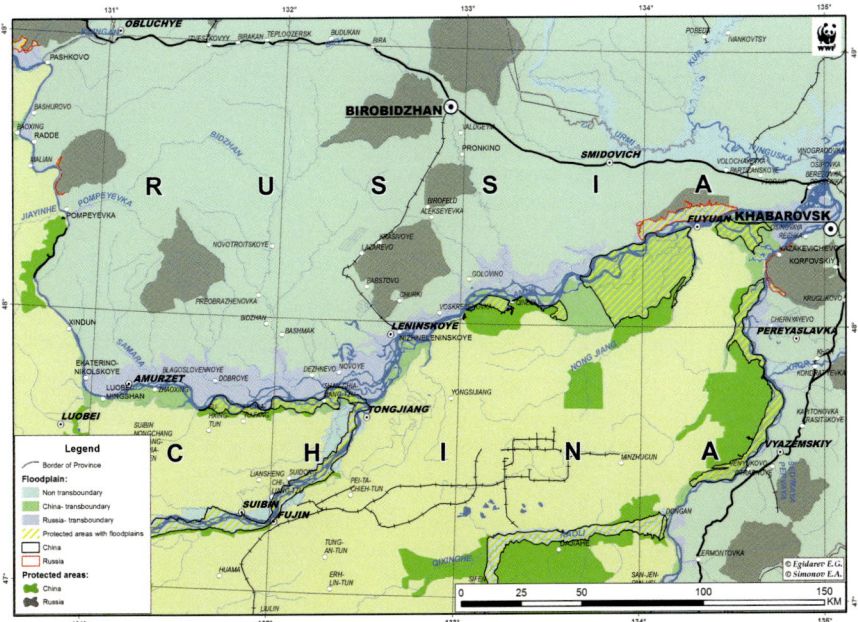

Fig. 7 Map of transboundary floodplains and nature reserves along Sanjiang Plain. Areas mapped as transboundary floodplain in China (including nature reserves) could be severely affected by antiflood dyke construction (E. Egidarev and E. Simonov © Rights remain with the author)

- Poaching, overfishing, and overhunting
- Industrial and municipal pollution
- Wildfires
- Expansion of intensive agriculture
- Competition for water
- Coal industry effects on water systems
- Logging in river valleys
- Cattle overgrazing in wetlands;
- Oil and gas extraction, processing, and transportation

On the positive, China and Russia in 2011 adopted the "Russian-Chinese Strategy for Development of Transboundary Network of Protected Areas in the Amur Basin for the period till 2020." Inventory and protection of wetlands is the highest priority in that strategy. The strategy has a special paragraph on the assessment of ecosystem services and could also be applied to protect natural floodplains in the scope of an integrated flood management strategy. It can also serve as the basis for improvement of existing collaboration between protected areas and conservation agencies and creation of new clusters of transboundary nature reserves at Ussuri/Wussuly Valley, Middle Amur/Sanjiang Plain, Hinggan Gorge, Korsakovski Gorge, and Cherpel Gorge in Amur River valley (Darman and Simonov 2015).

References

Abell R, Thieme ML, Revenga C, Bryer M, Kottelat M, Bogutskaya N, Coad B, et al. Freshwater ecoregions of the world: a new map of biogeographic units for freshwater biodiversity conservation. BioScience. 2008;58:403–14.

Bocharnikov VN, editor. Wetlands in Russia, volume 5. Wetlands in Southern Far-East Russia. Moscow: WWF and Wetlands International; 2005. p. 220.

Darman Y, Simonov E. Amur Heilong Ecoregion conservation program as a platform for Bio-Bridge initiative. 2015 Northeast Asia Peace and Cooperation Initiative Forum: New Horizons for Multilateral Cooperation in Northeast Asia, 2015 Oct 27–29, Seoul; 2015. p. 163–7.

Egidarev E, Simonov E. Wetlands of the Amur river basin. GIS maps of Amur-Heilong river basin. Vladivostok: WWF-Amur-Heilong Program; 2008. http://amur-heilong.net/http/gis_index.html.

Li XM. Wetlands of Heilongjiang basin and their protection. Monograph. Harbin: North East Forestry University Publishers; 2006. p. 254.

Markina A, Simonov E, Titova S, Minaeva T, Gafarov Y. Wetlands of the Amur river basin. Russia-China-Mongolia. Vladivostok: WWF Amur Branch; 2008. p. 20.

Ramsar Convention Secretariat. Designating Ramsar Sites: Strategic framework and guidance for the future development of the List of Wetlands in International Importance. Ramsar handbooks for the wise use of wetlands, 4th ed. vol. 17. Gland: Ramsar Convention Secretariat; 2010. p. 118.

Simonov EA. Protected areas network and transboundary conservation areas in Amur river Basin [dissertation]. Harbin: Northeast Forestry University; 2010.

Simonov EA, Dahmer T, editors. Amur-Heilong river basin reader. Hong Kong: Ecosystems Ltd; 2008. p. 426.

Simonov E, Nikitina O, Osipov P, Egidarev E. People and Amur floods: lesson (un)learned? Technical report submitted to UNECE Water Convention by the Pilot project on climate adaptation problems in Dauria and Amur River basins. 2015.

Sparling E, Simonov E, editors. Conservation investment strategy for the Russian Far East. San Francisco: Pacific Environment; 2014. p. 90.

Daurian Steppe Wetlands of the Amur-Heilong River Basin (Russia, China, and Mongolia)

124

Eugene Simonov, Oleg Goroshko, and Tatiana Tkachuk

Contents

Location and Climate	1500
Principal Wetlands of Eastern Dauria	1501
Biodiversity	1502
Ecosystem Services and Human Values	1503
Climate Change and Human Impacts	1503
Transboundary Wetland Management	1505
References	1507

Abstract

Dauria lies in the northern part of Central Asia and is ecologically strongly dependent on climate changes. Most of the Daurian steppe area is situated in Northeast China and East Mongolia; the Russian part is confined to Zabaikalsky Province and Buryat Republic. The area possesses a very high level of biodiversity for a steppe zone and is included in the Global 200 Ecoregions of the World as Dauria Steppe, which according to WWF covers the Nenjiang River grassland, the Daurian forest steppe, the Mongolian-Manchurian steppe, and the Selenge-Orkhon forest steppe ecoregions. These grassland areas are united by geographic

---

E. Simonov (✉)
Rivers without Boundaries Coalition, Dalian, China

Daursky NNR, Dalian, China
e-mail: esimonovster@gmail.com; Simonov@riverswithoutboundaries.org

O. Goroshko
Daursky Biosphere Reserve, Chita, Russia
e-mail: oleggoroshko@mail.ru

T. Tkachuk
Daursky Biosphere Reserve, Zabaikalsky State University, Chita, Russia
e-mail: tetkachuk@yandex.ru

© Springer Science+Business Media B.V., part of Springer Nature 2018
C. M. Finlayson et al. (eds.), *The Wetland Book*,
https://doi.org/10.1007/978-94-007-4001-3_170

location, annual and multiyear rhythms in ecological factors, and structure and composition of communities. The Daurian steppe's natural climate cycle with a span of 25–40 years is the major force shaping regional ecosystems and peoples' lifestyles. Each waterbody has its own drying and filling dynamics, depending on its depth, volume, hydrogeology, and location. In terms of freshwater ecosystems, Eastern Dauria is divided into three principal freshwater ecoregions, the Shilka River, the Argun River, and the endorheic basins of which the Torey Lakes/Uldz River basin is the most prominent. The principal wetlands of Eastern Dauria include the Argun River floodplain; the Hui and Moergol River floodplains; Dalai Lake and Ulan Lake; Buir Lake; and the Torey Lakes and Ulz River. The greatest management challenge is in ensuring proper water allocation to wetlands basin-wide. Despite the fact that bilateral agreements on transboundary waters exist between all three countries of the basin, they do not provide for wetland conservation, sustaining environmental flows, or adaptation to climate change. The lack of transboundary coordination in planning water use and regional development is leading toward drastic irreversible deterioration of Dauria's environment and loss of opportunities to adapt to a changing climate. Key measures required to reverse these negative trends are described.

Keywords

Dauria · Wetland dynamics · Steppe · Amur River basin · Migratory waterbirds

Location and Climate

Dauria lies in the northern part of Central Asia and is ecologically strongly dependent on climate changes. Most of the Daurian steppe area is situated in Northeast China and East Mongolia; the Russian part is confined to Zabaikalsky Province and Buryat Republic. The area possesses a very high level of biodiversity for a steppe zone and is included in the Global 200 Ecoregions of the World as *Dauria Steppe*, which according to WWF covers the Nenjiang River grassland, the Daurian forest steppe, the Mongolian-Manchurian steppe, and the Selenge-Orkhon forest steppe ecoregions. These grassland areas are united by geographic location, annual and multiyear rhythms in ecological factors, and structure and composition of communities (Kirilyuk et al. 2012). In terms of freshwater ecosystems, Dauria is divided into three principal freshwater ecoregions, Shilka River, Argun River, and Endorheic basins (Abell et al. 2008) of which the Torey Lakes/Uldz River basin is the most prominent (see Fig. 1).

Daurian steppe's natural climate cycle with a span of 25–40 years is the major force shaping regional ecosystems and peoples' lifestyles. Each waterbody has its own drying and filling dynamics, depending on its depth, volume, hydrogeology, and location. Giant Dalai Lake at maximum covers 2,300 km^2 but sometimes becomes a chain of shallow pools. "Pulsating" waterbodies provide higher but more uneven biological productivity than stable ones. The alternation of the wet

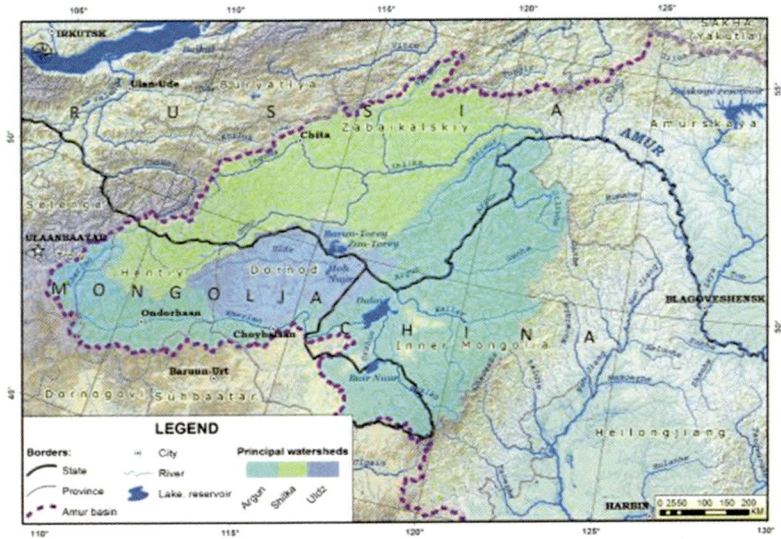

Fig. 1 Principal transboundary river basins of Eastern Dauria (E. Simonov, O. Gorpshoko, T. Tkachuk © Rights remain with the author)

and dry phases as well as the diversity in waterbodies creates a dynamic mosaic of habitats and triggers migration and changes in species populations. In 1999, the Torey Lakes yielded a thousand tons of fish annually, and in 2011, the meadow at Barun-Torey Lake bottom was a favorite pasture for Mongolian gazelle. River floodplains have much more frequent flooding events and thus preserve more stable habitat in times of drought. Once in 30 years during the most wet phase, thousands of ephemeral lakes scattered throughout the steppe provide the most productive habitat for birds and semiaquatic species. However, several of the most stable wetlands serve as life-support systems for wildlife and humans throughout all the phases of this climate cycle.

Principal Wetlands of Eastern Dauria

Argun River Floodplain. Argun (Erguna, Hailaer in Chinese) River is the largest watercourse in Daurian steppe, with a globally significant network of wetlands. For 940 km it serves as the Sino-Russian borderline, and the westernmost part has at least 200,000 ha of wide floodplain which is rich in biodiversity.

Hui and Moergol River Floodplains (approx 70,000 ha). These are small tributaries of the Argun River system in Hulunbeier Prefecture of Inner Mongolia in China, which possesses large floodplain wetlands with reedbeds known as breeding area for significant numbers of endangered cranes and geese.

Dalai Lake and Ulan Lake. Shallow Dalai (Hulun) Lake in Hulunbeier Prefecture receives waters of Kherlen and Wuershun rivers coming from Mongolia and is connected to the Argun River. The 750,000 ha Ramsar Site is a complex of lakes, rivers, marshes, shrublands, grasslands, and reedbeds typical of arid steppe wetlands, stretching north to south from the Russian to the Mongolian border. The lakes are important breeding, moulting, and stopover sites of waterbirds including endangered geese.

Buir Lake (61,500 ha) shared by Mongolia and China is fed by the Khalkh River, with its headwaters in China. This river forks at the 104,000 ha Ramsar Site in Mongolia and supplies water to Buir Lake and Dalai Lake via Wuershun River. The lake is an important breeding, moulting, and stopover site of waterbirds including endangered geese.

Torey Lakes and Ulz River. The endorheic Ulz River coming from Mongolia into Russia terminates into three large lakes of the Torey depression. The basin has many globally important biodiversity features and is protected by Daursky (172,500 ha) and Mongol Daguur (210,000 ha) Biosphere Reserves.

Biodiversity

Among vast arid steppe areas, the wetlands are nuclei of diversity especially for birds. Dauria's wetlands are globally important for the nesting of rare birds and huge numbers of migrating waterbirds. All sites are international Important Bird Areas (IBAs). Wetlands support globally significant breeding populations of many endangered bird species, including the red-crowned *Grus japonensis* and white-naped *G. vipio* cranes, swan goose *Anser cygnoides*, great bustard *Otis tarda*, and relict gull *Ichthyaetus relictus*, and are of international importance for the conservation of endangered migratory Siberian *G. leucogeranus* and hooded *G. monacha* cranes. More than 40 bird species registered here are listed both in the Red List of IUCN and the national Red Data Books of Russia, Mongolia, and China. The number of transitory migrants in the region's bird fauna is not less than 45%. Several million waterbirds pass through the wetlands in spring and autumn via the intracontinental branch of the East Asian-Australasian Flyway.

Dauria's wetlands have an amazing diversity of dynamic aquatic ecosystems, which have not been studied in depth yet. Buir Lake, the most species-rich lake in Mongolia, has 29 species of fish, including taimen *Hucho taimen*, lenok *Brachymystax lenok*, Amur grayling *Thymallus grubii*, Amur pike *Esox reichertii*, and Amur catfish *Silurus asotus*. The shore of Torey Lakes is a place of recovery for a herd of Mongolian gazelle *Procapra gutturosa*.

So far, just from Daursky Biosphere Reserve, 530 vascular plants, 50 mammals, 324 birds, three reptiles, three amphibians, and six fish species have been recorded.

The drought cycle dictates an unceasing succession in plant and animal communities, which increases the number of ecological niches and sustains a high diversity of species and habitats. Wildlife constantly moves between wetland sites. Therefore, the ecosystems' natural character dynamics must be preserved at many sites in Dauria to allow the long-term survival of regional biota.

Ecosystem Services and Human Values

These wetlands sustain the greatest fisheries in Dauria – for instance, at Dalai Lake the annual catch reaches 10,000 t in wet years. The upper Argun-Hailaer River is the source of water supply for southeast Zabaikalsky and western Hulunbeier. Farming communities depend on the Argun floodplain for cattle watering, pastures, hayfields, etc., most critically in dry years. The Khalkh River supports municipalities and irrigated agriculture in China and Mongolia. The grasslands around Dalai Lake's shores support two million livestock. The Uldz River provides water supply for cattle, mining, and settlements. In China, the Argun-Hailaer and Dalai Lake support numerous tourist camps and resorts. Altogether approximately two million people depend economically on the wetlands of the Argun and Ulz basins.

Main wetland ecosystem services include:

- Water retention in a semiarid region.
- Cyclical change in water levels sustains river floodplains and supports productivity and dynamic diversity of successional lake habitats.
- Important faunal refugia in times of drought.
- Important bird migration routes and stopover sites.
- High biological productivity, breeding areas for aquatic fauna and birds.
- Wetland groundwater recharge and discharge.
- Flood control, storm protection, flow regulation.
- Sediment retention and nutrient cycling, accumulation of organic matter.
- Climate regulation.

Cultural values of the wetland areas:

- Nomadic lifestyle of Mongolian tribes is the key cultural value and for centuries has been the most effective adaptation to climate fluctuations.
- Several regionally important sacred sites of the Buryats and Mongols attract a great number of pilgrims.
- Many areas in the Argun River basin are of cultural-historical interest, having been associated with Genghis Khan and his heirs.
- Buir Lake shores contain important memorials of Khalkhin-Gol Battle of 1939.
- Torey Lakes have a long history of research and monitoring since the eighteenth century.

Climate Change and Human Impacts

Within the last 55 years, Dauria's mean annual temperature has increased by two degrees C. In the past, during dry phases of the climatic cycle, populations of rare species have been especially vulnerable to human pressure. There will be more prolonged severe droughts within the natural cyclical pattern, resulting in low grass productivity, higher evaporation, and greater competition for remaining waterbodies between humans, cattle, and wildlife.

Mongolian nomadic tribes have been adapted to naturally occurring temporal and spatial changes in the availability of water, but modern rapid unsustainable development with stationary settlements and huge infrastructure is not. Traditional capacity for adaptation to climatic fluctuations is being rapidly decreased in the countries of Dauria, and risky projects like stabilizing the level of Dalai Lake by water transfers or massive tree planting in grasslands and wetlands are being presented as valid "adaptation to climate change." Poorly planned human activities initiated in anticipation of climate change may drastically hurt ecosystems much earlier and more severely than the consequences of actual global climate change.

The following factors have influenced wetlands for a long time:

- During drought periods, wetlands are impacted by concentrations of cattle to the extent that waterbirds may cease to breed (e.g., Ulz valley).
- Overgrazing is resulting in desertification in the area surrounding waterbodies (e.g., Dalai Lake, Argun valley).
- Uncontrolled wildfires have a negative impact on wetland biota (e.g., Argun River floodplain).
- Growing number of mining enterprises upstream from the sites' pumping water may affect water quality and quantity (e.g., small Ulz basin has 30 operations mining gold, tin, and fluorspar, and 70 more companies have been granted exploration licenses; Dalai BR has an oil field and multiple coal and copper mines in its immediate vicinity, etc.).
- Growing industrial and municipal sewage from sources in China (e.g., the Argun is considered the most polluted transboundary river in Russia).

Additional recently developed impacts on wetlands include:

- Massive water transfers: (1) Hailaer (Argun) River-Dalai Lake (in operation in China), (2) Kherlen River-Gobi Desert (planned in Mongolia), (3) Khalkh River-Xilingol (under EIA in China)
- Planned 1,000% water consumption increase from new reservoirs in Chinese tributaries, spurred by development of coal industry and agriculture
- Irrigation schemes along the Hailaer and Khalkh rivers (underway)
- Massive embankment construction along the Argun River in China and Russia (ongoing)

The cumulative impacts of the above activities may be enormous. For instance, several water infrastructure projects in China may reduce the upper Argun River flow by 50–60% and stop the flooding on which the well-being of riverine wetlands depends. And just one of those – the Hailaer-Dalai water transfer canal – may cause lake pollution, result in change in the ecological character of the lake by halting its dynamics, and provide grounds for large-scale industrial water supply to adjacent mines from this wetland of international importance designated under the Ramsar Convention.

Transboundary Wetland Management

Four out of seven sites are national-level protected areas (Table 1, Fig. 2). While all the major lakes of Dauria are Ramsar Sites and National Nature Reserves, only in Mongol Daguur and Hui River are there some protected floodplains. The Argun River floodplain is also nominally protected in China by local nature reserves. Since 1994, Daursky Biosphere Reserve in Russia, Dalai Lake in China, and Mongol Daguur in Mongolia have comprised Dauria International Protected Area (DIPA) with regular transboundary cooperation in research and environmental education. In 2006, the trilateral DIPA Committee approved a plan to expand DIPA reserves, including protection of the Argun River floodplain and Buir Lake.

Table 1 Principal wetlands of Argun and Ulz basins

	Name of transboundary wetland area	Countries sharing river/ wetland and location	Area and protection. International nominations
1	Argun River transboundary floodplains	Russia, Zabaikalsky Province, China, Inner Mongolia, Hulunbeier	200,000 ha, with 60% in China and 40% in Russia In China Erka, Huliyetu, and Ergunashidi nature reserves 2 IBAs in Russia and China IBA (not protected status)
2	Hui River floodplain	China, Inner Mongolia, Hulunbeier	Approx 50,000 ha, National Nature Reserve
3	Moergol River floodplain and Huh Lake	China, Inner Mongolia, Hulunbeier	Approx 10,000 ha
4	Dalai Lake National Nature Reserve	China, Inner Mongolia, Hulunbeier	750,000 ha. Covers Hulun (Dalai) Lake, Wuershun River valley, Ulan Lake and north shore of Buir Lake. Stretches north to south from vicinity of Russian border on Argun River to Mongolia border on Buir Lake. IBA, Ramsar Site
5	Lake Buir and its surrounding wetlands	Mongolia, Dornod Province	104,000 ha of which the lake covers 61,500 ha. Establishment of nature reserve planned by National Government. IBA, Ramsar Site
6	Torey Lakes, including "Daursky" National Nature Reserve	Russia, Zabaikalsky Province	172,500 ha Daursky Biosphere Reserve with 44,752 ha core area and 163,530 ha buffer zone. IBA, Ramsar Site
7	Mongol Daguur (Mongolian Dauria)	Mongolia, Dornod Province	210,000 ha territory of biosphere reserve in Uldz River valley and Barun-Torey lakeshore, IBA Adjacent Lake Huhnor also fed by Ulz River yet has to be protected by expansion of Biosphere Reserve. Ramsar Site

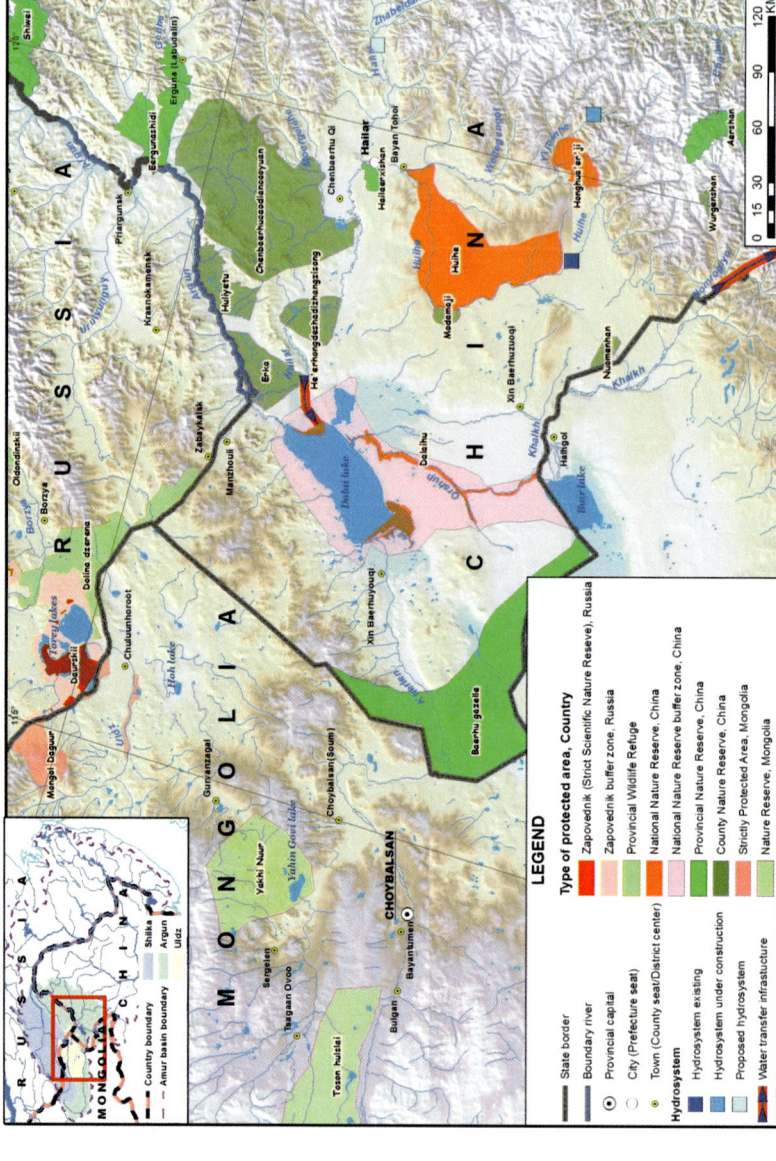

Fig. 2 Protected areas in Argun and Ulz River basins (E. Simonov, O. Gorpshoko, T. Tkachuk © Rights remain with the authors)

However, the greatest management challenge is in ensuring proper water allocation to wetlands basin-wide. Despite the fact that bilateral agreements on transboundary waters exist between all three countries of the basin, they do not provide for wetland conservation, sustaining environmental flows, or adaptation to climate change. Dialogues on transboundary waters have very limited scope and proceed with great difficulty, while major water infrastructure schemes are implemented unilaterally, without proper transboundary assessment.

The lack of transboundary coordination in planning water use and regional development leads to drastic irreversible deterioration of Dauria's environment and loss of opportunities to adapt to a changing climate. The globally important wetlands of Dauria may soon fall victim to uncoordinated water resource management.

The following measures are necessary to reverse negative trends:

1. Establishment of a Chinese-Russian-Mongolian intergovernmental task force on economic and ecological adaptation of management policies in Dauria to changing climate conditions
2. Agreement on environmental flow norms for transboundary rivers of the Argun basin and provisions for sustaining natural dynamics of water allocation to wetlands
3. Wetland monitoring system to measure the effects of climate change and human impacts
4. Wetland protected area network enhancement to provide for migration and breeding of species and preserve key hydrological features and all important refugia during drought periods (e.g., expanding DIPA to Argun floodplain and Buir Lake)
5. Awareness-raising program on climate adaptation in transboundary Dauria
6. Establishing specific basin-wide agreements for Torey-Ulz and Argun basin protection and management
7. Limiting the construction of embankments and dykes only to areas where dykes protect settlements

References

Abell RA, Thieme ML, et al. Freshwater ecoregions of the world: a new map of biogeographic units for freshwater biodiversity conservation. BioScience. 2008;58:403–14.

Greenpeace. Thirsty Coal: a water crisis exacerbated by China's new mega coal power bases. Beijing: Greenpeace; 2012a. http://www.greenpeace.org/eastasia/publications/reports/climate-energy/2012/thirsty-coal-water-crisis/

Greenpeace. Thirsty Coal: research on coal industrial bases and water resources. Institute of Geography and Natural Resources of China Academy of Science. Beijing: China Environmental Science Publishers; 2012b. Chinese. http://www.greenpeace.org/china/Global/china/publications/campaigns/climate-energy/2012/coalwest-report.pdf

Kirilyuk VE, Kirilyuk OK, Goroshko O. DIPA – 10 years of cooperation. "Express" Chita. 2006. http://www.dauriareserve.narod.ru/index_eng.htm

Kirilyuk VE, Obyazov VA, Tkachuk TE, Kirilyuk OK. Influence of climate change on vegetation and wildlife in the Daurian eco-region. In: Werger MJA, van Staalduinen MA, editors. Eurasian Steppes. Ecological problems and livelihoods in a changing world, Plant and vegetation, vol. 6. Dordrecht: Springer; 2012. p. 397–424. ISBN 978-94-007-3885-0 (Print) 978-94-007-3886-7 (Online).

Simonov E, Dahmer T. Amur-Heilong river basin reader. Hongkong: Ecosystems LTD; 2008. http://www.wwf.ru/resources/publ/book/299

Sanjiang Plain and Wetlands Along the Ussuri and Amur Rivers: Amur River Basin (Russia and China)

125

Thomas D. Dahmer

Contents

Introduction	1510
Geographic and Hydrologic Features	1511
The Ussuri-Wusuli River from Lake Khanka-Xingkaihu to Khabarovsk	1511
The Middle Amur River from Khabarovsk to Lake Bolon	1513
Wetland Features	1513
Threats	1514
The Ussuri-Wusuli River from Lake Khanka-Xingkaihu to Khabarovsk	1514
The Middle Amur River from Khabarovsk to Lake Bolon	1516
Conservation Status	1517
The Ussuri-Wusuli River from Lake Khanka-Xingkaihu to Khabarovsk	1517
Lake Khanka-Xingkaihu	1517
The Middle Amur River from Khabarovsk to Lake Bolon	1518
Future Challenges	1518
Research and Planning	1518
Conservation Projects	1519
New Protected Areas	1519
Transboundary Conservation	1519
References	1520

Abstract

The Amur-Heilong River basin is the world's eleventh largest by area. It covers 2,129,700 km^2 and drains the border areas of Russia, China, and Mongolia to the Tartar Straits of the Sea of Okhotsk. The basin is a study in contrasts, with nearly equal land areas shared by China and Russia but over 93% of the human population in China. A key wetland feature of the basin is the Sanjiang plain, a formerly forested but now mainly marsh wetland extending from northeast China

T. D. Dahmer (✉)
Ecosystems Ltd., Yau Tong, Kowloon, Hong Kong, China
e-mail: tdahmer@pacific.net.hk

© Springer Science+Business Media B.V., part of Springer Nature 2018
C. M. Finlayson et al. (eds.), *The Wetland Book*,
https://doi.org/10.1007/978-94-007-4001-3_30

downstream along the Amur-Heilong river into Siberia. Wetlands on the Sanjiang plain have declined in area by 86% from 108,900 km^2 prior to the 20th century to some 14,800 km^2 in 2000. Wetland loss is caused by conversion to farmland for exploitation of the rich black soils and abundant water resource that characterize the Sanjiang plain. The plain supports 28 species of globally threatened wildlife, 23 of which are birds and mammals. The region is rich in fish biodiversity and this economically valuable resource has been understudied and severely over-exploited. Protected areas in China (49) and Russia (11) aim to conserve wetlands but losses continue. Threats to wetlands include growing demand for farmland and irrigation water, and lack of transboundary cooperation in wetland and biodiversity research and conservation. The latter threat is most acute along the floodplains of the transboundary rivers Amur-Heilong and Wusuli-Ussuri where regulation is urgently needed to stop re-cultivation of fallow Russian farmlands.

Keywords

Amur · Biodiversity conservation · Heilongjiang · Marsh wetland · Sanjiang plain · Waterbird · Wetland conservation

Introduction

The Amur-Heilong River basin is the world's eleventh largest. It covers 2,129,700 km^2 and drains the border areas of Russia, China, and Mongolia to the Tartar Straits of the Sea of Okhotsk. Russia's Siberian portion of the basin covers just over one million km^2, accounting for 47% of the basin area. Northeast China's portion covers 905,700 km^2 or nearly 43% of the basin. Mongolia's northeast accounts for around 200,000 km^2 or just over 9% of the basin. Because Russia and China share the bulk of the basin area (90%), the basin is referred to here as the "Amur-Heilong" to combine the Russian and Chinese names. The name "Heilongjiang" (meaning "black dragon river") is the Chinese name for the Amur River and for Heilongjiang Province, which is bordered in the north by the Amur-Heilong River and in the east by the Ussuri River (in Chinese, Wusuli River). Similarly, the Ussuri River is referred to here as the Ussuri-Wusuli River to combine the names used on the two banks. Downstream from Khabarovsk, the river is bounded entirely by Russia and the reach is referred to only as the Amur River. Readers are referred to Simonov and Dahmer (2008) for a detailed description of the Amur-Heilong River basin as a whole.

Of WWF's 238 Global 200 ecoregions, the middle Amur-Sanjiang region is included in two, Russian Far East temperate forests (no. 71) and Russian Far East rivers and wetlands (no. 181). Global ecoregion no. 71 includes two forest types, of which Ussuri broadleaf and mixed forests are found in the middle Amur-Sanjiang region. These two global ecoregions are represented by five main habitat types: (i) Suifen-Khanka meadow and forest meadow, (ii) Ussuri broadleaf and mixed forest, (iii) Manchurian mixed forest, (iv) Amur meadow steppe, and (v) Okhotsk-

Manchurian taiga on the Primorsky (Russia) side of the basin. The wetlands in the middle Amur-Sanjiang region will be described in two sections. The first section and upstream reach covers the 900 km of the Ussuri-Wusuli River from Lake Khanka (Xingkaihu in Chinese) to the Ussuri-Wusuli confluence with the Amur-Heilong at Khabarovsk. The second and downstream reach covers the 260 km of the Amur River from Khabarovsk to Lake Bolon (Fig. 1).

Geographic and Hydrologic Features

In China, the Sanjiang is often considered as two plains separated by the Wanda mountains. In terms of physical geography, however, the Sanjiang is one plain extending northward from Lake Khanka-Xingkaihu on the Russia-China border and northeasterly from the confluence of the Songhua River with the Amur-Heilong to Lake Bolon on the Amur in Khabarovsky province.

The Sanjiang Plain upstream from Khabarovsk is low in elevation (<100 masl) and declines at a gradient of about 1:10,000 to the Ussuri-Wusuli and Amur-Heilong rivers. Prior to extensive agricultural development on the plains in the twentieth century, the plains covered 108,900 km^2, were largely unfragmented, and supported marshes with sparse forest that were seasonally submerged to depths of up to 2 m. Although written and photographic records are few, elderly villagers recall childhood memories of wildlife abundance that surely rivaled the most biodiverse regions of the world.

Average temperature on the Sanjiang Plain is $-22\ °C$ in January and $+22\ °C$ in July. The coldest recorded temperature is $-44.1\ °C$ and the warmest is over $+40\ °C$. Sunshine per year averages over 2,300 h. Rainfall averages 550 mm per year, with 50–70% falling during July–September. Annual evapotranspiration exceeds 1,100 mm. The annual ice period lasts 7 months in the north to 6 months in the south, and there is no permafrost. The depth of frost ranges from 80 to 160 mm. First frost occurs annually from 20–30 September, and spring thaw occurs from 10–20 May. The no-frost period lasts 114–150 days annually.

The Ussuri-Wusuli River from Lake Khanka-Xingkaihu to Khabarovsk

The Ussuri-Wusuli River demarcates the China-Russia border and is the second largest right-bank tributary of the Amur-Heilong after the Songhua River. The river drains Russia's western Primorsky Province, Russia's Khabarovsky Province, and eastern portions of China's Heilongjiang Province. The Ussuri-Wusuli River extends 897 km northward from its source in the southern Sikhote-Alin Ridge in southern Primorsky Province to its confluence with the Amur-Heilong at Khabarovsk. The watershed covers 193,000 km^2 of which 70% lies in Russia and 30% in China.

Fig. 1 Sanjiang Plain and wetlands along the Ussuri and Amur Rivers from Lake Khanka to Lake Bolon, Amur River Basin, Russian Federation and China. Notes: *Dashed black line* encircles subject wetlands; Portion of subject wetlands within China approximates the greater Sanjiang plain; Songhuajiang = Songhua River; Base map adapted with permission from Simonov and Dahmer (2008)

The Ussuri-Wusuli River and the middle Amur-Heilong cross the Sanjiang Plain or "Three Rivers" plain, named for the Amur-Heilong, Ussuri-Wusuli, and Songhua Rivers. The plain is divided by China's Wanda Range of low hills (<300 masl) that extend northeasterly some 230 km from Mishan city in the south to the Rao River

(Raohe) in the north, and continue from the confluence of the Ussuri-Wusuli and Bikin Rivers into the Sikhote-Alin range in Primorsky province in Russia. The conservation importance of this mountain range was highlighted in October 2012 by the establishment of Sredneussuriisky Wildlife Refuge in Russia to protect Siberian tiger habitats in the Sikhote-Alin range and connect them with those in the Wanda mountains.

The Middle Amur River from Khabarovsk to Lake Bolon

This reach of the Amur-Heilong River lies entirely in Russia. It begins at the confluence of the Amur-Heilong and Ussuri-Wusuli Rivers near Khabarovsk (42 masl) and extends downstream about 260 km to Lake Bolon (25 masl). The floodplain is wide, extending to over 50 km, and the river is braided.

Wetland Features

Wetland vegetation on the Sanjiang Plain is dominated by sedges and grasses, with sparsely distributed shrubs and dwarf trees on alluvial deposits not perennially inundated. Dominant vegetation over most of the uplands was mixed coniferous and broadleaf forest in prehuman times (Lu and Wang 1994). The Sanjiang Plain was more than 50% forested in the late 1800s (Liu and Ma 2002). This habitat can still be seen in Russia's nature reserves east of the Wusuli River and north of the Heilong River, but not on the China side where trees have been cut for use as fuel or in building construction. The remaining lowland forests are represented by small patches in protected areas and along the upstream reaches of the Wusuli River.

The Sanjiang Plain is not only noteworthy for its large wetland area but also because of its rich biodiversity. Sanjiang Plain wetlands are ranked as globally important in the Directory of Asian Wetlands (Scott 1989). According to Ni et al. (1999), there are about 1,000 species of plants and 37 ecosystem types in this area. The Sanjiang Plain also supports some 528 species of vertebrate fauna. Larger wild mammals have been virtually eliminated, but smaller mammals, birds, amphibians, reptiles, fish, and invertebrates are abundant and diverse. The Sanjiang is world-renowned for its waterbirds, most notable of which are cranes. Eight of the 15 species of cranes in the world regularly visit or occupy China wetlands (Harris and Mirande 2013, adjusted for loss of sarus crane), six have been recorded in the Songhua basin, and four species nest there (Ma and Jing 1987). Three species of cranes nest on the Sanjiang Plain, red-crowned *Grus japonensis*, white-naped *Grus vipio*, and hooded *Grus monacha*. The Heilong and Wusuli Rivers support the Huso (kaluga) sturgeon *Huso dauricus* and Amur sturgeon *Acipenser schrenckii*, both of which are globally endangered and trade restricted under Convention on International Trade in Endangered Species (CITES).

Species considered threatened according to IUCN World Conservation Union categories (critically endangered, endangered, vulnerable, near-threatened) and

Table 1 Number of fauna species of the Sanjiang Plain listed as globally threatened and trade restricted in 2010

	Global threat[a]				CITES listing[b]		
	Critical	Endangered	Vulnerable	Near-threatened	I	II	III
Mammals	1	1	3	2	4	2	2
Birds	1	7	12	8	7	37	
Reptiles			1				
Amphibians				2			
Fish		2				2	
Total	2	10	16	12	11	41	2

[a] According to IUCN World Conservation Union
[b] I = international trade not permitted under any circumstances
[b] II = international trade only with permit from authorized authority
[b] III = international trade permitted with quota

those species for which trade is restricted by the Convention on International Trade in Endangered Species are listed in Table 1.

Threats

The Ussuri-Wusuli River from Lake Khanka-Xingkaihu to Khabarovsk

An important difference between the Russian and Chinese sides of the middle Amur-Sanjiang region is the human population. Some 4–4.5 million Russians lived in the Amur-Heilong river basin in 2002, as compared to 65–70 million Chinese in 2005. Population densities were estimated in the early 2000s at 90/km^2 on the China side of the basin and 3.5/km^2 in the Russian Far East. The greater numbers of people on the China side have led to more extensive changes in land use, particularly for farming and livestock production. More intensive logging, hunting/poaching, and fishing on the China side have also taken a toll on wildlife habitats and populations.

Liu and Ma (2002) summarized five periods in the developmental history of the Sanjiang Plain in China. The first exploitation was in 1743 when 4.5 km^2 of marsh were converted to croplands to feed soldiers posted at the frontier. After this, slow conversion to agriculture continued through 1956. By 1949 only some 820 km^2 or less than 1% of the wetlands had been converted to croplands. At that time, carnivores were common, including Far Eastern leopard *Panthera pardus orientalis*, Siberian tiger *Panthera tigris altaica*, and bears (Asiatic black bear *Ursus thibetanus* and brown bear *Ursus arctos*), and waterbirds probably occurred in numbers that would defy belief today. Common carp *Cyprinus carpio* that sells today for over USD50 was priced at USD0.10 in local markets and free of charge to those willing to travel to the nearest river.

From 1956 through 1978, the second period of exploitation, the pace of wetland conversion accelerated. Some 100,000 military troops converted around

20,490 km² of marshland to cropland from 1956–1974, raising the farmland portion to over 18% of the plain area. Most of the six million people in northeast China in 1974 lived in urban areas distant from the Sanjiang Plain. Human impacts on the remote eastern portions of the plain were not yet severe. A 1976 survey of rare and endangered fauna counted several thousand cranes, including nearly 1,000 red-crowned cranes, and thousands of whooper swans *Cygnus cygnus* nesting on the Sanjiang. Over 100 oriental stork *Ciconia boyciana* nests were recorded in the area later gazetted as Honghe National Nature Reserve. Wild mammal skins sold in 1978 represented over 36,000 Siberian weasels *Mustela sibirica*, 562 sable *M. zibellina*, 120 deer (probably red deer *Cervus elaphus* and roe deer *Capreolus capreolus*), 102 wolves *Canis lupus*, and over 10 bears (probably *U. thibetanus*).

The third period of exploitation from 1978–1985 reduced the wetland area to less than half of that in the pre-1956 era, and forest cover was reduced by one third from the 1960s. The human population grew to over seven million by 1978 but was still concentrated in the few urban areas. Commercial markets for wildlife skins were active as late as 1983.

From 1985–1996 the fourth period of exploitation was the peak period of human impact. As the human population grew to over eight million, the environment was degraded by agricultural development, pollution, and unsustainable taking of wildlife. Sharp declines were observed in all wildlife populations but particularly for waterbirds. Oriental storks nearly disappeared and duck and goose numbers declined by 90%.

The fifth period of exploitation from 1996–2000 saw the beginning of protection and restoration of habitats and wildlife. The Wild Animal Protection Law of 1988 was implemented during this period and led to reduced pressure from hunting and egg collecting. Nature reserves were established to protect remaining key wetlands and wildlife. Nesting populations of oriental storks, red-crowned cranes, and white-naped cranes began to increase. Migrating flocks of waterbirds increased, with a total of over 12,000 greylag geese *Anser anser* seen in Yanwodao Nature Reserve in October 2000. Mammals, in particular, roe deer also began to increase in number but large carnivores did not return except for small numbers of Siberian tiger that occasionally explored and recently occupied areas in China along the Ussuri-Wusuli River.

Prior to the agricultural and urban developments of the late twentieth century, the Sanjiang Plain in China was the largest tract of wetlands in East Asia outside of Siberia. The area of wetlands in 1997 was reported to be 19,700 km² by ADB (Asian Development Bank/GEF Draft Project Concept Paper, 17 August 1999), or 18% of the historic 108,900 km². A more recent estimate by Changchun's Northeast Institute of Geography and Agricultural Ecology and the national Chinese Academy of Sciences was 14,800 km² (Liu and Ma 2002), or ±14% of the plain. If these estimates were accurate, they documented a 25% reduction in wetland area in the 3 years from 1997 to 2000. Song et al. (2014) used remote sensing to document a 38% loss of wetland area from 1976 to 1986, a further 16% loss from 1986 to 1995, and yet another 31% loss from 1995 to 2005. About 91% of wetland losses were caused by conversion to farmland. The value of ecosystem services during 1980–2000 declined for a total loss over the 20-year period of USD2.6 trillion, or around USD131 million per year (Chen et al. 2014).

In the twenty-first century, losses of wetlands have continued in spite of establishment of over 49 wetland nature reserves on the China side and 11 in Russia. Farming, livestock production, and transport are high priorities for economic development in northeast China. The rich soils (the Sanjiang is one of the three black soil regions of the world) and abundant water of the Sanjiang wetlands provide incentives for farming. Increasing availability of modern farming equipment and increased wealth of Chinese farmers have combined to exert greater pressure for conversion of wetlands lying outside of protected areas. This has extended in recent decades to the Russia side of the Sanjiang where immigrant farmers from China are increasingly active in commercial farming on previously abandoned farmlands leased from inactive Russian owners. Wetland water quality is degraded by the impacts of agricultural runoff and untreated domestic and industrial sewage as the region's population increases.

The Middle Amur River from Khabarovsk to Lake Bolon

The Tarabarovy-Heixiazi islands at the confluence of the Amur-Heilong and Ussuri-Wusuli Rivers are threatened with economic development following completion of the international border demarcation in 2009. Overfishing and pollution have led to loss of valuable fish resources. Since 2005, both countries have focused efforts on pollution monitoring in boundary rivers in response to industrial pollution incidents in Heilongjiang rivers during recent years. Anthropogenic wildfires have caused degradation of ecosystems and deforestation in Russia and continue to degrade wetland habitats in China.

Leasing of abandoned farmlands in Russia by Chinese farmers and resumption of intensive agriculture has not only caused loss of wetland habitats (abandoned floodplain farm plots) but has also led to increases in nonpoint source water pollution in the middle Amur and its tributaries. The loss of wetlands is not as severe downstream from Khabarovsk as in the upper reaches, but water pollution has an important downstream impact, particularly on fish, amphibians, and humans.

Although the Amur-Heilong main channel remains undammed, negative impacts have been documented on fish abundance and floodplain condition in the middle Amur from construction of the Zeya and Bureya hydropower plants upstream on these major Amur tributaries. Plans for a Taipinggou-Khingansky dam in the upstream Amur-Heilong main channel would degrade downstream wetland ecosystems, fish reproduction, fish migration, and riverbed and sedimentation processes. The potential impacts of this planned project are poorly understood and have been assessed in the absence of joint Sino-Russian cooperation.

Conservation Status

The Ussuri-Wusuli River from Lake Khanka-Xingkaihu to Khabarovsk

At year-end 2000, 43 nature reserves in Heilongjiang Province protected 16,889 km^2 of wetlands on the Sanjiang Plain. Twenty-one additional nature reserves were proposed for establishment during the 1998–2010 planning period to cover 6,101 km^2 of wetlands. At least six were established prior to 2011. In Primorsky and Khabarovsky Provinces, 11 protected areas are gazetted and one is proposed for wetlands along the Ussuri River. The acreage of protected areas on the Russia side is similar to that in China but in Russia, wetland acreage accounts for a smaller proportion.

Lake Khanka-Xingkaihu

Lake Khanka-Xingkaihu is an ancient lake with relict flora and fauna. It supports high levels of biodiversity including 523 algae species and 616 vascular plants (49 species are listed in the Red Data Book of Russia). The vertebrate fauna includes 454 species, including 333 birds. Freshwater biodiversity is among the highest in North Asia lakes with 69 species of fish (the Amur-Heilong basin in total supports over 110 species). The Ussuri-Wusuli River basin is part of the East Asian-Australasian Flyway and is critically important for waterbird breeding, staging, and migration: up to four million waterbirds migrate through the region annually. Six Sanjiang wetland nature reserves have been listed as Wetlands of International Importance by the Ramsar Convention (Ramsar Sites) in view of their importance for waterbird conservation. These are Xingkai Lake, Honghe, Sanjiang, Qixinghe, and Zhenbaodao National Nature Reserves in China and Lake Khanka in Russia (Table 2). China's five Sanjiang Plain Ramsar Sites represented 12% of the 41 Ramsar Sites in China by mid-2012 and 13% of the total Ramsar Site area in China (3,545,426 ha).

Significant gains have been made in the wetland protected area network during the early years of the twenty-first century. These include increases in nature reserve

Table 2 Sanjiang wetlands listed as Ramsar Sites

Ramsar site name	Surface area (ha)
Xingkaihu National Nature Reserve (transboundary)	222,488
Lake Khanka (transboundary)	310,000
Sanjiang National Nature Reserve	164,400
Honghe National Nature Reserve[a]	21,836
Qixinghe National Nature Reserve[a]	20,000
Zhenbaodao National Nature Reserve	44,364
Total Ramsar sites by 2012	**783,088**

[a]Outside the Ussuri-Wusuli River watershed but on the Sanjiang Plain

area, increased international recognition of important wetlands, enhanced capacity for effective conservation management, and increased funding for protected areas. But threats remain, including increasing demands for water supply to farms and cities, increasing water pollution, unsustainable fishing, fragmentation of wetlands by road and rail corridors, and continuing drainage of wetlands for conversion to farmlands and pastures.

The Middle Amur River from Khabarovsk to Lake Bolon

This region of the Amur wetlands has not received the attention focused on the upper Sanjiang Plain in terms of biodiversity and nature conservation. This is due in part to the paucity of international development aid (UNDP, ADB, World Bank, and others) targeted here as compared to northeast China. A bilateral agreement on "Environmental protection and development of Tarabarovy-Heixiazi islands" was signed by Khabarovsky and Heilongjiang Provinces in 2009, but it included no specific nature conservation measures. The only other wetland protected area in this reach of the Amur River is Bolonsky Nature Reserve (107,426 ha), which protects part of Lake Bolon Ramsar Site (53,800 ha) and plans to include adjacent wetlands by extending its boundaries.

Lake Bolon and the mouths of the Selgon and Simmi Rivers have been listed as a wetland of international importance by the Ramsar Convention. The Ramsar Site covers 53,800 ha and includes part of Bolonsky Nature Reserve. The wetland supports high levels of biodiversity including indigenous fish species and large flocks of breeding, migrating, and molting waterbirds.

Future Challenges

Simonov (2012) assessed the challenges facing the Sanjiang Plain and made recommendations for future work. These are adapted here with updates to reflect work completed since 2012.

Research and Planning

Although the Amur-Heilong is a water-rich basin, demand for water increases annually and aquifers on the China side of the basin have suffered depletion for decades. As a result, planners tend to seek options including additional water management infrastructure. It is critical to assess present and potential impacts of water infrastructure on floodplain ecosystems. Infrastructure includes upstream dams, dykes, and alterations of river channels such as those near Khabarovsk.

The distribution and area of high-quality wetlands remain largely undocumented, and protected area locations might not overlap wetlands of highest quality. Thus, it is important to assess wetland distribution and quality as a basis for ecological network planning. This began in 2004–2005 but covered only Russia and did not employ satellite imagery to support assessment of wetland condition.

Conservation and environmental objectives and safeguards are needed in urban planning for Khabarovsk city and Fuyuan county. Between the two lie the Tarabarovy-Heixiazi islands. The focus of conservation planning should be here to reduce/mitigate threats of urban and transport infrastructure development.

Conservation Projects

The outstanding fish resource of the Amur-Heilong basin is often neglected. Conservation is needed of key fish stocks, including sturgeon, salmon, and phytophilic fish species. The effectiveness of fish conservation measures needs assessment to identify sensitive areas and to plan and establish a network of nature reserves and protected areas for aquatic fauna (sturgeon spawning sites being top priority). Local people and border guard units should be involved to reduce illegal fishing.

Restoration of poor-quality farmland to wetlands complete with their necessary hydrologic connections (e.g., streams, canals) is desperately needed if biodiversity is to be restored and conserved. Huang et al. (2010) used remote sensing and GIS to identify land areas characterized by poor-quality farmland (lowest of three productivity categories) and medium to high wetness (top two of three wetness categories). The total area of farmland described by these combined criteria covered nearly 76,000 ha or about 1.3% of the total farmland area on the Sanjiang Plain. The restoration of these farmlands to wetlands could well follow the model developed on 4,000 ha of former farmland in the Naoli River catchment during the ADB-GEF Sanjiang Plains Wetland Protection Project of 2007–2011.

New Protected Areas

In Russia, Bolshekhekhtsyrskii Nature Reserve, Khekhtsyr Wildlife Refuge and adjacent monuments of nature should be connected in a system of protected areas covering 110,000 ha. This should be accompanied by establishment of "Podkhorenok" Provincial Wildlife Refuge in Khabarovskii Province (15,000 ha) and "Petrovskaya mar" crane and stork wildlife refuge in Evreyskaya Province (20,000 ha). Zabelovskii Wildlife Refuge should become an extension of Bastak Zapovednik in Evreyskaya Province.

In China, protected areas are needed for Chinese softshell (*Pelodiscus sinensis*, IUCN Red List vulnerable) breeding beaches near the mouth of Bidzhan River.

Transboundary Conservation

Policy and financial support is needed for cooperation between reserves in China and Russia to establish large transboundary protected areas on the middle Amur-Heilong floodplains as initially proposed in the Ussuri Sustainable Land-Use Program in

1996. An ecological corridor is needed along the Russia-China border as a buffer zone connecting nature reserves. Joint Russia-China activities in wetland research and conservation need support, including wildfire control; production of joint communications and scientific publications; stork, crane, fish, and turtle conservation; capacity building through training; and professional exchanges.

Extend wetland reserves on transboundary floodplains in China with focus on Tarabarovy-Heixiazi islands as the center of the reserve network. To supplement extension of reserves onto transboundary floodplains, transboundary species conservation plans are needed for oriental white stork, Chinese softshell, and kaluga sturgeon.

Transboundary exchanges are needed to develop wetland-friendly agriculture and prevent environmental damage from resumption of farming on the Russian side of the Amur-Heilong floodplain through work with village authorities, farmers, and environmental agencies.

References

Chen J, Sun B, Chen D, Wu X, Guo L, Wang G. Land use changes and their effects on the value of ecosystems services in the Small Sanjiang plain in China. Sci World J. 2014;752846. pub. online 2014 Mar 9. doi:10.1155/2014/752846.

Harris J, Mirande C. A global overview of cranes: status, threats and conservation priorities. Chinese Birds. 2013;4(3):189–209. doi:10.5122/cbirds.2013.0025.

Huang N, Wang Z, Liu D, Niu Z. Selecting sites for converting farmlands to wetlands in the Sanjiang plain, Northeast China based on remote sensing and GIS. Environ Manag. 2010; 46(5):790–800. doi:10.1007/s00267-010-9547-6. Epub 2010 Sep 7.

Liu X, Ma X. Natural environmental changes and ecological protection in the Sanjiang plain. Beijing: China Science Press; 2002. p. 355.

Ma Y, Jing L. The numerical distribution of the Red-crowned Crane in Sanjiang plain area of Heilongjiang province. Acta Zool Sin. 1987;33:82–7.

Ni Hongwei et al. Plant diversity of Honghe Nature Reserve in Sanjiang Plain. J. Territory & Natural Resources Study. 1999;(3):12–18. (in Chinese).

Scott DA. A directory of Asian Wetlands. Gland: IUCN; 1989. p. 1181 + maps.

Simonov EA. Unpublished manuscripts. Vladivostok: WWF Russian Far East; 2012.

Simonov EA, Dahmer TD, editors. Amur-Heilong river basin reader. Hong Kong: Ecosystems Ltd; 2008. p. 426.

Song K, Wang Z, Du J, Liu L, Zeng L, Ren C. Wetland degradation: its driving forces and environmental impacts in the Sanjiang plain. Environ Manag. 2014;54(2):255–71.

Zhalong Wetlands (China)

Liying Su

Contents

Introduction 1521
Hydrology 1523
Wetland Ecosystem 1524
Biodiversity 1525
Conservation Status 1526
Ecosystem Services 1527
Threats and Future Challenges 1527
References 1529

Abstract

Zhalong is a large freshwater wetland located in northeastern China, and currently has the largest breeding population of the endangered red-crowned crane. Zhalong has high biodiversity value with close to 100 waterbird species breeding there, and also has an irreplaceable role for migratory birds on the East Asian – Australasian Flyway with over 100 species staging in the wetland. Zhalong plays important roles in flood control, water purification, and income for local communities.

Keywords

Red-crowned crane · Reed marsh · Fresh water wetland

L. Su (✉)
International Crane Foundation, Baraboo, WI, USA
e-mail: 4_best@163.com; liying@savingcranes.org

Introduction

Zhalong Marsh is located in Heilongjiang Province, lying in the western part of Songnen Plain 24 km east of Qiqihar City on the eastern side of the Nen River between E 123°51.5′124°37.5′ and N 46°48′–47°31.5′ (Fig. 1).

Currently the largest wetland in Songnen Plain, Zhalong is located in the transition zone of the temperate continental climate in northeast China, which ranges from moist in the east to dry in the west (Simonov 2013). Zhalong has short, warm, and moist summers and long, cold, and dry winters. The annual rainfall is about 400 mm

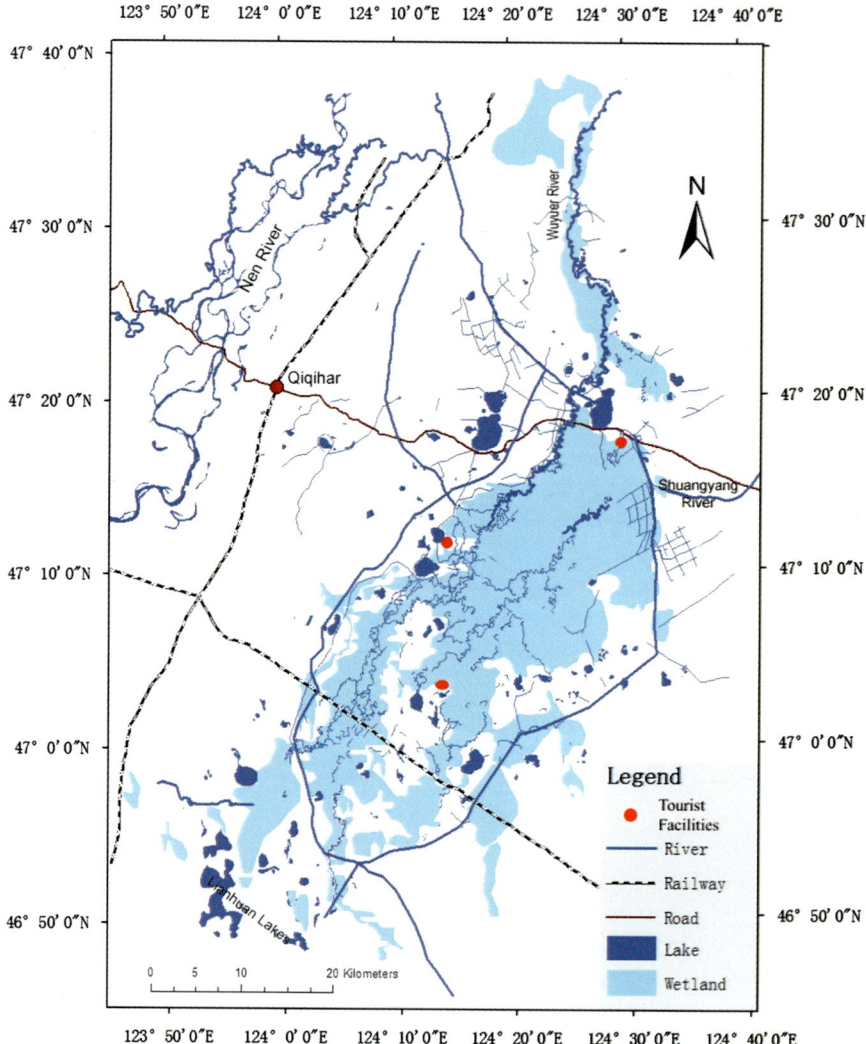

Fig. 1 Map of Zhalong Marsh

Fig. 2 The wide open landscape is dominated by reed marshes with shallow water at Zhalong (Photo by: Su Liying © Rights remain with the author)

with large variability among years (255–604 mm), mostly concentrated from June to August. The annual evaporation is usually between 1200 and 1600 mm. The wetland is surrounded mainly by farmlands and some dry grassland. The landscape is wide and flat, with lowest elevations in the middle parts of the wetland and a gentle slope from the northeast toward the southwest (Fig. 2). Most of the wetland lies between 140 and 160 m above mean sea level (Liu et al. 2007).

Hydrology

The wetland was formed by changes in the ancient Nen River channels and the inland delta of the Wuyuer and Shuangyang Rivers. Formerly water for Zhalong Marsh came from the Wuyuer River, the Shuangyang River, surface flows from surrounding areas, and occasional flood waters from the Nen River during wet years and rainfall (Zhang and Feng 2011). The Wuyuer River originates in the Lesser Xingan Mountains, while the Shuangyang River starts in rolling hills in the middle of Songnen Plain and flows west to enter Zhalong Marsh on the middle reaches of its eastern side. Under natural conditions, based on 46 years observations, the average annual flow into the wetland from the Wuyuer-Shuangyang Rivers was 4.08×10^8 m^3, representing over 85% of the total water input. In low water years (with below-average annual inflows), the flow was less than 2.0×10^8 m^3 (during the mid-late 1970s), and in high water years (with above-average annual inflows), the flow was more than 12.8×10^8 m^3 (in 1998) (Tong et al. 2008). The outlet of the wetland is located in the south in Duerbute, where several lakes (the Lianhuan Lakes) occur. There is no water inflow to the lakes when the water input at the upper end of Zhalong Marsh is less than 2.2×10^8 m^3. After a series of several dry years, even if the input is higher than 4.0×10^8 m^3, there will still be no water flowing into the lakes. With variable water supplies among years, the shallow water marsh area changes greatly from 130 to 1,240 km^2.

Because of increasing water demands from expanding agriculture, human population, and industry, water inflow from the Wuyuer River was greatly reduced and supply from the Shuangyang River almost completely cut off (Tong et al. 2011). There are several (Zhongyin, Eighth and Ninth) canals and dikes that have blocked and redirected surface water flow away from Zhalong Marsh (Su 2008). These canals are 30 m wide and 3–5 m in depth, with 1.5 m high dikes on each side. These canals eliminate surface water supply from adjacent areas and from the Nen River during rain or flood seasons. Currently, Zhalong is getting progressively drier, without artificial releases of water into the wetland except during very rich water years.

After several dry years and frequent fires, the Heilongjiang Government provided emergency water releases during the period 2001–2006 in order to restore the wetland. Based on water management plans developed during the UNEP/GEF Siberian Crane Wetland Project, in 2009 the Heilongjiang Government together with Qiqihar and Daqing City Governments set up a fund to support water releases for Zhalong when needed.

Wetland Ecosystem

The area is located within the eastern Daurian Steppe, with mountains on three sides: the Greater and Lesser Xingan and Changbai ranges. The plant flora is complex, with over 500 species present; most are wetland plants growing under moist conditions (Ni and Zhou 1999). In general, there are four main types of vegetation: open water with aquatic plants, reed and reed-sedge marshes, wet meadows, and grassland.

There are many small lakes in the marsh; most of these are < 1 m and a few are 2–3 m in depth. Submerged aquatic plants grow in the lakes, such as *Myriophyllum spicatum*, *Potamogetonaceae* spp., and *Ottelia alismoides*. Floating leaved plants include *Nymphoides peltatum*, *Nymphaea alba*, and *Trapa potaninii*. Emergent aquatic plants include reed *Phragmites australis*, *Typha angustifolia*, and *Scirpus tabernaemontani*.

The reed marsh and sedge-reed marsh, with shallow year-round water depth of 5–40 cm, cover about 46–60% of the total area. In the marshes, the most dominant species is reed accompanied by a variety of *Carex meyeriana*, *C. kirganica*, and *Sium suave*. When water is sufficient, *Caltha membranacea*, *Alisma orientale*, and *Scirpus fluviatilis* occur. In low places, such as the channels of river courses, there are some floating rafts of *Carex pseudocuraica*.

On the edges or higher ground (islands) in the marsh, there are some salty meadows, with complex plant communities and mixing with some scattered small patches of remnant grasslands. These meadows form the transition between the marshes and dry land. The soils are wet and salinized. The vegetation in these areas is very patchy and dynamic, constantly changing depending on the water conditions; thus it is wet meadow dominated by *Deyeuxia angustifolia* and *Puccinellia tenuiflora* during wet years, becoming grassland during dry years. The plant communities consist of salt-tolerant species, such as *Carex duriuscula*, *Potentilla anserine*, *Leymus chinensis*, and *Suaeda glauca*.

Biodiversity

These different types of vegetation provide habitats for large numbers of breeding and migratory birds (Su et al. 1987). Over 230 species have been recorded in the area, about one third of which are breeding birds and about 120 species are migrants. Among them, there are two critically endangered (CR), two endangered (EN), eight vulnerable (VU), and five near-threatened (NT) species. The most well known are the red-crowned *Grus japonensis* (EN), white-naped *G. vipio* (VU), and Siberian cranes *Leucogeranus leucogeranus* (CR), oriental stork *Ciconia boyciana* (EN), whooper and tundra swans *Cygnus cygnus and C. columbianus* (nationally protected), swan goose *Anser cygnoides* (VU), oriental white (or black-headed) ibis *Threskiornis melanocephalus* (NT), and Eurasian spoonbill *Platalea leucorodia* (Fei et al. 1985; Feng and Li 1985, 1986).

According to aerial surveys conducted in 2005 and 2008, there were over 200 red-crowned cranes breeding here, the largest breeding population in the world (Fig. 3). The white-naped crane also breeds at Zhalong, with less than 50 birds. The cranes arrive in mid-March and start nesting in early April (Ma 1981). The red-crowned cranes prefer to nest and brood in reed marshes, while the white-naped cranes prefer the shallower sedge marsh. Once their chicks are about 40–60 days old, they take them to open habitats – wet meadows, or hand-cut meadows, and grasslands.

Colonial breeders, namely, purple *Ardea purpurea* and gray herons *A. cinerea*, oriental white ibis (NT), Eurasian spoonbill, great egret *Ardea alba*, and great cormorant *Phalacrocorax carbo*, build ground nests on floating vegetation mats in the marsh. The nest areas are in deep water, of about 40–80 cm, beside or in the middle of the main streams of the marsh. These bird populations have greatly declined in recent decades because of water shortages impacting the wetland and egg collection.

Many species of geese and ducks breed in Zhalong. Among them, swan goose (VU), Baer's pochard *Aythya baeri* (CR), and falcated duck *Anas falcata* (NT) are globally threatened species. These birds are less common now than 20 years ago, and their population sizes are very small with most seen only during the migration seasons. Currently, common pochard, mallard, and spot-billed duck are commonly seen, but in low numbers.

During the migration seasons, large flocks of birds stage in the shallow lakes, marshes, and wet meadows. The critically endangered Siberian crane stops here for

Fig. 3 A pair of red-crowned cranes with their young chicks foraging in the wetland (Photo by Wang Keju © Rights remain with the author)

about 50 days in spring and forages in shallow water with sedges. Big flocks of these cranes were concentrated in the middle of the wetland in the 1980s. However, in more recent decades, the flock sizes have been much smaller, and the cranes have moved around a lot. Hooded cranes may gather in two or three large flocks at the edge of the wetland, with total numbers close to 3,000 birds. They forage in farm fields and grasslands during the day and roost in the marshes at night. Large flocks of geese and ducks usually stage in shallow lakes.

Conservation Status

The typical ecological system and rich biodiversity have ensured Zhalong Marsh an important position in the conservation of global biodiversity. Zhalong Provincial Nature Reserve was ratified by the Heilongjiang Provincial Government in 1979 and was promoted to a National Nature Reserve in 1987. Zhalong was listed under the Ramsar Convention as a Wetland of International Importance in 1992. The reserve covers 2,100 km^2 in area, which include adjacent areas from Tiefeng and Angangxi Districts of Qiqihar City, Duerbote Mongolian Autonomous Qi, Lindian, Tailai, and Fuyu Counties.

Ecosystem Services

Because rainfall in Songnen Plain is highly variable among years, one of the major values of Zhalong Marsh is flood control. Zhalong wetland provides an outlet for any local floods from Qiqihar (west and north) and Lindian (east) sides. During large-scale floods from the Nen River or from the upper reaches of the Hata and Wuyuer Rivers, large amounts of water are released into Zhalong. These floods usually occur following a cycle of about 4–15 years. Due to the large size of the wetland, Zhalong can absorb about 18×10^8 m^3 of flood water, a very important benefit for people in Qiqihar and surrounding areas (Hou and Chen 2003; Dong et al. 2008).

Besides storing flood water, Zhalong wetland purifies a large amount of water each year. It takes 2 months or longer for water to flow through the wetland from the inflow point in the north to the outlet lakes in the south. During this process, the dense vegetation slows down the flow, and suspended materials in the water column settle out as sediments, or are adsorbed on to the plants. Simultaneously, aquatic plants and animals can remove some harmful chemical substances from the water.

The fishery in Zhalong was formerly important to people living in the surrounding area, and production was large. For example, the annual production for Zhalong Village only was over 800 t in the 1960s, declining to 30 t in the 1980s, and declining further with very low production in recent years. The fish production declined sharply as the result of many years of overfishing. Currently, people can only catch very small fish used as feed material for domestic animals and with little economic value.

Reed harvesting has provided one of the main income sources for local people, especially during the 1980s. Local people would sell the reeds to paper mills, craft makers, mat makers, and other users. Reeds are also of importance for local subsistence uses such as building material for houses and fences and as fuel for cooking and warming houses (Fig. 4).

Tourism has become an important contributor to local incomes in the last decade after more than 20 years of promoting the cranes and the natural beauty of the wetland. There are three large tourism facilities in the wetland, situated at Qiqihar, Lindian, and Duerbote, together hosting at least ten million visitors/year. These visitors bring in millions of dollars to these sites and tourism-related services.

Threats and Future Challenges

Even though Zhalong Marsh is a well-known national level nature reserve whose ecological and biodiversity values have been recognized long ago, pressures on the wetland are enormous. In the past 20 years, development of the surrounding area, even in relatively distant places, has had strong impacts on Zhalong. In the early

Fig. 4 Reeds play an important part in local peoples' lives as a source of income and providing living materials (Photo by Su Liying © Rights remain with the author)

2000s, the wetland was facing serious threats and an ecological crisis – water shortage, fragmentation, frequent fires, and species and population declines.

Water shortage remains the most serious threat for the survival of the wetland ecosystem. There are three major aspects to this issue: first, the great reduction of natural water input to the wetlands; second, the large degree of channelization within the wetlands; and third, the conversion of wetland habitats for rice fields, fish ponds, and improved grazing lands. As result of these factors, the wetland is becoming increasingly fragmented, enabling easier and more frequent human access.

During the driest time of the year, in the months of April and May, there are frequent fires in the wetland during low water years (Cai et al. 2002; Xie 2002). Light fires occur as patches and only burn the top parts of new growing vegetation. However, occasional intensive fires burn into the deep roots of plants and cause serious damage to plant communities. Because of their timing in spring, these frequent fires not only have great impacts on plant communities but can also be devastating for nesting birds (Fig. 5).

The incidence of egg taking and poisoning of birds has been greatly reduced since the reserve was set up in the 1980s; however, these illegal acts have never completely stopped. With the greatly reduced sizes of local and regional waterbird populations now, these illegal activities have the potential to seriously impact these declining populations (Sha et al. 2011).

Declining water quality represents another serious potential threat for the future of the wetland (Li et al. 2007; Wang et al. 2012; Luo et al. 2013; Zhang et al. 2014). Recent studies indicate that the wetland receives polluted water from upstream and the surrounding area. The sources of pollution include nonpoint source pollution,

Fig. 5 Even if fires are only quick light fires, ground nesting birds still cannot escape, especially small birds. This is a burned common pochard's nest in a wet meadow in Zhalong (May 2014) (Photo by Su Liying © Rights remain with the author)

sewage, and industrial wastewater discharges. The types of pollution include both organic contamination and heavy metals which can be found in the sediments, water, plants, and animals in the wetland. This issue could become even more serious in the future due to the wetland's configuration – it acts effectively as an inland delta with almost no outlet. Once toxic substances have been allowed to accumulate in the wetland, they will be difficult to impossible to clean up.

References

Cai J, Wang Z, Wu Z, Xu L. Fire behavior in the wetland of Zhalong Nature Reserve and thinking. For Technol. 2002;6:30–1. [In Chinese with English abstract].

Dong L, Wu W, Ren C. 2008. Analyzing the capacity of flood storage of the Zhalong wetland. Northeast Water Utilization and hydropower. 2008;26(7):55,63. [in Chinese]

Fei D, Che R, Yang Z. A preliminary study on breeding habits of purple and Grey herons in Zhalong Nature Reserve. J Zool. 1985;2:12–6. [In Chinese]

Feng K, Li J. Aerial surveys of red-crowned cranes and other waterfowl in China. Journal of Northeastern Forestry University. 1985;13 (1):80–7. [In Chinese with English abstract]

Feng K, Li J. Reproductive ecology of the red-crowned crane. Journal of Northeast Forestry University. 1986;14(4):39–44. [In Chinese with English abstract]

Hou S, Chen Q. The function of Zhalong marshland to flood control and disaster relief. J Heilongjiang Hydraul Eng Coll. 2003;30:68–9 [In Chinese].

Li Feng, Zhang Weiwei, Liu Guang Ping. Bioaccumulation of heavy metal along food chain in the water of Zhalong wetland. J Northeast For Univ. 2007;35:44–6. [In Chinese with English abstract].

Liu J, Li Z, Qin X, editors. Natural history and study on human activity impacts in northeast (China). Beijing: Science Press; 2007. [In Chinese].

Luo J, Yin X, Ya Y, Wang Y, Zang S, Zhou X. Pb and Cd Bioaccumulations in the habitat and preys of red-crowned cranes (*Grus japonensis*) in Zhalong Wetland, Northeastern China. Biol Trace Elem Res. 2013. doi:10.1007/s12011-013-9837-y.

Ma G. Breeding habits of Red-crowned cranes. In: Lewis J, Masatomi H, editors. Crane research around the world. Baraboo: International Crane Foundation; 1981. p. 89–93.

Ni H, Zhou R. Flora and vegetation. In: Wu C, Li X, Ni H, Gao Z, editors. Studies on natural resources management of Zhalong Nature Reserve. Harbin: Northeast Forestry University Press; 1999. p. 20–108.

Sha J, Li F, Su L, Pang S, Gao Z. Numerical fluctuation of waterfowl in Zhalong wetland and its influencing factors. J Northeast For Univ. 2011;39(8):77–84. [In Chinese with English abstract].

Simonov EA, Dahmer TD, editors. Amur-heilong river basin reader. Hong King: Ecosystems Ltd.; 2008. p. 426.

Su L. Challenges for the Red-crowned crane conservation in China. In: Koga K, Hu D, Momose K, editors. The current status and issues of the Red-crowned crane. Hokkaido: Tancho Protection Group; 2008. p. 63–73.

Su L, Ma J, Xu J, Jiang X, Wu C. A preliminary study on avifauna in Zhalong Nature Reserve. J Northeast For Univ. 1987;15:62–72. [In Chinese with English abstract].

Tong S, Lv X, Su L, Jiang M, Yao Y. Changing process and impact factors of wetland ecosystem in Zhalong wetland. Wetl Sci. 2008;6:179–84.

Tong S, Lu X, Zhang M, Yao Y, Wang X. Water management and water supply restoration design for Zhalong wetlands. In: Prentice C, editor. Conservation of flyway wetlands in East and West/Central Asia. Proceedings of the project completion workshop of the UNEP/GEF Siberian crane wetland project, 14–15 October 2009, Harbin, China. Baraboo: International Crane Foundation; 2011.

Wang J, Zhang S, Wu B. Recent advances and future prospects of heavy metals pollution in Zhalong wetland. Nat Sci J Harbin Norm Univ. 2012;28:91–4, 90. [In Chinese with English abstract].

Xie Y. Fires in Zhalong wetland. J China Disaster Reduct. 2002;1:53–4. [In Chinese].

Zhang G, Feng H. Water monitoring program as tool for wetland restoration at Zhalong, China. In: Prentice C, editor. Conservation of flyway wetlands in East and West/Central Asia. Proceedings of the project completion workshop of the UNEP/GEF Siberian crane wetland project, 14–15 October 2009, Harbin, China. Baraboo: International Crane Foundation; 2011.

Zhang N, Zhang S, Sun Q. Health risk assessment of heavy metals in water environment of Zhalong wetland, China. Ecotoxicology. 2014;23:518–26. doi:10.1077/s10646-014-1183-0.

Highland Peatlands of Mongolia

127

Tatiana Minayeva, Andrey Sirin, and Chultemin Dugarjav

Contents

Introduction	1532
Highland Peatlands Under a Dry Climate	1533
Distribution of Peatlands in Mongolia	1534
Diversity of Mires	1535
Group of Mire Massif Types I: Transitional Peatlands (Ia) and Spring Fens (IIb) in Large Lowland Valleys	1535
Group of Mire Massif Types II: Mires of the Taiga Altitudinal Zone – Sphagnum-Sedge Kettle and Sloping Bogs (IIa); Transitional Open Sphagnum-Sedge Sloping Peatlands (IIb); Paludified Shallow Peat Forests and Coniferous Transition Peatlands (IIc); Forested Transitional Birch-Willow-Grass Peatlands (IId)	1536
Mire Massif Type III: Transitional Highland Mires on Permafrost	1541
Mire Massif Type IV: Highland Valley Meadow Fens	1542
Mire Massif of Highlands of Steppe Zone V: Highland Valley Sloping Fens (Va) Highland Valley Spring Fens (Vb)	1543
Mire Massif Type VI Highland Spring (VIb) and Blanket/Sloping Fen (VIa) on Permafrost	1544
Conservation	1546

T. Minayeva (✉)
Wetlands International, Ede, The Netherlands
e-mail: Tatiana.minaeva@wetlands.org; tania.minajewa@gmail.com

A. Sirin (✉)
Center for protection and restoration of peatland ecosystems, Institute of Forest Science, Russian Academy of Sciences, Moscow, Russia

Laboratory of Peatland Forestry and Amelioration, Institute of Forest Science, Russian Academy of Sciences, Moscow, Russia
e-mail: sirin@ilan.ras.ru

C. Dugarjav
Institute of General and Experimental Biology Mongolian Academy of Sciences, Ulaanbaatar, Mongolia
e-mail: chdugaa@yahoo.com

© Springer Science+Business Media B.V., part of Springer Nature 2018
C. M. Finlayson et al. (eds.), *The Wetland Book*,
https://doi.org/10.1007/978-94-007-4001-3_108

Threats and Future Challenges 1546
References 1548

Abstract

The geographical characteristics of Mongolia suggest the various conditions for peatlands development. Most of the country's territory is made up of highlands with pronounced taiga or forest tundra belts. The area underlain by permafrost used to comprise almost two thirds of Mongolia's territory; this area has dramatically decreased over the last three decades. The highlands in the steppe, semi-desert and desert zones were richer with precipitation during the last decades than the lowlands. Recent studies report that peatlands cover 27,000 km^2 or over 1.7% of the country. The article describes the diversity of peatlands in Mongolia by presenting seven basic mire massif types with variations which are typical for certain geographic conditions. Mire types demonstrate various natural values, contributions to human livelihoods and levels of resistance to climate change and anthropogenic impacts. Recent massive overgrazing in the steppe region, combined with permafrost degradation, has led to losses of peatland ecosystems. The key process contributing to this is peat mineralisation, which means the losses are largely irreversible. The mining industry in the highlands has also led to significant losses of peatlands upstream of key water sources both for people and downstream peatlands. The only solution to this problem is integrated water and ecosystem management.

Keywords

Peatlands · Highlands · Steppe · Dry climate · Desertification · Overgrazing · Degradation · Management

Introduction

Mongolia is a high elevation country dominated by a dry climate in which peatlands were previously not recognized as contributing to ecosystem diversity. This overview is based on a literature review and studies undertaken by the authors between 2002 and 2015. Recent studies report peatlands cover 27,000 km^2 or over 1.7% of the country and further demonstrated dramatic changes in these last decades. Peatlands in Mongolia occur under conditions of low annual precipitation and high summer temperatures and are thus susceptible to desertification processes. There are climatic differences in latitudinal and altitudinal zones with annual mean temperatures between approximately $-8\ °C$ and $6\ °C$, and the annual precipitation varies from 50 mm in the Gobi desert to 400 mm in the northern mountainous area. About 85% of the total precipitation falls from April to September (Dagvardorj 2010). Having originated under more favorable climatic conditions in the past, peatlands are progressively degrading during these last decades, especially those in the upland and forest steppe zones. Climate-driven desertification of mires is exasperated with degradation due to overpasturing and other human activities. Applying the principles for the wise use of peatlands could serve as a key measure for their adaptation to modern conditions.

Highland Peatlands Under a Dry Climate

Peatland ecosystems are especially important in a dry continental climate – storing water and contributing to hydrology regulation in headwaters and valleys of rivers and streams. Supporting specific flora and fauna, peatlands influence environmental conditions in adjacent habitats important to biodiversity and provide temporary habitats for "non-mire" species. Lands with wet conditions and relatively fertile peaty soils exhibit higher vegetation productivity that makes them attractive pastures especially under drying conditions. The vegetation is however much less resilient to the effects of livestock grazing compared to vegetation on mineral soils (Parish et al. 2008; Fig. 1).

Peatlands could indicate overall trends in the region. As peat accumulates, these wetlands preserve a unique record of their own development as well as of past changes in regional vegetation and climate. Plant macrofossils inform about shifts in vegetation cover and ecological condition of the mire including its hydrology and geochemistry. Spore-pollen data supported by radiocarbon dating provide important

Fig. 1 Highland valley peatland degradation is caused by both overgrazing and climate change; vegetation is destroyed by cattle and peat is exposed to mineralization during dry periods (Photo credit: T. Minayeva © Rights remain with the author)

background information for climate reconstruction. Temporal changes in peat increment growth and humification provide additional information on changes in the ecological conditions of these ecosystems. This is relevant to most highlands of Central Asia and possibly other areas with similar biogeographic and climatic conditions.

Distribution of Peatlands in Mongolia

Peatland ecosystems in Mongolia, as well as in other arid zones, are described more broadly than in temperate zones and include peatlands which are a phase of a rotating succession – with highly mineralized peat, sometimes partially transformed from peat to the turf-meadow soil, with peat as a buried soil layer. Alternatively, living peatlands – which are called mires – in the highlands can have a very shallow peat layer (e.g., 10 cm) especially in gravel bedrock. Nevertheless, they function as proper peatlands by accumulating peat, storing water, and hosting peat vegetation.

This extended concept is reflected in the classification of peatlands that is applied to describe peatlands in Mongolia. This classification is based on the traditional Russian integrated landscape approach (see Galkina 1946; Masing 1974; Tyuremnov 1976; Botch and Masing 1979; Yurkovskaya 1992) combining such features as position in the landscape, hydrological regime, and dominating vegetation. This classification approach best takes in consideration spatial heterogeneity at various levels which is a peculiar feature of most peatlands. The basic classification object is the mire massif – the entire mire/peatland within the enclosed continuous border of a peat deposit (Peat dictionary 1984). A mire massif is characterized by a singular period of origin, a defined position in the mesorelief, its vegetation, and hydrology (Galkina 1946; Ivanov 1953). The basic classification unit is mire massif type. For purpose of this publication, we use "mire massif type" and "mire type" as synonyms.

Peatlands are generally not recognized as specific ecosystems in many mostly dry countries. Mongolian peatlands were not mentioned in Global Peat Resources by Lappalainen (1996) and were poorly represented in all other global peatland reviews (see Minayeva et al. 2005). However, Tyuremnov's (1976) review of peat resources refers to Obruchev's (1863–1956) description of vast and impassable mires in the Onon and Shilka Valleys in eastern Mongolia, sedge fens on permafrost west of the Khingan Ridge, and on the lower plateau of the Large Khingan and in the Khangay Mountains.

An estimate of peatland distribution in Mongolia was undertaken within the framework of the Global Peatland Initiative Project (Minayeva et al. 2004). Using ecosystem and topographic maps and field surveys, it is estimated that peat-covered areas in Mongolia cover more than 27,000 km^2 (1.7% of the country). The distribution of peatlands and mires appear to be very similar to that predicted by Gunin and Vostokova (1995) in their *Map of the Ecosystems of Mongolia*. The peat areas are bound to foothills and mountains within forest steppe, taiga, and mountain tundra

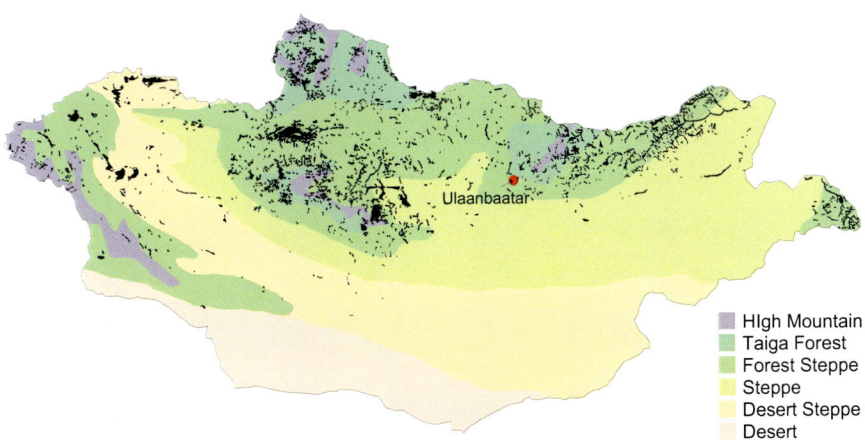

Fig. 2 Distribution of ecosystems which contain peatlands (in black) in Mongolia (© International Peat Society with permission; from Minayeva et al. 2004, Fig. 1, p. 44)

and are common in mountain river valleys in the steppe zone (Fig. 2). Sedge fens in highland valleys are the most widely dispersed mire type in Mongolia. These fens cover approximately 14,000 km^2, or 1% of the country's total area. These fens are a vital natural resource for Mongolia, as they provide highly productive pastures.

Diversity of Mires

Mire diversity in northern, central, and northeastern Mongolia is high and consists of several mire types in forest and steppe zones as well as in highlands, valleys, and permafrost. We are not including the alpine mires of the Altay Mountains which are influenced by a more continental temperate rather than arid climate. Peatlands of the Russian Altay Mountains are discussed in a number of publications (see Volkova et al. 2009).

The mire massif types can be found in different latitudinal and altitudinal zones as is reflected in Fig. 3 and Table 1. The species composition of these mire massif types is quite diverse and is not included in this overview. However, mire massif types do differ in their vegetation reflected in the composition of dominant species and have species that are characteristic to each type (Table 1).

Group of Mire Massif Types I: Transitional Peatlands (Ia) and Spring Fens (IIb) in Large Lowland Valleys

The Khentey Mountains are considered, together with Khovsgol Lake and Darchad Kettle, as part of the Sayan Ridge (Fig. 4). At this point, the Zabaikalie taiga penetrates into the Mongolian Gobi (Yunatov 1950). In the Khentey Mountains, at medium elevations (800–1,000 m asl), wide river valleys develop, such as the Huderijn-gol and Ero River Valleys (49°47′N, 107°29′E–49°49′N, 107°28′E).

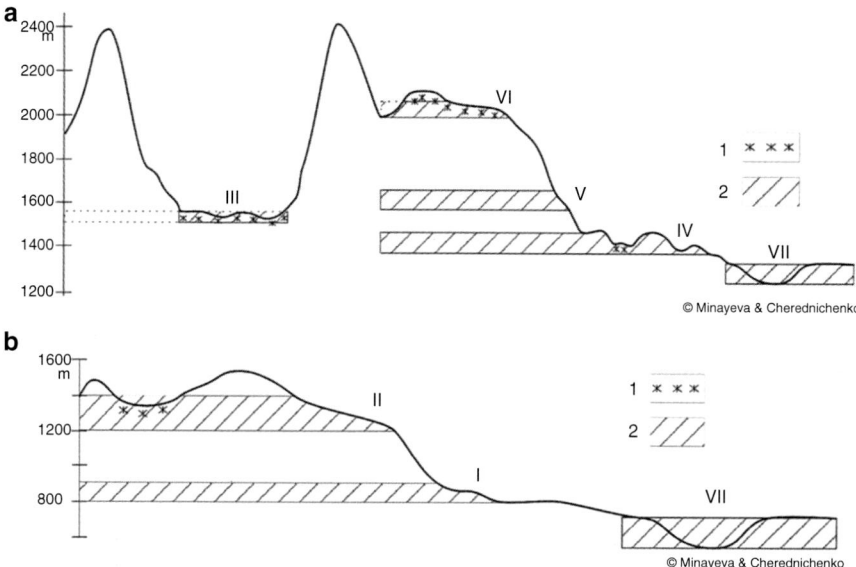

Fig. 3 (**a**) Steppe and forest steppe zone; (**b**) forest steppe and taiga forest zone. *1* – Permafrost; *2* – potential position of mire type

Vegetation at this elevation in Khentey is considered to be mountain forest steppe. The more humid climate, groundwater discharge, and temporary flooding by the river create good conditions for peat accumulation and mire vegetation development. Peat deposits up to 1–1.5 m depth are usually found directly in the valley alongside its gentlest slope.

Group of Mire Massif Types II: Mires of the Taiga Altitudinal Zone – Sphagnum-Sedge Kettle and Sloping Bogs (IIa); Transitional Open Sphagnum-Sedge Sloping Peatlands (IIb); Paludified Shallow Peat Forests and Coniferous Transition Peatlands (IIc); Forested Transitional Birch-Willow-Grass Peatlands (IId)

In the Khentey Mountains of northern Mongolia, the typical taiga belt is defined by the presence of coniferous forests and sphagnum-sedge bogs (Fig. 5). In its origin, the mountain massif belongs to the Sayan Mountains and forms the piedmonts along the Yablonovy Ridge. The vegetation structure and floristic composition of the highlands is very similar to the extensive Russian taiga to the north. At elevations ranging from 1,200 to 1,500 m, kettle (IIa) and sloping (IIb) transitional sphagnum-sedge open mires with peat deposits up to 5 m can be found in the watersheds of the Huderijn-gol and Ero Rivers.

Coniferous forests typical for the taiga zone, with sphagnum mosses and peat depths of 30–50 cm (IIc), are found in highland rivers and brook valleys and in

Table 1 Characteristics of Mongolian mire massif types

Type of mire massif		Zone	Elevation	Position in the landscape	Peat depth and age	Dominant and characteristic species
Transitional peatlands in large lowland valleys	Ia	Forest steppe, forest	800–1,100	Valley	1.0–1.5 m From 200 years BP near river to 1,500 years BP in valley	Dominant species: *Carex cespitosa, Equisetum palustre, Eriophorum polystachyon* Characteristic species: *Betula platyphylla Sukacz. Calamagrostis purpurea, Ligularia sibirica*
Spring fens in large lowland valleys	Ib	Forest steppe, forest	800–1,100	Gentle slopes	1.0–2.0 m	Dominant species: *Carex schmidtii, Carex cespitosa* Characteristic species: *Equisetum fluviatile, Eriophorum humile, Pedicularis karoi*
Sphagnum-sedge kettle and sloping bogs	IIa	Forest	1,200–1,600	Intermountain kettles, terraces	3–3.5 m 6,000 years BP	Dominant species: *Carex limosa, C. rostrata, Sphagnum flexuosum* Characteristic species: *Drosera anglica, Oxycoccus microcarpus, Scheuchzeria palustris*
Open transitional sedge sphagnum sloping peatlands	IIb	Forest, forest steppe	1,200–1,600	Plateaus and long gentle slopes	2–3 m	Dominant species: *Carex limosa, Ledum palustre, Sphagnum balticum* Characteristic species: *Parnassia palustris, Sphagnum subsecundum*
Paludified shallow peat forests and forested peatlands	IIc	Forest, forest steppe	900–1,600	Long moderate slopes	0.2–0.5 m	Dominant species: *Abies sibirica, Carex globularis, Linnaea borealis* Characteristic species: *Poa sibirica, Vaccinium vitis-idaea, Sphagnum* sp.
Forested transitional birch-willow-grass peatlands	IId	Forest, forest steppe	900–1,200	Kettles between gentle slopes close to the foot of the mountain ridge	1–1.5 m 900–1,500 years BP	Dominant species: *Betula platyphylla, Salix glauca, Agrostis mongolica, Calamagrostis purpurea* Characteristic species: *Carex lasiocarpa, Epilobium palustre, Filipendula palmata, Deschampsia koelerioides, Epilobium palustre*

(continued)

Table 1 (continued)

Type of mire massif		Zone	Elevation	Position in the landscape	Peat depth and age	Dominant and characteristic species
Transitional highland mires on permafrost	III	Steppe and forest steppe	1,500–1,600	Vast flat areas in plateau of the intermountain kettle	0.6 m of unfrozen peat – 2,000 years BP 0.3 m peat on gravel 2,400 years BP Estimated 1–2 m of frozen peat – no reliable data	Dominant species: *Carex coriophora, Triglochin maritimum, Angelica tenuifolia* Characteristic species: *Beckmannia syzigachne, Kobresia myosuroides, Pedicularis karoi*
Highland valley meadow fens, occur on permafrost	IV	Steppe and forest steppe	1,300–1,500			Dominant species: *Eleocharis palustris, Potamogeton perfoliatus, Carex duriuscula* Characteristic species: *Bromopsis inermis, Carex atherodes, Artemisia adamsii, Carex brunnescens*
Highland valley sloping fens	Va	Steppe and forest steppe	1,600–1,800	In the highland steep valleys on the slopes	0.2–1.0 m 1,500 years BP	Dominant species: *Iris lactea, Artemisia sericea*; on wet peat soils: *Agrostis mongolica, Angelica tenuifolia, Carex atherodes, Carex cespitosa, Carex cinerea, Carex coriophora, Carex curaica, Carex delicata*, in pools – *Batrachium divaricatum* Characteristic species: *Artemisia laciniata, Hordeum brevisubulatum, Sanguisorba officinalis*
Highland valley spring fens	Vb	Steppe and forest steppe	1,600–1,800	On the slope	0.5–1.0 m	Dominant species: *Bistorta alopecuroides, Carex reptabunda, Carex reptabunda, Betula rotundata* Characteristic species: *Carex enervis, Ligularia altaica, Salix glauca*

Highland blanket and sloping bogs	VIa	Steppe, forest steppe	1,800–2,200	On gentle slope and circus in the pass	0.6–1.0 m 900–1,500 years BP	Dominant species: *Alopecurus turczaninovii, Calamagrostis macilenta, Carex atrofusca* Characteristic species: *Carex enervis, Cirsium esculentum, Ranunculus repens*
Highland spring fens	VIb	Steppe, forest steppe	1,500–2,200	On the steep slopes	0.5–1.2 m	Dominant species: *Salix arctica, Potentilla fruticosa, Calliergon giganteum* Characteristic species: *Salix berberifolia, Fontinalis antipyretica, Calliergonella cuspidata, Melandrium apetalum*
Lacustrian brackish and saline marshes of large valleys	VII	Steppe, semidesert, desert	600–800	Lake kettles	No data	Not described yet

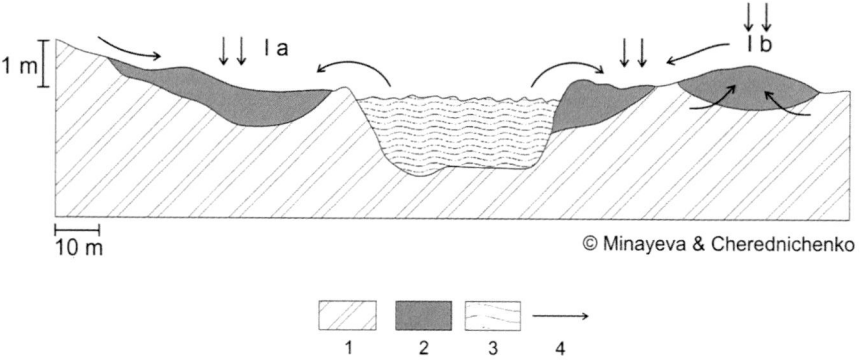

Fig. 4 Cross section of mire massifs of large river valleys in forest steppe and forest zone. Legend: *1* – bedrock; *2* – peat; *3* – water; *4* – water flows

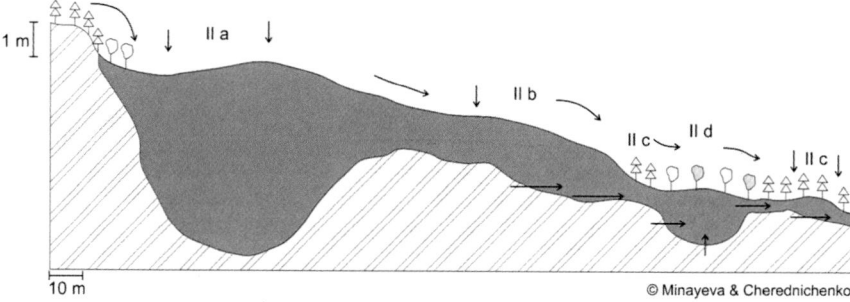

Fig. 5 Cross section of peatlands in taiga altitudinal belt. Legend the same as for Fig. 4

intermountain depressions. They are fed by surface flow, precipitation, and partly from groundwater discharge. Forested kettle mires with actively growing peat are sometimes found in the depressions between slopes (IId) and are mostly fed by groundwater and surface flow.

The Mire Nuur (49°38 N′, 107°48′E) is an example of the mire type IIa – in the Taiga altitudinal zone. Formed 6,000 years BP during the Atlantic Period, when the climate here was presumably wetter and warmer, the loamy deposits beneath the peat contain *Carex juncella* and algae macrofossils. The oldest peat is formed from the same sedge and horsetail. The sphagnum settled very soon (300 years after peat formation had begun), and the subsequent community can be classified as rich fen with relatively fast peat growth. Around 3,500 years ago, the site became more oligotrophic and more or less acquired its modern appearance, with the dominance of *Carex limosa, C. rostrata, Scheuchzeria palustris*, and *Sphagnum flexuosum*. The mire is fed by precipitations and surface flow.

Further east from Khentey, similar conditions can be found along the Ereen-Daba (Ereen-Nuuru) and Erman Ridges. Transitional valley fens were described from the

Bayan-gol (49°04′, 110°52′) and Dujche-gol (49°50′, 113°48′) Valleys and from the intermontane kettle (peat depth more than 70 cm) along the Ereen-Daba Ridge (49°18′, 112°35′). The mires here carry more shrubs, birch on drier patches, and willows in wetter areas.

Mire Massif Type III: Transitional Highland Mires on Permafrost

Mire development has a strong association with permafrost. The Darhad Kettle area is one area in Mongolia known for conditions conducive for permafrost (50°30′, 99°20′). In origin, this mountain area belongs to the Sayan Ridge. The Shishig-gol River, which runs through the kettle, is one of the main tributaries of the Yenisey River. The valley is located 1,500–1,600 m above sea level and is surrounded by mountain ridges with elevations of up to 2,500 m which serve as a barrier for climatic influences to the kettle environment.

In permafrost areas, the water level is always high and water temperature low; this creates constant habitat conditions for typical mire species and peat formation (Fig. 6). In the Shishig-gol Valley, relatively vast areas are occupied by very wet sedge fens with green mosses and shrubs. The density of willow shrubs is higher along the river and with a mixture of *Betula fusca* occurs on old peat deposits (2,360 years ±40 GIN–12829) underlain by gravel with ice at depths of 30–35 cm. Away from the riverbank, the peat increases in depth, is frozen beginning at 50 cm to an unknown depth and is aged at around 2,000 years BP. In this distant part, only dwarf willows are present and the mire surface is a complex of hummocks and hollows. Both hummocks and hollows carry transition vegetation and would be referred to as aapa mires in temperate areas.

From past descriptions and local knowledge, we know these mires had previously occupied a much larger area. Climate fluctuations create dramatic changes in the mires. Areas of the Darhad Kettle which previously contained mires now resemble a desert. The former hollows are still filled with peat, and the basal layer of peat dates back to 400 years BP (410 ± 40, GIN–12599). In these hollows, the same sedges as those encountered in the existing mire are found, but in a much suppressed living

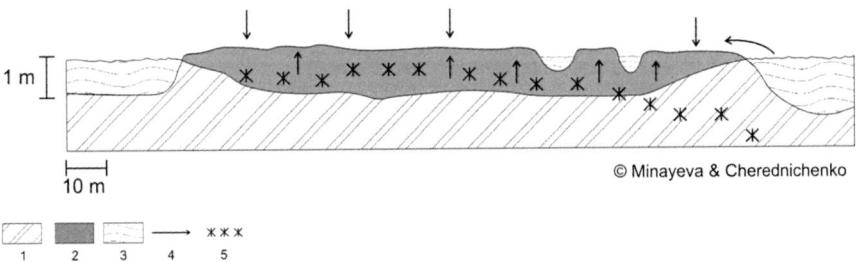

Fig. 6 The cross section of the transitional peatland on permafrost in Darhad Kettle. Legend: *1* – bedrock; *2* – peat; *3* – water; *4* – water flows; *5* – permafrost border

form. In between the hollows and on the ridges, vegetation is typical for steppe and gravel habitats.

Mire Massif Type IV: Highland Valley Meadow Fens

Valley meadow fens are found in steppe and forest steppe zones around mountain lakes and in large river valleys, such as the valleys of the Orkhon, Muren, Tesiin Gol, and Ider Rivers. In such valleys, trees are nearly absent, and steppe meadows neighbor sedge fens. Such landscapes might be called "lowlands," with some assumptions. These lake kettle mires and vast lowland valleys are found at an elevation of 1,300–1,500 m above sea level. The mires here primarily represent the remains of former large fens with tall sedges and presumably connected to the permafrost as well as fed by river flooding, surface flow from the surrounding mountains, and very limited precipitation (Fig. 7). This can be deduced from the hummock remains, from the composition and age of bottom peat deposits, from remains of natural patches where also permafrost is present non deeper than 1 m, and from stories of local nomadic people. Despite the fact that the peat surface can be extremely dry in the summertime, peat depth is still considerable, up to 1 m. These mires are sedge fens but are sometimes salty. Usually, a hollow-hummock structure is present. The hollows are filled with peat, which is still wet. Bottom peat layers vary significantly in age, depending on the history of the valley, and date back to 200–1,000 years before present (BP) (Carbon-14 noncalibrated dates). Pavlov (1929) described these lowland valley landscapes as vegetation complexes with pools, mires with hummocks, meadows on flat loamy sites, loamy-boulder ridges with meadow-steppe vegetation, and pure boulder sites with steppe-desert vegetation.

The Orkhon Valley (N 47°27′–47°44′, E 102°41′–102°48′) sedge fens are relatively young, with basal peat age not exceeding 250 years (240 ± 100 GIN–12811; 230 ± 70 GIN–12815). However, it is presently in a stage of peat mineralization. There are still pools with typical aquatic flora. The meadow and meadow-steppe sites are more extensive than they were in the past due to the fact that parts of mires had dried. Mire vegetation itself can be found in hollows, which are still filled with peat that is saturated by water. Sites with gravel and boulders host typical xerophytic

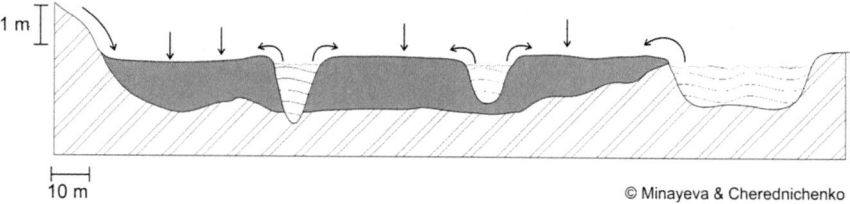

Fig. 7 Cross section of the transitional valley fen in the large river valleys in steppe latitudinal zone. The legend is the same as for Fig. 4

flora. All habitats are connected by species that can tolerate a broader range of conditions.

Sedge fens in the Tesiin Gol Valley (N 49°12′, E 98°12′) are similar in type, but they adjoin salty lakes and salt marshes. The "mire" history of that landscape is somewhat longer than that in the Orkhon Valley and dates back to 900 ± 60 years (GIN–12823). The primary peat-forming material is comprised of sedge macrofossils, in the beginning *Carex saxatilis*, replaced later by *C. schmidtii*. Green moss macrofossils are very sparse in the peat. The modern aquatic and mire flora is more diverse compared to that in the Orkhon Valley and also contains additional species. Halophytic flora was enriched by *Pedicularis dasystachys*, but *Triglochin palustre* does not occur. West Khangay is connected with the Large Lakes Kettle, which in the past, may have been a source of wet air masses (Yunatov 1950, p. 143). Now, the Booreg-Delig sands of that kettle, as well as the southern river valleys connected with the Gobi Desert, have become corridors to the West Khangay Valleys for desert species.

Sedge fens in the steppe zone of eastern Mongolia in the Ulz-gol and Onon Valleys are often connected with saucer-shaped lakes in valleys and oxbows fed by groundwater discharge. As a whole, the river valleys present a combination of steppe, meadow, dry peatland, mire, lake, and pond habitat types. The composition of vascular plant species is characterized by the relatively low occurrence of meadow and steppe species. The community structure is also slightly different; similar to the Orkhon River Valley, a significant part of the habitats are still wet with tall sedges dominating, but dry peat patches with steppe species do occur. Peat depths in living mires normally do not exceed 0.7 m. Another specific feature is the presence of the Dahuric biogeographical species group, including *Carex brunnescens*. Such mires are described from the kettle of Galutyn-Nur Lake in the Ulz-gol River Valley (N 49°43′–E 117°17′) and alongside small lakes in the Onon River Valley (N 48°41′–48°45′, E 110°10′–110°21′).

The described mires of large valleys still carry specific mire biodiversity and valuable paleoecological information in their peat deposits. Both the Orkhon and Tesiin Gol River Valleys have a significant portion of dry steppe flora species. Presumably, mires here had entirely lost their ability to maintain peat accumulation, water storage, and landscape integrity (against erosion). In eastern Mongolia, there are still some wet habitats among valley peatlands.

Mire Massif of Highlands of Steppe Zone V: Highland Valley Sloping Fens (Va) Highland Valley Spring Fens (Vb)

Sedge fens in highland valleys are the most widely dispersed mire type in Mongolia (Fig. 8). According to map analyses, these fens cover approximately 14,000 km^2, or 1% of the country's total area. These fens are a vital natural resource for Mongolia, as they comprise highly productive pastures.

In Khangay, at elevations ranging from 1,600 to 1,800 m, most small and medium highland valleys are filled with shallow peat. Mires are presented by sloping sedge

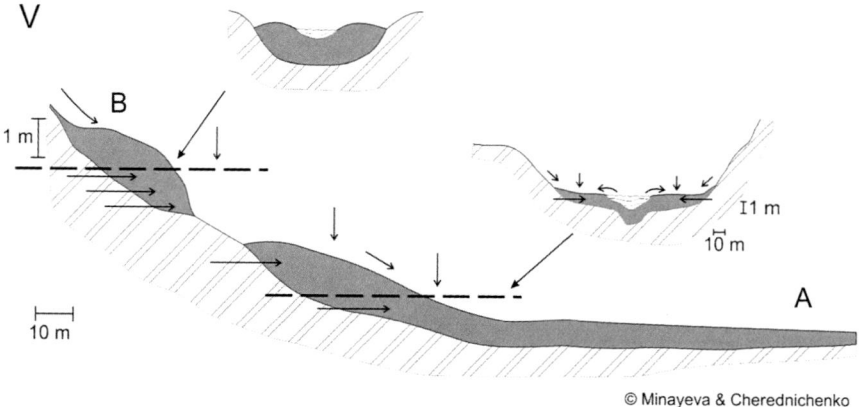

Fig. 8 The cross section of the sloping and spring highland fens of steppe zone. Legend the same as for Fig. 4

fens and are primarily used as pastures. During most of the year, the water level is low, but peat deposits with depths of 20–50 cm still retain enough water to support pastures. Occurring in the Uver-Teel River Valley – one of the sources of the Orkhon River (N 47°13′, E 101°51′), this type of fen has peat deposits dating back to 1,500 years BP. The inversion of peat layers in these deposits – older deposits (1,540 ± 90 GIN–12818) were found at depths of 25 cm overlying more recent deposits (830 ± 100 GIN–12819) at 42 cm – reflects the highly dynamic processes taking place in these highland mires. Occasionally, rainwater or mudflows from above move peat deposits downhill. Such "mire-burst" phenomena have also been described for the Alps.

Higher than sloping fens, very often one can find spring fens – fed by strong discharge of groundwater. That is also the source of water for laying down valley sloping fens together with moderate discharge and permafrost in the past.

Mire Massif Type VI Highland Spring (VIb) and Blanket/Sloping Fen (VIa) on Permafrost

Highland mires (Fig. 9) are the most intact peatlands in Mongolia. They cover small patches in watersheds at elevations around 2,200 m in the steppe zone, where steppe gives way to alpine tundra and/or slopes with water discharge or terraces. They are typical high-mountain sloping blanket bogs (Fig. 10). Annual precipitation is more than 600 mm. In most cases, these peatlands are associated with permanent or seasonally frozen bedrock. Typical bog vegetation is found including sphagnum mosses, dwarf shrubs, sedges, and cotton grass presented by two species.

These bog types are known from Pavlov (1929) and Yunatov (1950). Having examined the floristic composition, Pavlov had presumed that, in certain places, those mires remained above the last glaciation and served as refugia for various

Fig. 9 Spring and blanket bogs in highlands can survive because of permafrost and high precipitation (Photo credit: A. Sirin © Rights remain with the author)

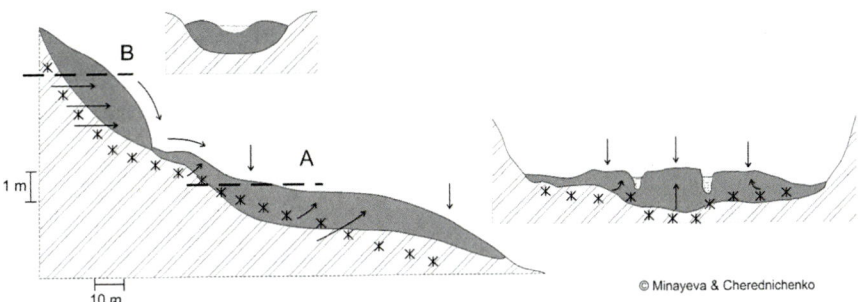

Fig. 10 Cross section of the highland sloping/blanket bog and spring fen on permafrost. The legend is the same as for Fig. 6

species groups. The floristic diversity is really very high here, and we have no other evidence to support the "refugia" theory. The peat deposits are not deep (up to 60 cm), date back to 1,300 years BP (1,310 ± 60 GIN–12566), and are underlain by frozen gravel. Peat is built up mainly from sedges – at the bottom *Carex saxatilis*, and in upper parts *C. atrofusca*. In the uppermost parts, cotton grass and sphagnum mosses are found as macrofossils. Macrofossils of green mosses – *Aulacomnium* sp., *Calliergon* sp., *Polytrichum* sp., *Tomenthypnum* sp. – are quite common in the peat. Willows are always present in the peat along the core. A general description of the vegetation structure is high ridges surrounding a small spring in the middle of one of the slopes and the valley of the small brook runs out of the spring. The slopes are covered by dwarf shrub-cotton grass-sphagnum vegetation. Some of the slopes have terraces on which hollow-hummock structures have begun to form. Sometimes, rock is on the surface, and then the group of species typical for gravel habitats settles there.

It is expected that a final mire type – lacustrian brackish and saline marshes of large valleys – will be found in the east of the country and its characteristics described.

Conservation

Mires are preserved within a number of nature reserves and Ramsar Sites in Mongolia. There are no special protected areas devoted exclusively to mire protection, nor is there special site management regarding peatlands. However, around 40% of peatlands are found within protected areas, but they are not recognized nor specially addressed in protected areas management plans.

The main threat to peatlands in Mongolia is the absence of detailed knowledge about their diversity, distribution, and natural functions. Peatlands use planning in Mongolia should be based on solid knowledge about their role in the landscape, including their catchment scale. For thousands of years, Mongolians have traditionally lived in harmony with nature. The principle of wise use and age-old Mongolian traditions of living with nature could help to facilitate the identification of a rational balance between the utilization and conservation of these unique ecosystems, which are so vitally important to people and to combat desertification and land degradation.

Threats and Future Challenges

Peatlands in Mongolia are typical landscape features that have developed over thousands of years. Peatlands constitute the last wet habitats in a considerable part of the country and are unique habitats that support special biodiversity. Vascular plant flora of Mongolian mires is reported as over 400 species, among them rare or newly found, and the broad spectrum of species within mire habitats is considered to be a response by plant communities to fluctuations in ecological conditions.

Peat deposits are not present in volumes necessary for industrial use, but they do serve another significant role in the Mongolian economy and social life. Peat stores water that maintains wet habitats and pastures, provides a source to feed rivers, prevents soil erosion, maintains groundwater levels necessary for forest and crop growth, and keeps wells full of water. Mires and peatlands are used mainly for livestock grazing and sometimes for arable lands. Mires' role as pastureland in Mongolia is really outstanding. Dry periods can continue for years, and during those periods, the moisture preserved in mires due to the unique nature of peat really becomes a source of life. The mires are living barriers on the road to desertification. Mires, which have been affected by people through overgrazing and drainage, can lose their ability to protect them from desertification.

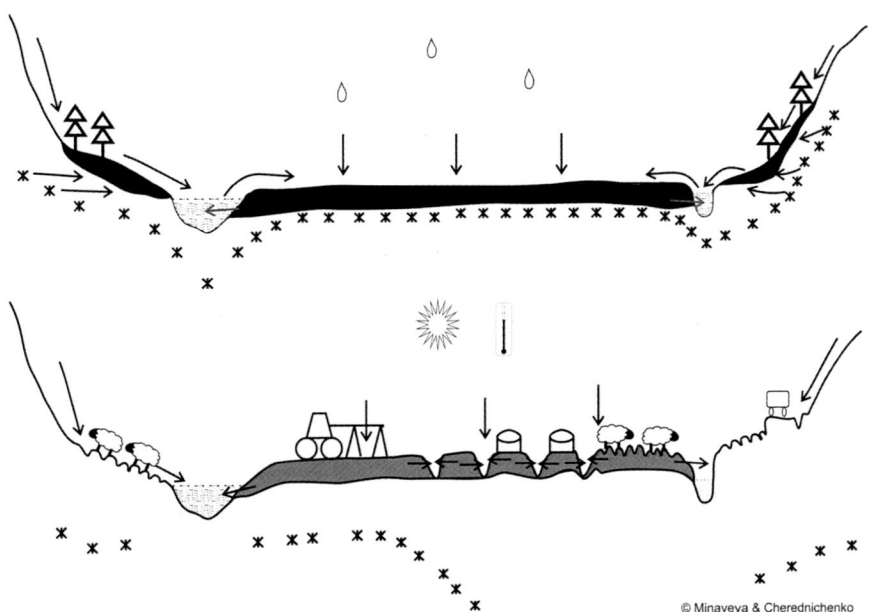

Fig. 11 Changes in ecosystems of Orkhon valley during last 1,000 years

A comparative analysis of existing peatlands to descriptions from the nineteenth and twentieth centuries provides support of dramatic changes in peatland landscapes during recent times. Paleoecological data based on peat macrofossil, decomposition rate, bulk density and ash content analyses, and radiocarbon dating demonstrates long-term changes in peatlands. Figure 11, based on numerous surveys and literature, is a representation of the changes in the ecosystems of the Orkhon River Valley due to climate change and human impacts during the last 1,000 years. Only the sloping mires in highlands, valley mires on permafrost, and sphagnum-sedge mires in taiga zone demonstrate stable conditions. The peatlands of steppe and forest steppe zone originated under more favorable climatic conditions and are progressively degrading during the last decades. This correlates with the overall trend of peat increment decrease in North Eurasia during warm periods within the last three millenniums (Klimanov and Sirin 1977).

Increasing temperatures and evapotranspiration rates will make soils much drier, and desiccation of peat surfaces will make them more susceptible to erosion. During droughts, the upper peat layers are eroded by wind and the surface becomes deflated, and during intense rainfall peat is moved down slope and lost in runoff. This has already occurred in many regions of Mongolia and is expected to expand in the future. Human activities can seriously increase the vulnerability of peatlands to climate change. The climate-driven desertification of mires is reinforced by overpasturing of livestock which leads to disturbance of plant and soil cover. Peat degradation and sometimes peat fires follow the drying up of the mires. Wise use of peatlands as pastures could serve as a key measure for their adaptation to climate change.

References

Botch MS, Masing VV. Ecosystems of mires in the USSR. Leningrad: Nauka; 1979. 183 p. (In Russian).
Dagvardorj D, editor. Mongolia Second National Communication under the UNFCCC. Ulaanbaatar: Ministry of Nature, Environment and Tourism, Mongolia; 2010. 159 p.
Galkina EA. Mire landscapes and the principles of their classification. In: Shishkin BK, editor. Collection of scientific works performed at the komarov botanical institute in leningrad during three years of world war II (1941–1943). Leningrad: Izdatelstvo Akademii Nauk SSSR; 1946. p. 129–56 (In Russian).
Gunin P, Vostokova EA, editors. Ecosystems of Mongolia. Moscow: Accord; 1995 (In Russian).
Ivanov K. Mire hydrology. Leningrad: Hydrometeoizdat; 1953. 300 p. (In Russian).
Joosten H, Clarke D. Wise use of mires and peatlands: background and principles including a framework for decision-making. Saarijärvi: International Mire Conservation Group/International Peat Society; 2002. 304 p.
Kats NJ. The mires of the globe. Moscow: Nauka; 1971. 295 p. (In Russian).
Klimanov VA, Sirin AA. The dynamics of peat accumulation by mires of Northern Eurasia during the last three thousand years. In: Trettin CC, editor. Northern forested wetlands, ecology and management. Boca Raton/London/Tokyo: Lewis Publishers/CRC Press; 1977. p. 319–30.
Lappalainen E, editor. Global peat resources. Jyskä: International Peat Society/Geological Survey of Finland; 1996. 368 p.
Lindsay RA, Charman DJ, Everingham F, O'Reilly RM, Palmer MA, Rowell TA, Stroud DA. The flow country: the peatlands of Caithness and Sutherland. Peterborough: Nature Conservancy Council; 1988. 174 p.
Markov WD, Olenin AS, Ospennikowa LA, Skobejewa EI, Khoroschew PI. Peat resources of the World. Reference book. Moscow, Nedra Publishing House; 1988. 384 p. (In Russian).
Masing VV. Some topical questions about classification and terminology in mire science. In: Abramova TG, Botch MS, Galkina EA, editors. Mire types of the USSR and principles of their classification. Leningrad: Nauka; 1974. p. 6–12 (In Russian).
Minayeva T, Gunin P, Sirin A, Dugardzhav C, Bazha S. Peatlands in Mongolia: the typical and disappearing landscape. Peatl Int. 2004;2:44–7.
Minayeva T, Sirin A, Dorofeyuk N, Smagin V, Bayasgalan D, Gunin P, et al. Mongolian Mires: from taiga to desert. In: Steiner GM, editor. Moore von Sibirien bis Feuerland. Stapfia, 85, zugleich Kataloge der OÖ, Landesmuseen Neue Serie. Linz, Austria: Land Oberösterreich, Biologiezentrum/Oberösterreichische Landesmuseen, vol. 35; 2005. p. 335–52.
Minayeva T, Bragg O, Cherednichenko O, et al. Peatlands and biodiversity. In: Parish F, Sirin A, Charman D, et al., editors. Assessment on peatlands, biodiversity and climate change: main report. Kuala Lumpur/Wageningen: The Netherlands Global Environment Centre/Wetlands International; 2008. p. 60–98.
Parish FA, Sirin D, Charman H, Joosten T, Minayeva M, Silvius, L. Stringer. Assessment on peatlands, biodiversity and climate change: main report. Global environment centre, Kuala Lumpur and Wetlands International, Wageningen. 2008. 179 pp.
Pavlov N. An introduction to Khangay Mountain area vegetation cover. A preliminary report from a botanical expedition to Northern Mongolia in 1926. Leningrad: USSR Academy of Sciences Publishing House; 1929. 72 p. (In Russian).
Peat Dictionary. Helsinki: International Pet Society; 1984. 595 p. (Russian-English-German-Finnish-Swedish).
Sirin A, Minayeva T, editors. Peatlands of Russia: towards an analysis of sectoral information. Moscow: Geos Publishing House; 2001 (In Russian).

Tyuremnov SN. Torfyanyje mestorozhdenija (Peat deposits). 3rd ed. Moscow: Nedra; 1976. 488 p. (In Russian).

Volkova I, Volkov I, Kuznetsova A. Mountain mires of South Siberia: biological diversity and environmental functions. Int J Environ Stud. 2009;66(4):465–72.

Yunatov A. The main features of vegetation cover of the Mongolian People's Republic. In: Proceedings of the Mongolian Commission of Academy of Sciences USSR – vol. 39. Moscow/Leningrad: Academy of Science USSR Publishing House; 1950, 224 p (In Russian).

Yurkovskaya. Geography and cartography of mire vegetation of European Russia and adjacent territories. Trudy BIN. Trans Komarov Bot Inst St. Petersburg. 1992;4:1–256 (In Russian).

Yangtze River Basin (China)

128

Cui Lijuan, Zhang Manyin, and Xu Weigang

Contents

Introduction	1552
Hydrology and Climate	1552
Important Wetlands in the Yangtze River Basin	1554
Wetlands in the Headstream	1554
Wetlands in the Upper Reaches	1555
Wetlands in the Middle and Lower Reaches	1556
Wetlands of the Yangtze River Estuary	1557
Current Status of Habitats and Threatened Species and Biodiversity	1558
Threats and Conservation Issues	1560
Changes in Hydrology and Runoff	1560
Sharp Decline in Wetland Biodiversity	1560
Aggravated Water Pollution from Pesticides	1561
Human Disturbance	1561
Future Challenges	1561
Influences of Future Climate Change	1561
Comprehensive Management of Nature Reserves	1562
References	1562

Abstract

The Yangtze River, originating in the Tanggula Mountains, is the longest river in China and the third longest in the world. It flows from west to east across 11 provinces (or autonomous regions and municipalities) for about 6,300 km before emptying into the East China Sea at Shanghai. The Yangtze River ranks first nationally in terms of water flow, with total water resources of 960 billion m^3, making up about 36% of the runoff volume in China, 20 times larger than that of the Yellow River. The wetlands along the Yangtze River Basin cover a total area

---

C. Lijuan (✉) · Z. Manyin (✉) · X. Weigang (✉)
Institute of Wetland Research, Chinese Academy of Forestry, Beijing, China
e-mail: lkyclj@126.com; cneco@126.com; xuweigang@foxmail.com

© Springer Science+Business Media B.V., part of Springer Nature 2018
C. M. Finlayson et al. (eds.), *The Wetland Book*,
https://doi.org/10.1007/978-94-007-4001-3_99

of 20.74 Mha, accounting for 38.69% of the national wetland area. This chapter introduces the wetlands distributed in the headstream, the upper reaches, the middle and lower reaches and the estuary of Yangtze River. Further, the chapter identifies the challenges faced by the wetlands along Yangtze River.

Keywords

Yangtze river · Wetland · Biodiversity · Human disturbance · Three gorges dam

Introduction

The Yangtze River, originating in the Tanggula Mountains, stands as the longest river in China and the third longest in the world. With its tributaries stretching across 8 provinces (or autonomous regions), it flows for about 6,300 km from west to east across 11 provinces (or autonomous regions and municipalities) before emptying into the East China Sea at Shanghai. The Yangtze River, with an average discharge of 960 billion m^3 per year (Ministry of Water Resources 2013), boasts a drainage area of about 1.8 million km^2, amounting to approximately 18.8% of the national territory. Along the river, abundant resources and multitudinous landforms can be found in the basin covering a total of 0.37 billion mu (246,666,666 ha) of fertile farmlands, accounting for one quarter of the total farmland in China. In addition, the grain output along the Yangtze River Basin constitutes as much as 40% of total national output, while the cotton output within the region comprises one third of the total.

Hydrology and Climate

The Yangtze River ranks first nationally in terms of water flow, with total water resources of 960 billion m^3, making up about 36% of the runoff volume in China, 20 times larger than that of the Yellow River. Also, it ranks third globally after the Amazon River and the Congo River (also known as the Zaire River) running through the tropical rainforests. The annual water vapor volume released into the atmosphere from the Yangtze has been averaged at $67,800 \times 10^8$ m^3 over the years, 25% of which has been transformed into rainfall with an annual average precipitation of approximately 1,100 mm. Of the annual 960 billion m^3 runoff volume of the Yangtze River, 47% originates from the upper reaches, 21% from the Dongting Lake, and 17% from the Poyang Lake (Yangtze River Water Resources Bulletin 2014).

Under the influences of the monsoon climate, the Yangtze River Basin has long been subject to an uneven temporal–spatial distribution of precipitation with a great interannual variation and concentrated intra-annual distribution, one of the most important factors with respect to the frequent floods. In recent years, it has seen obvious anomalous changes in the temporal–spatial distribution of precipitation compared with the past, which may very likely upset the established balance between the existing river runoff and flood control system and result in unexpected major disasters.

Relevant research studies show that the annual precipitation volume along the Yangtze River Basin is on the rise but has witnessed certain variations in temporal–spatial distribution, while the severe precipitations increased (Xu et al. 2006; Wang 2008).

The changes within the Yangtze River Basin in terms of hydrology and climate have exerted a great influence on the wetlands within the region. The alpine wetlands along the headwaters of the Yangtze River are an example. From the perspective of structure, these wetlands have been subject to decline with areas of the marshy grasslands, marshes, and lakes reduced to 1,843.76 km^2, 186.54 km^2, and 114.8 km^2, respectively, over the past 50 years; from the perspective of rate of change, the wetlands have decreased by 1.48%/a, of which the alpine peat marshes declined by 3.83%/a ranking first, while the marshy grasslands, marshes, and lakes declined by 2.72%/a, 0.85%/a, and 0.02%/a, respectively. In fact, the water in marshes averaged 20–40 cm in depth and could reach more than 1.0 m at its deepest before the 1930s; however, since 2004 the water depth only averaged 10–15 cm with a lot of marshy grounds merely in an over-wet status. The disappearance of the marshes has led to a decreased water vapor recharge within the region, increased desertification, and accelerated aridification (Liang et al. 2013).

The wetlands along the Yangtze River Basin cover a total area of 20.74 million hectares, accounting for 38.69% of the national wetland area. With great efforts over the past years, the State Forestry Administration has successfully established 164 wetland nature reserves and 51 national wetland parks along the Yangtze River. Some thirteen wetlands within the Basin have been listed as Ramsar Sites, marking the first achievement in building and improving a protection network of natural wetlands (Forest Administration 2010). Although wetlands can be found in all the provinces and autonomous regions throughout the Basin, they are distributed in an uneven manner, with most located in Qinghai, Hunan, and Jiangxi Provinces and with few in Guizhou and Yunnan Provinces, while still fewer are found in Gansu and Henan Provinces. Based on type, the common wetlands in the Yangtze River Basin can be classified into several categories including estuarial and coastal wetlands, lake and bottomland wetlands, marsh wetlands, and constructed wetlands, the distribution and area of which are presented in Fig. 1. and Table 1 (Yan et al. 2013).

The Yangtze River flows across southwest, central, and eastern China, stretching over plains, basins, hills, and plateaus with extremely complex geomorphic conditions and remarkably different geographic and hydrothermal conditions, which together contribute to a wide range of wetland types. The distribution of the wetlands across the basin is extensive and unbalanced. For example, marsh wetlands are widespread on the plateaus in the headwaters of the Yangtze River which constitute the largest marsh areas in China; plateau lakes and marsh wetlands can be found in the upper reaches of the river; the largest groups of freshwater lakes occur in the middle and lower reaches, forming the most concentrated distribution of freshwater lake wetlands in China; and large areas of tidal wetlands occur at the junction of the land and sea in the Yangtze River Delta (Wu et al. 2001; Ding et al. 2003). Dongting Lake and Poyang Lake had been listed under the Ramsar Convention as early as 1992, and presently, a total of 14 Ramsar Sites have been designated in the Yangtze River Basin from the river source to the Yangtze River Delta (Table 2).

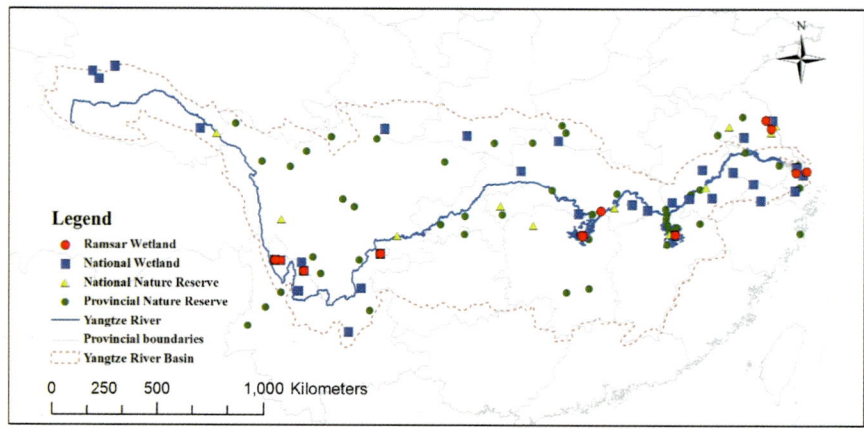

Fig. 1 The distribution of wetlands in the Yangtze River Basin

Table 1 The type and distribution of wetlands in the Yangtze River Basin

Type	Headwaters	Upper reaches	Middle and lower reaches	Estuary	Total (km²)
Estuarial and coastal lands	–	–	–	2,660	2,660
Lake wetlands	650	514.92	14,504.2	–	15,669.12
Marsh Wetlands	9,153	683.44	23,27.34	–	12,163.78
Flood plains	746.3	3,070.6	12,138.5	1,335.6	17,290.8
Subtotal (natural wetlands)	10,549.3	4,269	28,970	3,995.6	47,783.7
Paddy fields	–	38,738.4	87,919.2	–	126,657.5
Reservoirs	–	627.4	3,708.9	–	4,336.3
Artificial farms	–	–	28,666.7	–	28,666.7
Subtotal (constructed wetlands)	–	39,505.4	120,294.8	–	159,660.6
Total	10,549.3	39,936.76	145,555.9	3,995.6	207,444.3

Important Wetlands in the Yangtze River Basin

Wetlands in the Headstream

The headstream of the Yangtze River refer to the river source areas upstream of the Chumar Estuary of the Tongtian River, with a drainage area of about 0.11 million km². This area features marsh wetlands, lake wetlands, and lakeshore wetlands with large areas of marshes along the Tuotuo River, Chumar River, and Dangqu River, making it one of the largest distribution areas of natural marshes in the Yangtze River Basin, with a total area of 9,153 km². The main lakes here consist of

Table 2 Ramsar Sites in the Yangtze River Basin

No.	Name	Region	Area (km^2)
1	Bitahai Wetland	East of Shangri-La County, Yunnan Province	20
2	Napahai Wetland	Shangri-La, Yunnan Province	34.34
3	Lashihai Wetland	Lijiang County, Yunnan Province	14.43
4	Dashanbao	Zhaotong City, Yunnan Province	31.5
5	Dong Dongting Lake	Yueyang City, Hunan Province	1,900
6	Xi Dongting Lake Nature Reserve	Hanshou County, Hunan Province	356.8
7	Nan Dongting Wetland and Waterfowl Reserve	Dongting Lake	1,680
8	Poyang Lake	Nanchang City and Jiujiang City, Jiangxi Province	224
9	Hubei Honghu Wetlands	Jianli County and Honghu City, Hubei Province	414.1
10	Yancheng Wetland	Yancheng City, Jiangsu Province	2,841.79
11	Dafeng National Nature Reserve	Dafeng City, Jiangsu Province	780
12	Shanghai Yangtze Estuarine Wetland Nature Reserve For Chinese Sturgeon	Northeast of Shanghai City	37.6
13	Chongming Dongtan Nature Reserve	Chongming Island, Shanghai	326
14	Shengjin Lake	Chizhou City, Anhui	333.4

Duoergaicuo Lake, Yaxingcuo Lake, Quemocuo Lake, Telashi Lake, and Mazhangcuoqin Lake. Thanks to the low population density, these areas have been subject to little human disturbance, and it has been possible to maintain well-protected wetlands. Wider human activities, however, have indeed exerted some negative effects on these areas through global warming, resulting in increasing temperatures and glacial recession, which has led to the emergence of many runoffs, to evaporation losses exceeding snowmelt water runoff, and to the drying trend of marshes. Compounding the problem, it is very hard for the plateau wetlands, which have played an important role in water conservation in the Yangtze headwaters and in climate adjustment, to recover after destruction because of their own fragile ecological systems (Table 3).

Wetlands in the Upper Reaches

The wetlands in the upper reaches of the Yangtze River are mainly spread over the Yunnan-Guizhou Plateau and Sichuan Province. With high altitude and relatively low

Table 3 Wetlands in the headwaters of the Yangtze River Basin

Wetland level	Wetland name
Wetlands of International Importance	–
Wetlands of National Importance	Zhuonai Lake Wetland, Duoergaicuo Wetland, Kusai Lake Wetland, Longbaotan Nature Reserve Wetland
National Wetland Nature Reserve	Sanjiangyuan Reserve, Longbaotan Nature Reserve
Provincial Nature Reserve	–

precipitation, the marshes in the upper reaches are characterized by herbaceous marshes that mainly consist of reed and sedge *Carex spp.* marshes. The reed marshes are mainly clustered together in the rift lake basin in the northwest of Yunnan Province including Jian Lake. Mainly distributed in Guizhou Plateau are sedge marshes, cattail marshes, and club-rush marshes. Accompanying plants include *Trapa bispinosa* Roxb. and *Alisma plantago-aquatica* L. and *Juncus effusus* L. and *Polygonum hydropiper* L. A few small peat marshes such as Leigongze also occur in this region.

Some of the large lakes within this region are found in the fault zones or the watershed areas of different water systems, such as Dianchi Lake in the upper reaches of the Pudu River, a tributary of the Jinsha River, and in the source of the Nanpan River. Some of the lakes are of great depth and steep shores including Lugu Lake, Erhai Lake, and Chenghai Lake with average depths of over 10 m and poorly developed beaches compared with the lakes in the plains of the middle and lower reaches. These lakes are characterized by many inflowing tributaries and few outflows, some of which even have only a single outflow channel. With a high drop, the lakes provide rich hydroelectric resources and slow water exchange. For example, the water exchange periods of Lugu Lake, Erhai Lake, and Dianchi Lake are 2051, 891, and 485 days, respectively. Therefore, these lakes have a very vulnerable ecological system since it is difficult for them to recover quickly after losses of large amounts of water (Table 4).

Wetlands in the Middle and Lower Reaches

The wetlands in the middle and lower reaches represent the highest concentration of freshwater lakes in China, stretching across six provinces (or municipalities) including Hubei, Hunan, Jiangxi, Jiangsu, Anhui, and Shanghai with a total area of 780,000 km^2. This constitutes the largest complex wetland ecosystem in China, consisting of natural wetlands such as rivers (the main stream and tributaries of the Yangtze River), lakes, marshes, and coastal tidal flats and the artificial wetlands such as reservoirs and ponds. Based on the results of the 1995–2001 national wetland resources survey, the total area of the wetlands in the middle and lower reaches of the Yangtze River amounted to 5.8 million ha (excluding the man-made paddy fields), accounting for 15% of the national wetland area (Wang et al. 2006).

Table 4 Wetlands in the upper reaches of the Yangtze River Basin

Wetland level	Wetland name
Wetlands of International Importance	Bitahai Wetland, Napahai Wetland, Lashihai Wetland, Dashanbao, Shengjin Lake
Wetlands of National Importance	Bitahai Wetland, Napahai Wetland, Lashihai Wetland in Lijiang, Dianchi Wetland, Luguhu Wetland, Chenghai Wetland, Dashanbao Wetland in Zhaotong, Huize Black-Necked Cranes Habitat, Jiuzhaigou Marsh Wetland
National Wetland Nature Reserve	Haizishan Nature Reserve, Rare Fish Reserve in the Upper Reaches of the Yangtze River, Hanzhong Crested Ibis Reserve, Dashanbao Nature Reserve, Jiuzhaigou Nature Reserve
Provincial Nature Reserve	Bitahai Nature Reserve, Nashihai Nature Reserve, Luguhu Nature Reserve, Lashihai Plateau Wetland Reserve, Huize Black-necked Cranes Reserve, the North Sea Wetland Reserve, Dashahe Nature Reserve in Daozhen, Haifeng Reserve, Tianchi Reserve in Yunnan, Manzetang Wetland Reserve, Nanmoqie Wetland Reserve, Nuoshuihe Reserve, Xinluhai Reserve, Gongma Reserve in Changsha, Kashahu Reserve, Jianhu Wetland Reserve, Baicaopo Reserve, Yazui Reserve, Huanglong Reserve, Tianquanhe Reserve, Zhougonghe Reserve, Ertan Birds Reserve, Anlan Herons Reserve, Nantianhu Reserve

The famous "Five Great Lakes" in China, including Poyang Lake, Dongting Lake, Taihu Lake, Chaohu Lake, and Hongze Lake, are all clustered here, with five Ramsar Sites covering Dongdongting Lake, Xi Dongting Lake Nature Reserve, Nan Dongting Lake Wetland and Waterfowl Reserve, Poyang Lake, and Hubei Honghu Wetlands. This region has been listed as an International Wetland and Biodiversity Conservation Hotspot (http://www.shidi.org/lib/lore/ramsar.htm), being an important wintering ground for migratory waterbirds in the East Asian–Australasian Flyway, with the most abundant wetland resources in China. However, with agricultural development and industrial and urban progress, this region has been faced with great pressures with respect to wetland conservation, suffering from decreased areas of natural wetlands, weakened wetland functions, serious water pollution, and degraded wetland ecosystem. These problems have combined to pose a severe challenge to the stability and sustainability of the wetlands within this region (Table 5).

Wetlands of the Yangtze River Estuary

The silts and sediments brought by the Yangtze River act as the material base upon which the tidal flats are formed in the estuarial areas, which then gradually develop into widespread tidal flat wetlands. The tidal flat wetlands along the Yangtze River Estuary can be divided into two groups: the estuarial sandbank island wetlands and the tidal wetlands along the river and coast, most of which are located along the south bank of the Yangtze River Estuary (from the Liuhe Estuary in the west to Luchao Port in the east) with Nanhui Biantan Wetland as the most important. Some of the sandbank

Table 5 Wetlands along the middle and lower reaches in the Yangtze River Basin

Wetland level	Wetland name
Wetlands of International Importance	Dong Dongting Lake, Xi Dongting Lake Nature Reserve, Nan Dongting Lake Wetland and Waterfowl Reserve, Poyang Lake, Hubei Honghu Wetlands, Shenjinhu Wetland
Wetlands of National Importance	Liangzihuqun Wetland, Wanghu Wetland, Dajiuhu Wetland Park, Tianezhou Yangtze Gudaoqu Wetland, Dongtinghu Wetland, Poyang Lake Wetland, Chaohu Wetland, Taiping Lake Wetland, Taihu Wetland, Danjiangkou Reservoir Wetland, Shijiuhu Wetland, Chinese Alligator Nature Reserve, Gaoyouhu Wetland
National Wetland Nature Reserve	Zhangjiajie Andrias Davidianus Reserve, Yangtze Xinluoduan White-flag Dolphin Reserve, Yangtze Tianezhou White-flag Dolphin Reserve, Shishou Tianezhou Milu Reserve, Xingdoushan Reserve, Danjiang Wetland Reserve, Nanjishan Reserve, Shenjinhu Reserve, Poyang Lake Reserve, Anhui Chinese Alligator Reserve, Tongling Freshwater Dolphin Reserve, Hongzehu Wetland Reserve
Provincial Nature Reserve	Liangzihu Wetland Reserve, Zhongjianhe Reserve, Longganhu Reserve, Wanjianghe Andrias Davidianus Reserve, Chenhu Rare Fish Wetland Reserve, Liangzihu Reserve, Wanghu Wetland Reserve, Yichang Chinese Sturgeon Reserve, Jiangkou Niaozhou Reserve in Hunan, Qiyang Asiatic Salamander Reserve, Dong Dongtinghu Reserve, Xi Dongtinghu Reserve, Nan Dongtinghu Reserve, Henglinghu Reserve, Jichen Milu Reserve, Duchang Migratory Birds Reserve, Qinglanhu Reserve, Poyang Lake Silverfish Reserve, Poyang Lake Yangtze Neophocaena Reserve, Poyang Lake Clams Reserve, Poyang Lake Carp Crucian Fish Spawning Ground, Yuanyanghu Reserve, Honghu Wetland Reserve, Xixia Andrias Davidianus Reserve, Neixiang Tuanhe Wetland Reserve, Guichi Shibasuo Reserve, Dangtu Shijiuhu Reserve, Mingguang Nvshanhu Reserve, Anqing Yanjiang Fish Reserve, Zhenjiang White-flag Dolphin Reserve, Lianshui Egret Reserve

islands that rise permanently above water have been developed and inhabited such as Chongming Island, Changxing Island, and Hengsha Island, while some others that are covered by vegetation have not been populated, such as Jiuduansha and Qingcaosha. The main sandbank island wetlands consist of Dongtan and the northern tidal flat areas of Chongming Island, the western tidal flats of Hengsha Island, and the northern tidal flats of Changxing Island and Jiuduansha (Table 6).

Current Status of Habitats and Threatened Species and Biodiversity

The middle and lower reaches of the Yangtze River are endowed with a variety of ecosystems, attracting a great population of a diversity of waterbirds to overwinter here, including many rare species such as cranes, storks, and swans. A total of

Table 6 Wetlands of the Yangtze River estuary

Wetland level	Wetland name
Wetlands of International Importance	Shanghai Yangtze Estuarine Wetland Nature Reserve For Chinese Sturgeon, Chongming Dongtan Nature Reserve, Yancheng Wetland, Dafeng Wetland
Wetlands of National Importance	Changxingdao and Hengshadao Wetland, Chongmingdao Wetland, Xixi Wetland Park, Yancheng Coastal Wetland, Jinshan Sandao Wetland
National Wetland Nature Reserve	Chongming Dongtan Reserve, Jiuduansha Wetland Reserve, Rare Birds Reserve of the Coastal Tidal Flat Areas of Yancheng, Dafeng Milu Reserve
Provincial Nature Reserve	Qidong Yangtze Estuarine (North Branch) Wetland Reserve, Shanghai Yangtze Estuarine Wetland Nature Reserve, Dinghai Wuzhishan Birds Habitat and Reproduction Reserve, Jinshan Sandao Reserve, Drinking Water Resources Reserve of the Upper Reaches along the Huangpu River

762 bird species can be found in the Yangtze River Basin, belonging to 20 orders, 66 families, and 291 genera, respectively. These account for 61.12% of the total number of bird species in China, including 72 species endemic to China or mainly distributed in China, 26 species under state first-class protection, and 92 species under state second-class protection (Wu et al. 2004). The distribution of birds depends on their habitat requirements, with specific bird communities inhabiting different areas across the Yangtze River Basin. For instance, the areas with relatively smooth terrain, widespread lakes and marshes, and lush aquatic plants in the middle and lower reaches of the Yangtze River provide ideal habitat for waterfowl and waders, the most representative species of which include little egret *Egretta garzetta,* gray heron *Ardea cinerea,* mallard *Anas platyrhynchos,* Chinese spot-billed duck *A. zonorhyncha,* swans *Cygnus* spp., and cranes including the rare Siberian crane *Grus leucogeranus*. Other riparian specialist birds include the plumbeous water redstart *Rhyacornis fuliginosus*, white-crowned forktail *Enicurus leschenaulti*, and vinous-throated parrotbill *Paradoxornis webbiana*.

Since the last century, the freshwater lakes along the middle and lower reaches of the Yangtze River Basin have witnessed great changes in terms of number and area due to both natural and human factors. Lakes with an area larger than 10 km^2 have been selected from maps and remote sensing images in order to study trends of change and to analyze the causes and results of the variations. The results showed that the period from 1875 to 2011 has seen a remarkable decrease in both the number and area of lakes (Fig. 2), during which many lakes were transformed to paddy fields, fish ponds, or building land. Although some well-managed wetlands have stood the test of extreme climate, the wetlands along the Yangtze River are still on the decline in the long term with respect to their area and number, while the overall functions are in degeneration. Thus, the wetlands are still very vulnerable to extreme climate in terms of their adaptability and resiliency (Cui et al. 2013).

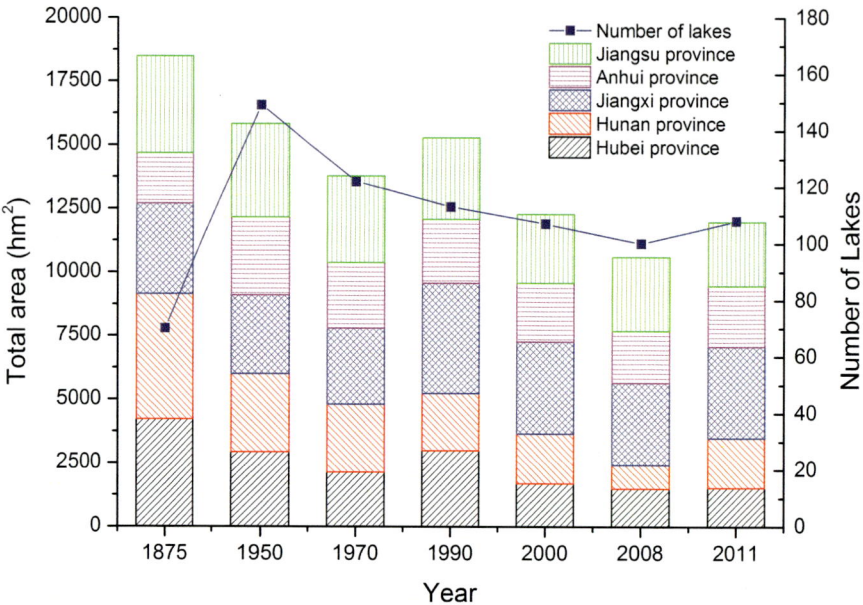

Fig. 2 Changes in the area of lakes in the middle and lower reaches of the Yangtze River Region at seven time intervals (Adapted from Cui et al. 2013; Ma et al. 2011)

Threats and Conservation Issues

Changes in Hydrology and Runoff

The completion and operation of Three Gorges Dam, which is the largest comprehensive water-control project spanning the Yangtze River, have not only generated tremendous social and economic development benefits in the region, but it has also exerted extensive and remarkable effects on the ecosystems of the Yangtze River. With runoff (flow) as a key factor determining the ecology of the river system, the adjustment of the hydrological regime by the operation of the Three Gorges Dam acts as a key driver in exerting effects on aquatic ecosystems. The Three Gorges Dam influence on the downstream water ecological environment mainly lies in the key aquatic organism protection, wetland ecosystems of lakes interlinked with rivers, groundwater and estuaries in the lake region on the plains, etc. These influences are closely related to the discharge altered by the Three Gorges Dam (Ran et al. 2011).

Sharp Decline in Wetland Biodiversity

The Yangtze River Basin has long been an important distribution region of wetlands in China and renowned for its biodiversity. However, excessive exploitation of the wetlands has brought great changes and destruction to the living environment of the animals and plants in the wetlands and brought increasing numbers of species, especially rare animals, into imminent danger of extinction, resulting in a sharp decline in wetland biodiversity in recent decades. For example, there has been a sharp decline in the number of fish species in Honghu Lake from about 100 species 50 years ago to over 50 at present and disappearance of large waterbirds such as the Eurasian crane *Grus grus* and Eurasian spoonbill *Platalea leucorodia*, which were once easy to find in the wetlands of Dianchi Lake in Yunnan Province (Duan et al. 2006; Li et al. 2006).

Aggravated Water Pollution from Pesticides

Pesticides, widely used in the farming areas along the middle and lower reaches of the Yangtze River, have posed a great threat to birds in the wetlands. Overuse of pesticides in these sensitive areas, which serve as important staging areas for migratory birds of the East Asian–Australasian Flyway, principal wintering grounds for huge concentrations of waterbirds in East Asia, and waterfowl breeding grounds, will not only put heavy pressures on the protection of birds in China but also exert far-reaching effects on the protection of birds across the globe (Qin et al. 2007; Chen et al. 2013). Key wetlands affected include Poyang Lake, the almost exclusive wintering ground for Siberian cranes; Yancheng Coastal Tidal Flat, the world's largest wintering ground for red-crowned cranes *Grus japonensis*; and the Chongming Dongtan in Shanghai, an important staging and wintering area for East Asian migratory shorebirds. The contaminants in the sediments (heavy metals, PAHs, DDT, HCH, solvents) have pernicious toxicological effects on the fish and amphibians (Wu 2007).

Human Disturbance

The wetlands of the Yangtze River Basin have been subject to the effects of both natural evolution and human activities simultaneously. Under pressure from human activities, these wetlands have seen a sharp reduction in the area and great increase in degree of fragmentation. The biodiversity of the region has also been severely threatened by the increase in water temperature and decrease in water level, leading not only to declines in the abundance and diversity of fish and birds but also to changes in the species composition and productivity of plant communities, bringing serious challenges to the wetland ecosystems (Mao et al. 2004).

Future Challenges

Influences of Future Climate Change

By the end of the twenty-first century, the global temperature will rise to 1.1–6.4 °C, while rainfall patterns will also see prominent changes (Solomon 2013). These future changes in temperature and the consequential changes in hydrological processes will work together to pose a serious threat to the stability and well-being of the wetland ecosystems. In the context of worsening wetland conditions arising from droughts, lack of water, declining areas, and degrading functions, great importance has been attached by scientists in China and internationally to the study of the effects of climate change on the ecosystems and hydrology of the wetlands of the Yangtze River Basin. The basin is considered relatively sensitive to climate change as it covers a wide area ranging from West China to East China. In addition, the changing trends of wetlands under current conditions of climate change also calls for further study and concern.

Comprehensive Management of Nature Reserves

There is a large number of nature reserves in the Yangtze River Basin, most of which are aimed mainly at the protection of single species. This approach, however, may result in the integrity of the whole ecosystems being overlooked. In fact, the uneven distribution of wetland nature reserves across the Basin has contributed to the concentration of nature reserves in certain areas. To protect the wetlands more efficiently, on the one hand, unified management can be carried out for nature reserves that are close to each other, together with strengthened cooperation between local governments and corresponding departments, as well as managerial and financial support given to less developed areas. On the other hand, the land ownership and boundaries of the wetland nature reserves should be cleared and protected, while updating the key nature reserves and reinforcing the ecological integrity of the wetland ecosystems in the nature reserves (Yan et al. 2013).

References

Chen J, Yuan SX, Qi SH. Distribution characteristics of organochlorine pesticides in tissues of Ardeidae and its risk assessment from Honghu Lake, southcentral China. Environ Chem. 2013;32(004):549–56.
Cui L, Gao C, Zhao X. Dynamics of the lakes in the middle and lower reaches of the Yangtze River basin, China, since late nineteenth century. Environ Monit Assess. 2013;185(5):4005–18.
Ding YJ, Yang JP, Liu SY. Exploration of eco-environment range in the Source Regions of the Yangtze and Yellow Rivers. Acta Geogr Sin. 2003;58(4):519–26.
Duan DX, Liu JH, Wu T. The current problems and protective measures of wetland resources in China. J Binzhou Univ. 2006;22(3):62–6.
Forestry Administration. Yangtze River Wetland Protection Network Annual Meeting. 2010. http://www.forestry.gov.cn//portal/main/s/72/content-444201.html
Li XG, Jiang N, Zhu XH. Study on lake surface area change of major lakes in the Taihu basin during the past 30 years. Trans Oceanol Limnol. 2006;4(17):17–23.
Liang C, Zhao LH, Zhang BX. Effects of climate change on hydrological environment in the extremely frigid zone of the source region of Yangtze River. South-North Water Transfers Water Sci Technol. 2013;1:81–6.
Ma RH, Yang GH, Duan HT. China's lakes at present: number, area and spatiual distribution. Sci China Earth Sci. 2011;41(3): 394–401.

Mao ZB, Peng WQ, Zhou HD. Ecological effects of dams to rivers and countermeasure study. China Water Resour. 2004;15(1):43–5.

Ministry of Water Resources of the People's Republic of China. Overview of the Yangtze River. 2013. http://www.mwr.gov.cn/ztpd/2013ztbd/2013kqjhkfzlbhxpz/lyxj/chly/ghyd/201305/t20130527_436300.htm

Qin WH, Shan ZJ, Wang Z. Threat of Carbofuran to wetland birds in China and the countermeasures. J Ecol Rural Environ. 2007;23(1):85–7.

Ran JJ, Chen M, Chen YB. Analysis on flow adjustment functions of the three gorges project which affect the downstream aquatic ecosystem and environment. J Hydroecology. 2011;32(1):1–6.

Solomon S. Contribution of working group I to the fourth assessment report of the intergovernmental panel on climate change, Climate Change 2007: the physical science basis. 2013.

Wang ML. Runoff changes and distributed hydrologic simulation in the upper reaches of Yangtze River. Institute of Geographic Sciences & Natural Resources Research, CAS: Beijing; 2008.

Wang XL, Xu HZ, Cai SM. Wetland protection and basin management in the middle and lower reaches of the YangTze River. Resour Environ Yangtze Basin. 2006;15(5):564–8.

Wu L. Distribution of two typical organic pollutants in the Yangtze River estuary and toxicology effects on fish. Shanghai: Tongji University; 2007.

Wu H, Yu XG, Jiang JH. The study means and important fields on the wetland eco-system in Changjiang River Valley. Ecol Econ. 2001;11:21–22,26.

Wu YM, Zhuang Y, Xu YG. Avifauna of Yangtze River Basin. Chin J Zool. 2004;39(4):81–4.

Xu JJ, Yang DW, Lei ZD. Precipitation and runoff in the Yangtze River basin long-term trend test. Yangtze River. 2006;37(9):63–7.

Yan RR, Cai XB, Wang XL. Distribution status of wetland nature reserves and the problems in Yangtze River Watershed. Wetl Sci. 2013;11(1):136–44.

Yangtze River Water Resources Bulletin. 2014. http://www.cjw.gov.cn/zwzc/bmgb. http://www.shidi.org/lib/lore/ramsar.htm.

Poyang Lake, Yangtze River Basin, China 129

James Harris

Contents

Introduction 1566
Hydrology 1568
Biodiversity 1568
Conservation Status 1570
Ecosystem Services 1570
Future Challenges 1571
References 1572

Abstract

Poyang Lake is the largest freshwater lake in China. Five main tributaries drain into the lake from the south, while the lake empties through a short outlet channel into the Yangtze River. Poyang is one of the two lakes in the mid Yangtze Basin that has retained its free connection to the river. In addition, Poyang has a seasonal, reverse-flow system that occurs in most years where waters from the Yangtze River flow into Poyang Lake during part of the summer. This reverse-flow system greatly contributes to the complexity of Poyang's yearly hydrological variation. In summer, the lake's surface area exceeds 4,000 km^2. Falling water levels during autumn expose extensive mudflats and leave behind isolated sublakes. Dramatic hydrological variations – water levels can fall as much as 11 m between summer floods and winter lows – drive ecological processes within the system and create a wide range of habitats supporting rich biodiversity. The

J. Harris (✉)
International Crane Foundation, Baraboo, WI, USA
e-mail: harris@savingcranes.org

© Springer Science+Business Media B.V., part of Springer Nature 2018
C. M. Finlayson et al. (eds.), *The Wetland Book*,
https://doi.org/10.1007/978-94-007-4001-3_34

seasonal changes in water levels create two separate ecological phases of Poyang Lake, supporting different sets of species and leading to near year-round productivity and rich diversity of life. Poyang is world famous for its concentrations of wintering waterbirds, including over 98% of the world population of the Siberian crane *Leucogeranus leucogeranus*. Poyang Lake has major value for flood control, with maximum storage of some 11.75 billion m^3, and provides valuable fish harvests. Changes in hydrology pose the greatest threat, while demand for water is growing for irrigation and industry. The Three Gorges Dam has reduced river flows during late summer and autumn. This change, combined with intensive sand dredging near the lake's outlet, has resulted in faster outflow from the lake into the Yangtze River from July to March, and reduced water storage in Poyang Lake. In addition, a dam has been proposed across Poyang Lake's outlet to stabilize water levels, which could transform the ecological character of the wetland and flood habitats used by Siberian cranes and many other waterbirds.

Keywords

Mid Yangtze Basin · Poyang Lake · Reverse-flow system · Sand dredging · Siberian crane

Introduction

Poyang Lake is the largest freshwater lake in China. Five main tributaries drain into the lake from the south, while the lake empties through a short outlet channel into the Yangtze River (see Fig. 1). Poyang is one of two lakes in the mid Yangtze Basin that has retained its free connection to the river. In addition, Poyang has a seasonal, reverse-flow system that occurs in most years where waters from the Yangtze River flow into Poyang Lake during part of the summer. This reverse-flow system greatly contributes to the complexity of Poyang's yearly hydrological variation.

In summer, the lake's surface area exceeds 4,000 km^2 (Shankman and Liang 2003). Falling water levels during autumn expose extensive mudflats and leave behind isolated sub-lakes. Dramatic hydrological variations at Poyang – water levels can fall as much as 11 m between summer floods and winter lows – drive ecological processes within the system and create a wide range of habitats supporting rich biodiversity. The seasonal changes in water levels create two, separate ecological phases of Poyang Lake: one dominated by subtropical vegetation most productive during the hot summers and another dominated by temperate vegetation primarily growing during the cool, wet winters (Zheng 2009). Different species rely on these separate ecological phases, leading to near year-round productivity and rich diversity of life. Poyang is world famous for its concentrations of wintering waterbirds, including over 98% of the world population of the charismatic and critically endangered Siberian crane *Leucogeranus leucogeranus* (Harris and Zhuang 2010).

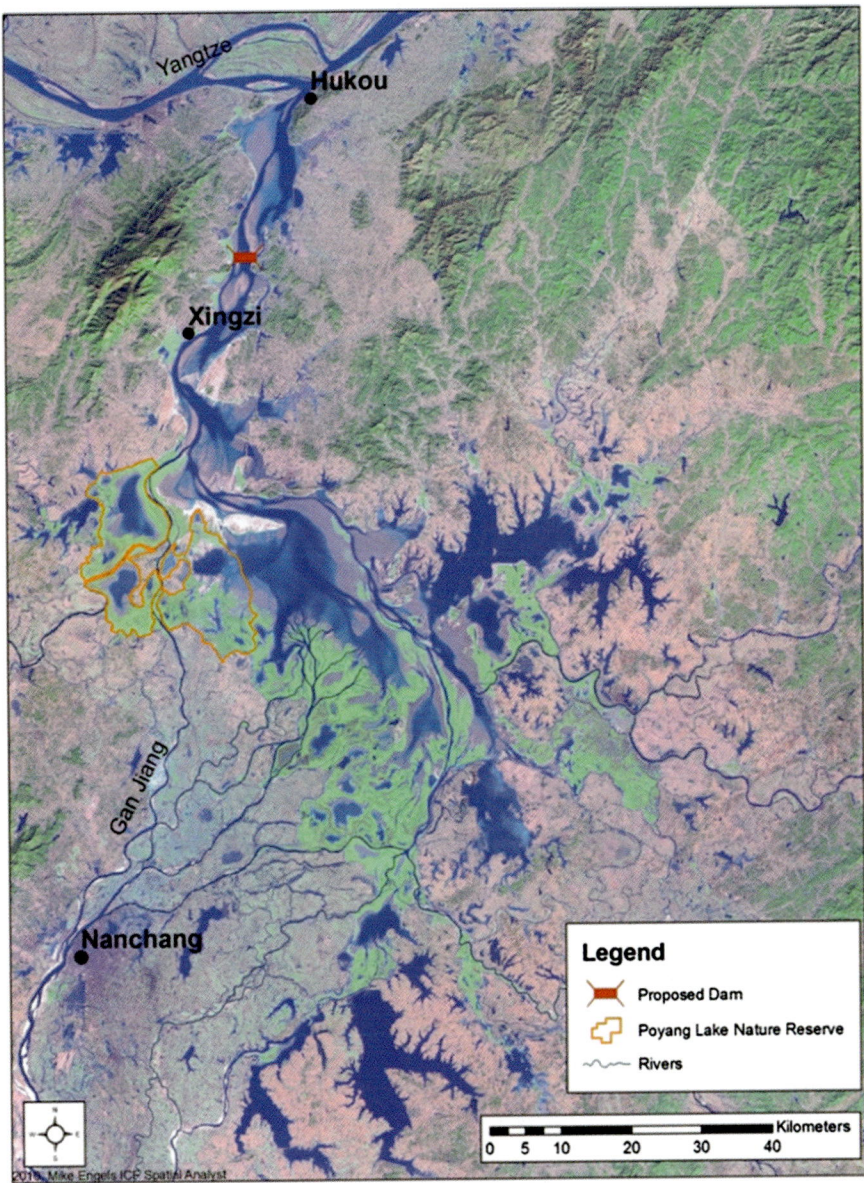

Fig. 1 Map of Poyang Lake showing boundaries of Poyang Lake National Nature Reserve, the location for the proposed dam, and hydrological station at Xingzi (Source: © Provided courtesy of International Crane Foundation)

The lake's catchment of 162,225 km^2 includes ~97.2% of Jiangxi Province (Editorial Committee of Studies on Poyang Lake 1988), while 18 million people live in the counties surrounding the lake.

Hydrology

The hydrological character of Poyang is unique among China's lakes. The flood peak caused by flooding in the lake's five tributaries usually happens between May and June. In late summer, the Yangtze flood can backflow into Poyang Lake, usually between July and October (Editorial Committee of Studies on Poyang Lake 1988; Harris and Zhuang 2010). After September, with the flood period ending, the water level in the lake gradually reduces until its lowest levels occur during December or January.

Water levels at Poyang in the low water season, as compared with the high water season, exhibit significant differences between the upper and lower reaches of Poyang. The lake surface slopes downwards from upstream areas in the south towards the north and its confluence with the Yangtze. The difference in water surface elevation between upper and lower parts of the lake can be >4 m. In spring, when the water level reaches about 15 m above sea level (Wu Song System – see Box 1 at end of chapter), however, the surface of the entire lake flattens out (i.e., the entire basin is flooded).

Average winter elevation of the water in Poyang Lake is 11.98 masl at the Gan River station at Wucheng near Poyang Lake National Nature Reserve (NNR), with minima fluctuating by +/− 1 m between and within years. There are over 60 sub-lakes in the system, so it is not easy to generalize about water depth – although the profile is relatively flat. Most of the sub-lakes (e.g., in Poyang Lake NNR) are very shallow in winter and can be more or less dry after being drained for fishing purposes, with just the channels having water remaining. Typically, winter depths will range from a bit more than 100 cm to less than 15 cm around the margins of the lake pans.

The mean annual runoff from Poyang into the Yangtze River is 143.6 million m^3. On average, the annual water exchange rate for Poyang Lake is 20.9 days (China Academy of Sciences unpublished data; Harris and Zhuang 2010). The waters thus exchange in the lake almost 18 times in one year, so that Poyang has the second fastest rate of exchange for large lakes in China. The rapid exchange helps keep the water clean within the lake, by flushing out pollutants. When the water level drops below 15 masl, the water develops a slope with water moving from south to north and the rate of flow increases to a relatively high rate. When the water level rises above 15 masl, the slope of the lake surface flattens and gravity flow weakens so that wind becomes an important factor. The higher the lake water level, the slower will be the flow rate. As a result, during high water the lake exchange period is longer than in winter (ignoring evaporation). Thus the natural process of flood and water drainage periods each year results in an overall rapid exchange of water for Poyang Lake, but at uneven rates among seasons.

Biodiversity

Poyang supports significant diversity of aquatic plants, with the different communities responsive to soil elevation and water depth (Cui et al. 2000; Yuan and Li 2008; Barzen et al. 2009). For a wetland as dynamic as Poyang Lake, the elevations at which particular plant communities grow, however, depends upon hydrological conditions in any one year. Submerged aquatic plants, for example, can grow in Poyang Lake at elevations as low as 8 or 9 masl during drought years or as high as 16 masl during flood years. On average, submerged aquatic plants grow at elevations from 13 to 14 masl, while the sedge zone dominates the 14–15 masl range (Barzen et al. 2009). At 15–16 masl reeds grow as do other grasses and forbs, though sedges are prevalent as well. Elevations above 16 masl tend to support upland vegetation composed mostly of grasses and shrubs.

The food resources provided by the aquatic plant diversity within the system are a major reason that hundreds of thousands of migratory birds travel to Poyang every winter (Li et al. 2005). Winter surveys at Poyang have recorded 425,000 waterbirds on average, with a peak count of 726,000 birds in 2005 (Barter et al. 2005; Li et al. 2005; Ji et al. 2007; Qian et al. 2011), making it the most important wetland for waterbirds in East Asia. In addition to the Siberian cranes, over 90% and 50% of the world's populations of the endangered oriental stork *Ciconia boyciana* and vulnerable swan goose *Anser cygnoides*, respectively, winters here. Formerly, half the world population of vulnerable white-naped cranes *Grus vipio* wintered at Poyang but the latter population has declined to roughly half the numbers present in the mid-1990s (Li et al. 2012), probably due in part to prolonged drought on the breeding grounds. In two years of winter waterfowl surveys in the middle and lower Yangtze River Basin, Poyang Lake supported more than 1% of the regional populations of 12–15 species of waterbird.

Poyang Lake provides habitat for numerous other aquatic animals dependent upon its wetlands, including the globally vulnerable finless porpoise *Neophocaena phocaenoides* and the vulnerable Chinese water deer *Hydropotes inermis inermis*. Zhang and Li (2007) report that 136 fish species have been recorded in Poyang Lake, although the critically endangered Chinese sturgeon *Acipenser sinensis* and critically endangered Chinese paddlefish *Psephurus gladius* are apparently gone from the lake.

Vegetation and hydrologic diversity at Poyang, together with the vast areas of shallow water and mudflats, offers conditions suitable for a variety of avian foraging guilds (Barzen 2008). Cranes, swan geese, and swans feed primarily on tubers of *Vallisneria* spp. and other submerged aquatic plants. Five species of geese, including the vulnerable lesser white-fronted goose *Anser erythropus*, feed in the sedge-grass zone. Large numbers of shorebirds forage for invertebrates exposed as waters recede from the muds, while spoonbills and avocets filter the water for zooplankton. Fish eaters are present, although in relatively small numbers.

Waterbirds move widely across different parts of the lake, responding not only to abundance of food (likely determined by summer water conditions) but also to availability of food that is determined by winter conditions such as water depth, vegetation patterns, human disturbance, and weather including the effects of wind. Thus any individual part of the lake does not consistently support populations of bird species of concern. During parts of some winters, for example, few if any cranes are present within Poyang Lake NNR as they move to other parts of Poyang's large basin.

Conservation Status

A total of 14 wetland nature reserves have been established at Poyang Lake, with a combined total area of over 200,000 ha or 50% of the lake basin (Wildlife Protection and Nature Reserve Management Department 2010). Of these, two are National Nature Reserves: Poyang Lake NNR (22,400 ha) and Nanjishan NNR (33,300 ha). Poyang Lake NNR was designated a Ramsar Site in 1992 and is on the site list for the Alliance for Zero Extinction (www.zeroextinction.org) because of its importance for the survival of the Siberian crane. Two of the reserves are provincial level: Duchang (41,100 ha) and Poyang (40,900 ha). The other reserves are municipal or county level reserves and have minimal staff or other resources.

Within Poyang Lake NNR are nine sub-lakes that separate from the main lake in winter and provide habitat for some of the greatest concentrations of waterbirds in the lake basin. The reserve has gained use rights to five of these lakes, thus having a greater degree of control over their aquatic resources and water management. Reserve control of sub-lakes is still constrained by contracts with outsiders for rights to fish and raise crabs. The protection status afforded the rest of the reserve and almost all other lands within the other nature reserves is limited to bird protection while resource rights belong to others.

After the 1998 floods, large numbers of people were moved out of the lowest areas of Poyang Lake to reduce flood risk and to restore a larger capacity for flood water storage. Many of these lowlands have been gradually resettled.

Ecosystem Services

Poyang Lake, due to its great extent, has major value for flood control. When water levels go from 8.33 masl (the 1998 lowest level within the outlet channel) to 16 masl, the storage is 11.75 billion m^3, equal to 50% of the Three Gorges Reservoir's adjusted storage capacity – a benefit provided without water infrastructure. This amount (11.75 billion m^3) mainly absorbs and controls the flood from Poyang's five tributaries and reduces the pressure from flooding on Poyang Lake and Jiangxi Province. During late summer flooding of the Yangtze, Poyang may absorb water from the Yangtze as well, buffering downstream communities.

Ecosystem services provided by the lake also include maintenance of water quality (see above) as well as important water supply and fishery production services. The lake supplies water to surrounding human communities, for urban, industrial, and agricultural purposes, and is an important navigation route, linking the city of Nanchang to the Yangtze River, other provinces, and eventually the sea. The lake provides valuable fish harvests, with an average annual catch of 198 kg/km^2 in the 1990s (Zhang and Li 2007). Portions of the lake support sand dredging, fish farming, and crab farming, but some of these practices may not be sustainable.

Future Challenges

Changes in hydrology pose the greatest threat to Poyang Lake's ecosystems and their biodiversity, but this is by no means the only threat facing the lake. As of 2001, 9,603 dams had been built on the five tributaries, with a total water storage capacity of 27.9 billion m^3 (Liu et al. 2009). Demand for water, for irrigation and industry, is growing. The Three Gorges Dam, although well upstream on the Yangtze, has reduced river flows during late summer and autumn, resulting in faster outflow from the lake into the Yangtze River from July to March, and reduced water storage in Poyang Lake (Guo et al. 2012). Numerous additional dams in the upper Yangtze Basin are in operation, under construction, or in the approval process that will further affect seasonal flows. Following the ban on sand dredging in the main channel of the Yangtze in 2000, a large part of this activity moved into Poyang Lake, where it has been particularly concentrated at the outlet channel. Aside from disturbance to substantial parts of the lake bottom and increasing turbidity, a deepening of the outlet channel could affect the rate of outflow to the Yangtze River (de Leeuw et al. 2010).

In the last six years four severe droughts and one severe flood have occurred. These extremes of weather appear to be compounding the impacts of dams and other human activities. During unusually low water periods, people encroach on the lake edges, construct dikes to separate fish ponds, and otherwise reduce the size of the natural wetland. Water quality has deteriorated, due to urban and industrial development within the neighboring counties. Lakes elsewhere in the mid Yangtze have seen major changes in vegetation as water quality declined, resulting in reduced food availability for tuber feeders including the Siberian cranes (Fox et al. 2010; Zhang et al. 2010). Poyang appears vital to their survival.

Over past decades, Jiangxi Provincial Government has sought to construct a dam across all or part of Poyang Lake to stabilize water levels. Recent hydrological changes have added wetland restoration to the objectives of projects designed to manage water. The current proposal calls for sluice gates to be built across the outlet of Poyang Lake; these gates would remain open and allow water to flow freely during high water periods but hold the water back from autumn through spring. In particular, the structure would prevent the rapid drop of lake levels in early autumn when water is needed for rice irrigation.

Yet many experts question whether the dam's negative impacts on ecosystem services of the wetland have been adequately evaluated and whether alternative strategies have been fairly compared. Raising winter water levels substantially from historic norms could transform the ecological character of the wetland and flood habitats used by Siberian cranes and many other waterbirds, pushing the birds to outer parts of the wetland where human disturbance is high (Barzen et al. 2009).

Other threats to waterbirds and their food supply include poaching, primarily by nets and poisons, and stocking of crabs into sub-lakes, consuming most or all submerged vegetation. The crab species are native to China but not naturally occurring in Poyang Lake.

Table 1 Water elevations measured in three datums used at Poyang Lake (Wu Song, National Vertical Elevation 1985, and Huang Hai) and their conversion factors to the Wu Song system. A water elevation of 12 m above sea level (Wu Song) approximates the average low water elevation at Poyang Lake (11.98 m Wu Song) as measured at Wucheng for the combined months of December, January, and February 1955–2006

Conversion table of analysis water levels for the three elevation datums (m)				
	Water level	Water level	Water level	Conversion
Wu Song	11.98[a]	14.00	16.00	
National 85	10.14	12.16	14.16	National 85 elevation = Wu Song elevation − 1.84 m
Huang Hai	9.72	11.74	13.74	Huang Hai elevation = Wu Song elevation − 2.26 m

[a]Historic Mean Low Water Level. Based on water elevation from Water Gauge Data of Gan and Xiu Rivers averaged for the months of December, January, and February 1955–2006

> **Box 1: Different Elevation Scales in Use in China**
> Three vertical elevation datums are commonly found in data obtained at Poyang Lake: Wu Song, Huang Hai, and National Vertical Datum 1985. Huang Hai elevation systems were used as the basis for reporting elevations on topographic maps whereas water levels were often recorded in Wu Song or National Vertical Datum 1985 elevation systems. In this account, any given elevation is measured in the Wu Song datum (Xu et al. 2007). "Wu Song" is named after a seaport near Shanghai (Table 1).

References

Barter M, Lei G, Cao L. Waterbird survey of the middle and lower Yangtze River floodplain in February 2005. Beijing: World Wildlife Fund-China and Chinese Forestry Publishing House; 2005.

Barzen J. Phase 1 report: How development projects may impact wintering waterbirds at Poyang Lake. Unpublished report submitted to Hydro-ecology Institute of the Yangtze Water Resources Commission. Baraboo: International Crane Foundation; 2008.

Barzen J, Engels M, Burnham J, Harris J, Wu G. Potential impacts of a water control structure on the abundance and distribution of wintering waterbirds at Poyang Lake. Unpublished report submitted to Hydro-ecology Institute of the Yangtze Water Resources Commission. Baraboo: International Crane Foundation; 2009.

Cui X, Zhong Y, Li W, Chen J. The effect of catastrophic flood on biomass and density of three dominant aquatic plant species in the Poyang Lake. Acta Hydrobiol Sinica. 2000;24(4):322–5.

de Leeuw J, Shankman D, Wu G, de Boer WF, Burnham J, He Q, Yesou H, Xiao J. Strategic assessment of the magnitude and impacts of sand mining in Poyang Lake, China. Reg Environ Chang. 2010;10:95–102.

Editorial Committee of Studies on Poyang Lake. Studies of Poyang Lake. Shanghai: Shanghai Scientific and Technical Publishers; 1988.

Fox A, Cao L, Zhang Y, Barter M, Zhao MJ, Meng FJ, Wang SL. Declines in the tuber-feeding waterbird guild at Shengjin Lake National Nature Reserve, China – a barometer of submerged macrophyte collapse. Aquatic Conserv Mar Freshw Ecosyst. 2010. doi:10.1002/aqc.1154. Wiley Online Library (wileyonlinelibrary.com), 10 pp.

Guo H, Hu Q, Zhang Q, Feng S. Effects of the Three Gorges Dam on Yangtze River flow and river interaction with Poyang Lake, China: 2003–2008. J Hydrol. 2012;416–417:19–27. doi:10.1016/j.jhydrol.2011.11.027.

Harris J, Zhuang H. An ecosystem approach to resolving conflicts among ecological and economic priorities for Poyang Lake wetlands. Baraboo/Gland: International Crane Foundation/IUCN; 2010.

Ji W, Zen N, Wang Y, Gong P, Xu B, Bao S. Analysis of the waterbirds community survey of Poyang Lake in winter. Ann GIS. 2007;13(1/2):51–64.

Li F, Ji W, Zeng N, Wu J, Wu X, Yi W, Huang Z, Shou F, Barzen J, Harris J. Aerial survey of Siberian cranes in the Poyang Lake Basin. In: Wang Q, Li F, editors. Crane Research in China. Yunnan: Crane and Waterbird Specialists Group of Chinese Ornithological Society, International Crane Foundation, Yunnan Educational Publishing House; 2005. p. 58–65.

Li F, Wu J, Harris J, Burnham J. Number and distribution of cranes wintering at Poyang Lake, China during 2011–2012. Chin Birds. 2012;3(3):180–90.

Liu J, Zhang Q, Xu C, Zhang Z. Characteristics of runoff variation of Poyang Lake watershed in the past 50 years. Trop Geogr. 2009;29(3):213–8.

Qian F, Yu C, Jiang H. Ground and aerial survey of wintering waterbirds in Poyang Lake basin. In: Prentice C, editor. Conservation of Flyway Wetlands in East and West/Central Asia. Proceedings of the Project Completion Workshop of the UNEP/GEF Siberian Crane Wetland Project, 14–15 October 2009, Harbin, China. Baraboo: International Crane Foundation; 2011. Paper #8, 9pp.

Shankman D, Liang QL. Landscape changes and increasing flood frequency in China's Poyang Lake region. Prof Geogr. 2003;55(4):434–45.

Wildlife Protection and Nature Reserve Management Department. 1999 annual report on nature reserves under the forestry system in China. Internal Report. Beijing: State Forestry Administration; 2010.

Xu J, Jiang B, Yao C. Overview on Wu Song Elevation System in main tributaries in lower and middle reaches of Yangtze River. Yangtze River. 2007;38(10):85–8.

Yuan L, Li W. Effects of water depths and substrate types on the distribution in winter buds of *Vallisneria spinulosa* in Poyang Lake. J Yangtze Univ (Nature Science Edit). 2008;5(1):55–8.

Zhang T, Li Z. Fish resources and fishery utilization of Poyang Lake. J Lake Sci. 2007; 19(4):434–44.

Zhang Y, Cao L, Barter M, Fox AD, Zhao M, Meng F, Shi H, Jiang Y, Zhu W. Changing distribution and abundance of Swan Goose *Anser cygnoides* in the Yangtze River floodplain: the likely loss of a very important wintering site. Bird Conserv Int. 2010;21:36–48.

Zheng Y. Prediction of the distribution of C3 and C4 plant species from a GIS-based model: a case study in Poyang Lake, China [MSc thesis]. Enschede: ITC; 2009.

Huang He (Yellow River) River Basin (China)

130

Cui Lijuan, Zhang Manyin, and Xu Weigang

Contents

Introduction	1576
Regional Climate and Runoff Features	1577
Overview of Wetlands in the Yellow River Basin	1577
Important Wetlands Along the River	1579
Wetlands in the Source Region and Upstream of the Yellow River	1579
Wetlands in Hetao Plain	1580
Wetlands in the Midstream and Downstream Reaches of the Yellow River	1581
Yellow River Delta Wetlands	1583
Distribution of Biodiversity in the Wetlands of the Lower Yellow River Basin	1584
Threats to the Wetlands of the Yellow River Basin	1584
Water Shortage of the Yellow River	1584
Sediment Deposition in the Yellow River	1585
Serious Water Pollution	1585
Fragile Ecological Environment and Soil Salinization in the Delta	1586
Future Challenges	1586
Influence of the Western Line of the South-to-North Water Diversion Project on Wetlands	1586
Influence of Reservoirs	1587
References	1587

Abstract

The Yellow River is the second largest river in China. It originates in the Yueguzonglie Basin, on the northern flank of the Bayan Har Mountains of the Qinghai-Tibet Plateau, at an altitude of 4500 m. Encompassing a drainage area of 795,000 km^2, the Yellow River's main channel is 5,464 km in length with an

---

C. Lijuan (✉) · Z. Manyin · X. Weigang
Institute of Wetland Research, Chinese Academy of Forestry, Beijing, China
e-mail: lkyclj@126.com; cneco@126.com; xuweigang@foxmail.com

© Springer Science+Business Media B.V., part of Springer Nature 2018
C. M. Finlayson et al. (eds.), *The Wetland Book*,
https://doi.org/10.1007/978-94-007-4001-3_98

average natural discharge of 53.48 billion m³. The Yellow River has both the highest sediment discharge and highest sediment concentration in the world. The total area of wetland in the Yellow River basin is 2.5135 Mha, primarily comprising of marsh (45.0%) and riverine wetlands (36.2%). This chapter describes the wetlands distributed in sections within the Yellow River basin, including the source and upstream, the Hetao Plain, midstream and downstream, and the Yellow River Delta. Further, the chapter identifies the challenges faced by the wetlands along the Yellow River.

Keywords
Yellow river basin · Wetland · Hetao plain · Yellow river delta · Ramsar · Water diversion

Introduction

The Yellow River is the second largest river in China. It originates in the Yueguzonglie Basin, on the northern flank of the Bayan Har Mountains of the Qinghai-Tibet Plateau, at an altitude of 4,500 m. It flows through nine provinces and autonomous regions including Qinghai, Sichuan, Gansu, Ningxia, Inner Mongolia, Shanxi, Shaanxi, and Henan, emptying into the Bohai Sea in Kenli county of Shandong province. The average natural discharge of the Yellow River is 53.48 billion m³, accounting for only 2% of the country's total. The volume of water per capita along the Yellow River is 23% of the national average, and its water resource is in relative shortage with great intra-annual and inter-annual variations in runoff. There is also uneven water distribution among different regions of the basin. More than 62% of the Yellow River's runoff comes from the section above Lanzhou city. At the same time, the Yellow River has both the highest sediment discharge and highest sediment concentration in the world. Its natural sediment discharge averages 1.6 billion tons, and its natural sediment concentration averages 35 kg/m³. The huge amounts of mineral-rich loess sediment impart a yellow color to the water and give the river its name.

The total length of the Yellow River's main channel is 5,464 km, and its drainage area covers 795,000 km² (including a 42,000 km² inland river basin), accounting for 8.3% of the national territory. The Yellow River Basin is the main distribution area of China's arable land, with 244 million mu (approx. 16.3 M ha) of arable land. The area of arable land per capita in rural areas along the Yellow River is 3.5 mu (0.23 ha), about 1.4 times the national average. The Huang-huai-hai Plain, Fen-Wei Plain, and Hetao irrigation area are major grain-producing areas accounting for 7.7% of the country's annual grain production. In 2010, the total population of the basin was about 113.68 million, 8.6% of the nation's total population (Ministry of Water Resources 2013).

Regional Climate and Runoff Features

The Yellow River Basin is located in arid and semi-arid regions, where the water resource is very sensitive to climate change. In recent decades, significant changes in temperature and rainfall have been observed in the Yellow River Basin. Since the mid-1980s, the temperature in the Yellow River Basin has risen obviously, mainly characterized by temperature increases in winter and the rise in the northern part of the basin. Rainfall in the Yellow River Basin decreased significantly in the 1990s but has increased slightly in this area since the beginning of the twenty-first century. The rise in temperature and decrease in rainfall are important reasons for the sharp drop in the Yellow River Basin's discharge. Based upon predictions of climate patterns, the temperature will rise distinctly and rainfall will increase slightly in the Yellow River Basin by 2050. The imbalance between the supply and demand for water resources will be further intensified, and adaptive countermeasures should be taken in order to reduce the adverse impacts of climate change on water resources (Liu et al. 2011; Zhang et al. 2013).

Although the Yellow River is the second largest river in China, it has an annual discharge of only 53.48 billion m^3, 90% of which originates from the upper and midstream sections. The volume of water in the main channel changes along its path. Water volume in the river section upstream of Lanzhou is characteristically abundant, greatly decreasing from Lanzhou to Hekou town because of the combined effects of diversion of river water for irrigation, high evaporation, and no tributaries to contribute additional flow. Moving downstream, the water volume increases substantially between Hekou town to Huayuankou due to tributary inflow from the Fen, Wei, and Jing rivers. Downstream of Huayuankou levee construction to confine the Yellow River has resulted in a "suspended river" as sediment deposition has elevated the channel bottom above the adjacent plain by 4–6 m to a maximum of 10–12 m along some sections (Leung 1996; Xu 2003; Zhai 2007). Water volume is again greatly reduced because of serious leakage, too much water diversion for irrigation, and lack of tributary inflow (Wen et al. 2012).

Overview of Wetlands in the Yellow River Basin

The distribution of wetlands in the Yellow River Basin is shown in Fig. 1, totaling an area of 2.5135 M ha (Table 1) and primarily comprising marsh (45.0%) and riverine wetlands (36.2%) and lesser amounts of lake wetlands (8.0%), artificial wetlands (6.9%), and coastal wetlands (3.9%). Based on differences in surface coverage and geomorphological and hydrological factors, marshes in the Yellow River Basin can be further characterized as alpine swamp meadow (23.5%), herbaceous swamp (19.3%), shrubby marsh (1.6%), and inland salt marsh (0.6%) (Huang et al. 2012).

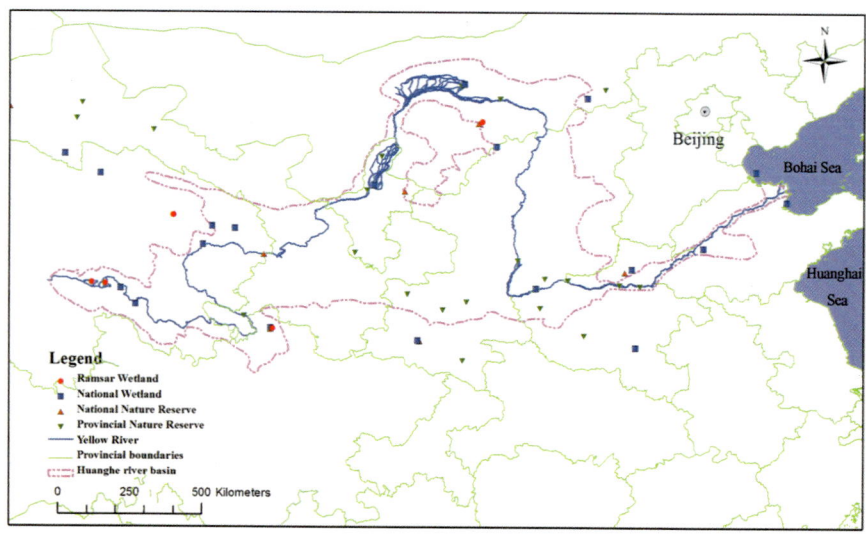

Fig. 1 The distribution of wetlands in the Yellow River Basin

Table 1 Types and distribution of wetlands in the Yellow River Basin

Types of wetlands (second level)	Area (ha)	Proportion of total wetland area (%)	Types of wetlands (third level)	Area (ha)	Proportion of total wetland area (%)
Riverine wetland	909,982	36.2	River channel	434,924	17.3
			Floodplain	475,059	18.9
Lake wetland	200,901	8.0	Lake	200,901	8
Marsh wetland	1,130,860	45	Herbaceous swamp	484,724	19.3
			Alpine herbaceous swamp	590,497	23.5
			Shrubby marsh	40,256	1.6
			Inland salt marsh	15,383	0.6
Coastal wetland	98,090	3.9	Shallow waters	40,296	1.6
			Estuarine waters	7,367	0.3
			Mud flat	36,004	1.4
			Coastal salt marsh	14,423	0.6
Artificial wetland	173,647	6.9	Pond	31,030	1.2
			Reservoir	142,617	5.7
Total	2,513,481	100		2,513,481	100

Table 2 Ramsar Sites in the Yellow River Basin

	Name	Region	Area (km^2)
1	Eling Lake wetland	Duo County in Qinghai	649
2	Bird Island wetland of Qinghai Lake	Gangcha County and Gonghe County in Qinghai	4,473
3	Zhaling Lake wetland	Maduo County in Qinghai	526
4	Ruoergai wetland	Ruoergai County in Sichuan	1,666
5	Erdos wetland	Dongsheng city and Yijinhuoluo County in Inner Mongolia	7,680
6	Zhangye Heihe wetland	Zhangye city in Gansu	412

Important Wetlands Along the River

The Bird Island Nature Reserve in Qinghai was included in the group of seven Wetlands of International Importance to be first designated under the Ramsar Convention in China. There are currently six Ramsar Sites in the Yellow River Basin (Table 2). These wetlands are mainly distributed in the upper and middle reaches of the Yellow River and its source area. In 2014 the Yellow River Basin included 154 of China's 569 National Wetland Parks.

Wetlands in the Source Region and Upstream of the Yellow River

The source of the Yellow River is located in the upper region of Duoshixia Valley in Maduo County, an area with a concentrated distribution of wetlands. The source encompasses three small basins in the form of a belt running from west to east within which the wetlands are distributed. The basins are dotted with many ponds with intervening meadows and abundant aquatic vegetation. The Yueguzonglie River flows through this area and in the downstream portion of the basin lies the vast fluvial marsh of the Xingxiu Sea. To the east of the Xingxiu Sea are the internationally important high altitude freshwater wetlands of Zhaling Lake (4,293 m asl) and Eling Lake (4,269 m asl) in the area known as the Double Lakes. Eling Lake is the larger of the two lakes, covering an area of 611 km^2 with an average depth of 17.6 m compared to Zhaling Lake's 526 km^2 and average depth of 8.9 m. Around the Double Lakes, there are ponds and marshes of various sizes (Table 3).

The Yellow River flows southeastwards out of Eling Lake between the Anyemaqen Mountains and Bayan Har Mountains. It then turns to the northwest at Minshan Mountain, meandering across the Nuoergai area and forming the vast prairie and famous first loop of the Yellow River. The Ruoergai prairie is gently sloping and is within the alpine zone at an elevation of 3,500–3,600 m asl. Its annual average temperature is around $-1.4\ °C$, and the rainfall ranges from 600 to 700 mm with an average annual evaporation of less than 800 mm. Rainfall occurs mostly in the summer and autumn across a large area but is of low intensity and long duration with floods

Table 3 Wetlands in the source region and upstream of the Yellow River

Level of wetlands	Name of wetlands
Ramsar Sites	Eling Lake, Bird Island of Qinghai Lake, Zhaling Lake, Ruoergai, Zhangye Heihe
Wetlands of National Importance	Maduo Lake, Gangnagemacuo, Duogeicuona Lake, Ruoergai plateau swamp area, Gahai Lake NR, Bosten Lake, Whale Lake, Hala Lake, Yirancuo, Guide Huanghexiaqing Wetland Park, Chakayan Lake, Big Sugan Lake and Small Sugan Lake wetland
National Wetland Nature Reserves (NR)	Mengda, Ruoergai, Gahaizecha, Dunhuang West Lake
Provincial Nature Reserves	Big Sugan Lake, Small Sugan Lake, Dunhuang South Lake, Guazhoushule River mid- and downstream NR, Changma River, Ganhaizi, Heihe River Basin, Yellow River Shouqu, Yellow River Three Gorges

consequently generally lasting for 20–30 days. Hills on the prairie undulate slightly, valleys spread vastly and are dotted with swamps, and water and grass are abundant and luxuriant. Favorable conditions for wetland development are created by the long persistence of water from precipitation and alpine snowmelt. Marshes, peatland wetland, and lakes are the main wetland types within this area holding the second most concentrated distribution of wetlands along the Yellow River. Ruoergai wetland is the world's largest and centralized distribution plateau peatland.

The source region of the Yellow River is known as "China's water tower." Wetlands in this region of the Yellow River perform a natural and important role in the regulation of water volume and disassociated sediments, water purification, flood prevention, water storage, as well as local climate regulation. These wetlands also provide important habitats for many rare fish and waterbirds of the plateau, including gulls, geese, ducks, and black-necked cranes *Grus nigricollis*. Birds are mostly migratory, with tens of thousands of wild geese and gulls arriving from their wintering grounds in the Indian Peninsula each spring to breed.

Wetlands in Hetao Plain

After the Yellow River has traversed the first step of China's landforms, it flows into the second step that includes the Inner Mongolian Plateau and Loess Plateau. Hetao Plain lies between the Yinshan Mountains and Erdos Plateau. The Yellow River flows through the Hetao Plain from west to east. Geologically, Hetao Plain belongs to the graben basin alluvial-proluvial plain. The pre-mountain diluvial plain accounts for one quarter of the whole plain, and the remaining three quarters comprise the Yellow River's alluvial plain where the wetlands are mainly distributed. There are three areas where concentrations of wetlands occur within this section, respectively Ningxia Plain, Houtao Plain, and Qiantao Plain.

As an alluvial plain on the wider Hetao Plain, the Ningxia Plain is located between the Helan Mountains and Erdos highlands, with an altitude of 1,100–1,200 m asl. This stretch of the Yellow River is more than 300 km from

Table 4 Wetlands associated with the Yellow River in Hetao Plain

Level of wetlands	Name of wetlands
Ramsar Sites	Erdos wetland
Wetlands of National Importance	Qingtongxia Reservoir, Hongjiannao, Crested Ibis habitat in Yang county, Daihai Lake, Wuliangsuhai Lake
National Wetland Nature Reserves	Yanchi county's Haba Lake, Erdos' *Larus relictus*, Hangzhong's Crested Ibis
Provincial Wetland Nature Reserves (NR)	Sand Lake, Qingtongxia Reservoir, Xiji county's Zhenhu Lake, Yinghu Lake, Qianhu Lake, Shaanxi's Yellow River, Jing-wei, Zhouzhi county's Heihe River, Nanhaizi, Daihai Lake, Wuliangsuhai Lake wetland waterfowl NR, Huangqihai Lake

north to south, from Zhongwei city to Shizuishan city. This area has a typical arid continental climate, with an annual average rainfall of 200–300 mm, while the average annual evaporation ranges from 1,200 to 2,000 mm. Due to the flat landscape along the river course, intensive branching streams and lakes were formed here. The total area of wetlands in Ningxia Hui Autonomous Region is about 256,000 ha, accounting for 4.9% of the region's total area. Among the provinces in Northwest China, Ningxia has the most abundant wetland resources with the greatest features. Wetlands here are mainly riverine and lakes, which played essential roles in flood mitigation and water regulation and storage.

Houtao Plain lies to the west of Qiantao Plain, running from Bayangol in the west to Xishanju of Ulateqianqi County in the east. It stretches 170 km from east to west and 75 km from north to south, covering an area of about 10,000 km^2. Qiantao Plain starts from the eastern foot of the Wula Mountains in the west to east of Hohhot. There are more wetlands on the Houtao Plain, and they are more widely distributed (Table 4).

The wetlands on Hetao Plain have not been investigated in detail. However, it is clear from remote-sensing images and topographic maps that the river network here is distributed extensively and that there is a high density of lakes and marshes. Because arable land is irrigated by water from the Yellow River, the drainage water from irrigation schemes maintains the existence of Wuliangsuhai Lake on Houtao Plain as a large multi-functional lake rarely seen in desert and semi-desert areas. Hasuhai Lake on Qiantao Plain is a large shallow freshwater lake of over 30 km^2 and only 2 m in depth that extends from a channel created by the Yellow River's movements. The wetlands on Hetao Plain in Inner Mongolia are mainly associated with the Yellow River, with their water mainly replenished by the river through channels or subsurface flow.

Wetlands in the Midstream and Downstream Reaches of the Yellow River

Wetlands in the midstream and downstream reaches of the Yellow River mainly include Maowusu sandy land, Xiaobei Main Stream wetlands, Sanmenxia Reservoir area wetlands, and downstream channel wetlands (Table 5).

Table 5 Wetlands in the midstream and downstream reaches of the Yellow River

Level of wetlands	Name of wetlands
Wetlands of National Importance	Sanmenxia Reservoir Area, Yellow River old riverway swamp area in Northern Henan, Suyahu Lake wetland, North Five Lake
National Wetland Nature Reserves	Yellow River, Xinxiang Yellow River wetland bird reserve
Provincial Wetland Nature Reserves	Sanmenxia Reservoir, Neixiang Tuan River, Lushi County giant salamander, Zhengzhou Yellow River, Kaifeng Liuyuankou, Yuncheng

The Maowusu sandy land undulates slightly and its surface is mainly characterized by shifting and fixed dunes and sand. Between the dunes and low-lying sand there are many salt, alkaline, and freshwater lakes. Hongjiannao Lake (1,200 m asl) is a typical lake, with seven seasonal inflowing rivers, and the water inflows equal evaporation losses. It has an area of 67 km^2 and an average water depth of 8.2 m. The groundwater of the Maowusu sandy land occurs at relatively shallow depths and has abundant reserves. Therefore, although it is located in an arid inland region, this area has many wetlands which differ from those in other parts of the Yellow River basin.

The river bed of Xiaobei Main Stream of the Yellow River is higher than the bottomlands (flat low land along a river) on either side and is still rising, with an elevation difference reaching 1.5 m. Many wetlands totaling some 255 km^2 have developed due to the constantly rising groundwater level on the bottomland. The wetland types are mainly salt marsh, catchment basin, swamp, wet grassland, and woodland wetland (Zhang et al. 2003). More than 50 species of First and Second-Grade State Protection birds live here. The main functions of the wetlands are flood detention storage and ecological balance.

The Sanmenxia Reservoir wetland (Mao et al. 2005) includes riverine, bottomlands, pond, and lake wetlands. The reservoir wetland is rich in animal and plant resources and provides a wintering ground and habitat for many rare waterbirds. In the area, there are 14 species of national and provincial key protected plants and 38 species of national key protected animals.

The downstream channel wetland mainly refers to wetlands distributed below Xiaolangdi to Dongpinghu Lake bottomland and on Dongpinghu Lake and its surroundings. From Baihe town of Mengjin County in Henan province to the mouth of the river in Kenli County of Shangdong province, there are more than 120 natural bottomlands, of which the area of the five bottomlands in Yuanyang County, Changyuan County, Puyang County, Dongming County, and Changqing County all exceed 200 km^2. The downstream channel wetland of the Yellow River is the by-product of flood sediment deposition, changing with shifts in channel morphology. Its formation, development, and shrinkage are closely related to the Yellow River's water and sediment conditions and the river course's boundary conditions. Consequently, it is characterized by instability, originality, vulnerability of ecological environment, and paucity of aquatic plants. Quite a few of the wetlands are seasonal, with their water mainly replenished by floods and groundwater. The ecological environment of the wetland is very complex, providing conditions that are suitable for the reproduction of a wide diversity of animals and plants as well as habitats for migratory waterbirds (Hao et al. 2005).

Yellow River Delta Wetlands

Wetlands in the Yellow River estuary are mainly distributed in the delta, with Ninghai at its vertex. The Yellow River delta is relatively recent having formed when the levee at Tongwaxiang broke and flood waters captured the Daqing River channel in 1855 (Chunhong et al. 2006). The delta is the youngest, most widely distributed, and most intact wetland complex covering the largest total area in China's warm zone. Its five main wetland types are shrub, meadow, marsh, riverine, and coastal wetlands.

The Yellow River delta is located in the two districts of Dongying and Hekou in Dongying city and the three counties of Guangrao, Lijin, and Kenli. It occurs in the North Temperate Zone and has a semi-humid continental climate with four distinctive seasons, moderate temperature, rain and heat in the same period, and sufficient sunlight. Its average annual rainfall and temperature is 533 mm and 12.2 °C, respectively. Due to the Yellow River's sediment deposition, new coastal land is formed at an average annual rate of 2,000–3,000 ha. The area of the delta is 8,053 km^2, of which the wetland area is 4,220 km^2, including the following wetland types:

Wetland type	Area (km^2)	% of total area
Shallow sea	1,680	41.2%
Tidal flat	1,020	25.4%
Open water	159	4.0%
Reservoir	136	3.4%
Pit-pond and reed marsh	420	10.4%
Canal and hydraulic structure	572	14.2%
Other wetlands	35	0.86%

The Yellow River delta has been gazetted as a National Nature Reserve (NNR) for protecting the new wetland ecosystem and rare and endangered birds. It includes 153,000 ha of wetlands, of which 79,000 ha is the core area, 11,000 ha is the buffer area, and 63,000 ha is the experimental area (Zong et al. 2009). However, intensifying human activities and decreasing Yellow River runoff and sediment resulted in a number of important changes to the delta (such as decrease in natural wetlands, increase in artificial wetlands, and degradation of regional wetlands) especially between 1973 and 2013. In contrast, a large area of natural wetland was constructed and developed in response to these factors during the past 40 years (Hong et al. 2016).

Distribution of Biodiversity in the Wetlands of the Lower Yellow River Basin

The wetlands in the Yellow River basin are staging areas and wintering and breeding grounds for migratory birds from northeast Asia and the western Pacific. The Yellow River Delta wetland and the National Nature Reserve for Yellow River wetlands in Henan are important regions for such birds.

The Yellow River Delta wetland is a Wetland of International Importance. It is the most intact, largest, and youngest coastal wetland ecosystem in China's warm temperate zone, occupying an important position in wetland biodiversity conservation practice and research in China. It is also rich in biological resources, where 116 phytoplankton, 271 angiosperm, two gymnosperm, four fern, 25 mammal, nine reptile, six amphibian, 108 freshwater fish, 583 terrestrial invertebrate, and 283 bird species (including nine species in the first grade of nationally protected birds and 42 in the second grade) have been found (Zhao 1995). In the nature reserve, the natural habitats consist of 33,000 ha of reeds, 18,000 ha of grasslands, 2,000 ha of willows, 8,100 ha of tamarisk shrubs, as well as 5,600 ha of artificial locust trees. At present, it is the best-preserved and largest area of natural vegetation in coastal North China. Every year more than one million snipe pass through the delta. It is an important breeding area for Saunders' Gulls Larus *saundersi/Saundersilarus saundersi*, a globally threatened species. In this area, the number of Saunders's gulls ranged from 1,254 to 4,012 (1998–2014), indicating the increasing trend (Li et al. 2013; Liu 2015).

The NNR for Yellow River wetland in Henan is located in the mid and downstream areas of the Yellow River, mainly protecting the wetland ecosystem and rare birds. The landscape here is complex, attracting a large number of wintering and passage birds. Fifteen globally threatened species have been recorded here, including two Critically Endangered species, namely the Siberian crane *Leucogeranus leucogeranus* and Baer's pochard *Aythya baeri*; two Endangered species, namely the oriental stork *Ciconia boyciana* and red-crowned crane *Grus japonensis*; and 10 Vulnerable species, such as Dalmatian pelican *Pelecanus crispus*, Chinese egret *Egretta eulophotes*, lesser white-fronted goose *Anser erythropus*, swan goose *Anser cygnoid,* Pallas' sea eagle *Haliaeetus leucoryphus*, white-naped crane *Grus vipio*, great bustard *Otis tarda*, etc. (Niu et al. 2009).

Threats to the Wetlands of the Yellow River Basin

Water Shortage of the Yellow River

Currently, the Yellow River's average natural runoff is 53.48 billion m^3. Due to water usage in agriculture and for water storage and the influence of other factors, it is predicted that the average natural runoff will decrease to 51.48 billion m^3 by 2030.

It is forecasted that the Yellow River's average natural runoff will further decrease for a long period of time. As a result of overexploitation of the Yellow River's water resources, a water cut-off (no flow at all) appeared for the first time in 1972 in the downstream reaches of the Yellow River. The water cut-off has now spread to the Yellow River's source region beginning in the 1990s. Since 1999, the state has authorized the Yellow River Conservancy Commission to implement unified management of the Yellow River's water resources. At present, the Yellow River has not seen any water cut-off for 13 consecutive years, and the water quality of the main stream and wetland's ecological environment have improved significantly. However, the water shortage situation in the Yellow River basin has not been relieved as the water volume in the outer and inner river is deficient by 2.66 billion m^3 and 860 million m^3, respectively, and the total water shortage is 3.52 billion m^3. Water shortage will worsen in dry years (Wang et al. 2013).

Sediment Deposition in the Yellow River

Water required to sustain the ecological conditions in the main stream is diverted for other usages, and the water-sediment relation is deteriorating. With the implementation of the unified management of water volume in the mainstream of the Yellow River, water cut-offs have been prevented in the downstream reaches. Yet there remains severe shortage of water resources and the supply and demand relationship is still in great tension. At the same time, the shortage of water to support ecological functions in the inner rivers also led to a series of problems, including river channel sedimentation, worsening of river bottom rising, and deterioration of the aquatic environment. After the 1980s and before the Xiaolangdi reservoir became operational, the water volume flowing into the Yellow River's downstream reaches decreased dramatically, worsening the disproportional relationship between water and sediments and compounding river channel sedimentation. Once the Xiaolangdi reservoir is fully filled with sediment, the sediment flowing into the downstream reaches will gradually increase again. Then the restored river channel will gradually shrink, and the Yellow River's downstream channel will once again be in a serious situation (Zhang et al. 2009a).

Serious Water Pollution

The situation of water pollution is very serious in the Yellow River. Only 8.2% of stream segments basin conform to the national water quality standards. The water quality of nearly 40% of the Yellow River's mainstream is worse than Class V, basically losing its functions as a water body. Most branches in the Yellow River's upstream are subject to different degrees of pollution. The water quality of almost all branches in the midstream and downstream is worse than Class V at all times of year. Within a 200 km stretch of the downstream reaches of the Weihe River, the largest branch of the Yellow River in Shaanxi, four cities (Baoji, Xianyang, Xi'an, and Weihe) discharge 600 million tons per year of industrial and sewage waste water. Nearly one third of the aquatic organisms in the Yellow River have disappeared. Many branches are now polluted overall, and fish and shrimp have fundamentally disappeared. Two rare species, *Cyprinus carpio haematoperus* and *Megalobrama amblycephala*, became extinct due to water pollution.

Fragile Ecological Environment and Soil Salinization in the Delta

The soil's original and secondary salinization has led to the degradation of land functions. The Yellow River Delta has only recently become land and has a high water table and high degree of mineralization. Coupled with a high evaporation-precipitation ratio, it is easy for salt to rise to the surface of the soil, causing soil salinization. The native herbaceous vegetation in the Yellow River Delta consists mostly of halophytic herbaceous plants, which can help to increase soil organic matter and inhibit evaporation. With the loss of this vegetative cover evaporation of the surface soil will increase and the salt will rise to the surface, leading to an increase in the soil's secondary salinization. Moreover, farmland development follows an unsustainable approach in the Yellow River Delta, which together with excessive exploitation and grazing seriously damage the consequent evolutionary changes of the land resources. Thus, a vicious cycle of "land reclamation- abandonment- reclamation again- abandonment again" has taken shape, leading to man-made secondary salinization (Wu et al. 2010).

Future Challenges

Influence of the Western Line of the South-to-North Water Diversion Project on Wetlands

The western line of the South-to-North Water Diversion Project will flood river valleys and arable land, necessitating the migration of human populations to new land from the inundated areas is difficult. The potential impacts of this diversion project are numerous and include deterioration of water quality, loss of meadows, water cut-off during the dry season, and expansion of dry river valleys. The already fragile natural balance may be further compromised, accelerating the evolutionary process of desertification. As an important site of water conservation in the upstream reaches of the Yellow River, the Ruoergai-Maqu plateau wetland is considered as the "water tower of the plateau" and the kidney of the Yellow River. Thirty percent of the water in the Yellow River originates from the Ruoergai wetland. Three water diversion lines planned in the western line of the South-to-North Water Diversion Project will eventually flow through the Ruoergai wetland into the Yellow River. This water diversion may affect the water level of wetlands, aggravating the aridification in the Ruoergai wetland (Shang et al. 2001).

Influence of Reservoirs

From 1950 to 1989, soil and water conservation projects in the middle and upstream of the Yellow River intercepted 1.8811 trillion tons of sediment. The sediment interception by reservoirs comprised 76.12% of the total sediment interception. It should be noted that the amount of sediment intercepted by these reservoirs is not equal to the overall decrease in sediment discharge in the downstream reaches. This is because after a reservoir retains the sediment, the flow and sediment conditions in the channel of the reservoir change, and the bed of river course adjusts itself, reaching a new balance. Due to the change in the distribution of sediment deposition, it is mainly deposited in the river main channels and narrow river courses of the lower section, resulting in the decline of its flood control capability. It can be said that reservoirs retaining sediment must have a certain positive role, but in reality it is unfavorable for water storage regulation, water conservancy, and flood control (Zhang et al. 2009b).

References

Chunhong H, Yangui W, Yanjing Z, Hongling S, Yuling T, Cheng L, Zhao F, Jayakumar R, Zhide Z. Case study on sediment management and wetland conservation at Yello River Mouth. Beijing (China): International Research and Training Center on Erosion and Sedimentation; 2006. 62 p. IRTCES Report −2006-2-01.

Hao FQ, Gao CD, Huang JH. Analysis on the wetland of the downstream river channel of the Yellow River. Yellow River. 2005;27(4):5–7.

Hong J, Lu XN, Wang LL. Quantitative analysis of the factors driving evolution in the Yellow River delta wetland in the past 40 years. Acta Ecol Sin. 2016;36(4):924–35.

Huang C, Liu GH, Wang XG. Features of wetland landscape in the Yellow River basin, controlling factors and protection. Geogr Res. 2012;31(10):1764–74.

Leung GY. Reclamation and sediment control in the middle Yellow River valley. Water Int. 1996;21:12–9.

Li CK, Zhao HP, Deng PY. Bird fauna and diversity at Zhengzhou Yellow River wetland's provincial natural reserve. J Henan Univ (Nat Sci). 2013;43(4):416–22.

Liu HF. Propagation study of Saunders's Gull at Shandong Yellow River delta. Shandong For Sci Technol. 2015;220(5):86–7.

Liu JF, Wang JH, Jiao MH. Response of China's Yellow River basin to the global climate change. Arid Zone Res. 2011;28(5):860–5.

Mao ZB, Peng WQ, Zhou HD. Research on the influence of Sanmenxia Reservoir's operating mode on wetland's ecosystem. Water Resour Dev Res. 2005;5(9):12–7.

Ministry of Water Resources of the People's Republic of China. Overview of the Yellow River basin; 2013. http://www.mwr.gov.cn/ztpd/2013ztbd/2013kqjhkfzlbhxpz/lyxj/huanghly/ghyd/201305/t20130527_436362.htm

Niu JY, Ma CH, Ma SZ. Survey of bird resources at Henan Yellow River wetland's national natural reserve. Sichuan J Zool. 2009;28(3):462–7.

Shang YM, Ding ZX, Tong HO. Analysis of the western route of the South-to-North water diversion project's influence on ecological environment. Yellow River. 2001;23(10):27–30.

Wang Y, et al. Prospect and countermeasures of Yellow River's development and protection. China Water Resour. 2013;13:30–2.

Wen W, et al. Analysis of eco-hydrology features change and rainfall in the upstream of the Yellow River. Urban Constr Theory Res. 2012;10(29):1–6.

Wu D, et al. Dynamic change simulation and situational analysis of the agricultural land at the Yellow River delta. Trans Chin Soc Agric Eng. 2010;26(4):285–90.

Xu JX. Sedimentation rates in the lower Yellow River over the past 2300 years as influenced by human activities and climate change. Hydrol Process. 2003;19:3359–71.

Zhai JR. Contemplation for floodplain problems in the lower Yellow River. In: Shang HQ, Luo XX, editors. Proceedings of the second international Yellow River forum, Vol. III. Zhengzhou (China): Yellow River Conservancy Press; 2007. p. 3–10

Zhang ZJ, Yuan BP, Hu TJ. Discussion on the wetland protection of the Yumenkou-Tongguan section of the yellow river. Yellow River. 2003;25(12):3–4.

Zhang JF, Wang Y, An CH. Yellow River's water and sediment controlling system overview from the point of maintaining Yellow River's healthy life. Yellow River. 2009a;31(12):6–10.

Zhang JY, Wang GQ, He RM. Hydrological variation tendency in the mid-stream of the Yellow River and its response to climate change. Adv Water Sci. 2009b;20(2):153–8.

Zhang GH, Wang XL, Guo MP. The Yellow River basin's overland runoff changing features and its relationship with the climate change. J Arid Land Res Environ. 2013;27(7):91–5.

Zhao YM. Scientific survey on Yellow River delta nature reserve. Beijing: China Forestry Publishing House; 1995.

Zong XY, Liu GH, Qiao YL. Analysis on Yellow River Delta's wetland landscape's dynamic changes. J Geogr-inf Sci. 2009;11(1):91–7.

Current Status of Seagrass Habitat in Korea

131

Kun-Seop Lee, Seung Hyeon Kim, and Young Kyun Kim

Contents

Introduction	1590
Seagrass Habitat in Korean Coastal Areas	1590
Seagrass Growth and Water Temperature	1593
Threatened Seagrasses in Korea	1593
Threats to Seagrass Habitat	1595
References	1595

Abstract

Nine seagrass species of four genera, including five *Zostera* species (*Z. marina*, *Z. asiatica*, *Z. caespitosa*, *Z. caulescens*, and *Z. japonica*), two *Phyllospadix* species (*P. iwatensis* and *P. jaonicus*), *Ruppia maritima*, and *Halophila nipponica* are distributed on soft sediments and rocky substrata from intertidal to approximately 15 m water depth in coastal waters of Korea. *Zostera marina* is the dominant seagrass species, and *H. nipponica*, a species of the tropical genus, has been observed recently on the coasts of Korea. *Zostera asiatica*, *Z. caespitosa*, *Z. caulescens*, and two *Phyllospadix* species are endemic in the northwestern Pacific. Most seagrasses in Korea have growth dynamics adapted to temperate environments, but *H. nipponica* exhibits a growth pattern similar to tropical/subtropical seagrass species. Six seagrass species including *Z. marina*, *Z. caespitosa*, *Z. caulescens*, *Z. asiatica*, *P. iwatensis*, and *P. japonicus* are protected as "Marine Organisms under Protection" in Korea. Seven of the 15 IUCN Red List seagrass species occur in coastal waters of Korea. Significant loss of seagrass habitats has been reported due to both anthropogenic and natural disturbances

---

K.-S. Lee (✉) · S. H. Kim · Y. K. Kim
Department of Biological Sciences, Pusan National University, Busan, South Korea
e-mail: klee@pusan.ac.kr; shkim0928@pusan.ac.kr; ykkim@pusan.ac.kr

© Springer Science+Business Media B.V., part of Springer Nature 2018
C. M. Finlayson et al. (eds.), *The Wetland Book*,
https://doi.org/10.1007/978-94-007-4001-3_264

such as land reclamation and coastal eutrophication in Korea. Since the majority of Korean seagrasses are cold water adapted, global climate change appears to be a new threat to seagrasses in Korea.

Keywords

Anthropogenic disturbances · Growth dynamics · Korea · Seagrasses · Seagrass decline · Threatened seagrasses

Introduction

The Republic of Korea is located in the temperate zone of the northwestern Pacific. It is enclosed by the Yellow Sea to the west, the East Sea to the east, and the Korean Strait to the south. The eastern coastline of Korea is simple and linear with well-developed sand dunes and lagoons, whereas the southern and western coastlines are highly complex and indented with many islands. The western coast has extremely high tidal amplitude (maximum tidal range of ~10 m), resulting in large tidal flats on the west and south coasts of Korea. Because each coast has a distinct structure, patterns of seagrass habitat also vary along the coasts (Fig. 1). The southern coastal areas are on the border between the temperate North Pacific and tropical Indo-Pacific bioregions (Short et al. 2007), and thus are sensitive to global climate change. One species (*H. nipponica*) of the tropical genus *Halophila* has been observed recently on the southern coast of Korea.

Nine seagrass species of four genera, including five *Zostera* species, two *Phyllospadix* species, *Ruppia maritima*, and *Halophila nipponica* are distributed on soft sediments and rocky substrata from the intertidal to a water depth of about 15 m in coastal waters of Korea (Table 1). *Zostera marina* is the most abundant seagrass species, accounting for nearly 90% of seagrass cover in Korea. Unlike most other seagrass species, which occur on soft bottoms, two *Phyllospadix* species occur on rocky substrata of high-energy open coasts. *Zostera japonica* is distributed primarily in intertidal areas. *Zostera asiatica, Z. caespitosa, Z. caulescens*, and two *Phyllospadix* species are endemic to Northeast Asia. *Halophila nipponica* was first observed in Korea in 2007, and several meadows of this species currently occur along the southern coast of Korea (Kim et al. 2009).

Seagrass Habitat in Korean Coastal Areas

The eastern coast of Korea, which is open and exposed to constant wave action, consists of sandy beaches, lagoons, rocky shores, and small ports. Of the nine Korean seagrass species, seven (five *Zostera* and two *Phyllospadix*) occur on the eastern coast (Fig. 1). Although *Z. marina* is the dominant seagrass species in coastal waters of Korea, two *Phyllospadix* species are dominant on the eastern coast. Due to the high wave energy and strong currents, *Z. marina* occurs only on the eastern coast

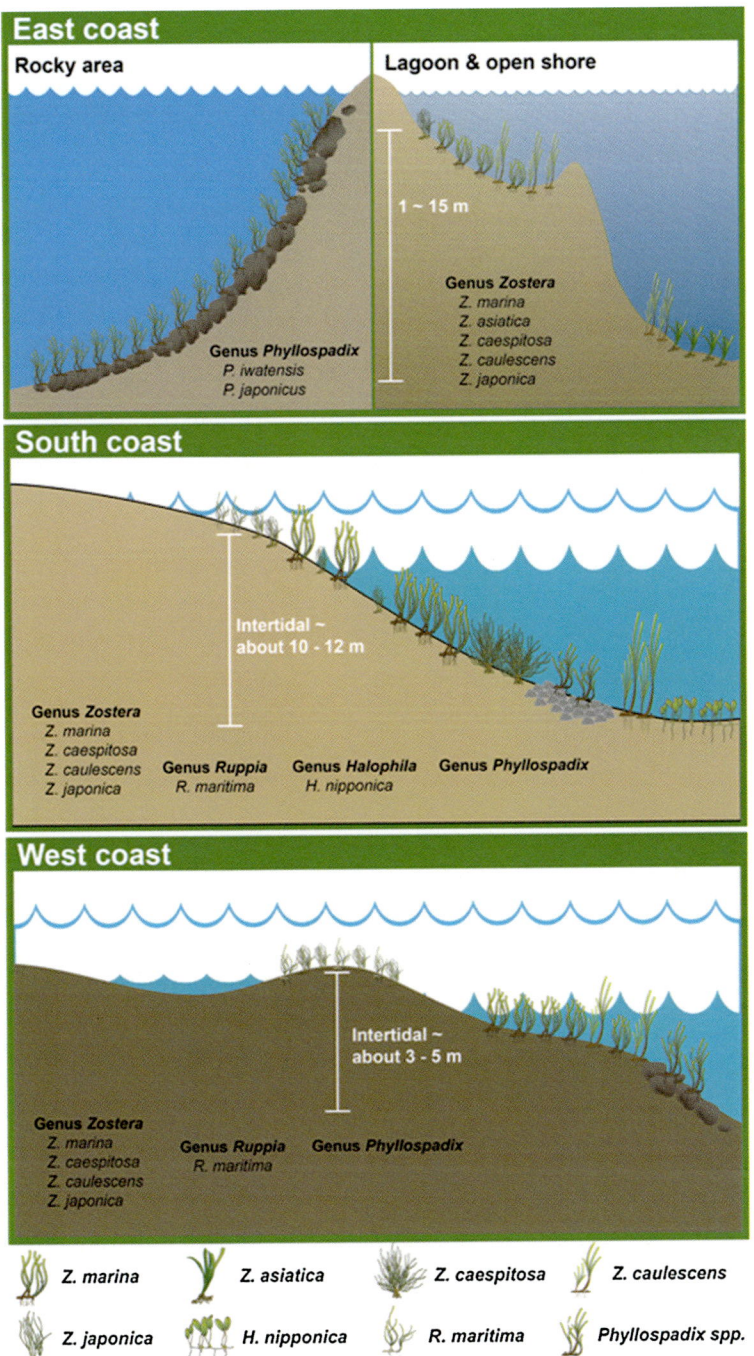

Fig. 1 Seagrass distribution in coastal areas of Korea (Kun-Seop Lee © Rights remain with the author)

Table 1 Seagrass species in coastal waters of Korea

Genus	Species
Zostera	Zostera marina
	Zostera asiatica
	Zostera caespitosa
	Zostera caulescens
	Zostera japonica
Phyllospadix	Phyllospadix iwatensis
	Phyllospadix japonicus
Ruppia	Ruppia maritima
Halophila	Halophila nipponica

Fig. 2 *Zostera asiatica* is endemic to northeastern Asia (Korea, China, Japan, and Russia). In Korea, *Z. asiatica* is distributed in relatively deep water (8–15 m) along open shores of the east coast. Few *Z. asiatica* meadows occur along the east coast, and they have been heavily impacted by construction of industrial areas, power plants, and ports. Thus, *Z. asiatica* has a high probability of disappearance from the Korean coast (Photo credit: Kun-Seop Lee © Rights remain with the author)

in isolated enclosed areas, such as inside small ports or lagoons, where it sometimes co-occurs with *Z. caulescens*. Although most *Zostera* species are distributed in sheltered areas with calm waters, *Z. asiatica* forms unique populations in relatively deep water (8–15 m) in open, high-energy zones on the eastern coast (Fig. 2).

The southern coast of Korea is heavily indented, and includes approximately 2,000 islands, creating a number of very diverse habitats for seagrasses. Four *Zostera* (*Z. marina, Z. caespitosa, Z. caulescens,* and *Z. japonica*), two *Phyllospadix*, *Ruppia maritima,* and *Halophila nipponica* are distributed along the southern coast (Fig. 1). *Zostera marina* forms relatively large meadows from the intertidal to approximately 5 m water depth. Although *Z. marina* typically has a perennial life history strategy in coastal waters of Korea, a few annual populations have been observed near the maximum depth limit of this species on the southern coast (Kim et al. 2014). *Zostera caespitosa* and *Z. caulescens* usually form small patches in slightly deeper water (3–12 m) than *Z. marina*. *Zostera japonica* is distributed primarily in intertidal areas, while *R. maritima* occurs only in estuarine areas on the southern coast. The Tsushima current, the northeast-flowing warm branch of the Kuroshio, passes through the South Sea to the East Sea. Because the water temperature on the southern coast is increasing due to global climate change, *H. nipponica*, which typically occurs in monotypic or mixed meadows with *Z. marina*, has been discovered recently on the southern coast. *Phyllospadix* species on the southern coast occur only around islands, which are away from the mainland and exposed to constant wave action or strong currents.

The western coast of Korea also has a very complex and indented coastline. Vast tidal flats with extremely high tidal range occur along the western coast. Because of the high tidal range, most seagrass beds on the western coast are located around islands. Four *Zostera* (*Z. marina, Z. caespitosa, Z. caulescens,* and *Z. japonica*), two *Phyllospadix,* and *Ruppia maritima* occur on the western coast (Fig. 1).

Seagrass Growth and Water Temperature

Because the coastal area of Korea belongs to the temperate North Pacific bioregion (Short et al. 2007), most Korean seagrasses have growth dynamics adapted to cold water (Fig. 3). The optimal growth temperatures for *Z. marina* and *P. japonicus* are 15–20 °C and 14 °C, respectively, and their growth is inhibited by water temperatures above 20 °C (Lee et al. 2005; Park and Lee 2009). However, *H. nipponica*, a recently discovered seagrass species in Korea, exhibited a growth pattern different from other Korean seagrass species but similar to tropical or subtropical seagrass species. The growth of *H. nipponica* was severely restricted at water temperatures below 15 °C, and it showed little growth until late spring (Fig. 3; Kim et al. 2012). The highest productivity of *H. nipponica* occurred at approximately 25 °C, and there was no growth inhibition in high water temperatures typical of summer. This result implies that *H. nipponica* in temperate waters of Korea possesses the tropical traits of the genus *Halophila*. If coastal water temperature increases steadily due to global warming, cold-water-adapted temperate seagrasses may be replaced by warm-water-adapted species such as *H. nipponica* in coastal waters of Korea.

Threatened Seagrasses in Korea

According to the "Conservation and Management of Marine Ecosystems Act," which was enacted in 2007, six seagrass species including *Z. marina, Z. caespitosa, Z. caulescens, Z. asiatica, P. iwatensis,* and *P. japonicus* are protected as "Marine Organisms under Protection" in Korea. In the Red List of Threatened Species produced by the IUCN (International Union for the Conservation of Nature),

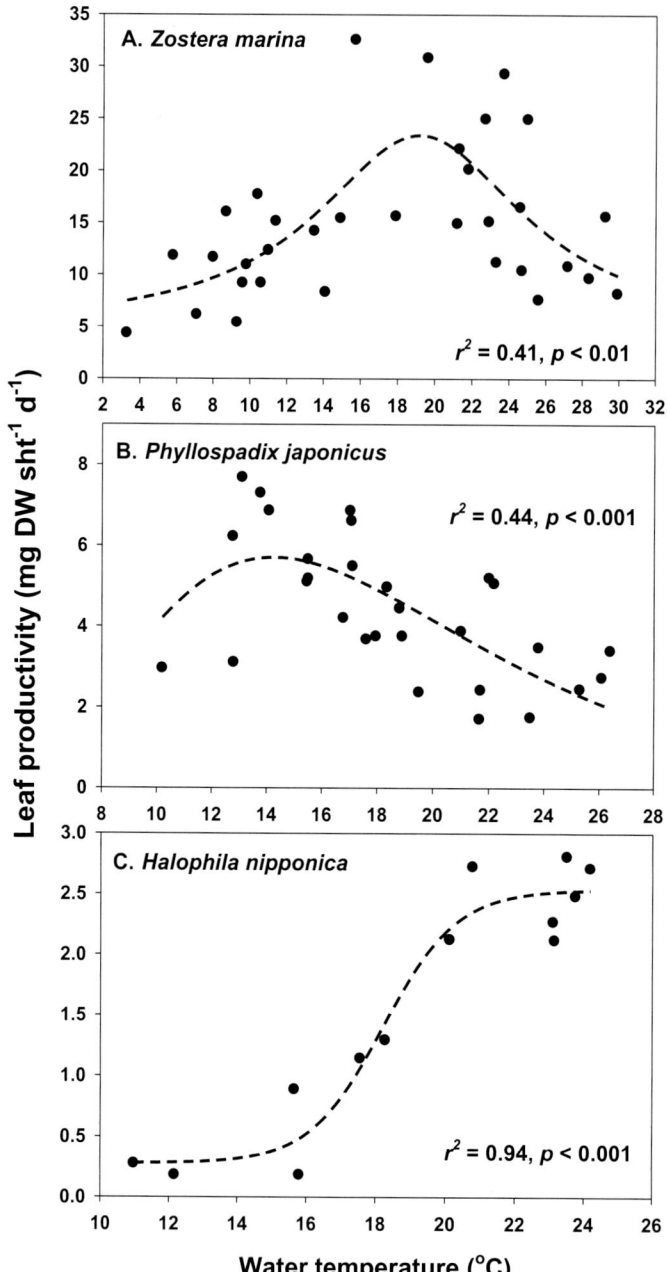

Fig. 3 Best-fit regressions of productivity for *Zostera marina* (**a**), *Phyllospadix japonicus* (**b**), and *Halophila nipponica* (**c**) against water temperature in temperate waters of Korea (© With kind permission from Springer Science + Business Media: Marine Biology, Growth dynamics of the seagrass *Halophila nipponica*, recently discovered in temperate coastal waters of the Korean peninsula, 159(2), 2012, 255–67 Kim SH, Kim YK, Park SR, Li W-T, Lee K-S, Fig. 10)

15 seagrass species are listed as threatened (Endangered or Vulnerable) or near threatened (Short et al. 2011). Seven of these 15 species occur in Korea: *Z. geojeensis* and *P. japonicus* are classified as Endangered; *Z. caespitosa* and *P. iwatensis* are classified as Vulnerable; and *Z. caulescens*, *Z. asiatica*, and *H. nipponica* are classified as Near Threatened. *Z. geojeensis* appears to be a synonym of *Z. caespitosa*. These threatened or near-threatened seagrass species have very limited distribution in Korea, Japan, and northern China and are experiencing severe human impacts. Thus, some threatened species in Korea may suffer extinction in the near future.

Threats to Seagrass Habitat

Although seagrass is a valuable component of estuarine and coastal ecosystems, significant declines of seagrass habitats have been reported worldwide (Waycott et al. 2009). In Korea, more than 70% of seagrass cover has been lost due to both anthropogenic and natural threats. Seagrass habitats in Korea have been severely impacted by land reclamation, aquaculture, and coastal eutrophication. Dredging, fisheries activity, and boat traffic also appear to be causes of seagrass decline in Korea. Many large *Z. marina* meadows in estuarine areas disappeared due to construction of ports or industrial complexes or large-scale pollutant input from upstream industrial areas (Park et al. 2009; Li et al. 2013). The intertidal seagrass, *Z. japonica*, has been severely affected by clam harvesting (Park et al. 2011). Nearly all *Z. japonica* shoots were uprooted or buried in sediment after clamming activity. Microalgal blooms lead to severely diminished underwater irradiance in coastal areas, and the frequency of bloom events has increased in recent decades along the Korean coast (Kim et al. 2015b). Disappearance of *Z. marina* has been reported due to red tide algal bloom on the southern coast of Korea (Lee et al. 2007). Severe light reduction lasted for several weeks, and a thick layer of mucilaginous substance secreted by algal cysts suffocated *Z. marina* for weeks during the bloom event.

Global climate change appears to be another threat to seagrass habitats in Korea. Extreme climatic events are predicted to increase in frequency and magnitude as a consequence of climate change. Already, the frequency and magnitude of typhoons passing over the Korean peninsula has increased during the last few decades. For example, three consecutive powerful typhoons over 3 weeks in the summer of 2012 caused the total destruction of a large *Z. marina* bed (about 4 km^2) on the southern coast of Korea (Kim et al. 2015a). Rising sea level may also alter the growth and species composition of seagrasses.

References

Kim JB, Park J-I, Jung C-S, Lee P-Y, Lee K-S. Distributional range extension of the seagrass *Halophila nipponica* into coastal waters off the Korean peninsula. Aquat Bot. 2009;90 (3):269–72.

Kim SH, Kim YK, Park SR, Li W-T, Lee K-S. Growth dynamics of the seagrass *Halophila nipponica*, recently discovered in temperate coastal waters of the Korean peninsula. Mar Biol. 2012;159:255–67.

Kim SH, Kim J-H, Park SR, Lee K-S. Annual and perennial life history strategies of *Zostera marina* populations under different light regimes. Mar Ecol Prog Ser. 2014;509:1–13.

Kim K, Choi J-K, Ryu J-H, Jeong HJ, Lee K, Park MG, Kim KY. Observation of typhoon-induced seagrass die-off using remote sensing. Estuar Coast Shelf Sci. 2015a;154:111–21.

Kim YK, Kim SH, Lee K-S. Seasonal growth responses of the seagrass *Zostera marina* under severely diminished light conditions. Estuar Coast. 2015b;38:558–68.

Lee K-S, Park SR, Kim J-B. Production dynamics of the eelgrass, *Zostera marina* in two bay systems on the south coast of the Korean peninsula. Mar Biol. 2005;147(5):1091–108.

Lee K-S, Park J-I, Kim YK, Park SR, Kim J-H. Recolonization of *Zostera marina* following destruction caused by a red tide algal bloom: the role of new shoot recruitment from seed banks. Mar Ecol Prog Ser. 2007;342:105–15.

Li W-T, Kim SH, Kim JW, Kim J-H, Lee K-S. An examination of photoacclimatory responses of *Zostera marina* transplants along a depth gradient for transplant-site selection in a disturbed estuary. Estuar Coast Shelf Sci. 2013;118:72–9.

Park J-I, Lee K-S. Peculiar growth dynamics of the surfgrass *Phyllospadix japonicus* on the southeastern coast of Korea. Mar Biol. 2009;156(11):2221–33.

Park SR, Kim J-H, Kang C-K, An S, Chung IK, Kim JH, Lee K-S. Current status and ecological roles of *Zostera marina* after recovery from large-scale reclamation in the Nakdong River estuary, Korea. Estuar Coast Shelf Sci. 2009;81:38–48.

Park SR, Kim YK, Kim J-H, Kang C-K, Lee K-S. Rapid recovery of the intertidal seagrass *Zostera japonica* following intense Manila clam (*Ruditapes philippinarum*) harvesting activity in Korea. J Exp Mar Biol Ecol. 2011;407:275–83.

Short F, Carruthers T, Dennison W, Waycott M. Global seagrass distribution and diversity: a bioregional model. J Exp Mar Biol Ecol. 2007;350:3–20.

Short FT, Polidoro B, Livingstone SR, Carpenter KE, Bandeira S, Bujang JS, Calumpong HP, Carruthers TJB, Coles RG, Dennison WC, Erftemeijer PLA, Fortes MD, Freeman AS, Jagtap TG, Kamal AHM, Kendrick GA, Judson Kenworthy W, La Nafie YA, Nasution IM, Orth RJ, Prathep A, Sanciangco JC, van Tussenbroek B, Vergara SG, Waycott M, Zieman JC. Exinction risk assessment of the world's seagrass species. Biol Conserv. 2011;144:1961–71.

Waycott M, Duarte CM, Carruthers TJB, Orth RJ, Dennison WC, Olyarnik S, Calladine A, Fourqurean JW, Heck Jr KL, Hughes AR, Kendrick GA, Kenworthy WJ, Short FT, Williams SL. Accelerating loss of seagrasses across the globe threatens coastal ecosystems. Proc Natl Acad Sci. 2009;106:12377–81.

Hokkaido Marshes (Japan)

132

Satoshi Kobayashi

Contents

Introduction 1598
Threats and Future Challenges 1601
References 1602

Abstract

Japan has lost 60% of its marshes in the twentieth century. About 86% of all marshes in Japan today can be found in Hokkaido, the northernmost main island of Japan. However, Hokkaido has lost over 100,000 ha of marshes in the past. Kushiro Marsh is the largest remaining marsh in Japan, providing habitats for endangered species such as red-crowned crane *Grus japonensis* and Siberian salamander *Salamandrella keyserlingii*. The other principal marsh areas in Hokkaido are Kiritappu and Bekambeushi marshes in the eastern part and Sarobetsu marsh in north. The increasing deer population in Hokkaido marshes has become a serious problem.

Keywords

Kushiro marsh · Kiritappu marsh · Bekambeushi marsh · Sarobetsu marsh · Red-crowned crane · Siberian salamander

---

S. Kobayashi (✉)
Faculty of Economics, Kushiro Public University, Kushiro, Hokkaido, Japan
e-mail: satoshi@kushiro-pu.ac.jp

© Springer Science+Business Media B.V., part of Springer Nature 2018
C. M. Finlayson et al. (eds.), *The Wetland Book*,
https://doi.org/10.1007/978-94-007-4001-3_35

Introduction

The Geospatial Information Authority of Japan carried out a survey of lakes and marshes from 1996 to 1999. According to the survey, there were marsh areas covering more than 210,000 ha in early twentieth century Japan. The present area of marshes is approximately 82,000 ha. Thus, Japan has already lost 60% of its marshes. About 86% of all marshes in Japan today can be found in Hokkaido, the northernmost main island of Japan. However, Hokkaido has lost over 100,000 ha of marshes in the past. The principal marsh areas in Hokkaido are described as follows.

Kushiro Marsh, 18,000–22,000 ha, 43°09'N 144°26'E

The majority of the remaining marsh areas in Hokkaido are located in the eastern part of the island. Among the various marshes in eastern Hokkaido, Kushiro Marsh, the first Ramsar Site designated in Japan, is the largest marsh area in the country. However, its size makes conservation measures more complicated and difficult. For example, Kushiro Marsh is shared by four municipalities. Besides, its status as a national park brings in the Ministry of the Environment, and Kushiro River as a major river system brings in the Ministry of Land, Infrastructure, Transport and Tourism for flood control. Various types of NGOs are represented among stakeholders.

The majority of Kushiro Marsh is fen dominated by reeds and sedges. Some areas are covered with *Salix*, and there are small portions of bog with *Sphagnum*; swamp alder *Alnus japonica* develops in drier areas. Kushiro Marsh is surrounded by hillsides except for its southern part. Kushiro Marsh is the largest remaining marsh area in Japan, although it has been shrinking. A survey of Kushiro Marsh in the 1970s reported its area as approximately 30,000 ha, declining to approximately 22,000 ha in the latest survey (1996–1999). This area reduction may be attributed to soil erosion. Between Kushiro Marsh and the Pacific Ocean, Kushiro City and other municipalities are expanding toward Kushiro Marsh. Although there has been less development on the hillsides compared with residential areas in the south, deforestation and other activities have led to unavoidable soil erosion. Soils from catchment areas may have accumulated in the marsh area and thus led to drier conditions in some marsh areas.

The Kushiro River runs through Kushiro Marsh from north to south and flows to the Pacific Ocean after flowing through Kushiro City. There are three major lakes in eastern Kushiro Marsh, and many tributaries run from west to east to join the Kushiro River (Ministry of the Environment 2015). Kushiro Marsh National Park includes four municipalities, and each municipality has its own visitor center together with long boardwalks onto marsh areas. Thus, many visitors can enjoy various parts of Kushiro Marsh throughout the seasons, although activities are limited in winter. Usually, visitors walk only on the boardwalks without damaging the fragile marsh areas. Apart from hiking, bird watching, and horse riding, people can enjoy canoeing in some parts of the Kushiro River and in the lakes. Sport fishing is allowed in many parts of the national park. The core area of the national park is also designated as a wildlife protection area and as a natural monument area, as well as being a Ramsar Site designated in 1980.

Red-crowned cranes *Grus japonensis* are not just the symbol of all conservation efforts in Kushiro Marsh but also the cutting-edge symbol of the new conservation era in Japan. The cranes were once thought to be extirpated throughout Japan. A remnant population of 20–30 cranes were accidentally discovered within Kushiro Marsh in the 1930s. Great efforts were initially made by local farmers to feed them, supported by citizens, municipalities, and people from all over Japan. According to recent census data, the total number of red-crowned cranes in eastern Hokkaido now exceeds 1,400 (Harris and Mirande 2013). The recovery of this population from near extirpation to such a number is a success story of the conservation efforts in this area (Sugimoto et al. 2015). In addition to the cranes, there are two other large bird species that were once completely extirpated in Japan. Oriental storks *Ciconia boyciana* (Anon 2010a) and Asian crested ibises *Nipponia nippon* have now been reintroduced into the central part of Japan (Anon 2010b; Nishimiya and Hayashi 2010; Nagata and Yamagishi 2013). They have been captive bred and released into the wild. Many conservationists believe that these reintroduced storks and ibises will fly wild like the cranes in the not-too-distant future. The people facilitating these reintroduction activities for storks and ibises have learned a lot from conservation efforts for cranes in the past and present.

The distribution of the Siberian salamander *Salamandrella keyserlingii* in Japan is limited to Kushiro Marsh. The salamander has a wide distribution in Russia, northern Kazakhstan, China, Korea, and Japan. There are only two salamander species reported in Hokkaido. The other salamander species is the Ezo or Hokkaido salamander *Hynobius retardatus* which is endemic to Hokkaido Island. The Hokkaido salamander has a wider distribution within Hokkaido Island; thus, the Siberian salamander has survived only in Kushiro Marsh due to interspecific competition (Sato and Matsui 2013).

While the number of red-crowned cranes has increased in Hokkaido, the habitat available for both the cranes and Siberian salamanders has decreased. Reintroduction of cranes into the other parts of Hokkaido has been discussed. The translocation of eggs of Siberian salamanders toward the core area of Kushiro Marsh has been carried out at an experimental level. But the question remains: does anybody know how to stop Kushiro Marsh from shrinking? The Japanese Ministry of the Environment and the Ministry of Land, Infrastructure, Transport and Tourism jointly organized the Council for Kushiro Marsh Restoration Project in 2005 and are seeking a solution (http://www.kushiro.env.gr.jp/saisei/english/top_e.html).

Kiritappu Marsh, 3,200 ha, 43°05'N 145°05'E

Along the Pacific Ocean coast and the Okhotsk Sea coast in the eastern part of Hokkaido, there are many medium- and small-sized marsh areas. Some six out of the thirteen Ramsar Sites found in Hokkaido Island are located in its eastern part. While Kushiro Marsh is well known throughout Japan as a national park dedicated to a wetland type, it is also familiar globally as the host city for the Ramsar Convention's Fifth Conference of the Contracting Parties (COP5) in 1993. For the residents of eastern Hokkaido, the name of Kiritappu Marsh is also well known.

Compared with Kushiro Marsh, citizens and officers in the Kiritappu area are fortunate in the sense that it has been a bit easier for them to reach consensus on conservation practices for Kiritappu Marsh. Officers at Hamanata Town Hall have been determined to promote conservation of Kiritappu Marsh (which lies in their jurisdiction). Citizens organized a local conservation NGO – Kiritappu Wetland National Trust (http://www.kiritappu.or.jp/- in Japanese) – and it has also facilitated conservation works. The NGO applied a national trust approach in securing some important areas for the conservation of Kiritappu Marsh. Furthermore, supporters from other parts of Japan contribute a lot to the local NGO. A large visitor center in the middle of Kiritappu Marsh with a café and restaurant functions as a cultural center for local citizens and attracts various research efforts necessary for the conservation of the marsh.

Bekambeushi Marsh, 8,300 ha, 43°03'N 144°54'E

Names like Kiritappu, Bekambeushi, and Kushiro came from the language of the indigenous Ainu people. Bekambeushi Marsh is a part of the larger wetland complex designated under the Ramsar Convention, known as "Lake Akkeshi and Bekambeushi Marsh." Along the Bekambeushi River, there is the second largest marsh area in eastern Hokkaido. Lake Akkeshi developed at the mouth of the Bekambeushi River, and the river water flows on to Akkeshi Bay.

There is a large restricted area for military drills by the Japanese Defense Forces. The area has been also formerly used by US Marines. Due to the upstream presence of this restricted area, Bekambeushi Marsh remains more or less intact. Only a few sport fishing people enter the marsh upstream. In some areas there is still the possibility to encounter a stray brown bear.

Downstream Lake Akkeshi is surrounded by residential areas, and the aquaculture of oysters and clams is active within the lake, which is well known for its oysters. Fisheries are the major industry in Akkeshi Town, especially the majority of oysters and clams come from Akkeshi Bay. Thus, local fishermen in Lake Akkeshi and Akkeshi Bay have been aware of the importance of conservation efforts in and around Bekambeushi Marsh.

The presence of a large dam to prevent soil erosion was only revealed to the public after its completion. Constructed upstream on the Bekambeushi River within the restricted area, only a few people knew of its construction. The local fishermen were upset, as preventing soil erosion also meant creating a barrier to the nutrients which the river brings downstream. Local conservationists were equally upset, since the river offers breeding sites for the endangered freshwater salmon species *Hucho perryi*, the largest freshwater fish species found in Japan. The dam was also surprisingly large: at the site of the dam, the river width was originally just three meters or so. However, the width of the dam was over 100 m.

When the concealed dam was finally made known to the public, it created a big controversy. This led to the conclusion by an ad hoc panel of experts to make an opening in the middle of the dam. Also, the local conservationists and fishermen, together with other citizens, now keenly watch what the municipal governments will do with the dam and for the conservation of their wetland.

Sarobetsu Marsh, 6,700 ha, 45°03'N 141°42'E

At least one marshland occurring outside eastern Hokkaido should be mentioned. Sarobetsu Marsh, 6,700 ha, is a part of the northernmost national park in Japan. Two offshore islands, namely Rishiri and Rebun islands, together with Sarobetsu Marsh comprise Rishiri, Rebun, and Sarobetsu National Park (more than 24,000 ha). The national park designation took place in 1974, before Kushiro Marsh National Park, which was established in 1987, although its Ramsar site designation was only obtained in 2005.

As in the case of Kushiro Marsh National Park in eastern Hokkaido, there has been a wetland restoration project in Sarobetsu Marsh (http://sarobetsu-saisei.jp/- in Japanese). The procedure of the project is also similar. The council with various stakeholders was set up and conducted negotiations, and the civil engineering works have been carried out.

Threats and Future Challenges

It is certain that the future of marshes in Hokkaido depends on collaboration between government officers and local citizens. Exchange of information is the key for such collaborations. Local conservation NGOs have often played very important roles in facilitating such exchange of information on marshes.

Another large problem emerged in Kushiro Marsh in recent years. Abundant deer cause damage to crops and natural vegetation throughout Japan. Over 20 national park authorities have reported serious damage caused by wild sika deer *Cervus nippon*, and Kushiro Marsh National Park is no exception. Wild deer have been observed within Kushiro Marsh for a few decades, but witnessing several hundred wintering deer in the central part of Kushiro Marsh is a recent phenomenon. Wolves, the natural predator of deer, were extirpated both in mainland Japan and in Hokkaido over 100 years ago; thus, their extirpation does not explain the recent increase in the deer population throughout the country. It is more likely that the expansion of pastoral land areas provides sufficient food resources for the deer, and recent changes in climate enable deer to survive otherwise severe winters, especially in Hokkaido (Uno et al. 2006; Inatomi et al. 2015).

The local office of the Ministry of the Environment has confirmed that damage has been caused by deer affecting some endangered plant species, e.g., *Chamaedaphne calyculata* and *Ledum palustre* subsp. *diversipilosum*, within Kushiro Marsh and concluded that a deer cull is necessary, including within the national park boundary. Effective countermeasures are difficult to identify as Kushiro Marsh is, however, large enough for deer to disperse and hide if hunters are encountered or gunshots are heard. Any culling program in Kushiro Marsh should also take into account the presence of other endangered species such as the cranes and salamanders. Wild deer also cause similar problems in Kiritappu and Sarobetsu Marshes.

References

Anon. Restoring rice paddy habitats to reintroduce the oriental white stork in Toyooka City. Tokyo: Ministry of Environment; 2010a. http://www.biodic.go.jp/biodiversity/shiraberu/policy/pes/en/satotisatoyama/satotisatoyama02.html

Anon. Reintroducing the crested ibis and rice production. Tokyo: Ministry of Environment; 2010b. http://www.biodic.go.jp/biodiversity/shiraberu/policy/pes/en/satotisatoyama/satotisatoyama03.html

Harris J, Mirande C. A global overview of cranes: status, threats and conservation priorities. Chin Birds. 2013;4(3):189–209.

Inatomi Y, Osa Y, Uno H, Ueno M, Kobayashi S, Hino T, Yoshida T. Changes in density and habitat selection of Sika deer during winter in Kushiro-Shitsugen (Marsh) National Park. Oral presentation at the Vth International Wildlife Management Congress, Sapporo; 2015.

Nagata H, Yamagishi S. Re-introduction of crested ibis on sado island, Japan. In: Soorae PS, editor. Global reintroduction perspectives: further case-studies from around the globe. Abu Dhabi: IUCN/SSC Re-introduction Specialist Group and Environment Agency; 2013. p. 58–62.

Nishimiya H, Hayashi K. Reintroducing the Japanese Crested Ibis in Sado, Japan. TEEBcase; 2010. https://www.cbd.int/financial/greenmarkets/japan-greenibis.pdf

Ministry of the Environment. Ramsar sites in Japan. Tokyo: Ministry of the Environment; 2015. http://www.env.go.jp/en/nature/npr/ramsar_wetland/pamph/index.html

Sato T, Matsui M. Salamanders of Hokkaido. Eco Network; 2013. [in Japanese].

Sugimoto T, Hasegawa O, Azuma N, Masatomi H, Sato F, Matsumoto F, Masatomi Y, Izumi H, Abe S. Genetic structure of the endangered Red-crowned cranes in Hokkaido, Japan and conservation implications. Conserv Genet. 2015;16(6):1395–401.

Uno H, Kaji K, Saitoh T, Matsuda H, Hirakawa H, Yamamura K, Tamada K. Evaluation of relative density indices for sika deer in eastern Hokkaido. Jpn Ecol Res. 2006;21:624–32.

Section X

Central and South Asia

High Altitude Wetlands of Nepal

133

Lalit Kumar and Pramod Lamsal

Contents

Introduction	1606
Hydrology	1608
Wetland Ecosystems	1608
Biodiversity	1609
Flora	1609
Fauna	1609
Conservation Status	1610
Ecosystem Services	1611
Threats, Future Challenges, and Remedial Measures	1611
References	1613

Abstract

High altitude wetlands (HAWs) have been described as "areas of swamp, marsh, meadow, fen, peat land, or water located at an altitude above 3,000 m, whether natural or artificial, permanent or temporary, with water that is static or flowing, fresh, brackish, or saline and are generally located at altitude between continuous natural forest border and the permanent snow." HAWs include different categories of water bodies, such as lakes, ponds, rivers, glaciers, and glacial lakes. They are characterized by a unique diversity of water sources, habitats, species, and communities and generally have not been subjected to rampant human interference compared to other wetland ecosystems. Most of the HAWs in South Asia, including Nepal, lie within the Hindu Kush Himalayan Region that extends over 3,500 km and covers approximately 3.5 million km^2, acting as a fresh water

L. Kumar (✉) · P. Lamsal (✉)
Ecosystem Management, School of Environmental and Rural Science, University of New England, Armidale, NSW, Australia
e-mail: lkumar@une.edu.au; plamsal@myune.edu.au

© Springer Science+Business Media B.V., part of Springer Nature 2018
C. M. Finlayson et al. (eds.), *The Wetland Book*,
https://doi.org/10.1007/978-94-007-4001-3_278

reservoir to the major river basins such as the Ganges, Indus, Yangtze, Mekong, Amu Darya, and Hilmand. Of the nine wetlands that have been declared as Wetlands of International Importance in Nepal, four are HAWs located in the Palearctic biogeographic region in central Himalaya. There are numerous other HAWs distributed across Nepal, including many smaller HAWs, either within or outside protected areas, that bear ecological, economic, and cultural significance; however, many of these have yet to be inventoried. Fragility and sensitiveness are the main characteristics of HAWs and any small change in the water chemistry, either naturally or through anthropogenic disturbances, could lead to large impacts on their ecosystems, directly affecting their flora and fauna. Three common pressures on all HAWs of Nepal are grazing; over extraction of fuel-wood, timber, and nontimber forest products; and pollution. Acid deposition during the spring season is a major threat to most of the HAWs in Nepal. Conservation effort targeting both Ramsar and Non-Ramsar HAWs is urgently needed to minimize ongoing anthropogenic pressures and preserve existing rare and endangered flora and fauna species found in such fragile wetland ecosystems.

Keywords

High altitude wetlands · Biodiversity · Ecosystem service · Conservation threat · Ramsar · Nepal

Introduction

Currently there is no precise definition available in the scientific literature for the term high altitude wetlands (HAWs), however Chatterjee et al. (2010) describe HAWs as "areas of swamp, marsh, meadow, fen, peat land, or water located at an altitude above 3,000 m, whether natural or artificial, permanent or temporary, with water that is static or flowing, fresh, brackish, or saline and are generally located at altitude between continuous natural forest border and the permanent snow." HAWs include different categories of water bodies, such as lakes, ponds, rivers, glaciers, and glacial lakes. They are characterized by a unique diversity of water sources, habitats, species, and communities and generally have not been subjected to rampant human interference compared to other wetland ecosystems. Nepal is blessed with the highest peak in the world, Mt. Everest, along with another ten of the fourteen highest peaks, all over 8,000 m. These mountains are the source of many glaciers and lakes in the high altitude regions across the country. Most of the high altitude wetlands in South Asia, including Nepal, lie within the Hindu Kush Himalayan Region that extends over 3,500 km and covers approximately 3.5 million sq. km., acting as a fresh water reservoir to the major river basins such as the Ganges, Indus, Yangtze, Mekong, Amu Darya, and Hilmand (Gujja 2005).

Of the nine wetlands that have been declared as Ramsar Sites (Wetlands of International Importance) in Nepal, four are high altitude wetlands (Fig. 1, Table 1). All four HAWs are located in the Palearctic biogeographic region in central Himalaya and were ratified in 2007.

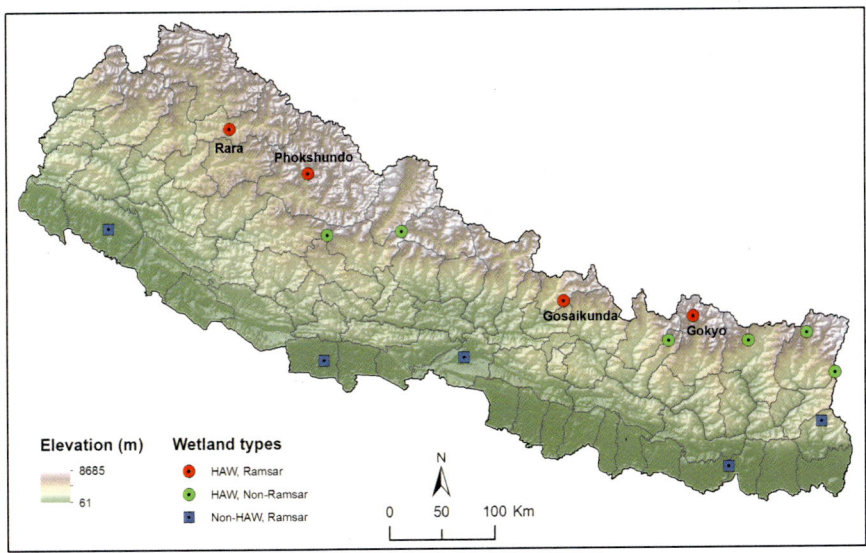

Fig. 1 Map of Nepal showing the locations of ten high altitude wetlands, four of which are Ramsar Sites, together with other Ramsar Sites

Table 1 Ramsar listed high altitude wetlands (HAWs) of Nepal

Name	Area (ha)	Location	Approx. Altitude (m)[a]	Headwaters
Gokyo and associated wetlands	7,770	Sagarmatha National Park, Solukhambu	4,710–4,950	Dudhkoshi River, a major tributary of Saptakoshi River
Gosaikunda and associated wetlands	1,030	Langtang National Park, Rasuwa	4,054–4,620	Trishuli River, a major tributary of Narayani River
Phoksundo wetland	494	Phoksundo National Park, Dolpa	3,610	Bheri River
Rara wetland	1,583	Rara National Park, Mugu	2,900	Karnali River

[a]The elevation range is for the series of wetlands that form the wetland complex

Apart from the abovementioned four Ramsar listed and six non-Ramsar HAWs shown in Fig. 1, there are numerous other HAWs distributed across Nepal. Some of these are Panch Pokhari, Dudh Pokhari, Surya Kunda, Bhairav Kunda, Salpa Pokhari, Titi Tal, Damodhar Kunda, Parbati Kunda, Sundaha, Kyangjing, and Khaptad Daha. There are many other smaller HAWs in the high altitude region, either within or outside of protected areas, that bear ecological, economic, and cultural significance; however, many of these have yet to be inventoried and hence little information exists. This article therefore focuses on the HAW Ramsar Sites that have been studied to some extent.

Hydrology

The Phoksundo wetland is an alpine freshwater oligotrophic lake that drains into the Phokhsundo River. The Lake is 4.8 km in length and 1.3 km in width, with a depth of 145 m. The main source of water for this wetland is from the mountains located on the northern side. The water flows through different tributaries, such as Chisa, Baulaha, Dekhutaichu, Jagatilumba, and Chollapu Rivers, before entering the wetlands. The water volume of the lake is approximately 408 M m^3 and the discharge rate is 3.7 cubic meters per second.

The Gosaikunda wetlands are a series of lakes formed by glacial water in the eastern and central Himalayas and remain frozen for at least 6 months of the year (Lacoul and Freedman 2005). The lake surface area is 13.8 ha, water volume nearly to 1.5 M m^3, and the maximum depth is around 24.1 m (Niraula 2011). It is one of the world's highest alpine oligotrophic freshwater lake systems, with a water discharge rate of 60 l per second.

Rara wetland is an alpine oligotrophic freshwater lake and is considered the largest lake system of Nepal, having a shoreline of 14 km and a depth of 167 m. Water volume of the lake is approximately 0.99 M m^3. It has 34 inlets, with only one outlet that discharges into the Karnali River, a tributary of the Ganges.

The Gokyo wetlands are a series of lakes that are moraine dammed glacier lakes with grassland, boulder/rocks, glaciers, moraines, and eroding debris present in the wider catchment (Sharma et al. 2010). The major source of water for this wetland series are the direct melting glacier snow from the Himalaya as well as small stream channels from the nearby surrounding forest lands (WWF Nepal 2010a).

Wetland Ecosystems

The most common characteristics of HAWs are the extreme cold, dry, and alpine climatic conditions due to low air temperatures and high ultraviolet radiation. Rara and Phokshundo wetlands lie in the Western Himalayan Temperate Forest ecoregion, as defined by WWF Global (2015a). This ecosystem comprises mainly Himalayan broadleaved and subalpine conifer forests with high altitude flora, such as blue pine *Pinus wallichiana*, spruce *P. smithiana*, and fir *Abies pindrow*, as well as fauna such as snow leopard *Uncia uncia*, Alpine musk deer *Muschus chrysogaster*, and red panda *Ailurus fulgens*. Similarly, Gokyo and Gosaikunda wetlands lie in the Eastern Himalayan Alpine Meadows ecoregion, as defined by WWF Global (2015b). This ecosystem is characterized by high elevation meadow and shrub habitats, suitable for flora such as Rhododendron *Rhododendron nivale* and different medicinal orchids and fauna such as snow leopard, Himalayan black bear *Ursus thibetanus*, and red panda *A. fulgens* as well as many high altitude birds.

Biodiversity

Flora

Most of the plant species found in Nepal's HAWs are endemic to each site, a characteristic of such wetlands. More than 80 species of flowering plants have been recorded from the Gokyo wetlands (WWF Nepal and DNPWC 2006); endemic species include *Kobresia fissiglumis*, *K. gandakiensis*, *Pedicularis poluninii*, and *P. pseudoregelina* while rare and vulnerable plant species include *Neopicrorhiza scrophulariifolia*, *Swertia multicaulis*, *Saussurea gossipiphora,* and *Meconopsis horridula*.

In Gosaikunda wetlands, around 100 species of flowering plants have been recorded (WWF Nepal and DNPWC 2006). Endemic plant species, such as *Meconopsis dhwojii, M. taylorii, Heracleum lallii, Primula aureata, P. sharmae, P. pseudoregelian,* and *Rhododendron cowanianum* are found in this wetland. Threatened species, based on the IUCN categories, include *Aconitum spicatum* (Vulnerable), *H. lallii* (Endangered), *Jurinea dolomiaea* (Near Threatened), *M. dhwojii* (Near Threatened), *Nardostachys grandiflora* (Vulnerable), *N. scrophulariifolia* (Vulnerable), *Rheum australe* (Vulnerable), and *R. moorcroftianum* (Near Threatened).

In Phoksundo wetlands, more than 150 plant species have been recorded (WWF Nepal and DNPWC 2006). Threatened species include *N. scrophulariifolia* (Vulnerable), *Dactylorhiza hatagirea* (Critically Threatened), *Dioscorea deltoidea* (Critically Threatened), *Aconitum spicatum* (Vulnerable), *N. grandiflora* (Vulnerable), *Podophyllum hexandrum* (Vulnerable), and *Megacarpea polyandra* (Vulnerable).

In Rara wetland, endemic plant species found are *Mecanopsis regia*, *Primula poluninii*, and *Cirsium flavisquamatum* (WWF Nepal and DNPWC 2006). The lake border is filled by reeds *Phragmites* spp., rushes *Juncus* spp., and sedges *Fimbristylis* spp., while *Polygonum* spp. and *Myriophyllum* spp. occupy shallow portions of the lake. Some of the threatened species include *D. hatagirea, N. grandiflora,* and *N. scrophulariflora.*

Fauna

While some mammals and birds are found in all HAWs in Nepal, others are recorded only at specific sites. Alpine musk deer is found in all four HAWs, while red panda is recorded only in Gosaikunda and Rara wetlands. Snow leopard and Tibetan Wolf *Canis lupus chanco* have been found in Phoksundo wetlands while smooth-coated otter *Lutra perspicillata* and Himalayan black bear *Selenarctos thibetanus laniger* are found in Rara wetland. Similarly, common coot *Fulica atra* has been recorded in

Gokyo, Phoksundo, and Rara wetlands, while ruddy shelduck *Tadorna ferruginea* and Eurasian wigeon *Anas penelope* are found in both Gokyo and Phoksundo wetlands. Common teal *Anas crecca*, tufted duck *Aythya fuligula*, and common merganser *Mergus merganser* are found in Gosaikunda and Rara wetlands. Globally threatened cheer pheasant *Catreus wallichii* is found in Rara wetland, while wood snipe *Gallinago nemoricola* is found in the catchment area of Gokyo and Rara wetlands. Bar-headed goose *Anser indicus* and brown dipper *Cinclus pallasii* are found in Gosaikunda and Phoksundo wetlands.

Apart from the above fauna recorded in more than one HAW, there are individual species found in each of the HAWs. For instance, Gokyo wetland harbors additional species such as northern pintail *Anas acuta*, common pochard *Aytha ferina*, Eurasian woodcock *Scolopax rusticola*, and great crested grebe *Podiceps cristatus*, while Gosaikunda has species such as ruddy shelduck *Tadorna ferruginea*, bar-headed goose, northern pintail, brown dipper, white-capped water redstart *Chaimarrornis leucocephalus*, and plumbeous water redstart *Rhyacornis fuliginosus*. Red-crested pochard *Netta rufina*, common moorhen *Gallinula chloropus*, white-throated dipper *C. cinclus*, and white-throated redstart *Phoenicurus schisticeps* are commonly seen birds in Phoksundo wetlands. Gadwall *Anas strepera*, mallard *Anas platyrhynchos*, northern shoveler *Anas clypeata*, common goldeneye *Bucephala clangula*, and solitary snipe *Gallinago solitaria* are bird species recorded in Rara wetland.

Conservation Status

All the HAWs designated as Ramsar Sites are located within national park boundaries and hence protected under various national acts within the jurisdiction of Ministry of Forest and Soil Conservation, Government of Nepal. Though they became party to the Ramsar Convention on Wetlands in 1988, the much needed Nepal National Wetland Policy was formulated and approved in the year 2003, and this provides a framework and direction for the conservation and sustainable utilization of all the wetlands in Nepal. As the country does not have a wetland related act, the National Wetland Policy (2003) is supported by the following acts: the National Conservation Strategy (1988), Master Plan for the Forestry Sector (1989), Nepal Environmental Policy and Action Plan (1993), Forestry Sector Policy (2000), Water Resource Strategy (2002), Wildlife Protection Act (1958), Aquatic Animal Protection Act (1961), National Parks and Wildlife Protection Act (1973), Soil and Watershed Management Act (1982), Electricity Act (1992), Water Resource Act (1992), Forest Act (1993), Environment Protection Act (1997), and Local Self-Governance Act (1999). Special conservation emphases are being given to these wetlands, from both government and non-government organizations, since they were listed as Ramsar Sites. Five-year site-specific management plans have been developed for all the Ramsar listed HAWs of Nepal. These plans will help in identifying field-based conservation and management problems and suggest remedial measures for sustainable use and conservation of wetlands.

Ecosystem Services

Human habitation is generally very low in all the HAWs of Nepal due to the difficult terrain and remote location. Therefore, most of the ecosystem services that such wetlands provide fall in the categories of regulating, cultural, and supporting compared to that of provisioning. Therefore, HAWs in Nepal are important for services such as climate and water regulation; spiritual, recreational, and aesthetic; as well as nutrient recycling. The local people residing in the high hill and mountainous areas of Nepal use the catchment and surrounding areas of HAWs for collecting yarsagumba *Ophiocordyceps sinensis*, a species of entomopathogenic fungi that parasitizes moth larvae known for its medicinal properties that fetches high value in the local market and is considered as a major source of livelihood, in the winter season.

The Gosaikunda wetlands bear cultural significance and are a shrine for Hindu and Buddhist pilgrims. In the event of Janaipurnima (full moon festival in August), thousands of devotees travel to these lakes, especially from Nepal, India, and some other South and South East Asian countries. Tourism is a vibrant industry in this area as thousands of trekkers visit this wetland series every year, helping to create employment opportunities for the local people of the area.

The human habitations near the Phoksundo wetland are very low compared to other Ramsar Sites in Nepal and are mostly Buddhist communities. The water in the wetland holds religious significance and is worshipped by the Buddhist communities. Resources harvested to sustain their livelihood include fuelwood, timber, and medicinal herbs. The pasture land around the wetland is used for cattle and sheep grazing. Tourism is another important service of Phoksundo wetland as thousands of national and international tourists visit the lake each year.

Rara wetland also bears cultural and religious significance and local people worship the lake during most of the religious festivals. The wetland complex includes dense forest and pasture land and local people depend on the extraction of fuelwood, fodder, timber, and medicinal herbs. Tourism is also a potential source of income but visitor numbers are low compared to other Ramsar-listed wetlands.

The Gokyo wetlands, containing ten lakes within a small periphery, bear cultural and religious significance, similar to the other HAWs discussed above.

Threats, Future Challenges, and Remedial Measures

Some of the international nongovernmental organizations in Nepal, such as the International Centre for Integrated Mountain Development (ICIMOD) and World Wildlife Fund (WWF) Nepal, are undertaking pioneer research work in the HAWs, especially from the perspective of climate change and community livelihood. They have inventoried and listed many HAWs in the high altitude region and have implemented adaptation and conservation programs. For example, WWF Nepal selected Gosaikunda and Gokyo wetlands as priorities under the Sacred Himalayan Landscape project. However, it is important to note that HAWs are the least studied ecosystems in Nepal due to their remoteness and limited provision of ecosystem services. Moreover, HAWs are sensitive to climate change because of their small catchment size, scant vegetative cover, low nutrient content in surface water, and shallow soil and low bedrock weathering rate (Strang et al. 2010).

Fragility and sensitiveness are the main characteristics of HAWs and any small change in the water chemistry, either naturally or through anthropogenic disturbances, could lead to large impacts on their ecosystems, directly affecting the flora and fauna in them. Acid deposition during the spring season is a major threat to most of the HAWs in Nepal (Bhuju et al. 2012); and the mercury and cadmium in the Gokyo wetlands is thought to result from long range transport by monsoon precipitation originating from industrial areas of lowland parts of the country, including that of India (Sharma et al. 2012). Water pollution, mainly from hotels and lodges and religious rituals, also affects water chemistry. Most of the wetlands located in the Himalayan region of Nepal are contaminated by fecal pollution, mainly from uncontrolled tourism activities (Sharma et al. 2005). Raut et al. (2012) reported the presence of chlorine ions in Gosaikunda wetlands, possibly due to discharge of waste water from the lodges in the area. Similarly Ghimire et al. (2013) reported the presence of nitrogen and phosphorous in the lake water of Gokyo wetlands from anthropogenic activities, such as excess agricultural fertilizer and human waste.

Karki et al. (2007) found three common pressures on all HAWs of Nepal, viz., grazing, over extraction of fuelwood, timber and non-timber forest products (NTFP), and pollution. Excessive pasture grazing by cattle and sheep around the wetlands has led to soil erosion and threatens siltation in HAWs. The heavy influx of trekkers and hikers, together with their recreational activities, such as camping, in the Himalayan region could be one potential reason for the degradation and depletion of forests and water resources, triggering soil erosion and landslides, shrinking of wildlife and medicinal plant habitat, and drying up of water sources in and around HAW catchments. Similarly, uncontrolled extraction of fuelwood for cooking, timber for rural building construction, and NTFP for medicinal and livelihood uses are common practices, endangering the existence of biodiversity.

Baral and Bhandari (2011) reported the importance of HAWs as breeding and staging areas for many globally threatened wetland birds, such as Baer's pochard *Aythya baeri*, Baikal teal *Anas formosa*, marbled teal *Marmaronetta angustirostris*, and the elegant black-necked crane *Grus nigricollis*, and emphasized the need for conservation of this fragile ecosystem from climatic and anthropogenic threats through in-depth scientific studies. However, a lack of understanding of such fragile ecosystems and climatic and anthropogenic disturbances on them could seriously threaten the overall HAW biodiversity of Nepal in the future.

The knowledge gap and a lack of available baseline information are the major issues that hinder conservation efforts of HAWs in Nepal. Most of the wetlands in Nepal are located in the lowland Terai and Mid-Hill regions that are easily accessible, and so have been inventoried. However, due to difficult terrain, harsh climatic conditions, and inaccessible locations of HAWs, researches have difficulty in gathering information about them. There are a range of issues concerning HAWS, such as status of biodiversity, climate change impacts on the hydrology and biodiversity, and identification and quantification of ecosystem services that demand immediate investigation from the scientific community. For instance, many of the HAWs that are not listed as Ramsar sites lack basic information, such as exact location and elevation, water quality and hydrology, biodiversity and ecosystem services. However, based on the limited information available, it can be assumed that these Non-Ramsar HAWs are important in terms of their ecosystem services and human welfare in the high altitudinal areas of Nepal. The WWF Nepal (2010b) reported that there is evidence to suggest the existence of Alpine musk deer, barking deer *Muntiacus muntjak*, ghoral *Naemorhedus goral*, snow leopard, common leopard *Panthera pardus*, red panda, and wild cat *Felis chaus*, in addition to numerous rare and endangered medicinal herbs and bird species, in these Non-Ramsar HAW catchment areas. Thus, this clearly indicates the immediate need for a coordinated effort from all conservation stakeholders to focus on landscape level conservation, corridor connectivity preservation, and site-specific management for such a fragile but valuable ecosystem.

References

Baral HS, Bhandari BB. Importance of high altitude wetlands for protection of avian diversity in the Hindu Kush Himalayas. Initiation. 2011;4:96–102.

Bhuju DR, Sharma S, Jha PK, Gaire NP. Scientific discourse of lakes in Nepal. Nepal J Sci Technol. 2012;13(2):147–58.

Chatterjee A, Blom E, Gujja B, Jacimovic R, Beevers L, O'Keeffe J, et al. WWF initiatives to study the impact of climate change on Himalayan High Altitude Wetlands (HAWs). Mountain Research and Development (Mt Res Dev) 2010;30(1):42–52.

Ghimire NP, Jha PK, Caravello G. Water quality of high altitude lakes in the Sagarmatha (Everest) National Park. Nepal J Environ Prot. 2013;4:22–44.

Gujja B. WWF International's regional approach to conserving high altitude wetlands and lakes in the Himalaya. Mt Res Dev. 2005;25(1):76–9.

Karki JB, Siwakoti M, Pradhan NS. High altitude Ramsar sites in Nepal: criteria and future ahead. Initiation. 2007;1:9–15.

Lacoul P, Freedman B. Physical and chemical limnology of 34 waterbodies along a tropical to alpine altitudinal gradient in Nepal. Int Rev Hydrobiol. 2005;90(3):254–76.

Niraula SR. Application of GIS in Bathymetric mapping of Gosaikunda Lake [Dissertation]. Dhulikhel: Kathmandu University; 2011.

Raut R, Sharma S, Bajracharya RM, Sharma CM, Gurung S. Physico-chemical characterization of Gosaikunda Lake. Nepal J Sci Technol. 2012;13(1):107–14.

Sharma S, Bajracharya RM, Situala BK, Merz J. Water quality in the Central Himalaya. Curr Sci. 2005;89(5):774–86.

Sharma CM, Sharma S, Gurung S, Juttner I, Bajracharya RM, Pradhan NS. Ecological studies within the Gokyo wetlands, Sagarmatha National Park. In: Jha PK, Khanal IP, editors. Contemporary research in Sagarmatha (Mt. Everest) region, Nepal; Nepal Academy of Science and Technology, Khumaltar, Lalitpur 2010. p. 139–54.

Sharma CM, Sharma S, Bajracharya RM, Gurung S, Juttner I, Kang S. et al. First results on bathymetry and limnology of high altitude lakes in the Gokyo Valley, Sagarmatha (Everest) National Park, Nepal. Limnology 2012;13:181–92.

Strang D, Aherene J, Shaw P. The hydrochemistry of high elevation lakes in the Georgia Basin, British Columbia. J Limnol. 2010;69:56–66.

WWF Global. Western Himalayan Temperate Forests. WWF Nepal, Kathmandu 2015a. http://wwf.panda.org/about_our_earth/ecoregions/westhimalayan_temperate_forests.cfm. On-line accessed on 14 July 2015.

WWF Global. Eastern Himalayan Alpine Meadows. WWF Nepal 2015b. http://wwf.panda.org/about_our_earth/ecoregions/easthimalayan_alpine_meadows.cfm. On-line accessed on 14 July 2015.

WWF Nepal. Gokyo: the importance of religion, culture and tradition for the conservation. 2010a.

WWF Nepal. Panch Pokhari: the importance of religion, culture and tradition for the conservation. 2010b.

WWF Nepal and Department of National Parks and Wildlife Conservation. Factsheet: wetlands of Nepal. Kathmandu: WWF Nepal and Department of National Parks and Wildlife Conservation; 2006.

Anzali Mordab Complex (Islamic Republic of Iran)

134

Masoud Bagherzadeh Karimi

Contents

Introduction	1616
Climate	1616
Hydrology	1616
Wetland Ecosystem	1618
Biodiversity	1619
Ecosystem Services and Land Use	1621
Conservation Measures	1621
Threats and Future Challenges	1623
References	1624

Abstract

Anzali Mordab Complex (Anzali wetland) located on the southwestern shore of the Caspian Sea, close to the city of Bandar-e-Anzali, is a good example of a natural wetland characteristic of the south Caspian lowlands. The 10 perennial flowing rivers draining the 3,610 km^2 watershed flow into the 15,000 ha Anzali wetland that supports an extremely diverse flora and fauna. A Ramsar Site, the Anzali wetland is internationally important for migratory waterbirds and provides important ecosystem services including a commercial fishery. The site is listed on Ramsar's Montreux Record requiring conservation action due to threats that include pollution from urban and agricultural waste water, spread of the exotic floating water-fern *Azolla sp.*, sedimentation from upstream deforestation, and increased hunting pressure.

M. B. Karimi (✉)
Department of Environment, Study Center for Environment, Wetlands and National Parks, Tehran, Iran
e-mail: mbkarimi@yahoo.com

© Springer Science+Business Media B.V., part of Springer Nature 2018
C. M. Finlayson et al. (eds.), *The Wetland Book*,
https://doi.org/10.1007/978-94-007-4001-3_231

Keywords

Caspian Sea · Ramsar Site · *Phragmites australis* · Wetland complex · Migratory waterbirds · Pollution

Introduction

Anzali Mordab Complex (Anzali Wetland) is one of 24 Ramsar Sites in Iran (site number 3 in Fig. 1). Situated in the province of Gilan, in northern Iran, Anzali wetland is located on the southwestern shore of the Caspian Sea, close to the city of Bandar-e-Anzali. It is considered a good example of a natural wetland characteristic of the south Caspian lowlands (Fig. 2).

Climate

The climate in the northern region of Iran where Anzali Wetland is located is referred to as the Caspian or Hyrcanian climate. Its influence on this thin coastal strip of land along the Caspian Sea, coupled with the close proximity of the Alborz Mountain Range to the south, results in a climate that is distinct from the arid climate typical of the rest of Iran. Regional rainfall varies between 400 and 2,000 mm per year and decreases gradually from west to east. Evaporation increases from west to east with a regional average of 800 mm/year. The temperature is mild, ranging between -0.8 °C and 37.3 °C with an annual average of 17 °C. The regional relative humidity averages 66% but varies between 24% and 100% depending on the location and season. The climate in the Anzali Wetland watershed is characterized by two distinct types. The lowland area to the north with elevations between 25 and 500 m is characterized by warm temperatures, high moisture, and abundant rainfall (Anzali station's rainfall is the highest along Iran's Caspian coast) during the summer with a mild climate during the winter. The climate between elevations 500 m and 3,000 m (further south, including the Alborz Mountains) is noticeably different from the lowland, characterized by cooler temperatures, drier conditions, and less rainfall.

Hydrology

The watershed of Anzali Wetland has an area of 3,610 km^2 encompassing 10 river systems (Fig. 3) with individual subcatchment between 100 and 700 km^2. The perennial flowing rivers originate in the Alborz Mountains to the south and provide a mean annual discharge to the wetland estimated at about 2,400 M m^3 (a mean flow rate of 76 m^3/s).

The water level of the Caspian Sea is measured at Anzali Port, where the wetland is connected to the sea. The maximum water level during the period from 1930 to

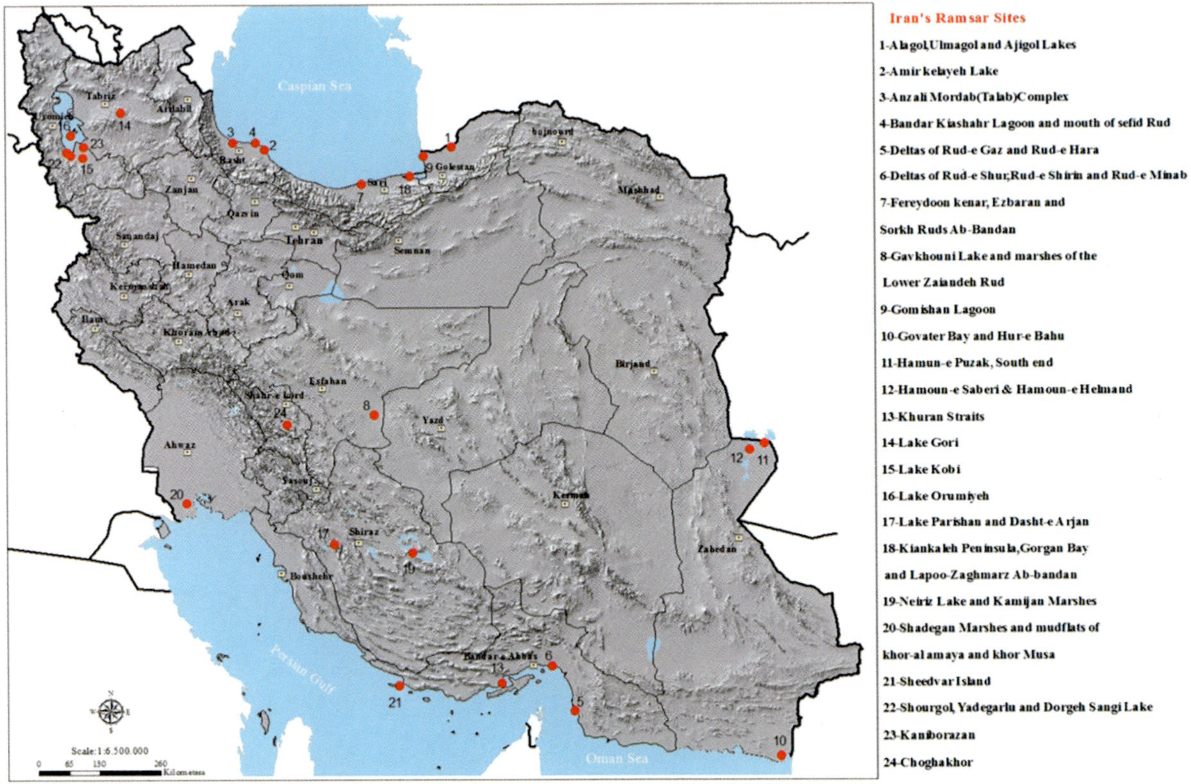

Fig. 1 Location of Ramsar Sites in Iran (© DoE 2010; reproduced with permission)

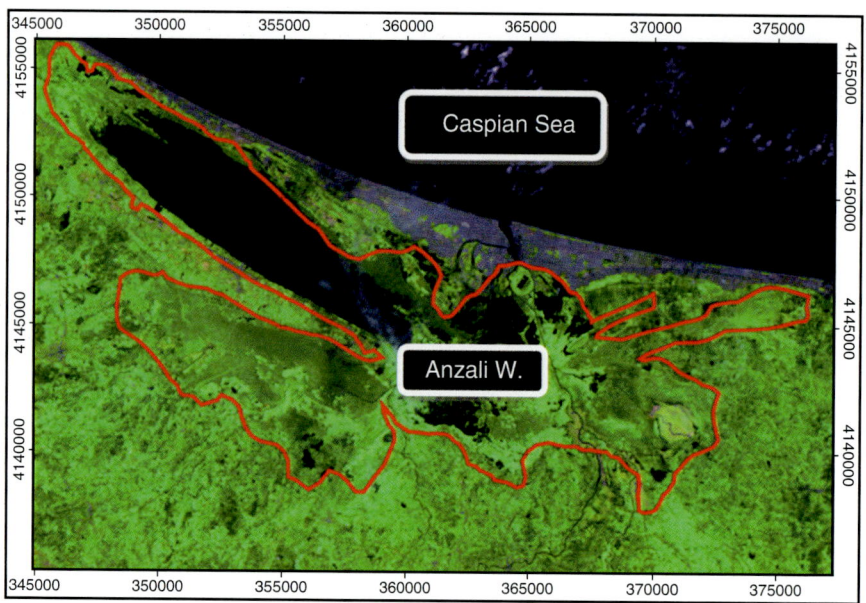

Fig. 2 Location of Anzali Wetland on the southern shore of the Caspian Sea (© DoE 2004; reproduced with permission)

2000 was 25.27 m in 1929 and the minimum was 28.44 m in 1977. In 1977, the water level began to rise, reaching a recent maximum of 26.10 m in 1994. However, the water level has receded since then (DoE and JICA 2005). The latest water level in September 2013 ranged from 26.10 m to 26.30 m.

Wetland Ecosystem

The Anzali Mordab Complex (15,000 ha) comprises large, shallow, eutrophic freshwater lagoons, shallow impoundments ("ab-bandans"), marshes, and seasonally flooded grasslands. It is separated to the north from the Caspian Sea by a sand dune barrier about 1 km wide, with open grassland and scrubby vegetation, and to the south by cultivated land (mainly rice) and patches of woodland. The main wetland covers about 11,000 ha, and comprises an open lagoon, 26 km long by 2.0–3.5 km wide, surrounded by reedbeds which extend its eastern limits a further 7 km. Several perpetual streams emanating in the nearby Talesh part of the Alborz Mountains feed into the Anzali Complex. The entire marsh and lagoon complex drains into the deepwater harbor of Bandar-e-Anzali and Caspian Sea through the main channel at the northeastern end of the main lagoon.

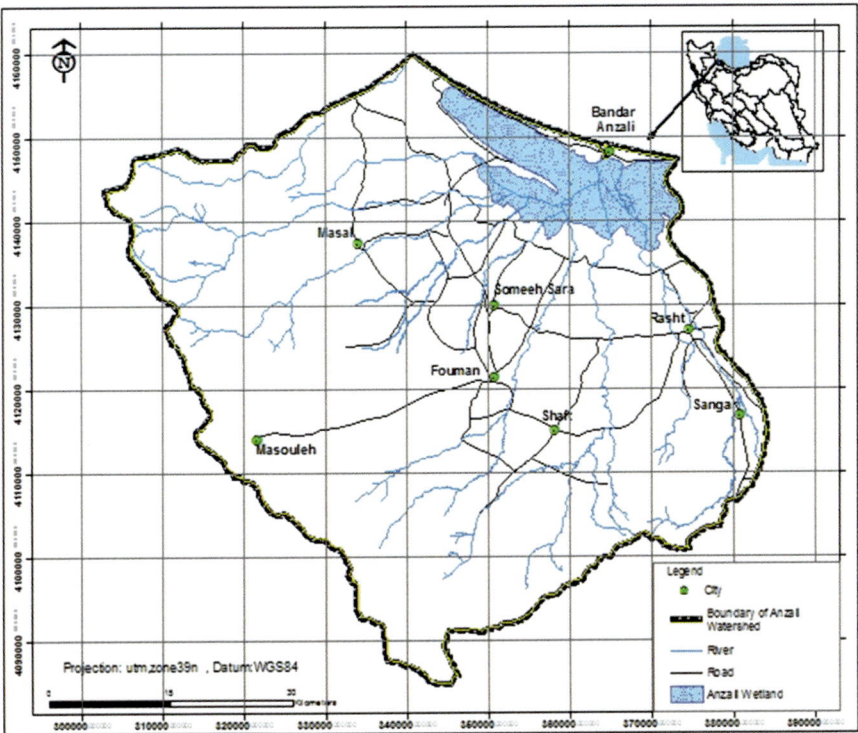

Fig. 3 Anzali river basin (© DoE and JICA 2005; reproduced with permission)

Biodiversity

Anzali Wetland supports an extremely diverse wetland flora and fauna (Fig. 4). The dominant vegetation throughout much of the Anzali Wetland consists of vast beds of *Phragmites australis* which in places grow to 6 m in height. Due to falling levels of the Caspian Sea in the late 1960s, a rapid expansion of the *Phragmites* reed began, and by the early 1980s, large parts of the main wetland were covered (Scott 1995). Together with increased pollution and eutrophication, the situation had become so serious that methods of control were considered. The recent rapid rise in water level in the wetland stopped the expansion of *Phragmites* and recreated open water areas. The new water areas support vast beds of lotus *Nelumbo nucifera* var. *caspica* and a very rich growth of other floating and submerged macrophyte vegetation.

Anzali Wetland and its satellite wetlands are extremely important for a wide variety of breeding, passage, and wintering waterfowl (Fig. 5), including huge numbers of wintering ducks, geese, swans, and coots. The wetlands support a large breeding colony of whiskered tern *Chlidonias hybridus*, colonies of six species of Ardeidae, and a resident population of purple swamphen *Porphyrio porphyrio*.

Fig. 4 Lotus at Anzali Wetland in summer 2004 (Photo credit: Masoud Bagherzadeh Karimi © Rights remain with the author)

Fig. 5 Anzali Wetland (Siakeshim) – migratory birds wintering (2009) (Photo credit: Masoud Bagherzadeh Karimi © Rights remain with the author)

Moustached warbler *Acrocephalus melanopogon* and great reed warbler *A. arundinaceus* are very common breeding birds in the reedbeds.

Anzali Wetland is internationally important for migratory waterbirds, supporting over 1% of the regional Middle Eastern wintering populations of several species, such as mallard *Anas platyrhynchos*, northern pintail *A. acuta*, common pochard *Aythya ferina*, and common coot *Fulica atra*. The wetland is the most important wintering area in Iran for pygmy cormorant *Phalacrocorax pygmeus*. Eurasian woodcock *Scolopax rusticola* is a common winter visitor to the surrounding damp woodlands and scrub. Many birds of prey also winter at the wetland, such as white-tailed eagle *Haliaeetus albicilla*, great spotted eagle *Aquila clanga*, golden eagle *A. chrysaetos*, peregrine falcon *Falco peregrinus*, saker falcon *F. cherrug*, merlin *F. columbarius*, short-eared owl *Asio flammeus*, and marsh harrier *Circus aeruginosus* are common throughout the year. It also provides wintering habitat for a diversity of rare and threatened bird species in addition to some of those mentioned above including white pelican *Pelecanus onocrotalus*, Dalmatian pelican *P. crispus*, lesser white-fronted goose *Anser erythropus*, white-headed duck *Oxyura leucocephala*, and imperial eagle *Aquila heliaca*.

Mammals include the golden jackal *Canis aureus*, Eurasian otter *Lutra lutra*, jungle cat *Felis chaus*, wild boar *Sus scrofa*, white-toothed shrew *Crocidura leucodon*, crested porcupine *Hystrix indica*, and wolf *Canis lupus* has been recorded. Anzali and Siakesheem are important spawning and nursery grounds for several fish species.

Ecosystem Services and Land Use

The site is mainly state owned, while some parts are privately owned. Anzali Wetland locally supports a major commercial fishery. The wetland and deeper rivers flowing into it are used for transportation of people, farm goods, and other materials to the various villages around the wetland and to Bandar-e-Anzali. Parts of the marsh and open wetlands in the south are heavily utilized by domestic livestock for grazing. Several villages cut the reeds for mat weaving, fencing, and building materials. Many of the ab-bandans are managed as duck-hunting areas throughout the winter months. At these sites, the duck hunters use a traditional dazzling and hand-netting technique (the "net, gong and flare" technique) to catch ducks and coots from a boat at night. Elsewhere in the wetland, hunting is mostly done by shotgun. The ab-bandans also provide a source of water for irrigation of rice fields during the dry summer months. In the summer time, the wetland is an important recreation center in the north of Iran (Figs. 6 and 7).

Conservation Measures

Anzali Mordab Complex (15,000 ha) was designated as a Ramsar Wetland of International Importance on 23/06/1975 on account of its values as a natural wetland characteristic of the South Caspian lowlands, its support of a diverse fauna and flora

Fig. 6 Anzali Wetland used for sport tourism (summer 2010) (Photo credit: Masoud Bagherzadeh Karimi © Rights remain with the author)

Fig. 7 Anzali Wetland is popular for ecotourism (summer 2010) (Photo credit: Masoud Bagherzadeh Karimi © Rights remain with the author)

including a range of threatened species, spawning and nursery grounds for fish, significant waterbird breeding colonies, more than 100,000 wintering waterbirds, and more than 1% of the regional populations of several waterbird species. Anzali is recognized as a significant site for preserving plant and animal genetic resources and diversity.

Three reserves have been established in the Anzali complex. The central portion of Siakesheem Marsh (3,515 ha) was first established as a Protected Area in 1967. The reserve was enlarged to 6,701 ha and upgraded to Wildlife Refuge in 1971 but

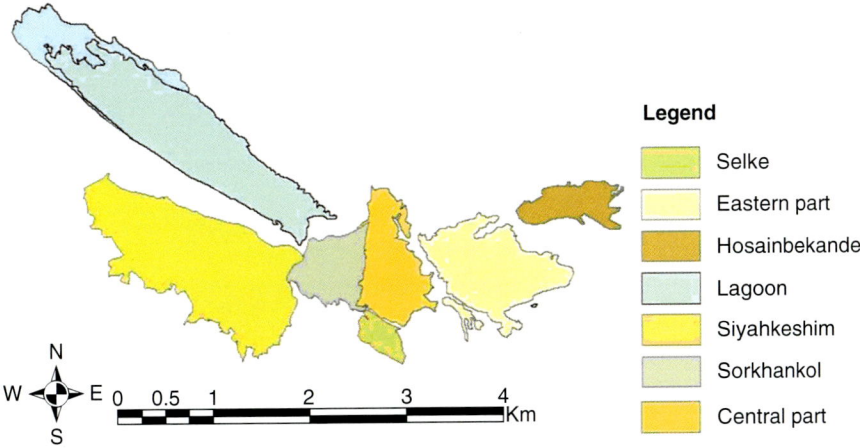

Fig. 8 Different parts of Anzali Wetland Complex (© Fathi 2009; reproduced with permission)

subsequently reduced to its present size of 4,500 ha and downgraded to a Protected Area in the 1980s. Selke Ab-bandan (360 ha) has been protected as a Wildlife Refuge since 1970. The Department of Environment (DOE) newly established Sorkan Kol wetland as a part of Anzali Wetland as a Wildlife Refuge (Fig. 8).

Annual mid-winter waterfowl censuses have been carried out by the Ornithology Unit of the DOE since 1967. Bird ringing concentrates on terns and passerines. The Department of Environment is currently undertaking a major research program which has included establishing 35 monitoring stations throughout the wetland to measure a variety of parameters including changes in water level, water quality, and physicochemical characteristics. Numerous limnological and hydrological studies have been conducted by the national fisheries organization, Shilat.

Threats and Future Challenges

The Ramsar Site encompasses the whole of the Anzali Wetland, Siakesheem Marsh, Selke Protected Area, and several other ab-bandans bordering the marshes. The site has been placed on the Ramsar Convention's Montreux Record of priority sites for conservation action since December 1993. Ramsar Advisory Missions visited the area in January 1992 and May 1997 (see: http://www.ramsar.org/cda/en/ramsar-documents-rams/main/ramsar/1-31-112_4000_0__). The site is on the Montreux Record for many reasons. The expanded reedbeds that developed due to water level changes have not started to die back yet, representing a continuing problem. Other problems include major eutrophication due to pollution from ricefields and urban waste water, sediment deposition in the wetland as a result of upstream deforestation and overgrazing, massive spread of the exotic floating water fern *Azolla* sp., increased hunting pressure (up to 200,000 birds per season), poaching (especially at Siahkesheem), and extensive disturbance of wintering waterbirds caused by hunting activities. In addition, development projects such as the Anzali Port ring road are threats for the wetland.

References

Department of Environment (DoE) Project Team, JICA Expert Team (Nippon Koei co., LTD). The study on integrated management for ecosystem conservation of the Anzali wetland. Tehran: DoE; 2005.

Department of Environment (DoE), GIS and Geographical Studies Group. DoE annual report, Tehran; 2004.

Department of Environment (DoE), Wetlands Group. DoE annual report, Tehran; 2010.

Fathi F. Determination of buffer for Anzali wetland. Tehran: Department of Environment (DoE); 2009.

Scott DA. A directory of wetlands in the Middle East. IUCN, WWF, IWRB, BirdLife International and Ramsar Convention Bureau; 1995, 560 pp.

Bujagh National Park (Islamic Republic of Iran)

Sadegh Sadeghi Zadegan

Contents

Introduction	1626
Hydrology	1627
Wetland Ecosystems	1627
Biodiversity	1628
Birds	1628
Mammals	1629
Fish	1630
Ecosystem Services	1630
Conservation Status and Activities	1631
Current Threats and Future Challenges	1632
References	1632

Abstract

Bujagh National Park (3,177 ha) is located in the South Caspian lowlands and encompasses both the largest delta in the South Caspian region (the Sefid Rud River delta – 1,350 ha) and one of the oldest lagoons in Gilan Province (Bandar Kiashahr Lagoon). Historically, fishing has been the common activity in the area. Villages are adjacent to the national park, and some farmland lies within its boundary. The site is important as spawning and nursery grounds for fish and as breeding, staging, and wintering areas for a wide variety of waterfowl. The coastal national park hosts more than 100,000 migratory waterbirds annually and is considered to be a potential secure release and wintering site for the Siberian crane, which traditionally has used Fereydoon Kenar, a short distance to the east.

S. Sadeghi Zadegan (✉)
Ornithological Unit, Wildlife Bureau, Department of Environment, Tehran, Iran
e-mail: sadegh64@hotmail.com

© Springer Science+Business Media B.V., part of Springer Nature 2018
C. M. Finlayson et al. (eds.), *The Wetland Book*,
https://doi.org/10.1007/978-94-007-4001-3_234

Keywords

Wetlands · Sefid rud river · Deltaic System · Grasslands · Rice paddies · Waterbirds · Ramsar convention · Iran · Gilan · Caspian sea

Introduction

Bujagh National Park (3,177 ha) (Fig. 1) is located in the South Caspian lowlands and encompasses both the largest delta in the South Caspian region (the Sefid Rud river delta – 1,350 ha) and one of the oldest lagoons in Gilan Province (Bandar Kiashahr Lagoon). Historically, fishing has been the common activity in the area. Villages are adjacent to the national park and some farmland lies within its boundary. The site is important as spawning and nursery grounds for fishes and as breeding, staging, and wintering areas for a wide variety of waterfowl. The coastal national park hosts more than 100,000 migratory waterbirds annually and is considered to be a potential secure release and wintering site for the Siberian crane *Leucogeranus leucogeranus*, which traditionally has used Fereydoon Kenar, a short distance to the east (see "Cross-References").

The national park comprises a shallow sea bay (formerly an enclosed lagoon), the mouth of the main channel of the Sefid Rud (or Sefid Rood) and its riverine marshes, extensive open grasslands, and dunes near the mouth of the river, and the associated

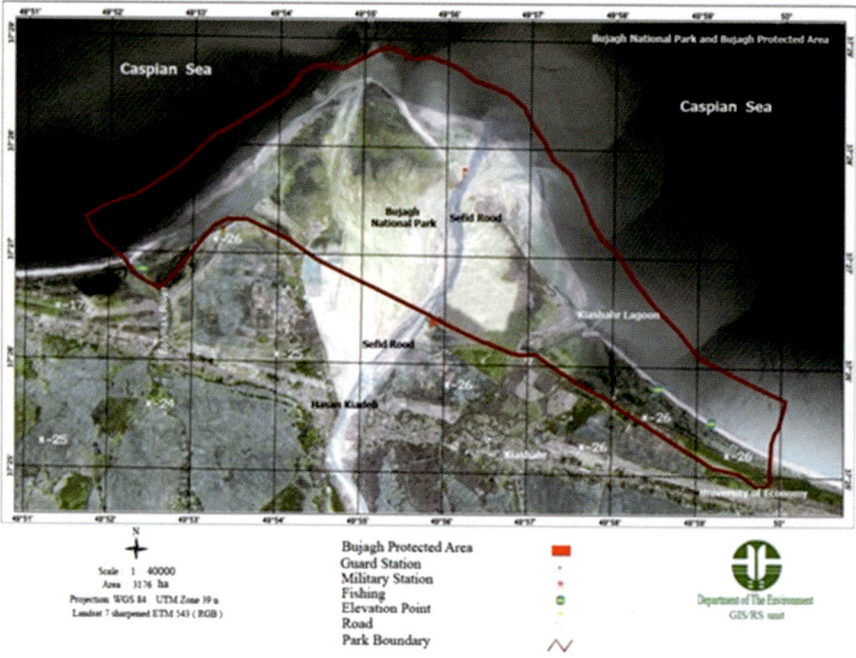

Fig. 1 Map of Bujagh National Park (© Convention on the Conservation of Migraory Species of Wild Animals, with permission)

fresh to brackish marshes of the Bandar Kiashahr Lagoon located adjacent to the shallow bay of the Caspian Sea. Land cover in the national park is 43% marine ecosystem (located in the north), 28% grassland (to the west and east of the Sefid Rud River), 15% water bodies and reedbeds, 9.4% sand dunes interspersed with sedge marshes, and 2.39% Sefid Rud river channels, and the remainder (about 0.44%) is small patches of forest and some farmlands.

Part of the Sefid Rud delta was declared a Ramsar Site (Bandar Kiashahr Lagoon and the mouth of Sefid Rud) on 23 June 1975, subsequently extended to include the whole of Bujagh National Park on 17 September 2009 (Ramsar Convention 2009). The national park is located between two other Ramsar sites, namely, Anzali and Amirkelayeh wetlands, to the west and east, respectively.

Hydrology

The Sefid Rood is the most important river contributing to the hydrology of Bujagh National Park. Originating in the Alborz and Zagros mountains, the Sefid Rood is the second largest river in Iran, winding through five provinces with a catchment area of over 54,000 sq. km before finally entering the Caspian Sea. The river divides into several distributary channels on the plains of Gilan, the main channel entering the Caspian Sea at Bandar Kiashahr. The national park encompasses the Sefid Rood's deltaic system and includes the outflows of an additional two small and local rivers with very small catchment areas. Within the national park, groundwater also enters the Sefid Rood River (Scott 1995; Ramsar Convention 2009).

Within the national park, the Sefid Rood divides the park's main wetland (Bujagh wetland) into two parts. The Sefid Rood has a natural flood discharge of 3,400–4,200 m^3s^{-1} that diminishes to less than 20 m^3s^{-1} during late summer (Scott 1995). The average depth of the wetland is 70 cm and the average volume of water within the wetland is 381,000 m^3. Dependent upon a freshwater supply from the Sefid Rood River, the depth and area of Bujagh wetland change during the flood and dry seasons. During flood season, freshwater enters Bujagh wetland from the deltaic system. When the wetland's water level is high, the wetland is connected to the sea by streams.

Inefficient management and fragile and unstable land result in erosion and huge amounts of sediments being carried by the river beginning in the mountains high in the catchment and deposited behind dams and the deltaic system. Erosion is higher during spring when there is more rain, the snow is melting, and grazing is taking place. Less sedimentation occurs in summer with low rainfall and low river flows.

Wetland Ecosystems

The national park covers the entire range of riverine, deltaic, and coastal habitats of the Sefid Rud delta, including shallow marine waters, coastal lagoons, river channels and streams, freshwater and brackish marshes, sand dunes, grasslands, and some rice fields. Sandy areas to the west and northwest are covered in shrub and grassland which give way to sand dune vegetation near the Caspian shore. Grassland along the banks of the Sefid Rood floods seasonally.

In the 1960s and 1970s, this wetland was a shallow, brackish coastal lagoon, 3.75 km long by 1.5 km wide, with fringing *Juncus* marshes and about 140 ha of *Phragmites* and *Typha* reedbeds at its west end. The lagoon was fed by two streams from the Sefid Rood and local runoff and drained northeast through a narrow channel into the Caspian Sea. The bottom was a mixture of sand and mud, and the waters were predominantly oligotrophic, except toward the marshy western extremity.

The Bandar Kiashahr Lagoon was formed in 1960 as a result of the falling level of the Caspian Sea and development of coastal sandspits. The 1.8 m rise in the level of the Caspian Sea since 1978 has obliterated the sand barrier between the lagoon and the sea, with the result that the wetland now constitutes a sea bay with broad entrance to the sea (similar to the situation in the 1950s). The extensive seasonally flooded grassland and sand dune areas at the mouth of the Sefid Rood have, however, remained more or less unchanged, while new wetland habitats have been created to the west of the river mouth (Scott 1995).

Biodiversity

Bandar Kiashahr Lagoon and the mouth of Sefid Rood within the national park are good representative examples of natural wetlands characteristic of the South Caspian lowlands, encompassing the entire range of riverine, deltaic, and coastal habitats associated with the Sefid Rud delta (see above for details). The national park has high ecological value and supports diverse plant and animal species. It is of particular importance for staging and wintering waterbirds, a range of globally threatened species, and as a breeding and nursery ground for a variety of fish species.

Birds

The South Caspian lowlands' greatest claim to fame is its avifauna. Its location at the crossroads of the western, central, and eastern Palearctic brings phenomenal numbers of migratory birds from three regions, making it one of the truly great overwintering areas in the world. Bujagh itself is an important staging and wintering area for a wide variety of migratory waterfowl, regularly supporting more than 100,000 waterbirds and more than 1% of the regional populations of several waterbird species, including wintering greater white-fronted goose *Anser albifrons*, whooper swan *Cygnus cygnus*, gadwall *Anas strepera*, mallard *A. platyrhynchos*, black-necked grebe *Podiceps nigricollis* and black-headed gull *Larus ridibundus*, and breeding greater cormorant *Phalacrocorax carbo*.

In addition to the abovementioned species, the site also supports ferruginous duck *Aythya nyroca*, little bustard *Tetrax tetrax* and corncrake *Crex crex*, pygmy cormorant *Phalacrocorax pygmeus* (up to 300), ducks, shorebirds, gulls and terns, and for

Table 1 Globally threatened species recorded at **Bujagh National Park** (Source: Ramsar Convention 2009)

English name	Scientific name	International status			National status
		IUCN Red List	CITES Appendix	CMS Appendix	
Caspian seal	*Phoca (Pusa) caspica*	EN			In danger of extinction
Dalmatian pelican	*Pelecanus crispus*	VU	I	I/II	In danger of extinction
Lesser white-fronted goose	*Anser erythropus*	VU		I/II	Protected
Sociable lapwing	*Vanellus gregarius*	CR		I/II	Protected
White-headed duck	*Oxyura leucocephala*	EN	II	I/II	In danger of extinction
Greater spotted eagle	*Aquila clanga*	VU	II	I/II	Protected
Stellate sturgeon	*Acipenser stellatus*	EN	II	II	In danger of extinction

IUCN Red List status: *CR* critically endangered, *EN* endangered, *VU* vulnerable

the raptors marsh harrier *Circus aeruginosus* and merlin *Falco columbarius*. A flock of Dalmatian pelican *Pelecanus crispus* (usually 30–40 birds) wintered at the mouth of the Sefid Rood in the 1970s but apparently disappeared by about 1980, probably because of increased disturbance. Lesser white-fronted goose *Anser erythropus* was also an occasional winter visitor to the area in the 1970s. The open grassy areas and dunes near the river mouth provide breeding habitat for 20–30 pairs of common pratincole *Glareola pratincola*, black-winged stilt *Himantopus himantopus*, and little ringed plover *Charadrius dubius*, while a small patch of woodland to the south of the lagoon supports a large breeding colony of herons and egrets (Evans 1994). The globally threatened and near-threatened species recorded at the park are listed in Table 1.

Mammals

The original mammalian fauna of the lower Caspian has largely disappeared along with its forested habitats. The modified agricultural landscape today supports few mammal species, all of which are adapted to the existing habitats, such as the golden jackal *Canis aureus*, which thrives in the shrub and treelined rice fields. Only the common otter *Lutra lutra* is of conservation significance, listed as near threatened by IUCN, on Appendix I of CITES, and totally protected in Iran. Records of otters are rare, and it is likely sightings are of individuals wandering down the streams and rivers from the foothills. Pesticide levels and food availability in the rice fields are unlikely to support a population of otters in Bujagh. The Caspian seal *Pusa caspica* is one of the smallest members of the earless seal family and unique in that it is found exclusively in the brackish Caspian Sea. Since the late 1990s, there have been several cases of many Caspian seals dying due to canine distemper virus. A century ago there were an estimated 1.5 million seals; by the 1980s, there were about 400,000.

Fish

The national park is also important for its role in supporting fish diversity. Stellate sturgeon *Acipenser stellatus* is the most important fish species of the national park because it is globally endangered. Also southern Caspian kutum *Rutilus frisii kutum* and vimba *Vimba vimba* are important fish species. These species are migratory, and they enter freshwater (Sefid Rood River) by the end of winter and beginning of spring for breeding and laying eggs; therefore they are in the area only during the breeding season and it is very important to control fishing during this time.

Ecosystem Services

As the delta of the Sefid Rud, is the second largest river in Iran with a catchment area of over 54,000 km^2, and significant flood discharge to the Caspian Sea. The Sefid Rud plays a significant role in coastal geomorphological processes, with shifts in its channel paths and deposition of huge quantities of sediments, providing inflows of freshwater and nutrients to the Caspian Sea.

It remains an important commercial fishery, providing breeding and nursery grounds for a range of Caspian species, some of which migrate up the river to breed further upstream (see above). A collective fishing company is based to the northwest of the site with around 50 members. In addition to such legal fishing, there is considerable illegal fishing activity both in the sea and in the river, mainly targeting caviar and kutum fish. A considerable body of fisheries research has been carried out by the National Fisheries Organization (Shilat). There is a fisheries research center based in the reserve, whose main activity is to breed kutum fish for restocking purposes. Currently, about 100 ha out of the planned 500 ha scheme is operational (including 27 breeding ponds with an area of 60 ha.)

The natural coastline attracts large numbers of visitors in summer, with the beach on the eastern side of the park receiving heavy usage. Kiashahr Lagoon is widely used for recreation, including some tourism resorts, attracting thousands of tourists mainly during the summer and Persian New Year (March–April) period. In addition, many people visit the western parts to use the sandy beach and riverside for fishing with hook and line. The extensive grasslands provide grazing lands for local communities, including dairy cattle, buffalos, and horses.

There is an area of rice fields within the national park. The farmers work in the rice fields from April to September and do other jobs for the rest of the year. Chemical use is low and the government is promoting biological pest control (*Trichodrama*). There are approximately 120 rice farmers at Bujagh, farming up to 1 ha each, in total 200 ha. The rice fields were established about 18 years ago (after the revolution), when the emphasis was on social support. No further expansion of agriculture will be allowed.

Conservation Status and Activities

The western part of the Sefid Rud delta was initially protected as a non-hunting area, declared in 1998 and covering 800 ha. The entire delta was subsequently designated as Bujagh National Park in 2002 (3,250 ha, IUCN protected area category II), with upgraded management facilities and staffing. There are around 20 staff and two guard stations.

Bandar Kiashahr Lagoon and the mouth of Sefid Rood were originally designated as a Ramsar Site on 23 June 1975. The existing Ramsar Site (500 ha) included the whole of the lagoon area, its associated marshes, and the marshes and sand flats at the mouth of the Sefid Rood to the west. It was subsequently extended to share the same boundaries as Bujagh National Park on 17 September 2009.

The site has been listed as an "Important Bird Area" by BirdLife International (BirdLife International 1994; Evans 1994). In view of its potential as a secure release and wintering site for the Siberian crane, it was also included in the Western/Central Asian Site Network for the Siberian Crane and Other Waterbirds at the launch of this network in May 2007 (Sadeghi-Zadegan and Fazeli 2007; UNEP/CMS 2008, 2011; Ilyashenko 2010).

Annual midwinter waterfowl censuses have been carried out by the Ornithology Unit, Department of Environment (DoE), since 1968. Many ornithological surveys have been undertaken at other times of the year, including comprehensive waterfowl censuses in mid-November in 1972, 1973, and 1974. Ringing of waterbirds has been conducted by Gilan Provincial DoE Office.

The DoE is developing comprehensive management plans for all the protected areas of Iran, including Bujagh National Park, for which the comprehensive management plan has been finalized (DoE 2007). Its implementation, through consultation and agreement with local stakeholders, is needed to address major problems at the site, including restoration and rehabilitation activities.

Under the UNEP/GEF project "Development of a Wetland Site and Flyway Network for Conservation of the Siberian Crane and Other Migratory Waterbirds in Asia," several outcomes were achieved at the national park (one of two main sites included in the project) to strengthen management capacity, which included the training of management staff, the establishment of a site management committee, stakeholder participation in management plan development, construction of a multipurpose building for site management, completion of boundary demarcation for the national park, GIS mapping, etc. (see Harris 2010; Mirande and Prentice 2011).

Current Threats and Future Challenges

Within the national park, there remains some illegal hunting pressure on waterfowl (mainly in winter) and heavy transport pressure by boats from the extensive commercial fisheries and associated industries. There is also considerable disturbance from recreational activities (tourism) during weekends and holidays, mainly during summer (Fallah Farbod 2007). The great decrease in the numbers of wintering birds in recent decades has been attributed to the increasing disturbance from fishing activities and previous heavy hunting pressure. The main activities currently affecting the site include illegal fishing (the local population is practicing a lucrative fishery, mostly taking sturgeon and white fish (mullet), for food and/or profit), grazing (there is very little control over grazing at present (the site is heavily grazed in summer, with c. 10,000 animals, but only a few horses graze in the area during the winter)), and illegal hunting (some local people continue to poach waterfowl for food and/or profit).

In the surrounding area, there is an army camp located next to the site. This has caused disturbance to the waterfowl, and impacts of agriculture and urban waste affect the site.

Concerns have been expressed by national and local groups regarding current threats to the site, including development and construction plans impacting wetland habitats, tourism development, pollution, and overfishing. There is particular concern regarding a development scheme to expand the current fishing harbor to a major trade harbor. Further investigations are required to assess the ecological changes which have occurred at the wetland, to mitigate the environmental impacts of planned developments, and to identify ways of reducing the disturbance to waterfowl from fishing activities. Monitoring of Caspian seal mortality is also important, in view of high mortality rates in recent years, including bycatch and killing by fishermen.

References

BirdLife International. Important Bird Areas (IBA's), IR017 Bandar Kiashahr lagoon and mouth of Sefid Rud; 1994. http://www.birdlife.org/datazone/sitefactsheet.php?id=8079

CMS. Western/Central Asian Site Network for the Siberian Crane and Other Waterbirds. Guidelines for the preparation of site nomination documents. Annotated list of nomination sites. Version: 13 March 2008. Bonn: CMS Secretariat, United Nations Campus; 2008. p. 221.

Department of Environment, Habitat and Protected Areas Bureau. Comprehensive management plan for Bujagh National Park. Tehran: Department of Environment; 2007.

Evans MI. Important Bird Areas in the Middle East. Cambridge, UK: BirdLife International; 1994.

Fallah Farbod SS. Sustainable recreational design of Bujagh National Park for tourism development. MS thesis. University of Tehran; 2007.

Harris J. Safe flyways for the Siberian crane. A flyway approach conserves some of Asia's most beautiful wetlands and waterbirds. Terminal report of the UNEP/GEF Siberian crane wetland project: development of a wetland site and flyway network for conservation of the Siberian crane and other migratory waterbirds in Asia, GF/2712-03-4627. International Wisconsin: Crane Foundation; 2010. http://www.scwp.info/final_report.shtml

Ilyashenko EI, editor. Atlas of key sites for the Siberian crane and other waterbirds in Western/Central Asia. Baraboo: International Crane Foundation; 2010. 116 pp.

Mirande CM, Prentice C. Conservation of flyway wetlands in Asia using the Siberian crane as a flagship species: an overview of the outcomes of the UNEP/GEF Siberian crane wetland project. In: Prentice C, editor. Conservation of flyway wetlands in East and West/Central Asia. Proceedings of the Project Completion Workshop of the UNEP/GEF Siberian Crane Wetland Project, 14–15 October 2009, Harbin. Baraboo: International Crane Foundation; 2011.

Ramsar Convention. Information sheet on Ramsar wetlands, Bujagh National Park; Apr 2009. https://rsis.ramsar.org/RISapp/files/RISrep/IR46RIS.pdf

Sadeghi-Zadegan S, Fazeli A. West & Central Asian site network, Bujagh National Park, Iran; 2007. http://sibeflyway.org/wp-content/uploads/2012/03/5-Budjagh-SIS-Part1-eng.pdf

Scott DA. A directory of wetlands in the Middle East. Gland/Slimbridge: IUCN/IWRB; 1995.

UNEP/CMS. Conservation measures for the Siberian crane, CMS Technical Series Publication No.16. 4th ed. Bonn: UNEP/CMS Secretariat; 2008.

UNEP/CMS. Conservation measures for the Siberian crane. 5th ed. CMS Technical Report Series No.25. Bonn: UNEP/CMS Secretariat; 2011. http://www.cms.int/species/siberian_crane/pdf/conservation_plans_all_pops_2010_12_e.pdf

Fereydoon Kenar, Ezbaran, and Sorkh Ruds Ab-Bandans

136

Sadegh Sadeghi Zadegan

Contents

Introduction	1636
Hydrology	1636
Wetland Ecosystems	1638
Biodiversity	1640
Mammals	1642
Fish	1642
Ecosystem Services	1642
Conservation Status	1643
Current Threats and Future Challenges	1644
References	1645

Abstract

Fereydoon Kenar Ramsar Site or "Fereydoon Kenar, Ezbaran & Sorkh Ruds Ab-Bandans" (FDK) occupies a total area of 5,427 ha, situated on the coastal plain of the South Caspian, in Mazandaran province. This site is also a nonshooting area (NSA). It includes four "damgahs" or duck-trapping areas and Fereydoon Kenar Wildlife Refuge (48 ha). Each damgah consists of shallow freshwater impoundments situated in harvested rice paddies, which have been developed as a duck-trapping area, surrounded by forest strips and reed enclosures. The area is the only known overwintering quarters of the western population of the Siberian crane (*Leucogeranus leucogeranus*) and is also an important wintering area for many other waterbirds, notably dabbling ducks (*Anas* spp.) and grey geese (*Anser* spp.). With the exception of small streams, the wetlands are almost entirely modified or man-made, consisting of a large expanse of rice fields surrounding four isolated

---

S. Sadeghi Zadegan (✉)
Ornithological Unit, Wildlife Bureau, Department of Environment, Tehran, Iran
e-mail: sadegh64@hotmail.com

© Springer Science+Business Media B.V., part of Springer Nature 2018
C. M. Finlayson et al. (eds.), *The Wetland Book*,
https://doi.org/10.1007/978-94-007-4001-3_233

damgah areas – shallow reservoirs that are flooded in winter with the dual purpose of providing irrigation water for the rice fields and attracting large numbers of waterfowl. Being located at the crossroads of the western, central, and eastern Palearctic and with a mild and wet winter climate, phenomenal numbers of migratory birds pass through and overwinter in the south Caspian lowlands, making it one of the truly great wintering sites in the world. Fereydoon Kenar NSA is not officially in the protected area network; therefore, the resolution of most threats to wintering Siberian cranes and other waterbirds requires a collaborative approach based on comanagement principles. The traditional duck-trapping practices used at the damgahs are slowly dwindling, and it is possible that in time they might cease altogether, replaced by hunting with shotguns as in other areas. High land prices have attracted local farmers to sell off their land for holiday accommodation and similar development; thus fragmentation and piecemeal loss of the rice field landscape are taking place, exacerbated by infrastructure development.

Keywords

Wetlands · Duck trapping area · Artificial wetlands · Siberian crane · Rice paddies · Waterbirds · Ramsar convention · Iran · Mazandaran · Caspian sea

Introduction

Fereydoon Kenar Ramsar Site or "Fereydoon Kenar, Ezbaran & Sorkh Ruds Ab-Bandans" (abbreviated to FDK) occupies a total area of 5,427 ha, situated on the coastal plain of the South Caspian, just south of the town of Fereydoon Kenar (Fig. 1) and 13 km southwest of Babolsar, in the province of Mazandaran. This site (36°40′N, 52°33′E), including the rice fields, is also a non-shooting area (NSA). The area includes four "damgahs" or duck-trapping areas (Fereydoon Kenar, Ezbaran, East and West Sorkh Ruds) and a small wildlife refuge (Fereydoon Kenar WR, 48 ha.) located in the northeastern part of the area. Each damgah consists of a complex of shallow freshwater impoundments situated in harvested rice paddies which have been developed as a duck-trapping area, surrounded by forest strips and reed enclosures. The area is of outstanding importance as the only known overwintering quarters of the western population of the Siberian crane *Leucogeranus leucogeranus*, but also extremely important as a wintering area for many other species of waterbird, notably dabbling ducks *Anas* spp. and gray geese *Anser* spp.

Hydrology

Mazandaran Province is humid, with a high rainfall averaging 700 mm per annum, resulting in a high water table throughout the year. The impenetrable underground base layer (Marney Argil) varies in depth and slope from the Alborz foothills to the Caspian's shore (the Mazandaran Plain), causing numerous

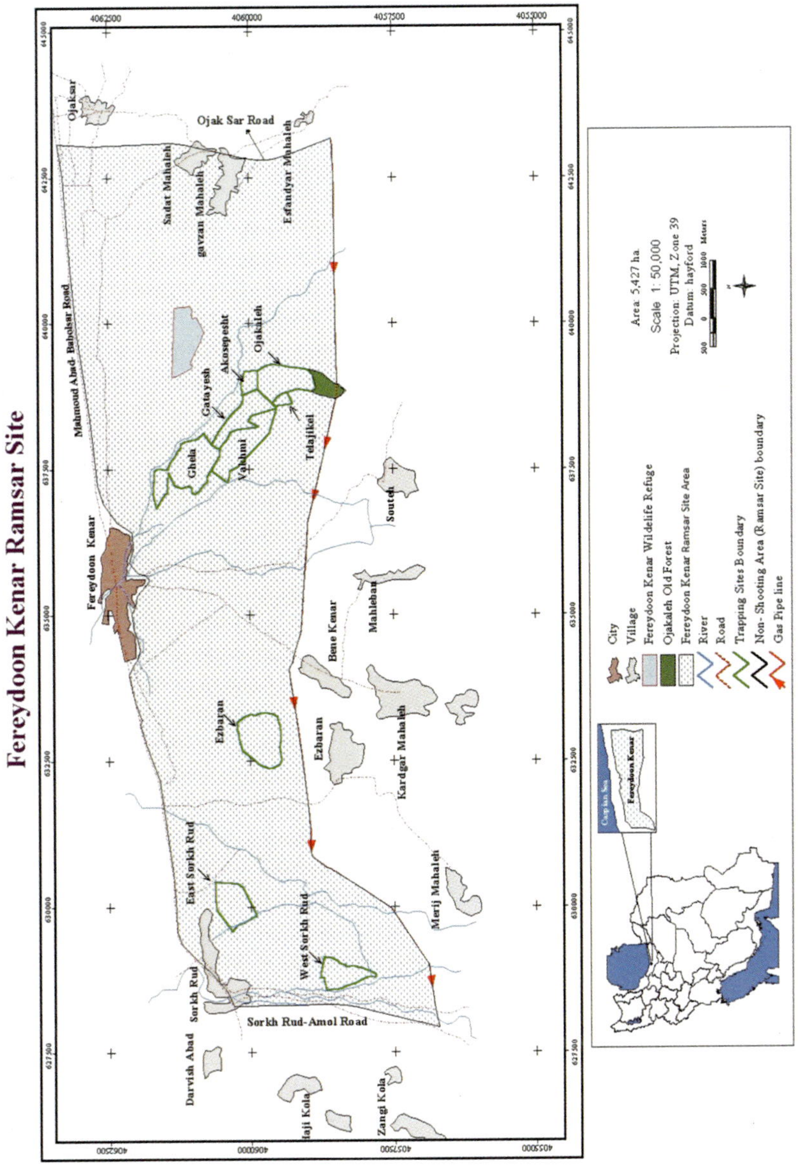

Fig. 1 Map of Fereydoon Kenar Ramsar Site (© Convention on the Conservation of Migraory Species of Wild Animals, with permission)

natural basins (reservoirs) to form. Most of these reservoirs have been altered with man-made dams to hold surface water for agricultural usage, especially during the dry season, and are usually situated in the upper (south) parts of villages. These reservoirs/wetlands are in fact agricultural fields that are cultivated during spring and summer and flooded in autumn and winter to become shallow reservoirs. The reservoirs of Babolsar are of greater importance than those associated with other cities of Mazandaran Province. Babolsar is the city most distant (on the Mazandaran Plain) from the Alborz mountain range, and with the area's gentle slope and nearness to the Caspian Sea, saline water intrusion with littoral groundwater can and does occur.

Babolsar has 80 reservoirs with total area of 1,800 ha, of which 11 and 34 are in Barikrood and Emamzadeh Abdollah villages, respectively, in Fereydoon Kenar district. Fereydoon Kenar NSA is located in the basin of the Haraaz River, flowing from the central parts of the Alborz range, and is delimited in the west by the Aleshrood basin, in the east by the Garmrood and Babolrood basins, and in the north by the Caspian Sea. Before reaching the plain and along its final 20 km in Mohammadabad rocky forest area, the Haraaz River divides into two main streams, one of called "Kari" that flows to the agricultural lands of east Amol region and into creeks on both sides which provide water to the surrounding farms, mostly rice fields. Along the south highlands of Alborz, Karirood (Henrood) flows to the east and after merging with Garmrood finally joins Babolrood. The river "Souteh" (Katkash or SouteKal) in the east and the river "Hakardeh Kel" in the west are boundaries of the damgah of Fereydoon Kenar. These two rivers join "Valikrood" river passing through this damgah and enters the Caspian Sea near Fereydoon Kenar city.

Wetland Ecosystems

With the exception of the small streams crossing the site, the wetlands of Fereydoon Kenar are almost entirely modified or man made, consisting of a large expanse of rice fields surrounding four isolated damgah areas – shallow reservoirs that are flooded in winter with the dual purpose of providing irrigation water for the rice fields and attracting large numbers of waterfowl. The damgahs are a system of small circular or strip forests including ponds and channels designed by villagers to catch ducks and geese. Rice fields are under cultivation during spring and summer (April–early September) and become flooded in autumn and winter (October–March) with a depth of 10–30 cm.

The major crop cultivated on the Mazandaran Plain is rice, fed by the waters of numerous streams flowing northward to the Caspian. This agricultural landscape is characterized by vast open spaces interspersed with rows of planted trees on bunds or lower stature hedgerows of dense shrubbery. The presence of water is cyclic, depending on the growing stages of the rice crop. In winter, most of the farmlands are dry, consisting of stubble. This farmland habitat is the primary reason for the large wintering waterbird population each winter. After crop harvesting in late autumn, some of the agricultural lands are flooded and maintained as reservoirs over the winter. These reservoirs are generally permanent in location with higher and stronger bunds to hold water after cropping is completed. There are some permanent reservoirs in the area, with six in the Amol area. These reservoirs are important breeding sites for resident waterbirds, but are also subjected to heavy trapping. Their fish fauna is inconsistent because of their seasonal nature.

Mazandaran Province is part of the Hyrcanian forests and was formerly entirely cloaked in the forest. The small forest patch of Ojakaleh in the NSA is all that is left of this habitat on-site. It harbors a depleted representation of the former flora and fauna of the area, including passerines and a few small mammals, and provides cover for local bird movements from the mountains down to the lowlands. Damgahs also have small forested margins along their boundaries, which serve as important habitat for resident and migratory birds and mammals. On the floor of shaded groves and forests are found herbaceous and perennial plants such as *Ranunculus*, wood sorrels *Oxalis acetosella*, pale yellow iris *Iris pseudacorus*, rush *Juncus effusus*, water plantain *Alisma plantago-aquatica*, wood horsetail *Equisetum sylvaticum*, and Arctic raspberry *Rubus arcticus*. There are some herbaceous types identified as different types of clovers, alfalfa, yellow pea, and sessile joyweed *Alternanthera sessilis*. Within the remnant forest area, 152 species representing 62 plant families have been recorded of which 91 are herbaceous plants, 16 are treelike (arborous), and 45 are hydrophytes; 31.2% of the site's vegetation comprises wetland plants.

Reed *Phragmites australis*, common cattail *Typha angustifolia*, flowering rush, and sedges are among the emergent species, while lesser duckweed *Lemna minor* is a floating hydrophyte found in the tranquil, still areas of water, including within the damgahs. The main submerged plants of the area are hornwort *Ceratophyllum* sp., longleaf pondweed *Potamogeton nodosus*, and water crowfoot *Ranunculus rionii*, which covers the wetland with white flowers in winter. The site's dominant wetland vegetation is marginal communities (24 species) with dwarf elderberry and knotgrass being the most dominant and widespread. Reed and nut grass *Cyperus rotundus* (Cyperaceae) are the dominant emergent plants. Common cattail and some simple-stemmed bur reed *Sparganium erectum* populations are present in some areas.

Trees are evident throughout the site, but few are native. Examples of the original endemic tree flora still exist in Ojakaleh forest patch, including five taxa that are endemic to the Caspian region:

Quercus castaneifolia castaneifolia, a subspecies of oak endemic to the Caspian lowlands.
Zelkova carpinifolia, the Siberian elm, also known as the Caucasian zelkova, of the family Ulmaceae. Confined to the lower northern slopes of the Alborz range and in two other small patches in the Zagros range.
The red elm *Ulmus carpinifolia*, also confined to the Caspian lowlands. Its Persian name is *Oja*, hence the Ojakaleh forest referring to the "Place of the *Oja*."
The Persian ironwood *Parrotia persica*, endemic to the lower Caspian.
The pea tree (or pea shrub) *Carpinus betulus*, endemic to the lower Caspian.

The most common tree species found in the central part of Fereydoon Kenar, East and West Sorkh Rud, and sides of Ezbaran Damgah is the European alder. There is also some Caucasian walnut *Juglans regia*, common ash *Fraxinus excelsior*, tree of heaven *Ailanthus altissima*, and white willow *Salix alba*. Spruce is mostly hand planted. There are also figs and climbing plants (family Liliaceae) which climb alders.

Biodiversity

Being located at the crossroads of the western, central, and eastern Palearctic and with a mild and wet winter climate, phenomenal numbers of migratory birds from three regions pass through and overwinter in the south Caspian lowlands, making it one of the truly great wintering sites in the world. BirdLife International lists this region as "a significant middle-eastern bird refuge," with 62.5% of its avifauna composed of migratory species. Local communities have capitalized on this abundant resource through a wide range of ingenious trapping methods, and this cultural tradition continues at Fereydoon Kenar, although it is dwindling due to socioeconomic changes. The complex of four damgahs in Fereydoon Kenar NSA, comprising Fereydoon Kenar, Ezbaran, East and West Sorkh Rud (FDK), and Lapoo-ye Fereydoon Kenar (the reservoir) is listed by BirdLife International as a significant bird refuge in the Middle East. A total of 109 species have been recorded in the NSA itself, 61 of which are migratory waterbirds belonging to 17 families. Waterbirds are the most significant feature of FDK, with an average of at least 155,000 birds counted each year during the annual waterbird census conducted in midwinter (January). Total numbers of waterbirds counted in the NSA are shown in Table 1 below. The most abundant species are common teal *Anas crecca*, northern pintail *Anas acuta*, greylag goose *Anser anser*, mallard *Anas platyrhynchos*, northern lapwing *Vanellus vanellus*, and black-tailed godwit *Limosa limosa*. In addition, there are over 40 other species of birds, with passerines such as buntings and Eurasian starlings forming large flocks during winter.

The artificially maintained shallow impoundments and extensive rice fields at Fereydoon Kenar provide excellent feeding and roosting habitat for large numbers of wintering waterfowl, notably *Phalacrocorax carbo* (maximum 1,560), dabbling ducks (maximum 200,000), *Anser albifrons* (maximum 1,700), *A. anser* (maximum 6,000), *Vanellus vanellus* (maximum 16,000), and *Limosa limosa* (maximum 5,000). Peak counts of dabbling ducks have included 14,500 *Anas penelope*, 20,000

Table 1 Annual midwinter count totals of waterbirds in Fereydoon Kenar

Year	2002	2003	2004	2005	2006	2007
Number of birds recorded	214,956	494,956	136,936	162,217	166,814	161,035
Year	2008	2009	2010	2011	2012	2013
Number of birds recorded	103,731	97,168	80,563	191,952	171,313	241,881

Source: DOE, Unpublished data from Midwinter Waterbird Census Annual Reports

A. strepera, 80,000 *A. crecca*, 80,000 *A. platyrhynchos*, 60,000 *A. acuta*, and 12,000 *A. clypeata*. A small flock of 11 *Anser erythropus* was present in January 1992 and occasional vagrants occur. Other wintering waterfowl have included up to 500 *Aythya ferina*, 330 *A. fuligula*, 900 *Fulica atra*, and smaller numbers of *Pluvialis apricaria* and *Gallinago gallinago*. These large concentrations of waterbirds attract a variety of wintering raptors including *Haliaeetus albicilla*, *Aquila heliaca*, *A. clanga*, and *Falco peregrinus*. Large concentrations of *Philomachus pugnax* (maximum 2,800) have been recorded on spring migration.

The wetland gained international fame in 1978 when ornithologists from the Department of Environment (DOE) discovered a tiny wintering population of the endangered Siberian crane *Leucogeranus leucogeranus* at the site. The local duck hunters were very familiar with the cranes and reported that they had been coming to this area for many years. The cranes arrive in October and depart in mid-March. Since the discovery of the cranes in mid-January 1978, their numbers fluctuated between 7 and 14 birds, before slowly dwindling to a single bird in recent years, attributed mainly to mortality from hunting along its migration route. The rediscovery of the Siberian crane in the South Caspian, after an absence of records for 60 years, has been described by Ashtiani (1987) and subsequently documented in reports to the CMS Memorandum of Understanding on Conservation Measures for the Siberian Crane (e.g., UNEP/CMS 2011).

Nine globally threatened bird species have been recorded in the NSA, listed in Table 2 below.

The site's significance is primarily associated with the presence of damgahs, which provides daytime roosts for waterbirds, thus attracting the large numbers each winter. Monitoring has not been conducted to determine the site's significance for birds on active migration, i.e., stopover sites during passage. However, all evidence points to this being a very significant passage site for waterbirds and raptors. Outside migration, the site is intensively cultivated and supports an avifaunal composition representative of an agricultural landscape. There are no significant species breeding in the area.

Table 2 Globally threatened birds recorded at Fereydoon Kenar NSA

	Common name	Scientific name	IUCN status[a]
1	Dalmatian pelican	*Pelecanus crispus*	VU
2	Lesser white-fronted goose	*Anser erythropus*	VU
3	Red-breasted goose	*Branta ruficollis*	EN
4	Ferruginous duck	*Aythya nyroca*	NT
5	Siberian crane	*Grus leucogeranus*	CR
6	Eurasian curlew	*Numenius arquata*	NT
7	Great snipe	*Gallinago media*	NT
8	Greater spotted eagle	*Aquila clanga*	VU
9	Imperial eagle	*Aquila heliaca*	VU

[a]IUCN Red List categories: *EN* endangered, *CR* critically endangered, *VU* vulnerable, *NT* near threatened, *LC* least concern. Source: http://www.iucnredlist.org/ Accessed 3 June 2014

Surveys by the DOE in the area since 1995 have shown a decline in diversity, especially Dalmatian pelican *Pelecanus crispus*, pygmy cormorant *Phalacrocorax pygmeus*, night heron *Nycticorax nycticorax*, Eurasian spoonbill *Platalea leucorodia*, greater flamingo *Phoenicopterus roseus*, mute swan *Cygnus olor*, lesser white-fronted goose *Anser erythropus*, red-breasted goose *Branta ruficollis*, ruddy shelduck *Tadorna ferruginea*, common shelduck *Tadorna tadorna*, common pochard *Aythya ferina*, and whimbrel *Numenius phaeopus*, all of which were formerly present in larger numbers.

Mammals

The original mammalian fauna of the lower Caspian has largely disappeared along with its forested habitats. The region's iconic symbol, the Caspian tiger *Panthra tigris virgata*, is now extinct. The modified agricultural landscape today supports few mammal species, all of which are adapted to the existing habitats. Eight species have been recorded, the largest being the golden jackal *Canis aureus*, which thrives in the shrub and treelined rice fields. Only the common otter *Lutra lutra* is of conservation significance, listed on Appendix I of CITES, and totally protected in Iran. Records of otters are rare, and it is likely sightings are of individuals wandering down the streams and rivers from the foothills. Pesticide levels and food availability in the rice fields are unlikely to support a population of otters in FDK.

Fish

The only water bodies that can support fish fauna are small creeks, canals, and reservoirs. However, most reservoirs are drained over the course of the year. The fish fauna is therefore poor, with ten species recorded.

Ecosystem Services

The wetlands at Fereydoon Kenar are highly modified or man made, located within a landscape dominated by rice cultivation. The shallow impoundments that both provide water for rice field irrigation and attract wintering waterfowl support abundant floating and submerged aquatic vegetation and are fringed by *Phragmites australis* and *Typha* sp. *Cyperus rotundus* (the principal food of the wintering Siberian cranes) is common. One interesting feature of the damgahs, appreciated by the local farmers, is that the huge concentrations of wintering waterbirds provide a natural source of fertilization for the rice crops through their abundant droppings. The harvest of ducks and geese from the damgahs also provides a supplementary source of income for the rice farmers during the fallow winter period.

Conservation Status

To ensure that the waterfowl are not disturbed, the duck trappers enforce a very strict ban not only on shooting activities in the area but also on all other unnecessary human activity. As a result, the damgah wetland and surrounding rice fields constitute one of the best protected and least disturbed wetlands in the South Caspian lowlands. Few birds other than *Anas platyrhynchos* and *A. crecca* are trapped, and thus for the many thousands of other ducks, geese, and shorebirds and for the cranes, conditions are ideal, although incidental catch of rare and endangered species constitutes a threat to their conservation. The site has been identified as an "Important Bird Area" by BirdLife International (Evans 1994). The site was designated as a non-shooting area in June 2001, covering Fereydoon Kenar, Ezbaran, East and West Sorkh Rud damgahs, and Fereydoon Kenar Wildlife Refuge, and includes a buffer zone around each of these areas. Fereydoon Kenar Wildlife Refuge (48 ha) is situated 2 km to the NE of the Fereydoon Kenar damgah. The 8 ha Ojakaleh patch of remnant Caspian forest was transferred to the Department of Environment for management as part of the NSA, and a guard station and interpretive building were constructed here to support site management.

Toward the end of each season, when duck netting becomes unprofitable, the area used to be opened up to hunting with guns in a massive "shootout." At the time, this represented the single greatest threat to the surviving Siberian cranes. This practice was forbidden after the Department of Environment designated a non-shooting area for the whole area of Fereydoon Kenar with a total area of 5,427 ha in 2001.

Annual midwinter waterfowl censuses have been carried out by the Ornithology Unit, Department of Environment (DOE), since 1974. During the annual mid-January waterfowl census in 1978, ornithologists of the DOE discovered a flock of about 11 Siberian cranes near the southeast Caspian town of Fereydoon Kenar. It was the first sighting after 60 years. According to the local villagers, these cranes were yearly visitors to the flooded fields near the town and, like their conspecifics in India, spent their time wading in shallow water and digging plant roots. This population has been monitored closely since its discovery in 1978, and the DOE established a long-term research and conservation project on the cranes, through two main instruments.

First, Iran and ten other countries "range states" joined an international effort that was initiated through the adoption, in 1993, of a "Memorandum of Understanding (MoU) on Conservation Measures for the Siberian Crane" under the auspices of the Convention on the Conservation of Migratory Species of Wild Animals (CMS) to help further protect and conserve this important endangered species (see http://www.cms.int/species/siberian_crane/sib_bkrd.htm). Under the CMS MoU (updated in 1998), the participating range states have committed to identify and conserve wetland habitats essential to the survival of Siberian cranes, to cooperate with international organizations and other range states, and to develop a long-term conservation plan.

The following activities concerning Fereydoon Kenar were assigned to Iran under the conservation plan (UNEP/CMS 2011), to be undertaken in 2010–2012:

Implement the Western/Central Asian Site Network (WCASN) for Siberian Cranes and Other Waterbirds Action Plan for 2010–2012;

Evaluate and specify the requirements for WSCAN site management plan development;

Involve communities in effective management of WCASN sites;

Maintain site management committees and encourage effective participation of local hunters in the committees;

Enforce protection of the site;

Support duck trapper associations and their trust funds;

Finalize, sign, and implement FDK Management Plan prepared under UNEP/GEF Siberian Crane Wetland Project;

Produce information materials for hunters;

Provide guidelines for raising the awareness of the key stakeholders, target groups, and audiences at national, provincial, and site levels about crane and wetland protection;

Facilitate and monitor livelihood activities in the area through intersectoral coordination;

Provide necessary equipment for FDK and Bujagh for guarding and monitoring; and

Support and maintain legal and sustainable traditional waterfowl trapping in damgah areas.

Secondly, the management of Fereydoon Kenar received significant technical assistance between 2003 and 2009 through the UNEP/GEF Siberian Crane Wetland Project (www.scwp.info) – a regional flyway conservation project spanning 16 wetland sites in four countries, including Fereydoon Kenar and Bujagh National Park in Iran. The project was led by the International Crane Foundation in collaboration with the governments of China, Iran, Kazakhstan, and the Russian Federation. The project focused on protecting a network of globally important wetlands in Eurasia that are of critical importance for migratory waterbirds and other wetland biodiversity, using the critically endangered Siberian crane as a flagship species, linking activities at the key wetlands along the species' western and eastern flyways. This project functioned at the site, national, and regional levels. At the site level, activities aimed to reduce external threats and strengthen site management effectiveness. At Fereydoon Kenar, activities included strengthening legal protection and enforcement, training nature reserve staff, involving local communities, developing a site management plan, conducting environmental education and public awareness programs, and piloting projects on sustainable livelihoods for local communities.

Current Threats and Future Challenges

Fereydoon Kenar NSA (except Fereydoon Kenar Wildlife Refuge) is not officially included in the protected area network. Therefore, the resolution of most threats to wintering Siberian cranes and other waterbird species, including the use of aerial nets, requires a collaborative approach based on co-management principles – a new concept for Iran's protected area system. The damgah has been maintained by the local community for the purpose of trapping ducks. The local duck trappers are very concerned about the disturbance of their quarry and prevent shooting in the area, which is probably the only reason the Siberian cranes have been able to survive through the years. The traditional use of captive ducks and baited ponds with clapnests is a legal trapping method. However, the extensive aerial nets used around the damgahs and in the surrounding areas are more of a problem because they are illegal. The options are to register them (under license from DOE, with negotiated conditions) or to phase them out over a period of time with the full agreement of the trappers. Some compensation or other benefits would be necessary for the second option. There have been no confirmed reports about Siberian cranes being taken with aerial nets. The incidence of lead poisoning in waterfowl is poorly known, but is likely to be significant given the level of hunting with shotguns in the area. Overhead power cables pose a hazard to large waterbirds in flight, including the Siberian cranes.

The emergence and rapid spread of H_5N_1 highly pathogenic avian influenza across Asia in the last decade also posed a threat to the large waterbird concentrations in this area, which mix with flocks of domestic ducks and geese in the rice fields and nearby wetlands and are handled by the duck trappers and hunters in large numbers. Consequently the Iranian authorities identified Fereydoon Kenar as a high-risk area and conducted public awareness campaigns in the past to raise awareness of the risks to human health.

The traditional duck-trapping practices used at the damgahs are slowly dwindling, and it is possible that in time they might cease altogether. This once widespread tradition in the South Caspian lowlands has largely been replaced by hunting with shotguns in other areas. The maintenance of the damgahs for waterfowl trapping may therefore give way in time to additional rice crops or other land uses that are more profitable to the farmers but detract from the area's conservation values. Another significant challenge is the rapidly increasing price of land and housing along the Caspian shoreline, associated with an increasing population. Such high prices have attracted local farmers to sell off their land for holiday accommodation and similar development; thus fragmentation and piecemeal loss of the rice field landscape are taking place, further exacerbated by the development of infrastructure such as roads, pipelines, and power lines. Given that the site is almost entirely under private land ownership, it is hard for the government authorities to intervene.

References

Archibald G. Ron Sauey and the Siberian Cranes. J Ecol Soc. 1992;5:41–8.
BirdLife International. Important Bird Areas, IR019 Fereydoonkenar Marshes. 1998. http://www.birdlife.org/datazone/sitefactsheet.php?id=8081
CMS. Western/Central Asian Site Network for the Siberian Crane and Other Waterbirds. Guidelines for the preparation of site nomination documents. Annotated list of nomination sites. Version: 13 March 2008. Bonn: CMS Secretariat, United Nations Campus; 2008. p. 221.

Farhadpour H. Capturing common crane with Alpha-Chloralose, 1st meeting of the working group on European cranes. 1985. Aquila. 1987;93–94:237.

Harris J. Safe flyways for the Siberian crane. A flyway approach conserves some of Asia's most beautiful wetlands and waterbirds. Terminal report of the UNEP/GEF Siberian crane wetland project: development of a wetland site and flyway network for conservation of the Siberian crane and other migratory waterbirds in Asia, GF/2712-03-4627. 2010.http://www.scwp.info/final_report.shtml

Ilyashenko EI, editor. Atlas of key sites for the Siberian crane and other waterbirds in Western/Central Asia. International Crane Foundation, Baraboo. 116 p. 2010. http://sibeflyway.org/wp-content/uploads/2012/03/Atlas_English-small.pdf

Kelly M. Project "Sterkh": summary of Siberian crane reintroduction program 1983–1998. International Crane Foundation. 1998.

Mirande CM, Prentice C. Conservation of flyway wetlands in Asia using the Siberian crane as a flagship species: an overview of the outcomes of the UNEP/GEF Siberian crane wetland project. In: Prentice C, editor. Conservation of flyway wetlands in East and West/Central Asia. Proceedings of the Project Completion Workshop of the UNEP/GEF Siberian Crane Wetland Project; 2009 Oct 14–15; Harbin. 2011. http://www.scwp.info/proceedings/Final%20Papers/1-%20SCWP%20overview%20paper.pdf

Neshat SN. Community participation and development of local environment programmes through community-based organizations at Fereydoon Kenar. In: Prentice C, editor. Conservation of flyway wetlands in East and West/Central Asia. Proceedings of the Project Completion Workshop of the UNEP/GEF Siberian Crane Wetland Project, 14–15 October 2009, Harbin. 2011. http://www.scwp.info/proceedings/l03_CommunityParticipation.html

Ramsar Convention Secretariat. Information sheet on Ramsar wetlands (RIS), Ramsar convention on wetlands, 2003. Fereydoon Kenar, Ezbaran & Sorkh Ruds Ab-Bandans. https://rsis.ramsar.org/RISapp/files/RISrep/IR1308RIS.pdf

Sadeghi-Zadegan S. An overview of the historical situation of the Siberian crane and common crane in Iran. Report of the third meeting of Siberian crane range states, 8–13 December 1998, Ramsar. In: UNEP/CMS, editor. Conservation measures for the Siberian crane. CMS Technical Series Publication No.1. Bonn: UNEP/CMS Secretariat; 1999. http://www.cms.int/sharks/sites/default/files/publication/TechSeries_SibCrane_3_0_0.PDF

Sadeghi-Zadegan S, Fazeli A. Fereydoon Kenar, Ezbaran & Sorkh Ruds Ab-Bandans, Iran. 2007. http://sibeflyway.org/wp-content/uploads/2012/03/3-FDK-SIS-Part1-eng.pdf

Sadeghi-Zadegan S, Ilyashenko E, Prentice C. Western flyway of the Siberian crane *Grus leucogeranus*: further releases of captive-reared birds in Iran. Sandgrouse J Ornithol Soc Middle East. 2009;31(2):112–21. http://www.osme.org/sites/default/files/sandgrouse/4-Zadegan_C_pp112-121_31_2_Sandgrouse.pdf

Scott DA. A directory of wetlands in the Middle East. Gland: IUCN/IWRB; 1995.

Siberian Crane Flyway Conservation Programme. Fereydoon Kenar, Ezbaran and Sorkh Rud Ab-Bandans. 2007. http://sibeflyway.org/wcsan/site-network/islamic-republic-of-iran/i-fereydoon-kenar-ezbaran-sorkh-rud-ab-bandans/. Accessed 21 May 2014.

UNEP/CMS. Conservation measures for the Siberian crane, 5th ed. CMS Technical Report Series No.25, UNEP/CMS Secretariat. 2011. http://www.cms.int/species/siberian_crane/pdf/conservation_plans_all_pops_2010_12_e.pdf

Lake Parishan (Islamic Republic of Iran)

Ahmad Lotfi

Contents

Introduction	1648
Location and Description	1648
Hydrology	1649
Ecological Attributes and Functioning	1652
Conservation Status of Lake Parishan	1654
Ecosystem Services	1654
Recent Developments	1654
Threats and Future Challenges	1655
References	1657

Abstract

Lake Parishan in Kazeroun Area is a protected wetland, a Ramsar site, and UNESCO Biosphere Reserve. It is one of the Important Bird Areas of Iran hosting a large number of migratory birds including rare and threatened species. It is a water body mainly recharged by ground water resources, i.e., seepage from karstic formations and spring flows. Evaporation is the sole discharging component of the lake. During the last decade, however, because of prolonged drought and increasing ground water exploitation for irrigated farmlands, the wetland is in crisis for unbalanced inflows vs evaporation. A joint Department of Environment and UNDP project conducted during 2004-2012 resulted in the development of an ecosystem based Management Plan which emphasized, among other conservation rules, improving on-farm water uses as well as alternative livelihood sources for local people to reduce pressure on wetland resources.

A. Lotfi (✉)
Conservation of Iranian Wetland Project, Department of Environment, Tehran, Iran

Senior Irrigation Engineers, Member of Board of Directors, Pandam Consulting Engineers, Tehran, Iran
e-mail: ah.lotfi.pandam2@gmail.com

© Springer Science+Business Media B.V., part of Springer Nature 2018
C. M. Finlayson et al. (eds.), *The Wetland Book*,
https://doi.org/10.1007/978-94-007-4001-3_228

Keywords

Biosphere Reserve · Bird Area · Climate change · Desiccation · Drought · Karstic aquifer · Parishan Lake · Ramsar site

Introduction

Lake Parishan in Kazeroun Area is a protected area, a Ramsar Site, and UNESCO Biosphere Reserve. It is one of the Important Bird Areas of Iran, hosting large numbers of migratory birds including rare and threatened species.

Location and Description

Lake Parishan is located in the Kazeroun valley and is formed in a natural depression along a foothill 15 km east of Kazeroun city in Fars Province. Geographically its center is defined as $51°, 20'$ E and $29°, 30'$ N (Fig. 1). It is about 820 m above mean sea level, about 10 km long and 1–5 km wide, and when full (a gauge reading of 3.1 m) has a water surface area of about 5,000 ha.

At its maximum surface area, the depth of water at the deepest point of the lake is about 5.5 m. Water depth in the lake varies considerably: relatively deep (2–5 m), medium deep (0.5–2 m), and shallow (<0.5 m). Vast shallow areas occur in the south

Fig. 1 Satellite image of Lake Parishan, Kazeroun City, and Arjan Wetland and their locations in Iran (*inset* image, *red spot*) (Source: Google Earth)

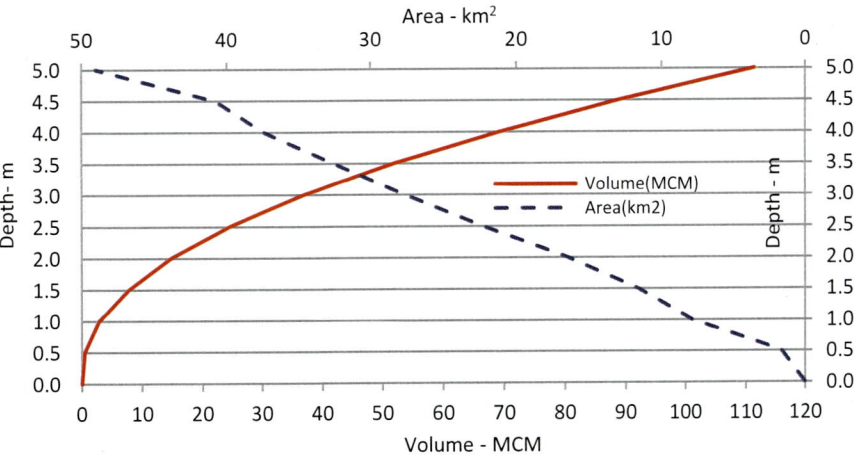

Fig. 2 Area-volume-depth curves of Lake Parishan (Original by author)

and southwest parts of the lake. The maximum volume of water in the lake is estimated to be about 115×10^6 m³ (Fig. 2).

The lake includes relatively large areas of *Phragmites australis* reed beds (Fig. 3) in its east and west ends as well as narrow discrete bands along its northern margins. Inside the wetland there are several very small patches of *Phragmites* reeds which form isolated islands. The latter provide breeding habitat for waterbirds. In the medium deep parts of the lake, particularly along its northern length, *Najas minor* is the only submerged vegetation.

Hydrology

The rather small catchment area of the lake (270 km²) is located in the semiarid region of Iran and has long warm summers and short temperate winters. Average annual temperature is 22 °C with a maximum of 48 °C and minimum of 0 °C. Average annual precipitation and evaporation in the lake area are 450 and 2,500 mm, respectively.

Much of the catchment area is covered by rather high elevated outcrops, mainly comprised of highly fractured karstic limestone and steep, very pervious piedmont alluvial fans northeast of the lake.

Because of the small catchment area, rather low precipitation rates, and porous alluviums just upstream of the lake, surface runoffs flowing into the lake are very low and have a minor role in recharging the lake. Indeed the lake is mainly recharged by ground water resources, i.e., excess flows from several karstic springs, and direct underground seepages particularly through northern, eastern, and western boundaries of the lake.

Fig. 3 *Phragmites australis* reed beds of Lake Parishan (Photo credit: Environment Conservation Office, Fars Province ©)

Ground water contours around the lake depict a generally converging movement of ground water towards the lake, while evaporation seems to be the only natural source of water loss from the lake.

Geological investigation has revealed that there are a number of sink holes to the south of Lake Arjan about 15 km northeast and 700 m higher than Lake Parishan (see Fig. 1) that are connected through karstic galleries with karstic aquifers in the Lake Parishan and Kazeroun areas (Karst Research Center 1995). These galleries have an important role in recharging karstic springs as well as alluvial aquifers in the Kazeroun area and around Lake Parishan (Petar and Aghili 1990).

About 1,000 shallow wells equipped with small diesel engines and centrifuge pumps are operating around the lake for irrigated agriculture. More than 70% of the wells are located only in the northeastern alluvial fan. These wells abstract more than 35×10^6 m^3/year (mcm/year) of ground water resources, which otherwise would discharge into the lake. This means that these wells and the irrigated agriculture they support represent significant competition for the lake's water resources. Although the first engine-equipped water wells were constructed in the early 1960s, the number increased tremendously after 1975. A long-lasting drought during the 2000s, which is still continuing, has also accelerated the construction of new wells.

Existing data indicate that during 1995–2005 while the total water withdrawal from the alluvial aquifer by water wells increased from about 20 to about 35 mcm/year the average discharges of springs have decreased from 45 to less than 30 mcm/year – a similar order of magnitude to the increase in abstraction by wells.

Fig. 4 Satellite image of Lake Parishan showing it in its almost dry state. Darker gray spots in the right and upper left edges of the wetland depict areas of plant growth, mainly reed beds (Source: Google Earth, image from 20 June 2011)

The electro-conductivity (EC) of the lake's water varies between 4 and 15 dS/m depending on the season, water level, and whether the lake is recharging or at the end of its evaporating stage. In parts of the lake where springs flow, and/or ground water seeps into the lake's water body, the quality of water around the inflowing point is higher, and fish fry usually gather in these areas. The quality of ground water in the northern alluvial aquifer is very high EC < 1 dS/m, while EC in the southern aquifer exceeds 8 dS/m.

As with many other places in Iran, the Lake Parishan area has been experiencing a long-lasting drought since the early 2000s that continued in 2014. As a result of this drought almost all the springs have dried up and/or are discharging very low flows. A similar event occurred in 1965, when the number of wells was much less and irrigation water use was much lower than at the present time. During the recent drought, huge pressure for ground water depletion has negatively affected the hydrological condition of the lake. Due to the recent desiccated state of the lake, potential evaporation and hence water requirement of irrigated crops has increased by more than 30%.

Since the early 2000s, the lake has not received enough water and has regularly lost its water reserves such that in many years during the last decade most of the lake was dry. Since 2009, however, the lake was almost entirely (more than 90%) dry (Fig. 4). This condition more or less remained unchanged until 2013.

In addition to the drought, the area is affected by impacts from climate change. Existing records indicate a distinct increasing trend in ambient temperature and decreasing trend in precipitation. During 2000–2012, the average annual ambient temperature increased by 1.16 °C. In the same period the average annual precipitation decreased by about 13% as compared with the average annual precipitation for the period 1980–2012 (Ranjbar 2012).

Ecological Attributes and Functioning

When not dry, Lake Parishan with its different physical features provides a range of diverse habitats for different fauna.

The water body supports habitat for several indigenous and introduced species of fish. The eastern and northern margins of the lake, which receive fresh groundwater seepage, constitute the freshest part of the water body and provide nursery habitat for fish larvae and fingerlings. Also, some fish species use the rather dense submerged aquatic plants (*Najas minor*) at the northern and particularly northeastern parts of the lake as nursery, feeding, and sheltering habitats. Water courses in the eastern and northeastern parts of the lake are spawning habitats for some fish species.

Migratory waterbirds use different parts of the wetland for feeding, sheltering and breeding. Deep and shallow water areas in the lake support a wide range of waterbirds including divers, surface feeders, and waders. The water body in its central deeper parts provides the main feeding habitat for piscivorous waterbirds. The dense reed beds at the eastern and western ends of the lake provide habitat for several species of migratory as well as resident waterbirds that use them as sheltering, nesting, and breeding sites. The lakeward margins of the eastern reed beds are particularly important because they provide habitat for nesting and breeding pelicans (Fig. 5). Vast areas of shallow wetlands provide important habitat for waders. Cereal farmlands in the south west of the lake are used by geese as feeding habitat. Several individual small patches of reed beds in the central as well as south western extension of the lake create isolated safe islands for some breeding birds.

Fig. 5 The globally threatened Dalmatian pelican *Pelecanus crispus* (Photo credit: H. Verdaat ©)

Fig. 6 Reed beds along the edge of Lake Parishan (Photo credit: Environmental Conservation Office, Fars Province ©)

High populations of waterbirds have been recorded for Lake Parishan, for example, an average of 25,000 ducks and 120,000 coot *Fulica atra* during four winters in the 1970s, and a marbled teal *Marmaronetta angustirostris* population of over 2,000 in the 1970s, and up to 5,500 in the late 1980s (Scott 1995). The increase in cultivated areas around the wetland has provided feeding areas for some species, i.e., greylag goose *Anser anser* and Eurasian crane *Grus grus*, while an increase in fishing activities and particularly disturbances due to motor boat traffic has caused a significant decrease in the populations of ducks.

More than 90 waterbird species have been reported in the lake, of which more than 37 species breed. Globally threatened waterbirds including the Dalmatian pelican *Pelecanus crispus*, marbled teal, white-headed duck *Oxyura leucocephala*, and ferruginous duck *Aythya nyroca* are among these breeding species (Scott 1995; Farhadpour 1997; Rahbar 2005).

Dense or moderately dense reed beds around the wetland and isolated islands (Fig. 6) inside the lake provide important habitats for breeding birds. The rather narrow strip of *Phragmites australis* reeds alongside the rocky shores at northern margins of the lake provide habitat for Eurasian otter *Lutra lutra*. Open grasslands in northeast of the lake are used for ranching horses. The shrub lands at the northeast part of the lake provide habitat for mammals, particularly wild boar *Sus scrofa*. Some of the villagers at the northeastern side of the lake breed water buffaloes that use the lake for their feeding, resting, and sheltering.

Wetlands in the eastern waterlogged areas provide habitat for wild daffodil *Narcissus tazetta* L., a well-known wetland flower with high marketing potential, while a small area of *Typha angustifolia* at the southeastern part of the lake seems to be a unique patch of this plant within the lake. The rural population harvests reeds mainly for feeding domestic animals.

Conservation Status of Lake Parishan

Management of Lake Parishan is under the jurisdiction of Kazeroun Environmental Conservation Office of the Fars Province Department of Environment (DOE). However, management of ground water resources of the area is the responsibility of Fars Water Authority under the Ministry of Energy.

Lake Parishan together with Arjan Wetland is protected under national laws as a National Park and is designated as a Ramsar site (1975) and a UNESCO Biosphere Reserve (1976). Any activity with potential impact on the lake needs permission from Fars Province Environment Conservation Office.

Ecosystem Services

Lake Parishan provides a range of ecosystem services to more than 18 villages located around the wetland with about 13,000 inhabitants. While irrigated cultivation and animal husbandry are the main activities of the majority of the rural population, some local families are partially dependent on the lake's resources for their livelihoods. The fish resource of the lake is an important source of livelihood for the local rural population which also harvests reeds for feeding domestic animals. The lake supports a small grazing area for horses and buffalos.

The importance of Lake Parishan in supporting large numbers of migratory waterbirds for wintering and for breeding has been described above. Moreover, the spectacular landscapes around the lake provide considerable potential for recreation, ecotourism, sport fishing, and hunting, while the considerable open water area of the lake has significant moderating effects on the summer climate of the surrounding areas.

Recent Developments

A project titled "Conservation of Iranian Wetlands Project" (CIWP) was undertaken during the last decade with support from the Global Environment Fund (GEF), UNDP, and DOE Iran. Despite the fact that the implementation period of the CIWP overlapped with a long-lasting drought in the area, seriously impacting its activities and results, its achievements included the development of a strategic plan for ecosystem-based management of the lake. Parishan Lake has no satellite wetlands. The main objectives of this plan include:

- Increasing the knowledge of decision-makers and awareness of stakeholders in relation to Lake Parishan;
- Developing capacity among stakeholders (training, participation, provision of equipment); and
- Establishing mechanisms for sustainable management of the wetland (intersectoral committees, ecosystem approach management plans, laws and regulations, monitoring).

Moreover, local and provincial committees have been established for management and decision-making for sustainable conservation of Lake Parishan. In addition to a monitoring plan, direct action includes establishing facilities for collection and disposal of garbage from villages around the lake, demonstration farms for on-farm water management, wetland habitat classification and zoning; marking the lake's boundaries; and organizing of the first Iranian wetland festival in March 2009.

Plans for alternative/supplementary livelihoods for villagers around the lake include management of fishing in the lake, a strategic plan for establishing ecotourism, and a wetland visitor center. Conservation planning for the lake's biodiversity includes protecting the endangered Eurasian otter *Lutra lutra,* restoring the *Typha angustifolia* community, and creation of an artificial pond by the local community to safeguard endemic turtles and fish during drought conditions.

A bilingual "Lake Parishan, concise baseline report" is accessible through: http://www.wetlandsproject.ir/en/publication/books/lp-bs-en.

Threats and Future Challenges

The lake has experienced intense pressure during the last decade and is currently in a state of ecological crisis, with major impacts on biodiversity and socioeconomic conditions. The current state of water inflows and outflows into and from the lake is unbalanced, and current inflows are much less than annual evaporation. The main reasons for this situation are:

- Reduced precipitation, apparently because of long-term climate change;
- Increased water abstraction from ground water resources;
- Trends of expanding cash crop cultivation which consume more groundwater; and
- Local farmers increasingly tend to expand early season cash crop production under plastic cover (Fig. 7), which generates higher income.

Fig. 7 Under plastic cultivation adjacent to wetland (Photo credit: A. Lotfi ©, Rights remain with author)

Contamination is another important threat to the lake. Sewage outflows from villages around the lake enter the lake directly or through seepage. Also, residual fertilizers and pesticides dissolved in irrigation return flows eventually discharge into the lake and increase its eutrophic level.

Unsustainable use of wetland's fish resources by the rural population is also considered a potential pressure on the lake's ecology.

The most effective solution to the present problems facing the lake is to establish a sustainable balance between groundwater inflows and evaporation from the lake. This implies a critical revision of groundwater management as well as crop production policies. Supported by CIWP (the Conservation of Iranian Wetlands Project), Fars Province Environmental Conservation Office in collaboration with Fars Province Jihad-Agriculture Organization have started a participatory program for establishing demonstration farms to instruct local farmers around the lake regarding opportunities for better on-farm water management, increasing water use efficiency through new irrigation technologies and a more adaptable cropping pattern so that they can effectively grow crops with less water (Sharifi Moghaddam 2013). At the same time, the water authority has started a challenging program to block unauthorized water wells with effective support from relevant governmental organizations. The early outcomes of these programs are promising. More ground water began flowing into the lake in 2014.

Promoting new sources of livelihood for the local rural population is another policy for reducing pressure on the wetland resources. There is good potential for nature tourism and handicraft promotion in the Lake Parishan area.

References

Farhadpour H. Birds in Lake Parishan. Internal report of Fars Province Environment Conservation Office, Shiraz. 1997. (In Farsi). unpublished.

Karst Research Center. Geo-hydrology of karsts in Arjan-Parishan zone. Tehran: Geological Survey Organization of Iran; 1995. (In Farsi).

Petar M, Aghili B. Hydrogeological characteristics and groundwater mismanagement of Kazeroun karstic aquifer, Zagros Iran. In: Proceedings of the Antalya Symposium and Field Seminar. IAHS Publication; 1990. p. 207.

Rahbar N. Bird population density and distribution in Lake Parishan as a tool to identify sensitive habitats [MSc dissertation]. Tehran: Islamic Azad University; 2005 (In Farsi).

Ranjbar A. Wetlands, climate change and drought. In: Wetlands, and a review on the factors affecting their sustainability. Tehran: Iranian Society of Consultant Engineers; 2012. p. 198–209 (In Farsi).

Scott DA, editors. A directory of wetlands in the Middle East. Gland/Slimbridge: IUCN/IWRB; 1995. http://www.wetlands.org/?TabId=56&mod=1570&articleType=ArticleView&articleId=1891

Sharifi Moghaddam M. Development of sustainable agricultural for applying ecosystem approach in wetlands management, Case studies on Lake Parishan, Conservation of Iranian Wetlands Project; 2013 (In Farsi).

Lake Uromiyeh (Islamic Republic of Iran)   138

Ahmad Lotfi

Contents

Introduction	1660
Location and Geography	1660
Hydrology (Surface and Groundwater)	1662
Ecological Functioning	1664
Conservation Status of Lake Uromiyeh and Satellite Wetlands	1668
Ecosystem Services	1669
Recent Developments	1671
Threats and Future Challenges	1672
References	1673

Abstract

Lake Uromiyeh is a vast hypersaline lake in northwestern Iran shared between West and East Azerbaijan provinces. It receives surface and ground water flows from a vast watershed. The lake is a National Park, a Ramsar Site and a UNESCO Biosphere Reserve and is the largest inland lake in Iran. Several fresh-brackish water satellite wetlands exist mainly to the south of the lake, a few of which are registered as Ramsar Sites. The lake and its satellite wetlands provide globally significant ecological services.

Since late 1990s the basin has been experiencing rather severe continuous drought which has dramatically affected the hydrological regime of the basin. During the same period, construction of several storage dams, significant increases in cultivated areas as well as increased abstraction of ground water resources has resulted in continuous reduction of water inflow into the lake.

A. Lotfi (✉)
Conservation of Iranian Wetland Project, Department of Environment, Tehran, Iran

Senior Irrigation Engineers, Member of Board of Directors, Pandam Consulting Engineers, Tehran, Iran
e-mail: ah.lotfi.pandam2@gmail.com

Presently the lake is under very severe pressure such that major parts of it are completely dry and vast salt beds are exposed to wind erosion with significant negative environmental impacts on human populations, settlement areas and croplands.

During 2004-12, Department of Environment of Iran (DOE) and the Global Environment Facility (GEF) conducted a joint "Conservation of Iranian Wetlands Project" aiming to develop and execute an ecosystem-based management plan of the lake. Also, since 2010, a national high level committee was organized to identify approaches and conduct programmes for the lake's survival.

Keywords

Biosphere Reserve · Desiccation · Drought · Hypersaline · Lake Uromiyeh · Ramsar site · satellite wetlands

Introduction

Lake Uromiyeh is a vast hypersaline lake in northwestern Iran shared between the provinces of East and West Azerbaijan (Fig. 1). The lake is a National Park (NP), one of the largest Iranian Ramsar Sites, a UNESCO Biosphere Reserve, and is the largest inland lake in Iran. The lake is surrounded by a number of important fresh-brackish water satellite wetlands, several of which are also of global significance for their biodiversity. Also the lake includes several islands that provide important habitats for birds and mammals.

Location and Geography

Lake Uromiyeh is formed in a natural depression at the lowest point within the closed Lake Uromiyeh basin, where water enters through several rivers but leaves only by evaporation. The Lake Uromiyeh basin, 51,876 km^2, is apportioned among the provinces of East Azerbaijan (39%), West Azerbaijan (51%), and the Southern Province of Kurdistan (10%). The basin is generally mountainous, containing two of the famous Iranian volcanic peaks (Sahand 3,707 m and Sabalan 4,810 m), and with several vast productive plains in the valleys and around the lake. The lake's water level varies between 1,270 and 1,280 m above mean sea level (AMSL).

Lake Uromiyeh's geographic coordinates lie between 37° 06' 15" and 38° 15' 15" North and 45° 00' 13" and 45° 55' 20" East. The lake has a maximum extent of 130 km × 48 km and covers an average area of ca. 5,000 km^2 when its water level is at 1,276 m AMSL. However the official marginal boundary of Lake Uromiyeh is the 1,278.1 contour lines, resulting in a maximum area close to 6,000 km^2. Under normal hydrological conditions, the lake has an average depth of about 5.4 m and a maximum depth around 15 m in the northern parts (Yekom 2005).

Numerous fresh- and brackish-water wetlands exist around the lake, supporting and complementing its ecological functioning. Sixteen of these, mainly located in

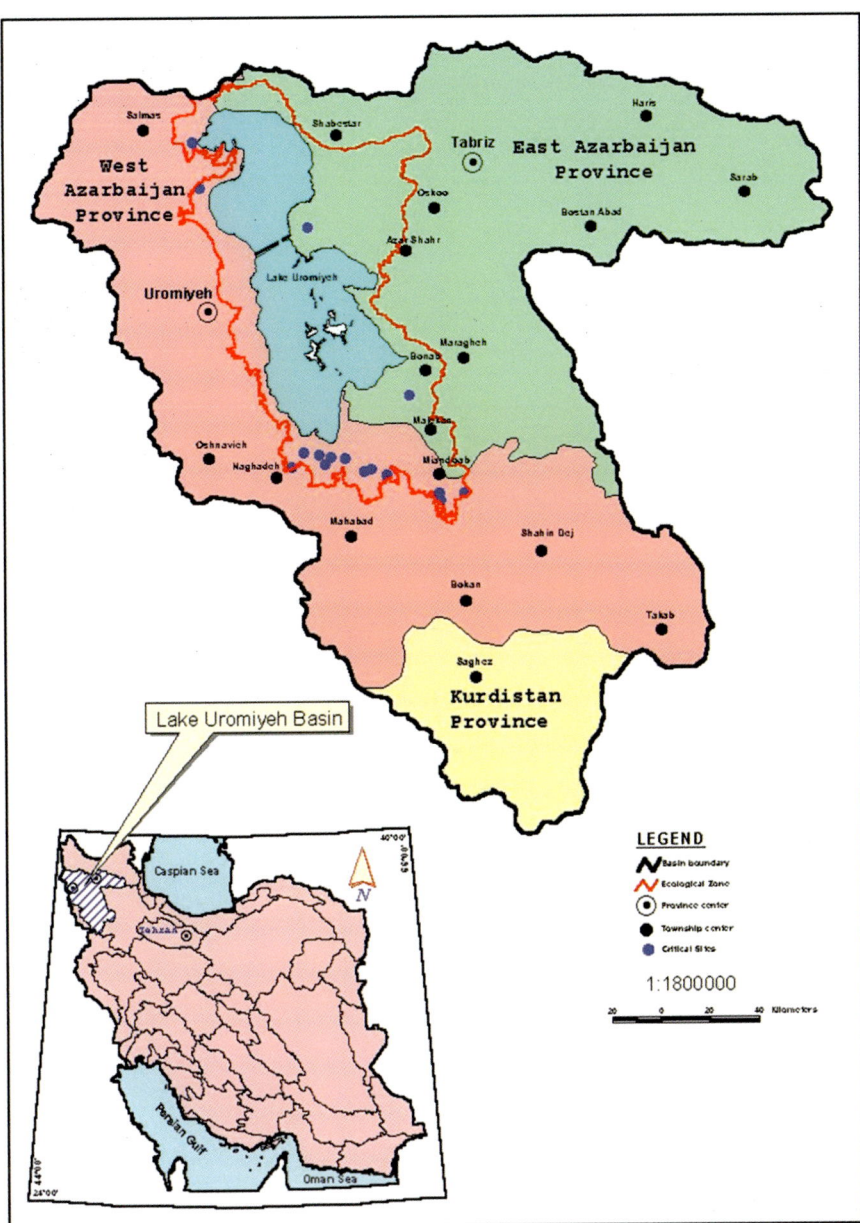

Fig. 1 Lake Uromiyeh basin is shared among three provinces (© Yekom Consulting Engineers 2002, with permission)

the south (red spots, Fig. 6), are sizeable, and some are designated as Ramsar Sites, i.e., Dorgeh Sangi, Hassanlou, Yadegarlou, Gopy Baba Ali, and Kani Brazan. These wetlands formerly received water from fresh river flows, spring flows, and/or seepage flows but presently are mainly fed from irrigation return flows (Pandam 2005).

Hydrology (Surface and Groundwater)

Hydrology, both quantitative and qualitative, is the most complicated physical attribute affecting the lake and its satellite wetlands. Its complications arise from several sources of surface and groundwater in the three provinces with different needs, policies, and strategies for their management.

The entire basin consists of 12 main subbasins (Fig. 2) with 17 permanent and 12 seasonal rivers. Most of the surface water is flowing in West Azerbaijan in the southern and western part of the basin. The Simineh Rud and Zarrineh Rud rivers produce about 51.6% of the long-term total surface water inflows into the lake (Figs. 2 and 3).

Long-term data (1965–2004) show that annual flow of the rivers into the lake averages 4,327 million cubic meters per year (mcm/year), varying between 12,574 mcm (in 1969) and less than 243 mcm (in 2000) (Yekom 2004; Yekom 2005; CIWP 2008).

Fig. 2 The major river systems that drain into Lake Uromiyeh are shown along with minor sub-basins (Data sources: Lehner and Döll 2004; Lehner and Grill 2013)

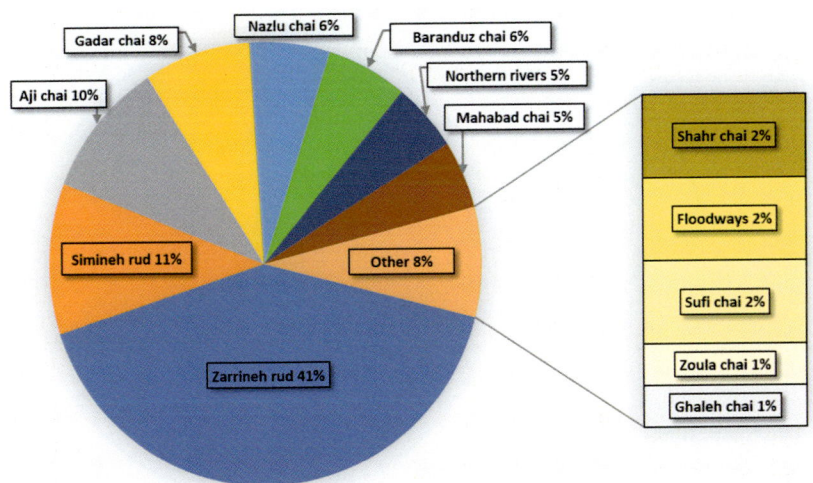

Fig. 3 Contribution of different rivers in water supply to Lake Uromiyeh period: 1965–2004 (WRRI 2005)

River flows are also one of the main sources of water supply for agricultural activities. In 2006, about 400,000 ha of cultivated lands in the Lake Uromiyeh basin were irrigated. Of the total water consumption for irrigation (5,600 mcm/year), about 59% is supplied from surface water resources (Yekom 2004; Hashemi 2008).

Within the Lake Uromiyeh basin, 25 groundwater basins are exploited of which 13 are located around the lake (Yekom 2004; Yekom 2005; Hashemi 2008). In 2008, close to 74,000 wells, 1,579 springs, and 5,747 qanats (underground galleries for collecting groundwater) produced about 2,440 mcm/year of which more than 90% was extracted from wells. The contribution of groundwater to the lake water resources is estimated at around 220 mcm/year, part of which evaporates in the wetlands around the lake before it reaches the main waterbody (Yekom 2004; Hashemi 2008). Generally with the exception of Aji Chai River in the northeastern part of the basin, all other rivers have good water quality suitable for almost all uses. Aji Chai, as a result of flowing across salt dome terrains, carries more than 1,270,000 t/year of dissolved salts into the lake, about half of the total annual salt intake of the lake (Yekom 2004; Yekom 2005; Hashemi 2008).

Long-term data collection (1969–2005) demonstrates the salinity of the lake's water and is closely related to the water level, i.e., volume of water in the lake, and its average electrical conductivity (EC) is 382 dS/m, close to 270 g/l (ranging annually between 577–279 dS/m and 400–200 g/l). In 2010 the salt concentration was 380 g/l. In 2011–2012 the lake's water level continued to decline, and salt concentration in the water increased to as high as 550 g/l (Tarbiat Modarres University 2012a, b). More than 90% of the salts occur as sodium chloride (Yekom 2004; Yekom 2005).

Because of the high erosion capacity of the surface geology, sparse vegetation cover, and special traditions of downslope plowing of steep foothills for rain-fed

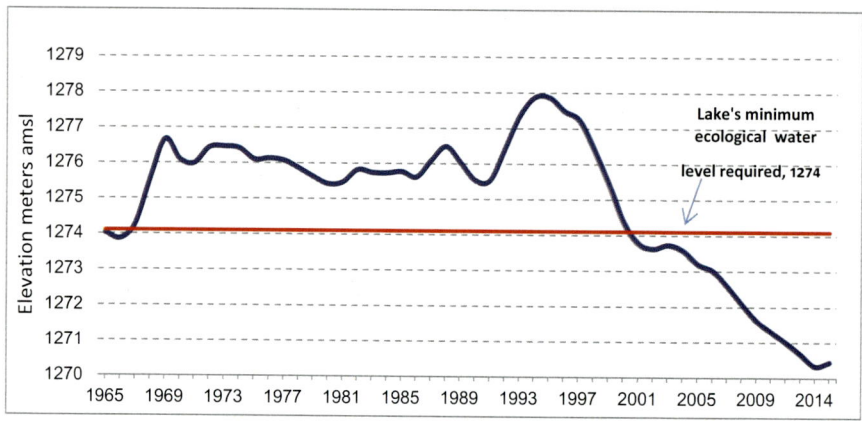

Fig. 4 Lake Uromiyeh water level records (Data Source: West Azerbaijan Water Authroity)

cultivation in the catchment area, the lake receives a rather high sediment load of about 5.8 million tons per year (Yekom 2004; Yekom 2005; Hashemi 2008).

Though drought is a common event in the region, beginning in the early 2000s, an exceptional continuous and long-lasting drought occurred in the basin which has continued through 2013 and into 2014. This severe drought condition was exacerbated by additional groundwater abstraction for irrigation in both provinces of Azerbaijan, resulting in acute pressure on the lake's water resources.

Up to about the mid-1990s, the maximum variation of water level between normal wet and dry years was about 2.7 m (Fig. 4). During the extended drought period, the lake's water level continued to decline and reached a minimum very close to 1,270 m AMSL. At the end of the 2013 summer season, almost the entire southern half of lake was dry such that a car could pass across it (Fig. 5).

Because of agricultural, industrial, and residential developments during the last three decades, there has been an enormous increase in the use of groundwater particularly during the recent drought. There is also a trend for the lake's salty water to intrude into freshwater aquifers. In order to control the overextraction of the groundwater resources, further groundwater-based irrigation development is banned in areas around the lake (Lotfi 2012).

Ecological Functioning

Species richness within the waterbody is low because of the hypersaline quality of Lake Uromiyeh water. *Artemia urmiana,* an endemic brine shrimp, is the most important species identified and can reach densities of 4,000 per liter when water quality is within its tolerance range. It is the main food source for the migratory birds that visit the lake. *Artemia urmiana* itself feeds on *Dunaliella* sp. and *Tetraselmis* sp. and some other phytoplankton (Yekom 2002, 2005).

16 Jul. 1995- Landsat5   22 Sept. 2014- Landsat 8

Fig. 5 Decline in extent of water coverage of Lake Uromiyeh (Landsat imagery courtesy of US Geological Survey)

Within Lake Uromiyeh basin, 239 species of birds have been recorded, while 212 species have been recorded within the National Park (the lake and its satellite wetlands). Of these, 71 species are protected under national law. More than 78 bird species have bred in the lake, while 170 species are migratory and use the lake and satellite wetlands for wintering, staging, or feeding as part of their migratory cycle. About 50 species are showing evidence of declines, and 11 globally threatened bird species have been recorded (Yekom 2002).

The lake (including its islands) has been extremely important for breeding American flamingo *Phoenicopterus ruber*, great white pelican *Pelecanus onocrotalus*, common shelduck *Tadorna tadorna*, ruddy shelduck *T. ferruginea*, pied avocet *Recurvirostra avosetta*, common redshank *Tringa totanus*, Armenian gull *Larus armenicus*, slender-billed gull *L. genei*, little egret *Egretta garzetta*, Eurasian spoonbill *Platalea leucorodia*, stone curlew *Burhinus oedicnemus*, marbled teal *Marmaronetta angustirostris*, and ferruginous duck *Aythya nyroca* (Scott 1995).

Over 425,000 waterfowl of at least 53 species were recorded in Lake Uromiyeh during an aerial survey in August 1973 (Scott 1995), compared to around 150,000 during another aerial survey in August 2001 (Yekom 2005).

The lake has been an important molting area for *T. tadorna* and flamingos and in midwinter may support large numbers of wintering waterfowl (Yekom 2002, 2005).

Under normal conditions, Lake Uromiyeh and its satellite wetlands regularly host waterbird populations exceeding 20,000 birds, meeting Ramsar Convention criterion 3a, and at least 14 species of waterbirds occur in numbers exceeding the Ramsar 1% criterion for international importance within the Ecological Zone (criterion 3c). Both of these criteria confirm that Lake Uromiyeh and several of its satellite wetlands are internationally important (Yekom 2002). However as a result of the increased salinity of the lake's waterbody during recent years, the densities of *Artemia* and other prey have crashed, leading to massive declines in the number of waterbirds breeding and feeding in the lake, with virtually no breeding of flamingos in recent years.

In some of the islands and under a successful conservation program, the Department of Environment (DOE) had introduced and propagated *Ovis ammon* and *Dama mesopotamica* (Persian fallow deer). However these populations have become extremely stressed and have faced noticeable losses due to shortage of freshwater and food on the islands and have therefore been relocated to a safer location.

While the lake supports no fish because of its high salinity, some of the satellite wetlands around the lake provide habitats for fishes. Fifteen fish species of four families (Balitoridae, Siluridae, Poeciliidae, and Cyprinidae) have been recorded. These include at least three species that are endemic to the Lake Uromiyeh basin, two species that are endemic to the Lake Uromiyeh and Caspian (Aras) basins, and three species that are endemic to Iran and the Lake Uromiyeh basin (Yekom 2002, 2005; Fig. 6).

There are about 546 plant species of 299 genera and 64 families within the ecological zone of Lake Uromiyeh. The majority of species are annual or perennial herbs of grass types, and only 11.5% are shrubs and trees. Halophytes constitute 17.8% of the total species distributed closer to the lake and salt marshes (Yekom 2002, 2005). Several of the islands, notably Ashk and Kaboodan, support close to 50 plant species including two endemic, seven native, and five vulnerable species (Yekom 2002). Satellite wetlands around the lake usually support a rich floristic diversity. Depending on the quantity and persistence of water, different vegetation communities have evolved around and inside the wetlands. In Solduz (Fig. 7), for example, a dense *Phragmites australis* reed bed has grown at the outlet of the drainage canal just upstream of the lake, while in Kani Brazan (Fig. 8) wetland, patches of reed have grown in the inner parts with a rather dense and uniform growth of *Schoenoplectus* in the eastern and northern parts of the wetland. In Gerdeh gheet wetland (Fig. 9), an extensive growth of *Tamarix* has created a very typical and spectacular landscape (Pandam 2005).

Dorgeh Sangi, Shorgol (Hassanlou), and Kani Brazan encompass considerable pasturelands beside and around the wetland which are normally supported by the groundwater table. These grazing lands, still considered as part of the wetland areas, have great economic importance for animal husbandry in the rural areas. Kani Brazan wetland seems to be the only satellite wetland south of the lake which supports intensive submerged vegetation of *Najas graminea*. Other wetlands do not contain such vegetation, most likely because of the quality of water. Growth of algae can be found in almost all the wetlands, indicating the eutrophic status of the water bodies. Nutrient-rich irrigation return flows recharging the wetlands are the main reasons for eutrophic status of wetlands (Pandam 2005).

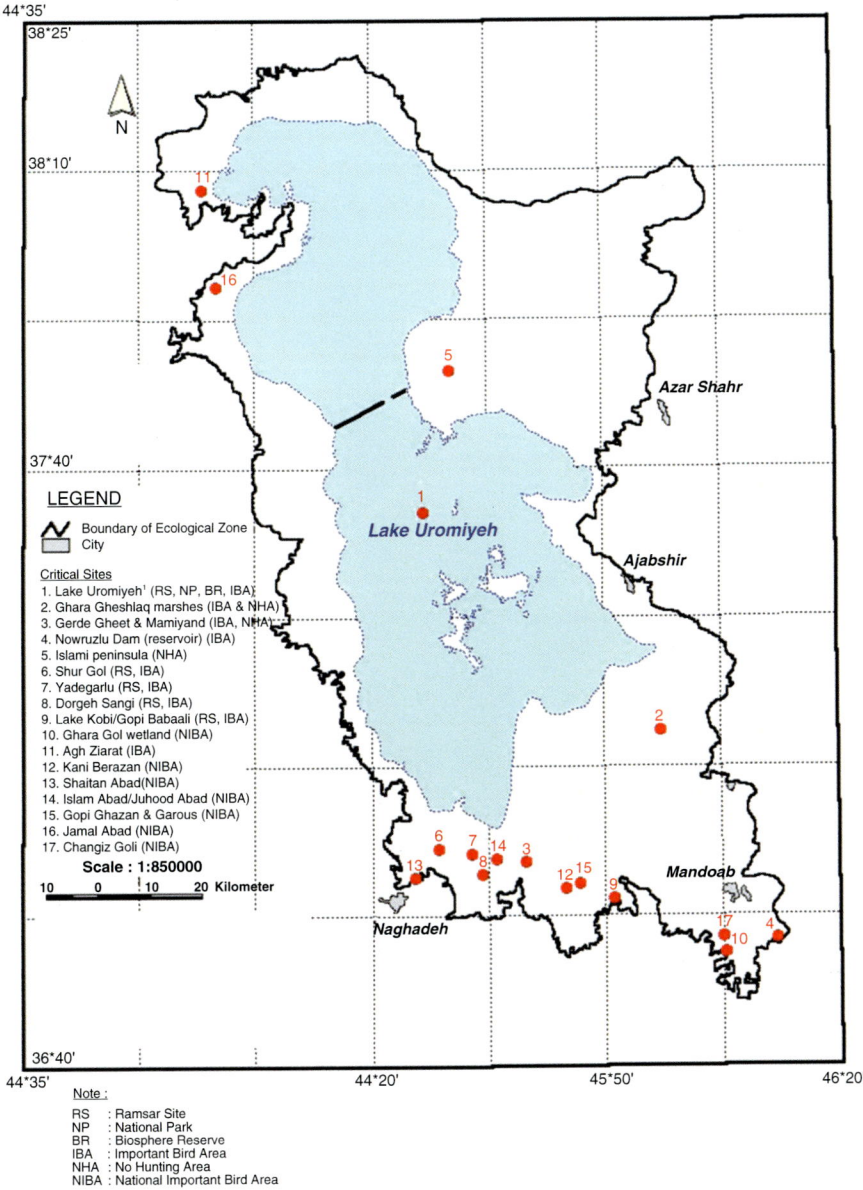

Fig. 6 Critical sites in the ecological zone of Lake Uromiyeh (© Yekom Consulting Engineers 2002, with permission)

Fig. 7 Dense reed beds just at the outlet of the drainage canal which extends toward Solduz Wetland at the right-hand side (Photo credit: A. Lotfi © Rights remain with the author)

Fig. 8 Extensive *Schoenoplectus litoralis* in Kani Brazan wetland, 2005 (Photo credit: A. Lotfi © Rights remain with the author)

Conservation Status of Lake Uromiyeh and Satellite Wetlands

Management of Lake Uromiyeh is under the sole jurisdiction of West Azerbaijan Environmental Conservation Office (WAECO) of the Department of Environment (DOE); while depending on their locations, the satellite wetlands are managed by the relevant provinces. However, management of water resources (rivers, groundwater resources) in the three provinces involved in Lake Uromiyeh is the responsibility of the water authorities in each of the provinces.

Fig. 9 Gerdeh gheet wetland, looking south toward Mamyand village, *Tamarix* woodlands and Gadar River (View from Gerdeh Sagav, July 2005) (Photo credit: A. Lotfi © Rights remain with the author)

The lake is protected under national laws as a National Park and is designated as a Ramsar Site (1975) and a UNESCO Biosphere Reserve (1976). Any activity with potential impact on the lake needs permission from WAECO.

Almost all the wetlands in each province are under the jurisdiction of the Environmental Conservation Offices of the relevant towns. Also most of the satellite wetlands are protected by DOE under the national laws of "Important National Bird Area" and "No Hunting Area." Five of the satellite wetlands are designated as Ramsar Sites.

Ecosystem Services

Lake Uromiyeh and its satellite wetlands support a range of ecosystem services that include providing habitats for large numbers of migratory waterbirds (Fig. 10) that visit the lake and its satellite wetlands for wintering and breeding.

Most of the satellite wetlands support noticeable grassland areas which are used by local herders for grazing considerable numbers of domestic animals including sheep, goats, cows, and buffalos (Fig. 11). Water buffalos are an important source of livelihoods in rural areas, particularly in West Azerbaijan. The lake also provides a flood storage function, with eleven large rivers around the lake discharging their

Fig. 10 Flamingos feeding in Goppy wetland, Lake Uromiyeh (Photo credit: A.Lotfi © Rights remain with the author)

Fig. 11 Grazing land at Dorgeh Sangi, one of several wetlands of Lake Uromiyeh (Photo credit: A.Lotfi © Rights remain with the author)

flood flows into the lake. Similarly, a large volume of salts delivered by the rivers and in particular the Aji Chai is collected in the lake.

The recent exposure of the lake's salt beds has revealed the crucial role of the lake in collecting these salts. Exposed salt beds of the lake cause salt dusts to spread on the agricultural lands and impact human health in the surrounding areas. Conversely, traditional salt harvesting is a noticeable source of income to many rural families particularly in East Azerbaijan. A few factories in this province also harvest salt from the lake.

The vast open water areas of the lake and satellite wetlands have significant moderating effects on the summer climate of the surrounding areas.

Mud flats around the lake are a resource for mud therapy, and large spectacular landscapes around the lake and satellite wetlands provide considerable potential for recreation, ecotourism, sport fishing, and hunting.

Recent Developments

The "Conservation of Iranian Wetlands Project" (CIWP) was undertaken between 2004 and 2012 with support from the Global Environment Facility (GEF), United Nations Development Programme (UNDP), and the Department of Environment (DOE), Iran. Key achievements of this project in relation to the lake and some of its satellite wetlands included developing a strategic plan for ecosystem-based management of the wetland which has as its main objectives:

- Raising knowledge of decision-makers and awareness of stakeholders in relation to Lake Uromiyeh and its satellite wetlands;
- Developing capacity among stakeholders (training, participation, provision of equipment); and
- Establishing mechanisms for sustainable management of the wetland (intersectoral committees, basin-wide management plans, laws and regulations, monitoring).

In addition to providing interprovincial coordination for land and water resources development and any other activity with potential effects on the lake, the following institutional arrangements were recently developed at national and provincial levels (CIWP 2008):

- National Executive Committee (NEC) consisting of first Vice President of Iran (chairman); Ministers of Energy, Agriculture Jihad, Interior, Roads, and Housing and Urbanization; Head of Planning and Budget Organization; Head of DOE (Secretariat); and Governors of East and West Azerbaijan and Kurdistan. The NEC is authorized to make the highest level decisions on any issue (particularly development programs) that may affect the lake or any of its satellite wetlands. This committee has organized a technical subcommittee to provide technical assistance to the NEC to decide on the particular threats which are facing or will face the lake; and
- Regional Level Management Council, consisting of the Provincial Governors and head of provincial offices from each of the three provinces involved. West Azerbaijan Environment Conservation Office was assigned as the secretariat for

this Regional Council. This council is capable to decide on any action within each province that has the potential to affect the lake (i.e., land and water resources development). This council in turn has organized three provincial committees to provide technical assistance to the council on different issues relevant to the lake. One of the crucial outcomes has been establishing a water right for the lake and the shares of provinces to provide the necessary water supplies.

In addition to developing a monitoring plan for Lake Uromiyeh, the project also prepared a strategic plan for establishing ecotourism, established demonstration farms for on-farm water management and sustainable agriculture, and undertook environmental research and feasibility studies for establishing alternative sources of livelihood for the local population in Gharagheshalgh hunting-prohibited area.

Threats and Future Challenges

The lake has faced intense pressure since the mid-1990s and is currently in a state of ecological crisis with major impacts on biodiversity and socioeconomic conditions. The water level has continuously decreased and the salt concentration increased. The main reasons for this situation are:

- Reduced precipitation, apparently because of long-term climate change;
- Increased water abstraction from rivers and groundwater resources; and
- Increasing trends for water and land development projects in the related provinces.

Yadegarlou wetland (a Ramsar Site) has been desiccated because of construction of a deep open drainage channel in its close vicinity. However WAECO has started attempts to restore it.

Shorgol wetland (also a Ramsar Site) has been converted into a reservoir for irrigation development, although a new wetland area is being created in its vicinity.

Most of the satellite wetlands are increasingly receiving irrigation return flows instead of fresh river/spring flows. Upstream irrigation developments are responsible for this conversion. As a result they are increasingly becoming eutrophic.

Considering the particular importance of the lake and serious concerns arising from its desiccation, in 2013 the Government of Iran organized a high-level committee (chaired by the former minister of agriculture) with special assignment for rescuing the lake as a priority task. Following several seminars, conferences, and consultation workshops, the committee focused, as a priority action, on managing water consumption particularly for agricultural uses in order to save water for the lake. Considering the nature of the work, the program would be a long-term one with huge budget requirements and organizational preparations. The program has started with large scale on-the-farm demonstration works to educate farmers in the involved provinces regarding different opportunities that could be adopted for optimizing water uses for crop production. At the same time, water resources development programs in the related provinces were subject to a critical revision such that some of the existing planned or even under construction programs for construction of storage dams were suspended, and existing water release plans of functioning dams were also reviewed, so that more water could be released toward the lake. Raising public awareness on the importance of the lake and the impacts arising from its desiccation is another critical component of the program.

References

Conservation of Iranian Wetlands Project (CIWP). Management plan for the Lake Uromiyeh ecosystem. Tehran: Department of Environment; 2008. http://www.wetlandsproject.ir/en/publication/books/lu-mp-en

Hashemi M. Review of the status of water resources in the Lake Uromiyeh Basin, conservation of Iranian wetland project. Tehran: Department of Environment; 2008. http://www.wetlandsproject.ir/MMCD/0DB/2-BS/INV/PROD/bs-inv-prod-lu-en-re-2008.pdf

Lehner B, Döll P. Development and validation of a global database of lakes, reservoirs and wetlands. J Hydrol. 2004;296;1–22. www.worldwildlife.org/pages/global-lakes-and-wetlands-database

Lehner B, Grill G. Global river hydrography and network routing: baseline data and new approaches to study the world's large river systems. Hydrol Process. 2013;27(15):2171–86. www.hydrosheds.org

Lotfi, A. Concise baseline report on Lake Uromiyeh, conservation of Iranian wetland project. Tehran: Department of Environment; 2012. http://www.wetlandsproject.ir/en/publication/books/lake-urmia-concise-baseline-report

Pandam Consultants. Integrated water resources management for the Lake Uromiyeh, module 3: water for ecosystem, Prepared for the Water Research Institute-Ministry of Energy. Tehran; 2005.

Scott DA, editor. A directory of wetlands in the Middle East. Gland/Slimbridge: IUCN/IWRB; 1995. http://www.wetlands.org/?TabId=56&mod=1570&articleType=ArticleView&articleId=1891

Tarbiat Modarres University. Drought risk management of Uromiyeh Lake, 10 volumes in Farsi, study conducted for conservation of Iranian wetlands project. Tehran: Department of Environment; 2012a.

Tarbiat Modarres University. Drought risk management of Lake Urmia Basin (Summary Report in English), Conservation of Iranian Wetlands Project. Tehran: Department of Environment; 2012b. http://www.wetlandsproject.ir/MMCD/0DB/5-IMP/DRM/PROD/imp-drm-prod-lu-pln-en-2012.pdf

Water Resources Research Institute. Integrated water resources management for Lake Uromiyeh Basin, module 1, water resources management, prepared for West Azerbaijan Water Authority. Tehran; 2005.

Yekom Consultants. Management plan for the Lake Uromiyeh ecosystem. Iran Irrigation Project, 4 volumes, prepared for the Ministry of Jihad Agriculture Iran. Tehran; 2002.

Yekom Consultants. Environmental impacts of water resources development on the Lake Uromiyeh, 9 volumes, prepared for West Azerbaijan Water Authority, Uromiyeh; 2004.

Yekom Consultants. Integrated water resources management for the Lake Uromiyeh Basin, module 2- water for food. Prepared for the Water Research Institute- Ministry of Energy. Tehran; 2005.

Shadegan Wetland (Islamic Republic of Iran)

139

Ahmad Lotfi

Contents

Introduction	1676
Location and Geography	1676
Hydrology	1676
Ecological Attributes and Functions	1678
Conservation Status	1681
Ecosystem Services	1681
Recent Developments	1682
Threats and Future Challenges	1682
References	1683

Abstract

Shadegan wetland at the most downstream reach of Jarrahi river is one the largest Iranian Ramsar Sites located in Khuzestan Province in the southwest of Iran. It includes a combination of vast areas of reed beds, open water, mudflats, estuaries, Khur musa bay, isolated small islands and shorelines along the Persian Gulf. Because of such habitat diversity, Shadegan wetland represents a unique wetland along the Persian Gulf and provides globally significant ecological services.

Shadegan wetland hosts large numbers of migratory birds including threatened and rare species. It is also important habitat for fish. A large number of rural people are resident around the wetland and dependent on its resources. Fishery in

---

A. Lotfi (✉)
Conservation of Iranian Wetland Project, Department of Environment, Tehran, Iran

Senior Irrigation Engineers, Member of Board of Directors, Pandam Consulting Engineers, Tehran, Iran
e-mail: ah.lotfi.pandam2@gmail.com

© Springer Science+Business Media B.V., part of Springer Nature 2018
C. M. Finlayson et al. (eds.), *The Wetland Book*,
https://doi.org/10.1007/978-94-007-4001-3_245

the Khur Musa bay and along the shorelines of Persian Gulf is an important source of income to the rural families. Buffalo and cow breeding is also a common industry in all the rural areas around the wetland.

The wetland which was originally fed by fresh flood flows of the Jarrahi River is now receiving major part of the water as irrigation return flows. Two storage dams in the upstream catchment control parts of the floods and several irrigation development projects abstract river flows for crop production.

Keywords
Shadegan · Ramsar site · Protected area · Habitat · Migratory birds

Introduction

Shadegan Wetland in Khuzestan Province of southwestern Iran is one of the largest Iranian Ramsar Sites and an Important Bird Area which hosts large numbers of migratory birds including rare and globally threatened species.

Location and Geography

Shadegan Wetland (537,700 ha) lies in the southwest of Iran in the most downstream reaches of the Jarrahi River catchment at the head of the Persian Gulf. It seems to be a remnant of a much larger marsh, which formerly extended to the Hawr Al Azim on the Iraqi border and is therefore an outlying component of the former Mesopotamian Marshes. The three cities of Shadegan, Abadan, and Mahshahr and several other towns and villages are located inside, around, or in close vicinity to the wetland. The Karoun River flows southwards just a few kilometers from the western boundary of the intertidal flats. Moving southwards, the wetland can broadly be divided into a large freshwater marsh, extensive intertidal flats, the Khur Musa Bay and associated islands, and the coastal fringes of the Persian Gulf (Map 1).

Hydrology

The rather large catchment area of the wetland is located in the semiarid region of southwestern Iran, where the long warm summers fade into short temperate winters. Average annual temperature of the wetland area is 25 °C with a maximum of 51 °C and minimum of -1.4 °C. Average annual precipitation in the catchment area varies between 200 mm in the wetland area in the south to about 800 mm in the northern higher altitudes of the catchment. Pan evaporation in the wetland area is about 3,500 mm/year corresponding to about 2,000 mm/year from the open water surface.

Shadegan Wetland is mainly (90%) supplied by freshwater from the Jarrahi River and by tidal influxes of seawater from the Persian Gulf. The wetland used to be also

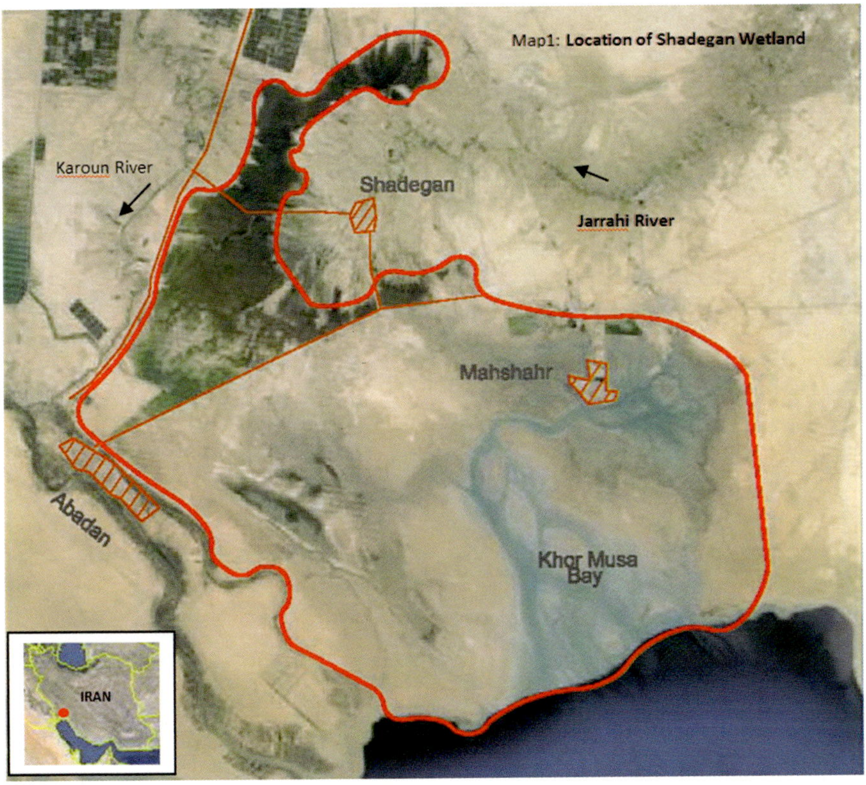

Map 1 Location (*inset map*) and remote sensing image showing the main features of Shadegan Wetland (*red line*) (© Pandam Consulting Engineers, with permission)

recharged by floodwater spills from the Karoun River. However, during recent decades, because of construction of several storage dams and irrigation development projects, these spills have been replaced by drainage flow from several irrigation development projects. Yet on-going irrigation development projects in the Jarrahi River catchment also will affect the water flow regime into the wetland. Due to these developments, the role of summer river and drainage flows in feeding the wetland will overshadow the role of winter fresh water flows from floods of the Jarrahi and Karoun Rivers, evidently with impacts on the wetlands' ecological condition.

According to historic data, before the recent water development projects were carried out, the average inflow from the Jarrahi River into the wetland was estimated to be $2,300 \times 10^6$ m^3/year (mcm/year) ranging from 300 to 6,000 mcm/year (Pandam Consulting Engineers 2002). Recent water developments include construction of two storage dams and irrigation developments of around 50,000 ha which will eventually use additional water totaling about 1,000–1,500 mcm/year (Pandam Consulting Engineers 2002; CIWP 2012).

Drainage flows from sugar cane and fish culture developments just upstream of the wetland (from Karoun River) add up to about 350 mcm/year (CIWP 2012).

The quality of the water in the Jarrahi River just upstream from the wetland ranges from 1.5 to 5 dS/m depending on the status of flow. Salinity of sugar cane drainage water ranges from 15 to 25 dS/m (Pandam Consulting Engineers 2002; CIWP 2012).

Ground water in the area is very shallow (<2.0 m) and with very low quality not suitable for any kind of use, and practically has no contribution in recharging the wetland.

There is very little documented information on the depth and volume of water in the wetland. The depth of water varies widely in different parts of the wetland and ranges from a few cm to a maximum of 3 m in the deeper water ways of the freshwater part of the wetland, with an overall average depth of less than 1 m (Pandam Consulting Engineers 2002). The quality of water in the wetland varies according to the season and availability of water and ranges between 5 and 10 dS/m in the wet years with good inflows and 25–30 dS/m in dry years with little or no flows.

As in other parts of Iran, the wetland is prone to periodic droughts. The last drought event started in the late 1990s and lasted until the mid-2000s. The effect of this drought was that almost the entire fresh water part of the wetland was desiccated. According to historic hydrological data, similar events have happened in the past. The area is also subject to impacts from climate change. Analyses indicate that there is an increasing trend in ambient air temperature and a decreasing trend in annual precipitation (Hemadi et al. 2011).

Ecological Attributes and Functions

Shadegan Wetland is the largest wetland in Iran and among the largest designated Ramsar Sites in the world. It is the largest coastal wetland in the Persian Gulf, and following the recent destruction of the Mesopotamian marshes has taken on particular conservation significance in the Middle East.

Despite several cases of human interventions, the wetland still displays a high degree of naturalness. Little wetland habitat has been lost to date, and much of the area remains in a near-natural condition. The most important modification to the naturalness of the freshwater part of the wetland is a trend in changing the hydrological regime from mainly winter fresh flood flows to summer irrigation return flows. Tidal mudflats are also relatively natural.

Shadegan Wetland is particularly distinguished by its habitat heterogeneity, i.e., extensive freshwater marsh with reed beds and open water areas, intertidal mud flats, several brackish water creeks and estuaries, Khur Musa Bay, offshore islands, and coastal areas in the Persian Gulf (Map 2). Because of this diversity of habitats, the wetland possesses a unique importance among the wetlands of the Persian Gulf.

The wetland supports a diverse flora, with 110 plant species in 17 plant communities identified during a special survey in 2000–2001. It also supports 40 species of mammals, 174 species of birds, 9 species of reptiles (including breeding marine

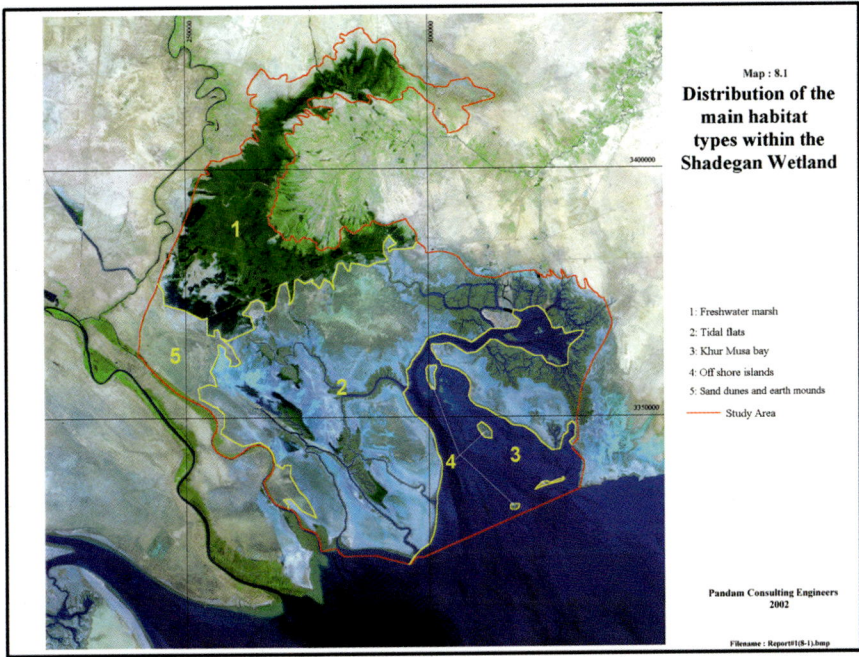

Map 2 Distribution of the main habitat types within Shadegan Wetland (© Pandam Consulting Engineers, with permission)

turtles), 3 species of amphibians, 4 species of shrimp, and over 90 species of fish (KFRC 1996; Pandam Consulting Engineers 2002; CIWP 2012). The birds and fish are the best studied and most important groups.

The freshwater part of the wetland has a unique extensive land cover of *Typha angustifolia* and *Scirpus maritimus* (Map 3). This land cover holds a high habitat value for breeding birds and displays excellent panoramic scenery.

Fishes are a very important component of the Shadegan Wetland, providing food for large populations of piscivorous birds as well as the human population around the wetland. Fishing is an important commercial occupation in the Khur Musa Bay and to lesser extent a sport activity in the fresh water marsh.

Marine, brackish, and freshwater fishes inhabit Shadegan Wetland. Freshwater species mainly inhabit the northern and central parts of the marsh; brackish-water fishes live in the southern part of the marsh where fresh and marine waters mix; and marine fishes inhabit the Khur Musa Bay and the creeks in the intertidal mudflats in the southern parts of the wetland. In winter and spring, the brackish water covers a wider area of the wetland and therefore the distribution of brackish water species increases. These patterns are also masked by extensive movements of marine and brackish species into the freshwater areas for breeding.

Different parts of the wetland including the extensive freshwater marsh in the north as well as creeks and estuaries in the south are important breeding and nursery habitats

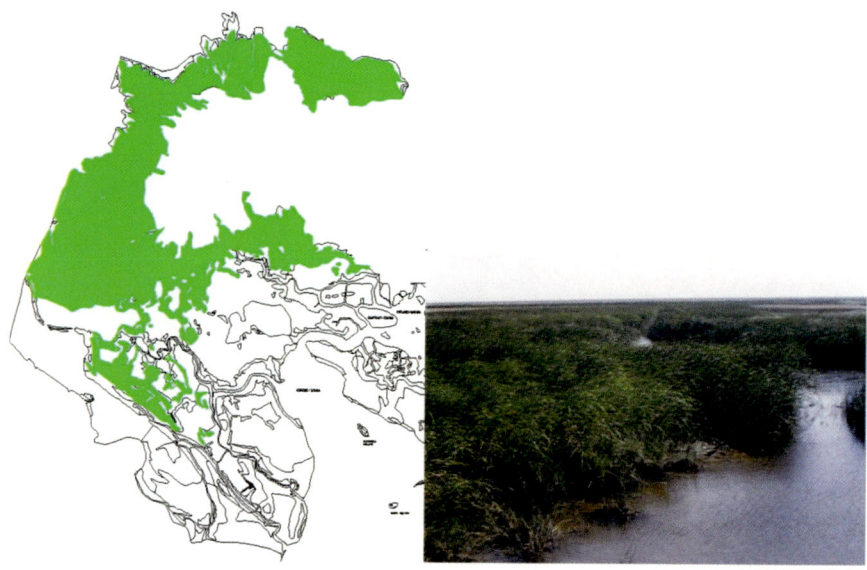

Map 3 Map of Shadegan Wetland indicating areas of *Typha angustifolia* and *Scirpus maritimus* (*green shading*) (Source: Pandam Consulting Engineers 2002)

for many fish species. In the Shadegan Marsh, 36 species of fish have been recorded upstream from the Abadan-Mahshahr highway (KFRC 1996). Seventeen of these are fresh water species, and the rest (21 species) are originally marine. Khur Musa and the creeks in the southern wetland support 45 species of fish (KECO 1995–1996).

Research by Khuzestan Fisheries Research Centre as well as several other case studies has estimated the existing fish resources of the marsh as being in the order of 15,000–25,000 t per year. Assuming the marsh benefits from permanent water recharge, the potential resources of the marsh have been estimated at 81,000 t per year (CIWP 2012; Makvandi 2011).

Forty species of mammals have been reported from Shadegan Wetland (Kiabi 1996). Wild boar *Sus scrofa* and golden jackal *Canis aureus* are abundant and mice (Muridae) are widely distributed throughout the region. Wild cats *Felis catus* and jungle cats *Felis chaus* are protected and occur as low populations within the limits and margins of the wetland with dense vegetation.

Khur Musa provides habitat for hump-backed dolphin *Sousa plumbea*. This species is registered in the data deficient group of the IUCN Red Book.

The wetland is internationally important for its wintering waterbirds (particularly ducks in the freshwater areas and waders in the tidal areas), with almost 55,000 birds counted during a survey in 2000–2001. Numbers of wintering waterbirds appear to have declined by almost one half since the early 1970s. The main losses seem to be among the herbivorous duck species, while increases in fish- and invertebrate-eating species may be indications of eutrophication of the wetland. Important colonies of terns and Ardeidae breed on the islands.

Birds are among the most important vertebrate groups of Shadegan Wetland because of their diversity and abundance. These were indeed the reasons that led to the designation of Shadegan Wetland as a Ramsar Site in 1975.

With the exception of 2000/2001 avifaunal surveys covering the entire wetland over a full year (Pandam Consulting Engineers 2002), our information on the birds of Shadegan Wetland comes from the annual midwinter counts that have been carried out since the mid-1970s in the freshwater part of the wetland. In summary, 32 species of breeding birds have been recorded in the wetland including the islands of Khur Musa Bay; and 13 species of globally threatened birds have been reported, although several not in recent years with white-headed duck *Oxyura leucocephala*, marbled teal *Marmaronetta angustirostris*, and Iraq babbler *Turdoides altirostris* the most frequently recorded.

Although covering only the freshwater part of the wetland, midwinter counts have almost always exceeded the 20,000 bird threshold used to identify wetlands of international importance under the Ramsar Convention. Moreover, 14 species have been recorded with regional populations above the Ramsar 1% threshold for international importance. However, there appears to have been a dramatic decline in the peak counts with several records of counts exceeding 100,000 waterbirds up to the mid-1980s (max of 684,145 in 1975) (Scott 1995) but no counts above 50,000 since 1992. The dramatic declines occurred in dabbling duck species dependent on aquatic macrophytes and seeds for food. This suggests there may have been significant ecological changes in the wetland since the 1970s affecting their food source, possibly an indication of severe eutrophication. In contrast, there have been increases in species which eat invertebrates or fish.

Conservation Status

The management of Shadegan Wetland is under the jurisdiction of Shadegan Environmental Conservation Office of Khuzestan province Department of Environment. However, management of surface water resources of the area is the responsibility of Khuzestan Water and Power Authority of Ministry of Energy.

Part of the northern freshwater marsh and a major part of the tidal mudflat totaling about 400,000 ha is protected under national laws as a wildlife refuge and is designated as a Ramsar Site (1975). Any activity with potential impact on the wetland needs permission from Khuzestan Province Environment Conservation Department. In the protected area, no human activity is admitted.

Ecosystem Services

Shadegan Wetland provides a range of ecosystem services. The expanse of open water area of the lake has significant moderating effects on summer climate to the surrounding areas. The vast area of the wetland absorbs flood flows of the Jarrahi River and provides protection for three important cities in close vicinity of the wetland, i.e., Shadegan, Abadan, and Mahshahr as well as several villages around the wetland. Large volume of sediments carried by the Jarrahi River, particularly during floods, settle within the wetland along with nutrients and chemicals originating from agricultural runoff which are bound in the sediments or assimilated by the expanses of vegetation. Indigenous wetland people use reed stems as source material for covering roofs and constructing shelters and in producing handicrafts such as reed mats and window sun shades. In addition, reeds are harvested by the rural population to feed domestic animals, and the wetland provides a grazing area for large numbers of cows and buffalo.

The importance of Shadegan Wetland as a fishery and in supporting large numbers of mammals and migratory waterbirds has been described above. Fish resources of the freshwater marsh as well as fish and shrimp harvest in Khur Musa are important sources of livelihood for the local population. Spectacular landscapes around the lake provide considerable potential for recreation, bird watching, ecotourism, sport fishing, and hunting.

Recent Developments

A project entitled "Conservation of Iranian Wetlands Project" (CIWP) was undertaken during the last decade with support from the Global Environment Facility (GEF), UNDP, and the Department of Environment (DOE), Iran. Key achievements of this project in relation to Shadegan Wetland included developing a strategic plan for ecosystem-based management of the wetland (CIWP 2008) which has as its main objectives:

- Raising knowledge of decision-makers and awareness of stakeholders in relation to Shadegan Wetland;
- Developing capacity among stakeholders (training, participation, provision of equipment); and
- Establishing mechanisms for sustainable management of the wetland (intersectoral committees, ecosystem approach management plans, laws and regulations, monitoring).

In addition to developing a monitoring plan, local and provincial committees were established for the management and decision-making required for the sustainable conservation of Shadegan Wetland. The project also prepared a strategic plan for establishing ecotourism and plans for establishing a wetland visitor center.

Threats and Future Challenges

The wetland is facing intense pressure from natural events and human activities in both the upstream river catchment and the wetland areas. Periodic droughts can last for several years and the construction of highways across the wetland are planned.

The wetland is also experiencing changes in its hydrological regime as a result of water demands for irrigation. There are changes in the timing of the main pulse of water inflow from fall/winter to spring/summer as the fall/winter floods are impounded behind dams and flows regulated. This is less favorable for migratory birds. Moreover, additional use of freshwater for irrigation developments (1,000–2,000 mcm/year) will reduce the volume of freshwater inflow into the wetland and increase the volume of nutrients and chemicals from agricultural runoff. Water quality concerns are identified with chemically contaminated agricultural runoff and the disposal of municipal wastes around and close to the wetland.

Shadegan has experienced industrial development with the construction of a thermal power plant within the wetland and several transmission lines that cross it. A steel factory is under construction and there are several petrochemical units inside and around the wetland and oil/gas transmission pipelines crossing it.

Encroachment by rural communities on wetland boundaries and the introduction of exotic species of fish originating from fish culture farms along the river and adjacent to the wetland is a pressure on the wetland's ecology. Unsustainable use of the fish resources and overgrazing by the rural population is also concern.

A challenge is the involvement of different organizations with contradictory policies/benefits in the management of the wetland. This has been to some extent resolved with the establishment of an ecosystem-based management plan for the wetland (see above).

References

CIWP. Ecosystem based management plan for the Shadegan wetland. Tehran: Conservation of Iranian Wetlands Project, Department of Environment; 2008. http://www.wetlandsproject.ir/en/publication/books/sw-mp-en

CIWP. Shadegan wetland baseline information. In: Hashemi M, Jamei N, editors. Synthesis report. Tehran: Conservation of Iranian Wetlands Project, Department of Environment; 2012.

Hemadi K, Jamei M, Houseini FZ. Climate change and its effect on agriculture water requirements in Khuzestan plain. J Food Agric Environ. 2011;9(1):624–8. http://world-food.net/wfl/download/journals/2011-issue_1/36(3).pdf

KECO. Limnological surveys of Shadegan Marsh. Ahvaz, Iran: Khuzestan Environment Conservation Office; 1996.

KFRC. Shadegan Marsh master plan. Ahvaz, Iran: Khuzestan Fishery Research Center; 1996.

Kiabi B. Zoology of Shadegan wetland. Ahvaz, Iran: Khuzestan Fishery Research Center; 1996.

Makvandi M. Collection of articles and dissertations on the Shadegan wetland. Ahvaz, Iran: Khuzestan Environment Conservation Office; 2011.

Pandam Consulting Engineers. Environmental management project for Shadegan wetland. Vol 1: the natural environment of the Shadegan wetland ecosystem. Tehran, Iran: Ministry of Jihad Agriculture, Environmental Component, Irrigation Improvement Project (EC-IIP); 2002.

Scott DA, editor. A directory of wetlands in the Middle East. Gland/Slimbridge: IUCN/IWRB; 1995. 560 p. http://www.wetlands.org/?TabId=56&mod=1570&articleType=ArticleView&articleId=1891

Mesopotamian Marshes of Iraq

140

Curtis J. Richardson

Contents

Introduction	1686
Hydrology	1688
Biodiversity	1688
Threats and Future Challenges	1692
Competing Water Issues	1692
References	1694

Abstract

The Mesopotamian marshes of Iraq were once the largest wetland in the Middle East and home to an ancient civilization of Marsh Dwellers know as the Madan. By 2000, after massive drainage by the Iraqi government only 7% remained. This environmental genocide resulted in the near extinction of numerous endemic species of birds and mammals as well as the livelihood of the Marsh Dwellers. Today efforts are underway to restore the hydrology of the marshes, but upstream retention of water by Turkey, Iran and Syria through a series of dams along with internal water reallocations for agriculture and urban use seriously reduce the water available for restoration. Fortunately, Iraq working with international agencies has created marsh restoration plans, protected Ramsar Sites, a National Park, and just recently a World Heritage Site in the marshes, conservation efforts that promise a better future for the marshes and the Madan.

Keywords

Hydrology · Transboundary · Marsh · Iraq · Madan

C. J. Richardson (✉)
Nicholas School of the Environment, Duke University Wetland Center, Durham, NC, USA
e-mail: curtr12@gmail.com; curtr@duke.edu

© Springer Science+Business Media B.V., part of Springer Nature 2018
C. M. Finlayson et al. (eds.), *The Wetland Book*,
https://doi.org/10.1007/978-94-007-4001-3_70

Introduction

Iraq's Mesopotamian marshes (Fig. 1) – considered by many to be the "cradle of western civilization" – have often been referred to as the Garden of Eden (Thesiger 1964; Nicholson and Clark 2002). The word Mesopotamia means "between rivers," a reference to the Tigris and the Euphrates rivers. During the Islamic Age, the lakes and marshes were called Al-Bataih, "the lands covered with torrents" (Al-Ansari et al. 2012). The marshlands are located in southeastern Iraq but also are found across the border in Iran and are located between 31° 01′ N 46° 14′ E and 30° 34′ N 47° 47′ E. These wetlands comprise three main areas, which are the Al-Hammar, the Central (Qurnah), and the Al-Hawizeh Marshes (Fig. 1).

For thousands of years, the marshes functioned essentially as a "River of Grass," as the Tigris and Euphrates, which were fed by the spring snowmelt from the mountains of Turkey and Iran, flooded the southern landscape of Iraq, and *Phragmites australis*, an aquatic grass better known as common reed, dominated the landscape. Excess springwater flow from these rivers was the lifeblood from which the marshes sprang and in turn nourished the early Samarian civilization.

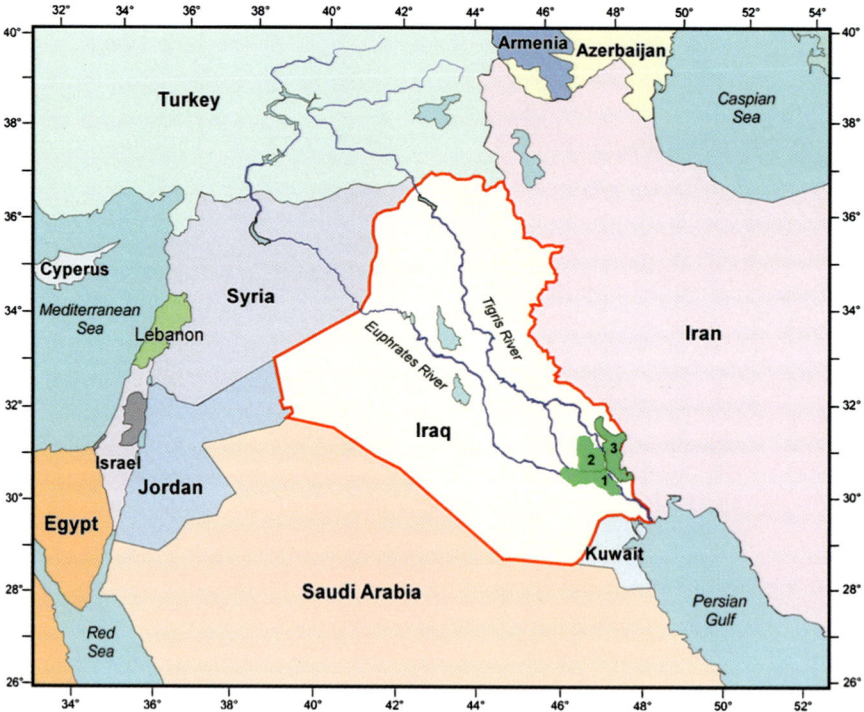

Fig. 1 Regional map of Middle East showing the location of the three main areas of the Mesopotamian Marshes (*1* Al-Hammar, *2* Central, and *3* Al-Hawizeh) southern Iraq (colored *dark green*). The main water sources for the marshes are the Euphrates and Tigris River (Figure by Richardson)

The water in the marshes lays over silt and clay soils where historically water depths were 0.5–2.0 m in the dry and wet seasons, respectively, while reaching 6.0 m depth in permanent lakes (Al-Ansari et al. 2012). These marshes, once the largest wetlands in Southwest Asia and the third largest in the world, once covered more than 15,000 km^2, an area nearly twice the size of the original Everglades (Richardson 2010). However, by the year 2000, less than 10% of the area remained as functional marshes due to drainage and water diversion projects carried out by the Hussein regime during the 1980s and 1990s (Partow 2001; Brasington 2002; Alwash and Alwash 2004; Richardson et al. 2005). In total almost 9,000 km^2 of primary wetlands disappeared between 1973 and 2011 (Zhang and Abed 2013). They reported the most serious losses were in the central marsh, which went from 3,121 km^2 to only 188 km^2 in 2011.

The drainage of these wetlands was also a major problem as they were once home to nearly 500,000 Marsh Arabs or Madan as they are locally know (Young 1977; Coast 2002). In the early 1990s, the Sunni-controlled Baghdad regime crushed an uprising by the Shi'a (the largest Muslim sect in Iraq). Their settlements were raided and destroyed killing thousands of Marsh Arabs; lakes were drained, date palm trees removed, and livestock killed, thereby wrecking the core of the local economy. In the early 2000s, the marsh dweller population living near the marshes was estimated to be 75,000–85,000 with the remaining people scattered in small villages throughout the desert or were refugees in larger cities (DAI 2004). More recent data are hard to obtain due to continued secular violence in the country, but reports continue to say that more and more villages are springing up in the marshes as they are restored (Alwash and Alwash 2004; UNEP 2007; Alwash 2012).

In April 2003, soon after the collapse of the Saddam Hussein regime, farmers and local populations blew up dikes and earthen dams releasing water back into many former marsh areas. Nearly 20% of the 15,000 km^2 of the former drained marshes had been re-flooded by early 2004 (Richardson and Hussain 2006). By late 2006 the Iraq Water Ministry and United Nations Environment Program satellite photos indicated that 58% of the destroyed marshes had standing water. A satellite-based trend analysis of vegetation regrowth from early 2003 until December 2006 indicated that plant cover was increasing at the rate of 800 km^2 per year (UNEP 2007). Importantly, habitat fragmentation (disconnected wetland patches) a commonly cited threat to species extinction and biological diversity (Wiens 1996) was found when a survey was carried out and the re-flooding patterns analyzed (Richardson and Hussain 2006). The lightly vegetated re-flooded areas were much more dispersed compared to the continuous wetland landscape found in 1973, and many former hydrologic connections between marsh patches were still obstructed by dikes and canals (Richardson et al. 2005; Hamdan et al. 2010). Further complicating marsh restoration was the erection of more than 30 dams and several thousand kilometers of dikes in Iraq during the past 40 years, resulting in the holding of most of the water in the central areas of Iraq for agricultural irrigation and city water supplies; concomitant with this was a serious reduction of new sediment accumulation in the marshes (Partow 2001; Nicholson and Clark 2002; Richardson et al. 2005).

Hydrology

Historically, the pulsed flow of water, nutrients, and sediments entering the marshes came in spring melt since rainfall in the region seldom exceeds 10–18 cm year^{-1} (Al-Ansari et al. 2012). If anything, substantial flooding was the norm during this period, with marsh expansion increasing from 15,000 to 20,000 km^2 followed by a decrease in marsh area due to high summer evaporation rates (>245 cm year^{-1}) and temperatures often exceeding 50 °C. It was following the spring melt water period that the Madan planted their rice and barley crops at the marsh edges and used the yearly revitalized marsh soils for their crops. Importantly, the water always continued flowing slowly toward the Persian Gulf, and it was this movement downstream that maintained low salinity concentrations, preventing the buildup of potentially toxic elements like selenium and salts, as found in the totally diked area of the Al-Sanaf (Richardson et al. 2005). However, it has been reported more recently after drainage, water diversions and diking, that water salinity in the Euphrates River south of Baghdad had increased from 1,080 ppm in 1979 to more than 4,500 ppm by 2001, which far exceeds standards for safe drinking and irrigation water for agricultural (Richardson et al. 2005; Rahi and Halihan 2010). High salinity also causes severe problems for the survival of native freshwater marsh plant species and further complicates restoration efforts (Richardson and Hussain 2006; Al-Ansari et al. 2012).

Biodiversity

The Mesopotamian marshes were once celebrated for both their cultural richness and biodiversity. They were the enduring habitat for hundreds of thousands of birds and a flyway for millions more migrating between Siberia and Africa (Maltby 1994; Evans 2002; Richardson and Hussain 2006). Nearly 100 bird species were found in the marshes during the last complete census in the 1970s (Evans 2002). Populations of rare species like the marbled duck *Marmaronetta marmaronetta* (40–60% of the world population) and the Basra reed warbler *Acrocephalus griseldis* (more than 90% of the world population) were declared nearly extinct (Evans 2002) but were again seen in a winter bird survey (Salim et al. 2005). Coastal fisheries in the Persian Gulf needed the marshlands for spawning migrations, and they functioned as nursery grounds for penaeid shrimp *Metapenaeus affinis* and numerous fish species. Fish catches initially decreased significantly after drainage, but re-flooding of nearly 50% of the marshes and reestablishment of some of the lakes have resulted in the return of most native species (Maltby 1994; Richardson and Hussain 2006; Hussain et al. 2009).

To assess recovery in the three historic marsh areas (Central, Al-Hawizeh, and Al-Hammar), Richardson and Hussain (2006) selected four marshes: Al-Hawizeh, the only natural remaining marsh on the Iranian border, eastern Hammar marsh, Abu Zarag (western Central marsh), and Suq Al-Shuyukh (western part of Al-Hammar) (Fig. 1). They monitored ecological indicators of plant and algal productivity and

Fig. 2 Water buffalo feeding on the native reed grass (*Phragmites australis*) in the Al-Hawizeh natural marsh in Iraq which borders Iran (Photo credit: C. J. Richardson © Rights remain with the author)

species recovery for birds and fish, in addition to making an assessment of macroinvertebrate populations for several years following re-flooding. This work was done in conjunction with Iraqi scientists to assess the dominant flora and fauna in the natural Al-Hawizeh as compared with the three marshes re-flooded in 2003. A year after re-flooding, restoration occurred in all three former marshland sites but with varying degrees of success and at different successional rates (Richardson et al. 2005). A 2005 survey from the Al-Hammar and Suq Al-Shuyukh marshes indicated that most macrophyte, macroinvertebrate, fish, and bird species had returned to the restored marshes, although densities were low compared to historical records. Al-Hawizeh, with only six species of macrophytes, had the lowest number of dominant species and was dominated by *Phragmites australis* stands and *Ceratophyllum demersum* in the open water areas, while Al-Hammar had nearly double the number of plant species due to the influence of salt-tolerant species (Fig. 2). Collectively, the restored sites had 15 species of macrophytes, a number close to historic values; however, a number of nondominant species found by Al-Hilli's (1977) extensive survey of the marshes had not been recorded. For example, *Polygonum salicifolium* was a common species in the wetlands before drainage and was an important food source for water buffalo (Fig. 2) that are central to the traditional livelihoods of local inhabitants, but it was not found in the re-flooded wetlands in a 2006 survey (Hamdan et al. 2010). Also, species known to have existed in these marshes like *Nymphaea*, *Nymphoides*, and *Utricularia* spp.,

Table 1 A comparison of ecological diversity indices for fish in the restored marshes of Hammar and Suq Al-Shuyukh compared to the natural Hawizeh (ARDI 2006, in public domain)

Indices	Suq Al-Shuyukh		Al-Hawizeh		East Hammar	
	2005	2006	2005	2006	2005	2006
Diversity	1.62	1.73	1.70	1.74	1.91	1.69
Richness	1.71	1.95	1.43	1.84	1.70	1.79
Evenness	0.61	0.62	0.68	0.64	0.60	0.64

great indicators of restoration success, had not reestablished. Over 100 species of macroinvertebrates were found in the three marshes, and surveys indicated the presence of the key species of Crustacea, Mollusca, and Arthropoda reported in the past (Hussain 1992; Scott 1995), with numbers of individuals per unit area being 1.5 times greater at the open water Suq Al-Shuyukh sites than at the thickly vegetated Al-Hawizeh (Richardson et al. 2005). Suq Al-Shuyukh has the highest number of species and the Al-Hammar the lowest. Macroinvertebrate abundance at all marsh sites was dominated by snails (*Lymnaea* spp.) and coleopterans (beetles).

The number of fish species found was very close to historic records (21 species) in Al-Hammar but was lowest in the Al-Hawizeh. A detailed survey of fish species and their size assessed in conjunction with local fish market data indicate that bunni *Barbus sharpeyi* – the most historically important endemic fish species with the highest commercial value – was present but greatly reduced in size and numbers in all the marshes (Richardson et al. 2005). An introduced carp species (*Carassius* sp.) from Iran comprised 20% of the catch in Suq Al-Shuyukh but up to 46% of summer 2004 catches in Al-Hawizeh. A survey of the fishermen (DAI 2004) indicated that fishing was extremely poor due to the small size of the edible fish. A hearty carnivorous catfish species (*Silurus triostegus*), 40–55 cm in length, comprised up to 60% of the catch but was not eaten by the local Shi'a population due to religious restrictions.

More recent indices that the marshes are continuing to recover include a recent comparison of the richness of aquatic fish and bird counts in Al-Hawizeh, Suq Al-Shuyukh, and Al-Hammar, in 2005 versus 2006 (Hussain et al. 2009). In general, biological indices of fish recovery (diversity, richness, and evenness) increased in value from 2005 to 2006, which indicated an improvement in the restored marshes (Table 1). Diversity was slightly improved in the Al-Hawizeh and Suq Al-Shuyukh marshes. In Al-Hammar, there was a slight decrease in diversity due to the reintroduction of saltwater from the Persian Gulf. Richness improved in all monitored marshes, revealing an increased number of individuals in comparison to the number of species collected. Evenness was also improved, but no species was dominant in the fish assemblage of the three marshes, and species were represented by a balanced number of individuals. The reappearance of merchantable fish to the marshes is essential to the livelihood of the Madan and the reestablishment of successful new villages in the interior of the marsh. The lack of good fishing is thus a key reason that village resettlement in the marshes has occurred in only a few locations like Suq Al-Shuyukh.

Table 2 A comparison of the 2005–2006 bird counts with previously collected data (2004–2005) in the monitored Iraqi southern marshes (ARDI 2006, in public domain)

Marsh	2004–2005		2005–2006	
	Species	Individuals	Species	Individuals
Al-Hawizeh	52	9,399	62	28,331
Suq Al-Shuyukh	37	1,975	53	4,443
East Hammar	29	1,998	53	8,465

Avian surveys in 2003 and 2004 in the three marshes found a total of 56 species compared to historical counts of 81 species (Scott 1995). The natural Al-Hawizeh marsh had the most species (51), nearly matching Richardson and Hussain's (2006) total count of 56. Al-Hammar had the lowest species number, which was nearly half of the total species identified. Summer counts of bird species were low (less than 50 birds per day), except for little egret *Egretta garzetta*), squacco heron *Ardeola ralloides*, and the threatened pygmy cormorant *Phalacrocorax pygmaeus* that was found in Al-Hawizeh; however, winter counts increased dramatically in 2005. Importantly, the endemic Iraq babbler *Turdoides altirostris* and the marbled teal *Marmaronetta angustirostris* were seen in the re-flooded wetlands but in low numbers. One suggestion that marsh recovery was happening was that nearly half of the species were recorded as breeding in the marshes during the summer of 2004 and 2005. Another indication of bird habitat restoration came from the Canadian International Development Agency's (CIDA) more complete survey of 28 marsh areas in the winter of 2005, which recorded 74 bird species, including ten rare and endangered species not seen in over 25 years in the Mesopotamian marshes (Salmin, personnel communication, August 8, 2005). The number of bird species counted and total number of individuals in the marshes in 2006 was also higher than in 2005 (Table 2).

This indicated that the marshes were providing an improved habitat for the bird populations. This information and the number of breeding birds found in the marsh provide evidence for an improvement in future bird populations in the restored marshes. The number of ducks also increased dramatically in Al-Hawizeh in 2006 compared to 2005. Duck counts also increased in the other marshes but not as significantly. A comparison of the number of bird species and individual birds with historical records indicates that in 2004–2005 there was a 65% recovery in species. In the 2005–2006 bird survey, the number of species increased and the recovery index was 88%. In other words, 88% of the historical recorded bird species have returned to the restored marshes, but the total number of individuals is much lower than recorded earlier. These variations in bird counts may in part be due to changes in winter weather conditions as well, which greatly influence southern migrations. The most recent surveys from 2005 to 2011 have greatly increased the information on the avian diversity of the Mesopotamian Marshes in southern Iraq (Salim and Porter 2015). A total of 264 species were identified in the marshes and their adjacent environments. Of the 77 breeding species found, 54 are resident in the marshes though some also have migratory populations (Salim and Porter 2015).

Threats and Future Challenges

Competing Water Issues

Available flow of high-quality freshwater supplies is the key to marsh sustainability and continued restoration of these wetlands. The major threat to the survival of these marshes are a series of major transboundary water projects which include the completion of the massive South-Eastern Anatolian Project (GAP) in Turkey, which when completed will create 22 dams to supply irrigation water to 1.7 million ha of agriculture land, and Syria's Tabqa dam projects that will supply water to 345,000 ha of irrigated land. In addition, the dike recently built by Iran to cut the Iranian water supply to the Iraqi portion of the Al-Hawizeh (Fig. 3) is now fully functioning (Richardson and Hussain 2006; UNEP 2007; Alwash 2012). The water from the Iranian project reportedly will be used to support increased agriculture and livestock production in the Karkheh basin in Iran (Ahmad et al. 2009) but at the expense of marsh survival. The Ataturk Dam built in 1998 can store more than the 30.7 billion cubic meters (BCM) of water, which flows from the Euphrates annually from Turkey into Iraq, and can almost alone dry up the Euphrates (Partow 2001).

Anticipated future demands on water for agriculture and urban use are huge, with Iraqi water needs estimated to be close to 95 BCM by 2020 (ARDI 2006). It is also estimated that to restore 10,000 km^2 of marshes will require 20–30 BCM of water or

Fig. 3 Drained portion of the Al-Hawizeh marsh cut off from river water flow due to an Iranian dike displays desert-type vegetation (Photo credit: C. J. Richardson © Rights remain with the author)

nearly 50% of Iraq's available water after the completion of the water projects and dams in Turkey and Syria (Partow 2001; ARDI 2006). These estimates may be high but suggest that there will not be enough water to meet the expected needs for Iraq's agriculture and growing population, and thus the marshes will be in direct competition for the same water resource. This will be heightened in drought years like that experienced during 2009. However, some of the water released from agricultural lands may be adequate for use in the marshes, but elevated salinity and long-term nutrient effects have not been analyzed. Another issue related to re-flooding is that re-flooding alone does not equal wetland restoration. While it is true that the addition of water is critical to marsh restoration, it is also true that the restoration of wetland functions requires proper water hydroperiod (period of time water is at or near the surface), hydropattern (distribution of water), and water quality. These conditions are complex in nature, and early successes can often result in major failures after a few years or limited restoration of important ecosystem functions (Zedler and Calloway 1999).

Competition for water among cities, agriculture, and the marshes indicates serious shortages will exist, especially in dry years. This suggests that the amount of water available for the marshes may be severely restricted in some years, and thus water should be directed only into former marshes with the most potential for maintenance of natural existing areas (e.g., Al-Hawizeh) or former marshes with the most potential for functional wetland ecosystems restoration like Suq Al-Shuyukh. Other sites need to be determined by ecological analyses and soils assessments to prevent the release of precious water into areas with lower restoration potential due to soil salinity or pollution. Another problem will be the release of water into areas that will become partially restored, but then water is restricted in drought years due to shortages as occurred in 2009. This alternating wet and dry set of conditions will result in the destruction of the soil structure and a severe loss of biota. To prevent this situation, minimum yearly allocations of water should be made to the most viable former marsh areas. In Iraq, the Center for Restoration of Iraqi Marshes (CRIM) will need to work closely with all ministries, especially the water and agriculture ministries, to maintain future water supplies for the marshes. The wildcard in this plan is the Ministry of Oil, which in the past had not fully participated in the marsh restoration program, probably because vast quantities of oil exist under former marsh areas in southern Iraq.

Water Restoration Guidelines

The restoration of the Mesopotamian marshes of southern Iraq is now underway. Release of water to former wetland areas has resulted in the return of native plants and animals, including rare and endangered species. Plans are now underway by the Eden Again Project in conjunction with the Iraqi government (Alwash and Alwash 2004; Alwash 2012) to try to restore as much as 80% of the marshes. Al-Ansari et al. (2012) report that 75% of the area covered by the marshes can be restored if 3,263, 5,495, and 4,128 $\times 10^6$ m^3 of water is supplied for the Al-Hammar, Al-Hawizeh, and Central marshes, respectively. Some say even this goal is way too high given the multiple demands on the water supply and the lower revised volume of remaining water projected to help restore the marshes. Nevertheless, the current rate of recovery is remarkable, considering that re-flooding occurred only about 10 years ago. While some areas are experiencing reduced recovery due to salinity and toxicity problems, many locations seem to be recovering well. The major unknowns are (a) whether the Marsh Arab culture can ever become established again in the restored marshes in any significant way; (b) how Iraq's multiple use issues and competition for water with Turkey, Syria, and Iran will affect the future water supplies needed for marsh restoration; and (c) whether or not landscape connectivity of the marshes can be reestablished to maintain species diversity. What is evident is that there is not a sufficient supply of water to fully restore all the marshes, and thus a series of marshes with connected habitats of adequate size to maintain a functioning wetland landscape needs to be established. Clearly, the long-term future of the former "Garden of Eden" depends on the willingness of the Iraqi government to commit sufficient water for marsh restoration and sustain vital areas designated as Ramsar Sites like the Al-Hawizeh and the recently (2013) dedicated Mesopotamia Marshland National Park, which is a unique wetlands complex rich in wildlife located in southern Iraq, North of the Euphrates River and West of the Tigris and Glory River. In 2016 Iraq put forward areas for consideration as a World Heritage Site and they were approved in July of 2016.

References

Ahmad MD, Islam A, Masih I, Muthuwatta L, Karimi P, Turral H. Mapping basin-level water productivity using remote sensing and secondary data in the Karkheh River Basin, Iran. Water Int. 2009;34:119–33.

Al-Ansari N, Knutsson S, Ali AA. Restoring the Garden of Eden, Iraq. J Earth Sci Geotech Eng. 2012;2:53–88.

Al-Hilli MR. Studies on the plant ecology of the Ahwar region in southern Iraq [dissertation]. Cairo: University of Cairo; 1977.

Alwash S. Eden again: hope in the marshes of Iraq. Huntington Beach: Tablet House; 2012.

Alwash A, Alwash S. Eden again: restoring Iraq's Mesopotamian marshes. Natl Wetl Newsl. 2004;26:1–15.

ARDI (Agriculture Reconstruction and Development Program for Iraq). Final report. Washington, DC: United States Agency for International Development; 2006.

Brasington J. Monitoring marshland degradation using multispectral remote sensed imagery. In: Nicholson E, Clark P, editors. The Iraqi marshlands: a human and environmental study. London: Politico's; 2002. p. 151–68.

Coast E. Demography of the Marsh Arabs. In: Nicholson E, Clark P, editors. The Iraqi marshlands: a human and environmental study. London: Politico's; 2002. p. 19–35.

DAI (Development Alternatives, Inc.). Iraq marshlands restoration program action plan. Washington, DC: United States Agency for International Development; 2004.

Evans MI. The ecosystem. In: Nicholson E, Clark P, editors. The Iraqi marshlands: a human and environmental study. London: Politico's; 2002. p. 201–19.

Hamdan MA, Asad T, Jassan FM, Warner BG, Douabul A, Al-Hilli MRA, Alwan AA. Vegetation response to re-flooding in the Mesopotamian wetlands, southern Iraq. Wetlands. 2010;30:177–88.

Hussain NA, editor. Ahwar of Iraq: an environmental approach. Baghdad: Marine Science Center, Basra University; 1992.

Hussain NA, Mohamed AM, Al-Noo SS, Mutlak FM, Abed IM, Coad BW. Structure and ecological indices of fish assemblages in the recently restored Al-Hammar Marsh, southern Iraq. BioRisk. 2009;3:173–86.

Maltby E, editor. An environmental and ecological study of the marshlands of Mesopotamia. Draft consultative bulletin. London: Wetland Ecosystems Research Group, University of Exeter; 1994.

Nicholson E, Clark P, editors. The Iraqi marshlands: a human and environmental study. London: Politico's; 2002.

Partow H. The Mesopotamian marshlands: demise of an ecosystem. Nairobi: United Nations Environment Programme, Division of Early Warning and Assessment; 2001. 46 p. Report No. UNEP/DEWA/TR.01–3.

Rahi KA, Halihan T. Changes in the salinity of the Euphrates River system in Iraq. Reg Environ Chang. 2010;10:27–35.

Richardson CJ, Hussain NA. Restoring the Garden of Eden: an ecological assessment of the marshes of Iraq. Bioscience. 2006;56:477–89.

Richardson CJ, Reiss P, Hussain NA, Alwash AJ, Douglas JP. The restoration potential of the Mesopotamian marshes of Iraq. Science. 2005;307:1307–11.

Salim MA, Porter RF. The ornithological importance of the southern marshes of Iraq. Marsh Bull. 2015;10(1):1–24.

Salim MA, DouAbul A, Porter RF, Rubec CD. Birds: a key element of a biodiversity survey in the marshes of Iraq. Paper presented at: The 9th annual meeting of the Ecological Society of America, special session 9: restoration of Mesopotamian marshes of Iraq; 9 Aug 2005; Montreal.

Scott D, editor. A directory of wetlands in the Middle East. Gland: IUCN/IWRB; 1995.

Thesiger W. The Marsh Arabs. London: Penguin; 1964.

UNEP (United Nations Environment Programme). UNEP project to help manage and restore the Iraqi Marshlands [internet]. 2007. Available from: http://marshlands.unep.or.jp

Wiens J. Wildlife in patchy environments: metapopulations, mosaics, and management. In: McCullough DR, editor. Metapopulations and wildlife conservation. Washington, DC: Island Press; 1996. p. 53–84.

Young G. Return to the marshes: life with the Marsh Arabs of Iraq. London: Collins; 1977.

Zedler JB, Calloway JC. Tracking wetland restoration: do mitigation sites follow desired trajectories. Restor Ecol. 1999;7:69–73.

Zhang H, Abed FH. Dynamics of land use/cover change in Iraq marshlands using remote sensing techniques. Environ Eng. 2013;12:1825–8.

Indus River Basin Wetlands

141

Rab Nawaz, Ali Dehlavi, and Nadia Bajwa

Contents

Introduction	1698
Hydrology	1699
Wetland Ecosystems	1699
Biodiversity and Conservation Status	1700
Threats and Future Challenges	1701
References	1702

Abstract

The Indus Basin extends over 1,120,000 km^2 in total, straddling four countries: Pakistan (47% of basin), India (39%), China (8%), and Afghanistan (6%). Among transboundary basins, the Indus Basin is notable for its sizable population of about 250 million in 2013. This heightens the importance of its wetlands to natural resource-dependent communities. Wetlands of the Indus Basin, spanning the Tibetan Himalaya to the Arabian Sea, may be categorized as upland (alpine and piedmont), midland (floodplain, terrace, and valley), and low-lying (lacustrine and riverine) wetlands. The wetlands run alongside the Indus River and its numerous tributaries, 27 of which lie above Guddu Barrage. They include significant natural and man-made wetlands supplying subsistence, commercial, and ecological services to human settlements. In the Pakistan portion of the Indus Basin alone, there exist about 225 nationally significant wetlands, of which 15 are Ramsar Sites. The Upper Indus Basin has two sub-basins divided by the Indus River, the Kohat sub-basin to the west and the Potwar sub-basin to the east. Similarly, the Lower Indus Basin comprises the central and southern sub-basins. A third of the Upper Indus Basin lies above 5,000 m a.s.l. Here, the Indus River

R. Nawaz · A. Dehlavi (✉) · N. Bajwa
Indus Ecoregion, WWF, Karachi, Pakistan
e-mail: rnawaz@wwf.org.pk; adehlavi@wwf.org.pk

© Springer Science+Business Media B.V., part of Springer Nature 2018
C. M. Finlayson et al. (eds.), *The Wetland Book*,
https://doi.org/10.1007/978-94-007-4001-3_219

has its source in the Tibetan Himalaya, before it crosses alluvial plains and reaches its delta on the Arabian Sea. The Basin's hydrology relies on snow, glacial melt, rainwater, and runoff. Its estimated annual flow is 207 billion m^3. The basin supplies a number of anthropogenic ecosystem services, among the most important of which are water supply to large cities, power generation, and food production. However, agricultural ecosystem functioning for food supply in the basin is threatened by rapid population growth, groundwater overexploitation, waterlogging, soil salinization, and inefficiencies in water transport and storage. Irrigated agriculture is estimated to account for 53% of the Indus Basin's total evapotranspiration.

Keywords

Wetlands · Pakistan · Indus River · Ecosystem services

Introduction

The Indus Basin extends over 1,120,000 km^2 in total (Fig. 1), straddling four countries between 24°38′–37°03′ N and 66°18′–82°28′ E: Pakistan (47% of basin), India (39%), China (8%), and Afghanistan (6%) (Hartman and Andresky 2013). Among transboundary basins, the Indus Basin is notable for its sizable population of about 250 million in 2013. This only heightens the value of its wetlands to natural resource-dependent communities.

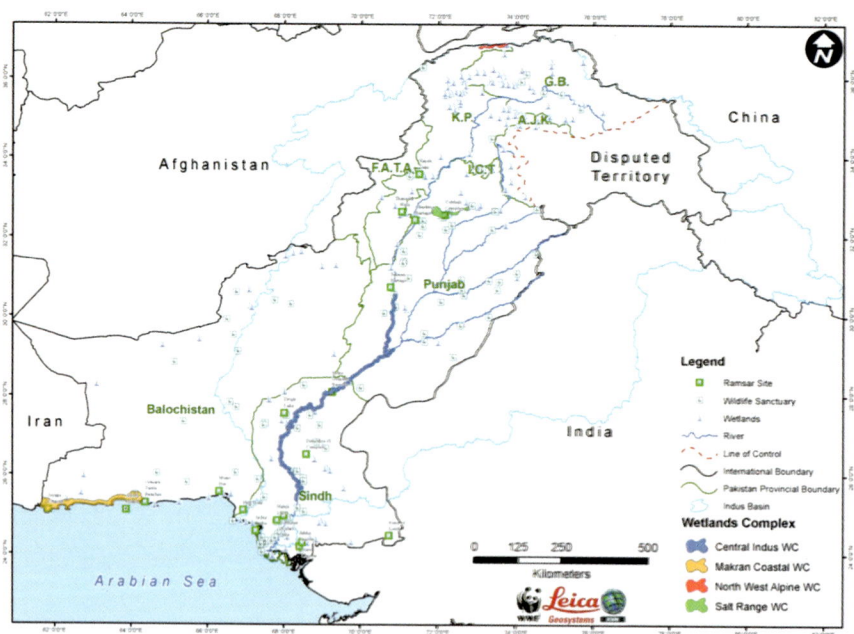

Fig. 1 The Indus river basin. Copyright WWF Pakistan, reproduced with permission

Wetlands of the Indus Basin run alongside the Indus River and its numerous tributaries, 27 of which lie above Guddu Barrage. They include significant natural and man-made wetlands supplying subsistence, commercial, and ecological services to human settlements. In the Pakistan portion of the Indus Basin alone, there exist about 225 nationally significant wetlands, of which 15 are Ramsar Sites (PWP 2009).

The Upper Indus Basin has two subbasins divided by the Indus River, the Kohat subbasin to the west, and the Potwar subbasin to the east. Similarly, the Lower Indus Basin comprises the central and southern subbasins (Ghouri and Kemal 1999). A third of the Upper Indus Basin lies above 5,000 masl. Here, the Indus River has its source in the Tibetan Himalaya, before it crosses alluvial plains and reaches its delta on the Arabian Sea. The Indus River's major left bank tributaries include the Jhelum, Chenab, Sutlej, Ravi, and Beas. The waters of the last three are allocated to India under the 1960 Indus Water Treaty.

Hydrology

The Indus Basin's hydrology relies on snow, glacial melt, rainwater, and runoff. The Indus River's headwaters emerge in the mountains and adjacent foothills of the Himalayan, Hindu-Kush, and Karakoram ranges. With the exception of a steep slope in Baltistan, the lower Indus is flat throughout its length and breadth and falls naturally at 1 ft per mile (Roberts 1999). Its estimated annual flow is 207 billion m^3, approximately twice that of the Nile River (Babel and Wahid 2008).

As the climate is arid and semiarid, the low reliance on rainfed agriculture has resulted in overexploitation of groundwater resources in the order of about 30 km^3 in 2007 (Karimi et al. 2013). The basin houses one of the world's largest contiguous irrigation schemes: Pakistan's 160,000 km^2 Indus Basin irrigation scheme, built by the British in the mid-nineteenth century and expanded in the 1960s and 1970s. In the present day, including both India and Pakistan's agricultural lands, as much as 22% of the total area of the basin is irrigated (Karimi et al. 2013).

The Indus River's discharge varies considerably over the calendar year, with a minimum expected in December-February, increased flows from March-June, and peak annual flows in July-December. Areas below the Sargodha High, which divides the Upper and Lower Indus Basin, are less reliable in terms of discharge availability. A 2005 Government of Pakistan report estimates a daily 5,000 cubic feet per second requirement below Kotri Barrage in the southern subbasin (Government of Pakistan 2005). However, the Indus Basin's deltaic ecosystem, host to significant mangrove and fish species, needs detailed investigation of species-specific hydrological requirements, e.g., spawning area inundation, flow velocity, sedimentary load, and water depth.

Wetland Ecosystems

Wetlands of the Indus Basin, spanning the Tibetan Himalaya to the Arabian Sea, following a geomorphic perspective and the Cowardin classification (Cowardin 1979), may be categorized as upland (alpine and piedmont), midland (floodplain, terrace, and valley), and low-lying (lacustrine and riverine) wetlands. There are 225 nationally significant wetlands in the Pakistan portion of the Indus Basin, including 11 Ramsar Sites, three Ramsar Sites in the Indian portion, and at least one wetland-dependent important bird area in the Afghan portion (WRI 2013).

Wetland types occurring among the upland wetlands in China, Afghanistan, India, and Pakistan include highland marshes, alpine lakes, glacier lakes, and midland lakes. Species common to the four riparian states are plain mountain finch *Leucosticte nemoricola*, white-winged redstart *Phoenicurus erythrogastrus*, little stoat *Mustela erminea*, and golden marmot *Marmota caudata*. Recharge to these marshes and lakes is restricted to the rainy seasons or summer snowmelt and thawing of permafrost (PWP 2012). The conservation management of both piedmont and alpine wetlands is highly onerous owing to low groundwater reserves, limited resources for water replenishment, and increased water loss due to climate change.

As concerns midland wetlands, India and Pakistan boast a large range of wetlands including desert lakes, lowland lakes, fresh water seepage lakes, oxbow lakes, riverine marshes, rice paddies, deep/shallow water dams, distributing link canals, and irrigation barrages. The cup-shaped catchment area at Soan Valley of the Upper Indus Basin hosts Uccali, Khabbeki, and Jahlar lakes that have been designated as a Wetland of International Importance under the Ramsar Convention (2010). Migratory and endemic species occur here, such as purple swamphen *Porphyrio porphyrio* and fat-tailed gecko *Hemitheconyx caudicinctus*, respectively. Swamps may be found in spatially expansive floodplains of the Indus and its tributaries, replenished by river water via overland or overbank flow. Keenjhar Lake (14,000 ha), is notable as a Ramsar Site, providing domestic/commercial water supply to Karachi's population of 20 million (2013), of which about 1 million are connected to the reticulation system (Dehlavi et al 2010).

In the categories of lacustrine and riverine wetlands, we find saline seepage marshes, estuaries and bays with/without mangroves, and the delta of the Indus which covers 30,000 km^2. The catchment of the Indus Delta is rich in biodiversity, including the world's largest arid climate mangrove forest stands.

Biodiversity and Conservation Status

As the basin corresponds to 65% of Pakistan's total land, but only 14% of India, 11% of Afghanistan, and 1% of China, the Indus Basin hosts a majority of Pakistan's total biodiversity (Hartman and Andresky 2013). This includes the following array of species (endemic species bracketed): 195 mammal species (6), 22 amphibians (9), 198 freshwater species (29), 5,700 species of flowering plants (400), over 5,000 species of invertebrates including insects, and 668 bird species including 25 endangered species (Government of Pakistan 2009). The low number of amphibians is unsurprising owing to the Indus Basin's arid and semiarid climate.

The Indus Basin's wetland ecology is divided across the Northern Alpine wetland complex; two midland wetland complexes, the Central Indus and Salt Range; and, at the basin's mouth, the fan-shaped Indus Delta. As much as 80% of flowering plant species in the Indus Basin are located in the Northern Alpine wetland complex. Naturally occurring grasses in the alpine pastures and meadows are *Festuca*, *Poa*, *Lolium*, *Eragrostis*, *Danthonia*, and *Phleum* species. There are also *Primula*, *Aremons*, *Fritillaria*, and *Gentiana* species, among forbs (Government of Pakistan 2009).

The Salt Range and Potwar subbasin are characterized by a cropland/natural vegetation mosaic which also narrowly flanks the Indus up to its delta. In the Indian portion of this region, three of the Indus Basin's 15 Ramsar Sites are located, each of which report invasive plant species. The other twelve Ramsar Sites span the length of the Indus River, with four clustered in the southernmost part of the basin. The large mammals include the endemic and endangered Indus river dolphin *Platanista gangetica minor* which is the second most endangered obligate freshwater dolphin species after the Yangtze River Dolphin (Akbar et al. 2013). The Indus river dolphin's population was estimated to be 1,452 in 2011 (WWF – Pakistan 2011).

Barren land is scattered in the Basin toward Sindh and the Himalayan, Hindu-Kush, and Karakoram ranges. The length of the Indus' right bank is mostly covered by shrubland. The Indus Delta hosts 147 fish species, of which 22 are endemic. *Avicennia Marina* is the principal mangrove species in the Indus Delta. Among marine fisheries resources, the Indus Basin has about 350 species, of which 240 are demersal fish, 50 small pelagics, 10 medium-sized pelagics, and 18 large pelagics, and there are 21 species of shrimp, twelve species of squid/cuttlefish/octopus, and five species of lobsters (Government of Pakistan 2009).

Threats and Future Challenges

The Indus Basin supplies a number of anthropogenic ecosystem services, among the most important of which are water supply to large cities, power generation, and food production. Its wetlands also help micro-climate regulation, groundwater recharge, nutrient cycling, bioremediation, and supply foraging, staging, and breeding grounds for different wildlife species.

Agricultural ecosystem functioning for food supply in the Indus Basin is threatened by rapid population growth, groundwater overexploitation, waterlogging, soil salinization, and inefficiencies in water transport and storage. Irrigated agriculture is estimated to account for 53% of the Indus Basin's total evapotranspiration, the majority of which is non-beneficial depletion in the form of excessive soil evaporation (Karimi et al. 2013). In the absence of long-term planning that incorporates expected variability in precipitation, temperatures (including night-time temperatures that especially affect cereal crops), and other climatic changes, it is anticipated that a high level of benefits to public agricultural investment will be foregone.

Wetland conservation and Indus River conservation work is in its infancy. Transboundary management schemes stand to regulate water usage for multiple competing uses. Among others, such schemes are needed if the Basin is meeting growing human needs such as national agricultural production, subsistence agriculture, domestic/ commercial water supply, power generation, and fisheries in the coming decades.

Ecological services provided by the wetlands of the Indus Ecoregion include nonuse values such as option use, bequest values, and existence values. Through these values, people seek to preserve biodiversity so that it may be passed on to future generations, for possible future recreational use, or simply for reasons of conscience. Indus Delta's biodiversity depends heavily on inflows of freshwater and sediments transported to flush its mangroves and wider marine ecology. In this regard, barrages and dam construction, in the presence of inappropriate fish ladders, have had the effect of shortening the length of riverine fish habitat. For example, the palla *Tenualosa ilisha* has had its riverine range cut from 1,200 to 300 miles (1,931–483 km) (Ahmad 1999). River regulation also presents challenges to the ecological requirements for species other than fish upon which the Delta's poorest coastal fishers and farmers are heavily dependent.

References

Ahmad MF. The wildlife of the Indus river and allied a. In: Meadows A, Meadows P, editors. The Indus river: biodiversity, resources, humankind. Proceedings of a symposium held at the Linnean Society, Burlington House, London; 13th–15th July 1994; Oxford: Oxford University Press; 1999. p. 4–11.

Akbar G, Arshad M, Chaudhry M, Pirzada S. Taunsa Barrage wildlife sanctuary: biodiversity profile. Lahore: WWF – Pakistan; 2013. p. 10–30.

Babel MS, Wahid SM. Freshwater under threat: South Asia vulnerability assessment of freshwater resources to environmental change – Ganges-Brahmaputra-Meghna River Basin, Helmand River Basin, and Indus River Basin Asian Institute of Technology and the United Nations Environment Programme; Nairobi: Kenya; 2008.

Cowardin LM. Classification of wetlands and deepwater habitats of the US. Washington, DC: Diane Publishing; 1979.

Dehlavi A, Groom B, Khan BN, Shahab A. In: Bennett J, Birol E, editors. Non-use values of ecosystems dependent on the Indus River, Pakistan: a spatially explicit, multi-ecosystem choice experiment in choice experiments in developing countries: implementation, challenges and policy Implications. Washington, DC: Edward Elgar; 2010.

Ghouri SS, Kemal A. Oil and gas resources in the Indus Basin: history and present status. In: Meadows A, Meadows P, editors. The Indus River: biodiversity, resources, humankind. Proceedings of a symposium held at the Linnean Society, Burlington House, London; 13th–15th July 1994; Oxford: Oxford University Press; 1999. p. 116–7.

Government of Pakistan, Ministry of Water and Power, Federal Flood Commission. Minimum water escapage needs downstream Kotri barrage to check sea water intrusion and environmental concerns – report of international panel of experts; November 2005.

Government of Pakistan, Ministry of Environment. Pakistan: fourth national report [to the convention on biological diversity]. Islamabad; 2009.

Hartmann H, Andresky L. Flooding in the Indus River Basin – a spatiotemporal analysis of precipitation records. Glob Planet Chang. 2013;107:25–35.

Karimi P, Bastiaanssen WGM, Molden D, Cheema MJM. Basin-wide water accounting based on remote sensing data: an application for the Indus Basin. J Hydrol Earth Syst Sci. 2013;17: 2473–86.

Pakistan Wetlands Programme (PWP) and Ministry of Environment, Government of Pakistan. Wetland policy: final draft. Islamabad: World Wide Fund for Nature - Pakistan; 2009.

Pakistan Wetlands Programme (PWP) and Ministry of Environment, Government of Pakistan. Four wetlands complex promo; 2012. http://pakistanwetlands.org/interavtive_app.php. Date accessed 5 Oct 2013.

Ramsar Convention Secretariat. Designating Ramsar sites: strategic framework and guidelines for the future development of the list of wetlands of international importance, Ramsar handbooks for the wise use of wetlands, vol. 17. 4th ed. Gland: Ramsar Convention Secretariat; 2010.

Roberts TJ. A Pictorial view of the Indus River and man's impact on its role and resources. In: Meadows A, Meadows P, editors. The Indus River: biodiversity, resources, humankind. Proceedings of a symposium held at the Linnean Society, Burlington House, London; 13th–15th July 1994; Oxford: Oxford University Press; 1999. p. 92–3.

World Resource Institute (WRI). Watersheds of the World CD. The World Conservation Union (IUCN), the International Water Management Institute (IWMI), the Ramsar Convention Bureau, and the World Resources Institute. http://multimedia.wri.org/watersheds_2003/as12.html. Date accessed 31 Oct 2013.

World Wide Fund for Nature – Pakistan (WWF – Pakistan). Dolphin survey. Lahore; 2011.

Wular Lake, Kashmir

142

Ritesh Kumar

Contents

Introduction 1706
Catchments and Hydrological Regimes 1706
Biodiversity and Ecosystem Services 1708
Threats 1708
Conservation Status and Management 1710
References 1710

Abstract

Lake Wular, the largest wetland of Kashmir Valley, is its natural flood buffer, and a rich source of fish and aquatic vegetation, along with being an important habitat for migratory waterbirds in Central Asian Flyway. Over the last century, the wetland has been subject to series of interventions for reclamation of shorelines and marshes for flood control, agriculture and willow plantation. Located at the trough of the valley, the wetland becomes a receptacle of waste generated in upstream towns leading to severe degradation in water quality. More recently, the spread of the invasive alligator weed *Alternanthera philoxeroides* has been a cause of concern. There has been a significant decline in wetland resources, as well as water regime moderation capability ultimately making the valley region vulnerable to floods and droughts. An integrated management plan for the wetland was formulated in 2007 and has been in implementation since 2011 through Wular Conservation and Management Authority.

Keywords

Flood buffer · Willow · Kashmir valley · Embankments · Alligator weed

R. Kumar (✉)
Wetlands International South Asia, New Delhi, India
e-mail: ritesh.kumar@wi-sa.org

© Springer Science+Business Media B.V., part of Springer Nature 2018
C. M. Finlayson et al. (eds.), *The Wetland Book*,
https://doi.org/10.1007/978-94-007-4001-3_250

Introduction

Located 34 km northwest of Srinagar City at an elevation of 1,530 masl, Wular is the largest wetland of the Kashmir Valley. The elliptical-shaped waterbody with a length and breadth of 16 km and 7.6 km, respectively, has an area of 160 km^2 with 18 km^2 as associated marshes. Wular is surrounded by high mountains on the northeast and northwest, which drain their runoff into the wetland through various streams (known locally as *nallahs*), the most prominent being Erin and Madhumati (Fig. 1). Low-lying areas surrounding the east, south, and west margins are cultivated mainly with paddy. In mid-fifteenth century, the Kashmiri Sultan, Zain-ul-Abidin, is reputed to have constructed the artificial island of Zaina Lank in middle of the wetland as a storm refuge for boats, for Wular often witnessed strong gales from the mountains of Erin and Bandipora.

Catchments and Hydrological Regimes

Wular Lake's hydrological regime is primarily linked to the River Jhelum and its tributaries. The River Jhelum, a tributary of the River Indus, enters Wular in a braided form at Banyari and flows downstream through Sopore. The basin covers an area of 33,300 km^2 forming a bowl-shaped elongated depression between the Great Himalayas in the northeast and the Pir Panjal ranges in the southwest. The highest mountain peaks enclosing the basin have an elevation of more than 5,300 masl on the Great Himalayan side and more than 5,500 masl on the Pir Panjal side. Of the total basin, an area of 12,777 km^2 drains directly into Wular.

Forests in the Wular Lake catchment cover an area of over 5,400 km^2 within the altitudinal range of 1650–3500 masl, with distinct species changes along an altitudinal gradient. Deodar *Cedrus deodara* and kail *Pinus wallichiana* forests are located within 1650–2600 masl, followed by the fir (Abies) forests found within 2600–3500 masl. The tree line on the upper fringes of southern and south western forests is dotted with alpine pastures (locally called *margs*), significant being Tangmarg, Gulmarg, Khilanmarg, and Sonmarg. During the spring when the snow melts, flowers of all colors appear in the pastures creating a stunning panorama. The rest of the catchment comprises the vast plains of the Kashmir valley and its four side valleys, namely, the Lolab, Lidder, Sind and Kishanganga. The Kashmir Valley has an area of 4,865 km^2 with altitudes ranging between 1400 and 1650 masl and is the demographic and economic hub of the basin, inhabited by more than 85% of its total population. Agriculture and horticulture account for 38% of the basin area. Rice is the primary food crop of the basin and accounts for 35% of the gross cultivated area. Area under orchards accounts for 18% of the gross cropped area and is the mainstay of the economy of the state. There are 31 settlements located around the wetland which derive livelihoods from aquatic vegetation and fisheries. The alpine pastures are dotted by Gujjar and Bakarwal communities, nomadic tribes raising livestock for sustenance.

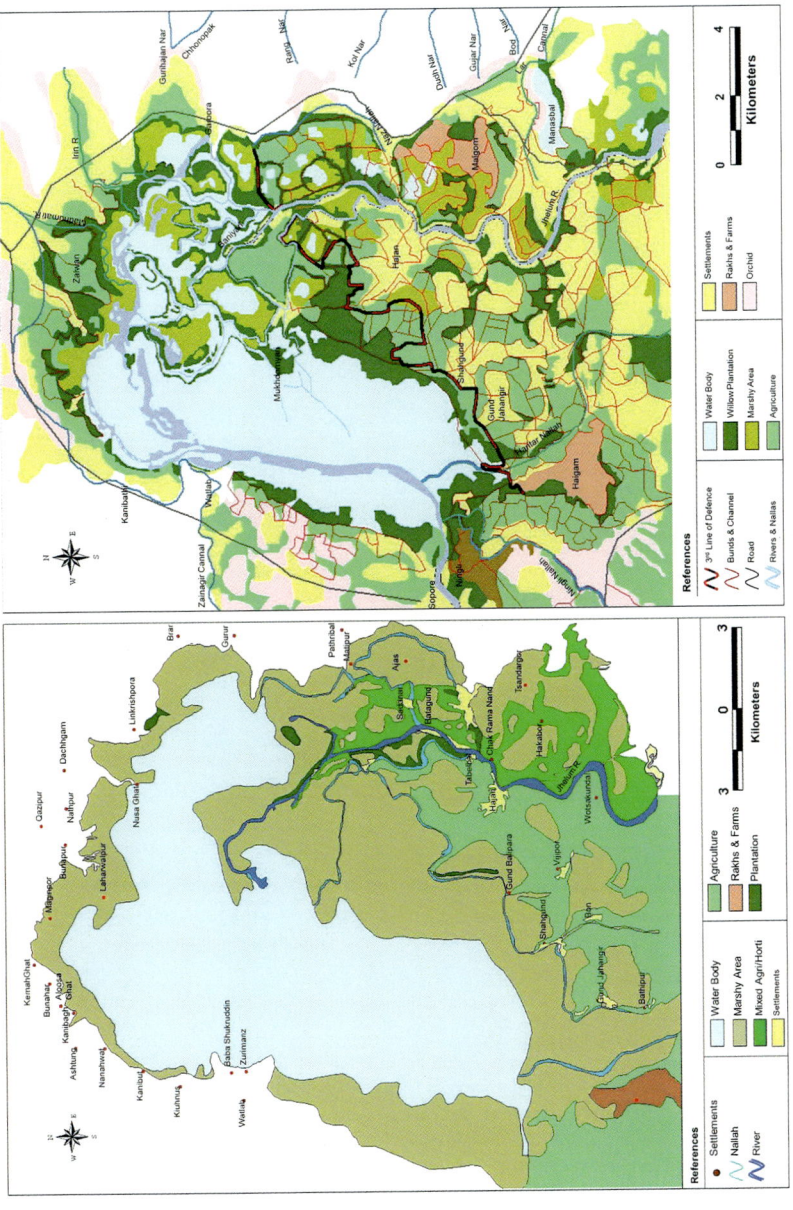

Fig. 1 Changes in Wular Lake from 1911 (*left*) to 2007 (*right*) reproduced with permission from Wetlands International South Asia

Biodiversity and Ecosystem Services

The wetlands of Kashmir Valley, including Wular Lake, are significant waterbird habitats for species migrating within the Central Asian Flyway. Twenty-two species, including the globally threatened ferruginous duck *Aythya nyroca*, are recorded to breed in the valley wetlands. In addition, a range of reed and tree nesting birds (e.g., warblers and raptors) also breed in these wetlands. During March–June, floating vegetation in Wular serves as breeding grounds for whiskered tern *Chlidonias hybridus*, pheasant-tailed jacana *Hydrophasianus chirurgus,* and little grebe *Tachybaptus ruficollis*. The peripheral shallow areas with longer emergent macrophytes serve as breeding sites for common moorhen *Gallinula chloropus*, little bittern *Ixobrychus minutus*, purple moorhen *Porphyrio porphyrio*, great reed warbler *Acrocephalus arundinaceus,* and others. The marshes associated with Wular, the Mukhdoomyari and Saderkote, are known to hold breeding grounds of mallard *Anas platyrhynchos*.

Surveys undertaken in 2000 indicated the presence of 13 fish species in Jhelum and associated wetlands, of which three, viz., *Schizothorax niger* (snow trout), *Triplophysa marmorata*, and *T. kashmiriensis*, are endemic to the Kashmir Valley (Anon 2000). Commercially important fish species in the wetland include *Schizothorax esocinus, S. curvifrons, S. micropogon, S. niger, S. longipinus, S. richardsonii, Nemacheilus* sp., *Cyprinus carpio communis*, and *C. c. specularis*. *Schizothorax* spp. constitute nearly 60% of the approximate 1400 MT annual fish catch.

Kundangar et al. (1992) reported 13 species of macrophytes from Wular, and Kaul and Trisal (1985) reported 24 species from associated marshes. The shorelines and marshes have been extensively planted with willow *Salix* sp. Water chestnut *Trapa* sp. and *Nelumbium* sp. (locally called *singhara* and *nadru*, respectively) are harvested for local consumption and trade. *Trapa* is distributed throughout the wetland, but the maximum concentration is found on the eastern shoreline. *Nelumbo* species is mainly found in some areas of Ashtangu to Kanusa in the southwest and Lunkershpura, Kolhama, and Garoora in the northeast. *Nymphoides* and *Nymphaea* form large belts in the Garoora-Laharwalpora portion. Apart from the direct livelihood benefits, a key value associated with Wular is its ability to regulate floods by providing storage for high volumes of floodwater received from the melting glaciers. The wetland is a natural food defence for the Kashmir valley.

Threats

The onset of the twentieth century witnessed a series of interventions for reclamation of Wular and its associated marshes. Wular was perceived to be a major factor contributing to floods in Srinagar City and interventions were planned for draining the wetland to as large an extent as possible. Dredging in the downstream reaches of Wular, from Sopore to Baramulla, was initiated in December 1902 to enable the lowering of water levels. However, a substantial reduction in water availability during the winter months was noticed immediately after this intervention, leading to a reconsideration of the effectiveness of dredging. Alternatively, a system of flood embankments was constructed in order to reclaim peripheral marshes for agriculture. In July 1949, a proposal for reclamation of marshes in Kashmir by embankments, drainage, and leveling was envisaged under the Grow More Food Campaign, which was duly considered by the state government. Marshes came under fresh introspection as potential areas for reclamation. The first series of embankments were constructed in the mid-fifties which is now known as first line of defense against flooding. The floods of 1957–1959 prompted the construction of a second line of defense in the mid-sixties. The third line of defense was constructed in 1975–1976, which marks the present inundation boundary. Subsequently the embankment heights were increased to above 1580 masl. A series of pumping stations were also constructed to drain the marshes for agriculture.

In parallel to the hydrological interventions was the introduction of willow plantations in the shallow water zones of Wular. The Ningli plantations in the south were created during 1916–1924 as a means to provide firewood for the region. The State Department of Rakhs and Farms, constituted to manage and administer the marshes reclaimed for agricultural, undertook willow plantation in a major way after the 1950s. The Department promoted plantations in shallower zones of the marshes and water bodies primarily to provide fuelwood and subsequently to support match and cricket bat manufacturing industries. As of 2006, nearly 35 km^2 of peripheral marshes had been converted into willow plantations.

Apart from the landscape transformation resulting from plantation and hydrological regime modifications, the wetland has also been subject to severe pollution. Being at the terminus of the Jhelum drainage system, Wular is a receptacle for pollutants flowing downstream from highly urbanized areas of Srinagar, Anantnag, Sopore, and Baramulla. Fertilizers and pesticides in agricultural fields and chemical sprays for pest control in orchards is ultimately washed into the Jhelum, subsequently flowing into Wular Lake. The water quality of the wetlands has significantly deteriorated due to uncontrolled dumping of sewage and solid waste by the adjoining settlements. The situation is most glaringly reflected in Wular, which due to its physiographic setting becomes the recipient of all wastes dumped into the river upstream and turns into a cesspool of wastewater leading to high incidence of waterborne diseases in the peripheral communities (WISA 2007).

Developmental interventions have resulted in extensive conversion of wetlands and loss of their valuable ecosystem services. A large segment of marshes along Bad Nambal, Rakh Ajas, Malgom, Haigam, and Nawgam have been converted for agriculture. Within Wular Lake alone, 71.55 km^2 has been converted for willow plantation and agriculture development (WISA 2007). The willow plantations act as barriers to silt laden waters of the Jhelum. The sediment load is discharged into the waterbody thereby reducing its water holding capacity. A key impact of these developments has been the impaired ability of the wetlands to moderate floods. Conversion of marshes has been identified as one of the major factors behind the extensive damages caused by the devastating floods of September 2014. Reduced connectivity of wetland complexes with the Jhelum River has accelerated the shrinkage of wetlands. More recently, the spread of the invasive alligator weed *Alternanthera philoxeroides* has been a cause of concern (Masoodi and Khan 2012), while communities have been impoverished by significant declines in the abundance of fish and lotus *Nelumbo nucifera*.

Conservation Status and Management

Wular was identified as a Wetland of National Importance in 1986 under the National Wetland Conservation Programme of the Ministry of Environment, Forest and Climate Change (MoEFCC), Government of India. In 1990, Wular was designated as a Wetland of International Importance under the Ramsar Convention by the Ministry. The initial focus of restoration efforts was on revegetation of catchments. Limited investments through programs under state government departments of revenue, social forestry, fisheries, rural development, irrigation, and flood control have also been made. However, concerns have been expressed regarding the need to coordinate various programs in order to tangibly address the drivers of wetland degradation. A comprehensive management action plan for lake restoration was formulated for the State Department of Wildlife Protection (WISA 2007), based on a comprehensive evaluation of wetland features and governing factors and was approved for implementation by the MoEFCC. Implementation is presently being coordinated by Wular Conservation and Management Authority (WUCMA) under the aegis of State Department of Wildlife Protection. A crucial part of the management plan is restoration of hydrological regimes by removal of willow plantations from the lake, revegetation of the catchments, and abatement of pollution. The Authority has delineated wetland boundary with geo-tagged pillars and made necessary inclusions in land records. Partial removal of willows has also been carried out. Based on review of management plan implementation during 2011–15, a follow up plan for the coming five years is under development.

References

Anon. Ecological status and conservation of River Jhelum. Srinagar: Jammu and Kashmir Lakes and Waterways Development Authority; 2000. Technical report, National Institute of Aquatic Ecology.
Kaul S, Trisal CL. Modelling nutrient dynamics in a Kashmir wetland. Pollut Res. 1985;4:1–6.
Kundangar MRD, Sarwar SG, Shah MA. Ecology and conservation of wetland of Wular Lake (Kashmir). Final report submitted to the Ministry of Environment and Forest. New Delhi: Government of India; 1992.
Masoodi A, Khan FA. Invasion of alligator weed (*Alternanthera philoxeroides*) in Wular Lake, Kashmir, India. Aquat Invasions. 2012;7:143–6.
Wetlands International South Asia. Comprehensive management action plan for Wular Lake, Kashmir. New Delhi: Wetlands International South Asia; 2007. Technical Report prepared for Department of Wildlife Protection, Government of Jammu and Kashmir. Available from http://ramsar.rgis.ch/pdf/wurc/wurc_mgtplan_india_wular.pdf. Accessed 23 Mar 2016.

Wetlands of the Ganga-Brahmaputra Basin

143

Ritesh Kumar and Kalpana Ambastha

Contents

The Ganga-Brahmaputra Basin	1712
Wetlands: Extent, Biodiversity, and Ecosystem Services	1713
Wetlands in the Himalayan Region	1713
Wetlands of the Terai Region	1715
Wetlands of Gangetic Plains	1716
Wetlands Within the Valley of Brahmaputra and its Distributaries	1717
Wetlands of the Central Highlands	1718
Wetlands of Gangetic Delta	1718
Conservation and Management	1719
Threats and Future Challenges	1720
References	1721

Abstract

The international basin of Rivers Ganga and Brahmaputra is endowed with a diverse wetland regime, which ranges from high altitude oligotrophic lakes in the Himalayas, the marshes and swamps of the Terai region, and floodplain and riverine wetlands in the Gangetic and Brahmaputra alluvial plains to coastal wetlands in the deltaic tracts. Besides, being key biodiversity habitats, these wetlands play an important role in providing water, food, and climate security to the basin's 630 million inhabitants. The basin countries, under the overarching principle of wise use, have evolved policy frameworks, regulatory regimes, and national programs for securing the health and ecological integrity of these wetlands. Yet, alteration of natural flow regimes, expansion and intensification of agriculture and settlements, pollution, unregulated tourism, and invasive species continue to stress these ecosystems. Integration of the full range of

R. Kumar (✉) · K. Ambastha (✉)
Wetlands International South Asia, New Delhi, India
e-mail: ritesh.kumar@wi-sa.org; kalpana.ambastha@wi-sa.org

© Springer Science+Business Media B.V., part of Springer Nature 2018
C. M. Finlayson et al. (eds.), *The Wetland Book*,
https://doi.org/10.1007/978-94-007-4001-3_93

ecosystem services and biodiversity values of basin wetlands, taking into account the ecological continuum between high altitude, planes, and deltaic wetlands, within an integrated water resources management framework can secure the future of these ecosystems.

Keywords

High altitude wetlands · Terai · Gangetic plains · Gangetic delta · Transboundary

The Ganga-Brahmaputra Basin

The basins of Rivers Ganga and Brahmaputra, the Ganga-Brahmaputra Basin (GBB), span over 1.7 million km^2 across India, China, Nepal, Bangladesh, Bhutan, and a small portion of Myanmar. Traversing diverse geographic regions as the cold dry plateau of Tibet, the rain-drenched Himalayan slopes, the landlocked alluvial plains of Assam, and the vast deltaic lowlands of Bangladesh, the transboundary basin serves as the lifeline for 630 million population (FAO 2012). The region has a monsoonal climate, with low precipitation in the north and higher precipitation along the coast (ibid).

The Ganga originates as Bhagirathi from the Gangotri Glaciers in the Himalayas and flows for a length of 2,525 km till its outfall in the Bay of Bengal, draining a basin area of 1.09 million km^2. The Brahmaputra originates in the Kailas ranges of southern Tibet and flows for over 2,900 km draining a basin of 0.58 million km^2. Upon reaching the Indian state of West Bengal, Ganga River splits into two main distributaries, the Hooghly, which continues the course toward the south into West Bengal, and the Padma, which flows eastward into Bangladesh. Similarly, the course of Brahmaputra River upon reaching Bangladesh is split into two major distributaries, the Jamuna and the Meghna. Brahmaputra River after meeting with the Tista River in Bangladesh is known as Jamuna. The Meghna River arises from the Manipur hills as River Barak with a drainage basin of 41,700 km^2 spread across India, Bangladesh, and Myanmar. The Jamuna coalesces with the Padma at Goalundo (District Rajbari of Bangladesh) and subsequently with Meghna River at Chandpur (Chittagong Division of Bangladesh), subsequently flowing toward the Bay of Bengal. Together with the Meghna, the Rivers Ganga and Brahmaputra form the Gangetic Delta at the mouth of the Bay of Bengal (Chowdhury and Ward 2004).

The geology of GBB is shaped by active faulting due to collision of continental plates containing India and Asia during mid-Oligocene (23–34 Ma BP) and formation of the Bengal Basin thereafter (Brown and Nicholls 2015). Based on physiography, the basin can be classed into four major divisions, namely, the Himalayas, the Indo-Gangetic Plains, the eastern hills, and plateau tracts of the south. The Himalayan region, comprising the three almost parallel fold ranges (the Greater Himalayas or the Himadri, the Middle Himalaya or the Himachal, and the Lower Himalayas or the Shivalik) interspersed with deep valleys and plateaus, forms the northern boundary of the GBB. At the region of the break of slope of Lower Himalayas, an 8–10 km narrow Bhabar belt is formed running parallel to the Shivaliks wherein streams and rivers tend to disappear beneath the deposited boulders and rocks. Such channels reappear in the Terai belt, south of Bhabar belt, creating marshy and swampy conditions with diffuse drainage patterns. South of the Terai belt is the surface of the basin filled by sediments derived from the Himalayas known as the Indo-Gangetic Plains. These plains exhibit mature fluvial erosional and depositional landforms as braided channels, sandbars, meanders, and oxbow lakes. The plain of Brahmaputra is aggradational, built up from sediments brought in by Brahmaputra and its major tributaries, with several riverine islands and sandbars. The drainages of Ganga, Brahmaputra, and Meghna merge to form the arcuate Gangetic Delta, spanning an area of 0.105 million km^2, approximately two-thirds of which is in Bangladesh.

Wetlands: Extent, Biodiversity, and Ecosystem Services

GBB is endowed with a diverse wetland regime, ranging from high-altitude oligotrophic lakes in the high-altitude Himalayas, marshes and swamps on the Himalayan foothills, and floodplain and riverine wetlands in the Gangetic Plains to coastal wetlands in the deltaic tracts. The wide ranging role of these wetlands as sources of major rivers, biodiversity habitats, backbone of highly productive agriculture and fisheries, buffers against floods and coastal storms, and cultural heritage of populations living in and around mark their critical role in food, water, ecological, and climate security of the GBB (Fig. 1).

Wetlands in the Himalayan Region

Wetlands in the Himalayan region are majorly of glacial or tectonic origin. Most of the high-altitude wetlands, located above 3,000 masl elevation, are formed in depressions caused by glacial erosion dammed by moraines and are deep, oligotrophic, clear with turquoise blue waters, and absent to very marginal macro-vegetation. At lower elevations, formation of wetlands is mostly related to tectonic processes.

Within GBB, high-altitude wetlands are located in China, India, Nepal, and Bhutan. High-altitude wetlands of the basin in China are mostly located at the fringes of the extensive peatlands of Qinhai-Tibetan Plateau. India has 4,699 high-altitude lakes spanning 0.12 million ha, spread across GBB and Indus Basins (Trisal and Kumar 2008; Panigrahy et al. 2012). Gurudongmar (in Sikkim) and wetland complexes of Nagula and Bhagajang (in Arunachal Pradesh) are some of the major high-altitude wetlands of the Indian part of GBB. Bhutan has over 3,000 high-altitude wetlands spanning 102 km^2, comprising supra-snow and glacial lakes, lakes in alpine meadows, and marshes (WWF undated). Drake Phangtsho, Nub Tsonapatra, Bothso, Tsheringma Lhatso, and Pemaling complexes are the major

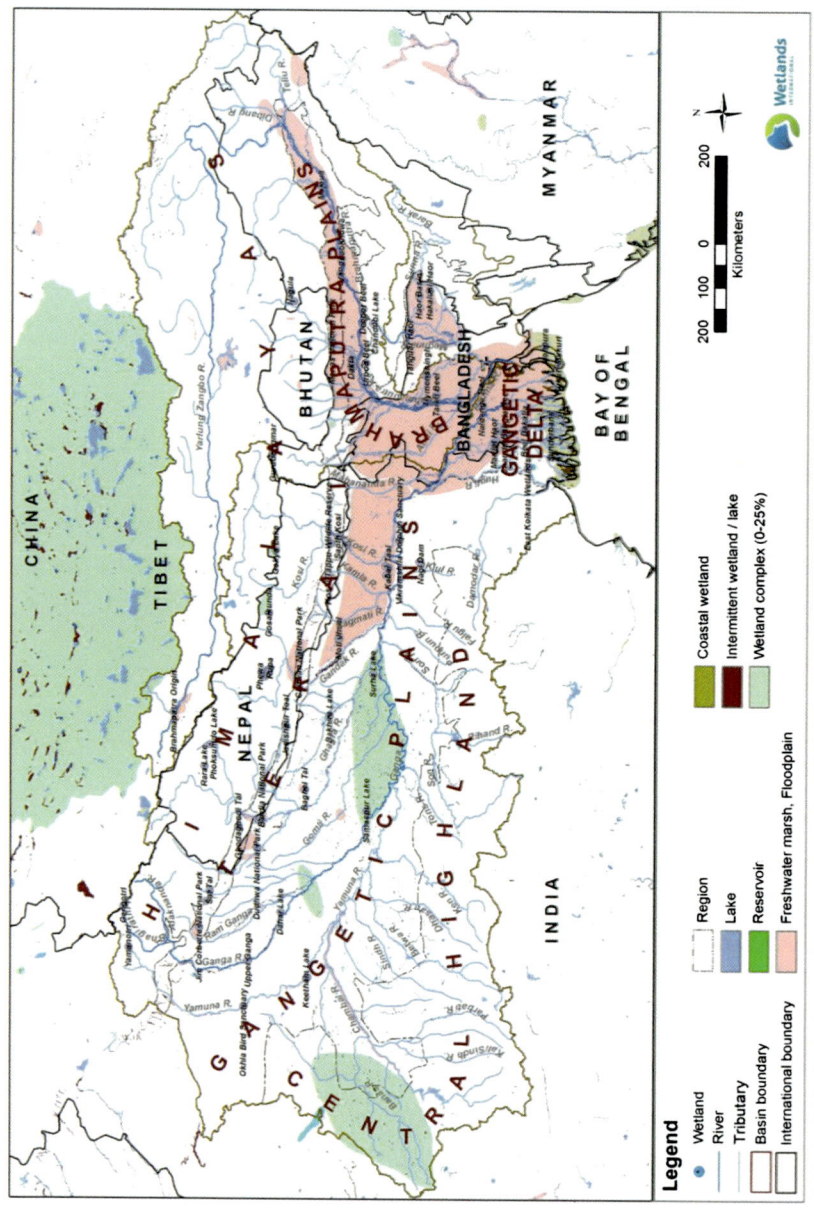

Fig. 1 Wetlands of the Ganga-Brahmaputra Basin (Prepared using GIS Datasets from Lehner and Döll 2004)

high-altitude wetlands of the country. Rara and wetland complexes of Gokyo, Phoksundo, and Gosaikunda are some of the major high-altitude wetlands of Nepal.

A number of wetlands at elevation below 3,000 masl in the Himalayan region of Nepal and India have been formed in depressions created by land subsidence due to tectonic movements and have major feeder streams aligned along fault lines and joints. Lakes Nainital and Sat Tal in the Indian state of Uttarakhand and the lakes of Pokhara Valley in Nepal (Phewa, Begnas, Rupa, Khaste, Dipang, Gunde, Kamalpokhari, and Maidi) are such examples. The average altitude of these lakes ranges from 750 m to 1,000 m. Phewa Taal is a clear, freshwater lake of associated marshes and adjacent rice paddies in the Pokhara Valley famous for its rich fish fauna including several native species of carps and two very popular sport fishes, the mahseer *Tor tor* and the asla *Schizothorax progastus*.

The high-altitude wetlands of Himalayas, located at the crest of GBB, play an important role in capturing and retaining snow and ice melt and rainfall, releasing water gradually during lean seasons, thereby acting as suppliers and regulators of water (Trisal and Kumar 2008). With the contribution of glacial melt to the mean discharge being nearly one-tenth of the flows of the Ganga and the Brahmaputra, Himalayan wetlands acquire a highly significant role (Eriksson et al. 2009). Besides water security, the wetlands of the Himalayas have an important role in securing biodiversity. Gokyo and Rara (Nepal) are important nesting and roosting sites for several trans-Himalayan migratory birds as the ruddy shelduck *Tadorna ferruginea* and bar-headed goose *Anser indicus*. Species of *Schizothorax*, *Orienus*, and *Tor* constitute the major coldwater fisheries of these wetlands (Swar 2002). Several Himalayan wetlands in India, Nepal, and Bhutan have been associated with religious and cultural values. The Buddhists hold high-altitude wetlands of Gosaikunda in high reverence as several of their teachers are believed to have obtained spiritual insights within these wetlands.

Wetlands of the Terai Region

The wetlands of Terai region primarily comprise marshes, swamps, and seasonally inundated grasslands. Most of these regions fall within protected area network of India and Nepal. Wetlands located within National Parks of Jim Corbett, Kishanpur, Dudhwa, Katarniaghat, and Manas are some representative examples of Terai wetlands in India. The marshes and swamps of Manas National Park are inhabited by the Indian one-horned rhinoceros *Rhinoceros unicornis*, tiger *Panthera tigris*, wild buffalo *Bubalus arnee*, swamp deer *Rucervus duvacelii*, and Asian elephant *Elephas maximus* among others. The region also forms the core of the 283,700 ha Manas Tiger Reserve. The critically endangered pygmy hog *Porcula salvania* is known to exist presently only in few locations around Manas National Park (Narayan et al. 2008). The swamps and lush grasslands of Katarniaghat connect the tiger habitats of Dudhwa and Kishanpur in India and the Bardia National Park in Nepal. Girwa River which flows through Katarniaghat is the natural habitat for a breeding population of the gharial *Gavialis gangeticus*.

Koshi Tappu Wildlife Reserve, Jagdishpur, and Gaidhwa harbour some of the major wetlands of Terai region in Nepal. Beeshazar Tal wetlands, located in the buffer zone of Royal Chitwan National Park, is an important habitat for marsh crocodiles and over 250 bird species. Water from these wetlands are used to irrigate adjoining agricultural fields. The Koshi Tappu Wildlife Reserve, a 24 km stretch of the Sapta Kosi and its adjacent floodplains, is a natural river system located close to the Koshi Tappu reservoir near the India-Nepal border. 485 species of birds, including at least 114 species waterbirds, are reported from the area. It is the only area in Nepal where water cock *Gallicrex cinerea* and Abbott's babbler *Malacocincla abbotti* are found. Nepal's last remaining population of wild buffalo can be seen here. Ghodaghodi Tal spanning 2,563 ha area is a complex of 13 oxbow lakes, marshes, and meadows and has been designated as a Ramsar site. The wetland hosts the shrine of local deity and is highly revered.

Wetlands of Gangetic Plains

The Gangetic Plains are a vast stretch of highly fertile alluvium, which is the basis of agriculture development within the GBB. The riverine and floodplain wetlands are intricately linked with flood pulses of the Ganga River and its tributaries. There is a significant exchange of water, sediments, nutrients, and species between the main river channel and associated wetlands. Thus the two form an interconnected system, with linked biodiversity and ecosystem services values.

From west to the east, the Gangetic Plains comprise the plains of Yamuna River within the northwestern states of Haryana and Delhi, with the major proportion lying within the north Indian states of Uttar Pradesh and Bihar. The Yamuna River floodplains play a major role in recharging groundwater for the City of Delhi, besides serving as habitat of birds in the linked Okhla Bird Sanctuary (Trisal et al. 2008). Some significant wetland areas in Uttar Pradesh are the Baghel Tal in Bahraich district which is a riverine wetland along Terhi River, the Bakhira Lake situated in Sant Kabir Nagar district spanning an area of 3,905 ha, the Surha Lake in Ballia district spanning an area of about 2,357 ha, the Dahar Lake in Hardoi district, the Keetham Lake located within Soor Sarovar Bird Sanctuary in Agra district on the bank of Yamuna River with an area of 403 ha, and the Samaspur Lake in Raebareli district. The Upper Ganga River stretch between Brijghat and Narora spanning 266 km^2, comprising shallow waters with intermittent deep-water pools and reservoirs upstream from barrages, is a designated Ramsar Site. Flooded rice paddies of districts Etawah and Mainpuri of southwestern Uttar Pradesh support the largest known population of sarus cranes *Grus antigone* in the world (Sundar 2009).

Wetlands within the state of Bihar are attributed mainly to the complex fluvial geomorphology of the Gangetic tributaries which have over a period of time created a number of natural depressions and cutoff meanders. Known variously as *maun*, *chaur, and taal*, these shallow wetlands with a maximum depth of 1.5 m are a characteristic feature of the interfluvial regime of Gangetic Plains, being completely inundated during the monsoon, and mostly dry by March–June. Kanwar Jheel (in Begusarai), Kusheshwarsthan (in Darbhanga), Bariella (in Vaishali), Vikramshila Dolphin Sanctuary (in Bhagalpur), and Moti Jheel (in East Champaran) are some of the major wetlands of the state. These wetlands are critical for water security due to their ability to buffer floods, provide water for drinking and irrigation, and recharge groundwater aquifers. With over 80–90% of the river runoff confined to only four monsoon months, the ability of wetlands to store water and regulate overall hydrological regimes is important for securing water availability in the region. The diverse and dynamic assemblage of fish, invertebrate, and crustaceans provides the basis of rich fishery which supports livelihoods of 4.9 million fishers of the state. Wild rice, makhana *Euryale ferox*, singada *Trapa natans*, and edible mollusk *Pila globosa* are some of the main wetland products harvested for local consumption by the dependent communities. Water and wetlands form an integral part of local culture, with several festivals (e.g., *Chhath*), local practices (e.g., *Jhaur Sheetal*), and folklores linked to these ecosystems.

Wetlands Within the Valley of Brahmaputra and its Distributaries

The valley of Brahmaputra and its distributaries Jamuna and Meghna have a number of riverine wetland formations. The State of Assam, in which major parts of the Brahmaputra Valley are located, has 873 oxbow lakes and cutoff meanders (14,173 ha) and 139 riverine wetlands (4,258 ha) (SAC 2011). These wetlands play an important role in buffering settlements from floods and are a rich source of fisheries. Deepor Beel, a Ramsar Site, located southwest of Guwahati City is a critical component of urban flood defense (Gogoi 2007). Majauli is a large inhabited riverine island in the Brahmaputra River, spanning 88,000 ha, with a number of oxbow lakes and sandbar formations (Islam and Rahmani 2004). The island has records of over 250 species of birds, including 14 of high conservation significance (ibid). Dhir-Dilpai-Dakra complex, Tamranga-Dalani complex, Urpod Beel, Chandubi Lake, Orang, and Laokhowa are some other important wetlands of the region, especially as habitats of waterbirds and freshwater fish.

The grassland wetland habitat of Kaziranga, a world heritage site, is an area of internationally famed wilderness, especially for Indian one-horned rhinoceros. The region is also inhabited by several other endangered species including tiger, wild buffalo, swamp deer, and Asian elephant among others (Islam and Rahmani 2008). As many as 490 bird species have been recorded from the Kaziranga National Park region (Chowdhury 2003). The region is famous for nesting colonies of spot-billed pelican *Pelecanus philippensis* and adjutant stork.

Within the Bangladesh segment of the valley, the Haor Basin of Sylhet and eastern Mymensingh is the most prominent wetland complex of over 400 riverine wetlands supporting fisheries, abundant aquatic vegetation used for domestic livestock, and seasonally flooding margins used for growing rice. Tanguar Haor within this complex is a designated Ramsar Site. Spanning 9,500 ha within the floodplains of Surma River at the base of Meghalaya Hills, this wetland has remnants of natural swamp forests, supports Bangladesh's largest commercial fishery, and provides habitat for at least 135 fish and 208 bird species (Ramsar 2000). The Hakaluki Haor (20,400 ha) is a complex of over 80 interconnected depressed lands (variously referred as beel and baor in India and Bangladesh), an important fisheries resource, and breeding ground for freshwater fish and habitat for migratory waterbirds.

Wetlands of the Central Highlands

The Central Highlands are of volcanic origin and studded with human-made wetlands in the form of tanks and reservoir. These wetlands play a critical role in providing water for irrigation and human uses while also serving as habitat for wetland-dependant species. Gandhisagar Reservoir, formed due to impoundment of Chambal River, draws a large number of migratory and nonmigratory waterbirds throughout the year. One can also witness over 100,000 waterbirds in Dihalia Jheel and adjoining wetlands, which are also an important water source for irrigation. Reservoirs of the Madhav National Park and Rangawa are other major wetlands of this region, located within the central Indian state of Madhya Pradesh.

Wetlands of Gangetic Delta

The Gangetic Delta is dotted with wetlands in the forms of river creeks, estuaries, mangrove swamps, marshes, oxbow lakes and depressed lands, fish ponds, and water storage structures. The East Kolkata Wetlands (12,500 ha) is an agglomerate of sewage-fed fish farms developed on marshes on the eastern periphery of Kolkata City (India). The wetlands are known to treat the city's wastewater and produce 10,500 tonnes of table fish and 150 t of vegetables, providing livelihoods to 50,000 people (Ramsar 2002). A number of deltaic wetlands in Bangladesh (as Marijat Haor, Atadanga Haor, Chanda-Bahgia Beel, Sowagram-Gopalpur Beel, Dakatia Beel, Naldanga Beel, and Tarail Beel) have been identified as priority management areas (Islam 2016). Most of these haor and beel areas are under intensive fisheries and margins cultivated with rice. During dry seasons, the wetlands are used as grazing grounds by domestic livestock (Khan et al. 1994). Kukuri Mukuri and Monpura are some prominent *char* lands (sediment deposits) in the estuarine region of Meghna River.

The Sundarbans located within the lower part of the Gangetic Delta covering about one million ha within Bangladesh and India is a mangrove swamp, crisscrossed by a network of streams, mudflats, and creeks. The region contains the world's largest contiguous area of mangroves and acts as a natural buffer for the Kolkata Metropolitan Region with 14 million inhabitants and adjoining settlements from the impacts of cyclones, sea level rise, and other adverse natural events (World Bank 2014). The biodiversity in these wetlands includes about 350 species of vascular plants, 250 fishes, and 300 birds, besides numerous species of high conservation significance (*Batagur baska, Pelochelys bibroni, Chelonia mydas*), especially the emblematic royal Bengal tiger *Panthera tigris* and several species of river dolphins (Gopal and Chauhan 2006).

Conservation and Management

All countries of GBB are signatories to the Ramsar Convention on Wetlands and have endorsed the "wise use of wetlands" as the guiding principle for wetland conservation and sustainable management. Countries have evolved policy frameworks, regulatory regimes, and programs for conservation and sustainable management of wetlands, including those of the GBB. In 2000, the Chinese government developed the China National Wetland Conservation Action Plan and approved the 2002–2030 plan establishing and setting a set of ambitious goals, including establishing 713 wetland reserves – with more than 90% of natural wetlands effectively protected by 2030; restoring 1.4×10^9 ha of natural wetlands; and building 53 national pilot zones for wetland protection and prudent use. High-altitude wetlands of Tibetan Plateau, the source of Brahmaputra River, have been identified as one of the pilot zones (Wang et al. 2012). The Government of India's National Environment Policy (2006) recommends prudent use and mainstreaming within development programming as core strategies for managing wetland ecosystems. The Ministry supports restoration of prioritized wetlands through a dedicated program entitled the National Plan for Conservation of Aquatic Ecosystems (NPCA). A regulatory framework for wetlands has also been put in place. There has been proactive effort to include Himalayan wetlands within the network of national priority wetlands. Policy and programming frameworks of water resources also include management of river floodplains and associated wetlands as means to ensure water security.

The Government of Nepal adopted the National Wetlands Policy in 2003. The policy aims to conserve and manage wetland resources wisely and in a sustainable way with local people's participation. The policy also aims to put the conservation and management aspects of wetland conservation within the framework of broader environmental management. The Government of Bhutan is using its accession to the Ramsar Convention in 2012 as an opportunity to shape a national policy framework and programming for wetlands in the country. Within Bangladesh, the National Water Policy (1999) sets the overarching principles for conservation and management of wetlands. The policy accords high priority to fisheries and wildlife in water resources planning, minimizes disruptions to natural regimes, and restricts drainage of wetlands, especially those of high ecological and socioeconomic values. The Bangladesh Water Act (2013) sets the overarching regulatory framework for the country's wetlands. Within the ambit of national policies and programs, several wetlands of GBB have been prioritized for conservation and sustainable management.

Given the transboundary characteristics of the wetlands of the GBB, there has been an effort since 2002 to establish a Himalayan Wetlands Initiative in the region (Harris et al. 2009). The initiative was endorsed as a regional initiative of the Ramsar

Convention in 2009; endorsed by Governments of India, Myanmar, Nepal, and Pakistan; and supported by International Center for Mountain Development (ICIMOD), Wetlands International, WWF International, and IUCN. However, the initiative is yet to materialize on account of lack of endorsement by all members of the region and clarity on financing arrangements.

Governments of India and Bangladesh have signed a Memorandum of Understanding on Conservation of the Sundarbans, agreeing to consider and adopt joint monitoring and management, implement conservation and protection efforts, and develop a long-term strategy for creating ecotourism opportunities.

Threats and Future Challenges

Wetlands of the GBB face a range of pressures emanating from anthropogenic and non-anthropogenic sources. Unregulated mountain tourism, seasonality of which corresponds with periods of high biological activity in high-altitude wetlands of Himalayas, poses high threat to these ecosystems. The impacts of such pressures are clearly evident in wetlands of the Lower Himalayas.

Alteration of natural hydrological regimes with hydraulic structures constructed to meet various human water resource requirements have been a major threat to wetlands of GBB. Reduction in river flows caused by diversion structures on the tributaries is implicated as one of the major causes of the decline in fish catch from Ganga River and reduced connectivity of the river with the floodplains (Vass et al. 2011). Construction of Farakka Barrage at the head of Gangetic Delta in 1975 to augment water supply at the Kolkata port is implicated in reduction in sediment transport and impeding delta development, along with altering natural ecological profiles of the region (Sinha 2004). The impact of hydrological regime alternation has been highly amplified in the Sundarbans region in the form of increased salinity, which is one of the causative factors for decline in dominant salinity-sensitive mangrove species of the region *Heritiera fomes* and *Ceriops decandra* (Gopal and Chauhan 2006; Islam 2016).

The plains of Ganga and Brahmaputra, being naturally of high soil fertility, have been the center of agriculture development. Traditional mixed cropping farming methods gradually transformed to intensive agriculture-based economies, supported by expansion of the irrigation network and high agriculture inputs (Abrol et al. 2002). An analysis of land use/land cover change in the Gangetic Plains over 1880–1980 indicated intensification of permanent agriculture, along with a 47% decline in area under forests and 27% decline in wetlands (Flint 2002). Besides conversion, development of irrigation infrastructure has also altered the natural hydrological regimes of several floodplain wetlands. Insufficient water treatment infrastructure in the population centers and excessive use of chemical fertilizers and pesticides are implicated in nutrient enrichment, promoting growth of the invasive water hyacinth *Eichornia crassipes* and ipomea *Ipomea aquatica*. In several cases, these invasives further compound problems due to fragmentation of water regimes.

The Gangetic Delta, inhabited by over 108 million people in 2011 and increasing at rate of 17.5% over the last two decades, is one of the most populous deltas of the world (Szabo et al. 2016). Such a rapid increase in population in an environmentally fragile landscape has created stresses for natural resources, particularly wetlands. Growth in urban centers as Kolkata (India) and Dhaka (Bangladesh) has been with a concomitant decline in wetlands areas, and increasing population vulnerabilities. Over 50% reduction in wetland area in Dhaka City during 1968–2001 is implicated for increased waterlogging and flood risks in one of the most rapidly expanding urban agglomerates in the basin (Sultana et al. 2009). The eastward expansion of Kolkata City has been through conversion of extensive salt marshes on the city's periphery (Wetlands International 2010). Large areas of the Sundarbans mangroves were converted into paddy fields over the past two centuries and more recently into shrimp farms (Gopal and Chauhan 2006). The annual cost of environmental damage due to degradation of Sundarbans has been estimated to be INR 6.7 billion (World Bank 2014).

With a growing population in the deltaic region spurred on by construction of ports, trade, and shipping, investments in dams and dykes have also been made for flood control, prevention of salinization, and expansion of agriculture (Islam 2016). In the 1960s, construction of embankments was intensified primarily to reclaim land for agriculture and secure the deltaic population from the threats of salinization and floods. However, such interventions only led to drainage congestion and siltation of the channels. Sea level rise in the Bay of Bengal has further complicated impacts of developmental activities, increasing the risk of submergence of mangroves and coastal erosion (Raha et al. 2012).

Wetlands of GBB underpin the hydrological and socioeconomic values of the river systems, yet their integration in water resources programming remains at the best marginal. Lack of consideration of the full range of biodiversity and ecosystem services values of these ecosystems in developmental programming has often led to wanton destruction, ultimately reducing the basin's resilience. There is a need to urgently upscale investments and programming for wise use of wetlands in all countries of the basin. There is also need to recognize the ecological continuum between high altitude, plains, and deltaic wetlands, so as to consider the impact of upstream developments on downstream ecosystems. Such a programming needs to be based on comprehensive inventories of basin wetlands, particularly on structural and functional interdependencies between different wetland types and between wetlands and river systems.

References

Abrol YP, Sangwan S, Dadhwal VK, Tiwari MK. Land use land cover in Indo-Gangetic Plains – history of changes, present concerns and future approaches. In: Abrol YP, Sangwan S, Tiwari MK, editors. Land use – historical perspectives: focus on Indo-Gangetic Plains. New Delhi: Allied Publishers Private Limited; 2002. p. 1–28.

Brown S, Nicholls RJ. Subsidence and human influences in mega deltas: the case of the Ganges–Brahmaputra–Meghna. Sci Total Environ. 2015;527-528:362–74.

Chowdhury AU. Birds of Kaziranga National Park: a checklist. Guwahati: Gibbon Books and the Rhino Foundation; 2003.

Chowdhury R, Ward N. Hydro-meteorological variability in the greater Ganges–Brahmaputra–Meghna basins. Int J Climatol. 2004;24(12):1495–508.

Eriksson M, Vaidya R, Jianchu X, Shrestha AB, Nepal S, Sandstrom K. The changing Himalayas: impact of climate change on water resources and livelihoods in the greater Himalayas. Kathmandu: International Center for Integrated Mountain Development (ICIMOD); 2009.

FAO. Irrigation in Southern and Eastern Asia in figures: AQUASTAT survey – 2011. FAO Water Reports 37. Rome: Food and Agriculture Organization of the United Nations (FAO); 2012.

Flint E. Historical reconstruction of changes in land use and land cover of vegetation in the Gangetic Plain 1880–1980: methodology and case studies. In: Abrol YP, Sangwan S, Tiwari MK, editors. Land use – historical perspectives: focus on Indo-Gangetic Plains. New Delhi: Allied Publishers Private Limited; 2002. p. 189–248.

Gogoi R. Conserving Deepor Beel–Ramsar Site, Assam. Curr Sci. 2007;93(4):445–6.

Gopal B, Chauhan M. Biodiversity and its conservation in the Sundarban Mangrove ecosystem. Aquat Sci. 2006;68:338–54.

Harris C, Bhandari BB, Hua O, Sharma E. Himalayan Wetlands Initiative – conservation and wise use of natural water storage in the HKH region. Sustainable Mountain Development. 2009;59. ICIMOD.

Islam SN. Deltaic floodplains development and wetland ecosystems management in the Ganges–Brahmaputra–Meghna Rivers Delta in Bangladesh. Sustain Water Resour Manag. 2016. doi:10.1007/s40899-016-0047-6.

Islam MZ, Rahmani AR. Important bird areas in India: priority sites for conservation. Mumbai: Indian Bird Conservation Network/Bombay Natural History Society and BirdLife International; 2004.

Islam MZ, Rahmani AR. Potential and existing Ramsar Sites in India. Indian Bird Conservation Network: Bombay Natural History Society, BirdLife International and Royal Society for the protection of birds. Oxford: Oxford University Press; 2008.

Khan SM, Haq E, Hug S, Rahman AA, Rashid SMA, Ahmed H. Wetlands of Bangladesh. Dhaka: Holiday Printers; 1994. p. 1–88.

Lehner B, Döll P. Development and validation of a global database of lakes, reservoirs and wetlands. J Hydrol. 2004;296(1-4):1–22. doi:10.1016/j.jhydrol.2004.03.028.

Narayan G, Deka P, Oliver W. *Porcula salvania*. The IUCN Red List of Threatened Species. 2008.

Panigrahy S, Patel JG, Parihar JS, editors. National wetland atlas: high altitude lakes of India. Ahmedabad: Space Applications Centre (SAC), ISRO, Government of India; 2012.

Raha A, Das S, Banerjee K, Mitra A. Climate change impacts on Indian Sundarbans: a time series analysis (1924–2008). Biodivers Conserv. 2012;20(1):1289–307.

Ramsar. Information Sheet on Ramsar Wetlands (RIS): Tanguar Haor. Gland: Ramsar Convention Secretariat; 2000.

Ramsar. Information Sheet on Ramsar Wetlands (RIS): East Calcutta Wetlands. Gland: Ramsar Convention Secretariat; 2002.

SAC. National Wetland Atlas. Ahmedabad: Space Applications Centre (SAC), ISRO, Government of India; 2011.

Sinha M. Farakka barrage and its impact on the hydrology and fishery of Hooghly estuary. Water Sci Technol Lib. 2004;49:103–24.

Sultana MS, Islam GMT, Islam Z. Pre and post-urban wetland area in Dhaka City, Bangladesh: a remote sensing and GIS analysis. J Water Resour Prot. 2009;1:414–21.

Sundar KSG. Are rice paddies suboptimal breeding habitat for Sarus Cranes in Uttar Pradesh, India? Condor. 2009;111(4):611–23.

Swar DB. The status of coldwater fish and fisheries in Nepal and prospects of their utilization for poverty reduction. In: Cold water fisheries in the trans Himalayan countries. Rome: FAO; 2002.

Szabo S, Brondizio E, Renaud FG, Hetrick S, Nicholls RJ, Matthews Z, Tessler Z, Tejedor A, Sebesvari Z, Foufoula-Georgiou E, da Costa S, Dearing JA. Population dynamics, delta vulnerability and environmental change: comparison of the Mekong, Ganges–Brahmaputra and Amazon delta regions. Sustain Sci. 2016;11:539–54.

Trisal CL, Kumar R. Integration of high altitude wetlands into river basin management in the Hindu Kush Himalayas: capacity building needs assessment for policy and technical support. New Delhi: Wetlands International South Asia; 2008.

Trisal CL, Tabassum T, Kumar R. Water quality of the River Yamuna in the Delhi stretch: key determinants and management issues. Clean-Soil Air Water. 2008;36(3):306–14.

Vass KK, Das MK, Tyagi RK, Katiha PK, Samanta S, Shrivastava NP, Bhattacharjya BK, Suresh VR, Pathak V, Chandra G, Debnath D, Gopal B. Strategies for sustainable fisheries in the Indian part of the Ganga-Brahmaputra river basins. Int J Ecol Environ Sci. 2011;37(4):157–218.

Wang Z, Wu J, Madden M, Mao D. China's wetlands: conservation plans and policy impacts. Ambio. 2012;41(7):782–6.

Wetlands International. Wetlands & water, sanitation and hygiene (WASH) – understanding the linkages. Wageningen: Wetlands International; 2010.

World Bank. Building resilience for sustainable development of the Sundarbans. Strategy Report. Washington, DC: World Bank: 2014.

Saline Wetlands of the Arid Zone of Western India

144

Malavika Chauhan and Brij Gopal

Contents

Introduction	1726
Location and Water Resources	1726
Distribution, Extent and Diversity of Wetland Types	1728
Significant Biodiversity	1728
Ecosystem Services	1729
Dependent People	1729
Conservation Status and Management	1730
Threats and Future Challenges	1730
References	1731

Abstract

The arid zone of western India hosts many natural wetlands which include the salt lakes such as Lake Sambhar- a Ramsar site in Rajasthan, and intertidal mud flats, saline marshes, and Avicennia-dominated mangroves in Gujarat. These wetlands have a relatively low but highly characteristic biodiversity that includes endemic and threatened species such as wild ass. Despite some conservation efforts and a few protected areas, arid zone wetlands are threatened by conflicting, multiple demands on their natural resources, changing hydrology, industry and anticipated climate change. Water management remains the most important future challenge in an already water-scarce region.

M. Chauhan (✉)
Himmotthan Society, Dehradun, Uttarakhand, India
e-mail: malavikachauhan@gmail.com

B. Gopal (✉)
Centre for Inland Waters in South Asia, Jaipur, Rajasthan, India
e-mail: brij.ciwsa@gmail.com

© Springer Science+Business Media B.V., part of Springer Nature 2018
C. M. Finlayson et al. (eds.), *The Wetland Book*,
https://doi.org/10.1007/978-94-007-4001-3_173

Keywords

Sambhar Lake · Mangroves · Indian Wild Ass Sanctuary · Kachchh Desert Wildlife Sanctuary · Great Indian Bustard Sanctuary · Banni Grasslands Reserve

Introduction

Arid regions are characterized by low annual rainfall (P) and high temperatures that cause very high evapotranspiration (PE). Generally, areas with a P/PE ratio of 0.03–0.2 (annual rainfall 100–300 mm) are considered arid, and those with a P/PE ratio 0.2–0.5 (annual rainfall 300–600 mm) are semiarid while those with P/PE <0.03 (annual rainfall below 100 mm) are hyper-arid (FAO 1989). Most of these arid zones lie within the tropical and subtropical belt, mostly to the west of the continents. Arid regions with extremely low precipitation occur also in cold climates at the higher altitudes (e.g., Tibet and Mongolia in Asia), known as cold deserts. Many areas in the rain shadow zones of mountains also receive low rainfall and hence are semiarid. Despite low precipitation however, many arid regions are rich in wetlands, which develop either because of geological factors, proximity to the sea, or because of the passage of rivers and streams. These wetlands are among some of the most spectacular in the world, and can be either temporary or permanent, are often brackish or saline, and host large populations of waterfowl and other characteristic biota (Kingsford 1997).

Location and Water Resources

Of the Indian subcontinent, most of the western part is arid. Geographically, the Indian arid zone includes the western parts of the states of Punjab, Haryana, Rajasthan (west of the Aravalli Range), and Gujarat (Mehta 2000; Fig. 1). It includes most of the Thar Desert which extends into Pakistan alongside the River Indus. It includes the Rann of Kachchh, a large (over 7,000 km^2) zone of seasonally flooded marsh and saline flats, which was formed by the receding sea level, neotectonic activity, and deposition of fluvial sediments. The Little Rann of Kachchh on the southeast of the Greater Rann is distinguished by extensive barren salt encrusted flats interspersed with vegetated islands called "bets" (Sathyapalan et al. 2014; Gupta 2015).

The region receives less than 400 mm of precipitation annually, which decreases westwards. Salinity is common throughout the region and is generally attributable to the influence of the sea (e.g., Godbole 1972). Several ephemeral rivers occur in this zone. The River Ghaggar arises in the Shivalik hills of Himachal Pradesh and flows through parts of arid Haryana and Rajasthan. The Luni, which rises on the western slopes of the Aravalli Range, flows southwest before dissipating into the Rann of Kachchh (= Kutch), while three ephemeral rivers, the West Banas, Saraswati, and

Fig. 1 Indian Arid Zone. *GRK* Greater Rann of Kachchh, *LRK* Little Rann of Kachchh. The arid zone lies west of the thin line that runs along the Aravalli ranges from which River Luni rises on the west. All rivers/streams are seasonal (M. Chauhan and B. Gopal © Rights remain with the authors)

Rupen, carry runoff from the southern flanks of the Aravallis into the Little Rann. There are 97 small ephemeral streams in Kachchh alone and several streams which flow from western Saurashtra into the Gulf of Kachchh (Stanley 2004). There are no natural freshwater lakes, but throughout the arid regions of Rajasthan and Gujarat, numerous reservoirs and tanks have been constructed over centuries to meet domestic and agricultural needs of water.

Distribution, Extent and Diversity of Wetland Types

Saline wetlands in Rajasthan consist almost exclusively of a few salt lakes, of which Lake Sambhar lies adjacent to the Aravalli hills (on the east) in the Districts of Nagaur and Jaipur. It covers an area of over 24,200 ha and has been divided by an artificial bund into a western natural shallow lake and an eastern part consisting of salt pans (Gopal and Sharma 1994). The lake is fed by two ephemeral rivers, the Mendha which comes in from the north and the Rupangarh which falls into the lake from the south, besides a few small streams on the northwest. Other major salt lakes lie to the west of the Aravalli hills and include Didwana (1,344 ha; Nagaur District), Phalodi (Jodhpur District), Pachpadra (Barmer District), and Lunkaransar (Bikaner District). There are hundreds of human-made salt pans, mostly around the salt lakes, which cover about 12,300 ha (SAC 2010). Intertidal flats are estimated to cover 18,900 ha in Barmer and Jalor districts along the border with Gujarat where the River Luni discharges its flow into the Rann of Kachchh (SAC 2010).

Gujarat contributes the largest area (34,749 km^2) to wetlands in India (SAC 2010). Most of these wetlands (28,083 km^2) are saline and are either in the Kachchh region or along the coast. They include 22,604 km^2 of intertidal mud flats (Rann of Kachchh), 144.3 km^2 of saline marshes, 90.5 km^2 of mangroves, and 172.2 km^2 of lagoons and creeks. Salt pans cover 922 km^2. Stanley (2004) identified 258 wetlands within Kachchh alone.

There is no clear evidence of a significant change in the area of these natural, saline wetlands either in Rajasthan or in Gujarat in recent decades, but their extent does fluctuate greatly between seasons and years depending upon the rainfall. The area under mangroves has increased through extensive plantations along the Gujarat coast (Vishwanathan et al. 2010; Pandey and Pandey 2012). At the same time the area of salt pans around Sambhar lake (Rajasthan) increased from 35 to 61 km^2 between 2003 and 2014, mainly because of salt-making activities using the subsurface brine (Gopal et al. unpublished).

Significant Biodiversity

The saline wetlands of Kachchh (Gujarat) have relatively low, but highly characteristic biodiversity. The inventory includes over 250 species of plants (Meena et al. 2005), 170 species of algae, numerous zooplankton and macroinvertebrates, as well as about 150 species of avifauna (Stanley 2004; Prusty 2009; Gajera et al. 2012). While Ishnava et al. (2011) record 104 plant species from the Little Rann of Kachchh alone, only a few species dominate. Among algae, *Spirulina* is dominant in the salt pans. The herbaceous vegetation of seasonally flooded areas comprises of typical halophytes such as *Salsola* sp., *Sesuvium portulacastrum*, *Salicornia brachiata*, *Suaeda fruticosa*, and *Cressa cretica*. The most abundant grasses and sedges include *Aelurops lagopoides* and species of *Scirpus* and *Juncus*. The mangroves are composed almost exclusively of *Avicennia marina*. Other occasional mangrove species are: *Aegiceras corniculatum, Avicennia officinalis, A. alba,*

Bruguiera gymnorhiza, Ceriops tagal, Rhizophora mucronata, and *Sonneratia apetala.*

The rich avifaunal diversity of Kachchh is dominated by the greater flamingo *Phoenicopterus roseus*, lesser flamingo *P. minor*, and great white pelican *Pelecanus onocrotalus* which breed here. At least two species, oriental darter *Anhinga melanogaster* and painted stork *Mycteria leucocephala*, are near threatened whereas the Dalmatian pelican *Pelecanus crispus* and sarus crane *Grus antigone* are in the vulnerable category (IUCN RedList 2015). Among other wildlife dependent on seasonal wetland grasses, the most noteworthy is the wild ass *Equus hemionus khur*, an endemic and threatened species. Another species of interest is the endemic reptile, *Cyrtopodion kachhense kachhensis*.

Plant biodiversity in saline wetlands of Rajasthan is similar to that of Gujarat. In the salt pans of Lake Sambhar, *Dunaliella salina* is the dominant alga. Both the greater and lesser flamingos occur in Lake Sambhar whereas the brackish and freshwater habitats around the lake support several other birds during the rainy season (Gopal and Sharma 1994).

Ecosystem Services

Arid areas are characterized by seasonal rainfall and often wetlands that retain water long after the rest of the landscape has dried out. These arid zone wetlands are vital sources of water in otherwise uninhabitable landscapes. They are critically important life-support systems for the survival of people as sources of water, food, and fiber. They help provide regular water supplies and fertile soils, improve water quality, recharge aquifers, and lessen the impact of seasonal floods.

While high salinity and a prolonged dry season coupled with frequent droughts restricts overall productivity of the area, the extensive grass cover over seasonally flooded areas provides valuable forage for sheep, goats, cattle, wild ass, and other wildlife. Wetlands further provide habitat and feeding and breeding grounds for a large population of birds. They are also an important source of edible and industrial grade salt. The mangroves not only protect the shoreline and support significant components of faunal biodiversity, they also support rural livelihoods across the region. A recent study estimated the total economic value (direct + indirect) of mangroves in Gujarat at Indian Rupees 7,731.3 million per year (Valuation Year = 2003; Hirway and Goswami 2007).

Dependent People

All local communities, many of which are old tribes, depend exclusively on the wetlands as grazing grounds and as a source for fodder for their herds of cattle, camel, sheep, and goats, which support their lives and livelihoods through milk and meat. Numerous people in both Rajasthan and Gujarat are engaged in the salt industry, a large employment generator in the region (Dave 2010; Sathyapalan et al. 2014). Most people also depend upon the scarce plant resources for food (e.g., from *Prosopis spicigera* and *Capparis decidua*), fuel (*Salvadora* and *Tamarix*), fiber (various grasses), thatch, weaving, and other uses (Ishnava et al. 2011).

Conservation Status and Management

Various wetlands in the region have been accorded different conservation status, and are under different management regimes. The southern part of the Gulf of Kachchh was declared as a Marine Sanctuary in 1980 (expanded to about 45,800 ha in 1982), and later, was designated as a Marine National Park by including islands and some of the intertidal areas for providing total protection over a consolidated area. Among other protected areas in Gujarat are the Gir, Black Buck, and Vansda National Parks, Indian Wild Ass Sanctuary, Kachchh Desert Wildlife Sanctuary, Narayan Sarovar Sanctuary, Kachchh Great Indian Bustard Sanctuary, Nalsarovar (also a Ramsar Site), Thol Lake, and Gaga Great Indian Bustard Wildlife Sanctuary (http://wiienvis.nic.in/Database/Gujarat_7821.aspx). The Wild Ass Sanctuary encompasses an area of 495,400 ha of the Little Rann of Kachchh and adjoining districts. In 2008, the Ministry of Environment and Forests designated Kachchh as a Biosphere Reserve (Pardeshi et al. 2010).

There are two conservation reserves – the Banni Grasslands Reserve and Chari-Dhand Wetland Conservation Reserve – of which the latter (located on the edge of the arid Banni grasslands and the Rann of Kachchh) receives water from the north-flowing rivers and runoff from the hills. Conservation reserves are government lands, usually adjacent to a protected area, identified by communities as important for biodiversity, as corridors for wildlife, or as buffer areas. The management of these reserves is similar to that of sanctuaries.

Within the arid region of Rajasthan there are several protected areas where salinity is low. For example, the Tal Chappar Sanctuary in Churu District is a unique arid, seasonal wetland which hosts the rare blackbuck *Antilope cervecapra*. The Keoladeo National Park, Bharatpur, is a World Heritage and Ramsar Site well known for its rich avifauna (over 350 species). It has large tracts of saline-alkaline soils and salt-tolerant vegetation (e.g., *Salvadora* species). Among the salt lakes, Sambhar lake is also a Ramsar Site.

Threats and Future Challenges

The major threats to natural wetlands stem from conflicting multiple demands on their natural resources, changing hydrology, industry, and anticipated climate change (Chauhan 2003). Widespread mining for minerals (bauxite, limestone, gypsum, etc.) and conversion into salt pans are serious threats. In 1999, the state government had identified a 58,900 ha area of the Wild Ass Sanctuary for the development of the salt industry. Water diversion for agriculture in areas upstream of wetlands is common threat throughout the region (Chauhan 2006; Sathyapalan et al. 2014). The growing need for freshwater, primarily for agriculture in coastal areas, has also resulted in the conversion of saline tidal marshes into shallow freshwater reservoirs by the construction of weirs at the mouths of estuaries to prevent tidal influx. This alters the nature of biodiversity and its functions, though its impacts have not yet been assessed. In Rajasthan, Sambhar and other such salt lakes receive reduced or even no inflows from seasonal rivers, as increasingly their water is used up in the catchment areas. Excessive abstraction of subsurface water for making salt in Rajasthan also impacts the salt lakes (Gopal et al. unpublished). It is further, worth mentioning that the provision of irrigation by the Indira Gandhi Canal in Rajasthan and the Narmada Canal system in Kachchh and Barmer (Rajasthan) has also impacted the native biodiversity.

The future of biodiversity of the Indian arid zone wetlands depends upon land and water management practices, to meet economic goals of development on the one hand and climate change on the other. This water-scarce region is witnessing large-scale changes in water resource management through extensive storages, diversions, and transfers from outside the region. An increasing frequency of extreme events, coupled with enhanced losses due to evapotranspiration as a result of rising temperatures, is a major challenge in the future management of these wetlands.

References

Chauhan M. Conserving biodiversity in arid regions – experiences with protected areas in India. In: Lemons J, Victor R, Schaffer D, editors. Conserving biodiversity in arid regions: best practices in developing nations. Dordrecht: Kluwer; 2003.

Chauhan M. Biodiversity vs irrigation – case of Keoladeo National Park. Econ Polit Weekly, 18 Feb 2006; p. 575–77.

Dave CV. Understanding conflicts and conservation of Indian wild ass around Little Rann of Kachchh, Gujarat, India. Final Technical Report. Rufford Small Grant Program; 2010. http://www.rufford.org/files/47.09.08%20Detailed%20Final%20Report_0.pdf.

FAO. Arid zone forestry: a guide for field technicians. Rome: FAO; 1989.

Gajera NB, Mahato AKR, Vijay Kumar V. Wetland birds of arid region-a study on their diversity and distribution pattern in Kachchh. Columban J Life Sci. 2012;13:47–51.

Godbole NN. Theories on the origin of salt lakes in Rajasthan, India. In: 24th International Geological Congress, 10. 1972; p. 354–7.

Gopal B, Sharma KP. Sambhar Lake. Ramsar sites of India. New Delhi: WWF; 1994.

Gupta V. Natural and human dimensions of semiarid ecology; a case of Little Rann of Kutch. J Earth Sci Clim Chang. 2015;6:311. https://doi.org/10.4172/2157-7617.1000311.

Hirway I, Goswami S. Valuation of coastal resources: the case of mangroves in Gujarat. New Delhi: Academic Foundation; 2007.

Ishnava K, Ramarao V, Mohan JSS, Kothari IL. Ecologically important and life supporting plants of Little Rann of Kachchh, Gujarat. J Ecol Nat Environ. 2011;3(2):33–8.

IUCN Red List of Threatened Species. Version 2015-4. www.iucnredlist.org. Downloaded on 15 January 2016.

Kingsford RT. Wetlands of the world's arid zones. In: Briefing paper at the First Conference of Parties of the UN Convention to Combat Desertification, vol. 29; 1997 Sept; Rome. Available from ftp://ftp.unccd.int/disk1/Library/Adlib_Catalogued_books/2_Loose_Leaf_Ramsar_World_Arid_Zones_UNCCD_1997.pdf. Accessed 9 Mar 2016.

Meena RL, Verma YL, Korvadiya VT, Pathak BJ, Kshatriya AR. Kachchh biosphere reserve: a management plan for protection, conservation research and development. Ahmedabad: Gujarat State Forest Department; 2005.

Mehta R. Status report on hydrology of arid zones of India. SR-2/1999-2000. Roorkee: National Institute of Hydrology; 2000. 59 p.

Pandey CN, Pandey R. Afforestation of mudflats in coastal Gujarat, India. In: Macintosh DJ, Mahindapala R, Markopoulos M, editors. Sharing lessons on mangrove restoration. Bangkok/Gland: Mangroves for the Future/IUCN; 2012. p. 123–31.

Pardeshi M, Gajera N, Joshi PN. Kachchh biosphere reserve: Rann and biodiversity. Res J For. 2010;4(2):72–6.

Prusty BAK. Need for conservation of wetlands in arid Kachchh region. Curr Sci. 2009;97:745–6.

SAC (Space Applications Centre). National wetland Atlas. SAC/EPSA/AFEG/NWIA/ATLAS/31/2010. Ahmedabad: Space Applications Centre (ISRO); 2010.

Sathyapalan J, Bhatt AM, Easa PS, Srinivasan JT, Shukla N, Jog P. Livelihoods of Agariyas and biodiversity conservation in the Little Rann of Kutch, Gujarat. Hyderabad: Centre for Economic and Social Studies; 2014.

Stanley OD. Wetland ecosystems and coastal habitat diversity in Gujarat, India. J Coastal Dev. 2004;7(2):49–64.

Vishwanathan PK, Pathak KD, Mehta I. Socio-economic and ecological benefits of mangrove plantation: a study of community based mangrove restoration activities in Gujarat. Gandhinagar: Gujarat Ecology Commission (GEC); 2010.

The Transboundary Sundarbans Mangroves (India and Bangladesh)

145

Brij Gopal and Malavika Chauhan

Contents

Introduction 1734
Location 1734
Extent 1735
Diversity of Wetland Types 1736
Significant Biodiversity 1736
Ecosystem Services 1737
Mangrove-Dependent People 1738
Conservation Status and Management 1738
Threats 1739
Future Challenges 1740
References 1740

Abstract

The Sundarban mangroves, the world's largest contiguous forested wetland system (estimated at about 10,000 sq km), lie in the deltas of the Ganga-Brahmaputra-Meghna rivers. Rivers Hooghly and Baleshwar form its western and eastern boundaries respectively whereas the river Harinbhanga (= Ichamati or Raimongal) demarcates the border between India and Bangladesh. The Sundarban delta has undergone rapid changes caused by neotectonic activity over the past millennium, and geomorphic processes of sediment accretion and erosion have influenced its extent. Habitat diversity of the Sundarban includes

B. Gopal (✉)
Centre for Inland Waters in South Asia, Jaipur, Rajasthan, India
e-mail: brij.ciwsa@gmail.com

M. Chauhan (✉)
Himmotthan Society, Dehradun, Uttarakhand, India
e-mail: malavikachauhan@gmail.com

© Springer Science+Business Media B.V., part of Springer Nature 2018
C. M. Finlayson et al. (eds.), *The Wetland Book*,
https://doi.org/10.1007/978-94-007-4001-3_26

freshwater, brackish and saline marshes, rice paddies and shrimp farms, besides the multilayered forest. The Sundarbans, extremely rich in its biodiversity are the only known mangrove habitat of the Bengal tiger (*Panthera tigris*). More than 12 million people live in and around the Sundarbans, of which 2.5 million depend almost entirely upon the mangroves for their livelihoods. Substantial efforts have been made for conservation of the Sundarbans by designating parts as sanctuaries, national parks, biosphere reserves and even as World Heritage in both India and Bangladesh. However, these mangroves are being impacted by water management strategies (including flow diversion) in the two countries, various human pressures – particularly conversion for agriculture and aquaculture, and are threatened by the sea level rise due to global climate change. The future of the Sundarban mangroves depends on our ability to manage efficiently the freshwater resources and effective adaptive responses to climate change.

Keywords
Ganga-Brahmaputra-Meghna basin · Bengal tiger · Sea level rise · Aquaculture · Cyclonic storms

Introduction

Mangroves are forested wetlands found along vast stretches of estuaries and tidal backwaters across tropical and subtropical coastal areas. They are adapted to the dynamic coastal environment, where a regular tidal flux and flushing flows of freshwater together with sediments and nutrients determine the distribution and patterns of biodiversity, ecosystems, and ecosystem functions. Approximately 42% of the world's mangrove areas lie in the river deltas of South and Southeast Asia (Giri et al. 2007, 2011), and these mangroves are amongst the richest in species worldwide (Ellison et al. 1999).

Location

Whereas most mangroves occur in specific, contained patches, the Sundarban mangroves which occupy the deltas of the Ganga-Brahmaputra-Meghna rivers (21° 30′ to 22° 40′N, 88° 05′ to 89° 55′ E) form the single largest contiguous forested wetland system in the world. Since 1947, the Sundarban mangroves have been shared between India and Bangladesh (formerly East Pakistan). The western and eastern limits of the Sundarbans are defined by the Rivers Hooghly (a distributary of the Ganga) and Baleshwar, respectively (Fig. 1). The River Harinbhanga (also known as the Ichamati or Raimongal in Bangladesh) demarcates the border between India and Bangladesh.

The Sundarbans essentially comprise numerous islands formed by sediments deposited by the three rivers and a dense network of smaller rivers, channels, and creeks. As the

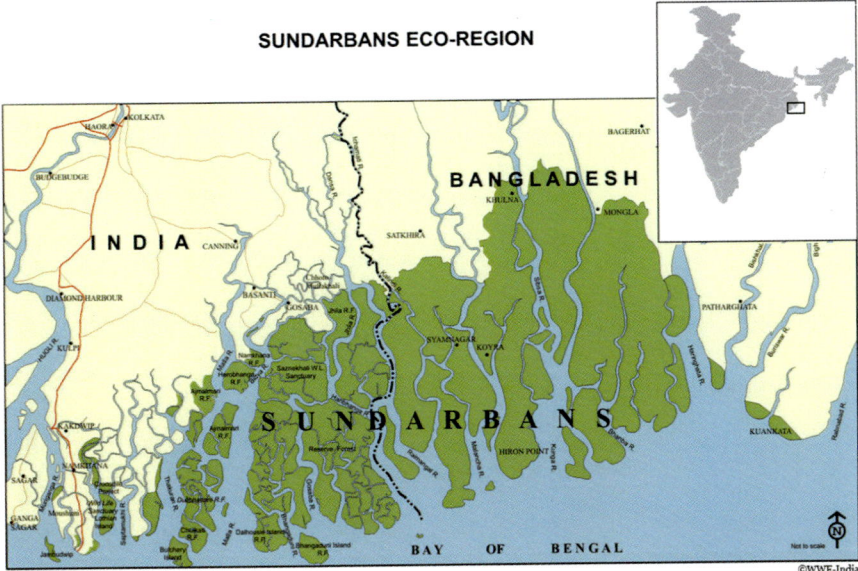

Fig. 1 The transboundary Sundarbans mangroves (India and Bangladesh) (© WWF-India, with permission)

maximum elevation within the Sundarbans is only 10 m above the mean sea level, tides ranging in height from 2.0 to 5.9 m flood large areas of the mangroves twice a day.

Extent

Geologically, the Sundarban delta developed relatively recently, following the filling of the Bengal basin with sediments carried down by Himalayan rivers. Neotectonic activity since the Tertiary period in north-western Punjab caused the River Ganga to flow southeast, and tectonic movements in the Bengal basin between the twelfth and fifteenth century AD resulted in an easterly tilt of the delta. During the sixteenth century, the Ganga again shifted its course eastwards and joined the Brahmaputra. By the mid-eighteenth century, the combined Ganga and Brahmaputra, now known as the Padma, tilted further eastwards to empty into the Meghna. Continuing tectonic activity shaped the hydrology of the deltaic region, through changes in sedimentation patterns and reduction in freshwater inflows (Snedaker 1991). Most rivers (distributaries) other than the Hooghly, that contributed in the formation of the Ganga Delta (from west to east: Muriganga, Saptamukhi, Thakuran, Matla, Gosaba, and Bidya), lost their original connections with the Ganga due to siltation, and their estuarine character is now maintained by monsoonal runoff alone. As a result of these changes, the delta-building process has nearly ceased in the west but has accelerated in the eastern part. An example of the high rates of sediment deposition in the Sundarbans is an increase in the land area by more than 80,000 ha during the period 1793 to 1870. Currently the mangroves are dominant geomorphic agents in the development of tidal shoals which contribute to the growth of the main landmass (Chakrabarti 1995).

It is estimated that the Sundarban mangroves originally covered more than 40,000 km^2 in coastal Bengal. Around 1980, the area was reported to be only about 10,000 km^2 (with an estimated 599,330 ha in 1978 in the Khulna District of Bangladesh (Rahman et al. 1979) and approximately 426,300 ha in the 24-Paragnas District of West Bengal, India (Sanyal 1983)). Through an analysis of the Geocover data set of Landsat images for the period 1973–2000 for the total Sundarbans covering 945,850 ha, Giri et al. (2007) observed that the mangrove forest area did not change significantly during the 30-year period, though it varied between 581,642 ha and 596,842 ha. In another study using Global Land Survey data of Landsat images, Giri et al. (2011) estimated the area of true mangroves at only 804,846 ha (436,570 ha in Bangladesh and 368,276 ha in India).

Diversity of Wetland Types

In common with other mangroves, the Sundarban mangroves are dominated by a tree layer. Shrubs, herbs, and grasses constitute the lower strata within the forest. Habitat diversity is promoted by the development of varied vegetation in response to salinity gradients (from fresh to highly saline), hydrological regimes, and the erosion and deposition of sediments. Within the Sundarbans one can recognize freshwater, brackish, and saline marshes as well as beds of submerged macrophytes in the creeks. The conversion of mangroves by villagers into rice paddies and shrimp farms has further resulted in expanses of these two kinds of man-made wetlands. Seaward of the Sundarban mangroves lie vast marine algal beds.

Significant Biodiversity

The Sundarban mangroves harbor a rich diversity of flora and fauna (Chaudhuri and Choudhury 1994; Hussain and Acharya 1994; GuhaBakshi et al. 1999). The inventory of species as total species richness, however, depends upon which areas are taken into consideration. Whereas only about 100 species of trees, shrubs, herbs, grasses, sedges, and ferns have been listed from the Indian Sundarban, the total flora of the Bangladesh Sundarban is estimated at 334 species (including many climbers, orchids, and ferns; Seidensticker and Hai 1983). Mangrove plants are usually divided into "true mangrove" (exclusively occurring in mangroves) and "mangrove associate" species. Of the 69 species of true mangroves worldwide, at least 30 species occur in the Sundarbans. Interestingly, the Sundarban mangroves have slightly fewer species than other mangroves on India's eastern coast. *Heritiera fomes* (or sundari, from which Sundarban derives its name) of the family Sterculiaceae and *Excoecaria* (Euphorbiaceae) are abundant in the Bangladesh part of the Sundarbans. Various studies show that in the Indian part, *H. fomes* has declined in recent decades due to an increase in salinity and is listed as an endangered species. *Sonneratia griffithii* is critically endangered, whereas *Ceriops decandra, Aegialitis rotundifolia, Phoenix paludosa*, and *Brownlowia tersa* are near threatened species (Polidoro et al. 2010; Barik and Chowdhury 2014; IUCN Red List 2015). The algal flora of the Sundarbans, though poorly investigated, comprises benthic, epiphytic, and planktonic forms from the freshwater to marine environments. A recent report from India listed 150 species, including a few red and brown algae. The Sundarbans also host 32 species of lichens (Santra 1998), 44 arbuscular mycorrhizal fungi (Kumar and Ghose 2008), and 62 macrofungal species (Dutta et al. 2013).

The Sundarbans are also known and valued for their faunal diversity. In addition to being the only known mangrove habitat of the Bengal tiger *Panthera tigris tigris* in the world, they support an enormous diversity and huge populations of fish and shrimp. Some 600 species of vertebrates in the Indian Sundarbans include 165 species of fish, 58 reptiles, 8 amphibians, 300 birds, and 40 mammals. Amongst the fish, the most valued is the hilsa *(Tenualosa) ilisha* which once migrated upstream in the Ganga, as far as its middle reaches. More species of fish (177) and birds (315) have been reported from the Bangladesh Sundarbans. Noteworthy are the spotted deer *Cervus axis*, wild boar *Sus scrofa*, three species of otters, wild cats *Felis bengalensis, F. chaus,* and *F. viverrina*, and the globally threatened Ganges river dolphin *Platanista gangetica*.

A few species of animals which were once common have disappeared locally. They include the Javan rhinoceros *Rhinoceros sondaicus*; water buffalo *Bubalus bubalis*; swamp deer *Rucervus duvauceli*; hog deer *Axis porcinus*; one-horned rhinoceros *Rhinoceros unicornis*; Indian bison (= gaur) *Bos gaurus*; and sambar deer *Rusa* (= *Cervus*) *unicolor*. The only primate, the rhesus macaque *Macaca mulatta* which still occurs in good numbers, is also disappearing (Gopal and Chauhan 2006). Some species are endangered (green sea turtle *Chelonia mydas*) or critically endangered (river terrapin *Batagur baska*) (IUCN Red List 2015).

The invertebrate fauna is dominated by crustaceans (240 sp.), insects (>200 sp.), molluscs (143 sp.), platyhelminthes (41 sp.), nematodes (68 sp.), and polychaetes (70 sp.). About 40 species of shrimp, crabs, and lobsters and 8 species of molluscs are commercially exploited.

Ecosystem Services

The Sundarban mangroves provide numerous ecosystem services. Provisioning services include the production of valuable timber (that once formed the sole source of railway sleepers in the region), fuelwood, fish and prawns, thatch, and honey. They provide habitats and food for a very high diversity of biota that includes the Bengal tiger. These mangroves contribute significantly to delta building and prevent erosion. They support enormous coastal fisheries, which depend on the mangroves for detritus and nutrients. Furthermore, the mangroves protect coastal human populations and assets from the frequent cyclonic storms of the region. The Sundarbans also have great cultural value, for services related to the life and society of local communities, which is closely interwoven with the resources of the mangroves. The Sundarbans form the central theme of many Bengali folk songs and dances, paintings and stories, arts and crafts, besides being associated with specific gods and goddesses. Sagar Island at the mouth of Hooghly estuary is visited by millions of believers who flock to the outflow of the River Ganga, where it meets the sea, for a holy bath.

Mangrove-Dependent People

More than 12 million people (2011 estimate; 4.5 million in India, 7.5 million in Bangladesh) live in and around the Sundarbans. Of these an estimated 2.5 million depend almost entirely upon the mangroves for their livelihoods (Kabir et al. 2007; Raha et al. 2013). They are engaged in rice paddy cultivation, shrimp farming, fishing, and fisheries related services. Collecting wild honey and wax from the forest is a specialized task of the *mawali* community. Similarly, the wood cutters (*bawalis*), leaf collectors, and fishermen (*jele*) are the main forest-dependent communities in Bangladesh. Many are engaged in grazing a large cattle population in the mangroves. Palm leaves (*Nypa fruticans* and *Phoenix paludosa*) are collected for thatch and a variety of other uses. Others are engaged in the transport of people and goods by boats. It is estimated that in Bangladesh about 50% of the forest-dependent households earn 75–100% of their total income from the mangrove forest resources and annually extract resources worth up to $1,000 per household for sale in the market.

Conservation Status and Management

During the past 60 years, several parts of the Sundarban mangroves have been protected in different ways. Soon after independence, India declared Lothian Island (3,800 ha) as a wildlife sanctuary, followed in 1960 with the Sajnakhali Wildlife Sanctuary (35,240 ha). A tiger reserve was established covering 258,500 ha in 1973 for the Bengal tiger, a globally endangered species, with an additional 24,100 ha demarcated as a subsidiary wilderness area. A core area of 133,000 ha was designated as a national park. In 1976, another wildlife sanctuary was established on Haliday Island (595 ha) to protect the spotted deer, wild boar, and rhesus macaque, which are major species in the forest dominated by *Ceriops decandra* (Gopal and Chauhan 2006).

Within Bangladesh, three wildlife sanctuaries – the Sundarban West (71,502 ha), Sundarban East (31,226 ha), and Sundarban South (36,970 ha) were created in 1977 under the Bangladesh Wildlife (Preservation) (Amendment) Act of 1974. These sanctuaries lie on disjunct deltaic islands in the Sundarban Forest Division of Khulna District, close to the border with India and just west of the main outflow of the three rivers (Gopal and Chauhan 2006).

Parts of the Sundarbans protected areas have also been inscribed on the World Heritage list in both India (1987) and Bangladesh (1997). Further, the entire Indian

Sundarban area south of the Dampier-Hodges line (demarcating the inward limit of tidal influence), including 536,600 ha of reclaimed lands, has been designated as the Sundarban Biosphere Reserve – with a core zone comprising of a national park, a tiger reserve, a manipulation zone (240,000 ha of mangrove forests), a restoration zone (24,000 ha of degraded forest and saline mud flats), and a development zone (including mostly the reclaimed areas). Only the core zone is under strict conservation measures. Collection of seeds of black tiger prawn *Penaeus monodon*, culturing of oysters and crabs, cultivation of mushrooms, and bee-keeping for honey production are allowed in the manipulation zone. Efforts are being made, however, to rehabilitate certain degraded areas through afforestation. Amongst the fauna, the estuarine crocodile *Crocodylus porosus* and the olive ridley sea turtle *Lepidochelys olivacea* are receiving some attention by way of captive breeding in the Sundarban Tiger Reserve in India (Roy Chowdhury et al. 2006).

The Bangladesh Sundarban is managed as a refuge where wildlife is protected in small sanctuaries located within the larger forest tract. Resource "hotspots" essential to the maintenance of wildlife populations are provided protection. Core areas surrounded by restricted-use buffer zones are expected to ensure the survival of large mammals and birds dependent upon the area for food or nesting and roosting sites. In India, larger areas are set aside for conservation, taking into consideration all ecological needs of wildlife. In both countries the current management focuses on silviculture, zoning, protection of existing natural forests, nonexploitive uses, and plantations. However, the management of mangroves requires an ecosystem approach (Iftekhar 2008) and a recognition of hydrology as the main driver of their specific attributes (Gopal 2013).

Threats

The Sundarbans have been exploited for centuries by clear felling the forests for timber, conversion of mangrove to paddy fields, and reclamation of land for various other uses. The former Turk sultan rulers of Bengal were the first to promote forest clearing and rice cultivation. Later, the British administration also actively promoted deliberate conversion to agriculture. The "waterlogged forests and swamps of the lower delta" were managed "to ensure that private landowners cleared, settled and reclaimed Sundarban forests and swamps for rice cultivation." Between 1776 and 1968, 428,000 ha of mangroves had been cleared (Ghosh et al. 2015). An additional 32,400 ha of forest was reclaimed within the next 20 years within India despite the Sundarban mangroves being declared as protected or reserve forest. The protected forests were available for clearance on leasing, while timber extraction was allowed in the reserve forests.

Major threats to the Sundarban mangroves continue to be various societal pressures, particularly the conversion for agriculture and aquaculture (Seidensticker et al. 1991). A major threat of a rather complex nature comes from human-induced hydrological alterations (Chauhan and Gopal 2013). While freshwater flows diverted upstream for irrigation, flood control, and hydropower have an effect on the Sundarbans, the diversion of flow of the river Ganga into the Hooghly River by the Farakka barrage (on the Indian side of the border with Bangladesh) to keep the channel navigable up to Calcutta Port has caused a noticeable change in salinity gradients and vegetation (Mirza 1997; Mirja and Mirza 2005). The hilsa fisheries have declined considerably within India as the fish are unable to migrate upstream to freshwater breeding sites. However, recent reports suggest that the hilsa has started recovering in freshwater areas of the Indian Sundarbans.

Within Bangladesh, the situation is compounded by extensive (about 3,700 km) embankments constructed between 1960 and 1970. The altered flooding regimes have increased salinity intrusion, promoted erosion, accelerated siltation, and reduced nutrient exchanges.

Hydrological alteration is also being experienced as a result of a rapid rise in sea level, resulting from human-induced climate change (Pethick and Orford 2013). The projected rise in sea level by about 30–50 cm by the end of the current century will submerge significant areas and impact the biodiversity, especially the tigers (Loucks et al. 2010). The consequent rise in salinity is seen as an important threat (Zaman et al. 2013). The Sundarbans are also facing an increasing frequency of tropical cyclonic storms. The cyclones Sidr (2007) and Aila (2009) caused widespread devastation in quick succession (Bhowmik and Cabral 2013; Saha 2015).

Among other threats, the most noteworthy is pollution from both the landward and seaward sides of the Sundarbans. The agrochemicals (fertilizers and pesticides) used extensively in the catchments of the Ganga and Brahmaputra rivers and their numerous tributaries reach and accumulate in the sediments and biota of the Sundarbans (Rahman et al. 2009; Manna et al. 2010). On the seaward side, oil spills on the sea as well as off-shore exploration for oil and gas by both Bangladesh and India threaten the aquatic biota (Saha et al. 2005).

Future Challenges

The mangroves have a long and continuing history of being managed as forests, based on the classical concepts of management of terrestrial forests and protected areas. Despite their recognition as wetlands, management policies and plans have yet to change to reflect their characteristic ecosystem attributes and driving variables (Gopal 2013). The future of the Sundarban mangroves hinges upon humanity's ability to manage efficiently the limited, available freshwater resources for meeting both human and environmental needs, coupled with effective adaptive responses to climate change that is causing increased spatial and temporal variability in precipitation as well as a rise in sea level (Gopal and Chauhan 2006).

References

Barik J, Chowdhury S. True mangrove species of Sundarbans delta, West Bengal, Eastern India. Check List. 2014;10(2):329–34.

Bhowmik AK, Cabral P. Cyclone Sidr impacts on the Sundarbans floristic diversity. Earth Sci Res. 2013;2(2):62–79.

Chakrabarti P. Evolutionary history of the coastal quaternaries of the Bengal Plain. Proc Indian Natl Sci Acad. 1995;61A:343–54.

Chaudhuri AB, Choudhury A. Mangroves of the Sundarbans. Volume 1: India. Gland: World Conservation Union; 1994. p. 247.

Chauhan M, Gopal B. Sundarban Mangroves: impact of water management in the Ganga River Basin. In: Sanghi R, editor. Our national river Ganga: lifeline of millions. Dordrecht: Springer; 2013.

Dutta A, Pradhan P, Basu SK, Acharya K. Macrofungal diversity and ecology of the mangrove ecosystem in the Indian part of Sundarbans. Biodiversity. 2013;14:196–206.

Ellison AM, Farnsworth EJ, Merkt RE. Origins of mangrove ecosystems and the mangrove biodiversity anomaly. Global Ecol Biogeogr. 1999;8:95–115.

Ghosh A, Schmidt S, Fickert T, Nüsser M. The Indian Sundarban mangrove forests: history, utilization, conservation strategies and local perception. Diversity. 2015;7:149–69.

Giri C, Pengra B, Zhu Z, Singh A, Tieszen LL. Monitoring mangrove forest dynamics of the Sundarbans in Bangladesh and India using multi-temporal satellite data from 1973 to 2000. Estuar Coast Shelf Sci. 2007;73:91–100.

Giri C, Ochieng E, Tieszen LL, Zhu Z, Singh A, Loveland T, Masek J, Duke N. Status and distribution of mangrove forests of the world using earth observation satellite data. Global Ecol Biogeogr. 2011;20:154–9.

Gopal B. Mangroves are wetlands, not forests: some implications for their management. In: Hanum F, Mohamad L, Hakeem KR, Ozturk M, editors. Mangrove ecosystem in Asia: current status, challenges and management strategies. Berlin: Springer; 2013.

Gopal B, Chauhan M. Biodiversity and its conservation in the Sundarban Mangrove Ecosystem. Aquat Sci. 2006;68:338–54.

GuhaBakshi DN, Sanyal P, Naskar KR, editors. Sundarbans Mangal. Calcutta: Naya Prokash; 1999. p. 771.

Hussain Z, Acharya G, editors. Mangroves of the Sundarbans. Vol. 2: Bangladesh. Gland: World Conservation Union; 1994. p. 257.

Iftekhar MS. An overview of mangrove management strategies in three South Asian countries: Bangladesh, India and Sri Lanka. Int For Rev. 2008;10(1):38–51.

Kabir DMH, Hossain J. Sundarban reserve forest: an account of people's livelihood and biodiversity conservation. Unpublished report. Dhaka: Unnayan Onneshan; 2007.

Kumar T, Ghose M. Status of arbuscular mycorrhizal fungi (AMF) in the Sundarbans of India in relation to tidal inundation and chemical properties of soil. Wetland Ecol Manag. 2008;16:471–83.

Loucks C, Barber-Meyer S, Hossain Md AA, Barlow A, Chowdhury RM. Sea level rise and tigers: predicted impacts to Bangladesh's Sundarbans mangroves. Clim Change. 2010;98:291–8.

Manna S, Chaudhuri K, Bhattacharyya S, Bhattacharyya M. Dynamics of Sundarban estuarine ecosystem: eutrophication induced threat to mangroves. Saline Systems. 2010;6:8. http://www.salinesystems.org/content/6/1/8. Accessed 28 Dec 2015.

Mirja EMK, Mirza MMQ, editors. The Ganges water diversion: environmental effects and implications. New York: Springer; 2005. p. 364.

Mirza MMQ. Hydrological changes in the Ganges system in Bangladesh in the post-Farakka period. Hydrol Sci J. 1997;42:613–31.

Pethick J, Orford JD. Rapid rise in effective sea-level in southwest Bangladesh: its causes and contemporary rates. Global Planet Change. 2013;111:237–45.

Polidoro BA, Carpenter KE, Collins L, Duke NC, Ellison AM, et al. The loss of species: mangrove extinction risk and geographic areas of global concern. PLoS One. 2010;5(4), e10095. doi:10.1371/journal.pone.0010095.

Raha AK, Zaman S, Sengupta K, Bhattacharya SB, Raha S, Banerjee K, Mitra A. Climate change and sustainable livelihood programmes: a case study from the Indian Sundarbans. J Ecol Photon. 2013;107:335–48.

Rahman N, Billah MM, Chaudhury MU. Preparation of an up to date map of Sundarban forests and estimation of forest areas of the same by using Landsat imageries. Second Bangladesh National Seminar on Remote Sensing (9–15 December), Dhaka; 1979.

Rahman MM, Chongling Y, Islam KS, Haoliang L. A brief review on pollution and ecotoxicologic effects on Sundarbans mangrove ecosystem in Bangladesh. Int J Environ Eng. 2009;1:369–83.

Roy Chowdhury B, Das BK, Ghose PS. Marine turtles of West Bengal. In: Shanker K, Choudhury BC, editors. Marine turtles of the Indian subcontinent. New Delhi: Universities Press (India); 2006. p. 107–16.

Saha CK. Dynamics of disaster-induced risk in southwestern coastal Bangladesh: an analysis on tropical Cyclone Aila 2009. Nat Hazards. 2015;75:725–54.

Saha SK, Roy K, Banerjee P, Al Mamun A, Rahman MA, Ghosh GC. Technological and environmental impact assessment on possible oil and gas exploration at the Sundarbans coastal region. Int J Ecol Environ Sci. 2005;31:255–64.

Santra SC. Mangrove lichens. Indian Biol. 1998;30(2):76–8.

Sanyal P. Mangrove tiger land: the Sundarbans of India. Tigerpaper. 1983;10(3):1–4.

Seidensticker J, Hai MA. The Sundarbans wildlife management plan. Conservation in the Bangladesh Coastal Zone. Gland: IUCN; 1983. p. 120.

Seidensticker J, Kurin R, Townsend AK, editors. The Commons in South Asia: societal pressures and environmental integrity in the Sundarbans of Bangladesh. Proceedings from a workshop held November 20–21, 1987, Washington, DC. Washington, DC: The International Center Smithsonian Institution; 1991.

Snedaker C. Notes on the Sundarbans with emphasis on geology hydrology and forestry. In: Seidensticker J, Kurin R, Townsend AK, editors. The commons in South Asia: societal pressures and environmental integrity in the Sundarbans. Washington, DC: The International Center, Smithsonian Institution; 1991.

The IUCN Red List of Threatened Species. Version 2015–4. www.iucnredlist.org. Downloaded on 15 Jan 2016.

Zaman S, Bhattacharyya SB, Pramanick P, Raha AK, Chakraborty S, Mitra A. Rising water salinity: a threat to mangroves of Indian Sundarbans. Community Environ Disaster Risk Manag. 2013;13:167–83.

Wetlands of Mahanadi Delta (India)   146

Ritesh Kumar and Pranati Patnaik

Contents

Introduction	1744
Wetland Distribution	1744
Biodiversity and Ecosystem Services	1747
Threats	1748
Management	1748
Future Challenges	1749
References	1749

Abstract

The Delta of Mahanadi, a major east flowing peninsular river of India, has a diverse regime of wetlands closely governed by landforms and hydrological regimes. The upper delta region, having a fanning distributary system with intervening alluvial plains, has wetlands mostly in the form of small fresh waterbodies developed as depressions on both flanks of the Rivers Mahanadi and Brahmani. The central delta, with predominantly clayey soil has wetlands primarily in the form of floodplains connected to active river channels as well as paleo-channels. Also located within the central delta are ephemeral wetlands located in the alluvial floodplain zones. Within the narrow coastal plains, wetland distribution is governed by the interaction between coastal and freshwater processes. Extensive hydrological fragmentation and landuse changes have significantly altered water and sediment distribution patterns within the delta, and loss of connectivity between river channels and floodplains. Owing to a range of pressures, the wetland area has undergone at least 32% decline during the period between 1975 and 2010. The State Government of Odisha constituted the State

R. Kumar (✉) · P. Patnaik (✉)
Wetlands International South Asia, New Delhi, India
e-mail: Ritesh.kumar@wi-sa.org; pranati.patnaik@wi-sa.org

© Springer Science+Business Media B.V., part of Springer Nature 2018
C. M. Finlayson et al. (eds.), *The Wetland Book*,
https://doi.org/10.1007/978-94-007-4001-3_28

Wetland Development Authority in 2011 for undertaking integrated management of the entire wetlands of the state, including those of the Mahanadi Delta. The government has initiated management planning for wetlands with high biodiversity value on a priority basis. Changing climate within the deltaic landscape further reinforces the urgency for securing wetland functioning within a rapidly developing delta.

Keywords

Delta · Floodplains · Ephemeral wetlands · Mangroves · Embankments

Introduction

The Delta of Mahanadi River spans 13,871 km^2 (including the area of Chilika lagoon and its direct catchment) between $19^0 40'$ – $20^0 35'$ N latitude and $85^0 40'$ – $86^0 45'$ E longitude around the confluence of River Mahanadi with the Bay of Bengal on the east coast of India. Mahanadi is one of the major east flowing peninsular rivers of the country flowing for 958 km and draining a basin of 139,681 km^2 across the states of Chattisgarh, Maharashtra, Jharkhand, and Odisha before meeting with the sea (CWC and NRSC 2014). The delta apex is located about 100 km from the Bay of Bengal coastline at Naraj wherein the River Mahanadi bifurcates into the Mahanadi and Kathjodi channels and further into several distributaries giving rise to 8 *doabs* (deltaic land between two distributary channels). Delta formation is attributed to a tectonic down wrap of the Gondwana graben (Jagannathan et al. 1983) with major delta building processes placed between 6,000 years BP and 800 years BP and the shoreline shifting seaward during the last 800 years (Mohanti 1993).

The River Mahanadi receives the flows of around 78 tributaries directly and 22 indirectly to build up a discharge of 66,640 Mm^3 while discharging into the Bay of Bengal. Nearly 30 million tons of sediments carried annually with this flow (GoO 1986) helps delta build up through its spread in the floodplains and along the shoreline by mangrove swamps and tidal flats. The deltaic environment is principally monsoon-dominated (receiving an annual rainfall of around 1,000 mm). Hirakud Dam, located 290 km upstream of Naraj with a gross storage capacity of 7,189 Mm^3 and a catchment area of 83,400 km^2, has a significant influence on the flows and sediment received within the delta. Highly fertile soils and abundance of water have made the Mahanadi Delta a hub of economic activities. As per population census of 2011 (Census of India 2011), 7.96 million people inhabit the delta constituting 19% of the total population of state while forming only 9% of its total geographical area.

Wetland Distribution

The distribution of wetlands in the Mahanadi Delta is closely related to the landforms and hydrological regimes (Fig. 1). The delta region can be classed in three subregions, namely upper delta, central delta, and coastal plains. The upper delta

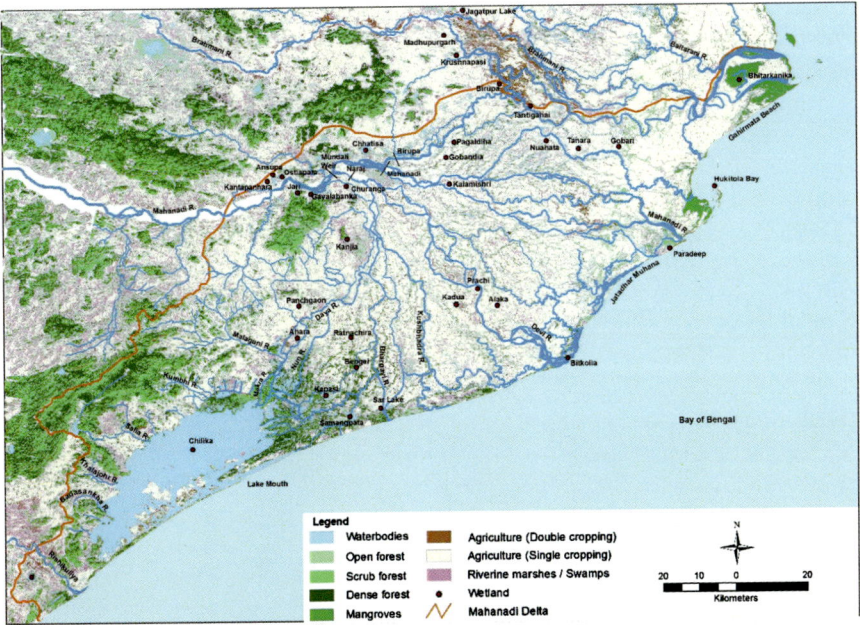

Fig. 1 Wetlands of Mahanadi Delta (© Provided courtesy of Wetlands International South Asia, based on Satellite imagery IRSP6, LISS 3 of January 24, 2010)

region has a fanning distributary system with intervening alluvial plains which are primarily used for agriculture. The region constitutes 29% of delta's area, has elevations ranging between 20 and 150 masl, with 43% area under forests and 41% under cultivation. Wetlands in this region are mostly in the form of small fresh waterbodies locally known as *pata* developed as depressions on both flanks of the Rivers Mahanadi and Brahmani. These receive runoff from the local catchments as well as from the rivers. Ansupa, Ostia, Kantapanhara, and Chattisa are important wetland formations to the north of the delta head, whereas Kanjia, Jari, Churanga, and Gayalabanka lie to the south. Madhupurgarh pata and Krushnapasi pata are major wetlands to the south of the Brahmani River, with the Jagatpur lakes on the northern side. Lake Kanjia also forms a part of the Nandankannan Zoological Park.

The central delta, constituting 53% of the total delta area is flat land with elevations ranging between 5 and 20 masl. This region is predominantly clayey with much of the area intensely cultivated. A dense network of embankments impede drainage creating waterlogging conditions during post monsoon and winter months. Wetlands in the central delta are primarily in the form of floodplains connected to active river channels as well as paleo-channels. Also located within the central delta are a number of ephemeral wetlands. Gobari, Hansua, Kadua, Bengai, and Ratnachira are floodplain wetlands connected to the fluvial zones. Pagaldiha, Nuahata, Gobandia, and Tanara are major ephemeral wetlands in this region.

The narrow coastal plains runs parallel to the Bay of Bengal coastline and constitutes 15% of the total delta area, a major part (68%) of which is accounted for by wetlands. The interaction between freshwater and coastal processes gives rise to some unique geomorphological features along the delta coastline. In the Mahanadi mouth, a complex-spit with a number of hooks, prominent being Hukitola Bay and Jatadhar Muhan, are formed due to the offshore long current and the strong longshore drift during the rainy season when the sediment load discharge in the Mahanadi is high. The high tidal prism keeps the mouths of the Devi, the Mahanadi, the Brahmani, the Baitarani, and the Rushikulya open to form estuaries. Lake Chilika, a brackish coastal lagoon spanning 1,165 km^2, marks the southwestern limit of the delta. Formation of Prachi and Bitkolia Muhana is attributed to changes in the fluvial distributary systems leading to development of paleo-channels and defunct river mouths.

Bhitarkanika, a tidal swamp at the opening of Rivers Brahmani and Baitarni to the sea, has a luxuriant growth of mangroves extending 202 km^2 (Ravishankar et al. 2004), the second largest contiguous mangrove area on the east coast of India. Mangroves are also found in small patches at the mouth of the Rivers Mahanadi and Devi.

Bhitarkanika and Lake Chilika have been designated as Wetlands of International Importance (Ramsar Sites) under the Ramsar Convention on Wetlands. The Bhitarkanika Conservation Area, which includes Bhitarkanika Wildlife Sanctuary and National Park covering 672 km^2, the Gahirmatha (Marine) Wildlife Sanctuary covering 1,435 km^2, and a buffer zone of 47.26 km^2, has been recommended by the Government of India for declaration as a World Heritage Site under UNESCO.

Vegetation in the delta wetlands is characterized by grasses and sedges (e.g., *Oryza rufipogon, Eleocharis dulcis, Cyperus platystylis, C. iria in association with Adenostemma lavenia, Ludwigia octovavis, L. adscendens, Limnophyton obtusifolium, Monochoria hastate, Sagittaria trifolia*) and emergents (dominant being *Typha angustifolia, Phragmites karka,* and *Scirpus grossus*). Patches of *Adenostemma lavenia, Polygonum glabrum, P. barbatum, Cyperus platystylis,* and *C.s cephalotes* are also present in some wetlands. Open water areas have submerged and floating vegetation forms as *Ottelia alismoides, Aponogeton natans, Vallisneria natans, Hydrilla verticillata, Najas minor,* and *Ceratophyllum demersum. Salvinia cucullata, S. natans, Utricularia stellaris, U. flexuosa, Azolla pinnata, Lemna globosa, Pistia stratiotes, and Eichhornia crassipes* are the major floating forms. Blooms of *A. pinnata* and *L. globosa* are common during the monsoon and post-monsoon periods, especially in central deltaic and delta-head wetlands. Salinity has major influence on vegetation of the coastal wetlands. *Potamogenon* sp. and *Najas* sp. are usually observed in low salinity areas of coastal wetlands, whereas *Ruppia* sp. and *Halophila* sp. are found in areas of high salinity conditions (Patnaik et al. 1990).

Biodiversity and Ecosystem Services

The wetlands of the Mahanadi Delta support rich floral and faunal diversity. Chilika, its fringe areas, and islands serve as habitat for 729 plants, 314 fish, and 224 bird species (Kumar and Pattnaik 2012). The avian diversity at Chilika includes 129 waterbirds of which 97 are intercontinental migrants from Arctic Russia, West Asia, Europe, Northeast Siberia, and Mongolia. The total population of waterbirds at this site is known to exceed one million. The flagship cetacean, Irrawaddy dolphin *Orcaella brevirostris* also inhabits the wetland with an increasing trend in population and habitat use (ibid). With 26 true mangrove species (Selvam 2003) and four mangrove-associate species (Badola and Hussain 2003), Bhitarkanika has the highest mangrove diversity in India. It is also known to support 220 bird species, 57 of which are winter visitors. Gahirmatha Beach, Devi Mouth, and Rishikulya Estuary support extensive rookeries of globally vulnerable olive ridley sea turtle *Lepidochelys olivacea*. Bhitarkanika is also famed for having the high density of saltwater crocodiles *Crocodylus porosus*. There are records of 31 species of mammals in Bhitarkanika including five species of marine dolphins, i.e., the humpbacked dolphin *Sousa* sp., Irrawaddy dolphin, pantropical spotted dolphin *Stenella attenuate*, common dolphin *Delphinus delphis*, and finless black porpoise *Neophocaena phocaenoides* (UNESCO 2009). A total of 253 species of fish have been recorded from the River Mahanadi, of which 160 species are found in the delta region (Chakrabarti 1999).

The rich biodiversity of the wetlands forms the basis of a range of ecosystem services. The total inland, brackish, and marine fish catch from the delta region in 2001 was 1.6 million MT, accounting 48% of the total Odisha State catch. Inland wetlands also constitute an important source for drinking water and irrigation. People attach high aesthetic, educational, cultural, and spiritual values to these wetlands. A majority of the places of tourist interest in Odisha State lie in the delta. The solitary hillocks within the periphery of the delta, coastal sands, sea shore, lakes, swamps, river courses, and river mouths have been sites of human interest since time immemorial. Some of the sites are associated with religious practices such as the spread of Buddhism in Odisha after the Kalinga war near Daya River. Chilika is visited annually by 0.3 million domestic and international tourists generating an economy of US$ 46 million annually (Kumar and Pattnaik 2012).

As the delta is frequently affected by floods and cyclones, the wetlands play an important role in buffering the impacts of extreme events. Assessments conducted after the super cyclone of 1999 highlighted the role of mangroves in providing protection against storms. Villages with wider mangrove belts between them and the coast experienced significantly fewer deaths than those with narrower belts or no mangroves (Das and Vincent 2009). The opportunity cost of saving a life by retaining mangroves was assessed to be Rs. 11.7 million per life saved (ibid).

Threats

The Mahanadi Delta has been subject to high degree of hydrological fragmentation by structures aimed at providing water for agriculture and domestic use. Historically, delta cultivators evolved farming practices which adequately distributed crop failure risks emerging through recurrent floods and droughts (D'Souza 2006). Cropping benefited immensely from the floods and ensuing inundation which led to natural fertilization of agriculture lands. The onset of nineteenth century marked extensive efforts for regulation of the water regimes, with an intent to raise revenues by controlling water supply to irrigators. Series of hydraulic structures were constructed to harness flows of various delta tributaries. The Naraj spur was constructed in 1856–1863 to maintain discharge distribution between various branches of Mahanadi River. Simultaneously, the Mahanadi anicut at Jobra (Cuttack) and Birupa anicut at Jagatpur were constructed along with canal systems to irrigate the deltaic land. Hirakud Dam was constructed in 1958 as a multipurpose project supporting irrigation, hydropower, and flood control. A weir was constructed at Mundali at the head of delta to support irrigation in lower parts of delta. The Naraj spur was subsequently converted to a barrage. Embankments were constructed to supply the stored water to agriculture farmers, against the wishes of communities apprehensive of implications of waterlogging and decline in agricultural productivity (D'Souza 2002). Later development in the delta emphasized on the extension of these activities without reviewing their long-term implications. Hydrological regime fragmentation was coupled in the later part of the twentieth century by loss of forest cover and extensive conversion of wetlands for permanent agriculture and settlements.

Extensive hydrological fragmentation and landuse changes have gradually converted a "flood dependent" landscape into being "flood vulnerable." The deltaic build-up processes have been impeded with as much as 67% reduction observed in the river sediments reaching downstream (Gupta et al. 2012) leading to delta's shrinkage. As much as 0.27 million ha of the delta area is recorded as being extensively waterlogged due to drainage congestion created by embankments which disconnect floodplains with river channels (Khatua and Patra 2004). Analysis of remote sensing imageries of 1975 and 2010 indicate that the total area of wetlands has undergone a 32% decline (WISA 2010). Along the coastline, extensive conversion of mangroves for aquaculture and settlements took place during the 1950s and 1970s (Ravishankar et al. 2004). In wetlands of delta head and central delta, nutrient enrichment due to discharge of untreated sewage and agricultural runoff and invasion by *Eichhornia sp.* and *Phragmites karka* are common.

Management

The State Government of Odisha constituted the State Wetland Development Authority in 2011 for undertaking integrated management of the all wetlands of the state, including those of the Mahanadi Delta. The government has initiated management planning for wetlands with high biodiversity value on a priority basis. Chilika

Development Authority (CDA) was constituted in 1991 for the restoration of Lake Chilika. The Authority initiated implementation of an integrated restoration program in 1997 leading to rejuvenation of the wetland system and rapid recovery of resource base. Chilika became the first wetland site in India to be removed from Ramsar's Montreux Record (in 2001) and the initiative recognized with a Ramsar Award to CDA in 2002. Since September 2010, an integrated coastal zone management project is also being implemented by the Government of Odisha in two stretches of the Mahanadi Delta (with financial support of the World Bank) to enable coordination of activities and programs of different government departments and agencies for sustainable management of coastal resources. A regional coastal processes study has led to mapping of the sediment cell and sediment budget to enable management of various developmental activities within the coastline. A detailed mapping of areas of coastal erosion and accretion has also been completed. The project will ultimately lead to development of an integrated coastal zone management plan for the state.

Future Challenges

The wetlands of Mahanadi Delta are located in an ecologically fragile, yet highly developing economic landscape. Restoration of Lake Chilika is an apparent demonstration of the importance of wetland restoration, both for biodiversity values as well as securing well-being of dependent communities. It is therefore of utmost importance that developmental programming in the delta takes into account the wide ranging ecosystem services and biodiversity values of its wetlands. Changing climate within the deltaic landscape further reinforce the urgency for securing wetland functioning within a rapidly developing delta. Odisha State level assessments of monsoonal patterns indicate declining dry season rainfall and intensifying monsoon rainfall with intensification of peaks (Ghosh and Mujumdar 2006). Variability in river flows, particularly increasing monsoon season flows, is also predicted along with high probability of extreme hydrological events (Gosain et al. 2006). The Bay of Bengal has recorded the maximum annual sea level rise within the Indian coast (Unnikrishnan and Shankar 2007). Such changes are likely to increase intensity as well as variability in extreme events. The value of wetlands as providers of water and food security and buffer of extreme events are therefore important components of climate change adaptation planning in this deltaic landscape.

References

Badola R, Hussain SA. Valuation of the Bhitarkanika mangrove ecosystem for ecological security and sustainable resource use. Study report. Deheradun: Wildlife Institute of India; 2003.
Census of India. Provisional population totals – Orissa. New Delhi: Office of the Registrar General and Census Commissioner, Ministry of Home Affairs, Government of India; 2011.
Central Water Commission (CWC) and National Remote Sensing Centre (NRSC). Mahanadi Basin. New Delhi/Hyderabad: Central Water Commission (CWC), Ministry of Water Resources/

National Remote Sensing Centre (NRSC), ISRO, Department of Space, Government of India; 2014.

Chakrabarti PK. River Mahanadi- environment and fishery. In: Sinha M, Khan MA, Jha BC, editors. Ecology, fisheries & fish-stock assessment of Indian rivers. Barrackpore: Central Inland Capture Fisheries Research Institute; 1999. p. 104–12.

D'Souza R. Colonialism, capitalism and nature: debating the origins of Mahanadi delta's hydraulic crisis (1803–1928). Econ Pol Wkly. 2002;37(13):1261–72.

D'Souza R. Drowned and dammed: colonial capitalism and flood control in Eastern India. New Delhi: Oxford University Press; 2006.

Das S, Vincent JR. Mangroves protected villages and reduced death toll during Indian super cyclone. Proc Natl Acad Sci U S A. 2009;106(18):7357–60.

Ghosh S, Mujumdar PP. Future rainfall scenario over Odisha with GCM projections by statistical downscaling. Curr Sci. 2006;90(3):396–404.

GoO–Government of Orissa. Delta development plan mahanadi delta command area: geology geomorphology and coast building, vol. IV. Irrigation Department, Government of Orissa; 1986. [Unpublished Report with Engineer in-Chief].

Gosain AK, Rao S, Basuray D. Climate change impact assessment on hydrology of Indian river basins. Curr Sci. 2006;90(3):346–53.

Gupta H, Kao S, Dai M. The role of Mega Dams in reducing sediment fluxes: a case study of large Asian rivers. J Hydrol. 2012;464–465:447–58.

Jagannathan CR, Ratnam C, Baishya NC, Dasgupta U. Geology of the offshore Mahanadi basin. Petroleum Asia J. 1983;IV(4):101–4.

Khatua KK, Patra PC. Management of high flood in Mahanadi and its tributaries below Naraj. Published in 49th Annual Session of IEI (India). Bhubaneswar: Orissa State Center; 2004.

Kumar R, Pattnaik A. Chilika- an integrated management planning framework for conservation and wise use. New Delhi/Bhubaneswar: Wetlands International- South Asia/Chilika Development Authority; 2012.

Mohanti M. Coastal processes and management of the Mahanadi River Deltaic complex, East Coast of India. In: Kay R, editor. Deltas of the world. New York: American Society of Civil Engineers; 1993.

Patnaik SS, Mahalik NK, Bhunya SP. Mapping and characterization of the wetlands along the Eastern coast of Orissa. New Delhi: Ministry of Environment and Forest; 1990.

Ravishankar R, Navamuniyammal T, Gnanappazham L, Nayak SS, Mahapatra GC, Selvam V. Atlas of mangrove wetlands of India. Chennai: M.S. Swaminathan Research foundation; 2004.

Selvam V. Environment classification of mangrove wetlands of India. Curr Sci. 2003;84(6):757–65.

UNESCO. Bhitarkanika conservation area. 2009. Available from: http://whc.unesco.org/en/tentativelists/5446/. Accessed 17 Oct 2012.

Unnikrishnan AS, Shankar D. Are sea-level-rise trends along the North Indian Ocean coasts consistent with global estimates? Global Planet Change. 2007;57(3–4):301–7.

Wetlands International South Asia. Wetlands and livelihood project technical report. New Delhi: Wetlands International South Asia; 2010.

Section XI
Southeast Asia

Tropical Peat Swamp Forests of Southeast Asia

147

Susan Page and Jack Rieley

Contents

Introduction	1754
The Peat Swamp Environment	1754
Peat Swamp Forest Vegetation and Fauna	1756
Ecosystem Service Provision	1759
Threats and Future Challenges	1759
References	1760

Abstract

Tropical peatlands of Southeast Asia occupy an estimated area of 248,000 km^2 (56% of the global resource). They store ~50 gigatonnes (Gt = t × 10^9) of carbon. Lowland tropical peat consists of organic matter from the partially decomposed remains of trees that accumulates to a thickness of 10 m or more. It is characterized by low ash and nutrient contents and high acidity. Under natural conditions, the water table is close to or at the peat surface for much of the year. The natural vegetation is forest that forms a continuum of different types from the edge to the centre of the peat dome. Many of the trees show adaptations to the waterlogged environment, for example, buttress and stilt roots and pneumatophores. Peat swamp forests have high biodiversity and are important habitats for many species of mammals, birds, fish and amphibians, some of which are endangered. In addition, they provide an array of other important ecosystem services. Land use changes are causing rapid loss and degradation of peat swamp forests driven by

S. Page (✉)
Department of Geography, University of Leicester, Leicester, UK
e-mail: sep5@le.ac.uk

J. Rieley
School of Geography, University of Nottingham, Nottingham, UK
e-mail: jack.rieley@btinternet.com

© Springer Science+Business Media B.V., part of Springer Nature 2018
C. M. Finlayson et al. (eds.), *The Wetland Book*,
https://doi.org/10.1007/978-94-007-4001-3_5

timber extraction and plantation establishment. Peat oxidation and fire have contributed substantially to global greenhouse gas emissions. Prospects for peat swamp forests are not promising, although initiatives such as the UN REDD+ programme may ensure that some remaining forests are safeguarded.

> **Keywords**
> Biodiversity · Ecosystem services · Fire · Land use change · Peat characteristics · Tropical peatland

Introduction

Tropical peatlands have a global area of around 441,000 km^2 (Page et al. 2011) and reach their greatest extent in Southeast Asia, where there are 248,000 km^2, representing 56% of the total global tropical peatland resource. Within the Southeast Asian region, Indonesia has the largest peatland extent (207,000 km^2), followed by Malaysia (26,000 km^2) and Papua New Guinea (11,000 km^2). Extensive peatlands are found along the east coast of Sumatra, in lowland areas of the island of Borneo (principally in Kalimantan and Sarawak), along the south coast of West Papua and around the coastal lowlands of Peninsular Malaysia (Page et al. 2011). The peat is mostly rain-fed and forms extensive domes that can extend for distances of 50 km or more between river valleys or inland behind coastal mangrove forests.

The natural vegetation cover of lowland tropical peatland in Southeast Asia is peat swamp forest. On most peatland domes, there is usually a continuum of at least two or three and sometimes up to seven forest communities from the edge to the center of the dome reflecting progressive increase in peat thickness and degree of waterlogging (Anderson 1983; Page et al. 1999). The substrate consists of partially decomposed organic matter derived from the remains of the forest vegetation, namely, branches, trunks, and tree roots (Brady 1997). Under the wet, acidic, and nutrient-poor conditions, decomposition of this organic material is incomplete and it accumulates to form deposits of woody peat up to 10 or even 20 m thick. The peat surface has a "hummock-hollow" topography. Hummocks (0.5–1.0 m in height) are formed from tree breathing roots and tree bases and other organic matter from leaves and small branches trapped among them. Hollows are usually devoid of woody vegetation other than protruding breathing roots from trees growing on adjacent hummocks (Lampela et al. 2016; Fig. 1); they can retain water during the wet season.

The Peat Swamp Environment

Lowland tropical peats are characterized by low ash and nutrient contents and high acidity; most contain much less than 5% inorganic constituents (i.e., the organic content is >95%). The pH of both peat and surface waters is highly acidic ranging from pH 2.3 to 4.5 (Page et al. 1999), and the availability of essential plant nutrients is very low. Carbon is the element present in largest amount, usually around 50–55%

Fig. 1 View inside low pole peat swamp forest, Riau, Sumatra. Trees grow on low hummocks separated by hollows which will be water-filled during the wet season (Photo credit: A. Hoscilo © rights remain with photographer)

of peat dry weight (Page et al. 2011), while nitrogen contents are much lower and usually less than 2%, with most present in organic compounds and thus largely unavailable to the vegetation.

The majority of lowland tropical peat swamps are completely rain-fed, and water flow from upland areas or adjacent rivers does not enter them. Under natural conditions, the water table is close to or at the peat surface for much of the year; it rises with rainfall and falls as a result of surface outflow and evapotranspiration. When the water table is low, rain water is absorbed by the surface layer of aerobic peat, but during the wet season the peat becomes saturated producing anaerobic conditions and the forest floor may be flooded to heights of 10–50 cm for several months (Lampela et al. 2016). The hummock-hollow topography of the peat surface and the low surface gradient across the peat dome ensure that water moves slowly, helping to maintain a high peat moisture content which is favorable to peat formation. The water in the streams and rivers emerging from peat swamp areas is characteristically transparent but with a dark brown coloration. These "blackwaters" contain a high concentration of tannic and other organic acids arising from the peat (Fig. 2; Moore et al. 2011).

Fig. 2 A "tea-stained" blackwater stream inside peat swamp forest, Central Kalimantan (Photo credit: S. Thornton © rights remain with author)

Peat Swamp Forest Vegetation and Fauna

The vegetation of peat swamp forest ranges from a tall, closed-canopy forest on shallow, marginal peat through to stunted, open-canopy pole forest on the thickest, most waterlogged peat occupying the central parts of the peat dome (Fig. 1; Anderson 1983; Page et al. 1999). Many of the trees show adaptations to the waterlogged environment: buttress and stilt roots provide stability in the soft ground; knee roots and pneumatophores (breathing roots) enable tree roots to obtain oxygen in the anaerobic, waterlogged peat; and mats of fine roots just beneath the surface enhance nutrient uptake from the upper peat layer where concentrations are highest. With increasing distance onto the peat dome, the trees show further evidence of adaptation to the nutrient-poor environment: average tree height, diameter, and leaf area all decrease, while leaf cuticle thickness and the levels of toxic compounds within the leaves increase, possibly as a defense strategy against herbivory (Yule and Gomez 2008).

Most of the tree families of lowland Dipterocarp forest are found in the peat swamp forests of Southeast Asia with members of the Anacardiaceae, Annonaceae, Burseraceae, Clusiaceae, Dipterocarpaceae, Euphorbiaceae, Lauraceae, Leguminosae,

Fig. 3 The sealing wax palm *Cyrtostachys renda* is a characteristic palm species of lowland peat swamp forests in Southeast Asia (Photo credit: A. Hoscilo © rights remain with photographer)

Myristicaceae, Myrtaceae, and Rubiaceae being well represented. In common with other types of forest within the region, the tree species composition shows considerable variation according to geographical location at both a regional and a local scale. Although there have been insufficient studies to determine clear biogeographical patterns, the peat swamps of Borneo support a greater diversity of both forest vegetation types and tree species than those elsewhere in the region. Several habitat endemic or near-endemic tree species have been identified in Southeast Asian peat swamps including *Archidendron clypearia, Dactylocladus stenostachys, Durio carinatus, Gonystylus bancanus, Horsfieldia crassifolia, Shorea albida* (specific to western and northern Borneo), *S. belangeran*, and *S. teysmanniana*. Most of these species are endangered and protected by law (CITES Appendix III). Various species of palms are found in peat swamp forests, including the brightly colored *Cyrtostachys renda*, which is used widely as an ornamental species (Fig. 3). Along blackwater streams and in wetter parts of the forest there is a dense growth of pandans (Pandanaceae), while climbing plants scramble up into the canopy, including several species of insectivorous pitcher plants (*Nepenthes* spp.) (Fig. 4).

Peat swamp forests provide an important habitat for several primates, including orangutan *Pongo pygmaeus* (Fig. 5), with some of the largest remaining populations

Fig. 4 Pitcher plants (*Nepenthes* spp.) are abundant in peat swamp forests. They are able to trap insects as a supplement to their mineral nutrition (Photo credit: A. Rieley © rights remain with photographer)

Fig. 5 Young orangutan in peat swamp forest in Central Kalimantan, Indonesia. Peat swamp forests are an important habitat for these endangered primates (Photo credit: A. Rieley © rights remain with photographer)

found in the peat swamps of Central Kalimantan (Morrogh-Bernard et al. 2003). Gibbons (*Hylobates* spp.) are also widespread (Cheyne et al. 2008), together with more localized occurrences of red langur *Presbytis rubicunda* and, on Borneo, the proboscis monkey *Nasalis larvatus*. Peat swamps also provide important habitat for cats: the Bornean clouded leopard *Neofelis diardi* has been recorded in Central Kalimantan; in Riau, Sumatra, there are records of the endangered Sumatran tiger *Panthera tigris sumatrae*, while marbled cat *Pardofelis marmorata*, leopard cat *Felis bengalensis* and flat-headed cat *Prionailurus planiceps* have also all been recorded. Among smaller mammals there is a high diversity of bats including several regionally rare and threatened species (Struebig et al. 2006) and of birds, with the remaining peat swamps providing a refuge for several highly endangered wetland species, including Storm's, milky, and lesser adjutant storks (*Ciconia stormi*, *Mycteria cinerea*, and *Leptoptilos javanicus*). The water-filled hollows on the forest floor and the "blackwater" streams and rivers flowing from the peat swamps are important habitats for endemic species of freshwater fish adapted to the highly acidic and low nutrient conditions. These include an unusually high number of species of miniature cyprinid fish (*Paedocypris* spp. and *Sundadanio* spp.) (Ng et al. 1994; Kottelat et al. 2006).

Ecosystem Service Provision

In addition to their role in maintaining regional biodiversity, the peat swamp forests of Southeast Asia provide an array of other important ecosystem services. Across the tropical regions of the world, peatland ecosystems store ~89 gigatonnes (Gt = t × 10^9) of carbon, with a pool of ~69 Gt in the Southeast Asian region alone (Page et al. 2011). In both Indonesia and Malaysia, the size of the peat carbon pool (at 57 Gt in Indonesia and 9 Gt in Malaysia) is more than twice that stored in the entire forest vegetation of these countries, i.e., 20 and 4 Gt carbon, respectively. Although a major part of the carbon pool in the peat swamp forest ecosystem is located below ground in the peat rather than in the aboveground forest vegetation, protection of the latter is essential in order to maintain the carbon sequestration and storage function of the peatland. The forest provides the dead organic matter (leaf litter, woody material, and roots) that is the source of carbon for the accruing peat deposit. Any disturbance of that vegetation, for example through logging, drainage, deforestation, or fire, will lead to a reduction in the supply of organic matter and ultimately to the cessation of peat formation.

Tropical peat swamps also play a role in regional and local hydrological regulation. Water flow over the surface of an intact peat dome is so slow that it can continue for several weeks after rainfall, possibly months in the case of very large peat domes, thereby slowly feeding into the streams and rivers that drain the peatland catchment and contributing to the maintenance of river flows during periods of low rainfall. When an intact peatland is drained, for example, by the construction of canals for agricultural development, the most immediate effects are that water moves more rapidly over the peat surface while surface water depths during wet periods are lowered. Lowering of the peatland water table leads to aerobic decomposition of the peat surface and hence oxidation of stored carbon that is emitted as the greenhouse gas carbon dioxide. Additional effects include peat surface compaction and subsidence, which, ironically, can over time increase the risk of the peat surface becoming flooded as it is lowered close to river or sea levels.

Threats and Future Challenges

Over the last two decades, land use changes involving deforestation, drainage, and fire have led to the rapid loss and degradation of the peat swamp forest ecosystem in Southeast Asia. The principal drivers of these changes have been timber extraction and the development of plantation estates for export crops (palm oil and paper pulp); the increasing occurrence of forest fires has also made a contribution, with some of the most extend areas (Page et al. 2002; Langner et al. 2007; Langner and Sination of forest loss and land drainage has destroyed the ction of the peat swamp ecosystem. Drainage has produced he peat that has favored microbial activity, resulting in eny peat oxidation, while an increased risk of fire has promo; of carbon from the peatland store.

In combina fire have contributed substantially to global greenhouse ga on peatlands are now an increasingly regular feature of mos ast Asia, their extent and severity is exacerbated during e infall associated with El Niño events. During the 2015 El Ni idental fires on peatland resulted in substantial smoke emi in air quality with consequences for human health and reg es. Indonesia has now signed the ASEAN Agreement on llution and is participating in national and regional peatla s which may result in improved peatland management in on, the aftermath of the 2015 fires has seen the establishm ration Agency by the Government of the Republic of In tions to prevent and suppress future forest and peatland fir

The prospec swamp forest surviving into the second half of the twenty-f nising, however, although recent initiatives such as the UN bined with a political willingness to limit further developt may ensure that at least some tracts of peat swamp forest ar

References

Anderson JAR. Th stern Malesia. In: Gore AJP, editor. Mires: swamp, bog, fen and m sterdam: Elsevier; 1983. p. 181–99.

Brady MA. Organi peat deposits in Sumatra. PhD Thesis; Vancouver: Faculty of Grad Forestry, University of British Columbia; 1997.

Cheyne SM, Thom RMC, Limin SH. Density and population estimate of gibbons (*Hyl* bangau catchment, Central Kalimantan, Indonesia. Primates. 2008;

Kottelat M, Britz R *ypris*, a new genus of Southeast Asian cyprinid fish with a remarkal prises the world's smallest vertebrate. Proc R Soc B. 2006;273:89

Lampela M, Jauhi en M, Tanhuanpää T, Valkeapää A, Vasaander H. Ground surf getation patterns in a tropical peat swamp forest. Catena. 2016;13

Langner A, Siegert rrence in Borneo over a period of 10 years. Glob Chang Biol. 200

Langner A, Miettin change 2002–2005 in Borneo and the role of fire derived from M Biol. 2007;13:2329–40.

Morrogh-Bernard H JO. Population status of the Bornean orang utan (*Pongo pygmae* amp forest, Central Kalimantan, Indonesia. Biol Conserv. 2003;1

Moore S, Gauci V, Evans CD, Page SE. Fluvial organic carbon losses from a Bornean blackwater river. Biogeosciences. 2011;8: 901–909.

Ng PKL, Tay JB, Lim KKP. Diversity and conservation of blackwater fishes in Peninsular Malaysia, particularly in the North Selangor peat swamp forest. Hydrobiologia. 1994;285:203–18.

Page SE, Siegert F, Rieley JO, Boehm H-DV, Jaya A, Limin S. The amount of carbon released from peat and forest fires in Indonesia in 1997. Nature. 2002;420:61–5.

Page SE, Rieley JO, Shotyk OW, Weiss D. Interdependence of peat and vegetation in a tropical peat swamp forest. Proc Trans R Soc B. 1999;345:1885–97.

Page SE, Rieley JO, Banks CJ. Global and regional importance of the tropical peatland carbon pool. Glob Chang Biol. 2011;17:798–818.

Struebig MJ, Galdikas BMF, Suatma. Bat diversity in oligotrophic forests of southern Borneo. Oryx. 2006;40:447–55.

Yule CM, Gomez LN. Leaf litter decomposition in a tropical peat swamp forest in Peninsular Malaysia. Wetland Ecol Manag. 2008;17:231–41.

Wetlands of the Mekong River Basin: An Overview

148

Peter-John Meynell

Contents

Introduction	1764
Climate and Hydrology	1767
Distribution and Diversity of Lower Mekong Wetlands	1768
Wetland Types	1768
Distribution by Bioregion	1769
Important Wetland Sites in the Lower Mekong Basin	1771
Biodiversity and Uniqueness of Mekong's Wetlands	1773
Dependent People	1779
Threats and Future Challenges	1780
Climate Change	1782
References	1783

Abstract

The Mekong River extends for 4,909 km from the Tibetan plateau, through to the Delta in Vietnam, passing through Yunnan province in China, Myanmar, Lao PDR, Thailand and Cambodia. The drainage basin of the Mekong River covers nearly 800,000 km^2 of which about 80% lies in the Lower Mekong Basin. Important wetland features include the Khone Falls and the braided river complex of Siphandone and Stung Treng and the Tonle Sap Great Lake. For much of its length, the Mekong flows through bedrock confined channels or old alluvium in the river bed and banks. Meanders, oxbow lakes, cut-offs and extensive floodplains are restricted to a short stretch around Vientiane, and below Kratie, where the river develops unrestrictedalluvial channels. Throughout its length, it is characterized by about 500 deep pools which are important dry season refuges for fish.

P.-J. Meynell (✉)
ICEM – International Centre for Environmental Management, Hanoi, Vietnam
e-mail: peterjohn.meynell@gmail.com

© Springer Science+Business Media B.V., part of Springer Nature 2018
C. M. Finlayson et al. (eds.), *The Wetland Book*,
https://doi.org/10.1007/978-94-007-4001-3_244

The climate of the Lower Mekong Basin is tropical monsoonal and the Mekong flows are characterized by an annual flood pulse which drives the reverse flow that occurs in the Tonle Sap in Cambodia, so that the area of wetlands in the Great Lake increase 6-fold at the peak of the flood season. About 42% of the total land area of the LMB has been classified as wetland, of which only 56,000 km^2 can be classified as natural wetlands; the majority of man-made wetland areas are paddy rice fields. The natural wetlands can be divided into wet "lands", open water, flowing water and coastal and estuarine wetlands. Most of the wetlands lie in the Central Indochina Dry Forests and Tonle Sap freshwater swamp forest Ecoregions. There are nearly 100 wetland sites of national and regional importance in the Lower Mekong Basin and of these 12 have been designated as Ramsar sites.

The biodiversity of the Mekong and its wetlands is second only to the Amazon and Congo rivers, especially its fish and mollusk diversity, with over 850 recorded fish species and over 83 species of mollusk. Key flagship species include the Mekong giant catfish, Irrawaddy dolphin, Eastern sarus crane and Siamese crocodile.

Most of the 60 million people that live in the LMB live in rural areas and most live near rivers, lakes and wetlands. The collection of fish and other wetland products are very important for their livelihoods. The Mekong Basin has the largest inland fishery in the world, estimated at 2.8 Mt per year of which 1.9 Mt comes from the capture fishery, i.e., from the wetlands.

Conversion of wetlands to agriculture and loss of habitat has been the most significant threat to wetlands in the LMB with an estimated 78% of the original natural wetland area being already converted. Other threats include the construction of hydropower dams and irrigation schemes which change the hydrology of the basin, block important fish migrations and trap the sediment, leading to a significant reduction in sediment transport to the Delta. Climate change is also expected to both increase the wet season flows and reduce the dry season flows, which will have impacts upon the extent of permanent and seasonal wetlands. In the Delta, the combination of sea level rise and changing flow patterns is expected to lead to erosion of the delta and saline intrusion, including permanent inundation of some important mangrove areas.

Keywords

Climate change · Fish and fisheries · Flood pulse · Hydropower · Mekong river · Wetland biodiversity · Wetland conversion to agriculture · Wetland distribution

Introduction

The Mekong River extends for 4,909 km from the Tibetan plateau to the Delta in Vietnam. On the way it passes through the province of Yunnan in the People's Republic of China (PRC), where it flows for 500 km parallel to two other of the region's great rivers and only 100 km apart – the Salween and Yangtze. Leaving China, it flows through a narrow valley for about 100 km between Lao People's

Table 1 The basin area, catchment, and flow distribution among countries of the Mekong River Basin (Mekong River Commission 2005)

Country	Basin area – km²	Catchment as % of MRB	Flow as % of MRB
PR China	165,000	21	16
Myanmar	24,000	3	2
Lao PDR	206,620	25	35
Thailand	203,060	23	18
Cambodia	156,435	20	18
Vietnam – Central Highlands Delta	32,400 34,373	8	11
Total	795,000		
Lower Mekong Basin	632,888	80	82

Source: Adapted from Mekong River Commission 2005, in public domain

Democratic Republic (PDR), Myanmar to the Golden Triangle with Thailand, and then into Lao PDR. Just upstream of Vientiane, the capital of Lao PDR, it forms the border between Thailand and Lao PDR for some 500 km, separating the drainage basin of the Khorat Plateau in NE Thailand from the Annamite mountains to the east. In the south of Lao PDR, there is a significant geological formation at the Khone Falls, the largest waterfall in southeast Asia where the river falls by some 17 m. This forms the border between Lao PDR and Cambodia, flowing through the wide, braided river channels of Siphandone and Stung Treng down to Kratie where the Mekong expands into the Cambodian floodplain. Just upstream of Phnom Penh, the Mekong mainstream is joined by the Tonle Sap River, linking the Tonle Sap Great Lake, a unique freshwater lake that expands in the wet season to three to six times its dry season size, driven by backflows from the Mekong. After Phnom Penh, the Mekong flows into the Delta in Vietnam, separating out into two main distributary channels and seven smaller ones and out into the South China Sea (Mekong River Commission 2010; Mekong River Commission 2011).

The drainage basin of the Mekong River covers nearly 800,000 km² of which about 80% lies in the Lower Mekong Basin (see Table 1).

For much of its length, the Mekong flows through channels that are confined by bedrock or old alluvium in the riverbed and banks. Geomorphological features such as meanders, oxbow lakes, cutoffs, and extensive floodplains are restricted to a short stretch of the Mekong around Vientiane and below Kratie, where the river develops alluvial channels unrestricted by the underlying bedrock (MRC 2010). Throughout its length, it is characterized by about 500 deep pools which are important dry season refuges for fish. The deepest of these pools may be up to 90 m deep and extend for almost 10 km. The median depth of deep pools in bedrock channels is about 22 m and 7.5 m in alluvial reaches. Other reaches include extensive beds of deposited sand, and in-channel wetland areas, seasonally exposed and flooded by the large changes between dry and wet seasons (MRC 2010).

The Mekong Basin contains 16 of the WWF Global 200 ecoregions and is one of the most biodiverse regions of the world, lying in the Indo-Burma biodiversity

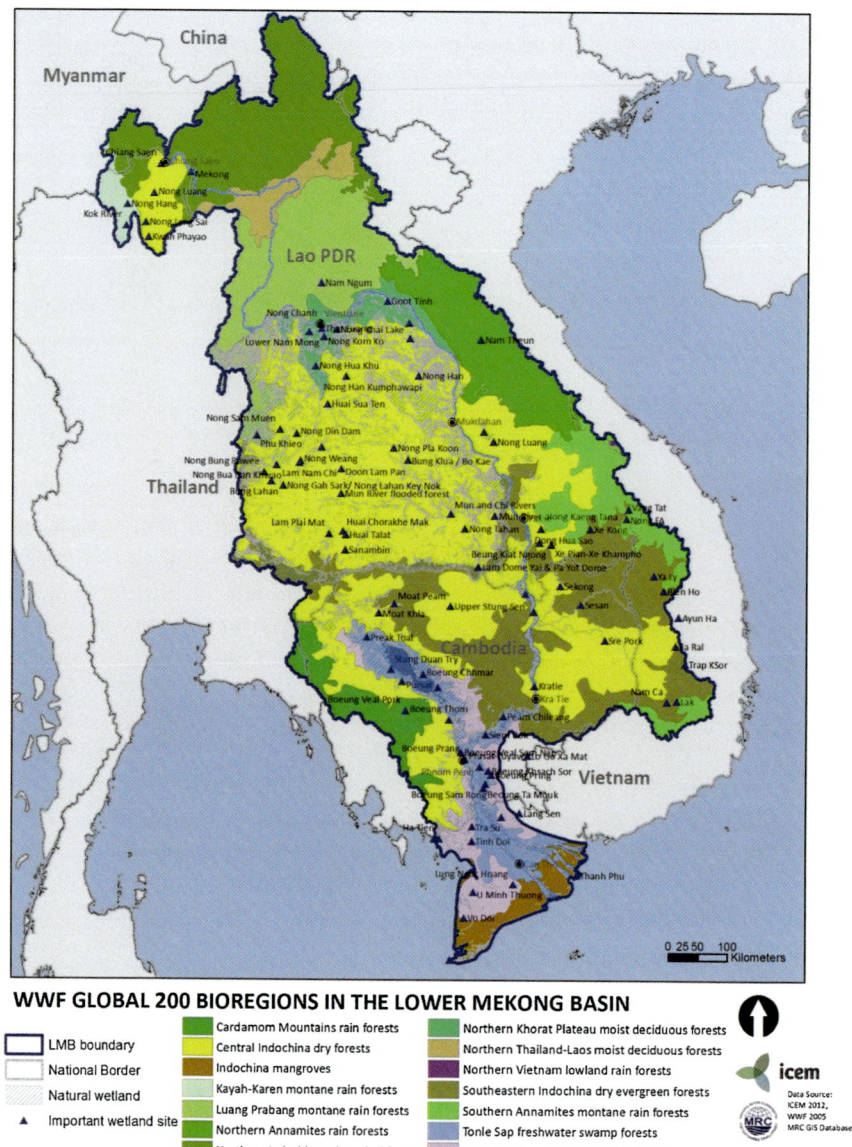

Fig. 1 Map of important wetlands in the Lower Mekong, showing the WWF Global 200 bioregions (Source: ICEM 2012, © with permission)

hotspot. Within the Mekong Basin, most of the wetlands occur in the Lower Mekong Basin, including the countries of Lao PDR, Thailand, Cambodia, and Vietnam (see Fig. 1). There are important areas of alpine and high-altitude wetlands occurring in Qinghai-Tibetan plateau which serves as the source region for Yellow, Yangtze, and

Mekong (Gao et al. 2013). In the upper Mekong the river basin is mountainous with a narrow and rocky valley, with little floodplain and opportunity for non-riverine wetland formation. This section focuses on the wetlands of the Lower Mekong Basin.

Climate and Hydrology

The climate of the Lower Mekong Basin is tropical monsoonal. Three seasons can be distinguished, the hot and wet season between mid-May and October driven by the southwest monsoon; the cooler dry season from December to February, characterized by the northeast monsoon; and the hot/dry season between February and May. There is also a shorter hot dry period at the end of the wet season through mid-December. Mean monthly temperatures rarely fall below 20 °C. Maximum summer temperatures at altitudes of less than 500 masl typically reach 35 °C, while minimum winter temperatures can fall as low as 15 °C. The highest rainfalls of 2,500 mm/year occur in the western montane regions of Lao PDR and the lowest which may be below 1,000 mm/year are found in the Khorat Plateau in NE Thailand. The late monsoon period also brings storms and typhoons from the east that can bring extreme rainfall events adding to the already high flows in the Mekong and causing large floods in Vietnam, Cambodia, and Lao PDR.

The Mekong is characterized by an annual flood pulse with a distinct seasonality in the hydrograph. The estimated mean annual flow of the Mekong is almost 460 km^3, but almost 75% of the flow occurs within the 4 months between July and October (July 14%, August 23%, September 23%, October 14%). While the Mekong is a highly predictable river in terms of timing of its hydrograph, there is considerable variation in the flows from year to year. The average discharge to the Delta is about 16,000 m^3/s with a maximum flow of 39,000 m^3/s.

One of the unique features of the Mekong system is the reverse flow that occurs in the Tonle Sap in Cambodia and to a much lesser extent in some of the tributaries, e.g., Songkhram River in NE Thailand. The Tonle Sap River links the Mekong with the Great Lake which lies in a low-relief, low-gradient floodplain landscape. During the dry season the Tonle Sap operates as a normal tributary draining the Cambodian floodplain into the Mekong. As water levels rise in the Mekong in the wet season, the flow reverses back up the Tonle Sap River. The area of the Great Lake increases sixfold from 2,500 to 15,000 km^3. As water levels fall at the end of the wet season, the flow in the Tonle Sap River returns to normal.

The sediments of the Mekong are an important flow feature, maintaining floodplain fertility and delta formation. The total sediment flow has been estimated at between 150 and 170 million tons per year. The two main sources of sediment in the Mekong are the Lancang subbasin in China and the Central Highlands of Vietnam (the "3S river basins" – the Sesan, Sre Pok, and Sekong rivers) which together contribute over 70% of the sediments (Mekong River Commission 2010).

Distribution and Diversity of Lower Mekong Wetlands

Wetland Types

The Mekong River Commission prepared a wetland database in 2006. From country reviews of wetlands prepared by the National Mekong Committees in 2003, about 94 important wetland sites have been identified. This database has been analyzed to show that 254,144 km^2 (about 42% of the total land area of the LMB) can be classified as wetland (seasonal and permanent), of which only 55,498 km^2 or 22% can be classified as natural wetlands. The large majority is man-made or converted wetlands, most of which are associated with agriculture, especially paddy rice fields (see Tables 2, 3, and 4 and Fig. 2) (ICEM 2012).

The natural wetland types can be subdivided into four groups:

(i) **Wet "lands"** (84% of the natural wetlands) which include flooded forest, swamps and woody scrub, peat swamps, marshes, and grasslands.
(ii) **Open water** (5.9%) including lakes, ponds, and saline lakes.
(iii) **Flowing water** (7.0%) rivers and streams, both mainstream Mekong and tributaries.
(iv) **Coastal and estuarine** wetlands (2.9%) including mangroves, salt marshes, and mudflats. Many of the wetland areas are seasonally or temporarily inundated – some 85% of flooded forests experience temporary flooding, as do 67% of grasslands and marshes, and 70% of the estuarine and coastal wetlands.

More than 98% of open water and flowing water wetlands are permanent (see Table 4). However, the wet "lands" vary significantly – peat swamps are almost exclusively permanent, while the majority of flooded forests (85%), swamp/woody scrub (98.5%), and some 67% of grasslands and marsh would be seasonally flooded. Further breakdown of the grassland/marsh category reveals that the majority of the permanently flooded wetlands under this category are marshes, while the grasslands would constitute the majority of the temporarily flooded areas of the category.

Table 2 Breakdown of man-made wetland types in the LMB

Man-made wetland	Area (ha)	%
Agriculture – rice	18,049,371	90.86
Agriculture – other wet crops	504,353	2.54
Aquaculture	253,268	1.27
Grasslands	5,552	0.03
Flooded forest plantations	10,578	0.05
Saltworks	9,218	0.05
Lakes and Ponds – irrigation, hydropower, drinking water	214,359	1.08
Urban lake/ponds and wetlands	368,451	1.85
Man-made artificial channels	449,451	2.26
Total	19,864,601	

Source: ICEM 2012, © with permission

Table 3 Areas and number of the main natural wetland types in the LMB

Wetlands	Total (ha)	%
Wet "lands"	**4,688,327**	**84**
Flooded forest	7,570.3	0.14
Swamp/woody scrub	3,966,671.0	71.21
Grassland and marsh	486,062.2	8.73
Peat swamp	228,023.2	4.09
Open water	**330,946.5**	**5.9**
Lakes >8 ha	322,764.8	5.79
Ponds <8 ha	1,536.0	0.03
Saline lakes >8 ha	6,645.7	0.12
Flowing water	**391,008.0**	**7.0**
Riverine and streams	391,008.0	7.02
Estuarine, coastal, and marine	**160,289.3**	**2.9**
Watercourse	64,202.0	1.15
Intertidal lagoon	335.4	0.01
Salt marsh	7,992.4	0.14
Non-vegetated, bare, sand	3,502.3	0.06
Mangrove	84,257.2	1.51
Total	**5,570,570.5**	100.00

Source: ICEM 2012, © with permission

Nearly 70% of the estuarine, coastal, and marine wetlands are temporarily flooded, while 30% are permanently flooded. This can be explained by the tidal nature of these wetlands. All of the mangroves would be temporarily flooded by the tide, and a smaller part of the non-vegetated mudflats and watercourses would remain underwater.

Distribution by Bioregion

As part of a process of strategic conservation planning, WWF considered that the Lower Mekong Basin could be classified into 11 main bioregions.

(i) Cardamom Mountains rain forests
(ii) Central Indochina dry forests
(iii) Indochina mangroves
(iv) Luang Prabang mountain rain forests
(v) Northern Annamites rain forests
(vi) Northern Indochina subtropical forests
(vii) Northern Khorat Plateau moist deciduous forests
(viii) Northern Thailand-Laos moist deciduous forests
(ix) Southeastern Indochina dry evergreen forests
(x) Tonle Sap freshwater swamp forests
(xi) Tonle Sap-Mekong peat swamp forests

Table 4 Proportions of temporary and permanent natural wetlands in the LMB

Wetland group	Wetland type		Ha	%
Wet "lands"	Flooded forest	Temp	17,034	85.0
		Perm	3,008	15.0
		Total	**20,042**	
	Swamp, woody scrub	Temp	3,946,569	98.5
		Perm	60,112	1.5
		Total	**4,006,681**	
	Grassland/marsh	Temp	336,846	67.2
		Perm	164,441	32.8
		Total	**501,287**	
	Peat swamp	Temp	–	0.0
		Perm	226,069	100.0
		Total	**226,069**	
Open water	Open water	Temp	5,983	1.8
		Perm	323,581	98.2
		Total	**329,564**	
Flowing water	Rivers and streams	Temp	4,841	1.2
		Perm	385,825	98.8
		Total	**390,666**	
Estuarine, coastal, and marine	Estuarine, coastal, and marine	Temp	999,469	69.8
		Perm	432,047	30.2
		Total	**1,431,516**	
	Non-vegetated		354,610	24.8
	Salt marsh		6,875	0.5
	Mangrove		977,169	68.3
	Intertidal lagoon		1,535	0.1
	Watercourse		91,327	6.4

Source: ICEM 2012, © with permission

It can be clearly seen in Table 5 and Fig. 3 that the vast majority of the Mekong wetlands lie in the Central Indochina dry forests (39%, 2.2 M ha) and Tonle Sap freshwater swamp forests (22%, 1.3 M ha), with very few lying in the mountainous areas. There is some freshwater woody scrub, marshes, and ponds and lakes in the Luang Prabang mountain rain forests and the Southeastern Indochina dry evergreen forests, but these are the only significant wetland types represented at these higher altitudes. There are nearly no wetlands in the Cardamom Mountains rain forests, the Northern Annamites rain forest, and the Northern Thailand-Laos moist deciduous forest bioregions.

In the Central Indochina dry forests, the large majority of wetlands are the freshwater woody scrub (84%), with some grassland and marshes (8.6%), and the riverine wetlands (5.9%) and lakes (1.3%). Saline lakes are also only found in the Central Indochina dry forest areas. 98% of peat swamps are found in the Tonle Sap freshwater swamp forests. As is to be expected, all of the estuarine wetlands, salt

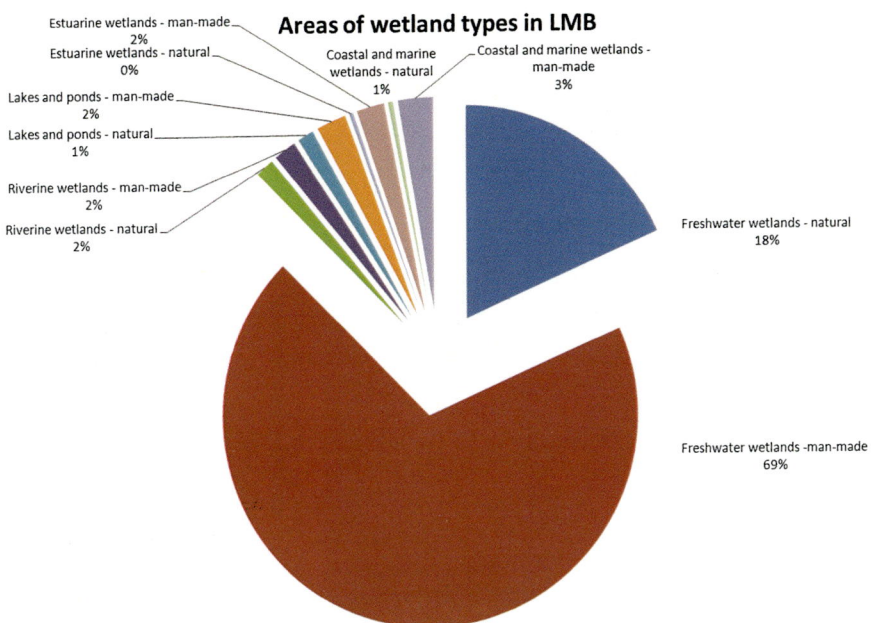

Fig. 2 Distribution of man-made and natural wetland types in the LMB (Source: ICEM 2012, © with permission)

marshes, marine and tidal lagoons, and mangroves are all found in the Indochina mangroves and Tonle Sap-Mekong peat swamp forests.

Swamp and woody scrub, lakes, grassland and marshes, and the riverine and stream wetlands are spread throughout most of the bioregions. The distribution of important wetland sites in the bioregions is shown in Fig. 1.

Important Wetland Sites in the Lower Mekong Basin

There are nearly 100 wetland sites which have been identified as nationally and regionally important, 24 sites in Cambodia, 16 in Lao PDR, 39 in Thailand, and 18 in Vietnam, of which 8 are located in the Central Highlands and 10 in the Mekong Delta. These are indicated in Fig. 1. Several of the most important wetlands in each country are summarized in Table 6. Each of the important wetland sites is made up of a mosaic of different wetland habitats, so a simple identification of the wetland sites into the different wetland types is not possible.

There are 12 designated Ramsar Sites in Thailand, of which three lie in the basin (Nong Bong Khai, Bung Khong Long, and Goot Ting); in Cambodia there are four designated Ramsar sites of which three lie in the Lower Mekong Basin. One of these lies on the Mekong mainstream (Stung Treng) and two in the Tonle Sap Great Lake (Boeng Chhmar and Prek Toal). Lao PDR acceded to the Ramsar Convention in 2010 and designated two sites (Beung Kiat Ngong and Xe Champhone), and

Table 5 Distribution of wetlands in the main bioregions of the Lower Mekong Basin

Wetland classification	Bioregions (WWF Global 200)																	Total
	Mixed forest[a]		Central Indochina dry forests		Indochina mangroves		Luang Prabang montane rain forests		Northern Indochina subtropical forests		Northern Khorat Plateau moist deciduous forests		Southeastern Indochina dry evergreen forests		Tonle Sap freshwater swamp forests		Tonle Sap-Mekong peat swamp forests		
	%	ha	%	ha	%	ha	%	ha	%	ha	%	ha	%	ha	%	ha	%	ha	
Flooded forest	–	–	39	2,969	0	7	–	–	–	–	–	–	3	264	20	1,547	37	2,783	7,570
Swamp/woody scrub	1	20,144	45	1,770,267	0	46	14	563,640	0	104	5	208,912	18	711,934	13	528,054	4	163,569	3,966,671
Grassland and marsh	0	6	49	239,638	0	750	1	3,020	–	–	10	46,513	3	15,119	16	77,261	21	103,756	486,062
Peat swamp	–	–	0	627	–	–	–	–	–	–	–	–	–	–	98	223,296	2	4,100	228,023
Lakes >8 ha	1	1,932	10	31,594	–	–	0	279	0	355	1	3,892	2	5,835	83	266,838	4	12,041	322,765
Ponds <8 ha	–	–	17	261	–	–	–	–	–	–	–	–	13	205	43	662	27	409	1,536
Saline lakes >8 ha	–	–	98	6,512	–	–	–	–	–	–	2	134	–	–	–	–	–	–	6,646
Rivers and streams	3	10,589	38	148,166	1	3,455	5	20,318	3	12,927	7	28,814	7	27,230	32	126,851	3	12,659	391,008
Watercourse (estuarine)	–	–	–	–	53	33,888	–	–	–	–	–	–	–	–	34	21,630	14	8,684	64,202
Intertidal lagoon	–	–	–	–	3	10	–	–	–	–	–	–	–	–	–	–	97	326	335
Non-vegetated, bare, sand (estuarine and marine)	–	–	–	–	92	3,225	–	–	–	–	–	–	–	–	–	–	8	277	3,502
Salt marsh	–	–	–	–	39	3,099	–	–	–	–	–	–	–	–	0	19	61	4,875	7,992
Mangroves	–	–	–	–	69	58,382	–	–	–	–	–	–	–	–	0	33	31	25,842	84,257
Grand total	1	32,671	39	2,200,033	2	102,861	11	587,257	0	13,386	5	288,265	14	760,588	22	1,246,191	6	339,319	5,570,570

[a]Combines: (a) Cardamom Mountains rain forests, (b) Northern Thailand-Laos moist deciduous forests, (c) Northern Annamites rain forests (Source: ICEM 2012, © with permission)

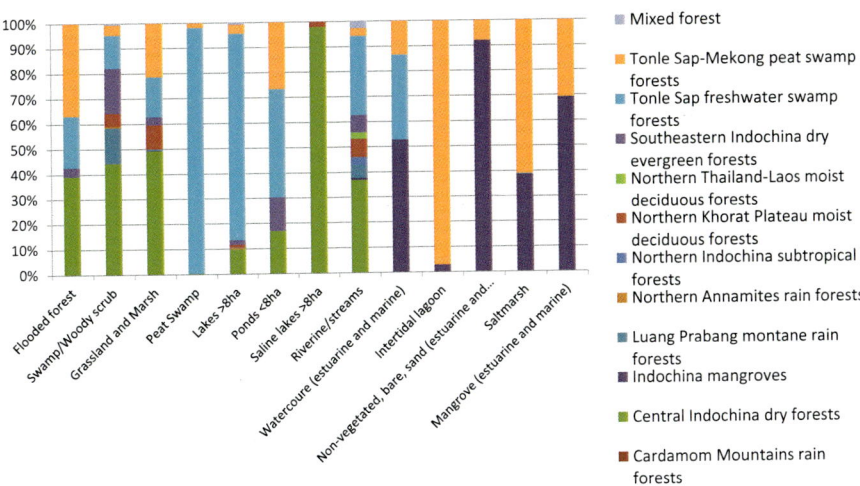

Fig. 3 Distribution of wetlands in WWF Global 200 bioregions of the LMB (Source: ICEM 2012, © with permission)

Vietnam has designated two Ramsar Sites in the (Delta – Tram Chim, 2012) and Mui Ca Mau (2013), and in 2015 two more wetland sites were designated in the Delta – U Minh Thuong and Lang Sen.

Biodiversity and Uniqueness of Mekong's Wetlands

The biodiversity of the Mekong and its wetlands is second only to the Amazon and the Congo rivers. The key endangered wetland species in the four countries of the Lower Mekong include one critically endangered, four endangered, and two vulnerable mammals; three critically endangered, six endangered, and ten vulnerable bird species; five critically endangered, four endangered, and five vulnerable reptile species; and one endangered and ten vulnerable species of amphibian.

Plant diversity has not been accurately assessed in the Mekong Basin. Wetland plants play an important part in people's livelihoods for food, medicines, firewood, construction material, and handicrafts. Popular plants are the lotus *Nelumbo nucifera* (tubers, stems, and seeds), water hyacinth *Eichhornia crassipes*, riang leaves (*Cissus hastata*), and morning glory (Convolvulaceae spp.). Mekong riverweed (*Cladophora* spp.) grows on exposed rocks in the Mekong and its tributaries in northern Lao PDR and is harvested by the tonne as a popular delicacy. Unique flooded forests occur both in the Tonle Sap and in some areas of the Mekong River, e.g., Stung Treng. The main trees species are *Barringtonia acutangula* and *Diospyros cambodiana*, with common in-channel wetland shrub species such as *Homonoia riparia*. Invasive species such as *Mimosa pigra* and water hyacinth can be especially problematic in some wetland areas. Cultivated *Melaleuca* spp. forests in the Mekong Delta provide fuel wood and construction timber.

Table 6 Character of some important wetland sites in the Lower Mekong Basin

Wetland site name	Importance	Area (ha)	Wetland habitat types	Key biodiversity information
Cambodia				
Middle stretches of the Mekong river, north of Stung Treng (Stung Treng province)	Ramsar Site	14,600	Flooded forest	Important breeding and feeding habitat for rare species of fish and globally threatened wetland-dependent birds, including CE white-shouldered ibis *Pseudibis davisoni*
Middle stretches of the Mekong river, between Stung Treng and Kratie towns (Stung Treng and Kratie provinces)	Rich biodiversity	33,808	Braided channels, islands and sandbars, flooded forest and grasslands, gallery forest, and channel woodland: deep pools, rocky rapids, and turbulent stretches	Variety of habitats supporting a wide range of aquatic species as well as water birds and mammals. Most of the Mekong population of the Irrawaddy dolphin is found here
Prek Toal (Battambang province)	Core Zone of Tonle Sap Biosphere reserve, Ramsar Site	39,873	Flooded forest – savanna forest with almost complete ground cover of *Sesbania javanica* with groups of trees and isolated trees of mainly *Barringtonia* (freshwater mangrove)	Supports internationally important colonies of globally threatened water birds, including breeding colonies of spot-billed pelicans *Pelecanus philippensis*, greater adjutant *Leptoptilos dubius*, and white-winged duck *Cairina scutulata*
Boeng Chhmar (Kampong Thom province)	Ramsar Site High biodiversity	28,000	Seasonally flooded forest and swamp	Supports large assemblage of plant, fish, and water bird species, many of which are listed as rare and endangered.
Bassac marshes (Kandal province)	Potential Ramsar Site (criteria 2,4, 6)	52,316	Seasonally inundated herbaceous, shrub, and savanna vegetation	Of great importance to water birds, including spot-billed pelican

(*continued*)

Table 6 (continued)

Wetland site name	Importance	Area (ha)	Wetland habitat types	Key biodiversity information
Lao PDR				
Siphandone wetland (Champasak province)	High biodiversity	6,000	Perennial wetlands, seasonal river channels, rapids and waterfalls, and seasonal flooded forest	At least 205 species of fish recorded at Siphandone; critical to the life cycle of many migratory fish species and home to most northerly group of Mekong Irrawaddy dolphins
Xepian – Xe Khampho (Attapeu and Champasak provinces)	High biodiversity	2,000	Freshwater lakes and ponds, freshwater marsh, seasonally flooded grassland	Critically important fish breeding and spawning grounds: endangered Siamese crocodiles *Crocodylus siamensis* confirmed
Xe Champhone (Savannakhet province)	High biodiversity, Ramsar Site	12,400	River meanders, oxbow lakes and ponds, freshwater marsh, seasonally flooded grassland	Critically important fish breeding and spawning grounds: breeding population of endangered Siamese crocodiles *Crocodylus siamensis*
Beung Kiat Ngong (Champasak province)	High biodiversity, Ramsar Site	2,360	Freshwater marshes and seasonally flooded forest	Rich fish fauna: important for a wide variety of resident and migratory waterfowl
That Luang marsh (Vientiane muncipality)	Natural urban wastewater purification, flood protection	2,000	Seasonal freshwater marsh, aquaculture ponds, seasonal flooded grassland/shrubland, irrigated floodplain wet rice	Fish and other aquatic organisms, snails, frogs, and invertebrates all collected by local people. Also aquatic plants, morning glory, water hyacinth, lotus
Thailand				
Nong Bong Khai nonhunting area (Chiang Rai province)	Ramsar Site	434	Riverine plain, seasonal and intermittent marsh	Includes 121 species of birds, 57 residents, including 53 water birds and 13 species of fish. Threatened Siamese fighting fish *Betta splendens*
Bung Khong Long nonhunting area (Nong Khai province)	Ramsar Site	2,214	Permanent natural water reservoir and associated forest and marshes	Includes 64 fish species (31 of economic importance) and five species of threatened birds

(*continued*)

Table 6 (continued)

Wetland site name	Importance	Area (ha)	Wetland habitat types	Key biodiversity information
Lower Songkhram river floodplains (Nong Khai, Udon Thani, Sakhon Nakhon, and Nakhon Phanom provinces)	Potential Ramsar Site	96,000	Free-flowing river and seasonally inundated floodplain	Supports 183 species of fish including 19 Mekong endemics and 18 species threatened in the wild
Goot Ting marshes (Nong Khai province)	Ramsar Site	2,600	Marshes, streams, evergreen forest, flooded grassland	Supports at least 123 species of fish (56 of economic importance) including Thai endemic *Boraras micros*, one of the smallest fish in the world and globally threatened *Probarbus jullieni*. 100 species of birds (60 wetland dependent). Large numbers of wigeon, falcated teal, and endangered Baer's pochard *Aythya baeri*
Vietnam				
Tram Chim National Park (Dong Thap province)	Ramsar Site, Plain of Reeds	7,588	Melaleuca swamp, seasonally flooded grassland, lotus swamp	Created in 1998 to protect several rare birds especially sarus crane *Grus antigone* and Bengal florican *Houbaropsis bengalensis* and typical ecosystems found in the Plain of Reeds: important site for the conservation of wild rice *Oryza rufipogon*
Lang Sen Wetland Reserve (Long An Province)	Ramsar Site, Plain of Reeds	4,802	Mosaic of seasonally flooded grassland, open swamp, and riverine Melaleuca and mixed forests	More than 20,000 water birds in dry season, including greater adjutant *Leptoptilos dubius* and sarus crane *Grus antigone*. Globally vulnerable reptiles such as the Indochinese spitting cobra *Naja siamensis* and Southeast Asian softshell turtle *Cuora amboinensis*. 27 of 87 fish species recorded in Lang Sen live only in the Lower Mekong Basin

(*continued*)

Table 6 (continued)

Wetland site name	Importance	Area (ha)	Wetland habitat types	Key biodiversity information
U Minh Thuong National Park (Kien Giang province)	National park, Ramsar site	8,514	Peat land, Melaleuca forest, swamp, grassland and open water, waterways	Supports on of the largest breeding colonies of water birds in the Mekong Delta – 187 bird species confirmed including spot-billed pelican and lesser adjutant; diversity of vascular plants
U Minh Ha National Park (Ca Mau province)	National park	8,286	Peat land, Melaleuca forest, swamp, grassland and open water, waterways	High abundance and species richness of water birds including small bitterns (*Ixobrychus sinensis* and *I. flavicollis*) bronze-winged jacanas *Metopidius indicus*, and purple swamp hen *Porphyrio porphyrio*; habitat for endangered mammals, e.g., hairy-nosed otter *Lutra sumatrana* and Asian small-clawed otter *Aonyx cinerea*
Mui Ca Mau National Park (Ca Mau province)	Ramsar site, UNESCO Biosphere Reserve	41,862	Subtidal and tidal coastal wetlands, mudflat, mangrove swamp, shrimp ponds	Excellent habitat for migrating shore birds, 93 species of birds including seven Red-listed species. Extensive shellfish harvesting

Sources: MRC 2010; Ramsar Sites Information Service website: https://rsis.ramsar.org/ris/2227; Tonle Sap, Tram Chim, and U Minh – this volume)

Invertebrates: The Mekong has one of the most diverse freshwater mollusk faunas in the world. In parts of the Mekong, catchment over 83 species found. Species are not evenly distributed across the basin, but most species are restricted to a relatively short stretch of the river between Pakse in southern Lao PDR and Kratie in Cambodia. Snails from the wetlands contribute significantly to livelihoods and nutrition. One of the invasive mollusk species commonly found in the region is the golden apple snail *Pomacea canaliculata*, which is a significant pest of paddy rice (Allen 2012).

The assessment of dragonflies of the Indo-Burma region covered about 473 odonate species of which 160 are thought to be endemic. New species are still being described. Two species were identified as critically endangered, two as endangered, and ten as vulnerable. The Mekong Basin appears to be somewhat less diverse than the Irrawaddy and Salween, but this may reflect lack of detailed studies in Cambodia and Lao PDR (Allen 2012).

Fish: The Mekong has a very diverse fish fauna, with over 850 recorded species in 65 families in the Cambodian Mekong and 50 families in the Lao/Thai Mekong, and although all are caught at some stage, only 50–100 are common in the fishery. The fish species are dominated by Cyprinidae with 54 cyprinid genera in the Cambodian Mekong and 75 cyprinid genera in the Lao/Thai Mekong. In general, they can be divided into three groups according to their ecology and migration patterns:

- Blackfish, which have limited lateral migrations from the river into the floodplains and no longitudinal migrations up- or downstream. These include Channidae (snakeheads), Clariidae and some Bagridae (catfish), and Anabantidae (climbing perch).
- Whitefish, which undertake long-distance migrations between lower floodplains and the Mekong mainstream, including many cyprinids and most of the Pangasiidae catfish.
- Grayfish, which do not spend the dry season in the floodplains nor undertake long-distance migrations. When the floods recede, they leave the floodplain and move to local tributaries. These include *Mystus* catfish.

Though fish catches have been quite steady, there has been a decline in large fish in the basin, especially some of the endangered species such as the Mekong giant catfish *Pangasianodon gigas* and *P. sanitwongsei*, the thicklip barb *Probarbus labeamajor*, and the giant barb *Catlocarpio siamensis*. Julien's golden carp *Probarbus jullieni* is also endangered. The recent IUCN Redlist Assessment of freshwater biodiversity in Indo-Burma indicated that seven of the fish species found in the Mekong and its tributaries are critically endangered, 14 species are endangered, and 20 species are vulnerable. There are at least 150 endemic species in the Mekong. There are 14 exotic species including tilapia, common carp, and mosquito fish (Meynell 2012; Allen 2012).

Amphibians and reptiles: 91 new species of amphibians have been described in the Mekong region since 1997, with 30 species of frogs documented in Cambodia, including 2 new species and 11 species recorded there for the first time. In Lao PDR the frog fauna is relatively poorly known, but scientists have been able to record at least 46 species. Thailand, Vietnam, and China have better described amphibian fauna (Mekong River Commission 2010). There are many endemics, including the Laos warty newt, *Paramesotriton laoensis*, which is only found in streams in northern and central Lao.

Reptiles in Mekong wetlands include the critically endangered Siamese crocodile *Crocodylus siamensis*. Wild populations are still found in the Cardamom Mountains and in Prek Toal in the Tonle Sap in Cambodia and in Xe Champhone Ramsar site in Lao PDR. They are also farmed intensively throughout the region. There is a high diversity of water snakes (22 species) in the Tonle Sap Great Lake and other wetlands, with a very large harvest for crocodile and human food. 19 species of freshwater turtles have been recorded in the Mekong region, with five species occurring in the Tonle Sap and the Mekong in northern Cambodia being an important remaining location for Cantor's giant softshell turtle *Pelochelys cantorii*. Turtles are collected and traded as favored foods throughout the region.

Birds: Most of the bird species associated with wetlands in the Mekong have declined under severe threat from hunting and habitat loss, especially the species that use sandbars and large rivers for breeding and feeding habitats such as plain martin *Riparia paludicola* and the endemic Mekong wagtail *Motacilla samveasnae* and Indochinese populations of river tern *Sterna aurantia*, wooly necked stork *Ciconia episcopus*, and pied kingfisher *Ceryle rudis*. The dry dipterocarp forests with their "trapeangs" or seasonally inundated grasslands and wetland pools, on the border between Lao and northern Cambodia, are breeding grounds for giant ibis *Pseudibis gigantea*, white-shouldered ibis *Pseudibis davisoni*, and Eastern sarus crane *Grus antigone*. Around the Tonle Sap, the grasslands are a stronghold for the Bengal florican *Houbaropsis bengalensis*, and the flooded forests of Prek Toal are important breeding grounds for spot-billed pelican *Pelecanus philippensis* and lesser and greater adjutant (*Leptoptilos javanicus* and *L. dubius*). The sarus cranes migrate to the Mekong Delta in the dry season, especially to remaining seasonally inundated grasslands such as at Tram Chim.

Mammals: A number of rare mammal species are associated with the river and its wetlands. These include the critically endangered subpopulation of the Irrawaddy dolphin *Orcaella brevirostris*, a flagship species and tourism attraction, which is found in the lower reaches of the Mekong below the Khone Falls in Lao PDR and Cambodia. There are three species of otter of special interest, the hairy-nosed otter *Lutra sumatrana*, the smooth-coated otter *Lutrogale perspicillata*, and the oriental small-clawed otter *Aonyx cinerea*. The first two are found in the Tonle Sap Great Lake, and all are under heavy threat due to hunting and loss of habitat. The flooded forests of the Tonle Sap are home to flying foxes and three species of primate. The Mekong mainstream in Cambodia between Stung Treng and Kratie is important for the conservation of three mammal species – hog deer *Axis porcinus*, silvered leaf monkey *Semnopithecus cristatus*, and otters (Mekong River Commission 2010).

Dependent People

The population of the Lower Mekong Basin has been estimated at about 60 million people, of whom about 85% live in the rural areas, at an overall population density of about 124 people/km^2. Most live near rivers, lakes, and wetlands with 25 million living within a 15 km corridor on either side of the Mekong mainstream (Mekong River Commission 2010). The population is predominantly young, e.g., 37–39% under 15 in Lao PDR and Cambodia, and has been growing at about 1.74% per annum. The rural population is very dependent upon agriculture, with fisheries and the collection of non-timber forest products as important contributors to their nutrition and livelihoods. There are more than 70 ethnic groups living in the Mekong Basin, especially in the upland areas.

While all four countries of the Lower Mekong are now classified as middle-income countries and there has been steady improvement in the Human Development Indices in recent years, significant proportions of the population are still considered to be poor. In 2010, the Human Poverty Index in Cambodia was 27.7% and 30.7% in Lao PDR, compared to 12.4% in Vietnam and less than 10% in Thailand (Mekong River Commission 2010).

The Mekong Basin has the largest inland fishery in the world. In the year 2008, the LMB fishery yield was estimated at 2.8 million tonnes (Mt) per year with a consumption per head ranging from over 43 kg/head/year around the Tonle Sap in Cambodia to about 25 kg/head/year in Lao PDR, excluding consumption of other aquatic animals. Of this figure, capture fisheries contributed about 1.9 Mt, with a first point of sale market value of US$ 3.9–7 billion per year, and making a significant contribution toward food security and nutrition. Other aquatic animals (insects, snails, frogs, snakes, etc.) make up about 20% of the total catch.

In 2003, it was reported that some 40 million people in the LMB were involved in fishing, at least part time or seasonally. In Lao PDR more than 70% of rural households depend upon fishing for subsistence livelihoods and additional cash income. In Cambodia 40% of the total population is dependent upon the Tonle Sap Great lake and its floodplains for their livelihoods, and over 1.3 million people reside in fishing communities around the Lake. Most of the rural households describe themselves as first of all rice farmers, with fishing and fish processing as secondary activities (Mekong River Commission 2010).

Apart from the important provisioning, other ecosystem services provided by the wetlands in the Lower Mekong Basin include the regulating services such as flood mitigation; without the wetlands of the Tonle Sap and other tributaries, large floods would be more regular and intense, especially in the Delta. Water purification is also important, with the dilution of the Mekong flows providing an important factor in limiting water pollution. The Mekong has contributed to the cultural heritage and diversity of the region, with many locations and festivals associated with the river, e.g., boat racing festivals throughout its length.

Threats and Future Challenges

The extent of wetland areas, both natural and converted, in the Lower Mekong Basin is taken as 254,144 km^2 with only 55,498 km^2 be classified as natural wetlands. If it is assumed that the total extent was the original cover of natural wetlands within the basin, then it can be seen that about 78% of the original wetland coverage has been lost or converted. In the Delta, less than 2% of the area's original wetlands remain. Conversion to agriculture has been the most significant change, although rice paddy fields still retain the character and some of the biodiversity of wetlands. The conversion of wetlands within urban areas has also accelerated, with dramatic urban expansion in all the countries of the region and many urban wetlands that remain, e.g., That Luang in Vientiane and Boeung Kak Lake in Phnom Penh are threatened or have been filled in.

Hydropower and irrigation dams have created large expanses of reservoirs, some which are important wetland areas in their own right, but are changing the hydrology of the both tributaries and the Mekong mainstream. Some 26 hydropower plants have been constructed in the Lower Mekong Basin, with another 134 in different stages of planning, design, and construction. In addition to the eight mainstream dams in China, a cascade of five dams is planned in northern Lao PDR and six more possible dams further downstream. The Xayaboury dam is under construction (due to be commissioned in 2019) and construction starting on the Don Sahong dam across one of the channels in the Siphandone area. Storage dams hold water back in the wet season, reducing the flood peak, and release it during the dry season, tending to balance out the flows and making the flood pulse less significant. This has implications for the overall ecological functioning of the Mekong, with changes in geomorphology and wetland habitats, including in the floodplain. Seasonally inundated in-channel wetland areas, upon which the productivity of the river depends, are likely to be reduced in extent. The seasonal shifts in hydrology and the barriers of dams can prevent or reduce fish migrations and ultimately will reduce fish populations.

Dams also tend to trap sediment, and even if they are fitted with functioning sediment flushing, it is likely that the quantities of bed load and fine sediments passing down the river and reaching the Delta will be significantly reduced in the future. Sand and gravel extraction in major tributaries and the Mekong mainstream have also contributed to a significant reduction in sediment transport; the suspended sediment budget in the Lower Mekong is now estimated at about 75 Mt per year, less than half of the original 160 Mt per year (Koehnken 2014) This will have implications for the nutrients in the floodplain and the state of the Delta which is already showing signs of regression.

Land use change within the catchment can also cause degradation of wetlands. In recent years there has been large-scale deforestation, especially in Cambodia and Lao PDR, as a result of both legal and illegal logging and conversion of natural forests to agroforestry, e.g., rubber, mining, and road construction, all of which tend to increase the amount of sediment in the tributaries, which smothers habitats and spawning sites, clogs fish gills, and reduces the amount of light penetrating the water, thus reducing plant productivity. Degradation of the riverine and flooded forests in the Mekong mainstream has also resulted in significant impacts on their biodiversity and productivity.

The wetlands have provided an important part of the food and livelihoods of many rural people, but with increasing populations and increasing effectiveness of fishing, hunting, and collection methods, including illegal methods such as the use of explosives and electrofishing, there is increasing pressure on these natural resources. In many cases this is shown by a decline in certain species and the larger fish and rarity of wetland mammals and birds.

Pollution and declining water quality is another major issue that is associated with increasing populations and industrialization. While the water quality in most of the river is considered to be acceptable, there are hotspots of high pollution levels around the major cities, e.g., Phnom Penh and in the Delta. Other forms of pollution such as mercury from mining activities are a source of concern, and the increasing use of agricultural chemicals is thought to be contributing to both eutrophication and toxicity in some parts of the river and its wetlands.

Climate Change

An MRC-sponsored study of the vulnerability of wetlands in the LMB to climate changes was carried out in 2011–2012. Direct climate threats include changes in the meteorological conditions including precipitation, as well as changes in temperature, hydrology, and sea levels, which will affect the source, transport, and fate of water in the wetland system (ICEM 2012).

Increases in wet season precipitation in wetland areas and upstream would result in expansion of the aquatic extent of wetlands. Conversely, an increase in dry season precipitation could prolong the presence of standing water during terrestrial phases, while both could change the timing of ephemeral components of a wetland and the duration of transition seasons. Mean annual precipitation is predicted to increase in the LMB except for isolated areas in the Cardamom ranges. At the basin level, mean annual precipitation may increase by 100–300 mm/year depending on the Global Circulation Model (GCM) used. The highest increases are predicted in the Central and Northern Annamites to the east of the basin. In these high elevation areas where rainfall is historically high, increases of over 1,000 mm are predicted. In the northern mid-elevation areas near the borders with China and Myanmar, the precipitation will increase by around 100–200 mm. Lower increases in precipitation will occur in the center of the basin, in the Khorat Plateau, Vientiane and Cambodian floodplains, and the Delta of Vietnam. In these areas, average annual precipitation will increase by approximately 50–150 mm. However, these areas are historically areas of low rainfall, and the predicted increase in temperature is proportionally significant (10–15%) compared to the baseline. This will result in greater variability in the Mekong moisture budget with the highest level of exposure correlated with increasing elevation.

There is considerable variability between changes in temperature throughout the LMB. The average annual maximum temperature will increase by between 2 °C and 3 °C with greater increases in the southern and eastern regions of the basin, with the largest change in temperature occurring in the 3S rivers catchment including a small area of the Srepok catchment with an increase of almost 5 °C. The expected increase in temperature gradually decreases toward the north along the Annamite ranges, and south east, toward the Mekong Delta. The historically warm areas of the Khorat Plateau are predicted to have a smaller increase in annual average maximum temperature of around 1–2 °C.

The hydrological regime plays a major role in shaping and maintaining wetland systems, through the seasonal input of water and by determining the morphological features of the wetland. Any alterations of these regimes will influence the biological, biogeochemical, and hydrological functions of wetlands. The flow regime incorporates consideration of variability, magnitude, frequency, duration, timing, and rate of change of the flow. This is particularly important in the riverine wetlands of Mekong Basin, where the annual flood pulse is central to the high levels of biodiversity and productivity. There would likely be an increase in flood magnitude and volume in all reaches of the Mekong between 15% and 25%, and climate change is likely to increase the duration of the flood season. The transition season between dry and wet seasons may also be shortened with implications for biological triggers, e.g., fish migration. In the dry season, climate change is expected to increase dry season flows, with increases of up to 20–30% in dry season water levels in some parts of the Mekong (e.g., between Vientiane and Pakse).

For the past 6,000 years, the Mekong Delta has been in a period of aggradation, in which catchment hydrology and sedimentation dynamics have resulted in the formation and net expansion of the Delta. Sea level rise will enhance the effectiveness of marine processes shifting the balance and resulting in permanent inundation, erosion, and salinization of a greater proportion of the Deltaic environment and an inland migration of coastal wetland environments where possible. For mean annual sea levels, the 2007 Intergovernmental Panel on Climate Change (IPCC) prediction of sea level rise (26–59 cm) would lead to a 36–63% inundation of the Ca Mau Peninsula.

References

Allen DK. The status and distribution of freshwater biodiversity in Indo-Burma. Cambridge, UK/Gland: IUCN; 2012.

Gao J, Li XL, Brierley G, Cheung A, Yang YW. Geomorphic-centered classification of Wetlands on the Qinghai-Tibet Plateau, Western China. J Mt Sci. 2013;10(4):632–42.

ICEM. Vulnerability report: Basin-wide climate change impact and vulnerability assessment for Wetlands in the Lower Mekong Basin for adaptation planning. Hanoi: Consultant report prepared for the Mekong River Commission, Hanoi; 2012.

Koehnken L. Discharge sediment monitoring project (DSMP) 2009–2013: summary and analysis of results, MRC, Office of the Secretariat in Vietnam, Vientiane; 2014.

Mekong River Commission. Overview of the hydrology of the Mekong Basin. Vientiane: Mekong River Commission; 2005.

Mekong River Commission. State of the Basin report. Vientiane: Mekong River Commission; 2010.

Mekong River Commission. Planning atlas of the Lower Mekong Basin, Basin Development Programme. Vientiane: Mekong River Commission; 2011.

Meynell P-J. Ecological significance of the tributaries of the Mekong – unpublished report to MRC. Vientiane: Mekong River Commission; 2012.

Tonle Sap Lake: Mekong River Basin (Cambodia)

149

Colin Poole

Contents

Introduction	1786
Extent (Current and Historical)	1786
Wetland Type Diversity	1786
Conservation Status	1787
Significant Biodiversity	1787
Management	1788
Ecosystem Services	1789
Dependent Peoples	1789
Threats and Future Challenges	1790
References	1791

Abstract

Located in the centre of Cambodia, Tonle Sap lake is the largest freshwater lake in South-east Asia and with a unique hydrology has for centuries been at the core of Cambodian life and culture. The flow of the Tonle Sap river is reversed when the level of the Mekong waters rise in the flood season (June-September) pushing water into the lake and increasing the inundated area up to five-fold. Vegetation of the floodplain is largely secondary in nature but the flooded forest supports populations of globally threatened species, is one of the region's most important areas for bird conservation, and supports a highly productive and diverse inland fishery and large human population. Listed as a Biosphere Reserve with two of three Core Areas designated as Ramsar Sites, Tonle Sap is threatened with the abolition of the traditional fishery, overharvesting of fish and wildlife, transfor-

C. Poole (✉)
Wildlife Conservation Society, Phnom Penh, Cambodia
e-mail: cpoole@wcs.org

mation of inundated grassland areas to irrigated dry season rice, and construction of hydropower dams on the mainstream of the Mekong River and major upstream tributaries.

Keywords

Tonle Sap lake · floodplain wetland · hydrology · fish diversity · globally threatened species · Biosphere Reserve · Ramsar Site · traditional fishery · hydropower dams

Introduction

The Tonle Sap Lake is located in the center of Cambodia and is the largest freshwater lake in Southeast Asia. It has for centuries been at the core of Cambodian life and culture, and its shores host the world renowned temple complex of Angkor, dating back to around 800 AD, one of the world's greatest early civilizations (Higham 2003), and one which was largely based on the productivity of the Tonle Sap Lake.

Extent (Current and Historical)

The formation of the lake itself is, in geological time, very young. The Tonle Sap lies on a northwest–southeast geological fault line and was created by the subsidence of the "Cambodian platform" around 5,700 years ago. The lake has a unique hydrology: When the level of the Mekong waters rise in the flood season (June–September), the flow of the Tonle Sap River which connects the lake with the Mekong is reversed so that water is pushed into the lake. In the dry season, the Tonle Sap Lake stretches for approximately 150 km in length and averages around 20 km in width; however, at the peak of the wet season it can expand to 250 km long and in places more than 100 km wide. The lake is shallow, measuring only 1–2 m at its deepest in the dry season, rising to more than 10 m in the wet season. As a result, as it floods, the total inundated area increases up to fivefold, from 2,500 km^2 to as much as 15,000 km^2 (Campbell et al. 2006).

Wetland Type Diversity

The vegetation of the Tonle Sap Lake floodplain has undergone a long history of alteration, mainly due to human disturbances, and is now largely secondary in nature. In its most pristine form, it consists of an almost closed canopy of small- to medium-sized trees – often referred to as "flooded forest." Most of the floodplain wetland, however, consists of scrubland with shrubs and stunted trees, interspersed with swards of herbaceous vegetation dominated by grasses (Campbell et al. 2006). It can largely be divided into a gradient of wetland types from the permanent open waters to the forested uplands, including flooded forest around the open water surface of the lake, estimated to cover about 3,000–3,600 km^2; seasonally inundated grasslands located beyond the flooded forest; and permanent rice-field agriculture, mainly flood recession rice (planting rice as the flood waters recede), and increasingly, irrigated dry season rice.

Conservation Status

The Tonle Sap Lake and floodplain was designated by the Royal Cambodian Government as a Multiple Use Area under the 1993 Royal Decree on Protected Areas. In October 1997, UNESCO accepted the nomination of the Tonle Sap as a biosphere reserve with a total area of 1,481,257 ha. This included a buffer zone of 510,768 ha and three core areas – Prek Toal (21,342 ha), Boeng Tonle Chmar (14,560 ha), and Stung Sen (6355 ha) – designated in April 2001 by a Royal Decree implementing the biosphere reserve. The core areas are devoted to "long term protection and conservation of natural resources and ecosystem, in order to preserve flooded forest, fish, wildlife, hydrological system, and natural beauty." Two of the core areas, Boeng Tonle Chmar and Prek Toal, are also designated as Ramsar Sites, the former in June 1999 and the latter in December 2015.

In addition, six Bengal Florican Conservation Areas were established by the Ministry of Agriculture Forestry and Fisheries (MAFF) in February 2010 totaling 31,159 ha of seasonally inundated grassland on the northeast side of the Tonle Sap Lake in Kampong Thom and Siem Reap Provinces. These are now managed by the Ministry of Environment under the Northern Tonle Sap Protected Landscape.

Significant Biodiversity

The Tonle Sap Lake is one of the most important areas for bird conservation in the region and is extremely important for gregarious large waterbirds. Seventeen IUCN globally threatened or near threatened species (IUCN 2013) are known to occur regularly around the Tonle Sap Lake, and a study to identify Important Bird Areas (IBAs), based on the global criteria of BirdLife International, identified ten IBAs within the inundation zone (Seng et al. 2003). These are divided between the flooded forest, particularly of the Prek Toal core area, and the inundated grassland.

Prek Toal is the only remaining breeding site in mainland Southeast Asia for two globally threatened species, spot-billed pelican *Pelecanus philippensis* and milky stork *Mycteria cinerea* and the largest colony in the region for six more globally threatened or near-threatened species, namely, oriental darter *Anhinga melanogaster*, lesser adjutant *Leptoptilos javanicus*, greater adjutant *Leptoptilos dubius*, black-headed ibis *Threskiornis melanocephalus*, painted stork *Mycteria leucocephala*, and gray-headed fish eagle *Ichthyophaga ichthyaetus* (Sun and Mahood 2012). The inundated grasslands support a unique community of birds, including the world's largest population of the critically endangered Bengal florican *Houbaropsis bengalensis*, and number of small grassland birds including the vulnerable Manchurian reed warbler *Acrocephalus tangorum* (Bird et al. 2012).

The flooded forest areas support populations of other globally threatened species, such as Siamese crocodile *Crocodylus siamensis*, Germain's silvered langur *Trachypithecus germaini*, and two species of otter- hairy-nosed *Lutra sumatrana* and smooth-coated *Lutrogale perspicillata* (Sun and Mahood 2012), also known to support seven species of watersnakes, one of which – Tonle Sap watersnake *Enhydris longicauda* – is believed to be endemic. It is not only one of the most productive but also one of the most diverse fish systems in the world. More than 2000 species occur in the Mekong system, and over 500 are recorded from Cambodia. In the Tonle Sap Lake itself there are around 200 species, of which more than 100 regularly occur in fish catches (Rainboth 1996; van Zalinge 2002).

Management

The Tonle Sap Lake (including the three core areas) was formally managed through a system of fishing concessions ("lots") auctioned every 2 years. The lot owners built lengthy bamboo fences around the fishing lot in order to maximize the return from fish which provided a high level of protection to the flooded forest and other biodiversity found in the lots. In 2002, half the fishing lots were canceled and converted into community fisheries. In March 2012, the remaining lots on the lake were also canceled and converted into either Community Fisheries Areas or 13 new Fisheries Conservation Areas totaling 81,921 ha around the Tonle Sap Lake. The most productive of the former fishing lots, Battambang Province fishing lot 2, which also included the Prek Toal core area, is the largest of the Fisheries Conservation Areas at 50,134 ha.

Management jurisdiction for the Tonle Sap Lake is divided between Royal Cambodian Government agencies as follows:

- The General Department Administration of Nature Conservation and Protection (GDANCP) of the Ministry of Environment is responsible for management of the three core areas (including the two Ramsar Sites) of the Tonle Sap Biosphere Reserve and the Northern Tonle Sap Protected Landscape.
- The Fisheries Administration of MAFF is responsible for management of all fisheries around the lake. Historically, this was largely the fishing lots but is now the Fish Conservation Areas and Community Fisheries Areas.
- The Forestry Administration of MAFF is responsible for management of forestry areas and wildlife.

- The Tonle Sap Authority, chaired by the Minister of Water Resources and Meteorology, has overall responsibility for regulation of the Tonle Sap Lake.

Ecosystem Services

In 1860, when French explorer Henri Mouhot crossed the Tonle Sap en route to Angkor, he wrote of the sheer numbers of fish hindering progress as they struck oars and the hull of the boat. The lake was, he commented, the center of a huge fishing system that supported thousands if not millions of people (Mouhot 2000). The annual consumption of fish and other aquatic animals in Cambodia is at least 720,000 tons, and the inland fisheries of Cambodia are among the most significant in the world (Hortle et al. 2004). Few countries in the world are so dependent on inland fisheries as is Cambodia where fish provide people with 80% of their animal protein and contribute to 16% of the country's GDP (Baird et al. 2003).

The reason for this is the magnitude and timing of the flood. Every year as the rising water inundates the floodplain, fish have access to an extensive environment rich in nutrients. The dependability of the system has allowed fish to evolve to take maximum advantage of the enormous quantities of food produced by this annual opportunity. The flooded forest areas also play a key role in the lake's fisheries as an important nursery habitat for fish species and are therefore also critical for the maintenance of ecosystem services.

Dependent Peoples

The Tonle Sap basin, including the catchment of the lake and the Tonle Sap river, has an estimated population of 4.5 million people (Leang 2003) giving an average population density of 53 people per km^2. Of these, more than 80,000 people have been estimated to live in about 170 floating villages on the Tonle Sap Lake (Keskinen 2003). These floating villages are also some of the most ethnically diverse in Cambodia; in addition to Khmer people, they contain significant Cham Muslim, Vietnamese, and Chinese communities. Traditionally, many floating villages have been inhabited by Vietnamese and about 12,000 Vietnamese are thought to live on the lake, but as many have no official status their true numbers are difficult to determine.

Although placed directly over Cambodia's richest natural resource, the floating villages are some of the country's poorest communities. A survey of eight floating villages around Prek Toal found that the major source of income for over 90% of people interviewed was family-scale fishing. It also revealed that the floating villages were generally worse off than their upland counterparts; villagers were less educated, with fewer livelihood options, no agricultural land, and a strong dependence on common property resources such as fisheries and flooded forests (Gum 1998).

Threats and Future Challenges

All three core areas are under threat because the abolition of the fishing lots has removed a crucial layer of protection. At the same time, the flooded forest is under increasing threat from upland development and man-made fires, which spread into the area during the dry season. Management and control of upland access to all three core areas in the dry season is extremely difficult, but loss of significant flooded forest habitat would seriously affect the productivity of the lake.

The inundated grassland areas of the Tonle Sap are undergoing a rapid transformation to irrigated dry season rice. 28% of the grasslands were lost from 2005 to 2007, and losses have continued at a high rate since then. Most of the loss is due to a recent wave of large-scale agricultural conversion. Earth dams of 100–1,000 ha are built to capture the floodwaters of the lake in the rainy season and then irrigate surrounding areas in the dry season. If conversion continues, it is likely that most of the remainder of this critical habitat will be lost in the coming years (Mahood et al. 2012).

Overharvesting of many species is of concern. There is little historical information on the sustainability of fish catches in Cambodia, but a growing human population has increased fishing pressure, and the catches of individual fishermen have undoubtedly decreased. Significant declines are known to have taken place in some larger species, particularly those that spawn later in life. Although no species is yet known to have gone extinct in the Mekong system, some, such as the Mekong giant catfish *Pangasianodon gigas*, are now coming close (van Zalinge 2002). Fish populations are threatened by the increasing use of nontraditional, and largely illegal, fishing technologies including small-mesh gill nets, electrofishing, and the use of small powerful pumps to pump dry streams and ponds, particularly in the flooded forest.

Commercial level harvesting of watersnakes has begun only in the last decade, much of it as a cheap food for farmed crocodiles. This has resulted in up to 6.9 million snakes being removed annually, representing the world's largest snake exploitation. Interviews with hunters suggest that snake catches could have declined by as much as 80% between 2000 and 2005, raising strong concerns about the sustainability of this exploitation (Brooks et al. 2007).

Harvesting of eggs and chicks for food was formerly the greatest threat to the large waterbird colonies, and in 1996 it was estimated that 26,000 eggs and close to 3,000 chicks were taken from the colonies of Prek Toal (Parr et al. 1996). In 2001, the GDANCP and the Wildlife Conservation Society (WCS) began a community-based program of ranger patrols, education, and monitoring in the Prek Toal core area as a deterrent to egg and chick collection. Since this began there has been a significant increase in the number of breeding waterbirds. Oriental darter, for example, has shown an increase from 241 nests in 2002 to 6,751 in 2011 (Sun and Mahood 2012). However, such collection is still of serious concern, particularly in unprotected areas away from the Prek Toal core area.

The greatest long-term threat to the Tonle Sap Lake is the significant number of hydropower dams that are completed, under construction, or planned on the mainstream Mekong River and its major tributaries upstream, particularly in China and Lao PDR, but also in Vietnam and Cambodia. The major impact of such development will be an increase in dry season flows and a decrease in wet season flows downstream, resulting in increasing dry season minimum water levels and decreasing wet season maximum water levels. This will produce a corresponding decrease in the seasonally inundated area that will in turn alter the plant communities and the productivity of the system. If the change in inundated area results in a similar average reduction in the fish harvest, it would have serious consequences for the livelihoods of those living around the lake and throughout Cambodia.

References

Baird IG, Flaherty MS, Bounpheng Phylavanh. Rhythms of the river: lunar phases and migrations of small carps (Cyprinidae) in the Mekong River. Nat Hist Bull Siam Soc. 2003;51:3–36.

Bird JP, Mulligan B, Rours V, Round PD, Gilroy JJ. Habitat associations of the Manchurian Reed Warbler *Acrocephalus tangorum* wintering on the Tonle Sap floodplain and an evaluation of its conservation status. Forktail. 2012;28:71–6.

Brooks SE, Allison EH, Reynolds JD. Vulnerability of Cambodian water snakes: initial assessment of the impact of hunting at Tonle Sap Lake. Biol Conserv. 2007;139(3–4):401–14.

Campbell IC, Poole CM, Giesen W, Valbo-Jorgensen J. Species diversity and ecology of Tonle Sap Great Lake, Cambodia. Aquat Sci. 2006;68(3):355–73.

Gum W. Natural resource management in the Tonle Sap biosphere reserve in Battambang Province. Phnom Penh: SPEC (European Commission); 1998.

Higham C. The civilization of Angkor. London: Phoenix Press; 2003.

Hortle KG, Lieng S, Valbo-Jorgensen J. Cambodia's Inland fisheries, Mekong development series, vol. 4. Phnom Penh: MRC and the Inland Fisheries Research and Development Institute; 2004.

IUCN. The IUCN red list of threatened species. Version 2013.1. 2013. http://www.iucnredlist.org. Downloaded on 02 July 2013.

Keskinen M. The great diversity of livelihoods? – socio-economic survey of the Tonle Sap Lake. Phnom Penh: WUP-FIN Socio-economic Studies on Tonle Sap 8, MRCS/WUP-FIN; 2003.

Leang P. Sub-area analysis, the Tonle Sap sub-area. Report for the basin development plan. Phnom Penh: MRC; 2003.

Mahood S, Son V, Hong C, Evans T. The status of Bengal Floricans in the Bengal Florican conservation areas, 2010/11 monitoring report. Phnom Penh: WCS Cambodia Program; 2012.

Mouhot H. Travels in Siam, Cambodia, Laos, and Annam. Bangkok: White Lotus; 2000 (Reprint).

Parr JWK, Eames JC, Sun H, Hong C, Som H, Pich VL, Seng KH. Biological and social aspects of waterbird exploitation and natural resource utilization of Prek Toal, Tonle Sap Lake, Cambodia. Cambridge, UK: IUCN/SSC; 1996.

Rainboth WJ. Fishes of the Cambodian Mekong. Rome: FAO; 1996.

Seng KH, Pech B, Poole CM, Tordoff AW, Davidson P, Delattre E. Directory of important bird areas in Cambodia: key sites for conservation. Phnom Penh: DFW, DNCP, BirdLife International and WCS Cambodia Program; 2003.

Sun V, Mahood S. Monitoring of large waterbirds at Prek Toal, Tonle Sap Great Lake, 2011. Phnom Penh: WCS Cambodia Program; 2012.

van Zalinge N. Update on the status of the Cambodian inland capture fisheries sector with special reference to the Tonle Sap Great Lake. Catch Culture. 2002;8(2):1–9.

Tram Chim: Mekong River Basin (Vietnam)

150

Triet Tran and Jeb Barzen

Contents

Introduction	1794
Physical Environment	1795
Wetland Ecosystems	1795
Biodiversity	1796
Ecosystem Services	1796
Conservation Status and Wetland Management	1797
Threats and Future Challenges	1798
References	1798

Abstract

Tram Chim National Park (7,313 ha) preserves the largest remnant of the Plain of Reeds, a freshwater floodplain once covered approximately one million hectares of the Mekong Delta in Cambodia and Vietnam. Wetlands in Tram Chim are of three main types: *Melaleuca* woodland, seasonally inundated marsh and permanently inundated swamp. The biodiversity of Tram Chim wetlands is very diverse, especially in fish and bird. Tram Chim is an important site for migrating birds and one of the most important wintering sites of the threatened Eastern sarus crane *Grus antigone sharpii* in the Mekong Delta. Wetland products from Tram Chim, especially fish and grazing grasses, are important for subsistent livelihoods of the local communities. Located downstream of the Mekong River, Tram Chim wetlands are vulnerable to abrupt environmental changes caused by upstream hydropower development.

T. Tran (✉)
International Crane Foundation, Baraboo, WI, USA

University of Natural Science, Vietnam National University, Ho Chi Minh City, Vietnam
e-mail: ttriet@gmail.com

J. Barzen
Private Lands Conservation, Spring Green, WI, USA

© Springer Science+Business Media B.V., part of Springer Nature 2018
C. M. Finlayson et al. (eds.), *The Wetland Book*,
https://doi.org/10.1007/978-94-007-4001-3_41

Keywords

Tram Chim · Mekong Delta · Plain of Reeds · Sarus crane · Hydropower

Introduction

Tram Chim National Park: Mekong River Basin (Vietnam) is located in Dong Thap Province, in the northeastern part of the Mekong Delta (Fig. 1). At 7,313 ha, Tram Chim is the largest remnant of the Plain of Reeds, which was once a vast, relatively closed basin dominated by flooding from sheet flow during the rainy season and slow recession during the dry season. The Plain of Reeds spans an area of approximately 1,000,000 ha in both Cambodia and Vietnam. Since the early 1980s, the Plain of Reeds has been rapidly converted into agricultural production. A dense network of irrigation canals was dug to drain wetlands, greatly altering the hydrology of the entire region, including the area which is now Tram Chim National Park. Restoring the natural hydrological regime became the basis for wetland management at Tram Chim. The water level inside the Park is now regulated by a system of dikes and sluice gates intended to reestablish the natural wet-dry cycle for the wetlands of this area (Beilfuss and Barzen 1994; Meynell et al. 2012).

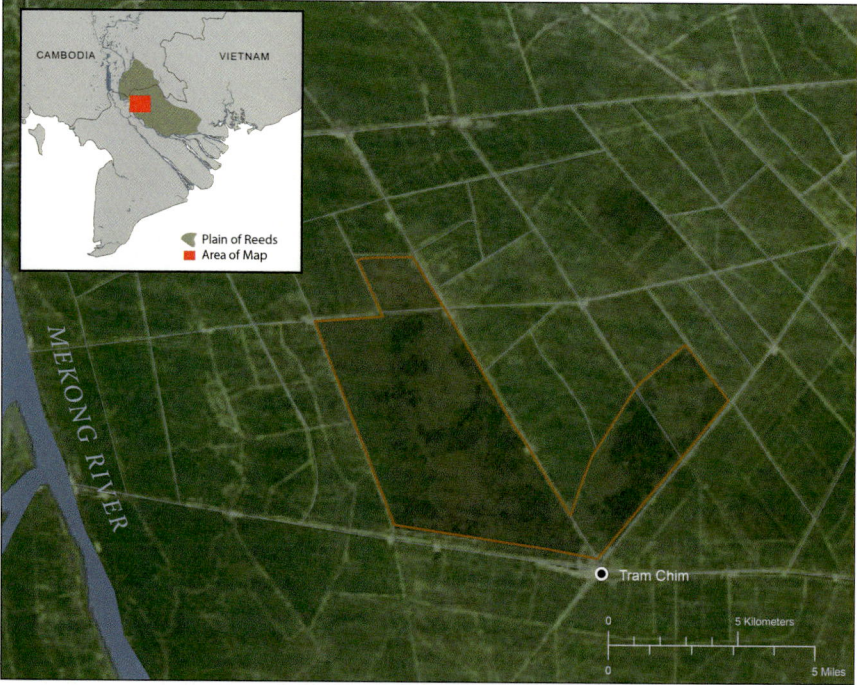

Fig. 1 Tram Chim National Park, a remnant of the Plain of Reeds, Mekong Delta, Vietnam (Source: © Provided courtesy of International Crane Foundation)

Physical Environment

Even though the entire Tram Chim National Park is located in a relatively flat basin, its microtopography varies greatly. Sand ridges (from old beaches) and old alluvium formations create relatively higher ground composed of different soil types, as high as 2.3 m, while the lower areas, consisting of mostly acid sulfate soils, have an average elevation of 1.0–1.2 m above the mean sea level. These variations in microtopography and soil conditions account for marked differences in the composition and structure of vegetation among the wetlands of Tram Chim and have important hydrological management implications (Beilfuss and Barzen 1994; Tran 2005). Tram Chim has a monsoonal climate with an average annual rainfall of 1,400 mm. During the rainy season (May to November), Tram Chim wetlands are inundated by overbank flows from the Mekong River and high rainfall. At the peak of the flood season (late September to October), Tram Chim wetlands can be submerged under 4–5 m of water. During the dry season (December to April), flood water recedes gradually through evaporation and natural drainage streams until there is standing water in only a few places at the onset of the following rainy season.

Wetland Ecosystems

Wetland ecosystems at Tram Chim are grouped into three broad types: *Melaleuca* woodlands, seasonally inundated marshes, and permanently inundated swamps (Tran 2005). *Melaleuca cajuputi* (Myrtaceae) forest is the only type of forest known to exist in the region since reliable scientific records were established. There is, however, evidence of more ancient forests dominated by other species of trees in the Plain of Reeds. Large tree stumps (1.8–2 m in diameter) with shallow-spreading root systems were discovered in the buffer zone of Tram Chim National Park (Le 1993) and were subsequently identified as a species of the genus *Syzygium*. The reasons why this forest type failed to reestablish remains unknown. At present, all forest stands at Tram Chim are replanted with the native *Melaleuca cajuputi*.

The seasonally inundated marshes at Tram Chim are the last extensive remnant of once immense freshwater marshes of the Plain of Reeds. The marsh plant communities form a continuum, closely following the gradient of soil surface elevation and water permanence. The *Panicum repens* (Poaceae) community is located on sand ridges and old-alluvium formations, with inundation time of 1–3 months a year. *Oryza rufipogon* (Poaceae) and *Eleocharis dulcis* (Cyperaceae) communities are located on the most depressed areas that may be flooded up to 9–10 months a year. The *Ischaemum rugosum* (Poaceae) community is located in areas of medium elevation, with average flooding time of 4–5 months a year.

Lotus *Nelumbo nucifera* (Nelumbonaceae) is abundant in permanently inundated swamps which are located on old riverbeds and shallow streams. Besides *Nelumbo nucifera*, many other aquatic plants are also found in lotus swamps such as *Nymphaea nouchali*, *Nymphaea tetragona* (Nymphaeaceae), *Polygonum tomentosum* (Polygonaceae), *Ludwigia adscendens* (Onagraceae*)*, *Nymphoides indica* (Menyanthaceae), and *Hymenachne acutigluma* (Poaceae).

Biodiversity

The avifauna of Tram Chim National Park is both diverse and abundant. To date, 231 bird species have been identified at Tram Chim (Nguyen 2006), of which 15 species are either endangered, threatened, or of special concern: comb duck *Sarkidiornis melanotos*, grass owl *Tyto capensis*, Bengal florican *Houbaropsis bengalensis*, Eastern sarus crane *Grus antigone sharpii*, greater spotted eagle *Aquila clanga*, Oriental darter *Anhinga melanogaster*, black-headed ibis *Threskiornis melanocephalus*, black-faced spoonbill *Platalea minor*, painted stork *Mycteria leucocephala*, Asian openbill stork *Anastomus oscitans*, lesser adjutant *Leptoptilos javanicus*, greater adjutant *Leptoptilos dubius*, black-necked stork *Ephippiorhynchus asiaticus*, Malaysian plover *Charadrius peronii*, and Asian golden weaver *Ploceus hypoxanthus*. Some of these birds, however, are infrequent visitors to Tram Chim wetlands or are represented by very few records. Tram Chim frequently hosts large flocks of water birds such as garganey *Anas querquedula*, common teal *Anas crecca*, and little cormorants *Phalacrocorax niger* and a variety of wading birds (e.g., *Ardea* spp., *Egretta* spp., *Ixobrychus* spp.) (Tordoff 2002).

Tram Chim National Park is well known in Vietnam for the presence of the Eastern sarus crane. The discovery of large flocks of sarus cranes in Tram Chim in the mid-1980s generated much excitement in the country and was the main reason for Tram Chim to be gazetted for conservation as a provincial nature reserve and then as a national park (Barzen 1991). Even though the number of cranes has declined at Tram Chim from the highest two decades ago, Tram Chim National Park is still one of the most important sites for sarus cranes in the Mekong Delta (Van Zalinge et al. 2011).

The wetlands of Tram Chim National Park provide important sources of food, spawning grounds, and migration paths for dependent fish stocks, both within and outside the wetlands (Duong et al.2007). Of the 130 fish species identified in Tram Chim (MWBP 2007), 5 species are globally threatened and 20 species are ranked as high or very high vulnerability in the FishBase 2004, an online-data base of fish species worldwide (WorldFish Center 2004).

Ecosystem Services

For people living in the buffer zone, fish from Tram Chim are probably the most desirable resource. Local people also harvest turtles, snakes, birds, Melaleuca for fuel wood, and some aquatic plants such as lotus and water lilies for food. Wetland plants, such as *Panicum repens* and *Eleocharis dulcis*, are an important source of mulching material for local vegetable gardeners. Tram Chim recently has become a popular ecotourism destination in Vietnam, serving over 20,000 visitors a year, many of whom are students.

Conservation Status and Wetland Management

Tram Chim was first established as a state-owned agro-forestry-fishery enterprise in 1985, became a district conservation area in 1988, a provincial protected area in 1993, and national park in 1998. In 2012, Tram Chim was designated the 2000th Wetland of International Importance by the Ramsar Convention on Wetlands. Since the beginning of wetland restoration and protection in Tram Chim, fire and water management has always been an important issue. Even though the initial water control structure was designed to create a hydrological condition that mimics the natural wet-dry cycle (Beilfuss and Barzen 1994), during 1995–2004 actual management practices mainly relied upon keeping high water levels inside Tram Chim to prevent forest fires. Prolonged flooding created unfavorable conditions for many wetland organisms that require seasonal water fluctuation and greatly diminished the diversity of Tram Chim wetland ecosystems. During 2005–2008, a comprehensive fire and water management plan was developed and tested at Tram Chim by a team of international experts (Meynell et al. 2012). The plan sought to minimize fire risks while taking into account ecological requirements of key plant and animal species of Tram Chim wetlands. Water drawdowns were implemented during dry seasons to recreate the wet-dry cycle. Monitoring data showed good recovery of wetland ecosystems following the implementation of the new fire and water management plan (Duong et al. 2007; Meynell et al. 2012). The plan was approved by Vietnamese authorities in 2011 to be applied permanently at Tram Chim.

The Tram Chim buffer zone comprises five villages and a town situated near the park, with a total population of approximately 42,000 people (Meynell et al. 2012). In addition, many seasonal migrants from nearby districts and provinces come to Tram Chim during flood season to fish (Nguyen 1997). Buffer zone management has been identified as the key element for the success of wetland conservation at Tram Chim and is integrally linked with the prevailing socioeconomic conditions that occur in the villages and town located near the park boundary (Duong 1997; Nguyen 1997). Poor people living in the surrounding area depend on fish and other wetland resources found at Tram Chim for their daily living. Following national regulations, Tram Chim National Park outlawed all forms of resource exploitation, creating much resentment from local communities. Still, illegal fishing for commercial purposes, especially those using harmful fishing methods such as electro-fish shocking, threatened the aquatic biodiversity of Tram Chim. After being tested in a pilot study during 2006–2007, a balanced approach towards wetland resource management has been applied in Tram Chim since 2008, which allows resource sharing in a sustainable manner (Meynell et al. 2012). Local people were organized into Resource User Groups, which together with park managers developed and implemented Sustainable Resource Use Plans designed to allow local people to harvest certain species of fish and wetland plants at a level that is thought to be sustainable.

Threats and Future Challenges

Besides the fire and water issues, invasive alien species is another important threat to Tram Chim. The giant sensitive plant *Mimosa pigra* (Fabaceae) and the golden apple snail *Pomacea canaliculata* are the two most serious invasive species at Tram Chim. *Mimosa pigra* once covered more than 3,000 ha of wetlands in Tram Chim (Tran et al. 2004). Recent eradication activities have reduced the area of *Mimosa pigra* infestation. Golden apple snails were first detected at Tram Chim during 2000–2002 and quickly became abundant by 2004. An effective invasive species management program needs to be implemented on a permanent basis at Tram Chim in order to keep harmful invasive species under control. Even though there has been good progress in managing resource sharing with local communities, illegal exploitation of wetland resources can still be a major threat to theTram Chim wetlands.

Located downstream of the Mekong river, Tram Chim National Park is susceptible to changes in the river hydrology and sedimentation caused by upstream development, especially hydropower development. Twelve hydropower dams, proposed to be constructed along the Mekong main channel in Laos and Cambodia, together with the existing and planned dams, in the upper Mekong basin in China, would cause tremendous changes in river hydrology and sedimentation in the delta area (ICEM 2010) and as a result would profoundly affect Tram Chim.

References

Barzen J. Restoration mixes science, people and luck in Vietnam. ICF Bugle. 1991;17(2):2–3.

Beilfuss R, Barzen J. Hydrological wetland restoration in the Mekong Delta, Vietnam. In: Mitsch WJ, editor. Global wetlands: old world and new. Amsterdam: Elsevier Science B.V.; 1994. p. 453–68.

Duong Van Ni. Sustainable development in the buffer zones of Tram Chim. In: Safford RJ, Duong Van Ni, Maltby E, Vo Tong Xuan, editors. Towards sustainable management of Tram Chim National Reserve, Vietnam: proceedings of a workshop on balancing economic development with environmental conservation. London: Royal Holloway Institute for Environmental Research, University of London; 1997.

Duong Van Ni, Shulman D, Thompson J, Triet T, Truyen T, van de Schans M. Integrated fire and water management strategy for Tram Chim National Park, Vietnam. Mekong Wetlands Biodiversity Conservation and Sustainable Use Programme (MWBP). Vientiane, Lao PDR: United Nations Development Programme (UNDP); 2007. 41 p.

ICEM – International Center for Environmental Management. MRC strategic environmental assessment of hydropower on the Mekong mainstream. Hanoi, Vietnam; 2010. 197 p.

Le Cong Kiet. Dong Thap Muoi, restoring the mystery forest of the Plain of Reeds. Restor Manage Notes. 1993;11:102–5.

Meynell PJ, Nguyen Huu Thien, Duong Van Ni, Triet T, Van der Schans M, Shulman D, Thompson J, Barzen J, Shepherd G. An integrated fire and water management strategy using the ecosystem approach: Tram Chim National Park, Vietnam. In: Gunawardena ERN, Gopal B, Kotagama H, editors. Ecosystems and integrated water resources management in South Asia. New Delhi: Routledge; 2012. p. 199–228.

MWBP (Mekong Wetland Biodiversity Programme). Inception report of landscape management and sustainable livelihoods in and around Tram Chim National Park. Hanoi, Vietnam: World Wildlife Fund for Nature – Greater Mekong Programme; 2007.

Nguyen Huu Thien. Winning support for conservation from local communities around Tram Chim. In: Safford RJ, Ni DV, Maltby E, Xuan VT, editors. Towards sustainable management of Tram Chim National Reserve, Vietnam: proceedings of a workshop on balancing economic development with environmental conservation. London: Royal Holloway Institute for Environmental Research, University of London; 1997. p. 17–46.

Nguyen Phuc Bao Hoa. Report on grassland birds survey & correlations between grassland birds and their habitat variables in Tram Chim National Park. Vientiane: Mekong Wetlands Biodiversity Conservation and Sustainable Use Programme; 2006.

Tordoff A, editor. Directory of important bird areas in Vietnam – key sites for conservation. Hanoi: Birdlife International Indochina Programme and Institute of Ecology and Biological Resources; 2002. 233 pp.

Tran T. An introduction to the biophysical environment and management of wetlands of Tram Chim National Park, Dong Thap Province, Vietnam. J Sci Technol Develop. 2005;8(6):31–9.

Tran T, Le Cong Kiet, Nguyen Thi Lan Thi, Pham Quoc Dan. The invasion of *Mimosa pigra* in wetlands of the Mekong Delta, Vietnam. In: Julien M, Flanagan G, Heard T, Hennecke B, Paynter Q, Wilson C, editors. Research and management of *Mimosa pigra*. Canberra: CSIRO Entomology; 2004. p. 45–51.

Van Zalinge R, Triet T, Evans T, Chamnan H, Seng Kim Hout, Jeb Barzen. Census of non-breeding Sarus Cranes in Cambodia and Vietnam, 2011. Phnom Penh: Wildlife Conservation Society Cambodia Program; 2011.

WorldFish Center. FishBase: a global information system on fishes. 2004. www.fishbase.org. Accessed 9 Nov 2012.

Transboundary Mekong River Delta (Cambodia and Vietnam)

151

Triet Tran

Contents

Introduction	1802
Physical Environment	1802
Channelization and Wetland Drainage	1804
Wetland Vegetation	1805
Biodiversity	1805
Fishes	1805
Birds	1806
Ecosystem Products and Services	1806
Conservation Status and Management	1807
Threats and Future Challenges	1808
References	1810

Abstract

With 5.5 million hectares of land area, the Mekong River Delta is one of the largest river deltas in the world. Mangrove forests, Melaleuca freshwater swamps and seasonally inundated marshes are the three main wetland vegetation types. Wetlands of the Mekong River Delta support a rich biodiversity, especially of fishes and birds. Ecosystem products and services provided by wetlands of the Mekong River Delta are enormous, supporting the livelihood of more than 18 million people in Cambodia and Vietnam. Rapid economic development during the past decades, however, have resulted in large-scale wetland loss. As of 2016, there are 14 protected wetlands, covering 84,000 ha, approximately 1.5% of the total land area of the Mekong River Delta. Climate change and upstream hydropower development are the two most important threats to wetlands of the Mekong

T. Tran (✉)
International Crane Foundation, Baraboo, WI, USA

University of Natural Science, Vietnam National University, Ho Chi Minh City, Vietnam
e-mail: ttriet@gmail.com

© Springer Science+Business Media B.V., part of Springer Nature 2018
C. M. Finlayson et al. (eds.), *The Wetland Book*,
https://doi.org/10.1007/978-94-007-4001-3_40

River Delta. The projected rates of sea level rise are among the highest of those of major world river deltas. There are as many as 130 hydropower projects, either existing, under construction or planned on the Mekong River main stem and its tributaries. The cumulative impacts of climate change and hydropower development to wetlands and people of the Mekong River Delta are still poorly understood.

Keywords

Climate change · Hydropower development · Mekong River Delta · Sea level rise · Wetland

Introduction

The Mekong Delta is one of the largest river deltas in the world. Beginning at Phnom Penh city in Cambodia, where the Mekong main channel bifurcates into the Mekong and the Bassac, the Mekong Delta encompasses 5.5 million hectares of land, 3.9 million of which are in Viet Nam (see Fig. 1). Except for small mountainous areas in the southwest, the delta is a low and flat plain that has an average elevation of 0.8 m above the mean sea level (Anonymous 2013). Even though major land developments in the Mekong Delta began relatively recently – since the seventeenth century (Li 1998), evidence of early human settlement dated back to 200 BC (Biggs 2003). The human population of the Mekong Delta in Vietnam is 17.4 million with a population density of 429 people per square kilometer (General Statistics Office of Vietnam 2012 Census), making it one of the most populated rural areas in the world. Thanks to its fertile soils, a large part of the Mekong Delta has been converted into agricultural lands. In Vietnam, 2.6 million hectares (or 65% of the Vietnamese Mekong Delta land area) are being used for agriculture, while forest land, including forest plantations, covers only 0.31 million hectares or 7.8% of the land area (General Statistics Office of Vietnam, Year 2012 data).

Physical Environment

The topography of the Mekong Delta consists of two parts: an upper (inner) delta plain dominated by fluvial processes and a lower (outer) delta plain mainly influenced by marine processes (Nguyen et al. 2000). Land elevation typically decreases outwards from the natural levees along major streams and rivers, forming large back swamps with elevations ranging from 0.5 to 1.5 m above the mean sea level (Ta et al. 2005). The largest such swamps are the Bassac Marsh in Cambodia and the Plain of Reeds and the Long Xuyen Quadrangle in Vietnam. The eastern coastal area is characterized by rows of sandy beach ridges, 3–10 m high, separated by inter-ridge swamps (Ta et al. 2005).

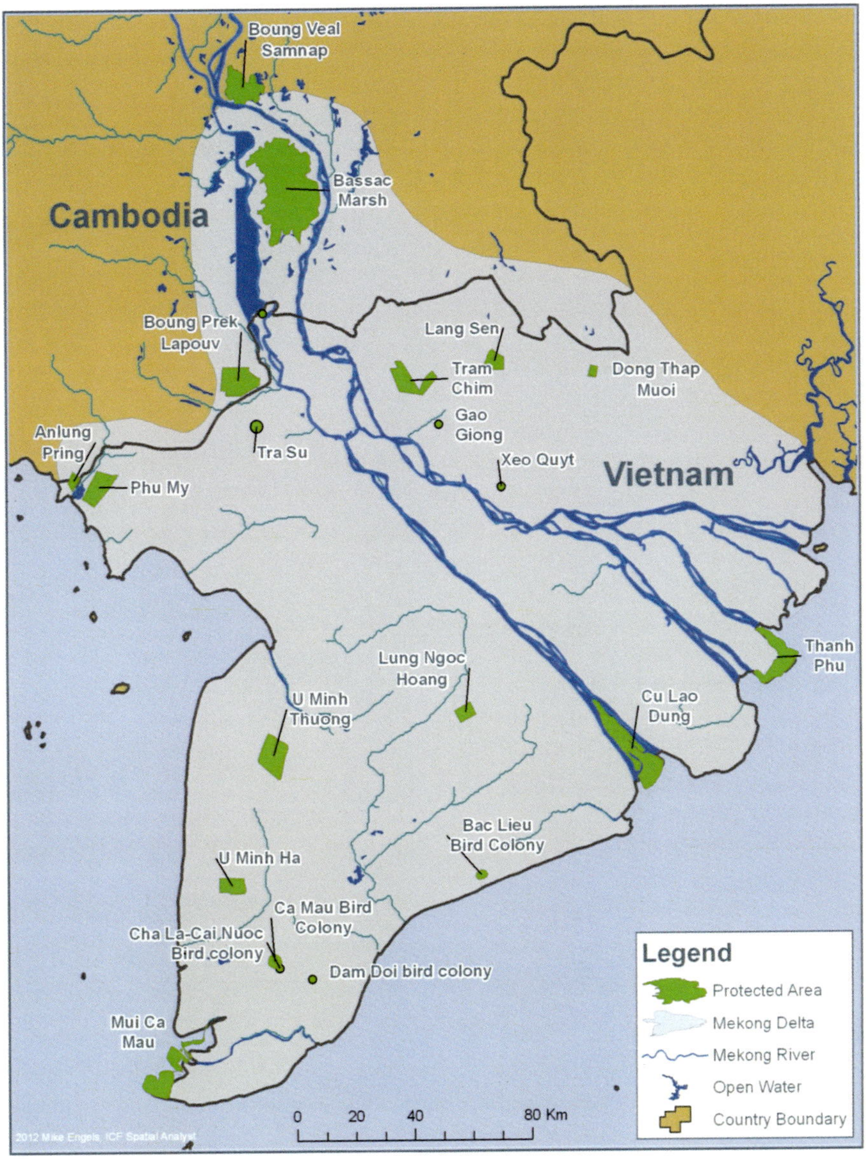

Fig. 1 Locations of some existing protected wetlands in the Mekong Delta

The Mekong Delta has a tropical monsoon climate, characterized by the alternation of one rainy and one dry season each year. The rainy season occurs from April to November in the west and from May to November in the rest of the Delta; mean annual rainfall varies across the Delta from 2,400 mm in the west to 1,300 mm in the center and 1,600 mm in the east. The rainy season is also the high-flow season of the Mekong

River. The combined effect of river flooding and local rainfall floods a significant part of the Mekong Delta during rainy seasons. Flooding can last for 4–5 months in the most low-lying areas. The maximum depth of inundation, ranging from 1.50 m to 5.50 m, varies in relation to the topography and the magnitude of flood (MRC 2005).

The Tonle Sap Lake in Cambodia plays an important role in the flood-pulse hydrology of the Mekong Delta. As the water level of the river rises during the wet season, at some point the Mekong river backs up the Tonle Sap river channel to fill the Tonle Sap Lake basin. At peak, the water depth in the Tonle Sap Lake can reach 8–10 m and the lake's water volume increases ten times. At the end of the wet season as the water level in the Mekong river subsides, water flows out of the Tonle Sap Lake downstream to the Mekong Delta area. Water flowing out of Tonle Sap Lake contributes on average 16% of the dry season flow of the Mekong river downstream of Phnom Penh (Campbell 2009).

The Mekong Delta is influenced by two tidal regimes: the semidiurnal tide from the South China Sea with tidal amplitudes of 3–3.5 m and the diurnal tide from the Gulf of Thailand with tidal amplitudes of 0.8–1.2 m (Nguyen Nghia Hung et al. 2012). High tide events, especially during river low-flow periods, can result in salinity intrusion over vast areas. During a normal dry season, salinity intrusion can cover 1.5–2.5 million hectares of the coastal Mekong Delta and can be as high as 2.85 million hectares during extreme drought events (MRC 2005).

Most of the soil surface in the delta is covered by alluvial sediments. Acid sulfate soils cover approximately 1.6 million ha, or 40% of the land surface area of the delta (Le Quang Minh et al. 1998). Pyrite (FeS_2), contained in acid sulfate soils (see box in ▶ Chap. 154, "Wetlands of Berbak National Park (Indonesia)"), can be quickly oxidized under aerobic conditions, resulting in high acid concentrations. The acidification of these soils not only lowers the soil pH but also releases large quantities of toxic elements, most importantly soluble aluminum, iron, and manganese (Dent 1992; Le Quang Minh et al. 1998).

Channelization and Wetland Drainage

Wetland drainage by means of canal digging, mainly for agricultural development, was the main force that transformed the Mekong Delta landscape. Efforts to drain wetlands in the delta began in the mid-nineteenth century by the French colonial government and were much more intensified after 1975 (Biggs 2003). To date, a dense network of canals has been developed, in total more than 50,000 km long and increasing (To 2006). The most dramatic effect of canal construction was the facilitation of rural development, resulting in large-scale wetland habitat loss.

Canals drain wetlands, supply irrigation water, provide settlement areas, and serve as a means of transportation. While the canal system promoted human settlement and agricultural development, it also altered the natural hydrological regime. Drainage shortened inundation periods, lowered the dry season water table, facilitated soil acidification, and changed the sedimentation patterns (Beilfuss and Barzen 1994; Le Quang Minh et al. 1998). Shorter periods of floodwater retention and lower water tables accelerated the desiccation of wetlands, triggering more frequent wild fires (Beilfuss and Barzen 1994).

Prior to channelization, flood water in the delta entered low areas primarily as sheet flow with the sediments being filtered out by the vegetation. Because of direct links to the Mekong River, channeled water carries higher sediment loads at the beginning of the rainy season and can significantly increase the turbidity of standing water during this time (Beilfuss and Barzen 1994; Nguyen Nghia Hung et al. 2012). Turbid water may limit the occurrence of many submerged plant species in the river and wetlands (Le Cong Kiet 1994).

Wetland Vegetation

Phung et al. (1989) classified the natural vegetation of the Mekong Delta into nine broad types: deciduous forests on limestone mountains, grasslands on high terraces, tropical evergreen forests on islands, vegetation on coastal sand dunes, mangrove forests, *Melaleuca* swamp forests, riparian vegetation, aquatic vegetation in permanent water bodies (lakes, reservoirs), and seasonally inundated marshes.

The wetland communities among these natural vegetation types exhibit tremendous variation in floristic composition and structure. The seasonally inundated freshwater marsh, mangrove forest, and *Melaleuca* swamp forest are the three most extensive wetland types in the delta.

Data collected in the 1950s reported roughly 200,000 ha of mangrove forests in the Mekong Delta (Phan Nguyen Hong and Hoang Thi San 1993). Large areas of mangrove forests in the Mekong Delta were destroyed during the Vietnam War by military-used defoliants (Tran et al. 2004) and after the war due mostly to shrimp farming development (Environment Justice Foundation 2003). A forest inventory conducted in 1999 showed that only 82,000 ha of mangrove forests remained (Do Dinh Sam et al. 2005). Mangrove forests in the Mekong Delta have very high floristic diversity, with the presence of more than 40 true mangrove plant species (Duke 2013).

Seasonally inundated grasslands are distributed predominantly within the freshwater zone of the delta (Le Cong Kiet 1994; Tran Triet et al. 2000; Tran Triet 2003). Recent biodiversity surveys (Safford et al. 1998; Tran Triet et al. 2000; Buckton and Safford 2004) found that these grasslands, which include permanent water bodies and *Melaleuca* woodlands, support the highest levels of species diversity of plants (and also birds) in the delta.

Biodiversity

Fishes

The Mekong River is the second richest river in the world in terms of fish biodiversity after the Amazon (Baran 2010; Baran et al. 2012). The Mekong Delta's fish biodiversity is particularly high because of the presence of marine fish species.

Vidthanyanon (2008) compiled a checklist of 460 fish species known to occur in the Mekong Delta. Carps, barbs, and minnows (cyprinids) and gobies (gobiids) are predominant groups, accounting for about 30% of all species. Twenty-eight species are endemic to the Mekong, of which four are restricted to the delta: dwarf catfish *Akysis filifer*, Mekong sea catfish *Hemipimelodus daugueti*, the Mekong goby *Stenogobius mekongensis*, and the Mekong blind sole *Typhlachirus elongates*. Seventy-five species are regular marine visitors. Among the recorded fishes, more than 250 species are economically important as food fish and 25 are common in aquarium trade.

Birds

A review of field records made during 1988–2000 (Buckton and Safford 2004) listed 247 bird species for the Mekong Delta in Vietnam, 50% of which are dependent on wetlands. Among the species recorded, twenty species are listed as globally threatened or near threatened. Wetlands of the Mekong Delta support internationally important populations of 21 waterbird species. Seasonally inundated wetlands in Tram Chim National Park and the Ha Tien Plain in Vietnam, Takeo and Kampot provinces in Cambodia provide habitats for the threatened Eastern sarus crane *Grus antigone sharpii*, which has a population of only around 1,000 individuals (Tran Triet and van Zalinge 2013). There are recent sightings of the critically endangered spoon-billed sandpiper (*Eurynorhynchus pygmeus*) in the coastal wetlands of the Mekong Delta (Nguyen Hoai Bao et al. 2013).

Ecosystem Products and Services

The value of ecosystem services and products in the Mekong Delta is enormous. The majority of the human population in the Mekong Delta lives in rural areas, where their livelihoods depend directly on products derived from natural ecosystems. The Mekong provides an ample, year-round flow of freshwater that nourishes life and makes economic activities such as agriculture, fishery, aquaculture, and transportation possible. Food sources and livelihood activities of rural people are heavily dependent on the Mekong flows and wetlands. Many species of aquatic plants and animals are directly collected from rivers and wetlands for use by rural people. Water-related occupations are mainly farming, aquaculture, and fishing, together with various types of jobs that are linked to those main occupations such as postharvest processing, marketing and trading, and equipment production and repair. Agricultural production, especially rice production, in the Mekong Delta is responsible for food security for millions of people who not only live in the Mekong Delta but also in other provinces of Cambodia and Vietnam. Rice produced in the Mekong Delta accounts for up to 90% of rice exported from Vietnam. Less known, but of no less importance, are supporting, regulating, and cultural services from the Mekong. Wetlands of the Mekong Delta, including inland freshwater, brackish, and coastal marine systems, perform many supporting and regulating functions such as providing habitats for biodiversity, flood control, erosion control, shoreline protection, and water purification.

Conservation Status and Management

Currently, there are 14 protected wetlands in the Mekong Delta, including 2 in Cambodia and 12 in Vietnam (Table 1, Fig. 1), with a total land area of 84,400 ha, roughly 1.5% of the total land area of the delta. Two large wetlands, Boeung Veal Samnap (11,286 ha) and Bassac Marsh (52,316 ha) in Kandal Province, Cambodia are listed as Important Bird Area of Cambodia (Seng Kim et al. 2003), but they are not protected areas. Tram Chim National Park and Lang Sen Nature Reserve are Ramsar Sites. As of January 2014, the Government of Vietnam announced a plan to establish nine new protected wetlands in the Mekong Delta by the year 2020 (Table 2) with a total land area of 51,264 ha.

Table 1 Existing protected wetlands and Important Bird Areas in the Mekong Delta (Sources: Buckton et al. 1999; Seng Kim et al. 2003; Government of Vietnam 2014)

Name	Major wetland types	Location	Area (hectares)	Protection status
Boung Prek Lapouv sarus crane sanctuary	Freshwater marshes	Takeo Province, Cambodia	9,276	Wildlife sanctuary
Anlung Pring sarus crane sanctuary	Brackish and freshwater marshes	Kampot Province, Cambodia	217	Wildlife sanctuary
Boeung Veal Samnap	Freshwater marshes and shallow lake	Kandal Province, Cambodia	11,286	Important Bird Area
Bassac Marsh	Freshwater marshes	Kandal Province, Cambodia	52,316	Important Bird Area
Lang Sen	Melaleuca swamp forest and freshwater marshes	Long An Province, Vietnam	5,030	Provincial protected area (Ramsar Site)
Dong Thap Muoi	Melaleuca swamp forest and freshwater marshes	Long An Province, Vietnam	633	Provincial protected area
Tram Chim	Melaleuca swamp forest and freshwater marshes	Dong Thap Province, Vietnam	7,300	National park (Ramsar Site)
Xeo Quyt	Melaleuca swamp forest and freshwater marshes	Dong Thap Province, Vietnam	50	Provincial nature conservation and historical site

(continued)

Table 1 (continued)

Name	Major wetland types	Location	Area (hectares)	Protection status
Tra Su	Melaleuca swamp forest and freshwater marshes	An Giang Province, Vietnam	850	Provincial protected area
Lung Ngoc Hoang	Melaleuca swamp forest and freshwater marshes	Hau Giang Province, Vietnam	790	Provincial protected area
U Minh Thuong	Peat swamp forest	Kien Giang Province, Vietnam	8,038	National park
U Minh Ha	Peat swamp forest	Ca Mau Province, Vietnam	7,926	National park
Thanh Phu	Mangrove forests	Ben Tre Province, Vietnam	2,584	Provincial protected area
Bac Lieu bird colony	Mangrove forest	Bac Lieu Province, Vietnam	385	Provincial protected area
Dam Doi bird colony	Mangrove forest	Ca Mau Province, Vietnam	130	Provincial protected area
Mui Ca Mau	Mangrove forest	Ca Mau Province, Vietnam	41,089	National park

Threats and Future Challenges

Climate change and upstream hydropower development are the two most important threats to the wetlands of the Mekong Delta. The Mekong Delta is particularly at risk of climate change-induced sea level rise (Wassmann et al. 2004; Eastham et al. 2008; Doyle et al. 2010). An analysis by the US Geological Survey (Doyle et al. 2010) estimated that sea levels would rise during the period 2010–2100 at the rate of 4.3 mm/year under scenario B1 (best case) to 5.5 mm/year under scenario A1FI (worst case). According to the study, all areas of the Mekong Delta that are less than 0.5 m above the current sea level are expected to begin inundation by 2035 and to be completely submerged by 2068. Climate change is expected to result in more erratic dry season rainfall and increased drought in the Delta region (Johnston et al. 2010). These changes could impact native vegetation communities of seasonally inundated wetlands such as those at Tram Chim National Park in the Mekong Delta and the flooded forests of the Tonle Sap Lake (ICEM 2012). Projected impacts of climate change on ecosystem services in the Mekong region include decreasing overall water availability, decreasing food production capacity – especially rice and aquaculture

Table 2 New protected wetlands to be established by 2020 by the Vietnam Government (Source: Government of Vietnam 2014)

Name	Major wetland types	Location	Area (hectare)	Protection status
Phu My	Melaleuca swamp forest and freshwater marshes	Kien Giang Province	1,106	Provincial protected area
Tri Ton protected forest	Melaleuca swamp forest	An Giang Province	1,900	Provincial protected area
Bung Binh Thien	Shallow freshwater lake	An Giang Province, Vietnam	500	Provincial fishery conservation area
Dong Ho Lagoon	Brackish water lagoon	Kien Giang Province	1,597	Provincial protected area
Thi Tuong Lagoon	Brackish water lagoon	Ca Mau Province	700	
Cu Lao Dung	Mangrove forests	Tra Vinh Province, Vietnam	25,333	Provincial protected area
Long Khanh protected forest	Mangrove forests	Tra Vinh Province	828	Provincial protected area
Ba Lai estuary	Estuary	Ben Tre Province	10,000	Provincial protected area
Ham Luong estuary	Estuary	Ben Tre Province	10,000	Provincial protected area

production – and increasing incidence of extreme floods and droughts, sea level rise, and land submersion in the Mekong Delta (Wassmann et al. 2004; Johnston et al. 2010; Mainuddin et al. 2010; ICEM 2012).

The Mekong River offers significant potential for hydropower generation. In the upper basin in Yunnan Province, China, a cascade is planned to consist of eight dams, six of which have a total generation potential of 15,600 MW, while the capacity of the other two is unreported (MRC 2009a). Three of the six upstream dams are completed (Xiaowan, Manwan, Dachaoshan) and the others are in various stages of implementation. The cascade, projected to be fully operational in 2025, will have a storage capacity of more than 23.2 km^3, corresponding to 28% of the mean annual flow that enters the lower Mekong basin. Cumulatively, the dams would regulate more than 20% of Mekong mean annual flows from the upper basin, significantly increasing dry season flows and reducing wet season flows. Substantial changes in downstream flow volumes and seasonal patterns are anticipated as far downstream as Kratie in Cambodia (MRC 2009a; Räsänen et al. 2012). In the lower Mekong basin, up to 12 dams have been proposed for construction on the main stem of the Mekong River with a total installed capacity of 13,000 MW (MRC 2009b). Two of these dams, the Xayaburi and the Don Sahong, are currently under construction. The total capacity of tributary dams, including those in operation, under construction, and proposed, is nearly 29,000 MW, bringing the total capacity of hydropower generation in the lower Mekong basin to nearly 42,000 MW from more than 130 hydropower projects (MRC 2009b).

The adverse effects of altered hydrological regimes due to large dams on ecosystem functions and services are well known: reduced sediment load and nutrient availability, channel degradation, loss of wetland habitats, altered food chain dynamics, habitat fragmentation, salt water intrusion, disruption of fish migration and reproduction, coastal erosion, and loss of mangroves (Beilfuss 2012). The reduction of river sediments that reach the Mekong Delta is projected to be severe. According to one of the most recent analyses (Kondolf et al. 2013), under the most possible scenario – with 38 upstream dams built or under construction – cumulative reduction of sediments to the Mekong Delta would be 51%; under the full scenario – with all planned dams built – cumulative sediment trapping will be 96%, which means only 4% of sediments would reach the Mekong Delta.

Finally, the combined impacts of climate change and upstream hydropower development on wetland ecosystems of the Mekong Delta are still poorly understood and need further study.

References

Anonymous. Mekong Delta Plan: long-term vision and strategy for a safe, prosperous and sustainable delta. Hanoi: The Socialist Republic of Vietnam and the Kingdom of the Netherlands. Hanoi, Vietnam; 2013; 126 pp.

Baran E. ICEM 2010. Mekong River Commission Strategic Environmental Assessment of hydropower on the Mekong mainstream. Hanoi: International Centre for Environmental Management; 2010. 145 p.

Baran E, Nith C, Fukushima M, Hand T, Hortle KG, Jutagate T, Kang B. Fish biodiversity research in the Mekong Basin. In: Nakano S, Yahara T, Nakashizuka T, editors. The biodiversity observation network in the Asia-Pacific Region: toward further development of monitoring, Ecological research monographs. Japan: Springer; 2012. p. 149–64.

Beilfuss R. A risky climate for Southern African hydro: assessing hydrological risks and consequences for Zambezi River Basin Dams. Berkeley: Int Rivers; 2012.

Beilfuss R, Barzen J. Hydrological wetland restoration in the Mekong Delta, Vietnam. In: Mitsch WJ, editor. Global wetlands: old world and new. Amsterdam: Elsevier Science B.V; 1994. Berkeley, California USA. 56 p. 453–68.

Biggs D. Problematic progress: reading environmental and social changes in the Mekong Delta. J Southeast Asian Stud. 2003;31:77–96.

Buckton ST, Safford RJ. The avifauna of the Mekong delta. Bird Conserv Int. 2004;14:279–322.

Buckton ST, Nguyen Cu, Nguyen Duc Tu, Ha Quy Quynh. The conservation of key wetland sites in the Mekong Delta. Hanoi: Birdlife International Vietnam Programme; 1999. 101 p.

Campbell, I. Development scenarios and Mekong River flows. In: Campbell I, editor. The Mekong: biophysical environment of an international river basin. Academic Press/Elsevier; 2009. p. 389–402.

Dent D. Reclamation of acid sulphate soils. In: Lal R, Stewart BA, editors. Advances in soil science, vol. 17. New York: Springer; 1992. p. 79–122.

Do Dinh Sam, Nguyen Ngoc Binh, Ngo Dinh Que, Vu Tan Phuong. A review of mangrove forests in Vietnam. Ha Noi: Agricultural Publishing House; 2005. 136 p. [in Vietnamese].

Doyle T, Day R, Michot T. Development of sea level rise scenarios for climate change assessment of the Mekong Delta Vietnam. U.S. Geological Survey Open File Report 2010–1165. 2010; 110p.

Duke N. Mangroves of the Kien Giang Biosphere Reserve Vietnam. Rach Gia: Deutsche Gesellschaft fur Internationale Zusammenarbeit (GIZ); 2013. 108 p.

Eastham J, Mpelasoka F, Mainuddin M, Ticehurst C, Dyce P, Hodgson G, Ali R, Kirby M. Mekong river basin water resources assessment: impacts of climate change. Canberra: CSIRO: Water for a Healthy Country National Research Flagship; 2008. 153 p.

Environment Justice Foundation. Risky business: Vietnamese shrimp aquaculture – impacts and improvements. London: Environmental Justice Foundation; 2003.

General Statistics Office of Vietnam. Area, population and population density in 2012 by province. Available from: http://www.gso.gov.vn/default_en.aspx?tabid=467&idmid=3&ItemID=14459. Access date 19 Mar 2014.

Government of Vietnam. Prime Minister Decree 45/QD-TTg 08 January 2014. Approval of the general plan for biodiversity conservation in Vietnam until 2020 with a vision to 2030. 2014. [in Vietnamese].

ICEM. Impact of climate change on Mekong wetlands and adaptation responses: synthesis paper on adaptation of Mekong wetlands to climate change. Prepared for the Mekong River Commission by ICEM in partnership with IUCN, World Fish and SEA Start. Hanoi; 2012. 110 p.

Johnston R, Lacombe G, Hoanh CT, Noble A, Pavelic P, Smakhtin V, Suhardiman D, Kam SP, Choo PS. Climate change, water and agriculture in the Greater Mekong Subregion (IWMI Research Report 136). Colombo: International Water Management Institute; 2010. 52p.

Kondolf GM, Rubin ZK, Minear JT. Dams on the Mekong: cumulative sediment starvation. Water Resour Res. 2013;50:1–12.

Le Cong Kiet. Native freshwater vegetation communities in the Mekong Delta. Int J Ecol Environ Sci. 1994;20:55–71.

Le Quang Minh, To Phuc Tuong, van Mensvoort MEF, Bouma J. Soil and water table management effects on aluminum dynamics in an acid sulphate soil in Vietnam. Agric Ecosyst Environ. 1998;68:255–62.

Li T. Nguyen Cochinchina: Southern Vietnam in the seventeenth and eighteenth centuries. Ithaca: Cornell University Press; 1998.

Mainuddin M, Hoanh CT, Jirayoot K, Halls AS, Kirby M, Lacombe G, Srinetr V. Adaptation options to reduce the vulnerability of Mekong water resources, food security and the environment to impacts of development and climate Change. Canberra: CSIRO: Water for a Healthy Country National Research Flagship; 2010. 151 p.

MRC (Mekong River Commission). Overview of the hydrology of the Mekong basin. Vientiane: Mekong River Commission; 2005. 73 p.

MRC. Economic, environmental and social impact assessment of basin-wide water resources development scenarios, Assessment methodology. Vientiane: Mekong River Commission (MRC) technical note. 2009a. Available from: http://www.mrcmekong.org/assets/Other-Documents/BDP/Tech-Note2-Scenarioassessment-methodology-complete-Report091104.pdf. Accessed Jan 2012.

MRC. Database of the existing, under construction, and planned/proposed hydropower projects in the Lower Mekong Basin. Vientiane: Mekong River Commission; 2009b.

Nguyen Hoai Bao, Nguyen Hao Quang, Tran Duc Thien. Spoon-billed Sandpiper survey in Mekong Delta 2013. Ho Chi Minh City: University of Science, Vietnam National University; 2013. 7 p. [unpublished report].

Nguyen Nghia Hung, Delgado JM, Vo Khac Tri, Le Manh Hung, Merz B, Bardossy A, Apel H. Floodplain hydrology of the Mekong Delta, Vietnam. Hydrol Process. 2012;26:674–86.

Nguyen VL, Ta TKO, Tateishi M. Late Holocene depositional environments and coastal evolution of the Mekong Delta, southern Vietnam. J Asian Earth Sci. 2000;18:427–39.

Phan Nguyen Hong, Hoang Thi San. Mangroves of Vietnam. Bangkok: IUCN; 1993. 173 p.

Phung Trung Ngan, Duong Tien Dung, Ngoa Thanh Loan, Tran Triet. Survey of the vegetation of the Mekong Delta. State Program 60 – B: Integrated Basis Studies of the Mekong Delta. Hanoi; 1989. [in Vietnamese].

Räsänen TA, Koponen J, Lauri H, Kummu M. Downstream hydrological impacts of hydropower development in the Upper Mekong Basin. Water Resour Manag. 2012;26:3495–513.

Safford RJ, Triet T, Malby E, Ni DV. Status, biodiversity and management of the U Minh wetlands, Vietnam. Trop Biodivers. 1998;5(3):217–44.

Seng Kim Hout, Pech Bunnat, Poole CM, Tordoff AW, Davidson P, Delattre E. Directory of important bird areas in Cambodia: key sites for conservation. Birds of the Mekong Delta. Department of Forestry and Wildlife, Department of Nature Conservation and Protection, BirdLife International in Indochina and the Wildlife Conservation Society Cambodia Program; 2003.

Ta TKO, Nguyen VL, Tateishi M, Kobayashi I, Saito Y. Holocene delta evolution and depositional models of the Mekong River Delta, Southern Vietnam, River Deltas – concepts, models, and examples. SEPM Special Publication, vol 83. SEPM (Society for Sedimentary Geology); 2005. p. 453–66.

Tran Triet. Boeung Prek Lapouv, Takeo Province, Cambodia: a rare wetland of the lower Mekong basin. ASEAN Biodivers. 2003;3(3–4):27–31.

Tran Triet, van Zalinge R. Census of non-breeding Sarus Cranes in Cambodia and Vietnam, 2013. Baraboo: International Crane Foundation; 2013. [unpublished report].

Tran Triet, Safford RJ, Tran Duy Phat, Duong Van Ni, Maltby E. Wetland biodiversity overlooked and threatened in the Mekong Delta, Viet Nam: grassland ecosystems in the Ha Tien Plain. Trop Biodivers. 2000;7(1):1–24.

Tran Triet, Barzen J, Le Cong Kiet, Moore D. War-time herbicides in the Mekong Delta and their implications on post-war wetland conservation. In: Furukawa H, Nishibuchi M, Kono Y Kaida Y, editors. Ecological destruction, health and development advancing Asian paradigms. Kyoto: Kyoto University Press/Trans Pacific Press; 2004. p. 199–211.

To VT. Identification, forecasting and control floods in the Mekong River Delta. Ha Noi, Vietnam: Agricultural Publishing House; 2006. 472 p [in Vietnamese].

Vidthayanon C. Field guide to fishes of the Mekong Delta. Vientiane: Mekong River Commission; 2008. 288 p.

Wassmann R, Hien NX, Hoanh CT, Tuong TP. Sea level rise affecting the Vietnamese Mekong Delta: water elevation in the flood season and implications for rice production. Clim Change. 2004;66:89–107.

U Minh Peat Swamp Forest: Mekong River Basin (Vietnam)

152

Triet Tran

Contents

Introduction	1813
Physical Environment	1815
Wetland Ecosystems	1815
Biodiversity	1816
Ecosystem Services	1816
Conservation Status	1816
Threats and Future Challenges	1817
References	1817

Abstract

The U Minh region contains the largest peat swamp areas in the Mekong River basin. U Minh peat swamp forests are currently protected in two national parks: U Minh Thuong (3,000 ha) and U Minh Ha (9,800 ha). Both national parks are Important Bird Areas of Vietnam. These peat swamps are important sources of freshwater and fish for local consumption. Forest fire is the main threat for U Minh peat swamp forests.

Keywords

Mekong Delta · Peat swamp forest · U Minh

Introduction

Located in the southwestern part of the Mekong Delta, U Minh has the most extensive peat swamp area in the Mekong river basin, containing two noncontiguous regions: U Minh Thuong (Kien Giang Province) and U Minh Ha (Ca Mau Province),

---

T. Tran (✉)
International Crane Foundation, Baraboo, WI, USA

University of Natural Science, Vietnam National University, Ho Chi Minh City, Vietnam
e-mail: ttriet@gmail.com

© Springer Science+Business Media B.V., part of Springer Nature 2018
C. M. Finlayson et al. (eds.), *The Wetland Book*,
https://doi.org/10.1007/978-94-007-4001-3_174

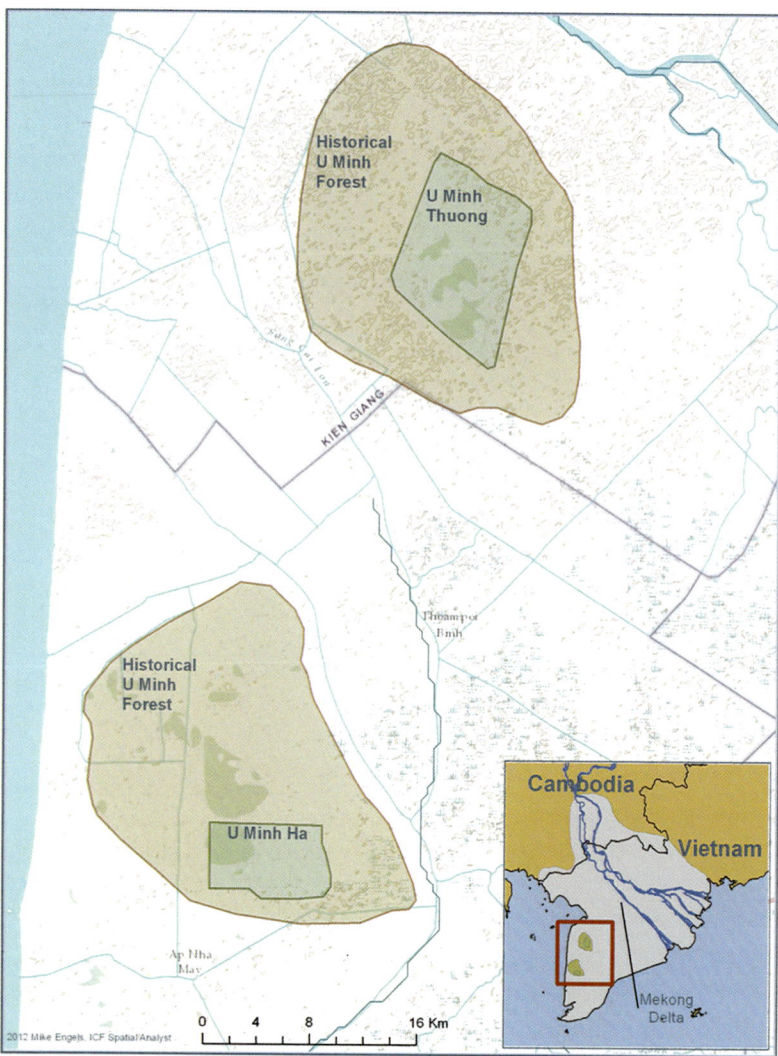

Fig. 1 Approximate historical extents of U Minh peat swamps forest and current boundaries of U Minh Thuong and U Minh Ha National Park, Mekong Delta, Vietnam (Source: © Provided courtesy of International Crane Foundation)

which literally mean upper and lower U Minh, respectively (Fig. 1). A survey conducted in 1976 by the Geological Survey Agency of Vietnam documented 12,400 ha of peatland in U Minh Thuong and 20,200 ha in U Minh Ha (Vo et al. 2010). Since then frequent fires, exacerbated by hydrological changes due to canals, have greatly reduced the extent of peat swamp forests and the thickness of peat layers.

Agricultural development in the area also claimed much of the degraded forests, further reducing the area of U Minh peat swamps. The current area of peat swamp is estimated at approximately 3,000 ha in U Minh Thuong and 9,800 ha in U Minh Ha, with the thickness of peat layers ranging from 0.4 to 1.2 m (Le 2010).

Physical Environment

U Minh Thuong and U Minh Ha peat swamps are located in the area that receives the highest rainfall in the Mekong Delta, with an average annual rainfall of 2,400 mm. Historically these two regions also had poor drainage through river channels so there was a tendency for inundated conditions to occur, allowing peat to accumulate. The region has a monsoonal climate, characterized by a December–April dry season and a May–November wet season in which 90% of rainfall occurs. Peat layers act like water sponges, absorbing rain water during wet seasons and gradually releasing it during dry seasons. Although located close to the sea, the U Minh peat swamps are a freshwater wetland ecosystem. In addition to altering the hydrology of the peat system, extensive canal systems developed recently for agricultural expansion allow saltwater intrusion in some parts of U Minh Thuong and U Minh Ha, especially during dry seasons.

Wetland Ecosystems

Melaleuca cajuputi (Myrtaceae) is the dominant tree species in the U Minh peat swamp forests. On the forest floor, the luxurious growth of ferns, mostly *Asplenium* sp. (Aspleniaceae) and *Stenochlaena palustris* (Blechnaceae), creates dense thickets that are often difficult to penetrate. The climax forest vegetation in U Minh peat swamps is, however, a type of mixed forest where several tree species codominate, most importantly *Alstonia spathulata* (Apocynaceae), *Ilex cymosa* (Aquifoliaceae), *Syzygium cumini* (Myrtaceae), *Acronychia pedunculata*, and *Euodia lepta* (both Rutaceae). *Alstonia spathulata* trees can reach more than 30 m in height, forming an above-canopy stratum in the forest (Safford et al. 1998; Tran 2005). This mixed forest type occurred only on peat domes which were 2–4 m higher than the surrounding area and were often not flooded even at the peak of the rainy season. Forest fires have destroyed most of the mixed peat swamp forest in U Minh. When peat layers were greatly reduced or disappeared because of fires, *Melaleuca* became the only tree species in the forest and ferns also largely disappeared from the forest floor, replaced by *Phragmites vallatoria* (Poaceae) or *Eleocharis dulcis* (Cyperaceae). Severe fires could eradicate *Melaleuca* forests entirely, giving way to Phragmites/Eleocharis marshes or even open water bodies covered by floating plants such as *Pistia stratiotes* (Araceae), *Salvinia cucullata* (Salviniaceae), and *Eichhornia stratiotes* (Pontederiaceae) where water levels are artificially maintained to prevent further fire (Tran 2005).

Biodiversity

Both U Minh Thuong and U Minh Ha are considered to be Important Bird Areas in Vietnam (Tordoff 2002), with 187 bird species recorded from the two sites. The region hosted some of the largest waterbird colonies in the Mekong Delta (Buckton et al. 1999) even though the size of these bird colonies has been reduced due to frequent fires. Birds of conservation importance are greater spotted eagle *Aquila clanga*, spot-billed pelican *Pelecanus philippensis*, lesser adjutant *Leptoptilos javanicus*, gray-headed fish eagle *Ichthyophaga icthyaetus*, oriental darter *Anhinga melanogaster*, black-headed ibis *Threskiornis melanocephalus*, painted stork *Mycteria leucocephala*, and Asian golden weaver *Ploceus hypoxanthus* (Safford et al. 1998; Buckton et al. 1999; Tordoff 2002). Mammal species of special concern are small-clawed otter *Aonyx cinerea* and hairy-nosed otter *Lutra sumatrana*. Up until the late 1980s, Vo Doi (literally meaning "Bat Hill"), once the largest peat dome in U Minh Ha, was home to the largest fruit bat *Pteropus vampirus* colony in the Mekong Delta (Phung and Chau 1980). Even though none of the plant species are considered endemic, some plants are rarely found elsewhere in the Mekong Delta, e. g., *Alstonia spathulata*, *Lemna tenera*, *Nepenthes mirabilis*, *Asplenium confusum*, *Licuala spinosa*, *Hydnophytum formicarum*, and two orchids *Eulophia graminea* and *Spiranthes sinensis* (Tran 2005).

Ecosystem Services

U Minh peat swamps play an important role in freshwater supply for U Minh and the surrounding area. Peat layers absorb a large amount of rain water in the wet season and gradually release it during the dry season, supplying potable freshwater for local people as well as water for wildlife and agriculture. U Minh peat swamps also provide large amounts of freshwater fish for local consumption and commercial trade. Recent reduction in the area of peat swamps, however, has diminished much of these services.

Conservation Status

To preserve the remaining peat swamp forests, the Government of Vietnam established U Minh Thuong National Park in 2002 and U Minh Ha National Park in 2006. The core zone of U Minh Thuong National Park covers an area of 8,038 ha, of which 3,000 ha still have peat soils. The buffer zone of U Minh Thuong includes 13,000 ha, most of which are farm lands. U Minh Ha National Park has an 8,256 ha core zone, of which 6,300 ha still have peat soils (Le 2010). The buffer zone of U Minh Ha National Park covers a 25,000 ha mosaic of Melaleuca forests and farm lands, within which almost 3,500 ha of peat soils, composed of small, fragmented patches, still exist in the buffer zone (Le 2010). U Minh Thuong National Park was included in the Kien Giang Biosphere Reserve (recognized by UNESCO in 2006), and U Minh Ha was included in the Mui Ca Mau Biosphere Reserve (recognized by UNESCO in 2009).

Threats and Future Challenges

Forest fire is the most important threat to the U Minh peat swamps. Extensive canal digging in the surrounding areas has lowered the water table, dried peat domes, and increased the risk of uncontrollable fires. The last catastrophic fire happened in March and April 2002. This fire burned as much as 90% (2,800 ha) of the Melaleuca forest in the core zone of U Minh Thuong National Park and 3,300 ha of the Melaleuca forest in U Minh Ha National Park (Le 2010). Currently, the core zones of U Minh Thuong and U Minh Ha are protected by dikes that help store water to recreate wet conditions for peats and to prevent fire. Too much water being stored, however, has turned many areas of peat swamps into permanent water bodies. The growth of Melaleuca forests has been suppressed as a result. Fire and water management plans which reduce the risk of catastrophic fires and simultaneously take into account the ecological requirements of wetland organisms are needed for the sustainable management of the U Minh peat swamps. Besides fire and water issues, large-scale illegal hunting and fishing are other important threats to U Minh's wildlife.

References

Buckton ST, Nguyen C, Nguyen Duc Tu, Ha Quy Quynh. The conservation of key wetland sites in the Mekong Delta. Hanoi: Birdlife International Vietnam Programme; 1999.

Le Phat Quoi. Inventory of peatlands in U Minh Ha region, Ca Mau Province, Vietnam. SNV Netherlands Development Organisation REDD+ Programme; Hanoi, Vietnam, 2010. 24 pp.

Phung Trung Ngan, Chau Quang Hien. Flooded forests of Vietnam. Education Publisher; Hanoi, Vietnam, 1980. 138 pp. [in Vietnamese].

Safford RJ, Tran T, Maltby E, Duong Van Ni. Status, biodiversity and management of the U Minh Wetlands, Vietnam. Trop Biodivers. 1998;5(3):217–44.

Tordoff A, editor. Directory of important bird areas in Vietnam – key sites for conservation. Hanoi: Birdlife International Indochina Programme and Institute of Ecology and Biological Resources; 2002. 233 pp.

Tran T. Flora and vegetation of U Minh Thuong National Park. In: Sage NS, Kutcher NX, Vinh WP, Dunlop J, editors. Biodiversity of U Minh Thuong National Park. Ho Chi Minh City: Agricultural Publishing House; Ho Chi Minh City, Vietnam, 2005. p. 7–19.

Vo Dinh Ngo, Nguyen Sieu Nhan, Tranh Manh Tri. Peats and their uses. Hanoi: Science and Technology Publisher; 2010. 272 pp. [in Vietnamese].

Sembilang National Park: Mangrove Reserves of Indonesia

153

Marcel J. Silvius, Yus Rusila Noor, I. Reza Lubis, Wim Giesen, and Dipa Rais

Contents

Introduction	1820
Hydrology	1820
Wetland Ecosystems	1822
Peat Carbon Stock	1823
Biodiversity	1824
Conservation Status	1826
Current Threats and Future Challenges	1827
References	1828

Abstract

Sembilang National Park with Berbak National Park (a Ramsar Site) to the north is part of the Greater Berbak-Sembilang Ecosystem on the Indonesian island of Sumatra and comprises the largest mangrove area (77,500 ha) of the Indo-Malayan region and the only mangrove area that still has an intact natural transition into adjacent freshwater and peatswamp forest. It is an important breeding and nursery area for fish and shrimp and one of the most important areas for resident and migratory waterbirds in Southeast Asia. The peatland area in Sembilang NP is approximately 31% of total park area and constitutes a huge carbon store estimated at around 164 million tons. Officially designated in 2011

M. J. Silvius (✉)
Wageningen, The Netherlands
e-mail: msilvius58@gmail.com

Y. R. Noor (✉) · I. R. Lubis (✉) · D. Rais (✉)
Wetlands International Indonesia, Bogor, Indonesia
e-mail: noor@wetlands.or.id; rezalubis@wetlands.or.id; elitelongbowman77@yahoo.com

W. Giesen
Euroconsult Mott MacDonald, Arnhem, AK, The Netherlands
e-mail: wim.giesen@mottmac.com

© Springer Science+Business Media B.V., part of Springer Nature 2018
C. M. Finlayson et al. (eds.), *The Wetland Book*,
https://doi.org/10.1007/978-94-007-4001-3_213

as one of Indonesia's Ramsar Sites, the most serious threat to the area is rapid degradation of the peat swamp forest outside of the park boundaries, involving illegal logging as well as the conversion of large areas to cultivation of oil palm or Acacia (for pulp wood for paper production). The peat domes that support the peat swamp forests in the national park are only partly covered by the protected area, and drainage (including legal drainage) outside of the park may strongly impact the eco-hydrology of the protected area. About 3,000 ha of mangrove in the eastern part of the park converted for aquaculture since 1995 has been subsequently included in the park's restoration zone, and in 2013 a mangrove restoration project was initiated. The park is still threatened by external developments including a harbor project and plantations.

Keywords
Mangrove · Peatswamp forest · Mudflats · Waterbirds · Sumatra

Introduction

Sembilang National Park (104°14′ – 104°54′ E, 1°53′ – 2°27′ S) comprises the largest mangrove area (77,500 ha) of the Indo-Malayan region and the only mangrove area that still has an intact natural transition into adjacent freshwater and peat swamp forest. The Park covers a total of 202,896 ha, excluding extensive mudflats along the coast. It forms part of the vast coastal plain of south-eastern Sumatra. The area is flat and swampy reaching an elevation of around 15 masl in the west and is bounded by the Benue River and the provincial boundary with Jambi province in the north and the Lalan and Banyuasin rivers in the south. It is bisected by the Sembilang River. Administratively it belongs to the province of South Sumatra and Banyuasin District, but day-to-day management is the responsibility of the Ministry of Forestry. Established in 2003, it is one of the most recently established national parks of Indonesia, especially founded in recognition of the presence of a unique estuarine environment. Not only is it the largest mangrove area in western Indonesia, but also an important breeding and nursery area for fish and shrimp and one of the most important staging areas for migratory waterbirds in the East Asia–Australasian flyway. It forms one contiguous protected area with the Berbak National Park (a Ramsar Site) to the north, forming the Greater Berbak-Sembilang Ecosystem (Fig. 1).

Hydrology

Annual rainfall of SNP is about 2,466 mm with minimum and highest records being 933 mm and 3,972 mm. Two monsoons strongly affect the climate in the area: the North-western monsoon carries humid air and causes heavy rainfall that lasts from November to March, while the South-eastern monsoon lasts from May to September, carries dry air, and causes a dry season during June–September

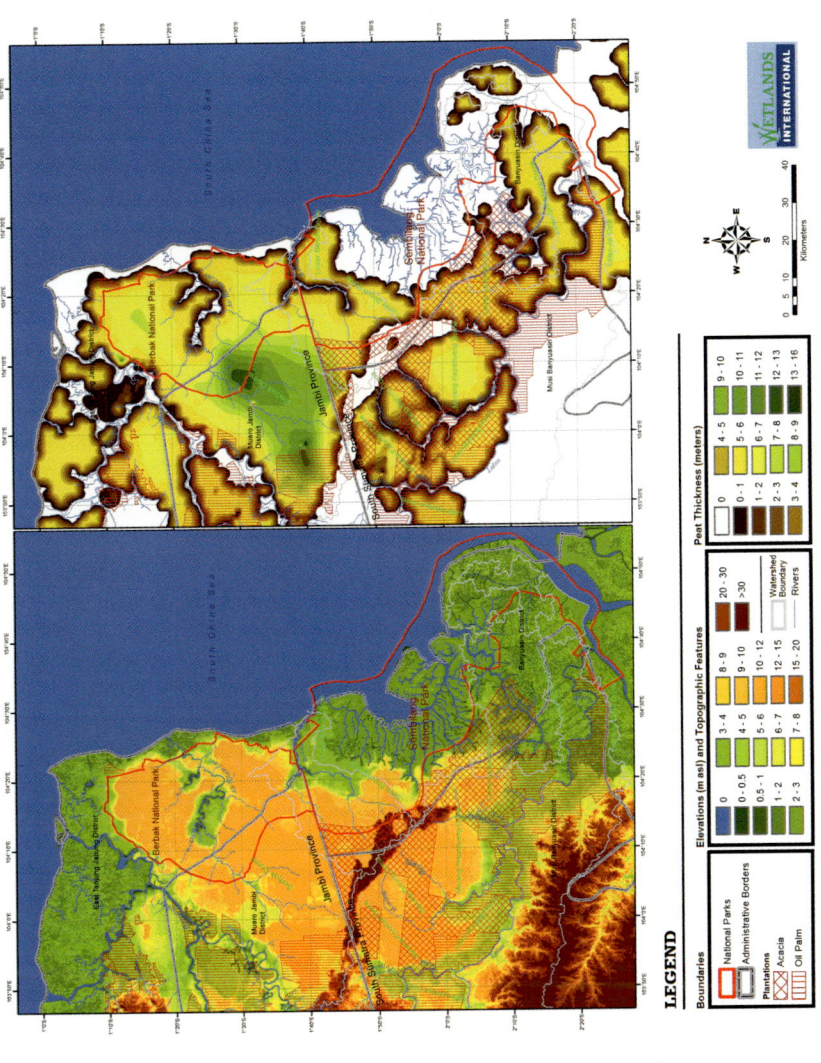

Fig. 1 Location and topography of the Sembilang and Berbak National Parks and surrounding areas (*left*) and peat thickness distribution (*right*), including location of plantation concessions

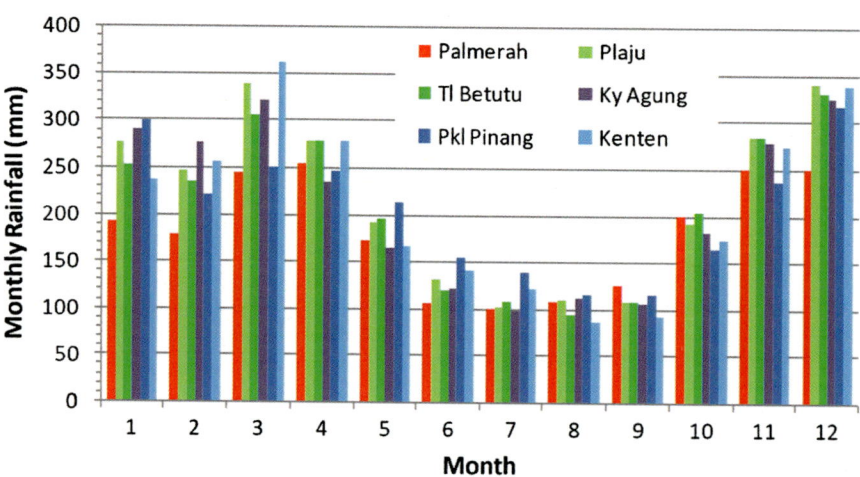

Fig. 2 Monthly rainfall of recorded at six stations closest to SNP (Source: National Bureau of Meteorology, Climatology and Geophysics)

(Fig. 2). The hydrology of the Sembilang NP is characterized by the smooth transition of freshwater and brackish water habitats. The mangrove belt that can exceed 12 km in width is regularly and largely inundated by black, tannin-colored freshwater from adjacent peat swamp forests. During high tides, freshwater is pushed back to some extent, but only during the dry season will brackish water reach the landward parts of the mangrove zone. The mangrove area is bisected by a myriad of creeks stretching into the back swamps. The Benue River on the border of Berbak NP is a black water river, originating in a large peat dome shared with Berbak National Park. The Sembilang River originates within the national park at about 30 km inland in shallow peat and clay-dominated habitats and carries more silt. The Lalan River originates from peat swamp forest on deep peat outside of the reserve, over 100 km inland, and has a catchment of about 14,100 km^2. The freshwater influx from the peat swamp forests inland is crucial for the functioning of the mangrove area, but over the last decade extensive peat swamp forest areas in the buffer zone of the park have been clear-felled, and other areas illegally logged, burnt, and partly converted to oil palm plantations, resulting in substantial change of vegetation and hydrology in the catchment of the park. As peatland drainage inevitably causes land subsidence of around 5 cm per year (Hooijer et al. 2012), this poses a long-term threat to the hydrological integrity of the park.

Wetland Ecosystems

Sembilang NP has six main habitat types: freshwater swamp forest (on mineral soil, 10%), peat swamp forest (30%), mangrove forest (40%), swamps with freshwater herbaceous vegetation (10% – mainly on the Banyuasin Peninsula), and waterbodies and mudflats (10%) [Note that percentages are rough estimates]. Intertidal mudflats are a striking feature of the coastal belt of South Sumatra. During neap tides, an area of up to 2 km from the coast can be exposed (Danielsen and Verheugt 1990). Most of these, however, are not included in the national park, but do contribute significantly to its importance for waterbirds and marine life.

Riparian vegetation attains a height of 15–25 m with emergent trees of up to 40 m, while the freshwater swamp forest is usually 35–45 m tall, dominated by many *Myrtaceae* such as *Eugenia* and *Syzygium* species, *Alstonia pneumatophora, Baccaurea* species, *Elaeocarpus* species, *Flacourtia rukam, Koompassia malaccensis,* and a wide range of *Ficus* species, and with many emergent trees of 50–55 m (e.g., *Alstonia pneumatophora, Koompassia malaccensis*). The peat swamp forests are tall (30–40 m) and relatively diverse, lacking the so-called Padang (pole-) forest type, with key species similar to those of Berbak NP. Danielsen and Verheugt (1990) mention a variety of mangrove communities, dependent especially on different edaphic conditions, salinity, drainage, and water currents. Along or near accreting saline coastlines, 12–15 m tall vegetation of *Avicennia marina* dominates and sometimes *Sonneratia alba,* near river mouths *Avicennia alba* and *Rhizophora apiculata.* Further inland on more firm soil and less saline conditions, up to 30 m high communities dominated by a mixture of the genera *Rhizophora* (mainly *R. apiculata*), *Bruguiera* (mainly *B. gymnorrhiza* and *B. cylindrica*), and *Xylocarpus* (*X. granatum, X. moluccensis*) are found with basal areas of up to 45–85 m^2, with tree densities varying between 700 and 2,000 trees per ha. This forest has a relatively open understory with *Acrostichum* ferns on mud-lobster hills. Along the Banyuasin River, mangroves are dominated by *Rhizophora* spp. and *Sonneratia caseolaris.* The rivers and parts of the coastline are fringed by 8–10 m tall *Nypa fruticans* vegetation, sometimes forming patches with a density of around 1.500 palms per ha. Further inland the mangroves are more species-rich, including *Bruguiera cylindrica, Cerbera odollam, Hibiscus tiliaceus, Thespesia populnea, Excoecaria agallocha, Sonneratia ovata, Lumnitzera racemosa, and Xylocarpus* spp. In the back swamp where the mangrove merges into freshwater swamp forest, *Ficus microcarpa, Cerbera odollam,* and *Pandanus furcatus* occur increasingly, and on ridges the valuable palm *Oncosperma tigillarium.* Silvius (1986) reports a total of 28 true mangrove species. Including the freshwater habitats, the lowland areas of South Sumatra hold over 200 tree species (Danielsen and Verheugt 1990).

Peat Carbon Stock

The total peatland area in SNP is approximately 81,272 ha (about 31% of total SNP area), with an average peat thickness of about 4 m. The thickest and most extensive peat is located in the northwestern part, reaching 7.5 m, while in the southeast smaller peatland areas with thinner peat layers are located (Fig. 1). The peat is most likely hemic or fibric, similar to the peat found in Berbak National Park and in the Merang-Kepahyang area further inland from Sembilang NP (Rosalina et al. 2004). The peat in Sembilang National Park constitutes a huge carbon store. The

Table 1 Summary of peatland and carbon stock distributions in Sembilang National Park (values determined by MJ Silvius based on primary data compiled by Wetlands International – Indonesia Programme)

Peat thickness range (m)	Area (ha)	Peat carbon stock (t)
<1	9,910	2,532,110
1–2	8,448	6,234,125
2–3	7,593	9,471,985
3–4	7,476	13,068,630
4–5	9,513	21,605,625
5–6	26,214	72,368,390
6–7	11,434	36,763,520
7–7.4	684	2,436,760
Total	**81,272**	**164,481,145**
Average		**2,024 t/ha**

total carbon stock is estimated at around 164 million tons, with an average C stock per hectare of around 2,024 t (Table 1).

Biodiversity

Sembilang NP is particularly important for its migratory waterbird fauna and ranks among the most important waterbird areas in South-east Asia, both for resident waterbird populations and a stop-over and wintering site for migratory waders and terns. First surveys of the area were done in 1984 and 1986 by Silvius, Verheugt, and Iskandar (Silvius et al. 1986). They found flocks of tens of thousands of waterbirds, including over 1,500 milky stork *Mycteria cinerea*, over 500 lesser adjutant storks *Leptoptilos javanicus*, and over 100 black-headed ibis *Threskiornis melanocephalus*, besides thousands of herons (ten species). Most common was the great egret *Egretta alba*. Rarest were the great-billed heron *Ardea sumatrana*, black-crowned night-heron *Nycticorax nycticorax*, and Chinese egret *E. eulophotes*. They also discovered the occurrence of a small population of spot-billed pelicans *Pelecanus philippensis*, including some immature, indicating possible breeding. In 1988, a breeding colony of milky storks and herons with over 280 nests was discovered in mangrove back swamps with herbaceous swamp vegetation, some 3–4 km inland. The milky stork nests were located in up to 15 small bushes (Danielsen et al. 1991a), while the egrets had nests build on the ground. Other colonies of milky storks were found south of Sungsang at Tanjung Koyan and Tanjung Selokan the other side of the Banyuasin River, but these areas have since been reclaimed and it is doubtful that colonies remain there. Two later milky stork surveys during 2001–2005 and 2008 observed a total population of 500 and 322 individual, respectively, indicated the decline of about 70% in 22 years (Iqbal et al. 2012). In 1990, a nest of the rare Storm's Stork was found by Danielsen and Verheugt (1990) in the by then proposed Sembilang NP.

The extensive mudflats in front of the mangrove coast serve as foraging areas for the storks, ibises, and herons, as well as for thousands of migratory waders. The area is one of the most important wintering sites for the Asian dowitcher *Limnodromus semipalmatus* (Silvius 1986, 1988; Danielsen and Skov 1987) including a count in

April 1986 of 1,780 and in March 1989 of up to 10,000 individuals at the Banyuasin Peninsula. The world population of that species until then had been estimated at less than 1,000 individuals (Liedel 1982). More waterbird surveys followed between 1986 and 1990 by the Asian Wetland Bureau, now Wetlands International (Verheugt et al. 1991, 1993; Danielsen and Verheugt 1990; Danielsen et al. 1991a, b), finding peak numbers of up to 80,000 waders during October–November, and up to 10,000–30,000 in other months (January–March, August). It is estimated that the total number of waders that use the area at during migration may number up to 0.5–1 million birds, or 8–20% of the total East Palearctic flyway (Danielsen and Verheugt 1990). The wader counts revealed a total of 30 species, including some very rare and endangered species, such as Nordmann's greenshank *Tringa guttifer*, Far Eastern curlew *Numenius madagascariensis,* and oriental plover *Charadrius veredus*. The area also supports an appreciable number of terns, with peak counts of between 2,500 and 3,000 terns. Of the total of 112 species of nonwaterbirds, 44 species occur in the mangroves and some are more or less confined to the mangroves, including *Ducula bicolor, Batrachostomus javensis, Alcedo euryzona, Halcyon coromanda, H. concreta, Merops philippinus, Berenicornis comatus, Picus canus, Parus major, Zoothera sibirica, Phylloscopus borealis, Muscicapa sibirica,* and *Cyornis rufigastra*.

The area is believed to still hold a remnant population of the estuarine crocodile *Crocodylus porosus*, severely diminished and threatened by hunting. The black water rivers in the Merang area, further inland from the national park, are known to hold a population of the endangered false gharial *Tomistoma schlegelii*, which therefore may also occur within the park. In addition, giant freshwater turtles may occur, including the critically endangered three-striped Batagur *Batagur borneoensis* and the vulnerable Asian giant softshell turtle *Pelochelys bibroni* as well as the near-threatened brown stream terrapin *Cyclemys dentata*. Mangroves and *Nypa* fringes are a favored habitat of the mangrove snake *Boiga dendrophila*, while reticulated python *Python reticulatus* is common in swamp forests. Monitor lizards (esp. *Varanus salvator*) are common throughout the area.

The area harbors a rich mammalian fauna, including along the coast the hump-backed dolphin *Sousa chinensis,* and the estuarine system has been frequented by the vulnerable Irrawaddy dolphin *Orcaella brevirostris* with groups of 6–10 individuals (Danielsen and Verheugt 1990). New surveys are needed to establish the current situation. The area holds at least three species of otter, including the smooth-coated otter *Lutra perspicillata*, small-clawed otter *Aonyx cinerea*, and the very rare and endangered hairy-nosed otter *L. sumatrana* (Danielsen and Verheugt 1990). The moonrat *Echinosorex gymnurus* is common. The swamp forests hold healthy populations of the siamang *Symphalangus syndactylus* and white-handed gibbon *Hylobates agilis*. Monkeys include the long-tailed macaque *Macaca fascicularis,* pig-tailed macaque *M. nemestrina*, and the silvered leaf-monkey *Trachypithecus cristatus*. Together with Berbak NP, Sembilang NP is the most important lowland reserve in Indonesia for the critically endangered Sumatran tiger *Panthera tigris sumatrae* (Silvius et al. 1984; Olviana 2011) and also supports several other cat species, including the near-threatened Asiatic golden cat *Catopuma temminckii* and the common Asian leopard cat *Felis bengalensis*. The presence of Sumatran tiger was confirmed during direct observation as well as tracks in 2011 and 2012, including a videoed observation of two immature tigers. In addition, reports on attacks on humans by tigers in the local communities appeared in local media in 2011. At least one tiger caught attacking villagers was reported captured and released in the adjacent Betet island in the Park.

Tracks of Asian tapir *Tapirus indicus* can be commonly encountered, especially in muddy areas along rivers in the freshwater swamp forest. From the Tragulidae, the Java mouse deer *Tragulus javanicus* and *Tragulus napu* both occur. The sambar deer *Cervus unicolor* frequents all habitats, as do the bearded and the common pig *Sus barbatus, S. scrofa*. Critically endangered Sumatran elephants *Elephas maximus sumatranus* have also been observed traveling between the Park and adjacent disturbed forests. The area is known to occasionally harbor huge congregations of flying foxes that migrate to and from Malaysia across the Malacca Straits.

The Park and its water areas have a rich fish and crustacean fauna. At least 142 fish species including the Sembilang fish *Plotosus canius*, 38 crab species, and 13 shrimp species have been recorded.

Conservation Status

Silvius (1986) recommended a 70,000 ha area in the southeast, the Banyuasin Peninsula, for conservation. Under projects carried out by the Asian Wetland Bureau (predecessor of Wetlands International) in the early 1990s, increasing evidence was documented on the high conservation values of the area, including also for fisheries. The area is estimated to support 70% of the significant coastal fisheries of South Sumatra in terms of breeding, spawning and nursery areas. This captured the attention of Sriwijaya University (in Palembang) and provincial authorities, with even a World Bank (Forestry II) project implemented in the early 1990s to enhance the conservation management of the area in conjunction with the Berbak NP. In 2000–2004, a project administered by Wetlands International under grant from GEF-World Bank succeeded to prepare and establish the area as a National Park (Wibowo 2004). In 2003 an area comprising the entire mangrove area and adjacent peat swamp forests was finally gazetted as a National Park, but covering unfortunately a much smaller area of the peat and freshwater swamp forests than had been proposed.

The National Park consists of the following zones:

- Core Zone, still in very good condition. Its physical features are still in their original state and the forest has not yet been exploited. It is designated to be fully protected.
- Wilderness Zone: another protection zone around the core zone, to protect the core zone.
- Utilization Zone, designated for tourism and other environmental services for local communities.

- Traditional Zone, mostly a web of rivers and creeks which serves to accommodate local transportation.
- Rehabilitation/Restoration Zone: specifically designated for rehabilitation/restoration activities.
- Specific treatment zone, which contains local villages that existed before the national park was created.

Sembilang National Park was officially designated as one of Indonesia's six Ramsar Sites (6 March 2011). In 2012, the National Park was also listed as one of the two designated sites under the East Asian Australasian Flyway Partnership.

Current Threats and Future Challenges

The most serious threat to the area is rapid degradation of the peat swamp forest outside of the park boundaries, involving illegal logging as well as the conversion of large areas to cultivation of oil palm or Acacia (for pulp wood for paper production). Over the last 10 years (since 2005), most of the inland area has been allocated to such concessions, and areas outside of official concessions were subject to illegal logging. Illegal loggers make use of small channels to float in equipment and float out the logs, thus creating a dense drainage pattern that desiccates the peat swamp, making them very vulnerable to fire. This is the single most severe threat to the area, especially as the peat domes that support the peat swamp forests in the national park are only partly covered by the protected area, and as such drainage (including legal drainage) outside of the park may strongly impact the eco-hydrology of the protected area. The peat swamp forest in the National Park is still in a relatively good state, but further inland huge fires have raged through this habitat. The conservation and rehabilitation of the peat swamp forests inland may be crucial to the eco-hydrological functioning of the coastal habitats, which will succumb if the freshwater influx is diverted as a result of inland drainage and bund construction.

About 3,000 ha of mangrove in the eastern part of the Park has been converted for aquaculture since 1995 as part of a misplaced government-led transmigration scheme which should have been located at some distance east of the current Park borders. Various consultations were made between communities involved, park managers, and local government, in order to relocate the households. In 2004 a consensus was reached with the fish famers to halt expansion of the existing fishponds and gradually restore any abandoned ones. The area has been subsequently included in the park's restoration zone and a mangrove restoration project started in 2013 with support of Japanese International Cooperation Agency (JICA) and Sriwijaya University.

In 2004, the local government of South Sumatra decided to build a harbor to the east of the park (at a distance of 5 km) near the mouth of the Banyuasin River. This is located in the vicinity of extensive mudflats and an extensive mangrove belt with the status of Protection Forest (Hutan Lindung). The harbor and roads were to result in the conversion of over 600 ha of mangroves. The plan received much protest from environmentalists as it would threaten the National Park as well as estuarine biodiversity and the mangrove habitat. Over the years, the harbor project plan has been repeatedly proposed and withdrawn, and while it is currently halted under the Presidential Decree on the Moratorium on new Concession Licenses in Forests and Peatlands (which has been extended twice since 2011, with the current end date in 2017), it may just be a matter of time before the development will be proceed and the national park and its surroundings will suffer predicted impacts.

The rare species harbored by the park remain very vulnerable to poaching, especially the Sumatran tiger. The freshwater habitats harbor many valuable fish species, both species for consumption and many small but attractive species for the aquarium industry. Already in the early 1990s, it was noted that the belida fish *Notopterus chitala* was suffering from overharvesting. The park may provide the last suitable area for supporting large breeding colonies of resident storks, herons, and ibises. The rapid decrease of colonies elsewhere in the province and in Jambi (Danielsen et al. 1991b) shows, however, how vulnerable these are.

Government funds and park staffing are inadequate, and alternative sources of income such as from ecotourism are almost nonexistent as the park is simply too inaccessible and lacks infrastructure. It is now hoped that payment for environmental service programs such as REDD+ that are being developed (2011–2013) will provide a longer term solution in securing funds for conservation and sustainable livelihood programs. Another avenue that is being pursued is formally linking the Sembilang NP with Berbak NP and its adjacent peatland buffer zone areas. A buffer zone management plan for Berbak NP exists, but this is not the case for Sembilang NP. Hydrologically both national parks come together in the small blackwater Benuh River and in a large peat dome in the west that forms the upper catchment of the Benuh River. They share major wildlife populations and together form what is probably the last viable habitat for the Sumatran tiger along the island's east coast. There have been proposals for extending a protected area status to areas in the west of the park, including the Merang area, but so far this has not happened. Instead, a large concession was granted and clear-felled for establishment of plantations by one of the subsidiaries or supply companies belonging to the Sinar Mas Group.

References

Danielsen F, Skov H. Waterbird study results from South East Sumatra. Oriental Bird Cl. Bull. 1987;6:8–11

Danielsen F, Verheugt WJM. Integrating conservation and land-use planning in the coastal region of South Sumatra. Bogor: PHPA/AWB-Indonesia; 1990.

Danielsen F, Purwoko A, Silvius MJ, Skov H, Verheugt W. Breeding colonies of Milky Stork in South Sumatra. Kukila. 1991a;5(2):133–5.

Danielsen F, Skov H, Suwarman U. Breeding colonies of waterbirds along the coast of Jambi Province, Sumatra, August 1989. Kukila. 1991b;5(2):135–8.

Hooijer A, Page S, Jauhiainen J, Lee WA, Lu XX, Idris A, Anshari G. Subsidence and carbon loss in drained tropical peatlands. Biogeosciences. 2012;9:1053–71.

Iqbal M, Mulyono H, Riwan A, Takari F. An alarming decrease in the Milky Stork *Mycteria cinerea* population on the east coast of South Sumatra Province. Indones Birding Asia. 2012;18:68–70.

Liedel K. Verbreitung und Okologie des Steppenschlammläufers, Limnodromus semipalmatus (Blyth). Ergebnisse der Mongolischn Gemeinschaftsreise von Ornithologen aus der DDR 1979. XI. Mitt. Zool. Mus. Berlin. Bd. 58, Suppl: Ann Orn. 1982; 6:147–162.

Olviana EK. Pendugaan populasi harimau Sumatra Pantera tigris sumatrae, Pocock 1929 menggunakan metode kamera jebakan di Taman Nasional Berbak. Bogor Agricultural University (IPB); Bogor, Indonesia, 2011.

Rosalina U, Rahayu S, Permana E, Suryawan SI, Hidayat A, Waluyo. Laporan Final Delineasi Potensi Proyek Carbon dan Pendugaan Cadangan Carbon pada areal kegiatan proyek (Carbon Stocks in Pilot Sites at Project Commencement). Fakultas Kehutanan IPB – Wetlands International; Bogor, Indonesia, 2003/2004.

Silvius MJ. On the importance of Sumatra's east coast for waterbirds, with notes on the Asian Dowitcher *Limnodromus semipalmatus*. Kukila. 1988;3(3/4):117–37.

Silvius MJ. Survey of coastal wetlands in Sumatra Selatan and Jambi, Indonesia. March/April 1986. Kuala Lumpur: PHPA- Interwader Report No.:1; 1986.

Silvius MJ, Verheugt WJM, Simons HW. Soils, vegetation, fauna and nature conservation of the Berbak Game Reserve, Sumatra, Indonesia, RIN contributions to Research on Management of Natural Resources, vol. 1984-3. Arnhem: Research Institute for Nature Management; 1984.

Silvius MJ, Verheugt WJM, Iskandar J. Coastal wetlands inventory of Southeast Sumatra. Report of the Sumatran Waterbird Survey Oct–Dec 1984. Cambridge: ICBP study report No.: 9; 1986.

Verheugt WJM, Purwoko A, Danielsen F, Skov H, Kadarisman R. Integrating mangrove and swamp forests conservation with coastal lowland development; the Banyuasin Sembilang swamps case study, South Sumatra Province, Indonesia. Landscape Urban Ecol. 1991;20(1–3):85–94.

Verheugt WJM, Skov H, Danielsen F. Notes on the birds of the tidal lowlands and floodplains of South Sumatra Province Indonesia. Kukila. 1993;6(2):53–84.

Wibowo P. The greater Berbak Sembilang integrated coastal wetland conservation project. Project final report. Wetlands International Indonesia Programme, Bogor; 2004.

Wetlands of Berbak National Park (Indonesia)

154

Wim Giesen, Marcel J. Silvius, and Yoyok Wibisono

Contents

Introduction	1832
Hydrology	1832
Wetland Ecosystems	1834
Biodiversity	1835
Conservation Status	1836
Ecosystem Services	1836
Current Threats and Future Challenges	1836
References	1838

Abstract

Berbak National Park extends over 162,700 ha and is the largest relatively intact swamp forest reserve on the Indonesian island of Sumatra. One third of the park consists of freshwater swamp forests and the remainder consists of peat swamp forests. Berbak is highly important for biodiversity conservation, especially for species such as false gharial *Tomistoma schlegelii*, painted terrapin *Batagur borneoensis*, Storm's stork *Ciconia stormi*, white-winged duck *Cairina scutulata* and Sumatran tiger *Panthera tigris sumatrae*. Although officially protected since

---

W. Giesen (✉)
Euroconsult Mott MacDonald, Arnhem, AK, The Netherlands
e-mail: wim.giesen@mottmac.com

M. J. Silvius (✉)
Wageningen, The Netherlands
e-mail: msilvius58@gmail.com

Y. Wibisono (✉)
Wetlands International – Indonesia Programme, Bogor, Indonesia
e-mail: wibisono_itc@yahoo.com

© Springer Science+Business Media B.V., part of Springer Nature 2018
C. M. Finlayson et al. (eds.), *The Wetland Book*,
https://doi.org/10.1007/978-94-007-4001-3_43

1935 it is under severe threat from illegal logging, encroachment by agriculture (oil palm) and fires, the latter mainly occurring in peatland areas drained by canals excavated for agriculture.

Keywords

Berbak · Swamp forest · Peatland · Sumatra · Indonesia

Introduction

Berbak National Park (104° 20'E, 1° 10'S) is one of the last examples of the formerly extensive lowland wetland forest habitats in Sumatra, Indonesia, comprising of peat swamp and freshwater swamp forests and covering 162,700 ha. It forms part of the vast coastal plain of eastern Sumatra. The area is flat and swampy reaching an elevation of 16 masl in the west and is bounded by the Berbak River in the north, the Benu River in the south, and the largely cultivated narrow alluvial coastal fringe in the east. It is one of the oldest protected areas in Indonesia, first established in 1935 as a wildlife reserve and designated as a national park and as Indonesia's first Ramsar Site in 1992 (Fig. 1). In the south, it is connected to Sembilang National Park (established in 2003), which holds the most extensive protected mangrove area in the Indo-Malayan region and with which it forms the Greater Berbak-Sembilang Ecosystem.

Hydrology

Berbak is bisected by one of the longest blackwater (see Box 1 in Chap. 155, "▶ Danau Sentarum National Park (Indonesia)") streams in Sumatra, the Air Hitam Laut River that has its upper catchment about 90 km inland in a peat dome partly shared with Sembilang NP and partly with unprotected areas. The Benu River on the border of Berbak and Sembilang NP is a similar blackwater river, but starts at only 30 km inland. Both rivers have part of their upper catchment located on a 20 m deep unprotected peat dome, and this status poses a major threat to the long-term viability of Berbak. Southern tributaries of the Air Hitam Laut, such as the Simpang Kubu and Simpang Gadjah, and the northern tributaries of the Benu River such as the Simpang Kanan share the same peat domes in the park. The main northern tributary of the Air Hitam Laut River is the Simpang Melaka. West of the Berbak NP run the Kumpeh and Berbak Rivers (a distributary of the Batanghari River) and very close to its northwestern boundary the Sungai Air Hitam Dalam (a tributary of the Berbak River). Along the coast there are a number of small natural streams that originate in the peat domes. Berbak's peat swamps are dome shaped and in some places attain over 15 m depth. Their water and nutrient supply is derived entirely from rainfall (ombrogenous), and both waters and peat layer are very acidic (pH 3.5–3.9) and nutrient deficient (Andriesse 1986).

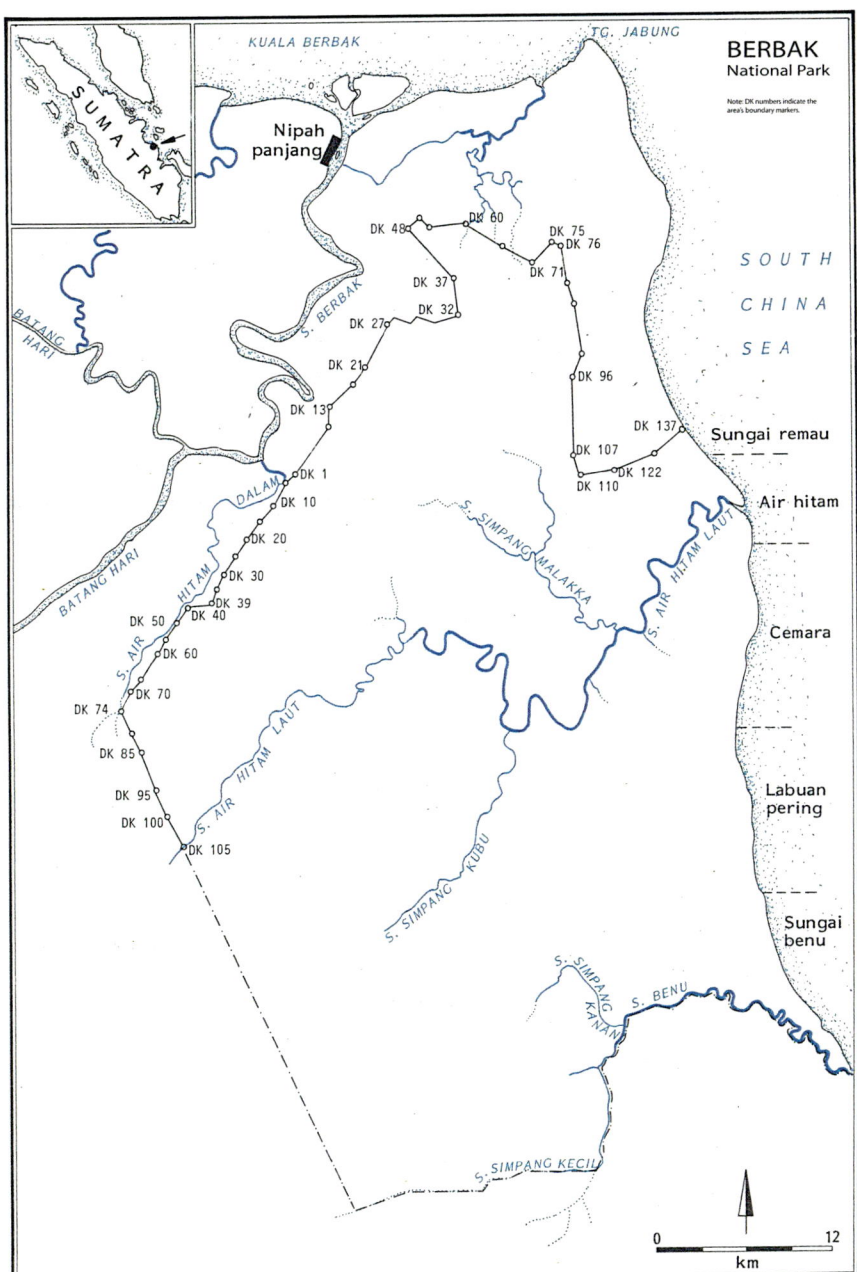

Fig. 1 Location map of Berbak National Park (© MJ Silvius, Rights remain with author) (The original map came from Silvius et al (1983))

Wetland Ecosystems

Berbak NP originally had two main habitat types – freshwater swamp forest on mineral soil and peat swamp forest – that extend over 32% and 58% of the park, respectively, with the balance (10%) consisting of riparian forest and coastal scrub and mangrove. However, coastal scrub and mangroves have since been excised from the park, and a large (25,000 ha) central portion was degraded by illegal logging and fires in the late 1990s. However, the most extensive mangroves of Southeast Asia are located in adjacent Sembilang NP.

Riparian vegetation attains a height of 15–25 m with emergent trees of up to 40 m, while swamp forest is usually 35–45 m tall, with many emergent trees of 50–55 m (e.g., *Alstonia pneumatophora*) and occasionally up to 60 m (especially *Koompassia malaccensis*). In the central Air Hitam Laut region, the riparian forest is fringed along the river by a zone of *Pandanus helicopus* (*rasau*), which may vary from few meters to the entire width of the river preventing river transport. It is often backed by vegetation dominated by *Syzygium* species mixed with other riparian species including *Gardenia tubifera* and *Barringtonia racemosa*. Aquatic vegetation is generally absent in the blackwater streams, apart from free-floating *Hanguana malayana*, which forms dense mats that – as with the pandans – may choke sections of the river and provide important habitat for crocodiles. In some areas small patches of bladderwort *Utricularia exoleta* occur, while in burnt areas, sedges (mainly *Thoracostachyum* and *Scleria* species) and grasses may abound. *Nypa fruticans* occurs along the Air Hitam Laut until about 600 m upstream of the confluence with the Simpang Melaka, where it gives way quite abruptly to *Pandanus helicopus*. To the north of Berbak, streams such as Air Hitam Dalam have riparian vegetation typical of mineral soils, with species including *Cerbera odollam, Ficus microcarpa, Flacourtia rukam, Gluta renghas, Hibiscus tiliaceus, Kleinhovia hospita,* and *Lagerstroemia speciosa*. On the inland beach ridges and some higher river levels near the coast, the valuable palm *Oncosperma tigillarium* abounds.

Peat swamp forests at Berbak are tall (30–50 m) and relatively diverse, and even centers of main peat domes have a species-rich peat swamp forest vegetation that can attain a height of over 35 m, with emergents of up to 40–45 m (Franken and Roos 1981; Silvius et al. 1984; Giesen 1991). Key species include *Alstonia pneumatophora, Archidendron clypearia, Cyrtostachys renda* (*C. lakka*), *Diospyros areolata, D. lanceifola, Durio carinatus, Dyera polyphylla, Gluta* (*Melanorrhoea*) *wallichii, Eleiodoxa* (*Salacca*) *conferta, Knema conferta, Koompassia malaccensis, Madhuca* (*Ganua*) *motleyana, Mangifera foetida, Mussaendopsis beccariana, Myristica elliptica, M. iners, Parastemon urophyllus, P. lateriflora, P. sumatrana, Santiria griffithii, Shorea platycarpa, S. teysmanniana, S. uliginosa, Tetramerista glabra, Teijsmanniodendron hollrungii, T. pteropodum,* and *Urophyllum arboreum*. Freshwater swamps on mineral soils attain heights of 35–55 m and are dominated by many Myrtaceae such as *Eugenia* and *Syzygium* species, *Alstonia pneumatophora, Baccaurea* species, *Elaeocarpus* species, *Flacourtia rukam, Koompassia malaccensis,* and a wide range of *Ficus* species.

Biodiversity

A total of 261 flowering plant species have been recorded at Berbak, of which 67% are trees and shrubs, with climbers accounting for a further 17% and herbs and epiphytes 8% each (Giesen 1991). The flora is still poorly studied, but includes a first record in Sumatra for the rare mistletoe *Lepidaria kingii* and at least 23 palm species, among them rare species such as *Johannesteijsmannia altifrons* and *Cyrtostachys renda* (Dransfield 1974; Giesen 1991).

Berbak supports a rich reptilian fauna, including estuarine crocodile *Crocodylus porosus*, false gharial *Tomistoma schlegelii* (Stuebing et al. 2006) and giant freshwater turtles including in the blackwater habitats the critically endangered painted terrapin *Batagur borneoensis* and the endangered Asian giant softshell turtle *Pelochelys cantorii* in the Berbak River, as well as the near-threatened common leaf turtle *Cyclemys dentata* (in the freshwater swamp forest; Silvius et al. 1984). Mangroves and *Nypa* fringes are a favored habitat of the mangrove snake *Boiga dendrophila*, while reticulated python *Python reticulatus* is common in swamp forests.

More than 230 bird species have been recorded at Berbak (Silvius et al. 1984; Hornskov 1987), although more than 350 species are expected if all migratory species are included (Perbatakusuma et al. 2010). This area is especially important for swamp bird species such as Storm's stork *Ciconia stormi*, lesser adjutant *Leptoptilos javanicus*, milky stork *Mycteria cinerea*, and white-winged duck *Cairina scutulata* (Silvius et al. 1984; Silvius and de Iongh 1989; Perbatakusuma et al. 2010), but also forest species such as Jerdon's baza *Aviceda jerdoni*, Wallace's hawk-eagle *Spizaetus nanus*, black partridge *Melanoperdix niger*, and six hornbill species including several near-threatened species. The coastal mudflats include some of the most important staging areas of migratory waterbirds in Southeast Asia, with flocks of tens of thousands of waders congregating including some rare species such as the endangered Nordmann's greenshank *Tringa guttifer* and on the sandy beach of Cemara also the rare white-faced plover *Charadrius alexandrinus dealbatus*. The Berbak-Sembilang coastal mudflats represent the key wintering area of the near-threatened Asian dowitcher *Limnodromus semipalmatus* with a top count of over 10,000 individuals. The mudflats also serve as a feeding area for thousands of local and migratory herons (including the vulnerable Chinese egret *Egretta eulophotes*) and storks, (milky stork and lesser adjutant) and hundreds of black-headed ibis *Theskiornis melanocephalus* (Silvius and Verheugt 1986; Silvius 1987, 1988; Perbatakusuma et al. 2010). Recent (2013) surveys by Zoological Society of London suggest that the number of species may have dropped and that certain groups such as hornbills may have been affected by hunting and habitat degradation.

Berbak and the adjacent Sembilang area are the most important lowland reserves in Indonesia for the Sumatran tiger *Panthera tigris sumatrae* (Silvius et al. 1984; Olviana 2011). Other characteristic species include the clouded leopard *Neofelis nebulosa*, Malayan tapir *Tapirus indicus*, and Malaysian sun bear *Helarctos malayanus*. Early mornings resound with the calls of the white-handed gibbon *Hylobates lar* and siamang *Symphalangus syndactylus*. Monkeys include the long-tailed macaque *Macaca fascicularis*, pig-tailed macaque *Macaca nemestrina*, and the silvered leaf monkey *Trachypithecus cristatus*. Notewothy are also the smooth-coated otter *Lutra perspicillata*, fishing cat *Prionailurus viverrinus*, binturong *Arctictis binturong*, flying lemur *Cynocephalus variegatus*, and the many squirrel species, including the red giant flying squirrel *Petaurista petaurista*. In 1983, Silvius et al. (1984) found evidence of local migration of a large group of bearded pig *Sus barbatus* that had left a long track through the freshwater swamp forest of over 10 m wide. The Java mouse deer *Tragulus javanicus* is very common, while sambar deer *Rusa unicolor* is rare. In the early 1990s, there was a rare record of a footprint of the Sumatran rhino *Dicerorhinus sumatrensis* in the upper catchment of the Simpang Melaka.

Conservation Status

Berbak was first gazetted as a protected area in 1935, when it was proclaimed a "wildreservaat" (i.e., game reserve) by a decree by the Governor-General of the Netherlands Indies on 29 October 1935. The (then) Indonesian Directorate General of Forest Protection and Nature Conservation (PHPA; KSDA at provincial level) began management activities in Jambi in 1972, and boundary demarcation of Berbak as a wildlife reserve *Suaka Margasatwa* was completed by 1974. It was designated as Indonesia's first Wetland of International Importance under the Ramsar Convention on 8 April 1992, and its status upgraded to National Park in 1997 (No. 185/Kpts-II/1997), with management handed over from KSDA to the Provincial National Park Unit *Taman Nasional* section of the Forestry Department in 1998.

Ecosystem Services

The peat domes that largely (but not entirely) lie in Berbak NP are sources of freshwater in the drier months, ensuring that waters in rivers such as the Air Hitam Laut, Simpang Melaka, and Benu remain fresh, thereby protecting important riverine biodiversity and providing drinking water for coastal communities. Berbak is one of the last remaining areas of reasonably intact peat swamp forest in Sumatra, providing a habitat for important biodiversity, including many rare and endangered species. The vast peat resources are an important repository of carbon reserves.

Current Threats and Future Challenges

The main ongoing threats to Berbak NP are illegal conversion and logging, fires, and changes in hydrology, all of which are interlinked. Conversion was originally along the coast, where Buginese settlers already established themselves in the 1960s, excavating small drainage channels (*parit*) and cultivating coconut palms. From the early 1990s onward, illegal logging from the west and northwest led to direct loss of forest, but also to subsequent drying out of peatland (due to canals dug to extract logs) and subsequent fires, especially in 1997–1998 when 17,000 ha of peatland burnt in the center of the park (Giesen 2004; van Eijk et al. 2009). Large parts of the peat domes have since been affected by drainage from illegal logging channels and – mainly along the coast – the Buginese *parit* drainage and irrigation system used for their coconut plantations (Silvius et al. 1984).

The upper catchment of the Air Hitam Laut River was until recently relatively undisturbed, and the main economic activity there was forestry based on selective logging. However, in the 1990s the area was opened up for oil palm plantations under an official transmigration program, and both the newly established cluster of villages (Desa Sungai Gelam) and the plantations are located on >10 m deep peat. The development scheme was accompanied by spontaneous settlers, who are mainly occupied with illegal logging of adjacent forests in the park. Drainage for oil palm is often deep (>100 cm), which inevitably leads to peat soil compaction, shrinkage, and oxidation, resulting in soil subsidence. A hydrological modeling study by Wösten et al. (2006a, b) predicts major impacts on the natural drainage of the Air Hitam Laut River and a reduction of its natural extent. This in turn will lead to a lowered river discharge with a detrimental effect on the sustainability of the Berbak NP in the center of the catchment and potential negative impacts for agriculture and fisheries in the coastal zone, especially in relation to potential acid sulfate soils (see Box 1 below), saltwater intrusion, and lack of irrigation water. Conversions have continued since 2000, affecting much of the upper Air Hitam Laut catchment. Fortunately, for now the main and very deep peat dome in the southwest of the park remains relatively intact as it has been part of logging concessions licensed for selective logging only.

Issues such as fires, clearing, conversion, and loss of adjacent peatland have been repeatedly emerged over the past decades, and attempts at addressing these included the establishment of firefighting brigades, development of community livelihood programs, and attempts at peat swamp forest restoration in burnt areas. However, as these attempts were invariably project based, they have not had lasting success or a major impact. Government funds and park staffing are inadequate, and alternative sources of income such as from ecotourism are almost nonexistent as the park is simply too inaccessible and lacks infrastructure. It is now hoped that payment for environmental service programs such as REDD+ that are being developed (2011–2013) will provide a longer term solution in securing funds for conservation and sustainable livelihood programs. Another avenue that is being pursued is formally linking Berbak with the Sembilang National Park, a large mangrove and peat swamp forest reserve adjacent to the south of the park located in South Sumatra Province. Hydrologically, both national parks come together in the small blackwater river – Sungai Benu – and large peat domes in the west that form the upper catchment of the Air Hitam Laut and Sungai Benu Rivers. While habitats are different, they share major wildlife species and together form what is probably the last viable habitat for the Sumatran tiger along the island's east coast.

> **Box 1: Potential Acid Sulfate Soils**
>
> A common feature of (near) coastal areas is the occurrence of iron sulfide (or iron pyrite, FeS_2) in the wetland soil. These iron pyrite-containing soils are known as "potential acid sulfate" (PAS) soils or "cat clays." Such soils can occur in all geographic regions, from the tropics to temperate zones. Formation of PAS soils typically occurs in estuaries because of the presence of iron (scarce in seawater, but abundant in river water), sulfates (in seawater), and organic matter and a lack of oxygen in the soil (leading to reducing conditions). Hence, PAS soils are common in (former) mangrove and salt-marsh areas, but in some regions (e.g., Sumatra and Borneo), they may also underlie coastal peat.
>
> Metallic sulfides such as iron sulfide are highly insoluble and remain in situ unless disturbed. Upon development (e.g., typically involving soil drainage by human activity for agriculture) and exposure to air, they will turn highly acidic (pH often <1–2) due to the reaction of iron pyrites with oxygen, resulting in the production of sulfuric acid (Dent 1986; Craswell and Pushparajah 1989; Hardjowigeno 1989; Konsten and Klepper 1992; Giesen et al. 2007). Mobilization of toxic aluminum ions due to a lowering of the pH seems to be one of the major additional problems associated with these soils (Dent 1986). Plants growing on drained PAS soils may therefore suffer from both aluminum poisoning and high acidity. The degree to which problems arise depends on a complex interplay between pyrite concentrations, (depth) distribution of pyrite in the soil, rate of oxidation (how much is exposed to oxygen), and possibilities for flushing a soil (e.g., with freshwater). These factors may vary considerably over short distances and are one of the challenges facing the development of such areas (Wösten et al. 2013).

References

Andriesse JP. Characteristics and management of tropical peat soils. Amsterdam: Royal Tropical Institute; 1986.

Craswell ET, Pushparajah E. Management of acid soils in the Humid Tropics of Asia. Canberra: Australian Centre for International Agricultural Research (ACIAR)/International Board for Soil Research and Management; 1989. 118 pp.

Dent D. Acid sulphate soils: a baseline for research and development, ILRI Publication No. 39. Wageningen: International Institute for Land Reclamation and Improvement/ILRI; 1986. 204 pp.

Dransfield J. Notes on the palm flora of Central Sumatra. Reindwardtia. 1974;8:519–31.

Franken NAP, Roos MC. Studies in lowland equatorial forest in Jambi Province, Central Sumatra. Bogor: BIOTROP/SEAMEAO Regional Centre for Tropical Biology; 1981.

Giesen W. Berbak Wildlife Reserve, Jambi, Sumatra. Final Draft Survey Report. PHPA/AWB Sumatra Wetland Project Report No. 13, Bogor; 1991.

Giesen W. Causes of peatswamp forest degradation in Berbak NP, Indonesia, and recommendations for restoration. Water for food and ecosystems programme project on "Promoting the river basin and ecosystem approach for sustainable management of SE Asian lowland peatswamp forests: Case study Air Hitam Laut River basin, Jambi Province, Indonesia." Arnhem: ARCADIS Euroconsult; 2004.

Giesen W, Wulffraat S, Zieren M, Scholten L. Mangrove guidebook for Southeast Asia, RAP Publications 2006/07. Bangkok: FAO & Wetlands International; 2007. 769 pp. ISBN 974-7946-85-8.

Hardjowigeno S. Mangrove soils of Indonesia. In: Soerianegara I et al., editors. In: Proceedings Symposium on Mangrove Management: its Ecological and Economic Considerations, August 9–11, 1988, Bogor; 1989. p. 257–65.

Hornskov J. More birds from Berbak Game reserve, Sumatra. Kukila. 1987;3:58.

Konsten CJM, Klepper O. Pyrite in coastal wetlands: a natural chemical time bomb. Paper presented at the European state of the art conference on delayed effects of chemicals in soils and sediments (Chemical Time Bombs), 2 5 September 1992, Veldhoven; 1992, 14 pp.

Olviana EK. Pendugaan Populasi Harimau Sumatera. *Panthera tigris sumatrae*, Pocock 1929 Menggunakan Kamera Jebakan di Taman Nasional Berbak. Dep. Konservasi Sumberdaya Hutan dan Ekowisata, Fak. Kehutanan, Institut Pertanian Bogor; 2011.

Perbatakusuma EA, Rachman D, Collins M. Bird species diversity and reducing emissions from deforestation and degradation in Berbak Peat Swamp Forest Jambi Province Indonesia. Technical Report. Bird and REDD Report. Zoological Society London – Ministry of Forestry. Jambi; 2010.

Silvius M. Notes on new wader records for Berbak Game Reserve, Sumatra. Kukila. 1987;3:59.

Silvius MJ. On the importance of Sumatra's east coast for waterbirds, with notes on the Asian Dowitcher *Limnodromus semipalmatus*. Kukila. 1988;3:117–38.

Silvius MJ, de Iongh H. White-winged Wood Duck, a new site for Jambi province. Kukila. 1989;4(3–4):150.

Silvius MS, Verheugt WJM. The birds of Berbak Game Reserve, Jambi Province, Sumatra. Kukila. 1986;2:76–85.

Silvius MJ, Simons HW, Verheugt WJM. Soils, vegetation, fauna and nature conservation of the Berbak Game Reserve, Sumatra, Indonesia, RIN Contributions to research on management of natural resources. Arnhem: Research Institute for Nature Management; 1984.

Silvius MJ, Verheugt WJM, Iskandar J. Coastal wetlands inventory of Southeast Sumatra. Report of the Sumatra Waterbird Survey Oct–Dec 1984. ICBP Study Report No 9. Cambridge; 1986.

Stuebing RB, Bezuijen MR, Auliya M. The current and historic distribution of *Tomistoma schlegelii* (The False Gharial) (Müller, 1838) (Crocodylia, Reptilia). Raffles Bull Zool. 2006;54:181–97.

Van Eijk P, Leenman P, Wibisono ITC, Giesen W. Regeneration and restoration of degraded peat swamp forest in Berbak NP, Jambi, Sumatra, Indonesia. Malay Nat J. 2009;61:223–41.

Wösten H, Hooijer A, Siderius C, Dipa SR, Idris A, Rieley J. Tropical peatland water management modeling of the Air Hitam Laut catchment in Indonesia. Int J River Basin Manag. 2006a;4:233–44.

Wösten JHM, van den Berg J, van Eijk P, Gevers GJM, Giesen WBJT, Hooijer A, Idris A, Leenman PH, Rais DS, Siderius C, Silvius MJ, Suryadiputra N, Wibisono ITC. Interrelationships between hydrology and ecology in fire degraded tropical peat swamp forests. Water Resour Dev. 2006b;22:157–74.

Wösten H, Haag A, Boissevain W. Soil and water management for sustainable agriculture in lowlands. Quick Assessment and Nationwide Screening (QANS) of Peat and Lowland Resources and Action Planning for the Implementation of a National Lowland Strategy. PVW3A10002 Agentschap NL 6201068 QANS Lowland Development. For Indonesian Ministry of Public Works, Jakarta; 2013, 77 pp.

Danau Sentarum National Park (Indonesia)

155

Wim Giesen and Gusti Z. Anshari

Contents

Introduction	1842
Hydrology	1843
Wetland Ecosystems	1843
Biodiversity	1844
Conservation Status	1846
Ecosystem Services	1846
Threats and Future Challenges	1847
References	1849

Abstract

Danau Sentarum National Park (>132,000 ha) is located on an average elevation of 35 m ASL in the floodplain of the upper Kapuas River in West Kalimantan, Indonesian Borneo. The area consists of a series of interconnected seasonal lakes (ca. 25%), interspersed with swamp forest (ca. 32%), inland ombrogenous peat swamp forest (ca. 16%; dated at >30,000 years BP, the oldest peat swamp in Indonesia) with very high carbon stocks, dry lowland forest on isolated hills, shifting agriculture, and settlements. It is a hot spot for endemism of wetland flora and fauna, populations of threatened species, and is highly important for productive fisheries that form the main livelihood for local communities. Declared Indonesia's second Ramsar Site in 1999, its status was upgraded to that of *Taman Nasional* (i.e., National Park) in 1999, which includes the 132,000 ha

---

W. Giesen (✉)
Euroconsult Mott MacDonald, Arnhem, AK, The Netherlands
e-mail: wim.giesen@mottmac.com

G. Z. Anshari
Center for Wetlands People and Biodiversity, University Tanjungpura, Pontianak, W. Kalimantan, Indonesia

© Springer Science+Business Media B.V., part of Springer Nature 2018
C. M. Finlayson et al. (eds.), *The Wetland Book*,
https://doi.org/10.1007/978-94-007-4001-3_44

core area, along with a 65,000 ha buffer zone proposed in 1997 that is disputed and partly earmarked for oil palm estate development. The main threats to the integrity of DSNP are fires, habitat conversion along the periphery, direct overexploitation of resources, and a recurrent plan to construct a dam on the Tawang River at the outlet of the lakes.

Keywords

Danau Sentarum · Indonesia · Lakes · Oldest Indonesian Peat swamp · Endemism

Introduction

Danau Sentarum National Park (DSNP) extends over 132,000 ha and is located in the floodplain of the upper Kapuas River in West Kalimantan, Indonesian Borneo (Fig. 1). The Park lies between the Kapuas River and the border with Sarawak and is located between 0°40'–0°55' N and 112°00'–112°25' E at an average elevation of 35 masl.

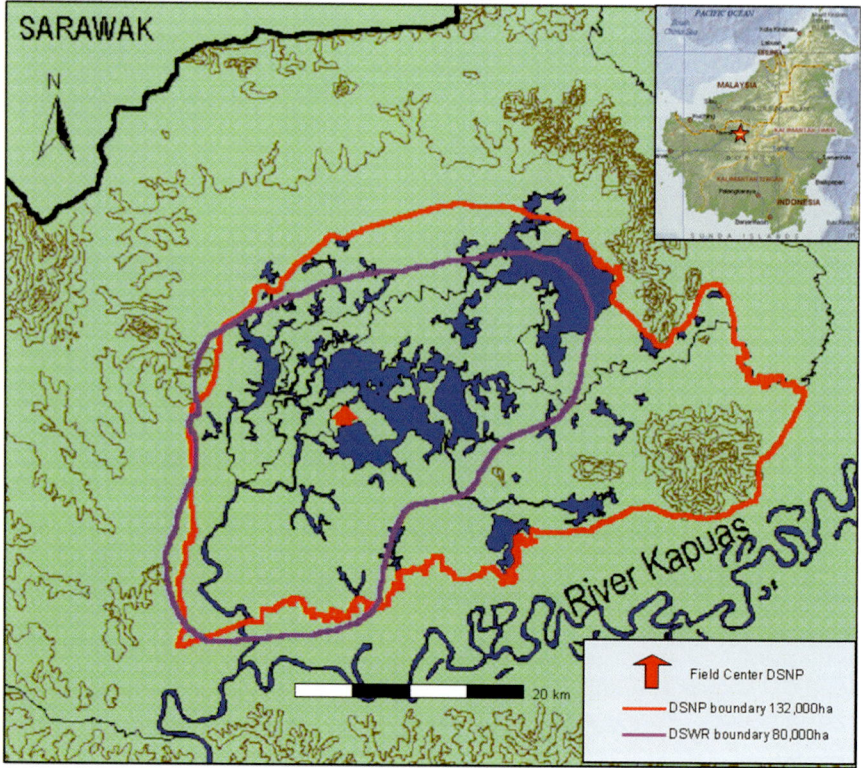

Fig. 1 Location map of Danau Sentarum National Park (Adapted from Giesen and Aglionby 2000, with permission)

The area consists of a series of interconnected seasonal lakes (*danau* in Bahasa Indonesia), interspersed with swamp forest, inland ombrogenous (receiving water and nutrients only through precipitation) peat swamp forest, and dry lowland forest on isolated hills. It is a hot spot for endemism of wetland flora and fauna and is highly important for productive fisheries that form the main livelihood for local communities.

Hydrology

The upper Kapuas basin is very flat, and waters of the Kapuas River tend to pond upstream of a natural "bottleneck." Because of high precipitation levels (>3,000 mm/year), most of the low-lying areas in the basin are flooded in the wetter months. As a result, DSNP consists of a mosaic of various types of swamp forest, along with seasonal lakes that are darkly colored due to dissolved organic acids from peat oxidation and have a pH of 4.5–5.5 and a conductivity of 9–24 µS. According to a model developed by Klepper et al. (1995), the lakes and swamp forests act as a buffer for the Kapuas system, mitigating floods (absorbing ¼ of peak flood) and maintaining water levels in the dry season by providing 50% of dry season flow in the upper Kapuas River. The ecosystem is characterized by an annual rise and fall in water levels of up to 12 m. The lakes are 6–8 m deep and are connected to the Kapuas River via the Tawang River that reverses flow when water levels in the Kapuas drop. When monthly rainfall in the upper Kapuas is below 250 mm for three consecutive months (usually June–August), water levels in the lakes start to drop, and lakes may dry out, exposing the beds for 1–3 months (Giesen 1987). In the prolonged dry season associated with the strong 1997 El Niño, DSNP lakes dried out for 5–6 months. In addition to the Tawang River, other major streams in the Park are the Leboyan, Bunut, and Sekulat rivers.

Wetland Ecosystems

About half of DSNP consists of swamp forest, while a quarter consists of lakes and the remainder of hill forest and (wet) heath forest (Table 1). Three major types of swamp forest occur: peat, stunted, and dwarf swamp forests, which have average canopy height of 22–30 m, 8–15 m, and 5–8 m, respectively. Almost two-thirds of the swamp forest consists of stunted swamp forest, while one-third consists of peat swamp forest. Dwarf swamp forest forms a minor element only. A large part of the peat swamp forest has developed on (deep) ombrogenous peat soil, while the stunted and dwarf swamp forests are on mineral soils.

Peat swamp forests are characterized by several phases of peat accumulation that has occurred over the past 30,000 years (Anshari et al. 2001, 2004, 2012). Peat depths are variable, depending on the dynamics of river courses. Shallow peat (0.5–2.0 m) occurs near river edges and may be frequently flooded by 1–2.5 m for 2–3 months (Giesen 1987, 1996, 2000). The average peat depth is about 7.7 m, and a maximum peat depth of 12 m was recorded in Nung peat swamp forest in the central

Table 1 Land cover analysis for Danau Sentarum National Park

Land cover	Area (ha)	Land cover	Area (ha)
Dryland:		**Swamps:**	
Lowland hill forest	6,767	Peat swamp forest	21,915
Heath forest	2,008	Stunted swamp forest	39,469
Shifting cultivation	4,603	Dwarf swamp forest	2,362
		Burnt swamps	23,084
		Floating grass	256
		Miscellaneous:	
Open water:		Settlements	32
Lakes, streams, and rivers	30,095	Others (e.g., cloud cover)	2,000

Note: Adapted from Jeanes (1997): © Wetlands International – Indonesia Programme, with permission; total area = 132,590 ha

part of DSNP, which is managed as a customary forest by local communities (Anshari et al. 2012).

Dwarf swamp forest develops in deeply flooded areas that may be flooded by up to 4–5.5 m for 8–12 months per year. Stunted swamp forest is intermediate between peat and dwarf swamp forest in terms of flooding depth and duration. Herbaceous submerged and emergent aquatic herbs are rare, as the extreme annual fluctuation in water level limits their growth.

Danau Sentarum's swamp forests are prone to fires (see Table 1), possibly due to the accumulation of large amounts of organic matter in the wet months, and repeated fires appear to be leading to an expansion of dwarf swamp forest, at the expense of stunted swamp forest (Giesen 1996; Dennis et al. 2000a, b).

Biodiversity

More than 500 plant species have been recorded at DSNP, of which 262 species occur in the swamp forests and three-quarters of which are trees and shrubs. DSNP harbors many novel and interesting plant species, such as the unusual tree *Dichilanthe borneensis,* which was collected at DSNP by Beccari in 1867 and is not found elsewhere. This unique species represents a link between the coffee family (Rubiaceae) and the figwort family (Scrophulariaceae), incorporating characteristics from both families. A new species of *Rhodoleia* collected in 1993 belongs to the witch-hazel family (Hamamelidaceae), a family poorly represented in Asia, with only seven genera occurring in Southeast Asia, each represented by only one species. Other rare species include the small tree *Dicoelia beccariana*, the sedge *Hypolytrum capitulatum,* the stemless palm *Eugeissona ambigua*, and the rattan *Plectocomiopsis triquetra* (Giesen 1987, 1996, 2000).

The lakes and **blackwaters** (see Box 1 below) of the swamp forests of Danau Sentarum are remarkable for their fish diversity, and 212 fish species have been identified in and around the Park, including 11 species new to science (*Akysis fuscus, Betta enisae, B. pinguis, Gastromyzon embalohensis, Homaloptera yuwonoi, Hyalobagrus leiacanthus, Osteochilus partilineatus, Parachela cyanea, Puntius trifasciatus, Rasbora tuberculata, Sundasalanx platyrhynchus*), while a further nine additional species are either new or require further study before their identity can be cleared (Kottelat and Widjanarti 2005). [Note: Jeanes and Meijaard (2000) recorded 266 species (>70% of all fish species found on Borneo), but not all of these are taxonomically confirmed.] The DSNP fish fauna includes two highly popular aquarium fish: the rare and valuable red variety of the endangered Asian arowana *Scleropages formosus* and the clown loach *Botia macracanthus,* known from only several localities in Kalimantan and Sumatra. A large number of fish species migrate upriver to headwaters or downriver to the Kapuas River at some time of the year, while in addition there are lateral movements between the rivers and lakes and the flooded forest during the wet season (Kottelat and Widjanarti 2005).

Three species of crocodile occur at DSNP, including the rare and vulnerable false gharial *Tomistoma schlegelii*, the estuarine crocodile *Crocodylus porosus*, and an enigmatic third species (possibly of the *C. raninus* group; Frazier 2000). Eleven turtle and tortoise species have been recorded, including the endangered keeled box turtle *Cuora mouhotii,* the vulnerable Asian softshell turtle *Amyda (Trionyx) cartilaginea,* vulnerable Asian giant softshell turtle *Pelochelys bibroni,* endangered Asian brown/forest tortoise *Manouria emys,* vulnerable black marsh turtle *Siebenrockiella crassicollis,* endangered spiny turtle/spiny terrapin *Heosemys spinosa,* and the vulnerable rice field/Malayan snail-eating turtle *Malayemys subtrijuga* (Walter 2000).

DSNP's birdlife has been relatively well studied and has been found to include 237 confirmed species, which is half of the species recorded on Borneo (Van Balen and Dennis 2000). These include 9 threatened and 22 near-threatened species, including the great argus pheasant *Argusianus argus* and the endangered Storm's stork *Ciconia episcopus stormi.* Storm's stork, which breeds in these inland swamp forests, is listed as rare and is considered to be the world's rarest stork. The vast majority of bird species are forest dwellers, and waterfowl are relatively rare, probably because of a lack of herbaceous aquatic vegetation cover. Colonial water birds such as egrets and herons have been wiped out due to hunting and egg collecting (Giesen 1987), and the area has probably never had many ducks or waders.

DSNP has the largest inland population of endangered proboscis monkeys *Nasalis larvatus* (Sebastian 2000; Meijaard and Nijman 2000), but they are elusive, probably due to past hunting pressures, and unlike other populations of this species, they venture far from waterways frequented by fisherfolk. A remarkable discovery has been that the swamp forests and peat swamp forests around Danau Sentarum harbor what may be one of Borneo's largest populations of orangutan *Pongo pygmaeus* (Russon et al. 2000, 2001). In addition to these primates, a total of possibly 143 mammal species have been recorded at DSNP, which is 65% of Borneo's nonmarine mammal species (Jeanes and Meijaard 2000). Species include sun bear *Helarctos malayanus,* clouded leopard *Neofelis nebulosa,* bay cat *Catopuma badia,* smooth-coated otter *Lutra perspicillata,* and otter civet *Cynogale bennettii.*

Conservation Status

Danau Sentarum was first gazetted as a *Suaka Margasatwa* (Wildlife Reserve) in 1982 by Decree SK No. 757/Kpts/Um/10/1982, when it extended over 80,000 ha, with just under one-third consisting of open water (see Fig. 1). In 1994, it was enlarged to 132,000 ha to include extensive tracts of peat swamp forest and several hill ranges with dry lowland and heath forest. In April 1994, Danau Sentarum was declared Indonesia's second Wetland of International Importance under the Ramsar Convention, and on 4 February 1999, its status was upgraded to that of *Taman Nasional* (i.e., National Park) by Decree SK 34/Kpts-II/1999, which includes the 132,000 ha core area, along with a 65,000 ha buffer zone proposed in 1997. The latter is disputed and has been partly earmarked for oil palm estate development (see Wadley et al. 2000).

Ecosystem Services

Danau Sentarum is one of the most important areas on Borneo in terms of biodiversity, supporting not only many diverse species but also a high degree of endemism (esp. plants and fish) and important populations of threatened species (e.g., Asian arowana, orangutan, proboscis monkey).

The lakes support a large traditional fishing industry, utilized by over 8,500 fisherfolk (85% Melayu) inhabiting 39 villages in and adjacent to the Park (Giesen and Aglionby 2000; Dennis et al. 2000b). The Melayu economy revolves around fishing, which is the major source of protein and provides most of the Melayu income (Colfer et al. 2000). When water levels are high, fishing activity is at an ebb and carried out for subsistence only. During the onset of the dry season (usually June), as water levels drop, fishing activity picks up, and when the lakes have almost dried out, fishing activity peaks. Fishing practices include the use of a wide range of cast nets, gill nets, fixed nets, funnel nets, lift nets, traps, barriers, hooks and lines, and even excavated pits (Giesen 1987; Dudley 2000). Most fish are sun dried and salted, as the remote location excludes the possibility of marketing fresh fish, with the exception of several high-value species. The latter include ornamentals, such as the clown loach and the Asian arowana, but also food fish such as marbled goby *Oxyeleotris marmorata* or *ikan lemas*, sultan fish *Leptobarbus hoevenii*, featherback *Chitala lopis*, and giant snakehead *Channa micropeltes*. The estimated total annual catch ranges from 7,800 to 13,000 t (97.5–162.5 kg/ha; Giesen 1987; Dudley 2000).

Compared to Melayu people along the Borneo coast and in Peninsular Malaysia, however, the Melayu in the DSNP area still maintain strong links with the forests, as they still harvest timber and minor forest products, and practice some form of shifting cultivation. Forests are heavily utilized, both for construction timber and for a wide variety of non-timber forest products. Most (87%) of forest plant species are utilized locally for traditional medicines and other daily uses (Giesen 1987; Giesen and Aglionby 2000).

The lakes and swamp forests act as a buffer for the Kapuas River mitigating floodwaters (absorbing ¼ of peak floods) and maintaining water levels in the dry season by providing 50% of dry season flow in the upper Kapuas River. The larger rivers in DSNP (Tawang, Leboyan, Sekulat, Bunut) serve as a means of transportation, uniting lakes in the Park and connecting wetland ecosystems with the Kapuas River.

Carbon stocks are very high in DSNP's peat swamp forests. Aboveground carbon stock (AGC) in Nung peat forest is well correlated with tree diameter, and in an analysis of >1,000 individual trees from 31 families, this was found to average at about 85 tCha^{-1} (Anshari et al. 2012). The belowground carbon stock in peat forest is much higher than the AGC and depends on peat depths, carbon fraction, and bulk density. At Nung peat swamp forest at DSNP, it was found to range from 2,000 to 10,000 t C ha^{-1} (Anshari 2010; Anshari et al. 2012).

Threats and Future Challenges

The main threats to the integrity of DSNP are fires, habitat conversion along the periphery, and direct over-exploitation of resources. Land use change in the upper Kapuas region is still occurring at a rapid rate, and after the timber boom (which ended in the 1980s), the government widely supported the development of oil palm plantations. This also occurred around DSNP where at present there are about 20 companies that extend over 300,000 ha (Wadley et al. 2000; Giesen 2012). These large-scale oil palm plantations form the greatest threat to the Park. Pollutants from oil palm estates around DSNP flow into the Park's waterways and end up in swamp forests, lakes, fish, and wildlife. Also, land cover change due to plantation development has contributed to significant hydrological changes at DSNP over the past decade. Instead of a single 1–3-month period during which the lakes fall dry, this drying out now occurs more often (twice) and for briefer periods. As seasonal water dynamics are important for fish breeding, rapid water fluctuation not only disturbs seasonal fishing activities but also jeopardizes fish reproduction.

Danau Sentarum has a long history of burning that goes back to >30,000 years BP (Anshari et al. 2001), and most fires are caused by human interventions (Giesen 1987, 1996). However, a marked increase can be noted since 1990 (Anshari et al. 2001; Dennis et al. 2000a, b), with burn scars increasing from 5,483 ha in 1973 to 18,905 ha in 1997 (Dennis et al. 2000b). As a result, recently burnt areas and swamp forest regenerating after fires now together account for about 20% of the Park. The main direct causes are fishing and forest exploitation, and the increase is explained by increasing human pressures and conflicts over resources (Dennis et al. 2000b).

Except for colonial waterfowl, exploitation levels of various resources were fairly sustainable until about two to three decades ago. Since then, however, the resource base appears to have been steadily eroding, with fish catches declining and tall forest area dwindling. The reasons for this are complex, involving the influx of migrants, increased nonadherence to local customary law, population increase, increased access to external markets (road construction), and a steady development of adjacent areas (e.g., by large-scale logging and plantation companies; Giesen 2012).

A recurrent potential threat is the plan to construct a dam on the Tawang River at the outlet of the lakes, on the misguided concept that this would improve fisheries. This plan has resurfaced repeatedly over the past decades, but if implemented, it would certainly be a major disaster for the Park, negatively affecting its biodiversity as well as its ecosystem services, including fisheries, transportation, and Kapuas River buffering capacity.

> **Box 1: Blackwater**
> By Wim Giesen
>
> Blackwater – in an ecological sense[a] – refers to surface waters that are darkly colored by humic acids (tannins and other polyphenols) leaching from decaying plant matter, the result being water that looks black and tea colored from above, but appears clear when held in a glass. Transparency is often high. Blackwaters usually drain from catchments such as peat domes, podzols, or sandstone hills, which are poor in nutrients and minerals. As a result, blackwaters are themselves nutrient poor and deficient in ions, especially of magnesium and calcium, but also others such as sodium and potassium. Electroconductivity is low (commonly 8–15 µS) and close to that of rainwater, while the pH is commonly acidic and around pH 4–5. Blackwater rivers may be biodiverse, especially of certain taxa such as rotifers, but with lower densities than nutrient-rich waters. Taxa requiring calcium for shells and exoskeletons, for example, such as crustaceans and mollusks, are much less common, as are fish that depend on these for food.
>
> The term has been especially used in South America, where major rivers are classified as blackwater (e.g., Rio Negro), white water (e.g., Solimões River, main course Amazon River), or clear water (e.g., Branco River, Xingu River; Junk et al. 2011; Goulding et al. 2010). A comparison between these is summarized below in Table 2. The term blackwaters has also been applied to Southeast Asian waters, especially those draining peatlands (Janzen 1974; Giesen 1987; Moore et al. 2011). Janzen (1974) proposes that blackwaters are rich in humic acids because the vegetation growing on these poor soils is exceptionally rich in secondary compounds[a] especially those that act as a deterrent for herbivores. Dark-colored humic acids may only be the most conspicuous of the secondary compounds present in blackwaters. This is supported by other studies, such as Yule and Gomez (2009), who found in Peninsular Malaysia that sclerophyllous, toxic leaves of endemic peat forest plants (*Macaranga pruinosa, Campnosperma coriaceum, Pandanus atrocarpus, Stenochlaena palustris*) were barely decomposed by bacteria and fungi (decay rates of only 0.0006–0.0016 k day − 1), while leaves of *M. tanarius*, a secondary forest species, were almost completely decomposed (decay rates of 0.0047–0.005 k day − 1) after 1 year.

Table 2 Ecological attributes of Amazonian white water, blackwater, and clear water rivers (Adapted from Junk et al 2011, Table 1; Wetlands, A classification of major naturally-occurring Amazonian lowland wetlands, 31, 2011, 623–40, Junk WJ, Fernandez Piedade MT, Schöngart J, Cohn-Haft M, Adeney JM, Wittmann F; © 2011, Society of Wetland Scientists, with permission of Springer Science + Business Media)

Ecological attributes	White water	Blackwater	Clear water
pH	Near neutral	Acidic, <5	Variable, 5–8
Electroconductivity (µS)	40–100	<20	5–40
Transparency (Secchi depth)	20–60 cm	60–120 cm	>150 cm
Water color	Turbid	Brown	Greenish
Humic substances	Low	High	Low
Inorganic suspensoids	High	Low	Low
Relationship of alkali earth (Ca, Mg) and alkali (Na, K) cations	Ca, Mg > Na, K	Na, K > Ca, Mg	Variable
Dominating anions	CO3-	SO4 –, Cl-	Variable
Fertility of substrate and water	High	Low	Low to intermediate

^aIn an urban, residential, or industrial setting, blackwater refers to wastewater containing fecal matter and urine and is a synonym for foul water or sewage

References

Anshari G. Carbon content of the freshwater peat swamp forests of Danau Sentarum. Borneo Res Bull. 2010;41:62–73.

Anshari G, Kershaw AP, Van der Kaars S. A late Pleistocene and Holocene pollen and charcoal record from peat swamp forest, Lake Sentarum Wildlife Reserve, West Kalimantan, Indonesia. Palaeogeogr Palaeoclimatol Palaeoecol. 2001;171:213–28.

Anshari G, Kershaw AP, Van der Kaars S, Jacobsen G. Environmental change and peatland forest dynamics in the Lake Sentarum area, West Kalimantan, Indonesia. J Quat Sci. 2004;19:637–55.

Anshari G, Gusmayanti E, Afifudin M, Widhanarto G. A study of carbon balance in customary peat forest (Hutan Nung) in Taman Nasional Danau Sentarum. Pontianak: Universitas Tanjungpura; 2012. Final Report. Submitted to Center for International Forestry Research (CIFOR). Project code: ENV01100-USA19-DM3.

Colfer CJP, Salim A, Wadley RL, Dudley RG. Understanding patterns of resource use and consumption. Borneo Res Bull. 2000;31:29–88.

Dennis R, Erman A, Meijaard E. Fire in the Danau Sentarum landscape: historical, present perspectives. Borneo Res Bull. 2000a;31:123–37.

Dennis R, Erman E, Stolle F, Applegate G. The underlying causes and impacts of fires in South-east Asia Site 5. Danau Sentarum, West Kalimantan Province, Indonesia. Bogor: Center for International Forest Research (CIFOR); 2000b. 58 p.

Dudley R. The fishery of Danau Sentarum. Borneo Res Bull. 2000;31:261–306.

Frazier S. Crocodiles of Danau Sentarum. Borneo Res Bull. 2000;31:307–22.

Giesen W. Danau Sentarum wildlife reserve-inventory, ecology and management guidelines. Bogor: World Wildlife Fund Report for the Indonesian Directorate General of Forest Protection and Nature Conservation (PHPA); 1987. 284 p.

Giesen W. Habitat types and their management: Danau Sentarum wildlife reserve, West Kalimantan, Indonesia. Bogor: Wetlands International-Indonesia Programme/PHPA; 1996. 97 p.

Giesen W. Flora and vegetation of Danau Sentarum: unique lake and swamp forest ecosystem of West Kalimantan. Borneo Res Bull. 2000;31:89–122.

Giesen W. The heart of Borneo's swamp forests. In: Wulffraat S, editor. The environmental status of the heart of Borneo. WWF Indonesia, Jakarta: Heart of Borneo Programme; 2012. p. 32–3.

Giesen W, Aglionby J. Introduction to Danau Sentarum National Park, West Kalimantan. Borneo Res Bull. 2000;31:5–28.

Goulding M, Barthem R, Ferreira E. Atlas of the Amazon. Washington, DC: The Smithsonian Institution; 2010. 253 p.

Janzen DH. Tropical blackwater rivers, animals and mast fruiting of the Dipterocarpaceae. Biotropica. 1974;6(2):69–103.

Jeanes KW. A biophysical profile of Danau Sentarum wildlife reserve. Indonesia – UK-Tropical Forest Management Programme: Project 5 – Conservation, Wetlands International – DFID; 1997.

Jeanes K, Meijaard E. Danau Sentarum's wildlife, part 1: biodiversity value and global importance. Borneo Res Bull. 2000;31:150–229.

Junk WJ, Fernandez Piedade MT, Schöngart J, Cohn-Haft M, Adeney JM, Wittmann F. A classification of major naturally-occurring amazonian lowland wetlands. Wetlands. 2011;31:623–40.

Klepper O, Asmoro PB, Suyatno N. A hydrological model of the Danau Sentarum floodplain lakes, Kalimantan (Borneo), Indonesia. Tropical limnology, vol. 2. Tropical lakes and reservoirs; 1995. p. 51–8.

Kottelat M, Widjanarti E. The fishes of Danau Sentarum National Park and the Kapuas Lakes Area, Kalimantan Barat, Indonesia. Raffles Bull Zool. 2005;18(Suppl. No. 13):139–73.

Meijaard E, Nijman V. Distribution and conservation of the proboscis monkey (*Nasalis larvatus*) in Kalimantan, Indonesia. Biol Conserv. 2000;92:15–24.

Moore S, Gauci V, Evans CD, Page SE. Fluvial organic carbon losses from a Bornean blackwater river. Biogeosciences. 2011;8:901–9.

Russon AE, Meijaard E, Dennis R. Declining orangutan populations in and around Danau Sentarum. Borneo Res Bull. 2000;31:372–84.

Russon AE, Erman A, Dennis R. The population and distribution of orangutans (*Pongo pygmaeus pygmaeus*) in and around the Danau Sentarum Wildlife Reserve, West Kalimantan, Indonesia. Biol Conserv. 2001;97:21–8.

Sebastian AC. Proboscis monkeys in Danau Sentarum National Park. Borneo Res Bull. 2000;31:359–71.

Van Balen S, Dennis R. Birds of Danau Sentarum. Borneo Res Bull. 2000;31:336–58.

Wadley RL, Dennis RA, Meijaard E, Erman A, Valentinus H, Giesen W. After the conservation project: Danau Sentarum National Park and its vicinity – conditions and prospects. Borneo Res Bull. 2000;31:385–401.

Walter O. A study of hunting and trade of freshwater turtles and tortoises. Borneo Res Bull. 2000;31:323–35.

Yule CM, Gomez LN. Leaf litter decomposition in a tropical peat swamp forest in Peninsular Malaysia. Wetl Ecol Manag. 2009;17:231–41.

Wetlands of Tasek Bera (Peninsular Malaysia)

156

R. Crawford Prentice

Contents

Introduction	1852
Location	1853
Physical Environment	1853
Formation of Tasek Bera	1856
Habitats and Flora	1856
Flora	1859
Fauna	1859
Key Species	1860
Local History and Cultural Heritage	1861
Land Use	1861
Factors Influencing Site Management	1862
Future Challenges	1863
References	1863

Abstract

Tasek Bera is an alluvial peat swamp ecosystem including a range of freshwater and lowland forest habitats located in the central lowlands of Peninsular Malaysia. The wetland system consists of a dendritic complex of inflowing streams and swamps whose water levels fluctuate markedly, rising by 1-5m during the monsoon periods. This dynamic flooding regime is an important underlying factor in the ecology of the wetland. The catchment area is around 61,380 ha, while the Ramsar Site of 38,446 ha includes 6,830 ha of wetland habitats supporting diverse flora and fauna including 94 fish and 10 turtle species and many globally threatened species. Its catchment has been occupied for over 600 years by the Semelai Orang Asli (aboriginal people), who traditionally practice

R. C. Prentice (✉)
Nature Management Services, Histon, Cambridge, UK
e-mail: Crawford.prentice@gmail.com

© Springer Science+Business Media B.V., part of Springer Nature 2018
C. M. Finlayson et al. (eds.), *The Wetland Book*,
https://doi.org/10.1007/978-94-007-4001-3_196

shifting cultivation of hill rice combined with collection of forest and wetland products, although their livelihoods are changing. The site was relatively unknown until the early 1970s, when a detailed ecological study was conducted before major clearance of lowland forest took place for the establishment of rubber and oil palm plantations. During the last 40 years, the surface layer of sediments has been enriched by clay minerals in response to disturbance of the wetland environment, especially plantation development, which may eventually lead to the termination of peat accumulation processes. Tasek Bera was designated as Malaysia's first Ramsar Site in 1994, with its management subsequently established with assistance from an international project from 1996 to 1999. While the area has remained protected, water levels have reportedly declined with more prolonged dry periods in recent years, likely attributable to the continuing development of the catchment area. In order to stabilize or reverse these trends, improved catchment management is needed based on systematic monitoring and research on the wetland's hydrological regime and water quality.

Keywords
Peat swamp forests · Semelai orang asli · Sedge · Pandanus · Wetland fauna · Oil palm plantation

Introduction

Tasek Bera is an alluvial peat swamp ecosystem including a range of freshwater and lowland forest habitats located in the central lowlands of Peninsular Malaysia. Its catchment has been occupied for over 600 years by the Semelai group of Orang Asli (aboriginal people), who traditionally practice shifting cultivation of hill rice combined with collection of forest and wetland products, and Tasek Bera remains the center of their population today, although their livelihoods are changing (Gianno 1990; Gianno and Bayr 2009).

Tasek Bera was relatively unknown until the early 1970s, when a detailed ecological study of the wetland was conducted under the auspices of the International Biological Program as a collaborative effort by Malaysian and Japanese scientists (Furtado and Mori 1982). This provided valuable baseline ecological information on Tasek Bera, before major clearance of lowland forest took place in the catchment for the establishment of rubber and oil palm plantations, with consequent impacts on the wetland.

Tasek Bera was designated as Malaysia's first Wetland of International Importance (Ramsar Site 712), when Malaysia joined the Ramsar Convention on 10 November 1994 (Ramsar Convention Secretariat 1998). An international project supported financially by Danish Cooperation for Environment and Development (DANCED), *The Integrated Management of Tasek Bera – Support for the*

Implementation of Obligations under the Ramsar Convention, was executed from 1996 to 1999 by the Asian Wetland Bureau (now part of Wetlands International) and the State Government of Pahang. A wide range of baseline surveys were conducted to support the formulation of a site management plan. Subsequently, the area has been managed by the government, with some small-scale projects aiming to assist ecotourism and other livelihoods among the Semelai community.

Location

Tasek Bera (02°58′00″N 102°36′00″E; elevation 30–35 m) is located in the central lowlands of Peninsular Malaysia, within the catchment of the Pahang River, the Peninsula's largest river (Fig. 1). The wetland lies entirely within Bera District, in the state of Pahang Darul Makmur. The wetland system consists of a dendritic complex of inflowing streams and swamps, measuring 34.6 km long by 25.3 km wide. The catchment area of Tasek Bera is around 61,380 ha, while the Ramsar Site of 38,446 ha includes 6,830 ha of wetland habitats (WIAP 1999).

Physical Environment

The climate is humid tropical, with two monsoon periods. Annual rainfall varies between 1,200 and 2,500 mm, following a bimodal pattern with wetter periods during the southwest monsoon in March to May and the more extensive northeast monsoon during September to January.

Tasek Bera lies in a subcatchment of the Bera River basin (Fig. 2), although its southern reaches drain southward into the Muar River basin. Water from Tasek Bera generally drains northward along a main channel to a single confluence with the Bera River (which originates to the northeast of the site), which then flows northward into the Pahang River and on eastward to the South China Sea. Water levels average 0.8 m in the littoral region, 2.0 m in lakes, and 2.5 m in the main channel (with a maximum depth of 7.0 m near the outlet). However, water levels fluctuate markedly, rising by 1–5 m during the monsoon periods in response to local rainfall. This dynamic flooding regime is an important underlying factor in the ecology of Tasek Bera (Furtado and Mori 1982; Giesen 1998; WIAP 1999). However, in recent years water levels have reportedly become lower with more prolonged dry periods, likely attributable to the continuing development of plantations in the catchment (Chong 2007).

Furtado and Mori (1982) provide a detailed assessment of the wetland's water quality characteristics while it was still in a relatively undisturbed condition during 1970–1974. Features included relatively low pH, alkalinity, conductivity, and sodium, calcium, and magnesium content. Phosphorus and nitrogen content was strongly influenced by runoff during the seasonal monsoon periods. No such detailed work has since been conducted, but the limited information available points toward a degradation of water quality associated with development of plantations in the catchment (Chong 2007).

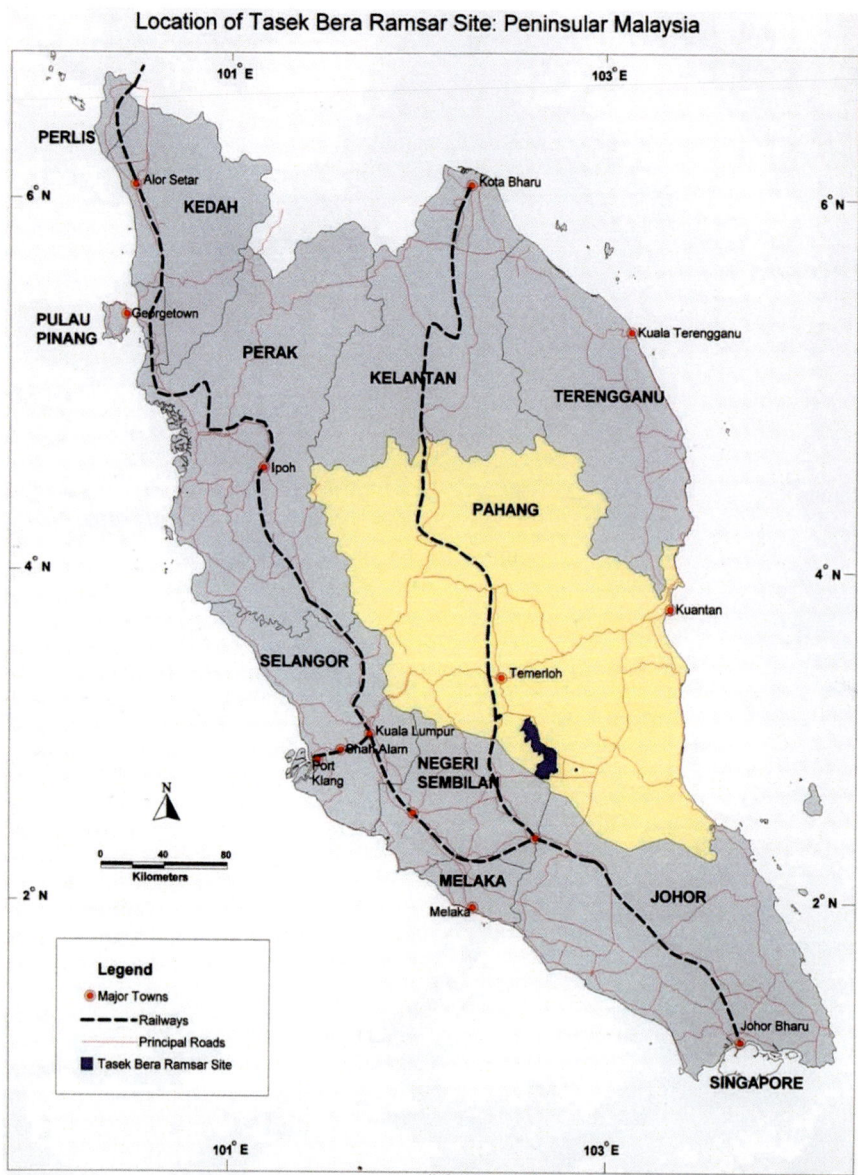

Fig. 1 Location of Tasek Bera Ramsar Site (Published with permission: Wetlands International – Asia Pacific 1999)

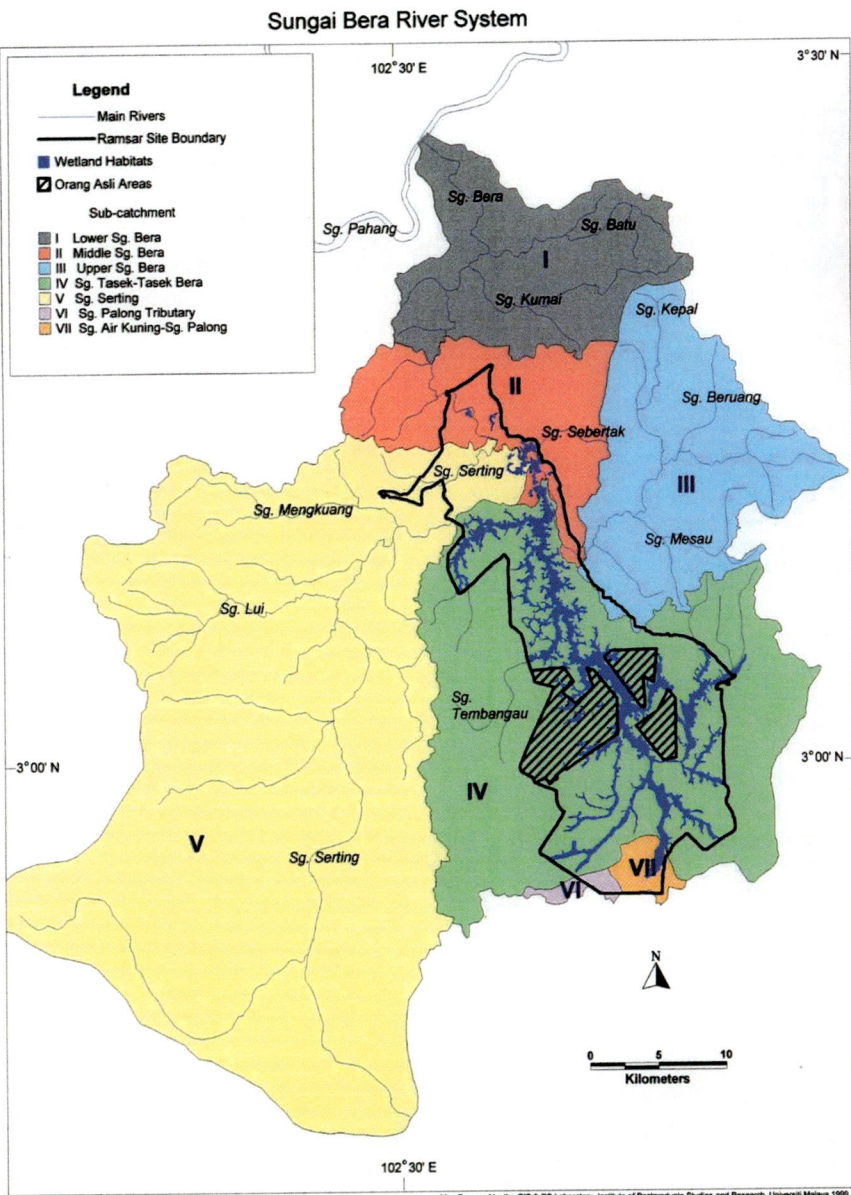

Fig. 2 The Bera River system showing Tasek Bera Ramsar Site (Published with permission: Wetlands International – Asia Pacific 1999)

Formation of Tasek Bera

Tasek Bera lies on the historical course of the Pahang River, which flowed south and westward to the Malacca Straits. This river was "captured" by the Lower Pahang River which flows eastward to the South China Sea. Subsequent blocking of drainage resulted in the formation of the alluvial peat swamp we now know as Tasek Bera. There is a mosaic of peat and mineral soils underlying Tasek Bera. Peat dominates in the central region and along the main channel, while mineral soils dominate the margins. These peat soils vary greatly in depth, averaging 1–3 m, with a maximum depth of 7 m. Underneath the peat, there are clay, silts, and silty sand.

The development of the peat swamp and its soils are described in Furtado and Mori (1982) and more recently by Phillips and Bustin (1995) and Wust and Bustin (1999, 2004). Wet climatic conditions in SE Asia around 6000 years BP favored organic matter accumulation in poorly drained areas. The formation of the Tasek Bera peat deposits started with the deposition of impervious fine clays that favored the formation of a lake system, followed by initial damming of the basin by plants such as *Pandanus* spp. This led to the development of various peat sediments and the formation of peat swamp forest. During the last 600 years or so, the area of *Pandanus–Lepironia* swamp has greatly expanded, probably due to the arrival of the Semelai in the area. During recent decades, there is evidence that burning of peat swamp forest has resulted in the present mixture of open water, *Pandanus* clumps, extensive sedge beds, and swamp forest patches, likely the same process as has been occurring over the 600 years, accounting for the loss of one third of the original swamp forest area. This process of change is still ongoing due to continued use of fire for turtle hunting and to keep navigation channels open (Giesen 1998). During the last 40 years, the surface layer of sediments has been enriched by clay minerals in response to disturbance of the wetland environment, especially the development of plantations in the catchment, which may eventually lead to the termination of peat accumulation processes.

Habitats and Flora

Tasek Bera is an inland riverine swamp which supports a biological community unique within Malaysia and possibly represented nowhere else. It is one of the few major natural bodies of freshwater in Peninsular Malaysia. The peat swamp forest here is unique in Peninsular Malaysia (Giesen 1998), growing on topogenous peat formations (geogenous peat with a virtually horizontal water table, located in basins; peat that has originated as a result of the features of an area (Joosten and Clark 2002)), while the large coastal peatlands have ombrogenous peat (receiving water and nutrients only from the atmosphere through precipitation).

Of the 6,830 ha of wetlands inside the Ramsar Site, the main habitats comprise freshwater and peat swamp forests (about 5,400 ha, 79%), transitional open-forested swamps (about 510 ha, 7%), *Pandanus helicopus* swamps and *Lepironia articulata* reed beds (about 800 ha, 12%), and open water (about 120 ha, 2%) with beds of a

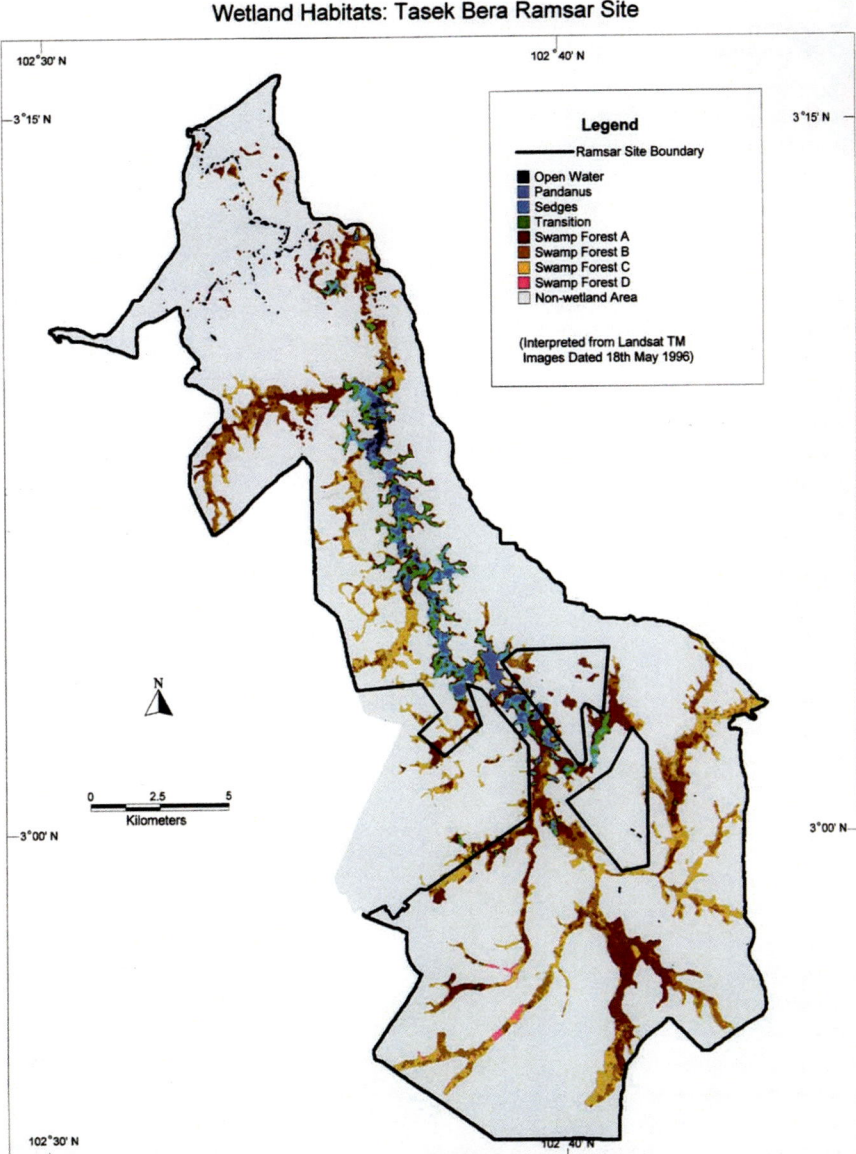

Fig. 3 Wetland habitats at Tasek Bera (Published with permission: Wetlands International – Asia Pacific 1999)

highly diverse algal community and submerged macrophytes (Fig 3). Lowland forests cover most of the immediate surroundings of the wetland and lie largely within the Ramsar Site (Fig. 4). The only lowland forest remaining in the catchment outside the Ramsar Site occurs in the headwaters of Sungai Bera to the northeast.

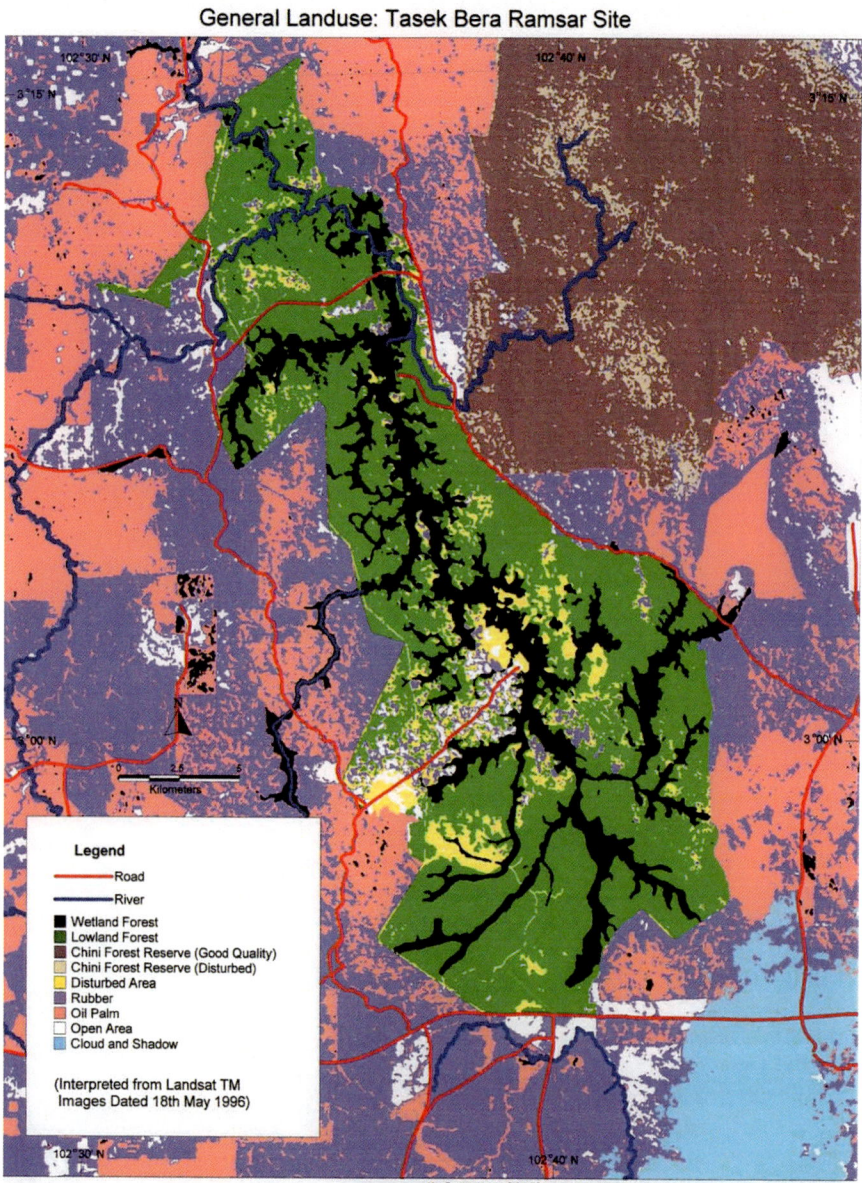

Fig. 4 Map showing land use at Tasek Bera (Published with permission: Wetlands International – Asia Pacific 1999)

The lowland forests have been extensively disturbed by shifting cultivation and commercial logging (WIAP 1999).

Flora

A total of 807 species of vascular plants in 400 genera and 128 families have been recorded from Tasek Bera (Giesen 1998; Rafidah et al. 2010). Of these, at least 11 species are endemic to Peninsular Malaysia, including the purple water trumpet *Cryptocoryne purpurea* which is only likely to exist at Tasek Bera and the pandan *Pandanus immersus* which is only known from one other site. In Malaysia, the rare sedge *Scirpus confervoides* is only known from Tasek Bera and one other site. Twelve species of aquatic plants have been recorded, including three species of bladderwort *Utricularia*. One of these, *U. punctata*, has disappeared at Tasek Chini due to the introduction of *Cabomba furcata* and its survival may now depend on Tasek Bera.

Previous studies in the 1970s (Furtado and Mori 1982) recorded some 328 species, varieties, and forms of algae at the site, although more recent studies found fewer species, indicating a shift in species composition related to changes in water quality.

Tasek Bera is largely free of exotic aquatic weeds, although water hyacinth *Eichhornia crassipes* and the invasive shrub *Mimosa pigra* have been recorded in the catchment area, and the aquatic plant *Hydrilla verticillata* occurs in the wetlands.

Fauna

Studies in the wetlands and surrounding lowland forests of Tasek Bera Ramsar Site have revealed a great diversity of animal life, although further studies will certainly extend these findings (see Ramsar Convention Secretariat 1998; WIAP 1999). Recorded representation of selected taxonomic groups is given below.

Group	No. species recorded
Mammals	67
Birds	230
Crocodiles	1
Turtles	10
Lizards	15
Snakes	18
Amphibians	19
Fish	94
Dragonflies and damselflies (Odonata)	78
Crabs	2
Shrimps	2
Zooplankton (taxa)	64

The relatively high diversity of some groups (e.g., fish, turtles, dragonflies, and damselflies) is a direct reflection of the diversity of wetland habitats, while the composition of other groups (e.g., mammals, birds, snakes, lizards) is more reflective of the lowland forest but includes some wetland specialists. Some aquatic groups are poorly represented (e.g., zooplankton, molluscs) due to the low pH value, softness, and low calcium content of the water (Furtado and Mori 1982).

Key Species

Tasek Bera supports a large number of globally threatened species*, including the Asian elephant *Elephas maximus (E)*, Sunda otter civet *Cynogale bennettii* (E), Malayan tapir *Tapirus indicus* (E), tiger *Panthera tigris* (E), clouded leopard *Neofelis nebulosa* (V), flat-headed cat *Prionailurus planiceps* (E), smooth-coated otter *Lutrogale perspicillata* (V), Asian small-clawed otter *Aonyx cinerea* (V), sun bear *Helarctos malayanus* (V), lar gibbon *Hylobates lar* (E), siamang *Symphalangus syndactylus* (E), Sunda pangolin *Manis javanica* (C), lesser adjutant stork *Leptoptilos javanicus (V)*, masked finfoot *Heliopais personatus* (E), Malay crestless fireback *Lophura erythrophthalma* (V), Southeast Asian narrow-headed softshell turtle *Chitra chitra* (C), Southeast Asian box turtle *Cuora amboinensis* (V), black pond turtle *Siebenrockiella crassicollis* (V), Malaysian giant turtle *Orlitia borneensis* (E), sunburst turtle *Heosemys spinosa* (E), Burmese mountain tortoise *Manouria emys* (E), Southeast Asian softshell turtle *Amyda cartilaginea* (V), Asian giant softshell turtle *Pelochelys Cantorii* (E), king cobra *Ophiophagus hannah* (V), and Asian arowana *Scleropages formosus* (E). At least three species have become locally extinct, including the banteng *Bos javanicus* (E), dhole *Cuon alpinus* (E), and silver shark *Balantiocheilos melanopterus* (E), and the Malayan false gharial *Tomistoma schlegelii* (V) is now extremely rare. In addition, the area supports species that are endemic to Peninsular Malaysia, including the dusky leaf monkey *Presbytis obscura* and Malaysian peacock-pheasant *Polyplectron malacense* (V) as well as a significant number of fish and invertebrate species (Ramsar Convention Secretariat 1998; WIAP 1999; Sim et al. 2002; Yeap et al. 2004).[1]

The great diversity of freshwater fish (94 species, almost all endemic to Peninsular Malaysia) is a primary value of Tasek Bera (e.g., see Sim 2002), and the wetland appears to play a role in providing breeding, nursery, or feeding grounds for fish from the Pahang River system (e.g., the giant catfish *Wallagonia leerii*). Some 30 species are considered food fish and 50 species as aquarium fish (including the Asian arowana, tiger barb *Puntius partipentazona*, harlequin *Rasbora*

---

[1] Critically endangered (C), endangered (E), and vulnerable (V) species are listed here, following www.iucnredlist.org.

heteromorpha, etc.). Similarly, the occurrence of ten species of turtles is of great conservation importance, with a need for monitoring and control of exploitation.

Local History and Cultural Heritage

Tasek Bera has been inhabited for over 600 years by indigenous peoples, currently the Semelai. This *Orang Asli* group has a strong cultural attachment to Tasek Bera and still partially depends on its natural resources. Sites of cultural and historical interest are distributed throughout the Ramsar Site (Zaiton 1997), reflecting the long history of occupation by the Semelai. Their culture, traditions, and socioeconomy have been studied by various authors (e.g., Hood 1978; Gianno 1990; Mohd Shawahid 1997; Gianno and Bayr 2009; Mohamad 2010).

There are at present about 4,000 Semelai and about 1,500 of them living at Tasek Bera, mostly in settlements around Pos Iskandar, the main settlement established during the Communist Emergency in Peninsular Malaysia in 1948–1960. Before the Emergency, the Semelai lived in scattered communities throughout the area and practiced swidden cultivation, hunting, forest product collection, and fishing, according to established cultural traditions. Following congregation at Pos Iskandar, they have had to commute to swidden fields and rubber smallholdings or live away from the facilities of the main settlements. Collection of natural resources such as tapping resin trees and collection of rattans, fishing, and handicraft production have declined as main economic activities, but are still supplementary livelihoods, especially in outlying villages.

The Government has encouraged the resettlement of outlying villages to the Pos Iskandar Resettlement Scheme in order to provide development and welfare assistance to the community. At present, most of the Semelai population at Tasek Bera is resident in the three areas covered by this scheme, a systematic agricultural land development program, where basic infrastructure and facilities such as clinics and schools as well as a smallholders' rubber plantation have been provided. Rubber tapping now provides a basic income for many households, reducing dependence on natural resources.

Land Use

The Tasek Bera area is almost entirely under State ownership. The Ramsar Site (38,446 ha) is gazetted as a forest reserve, while the buffer zone (71,500 ha) consists mostly of rubber and oil palm plantations on government-owned land (see Fig. 4). The buffer zone also includes the upper Bera River catchment that lies within adjacent forest reserves, which have been logged in the past and remain production forest. The remainder of the buffer zone consists of three Orang Asli areas within the Ramsar Site (see Fig. 1, totaling 5,120 ha). The Semelai continue to clear small blocks of forest in and around the Ramsar Site for shifting cultivation and rubber smallholdings (Ramsar Convention Secretariat 1998; WIAP 1999).

The Ramsar Site is managed by the federal Department of Wildlife and National Parks, in collaboration with the state Departments of Forestry, Environment, Fisheries, and Orang Asli Affairs. The site has received a lot of attention regarding tourism development. Currently one resort is operating, and a visitor center provides information on Tasek Bera's ecology, culture, and services. Ecotourism activities such as homestays, guided jungle trekking, camping, canoeing, etc., have been developed with external assistance, operated by the Semelai Association for Boating and Tourism. The area is also used for environmental education and research.

Factors Influencing Site Management

There are a number of factors that influence site management, including natural and human-induced factors and management constraints which were recognized in the first site management plan (WIAP 1999) informed by studies such as Giesen (1998). Natural processes include the evolution of the peat swamp through peat development (see above), which in turn is dependent upon suitable climatic and hydrological conditions, as well as the quality of surface water runoff from the catchment. Natural vegetation succession in the long term should result in the development of climax forest across the entire site, including recovery of degraded forest areas. More extreme dry periods due to climate change may result in lower water levels in the wetland and increased fire incidence.

Human-induced factors most critically concern the gross modification of the catchment area by forest clearance and the establishment and management of oil palm and rubber plantations, including periodic replanting of crops. An overall accumulation rate of eroded soils within the wetlands and open waters has been determined to be 1.025 cm year^{-1} since 1995 (Gharibreza et al. 2013). Impacts on the wetland include eutrophication due to fertilizer runoff, pollution by pesticides, changes to the seasonal flooding pattern, and severe soil erosion causing sediment deposition on the lake bed (e.g., Gharibreza and Ashraf 2014), accelerating the spread of *Pandanus* and *Lepironia* reed beds. Sediment loading and eutrophication also affect vegetation in the river channels, directly by smothering with silt or by encouraging algal growth which out-competes submerged plants.

Fire is a second major factor, traditionally used during the shifting cultivation cycle. However, in very dry conditions such as *El Niño*-induced droughts, such fires have potential to get out of control and destroy large tracts of forest (the El Niño–Southern Oscillation (ENSO) is an irregular periodic climate pattern that occurs across the tropical Pacific Ocean roughly every 5 years. Its extreme oscillations, El Niño and La Niña, cause floods and droughts in many regions of the world). Fires are also set to clear *Pandanus* and *Lepironia* swamp vegetation in order to hunt turtles and clear navigation routes. These fires can also spread and are considered responsible for the progressive replacement of swamp forest with open *Pandanus–Lepironia* swamp habitats at Tasek Bera.

Cutting of trees, poles, and bamboo and some illegal logging cause localized habitat disturbance, and selective removal of trees (e.g., *Aquilaria* spp.) may lead to their local extinction. Hunting, fishing, and collection may reduce populations of economic species such as mouse deer, wild pig, pythons, monitor lizards, and turtles. Illegal hunting by outsiders may deplete populations of prey species, e.g., mouse deer, barking deer, and possibly sambar, and consequently affect their predators (e.g., tiger). Illegal collection of fish (particularly *Scleropages formosus*) for the aquarium trade has depleted populations. Certain indiscriminate fishing techniques pose a threat to fish and crocodile populations, including fish poisoning carried out by outsiders. Finally, the introduction of alien invasive species of plants is a significant risk, and the culture of *Tilapia* spp. in cages along the Bera and Pahang rivers may result in colonization of Tasek Bera, with potential damage to valuable native fish communities.

Future Challenges

The factors described above are progressively impacting the ecological character of Tasek Bera's wetland and forest habitats, reducing its biological diversity (e.g., Chong 2007). In order to stabilize or reverse these trends, systematic monitoring and research on the wetland's hydrological regime and water quality are required to provide a scientific basis for planning and assessing remedial measures, such as improved watershed management involving the surrounding plantations.

The challenges of integrated management remain, with the need to involve a range of stakeholders in the development and implementation of site management plans, including catchment land managers, the indigenous Semelai communities, and agencies responsible for forest management, wildlife conservation, fisheries, environmental quality, water resource management, tourism, and district administration. While the Tasek Bera Ramsar Site would benefit from stronger legal protection than its existing forest reserve status to ensure its future integrity, the rights of the Semelai people for continued traditional use of its natural resources need to be maintained through formalized co-management mechanisms.

References

Chong G. Tasek Bera: past, present and future. In: Colloquium on lakes and reservoir management: status and issues, 2–3 August 2007, Putrajaya: NAHRIM; 2007. p35–40.

Furtado JI, Mori S, editors. Tasek Bera: the ecology of a freshwater swamp. The Hague/Boston/London: Dr. W. Junk Publ; 1982.

Gharibreza M, Ashraf MA. Applied limnology: comprehensive view from watershed to lake. Berlin: Springer; 2014.

Gharibreza M, Raj JK, Yusoff I, Othman Z, Tahir WZWM, Ashraf MA. Land use changes and soil redistribution estimation using 137 Cs in the tropical Bera Lake catchment, Malaysia. Soil Tillage Res. 2013;131:1–10.

Gianno R. Semelai culture and resin technology. Connecticut Acad Arts Sci Mem. 1990;22:1–238.

Gianno R, Bayr KJ. Semelai agricultural patterns: toward an understanding of variation among indigenous cultures in southern peninsular Malaysia. J Southeast Asian Stud. 2009;40(1):153–85.

Giesen W. The habitats and flora of Tasek Bera, Malaysia: an evaluation of their conservation value and management requirements. Integrated management of Tasek Bera Technical Report Series. Kuala Lumpur: Wetlands International – Asia Pacific; 1998.

Hood S. Semelai rituals of curing [dissertation]. Oxford: St Catherine's College, University of Oxford; 1978. 446p.

IUCN Red List of Threatened Species. http://www.iucnredlist.org. Accessed 10 Nov 2015.

Joosten H, Clark D. Wise use of mires and peatlands. Background and principles including a framework for decision-making. International Mire Conservation Group and International Peat Society; 2002. Available from http://www.gret-perg.ulaval.ca/fileadmin/fichiers/fichiersGRET/pdf/Doc_generale/WUMP_Wise_Use_of_Mires_and_Peatlands_book.pdf. Accessed 10 Nov 2015.

Mohamad S. The ethnobotany of the Semelai community at Tasek Bera, Pahang, Malaysia: an ethnographic approach for re-settlement [dissertation]. Adelaide: School of Architecture, Landscape Architecture and Urban Design, University of Adelaide; 2010. Accessed from https://digital.library.adelaide.edu.au/dspace/handle/2440/68557. Accessed 10 Nov 2015.

Mohd Shahwahid HO. Economics of fishing, natural product exploitation and shifting cultivation at Tasek Bera. Integrated management of Tasek Bera Technical Report Series. Kuala Lumpur: Wetlands International – Asia Pacific; 1997.

Phillips S, Bustin RM. Accumulation of organic rich sediments in a dendritic fluvial/lacustrine mire system at Tasik Bera, Malaysia: implications for coal. Department of Earth & Ocean Sciences, University of British Columbia, Vancouver; 1995. Available from http://cat.inist.fr/?aModele=afficheN&cpsidt=2231772. Accessed 10 Nov 2015.

Rafidah AR, Chew MY, Ummul-Nazrah AR, Kamarudin S. The flora of Tasek Bera, Pahang, Malaysia. Malay Nat J. 2010;62(3):249–306.

Ramsar Convention Secretariat. Ramsar wetland information sheet for Tasek Bera. 1998. Available from http://www.ramsar.wetlands.org/Database/Searchforsites/tabid/765/Default.aspx. Accessed 15 Jan 2013.

Ramsar Convention Secretariat. The annotated Ramsar list: Malaysia. Available from http://www.ramsar.org/cda/en/ramsar-pubs-notes-anno-malaysia/main/ramsar/1-30-168%5E16529_4000_0__. Accessed 15 Jan 2013.

Sim CH. A field guide to the fish of Tasek Bera Ramsar site, Pahang Malaysia. Kuala Lumpur: Wetlands International – Malaysia Programme; 2002. 104p.

Sim CH, Murugadas TL, Sundari R. A guide to the endangered and endemic flora and fauna of Tasek Bera Ramsar site, Pahang, Malaysia. Kuala Lumpur: Wetlands International – Malaysia Programme; 2002. 64p.

Wetlands International – Asia Pacific. Tasek Bera Integrated Management Plan (1999–2004). Kuala Lumpur: Wetlands International – Asia Pacific; 1999.

Wust RAJ, Bustin RM. Geological and ecological evolution of the Tasek Bera (Peninsular-Malaysia) wetland basin since the Holocene: evidences of a dynamic system from siliciclastic and organic sediments. Integrated management of Tasek Bera Technical Report Series. Kuala Lumpur: Wetlands International; 1999.

Wüst RAJ, Bustin RM. Late Pleistocene and Holocene development of the interior peat-accumulating basin of tropical Tasek Bera, Peninsular Malaysia. Palaeogeogr Palaeoclimatol Palaeoecol. 2004;211(3–4):241–70.

Yeap CA, Zubaid A, Prentice C, Lopez A, Davison GWH. Avifauna in a peat swamp forest at Tasek Bera, Malaysia's first Ramsar Site. In: Mansor M, Ali A, Rieley J. Ahmad AH, Mansor A, editors. Tropical peat swamps – safeguarding a global natural resource. Proceedings of the International Conference & Workshop on Tropical Peat Swamps, 1999, Pulau Pinang, Malaysia. Pulau Pinang: University Sains Malaysia; 2004. p. 59–67.

Zaiton S. Sites of cultural and historical interest at Tasek Bera. Integrated management of Tasek Bera Technical Report Series. Kuala Lumpur: Wetlands International – Asia Pacific; 1997.

Intertidal Flats of East and Southeast Asia

157

John MacKinnon and Yvonne I. Verkuil

Contents

Extent and Location	1866
Diversity of Wetland Types	1866
Conservation Status	1867
Management	1867
Threats	1868
Ecosystem Services	1869
Significant Biodiversity	1870
Human Dependence	1871
International Initiatives	1872
Future Challenges	1872
References	1872

Abstract

A recent rise in economic prosperity in Asia, the most densely populated region of the world, has created a shortage of land for industry, housing developments and aquaculture. Consequently, large extents of tidal flat habitat in East and Southeast Asia, and especially in the Yellow Sea, have been lost since 1980, some through sediment inflow reduction, some through reclamation to satisfy demand for land. Throughout the East Asian–Australasian Flyway (EAAF), over 600,000 ha of tidal flats were the subject of further proposed land claims in 2012; in the Yellow Sea, planned conversions of >300,000 ha would amount to

---

J. MacKinnon (✉)
University of Kent, Canterbury, UK
e-mail: arcbcjrm@gmail.com

Y. I. Verkuil (✉)
Conservation Ecology Group; Groningen Institute for Evolutionary Life Sciences (GELIFES), University of Groningen, Groningen, The Netherlands
e-mail: y.i.verkuil@rug.nl

© Springer Science+Business Media B.V., part of Springer Nature 2018
C. M. Finlayson et al. (eds.), *The Wetland Book*,
https://doi.org/10.1007/978-94-007-4001-3_51

a further loss of 40% of the remaining habitat. Here we articulate five arguments to contribute to convincing governments and other stakeholders in the EAAF that the current rate of loss is a disaster which must be urgently addressed. (1) Global responsibility: the EAAF is a large flyway supporting 176 waterbird species, of which 34 (19%) are globally threatened or Near Threatened. Nine more species are under consideration for such listing. Other flyways have 5–13 threatened species, amounting to 4–12%. (2) Regional responsibility: migratory shorebird species essentially make a single stop, or very few stops, when moving between non-breeding and breeding sites. In the EAAF, most of these critical sites where birds refuel for a few weeks are in the Yellow Sea. (3) Regional effects: shorebird population trends in Japan, and at a single wintering site in Australia showed that shorebirds dependent on the Yellow Sea during migration show the strongest population declines. (4) Local effects: migratory shorebirds that lost their fuelling site due to the largest land claim projects in the Yellow Sea (Saemangeum and Bohai Bay) did not all redistribute to the adjacent tidal flats, resulting in a net population decline. (5) Self-interest: Tidal flats and associated coastal ecosystems provide critical ecosystem services including protection from storm surges and sea level rise. This information was summarized in a 2012 IUCN report and subsequently EAAF governments have committed via IUCN Resolution 28 to protect the EAAF.

Keywords

East Asian-Australasian Flyway · Migratory waterbird · Habitat loss · Reclamation · Coastal protection · Coastal fishery · Population decline · Migratory species agreement

Extent and Location

The intertidal zone of East and Southeast Asia extends for 34,000 km from China and Korea, down along the coasts of Vietnam, Cambodia, Thailand, Peninsular Malaysia, and north around the coast of Myanmar to Bangladesh. An even greater length, 128,000 km, of coasts surrounds the islands of Japan and nations of Association of Southeast Asian Nations (ASEAN) – Philippines, Malaysia (East), Indonesia, Brunei, and Singapore. The intertidal zone is narrow – a few meters to a few hundred meters wide; the total area involved is very small, fragile, and rapidly vanishing. But this small area is of disproportionate significance in terms of biodiversity, ecosystem services, and human livelihoods.

Diversity of Wetland Types

The coasts of Asia range from cold temperate to tropical and offer a range of habitats from muddy mangrove to marshes and sand flats. Of particular value to wildlife are the tidal estuaries of some of Asia's great rivers. Sixteen key areas emerged from analysis as the most important for endangered waterbirds of which six in the Yellow Sea (MacKinnon et al. 2012; see also Battley et al. 2008; Barter 2006).

The East Asian coastline serves as a migration flyway (East Asian–Australasian Flyway (EAAF)) for the many species that nest in the northeastern Russia and Alaska. Large numbers of birds migrate annually through this flyway, to the nonbreeding areas in, for example, the Yangtze valley of China, or south to Australasia, e.g., Indonesia, New Zealand, and Australia.

Conservation Status

Current estimates of intertidal habitat loss in Asia equate to loss rates greater than or equal to global rates of declines of mangroves, tropical forests, and coral reefs (MacKinnon et al. 2012). Losses of up to 51% of coastal wetlands have occurred in China over the past 50 years, and in Singapore 76% of coastal wetlands have been reported lost (An et al. 2007). Loss of staging areas within migratory pathways, where birds must replenish their energy stores during migration, can have extreme consequences for shorebird populations (Conklin et al. 2016; Piersma et al. 2016; Myers et al. 1987; Buehler and Piersma 2008; Verkuil et al. 2012). For the millions of shorebirds that migrate through the East Asian–Australasian Flyway (EAAF), the intertidal areas of Asia are a crucial migratory bottleneck, and extreme habitat losses are driving major population declines in many of these species (Barter 2002; Rogers et al. 2010; Moores et al. 2016).

There are indications of serious problems along the migration flyway. Monitoring on beaches of Australia shows declines in the numbers of most flyway migrant shorebirds wintering there (Rogers et al. 2008; Wilson et al. 2011; Clemens et al. 2016). Analysis of monitoring data of Japanese shorebirds between 1975 and 2008 shows declines in most species but interestingly a much higher proportion among species that are dependent on Yellow Sea stopover sites (Amano et al. 2010).

Management

All countries of the region have well-developed protected area systems. Most countries are well over the 10% target proposed by the UN Convention on Biological Diversity (CBD). More specifically, most of the countries are parties to the Ramsar Convention of Internationally significant wetlands and have established country focal points for the protection and monitoring of wetlands, especially Ramsar Sites (BirdLife International 2005). So why cannot adequate intertidal zone habitat be acquired and protected?

Analysis of the impressive protected areas of the region (BirdLife International and IUCN 2007) reveals that there is a bias in establishment toward mountain reserves and inland wetlands but a significant lack of representation of lowlands and coastal and marine areas. Awareness of conservation needs is lowest, and due to reasons of demography and access, competition for coastal lands is greatest. Conservation agencies have low financing, limited resources resulting in weak protection, and management of existing sites combined with generally low political influence.

Threats

The fast pace and nature of human developments affecting this zone is not in harmony with the natural environment and jeopardizes both those species that depend on this zone and the valuable ecological services that intertidal zone ecosystems deliver. Shoreline viability and the health of bird populations are negatively affected by a wide range of threats and destructive processes.

- *Loss and fragmentation of habitat.* According to the China National Wetland Conservation Action Plan (2000), some 1.19 million ha of coastal tidal flats has been lost and 1 million ha of coastal wetlands has been urbanized or used for mining. This constitutes a loss of 50% of all China's coastal wetlands (CCICED 2010). Mangroves had decreased from ~50,000 ha in 1950 to 22,000 ha by 2000 – a 44% loss (Chen et al. 2009). Loss of coastal wetlands has continued and indeed accelerated during the following decade (see MacKinnon et al. 2012 for overview).
- *Damming of the major rivers* of the region leads to changes in silt discharge, seasonality and quality of freshwater discharge (Wang et al. 2010a; CCICED 2010).
- *Overuse of chemicals in agriculture* leads to excessive nitrogen in freshwater systems and growing threats from toxic algal blooms in many coastal reaches (CCICED 2010).
- *Pollution* due to industrial emissions, oil spills, wastewater, and sewage discharges both directly into the coastal zone and also into the rivers that flow into it (Li and Daler 2004).
- *Tidal energy developments*, which involve the construction of sea walls and tidal barrages, lead to direct loss of tidal flats. These developments also change nearshore tidal flows, which leads to increased impacts to siltation dynamics and damage to nearshore areas (Gill 2005).
- *Overharvesting* and overuse of intertidal resources, including fish, mollusks, sea cucumber, sea urchins, and seaweeds. The recent industrialization of harvesting methods has resulted in far greater harvests with less manual labor required, which is undoubtedly impacting ecosystem processes throughout the intertidal zone (CCICED 2010).
- *Hunting using* mist nets, fine fishnets, snares, poison, and guns is used on or adjacent to beaches throughout the region (e.g., Zöckler et al. 2010a).
- *Competition for food* by human fishermen together with associated disturbance by humans, boats, and dogs.
- *Anthropogenic climate change* leads to raised temperatures, sea levels, acidity, and reduced oxygen. Tropical cyclones and floods are becoming more frequent resulting in loss of many beaches and intertidal habitats and seasonal mismatch between migration times and habitat productivity.

Ecosystem Services

The intertidal zone has long provided a wealth of services to humans. Shorelines function as physical collecting zones of sand, mud, pebbles, and fringe vegetation that help slow and break the action of waves. Gentle beaches tame ocean waves providing safe places for villages, harbors, and towns and the protection of adjacent agricultural areas. The binding of sand, mud, and other sediments helps keep seas clean and productive and removes many pollutants from the air and water. Increasingly, these habitats are being recognized for their ability to store carbon (blue carbon) (Decho 2000).

Healthy strand vegetation, sea grass beds, algal beds, and mangroves provide significant shelter in the face of typhoons and storms and against the tsunamis that are frequent in a zone prone to devastating earthquakes. Coastal damage seen after the great tsunami in Aceh, Indonesia, in 2004 and again in Japan in 2011 revealed that sites protected by intact healthy coral, mangrove, or other coastal vegetation were dramatically less damaged than sites where these same habitats had been destroyed (Chang et al. 2006).

Intertidal habitats are among the most productive ecosystems on earth. Intertidal habitats, including tidal mudflats, tidal marshes, and mangroves, provide safe spawning areas and nurseries for countless species of fish and crustaceans on which coastal fisheries depend.

Constanza et al. (2014) valued the ecosystem services of the tidal marshes and mangroves globally at US$ 24.8 trillion per annum (or 20% of all global ecosystem services), and a 10 fold increase in value since 1997, due to factoring in their role in safeguarding human lives, property and crops from storms, tsunami and the like). More precise economic assessments of the values of these services need to be undertaken regionally. One preliminary study by the Korean Ocean Research and Development Institute (KORDI 2006) came up with the following estimates: annual value of a hectare of the ROK's intertidal habitats (US$ 32,660), which includes marine products (US$ 9,993), ecosystem preservation (US$ 8,548), habitat (US$ 7,533), water purification (US$ 3,702), recreation (US$ 1,443), and disaster prevention (US$ 1,442). Ecosystem service values for 170 km^2 of intertidal flats planned for reclamation in Xinghua Bay, Fujian, China, were estimated at US$ 0.65 billion/annum or US$ 38,235/ha/annum with an estimated loss of value of US$ 8,250/ha/annum if the land were reclaimed for agriculture or ponds (Yu et al. 2008). Given that there are more than 1 million ha of intertidal habitats in the Yellow Sea (including the Bohai Sea), these estimates point toward service values exceeding at least US$ 30 billion per annum.

An et al. (2007) estimate that the historical loss of 51% of China's coastal wetlands (not all intertidal) resulted in an annual loss of US$ 46 billion. The loss of ecosystem services caused by sea enclosures and land reclamation in China has been estimated at US$ 27.76 billion/annum (CCICED 2010).

Mangroves not only deliver a huge boon of services, but they also provide livelihood to large numbers of people. There are high social and financial losses when intertidal habitat is destroyed. These values are often not appreciated until they are lost (Wang et al. 2010b).

Significant Biodiversity

Asian intertidal habitats are vital for the survival of tens of millions of birds of more than a hundred species, as well as nesting beaches for endangered sea turtles, breeding areas for seals, spawning grounds for important economic fisheries, and homes of thousands of species of invertebrate crustaceans, worms, and mollusks. Many species which rely on intertidal habitats are in trouble; five regional species of intertidal sea grasses are globally threatened (Short et al. 2011), and the estuarine Indo-Pacific humpback dolphin is critically endangered. The clearest evidence of the high number of globally threatened species dependent on these habitats is among the birds, particularly waterbirds, with 24 globally threatened species among the shorebirds, waterfowl, spoonbills, cranes, seabirds, and pelicans (BirdLife 2005) that use Asian intertidal habitats and 9 more shorebird species under review to be listed.

At least 33 globally threatened/near-threatened birds occur (of which 24 depend on the tidal zone) with as many as nine additional shorebirds that may be added to these lists soon (see Appendix 1, in MacKinnon et al. 2012). The East Asian - Australasian Flyway is characterized by more threatened waterbird species than any other major migratory flyways, with less waterbird species listed as least concern and far more waterbird species listed as near threatened or globally threatened (Kirby 2010) (Fig. 1).

The fastest declining migratory shorebirds in the EAAF are the long-distance, Arctic-breeding migrants, such as the spoon-billed sandpiper *Eurynorhynchus pygmeus* (Amano et al. 2010; Zöckler et al. 2010b) and the red knot *Calidris canutus* (Wilson et al. 2011). At the current rate of decline (26% per annum), spoon-billed sandpipers could be extinct within the decade despite ongoing conservation action. Similarly, with the current rates of decline, for every 100 red knots migrating along the EAAF in 1992, only 7 will be left in 2020.

Of the 155 waterbird species in the flyway, at least *50* species of migratory shorebirds and *21* migratory gulls and terns in the flyway are strongly dependent on intertidal habitats. Fifteen globally threatened or near threatened migratory intertidal species, including the endangered spotted greenshank *Tringa guttifer* and the critically endangered spoon-billed sandpiper and Chinese crested tern *Sterna bernsteini*, have more than 95% of their entire global population in the EAAF; at least one species, entirely confined to the EAAF and currently listed as least concern, gray-tailed tattler *Heteroscelus brevipes*, is likely to be listed as threatened in the near future (Appendix 1 in MacKinnon et al. 2012). A further six migratory shorebird species, currently listed as least concern, also have more than 95% of their entire global population in the EAAF (sharp-tailed sandpiper *Calidris acuminata*, red-necked stint *C. ruficollis*, long-toed stint *C. subminuta*, Pacific golden plover *Pluvialis fulva*, oriental pratincole *Glareola maldivarum*, and Swinhoe's snipe *Gallinago megala*.

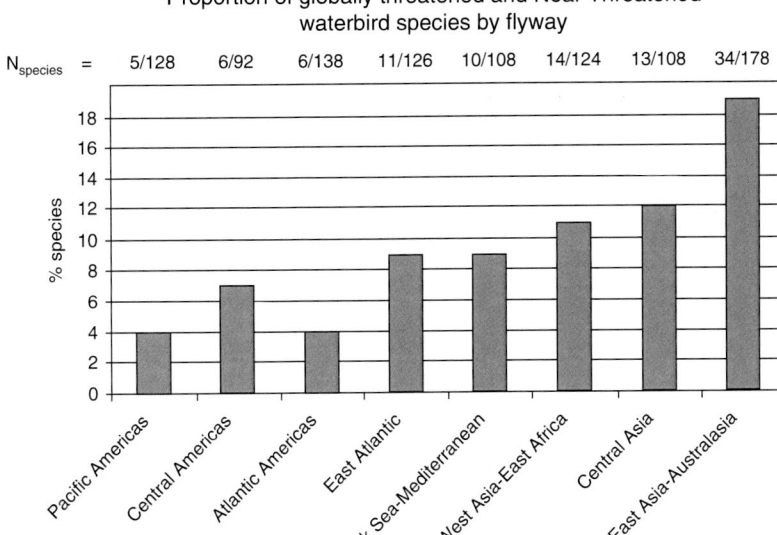

Fig. 1 Total number and proportion of globally threatened and near threatened waterbirds in the flyways of the world (Analysis and graphs reproduced from Kirby (2010))

Human Dependence

The loss of and degradation to intertidal flats are more than just the sad loss of our fascinating natural heritage but constitute a threat to the lives, health, safety, and welfare of hundreds of millions of our fellow humans, a threat to trillions of dollars worth of land and property and a risk to the health of the great oceans on which we all depend. The intertidal zone, with its sand and mud bars, beaches, and mangroves, delivers a vast list of valuable ecological services that are being discarded too causally in favor of nearsighted development goals (MacKinnon et al. 2012).

The bird declines are a sure warning that the productivity and health of the intertidal zone face an urgent crisis. Confirmation of these trends is paralleled in other taxa (WWF et al. 2006) and the growing frequency and scale of ecological disasters. Important stocks of economically important fish, crustaceans, shellfish, and cephalopods are all collapsing with associated loss of livelihood for hundreds of thousands of traditional fishermen. Sea mammals and turtles also show sharp declines. The frequency of toxic algal blooms increases dramatically; temperature, acidity, and water levels are rising (Nicholls and Cazenave 2010); the frequency of catastrophic storms increases and the damage to coastlines from cyclones and tsunamis becomes more serious wherever the natural coastline has been destroyed.

International Initiatives

In recognition of the fact that the problems transcend national boundaries, conservation efforts along the intertidal zone have been encouraged through the Convention on Migratory Species. Several bilateral migratory species agreements have been signed such as between Australia and China, Australia and Japan, and China and Japan. More recently, the activities of 13 countries and a number of international agencies (IUCN, WWF, Birdlife International, Royal Society for Protection of Birds (RSPB), Wildfowl and Wetlands Trust (WWT), and Wetlands International (WI)) have become coordinated through the establishment of the East Asian–Australasian Flyway Partnership (EAAFP) supported by the South Korean government. A recent situation analysis of the issue was commissioned by International Union for Conservation of Nature (IUCN), and a special motion was proposed and approved at the 2012 Jeju World Conservation Conference, urging countries along the flyway and especially around the Yellow Sea convergence area to restrain further land reclamation and undertake more detailed studies of the impacts to and conservation needs of these fragile mudflats.

Future Challenges

The challenge of harmonizing the conservation efforts and land reclamation policies of so many countries is big enough, but in the face of additional threats of increased pollution, poaching, decreased silt discharge, and climate change offer a bleak prospect. There will undoubtedly be more biodiversity losses, but it is important to acquire protection for the most important sites to salvage as much as one can.

References

Amano T, Székely T, Koyama K, Amano H, Sutherland W. A framework for monitoring the status of populations: an example from wader populations in the East Asian– Australasian Flyway. Biol Conserv. 2010;143:2238–47.

An S, Li H, Guan B, Zhou C, Wang Z, Deng Z, Zhi Y, Liu Y, Xu C, Fang S, Jiang J, Hongli LH. China's natural wetlands: past problems, current status, and future challenges. Ambio. 2007;34:335–42.

Barter MA. The Yellow Sea – a vitally important staging region for migratory shorebirds. In: Boere GC, Galbraith CA, Stroud DA, editors. Waterbirds around the world. Edinburgh: The Stationery Office; 2006. p. 663–7.

Barter M. Shorebirds of the Yellow Sea – importance, threats and conservation status. Global Series 9, International Wader Studies 12, Canberra; 2002. http://www.deh.gov.au/biodiversity/migratory/waterbirds/yellow-sea/index.html.

Battley P, McCaffery B, Rogers D, Hong J-S, Moores N, Ju Y-K, Lewis J, Piersma T, van de Kam J. Invisible connections. Why migrating shorebirds need the Yellow Sea. Collingwood: CSIRO Publishing; 2008.

BirdLife International. Important bird areas and potential Ramsar Sites in Asia. Cambridge: BirdLife International; 2005.

BirdLife International and IUCN-WCPA South-East Asia. Gap analysis of protected areas coverage in the ASEAN countries. Cambridge, UK: BirdLife International; 2007.

Buehler DM, Piersma T. Travelling on a budget: predictions and ecological evidence for bottlenecks in the annual cycle of long-distance migrants. Philos Trans R Soc B Biol Sci. 2008;363:247–66.

CCICED. Ecosystem Issues and policy options addressing sustainable development of China's ocean and coast. In: Report of Marine Ecosystems Task Force to CCICED AGM. Beijing; 2010. pp. 264–316 http://www.cciced.net/encciced/policyresearch/report/201205/P020120529358302221866.pdf.

Chang SE, Eeri M, Adams BJ, Alder J, Berke PR, Chuenpagdee R, Ghosh S, Wabnitz C. Coastal ecosystems and tsunami protection after the December 2004 Indian Ocean tsunami. Earthquake Spectra. 2006;22(S3):S863–87.

Chen L, Wang W, Zhang Y, Lin G. Recent progresses in mangrove conservation, restoration and research in China. J Plant Ecol. 2009;2:45–54.

Clemens RS, Rogers DI, Hansen BD, Gosbell K, Minton CDT, Straw P, Bamford M, Woehler EJ, Milton DA, Weston MA, Venables B, Weller D, Hassell C, Rutherford B, Onton K, Herrod A, Studds CE, Choi C-Y, Dhanjal-Adams KL, Murray NJ, Skilleter GA, Fuller RA. Continental-scale decreases in shorebird populations in Australia. Emu. 2016;116:119–35.

Conklin JR, Lok T, Melville DS, Riegen AC, Schuckard R, Piersma T, Battley PF. Declining adult survival of New Zealand Bar-tailed Godwits during 2005–2012 despite apparent population stability. Emu. 2016;116:147–57.

Constanza R, de Groot R, Sutton P, van der Ploeg S, Anderson SJ, Kubiszewski I, Farber S, Turner RK. Changes in the global value of ecosystem services. Glob Environ Chang. 2014;26:152–8.

Decho AW. Microbial biofilms in intertidal systems: an overview. Cont Shelf Res. 2000;20:1257–73.

Gill AB. Offshore renewable energy – ecological implications of generating electricity in the coastal zone. J Appl Ecol. 2005;42:605–15.

Kirby J. Review of current knowledge of bird flyways, principal knowledge gaps and conservation priorities. 2010. www.cms.int/sites/default/files/document/ScC16_Doc_10_Annex_2b_Flyway_WG_Review2_Report_Eonly_0.pdf.

KORDI – Korean Ocean Research and Development Institute. Coastal Wetlands conservation plan. Presented at the Symposium for Intertidal habitat Conservation and Sustainable Use, Gochang, Republic of Korea, 28 Sept 2006. 2006.

Li DJ, Daler D. Ocean pollution from land-based sources: East China Sea. Ambio. 2004;33:107–13.

MacKinnon J, Verkuil YI, Murray N. IUCN situation analysis on East and Southeast Asian intertidal habitats, with particular reference to the Yellow Sea (including the Bohai Sea), Occasional Paper of the IUCN Species Survival Commission No. 47. Gland: IUCN; 2012. ii + 70 pp. 2012. Available at: www.iucn.org/asiancoastalwetlands.

Melville DS, Chen Y, Ma Z. Shorebirds along the Yellow Sea coast of China face an uncertain future – a review of threats. Emu. 2016;116:100–10.

Moores N, Rogers DI, Rogers K, Hansbro PM. Reclamation of tidal flats and shorebird declines in Saemangeum and elsewhere in the Republic of Korea. Emu. 2016;116:136–46.

Myers JP, Morrison RIG, Antas PZ, Harrington BA, Lovejoy TE, Sallaberry M, Senner SE, Tarak A. Conservation strategy for migratory species. Am Sci. 1987;75:18–26.

Nicholls RJ, Cazenave A. Sea-level rise and its impact on coastal zones. Science. 2010;328:1517–20.

Piersma T, Lok T, Chen Y, Hassell CJ, Yang HY, Boyle A, Slaymaker M, Chan YC, Melville DS, Zhang ZW, Ma Z. Simultaneous declines in summer survival of three shorebird species signals a flyway at risk. J Appl Ecol. 2016;53:479–90.

Rogers D, Hassell C, Oldland J, Clemens R, Boyle A, Rogers K. Monitoring Yellow Sea migrants in Australia (MYSMA). North-western Australian shorebird surveys and workshops, December 2008. 2009.

Rogers DI, Yang H-Y, Hassell CJ, Boyle AN, Rogers KG, Chen B, Zhang ZW, Piersma T. Red knots (*Calidris canutus piersmai* and *C.c.rogersi*) depend on a small threatened staging area in Bohai Bay, China. EMU. 2010;110:307–15.

Short FT, Polidoro B, Livingstone SR, Carpenter KE, Bandeira S, Sidik Bujang J, Calumpong HP, Carruthers TJB, Coles RG, Dennison WC, Erftemeijer PLA, Fortes MD, Freeman AS, Jagtap TG, Kamal AHM, Kendrick GA, Kenworthy WJ, La Nafie YA, Nasution IM, Orth RJ, Prathep A, Sanciangco JC, van Tussenbroek B, Vergara SG, Waycott M, Zieman JC. Extinction risk assessment of the world's seagrass species. Biol Conserv. 2011;144:1961–71.

Verkuil YI, Karlionova N, Rakhimberdiev EN, Jukema J, Wijmenga JJ, Hooijmeijer JCEW, Pinchuk P, Wymenga E, Baker AJ, Piersma T. Losing a staging area: Eastward redistribution of Afro-Eurasian ruffs is associated with deteriorating fuelling conditions along the western flyway. Biol Conserv. 2012;149:51–9.

Wang H, Bi N, Saito Y, Wang Y, Sun X, Zhang J, Yang Z. Recent changes in sediment delivery by the Huanghe (Yellow River) to the sea: causes and environmental implications in its estuary. J Hydrol. 2010a;391:302–13.

Wang X, Chen W, Zhang L, Jin D, Lu C. Estimating the ecosystem service losses from proposed land reclamation projects: a case study in Xiamen. Ecol Econ. 2010b;69:2549–56.

Wilson HB, Kendall BE, Fuller RA, Milton DA, Possingham HP. Analyzing variability and the rate of decline of migratory shorebirds in Moreton Bay, Australia. Conserv Biol. 2011;4:758–66.

WWF, Korea Ocean Research and Development Institute (KORDI), Korea Environment Institute (KEI). The Yellow Sea ecoregion – a global biodiversity treasure. 2006. Available at: http://www.wwf.or.jp/activities/lib/pdf/200710y-seamap01e.pdf.

Yu WW, Chen B, Zhang LP. Cumulative effects of reclamation on ecosystem services of tidal flat wetland – a case in the Xinghua Bay, Fujian, China. Mar Sci Bull. 2008;1:88–94 (In Chinese).

Zöckler C, Htin Hla T, Clark N, Syroechkovskiy EE, Yakushev N, Daengphayon S, Robinson R. Hunting in Myanmar is probably the main cause of the decline of the spoon-billed sandpiper *Calidris pygmeus*. Wader Study Group Bull. 2010a;117:1–8.

Zöckler C, Syroechkovskiy EE, Atkinson PW. Rapid and continued population decline in the spoon-billed sandpiper *Eurynorhynchus pygmeus* indicates imminent extinction unless conservation action is taken. Bird Conserv Int. 2010b;20:95–111.

Seagrass in Malaysia: Issues and Challenges Ahead

158

Japar S. Bujang, Muta H. Zakaria, and Frederick T. Short

Contents

Introduction	1876
The Significance of Seagrass Beds and Their Resources	1877
The Decline of Seagrass Beds and Their Resources	1879
Threats and Future Challenges	1881
References	1882

Abstract

Seagrasses are submerged monocotyledonous angiosperms living in marine and estuarine habitats. They are plants differentiated into distinct segments: rhizomes, roots, and leaves. Seagrasses vary in morphology and size, ranging from the tropical eelgrass *Enhalus acoroides* with strap-like leaves that reach 1 m or more in height to shorter ovate leaved spoongrass *Halophila ovalis* that grows to only a few centimeters tall. Seagrasses produce flowers and seeds, disperse seeds and propagate vegetatively to maintain meadows. There are 16 species of seagrasses in Malaysia comprising *Enhalus acoroides*, *Halophila beccarii*, *Halophila decipiens*, *Halophila ovalis*, *Halophila major*, *Halophila minor*, *Halophila*

J. S. Bujang (✉)
Department of Biology, Faculty of Science, Universiti Putra Malaysia, Serdang, Selangor Darul Ehsan, Malaysia
e-mail: japar@upm.edu.my

M. H. Zakaria
Department of Aquaculture, Faculty of Agriculture, Universiti Putra Malaysia, Serdang, Selangor Darul Ehsan, Malaysia
e-mail: muta@upm.edu.my

F. T. Short
Department of Natural Resources and the Environment, Jackson Estuarine Laboratory, University of New Hampshire, Durham, NH, USA
e-mail: fredtshort@gmail.com

© Springer Science+Business Media B.V., part of Springer Nature 2018
C. M. Finlayson et al. (eds.), *The Wetland Book*,
https://doi.org/10.1007/978-94-007-4001-3_268

spinulosa, *Halophila* sp., *Halodule pinifolia*, *Halodule uninervis*, *Cymodocea rotundata*, *Cymodocea serrulata*, *Thalassia hemprichii*, *Syringodium isoetifolium*, *Thalassodendron ciliatum*, and *Ruppia maritima*. Healthy seagrasses may grow dense and form an extensive beds or meadows. Their characteristics and interactive community within and from outside account for the high diversity and enable survival of diverse invertebrates (shrimps, sea cucumbers, starfishes, bivalves, gastropods), vertebrates (dugongs, green sea turtles, fishes) and macroalgae. Seagrasses provide conditions for the growth and abundance of invertebrates and fish that many local coastal communities collect and catch for their livelihood. Seagrass ecosystems are sources of food and continually facing threats by natural events and, human activities, e.g., coastal development causing their fast degradation and possible habitat loss.

Keywords
Seagrass · diversity · significance · Malaysia · threats

Introduction

Seagrasses are submerged monocotyledonous angiosperms living in marine and estuarine habitats (den Hartog 1970). They are plants differentiated into distinct segments: rhizomes, roots, and leaves. Seagrasses produce flowers and seeds, disperse seeds and propagate vegetatively to maintain meadows (Muta Harah et al. 2002). Seagrasses vary in morphology and size, ranging from the tropical eelgrass *Enhalus acoroides* with strap-like leaves that reach 1 m or more in height to shorter ovate leaved spoongrass *Halophila ovalis* that grows to only a few centimeters tall (Fig. 1). There are 16 species of seagrasses in Malaysia comprising *Enhalus acoroides*, *Halophila beccarii*, *Halophila decipiens*, *Halophila ovalis*, *Halophila major*, *Halophila minor*, *Halophila spinulosa*, *Halophila* sp., *Halodule pinifolia*, *Halodule uninervis*, *Cymodocea rotundata*, *Cymodocea serrulata*, *Thalassia hemprichii*, *Syringodium isoetifolium*, *Thalassodendron ciliatum*, and *Ruppia maritima*.

In Malaysia, along its 4800 km coastline, that stretches along the Malay Peninsula and on the island of Borneo in both Sabah and Sarawak and bounds much of the southern part of the South China Sea, are coastal environments harboring mangroves, coral reefs, and seagrasses. The majority of seagrasses are found in sheltered shallow intertidal associated ecosystems, with mangroves, coral reefs, semi-enclosed lagoons, shoals, and subtidal zones. They sometimes form diverse extensive communities (Japar Sidik and Muta Harah 2003, Fig. 2). Historically, seagrasses (*E. acoroides*, *H. ovalis*) were common all around the coast of Peninsular Malaysia, on muddy shores and areas exposed at low tide (Ridley 1924; Henderson 1954). Since these early reports of extensive seagrass meadows, most have deteriorated due to coastal development. Moreover, seagrass habitats in Sabah, East Malaysia were described by Norhadi (1993) as already degraded by human activities, including deforestation (Short et al. 2014). This would explain the present patchy and no longer extensive distribution along the Malaysian coastline (Fig. 3).

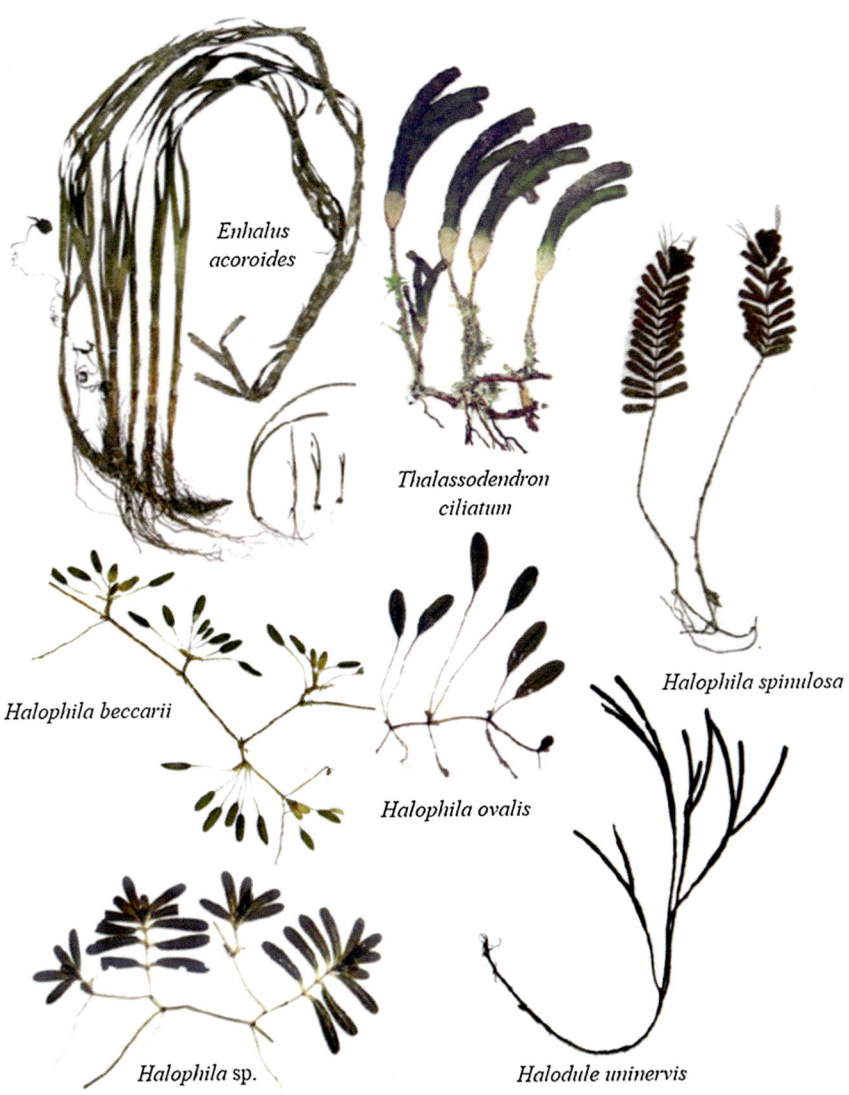

Fig. 1 Selected seagrass species of Malaysia (© JSB Bujang, Rights remain with the author)

The Significance of Seagrass Beds and Their Resources

Seagrasses are primary producers and through photosynthesis release dissolved oxygen into the water for use by marine and estuarine animals. Seagrasses are also primary food source for large species, such as dugongs and green sea turtles, and consumed by sea urchins. Algae and animals attached to seagrass leaves are an

Fig. 2 Merambong shoal during low tide is one of several areas that support a well-developed multi-species seagrass community (Photo credit: JS Bujang © Rights remain with the author)

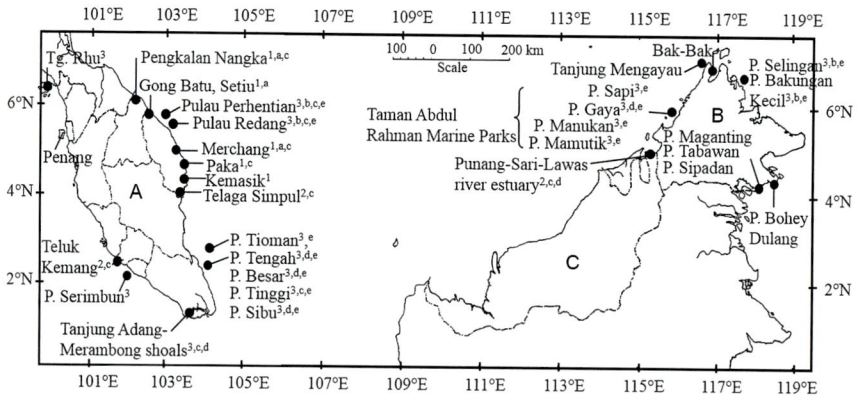

Fig. 3 The major seagrass areas, associated habitats, utilization by coastal communities and other users in Peninsular Malaysia (*A*), east Malaysia-Sabah (*B*), and Sarawak (*C*). ∆-Lagoon, o-intertidal, -subtidal. Aquaculture[1], turtle sanctuary[2], traditional capture fisheries[3], dugong feeding ground[4], and marine park[5] (© JSB Bujang, Rights remain with the author)

important food resource for many grazing animals, e.g., juvenile and adult fishes, and its disappearance could be detrimental to their survival.

Seagrass meadows are subject to tide and wave action that could erode and wash away sediments if not for the plants' underground rhizomes and roots forming dense mats that trap, hold, and stabilize the soft sediments. The seagrass meadows provide habitat and shelter for many organisms. Bivalves and other buried organisms burrow

Fig. 4 Fisherman with his catch from Sungai Pulai estuary seagrass meadow (Photo credit: JS Bujang © Rights remain with the author)

into the sediments to escape predation. In addition, many fish (grouper, seahorses), shellfish, and crustaceans found in seagrass meadows are a valuable economic entity (Sasekumar et al. 1989; Arshad et al. 2001). For example, local coastal communities around Sungai Pulai estuary harvest invertebrates and fishes in and around seagrass meadow for their livelihood (Fig. 4). While recognized as sources of food and an important reservoir of coastal biodiversity, seagrass meadows are continually threatened by human activities causing their degradation and habitat loss (Muta Harah and Japar Sidik 2011).

The Decline of Seagrass Beds and Their Resources

Seagrasses and their resources have declined due to natural causes and human-induced activities. Seagrasses can tolerate moderate disturbance through morphological and physiological adaptation, while heavy disturbances can result in seagrass loss (Hemminga and Duarte 2000). Along the east coast of Peninsular Malaysia, increased wave activity from storms during the northeast monsoon threatens meadows (e.g., Gong Batu, Merchang, Paka shoal, Terengganu) in semi-enclosed lagoons by either exposing or burying the plants with shoals of sand, silt, or mud. These seagrass meadows are also impacted from reduced salinity by frequent inputs of freshwater (high rain events) or increased salinity due to drought. Seagrass meadows can also be heavily impacted by flood events such as occurred in December 2014 in Paka shoal, Terengganu, destroying almost all seagrasses. Extremely high and continuous rainfall for a week contributed to the flood event that transported and deposited sand that covered the seagrass meadow (Fig. 5). Recovery may take years to occur or may not at all, as a result of a single monsoon flooding event. In Sabah, deforestation associated with palm oil plantations has resulted in extensive decline and loss of seagrass from reduced water clarity and

Fig. 5 (**a**) Dense *Halophila beccarii* bed in Paka shoal in 2004 (Photo credit: JS Bujang © Rights remain with the author) (**b**) The same Paka shoal buried by shoal of sand, devoid of seagrass after monsoon and flood in December 2014 (Photo credit: MH Zakaria © Rights remain with the author)

Fig. 6 Algal bloom, *Amphiroa fragilissima* at seagrass bed of Tanjung Adang Laut, Sungai Pulai estuary, Johor (Photo credit: MH Zakaria © Rights remain with the author)

sedimentation (Freeman et al. 2008), followed by limited recovery by a pioneering seagrass species (Short et al. 2014).

The human population utilizing the coasts and adjacent waters for numerous activities is increasing, and many of these activities are accompanied by major coastal developments. Activities attributed to seagrass declines are sediments in runoff from cleared lands flowing into coastal areas covering meadows; and discharge of untreated domestic sewage effluent, runoff of fertilizer, and outflow of animal waste; all rich in nutrients promoting large algal blooms that cover the seagrasses (Japar Sidik et al. 2006, Fig. 6) decreasing the plants' photosynthesis capability. Sand mining, dredging, bottom sediment removal, and land reclamation destroy seagrass meadows through direct removal and burial. Sourcing for sand is

Fig. 7 Massive earthwork of land reclamation across Merambong seagrass shoal (Photo credit: MH Zakaria © Rights remain with the author)

commonly undertaken in shallow coastal lagoons with seagrasses (e.g., east coast of Peninsular Malaysia) for land fill and shoreline stabilization projects. Coastal development (land reclamation) has reduced the seagrass area of Merambong shoal (Fig. 7), Johor from 26.3 ha to 21.1 ha. Similar land reclamation totally destroyed a Tanjung Adang Darat seagrass shoal in 2003 (Japar Sidik et al. 2007).

Threats and Future Challenges

By the mid-to-late 1990s, inland areas were extensively developed, and beginning in the early 2000s there has been a shift in the focus of development towards coastal reclamation and constructed island systems as the primary region for economic growth and development. Both systems have been recognized in Malaysia as an increasingly important source of foreign exchange earnings as well as a means to diversify the country's economy. The tourism industry can play a vital role to boost the economy drawing upon the scenic attraction of the aquatic environment, white sandy beaches, and underwater marine resources such as live corals. The government has allocated substantial funds to finance projects, basic infrastructure, public, and other tourist facilities. In addition, development of tourist related facilities, resorts, restaurants, and recreational centers are being undertaken by the private sector in line with the government's privatization policy. Many coastal zones are now being used intensively by a variety of activities, and these developments have not been without problems.

The Sungai Pulai estuary, Johor, is an example of issues that have arisen with the new focus on development of the coastal zone and islands. The estuary harbors mangrove and seagrass meadows used by local communities for their fisheries and transportation. Sungai Pulai estuary's seagrass ecosystem provides multiple benefits to local coastal inhabitants and other users including provisioning services (food, genetic resources, biochemical, and medicines); regulatory services (erosion regulation, water purification, and waste treatment); and cultural services (educational values) (Nakaoka et al. 2014). The most important provisioning services support the fisheries and any disturbance will directly affect the earnings and protein supply of coastal communities depending on the ecosystem.

Port and facilities development in Sungai Pulai estuary began in 2003, and development of a mega resort and city commenced in 2014 for the next 30 years. The development involves sand mining, filling, and land reclamation which has had an immediate and significant impact to the marine environment and associated resources of the seagrass beds. Land reclamation resulted in heavy loads of suspended solids in the water column, depositing thick glutinous silt often many centimeters deep over seagrasses and benthic communities (Japar Sidik et al. 2007). The effect of habitat change and altered ecology has been realized by the fishermen. They claim reclamation and pollution are severely affecting fish yields. Some fishermen have ceased this activity and are seeking alternative jobs elsewhere to generate income lost from fishing activities. It is expected that adjacent seagrass meadows of Merambong-Tanjung Adang shoals will face a similar fate as further expansion is planned for the port, mega resort, and city.

With the development and anticipated population growth, regulating domestic waste disposal and waste treatment is a big issue. Domestic waste is the largest contributor to the water quality problem and will worsen with growth in the population and visiting tourists (Sungai Pulai estuary has been earmarked for "community tourism"). There will be increased discharges of human waste and household and commercial compounds (e.g., detergents and sullage) directly into the watercourses. Inadequacy of sewage treatment systems will further aggravate the problem. Discharge wastes eventually reach the seagrass and coastal areas, placing great stress (causing algal bloom) to the ecosystem and associated resources. Maintenance and protection of seagrass habitats will conserve the water quality and other resource values that are, in fact, able to attract tourism.

Malaysian seagrass meadows are facing serious threats from external human sources and coastal development activities. The optimal and sustainable use of coastal habitats is a high priority due to their important ecological functions and socioeconomic benefits to coastal human population. However, the protection and conservation of seagrass ecosystems is only afforded to those within protected areas, e.g., marine parks under Fisheries Act 1985 and not for those outside their boundaries (Japar Sidik and Muta Harah 2003).

References

Arshad A, Japar Sidik B, Muta Harah Z. Fishes associated with seagrass habitat. In: Japar Sidik B, Arshad A, Tan SG, Daud SK, Jambari HA, Sugiyama S, editors. Aquatic resource and environmental studies of the straits of Malacca: current research and reviews. Serdang: Universiti Putra Malaysia Press; 2001. p. 151–62.

Den Hartog C. Seagrasses of the world. Amsterdam: North-Holland Publishing; 1970.

Freeman AS, Short FT, Isnain I, Razak FA, Coles RG. Seagrass on the edge: land-use practices threaten coastal seagrass communities in Sabah, Malaysia. Biol Conserv. 2008;141:2993–3005.

Hemminga MA, Duarte CM. Seagrass ecology. Cambridge: Cambridge University Press; 2000.

Henderson MR. Malayan wild flowers: monocotyledon. Kuala Lumpur: The Malayan Nature Society; 1954.

Japar Sidik B, Muta Harah Z. Chapter 14 Seagrasses in Malaysia. In: Green EP, Short FT, Spalding MD, editors. World atlas of seagrasses. Berkeley/Los Angeles: California University Press; 2003. p. 166–76.

Japar Sidik B, Muta Harah Z, Arshad A. Distribution and significance of seagrass ecosystems in Malaysia. Aquat Ecosyst Health Manag. 2006;9:1–12.

Japar Sidik B, Muta Harah Z, Arshad A. Seagrass resources and their declining in Malaysia. MIMA Bull. 2007;14(2):36–42.

Japar Sidik B, Muta Harah Z. Seagrasses in Malaysia. In: Ogawa H, Japar Sidik B, Muta Harah Z, editors. Seagrasses: resource status and trends in Indonesia, Japan, Malaysia, Thailand and Vietnam. Tokyo: Japan Society for the Promotion of Science (JSPS) and Atmosphere and Ocean Research Institute (AORI), The University of Tokyo: Seizando-Shoten Publishing Co., Ltd.; 2011. p. 22–37.

Muta Harah Z. Japar Sidik B. Disturbances in seagrass ecosystem in Malaysia. In: Ogawa H, Japar Sidik B, Muta Harah Z, editors. Seagrasses: resource status and trends in Indonesia, Japan, Malaysia, Thailand and Vietnam. Tokyo: Japan Society for The Promotion of Science (JSPS) and Atmosphere and Ocean Research Institute (AORI), The University of Tokyo: Seizando-Shoten Publishing Co., Ltd.; 2011. p. 67–78.

Muta Harah Z, Japar Sidik B, Arshad A. Flowering, fruiting and seedling of annual *Halophila beccarii* Aschers in Peninsular Malaysia. Bull Mar Sci. 2002;71(3):1199–205.

Nakaoka M, Lee Kun-Seop, Huang Xiaoping, Almonte Tutu, Japar Sidik B, Kiswara W, Rohani AR, Siti Maryam Y, Prabhakaran MP, Abu Hena MK, Masakazu H, Zhang P, Prathep A, Fortes MD. Regional comparison of the ecosystem services from seagrass beds in Asia. In: Shin-ichi Nakano, Tetsukazu Yahara, Tohru Nakazhizuka, editors. Integrative observations and assessments. Tokyo: Springer; 2014. p. 368–90.

Norhadi I. Preliminary study of seagrass flora of Sabah, Malaysia. Pertanika J Trop Agric Sci. 1993;16(2):111–8.

Ridley HN. The flora of the Malay Peninsula: monocotyledons. The authority of the government of the straits settlements and federated Malay States, vol. IV. Amsterdam/Holland: A. Asher & Co. L. Reeve & Co. Brook Nr. Ashford, Great Britain; 1924.

Sasekumar A, Leh CMU, Chong VC, Rebecca D, Audery ML. The Sungai Pulai (Johore): a unique mangrove estuary. In: Phang SM, Sasekumar A, Vickineswary S, editors. Malaysian society of marine science. Kuala Lumpur: Universiti Malaya Press; 1989. p. 191–211.

Short F, Coles R, Fortes M, Victor S, Salik M, Isnain I, Andrew J, Seno A. Monitoring in the Western Pacific Region shows evidence of seagrass decline in line with global trends. Mar Pollut Bull. 2014;83:408–16.

Section XII

Australia, New Zealand, and Pacific Islands

Murray-Darling River Basin (Australia)

159

Jamie Pittock

Contents

Introduction 1888
The Basin's Wetlands 1888
Climate and Hydrology 1890
Threats to the Wetlands 1891
Management Responses 1891
Environmental Flows 1892
Catchment Management 1894
Threatened Biota 1894
Protected Areas 1895
Future Challenges 1895
References 1895

Abstract

The wetlands of the Murray–Darling Basin (the Basin) in Australia are the focus of this article. Beginning with a description of the Basin's wetlands, the climate and hydrology are outlined as a point for discussing the threats to wetlands conservation, management responses, and future challenges.

Keywords

Great artesian basin · Murray-Darling Basin · Climate change · Environmental flows · Legislation · Protected areas · Ramsar Convention on Wetlands · Threatened species

---

J. Pittock (✉)
Fenner School of Environment and Society, The Australian National University, Canberra, Australia
e-mail: Jamie.pittock@anu.edu.au

© Springer Science+Business Media B.V., part of Springer Nature 2018
C. M. Finlayson et al. (eds.), *The Wetland Book*,
https://doi.org/10.1007/978-94-007-4001-3_102

Introduction

The Basin has the longest rivers in Australia and covers about one million square kilometers, around one seventh of the continent's land mass (Fig. 1). The Basin is estimated to contain around 23% of Australia's wetlands (Finlayson et al. 2011). The Basin is located in a region subject to great hydroclimatic variability and change, and with innovative policies and considerable resources being devoted to conservation, management of these wetlands may hold lessons for other regions of the world.

The Basin's Wetlands

There are different estimates for the extent of wetlands in the Basin. Some 440,000 km of rivers and 30,000 other wetlands have been identified. Based on paper maps, Geoscience Australia has identified 5.3 million hectares of wetlands (Table 1; MDBA 2010; Finlayson et al. 2011).

Another analysis based on satellite images concludes that wetlands occur over 5.7 million hectares (Table 2; Kingsford et al. 2004). A more recent estimate is 6.2 million hectares covering 5.87% of the land area of the Basin (Nairn and Kingsford 2012).

The types of wetlands in the Basin vary considerably (SEWPAC 2012). Two broad classifications of wetlands by area are shown in Tables 1 and 2, although these do not indicate the breadth of smaller and biologically diverse wetland types. In the southeastern highlands, the rivers start amid alpine peatlands, a few glacial lakes, and, in some localities, karst wetlands. The rivers spill from the eastern rim of the Basin onto low-lying plains through current and many prior streams, forming extensive floodplains and inland deltas that comprise the overwhelming majority of wetlands by area. Many parts of the Basin have limited hydrological connectivity due to aeolian sand deposits, low gradients, and river flows. Some large freshwater lakes form in wetter times. High levels of salinity have also resulted in the formation of extensive saline lakes in the western portions of the Basin.

The huge Great Artesian Basin underlies much of the northwestern portion of the area and is expressed as mound springs. There are extensive areas of groundwater-dependent ecosystems, often associated with the extensive floodplains (BoM 2012). Knowledge of these ecosystems is growing, as are attempts to conserve them.

The River Murray flows through the large freshwater lakes Alexandrina and Albert before reaching the Coorong, a back-barrier lagoon, adjacent to the Murray Mouth. Historically in severe droughts, the river would cease flowing to the sea around 1 year in a hundred; however, the mouth is now closed in 36% of years due to excessive water diversions (CSIRO 2008; MDBA 2010).

The ecosystem services generated from healthy wetlands in the Basin have been estimated in different studies. Most notably, increasing environmental flows by 2,800 GL/year was estimated to increase Basin-wide value of enhanced habitat ecosystem services arising from floodplain vegetation, water bird breeding, native fish and the Coorong, Lower Lakes, and Murray Mouth worth between AUD $3

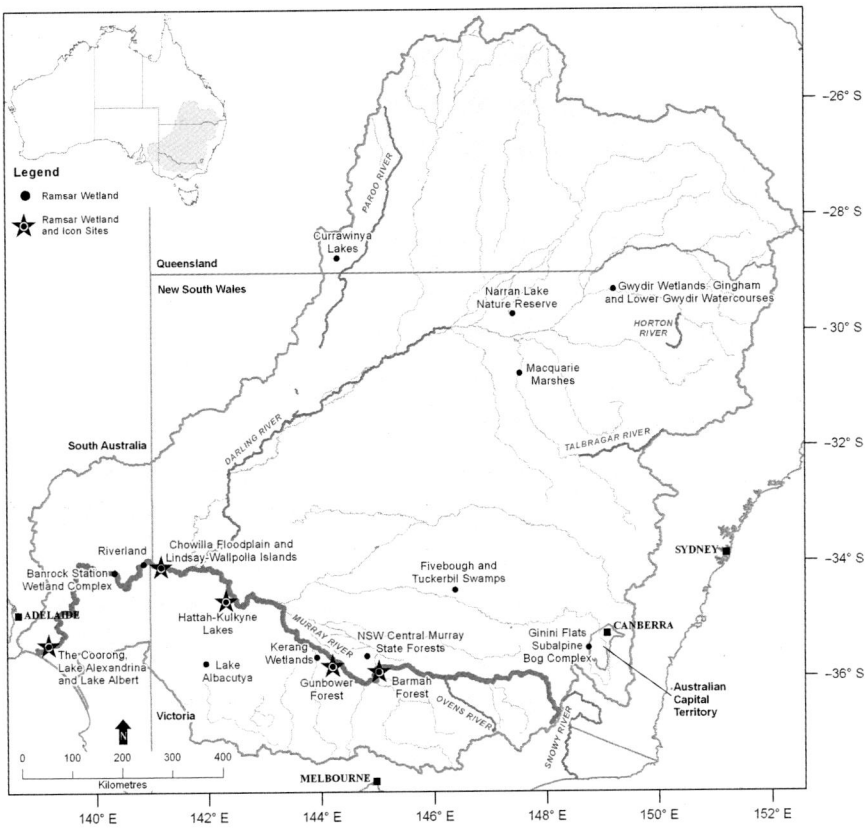

Fig. 1 Location of the Murray–Darling Basin and the 16 designated Ramsar wetlands and icon wetlands (the focus of government conservation efforts from 2003 to 2012) (Map credit: © Charles Sturt University, with permission)

Table 1 One estimate of the area of wetlands in the Murray–Darling Basin (Derived from Finlayson et al. 2011: Table 2; Aquatic Sciences – Research Across Boundaries, The status of wetlands and the predicted effects of global climate change: the situation in Australia, 75, 2011, 73–93, Finlayson CM, Davis JA, Gell PA, Kingsford RT, Parton KA; © with permission of Springer Science + Business Media)

Wetland type	Area (ha)	Portion (%)
Lakes	1,006,629	18.8
Floodplains	3,950,009	73.9
Swamps	298,575	5.6
Marine wetlands	1,607	0.0
Rivers	84,838	1.6
Anthropogenic wetlands	199,623	3.7
Total Basin wetland area	5,341,656	100.0

Table 2 A second estimate of the area of wetlands in the Murray–Darling Basin (Reproduced from Kingsford et al. 2004: Table 3; with permission of CSIRO Publishing)

Wetland type	Number	Area (ha)	Portion (%)
Saline lake	205	16,476	0.3
Freshwater lake	367	275,281	4.9
Floodplain	n/a	5,383,820	94.9
Total natural wetlands	n/a	5,675,577	100.0
Anthropogenic reservoirs	2,232	101,142	n/a

billion and AUD $8 billion, carbon sequestration by AUD $120 million to AUD $1 billion, aesthetic appreciation by more than AUD $330 million, avoided damage and treatment of freshwater of AUD $30 million, and tourism benefits of up to AUD $160 million annually (AUD $1.00 = USD $0.76; CSIRO 2012). This compares to an estimated annual AUD $542 million reduction in gross irrigated agricultural production due to water reallocation.

There has been no systematic classification of wetlands within the basin. Wetland management has been the primary responsibility of each of the four states and the Australian Capital Territory that have jurisdiction in the Basin, and they have each adopted different classification systems (Pressey and Adam 1995; Finlayson et al. 2011). A *National Directory of Important Wetlands* publishes information on those significant wetlands for which there are data (SEWPAC 2012). The Murray–Darling Basin Authority (the Authority) has assessed 20,000 wetland records held by state agencies and designated 2,442 as "key environmental assets," including a subset of 18 as "hydrologic indicator" sites for these assets (MDBA 2010).

Climate and Hydrology

Wetland health depends on adequate water inflows. Average annual depth of runoff in the Basin is only 21 mm/year, but this varies considerably from the high rainfall Australian Alps in the southeast to the semiarid Paroo River catchment in the northwest (Pressey and Adam 1995). The northern Darling River Basin is summer rainfall dominated but only provides 10% of the inflows into the river system. The southern River Murray Basin is winter rainfall dominated and receives 90% of system inflows.

The Basin has a long history of hydroclimatic variability (Finlayson et al. 2011). Long droughts and large floods linked to the El Niño/La Niña Southern Oscillation and other climate influences make the Basin among the regions with the greatest variability in the world. Wetland biota has adapted to the boom and bust cycle, but has been severely impacted as the hydrological variability has been smoothed out to supply water for irrigation and other human consumption (Kingsford 2006). A remarkable expression of this adaptation is given by the colonial nesting water birds that congregate at Basin wetlands after big floods provide the specific conditions that they require to successfully breed.

Table 3 Scenarios for average water availability in the Murray–Darling Basin compared to historical flows under climate change compared with the short-term situation in 2009

CSIRO scenario (CSIRO 2008)	Average surface water availability in 2030 (%)	End of system flows in 2030 (%)
Extreme wet	7	20
Median	−12	−24
Extreme dry	−37	−69
Situation in 2009 (Pittock and Finlayson 2011)	−66	No outflows 2002–2010

The Basin falls within the mid-latitude zone of "difficult hydrology" anticipated to be most impacted by climate change. Modeling of prospective climate change impacts on the Basin to 2030 (Table 3) suggest that long-term average conditions may range from a little wetter through to substantially drier, and as these impacts are magnified downriver, the consequences for wetlands lower in the system may be severe (CSIRO 2008). When short-term hydroclimatic variability is considered in addition to the expected long-term change, such as the extreme 2002–2010 drought in the Basin, the impacts on wetland biota may be substantial.

Threats to the Wetlands

The waters of the Basin are being extensively exploited for human use. Median annual end of system flows have fallen to 29% of levels before development, and the river mouth was closed from 2002 to 2010 due to over-extraction of water and climatic variability. Tens of thousands of hectares of wetlands have suffered from reduced water flows, salinization, acidification, and desiccation with consequential loss of floodplain forests and other wetlands (Fig. 2; Pittock et al. 2010). Conversion of habitat, levees, and other diversions of water for irrigated agriculture is the primary impact on the extensive floodplain wetlands (Kingsford 2006). Lack of inflows also drove closure of the Murray Mouth, hypersalinity in the Coorong, and desiccation of lakes Albert and Alexandrina, impacting on migratory water bird populations and other biota (Kingsford et al. 2011). While the impacts of reduced flows may be exacerbated by climate change, in the medium term, the primary impact is due to diversion of water for human consumption (Pittock et al. 2010). The Basin's wetlands are also significantly impacted by invasive species, loss of riparian vegetation, barriers to migration of aquatic fauna, and thermal pollution resulting from water infrastructure (Pittock and Finlayson 2011).

Management Responses

Australia has experimented with a number of policy innovations to restore and conserve the health of wetlands in the Basin, and those in the area of environmental water, catchment management, conservation of threatened biota, and protected areas

Fig. 2 Floodplain wetlands at Bottle Bend (*left*) and Psyche Bend (*right*) along the River Murray in 2009 showing salinization, acidification, and floodplain forest death as a result of inadequate flows (Photo credit: J Pittock © Rights remain with the author)

are briefly summarized here. Australia's federal constitution left primary responsibility for management of natural resources like water and biodiversity with the states and territories. However, in the past two decades, the Federal Government has increasingly used its powers to implement treaties like the Convention on Biological Diversity, the Ramsar Convention on Wetlands, and those for migratory birds, to regulate management of water and biodiversity – especially in the Murray–Darling Basin (Pittock et al. 2010).

Environmental Flows

The ecological health of wetland habitats in the Basin depends on delivery of adequate water that mimics the volume, timing, and quality of natural flows. As a water-scarce region, the contested management of this resource has dominated Basin management institutions (Connell 2007). In 2007–2008, a new national Water Act was adopted that extended Federal Government control over water use in the Basin and requires the development of a Basin Plan with sustainable diversion limits for water extraction. The Act largely draws its constitutional mandate on Australia's obligations to implement the Ramsar Convention on Wetlands and thus requires conservation of key wetland environmental assets, ecosystem functions, and services (MDBA 2010; Pittock et al. 2010).

The Authority has assessed available data on 20,000 wetlands in the Basin, and a subset of 18 hydrologic indicator sites for environmental assets has been used to propose environmental flows. While welcome, the Authority's assessment did not apply all of the Ramsar criteria for identifying wetlands of international importance, omitting representativeness for instance, such that a number of wetland types in the Basin are not targeted for conservation under the proposed plan. Further, this systems approach to allocating water may not always adequately prioritize the habitat of particular threatened or migratory wetland species.

On average, 11,164 GL/year of water is diverted from the Basin, and it has been estimated that as much as 7,600 GL/year of this needs to be returned to sustain a health river system (MDBA 2010). In 2012, the Federal Government adopted a Basin Plan (Commonwealth of Australia 2012) and associated programs that may reallocate up to an average 3,200 GL/year (29% of consumptive water), and while this may improve some wetland health, it would meet only 66% of the Authority's 112 wetland environmental targets (WGCS 2012). Further, full implementation of the proposed Plan is not scheduled until 2019.

There are other concerns with the Plan, including increased groundwater extraction when there is little available science to judge its impacts on wetlands (WGCS 2012). No specific reallocation to reduce the impact of climate change impacts on water availability has been proposed when climate-induced reductions in surface water availability of up to 37% by 2030 have been modeled (Pittock 2013). Reduced allocations of water for the environment are being justified in part on the beguiling notion of "environmental works and measures" or "environmental water demand management," engineering measures using pumps and levees and weirs that aim to save more floodplain wetland biodiversity with less water. This has great risks, for instance, in breaking habitat connectivity and in concentrating salt on wetlands (Pittock et al. 2012). While instituting adequate environmental flows is a primary measure for conserving the Basin's wetland habitats, there are other important actions that lie outside the Authority's mandate or have been overlooked. These include restoration of riparian forests, protection of remaining free-flowing rivers (such as the Paroo and Ovens rivers; Fig. 1), and reengineering the thousands of dams and weirs that fragment the Basin's rivers so as to eliminate cold water pollution and provide fish passage (Lukasiewicz et al. 2013).

The ongoing acquisition of water entitlements from irrigators for environmental flows will enhance the health of many wetlands. As of 30 June 2016, the Federal Government had acquired entitlements for 1,981.4 GL in long-term average annual yield. Significantly these entitlements, which may eventually comprise a quarter of the total environmental water, are now owned and independently managed for conservation by the Commonwealth Environmental Water Holder. This makes this Commonwealth environmental water a legal entitlement equal to commercial holdings rather than the previous approach by the states where the environment was supposed to be supplied by "rule-based" (leftover) water that largely disappeared through administrative reallocation in times of scarcity (Connell 2011). Further, the Basin Plan is to be revised periodically, enabling revisions to incorporate experience and new knowledge (Pittock and Finlayson 2011).

Catchment Management

In an effort to develop common visions for natural resource management and coordinate action across different scales of government and between agencies, Australia's governments have established catchment management agencies. These institutions are also called regional natural resource management organizations and equate elsewhere in the world to watershed and river basin management organizations. There have been up to 21 in the Basin, where they have formed a de facto fourth tier of government. While their mandate varies by state, generally they are charged with preparing regional plans to conserve water, vegetation, soils, biodiversity, and other natural resources and recommend or oversee government grants to implement the selected priorities. These institutions have been established to seek cooperation between stakeholders, including farmers, environmentalists, and governments. At their best they are a powerful institution for wetland conservation; however, their mandate has often overlapped with those of state government departments, and consequently their functions and funding have been repeatedly built up and taken down with changes in government policies, diminishing their effectiveness (Robins and Kanowski 2011). As an example, in 2013 the State of New South Wales abolished its catchment management authorities, merging their functions into larger Local Land Services regional agencies.

Threatened Biota

Australian state and federal governments each have laws to conserve threatened species and, in some cases, threatened ecological communities. Under the Federal Government's Environment Protection and Biodiversity Conservation Act, threatened species and ecological communities can be nominated for assessment and listing for legal protection, which is then intended to trigger the preparation of a recovery plan for their conservation (DEWHA 2008). Listing also means that the Federal Minister for the Environment must assess and approve any new proposed action that may significantly impact on threatened biota. A great many wetland species and three wetland ecological communities in the Basin are currently listed under this Act. "Key threatening processes" can also be designated for preparation of threat abatement plans. Similar site protection measures exist under the Act for Ramsar and World Heritage wetlands, as well as for migratory water bird species.

Despite this mechanism for legal protection, large numbers of threatened wetland ecological communities are not yet designated under the Act. Lists of threatened biota under state and federal laws are poorly coordinated, thus diminishing the effectiveness of government responses. Regrettably, few recovery plans have been prepared, let alone implemented, diminishing the benefits from these conservation laws.

Protected Areas

Sixteen sites in the Basin covering 636,300 ha have been designated as Wetlands of International Importance under the Ramsar Convention (Fig. 1; Pittock et al. 2010; SEWPAC 2012). These designations have been made in an ad hoc manner, and many more wetlands in the Basin would also qualify for designation. Since the adoption of federal law to conserve Ramsar wetlands in 1999, few new sites have been proposed for designation by the state governments.

An estimated 5.44% of the Basin's wetlands are in protected areas or reserves (Nairn and Kingsford 2012), and this area has been increasing rapidly. However, these reserves are unrepresentative of the diversity of wetland ecosystems even though the Federal Government has made commitments to build a more representative reserve system (Fitzsimons and Robertson 2005). Wetland reserves are largely held and managed by state governments, but in the past two decades, there has been a growing trend for non-government institutions to own and manage wetland reserves. In the Basin there are now protected areas established by private landholders applying conservation agreements and covenants, non-government conservation land trust organizations, as well as Indigenous communities (Thackway and Olsson 1999). This diversity of managers is bringing more resources and greater innovation into wetland conservation in the Basin, for instance, in terms of experimenting with environmental flows.

Future Challenges

Efforts to conserve these 6 million hectares of wetlands will hold important lessons for the rest of the world, owing to the location of the Murray–Darling Basin in a region subject to great hydroclimatic variability and change and due to the innovative policies and considerable resources being devoted to this cause. Successfully reversing the overallocation of water to farming and instituting effective environmental flows are the major challenges. Associated with these approaches is the need to reengineer water infrastructure to reduce impacts on wetlands. Stronger institutions are required for effective conservation, including catchment management agencies, programs for threatened species and ecosystems, as well as a more representative protected area network. The key elements have been established and only need the political will, further institutional reform, and more resources to set an example for effective wetland conservation.

References

BoM. Atlas of groundwater dependent ecosystems. Melbourne: Bureau of Meteorology; 2012.
Commonwealth of Australia. Basin plan. Canberra: Commonwealth of Australia; 2012.
Connell D. Water politics in the Murray-Darling Basin. Leichardt: The Federation Press; 2007.
Connell D. The role of the commonwealth environmental water holder. In: Connell D, Grafton RQ, editors. Basin futures: water reform in the Murray-Darling basin. Canberra: ANU E Press; 2011. p. 327–38.

CSIRO. Water availability in the Murray-Darling Basin. Canberra: CSIRO; 2008.

CSIRO. Assessment of the ecological and economic benefits of environmental water in the Murray–Darling Basin. Canberra: CSIRO; 2012.

DEWHA. Independent review of the Environment Protection and Biodiversity Conservation Act 1999. Discussion paper. Canberra: Department of the Environment, Water, Heritage and the Arts; 2008.

Finlayson CM, Davis JA, Gell PA, Kingsford RT, Parton KA. The status of wetlands and the predicted effects of global climate change: the situation in Australia. Aquat Sci. 2011;75:73–93.

Fitzsimons J, Robertson H. Freshwater reserves in Australia: directions and challenges for the development of a comprehensive, adequate and representative system of protected areas. Hydrobiologia. 2005;552:87–97.

Kingsford R. Impacts of dams, river managemnent and diversions. In: Kingsford R, editor. Ecology of desert rivers. Cambridge: Cambridge University Press; 2006. p. 203–47.

Kingsford RT, Brandis K, Thomas RF, Crighton P, Knowles E, Gale E. Classifying landform at broad spatial scales: the distribution and conservation of wetlands in New South Wales, Australia. Mar Freshw Res. 2004;55:17–31.

Kingsford RT, Walker KF, Lester RE, Young WJ, Fairweather PG, Sammut J, Geddes MC. A ramsar wetland in crisis – the Coorong, lower lakes and Murray mouth, Australia. Mar Freshw Res. 2011;62:255–65.

Lukasiewicz A, Finlayson CM, Pittock J. Identifying low risk climate change adaptation in catchment management while avoiding unintended consequences. Gold Coast: National Climate Change Adaptation Research Facility; 2013.

MDBA. Guide to the proposed basin plan: overview. Canberra: Murray-Darling Basin Authority; 2010.

Nairn LC, Kingsford RT. Wetland distribution and land use in the Murray-Darling Basin: Australian wetlands, rivers and landscapes centre. Sydney: University of NSW; 2012.

Pittock J. Lessons from adaptation to sustain freshwater environments in the Murray–Darling Basin, Australia. Wiley Interdiscip Rev Clim Chang. 2013;4:429–38.

Pittock J, Finlayson CM. Australia's Murray-Darling Basin: freshwater ecosystem conservation options in an era of climate change. Mar Freshw Res. 2011;62:232–43.

Pittock J, Finlayson CM, Gardner A, McKay C. Changing character: the ramsar convention on wetlands and climate change in the Murray-Darling Basin, Australia. Environ Plan Law J. 2010;27:401–25.

Pittock J, Finlayson CM, Howitt JA. Beguiling and risky: "Environmental works and measures" for wetlands conservation under a changing climate. Hydrobiologia. 2012: online.

Pressey RL, Adam P. A review of wetland inventory and classification in Australia. Vegetatio. 1995;118:81–101.

Robins L, Kanowski P. 'Crying for our Country': eight ways in which 'Caring for our Country' has undermined Australia's regional model for natural resource management. Australas J Environ Manag. 2011;18:88–108.

SEWPAC. Australian wetlands database. Canberra: Department of Sustainability, Environment, Water, Population and Communities; 2012.

Thackway R, Olsson K. Public/private partnerships and protected areas: selected Australian case studies. Landsc Urban Plan. 1999;44:87–97.

WGCS. Does a 3,200Gl reduction in extractions comnined with the relaxation of eight constraints give a healthy working Murray-Darling Basin River system? Sydney: Wentworth Group of Concerned Scientists; 2012.

Macquarie Marshes: Murray-Darling River Basin (Australia)

160

Rachael F. Thomas and Joanne F. Ocock

Contents

Introduction	1898
Location	1898
Hydrology	1900
Wetland Ecosystems	1900
Biodiversity	1901
Conservation Status	1904
Ecosystem Services	1904
Threats and Impacts	1904
Environmental Water Management	1905
Future Challenges	1906
References	1907

Abstract

The Macquarie Marshes are a large Australian floodplain wetland within the semi-arid region of the Murray-Darling River Basin, a Ramsar listed wetland of international importance recognised globally for its high biodiversity and conservation values. Unique flooding regimes from variable river flows of the Macquarie River are vital for a flourishing mosaic of diverse wetland habitats: open water lagoons and vast common reed-beds to lush water couch meadows to extensive stands of river red gum forests and woodlands. They are renowned for

---

R. F. Thomas (✉) · J. F. Ocock (✉)
Water and Wetlands Team, Science Division, NSW Office of Environment and Heritage, Sydney, NSW, Australia

Centre for Ecosystem Science, School of Biological, Earth and Environmental Sciences, UNSW, Sydney, NSW, Australia
e-mail: rachael.thomas@environment.nsw.gov.au; joanne.ocock@environment.nsw.gov.au

© Springer Science+Business Media B.V., part of Springer Nature 2018
C. M. Finlayson et al. (eds.), *The Wetland Book*,
https://doi.org/10.1007/978-94-007-4001-3_209

high waterbird abundance and diversity, especially colonial nesting waterbirds which can breed in amazingly large numbers, up to hundreds of thousands at a time. Fish, frogs, turtles and a diversity of other fauna species rely on the Marshes. The Macquarie Marshes also support a rich Aboriginal and European cultural heritage. But river regulation by large dams and river flow extraction have altered river flows changing flooding regimes. A concomitant decline in waterbird abundance and breeding events, fish populations, and the health of river red gum trees occurred. Recent concerns about degradation in the Murray-Darling River Basin initiated an Australian governments' recovery program, significantly increasing the water allocated to the Macquarie River for environmental purposes. Evaluation of environmental water use under the Basin Plan is vital for effective adaptive management, but will be contingent on a commitment to long-term ecological research and monitoring.

Keywords

Floodplain wetlands · Environmental Flows · Waterbirds · Wetland vegetation · Adaptive management · Basin plan

Introduction

The Macquarie Marshes ("the Marshes") are an iconic Australian dryland floodplain wetland recognized globally for its ecological and conservation significance. A biodiversity hotspot and complex socio-ecological system, it has been at the center of a long debate about competing demands for water and the balance between environmental, social, and economic requirements. Future management challenges remain to restore degraded systems and to ensure the critical ecological structure and functions of the Marshes are maintained.

Location

The Macquarie Marshes (30°50′ S, 147 °30′ E, Fig. 1) are a large ($\sim$2,000 km^2, Thomas et al. 2015) freshwater floodplain wetland of the Murray-Darling River Basin, located on the lower reaches of the Macquarie River in the Macquarie-Bogan River catchment (74,700 km^2) in south-eastern Australia (Fig. 1a). The Macquarie River originates west of the Great Dividing Range and flows north-west towards the Barwon-Darling River (Fig. 1a). The Marshes are situated on a complex dryland alluvial floodplain north of Marebone Weir where the Macquarie River breaks down into an intricate network of braided channels and creeks with interconnected wetland habitats (Ralph and Hesse 2010) (Fig. 1b). Wetlands in the south and north marsh regions form along the Macquarie River, and wetlands of the east marsh are formed along the Gum Cowal-Terrigal Creek (Fig. 1b). The creeks re-form to one main

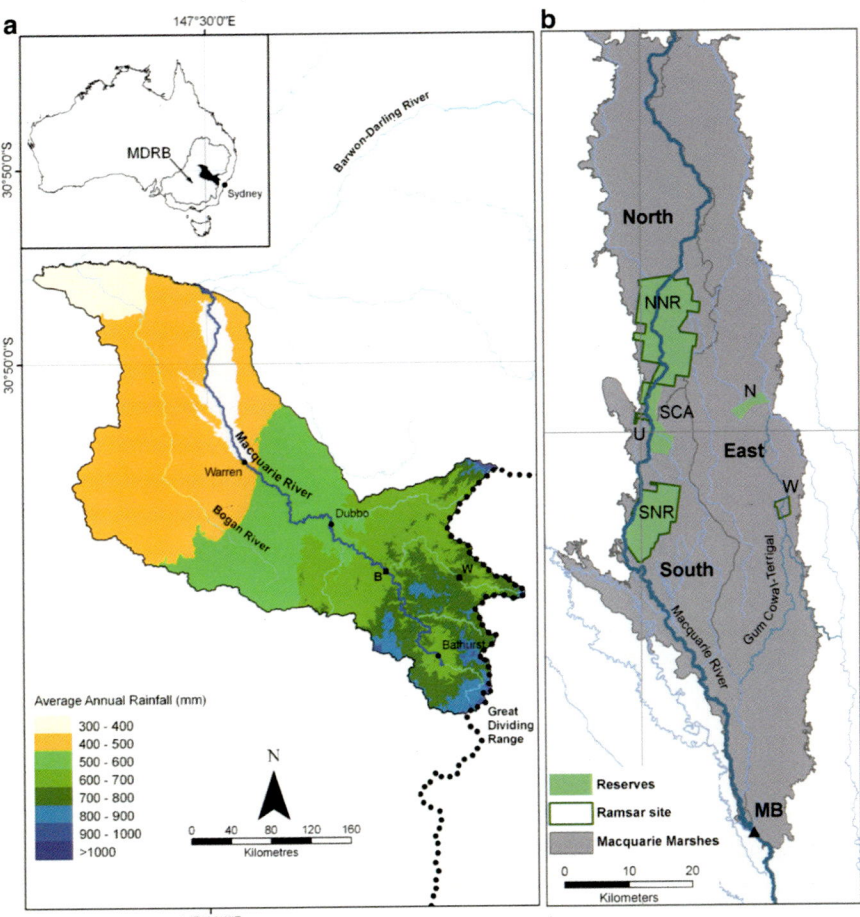

Fig. 1 (a) Location of the Macquarie Marshes floodplain and the major reservoirs of Burrendong (*B*) and Windamere (*W*) Dams in the Macquarie-Bogan River Catchment of the Murray-Darling River Basin (*MDRB*) and (b) the Macquarie Marshes located north of Marebone Weir (*MB*) with wetland habitats forming along the Macquarie River in the north and south regions (*grey boundary*) and in the east forming along the Gum Cowal-Terrigal Creek. Protected areas include: the Macquarie Marshes Nature Reserve (*NR*) in three sections – Northern NR (*NNR*), Southern NR (*SNR*), and Ninia (*N*); the State Conservation Area (*SCA*); and the Macquarie Marshes Ramsar Site (*thick outline*) including the NNR and SNR, Wilgara (*W*) and U-Block (*U*)

channel before joining the Barwon-Darling River (Fig. 1a). The Marshes are in a semi-arid region with low mean annual rainfall (400–500 mm, Fig.1a) and high average annual pan-evaporation rates (1,800–2,400 mm). Average annual temperatures range from minima of 3–6 °C during winter to maxima of 30–36 °C during summer.

Hydrology

The Marshes rely on river flows from the Macquarie River which are naturally highly variable, as is typical of most Australian dryland rivers (Ren and Kingsford 2011) (Fig. 2). The Macquarie River is highly regulated, predominantly by Burrendong Dam (operational: 1967) and Windamere Dam (operational: 1984) (Fig. 1a), as well as another eight large dams, weirs, regulators, channels, and off-river storages (Kingsford 2000).

Flooding is spatially variable and closely linked to river flows mirroring their high temporal variability (Ren et al. 2010; Thomas et al. 2015). There is a gradient of flood frequency across the floodplain with core wetland areas inundated across a range of frequencies: moderate (once every 4–6 years) to high (once every 1–3 years) and the outer floodplain confined to low inundation frequencies (once every 10–20 years) (Thomas et al. 2011).

Wetland Ecosystems

Flooding is the most influential ecological driver of the Marshes and triggers a series of dynamic ecological processes which produce a "boom" in species biodiversity and abundance. Fundamental to this biodiversity is the mosaic of diverse wetland habitats that reflect the variable inundation patterns in accordance with the flood dependencies of different wetland vegetation types (Fig. 3). Open water lagoons with varying depths are interspersed within vast beds of common reed *Phragmites australis* fringed with cumbungi *Typha domingensis* forming important habitat for waterbirds, fish, and frogs.

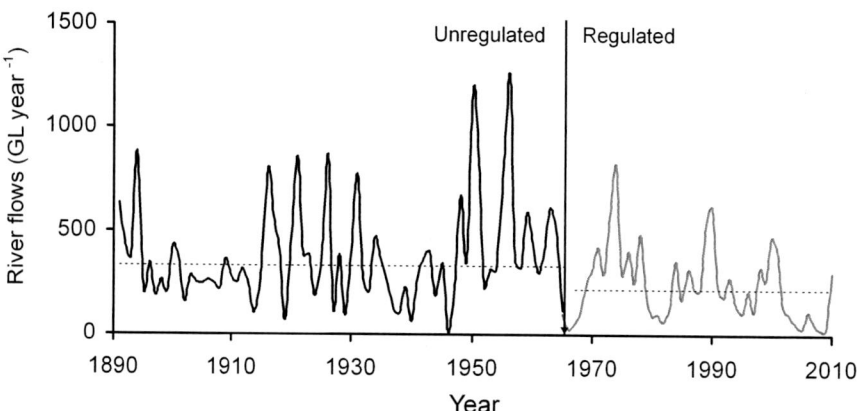

Fig. 2 Unregulated (median = 331GL) and regulated (median = 220GL) river flows to the Macquarie Marshes) measured at Oxley, Fig. 1b) (Modeled data sourced from Ren and Kingsford 2011; Environmental Management, Statistically integrated flow and flood modelling compared to hydrologically integrated quantity an quality model for annual flows in the regulated Macquarie River in arid Australia, 48, 2011, 177–88, Ren SQ, Kingsford RT; © 2011, with permission of Springer Science + Business Media)

Fig. 3 Wetland vegetation types of the Macquarie Marshes: (**a**) Common reed beds and open water lagoons, (**b**) Water couch marsh, (**c**) Mixed marsh sedgelands, and (**d**) River red gum forests and woodlands (Photo credits: © Rights remain with the photographer shown with the image)

Water couch *Paspalum distichum* forms lush grassland swamps. Mixed marsh sedgelands are a dynamic combination of emergent macrophytes that vary with flooding regimes and can include *Eleocharis* and *Juncus* spp. and nardoo *Marsilea drummondii*. Lignum shrublands *Duma florulenta* and river cooba *Acacia stenophylla* form important colonial-nesting waterbird habitat. There are extensive river red gum *Eucalyptus camaldulensis* forests and woodlands, often associated with coolibah *E. coolabah* and black box *E. largiflorens* woodlands that occur at the outer limits of flooding. At higher elevations, dry land floodplain woodlands, shrublands, and grasslands can occasionally receive flooding (Paijmans 1981; DECCW 2010; OEH 2012).

Biodiversity

The Marshes are renowned for high waterbird species richness and abundance. At least sixty different species have been recorded with total count estimates of up to 300,000 waterbirds (Kingsford and Thomas 1995). They are also a globally significant site for waterbird breeding (Fig. 4) as some of Australia's largest waterbird breeding events have been recorded in the Marshes (Kingsford and Johnson 1998). Forty-four species are known to breed in the Marshes, of which 16 are colonial-nesting waterbirds (Kingsford and Auld 2005). The Marshes are an especially

Fig. 4 Waterbirds of the Macquarie Marshes: (**a**) Australian white ibis chicks, (**b**) Straw-necked ibis, (**c**) Egrets, and (**d**) Plumed whistling ducks (Photo credits: © Rights remain with the photographer shown with the image)

important breeding site for ibis (e.g., Fig. 4a Australian white ibis *Threskiornis molucca*, Fig. 4b straw-necked ibis *Threskiornis spinicollis*, and glossy ibis *Plegadis falcinellus*) and egrets (e.g., Fig. 4c intermediate egret *Ardea intermedia*, eastern great egret *Ardea modesta*, little egret *Egretta garzetta*, and cattle egret *Ardea ibis*). Significant breeding colonies also occur for herons (e.g., nankeen night heron *Nycticorax caledonicus* and the white-necked heron *Ardea pacifica*), and cormorants (e.g., pied cormorant *Phalacrocorax varius*, little pied cormorant *Microcarbo melanoleucos*, great cormorant *Phalacrocorax carbo*, and little black cormorant *P. sulcirostris*) (Kingsford and Auld 2005; OEH 2012).

Several waterbird species are legislated as threatened at state, national, and international levels, including the Australasian bittern *Botaurus poiciloptilus*, black-necked stork *Ephippiorhynchus asiaticus*, brolga *Grus rubicundus*, and freckled duck *Stictonetta naevosa*. Migratory waterbird species that utilize the Marshes habitats are listed under international migratory bird agreements (Kingsford and Auld 2005; OEH 2012).

A large number of bird species (>130) other than waterbirds occur in the Marshes area. Mammals such as the red kangaroo *Macropus rufus* (Fig. 5a), eastern grey kangaroo *M. giganteus*, the common water-rat *Hydromys chrysogaster*, the short-beaked echidna *Tachyglossus aculeatus*, bats, and the brush tail possum *Trichosurus*

Fig. 5 Fauna of the Macquarie Marshes: (**a**) Red kangaroo, (**b**) Inland carpet python, (**c**) Barking marsh frog, and (**d**) Inland crab (Photo credits: © Rights remain with the photographer shown with the image)

vulpecula are common, whereas the Gould's long-eared bat *Nyctophilus gouldii* is at the western limit of its distribution (OEH 2012).

Fifteen species of fish, four species of turtle, 30 species of lizards, and 14 species of snakes (Fig. 5b) have also been recorded (OEH 2012). All native fish species in the Macquarie Marshes and lower Macquarie River typically recruit during spring and early summer. Murray cod *Maccullochella peeli* and silver perch *Bidyanus bidyanus* are listed as vulnerable species (Rayner et al. 2009).

Fifteen species of frogs have been detected in the Marshes, including striped-salmon frog *Limnodynastes salmini*, barking marsh frog *Limnodynastes fletcheri* (Fig. 5c), crucifix frog *Notaden bennetti*, and green tree frog *Litoria caerulea* (DECCW 2010; Ocock et al. 2013). This is among the highest of any wetland in the Murray-Darling River Basin. They are found around all types of wetland habitats including common reed beds, water couch marshes, flooded woodland, and rain-fed ephemeral ponds. Frogs and tadpoles are an important source of food for a number of animals, including snakes and water birds.

The Marshes support over 160 aquatic invertebrate species. Extremely high densities of microinvertebrates (<250 mm) can occur during flooding (Jenkins et al. 2009), and the inundation of diverse wetland habitats drives planktonic diversity, respiration, and primary productivity (Kobayashi et al. 2015). Crayfish, freshwater crabs (Fig. 5d), and mussels are also often found throughout the Marshes.

Conservation Status

The Macquarie Marshes have a long history of protection (Kingsford and Thomas 1995; OEH 2012). In 1971, parts of the Marshes were formally legislated as a Nature Reserve (IUCN category IV) in three sections: Northern Nature Reserve (12,260 ha), Southern Nature Reserve (6,292 ha), and Ninia (923 ha) (Fig. 1b). The Northern and Southern sections of the Nature Reserve and Wilgara wetland, a privately owned property, were listed under the Ramsar Convention as a Wetland of International Importance in 1986 and 1999, respectively. In recent years, there have been further additions to the Marshes reserve network (State Conservation Area) and to the Ramsar site (U-Block) (OEH 2012) (Fig. 1b).

Ecosystem Services

The ecosystem services and benefits that the Macquarie River and the Macquarie Marshes provide, either directly or indirectly, are economically, socially, and ecologically wide ranging (OEH 2012). The Macquarie River provides water to regional townships and agricultural economy, and the floodplain provides fertile land for primary production. The main agricultural industries of the Marshes region are cattle and sheep grazing and cropping (predominantly dryland wheat and irrigated cotton) (Herron et al. 2002). The Macquarie River was developed largely for the purposes of irrigation (using 89% of all water diverted) (Kingsford 2000).

The Marshes provide important cultural services and many people have significant links through historic connections. They were an important cultural and spiritual focus for the Wailwan people, the traditional Aboriginal custodians. Contemporary Aboriginal communities value the Marshes through their continued custodial connection. Pastoral graziers also have deep connections to the Marshes through multi-generational family history of working the land. Graziers on the Marshes benefit from the flooding of native pasture which enhances livestock productivity (DECCW 2010; OEH 2012).

Wetland hydrological cycles are regulated and maintained by the Macquarie Marshes, and they may play a role in balancing the local climate. The Marshes support waterbirds, frogs, and fish that biologically regulate disease vectors (e.g., mosquitoes) and agricultural pests (e.g., grasshoppers). The Marshes are critical for nutrient cycling, trapping and stabilization of sediments, and accumulation of organic matter for the formation of fertile soil (OEH 2012).

Threats and Impacts

River regulation and water diversions have altered river flows to the Marshes (Ren and Kingsford 2011), with flows reduced by approximately 45% and subsequent flooding by 41% (Ren et al. 2010). Changed flooding patterns have reduced the area of frequently flooded core wetland by up to 50% (Thomas et al. 2011). Earthworks associated with water resource development upstream and adjacent to the Marshes also affect flow paths (Steinfeld and Kingsford 2011).

The ecological impacts of river regulation are well established. Reduced flows have led to declines in waterbird populations (Kingsford and Thomas 1995), breeding events of colonial-nesting waterbirds (Kingsford and Johnson 1998; Kingsford and Auld 2005), and fish populations (Rayner et al. 2009). Changes in flooding patterns have extended dry periods which in turn have reduced microinvertebrate productivity and biodiversity (Jenkins et al. 2009) and altered the structure and composition of flood dependent vegetation in the Marshes (Thomas et al. 2010; Bino et al. 2015). The impact of floodplain earthworks on flooding have contributed to the decline in health of river red gums (Steinfeld and Kingsford 2011). Social changes in the local community may be related to a reduced ability of the Marshes to support grazing with many families leaving the area due to a decline in capacity to sustainably raise cattle (Fazey et al. 2006).

Wetland habitat and river red gum forests and woodlands in the Marshes are vulnerable to fire especially when they are dry (DECCW 2010). Fire management strategies for the Marshes Reserves consider the optimal fire regimes for vegetation community persistence, for the protection of endangered ecological communities, and for threatened fauna and waterbird rookery sites. Fire is also managed to protect heritage sites of Aboriginal cultural and historic significance. Management strategies are in place to control invasive pest species including wild pigs and foxes (DECCW 2010). Fish communities are dominated by alien species, particularly European carp *Cyprinus carpio* (Rayner et al. 2009). Lippia, *Phyla canescens*, is a weed of significant risk to the ecological and agricultural values of the marshes (DECCW 2010).

Water availability projections for the Macquarie catchment under climate change are wide ranging with the best estimate predicting an 8% decrease in 2030, and at the extremes (low and high warming scenarios) a 25% decrease and 25% increase (CSIRO 2008). Reduced water availability will impact water allocations to agriculture and the environment. Flood frequency is predicted to decrease and dry periods will be extended, exacerbating impacts on species already vulnerable to reduced water availability and other threats such as fire. However, the ecological impacts of these predictions are unlikely to be as severe as the effects of river regulation (CSIRO 2008; Kingsford 2011).

Environmental Water Management

There have been decades of environmental water management in the Macquarie Marshes. The need to mitigate the effects of Burrendong Dam on river flows was recognized prior to its construction. An environmental water allocation to the Marshes was recommended, however not granted until 1980 (18.5GL). Water management plans recognized that the subsequent increases in environmental water (50GL in 1985, 125GL in 1996) were not adequate due to over-allocation of water to extractive users (OEH 2012).

The current Water Sharing Plan (WSP) governing water to the Marshes apportions a 160GL share to the environment (the Environmental Water Allowance) out of 900 shares available in Burrendong Dam. Only a proportion of the 160GL share is annually allocated to the environment based on dam water availability using a series of planning rules. The WSP also includes recommended restrictions to water access throughout the region (DECCW 2010).

Concerns about degradation of river systems in the Murray-Darling River Basin led Australian governments' to embark on a major rehabilitation program to restore flows to the environment. Through a buyback of water licenses and improved irrigation efficiencies, there is the provision of additional environmental water, held as Adaptive Environmental Water in New South Wales (NSW) and as Commonwealth Environmental Water Holder (CEWH) entitlements, bringing the total share to almost 350GL. The Office of Environment and Heritage (OEH) manages NSW environmental water on behalf of the NSW Government in collaboration with the Macquarie Cudgegong Environmental Flows Reference Group (OEH 2014).

With the considerable public investment in environmental water for the Macquarie Marshes and other wetlands in the Murray-Darling River Basin comes the demand to demonstrate its environmental, economic, and social benefits. Recently the Australian government enacted the Basin Plan 2012 (Australian Government 2012) to provide comprehensive integrated management of the Murray-Darling River Basin's water resources across the state government jurisdictions. By balancing water access for human consumption and environmental needs, the Basin Plan strives to protect and restore key iconic wetlands and functions, setting new limits on the amount of water that can be taken for consumptive use (referred to as long-term average sustainable diversion limits). The Basin Plan offers an adaptive governance framework that is necessarily dynamic for planning at such a large scale and high level of complexity. It is currently being implemented through a long-term environmental water planning process (OEH 2014).

Future Challenges

While water remains scarce and in high demand, the protection and recovery of river flows is critical for the long-term resilience of the Macquarie Marshes. Essential to this is recognizing that the Marshes are connected to the rivers and other wetland systems within the Murray-Darling River Basin, presenting unique organizational challenges when managing at large spatial scales within the context of river catchments and basins (Kingsford 2011). The dynamic ecological processes of the Marshes are not well understood, introducing multiple uncertainties for environmental managers. Further complicating this understanding is the difficulty in separating anthropogenic impacts from the naturally high variability of river flows and climate which are only partially predictable over time.

Recognition of key uncertainties and accommodating them into environmental water management brings a unique opportunity to embrace an active adaptive management approach (Kingsford et al. 2011; OEH 2014). Crucial to the success of adaptive decision-making starts with stakeholder collaboration in the process (OEH 2014). Clear, measurable, and agreed objectives of management relevance and scales are critical for evaluating performance. Models that link potential management actions to ecosystem responses will play a key role in informed decision-making and in representing uncertainty. Evaluation is a central tenet of adaptive management: an iterative process of progress assessment towards achieving objectives and feedback for improved decision-making (Williams 2011). However, learning can only be effective through an adaptive long-term ecological research and monitoring approach (Lindemayer and Likens 2009). One of the biggest challenges will be a long-term commitment to learning through simultaneous management and scientific inquiry, a necessity for the protection of the Marshes into a future that remains uncertain.

References

Australian Government. Basin Plan 2012. Water Act 2007. Canberra; 2012. Available from: https://www.legislation.gov.au/Details/C2016C00469. Accessed 2 Sep 2016.

Bino G, Sisson SA, Kingsford RT, Thomas RF, Bowen S. Developing state and transition models of floodplain vegetation dynamics as a tool for conservation decision-making: a case study of the Macquarie Marshes Ramsar wetland. J Appl Ecol. 2015;52:654–64.

CSIRO. Water availability in the Macquarie-Castlereagh. A report to the Australian Government from the CSIRO Murray-Darling Basin Sustainable Yields Project. Australia: CSIRO; 2008. 144 p.

DECCW. Macquarie Marshes adaptive environmental management plan. Synthesis of information projects and actions. Sydney: Department of Environment, Climate Change and Water NSW; 2010. 100 p. Available from: http://www.environment.nsw.gov.au/resources/environmentalwater/100224-aemp-macquarie-marsh.pdf. Accessed 2 Sep 2016.

Fazey I, Proust K, Newell B, Johnson B, Fazey JA. Eliciting the implicit knowledge and perceptions of on-ground conservation managers of the Macquarie Marshes. Ecol Soc 2006;11. http://www.ecologyandsociety.org/vol11/iss1/art25/. Accessed 2 Sep 2016.

Herron N, Davis R, Jones R. The effects of large-scale afforestation and climate change on water allocation in the Macquarie River catchment, NSW, Australia. J Environ Manag. 2002;65:369–81.

Jenkins K, Kingsford R, Ryder D. Developing indicators for floodplain wetlands: managing water in agricultural landscapes. Chiang Mai J Sci. 2009;36:224–35.

Kingsford R. Ecological impacts of dams, water diversions and river management on floodplain wetlands in Australia. Austral Ecol. 2000;25:109–27.

Kingsford RT. Conservation management of rivers and wetlands under climate change – a synthesis. Mar Freshw Res. 2011;62:217–22.

Kingsford RT, Auld KM. Waterbird breeding and environmental flow management in the Macquarie Marshes, arid Australia. River Res Appl. 2005;21:187–200.

Kingsford RT, Johnson WJ. The impact of water diversions on colonially nesting waterbirds in the Macquarie Marshes in arid Australia. Colon Waterbirds. 1998;21:159–70.

Kingsford R, Thomas R. The Macquarie Marshes in arid Australia and their waterbirds: a 50-year history of decline. Environ Manag. 1995;19:867–78.

Kingsford RT, Biggs HC, Pollard SR. Strategic adaptive management in freshwater protected areas and their rivers. Biol Conserv. 2011;144:1194–203.

Kobayashi T, Ralph T, Ryder D, Hunter S, Shiel R, Segers H. Spatial dissimilarities in plankton structure and function during flood pulses in a semi-arid floodplain wetland system. Hydrobiologia. 2015;747:19–31.

Lindemayer DB, Likens GE. Adaptive monitoring: a new paradigm for long-term research and monitoring. Trends Ecol Evol. 2009;24:482–6.

Ocock J, Rowley JL, Penman T, Rayner T, Kingsford R. Amphibian chytrid prevalence in an amphibian community in arid Australia. Ecohealth. 2013;10:77–81.

OEH. Environmental water use in New South Wales. Outcomes 2013–14. Sydney: Office of Environment and Heritage; 2014. 30 p. Available from: http://www.environment.nsw.gov.au/resources/environmentalwater/140811-env-water-outcomes-1314.pdf. Accessed 2 Sep 2016.

OEH. Macquarie Marshes Ramsar site. Ecological character description – Macquarie Marshes and U-block components. Sydney: Office of Environment and Heritage, NSW Department of Premier and Cabinet; 2012. 144 p. Available from: http://www.environment.nsw.gov.au/resources/water/120517MMECDPt1.pdf. Accessed 2 Sep 2016.

Paijmans K. The Macquarie Marshes of Inland Northern New South Wales, Australia. CSIRO Australian Division of Land Use Research Technical Paper No. 1981;41:1–22.

Ralph TJ, Hesse PP. Downstream hydrogeomorphic changes along the Macquarie River, southeastern Australia, leading to channel breakdown and floodplain wetlands. Geomorphology. 2010;118:48–64.

Rayner TS, Jenkins KM, Kingsford RT. Small environmental flows, drought and the role of refugia for freshwater fish in the Macquarie Marshes, arid Australia. Ecohydrology. 2009;2:440–53.

Ren SQ, Kingsford RT. Statistically integrated flow and flood modelling compared to hydrologically integrated quantity and quality model for annual flows in the regulated Macquarie River in arid Australia. Environ Manag. 2011;48:177–88.

Ren S, Kingsford RT, Thomas RF. Modelling flow to and inundation of the Macquarie Marshes in arid Australia. Environmetrics. 2010;21:549–61.

Steinfeld CMM, Kingsford RT. Disconnecting the floodplain: earthworks and their ecological effect on a dryland floodplain in the Murray-Darling Basin. Aust River Res Appl. 2011;29:206–18.

Thomas RF, Bowen S, Simpson SL, Cox SJ, Sims NC, Hunter SJ, et al. Inundation response of vegetation communities of the Macquarie Marshes in semi-arid Australia. In: Saintilan N, Overton I, editors. Ecosystem response modelling in the Murray Darling Basin. Melbourne: CSIRO Publishing; 2010. 18 p.

Thomas RF, Kingsford RT, Lu Y, Hunter SJ. Landsat mapping of annual inundation (1979–2006) of the Macquarie Marshes in semi-arid Australia. Int J Remote Sens. 2011;32:4545–69.

Thomas RF, Kingsford RT, Lu Y, Cox SJ, Sims NC, Hunter SJ. Mapping inundation in the heterogeneous floodplain wetlands of the Macquarie Marshes, using Landsat Thematic Mapper. J Hydrol. 2015;524:194–213.

Williams BK. Adaptive management of natural resources-framework and issues. J Environ Manage. 2011;92:1346–53.

The Coorong: Murray-Darling River Basin (Australia)

161

Peter Gell

Contents

Introduction	1910
A Ramsar Wetland	1911
Wetland Change	1913
Future Challenges	1916
References	1918

Abstract

The Coorong is a long, narrow back-barrier lagoon near the mouth of the River Murray, Australia. It was accorded the status of a Wetland of International Importance under the Ramsar Convention in 1985 when it was described as a shallow, brackish-to-hypersaline lagoon. Historically the lagoon has played an important role as habitat for waterbird and fish populations, in particular migratory wading bird species covered under international agreements, underpinned by extensive seagrass beds. Evidence of long term change reveals the Coorong to have been a highly tidal system for several thousand years but was substantially affected by water diversions in the catchment and the construction of end-of-system barrages in 1940. These changes have seen declines in seagrass cover and associated birds and fish, exacerbated by an extended dry period in recent years with the Coorong experiencing extreme hypersalinity and consequent change in its perceived natural ecological character. The extended record of change reveals

P. Gell (✉)
Water Research Network, Federation University Australia, Ballarat, VIC, Australia
e-mail: p.gell@federation.edu.au

© Springer Science+Business Media B.V., part of Springer Nature 2018
C. M. Finlayson et al. (eds.), *The Wetland Book*,
https://doi.org/10.1007/978-94-007-4001-3_210

the Coorong to be outside its historical range of variability and the challenge to restore its condition is considerable.

Keywords

Ramsar · Estuary · Water diversion · Migratory waders · *Ruppia* · Paleolimnology

Introduction

The Coorong and Lower Murray Lakes Wetland complex comprises a long, narrow, back-barrier lagoon connected by a series of channels to Lakes Alexandrina and Albert, two large terminal lakes at the end of the Murray-Darling Basin, referred to collectively as the Lower Lakes (Fig. 1). The Basin, in excess of 1 million km^2, is Australia's largest covering 14% of the continent. The watershed to the Basin is mostly the inner ramparts of the Great Dividing Range extending from the subtropical climates of southern Queensland to the alpine zone (>2,000 m above sea level) in temperate Victoria.

The Coorong is a large, coastal lagoon complex situated between a Holocene barrier dune, the Younghusband Peninsula, and, on its north-eastern margin, a fossil shoreline formed during the last interglacial period ~120,000 years ago, the Woakwine Range. Naturally, while fresh and tidal water was exchanged between the Coorong, the Murray Mouth, and the Lower Lakes, the fossil dune and Hindmarsh Island limited this exchange. East of the Coorong is evidence of a sequence of dunes representing each high sea level stand of the last 700,000 years. The swales, depressions between the dunes, became inundated from winter rains and these waters slowly made their way to the Coorong via Salt Creek. The Coorong is comprised of the south and north lagoons with exchange limited by a narrow channel at Parnka Point. Being closer to the mouth, the north lagoon was more strongly influenced by tidal water but historically this too reached well into the south lagoon.

Presently the Coorong is separated from the lakes by a series of five barrages. The first barrage was constructed of sand bags emplaced near the site of the present Mundoo Barrage in 1914 (Sim and Muller 2004) in order to maintain water quality at a level suitable for irrigation by reducing the intrusion of tidal water into Lake Alexandrina. These were replaced with a timber barrage in 1919 and the final construction was in place by 1939. The shelly calcarenite reef, the remnant of the last interglacial dune, was used as the foundation for the Tauwitchere, Ewe Island, Mundoo, and Boundary Creek Barrages (Bourman et al. 2000), while the Goolwa Barrage, constructed across the main channel, was built on piles driven into soft sediments. The entire barrage system was completed by 1940. The seasonal flooding of the interdune swales east of the Coorong interrupted agricultural production so, from 1865, an extensive network of channels was excavated to shed this fresh water to the ocean.

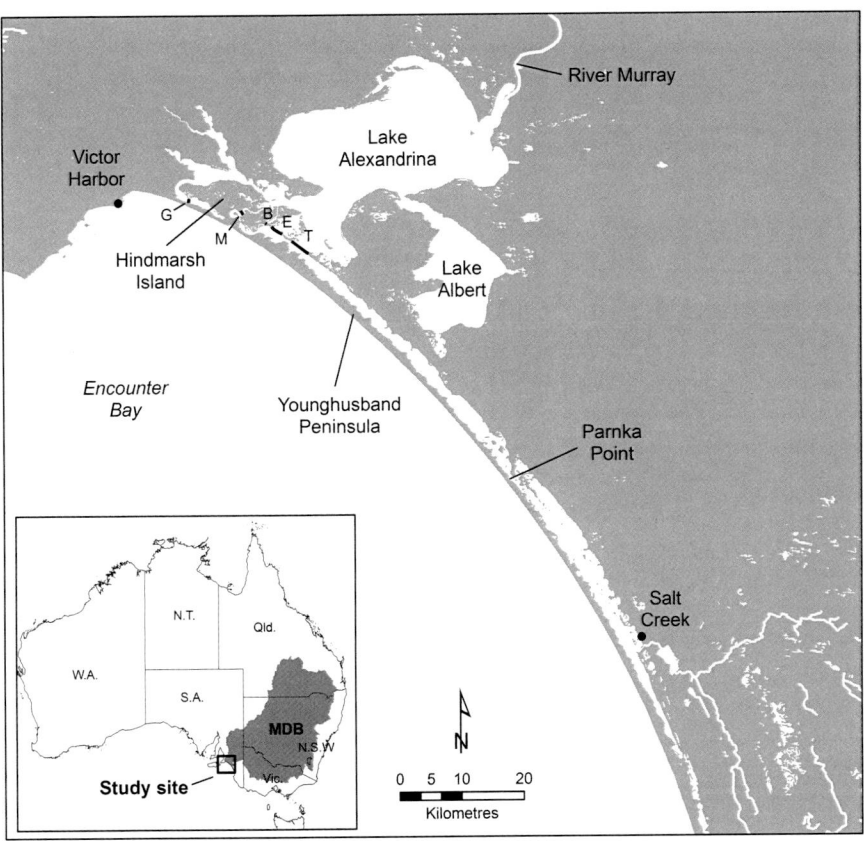

Fig. 1 The Coorong, Lower Lakes, and River Murray mouth region in South Australia. *G* Goolwa Barrage; *M* Mundoo Barrage; *B* Boundary Creek Barrage; *E* Ewe Island Barrage; *T* Tauwitchere Barrage (Figure credit: P Gell © Rights remain with the author)

A Ramsar Wetland

In 1985 the wetland complex was accorded status as a Wetland of International Importance under the Ramsar Convention. The site listing (https://rsis.ramsar.org/ris/321) states:

> **The Coorong, and Lakes Alexandrina & Albert Wetland.** 01/11/85; South Australia; 142,530 ha; 35°56′S 139°18′E. National Park, Game Reserves and Crown Land; Shorebird Network Site. The site is located at the mouth of the River Murray, south east of the city of Adelaide. It consists of two lakes forming a wetland system at the river's mouth and a long, shallow brackish-to-hypersaline lagoon which they feed into, separated from the ocean by a narrow sand dune peninsula. The lakes contain water of varying salinity and include a unique mosaic of 23 wetland types including intertidal mud, sand and salt flats, coastal brackish/saline lagoons and permanent freshwater lakes. The site is of international

importance for migratory waterbirds and supports the greatest wealth of waterbird species in the Murray-Darling Basin. It hosts important nesting colonies of cormorants, plovers, ibises and terns, and also supports globally endangered species such as the orange-bellied parrot (*Neophema chrysogaster*) and the Murray cod (*Maccullochella peelii peelii*). The site is popular for recreation activities include camping, boating, regulated duck hunting, and supports a range of commercial activities related to tourism, irrigated agriculture, and commercial fishing. The area is central to aboriginal culture and spiritual beliefs, and it is noted for its extensive sites of historic and geological importance. Ramsar site no. 321.

Water levels in the Coorong vary seasonally with dry conditions in summer exposing mudflats that provide habitat to many migratory and resident waterbirds. Over 80 species of waterbirds have been recorded in the region (DEH 2000), and the site is considered a major site for the conservation of 30 species. It is particularly important as wintering habitat for red-necked stints *Calidris ruficollis*, sharp-tailed sandpipers *C. acuminata*, and curlew sandpipers *C. ferruginea* that are protected under the Japan (JAMBA) and China (CAMBA) Australia Migratory Bird Agreements. An excess of 50,000 grey teal *Anas gracilis* have been recorded from the site, and the region hosts about 2,000 Cape Barren geese *Cereopsis novaehollandiae*, one of the world's rarest.

The orange-bellied parrot *Neophema chrysogaster* is a critically endangered, endemic species that breeds in south west Tasmania and overwinters in the coastal marshes of south eastern Australia. While there are over 200 captive birds in the species recovery program, its wild population is estimated to be as few as 30 individuals. A few parrots were recorded in the Coorong in 2011. However, despite modest recruitment in the summer of 2011–2012, the species is projected to go extinct within the decade. Its future hinges on the success of a translocation strategy and feeding and nestbox support for breeding pairs in Tasmania. The Ramsar area also includes the high energy surf zone on the south side of the Younghusband Peninsula. This area represents important habitat for the endemic and nationally and internationally vulnerable hooded plover *Thinornis rubricollis* that nests above the high water mark but occasionally uses the Coorong lagoon. The rare fairy tern *Sternula nereis*, and vulnerable little tern *Sternula albifrons*, also nest on the islands within the Coorong.

The Coorong and Lower lakes support 59 fish species (DEH 2000) of which 33 are principally marine, 12 estuarine, and 23 largely freshwater. The site is a renowned fishery for mulloway *Argyrosomus hololepidotus* providing 80% of the South Australian catch in the 1990s. This species moves into the system from the ocean after flooding and this is now limited due to reduced flows. The large mulloway fishery in Lake Alexandrina was lost after the barrages limited the tidal extent of the system. The region is also known for the "Coorong (yellow-eye) mullet" *Aldrichetta forsteri* catch of which 150 t is harvested annually, and the freshwater lakes support murray cod *Maccullochella peelii*, silver perch *Bidyanus bidyanus,* and bony bream *Nematolosa erebi*. The value of the annual commercial fishery was estimated in 1997 at $4.6 million.

The latest management plan for the wetland complex cites, as part of its vision,

> …conservation of the Coorong…by incorporating the world's best practice in integrated natural resource management to:
>
> - conserve the … ecological attributes of the wetlands for the benefit of future generations;
> - protect and restore natural habitats;
> - restore viable populations of native species; and
> - improve water quality and increase flows through the wetlands. (DEH 2000)

Part of this commitment is to maintain the ecological character of the Coorong and Lower Lakes. The ecological character of a wetland, as defined by the Ramsar Convention Bureau, is: "…the structure and the interrelationships between the biological, chemical, and physical components of the wetland." However, the natural attributes of the Coorong are not clear and are presently the subject of considerable national debate. Issues central to this debate include the volume of natural flows of the River Murray into the system, the dynamics of the River mouth that required dredging to remain open through the recent drought, the impact of the barrages that separate Lake Alexandrina fresh water from the Coorong (DEH 2000), and the contribution of freshwater from the ephemeral, interdune wetlands of the Upper South East (USE) of South Australia.

Wetland Change

The Murray mouth closed for the first time in memory in 1981, associated with low river flows under drought. At this time, and at the time of the Ramsar listing, the wetland was recognized as a saline to hypersaline lagoon influenced by river flows. The lakes were considered to be mostly fresh. Analysis of sediment records that accumulate in the wetlands continuously over time reveal a different history to that based on memory and monitoring. The pre-barrage sediments in the Coorong, particularly in the sections more strongly influenced by tidal waters, consist of shelly marls, attesting to a marine past. Evidence for long-term change from microscopic indicators within these sediments reveals that, for most of the last 7,000 years, the Coorong was indeed an active tidal system with substantial contributions of fresh water from the USE via Salt Creek and relatively little influx of river water (Fluin et al. 2007). Preserved remains of diatom algae, foraminifers, and ostracod crustacea reveal that estuarine species predominated until after European settlement, most likely until the barrages were commissioned (Reeves et al. 2014). The Lower Lakes, more directly influenced by river flows, were largely fresh but did experience tidal intrusion. In the sediments in the south of Lake Alexandrina 5–30% of the diatoms preserved had marine affinities indicating a persistent tidal influence (Battarbee et al. 2012).

The ecological character description for the Ramsar listing of the Coorong identifies its connection with the lower lakes. Bourman et al. (2000), however, describe the flow restrictions of the last interglacial shoreline and reveal the differential depths of the respective channels. The Mundoo Channel is naturally shallow and sinuous and the broad Tauwitchere Channel varies in depth between 0.5 and 2 m. The Goolwa Channel by contrast is much larger, more efficient hydrologically, and is estimated to have carried 70% of the flow from the lower River Murray into Encounter Bay. So, naturally, the connection between Lake Alexandrina and the Coorong was relatively small. The interrelationship between the River Murray and the Coorong was related mostly to the importance of the flows through the Goolwa Channel on the location and functioning of the "Murray Mouth." These flows, therefore, allowed for an open mouth and the formation of an extensive tidal prism to the full length of the Coorong and into the lakes when river flows were low.

The barrages have created permanently elevated water levels in the lakes, and, in concert with declining flows through increasing abstraction upstream, limited the influence of river flows on the mouth. The effectiveness of the barrages in limiting marine water intrusion into Lake Alexandrina is illustrated by the maximum recorded salinity level of 833 ppm in the 1945 drought, compared to pre-barrage salinities of 6,929 ppm at Murray Bridge, 30 km upstream, during the 1914 drought (Bourman and Barnett 1995). Sedimentary evidence for the sudden loss of estuarine foraminifera behind the Goolwa channel barrage, from immediately after its construction, attest to an abrupt shift from estuarine to fresh conditions.

The morphology of the Murray Mouth has always been dynamic and has likely followed a natural, evolutionary trend, typical of drowned valleys, from an open system to an estuary that would ultimately progress to a stranded coastal lagoon. While this would occur naturally over centuries, geomorphological research within the region of the Coorong and lower lakes suggests that sediment has recently been accumulating at unprecedented rates. The mouth of the Murray migrated 1.3 km to the north-west between 1965 and 1995 and the sediments in the lagoon inside the mouth have been accreting rapidly. The accumulation of fine sediments across the system increased greatly after European settlement, but particularly after barrage construction, and the lakes have become more turbid over time (DEH 2000; Gell et al. 2009). Artificially elevated lake levels since barrage construction have led to shoreline progradation in sheltered and vegetated areas around Lake Alexandrina, whereas in exposed and vulnerable localities accelerated shoreline erosion of up to 12 m/year has occurred. Whether the principal source of increased sediment accumulation is shoreline erosion or increased transport by the river is uncertain. It is clear, however, that the volume of fine clays accumulating in the lower lakes and the Coorong has increased substantially, at rates unprecedented in the history of the system.

Throughout much of the system, the sediments that have accumulated since the commissioning of the barrages consist of fine, organic muds. These have accumulated sulfate salts that have, with permanent inundation, oxidized into sulfides. In the extended drought from 1997 to 2010 water levels were critically low at 0.75 m below mean annual sea level (MASL). The exposed sulfidic sediments risked a major acidification event in Lake Alexandrina but particularly in Lake Albert. Freshwater in Lake Alexandrina was pumped into Lake Albert to avert the risk of oxidation. With river flows at historic lows, there was little resource available to limit exposure of the sediments. A major acidification event was averted with the onset of a substantial wet period in 2010–2011 which returned substantial flows to the river. Sulfidic sediments also cover the Coorong lagoon for most of its 140 km length. This, and increasing salinity since the commissioning of the barrages, has seen a decline in populations of the aquatic plant *Ruppia*. Historically, it was thought that *Ruppia megacarpa* was the dominant species in the north lagoon and the more salt tolerant *Ruppia tuberosa* dominant in the south. Evidence from fossil seeds (Dick et al. 2011) shows that the south lagoon transitioned through *R. megacarpa* to *R. tuberosa* as water quality declined since the barrages, but now, with toxic sulfides widespread, both species are largely lost to the Coorong. Stable isotope and diatom evidence reveals the widespread shift from a food web driven by a structurally complex aquatic macrophytes to one driven by phytoplankton (Krull et al. 2009). While this trend commenced with the commissioning of the barrages, the high oxygen demand of the sediments continues to impact on aquatic fauna despite the respite afforded by the recent wet period.

The population of waterbirds in the area in 1981 was estimated at 122,000 which, at the time, was thought to represent more than 25% of the Australian total. Improved national waterbird population assessments using aerial surveys estimated the national total to be 4.65 million waterbirds in 2008, with the Coorong and Lower Lakes supporting 172,000 birds making it the 4th ranked Australian wetland complex for waterbirds (Kingsford et al. 2012). This reflects the importance of the site during drought when other more important wetlands in inland Australia are dry. When comparing waterbird numbers from 1985 with a series of counts from 2000 to 2007, Rogers and Paton (2009) observed a decline in the key species that characterized the South Lagoon in 1985, including fairy tern, common greenshank *Tringa nebularia*, grey teal, masked lapwing *Vanellus miles*, red-capped plover *Charadrius ruficapillus*, and white-faced heron *Egretta novaehollandiae*, with the lowest numbers of grey teal (~5,000) on record in 2007. They attribute the loss of black swan *Cygnus atratus* numbers in the south lagoon between 1985 and 2000 as due to the loss of *Ruppia tuberosa*. Similar correlations were detected in the South Lagoon between smallmouth hardyhead fish *Atherinosoma microstoma* and the two piscivorous bird species investigated, fairy tern and Australian pelican *Pelecanus conspicillatus*.

They do acknowledge that these observations are limited by the fact that comparisons have been made between a series of years (2000–2007) and a single year (1985) that may not be representative of the period. Even so, while numbers of sharp-tailed sandpiper and red-necked stint appear not to show a trend across the period 2000–2007, their numbers have declined by 60% relative to the isolated surveys undertaken in the 1980s. Curlew sandpipers did decline through 2000–2007 to populations around 10,000 and this continued a trend from the 1980s when 40,000 birds were regularly seen. Where they feed has also changed, with 87.8% of curlew sandpipers recorded nearer the Murray Mouth in 2007 compared to 17.9% in 2000 and 8.3% in 2001. The distribution of these species appears to be contracting in response to the increasingly limited tidally mixed zone as the habitat suitability of the Coorong declines.

On the other hand, banded stilt *Cladorhynchus leucocephalus*, a species strongly associated with ephemeral lakes, has increased in abundance dramatically over the

2000–2007 period, from 2,354 to 64,552 individuals. They became especially abundant in the Coorong with the appearance of large numbers of Australian brine shrimp *Parartemia zietziana* in the system in 2005. At the time, all of this increase was recorded in the South Lagoon; however, Thiessen (2011) observed 142,000 stilt in the north lagoon in 2008/9. This population left when floodwaters reduced salinity levels in 2010/11. By the end of the drought, when the Coorong was at its most saline, this species had become the most abundant with almost 300,000 birds.

In fact much of the assessment of trends in waterbird numbers is confounded by the hydroclimatic variability in the system. For example, Thiessen (2011) reported a 74% decline in waterbird numbers from nearly 540,000 in 2008/09 to less than 140,000 in 2010/11 when inland flooding reduced the role of the site as a drought refuge. Locally, the extensive mudflats were inundated after the 2010 floods resulting in a significant decline in the number of wading birds. Also, the freshening of the south lagoon reduced the food source for banded stilt, whose abundance declined from over 270,000 to less than 10,000 after the floods. In contrast, the flooding prompted fish breeding which attracted large numbers of piscivorous birds such as the Australian pelican.

Little is known of the long-term changes in the fish populations of the Coorong. The abundance of estuarine species on the seaward side of the barrages declined as the drought deepened through 2006–2009 as reduced flows led to elevated salinity levels (Zampatti et al. 2010). This lack of flow allowed easterly winds to drive hypersaline south lagoon waters to the north leading to a collapse in the numbers of the previously abundant smallmouth hardyhead and sandy sprat *Hyperlophus vittatus*. These were confined to the northern Goolwa Channel that remained subsaline while the hypersaline Tauwitchere populations were replaced with marine immigrants such as yellow-eyed mullet.

The sedimentary record has documented substantial change in the Coorong region since the barrages were put in place, in contrast to considerable stability, or gradual geomorphic evolution, over the previous 7,000 years. The ecological observations that have underpinned the Ramsar listing commenced when the wetlands were already under rapid transition. Despite much variability the trends in bird numbers are clear and constitute significant end points to a phase of wetland change that has been in place for many decades.

Future Challenges

The decline in the condition of both the Coorong, and the Lower Lakes, is widely acknowledged and was highlighted by the acute risk of widespread acidification in 2009 during the drought. Kingsford et al. (2011) identify the system as a Ramsar Site in crisis. The ecological character of the Coorong was identified, at the time of listing, as being a saline to hypersaline lagoon. An embargo was placed on the release of fresh water from the upper south-east to the Coorong to preserve its presumed natural character. Evidence from sediments records revealed the Coorong to be strongly tidal with freshening from Salt Creek being a frequent phenomenon.

Denied freshwater and sea water due to the increasingly restricted mouth, the salinity of the Coorong rose. Through the extended drought the salinity of the south lagoon rose to greater than five times that of seawater driving its ecosystem into one of phytoplankton, brine shrimp, and banded stilt and red-necked avocet *Recurvirostra novaehollandiae*. The native fish communities have declined, as have the nationally significant populations of migratory wading birds.

This long-term decline in condition, partly through drought and declining surface water inflow, and in part through deliberate intervention to maintain a 1985 baseline, has elevated the challenge of rehabilitating the Coorong and lower lakes system. Short-term solutions directed at maintaining key populations have been posed. Thiessen (2011) suggests that water be released in summer to lower water levels to enhance habitat availability for migratory waders, while Rogers and Paton (2009) suggested that the hypersaline condition of parts of the Coorong be maintained, since these supported unique waterbird communities that supplemented the site's ecological diversity. At the same time, they advocate mitigating salinity levels to provide for populations of smallmouth hardyhead, key prey for small terns. Ultimately they suggest that managers need to provide water level regimes that support keystone species such as *Ruppia* but also provide mudflats that are inundated at the right depths and frequencies to provide habitat for wading birds. These solutions, while appropriate from a local bird conservation perspective, underestimate the temporal and spatial scale of the drivers of changes observed in the system revealed in part by long-term monitoring but extended by the paleoenvironmental record.

This longer record has revealed that the system has undergone significant change since the commissioning of the barrages and that lack of flow, both fresh and tidal, has permitted the accumulation of high acid-potential sediments. Lakes Alexandrina and Albert faced ecological collapse when these sediments were exposed through drought. The lakes were considered to be permanently fresh (Sim and Muller 2004) and were managed to remain so. This was despite anecdotal evidence for explorers considering the water's quality as too poor for livestock to drink. Significantly, in the argument for water entitlements under the Murray-Darling Basin agreement, a permanent freshwater condition is the preferred ecological character despite sedimentary evidence revealing an increasing influence of tidal waters across Lake Alexandrina from the River Murray towards the barrages. While it is likely that sea water may have been let in to avoid acidification had the drought persisted, measures to resolve a repeated low flow phase, when the water levels in the lakes are again below sea level, remain to be resolved. A strong case has been made to reserve certain water volumes at times of low river flow to avoid salinization and ensure a freshwater-driven solution to the lower lakes (Kingsford et al. 2011). If a variable solution is considered, then periodic tidal inflow, akin to the pre-barrage state, could be permitted, and the environmental flow reserve could be applied if the interval to natural flooding was prolonged.

The waters of the Murray-Darling Basin have been greatly over-allocated and only recently (1995) have measures to cap abstraction and return environmental flows to the system in the form of the Murray-Darling Basin Plan (MDBA 2012) been implemented. Water buy-back schemes and water use efficiency measures across the basin are proposed to return 3,200 GL of flow, ~25% of the 14,000 GL mean annual flow, to the rivers and wetlands of the basin. The Coorong and Lower Lakes are identified as "Icon Sites" in the Living Murray Plan, special sites deserving of water allocation for wetland restoration (MDBA 2011). The basin lies in a temperate zone expected to receive declining winter rains under climate change scenarios for this century with 25–50% declines in runoff predicted across the region (Finlayson et al. 2013). Further, with sea levels rising 3.2 mm/a presently, management measures to retain good function to retain the vital ecological attributes that afforded it international significance in the first place will need a flexible approach and the capacity to accommodate ongoing climatic variation and estuarine evolution (Lester et al. 2009).

References

Battarbee R, Bennion H, Rose N, Gell P. Chapter 27, Human impact on freshwater ecosystems. In: Matthews JA, editor. The SAGE handbook of environmental change. London: SAGE Publications; 2012.

Bourman RP, Barnett EJ. Impacts of river regulation on the terminal lakes and mouth of the River Murray, South Australia. Aust Geogr Stud. 1995;33:101–15.

Bourman RP, Murray-Wallace CV, Belperio AP, Harvey N. Rapid coastal change in the River Murray estuary of Australia. Mar Geol. 2000;170:141–68.

Department of Environment and Heritage. Coorong, and Lakes Alexandrina and Albert Ramsar management plan. Adelaide: South Australian Department of Environment and Heritage; 2000. 63 p.

Dick J, Haynes D, Tibby J, Garcia A, Gell P. A history of aquatic plants in the Ramsar-listed Coorong wetland, South Australia. J Paleolimnol. 2011;46:623–35.

Finlayson M, Davis J, Gell P, Kingsford R, Parton K, Smith P. The status of wetlands and the predicted effect of global climate change: the situation in Australia. Aquat Sci. 2013;75:73–93.

Fluin J, Gell P, Haynes D, Tibby J. Paleolimnological evidence for the independent evolution of neighbouring terminal lakes, the Murray Darling Basin, Australia. Hydrobiologia. 2007;591:117–34.

Gell P, Fluin J, Tibby J, Hancock G, Harrison J, Zawadzki A, Haynes D, Khanum S, Little F, Walsh B. Anthropogenic acceleration of sediment accretion in lowland floodplain wetlands, Murray-Darling Basin, Australia. Geomorphology. 2009;108:122–6.

Kingsford RT, Walker KF, Lester RE, Young WJ, Fairweather PG, Sammut J, Geddes MC. A Ramsar wetland in crisis – the Coorong, Lower Lakes and Murray Mouth, Australia. Mar Freshw Res. 2011;62:255–65.

Kingsford RT, Porter JL, Halse SA. National waterbird assessment. Canberra: National Water Commission; 2012. 170 p. Waterlines report no. 74.

Krull E, Haynes D, Lamontagne S, Gell P, McKirdy D, Hancock G, McGowan J, Smernik R. Changes in the chemistry of sedimentary organic matter within the Coorong over space and time. Biogeochemistry. 2009;92:9–25.

Lester RE, Webster IT, Fairweather PG, Langley RA. Predicting the future ecological condition of the Coorong – the effect of management actions and climate change scenarios. Canberra: CSIRO: Water for a Healthy Country National Research Flagship; 2009. 94 p.

Murray Darling Basin Authority. The Living Murray story – one of Australia's largest river restoration projects. Canberra: MDBA; 2011. 108 p.

Murray Darling Basin Authority. Basin plan. Canberra: MDBA; 2012. 245 p.

Reeves JM, Haynes D, Garcia A, Gell PA. Hydrological change in the Coorong estuary, Australia, Past and Present: evidence from fossil invertebrate and algal assemblages. Estuar Coasts. 2014;38(6):2101–16.

Rogers DJ, Paton DC. Spatiotemporal variation in waterbird communities in the Coorong. Canberra: CSIRO: Water for a Healthy Country National Research Flagship; 2009. 49 p.

Sim T, Muller K. A fresh history of the lakes: Wellington to Murray Mouth, 1800s to 1935. Strathalbyn: River Murray Catchment Water Management Board; 2004. 75 p.

Thiessen J. Comparison of bird abundance, diversity and distribution observed in 2008/09 and 2010/11 in the Coorong Murray Mouth and Lower Lakes. Adelaide: Department for Environment and Natural Resources; 2011. 33p.

Zampatti BP, Bice CM, Jennings PR. Temporal variability in fish assemblage structure and recruitment in a freshwater-deprived estuary: the Coorong, Australia. Mar Freshw Res. 2010;61:1298–312.

Kati Thanda: Lake Eyre (Australia)

162

Richard T. Kingsford

Contents

Introduction 1922
Hydrology 1922
Aquatic Ecology 1925
Future Challenges 1926
References 1927

Abstract

Kati Thanda-Lake Eyre is a dry salt lake much of the time, in the middle of the arid zone in Australia (<250 mm annual rainfall), with the lowest natural point in Australia at about 15 m below sea level. Located in the southern portion of the endorheic lake Eyre Basin, it is the fifth largest terminal lake in the world, covering 9,690 km^2. Consisting of two parts, connected during large floods by the 15 km long Goyder channel (a large Lake Eyre North and the small Lake Eyre South), the lake's aquatic ecology is defined by interaction between highly variable inflowing rivers and the layer of salt that varies in thickness across the lake, reflecting the lake's flooding patterns. Few animals tolerate the extreme high salinities of Lake Eyre, but with sufficient freshwater delivered by inflowing rivers there is a diverse and abundant aquatic life. The most serious threat to the lake's ecology is the development of water resources upstream and mining affecting flow patterns.

Keywords

Salt lake · Flooding · Waterbirds · Shorebirds · Cooper Creek · Diamantina River

---

R. T. Kingsford (✉)
Centre for Ecosystem Science, School of Biological, Earth and Environmental Sciences, UNSW Australia, Sydney, NSW, Australia
e-mail: richard.kingsford@unsw.edu.au

© Springer Science+Business Media B.V., part of Springer Nature 2018
C. M. Finlayson et al. (eds.), *The Wetland Book*,
https://doi.org/10.1007/978-94-007-4001-3_175

Introduction

Lake Eyre was renamed Kati Thanda-Lake Eyre in December 2012, incorporating its aboriginal name. The lake is a dry salt lake much of the time, in the middle of the arid zone in Australia (<250 mm annual rainfall), with the lowest natural point in Australia at about 15 m below sea level. Located in the southern portion of the endorheic lake Eyre Basin, it is the fifth largest terminal lake in the world, covering 9,690 km^2 but was once much larger. About 35,000 years ago, rivers in the Lake Eyre Basin catchment flowed constantly, forming an extensive and deep freshwater lake about three times the size of the current Lake Eyre, Lake Dieri. Then, the lake and its surroundings supported thick lush vegetation, providing habitat for many of Australia's extinct megafauna, including giant kangaroos, wombats, goannas, flightless birds, and pythons and their predators, marsupial lions and thylacines. The lakes and rivers supported platypus, dolphins, and a range of waterbird species. Many species were driven to extinction as the climate became drier and the rivers flowed less often, about 20,000 years ago. Today's rivers are highly variable in their flow patterns into the lake.

When dry, a white layer of salt (400 t) varies in thickness across the lake, reflecting the lake's flooding patterns. In the south, the salt crust is hard and thick, but it remains thin in the north where the large rivers enter (Fig. 1), giving way to soft black mud. The lake consists of two parts, connected during large floods by the 15 km long Goyder channel: a large Lake Eyre North and the small Lake Eyre South (Fig. 1). The lake's aquatic ecology is defined by interaction between highly variable inflowing rivers and the salt. Few animals tolerate the extreme high salinities of Lake Eyre that sometimes exceed that of seawater, particularly as the lake is drying. With sufficient freshwater delivered by inflowing rivers (Fig. 1), there is a diverse and abundant aquatic life. The aquatic ecology of the lake is highly dependent on the variable flow patterns of its inflowing rivers (Puckridge et al. 1998).

Hydrology

Periodically, about every 2 years, one or several rivers flow into the lake (Fig. 1) (Kotwicki 1986). In 23 of the 33 years since 1979, water flowed into the lake; there were more years that it received water than years that it did not (Kotwicki 2009). Often much of the surface area of the lake can be covered by shallow water, but this is insufficient to trigger widespread biological activity. For example, in 2009 and 2010 (Fig. 2), more than half the surface area of the lake was flooded but few waterbirds used this area, partly reflecting the absence of invertebrates or fish. At times of rare high flooding, only achieved during a series of large floods in the 1950s, 1970s, and 1980s, the lake fills to 4 m. The 1970 filling was one of the largest filling events for a hundred years. If the lake fills, it can hold water for 4–5 years, as in the 1970s.

The rivers of the Lake Eyre Basin flow in response to large tropical cyclones and large rainfall depressions in the upper (northern) reaches of the basin (Fig. 1). Floods

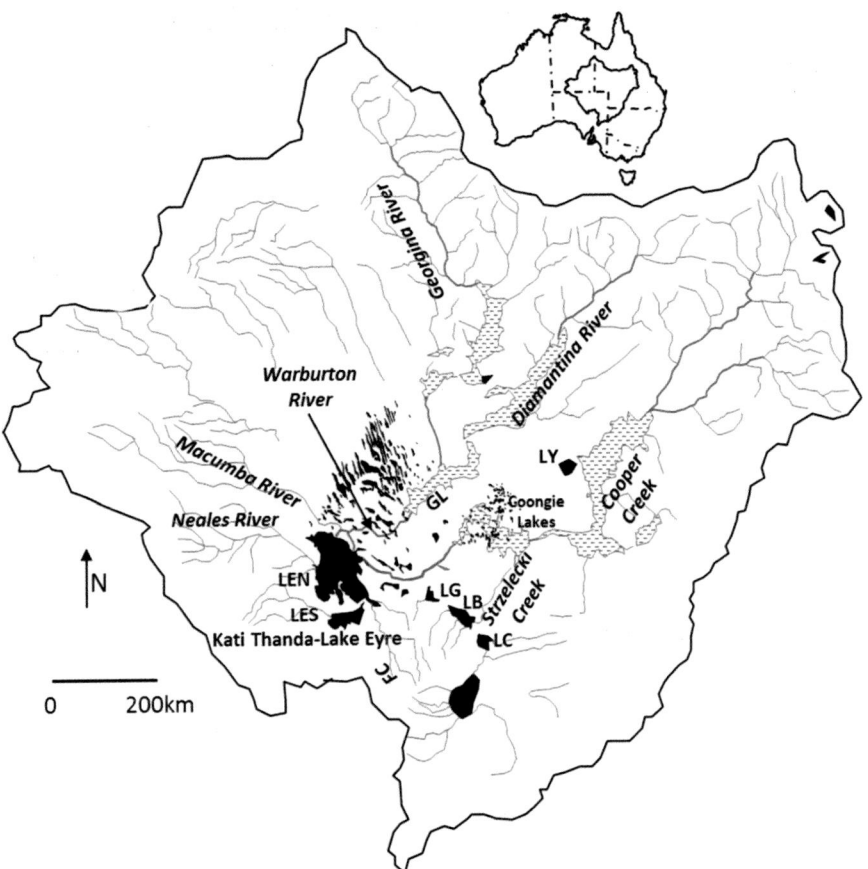

Fig. 1 Kati Thanda-Lake Eyre (*E*) consists of Lake Eyre North (*LEN*) and Lake Eyre South (*LES*), supplied by the major rivers of the Lake Eyre Basin in central Australia (*inset*), including the Georgina River, the Diamantina River and Cooper Creek (also supplies Lake Yamma Yamma – *LY*) which flows into Strzelecki Creek which fills Lakes Gregory (*LG*), Blanche (*LB*) and Callabonna (*LC*). Frome Creek (*FC*) flows into the southeastern part of Lake Eyre North. Rivers and creeks inundate floodplains (stippled), including Goyders Lagoon (*GL*) on the Diamantina River and lakes (filled)

are often dependent on La Niña phases of the El Niño–Southern Oscillation (ENSO) which produce a series of floods over several years which flow down the large rivers, progressively filling the large wetland systems upstream of Lake Eyre (Kingsford et al. 1999; Puckridge et al. 2000). Until these are filled, the floods will not reach Lake Eyre. About 64% of Lake Eyre water comes from the Georgina-Diamantina rivers (Kotwicki 1986); the Georgina River flows from the north and the Diamantina River flows from the northeast (Fig. 1). On the way to Lake Eyre, the Diamantina River floods the extensive floodplain of Goyders Lagoon (~120 × 30 km) where it

Fig. 2 Lake Eyre drying back after the 2010 flood (Photo credit: RT Kingsford © Rights remain with the author)

is joined by the Georgina River system which has become Eyre Creek and filled large lakes and floodplains before reaching the lagoon (Fig. 1). Goyders Lagoon floodplain forms a channel at its most downstream point which becomes the Warburton River. This river flows into Lake Eyre (Fig. 1), carving out the Warburton Groove (~80 km), down the north–south axis of the lake and easily distinguishable from satellite imagery. Water flows down this groove, filling the southern parts of Lake Eyre North, Belt Bay, Jackboot Bay, and Madigan Gulf before spreading more widely.

Cooper Creek is the next most important source of water for Lake Eyre, entering the lake from the northeast (Fig. 1) and contributing about 17% of flows to Lake Eyre (Kotwicki 1986). Cooper Creek originates in the northeast of the Lake Eyre Basin, with flows from the Great Dividing Range forming two main tributaries: the Thomson River, flowing from the northeastern part of the catchment, and Barcoo River, flowing from the eastern part of the catchment. In the lower part of the Cooper Creek catchment, the river forms extensive floodplains and fills large freshwater and salt lakes (e.g., Coongie Lakes, Lake Yamma Yamma). In large floods, there is also sufficient water to flow south into Strzelecki Creek which eventually runs into Lakes Callabonna, Blanche, and Gregory on the northern edge of the Flinders Ranges (Fig. 1). Strzelecki Creek only reaches Lake Blanche about every 10–15 years (Kingsford et al. 1999) while Cooper Creek flows into Lake Eyre about every 13 years (Kingsford et al. 1999). The other river and creek systems from the west and south are considerably smaller and contribute the remaining flow into Lake Eyre (Fig. 1). These include the Neales and Macumba Rivers which rarely flow in from the northwest. Frome River or Creek flows into the southeastern part of Lake Eyre North from the Flinders Ranges. In 1984 and 2011, this almost filled Lake Eyre South. Finally, the Warriner and Margaret Creeks flow into Lake Eyre South from the Stuart Range in the southwest. The network of branching creeks, rivers, and floodplains is more often dry than not with rare flooding years.

Aquatic Ecology

When sufficient freshwater flows into the lake, flooding triggers significant biological productivity from the microscopic organisms up to the top of the food web occupied by fish and waterbirds. Freshwater invertebrates, particularly crustaceans, hatch from sediment or are transported by floodwaters (Jenkins and Boulton 2003) into Lake Eyre. Eggs remain viable for decades, although their viability declines with time. Within days and weeks, these animals reach adulthood and breed, providing abundant food for higher order predators. An incredible diversity and abundance of invertebrates provide prey for fish and waterbirds when Lake Eyre floods, capitalizing on the boom in food resources.

Fish build up in extraordinary numbers colonizing the lake from waterholes upstream where they have bred; after the large floods of the 1970s, there were an estimated 40 million dead fish lying in contours around the edge of Lake Eyre, including equal numbers of bony bream *Nematalosa erebi* and Lake Eyre hardyhead *Craterocephalus eyresii* (Ruello 1976). Waterbirds follow the rivers as they flood Lake Eyre. In small floods, waterbirds remain largely confined to the central Warburton Groove, but even in moderate floods Lake Eyre can support large concentrations of waterbirds. In the 1990 flood of Lake Eyre, there were estimated to be more than half a million waterbirds, including more than 100,000 migratory shorebirds on a relatively small part of Lake Eyre, where Cooper Creek flowed into the lake (Kingsford and Porter 1993). At such times, breeding colonies of some species of waterbirds also establish on the lake. These principally include Australian pelicans *Pelecanus conspicillatus*, silver gulls *Larus novaehollandiae*, and banded stilts *Cladorhynchus leucocephalus*. In 1990, a hundred thousand pelicans raised an estimated 90,000 chicks on Lake Eyre (Waterman and Read 1992). At least 36 species of waterbirds have been recorded on Lake Eyre (Kingsford and Porter 1993).

The "boom" and "bust" cycles of flooding in the Lake Eyre Basin drive waterbird movements across a large part of the continent (Roshier et al. 2001, 2002; Kingsford et al. 2010). They can result in waterbirds moving to productive wetlands and waters, including Lake Eyre from southeastern parts of the continent (Kingsford and Norman 2002). The extensive and highly productive habitat supports large numbers of waterbirds, among some of the highest densities on the continent (Kingsford and Porter 1993). The mechanisms through which these waterbirds are able to exploit productive habitats of the Lake Eyre Basin are just beginning to be understood, with evidence that landscape structure combined with memory and weather patterns are important (Roshier et al. 2008; Reid 2009; Kingsford et al. 2010). Satellite tracking of waterfowl has further contributed to improved understanding. One individual grey teal *Anas gracilis* flew 1,268 km, including traveling 502 km in the first 6.6 h, over about a 10 day period, and 4,800 km in the year it was tracked, mostly at night (Roshier et al. 2008). Similarly, black swans *Cygnus atratus* and straw-necked ibis *Threskiornis spinicollis* make long flights and then remain in an area (Kingsford et al. 2010). These extraordinary movement abilities allow waterbirds to exploit remote and productive wetland habitats in the Lake Eyre Basin. For example, waterbirds such as banded stilts move to Lake Eyre from perennial habitats near the coast (e.g., Coorong, mouth of the River Murray) to breed (sometimes laying more than one clutch, Minton et al. 1995), but their life history is highly dependent on the availability of abundant crustacean prey (Burbidge and Fuller 1982).

Similarly, Australian pelicans travel long distances to the flooding on the rivers of the Lake Eyre Basin (Reid 2009). The ability to move to wetland habitat as the wetlands of Lake Eyre dry is equally important. When the rivers stop flowing, fresh water no longer reaches Lake Eyre and high summer temperatures ($>40\ °C$) drive high evaporation rates which result in the drying up of productive wetlands. Vertebrate animals must move before it is too late and there is not enough food to provide them with the energy for escape. After the 1974–1976 floods, pelicans hatched and banded on Lake Eyre moved considerable distances, including to Palau, Christmas Island, and New Zealand. Inevitably, some waterbirds delay too long, as food resources rapidly diminish, affecting breeding success. Some pairs of Australian pelicans can nest too late, when food resources are low. They will subsequently abandon eggs and even almost fully grown chicks (Kingsford and Porter 1993; Kingsford et al. 2010). Chicks unable to fend for themselves die in their hundreds. Even some adult waterbirds fail to escape the "bust" period of insufficient food resources. After the big floods of Lake Eyre in the mid-1970s, a thousand cormorants were found dead in a concrete water tank with only 30 cm of water (Kingsford et al. 1999).

Future Challenges

Lake Eyre's aquatic ecology, like all wetlands, is fundamentally dependent on inflowing rivers and their supply of freshwater. The most serious threat to this ecology is the development of water resources upstream and mining affecting flow patterns, particularly in Queensland (Kingsford et al. 1998; Kingsford et al. 2014). Periodically, governments consider the opportunity to stimulate development of water resources in the catchment. Unlike many other wetlands, river flow is seldom sufficient to fill Lake Eyre, but small floods provide extensive habitat for waterbirds throughout the Lake Eyre Basin, along the rivers that run into Lake Eyre. Development will affect this flooding regime. If the floods are large and long enough to reach Lake Eyre, this triggers not only immigration of waterbirds from many different parts of the continent but often breeding in spectacular numbers. Boom and bust ecology drives the ecology of Lake Eyre, one of the world's most extreme wetlands.

References

Burbidge AA, Fuller PJ. Banded stilt breeding at Lake Barlee, Western Australia. Emu. 1982;82:212–6.

Jenkins KM, Boulton AJ. Connectivity in a dryland river: short-term aquatic microinvertebrate recruitment following floodplain inundation. Ecology. 2003;84:2708–23.

Kingsford RT, Norman FI. Australian waterbirds – products of a continent's ecology. Emu. 2002;102:1–23.

Kingsford RT, Porter JL. Waterbirds of Lake Eyre. Aust Biol Conserv. 1993;65:141–51.

Kingsford RT, Boulton AJ, Puckridge JT. Challenges in managing dryland rivers crossing political boundaries: lessons from Cooper Creek and the Paroo River, central Australia. Aquat Conserv Mar Freshwat Ecosyst. 1998;8:361–78.

Kingsford RT, Curtin AL, Porter J. Water flows on Cooper Creek in arid Australia determine "boom" and "bust" periods for waterbirds. Biol Conserv. 1999;88:231–48.

Kingsford RT, Roshier DA, Porter JL. Australian waterbirds – time and space travellers in dynamic desert landscapes. Mar Freshw Res. 2010;61:875–84.

Kingsford RT, Costelloe J, Sheldon F. Lake Eyre Basin – challenges for managing the world's most variable river system. In: Squires VR, Milner HM, Daniell KA, editors. River basin management in the twenty-first century. Boca Raton: CRC Press/Taylor Francis Group; 2014. p. 346–67.

Kotwicki V. The floods of Lake Eyre. In: Government Printer. Adelaide: South Australian Engineering and Water Supply Department; 1986. 99 p.

Kotwicki V. Floods of Lake Eyre. 2009. Available from: http://www.k26.com/eyre/index.html. Accessed 22 Nov 2015.

Minton C, Pearson G, Lane J. History in the mating: banded stilts do it again. Wingspan. 1995;13–15.

Puckridge JT, Sheldon F, Walker KF, Boulton AJ. Flow variability and ecology of large rivers. Mar Freshw Res. 1998;49:55–72.

Puckridge JT, Walker KF, Costelloe JF. Hydrological persistence and the ecology of dryland rivers. Regul Rivers: Res Manage. 2000;16:385–402.

Reid J. Australian pelican: flexible responses to uncertainty. In: Robin L, Heinsohn R, Joseph L, editors. Boom and bust-bird stories for a dry continent. Melbourne: CSIRO Publishing; 2009. p. 95–120.

Roshier DA, Robertson AI, Kingsford RT, Green DG. Continental-scale interactions with temporary resources may explain the paradox of large populations of desert waterbirds in Australia. Landsc Ecol. 2001;16:547–56.

Roshier DA, Robertson AI, Kingsford RT. Responses of waterbirds to flooding in an arid region of Australia and implications for conservation. Biol Conserv. 2002;106:399–411.

Roshier DA, Asmus M, Klaassen M. What drives long-distance movements in nomadic Grey Teal *Anas gracilis* in Australia? Ibis. 2008;150:474–84.

Ruello NV. Observations on some massive fish kills in Lake Eyre. Aust J Mar Freshw Res. 1976;106:667–72.

Waterman MH, Read JL. Breeding success of the Australian pelican *Pelecanus conspicillatus* on Lake Eyre South in 1990. Corella. 1992;16:123–6.

Myall Lakes (Australia)

Brian G. Sanderson and Anna M. Redden

Contents

Introduction 1930
Hydrology 1931
Macrophytes and Gyttja 1933
Nutrients and Plankton 1935
Threats and Future Challenges 1937
References 1938

Abstract

The Myall Lakes comprise four connected basins: Bombah Broadwater, Two Mile Lake, Boolambayte Lake, and Myall Lake, and is the biggest lagoon system in New South Wales, Australia. Bombah Broadwater receives most of the catchment discharge and is weakly connected to the ocean via Myall River. Myall Lake is the ultimate backwater, receiving very little catchment discharge and having a constricted, shallow connection to Boolambayte Lake which, in turn, connects to Bombah Broadwater via the narrow Two Mile Lake. Infrequent flood events and droughts cause large salinity changes in Bombah Broadwater but Myall Lake remains stable. Plankton assemblages vary in Bombah Broadwater in ways that are related to salinity. In Myall Lake chlorophyll is stable and low. Meadows of charophytes and *Najas marina* are widespread in Myall Lake and these macrophytes are associated with high water clarity and a thick layer of ammonium-rich gyttja. Rising sea level, associated with climate change, has the potential to disrupt the hydrologic stability of Myall Lake and induce an ecological change of state.

B. G. Sanderson (✉) · A. M. Redden (✉)
Acadia Centre for Estuarine Research, Acadia University, Wolfville, NS, Canada
e-mail: bxs@eastlink.ca; anna.redden@acadiau.ca

Keywords

Lagoon · Hydrological stability · Salinity · Hypsometry · Water clarity · Charophytes · Gyttja

Introduction

The Myall Lakes comprise four connected basins: Bombah Broadwater, Two Mile Lake, Boolambayte Lake, and Myall Lake, located on the coast of New South Wales (NSW), approximately 75 km north of Newcastle, Australia, at longitude 152° 22′ E and latitude 32° 26′ S (Fig. 1). Collectively, the lakes have a waterway area of 105 km^2 and a 490 km^2 catchment. Although these water bodies are named "lakes," one should consider them to be lagoons or even an estuary because they have an open connection to the ocean. Myall Lakes constitute the biggest lagoon system in New South Wales, and they are unique in that they have a hydrodynamic connection

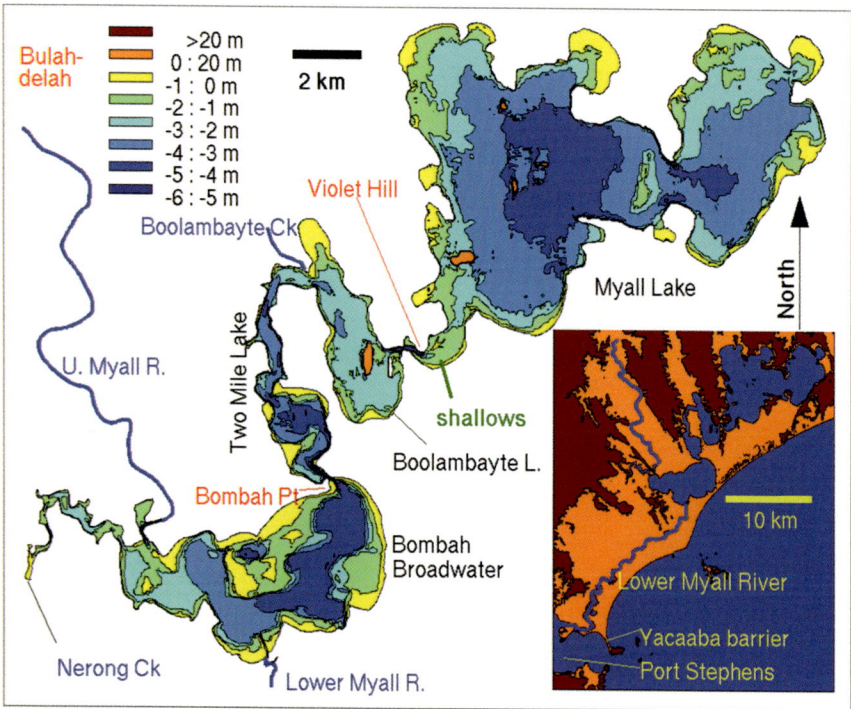

Fig. 1 The waterbodies and bathymetry of the Myall Lakes, NSW, Australia (Image credit: Brian Sanderson © Rights remain with the author)

to the ocean that is both highly restrictive in tidal exchange and yet remains permanently open without human intervention. In Australia four out of five people live within 50 km of the coast (ABS 1998). Coastal development has impacted the ecology of many coastal lakes in NSW (HRC 2002) but has been relatively limited in the Myall Lakes area.

The Worimi people were the aboriginal inhabitants in the vicinity of the Myall Lakes. European settlement first occurred during 1800–1830 to harvest timber and develop agriculture. The town of Bulahdelah (present population about 1,500) was established on the upper Myall River in 1840. The timber industry declined as the resource was depleted. Beef and dairy farming became established in the early 1900s and continue to this day. In 1968, Bulahdelah established a sewage treatment plant (STP) that discharged into a tributary of upper Myall River. The STP was upgraded to tertiary treatment in 1996 (Drew et al. 2008). Much of the catchment is now a designated National Park or managed by State Forests. In 1999, the Myall Lakes, a nearly pristine ecosystem, was listed as a Ramsar Site (OEH 2012).

The relatively large uppermost lake, Myall Lake, is the ultimate backwater, being weakly connected to both the ocean and the catchment and with little human activity. As there are no rivers that directly enter Myall Lake, the waters are both low in salinity (usually 2–3 ppt) and distinctively clear. Myall Lake supports extensive growth and high biomass of charophytes, *Chara fibrosa* and *Nitella hyalina*, which are limited to waters with low salinity (Garcia and Chivas 2006). The combination of clear water and abundant charophytes surrounded by an undeveloped catchment marks Myall Lake as a place with rare beauty and natural serenity. This remarkable ecosystem is on the one hand accessible and on the other hand made remote from human disturbance due to an unusual degree of hydrodynamic isolation.

Given the limited catchment inputs and low levels of human use of the system, a cyanobacterial bloom observed in the Myall Lakes during April 1999 was unexpected. Prior to the bloom there had been one ecological study (Atkinson et al. 1981) and no long-term water quality monitoring in the Myall Lakes. A flurry of scientific investigations was undertaken following the bloom (Wilson 2008).

Hydrology

The upper Myall River and Nerong Creek have catchment areas 240 km^2 and 88 km^2, respectively, and they both drain into the northwestern end of Bombah Broadwater. Bombah Broadwater is the southernmost basin and receives most (about 65%) of the total catchment discharge into Myall Lakes. The southern side of Bombah Broadwater is connected to the outer harbor of Port Stephens by the 26 km long lower Myall River (inset in Fig. 1). Morphology of the outer harbor of Port Stephens is rapidly changing (Vila-Concejo et al. 2007), and although the morphology of the entrance to lower Myall River has also changed, its connection to the ocean remains open. Nevertheless the connection to the ocean is strongly

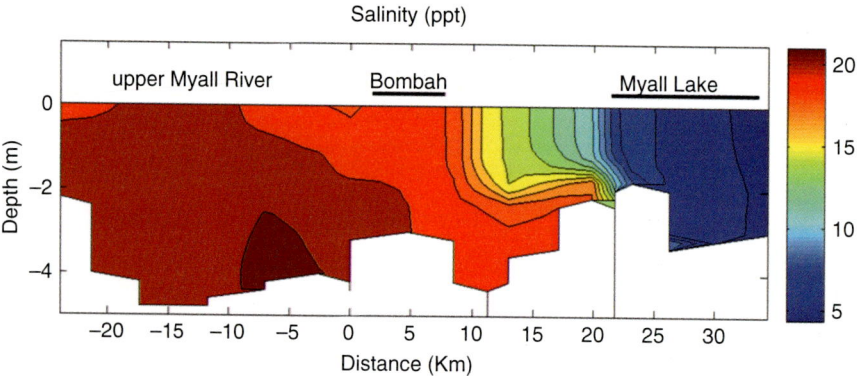

Fig. 2 Salt intrusion into upper Myall River and Myall Lakes during drought conditions. The origin for distance is where lower Myall River enters Bombah Broadwater. Negative distances are up lower Myall River, positive along a thalweg through the lakes. Measurements in upper Myall River were made on 15 June 2006. Salinity was measured on 28 June 2006 along the thalweg through Bombah Broadwater, Two Mile Lake, Boolambayte Lake, and Myall Lake (Image credit: Brian Sanderson © Rights remain with the author)

choked, with the 1.5 m tidal range in Port Stephens being reduced to 1–2 cm in Bombah Broadwater. The oscillating tide loses amplitude as it travels up lower Myall River, and its momentum is converted into a stress which raises the mean water level, a phenomenon known as tidal setup. Tidal setup is about 0.2 m and oceanic processes with sub-tidal frequencies can sometimes raise the water level of the lakes by a further 0.2 m.

Catchment discharge is usually very low, often being less than the loss by evaporation from the extensive waterways. At these times, lower Myall River functions as an estuary, with the choked tides slowly transporting salt toward the headwaters, slowly raising the salinity of Bombah Broadwater to levels that have been measured to be as high as 27 ppt under drought conditions. Upper Myall River is deep so there is no obstruction to the gravitational intrusion of saline water from Bombah Broadwater into upper Myall River. Observations confirm intrusion into the river with salinity as high as 21 ppt during drought conditions (Fig. 2). Catchment discharge is highly intermittent and most of the time is negligible. Salinity of Bombah Broadwater is lowered, from time to time, by small catchment discharge events. Infrequently, big discharge events totally flush Bombah Broadwater with freshwater, raising lake levels by as much as 2.7 m, perhaps higher (DPW 1980; OEH 2012). At such times, lower Myall River functions as a river for a few weeks, as the water drains.

Being broad and shallow (average depth 2.4 m), with a weak connection to the ocean, Bombah Broadwater is usually well mixed. At Bombah Point, a constricted channel connects Bombah Broadwater to the southern end of Two Mile Lake. Two Mile Lake is deep (average 2.7 m) and narrow with thalweg aligned north–south. Roughly 25% of the catchment (100 km^2) discharges from Boolambayte Creek into the northern end of Two Mile Lake. Under drought conditions, saltwater intrudes from Bombah Broadwater, and the water column can stratify in Two Mile Lake, as well as in the deeper channels of Boolambayte Lake (Fig. 2).

The northern end of Two Mile Lake opens onto Boolambayte Lake which is shallow (average depth 2 m) and has no significant rivers. Myall Lake connects to Boolambayte Lake via a deep, narrow passage at Violet Hill. Shallow water at the southwestern corner of Myall Lake (Fig. 1) restricts intrusion of deep, saline water. These shallows are critical for maintenance of low salinity in Myall Lake. Myall Lake has a small catchment (68 km^2) with no river discharge.

Myall Lake is remotely connected to catchment discharge and even more remotely connected to the ocean. Remarkably, salinity is sometimes observed to increase in Myall Lake after a catchment discharge event. This happens because salinity can slowly build in the three lower lakes (Bombah, Two Mile, and Boolambayte) during a long dry period. A large discharge event then drives freshwater into Bombah Broadwater causing water level to rise due to constricted drainage to the ocean. As water level rises, the more salty water of Boolambayte Lake can only be driven into Myall Lake. Given its large volume and weak coupling to the ocean, the salinity in Myall Lake (typically 2–3 ppt, measured at 6 ppt during a severe drought) is very stable compared to the other basins and compared to other coastal lakes and lagoons in NSW.

With an average depth of 2.9 m, Myall Lake is the deepest of the four interconnected basins and has the largest surface area, 65 km^2, greater than the other basins combined. Displacing the combined volume of the other three lakes into Myall Lake would raise water level in Myall Lake by 1.5 m. Catchment discharge into Bombah Broadwater infrequently raises water level to that extent (Sanderson 2008).

Myall Lake is hydrodynamically isolated from the modified Bombah Broadwater catchment, but up to 40% of discharge from the Boolambayte catchment could be washed into Myall Lake (Sanderson 2008). The Boolambayte catchment is in a relatively natural state, so Myall Lake can be considered to be a large, pristine ecosystem with only weak coupling to ecosystems that have been modified by human activity.

Macrophytes and Gyttja

In the Myall Lakes region, land has been rising relative to sea level for the last 6,000 years (Lambeck 1996). Skilbeck et al. (2005) examined cores which showed that Myall Lake made the transition from marine estuary to charophyte lake about 1,100 years BP. A thick layer of hydrous organic sediments known as "gyttja" has accumulated since then. Samples obtained from shallow sites in Myall Lake show a continuous transition from growth at the charophyte apex to a decaying base and then to gyttja (Sanderson et al. 2008).

The hypsometry of Myall Lake is deep and flat bottomed compared to the other basins. Shallow flats are found in sheltered bays, mostly along the north–west shoreline which is in the lee of the strongest winds. These shallow flats are formed by accumulations of thick layers of gyttja. The southeastern shorelines are sandy, consistent with mechanical stress by wind waves and currents driven by the strongest winds.

The top few cm of the gyttja layer is soupy, floc-like, and easy to disturb when without overlying macrophytes. Sometimes flocculent gyttja is swept onto sandy areas, for a time, before being washed away by waves and currents. Beneath this surface layer the gyttja sediment is more gelatinous and more coherent. From 5 to 30 cm beneath the sediment surface, the density of gyttja is not particularly high ($1,028 \pm 7$ kg/m^3), but its static strength (11 ± 1 N/m^2) is sufficient to resist erosion by the typically small wind-driven currents in Myall Lake. When disturbed, it settles quickly.

Conditions are calm on shallow flats that have charophytes growing on a gyttja substrate, even when onshore winds have raised a surface chop offshore. In part the dense vegetation might damp wind-driven waves as they propagate into shallow water (Granata et al. 2001). Additionally, gyttja may deform with the pressure of passing waves in which case it might be approximated as a viscous fluid which damps wave energy (Dalrymple and Liu 1978). Measurements suggest that dynamic viscosity is 1 ± 0.6 kg/m/s. Given this value for viscosity, the available theory is broadly consistent with qualitatively observed wave damping.

Physical stability, low salinity, decoupling from catchment nutrient discharge, and relatively deep and flat-bottomed hypsometry of Myall Lake are all factors that favor charophytes over angiosperms (Andrews et al. 1984). In Myall Lake, *Chara fibrosa* dominates biomass and is abundant over the depth range 0.5–4 m. Ultraviolet radiation may restrict charophytes from more shallow waters (Asaeda et al. 2007). The other commonly found charophyte, *Nitella hyalina*, is found over the same depth range as *Chara fibrosa* but is patchy in distribution and less abundant where depth is greater than 2 m. Both species show little seasonal variability, consistent with the small annual range of water temperature (13–28 °C).

The dominant angiosperm, *Najas marina*, has great seasonal variability and can achieve very high biomass in waters 1.5–2.7 m deep. Water temperature and light can support two growing seasons for *Najas marina*. This species requires soft sediment in order to establish roots (Handley and Davy 2002); mechanical disturbance by wind-driven waves and currents can be high in the spring, so the greatest biomass is often observed in the fall. The distribution of *Najas marina* is very patchy. Shading by dense stands of *Najas marina* is associated with the patchy distribution and reduced biomass of the charophyte *Nitella hyalina* in deeper water (Sanderson et al. 2008).

Charophyte meadows increase water clarity by stabilizing the sediment and trapping particulate material (Blindow et al. 2002). The coefficient of light attenuation is 0.5 m^{-1} in charophyte-dominated Myall Lake. Light attenuation increases and charophyte abundance diminishes progressing through Boolambayte Lake and into Two Mile Lake. There are no charophytes in Bombah Broadwater where the coefficient of light attenuation is typically 2 m^{-1} and can be much higher near upper Myall River and Nerong Creek.

While hydrology makes Myall Lake favorable for charophytes, it is the charophytes that do much to determine the physical and chemical characteristics of Myall Lake.

Nutrients and Plankton

Catchment discharge from upper Myall River into Bombah Broadwater is the most obvious nutrient source for Myall Lakes (Fig. 3). The muddy sediment of Bombah Broadwater contains fecal matter from upstream pasture lands and a predominance of terrestrial organic matter (Heggie et al. 2008).

Calculation of nutrient budgets from catchment loads (Sanderson 2008) demonstrated that Myall Lake should have a nitrogen source and a phosphorus sink. Chemical analyses of gyttja cores have shown ammonium (NH_4^+) concentrations of 10 mg/L in lower gelatinous layer, reducing to 1–4 mg/L near the top flocculent layer (Sampaklis 2003). The large accumulation of NH_4^+ in gyttja suggests that Myall Lake could be an overall nitrogen sink while still having the capacity to be a nitrogen source when disrupted or when ecological state changes. Sampaklis demonstrated that physical disturbance of gyttja released high concentrations of NH_4^+

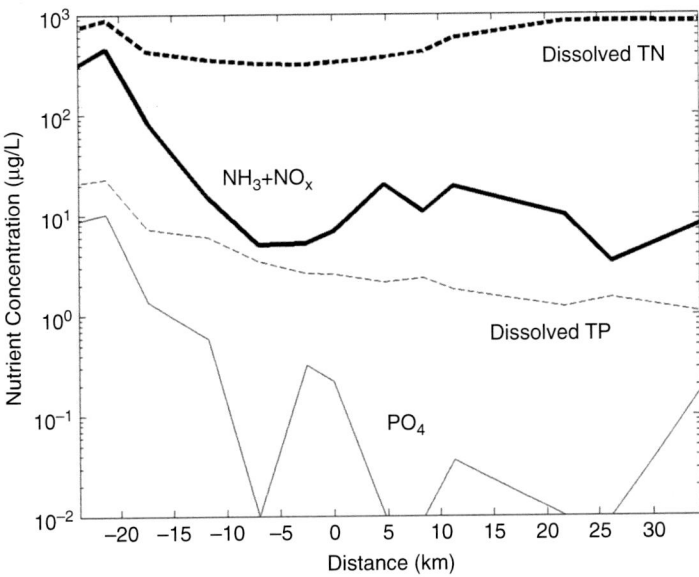

Fig. 3 Nutrients in upper Myall River and Myall Lakes during drought conditions. The origin for distance is where lower Myall River enters Bombah Broadwater. Negative distances are up lower Myall River, positive along a thalweg through the lakes. Measurements in upper Myall River were made on 15 June 2006. Salinity was measured on 28 June 2006 along the thalweg through Bombah Broadwater, Two Mile Lake, Boolambayte Lake, and Myall Lake (Image credit: Brian Sanderson © Rights remain with the author)

and silicate into the water column. Siong and Asaeda (2008) find that incomplete decomposition of charophytes results in burial of phosphorus in the gyttja.

Little is known about the microphytobenthos community in Myall Lakes. Biomarkers indicate that diatoms and dinoflagellates contribute to the organic matter in the sediments of Bombah Broadwater, and *Microcystis* filaments have been reported to emerge from surface sediments (Heggie et al. 2008). The gyttja in Myall Lake provides substrate for large benthic diatoms and abundant Cyanophyceae (mostly Chroococcales and Nostocales). Potentially toxic cyanobacteria, *Microcystis flos-aquae*, is also found in the gyttja.

Bombah Broadwater is strongly and variably influenced by both catchment discharge and marine input. Chlorophyll-*a* measurements in 2003 varied with respect to both time and location (Cohen 2004). Average chlorophyll-a was 4.1 µg/L ($\pm$3.4 µg/L) and range 1.6 to 15.6 µg/L. *Anabaena* blooms were observed during freshwater periods in 1999 and early 2000 (Ryan et al. 2008). Redden and Rukminasari (2008) demonstrated that small increases in salinity (1.5–5.5 ppt) negatively affected growth of *Anabaena circinalis* collected from Bombah Broadwater.

The composition of the plankton assemblage in Bombah Broadwater might change for many reasons. Salinity is one factor, and injection of either marine species from lower Myall River or freshwater species from upper Myall River is another. Bombah Broadwater was dominated by marine dinoflagellates in the 2003 autumn when salinity was 8–9 ppt. By winter the salinity reduced to 2–3 ppt and Chlorophyceae dominated. Redden and Rukminasari (2008) incubated natural phytoplankton assemblages and found that elevating salinity levels from 4 ppt to 8 ppt caused chlorophytes, particularly *Scenedesmus* sp., to decline. Salinity of Bombah Broadwater rose to 5 ppt in spring and 6 ppt in summer, and this heralded a switch to dominance by Cyanophyceae (Cohen 2004).

The stable physical and chemical environment of Myall Lake is reflected by the stability of its phytoplankton community. Chlorophyll-*a* is low and stable in the water column of Myall Lake. Measurements spanning 2003 at multiple sites (Cohen 2004) showed the average value for Chlorophyll-*a* was 2.3 µg/L ($\pm$0.7 µg/L) and range 1.4 to 3.8 µg/L. Cohen (2004) found that the Cyanophyceae consistently dominated throughout 2003 in Myall Lake. Similarly, Ryan et al. (2008) report that *Chroococcus*, *Merismopedia*, and chlorophyte taxa were most common from 2000 through 2002. Two Mile Lake and Boolambayte Lake have phytoplankton assemblages that might be characterized as a mix of those found in Bombah Broadwater and Myall Lake.

Copepods dominate mesozooplankton assemblages in both Bombah Broadwater and Myall Lake, but counts are both higher and more variable with respect to time and site within Bombah Broadwater than within Myall Lake (Cohen 2004). Large numbers of very small rotifers are also found in both lakes. Nevertheless, the zooplankton community of stable Myall Lake has clear differences from more variable Bombah Lake. Myall Lake is distinguished by large populations of the rotifer *Brachionus baylyi* (Timms 1976a). Cohen (2004) observed that Bombah Broadwater was set apart by abundant cyclopoids in the fall of 2003. In 2001,

Muschal (2003) found that the two lakes differed due to high densities of polychaete larvae in Bombah Broadwater. Vertical migration causes zooplankton to become more abundant at nighttime in Myall Lake but not so in Bombah Broadwater (Timms 1976b).

Microzooplankton dilution studies (Cohen 2004) shed light upon the growth and daytime grazing dynamics of the plankton communities in the water columns of Myall Lake and Bombah Broadwater. In Myall Lake both zooplankton grazing and phytoplankton growth rate were lower and much less variable than in Bombah Broadwater. Phytoplankton growth rate of samples from Myall Lake was not stimulated by the addition of phosphorus and/or nitrogen, whereas samples from Bombah Broadwater were often found to be nitrogen limited and sometimes phosphorus limited. Mechanisms that limit phytoplankton growth in Myall Lake remain unknown.

Threats and Future Challenges

Myall Lake and Boolambayte Lake have expansive charophyte meadows in shallow, calm water. Both the charophytes and gyttja are easily disturbed by motorized craft. Gyttja entombs high concentrations of NH_4^+ which are released into the water column when sediment is disturbed. A recent map (RMS 2013) designates some bays as being restricted to paddle craft and others to idle speeds. Interestingly, some of the restricted areas are sandy beaches, and many charophyte meadows (including shallow and easily disturbed Bibby Harbour) are unprotected.

The modern-day ecology of Myall Lakes is believed to have begun about 1,000 years ago as a consequence of 6,000 years of land rising relative to the level of the ocean. Now the long-term trend has turned for water level. Sea level is rising, at an accelerating rate, as anthropogenic emissions warm the planet. The entrance of lower Myall River is protected by Yacaaba barrier where sand presently accretes, although erosion is severe at neighboring locations (Vila-Concejo et al. 2007). Should further sea level rise shift the pattern of erosion, the entrance might become exposed and more susceptible to closure. Alternatively, perhaps inevitably, continued rising sea levels and increased storminess might breach sand barriers, opening direct communication of Bombah Broadwater with the nearby ocean (inset of Fig. 1).

The apparent stability of vast charophyte meadows and the lack of response of phytoplankton to nutrient enrichment experiments should not be construed as indicating that Myall Lake will be robust to anthropogenic disturbance. Eutrophication and/or changed hydrology can be expected to be followed by invasion of other species of phytoplankton which might grow and fundamentally change the ecosystem. It has happened before.

The ecological stability of Myall Lake is predicated upon physical stability and isolation from nutrient loading by the catchment. Shallow water in the southwestern corner of Myall Lake plays a critical role, preventing rapid intrusion of saline water during drought. The dominant submerged vegetation – charophytes and *Najas marina* – seems to grow best when salinity is low (Garcia and Chivas 2006). To maintain this unique, pristine ecosystem, it is critical to preserve the shallow barrier. Presently this area is mapped as a "channel," and there are no restrictions placed upon the draft of motorized vessels passing through it (RMS 2013).

References

Andrews M, Davidson IR, Andrews ME, Raven JA. Growth of *Chara hispida* I. Apical growth and basal decay. J Ecol. 1984;72:873–84.

Asaeda T, Rajapakse L, Sanderson BG. Morphological and reproductive acclimation to growth of two charophyte species in shallow and deep water. Aquat Bot. 2007;86:393–401.

Atkinson G, Hutchings P, Johnson M, Johnson WD, Melville MD. An ecological investigation of the Myall Lakes region. Aust J Ecol. 1981;6:299–327. https://doi.org/10.1111/j.1442-9993.1981.tb01580.x.

Australian Bureau of Statistics. Most Australians still live near the coast – ABS. Media Release, 54/98, May 27, 1998. http://www.abs.gov.au/ausstats/abs@.nsf/mediareleasesbytitle/D1D3980B1944DAC6CA2568A900136291?OpenDocument

Blindow I, Hargeby A, Andersson G. Seasonal changes of mechanisms maintaining clear water in a shallow lake with abundant Chara vegetation. Aquat Bot. 2002;72:315–34.

Cohen D. An examination of planktonic processes in the Myall Lakes. B.Sc. (Hons.) thesis, School of Applied Sciences, University of Newcastle, Australia; 2004.

Dalrymple RA, Liu PL-F. Waves over soft muds: a two layer fluid model. J Phys Oceanogr. 1978;8:1121–31.

Department of Public Works, NSW. Lower Myall River flood analysis. Report prepared by PWD for Great Lakes Shire Council; 1980.

Drew S, Flett I, Wilson J, Heijnis H, Skilbeck CG. The trophic history of Myall Lakes, New South Wales, Australia: interpretations using δ13C and δ15N of the sedimentary record. Hydrobiologia. 2008;608:35–47. https://doi.org/10.1007/s10750-008-9383-3.

Garcia A, Chivas A. Diversity and ecology of extant and Quaternary Australian charophytes (Charales). Crytpogam Algol. 2006;27:323–40.

Granata TC, Serra T, Colomer J, Casamitjana X, Duarte CM, Gacia E. Flow and particle distributions in a nearshore seagrass meadow before and after a storm. Mar Ecol Prog Ser. 2001;218:95–106.

Handley RJ, Davy AJ. Seedling root establishment may limit *Najas marina* L. To sediments of low cohesive strength. Aquat Bot. 2002;73:129–36.

Healthy Rivers Commission. Independent inquiry into Coastal Lakes – final report, Sydney; 2002.

Heggie DT, Logan GA, Smith CS, Fredericks DJ, Palmer D. Biochemical processes at the sediment-water interface, Bombah Broadwater, Myall Lakes. Hydrobiologia. 2008;608:49–67. https://doi.org/10.1007/s10750-008-9378-0.

Lambeck A. Limits of the areal extent of the Barents Sea Ice sheet in Late Weischelian time. Global Planet Change. 1996;12:41–51.

Muschal M. Zooplankton assemblages of the Myall Lakes; a unique coastal lake system of New South Wales, Australia. Wetlands. 2003;21:16–28.

Office of the Environment and Heritage. Myall Lakes Ramsar site, ecological character description. 2012. ISBN 978 1 74293 657 4, OEH 2012/0423

Redden AM, Rukminasari N. Effects of increases in salinity on phytoplankton in the Broadwater of the Myall Lakes, NSW, Australia. Hydrobiologia. 2008;608:87–97.

Roads and Maritime Services. Map of Myall Lakes and upper Myall River area and inset of Seal Rocks Area. 2013. http://www.rms.nsw.gov.au/documents/maritime/usingwaterways/maps/boating-maps/7b-myall-river-seal-rocks.pdf

Ryan NJ, Mitrovic SM, Bowling L. Temporal and spatial variability in the phytoplankton community of Myall Lakes, Australia, and influences of salinity. Hydrobiologia. 2008;608:69–86. https://doi.org/10.1007/s10750-008-9375-3.

Sampaklis A. Nutrient content of hydrous organic sediment (gyttja) in Myall Lake and the effects of its injection into overlying water. B.Sc. (Hons.) thesis, School of Applied Sciences, University of Newcastle, Australia; 2003.

Sanderson BG. Circulation and the nutrient budget in Myall Lakes. Hydrobiologia. 2008;608:3–20. https://doi.org/10.1007/s10750-008-9380-6.

Sanderson BG, Redden AM, Asaeda T. Mechanisms affecting biomass and distribution of charophytes and *Najas marina* in Myall Lake, New South Wales, Australia. Hydrobiologia. 2008;608:99–119. https://doi.org/10.1007/s10750-008-9373-5.

Siong K, Asaeda T. Effect of magnesium on charophytes calcification: implications for phosphorus speciation stored in plant biomass and sediment of Myall Lake, NSW, Australia. In: Sengupta M, Dalwani R, Editors. Proceedings of Taal2007: The 12th World Lake Conference; 2008. p. 264–74.

Skilbeck GCT, Rolph TC, Hill N, Woods J, Wilkens RH. Halocene millennial/centenial-scale multiproxy cyclicity in temperate eastern Australian estuary sediments. J Quat Sci. 2005;20:327–47.

Timms BV. New and rare aquatic invertebrates of the Myall Lakes. Hunter Nat Hist. 1976a;8:204–7.

Timms BV. Salinity regime and zooplankton of Myall Lakes. Hunter Nat Hist. 1976b;8:6–13.

Vila-Concejo A, Short AD, Hughes MG, Ranasinghe R. Flood-tide delta morphodynamics and management implications, Port Stephens, Australia. J Coast Res Spec Issue. 2007;50:705–9.

Wilson J. Preface, The Myall Lakes: patterns and processes in an unusual coastal lake system in eastern Australia. Hydrobiologia. 2008;608:1–2.

Australia's Wet Tropics Streams, Rivers, and Floodplain Wetlands

164

Richard G. Pearson

Contents

Introduction	1942
Landscape, Climate, and Hydrology	1942
Wetland Ecosystems	1944
Biodiversity	1945
Conservation Status	1947
Ecosystem Services and Values	1947
Threats and Future Challenges	1948
References	1949

Abstract

The Australian Wet Tropics biogeographic region is known for its ancient World Heritage rainforests and its proximity to the Great Barrier Reef. It includes ancient mountains and floodplains, and diverse wetlands that support high biodiversity and have important socio-economic and ecological values. Aboriginal people have maintained strong cultural ties to wetlands over 50,000 years. Wetlands include streams, rivers, estuaries, crater lakes, floodplain lagoons and swamps, variously lined by rainforest, paperbark and mangroves. The wetland invertebrate and fish fauna is the most diverse in Australia and includes many endemics such as mountain mayflies and crayfish, and several fish species. Wetlands attract a rich amphibian, reptile and bird fauna, and the platypus is common in the uplands. Wetlands are of value in providing water for agriculture and other industries as well as supporting commercial and recreational fisheries and rich biodiversity and other ecosystem services. Wetlands are adversely affected by extensive use of the region, especially for agriculture, water harvesting and other human activity. Many streams and some coastal wetlands are in protected areas, but protection

R. G. Pearson (✉)
College of Science and Engineering, James Cook University, Townsville, QLD, Australia
e-mail: richard.pearson@jcu.edu.au

© Springer Science+Business Media B.V., part of Springer Nature 2018
C. M. Finlayson et al. (eds.), *The Wetland Book*,
https://doi.org/10.1007/978-94-007-4001-3_45

of floodplain rivers and associated wetlands is limited, and they remain threatened by human activity, especially agriculture and invasive species, and in the long term, by climate change and sea-level rise. Management of these issues is progressing because of the downstream impacts on the Great Barrier Reef, rather than for the inherent values of the wetlands.

Keywords

Tropical wetlands · Biodiversity · Conservation · Management · Ecosystem services · Queensland · Rainforest

Introduction

The Wet Tropics biogeographic region of Queensland is best known for its ancient World Heritage rainforests and its proximity to the Great Barrier Reef World Heritage Area (Fig. 1). It is one of only a few World Heritage areas that meet all four natural criteria for listing: it represents a major stage of the earth's evolutionary history, is an outstanding example of ecological and biological processes, contains superlative natural phenomena, and contains the most important natural habitats for conservation of biological diversity. The region includes diverse wetlands that reflect these criteria: they support high biodiversity and have important socioeconomic and ecological values, including biota of conservation significance, as well as being integral to water quality maintenance and flood detention (Environment Australia 2001). Wetlands are affected by extensive use of the region for agriculture (sugar cane, cattle grazing, horticulture, and forestry), urban development, fishing, and tourism. The region has a ~50,000-year history of Indigenous occupation, and Aboriginal people maintain strong cultural ties to wetlands, which include hunting and ceremonial sites. Many wetland plants, crustaceans, fish, and other vertebrates were harvested traditionally. Wet Tropics wetlands have received substantial research attention because of their scientific and conservation values and because of their role in transporting agricultural contaminants to the Great Barrier Reef (Pearson and Stork 2008).

Landscape, Climate, and Hydrology

The Wet Tropics landscape includes rainforest-clad, weathered granitic mountains to about 1,700 m, tablelands with some basalt flows, gorges, and alluvial floodplains. The climate is warm monsoonal, with highest rainfall in the summer wet season (November–March) and occasional cyclones producing the highest rainfall in Australia (annual maximum of 11,000 mm on Mt. Bellenden Ker), but with less extreme seasonality than the rest of tropical Australia. The region covers only 0.26% of Australia's land area but contributes 7% of its annual runoff. Monthly stream flows peak in the late wet season and are highly predictable. In the dry season, orographic rainfall and forest cloud capture provide for perennial flow (McJannet et al. 2007). Water storage in areas of basalt or alluvium also boosts base flows and

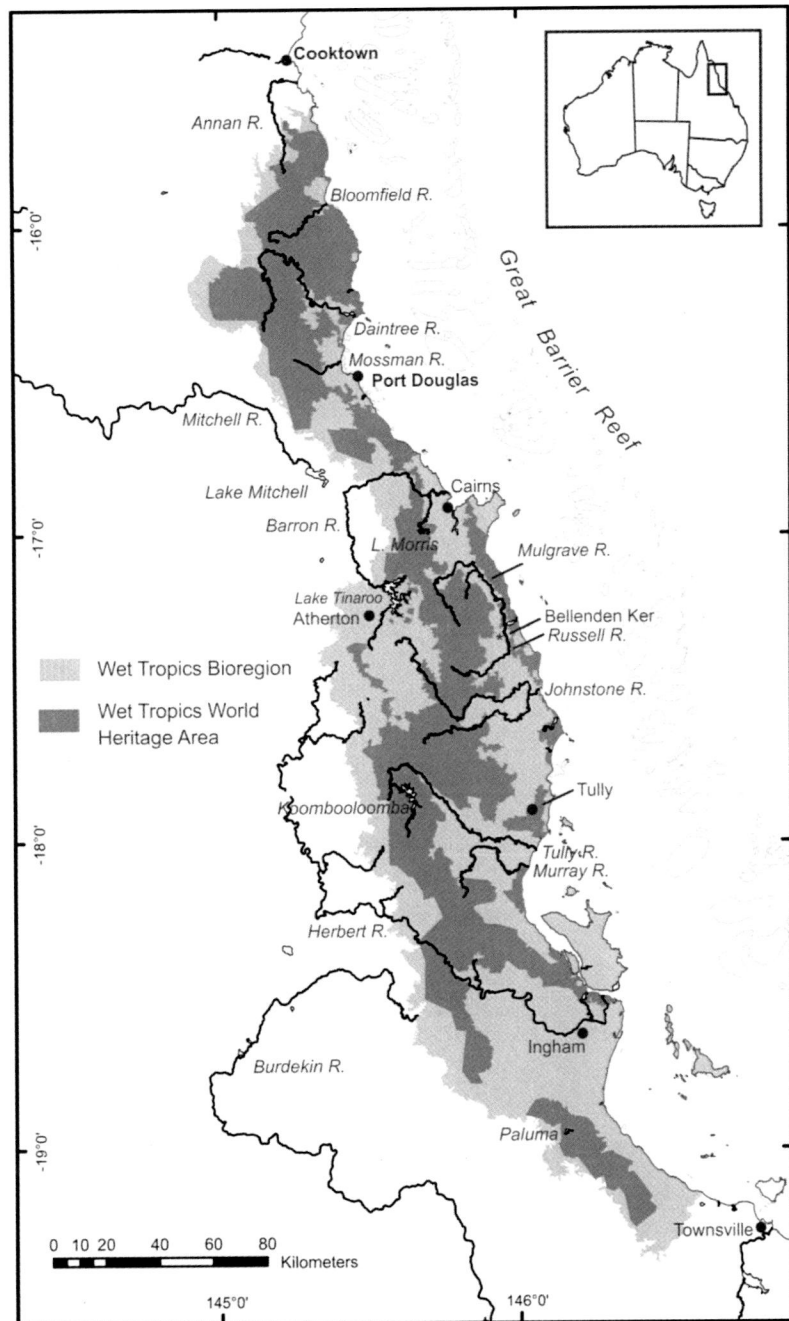

Fig. 1 Location of the Australian Wet Tropics and its major rivers, impoundments (Lakes Morris, Tinaroo, Koombooloomba, and Paluma), and towns (Image credit: RG Pearson © Rights remain with the author (Drawn by Adelle Edwards))

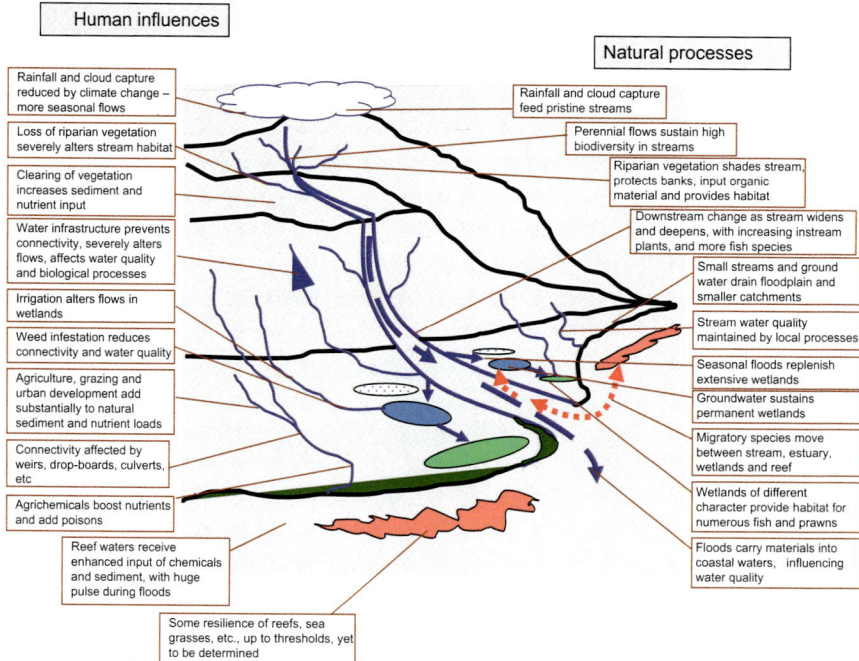

Fig. 2 Outline of natural and human-influenced processes in Wet Tropics catchments. Floodplains include wetlands that are temporary (*stippled*), freshwater (*blue*), and saline (*light green*). Saline areas are characterized by extensive mangroves (*dark green*) (Modified from Pearson and Stork (2008); reproduced with permission, © 2008 by Blackwell Publishing Ltd)

sustains water levels in floodplain lagoons during drier periods. Major habitats include streams, rivers, crater lakes, and extensive freshwater and saline floodplain wetlands (Fig. 2). Most rivers discharge into the Great Barrier Reef lagoon (Fig. 1). During heavy monsoonal or cyclonic rainfall, flows exceed the rivers' carrying capacity and spill out on to the floodplains, replenishing distributaries and wetlands, but levee banks and agricultural drains have generally reduced the extent and duration of flooding (Karim et al. 2011).

Wetland Ecosystems

Wetlands comprise about 6% of the area of the Wet Tropics, of which 38% are estuarine, 28% riverine, 26% palustrine, 7% artificial, and less than 1% lacustrine (DERM 2012). Because mountains are ancient and eroded, streams typically rise with shallow gradient, before descending steeply down mountainsides or escarpments. Sandy and rocky habitats are characteristic, with cascade, riffle, and pool sequences. Rocky gorges and waterfalls are prominent in the middle sections of rivers: for example, Wallaman Falls, in the Herbert catchment, is the highest fall in Australia (268 m). Unusual systems include hot springs in the upper Herbert catchment. The lower meandering sections on the floodplain are relatively short. In larger rivers, deeper reaches are interspersed with riffles and sand bars, with more consistently deep water in tidal reaches. Natural riparian vegetation is dominated by rainforest or paperbarks (*Melaleuca* spp.), with mangroves in saline reaches.

Natural lentic waters in the uplands are scarce, comprising swamps and deep lakes in volcanic craters from ~12,000 years ago. Two rivers and two smaller streams have dams in their upper reaches, creating substantial lakes (Fig. 1). Floodplain wetlands are diverse and include lagoons, swamp forests, and seasonally inundated plains, within the largely agricultural landscape (Environment Australia 2001). Extensive mangrove forests occur toward the coast, including the most complex area of mangroves in Australia in and adjacent to the Herbert delta. Regular rainfall precludes development of the extensive salt flats. The extent of floodplain wetlands, other than drains, has been greatly reduced with the development of agriculture (Johnson et al. 1999).

Biodiversity

Despite its small area, the Wet Tropics is remarkable for having the greatest concentration of biodiversity in Australia – for example, it supports 30% of Australian terrestrial vertebrates (Williams et al. 1996) and over 40% of its freshwater fishes (Pusey et al. 2008). Aquatic plants dominate many wetlands – for example, swamps with trees (*Melaleuca* and *Licuala*), sedges (*Cyperus* and *Eleocharis*), and lilies (*Nymphaea* and *Ottelia*) and streams with diverse bryophytes – and, like the surrounding forests, diversity is generally high (Mackay et al. 2010).

The invertebrate fauna is mainly Gondwanan, with some taxa derived from Southeast Asia, and is diverse. For example, the Ephemeroptera of the region comprises 58 species, higher than in comparable regions in Australia, and includes a high proportion of endemics (Pearson et al. 2015); and the Trichoptera, with 78 species recorded at one site (Pearson et al. 1986), is much more diverse than elsewhere in Australia. Patterns of diversity vary: for example, the Trichoptera are homogeneously distributed, while many Ephemeroptera are restricted to particular subregions (Connolly et al. 2008), and some species have restricted distributions – thus, different species of *Euastacus* crayfish, living in streams above 700 m and growing to 30 cm body length, are endemic to different mountains (Short and Davie 1993). Smaller crayfish (*Cherax*) and shrimps (Atyidae and *Macrobrachium*) are common in many waterways, but some are restricted to downstream reaches because they require saline water for larval development.

Although Australia is depauperate in freshwater fishes, about 104 native species and six alien species are recorded from the Wet Tropics, but only eight species are endemic (Pusey et al. 2008). Some endemics have very restricted distributions – for example, *Guyu wujalwujalensis* is recorded only from the Bloomfield River, with its nearest relatives occurring ~2,000 km south (Pusey et al. 2008). Many fishes return to saline waters to breed, including prized angling species such as barramundi *Lates calcarifer* and jungle perch *Kuhlia rupestris*, and this requirement precludes their colonization upstream of falls. For this reason, and because of an increase in habitat diversity, fish diversity increases downstream. Eels *Anguilla reinhardtii* are also marine breeders, but juveniles manage to negotiate barriers during their return migration, such that in many upland streams fishes are represented by only eels and perhaps the purple-spotted gudgeon *Mogurnda adspersa*, which also appears adept at surmounting obstacles.

The Wet Tropics supports about 54 species of frogs (28% of the Australian total, including eight endemics), 42 of which are dependent on wetlands for breeding (Williams et al. 1996). Tadpoles are particularly conspicuous in small streams and temporary pools. Over the last couple of decades, there have been widespread population declines and extinctions of upland stream-dwelling frogs in the Wet Tropics and elsewhere globally, as a result of infection by the fungus *Batrachochytrium dendrobatidis*.

The Wet Tropics supports about 131 species of reptiles (21% of the Australian total), 14 of which occur in freshwater habitats and 14 in mangroves (Williams et al. 1996). These include tortoises, e.g., the saw-shell tortoise *Elseya latisternum*, common in many habitats; lizards, e.g., the water dragon *Physignathus lesueurii*; snakes, e.g., the water python *Liasis fuscus*; and the estuarine and freshwater crocodiles, e.g., *Crocodylus porosus* and *C. johnstoni*. The estuarine crocodile occurs in rivers and breeds in freshwater wetlands. It grows to around 6 m and preys on fish and other large vertebrates and is dangerous to humans. In the Wet Tropics, the freshwater crocodile occurs only in western-draining rivers (e.g., the Mitchell) and the Herbert River. It grows to 3 m and eats fish.

About 316 bird species (43% of the Australian total) occur in the Wet Tropics, including 73 species in fresh water and 63 in mangroves (Williams et al. 1996). Small streams are characterized by kingfishers, which take their prey from the water, but also are used by forest species, which feed on emerging insects. Swallows and martins frequent open waters, again feeding on insects emerging from the water. Crakes and rails inhabit reed beds, and jacanas *Irediparra gallinacea* are characteristic of floating vegetation. There are many species of ibis, heron, and egret found on river margins and in shallow swamps, along with spoonbills, stilts, geese (*Anseranas semipalmata*), ducks, cranes (brolga *Grus rubicundus* and sarus crane *G. antigone*), and black-necked stork *Ephippiorhynchus asiaticus*. In open water, pelicans, grebes, cormorants, darter *Anhinga melanogaster*, terns, osprey *Pandion haliaetus*, and other birds of prey feed. Along rivers and on mudflats, migratory sandpipers, plovers, etc. spend the southern summer, having traveled from their subarctic breeding grounds.

Only three of the 109 nonmarine mammal species (45% of the Australian total) that occur in the Wet Tropics are freshwater specialists (Williams et al. 1996). They include the native water rat *Hydromys chrysogaster*, which occurs in upland streams through to coastal areas, and the platypus *Ornithorhynchus anatinus*, which occurs mainly in upland streams and impoundments and whose northern distributional limit is in the Wet Tropics.

Conservation Status

Many smaller streams and some coastal wetlands are protected in World Heritage areas, national parks, and designated fish habitat areas, which allow no development apart from minor tourism infrastructure. The Wet Tropics Management Authority aims to rehabilitate wetlands "where restoration of World Heritage values is achievable," and apply legislation to "conserve and maintain threatened aquatic ecosystems" (WTMA 2004). It prioritizes ecosystems that are of particularly high biodiversity value, including: 35 "wetlands of national importance" (Environment Australia 2001), entire catchments such as the Bloomfield River, rivers important for fish diversity (Johnstone, Russell, Mulgrave), wild rivers and gorges (e.g., North Johnstone and Herbert), crater lakes, remnant upland wetlands, coastal lowland wetlands, and selected estuarine systems. However, protection of large areas of floodplain wetland and larger streams and rivers, including those in the cleared areas, is limited; moreover, well-protected terrestrial ecosystems do not adequately include fresh waters (Januchowski-Hartley et al. 2011). The values of wetlands are now receiving management attention, both as an integral part of the reef catchment (e.g., Brodie et al. 2008) and in their own right (Pearson et al. 2013).

Ecosystem Services and Values

By virtue of its climate, the Wet Tropics has abundant fresh water, which is harvested via direct extraction from basalt aquifers and alluvial deposits, and via dams and weirs. Water is used for irrigation on parts of the Atherton Tableland, for the aquaculture industry, for domestic supply, and to service the tourism industry. Other contributions to provisioning services include maintenance of normal, environmental conditions in rivers and estuaries, which are the basis of important recreational and commercial fisheries and biodiversity.

Wetlands may provide important regulating services: for example, forested wetlands can reduce water flows, and wetland vegetation can remove nutrients and sediments that would otherwise reach the reef lagoon. Such processes are so highly valued that some farmers build their own retention lagoons. However, despite general belief that wetlands are the "kidneys of the landscape," evidence has been scant, and some wetlands do not perform this role because of short retention times (McJannet et al. 2012). It is unclear whether regular flooding negates any pollution-reducing effects of wetlands.

The wetlands provide important supporting services, such as breeding and nursery habitat for the rich biodiversity, especially waterbirds, fish, and crustaceans. They also provide cultural services for Indigenous people and opportunities for recreation and tourism, including nature-based activities, white-water rafting, angling, and hiking.

Threats and Future Challenges

The major human impact on Wet Tropics wetlands has been land clearing and associated agriculture (Fig. 2). Agricultural impacts on wetlands stem from the input of nutrients, sediments, organic material, and pesticides. Fertilizers, especially nitrate, are widely applied in the Wet Tropics, and large quantities end up in waterways via surface runoff or infiltration. Nutrients promote eutrophication, with excessive growth of algae, phytoplankton, or macrophytes, especially invasive weeds. The result is clogging of waterways, impeding of fish movements, and severe depletion of dissolved oxygen, making the habitat unavailable for many species (Pearson and Stork 2008).

Sedimentation occurs as a result of erosion and runoff from agriculture and unprotected stream banks. Cattle grazing exacerbates soil and bank erosion, while cropping increases erosion by exposing soils to heavy rain and by removal of riparian vegetation that might otherwise arrest water flow and sediment transport. These processes are evident in Wet Tropics waterways to a variable extent, depending on land and riparian management practices and on the mitigating effects of high flows (e.g., Connolly et al. 2007).

Organic inputs, such as effluents from sewage works and sugar mills, typically cause oxygen depletion through bacterial respiration of organic materials, with subsequent loss of hypoxia-intolerant fauna (Pearson and Penridge 1987). In the Wet Tropics, there has been substantial effort to remove or clean up discharges to waterways. Sources of organic materials include decaying vegetation such as sugar trash on fields and dead aquatic plants. When trash sits in wet fields, fermentation can take place, deoxygenating the water and producing a toxic solution that can find its way into waterways and cause fish kills. The effect on the oxygen content of the waterway is largely dependent on the flow regime: fast-flowing streams are generally well flushed, whereas still lagoons are particularly vulnerable to hypoxia and fish kills. Problems are exacerbated by high temperatures, allowing high rates of biological activity and hypoxic conditions to continue throughout the year (Pearson and Stork 2008).

Major pesticides reaching waterways in the Wet Tropics include weed-control agents such as atrazine and diuron. Some monitoring of pesticide concentrations has been done in the region, but there is very little information on the impacts of pesticides on native biota (Pearson et al. 2013). Invasive weeds and feral animals are an important threat to habitats and biodiversity in the Wet Tropics wetlands. For example, nonnative pasture grasses, such as para grass *Brachiaria mutica*, proliferate wherever there is sufficient light and appropriate substrate. Weed growth is enhanced by dissolved nutrients and loss of riparian shade. Para grass has severe impact on drainage and habitats and is a nuisance to farmers, who control it through mechanical or chemical means. Water hyacinth *Eichhornia crassipes* is another major weed that forms a thick mat over slow-flowing waterways, preventing gas exchange and rendering the waterway hypoxic and uninhabitable for native plants and animals (Pearson and Stork 2008).

Feral animals include pigs, which severely disturb the sediments and fauna of shallow wetlands, and several species of fish. Invasive fishes of major concern include tilapia *Oreochromis mossambicus*, which can displace native species. Translocation of native species can also have serious effects on local faunas (Burrows 2004). For example, introduction of native species to a crater lake (Lake Eacham) resulted in local extinction of the rare endemic Lake Eacham rainbowfish *Melanotaenia eachamensis*, and the crayfish *Cherax quadricarinatus*, introduced from its normal range further west, is likely to have adverse consequences for other crustaceans and plants.

Water abstraction occurs across the Wet Tropics from basalt and alluvial aquifers and pristine streams and via major impoundments (Fig. 1), for which there is some provision for environmental flows. Flow management may threaten habitat integrity and biodiversity, but the environmental effects of water abstraction in the Wet Tropics are largely unstudied. Drainage works on floodplains have variable effects on wetland integrity and can expose acid-sulfate soils, causing high levels of acidity and reduced biodiversity in some lowland systems.

While the general effects of human impact on Wet Tropics wetlands are broadly described (Fig. 2), the major challenges are in implementation of appropriate land and water management to achieve good conservation outcomes while sustaining socio-economic benefits. Management has improved substantially in recent years, especially through rural industries developing "best practice" guidelines regarding stock access to streams, fertilizer application, etc. But changes in the climate and the economics of particular crops may bring new land uses and new problems. For example, a change in sugarcane harvesting, from the old method of burning to remove trash to the current approach of green cane harvesting, had unpredicted impacts: leaving the trash on the land had the benefits of retaining organic material, removing smoke, and protecting the soil against erosion; but the interaction of rainfall with trash can produce organic pollution in streams and fish kills (Pearson and Stork 2008). Restoration of riparian vegetation would greatly improve wetland ecosystem health but is a huge task, and even well-managed lands continue to produce substantial contaminant input to waterways (Connolly et al. 2007). Additionally, while there is little that can be done regionally to avert the local effects of climate change, addressing other issues such as pollution, invasive species and habitat damage can improve the resilience of systems in the long term. Australia is well placed to deal with such issues and can take a major lead in good ecosystem management of its wetlands, just as it has for the Great Barrier Reef (Pearson and Stork 2008).

References

Brodie JE, Binney J, Fabricius K, Gordon I, Hoegh-Gouldberg O, Hunter H, O'Reagain P, Pearson RG, Quirk M, Thorburn P, Waterhouse J, Webster I, Wilkinson S. Synthesis of evidence to support the scientific consensus statement on water quality in the Great Barrier Reef. The State of Queensland (Department of the Premier and Cabinet). Reef Water Quality Protection Plan Secretariat; 2008.

Burrows DW. Translocated fishes in streams of the Wet Tropics region, North Queensland: distribution and potential impact. Cairns: James Cook University; 2004. Rainforest CRC Report No 30.

Connolly NM, Pearson BA, Pearson RG. Macroinvertebrates as indicators of ecosystem health in wet tropics streams. In: Arthington AH, Pearson RG, editors. Biological indicators of ecosystem health in wet tropics streams. Townsville: James Cook University; 2007. p. 128–75. CRC for Rainforest Ecology and Management and CRC for the Great Barrier Reef. ISBN 9780864437921.

Connolly NM, Christidis F, McKie B, Boyero L, Pearson RG. Diversity of invertebrates in Wet Tropics streams: patterns and processes. In: Stork NE, Turton S, editors. Living in a dynamic tropical forest landscape. Blackwells Publishing; 2008. p. 166–77.

DERM (Queensland Department of Environment and Resource Management) WetlandInfo Wetlandinfo.derm.qld.gov.au/wetlands/; 2012.

Environment Australia. A directory of important wetlands in Australia. 3rd ed, Canberra. Available from https://www.environment.gov.au/water/wetlands/australian-wetlands-database/directory-important-wetlands (2001). Accessed 28 Nov 2015.

Januchowski-Hartley SR, Pearson RG, Puschendorf R, Rayner T. Fresh waters and fish diversity: distribution, protection and disturbance in tropical Australia. PLoS One. 2011;6:e25846.

Johnson AKL, Ebert SP, Murray AE. Distribution of coastal freshwater wetlands and riparian forests in the Herbert River catchment and implications for management of catchments adjacent the Great Barrier Reef Marine Park. Environ Conserv. 1999;26:229–35.

Karim F, Kinsey-Henderson A, Wallace J, Arthington AH, Pearson RG. Modelling wetland connectivity during overbank flooding in a tropical floodplain in north Queensland, Australia. Hydrol Process. 2011;26:2710–23.

Mackay SJ, James CS, Arthington AH. Macrophytes as indicators of stream condition in the wet tropics region, Northern Queensland, Australia. Ecol Indic. 2010;10:330–40.

McJannet D, Wallace J, Reddell P. Precipitation interception in Australian tropical rainforests: II. Altitudinal gradients of cloud interception, stemflow, throughfall and interception. Hydrol Process. 2007;21:1703–18.

McJannet D, Wallace J, Keen R, Hawdon A, Kemei J. The filtering capacity of a tropical riverine wetland: II Sediment and nutrient balances. Hydrol Process. 2012;26:53–72.

Pearson RG, Penridge LK. The effects of pollution by organic sugar mill effluent on the macro-invertebrates of a stream in tropical Queensland, Australia. J Environ Manage. 1987;24:205–15.

Pearson RG, Stork NE. Catchment to Reef: water quality and ecosystem health in tropical streams. In: Stork NE, Turton S, editors. Living in a dynamic tropical forest landscape. Melbourne: Blackwells Publishing; 2008. p. 557–76.

Pearson RG, Benson LJ, Smith REW. Diversity and abundance of the fauna in Yuccabine Creek, a tropical rainforest stream. In: de Deckker P, Williams WD, editors. Limnology in Australia. Melbourne: CSIRO; 1986. p. 329–42.

Pearson RG, Godfrey PC, Arthington AH, Wallace J, Karim F, Ellison M. Biophysical status of remnant freshwater floodplain lagoons in the Great Barrier Reef catchment: a challenge for assessment and monitoring. Mar Freshw Res. 2013;64:208–22.

Pearson RG, Connolly NM, Boyero L. Ecology of streams in a biogeographic isolate – the Queensland Wet Tropics, Australia. Freshw Sci. 2015;34:797–819.

Pusey B, Kennard M, Arthington AH. Origins and maintenance of freshwater fish biodiversity in the Wet Tropics region. In: Stork NE, Turton S, editors. Living in a dynamic tropical forest landscape. Melbourne: Blackwells Publishing; 2008. p. 150–60.

Short JW, Davie PJF. Two new species of freshwater crayfish (Crustacea: Decapoda: Parastacidae) from northeastern Queensland rainforest. Mem Qld Museum. 1993;34:69–80.

Williams SE, Pearson RG, Walsh PJ. Distributions and biodiversity of the terrestrial vertebrates of Australia's Wet Tropics: a review of current knowledge. Pac Conserv Biol. 1996;2:327–62.

WTMA. (Wet Tropics Management Authority) Wet tropics conservation strategy. Cairns: WTMA; 2004. ISBN 0-9752202-0-9.

Wetlands of Kakadu National Park (Australia)

165

C. Max Finlayson

Contents

Introduction	1952
Biodiversity	1952
Conservation Status	1955
Ecosystem Services	1956
Threats and Future Challenges	1956
References	1957

Abstract

Kakadu National Park is located to the east of Darwin in the Northern Territory of Australia. The main wetlands in the park include mangroves, salt flats, freshwater flood plains, and small permanent lakes (billabongs) as well as springs and pools along the many streams. The wetlands contain a high diversity of plants and animals that have adapted to the marked seasonal flooding and drying and/or to the tidal range. Threats include invasive species, salinisation of freshwater wetlands, recreational activities, and potential pollution from uranium mining. The park is managed jointly by the federal government and traditional Indigenous owners.

Keywords

Floodplain wetlands · Mangroves · Invasive species · Salinisation · Indigenous owners

---

C. M. Finlayson (✉)
Institute for Land, Water and Society, Charles Sturt University, Albury, NSW, Australia

The Institute for Water Education UNESCO-IHE, Delft, The Netherlands
e-mail: mfinlayson@csu.edu.au

© Springer Science+Business Media B.V., part of Springer Nature 2018
C. M. Finlayson et al. (eds.), *The Wetland Book*,
https://doi.org/10.1007/978-94-007-4001-3_47

Introduction

Kakadu National Park is located to the east of Darwin in the Northern Territory of Australia (Fig. 1). The park contains large parts of the catchments of the West Alligator, South Alligator, and East Alligator Rivers, as well as small parts of the Mary and Katherine River catchments. The main wetlands in the park include mangroves, salt flats, freshwater flood plains, and small permanent lakes (billabongs) in the coastal plains, as well as springs and pools along the streams that dissect the sandstone plateau in the eastern part of the park (Finlayson and Woodroffe 1996).

The freshwater wetlands are subject to an annual flood pulse during a wet season with approximately 1,560 mm rainfall over 3–4 months (Fig. 2). During the wet months the floodplains are inundated to a depth of several meters, while they may dry completely during the dry months. The coastal wetlands are influenced by a diurnal tide of 5–6 m in addition to the annual flooding.

Biodiversity

The extensive literature covering the occurrence and distribution of the main plants and animals is largely summarized in Finlayson et al. (2006).

Flora The vegetation of the wetlands has been described in a general sense with summaries in Finlayson (2005) and Bayliss et al. (2012). The freshwater wetlands are characterized by highly productive grasses, including *Pseudoraphis spinescens*, *Hymenachne acutigluma*, and *Oryza meridionalis*; *Melaleuca* trees; and a mix of waterlilies, sedges, and herbs. Mangroves fringe the northern coast and estuarine reaches of the rivers. The vegetation of the freshwater wetlands has undergone considerable change as a consequence of invasion by feral animals, especially buffalo *Bubalus bubalis* and pigs *Sus scrofa*, as well as by weeds, especially mimosa *Mimosa pigra*, salvinia *Salvinia molesta*, and paragrass *Urochloa mutica* with the latter having expanded greatly in recent years, and changes in fire regimes and saline intrusion (driven by landward advancement of tidal streams carrying salt water into formerly fresh water areas) from coastal areas.

Aquatic invertebrates include approximately 300 microinvertebrate (Copepoda, Cladocera, and Rotifera) and over 600 macroinvertebrate taxa (insects, worms, mites, larger crustaceans, mollusks, and sponges) from freshwater habitats. At family level, the fauna has a high year-to-year persistence as a consequence of the low interannual variability of stream flow.

Fishes include 62 species, represented by 44 entirely freshwater species, 4 species that reproduce in estuarine or marine waters (catadromous) and 14 marine or estuarine species (marine vagrants). This level of diversity is high by comparison to other regions within Australia. About half the fish species are small to medium in size, generally less than 30 cm in length. Marine vagrants may be found in freshwater bodies close to the limit of tidal influence. Upstream there is little variation in

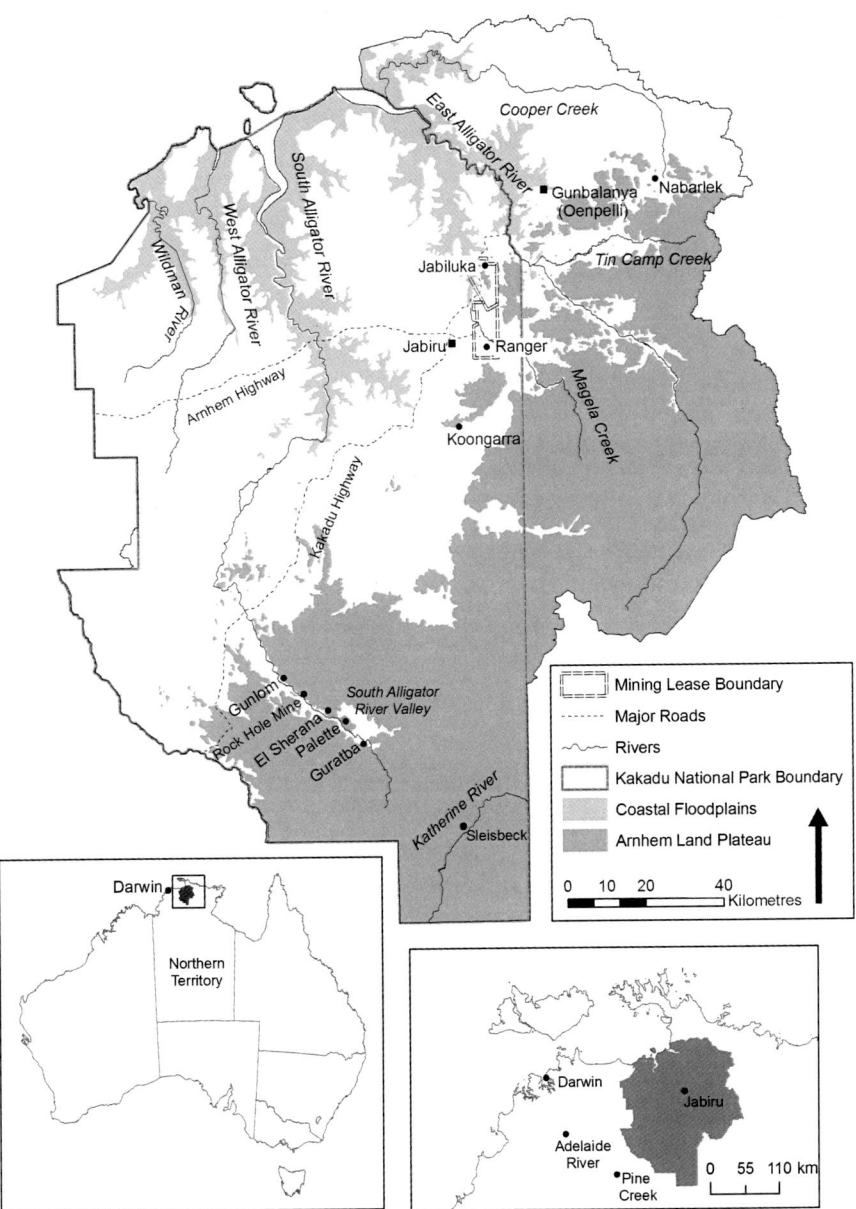

Fig. 1 Map of Kakadu National Park (Courtesy, Supervising Scientist Division, Darwin, Australia)

the species composition until the first dispersal barriers occur near the escarpment of the sandstone plateau.

Amphibians include 25 frog species from three families, the Hylidae, the Myobatrachidae, and the Microhylidae. The cane toad *Bufo marinus* is a recent

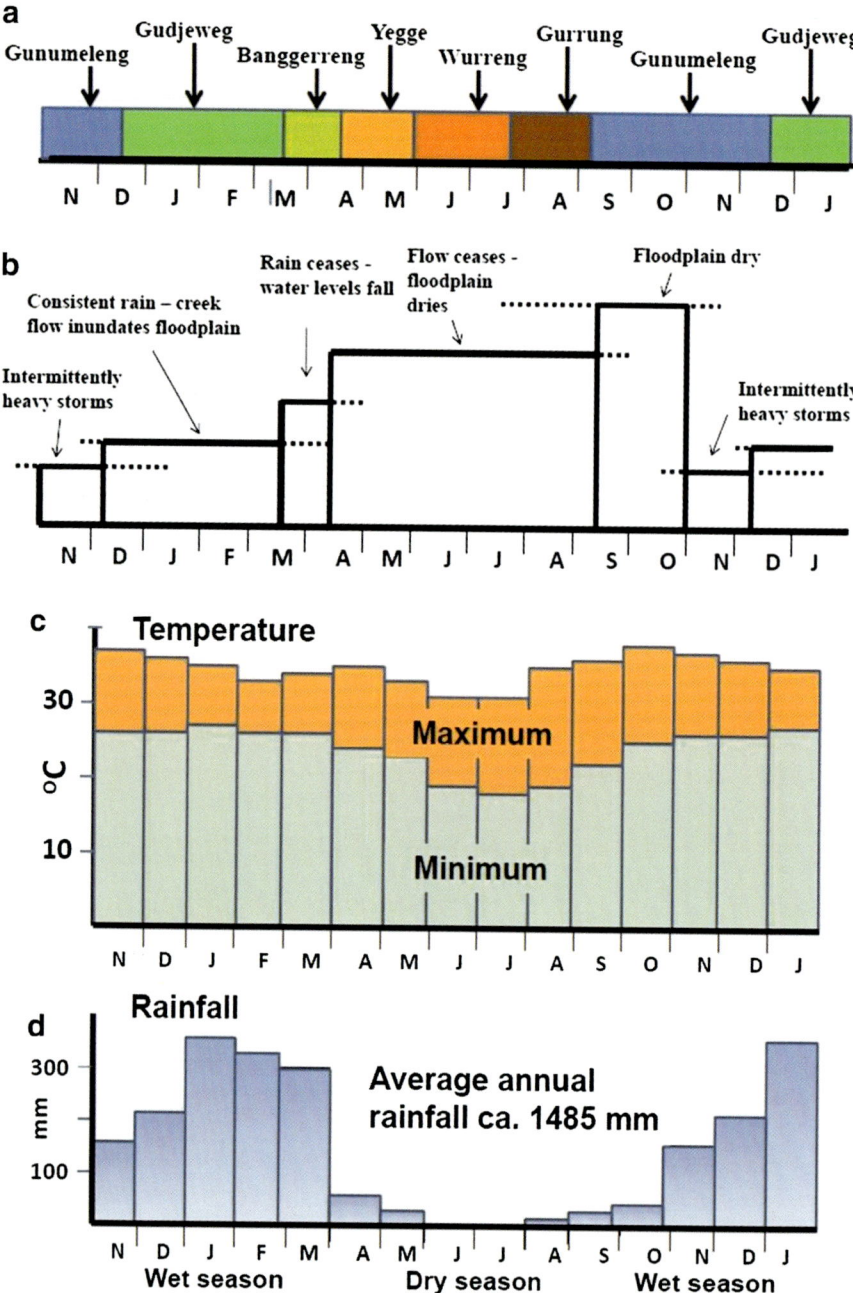

Fig. 2 The Aboriginal calendar for the Kakadu region (a) compared to the seasonal flooding cycle (b) and annual temperature (c) and rainfall patterns (d) (Adapted from information summarized in Finlayson et al. 2005)

invader. Most of the frog species are inactive during the dry season with at least nine species estivating underground. During the wet season, frogs are very obvious and abundant. The highest diversity of frogs occurs on lowland wet clay flats, with open vegetation and tussock grass.

Reptiles include 18 large species that use the riparian and floodplain habitats with a further six water-dependent species. The latter includes the pig-nosed turtle *Carettochelys insculpta*; the mangrove monitor *Varanus indicus*, restricted to mangrove communities; the little file snake *Acrochordus granulatus*; and three species of colubrid snakes. The estuarine crocodile *Crocodylus porosus* inhabits coastal rivers and rivers and billabongs on the floodplains, while the freshwater crocodile *Crocodylus johnstoni* inhabits permanent freshwater rivers and billabongs. The characteristic floodplain turtle is the carnivorous northern long-necked turtle *Chelodina rugosa*.

Waterbirds are a predominant feature of the wetlands, especially the floodplains. During the dry season on the floodplains there may be as many as two million waterbirds, including large concentrations of geese and ducks, such as the magpie goose *Anseranas semipalmata*, the wandering whistling duck *Dendrocygna arcuata*, and plumed whistling duck *Dendrocygna eytoni*. There are numerous roosts of shorebirds, containing 2,000 or more birds, along the coastline. The more numerous species included whimbrel *Numenius phaeopus*, Far Eastern curlew *Numenius madagascariensis*, common greenshank *Tringa nebularia*, and black-tailed godwit *Limosa limosa*. The mangroves provide habitat for large multispecies colonies of egrets and herons (*Ardea ibis, A. alba, A. picata, Egretta intermedia*, and *E. garzetta*), cormorants (*Phalacrocorax melanoleucos* and *P. sulcirostris*), and Australian white ibis *Threskiornis molucca*.

Mammals include the dugong *Sirenia artiodactyla* in the estuaries. A number of mammals make some use of the floodplains, including two species of rodents, two dasyurid carnivorous marsupials, three macropods, a number of insectivorous bats, two flying foxes, and the dingo. The dusky rat *Rattus colletti* shelters in cracks in the floodplain soils during the dry season and reaches high densities. Flying foxes (*Pteropus scapulatus* and *P. alecto*) roost in the fringing vegetation. Introduced mammals that have invaded the wetlands include the Asian water buffalo, pig, cattle *Bos taurus*, and horse *Equus caballus*.

Conservation Status

The wetlands of the park have been designated as internationally important under the Ramsar Convention on Wetlands due to their biogeographical importance, the diversity of their plant communities, and their role in conserving large numbers of waterfowl that congregate during the dry season. World Heritage listing of the park also refers to the natural heritage value of the tidal flats and flood plains, and the floristic diversity and endemism of the wetland vegetation. The waterbirds are protected through the Japan Australia Migratory Bird Agreement and the China Australia Migratory Bird Agreement.

The park has a management plan that supports joint management between traditional landowners and the Australian Government. The management of the park is widely regarded as a model of wetland wise use, incorporating traditional knowledge and providing a balance between multiple interests, such as conservation and tourism. Traditional people harvest plants and animals for food, art and craft, and medicinal purposes, and play a major role in managing the wetlands. The joint management arrangements have been built on successful partnerships between the traditional Indigenous owners, governments and park users to ensure the cultural and natural values of the Park are maintained. The current plan of management runs from 2016–26 (Director of National Parks 2016).

Ecosystem Services

The main ecosystem services from the wetlands include: (i) the maintenance of global important biodiversity, including critical habitat for globally and nationally threatened species, and for locally endemic species; (ii) fisheries which has a strong cultural and recreational element, the latter being largely based around the barramundi *Lates calcarifer*; and (iii) maintenance of a contemporary living culture. The biodiversity and fisheries services are largely dependent on maintenance of the important habitats within the park, as well as their interconnectedness, including with the catchments of the rivers. Maintenance of the living culture is dependent on land ownership, access to land and resources, transmission of cultural knowledge and practices to younger generations, protection of sites, and documentation of cultural heritage.

The park is an iconic destination for visitors, attracted by the wildlife and landscape, as well as the cultural heritage, including Aboriginal rock paintings. People from nearby places regularly use the park for recreation, including bushwalking, swimming, boating, and fishing. Tourism is a major source of employment. The wetlands also provide many opportunities for scientific research and have yielded archaeological materials that provide an exceptional record of the lifestyle of the indigenous people.

Aboriginal communities in the park have maintained a strong relationship with the wetlands, particularly through the provision of traditional food items, such as geese eggs, mussels, fish, turtles, and crocodiles; plants, including several species of water lily *Nymphaea* spp. and water chestnut *Eleocharis dulcis*; and for customary uses such as medicine, craft, and utensils.

Threats and Future Challenges

A number of analyses and reviews over the past few decades have identified the threats and management issues faced by floodplain wetlands in northern Australia (Finlayson et al. 2006). As elsewhere in northern Australia these issues have been assessed in site-based analyses, and some comprehensive databases now exist (e.g., for the management of specific invasive species). Information on many threats is increasingly being collected through the adoption of structured risk assessments.

Weeds present an ongoing threat to the wetlands as they can outcompete native plants, leading to their displacement, or change the structure of vegetation communities, or alter the fire regimes. Introduced pasture grasses, in particular paragrass and olive hymenachne *Hymenachne amplexicaulis*, may displace native vegetation and lead to changes in the fire regimes. Salvinia has reportedly led to a decrease in hunting and fishing grounds. Giant mimosa is largely controlled but is an ever-present threat.

Introduced animals also pose a threat to the wetlands. Ongoing efforts are made to control feral pigs, cattle, and buffalo, although they can reinvade from neighboring areas. Cane toads are a relatively recent arrival in 2001, but having since established, and are likely to colonize all wetland habitats. They are a threat to many native animals from direct consumption, competition for resources, and toxic effects on toad predators.

The implications of climate change and sea level rise for the wetlands have been assessed. The principal threats to the wetlands include increased rate and extent of saltwater inundation into freshwater environments, response of mangrove communities to rising sea level, and possibly more intensive fire regimes that eventuate due to hotter dry seasons. Bayliss et al. (1997) indicated that all wetland areas in the region below four meters in elevation are assessed as being vulnerable to climate induced changes. The resulting impacts may be drastic, particularly if the tides overtop the coastal levees and river banks. A comprehensive and multiple scale monitoring and assessment program for climate change was outlined by Finlayson et al. (2009). It is also expected that the expansion of mangroves that has been observed over the past 50 years will increase under sea level rise scenarios.

Pressure from recreational activity and tourism is also considered a threat, but under the existing management regime the risk of adverse impacts is low.

The continued mining and processing of uranium ore in a mining lease adjacent to the park poses an ongoing threat to rivers and wetlands from the potential dispersion of mine waste waters to streams and shallow wetlands, including contamination with radioactive substances (Bayliss et al. 2012). In high rainfall years low-level contaminated runoff is released to the nearby creek that drains into the park. This is accompanied by an intensive program containing biological, chemical, and radiological analyses to assess and monitor potential impacts upon ecosystems and humans.

References

Bayliss BL, Brennan KG, Eliot I, Finlayson CM, Hall RN, House T, Pidgeon RWJ, Walden D, Waterman P. Vulnerability assessment of the possible effects of predicted climate change and sea level rise in the Alligator Rivers Region, Northern Territory, Australia. Supervising Scientist Report. 1997;123:134 pp.

Bayliss P, van Dam R, Bartolo R. Quantitative ecological risk assessment of Magela Creek floodplain on Kakadu National Park: comparing point source risks from Ranger uranium mine to diffuse landscape-scale risks. Hum Ecol Risk Assess. 2012;18:115–51.

Director of National Parks. Kakadu National Park Plan of Management 2016–2026. Canberra, Australia. 2016.

Finlayson CM. Plant ecology of Australia's tropical floodplain wetlands: a review. Ann Bot. 2005;96:541–55.

Finlayson CM, Woodroffe CD. Wetland vegetation. In: Finlayson CM, von Oertzen I, editors. Landscape and vegetation ecology of the Kakadu Region, Northern Australia, Geobotany, vol. 23. Dordrecht: Kluwer; 1996. p. 81–112.

Finlayson CM, Bellio MG, Lowry JB. A conceptual basis for the wise use of wetlands in northern Australia – linking information needs, integrated analyses, drivers of change and human well-being. Marine & Freshwater Research. 2005;56:269–277.

Finlayson CM, Lowry J, Bellio MG, Walden D, Nou S, Fox G, Humphrey CL, Pidgeon R. Comparative biology of large wetlands: Kakadu National Park, Australia. Aquat Sci. 2006;68:374–99.

Finlayson CM, Eliot I, Eliot M. A strategic framework for monitoring coastal change in Australia's wet-dry tropics- concepts and progress. Aust Geogr. 2009;47:109–23.

Groundwater Dependent Wetlands of the Gnangara Groundwater System (Western Australia)

Ray H. Froend, Pierre Horwitz, and Bea Sommer

Contents

Introduction	1959
Biophysical Characteristics	1960
Change to the Ecological Character of Gnangara Mound GDWs	1961
Future Challenges	1965
References	1965

Abstract

This case study focuses on the inland, freshwater ecosystems of the Gnangara Groundwater System, principally those ecosystems with an interaction between surface water and groundwater or a dependence upon groundwater. It discusses the hydrology, water chemistry, soils and sediments, and plant and animal communities of groundwater dependent ecosystems, and the effects of climate change on them over the millenia. It also presents a synopsis of other changes to the ecosystems such as urbanization (and land use change), eutrophication, acidification and the effects of fire, and provides a prognosis for the ecosystems in the face of groundwater declines associated with climatic drying.

Keywords

Hydrology biology linkages · climate change · fire · acidification · phreatophytic vegetation · stygofauna · tumulus mound springs

---

Alphabetical authorship

R. H. Froend · P. Horwitz (✉) · B. Sommer
Centre for Ecosystem Management, School of Science, Edith Cowan University, Joondalup, WA, Australia
e-mail: r.froend@ecu.edu.au; p.horwitz@ecu.edu.au; b.sommer@ecu.edu.au

© Springer Science+Business Media B.V., part of Springer Nature 2018
C. M. Finlayson et al. (eds.), *The Wetland Book*,
https://doi.org/10.1007/978-94-007-4001-3_110

Introduction

This case study focuses on the inland, freshwater ecosystems of the Gnangara Groundwater System, principally those ecosystems with an interaction between surface water and groundwater or a dependence upon groundwater. The focus ranges from nontidal surface water systems to other groundwater dependent ecosystems (GDEs), communities, and components such as caves, stygobiota, phreatophytic vegetation, and areas of groundwater discharge including coastal areas. The seasonal rainfall patterns, temperature regimes, landforms, and soil types of the area create a profound, bioregionally idiosyncratic, and significant water dependence for the biota (Table 1).

The system occurs wholly within the Swan Coastal Plain (SCP) bioregion of Western Australia. It is bounded by the Darling Fault and Gingin Scarp on the eastern side and the coast on the western side (Fig. 1). This work is derived from Horwitz et al. (2009a, b), and for full referencing see work cited therein.

Biophysical Characteristics

Ultimately, biophysical characteristics of GDWs are dependent on their water regime. Wetlands usually occur in depressions (Fig. 1). The majority of permanently inundated wetlands, or lakes, are shallow expressions of the groundwater table which fill following winter rainfall and dry during summer as groundwater levels fall, rainfall decreases, and surface evaporation increases. Sumplands, holding water only during winter and spring, and damplands, seasonally waterlogged areas, can form as expressions of underlying groundwater. Floodplains hold water when streams like Ellen Brook, or Gingin Brook, overflow in winter or spring, and palusplains (seasonally waterlogged flats) become saturated from seasonally elevated groundwater levels. Sumplands and palusplains can also form from rainfall perching over more impermeable soils (like diatomaceous earths which have increased water holding capacities).

Strongly related to the hydrological regimes are the soils and sediments of the Swan Coastal Plain. On the Gnangara Groundwater System the large scale landforms run roughly parallel with the Darling Fault and the coast, corresponding with sedimentary formations. Landforms are dunal systems and groundwater dependent wetlands occur in a sequence east to west (Davidson 1995); the sequence also corresponds to geomorphological age, with the dunes closest to coast being the most recent, and those furthest being the oldest.

The work of Semeniuk and Semeniuk (2004, 2006) has advanced our understanding of the types and distribution of sediments on the Swan Coastal Plain. It describes the main wetland sediment types as peat, diatomite, and calcilutite (carbonate mud); wetlands can also have intermediaries with proportional mixtures of these. These wetland sediments accumulate in depressions under conditions of permanent, seasonal, or intermittent saturation, from mainly internal biological processes (biogenic: from sponges, diatoms, shells, charophytes, and so on) or from basin setting (internal) or extrabasinal (washed in) (terrigenic) processes. These accumulated sediments are then further subjected to biological, chemical, and physical (diagenic) processes in

Table 1 Hydrology–biology linkages to demonstrate those biotic components (from individuals, populations, communities, and ecosystems) that have a hydrological requirement for their continued survival (Adapted from Horwitz et al. 2008, with permission Royal Society of Western Australia; for full explanation see ▶ Chap. 24, "Groundwater Dependent Wetlands" by Froend et al. this volume)

Hydrology–biology linkage	Relevance to the Gnangara Groundwater System (with examples)
Requirement for seasonal soil moisture	All vegetation, to different degrees
Requirement for seasonally moist habitat for aestivation/drought avoidance	Burrowing crayfish *Cherax quinquecarinatus*, burrowing frogs *Heleioporus eyrei*, or aestivating fish *Galaxiella nigrostriata*
Requirement for a seasonal or intermittent surface saturation	Microcrustacean faunas with drought resistant eggs. Waterbirds that feed on them (i.e., stilts and sandpipers). Palusplain wetlands on eastern part of the Mound associated with Ellen Brook
Terrestrial requirement (obligate and facultative) for access to groundwater table	Phreatophytic vegetation (i.e., Banksia woodlands or tuart *Eucalyptus gomphocephalus*)
Requirement for groundwater discharge to maintain a particular quality of surface water or habitat	Tumulus mound springs: groundwater discharge enables accumulation of organic material, creating well-vegetated well-shaded habitats for other organisms
Requirement for permanent surface or subsurface saturation	Any lake with permanent water (e.g., Loch McNess, Yonderup, Goolellal Lakes, etc.). Permanent subsurface saturation required to prevent acidification
Requirement for an exchange between surface/subsurface flows and groundwater	Baseflows in Ellen Brook and Gingin Brook. Flow through lakes like Lake Jandabup and Lake Nowergup
Requirement for saturated hypogean (interstitial) spaces (aquifer)	Stygofauna in spaces between sand grains. Fauna dependent on saturation of tuart root mats in caves

situ. Consanguineous suites of wetlands and those wetlands on major landforms, and in interbarrier depressions, tend to have characteristic types of sedimentary fill. Water chemistry follows these characteristics; for example, wetlands with intrabasinal peat or diatomaceous peat sediments tend to be colored, tannin-rich, acidic (to alkaline), and cation poor, whereas wetlands with a limestone basement setting and calcilutite sediments tend to be clear and uncolored, tannin-poor, alkaline, and cation enriched.

Change to the Ecological Character of Gnangara Mound GDWs

Wetlands have experienced periods of higher water levels and lower water levels over the last 10,000 years but also in more recent times. Wetlands have undergone many changes over the last 200 years, including drainage of low-lying areas on the SCP, infilling of wetlands, groundwater abstraction, and a more recent decline in

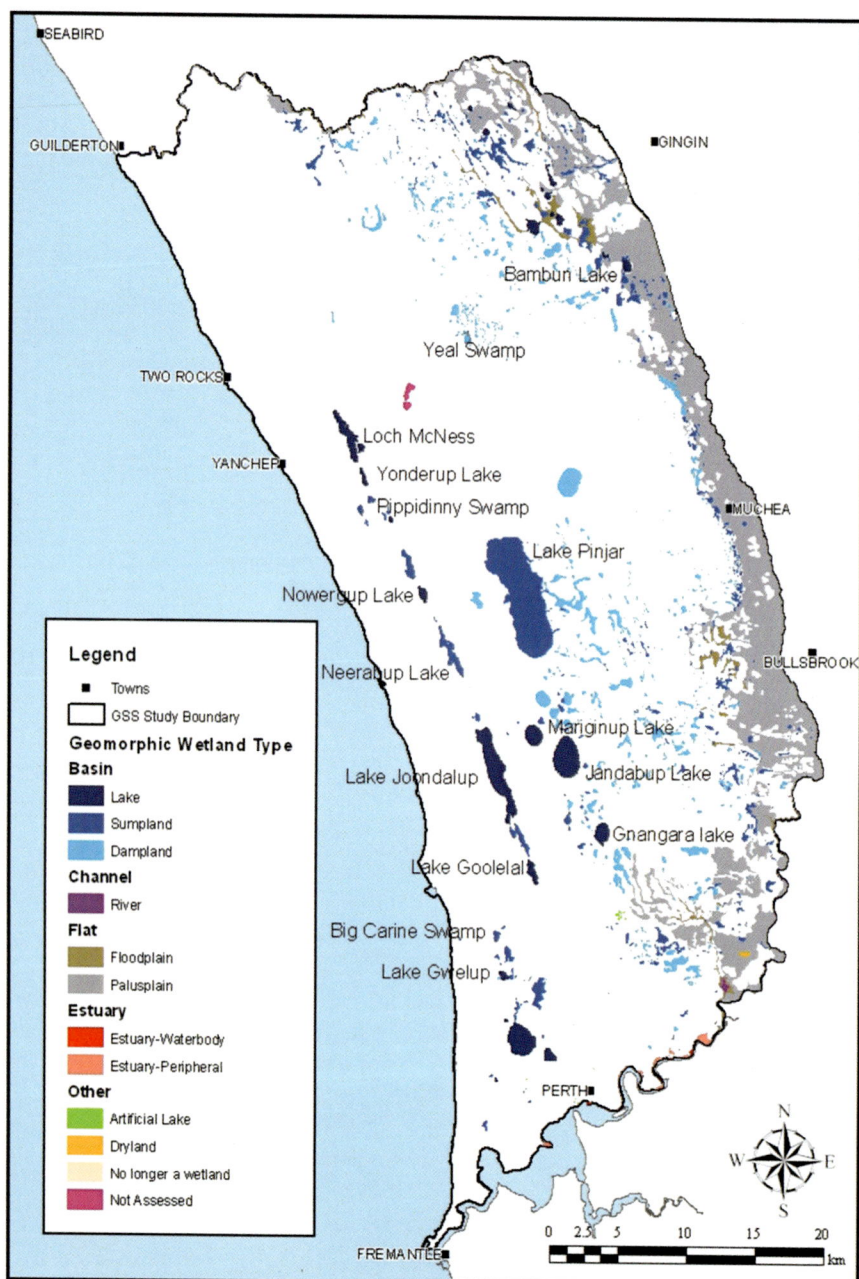

Fig. 1 Wetland mapping for the Gnangara Groundwater System area. Mapped wetlands are assigned geomorphic attributes in the database held by the Western Australian Department of Parks and Wildlife (formerly Department of Environment and Conservation) (Map reproduced from Horwitz et al (2009a), with permission Department of Biodiversity Conservation and Attractions, Perth)

rainfall. Together these factors have produced current groundwater declines across the Gnangara Mound and associated declines in water levels within the wetlands. In fact, processes that are involved in the degradation and loss of groundwater-dependent wetlands cover the full spectrum discussed by Froend et al. (▶ Chap. 24, "Groundwater Dependent Wetlands," this volume).

Rainfall has had the greatest impact on groundwater and wetland depth; however, other factors have influenced long-term groundwater levels, such as urbanization, which has increased the amount of water abstracted for public and private use. In response to a drying hydrological regime, changes in vegetation that are dependent on groundwater have varied, depending on their tolerance of the amount of moisture in the soil. Wetland sediments have also been affected by a drying hydrological regime. Drying of permanently inundated wetland sediments has had in some instances irrevocable consequences such as acidification, eutrophication, or irreversible changes to sediment-water relationships.

Land-use changes include clearing, filling, and fragmentation of wetlands habitats. A rise in the water levels of wetlands has occurred in some areas with clearing of native vegetation for urbanization, while excessive groundwater abstraction has reduced water levels in other areas.

Wetlands are parts of landscapes where nutrients generated in the catchment can accumulate, where they are involved in biological metabolic processes, biomass growth, decomposition, and deposition. An oversupply of nutrients has in places resulted in an altered ecology and diminished ecosystem services, such as diminished water quality, deoxygenation of the water body, which in turn has affected wetland invertebrates and vertebrates, and build-up of toxins that could have consequences for the health of human and birdlife.

Acidification (oxidation of iron sulfides, which can be drought-induced or has occurred as a result of dewatering or direct disturbance by excavation) produces a significant change in water chemistry. Its consequences have been shifts in aquatic plant and algal communities, declines of sensitive species (i.e., fish, crustaceans, and molluscs), and characteristic reductions in macroinvertebrate richness and abundance.

While fire is part of the ecology of wetlands, and some riparian species depend on fire for recruitment, wetlands are also places where fire sensitive species and processes occur, and changed fire regimes particularly fires that are intense or frequent can alter wetland ecologies. Groundwater dependent wetlands have incurred altered hydrologies, as a result water chemistry can change (even toward acidification), the sediment load in the water can increase, species of flora and fauna can disappear (or dominate, as is the case with some opportunistic or weedy species), and organic rich sediment can be consumed by fire, releasing carbon into the atmosphere and toxic materials into the groundwater. These impacts have been exacerbated by, and perhaps linked to, recent declines in rainfall, groundwater levels, and reduced saturation of sediments, such as peat.

Seven threatened ecological communities are either wetland in nature or are groundwater dependant ecosystems. Threats to these communities include altered hydrological regimes, clearing, fire, climate change, and invasive weeds.

Table 2 Broad classes of Hydrology–Biology linkages for inland aquatic and terrestrial species, communities and/or ecosystems on the Gnangara Groundwater System with a prognosis under drying scenarios, examples from the literature where such effects have been detected or predicted, and possible irreversibilities and thresholds of change (Adapted from Horwitz et al. 2008, with permission Royal Society of Western Australia; for full explanation see ▶ Chap. 24, "Groundwater Dependent Wetlands" by Froend et al. this volume)

Hydrology–Biology linkages	Prognosis under scenarios of declining groundwater levels and decreasing rainfall	Possible irreversibilities and thresholds of change
Requirement for seasonal soil moisture	Reduced vigor as a result of lower water availability in summer, altered rates of surface soil carbon and nutrient cycling. Potentially reduced seed set and shift in population distribution and persistence and community composition	Transitional states with drying-more research required
Requirement for seasonal availability of moist habitat for aestivation/drought avoidance	Survival provided moisture levels are sustained; however, if otherwise moist habitats dry, less frequent reemergence and probably reduced reproduction may result (examples include burrowing crayfish and frogs, aestivating fish)	Once-off severe drying (of one or more years) may eliminate local and regional populations
Requirement for a seasonal or intermittent surface saturation	Inundation less frequent (i.e., inundation once every 5 years to once every 10 years) or seasonality of inundation changes (decreasing winter-spring inundation and possible incidence of summer inundation) Decreased areal extent and duration of inundation can result in reduced frequency of plant recruitment events and reduced richness of wetland invertebrates	Transitional states with drying for vegetation-more research required
Terrestrial requirement for access to groundwater table	Acute drawdown and low recharge can result in loss of adult individuals of overstorey and understorey species or local extinction of susceptible species. Less severe circumstances can result in reduced vigor of adults and a shift in the distributions of established juveniles	Transitional states with drying for vegetation – more research required
Requirement for groundwater discharge to maintain a particular quality of surface water	Extinction of discharge means loss of suitable habitat. Declined volumetric discharge can mean local contraction of habitat, altered water temperature regimes and chemical characteristics	Local and regional extinction of endemic and other forms wedded to discharge habitats like mound springs
Requirement for permanent surface or subsurface saturation	Change from permanent to temporary lake or stream systems flora and fauna dependant on permanent water lost from location Sediments exposed to more frequent drying, potentially displacing biota. Most severe will be drying, heating, and cracking of sediments that have never been so, changing sediment structure and biogeochemistry; acidification under certain conditions	Local and regional extinction of endemic forms wedded to permanent surface water Permanent change in sediment structure from severe drying; perpetual acidification from continued drying. Permanent loss of calcium from system. Permanent loss of sediment from burning

(continued)

Table 2 (continued)

Hydrology–Biology linkages	Prognosis under scenarios of declining groundwater levels and decreasing rainfall	Possible irreversibilities and thresholds of change
Requirement for an exchange between surface/subsurface flows and groundwater	Altered patterns of carbon and nutrient cycling. Reduced ability to retreat or emerge according to life history requirement. Potential loss of habitat. Seasonal switches between surface water recharging to groundwater discharging, shifts to surface water recharging only	If shift to recharge only, subsurface ecologies change permanently
Requirement for saturated hypogean (interstitial) spaces (aquifer)	Where habitat is fixed at a certain stratigraphic level then declines in the saturated zone will strand dependent biota resulting in local extinctions. Otherwise distributions of short-range endemics may change according to the extent of groundwater level reduction	Local extinctions of biota wedded to particular stratigraphic zones

Future Challenges

Hydrological change on the SCP is therefore the underlying feature of wetland change, but this is driven by several processes, mostly anthropogenic: climate variability, patterns of land-use change, patterns of water regulation, patterns of groundwater extraction, water infrastructure, and distal societal drivers (e.g., lack of adequate water pricing) (see McFarlane et al. 2012). Prognoses for wetland biota experiencing the drying scenarios that result from these processes are given in Table 2.

Management interventions for wetland restoration and to prevent permanent losses or changes in groundwater dependent wetland systems include shutting off nearby extraction bores to reduce local drawdown effects. Artificial water supplementation is an important short-term measure to reinstate hydrological regimes, reinstate saturation of sediments, and provide water for hydrology-biology linkages. Rehabilitation of inner urban wetlands, particularly to reduce nutrients flowing into wetlands with better riparian buffers and a change in social behavior in catchments, will address the effects of hydrological change, eutrophication, and habitat alteration.

References

Davidson WA. Hydrogeology and groundwater resources of the Perth Region, Bulletin 142. Perth: Geological Survey of Western Australia; 1995.

Horwitz P, Bradshaw D, Hopper SD, Davies PM, Froend R, Bradwhaw F. Hydological change escalates risk of ecosystem stress in Australia's threatened biodiversity hotspot. J R Soc West Aust. 2008;91:1–11.

Horwitz P, Sommer B, Froend R. Biodiversity of wetlands and groundwater dependent ecosystems, Chapter 4. In: Wilson B, Valentine LE, editors. Biodiversity values and threatening processes of the Gnangara Groundwater System. Perth: Department of Environment and Conservation; 2009a.

Horwitz P, Sommer B, Hewitt P. Wetlands: changes, losses and gains, Chapter 5. In: Wilson B, Valentine LE, editors. Biodiversity values and threatening processes of the Gnangara Groundwater System. Perth: Department of Environment and Conservation; 2009b.

McFarlane D, Strawbridge M, Stone R, Paton A. Managing groundwater levels in the face of uncertainty and change: a case study from Gnangara. Water Sci Technol Water Supply. 2012;12:321–8.

Semeniuk V, Semeniuk CA. Sedimentary fill of basin wetlands, central Swan Coastal Plain, southwestern Australia. Part 1: sediment particles, typical sediments, and classification of depositional systems. J R Soc West Aust. 2004;87:139–86.

Semeniuk V, Semeniuk CA. Sedimentary fill of basin wetlands, central Swan Coastal Plain, southwestern Australia. Part 2: distribution of sediment types and their stratigraphy. J R Soc West Aust. 2006;89:185–220.

Seagrass Meadows of Northeastern Australia

167

Robert G. Coles, Michael A. Rasheed, Alana Grech, and Len J. McKenzie

Contents

Introduction	1968
Seagrass Species	1970
Current Status	1971
Seagrass Area	1971
Threat Mapping	1972
Monitoring	1972
Threats and Future Challenges	1973
References	1974

Abstract

The enormous seagrass meadows that stretch along the shallow coastal waters of Northeastern Australia from intertidal banks to about 60 m deep are of great importance to ecosystem functionality. A large proportion of these meadows are remote from human populations and anthropogenic impacts, and many are inaccessible and rarely visited by people. The meadows provide a range of important ecological services and support economically valuable fish and shrimp populations, and some of the largest remaining populations of the dugong (*Dugong dugon*) and green sea turtles (*Chelonia mydas*). Northeastern Australia's fifteen seagrass species represent just over 20% of the world's 72 species. The greatest impact to seagrasses along the Great Barrier Reef coast comes from agricultural runoff, followed by urban and industrial runoff, urban port and

---

R. G. Coles (✉) · M. A. Rasheed · L. J. McKenzie
Centre for Tropical Water and Aquatic Ecosystem Research, James Cook University, Cairns and Townsville, QLD, Australia
e-mail: rob.coles@jcu.edu.au; michael.rasheed@jcu.edu.au; len.mckenzie@jcu.edu.au

A. Grech
Department of Environmental Sciences, Macquarie University, Sydney, NSW, Australia
e-mail: Alana.grech@mq.edu.au

© Springer Science+Business Media B.V., part of Springer Nature 2018
C. M. Finlayson et al. (eds.), *The Wetland Book*,
https://doi.org/10.1007/978-94-007-4001-3_266

infrastructure development, dredging, shipping accidents, bottom trawling, boat damage, and other fishing methods. Even in the areas of lower anthropogenic disturbance such as the Gulf of Carpentaria, the influence of predicted climate change on seagrasses may lead to substantial future impacts. The challenge in the future will be to mitigate the impacts of development from urban and agricultural activity where these may interact synergistically with increasing ecological pressure from climate change.

Keywords

Seagrass · Great Barrier Reef · Dugong · Turtles

Introduction

The northeastern Australian coastline from the tropics (10°S) to the subtropical zone of the southern Great Barrier Reef (GBR) and its World Heritage Area (25°S) is famous for coral reefs, a huge drawcard for the Australian tourist industry. Less appreciated but of at least equal importance to ecosystem functionality are the enormous seagrass meadows that stretch along these shallow coastal waters from

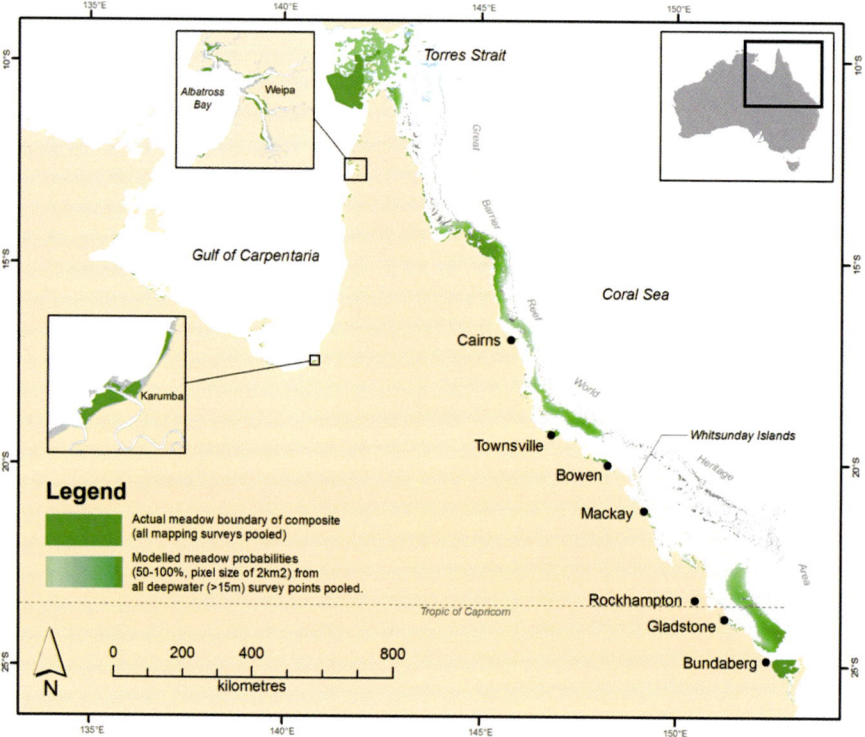

Fig. 1 Distribution of northeastern Australian seagrass meadows (Compiled from McKenzie et al. 2010 and Coles et al. 2004)

Fig. 2 Vast tropical wetlands adjacent to the Gulf of Carpentaria coastline are largely inaccessible ensuring the seagrasses in the shallow coastal water are rarely visited (Photograph by TropWATER James Cook University © Rights remain with the author)

intertidal banks to about 60 m deep (Fig. 1). Unique for a developed country such as Australia, a large proportion of these meadows are remote from human populations and anthropogenic impacts. Many meadows, such as those in the Gulf of Carpentaria, are inaccessible and rarely visited by people (Fig. 2).

Of global ecological importance, northeastern Australia's seagrass meadows support economically valuable fish and shrimp populations and some of the largest remaining populations of the marine mammal, the dugong *Dugong dugon* (Fig. 3) and green sea turtles *Chelonia mydas* (Coles et al. 1993; Watson et al. 1993; Marsh et al. 2012). Seagrasses are critical to the survival of these animals. The meadows also provide a range of other important ecological services. They stabilize the substrate, filter the water of organic matter and recycle nitrogen, baffle wave and tidal energy, and incorporate large amounts of carbon (Kenworthy et al. 2006).

Northeastern Australia includes coastal areas with diverse physical characteristics. The tropical north coast and Torres Strait are influenced by monsoonal rains and associated pulses of turbid waters draining from adjacent catchments. The coasts are mostly lined with muddy sediments, extensive intertidal flats, and shallow inshore reefs. The tropical and a portion of the subtropical Queensland east coast are sheltered by the Great Barrier Reef which effectively encloses a long lagoon. It is within this lagoon that seagrass meadows flourish. Four general seagrass habitat types have been identified: estuarine (and inlet), coastal, deep water (>15 m depth), and reef, with dominant drivers of each habitat being terrigenous runoff, physical disturbance, light, and low nutrients, respectively (Carruthers et al. 2002). Inshore meadows are influenced by coastal topography and shelter, with most meadows

Fig. 3 Trails left by feeding dugong in seagrass near Magnetic Island (Photograph by TropWATER James Cook University © Rights remain with the author)

occurring in north facing bays and estuaries that are protected from the dominant south-easterly winds. The most extensive meadows, however, cover much of the sea floor of the Torres Strait and GBR lagoon.

Seagrass Species

Northeastern Australia has fifteen seagrass species, representing just over 20% of the world's 72 species (Table 1). All species are important for the ecological health of the region, and none are listed as Endangered, Vulnerable, Near Threatened, or Data Deficient under the IUCN Red List criteria (Short et al. 2011). One species is potentially endemic to northern Australia, *Halophila tricostata* (Kuo et al. 1992); most other species are widely distributed or common across northern Australia and the Western Pacific region. The tropical species *Enhalus acoroides* and *Thalassodendron ciliatum* as well as *Halophila tricostata*, *Thalassia hemprichii*, and *Cymodocea rotundata* are found mainly in the north with small populations at least as far south as the Whitsunday Islands (approximately 22°S, Lee Long et al. 1992). Except for *Halophila capricorni*, all species are present in water shallower than 10 m. Only seagrass of the genus *Halophila* are found in water deeper than 15 m (Lee Long et al. 1992; Coles et al. 2009).

Most species in the region are capable of rapid recovery from losses due to fast asexual growth rates and capacity for generating large seed banks (Rasheed 1999, 2004). Previous large scale impacts have recovered within 3–5 years (Preen et al. 1995). However, recently this resilience of seagrasses has been tested by sustained periods of unfavorable climate which have resulted in seagrasses not recovering as expected (McKenna et al. 2015; Rasheed et al. 2014).

Table 1 Seagrass species in the Great Barrier Reef World Heritage Area (Based on Carter et al. 2016. © James Cook University licensed for use under a Creative Commons Attribution 4.0 Australia license)

FAMILY – Species	Habitat
CYMODOCEAE	
Cymodocea rotundata	Reef
Cymodocea serrulata	Estuarine, coastal, deep-water, reef
Halodule pinifolia[a]	Estuarine, coastal, reef
Halodule uninervis	Estuarine, coastal, reef
Syringodium isoetifolium	Coastal, reef
Thalassodendron ciliatum	Reef
HYDROCHARITACEAE	
Enhalus acoroides	Estuarine, coastal
Halophila capricorni	Deep-water
Halophila decipiens	Coastal, deep-water, reef
Halophila minor	Estuarine, coastal
Halophila ovalis	Estuarine, coastal, deep-water, reef
Halophila spinulosa	Coastal, deep-water
Halophila tricostata	Deep-water
Thalassia hemprichii	Reef
ZOSTERACEAE	
Zostera muelleri	Estuarine, coastal
POTAMOGETONACEAE	
Ruppia spp.[b]	Estuarine

[a]Taxonomy of *H. pinifolia* is uncertain. It is considered synonymous with *H. uninervis* by some taxonomists
[b]*Ruppia* spp. is sometimes excluded from seagrass lists as it is a brackish species that also grows in freshwater

Current Status

Managing, monitoring, and providing advice on seagrass ecology and seagrass conservation and development approvals at the scale of over 4,000 km of coast line presents many logistic challenges. Three key approaches have been used to meet this challenge. Maps have been compiled to form a composite spatial (geographic information system) layer of all recorded seagrass locations and modeling techniques used to fill in gaps in our spatial knowledge. Threats to seagrass have been investigated and mapped to enable priorities to be established for areas requiring detailed studies and long-term monitoring projects have been developed to follow trends through time in key locations.

Seagrass Area

Mapping estimates suggest that there is as much as 900 km^2, 13,413 km^2, and 3,063 km^2 of seagrass in water shallower than 15 m in the Gulf of Carpentaria, the Torres Strait, and

GBR, respectively. There is an additional 35,000 km^2 of meadows in the Torres Strait and GBR deeper than 15 m (Coles et al. 2004, 2009; McKenzie et al. 2010) (see Fig. 1). The highest species diversity (14 species) of seagrass is found in the waters north of Mackay (Fig. 4) in the central region of the GBR (Coles et al. 2003). But sites with diverse topography, mud and sand flats, reef platforms, and inter-reef habitats such as the Torres Strait (Fig. 5) commonly have 10 or more species (Carter et al. 2014).

Threat Mapping

The greatest impact to seagrasses along the GBR coast comes from agricultural runoff, followed by urban and industrial runoff, urban port and infrastructure development, dredging, shipping accidents, bottom trawling, boat damage, and other fishing methods. The seagrass meadows identified as at greatest risk of cumulative impact are all adjacent to population centers or areas of farmland, in sheltered north facing bays, and all in the populated southern half of the GBR. Industrial ports are also located in sheltered bays and although heavily regulated contribute to pressures on seagrass meadows (Grech and Coles 2010; Grech et al. 2011).

Monitoring

To include as large a geographic spread as possible, monitoring data for this region relies on several sources: the GBR has a long-term Marine Monitoring Program as

Fig. 4 Seagrass meadows exposed at low tide on the Green Island reef flat Cairns. There are around nine species of seagrass accessible at low tide (Photograph by TropWATER James Cook University © Rights remain with the author)

Fig. 5 Extensive mixed species intertidal seagrass meadows near Thursday Island in the Torres Strait (Photograph by TropWATER James Cook University © Rights remain with the author)

part of the Commonwealth and Queensland State Governments Reef Plan, collecting seagrass metrics from 45 sites spread along the east coast (Coles et al. 2015); the Seagrass-Watch program which collects data from mostly intertidal sites and includes monitoring with indigenous ranger groups along the east coast and in the Torres Strait (McKenzie et al. 2000; Mellors et al. 2008); the SeagrassNet monitoring project included a site at Green Island near Cairns with data extending over several years (Short et al. 2014); and the Queensland Ports Seagrass Monitoring Program currently monitors seagrass status in every major port at least annually including data sets spanning 20 years (Coles et al. 2015).

Threats and Future Challenges

Because there are no major human population centers along the northeastern coast of Australia and north of 17°S, seagrass meadows in this region are spared many of the anthropogenic hazards that have resulted in losses in many other parts of the world (Halpern et al. 2008). The southern tropical and subtropical east coast includes the larger (by Australian standards) population centers and ports of Cairns, Townsville, Mackay, and Gladstone, which have a standard suite of associated anthropogenic disturbances (dredging, land reclamation, inshore turbidity, and pesticide runoff). Seagrass meadows in these locations show indications of stress with declines in many places, some quite dramatic. These declines are almost certainly the result of multiple stressors, exacerbated by severe tropical storms and floods that have been a feature of the coastline since 2006 (McKenna et al. 2015; Rasheed et al. 2014).

Even in the areas of lower anthropogenic disturbance such as the Gulf of Carpentaria, the influence of predicted climate change on seagrasses may lead to substantial future impacts (Rasheed and Unsworth 2011).

Seagrasses, along with other marine plants, are protected in Queensland waters and any more than incidental damage can only occur with legislative approval. Almost a third of the GBR is zoned as highly protected areas, and activities such as bottom trawling have been reduced in recent years to a level unlikely to cause serious damage to seagrass (Grech and Coles 2011). Maintaining this level of protection requires a constant level of advocacy, and there are no guarantees it will be maintained. The challenge in the future will be to mitigate the impacts of development from urban and agricultural activity where these may interact synergistically with increasing ecological pressure from climate change.

Seagrasses as flowering plants have adapted to live submerged in a marine environment. However, there is surprisingly little research on how tropical meadows colonize or regrow from propagules and how that process interacts with the complex set of constraints in the marine environment both physical (light and light spectra, sediment type, current stress, etc.) and biological (propagule buoyancy and viability, herbivory, seed and fruit predation, etc.) that determine survival. If the seagrass meadows that are such an important feature of this coastline are to survive there will be a continuing need to provide credible science to understand the ecological processes occurring and the values of these seagrass meadows.

References

Carruthers TJB, Dennison WC, Longstaff BJ, Waycott M, Abal EG, McKenzie LJ, Lee Long WJ. Seagrass habitats of Northeast Australia: models of key processes and controls. Bull Mar Sci. 2002;71:1153–69.

Carter AB, Taylor HA, Rasheed MA. Torres strait mapping: seagrass consolidation, 2002 – 2014, JCU publication, report no. 14/55. Cairns: Centre for Tropical Water & Aquatic Ecosystem Research; 2014.

Carter AB, McKenna SA, Rasheed MA, McKenzie L, Coles RG. Seagrass mapping synthesis: A resource for coastal management in the Great Barrier Reef World Heritage Area. Report to the National Environmental Science Programme. Cairns: Reef and Rainforest Research Centre Limited; 2016. p. 22.

Coles RG, Lee Long WJ, Watson RA, Derbyshire KJ. Distribution of seagrasses, and their fish and penaeid prawn communities, in Cairns Harbour, a tropical estuary, northern Queensland, Australia. Aust J Mar Freshwat Res. 1993;44:193–210.

Coles RG, McKenzie LJ, Campbell SJ. The seagrasses of eastern Australia. In: Green EP, Short FT, Spalding MD, editors. The world atlas of seagrasses: present status and future conservation. Berkeley Los Angeles London: University of California Press; 2003. p. 131–47.

Coles RG, Smit N, McKenzie LJ, Roelofs A, Haywood, Kenyon R. Seagrasses. In: National oceans office. Key species. A description of key species groups in the Northern Planning Area. Hobart: National Oceans Office; 2004. p. 9–16.

Coles RG, McKenzie LJ, De'ath G, Roelofs AJ, Lee LW. Spatial distribution of deepwater seagrass in the inter-reef lagoon of the Great Barrier Reef World Heritage Area. Mar Ecol Prog Ser. 2009;392:57–68.

Coles R, Rasheed M, McKenzie L, Grech A, York P, Sheaves M, McKenna S, Bryant C. The Great Barrier Reef World Heritage Area seagrasses: managing this iconic Australian ecosystem for the future. Estuar Coast Shelf Sci. 2015;153:A1–12.

Grech A, Coles RG. An ecosystem-scale predictive model of coastal seagrass distribution. Aquat Conserv. 2010;20:437–44.

Grech A, Coles RG. Interactions between a Trawl fishery and spatial closures for biodiversity conservation in the Great Barrier Reef World Heritage Area, Australia. PLoS One. 2011;6(6): e21094. doi:10.1371/journal.pone.0021094.

Grech A, Coles RG, Marsh H. A broad-scale assessment of the risk to coastal seagrasses from cumulative threats. Mar Policy. 2011;35(5):560–7.

Halpern BS, Walbridge S, Selkoe KA, Kappel CV, Micheli F, D'Agrosa C, Bruno JF, Casey KS, Ebert C, Fox HE, Fujita R, Heinemann D, Lenihan HS, Madin EMP, Perry MT, Selig ER, Spalding M, Steneck R, Watson R. A global map of human impact on marine ecosystems. Science. 2008;319:948–52.

Kenworthy WJ, Wyllie-Echeverria S, Coles RG, Pergent G, Pergent C. Seagrass conservation biology: an interdisciplinary science for protection of the seagrass biome. In: Larkum AWD, Orth RJ, Duarte CM, editors. Seagrass biology. Dordrecht: Springer; 2006. p. 595–623.

Kuo J, Lee Long W, Coles RG. Fruits and seeds of *Halophila tricostata* greenway (hydrocharitacea) with special notes on its occurrence and biology. Aust J Mar Freshwat Res. 1992;44:43–59.

Lee Long WJ, Mellors JE, Coles RG. Seagrasses between Cape York and Hervey Bay, Queensland, Australia. Aust J Mar Freshwat Res. 1992;44:19–33.

Marsh H, O'Shea TJ, Reynolds III JE. Ecology and conservation of the Sirenia – Dugongs and Manatees, Conservation Biology No. 18. Cambridge, UK: Cambridge University Press; 2012.

McKenna S, Jarvis J, Sankey T, Reason C, Coles R, Rasheed M. Declines of seagrasses in a tropical harbour, North Queensland Australia – not the result of a single event. J Biol Sci. 2015;40:389–98.

McKenzie LJ, Lee Long WJ, Coles RG, Roder CA. Seagrass-watch: community based monitoring of seagrass resources. Biol Mar Mediterr. 2000;7:393–6.

McKenzie LJ, Yoshida RL, Grech A, Coles R. Queensland seagrasses. Status 2010 – Torres Strait and East Coast. Cairns: Fisheries Queensland (DEEDI); 2010. 6 pp, hdl:10013/epic.42902.d001.

Mellors J, McKenzie LJ, Coles RG. Seagrass-watch: engaging Torres Strait Islanders in marine habitat monitoring. Cont Shelf Res. 2008;28:2339–49.

Preen AR, Lee Long WJ, Coles RG. Flood and cyclone related loss, and partial recovery, of more than 1,000 km^2 of seagrass in Hervey Bay, Queensland, Australia. Aquat Bot. 1995;52:3–17.

Rasheed MA. Recovery of experimentally created gaps within a tropical *Zostera capricorni* (Aschers.) seagrass meadow, Queensland Australia. J Exp Mar Biol Ecol. 1999;235(2):183–200.

Rasheed MA. Recovery and succession in a multi-species tropical seagrass meadow following experimental disturbance: the role of sexual and asexual reproduction. J Exp Mar Biol Ecol. 2004;310:13–45.

Rasheed MA, Unsworth RKF. Long-term climate-associated dynamics of a tropical seagrass meadow: implications for the future. Mar Ecol Prog Ser. 2011;422:93–103.

Rasheed MA, McKenna S, Carter A, Coles RG. Contrasting recovery of shallow and deep water seagrass communities following climate associated losses in tropical north Queensland, Australia. Mar Pollut Bull. 2014;83:491–9.

Short FT, Polidoro B, Livingstone SR, Carpenter KE, Bandeira S, Bujang JS, Calumpong HP, Carruthers TJB, Coles RG, Dennison WC, Erftemeijer PLA, Fortes MD, Freeman AS, Jagtap TG, Kamal AHM, Kendrick GA, Judson Kenworthy W, La Nafie YA, Nasution IM, Orth RJ, Prathep A, Sanciangco JC, Tussenbroek BV, Vergara SG, Waycott M, Zieman JC. Extinction risk assessment of the world's seagrass species. Biol Conserv. 2011;144:1961–71.

Short F, Coles R, Fortes M, Victor S, Salik M, Isnain I, Andrew J, Aganto S. SeagrassNet monitoring in the western Pacific shows evidence of seagrass declines in line with global trends. Mar Pollut Bull. 2014;83:408–16.

Watson RA, Coles RG, Lee Long WJ. Simulation estimates of annual yield and landed value for commercial penaeid prawns from a tropical seagrass habitat, northern Queensland, Australia. Aust J Mar Freshwat Res. 1993;44(1):211–20.

Wetlands of New Zealand

168

Karen Denyer and Hugh Robertson

Contents

Introduction	1978
Wetland Types and Distribution	1978
Flora and Fauna	1978
Cultural Values and Management	1982
New Zealand's Ramsar Wetlands	1983
Threats to Wetlands	1984
Wetland Policy and Strategies	1986
Wetland Restoration	1987
Future Challenges	1988
References	1989

Abstract

New Zealand is an archipelago located in the South Pacific Ocean comprising two large islands and numerous smaller islands spread over 23° of latitude. It has a relatively wet climate with mean annual rainfall of around 600–4,000 mm per year depending on geographic region. The landscape is geologically active and varied, dominated in different regions by mountains, glacial valleys, volcanic plateaus, broad floodplains, and lowlands, with an extensive coastline. Here we provide an overview of New Zealand wetlands, describing the key wetland types, flora and fauna, progress with implementing the Ramsar Convention relevant wetland policy, approaches to restoration, and cultural connections. The focus is

---

K. Denyer (✉)
National Wetland Trust, Cambridge, Waikato, New Zealand
e-mail: karen.denyer@papawerageological.co.nz; karen.denyer@wetlandtrust.org.nz

H. Robertson (✉)
Freshwater Section, Department of Conservation, Nelson, New Zealand
e-mail: harobertson@doc.govt.nz

© Springer Science+Business Media B.V., part of Springer Nature 2018
C. M. Finlayson et al. (eds.), *The Wetland Book*,
https://doi.org/10.1007/978-94-007-4001-3_176

on freshwater and estuarine systems, particularly palustrine wetlands such as bogs, fens, swamps, and marshes.

Keywords

New Zealand · wetlands · pressures · management · threatened species

Introduction

New Zealand is an archipelago located in the South Pacific Ocean comprising two large islands and numerous smaller islands spread over 23° of latitude. It has a relatively wet climate with mean annual rainfall of around 600–4,000 mm per year depending on geographic region. The landscape is geologically active and varied, dominated in different regions by mountains, glacial valleys, volcanic plateaus, broad floodplains, and lowlands, with an extensive coastline.

Here we provide an overview of New Zealand wetlands, describing the key wetland types, flora and fauna, progress with implementing the Ramsar Convention, relevant wetland policy, approaches to restoration, and cultural connections. The focus is on freshwater and estuarine systems, particularly palustrine wetlands such as bogs, fens, swamps, and marshes.

Wetland Types and Distribution

New Zealand contains a variety of temperate wetland systems. Most contemporary wetlands formed at or after the end of the last glaciation, about 18,000 years ago (McGlone 2009). Prior to the arrival of humans around 800 years ago about 10% of the landscape was freshwater wetlands (Ausseil et al. 2011).

Wetland classification applied in New Zealand identifies nine hydrosystems and nine wetland classes (Johnson and Gerbeaux 2004), differentiated by water source, substrate, and nutrient status (Table 1). Bogs include the raised, blanket, and string mires, while swamps and marshes are often associated with river and lake systems. Less common, or naturally rare, wetland types include ephemeral "kettleholes" formed on glacial moraines, and geothermal wetlands.

The current extent of wetlands accounts for only 10% of the historic extent, and in some regions less than 1% of wetlands remain (Ausseil et al. 2011). Wetland loss is primarily due to wetland drainage and clearing as land was converted to farming. Of the remaining ~250,000 ha of wetlands, approximately 60% are located on public conservation land (Robertson 2016). Key wetland regions are the West Coast, Southland (South Island), and Northland and Waikato (North Island) (Fig. 1).

Flora and Fauna

Being isolated from other land masses for over 80 million years, New Zealand has a distinctive biota, with a high proportion of endemic species and a vertebrate fauna dominated by birds, fish, and small terrestrial reptiles.

Table 1 Key features of New Zealand wetlands (Adapted from Johnson and Gerbeux 2004; © Department of Conservation, with permission)

Wetland class	Water origin	Water fluctuation	Substrate	Nutrient status
Bog	Rain only	Slight	Peat	Very low to low
Fen	Rain + groundwater	Slight to moderate	Mainly peat	Low to moderate
Swamp	Mainly surface water	Moderate to high	Peat and/or mineral	Moderate to high
Marsh	Groundwater + surface water	Moderate to high	Usually mineral	Moderate to high
Seepage	Surface and/or groundwater	Nil to moderate	Peat, mineral, or rock	Low to high
Shallow water	Surface and/or groundwater	Nil to high	Usually mineral	Moderate
Ephemeral	Rain + groundwater	Marked wet/dry alteration	Mineral	Moderate
Pakihi and gumland	Mainly rain	Slight to moderate	Peat or mineral	Very low to low
Saltmarsh	Seawater, brackish water	Tidal or slight (supratidal zone)	Mainly mineral	Moderate

Fig. 1 Example of a swamp wetland type in an inter-montane region of the South Island. Tall *Carex secta* growing on pedestals over water (Photo credit: H. Robertson © Rights remain with the author)

The vegetation cover prior to human arrival was largely evergreen temperate rainforest but with wetlands distributed widely across the long, narrow island chain. The wetland flora exhibits a lower level of endemism than in other ecosystems, e.g., bogs (65%), open water and swamps (34%) (McGlone 2009), compared with 80% for the total vascular indigenous flora (Taylor and Smith 1997).

By the time humans arrived, most of the postglacial wetlands had evolved from herbaceous fens into wooded swamps or bogs (McGlone 2009). Dominant woody species were typically conifers including *Dacrycarpus dacrydioides*, *Manoao colensoi*, *Lepidothamnus intermedius*, *Phyllocladus alpinus*, and *Halocarpus biformis*. Angiosperms included manuka *Leptospermum scoparium*, turpentine bush *Dracophyllum* species, swamp maire *Syzygium maire*, mingiminigi *Coprosma* species, and one of the world's largest lilies, the cabbage tree *Cordyline australis*. Extensive bogs were, and remain, a characteristic feature of New Zealand wetlands, where jointed rushes or restiads (*Sporadanthus* and *Empodisma* species) along with *Sphagnum* species are important peat-formers.

Extensive and repeated burning of the native vegetation and conversion of land to pasture drastically altered catchment hydrology (Woodward et al. 2014), resulting in wetter, herbaceous wetlands of flax *Phormium tenax*, raupo reeds *Typha orientalis*, and rush and sedge species (e.g., *Carex*, *Juncus*, *Cyperus*, *Machaerina*). Drainage later caused the fragmentation of intact wetlands, and many of the remnants are becoming invaded by introduced species including willow (*Salix* species).

While wetlands now occupy just 1% of New Zealand's land cover (Ausseil et al. 2011), around 23% of the total vascular flora (559 species) is either restricted to or frequently encountered in freshwater or saline wetlands (Table 2). Almost one third of wetland plants are threatened or at risk, and 20% of all threatened

Table 2 Indigenous plant, bird, and fish species that occupy freshwater or estuarine habitats in New Zealand

Taxonomic group (total for all native species for all ecosystem types in brackets)	Extinct	Extant	Threatened or at risk	Percentage of taxa threatened or at risk
Vascular plants	2 (6)	559[a] (2415[b])	167[c] (831)	30 % (34 %)
Nonpelagic resident/breeding birds, coastal[d] and freshwater inhabitants	10 (57)	55 (137)	45 (107)	81 % (78 %)
Nonmarine resident fish	1	53	40	75 %

Values in brackets are totals for all ecosystem types. Data not available for invertebrate fauna. Plants, after de Lange et al. 2009, birds after Table 4 of Miskelly et al. 2008, fish after Goodman et al. 2014
[a]Data calculated for obligate wetland, facultative wetland, facultative, and facultative upland species from Clarkson et al. 2013
[b]Total extant plant species numbers from de Lange et al. 2009
[c]Threatened wetland plant data aggregate of threatened and at risk species from the following habitats: Aquatic, Estuary, Eu-/mesotrophic wetland, Flush and seepages, Oligotrophic wetland, River bed, Wetland margin, Inland saline, Turf and cushion (de Lange et al. 2009)
[d]May include some nonwetland species; however, most would utilize estuarine or intertidal areas

vascular plant species inhabit wetlands. Over 200 plant species have become naturalized in New Zealand wetlands, some having major impacts on wetland form and function.

As with terrestrial systems, the native wetland fauna is dominated by birds and invertebrates. Forty percent of the nation's resident native bird species (excluding oceanic species) inhabit or utilize freshwater or coastal wetlands. These include southern crested grebe *Podiceps cristatus australis*, Australasian bittern *Botaurus poiciloptilus*, herons (Ardeidae), royal spoonbill *Platalea regia*, black swan *Cygnus atratus*, ducks (e.g., *Tadorna variegata* and *Anas* species), a rail *Gallirallus philippensis assimilis*, crakes (*Porzana* species), and a swamp hen *Porphyrio melanotus*. New Zealand's avifauna is renowned for the many flightless species, which includes two species of teal confined to sub-Antarctic Islands, one of them, *Anas nesiotis*, is the nation's most threatened endemic duck (Taylor and Smith 1997, Robertson et al 2013). Migratory bird species such as red knot *Calidris canutus* and bar-tailed godwit *Limosa lapponica* also utilize New Zealand wetlands, which lie at the southern end of the East Asian-Australasian Flyway.

New Zealand's native mammals, reptiles, and amphibians largely inhabit terrestrial environments, although long-tailed bats will forage over water margins, and skinks and geckos also occupy woody wetland vegetation. No snakes or crocodilian species are present in New Zealand. Native amphibians are limited to four ancient species of frog that are typically found in forest litter or riparian zones. Introduced mammals and frogs have adapted to New Zealand wetlands, with predatory mammals responsible, along with loss of habitat, for the high number of threatened or at risk native wetland bird species (81%, see Table 2).

The freshwater fish fauna, with 53 extant resident native species described (Goodman et al. 2014), is similar in richness to the United Kingdom but includes a high level of endemism (about 90%, Taylor and Smith 1997), largely induced by speciation of land-locked populations of galaxiid fish. Many native fish reside in swampy areas, and a group of mudfish *Neochanna* species are adapted to living in dried up pools and tributaries, capable of burrowing into mud for extended periods and respiring through their mucus-coated skin (O'Brien and Dunn 2007). Galaxiid species are still being discovered and described, including the Smeagol galaxias *Galaxias* aff. *gollumoides* named after the swamp-dwelling creature in the Lord of the Rings Trilogy. Migration between inland swamps, lakes, and streams and the sea (diadromy) is a key feature for a large proportion of New Zealand's freshwater fish. Many species mature and spawn in freshwater, with their young spending a period of development at sea.

The aquatic invertebrate fauna of lowland wetlands is dominated by midges (Chironomidae), aquatic mites (Acarina), Copepoda, Nematoda, and Ostracoda (Suren and Sorrell 2010). The terrestrial invertebrate taxa of New Zealand wetlands are known to be relatively depauperate. The fauna is dominated by Diptera (flies), Hymenoptera (wasps, bees, and ants), Coleoptera (beetles), and Hemiptera (bugs) (Clarkson et al. 2008). There are many species new to science including the recently discovered species of stem boring moth, *Houdinia flexilissima*, which is host-specific to the nationally threatened giant cane rush *Sporadanthus ferrugineus*.

Cultural Values and Management

Maori people travelled from Polynesia to New Zealand (Aotearoa) 800–1,000 years ago and developed strong connections with freshwater and coastal wetlands. For example, wetlands were essential places for collection of food (mahinga kai) such as fishing for freshwater eels/tuna (*Anguilla* species), harvest of plants like raupo *Typha orientalis* and harakeke *Phormium tenax*, and collection of swamp mud used for staining flax.

Present day environmental management is increasingly learning from, and applying, Maori knowledge systems (matauranga maori) in wetland conservation. Maori also have an important role in setting wetland management objectives, especially for sites like Waituna Lagoon recognized in a Deed of Settlement with Ngāi Tahu, the largest iwi (tribe) in the South Island. Co-management of a number of waterways is occurring under Treaty of Waitangi settlement processes and through joint management agreements between Maori and the government, as is the case with Waikato-Tainui iwi for wetlands associated with the Waikato River.

Wetlands are enjoyed by a cross section of New Zealanders and international visitors for recreation, walking, bird watching, kayaking (Fig. 2), fishing, and gamebird hunting. Many wetlands are managed, or even created, as reserves for fishing and gamebirds by the Fish and Game Authority, while the Department of Conservation manages a number of Wildlife Reserves and National Parks with community use of wetlands as a key management objective. The perceived lack of suitable native species for hunting lead to introduction of northern hemisphere game

Fig. 2 Recreational enjoyment of the Awarua Wetland Ramsar site (Photo credit: S. Chesterfield © Rights remain with the photographer)

birds (e.g., mallards, Canadian geese) and coarse fish (e.g., rudd, perch, carp) by early acclimatization societies. Some of these have had negative impacts on native species.

New Zealand's Ramsar Wetlands

New Zealand became a signatory to the Convention on Wetlands of International Importance (Ramsar Convention) in 1976. Contracting parties are obliged to designate wetlands within their territory of international importance, based on a suite of nine criteria related to ecology, botany, zoology, and hydrology. New Zealand currently has six Ramsar Sites (Fig. 3) covering a surface area of 55,512 ha. Table 3 lists the key features of each site.

The convention urges contracting parties to nominate sites that fully represent the diversity of wetlands and their key ecological and hydrological functions. New Zealand's current suite of Ramsar wetlands is not fully representative. Of the six listed sites, four are coastal/estuarine systems, three are predominantly peatlands, three are in one administrative region, and all are below 20 m in elevation. The Department of Conservation (the designated Ramsar authority in New Zealand) has developed a set of guidelines to encourage designation of under-represented wetland types, including nival (alpine), geothermal, plutonic, riverine, marine, inland saline, and lacustrine wetland systems (Denyer and Robertson 2016).

Nevertheless, the current suite of Ramsar sites includes many special features including:

- Seasonal habitat for tens of thousands of migratory birds
- New Zealand's largest unaltered raised bog and most extensive stand of the peat-forming giant cane rush (Kopuatai)
- An accreting sand bar, gaining 14 million cubic meters of sand every year (Farewell Spit)
- The nation's largest population of Australasian bittern and the only remaining New Zealand location for the swamp helmet orchid *Corybas carsei* (Whangamarino)
- Rare coastal populations of typically alpine plants (Awarua-Waituna)

All of the nation's Ramsar Sites face a range of management issues, including invasive plants and animals, threats to water quality, and tensions between the hydrological needs of the wetlands and the land use requirements of the predominately farming community in their catchments (Robertson and Funnell 2012).

While there has been little overall change in the ecological state of the six sites since 2009, some Ramsar wetlands have showed improvement, for example, through wide-scale invasive plant control, while others (Whangamarino, Waituna Lagoon) are under threat from declining water quality (Department of Conservation 2011). Initiatives such as the Department of Conservation Arawai Kakariki wetland restoration programme and the efforts of community groups continue to help maintain these internationally significant ecosystems.

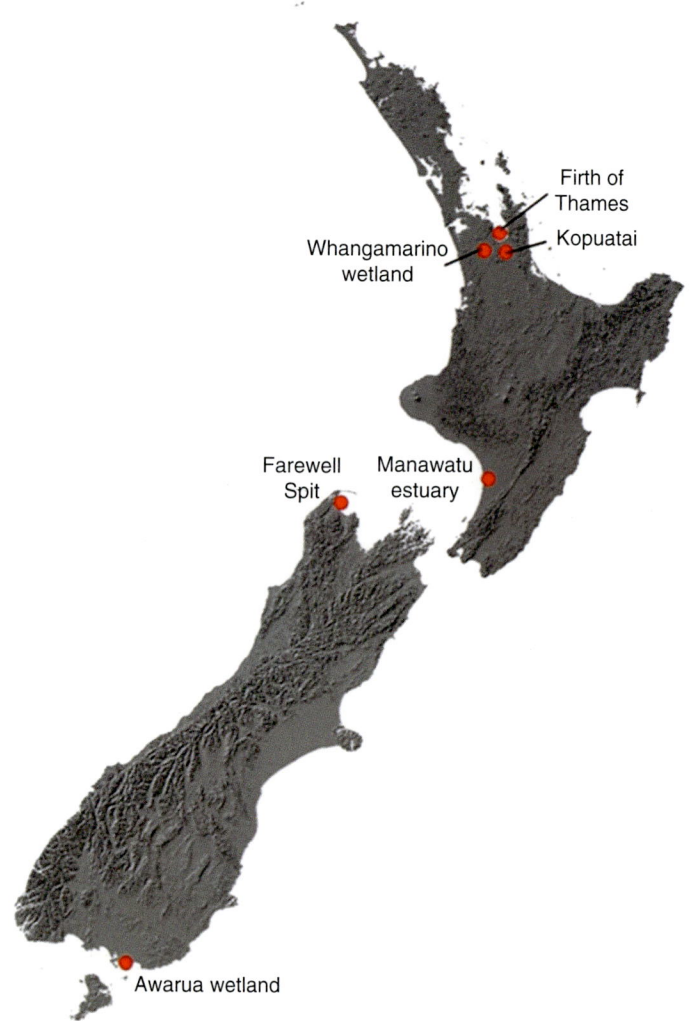

Fig. 3 Ramsar Sites in New Zealand as of October 2013 (Courtesy of the New Zealand Department of Conservation)

Threats to Wetlands

New Zealand's ecosystems evolved for millennia in the absence of land mammals, including humans, other than a handful of bat species. When humans arrived, they brought alien plants and animals to which the native biota were ill-adapted to coexist. They also brought fire, and, later, technology to dramatically transform the landscape via drainage of wetlands and conversion of the largely forest-covered land

Table 3 Key features of New Zealand's Ramsar Sites (Adapted from Robertson 2013)

Ramsar Site	Year listed	Area (ha)	Distinctive features	Ramsar criteria met
Awarua-Waituna Lagoon	1976	18,900	Extensive, intact peatlands, estuary, coastal lake	1, 2, 3, 4, 6, 7
Farewell Spit	1976	11,400	Expansive mudflats and sandspit, high bird diversity, migratory species	1, 2, 3, 4, 5, 6
Firth of Thames	1990	7,800	Shell banks, tidal mud, and sand flats offer extensive feeding for wading birds and waterfowl	1, 2, 3, 5
Kopuatai	1989	10,200	Largest raised peat dome in North Island, peat-forming *Sporadanthus* (rare bog plant)	1, 2, 3, 4
Manawatu Estuary	2005	200	Important feeding ground for migratory birds	1, 2, 3, 4, 6, 8
Whangamarino	1989	5,900	Extensive raised peat dome/swamp complex. Australasian bittern stronghold	1, 2, 3, 5, 6

Note: Criterion 9 was added to the criteria after most of the New Zealand Ramsar Sites were listed

to open ryegrass and clover pasture. Many wetlands that remain, particularly in lowland areas, are degraded by animal pests, weeds, livestock, drains, weirs, culverts, and excessive inputs of nutrients, sediments, and urban pollution.

The loss of freshwater swamps and bogs, much of it during the last 150 years, has been almost total, with just 10% of the prehuman extent remaining today. While national and local policies aim to stem further loss, ongoing drainage of smaller wetlands, either consented or undertaken unlawfully, continues (e.g., Myers et al. 2013).

Over 200 introduced plant species now inhabit wetlands in New Zealand, compared with a total native wetland flora of some 460 species (Clarkson et al. 2013). Some introduced plants are capable of completely transforming the wetland structure and functioning, e.g., woody species such as willow (*Salix* sp.) that invade sedgelands, and deciduous species that seasonally contribute nutrients and smothering litter.

An array of introduced mammals (possums, hedgehogs, rodents, mustelids), fish, and birds have adapted to wetlands where they prey on or compete with native species. Three quarters of New Zealand's wetland birds are considered threatened, and the native grey duck subspecies *Anas superciliosa superciliosa* is predicted to become extinct within a decade through hybridization with introduced mallards (Miskelly et al. 2008).

Large herbivorous birds (moa, geese, swans) that wetlands evolved with have been replaced with domestic ungulates (cattle, sheep, goats, deer, pigs). These farm animals, with their heavy body mass and high-nutrient diets, alter soil structure, browse and trample seedlings and ground-level nests, distribute weed seeds, and increase soil and water fertility. Dairy farming has become intensified in New Zealand in recent years, and the national herd has doubled since 1986 to 4.6

Fig. 4 Introduced species such as brushtailed possums *Trichosurus vulpecula* and grey willow *Salix cinerea* pose a threat to the health and functioning of New Zealand wetlands (Photo: Karen Denyer © Rights remain with the author)

million. In addition to the many small wetlands at risk of eutrophication from agricultural runoff, two of the nation's Ramsar designated wetlands (Whangamarino and Awarua) are at risk of changes to their ecological character because of increased sediment and nutrient input (Robertson and Funnell 2012).

Wetlands in urban areas are subject to contamination from industrial spills, sewage leaks, and road pollutants from stormwater. They are often also infested with alien plants from residential gardens.

Climate change poses new threats, including coastal erosion, increased extreme weather events, and increased vulnerability to invasive species and eutrophication. While there are substantial pressures on smaller and fragmented wetlands, many large systems and high country wetlands have retained significant biodiversity values, and an increasingly engaged community is involved in wide-ranging restoration and conservation projects (Fig. 4).

Wetland Policy and Strategies

New Zealand introduced the Resource Management Act (RMA) in 1991, comprehensive legislation that provides for a river basin management approach to water and its sustainable use. This Act includes provisions to protect the natural character of wetlands and identifies the cultural importance of lands and water to Maori.

Although significant improvements in the management of wetlands have been achieved, it is recognized that further improvements are still needed due to increasing demands for water and the intensification of land use (Myers et al. 2013).

Regional and district councils develop policy and plans under the RMA, relating to maintaining the natural character of wetlands and protecting significant indigenous vegetation and habitat. Both regulatory mechanisms and voluntary incentives are employed to encourage protection and restoration. However, the strength of wetland policy and enforcement is not consistent between regions (Myers et al. 2013). The release of the National Policy Statement for Freshwater Management in 2014 aimed to provide a more consistent approach to the management of freshwater resources in New Zealand, including protection of the significant values of wetlands.

The New Zealand Biodiversity Strategy released in 2000 made specific reference to wetlands, with a 2020 goal that the "extent and condition of remaining natural freshwater ecosystems and habitats are maintained" (Department of Conservation and Ministry for the Environment 2000). While some progress has been made, a review of the strategy in 2006 called for the protection and sustainable management of freshwaters to be accorded higher priority.

New Zealand became signatory to the Ramsar Convention in 1976, but this has not translated into any specific national legislation for wetlands, with the exception of the Crown Minerals Act which lists most Ramsar sites as being closed to mining (Robertson 2013).

Wetland Restoration

Restoration, and in some cases wholesale re-creation, has become a vital tool in maintaining and regaining wetlands across New Zealand. Projects encompass individual efforts by landowners on farms, corporate reconstructions to mitigate impacts of land developments, agency-led projects, and community-led projects.

Limited data are available on the extent of wetlands subject to active restoration, but of 270 community ecological restoration groups nationwide surveyed in 2013, 33% were engaged in restoring freshwater wetlands, 10% were restoring lake ecosystems, and 17% were restoring estuarine wetlands (Peters et al, 2015).

Innovative techniques are being employed to restore and manage wetlands including:

- Eradication of introduced mammalian predators from offshore islands and from within New Zealand-designed pest fence enclosures to create wildlife sanctuaries.
- Construction of safe spawning sites for native fish, and retrofitting drain culverts to aid passage for diadromous fish.
- Floating wetlands on shallow lakes designed to enhance uptake of nutrients.
- Combining scientific methods with traditional Maori methods for observing and monitoring changes in wetland health.

- Development of WETMAK – wetland monitoring methods targeted at community groups.
- Bi-annual symposia led by the National Wetland Trust that bring together researchers, wetland managers, and enthusiastic amateurs to share restoration experience.

Notable examples of wetland restoration projects in New Zealand include:

The Arawai Kakariki Programme. Translated as "green pathway," this programme led by the Department of Conservation aims to work with the community to enhance the ecological character of three outstanding wetlands while promoting and documenting research into wetland restoration techniques. The sites include two Ramsar Sites (Awarua and Whangamarino) and a montane wetland complex (O Tu Wharekai/Ashburton Lakes).

Travis Wetland. One of the largest urban restoration projects in the country (116 ha). Previously farmed and drained, the lowland freshwater wetland has been re-instated and area is now managed by the Christchurch City Council as a Nature Heritage Park. Travis Wetland is one of the few surviving fragments of the once extensive wetland habitat that covered much of Christchurch prior to European settlement.

Rotopiko/National Wetland Centre. A community-led restoration project at the site of the proposed National Wetland Centre in the Waikato region. A pest fence has been constructed around a peat lake and swamp forest complex to enable pest eradication and reintroduction of a range of native species.

Redcliff and Rakatu Wetlands. Funded by compensation for a hydroelectric power scheme, this project led by the Waiau Fisheries and Wildlife Habitat Trust has involved restoring nearly 300 ha of farmland on the floodplain of the Waiau River to re-create fish and waterfowl habitat (Fig. 5).

Future Challenges

Improving public perception of the values of healthy, functioning wetlands remains a challenge in New Zealand. The National Wetland Trust and other community based advocacy groups such as Fish & Game and Forest & Bird will play a vital role in encouraging public participation in wetland conservation.

Implementation of a longer-term strategy for freshwater management is also required, which considers the relationship between ecosystem services and human well-being. This includes obtaining a better understanding on specific water use limits to maintain wetland values and function (e.g., Clarkson et al. 2015).

Given New Zealand's isolation, it retains a high degree of endemism with flora and fauna communities dominated by native species. Biosecurity programs to limit the arrival and spread of exotic invasive species remain a priority as is controlling the impacts of species that have already arrived.

Future consequences of climate change on water regimes, temperature, and sea level are not fully recognized, even though they have potential to disrupt ecosystem

Fig. 5 Restoration at Rotopiko, site of proposed National Wetland Centre (Photo: Karen Denyer © Rights remain with the author)

processes. Inland and coastal wetlands will play a role in adapting to climate change through reducing flood impacts, restricting eutrophication, and provision of refugia, while functioning bogs have the potential to lock up significant amounts of carbon in peat soils, and their protection is a high priority.

Addressing the legacy of the 90% loss of palustrine wetlands due to historical land use change will require novel approaches to reconnect aquatic habitats and achieve a comprehensive network of wetlands within public and private protected areas. While approximately a third of New Zealand's land mass is public conservation land, half of the remaining wetland extent occurs in areas outside conservation reserves. Further support for community based initiatives to restore and protect vulnerable wetlands will therefore be increasingly important.

References

Ausseil A, Chadderton L, Gerbeaux P, Stephens RT, Leathwick JR. Applying systematic conservation planning principles to palustrine and inland saline wetlands of New Zealand. Freshw Biol. 2011;56:142–61.

Clarkson B, Watts C, Sorrell B, Bartlam S, Thornburrow D, Fitzgerald N, Chague-Goff C, Bodmin K, Champion P. Biotic composition of New Zealand lowland wetlands: I vegetation and II invertebrates. Landcare Research Contract Report [LC0708/142]. Hamilton: Landcare Research; 2008.

Clarkson BR, Champion PD, Rance BD, Johnson PN, Bodmin KA, Forester L, Gerbeaux P, Reeves PN. Wetland indicator status ratings for New Zealand species. Unpublished report. Hamilton: Landcare Research; 2013. Available from: www.landcareresearch.co.nz/__data/assets/pdf_file/0014/64400/Wetland_indicator_status_July13.pdf. Accessed 30 Dec 2105.

Clarkson BR, Overton JM, Ausseil AGE, Robertson HA. Towards quantitative limits to maintain the ecological integrity of freshwater wetlands: interim report. Landcare Research Contract Report [LC1933]. Hamilton: Landcare Research; 2015.

de Lange PJ, Norton DA, Courtney SP, Heenan PB, Barkla JW, Cameron EK, Hitchmough R, Townsend AJ. Threatened and uncommon plants of New Zealand (2008) revision. N Z J Bot. 2009;47:61–96.

Denyer K, Robertson H. National guidelines for the assessment of potential Ramsar wetlands in New Zealand. 2016. Department of Conservation, Wellington.

Department of Conservation. National report on the implementation of the Ramsar Convention on Wetlands. Wellington: Department of Conservation; 2011.

Department of Conservation, Ministry for the Environment. New Zealand biodiversity strategy. Wellington: Department of Conservation/Ministry for the Environment; 2000.

Goodman JM, Dunn NR, Ravenscroft PJ, Allibone RM, Boubee JAT, David BO, Griffiths M, Ling N, Hitchmough RA, Rolfe JR. Conservation status of New Zealand freshwater fish, 2013. Wellington: Department of Conservation; 2014. 12 p.

Graeme A. Taylor. 2013. Conservation status of New Zealand birds, 2012. New Zealand Threat Classification Series 4. Department of Conservation, Wellington. p.9.

Hugh A. Robertson, John E. Dowding, Graeme P. Elliott, Rodney A. Hitchmough, Colin M. Miskelly, Colin F.J. O'Donnell, Ralph G. Powlesland, Paul M. Sagar, R. Paul Scofield.

Johnson P, Gerbeaux P. Wetland types in New Zealand. Wellington: Department of Conservation; 2004.

McGlone MC. Postglacial history of New Zealand wetlands and implications for their conservation. N Z J Ecol. 2009;33:1–23.

Miskelly CM, Dowding JE, Elliott GP, Hitchmough RA, Powlesland RG, Robertson HA, Sagar PM, Scofield RP, Taylor GA. Conservation status of New Zealand birds. Notornis. 2008;55(3):117–35.

Myers SC, Clarkson BR, Reeves PN, Clarkson BD. Wetland management in New Zealand: are current approaches and policies sustaining wetland ecosystems in agricultural landscapes? Ecol Eng. 2013;56:107–20.

O'Brien L, Dunn NR. Mudfish (*Neochanna galaxiidae*) literature review, Science for conservation 277. Wellington: Department of Conservation; 2007. p. 88.

Peters MA, D Hamilton, C Eames. Action on the ground: A review of community environmental groups' restoration objectives, activities and partnerships in New Zealand. 2015. New Zealand Journal of Ecology (2015) 39(2):179–189.

Robertson HA. Ramsar wetlands in NZ: why are they important and where are we going? Waiology; 2013. Available from: http://sciblogs.co.nz/waiology/2013/02/05/ramsar-wetlands-in-nz-why-are-they-important-and-where-are-we-going/. Accessed 30 Dec 2015.

Robertson HA. Wetland reserves in New Zealand: the status of protected areas between 1990 and 2013. N Z J Ecol. 2016;40(1):1–11.

Robertson HA, Funnell EP. Aquatic plant dynamics of Waituna Lagoon, New Zealand: trade-offs in managing opening events of a Ramsar site. Wetl Ecol Manag. 2012;20(5):433–45.

Suren A, Sorrell B. Aquatic invertebrate communities of lowland wetlands in New Zealand. Characterising spatial, temporal and geographic distribution patterns, Science for conservation 305. Wellington: Department of Conservation; 2010.

Taylor R, Smith I (Principal authors). The state of New Zealand's environment. Wellington: NZ Government Printer; 1997.

Woodward C, Shulmeister J, Larsen J, Jacobsen GE, Zawadzki A. The hydrological legacy of deforestation on global wetlands. Science. 2014;346(6211):844–7.

New Zealand Restiad Bogs

Beverley R. Clarkson

Contents

Introduction	1992
Restiad Bog Species	1992
Restiad Bog Types	1994
Empodisma robustum Bogs	1994
Empodisma minus Bogs	1994
Sporadanthus ferrugineus Bogs	1994
Sporadanthus traversii Bogs	1995
Ecosystem Services	1996
Threats, Conservation, and Restoration	1998
References	1999

Abstract

Restiad bogs, dominated by species in the predominantly Southern Hemisphere family Restionaceae, are most extensively developed on the temperate oceanic islands of New Zealand. They form raised and blanket bogs on mainly flat, poorly drained lowlands, but also occur in the montane and subalpine zones. The main peat forming species are *Empodisma robustum* (northern North Island), *E. minus* (central and southern North Island, and most of South and Stewart Islands), *Sporadanthus ferrugineus* (northern North Island), and *S. traversii* (Chatham Island). *Empodisma minus* also grows in south-eastern Australia. Since 1840, restiad bogs in the lowland zone have been much reduced by widespread agricultural development, associated with European settlement. As a result, some are now classified as threatened ecosystems, and most contain threatened plant and

B. R. Clarkson (✉)
Landcare Research, Hamilton, New Zealand
e-mail: clarksonb@landcareresearch.co.nz

© Springer Science+Business Media B.V., part of Springer Nature 2018
C. M. Finlayson et al. (eds.), *The Wetland Book*,
https://doi.org/10.1007/978-94-007-4001-3_222

animal species. Recognition of the magnitude of this loss, ongoing threats to ecological integrity, and the important role of restiad bogs in providing ecosystem services has led to increased efforts to protect, manage, and restore them.

Keywords

Empodisma minus · *Empodisma robustum* · Peatlands · Restionaceae · *Sporadanthus ferrugineus* · *Sporadanthus traversii*

Introduction

Restiad bogs are peatlands characterized by species in the predominantly Southern Hemisphere angiosperm family Restionaceae. The family comprises jointed rush-like or bamboo-like herbs with rigid or flexuose photosynthetic stems and reduced scale-like sheathing leaves that grow on low fertility, seasonally moist or wet soils. While the distribution of Restionaceae centers on Australia and southern Africa, restiad bogs are most extensively developed in New Zealand (North, South, Stewart, and Chatham Islands; Clarkson and Clarkson 2013). New Zealand restiad bogs are most common on flat, poorly drained lowlands, but also occur in montane and subalpine settings. They are considered as ecological equivalents of the moss-dominated Northern Hemisphere *Sphagnum* bogs (Hodges and Rapson 2011), as functional processes such as bog development and litter decomposition have been shown to be similar. *Sphagnum* bogs are of limited extent in New Zealand but may be of regional importance in wetter and cooler climates such as in southern South Island and mountainous areas. *Sphagnum* species are usually present in restiad bogs, but the plants do not thrive in the shade of the taller restiads.

Restiad Bog Species

There are four restiad bog species: *Empodisma robustum* (northern North Island), *E. minus* (central and southern North Island, and most of South and Stewart Islands), *Sporadanthus ferrugineus* (northern North Island), and *S. traversii* (Chatham Island). All are endemic (confined to New Zealand) except *Empodisma minus*, which also grows in southeastern Australia.

The *Empodisma* are the least robust species, having fine intertwining wiry stems, averaging 0.7 m (*E. robustum*) and 0.4 m in height (*E. minus*). They are vigorous peat formers, producing dense masses of fine roots and root hairs (cluster roots) at the bog surface, which form the bulk of the peat (Fig. 1). These cluster roots have similar water-holding properties to *Sphagnum* moss, similarly extremely slow decomposition rates (Clarkson et al. 2014), and are very efficient at absorbing nutrients from rainfall. *Empodisma robustum* and *E. minus* have adaptations to conserve water, often allowing bogs to form in areas of high seasonal water deficits

Fig. 1 Profile of the surface layers of *Empodisma robustum* peat showing the dense growth of fine roots (Photo credit: EWE Butcher © Rights remain with the author)

where typically bogs do not occur. Strict stomatal control of plant transpiration and a dense, mulch-like mattress of decay-resistant stems contribute to exceptionally low daily evaporation rates (Thompson et al. 1999). Both *Empodisma robustum* and *E. minus* are mid- to late-successional species, establishing early in minerotrophic (groundwater-fed, relatively high nutrient) wetlands to initiate restiad bog development and persisting in significant amounts through to late ombrotrophic (rain-fed, low nutrient) phases (Clarkson et al. 2004b). *Empodisma robustum* is killed by fire, with recruitment occurring relatively quickly from seed, whereas *E. minus* usually resprouts after fire.

The *Sporadanthus* are more robust and more erect bamboo-like species (de Lange et al. 1999). They are also important peat formers; however, their cluster roots are not as well developed as the *Empodisma* species and remain below the bog surface. *Sporadanthus ferrugineus* is a late-successional species, establishing after an initial *Empodisma robustum* phase to overtop existing vegetation and ultimately become the physiognomic dominant up to 3 m tall. *Sporadanthus ferrugineus* is not fire resistant, and as seedling establishment is slow, it can take several years for *S. ferrugineus* cover to reestablish after being killed by fire. It is tolerant of very low nutrient conditions, having high nutrient use efficiency, low tissue nutrient concentrations, and other features typical of stress-tolerant species. *Sporadanthus traversii* has dense, somewhat flexuose stems reaching 1.5–2 m in height. Its ecological role in restiad bog development on Chatham Island is similar to that of *Empodisma robustum* on mainland New Zealand, being the main peat former, having a dense decay-resistant litter layer, and reestablishing relatively rapidly by seed after fire.

Restiad Bog Types

Empodisma robustum Bogs

Empodisma robustum bogs are uncommon and confined to lowland zones between latitudes 35° and 38° S on the northern North Island. This is the warmest region in New Zealand, with a mean annual temperature of 13–15 °C and mean annual rainfall of 1,100–1,400 mm. *Empodisma robustum* dominates fens and young bogs where it forms a dense layer of sprawling intertwined wiry stems up to 0.9 m tall. Associated species include the fern *Gleichenia dicarpa*; sedges *Machaerina teretifolia* and *Schoenus brevifolius*; heath shrubs *Leptospermum scoparium*, *Dracophyllum lessonianum*, and *Epacris pauciflorus*; and forked sundew *Drosera binata*. In older raised bogs, *Empodisma robustum* becomes subdominant to the taller restiad, *Sporadanthus ferrugineus* (see section "*Sporadanthus ferrugineus* Bogs" below).

Empodisma minus Bogs

Empodisma minus bogs are relatively common south of latitude 38° S on North Island, through South Island, to Stewart Island at around latitude 47° S. The stature of *Empodisma* becomes shorter and stems finer with increasing latitude and elevation, and it is subdominant to *Sphagnum* (mainly *S. cristatum*, *S. australe*, and *S. novo-zelandicum*) in wetter areas. Strongholds for large blanket and raised bogs dominated by *Empodisma minus* are lowlands on the west coast of the South Island and southern, cooler areas, including Stewart Island. The species is common also in mountain bogs up to 1,350 m a.s.l. Typically, dense springy carpets of *Empodisma minus* averaging 40–50 cm high cover large expanses, associated with heath shrubs (*Leptospermum scoparium*, *Dracophyllum* spp.), *Gleichenia dicarpa*, sedges (e.g., *Machaerina teretifolia*, *M. tenax*), sundews (*Drosera binata*, *D. spatulata*), bladderworts (*Utricularia* spp.), and mosses (*Sphagnum cristatum*, *Dicranoloma robustum*). Regular burning since human settlement has reduced the woody content of these bogs, promoting wetter, more herbaceous systems (McGlone 2009; Fig. 2). The peats were initiated in the postglacial period (after about 14,000 years BP) and are typically shallower (usually 1–5 m deep) than in northern North Island late-successional raised bogs.

Sporadanthus ferrugineus Bogs

Late-successional bogs dominated by *Sporadanthus ferrugineus* characterize the mild climates north of latitude 38° S on the North Island (Clarkson et al. 2004b). *Sporadanthus ferrugineus* forms an upper tier, 2–3 m tall, above a dense understorey of sprawling *Empodisma robustum* (Fig. 3). The species are able to coexist by accessing nutrients from different sources: *Empodisma* from the atmosphere via rainfall and *Sporadanthus* from the peat substrate (Clarkson et al. 2009). Heath

Fig. 2 *Empodisma minus* and *Sphagnum cristatum* dominate foreground of Centre Burn bog, southern South Island, with a woody remnant of *Dracophyllum oliverii* and *Halocarpus bidwillii*, which has avoided recent burning, in the background (Photo credit: BR Clarkson © Rights remain with the author)

shrubs are usually also present, mainly *Leptospermum scoparium* and *Epacris pauciflora*. The fern *Gleichenia dicarpa* is often abundant, along with a limited number of sedges (e.g., *Machaerina teretifolia*, *Schoenus brevifolius*), ground-cover herbs (e.g., *Utricularia delicatula*), mosses (e.g., *Campylopus acuminatus* var. *kirkii*), and liverworts (e.g., *Riccardia crassa*, *Goebelobryum unguiculatum*). *Sphagnum cristatum* also occurs but is usually not common in the densely vegetated restiad bogs. Initiated early in the postglacial period, these systems typically formed extensive raised bogs or domes covering up to 15,000 ha, with peat 10–12 m deep.

Sporadanthus traversii Bogs

Sporadanthus traversii blanket and raised bogs occur on Chatham Island, a sparsely populated, isolated island 90,000 ha in area, lying 870 km east of the South Island at latitude 44° S (Clarkson et al. 2004a). The island has comparatively low relief, with a cool and windy climate (mean annual temperature 11 °C), low rainfall (mean annual 700–1,000 mm), and frequent cloud cover. Extensive areas are peat-covered, with 60% of soils being peat or derived from peat. *Sporadanthus traversii* dominates the raised bogs and many blanket peatlands (Fig. 4), in association with the heath shrub *Dracophyllum scoparium*, and a shrub daisy, *Olearia semidentata*. The three

Fig. 3 Canopy of flowering *Sporadanthus ferrugineus* overtopping *Empodisma robustum* and the fern *Gleichenia dicarpa* at Kopuatai Peat Dome, Waikato, North Island, New Zealand (Photo credit: BR Clarkson © Rights remain with the author)

mainland New Zealand restiad bog species are absent; however, several species in common include *Gleichenia dicarpa*, *Machaerina tenax*, *Utricularia delicatula*, *Goebelobryum unguiculatum*, *Riccardia crassa*, *Sphagnum australe*, *S. falcatulum*, and *S. cristatum*. Compared with mainland New Zealand peats, Chatham Island peats are deeper (>10 m thick) and older, often dating back to the last previous interglacial period (40,000–30,000 years ago). They have lower pH and generally higher nutrient contents than mainland New Zealand peats, which may be partly due to oceanic influence and a long history of seabird nutrient inputs.

Ecosystem Services

Restiad peatlands, as with wetlands in general, provide a wide range of economic, social, environmental, and cultural benefits known as ecosystem services. These include improving water quality, abating floods, and providing habitat to sustain vital ecosystem functions and unique biodiversity. In addition, peatlands play an increasingly recognized role in sequestering and storing carbon, which in the long term creates a net climate cooling effect (Frolking and Roulet 2007).

Fig. 4 *Sporadanthus traversii*-dominated restiad blanket bog at Wharekauri, Chatham Island (Photo credit: BR Clarkson © Rights remain with the author)

A detailed economic evaluation has been undertaken for Whangamarino Wetland, a large restiad bog, fen, swamp, and open-water complex. The wetland provides both use (direct use of a wetland's goods) and nonuse (existence or society's willingness to pay for preservation) values, estimated at US$$_{2003}$9.9 million per year (Kirkland 1988 in Schuyt and Brander 2004). The main nonuse values center on the high diversity of habitats and species contained within the wetland. It is home to several threatened plant species, including the swamp helmet orchid *Anzybas carsei*, which is found only at Whangamarino, as well as the more widely distributed water milfoil *Myriophyllum robustum*, fern *Cyclosorus interruptus*, bladderwort *Utricularia delicatula*, clubmoss *Lycopodiella serpentina*, and liverwort *Goebelobryum unguiculatum*. Whangamarino provides habitat for one fifth of New Zealand's population of Australasian bittern *Botaurus poiciloptilus*, as well as other threatened birds such as the grey teal *Anas gibberfrons*, spotless crake *Porzana tauensis plumbea*, and North Island fernbird *Bowdleria punctata vealeae*. The wetland contains a key population of the threatened black mudfish *Neochanna diversus*, which survive dry periods by burying themselves in moist mud or under logs until the water returns.

The main use values of Whangamarino are flood control, gamebird hunting, recreation, commercial fishing of eels, and carbon storage. Of increasing economic significance is the wetland's role as part of the substantial flood control scheme on the lower Waikato River (Waugh 2007), which resulted in lowered regional water levels. The scheme reproduces the natural water storage function of Whangamarino

Wetland and adjoining Lake Waikare, but in a more controlled way, to depress flood peaks in the Waikato River (Department of Conservation 2007). Water storage in the wetland has reduced public works costs (e.g., stopbank construction) and damage to farmland during regularly occurring flood events. This includes a saving of NZ$5.2 million in flood control costs during a single 1-in-100-year flood event in 1998 (Waugh 2007).

Threats, Conservation, and Restoration

Restiad peat bogs throughout the lowland zone have been destroyed by widespread drainage and conversion to pasture since 1840, as European settlement gathered pace. The *Sporadanthus ferrugineus* type, in particular, has been severely reduced in extent (from more than 100,000 to 3,000 ha; de Lange et al. 1999) and is now confined to three sites in the Waikato region of the North Island between 37° and 38° S latitude. Other threats to the ecological integrity of restiad bogs include introduced weeds and pests, domestic livestock, nutrient inputs, fire, peat mining, and blueberry (*Vaccinium* spp.) horticulture. The restiad bogs provide habitat for several threatened plants and animals, and efforts are now concentrated on protection, management, and restoration. For instance, three large Ramsar Sites – Kopuatai Peat Dome (10,000-ha raised bog) and Whangamarino Wetland (7,300-ha wetland complex) in the northern North Island, and Awarua Wetland (19,000 ha of estuary, lagoon, and blanket bog) in southern New Zealand – have statutory protection and contain representative examples of *Sporadanthus ferrugineus* (at Kopuatai), *Empodisma robustum* (Kopuatai, Whangamarino), and *E. minus* (Awarua) bog types. Smaller areas of *Sporadanthus ferrugineus*, *Empodisma robustum*, and *E. minus* elsewhere on mainland New Zealand and *S. traversii* on Chatham Island are also protected.

Most New Zealand wetlands that have survived the human settlement phase are modified to some degree, particularly those in agricultural landscapes or urban environments. As awareness of wetland values spreads, the number of private individuals, community groups, iwi (Māori tribes), and organizations restoring the full range of wetlands is rapidly increasing. The following case study centers on restoration of a priority restiad cut-over bog and "reconstruction" of populations of the bog type at three sites where the bog originally occurred.

Torehape, a *Sporadanthus ferrugineus* bog on the Hauraki Plains, northern North Island, provides a rare example of an attempt to harvest peat sustainably and restore biodiversity. It consists of 180 ha of privately owned bog, which is currently being mined for horticultural peat, adjoining 350 ha of wetland management reserve administered by the Department of Conservation. The mining company has resource consent to mine the top meter of a 4–6 m depth of peat on private land and is then required to block drainage ditches and restore the bare surface to original bog vegetation. A patch dynamic approach to restoration (Schipper et al. 2002; Clarkson et al. 2013) has been developed for after peat harvesting whereby small "islands" of milled peat scattered over the mine surface are seeded with the early successional heath shrub *Leptospermum scoparium*. The developing shrubland functions as a nurse, providing suitable environmental conditions for seeds and propagules of later successional bog species (*Sporadanthus ferrugineus*, *Empodisma robustum*, *Sphagnum cristatum*) that are blown in from the adjoining intact peatland. Recovery of invertebrate communities follows relatively quickly (Watts et al. 2008a); however, reinstatement of ecosystem processes such as litter decomposition and microbial respiration will take much longer (Watts et al. 2008b). The project has provided techniques and resources to allow the successful translocation of three new populations of the restiad bog type – at Lake Serpentine, Lake Komakorau, and Waiwhakareke Natural Heritage Park (Watts et al. 2013). These are important for educational purposes, with the Lake Serpentine population being showcased within a predator-proof fence as part of the proposed National Wetland Trust interpretation center (http://www.wetlandtrust.org.nz/Site/National_Wetland_Centre.ashx, accessed 13 January 2016).

References

Clarkson BR, Ausseil A-GE, Gerbeaux P. Wetland ecosystem services. In: Dymond JR, editor. Ecosystem services in New Zealand – conditions and trends. Lincoln: Manaaki Whenua Press; 2013. p. 192–202.

Clarkson BR, Clarkson BD. Restiad bogs in New Zealand. In: Rydin H, Jeglum JK, editors. The biology of peatlands. 2nd ed. Oxford: Oxford University Press; 2013. p. 241–8.

Clarkson BR, Moore TR, Fitzgerald NB, Thornburrow D, Watts CH, Miller S. Water table regime regulates litter decomposition in restiad peatlands, New Zealand. Ecosystems. 2014;17:317–26.

Clarkson BR, Schipper LA, Clarkson BD. Vegetation and peat characteristics of restiad bogs on Chatham Island (Rekohu), New Zealand. N Z J Bot. 2004a;2:293–312.

Clarkson BR, Schipper LA, Lehmann A. Vegetation and peat characteristics in the development of lowland restiad peat bogs, North Island, New Zealand. Wetlands. 2004b;24:133–51.

Clarkson BR, Schipper LA, Silvester WB. Nutritional niche separation in co-existing bog species demonstrated by ^{15}N-enriched simulated rainfall. Aust Ecol. 2009;34:377–85.

de Lange PR, Heenan PB, Clarkson BD, Clarkson BR. *Sporadanthus* in New Zealand. N Z J Bot. 1999;37:413–31.

Department of Conservation. The economic values of Whangamarino Wetland. 2007. DOCDM-141075. Available from: http://www.doc.govt.nz/Documents/conservation/threats-and-impacts/benefits-of-conservation/economic-values-whangamarino-wetland.pdf. Accessed 13 Jan 2016.

Frolking S, Roulet NT. Holocene radiative forcing impact of northern peatland carbon accumulation and methane emissions. Glob Chang Biol. 2007;13:1079–88.

Hodges TA, Rapson GL. Is *Empodisma* the ecosystem engineer of the FBT (fen-bog transition zone) in New Zealand? J R Soc. 2011;40:181–207.

McGlone MS. Postglacial history of New Zealand wetlands and implications for their conservation. N Z J Ecol. 2009;33:1–23.

Schipper LA, Clarkson BR, Vojvodic-Vukovic M, Webster R. Restoring cut-over peat bogs: a factorial experiment of nutrients, seeds and cultivation. Ecol Eng. 2002;19:29–44.

Schuyt K, Brander L. The economic value of the world's wetlands. Gland/Amsterdam: World Wildlife Fund; 2004.

Thompson MA, Campbell DI, Spronken-Smith RA. Evaporation from natural and modified raised peat bogs in New Zealand. Agric For Meteorol. 1999;95:85–98.

Watts CH, Clarkson BR, Didham RK. Rapid beetle community convergence following experimental habitat restoration in a mined peat bog. Biol Conserv. 2008a;141:568–79.

Watts CH, Vojvodic-Vukovic M, Arnold GC, Didham RK. A comparison of restoration techniques to accelerate recovery of litter decomposition rates and microbial activity in an experimental peat bog restoration trial. Wetl Ecol Manag. 2008b;16:103–22.

Watts C, Thornburrow D, Clarkson B, Dean S. Distribution and abundance of a threatened stem-boring moth, *Houdinia flexilissima*, (Lepidoptera: Batrachedridae) in New Zealand peat bogs. J Res Lepid. 2013;46:81–9.

Waugh J. Report on the Whangamarino Wetland and its role in flood storage on the lower Waikato River. Hamilton: Department of Conservation; 2007.

Atolls of the Tropical Pacific Ocean: Wetlands Under Threat

170

Randolph R. Thaman

Contents

Introduction	2002
Pacific Atolls Described	2003
Distribution	2006
Atoll Biodiversity	2006
Ecosystem and Habitat Diversity	2007
Species and Taxonomic Diversity	2010
Genetic Diversity	2013
Ecosystem Goods and Services	2013
Threats and Future Challenges	2013
Conservation of Atoll Ecosystems	2019
Conclusion	2022
References	2023

Abstract

Atolls are small, geographically isolated, resource-poor islands scattered over vast expanses of ocean. There is little potential for modern economic or commercial development, and most Pacific Island atoll countries and communities depend almost entirely on their limited biodiversity inheritances for ecological, economic, and cultural survival in a rapidly globalizing world. Atolls rarely have elevations over 2 or 3 m above sea level and commonly have extensive areas of intertidal flats, mangroves, shallow lagoons, coral reefs, and limited areas of brackish water marshes or landlocked fossil lagoons and are subject to periodic tidal inundation during extreme weather and tidal events, such as "king tides." Under the Ramsar Convention definition, atolls and their nearshore waters are essentially "wetlands." Although arguably among the Earth's "biodiversity

R. R. Thaman (✉)
The University of the South Pacific, Suva, Fiji
e-mail: Thaman_r@usp.ac.fj; randolph.thaman@usp.ac.fj

© Springer Science+Business Media B.V., part of Springer Nature 2018
C. M. Finlayson et al. (eds.), *The Wetland Book*,
https://doi.org/10.1007/978-94-007-4001-3_270

coolspots" with the poorest and highly threatened terrestrial biodiversity inheritances on Earth, they are among the last remaining sanctuaries for extensive, but highly threatened populations of breeding seabirds and coral reef-associated biodiversity. Fortunately, under Ramsar, UNESCO World Heritage, and other relevant conventions and initiatives, the conservation, restoration and sustainable use of atoll biodiversity and associated ethnobiodiversity (the uses, knowledge, beliefs, management systems, taxonomies and language that traditional and the scientific communities have for biodiversity) is now clearly on the conservation agenda.

Keywords

Atolls · Biodiversity cool spots · Ethnobiodiversity · Coastal plants · Coral reef diversity · Food and livelihood security · Global change · Lagoons · Mangroves · Marine biodiversity · Pacific Islands · Sea birds

Introduction

Atolls are small, geographically isolated, resource-poor islands scattered over vast expanses of ocean. There is little potential for modern economic or commercial development, and most Pacific Island atoll countries and communities depend almost entirely on their limited biodiversity inheritances for ecological, economic, and cultural survival. Atolls rarely have elevations over 2 or 3 m above sea level and commonly have extensive areas of intertidal flats, mangroves, shallow lagoons, coral reefs, and limited areas of brackish water marshes or landlocked fossil lagoons and are subject to periodic tidal inundation during extreme weather and tidal events, such as "king tides" (Thaman 2008). Under the Ramsar Convention definition, atolls and their nearshore waters are essentially "wetlands."

Atolls are the opposite of "biodiversity hot spots" – areas such as Amazonia, Southeast Asia, Australia, and many Pacific Island areas such as New Guinea, New Caledonia, Solomon Islands, Hawaii, and the Galapagos – which have very high species and ecosystem diversity and high levels of endemism that are under threat of extinction and degradation (Whittaker 1998). Atolls are among the Earth's "biodiversity cool spots" because they have few, if any, endemic plants and animals and among the most impoverished and highly threatened terrestrial and freshwater biodiversity inheritances on Earth, with a high proportion of all economically, culturally, and ecologically important terrestrial plants and animals in danger of extirpation (local extinction) (Thaman 1992a, 2008). Although not as impoverished, atoll marine biodiversity is also under threat and in danger of extirpation, especially on inhabited and urbanized atolls.

Despite the poverty, fragility, threatened status, and the obligate dependence of atoll peoples on biodiversity, atoll biodiversity has received only limited attention from the international conservation community, which has focused mainly on the Earth's "biodiversity hot spots" (Thaman 2008). Fortunately, under Ramsar and a number of other conservation initiatives, Kiribati, the Marshall Islands, and a number of other Pacific Island atoll countries have designated atolls or atoll islets as "conservation areas" or initiated other initiatives to conserve atoll biodiversity. This chapter discusses the nature of atolls, atoll biodiversity, its value to atoll countries and communities, the threats to and conservation status of atoll biodiversity, and some conservation initiatives, including Ramsar initiatives, that have catalyzed the conservation of atoll biodiversity. Although there are atolls elsewhere, such as the Maldives in the Indian Ocean, the focus is on all atoll nations and atolls and low-lying limestone reef islands of the cultural regions of Melanesia, Polynesia, and Micronesia in the tropical Pacific Ocean (Fig. 1).

Pacific Atolls Described

The word atoll comes from the Malayalam word *atolu* or "reef" or *atollon*, the native name for the Maldives Archipelago (Newhouse 1980). Adapting the definitions of Bryan (1953) and Wiens (1962), the term "atoll" refers here to all low-lying oceanic limestone reef islands, with or without lagoons, that have formed on barrier reefs or in the shallow lagoons along the coastlines, or encircle long-submerged ancient volcanoes, which are not associated with nearby high islands or continents (Fig. 2). The term "islet" refers to the individual smaller islands or "motu" (a Polynesian name for reef islets) that are found on the reefs or in the lagoons of the main atoll islands. In other words, "atolls" include both "true atolls," the islets of which encircle, border, or are found within a lagoon, and individual, separate low-lying limestone reef islands that have no lagoon or may have "secondary" or remnant "fossil" lagoons on the actual limestone island or islets (Thaman 2008).

Most atolls have maximum elevations below 3–4 m above sea level, although some have limited areas of coral rubble ramparts deposited over time by high storm waves, limestone pinnacles (e.g., the raised limestone pinnacles on Tikehau Atoll in French Polynesia (Fig. 3), or windblown sand "dunes" that can reach elevations of over 10 m (e.g., Joe's Hill which attains 13 m on Kiritimati Atoll in Kiribati). Excluded from this definition of atolls are raised limestone islands or "raised atolls" that have average elevations much higher than 5 m, such as the main islands of the Tongatapu, Ha'apai, and Vava'u groups of Tonga and Ouvea, Lifou, and Mare in the Loyalty Islands of New Caledonia; raised phosphatic limestone islands, such as Nauru and Banaba (Ocean Island) and Makatea in the Tuamotus; and barrier reefs and associated islets surrounding high islands or continents, such as the "almost atolls" of the main Chuuk group and Bora Bora in French Polynesia and islets and reef structures with associated islands such as those on the Great Barrier Reef off northeastern Australia. Also excluded from this definition are "sunken atolls" without dry land, such as Middleton and Elizabeth Reefs in the Coral Sea off Australia (Thaman 2008).

The above definition of atoll is relatively clear for most of the well-studied "atolls" of Polynesia and Micronesia, although the status of many of the small

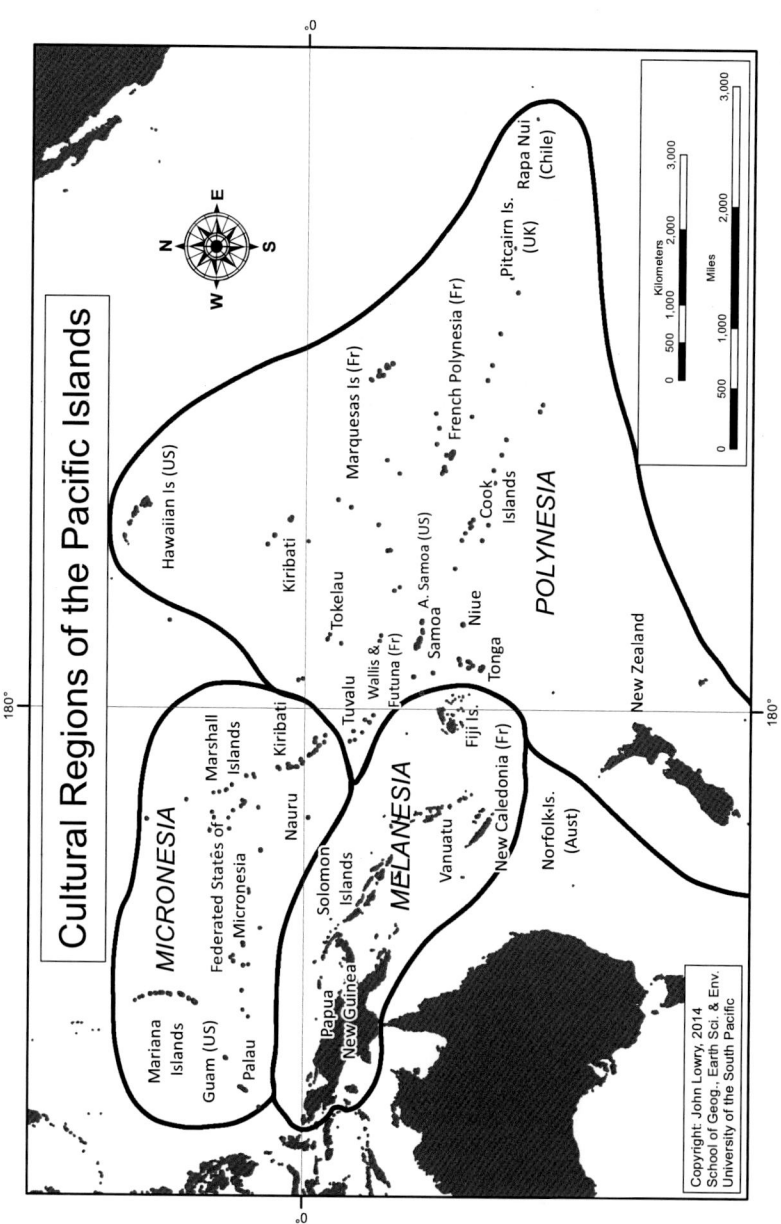

Fig. 1 Distribution of the main atoll countries and territories in the Pacific Ocean in the cultural regions of Micronesia, Melanesia, and Polynesia (J. Lowry © Rights remain with the author)

Fig. 2 Wailagi Lala Atoll, Fiji, partially obscured by cloud (Photo credit: R.R. Thaman © Rights remain with the author)

Fig. 3 Raised coastal karstified limestone pinnacles, Tikehau Atoll (Photo credit: R.R Thaman © Rights remain with the author)

islands included as atolls in the islands of Melanesia is uncertain due to the lack of detailed published information. For example, Bryan (1953) listed the island of Ouvea in the Loyalty Islands to the east of New Caledonia as the "world's largest atoll," although by his own definition, it is clearly not, but rather a large raised limestone island rising to a maximum elevation of 46 m, much more similar to some of the islands of Tonga, mentioned above (Thaman 2008).

Distribution

Globally, the Earth's oceans contain more than 400 "atolls" composed of thousands of individual islets! Also included under the definition are numerous small isolated individual low-lying limestone reef islands (Bryan 1953). Most are found in the tropics, especially in the tropical Pacific Ocean, where ocean water temperatures are suitable for coral reefs.

The main atoll "groups" in the "cultural areas" of Melanesia, Polynesia, and Micronesia (Fig. 1) include the Marshall, Gilbert, Phoenix, and Line Islands; the Tuvalu, Tokelau, and Northern Cook Islands; and the Tuamotu Archipelago in French Polynesia, which has 77, the most of any single group. There are also atolls in most of the other countries or island groups in the Pacific Islands, with Papua New Guinea, New Caledonia, Solomon Islands, Fiji, Palau, and the Federated States of Micronesia, all having atolls. Some of the Northwestern Hawaiian Islands, including the Midway Islands, are also atolls. There is also Clipperton Atoll, a French possession located in the Eastern Pacific about 1,080 km southwest of Mexico. Pacific countries and territories with no reported true "atolls" include Samoa, Tonga, Vanuatu, Wallis and Futuna, Niue, Nauru, and Kosrae (Thaman 2008) (Fig. 1). The largest group of atolls outside the Pacific Islands is the Maldives Archipelago in the Indian Ocean to the southwest of India.

Atoll Biodiversity

Atoll biodiversity encompasses (1) ecosystem and habitat diversity, including the staggering diversity of different atolls, atoll islets, and lagoon shapes and sizes, (2) species and taxonomic diversity, (3) genetic diversity, and (4) "ethnobiodiversity" (the knowledge, uses, beliefs, resource-use systems and conservation practices, taxonomies, and language that a given society or community, including the modern scientific community, has for its ecosystems, species, taxa, and genetic diversity). Ethnobiodiversity is here considered an integral component of atoll biodiversity because atoll people and their knowledge, traditions, and spirituality are seen as inseparable from their terrestrial, freshwater, and marine ecosystems, rather than separate external entities. This holistic view is embodied in the Melanesia pidgin concepts of *kastom*/custom or *ples*/place; the all-encompassing pan-Polynesian concept of land/*fonua, fanua, fenua, henua*, or *'enua*; or the concepts of *te aba* in Kiribati and *bwirej* in the Marshall Islands (Thaman 2004a, 2008).

Ecosystem and Habitat Diversity

In terms of ecosystem diversity, each "atoll," whether a true atoll with a central lagoon or lagoons and associated islets or a single low-lying reef island, is somewhat unique, with there being a staggering diversity of atoll island and lagoon types and associated smaller individual islets, lagoons, and ecosystem or habitat types. Depending on the size and type of atoll, ecosystem diversity can include coastal inland, coastal littoral, and mangrove forests; scrublands, grasslands, and herblands; brackish swamps, marshes and ponds (Fig. 4), maricultural areas, and freshwater (groundwater) lenses and wells; swamp taro gardens; agroforests; towns, villages, and houseyard gardens; beaches, rocky limestone shores, terraces, and limestone reef rock; and reefs, seagrass beds, lagoons, open ocean, sea mounts, and ocean floor.

Indigenous atoll vegetation is composed almost exclusively of widespread, ocean-dispersed or, less commonly, wind- or bird-dispersed, salt-tolerant pan-Pacific or pan-tropical coastal plants and mangroves. There are no endemic plant species.

Relatively undisturbed indigenous inland atoll forest is now absent on most atolls. Although remnants are found on some inhabited atolls, they are almost exclusively on uninhabited atolls and less accessible uninhabited atoll islets. There remains, however, a significant amount of coastal shoreline forest and scrub vegetation in various stages of disturbance on many atolls, again, more commonly found on isolated uninhabited islets, many of which are globally among the world's most

Fig. 4 Inland coastal marsh, Temaiku, S. Tarawa, Kiribati (Photo credit: R.R. Thaman © Rights remain with the author)

Fig. 5 *Sonneratia alba* mangrove forest, Butaritari Atoll, Kiribati (Photo credit: R.R. Thaman © Rights remain with the author)

important seabird reserves. On some of the drier atolls, such as Kiritimati which lies in the tropical dry belt, there are also extensive areas of grassland and low open scrubland.

It should be stressed that on atolls there is no surface water in the form of rivers and lakes, with the only real surface water being in the form of limited areas of freshwater marshes and brackish ponds that are found on some atolls, such as the extensive system of landlocked hypersaline ponds on Kiritimati (Christmas) Atoll in the Line Islands of Kiribati. Many of these are polluted, affected by saltwater incursion, decreasing in size or being reclaimed, and are under threat as critical "wetland habitats." Larger wetter atolls, such as Jaluit in the Marshall Islands (399 cm $year^{-1}$ rainfall), Funafuti in Tuvalu (338 cm $year^{-1}$ rainfall), Butaritari in Kiribati (310 cm $year^{-1}$ rainfall), and Ontong Java in Solomon Islands, have the most extensive areas of most of the major atoll ecosystems, such as mangroves (Fig. 5), saline swamps, inland atoll forest, and areas of coral reef, and the richest species diversity (Thaman 2008).

Although mangroves are found on most of the larger true atolls, almost always along protected lagoon shores or in back-beach basins in brackish ponds, they are normally not found on lagoon-less reef islands and the atolls of the Central Pacific (Thaman 2008). On many atolls, such as on Funafuti in Tuvalu, extensive areas of mangroves have been reclaimed or destroyed, in the Tuvalu case to build airstrips during World War II (Thaman et al. 2012).

Fig. 6 Giant swamp taro (*Cyrtosperma chamissonis*) excavated pit cultivation, Funafuti, Tuvalu (Photo credit: R.R. Thaman © Rights remain with the author)

The most widespread vegetation type on most atolls now consists of coconut-dominated agroforests that are dominated almost exclusively by the coconut palm, often on the best soils, but will include other useful indigenous trees, shrubs and other plants, depending of the level of maintenance of former coconut plantings and regeneration of native plants and introduced weedy species over time. Pandanus *Pandanus* spp., breadfruit (*Artocarpus altilis* and *A. mariannensis*), and sometimes bananas *Musa* spp. or other useful trees are also planted, sometimes as small tree groves in more favorable sites, usually near villages or residences or wetter sites (Thaman 1990, 2008; Thaman and Whistler 1996).

Excavated taro pits are a unique and specialized wetland agricultural ecosystem found in the central parts of the larger islets of many atolls and in and around villages. These pits have been excavated to the level of the freshwater lens through the limestone bedrock to depths of 1.5–3 m. The artificially enriched soils in these pits, known as taro mucks, are fertile, swampy, and very rich in organic material. The main crop planted in taro pits is giant swamp taro *Cyrtosperma chamissonis* (Fig. 6), although common taro *Colocasia esculenta*, giant taro *Alocasia macrorrhiza*, and bananas *Musa* cultivars often using intensive traditional mulching systems are also occasionally planted. Coconut palms, pandanus, breadfruit, papaya *Carica papaya*, native fig *Ficus tinctoria*, and other multipurpose plants are also planted in or near pits. Shrubby species are also found on the margins of the pits; and other multipurpose native trees, including *Tournefortia argentea*, *Guettarda speciosa*,

Fig. 7 Fish harvest from traditional fish weirs, Tikehau Atoll, Tuamotus, French Polynesia (Photo credit: R.R. Thaman © Rights remain with the author)

and *Pipturus argentea*, the leaves of which are an important component of the fertilizer or mulch, are also present (Thaman 1990, 2008).

There is also limited mariculture practiced on atolls. This includes the culture of black-lip pearl oysters *Pinctada margaritifera* in the Northern Cook Islands and Tuamotus; milkfish *Chanos chanos* mariculture in Kiribati on both Tarawa and Kiritimati Atolls; limited brackish water aquaculture of tilapia *Oreochromis* spp. in Kiribati and on some other atolls, although tilapia is seen as a pest and a hindrance to the mariculture of milkfish in Kiribati; and the mariculture of seaweed *Kappaphycus alvarezii* in Kiritimati and a number of other atolls, although ventures have often failed (Gillett 1989; Thaman 2008). There has also been the commercial maricultural production of brine shrimp *Artemia salina* on Kiritimati Atoll, beginning in the 1970s, but abandoned in 1978 (Teeb'aki 1993). Also of note is the use of intricate systems of fish weirs or traps on the intertidal and shallow subtidal areas of atolls as a management strategy for nearshore intertidal finfish resources (Fig. 7).

Species and Taxonomic Diversity

There are four generalizations that apply to the Pacific Islands as a whole for species and taxonomic diversity. These are (1) the "western affinity" of most taxa; (2) a diversity gradient or attenuation of diversity from west to east, as one moves away from the Western Indo-Pacific source area of most taxa; (3) a gradual elimination of major groups (higher taxa) of plants and animals from west to east; and (4) a range from very high endemism for high isolated islands to virtually no endemism for the smaller, low-lying limestone islands and atolls and sand cays. The "Theory of Island Biogeography" (MacArthur and Wilson 1967) suggests that species diversity on islands is controlled mainly by island area and distance from the source of colonizing propagules, although as stressed by Fosberg (1974) and Stoddart (1992), species richness on atolls is related more to moisture availability, rainfall extremes, and susceptibility to drought, rather than to island size or to the closeness of source areas of colonizing organisms in the Western Pacific. Atolls in the far west, such as Ontong Java, do however have richer floras and marine biotas. For example, some of the most extensive areas of mangroves and freshwater swamps in Kiribati are found on Butaritari, which is the wettest and most westerly atoll in the main Gilbert group (Thaman 2008).

In terms of western affinity of atoll taxa, studies show that almost all terrestrial, freshwater, and marine plant and animal taxa on atolls (e.g., ferns, algae, sea grasses, corals, echinoderms, marine and terrestrial mollusks, insects, birds, bats, etc.) are mainly of Asian or Western Indo-Pacific origin.

The decrease in the total number of species, genera, and families with increasing distance from the Western Indo-Pacific is considerable, and smaller and lower islands have fewer species than larger higher islands. This is due to the differential dispersability of the different organisms, some of which were never able to reach the more distant islands and atolls of the Central Pacific. It is also due to the greater habitat diversity on larger islands, the greater chance of extinction or extirpation (local extinction) among smaller populations on atolls and small islands (often due to prolonged drought or other extreme events, such as "king tides" or tropical cyclones and associated saltwater inundation), and the associated lower probability of initial colonization or recolonization after extirpation or extinction. For example, a large number of marine finfish families absent from the more distant islands on the Pacific Plate in the Central Pacific have shorter larval stages and are unable to disperse over the great distances of open water between oceanic islands. Also estuarine and freshwater habitats are limited to high islands, with such species never being found on atolls (Myers 1991).

For whatever reasons, regardless of their distance from the Western Indo-Pacific "hot spot," as "biodiversity cool spots," atolls have the most limited species diversity for most taxa, although this is not as dramatic in the marine environment. For example, the number of indigenous fern species drops from 230 in Fiji to 215 in Samoa, 150 in the Society Islands, and to only 9, 6, 5, and 5 species, respectively, in atolls of the Marshall Islands, Tuvalu, Kiribati, and Tikehau Atoll in the Tuamotus. Even among orchids, which are famous for the high dispersability of their very small light seeds, the same attenuation and almost absolute poverty on atolls are reflected, with the number of orchid species dropping from over 3,000 for Papua New Guinea, which has one of the richest orchid floras in the world, to 164 for Fiji, 100 for Samoa, only 3 for Hawai'i, and none for the low-lying atolls of Micronesia and Polynesia, although raised limestone islands, such as Makatea,

Fig. 8 Surgeonfishes (Acanthuridae) on fringing reef, Tikehau Atoll, Tuamotu Archipelago, French Polynesia (Photo credit: R.R. Thaman © Rights remain with the author)

which is located only a short distance from the atolls of Tikehau and Rangiroa in the Tuamotus, has two orchids. The number of native angiosperm genera drops from 654 in Solomon Islands, to 476 in Fiji, 302 in Samoa, 263 in Tonga, 201 in the Society Islands, and only 57, 56, 50, and 45 for the Marshall Islands, Gilbert Islands, Tuvalu, and Tikehau Atoll, respectively, in French Polynesia species (Sekhran and Miller 1996; Kores 1991; Whistler 1992; Wilder 1934; Carlquist 1980; Thaman 2008).

The disparity is not as great in the marine environment, with benthic marine algae (including green, brown, and red algae) species numbers dropping from 1,185 in North Australia, to 520 for all of Micronesia, 336 for New Caledonia, 302 for Fiji, 219 for the Solomon Islands, 151 for Tahiti, 90 for Samoa, 40 for Nauru, and 268 for the Marshall Islands, one of the larger more westerly atoll groups (N'Yeurt and South 1997). The diversity of corals and reef related finfish diversity is also considerable around atolls relative to terrestrial and freshwater diversity, with atolls considered to be among the best dive sites (Fig. 8).

Despite its relative poverty, the conservation of atoll species diversity is critical to the heath and sustainable livelihoods of atoll people. This is reflected in the indigenous knowledge and taxonomy (ethnobiodiversity) that atolls' peoples have for their biodiversity. For example, the I-Kiribati (people of Kiribati) have local vernacular names for approximately 144 small and 115 large finfishes, 20 sharks, 9 rays, 25 eels, 5 whales or dolphins, 5 sea turtles, 6 sea snakes or snakelike animals, 16 seabirds, and 74 marine mollusks (Thaman 2008).

Genetic Diversity

The very limited species diversity is, however, enriched by the genetic diversity of cultivars of the limited number of crops that can be cultivated on atolls. In Kiribati, for example, there are reportedly over 200 named cultivars of pandanus *Pandanus* spp., over 30 cultivars of giant swamp taro *Cyrtosperma chamissonis*, and at least ten named coconut *Cocos nucifera* cultivars (Small 1972), many of which are now threatened or lost.

Ecosystem Goods and Services

The ecosystem goods and services provided by the highly impoverished terrestrial and freshwater and the richer marine biodiversity of atolls constitute a critical ecological and cultural resource. In the case of plants, this is particularly true, virtually all of which have wide cultural utility, and despite the importance of the sea, plants continue to provide the majority of foods, medicines, handicrafts, construction materials, and other material and nonmaterial needs and ecosystem services required by atoll peoples (Table 1).

In terms of cultural utility, 140 species of widespread indigenous coastal littoral and mangrove vascular plants, almost all of which are found on Pacific atolls, were shown to have 75 different purpose/use categories (Thaman 1992b). Collectively 1,024 uses were recorded for these 140 species, ranging from no reported uses (two species) to as many as 125 for the coconut if distinct uses are included (e.g., tools with distinct functions). Another 17 species have 20 or more reported uses, and 29 species have at least 7 uses each. Moreover, the list does not include the more strictly ecological services provided by coastal plants listed in Table 1. For example, for the Marshall Islands, studies showed that there were 168 use categories for 37 of 58 indigenous species found there and another 303 uses for 283 introduced species, the majority of which are ornamentals, food plants, or "weeds" (Thaman 2008).

It should be noted that although "plants," including sea grasses and marine algae, are particularly critical to atoll peoples, coral reefs, currents, areas of upwelling, waves, tidal flows and flushing, and other components of marine biodiversity also provide critical ecosystem goods and services that are disproportionately more important to atoll people due to the poverty of terrestrial biodiversity.

Threats and Future Challenges

As stressed above, atolls have among the most impoverished and highly threatened biodiversity inheritances on Earth, with a high proportion of all economically, culturally, and ecologically important terrestrial plants and animals in danger of extirpation (Thaman 2008).

Atoll communities are clearly on the frontline in the battle against climate change, sea-level rise, extreme weather and tidal events, loss of biodiversity, environmental

Table 1 Ecological and cultural goods and services provided by terrestrial, freshwater, and marine biodiversity on atolls (Adapted from Thaman and Clarke 1993, permission provided by Creative Commons Attribution-NonCommercial-ShareAlike 3.0 IGO License)

Ecological			
Shade/sun protection	Soil/substrate improvement	Animal/plant habitats	
Protection from waves/tides	Sand/sediment provision	Pollution control	
Erosion control	Spawning/breeding grounds	Flood/runoff control	
Oxygenation	Moisture regulation	Dispersal facilitation	
Wind protection	Wild animal food	Weed/disease control	
Coastal reinforcement	Water purification	Protection from salt spray	
Climate regulation	Current maintenance		
Cultural/economic			
Timber/wood	Brooms	Prop or nurse plants	
Boatbuilding (canoes)	Parcelization/wrapping	Food crops	
Housing/shelter	Abrasives/sandpaper	Wild/emergency foods	
Fuel/firemaking	Illumination/torches	Spices/sauces	
Woodcarving	Insulation	Drinks/teas/coffees	
Tools/weapons	Decoration	Alcoholic beverages	
Containers	Body ornamentation	Masticants/gum	
Cordage/lashing	Fiber/fabric	Meat tenderizer	
Caulking	Plaited ware	Preservatives	
Fishing equipment	Handicrafts	Stimulants/narcotics	
Floats	Clothing/hats	Medicines	
Toys	Dyes/paints	Aphrodisiacs	
Musical instruments	Paint brushes	Fertility control	
Cages/roosts	Tannin/preservative	Abortifacients	
Sails	Rubber	Ritual exchange	
Baskets	Glues/adhesives	Magic/sorcery	
Religious objects	Scents/perfume	Totems	Deodorants
Oils/lubricants	Subjects of mythology		
Toothbrush	Poisons	Recreation/sport	
Paper/toilet paper	Sunscreens	Secret meeting sites	
Religious objects	Insect repellents	Commercial/export	
Switch for children	Embalming corpses	Products	
Discipline	Cosmetics	Tourism attractions	

degradation, invasive alien species (IAS), and economic uncertainty. Particularly impacted has been Takuu Atoll, located in Papua New Guinea northeast of Bougainville, in which, due to rising sea levels, waves have "savagely eroded the coastline" and tidal incursion has contaminated their limited groundwater supplies and taro gardens. As a result, the Polynesian people of Takuu are considered by some authorities to be the first "climate change refugees," with many having been resettled

Fig. 9 Takuu Atoll (October 2004), Papua New Guinea, which has been heavily impacted by sea-level rise (Photo credit: R.R. Thaman © Rights remain with the author)

on Bougainville and elsewhere in PNG with predictions being made that "the sea's relentless advance will extinguish the atoll's ability to sustain life within the next 2–3 years" (Wane 2005) (Fig. 9). Atoll biodiversity and associated ethnobiodiversity are also threatened by increasing population, urbanization, modern education, commercialization, and overexploitation.

The highly threatened status of atoll biodiversity is an accelerating phenomenon that began long before European contact when the early Pacific Islanders severely deforested their islands and brought many birds and other species, including some shellfish to extinction or extirpation throughout many areas of Polynesia and Micronesia (Kirch 1982; Steadman 1995). Losses are greatest on more urbanized and densely inhabited atolls. Although biodiversity is also threatened on larger high volcanic and raised limestone islands, the need for protection is far greater on atolls where there is far less ecosystem and biotic diversity; smaller, genetically less diverse populations; and higher human population densities (Thaman 2008).

Surveys of plants used for specific purposes in Kiribati, Tuvalu, and the Marshall Islands show that there is widespread concern over the loss or scarcity of a wide range of trees and plants used for housebuilding, woodcarving, medicine, body ornamentation, and sacred and other purposes. Even common fruit trees, such as coconut palm and pandanus cultivars, papaya, and breadfruit are reportedly increasingly scarce or disappearing; and traditional cultivars of important tree crops and taros are rare or are no longer cultivated because of overemphasis on monoculture,

Fig. 10 Frigatebird (*Fregata minor*) with chick, Ngaon te Taake reserve, Kiritimati Atoll, Kiribati (Photo credit: R.R. Thaman © Rights remain with the author)

diseases, tropical cyclones, drought, failure to replant, lack of planting material, and increasing dependence on or taste for imported foods (Thaman and Whistler 1996). Results from studies in the Marshall Islands showed that 38 of a total 61 medicinal plants were considered to be rare or in short supply in some areas by 26 local experts (Taafaki et al. 2006). The loss of medicinal plant biodiversity on atolls is a serious problem because few if any modern medicines are available in rural villages and urban communities.

A large number of indigenous birds, reptiles (e.g., lizards and turtles), and a number of terrestrial invertebrates, such as land crabs, are also threatened on atolls. Land crabs, especially the coconut crab *Birgus latro* and larger land hermit crabs *Coenobita* spp., are now extirpated or rare on most inhabited atoll islets (Thaman 1999a, 2008).

Land birds most often mentioned as being rare or endangered include larger land birds such as doves, pigeons, and lorikeets, which are found on some atolls. Also seriously endangered are seabirds and migratory birds that used to be abundant throughout the atoll Pacific. These include noddies, terns, tattlers, godwits, plovers, frigate birds, boobies, tropic birds, petrels, and shearwaters (Fig. 10). Many of these birds and their eggs, which are important traditional foods and considered delicacies, are now rare or endangered on most atolls. Some species, such as frigate birds and noddies, are of critical importance to atoll peoples as a sign to fishermen of the location and identification of schools of fish (Thaman 2008); and studies in the early 1990s for the 1992 Rio Summit report identified the loss of seabirds as the second most serious environmental concern in Tuvalu (Thaman and Neemia 1991).

Almost all indigenous birds, the focus of many Ramsar wetland conservation programs, should be given some form of protected status on atolls and their preferred habitats, tree groves, remaining areas of coastal, mangrove, and inland forest and uninhabited islets given protected status. Of particular concern, especially on atolls, has been the widespread removal of groves of *Pisonia grandis*, the preferred rookery species for a wide range of sea birds. In many cases *Pisonia* groves have been cleared, beginning in the early colonial period, for the expansion of copra plantations onto the fertile, guano-rich soils resulting from the thousands of generations of seabirds that had occupied the groves prior to their removal (Thaman 2008).

Although forest removal, habitat degradation, and hunting have historically been the main causes of the loss of birds, land crabs, and reptiles, the introduction of invasive alien species (IAS), such as pigs, goats, rats, dogs, cats, and ants, may now be the main drivers of the loss of a wide range of birds, crabs and other invertebrates, natural vegetation, farming systems, and human health. The situation is most serious on Guam (which, although not an atoll, has regular air connections with atoll countries) where the accidental introduction of the brown tree snake *Boiga irregularis* near the end of World War II has led to the near extinction of almost all indigenous land birds and the serious endangerment of flying foxes and lizards (Rodda and Fritts 1993; Quammen 1996). In Kiribati, Tuvalu, Tokelau, the Tuamotus, the atolls of Palau and New Caledonia, and possibly other areas, introduced ants, such as the yellow crazy ant *Anoplolepis gracilipes* and the little fire ant *Wasmannia auropunctata*, both considered to be among the world's 100 worst IAS, threaten indigenous birds, invertebrates, crops, and human health (CGAPS *c.* 1996; Wetterer et al. 1997). Although not currently affected by the brown tree snake and some of the more destructive ant species, atoll countries must strengthen their quarantine services to ensure that these pests are not introduced or do not spread to outer islands or islets.

There are some serious exotic weeds on atolls that have outcompeted culturally and important indigenous plants. Of particular concern is wedelia or trailing daisy *Sphagneticola/Wedelia trilobata*, an introduced ornamental ground cover of tropical American origin considered one of the world's 100 worst invasive species. Wedelia has spread out of control on most of the main inhabited atoll islets of Kiribati, Tuvalu, Marshall Islands, and Tikehau Atoll in the Tuamotus, where it is encroaching on interior marshland habitats and coastal herblands. Despite concerted attempts, it has been almost impossible to eradicate due to its ability to spread vegetatively, often via discarded cuttings (Thaman 1999b, 2009).

Nutrient pollution from human waste (defecation on the reef is a common practice), sewerage, and pigpen waste, all of which are discharged near the sea or close to the freshwater table, also constitutes a serious threat to the health of atoll wetlands, irrigated taro patches, coral reefs, lagoons, and humans on atolls. Note that there are no larger grazing animals, such as cows, horses, and goats on Pacific atolls because of lack of water, space, and fodder (Thaman 2008).

Many nearshore marine species are overfished for export or local sale. Because of low cash incomes and scarcity of foreign exchange on atolls coupled with the increasing market demand for many marine products, there has been an increasing pressure to market shark fin, bêche-de-mer (sea cucumbers), giant clams, large coral reef fish, aquarium fish, live coral, and a number of other marine products, many of which are endangered and listed on the IUCN Red List or on the CITES list of restricted exports (Thaman 2004b). Marine vertebrates that are increasingly threatened include whales and dolphins, marine turtles, sharks and rays, eels, a number of smaller reef and lagoon fish, and a number of larger, commercially important finfish species commonly targeted by spearfishers, hook-and-line fishers, deepwater line fishers, and the live-fish export market, the most common being large rock cods, coral trout, or groupers (Epinephelidae, Serranidae); large parrotfishes (Scaridae); large wrasses (Labridae), particularly the humphead wrasse *Cheilinus undulatus*; and trevallies (Carangidae) (Thaman 2008).

Many culturally important mollusks are also rare or extirpated on most inhabited atolls, the most overexploited of which include giant clams *Tridacna* spp., turban snails *Turbo* spp., black-lip pearl oysters *Pinctada margaritifera*, conches (*Lambis* and *Strombus* spp.), triton trumpet shell *Charonia tritonis*, ark shells *Anadara* spp., topshells or trochus (*Trochus* and *Tectus* spp.), Venus clams *Periglypta* spp., and mussels (Mytilidae). Crustaceans, including reef and shore crabs (Calappidae, Carpilidae, Grapsidae, Ocipodidae, Portunidae, and Xanthidae) and mantis shrimps (Stomatopoda), are also rare due to overexploitation. Their declining yields constitute a serious nutritional and economic problem as they are one of the most easily accessible nutritional and commercial resources for low-income coastal communities. For some of these species, there is a need for a total ban on exploitation, whereas for others, there is a need for the establishment of local reserves or commercial or seasonal protection, until such times as stocks recover.

Holothurians or bêche-de-mer are overexploited in most atoll countries due to pressure in the past 30 years to export to Asian markets. All species should be given some protected status until stocks recover, with some species being reserved for local consumption or limited local sale. Other marine invertebrates considered to be threatened or overharvested in some areas include lobsters, octopus, and squid, all of which are in need of some form of protection, at least the designation of some local marine reserves or the enforcement of seasonal or size restrictions on their exploitation (Thaman 2008).

Finally, and most worrying, is that like biodiversity itself, atoll ethnobiodiversity is highly threatened. Many of the current generation, schooled in the modern educational system and living in the cash economy in urbanized, highly populated overexploited areas, often know few of the traditional uses and services provided by atoll biodiversity, let alone the local vernacular names of plants and animals and the places they live. This loss of knowledge has undoubtedly contributed to a loss of appreciation for, and is indirectly associated with, the degradation and loss of biodiversity on atolls (Thaman 2008). The conservation and re-enrichment of this knowledge must be an integral part of attempts at atoll biodiversity conservation.

Conservation of Atoll Ecosystems

Although the predominant focus of biodiversity conservation has been on endemic, charismatic, and officially threatened species in biodiversity hot spots, there is an increasing recognition of the uniqueness, fragility, and need for conservation of atoll biodiversity and other biodiversity cool spots, where seriously threatened very limited biodiversity inheritances constitute the foundation for the sustainability of local communities. This has resulted in an increasing range of collaborative initiatives under Ramsar, the UNESCO Man and the Biosphere Programme, World Heritage Convention, and other initiatives that have focused on the conservation, restoration, and sustainable use of atoll wetland ecosystems and their surrounding marine areas.

Firstly, three Ramsar Wetlands of International Significance have been designated on atolls in the Pacific. These are Jaluit and Namdrik Atolls in the Marshall Islands and Nooto Islet, which is part of North Tarawa Atoll, in Kiribati.

Inscripted in 2004, Jaluit is a large atoll covering an area of 69,000 ha comprised of some 91 islets with a land area of 700 ha enclosing a large lagoon. It has relatively pristine areas of coral reefs, intertidal flats, seagrass beds, and mangroves and is an important nesting, nursery, or spawning site for sea turtles, seabirds, and a wide range of finfish and marine invertebrates. About 2,000 local residents practice a reasonably sustainable subsistence lifestyle, although some of resources are under threat due to overharvesting, in the case of marine resources for off-island sale (http://www.ramsar.org/jaluit-atoll-conservation-area).

Inscripted in 2012, Namdrik Atoll is located about 390 km southwest from the capital atoll of Majuro. It has an area of 1,119 ha consisting of two wooded islets with an extensive reef flat lying between them and is unusual because there are no navigable waterways into the central lagoon. Being relatively isolated, the atoll is in near pristine condition, supports an extensive mangrove forest, is home to some 150 species of fish, and supports breeding populations of the endangered hawksbill *Eretmochelys imbricata* and green sea turtles *Chelonia mydas*. The wetland provides many resources for local people, although unsustainable harvesting practices place considerable pressure on the atoll's unique biodiversity (http://www.ramsar.org/namdrik-atoll).

Inscripted in 2013, Nooto is a relatively pristine northern islet of Tarawa Atoll, the capital atoll of Kiribati. It has an area of 1,033 ha and a wide range of coastal habitats including coral reefs, an extensive lagoon, intertidal flats, mangroves, and nesting, spawning, and nursery sites for sea turtles, bonefish *Albula* sp., and other threatened organisms. It is also an important resource island for local North Tarawa Atoll communities (Thaman et al. 1995).

Noteworthy international atoll conservation initiatives under the UNESCO Man and the Biosphere (MAB) Programme include the designation in 1977 of Taiaro Atoll in the Tuamotus 540 km to the northeast of Tahiti, as the first atoll ecosystem under MAB, and its redesignation in 2006, after close consultation with local communities, as the Tuamotu Biosphere Reserve to encompass seven additional atolls, including Fakarava, the largest and most economically developed atoll; Aratika, Kauehi, Raraka, and Toau Atolls that are inhabited and with navigable passages to interior lagoons; and Taiaro and Niau, two uninhabited closed atolls with interior lagoons but no navigable passages (www.unesco.org/mabdb/br/brdir/directory/biores.asp?; Jacques-Bourgeat 2015).

In 2007, Ant Atoll, the only uninhabited atoll in Pohnpei State of the Federated States of Micronesia, was designated as a MAB Reserve and as an official state protected area in 2010 in efforts to rehabilitate the island, control unsustainable fishing, and protect seabird and coconut crab populations and turtle nesting sites (CSP 2007; Cohen 2015).

In 2006 the Papahānaumokuākea Marine National Monument (PMNM) (originally Northwestern Hawaiian Islands Marine National Monument) was established and subsequently inscribed as a World Heritage Site in 2008. PMNM is one of the largest marine protected areas in the world, encompassing an area of 362,000 km^2, including ten islands and atolls of the Northwestern Hawaiian Islands. It is internationally recognized for its cultural and natural significance and supports 7,000 marine species, one quarter of which are endemic to the Hawaiian Islands. The islands and shallow water environments are important habitats for threatened species such as the green sea turtle and the Hawaiian monk seal *Monachus schauinslandi*, one of the rarest marine mammals in the world, as well as the 14 million seabirds representing 22 species that breed and nest there. The area has four species of bird found nowhere else in the world, including the world's most critically endangered duck, the Laysan duck *Anas laysanensis*, Laysan albatross *Phoebastria immutabilis*, Nihoa millerbird *Acrocephalus familiaris*, and Nihoa finches *Telespiza ultima*, the latter two which are found only on the small rocky island of Nihoa. Although the Nihoa millerbird used to be present with numbers up to 1,500 in 1915 on Laysan Atoll, it was brought to extinction between 1916 and 1923. There are also interesting species of plants including *Pritchardia* palms and many arthropods and crustaceans. Many marine species seriously overfished in the past are now under protection (PMNM 2014; Birdlife International http://www.birdlife.org/; Morin et al. 1997).

In 2007 Kiribati designated the Phoenix Islands Protected Area (PIPA) as the world's third largest marine protected area with an area of 408,250 km^2. PIPA includes eight atolls and two submerged coral reefs (PIPA 2007) and has extensive areas of pristine reefs, lagoons, intertidal flats, marshes, seabird, turtle, and land crab nesting areas and rich fisheries resources, many of which are threatened by overexploitation (e.g., illegal fishing) and invasive alien species (IAS), including weeds, rats, cats, rabbits, dogs, and ants, some of which have been the focus of recent successful eradication efforts. These include the successful eradication of Asian rats from McKean Atoll and rabbits from Rawaki Atoll, under the New Zealand-supported Pacific Invasives Initiative (PII), important for many threatened bird populations which had almost disappeared, and spectacular recovery of the vegetation (Koszler 2010). In 2009, the PIPA was inscribed as a World Heritage Site.

In 2008, the New Caledonia barrier reef was formally inscribed as a World Heritage Site under the name The Lagoons of New Caledonia: Reef Diversity and

Associated Ecosystems. The area includes the central lagoon of Ouvea, a raised limestone island which is sometimes incorrectly referred to as the largest atoll in the world, as well as the Entrecasteaux Atoll group to the north of New Caledonia's main island (CI 2014).

Bikini Atoll Nuclear Test Site in 2010 was inscribed as the first World Heritage Site in the Marshall Islands due mainly to its long history as a nuclear test site and as testimony and memorial to the destruction caused by nuclear testing. The atoll has a total area of 73,500 ha. Alinginae Atoll, another uninhabited atoll, is in the process of applying for inscription as the country's 2nd World Heritage Site (http://whc.unesco.org/en/list/1339).

Other initiatives include the establishment in 2012 by the Cook Islands of the 1.1 million km^2 Cook Islands Marine Park, which encompasses the atolls of the Northern Cook Islands and includes Suwarrow Atoll which has been designated as a Birdlife International Important Bird and Biodiversity Area (IBA) and has just been the focus of a rat eradication program to restore seriously threatened seabird populations (Leahy 2012). And, in 2014 New Caledonia formally created the world's largest marine park, the Natural Park of the Coral Sea, bringing under management a multiuse marine area of 1.3 million km^2, an area rich in atolls, coral reefs, marine mammals, marine turtles, and nesting seabirds (Vorrath 2014). This area includes much of the lagoon of New Caledonia World Heritage Site mentioned above. Almost immediately following the announcement of the establishment of the Natural Park of the Coral Sea, on June 17, 2014, US President Barack Obama used his executive powers to create an even larger marine park in the South-Central Pacific, known as the Pacific Remote Islands Marine National Monument, which protects two million km^2 of ocean and the reefs and atolls between Hawaii and American Samoa from commercial fishing (Neubauer 2014). It must be stressed, however, that the conservation success of all of these large marine protected areas will depend on continued political commitment and the ability to police them.

Of particular interest from a wetland conservation perspective have been the continuing conservation initiatives to protect the extensive bird populations on Kiritimati Atoll in Kiribati, the largest true atoll in the world (388 km^2). The bird populations include over 30 species, about 18 of which are seabirds, with the atoll being one of the most important breeding grounds on Earth for some of these species (Streets 1877; Teeb'aki 1993; Thaman et al. 1997) (Fig. 11). These populations have significantly declined and remain seriously threatened by poaching of birds and eggs for food, rats, ants, pigs, and especially feral cats which, since the late nineteenth century, have driven 60% of the seabirds from the mainland to offshore islets. Initiatives have included the gazetting of the atoll as a bird conservation area by the British in 1960 (when Kiribati was still a colony) and declaring it a wildlife sanctuary in 1975. This included restricting access to the five most important bird nesting areas, the lagoon islets, Cook Island, Motu Tabu, Motu Upua, and Ngaon te Taake, and the sooty tern nesting areas at Southeast Point. The atoll was also the focus of the South Pacific Biodiversity Conservation Programme in the late 1990s, and the conservation of the island is currently carried out by the area by Kiribati national authorities and part of the atoll is being considered as Kiribati's second

Fig. 11 Sooty terns *Onychoprion fuscatus* nesting on north coast, Cook Island (Islet), Kiritimati Atoll, Line Islands, Kiribati (Photo credit: R.R Thaman © Rights remain with the author)

Ramsar Site. Beginning in 2008 under the New Zealand-funded Pacific Invasives Initiative (PII), IAS were eradicated from 23 of Kiritimati's lagoon islets, which more than doubled the pest-free area within the lagoon (Koszler 2010).

Other initiatives to control IAS on atolls include the eradication of rats from islets on Kayangel Atoll in northern Palau, an area with the biggest population of the Micronesian megapode *Megapodius laperouse*, an IUCN red-listed species. This project has been a collaborative effort since 2008 of Birdlife International, PII, and the Palau Conservation Society. The eradication of pigs and rats is the basis for a spectacular recovery of seabird populations on Clipperton Atoll, a French territory to the southwest of Mexico by Island Conservation and partners (Pitman et al. 2005; Birdlife International 2012).

On a possibly negative note, Wailagi Lala, Fiji's only true atoll, which has one of Fiji's largest colonies of brown noddy *Anous stolidus* and several other seabirds, has been leased to a tourism developer without any provision for the nesting birds (Watling 2013).

Conclusion

In conclusion, the low-lying atolls of the Pacific Ocean and their lagoons and nearshore areas are, under the Ramsar definition, essentially "wetlands." They are unique wetlands with very limited but highly threatened biodiversity inheritances, most notable of which are extensive populations of breeding seabirds and other birds, for which atolls remain among their most important sanctuaries. As "biodiversity cool spots," the highly threatened terrestrial and nearshore marine biodiversity of atolls constitutes the only biodiversity that atoll nations and peoples have a basis for food and livelihood security and building resilience in the face of global change. Fortunately, under Ramsar, UNESCO World Heritage, and other relevant initiatives, such as the inscription as Ramsar sites of the atolls of Jaluit and Namdrik Atolls in the Marshall Islands and Nooto Islet of North Tarawa Atoll in Kiribati, the conservation and restoration of atoll ecosystems are now firmly on the global conservation agenda.

References

Birdlife International. Eradicating introduced mammals from Clipperton Island led to dramatic recovery of seabirds. Presented as part of the BirdLife State of the world's birds website. 2012. Available from http://www.birdlife.org/datazone/sowb/casestudy/261. Accessed 4 Nov 2014.

Bryan Jr EH. Check list of atolls. Atoll Res Bull. 1953;19:1–38.

C.I. (Conservation International). New Caledonia: home of the world's largest marine park. 2014. http://www.conservation.org/projects/Pages/New-Caledonia-Home-of-the-Worlds-Largest-Marine-Park.aspx. Accessed 21 Feb 2015.

Carlquist S. Hawaii: a natural history. Kauai/Lawai: Pacific Tropical Botanical Garden; 1980.

CGAPS. The silent invasion. Honolulu: Coordinating Group on Alien Pest Species; 1996.

Cohen J. Micronesia: keeping Ant Atoll safe. Nature e-News. 2015. http://www.nature.org/ourinitiatives/regions/asiaandthepacific/micronesia/explore/keeping-ant-safe.xml. Accessed 22 Feb 2015.

CSP (Conservation Society of Pohnpei). Ant Atoll declared a UNESCO Biosphere Reserve, vol. 7 (26). Kolonia, Pohnpei: Kaselehlie Press; 2007. p. 12.

Fosberg (1974) Proceedings of the Second International Coral Reef Symposium, vol. 1. Brisbane, Australia: The Great Barrier Reef Committee.

Gillett R. Tilapia in the Pacific Islands: are there lessons to be learned? Suva: FAO/UNDP Fishery Support Programme; 1989.

Jacques-Bourgeat C. Kauehi, un air de paradis. Air Tahiti. 2015;1:14–33.

Jaluit Conservation Area: Ramsar sites and countries. http://www.ramsar.org/jaluit-atoll-conservation-area. Accessed 4 Nov 2014.

Kirch PV. Ecology and adaptation of Polynesian agricultural systems. Archaeol Ocean. 1982;17:1–6.

Kores PJ. Family 32. Orchidaceae. In: Smith AC, editor. Flora Vitiensis nova: a new flora of Fiji (spermatophytes only). Comprehensive indices. Lawai/Kauai: Pacific Tropical Botanical Garden; 1991. p. 322–576.

Koszler N. Restoring biodiversity, one island at a time. Currents. 2010;21:8–13.

Leahy S. Tiny Pacific nations create world's largest marine parks. The Guardian. 2012; 30 August. http://www.theguardian.com/environment/2012/aug/30/pacific-marine-park. Accessed 4 Nov 2014.

MacArthur RH, Wilson EO. The theory of island biogeography. Princeton: Princeton University Press; 1967.

Morin MS, Conant S, Conant P. Laysan and Nihoa Millerbird (*Acrocephalus familiaris*). In: Poole A, Gill F, editors. The birds of North America. Philadelphia and Washington: The Academy of Naural Sciences and The American Ornithologists Union; 1997. No. 302, p. 1–19.

Myers RF. Micronesian reef fishes: a practical guide to the coral reef fishes of the tropical central and western Pacific. Barrigada: Coral Graphics; 1991.

N'Yeurt ADR, South GR. Biodiversity and biogeography of benthic marine algae in the southwest Pacific, with specific reference to Rotuma and Fiji. Pac Sci. 1997;51:18–28.

Neubauer IL. Is a vast marine sanctuary any use if you can't police it? Time Magazine. 2014; June 30. http://time.com/2910469/natural-park-of-the-coral-sea-new-caledonia. Accessed 21 Feb 2015.

Newhouse J. Marine and terrestrial flora of atolls. South Pac Bull. 1980;30:25–30.

Nooto-North Tarawa: Ramsar sites and countries. http://www.ramsar.org/nooto-north-tarawas. Accessed 4 Nov 2014.

PIPA (Phoenix Islands Protected Area). 2007. http://www.phoenixislands.org

Pitman RL, Balance LT, Bost C. Clipperton Island: pig sty, rat hole and booby prize. Mar Ornithol. 2005;33:193–4.

PMNM (Papahānaumokuākea Marine National Monument). 2014. http://www.papahanaumokuakea.gov/wheritage/measures.html. Accessed 4 Feb 2015.

Quammen D. The song of the Dodo: Island biogeography in an age of extinction. New York: Touchstone (Simon & Schuster); 1996.

Rodda GH, Fritts TH. The brown tree snake on Pacific islands. Pac Sci Assoc Inf Bull. 1993;45(3–4):1–3.

Sekhran N, Miller S. Papua New Guinea country study on biological diversity. Port Moresby: Department of Environment and Conservation; 1996.

Small CA. Atoll agriculture in the Gilbert and Ellice Islands. Tarawa: Department of Agriculture; 1972.

Steadman DW. Prehistoric extinctions of Pacific Island birds: biodiversity meets zooarchaeology. Science. 1995;267:1123–31.

Stoddart DR. Biogeography of the tropical Pacific. Pac Sci. 1992;46:276–93.

Streets TH. Some account of the natural history of the Fanning Group of islands. Am Nat. 1877;11:65–72.

Taafaki IJ, Fowler MK, Thaman RR, editors. Traditional medicine of the Marshall Islands: the women, the plants, the treatments. Suva: IPS Publications, University of the South Pacific; 2006.

Teeb'aki K. Republic of Kiribati. In: Scott DA, editor. A directory of wetlands in Oceania. Slimbridge/Kuala Lumpur: International Waterfowl and Wetlands Research Bureau/Asian Wetland Bureau; 1993. p. 199–228.

Thaman RR. Kiribati agroforestry: trees, people and the atoll environment. Atoll Res Bull. 1990;333:1–29.

Thaman RR. Vegetation of Nauru and the Gilbert Islands: case studies of poverty, degradation, disturbance and displacement. Pac Sci. 1992a;46:128–58.

Thaman RR. Batiri kei Baravi: the ethnobotany of Pacific Island coastal plants. Atoll Res Bull. 1992b;361:1–62.

Thaman RR. Pacific Island biodiversity on the eve of the 21st century: current status and challenges for its conservation and sustainable use. Pac Sci Assoc Inf Bull. 1999a;51(1–4):1–37.

Thaman RR. *Wedelia trilobata*: Daisy invader of the Pacific Islands, IAS Technical Report 99/2. Suva: Institute of Applied Science, University of the South Pacific; 1999b.

Thaman RR. Sustaining culture and biodiversity in Pacific Islands with local and indigenous knowledge. Pac Ecol. 2004a;7–8:43–8.

Thaman RR. Cool spots under threat: the conservation status of atoll biodiversity and ethnobiodiversity in the Pacific Islands. In: Lee KJ, Tsai H-M, editors. Changing islands – changing worlds: proceedings of Islands of the world VIII. International Small Islands Studies Association (ISISA); 2004b. p. 60–4.

Thaman RR. Atolls – the "biodiversity cool spots" vs "hot spots": a critical new focus for research and conservation. Micronesica. 2008;40(1/2):33–61.

Thaman RR, Fihaki E, Fong T. Plants of Tuvalu: Lākau mo mouku o Tuvalu. Suva: University of the South Pacific, Press; 2012. p. 259.

Thaman R. Agrodeforestation and the loss of agrobiodiversity in the Pacific Islands: a call for conservation. In: Jupiter S, Kingsford R, editors. Pacific conservation biology. Special issue on Biodiversity conservation in the Pacific Islands of Oceania, vol. 20(2). 2014. p. 180–92. http://pcb.murdoch.edu.au

Thaman RR. Wedelia (*Sphagneticola trilobata*) – daisy invader of the Pacific Islands: the worst weed in the Pacific? In: Frogier P, Mery P, editors. Proceedings of the 11th Pacific Science Intercongress and 2nd Symposium on French Research in the Pacific, Papeete/Tahiti; 2009. http://webistem.com/psi2009/output_directory/cd1/Data/articles/000568.pdf. Accessed 17 Mar 2015.

Thaman RR, Clarke WC. Pacific island agroforestry: functional and utilitarian diversity. In: Clarke WC, Thaman RR, editors. Pacific Island agroforestry: systems for sustainability. Tokyo: United Nations University Press; 1993. p. 17–33.

Thaman RR, Whistler WA. A review of uses and status of trees and forests in land-use systems in Samoa, Tonga, Kiribati and Tuvalu with recommendations for future action, Working Paper 5, June 1996 (RAS/92/361). Suva: South Pacific Forestry Development Programme; 1996.

Thaman RR, Neemia U. Tuvalu: National Report for the United Nations Conference on Environment and Development (UNCED), Rio de Janeiro, June 1992. Office of the Prime Minister, Vaiaku, Funafuti, Tuvalu and South Pacific Regional Environment Programme (SPREP), Noumea; 1991.

Thaman RR, Clarke WC, Tebano T, Taniera T, Eritaia B. North Tarawa conservation area: South Pacific biodiversity programme project preparation document. Bikenibeu/Tarawa: Environment Unit, Ministry of Environment and Social Development; 1995.

Thaman RR, Tuxson R, Eritaia B. Kiritimati Atoll conservation area: South Pacific biodiversity programme project preparation document. Bikenibeu/Tarawa: Environment Unit, Ministry of Environment and Social Development; 1997.

Vorrath S. New Caledonia creates world's largest marine park. Renew economy; 2014. http://reneweconomy.com.au/2014/new-caledonia-creates-worlds-largest-marine-park-94028. Accessed 4 Nov 2014.

Wane J. Before the flood. Ingenio et Labore: Magazine of the Auckland University; 2005. spring: 10, 12.

Watling D. Fiji: state of birds. Suva: NatureFiji-MareqetiViti; 2013.

Wetterer JK, Banks PC, Laniawe L, Slotterback JW, Brenner G. Exotic ants at high elevations on Mauna Kea, Hawaii. Pac Sci. 1997;52:228–36.

Whistler AW. Vegetation of Samoa and Tonga. Pac Sci. 1992;46:159–78.

Whittaker RJ. Island biogeography: ecology, evolution, and conservation. Oxford: Oxford University Press; 1998.

Wiens HJ. Atoll environment and ecology. New Haven: Yale University Press; 1962.

Wilder GP. The flora of makatea, Bulletin 120. Honolulu: Bernice P. Bishop Museum; 1934.

Index of Keywords

A
Aammiq wetland, 1156
Aapa mire, 261, 1541
  peat layer, 284
  vs. raised bog, 283
Abay in Ethiopia, 1247
Aberdare Range, 1389
Abies, 938
  A. pindrow, 1608
  A. sibirica, 1537
Abramis brama, 939, 953, 1443
Abu Dhabi Gulf coast, 1175
Abutilon
  A. longicuspe, 206
  A. mauritanum, 206
Acacia, 207, 1302
  A. caven, 816
  A. dealbata, 1313
  A. fistula, 1294
  A. mearnsii, 1313
  A. mellifera, 1385
  A. nubica, 1272
  A. polyacantha, 1236
  A. seyal, 206, 1262, 1272, 1293, 1294, 1302
  A. sieberiana, 206, 1236, 1294, 1302
  A. stenophylla, 1901
  A. thailandica, 208
  A. tortilis, 1353, 1368, 1385, 1386
  A. xanthophloea, 201, 1236, 1357, 1386
Acacia savanna, 1236
Acanthobrama hulensis, 1161, 1169
Accipter brevipes, 1255
Accretion, 162, 167, 411, 709, 714, 840, 969, 1019, 1102, 1109, 1120, 1258, 1749
Acer, 938
  A. negundo, 817
  A. rubrum, 177, 581, 672, 673
  A. saccharinum, 525

Acidification, 361, 1804, 1891, 1892, 1914, 1916, 1917, 1963
  drainage basin, 1409
  lakes, 139
  ocean, 12, 82, 451, 456–458, 502
Acinonyx jubatus, 1303, 1372
Acioa deweverei, 207
Acipenser
  A. baerii, 1444, 1455, 1474, 1480, 1481
  A. baerii baicalensis, 1481
  A. brevirostrum, 633
  A. fulvescens, 428
  A. gueldenstaedtii, 900, 918, 939, 953
  A. medirostris, 569
  A. nudiventris, 905, 918, 939, 953
  A. oxyrinchus, 633
  A. persicus, 940
  A. ruthenus, 900, 905, 940, 953, 1480
  A. schrenckii, 1513
  A. sinensis, 1569
  A. stellatus, 900, 918, 939, 953, 1629, 1630
  A. sturio, 928, 1131
  A. transmontanus, 569, 642
Aconitum spicatum, 1609
Acrocephalus
  A. arundinaceus, 1621, 1708
  A. familiaris, 2020
  A. griseldis, 1688
  A. melanopogon, 1621
  A. rufescens, 192
  A. scirpaceus, 1337
  A. tangorum, 1788
Acrochordus granulatus, 1955
Acronychia pedunculata, 1815
Acrostichum, 1823
  A. aureum, 96, 750, 765
Acrotelm, 234, 296, 309
Actitis hypoleucos, 1336, 1385
Active blanket bog, 290

Active raised bog, 290
Adaptive management, 351, 720, 1906, 1907
  Kakagon, 434
  Lake Chilika, 397–402
  Lake Fuquene, 781
Addax nasomaculatus, 1303
Adenostemma lavenia, 1746
Aechmophorus occidentalis, 597
Aegialitis rotundifolia, 1737
Aegiceras corniculatum, 1728
Aegopis, 339
Aeluropus
  A. lagopoides, 1728
  A. littoralis, 916, 1117
Aepyceros
  A. melampus, 1358, 1405
  A. melampus rendilis, 1385
Aerial roots, 94, 100, 217
Aeschynomene, 194
  A. indica, 205
  A. pfundii, 1353, 1354
Aeshna
  A. isosceles, 1030
  A. sitchensis, 571
  A. subarctica, 571
Africa, 9, 11, 28, 29, 59, 81, 96, 113, 116, 131, 145, 146, 150, 171, 175, 176, 1156, 1160
  Bahr el Ghazal, 1270, 1274
  Bijagós archipelago, 1334
  Congo River Basin, 1200, 1206, 1210–1212
  Cyperus papyrus swamps, 204, 205
  Itezhi-Tezhi dam, 1230
  Kilombero floodplain, 1342, 1343
  Lake Turkana, 1362
  Mimosa pigra, 385
  Nile Delta, 1254
  Nile River Basin, 1244, 1249
  Papyrus marshes, 184–186, 190–193, 195
  peatland, 229, 232, 1413–1421
  Rift Valley soda lakes, 153
  salt lakes, 147
  Sudd, 1300, 1302
  tropical freshwater wetlands, 202
  wetland loss, 372, 373
  Zambezi Delta, 1234
  Zambezi River Basin, 1222
African buffalo, 1227, 1236, 1237, 1240
Afro-Syrian rift valley, 1168
Agabus, 660, 661
  A. labiatus, 1071
Agamia agami, 750
Agarum cribrosum, 626

Agathis dammara, 239
Agave cundinamarcensis, 777
Agelaius phoeniceus, 612
Ageneiosus caucanus, 763
Aggressive urbanization, 1146
Agmon restoration project, 1171
Agriculture, 12, 31, 58, 61, 62, 89, 135, 138, 146, 152, 1156, 1163, 1193–1194, 1254
  Azraq Oasis wetland, 462, 466, 467
  Congo Basin, 1209
  Danube Delta, 890, 891, 917, 919–921
  Ebro Delta, 119, 1114, 1118
  food recession, 118, 1229, 1239
  Gangetic-Brahmaputra Plains, 1713, 1716, 1720, 1721
  Indus Basin, 1699, 1701, 1702
  karst wetland, 316, 325, 326, 328
  Kilombero Valley Floodplain, 1345–1347
  Lake Seyfe, 1108, 1193–1194
  Lake Wular, 1706, 1709
  New Zealand, 114
  Nile Delta, 1254–1257
  Nile River Basin, 1248, 1249
  peatlands for, 237, 238, 240, 938
  pocosin, 675
  Prairie Pothole Region, 685, 686
  Rhine River Basin, 927, 928
  Rhone River Delta, 1102, 1103, 1106, 1109
  rice, 407, 584, 587, 1114
  Rugezi Marsh, 1309, 1310, 1316
  on San Pablo Bay, 644, 647
  Volga River Basin, 934, 940, 942
  wetland conversion, 362–364, 366, 534
  Zambezi River Basin, 1222, 1225, 1227, 1229
Agropyron pungens, 964
Agropyrum spp., 851
Agrostis
  A. mongolica, 1537, 1538
  A. stolonifera, 916, 964, 1130
Ailanthus altissima, 1640
Ailurus fulgens, 1608, 1609
Air Hitam Laut River, 1832, 1836, 1837
Aix sponsa, 583
Akysis
  A. filifer, 1806
  A. fuscus, 1845
Alaria esculenta, 626
Alaska National Interest Lands Conservation Act, 548, 562
Alaskan boreal zone, 519, 522, 527, 529, 531, 532, 537, 539

Alas lake, 1446, 1467
Alauda
  A. alauda, 965
  A. arvensis, 1083, 1084
Al Bahr al-Azraq in Sudan, 1247
Alberta Provincial Lands, 553
Albizia
  A. amara, 1272
  A. inundata, 816
Albúferas, 827
Alburnoides bipunctatus thessalicus, 1143
Alburnus chalcoides, 939
Alca torda, 1094
Alcedo
  A. atthis, 905
  A. cristata, 1345
  A. euryzona, 1825
  A. semitorquata, 1345
Alcelaphus
  A. buselaphus, 1282
  A. lichtensteinii, 1227, 1236
Alces alces, 532, 553, 561, 1444
Alcolapia grahami, 1386–1388
Aldan and Amga rivers, 1449, 1472, 1473
Alder and Willow communities, 560
Aldrichetta forsteri, 1912
Alectoria, 279
Alestes stuhlmanni, 1344
Aleuropus littoralis, 916, 1117
Al-Hammar, 1686, 1688–1691, 1693
Al-Hawizeh, 1686, 1688, 1693, 1694
  bird counts, 1691
  drained portion, 1692
  ecological diversity indices for fish, 1690
  water buffalo feeding, 1689
Alien plants in wetlands
  Aeschynomene sp., 194
  Azolla filiculoides, 1133, 1140, 1237
  Butomus umbellatus, 603, 952
  control methods, 388
  Cyperus papyrus, 170, 177, 184, 190, 204,
    205, 209–211, 1156, 1168–1169,
    1210, 1237, 1244, 1265, 1272, 1302,
    1313, 1350, 1357, 1398, 1427
  definitions, 384
  Egeria densa, 385, 777
  Eichhornia crassipes, 117, 134, 137, 189,
    209, 218, 220, 384, 385, 713, 765,
    777, 779, 790, 809, 1210, 1237,
    1249, 1255, 1272, 1301, 1305, 1357,
    1368, 1720, 1746, 1773, 1859, 1948
  Elodea canadensis, 384, 914
  Halophila stipulacea, 81
  Hydrilla verticillata, 1746, 1859
  Hymenachne amplexicaulis, 1957
  impact assessment, 387
  Impatiens glandulifera, 164
  Ipomoea spp., 194
  invasion methods, 386–387
  Juncus acutus, 64, 1116, 1117
  Lasimorpha senegalensis, 1210
  Lyngbya wollei, 603
  Lythrum salicaria, 432, 597, 613, 915
  Mimosa pigra, 137, 206, 220, 366, 385,
    386, 1773, 1798, 1859, 1952
  Myriophyllum spicatum, 480, 613,
    914, 952, 1105, 1106, 1117, 1460,
    1524
  Phalaris arundinacea, 432, 613
  Phragmites australis, 58, 67, 149, 159, 163,
    164, 170, 385, 386, 431–432, 471,
    603, 612, 914, 915, 949, 965, 1105,
    1116, 1117, 1140, 1156, 1171, 1190,
    1237, 1398, 1416, 1427, 1524, 1619,
    1639, 1642, 1649, 1650, 1653, 1666,
    1686, 1689, 1900
  Phragmites karka, 204, 1272, 1302,
    1746, 1748
  Pistia stratiotes, 218, 220, 713, 765, 1210,
    1237, 1353, 1746, 1815
  population growth and migration, 384–385
  Potomageton crispus, 613, 914
  risk assessment, 387
  Salix cinerea, 915, 1029, 1986
  Salvinia molesta, 218, 220, 384, 1237,
    1357, 1409, 1952
  Spartina alterniflora, 58, 384, 969, 988
  Tamarix, 64, 1140, 1666, 1669, 1730
  Typha angustifolia, 163, 428, 613, 777, 914,
    915, 952, 1190, 1639, 1654, 1655,
    1679, 1680, 1746
  Typha domingensis, 190, 713, 1156, 1171,
    1221, 1265, 1272, 1302, 1353,
    1354, 1900
  Typha species, 26, 159, 163, 189, 204, 385,
    386, 612, 712, 715, 851, 915, 1117,
    1255, 1642
  Urochloa mutica, 765, 1952
Alien species, 179, 193, 220, 324, 327,
    384–387, 470, 638, 818, 1798, 1905,
    1945, 2014, 2017, 2020
Alisma
  A. orientale, 1524
  A. plantago-aquatica, 949, 1556, 1639
Alkaline, 146, 518, 706, 709, 1362, 1382,
    1384–1386, 1426, 1582, 1961

Alkaline (cont.)
  lakes, 1365, 1382, 1445, 1486
  swamps, 1386
Alkaliphilic bacteria, 1389
Alligator mississippiensis, 675
Alligator weed, 1710
Allium
  A. angulosum, 951
  A. chamaemoly, 1106
Allotropic riverine lake, 1365
Alluvial fan, 698, 1395, 1401, 1649, 1650
Alluvial terrace, 1453
Alnus, 525, 545, 560, 1043
  A. acuminata, 1313
  A. glutinosa, 915
  A. incana, 612
  A. japonica, 1598
  A. rubra, 571
Alocasia macrorrhiza, 2009
Alopecurus turczaninovii, 1539
Alosa
  A. fallax, 165
  A. kessleri, 939, 953
  A. kessleri volgensis, 953
  A. pontica, 918
  A. sapidissima, 165, 629
Alpine fens, 1420
Alpine wetlands, 130, 297, 888, 1414, 1445, 1535, 1553, 1699–1701
Alstonia
  A. pneumatophora, 1823, 1834
  A. scholaris, 208
  A. spathulata, 1815, 1816
Alternanthera
  A. philoxeroides, 816, 1710
  A. sessilis, 1639
Althaea officinalis, 951
Althea rosea, 916
Althenia filiformis, 1104
Amazonian large-river floodplains, 729–731, 733, 737
Amazonian white-sand savannas, 730, 732, 734, 738
Amazon of Europe, see Danube River; Drava River; Mura River
Amazon River Basin
  biodiversity, 733–736
  coastal wetlands, 733
  conservation status, 737
  deforestation, 219, 739
  ecosystem degradation, 739
  ecosystem services, 738–739
  fauna, 735–736
  flora, 733–735
  future challenges and developments, 739–741
  interfluvial wetlands, 732
  large-river floodplains, 729–731
  location, 728
  pollution, 740
  river damming for hydropower generation, 217, 739–740
  unsustainable use of resources, 740
  wetland types and extent, 204–207, 729–733
Ambystoma, 658
  A. jeffersonianum, 658
  A. laterale, 658
  A. maculatum, 658
  A. opacum, 658
  A. tigrinum, 612
  A. tigrinum tigrinum, 658
  A. tremblayi, 658
Ameiurus melas, 1133
Amietophrynus reesi, 1344
Ammannia, 1117
Ammodramus bairdii, 683
Ammophila arenaria, 1104
Amoria fragifera, 951
Amorpha fruticosa, 915, 918, 1107
Amphibacillus
  A. fermentum, 1388
  A. tropicus, 1388
Amphibians, 100, 101, 113, 149, 165, 176, 191, 192, 214, 236, 323, 352, 377, 584, 611, 612, 656, 657, 660, 662, 682, 692, 735, 736, 750, 805, 817, 824, 828, 830, 831, 852, 899, 918, 1067, 1105, 1107, 1118, 1119, 1142, 1228, 1238, 1315, 1368, 1406, 1469, 1474, 1502, 1513, 1516, 1561, 1584, 1679, 1700, 1737, 1859, 1981
  in Doñana wetlands, 1128, 1131
  egg clutches, 471
  in Kakadu National Park, 1953
  Lake Fúquene biodiversity, 776, 777
  in Lake Seyfe, 1190
  in Mekong River Basin, 1773, 1778
  in papyrus wetlands, 193
  Sanjiang Plain, 1514
  in Tasek Bera, 1859
  in tropical freshwater swamps, 212
  in turloughs, 1071
  in vernal pools, 652–654, 658–659
  in Volga Basin, 939

Index of Keywords

Amphibolis
   A. antarctica, 75
   A. griffithii, 75
Amphiroa fragilissima, 1880
Amplifier lake, 1363
Amur, 280, 1486–1489, 1496, 1497, 1502, 1509–1520
Amur-Heilong River Basin, 13, 1510, 1514
   corridors and regions, 1495
   Daurian steppe wetlands, 1499–1507
   distribution of wetlands, 1489
   dyke construction and reinforcement, 1496
   freshwater ecoregions, 1486
   future challenges and developments, 1496–1497
   Ramsar sites, 1486, 1488
   regions, 1494
   Russian-Chinese Strategy for Development, 1497
   threats, 1496
   transboundary floodplain in China, 1496, 1497
   types, 1488
   wetland-dominated regions, 1487, 1490
   Zeya River headwaters, 1494
Amur River Basin, 1486–1489, 1496, 1509–1520
Amyda cartilaginea, 1845, 1860
Anabaena, 1936
   A. circinalis, 1936
   A. flos-aquae, 1273
Anabaenopsis, 1385, 1387
   A. arnoldii, 1385
   A. magna, 1386
Anacardium excelsum, 765
Anadara
   A. grandis, 752
   A. senilis, 1325
   A. similis, 752
   A. tuberculosa, 752
Anadia bogotensis, 777
Anarhichas
   A. lupus, 633
   A. minor, 633
Anas, 1636, 1981
   A. acuta, 570, 682, 702, 989, 1016, 1481, 1610, 1621, 1640, 1641
   A. americana, 570
   A. clypeata, 166, 472, 583, 1029, 1610, 1641
   A. crecca, 472, 570, 978, 1029, 1083, 1190, 1481, 1610, 1640, 1641, 1643, 1796
   A. cyanoptera, 777

   A. cyanoptera borreroi, 776
   A. discors, 597, 777
   A. falcata, 1444, 1474, 1481, 1525
   A. formosa, 1444, 1474, 1612
   A. georgica niceforoi, 776
   A. gibberfrons, 1997
   A. gracilis, 1912, 1926
   A. laysanensis, 2020
   A. penelope, 978, 1003, 1029, 1083, 1092, 1481, 1610, 1640
   A. platalea, 859
   A. platyrhynchos, 570, 597, 965, 1073, 1095, 1559, 1610, 1621, 1628, 1640, 1641, 1643, 1708
   A. querquedula, 472, 1796
   A. rubripes, 629
   A. strepera, 472, 1031, 1448, 1469, 1610, 1628, 1641
   A. superciliosa, 1985
   A. zonorhyncha, 1559
Anastomus
   A. lamelligerus, 213, 1227, 1238, 1345
   A. oscitans, 1796
Anaxyrus
   A. americanus, 658
   A. fowleri, 658
Anchoa
   A. hepsetus, 763
   A. marinii, 850
   A. parva, 763
   A. trinitatis, 763
Anchovia clupeoides, 763
Andean wetlands, 781, 825, 827–829, 834
Andes, 130, 728, 731, 736, 774, 786, 798, 799, 825, 827, 828, 874, 876, 877, 879, 880
Andromeda, 517
Andropogon
   A. angustatus, 206
   A. canaliculatus, 205
Angelica
   A. lucida, 570
   A. tenuifolia, 1538
Angiosperm, 398, 1584, 1934, 1980, 1992, 2012
   salt tolerant, 149
   submerged, 1357, 1876
Angitis River, 1150, 1153
Anguilla, 1982
   A. anguilla, 118, 980, 989, 1002, 1104, 1107, 1131
   A. reinhardtii, 1946
   A. rostrata, 633

Anguillicola crassus, 1107
Anhinga
   A. melanogaster, 213, 1729, 1787, 1796, 1816, 1946
   A. rufa, 1274, 1354
Anillo de Cenotes, 323
Aniseia martinicensis, 205
Anisolepsis undulatus, 817
Ankistrodesmus, 1385
Annona glabra, 719
Annual precipitation
   Amazon Basin, 728
   in blanket mire regions, 298
   in Boreal wetlands, 530
   in Copper River, 559
   Dinder and Rahad River basins, 1290
   Lake Parishan, 1649, 1651
   Maputaland Coastal Plain, 1424
   Mekong River Basin, 1782
   in Mississippi Alluvial Valley, 579
   in Mongolia, 1532, 1544
   in Patagonian Andes, 876
   pocosin, 671
   prairies, 681
   Rugezi Marsh, 1308
   ShadeganWetland, 1676, 1678
   Yangtze River Basin, 1553
Anogeissus schimperii, 1294
Anomalocardia brasiliana, 844
Anoplolepis gracilipes, 220, 2017
Anous stolidus, 2022
Anser, 1636
   A. albifrons, 547, 552, 966, 1030, 1190, 1480, 1628, 1640
   A. albifrons elgasi, 702
   A. albifrons flavirostris, 1044
   A. anser, 236, 472, 1091, 1443, 1515, 1640, 1653
   A. brachyrhynchus, 980, 1015, 1016, 1029
   A. caerulescens, 552
   A. cygnoides, 1481, 1502, 1525, 1569, 1584
   A. erythropus, 953, 1480, 1569, 1584, 1621, 1629, 1641, 1642
   A. fabalis, 1480
   A. fabalis middendorffii, 1444, 1474
   A. indicus, 1610, 1715
Anseranas semipalmata, 1946, 1955
Anseriformes, 1454
Antelopes, 192, 211, 948, 1225, 1228, 1236, 1237, 1240
   Nile Basin, 1270
   Rugezi Marsh, 1315
   Sudd region, 1302

Anthropogenic activities, 8, 82, 193, 1358
   Gokyo wetlands, 1612
   Kakagon Sloughs, 428, 434
   Kenyan Soda lakes, 1389
   Lake Turkana, 1375
   Magdalena River, 762
   papyrus marsh, 193
   Volga River Basin, 940
   wetlands in Chile, 832
Anthropogenic change, 362
   intertidal zone, 1868
   water level management, 365
Anthropogenic disturbances, 12
   high altitude wetlands, Nepal, 1612
   Lake Superior, 606
   Myall Lake, 1937
   seagrass habitat, 1595
   seagrass meadow, 1973–1974
Anthropogenic effects/impacts, 8–9, 12, 83, 623, 1327, 1414, 1421, 1484, 1906, 1969
Anthus, 1084
   A. pratensis, 965
   A. richardi, 1448, 1469
   A. spragueii, 683
Antilope cervecapra, 1730
Anzali Mordab Complex, 12
   biodiversity, 1619–1621
   climate, 1616
   conservation measures, 1621–1623
   ecosystem services and land use, 1621
   future challenges and developments, 1623–1624
   hydrology, 1616–1618
   location, 1616
   threats, 1623
   wetland ecosystem, 1618–1621
Aonyx, 192
   A. capensis, 1238, 1315, 1337
   A. cinerea, 1777, 1779, 1816, 1825, 1860
Apera spica-venti, 916, 917
Aphanius
   A. baeticus, 1131
   A. chantra, 1190
   A. fasciatus, 1142
   A. iberus, 1116
   A. sirhani, 1161
Aphanizomenon, 478, 479
Apis
   A. ligustica, 809
   A. mellifera, 809
   A. scutellata, 809
Aplocheliches, 1354

Aponogeton natans, 1746
Aquaculture, 38, 49, 53, 62, 66, 89, 96, 102, 105, 107, 116, 119, 135, 136, 153, 177, 326, 370, 398, 454, 476, 477, 780, 1359, 1595, 1600, 1739, 1748, 1768, 1806, 1808, 1827, 1947, 2010
   channel catfish, 587
   Ebro Delta, 1118
   kelp, 83, 88
   for marine plants, 634
   pond, 103, 580, 583–584
   shrimp, 85, 103, 364
Aquatic diversity, Sudd region, 1249, 1305
Aquatic estuarine communities, 1002
Aquatic invertebrates, 191, 342, 581, 584, 641, 656, 659–661, 717, 735, 1903, 1952, 1981
Aquatic macroinvertebrates, 652, 656, 692, 727, 1131
Aquifer, 120, 316, 320, 322–324, 337, 346–348, 351, 463, 643, 690–692, 694, 825, 827, 836, 861, 1070, 1073, 1124, 1129, 1133, 1150, 1389, 1426, 1518, 1650, 1651, 1664, 1717, 1729, 1947, 1949, 1961, 1965
   freshwater aquifer, 364, 1664
   Mississippi River Valley, 588
   Prairie wetlands, 683
Aquila
   A. adalberti, 1130
   A. chrysaetos, 1047, 1621
   A. clanga, 953, 1141, 1142, 1188, 1621, 1629, 1641, 1796, 1816
   A. heliaca, 953, 1188, 1621, 1641
   A. nipalensis, 1255
   A. rapax, 1385
Aquilaria spp., 1862
Arabian Gulf, 1173–1182
Aramides ypecaha, 817
Aratus
   A. penii, 763
   A. pisonii, 763
Araucaria, 828
   A. patagioenas, 830
Arawai Kakariki programme, 1988
Archidendron clypearia, 1757, 1834
Archontophoenix alexandrae, 208
Arctanthemum hultenii, 285
Arctic peatlands, 10–13, 276–277
   challenges, 286–287
   coastal wetlands, 285
   distribution and diversity, 277–279
   drained depressions, 284
   paludified shallow, 279
   patterned string fen, 283–284
   peat plateau and palsa mires, 280–283
   polygon landscapes, 279–280
   riparian mires, 285
   thermokarst kettle hole peatlands, 284
Arctictis binturong, 1836
Arctic wetlands, 284, 391–392
   colonial birds, 392
   contaminant biotransport, 392–393
   distribution and diversity, 278
   nutrient biotransport, 392–393
Arctophila fulva, 546
Arctophylla, 560
Arctosa, 1082
Arctostaphylos uva-ursi, 1045
Ardea, 1274, 1796
   A. alba, 583, 751, 1354, 1525, 1955
   A. cinerea, 192, 905, 1238, 1385, 1525, 1559
   A. cocoi, 751
   A. goliath, 1354
   A. herodias, 583, 612
   A. herodias fannini, 568
   A. ibis, 1902, 1955
   A. intermedia, 1902
   A. modesta, 1902
   A. monicae, 1326
   A. pacifica, 1902
   A. picata, 1955
   A. purpurea, 1525
   A. sumatrana, 1824
Ardeola ralloides, 1131, 1141, 1238, 1338, 1354, 1691
Aremons, 1701
Arenaria
   A. interpres, 547
   A. melanocephala, 547
Arenicola marina, 569, 988, 1002
Argenteohyla siemersi, 817
Argentina, 29, 150, 160, 211, 814, 829, 830
   Paraná-Paraguay Corridor, 785–794
   Patagonian peatland, 873–881
   Rio de la Plata system, 847–854
   Tierra del Fuego, 268, 301, 307
Argusianus argus, 1845
Argyrosomus
   A. hololepidotus, 1912
   A. regius, 1325

Arid, 6, 56, 61, 119, 131, 132, 144, 146, 147, 152, 217, 319, 337, 350, 351, 372, 462, 516, 759, 766, 824, 1156, 1159, 1172, 1174, 1248, 1323, 1353, 1362, 1363, 1468, 1486, 1502, 1534, 1535, 1577, 1581, 1582, 1616, 1699, 1700, 1725–1731, 1900, 1922
Aristichthys nobilis, 477
Aristida
    A. plumosa, 1294
    A. setifolia, 206
Arius proops, 763
Armeria maritima, 279, 964
Artemia, 149, 153, 1104, 1666
    A. salina, 2010
    A. urmiana, 1664
Artemisia, 938
    A. adamsii, 1538
    A. laciniata, 1538
    A. santonicum, 916
    A. sericea, 1538
Artesian spring mire/fen, 254, 256
Arthrocnemum, 1117, 1140
    A. macrostachyum, 1105, 1117, 1130
Arthrospira fusiformis, 1384–1387
Artificial wetlands, 29, 375, 395, 1117–1119, 1556, 1577, 1578, 1583, 1640
Artiodactyla, 939
Artisanal fishery, 74, 767, 852, 1226
    Congo Basin, 1208
    seagrass, 437–443
Artocarpus
    A. altilis, 2009
    A. mariannensis, 2009
Arvicola
    A. amphibius, 1047
    A. sapidus, 1131
    A. terrestris, 1443, 1469, 1474
Ascophyllum, 48
    A. nodosum, 626
Asio flammeus, 570, 1031, 1045, 1084, 1621
Asplenium, 1815
    A. confusum, 1816
Astagobius, 340
Astelia, 878, 879
    A. pumila, 833, 876
Astelia–Donatia cushion bogs, 878, 879
Aster
    A. douglasii, 570
    A. tripolium, 963, 1082
    A. tripolium ssp. *pannonicus*, 482
Astronium graveolens, 206
Atbara River, 1248

Atherina boyeri, 1104, 1131
Atherinops affinis affinis, 642
Atherinosoma microstoma, 1915
Athyrium filix-femina, 560
Atlantoraja castelnaui, 850
Atolls
    biodiversity, 2006
    description, 2002
    distribution, 2004, 2006
    ecosystem conservation, 2019–2022
    ecosystem diversity, 2007–2010
    ecosystem goods and services, 2013, 2014
    genetic diversity, 2013
    global distribution, 2006
    habitat diversity, 2007–2010
    species and taxonomic diversity, 2010–2012
    threats and future challenges, 2013–2018
    tropical pacific ocean, 10, 2001–2023
Atractaspis microlepidota, 1274
Atractus, 777
Atriplex, 61, 963
    A. calotheca, 951
    A. portulacoides, 1082, 1085
    A. tatarica, 956
Aulacomnium, 1545
Aurelia aurita, 153
Australia, 13
    Atoll, 2002
    Burdekin Falls Dam, 128
    Coorong, 1909–1918
    Fraser Island, 247, 260, 262
    Great Barrier Reef, 447–458
    groundwater dependent ecosystems, 347, 350, 352
    Kakadu National Park, 1951–1957
    karst wetlands, 327
    Kati Thanda-Lake Eyre, 1921–1926
    Lake Eyre, 130, 145, 1921–1926
    Macquarie Marsh, 1897–1907
    mangrove, 94, 96, 98, 99
    Murray–Darling Basin, 1887–1895
    Myall Lakes, 1929–1938
    salt lake, 147, 149, 151, 152
    seagrass, 77, 81–83
    subterranean fauna, 336, 337
    tropical floodplain wetlands, 201, 202
Australian wet tropics
    biodiversity, 1945–1946
    climate, 1942
    conservation status, 1947
    ecosystem services and values, 1947
    geographical location, 1942, 1943

hydrology, 1942
landscape, 1942
natural and human-influenced processes, 1944
threats, 1948–1949
wetland ecosystems, 1944–1945
Avellara fistulosa, 1130
Aviceda jerdoni, 1835
Avicennia
  A. africana, 1322, 1323
  A. alba, 1728, 1823
  A. germinans, 750, 764, 766, 1336
  A. marina, 1174, 1180, 1182, 1237, 1701, 1728, 1823
  A. officinalis, 1728
Axios, Aliakmon and Gallikos Delta complex, northern Greece
  conservation status, 1143–1145
  ecosystem services, 1145–1146
  fauna, 1141–1143
  history, 1139–1140
  hydrology, 1139–1140
  location, 1138, 1139
  threats and challenges, 1146
  vegetation types, 1140–1141
Axis
  A. axis, 220, 818
  A. porcinus, 1737, 1779
Axonopus, 205
  A. aureus, 206
Aythya, 642
  A. affinis, 583, 777
  A. americana, 597
  A. baeri, 1525, 1584, 1612, 1776
  A. ferina, 1610, 1621, 1641, 1642
  A. fuligula, 472, 1094, 1610, 1641
  A. nyroca, 1141, 1628, 1641, 1653, 1665, 1708
  A. valisineria, 643
Azima tetracantha, 206
Aznalcóllar mine spill disaster, 1129
Azolla, 191, 1134, 1623
  A. caroliniana, 914
  A. filiculoides, 1133, 1140, 1237
  A. nilotica, 1272, 1301, 1353
  A. pinnata, 1353, 1746
Azraq Oasis, 9, 12, 364, 462
  Highland Water Forum, 466
  location, 462
  management, 463–466
  Water Resource Management Plan, 466
Azraq wetland, 346, 462, 1163

B

Ba Be National Park, 325
Babolsar, 1636, 1638
Baccaurea, 1823, 1834
Baccharis halimifolia, 1107, 1109
Bacopa caroliniana, 717
Badanloch Bogs, eastern Sutherland, 1045
Bahía Lomas, Chile, 834, 835
  future challenges and developments, 870
  location, 866
  management plan, 869–870
  shorebirds, 867–868
  studies and monitoring programs, 868–869
  threats, 870
Bahr el Ghazal, 12, 185, 1246, 1247, 1300
  biodiversity, 1271–1274
  conservation status, 1274
  ecosystem services, 1274
  future challenges and developments, 1275–1276
  grazing range, 1275
  hydrology, 1270
  oil exploration and production operations, 1276
  produced water, 1275
  waste management practices, 1276
  wetland ecosystems, 1270
Baikal lake, *see* Lake(s), Baikal
Balaeniceps rex, 192, 1265, 1274, 1302, 1304
Balaenoptera
  B. acutorostrata, 569, 629, 851
  B. borealis, 629
  B. musculus, 851
  B. physalis, 629
  B. physalus, 633, 851
Balanites aegyptiaca, 206, 1262, 1272, 1293, 1294, 1302, 1385
Balantiocheilos melanopterus, 1860
Balanus improvisus, 850, 918
Baldellia ranunculoides, 1130
Balearica
  B. pavonina, 213, 1274, 1304
  B. regulorum, 192, 1222, 1227, 1238, 1315
Balkans peatlands, 1156
Balneotherapeutic treatment, 1153
Banc d'Arguin, 12
  coastal and marine biodiversity, 1324–1328
  coastal and marine ecosystems, 1323
  coastal shoals, 1320
  conservation status, 1328
  ecosystem services, 1328–1329
  future challenges and developments, 1329–1330

Banc d'Arguin (*cont.*)
  hydrology, 1322–1323
  Mauritanian coastal zone, 1321
  offshore oil prospecting, 1322
  threats and future challenges, 1329–1330
Bandar Kiashahr Lagoon, 1626–1628, 1631
Bangladesh Sundarbans, 1736, 1737, 1739
Bangladesh Water Act, 1719
Bangweulu grassy swamps, 1201, 1203
Banni Grasslands Reserve, 1730
Baradla-Domica Cave System, 321
Bar-built estuaries, 39, 998, 1000
Barbus
  B. amphigramma, 1358
  B. brachycephalus caspicus, 953
  B. intermedius australis, 1354, 1355
  B. lineomaculatus, 1354
  B. neumayeri, 192
  B. sharpeyi, 1690
Baro-Akobo Basin
  extent and character, 1262–1263
  location, 1262
  lowland wetlands, 1265–1266
  threats, 1266
  uses and beneficiaries, 1263
  wetlands challenges, 1266–1267
Baro-Akobo River, 1247, 1248
Baro Akobo Sobat sub-basin (BAS), 1280–1282, 1284
Baro-Akobo system, 1262
Barotse floodplain, 1220, 1222, 1229
Barren land, 1701
Barrier Island, 56, 559, 595, 984, 986, 993, 1000, 1012, 1013
Barrier-protected, 593–595, 608, 614
Barringtonia, 200, 208, 1774
  B. acutangula, 205–208, 1773
  B. racemosa, 1237, 1433, 1834
Basin
  Amazon, 130, 201, 214, 221, 727–741, 792, 817
  Amur-Heilong River, 13, 1485–1497, 1499–1507, 1510–1520
  Anzali River, 1619
  Arguin, 1322
  Baro-Akobo River, 12, 1247, 1261–1267
  Baro-Akobo-Sobat, 1280, 1282, 1283
  catchment, 779, 1922
  Congo River, 12, 1199–1212, 1214
  Danube River, 11, 885–895, 898, 900, 908
  Dinder and Rahad, 12, 1287–1298
  Drama, 1150–1152
  fen, 229, 518
  Ganga-Brahmaputra, 13, 1711–1721
  Indus River, 13, 1697–1702
  infilled, 263–264
  Kalahari sand, 1395, 1399
  Lake Turkana, 1361–1378
  Lake Uromiyeh, 1659–1673
  Lena River, 13, 1439–1449, 1457–1460, 1463–1470
  local, 552
  Mekong River, 1763–1783, 1785–1791, 1813–1817
  Middle Aldan River, 13, 1471–1476
  Murray-Darling River, 1887–1895, 1897–1907, 1909–1918
  Nile River, 12, 184, 1243–1249, 1269–1276, 1280–1286
  Okavango drainage, 1396, 1397
  perched, 551, 552
  plan, 1892, 1893, 1906
  raised bog, 265–266
  Rhine River, 887, 923–930
  steep-sided, 252, 266, 310
  Upper Paraguay, 798, 799
  Volga River, 887, 933–942
  wetlands, 1219–1221
  Yangtze River, 1551–1562, 1565–1572
  Yenisei River, 13, 1477–1484
  Yueguzonglie, 1576–1587
  Zambezi River, 12, 1217–1230, 1240
Batagur
  B. baska, 1718, 1737
  B. borneoensis, 1825, 1835
Batis, 61
  B. maritima, 58, 766
Batrachium
  B. baudotii, 482, 977
  B. divaricatum, 1538
Batrachochytrium dendrobatidis, 1946
Batrachostomus javensis, 1825
Batrachyla antartandica, 831
Bauhinia cunninghamii, 208
Bay of Fundy, 10, 39, 41, 45, 47–50, 999
  Avon River estuary, 52
  biodiversity, 625–629
  conservation status, 632–633
  future challenges and developments, 634–635
  marshes and tidal flats, 629–632
  origin, 622
  physical characteristics, 622–625
  regions and tidal ranges, 623
  threats, 634

Beach rock reefs, 840, 843
Becheropsis uniseta, 1294
Beckmannia syzigachne, 1538
Bed of the Vixen Goddess, 777
Beetles, 149, 165, 336–340, 342, 656,
　　660–662, 1031, 1047, 1071, 1073,
　　1082, 1085, 1690, 1981
Bekambeushi marsh, 1600
Belgrandiella, 340
Bellis annua, 1105
Bengal tiger, 1719, 1737, 1738
Benthos, 44, 46, 78, 159, 988, 1092, 1324,
　　1460, 1466
　Bahr el Ghazal, 1273
　Banc d'Arguin, 1324–1325
　Danish waters, 1092
　macrofauna, 1324
　Tipperne Reserve, 977–978
　Wadden Sea, 988
Berbak National Park, 1820–1823,
　　1825, 1826, 1828
　biodiversity, 1835–1836
　conservation status, 1836
　ecosystem services, 1836
　future challenges and developments,
　　1836–1837
　hydrology, 1832
　location, 1833
　potential acid sulphate soils,
　　1837, 1838
　threats, 1836–1837
　wetland ecosystems, 1834
Berenicornis comatus, 1825
Berga mire, 1420
Bergia, 1117
Berlinia grandiflora, 207
Berosus signaticollis, 1071
Berula erecta, 1171
Best management practices (BMP), 457, 538,
　　662, 712
Betta
　B. enisae, 1845
　B. pinguis, 1845
　B. splendens, 1775
Betula, 518, 524, 1043
　B. fusca, 1541
　B. nana, 280, 1045
　B. papyrifera, 545
　B. platyphylla, 1537
　B. rotundata, 1538
Bidens, 163
　B. laevis, 159
Bidyanus bidyanus, 1903, 1912

Bijagós Archipelago, 12, 1334
　conservation status, 1338
　ecosystem services, 1338
　future challenges and developments,
　　1339–1340
　hydrology, 1335–1336
　threats, 1339–1340
　wetland ecosystem, 1336–1337
Biodiversity
　Amazon River Basin, 733–736
　amphibians, 1190
　Anzali Wetland, 1619–1621
　atoll encompasses, 2006
　Australian wet tropics, 1945–1946
　Bandar Kiashahr Lagoon, 1628
　BAS, 1281–1286
　Bay of Fundy, 625–629
　benthos, 977–978, 1273
　Berbak National Park, 1835–1836
　Bijagos Archipelago, 1337
　birds, 978–980, 1190–1192
　breeding birds, 989
　British estuaries, 1001–1004
　Camargue, Rhône River Delta, 1106–1107
　coastal and marine, 1324–1328
　Colombian mangrove wetlands, 750–751
　Conchalí Lagoon, 859–861
　Congo River Basin, 1205–1207
　conservation, 455, 998, 1034, 1118, 1119,
　　1237, 1240, 1294, 1296, 1329, 1557,
　　1584, 1894, 2018, 2019
　cool spots, 2002, 2011, 2019, 2023
　Copper River Delta, 561–562
　corridor, 792
　Danube Delta, 917–918
　Danube River Basin, 889–890
　Daurian Steppe Wetlands, 1502
　Doñana wetlands, 1130–1132
　dry phase, 1072–1073
　DSNP, 1844–1845
　Ebro Delta, 1115–1117
　estuaries, 44–48
　Europe's coastal saltmarshes, 961–966
　fauna, 1140–1143, 1609–1610
　Fereydoon Kenar, 1640–1642
　fish, 817, 977–978, 988–989, 1190, 1273
　flooded phase, 1071
　floodplain, 1344–1345
　flora, 1140, 1189–1190, 1609
　Fluvial Corridor, 792
　habitats, 1027–1030, 1189
　Hula Wetland, 1171–1172
　Indus basin's wetland, 1700–1701

Biodiversity (*cont.*)
  insects, 1272
  intertidal habitats, 1870–1871
  Kakadu National Park, 1952–1955
  Lake Baringo, 1353–1355
  Lake Fúquene, 776–777
  Lake Naivasha, 1357–1358
  Lake Seyfe, 1189–1192
  Lake Turkana, 1371–1372
  Lena River Basin, 1443–1444, 1454
  Lower Danube's floodplains, 898–899
  Mahanadi Delta, 1747
  mammals, 1192
  marine, 440
  Marshes, 1901–1903
  MAV, 584–585
  mayas wetlands, 1293–1294
  Mekong River, 1805–1806
  Mekong River Basin, 1773–1779
  Mesopotamian marshes, 1688–1691
  middle Aldan Basin, 1474–1475
  migratory birds, 989–992
  Mississippi Alluvial Valley, 584–585
  Nidjili Lake, 1459–1460
  Nile Delta, 1254–1255
  Nile River Basin, 1244, 1249
  Okavango Delta, 1406–1408
  Pantanal, 804–806
  in papyrus marshes, 190–192
  patterns, 849–852
  Peace-Athabasca Delta, 552
  peatlands of Mediterranean, 1159–1162
  plankton, 1273
  plants and benthos, 988
  playa wetlands, 692
  pocosins, 674–675
  Poyang Lake, 1568–1569
  prairie pothole, 682–683
  reptiles, 1190, 1337
  river basins, 112–115
  Rugezi Marsh, 1313–1315
  saline wetlands of Kachchh, 1728–1729
  San Francisco Bay Estuary, 641–643
  Sembilang National Park, 1824–1826
  species, 1030–1032
  submerged vegetation, 977
  Sudd, 1302–1303
  Sundarban mangroves harbor, 1736–1737
  swamps, 1271–1272
  Taiga, 1443, 1468–1469
  threats of, 1209–1211
  Tonle Sap Lake, 1787–1788
  Tram Chim National Park, 1796
  tropical freshwater swamps, 210–213
  U Minh Peat Swamp Forest, 1816
  vegetation, 750, 1140
  Volga River Basin, 939–940, 949–953
  Wash and North Norfolk Coast, 1015–1017
  of water birds, 471–473
  wetland ecosystems, 1140
  wildlife, 817, 1273–1274
  Wular Lake, 1708
  Yangtze River Basin, 1558–1560
  Yellow River Basin, 1584
  Yenisei River Basin, 1480
  Yukon and Kuskokwim, 547
  Zambezi Delta, 1237–1238
  Zhalong Wetlands, 1525–1526
Biofuels, 363, 598, 1174, 1181
Biological invasions, 139, 1133
Biological productivity, 131, 624, 869, 1400, 1403, 1410, 1500, 1503, 1925
Biomagnification, 392, 393
Biosphere Reserve, 317, 480, 494, 907, 954, 967, 1295, 1296, 1334, 1335, 1338
Biosphere Reserve concept, 907, 1295, 1296, 1334
Biota, 6, 31, 50, 52, 217, 286, 341, 360, 392, 395, 476, 482, 502, 532, 613, 653, 655, 656, 663, 739, 804, 810, 853, 1244, 1294, 1382, 1386, 1388, 1390, 1404, 1405, 1502, 1504, 1693, 1726, 1737, 1740, 1890, 1891, 1894, 1942, 1948, 1960, 1965, 1978, 1984, 2011
  in freshwater lakes, 133–135
  in freshwater swamps and marshes, 173–177, 180
  mangrove, 100–101
  peatlands, 235–236
  salt lakes, 148–151, 153
  seagrass, 85, 90
Biotransport, arctic wetlands
  of contaminants and effects, 393–395
  of nutrients and effects, 392–393, 395
Bird(s)
  bird area, 325, 480, 482, 490, 572, 595, 599, 645, 994, 1118, 1186, 1354, 1371, 1386, 1502, 1631, 1643, 1648, 1669, 1676, 1700, 1787, 1807, 1816, 1824
  breeding, 926, 969, 981, 989, 994, 1031, 1032, 1040, 1049, 1134, 1326, 1525, 1621, 1652, 1653, 1679, 1681, 1691, 1980
  fauna, 1386, 1502, 1824
  in Golfe d'Arguin, 1326–1328
  marine biodiversity, 1326–1328

Mekong Delta, 1806
Ringkøbing Fjord, 978–980
seabird colonies, 391–395
shorebirds, 44, 45, 47, 50, 51, 64, 133, 150, 377–380, 407–410, 533, 544, 547, 552, 561, 562, 568–571, 583–585, 630, 631, 633, 641, 642, 645, 647, 692, 693, 805, 867–868, 870, 965, 966, 970, 980, 1002, 1320, 1324, 1328, 1480, 1481, 1561, 1569, 1628, 1643, 1867, 1870, 1925, 1955
wading birds, 64, 86, 717, 718, 805, 1388, 1796, 1916, 1917, 1985
waterbirds (*see* Waterbirds)
Bird Island Nature Reserve, 1579
Birdlife International, 1444, 1631, 1640, 1643, 1787, 1867, 1872, 2020, 2022
Birgus latro, 2016
Bison bison athabascae, 552
Bistorta alopecuroides, 1538
Bitis arietans, 1274
Black Buck, 1730
Blackwater, 731, 735, 1203, 1207, 1755–1758, 1828, 1832, 1834, 1835, 1837, 1844, 1848, 1849
Black-water rivers, 207, 729, 731, 732, 1822, 1825
Blanket bog(s), 21, 230, 267, 269–271, 1043, 1045, 1054, 1159, 1446, 1544, 1545, 1997, 1998
  active, 290
  definition, 297, 304
  distinctive features, 304
  Isle of Lewis, Scotland, 304
  mesotopes, 305–308
  ombrotrophic, 304, 305, 867
  open-water pool complex, 307
  surface-water seepage, characteristic patterns, 306
  watershed plateau, 307
Blanket mires, 258, 264, 268–270, 296–297, 306, 415, 421
  alpine and sub-alpine zones, 297
  of Caithness and Sutherland, 1039–1054
  landscapes, 296–301, 304, 305, 417–420, 422, 423
  patterned "ladder" fen, 261
Blastocerus dichotomus, 817
Blechnum indicum, 205
Blepharocalyx, 829
  B. cruckshanksii, 825
  B. salicifolius, 816
Blue carbon, 74, 88, 502, 754, 1869

Blue crab, *see Callinectes sapidus*
Blue-green algae, 86, 483, 629, 712, 1273, 1353, 1371
Blue Nile, 186, 1244, 1247–1248, 1270
Blue Nile Basin, 1287–1298
Bog butter, 247
Bogs, 21, 230, 290, 291, 300, 310, 419, 516–519, 573, 827, 828, 880, 1044–1047, 1156, 1544, 1979
  Astelia–Donatia cushion, 876, 878, 879
  boreal wetlands, 523, 524
  Burns, 568, 571, 573
  Empodisma minus, 1994
  Empodisma robustum, 1994
  lake, 1044, 1446, 1466, 1467
  ombrotrophic mires, 264, 1159, 1978
  pocosin, 667–676
  restiad, 1991–1999
  Sphagnum, 247, 291, 877, 879, 1598, 1992
  Sporadanthus ferrugineus, 1994–1995, 1998
  Sporadanthus traversii, 1995–1996
  treed domed, 229
  valley bog, 254–256 (*see also* Blanket bog; Raised bog)
Boiga
  B. dendrophila, 1825, 1835
  B. irregularis, 2017
Bolama Bijagós Archipelago Biosphere Reserve, 1335, 1338
Bolax caespitosa, 876
Bolboschoenus maritimus, 951, 1105, 1130, 1140
Bolitoglossa yucatana, 323
Boloria spp., 612
Bombina bombina, 918
Boraras micros, 1776
Borassus aethiopum, 1236
Borassus palm savanna, 1236
Boreal wetlands, 10, 12, 231, 264, 277, 283, 519
  basin fens, 518
  beavers, 527, 531
  best management practices, 538
  bogs, 523
  climate change, 530, 536
  damming and water redistribution, 534
  direct loss, 534
  diversity, 527–531
  ecosystem services, 531–533
  of Eurasia, 239
  federal, provincial and territorial policies, 539

Boreal wetlands (*cont.*)
  fens, 523
  fragmentation, 534
  glaciers, 529
  humans, 531
  location, 522, 523
  major and minor classes, 527
  marshes, 525
  mineral wetlands, 529
  North America, 516, 522
  Permafrost, 530, 536
  protected area, 537
  resource development, 539
  restoration/reclamation, 538
  shallow open waters, 525
  soil compaction, 534
  surficial geology, 529
  swamps, 525
  topography, 529
  water contamination, 535
  water removal, 535
  water yield, 536
Borneo, 113, 201
  Danau Sentarum, 201, 1846
  DSNP's birdlife, 1845
  karst wetlands, 326
  potential acid sulfate soil, 1838
  seagrass, 1876
  Shorea balangeran swamp forests, 203, 219
  tropical peatswamp forest, 374, 1754, 1757, 1758
Bos
  B. gaurus, 1737
  B. javanicus, 1860
  B. taurus, 1955
Boscia senegalensis, 1272
Boswellia papyrifera, 1272
Botaurus
  B. lentiginosus, 612, 683
  B. poiciloptilus, 327, 1902, 1981, 1997
  B. stellaris, 1030
Bothriocephalus acheilognathi, 603
Botia macracanthus, 1845
Botriochloa ischaemum, 916
Botryococcus braunii, 1371
Bottomland hardwood forests, 177
  Mississippi Alluvial Valley, 580–583
  restoration in Grand Prairie region, 586
Boundary bay, 567–572
Bowdleria punctata vealeae, 1997
Brachiaria mutica, 1948
Brachiodontes darwinianus, 850

Brachionus, 1357
  B. baylyi, 1936
  B. plicatilis, 149, 1385
Brachydesmus spp., 339
Brachymystax lenok, 1448, 1502
Brachypodium sylvaticum, 1106
Brachyramphus marmoratus, 1444
Brackish marsh(es), 30, 56–59, 61, 167, 278, 323, 546, 851, 1019, 1116, 1117, 1486, 1627
Bradypterus
  B. carpalis, 1315
  B. graueri, 1314
Branta
  B. bernicla, 570
  B. bernicla bernicla, 966, 980, 990, 1003, 1015, 1016, 1083
  B. bernicla hrota, 990, 1003
  B. bernicla nigricans, 547
  B. canadensis, 552, 612
  B. canadensis leucopareia, 702
  B. canadensis occidentalis, 562
  B. hutchinsii, 547
  B. leucopsis, 966, 980, 990, 1003
  B. ruficollis, 889, 899, 918, 953, 1480, 1641, 1642
Brasenia schreberi, 612
Brazil
  aerial view of floodplain swamp and marsh vegetation, 172
  Amazon Basin, 732, 737, 740, 741
  Amazon State of, 739
  freshwater lakes and reservoirs, 131
  mangroves, 96, 98
  Pantanal, 23, 26, 798, 799, 801, 802, 810
  Paraná-Paraguay Corridor, 786
  peatlands, 232
  plans, 740
  seagrasses, 12, 839–846
  tidal marsh, 58
  tropical floodplain wetlands, 201, 202, 210, 213
Breeding birds, 926, 969, 981, 989, 994, 1031, 1032, 1040, 1049, 1134, 1326, 1525, 1621, 1652, 1653, 1679, 1681, 1691, 1980
Brevoortia, 46
  B. aurea, 850, 852
Breynia rhamnoides, 208
Bridelia cambodiana, 205
Britain, 267, 304, 418, 966, 998, 1000, 1001, 1003, 1004, 1019, 1027, 1047, 1049
See also Great Britain

British blanket mires, 299
British Estuaries, 371, 997–1008
Broadland rivers catchment area, 1026
Bromopsis inermis, 1538
Bromus, 1106
   B. catharticus, 816
Brosme brosme, 633
Brownlowia
   B. paludosa, 205
   B. tersa, 1737
Brown mosses, 518, 523, 524
Bruguiera
   B. cylindrica, 1823
   B. gymnorhiza, 1237, 1729, 1823
Brycinus
   B. ferox, 1371
   B. minutus, 1371
Bryophyte species, 236, 1313
Bryum, 285
Bubalus
   B. arnee, 1715
   B. bubalis, 809, 1737, 1952
Bubulcus ibis, 213, 1107
Bucephala
   B. clangula, 472, 1083, 1092, 1448, 1460, 1475, 1610
   B. islandica, 633
Buffer zone management, 1797
Bufo, 939
   B. bufo, 918, 1047
   B. calamita, 1017, 1128
   B. chappuisi, 1372
   B. marinus, 220, 1953
   B. turkanae, 1372
   B. viridis, 918
Bugeranus carunculatus, 1222, 1225, 1407
Bugs, water, 656, 660–662
Bujagh National Park, 12
   biodiversity, 1628–1630
   conservation status, 1631
   ecosystem services, 1630–1631
   future challenges and developments, 1632
   hydrology, 1627
   location, 1626
   threats, 1632
   wetland ecosystems, 1627–1628
Burhinus oedicnemus, 1665
Burns bog, 568, 571–573
Bursera simaruba, 719
Busanga wetland, 1220, 1223, 1224, 1228
Buteo
   B. buteo, 418, 1255
   B. lagopus, 570

Butomus umbellatus, 603, 915, 952
Butorides striatus, 1338
Bythotrephes cederstroemi, 603

C

Cabomba furcata, 1859
Cacajao spp., 736
Cadaba farinosa, 206
Caddisflies, 612, 657, 660
Caenis spp., 612
CAFF approach, 278
Caiman crocodilus, 735
Cairina scutulata, 1774, 1835
Caithness, 11, 297, 298
   animal life, 1047
   climate, 1041–1043
   drought-sensitive pool, 1046
   landforms, 1043
   land-use pattern, 1047–1048
   locations and features, 1041
   mire development, 1043
   Pool, ridge, and hummock system, 1045
   range of climatic conditions, 1043
   scale and significance, 1049
Caithness and Sutherland, northern Scotland, 11, 297, 1039–1054
Calamagrostis, 545, 552
   C. canadensis, 560
   C. deschampsioides, 285, 546
   C. epigeios, 951
   C. macilenta, 1539
   C. neglecta, 285
   C. purpurea, 1537
Calamus
   C. leptospadix, 207
   C. tenuis, 207
Calandrella brachydactyla, 1141
Calcarius
   C. lapponicus, 1084
   C. ornatus, 683
Calcium, 133, 148, 162, 290, 517, 654, 690, 712, 1070, 1402, 1848, 1853, 1860, 1964
Calidris
   C. acuminata, 410, 1870, 1912
   C. alba, 410, 1015, 1016
   C. alpina, 410, 547, 562, 570, 630, 642, 980, 989, 1003, 1016, 1030, 1044, 1073
   C. alpina schinzii, 965, 980
   C. canutus, 630, 990, 1003, 1015, 1328, 1337, 1870, 1981

Calidris (cont.)
  C. canutus islandica, 1003, 1016, 1017
  C. canutus rufa, 867–869
  C. ferruginea, 1337, 1481, 1912
  C. fuscicollis, 867
  C. mauri, 547, 562, 569, 642
  C. minuta, 1337
  C. minutilla, 630
  C. pusilla, 45, 630
  C. ruficollis, 1870, 1912
  C. subminuta, 1444, 1870
  C. temminckii, 1448, 1469, 1481
  C. tenuirostris, 410
California, 10, 51, 64, 81, 150, 152, 638, 641–643, 652, 671, 683, 702
California's Central Valley, 10, 638
  future challenges and developments, 702
  historic and current extent of wetlands, 698–700
  hydrologic modifier, 700
  hydrology, 698, 700
  location, 698
  waterfowl use, 702
  wetlands, 700
Callianassa sp., 569
Calliergon, 518, 1545
  C. giganteum, 1539
  C. stramineum, 284
Calliergonella cuspidata, 1539
Callinectes, 763, 1336
  C. sapidus, 844
Callitriche, 510, 561, 1105
  C. obtusangula, 1130
Calluna vulgaris, 255, 516, 1044, 1048
Calocephalus platycephalus, 205
Calophyllum, 208, 228, 290
  C. brasiliense, 207
Calophysus macropterus, 741
Caltha
  C. dioneifolia, 833
  C. membranacea, 1524
Calycophyllum spruceanum, 207
Calystegia sepium, 159, 916, 952
Camargue, Rhone River Delta, 1130
  biodiversity, 1106–1107
  coastal dunes, 1104
  conservation status, 1108
  ecosystem services, 1108
  freshwater lagoons, 1106
  future challenges and developments, 1109–1110
  grasslands, 1105
  human populations, 1102
  hydrology, 1102
  location, 1102
  Mediterranean coastal lagoons, 1104
  riparian forests, 1106
  salt marshes, 1105
  temporarily flooded marshes, 1105
  threats, 1109–1110
  wetland ecosystems, 1104–1106
Campanian Province, 1152
Campephilus principalis, 584
Camphorosma
  C. annua, 916
  C. monspeliaca, 1189
Campnosperma, 207
  C. brevipetiolata, 208
  C. coriaceum, 1848
  C. panamensis, 750
Campylopus
  C. acuminatus var. *kirkii*, 1995
  C. atrovirens, 1046
Canadian Boreal Forest Agreement (CBFA), 539
Canadian boreal zone, 519, 522, 527, 529–532, 537, 539
Canary current upwelling (CCU), 1320, 1322, 1323
Canis, 1274
  C. aureus, 1629, 1642, 1680
  C. lupis occidentalis, 552
  C. lupus, 561, 1107, 1142, 1515, 1621
  C. lupus chanco, 1609
  C. mesomelas, 1294, 1372
  C. rufus, 675
Capparis
  C. decidua, 1730
  C. erythrocarpos, 206
  C. micrantha, 205
  C. sepiaria, 206
Capreolus capreolus, 1047, 1515
Caquetaia krausii, 763
Caracal caracal, 1274
Caraipa, 207
Carallia
  C. brachiata, 207, 208
  C. bracteata, 205
Caranx hippos, 1336
Carassius, 1690
  C. auratus, 477, 479, 603, 776
  C. carassius, 1104, 1448, 1459, 1469
Carbon, 6, 82, 99, 100, 126, 134, 233, 236, 241, 242, 249–250, 304, 1311, 1313, 1759, 1760, 1963, 1969, 1989, 1996
  accumulation, 714, 752

balance, 304, 414, 423
emissions, 51, 232, 414, 1421
inorganic, 478
organic, 476, 684
sequestration, 7, 49, 63, 213, 892, 900, 908,
    1162, 1294, 1408, 1759, 1760
sinks, 88, 210, 237, 516, 519
stocks, 738, 741, 1823, 1824, 1847
storage, 62, 74, 88, 102, 167, 189, 190, 213,
    230, 232, 237, 249–250, 272, 293,
    300, 516, 519, 532, 534, 536, 714,
    752, 880, 1053, 1316, 1823–1824,
    1847, 1869, 1997
targets and environmental impacts, 413–423
Carbonate, 81, 133, 148, 314, 315, 318, 320,
    324, 690, 1070, 1323, 1960
Cardamine hirsuta, 1105
Cardisoma guanhumi, 763
Carduelis
    C. cannabina, 1083
    C. flammea, 1029
Caretta caretta, 324, 453, 584
Carettochelys insculpta, 211, 1955
Carex, 26, 61, 234, 510, 517, 524, 552, 612,
    828, 915, 1072, 1152, 1159, 1556,
    1980
    C. acuta, 951
    C. acutiformis, 915, 952, 1029
    C. aquatilis, 284, 518, 545, 546
    C. atherodes, 552, 1538
    C. atrofusca, 1539, 1545
    C. brunnescens, 1538, 1543
    C. cespitosa, 1537, 1538
    C. cinerea, 1538
    C. colchicum, 916
    C. coriophora, 1538
    C. curaica, 1538
    C. delicata, 1538
    C. duriuscula, 1525, 1538
    C. elata, 915
    C. enervis, 1538, 1539
    C. flacca, 1072
    C. glareosa, 285
    C. globularis, 1537
    C. hostiana, 1072
    C. juncella, 1540
    C. kirganica, 1524
    C. lasiocarpa, 1537
    C. limosa, 1046, 1537, 1540
    C. lyngbyei, 560, 561, 569
    C. meyeriana, 1524
    C. microglochin, 511
    C. panicea, 1072
    C. paniculata, 1029
    C. pauciflora, 1045
    C. pseudocuraica, 1524
    C. ramenskii, 546
    C. rariflora, 284, 511, 546
    C. reptabunda, 1538
    C. riparia, 915
    C. rostrata, 1537, 1540
    C. rotundata, 284
    C. saxatilis, 511, 1543, 1545
    C. schmidtii, 1537, 1543
    C. secta, 1979
    C. sitchensis, 560
    C. stans, 511
    C. subspathacea, 285, 510, 546
    C. utriculata, 546
Carica papaya, 2009
Caridina nilotica, 1273, 1303
Carix
    C. aquatilis, 518
    C. lasiocarpa, 518
Carolina bays, 668, 669
Caropsis verticillato-inundata, 1130
Carpha alpina, 876
Carpinus betulus, 1639
Carr, 21, 290, 480, 1027, 1029, 1072
Carya spp., 581, 1272
Caspian sea, 110, 118, 129, 145–149, 151,
    152, 934, 936–939, 946–949, 952,
    954–957, 1616, 1618, 1619,
    1627–1630, 1632, 1638
Cassia, 206
Cassis cornuta, 442
Castilla elastica, 750
Castor
    C. canadensis, 166, 527, 554, 561,
        612, 831
    C. fiber, 482, 889, 905,
        926, 1443
Casuarina cunninghamiana, 207, 208
Catharus minimus, 547
Cathorops spixii, 763
Catlocarpio siamensis, 1778
Catopuma
    C. badia, 1845
    C. temminckii, 1825
Catotelm, 234
Catreus wallichii, 1610
Cattails, 432, 525, 612, 777, 1117
Cattle ranching, 214, 738, 739, 793,
    802, 807, 808
Caulerpa, 86, 1323, 1336
Cavanilesia okantifolia, 765

Caves, 20, 110, 253, 314, 316, 317, 319–327, 332–343, 347, 1058, 1059, 1061–1063, 1960
Cavia pamparum, 852
Cebus albifrons, 751
Cedrela odorata, 741
Cedrus deodara, 1706
Ceiba pentandra, 741, 750
Celtis
    C. laevigata, 581
    C. wrightii, 206
Cenotes, 322–324
Centaurea
    C. arenaria, 916, 917
    C. pergamacea, 1189
Central Anatolia, 1188
Central Highlands, 1718
Central valley, *see* California's Central Valley
Central Volga Basin, 938
Central Yakutian lowland, 1442, 1446, 1458, 1463–1470
Centrolene buckleyi, 777
Centropomus
    C. latus, 763
    C. undecimalis, 763
Cephalanthus occidentalis, 164, 581
Cerambyx cerdo, 1142
Cerastoderma edule, 988, 1002
Ceratophyllum, 1301, 1639
    C. demersum, 612, 914, 952, 1272, 1353, 1354, 1368, 1689, 1746
Ceratotherium
    C. simum, 1408
    C. simum cottoni, 1386
Cerbera odollam, 1823, 1834
Cercopithecus pygerythrus, 1385
Cereopsis novaehollandiae, 1912
Ceriops
    C. decandra, 1720, 1737, 1738
    C. tagal, 1237, 1729
Ceritium atratum, 843
Cervus
    C. axis, 1737
    C. elaphus, 1047, 1107, 1515
    C. nippon, 1601
    C. timorensis, 220
    C. unicolor, 1826
Ceryle rudis, 192, 1345, 1779
Cetengraulis edentulous, 763
Cetraria, 279
Ceyx pictus, 1345
Chaetomorpha, 842
    C. linum, 1104

Chaimarrornis leucocephalus, 1610
Chamaecyparis thyoides, 177, 669
Chamaedaphne calyculata, 1601
Channa micropeltes, 1846
Chanos chanos, 2010
Chara, 952, 1105
    C. fibrosa, 1931, 1934
    C. intermedia, 1027
Charadriiformes, 1454, 1468
Charadrius
    C. alexandrinus, 1084, 1141, 1142
    C. alexandrinus dealbatus, 1835
    C. dubius, 1629
    C. falklandicus, 867
    C. hiaticula, 965, 1016, 1337
    C. melodus, 584, 633
    C. melodus melodus, 633
    C. modestus, 833
    C. mongolus, 410
    C. pallidus, 1388
    C. peronii, 1796
    C. ruficapillus, 1915
    C. semipalmatus, 630
    C. veredus, 1825
Chari-Dhand Wetland Conservation Reserve, 1730
Charonia tritonis, 2018
Charophytes, 1106, 1931, 1934–1937, 1960
Chasmagnathus granulata, 851
Chauna chavaria, 764
Cheilinus undulatus, 2018
Chelodina rugosa, 1955
Chelonia
    C. midas, 843
    C. mydas, 86, 324, 448, 1325, 1337, 1718, 1737, 1969
Chelon labrosus, 965
Chelus fimbriatus, 735
Chelydra serpentina, 612
Chemistry, water, 119, 230, 256, 360, 608, 614, 652, 654, 657, 660, 683, 739, 1129, 1248, 1410, 1612, 1961, 1963
Chen
    C. caerulescens, 547, 562, 1444
    C. caerulescens caerulescens, 569–570
    C. canagica, 547
Chenopodium rubrum, 956
Cherax, 1945
    C. quadricarinatus, 1949
    C. quinquecarinatus, 1961
Chile, 29, 268, 385, 874, 879
    Andean wetlands, 827–829

coastal forest, 829–830
coastal wetlands, 829–830
Conchalí Lagoon, 857–864
disturbances and threats, 836
locations, 10
national inventory and environmental track, 832–836
peatlands, 830–832, 873–881
Ramsar sites in, 834–835, 866–870
wetland types, 825–829
Chilia (Kilia) branch, Danube River, 912
Chilika Development Authority (CDA), 400
China's water tower, 1580
Chiridotea coeca, 630
Chironomids, 165, 1460
Chitala lopis, 1846
Chitra chitra, 1860
Chlidonias
    C. hybridus, 1619, 1708
    C. leucopterus, 1274
    C. niger, 683, 1338
Chloris
    C. gayana, 1386
    C. virgata, 1385
Chloropeta gracilostris, 192
Cholephaga picta, 831
Chondrus, 86
    C. crispus, 626
Chroicocephalus
    C. cirrocephalus, 1338
    C. genei, 1338
    C. ridibundus, 472, 1073
Chroococcus, 1936
Chrysemys picta, 612
Chrysomus icterocephalus, 777
Chthonius, 339
Cichla ocellaris, 809
Cicindela, 1082
Ciconia
    C. boyciana, 1515, 1525, 1569, 1584, 1599
    C. ciconia, 482, 918, 1127, 1188, 1254, 1274
    C. episcopus, 1779
    C. episcopus stormi, 1845
    C. nigra, 889, 905, 1254, 1481
    C. stormi, 1758, 1835
Cicuta, 612
    C. virosa, 915
Ciénaga Grande de Santa Marta (CGSM), 749, 751, 752, 754
    birds, reptiles, mammals, 764
    ecosystems services, 767–768
    environmental impact, conservation and management, 766–767
    environmental setting, 759–761
    fishes, 763–764
    future challenges and developments, 768–769
    invertebrates, 762–763
    location of, 758, 759
    plankton, 762
    threats and challenges, 768–769
    vegetation, 764–766
    water quality characteristics, 761–762
Cinclodes oustaleti, 833
Cinclus
    C. cinclus, 1610
    C. pallasii, 1610
Circaetus gallicus, 1255
Circus
    C. aeruginosus, 1030, 1084, 1621, 1629
    C. bofoni, 852
    C. cyaneus, 166, 418, 570, 1045, 1084
    C. macrourus, 1354
Cirsium
    C. esculentum, 1539
    C. flavisquamatum, 1609
Cissus, 207
    C. hastata, 1773
    C. hexangularis, 205
    C. ibuensis, 1272
Cisticola carunthesi, 192
Cistothorus apolinari, 777
Citharinus congicus, 1344
Cladium, 1116, 1418
    C. jamaicense, 715
    C. mariscus, 163, 1074, 1116, 1117, 1152, 1427
Cladonia, 279
    C. rangiferina, 571
Cladophora, 478, 842, 1773
Cladophoropsis, 842
Cladorhynchus leucocephalus, 150, 1915, 1925
Clanga pomarina, 1255
Clangula hyemalis, 1083, 1094
Clarias, 192, 1273
    C. gariepinus, 1254, 1354, 1355
"Class II" lake, 1365
Clear-water rivers, 729–731, 1849
Cleistanthus sumatranus, 208
Clematis vitalba, 916

Climate change, 6, 7, 12, 49, 62, 64, 67, 82, 105, 106, 118, 119, 137, 138, 151, 153, 167, 179, 180, 193, 194, 202, 220–221, 237, 241, 242, 277, 286, 327, 343, 349, 367, 434, 451, 454, 456–458, 466, 492, 501, 503, 519, 531, 532, 536, 537, 585, 613, 614, 632, 634, 646–648, 714, 715, 754, 755, 774, 780–782, 864, 880, 890, 900, 902, 929, 930, 969, 1032, 1034, 1035, 1053, 1109, 1110, 1120, 1133, 1146, 1163, 1172, 1211, 1212, 1230, 1249, 1258, 1323, 1330, 1358, 1359, 1375, 1378, 1382, 1389, 1390, 1500, 1503–1504, 1507, 1533, 1547, 1561–1562, 1577, 1590, 1593, 1595, 1611–1613, 1651, 1655, 1672, 1678, 1700, 1710, 1730, 1731, 1740, 1749, 1782–1783, 1808, 1810, 1862, 1868, 1872, 1889, 1891, 1893, 1905, 1918, 1949, 1957, 1963, 1974, 1986, 1988, 1989, 2013, 2014
  boreal wetlands, 530
  Congo River, 1211
  effect on estuarine marsh, 62
  freshwater swamps, 220–221
  impacts, 754, 1330, 1613, 1891, 1893
  Volga River Basin, 935–936
Climate driven desertification, 1532, 1547
Cloud forest, 231, 269, 272, 300, 301, 308, 420
Clupea, 46, 1083
  C. harengus, 629, 980
  C. pallasii, 641
Clupeonella cultriventris, 939
Clymenella torquata, 630
Coastal change, 96
Coastal dunes, 1104, 1239, 1255, 1434
Coastal Eurasian tundra, 285
Coastal fishery, 1912
Coastal habitat, 502, 547, 607, 1445, 1627, 1628, 1827, 1882, 2019
Coastal herbaceous communities, 546
Coastal lagoon, 39, 364, 398, 733, 766, 815, 836, 859, 974, 1001, 1102, 1104–1105, 1116–1118, 1120, 1144, 1322, 1627, 1628, 1746, 1881, 1910, 1914
Coastal peatlands, 1156, 1856
Coastal plains, 12, 38, 58, 65, 171, 200, 280, 544, 545, 585, 668, 669, 673, 675, 758, 840, 848, 998, 1000, 1230, 1414, 1433, 1434, 1636, 1744, 1746, 1820, 1832, 1952, 1960
  MCP, 1424–1434

Coastal protection, 49, 101, 115, 116, 968
Coastal squeeze, 68, 969, 1004, 1006, 1019
Coastal wetlands, 5, 7, 8, 21, 22, 27, 29, 56, 59, 62, 68, 105, 285, 322, 348, 361, 370, 372–375, 377, 380, 409, 428, 431, 432, 456, 558, 563, 593–594, 825, 827, 829, 830, 833, 984, 1086, 1320, 1553, 1559, 1577, 1583, 1584, 1678, 1713, 1746, 1768, 1783, 1806, 1867–1869, 1947, 1952, 1981, 1982, 1989
  Amazon River Basin, 733
  characteristic types of, 748
  Conchali Lagoon, 858–864
  distribution of, 607
  ecosystems, 7
  ecotypes, 827
  Lake Superior's South Shore, 605–617
  Laurentian Great Lakes, 606
  Manitoba's Great Lakes, 591–603
  wetlands of Chile, 829–830
Coccoceras anisopodum, 208
Cocconeis, 86
Cochlearia officinalis, 964
Coconut-dominated agroforests, 2009
Cocos nucifera, 2013
Coenobita, 2016
Cold deserts, 1726
Coleanthus subtilis, 480
Coleataenia prionitis, 791, 816
Collapse lake, 1465, 1466
Colocasia esculenta, 1263, 2009
Colombia, 29, 206, 732, 740, 767
  central Caribbean coast, 758
  Lake Fuquene, 774–782
  locations of wetlands, 10
  mangroves of, 748–755
Colombian law, 762
Colombian mangrove-associated fauna, 750
Colombian mangrove ecosystems
  climate, 749
  conservation status and management, 752–754
  control and management, 755
  distribution, 749
  ecosystem services, 751–752
  temperature, 748
  threats and challenges, 754–755
  tropical climate, 748
  vegetation, 750
Comarum palustre, 285
Combretum, 1294, 1385
  C. fragans, 1344
  C. glutinosum, 1272

C. hartmannianum, 1294
C. trifoliatum, 208
Comephorus
  C. baicalensis, 1481
  C. dybowskii, 1481
Commelina, 191
Committee on the Status of Endangered Wildlife in Canada (COSEWIC), 632, 633
Common goldeneye, 1083, 1092–1094, 1448, 1460, 1475, 1610
Common Saltmarsh Grass, 963, 964
Community Based Organizations (CBOs), 465, 493, 1240
Community institutions, 1264–1265
Community succession, 194, 519
Conchalí Lagoon
  aerial view, 858
  biodiversity, 859–861
  future challenges and developments, 864
  landscape process, 862–863
  locations of wetlands, 10
  physical process, 861–862
  threats and future challenges, 864
Condensation, 230, 271, 1384
Conepatus chinga, 852
Conflict resolution, 400
Conger orbignyanus, 850
Congo River Basin, 1552
  biodiversity, 1205–1207
  climate change, 1211
  conservation and management priorities, 1212
  conservation priorities, 1212
  Cuvette Centrale, 1204
  dams, 1210
  deforestation, 1209
  ecosystem services, 1207–1209
  forest conversion, 1209–1210
  freshwater ecoregions, 1201, 1202
  geographical map, 1201
  habitat disturbance, 1210
  location, 1200
  Malebo Pool, 1205
  management priorities, 1212
  migration, 1211
  mining, 1209
  overfishing, 1211
  pollution, 1210
Conifer, 299, 300, 528, 1040, 1049, 1051, 1052, 1480, 1608, 1980
Coniferous forests, 238, 480, 525, 685, 1536

Connochaetes
  C. gnou, 1222
  C. taurinus, 1358, 1405
Conocarpus erectus, 750, 764, 1336
Conservation, 14, 28, 51, 53, 90, 112, 115, 117, 153, 195, 240, 249, 287, 328, 377, 438, 440, 441, 473, 582, 588, 622, 653, 654, 700, 766, 801, 802, 834, 845, 859, 864, 868–870, 921, 955, 969, 1026, 1031, 1034, 1052, 1054, 1069, 1071, 1102, 1109, 1118–1120, 1129, 1146, 1161, 1171, 1172, 1182, 1212, 1226, 1229, 1237, 1240, 1264, 1284, 1290, 1291, 1294, 1316, 1325, 1329, 1334, 1338, 1346, 1409, 1444–1445, 1448–1449, 1487, 1489–1493, 1496, 1497, 1502, 1507, 1513, 1546, 1555, 1586, 1587, 1598–1601, 1611–1613, 1629, 1642, 1717, 1718, 1769, 1776, 1779, 1796, 1807, 1809, 1816, 1837, 1861, 1863, 1867, 1870, 1872, 1882, 1888, 1889, 1891–1895, 1898, 1912, 1913, 1917, 1942, 1949, 1956, 1971, 1979, 1988, 1989, 1998–1999, 2002, 2012
  atoll biodiversity, 2003
  atoll ecosystems, 2019–2022
  biodiversity, 1295, 1296
  challenges, 1239
  designations, 571–572, 1187
  ecological restoration and adaptive management, 397–402
  freshwater habitats, 121, 1207, 1208
  global biodiversity, 1526
  global concerns, 675–676, 764, 1237
  grasslands, 685–686
  Great Barrier Reef, 454, 455
  habitats and species, 342
  initiatives, 993–994
  inland waters, 120
  issues, 4, 409, 522, 652, 1051, 1560–1561
  Lake Parishan, 1654, 1655
  Lake Uromiyeh and Satellite Wetlands, 1668–1669, 1678
  Makgadikgadi wetlands, 487–494
  management, 117, 480–482, 511–512, 752–754, 769, 779–781, 901–902, 906–908, 1118–1119, 1375, 1491–1493, 1518, 1700, 1710, 1730, 1738–1739, 1807–1808, 1826, 1888
  measures, 1621–1623
  Migraory Species, 1626, 1637
  peatlands, 241, 1164

Conservation (*cont.*)
  peatland use, 879–880
  Ramsar wetland, 2017
  Shadegan Wetland, 1682
  status, 4, 480–482, 511–512, 537–539,
    547–548, 553, 562–563, 585–586,
    632–633, 638, 645–646, 684–686,
    692–693, 737, 752–754, 768,
    779–781, 792, 806–807, 818, 830,
    901–902, 906–908, 918–919, 954,
    966–968, 981, 1004, 1017–1018,
    1032, 1049–1051, 1058–1059, 1075,
    1086, 1094, 1108, 1118–1119, 1132,
    1143–1144, 1187–1188, 1206, 1238,
    1274, 1294, 1303, 1328, 1338, 1346,
    1375, 1386, 1454, 1460, 1469, 1472,
    1475, 1482, 1517–1518, 1526, 1570,
    1610, 1631, 1643–1644, 1654,
    1668–1669, 1681, 1700–1701, 1710,
    1730, 1738–1739, 1787, 1797,
    1807–1808, 1816, 1826–1827, 1836,
    1846, 1867, 1904, 1947, 1955–1956,
    2003
  for sustainable use, 833
  threat, 694, 768, 845–846
  transboundary, 1519–1520
  values or uniqueness, 317
  vernal pools, 662–664
  wetland, 7, 180, 409, 684–685, 741, 1557,
    1719, 1982
  wetland biodiversity, 957, 998, 1584
  Yukon River, 547–549
Conservation of Iranian Wetlands Project
  (CIWP), 1654, 1656, 1662, 1671,
  1677–1680, 1682
Conservation Reserve Program's Continuous
  Signup, 694
Conspicuous invertebrate animals, 149
Contaminant, 12, 43, 49, 391–395, 535, 554,
  683, 853, 1561, 1942, 1949
Conus, 86
Convention on Biological Diversity (CBD), 120,
  121, 287, 443, 1295, 1867, 1892
Convention on Conservation of Migratory
  Species of Wild Animals (CMS),
  443, 1641, 1643
Convolvulus
  C. cantabrica, 916
  C. persicus, 916, 917
Cooper Creek, 1923–1925
Coorong, 1888, 1891, 1926
  future challenges and developments,
    1916–1918
  location, 1910
  and Lower Murray Lakes Wetland complex,
    1910
  wetland change, 1913–1916
The Coorong, and Lakes Alexandrina & Albert
  Wetland, 1911
Copernicia
  C. alba, 206
  C. australis, 206
Copper River Delta
  future challenges and developments, 563
  hydrology, 559
  land cover, 560
  succession, 561
  tectonics, 559
  threats, 563
  wetlands, 560
Coprosma, 1980
Corallus
  C. cropanii, 764
  C. hortulanus, 764
Corals, 78, 79, 86, 326, 377, 440, 442, 456,
  457, 1881, 2011, 2012
  reefs, 27, 79, 82, 85, 87, 322, 439–443, 448,
    451, 458, 754, 825, 845, 1867, 1869,
    1876, 1968, 2002, 2006, 2008, 2013,
    2017–2021, 4450
  Sea, 2003, 2021
Coral Triangle, in Southeast Asia, 440
Corbicula
  C. fluminea, 854
  C. langilleri, 854
Cordyla Africana, 1236
Cordyline australis, 1980
Coregonus
  C. autumnalis, 1447
  C. lavaretus, 476, 980, 1448, 1475
  C. migratorius, 1443, 1483
  C. muksun, 1447, 1454
  C. oxyrhynchus, 989
  C. peled, 476, 1447
  C. sardinella, 1447
  C. tugun, 1448, 1475
Cornus, 525
  C. mas, 916
  C. sanguinea, 916
Corophium, 630
  C. volutator, 630, 977, 988
Corosium insidiosum, 851
Corylus avellana, 916
Cotton grass, 26, 228, 290, 1048, 1544, 1545
Coturnicops noveboracensis, 633
Coturnix coturnix, 1084

Cover moss, 304
Cracking, 278, 279, 354, 421, 1964
Craspedacusta sowerbyi, 603
Crassocephalum, 1313
Crassostrea
  C. gigas, 995
  C. rhizophorae, 763
Crataegus monogyna, 916, 1106
Crataeva
  C. nurvala, 208
  C. roxburghii, 208
Craterocephalus
  C. eyresii, 1925
  C. marianae, 211
Cratoxylum spp., 228, 290
Crepis spp., 1105
Cressa, 61
  C. cretica, 1728
Crex crex, 1628
Cricetulus migratorius, 1192, 1443, 1483
Critically endangered species, 633, 1472, 1584, 1806, 1870
Crocidura
  C. leucodon, 1142, 1621
  C. suaveolens, 1142
Crocodylus
  C. acutus, 324, 750, 764
  C. johnstoni, 1946, 1955
  C. moreletii, 324
  C. niloticus, 1238, 1274, 1354, 1368, 1369, 1371
  C. porosus, 1739, 1747, 1825, 1835, 1845, 1946, 1955
  C. raninus, 1845
  C. siamensis, 1775, 1778, 1788
  C. suchus, 1337
Crocuta
  C. crocuta, 1274
  C. crocuta fortis, 1294
Crotalus adamanteus, 675
Croton
  C. cf. *ensifolius*, 205
  C. krabas, 205
  C. mekongensis, 205
Crown of Thorns Starfish (COTS), 451, 456–458
Crucigenia, 1385
Crudia
  C. amazonica, 207
  C. teysmannii, 208
Crustaceans, 45, 50, 51, 86, 102, 149, 153, 165, 176, 337, 340, 569, 630, 654, 657, 659–661, 767, 843, 844, 917, 978, 1082, 1105, 1107, 1109, 1228, 1324, 1337, 1387, 1717, 1737, 1848, 1869–1871, 1879, 1925, 1942, 1947, 1949, 1952, 1963, 2018, 2020
  fairy shrimp, 654, 656, 659–661
Cryo-arid conditions, 1464
Cryolithozone, 1442, 1464, 1472
Crypsis
  C. aculeata, 951
  C. schoenoides, 951
Cryptocoryne purpurea, 1859
Ctenomys talarum, 852
Ctenopharyngodon
  C. idella, 476
  C. idellus, 1254
Ctenopoma muriei, 192
Cuculus canorus, 1030
Culex, 338
Cultivation, 102, 103, 239, 443, 1014, 1103, 1104, 1117–1119, 1145, 1151, 1153, 1174, 1176, 1180, 1181, 1193, 1263–1265, 1309, 1310, 1316, 1347, 1357, 1390, 1414, 1417, 1419, 1424, 1433, 1434, 1638, 1642, 1654–1656, 1664, 1738, 1739, 1745, 1827, 1846, 1852, 1859, 1861, 1862, 2009
Cultural ecosystem services, boreal wetlands, 532
Cultural eutrophication, 496
Culturally-accelerated sediment accumulation, 693
Cuon alpinus, 1860
Cuora
  C. amboinensis, 1776, 1860
  C. mouhotii, 1845
Cuscuta salina, 570
Cuvette centrale, 1202–1204, 1207, 1208, 1211
Cyanocorax yucatanicus, 324
Cybister tripunctatus africanus, 1131
Cyclarhis gujanensis, 851
Cyclemys dentata, 1825, 1835
Cyclestheria hislopi, 1273, 1303
Cyclonic storms, 1737, 1740
Cyclosorus interruptus, 1997
Cygnus, 1559
  C. atratus, 1915, 1926, 1981
  C. bewickii, 1481
  C. buccinator, 562, 569
  C. columbianus, 547, 552, 1030, 1525
  C. columbianus bewickii, 1480
  C. cygnus, 1515, 1525, 1628
  C. olor, 472, 978, 1091, 1642

Cymodocea
 C. angustata, 75
 C. nodosa, 75, 1323
 C. rotundata, 75, 84, 1876, 1970, 1971
 C. serrulata, 75, 1876, 1971
Cynocephalus variegatus, 1836
Cynocion guatucupa, 853
Cynodon, 191
 C. dactylon, 205, 816, 951, 1130, 1171, 1385
Cynogale bennettii, 1845, 1860
Cynorkis anacamptoides, 1313
Cynoscion guatucupa, 850, 853
Cyornis rufigastra, 1825
Cyperaceae, 187, 205, 510, 735, 876, 879, 1074, 1152, 1418, 1639, 1795, 1815
Cyperus, 189, 204, 1316, 1945, 1980
 C. articulatus, 205
 C. cephalotes, 1746
 C. digitatus, 1236, 1237
 C. dives, 191
 C. exaltatus, 1236
 C. giganteus, 791, 816
 C. haspan, 205, 206
 C. involucratus, 205
 C. iria, 1746
 C. laevigatus, 1385
 C. latifolius, 1248, 1262, 1313
 C. nitidus, 205
 C. papyrus, 6, 170, 177, 184, 190, 204, 205, 209–211, 1156, 1168–1169, 1210, 1237, 1244, 1265, 1272, 1302, 1313, 1350, 1357, 1398, 1427
 C. platystylis, 1746
 C. polystachyos, 205
 C. prolixus, 205
 C. rotundus, 1639, 1642
Cypraea, 86
Cyprinids, 939
Cyprinus
 C. carpio, 476, 597, 612, 776, 818, 850, 853, 939, 953, 1104, 1131, 1357, 1514, 1905
 C. carpio communis, 1708
 C. carpio haematoperus, 1586
 C. carpio specularis, 1708
 C. linnaeus, 1443
Cyrilla racemiflora, 672, 674
Cyrtograpsus angulatus, 850, 851
Cyrtopodion kachhense kachhensis, 1729
Cyrtosperma chamissonis, 2009, 2013
Cyrtostachys
 C. lakka, 1834
 C. renda, 1757, 1834, 1835

Czech Republic, fishponds, 470
 biodiversity, 471–473
 conservation status and management, 480
 ecosystem services, 473–476
 history, 470
 management, 476–477
 threats, 483
 Tøeboò Basin Biosphere Reserve area, 471
 vegetation management, 477–479

D

Dacrycarpus dacrydioides, 1980
Dacryodes edulis, 207
Dactylocladus stenostachys, 1757
Dactyloctenium aegyptium, 1385
Dactylorhiza hatagirea, 1609
Dalbergia
 D. louisii, 207
 D. pinnata, 205
Dama
 D. dama, 220
 D. mesopotamica, 1666
Damaliscus
 D. korrigum, 1274, 1371
 D. korrigum tiang, 1293
 D. lunatus, 1282, 1405
Dambos, 1219, 1221, 1229
Damgahs, 1636, 1638–1645
Damplands, 350, 1960
Dams, 12, 53, 65, 105, 117–119, 126–128, 131, 132, 136, 146, 151, 179, 217, 466, 470, 531, 534, 551, 561, 583, 640, 702, 731, 737, 739, 740, 786, 808, 818, 831, 887, 900, 904–906, 920, 924, 926, 939, 941, 1062, 1208–1210, 1218, 1219, 1225, 1229, 1240, 1249, 1266, 1284, 1377, 1409, 1410, 1445, 1483, 1518, 1571, 1627, 1638, 1673, 1677, 1683, 1687, 1692, 1693, 1700, 1721, 1798, 1809, 1810, 1893, 1899, 1900, 1945, 1947
 Gibe, 1376, 1377
 hydroelectric, 51, 789, 1086
 hydropower, 131, 907, 930, 1230, 1375, 1781, 1790
 irrigation, 1781
 Volga River Basin, 941
Damselflies, 176, 612, 660, 1859, 1860
Danau Sentarum, 13, 201, 203, 205, 208, 210, 212, 214, 216, 219
Danau Sentarum National Park (DSNP)
 biodiversity, 1844–1845

black water, 1848
conservation status, 1846
ecosystem services, 1846–1847
future challenges and developments, 1847–1848
geographical map, 1842
hydrology, 1843
land cover analysis, 1844
location, 1842
threats, 1847–1848
wetland ecosystems, 1843–1844
Danish Cooperation for Environment and Development (DANCED), 1852
Danish Wadden Sea
biodiversity, 988–993
biological characteristics, 984
conservation status, 993–994
ecosystem services, 993
habitat types, 984, 985
threats and future challenges, 994–995
tidal ecosystem, 984–987
Danthonia, 1701
Danube Delta, 886–888, 890
biodiversity, 917–918
challenges, 921
conservation status, 918–919
ecosystem services, 919–920
features, 889
future challenges and developments, 921
hydrology, 912–914
location, 912
threats, 920
wetland ecosystems, 914–917
Danube River, 476, 893, 894, 901, 912, 913, 919, 921
branches, 912, 913
channelling, 906
conservation status and management, 906–908
ecosystem service, 908
location, 904
threats, 905
wetland types and diversity, 904–905
Danube River Basin (DRB), 11, 900, 908
assessment of restoration potential, 894
biodiversity, 889
biogeographical regions, 888
climate, 887
Danube River Basin Management Plans, 890, 892, 894
delta features, 889
Drava-Mura wetlands, 889
eutrophication, 891

future challenges and developments, 892–895
hazardous substances, pollution by, 891
hydrology, 887
hydromorphological alterations, 891
International Commission for the Protection of the Danube River, 892, 893
Joint Programme of Measures (JPM), 892
location, 886
lower basin, 887
Lower Danube Green Corridor, 893
lower Danube wetlands, 888
middle basin, 887
morphological floodplain, 888
organic pollution, 891
population, 890
transboundary Danube Delta, 888
upper basin, 887
Urban Wastewater Treatment Directive, 893
values, 892
Daphnia spp., 478, 479, 612, 1357
Daqing River channel, 1583
Darcy's law, 233
Darhad Kettle, 1541
Dasyurus
D. albopunctatus, 211
D. spartacus, 211
Dauria steppe wetlands, 13
biodiversity, 1502
climate, 1500–1501
climate change and human impacts, 1503
cultural values, 1503
ecosystem services and human values, 1503
location, 1500–1501
principal wetlands, eastern Dauria, 1501
principal wetlands of Argun and Ulz basins, 1505
transboundary wetland management, 1505–1507
Declines, 7, 46, 50, 52, 59, 60, 82, 89, 113, 354, 377, 385, 410, 432, 442, 451, 453, 457, 472, 496, 500, 547, 645, 647, 648, 654, 681, 682, 1003, 1019, 1032, 1052, 1091, 1384, 1511, 1515, 1528, 1561, 1595, 1665, 1666, 1681, 1710, 1790, 1867, 1871, 1880, 1905, 1915, 1918, 1925, 1946, 1963, 1965, 1973
Degradation, 5, 14, 32, 46, 66, 105, 115, 117, 180, 214, 219, 241, 276, 286, 346, 366, 398, 411, 419, 441, 442, 457, 462, 490, 537, 538, 638, 647, 737, 752, 754, 817, 880, 900, 907, 1074, 1118, 1200, 1229, 1249, 1257, 1258,

1389, 1390, 1419, 1469, 1532, 1546, 1547, 1612, 1721, 1810, 1835, 1871, 1879, 1963, 2002, 2014, 2017, 2018
 diverse activity, 1316
 ecosystem, 116, 466, 496, 739, 1516
 environmental, 64, 779, 1293
 estuarine environments, 53
 estuary conditions, 49
 by human activities, 148, 151
 intertidal flats, 1871
 lagoons, 1120
 Lake Chilika, 364
 lake habitats, 135
 land functions, 1586
 marsh system, 1309
 of organic matter, 46
 of palsas, 283
 of peatlands, 237, 238, 241, 242, 1533, 1547
 quality of water, 217, 1853
 river and floodplain areas, 908, 1781
 riverine landscape, 907
 river systems, 1906
 seagrass, 438
 seagrass biodiversity, 85
 soil, 1433
 swamp forest, 1433, 1759, 1827
 of wetland, 8, 12, 117, 361, 362, 365, 370, 376, 956, 1264, 1265, 1583, 1710, 1781
De la Plata River Basin, 786
Delphinus delphis, 918, 1747
Delta, 38, 56, 186, 218, 277, 285, 364, 398, 639, 642, 643, 758, 765, 786, 789–792, 794, 840, 924, 929, 968, 976, 998, 1139, 1323, 1335, 1337, 1345, 1440, 1442–1446, 1486, 1523, 1529, 1623, 1699, 1700, 1734, 1764, 1765, 1767, 1780–1783
 Amur River, 280
 archipelago, 1334
 Axios, Aliakmon, and Gallikos, 1137–1146
 Axios–Loudias–Aliakmon, 1144
 Canadian Mackenzie, 280
 Ciénaga Grande de Santa Marta, 749
 Copper River, 558–563
 Danube, 886–890, 893, 898–901, 911–921
 Dique canal, 749
 drowned rivers/river mouth, 608, 609
 Ebro, 1114–1120
 Egypt's Nile, 1156
 Fraser River, 566–573
 Ganga-Brahmaputra-Meghna rivers, 1734
 Gangetic, 1712, 1713, 1718–1719, 1721
 Herbert, 1945
 Indus, 105, 172, 1700, 1701
 Inner Niger, 217
 Kerio/Turkwel, 1369
 Lena River, 1445, 1449, 1452–1455
 locations, 10–13, 26
 Mahanadi, 1744–1749
 Marsh, 26, 595, 598–602
 Mekong River, 96, 365, 1771, 1773, 1777, 1779, 1782, 1783, 1793–1798, 1802–1810, 1813–1816
 Mississippi, 116, 160, 587, 588
 Mkuze, 1426
 Neretva River, 1058, 1059
 Niger, 113, 217
 Nile, 185, 1156, 1160, 1248, 1249, 1252–1258
 Okavango, 110, 176, 201, 211, 488, 490, 1343, 1394–1410, 1414, 1415, 1418
 Paraná River, 814–818, 848
 Peace-Athabasca, 550–555
 region, 789–790
 Rhone River, 1102–1110, 1130
 Riparian mires, 285
 Rufiji, 1342
 Sacramento-San Joaquin, 698, 700
 Sefid Rud, 1626–1628, 1630, 1631
 Selenge River, 1481, 1482
 Sinú and Atrato River, 749
 Sundarban, 1735
 Volga River, 934, 937, 938, 940, 946–957
 wetlands, 160, 548, 554, 817, 920, 947, 957, 1583 (*see also* Omo Delta wetland)
 Yangtze River, 1553
 Yellow River, 1583–1586
 Yukon-Kuskokwim, 160, 544–548
 Zambezi, 1219, 1221, 1227, 1228, 1234–1240 (*see also* Axios, Aliakmon and Gallikos Delta complex, northern Greece; Omo Delta wetland)
Deltaic archipelago, 1334
Deltaic system, 1627
Delta Marsh, 26, 595, 598–602
Dendroaspis angusticeps, 1274
Dendrobranchiata, 570
Dendrocygna
 D. arcuata, 1955
 D. eytoni, 1955
 D. viduata, 1385
Dendropsophus labialis, 777

Denmark, 58, 83, 248, 967, 978, 981, 1000, 1092, 1094, 1095
　Ringkøbing Fjord, 974–975
　tidal vegetated marshes, geographic distribution, 960
　Wadden Sea, 984–995
Density fingering, 1402
Dependent communities, 1371, 1373–1374, 1717, 1749
Depressional wetlands, 680, 683, 684
Dermochelys coriacea, 324, 453
Desalination, 118, 1175, 1400–1403, 1409, 1410
Deschampsia koelerioides, 1537
Desertification, 238, 462, 1504, 1532, 1546, 1547, 1553, 1586
Desert Lake, 1362–1378, 1700
Desiccation, 148, 150, 152, 153, 200, 204, 209, 656, 661, 1130, 1316, 1433, 1547, 1672, 1673, 1805, 1891
Desmana moschato, 953
Detritus, 441, 569, 632, 652, 657–658, 843, 1176, 1273, 1737
Dew, 265, 271, 298
Deyeuxia angustifolia, 1525
Diadromous fish species, 7, 52, 165
Dialium, 207
Dialulinopsis deltaicus, 918
Dialyanthera spp., 750
Diamantina River, 1923
Dianthus pontederae, 916
Diapensia lapponica, 279
Diaphanosoma, 1357
Diaptomus castor, 1071
Dicentrarchus labrax, 965, 1002, 1083, 1104
Dicerorhinus sumatrensis, 1836
Diceros bicornis, 1408
Dichilanthe borneensis, 208, 210, 1844
Dichirotrichus gustavii, 1082
Dichrostachys
　D. cinerea, 1294
　D. glomerata, 1272
Dicoelia beccariana, 210–211, 1844
Dicranoloma robustum, 1994
Dicranum elongatum, 282
Dicrocephala, 1313
Didelphys albiventris, 852
Digital Chart of the World (DCW), 111
Digitaria, 204
　D. velutina, 1385
Dillenia
　D. alata, 208
　D. indica, 207

Dinaric Karst, 319, 320, 336–342
　blind and pocket valleys, 1061
　conservation status, 1058–1059
　future challenges and developments, 1063–1064
　lakes, 1062
　location, 1058
　water caves, 1062–1063
Dinder and Rahad basins, 1291
Dinder National Park (DNP)
　biodiversity, 1293–1294
　conservation status, 1294–1296
　ecosystem services, 1294
　features, 1288
　location, 1289
　threats, 1296
　topography, 1289
Dinder National Park Project (DNPP), 1295
Dinder River, 1288, 1290, 1292, 1294, 1297
Dinka communities, Sudd region, 1274, 1281, 1282, 1304
Dionaea muscipula, 674
Dioscorea deltoidea, 1609
Diospyros
　D. areolata, 1834
　D. cambodiana, 208, 1773
　D. coriacea, 208
　D. lanceifola, 1834
Dipsacus gmelinii, 952
Disa stairsii, 1313
Discharge regulation, Volga River Basin, 937, 939–941, 956
Discopyge tschudii, 850
Distichlis, 61
Distichodus spp., 1273
Distribution, 5, 46, 47, 78, 85, 88, 89, 99, 100, 107, 127, 138, 187, 221, 228, 266, 268, 277, 340, 342, 352–354, 384, 387, 453, 462, 483, 491, 510, 513, 579, 648, 654, 656, 657, 668, 680, 686, 714, 777, 786, 789, 792, 803, 806, 817, 849, 851, 868, 874, 914, 917, 980, 987, 998, 1002, 1003, 1116, 1117, 1125, 1140, 1156–1157, 1212, 1219, 1229, 1320, 1322, 1323, 1344, 1366, 1472, 1546, 1553, 1559, 1560, 1587, 1728, 1734, 1748, 1765, 1768–1769, 1838, 1876, 1903, 1915, 1934, 1945, 1946, 1952, 1960, 2004, 2006
　arctic peatland, 277–279
　areas, 202–203
　atoll countries and territories, 2004, 2006

Distribution (cont.)
  azonal patterns, 235
  benthic species, 988
  biodiversity, 1584
  bioregion, 1769–1771
  boreal wetland, 539
  brackish water species, 1679
  coastal wetlands, 607
  Danube River Basin, 888–889
  Estuarine Marsh, 58–60
  flooded areas, 1403
  freshwater lakes and reservoirs, 130–131
  freshwater swamps, 202–203
  geographic, 6, 58–60, 130–131, 133, 690, 960–961, 1093
  global, 7, 79–82, 94–98, 145–147, 149, 156, 230–233, 315
  habitat types, 1679
  historic, 643, 1408
  inland water systems, 171–172
  landscape, 652–653
  Lena River Basin, 1445–1448
  location climate, 748–749
  managed seasonal and semipermanent wetland areas, 701
  mangroves, 95–98, 748–749
  map, 1028
  Mediterranean drainage basin, 1156–1159
  Mekong wetlands, 1769–1771
  Najas marina, 1934
  New Zealand wetland, 1978
  northeastern Australian seagrass meadows, 1968
  Papyrus, 186
  Pareto, 128
  Patagonian peatlands, 875
  peatlands, 230–233, 277–279, 531, 876–879, 1415–1417, 1534–1535
  plant and animal species, 899
  playas, 690
  pocosin bogs and Carolina bays, 669
  pre-and post-construction, 1095
  primary production, 627
  raised bog, 265
  saline wetlands, 1728
  salinity gradients, 761
  salt lakes, 145–147
  saltmarsh area, 962
  seagrasses, 79–82, 840–842, 1591, 1595
  seasonal, 949
  Siberian salamander *Salamandrella keyserlingii*, 1599
  spatial, 527, 663, 715, 1081, 1082, 1091, 1132, 1249, 1258
  swamps and marshes, 171–172
  terrestrial taxa, 737
  tidal freshwater wetlands, 160–161
  tidal saltmarshes, 960–961
  vascular plant species, 1071
  vegetation community and species, 1070
  vernal pools, 652–653
  vertical salinity, 40
  water and sediment discharge, 913
  wetland articles, 9–13
  wetlands, 888–889, 1489, 1553, 1554, 1562, 1577–1580, 1744–1746, 1772, 1773
  Wetlands of Mahanadi Delta, 1744–1746
  and wetland types, 20–32, 1445–1448, 1978
  wintering, 1093, 1094
  Yangtze River Basin, 1554
  Yellow River Basin, 1578, 1584
  zonation of vegetation communities, 1071, 1073, 1075
  zooplankton, 762
Diurnal fluctuations, 162, 655
Diurnal tides, 42, 162, 1804, 1952
Diversity, 21, 46, 48, 74, 80, 101, 106, 107, 175, 191, 200–201, 228, 314, 337, 387, 393, 442, 471, 473, 480, 483, 544, 551, 553, 569, 585, 660, 693, 700, 721, 728, 733, 734, 767, 799, 805, 817, 824, 825, 843, 845, 867, 869, 915, 926, 1029, 1030, 1074, 1106, 1108, 1115, 1171, 1200, 1206, 1207, 1223, 1237, 1249, 1255, 1271, 1273, 1302, 1303, 1315, 1338, 1345, 1382, 1386, 1387, 1414, 1418, 1421, 1472, 1480, 1501–1503, 1558, 1561, 1566, 1568, 1583, 1606, 1621, 1622, 1642, 1666, 1678, 1687, 1690, 1691, 1694, 1728, 1729, 1737, 1778, 1780, 1797, 1859, 1860, 1863, 1903, 1917, 1925, 1952, 1955, 1972
  animal, 46, 1082–1084
  aquatic, 1305
  aquatic macroinvertebrates, 1131
  arctic peatland, 277–279
  biological, 210, 212, 326, 448, 818, 1161, 1309, 1942
  of biota, 173
  biotic, of peatlands, 6, 235–236, 335
  Boreal wetlands, 527–531
  British Estuaries, 998–1000

Colombian Mangrove Associated Fauna, 750–751
corals and reef, 451, 2012
cultural, 326
Danube River Basin, 888–889
Drava and Mura rivers, 904–905
ecosystem, 276, 319–320, 1104, 2002, 2006–2010
Europe vegetation, 159
exotic species, 853
faunal, 7, 165, 190, 571, 706, 939–940, 1124, 1747, 1757, 1758, 1763, 1768
fish, 192, 1630, 1788, 1844, 1946, 1947
floral, 190, 191, 1747
freshwater swamps, 200
fresh water wetlands, 510–511
functional, 818
genetic, 187, 2006
global occurrence/genetic, 79
habitat, 597, 626, 629, 632, 642, 1736, 2006–2010
hydrologic, 1569
hydroperiods and vegetation types, 1129
invertebrate, 236, 1076
Karst wetlands, 319–320
macrophytes, 149
mammalian species, 806
Mekong River, 1769–1771
microbial, 193
of mire, 1535–1546
in papyrus marshes, 190
peatlands, 235
phytoplankton, 1371
plant, 60–61, 157, 165, 166, 176, 717, 1074, 1081–1082, 1773
saline wetlands, 1728
saltmarsh zone, 963
seagrasses, 80–82, 85
taxonomic, 692, 2006, 2010–2011
terrestrial ground beetle, 1073
of tidal freshwater wetlands, 165
tropical black water fish, 240
vegetation, 750, 1569
vegetation communities, 716
waterbird habitats, 1447
wetlands, 833, 888–889, 1035, 1983
ecosystems, 1366–1371, 1895
types, 4, 5, 12, 20, 510–511, 904–905, 1736, 1786–1787, 1866–1867
Dolichopoda, 338
Dolichopus nigripes, 1030
Dolomedes plantarius, 1031
Doñana Biosphere Reserve, 1132
Doñana wetlands, 1133
biodiversity, 1130–1132
conservation status, 1132
distribution, 1125
ecosystem services, 1132
hydrology, 1124–1130
location, 11, 1124
threats and challenges, 1133–1134
Donatia, 878, 879
D. fascicularis, 265, 833, 876
Dongpinghu Lake, 1582
Dosinia, 1336
D. isocardia, 1324
Double lakes, 1579
Dracophyllum, 1980, 1994
D. lessonianum, 1994
D. oliverii, 1995
D. scoparium, 1995
Dragonflies, 176, 249, 571, 612, 656, 660, 661, 926, 1071, 1408, 1777, 1859, 1860
Drainage, 8, 12, 21, 22, 28, 38, 52, 62, 65, 111–113, 146, 153, 194, 228, 237, 238, 240, 241, 272, 284, 286, 299, 349, 363, 370, 395, 414, 418, 421–423, 470, 476, 480, 518, 519, 529, 531, 545, 567, 571, 578, 592, 594, 638, 643, 668, 671–673, 675, 681–685, 698, 709, 710, 712, 713, 717, 728, 739, 760, 779, 781, 786, 789, 808, 828, 836, 876–878, 881, 887, 917, 920, 924, 925, 934, 937, 938, 976, 994, 1026, 1031, 1043, 1052, 1060, 1063, 1064, 1073, 1083, 1106, 1109, 1114, 1117, 1119, 1129, 1133, 1138, 1140, 1145, 1150, 1152, 1153, 1156, 1162, 1163, 1168–1171, 1180, 1192, 1204, 1209, 1218, 1222, 1229, 1239, 1240, 1249, 1255–1257, 1263–1265, 1275, 1282, 1283, 1291, 1293, 1315, 1363, 1364, 1384, 1389, 1396–1400, 1409, 1410, 1419, 1421, 1428, 1433, 1442, 1464, 1472, 1518, 1546, 1552, 1554, 1568, 1576, 1581, 1666, 1668, 1672, 1677, 1678, 1687–1689, 1709, 1712, 1713, 1719, 1721, 1745, 1748, 1759, 1760, 1765, 1795, 1804–1805, 1822, 1823, 1827, 1836–1838, 1856, 1933, 1948, 1949, 1961, 1978, 1980, 1984, 1985, 1998
Drama Basin, 1150–1152
Drava-Mura wetlands, 889

Drava River
  channelling, 906
  conservation status and management, 906–908
  ecosystem service, 908
  hydropower dams, 906
  location, 904
  threats, 905
  wetland types and diversity, 904–905
Drawdowns, 187, 353, 431, 477, 480, 482, 525, 526, 652, 655, 656, 660, 661, 700, 701, 1797, 1964, 1965
Dreissena
  D. bugensis, 139
  D. polymorpha, 139, 603
Drepanocladus, 285
Drimycarpus racemosa, 207
Drimys winteri, 878
Drivers of change, 8, 68, 362–363, 385, 402, 492, 1917
  infrastructure, 364–365
  land conversion, 363
  pollution, 361, 366
  water level management, 365
  water withdrawals, 363–364
Drosera
  D. anglica, 428, 1537
  D. binata, 1994
  D. intermedia, 1046
  D. rotundifolia, 571
  D. spatulata, 1994
Drought, 65, 132, 173, 220, 221, 238, 286, 350–352, 360, 462, 531, 532, 652, 654, 655, 672, 674, 681, 682, 686, 702, 709, 710, 714, 719, 736, 789, 799, 803–805, 900, 902, 930, 1027, 1035, 1146, 1192–1195, 1207, 1218, 1230, 1235, 1240, 1257, 1271, 1275, 1294, 1386, 1410, 1421, 1501–1504, 1507, 1547, 1562, 1569, 1571, 1650, 1651, 1654, 1655, 1664, 1678, 1682, 1693, 1729, 1748, 1804, 1808, 1809, 1862, 1879, 1888, 1890, 1891, 1913–1917, 1932, 1933, 1935, 1937, 1961, 2011, 2016
Drought-sensitive pool, 1046, 1047
Drowned River estuaries, 39, 41
Dry climate, 876, 1180, 1532–1534
Drying, 12, 132, 136, 177, 179, 216, 221, 239, 284, 352–354, 365, 419, 421, 531, 536, 551, 655–661, 682, 698, 710, 1170, 1176, 1390, 1446, 1467, 1500, 1533, 1547, 1555, 1612, 1837, 1847, 1922, 1924, 1926, 1963–1965

Dryobalanops abnormis, 208
Dryomys nitedula, 1192
Dryopteris
  D. cristata, 1027
  D. gongylodes, 1272
Dubravius ponds, 477
Duck-trapping, 1636, 1645
Ducula bicolor, 1825
Dugong, 74, 80, 85–87, 448, 453, 457, 843, 1877, 1878, 1955, 1969, 1970
Dugong dugon, 86, 448, 1969
Duma florulenta, 1901
Dunaliella, 153, 1664
  D. salina, 149, 482, 1174, 1180–1182, 1729
Dune ponds, 1124, 1127–1130, 1133
Dunlin, 410, 547, 562, 570, 571, 630, 642, 965, 980, 989, 1003, 1016, 1030, 1044, 1045, 1052, 1073
Duoshixia Valley, 1579
Durio carinatus, 1757, 1834
Dwarf shrub, 94, 279, 280, 282, 284, 300, 510, 545, 876, 1044, 1544, 1546
Dwarf swamp forest, 203, 1843, 1844
Dyera polyphylla, 1834
Dynamic Interactive Vulnerability Assessment Wetland Change Model, 68

E
Earthcover mapping project, 545
East Asian-Australasian Flyway (EAAF), 410, 1444, 1502, 1517, 1557, 1561, 1820, 1827, 1867, 1870, 1872, 1981
Eastern Caprivi floodplain, 1222
Eastern Nile Technical Regional Office (ENTRO), 1284, 1296
Ebro Delta, 11
  biodiversity, 1115–1117
  conservation status and management, 1118–1119
  ecosystem services, 1117–1118
  environmental crisis, 1119
  extension of tourism, 1120
  fragmentation, 1119
  future challenges and developments, 1119–1120
  habitats, 1116
  location, 1114
  river, 1114
  sediment transport reduction, 1120
  wetland land use and habitats, 1115
  wetland types, 1115–1117
  wetland use, 1117–1118

Echinochloa, 177, 210, 580, 816, 1117, 1302
  E. colona, 1344
  E. colonum, 204
  E. haploclada, 1302
  E. polystachya, 210, 211, 213, 765
  E. pyramidalis, 1236, 1265, 1302
  E. stagnina, 205
Echinometra lucunter, 843
Echinosorex gymnurus, 1825
Echis spp., 1274
Ecological diversity indices for fish, 1690
Ecological gradients, 874, 878, 879
Ecological network, 906, 907, 1489, 1491, 1518
Ecoregion, 171, 173, 545, 585, 685, 694, 794, 898, 1201–1205, 1208–1211, 1235, 1237, 1343, 1478, 1480, 1486–1493, 1500, 1510, 1608, 1702, 1765
Ecosystem-based adaptation (EBA), 754, 755
Ecosystem degradation, 466, 496, 739
Ecosystem services, 4, 7, 12, 52, 61–64, 74, 94, 107, 139, 145, 162, 166, 180, 250, 300, 323, 325, 327, 328, 361, 438, 483, 492, 522, 534, 536, 537, 539, 594, 613, 662, 728, 737, 741, 774, 780–782, 799, 845, 854, 859, 890, 892, 900, 901, 905, 908, 957, 998, 1005, 1230, 1248, 1249, 1264, 1266, 1267, 1275, 1282, 1290, 1375, 1378, 1382, 1389, 1479, 1516, 1571, 1612, 1613, 1701, 1709, 1716, 1721, 1749, 1780, 1808, 1848, 1866, 1888, 1963, 1988, 2013
  Amazonian wetlands, 738–739
  Anzali Mordab Complex, 1621
  arid areas, 1729
  Axios, Aliakmon, and Gallikos Delta Complex, 1144–1146
  Banc d'Arguin, 1328–1329
  Berbak National Park, 1836
  Bijagos Archipelago, 1338–1339
  boreal wetlands
    cultural, 532
    provision, 532
    regulation, 531
    supporting, 533
    valuation, 533
  Broads, 1034–1035
  Bujagh National Park, 1630–1631
  Camargue, Rhône River Delta, 1108–1109
  Ciénaga Grande de Santa Marta, 767
  Congo Basin, 1207–1211
  Danau Sentarum, 1846–1847
  Danube Delta, 919–920
  Daurian Steppe Wetlands, 1503
  Dinder National Park, 1294, 1295
  Doñana wetlands, 1132–1133
  Ebro Delta, 1117–1118
  estuaries, 48–50
  Fereydoon Kenar, 1642
  fishponds, 473–476
  freshwater lakes, 135–137
  freshwater marshes and swamps, 177–178
  Ganga-Brahmaputra Basin, 1713
  high altitude wetlands, 1611
  Indus Basin, 1701
  intertidal zone, 1869–1870
  Kakadu National Park, 1956
  Karst systems, 316–317, 319–320
  Kilombero Valley Floodplain, 1345–1346
  Lake Fúquene, 777–778
  Lake Parishan, 1654
  Lakes Baringo and Naivasha, 1355, 1358
  Lake Turkana, 1372–1373
  Lake Uromiyeh, 1669–1671
  Lena River Basin, 1443–1444, 1455
  Macquarie Marshes, 1904
  Mahanadi Delta, 1747
  mangrove wetlands, 101–102, 751–752
  marshes, 190, 191
  Mekong River Delta, 1806–1807
  Messerya cattle herdsmen, 1274
  Middle Aldan Basin, 1475
  Mississippi Alluvial Valley, 586–587
  Nidjili Lake, 1460
  Okavango Delta, 1408–1409
  Pantanal, 807
  papyrus wetlands, 190–191
  Paraná-Paraguay Fluvial Corridor, 793
  Peace-Athabasca Delta, 553–554
  peatlands, 237–240
  peatlands of the Mediterranean, 1159–1162
  Poyang Lake, 1570
  Prairie Pothole Region, 683–684
  Restiad peatlands, 1996–1998
  Ringkøbing Fjord, 980–981
  Río de la Plata River, 852–853
  River basins, 115–117
  Rugezi Marsh, 1315–1316
  salt lakes, 153
  salt marsh functioning and potential loss, 1084–1086
  San Francisco Bay Estuary, 643–645
  seagrasses, 85–88
  Shadegan Wetland, 1681–1682
  Sudd, 1304

Ecosystem services (*cont.*)
　Sundarban mangroves, 1737–1738
　Taiga-Alas Landscape, 1469
　tidal marshes, 61–64
　Tonle Sap lake, 1789
　Tram Chim, 1796–1797
　tropical peat swamps, 1759
　U Minh peat swamps, 1816
　Volga-Akhtuba floodplain, 955
　Volga Delta, 955
　Wadden Sea, 993
　water levels, 208–209
　water quality, 209
　Wet Tropics, 1947
　Wular lake, 1708
　Zambezi Delta, 1239
　Zhalong wetland, 1527
Ecotourism, 62, 107, 191, 325, 326, 364, 365, 462, 493, 595, 738, 807, 908, 1118, 1120, 1133, 1164, 1171, 1239, 1372, 1396, 1622, 1654, 1655, 1671, 1672, 1682, 1720
Edge effects, boreal wetlands, 534–535
EEC-Birds Directive, 994
Eelgrass, *see Zostera marina*
Egeria densa, 385, 777
Egna, 191
Egretta, 1238, 1274, 1796
　E. alba, 1824
　E. eulophotes, 1584, 1824, 1835
　E. garzetta, 1141, 1190, 1338, 1385, 1559, 1665, 1691, 1902, 1955
　E. gularis, 1327, 1338
　E. intermedia, 1955
　E. novaehollandiae, 1915
　E. thula, 859
　E. vinaceigula, 1407
Egypt, 152, 184, 185, 193, 1102, 1160, 1248, 1275
　Nile Delta, 1156, 1251–1258
Eichhornia, 400, 816, 1748
　E. azurea, 790, 809
　E. crassipes, 117, 134, 137, 189, 209, 218, 220, 384, 385, 713, 765, 777, 779, 790, 809, 1210, 1237, 1249, 1255, 1272, 1301, 1305, 1357, 1368, 1720, 1746, 1773, 1859, 1948
　E. stratiotes, 1815
Elaeocarpus, 1823, 1834
Elanus leucurus, 852
Eleiodoxa conferta, 1834
Eleocharis, 205, 210, 211, 612, 715, 717, 828, 851, 1815, 1901, 1945

E. acutangula, 1236
E. dulcis, 1236, 1746, 1795, 1796, 1815, 1956
E. minima, 205
E. multicaulis, 1045, 1130
E. palustris, 949, 1130, 1538
E. plicarhachis, 205
Elephas
　E. maximus, 1715, 1860
　E. maximus sumatranus, 1826
Elfin forests, 232, 300
Eling lake, 1579
Eliomys quercinus, 1107
El Niño–Southern Oscillation (ENSO), 147, 790, 1862, 1923
Elodea
　E. canadensis, 384, 914
　E. densa, 779
Elops saurus, 763
El Rocío pilgrimage, 1133
Elseya latisternum, 1946
Elymus
　E. athericus, 1082, 1084, 1085
　E. giganteus, 916, 917
　E. mollis, 570
Elytrigia repens, 951, 964
Embankments, 58, 285, 364, 365, 371, 398, 769, 905, 908, 1004, 1102, 1107, 1110, 1139, 1235, 1496, 1504, 1507, 1709, 1721, 1740, 1745, 1748
Emberiza schoeniclus, 965
Emerita brasiliensis, 850
Emissions, 28, 63, 232, 237, 242, 286, 415, 457, 502, 503, 684, 754, 891, 893, 894, 1181, 1182, 1868, 1937
　carbon, 51, 414, 423, 1421
　greenhouse gas, 740, 1760
　methane, 213, 214, 532, 684
Empetrum, 545, 876
　E. hermaphroditum, 285
　E. nigrum, 280, 291
Emphanes, 1082
Empodisma, 260, 1980, 1992–1994
　E. minus, 26, 247, 259, 263, 1992–1995, 1998
　E. robustum, 1992–1994, 1996, 1998, 1999
Emys orbicularis, 939, 1131, 1142, 1190
Enallagma spp., 612
Endangered species, 213, 324, 364, 365, 480, 519, 638, 643, 648, 817, 905, 906, 940, 952, 1116, 1303, 1316, 1344, 1472, 1584, 1601, 1643, 1691, 1693, 1700, 1717, 1737, 1738, 1778, 1825, 1836, 1912

Endemism, 6, 113, 135, 148, 317, 490, 739, 824, 828, 1244, 1408, 1480, 1481, 1843, 1846, 1955, 1980, 1981, 1988, 2002, 2011
Endorheic freshwater, 1349–1359
Energy, 5, 6, 8, 21, 49–52, 81, 94, 100, 115, 116, 119, 149, 157, 184, 188, 333, 336, 348, 367, 414, 415, 423, 448, 535, 554, 559, 581, 584, 608, 629, 631, 641, 663, 675, 694, 702, 740, 769, 793, 808, 818, 921, 926, 928, 941, 942, 969, 1005, 1035, 1200, 1282, 1324, 1369, 1372, 1376, 1382, 1400, 1419, 1425, 1428, 1443, 1445, 1590, 1592, 1654, 1671, 1681, 1867, 1868, 1912, 1926, 1934, 1969
   alien plants, 384
   amphibians, 658
   karst ecosystems, 327
   peatlands, 239
   production, 5, 49, 116, 119, 417, 423, 740, 808, 941, 942
   renewable energy, 634, 941
   River Rhine, 927
   Volga Basin, 940
Energy development, 51, 554, 1868
Enhalus acoroides, 75, 82, 1876, 1877, 1970, 1971
Enhydris
   E. enhydris, 219
   E. longicauda, 1788
Enicurus leschenaulti, 1559
Ensis, 1002
   E. americanus, 1092
Environmental flows, 120, 808, 1212, 1230, 1240, 1376, 1490–1492, 1507, 1888, 1892–1893, 1895, 1906, 1917, 1949
Environmental monitoring system, 832
Environmental planning, Congo River, 1212
Environment-development framework (EDF), 489
Eolian sand, 1124, 1322, 1447, 1459, 1888
   dunes, 1447
Eophreatoicus, 211
Epacris
   E. pauciflora, 1995
   E. paucifloris, 1994
Ephedra distachya, 916
Ephemeral, 6, 173, 285, 372, 551, 655, 658, 690, 733, 1105, 1414, 1727, 1782, 1903, 1913, 1979
   kettleholes, 1978
   lakes, 130–133, 136, 146, 1501
   rivers, 1726, 1728
   wetlands, 278, 350, 662, 683, 1069, 1745
Ephippiorhynchus
   E. asiaticus, 1796, 1902, 1946
   E. senegalensis, 1354
Ephydra, 149
Epilobium
   E. angustifolium, 560
   E. palustre, 285, 1537
Epinetrum villosum, 207
Equisetum, 552, 560, 612
   E. arvense, 560
   E. fluviatile, 1537
   E. palustre, 915, 1537
   E. sylvaticum, 1639
Equus
   E. burchelli, 1405
   E. caballus, 1955
   E. grevyi, 1303, 1371
   E. hemionus khur, 1729
   E. quagga, 1358
   E. quagga burchellii, 1371
   E. quagga crawshayi, 1238
Eragrostis, 205, 1701
Eremias, 939
Eremophilus mutisii, 776, 778, 779
Eretmochelys imbricata, 324, 2019
Erica, 1130
   E. rugegensis, 1313
   E. tetralix, 1044
Erignathus barbatus, 1443
Erigone, 1082
Erinaceus concolor, 1142
Eriocoelum microspermum, 207
Eriophorum, 228, 290, 517, 545, 560
   E. angustifolium, 256, 511, 546, 1044
   E. humile, 1537
   E. polystachyon, 1537
   E. scheuchzeri, 512
   E. triste, 511
   E. vaginatum, 256, 517, 1045, 1048
Erodona mactroides, 850
Erosion, freshwater swamps, 218
Erythrina, 208
   E. crista-galli, 164, 816
Erythrocebus patas, 1294
Erythrococca bongensis, 206
Erytroxylum cartagenense, 765
Eschrichtius robustus, 569
Esox
   E. lucius, 476, 597, 612, 939, 953, 980, 1459, 1469
   E. reichertii, 1502

Estuarial sandbank island wetlands, 1557
Estuaries, 5, 7, 10, 11, 13, 22, 26, 37, 56, 60, 64,
    81, 101, 102, 106, 117, 156, 161, 162,
    165, 167, 174, 231, 348, 371, 406–408,
    623–625, 638, 641–644, 647, 675,
    700, 701, 733, 749, 753, 786, 787,
    790–792, 794, 814–817, 824, 825,
    827, 829, 830, 836, 841, 848–853, 864,
    887, 888, 930, 960, 963, 965, 968–971,
    986, 988, 1024, 1025, 1029, 1063,
    1080, 1090, 1124, 1129, 1142, 1144,
    1156, 1337, 1424, 1426, 1482, 1488,
    1554, 1557–1560, 1583, 1678–1700,
    1718, 1731, 1734, 1738, 1746, 1747,
    1809, 1838, 1866, 1879–1882, 1914,
    1930, 1932, 1933, 1947, 1955, 1970,
    1980, 1984, 1985, 1998
  barbuilt estuaries, 39
  British, 1007–1008
  common names, 21
  definition, 38
  dredging, 50
  drowned River estuaries, 39
  dynamics, 41–44
  ecosystem services, 48–50, 61–64,
    643–645, 852–853, 1034–1035
  fjords, 39
  fresh part of, 157–158
  of Great Britain, 997–1008
  in healthy, 49
  intertidal areas, 50
  macrotidal estuary, 41
  negative, 39
  nutrient enrichment, 50
  oceanic water movements, 38
  plankton of, 44
  population growth and associated socio-
    economic drivers, 52
  positive, 39
  productivity and biodiversity, 44–48
  salt wedge estuary, 40, 41, 45, 46, 570–571,
    849
  shoreline modifications, 50
  Sungai Pulai, 1881–1882
  tectonic estuaries, 39
  threats and future challenges, 50–53
  tidal barriers/seawalls, 50
  vertically homogeneous estuary, 41
  wash, 1011–1021
  waste disposal, 50
  Yangtze River, 1557–1558
  Yellow River, 1583 (*see also* San Francisco
    Bay Estuary (SFBE))

Estuarine marsh, 5, 7, 49, 56, 57
  ecosystem services, 61–64
  formation, 58
  Fraser River Delta, 569–570
  geographic distribution and extent, 58–60
  plant species diversity, 60–61
  threats and challenges, 64–68
Estuarine raised bog, 265, 266
Estuarine vegetation, 74, 78, 81
Ethiopia, 184–186, 202, 1247–1249, 1290,
    1294, 1362, 1367, 1375, 1376, 1414,
    1415
  Baro-Akobo system, 1262–1267
  Berga mire in, 1420
  Lake Tana, 1244
  map of water systems, 1363
Ethmalosa, 1336
Ethnobiodiversity, 2006, 2012, 2015, 2018
Euastacus, 1945
Eubalaena glacialis, 629, 633
Eubosmina coregoni, 603
Eubranchipus spp., 659
Eucalyptus, 1433
  E. camaldulensis, 208, 1901
  E. coolabah, 1901
  E. gomphocephalus, 1961
  E. grandis, 1313
  E. largiflorens, 1901
Eugeissona ambigua, 211, 1844
Eugenia, 1823, 1834
  E. formosum, 207
Eugerres plumieri, 763
Euglena spp., 1385
Eulophia graminea, 1816
Eunapius subterraneus, 341
Eunectes murinus, 735
EUNIS habitat classification, 961, 963, 964
Euodia lepta, 1815
Euonymus europaea, 916
Eupatorium tremulum, 164
Euphagus carolinus, 547
Euphorbia
  E. candelabrum, 1386
  E. palustris, 951
  E. peplis, 1104
Euphrates River, 1686, 1688, 1694
Euplectes jacksoni, 1387
Europe, 9, 11, 22, 28, 29, 58, 59, 64, 113, 115,
    121, 150, 156, 159–161, 163–166,
    175, 180, 184, 238, 240, 311, 317,
    319–322, 335, 337, 340, 370,
    372–377, 380, 385, 476, 622, 629,
    886, 888, 905, 912, 918, 920, 924,

935, 960–966, 968–970, 980, 987, 1000, 1003, 1013, 1047, 1058, 1081, 1094, 1107, 1124, 1131, 1133, 1134, 1140, 1142, 1143, 1156, 1161, 1163, 1164, 1171, 1200, 1209, 1254, 1302, 1325, 1337, 1747
  Amazon of, 904–908
  coastal saltmarsh, 960–961
  Danube River Basin, 887
  Doñana, 1130
  finest wetlands, 1032
  peatlands in, countries, 1158
  salt lakes distribution, 147
  temporal patterns of rates, 374
  Volga Delta, 946–947
  Volga River, 934
  Zostera noltii, 81
European alder, 1640
European Commission, 252, 263, 290
European tidal saltmarshes, *see* Tidal saltmarshes
Euryale ferox, 1717
Eurycercus glacialis, 1071
Euryhaline species, 101, 139, 333, 850, 918
Eurynorhynchus pygmeus, 410, 1806, 1870
Eustatic sea-level rise, 67
Euterpe oleracea, 206, 734, 750
Eutrophication, 8, 66, 84, 89, 105, 139, 353, 361, 366, 496, 502, 587, 598, 600, 602, 684, 740, 766, 779, 830, 864, 891, 976–978, 981, 1027, 1085, 1091, 1107, 1109, 1119, 1133, 1134, 1145, 1153, 1162, 1240, 1258, 1297, 1377, 1390, 1409, 1459, 1595, 1619, 1623, 1680, 1681, 1782, 1862, 1937, 1948, 1963, 1965, 1986, 1989
  fishponds, 478, 480, 483
  of Lake Winnipeg., 603
  land-use change, 138
  nutrient pollution, 891
  pollution, 218
  seagrasses in southeastern Brazil, 845
Evan's Line, 1175–1177
Everglades, 10, 26, 171, 174, 176, 362, 707, 708, 1687
  carbon storage and accumulation, 714
  climate, 709–710
  forest types, 718–719
  future challenges and developments, 720–721
  hydrology, 710–711
  location, 706
  mire/peatland, 709
  nutrients, 711–713
  periphyton, 719
  ponds, 718
  sawgrass in, 706, 715
  sloughs, 716–718
  tree islands, 718
  vegetation and plant communities, 714–720
  wet prairies, 715
Everglades Agricultural Area (EAA), 706–708, 712, 713, 716
Everglades National Park (ENP), 706, 707, 709–712, 714, 716, 717, 720
Excoecaria, 1736
  E. agallocha, 1823
Extent, 4–7, 20, 26–31, 43, 47, 51, 62, 64, 98, 102, 111, 118, 121, 130, 136, 152, 157, 161, 176, 177, 179, 180, 186, 202, 231, 233, 246, 250, 264, 266, 267, 347, 352–354, 363, 370, 372, 376, 384–387, 395, 418, 419, 423, 440, 451, 462, 472, 480, 498, 500, 501, 510, 528, 530, 537, 552, 559, 593, 601, 606, 625, 654, 663, 668, 685, 699, 712, 714, 728, 732, 733, 755, 765, 779, 786, 793, 799, 802, 805, 862, 877, 879, 880, 887, 888, 894, 905, 970, 998, 1014, 1041, 1043, 1049, 1059, 1095, 1102, 1104, 1106, 1129, 1159, 1162, 1210, 1212, 1221, 1222, 1228, 1235, 1245, 1248, 1271, 1283, 1300, 1310, 1311, 1343, 1404, 1418, 1426, 1441, 1442, 1482, 1504, 1570, 1608, 1660, 1679, 1683, 1708, 1760, 1764, 1767, 1780–1781, 1814, 1822, 1837, 1888, 1912, 1933, 1944, 1945, 1948, 1957, 1964, 1965, 1978, 1985, 1987, 1989, 1992, 1998
  Amazonian wetland types, 729–731
  Baro-Akobo Basin, 1265
  boreal zone, 522–523
  of coastal sabkha, 1179
  of coastal zone, 1178
  Congo River, 1200–1205
  estuarine marshes, 58–60
  freshwater swamps, 171
  Ganga-Brahmaputra Basin, 1713
  highland wetlands, 1262–1263
  intertidal zone of East and Southeast Asia, 1754, 1785
  Lower Danube Green Corridor, 898
  lowland wetlands, 1265
  mangroves, 95–96
  Norfolk and Suffolk Broads, 1024–1027

Extent (*cont.*)
  saline wetlands, in Rajasthan, 1728
  San Francisco Bay Estuary, 642
  seagrass meadows, 450
  subterranean space, 334
  Sundarban Delta, 1735–1736
  tidal saltmarshes, 960
  Tonle Sap lake, 1786
  water coverage of, 1665
  wetland dysfunction, 384
  wind farms, 415–417

F

Faidherbia albida, 1272
Fairy shrimp, 654, 656, 659–661
Falcaria vulgaris, 916
Falco
  F. cherrug, 953, 1621
  F. columbarius, 1045, 1084, 1142, 1621, 1629
  F. naumanni, 953, 1354
  F. peregrinus, 980, 1047, 1084, 1142, 1621, 1641
  F. peregrinus anatum, 633
  F. tinnunculus, 1084, 1186
Falkland Islands, 298, 299
Farm Bill, 581, 685, 694
Farmland habitat, 1638
Fauna, 20, 47, 66, 100, 119, 137, 148, 156, 159, 162, 191, 210, 211, 235, 316, 317, 320, 321, 323, 324, 326, 327, 350, 353, 354, 392, 395, 441, 442, 457, 462, 498, 586, 587, 625, 626, 629, 638, 642, 647, 657, 674, 684, 733, 738, 750, 755, 762, 767, 803, 805, 808, 829, 833, 845, 858, 859, 880, 889, 914, 917, 926, 938, 940, 967, 977, 978, 981, 1002, 1060, 1082, 1092, 1107, 1115, 1118, 1119, 1124, 1129, 1130, 1133, 1153, 1206, 1223, 1265, 1272–1274, 1281, 1288, 1315, 1336, 1344, 1358, 1386, 1387, 1395, 1443, 1482, 1502, 1503, 1513–1515, 1517, 1519, 1533, 1608, 1612, 1619, 1621, 1639, 1642, 1652, 1689, 1715, 1736, 1737, 1739, 1775, 1778, 1824–1826, 1835, 1843, 1845, 1891, 1905, 1915, 1948, 1952, 1961, 1963, 1964, 1980, 1988
  aquatic invertebrate, 1981
  aquatic vertebrates, 735
  Axios, Aliakmon, and Gallikos Delta Complex, 1140–1143
  benthic, 977–978, 1336
  biodiversity of, high altitude wetlands, 1609–1610
  Colombian mangrove wetlands, 750–751
  composition of subterranean, 336–337
  diversity, 7, 939
  fish of, Danube Delta, 918
  freshwater fish, 1981
  invertebrate, 1945
  of Macquarie Marshes, 1903
  mammalian, 1629–1630
  mangrove, 101
  native wetland, 1981
  papyrus marshes diversity, 190
  peat swamp forest vegetation, 1756–1759
  species of, 1514
  subterranean wetlands, 331–343
  Tasek Bera Ramsar Site, 1859–1860
  terrestrial vertebrate species, 735–736
  tidal freshwater wetlands diversity, 165–166
  vertebrate, 1978
Felis
  F. bengalensis, 1737, 1758, 1825
  F. catus, 1680
  F. chaus, 1613, 1621, 1680, 1737
  F. silvestris, 918
  F. viverrina, 1737
Fens, 5, 20–22, 229, 230, 233–235, 240, 262–266, 278–280, 282, 290, 296–298, 304, 327, 430, 432, 493, 510–512, 516, 518, 528, 529, 534, 536, 608, 609, 706, 708, 709, 715, 717, 720, 827, 876, 877, 879, 880, 938, 1014, 1025–1027, 1029–1032, 1034, 1035, 1043, 1069, 1072, 1156, 1158–1163, 1311, 1316, 1417–1420, 1426–1428, 1434, 1445, 1486, 1534–1541, 1598, 1606, 1978, 1980, 1994, 1997
  artesian spring mire, 256
  blanket/sloping, 1544–1546
  boreal wetlands, 523–524
  floodplain, 254, 256, 264–266, 270, 285, 310
  highland valley meadow, 1542–1543
  lagg, 309–311
  management, 1032–1034
  Mfabeni sedge, 1428
  open-water transition, 252–254
  patterned mires, 259, 260, 283–284
  percolation mire, 257–258
  polygonal fens, 261
  Schwingmoor, 252, 254, 263, 264, 266

sloping sedge, 1543–1544
snowmelt water flow, 261
surface-flow spring mire, 256–257
surface-water flow, 258–261
valley mire, 254–256, 258, 270, 938, 1159, 1540, 1542, 1547
Feral horses, 1126
Fereydoon Kenar Ramsar Site, 12, 1626
  biodiversity, 1640–1642
  conservation status, 1643–1644
  ecosystem services, 1642
  future challenges and developments, 1644–1645
  hydrology, 1636–1638
  location, 1636
  threats, 1644–1645
  wetland ecosystems, 1638–1640
Ferguson's Gulf, Kenya, 1364–1366, 1370, 1377
Festuca, 828, 938, 1701
  F. arenicola, 916
  F. callieri, 916
  F. ovina, 279
  F. richardsonii, 285
  F. rubra, 510, 964, 1082, 1084
Ficopomatus enigmaticus, 340
Ficus, 1385, 1418, 1823, 1834
  F. aurea, 719
  F. carica, 1106
  F. coronata, 207, 208
  F. microcarpa, 1823, 1834
  F. racemosa, 208
  F. tinctoria, 2009
  F. trichopoda, 1427
Filinia, 1357
Filipendula palmata, 1537
Fimbristylis, 205, 1609
  F. dichotoma, 205
  F. longiculmis, 1427
Fine sediments, 45, 451, 622, 848, 1781, 1914
Fine-textured mineral sediments, 282
Fire, 65, 203, 206–208, 221, 237, 310, 349, 352, 516, 517, 519, 581, 672–674, 711, 714, 718, 719, 721, 768, 791, 792, 802, 803, 877, 938, 956, 1048, 1049, 1163, 1240, 1271, 1275, 1297, 1404, 1408, 1421, 1433, 1524, 1528, 1529, 1547, 1759, 1760, 1790, 1797, 1798, 1805, 1814–1817, 1827, 1834, 1836, 1837, 1844, 1847, 1856, 1905, 1952, 1957, 1963, 1984, 1993, 1998, 2017

Everglades, 715
factors influencing site management, 1862
fishponds, 476
forest, 324, 768, 877, 1759, 1797, 1815, 1817
peatlands, 1162
tropical freshwater swamps, 219
First Nations, 548, 553
Fishery, 5, 51, 53, 64, 85, 87, 116, 118, 119, 126, 136–138, 153, 177, 214, 216–218, 328, 364, 365, 398, 400–402, 406, 407, 410, 428, 450, 452, 456, 476, 538, 553, 570, 585, 588, 597, 625, 628, 629, 634, 644, 737, 738, 752, 764, 767, 779, 780, 793, 829, 830, 844, 852, 853, 900, 901, 918, 940–942, 1014, 1019, 1095, 1118, 1208, 1211, 1222, 1226, 1227, 1229, 1239, 1248, 1249, 1256, 1258, 1275, 1296, 1304, 1325, 1328, 1329, 1338, 1339, 1345, 1346, 1355, 1364, 1377, 1443–1445, 1460, 1570, 1595, 1600, 1623, 1630, 1632, 1688, 1690, 1701, 1702, 1706, 1710, 1713, 1715, 1717–1719, 1737, 1738, 1740, 1778, 1788, 1789, 1797, 1806, 1809, 1826, 1837, 1843, 1848, 1862, 1863, 1869, 1870, 1878, 1881, 1882, 1912, 1947, 1956, 1988, 2020
  Amazon Basin, 741
  Anzali wetland, 1621
  artisanal, 438–443
  Cambodia, 1789
  Ciénaga Grande de Santa Marta, 767
  Dalai Lake, 1503
  Danube Delta, 919–920
  Lake Baikal, 1483
  Lake Tonlé Sap, 210
  Lake Turkana, 1370–1373
  Lena Delta, 1452, 1455
  LMB yield, 1780
  management, 398, 400–401, 440, 441, 443, 482, 642, 1375
  Nile Delta, 1254
  Saemangeum Estuarine System, 406, 410
  San Francisco Bay Estuary, 645
  seagrass, 438–442
  Shadegan Wetland, 1682
  yield for Lake Turkana, 1373
  in Zhalong, 1527
Fishes, 50, 52, 86, 100, 101, 113–115, 166, 176, 337, 398, 652, 656, 682, 738, 750, 761, 776, 816, 817, 930, 1107, 1118, 1119, 1207, 1254, 1273, 1323, 1325,

1336, 1337, 1357, 1366, 1377, 1403,
1405, 1406, 1447, 1469, 1474, 1626,
1666, 1679, 1715, 1718, 1765,
1774–1776, 1778, 1878, 1879, 1945,
1946, 1948, 1952–1953
annual yields, 477
assemblage variation, 850
biodiversity, 977–978
British estuaries, 1002
British fish species, 1002
Bujagh National Park, 1630
Ciénaga Grande de Santa Marta, 763–764
Congo Basin, 1206, 1208
DanishWadden Sea, 988–989
diverse and dynamic assemblage, 398, 1717
fauna, 1642
freshwater, 113
Golfe d'Arguin, 1325
harvesting, 478, 2010
Lake Seyfe, 1190
Mekong Delta, 1805–1806
Mekong River, 1805–1806
Nile River Basin, 1273
Paraná-Paraguay River, 817
production management, 476–477, 900
sawfishes, 1339
stocks, 161
Fishing club, 1339
Fishpond(s), 11, 103, 583, 740, 1528, 1559,
1571, 1718, 1827
aquaculture, 476
biodiversity of water birds, 471–473
conservation status and management,
480–482
ecosystem services, 473–476
fish production management, 476–477
future challenges and developments, 483
Lednice, 481
management, 472
vegetation management, 477–479
Fjords, 38–40, 46, 48, 510, 829, 879, 974, 977,
980, 982, 998, 1000, 1090–1092
Flacourtia rukam, 1823, 1834
Flagship species, 928, 1086, 1229, 1472, 1644,
1747, 1779
Floating reed beds, 888, 915
Floating roads, 421, 422
Flood/flooding, 39, 41, 43, 45, 62, 63, 102,
115–119, 126, 153, 161, 162, 178,
201, 207, 215, 238, 246, 254, 265,
277, 279, 316, 321, 325, 365,
473–475, 533, 552, 578, 586, 594,
599, 601, 647, 652, 653, 706, 729,

732, 733, 735–737, 739, 760,
788–791, 802–804, 807, 808, 815,
816, 830, 831, 880, 890–893,
905–908, 915, 920, 929, 951, 1026,
1031, 1032, 1034, 1060, 1064, 1070,
1071, 1074, 1109, 1145, 1162, 1202,
1208, 1218–1220, 1222, 1223, 1225,
1228–1230, 1239, 1246, 1263, 1265,
1266, 1270, 1271, 1275, 1280–1283,
1294, 1300, 1302, 1305, 1316, 1366,
1370, 1373, 1376, 1418, 1426, 1428,
1441, 1442, 1490, 1491, 1493, 1503,
1568–1571, 1579–1583, 1586, 1587,
1598, 1669, 1670, 1677, 1678, 1682,
1683, 1710, 1713, 1717, 1721, 1729,
1739, 1747, 1748, 1767, 1775, 1778,
1780, 1781, 1789, 1804, 1805, 1807,
1809, 1843, 1847, 1862, 1868, 1879,
1880, 1890, 1905, 1916, 1922–1924,
1942, 1952, 1955, 1973, 1989,
1996–1998, 2014
Amazon River, 706, 729, 732–735
Amur River, 1496
Bad River, 433
Bay of Fundy, 622, 624, 625
bottomland hardwood forest, 581
buffer, 1717
control, Zhalong marsh, 1527
forest, 1786
freshwater swamps, 216–217
Ganga River, 1716
Great Barrier Reef, 452, 454, 457
intertidal flat, 21
Kashmir valley, 1708, 1709
Kilombero Valley, 1342
Lake Eyre, 1925, 1926
Lower Danube Green Corridor, 898–902
Mekong Basin, 1783, 1786
mires, 1418
Mississippi Alluvial Valley, 578, 580,
583–585
morphological floodplain, 888
Nen River, 1523, 1524
Okavango River, 1398–1402
Pantanal, 798, 799, 801
Peace River, 551
Poyang Lake, 1556
prairie wetlands, 683
prevention, 926, 927
reproductive adaptations, 204
safety and nature development, 928
salt marsh, 22
San Francisco Bay Estuary, 638, 643, 644

savanna, 26, 174, 798
Sefid Rood, 1627, 1628
Sefid Rud Delta, 1630
tidal freshwater wetlands, 167
Tonle Sap Lake, 1787
Tram Chim, 1795
wetland vegetation types, 1900
Yangtze River Basin, 1552
Zambezi River, 1235
Zhalong Marsh, 1527
Floodplain(s), 12, 21, 26, 27, 110, 114, 116, 117, 120, 128, 130, 132, 135, 137, 173, 174, 177, 178, 194, 200–203, 205, 208, 209, 211, 214, 215, 219, 246, 348, 350, 373, 385, 400, 481, 546, 551, 560, 578, 580, 587, 653, 698, 728, 734–739, 741, 750, 760, 786, 789, 790, 793, 794, 799, 803–805, 810, 889–902, 906, 908, 920, 924, 927, 928, 937–940, 951, 1208, 1220–1222, 1225–1230, 1235, 1238–1240, 1246, 1262, 1263, 1270, 1281, 1282, 1298, 1303, 1305, 1398, 1400, 1402–1407, 1414, 1426, 1432, 1433, 1445–1448, 1453, 1454, 1473, 1475, 1482, 1486, 1488, 1489, 1491, 1496, 1497, 1516, 1518–1520, 1554, 1578, 1699, 1700, 1713, 1716, 1717, 1719, 1720, 1744, 1745, 1748, 1775, 1776, 1778, 1780–1782, 1789, 1904, 1905, 1923–1925, 1952, 1955, 1956, 1960, 1978, 1988
    Argun River, 1501–1505, 1507
    Bahr el Ghazal wetlands, 1274
    channels, 791
    Congo River, 1201
    Danau Sentarum National Park, 1842
    Danube-Drava National Park, 905
    Danube River Basin, 888–895
    fen, 254–256, 270, 285, 286, 310
    fisheries, 1208
    former, 888
    freshwater swamps in, 210
    grassland and swamp communities, 1236–1237
    habitats, 698
    Kilombero, 1341–1347
    Lake Bolon, 1513
    Lake Chilika, 398, 399
    lakes, 791
    Lake Turkana, 1372
    large-river, 729–731
    Lower Danube's floodplains, 898
    Lužnice River, 480
    Macquarie Marshes, 1898–1901
    Mekong River Basin, 1765, 1767
    mires, 1428, 1431
    Mississippi Alluvial Valley, 578, 580, 587
    morphological, 888
    Odra River, 482
    papyrus wetlands, 184–186
    Paraná River Delta, 814
    productive systems, 210
    raised bog, 264–266
    restoration, 895, 900–902
    River Mahanadi, 1744
    River Murray, 1888–1893
    San Francisco Bay Estuary, 647
    Savanna Communities, 1236
    Sudd, 1300
    swamp and marsh vegetation, 171, 172
    Tonle Sap Lake, 1786, 1787
    tropical wetlands, 201
    Volga Basin, 936, 951, 956–958
    wetland, 1786, 1892, 1898, 1942–1949, 1956 (*see also* Australian wet tropics)
    Wet Tropics landscape, 1942
    woodland, 1302
    Yukon River, 545
    Zambezi River Basin, 1219
Flood pulse, 204, 365, 551, 729, 733, 734, 736, 739, 788–790, 799, 801, 804, 816, 888, 1398–1401, 1403, 1405–1406, 1716, 1767, 1781, 1783, 1804, 1952
Flood switching, 1399–1401, 1403–1410
Flora, 45, 66, 96, 156, 235, 317, 324, 326, 327, 353, 392, 395, 462, 480, 518, 524, 586, 587, 626, 638, 642, 647, 684, 731, 738, 761, 767, 776, 803, 804, 808, 845, 858, 859, 889, 914, 926, 940, 967, 1002, 1060, 1081, 1129, 1130, 1206, 1223, 1265, 1281, 1288, 1386, 1443, 1524, 1542, 1543, 1612, 1619, 1621, 1639, 1678, 1689, 1843, 1952, 1963, 1964, 1985, 1988
    Amazon River Basin, 733–735
    Axios, Aliakmon and Gallikos Delta Complex, 1140
    Berbak National Park, 1835
    biodiversity, 1344
    Central Yakutian region, 1468
    high altitude wetlands, Nepal, 1608, 1609
    highland peatlands, 1533
    Lake Baikal, 1482
    Lake Chilika, 398

Flora (*cont.*)
  Lake Elementeita, 1387
  Lake Khanka-Xingkaihu, 1517
  Lake Naivasha, 1358
  Lake Seyfe, 1189–1190
  middle Aldan Basin, 1474
  Mongolian mires, 1546
  New Zealand wetlands, 1978–1981
  Sundarban mangroves, 1736, 1737
  Tasek Bera, 1856–1859
  tidal freshwater wetlands, 163
  Volga Basin, 939
Flow, 5, 38, 39, 41–43, 51, 99–101, 110–112, 117–119, 131, 135, 138, 152, 157, 158, 163, 167, 204, 209, 213, 214, 217, 233, 284–286, 314, 323, 325, 350, 351, 353, 360, 421–423, 431, 433, 474, 519, 523, 524, 528–532, 534, 545, 551, 552, 554, 559, 568, 572, 594, 598, 626, 628, 629, 634, 640, 647, 694, 698, 710, 711, 719–721, 728, 738, 760, 761, 766, 777, 780, 786, 789–791, 793, 807, 808, 888, 891, 893, 924, 929, 940, 949, 951, 986, 1014, 1062, 1070, 1074, 1080, 1114, 1133, 1140, 1205, 1210, 1212, 1218, 1219, 1221–1223, 1225, 1226, 1228–1230, 1239, 1249, 1266, 1270, 1271, 1274, 1275, 1288, 1297, 1302, 1305, 1310, 1315, 1376, 1397, 1399–1402, 1410, 1418, 1426–1248, 1440, 1441, 1458, 1490–1492, 1503, 1507, 1540, 1542, 1571, 1577, 1581, 1586, 1587, 1651, 1677, 1678, 1686, 1688, 1692, 1700, 1735, 1739, 1744, 1755, 1759, 1783, 1803–1806, 1809, 1868, 1900, 1905, 1913, 1916–1918, 1948, 1949, 1952, 1961
  annual of dinder river at Dinder station, 1292
  Anzali Wetland, 1616
  Argun River, 1504
  Baro-Akobo system, 1262
  Baro River, 1282
  channeled surface-water, 258–261
  Conchalí Lagoon, 861
  Danube River, 887
  Dinder River, 1290–1292
  of floodwaters, 254
  high-quality freshwater, 1692
  Indus Basin, 1699
  Kapuas River, 1843, 1847
  Kilombero River, 1342
  Lake Bogoria, 1384
  Lake Eyre Basin, 1922–1926
  Lake Manitoba, 592
  Lake Uromiyeh, 1662
  management, 1949
  Mekong River, 1765, 1767
  Nile Basin, 1248
  Okavango River, 1398
  Omo River, 1377
  Paraná River Delta wetlands, 817
  Rann of Kachchh, 1727, 1728
  regulation, 940, 949, 951, 1230
  Río de la Plata system, 848
  River Murray, 1914
  Sudd, 1304
  through mires, 1426, 1429
  Tonle Sap River, 1786
  Tram Chim National Park, 1794
  Volga River, 937
  wet tropics landscape, 1942
  White Nile, 1246, 1247
  Yangtze River, 1552, 1560, 1566, 1568
  Yenisei River Basin, 1480, 1484
  Yukon River, 547
  Zhalong Marsh, 1523, 1524
Flow Country, northern Scotland, 261, 297, 298, 300, 307, 1040–1054
  animal life, 1047
  bog systems, 1045
  climate, 1041–1043
  conservation status, 1049–1051
  land-use pattern, 1047–1048
  location, 1040
  mire development, 1043
  physical setting, 1040–1041
  surface patterns, 1043
  threats, 1051–1054
  vegetation, 1043
Flush, 105, 258–259, 647, 709, 842, 1180, 1433, 1568, 1702, 1734, 1781, 1932, 1948, 1980, 2013
Fluvial wetlands, 728, 732, 786, 834
Fog, 298, 300, 304
  basin raised bog, 265
  blanket mire, 297
  occult precipitation mires, 271
  watershed blanket bog, 270
Fontinalis antipyretica, 1539
Food and livelihood security, 2023
Food chain, 46, 103, 393, 477, 631, 843, 1324, 1404, 1810
Food quality, 657

Index of Keywords   2067

Food security, 7, 115, 177, 438, 440–441, 443, 778, 779, 1229, 1240, 1258, 1338, 1373, 1433, 1749, 1780, 1806
Foraminifera, 86, 1914
Fordia splendissima, 208
Forest(s), 6, 21, 22, 58, 63, 85, 99, 138, 158, 166, 174, 193, 216–218, 220, 231–233, 238, 241, 249, 260, 277, 285, 310, 323–326, 339, 340, 372, 374, 400, 401, 417, 480, 516, 523, 533, 534, 538, 539, 545, 559–561, 563, 578, 579, 615–617, 652, 657–660, 663, 668, 670–674, 685, 733–740, 767, 777, 779, 780, 782, 791, 801, 802, 804–807, 810, 816, 817, 825, 836, 874, 876–879, 889, 891, 899, 901, 904, 905, 919, 920, 928, 934, 937, 938, 940, 952, 1040, 1050, 1082, 1116, 1192, 1203, 1204, 1207–1209, 1226, 1235, 1246, 1262, 1282, 1293, 1296, 1302, 1316, 1334, 1338, 1343, 1345, 1347, 1359, 1386, 1389, 1390, 1406, 1408, 1465, 1469, 1480, 1482, 1488, 1491, 1493, 1510, 1511, 1513, 1515, 1532, 1534, 1542, 1546, 1547, 1606, 1608, 1610–1612, 1627, 1636, 1638, 1639, 1643, 1700, 1710, 1717, 1720, 1748, 1776, 1777, 1779, 1781, 1795, 1797, 1845, 1847, 1848, 1852, 1857–1863, 1867, 1891, 1892, 1901, 1905, 1945–1947, 1981, 1984, 1988, 2009, 2017
amphibians, 658
boreal forest, 10, 26, 523, 533, 535, 539, 545, 559, 592, 673, 681, 1480
bottomland hardwood, 580–583
Canadian and Alaskan boreal, 533–534
cloud, 231, 269, 272, 300, 301, 308, 420
coastal wetlands, 829–830
in Congo River Basin, 1200
Danube Delta, 915–917
Dauria, 1500
dipterocarp, 1756, 1779
dwarf swamp, 1843, 1844
elfin, 232, 300
flooded forests, 174, 207, 209, 372, 734, 735, 1208, 1445, 1768–1770, 1772–1775, 1779, 1781, 1786–1790, 1808, 1845
fluvial, 792
freshwater swamp, 170, 171, 207–208
Hyrcanian forests, 1639
Mahanadi Delta, 1744
mangroves, 26, 93–107, 323, 326, 400, 440, 452, 750–754, 764–766, 768, 1221, 1227, 1235, 1237, 1337, 1700, 1734, 1736, 1738–1740, 1754, 1805, 1808, 1809, 1822, 1945, 2007, 2008, 2019
Melaleuca, 208, 219, 220, 1777, 1815–1817
Mississippi Alluvial Valley, 585–588
Mongolia, 1534–1540
Nile Basin, 1270
nothofagus, 879
peat swamp, 26, 230, 234, 239, 240, 265, 1415, 1426, 1428–1432, 1434, 1754–1760, 1768–1772, 1808, 1813–1817, 1820, 1822, 1823, 1826, 1827, 1832, 1834, 1836, 1837, 1843–1847, 1852, 1856, 1862
pine forests, 938, 1124, 1132
plant communities, 657
ponds, 653
rainforest, 79, 208, 219, 230, 236, 323, 326, 737, 738, 741, 825, 840, 876, 877, 1200, 1205, 1208, 1209, 1552, 1769, 1770, 1772, 1942, 1945, 1980
riparian, 164, 578, 585, 588, 765, 1106, 1109, 1121, 1140, 1142, 1226, 1359, 1834, 1893
stunted swamp, 208, 1844
Swamp Communities, 1237
swamp forest, 114, 171, 177, 200, 203, 204, 207–208, 211, 214, 216, 219, 221, 228, 718–719, 734, 878, 1237, 1415, 1418, 1419, 1426–1428, 1433, 1434, 1717, 1843
tall swamp, 200, 208
temperate riverine, 916
tropical peat swamp, 240, 264, 269, 301, 1424, 1753–1760
tropical swamp, 1414, 1423–1434
U Minh peat swamp, 1813–1817
vernal pools, 653
wetlands of Chile, 828–830
Wular Lake, 1706
Zambezi Delta, 1239
Forested peatlands, 238, 1445, 1537, 1760
Forestiera acuminate, 583
Forestry, 107, 138, 170, 237, 238, 321, 349, 534, 538, 554, 607, 668, 675, 741, 891, 917, 938, 1049–1054, 1188, 1553, 1610, 1710, 1781, 1787, 1788, 1820, 1836, 1837, 1862, 1942
Forsinard, 1051, 1053
Fothergilla gardenii, 674
Fragaria chiloensis, 560

Frankenia
 F. hirsuta, 1189
 F. laevis, 965
Fraser Delta, 566–573
 bog, 571
 Boundary Bay, 570
 conservation designations, 571
 ecological importance, 568–569
 estuarine marshes, 569
 intertidal, 569
 location, 566
 ocean, 570
 salt marshes, 570
 salt wedge and plume, 570
 threats, 572–573
 upland farmland, 571
 wetland diversity, 569, 572
Fraser estuary, 572
Fraser River
 burns bog, 571
 conservation designations, 571–572
 diversity, 569
 ecological importance, 568–569
 estuarine marshes, 569–570
 formation of delta, 567–568
 future challenges and developments, 572–573
 salt marshes, 570
 salt wedge and plume, 570–571
 threats, 572–573
Fraxinus, 1072
 F. angustiflora, 916
 F. angustifolia, 915, 1106
 F. excelsior, 1072, 1640
 F. nigra, 612
 F. pallisiae, 916
 F. pennsylvanica, 581
Fregata minor, 2016
Freshwater, 29, 30, 41, 44, 46, 50, 53, 94, 101, 103, 105, 110, 112–121, 148, 150, 151, 153, 238, 278, 324, 327, 333, 335, 340, 341, 350, 361, 364, 373, 546, 559, 561, 567, 569, 571, 603, 623, 674, 675, 721, 730, 731, 733, 737, 830, 834, 862, 864, 880, 889, 915, 918, 948, 977, 980, 998, 1004, 1006, 1017, 1021, 1025, 1027, 1032, 1047, 1081, 1082, 1107–1109, 1116, 1117, 1119, 1130, 1133, 1146, 1160, 1170, 1181, 1188, 1190, 1238, 1239, 1273, 1303, 1320, 1322, 1337, 1345, 1368, 1556, 1600, 1664, 1666, 1683, 1688, 1692, 1700, 1718, 1727, 1731, 1759, 1774, 1775, 1778, 1779, 1825–1828, 1838, 1868, 1879, 1888, 1890, 1903, 1911, 1917, 1932, 1933, 1936, 1955, 1957, 1971, 1988, 2002, 2006–2009, 2011–2014, 2017
 agricultural activities, 1145
 Amazon Basin, 728
 Amur River, 1486–1493
 Anzali Mordab Complex, 1618
 Axios, Aliakmon, and Gallikos Delta Complex, 1138
 Azraq Oasis, 462
 Berbak National Park, 1832, 1834–1836
 Camargue, Rhône River, 1103–1105
 Central Valley, 700
 Ciénaga Grande de Santa Marta, 759–761, 764–766
 coastal marshes, 285
 coastal wetlands, 593
 Conchalí Lagoon, 858
 Congo River, 1200–1202, 1204–1212
 Cooper Creek, 1924
 Coorong, 1912, 1913
 Dauria, 1500
 distribution of peatlands, 231
 ecoregions, 1201, 1202, 1204, 1487, 1489
 ecosystems, 5, 110, 112, 118, 121, 232, 949–953, 955
 emergent marsh, 546
 endorheic, 1350–1359
 estuarine environments, 37–39
 estuarine marshes, 56
 Everglades, 710–711, 714
 Fereydoon Kenar Ramsar Site, 1636
 fish, 113–115, 583, 584, 817, 953, 1239, 1584, 1600, 1679, 1717, 1718, 1759, 1816, 1860, 1945, 1981
 fisheries, 1345
 Fraser River, 570
 Ganga-Brahmaputra Basin, 1715, 1717
 Gnangara Groundwater System, 1960
 hydrology, 157
 hydrophytic vegetation, 1140
 Indus River, 1701, 1702
 Kakadu National Park, 1952
 Kakagon/Bad River Sloughs, 428
 lagoons, 827, 1106, 1618
 Lake Alexandrina, 1914
 Lake Baikal, 1479–1482
 Lake Bogoria, 1384
 Lake Chilika, 398
 Lake Eyre, 1922, 1924–1926
 Lake Fúquene, 774, 776, 777

Lake Khanka-Xingkaihu, 1517
Lake Magadi, 1388
Lake Sambhar, 1729
lakes and reservoirs, 126–139
Lena River Basin, 1443–1445
lentic ecosystems, 914
Living Planet Index, 377
Lower Danube Green Corridor, 898
Macquarie Marshes, 1898
Mahanadi Delta, 1746
marshes, 1117
Mekong Basin, 1777, 1805–1809
Mont-Saint-Michel Bay, 1080
New Zealand wetlands, 1978, 1980–1982, 1985, 1987
North Norfolk Coast, 1018, 1019
Okavango Delta, 1395, 1400, 1408
Omo River, 1362
Paraná-Paraguay Fluvial Corridor, 790, 793
Peace-Athabasca Delta, 550
Phoksundo wetland, 1608
Poyang Lake, 1566
Ramsar wetland, 5
Río de la Plata system, 848, 850, 851, 853
Saemangeum, 406, 410
San Francisco Bay Estuary, 639–641, 647
sawgrass, 715
Sefid Rood River, 1627, 1630
Sembilang National Park, 1820–1822
ShadeganWetland, 1676–1681
soda lakes, 1382
Sundarban mangroves, 1734–1737, 1739, 1740
swamps and marshes, 6, 170–180
swamps on mineral soils, 200–221
Tasek Bera, 1852, 1856, 1860
tidal freshwater wetlands, 156–167
Tonle Sap River, 1765, 1786
Tram Chim, 1795
turloughs, 1069
U Minh peat swamps, 1815, 1816
Volga Delta, 940, 950, 952, 953
Wadden Sea, 986–989
wetlands (*see* Tidal freshwater wetlands)
wet prairies, 715
wet tropics, 1944–1946
Yangtze River, 1553, 1559
Zambezi River, 1228, 1235
Zhaling Lake, 1579
Freshwater Ecoregions of the World (FEOW), 1480
Freshwater lakes and reservoirs, 30, 50, 144, 147, 150, 152, 175, 184, 364, 406, 410, 462, 828, 940, 1006, 1322, 1445, 1486, 1553, 1556, 1559, 1581, 1582, 1608, 1715, 1727, 1731, 1765, 1775, 1809, 1890, 1911, 1912, 1922
area, 127–128
Azraq Oasis, 462
biota, 133–135
ecological features, 131–133
ecosystem services, 135–137
Estuarine environments, 37
geographic distribution, 130–131
glacier ice movement, 126
global biogeochemical cycles, 126
Golfe d'Arguin, 1320
Lake Baikal, 1479
Lake Baringo, 1351
Lake Manitoba, 592
lake volume and residence time, 128–129
Manitoba Great Lakes, 592
papyrus marshes, 184
Poyang Lake, 1566
River Murray, 1888
threats and future challenges, 137–139
Tonle Sap Lake, 1786
Freshwater swamps, 326, 733, 1190, 1445, 1826, 1985, 2011
Berbak National Park, 1832, 1834–1836
biota, 173–177
ecological features, 173
ecosystem services, 177–178
global biogeographic regionalization, 171
global distribution, 171–173
mineral soil-based, 6
papyrus swamps, 170
Sembilang National Park, 1822, 1823
threats and trends, 178–180, 215–216
variety, 170
Fringillidae, 939
Fritillaria, 1701
Frost heaving, 278, 279, 282
Frozen peatlands, 277, 278, 284, 286
Fucus, 48
 F. edentatus, 626
 F. serratus, 626
 F. spiralis, 626
 F. vesiculosus, 626
Fuirena umbellata, 205
Fulica
 F. americana, 583, 683
 F. armillata, 859
 F. atra, 971, 1073, 1092, 1190, 1469, 1609, 1621, 1641, 1653
 F. cristata, 1131

Fulica (cont.)
   F. leucoptera, 859
   F. rufifrons, 859
Fumana procumbens, 916
Functional landscape approach, 1266
Fundy Basin, 622
Fuquene lake, 774–782

G

Gadus morhua, 629, 633, 1016
Galaxias gollumoides, 1981
Galaxiella nigrostriata, 1961
Galeocerdo cuvier, 1325
Galictis cuja, 852
Galium palustre, 915
Gallesia, 206
Gallicrex cinerea, 1716
Gallinago
   G. gallinago, 472, 980, 1073, 1641
   G. media, 1641
   G. megala, 1870
   G. nemoricola, 1610
   G. solitaria, 1610
Gallinula chloropus, 1610, 1708
Gallirallus philippensis assimilis, 1981
Gambusia, 1358
   G. affinis, 1109
   G. holbrooki, 1131, 1134
Gammarus spp., 612
Ganga-Brahmaputra Basin (GBB), 1712–1721
   challenges, 1720–1721
   conservation and management, 1719–1720
   extent, biodiversity and ecosystem services, 1713–1719
   future challenges and developments, 1720–1721
   geology of, 1712
Ganga-Brahmaputra-Meghna Basin, 1734
Gangetic Delta, 1712, 1713, 1718–1721
Gangetic plains, 1713, 1716–1717, 1720
Garcinia
   G. bancana, 208
   G. borneensis, 205
   G. dulcis, 208
Gardenia
   G. kambodiana, 205
   G. tentaculata, 205
   G. tubifera, 1834
Gastromyzon embalohensis, 1845
Gavia
   G. arctica, 236, 547, 1047, 1094

   G. immer, 612
   G. stellata, 418, 547, 1044, 1094
Gavialis gangeticus, 1715
Gazella
   G. dama, 1303
   G. dorcas, 1303, 1324
   G. leptoceros, 1303
   G. rufifrons, 1303
   G. soemmerringii, 1303
Gelochelidon nilotica, 1338
Gemenc-Béda-Karapancsa wetlands, 889
General Department Administration of Nature Conservation and Protection (GDANCP), 1788, 1790
Genetta genetta, 1107, 1131
Genidens barbus, 849, 853
Gentiana, 1701
Geothlypis
   G. aequinoctialis, 851
   G. trichas, 648
Ghawarna, 1170
Gibbula umbilicalis, 1324
Gilan, 1616, 1626, 1627, 1631
Giraffa
   G. camelopardalis, 1358, 1405
   G. camelopardalis rothschildi, 1386
Glacicavicola, 339
Glareola
   G. maldivarum, 1870
   G. pratincola, 1141, 1142, 1629
Glaux, 61
   G. maritima, 964
Gleichenia dicarpa, 26, 1994–1996
Global change, 98, 501, 1110, 2023
Globally threatened species, 325, 377, 953, 1130, 1188, 1190, 1525, 1584, 1628, 1629, 1788, 1860, 1870
   birds, 1641
   Bujagh National Park, 1629
   Shadegan Wetland, 1676
   Tonle Sap Lake, 1787
Global Peatland Initiative Project, 1534
Global warming, 53, 67, 85, 167, 286, 367, 456, 476, 632, 845, 1035, 1194, 1390, 1555, 1593
Globicephala melaena, 629
Glossina morsitans, 1385
Gluta
   G. pubescens, 208
   G. renghas, 1834
   G. wallichii, 208, 1834
Glycera, 630
   G. dibranchiata, 47

Glyceria aquatica, 915
Glycyrrhiza
  G. echinata, 951
  G. glabra, 951
Glyptemys insculpta, 612
Gmelina, 208
  G. asiatica, 205
Gnangara groundwater system, 1960
  biophysical characteristics, 1960–1961
  change to ecological character, 1961
  wetland mapping, 1962
Gobio sibiricus, 1480
Godwin, H., 290, 304
Goebelobryum unguiculatum, 1995–1997
Gokyo wetlands, 1608, 1612
  cultural and religious significance, 1611
  fauna, 1610
  flora, 1609
Golfe d'Arguin, 1320, 1322, 1325, 1326, 1328, 1329
Gonystylus bancanus, 239, 1757
Goolwa channel, 1914, 1916
Gordonia lasianthus, 672–674
Gorsachius magnificus, 325
Gosaikunda wetlands, 1607, 1608, 1612, 1715
  cultural significance, 1611
  fauna, 1609–1610
  flora, 1609
Governance, 195, 464, 467, 769, 854, 1610, 1906
  Azraq Oasis management, 465–467
  Pantanal, 809
Gracilaria, 86
Graphoderus bilineatus, 1071
Grasslands, 171, 200, 215, 217, 220, 233, 510, 552, 568, 641, 653, 701, 791, 793, 816, 856, 879, 880, 1001, 1004, 1013, 1069, 1072–1075, 1082, 1105–1106, 1144, 1189, 1220, 1223, 1225, 1227, 1228, 1230, 1235–1237, 1239, 1247, 1265, 1272, 1281, 1283, 1293, 1297, 1302, 1344, 1357, 1386, 1399, 1402, 1404, 1405, 1416, 1493, 1502–1504, 1524–1526, 1553, 1630, 1653, 1669, 1715, 1730, 1768–1772, 1774–1777, 1779, 1787, 1788, 1790, 1805, 1901, 2007, 2008
  Anzali Mordab Complex, 1618
  Baro-Akobo Basin, 1262
  Central Valley, 698
  Gokyo wetlands, 1608
  Kaziranga National Park, 1717
  Kilombero floodplain, 1343

Lake Naivasha, 1357
Nenjiang River, 1500
North Norfolk Coast, 1018
Prairie Pothole Region, 680–683, 685–686
rain floodplain, 1236, 1302
and riparian habitats, 698
river floodplain, 1236
Sefid Rood, 1626–1628
Sudd, 1300
Terai region, 1715
upper Zambezi, 1221
Wadden Sea, 994
Yangtze River Basin, 1553
Yellow River, 1582, 1584
Zhalong marsh, 1523, 1525
Gratiola officinalis, 951
Grauer's swamp warbler, 1314
Grazing, 58, 62, 65, 66, 85, 118, 136, 152, 178, 187, 194, 214, 217, 239, 240, 327, 349, 442, 452, 630, 631, 647, 681, 843, 919, 926, 928, 929, 954, 956, 968, 980, 994, 1026, 1027, 1029, 1031, 1032, 1043, 1048, 1069, 1082, 1083, 1086, 1140, 1162, 1163, 1175, 1226, 1227, 1239, 1249, 1263, 1264, 1266, 1274, 1275, 1296, 1297, 1302, 1309, 1419, 1420, 1528, 1533, 1612, 1621, 1627, 1630, 1632, 1654, 1666, 1669, 1670, 1682, 1718, 1729, 1738, 1878, 1942, 1948, 2017
  arctic peatlands, 287
  blanket mire, 299
  British estuaries, 1004
  Camargue wetlands, 1108, 1109
  Doñana wetlands, 1133
  Ebro Delta, 1117
  European saltmarshes, 969
  high altitude wetlands of Nepal, 1611, 1612
  highland peatlands of Mongolia, 1546
  karst systems, 316
  Machar marshes, 1281, 1282, 1284
  Macquarie Marshes, 1904
  Makgadikgadi wetlands, 488
  Myall Lake, 1937
  North Norfolk Coast, 1013, 1018
  Okavango Delta, 1403, 1405, 1407
  Patagonian peatlands, 879
  Ringkøbing Fjord, 974
  Rugezi Marsh, 1309
  saltmarsh, 452
  Sudd, 1304, 1305
  turloughs, 1073–1075
  Yellow River Delta, 1586

Grazing regime, turloughs, 1069, 1074, 1075
Great Artesian Basin, 1888
Great barrier reef (GBR), 9, 13, 366, 1942,
    1944, 1949, 1972, 1974, 2003
  biological diversity, 448
  coral reefs, 451
  future aspects, 457
  GBRMP and GBRWHA, 454–456
  geographical map, 449
  landscape, 448
  managemet status, 456–457
  mangroves, 452
  megafauna, 453–454
  northeastern Australian coastline,
      1968–1970
  saltmash, 452–453
  satellite image, 448, 450
  seagrass, 451–452
  types, 451
Great Barrier Reef Marine Park (GBRMP), 454,
    455, 457
Great Barrier Reef Marine Park Authority
    (GBRMPA), 450, 454, 455
Great Barrier Reef World Heritage Area
    (GBRWHA), 450, 454, 455, 1971
Great Britain, 11, 960, 965, 966, 968, 998,
    1029, 1049
Great Britain estuaries
  biodiversity, 1001
  conservation status, 1004
  fish, 1002
  future aspects, 1005
  habitats and communities, 1001
  land claim, 1004
  location and area, 998, 999
  Morecambe Bay, 999, 1000
  types of, 998, 1000
  waterbirds, 1002
Greater flamingo, 490, 1104, 1106, 1186, 1188,
    1190, 1191, 1324, 1326, 1386, 1444,
    1642, 1729
Great Indian Bustard Sanctuary, 1730
Great plains, 147
  biodiversity, 692
  conservation status, 692–693
  hydrology, 690–692
  threats and future challenges, 693–694
  water dissolution hypothesis, 690
  wind erosion hypothesis, 690
Great Rift Valley, 1350, 1354, 1414
Great River, *see* Lena River Basin
Greece, 1138, 1139, 1142, 1145, 1150, 1153,
    1156, 1158, 1162, 1163

Greenhouse gases (GHG) emissions, 63,
    242, 286, 415, 457, 502, 684, 740,
    752, 1760
Greenland's Ramsar sites, 513
Grewia
  G. bicolor, 1385
  G. densa, 206
  G. tenax, 206, 1385
Greylag Geese, post-breeding concentration,
    472
Grindelia integrifolia, 570
Groundwater, 22, 66, 103, 105, 115, 118, 131,
    132, 135, 146, 151, 157, 163, 178,
    201, 238, 252, 254, 256, 258, 259,
    261–263, 266, 267, 309–311, 327,
    336, 343, 363, 364, 366, 464, 467,
    491, 492, 525, 536, 655, 674, 683,
    732, 789, 791, 807, 828, 829, 900,
    905, 1035, 1061, 1069–1071, 1073,
    1171, 1175, 1180, 1235, 1239, 1258,
    1290, 1291, 1294, 1305, 1311,
    1426–1428, 1433, 1434, 1503, 1536,
    1540, 1543, 1544, 1546, 1560, 1638,
    1652, 1655, 1656, 1662, 1666, 1668,
    1672, 1700, 1701, 1717, 1888, 1979,
    1993, 2014
  abstraction, 347, 349, 1664, 1961, 1963
  Andean wetlands, 827
  aquifers, 120
  Axios, Aliakmon, and Gallikos Delta
      Complex, 1145
  Azraq Basin, 462
  boreal wetlands, 523, 524
  calcareous mires, 1426–1427
  Conchalí Lagoon, 859
  Danube River, 888, 890
  Doñana wetlands, 1133
  fens waterlogged by level, 252–258
  fishponds, 476
  floodplain mires, 1428
  freshwater swamps, 200
  Hula wetland, 1171
  hydrology, 1662–1664
  Indus River, 1699
  karst wetlands, 314
  Lake Magadi, 1388
  Lake Okeechobee, 709–711
  Lake Seyfe, 1188, 1189, 1192, 1194
  Lake Turkana, 1363
  Lake Uromiyeh, 166, 1663, 1664
  Okavango River, 1398, 1400, 1402
  peatlands, 230, 233–235
  peatlands in Africa, 1414, 1417, 1418

recharge, 683
Rugezi Marsh, 1316
sabkha landforms, 1176
salt lakes, 153
Sefid Rood, 1627
sloping mires, 1426
terrestrializing mires, 1427
through flow mires, 1426
turloughs, 1069–1071
Volga River Basin, 937
Yamuna River, 1716
Yellow River, 1582, 1583
Zambezi River, 1223
Groundwater-dependent wetlands
    definition, 346
    ecosystems, 347–349
    Gnangara Groundwater System, 1959–1965
    threats, 349–355
    types, 347–349
Groundwater discharge, 347, 349–351, 353, 683, 1311, 1536, 1540, 1543, 1960, 1961, 1964
    karst wetlands, 327
    peatlands in Africa, 1414
Growth dynamics, 1593, 1594
Grundulus bogotensis, 776, 779
Grus
    G. antigone, 1716, 1729, 1776, 1779, 1946
    G. antigone sharpii, 1796, 1806
    G. canadensis, 547, 692
    G. carunculatus, 1236
    G. grus, 482, 1186, 1190, 1254, 1561, 1653
    G. japonensis, 1502, 1513, 1525, 1561, 1584, 1599
    G. leucogeranus, 953, 1444, 1472, 1481, 1502, 1559, 1641
    G. monacha, 1444, 1474, 1502, 1513
    G. nigricollis, 325, 1580, 1612
    G. rubicundus, 1902, 1946
    G. vipio, 1493, 1502, 1513, 1525, 1569, 1584
Guadalquivir River, 1124, 1129, 1133, 1134
Gubernatrix cristata, 817
Guettarda speciosa, 2009
Gustavia augusta, 207
Guyu wujalwujalensis, 1945
Gymnarchus niloticus, 1368
Gymnocephalus cernuus, 612
Gymnomitrion concinnatum, 279
Gypohierax angolensis, 1336–1337
Gypsophila perfoliata, 916, 917, 1189
Gyrinus opacus, 1047
Gyrosigma, 630
Gyttja, 1933–1935

H
Habitat, 5, 6, 20, 26, 51, 53, 60, 64, 66, 67, 79, 81, 89, 90, 100–102, 110, 113, 115, 117, 119, 133, 135, 137, 151, 153, 165, 166, 174, 176, 184, 191, 194, 210, 211, 219, 220, 237, 240, 268, 269, 287, 290–293, 297, 310, 316, 320, 321, 323, 325, 326, 347, 350, 352–355, 361, 364–366, 384, 386–388, 395, 398, 401, 411, 415, 417, 418, 423, 428, 432, 440–443, 448, 452, 453, 455, 462, 482, 496, 502, 532, 533, 535, 547, 552, 553, 594, 595, 597, 598, 600, 607, 608, 611, 613, 654, 657–663, 674, 692, 693, 698, 702, 738, 740, 750, 776, 777, 779, 780, 791, 793, 794, 799, 802, 804, 805, 809, 816, 825, 828, 830, 831, 836, 850, 852, 853, 861–864, 867, 870, 880, 889–891, 893, 917, 920, 921, 926–928, 936, 938, 961, 963, 964, 967–969, 981, 985, 987, 988, 994, 998, 1014, 1043, 1047, 1049, 1051, 1052, 1054, 1069, 1071, 1075, 1082, 1084, 1086, 1091, 1092, 1190, 1491, 1492, 1557, 1678–1680, 1687, 1688, 1691, 1694, 1702, 1729, 1757–1759, 1961, 1963–1965
Amazon of Europe, 904–908
Amazon River Basin, 731, 733–736
Anzali Mordab Complex, 1621
aquatic, 110, 120–121, 580, 611, 658, 660, 663, 780, 791, 805, 889, 955, 1130, 1403, 1989
Australia's Wet Tropics, 1942, 1944, 1947, 1948
Axios, Aliakmon, and Gallikos Delta complex, 1139, 1143, 1144, 1146
Bahr el Ghazal, 1272, 1274, 1275
Banc d'Arguin, 1323, 1325
Baro-Akobo River Basin, 1265
Bay of Fundy, 622, 625, 626, 629, 630, 632
Berbak National Park, 1832, 1834–1837
Bijagos Archipelago, 1336, 1337
Broads area, 1027–1030
Bujagh National Park, 1627–1629, 1632
Camargue, 1102, 1104–1109
Ciénaga Grande of Santa Marta, 752, 763–765
complexes, 585
Congo River Basin, 1200, 1207–1210, 1212
Coorong, 1912, 1915, 1917

Habitat (cont.)
 Copper River Delta, 560–563
 current status, 1558–1560
 Czech Republic, 471, 472
 Danau Sentarum National Park, 1847
 Danube floodplains, 898–900
 Dinder and Rahad rivers, 1294, 1295, 1297
 disturbance, 1210
 diversity, 2007–2010
 Doñana wetlands, 1124, 1130–1132
 Ebro Delta, 1115–1120
 estuarine, 45–48, 1001–1002, 1005, 1006
 EUNIS classification, 963
 Everglades, 709, 715, 717, 718, 720, 721
 Fereydoon Kenar Ramsar Site, 1638–1640, 1642, 1643
 floodplain, 698
 Fraser Delta, 566, 568, 570, 571, 573
 freshwater marshes and swamps, 173, 174
 Ganga-Brahmaputra Basin, 1713, 1715–1718
 high altitude wetlands, Nepal, 1606, 1611, 1612
 highland peatlands of Mongolia, 1533, 1541, 1543, 1546
 Hula wetland, 1170, 1171, 1176
 intertidal, 1869
 intertidal zone of East and Southeast Asia, 1869, 1870
 Kakadu National Park, 1952, 1955–1957
 Kakagon Sloughs, 434
 Kati Thanda, 1925
 Kilombero Valley Floodplain, 1344, 1345
 Lake Eyre, 1922
 Lake Naivasha, 1358
 Lake Parishan, 1649, 1652–1655
 Lake Seyfe, 1189
 Lake Turkana, 1366, 1370, 1371, 1377
 Lake Uromiyeh, 1660, 1666, 1669
 loss, 1209
 Machar marshes, 1281
 Macquarie Marshes, 1898–1905
 Mahanadi Delta, 1747
 mangroves, 100–102
 Mekong River Basin, 1771, 1774–1777, 1779, 1781, 1803, 1807, 1810
 Mesopotamian Marshes of Iraq, 1688, 1691, 1694, 1987
 middle Aldan River, 1472
 Mississippi Alluvial Valley, 580, 583–586
 Murray-Darling River Basin, 1888, 1891–1893
 Natura 2000, 1144
 New Zealand restiad bogs, 1996–1998
 Nidjili Lake, 1460, 1468, 1470
 Nile Delta, 1252, 1256, 1257
 Nile River Basin, 1249
 Norfolk and Suffolk Broads, 1024–1032, 1035
 Okavango Delta, 1403, 1407
 papyrus marshes, 191
 papyrus wetlands, 184
 peatlands of Africa, 1445–1447
 Poyang Lake, 1566, 1569–1571
 Prairie Pothole Region, 682, 683, 686
 productivity and diversity, 632
 protection, 342, 646, 648
 Ramsar Site, 1189
 riparian, 547, 698
 Rugezi Marsh, 1309, 1315, 1316
 Sabkhas, 1180
 salt lakes, 151, 153
 San Francisco Bay Estuary, 638, 641, 644–648
 Sanjiang Plain, 1513–1516
 seagrass, 78, 80, 85, 87, 90, 440, 453, 496, 502, 845, 1589–1595, 1876, 1882, 1969
 seagrass in Malaysia, 1876, 1878, 1882
 seagrass meadows, 438
 seagrass meadows of Northeastern Australia, 1969, 1971, 1972
 Sembilang National Park, 1822, 1825–1828
 Shadegan Wetland, 1678–1680
 soda lakes, 1382, 1383, 1386, 1387, 1389
 southeast Brazil, 840
 subterranean karst systems, 331–343
 Sudd, 1300, 1302, 1303, 1305
 Sundarban mangroves, 1736–1737
 Tampa Bay, 501
 Tasek Bera, 1852, 1856–1859
 tidal freshwater wetlands, 165
 Tonle Sap Lake, 1789, 1790
 tropical freshwater swamps, 213–216
 vernal pools, 652
 Volga Delta, 955–957
 wetlands of Chile, 833
 wetlands of New Zealand, 1980, 1983, 1987–1989
 wild rice, 431
 Wular Lake, 1708
 Yangtze River Basin, 1558, 1559
 Yellow River, 1580–1584
 Yenisei River Basin, 1480
 Zambezi River Basin, 1129, 1222, 1223, 1225
 Zhalong wetlands, 1525, 1528

Habmonas campisalis, 1388
Hadziella, 340
Haematobia irritans, 809
Haematopus
  H. leucopodus, 867
  H. ostralegus, 965, 989, 1003, 1016, 1083, 1141
Halcyon
  H. albiventris, 1345
  H. chelicuti, 1345
  H. concreta, 1825
  H. coromanda, 1825
Haliaeetus
  H. albicilla, 418, 473, 889, 899, 904, 918, 1092, 1142, 1444, 1460, 1474, 1621, 1641
  H. leucocephalus, 568, 612
  H. leucoryphus, 953, 1584
  H. pelagicus, 1444
  H. vocifer, 1337, 1357, 1385
Halichoerus grypus, 629, 993, 1015, 1094
Halimeda, 86, 448
Halimione
  H. portulacoides, 964, 971, 1001, 1140
  H. verrucifera, 916
Haliplus andalusicus, 1131
Haloalkaliphilic archaea, 1388
Halocarpus
  H. bidwillii, 1995
  H. biformis, 1980
Halocnemum strobilaceum, 916, 1140, 1189
Halodule, 841, 842
  beds, 843
  H. beaudettei, 75
  H. bermudensis, 75
  H. ciliata, 75
  H. emarginata, 75, 840, 841, 845
  H. pinifolia, 75, 1876, 1971
  H. uninervis, 75, 82, 1876, 1877, 1971
  H. wrightii, 75, 81, 840–844, 1323, 1336
Haloleptolyngbya alcalis, 1385
Halonatronum saccharophilum, 1389
Halophila, 74, 76, 1592, 1593, 1746, 1876, 1877, 1970
  H. australis, 75, 77
  H. baillonii, 75
  H. beccarii, 75, 85, 1876, 1877, 1880
  H. capricorni, 75, 1970, 1971
  H. decipiens, 75, 840–842, 1876, 1971
  H. engelmannii, 75
  H. euphlebia, 75
  H. hawaiiana, 75

H. johnsonii, 75, 81
H. major, 1876
H. minor, 75, 1876, 1971
H. nipponica, 75, 1590–1595
H. ovalis, 75, 81, 1876, 1877, 1971
H. ovataa, 75
H. spinulosa, 75, 1876, 1877, 1971
H. stipulacea, 75, 81
H. sulawesii, 75
H. tricostata, 75, 1970, 1971
Halophytic vegetation, 960, 1140
Hamatocaulis, 518
Hanguana malayana, 205, 1834
Haplochromis turkanae, 1371
Haraaz River, 1638
Harbour Seal, 993
Hauffenia, 340
Hedera helix, 916
Hediste diversicolor, 977, 978
Heilongjiang, 26, 1488, 1496, 1510, 1511, 1516–1518, 1522, 1524, 1526
Helarctos malayanus, 1835, 1845, 1860
Heleioporus eyrei, 1961
Helichrysum arenarium, 917
Helicops spp., 735
Heliopais personatus, 1860
Hemichromis exsul, 1371
Hemidactylium scutatum, 612
Hemigrapsus sanguineus, 995
Hemipimelodus daugueti, 1806
Hemipodus olivieri, 850
Hemitheconyx caudicinctus, 1700
Heosemys spinosa, 1845, 1860
Heracleum, 560
  H. lallii, 1609
  H. lanatum, 570
Herbaceous swamps, 177, 204–205, 214, 1577, 1578, 1824
Heritiera fomes, 1720, 1736
Hermathia, 205
Hetao plain, 1580–1581
Heteranthera spp., 1107
Heteromastis, 630
Heteromastus similis, 851
Heteroscelus brevipes, 1870
Heterotis niloticus, 1273, 1368
Hetrotis, 1273
Hexarthra, 1357
  H. jenkinae, 1385
Hibiscus
  H. calophyllus, 206
  H. moscheutos, 159
  H. tiliaceus, 1237, 1823, 1834

Hieraaetus pennatus, 1255
Hierochloe odorata, 951
High altitude wetlands (HAWs), Nepal
  characteristics, 1608
  conservation status, 1610
  description, 1606
  ecosystem services, 1611
  fauna, 1609–1610
  flora, 1609
  fragility and sensitiveness, 1612
  hydrology, 1608
  issues in conservation efforts, 1612
  Ramsar Sites, 1606, 1607
  threats, 1612
  water bodies, categories of, 1606
High arctic, 286
  in Canada, 394
  in Greenland, 510, 511
  North America, 280
High-centred polygon and low-centred polygon mire, 281
High endemism, 824, 2011
Higher Council for Environment and Natural Resources (HCENR), 1295
High ion lake, 1365
Highlands, 26, 230, 279, 337, 824, 825, 1048, 1053, 1265, 1266, 1271, 1272, 1280, 1284, 1290, 1292, 1293, 1304, 1309, 1342, 1355, 1398, 1580, 1638, 1700, 1765, 1767, 1771, 1888
  Andean wetlands, 827, 828
  Baro-Akobo Basin, 1642
  Blue Nile, 1247
  central, 1718
  extent and character, 1262–1263
  local use and knowledge, 1263–1264
  Main Nile, 1248
  Omo River, 1362, 1364
  peatlands of Mongolia, 1531–1547
  saddle raised bogs, 268, 269
  of Scotland, 257
  sedge fens, 1543–1544
  spring and blanket bogs, 1545
  valley meadow fens, 1542–1543
  water forum, 466
  White Nile, 1246, 1247
High plains acquifer
  geographic distribution, 690, 691
  source of water recharge, 692
High productivity, 50, 52, 87, 490, 594, 622, 626, 735, 738, 752, 761, 817, 843, 986, 1105, 1117, 1249, 1257, 1404, 1405, 1535, 1543, 1593, 1713, 1925, 1952

Bahía Lomas, 867
  of Ciénaga Grande de Santa Marta, 753
  Colombian mangroves, 753
  estuaries, 50
  Fundy salt marshes, 634
  Okavango Delta, 1401
  papyrus marshes, 190
  seagrass, 74, 87
  tropical freshwater swamps, 210
  Volga River, 953
Himalayan region, 1612, 1712–1715
Himantopus himantopus, 1141, 1190, 1629
Hiodon alosoides, 553, 597
Hippoglossus hippoglossus, 629
Hippolais polyglotta, 1337
Hippophae rhamnoides, 916, 917
Hippopotamus amphibius, 192, 1238, 1274, 1337, 1338, 1354, 1358, 1372, 1405
Hippotragus
  H. equinus bakeri, 1293
  H. niger, 1227, 1236
Hippuris lanceolata, 510, 511
Histosol, 246, 1160, 1161
  peat soil, 246
  Pocosin vegetation, 671
Histrionicus histrionicus, 633, 1444
Hokkaido marshes, 13
  Bekambeushi marsh, 1600
  future aspects, 1601
  future challenges and developments, 1601
  Geospatial Information Authority of Japan, 1598
  Kiritappu marsh, 1599
  Kushiro marsh, 1598–1599
  Sarobetsu marsh, 1601
  threats, 1601
Hollows, 259, 261, 280, 282–284, 305, 524, 653, 877, 1044–1046, 1541, 1542, 1546, 1754, 1755, 1758
Holocene, 529, 815, 874, 947, 1252, 1443
  Ciénaga Grande de Santa Marta, 758
  Coorong, 1910
  Lena Delta, 1453
  Paraná River Delta, 815
  in western Canada, 517
Homalium
  H. caryophyllaceum, 208
  H. molle, 207
Homaloptera yuwonoi, 1845
Homarus americanus, 628
Homonoia riparia, 1773
Homoporus deltaicus, 918
Hongjiannao lake, 1582

Hopea
 H. mengerawan, 208
 H. novoguineensis, 208
Hoplias malabaricus, 763
Hordeum, 61
 H. brevisubulatum, 1538
Horsfieldia crassifolia, 1757
Houbaropsis bengalensis, 1776, 1779, 1788, 1796
Houdinia flexilissima, 1981
Houting, 989
Hualves, 825, 827, 829
Hucho
 H. perryi, 1600
 H. taimen, 939, 1444, 1475, 1502
Hudson Bay Lowland, 516, 517
Hula Drainage Project, 1170
Hula Nature Reserve, 1161, 1170–1172
Hula peatland, 1156, 1161, 1163
Hula wetland, 1160
 biodiversity, 1171
 climate, 1168
 drainage and restoration, 1168–1171
 future challenges and developments, 1172
 hydrology, 1168
 location, 1168
Human disturbance, 64, 191, 558, 647, 1091, 1327, 1555, 1561, 1569, 1571, 1786, 1931
Human impacts, 7, 14, 66, 765, 1159, 1469, 1507, 1547, 1595
 Bay of Fundy, 634
 Dauria's wetlands, 1503–1504
 and distribution, 160
 Drava and Mura rivers, 905
 Greenland's Ramsar Sites, 513
 Laurentian Great Lakes, 606
 Mediterranean peatlands, 1162–1163
 Sanjiang Plain, 1515
 seagrasses, 83, 85, 89
 tidal freshwater wetlands, 161–162
 Volga Basin, 940–942
 Wet Tropics wetlands, 1948, 1949
Humulus lupulus, 916
Hura crepitans, 741
Huso
 H. dauricus, 1513
 H. huso, 900, 918, 939, 953
Hutchinsia procumbens, 1105
Hyaena
 H. hyaena, 1274, 1372
 H. hyaena dubbah, 1294
Hyalobagrus leiacanthus, 1845

Hydnophytum formicarum, 1816
Hydrictis maculicollis, 1274
Hydrilla verticillata, 1746, 1859
HydroBASINS, 112
Hydrobia spp., 977
Hydrocharis morsus-ranae, 952, 1130, 1171
Hydrocharition, 1144
Hydrochemistry, 1075
Hydrochoerus hydrochaeris, 735, 809, 817, 843, 852
Hydrocotyle, 1368
HYDRO 1 K, 111, 112
Hydrological stability, 1960, 1963
Hydrology, 6, 8, 56, 65, 66, 157, 158, 162, 172, 191, 217, 241, 277, 283, 287, 304, 322, 323, 349–351, 354, 365, 419, 489, 526, 532, 534, 558, 652, 655–656, 662, 668, 670, 693, 694, 701, 728, 733, 739, 766, 781, 791, 808, 810, 831, 879, 887–889, 906, 920, 955, 1058, 1069, 1073, 1075, 1086, 1104, 1107, 1109, 1119, 1133, 1146, 1175, 1210, 1225, 1228–1230, 1248, 1249, 1281, 1284, 1298, 1304, 1305, 1359, 1384, 1410, 1493, 1533, 1534, 1553, 1562, 1571, 1613, 1730, 1735, 1739, 1781–1783, 1786, 1794, 1798, 1804, 1815, 1827, 1836, 1891, 1935, 1937, 1961, 1964, 1965, 1980, 1983
 Anzali Wetland, 1616–1618
 archipelago, 1335–1336
 Axios, Aliakmon, and Gallikos Delta Complex, 1138–1139
 Bahr el Ghazal, 1270
 Banc d'Arguin, 1322–1323
 Berbak National Park, 1832
 biology linkages, 1961, 1964, 1965
 Bujagh National Park, 1627
 Camargue, 1102–1103
 Copper River, 559
 Danau Sentarum National Park, 1843
 Danube River Basin, 887–888
 Danube River branches, 912–914
 Doñana, 1124–1130
 drivers of wetland degradation, 362–363
 ecological features of lakes, 131
 Fereydoon Kenar, 1636–1638
 Hula Wetland, 1168
 Indus River Basin Wetlands, 1699
 Kakagon Sloughs, 434
 karst systems, 314–316
 Kati Thanda, 1922–1925

Hydrology (cont.)
 Kilombero floodplain, 1342–1343
 lagg fen, 311
 Lake Bogoria, 1384
 Lake Okeechobee, 710
 Lake Parishan, 1649–1651
 Lake Turkana, 1363–1365
 Lake Uromiyeh, 1662–1664
 Lena River Delta, 1452–1453
 Mackenzie River watershed, 550–551
 Macquarie Marshes, 1900
 Macrotidal Bay, 1080–1081
 mangroves, 103
 mayas, 1290–1291
 Mekong Basin, 1767
 Mesopotamian Marshes, 1688
 Middle Aldan River Basin, 1472–1473
 Mississippi Alluvial Valley, 578–580
 Murray-Darling River Basin, 1890–1891
 Myall River and Nerong Creek, 1931–1933
 Nidjili Lake, 1458
 Paraná-Paraguay Fluvial Corridor, 788–790
 Paraná River Delta, 815–816
 peatland, 1150
 Phoksundo wetland, 1608
 Playas, 690–692
 Pocosins, 673–674
 Poyang Lake, 1568
 Prairie Pothole Region, 681–682
 Rugezi Marsh, 1310–1311
 Sembilang National Park, 1820–1822
 Shadegan Wetland, 1676–1678
 significant changes, 362–363
 Sudd, 1300
 Taiga-Alas Landscape, 1464–1465
 tidal freshwater wetlands, 157
 turloughs, 1070–1071
 Volga River Basin, 936–937, 949
 wet tropics, 1942–1944
 Yangtze River, 1552–1554
 Yukon River Basin, 544–545
 Zambezi River Delta, 1235
 Zhalong wetland, 1523–1524
Hydromys chrysogaster, 1902, 1946
Hydroperiod, 209, 668, 1071, 1075, 1129, 1133, 1693
 Doñana wetlands, 1127
 Everglades, 714
 playa wetlands, 693
 pocosin, 673
 vernal pools, 652, 654–660
Hydrophasianus chirurgus, 1708
Hydrophytic vegetation, 828, 1140

Hydroporus lucasi, 1131
Hydropotes inermis inermis, 1569
Hydropower, 116, 126, 127, 135, 139, 474, 533, 534, 731, 740, 793, 891, 920, 941, 1102, 1249, 1309–1311, 1421, 1445, 1516, 1739, 1748
 Amazonian wetlands, 739
 Amur River, 1496
 Congo Basin, 1208
 dams, 131, 217, 905–907, 930, 1230, 1375, 1790, 1798
 development, 893, 1266, 1496, 1798, 1808, 1810
 development, Amur-Heilong River Basin, 1496
 freshwater swamps, 217
 generation, 731, 737, 739, 741, 891, 1219, 1809, 1810
 Lower Mekong Basin, 1768, 1781
 Yenisei River, 1483
 Zambezi River Basin, 1219, 1229, 1230, 1235, 1241
Hydrops spp., 735
Hygrotus
 H. lagari, 1131
 H. quinquelineatus, 1071
Hyla, 735
 H. andersonii, 674
 H. arborea, 918
 H. chrysoscelis, 658
 H. versicolor, 658
Hylobates, 1758
 H. agilis, 1825
 H. lar, 1835, 1860
Hyloxalus subpunctatus, 777
Hymenachne, 816
 H. acutigluma, 215, 219, 1796, 1952
 H. amplexicaulis, 1957
Hynobius retardatus, 1599
Hyparrhenia, 1294, 1302
 H. dichroa, 1236
 H. diplandra, 205
 H. hirta, 1386
 H. rufa, 1236, 1272, 1302
Hyperlophus vittatus, 1916
Hyperoodon ampullatus, 633
Hyperopisus bebe, 1368
Hypersaline, 81, 105, 147, 149, 151, 840, 1002, 1174, 1664, 1911, 2008
 Abu Dhabi Emirate, 1180
 Coorong, 1916, 1917
 Lake Bogoria, 1385
 Lake Uromiyeh, 1660, 1664

River Murray, 1913
    saline mudflats, 1237
    salt lakes, 144, 145
Hypersalinization, 766
Hyperseasonal savanna, 804
Hypertidal ecosystem, 624, 625
Hypertidal estuaries, 41
Hyphaene, 1236
    H. coriacea, 1368
    H. thebaica, 1294, 1368
Hyphaene palm savanna, 1236
Hypnea, 842
Hypnum, 285
Hypogean, 348, 351, 354, 1961, 1965
    classification, 332
    composition, 336–337
    environment, 334–335
    inhabitants, 335
    karst, 332
    karst hydrology, 316
    threats, 342–343
Hypolytrum capitulatum, 1844
Hypomesus transpacificus, 642
Hypophthalmichthys molitrix, 476
Hyporheic zone, 347, 348, 351
Hypsometry, 1933, 1934
    of Myall lake, 1933
Hysterocarpus traskii traskii, 642
Hystrix indica, 1621

I

Ice, 65, 67, 110, 126, 138, 261–263, 277, 279–284, 305, 319, 322, 340, 402, 431, 517–519, 530, 536, 551, 554, 562, 567, 594, 614, 625, 628, 631, 632, 653, 658, 673, 681, 714, 874, 924, 936, 987, 1040, 1090, 1102, 1442, 1443, 1452, 1453, 1458, 1459, 1464, 1465, 1473, 1479, 1511, 1541, 1715
Ichthyaetus
    I. audouinii, 1131
    I. relictus, 1502
Ichthyophaga ichthyaetus, 1788, 1816
Icmadophila ericetorum, 283
Ictalurus punctatus, 583
Icterus cayanensis, 851
Iglica, 340
Ilex, 1029
    I. cassine, 719
    I. cymosa, 208, 1815
    I. glabra, 672

Ilisha, 1336
Ilyanassa, 630
    I. obsoleta, 630
IMCG, *see* International Mire Conservation Group (IMCG)
Immersion mire, 253, 254
Impatiens, 205
    I. glandulifera, 164
Imperata cylindrica, 1236
Important Bird Area (IBA), 325, 480, 482, 490, 572, 595, 599, 645, 994, 1118, 1186, 1265, 1354, 1386, 1502, 1505, 1631, 1643, 1648, 1676, 1700, 1787, 1807, 1816, 2021
Imraguen fishermen, 1328, 1329
Indian arid zone, 1726, 1727, 1731
    conservation status and management, 1730
    dependent people, 1729–1730
    distribution, 1728
    diversity, 1728
    ecosystem services, 1729
    extent, 1728
    location, 1726–1727
    significant biodiversity, 1728–1729
    threats, 1730–1731
    water resources, 1726–1727
Indian Sundarban, 1736, 1737, 1739, 1740
Indian Wild Ass Sanctuary, 1730
Indigenous atoll vegetation, 2007
Indigenous owners, 1956
Indira Gandhi Canal, 1731
Indonesia, 83, 96, 98, 103, 201–203, 206, 209, 216, 219, 220, 232, 237, 240, 301, 440–442, 522, 1754, 1758–1760, 1804, 1819–1828, 1831–1838, 1841–1849, 1866, 1867, 1869
Indus River Basin, 13
    biodiversity and conservation status, 1700–1701
    future challenges and developments, 1701–1702
    hydrology, 1699
    location, 1698
    threats, 1701–1702
    wetland ecosystems, 1699–1700
Infilled basin mires, 263–264
Inia geoffrensis, 735
Inner Danish waters, 1089–1095
    benthos, 1092
    birds and mammals, 1092–1094
    conservation status, 1094
    physio-chemical conditions, 1091
    threats, uses and challenges, 1095

Insect, 65, 100, 101, 119, 149, 165, 166, 176, 191, 192, 236, 336–338, 352, 393, 395, 532, 571, 630, 631, 660–662, 791, 899, 918, 1073, 1272, 1273, 1296, 1371, 1385, 1405, 1700, 1737, 1758, 1780, 1946, 1952, 2011, 2014
Intada sudanica, 1294
Integrated planning stakeholder participation, 488–490, 1631
Interfluvial wetlands, Amazon River, 728, 732
Intergovernnmetal Panel on Climate Change (IPCC), 63, 67, 119, 242, 286, 676, 714, 803, 1230, 1783
Intermediate raised/blanket mire, 268–269
Intermediate raised bogs, 267–269
International Commission for the Protection of the Danube River (ICPDR), 887–894, 926
International cooperation, 14, 53, 111, 241, 370, 926, 1827
International Mire Conservation Group (IMCG), 230–232, 290–292, 1050, 1160, 1169, 1311
Interstitial, 187, 332–334, 336, 343, 351, 354, 1109, 1961, 1965
Intertidal
   flats, 21, 47, 49, 371, 375, 406, 622, 625, 629, 633, 834, 963, 998, 1013, 1323, 1324, 1336, 1676, 1728, 1865–1872, 1969, 2002, 2019, 2020
   habitats, 51, 165, 625, 626, 1001, 1014, 1867–1870
   mudflats, 544, 631, 1001, 1320, 1325, 1679, 1823
   zone, definition, 7, 22, 45, 47, 48, 56, 58, 61, 622, 625, 626, 748, 749, 841, 851, 961, 967, 1013, 1080, 1081, 1323, 1866–1869, 1871, 1872
Intertidal zone, East and Southeast Asia, 51, 60, 318, 1865–1872
   agricultural chemical overuse, 1868
   anthropogenic climate change, 1868
   biodiversity, 1870
   competition for food, 1868
   conservation status, 1867
   damming major rivers, 1868
   diversity, 1866–1867
   ecosystem services, 1869–1870
   extent and location, 1866
   future aspects, 1872
   habitat loss and fragmentation, 1868
   human dependence, 1871
   hunting, 1868
   international agencies, 1872
   management, 1867
   overharvesting, 1868
   pollution, 1868
   tidal energy developments, 1868
Intertropical Convergence Zone (ICZ), 748, 1218, 1235
Intra-montane basin fens, 1156, 1158
Inula britannica, 951
Invasive alien species (IAS), 179, 220, 324, 327, 384, 818, 1798, 2014, 2017, 2020, 2022
Invasive species, 8, 66, 137, 194, 220, 360, 361, 365, 384–388, 432, 434, 603, 613, 614, 646, 648, 713, 853, 918, 978, 995, 1035, 1109, 1133, 1210, 1249, 1272, 1305, 1358, 1359, 1389, 1773, 1798, 1863, 1891, 1949, 1957, 1986, 1988, 2017
Inventory, 5, 20, 22, 27, 28, 31–32, 98, 120, 147, 230, 232, 277, 595, 607, 663, 755, 790, 816, 832–835, 892, 998, 1163, 1324, 1337, 1446, 1497, 1728, 1736, 1805
   Lake Superior coastal wetlands, 607
   national, 832
Invertebrate seagrass fisheries, 442
Ipomoea, 194
   I. aquatica, 1272, 1720
   I. cairica, 191, 1272
Iran, 120, 132, 135, 1615–1632, 1642–1645, 1647–1673, 1675–1683, 1686, 1689, 1690, 1692, 1694
Iraq, 363, 1174, 1676, 1681, 1685–1694
Irediparra gallinacea, 1946
Ireland, 239, 253, 267, 268, 304, 415, 417, 419, 960, 967, 998, 1004, 1067–1075
Iris
   I. lactea, 1538
   I. pseudacorus, 818, 915, 1190, 1639
   I. spuria, 1106
Irish blanket mires, 299
Irrigation schemes, 135, 1229, 1297, 1375, 1377, 1410, 1504, 1581, 1699, 1764
Irrigation threats, 1409–1410
Ischaemum
   I. afrum, 1236
   I. rugosum, 1795
Island growth, 1402, 1403, 1409
Islet, 899, 1140, 1142, 1190, 1320, 1334, 1338, 1339, 2003, 2006, 2007, 2009, 2016, 2017, 2019, 2021–2023
Isolepis fluitans, 1130

Isophya dobrogensis, 918
Israel, 151, 184, 1156, 1160, 1161, 1163, 1167–1172
Itezhi-tezhi dam, 217, 1220, 1230
Ixobrychus, 1796
    I. exilis, 633, 777
    I. flavicollis, 1777
    I. minutus, 1141, 1142, 1708
    I. sinensis, 1777
Ixora mentanggis, 205

J
Jade Sea, 1363
Jaluit atoll, 2019
Jameson land, 510, 513
Jania, 842
Johannesteijsmannia altifrons, 1835
Joint Nature Conservation Committee (JNCC), 306, 371, 999, 1006, 1013, 1019, 1020, 1041, 1052
Joint venture, 493, 548, 579, 585, 635, 645, 646, 648, 686, 690, 691, 702, 1230
Jonglei Canal, Sudd region, 216, 1249, 1275, 1305
Juglans regia, 1640
Juncetalia maritimi, 1144
Junco, 750
Juncus, 61, 284, 828, 851, 967, 1104, 1140, 1159, 1609, 1628, 1728, 1901, 1980
    J. acutus, 64, 1116, 1117
    J. biglumis, 511
    J. castaneus, 511
    J. effusus, 1556, 1639
    J. emmanuelis, 1130
    J. gerardii, 630, 916, 964
    J. heterophyllus, 1130
    J. kraussii, 64
    J. maritimus, 965, 967, 1105, 1116, 1117, 1130
    J. oxycarpus, 1313
    J. roemerianus, 58
    J. subulatus, 1130
    J. trifidus, 279
    J. triglumis, 511
Juniperus, 1043
    J. turbinata, 1104
Jurinea dolomiaea, 1609

K
Kachchh Desert Wildlife Sanctuary, 1730
Kafue River, 201, 217, 1219, 1220, 1223, 1225
Kakaducaris, 211
Kakadu National Park, 13, 26, 1951–1957
    biodiversity, 1952–1955
    catchments, 1952, 1956
    conservation status, 1955–1956
    ecosystem services, 1956
    freshwater wetlands, 1952
    future challenges and developments, 1956–1957
    threats, 1956–1957
Kakagon/Bad River Sloughs, 10, 427–435, 616
    ecology of Manomin, 431
    geologic setting, 429–430
    invasion of exotics, 431–432
    location, 428
    sediment and nutrient loading, 432–434
    water level dynamics, 431
    wetland setting, 430–431
Kalmia, 517
    K. cuneata, 674
    K. microphylla, 571
Kamiranzovu valley, 1309, 1311
Kappaphycus alvarezii, 2010
Kariba dam, 1223, 1229, 1235
Karirood stream, 1638
Kari stream, 1638
Karstic aquifer, 1150, 1650
Karstic limestone, 1070, 1649
Karst wetlands, 11, 26, 313–328, 1057–1064, 1888
    complexity, 317
    conservation and wise use, 317–319
    diversity, 319–327
    ecosystem services, 316–317
    hydrology, 314–316
    immersion mire, 253
    landscape, 22, 319, 324–326, 332, 1062, 1069, 1070, 1414
    location, 314
    phenomena, 314, 316, 317, 319–321, 324, 332, 1058, 1060–1062
    in temperate zone, 320–322
    threats, 327–328
    in tropics/subtropics, 322–327
Kasai, 1202–1204, 1206, 1209–1211
Kashmir valley, 1706, 1708
Kati Thanda-Lake Eyre, Australia, 6, 13, 1921–1926
    aquatic ecology, 1925–1926
    hydrology, 1922–1925
    location, 1923
Kekopey hot springs, 1387

Kenya, 150, 152, 186, 193, 202, 211, 213, 217, 218, 1349–1359, 1361–1378, 1381–1390, 1419
Keratella, 1357
Kerio/Turkwel deltas, 1362, 1366, 1369, 1377
Kerivoula muscina, 211
Kettle-hole fens, 518
Key threatening processes, 1894
Khentey Mountains, 1535, 1536
Kiashahr Lagoon, 1626–1628, 1630, 1631
Kigelia africana, 1236
Kilombero floodplain, 1342, 1343, 1345–1347
    biodiversity, 1344–1345
    conservation status, 1346
    ecosystem services, 1345–1346
    future challenges and developments, 1346–1347
    hydrology, 1342–1343
    location, 1342
    threats, 1346–1347
    wetland ecosystem, 1343–1344
Kinosternon scorpioides, 735
Kiritappu marsh, 1599, 1600
Kirsehir Museum, 1192
Kissimmee Basin, 710
Kleinhovia hospita, 1834
Knema conferta, 1834
Kniphofia caulescens, 1417
Knockfin Heights, 1041, 1042
Kobresia
    K. fissiglumis, 1609
    K. gandakiensis, 1609
    K. myosuroides, 510, 1538
    K. simpliciuscula, 511
Kobus
    K. defassa harnieri, 1293
    K. ellipsiprymnus, 1227, 1237, 1274
    K. ellipsiprymnus defassa, 1274
    K. kob, 1265
    K. kob ssp. *leucotis*, 1282
    K. leche, 192, 211, 217, 1225, 1405
    K. leche kafuensis, 211, 217, 1225
    K. megaceros, 211, 1265, 1274, 1303
    K. vardonii, 1345
Koompassia malaccensis, 1823, 1834
Korea, 13, 50, 81, 380, 405–411, 1589–1595, 1599, 1866, 1869, 1872
Krascheninnikovia ceratoides, 1189
Kuhlia rupestris, 1945
Kultuk zone, Volga Delta, 947, 952, 954
Kushiro marsh, 1598–1601
Kuskokwim River, 544–546, 548
Kutum Fish, 1630

L

Labeo, 192
    L. cylindricus, 1354
Labyrinthula, 82, 86
    L. zosterae, 82
Lacerta, 939
    L. clarkorum, 1190
Lacustrine, 6, 27, 126, 184, 186, 279, 525, 580, 583–584, 593, 607, 608, 614, 698, 825, 1159, 1223, 1388, 1452, 1454, 1458, 1468, 1700, 1944, 1983
Ladder fen, 259, 261, 297
Laetia corymbulosa, 207
Lagerstroemia speciosa, 1834
Lagg fen, 21, 265, 266, 304, 309–311, 571
    minerotrophic mire, 252–264, 310
    zone, 266, 310, 311
Lagomorphs, 939
Lagoon Ciénaga Grande de Santa Marta, Colombia, *see* Ciénaga Grande de Santa Marta (CGSM)
Lagoons, 21, 38, 81, 90, 94, 120, 185, 364, 392, 398, 758, 760, 761, 763–766, 1138, 1139, 1141, 1143, 1144, 1454, 1930, 1933
    Bandar Kiashahr, 1626–1628, 1631
    coastal, 39, 364, 398, 733, 766, 815, 836, 859, 974, 1001, 1104, 1116–1118, 1120, 1144, 1322, 1627, 1628, 1746, 1881, 1910, 1914
    Conchalí, 10, 857–864
    geomorphic type, 614
    physiochemical conditions in, 974–976
Lagopus lagopus scoticus, 1048
Laguncularia racemosa, 750, 764–766, 1336
Lake(s)
    Agassiz, 592, 599
    Athabasca, 536, 550, 551
    Baikal
        biodiversity, 1480–1481
        conservation status, 1482
        description, 1479
        human uses, 1482–1483
        threats and future challenges, 1484
    Baringo, 12, 186, 1349–1359
    Bogoria, 186, 1382–1385, 1389, 1390
    Borullus, 1255, 1257
    Chilika, India
        adverse change, ecological character, 398–400
        ecological restoration, 400–401
        integrated management planning, 402
        location, 398

Montreux Record, removal from, 401–402
restoration, 1749
Edku, 1255, 1256, 1258
Elementeita, 1382, 1383, 1387, 1389, 1390
Eyre, 6, 13, 130, 145, 1921–1926, 1961
freshwater (see Freshwater lakes and reservoirs)
Fuquene, 10, 773–782
  biodiversity, 776–777
  climate change, 781, 782
  conservation status, 779–781
  ecosystem services, 777–778
  geographical position, 775
  management, 779–780
  water quality, 774, 776
Khanka-Xingkaihu, 1511–1518
Kinneret, 1168, 1170
Magadi, 1382, 1383, 1388–1390
Manitoba, 10, 591–603
Manzalah, 1255–1257
Mariut, 1255, 1256
Naivasha, 12, 186, 187, 189, 190, 1349–1359
Nakuru, 150, 1382, 1383, 1385–1387, 1390
Parishan, Iran
  conservation status, 1654
  developments, 1654–1655
  ecological attributes and functioning, 1652–1654
  ecosystem services, 1654
  hydrology, 1649–1651
  location, 1648
  Phragmites australis reed beds, 1649, 1650, 1653
  threats, 1655–1656
Sambhar, 1728–1731
Seyfe, 1194–1195
  biodiversity, 1189–1192
  conservation status, 1187–1188
  cultural and social aspects, 1192–1193
  geology, 1189
  hydrology, 1188–1189
  location, 1186, 1187
  management structure, 1188
  natural resources, 1193–1195
  threats, 1194
Superior coastal wetlands
  ecology, 610–613
  future challenges and developments, 613–614
  inventory of, 607
  threats, 613–614
  types of, 607–610
Turkana, 6, 12, 135, 145, 150, 151, 1362, 1375–1378, 1382, 1383
  biodiversity, 1371–1372
  conservation and management, 1375
  ecosystem services, 1372–1373
  geochemistry, 1365–1366
  hydrology, 1363–1364
  threats and future challenges, 1375–1378
  water levels, 1364–1366
  wetland ecosystem, 1366–1371
Uromiyeh, 12, 1672–1673
  conservation status, 1668–1669
  developments, 1671–1672
  ecological functioning, 1664–1668
  ecosystem services, 1669–1671
  geography, 1660–1662
  hydrology, 1662–1664
  location, 1660–1662
  threats, 1672–1673
Winnipeg, 592–598, 602, 603
Winnipegosis, 592, 595, 596
Lama guanicoe, 831
Laminaria
  L. digitata, 626
  L. longicruris, 626
Lamna nasus, 633
Lampetra fluviatilis, 939, 1002
Lamprothamnium papulosum, 1104, 1106
Land-claim, British estuaries, 1004–1005
Land conversion, 6–8, 12, 104, 179, 240, 376, 380, 539, 623, 633, 634, 685, 686, 1019, 1209, 1514
  ambitious program, 363
  wetland hydrology, 363
Landscape
  change, 152, 533, 535, 537
  conservation, 470, 473, 955, 1017
  patterns, 694
  vernal pools in, 652–653, 663–664
Land use, 6, 50, 51, 103, 111, 195, 215, 239, 277, 287, 362, 363, 370, 407, 433, 449, 455, 472, 476, 483, 488, 489, 493, 497, 498, 537–539, 553, 563, 571, 573, 647, 654, 660, 686, 754, 755, 794, 833, 840, 853, 881, 920, 953–954, 980, 994, 1005, 1060, 1069, 1075, 1102, 1108, 1110, 1621
  change, 6, 138, 152, 193, 217, 218, 794, 1052, 1054, 1146, 1759, 1781, 1847, 1963, 1965, 1989
  coordination of competing, 1266
  past and present, 1192–1193

Land use (*cont.*)
    pattern, 1047–1048, 1274
    Ramsar Site, 1861–1862
    wetland change, 361–362
La Niña phases, 1923
La Plata River system, 848
Large-River floodplains, 114, 729–731, 733, 737, 1446
Largest lake, 129, 145, 592, 1062, 1458, 1608
Larix, 1486
    L. cajanderi, 1460
    L. gmelinii, 1444
    L. laricina, 518, 524, 612
Larus
    L. armenicus, 1665
    L. californicus, 150, 683
    L. canus, 1047
    L. dominicanus, 1327
    L. genei, 1141, 1142, 1665
    L. marinus, 629
    L. melanocephalus, 1141–1142, 1190
    L. novaehollandiae, 1925
    L. pipixcan, 683
    L. ridibundus, 1192, 1628
Lasimorpha senegalensis, 1210
Lates, 1371
    L. calcarifer, 1945, 1956
    L. niloticus, 139, 1273, 1371
Latonia nigriventer, 1170, 1171
Laurentian great lakes (LGL), 129, 137, 593, 595, 603, 606, 611
Lecane, 1357
Lechwe in Kafue Flats, 211, 217, 1225
Ledum, 545
    L. groenlandicum, 516, 517, 571
    L. palustre, 1537
    L. palustre ssp. *decumbens*, 280
    L. palustre ssp. *diversipilosum*, 1601
Leersia hexandra, 191, 204, 205, 214, 816, 1236, 1237
Legislation, 105, 407, 453, 454, 456, 462, 483, 498, 538, 539, 685, 753, 810, 845, 892, 906, 919, 940, 1004, 1017, 1018, 1050, 1068, 1075, 1091, 1143, 1338, 1947, 1986, 1987
Lemna, 952, 1140
    L. gibba, 191, 914, 1301, 1368
    L. globosa, 1746
    L. minor, 765, 777, 914, 1639
    L. tenera, 1816
Lemnis, 552
Lena Delta, 280, 1445, 1449
    biodiversity, 1454
    conservation status, 1454
    ecosystem services, 1455
    hydrology, 1452
    location, 1452
    threats, 1455
    wetland ecosystems, 1453–1454
Lena River Basin
    area, 1441
    biodiversity, 1443–1444, 1459–1460, 1468–1469
    changes in ecological character, 1444–1445
    coastal habitats, 1445
    conservation, 1448–1449, 1460, 1469
    ecosystem services, 1443–1444, 1460, 1469
    eolian sand dune, 1447
    factors, 1442–1443
    future challenges and development, 1460
    hydrology, 1458, 1464–1465
    location, 1440, 1442
    natural inland wetlands, 1445
    plateaus and piedmont areas, 1448
    regional classification of waterbird habitats, 1446
    threats, 1460, 1469–1470
    types, 1445–1448
    wetland structure, 1458–1459, 1465–1468
Lena River Delta, 13, 1451–1455
Lentic ecosystem, 350, 914
Leontodon autumnalis, 1073
Leopardus
    L. geoffroyi, 817
    L. wiedii, 736
Lepidaria kingii, 1835
Lepidium
    L. caespitosum, 1189
    L. cartilagineum, 916
    L. latifolium, 951
Lepidochelys
    L. kempii, 584
    L. olivacea, 1739, 1747
Lepidopyga lilliae, 764
Lepidothamnus intermedius, 1980
Lepilaena, 149
    L. australis, 76
    L. marina, 76
Lepironia, 1856, 1862
    L. articulata, 1856
Lepomis
    L. gibbosus, 612
    L. macrochirus, 612
Leporinus obtusidens, 817
Leptadenia pyrotechnica, 1272
Leptobarbus hoevenii, 1846

Leptochloa
  L. chinensis, 204
  L. fusca, 205
Leptodacylus, 735
  L. latrans, 817
  L. ocellatus, 852
Leptodirus, 339
  L. hochenwartii, 339
Leptomys signatus, 211
Leptopalaemon, 211
Leptoptilos
  L. dubius, 213, 1774, 1776, 1779, 1787, 1796
  L. javanicus, 213, 1758, 1779, 1787, 1796, 1816, 1824, 1835, 1860
Leptospermum, 208
  L. scoparium, 1980, 1994, 1995, 1998
Lepus
  L. capensis, 1177
  L. europaeus, 220, 1084
Lesser Flamingo, 490, 1354, 1371, 1384–1386, 1729
Lestes
  L. dryas, 1071, 1073
  L. macrostigma, 1127, 1131
Leuciscus
  L. cephalus, 953
  L. idus, 1459
  L. leuciscus, 1459
Leucogeranus leucogeranus, 1525, 1566, 1584, 1626, 1636, 1641
Leucophaeus pipixcan, 596
Leucoraja ocellata, 633
Leucosticte nemoricola, 1700
Leuzea salina, 916
Levant, 462, 1156, 1244, 1255
Leymus
  L. chinensis, 1525
  L. mollis, 546, 560
Liasis fuscus, 1946
Licuala, 1945
  L. ramsayi, 208
  L. spinosa, 1816
Light, 31, 44, 45, 48, 66, 74, 78, 82, 89, 99, 133, 184, 218, 332, 334, 335, 391, 434, 451, 452, 466, 482, 528, 573, 614, 629, 734, 761, 1003, 1035, 1052, 1071, 1210, 1297, 1329, 1444, 1470, 1528, 1529, 1595, 1781, 1934, 1937, 1944, 1948, 1969, 1974, 2011
  fires, 1297, 1528, 1529
  limitations, 48, 1071
  shading, 654

Ligularia
  L. altaica, 1538
  L. sibirica, 1537
Ligustrum
  L. lucidum, 817
  L. sinense, 817, 818
Liman, 145, 888, 912
Limicola falcinellus, 1448, 1469
Limnanthemum peltata, 949
Limnodromus semipalmatus, 1481, 1824, 1835
Limnodynastes
  L. fletcheri, 1903
  L. salmini, 1903
Limnoperna fortunei, 809, 818, 854
Limnophilous, 940
Limnophyton obtusifolium, 1746
Limnornis curvirostris, 851
Limnothrix, 479
Limonia, 338
Limoniastrum spp., 1116
Limonium, 61, 1001, 1104, 1106, 1116, 1140
  L. bellidifolium, 965, 1017
  L. binervosum, 965, 1017
  L. globuliferum, 1189
  L. gmelinii, 916
  L. iconicum, 1189
  L. vulgare, 964, 965, 1018
Limosa
  L. fedoa, 683
  L. haemastica, 833, 867
  L. lapponica, 547, 980, 990, 1003, 1015, 1016, 1337, 1981
  L. limosa, 980, 989, 1015, 1016, 1030, 1141, 1444, 1640, 1955
Limprichtia revolvens, 284
Lindenia tetraphylla, 1142
Lindera melissifolia, 584
Lindernia, 1117
Line transect, 1091
Liniarius mufumbiri, 192
Linnaea borealis, 1537
Linum maritimum, 1104
Liophis epinephelus, 777
Liopsetta glacialis, 1480
Liparis loeselii, 1027
Lipotes vexillifer, 115
Liquidambar styraciflua, 581
Lissotriton
  L. boscai, 1131
  L. vulgaris, 1047
Listera cordata, 1044

Lithobates
  L. catesbeiana, 656
  L. clamitans, 612, 656
  L. sylvaticus, 658
Litoria caerulea, 1903
Little Ice Age, 519, 1102
Littoridina australis, 850
Livelihoods, 62, 85, 101, 118, 178, 180, 190, 192–194, 214, 217, 237, 240, 316, 325, 326, 398, 400, 407, 410, 438, 443, 462, 463, 466, 488–492, 752, 758, 782, 786, 900, 954, 1208, 1222, 1229, 1230, 1240, 1249, 1261–1267, 1282, 1296, 1305, 1316, 1373, 1386, 1390, 1396, 1421, 1611, 1612, 1644, 1654–1656, 1669, 1672, 1682, 1689, 1690, 1706, 1708, 1717, 1718, 1729, 1738, 1764, 1773, 1777, 1779–1781, 1789, 1791, 1806, 1828, 1837, 1843, 1852, 1853, 1861, 1866, 1870, 1871, 1879, 2012, 2023
Living rivers, 927, 928
Livistona saribus, 207
Liza, 1083
  L. aurata, 965, 1254
  L. ramada, 965, 1254
Local knowledge, 1541
Local water resource management plan, 466
Loiseleuria procumbens, 279
Lolium, 1701
  L. multiflorum, 851
  L. perenne, 1068, 1073
Long-term monitoring, 453, 503, 1917, 1971
Long-term wetland losses, 372–373
Lontra
  L. canadensis, 166, 612
  L. enudris, 735
  L. longicaudis, 764, 817
  L. provocax, 825, 830
Lophodytes cucullatus, 584
Lophura erythrophthalma, 1860
Loripes lucinalis, 1324
Loudetia phragmitoides, 205
Low arctic, 511
Lower Amur-Ussuri freshwater ecoregion, 1489
Lower Congo Rapids, 1202, 1203, 1205, 1207–1210
Lower Danube Green Corridor (LDGC), 11, 886, 889, 893, 897–902, 921
  biodiversity, 898–899
  conservation status and management, 901–902
  ecosystem service, 900
  fish production, 900
  geographical maps, 899
  inhabitants, 901
  institute for European Environmental Policy, 900
  location, 898
  wetland types, 898–899
Lower Mekong Basin, 1809, 1810
  biodiversity of, 1773–1779
  climate change, 1782–1783
  climate of, 1767
  distribution and diversity of, 1768–1773
  hydrology, 1767
  population, 1779
  threats, 1780–1782
  uniqueness of, 1773–1779
  wetland sites, 1771–1773
Lower Mekong wetlands, 1768–1769
Lower Mississippi Valley Joint Venture (LMVJV), 579, 585
Lower Volga Basin, 938
Lower Zambezi, 1218, 1219, 1221, 1226–1228, 1230, 1235, 1238
Lowland Dipterocarp forest, 1756
Lowland tropical peats, *see* Peat, swamp forests
Low marsh, 61, 163, 165, 963, 964, 1082, 1084, 1117
Loxodonta africana, 1227, 1236, 1274, 1405
Luangwa River, 1226
Lucipimelodus pati, 850
Ludwigia, 205, 1105, 1107, 1109
  L. adscendens, 1746, 1796
  L. elegans, 816
  L. hyssopifolia, 205
  L. octovavis, 1746
  L. peruviana, 718
Luffa cylindrica, 1272
Lukanga wetland, fishing camps, 1223, 1224
Luma, 829
  L. chequen, 825
Lumnitzera racemosa, 1823
Lupinus spp., 560
Luscinia svecica, 547, 1084
Lutjanus cyanopterus, 763
Lutra
  L. lutra, 473, 889, 905, 918, 926, 980, 1047, 1107, 1131, 1142, 1192, 1443, 1474, 1621, 1629, 1642, 1653, 1655
  L. perspicillata, 1609, 1825, 1836, 1845
  L. sumatrana, 1777, 1779, 1788, 1816, 1825
Lutrogale perspicillata, 1779, 1788, 1860

Luziola peruviana, 816
Lycaena dispar, 1142
Lycaon pictus, 1222, 1238, 1303, 1372
Lycengraulis grossidens, 817, 849
Lycium depressum, 1189
Lycopodiella
    L. cernua, 1313
    L. serpentina, 1997
Lymnaea spp., 1690
Lynceus brachyurus, 659
Lyngbya
    L. limnetica, 1273
    L. wollei, 603
Lynx
    L. canadensis, 554
    L. pardinus, 1130, 1131
    L. rufus, 675
Lyonia lucida, 672
Lyrurus tetrix, 236
Lysapsus spp., 735
Lysimachia asperulaefolia, 674
Lystrophis dorbignyi, 852
Lytechinus variegatus, 843
Lythrum salicaria, 432, 597, 613, 915

M
Macaca
    M. fascicularis, 1825, 1835
    M. mulatta, 1737
    M. nemestrina, 1825, 1835
Macaranga
    M. pruinosa, 1848
    M. tanarius, 1848
Maccullochella peelii, 118, 1903, 1912
Machaerina, 1980
    M. tenax, 1994, 1996
    M. teretifolia, 1994, 1995
Machar marshes, 12, 185, 1247, 1270, 1279–1286
    future challenges and developments, 1284–1286
    location, 1280
    resources, 1281–1284
Machar wetland, 1280, 1283
Machilus gamblei, 207
MacKenzie River Basin, 549–555
    biodiversity, 552
    conservation status, 553
    ecosystems, 552
    ecosystem services, 553–554
    future challenges and developments, 554–555
    hydrology, 550–551
    location, 550
    threats, 554–555
Mackenzie River watershed, 550
    in Alberta, 516
    in North America, 550
Macoma balthica, 630, 1002
Macquarie Marshes, 1897–1907
    biodiversity, 1901–1903
    conservation status, 1904
    ecosystem services, 1904
    environmental water management, 1905–1906
    fauna, 1903
    future challenges and developments, 1906–1907
    hydrology, 1900
    location, 1898–1899
    threats, 1904–1905
    wetland ecosystems, 1900–1901
Macroalgae mats, 498
Macrobenthic species, 988
Macrobrachium, 1945
Macrodon ancylodon, 850, 852
Macrohabitat, 802–804, 806
Macrophytes, 48, 132, 149, 176, 184, 185, 187, 189–191, 193, 211, 311, 327, 353, 400, 401, 471, 478–480, 482, 483, 525, 602, 629, 630, 720, 765, 791, 804, 899, 988, 1105, 1116, 1117, 1128, 1130, 1257, 1301, 1305, 1350, 1353, 1354, 1357, 1359, 1367, 1369, 1370, 1372, 1384, 1398, 1409, 1619, 1681, 1689, 1708, 1736, 1857, 1901, 1915, 1933–1935, 1948
Macropus
    M. giganteus, 1902
    M. rufus, 1902
Mactra isabelleana, 850
Madan, 1687, 1688, 1690
Madhuca motleyana, 1834
Madoqua, 1387
Maerua kirkii, 206
Maerua subcordata, 206
Magellanic moorland, 876, 878
Magellanic tundra, 878
Magnolia
    M. griffithii, 207
    M. virginiana, 673, 719
Magnopotamion, 1144
Mahanadi River Delta
    biodiversity, 1747
    challenges, 1749

Mahanadi River Delta (*cont.*)
  ecosystem services, 1747
  future challenges and developments, 1749
  geographical description, 1744
  management, 1748–1749
  threats, 1748
  wetland distribution, 1744–1746
Makgadikgadi Framework Management Plan (MFDP), 488
Makgadikgadi wetlands, 9, 12
  development scenario, 492–493
  future challenges and developments, 492–493
  livelihoods, 490
  location, 488
  opportunities and challenges, 493–494
  principles and approach, 488–490
  resource base, 490
  resource value, 491–492
Malacocincla abbotti, 1716
Malayemys subtrijuga, 1845
Malaysia, 13, 26, 83, 98, 102, 202, 216, 232, 239, 324, 326, 440, 1754, 1759, 1796, 1826, 1846, 1848, 1852–1863, 1866, 1875–1882
Malaysian seagrass, 13
  decline, 1879–1881
  selected species, 1876, 1877
  significance, 1877–1879
  threats and challenges, 1881–1882
Malebo Pool, 1202, 1203, 1205, 1207, 1210, 1211
Mallines, 876, 877
Malus silvestris, 916
Malvinas, 298, 299
Mamawi Lake, 551
Mammals, 46–48, 51, 53, 86, 100, 113, 134, 149, 166, 176, 191–193, 211, 212, 214, 219, 220, 236, 287, 366, 512, 552, 569–571, 584, 598, 612, 628, 629, 632, 662, 675, 682, 692, 733, 735, 736, 750, 764, 766, 792, 806, 817, 818, 831, 852, 859, 889, 899, 918, 926, 939, 952, 953, 955, 980, 989, 993, 995, 1029, 1047, 1084, 1092–1094, 1106, 1107, 1109, 1131, 1142, 1192, 1206, 1228, 1229, 1237, 1265, 1302, 1315, 1328, 1337, 1338, 1345–1347, 1358, 1371, 1382, 1385–1387, 1395, 1405–1407, 1443, 1444, 1454, 1460, 1474, 1502, 1513–1515, 1584, 1609, 1621, 1629–1630, 1639, 1642, 1653, 1678, 1680, 1682, 1700, 1701, 1737, 1739, 1747, 1758, 1773, 1774, 1777, 1779, 1781, 1816, 1825, 1845, 1859, 1860, 1871, 1902, 1946, 1955, 1969, 1981, 1984, 1985, 1987, 2020, 2021
Managed realignment, 969–971, 1006, 1019, 1021
Management plan, 107, 324, 400, 402, 428, 443, 463–466, 480, 501, 585, 601, 663, 710, 720–721, 754, 779, 781, 792, 818, 833, 869–870, 895, 917, 918, 1035, 1143, 1194, 1240, 1265, 1296, 1313, 1375, 1496, 1524, 1546, 1610, 1631, 1644, 1655, 1671, 1682, 1683, 1710, 1748, 1749, 1797, 1817, 1828, 1853, 1862, 1863, 1905, 1913, 1956
  bahía lomas, 869–870
  integrated process, 402
  local water resource, 466
  specific and appropriate, 443
  water, 466
Man & Biosphere Reserve, 1108
Mangal, 58, 94, 100, 1281
Mangifera foetida, 1834
Mangrove(s), 452, 1336, 1728, 1819–1828
  aquaculture, 96, 102–105, 107
  area in individual countries, 96, 98
  biodiversity, 750–751, 1736–1737, 1824–1826
  biotic components, 100–101, 763–766
  climate change, 105, 106
  Colombia, 747–755
  conservation status, 752–754, 1738–1739, 1826–1827
  dependent people, 1738
  diversity of, 1736
  ecophysiological adaptations, 94
  ecosystem and ecosystem services, 101–102, 751–752, 766, 1737–1738, 1822–1823
  fauna, Nagelkerken, 100–101, 750
  food webs, 99, 100
  future challenges and developments, 102–107, 754–755, 1740, 1827–1828
  global distribution, 30, 59, 94–98
  hydrological changes, 99, 103
  hydrology, 1820–1822
  location, 748–749, 1734–1735
  management, 752–754, 1738–1739
  Mekong Delta, 1805
  natural events, 103, 1718

origin, 94, 102, 1822, 1826
peat carbon stock, 1823–1824
pollutants, 99, 105
within protected areas, 102, 103, 107, 752, 1737–1740, 1820, 1826–1828
and swamp forest communities, 1237
threats, 104, 754–755, 1738–1740, 1827–1828
in Zambezi Delta, 1227
zonation pattern, 99 (*see also* Colombian mangrove ecosystems)
Manicaria saccifera, 206
Manis javanica, 1860
Manitoba's Great Lakes, 10
coastal wetlands, 593–595
delta marsh, 598–602
future challenges and developments, 602–603
location, 592
threats, 602–603
water levels on, 600–601, 603
Manoao colensoi, 1980
Manouria emys, 1845, 1860
Maowusu sandy land, 1581, 1582
Maputaland coastal plain (MCP), 12, 1419
future challenges and developments, 1433–1434
geomorphological history, 1424–1426
location, 1424, 1425
mire types, 1426–1428
peat swamp forests, 1428–1432
swamp forest on mineral soils, 1433
threats, 1433–1434
Marañon-Ucayali Basins, 732
Marcusenius
  M. macrolepidotus, 1368
  M. stanleyanus, 1368
  M. victoriae, 1368
Marenzelleria viridis, 978
Margaritifera margaritifera, 1047
Margin, 68, 136, 149, 194, 255, 256, 270, 271, 304, 309, 311, 316, 422, 480, 586, 643, 671, 787, 789, 791, 824, 914–916, 1031, 1044, 1046, 1060, 1074, 1150, 1153, 1161, 1163, 1208, 1227, 1230, 1236, 1237, 1255, 1310, 1342, 1344–1347, 1366, 1386, 1568, 1639, 1649, 1652, 1653, 1660, 1680, 1706, 1717, 1718, 1856, 1910, 1946, 1980, 1981, 2009
Marifugia, 340–343
  M. cavatica, 340
Mari lake, 1446, 1447, 1459, 1473

Marine biodiversity, 85, 440, 1329, 2002, 2013, 2014, 2023
benthos, 1324–1325
birds, 1326–1328
fishes, 1325
marine mammals, 1328
sea turtles, 1325–1326
Marine mammals, 46, 51, 569, 570, 629, 1094, 1328, 1443, 1454, 1845, 1946, 1969, 2020, 2021
Marine transgression, 58, 265
Marmaronetta
  M. angustirostris, 953, 1130, 1612, 1653, 1665, 1681, 1691
  M. marmaronetta, 1688
Marmosa xerophila, 750
Marmota caudata, 1700
Marshes
boreal wetlands, 525
delta, 26, 595–602
estuarine, 7, 49, 56–58, 569–570
freshwater, 159–162, 169–180, 285, 341, 371, 698, 765, 1021, 1105, 1108, 1117, 1130, 1160, 1188, 1445, 1775, 1795, 1807–1809, 2008
Netley-Libau, 595–597, 599, 600, 602
salt, 48, 56, 58, 60–62, 64, 68, 78, 156, 170, 232, 407, 481, 510, 568, 570, 622, 629, 630, 634, 638, 833, 967, 984, 989, 994, 1079–1087, 1105, 1116, 1140, 1142, 1189, 1190, 1543, 1666, 1721, 1768
vegetation, 21, 58, 162, 172, 598, 644, 969
wetland, 525, 1553, 1554, 1557
Marsilea
  M. drummondii, 1901
  M. minuta, 1171
Marsippospermum gradiflorum, 876
Martes
  M. americana, 554
  M. foina, 1142
  M. martes, 1106, 1142
  M. zibellina, 1444, 1515
Mathenge, 1390
Mato Grosso National Park, 806
Matteuccia struthiopteris, 532
Mau Escarpment, 1389
Mauremys
  M. caspica, 1142
  M. leprosa, 1131
Mauritia, 219
  M. armata, 206
  M. flexuosa, 206, 732, 735

MAV, *see* Mississippi Alluvial Valley (MAV)
Mayas ecosystem, 1290, 1293
Mayas wetlands
  biodiversity, 1293–1294
  conservation status, 1294–1296
  ecosystem services, 1294
  future challenges and developments, 1296–1298
  hydrology, 1290–1293
  location, 1289
  threats, 1296–1298
Maytenus undata, 206
Mazandaran, 1636, 1638, 1639
Meadow, 22, 74, 77, 372, 401, 430, 442, 452, 480, 545, 552, 560, 598, 608, 612, 825, 826, 828, 875–877, 880, 904, 916, 938, 949, 951, 954–956, 967, 974, 976, 980, 986, 989, 1082, 1084, 1086, 1092, 1116, 1140, 1144, 1171, 1194, 1445, 1446, 1454, 1467–1469, 1473, 1480, 1482, 1490, 1491, 1501, 1510, 1524, 1525, 1529, 1534, 1538, 1542–1543, 1577, 1579, 1583, 1586, 1590, 1592, 1593, 1595, 1606, 1608, 1701, 1713, 1716, 1880, 1934, 1937
  atlantic salt, 967
  Mediterranean salt, 967, 1144
  seagrass, 13, 74, 87, 90, 438–443, 448, 450, 453, 841, 843, 845, 1876, 1878–1882, 1967–1974
  Volga Delta, 951
Mecanopsis regia, 1609
Meconopsis
  M. dhwojii, 1609
  M. horridula, 1609
  M. taylorii, 1609
Medicago spp., 1106
Mediterranean coastal lagoons, 1104
Mediterranean peatlands, 1156, 1158, 1161–1164
  biodiversity and ecosystem services, 1159
  distribution, 1156–1159
  future challenges and developments, 1163–1164
  human activity, 1162–1163
  types, 1156
Meesia triquetra, 284
Megacarpea polyandra, 1609
Megaceryle
  M. alcyon, 612
  M. maxima, 1345
Megalobrama amblycephala, 1586
Megalops atlanticus, 750

Megapodius laperouse, 2022
Megaptera novaeangliae, 448, 569, 629
Mekong River Basin, 13, 217, 1765, 1785–1791, 1793–1798, 1813–1817
  biodiversity, 1773–1779
  climate, 1767
  distribution, 1768–1773
  drainage basin, 1765
  future challenges and developments, 1780–1783
  hydrology, 1767
  threats, 1780–1782
Mekong River Delta, 13
  biodiversity, 1805–1806
  channelization and wetland drainage, 1804–1805
  climate change, 1808
  conservation status and management, 1807
  ecosystem products and services, 1806–1807
  human population, 1802
  locations, 1802, 1803
  physical environment, 1802–1804
  threats and future development, 1808–1810
  upstream hydropower development, 1809
  wetland vegetation, 1805
Mekong's wetlands, 1773–1779
Melaleuca, 207, 208, 211, 215, 219, 220, 1773, 1795, 1805, 1815, 1945, 1952
  M. argentea, 208
  M. bracteata, 208
  M. cajuputi, 206–208, 1795, 1815
  M. leucadendron, 206–208
  M. nervosa, 206, 208
  M. quinquenervia, 210, 211
  M. viridiflora, 206–208
Melaleuca forests, 208, 219, 220, 1777, 1815–1817
Melampus coffeus, 763
Melampyrum arvense, 916
Melandrium apetalum, 1539
Melanitta, 642
  M. deglandi, 1444, 1460
  M. fusca, 1093
  M. nigra, 1091
  M. perspicillata, 642
Melanogrammus aeglefinus, 629
Melanoperdix niger, 1835
Melanoptila glabirostris, 324
Melanosuchus niger, 735
Melanotaenia eachamensis, 1949
Meles meles, 1106, 1142
Melica ciliata, 916

Mellanitta deglandi, 1444, 1460
Melochia crenata, 766
Melongena melongena, 763
Melosira granulata, 1273
Melospiza melodia, 648
Memecylon edule, 205
Menippe nodifrons, 844
Mentha
  M. aquatica, 915
  M. pulegium, 1130
Menticirrhus americanus, 850
Menyanthes trifoliata, 510, 560, 1046, 1072
Mergellus albellus, 1094
Mergus, 1091
  M. merganser, 1094, 1610
  M. serrator, 1092
Merismopedia, 1936
Merops philippinus, 1825
Merremia hederacea, 205
Mesocyclops, 1357
Mesodesma mactroides, 850
Mesohaline, 157
Mesoplodon bidens, 633
Mesopotamia Marshland National Park, 1694
Mesopotamian marshes, Iraq, 363
  biodiversity, 1688–1691
  future challenges and developments, 1692–1694
  hydrology, 1688
  location, 1686
  threats and challenges, 1692–1693
  water restoration guidelines, 1693
Mesosetum liliiforme, 206
Mesotidal estuaries, 41
Mesotopes, 305–308, 1052
Mesua hexapetalum, 208
Metacarcinus magister, 570
Metallina, 1082
Metamorphic rocks, 1189
Metapenaeus affinis, 1688
Methane production, 235, 684
Methanogenesis, 157, 163, 167
Methylohalomonas lacus, 1388
Methylonatrum kenyese, 1388
Metopidius indicus, 1777
Metroxylon sagu, 208
Mexico's Yucatan Peninsula, 322
Microcarbo
  M. africanus, 1274, 1338
  M. melanoleucos, 1902
  M. pygmaeus, 918
Microcos cf. *stylocarpa*, 208

Microcystis, 479, 1936
  M. aeruginosa, 1371
  M. flosaquae, 1385, 1936
Micropogonias furnieri, 844, 850, 852, 853
Micropterus
  M. dolomieu, 612
  M. salmoides, 612, 1357, 1358
Micropyropsis tuberosa, 1130
Microtidal estuaries, 41
Microtus townsendii, 568
Microzooplankton dilution studies, 1937
Middle Aldan River, 13
  biodiversity, 1474
  conservation status, 1475
  ecosystem services, 1475
  future challenges and development, 1475
  hydrology, 1472–1473
  and territories, 1474
  threats, 1476
  wetland ecosystems, 1473
Middle Amur River
  conservation status, 1518
  from Khabarovsk to Lake Bolon, 1513
  threat with economic development, 1516
Middle East, 9, 11, 27, 31, 103, 104, 116, 121, 184, 318, 340, 346, 1164, 1168, 1171, 1621, 1640, 1678, 1686
Middle Zambezi, 1218, 1222–1226
Mid Yangtze Basin, 1566
Migratory birds, 52, 392, 398, 572, 585, 586, 682, 683, 733, 816, 829, 833, 859, 868, 930, 948, 952, 957, 989–992, 1030, 1085, 1115, 1141, 1161, 1169, 1192, 1371, 1525, 1558, 1561, 1569, 1584, 1620, 1628, 1639, 1640, 1648, 1664, 1676, 1683, 1715, 1892, 1902, 1912, 1955, 1981, 1983, 1985, 2016
Migratory shorebirds, 150, 377, 379, 407, 408, 630, 633, 641, 1561, 1870, 1925
Migratory species agreement, 1872
Migratory waders, 1324, 1327, 1337, 1824, 1917
Migratory waterbirds, 151, 153, 364, 406, 409, 989, 994, 1002, 1095, 1130, 1557, 1583, 1621, 1626, 1631, 1640, 1644, 1652, 1654, 1669, 1682, 1718, 1820, 1824, 1835, 1902, 1912
Mikania, 191
Millennium Ecosystem Assessment, 8, 48, 101, 177–179, 361, 531, 532, 859
Millingerwaard, 928, 929
Milvus migrans, 1106

Mimosa pigra, 137, 206, 220, 366, 385, 386, 1773, 1798, 1859, 1952
Minas Passage, 628, 634
Mineral wetlands, 529
Minerotrophic fen, 230, 252, 261–263, 270, 296, 304, 524, 525, 674, 880, 1156, 1158, 1159, 1163, 1993
Minerotrophic mires (fens)
  lagg fen, 310
  open-water transition fen, 252, 253
  Schwingmoor fen, 252, 253
Mining, 51, 105, 117, 151, 442–443, 454, 457, 491, 512, 534–536, 538, 539, 634, 737, 739, 741, 830, 834, 859, 870, 891, 1200, 1209, 1210, 1223, 1339, 1397, 1416, 1419, 1445, 1483, 1493, 1503, 1504, 1730, 1781, 1782, 1868, 1926, 1957, 1987
  artisanal, 754
  clay, 191
  gold, 218, 548, 740, 1491, 1496
  hydraulic, 640
  industrial, 754
  open-pit, 881
  peat, 363, 370, 573, 879, 938, 1998
  salt, 147, 152, 488
  sand, 1880, 1882
  silver, 548
Miniopterus schreibersii, 1192
Minister of Water Resources and Meteorology, 1789
Ministry of Agriculture Forestry and Fisheries (MAFF), 1787, 1788
Ministry of Water Resources and Electricity (MoWRE), 1296
Mire, 21, 26, 228, 230, 238, 247, 250, 259, 292, 311, 480, 708, 709, 712, 720, 879, 1150–1152, 1158, 1309, 1310, 1414–1417, 1429, 1430, 1442, 1443, 1446, 1454, 1486, 1547, 1978
  alpine, 1414, 1535
  artesian spring, 256
  blanket, 269–270, 295–301, 304–306, 415, 417–423, 1039–1054
  cloud-affected forests, 272
  condensation, 271
  definition, 252, 290–291
  diversity, 1535–1546
  Elatia, 1162
  floodplain, 1428, 1431
  karst immersion, 253
  Maputaland Coastal Plain, 1426–1428, 1434

Mfabeni, 1426
minerotrophic, 252–254, 310
Nuur, 1540
occult precipitation, 271
ombrotrophic, 233, 235, 264–271, 309
open-water transition, 253, 263
palsa, 262–263, 279–283
percolation, 257–258
polygon, 279–280
riparian, 285
Rugezi, 1421
Schwingmoor, 263
snowmelt-patterned mixed, 261
surface-flow spring, 256–257
tidal immersion, 254
transition, 263–264
tropical and subtropical regions, 262
types, 1417–1419
valley, 254–256, 258, 1159
Miscanthidium violaceum, 205, 1313
Miscanthus, 189, 1418
  M. violaceus, 190, 1311
Mississippi Alluvial Valley (MAV), 12, 113, 578, 580, 583, 587, 588, 683, 698
  biodiversity, 584–585
  conservation status, 585–586
  ecosystem services, 586–587
  future challenges and developments, 587–588
  hydrology, 578–580
  location, 578
  wetland and aquatic systems, 580–584
Mist, 271, 298, 304, 1343, 1868
Mitoura hesseli, 674
Mitragyna, 207
  M. ciliata, 207
  M. stipulosa, 207
Mixed mire, 247, 262, 263
  palsa mires, 262–263
  snowmelt-patterned, 261
  transition mire, 263–264
  waterlogging, 261–264
  wet-season subtropical/tropical patterned, 262, 263
Mnemiopsis leidyi, 152–153, 918, 995
Mnium, 284
Mobutu-Sese Seko, 1246
Modiolus modiolus, 628
Mogurnda adspersa, 1946
Moina, 1357
Molinia, 1046
  M. caerulea, 256, 261, 1045, 1046
Mollusca, 569

Molluscs, 410, 440, 628, 630, 660, 661, 752, 917, 918, 1273, 1336, 1737, 1860, 1963
Molothrus aeneus, 764
Molt, 1093
Monachus monachus, 1320, 1328
Mongolia, 13, 26, 147, 232, 1478, 1479, 1487–1493, 1499–1507, 1510, 1547, 1576, 1579, 1581, 1726, 1747
    highland valley peatland degradation, 1533
    mire conservation, 1546
    mire diversity, 1535–1546
    mire massif types, 1537
    peatland distribution, 1534–1535
Monochoria hastate, 1746
Monolistra spp., 340
Monoraphidium minutum, 1385
Montreux record, 400, 480, 482, 1623, 1749
    restoration initiatives, 401–402
Montrichardia arborescens, 206, 750
Mont-Saint-Michel bay, 11
    animal diversity, 1082–1084
    conservation status, 1086
    ecosystem services, 1084–1086
    hydrology and geomorphology, 1080–1081
    location, 1080
    plant diversity, 1081
    salt marshes, 1082, 1084
Moor, 21, 290, 417, 422, 1046, 1047
Moose, 240, 532, 553, 554, 561, 562
Mora oleifera, 750
Morecambe Bay, 999, 1000, 1003, 1008
Morimus funereus, 1142
Mormyrus, 1368
    M. anguilloides, 1368
    M. caschive, 1273
    M. kannume, 1368
    M. longirostris, 1368
Morone
    M. chrysops, 603
    M. saxatilis, 633
Morphological floodplain, 888, 893, 898
Moschus moschiferus, 1444
Moss/Graminoid class, 546
Motacilla
    M. flava, 965, 1105
    M. samveasnae, 1779
    M. tschutschensis, 547
Mudflats, 50, 51, 64, 158, 462, 544, 561, 569, 580, 583–585, 630–632, 641, 693, 869, 969, 970, 978, 1001, 1081, 1083, 1140, 1236, 1237, 1320, 1322, 1323, 1325, 1345, 1566, 1569, 1678, 1679, 1681, 1718, 1768, 1769, 1777, 1820, 1823, 1824, 1827, 1835, 1869, 1872, 1912, 1916, 1917, 1946, 1985
Mugil, 763, 852, 1083
    M. cephalus, 859, 1131, 1254, 1325
    M. incilis, 763
    M. platanus, 844
Multidisciplinary approach, 4, 489, 755
Muntiacus muntjak, 1613
Mura-Drava-Danube, 908
Mura River, 11, 886
    channelling, 906
    conservation status and management, 906–908
    ecosystem service, 908
    hydropower dams, 906
    location, 904
    threats, 905
    wetland types and diversity, 904–905
Murray-Darling Basin, 113, 118, 1889–1892, 1895, 1897–1907, 1909–1918
    catchment management, 1894
    climate and hydrology, 1890–1891
    environmental flows, 1892–1893
    future challenges and developments, 1995
    management responses, 1891–1892
    protected areas, 1895
    threatened biota, 1894
    wetlands, 1888–1890
Murray mouth, 1888, 1891, 1910, 1911, 1913–1915
Musa, 2009
Muschus chrysogaster, 1608
Muscicapa
    M. aquatica, 192
    M. sibirica, 1825
Mussaendopsis beccariana, 1834
Mussel culture, 1144, 1145
Mustela
    M. erminea, 570, 918, 1700
    M. eversmanni, 1443
    M. lutreola, 918, 953
    M. nivalis vulgaris, 1142
    M. sibirica, 1515
    M. vison, 1443, 1474
    M. zibellina, 1515
Mustelus schmitti, 850, 853
Mya arenaria, 630, 918, 978, 1092
Myall Lakes, Australia
    catchment discharge, 1932, 1933
    future challenges and developments, 1937–1938
    gyttja, 1933–1935

Myall Lakes, Australia (*cont.*)
  hydrology, 1931–1933
  hypsometry, 1933
  location, 1930
  macrophytes, 1933–1935
  nutrients and plankton, 1935–1937
  waterbodies and bathymetry, 1930
Mycteria
  M. cinerea, 213, 1758, 1787, 1824, 1835
  M. ibis, 1385
  M. leucocephala, 213, 1729, 1787, 1796, 1816
Myliobatis goodei, 850
Myocastor coypus, 817, 852, 1109, 1142
Myodes gapperi occidentalis, 571
Myosurus minimus, 1105
Myotis
  M. capaccinii, 1192
  M. grisescens, 584
  M. myotis, 1192
Myr, 290
  See also Mire
Myrceugenia, 829
  M. exsucca, 825
Myrica, 1043
  M. cerifera, 719
  M. gale, 545, 560, 571
Myriophyllum, 510, 561, 612, 1609
  M. alterniflorum, 1130
  M. robustum, 1997
  M. spicatum, 480, 613, 914, 952, 1105, 1106, 1117, 1460, 1524
  M. verticillatum, 914
Myristica, 207
  M. elliptica, 1834
  M. hollrungii, 208
  M. iners, 1834
Myrmecophaga tridactyla, 220
Mystus, 1778
Mytella charruana, 850
Mytilopsis, 341, 343
  M. kusceri, 341
Mytilus
  M. edulis, 988, 1092
  M. galloprovincialis, 1145

N
Ñadis, 825
Naemorhedus goral, 1613
Najas, 613, 1274, 1746
  N. graminea, 1666
  N. horrida, 1353
  N. marina, 952, 1031, 1105, 1934, 1937
  N. minor, 1272, 1649, 1652, 1746
  N. pectinata, 1272, 1357
Naja siamensis, 1776
Nakaikemi-shicchi peatland, 249
Nalsarovar, 1730
Namdrik atoll, 2019, 2023
Nanger granti, 1371
Narcissus tazetta, 1654
Nardostachys
  N. grandiflora, 1609
  N. scrophulariifolia, 1609
Narmada Canal system, 1731
Narthecium ossifragum, 1044
Nasalis larvatus, 210, 1758, 1845
Nasua nasua, 809
Natator depressus, 453
National inventory, 31, 607, 824, 832–836
National Nature Reserve (NNR), 480, 482, 1018, 1090, 1490–1493, 1568–1570, 1583, 1584
  Camargue, 1108
  Dafeng, 1555
  Dalai Lake, 1505
  Faioleag, 307
  Honghe, 1515, 1517
  Hui River, 1492
  Poyang Lake, 1567, 1568
  Qixinghe, 1517
  Sanjiang, 1517
  Thursley, 256
  Vigueirat, 1108
  Xingkaihu, 1517
  Zhenbaodao, 1517
National Park of Banc d'Arguin, *see* Banc d'Arguin
National Water Act, 1892
National Wetland Policy (2003), 1610
Natrionella acetigena, 1389
Natrix
  N. maura, 1131
  N. natrix, 926, 1131
  N. tessellata, 1142
Natronococcus amylolyticus, 1388
Natronoincola histidinovorans, 1389
Natura 2000, 480, 482, 889, 906, 917, 967, 1032, 1059, 1086, 1108, 1130, 1132, 1141, 1144, 1145
  sites, 906, 967, 1004, 1027, 1032, 1059, 1108, 1141, 1143
Natural ebb, 360, 1239
Natural effects, 4, 8–9, 12, 359–367, 623, 791, 834

Natural National Park, 1086–1087
Natural resource(s), 4, 9, 62, 63, 178, 180, 214, 286, 355, 414, 441, 463, 490, 491, 494, 512, 513, 522, 533, 534, 537, 539, 585, 753, 754, 781, 818, 836, 898, 901, 905, 919, 921, 942, 946, 955–957, 1014, 1019, 1034, 1035, 1117, 1118, 1133, 1144, 1193, 1211, 1212, 1226, 1264, 1275, 1281–1286, 1296, 1304, 1329, 1334, 1340, 1375, 1386, 1482, 1535, 1543, 1721, 1730, 1781, 1787, 1789, 1861, 1863, 1892, 1894, 1913
- agriculture, 1193–1194
- biodiversity, 1281–1282
- of coastal environments, 51
- exploitation of, 512–513, 956
- lake, 1194–1195
- livestock, 1193–1194
- management, 9, 441, 450, 463–465, 488, 492, 1296, 1894, 1913
- recreation, 1194
- tourism, 1194
- Volga Delta wetlands, 955, 957
- water resources, 1282–1284
- wetland management plan, 1194

Nature Conservancy Council (NCC), 259, 1049, 1050
Nature conservation, 249, 482, 906, 946, 967, 969, 1004, 1050, 1109, 1132, 1143, 1171, 1172, 1188, 1518, 1836
Nauclea, 207
- *N. coadunata*, 207
- *N. orientalis*, 208
Navicula, 630, 1385
Neanthes, 630
- *N. succinea*, 850
Neap tides, 41, 1322, 1823
Nectophrynoides asperginis, 1344
Necturus maculosus, 612
Negative estuary, 39
Nelumbium, 1708
Nelumbo, 1708
- *N. caspica*, 952
- *N. nucifera*, 134, 204, 1710, 1773, 1795
- *N. nucifera* var. *caspica*, 1619
Nemacheilus, 1708
Nematalosa erebi, 1912, 1925
Nemotaulius punctatolineatus, 1047
Neobisium, 339
Neochanna, 1981
- *N. diversus*, 1997

Neofelis
- *N. diardi*, 1758
- *N. nebulosa*, 1835, 1845, 1860
Neogobius melanostomus, 612
Neophema chrysogaster, 327, 1912
Neophocaena phocaenoides, 1569, 1747
Neopicrorhiza scrophulariifolia, 1609
Neostrengeria macropa, 776, 778, 779
Neotropics, 28, 29, 31, 171, 372, 373, 758
Neovison vison, 612, 995
Nepal, 1605–1613, 1712, 1713, 1715, 1716, 1719, 1720
Nepenthes, 1757, 1758
- *N. mirabilis*, 1816
Nephtys, 630
Nereis, 630
- *N. diversicolor*, 988
Nerodia sipedon, 612
Netley Cut, 598
Netley-Libau marsh, 595–600, 602
Netta
- *N. erythrophthalma*, 764, 777
- *N. rufina*, 472, 1190, 1610
New Forest Code, 798, 809, 810
Newtonia devredii, 207
New Zealand wetlands, 1978–1980, 1998
- cultural values and co-management, 1982–1983
- flora and fauna, 1978–1981
- location, 1978
- policy and strategies, 1986–1987
- Ramsar sites, 1983–1984
- restoration, 1987–1988
- threats to, 1984–1986
- types and distribution, 1978
Nidjili Lake, Lena River Basin
- biodiversity, 1459–1460
- conservation status, 1460
- ecosystem services, 1460
- future challenges and developments, 1460
- hydrology, 1458
- location, 1458
- threats, 1460
- wetland structure, 1458–1459
Nilaus afer, 1385
Nile, 38, 53, 111, 113, 185, 186, 200, 1102, 1218, 1238, 1246–1249, 1252, 1255, 1257, 1262, 1265, 1270, 1274, 1279–1286, 1296, 1297, 1304, 1305, 1699
Nile Delta, 185, 1102, 1156, 1160, 1248, 1249
- biodiversity, 1254–1255
- characteristics, 1253–1254

Nile Delta (cont.)
    Egypt, 1251–1258
    future challenges and developments,
        1257–1258
    location, 1252
    north coast wetlands, 1255
    North Delta Lakes, 1255–1257
    threats, 1257–1258
Nile River Basin, 184, 1269–1276, 1280
    Atbara River, 1248
    Blue Nile, 1247
    future challenges and developments, 1249
    location, 1244
    role of wetlands, 1248
    spatial distribution, 1245
    Tekeze and Bahr el Salam, 1248
    threats, 1249
    tributaries, 1244
    White Nile, 1246–1247
Nilotes, Sudd catchment, 1304
Ningxia Plain, 1580
Niphargus, 321, 340
Nipponia nippon, 1599
Nitella hyalina, 1931, 1934
Nitrification, 162, 587, 738
Nitrogen, 8, 44, 66, 83, 89, 100, 133, 162–165,
        184, 188, 189, 209, 235, 366, 392,
        394, 433, 434, 456, 498, 499, 503,
        516, 519, 586, 587, 684, 712, 754,
        779, 830, 894, 901, 987, 1085, 1091,
        1103, 1353, 1357, 1365, 1366, 1394,
        1403, 1612, 1755, 1853, 1868, 1935,
        1937, 1969
Nitzschia, 86, 630, 1385
Nomonyx dominicus, 777
Non-forested peatlands, 1445
Norfolk and Suffolk Broads, UK, 11
    biodiversity, 1027–1032
    catchment area, 1026
    conservation status, 1032
    dependent peoples, 1035
    ecosystem services, 1034–1035
    future challenges and developments,
        1035–1036
    habitats, 1026, 1027
    location, 1024
    management, 1032–1034
    species, 1030–1032
North Africa, 116, 1156, 1164
North American tidal freshwater wetlands, 160
Northeastern Australian seagrass meadows
    distribution, 1968
    fish and shrimp populations, 1969
    monitoring, 1972–1973
    seagrass area, 1971–1972
    species, 1970–1971
    threat mapping, 1972
    threats and challenges, 1973–1974
Northeastern North America, Vernal pools,
    651–664
Northern Andes, 774
Notaden bennetti, 1903
Nothofagus, 307, 828, 879
    N. antarctica, 876, 878
    N. betuloides, 876, 878
    N. nitida, 876
Notophthalmus viridescens, 612, 656
Notopterus chitala, 1828
Notropis spp., 597
Ntaruka hydropower plant, 1309, 1311
Nuer communities, Sudd region, 1303, 1304
Numenius
    N. arquata, 980, 1016, 1073, 1083, 1141,
        1324, 1641
    N. madagascariensis, 1444, 1474, 1825,
        1955
    N. minutus, 1444
    N. phaeopus, 833, 1324, 1336, 1337, 1448,
        1642, 1955
    N. tenuirostris, 953, 1141
Nuphar, 526, 612, 718
    N. lutea, 159, 163, 717, 914, 952
    N. luteum, 1106
    N. polysepalum, 561
Nutrient pollution, 603, 891, 893, 894, 900,
    2017
Nutrients, 12, 20, 43, 44, 49, 85, 87–89, 99, 100,
        102, 103, 105, 115, 117, 126, 132, 133,
        135, 137, 138, 157, 158, 161, 162, 165,
        166, 178, 187, 201, 209, 218, 238, 353,
        364, 392–395, 432, 441, 451, 473,
        478, 483, 510, 524, 525, 532, 551, 553,
        597, 598, 602, 614, 624, 626, 643,
        671–673, 683, 711–714, 720, 731,
        739, 761, 817, 829, 840, 890, 900, 901,
        940, 974, 976, 981, 982, 987, 1074,
        1085, 1117–1119, 1145, 1170, 1208,
        1239, 1316, 1322, 1334, 1336, 1345,
        1358, 1365–1366, 1403–1405, 1433,
        1600, 1630, 1682, 1683, 1688, 1716,
        1734, 1737, 1754, 1781, 1789, 1843,
        1848, 1856, 1880, 1935–1937, 1947,
        1948, 1963, 1965, 1969, 1985, 1987,
        1992, 1994
    biotransport of, 392–393
    in Lake Turkana, 1365, 1366

in Myall Lakes, 1935–1937
and sediment, 432–434
total nitrogen, 499–500
in Wadden Sea, 987
Nuttallia obscurata, 573
Nyctereutes procyonoides, 995, 1134
Nycticorax
   N. caledonicus, 1902
   N. nycticorax, 597, 1141, 1642, 1824, 1902
Nyctophilus gouldii, 1903
Nymphaea, 134, 189, 612, 718, 1368, 1398, 1689, 1708, 1945, 1956
   N. alba, 914, 949, 1072, 1117, 1171, 1524
   N. ampla, 765
   N. caerulea, 1237, 1357
   N. candida, 480, 952
   N. gardneriana, 735
   N. lotus, 1272, 1301, 1353, 1354
   N. micrantha, 1272
   N. nouchali, 1313, 1795
   N. nouchali var. *caerulea*, 1314
   N. odorata, 717, 718
   N. tetragona, 1795
Nymphoides, 1689, 1708
   N. indica, 1237, 1796
   N. nilotica, 1237
   N. peltata, 482, 717, 914, 952
   N. peltatum, 1524
Nypa, 1825, 1835
   N. fruticans, 1738, 1823, 1834
Nyssa
   N. aquatica, 177, 581
   N. sylvatica var. *biflora*, 177, 669

O

Occult precipitation mires, 270–272, 298
Ochrolechia, 279
   O. frigida, 283
Ocotea, 206
   O. cymbarum, 207
Ocypode cuadrata, 850
Odobenus rosmarus, 1443, 1454
Odocoileus virginianus, 675
Odontesthes, 849, 853
   O. bonariensis, 817
   O. brevianalis, 859
Odontophrynus americanus, 852
Oecetis spp., 612
Oedothorax, 1082
Oenanthe
   O. aquatica, 915, 949
   O. oenanthe, 547

Ogallala acquifer, 690
Ogilbia pearsei, 323
Oil palm plantation, 215, 241, 768, 1822, 1837, 1847, 1852, 1861
Ojakaleh Forest, 1639
Okavango Delta, 11, 12, 110, 176, 201, 211, 488, 490, 1343, 1393–1410, 1414, 1415, 1418
   biodiversity, 1406–1408
   branching stream network, seasonal and permanent swamps, 1396
   desalination, 1401–1403
   ecosystems services, 1408–1409
   flood switching, 1400–1401, 1403–1405
   future challenges and developments, 1409–1410
   landscape, 1397–1399
   location, 1395
   seasonal flood pulse, 1405–1406
   site specific features, 1399–1400
   social and economic factors, 1396–1397
   threats and future challenges, 1409–1410
Oldest Indonesian Peat swamp, 1758, 1759, 1836, 1843–1847
Olea hochstetteri, 1386
Olearia semidentata, 1995
Oligohaline, 56, 157, 482
Oligotrophic peatlands, 672
Ombrotrophic bog(s), 252, 256, 261, 270, 296, 304, 310, 876, 1156, 1159
Ombrotrophic mires, 233, 235, 264
   Blanket bogs, 269–271, 304, 305
   intermediate raised bogs, 267–269
   occult precipitation mires, 271–272
   raised bogs, 264–269
Omo Delta wetland, 1367–1369
Omphalina hudsoniana, 283
Omphiscola glabra, 1071
Oncorhynchus
   O. clarkii, 561
   O. clarkii clarkii, 569
   O. gorbuscha, 569
   O. keta, 569
   O. kisutch, 569
   O. mykiss, 561, 569, 642, 1358
   O. nerka, 569
   O. tshawytscha, 569, 642
Oncosperma tigillarium, 1823, 1834
Ondatra zibethicus, 166, 552, 597, 1443, 1460, 1469, 1474
Onon valley, 1543
Ontragus megaceros, 1304

Onychoprion
   O. anaethetusi, 1327
   O. fuscatus, 2022
Open water vegetation, 1237
   Sudd region, 1301
Ophiocordyceps sinensis, 1611
Ophiophagus hannah, 1860
Ophisternon infernale, 323
Opuntia, 1385, 1390
Orbignya martiana, 206
Orcaella brevirostris, 398, 1747, 1779, 1825
Orchestia gammarella, 1082, 1085
Orcinus orca, 569, 1328
Orconectes, 612
   O. rusticus, 603
Oreobulus obtusangulus, 876
Oreochromis, 192, 2010
   O. alcalicus grahami, 150
   O. leucostictus, 1357
   O. mossambicus, 1949
   O. niloticus, 763, 764, 853, 1254, 1354, 1357, 1370, 1371
   O. niloticus baringoensis, 1354, 1355
Oreodytes alpinus, 1047
Oreotragus oreotragus, 1387
Organic pollution, 891, 893, 894, 1949
Organic wetlands, 529
Oriental darter, 1729, 1787, 1790, 1796, 1816
Orienus, 1715
Orkhon River valley, 1543, 1547
Orlitia borneensis, 1860
Ornithorhynchus anatinus, 1946
Orographic, 266, 296, 297, 1442, 1443, 1942
Oryctolagus cuniculus, 1107
Oryx beisa, 1371
Oryza, 1302
   O. longistaminata, 1265, 1302
   O. longistaminus, 1236
   O. meridionalis, 205, 1952
   O. rufipogon, 214, 1746, 1776, 1795
   O. sativa, 214
Osmerus
   O. eperlanus, 980
   O. mordax, 603
Osmunda regalis, 1313
Osteochilus partilineatus, 1845
Osteolaemus tetraspis, 1337
Ostrea puelchana, 845
Osvald, H., 304
Otis tarda, 953, 1131, 1186, 1493, 1502, 1584
Ottelia, 1945
   O. alismoides, 1272, 1524, 1746
   O. brachyphylla, 1272
   O. ovalifolia, 1368
Ourebia
   O. ourebi, 1274, 1282
   O. ourebia montana, 1293
Overgrazing, 325, 828, 880, 956, 1074, 1367, 1420, 1470, 1497, 1504, 1533, 1546, 1623, 1683
Overharvesting, 66, 117, 179, 361, 365, 366, 1211, 1455, 1790, 1828, 1868, 2018, 2019
Ovibos moschatus, 1443
Ovibus moschatus, 512
Ovis ammon, 1666
Oxalis acetosella, 1639
Oxidation, 133, 237, 343, 354, 419, 421, 673, 1153, 1162, 1163, 1433, 1445, 1759, 1760, 1837, 1838, 1843, 1914, 1963
Oxidoras kneri, 817
Oxycaryum cubense, 809
Oxycoccus
   O. microcarpus, 1537
   O. palustris, 511
Oxyeleotris marmorata, 1846
Oxyrhopus rhombifer, 852
Oxyura
   O. jamaicensis, 583, 777, 1134
   O. leucocephala, 953, 1131, 1621, 1629, 1653, 1681
Ozotoceros bezoarticus, 852

P
Pachygnatha, 1082
Pacific atolls, 2003–2006
Pacific islands, 10, 13, 67, 94, 236, 2002, 2003, 2006, 2010
Paddle grass, *see Halophila decipiens*
Padina, 86
Paedocypris spp., 1759
Painted stork, 1729, 1787, 1796, 1816
Pakistan, 105, 202, 217, 1698–1700, 1720, 1726, 1734
Palaemon nilotica, 1273, 1303
Palearctic waders, 1320
Paleo-deltas, 799
Paleo-dikes, 799
Paleolimnology, 1917
Palestine, 1170
Palmaria palmata, 626
Palsa, 230, 263, 279–283, 286
   degradation, 283
   mires, 26, 262–263, 278–283

Paludella squarrosa, 284
Paludification, 267, 286, 296, 310, 311, 517, 529, 673, 1041
Paludified shallow peatland, 276, 278, 279
Palustrine emergent wetlands, 561, 580, 581
Palustrine scrub-shrub, 583
Pancratium maritimum, 1104
Pandanus, 2009, 2013, 2105
Pandanus, 207, 208, 1856, 1862, 2009, 2013, 2015
   P. atrocarpus, 1848
   P. furcatus, 1823
   P. helicopus, 26, 1834, 1856
   P. hollrungii, 208
   P. hysterix, 208
   P. immersus, 1859
   P. kaernbachii, 208
   P. lauterbachii, 208
   P. leiophyllus, 208
   P. scabribracteatus, 208
   P. tectorius, 208
Panderia pilosa, 1189
Pandion haliaetus, 166, 612, 777, 1337, 1444, 1460, 1474, 1946
Pangasianodon
   P. gigas, 118, 1778, 1790
   P. sanitwongsei, 1778
Panicum, 205, 214, 580, 715, 851
   P. coloratum, 1344
   P. conjugatum, 204
   P. elephantipes, 791, 816
   P. fluviicola, 1344
   P. grumosum, 163
   P. hemitomon, 715
   P. maculatum, 1236
   P. maximum, 1236
   P. parvifolium, 205
   P. repens, 204, 1795, 1796
   P. subalbidum, 205
Pantanal, 10, 26, 171, 172, 174, 176, 200, 201, 205, 206, 210, 212, 218, 736, 786, 787, 789, 793
   biodiversity, 804–806
   conservation status, 806–807
   ecosystem services, 807
   future challenges and developments, 810
   geographic and ecological setting, 799–804
   herpetofauna, 805
   human occupation, 798
   landscape and dynamic vegetation, 803
   mammal species, 806
   map and catchment area, 800
   paleo-climatic history, 803
   temporary wetland, 801
   threats, 807–810
   vertebrates, 805
Panthera
   P. leo, 1238, 1274, 1372
   P. leo leo, 1294
   P. onca, 324, 736
   P. pardus, 1238, 1372, 1613
   P. pardus orientalis, 1514
   P. tigris, 1715, 1719, 1860
   P. tigris altaica, 1514
   P. tigris sumatrae, 236, 1758, 1825, 1835
   P. tigris tigris, 1737
   P. tigris virgata, 1642
Panulirus
   P. argus, 844
   P. laevicauda, 844
Panurus biarmicus, 1030
Papahanaumokuakea Marine National Monument (PMNM), 2020
Papaver maeoticum, 916
Papilio
   P. machaon, 1030
   P. palamedes, 674
   P. zelicaon, 570
Papio anubis, 1294
Papyrus, 170, 171, 200, 209, 1170, 1171, 1220, 1226, 1229, 1237, 1244, 1281, 1309, 1400, 1418, 1419, 1426–1428
   fen, 1419
   marshes, 184
      biodiversity, 190–192
      conservation and management status, 190
      ecosystem services, 190
      features of, 186–189
      human dependence, 192–193
      occurrence of, 184–186
      threats, 193–195
   swamps, 170, 209, 213, 1227, 1235, 1237, 1247, 1418
   zonation, 189
Parachela cyanea, 1845
Paradoxornis webbiana, 1559
Paraguay River, 26, 786, 789, 793, 794, 798, 799, 801, 807, 808, 810, 814, 816, 817
Paralichthys
   P. orbignyanus, 850
   P. patagonicus, 850
Paralonchurus brasiliensis, 850
Paramesotriton
   P. deloustali, 325
   P. laoensis, 1778

Paraná Delta region
  conservation status, 818
  ecosystem functions and values, 817–818
  fish and wildlife biodiversity, 817
  geomorphology and hydrology, 815–816
  landscapes, 816–817
  location, 814
  threats, 818
Paraná-Paraguay Fluvial Corridor
  biodiversity, 792
  conservation status, 792
  distributaries, 791
  ecosystem services, 793
  floodplain channels, 791
  floodplain lakes, 791
  fluvial forests, 792
  future challenges and developments, 793–794
  hydrology, 788–790
  location, 787
  main channel and anabranches, 790
  marshlands, flooded grasslands and savannas, 791
  threats and future challenges, 793
  wetland complexes, 788
Parapholis strigosa, 61, 964
Parapimelodus valenciennsi, 850
Parartemia zietziana, 1916
Parastacus nicoletti, 825
Parastemon
  P. lateriflora, 1834
  P. sumatrana, 1834
  P. urophyllus, 1834
Pardofelis marmorata, 1758
Pardosa, 1082
Parinari glabra, 207
Parishan Lake, see Lake(s), Parishan, Iran
Parnassia palustris, 285, 1537
Parona signata, 850
Parrotia persica, 1639
Partially mixed estuary, 41, 44
Parula pitiayumi, 852
Parus major, 1825
Paspalidium
  P. geminatum, 1353, 1354, 1368
  P. paludiphalus, 851
  P. paludivagum, 205
Paspalum, 61, 851
  P. commersonii, 205
  P. distichum, 1901
  P. fasiculatum, 765
  P. melanospermum, 206
  P. paspalodes, 1130
  P. repens, 204, 205, 791
  P. scrobiculatum, 1344
  P. setllatum, 206
  P. strigosum, 205
  P. vaginatum, 816, 1117
  P. virginatum, 205
Passerculus sandwichensis, 570
Pastoral sheep activities, 1086
Patagonia peatlands
  area and origin, 874–876
  distribution, 876–879
  future challenges and developments, 881
  threats and challenges, 881
  values and functions, 880
Patterned fen, 259–262, 519
Patterned string fen, 276, 278, 283–284
Pavona, 86
Pavonia zeylanica, 206
Peace-Athabasca Delta (PAD), 550
  biodiversity, 552
  conservation status, 553
  ecosystem services, 553–554
  future challenges and developments, 554–555
  hydrology, 550–551
  wetland ecosystem, 552
Peace-Athabasca Ecological Monitoring Program, 553
Peace River, 544, 551, 553, 554
Peat, 6, 21, 26, 170, 173, 190, 200, 228–241, 255, 264, 266, 877, 878, 1040, 1041, 1044, 1047, 1048, 1052, 1068, 1070, 1071, 1151–1153, 1158–1160, 1163, 1170, 1171, 1309, 1312, 1316, 1408, 1415, 1416, 1418, 1419, 1421, 1427, 1428, 1433, 1434, 1446, 1448, 1533, 1534, 1537, 1538, 1540–1547, 1553, 1556, 1755, 1756, 1759, 1760, 1814, 1815, 1821–1823, 1826, 1832, 1836–1838, 1843, 1847, 1848, 1856, 1862, 1960, 1961, 1979, 1985, 1988, 1992–1996
  accumulation, 236, 254, 264, 267, 285, 286, 671, 877, 1428, 1433, 1536, 1543, 1843, 1856
  bog, 248, 309, 938, 1998
  characteristics, 1755, 1757
  cutting, 267, 299, 879, 1025, 1162, 1163
  deposits, 246, 249, 250, 285, 415, 516, 671, 1150, 1168, 1310, 1424, 1425, 1534, 1536, 1541–1546, 1759, 1856
  digging, 1163

dome, 1755, 1756, 1759, 1816, 1822, 1828, 1832, 1837, 1985, 1996, 1998
formation, 228, 246, 256, 257, 268, 271, 280, 284, 290, 291, 305, 310, 415, 517, 673, 732, 1151, 1163, 1408, 1416, 1427, 1540, 1541, 1755, 1759, 1856
vs. freshwater swamps, 170, 201
genesis, 230, 251–272
mining, 363, 370, 573, 879, 938, 1998
plateaus, 276, 278–283, 518, 545
of Rugezi Marsh, 1311–1313
soils, 246–248, 250, 415, 422, 525, 546, 675, 1019, 1025, 1052, 1171, 1433, 1816, 1837, 1843, 1856, 1989
swamp forests, 26, 230, 234, 239, 240, 264, 265, 269, 301, 1415, 1424, 1426, 1428–1434, 1753–1760, 1769, 1771, 1808, 1813–1817, 1820, 1822, 1823, 1826, 1827, 1834, 1836, 1837, 1843–1847, 1856
   adaptations to waterlogged environment, 1756
   Cyrtostachys renda, 1757, 1834, 1835
   ecosystem services, 1759
   environment, 1754–1756
   hummock-hollow topography, 1754, 1755
   landscape, 1754
   orangutan, 236, 1757, 1758, 1845
   pitcher plants, 1757, 1758
   regional and local hydrological regulation, 1759
   threats, 1759
   vegetation, 1756, 1834
wind farms, 413–423
Peatland(s), 5, 6, 10–13, 20, 21, 26, 28–30, 113, 170, 171, 247, 249, 250, 276–278, 280, 284, 286, 287, 290–292, 296, 300, 301, 304–306, 351, 363, 373, 423, 430, 522, 529–536, 538, 539, 545, 592, 606, 608, 670, 672, 673, 675, 706, 708, 709, 715, 720, 825–827, 830–832, 938, 940, 1047, 1049, 1050, 1117, 1169–1171, 1307–1316, 1423–1434, 1445, 1446, 1532–1541, 1543, 1544, 1546, 1547, 1580, 1713, 1754, 1759, 1760, 1814, 1822–1824, 1828, 1837, 1848, 1856, 1888, 1983, 1985, 1992, 1995, 1996, 1999, 1052–1054
Africa, 229, 1413–1421
biota, 6, 235–236
carbon storage, 249–250
characterization, 228
Chilean, 830–833
classification, 230, 240, 241, 1534
complex, 256, 516, 518, 608, 720, 878
in continental North America, 515–519
definition, 228
distribution, 277–279, 531, 876–879, 1415–1417, 1534
ecological features, 233–235
ecosystem services, 237
fens waterlog
   channeled surface-water flow, 258–261
   groundwater moving downslope, 254–258
functions, 880, 1054
global distribution, 230–233
Karst Immersion Mire, 253–254
landscapes, 518–519, 1547
management, 240, 241, 879–880, 1760
Mediterranean Region, 1155–1164
minerotrophic mires (fens), 252–253
mires, 415
mixed mires waterlog, 261–264
in Mongolia (*see* Mongolia)
occult precipitation mires, 271–272
ombrotrophic mires (bogs), 264–271
Patagonian, 873–881
Philippi, 1149–1153
production services, 238–240
regulation services, 237–238
threats and challenges, 240–242
tidal immersion mire, 254
turbine towers and turbine bases, 418–419
Volga River Basin, 938
wind farms, 415–418
Pedicularis
   P. dasystachys, 1543
   P. karoi, 1537, 1538
   P. poluninii, 1609
   P. pseudoregelina, 1609
Pelagic fish species, 165, 1323
Pelecanus
   P. conspicillatus, 1915, 1925
   P. crispus, 889, 899, 918, 953, 1141, 1584, 1621, 1629, 1641, 1642, 1652, 1653, 1729
   P. erythrorhynchos, 683
   P. onocrotalus, 918, 1190, 1227, 1238, 1255, 1274, 1323, 1327, 1387, 1444, 1621, 1665, 1729
   P. philippensis, 213, 1717, 1774, 1779, 1787, 1816, 1824
   P. rufescens, 1338

Pelliciera rhizophorae, 750, 751
Pelobates
   P. cultripes, 1131
   P. fuscus, 918
Pelochelys
   P. bibroni, 1718, 1825, 1845
   P. cantorii, 1779, 1835, 1860
Pelodiscus sinensis, 1519
Pelomedusa subrufa, 1274
Peltandra
   P. sagittaefolia, 674
   P. virginica, 159, 163
Peltocephalus tracaxa, 735
Pelusios
   P. broadleyi, 1372
   P. castanoides, 1238
Penaeus
   P. brasiliensis, 844
   P. monodon, 1739
   P. paulensis, 844
Penelope obscura, 817
Pennisetum
   P. clandestinum, 1313
   P. purpureum, 189
Perca
   P. flavescens, 597, 612
   P. fluviatilis, 236, 980, 1459, 1469
Perched basin, 551, 552
Percids, 939
Percolation mire/fen, 257, 258, 1426
Percophis brasiliensis, 850
Periglypta spp., 2018
Periploca graeca, 916
Permafrost, 230, 238, 261–263, 276–280, 282–284, 286, 287, 349, 518, 519, 530–532, 536, 538, 545, 552, 1041, 1442, 1446, 1447, 1453, 1464, 1486, 1511, 1534–1536, 1538, 1541–1542, 1700
   arctic frozen peatlands, 278
   highland sloping/spring, 1544–1546
   polygonal fens of, 261
   thaw, 276, 278, 286, 519, 531, 532, 1464
   thermokarst lake, 284, 1446, 1447, 1464–1466
   transitional peatland, 1536, 1541
   wetland alteration, 536
Permanent swamps, Amazon River, 732
Pernis apivorus, 1255
Persea borbonia, 672–674, 719
Persicaria amphibia, 951
Pertusaria, 279
Petasites frigidus, 545

Petaurista petaurista, 1836
Petromyzon marinus, 139, 1002
Petrosimonia brachiata, 1189
Phacochoerus
   P. aethiopicus, 1405
   P. aethiopicus aelinani, 1293–1294
Phaeocystis, 218
Phagmites australis, 1117
Phalacrocorax
   P. africanus, 1327
   P. auritus, 583
   P. brasilianus, 765
   P. carbo, 905, 1091, 1327, 1525, 1628, 1640, 1902
   P. lucidus, 1338
   P. melanoleucos, 1955
   P. niger, 1796
   P. pygmaeus, 1141, 1142, 1621, 1628, 1642, 1691
   P. sulcirostris, 1902, 1955
   P. varius, 1902
Phalaris arundinacea, 432, 613
Phalaroides arundinacea, 949
Phalaropus
   P. fulicarius, 547
   P. lobatus, 547
   P. tricolor, 150
Philippi peatland, 12, 1156
   location, 1150
   maize, tobacco and sugar beet cultivation, 1150
   peat alternation, 1151
   threats, 1153
   uses, 1153
Phillyrea angustifolia, 1106
Philochthus, 1082
Philomachus pugnax, 980, 989, 1030, 1448, 1469, 1641
Phleum, 1701
Phoca
   P. caspica, 1629, 1630
   P. vitulina, 569, 629, 993, 1015, 1094
Phocoena, 1094
   P. dispotica, 850
   P. phocoena, 629, 633, 993
Phoebastria immutabilis, 2020
Phoeniconaias minor, 490, 1354
Phoenicopterus, 150
   P. andinus, 829
   P. chilensis, 829, 833, 834
   P. minor, 1384, 1729

P. roseus, 490, 1104, 1186, 1324, 1386, 1444, 1642, 1729
P. ruber, 751, 764, 1126, 1131, 1387, 1665
Phoenicurus
   P. erythrogastrus, 1700
   P. schisticeps, 1610
Phoenix, 207
   P. paludosa, 1737, 1738
Phoenix Islands Protected Area (PIPA), 2020
Phoksundo wetlands, 1607, 1608, 1610
   Buddhist communities, 1611
   fauna, 1609–1610
   flora, 1609
Pholidocarpus sumatranus, 207
Phormium tenax, 1980, 1982
Phosphorus concentration, in turlough water, 1070
Phoxinus
   P. percnurus, 1459, 1469
   P. phoxinus, 939
Phragmites, 22, 26, 163, 185, 189, 220, 234, 1220, 1223, 1255, 1302, 1418, 1609, 1628, 1649
   P. australis, 58, 67, 149, 159, 163, 164, 170, 385, 386, 431–432, 471, 603, 612, 914, 915, 949, 965, 1105, 1116, 1140, 1156, 1171, 1190, 1237, 1398, 1416, 1427, 1524, 1619, 1639, 1642, 1649, 1650, 1653, 1666, 1686, 1689, 1900
   P. communis, 1447, 1460
   P. karka, 204, 1272, 1302, 1746, 1748
   P. mauritianus, 190, 1221, 1343
   P. vallatoria, 1815
Phragmites reedswamps and saline grasslands, 1237
Phreatophytes, 353
Phreatophytic vegetation, 350, 1960, 1961
Phrynobatrachus zavattari, 1372
Phrynops hilarii, 817, 852
Phrynops spp., 735
   P. hilarii, 817
Phyla canescens, 1905
Phyllanthus
   P. ovalifolius, 206
   P. taxodiifolius, 205
Phyllocladus alpinus, 1980
Phylloscartes ventralis, 852
Phylloscopus
   P. borealis, 1825
   P. trochilus, 1337
Phyllospadix, 74, 78, 81, 1590, 1593
   P. iwatensis, 76, 1591–1593, 1595

P. japonicus, 76, 88, 1591–1595
P. scouleri, 76, 641
P. serrulatus, 76
P. torreyi, 76, 641
Physeter catodon, 324
Physignathus lesueurii, 1946
Phytophthora cinnamomi, 1134
Pibor River, 1247
Picea, 938
   P. glauca, 525
   P. mariana, 26, 238, 516, 517, 523, 612
   P. sitchensis, 560
Picoides borealis, 584, 675
Picus canus, 1825
Pila globosa, 1717
Pilgerodendron uviferum, 877, 878
Piliostigma reticulate, 1272
Pimelodus albicans, 850
Pinctada margaritifera, 2010, 2018
Pingalla midgleyi, 211
Pinus, 669, 938, 1043, 1104, 1433
   P. contorta var. *contorta*, 571
   P. serotina, 668, 672–674
   P. smithiana, 1608
   P. sylvestris, 236, 238, 516, 1460
   P. taeda, 668
   P. wallichiana, 1608, 1706
Pioneer marshes, 963, 969
Pipistrellus nathusii, 1142
Pipturus argentea, 2010
Pisonia grandis, 2017
Pistia stratiotes, 218, 220, 713, 765, 1210, 1237, 1353, 1746, 1815
Pityrogramma aurantiaca, 1313
Placopecten magellanicus, 628
Plain of Reeds, 1776, 1794, 1795, 1802
Planarians, flatworms, 654, 660
Planchonia caeya, 208
Plankton, 44–46, 472, 626, 762, 790, 1273, 1371, 1935–1937
Planktothrix, 479
Planning
   integrated management process, 402
   oasis management, 463
   water management, 720–721, 1496
Plant(s), 7, 21, 38, 44, 56, 58, 65–67, 78–80, 82, 85, 94, 96, 110, 119, 132, 133, 138, 160–164, 167, 173, 187–190, 194, 218, 228, 229, 231, 233, 235, 239, 246, 247, 249, 264, 277, 280, 323, 352–354, 366, 384, 385, 387, 418, 431, 443, 483, 510–512, 518, 519, 523, 532, 533, 552, 558, 561, 569,

571, 579, 580, 584, 598, 600, 608,
638, 641, 654, 657, 672, 674, 682,
692, 693, 709, 710, 712–715, 717,
728, 733, 734, 741, 750, 754, 767,
777, 801, 803, 804, 809, 825, 889,
891, 899, 915, 916, 920, 926, 928,
955, 964, 969, 970, 988, 1006, 1027,
1029, 1032, 1051, 1068, 1069, 1071,
1072, 1075, 1085, 1086, 1134, 1140,
1153, 1164, 1171, 1174, 1175, 1182,
1189, 1203, 1206, 1210, 1296, 1300,
1309, 1311, 1313, 1375, 1385, 1386,
1390, 1401, 1403, 1408, 1444, 1467,
1480, 1481, 1483, 1484, 1493, 1524,
1533, 1543, 1546, 1547, 1569, 1582,
1601, 1609, 1612, 1628, 1639, 1643,
1651, 1654, 1666, 1678, 1683,
1687–1689, 1700, 1701, 1728–1730,
1754, 1774, 1781, 1797, 1798, 1805,
1816, 1835, 1844, 1846, 1848, 1859,
1915, 1931, 1955, 1964, 1980, 1981,
1983, 1985, 1993, 1997, 2007, 2011
and animal diversity, 61, 175, 483, 523, 825,
1081–1084, 1569, 1773
aquatic, 113, 159, 161, 163, 483, 511, 597,
601, 641, 776, 777, 779, 858, 864,
1130, 1189, 1210, 1237, 1302, 1467,
1524, 1527, 1559, 1568, 1569, 1572,
1583, 1652, 1775, 1795, 1796, 1806,
1859, 1915, 1945, 1948, 1963
biodiversity, 988, 1729, 2016
coastal, 2007, 2013
community, 58, 65, 67, 552, 579, 600, 654,
715, 851, 964, 970, 988, 1082
diversity, 61, 483, 523, 1081–1082, 1569,
1773
diversity-estuarine marsh, 61
macrofossils, 1533
medicinal, 1239, 1248, 1263, 1612, 2016
species diversity, 60–61, 176, 717, 969
succession, 558, 561
vascular, 176, 236, 279, 280, 285, 512, 525,
552, 611, 612, 733, 816, 824, 880,
1071, 1130, 1206, 1443, 1454, 1469,
1502, 1517, 1543, 1546, 1718, 1777,
1859, 1980, 1981, 2013
vegetation, 657
Plantago, 61
P. coronopus, 61, 916, 917
P. maritima, 630, 916, 917, 964
P. schrenkii, 285
Plantations, 160, 215, 238, 241, 299, 300, 349,
668, 675, 740, 768, 792, 810, 914,
917, 1049–1051, 1053, 1054, 1180,
1238, 1255, 1296, 1310, 1345, 1347,
1376, 1424, 1433, 1709, 1710, 1728,
1739, 1759, 1768, 1802, 1822, 1828,
1837, 1847, 1848, 1852, 1853, 1856,
1861–1863, 1879, 2017
palm oil, 215, 238, 241, 768, 1822, 1837,
1847, 1852, 1861, 1879
poplar, 914, 917
sugar-cane, 1255, 1345
willow, 160, 1709, 1710
Plasmodiophora, 86
Platalea
P. alba, 1238, 1338
P. leucorodia, 918, 1131, 1141, 1142, 1190,
1324, 1326, 1525, 1561, 1642, 1665
P. leucorodia balsaci, 1326
P. minor, 1796
P. regia, 1981
Platanista
P. gangetica, 1737
P. gangetica minor, 1701
Platanus occidentalis, 581
Platemys platycephala, 735
Platichthys flesus, 980, 1083
Platysternon megacephalum, 325
Playa wetlands
biodiversity, 692
conservation planning, 693
Conservation Reserve Program's
Continuous Signup, 694
conservation status, 692–693
culturally-accelerated sediment
accumulation, 693
distribution, 690
ecosystem, 692
'Farm Bill,', 694
future challenges and developments,
693–694
geographic distribution, 690
hydrological modifications, 693
physical structure, 692
probable playas and high plains, 691
water dissolution hypothesis, 690
water recharge, 692
Wetlands Reserve Program, 694
wind erosion hypothesis, 690
Plectocomiopsis triquetra, 211, 1844
Plectrophenax nivalis, 395, 1084
Plectropterus gambensis, 213
Plegadis
P. chihi, 830
P. falcinellus, 213, 1131, 1142, 1902

Pleurochrysis carterae, 1181
Pleuronectes platessa, 1016, 1083
Pleuropogon sabinei, 511
Pleurosigma, 630
Pleurozia purpurea, 1045
Ploceus
   P. burnieri, 1344
   P. cucullatus, 192
   P. hypoxanthus, 1796, 1816
Plotosus canius, 1826
Pluvialis
   P. apricaria, 418, 980, 1016, 1029, 1045, 1641
   P. dominica, 630
   P. fulva, 1481, 1870
   P. squatarola, 570, 630, 1015, 1016, 1337
Pluvianellus socialis, 867
Pneumatophores, 26, 95, 750, 1756
Poa, 1701
   P. eminens, 546
   P. sibirica, 1537
Pocillopora, 86
Pocosin (USA), 10
   biodiversity, 674–675
   characterization, 668
   classification, 668
   definition, 668
   distribution, 669
   geology, 673
   hydrology, 673–674
   preservation activity, 675
   wetland ecosystems, 669–673
Podica senegalensis, 1345
Podiceps
   P. andinus, 776
   P. cristatus, 1354, 1387, 1443, 1610
   P. cristatus australis, 1981
   P. nigricollis, 150, 1628
   P. occipitalis, 831
Podilymbus podiceps, 683
Podocnemis spp., 735, 736
Podophyllum hexandrum, 1609
Poecilia
   P. eticulata, 1358
   P. velifera, 323
Pogonias cromis, 850, 852
Pogonus, 1082
Polemaetus bellicosus, 1354
Poljes, 314, 320, 321, 341, 343, 1058–1060, 1062–1064
Pollachius virens, 629
Pollen, 82, 219, 247, 673
   analysis, 673

Pollution, 7, 8, 66, 79, 83, 117, 138, 139, 151, 152, 161, 179, 193, 216, 218, 240, 299, 325, 327, 342, 343, 346, 347, 349, 354, 361, 366, 392, 406, 441, 455, 456, 548, 573, 587, 603, 638, 739–741, 793, 798, 840, 870, 890–894, 900, 915, 921, 937, 942, 955, 956, 1027, 1043, 1063, 1145, 1209, 1210, 1240, 1258, 1297, 1330, 1346, 1358, 1389, 1390, 1455, 1484, 1490, 1492, 1497, 1515, 1516, 1518, 1528, 1529, 1557, 1561, 1585–1586, 1612, 1619, 1623, 1632, 1693, 1709, 1710, 1740, 1760, 1780, 1782, 1862, 1868, 1872, 1882, 1891, 1893, 1947, 1949, 1985, 2014, 2017
   air, 6, 240
   chemical, 218, 1114, 1119
   Congo River, 1210
   freshwater swamps, 179, 218
   intertidal zone, 7, 1868, 1872
   thermal, 216, 1891
   urban, 1133, 1985
   water pollution, 66, 115, 322, 323, 942, 1035, 1145, 1455, 1516, 1518, 1557, 1561, 1585–1586, 1612, 1780, 1893
Poludora ligni, 851
Polyborus
   P. chimango, 852
   P. plancus, 852
Polygala polygonifolius, 1045
Polygon, 111, 230, 261, 276, 278–282
   landscapes, 279–280
   mires, 276, 278–281
Polygonal fens of permafrost regions, 261
Polygonum, 163, 164, 580, 791, 816, 1272, 1609
   P. acuminatum, 765
   P. amphibium, 915, 1073
   P. arifolium, 159
   P. barbatum, 205, 1746
   P. celebicum, 205
   P. glabrum, 1746
   P. hydropiper, 915, 1556
   P. punctatum, 159
   P. salicifolium, 191, 1689
   P. tomentosum, 1795–1796
Polyhaline, 157
Polymesosa solida, 763
Polyplectron malacense, 1860
Polypogon elongatus, 851
Polypterus, 1273
   P. senegalus, 1368, 1369

Polytrichum, 1545
  P. juniperinum, 282
  P. piliferum, 279
Pomacea canaliculata, 1777, 1798
Pomatoschistus, 1083
  P. microps, 965
Pomoxis nigromaculatus, 612
Pomponal, 877
Pongo
  P. abelii, 236
  P. pygmaeus, 236, 1757, 1845
Ponors, 314, 321, 335, 1058, 1060–1063
Pontederia
  P. cordata, 163, 612, 715, 718
  P. lanceolata, 205
Pontoporia blainvillei, 850
Poodří–Odra River floodplain ramsar site, 482
Pools, 21, 63, 81, 191, 259, 261–263, 297, 298, 305, 307, 320, 322, 440, 488, 518, 530, 652–664, 690, 738, 843, 1042, 1044–1047, 1050, 1068, 1072, 1074, 1205, 1220, 1221, 1223, 1228, 1302, 1403, 1406, 1445, 1486, 1500, 1538, 1542, 1709, 1716, 1779, 1944, 1946, 1952
  bog, 1046, 1047
  carbon, 237, 1759
  deep, 1765, 1774
  long-hydroperiod, 655, 656, 660
  Malebo, 1202, 1203, 1205, 1207, 1210, 1211
  oil, 536
  saline, 488, 1176
  saltwater, 322
  shallow, 56, 260, 659, 1045, 1072, 1500
  short-hydroperiod, 655–657, 660
  vernal, 10, 651–664
Popowia diospyrifolia, 205
Population decline, 457, 472, 547, 682, 867, 868, 980, 1019, 1528, 1867, 1946
Populus, 816
  P. alba, 915, 916, 1106, 1144
  P. balsamifera, 525
  P. canescens, 916
  P. deltoides, 581
  P. tremula, 916
  P. tremuloides, 681
  P. trichocarpa, 560
Porcula salvania, 1715
Porites, 86
Porphyrio
  P. melanotus, 1981
  P. porphyrio, 1107, 1619, 1700, 1708, 1777

Porphyriops melanops, 777
Portsmouth Harbour, southern England, 371, 1008
Porzana, 1981
  P. carolina, 612, 683
  P. pusilla, 1354
  P. tabuensis plumbea, 1997
Posidonia, 81
  P. angustifolia, 75
  P. australis, 75
  P. coriacea, 75
  P. denhartogii, 75
  P. kirkmanii, 76
  P. oceanica, 76, 81, 82
  P. ostenfeldii, 76
  P. sinuosa, 76
Positive feedback, 263, 276, 286
Potamogeton, 149, 211, 480, 510, 526, 552, 561, 612, 1072, 1106, 1117, 1140, 1272, 1301, 1357, 1367, 1746
  P. australis, 1105, 1117
  P. crispus, 914
  P. fluitans, 914
  P. lucens, 949
  P. natans, 914, 952
  P. nodosus, 1639
  P. octandrus, 1357
  P. pectinatus, 149, 641, 914, 977, 1105, 1117, 1190, 1357, 1370
  P. perfoliatus, 914, 949, 1538
  P. pusillus, 1105
  P. schweinfurthii, 1357
Potamogetonaceae spp., 1524
Potamonautes, 1273
Potentilla
  P. anserina, 1072, 1073, 1525
  P. egedii, 285
  P. erecta, 1045
  P. fruticosa, 1539
  P. palustris, 560
  P. reptans, 915
Potomageton crispus, 613
Pottia heimii, 61
Potwar subbasin, 1699, 1701
Pouteria glomerata, 207
Power generation, 51, 117, 891, 942, 1006, 1153, 1175, 1208, 1701, 1702
Poyang Lake, 13, 1552, 1553, 1555, 1557, 1558, 1561
  biodiversity, 1568, 1569
  catchment, 1567
  conservation status, 1570
  ecosystem services, 1570

elevation scales, 1571–1572
future challenges and developments, 1571–1572
geological map, 1567
hydrology, 1568
origin, 1566
surface area, 1566
threats, 1571
Poyang Lake National Nature Reserve (NNR), 1567–1570
Prairie Pothole Region (PPR)
  biodiversity, 682–683
  conservation status and threats, 684–686
  ecosystem services, 683–684
  hydrology, 681–682
  location, 680
Precipitation, 38, 39, 41, 67, 131, 133, 138, 144, 146, 179, 233, 234, 252, 262–266, 286, 296, 300, 304, 305, 309, 332–334, 360, 366, 415, 433, 516, 519, 523, 525, 530, 536, 559, 579, 655, 671, 674, 681, 682, 691, 702, 709, 728, 739, 759, 762, 788, 789, 801, 861, 875, 876, 878, 879, 887, 935, 937, 947, 1041, 1042, 1150, 1157, 1163, 1168, 1188, 1211, 1221, 1223, 1290, 1308, 1322, 1355, 1362, 1384, 1387, 1424, 1441, 1442, 1446, 1458, 1464, 1468, 1532, 1540, 1542, 1544, 1545, 1552, 1553, 1556, 1580, 1612, 1649, 1651, 1655, 1672, 1676, 1678, 1701, 1712, 1726, 1740, 1782, 1843, 1856
  continental-scale variations, 955
  Dinaric karst, 1063
  hidden mires, 271
  karst evolution, 319
  in mangrove ecosystems, 107
  meteorological drought, 1194
  occult mires, 270–272, 298
  surface water storage and flows, 683
  water sources, 655
Predation/predators, 46, 48, 51, 99, 101, 110, 119, 442, 477, 535, 547, 552, 583, 647, 652, 656, 660, 661, 762, 809, 817, 843, 939, 965, 980, 989, 994, 995, 1071, 1131, 1273, 1274, 1294, 1323, 1336, 1403, 1404, 1408, 1601, 1863, 1879, 1922, 1925, 1957, 1974, 1987
Prek Toal, 1771, 1774, 1778, 1779, 1787–1790
Presbytis
  P. obscura, 1860
  P. rubicunda, 1758

Preservation, 492, 662, 675, 753, 754, 761, 767, 780, 834, 1108, 1320, 1328, 1372, 1486, 1613, 1738, 1869, 1997
Pressures, 7, 64, 79, 120, 138, 151, 163, 179, 180, 193, 215, 240, 299, 325, 327, 328, 349, 361, 366, 377, 398, 400, 417, 423, 441, 452, 478, 498, 536, 551, 573, 594, 606, 630, 634, 648, 652, 656, 753, 768, 769, 808, 832, 859, 891, 894, 955, 956, 968, 989, 993–995, 998, 1035, 1043, 1071, 1074, 1075, 1104, 1107, 1109, 1146, 1163, 1172, 1189, 1208, 1211, 1229, 1230, 1249, 1263, 1266, 1296, 1304, 1329, 1345, 1420, 1433, 1503, 1515, 1516, 1527, 1557, 1561, 1570, 1612, 1623, 1632, 1651, 1655, 1656, 1664, 1672, 1682, 1683, 1720, 1739, 1781, 1790, 1845, 1847, 1934, 1957, 1974, 1986, 2018, 2019
  anthropogenic, 845, 1087
  biotic, 389
  coastal ecosystems, 8
  coastal swamp forests, 1434
  coastal wetlands, 830
  Dinder National Park, 1296
  estuaries, 50, 52
  freshwater ecosystems, 116
  mangroves, 105
  natural ecosystems, 942
  seagrass meadows, 441, 1972
  urban land development, 708
  Wadden Sea ecosystem, 993
Primula, 1701
  P. aureata, 1609
  P. poluninii, 1609
  P. pseudoregelian, 1609
  P. sharmae, 1609
Prince William Sound, 83, 563
Prionailurus
  P. planiceps, 1758, 1860
  P. viverrinus, 1836
Prionops poliolophus, 1387
Prionotus punctatus, 843, 850
Pristimantis bogotensis, 777
Pristine mires, 1419
Pristis, 1325
  P. pectinata, 1337
  P. pristis, 1337
Private Natural Heritage Reserve, 806, 807
Probarbus
  P. jullieni, 1776, 1778
  P. labeamajor, 1778

Procambarus clarkii, 585, 1107, 1109, 1133, 1358
Procapra gutturosa, 1502
Prochilodus lineatus, 790, 817, 850, 852
Productivity, 6, 7, 38, 51, 66, 67, 131, 132, 137, 177, 184, 192, 201, 210, 218, 231, 287, 348, 431, 490, 534, 552, 594, 626, 632, 694, 728, 733, 735, 738, 739, 752, 753, 755, 761, 782, 829, 876, 936, 1026, 1031, 1108, 1120, 1181, 1182, 1207, 1208, 1228, 1258, 1310, 1335, 1366, 1375–1377, 1401, 1404, 1433, 1460, 1503, 1504, 1519, 1533, 1561, 1566, 1593, 1594, 1688, 1729, 1748, 1781, 1783, 1786, 1790, 1791, 1868, 1871, 1903–1905
  biological, 624, 869, 1400, 1403, 1410, 1500, 1925
  ecosystems, 74, 78
  estuaries, 43–48, 53
  fisheries, 438–440, 1249, 1338
  Fundy salt marshes, 634
  mangrove forests, 100, 105, 107, 763, 1334
  papyrus biomass, 188–190
  peatlands, 238
  prairie wetlands, 682
  seagrasses, 87
  tropical swamp vegetation, 211
  vascular plants, 525
Proleptonchus deltaicus, 918
Prop-roots, 95, 101
Prosopis
  P. juliflora, 1359, 1369, 1370, 1385, 1390
  P. spicigera, 1730
Prosopium cylindraceus, 1448
Protected areas, 5, 53, 102, 325, 326, 443, 454, 455, 488, 494, 537–538, 633, 737, 741, 779, 780, 792, 800, 806, 807, 901, 906, 907, 918, 919, 940, 956, 957, 1075, 1104, 1107–1109, 1124, 1125, 1132, 1139, 1143, 1144, 1212, 1295, 1303, 1322, 1325, 1328, 1329, 1334, 1338, 1345, 1370–1371, 1375, 1448, 1449, 1452, 1482, 1487, 1489, 1505–1507, 1513, 1516–1519, 1525, 1546, 1607, 1631, 1644, 1645, 1681, 1715, 1730, 1738, 1740, 1867, 1882, 1891–1892, 1895, 1974, 1989, 2020
  Amazon Basin, 737
  Amur River Basin, 1487, 1489
  Berbak National Park, 1832, 1836
  conservation for sustainable use, 833
  Karukinka Natural Park, 830, 831
  Lake Parishan, 1648
  Lower Danube Green Corridor, 899
  mangroves, 103, 107
  Murray–Darling Basin, 1895
  in Patagonia, 880
  Shadegan Wetland, 1681
  Siakesheem Marsh, 1622–1623
  Sibiloi/Koobi Fora, 1370
  terrestrial, 120, 121
  Tipperne Reserve and Ringkøbing Fjord, 981
  Tram Chim, 1797
  wetland conservation efforts, 537–538
Protection, 12, 74, 85, 115, 317, 319, 324, 328, 342, 343, 362, 443, 454, 456, 457, 473, 494, 538, 539, 553, 571, 572, 586, 599, 613, 632, 633, 662, 663, 685, 692, 693, 700, 706, 737, 738, 741, 754, 781, 782, 792, 794, 818, 830, 831, 833, 834, 889, 895, 901, 902, 906, 919, 956, 994, 1029, 1030, 1049–1051, 1086, 1107, 1109, 1118, 1119, 1132, 1139, 1143, 1212, 1239, 1316, 1322, 1328, 1334, 1404, 1406, 1449, 1460, 1472, 1487, 1488, 1497, 1503, 1505, 1507, 1515, 1518, 1546, 1553, 1559–1562, 1570, 1644, 1681, 1719, 1720, 1730, 1739, 1747, 1759, 1787, 1788, 1790, 1797, 1807, 1826, 1863, 1867, 1869, 1872, 1882, 1893, 1894, 1904–1907, 1947, 1956, 1974, 1987, 1989, 1998, 2015, 2018, 2020
  aquatic habitats, 120–121
  coastal, 49, 101, 115, 116, 968
  flood, 63, 126, 578, 638, 644, 706, 888, 890–893, 898, 905, 1235
  habitat, 342, 646, 648
  mangroves, 102, 452
  of Pantanal, 797–810
  peatland, 242, 291, 1164
  of rivers and aquatic habitats, 120–121
  Santuario de la Naturaleza, 870
  seagrass, 498
  shoreline, 50, 63, 613, 1807
  storm, 63, 1503
Proteus, 341, 342
  P. anguinus, 337, 340
Protopterus aethiopicus, 192, 1354, 1355
Prunus spinosa, 916
Psammocora, 86
Psephurus gladius, 1569
Pseudacris, 658
  P. crucifer, 658
Pseudamnicola confuse, 1030

Pseudibis
 P. davisoni, 1774, 1779
 P. gigantea, 1779
Pseudobombax munguba, 207
Pseudocolopteryx acutipennis, 777
Pseudopimelodus pati, 817
Pseudoplatystoma
 P. corruscans, 790, 817
 P. reticulatus, 817
Pseudoraphis spinescens, 205, 1952
Pseudorasbora parva, 476–477, 479
Psiadia arabica, 1386
Psittacus timneh, 1338
Psyllospadix
 P. scouleri, 641
 P. torreyi, 641
Pternandra teysmanniana, 205
Pterocarpus officinalis, 765
Pterodoras granulosus, 817
Pterois volitans, 764
Pteronura brasiliensis, 735
Pteropus
 P. alecto, 1955
 P. scapulatus, 1955
 P. vampirus, 1816
Pterygoplites undecimalis, 763
Puccinellia, 61
 P. distans, 916
 P. maritima, 963, 1082, 1083, 1085
 P. phryganodes, 285, 510
 P. sellerietum radicantae, 833
 P. tenuiflora, 1525
Puerto Princesa Subterranean River National Park, 326
Pulse flood, 204, 365, 551, 729, 733, 734, 736, 739, 788–790, 799, 801, 804, 816, 888, 1398–1401, 1403, 1405–1406, 1716, 1767, 1781, 1783, 1804, 1952
Puma concolor, 324
Puntius
 P. partipentazona, 1860
 P. trifasciatus, 1845
Pusa
 P. caspica, 1629, 1630
 P. hispida, 1444
 P. sibirica, 1481
Pygidium bogotense, 776
Pyncbostacbys defixxifolia, 191
Pyrodinium bahamense, 500
Pyrus pyraster, 916
Python
 P. reticulatus, 817, 1825, 1835
 P. sebae, 192, 1274

Q
Qa'a Azraq, *see* Azraq Oasis
Queensland, 259, 260, 262, 452–456, 1910, 1926, 1942, 1969, 1974
Queensland Ports Seagrass Monitoring Program, 1973
Quercus, 177, 581, 938, 1106
 Q. castaneifolia castaneifolia, 1639
 Q. humboldtii, 777
 Q. lyrata, 581
 Q. pedunculiflora, 916
 Q. robur, 916
 Q. virginiana, 719
Quisqualis indica, 205

R
Racomitrium lanuginosum, 279, 1045, 1046
Rahad River, 1288, 1290, 1293, 1297
Rain, 110, 150, 208, 219, 221, 253, 300, 304, 314, 326, 681, 738, 741, 765, 789, 791, 825, 876, 877, 928, 1200, 1254, 1271, 1342, 1355, 1398, 1409, 1524, 1583, 1627, 1726, 1755, 1769, 1770, 1772, 1815, 1816, 1879, 1948, 1979
 acid rain, 114, 535, 654
 agriculture, 1193, 1297
 Amazon Basin, 10, 728–741
 days, 298, 415, 1042
 flooded grasslands, 1236, 1302
Rainfall, 6, 99, 107, 112, 119, 133, 137, 138, 145, 146, 153, 230, 235, 272, 284, 286, 347, 349, 415, 421, 451, 500, 531, 554, 571, 580, 638, 673, 709–712, 714, 732, 734, 739, 748, 749, 759, 761, 763, 767, 774, 789, 799, 801, 816, 828, 836, 930, 937, 947, 1027, 1041, 1042, 1070, 1102–1105, 1114, 1127, 1129, 1134, 1159, 1162, 1168, 1172, 1175, 1194, 1207, 1218, 1225, 1230, 1235, 1262, 1283, 1284, 1292, 1300, 1304, 1311, 1342, 1343, 1353, 1355, 1364, 1384, 1386, 1390, 1396, 1398, 1414, 1420, 1424, 1433, 1434, 1479, 1511, 1522, 1523, 1527, 1547, 1552, 1561, 1577, 1581, 1583, 1616, 1627, 1688, 1715, 1726, 1729, 1744, 1749, 1755, 1759, 1760, 1767, 1782, 1795, 1803, 1804, 1808, 1815, 1820, 1832, 1843, 1853, 1879, 1890, 1942, 1949, 1952, 1954, 1957, 1960, 1963, 1978, 1994, 1995
 arid regions, 1726
 Baro and Akobo rivers, 1262, 1265

Rainfall (*cont.*)
    blanket mire development, 298
    Blue Nile drains, 1247
    Caithness, 1041
    Dinder and Rahad Basin, 1291, 1292
    Doñana wetlands, 1124
    Everglades, 709
    fishponds, 473
    freshwater swamps, 203
    Hula Basin, 1168
    Lake Bogoria, 1384–1385
    Lake Fúquene, 774–775
    Lake Nakuru, 1385–1387
    Macquarie Marshes, 1899
    Mazandaran Province, 1636
    Mekong Delta, 1815
    New Zealand, 1978
    pocosins, 674
    in San Francisco Bay Estuary, 638
    Sanjiang Plain, 1511
    Sembilang National Park, 1820
    Tasek Bera, 1853
    Volga River Basin, 937
    Yangtze River, 1552
    Yellow River, 1579
    Zambezi River Basin, 1218
    Zhalong Marsh, 1522–1523
Rain-fed agriculture, 1193, 1297
Rain flooded grasslands, 1236
    Sudd region, 1302
Rainforest, 79, 230, 236, 323, 326, 840, 1205, 1208, 1209, 1552, 1942, 1945, 1980
Raised atolls, 2003
Raised bog, 21, 230, 236, 258, 264–269, 276, 278, 283–284, 290, 304–306, 310, 311, 373, 1044, 1159, 1983, 1994, 1995, 1998
Raised limestone islands, 2003, 2006, 2011, 2015, 2021
Raised mire, 238, 266, 267, 298, 301, 309–311, 415, 417
Raja, 1083
Rallus
    R. aquaticus, 1073
    R. obsoletus, 641
    R. semiplumbeus, 777
Ramsar convention, 26, 28, 30–32, 117, 120, 180, 241, 287, 328, 346–348, 372, 376, 377, 401, 406, 409, 470, 480, 488, 510, 741, 825, 869, 918, 946, 966, 978, 1094, 1131, 1187, 1238, 1328, 1338, 1346, 1384, 1482, 1486, 1504, 1517, 1518, 1526, 1553, 1599, 1600, 1623, 1627, 1629, 1666, 1681, 1700, 1710, 1836, 1846, 1852, 1867, 1895
    bureau, 1913
    in China, 1579
    definition, 21, 2002
    Everglades, 706
    on wetlands, 4, 5, 14, 20, 68, 277, 370, 428, 553, 572, 633, 638, 737, 781, 870, 1004, 1017, 1032, 1353, 1357, 1610, 1719, 1746, 1771, 1797, 1892, 1895, 1904, 1911, 1955, 1978, 1983, 1987
    classification of, 21
    wise use of, 317
Ramsar site, 10, 12, 51, 58, 190, 317, 321–327, 370, 398, 462, 480, 482, 513, 572, 792, 827, 849, 851, 864, 880, 888, 889, 918, 954, 966, 974, 975, 981, 1004, 1006, 1017, 1059, 1094, 1108, 1118, 1143, 1159, 1188, 1189, 1238, 1248, 1257, 1334, 1346, 1375, 1386, 1387, 1397, 1434, 1490–1493, 1502, 1505, 1517, 1546, 1553, 1557, 1570, 1598, 1601, 1606–1608, 1610, 1611, 1613, 1616, 1623, 1627, 1631, 1636, 1648, 1654, 1660, 1662, 1669, 1678, 1694, 1699–1701, 1716, 1717, 1730, 1746, 1771, 1773–1778, 1787, 1788, 1807, 1832, 1852, 1853, 1856, 1857, 1861–1863, 1867, 1899, 1904, 1916, 1931, 1983, 1987, 1988, 2022, 2023
    Amazon, 737
    Amur-Heilong River Basin, 1488, 1494
    Andean wetlands, 834–835
    Anzali Mordab, 1616, 1617
    AwaruaWetland, 1982
    Azraq Wetland Reserve, 462
    Bahía Lomas, 866–870
    Berbak National Park, 1820, 1832
    in Chile, 834–535
    definition, 30
    Dinaric karst, 1058
    in England, 1004
    Fereydoon Kenar, 11, 12, 1636, 1637
    Hokkaido Island, 1599
    in Iran, 1617, 1676
    karst wetlands, 317–328
    Kilombero Valley Floodplain, 1343
    Kopuatai Peat Dome, 1998
    Lake Baikal, 1482
    Lake Bolon, 1518
    Lake Elementeita, 1387
    Lake Nakuru, 1386

Lake Seyfe, 1186, 1187
Lake Uromiyeh, 1660
Mato Grosso National Park, 806
Mekong Basin, 1771
Mont-Saint-Michel Bay, 1086
Napahai Wetland, 325
Nepal, 1607
in New Zealand, 1984, 1985
Ringkøbing Fjord, 974, 976
Rugezi Marsh, 1314
San Francisco Bay Estuary, 645
Sanjiang Plain, 1517
Sefid Rood, 1631
Sefid Rud Delta, 1627
Sembilang National Park, 1821, 1827
Shadegan Wetland, 1676, 1681
Southern Bight Minas Basin, 26
Tasek Bera, 1589, 1854, 1855
Třeboň fishponds, 480
Wetlands of International Importance, 1482, 1517
Yadegarlou wetland, 1672
Yangtze River Basin, 1555
Yellow River Basin, 1579
Yucatan Peninsula, 319
Ramsar wetlands, 5, 68, 387, 402, 599, 767, 828, 859, 866, 1132, 1222, 1228, 1229, 1303, 1486, 1621, 1889, 1895, 1911–1913, 2017
Ramsar Wetland Steering Committee, 767
Rana, 1354
   R. catesbeiana, 612
   R. palustris, 612
   R. pipiens, 612
   R. ridibunda, 918
   R. septentrionalis, 612
   R. temporaria, 1047
Rangifer
   R. tarandus, 1455
   R. tarandus caribou, 519, 532
   R. tarandus groenlandicus, 512
   R. tarandus sibiricus, 1480
Ranidae, 939
Rann of Kachchh, 1726–1728, 1730
Ranunculus, 1105, 1117, 1639
   R. confervoides, 511
   R. flammula, 1073
   R. lingua, 915
   R. pallasii, 284
   R. peltatus, 1128, 1130
   R. repens, 1072, 1073, 1539
   R. rionii, 1639
Rapana venosa, 854, 918

Raphanus raphanistrum, 915
Raphia, 207
   R. taedigera, 206
Raphidiopsis, 1385
Rapid
   attenuation of light, 78
   coastal development, 90, 1749, 1868
      industrial development, 103, 455, 457, 1175
   coastal wetlands loss, rates of, 8, 60, 373, 675, 1867
   ecological succession, 561
   eggs and larvae development, 658
   fishery declines, 118
   human population growth, 440, 498, 1230, 1240, 1249, 1345, 1358, 1645, 1701, 1721
   impoundment
      construction, rate of, 126, 810, 1249, 1809
      lake effects, 151
      sediment deposition, 634
   in situ peat decomposition, 276, 286
   lake response and monitoring, 138
   lake warming trends, 138
   mangrove canopy production, rates of, 94
   organic matter
      decomposition, 483, 1760
      turnover, 177
   peat swamp forest degradation, 1759, 1827
   small plankton, abundance declines, 46
   seagrass declines, 82, 85
   seagrass nutrient uptake, 87
   water seepage un-humified peat, 258
   water flow changes, 891, 906, 1473, 1847
   wetland
      biodiversity deterioration, 361, 377, 380
      loss rates of, 28, 167, 363, 377, 1794
Rapids (river condition), 1204, 1207, 1246, 1774, 1775
Rara wetlands, 1607, 1608, 1610
   cultural and religious significance, 1611
   fauna, 1609
   flora, 1609
Rasbora
   R. heteromorpha, 1860–1861
   R. tuberculata, 1845
Rate of loss, 59, 98, 121, 372–376
Rattus colletti, 1955
Reclamation, 38, 50, 58, 539, 797, 917, 920, 1064, 1118, 1138, 1181, 1251, 1257, 1258, 1389, 1595, 1708, 1709, 1739, 1869, 1872, 1880–1882, 1973

Reclamation (cont.)
  and conversion, 202, 215, 216, 219
  East Asian-Australasian Flyway, 410
  land, 50, 58, 219, 920, 1064, 1138, 1257, 1258, 1595, 1869, 1872, 1880–1882, 1973
  Project Rationale, 407
  restoration, 538
  Saemangeum, 410, 411
    coastal wetlands, 409
    Estuarine System, 406
  tropical freshwater swamps, 216
Recurvirostra
  R. avosetta, 980, 1029, 1141, 1190, 1665
  R. novaehollandiae, 1917
Redcliff and Rakatu wetlands, 1988
Red-crowned crane, 1515, 1525, 1526, 1561, 1584, 1599
  breeding, 1525, 1526
Red knots *(Calidris canutus)*, 867, 868, 990, 1003, 1015–1017, 1019, 1328, 1337, 1870, 1981
Redunca
  R. arundinum, 1227, 1237
  R. bohor cottoni, 1293
Reed and sedge-reed marshes, Zhalong marsh, 1524
Reed bed(s), 228, 791, 862, 863, 888, 901, 915, 919, 920, 1116, 1117, 1140–1142, 1171, 1193, 1194, 1394, 1649–1653, 1666, 1668, 1678, 1856, 1862, 1901, 1903, 1946
Reed harvesting, Zhalong marsh, 1527
Reed marsh, 166, 1523–1525, 1556, 1583
Regional Autonomous Environmental Authority (CAR), 491, 494, 752, 753, 779
Regional natural resource management organizations, 1894
Regulated flow, 937, 949
Reithrodontomys raviventris, 641
Relative sea-level rise, 7, 67, 68, 675
Reptiles, 86, 100, 101, 149, 165, 176, 191–193, 212, 214, 236, 323, 584, 612, 735, 736, 750, 764, 776, 777, 792, 805, 817, 818, 852, 899, 918, 939, 1107, 1128, 1131, 1142, 1190, 1238, 1274, 1337, 1368, 1372, 1406, 1407, 1474, 1502, 1513, 1514, 1584, 1678, 1729, 1737, 1773, 1776, 1778–1779, 1946, 1955, 1978, 1981, 2016, 2017
Reserva de la Biosfera Los Petenes, 323

Reserve, 325, 473, 475, 568, 752, 801, 908, 919, 974, 976–981, 994, 1018, 1029, 1086, 1108, 1118, 1328, 1338, 1346, 1375, 1376, 1384, 1452, 1454, 1490, 1492, 1493, 1517, 1520, 1526–1528, 1570, 1584, 1622, 1630, 1644, 1730, 1738, 1739, 1787, 1796, 1822, 1825, 1832, 1836, 1837, 1861, 1863, 1895, 1904, 1917, 1998, 2016
Reservoirs, 29, 30, 172, 186, 217, 327, 375, 406, 410, 470, 519, 580, 662, 702, 713, 720, 793, 808, 934, 936–939, 941, 942, 949, 956, 1064, 1102, 1120, 1145, 1146, 1171, 1222, 1226, 1230, 1248, 1255, 1275, 1376, 1419, 1445, 1448, 1482, 1483, 1486, 1489, 1495, 1504, 1556, 1581, 1582, 1585, 1606, 1638–1640, 1642, 1672, 1716, 1718, 1727, 1731, 1781, 1879, 1899
  Amazon, 740
  Central Valley, 702
  construction of, 128
  freshwater lakes and, 50, 126–139
  global distribution of, 131
  hydroelectric, 592, 891
  influence of, 1587
  Mississippi, 583
  Nile Delta, 1255
  Volga River Basin, 941, 942
  Zeya Hydro, 1495
Residence time, 105, 126, 128–129, 209, 497, 1070, 1322, 1364
Resistance, 163, 209, 287, 464, 808, 1142
Resource Management Act (RMA), 1986, 1987
Restiad bogs
  conservation, 1998–1999
  ecosystem services, 1996–1998
  Empodisma minus, 1994
  Empodisma robustum, 1994
  location, 1992
  New Zealand, 1992
  restoration, 1998–1999
  species, 1992–1993
  Sporadanthus ferrugineus, 1994–1995
  Sporadanthus traversii, 1995–1996
  threats, 1998–1999
  types, 1994–1996
Restionaceae, 236, 265, 1992
Restoration, 8, 22, 26, 107, 116, 139, 152, 166, 180, 241, 242, 285, 287, 311, 362, 363, 365, 377, 431, 435, 456, 458, 496, 498, 501, 503, 538, 539, 585, 602, 613, 638, 644–648, 662, 684,

693, 694, 709, 711, 720, 753–755,
759, 767, 780–782, 888, 892–895,
900–902, 907, 908, 917, 920, 921,
928–930, 989, 1006, 1027, 1030,
1032, 1050, 1052–1054, 1119, 1120,
1129, 1130, 1310, 1316, 1421, 1515,
1519, 1571, 1601, 1631, 1687–1694,
1710, 1719, 1739, 1797, 1827, 1837,
1893, 1918, 1947, 1949, 1965, 1978,
1983, 1986–1989, 1998–1999
  Azraq Oasis, 466
  bottomland hardwood forest, 586
  coastal wetland, 858–864
  drainage and, 1168–1170
  dynamic flooding, 7
  ecological, 398–402
  integrated management planning, 402
  Lake Chilika, 1749
  Lake Hula, 1171
  Mesopotamian marshes, 1693
  papyrus vegetation, 193
  peatlands, 6, 414
  Rhine River, 926–927
  salt marsh, 63
  seagrass, 89
  stakeholder representation, 462–467
  wetland ecosystem, 401–402
Restoring the Tradition at Delta Marsh, 602
Reverse-flow system, 1566
Review, 6, 8, 31, 32, 60, 85, 94, 100, 120, 145,
  156, 178, 235, 290, 336, 349, 388,
  455, 463, 493, 522, 654, 658, 661,
  662, 712, 734, 798–810, 998, 1006,
  1069, 1132, 1443, 1487, 1532, 1534,
  1710, 1768, 1806, 1870, 1956, 1987
Rhabdadenia biflora, 765
Rhamnus, 1029, 1072
  R. cathartica, 916, 1072
  R. frangula, 916
Rhantus hispanicus, 1131
Rhea americana, 817
Rheedia brasiliensis, 207
Rheum
  R. australe, 1609
  R. moorcroftianum, 1609
Rhinella arenarum, 852
Rhine River Basin, 11, 887
  course and drainage system, 925
  future aspects and developments, 929–930
  historical background, 924–926
  restoration, 926–929
Rhinobatos cemiculusi, 1325
Rhinocerophis alternatus, 817

Rhinoceros
  R. sondaicus, 1737
  R. unicornis, 1715, 1737
Rhinoclemmys punctularia, 735
Rhinolophus hipposideros, 1192
Rhinoptera marginatai, 1325
Rhino sanctuary, 1386
Rhizoclonium, 842
Rhizophora, 1823
  R. apiculata, 1823
  R. harrisonii, 750, 1336
  R. mangle, 750, 752, 764–766, 1336
  R. mucronata, 1237, 1729
  R. racemosa, 1336
Rhodeus sericeus amarus, 1142
Rhodiola rosea, 285
Rhododendron
  R. cowanianum, 1609
  R. nivale, 1608
Rhodoleia, 210, 1844
Rhogeessa minutilla, 750
Rhus taitensis, 208
Rhyacornis fuliginosus, 1559, 1610
Rhynchanthera grandiflora, 206
Rhynchops flavirostris, 1344
Rhynchospora, 205, 715
  R. alba, 674
  R. corymbosa, 204
  R. fusca, 1046
  R. subsubquadrata, 205
Riama estriata, 777
Riccardia crassa, 1995, 1996
Rice farming, in Ebro Delta, 1118
Rice fields, 174, 408, 578, 580, 584, 585, 702,
  1114–1119, 1140–1142, 1528, 1621,
  1629, 1631, 1636, 1638, 1640, 1642,
  1643, 1645, 1768, 1787, 1845
Rice paddies, 29, 249, 375, 1118, 1636, 1700,
  1715, 1716, 1736, 1738, 1780
Ridges, 22, 259–263, 267, 268, 270, 305, 307,
  308, 334, 420, 518, 560, 610, 622,
  628, 653, 840, 915, 916, 1013, 1044,
  1045, 1082, 1201, 1257, 1426, 1440,
  1442, 1472, 1473, 1537, 1540–1542,
  1545, 1795, 1802, 1823, 1834
Ridgeway's rail, 644
Rift valley, 12, 113, 130, 135, 147, 152, 153, 186,
  1168, 1202, 1208, 1246, 1349–1359,
  1362, 1381–1390, 1414, 1479
Ringkøbing Fjord, 11
  benthos and fish, 977
  biodiversity, 977–980
  birds, 978–980

Ringkøbing Fjord (*cont.*)
  conservation status, 981
  ecosystem services, 980
  location, 974
  map of, 975
  physio-chemical conditions, 974–976
  threats, 981–982
  Tipperne Reserve, 976
Río de la Plata system
  biodiversity patterns, 849–852
  characteristics, 848–849
  crab community, 851
  ecosystem services, 852–854
  fishing boats, 852
  future aspects, 853–854
  location, 849
  threats, 853–854
  water circulation, 850
Rio Negro National Park, 807
Riparia
  R. paludicola, 1779
  R. riparia, 906
Riparian, 350, 539, 698, 765, 836, 1140
  areas, 457, 533, 545, 608, 728, 1359, 1389, 1390
  buffers, 121, 1965
  communities, 1373, 1390
  countries, 1218, 1229, 1230, 1249
  cultivation, 1257
  Danubian countries, 921
  ecosystems, 177, 185, 217, 876
  edge vegetation, 828
  and floodplain habitats, 698, 736, 1955
  forests, 164, 578, 585, 588, 926, 1106, 1121, 1140–1142, 1211, 1226, 1359, 1834, 1893
  habitats, 547, 698, 889, 906, 1209
  management, 538, 1948
  mires, 276, 278, 285
  nations, 111
  peatlands, 238
  reed swamps, 1221
  settlements in Lake Turkana, 1374
  species, 1963
  trees, 1048
  vegetation, 138, 859, 950, 1116, 1209, 1805, 1823, 1834, 1891, 1945, 1948, 1949
  willow formations, 888, 915
  woody wetlands, 698, 700, 1398, 1399, 1402–1404
  zones, 350, 608, 730, 732, 734, 735, 737–739, 741, 1358, 1359, 1981

Risk assessment, 385, 387, 389, 794, 1957
River, 7, 12, 20–22, 37, 38, 43, 45, 50, 53, 56, 61, 66, 110–113, 116, 132, 133, 137, 150–152, 156, 157, 160, 161, 165, 171, 173–175, 177, 185, 186, 189, 194, 200, 201, 207, 209, 217, 254, 277, 280, 296, 314, 320, 322, 326, 327, 333–335, 348, 349, 351, 360, 364, 430, 452, 474, 510, 547, 551, 554, 580, 642, 712, 731, 740, 760, 824, 836, 880, 924, 986, 987, 1035, 1061–1063, 1080, 1129, 1139, 1200, 1207, 1221, 1247, 1281, 1343, 1346, 1399, 1426, 1444, 1446, 1448, 1454, 1464, 1465, 1472, 1473, 1475, 1486, 1496, 1501, 1502, 1507, 1516, 1533, 1536, 1546, 1660, 1686, 1713, 1779, 1802, 1806, 1827, 1848, 1900, 1942–1949
  avulsion, 1399, 1401, 1410
  basins, 5, 9, 12, 13, 110, 186, 854, 880, 1288, 1290, 1344, 1396, 1397, 1455, 1606, 1619, 1853, 1986
  Amazon, 10, 728–741
  Amur, 1510–1520
  Amur-Heilong, 1485–1497, 1500–1507
  Baro-Akobo, 1262–1267, 1270–1276
  biodiversity, 112–115
  categories, 111
  Congo, 1200–1212
  Copper, 559
  Danube, 886–895, 898, 900, 908
  De la Plata, 786
  Huang He, 1576–1587
  human well-being, 115–117
  Indus, 1698–1702
  Lena, 1440–1449, 1458–1460, 1464–1470, 1472–1476
  MacKenzie, 550–555
  Mahanadi, 402
  Mekong, 217, 1764–1783, 1786–1791, 1794–1798, 1813–1817
  Mississippi, 588
  Murray-Darling, 1888–1895, 1898–1907, 1910–1918
  Nile, 1244–1249, 1270–1276
  Tara, 1059
  threats to rivers, 117–119
  Ubaté, 774
  Volga, 934–942
  White Nile, 184
  of world, 110–121
  Yangtze, 1552–1562, 1566–1572

Yenisei, 1478–1484
Yukon, 544–548
Zambezi, 1218–1230, 1240
catchments, 113, 162, 325, 327, 481, 1085, 1168, 1519, 1676, 1677, 1682, 1861, 1890, 1898, 1899, 1906, 1952
channels, 186, 254, 430, 550, 551, 568, 598, 736, 739, 789, 799, 808, 899, 906, 936, 1117, 1138, 1223, 1225, 1226, 1247, 1282, 1288, 1300, 1368, 1399, 1479, 1518, 1523, 1578, 1583, 1585, 1627, 1716, 1745, 1748, 1765, 1775, 1804, 1805, 1815, 1862
discharge, 7, 112, 160, 167, 451, 452, 570, 761, 841, 848, 926, 930, 949, 950, 955, 957, 1297, 1484, 1837, 1933
ecosystem, 157, 890, 905, 1293, 1371
fisheries, 1208
flooded grasslands, 1236, 1302
grasslands, 1493, 1500
Jhelum, 1706
Jordan, 1168, 1170, 1171, 1244
levels, 157, 581, 1281, 1834
Malewa, 1355
networks, 110–112, 117, 121, 126, 129, 1027, 1208, 1581
protection, 120, 121
Ramsar wetland definition, 5
regulation, 905, 907, 1235, 1240, 1702, 1904, 1905
restoration approaches, 904, 907, 927–929, 989
Restoration Programme, 907
Rhine, 11, 161, 887, 924–930
storage, 1900
systems, 5, 53, 110, 111, 186, 326, 419, 568, 728, 906–908, 926–928, 930, 988, 1228, 1234, 1248, 1262, 1284, 1300, 1371, 1426, 1443, 1466, 1472, 1482, 1501, 1560, 1598, 1616, 1662, 1716, 1721, 1855, 1860, 1890, 1893, 1906, 1924
threats to, 117–119
transport, 433, 912, 1834
valleys, 38, 58, 167, 238, 325, 337, 588, 1048, 1247, 1446, 1493, 1497, 1505, 1535, 1540, 1542–1544, 1547, 1586, 1754
Varde, 986, 988
wetlands, 186, 1395, 1486, 1496, 1505, 1582, 1584
Yellow, 1552, 1576–1587

Riverine, 26, 27, 110, 118, 126, 135, 160, 184–186, 265, 279, 363, 431, 441, 525, 580, 585, 593–595, 607–610, 614–617, 682, 807, 848, 850, 853, 892, 904, 905, 907, 916, 926–930, 948, 1221–1223, 1239, 1288, 1293, 1365, 1367–1369, 1433, 1452, 1458, 1473, 1504, 1577, 1578, 1581–1583, 1626–1628, 1700, 1702, 1713, 1716, 1717, 1767, 1770, 1771, 1775, 1776, 1781, 1783, 1836, 1856, 1944, 1983
Rizonopriodon acutus, 1325
Roads, 105, 364, 407, 533, 536–539, 554, 610, 614–617, 683, 693, 739, 740, 766, 769, 794, 802, 808, 818, 836, 956, 1041, 1070, 1081, 1129, 1134, 1146, 1151, 1152, 1210, 1235, 1257, 1275, 1303, 1305, 1309, 1316, 1329, 1384, 1434, 1518, 1546, 1645, 1671, 1781, 1827, 1848, 1986
 Anzali Port ring, 1624
 boreal wetlands, 534, 535
 floating, 421, 422
 ice, 554
 infrastructure, 111
 network, 365, 1049
 and railroad crossings, 65
 Tampa Bay, 502
 wind farm, 417, 419–423
Rocky shores, 47, 48, 448, 653, 840, 843, 1001, 1013, 1370, 1590, 1653
Rollandia rolland, 831
Rorippa
 R. amphibia, 915, 1171
 R. palustris, 915
 R. valdes-bermejoi, 1130
Rosa canina, 916
Rostkovia, 876
Rotopiko/National Wetland Centre, 1988, 1989
Royal Society for the Protection of Birds (RSPB), 1018, 1049–1051, 1872
Rubus
 R. arcticus, 1639
 R. caesius, 915
 R. chamaemorus, 517, 545, 571
 R. ulmifolius, 818
Rucervus duvaucelii, 1715, 1737
Rugezi Marsh, 11, 12
 biodiversity, 1313–1315
 ecosystem services, 1315–1316
 future aspects, 1316
 hydrology, 1310–1311
 location, 1308

Rugezi Marsh (*cont.*)
   peat, 1309, 1311–1313
   in Rwanda, 1415, 1417
   threats, 1316
   wetlands in catchment area, 1311
Rugezi mire, 1421
Rumex
   R. crispus, 1073
   R. hydrolapatum, 915
Ruoergai-Maqu plateau wetland, 1586
Ruppia, 81, 149, 151, 977, 1092, 1117, 1746, 1915, 1917, 1971
   R. cirrhosa, 76, 977, 1104
   R. filifolia, 76
   R. maritima, 76, 79, 81, 641, 840–843, 1104, 1140, 1590, 1592, 1593, 1876
   R. megacarpa, 76, 1915
   R. polycarpa, 76
   R. tuberosa, 76, 1915
Rusa unicolor, 1737, 1826, 1836
Rush *(Juncus)* meadows, 1140
Russian integrated landscape approach, 1534
Rusumo stream, 1310
Rutilus
   R. frisii kutum, 1630
   R. rutilus, 939, 953, 980, 1459, 1469
   R. rutilus caspicus, 939, 953
Ruwenzori, 297, 1246
Rynchops
   R. flavirostris, 1238, 1354
   R. niger, 851

S

Saaminus brasiliensis, 850
Sabal palmetto, 719
Sabkha, 22, 26
   coastal, 11, 12, 1174–1182
   landforms, 1176–1179
   location, 1177
   sustainable saline agro-systems, 1180–1181
Saccharum spontaneum, 204
Sacramento-San Joaquin Delta, 698, 700
Sacramento valley, 698, 701
Saddle, 267–270, 305
   blanket bog, 270
   raised bog ("Sattel Hochmoor"), 268
Saemangeum estuarine system (SES), 13, 50, 406
   civil society opposition, 409
   human use, 407–408
   location, 406–407
   opposition history, 409
   physical geography, 407–408
   reclamation project, 407
   Saemangeum shorebird monitoring program, 409–410
   seawall closure, 408–409
   threats, 410–411
   waterbirds, 408
Saemangeum reclamation project, 407
Saemangeum Shorebird Monitoring Program (SSMP), 408–410
Sagina maritima, 61
Sagittaria, 205, 612
   S. lancifolia, 715
   S. natans, 1447
   S. sagittifolia, 915, 952
   S. trifolia, 1746
Saiga tatarica, 948
Salacia erecta, 206
Salamanders, 323, 325, 337, 338, 340, 342, 612, 658, 659, 661, 1558, 1582, 1599, 1601
Salamandrella keyserlingii, 1599
Salicornia, 58, 61, 963, 1014, 1083, 1106, 1116, 1144, 1237
   S. ambigua, 851
   S. brachiata, 1728
   S. dolichostachya, 1082
   S. europaea, 951, 988, 1140
   S. fragilis, 1082
   S. fruticosa, 1105
   S. gr. europaea, 1105
   S. perennis, 1105
   S. prostrata, 1189
   S. virginica, 570
Salinas, 56, 58, 61, 174, 643
Saline, 22, 26, 28, 37, 58
   agro-systems, 1174, 1180–1182
   algae, 1176, 1181
   brackish lakes, 488, 1445, 1486
   brine, 1174–1176, 1182
   coastlines, 1823
   conditions, 99, 149, 728, 1032, 1823
   environments, 1140
   flats, 26, 144, 1726
   floodplain, 1944
   gradient, 335, 761, 849, 1104, 1736, 1740
   grasslands, 1230, 1237
   groundwater, 22, 153, 1175, 1176, 1180, 1400
   inland waters, 173
   intrusion, 238, 1035, 1952
   lagoons, 827, 1002, 1013, 1108, 1911

lakes, 144, 145, 148, 149, 152, 153, 1365, 1368, 1660, 1768–1770, 1772, 1888
land drainage, 349
marshes, 278, 1539, 1546, 1728, 1736
mudflats, 1237, 1739
pan, 145, 1395
pastures, 826
ponds, 1180, 1181
pools, 488
reedbeds, 961
soils, 58, 61, 149, 851, 916, 960, 1105, 1237, 1470
swamps, 2008
tides, 1032
tolerant algae, 1180, 1181
waters, 37, 147, 157, 160, 162, 228, 462, 481, 641, 998, 1021, 1117, 1174, 1182, 1638, 1932, 1933, 1937, 1945
wetlands (see Indian arid zone)
Salinisation, 151, 152, 220, 221, 353, 361, 916, 1133, 1258, 1468, 1586, 1701, 1721, 1783, 1891, 1892, 1917
Salinity, 6, 7, 20, 38, 39, 41, 46, 47, 56, 61, 65, 66, 89, 99, 105, 106, 131, 133, 144, 146, 148, 149, 151–153, 157, 165, 335, 350, 400, 401, 462, 481, 571, 623, 638, 647, 733, 748, 761, 763–767, 769, 789, 828, 829, 848–850, 858, 859, 949, 954, 974, 1002, 1081, 1082, 1091, 1092, 1103–1105, 1107, 1109, 1116–1119, 1140, 1175, 1176, 1180, 1256, 1335, 1365, 1366, 1377, 1390, 1402, 1467, 1663, 1666, 1678, 1688, 1693, 1694, 1720, 1726, 1729, 1730, 1736, 1737, 1740, 1746, 1804, 1823, 1879, 1888, 1891, 1911, 1914–1917, 1931, 1933–1937
  biogeochemical processes, 167
  Bombah Broadwater, 1932
  CGSM system, 762
  lagoon, 976, 981
  structure, 40
  sugar cane drainage water, 1678
  wetland ecosystems, 1104
Salix, 160, 161, 164, 300, 518, 524, 545, 560, 612, 816, 1043, 1072, 1144, 1598, 1708, 1980, 1985
  S. alba, 915, 916, 952, 1144, 1640
  S. arctica, 1539
  S. arctophila, 511
  S. aurita, 915
  S. berberifolia, 1539

S. caroliniana, 718
S. cinerea, 915, 1029, 1986
S. fragilis, 915
S. glauca, 1537, 1538
S. humboldtiana, 791, 816
S. nigra, 581
S. repens, 1072
S. reptans, 285
S. triandra, 952
Salminus brasiliensis, 790, 817
Salmo
  S. salar, 138, 633, 928, 989, 1002, 1052
  S. trutta, 939, 1002
  S. trutta caspius, 939, 953
Salmon, 46, 118, 138, 165, 547, 561, 569, 617, 633, 634, 641–643, 928, 930, 939, 989, 1002, 1052, 1053, 1444, 1447, 1455, 1491, 1519, 1600
Salmonid, 119, 566, 569, 926, 939, 1444, 1455
Salsola, 1728
  S. inermis, 1189
Salt industry, 1729, 1730
Salt lakes, 1190, 1728, 1730, 1731, 1922, 1924
  biota, 149–150
  continental basis, distribution, 147
  defined, 6, 144
  ecological features, 148–149
  ecosystem services, 153
  global distribution, 145–147
  hypersaline lakes, 145
  south-western Australia, 145
  Tanzania, 144
  threats and future aspects, 151–153
Salt marshes, 22, 26, 44, 48, 49, 52, 56, 61, 64, 67, 68, 78, 81, 107, 156, 170, 232, 407, 452–453, 481, 482, 510, 546, 561, 568, 622, 629, 631, 634, 833, 967, 984, 988, 989, 994, 1006, 1105, 1116–1118, 1130, 1140, 1142, 1189, 1190, 1410, 1543, 1577, 1582, 1666, 1721, 1768–1770, 1772, 1838
  animal diversity, 1082–1084
  Bay of Fundy, 622, 629–631, 634
  for birds, 1190
  Boundary Bay, 570
  British estuaries, 1001
  communities, 1001
  estuarine marshes, 57
  Fraser Delta, 570
  functions, 1082–1086
  geomorphology, 1080–1081
  Great Barrier Reef, 452–453
  harvest mouse, 641, 643

Salt marshes (cont.)
  hydrology, 1080–1081
  loss, 60
  marine transgression, 58
  Mont-Saint-Michel Bay, 1080, 1082, 1084
  plant communities, 988
  plant diversity, 1081–1082
  primary production, 1081–1082
  San Francisco Bay Estuary, 638, 641, 643
  soils, 63
  structure, 1081–1082
  tidal, 62, 63, 638
  US west coast, 64
  vegetation diversity, 963
  woody vegetation, 62
Salt Range Basin, 1700, 1701
Salt wedge, 41, 46,
      570–571, 849
  estuary, 40, 45, 46
Salvadora, 1730
  S. persica, 206
Salvelinus
  S. alpinus, 1448
  S. jakuticus, 1454
  S. malma, 561
Salvinia
  S. cucullata, 1746, 1815
  S. molesta, 218, 220, 384, 1237, 1357,
      1409, 1952
  S. natans, 914, 952, 1746
Sambhar lake, 1728, 1730
Sampling method, 233, 433, 452, 978, 1091,
      1309, 1312, 1324, 1335, 1337
Sand dredging, 1570, 1571
Sand dunes, 262, 559, 888, 1001, 1013, 1017,
      1399, 1425, 1447, 1459, 1590, 1618,
      1627, 1628, 1805, 1911
Sander
  S. canadensis, 597
  S. vitreus, 553
San Francisco Bay Estuary (SFBE), 10, 639
  biodiversity, 641–643
  conservation status, 645–646
  defined, 638–640
  ecosystem services, 643–645
  fisheries, 645
  future aspects, 647–648
  location, 638
  rainfall, 638
  salinity, 638
  sediment, 640
  threats, 647
  water depth, 640

San Francisco bay joint venture (SFBJV), 645,
      646, 648
Sanguisorba officinalis, 1538
Sanionia uncinata, 284
Sanjiang Plain, 13
  Amur-Heilong River, 1510–1512
  conservation projects, 1519
  conservation status, 1517–1518
  fauna species, 1514
  future aspects, 1518–1520
  geographical map, 1511, 1512
  geographic features, 1511
  habitat types, 1510
  hydrologic features, 1511
  middle Amur River, 1513
  new protected areas, 1519
  Ramsar sites, 1517
  research and planning, 1518–1519
  threats, 1514–1516
  transboundary conservation, 1519–1520
  transboundary floodplains, 1497
  Ussuri-Wusuli River, 1510–1513
  wetland vegetation, 1513–1514
  wildlife, 366
San Joaquin valley, 698, 701
Sanmenxia Reservoir wetland, 1581, 1582
Santiria griffithii, 1834
Sarcocornetea fruticosi, 1144
Sarcocornia, 61, 1116
  S. fruticosa, 1117
  S. pacifica, 641
  S. perennis, 1117
Sardina, 1083
Sardinella, 1336
Sargassum, 1336
Sarkidiornis melanotos, 1796
Sarobetsu marsh, 1601
Sarotherodon melanotheron, 1322
Sarracenia
  S. purpurea, 612
  S. rubra, 674
Sarus crane, 1513, 1716, 1729, 1776, 1779,
      1796, 1806, 1807, 1946
Satellite wetlands, 1619, 1655, 1660, 1662,
      1665, 1666, 1668–1669, 1671, 1672
Satyrium crassicaule, 1313
Saundersilarus saundersi, 1584
Saussurea gossipiphora, 1609
Savanna swamps, 206–207
Sawgrass, 706, 710, 714, 715, 717–719
Scapharca inaequivalvis, 918
Scaphiopus holbrookii, 658
Scardinius erythrophthalmus, 953

Index of Keywords

Scenedesmus, 1936
Schenoplectus californicus, 791
Scheuchzeria palustris, 1537, 1540
Schizachyrium brevifolium, 206
Schizodon borelli, 817
Schizothorax, 1708, 1715
    S. curvifrons, 1708
    S. esocinus, 1708
    S. longipinus, 1708
    S. micropogon, 1708
    S. niger, 1708
    S. progastus, 1715
    S. richardsonii, 1708
Schoemus antarcticus, 876
Schoenoplectus, 159, 161, 163, 164, 607, 1666
    S. californicus, 791, 816
    S. giganteus, 164
    S. litoralis, 1130, 1668
Schoenus
    S. brevifolius, 1994, 1995
    S. ferrugineous, 290, 292
    S. nigricans, 256, 1045
Schwingmoor, 252–254, 263, 264
    raised bog, 266–267
Scirpo-cotuletum coronopifoliae, 833
Scirpus, 204, 517, 552, 641, 1104, 1728
    S. californicus, 205, 777, 851
    S. confervoides, 1859
    S. fluviatilis, 1524
    S. giganteus, 205, 211, 816
    S. grossus, 1746
    S. inclinatus, 205
    S. lacustris, 915, 1447
    S. maritimus, 149, 963–965, 1117, 1130, 1679, 1680
    S. paludosus, 569
    S. radicans, 915
    S. tabernaemontani, 1524
Sciurus
    S. anomalus, 1192
    S. carolinensis, 675
    S. vulgaris, 1106
Scleria, 1834
Sclerocarya birrea, 1272
Scleropages formosus, 1845, 1860, 1863
Scoliopteryx, 338
Scolopax rusticola, 1610, 1621
Scomberomorus
    S. brasiliensis, 763
    S. tritor, 1336
Scophthalmus rhombus, 1083
Scopus umbretta, 192, 1385
Scotland, *see* Flow Country, northern Scotland
Scotopelia peli, 1345
Scutellaria
    S. altissima, 916
    S. galericulata, 1171
Scylla serrata, 1228
Seabirds, 366, 392–395, 626, 843, 1131, 1326, 1328, 1870, 1996, 2008, 2012, 2016, 2017, 2019–2022
    colonies, 393, 395
Sea couch grass, 964, 1084, 1085
Seagrass(es), 7, 8, 10, 29, 30, 438, 442, 443, 451–452, 505, 1592, 1593, 1876, 1877, 1879–1882, 1969, 1970, 1972, 1974
    areas, 74, 79, 80, 86, 87, 89, 90, 442, 501, 502, 845, 1878, 1881, 1971–1972
    beds, 74, 78, 87, 323, 442, 448, 498, 841–843, 846, 1320, 1323, 1325, 1593, 1877–1882, 2007, 2019
    biodiversity, 85, 443
    biology, 78, 80, 85, 443
    bioregions, 75–76
    communities, 442, 1878
    conservation, 1971
    decline, 82, 89, 443, 496, 1595, 1880
    degradation, 85, 438
    description, 74, 1876
    detrital food web, 87
    diversity, 80–82
    ecological value, 76, 1971
    ecosystem services, 85–88
    extraction, 443
    families, 75–76
    fauna, 442
    global distribution, 79–82
    Halophila australis, 77
    improvements, 501
    Korean coastal areas
        distribution, 1591
        growth, 1593
        habitat, 1590–1593
        species, 1590
        threats, 1593–1595
        water temperature, 1593
    leaf reddening phenomenon, 82, 85
    losses, 79, 83, 84, 89, 442, 1879
    meadows, 13, 74, 87, 90, 439, 441–443, 841, 843, 845, 1876, 1878–1881
    in Australia, 453
    Great Barrier Reef, 448
    Malaysian, 1882
    mapped extent, 450
    Merambong-Tanjung, 1882

Seagrass(es) (*cont.*)
    Northeastern Australia, 1968–1974
    seascape forming, 440
    Southeast Asia, 438
    Sungai Pulai estuary, 1879
monitoring, 845
organic biomass, 78, 80, 85
photosynthesis, 88, 89
Phyllospadix japonicus, 88
plants, 89
populations, 80
protection, 498
recovery, 84, 89, 495–505
restoration, 89, 503
root-rhizome structure, 87
in Southeast Asia, 438
    challenges, 443
    coral reef conservation, 441
    destructive fishing, 442
    extraction, 443
    fisheries and food security, 438–441
    mining, 442
    overfishing, 441–442
    stakeholder engagement, 441
in Southeastern Brazil
    beds, 841, 842
    biology, 842–845
    conservations, 845
    ecology, 842–845
    geographical regions, 840
    locations, 840
    model of species interactions, 843–845
    oceanographic regions, 840
    paddle grass, 842
    predators, 843
    seasonal changes, 842–843
    sexual reproduction, 842
    shoal grass, 842
    species and distribution, 840–842
    threats, 845–846
    widgeon grass, 842
species, 75–76, 79–82, 85, 89, 840–842, 1323, 1590, 1595, 1877, 1880
    in coastal waters of Korea, 1592, 1593
    Great Barrier Reef World Heritage Area, 1970, 1971
    Halophila, 76, 1970
    International Union for the Conservation of Nature (IUCN), 78
taxonomic groups, 86
threats and challenges, 88–90
volumes, 78
Zostera marina, 76, 77, 81, 82, 88, 89

SeagrassNet monitoring project, 1973
Seagrass-Watch program, 1973
Sea-level rise (SLR), 7, 12, 38, 53, 65, 105, 106, 156, 349, 367, 502, 623, 647, 648, 675, 706, 714, 755, 952, 955, 956, 987, 1034, 1109, 1133, 1146, 1253, 1258, 1718, 1749, 1783, 1937, 2013
accelerated, 676
in Bay of Bengal, 1721
Camargue, 1109
Caspian, 949
climate change, 220–221, 349, 1109, 1808, 1809, 1957
coastal dunes, 1104
coastal plain subsidence, 65
coastal wetlands, 68
Ebro River, 1120
estuarine marshes, 58, 454, 502, 647
eustatic, 67
Everglades, 706, 714
global warming, 167, 1035
mangrove ecosystems, 105, 106
Mekong Delta, 1783
negative relative, 68
post-glacial, 1322
relative, 67, 675, 1120
salinization, 221
saltmarsh with, 454
San Francisco Bay Estuary, 647, 648
Takuu Atoll, Papua New Guinea, 2015
Tampa Bay, 501
tidal freshwater wetlands, 156, 167
tropical freshwater swamps, 220
Seals, 149, 530, 569, 570, 629, 692, 993–995, 1015, 1094, 1175, 1180, 1320, 1328, 1443, 1444, 1481, 1629, 1630, 1632, 1757, 1870, 2020
Sea mammals, 993, 1871
Seasonal flood pulse, 1401, 1405–1406
Seasonally flooded palustrine emergents, 581, 700
Seasonal pools/ponds, 22, 26, 652–664, 1133
Seasonal wetland, 110, 584, 585, 645, 681, 682, 685, 698, 700–702, 825, 1219, 1262, 1265, 1284, 1434, 1729, 1730
Sea turtles, 66, 74, 80, 85, 87, 584, 1325–1326, 1338, 1339, 1737, 1739, 1747, 1870, 1877, 1969, 2012, 2019, 2020
Secchi disk depth, 498, 500
Securinegion tinctoriae, 1144
Sedges, 173, 191, 192, 205, 206, 228, 258, 279, 284, 285, 290, 432, 517–519, 523, 525, 560, 561, 581, 586, 612, 735,

791, 805, 1026, 1029, 1032, 1068,
1069, 1074, 1117, 1151, 1156, 1221,
1223, 1236, 1302, 1316, 1402, 1406,
1418, 1426, 1427, 1460, 1465, 1467,
1513, 1526, 1534, 1535, 1540–1542,
1544, 1545, 1569, 1598, 1609, 1627,
1639, 1728, 1736, 1746, 1834, 1945
Aelurops lagopoides, 1728
alkaline swamps, 1386
beds, 961, 1856
Carex aquatilis, 546, 1072
Carex flacca, 1072
Carex hostiana, 1072
Carex lyngbyei, 569
Carex panicea, 1072
Carex spp., 524, 1556
Cladium mariscus, 1427
Cyperus haspan, 206
Cyperus latifolius, 1248
Cyperus papyrus, 170, 184
Eriophorum angustifolium, 546
estuarine marshes, 569
fens, 1428, 1534, 1535, 1541–1543
Fimbristylis longiculmis, 1427
flooded/semi-flooded, 828
Hypolytrum capitatum, 1844
Machaerina teretifolia, 1994, 1995
macrofossils, 1543
marshes, 1445, 1525, 1556
meadows, 430
Mfabeni, 1428
mud, 1046
peat, 1311
Schoenus brevifolius, 1994, 1995
Scirpus confervoides, 1859
seasonal floodplains, 1398
slender-tufted sedge, 951
social/ceremonial uses, 1263
species, 1980
swamps, 952
vegetation, 1264
Sediments, 7, 8, 38, 43–45, 47–51, 56, 58,
64–66, 68, 74, 83, 87–90, 100, 101,
105–107, 110, 115, 117, 126, 133,
135, 138, 149, 152, 157, 159, 166,
167, 178, 210, 218, 256, 279, 285,
347, 348, 350, 364, 366, 392, 394,
398, 400, 407, 451, 456, 457, 473,
476, 530, 532, 538, 553, 559, 561,
567, 568, 570, 572, 594, 622, 625,
626, 629, 630, 634, 641, 644, 646,
647, 655, 660, 661, 673, 683, 684,
693, 694, 712, 714, 721, 731, 734,
735, 739, 740, 750, 752, 754, 758,
761, 763, 780, 789–791, 808, 829,
842, 848, 849, 853, 869, 876, 888,
890, 905, 907, 912–914, 920, 937,
938, 974, 986–988, 1032, 1060,
1070, 1071, 1081, 1102, 1104, 1109,
1114, 1117–1120, 1129, 1138, 1139,
1150–1152, 1156, 1175, 1204, 1209,
1247, 1252, 1258, 1288, 1291, 1298,
1315, 1316, 1322, 1323, 1334, 1336,
1345, 1353, 1366, 1367, 1384,
1387–1389, 1399–1401, 1406, 1409,
1410, 1426, 1428, 1441, 1443, 1465,
1479, 1484, 1503, 1527, 1529, 1557,
1561, 1576, 1577, 1580, 1587, 1590,
1595, 1627, 1630, 1664, 1682, 1687,
1688, 1699, 1702, 1709, 1713, 1716,
1718, 1720, 1734–1736, 1740, 1744,
1746, 1748, 1749, 1781, 1805, 1810,
1856, 1862, 1869, 1879, 1880, 1904,
1910, 1913, 1914, 1916, 1925, 1933,
1934, 1937, 1947, 1948, 1960, 1961,
1963–1965, 1985, 1986, 2014
accumulation, 85
acid-potential, 1917
alluvial, 21, 254, 265, 1804
anaerobic bottom, 483
anoxic, 78
Bad River, 432
Bombah Broadwater, 1936
Ciénaga Grande de Santa Marta, 758
clay, 610
colluvial, 1061
Copper River, 559
Danube River, 912
deficit, 1120
deposition, 799, 803, 809, 1129, 1210,
1264, 1297, 1417, 1582, 1583, 1585,
1623
Ebro River, 1120
fine-textured mineral, 282
fluvial, 1726
Fraser Delta, 567
glacial, 559
land-use change, 138
marine, 52
Mekong, 1767
mineral soil, 246
muddy, 162, 1935, 1969
nonconsolidated, 333, 334, 346
nutrients and, 201, 428, 432–434
organic matter in, 776
Paraná River, 848

Sediments (*cont.*)
    phosphorus in, 776
    playa basin, 693
    Polje, 1060
    River-deposited, 599
    San Francisco Bay Estuary, 640
    seagrass meadows, 1878
    SFBE, 640
    soils, 511, 719
    sulfidic, 1915
    unvegetated wetlands, 30, 59
    Volga River, 937
    Wadden Sea, 986–987
    The Wash, 1013
    Yellow River, 1585
Sefid Rood, 1626–1631
Sefid Rud Delta, 1627, 1628, 1631
Sefid Rud River, 1626, 1627
Seiche, 431, 551, 593, 594, 606
Selenarctos thibetanus laniger, 1609
Sembilang National Park, 13, 1820, 1832, 1837
    biodiversity, 1824–1826
    conservation status, 1826–1827
    future aspects, 1827–1828
    hydrology, 1820–1822
    location, 1821
    peat carbon stock, 1823
    threats, 1827
    topography, 1821
    wetland ecosystems, 1822–1823
Semelai orang asli, 1852, 1861, 1862
Semi-permanent and permanent wetlands, 113, 350, 681, 682, 700, 701, 1262, 1263, 1281, 1944
Semi-permanently flooded palustrine emergents, 655, 700
Semnopithecus cristatus, 1779
Senecio paludosa, 159
Senilia, 1325
    S. senilis, 1324, 1325, 1336
Senna reticulata, 207
Sentinel wetland ecosystem, 706
Serinus
    S. citrinelloides, 1385
    S. koliensis, 192
Sesbania, 200
    S. javanica, 205, 1774
Sesheke Maramba Floodplain, 1220, 1222
Sesuvium, 61
    S. portulacastrum, 766, 1728
Setaria, 1236, 1302
    S. incrassate, 1272
Setophaga striata, 547

Settlement, 45, 50, 60, 62, 64, 105, 110, 117, 161, 193, 321, 326, 363, 366, 419, 422, 431–433, 452, 462, 531, 532, 534, 548, 578, 608, 669, 681, 684, 685, 768, 802, 808, 824, 901, 906, 914, 917, 1060, 1192, 1210, 1265, 1274, 1295, 1309, 1310, 1323, 1346, 1358, 1371, 1374, 1389, 1390, 1445, 1469, 1472, 1473, 1503, 1504, 1507, 1687, 1690, 1699, 1706, 1709, 1717, 1718, 1748, 1802, 1804, 1844, 1861, 1913, 1914, 1931, 1982, 1988, 1994, 1998
Seyfe Lake, *see* Lake(s), Seyfe
Shadegan Wetland, Iran
    conservation status, 1681
    ecological attributes and functions, 1678–1681
    ecosystem services, 1681–1682
    future challenges and developments, 1682–1683
    geography, 1676
    hydrology, 1676–1678
    location, 1676
    threats, 1682
Shallow open waters, boreal wetlands, 525, 526, 529, 537, 606, 608, 1235
Shallow peatlands, 277–279
Shilluk communities, Sudd region, 1304
Shire River, 1218, 1220, 1221, 1226
Shishig-gol valley, 1541
Shoal grass, *see Halodule wrightii*
Shorea, 240
    S. albida, 1757
    S. balangeran, 203, 208, 219, 1757
    S. platycarpa, 1834
    S. teysmanniana, 1757, 1834
    S. uliginosa, 1834
Shorebirds, 44, 51, 64, 133, 150, 377, 407, 533, 544, 547, 552, 561, 568–571, 583, 585, 630, 631, 633, 641, 645, 647, 692, 693, 805, 867–868, 870, 970, 1002, 1320, 1324, 1328, 1480, 1481, 1561, 1569, 1628, 1643, 1870, 1911, 1925, 1955
    Bahía Lomas, 867–868
    Bay of Fundy, 630
    breeding, 965, 966, 980
    Copper River Delta, 562
    East Asian–Australasian Flyway, 1870
    Lake Eyre, 1925
    migration, 45, 47, 408
    Mississippi Alluvial Valley, 584

populations, 50, 378–380, 1867
Saemangeum Estuarine System, 408–409
Saemangeum Shorebird Monitoring
    Program, 409–410
San Francisco Bay Estuary, 642
Short grass-dominated saltmarshes, 966
Short Pocosin peatlands, 671–674
Shrub-dominated wetlands, 1445
Shrubland swamps, 205–206
Shrubs and herbaceous vegetation, 914, 916
Sian Ka'an, 324
Siberia, 161, 319, 516, 1002, 1047, 1479, 1480,
    1482, 1515, 1688, 1747
Siberian crane, 953, 1444, 1448, 1471–1476,
    1481, 1524, 1525, 1559, 1561, 1566,
    1569–1571, 1584, 1626, 1631, 1636,
    1641–1645
Siberian salamanders, Kushiro marsh, 1599
Siberian sturgeon, 1444, 1447, 1448, 1455,
    1474, 1475, 1480, 1481
Sibiloi/Koobi Fora protected area, 1370
Siderastrea, 86
Siebenrockiella crassicollis, 1845, 1860
Sierra Nevada de Santa Marta (SNSM),
    760–762
Silene thymifolia, 916
Silurus
    S. asotus, 1502
    S. glanis, 953, 1109
    S. triostegus, 1690
Simocephalus, 1357
Sinkholes, 253, 314, 323, 335
Sipunculus, 86
Sirenia artiodactyla, 1955
Site of Special Scientific Interest (SSSI), 1049,
    1051
Sitka Spruce, 560
Sium
    S. latifolium, 1027
    S. suave, 1524
Skallingen peninsula, 984, 986, 988, 989
Sloping and spring highland fens, 1544
Small ruminant husbandry, 1194
Smart rivers, 927
Snowmelt patterned mixed mires, 261
Sobat River, 185, 1247
Socio-ecological conflicts, 361, 777–778, 780
Soda lakes, 11, 12, 146, 1383
    Bogoria, 1384–1385
    Elementeita, 1387
    fish of East Africa, 150
    Magadi, 1388–1389
    Nakuru, 1385–1387
    Rift Valley, 153, 1383
    threats and challenges, 1389
Soft-shore communities, 1002
Soil(s), 20, 58, 61, 63, 67, 94, 103, 117, 132,
    149, 163, 170, 173, 187, 188, 242,
    310, 430, 462, 471, 481, 523, 525,
    528–530, 534, 536, 546, 559, 566,
    610, 654, 657, 668, 673, 685, 686,
    690, 692, 712, 731, 732, 738, 765,
    766, 782, 802, 825, 851, 878, 916,
    926, 1070, 1074, 1105, 1116, 1236,
    1272, 1294, 1323, 1396, 1433, 1460,
    1468, 1470, 1516, 1525, 1533, 1547,
    1598, 1693, 1848, 1862, 1894, 1948,
    1960, 1995, 2009, 2017
    acidification, 1804
    acid sulfate, 103, 353, 354, 1795, 1804,
        1837, 1838, 1949
    agricultural, 566
    archaic peat, 250
    biodiversity, 230, 1894
    chemical properties, 879
    clay, 163, 608, 673, 926, 1302, 1687
    compaction, boreal wetlands, 534
    conditions, 61, 157, 851, 1795
    conservation, 1264
    degradation, 1433
    deposits, 529, 536
    drainage, 836, 1838
    dunes, 690
    environments, 1314
    erosion, 89, 107, 138, 434, 451, 739, 880,
        956, 1297, 1359, 1389, 1546, 1598,
        1600, 1612, 1862
    evaporation, 1701
    fens, 510
    fertile, 573, 1397, 1729, 1744, 1802, 1904
    fertility, 728, 738, 1264, 1720
    floodplain, 1955
    formation, 62, 1294
    humidity, 1140
    hydric organic, 187
    inorganic, 1169, 1445
    layers, 229, 272, 279, 343, 693, 1470, 1534
    leached, 415
    marshes, 1019
    mineral content, 671
    mineral soils, 21, 173, 200–221, 229, 246,
        279, 280, 282, 290, 517, 519, 525,
        529, 671, 672, 674, 1433, 1533,
        1822, 1834, 1843, 1856
    mineral wetlands, 525
    mobilization, 138

Soil(s) (cont.)
  moisture, 187, 350, 352, 353, 400, 530, 1236, 1291, 1961, 1964
  nitrogen, 516
  non-peat, 246
  nutrient-poor, 391, 511
  nutrients, 99, 669, 671, 673, 715, 734
  organic, 63, 187, 242, 272, 279, 285, 671, 698, 825, 1160, 1161, 1169, 1414
  organic carbon, 476, 684
  organic matter, 66, 1586
  oxidation, 1433
  peat, 208, 246–250, 415, 422, 471, 525, 546, 675, 1019, 1025, 1052, 1171, 1433, 1533, 1538, 1816, 1837, 1843, 1856, 1989
  pH, 1804
  physical compaction, 534
  potential acid sulfate, 1838
  properties, 360
  quality, 685
  saline-alkaline, 1730
  salinity, 764, 949, 1116–1118, 1140, 1467, 1693
  salinization, 1258, 1468, 1586, 1701
  sandy/loamy, 279
  saturation, 61, 361
  sediments, 246, 350, 719
  storage capacity, 674
  structure, 1074, 1693, 1985
  transportation, 364, 400
  types, 99, 246, 528, 681, 715, 1302, 1469, 1486, 1795, 1960
  volume, 1236
  wet/damp, 510
Solanum, 1313
  S. dulcamara, 915
Solea solea, 1016, 1083
Solidago
  S. canadensis, 570
  S. verna, 674
Solifluction, 510
Soligenous, 252, 254–258, 519
Somateria
  S. mollissima, 394, 1091
  S. mollissima borealis, 394
Sonneratia
  S. alba, 1237, 1823, 2008
  S. apetala, 1729
  S. caseolaris, 1823
  S. griffithii, 1737
  S. ovata, 1823
Sorbus, 300, 1029, 1043

Sorghum, 1302
  S. sudanense, 1294
  S. sudanica, 1302
Sorubium lima, 763
Sotalia fluviatilis, 735
Sousa, 1747
  S. chinensis, 1825
  S. plumbea, 1680
  S. teuszii, 1328, 1338
South East Asia, 9, 13, 131, 319, 322, 324, 1611, 1824
South-Eastern Anatolian Project, 1692
Southern Africa, 186, 370, 1218, 1234, 1393–1410, 1414, 1418, 1420, 1424, 1992
Southern and western Arabian Gulf coast, 1174
  Sabkha landforms, 1176–1179
  sustainable saline agro-systems, 1180–1181
Southern Barotse floodplain, 1222
South Florida Water Management District (SFWMD), 707, 711
South Sudan, 201, 202, 210, 216, 220, 1246, 1262
  Bahr el Ghazal, 1269–1276
  Machar Marshes, 1279–1286
  Nile River, 171, 185, 200, 1270–1276
  Sudd, 189, 1265, 1299–1305
South-to-North Water Diversion Project, 1586
Spalax
  S. giganteus, 953
  S. leucodon, 1192
Sparganium, 510, 612
  S. erectum, 949, 1190, 1639
  S. hyperboreum, 511
Spartina, 61, 64, 573, 631, 851, 963, 969
  S. alterniflora, 26, 49, 52, 57, 58, 384, 630, 631, 969
  S. alterniflora x maritime, 988
  S. anglica, 969, 970, 1082
  S. foliosa, 641
  S. maritima, 969, 1323
  S. patens, 57, 67, 630
  S. townsendii, 1001
Sparus auratus, 1104
Spatial data, 111, 1363
Special Area for Conservation (SAC), 981, 1051, 1052, 1143, 1717, 1728
Special Protection Areas (SPA), 975, 976, 981, 1004, 1017, 1094–1095, 1118, 1132, 1143
Species composition, 101, 106, 167, 176, 510, 660, 767, 768, 988, 1105, 1107, 1302, 1330, 1459, 1535, 1561, 1595, 1757, 1859, 1953

Spergularia, 851
  S. media, 964
Spermatophyta, 804
Spermophilus citellus, 1142
Sphaenorhynchus spp., 735
Sphaeroma quoyanum, 64
Sphaerophorus globosus, 279
Sphagniana sphagnorum, 533
Sphagnum, 21, 26, 218, 228, 234–236, 247,
        256, 258–260, 263–265, 271, 279,
        280, 282, 284, 285, 291, 310,
        517–519, 523, 524, 532, 533, 546,
        612, 831, 876, 877, 879, 1044, 1045,
        1158, 1159, 1163, 1311, 1313, 1314,
        1418, 1537, 1598, 1980, 1992–1994
  bog, 247, 291, 877, 879, 1992
  bog moss, 247, 291
  distribution, 236
  peats, 234, 280, 284, 532, 533, 879, 1159,
      1311, 1418
  S. angustifolium, 284, 517, 518
  S. auriculatum, 1045, 1046, 1160
  S. australe, 1994, 1996
  S. balticum, 1537
  S. capillifolium, 291, 517, 1044
  S. cristatum, 1994–1996, 1999
  S. cuspidatum, 517, 1044, 1045, 1313
  S. eriophorum vaginatum, 1045
  S. falcatulum, 1996
  S. fallax, 264, 291, 518
  S. fimbriatum, 284
  S. flexuosum, 571, 1537, 1540
  S. fuscum, 261, 282, 517, 1045
  S. girgensohnii, 284
  S. jensenii, 518
  S. lindbergii, 284
  S. magellanicum, 291, 517, 518, 877
  S. majus, 518
  S. novo-zelandicum, 1994
  S. papillosum, 256, 1044, 1045
  S. perichaetiale, 1313
  S. riparium, 518
  S. skyense, 1045
  S. squarrosum, 284
  S. subsecundum, 1537
  S. tenellum, 256
  S. teres, 518
  S. warnstorfii, 284, 518
Sphyrna, 1325
Spider population, 1082
Spilanthes, 1313
Spiophanes spp., 630
Spiranthes sinensis, 1816

Spirlinus, 1190
Spirochaeta
  S. africana, 1389
  S. alkalica, 1389
  S. asiatica, 1389
Spirulina, 150, 153, 1728
Spisula subtruncata, 1094
Spizaetus nanus, 1835
Spondianthus, 207
Sporadanthus, 1980, 1985, 1993, 1994
  S. ferrugineus, 1981, 1992–1996, 1998, 1999
  S. traversii, 1992, 1993, 1995–1998
Sporobolum virginicus, 766
Sporobolus, 61
  S. ioclados, 1385
  S. pyramidalis, 1302
  S. spicatus, 1368, 1385, 1386
Sporophila palustris, 817
Sprattus, 1083
Spring(s), 53, 67, 110, 119, 121, 254, 327, 364,
        434, 472, 478, 551, 562, 570, 580,
        602, 641, 643, 647, 652, 653,
        655–661, 674, 683, 700, 789, 901,
        912, 914–916, 918, 937, 940, 942,
        949, 956, 966, 980, 990, 1035,
        1061–1063, 1073, 1080, 1081, 1092,
        1102, 1104, 1105, 1124, 1128, 1142,
        1170, 1171, 1255, 1312, 1322, 1324,
        1327, 1330, 1384, 1426, 1441, 1442,
        1452, 1454, 1459, 1472–1475, 1479,
        1502, 1511, 1526, 1528, 1535–1536,
        1544, 1545, 1568, 1571, 1580, 1593,
        1612, 1627, 1630, 1638, 1641, 1662,
        1672, 1679, 1683, 1686, 1688, 1706,
        1903, 1934, 1936, 1960
  and blanket bogs, 1545
  fens, 256, 1535–1539, 1543–1545
  mire, 254, 256–258, 290, 292, 1426
  Peace River, 551
  periods, 45
  surface water storage and flows, 683
  tides, 41, 98, 159, 407, 1035, 1080, 1081,
      1322, 1330, 1334, 1335
Spur, 267–268, 270, 305, 1080, 1504, 1748
  blanket bog, 270
Squatina guggenheim, 850
Stachys
  S. maritima, 916, 917
  S. palustris, 915, 952
Stelgidopteryx ridgwayi, 323
Stellaria
  S. aquatica, 915
  S. humifusa, 510

Stemonurus, 208
Stenella attenuate, 1747
Stenocaulon kleinii, 205
Stenochlaena palustris, 205, 1815, 1848
Stenodus
  S. leucichthys, 1444
  S. leucichthys leucichthys, 953
Stenogobius mekongensis, 1806
Steppe, 824, 827, 828, 876, 916, 934, 937, 938, 948, 954, 957, 1130, 1188–1190, 1192, 1255, 1446, 1448, 1467–1469, 1480, 1492, 1493, 1500–1507, 1510, 1524, 1532, 1534–1540, 1542–1544, 1547
Sterculia apetala, 765
Sterna
  S. albifrons, 905, 1141, 1142
  S. aurantia, 1779
  S. bernsteini, 1870
  S. dougallii, 633
  S. forsteri, 596
  S. hirundo, 851, 1030, 1142
  S. maxim, 1324
  S. nilotica, 1141, 1142, 1192
  S. paradisaea, 394, 980
  S. sandvicensis, 1142
Sternula
  S. albifrons, 1338, 1912
  S. antillarum athalassos, 584
  S. caspia, 1338
  S. hirundo, 1338
  S. maxima, 1338
  S. nereis, 1912
  S. sandvicensis, 1338
Stictonetta naevosa, 1902
Stikine River, 559
Stipa, 938
  S. borysthenica, 916
  S. capillata, 916
  S. pulcherrima, 916
Stizostedion
  S. lucioperca, 476, 953
  S. vitreum, 612
Storage Treatment Areas, 707
Stormwater Treatment Areas (STAs), 712, 713
Stratiotes aloides, 914
Stressors, 9, 82, 89, 138, 139, 361, 366, 434, 585, 594, 603, 607, 613, 755, 845, 1175, 1375, 1973
Stromateus brasiliensis, 850
Strombus, 86, 2018
Struthio camelus, 1385
Sturnus vulgaris, 1083

Stygobitic (aquatic cave-adapted) species, 322, 323
Stygofauna, 348, 351, 1961
Suaeda, 61, 1117
  S. confusa, 951
  S. fruticosa, 1001, 1728
  S. glauca, 1525
  S. japonica, 407
  S. maritima, 916, 1082, 1083
  S. vera, 965
Submerged aquatic plants, 641, 1237, 1569, 1652
Zhalong marsh, 1524
Submerged marine flowering plants, 74
Subsidence, 41, 65–67, 238, 311, 352–354, 421, 422, 536, 690, 698, 1060, 1109, 1120, 1138, 1151, 1153, 1163, 1170, 1255, 1258, 1400, 1715, 1759, 1786, 1822, 1837
Subsistence cultivation, 1433
Subsurface conduits, 363
Subterranean (hypogean) habitats
  caves, 332–334
  classification, 332
  composition, 336–337
  Dinaric Karst, 338–342
  environment, 334–335
  future aspects, 342–343
  inhabitants, 335–336
  interstitial, 332–334
  karst, 332–334
  karst hydrology, 316
  threats, 342–343
Subtidal zone, 841, 1876
  definition, 47
  intertidal and, 1083
  Outer Bay, 626
Sudan Community Watershed Management Project (SCWMP), 1296
Sudan Wildlife Provisions Act of 2003, 1303
Suddia sagittifolia, 211
Sudd, South Sudan, 12, 171, 185, 189, 200, 201, 204, 206, 209–214, 216, 1244, 1246, 1247, 1249, 1265, 1270–1272, 1280, 1284, 1414, 1415
  aquatic diversity, 1305
  biodiversity, 1302–1303
  conservation status, 1303
  ecosystem services, 1304
  future aspects, 1304–1305
  hydrology, 1300
  location, 1300
  threats, 1304–1305
  wetland ecosystems, 1300–1302

Sueda
    S. fruticosa, 1001
    S. maritima, 963
Suisun Basin, 701
Sumatra, 113, 203, 216, 1754, 1755, 1758, 1820, 1823, 1826, 1827, 1832, 1835–1838, 1845
Sumphonia, 206
Sumplands, 1960
Sundadanio spp., 1759
Sundarban Biosphere Reserve, 1739
Sundarban mangroves
    biodiversity, 1736–1737
    challenges, 1740
    conservation status and management, 1738–1739
    ecosystem services, 1737–1738
    flora and fauna, diversity of, 1736
    geological description, 1735–1736
    location, 1734–1735
    mangrove-dependent people, 1738
    threats, 1739–1740
    wetland types, 1736
Sundasalanx platyrhynchus, 1845
Sungai Pulai estuary, 1879–1882
Suo, 290
    See also Swamps
Suq Al-Shuyukh, 1688–1691, 1693
Surface ditches, 363
Surface-flow 'flush,', 258–259
Surface-flow spring mire/fen, 256–258
Surface patterning, 230, 264, 268, 270, 271, 305, 306, 308, 524, 1043–1047
Sus
    S. barbatus, 1826, 1836
    S. scrofa, 220, 1084, 1134, 1621, 1653, 1680, 1737, 1826, 1952
Sustainable development, 178, 400, 492, 737, 767, 810, 1340, 1504
Sustainable wetland management, 1264
Sutherland, 11, 253, 261, 297, 305, 1039–1054
Swamps, 7, 20, 22, 27, 137, 157, 159, 160, 163, 165, 166, 170–180, 184–186, 189, 192, 200–221, 326, 473, 480, 525, 529, 531, 538, 671, 709, 728, 730, 734, 735, 738, 739, 750, 752, 789, 791, 808, 827, 829, 938, 1138, 1188, 1201, 1203, 1204, 1219–1221, 1223, 1226–1228, 1235, 1246, 1254, 1262, 1270–1272, 1274, 1283, 1284, 1300, 1301, 1303–1305, 1342, 1384, 1386, 1396, 1398, 1403, 1445, 1486, 1489, 1491, 1580, 1700, 1713, 1715, 1739, 1744, 1747, 1768, 1795, 1802, 1822, 1824, 1844, 1853, 1856, 1889, 1901, 1945, 1946, 1978, 1980, 1981, 1985, 2007, 2008
    Amazonian permanent, 732
    boreal wetlands, 525
    Central Africa, 207
    communities, 1236–1237
    Danau Sentarum National Park, 1843
    deepwater bald cypress–tupelo, 177
    defined, 6
    depressions, 1237
    ecosystems, 173, 1760, 1852
    environment, 1754–1756
    floodplain, 172, 200, 203, 208, 209, 211, 1220
    freshwater, 6, 30, 170–173, 175, 177–180, 200–221, 326, 733, 1190, 1445, 1769, 1770, 1772, 1822, 1823, 1826, 1832, 1834–1836, 1985, 2011
    grasses, 220, 1777
    habitats, 1862
    herbaceous, 177, 204, 205, 214, 1577, 1578, 1824
    inland mangrove, 1237
    mangrove, 27, 57, 58, 114, 174, 246, 322, 758, 1718, 1744, 1777
    meadow, 1577
    mineral soil-based freshwater, 6
    mineral soils, 1433, 1834
    mud, 1982
    papyrus, 170, 177, 213, 1227, 1235, 1237, 1247, 1418
    peat, 26, 201, 230, 234, 238–240, 264, 265, 269, 301, 1415, 1424, 1426, 1428–1432, 1434, 1753–1760, 1768–1773, 1808, 1813–1817, 1820, 1822, 1823, 1826, 1827, 1832, 1834, 1836, 1837, 1843–1847, 1852, 1856, 1862
    pond cypress–black gum, 177
    reed, 22, 170, 1221, 1226
    restoration, 26
    shallow, 22, 1220, 1946
    shrub, 177, 204, 609, 1445
    South America, 207
    Sudd, South Sudan, 1300
    supratidal, 159
    tree islands, 718–719
    tropical, 192, 211, 213, 1414, 1423–1434
    valley, 185, 186
    vegetation, 209, 211, 955, 1246, 1824, 1862

Swamps (*cont.*)
  wetland type, 1979
  woody scrub, 1768–1770, 1772
Swan coastal plain (SCP), 1960, 1961
  hydrological change, 1965
  sediments, 1960
Swertia multicaulis, 1609
Sylvia
  S. cantillans, 1337
  S. conspicillata, 1105
Sylviidae, 939
Sylvilagus floridanus, 675
Symmeria paniculata, 207
Sympetrum sanguineum, 1073
Symphalangus syndactylus, 1825, 1835, 1860
Symphonia, 207
  S. globulifera, 207
Symphytum
  S. officinale, 164, 915
  S. officinalis, 915
Sympterigia bonapartei, 850
Synallaxis
  S. cinerascens, 851
  S. frontalis, 851
Syncerus
  S. caffer, 1227, 1236, 1274, 1405
  S. caffer aequinoctialis, 1293
Syncomistes butleri, 211
Syringodium, 74
  S. filiforme, 75, 81
  S. isoetifolium, 75, 82, 1876, 1971
Syzygium, 207, 1221, 1418, 1795, 1823, 1834
  S. claviflora, 205
  S. cordatum, 1427
  S. cumini, 207, 1815
  S. maire, 1980

T

Tabebuia barbata, 207
Tabernaemontana juruana, 207
Tachybaptus ruficollis, 1073, 1708
Tachyglossus aculeatus, 1902
Tachyphonus rufus, 851
Tachys, 1082
Tadorna
  T. ferruginea, 1186, 1190, 1610, 1642, 1665, 1715
  T. tadorna, 1003, 1016, 1030, 1141, 1142, 1190, 1642, 1665
  T. variegata, 1981
Tagelus
  T. adamsonii, 1336

T. gibbus, 851
T. plebeius, 844–845, 850
Taiga-alas landscape, 13, 1463–1470
Takuu Atoll, 2014, 2015
Tall-grass prairies, 681
Tall/low shrub, 545
Tall Pocosin peatlands, 671, 672
Tamandua tetradactyla, 220
Tamarix, 64, 1140, 1666, 1669, 1730
  scrubland, 1140
  T. gallica, 1106
  T. nilotica, 206
  T. ramosissima, 916, 951
Tampa Bay, 10, 89
  blue carbon benefits, 502
  characteristics, 496
  habitat modeling, 502
  historical loss and recovery stage, 498–499
  nutrient loading, 499–500
  pH conditions, 501
  seagrass cover, 501
  seagrass loss, 83, 84
  water quality and clarity, 500
Tampa Bay Estuary Program (TBEP), 497–500, 502, 504
Tampa Bay Nitrogen Management Consortium (TBNMC), 498, 499, 503
Tansley, A. G., 290, 296, 298, 304, 305, 1049
Tanymastrix stagnalis, 1071
Tanzania, 130, 144, 186, 1218, 1341–1347, 1414
Tapirus
  T. bairdii, 324
  T. indicus, 236, 1826, 1835, 1860
  T. terrestris, 735
Taraxacum bessarabicum, 916
Tarchonanthus camphoratus, 1386
Taro pits, 2009
Tarpon atlanticus, 763
Tasek Bera, 13, 26
  cultural heritage, 1861
  fauna, 1859
  flora, 1859
  formation, 1856
  future challenges and developments, 1863
  habitats, 1856–1859
  key species, 1860–1861
  land use, 1861–1862
  local history, 1861
  location, 1852, 1853
  physical environment, 1853
  site management, factors influencing, 1862

Taurotragus
　T. derbianus, 1387
　T. oryx, 1227, 1238
Taxodium
　T. ascendens, 719
　T. distichum, 164, 177, 581,
　　　668, 669
　T. distichum var. *nutans*, 177
Tayassu
　T. pecari, 736, 750
　T. tajacu, 220
Tea plantation project, 1310
Tea-stained blackwater stream, 1756
Tectonic, 67, 126, 201, 332, 559, 622, 786, 828,
　　　1186, 1387, 1445, 1713, 1715, 1735,
　　　1744
Tectonic estuaries, 39
Tees estuary, 371, 375, 1004, 1005
Teijsmanniodendron
　T. hollrungii, 1834
　T. pteropodum, 1834
Telespiza ultima, 2020
Temperate riverine forests, 916
Temperature, 38, 41, 44, 46, 67, 82, 89, 105,
　　　117, 119, 132, 133, 138, 144, 146,
　　　153, 238, 266, 271, 281, 286, 298,
　　　335, 349, 391, 442, 457, 519, 531,
　　　559, 630, 647, 652, 655, 658, 660,
　　　709, 748, 749, 774, 799, 827, 956,
　　　1042, 1090, 1094, 1124, 1150, 1157,
　　　1163, 1168, 1211, 1254, 1281, 1293,
　　　1294, 1365, 1384, 1390, 1441, 1442,
　　　1444, 1464, 1479, 1511, 1532, 1547,
　　　1555, 1561, 1577, 1579, 1583, 1616,
　　　1651, 1676, 1678, 1688, 1701, 1726,
　　　1767, 1782, 1871, 1899, 1926, 1948,
　　　1954, 1960, 1988
　air, 106, 476, 760, 799, 935, 947, 1114,
　　　1355, 1387, 1608, 1678
　mean annual temperature, 530, 536, 1293,
　　　1442, 1503, 1994, 1995
　ocean, 89
　water, 119, 256, 434, 479, 614, 625, 640,
　　　655, 660, 739, 761, 1091, 1175,
　　　1366, 1541, 1561, 1593, 1934
Temporarily flooded marshes, 1105
Temporary marshes, 1127
Temporary ponds, pools or waters, 652–654,
　　　656, 657, 1071, 1124, 1127–1131,
　　　1133, 1445, 1946
Tenualosa ilisha, 1702, 1737
Tepuales, 877, 878
Tepualia, 829
　T. stipularis, 825, 878

Terai, 1612, 1713, 1715–1716
Terminalia, 1385
　T. cambodiana, 205, 208
　T. copelandii, 208
　T. laxiflora, 1272
Terrapene carolina yucatana, 323
Terrestrializing fens, 1418
Tessaria integrifolia, 791
Testudo graeca, 1142, 1190
Tetramerista glabra, 1834
Tetrao parvirostris, 1444
Tetraselmis, 1664
Tetrax tetrax, 1628
Tevreden Pan, 1416
Teysmanniodendron sarawakanum, 208
Thais haemastoma, 845
Thalasseus
　T. maximus, 851
　T. sandvicencis, 851
Thalassia
　T. hemprichii, 75, 82, 1876, 1970, 1971
　T. testudinum, 75, 80–82
Thalassodendron
　T. ciliatum, 75, 82, 1876, 1877,
　　　1970, 1971
　T. pachyrhizum, 75
Thalassornis leuconotus, 1354
Thaleichthys pacificus, 569
Thalia
　T. geniculata, 791
　T. multiflora, 791
Thalictrum flavum, 951
Thamnophis sirtalis, 612
Thelypteris
　T. confluens, 1313
　T. palustris, 915
　T. striata, 191
Thermaikos Gulf, 1138, 1139,
　　　1142, 1145
Thermal corridor, 786
Thermocyclops, 1357
Thermo-isolation, 286
Thermokarst, 545, 1443, 1454, 1458
　depression, 1466, 1468, 1469
　kettle hole peatlands, 284
　lakes, 284, 1446, 1447, 1464–1466
　ponds, 278–280, 283, 286
Thespesia populnea, 1823
Thessaloniki, 1138, 1139, 1143, 1145
Thinocorus rumicivorus, 867
Thinornis rubricollis, 1912
Thoracostachyum, 1834
　T. sumatranum, 204
Threatened seagrass, in Korea, 1593–1595

Threatened species, 53, 164, 323, 325, 377,
   455, 490, 817, 889, 953, 1071, 1073,
   1115, 1130, 1131, 1188, 1354, 1525,
   1558–1560, 1584, 1595, 1622, 1729,
   1737, 1758, 1846, 1860, 1956, 2019,
   2020
 amphibians and reptiles, 1190
 biodiversity, 1628, 1787–1788, 1870
 Bujagh National Park, 1629
 Dactylorhiza hatagirea, 1609
 ecosystem services, 1956
 flooded forest areas, 1788
 Nardostachys grandiflora, 1609
 Nardostachys scrophulariflora, 1609
 Paraná River Delta, 817
 Tasek Bera, 1860
 threatened biota, 1894
 tidal freshwater wetlands, 164
 waterbirds, 377
 Yangtze River, 1158–1560
Three Gorges Dam, 1560, 1571
Threskiornis
 T. aethiopicus, 1109, 1238, 1338
 T. melanocephalus, 213, 1444, 1525, 1787,
   1796, 1816, 1824, 1835
 T. molucca, 1902, 1955
 T. spinicollis, 1902, 1926
Thrombolites, 148, 149
Thryonomys, 1315
Thuja
 T. occidentalis, 612
 T. plicata, 571
Thylogale bruinii, 211
Thymallus
 T. arcticus, 1448
 T. grubii, 1502
 T. nigrescens, 1480
 T. thymallus, 939
Thymus zygoides, 916
Tidal flats, 56, 406, 407, 410, 562, 569, 572,
   829, 866–868, 1001, 1006, 1016,
   1019, 1324, 1325, 1338, 1556, 1590,
   1593, 1744
 biodiversity, 641
 conservation status, 645, 994, 1955
 hydrology, 1335
 marshes and, 629–632
 plants and benthos, 988, 1324
 sediments, 986, 987
 Yangtze River Estuary, 1557–1558
Tidal freshwater wetlands (TFWs), 7
 description, 156
 distribution, 160–161
 fauna, 165–166
 flora, 163–165
 fresh water hydrology, 157
 global warming and sea level rise, 167
 gradient water–land, 158–160
 human impacts, 160–161
 physical and chemical processes, 162–163
 survival, 157
 in USA, 159
Tidal immersion mire, 254
Tidal saltmarshes, 960, 1019, 1021
 biodiversity, 961–966
 conservation status, 966–968
 geographic distribution, 960
 threats and challenges, 968–971
Tidal wetlands, 56, 59, 60, 62, 63, 65–68, 156,
   158, 160, 164–166, 406, 407, 410,
   502, 560, 638, 643, 646, 647, 668,
   733, 816, 833, 1553, 1557
Tierra del Fuego, 268, 299, 301, 307, 830, 831,
   834, 866–868, 874, 876–878, 880,
   1043
Tiger reserve, 1738, 1739
Tilapia, 1863
 T. zillii, 1357
Tilia, 938
 T. cordata, 916
 T. tomentosa, 916
Tillandsia usneoides, 585
Timonius, 207
 T. salicifolius, 205
Tinca tinca, 476, 953, 1480
Tindallia magadii, 1388
Tipperne Reserve, 974–976,
   978–981
Titanethes, 339
Tivela mactroides, 845
Tollund Man, 248
Tolypella, 1105
 T. salina, 1104
Tomenthypnum, 518
Tomistoma schlegelii, 1825, 1835, 1845, 1860
Tonle Sap Lake, 13, 205, 1808
 biodiversity, 1787
 conservation status, 1787
 damming (off-site) and LU changes
   upstream, 217
 dependent peoples, 1789
 ecosystem services, 1789
 extent of landscape, 1786
 flood-pulse hydrology, 1804
 future challenges and developments,
   1790–1791

location, 1786
management, 1788–1789
swamp forests, 207
threats, 1790–1791
wetland diversity, 1786
Tope System, 230
Topogenous, 252, 254, 518, 1856
Tor, 1715
   T. tor, 1715
Tourism, Zhalong marsh, 1527
Tournefortia argentea, 2009
Tower karst, 326
Trachemys
   T. scripta, 1134
   T. scripta ssp. *elegans*, 1107
Trachycardium muricatum, 845
Trachypithecus
   T. cristatus, 1825, 1836
   T. francoisi, 325
   T. germaini, 1788
Traditional fishery, 364, 400, 764, 1208, 1871
Tragelaphus
   T. scriptus bor, 1294
   T. spekii, 192, 211, 1274, 1315, 1405
   T. strepsiceros, 1371, 1384,
       1385, 1405
Tragulus
   T. javanicus, 1826, 1836
   T. napu, 1826
Tram Chim National Park, 365, 366,
      1806–1808
   biodiversity, 1796
   conservation status, 1797
   ecosystem services, 1796
   future challenges and developments, 1798
   location, 1794
   physical environment, 1795
   threats and challenges, 1798
   wetland ecosystems, 1795
Transboundary Biosphere Reserve (TBR),
      906–908, 921
Transboundary cooperation, 921,
      926, 1505
Transboundary Mekong River Delta,
      see Mekong River Delta
Transboundary Sundarban mangroves,
      see Sundarban mangroves
Transboundary UNESCO Biosphere Reserve
      "Mura-Drava-Danube" (TBR
      MDD), 906–908
Transboundary Wetland, 911–921, 1496,
      1505–1507
Transdisciplinary approach, 4, 14

Transition, 132, 156, 252–254, 263, 270, 534,
      536, 608, 681, 798, 804, 984, 988,
      1017, 1069, 1075, 1193, 1525,
      1536–1541, 1552, 1822, 1933
Transitional (driftline) marsh, 964
Trapa, 1272, 1301, 1708
   T. bispinosa, 1556
   T. natans, 482, 914, 1443, 1717
   T. potaninii, 1524
Travis wetland, 1988
Trebon Biosphere Reserve, 473
Treed domed bog, 229
Tribulus terrestris, 952
Trichechus
   T. inunguis, 735
   T. manatus, 86, 87, 324, 584, 750, 764
   T. senegalensis, 86, 1337, 1338
Trichodrama, 1630
Trichophorum cespitosum, 1045
Trichosurus vulpecula, 1902–1903, 1986
Tridacna spp., 2018
Trifolium, 1105
   T. dubium, 916
   T. fragiferum, 916
Triglochin, 61
   T. maritima, 964
   T. maritimum, 570, 1538
   T. palustre, 1543
Trigonoceps occipitalis, 1354
Trilateral Wadden Sea Cooperation, 994
Tringa
   T. erythropus, 980
   T. guttifer, 1825, 1835, 1870
   T. nebularia, 980, 1044, 1915, 1955
   T. stagnatilis, 1448, 1469, 1481
   T. totanus, 965, 989, 1003, 1016, 1030,
       1073, 1105, 1337, 1665
Trionyx triunguis, 1274
Triphosa, 338
Triplophysa
   T. kashmiriensis, 1708
   T. marmorata, 1708
Tripolium pannonicum, 951
Triportheus magdalenae, 763
Trisopterus luscus, 1083
Tristania, 207
   T. suaveolens, 208
Tristaniopsis obovata, 208
Triturus, 939
   T. carnifex, 1142
   T. dobrogicus, 918
   T. pygmaeus, 1131
   T. vulgaris, 918

Troglocaris, 341
  T. anophthalmus, 342
Troglogloea, 336
Troglohyphantes, 339
Troglophilus, 338
Tropical freshwater swamps, 6, 171, 175, 177
  biodiversity, 210–213
  buffering water levels, 208–209
  carbon storage, 213
  climate change, 220–221
  damming and flooding, 216–217
  dependent peoples, 213–215
  distribution and area, 202
  diversity, 200–201
  erosion, 218
  fires, 219
  forests, 207–208
  herbaceous swamps, 204–205
  invasive species, 220
  land use changes upstream and damming, 217–218
  phenological adaptations, 204
  physiological adaptations, 204
  pollution, 218
  productive systems, 210
  reclamation and conversion, 216
  reproductive adaptations, 204
  resource over-utilisation, 219–220
  savanna swamps, 206–207
  shrubland swamps, 205–206
  structural adaptations, 204
  water quality, 209
  woody/shrubby vegetation, 203
Tropical Pacific Ocean, 10, 1862, 2001–2023
Tropical peatland, 231, 236, 238, 242, 1754
  Rugezi Marsh, 1307–1316
Tropical swamp forests, 1414
  future challenges and developments, 1759–1760
  Maputaland coastal plain (*see* Maputaland coastal plain (MCP))
  peat, 301 (*see also* Peat, swamp forests)
Tropical wetlands, 184, 221, 706, 733, 1270, 1300, 1969
Trumpeter swan, 562, 569, 571
Trypanosoma
  T. evansi, 809
  T. vivax, 809
Tsuga heterophylla, 560
Tubifex, 977
  T. costatus, 977
Tubificids, 165
Tubificoides benedii, 977

Tumulus mound springs, 1961
Tundra, 26, 276–279, 282, 283, 285, 392, 547, 552, 825, 878, 1040, 1041, 1052, 1443, 1445, 1447, 1448, 1453, 1469, 1480, 1482, 1525, 1534, 1544
Tupinambis
  T. merianae, 817
  T. teguxin, 852
Turbidity front, 848, 853, 854
Turbine, 52, 417–420, 423, 1052, 1095
Turbine towers, 418–419
Turbo spp., 2018
Turdidae, 939
Turdoides altirostris, 1681, 1691
Turf-cutting, 239
Turkana Lake, *see* Lake(s), Turkana
Turloughs, 11, 253
  conservation status, 1075
  description, 1069
  dry phase, 1069, 1072–1073
  flooded phase, 1071
  hydrology, 1070–1071
  inundated phase, 1069
  threats and management challenges, 1073–1075
  wetland description, 1069–1070
Turnix sylvaticus, 1131
Tursiops
  T. gephyreus, 850–851
  T. truncatus, 850–851, 918, 1328, 1338
Turtles, 46, 82, 118, 166, 219, 323–325, 448, 453, 454, 456, 610, 612, 735, 741, 750, 843, 939, 1131, 1206, 1336, 1337, 1655, 1679, 1796, 1845, 1856, 1860–1863, 1871, 1956, 2016, 2018, 2021
  freshwater, 113, 115, 1778–1779, 1825, 1835
  sea, 74, 80, 85, 87, 584, 1325–1326, 1338, 1339, 1737, 1739, 1747, 1870, 1877, 1969, 2012, 2019, 2020
Tympanuchus cupido attwateri, 584
Typha, 26, 159, 185, 189, 204, 385, 386, 552, 612, 712, 713, 851, 915, 1117, 1223, 1255, 1628, 1642
  T. angustifolia, 163, 428, 613, 777, 914, 915, 952, 1190, 1524, 1639, 1654, 1655, 1679, 1680, 1746
  T. australis, 205
  T. domingensis, 190, 713, 750, 1156, 1171, 1221, 1265, 1272, 1302, 1353, 1354, 1900
  T. dominguensis, 750, 765

T. latifolia, 163, 432, 569, 915, 1140, 1237, 1447
T. laxmannii, 952
T. orientalis, 1980, 1982
T. x glauca, 432, 598
Typhlachirus elongates, 1806
Typhlogammarus mrazeki, 341
Tyto capensis, 1796

U
Uapaca, 207
   U. guineensis, 207
Ubangi, 1204, 1206–1211
Uca
   U. rapax, 763
   U. tangeri, 1324, 1336
   U. vocator, 763
Ulmus, 581
   U. carpinifolia, 1639
   U. foliacea, 916
   U. minor, 1106
Ulva, 86, 498
Ulz-gol valley, 1543
U Minh peat swamp forest
   agriculture, 1815
   biodiversity, 1816
   conservation status, 1816
   ecosystem services, 1816
   future challenges and developments, 1817
   location, 1813
   physical environment, 1815
   threats, 1817
   wetland ecosystems, 1815
Uncaria, 207
Uncia uncia, 1608
UNEP/GEF Siberian Crane Wetland Project, 1524, 1644
Unique climatic conditions, 1442
Unique Trucial Coast Sabkha, 1175
United States Prairie Pothole Region, 363, 685
Upland fens, 1156
Upper Congo rapids, 1202, 1203, 1207
Upper Indus Basin, 1699, 1700
Upper (high) marsh, 964
Upper Paraguay Basin, 799
Upper Volga Basin, 938, 939
Upper Zambezi, 1218, 1221–1222, 1408
Upwelling, 138, 448, 624, 628, 629, 640, 840, 1322, 1323, 1334, 1335, 2013
Urban Wastewater Treatment Directive (UWWTD), 893–894
Uria aalge, 1094

Urochloa mutica, 765, 1952
Uromiyeh Lake, *see* Lake(s), Uromiyeh
Urophyllum arboreum, 1834
Ursus
   U. americanus, 675
   U. americanus luteolus, 584
   U. arctos, 1514
   U. maritimus, 1444, 1454
   U. thibetanus, 1514, 1609
1948 U.S. Congressional Flood Control Act, 706
U.S. Farm Bill, 685, 694
Ussuri-Wusuli River, 1510
   conservation status, 1517
   developmental history, 1514–1516
   from Lake Khanka-Xingkaihu to Khabarovsk, 1511
   Ramsar sites, 1517
Utricularia, 613, 717, 914, 1237, 1272, 1689, 1859, 1994
   U. australis, 1171
   U. delicatula, 1995–1997
   U. exoleta, 1834
   U. flexuosa, 1746
   U. punctata, 1859
   U. stellaris, 1746
   U. vulgaris, 952, 1046
Uvalas and dolines, 1060–1061
Uvaria welwitschii, 206

V
Vaccinium, 545, 612, 1998
   V. microcarpum, 1044
   V. stanleyi, 1313
   V. vitis-idaea, 1537
Vajgar fishpond, 473, 474
Valencia hispanica, 1116
Valikrood River, 1638
Valley, 12, 38, 58, 167, 185, 186, 238, 277, 280, 285, 325, 337, 473, 529, 717, 786, 825, 879, 880, 912, 924, 928, 938, 1048, 1223, 1226, 1247, 1262, 1264, 1308, 1310, 1398, 1400, 1401, 1416, 1419, 1420, 1425, 1426, 1442, 1446, 1454, 1473, 1479, 1533, 1535–1536, 1541, 1544–1546, 1580, 1586, 1660, 1712, 1717–1718, 1754, 1764, 1767, 1914, 1978
   blind and pocket valleys, 1061
   bog, 254–256
   bottom wetlands, 348
   California's Central Valley, 638, 697–702

Valley (*cont.*)
   Doon valley, 218
   fen, 254–256, 938, 1540, 1542–1543
   Hula Valley, 1168, 1170
   Kamiranzovu valley, 1309, 1311
   Kashmir valley, 1706, 1708
   Kazeroun valley, 1648
   Kilombero Valley, 1341–1347
   MAV, 577–588
   meadow-fens, 1542
   mire, 254–256, 258, 270, 1547
   Nile valley, 184, 1248, 1253
   Orkhon valley, 1547
   Pokhara Valley, 1715
   Rhône valley, 1103
   Rift Valley, 152, 153, 1202, 1208, 1246,
      1349–1359, 1381–1390, 1479
   Rugezi valley, 1309, 1311, 1313
   Sacramento Valley, 698
   San Joaquin Valley, 698, 701
   side, 255, 256, 267, 268, 270–271
Valleyside, 255, 256, 267, 305
   blanket bog, 270–271
   eccentric raised bog, 268
Vallisneria, 1272, 1569
   V. americana, 613
   V. natans, 1746
   V. spiralis, 952
Values, 22, 29, 32, 50, 62, 74, 76, 85, 116, 117,
   120, 136, 137, 152, 153, 156, 158,
   160, 161, 165, 173, 177, 178, 189,
   191, 192, 195, 217, 234, 240, 241,
   287, 316, 317, 321, 323, 326–328,
   335, 370, 372, 375, 377, 394, 398,
   401, 417, 428, 438, 440, 441, 443,
   451, 454, 455, 462, 471, 478, 531,
   533, 538, 553, 571, 583, 586, 587,
   594, 634, 644, 645, 654, 682, 711,
   720, 738, 761, 762, 807, 817–818,
   848, 880, 881, 890, 892, 900, 902,
   905, 906, 908, 929, 937, 953, 955,
   987, 1032, 1034, 1035, 1054, 1068,
   1075, 1081, 1083, 1086, 1110, 1114,
   1116, 1118, 1132, 1133, 1138, 1142,
   1144–1146, 1152, 1171, 1174–1176,
   1181, 1182, 1194, 1203, 1208, 1226,
   1229, 1239, 1266, 1267, 1304, 1324,
   1328, 1329, 1334, 1337, 1338, 1340,
   1346, 1357, 1372, 1382, 1387, 1455,
   1460, 1467, 1482, 1487, 1503, 1516,
   1527, 1570, 1611, 1621, 1628, 1645,
   1679, 1689, 1690, 1698, 1702, 1708,
   1715, 1716, 1719, 1721, 1729, 1737,
   1747–1749, 1780
   1860, 1866, 1869
   1904, 1905, 1912
   1955, 1956, 1974
   1998, 2003
   cultural values, 1982–19
   drought, 1194
   ecosystem service, 85, 1
   electrical conductivity, 8
   freshwater ecosystems, 1
   Hula wetland, 1171
   Macquarie Marshes, 190
   natural resources, 63
   of papyrus marshes, 191
   Patagonian peatlands, 88
   resource, 491–492
   swamps/floodplains, 137
Vanellus
   V. albiceps, 1344
   V. gregarius, 953, 1629
   V. miles, 1915
   V. spinosus, 1142, 1190
   V. vanellus, 965, 980, 98
      1073, 1448, 1469
Varanus, 192, 1354
   V. indicus, 1955
   V. niloticus, 1238, 1274
   V. salvator, 1825
Vascular plants, 176, 236, 2
   525, 552, 584, 61
   880, 1071, 1130,
   1469, 1502, 1517,
   1859, 1980–1981
Vatica
   V. cf. umbronata, 208
   V. lancaefolia, 207
   V. papuana, 208
   V. ressak, 208
Vaucheria, 1323
Vegetation, 6, 49, 60, 138, 1
   186, 192, 193, 20
   232–234, 237, 24
   267, 280, 282–28
   395, 414, 422, 46
   523–527, 534, 60
   674, 681, 700, 70
   732, 748, 750, 76
   808, 825, 828, 83
   887, 890, 914, 93
   1043–1047, 1082
   1156, 1159, 1171,
   1281, 1291, 1293
   1315–1316, 1324,

1401–1403, 1406, 1409, 1419, 1446,
1468, 1469, 1525, 1527, 1528,
1533–1536, 1541, 1558, 1566, 1569,
1571, 1619, 1628, 1639, 1666, 1682,
1687, 1736, 1740, 1746, 1815, 1822,
1848, 1862, 1869, 1888, 1894, 1905,
1922, 1934, 1946, 1960, 1963, 1980,
1993, 2007, 2009, 2020
acid rain, 535
aquatic vegetation, 133, 600, 646, 660, 779,
780, 851, 917, 1107, 1258, 1466,
1579, 1642, 1706, 1717
associations, 1081
bog, 971, 1544, 1998
characteristics, 569–570
Ciénaga Grande de Santa Marta, 764–766
Colombia, 750
composition and soil nutrient status, 669
decision tree, 559
delta wetlands, 1746
emergent vegetation, 151, 187, 191, 200,
209, 552, 561, 598, 607, 608,
915, 1445
estuarine marshes, 58, 569
Everglades, 714–720
fauna, 735
features of, 20, 1271
fishpond management, 477–479
flora, 163, 1952
foraging in floodplain, 214
fresh and saline water, 228
fringing, 191, 1273, 1274, 1357, 1358
growth of, 928, 935
halophilous vegetation, 1115
herbaceous vegetation, 6, 171, 175, 184,
200, 203, 393, 561, 914, 916, 1227,
1586, 1728, 1786, 1822
littoral vegetation, 472, 473, 480
marginal, 915
marsh, 162, 172, 598, 641, 644, 960, 1313
mires, 252, 264, 1043, 1542
natural vegetation, 7, 152, 166, 799, 956,
1309, 1584, 1601, 1701, 1754, 1805,
1862, 2017
peat-forming, 252, 256, 263, 266, 290, 291,
305, 1043, 1415
peat swamp forest, 1756–1759
Pocosin, 671, 674
removal of, 534, 536
riparian forest, 1834
riparian vegetation, 859, 956, 1209, 1823,
1834, 1891, 1948
saltmarsh, 969, 971, 1006

structure, 7, 179, 1074, 1545, 1795, 1957
submerged vegetation, 977–979, 981, 1649
tropical freshwater swamps, 177, 211
types, 173, 174, 177, 203–208, 246, 1454,
1524
wetland, 194, 365, 517, 657, 1367, 1513,
1805, 1900, 1901, 1947, 1955, 1981
woody vegetation, 62, 203, 204, 206, 1754
zonation, 1069, 1071, 1072, 1140
Velkovrhia enigmatica, 341
Vermivora bachmanii, 584
Vernal temporary pools, 656
Veronica anagallis-aquatica, 949, 951
Vertically homogeneous estuary, 41
Vetiveria nigritana, 1236, 1272
Viburnum
  V. opulus, 1029
  V. tinus, 1106
Victoria amazonica, 735
Vigna nilotica, 1272
Vimba
  V. vimba, 1630
  V. vimba persa, 953
Vipera ammodytes, 1142
Virola, 206, 750
  V. calophylla, 207
Vitex holoadenon, 205
Vitis sylvestris, 916
Voacanga, 207
  V. thouarsii, 1427
Vochysia, 750
  V. divergens, 206
Volga Delta, 912, 934, 937, 938, 940
  biodiversity, 949
  climate, 947
  conservation status, 954
  discharge regulation, 956
  ecosystem services, 955
  future aspects, 956
  hydrology, 949
  land usage, 953
  location, 946
  meadows, 949
  MSR image, 948
  natural environmental variability, 955
  sub regions, 947
  unsustainable natural resource exploitation,
    956
  wetland natural resources, 955
Volga River Basin, 887
  biodiversity, 939–940
  biomes, 934
  central basin, 938

Volga River Basin (*cont.*)
  climate, 935–936
  future aspects, 942
  future challenges and developments, 942
  human impacts, 940–942
  hydrological network, 936–937
  landscapes, 937
  location, 934
  lower basin, 938
  natural lakes, 938
  origin, 934
  threats, 942
  upper basin, 938
  Volume, 31, 78, 110, 127, 131, 147, 151, 158, 160, 175, 179, 206, 209, 216, 218, 234, 238, 310, 355, 364, 388, 421, 473, 536, 568, 592, 623, 624, 652, 655, 691, 710, 711, 774, 886, 887, 937, 940, 947, 956, 1062, 1145, 1175, 1236, 1239, 1257, 1270, 1275, 1366, 1377, 1382, 1386, 1400–1402, 1414, 1479, 1500, 1546, 1552, 1553, 1670, 1682, 1683, 1694, 1708, 1783, 1892, 1913, 1914, 1933
  discharge, 113, 545, 559, 949, 950, 1300
  flow volume, 111, 112, 1809
  hydric soil-defined volume, 693, 694
  lakes, 128–129, 146
  peat, 1311, 1313
  of seawater, 1080
  water volume, 128, 473, 720, 887, 912, 1188, 1204, 1244, 1281, 1311, 1376, 1458, 1576, 1577, 1580, 1585, 1608, 1627, 1649, 1663, 1678, 1804, 1917
Voluta ebraea, 845
von Post, 246
von Post humification scales, 246, 1311, 1312
Vormela peregusna, 953
Vossia cuspidata, 189, 205, 1265, 1272, 1302
Vulpes vulpes, 547, 980, 989, 1106, 1142
Vulpia spp., 1104

W
W.A.C. Bennett dam, 553, 554
Wadden Sea (Denmark), 960, 984–995, 1000
Wading birds, 64, 86, 717, 718, 805, 1388, 1796, 1916, 1917
Wailagi Lala Atoll, 2005
Wallagonia leerii, 1860
Warnstorfia exannulata, 284
Wash and North Norfolk Coast, UK
  biodiversity importance, 1015
  conservation status, 1017
  intertidal mud and sand flats, 1014
  land-claim and conversion, 1018
  maps of, 1013
  sediment, 1013
  threats, 1019
  tidal embayment, 1012
  tourism and recreation, 1015
Wash estuary, 11, 1003, 1011–1021
Wasmannia auropunctata, 2017
Wastewater treatment, 49, 89, 117, 162, 190, 434, 498, 503, 639, 740, 780, 894, 942, 1445
Water
  Anillo de Cenotes, 323
  biota, 100, 133, 149, 286
  consumption, 115, 1504, 1663, 1672
  distribution, 145–146, 160, 777, 1576
  estuaries, 39, 43
  HWF, 466
  hydroelectric production/ irrigation, 53
  hydrological cycle, 314
  levels (*see* Water levels)
  management (*see* Water management)
  open-water transition mire, 253
  peatland, 233, 234
  physical and chemical processes, 162–163
  quality (*see* Water quality)
  regimes, 20, 31, 32, 136, 151, 173, 177, 179, 277, 360, 363, 365, 534, 682, 698, 718, 780, 818, 828, 917, 955, 1720, 1748, 1960
  withdrawals, 5, 8, 116, 117, 137, 179, 360, 362–364, 554, 1258, 1650
Waterbirds, 20, 135, 136, 150, 151, 366, 377, 395, 547, 583, 584, 682, 692, 764, 889, 918, 954, 976, 981, 984, 1002–1003, 1089–1095, 1106, 1131, 1132, 1188, 1190, 1192, 1223, 1236, 1238, 1255, 1344, 1448, 1469, 1480, 1502, 1504, 1513–1515, 1518, 1558, 1561, 1569–1571, 1580, 1582, 1628, 1639–1641, 1645, 1681, 1787, 1790, 1816, 1823, 1866, 1871, 1916, 1917, 1922, 1925, 1947, 1955
  Anzali Wetland, 1621
  breeding, 395, 600, 682, 1338, 1517, 1622, 1790, 1901
  British estuaries, 1002, 1003
  colonial waterbirds, 133, 395, 600, 950, 1131
  conservation status, 377, 954, 1094, 1338
  Danish Wadden Sea, 992

ecosystem services, 1132
Fereydoon Kenar, 1640
habitats, 134, 135, 408, 1104, 1446, 1447, 1468, 1708, 1717, 1900, 1901, 1926
inner Danish waters (*see* Inner Danish waters)
international importance, 683, 1016
Lake Parishan, 1653
lakes, 150, 153, 583
Macquarie Marshes, 1902, 1904
migration, 133, 134, 151, 153, 323, 364, 406, 682, 692, 994, 1002, 1095, 1130, 1557, 1583, 1621, 1626, 1640, 1644, 1652, 1654, 1669, 1682, 1718, 1820, 1824, 1835, 1912
Mississippi Alluvial Valley, 578, 583, 584, 600
nonbreeding, 1015, 1016
populations, 136, 366, 377, 409, 1002, 1003, 1015, 1016, 1019, 1107, 1469, 1481, 1528, 1638–1639, 1653, 1666, 1747, 1824, 1905, 1915
Saemangeum estuarine system, 408
species, 984, 987, 989, 990, 992, 1015, 1095, 1227, 1238, 1622, 1628, 1644, 1653, 1806, 1825, 1870, 1901, 1902, 1912, 1922
staging, 981, 1094, 1628
Ussuri-Wusuli River Basin, 1517
wetland, 365, 547
wintering, 364, 641–642, 692, 1091, 1094, 1131, 1566, 1622, 1624, 1628, 1642, 1680
Yenisei River Basin, 1480
Yukom and Kuskokwim, 547
Water clarity, 48, 49, 78, 83, 84, 89, 138, 327, 496, 503, 610, 840, 1879, 1934
Water Conservation Areas (WCAs), 706–708, 712, 716, 720
Water dissolution hypothesis, 690
Water diversion, 146, 217, 361, 1073, 1218, 1350, 1577, 1586, 1687, 1688, 1730, 1888, 1904
Water-divide, 267, 268, 270, 304, 305, 307, 420
Waterfowl, 47, 48, 64, 74, 166, 532, 533, 544, 547, 554, 562, 563, 568, 571, 630, 641, 642, 645, 681, 682, 686, 692, 702, 804, 806, 905, 955, 978, 982, 1029, 1073, 1133, 1445, 1469, 1470, 1475, 1561, 1626, 1628, 1631, 1632, 1638, 1643–1645, 1726, 1845, 1847, 1870, 1925, 1955
abundance of, 472

conservation status, 585
habitat, 74, 86, 161, 166, 472, 533, 552, 563, 568, 583–585, 595, 612, 681–682, 692, 1559, 1988
human impacts, 161, 370
hunting, 480, 582, 587, 600, 601
populations, 285, 370
SFBE, 642
use, 702
wetland impacts, 285
wintering, 581, 584, 585, 698, 702, 1105, 1619, 1640–1642, 1665
Water Framework Directive (WFD), 892, 893, 895, 907, 1027, 1069
Water hyacinth, *see Eichhornia crassipes*
Water levels, 41, 131, 134, 137, 145, 146, 157, 160, 167, 188, 192, 217, 314, 349, 351, 360, 364, 432, 526, 550, 572, 580, 583, 593–595, 598–601, 603, 606, 614, 624, 682, 698, 715–717, 732, 735, 736, 765, 780, 790, 799, 801, 877, 878, 906, 914, 920, 942, 949, 952, 955, 974, 986–988, 1025, 1027, 1061, 1070, 1072, 1074, 1103, 1117, 1124, 1129, 1150, 1207, 1235, 1290, 1302, 1305, 1309–1311, 1342, 1344, 1350, 1352, 1353, 1356, 1364, 1366, 1370, 1377, 1384, 1442, 1541, 1544, 1561, 1566, 1570, 1571, 1586, 1616, 1618, 1619, 1623, 1651, 1660, 1663, 1672, 1708, 1767, 1783, 1791, 1794, 1797, 1804, 1815, 1843, 1844, 1846, 1847, 1853, 1862, 1871, 1912, 1914, 1917, 1932, 1933, 1937, 1944, 1961, 1963, 1997
analysis, 1572
buffer, 208–209
Caspian Sea, 1616
Coorong, 1912
dynamics, 431
fluctuations, 133, 157, 320, 431, 434, 482, 734, 779, 802, 920, 1061, 1364, 1370, 1376, 1453, 1458, 1465
lagg fen, 311
Lake Manitoba, 600, 603
Lake Turkana, 1365, 1376
Lake Uromiyeh, 1664
LakeWinnipeg, 592, 603
management, 365, 431
Myall River, 1932
Peace River, 551
Poyang, 1568
Rift Valley, 1353

Water levels (*cont.*)
    salt lakes, 151
    short and long term, 431
    Tasek Bera, 1853
    wetland biodiversity, 365
Waterlogging, 230, 238, 246, 254, 256, 257, 261, 270, 305, 310, 531, 792, 1701, 1721, 1745, 1748, 1754
Water management, 119, 324, 364, 365, 464, 482, 701, 708, 710, 720–721, 900, 1104, 1162, 1263, 1266, 1296, 1496, 1518, 1524, 1570, 1655, 1656, 1672, 1731, 1797, 1817, 1905–1906, 1949, 1988
Water provision, 116, 126, 807, 892, 900, 908, 1372
Water quality, 6, 20, 63, 78, 103, 137, 151, 161, 217, 238, 352, 361, 362, 411, 414, 443, 453, 455, 480, 483, 496, 498, 499, 505, 532, 548, 586, 587, 601–603, 612–614, 638, 643, 647, 663, 684, 754, 767, 774, 779, 780, 817, 830, 836, 880, 891, 919, 926, 976, 981, 987, 1027, 1032, 1035, 1053, 1103, 1130, 1170, 1212, 1249, 1255, 1258, 1297, 1305, 1390, 1410, 1504, 1516, 1528, 1571, 1585, 1586, 1613, 1623, 1663, 1664, 1683, 1693, 1709, 1853, 1882, 1910, 1915, 1942, 1963
    buffering, 209
    changes in, 434, 501, 1859
    characteristics, 761–762
    deforestation and reduction, 1209, 1210
    ecosystem services, 1304, 1570, 1729, 1996
    geochemistry, 1353, 1365–1366
    management, 456, 457
    pollution and declining, 1782, 1983
    River Rhine restoration, 927–928
    swamps and marshes, 173, 174
    targets, 503
    vernal pools, 654–655
Water-related vertebrates, 1469, 1474
Water retention, 247, 348, 476, 483, 532, 712, 817, 898, 902, 1309, 1503
Watershed-based nutrient reduction, 500, 503
Watershed (water-divide) blanket bog, 270
Watershed or spur raised bog, 267–268
Watershed water balance, 876
Wattled crane, 1222, 1225, 1227, 1229, 1236, 1238, 1407
Weed control, 388, 1948
Weeds, 364, 384, 385, 388, 580, 584, 777, 1304, 1368, 1394, 1710, 1859, 1905, 1948, 1952, 1957, 1963, 1985, 1998, 2013, 2017, 2020
Western Alaska, 544
Western Hemisphere Shorebird Reserve Network (WHRSN), 562, 633, 641, 867
Western sandpiper, 547, 562, 569, 642
Wet alpine heath, 297
Wet Graminoid communities, 546
Wetland(s), 509–513
    alteration, 533–537
    Amur River Basin, 1485–1497, 1499–1507
    assessing change, 361–362
    atolls (*see* Atolls)
    Baro-Akobo system, 1261–1267
    Bay of Fundy, 621–635
    Berbak National Park, 1831–1838
    biodiversity, 5, 6, 28, 178, 361, 364–366, 818, 948, 954, 955, 957, 998, 1560, 1561, 1584, 1644, 1893
    Central Valley of California, 697–702
    challenges, 14
    changes, 1913–1916
        natural and anthropogenic drivers, 8–9, 359–367
    Chile, 823–836
    common names in English, 21
    complex, 4, 6, 9–11, 13, 26, 173, 174, 176, 428, 429, 529, 548, 558, 709, 752, 769, 774, 782, 786, 876, 878, 888, 889, 920, 1143, 1145, 1168, 1408, 1454, 1482, 1600, 1611, 1623, 1700, 1701, 1713, 1715, 1717, 1911, 1913, 1988
    conservation, 7, 180, 317, 402, 409, 537, 737, 1102, 1119, 1284, 1291, 1719, 1797, 1894, 1895, 1968, 1982, 2017, 2021
    conversion to agriculture, 1780
    Danube Delta, 911–921
    definition, 5, 20
    degradation, 8, 12, 361, 362, 365, 1120, 1229, 1264
    Dinaric karst, 1057–1064
    distribution, 20, 539, 888–889, 1518, 1744–1746
    diversity, 5–7, 20, 21, 278, 569, 750
    Doñana wetlands, 1123–1134
    dynamics, 803, 1502
    ecosystems (*see* Wetland ecosystems)
    extent and distribution of, 26–31

fauna, 1859–1860, 1981
functions, estuarine marsh, 62, 347, 527, 534, 537, 1557, 1693, 1749
Ganga-Brahmaputra Basin, 1711–1721
global area, Ramsar regions, 29
global distribution of, 30
Great Barrier Reef, 447–458
HAWs, 1605–1613
Hula wetland, 1167–1172
hydrology, 346, 362–365, 535, 1230, 1249, 1904
Indus River Basin, 1697–1702
infrastructure development, 364
inputs of water, 366
of International Importance, 1710
inventory, 31–32
Kakadu National Park, 1951–1957
land conversion, 363
locations of, 10, 11, 13
loss(es), 8, 12, 28, 32, 68, 180, 193, 363, 369–380, 534, 537, 539, 683–685, 698–700, 741, 1516, 1978
    long-term, 372–373
    natural, 375
    in 20th and early 21st centuries, 373–375
Mahanadi Delta, 1743–1749
Makgadikgadi wetlands, 487–494
management, 779–781, 1194, 1264, 1797, 1890, 1982, 1998
management plan, Lake Seyfe, 1194
man-made, 1771
Mekong River Basin, 1763–1783, 1785–1791, 1793–1798
natural, 360, 1771
natural processes and anthropogenic activities, 8–9
New Zealand, 1977–1989
Norfolk and Suffolk Broads, 1023–1036
Playas, 689–694
regional compilations, 9
restoration, 644, 684, 857–864, 894–895, 920, 921, 1120, 1571, 1601, 1693, 1749, 1797, 1918, 1965, 1987–1988
saline wetlands, 1725–1731
Shadegan wetland, 1675–1683
species, 804
Sudd, South Sudan, 1300–1302
Tasek Bera, 1851–1863
of Tram Chim National Park, 365
types (see Wetland types)

vegetation, 194, 350, 365, 517, 593, 657, 1236, 1367, 1459, 1513, 1639, 1805, 1900, 1901, 1947, 1955, 1981
water level management, 365
water withdrawals, 363–364
Zhalong wetlands, 1521–1529
Wetland-dependent species, 52, 377–380
Wetland ecosystems, 7, 14, 21, 112, 156, 171, 284, 346, 347, 402, 476, 552, 602, 706, 816–817, 833, 888, 1119
    Anzali Mordab Complex, 1618–1619
    Australia, 1944–1945
    Berbak National Park, 1834
    Bijagós archipelago, 1336–1337
    biodiversity, 949–953
    Bujagh National Park, 1627–1628
    Camargue, 1104–1106
    cryo-arid conditions, 1465–1468
    Danau Sentarum National Park, 1843–1844
    Danube Delta, 914–917
    distributaries, 791
    disturbances and threats, 836
    Fereydoon Kenar Ramsar, 1638–1640
    floodplain channels, 791
    floodplain lakes, 791
    fluvial forests, 792
    Greek, 1140
    high altitude wetlands, 1608–1609
    Indus Basin, 1699–1700
    Kilombero floodplain, 1343–1344
    Lake Baringo, 1353–1355
    Lake Naivasha, 1357–1358
    Lake Turkana, 1366–1367
    Lena Delta, 1453–1454
    Macquarie Marshes, 1900–1901
    main channel and anabranches, 790–791
    marshlands, flooded grasslands, and savannas, 791
    middle Aldan River, 1473–1474
    Nile, 1270–1271
    Paraná-Paraguay Fluvial Corridor, 790–792
    Peace-Athabasca Delta, 552
    pocosin, 669–670
    restoration, 401
    Sembilang National Park, 1822–1823
    soil humidity and salinity, 1140
    Sudd, 1300–1302
    thermokarst kettle hole peatlands, 284
    Tram Chim, 1795–1796
    U Minh, 1815
    Volga Delta, 949–953
    water bodies, 914–915

Wetland ecosystems (*cont.*)
  Zambezi Delta, 1235
  Zhalong Marsh, 1524–1525
Wetlands International (WI), 319, 1487, 1720, 1825, 1826, 1872
Wetlands of Chile, 823–836
Wetlands Reserve Easement, 694
Wetland trends-estuarine marsh, 64–68
Wetland types, 12, 19–32, 59, 68, 163, 173, 211, 213, 214, 277, 326, 327, 349, 373, 430, 431, 482, 488, 669, 681–682, 694, 728, 734, 738, 790, 801, 825, 1228, 1414, 1424, 1444, 1488, 1553, 1580, 1582, 1583, 1599, 1700, 1721, 1805, 1888, 1893, 1911, 1979, 1983
  Amazon of Europe, 904–905
  boreal wetlands, 523–528
  Central Valley, 700
  coastal wetlands, 7, 733
  definition and classification, 21–22
  distribution, 1728, 1978
  diversity, 510–511, 1728, 1736
  East and Southeast Asia, 1866–1867
  Ebro Delta, 1115–1117
  estimates of area, 30
  extent, 1728
  Greenland, 510–511
  interfluvial wetlands, 732
  large-river floodplains, 729–731
  Lena River Basin, 1445–1448
  Lower Danube, 898–899
  Mekong River, 1768–1769
  New Zealand, 1978
  Paludified Shallow Peatland, 279
  patterned string fen (aapa mire) and raised bogs, 283
  peatlands, 6
  Ramsar classification, 278
  Ramsar wetland, 5
  riparian mires, 285
  Riverine, 607, 608
  Saline wetlands, 1728
  Sundarban mangroves, 1736
  Tidal freshwater wetlands, 7
  Tonle Sap lake, 1786–1787
Wet meadows, Zhalong marsh, 1524
Wet-season sub-tropical/tropical patterned mixed mires, 262, 263
Wet Tropics, 1941–1949
Whitelees Wind Farm, 417
White Nile, 1244, 1246–1249, 1262, 1281, 1283, 1300, 1305

White Storks, 482, 918, 1127, 1188, 1192, 1254, 1274, 1520
White-water rivers, 729, 731, 738, 1848
Widgeon grass, *see Ruppia maritima*
Wildfires, Amur-Heilong River Basin, 1497
Wildfowl and Wetlands Trust (WWT), 1872
Wildlife, 10, 153, 184, 365, 366, 370, 398, 400, 428, 462, 488, 490–493, 513, 519, 534, 535, 539, 548, 553, 561, 562, 568, 571, 573, 594, 595, 597, 600, 613, 633, 648, 682, 686, 694, 702, 767, 814, 816–818, 994, 1000, 1014, 1015, 1017, 1027, 1029, 1035, 1041, 1049, 1050, 1143, 1176, 1229, 1235, 1239, 1240, 1249, 1265, 1266, 1273–1274, 1281, 1290, 1291, 1293–1297, 1302–1305, 1315, 1345, 1358, 1359, 1375, 1376, 1384, 1389, 1390, 1396, 1409, 1501, 1502, 1504, 1511, 1515, 1519, 1598, 1612, 1636, 1681, 1694, 1701, 1719, 1729, 1730, 1738, 1739, 1787, 1788, 1816, 1828, 1832, 1836, 1837, 1847, 1863, 1866, 1956, 1987, 2021
Wildlife Conservation Society (WCS), 1790
Wildlife Management Area (WMA), 599
Wild rice, 159, 428, 430–432, 434, 532, 533, 553, 611, 612, 1302, 1717, 1776
Willow, 160, 279, 285, 524, 525, 545, 560, 561, 581, 717–719, 791, 888, 915, 952, 1029, 1072, 1140, 1171, 1193, 1194, 1337, 1482, 1536, 1541, 1545, 1584, 1640, 1708–1710, 1980, 1986
Wind, 38, 49, 132, 160, 391, 536, 551, 559, 593, 595, 599, 606, 759, 761, 874, 879, 915, 938, 981, 984, 1294, 1323, 1330, 1353, 1359, 1369, 1375, 1378, 1390, 1547, 1568, 1569, 1934
  dissolution hypothesis, 690
  Eichhornia crassipes, 134
  erosion hypothesis, 690
  farms, 413–423, 1040, 1052, 1095
  lake, 1365
  peat, 415, 417
  seed transport, 414
  water movements, 43
Windfarms on peatlands
  carbon balance, 423
  construction, 417–418
  global wind resources map, 415, 416
  landscapes, 415–417
  mires and climate, 415

roads, 419–423
turbine tower, 418
Wintering waterfowl, 581, 584, 585, 698, 702, 1105, 1619, 1640–1642, 1665
Wood Buffalo National Park, 553
Wood frog, 658, 659
Working with nature, 927
World Heritage Site, 317, 320, 321, 324–326, 553, 888, 994, 1054, 1086, 1132, 1222, 1223, 1363, 1386, 1397, 1482, 1694, 1717, 1746, 2020, 2021
World Wildlife Fund (WWF), 1486
Worms, 191, 917, 918, 1002, 1481, 1870, 1952
 annelid, 1324
 aquatic, 612
 bait, 47
 polychaete, 630, 843, 867, 1002
 rag, 977–978, 988
Wular Lake, Kashmir
 biodiversity and ecosystem services, 1708
 catchments, 1706
 conservation status, 1710
 hydrological regimes, 1706
 location, 1706
 management plan, 1710
 threats, 1708–1710

X

Xanthium strumarium, 956
Xanthophyllum affine, 208
Xanthopsar flavus, 817
Xolmis dominicanus, 817
Xylocarpus, 1823
 X. granatum, 1823
 X. moluccensis, 1823
Xyris validus, 1313

Y

Yangtze River Basin, 13, 115, 325, 1565–1572, 1701
 challenges, 1561–1562
 future challenges and developments, 1561–1562
 habitats and threatened species, 1558–1559
 headstream, 1554–1555
 human disturbances, 1561
 hydrology and climate, 1552–1553
 water pollution, 1561
 wetlands in middle and lower reaches, 1556–1557
 wetlands in upper reaches, 1555–1556

Yellow River Basin, 1582
 biodiversity, 1584
 distribution, 1576
 future challenges and developments, 1586–1587
 Ramsar sites, 1579
 regional climate and runoff features, 1577
 reservoirs, 1587
 sediment deposition, 1585
 soil salinization, 1586
 types and distribution of wetlands, 1578
 water pollution, 1585
Yellow River Delta, 1583, 1584, 1586
Yenisei River Basin, 13
 biodiversity, 1480
 conservation status, 1482
 description, 1479
 FEOW definition, 1480
 future challenges and developments, 1484
 human uses, 1482–1483
 location, 1479
 threats and future challenges, 1484
Yueguzonglie River, 1579
Yukon-Charley Rivers National Preserve, 547
Yukon-Kuskokwim Delta, 10
 biodiversity, 547
 conservation status, 548
 hydrology, 544–545
 wetlands, 160, 545–546
Yukon-Kuskokwim Delta National Wildlife Refuge, 544, 548
Yukon River, 543–548, 559
Yukon River Inter-tribal Watershed Council, 548

Z

Zaire River, 1552
Zambezian Phytochoria, 1408
Zambezi Delta, 1219, 1227
 biodiversity, 1237–1238
 conservation status, 1238
 ecosystem services, 1239
 hydrology, 1235
 location, 1234
 threats, 1239–1240
 wetland ecosystems, 1235–1237
Zambezi River Basin, 12, 186, 1233–1240, 1408
 biodiversity and productivity, 1228
 climate change, 1230
 dambos, 1219–1221
 dams and wetland areas, 1219

Zambezi River Basin (*cont.*)
  future challenges and developments, 1230, 1239–1240
  lower Zambezi, 1226
  major wetlands, 1220
  middle Zambezi, 1222–1226
  origin, 1218
  population, 1230
  Ramsar wetlands, 1229
  Riparian reed swamps, 1221
  role in hydrology, 1228
  role in livelihoods and wellbeing, 1229
  swamps, marshes and foodplains, 1219–1221
  threats, 1229–1230
  tributary dams, 1219
  upper Zambezi, 1221–1222
Zannichellia, 1105, 1117
Zapovednik, 1449, 1452, 1454, 1490–1492, 1519
Zelkova carpinifolia, 1639
Zeya Hydro Reservoir, 1495
Zhalong Marsh, 26, 1522–1524, 1526–1529
Zhalong wetlands, 13
  biodiversity, 1525–1526
  conservation status, 1526
  ecosystem, 1524–1525
  fishery, 1527
  flood control, 1527
  frequent fires, 1528
  geographical map, 1522
  hydrology, 1523–1524
  location, 1522
  population declines, 1528
  reed harvesting, 1527
  reed marshes with shallow water, 1523
  tourism, 1527
  water quality decline, 1528
  water shortage, 1528

Zizania, 164, 532
  Z. aquatica, 159, 163
  Z. palustris, 428, 611
Zizaniopsis, 791
  Z. bonariensis, 163, 851
  Z. miliacea, 163
Zonation, 46, 47, 98, 99, 102, 163, 164, 189, 207, 510, 561, 692, 963, 1068, 1069, 1071, 1072, 1081, 1082, 1085, 1237
Zoning categories, Colombian mangrove legislation, 753
Zonotrichia atricapilla, 547
Zooplankton, 44, 45, 100, 165, 394, 471, 472, 477, 479, 625, 762, 1105, 1131, 1273, 1303, 1353, 1357, 1371, 1385, 1406, 1569, 1728, 1859, 1860, 1936, 1937
  of turloughs, 1071
Zoothera sibirica, 1825
Zootoca vivipera, 1047
Zospeum spp., 339
Zostera, 48, 80, 81, 1590, 1592
  Z. asiatica, 76, 1590–1592, 1595
  Z. caespitosa, 76, 1590–1593, 1595
  Z. capensis, 76
  Z. caulescens, 76, 1590–1593, 1595
  Z. chilensisa, 76
  Z. geojeensis, 76, 1595
  Z. japonica, 76, 569, 573, 1590–1593, 1595
  Z. marina, 76–78, 81, 82, 88, 89, 569, 641, 988, 1091, 1590–1595
  Z. muelleri, 76, 1971
  Z. nigricaulis, 76
  Z. noltei, 1104, 1106
  Z. noltii, 76, 81, 988, 1323
  Z. pacifica, 76
  Z. polychlamys, 76
  Z. tasmanica, 76